HANDBOOK OF SINGLE-PHASE CONVECTIVE HEAT TRANSFER

HANDBOOK OF SINGLE-PHASE CONVECTIVE HEAT TRANSFER

Edited by

Sadık Kakaç

Department of Mechanical Engineering
University of Miami
Coral Gables, Florida

Ramesh K. Shah

Harrison Radiator Division
General Motors Corporation
Lockport, New York

Win Aung

National Science Foundation
Washington, D.C.

A Wiley-Interscience Publication

JOHN WILEY & SONS

New York • Chichester • Brisbane • Toronto • Singapore

Library of Congress Cataloging in Publication Data:

Handbook of single-phase convective heat transfer.

"A Wiley-Interscience publication."
Includes bibliographies and index.
1. Heat.—Transmission.—Handbooks, manuals, etc.
2. Heat.—Convection.—Handbooks, manuals, etc.
I. Kakaç, S. (Sadık) II. Shah, R. K. III. Aung, W.

QC320.4.H37 1987 536′.25 87-2155
ISBN 0-471-81702-3

Printed in the United States of America

10 9 8 7 6 5 4 3 2 1

CONTRIBUTORS

B. F. Armaly, Department of Mechanical Engineering, University of Missouri–Rolla, Rolla, Missouri

W. Aung, Division of Chemical, Biochemical and Thermal Engineering, National Science Foundation, Washington, DC

A. Bejan, Department of Mechanical and Materials Science, Duke University, Durham, North Carolina

M. S. Bhatti, Harrison Radiator Division, General Motors Corporation, Lockport, New York

T. S. Chen, Department of Mechanical Engineering, University of Missouri–Rolla, Rolla, Missouri

Jane H. Davidson, Department of Mechanical Engineering, College of Engineering, Colorado State University, Fort Collins, Colorado

P. F. Dunn, Department of Aerospace and Mechanical Engineering, University of Notre Dame, Notre Dame, Indiana

T. F. Irvine, Jr., Mechanical Engineering Department, State University of New York, Stony Brook, New York

Y. Jaluria, Department of Mechanical Engineering, Rutgers University, Piscataway, New Jersey

S. D. Joshi, Phillips Research Center, Phillips Petroleum Company, Bartlesville, Oklahoma

S. Kakaç, Mechanical Engineering Department, University of Miami, Coral Gables, Florida

J. Karni, Mechanical Engineering Department, State University of New York, Stony Brook, New York

F. A. Kulacki, College of Engineering, Colorado State University, Fort Collins, Colorado

P. E. Liley, School of Mechanical Engineering, Purdue University, W. Lafayette, Indiana

W. J. Marner, Jet Propulsion Laboratory, California Institute of Technology, Pasadena, California

M. N. Özişik, Mechanical & Aerospace Engineering, North Carolina State University, Raleigh, North Carolina

R. H. Pletcher, Mechanical Engineering Department, Iowa State University, Ames, Iowa

C. B. Reed, Advanced System Section, Engineering Division, Argonne National Laboratory, Argonne, Illinois

K. Rehme, Kernforschungszentrum Karlsruhe, Institut für Neutronenphysik und Reaktortechnik, Karlsruhe Federal Republic of Germany

R. K. Shah, Harrison Radiator Division, General Motors Corporation, Lockport, New York

J. W. Suitor, Jet Propulsion Laboratory, California Institute of Technology, Pasadena, California

R. L. Webb, Mechanical Engineering Department, Pennsylvania State University, University Park, Pennsylvania

K. T. Yang, Aerospace and Mechanical Engineering, University of Notre Dame, Notre Dame, Indiana

Y. Yener, Mechanical Engineering Department, Northeastern University, Boston, Massachusetts

A. A. Žukauskas, Academy of Sciences of the Lithuanian, U.S.S.R.

PREFACE

The field of heat transfer has grown enormously in the last 20 years with the explosion of scientific and engineering research. It has increased tremendously the depth of our understanding. It is no longer possible for a single individual to be intimately familiar with and/or be an expert in even some major subfields of heat transfer. One such subfield of great industrial importance is single-phase convective heat transfer. This is the subject that we have tried here in considerable depth with the dedicated effort of 25 specialists.

This handbook is intended to furnish the latest design and research information in the area of single-phase convective heat transfer to practicing engineers, researchers, academicians, and students. It consists of 22 chapters, a brief description of which is provided next.

Chapter 1. This chapter provides the reader with basic concepts and fundamentals of heat transfer. Four general laws are stated in terms of a system, and then control-volume forms are given. Particular laws of heat transfer are stated. Governing equations of convective heat transfer are formulated and are presented in tabular form for rectangular, cylindrical, and spherical coordinates. Boundary-layer approximations for laminar and turbulent flow are presented.

Chapter 2. This chapter deals with external flow forced convection. Fundamental equations in general form and definitions are presented first. Then, Reynolds equations for turbulent flow are described; reduced forms for inviscid flow and viscous flow are given. Summaries of the common turbulent models are presented. The flow over a flat plate and in other geometries is discussed; important correlations for heat transfer and friction coefficients are provided. General formulas and data correlations for use in preliminary design and as benchmark checks for computer codes are also discussed in this chapter. In addition, the capabilities of computational procedures for forced convection over external surfaces are also discussed.

Chapter 3. This chapter presents an up-to-date compilation of analytical solutions for laminar fluid flow and forced convection heat transfer in circular and noncircular ducts. The solutions are presented for four types of laminar flows of Newtonian fluids, viz., fully developed, hydrodynamically developing, thermally developing, and simultaneously developing flows. The most solutions are given in terms of mathematical expressions and in graphical form to elicit the general trends. In all, results are presented for 70 duct geometries covering a variety of thermal boundary conditions.

Chapter 4. In this chapter, the heat transfer and fluid flow characteristics of turbulent flows in ducts are considered. The turbulent pressure drop and heat transfer for various entry and wall surface conditions are presented. Other effects such as the influences of thermal boundary conditions, entrance shape, high velocity, porous walls, Prandtl number, body force, and internal heat generation on turbulent forced convection are discussed. Turbulent flows in planar ducts, rectangular ducts, and other-shaped ducts are presented. Various turbulence models are discussed. Correlations covering a number of geometries and conditions for turbulent flow in ducts are provided, along with constraints, if any, on their practical application.

Chapter 5. Based on an extensive literature search, dimensionless heat transfer and flow friction design data are presented for curved ducts with circular, square, and rectangular cross sections. These design data are based on the theoretical analyses and experimental measurements on laminar and turbulent flows of Newtonian and non-Newtonian fluids through curved ducts. The results are presented in terms of design correlations, graphs, or tables.

Chapter 6. This chapter deals with convective heat transfer in cross flow. Correlations for the local and average heat transfer coefficients of single tubes and bodies for cross flow are presented, and the factors influencing heat transfer are discussed. Heat transfer from smooth and rough tube bundles, and the drag of smooth and yawed tube bundles are treated, as well as heat transfer from finned tube bundles. Heat transfer correlations are presented in tabular forms. Extensive design information for convective heat transfer in cross flow is also provided in graphical form.

Chapter 7. Longitudinal flow over tube or rod bundles is common in most fuel elements of nuclear power reactors. Other applications of this geometry are encountered in shell-and-tube heat exchangers, boilers, condensers, etc. This chapter provides information on heat transfer and friction coefficients for laminar, transitional, and turbulent flow over rod bundles for various conditions. Various correlations useful to the designers are presented, and effects of spacers are discussed. Valuable information is provided in graphical forms.

Chapter 8. This chapter provides the reader with a short introduction to the fundamentals of liquid metal heat transfer, followed by detailed discussions of laminar and turbulent liquid metal heat transfer correlations. Thermal entry lengths, variable fluid properties, and natural convection are also treated in this chapter. Flows in various geometries, including round pipes, annuli, parallel plates, and various tube-bank geometries, are also covered. Both laminar and turbulent entry-length correlations are presented. Correlations covering a number of plate geometries in natural convection heat transfer are presented, as well as the correlation for heat transfer from horizontal cylinders.

Chapter 9. This chapter discusses the convective heat transfer with electric and magnetic fields. The important basic concepts of electrohydrodynamics (EHD) and of magnetohydrodynamics (MHD) are presented. EHD in external boundary layers and in confined flows is treated. The experimental and mathematical limitations of the existing literature have been emphasized. Governing equations and dimensionless groups are presented. The basic physics of magnetic field effects in electrically conducting liquids is discussed. MHD in confined flows, in external flows, and in natural convection are presented.

Chapter 10. Bends and fittings are most commonly used in pipelines, for which considerable pressure drop information has been summarized in the literature. A first attempt is made here to compile the available heat transfer information for bends with 90°, 180°, and other angles. Experimental and theoretical results for laminar and turbulent flow friction factors and Nusselt numbers are presented for bends having circular, square, and rectangular cross sections.

Chapter 11. This chapter is mainly concerned with the transient response of duct flows. The parallel-plate channel and circular duct, which are the two commonly encountered geometries in practice, are considered with both laminar and turbulent flows. The results of transient laminar forced convection in ducts for a step change in wall temperature and in wall heat flux are presented. Transient laminar forced convection in circular ducts with arbitrary time variations in wall temperature and with unsteady flow is also considered. The chapter also discusses transient turbulent forced convection in circular ducts with a step change in wall temperature, and in parallel-plate channels for a step change in wall temperature and wall heat flux.

Chapter 12. This chapter presents the basic considerations for the study of natural-convection heat transfer. The governing equations, along with their important simplifications, are presented to indicate the dimensionless parameters that arise and the basic nature of the transport process. Laminar natural convection over flat surfaces, cylinders, and spheres is discussed in detail, and the resulting heat transfer expressions presented. Transient effects and turbulent flow are outlined, since many practical problems involve these effects. Recommended empirical correlations for a variety of external natural convection heat transfer processes are given, along with the constraints, if any, on their application to physical problems.

Chapter 13. This chapter provides important basic information on the physics of the natural convection phenomena in enclosures; the formalism of the mathematical formulation of the natural convection problem; the available solution techniques; some significant results in the field, including both theoretical and experimental data and their correlation; and a brief description of recent studies of basic natural convection phenomena interacting with other heat transfer processes in an enclosure. Some emphasis is placed on the modern development in this field, including numerical and experimental techniques, laminar and turbulent flows, and two-dimensional and three-dimensional phenomena. Areas of future research are also delineated.

Chapter 14. This chapter deals with mixed convection in external flows. Results on the local Nusselt number are presented for various flow configurations, and instability studies conducted for these flows are described. This chapter summarizes and presents comprehensive correlations for the local and average Nusselt numbers that cover the entire mixed-convection regime, from pure forced convection to pure free convection, for various flow configurations of engineering interest, and for laminar as well as turbulent flows under the heating conditions of uniform wall temperature (UWT) and uniform surface heat flux (UHF). The instability characteristics of laminar mixed convection along flat plates, with regard to both wave and vortex instability, are also summarized.

Chapter 15. This chapter discusses combined free and forced convection in internal flow. The best available information is summarized, and correlations are presented for ducts in the vertical and horizontal orientations. Results are categorized according to laminar, transitional, and turbulent flow, and according to duct geometry, heating conditions, buoyancy-assisted or buoyancy-opposed flow, and whether or not the flow is hydrodynamically or thermally fully developed. The effects of secondary flow in horizontal ducts are indicated, and the contrasting influences of buoyancy forces on mixed convection in laminar flow and in turbulent flow are discussed.

Chapter 16. This is an up-to-date review of the literature on convective heat transfer in porous media. It emphasizes *scale analysis* as a means of identifying and sorting out the proper heat transfer scales of forced and natural convection through porous media. The engineering heat transfer correlations assembled in this chapter are all scaling-correct correlations, i.e., their analytical forms are the ones recommended by the appropriate scale analysis. In many instances, classical experimental and numerical results are here condensed into scaling-correct correlations.

Chapter 17. This chapter discusses special heat transfer surface geometries that yield higher heat transfer coefficients than "plain" surfaces do. The major emphasis of the chapter is on forced convection of gases and liquids. The enhancement geometries covered include finned surfaces for gases and tube-side enhancement for laminar and turbulent flow of gases. Design equations are provided to calculate the heat transfer coefficient and friction factor for all of the enhancement geometries discussed. Performance Evaluation Criteria are described, which are used to calculate the performance improvement provided by an enhanced surface, relative to that of a plain surface for specific design objectives and operating constraints.

Chapter 18. This chapter deals with the effect of temperature-dependent fluid properties on convective heat transfer. Correlations for heat transfer and friction coefficients which take into account temperature-dependent properties are described for viscous liquids and gases both for laminar and turbulent flow. Many solutions are surveyed which have been proposed to describe these effects as they occur in practical applications. Tabular forms of correlations for heat transfer and friction coefficients for turbulent flow in ducts with variable physical properties are provided. The particular characteristics of fluids at supercritical pressure are described, the role of property variations in influencing flow and heat transfer is discussed, and correlations of supercritical forced convection are presented. The chapter also surveys and summarizes the available solutions and experimental studies on temperature-dependent effects as they occur in natural convection.

Chapter 19. When heat transfer by radiation is of the same order of magnitude as convection, radiation and convection need to be treated simultaneously. In such situations, in the analysis of the problem distinction should be made between cases involving a completely transparent fluid and a fluid that absorbs, emits, and perhaps scatters radiation. In this chapter, radiation transfer in nonparticipating and participating media is discussed for such cases, and typical heat transfer results are presented on the effects of radiation parameters such as the conduction-to-radiation parameter, optical thickness, surface reflectivity, and single-scattering albedo on the Nusselt number and temperature distribution for forced convection over a flat plate and inside a parallel-plate duct.

Chapter 20. The basic definitions of non-Newtonian fluids and their rheological properties are presented. Methods of measuring these rheological properties are described. Analyses are made of the flow and heat transfer characteristics of non-Newtonian flows in both ducts and over external surfaces in laminar and turbulent flows. Both free and forced convection are considered. Emphasis has been placed on the presentation of results in a form suitable for engineering design calculations.

Chapter 21. This chapter is concerned with the fouling of heat transfer surfaces; a reasonable balance between gas-side and liquid-side fouling is maintained throughout. The treatment is design-oriented and includes tabulated values of liquid-side and gas-side fouling factors, along with properties of representative fouling deposits. Detailed procedures are presented for taking into account the effects of fouling on both pressure drop and heat transfer. A considerable amount of material on techniques available to combat fouling is also presented and discussed.

Chapter 22. Tables are given of the thermophysical properties: specific volume, specific enthalpy, specific entropy, specific heat at constant pressure, viscosity, thermal conductivity, and Prandtl number as a function of temperature. For saturated liquid and vapor, these are given for air, carbon dioxide, cesium, lithium, mercury, potassium, Refrigerant 12, Refrigerant 22, rubidium, sodium, and steam. Similar saturated tables for ice-water-steam are given as a function of pressure. Wherever possible, the different tables are presented at equal temperatures and temperature increments to facilitate comparisons in design and also computer programming of the data.

Most of the results of engineering utility are presented in terms of equations, tables, and/or figures. Where appropriate, most results are presented in nondimensional form; and the dimensional results are presented in two unit systems—the International System (SI) and the U.S. Customary System (USCS)—to allow for the worldwide use of this handbook. Although the nomenclature is listed at the end of each chapter, the editors have made a diligent effort to make most of the symbols common throughout the handbook.

The success of this handbook rests with the quality of the information provided by the contributors. We are grateful to them for providing excellent material in a timely manner and within the length limitations. The editorial staff at John Wiley has provided superb cooperation and continued support throughout the compilation of this handbook, from the inception of the idea to the final production. In particular, we sincerely appreciate the cordial and prompt support of Mr. Frank Cerra on all matters of concern. We also gratefully acknowledge the outstanding editorial work of Mr. Joseph Fineman, the efficient work and outstanding cooperation of Lisa VanHorn during the production and the excellent figures prepared by Mr. Ali Akgüneş of the Middle East Technical University, Ankara, Turkey. We also wish to thank the professional staff of John Wiley & Sons, Inc. who were involved with the publication of this Handbook. The first editor would like to express his appreciation to Norman Einspruch, Dean of the College of Engineering at the University of Miami for suggesting the idea of preparing a handbook.

Every effort has been made by the editors to minimize typographical errors. Each chapter has been independently reviewed by other experts in the field to enhance the quality and the correctness. If any errors come to the attention of readers, we would greatly appreciate being informed of them so that they can be eliminated in the subsequent printing. Of course, we would also appreciate any more general comments related to any chapter.

S. Kakaç
R. K. Shah
W. Aung

Coral Gables, Florida
Lockport, New York
Washington, D.C.
June 1987

CONTENTS

1

BASICS OF HEAT TRANSFER

S. Kakaç

University of Miami
Coral Gables, Florida

Y. Yener

Northeastern University
Boston, Massachusetts

1.1 INTRODUCTION

Convective heat transfer is the study of heat transport processes between the layers of a fluid when the fluid is in motion and/or between a fluid in motion and a boundary surface in contact with it when they are at different temperatures.

Heat is that form of energy which crosses the boundary of a thermodynamic system by virtue of a temperature difference existing between the system and its surroundings. *Heat flow* is vectorial in the sense that it is in the direction of negative temperature gradients, i.e., from higher toward lower temperatures.

The science of heat transfer is based upon foundations comprising both theory and experiment. As in other engineering disciplines, the theoretical part is constructed from one or more *physical* (or *natural*) *laws*. The physical laws are statements, in terms of various concepts, which have been found to be true through many years of experimental observations. A physical law is called a *general law* if its application is independent of the medium under consideration. Otherwise, it is called a *particular law*. There are, in fact, the following four general laws among others upon which all the analyses concerning heat transfer, either directly or indirectly, depend:

1. The law of conservation of mass
2. The first law of thermodynamics
3. The second law of thermodynamics
4. Newton's second law of motion

In addition to the general laws, it is usually necessary to bring certain particular laws into an analysis. Examples are Fourier's law of heat conduction, Newton's law of cooling, the Stefan-Boltzmann law of radiation, Newton's law of viscosity, the ideal-gas law, etc.

1.2 MODES OF HEAT TRANSFER

The mechanism by which heat is transferred in a heat exchange or an energy conversion system is, in fact, quite complex. There appear, however, to be three rather basic and distinct *modes* of heat transfer. These are *conduction*, *convection*, and *radiation*.

Conduction is the process of heat transfer by molecular motion, supplemented in some cases by the flow of free electrons, through a body (solid, liquid, or gaseous) from a region of high temperature to a region of low temperature. Heat transfer by conduction also takes place across the interface between two bodies in contact when they are at different temperatures.

The mechanism of heat conduction in liquids and gases has been postulated as the transfer of kinetic energy of the molecular movement. Transfer of thermal energy to a fluid increases its internal energy by increasing the kinetic energy of its vibrating molecules, and is measured by the increase of its temperature. Heat conduction is thus the transfer of kinetic energy of the more energetic molecules in the high-temperature region by successive collisions to the molecules in the low-temperature region.

Heat conduction in solids with crystalline structure, such as quartz, depends on energy transfer by molecular and lattice vibrations and free-electron drift. In general, energy transfer by molecular and lattice vibrations is not as large as the transfer by free electrons. This is the reason why good electrical conductors are always good heat conductors, and electrical insulators are usually good heat insulators. In the case of amorphous solids, such as glass, heat conduction depends only on the molecular transport of energy.

Thermal radiation, or simply *radiation*, is heat transfer in the form of electromagnetic waves. All substances, solid bodies as well as liquids and gases, emit radiation as a result of their temperature, and they are also capable of absorbing such energy. Furthermore, radiation can pass through certain types of substances (called *transparent* and *semitransparent* materials) as well as through vacuum, whereas for heat conduction to take place a material medium is absolutely necessary.

Conduction is the only mechanism by which heat can flow in *opaque* solids. Through certain transparent or semitransparent solids, such as glass and quartz, energy flow can be by radiation as well as by conduction. With gases and liquids, if there is no observable fluid motion, the heat transfer mechanism will be conduction (and radiation). However, if there is macroscopic fluid motion, energy can also be transported in the form of internal energy by the movement of the fluid itself. The process of energy transport by the combined effects of heat conduction (and radiation) and the movement of fluid is referred to as *convection* or *convective heat transfer*.

1.3 STATEMENTS OF GENERAL LAWS

In the following sections, the four general laws referred to in Sec. 1.1 are stated first in terms of a system, and then the control-volume forms are given.

1.3.1 Law of Conservation of Mass

A *system* is any arbitrary collection of matter of fixed identity bounded by a closed surface, which can be a real or an imaginary one. All other systems that interact with the system under consideration are known as its surroundings. The law of conservation of mass simply states that, in the absence of any mass-energy conversion, the mass of a system remains constant. Thus, for a system

$$\frac{dm}{dt} = 0, \qquad \text{or} \quad m = \text{constant} \tag{1.1}$$

where m is the mass of the system.

A *control volume* is any defined region in space, across the boundaries of which matter, energy, and momentum may flow, within which matter, energy, and momentum storage may take place, and on which external forces may act. Its position and/or size may change with time. Consider now a control volume fixed in space and of fixed size and shape, as illustrated in Fig. 1.1. Matter (e.g., a fluid) flows across its boundaries. The law of conservation of mass for this control volume can then be expressed as

$$\frac{\partial m_{\text{c.v.}}}{\partial t} = \dot{m}_{\text{in}} - \dot{m}_{\text{out}} \tag{1.2a}$$

where $m_{\text{c.v.}}$ is the instantaneous mass inside the control volume, and $\dot{m}_{\text{in}}$ and $\dot{m}_{\text{out}}$ are the instaneous mass flow rates into and out of the control volume, respectively. Equation (1.2a) can also be written as [1,2]

$$\frac{\partial}{\partial t}\int_{\text{c.v.}} \rho \, d\mathrm{V} = -\int_{\text{c.s.}} \rho \mathbf{V} \cdot \hat{\mathbf{n}} \, dA \tag{1.2b}$$

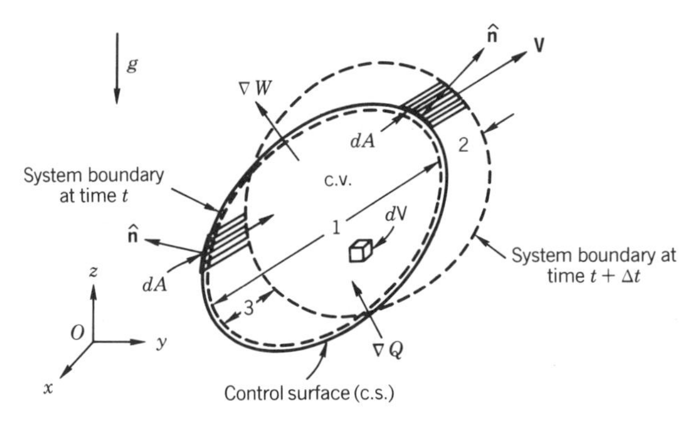

Figure 1.1. Flow through a control volume.

where $d\mathsf{V}$ is an element of the control volume, ρ is the local density of that element, and c.v. designates the control volume fixed in space and bounded by the control surface (c.s.). Finally, $\mathbf{V}$ is the velocity vector, and $\hat{\mathbf{n}}$ is the outward-pointing unit vector normal to the control surface.

The control-volume form of the law of conservation of mass states that the net rate of mass flow into a control volume is equal to the time rate of change of mass within the control volume.

1.3.2 Newton's Second Law of Motion

Newton's second law of motion states that the net force $\mathbf{F}$ acting on a system in an inertial coordinate system is equal to the time rate of change of the total linear momentum $\mathbf{M}$ of the system; that is,

$$\mathbf{F} = \frac{d\mathbf{M}}{dt} \tag{1.3}$$

which, for the control volume shown in Fig. 1.1, reduces to [1, 2]

$$\mathbf{F} = \frac{\partial}{\partial t}\int_{\text{c.v.}} \mathbf{V}\rho \, d\mathsf{V} + \int_{\text{c.s.}} \mathbf{V}\rho\mathbf{V} \cdot \hat{\mathbf{n}} \, dA \tag{1.4}$$

This result is usually called the *momentum theorem* or the *law of conservation of linear momentum* and states that the net force acting instantaneously on a control volume is equal to the time rate of change of linear momentum within the control volume plus the net flow rate of linear momentum out of the control volume.

Equation (1.4) is a vector relation. Referred to the rectangular coordinates x, y, and z, the component in the x direction, for example, can be written as

$$F_x = \frac{\partial}{\partial t}\int_{\text{c.v.}} u\rho \, d\mathsf{V} + \int_{\text{c.s.}} u\rho\mathbf{V} \cdot \hat{\mathbf{n}} \, dA \tag{1.5}$$

where u and F_x are the x components of the velocity vector $\mathbf{V}$ and the force vector $\mathbf{F}$, respectively.

1.3.3 First Law of Thermodynamics

The *thermodynamic state* of a system is its conditions, as described by a list of the values of all its properties. A *property* of a system is either a directly or an indirectly observable characteristic of the system and can, in principle, be quantitatively evaluated. Volume, mass, pressure, temperature, etc., are all properties. If all the properties of a system remain unchanged, then the system is said to be in an equilibrium state. A *process* is a change of state and is described in part by the series of states passed through by the system. A *cycle* is a process wherein the initial and final states of a system are the same.

When a system undergoes a cyclic process, the first law of thermodynamics can be expressed as

$$\oint \delta Q = \oint \delta W \tag{1.6}$$

where the cyclic integral $\oint \delta Q$ represents the net heat transfer *to* the system and the cyclic integral $\oint \delta W$ is the net work done *by* the system during the process. Both heat and work are *path* functions; that is, the amount of heat transferred or the amount of work done when a system undergoes a change of state depends on the path the system follows during the change of state. This is why the differentials of heat and work are inexact differentials, denoted by the symbols δQ and δW. For a process which involves an infinitesimal change of state during a time interval dt, the first law of thermodynamics is given by

$$dE = \delta Q - \delta W \tag{1.7}$$

where δQ and δW are the small amounts of heat added *to* the system and the work done *by* the system, respectively, and dE is the corresponding increase in the total energy of the system during the time internal dt. The *energy* E is a property of the system and, like all other properties, is a *point* function. That is, dE depends upon the initial and final states only, and not on the path followed between the two states. For a more complete discussion of point and path functions, the reader is referred to Refs. 3, 4. The property E represents all energies of a system and is customarily separated into three parts: bulk kinetic energy, bulk potential energy, and internal energy; that is,

$$E = \mathrm{KE} + \mathrm{PE} + \mathscr{U} \tag{1.8}$$

The internal energy $\mathscr{U}$ represents the energy associated with molecular and atomic structure and behavior of a system.

Equation (1.7) can also be written as a rate equation,

$$\frac{dE}{dt} = \frac{\delta Q}{dt} - \frac{\delta W}{dt} \tag{1.9a}$$

or

$$\frac{dE}{dt} = q - \mathsf{P} \tag{1.9b}$$

where $q = \delta Q/dt$ represents the rate of heat transfer to the system and $\mathsf{P} = \delta W/dt$ is the rate of work done (power) by the system.

Consider now the control volume illustrated in Fig. 1.1, and define a system whose boundary at time t happens to correspond exactly to that of the control volume. At some later time $t + \Delta t$ the system moves to another location and occupies a different volume in space. The first law of thermodynamics for this change is

$$\Delta E = \nabla Q - \nabla W \tag{1.10}$$

where ∇Q is the heat transferred to the system and ∇W is the work done by the system, and ΔE is the corresponding increase in the energy of the system during time interval Δt. Dividing Eq. (1.10) by Δt, one obtains

$$\frac{\Delta E}{\Delta t} = \frac{\nabla Q}{\Delta t} - \frac{\nabla W}{\Delta t} \tag{1.11}$$

As $\Delta t \to 0$, the left-hand side of Eq. (1.11) becomes [1,2]

$$\lim_{\Delta t \to 0} \frac{\Delta E}{\Delta t} = \frac{\partial}{\partial t} \int_{\text{c.v.}} e\rho \, d\mathsf{V} + \int_{\text{c.s.}} e\rho \mathbf{V} \cdot \hat{\mathbf{n}} \, dA \tag{1.12}$$

where e represents the energy per unit mass.

The first term on the right-hand side of Eq. (1.11) represents, as $\Delta t \to 0$, the rate of heat transfer across the control surface; that is,

$$\lim_{\Delta t \to 0} \frac{\nabla Q}{\Delta t} = \left(\frac{\delta Q}{dt} \right)_{\text{c.s.}} = q_{\text{c.s.}} \tag{1.13}$$

Similarly, the second term becomes

$$\lim_{\Delta t \to 0} \frac{\nabla W}{\Delta t} = \frac{\delta W}{dt} = \mathsf{P} \tag{1.14}$$

which is the rate of work done (power) by the matter in the control volume (i.e., the system) on its surroundings at any time t. Hence, as $\Delta t \to 0$, Eq. (1.11) becomes

$$\frac{\partial}{\partial t} \int_{\text{c.v.}} e\rho \, d\mathsf{V} + \int_{\text{c.s.}} e\rho \mathbf{V} \cdot \hat{\mathbf{n}} \, dA = q_{\text{c.s.}} - \mathsf{P} \tag{1.15}$$

which is the control-volume form of the first law of thermodynamics. However, a final form of this expression can be obtained after further consideration of the power term P. Work can be done by the system against its surroundings in a variety of ways. In the following discussion, only the work done against normal (hydrostatic pressure) and tangential (shear) stresses, the work done by the system that could cause a shaft to rotate (shaft work), and the power drawn to the system from an external electric circuit are considered. Capillary and magnetic effects will be neglected.

The net rate of work done by the system against normal stresses (pressure) is called *flow work* and can be written as [2,4]

$$\mathsf{P}_{\text{normal}} = \int_{\text{c.s.}} p \mathbf{V} \cdot \mathbf{n} \, dA \tag{1.16}$$

where p is the pressure at the control surface.

Let $\mathsf{P}_{\text{shaft}}$ be the rate at which the system does shaft work, and $\mathsf{P}_{\text{shear}}$ be the rate at which the system does work against shear stresses. The rate of work done on the system due to power drawn from an external electric circuit can be written as $\int_{\text{c.v.}} q_e''' \, d\mathsf{V}$, where q_e''' is the rate of internal energy generation per unit volume due to the power drawn to the system. Hence, Eq. (1.15) reduces to

$$\frac{\partial}{\partial t}\int_{\text{c.v.}} e\rho \, d\mathsf{V} + \int_{\text{c.s.}} \left(e + \frac{p}{\rho}\right)\rho\mathbf{V} \cdot \hat{\mathbf{n}} \, dA$$

$$= q_{\text{c.s.}} - \mathsf{P}_{\text{shaft}} - \mathsf{P}_{\text{shear}} + \int_{\text{c.v.}} q_e''' \, d\mathsf{V} \tag{1.17}$$

The energy per unit mass may be written as

$$e = u + \tfrac{1}{2}V^2 + gz \tag{1.18}$$

where u, $\frac{1}{2}V^2$ and gz are the internal, bulk kinetic and bulk potential energies per unit mass, respectively. Hence, Eq. (1.17) becomes

$$\frac{\partial}{\partial t}\int_{\text{c.v.}} e\rho \, d\mathsf{V} + \int_{\text{c.s.}} \left(i + \tfrac{1}{2}V^2 + gz\right)\rho\mathbf{V} \cdot \hat{\mathbf{n}} \, dA$$

$$= q_{\text{c.s.}} - \mathsf{P}_{\text{shaft}} - \mathsf{P}_{\text{shear}} + \int_{\text{c.v.}} q_e''' \, d\mathsf{V} \tag{1.19}$$

where

$$i = u + \frac{p}{\rho} \tag{1.20}$$

is the *enthalpy* per unit mass. Equation (1.19) is the first law of thermodynamics for the control volume.

1.3.4 Second Law of Thermodynamics

The first law of thermodynamics, which embodies the idea of conservation of energy, gives means for quantitative calculation of changes in the state of a system due to interactions between the system and its surroundings, but tells nothing about the direction that a process might take. On the other hand, observations concerning the unidirectionality of naturally occurring processes have led to the formulation of the second law of thermodynamics, which gives a sense of direction to energy-transfer processes.

The second law of thermodynamics leads to a thermodynamic property—*entropy*. For any *reversible* process that a system undergoes during a time interval dt, the change in the entropy S of the system is given by

$$dS = \left(\frac{\delta Q}{T}\right)_{\text{rev}} \tag{1.21}$$

For an *irreversible* process, the change, however, is

$$dS > \left(\frac{\delta Q}{T}\right)_{\text{irr}} \tag{1.22}$$

TABLE 1.1 Summary of General Laws [1, 5]

Law	For a System	For a Control Volume
Law of conservation of mass	$\frac{dm}{dt} = 0$	$\frac{\partial}{\partial t}\int_{\text{c.v.}} \rho\, dV = -\int_{\text{c.s.}} \rho \mathbf{V} \cdot \hat{\mathbf{n}}\, dA$
Newton's second law of motion	$\mathbf{F} = \frac{d\mathbf{M}}{dt}$	$\mathbf{F} = \frac{\partial}{\partial t}\int_{\text{c.v.}} \rho \mathbf{V}\, dV + \int_{\text{c.s.}} \mathbf{V}\rho\mathbf{V} \cdot \hat{\mathbf{n}}\, dA$
First law of thermodynamics	$\frac{dE}{dt} = q - P$	$\frac{\partial}{\partial t}\int_{\text{c.v.}} e\rho\, dV + \int_{\text{c.s.}} \left(i + \frac{1}{2}V^2 + gz\right)\rho\mathbf{V} \cdot \hat{\mathbf{n}}\, dA = q_{\text{c.s.}} - P_{\text{shaft}} - P_{\text{shear}} + \int_{\text{c.v.}} q_e'''\, dV$
Second law of thermodynamics	$\frac{dS}{dt} \geq \frac{1}{T}\frac{\delta Q}{dt}$	$\frac{\partial}{\partial t}\int_{\text{c.v.}} s\rho\, dV + \int_{\text{c.s.}} s\rho\mathbf{V} \cdot \hat{\mathbf{n}}\, dA \geq \int_{\text{c.s.}} \frac{1}{T}\frac{\delta Q}{dt}$

where δQ is the small amount of heat added to the system during the time interval dt, and T is the temperature of the system at the time of heat transfer. Equations (1.21) and (1.22) may be taken as the mathematical statement of the second law, which can also be written in rate form as

$$\frac{dS}{dt} \geq \frac{1}{T}\frac{\delta Q}{dt} \tag{1.23}$$

The control-volume form of the second law is given by [1, 2]

$$\frac{\partial}{\partial t}\int_{\text{c.v.}} s\rho\, dV + \int_{\text{c.s.}} s\rho\mathbf{V} \cdot \mathbf{n}\, dA \geq \int_{\text{c.s.}} \frac{1}{T}\frac{\delta Q}{dt} \tag{1.24}$$

where s is the entropy per unit mass, and the equality applies to reversible processes and the inequality to irreversible processes.

The efficient performance of systems in industrial applications involving heat transfer processes corresponds to the least generation of entropy; that is, the rate of loss of useful work in a process is directly proportional to the rate of entropy production during that process.

Table 1.1 summarizes the general laws discussed in this section.

1.4 STATEMENTS OF PARTICULAR LAWS

In the following sections the particular laws of heat transfer are reviewed.

1.4.1 Fourier's Law of Heat Conduction

Since the mechanism of heat conduction on the molecular level is thought to be the exchange of kinetic energy between the molecules, the most fundamental approach in analyzing heat conduction in a substance would be to apply the laws of motion to each individual molecule or a statistical group of molecules, subsequent to some initial state of affairs. In most engineering problems, however, primary interest does not lie in the molecular behavior of a substance, but rather in how the substance behaves as a continuous medium. In heat conduction (and, therefore, heat convection) studies, the molecular structure is neglected and the matter is considered to be a continuous medium (*continuum*), which fortunately is a valid approximation to many practical problems where only macroscopic information is of interest. Such a model may be used provided that the size and the mean free path of molecules are small compared with other dimensions existing in the medium. This approach, also known as the *phenomenological* approach to heat transfer problems, is simpler than microscopic approaches and usually provides the answers required in engineering.

Fourier's law of heat conduction, which is the basic law governing heat conduction based on the continuum concept, states that the heat flux (i.e., the rate of heat transfer per unit area) due to conduction in a given direction at a point within a medium (solid, liquid or gaseous) is proportional to the temperature gradient in the same direction at the same point. For heat conduction in any direction n, this law is given by

$$q_n'' = -k\frac{\partial T}{\partial n} \tag{1.25}$$

where q_n'' is the magnitude of the heat flux in the n direction, and $\partial T/\partial n$ is the temperature gradient in the same direction. Here k is a proportionality constant known as the *thermal conductivity* of the material of the medium under consideration, and is a positive quantity. The minus sign is included so that heat flow is positive in the direction of a negative temperature gradient.

Thermal conductivity is a thermophysical property and has the units $W/(m \cdot K)$ in the SI system. A medium is said to be *homogeneous* if its thermal conductivity does not vary from point to point within the medium, and *heterogeneous* if there is such a variation. Further, a medium is said to be *isotropic* if its thermal conductivity is the same in all directions, and *anisotropic* if there exists directional variation in thermal conductivity.

In anisotropic media the heat flux due to conduction in a given direction may also be proportional to the temperature gradients in other directions, and therefore Eq. (1.25) may not be valid [6].

In an isotropic medium, there is an equation like Eq. (1.25) in each coordinate direction. For example, in rectangular coordinates, the heat-flux relations can be written as

$$q_x'' = -k\frac{\partial T}{\partial x}, \qquad q_y'' = -k\frac{\partial T}{\partial y} \quad \text{and} \quad q_z'' = -k\frac{\partial T}{\partial z} \tag{1.26a, b, c}$$

These are, in fact, the three components in the x, y, and z directions of the *heat flux vector*

$$\mathbf{q}'' = -k\nabla T \tag{1.27}$$

which is the vector form of Fourier's law in isotropic media.

1.4.2 Newton's Law of Cooling

Convection has already been defined in Sec. 1.2 as the process of heat transport in a fluid by the combined action of heat conduction (and radiation) and fluid motion. As a mechanism of heat transfer it is important not only between the layers of a fluid but also between a fluid and a solid surface when they are in contact.

When a fluid flows over a solid surface as illustrated in Fig. 1.2, it is an experimental observation that the fluid particles adjacent to the surface stick to it and therefore have

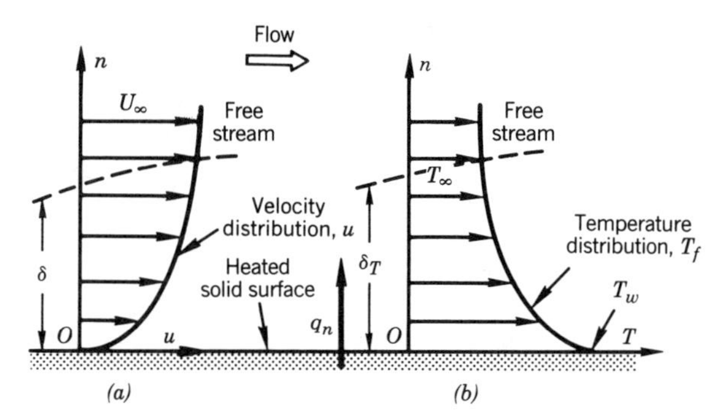

Figure 1.2. Velocity and thermal boundary layers along a solid surface.

zero velocity relative to the surface. Other fluid particles attempting to slide over the stationary ones at the surface are retarded as a result of viscous forces between the fluid particles. The velocity of the fluid particles thus asymptotically approaches that of the undisturbed free stream over a distance δ (*velocity boundary-layer thickness*) from the surface, with the resulting velocity distribution shown in Fig. 1.2.

As illustrated in Fig. 1.2, if $T_w > T_\infty$, then heat will flow from the solid to the fluid particles at the surface. The energy thus transmitted increases the internal energy of the fluid particles (*sensible* heat storage) and is carried away by the motion of the fluid. The temperature distribution in the fluid adjacent to the surface will then appear as shown in Fig. 1.2, asymptotically approaching the free-stream value T_∞ in a short distance δ_T (*thermal boundary-layer thickness*) from the surface.

Since the fluid particles at the surface are stationary, the heat flux from the surface to the fluid will be

$$q_n'' = -k_f\left(\frac{\partial T_f}{\partial n}\right)_w \tag{1.28}$$

where k_f is the thermal conductivity of the fluid, T_f is the temperature distribution in the fluid, the subscript w means the derivative is evaluated at the surface, and n denotes the normal direction from the surface.

In 1701, Newton expressed the heat flux from a solid surface to a fluid by the equation

$$q_n'' = h(T_w - T_\infty) \tag{1.29}$$

where h is called *heat transfer coefficient*, *film conductance* or *film coefficient*. In the literature, Eq. (1.29) is known as *Newton's law of cooling*. In fact, it is not a law, but the defining equation for the heat transfer coefficient; i.e.,

$$h \equiv \frac{q_n''}{T_w - T_\infty} = \frac{-k_f(\partial T_f/\partial n)_w}{T_w - T_\infty} \tag{1.30}$$

The heat transfer coefficient has the units $W/(m^2 \cdot K)$ in the SI system. It should be noted that h is also given by

$$h = \frac{-k_s(\partial T_s/\partial n)_w}{T_w - T_\infty} \tag{1.31}$$

TABLE 1.2 Approximate Values of the Heat Transfer Coefficient *h*

	h, W/(m$^2 \cdot$ K)	
Fluid	Free Convection	Forced Convection
Gases	5–30	30–300
Water	30–300	300–10,000
Viscous oils	5–100	30–3,000
Liquid metals	50–500	500–20,000
Boiling water	2,000–20,000	3,000–100,000
Condensing water vapor	3,000–30,000	3,000–200,000

where k_s is the thermal conductivity of the solid and T_s is the temperature distribution in the solid.

If the fluid motion involved in the process is induced by some external means such as a pump, blower, or fan, then the process is referred to as *forced convection*. If the fluid motion is caused by any body force within the system, such as those resulting from the density gradients near the surface, then the process is called *natural* (or *free*) *convection*.

Certain convective heat transfer processes, in addition to sensible heat storage, may also involve *latent* heat storage (or release) due to a phase change. Boiling and condensation are two such cases.

The heat transfer coefficient is actually a complicated function of the flow conditions, thermophysical properties (viscosity, thermal conductivity, specific heat, density) of the fluid, and geometry and dimensions of the surface. Its numerical value, in general, is not uniform over the surface. Table 1.2 gives the order of magnitude of the range of values of the heat transfer coefficient under various conditions.

1.4.3 Stefan-Boltzmann Law of Radiation

As mentioned in Sec. 1.2, all substances emit energy in the form of electromagnetic waves (i.e., thermal radiation) as a result of their temperature, and are also capable of absorbing such energy. When thermal radiation is incident on a body, part of it is reflected by the surface as illustrated in Fig. 1.3. The remainder may be absorbed as it travels through the body. If the material of the body is a strong absorber of thermal radiation, the energy that penetrates into the body will be absorbed and converted into internal energy within a very thin layer adjacent to the surface. Such a body is called *opaque*. If the thickness of the material required to substantially absorb radiation is large compared to the thickness of the body, then most of the radiation will be transmitted through the body. Such a body is called *transparent*.

When radiation impinges on a surface, the fraction that is reflected is defined as the *reflectivity* $\tilde{\rho}$, the fraction absorbed as the *absorptivity* $\tilde{\alpha}$, and the fraction transmitted as the *transmissivity* $\tilde{\tau}$. Thus,

$$\tilde{\rho} + \tilde{\alpha} + \tilde{\tau} = 1 \tag{1.32}$$

For opaque substances, $\tilde{\tau} = 0$, and therefore Eq. (1.32) reduces to

$$\tilde{\rho} + \tilde{\alpha} = 1 \tag{1.33}$$

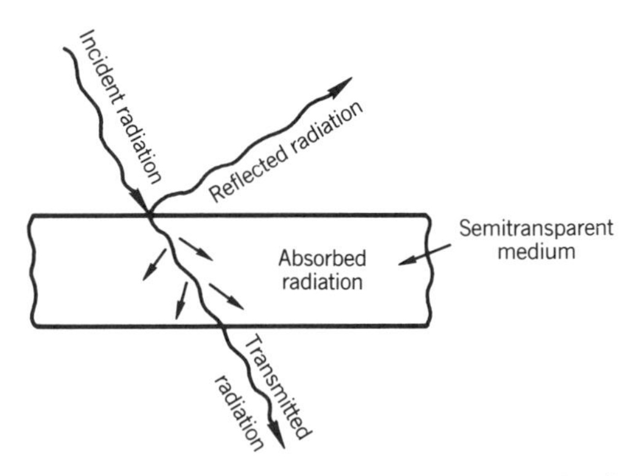

Figure 1.3. Absorption, reflection, and transmission of radiation.

An ideal body which absorbs all the impinging radiation energy without reflection and transmission is called a *blackbody*. Therefore, for a blackbody Eq. (1.32) reduces to $\tilde{\alpha} = 1$. Only a few materials, such as carbon black and platinum black, approach the blackbody in their ability to absorb radiation energy. A blackbody also emits the maximum possible amount of thermal radiation [7]. The total emission of radiation per unit surface area per unit time from a blackbody is related to the fourth power of the absolute temperature T of the surface by the *Stefan–Boltzmann law of radiation*, which is

$$q''_{r,b} = \tilde{\sigma} T^4 \tag{1.34}$$

where $\tilde{\sigma}$ is the *Stefan–Boltzmann constant* with the value 5.6697×10^{-8} W/(m$^2 \cdot$ K^4) in the SI system.

Real bodies (surfaces) do not meet the specifications of a blackbody, but emit radiation at a lower rate than a blackbody of the same size and shape and at the same temperature. If q'' is the radiative flux (i.e., radiation emitted per unit surface area per unit time) from a real surface at the absolute temperature T, then the *emissivity* of the surface is defined as

$$\epsilon = \frac{q''_r}{\tilde{\sigma} T^4} \tag{1.35}$$

Thus, for a blackbody $\epsilon = 1$. For a real body exchanging radiation only with other bodies at the same temperature (i.e., for thermal equilibrium) it can be shown that $\tilde{\alpha} = \epsilon$, which is a statement of Kirchhoff's law in thermal radiation [7]. The magnitude of emissivity depends upon the material, its state, and the surface conditions.

If two isothermal surfaces A_1 and A_2, having emissivities ϵ_1 and ϵ_2 and absolute temperatures T_1 and T_2, respectively, exchange heat by radiation only, then the net rate of heat exchange between these two surfaces is given by

$$q_r = \tilde{\sigma} A_1 \mathscr{F}_{12} \left(T_1^4 - T_2^4 \right) \tag{1.36}$$

where Kirchhoff's law is assumed to be valid. If A_1 and A_2 are two large parallel surfaces with negligible losses from the edges as shown in Fig. 1.4a, then the factor $\mathscr{F}_{12}$

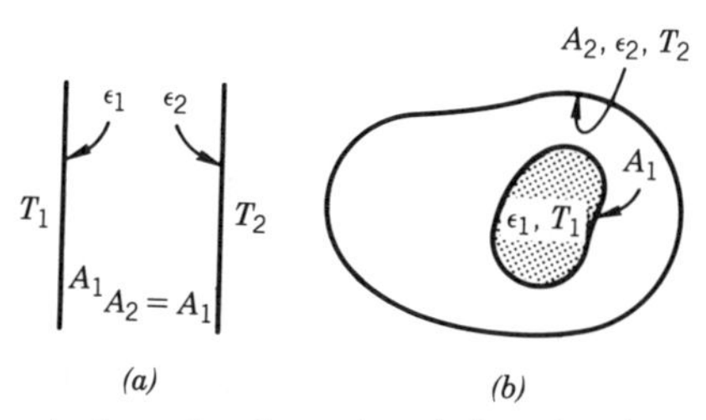

Figure 1.4. Two isothermal surfaces A_1 and A_2 exchanging radiation energy.

in Eq. (1.36) is given by

$$\frac{1}{\mathscr{F}_{12}} = \frac{1}{\epsilon_1} + \frac{1}{\epsilon_2} - 1 \tag{1.37}$$

If A_1 is completely enclosed by the surface A_2 as shown in Fig. 1.4*b*, then

$$\frac{1}{\mathscr{F}_{12}} = \frac{1}{F_{12}} + \frac{1}{\epsilon_1} - 1 + \frac{A_1}{A_2}\left(\frac{1}{\epsilon_2} - 1\right) \tag{1.38}$$

where F_{12} is a purely geometric factor called *radiation shape factor* or *configuration factor* between the surfaces A_1 and A_2, and is equal to the fraction of the radiation leaving surface A_1 that directly reaches surface A_2. Radiation shape factors are given in the form of equations and charts for several configurations in the literature [7, 8]. For surfaces A_1 and A_2 it is clear that

$$\sum_{j=1}^{2} F_{ij} = 1, \qquad i = 1,2 \tag{1.39}$$

Obviously, if the surface A_1 is a completely convex or a plane surface, then $F_{11} = 0$.

In certain applications it may be convenient to define a radiation heat transfer coefficient h_r as

$$q_r = h_r A_1 (T_1 - T_2) \tag{1.40}$$

When this concept is applied to Eq. (1.36), h_r becomes

$$h_r = \tilde{\sigma}\mathscr{F}_{12}(T_1 + T_2)(T_1^2 + T_2^2) \tag{1.41}$$

The particular laws of heat transfer are summarized in Table 1.3.

TABLE 1.3 Summary of the Particular Laws of Heat Transfer

Mode	Mechanism	Particular Law
Conduction	Diffusion of thermal energy	$q_n'' = -k\,\partial T/\partial n$
Convection	Diffusion and transport of thermal energy	$q'' = h(T_w - T_\infty)$
Radiation	Heat transfer by electromagnetic waves	$q_r'' = h_r(T_1 - T_2)$

1.5 GOVERNING EQUATIONS OF CONVECTIVE HEAT TRANSFER

In this section, the governing equations of convective heat transfer will be given for fluids which behave as a continuum. Ordinary fluids, such as air, water, and oils, behave as a continuum at atmospheric pressures and temperatures, and also exhibit a linear relation between the applied shear stress and the rate of strain. Such fluids are called *Newtonian* fluids. The expression relating the shear stress to the rate of strain (velocity gradient) for a Newtonian fluid in *simple shear flow*, where only one velocity component is different from zero, is given by

$$\tau = \mu \frac{du}{dy} \tag{1.42}$$

where the proportionality constant μ is called the *dynamic viscosity* or, more simply, the *viscosity* of the fluid. The viscosity is constant for each Newtonian fluid at a given temperature and pressure. For non-Newtonian fluids, the viscosity, at a given pressure and temperature, is also a function of the velocity gradient. Colloidal suspensions and emulsions are examples of non-Newtonian fluids. The above relation (1.42) is a particular law in fluid mechanics and also known as *Newton's law of viscosity*. For two- and three-dimensional flows the expressions relating the stresses to the strain rates are more complicated and are introduced later in Section 1.5.2.

The main objective of convective heat transfer studies is to determine the temperature distribution in a fluid, so that heat fluxes between the fluid and solid boundaries in contact with it can be calculated. Although it is desirable that such calculations should be possible for any boundary, initial and inlet conditions, there are certain mathematical difficulties in finding the temperature distribution in a fluid. In a given flow field, using Newton's second law of motion, together with the law of conservation of mass and the first law of thermodynamics, one can set up a system of five simultaneous partial differential equations for the three velocity components (e.g., u, v, and w in the x, y, and z directions in rectangular coordinates), the pressure p, and the temperature T. If the density of the fluid changes with pressure and temperature, that is, if the fluid is compressible, then a sixth equation has to be introduced to relate density to temperature and pressure, such as the equation of state for a perfect gas. Finally, if there are large temperature differences within the fluid and between the fluid and the bounding surfaces, then additional information for the variation of other thermophysical properties with temperature is also required.

1.5.1 Continuity Equation

Consider the flow of a single-phase and single-component fluid (invariant in composition), and define an elemental control volume with dimensions Δx, Δy, and Δz at a location (x, y, z) in the flow field as shown in Fig. 1.5. Let $\mathbf{V}$ $(= u\hat{\mathbf{i}} + v\hat{\mathbf{j}} + w\hat{\mathbf{k}})$ be the velocity vector at (x, y, z). In Fig. 1.5, the mass flow rates entering and leaving the control volume are also indicated. Hence, the net rate of mass entering the control volume is given by

$$-\left[\frac{\partial(\rho u)}{\partial x} + \frac{\partial(\rho v)}{\partial y} + \frac{\partial(\rho w)}{\partial z}\right] \Delta x \, \Delta y \, \Delta z \tag{1.43}$$

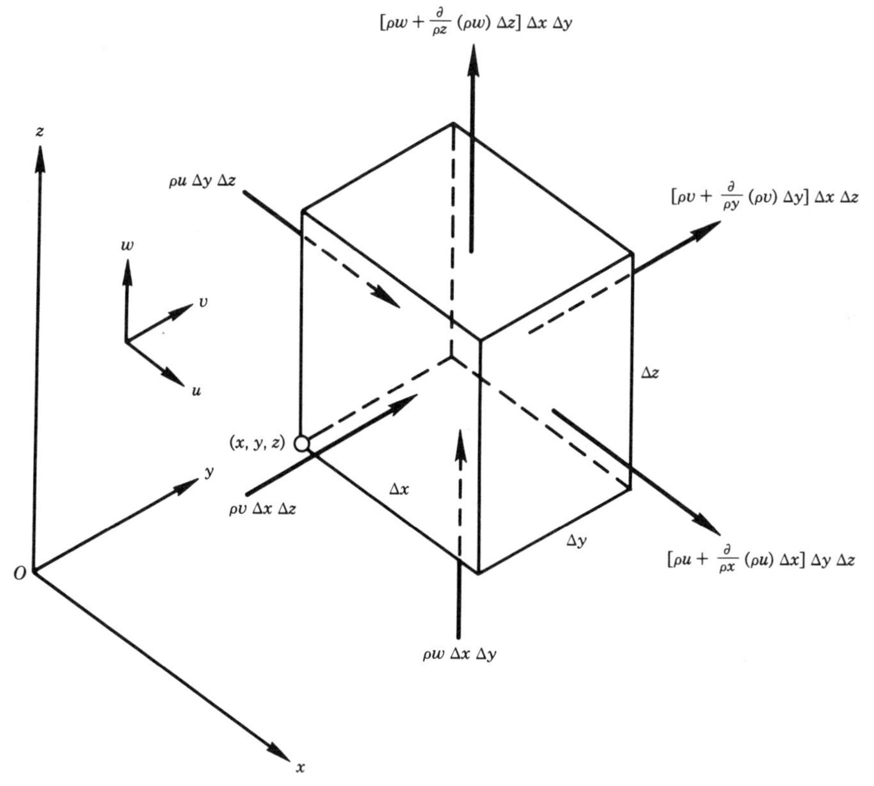

Figure 1.5. Elemental control volume in a flow field for the derivation of the continuity equation. The mass flow rates in and out of the control volume in the x, y, and z directions are also indicated.

The rate of increase of mass within the control volume is

$$\frac{\partial \rho}{\partial t} \Delta x \Delta y \Delta z \tag{1.44}$$

The law of conservation of mass, Eq. (1.2b), therefore leads to

$$\frac{\partial \rho}{\partial t} + \frac{\partial(\rho u)}{\partial x} + \frac{\partial(\rho v)}{\partial y} + \frac{\partial(\rho w)}{\partial z} = 0 \tag{1.45}$$

which is called the *continuity equation* and is the mathematical expression of the law of conservation of mass for an elemental control volume within a fluid flow field. The continuity equation (1.45) can also be written as

$$\frac{\partial \rho}{\partial t} + \nabla \cdot (\rho \mathbf{V}) = 0 \tag{1.46}$$

where

$$\nabla \cdot (\rho \mathbf{V}) = \frac{\partial(\rho u)}{\partial x} + \frac{\partial(\rho v)}{\partial y} + \frac{\partial(\rho w)}{\partial z} \tag{1.47}$$

or it can be rearranged as

$$\frac{D\rho}{Dt} + \rho\nabla \cdot \mathbf{V} = 0 \tag{1.48}$$

where

$$\frac{D}{Dt} \equiv \frac{\partial}{\partial t} + \mathbf{V} \cdot \nabla = \frac{\partial}{\partial t} + u\frac{\partial}{\partial x} + v\frac{\partial}{\partial y} + w\frac{\partial}{\partial z} \tag{1.49}$$

The differential operator of Eq. (1.49) is often called the *substantial derivative*, or sometimes the *derivative following the motion of the fluid*.

For *steady* flows, $\partial\rho/\partial t = 0$, and the continuity equation (1.46) reduces to

$$\nabla \cdot (\rho\mathbf{V}) = 0 \tag{1.50}$$

For *incompressible* fluids, ρ = constant, and Eq. (1.46) becomes

$$\nabla \cdot \mathbf{V} = 0 \tag{1.51a}$$

or

$$\frac{\partial u}{\partial x} + \frac{\partial v}{\partial y} + \frac{\partial w}{\partial z} = 0 \tag{1.51b}$$

which is valid for steady as well as unsteady flows.

The continuity equation in any other coordinate system can be derived similarly, or it can be obtained from Eq. (1.45) for rectangular coordinates by coordinate transformation. In Table 1.4, the continuity equation is tabulated in rectangular, cylindrical

TABLE 1.4 The Continuity Equation in Several Coordinate Systems

General	Compressible	$\frac{\partial\rho}{\partial t} + \nabla \cdot (\rho\mathbf{V}) = 0$
	Incompressible	$\nabla \cdot \mathbf{V} = 0$
Rectangular coordinates (x, y, z)	Compressible	$\frac{\partial\rho}{\partial t} + \frac{\partial}{\partial x}(\rho u) + \frac{\partial}{\partial y}(\rho v) + \frac{\partial}{\partial z}(\rho w) = 0$
	Incompressible	$\frac{\partial u}{\partial x} + \frac{\partial v}{\partial y} + \frac{\partial w}{\partial z} = 0$
Cylindrical coordinates (r, θ, z)	Compressible	$\frac{\partial\rho}{\partial t} + \frac{1}{r}\frac{\partial}{\partial r}(\rho r v_r) + \frac{1}{r}\frac{\partial}{\partial \theta}(\rho v_\theta) + \frac{\partial}{\partial z}(\rho w) = 0$
	Incompressible	$\frac{\partial v_r}{\partial r} + \frac{v_r}{r} + \frac{1}{r}\frac{\partial v_\theta}{\partial \theta} + \frac{\partial w}{\partial z} = 0$
Spherical coordinates (r, θ, ϕ)	Compressible	$\frac{\partial\rho}{\partial t} + \frac{1}{r^2}\frac{\partial}{\partial r}(\rho r^2 v_r) + \frac{1}{r\sin\theta}\frac{\partial}{\partial \theta}(\rho v_\theta \sin\theta) + \frac{1}{r\sin\theta}\frac{\partial}{\partial \phi}(\rho v_\phi) = 0$
	Incompressible	$\frac{1}{r}\frac{\partial}{\partial r}(r^2 v_r) + \frac{1}{\sin\theta}\frac{\partial}{\partial \theta}(v_\theta \sin\theta) + \frac{1}{\sin\theta}\frac{\partial v_\phi}{\partial \phi} = 0$

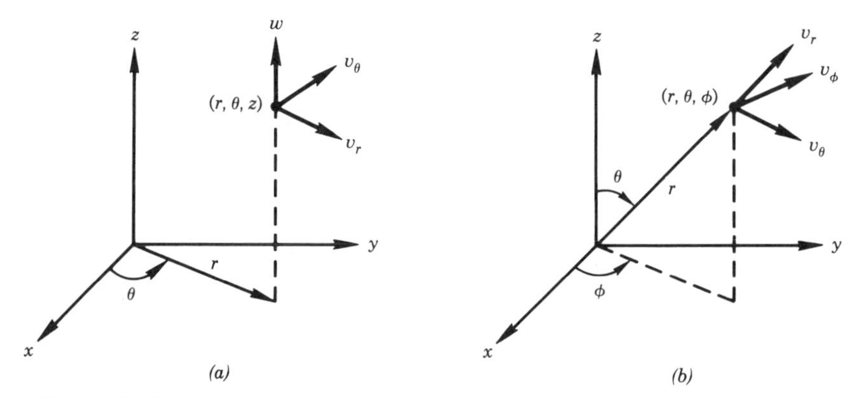

Figure 1.6. Velocity components in (a) cylindrical and (b) spherical coordinate systems.

and spherical coordinates. In addition, the velocity components in cylindrical and spherical coordinates are shown in Fig. 1.6.

1.5.2 Equations of Motion

The dynamic behavior of fluid motion is governed by a set of equations called the *momentum equations* or the *equations of motion*. These equations are obtained by applying either Newton's second law of motion (1.3) to an elemental fluid particle or the law of conservation of linear momentum, Eq. (1.4), to an elemental control volume in the flow field. In distinction to the approach followed in the derivation of the continuity equation, where an elemental volume element was used, consider here an elemental fluid particle and follow its motion. Newton's second law of motion, Eq. (1.3), for a fluid particle of mass m may be rewritten as

$$\mathbf{F} = \frac{d\mathbf{M}}{dt} = \frac{d(m\mathbf{V})}{dt} = m\mathbf{a} \tag{1.52}$$

where $\mathbf{F}$ is the net force acting on the fluid particle, and $\mathbf{a}$ is its acceleration.

A fluid particle situated at (x, y, z) at any time t will be at the new location $(x + \Delta x, y + \Delta y, z + \Delta z)$ at $t + \Delta t$. The total change in the velocity of the particle can be written as

$$\Delta\mathbf{V} = \frac{\partial\mathbf{V}}{\partial t}\Delta t + \frac{\partial\mathbf{V}}{\partial x}\Delta x + \frac{\partial\mathbf{V}}{\partial y}\Delta y + \frac{\partial\mathbf{V}}{\partial z}\Delta z \tag{1.53}$$

The acceleration of the particle situated at (x, y, z) at the instant t, therefore, becomes

$$\mathbf{a} = \lim_{\Delta t \to 0}\frac{\Delta\mathbf{V}}{\Delta t} = \frac{\partial\mathbf{V}}{\partial t} + u\frac{\partial\mathbf{V}}{\partial x} + v\frac{\partial\mathbf{V}}{\partial y} + w\frac{\partial\mathbf{V}}{\partial z} = \frac{D\mathbf{V}}{Dt} \tag{1.54}$$

The forces acting on a fluid particle can be of two types; namely, *body forces* such as forces of gravitational, electrical or magnetic origin, and *surface* (*contact*) forces. Let $\mathbf{f} = f_x\hat{\mathbf{i}} + f_y\hat{\mathbf{j}} + f_z\hat{\mathbf{k}}$ be the body force per unit mass acting on the fluid particle at (x, y, z). In addition, denote the surface stresses (surface forces per unit area) which lie

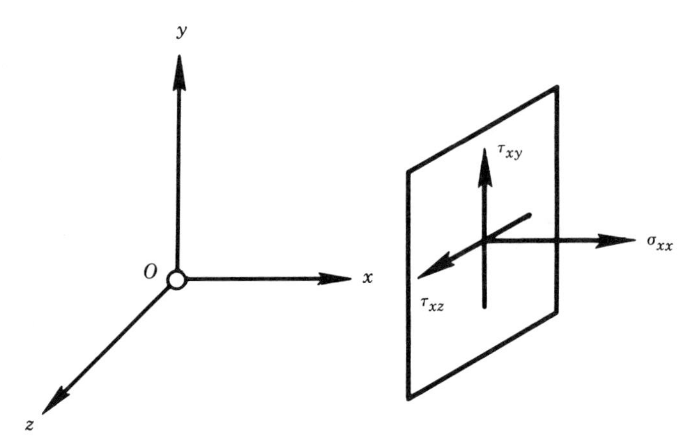

Figure 1.7. Normal and shear stresses acting on a surface element whose normal is in the x direction.

in the plane of the surface by the symbol τ (shear stress), and the stresses normal to the plane of the surface by σ (normal stress). Two subscripts are attached to each of the stress symbols: the first indicates the direction of the normal to the surface on which the stress acts, and the second indicates the direction in which the stress acts. The normal and shear stresses on a surface are reported in terms of a right-handed coordinate system in which the outwardly directed surface normal indicates the positive direction as illustrated in Fig. 1.7.

The state of stress at a point within a fluid is determined when each element of the following stress tensor is known:

$$\begin{bmatrix} \sigma_{xx} & \tau_{xy} & \tau_{xz} \\ \tau_{yx} & \sigma_{yy} & \tau_{yz} \\ \tau_{zx} & \tau_{zy} & \sigma_{zz} \end{bmatrix}$$

Consider now an elemental fluid particle situated at the location (x, y, z) at time t in a flow field as shown in Fig. 1.8. The acceleration of this particle in the x direction is the x component of the acceleration vector $\mathbf{a}$ given by Eq. (1.54), that is,

$$\frac{Du}{Dt} = \frac{\partial u}{\partial t} + u\frac{\partial u}{\partial x} + v\frac{\partial u}{\partial y} + w\frac{\partial u}{\partial z} \tag{1.55}$$

Referring to Fig. 1.8, the net force acting on this fluid particle in the x direction is

$$F_x = \left(\rho f_x + \frac{\partial \sigma_{xx}}{\partial x} + \frac{\partial \tau_{yx}}{\partial y} + \frac{\partial \tau_{zx}}{\partial z}\right) \Delta x \, \Delta y \Delta z \tag{1.56}$$

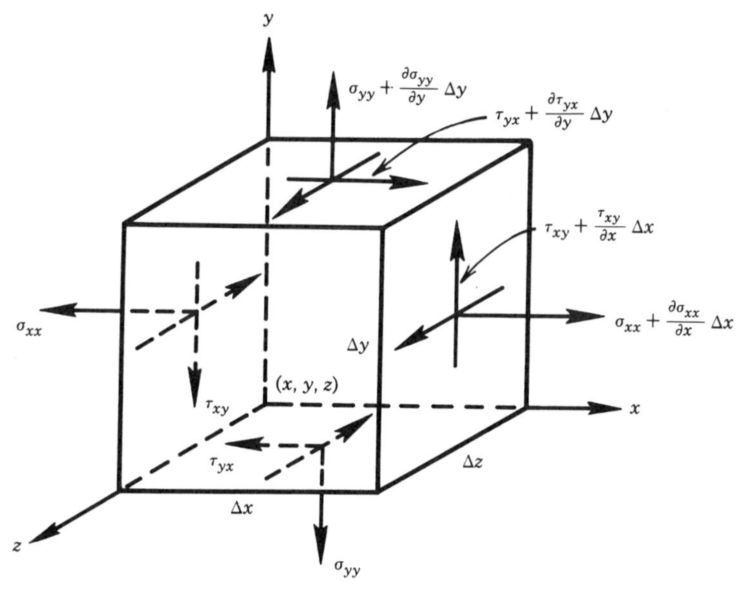

Figure 1.8. Normal and shear stresses acting on an element of fluid. Only the stresses on the surfaces with normals in the x and y directions (both positive and negative) are shown.

Thus, the x component of Newton's second law of motion (1.52) gives

$$\rho \frac{Du}{Dt} = \rho f_x + \frac{\partial \sigma_{xx}}{\partial x} + \frac{\partial \tau_{yx}}{\partial y} + \frac{\partial \tau_{zx}}{\partial z} \tag{1.57a}$$

Similar considerations in the y and z directions result in

$$\rho \frac{Dv}{Dt} = \rho f_y + \frac{\partial \tau_{xy}}{\partial x} + \frac{\partial \sigma_{yy}}{\partial y} + \frac{\partial \tau_{zy}}{\partial z} \tag{1.57b}$$

and

$$\rho \frac{Dw}{Dt} = \rho f_z + \frac{\partial \tau_{xz}}{\partial x} + \frac{\partial \tau_{yz}}{\partial y} + \frac{\partial \sigma_{zz}}{\partial z} \tag{1.57c}$$

Equations (1.57a), (1.57b) and (1.57c) are called the momentum equations or the *equations of motion*. In order to make use of them, the relations between the stresses and the deformation of the fluid particle must be known. It has been found experimentally that, to a high degree of accuracy, stresses in many fluids are related linearly to rates of strain (derivatives of the velocity components). It can be shown [9, 10] that for

Newtonian fluids the expressions are

$$\sigma_{xx} = -p + 2\mu\frac{\partial u}{\partial x} - \tfrac{2}{3}\mu\nabla\cdot\mathbf{V} \tag{1.58a}$$

$$\sigma_{yy} = -p + 2\mu\frac{\partial v}{\partial y} - \tfrac{2}{3}\mu\nabla\cdot\mathbf{V} \tag{1.58b}$$

$$\sigma_{zz} = -p + 2\mu\frac{\partial w}{\partial z} - \tfrac{2}{3}\mu\nabla\cdot\mathbf{V} \tag{1.58c}$$

$$\tau_{xy} = \tau_{yx} = \mu\left(\frac{\partial v}{\partial x} + \frac{\partial u}{\partial y}\right) \tag{1.58d}$$

$$\tau_{xz} = \tau_{zx} = \mu\left(\frac{\partial w}{\partial x} + \frac{\partial u}{\partial z}\right) \tag{1.58e}$$

$$\tau_{yz} = \tau_{zy} = \mu\left(\frac{\partial w}{\partial y} + \frac{\partial v}{\partial z}\right) \tag{1.58f}$$

where μ is the dynamic viscosity of the fluid and p the pressure.

Substitution of Eqs. (1.58a) through (1.58f) into the momentum equations (1.57a), (1.57b), and (1.57c) gives

$$\rho\frac{Du}{Dt} = \rho f_x - \frac{\partial p}{\partial x} + \frac{\partial}{\partial x}\left[\mu\left(2\frac{\partial u}{\partial x} - \tfrac{2}{3}\nabla\cdot\mathbf{V}\right)\right] + \frac{\partial}{\partial y}\left[\mu\left(\frac{\partial v}{\partial x} + \frac{\partial u}{\partial y}\right)\right] + \frac{\partial}{\partial z}\left[\mu\left(\frac{\partial w}{\partial x} + \frac{\partial u}{\partial z}\right)\right] \tag{1.59a}$$

$$\rho\frac{Dv}{Dt} = \rho f_y - \frac{\partial p}{\partial y} + \frac{\partial}{\partial x}\left[\mu\left(\frac{\partial v}{\partial x} + \frac{\partial u}{\partial y}\right)\right] + \frac{\partial}{\partial y}\left[\mu\left(2\frac{\partial v}{\partial y} - \tfrac{2}{3}\nabla\cdot\mathbf{V}\right)\right] + \frac{\partial}{\partial z}\left[\mu\left(\frac{\partial w}{\partial y} + \frac{\partial v}{\partial z}\right)\right] \tag{1.59b}$$

$$\rho\frac{Dw}{Dt} = \rho f_z - \frac{\partial p}{\partial z} + \frac{\partial}{\partial x}\left[\mu\left(\frac{\partial w}{\partial x} + \frac{\partial u}{\partial z}\right)\right] + \frac{\partial}{\partial y}\left[\mu\left(\frac{\partial w}{\partial y} + \frac{\partial v}{\partial z}\right)\right] + \frac{\partial}{\partial z}\left[\mu\left(2\frac{\partial w}{\partial z} - \tfrac{2}{3}\nabla\cdot\mathbf{V}\right)\right] \tag{1.59c}$$

These relations are the famous Navier-Stokes equations, and nearly all analytical investigations involving viscous fluids are based on them. They are general in the sense that they are valid for compressible Newtonian fluids with varying viscosity.

When the density and viscosity are constant—that is, when the fluid is incompressible and the temperature variations are small—the Navier-Stokes equations (1.59) simplify to

$$\rho \frac{Du}{Dt} = \rho f_x - \frac{\partial p}{\partial x} + \mu \nabla^2 u \qquad (1.60a)$$

$$\rho \frac{Dv}{Dt} = \rho f_y - \frac{\partial p}{\partial y} + \mu \nabla^2 v \qquad (1.60b)$$

$$\rho \frac{Dw}{Dt} = \rho f_z - \frac{\partial p}{\partial z} + \mu \nabla^2 w \qquad (1.60c)$$

These equations may conveniently be summarized in vector notation as

$$\frac{D\mathbf{V}}{Dt} = \mathbf{f} - \frac{1}{\rho} \nabla p + \nu \nabla^2 \mathbf{V} \qquad (1.61)$$

where $\nu = \mu/\rho$ is the *kinematic viscosity* of the fluid, **f** is the body force vector per unit mass, and ∇^2 is the Laplacian operator given by

$$\nabla^2 = \nabla \cdot \nabla = \frac{\partial^2}{\partial x^2} + \frac{\partial^2}{\partial y^2} + \frac{\partial^2}{\partial z^2} \qquad (1.62)$$

Rectangular coordinates may not always be the most useful coordinate system. In problems involving flows through circular tubes, for example, cylindrical coordinates are the most convenient. Similarly, for problems with flows around spheres the use of spherical coordinates is more appropriate. The Navier-Stokes equations in cylindrical and spherical coordinates, however, may be obtained from the above results by coordinate transformation, which is a straightforward but tedious procedure. In Table 5.1, the Navier-Stokes equations for an incompressible fluid with constant viscosity are tabulated in rectangular as well as in cylindrical and spherical coordinates.

Any fluid flow problem which involves the determination of the velocity components and the pressure distribution as a function of space coordinates and time requires the simultaneous solution of the continuity and the Navier-Stokes equations under suitable boundary and initial conditions. Although these equations are, in general, too complicated to be solved analytically, they may be solved by numerical methods. In many cases, however, the nature of the flow is such that they can be simplified considerably for an analytical solution.

1.5.3 Energy Equation

Consider a fluid element of mass $\rho \Delta x \Delta y \Delta z$ situated at the location (x, y, z) at time t in a flow field as shown in Fig. 1.8. The first law of thermodynamics, Eq. (1.9b), when applied to this fluid element, states that the net rate of heat transfer to the element minus the net rate of work done by the element must be equal to the rate of increase of energy of the element. The net rate of heat transfer to the element (ignoring radiation effects) is given by

$$\left[\frac{\partial}{\partial x}\left(k\frac{\partial T}{\partial x}\right) + \frac{\partial}{\partial y}\left(k\frac{\partial T}{\partial y}\right) + \frac{\partial}{\partial z}\left(k\frac{\partial T}{\partial z}\right)\right] \Delta x \Delta y \Delta z \qquad (1.63a)$$

TABLE 1.5 The Equations of Motion for an Incompressible Newtonian Fluid with Constant Viscosity in Several Coordinate Systems

RECTANGULAR COORDINATES (x, y, z)

x component
$$\frac{\partial u}{\partial t} + u\frac{\partial u}{\partial x} + v\frac{\partial u}{\partial y} + w\frac{\partial u}{\partial z} = f_x - \frac{1}{\rho}\frac{\partial p}{\partial x} + \nu\left(\frac{\partial^2 u}{\partial x^2} + \frac{\partial^2 u}{\partial y^2} + \frac{\partial^2 u}{\partial z^2}\right)$$

y component
$$\frac{\partial v}{\partial t} + u\frac{\partial v}{\partial x} + v\frac{\partial v}{\partial y} + w\frac{\partial v}{\partial z} = f_y - \frac{1}{\rho}\frac{\partial p}{\partial y} + \nu\left(\frac{\partial^2 v}{\partial x^2} + \frac{\partial^2 v}{\partial y^2} + \frac{\partial^2 v}{\partial z^2}\right)$$

z component
$$\frac{\partial w}{\partial t} + u\frac{\partial w}{\partial x} + v\frac{\partial w}{\partial y} + w\frac{\partial w}{\partial z} = f_z - \frac{1}{\rho}\frac{\partial p}{\partial z} + \nu\left(\frac{\partial^2 w}{\partial x^2} + \frac{\partial^2 w}{\partial y^2} + \frac{\partial^2 w}{\partial z^2}\right)$$

CYLINDRICAL COORDINATES (r, θ, z)

r component
$$\frac{\partial v_r}{\partial t} + v_r\frac{\partial v_r}{\partial r} + \frac{v_\theta}{r}\frac{\partial v_r}{\partial \theta} - \frac{v_\theta^2}{r} + w\frac{\partial v_r}{\partial z}$$
$$= f_r - \frac{1}{\rho}\frac{\partial p}{\partial r} + \nu\left\{\frac{\partial}{\partial r}\left[\frac{1}{r}\frac{\partial}{\partial r}(rv_r)\right] + \frac{1}{r^2}\frac{\partial^2 v_r}{\partial \theta^2} - \frac{2}{r^2}\frac{\partial v_\theta}{\partial \theta} + \frac{\partial^2 v_r}{\partial z^2}\right\}$$

θ component
$$\frac{\partial v_\theta}{\partial t} + v_r\frac{\partial v_\theta}{\partial r} + \frac{v_\theta}{r}\frac{\partial v_\theta}{\partial \theta} + \frac{v_r v_\theta}{r} + w\frac{\partial v_\theta}{\partial z}$$
$$= f_\theta - \frac{1}{\rho r}\frac{\partial p}{\partial \theta} + \nu\left\{\frac{\partial}{\partial r}\left[\frac{1}{r}\frac{\partial}{\partial r}(rv_\theta)\right] + \frac{1}{r^2}\frac{\partial^2 v_\theta}{\partial \theta^2} + \frac{2}{r^2}\frac{\partial v_r}{\partial \theta} + \frac{\partial^2 v_\theta}{\partial z^2}\right\}$$

z component
$$\frac{\partial w}{\partial t} + v_r\frac{\partial w}{\partial r} + \frac{v_\theta}{r}\frac{\partial w}{\partial \theta} + w\frac{\partial w}{\partial z}$$
$$= f_z - \frac{1}{\rho}\frac{\partial p}{\partial z} + \nu\left\{\frac{1}{r}\frac{\partial}{\partial r}\left(r\frac{\partial w}{\partial r}\right) + \frac{1}{r^2}\frac{\partial^2 w}{\partial \theta^2} + \frac{\partial^2 w}{\partial z^2}\right\}$$

SPHERICAL COORDINATES (r, θ, ϕ)[a]

r component
$$\frac{\partial v_r}{\partial t} + v_r\frac{\partial v_r}{\partial r} + \frac{v_\theta}{r}\frac{\partial v_r}{\partial \theta} + \frac{v_\phi}{r\sin\theta}\frac{\partial v_r}{\partial \phi} - \frac{v_\theta^2 + v_\phi^2}{r}$$
$$= f_r - \frac{1}{\rho}\frac{\partial p}{\partial r} + \nu\left(\nabla^2 v_r - \frac{2v_r}{r^2} - \frac{2}{r^2}\frac{\partial v_\theta}{\partial \theta} - \frac{2}{r^2}v_\theta\cot\theta - \frac{2}{r^2\sin\theta}\frac{\partial v_\phi}{\partial \phi}\right)$$

θ component
$$\frac{\partial v_\theta}{\partial t} + v_r\frac{\partial v_\theta}{\partial r} + \frac{v_\theta}{r}\frac{\partial v_\theta}{\partial \theta} + \frac{v_\phi}{r\sin\theta}\frac{\partial v_\theta}{\partial \phi} + \frac{v_r v_\theta}{r} - \frac{v_\phi^2\cot\theta}{r}$$
$$= f_\theta - \frac{1}{\rho r}\frac{\partial p}{\partial \theta} + \nu\left(\nabla^2 v_\theta + \frac{2}{r^2}\frac{\partial v_r}{\partial \theta} - \frac{v_\theta}{r^2\sin^2\theta} - \frac{2\cos\theta}{r^2\sin^2\theta}\frac{\partial v_\phi}{\partial \phi}\right)$$

ϕ component
$$\frac{\partial v_\phi}{\partial t} + v_r\frac{\partial v_\phi}{\partial r} + \frac{v_\theta}{r}\frac{\partial v_\phi}{\partial \theta} + \frac{v_\phi}{r\sin\theta}\frac{\partial v_\phi}{\partial \phi} + \frac{v_\phi v_r}{r} + \frac{v_\theta v_\phi}{r}\cot\theta$$
$$= f_\phi - \frac{1}{\rho r\sin\theta}\frac{\partial p}{\partial \phi} + \nu\left(\nabla^2 v_\phi - \frac{v_\phi}{r^2\sin^2\theta} + \frac{2}{r^2\sin\theta}\frac{\partial v_r}{\partial \phi} + \frac{2\cos\theta}{r^2\sin^2\theta}\frac{\partial v_\theta}{\partial \phi}\right)$$

[a]In these relations, $\nabla^2 \equiv \frac{1}{r^2}\frac{\partial}{\partial r}\left(r^2\frac{\partial}{\partial r}\right) + \frac{1}{r^2\sin\theta}\frac{\partial}{\partial \theta}\left(\sin\theta\frac{\partial}{\partial \theta}\right) + \frac{1}{r^2\sin^2\theta}\frac{\partial^2}{\partial \phi^2}$

which can also be written as

$$\nabla \cdot (k \nabla T)\, \Delta x\, \Delta y \Delta z \tag{1.63b}$$

where k is the thermal conductivity of the fluid. The net rate of work done by the fluid element against the surface and body forces is

$$-\left[\frac{\partial}{\partial x}(u\sigma_{xx} + v\tau_{xy} + w\tau_{xz}) + \frac{\partial}{\partial y}(u\tau_{yx} + v\sigma_{yy} + w\tau_{yz}) + \frac{\partial}{\partial z}(u\tau_{zx} + v\tau_{zy} + w\sigma_{zz}) + \rho \mathbf{V}\cdot\mathbf{f}\right] \Delta x\, \Delta y \Delta z \tag{1.64}$$

The rate of increase of internal and kinetic energies of the element can be written as

$$\rho\, \Delta x\, \Delta y \Delta z \frac{D}{Dt}\left[\mathscr{u} + \tfrac{1}{2}(u^2 + v^2 + w^2)\right] \tag{1.65}$$

where $\mathscr{u}$ is the internal energy of the fluid per unit mass. Noting the fact that the change in potential energy has already been included in the work of term of Eq. (1.64) by considering the work done against the body forces, the first law of thermodynamics for the element under consideration becomes

$$\rho \frac{D}{Dt}\left[\mathscr{u} + \tfrac{1}{2}(u^2 + v^2 + w^2)\right] = \nabla \cdot (k \nabla T) + \frac{\partial}{\partial x}(u\sigma_{xx} + v\tau_{xy} + w\tau_{xz}) + \frac{\partial}{\partial y}(u\tau_{yx} + v\sigma_{yy} + w\tau_{yz}) + \frac{\partial}{\partial z}(u\tau_{zx} + v\tau_{zy} + w\sigma_{zz}) + \rho \mathbf{V}\cdot\mathbf{f} \tag{1.66}$$

This result is also known as the *total energy equation* because it comprises both thermal and mechanical energies.

If the momentum equations (1.57a), (1.57b), and (1.57c) are multiplied by u, v, and w, respectively, then the resulting expressions can be summed to yield

$$\rho \frac{D}{Dt}\left[\tfrac{1}{2}(u^2 + v^2 + w^2)\right] = u\left(\frac{\partial \sigma_{xx}}{\partial x} + \frac{\partial \tau_{yx}}{\partial y} + \frac{\partial \tau_{zx}}{\partial z}\right) + v\left(\frac{\partial \tau_{xy}}{\partial x} + \frac{\partial \sigma_{yy}}{\partial y} + \frac{\partial \tau_{zy}}{\partial z}\right) + w\left(\frac{\partial \tau_{xz}}{\partial x} + \frac{\partial \tau_{yz}}{\partial y} + \frac{\partial \sigma_{zz}}{\partial z}\right) + \rho \mathbf{V}\cdot\mathbf{f} \tag{1.67}$$

which is an energy equation obtained directly from the laws of mechanics and is appropriately called the *mechanical energy equation*.

Subtraction of the mechanical energy equation (1.67) from the total energy equation (1.66) gives

$$\rho \frac{Du}{Dt} = \nabla \cdot (k \nabla T) + \sigma_{xx} \frac{\partial u}{\partial x} + \sigma_{yy} \frac{\partial v}{\partial y} + \sigma_{zz} \frac{\partial w}{\partial z} + \tau_{xy} \left(\frac{\partial v}{\partial x} + \frac{\partial u}{\partial y} \right) + \tau_{yz} \left(\frac{\partial w}{\partial y} + \frac{\partial v}{\partial z} \right) + \tau_{zx} \left(\frac{\partial u}{\partial z} + \frac{\partial w}{\partial x} \right) \tag{1.68}$$

which is called the *thermal energy equation* or, for short, the *energy equation*.

Equations (1.58), which are the relations between stresses and strains for Newtonian fluids, yield

$$\sigma_{xx} \frac{\partial u}{\partial x} + \sigma_{yy} \frac{\partial v}{\partial y} + \sigma_{zz} \frac{\partial w}{\partial z} = -p \nabla \cdot \mathbf{V} - \tfrac{2}{3} \mu (\nabla \cdot \mathbf{V})^2 + 2\mu \left[\left(\frac{\partial u}{\partial x} \right)^2 + \left(\frac{\partial v}{\partial y} \right)^2 + \left(\frac{\partial w}{\partial z} \right)^2 \right] \tag{1.69}$$

and

$$\tau_{xy} \left(\frac{\partial v}{\partial x} + \frac{\partial u}{\partial y} \right) = \mu \left(\frac{\partial v}{\partial x} + \frac{\partial u}{\partial y} \right)^2 \tag{1.70a}$$

$$\tau_{yz} \left(\frac{\partial w}{\partial y} + \frac{\partial v}{\partial z} \right) = \mu \left(\frac{\partial w}{\partial y} + \frac{\partial v}{\partial z} \right)^2 \tag{1.70b}$$

$$\tau_{zx} \left(\frac{\partial u}{\partial z} + \frac{\partial w}{\partial x} \right) = \mu \left(\frac{\partial u}{\partial z} + \frac{\partial w}{\partial x} \right)^2 \tag{1.70c}$$

When these relations are substituted into the energy equation (1.68), it reduces to

$$\rho \frac{Du}{Dt} = \nabla \cdot (k \nabla T) - p \nabla \cdot \mathbf{V} + \mu \Phi \tag{1.71}$$

where

$$\Phi = 2 \left[\left(\frac{\partial u}{\partial x} \right)^2 + \left(\frac{\partial v}{\partial y} \right)^2 + \left(\frac{\partial w}{\partial z} \right)^2 \right] + \left(\frac{\partial v}{\partial x} + \frac{\partial u}{\partial y} \right)^2 + \left(\frac{\partial w}{\partial y} + \frac{\partial v}{\partial z} \right)^2 + \left(\frac{\partial u}{\partial z} + \frac{\partial w}{\partial x} \right)^2 - \tfrac{2}{3} (\nabla \cdot \mathbf{V})^2 \tag{1.72}$$

which is called the *dissipation* function. The first term on the right-hand side of Eq. (1.71) represents the net rate of heat conduction to the fluid particle per unit volume; the second term, $-p \nabla \cdot \mathbf{V}$, is the rate of reversible work done on the fluid particle per unit volume; and the last term, $\mu \Phi$, is the rate at which viscous forces do irreversible work (i.e., viscous dissipation, or viscous heating) per unit volume.

Finally, the energy equation (1.71) may also be written in terms of the fluid enthalpy ($i = u + p/\rho$) as

$$\rho \frac{Di}{Dt} = \nabla \cdot (k \nabla T) + \frac{Dp}{Dt} + \mu \Phi \tag{1.73}$$

For a *perfect gas* $du = c_v\, dT$ and $di = c_p\, dT$, where c_v and c_p are the specific heats at constant volume and constant pressure, respectively. Hence, for a perfect gas, the energy equation (1.73) takes the form

$$\rho c_p \frac{DT}{Dt} = \nabla \cdot (k \nabla T) + \frac{Dp}{Dt} + \mu \Phi \tag{1.74}$$

In Eqs. (1.73) and (1.74), the term Dp/Dt is usually negligible except above sonic velocities. Therefore, for low-speed flows with constant thermal conductivity, the energy equation (1.74) reduces to

$$\frac{DT}{Dt} = \alpha \nabla^2 T + \frac{\mu}{\rho c_p} \Phi \tag{1.75}$$

where $\alpha = k/\rho c_p$ is the *thermal diffusivity* of the fluid and

$$\nabla^2 T = \nabla \cdot \nabla T = \frac{\partial^2 T}{\partial x^2} + \frac{\partial^2 T}{\partial y^2} + \frac{\partial^2 T}{\partial z^2} \tag{1.76}$$

For an *incompressible* fluid $du = c\, dT$, where $c = c_v \simeq c_p$. Thus, the energy equation (1.71) for an incompressible fluid takes the form

$$\rho c \frac{DT}{Dt} = \nabla (k \nabla T) + \mu \Phi \tag{1.77}$$

with

$$\Phi = 2\left[\left(\frac{\partial u}{\partial x}\right)^2 + \left(\frac{\partial v}{\partial y}\right)^2 + \left(\frac{\partial w}{\partial z}\right)^2\right] + \left(\frac{\partial v}{\partial x} + \frac{\partial u}{\partial y}\right)^2 + \left(\frac{\partial w}{\partial y} + \frac{\partial v}{\partial z}\right)^2 + \left(\frac{\partial u}{\partial z} + \frac{\partial w}{\partial x}\right)^2 \tag{1.78}$$

When the thermal conductivity is constant, the energy equation (1.77) for incompressible fluids reduces to

$$\frac{DT}{Dt} = \alpha \nabla^2 T + \frac{\mu}{\rho c} \Phi \tag{1.79}$$

The energy equation in cylindrical and spherical coordinates can be derived by following an approach similar to the one above, or they can be obtained from the equation in rectangular coordinates by coordinate transformations. The energy equation for Newtonian fluids in rectangular, cylindrical and spherical coordinates is listed in Table 1.6. In this table, an additional term q'''—representing the rate of internal

TABLE 1.6 The Energy Equation for Newtonian Fluids in Several Coordinate Systems

RECTANGULAR COORDINATES

$$\rho \frac{Di}{Dt} = \nabla \cdot (k \nabla T) + \frac{Dp}{Dt} + q''' + \mu\Phi$$

$$\frac{D}{Dt} \equiv \frac{\partial}{\partial t} + u\frac{\partial}{\partial x} + v\frac{\partial}{\partial y} + w\frac{\partial}{\partial z}$$

$$\nabla \cdot (k \nabla T) = \frac{\partial}{\partial x}\left(k\frac{\partial T}{\partial x}\right) + \frac{\partial}{\partial y}\left(k\frac{\partial T}{\partial y}\right) + \frac{\partial}{\partial z}\left(k\frac{\partial T}{\partial z}\right)$$

$$\Phi = 2\left[\left(\frac{\partial u}{\partial x}\right)^2 + \left(\frac{\partial v}{\partial y}\right)^2 + \left(\frac{\partial w}{\partial z}\right)^2\right] + \left(\frac{\partial v}{\partial x} + \frac{\partial u}{\partial y}\right)^2 + \left(\frac{\partial w}{\partial y} + \frac{\partial v}{\partial z}\right)^2$$
$$+ \left(\frac{\partial u}{\partial z} + \frac{\partial w}{\partial x}\right)^2 - \frac{2}{3}\left(\frac{\partial u}{\partial x} + \frac{\partial v}{\partial y} + \frac{\partial w}{\partial z}\right)^2$$

RECTANGULAR COORDINATES (PERFECT GAS)

$$\rho c_p \frac{DT}{Dt} = \nabla \cdot (k \nabla T) + \frac{Dp}{Dt} + q''' + \mu\Phi$$

$$\frac{D}{Dt} \equiv \frac{\partial}{\partial t} + u\frac{\partial}{\partial x} + v\frac{\partial}{\partial y} + w\frac{\partial}{\partial z}$$

$$\nabla \cdot (k \nabla T) = \frac{\partial}{\partial x}\left(k\frac{\partial T}{\partial x}\right) + \frac{\partial}{\partial y}\left(k\frac{\partial T}{\partial y}\right) + \frac{\partial}{\partial z}\left(k\frac{\partial T}{\partial z}\right)$$

$$\Phi = \left[\left(\frac{\partial u}{\partial x}\right)^2 + \left(\frac{\partial v}{\partial y}\right)^2 + \left(\frac{\partial w}{\partial z}\right)^2\right] + \left(\frac{\partial v}{\partial x} + \frac{\partial u}{\partial y}\right)^2 + \left(\frac{\partial w}{\partial y} + \frac{\partial v}{\partial z}\right)^2$$
$$+ \left(\frac{\partial u}{\partial z} + \frac{\partial w}{\partial x}\right)^2 - \frac{2}{3}\left(\frac{\partial u}{\partial x} + \frac{\partial v}{\partial y} + \frac{\partial w}{\partial z}\right)^2$$

RECTANGULAR COORDINATES (INCOMPRESSIBLE FLUID)

$$\rho c \frac{DT}{Dt} = \nabla \cdot (k \nabla T) + q''' + \mu\Phi$$

$$\frac{D}{Dt} \equiv \frac{\partial}{\partial t} + u\frac{\partial}{\partial x} + v\frac{\partial}{\partial y} + w\frac{\partial}{\partial z}$$

$$\nabla \cdot (k \nabla T) = \frac{\partial}{\partial x}\left(k\frac{\partial T}{\partial x}\right) + \frac{\partial}{\partial y}\left(k\frac{\partial T}{\partial y}\right) + \frac{\partial}{\partial z}\left(k\frac{\partial T}{\partial z}\right)$$

$$\Phi = 2\left[\left(\frac{\partial u}{\partial x}\right)^2 + \left(\frac{\partial v}{\partial y}\right)^2 + \left(\frac{\partial w}{\partial z}\right)^2\right] + \left(\frac{\partial v}{\partial x} + \frac{\partial u}{\partial y}\right)^2 + \left(\frac{\partial w}{\partial y} + \frac{\partial v}{\partial z}\right)^2 + \left(\frac{\partial u}{\partial z} + \frac{\partial w}{\partial x}\right)^2$$

TABLE 1.6 (continued)

CYLINDRICAL COORDINATES (INCOMPRESSIBLE FLUID)

$$\boxed{\rho c \frac{DT}{Dt} = \nabla \cdot (k \nabla T) + q''' + \mu \Phi}$$

$$\frac{D}{Dt} = \frac{\partial}{\partial t} + v_r \frac{\partial}{\partial r} + \frac{v_\theta}{r}\frac{\partial}{\partial \theta} + w \frac{\partial}{\partial z}$$

$$\nabla \cdot (k \nabla T) = \frac{1}{r}\frac{\partial}{\partial r}\left(rk\frac{\partial T}{\partial r}\right) + \frac{1}{r^2}\frac{\partial}{\partial \theta}\left(k\frac{\partial T}{\partial \theta}\right) + \frac{\partial}{\partial z}\left(k\frac{\partial T}{\partial z}\right)$$

$$\Phi = 2\left\{\left(\frac{\partial v_r}{\partial r}\right)^2 + \left[\frac{1}{r}\left(\frac{\partial v_\theta}{\partial \theta} + v_r\right)\right]^2 + \left(\frac{\partial w}{\partial z}\right)^2\right\} + \left(\frac{\partial v_\theta}{\partial z} + \frac{1}{r}\frac{\partial w}{\partial \theta}\right)^2 + \left(\frac{\partial w}{\partial r} + \frac{\partial v_r}{\partial z}\right)^2 + \left[\frac{1}{r}\frac{\partial v_r}{\partial \theta} + r\frac{\partial}{\partial r}\left(\frac{v_\theta}{r}\right)\right]^2$$

SPHERICAL COORDINATES (INCOMPRESSIBLE FLUID)

$$\boxed{\rho c \frac{DT}{Dt} = \nabla \cdot (k \nabla T) + q''' + \mu \Phi}$$

$$\frac{D}{Dt} = \frac{\partial}{\partial t} + v_r \frac{\partial}{\partial r} + \frac{v_\theta}{r}\frac{\partial}{\partial \theta} + \frac{v_\phi}{r \sin\theta}\frac{\partial}{\partial \phi}$$

$$\nabla \cdot (k \nabla T) = \frac{1}{r^2}\frac{\partial}{\partial r}\left(r^2 k\frac{\partial T}{\partial r}\right) + \frac{1}{r^2 \sin\theta}\frac{\partial}{\partial \theta}\left(k \sin\theta\frac{\partial T}{\partial \theta}\right) + \frac{1}{r^2 \sin^2\theta}\frac{\partial}{\partial \phi}\left(k\frac{\partial T}{\partial \phi}\right)$$

$$\Phi = 2\left[\left(\frac{\partial v_r}{\partial r}\right)^2 + \frac{1}{r^2}\left(\frac{\partial v_\theta}{\partial \theta} + v_r\right)^2 + \frac{1}{r^2}\left(\frac{1}{\sin\theta}\frac{\partial v_\phi}{\partial \phi} + v_r + v_\theta \cot\theta\right)^2\right] + \left[r\frac{\partial}{\partial r}\left(\frac{v_\theta}{r}\right) + \frac{1}{r}\frac{\partial v_r}{\partial \theta}\right]^2 + \left[\frac{1}{r \sin\theta}\frac{\partial v_r}{\partial \phi} + r\frac{\partial}{\partial r}\left(\frac{v_\phi}{r}\right)\right]^2 + \left[\frac{\sin\theta}{r}\frac{\partial}{\partial \theta}\left(\frac{v_\phi}{\sin\theta}\right) + \frac{1}{r \sin\theta}\frac{\partial v_\theta}{\partial \phi}\right]^2$$

thermal energy generation per unit volume within the fluid due to chemical, nuclear, electrical, etc., sources—has been included for completeness.

The continuity, Navier-Stokes, and energy equations presented in the preceding sections provide a comprehensive description of the thermal energy transfer in a flow field. These equations, however, present insurmountable mathematical difficulties due to the number of equations to be simultaneously satisfied and the presence of nonlinear terms such as $u\, \partial u/\partial x$. Because of the nonlinearities, the superposition principle is not applicable and complex flows may not be compounded from simple flows.

Exact solutions to these equations have been obtained for some simple cases [10]. In some of these cases, the troublesome nonlinear terms are either extremely small or identically zero. The flows represented by these solutions are referred to as *slow motions* or *creeping flows*. These solutions are important in the theory of lubrication and in the investigation of the settling of small particles in fluids. In most of the practical applications, however, flows of ordinary fluids, such as air and water, are generally quite different than creeping flows. In such flows the nonlinear terms are most often of greater magnitude than other terms in the Navier-Stokes equations.

The Reynolds number is a dimensionless quantity which measures the ratio of the inertia effects to the viscous effects in a fluid, and is defined as

$$\mathrm{Re} = \frac{\rho V L}{\mu} \tag{1.80}$$

where ρ is the fluid density, V is the fluid velocity, L represents a characteristic dimension in the flow field, and μ is the fluid viscosity.

Creeping flows are therefore characterized by small Reynolds numbers, whereas practical flows have Reynolds numbers which are most often large compared to unity. For example, experiments have indicated that the theory of slow motions may be used to predict the drag force exerted on a sphere moving at constant speed relative to a fluid when the Reynolds number (with the sphere diameter as the characteristic dimension) is less than about 1.

Finally, two important observations are worth mentioning. First, the velocity and temperature fields will be coupled if the fluid has temperature-dependent density and/or viscosity. Secondly, the temperature field can become similar to the velocity field under certain conditions. It can be seen from Eqs. (1.61) and (1.75), for example, that it is the terms in ∇p, $\mathbf{f}$, and Φ that prevent similarity between these two equations. Further, the viscosity μ and the thermal conductivity k may be different functions of temperature. If ∇p, Φ, and $\mathbf{f}$ are zero and if the Prandtl number $\mathrm{Pr} = \nu/\alpha = 1$, then the solutions for the velocity and temperature fields will be similar, provided that the corresponding boundary conditions are also similar.

1.6 BOUNDARY-LAYER APPROXIMATIONS — LAMINAR FLOW

L. Prandtl, in 1904, made a significant contribution to the field of fluid mechanics (and, therefore, to heat transfer) when he introduced the boundary-layer concept, which allowed flows at high Reynolds numbers to be studied mathematically. According to his theory, under certain conditions viscous forces in a flow field are of importance only in the immediate vicinity of the boundary surface, where the velocity gradient normal to the surface is large (see Fig. 1.2). In regions away from the boundary surface, the fluid motion may be considered frictionless (i.e., potential flow), because of negligible velocity gradients. There is, in fact, no precise division between the potential flow and boundary-layer regions, because the velocity component parallel to the surface approaches asymptotically its free-stream value (i.e., its value away from the surface). However, it is customary to define the boundary layer as that region where the velocity component parallel to the surface is less than 99% of its free-stream value.

Consider, for example, the case of a steady, two-dimensional incompressible laminar boundary-layer flow over a surface with a free-stream velocity $U_\infty(x)$. Assume that the body forces are negligible and the viscosity is constant. By the use of Prandtl's boundary-layer approximations, the continuity, Navier-Stokes, and energy equations reduce to [5,10]

$$\frac{\partial u}{\partial x} + \frac{\partial v}{\partial y} = 0 \tag{1.81}$$

$$u\frac{\partial u}{\partial x} + v\frac{\partial u}{\partial y} = U_\infty \frac{dU_\infty}{dx} + \nu \frac{\partial^2 u}{\partial y^2} \tag{1.82}$$

$$\rho c_p \left(u\frac{\partial T}{\partial x} + v\frac{\partial T}{\partial y} \right) = \frac{\partial}{\partial y}\left(k\frac{\partial T}{\partial y} \right) + \mu \left(\frac{\partial u}{\partial y} \right)^2 \tag{1.83}$$

where x denotes the direction parallel to the surface and y the direction normal to the surface. Equations (1.81), (1.82) and (1.83) are called Prandtl's *laminar boundary-layer equations*, for which the following boundary conditions apply:

$$\text{at} \quad y = 0: \qquad u = v = 0 \text{ and } T = T_w \text{ or } -k\frac{\partial T}{\partial y} = q''_w$$

$$\text{as} \quad y \to \infty: \qquad u \to U_\infty(x) \text{ and } T \to T_\infty$$

In addition, the initial velocity and temperature distributions at $x = 0$ must be specified.

1.7 TURBULENT FLOW

A turbulent flow is characterized by disorderly displacement of individual fluid particles within the flow field. Most flows of practical importance are turbulent. The continuity, Navier-Stokes, and energy equations, which were described in Sec. 1.5, are also valid for turbulent flows. It must be noted, however, that the velocity components, pressure, and temperature in these equations would have to be the instantaneous values. In turbulent flow, the instantaneous values always vary with time, and the variations are completely random with minor fluctuations about the mean values (Fig. 1.9). These random fluctuations are so complex that any direct mathematical treatment of the governing equations becomes impossible.

Methods of analysis of heat transfer in turbulent flows are by no means complete at present because of our limited understanding of the mechanism of turbulence. However, many analytical procedures and empirical correlations have been proposed by various investigators. In order to attack problems involving turbulence, analytically or numerically, it is convenient to define mean and fluctuating components of velocity, pressure, temperature, etc., as

$$\eta = \bar{\eta} + \eta' \tag{1.84}$$

where η may represent u, v, w, p, T, etc. The fluctuating components η' can have both positive and negative values, and the mean values are defined according to

$$\bar{\eta} = \frac{1}{\Delta t} \int_{t_0}^{t_0 + \Delta t} \eta \, dt \tag{1.85}$$

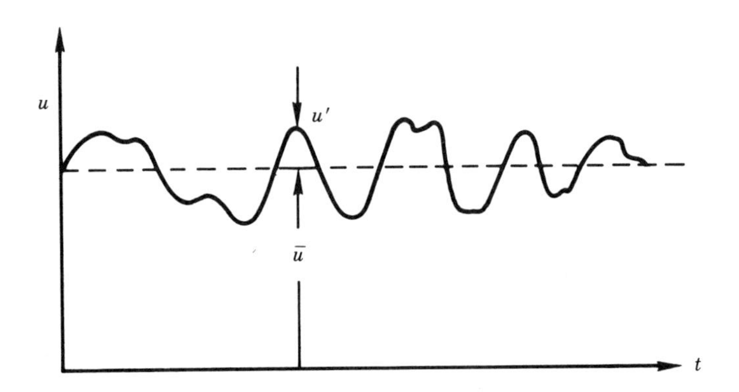

Figure 1.9. Variation of the velocity component u with time.

where Δt is sufficiently large to obtain a true average; that is, large enough for recording turbulent fluctuations, but sufficiently small for the quantity to be unaffected by external disturbances on the flow. Hence, it is obvious that

$$\int_{t_0}^{t_0+\Delta t} \eta' \, dt = 0 \tag{1.86}$$

A turbulent flow is called *steady* if the mean values $\bar{u}$, $\bar{v}$, $\bar{w}$, $\bar{p}$, $\bar{T}$, etc., do not change with time. This is reasonable, because instruments with long enough response times (e.g., a thermocouple), when placed in a turbulent stream, show readings that are entirely stable with time.

In some cases, thermophysical properties such as density, viscosity, specific heat, and thermal conductivity may also fluctuate, but such fluctuations are usually neglected.

Consider now a steady, two-dimensional turbulent flow with constant thermophysical properties implying that $\bar{u} = \bar{u}(x, y)$, $\bar{v} = \bar{v}(x, y)$, $\bar{w} = 0$, $\bar{p} = \bar{p}(x, y)$ and $\bar{T} = \bar{T}(x, y)$. Even in this case, $u' = u'(x, y, z, t)$, $v' = v'(x, y, z, t)$, $w' = w'(x, y, z, t)$, $p' = p'(x, y, z, t)$ and $T' = T'(x, y, z, t)$. If one, however, assumes that

$$u' = u'(x, y, t) \qquad p' = p'(x, y, t)$$

$$v' = v'(x, y, t) \qquad T' = T'(x, y, t)$$

$$w' = 0$$

then substitution of the instantaneous velocity components, pressure, and temperature as defined by Eq. (1.84) into Eqs. (1.51a), (1.60), and (1.79), and time-averaging the resulting equations, yield the following continuity, momentum, and energy equations for turbulent flows [5]:

Continuity equation.

$$\frac{\partial \bar{u}}{\partial x} + \frac{\partial \bar{v}}{\partial y} = 0 \tag{1.87a}$$

and

$$\frac{\partial u'}{\partial x} + \frac{\partial v'}{\partial y} = 0 \qquad (1.87b)$$

Momentum equations.

x component:

$$\bar{u}\frac{\partial \bar{u}}{\partial x} + \bar{v}\frac{\partial \bar{u}}{\partial y} = f_x - \frac{1}{\rho}\frac{\partial \bar{p}}{\partial x} + \nu \nabla^2 \bar{u} - \left[\frac{\partial}{\partial x}\left(\overline{u'u'}\right) + \frac{\partial}{\partial y}\left(\overline{v'u'}\right)\right] \qquad (1.88)$$

y component:

$$\bar{u}\frac{\partial \bar{v}}{\partial x} + \bar{v}\frac{\partial \bar{v}}{\partial y} = f_y - \frac{1}{\rho}\frac{\partial \bar{p}}{\partial y} + \nu \nabla^2 \bar{v} - \left[\frac{\partial}{\partial x}\left(\overline{u'v'}\right) + \frac{\partial}{\partial y}\left(\overline{v'v'}\right)\right] \qquad (1.89)$$

Energy equation.

$$\bar{u}\frac{\partial \bar{T}}{\partial x} + \bar{v}\frac{\partial \bar{T}}{\partial y} = \alpha \nabla^2 \bar{T} - \left[\frac{\partial}{\partial x}\left(\overline{u'T'}\right) + \frac{\partial}{\partial y}\left(\overline{v'T'}\right)\right] \qquad (1.90)$$

where the dissipation term in Eq. (1.79) has been neglected.

Prandtl's order-of-magnitude analysis can also be applied to the above equations in the same way as for laminar flow, and the two-dimensional turbulent boundary-layer equations for steady, constant-property flow with negligible body forces and heat dissipation become [5]

$$\frac{\partial \bar{u}}{\partial x} + \frac{\partial \bar{v}}{\partial y} = 0 \qquad (1.91)$$

$$\bar{u}\frac{\partial \bar{u}}{\partial x} + \bar{v}\frac{\partial \bar{u}}{\partial y} = -\frac{1}{\rho}\frac{d\bar{p}}{dx} + \frac{\partial}{\partial y}\left[(\nu + \epsilon_m)\frac{\partial \bar{u}}{\partial y}\right] \qquad (1.92)$$

$$\bar{u}\frac{\partial \bar{T}}{\partial x} + \bar{v}\frac{\partial \bar{T}}{\partial y} = \frac{\partial}{\partial y}\left[(\alpha + \epsilon_h)\frac{\partial \bar{T}}{\partial y}\right] \qquad (1.93)$$

where ϵ_m and ϵ_h are called *eddy diffusivity of momentum* and *eddy diffusivity of heat*, respectively, which were introduced by Boussinesq [10] as

$$-\overline{v'u'} = \epsilon_m \frac{\partial \bar{u}}{\partial y} \qquad (1.94)$$

and

$$-\overline{v'T'} = \epsilon_h \frac{\partial \bar{T}'}{\partial y} \qquad (1.95)$$

Both ϵ_m and ϵ_h are flow parameters, and not fluid properties.

The time-averaged two-dimensional boundary-layer equations for mass, momentum, and energy in cylindrical coordinates for steady flows with constant thermophysical properties and with negligible body forces and heat dissipation can be obtained as

$$\frac{\partial \bar{u}}{\partial x} + \frac{1}{r}\frac{\partial}{\partial r}(r\bar{v}_r) = 0 \tag{1.96}$$

$$\bar{u}\frac{\partial \bar{u}}{\partial x} + \bar{v}_r\frac{\partial \bar{u}}{\partial r} = -\frac{1}{\rho}\frac{d\bar{p}}{dx} + \frac{1}{r}\frac{\partial}{\partial r}\left[r(\nu + \epsilon_m)\frac{\partial \bar{u}}{\partial r}\right] \tag{1.97}$$

$$\bar{u}\frac{\partial \bar{T}}{\partial x} + \bar{v}_r\frac{\partial \bar{T}}{\partial r} = \frac{1}{r}\frac{\partial}{\partial r}\left[r(\alpha + \epsilon_h)\frac{\partial \bar{T}}{\partial r}\right] \tag{1.98}$$

As seen from these equations, there are five unknowns ($\bar{u}$, $\bar{v}$, $\bar{T}$, ϵ_m, and ϵ_h) but only three equations. Additional information can, however, be obtained from *turbulence modeling*, which provides generally applicable expressions for ϵ_m and ϵ_h.

1.8 FINAL REMARKS

In this chapter, the basics of heat transfer are reviewed. The intention is to present the fundamental concepts and working relations to be referred to later in the following chapters, where various topics of single-phase convection heat transfer are discussed in depth. The material in this chapter can also be used by engineering students, scientists, and practicing engineers who have interest in heat transfer problems. This chapter, being a short review, does not cover all the details, but the references at the end of this as well as the following chapters can be consulted for further information.

NOMENCLATURE

A	surface area, heat transfer area, m^2, ft^2
a	acceleration, m/s^2, ft/s^2
c_p	specific heat at constant pressure, J/(kg · K), Btu/(lb$_m$ · °F)
c_v	specific heat at constant volume, J/(kg · K), Btu/(lb$_m$ · °F)
E	energy, J, Btu
e	energy per unit mass, J/kg, Btu/lb$_m$
F, **F**	force, N, lb$_f$
F_{ij}	radiation shape factor
f	body force per unit mass, N/kg, lb$_f$/lb$_m$
g	gravitational acceleration, m/s^2, ft/s^2
h	heat transfer coefficient, W/(m^2 · K), Btu/(hr · ft^2 · °F)
h_r	radiation heat transfer coefficient, W/(m^2 · K), Btu/(hr · ft^2 · °F)
i	enthalpy per unit mass, J/kg, Btu/lb$_m$
$\hat{\mathbf{i}}$	unit vector in x direction
$\hat{\mathbf{j}}$	unit vector in y direction
KE	kinetic energy, J, Btu
k	thermal conductivity, W/(m · K), Btu/(hr · ft · °F)
k_s	thermal conductivity of solid, W/(m · K), Btu/(hr · ft · °F)
$\hat{k}$	unit vector in z direction

$\mathbf{M}$	linear momentum, kg · m/s, lb_f · ft/s
m	mass, kg, lb_m
$\dot{m}$	mass flow rate, kg/s, lb_m/hr
n	distance, m, ft
$\hat{\mathbf{n}}$	unit vector in n direction
P	power, W, Btu/hr
PE	potential energy, J, Btu
Pr	Prandtl number $= \mu c_p/k = \nu/\alpha$
Pr_t	turbulent Prandtl number
p	pressure, Pa, lbf/ft^2
Q	cumulative heat, J, Btu
q	heat transfer rate, W, Btu/hr
$q'', \mathbf{q}''$	heat flux, W/m^2, Btu/(hr · ft^2)
q''_w	wall heat flux, W/m^2, Btu/(hr · ft^2)
q'''	volumetric heat generation rate, W/m^3, Btu/(hr · ft^3)
q'''_e	volumetric heat generation rate due to electrical sources, W/m^3, Btu/(hr · ft^3)
Re	Reynolds number $= \rho VL/\mu$
r	radial coordinate, m, ft
s	entropy per unit mass, J/(kg · K), Btu/(lb_m · °R)
T	temperature, °C, K, °F, °R
T_s	temperature of solid, °C, K, °F, °R
T_w	wall temperature, °C, K, °F, °R
T_∞	free-stream temperature, °C, K, °F, °R
t	time, s
$\mathcal{U}$	internal energy, J, Btu
U_∞	free-stream velocity, m/s, ft/s
$\mathscr{u}$	internal energy per unit mass, J/kg, Btu/lb_m
u	velocity component in x direction, m/s, ft/s
$V, \mathbf{V}$	velocity, m/s, ft/s
V	volume, m^3, ft^3
v	velocity component in y direction, m/s, ft/s
v_θ	velocity component in θ direction, m/s, ft/s
v_ϕ	velocity component in ϕ direction, m/s, ft/s
W	work, J, Btu
w	velocity component in z direction, m/s, ft/s
x	rectangular coordinate, distance parallel to surface, m, ft
y	rectangular coordinate, distance parallel to surface, m, ft
z	rectangular coordinate, m, ft

Greek Symbols

$\tilde{\alpha}$	absorptivity
α	thermal diffusivity $= k/\rho c_p$, m^2/s, ft^2/s

δ	velocity boundary-layer thickness, m, ft
δ	inexact differential operator
δ_T	thermal boundary-layer thickness, m, ft
Δ	finite increment
ϵ	emissivity
ϵ_h	eddy diffusivity of heat, m^2/s, ft^2/s
ϵ_m	eddy diffusivity of momentum, m^2/s, ft^2/s
η	general variable
θ	latitude angle in cylindrical coordinates, rad, deg
μ	dynamic viscosity, Pa · s, $lb_m/(hr \cdot ft)$
ν	kinematic viscosity, m^2/s, ft^2/s
ρ	density, kg/m^3, lb_m/ft^3
$\tilde{\rho}$	reflectivity
$\tilde{\sigma}$	Stefan-Boltzmann constant = 5.6697×10^{-8} $W/(m^2 \cdot K^4)$ = 0.17×10^{-8} $Btu/(hr \cdot ft^2 \cdot R^4)$
σ	shear stress, Pa, lb_f/ft^2
τ	shear stress between fluid layers, Pa, lb_f/ft^2
$\tilde{\tau}$	transmissivity
ϕ	azimuth angle in spherical coordinates, rad, deg
Φ	viscous dissipation function, W/m^3, $Btu/(hr \cdot ft^3)$
$\mathscr{F}$	geometric shape and emissivity factor for radiation from one gray body to another

Subscripts

b	blackbody
c.v.	control volume
c.s.	control surface
f	fluid
irr	irreversible
n	normal direction
r	radiation
rev	reversible
w	wall condition
x	x direction
y	y direction
z	z direction
θ	θ direction
ϕ	ϕ direction
∞	free-stream conditions

Superscripts and Accents

$\bar{}$	mean value
$'$	randomly fluctuating value

$\dot{}$ rate

$\hat{}$ unit vector

REFERENCES

1. A. H. Shapiro, *The Dynamics and Thermodynamics of Compressible Fluid Flow*, Vol. 1, Ronald Press, New York, 1953.
2. S. Kakaç and Y. Yener, *Heat Conduction*, 2nd ed., Hemisphere, New York, 1985.
3. J. H. Keenan, *Thermodynamics*, Wiley, New York, 1941.
4. G. J. Van Vylen and R. E. Sonntag, *Fundamentals of Classical Thermodynamics*, 2nd ed., revised printing, SI version, Wiley, New York, 1978.
5. S. Kakaç and Y. Yener, *Convective Heat Transfer*, METU, Ankara, Turkey, distributed by Hemisphere, New York, 1980.
6. M. N. Özışık, *Heat Conduction*, Wiley, New York, 1980.
7. R. Siegel and J. R. Howell, *Thermal Radiation Heat Transfer*, 2nd ed., Hemisphere, New York, 1980.
8. M. N. Özışık, *Radiative Transfer and Interactions with Conduction and Convection*, Wiley, New York, 1978.
9. H. Lamb, *Hydrodynamics*, Dover, New York, 1945.
10. H. Schlichting, *Boundary Layer Theory*, 7th ed., McGraw-Hill, New York, 1979.
11. R. B. Bird, W. E. Steward, and E. N. Lightfoot, *Transport Phenomena*, Wiley, New York, 1960.
12. A. Bejan, *Convection Heat Transfer*, Wiley, New York, 1984.

2

EXTERNAL FLOW FORCED CONVECTION

R. H. Pletcher

Iowa State University
Ames, Iowa

2.1 INTRODUCTION

In the most common example of external flow forced convection one is called upon to compute the heat transferred to or from a body moving through a large (for practical purposes, infinite) ambient medium, uniform in properties and at rest. In the analysis of such problems, it is found that it is only the relative motion that matters, and it is frequently convenient to attach the coordinate reference frame to the body and consider the fluid to be moving past the body (now at rest) with uniform velocity. Such circumstances arise very frequently in applications, a very common example being the convective heat transfer between vehicles (or other objects of all sorts) and the earth's atmosphere.

The essential difference between the formulation of the convective problem for internal and external flows arises through the boundary conditions to the governing equations. In the external flow formulation, the fluid velocity and temperature approach known "free-stream" values at large distances from the convective surface. Sufficiently far from the surface, the flow can almost always be treated as inviscid, and frequently as irrotational. The designation "internal flow" on the other hand indicates that the flow is confined within a finite-size passage or device. Flow generally enters or leaves the device, of course, but inside, boundary conditions on velocity, temperature, or fluxes are imposed at solid surfaces or symmetry lines or surfaces. Parts of the flow may occasionally be treated as inviscid, but this is not the most common approach in analysis. In steady flow, the mass flux through the passage is constant and usually known. The mass flux is generally not specified or known for external flow problems.

External convection problems can be relatively simple, as in the case of low speed laminar flow over an isothermal flat plate, or very complex, as in the flow about the space-shuttle orbiter during atmospheric reentry. Features which tend to complicate the analysis of convective flows include turbulence, the presence of two or more phases (gas-liquid, gas-solid, liquid-solid flows), unsteady effects, and complex geometries which cause flows to be three-dimensional or cause regions of flow reversal and recirculation to occur.

In the simplest cases of external flow, the effects of the presence of a body in a large fluid stream are confined to a thin layer of fluid immediately adjacent to the body surface. This region is the well-known Prandtl "boundary layer" [1]. As the distance normal to the surface is increased, the fluid properties approach those of the external stream. As long as the flow does not separate, the Reynolds number is moderate to large, and the Prandtl number is of the order of 1 or larger, the boundary layer remains thin and has a negligible effect on the external flow. In this case, the heat transfer to the body can be obtained from a solution to the boundary-layer form of the governing equations. However, in order to solve these equations, the velocity and temperature at the outer edge of the boundary layer are required as boundary conditions. These can be obtained from the solution for the inviscid flow about the body. It is possible to obtain an improved inviscid flow solution by augmenting the physical thickness of the body by the boundary-layer displacement thickness. The improved edge conditions from the inviscid solution can then be used to obtain yet another viscous flow solution. This iterative procedure can account for "viscous-inviscid interaction." However, interaction should not be necessary for fully attached flows except at very small Reynolds numbers. In this simplest case, the flow problem has been divided into two distinct regions and the solutions for the two regions can be obtained independently, although the boundary conditions for the viscous region must be obtained from the solution for the inviscid flow. In the case of a thin flat plate placed parallel to a uniform stream, the edge conditions are constant at the stream values; i.e., the thin plate causes no disturbance in the inviscid flow, and no further attention need be given to obtaining the solution for the inviscid flow.

In many important external flows, the effects of viscosity are not confined to a thin layer next to the solid body. This situation occurs when the flow *separates* from the body due to an adverse pressure gradient or an abrupt change in the geometry. Then the strategy of solving the flow problem in two independent parts, one viscous and one inviscid, in a noniterative fashion fails. The major cause of the failure is that the inviscid flow solution over the solid body is no longer a good approximation. The "displacement effect" of the separated regions locally alters the pressure distribution in a significant manner. Even in this situation it is often possible to treat the viscous flow using the boundary-layer approximation. However, the displacement effect of the viscous region must be taken into account in the inviscid solution procedure. Because of this, the two solutions are no longer independent, and an iterative viscous-inviscid interaction procedure is required if the problem is to be modeled as having viscous and inviscid parts. Alternatively, a solution can be obtained by solving a single set of more complex governing equations, valid in all regions of the flow. In addition to flow separation, the other effects mentioned above (turbulence, multiphase flow, three-dimensional effects, etc.) force the use of more complex governing equations or recourse to empirically based approximations, or both.

Traditionally, both experimental and analytical methods have been used to obtain heat transfer information needed for design purposes. The traditional analytical method made use of simplifying assumptions in order to obtain closed-form solutions to problems. Correlations based on experimental measurements were sometimes incorporated into an analytical method for especially complicated problems. With the advent of the digital computer, a third method, the numerical or computational approach, has become available.

In the computational approach, a fairly complete mathematical description of the heat transfer phenomena is retained, and the governing equations, usually in partial differential form, are solved numerically. If the mathematical description is complete and involves few assumptions, the numerical solution provides a "computer simulation" for the physical process.

Although experimentation continues to be important, and in some instances essential, the trend is clearly toward greater reliance on computer-based predictions in design. The trend can be explained by economics. Over the years, computer speed has increased much more rapidly than computer costs. The net effect has been a significant decrease in the cost of performing a given calculation. On the other hand, the costs of performing experiments have been steadily increasing. The result of these trends has been to encourage the maximum use of computational tools to reduce the range of conditions over which testing is required.

Both experimentation and computer simulation have limitations. It should be clear that neither approach is capable of providing all of the information of interest to designers in every application. Existing test facilities are not always capable of simulating the severe operating conditions occurring in some applications. This may present no difficulty for a computer simulation, but computer storage and speed may limit the usefulness of the computer in some applications. Other limitations arise from our inability to understand and mathematically model certain complex phenomena.

While an ever increasing number of heat transfer engineers are using computational methods to solve problems in convective heat transfer, a need still exists for general formulas and data correlations that can be used in preliminary design, and as benchmark checks for computer codes. These will be included in this chapter. In addition, the capabilities of computational procedures for forced convection over external surfaces will be discussed. Details of computer codes can be found elsewhere in the technical literature and will not be included in this chapter.

2.2 FUNDAMENTAL EQUATIONS AND DEFINITIONS

2.2.1 The Basic Conservation Equations

The governing equations for single-phase external forced convection are based on the fundamental conservation laws for mass, momentum, and energy. It will be assumed that the fluid behaves as a continuum and can be treated as a single chemical specie. Gravity and other body forces will be neglected. Flows in which body forces become important are treated in other chapters in this handbook. Detailed derivations of the equations will not be presented here. They can be found in [1, 2].

Continuity Equation. Conservation of mass applied to a fluid in motion yields the following continuity equation:

$$\frac{\partial \rho}{\partial t} + \nabla \cdot (\rho \mathbf{V}) = 0 \tag{2.1}$$

It is often convenient to use the substantial derivative to write the continuity equation in the form

$$\frac{D\rho}{Dt} + \rho(\nabla \cdot \mathbf{V}) = 0 \tag{2.2}$$

For a Cartesian coordinate system, where u, v, w represent the components of the velocity in the x, y, z directions, Eq. (2.1) becomes

$$\frac{\partial \rho}{\partial t} + \frac{\partial(\rho u)}{\partial x} + \frac{\partial(\rho v)}{\partial y} + \frac{\partial(\rho w)}{\partial z} = 0 \tag{2.3}$$

For a flow in which the density remains constant, we find

$$\frac{D\rho}{Dt} = 0 \tag{2.4}$$

which reduces Eq. (2.2) to $\nabla \cdot \mathbf{V} = 0$ or, in the Cartesian coordinate system,

$$\frac{\partial u}{\partial x} + \frac{\partial v}{\partial y} + \frac{\partial w}{\partial z} = 0 \tag{2.5}$$

Momentum Equations. Application of Newton's second law to a fluid in which body forces are negligible yields

$$\frac{\partial}{\partial t}(\rho \mathbf{V}) + \nabla \cdot \rho \mathbf{V}\mathbf{V} = \nabla \cdot \Pi \tag{2.6}$$

The term $\nabla \cdot \rho\mathbf{VV}$ can be expanded as

$$\nabla \cdot \rho\mathbf{VV} = \rho\mathbf{V} \cdot \nabla \mathbf{V} + \mathbf{V}(\nabla \cdot \rho\mathbf{V}) \tag{2.7}$$

When this expression is used with the continuity equation (2.6), the momentum equation becomes

$$\frac{D\mathbf{V}}{Dt} = \nabla \cdot \Pi \tag{2.8}$$

where Π is the stress tensor. For all gases which can be treated as a continuum and most liquids, it has been observed that the stresses are linearly dependent on the rates of deformation of the fluid. Such a fluid is known as a Newtonian fluid. For Newtonian fluids it is possible to derive a relationship between the stress tensor and the pressure and velocity components [1, 2]. In Cartesian tensor notation, this relationship becomes

$$\Pi_{ij} = -p\delta_{ij} + \mu\left(\frac{\partial u_i}{\partial x_j} + \frac{\partial u_j}{\partial x_i}\right) + \delta_{ij}\mu'\frac{\partial u_k}{\partial x_k} \tag{2.9}$$

where δ_{ij} is the Kronecker delta function ($\delta_{ij} = 1$ if $i = j$ and $\delta_{ij} = 0$ if $i \neq j$); u_1, u_2, u_3 are the three components of the velocity vector in the x_1, x_2, x_3 coordinate directions; μ is the dynamic viscosity; and μ' is the second coefficient of viscosity. For an incompressible fluid, the term involving the second coefficient of viscosity vanishes. The combination $\frac{2}{3}\mu + \mu'$ is known as the bulk viscosity and is generally believed to be negligible whenever the time characterizing global processes in the flow is large compared to that of the molecular relaxation time. The assumption that the bulk viscosity is zero (known as the *Stokes hypothesis*) allows μ' to be evaluated as $\mu' = -\frac{2}{3}\mu$, and the stress tensor can be written as

$$\Pi_{ij} = -p\delta_{ij} + \mu\left[\left(\frac{\partial u_i}{\partial x_j} + \frac{\partial u_j}{\partial x_i}\right) - \frac{2}{3}\delta_{ij}\frac{\partial u_k}{\partial x_k}\right] \tag{2.10}$$

The stress tensor is frequently split in the following manner:

$$\Pi_{ij} = -p\delta_{ij} + \tau_{ij} \tag{2.11}$$

where τ_{ij} represents the viscous stress tensor given by

$$\tau_{ij} = \mu\left[\left(\frac{\partial u_i}{\partial x_j} + \frac{\partial u_j}{\partial x_i}\right) - \frac{2}{3}\delta_{ij}\frac{\partial u_k}{\partial x_k}\right] \tag{2.12}$$

Upon substituting Eq. (2.10) into Eq. (2.8), the Navier-Stokes equation is obtained:

$$\rho\frac{Du_i}{Dt} = -\frac{\partial p}{\partial x_i} + \frac{\partial}{\partial x_j}\left[\mu\left(\frac{\partial u_i}{\partial x_j} + \frac{\partial u_j}{\partial x_i}\right) - \frac{2}{3}\delta_{ij}\mu\frac{\partial u_k}{\partial x_k}\right] \tag{2.13}$$

Equation (2.13) can be separated into the following three scalar Navier-Stokes equations:

$$\rho\frac{Du}{Dt} = -\frac{\partial p}{\partial x} + \frac{\partial}{\partial x}\left[\tfrac{2}{3}\mu\left(2\frac{\partial u}{\partial x} - \frac{\partial v}{\partial y} - \frac{\partial w}{\partial z}\right)\right]$$

$$+\frac{\partial}{\partial y}\left[\mu\left(\frac{\partial u}{\partial y} + \frac{\partial v}{\partial x}\right)\right] + \frac{\partial}{\partial z}\left[\mu\left(\frac{\partial w}{\partial x} + \frac{\partial u}{\partial z}\right)\right] \tag{2.14}$$

$$\rho\frac{Dv}{Dt} = -\frac{\partial p}{\partial y} + \frac{\partial}{\partial x}\left[\mu\left(\frac{\partial v}{\partial x} + \frac{\partial u}{\partial y}\right)\right]$$

$$+\frac{\partial}{\partial y}\left[\tfrac{2}{3}\mu\left(2\frac{\partial v}{\partial y} - \frac{\partial u}{\partial x} - \frac{\partial w}{\partial z}\right)\right] + \frac{\partial}{\partial z}\left[\mu\left(\frac{\partial v}{\partial z} + \frac{\partial w}{\partial y}\right)\right] \tag{2.15}$$

$$\rho\frac{Dw}{Dt} = -\frac{\partial p}{\partial z} + \frac{\partial}{\partial x}\left[\mu\left(\frac{\partial w}{\partial x} + \frac{\partial u}{\partial z}\right)\right] + \frac{\partial}{\partial y}\left[\mu\left(\frac{\partial v}{\partial z} + \frac{\partial w}{\partial y}\right)\right]$$

$$+\frac{\partial}{\partial z}\left[\tfrac{2}{3}\mu\left(2\frac{\partial w}{\partial z} - \frac{\partial u}{\partial x} - \frac{\partial v}{\partial y}\right)\right] \tag{2.16}$$

In some applications it is convenient to write the Navier-Stokes equations in divergence (or conservation-law) form as

$$\frac{\partial \rho u}{\partial t} + \frac{\partial}{\partial x}\left(\rho u^2 + p - \tau_{xx}\right) + \frac{\partial}{\partial y}\left(\rho uv - \tau_{xy}\right)$$

$$+\frac{\partial}{\partial z}\left(\rho uw - \tau_{xz}\right) = 0 \tag{2.17}$$

$$\frac{\partial \rho v}{\partial t} + \frac{\partial}{\partial x}\left(\rho uv - \tau_{xy}\right) + \frac{\partial}{\partial y}\left(\rho v^2 + p - \tau_{yy}\right)$$

$$+\frac{\partial}{\partial z}\left(\rho vw - \tau_{yz}\right) = 0 \tag{2.18}$$

$$\frac{\partial \rho w}{\partial t} + \frac{\partial}{\partial x}\left(\rho uw - \tau_{xz}\right) + \frac{\partial}{\partial y}\left(\rho vw - \tau_{yz}\right)$$

$$+\frac{\partial}{\partial z}\left(\rho w^2 + p - \tau_{zz}\right) = 0 \tag{2.19}$$

where the components of the viscous stress tensor τ_{ij} are given by

$$\tau_{xx} = \tfrac{2}{3}\mu\left(2\frac{\partial u}{\partial x} - \frac{\partial v}{\partial y} - \frac{\partial w}{\partial z}\right)$$

$$\tau_{yy} = \tfrac{2}{3}\mu\left(2\frac{\partial v}{\partial y} - \frac{\partial u}{\partial x} - \frac{\partial w}{\partial z}\right)$$

$$\tau_{zz} = \tfrac{2}{3}\mu\left(2\frac{\partial w}{\partial z} - \frac{\partial u}{\partial x} - \frac{\partial v}{\partial y}\right)$$

$$\tau_{xy} = \mu\left(\frac{\partial u}{\partial y} + \frac{\partial v}{\partial x}\right) = \tau_{yx}$$

$$\tau_{xz} = \mu\left(\frac{\partial w}{\partial x} + \frac{\partial u}{\partial z}\right) = \tau_{zx}$$

$$\tau_{yz} = \mu\left(\frac{\partial v}{\partial z} + \frac{\partial w}{\partial y}\right) = \tau_{zy}$$

For an incompressible flow in which viscosity variations can be neglected, the Navier-Stokes equations reduce to the much simpler form

$$\rho\frac{D\mathbf{V}}{Dt} = -\nabla p + \mu\nabla^2\mathbf{V} \tag{2.20}$$

Energy Equation. In terms of the total enthalpy $H = i + u_i u_i/2$, the energy equation for a fluid in motion can be written as

$$\frac{\partial}{\partial t}(\rho H) + \frac{\partial}{\partial x_j}(\rho u_j H) = -\frac{\partial p}{\partial t} + \frac{\partial}{\partial x_j}(u_i\tau_{ij} - q_j) \tag{2.21}$$

where the heat flux q_j can be evaluated from Fourier's law,

$$q_j = -k\frac{\partial T}{\partial x_j} \tag{2.22}$$

Utilizing the substantial derivative and making use of the continuity equation, we can write the left-hand side of Eq. (2.21) as $\rho\, DH/Dt$.

It is sometimes more convenient to express the energy equation in terms of the static enthalpy rather than the total enthalpy. Then Eq. (2.21) becomes

$$\frac{\partial}{\partial t}(\rho i) + \frac{\partial}{\partial x_j}(\rho i u_j) = \frac{\partial p}{\partial t} + u_j\frac{\partial p}{\partial x_j} + \tau_{ij}\frac{\partial u_i}{\partial x_j} - \frac{\partial q_i}{\partial x_j} \tag{2.23}$$

where $\tau_{ij}\,\partial u_i/\partial x_j$ is commonly identified as the dissipation function ϕ. The value of the dissipation function represents the rate at which mechanical energy is converted to thermal energy, per unit volume. Utilizing the definition of the substantial derivative permits Eq. (2.23) to be written in the form

$$\rho \frac{Di}{Dt} = \frac{Dp}{Dt} - \nabla \cdot \mathbf{q} + \phi \tag{2.24}$$

Equation (2.23) can also be written in terms of the internal energy u as

$$\rho \frac{Du}{Dt} = -\nabla \cdot \mathbf{q} - p\nabla \cdot \mathbf{V} + \phi \tag{2.25}$$

It is frequently convenient to treat the temperature as the primary thermal dependent variable. In this case, Eq. (2.24) can also be written as

$$\rho c_p \frac{DT}{Dt} = \beta T \frac{Dp}{Dt} - \nabla \cdot \mathbf{q} + \phi \tag{2.26}$$

and Eq. (2.25) as

$$\rho c_v \frac{DT}{Dt} = -T\left(\frac{\partial p}{\partial T}\right)_v (\nabla \cdot \mathbf{V}) - \nabla \cdot \mathbf{q} + \phi \tag{2.27}$$

In the above, $\beta = -(1/\rho)(\partial \rho/\partial T)_p$ is the coefficient of thermal expansion (equal to $1/T$ for an ideal gas). Thc dissipation function ϕ is given by

$$\phi = \mu \left[2\left(\frac{\partial u}{\partial x}\right)^2 + 2\left(\frac{\partial v}{\partial y}\right)^2 + 2\left(\frac{\partial w}{\partial z}\right)^2 + \left(\frac{\partial v}{\partial x} + \frac{\partial u}{\partial y}\right)^2 + \left(\frac{\partial w}{\partial y} + \frac{\partial v}{\partial z}\right)^2 + \left(\frac{\partial u}{\partial z} + \frac{\partial w}{\partial x}\right)^2 - \frac{2}{3}\left(\frac{\partial u}{\partial x} + \frac{\partial v}{\partial y} + \frac{\partial w}{\partial z}\right)^2 \right] \tag{2.28}$$

When the flow can be treated as incompressible with constant thermal conductivity, Eq. (2.27) reduces to

$$\rho c_v \frac{DT}{Dt} = k \nabla^2 T + \phi \tag{2.29}$$

and the last term in the dissipation function, equal to $-\frac{2}{3}u(\nabla \cdot \mathbf{V})^2$, can be neglected. In many low-speed flows, the dissipation function can be neglected entirely.

2.2.2 Coordinate Systems

The basic equations for convection are derived from fundamental conservation principles. Thus, a form of these equations can be formulated for any coordinate system. The equations have been expressed in terms of a Cartesian coordinate system thus far in this chapter. For many applications it is more convenient to use a different coordinate system. Whenever possible, a coordinate line should lie on the boundary of the body undergoing convection. It is also convenient if the coordinate lines are orthogonal, because in this case the equations appear in their simplest form. This is not essential, however. The conservation principles can be formulated in coordinate systems which are not orthogonal. Nonorthogonal but body-conforming coordinate systems are frequently used in the numerical solution of convective flows. The generation of such body-conforming coordinate systems is discussed in considerable detail in [3].

Here the conservation equations will be formulated in a generalized orthogonal curvilinear coordinate system. Utilizing the development in [4], x_1, x_2, x_3 are defined as a set of generalized orthogonal curvilinear coordinates with $\hat{\imath}_1, \hat{\imath}_2, \hat{\imath}_3$ the corresponding unit vectors. The rectangular Cartesian coordinates are related to the generalized curvilinear coordinates by

$$x = x(x_1, x_2, x_3)$$

$$y = y(x_1, x_2, x_3)$$

$$z = z(x_1, x_2, x_3)$$

so that if the Jacobian $\partial(x, y, z)/\partial(x_1, x_2, x_3)$ is nonzero, then

$$x_1 = x_1(x, y, z)$$

$$x_2 = x_2(x, y, z)$$

$$x_3 = x_3(x, y, z)$$

The elemental arc length in Cartesian coordinates is

$$(ds)^2 = (dx)^2 + (dy)^2 + (dz)^2$$

and, in terms of the curvilinear coordinates,

$$(ds)^2 = (h_1\, dx_1)^2 + (h_2\, dx_2)^2 + (h_3\, dx_3)^2$$

where the metric coefficients h_1, h_2, and h_3 are given by

$$(h_1)^2 = \left(\frac{\partial x}{\partial x_1}\right)^2 + \left(\frac{\partial y}{\partial x_1}\right)^2 + \left(\frac{\partial z}{\partial x_1}\right)^2$$

$$(h_2)^2 = \left(\frac{\partial x}{\partial x_2}\right)^2 + \left(\frac{\partial y}{\partial x_2}\right)^2 + \left(\frac{\partial z}{\partial x_2}\right)^2$$

$$(h_3)^2 = \left(\frac{\partial x}{\partial x_3}\right)^2 + \left(\frac{\partial y}{\partial x_3}\right)^2 + \left(\frac{\partial z}{\partial x_3}\right)^2$$

If ϕ is an arbitrary scalar and **A** is an arbitrary vector, the expressions for the gradient, divergence, curl, and Laplacian operator in the generalized curvilinear coordinates become

$$\nabla\phi = \frac{1}{h_1}\frac{\partial\phi}{\partial x_1}\hat{\mathbf{i}}_1 + \frac{1}{h_2}\frac{\partial\phi}{\partial x_2}\hat{\mathbf{i}}_2 + \frac{1}{h_3}\frac{\partial\phi}{\partial x_3}\hat{\mathbf{i}}_3 \tag{2.30}$$

$$\nabla \cdot \mathbf{A} = \frac{1}{h_1 h_2 h_3}\left[\frac{\partial}{\partial x_1}(h_3 h_3 A_1) + \frac{\partial}{\partial x_2}(h_3 h_1 A_2) + \frac{\partial}{\partial x_3}(h_1 h_2 A_3)\right] \tag{2.31}$$

$$\begin{aligned}\nabla \times \mathbf{A} = \frac{1}{h_1 h_2 h_3}\Bigg\{ & h_1\left[\frac{\partial(h_3 A_3)}{\partial x_2} - \frac{\partial(h_2 A_2)}{\partial x_3}\right]\hat{\mathbf{i}}_1 + h_2\left[\frac{\partial(h_1 A_1)}{\partial x_3} - \frac{\partial(h_3 A_3)}{\partial x_1}\right]\hat{\mathbf{i}}_2 \\ & + h_3\left[\frac{\partial(h_2 A_2)}{\partial x_1} - \frac{\partial(h_1 A_1)}{\partial x_2}\right]\hat{\mathbf{i}}_3\Bigg\}\end{aligned} \tag{2.32}$$

$$\begin{aligned}\nabla^2\phi = \frac{1}{h_1 h_2 h_3}\Bigg[& \frac{\partial}{\partial x_1}\left(\frac{h_2 h_3}{h_1}\frac{\partial\phi}{\partial x_1}\right) + \frac{\partial}{\partial x_2}\left(\frac{h_3 h_1}{h_2}\frac{\partial\phi}{\partial x_2}\right) \\ & + \frac{\partial}{\partial x_3}\left(\frac{h_1 h_2}{h_3}\frac{\partial\phi}{\partial x_3}\right)\Bigg]\end{aligned} \tag{2.33}$$

The expression $\mathbf{V} \cdot \nabla \mathbf{V}$ which is contained in the momentum-equation term $D\mathbf{V}/Dt$ can be evaluated as

$$\begin{aligned}\mathbf{V} \cdot \nabla \mathbf{V} = & \left(\frac{u_1}{h_1}\frac{\partial u_1}{\partial x_1} + \frac{u_2}{h_2}\frac{\partial u_1}{\partial x_2} + \frac{u_3}{h_3}\frac{\partial u_1}{\partial x_3} + \frac{u_1 u_2}{h_1 h_2}\frac{\partial h_1}{\partial x_2}\right. \\ & \left. + \frac{u_1 u_3}{h_1 h_3}\frac{\partial h_1}{\partial x_3} - \frac{u_2^2}{h_1 h_2}\frac{\partial h_2}{\partial x_1} - \frac{u_3^2}{h_1 h_3}\frac{\partial h_3}{\partial x_1}\right)\hat{\mathbf{i}}_1 \\ & + \left(\frac{u_1}{h_1}\frac{\partial u_2}{\partial x_1} + \frac{u_2}{h_2}\frac{\partial u_2}{\partial x_2} + \frac{u_3}{h_3}\frac{\partial u_2}{\partial x_3} - \frac{u_1^2}{h_1 h_2}\frac{\partial h_1}{\partial x_2}\right. \\ & \left. + \frac{u_1 u_2}{h_1 h_2}\frac{\partial h_2}{\partial x_1} + \frac{u_2 u_3}{h_2 h_3}\frac{\partial h_2}{\partial x_3} - \frac{u_3^2}{h_2 h_3}\frac{\partial h_3}{\partial x_2}\right)\hat{\mathbf{i}}_2 \\ & + \left(\frac{u_1}{h_1}\frac{\partial u_3}{\partial x_1} + \frac{u_2}{h_2}\frac{\partial u_3}{\partial x_2} + \frac{u_3}{h_3}\frac{\partial u_3}{\partial x_3} - \frac{u_1^2}{h_1 h_3}\frac{\partial h_1}{\partial x_3}\right. \\ & \left. - \frac{u_2^2}{h_2 h_3}\frac{\partial h_2}{\partial x_3} + \frac{u_1 u_3}{h_1 h_3}\frac{\partial h_3}{\partial x_1} + \frac{u_2 u_3}{h_2 h_3}\frac{\partial h_3}{\partial x_2}\right)\hat{\mathbf{i}}_3\end{aligned} \tag{2.34}$$

where u_1, u_2, u_3 are the velocity components in the x_1, x_2, x_3 coordinate directions.

The components of the stress tensor given by Eq. (2.10) can be expressed in terms of the generalized curvilinear coordinate as

$$\Pi_{x_1 x_1} = -p + \tfrac{2}{3}\mu(2e_{x_1 x_1} - e_{x_2 x_2} - e_{x_3 x_3})$$

$$\Pi_{x_2 x_2} = -p + \tfrac{2}{3}\mu(2e_{x_2 x_2} - e_{x_1 x_1} - e_{x_3 x_3})$$

$$\Pi_{x_3 x_3} = -p + \tfrac{2}{3}\mu(2e_{x_3 x_3} - e_{x_1 x_1} - e_{x_2 x_2})$$

$$\Pi_{x_2x_3} = \Pi_{x_3x_2} = \mu e_{x_2x_3}$$

$$\Pi_{x_1x_3} = \Pi_{x_3x_1} = \mu e_{x_1x_3}$$

$$\Pi_{x_1x_2} = \Pi_{x_2x_1} = \mu e_{x_1x_2} \tag{2.35}$$

where the expressions for the strains are

$$e_{x_1x_1} = \frac{1}{h_1}\frac{\partial u_1}{\partial x_1} + \frac{u_2}{h_1h_2}\frac{\partial h_1}{\partial x_2} + \frac{u_3}{h_1h_3}\frac{\partial h_1}{\partial x_3}$$

$$e_{x_2x_2} = \frac{1}{h_2}\frac{\partial u_2}{\partial x_2} + \frac{u_3}{h_2h_3}\frac{\partial h_2}{\partial x_3} + \frac{u_1}{h_1h_2}\frac{\partial h_2}{\partial x_1}$$

$$e_{x_3x_3} = \frac{1}{h_3}\frac{\partial u_3}{\partial x_3} + \frac{u_1}{h_1h_3}\frac{\partial h_3}{\partial x_1} + \frac{u_2}{h_2h_3}\frac{\partial h_3}{\partial x_2}$$

$$e_{x_3x_3} = \frac{h_3}{h_2}\frac{\partial}{\partial x_2}\left(\frac{u_3}{h_3}\right) + \frac{h_2}{h_3}\frac{\partial}{\partial x_3}\left(\frac{u_2}{h_2}\right)$$

$$e_{x_1x_3} = \frac{h_1}{h_3}\frac{\partial}{\partial x_3}\left(\frac{u_1}{h_1}\right) + \frac{h_3}{h_1}\frac{\partial}{\partial x_1}\left(\frac{u_3}{h_3}\right)$$

$$e_{x_1x_2} = \frac{h_2}{h_1}\frac{\partial}{\partial x_1}\left(\frac{u_2}{h_2}\right) + \frac{h_1}{h_2}\frac{\partial}{\partial x_2}\left(\frac{u_1}{h_1}\right) \tag{2.36}$$

The components of $\nabla \cdot \Pi$ are

$$x_1: \quad \frac{1}{h_1h_2h_3}\left[\frac{\partial}{\partial x_1}\left(h_2h_3\Pi_{x_1x_1}\right) + \frac{\partial}{\partial x_2}\left(h_1h_3\Pi_{x_1x_2}\right) + \frac{\partial}{\partial x_3}\left(h_1h_2\Pi_{x_1x_3}\right)\right]$$

$$+\Pi_{x_1x_2}\frac{1}{h_1h_2}\frac{\partial h_1}{\partial x_2} + \Pi_{x_1x_3}\frac{1}{h_1h_3}\frac{\partial h_1}{\partial x_3} - \Pi_{x_2x_2}\frac{1}{h_1h_2}\frac{\partial h_2}{\partial x_1} - \Pi_{x_3x_3}\frac{1}{h_1h_3}\frac{\partial h_3}{\partial x_1}$$

$$x_2: \quad \frac{1}{h_1h_2h_3}\left[\frac{\partial}{\partial x_1}\left(h_2h_3\Pi_{x_1x_2}\right) + \frac{\partial}{\partial x_2}\left(h_1h_3\Pi_{x_2x_2}\right) + \frac{\partial}{\partial x_3}\left(h_1h_2\Pi_{x_2x_3}\right)\right]$$

$$+\Pi_{x_2x_3}\frac{1}{h_2h_3}\frac{\partial h_2}{\partial x_3} + \Pi_{x_1x_2}\frac{1}{h_1h_2}\frac{\partial h_2}{\partial x_1} - \Pi_{x_3x_3}\frac{1}{h_2h_3}\frac{\partial h_3}{\partial x_2} - \Pi_{x_1x_1}\frac{1}{h_1h_2}\frac{\partial h_1}{\partial x_2}$$

$$x_3: \quad \frac{1}{h_1h_2h_3}\left[\frac{\partial}{\partial x_1}\left(h_2h_3\Pi_{x_1x_3}\right) + \frac{\partial}{\partial x_2}\left(h_1h_3\Pi_{x_2x_3}\right) + \frac{\partial}{\partial x_3}\left(h_1h_2\Pi_{x_3x_3}\right)\right]$$

$$+\Pi_{x_1x_3}\frac{1}{h_1h_3}\frac{\partial h_3}{\partial x_1} + \Pi_{x_2x_3}\frac{1}{h_2h_3}\frac{\partial h_3}{\partial x_2} - \Pi_{x_1x_1}\frac{1}{h_1h_3}\frac{\partial h_1}{\partial x_3} - \Pi_{x_2x_2}\frac{1}{h_2h_3}\frac{\partial h_2}{\partial x_3} \tag{2.37}$$

In generalized curvilinear coordinates, the dissipation function becomes

$$\Phi = \mu\Big[2\big(e_{x_1x_1}^2 + e_{x_2x_2}^2 + e_{x_3x_3}^2\big) + e_{x_2x_3}^2 + e_{x_1x_3}^2 + e_{x_1x_2}^2 - \tfrac{2}{3}\big(e_{x_1x_1} + e_{x_2x_2} + e_{x_3x_3}\big)^2\Big] \tag{2.38}$$

The above expressions can now be used to derive the conservation equations in any orthogonal curvilinear coordinate system. Examples of the most common orthogonal coordinate systems are given below.

1. Cartesian coordinates:

$$x_1 = x, \quad h_1 = 1, \quad u_1 = u$$
$$x_2 = y, \quad h_2 = 1, \quad u_2 = v$$
$$x_3 = z, \quad h_3 = 1, \quad u_3 = w$$

2. Cylindrical coordinates:

$$x_1 = r, \quad h_1 = 1, \quad u_1 = u_r$$
$$x_2 = \theta, \quad h_2 = r, \quad u_2 = u_\theta$$
$$x_3 = z, \quad h_3 = 1, \quad u_3 = u_z$$

3. Spherical coordinates:

$$x_1 = r, \quad h_1 = 1, \quad u_1 = u_r$$
$$x_2 = \theta, \quad h_2 = r, \quad u_2 = u_\theta$$
$$x_3 = \phi, \quad h_3 = r\sin\theta, \quad u_3 = u_\phi$$

4. 2D or axisymmetric body intrinsic coordinates:

$$x_1 = \xi, \quad h_1 = 1 + K(\xi)\eta, \quad u_1 = u$$
$$x_2 = \eta, \quad h_2 = 1, \quad u_2 = v$$
$$x_3 = \phi, \quad h_3 = [r(\xi) + \eta\cos\alpha(\xi)]^m, \quad u_3 = w = 0$$

where $K(\xi)$ is the local body curvature, $r(\xi)$ is the cylindrical radius, and

$$m = \begin{cases} 0 & \text{for 2D flow} \\ 1 & \text{for axisymmetric flow} \end{cases}$$

These coordinate systems are illustrated in Fig. 2.1.

2.2.3 Reynolds Equations for Turbulent Flow

The unsteady Navier-Stokes equations are generally considered to be valid for turbulent flows in the continuum regime. However, the complexity and random nature of the turbulent motion has so far made exact analytical solutions of the Navier-Stokes

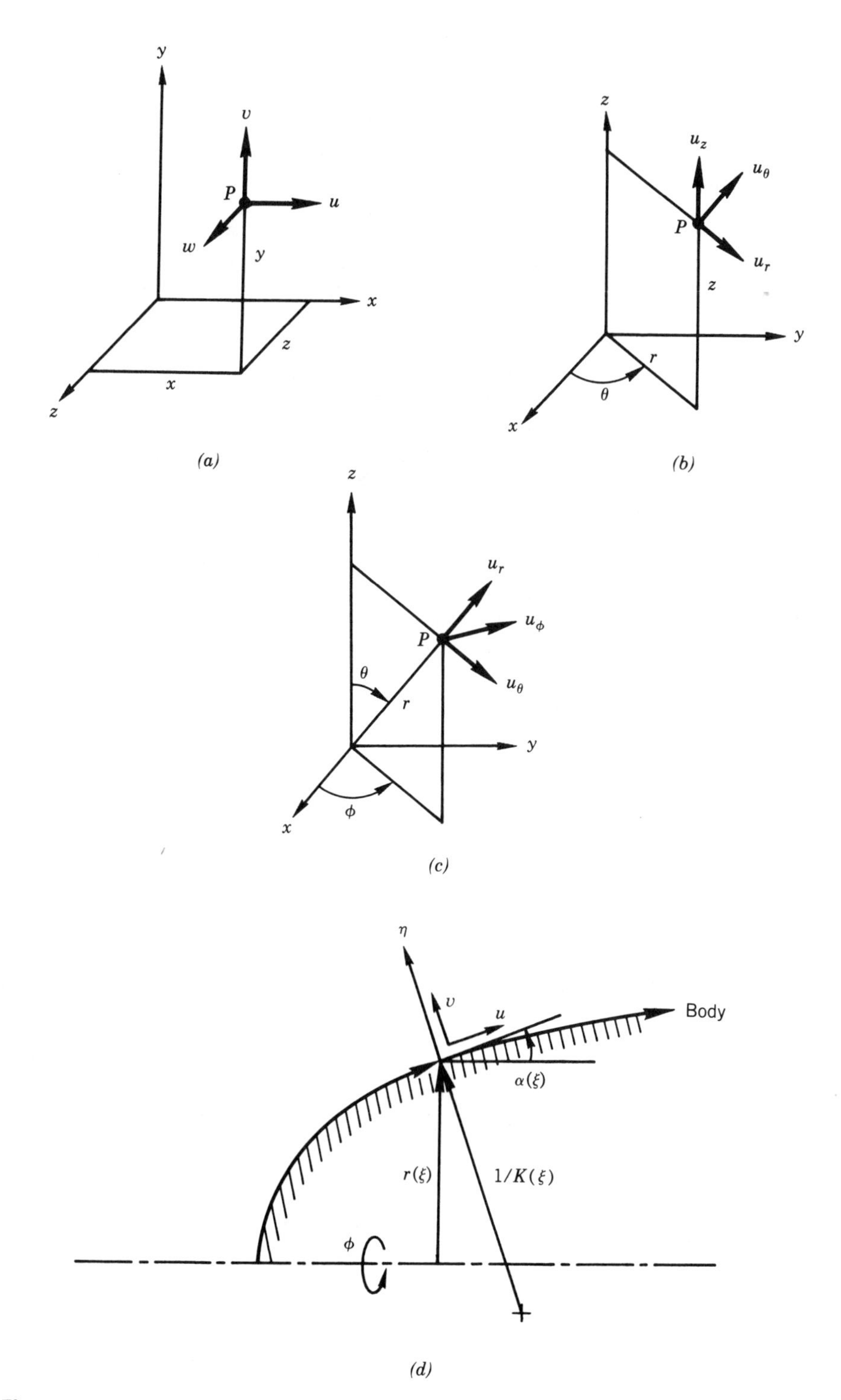

Figure 2.1. Orthogonal coordinate systems: (*a*) Cartesian coordinates (x, y, z); (*b*) cylindrical coordinates (r, θ, z); (*c*) spherical coordinates (r, θ, ϕ); (*d*) 2D or axisymmetric body intrinsic coordinates (ξ, η, ϕ). (Adapted from [4] by permission.)

equations for these flows impossible. Even computer simulations based on the Navier-Stokes equations are beyond reach at present for practical problems. This is because extremely large numbers of grid points are required to resolve the small time and space scales of the turbulent motion. If computer technology continues to advance at about the present rate, it is conceivable that at some time during the next century, computer simulations of flows of practical interest will be possible by solving the unsteady Navier-Stokes equations. However, some authorities believe this will never be possible.

Most of the present-day prediction methods for turbulent flows are based on the time-averaged Navier-Stokes equations. These equations are also referred to as the Reynolds equations of motion or the Reynolds-averaged equations. Time-averaging the equations of motion gives rise to new terms which can be interpreted as "apparent" stress gradients and heat-flux quantities associated with the turbulent motion. These new quantities must be related to the mean flow variables through turbulence models. This process introduces further assumptions and approximations. Thus, this attack on the turbulent flow problem through solving the Reynolds equations of motion does not follow entirely from first principles, since additional assumptions must be made to "close" the system of equations.

The Reynolds equations are derived by decomposing the dependent variables in the conservation equations into time-mean (obtained over an appropriate time interval) and fluctuating components and then time-averaging the entire equation. Two types of averaging are presently used, the classical Reynolds averaging and the mass-weighted averaging suggested by Favre [5]. For flows in which density fluctuations can be neglected, the two formulations become identical.

In the conventional averaging procedure, following Reynolds, we define the time average $\bar{f}$ of a quantity f as

$$\bar{f} \equiv \frac{1}{\Delta t} \int_{t_0}^{t_0+\Delta t} f \, dt \tag{2.39}$$

We require that Δt be large compared to the period of the random fluctuations associated with the turbulence, but small compared to the time constant for any slow variations in the flow field associated with ordinary unsteady flows. The interval Δt is sometimes taken to approach infinity as a limit, but this should be interpreted as being relative to the characteristic fluctuation period of the turbulence. For practical measurements, Δt must be finite.

In the conventional Reynolds decomposition, the randomly changing flow variables are replaced by time averages plus fluctuations about the average. For a Cartesian coordinate system this gives

$$\begin{aligned} &u = \bar{u} + u', \quad v = \bar{v} + v', \quad w = \bar{w} + w', \quad \rho = \bar{\rho} + \rho' \\ &p = \bar{p} + p', \quad i = \bar{i} + i', \quad T = \bar{T} + T', \quad H = \bar{H} + H' \end{aligned} \tag{2.40}$$

where the total enthalpy H is defined by $H = i + u_i u_i/2$. Fluctuations in other fluid properties such as viscosity, thermal conductivity, and specific heat are usually small and will be neglected here.

By definition, the time average of a fluctuating quantity is zero:

$$\overline{f'} = \frac{1}{\Delta t} \int_{t_0}^{t_0+\Delta t} f' \, dt \equiv 0 \tag{2.41}$$

It should be clear from these definitions that for symbolic flow variables f and g, the following relations hold:

$$\overline{\bar{f}\,g'} = 0, \qquad \overline{\bar{f}\,g} = \bar{f}\,\bar{g}, \qquad \overline{f + g} = \bar{f} + \bar{g} \tag{2.42}$$

It should also be clear that, whereas $\overline{f'} \equiv 0$, the time average of the product of two fluctuating quantities is, in general, not equal to zero, i.e., $\overline{f'f'} \neq 0$. In fact, the root-mean-square of the velocity fluctuations is known as the turbulence intensity.

For treatment of compressible flows and mixtures of gases in particular, the mass-weighted averaging is convenient. In this approach, mass-averaged variables are defined according to $\tilde{f} = \overline{\rho f}/\bar{\rho}$. This gives

$$\tilde{u} = \frac{\overline{\rho u}}{\bar{\rho}}, \quad \tilde{v} = \frac{\overline{\rho v}}{\bar{\rho}}, \quad \tilde{w} = \frac{\overline{\rho w}}{\bar{\rho}}, \quad \tilde{\imath} = \frac{\overline{\rho i}}{\bar{\rho}}, \quad \tilde{T} = \frac{\overline{\rho T}}{\bar{\rho}}, \quad \tilde{H} = \frac{\overline{\rho H}}{\bar{\rho}} \tag{2.43}$$

It should be noted that only the velocity components and thermal variables are mass-averaged. Fluid properties such as the density and pressure are treated as before.

To substitute into the conservation equations, new fluctuating quantities are defined by

$$u = \tilde{u} + u'', \quad v = \tilde{v} + v'', \quad w = \tilde{w} + w'', \quad i = \tilde{\imath} + i'', \quad T = \tilde{T} + T'', \quad H = \tilde{H} + H'' \tag{2.44}$$

It is important to note that the time averages of the doubly primed fluctuating quantities ($\overline{u''}$, $\overline{v''}$, etc.) are not equal to zero, in general, unless $\rho' = 0$. In fact, it can be shown that $\overline{u''} = -\overline{\rho' u'}/\bar{\rho}$, $\overline{v''} = -\overline{\rho' v'}/\bar{\rho}$, etc. Instead, the time average of the doubly primed fluctuation multiplied by the density is equal to zero:

$$\overline{\rho f''} \equiv 0 \tag{2.45}$$

The above identity can be established by expanding $\overline{\rho f} = \overline{\rho(\tilde{f} + f'')}$ and using the definition of $\tilde{f}$.

Reynolds Form of the Continuity Equation. Starting with the continuity equation written in the Cartesian coordinate system, Eq. (2.3), the variables are first decomposed into conventional time-averaged variables plus fluctuating components as given by Eq. (2.40). The entire equation is then time-averaged. Several of the terms are identically zero because of the identity of Eq. (2.41). Finally, the Reynolds form of the continuity equation is conventionally averaged variables can be written

$$\frac{\partial \bar{\rho}}{\partial t} + \frac{\partial}{\partial x_j}\left(\bar{\rho}\bar{u}_j + \overline{\rho' u_i'}\right) = 0 \tag{2.46}$$

To develop the Reynolds form of the continuity equation in mass-weighted variables, the variables in Eq. (2.3) are decomposed as indicated by Eq. (2.44) except for the density, which is decomposed according to Eq. (2.40). When the entire continuity equation is time-averaged, several terms are observed to be identically zero because of Eqs. (2.41) and (2.45). Finally, the continuity equation in mass-weighted variables can be written as

$$\frac{\partial \rho}{\partial t} + \frac{\partial}{\partial x_j}(\bar{\rho}\tilde{u}_j) = 0 \tag{2.47}$$

It is noted that Eq. (2.47) is more compact in form than Eq. (2.46). For incompressible flows, $\rho' = 0$ and the differences between the conventional and mass-weighted variables vanish, so that the continuity equation can be written as

$$\frac{\partial \rho \bar{u}_j}{\partial x_j} = 0 \tag{2.48}$$

Reynolds Form of the Momentum Equations. Working first with the conventionally averaged variables, the dependent variables in Eqs. (2.17) to (2.19) are replaced with the time averages plus fluctuations according to Eq. (2.40). Next the entire equation is time-averaged. Terms which are linear in fluctuating quantities become zero when time-averaged, as they did in the continuity equation. Several terms disappear in this manner, while others can be grouped together and found equal to zero through use of the continuity equation. The resulting Reynolds momentum equation (all three components) can be written

$$\frac{\partial}{\partial t}\left(\bar{\rho}\bar{u}_i + \overline{\rho' u_i'}\right) + \frac{\partial}{\partial x_j}\left(\bar{\rho}\bar{u}_i\bar{u}_j + \bar{u}_i\overline{\rho' u_j'}\right)$$

$$= -\frac{\partial \bar{p}}{\partial x_i} + \frac{\partial}{\partial x_j}\left(\bar{\tau}_{ij} - \bar{u}_j\overline{\rho' u_i'} - \bar{\rho}\overline{u_i' u_j'} - \overline{\rho' u_i' u_j'}\right) \tag{2.49}$$

where

$$\bar{\tau}_{ij} = \mu\left[\left(\frac{\partial \bar{u}_i}{\partial x_j} + \frac{\partial \bar{u}_j}{\partial x_i}\right) - \frac{2}{3}\delta_{ij}\frac{\partial \bar{u}_k}{\partial x_k}\right] \tag{2.50}$$

To develop the Reynolds momentum equation in mass-weighted variables, the decomposition indicated by Eq. (2.44) is used to represent the instantaneous variables in Eqs. (2.17)–(2.19). Next, the entire equation is time-averaged and the identity (2.45) is used to eliminate terms. The complete Reynolds momentum equation in mass-weighted variables becomes

$$\frac{\partial}{\partial t}(\bar{\rho}\tilde{u}_i) + \frac{\partial}{\partial x_j}(\bar{\rho}\tilde{u}_i\tilde{u}_j) = -\frac{\partial \bar{p}}{\partial x_i} + \frac{\partial}{\partial x_j}\left(\tau_{ij} - \overline{\rho u_i'' u_j''}\right) \tag{2.51}$$

where, neglecting viscosity fluctuations, $\bar{\tau}_{ij}$ becomes

$$\bar{\tau}_{ij} = \mu\left[\left(\frac{\partial \tilde{u}_i}{\partial x_j} + \frac{\partial \tilde{u}_j}{\partial x_i}\right) - \frac{2}{3}\delta_{ij}\frac{\partial \tilde{u}_k}{\partial x_k}\right] + \mu\left[\left(\frac{\partial \overline{u_i''}}{\partial x_j} + \frac{\partial \overline{u_j''}}{\partial x_i}\right) - \frac{2}{3}\delta_{ij}\frac{\partial \overline{u_k''}}{\partial x_k}\right] \tag{2.52}$$

The momentum equation (2.51) in mass-weighted variables is simpler in form than the corresponding equation using conventional variables. It is noted, however, that even when viscosity fluctuations are neglected, $\bar{\tau}_{ij}$ is more complex in Eq. (2.52) than $\bar{\tau}_{ij}$ which appeared in the conventionally averaged equation (2.50). In practice, the viscous terms involving the doubly primed fluctuations are expected to be small and are likely candidates for being neglected on the basis of order-of-magnitude arguments.

For incompressible flows the momentum equation can be written in the simpler form

$$\frac{\partial}{\partial t}(\rho\bar{u}_i) + \frac{\partial}{\partial x_j}(\rho\bar{u}_i\bar{u}_j) = -\frac{\partial\bar{p}}{\partial x_i} + \frac{\partial}{\partial x_j}\left(\bar{\tau}_{ij} - \rho\overline{u_i'u_j'}\right) \tag{2.53}$$

where $\bar{\tau}_{ij}$ takes on the reduced form

$$\bar{\tau}_{ij} = \mu\left(\frac{\partial u_i}{\partial x_j} + \frac{\partial u_j}{\partial x_i}\right) \tag{2.54}$$

As noted in connection with the continuity equation, there is no difference between the mass-weighted and conventional variables for incompressible flow.

Reynolds Form of the Energy Equation. The thermal variables H, i, and T are all related, and the energy equation takes on different forms depending upon which one is chosen to be the transported thermal variable. To develop one common form, the energy equation as given by Eq. (2.21) is used as a starting point. To obtain the Reynolds energy equation in conventionally averaged variables, the dependent variables in Eq. (2.21) are replaced with the decomposition of Eq. (2.40). After time averaging, the equation becomes

$$\frac{\partial}{\partial t}\left(\bar{\rho}\bar{H} + \overline{\rho'H'}\right) + \frac{\partial}{\partial x_j}\left(\bar{\rho}\bar{u}_j\bar{H} + \bar{\rho}\overline{u_j'H'} + \overline{\rho'u_j'}\bar{H} + \overline{\rho'u_j'H'} + \bar{u}_j\overline{\rho'H'} - k\frac{\partial\bar{T}}{\partial x_j}\right)$$
$$= \frac{\partial\bar{p}}{\partial t} + \frac{\partial}{\partial x_j}\left\{\bar{u}_i\left(-\tfrac{2}{3}\mu\delta_{ij}\frac{\partial\bar{u}_k}{\partial x_k}\right) + \mu\bar{u}_i\left(\frac{\partial\bar{u}_j}{\partial x_i} + \frac{\partial\bar{u}_i}{\partial x_j}\right)\right.$$
$$\left. -\tfrac{2}{3}\mu\delta_{ij}\overline{u_i'\frac{\partial u_k'}{\partial x_k}} + \mu\left(\overline{u_i'\frac{\partial u_j'}{\partial x_i}} + \overline{u_i'\frac{\partial u_i'}{\partial x_j}}\right)\right\} \tag{2.55}$$

It is frequently desirable to utilize the static temperature as the primary thermal variable in the energy equation. Letting $i = c_pT$ in Eq. (2.23), replacing the variables in Eq. (2.23) with the decomposition of Eq. (2.40) and time averaging gives

$$\frac{\partial}{\partial t}\left(c_p\bar{\rho}\bar{T} + c_p\overline{\rho'T'}\right) + \frac{\partial}{\partial x_j}\left(\bar{\rho}c_p\bar{T}\bar{u}_j\right)$$
$$= \frac{\partial\bar{p}}{\partial t} + \bar{u}_j\frac{\partial\bar{p}}{\partial x_j} + \overline{u_j'\frac{\partial p'}{\partial x_j}} + \frac{\partial}{\partial x_j}\left(k\frac{\partial\bar{T}}{\partial x_j} - \bar{\rho}c_p\overline{T'u_j'} - c_p\overline{\rho'T'u_j'}\right) + \bar{\phi} \tag{2.56}$$

where

$$\bar{\phi} = \overline{\tau_{ij}\frac{\partial u_i}{\partial x_j}} = \bar{\tau}_{ij}\frac{\partial\bar{u}_i}{\partial x_j} + \overline{\tau_{ij}'\frac{\partial u_i'}{\partial x_j}} \tag{2.57}$$

The $\bar{\tau}_{ij}$ in Eq. (2.57) should be evaluated as indicated by Eq. (2.50).

To develop the Reynolds form of the energy equation in mass-weighted variables, the dependent variables in Eq. (2.21) are replaced with the decomposition of Eq. (2.44) and the entire equation is time-averaged. The result can be written

$$\frac{\partial}{\partial t}(\bar{\rho}\tilde{H}) + \frac{\partial}{\partial x_j}\left(\bar{\rho}\tilde{u}_j\tilde{H} + \overline{\rho u_j''H''} - k\frac{\partial \bar{T}}{\partial x_j}\right) = \frac{\partial \bar{p}}{\partial t} + \frac{\partial}{\partial x_j}\left(\tilde{u}_i\bar{\tau}_{ij} + \overline{u_i''\tau_{ij}}\right) \quad (2.58)$$

where $\bar{\tau}_{ij}$ can be evaluated by means of Eq. (2.52) in terms of mass-weighted variables.

In terms of static temperature, the Reynolds energy equation in mass-weighted variables becomes

$$\frac{\partial}{\partial t}(\bar{\rho}c_p\tilde{T}) + \frac{\partial}{\partial x_j}(\bar{\rho}c_p\tilde{T}\tilde{u}_j) = \frac{\partial \bar{p}}{\partial t} + \tilde{u}_j\frac{\partial \bar{p}}{\partial x_j} + \overline{u_j''\frac{\partial p}{\partial x_j}}$$

$$+ \frac{\partial}{\partial x_j}\left(k\frac{\partial T}{\partial x_j} + k\frac{\overline{\partial T''}}{\partial x_j} - c_p\overline{\rho T''u_j''}\right) + \bar{\phi} \quad (2.59)$$

where

$$\bar{\phi} = \overline{\tau_{ij}\frac{\partial u_i}{\partial x_j}} = \bar{\tau}_{ij}\frac{\partial \tilde{u}_i}{\partial x_j} + \overline{\tau_{ij}\frac{\partial u_i''}{\partial x_j}} \quad (2.60)$$

For incompressible flows, the energy equation can be written in terms of the total enthalpy as

$$\frac{\partial \rho \bar{H}}{\partial t} + \frac{\partial}{\partial x_j}\left(\rho u_j\bar{H} + \overline{\rho u_j'H'} - k\frac{\partial \bar{T}}{\partial x_j}\right)$$

$$= \frac{\partial \bar{p}}{\partial t} + \frac{\partial}{\partial x_j}\left\{\mu\bar{u}_i\left(\frac{\partial \bar{u}_j}{\partial x_i} + \frac{\partial \bar{u}_i}{\partial x_j}\right) + \mu\left(\overline{u_i'\frac{\partial u_j'}{\partial x_i}} + \overline{u_i'\frac{\partial u_i'}{\partial x_j}}\right)\right\} \quad (2.61)$$

and in terms of the static temperature as

$$\frac{\partial}{\partial t}(\rho c_p\bar{T}) + \frac{\partial}{\partial x_j}(\rho c_p\bar{T}\bar{u}_j)$$

$$= \frac{\partial \bar{p}}{\partial t} + \bar{u}_j\frac{\partial \bar{p}}{\partial x_j} + \overline{u_j'\frac{\partial p'}{\partial x_j}} + \frac{\partial}{\partial x_j}\left(k\frac{\partial \bar{T}}{\partial x_j} - \rho c_p\overline{T'u_j'}\right) + \bar{\phi} \quad (2.62)$$

where $\bar{\phi}$ is reduced slightly in complexity due to the vanishing of the volumetric dilation term in $\bar{\tau}_{ij}$ for incompressible flow.

Comments on the Reynolds Equations. The Reynolds equations enforce the conservation principles in terms of time-averaged variables. In comparing the Reynolds equations with the Navier-Stokes equations, some differences are noted. The Reynolds equations contain terms involving velocity, density, and temperature fluctuations. Among these new terms can be identified "apparent" or turbulent stress and heat-flux quantities. For example, all the terms containing fluctuations on the right-hand side of Eqs. (2.49) and (2.51) represent apparent turbulent stresses. Similarly, terms in the energy equation containing the time average of the product of a temperature fluctuation and a velocity fluctuation can be identified in turbulent heat-flux terms. These new quantities originated from the momentum and energy flux terms of the Navier-Stokes equations.

The Reynolds equations cannot be solved in the form given, because the apparent turbulent stresses and heat-flux quantities must be viewed as new unknowns. To proceed further, it is necessary to find additional equations involving the new unknowns or make assumptions regarding the relationship between the apparent turbulent quantities and the time-mean flow variables. This is known as the closure problem. The process of obtaining closure, commonly called "turbulence modeling," will be discussed in a later section.

2.2.4 Reduced Forms of the Equations

Inviscid Flow. As was pointed out in the introduction to this chapter, there are many flows in which the important effects of viscosity and heat conduction can be limited to a thin boundary layer near solid surfaces. If this boundary layer is very thin compared to the characteristic length of the flow field, the inviscid (nonviscous, nonconducting) portion of the flow can be solved independently of the boundary layer. The appropriate equations for the inviscid flow are obtained by dropping both viscous and heat transfer terms from the complete Navier-Stokes equations. These simplified equations are generally known as the Euler equations. The solution of these equations provides the edge (or boundary) conditions needed for the solution to the boundary-layer form of the conservation equations, from which heat transfer information can be obtained.

The continuity equation contains no viscous or heat conduction terms, so that the various forms of the continuity equation given previously apply to the inviscid flow. However, if the steady form of the continuity equation reduces to two terms for a given coordinate system, it becomes possible to discard the continuity equation by introducing the so-called stream function. This can be done whether the flow is viscous or nonviscous. If the stream function ψ is defined such that

$$\rho u = \frac{\partial \psi}{\partial y} \qquad \rho v = -\frac{\partial \psi}{\partial x} \tag{2.63}$$

it can be seen by substitution that Eq. (2.3) is satisfied for steady flow. Hence, the continuity equation does not need to be solved, and the number of dependent variables is reduced by one. The disadvantage is that the velocity derivatives in the remaining equations are replaced using Eq. (2.63), so that these remaining equations will now contain derivatives which are one order higher. For a steady, incompressible flow, the density can be eliminated from the continuity equation and in Cartesian coordinates the stream function is defined by

$$u = \frac{\partial \psi}{\partial y} \qquad v = -\frac{\partial \psi}{\partial x} \tag{2.64}$$

For a steady, axisymmetric, compressible flow in cylindrical coordinates, the continuity equation is given by

$$\frac{1}{r}\frac{\partial}{\partial r}(r\rho u_r) + \frac{\partial}{\partial z}(\rho u_z) = 0 \tag{2.65}$$

and the stream function is defined by

$$\rho u_r = \frac{1}{r}\frac{\partial \psi}{\partial z}$$
$$\rho u_z = -\frac{1}{r}\frac{\partial \psi}{\partial r} \tag{2.66}$$

For the case of three-dimensional flows, it is possible to use two stream functions to replace the continuity equation. However, the complexity of this approach usually makes it less attractive than using the continuity equation in its original form.

When the viscous terms are dropped from the Navier-Stokes equations (2.8), the Euler equation is obtained:

$$\rho\frac{D\mathbf{V}}{Dt} = -\nabla p \tag{2.67}$$

For steady flow, this can be written

$$\mathbf{V}\cdot\nabla\mathbf{V} = -\frac{1}{\rho}\nabla p \tag{2.68}$$

Integrating this along a streamline in the flow gives

$$\frac{V^2}{2} + \frac{dp}{\rho} = \text{constant} \tag{2.69}$$

The integral in this equation can be evaluated if the flow is assumed barotropic. A *barotropic* fluid is one in which ρ is a function only of p (or a constant), i.e., $\rho = \rho(p)$. Incompressible flows and isentropic flows are examples of barotropic flows. For an incompressible flow, Eq. (2.69) can be integrated to give

$$p + \tfrac{1}{2}\rho V^2 = \text{constant} \tag{2.70}$$

which is Bernoulli's equation. For an isentropic, compressible flow, $\rho = \text{constant}\cdot p^{1/\gamma}$ and Eq. (2.69) can be integrated to give

$$\frac{V^2}{2} + \frac{\gamma}{\gamma-1}\frac{p}{\rho} = \text{constant} \tag{2.71}$$

which is sometimes referred to as the Bernoulli equation for compressible flow.

It is important to remember that the integration resulting in Eqs. (2.70) and (2.71) was carried out along a specific streamline. The constants appearing in the equations can vary from streamline and streamline. However, the integration can be carried out between arbitrary points in a flow field which is irrotational, that is, in a flow in which the particles do not rotate. In this case, Eqs. (2.70) and (2.71) remain valid throughout the flow with the same constants on the right-hand side. A proof of this result can be found in standard works on fluid mechanics such as Owczarek [6]. The inviscid portions of many flows arising in applications are, in fact, irrotational.

For flows which can be treated as irrotational, further useful results can be developed. The condition of irrotationality,

$$\nabla \times \mathbf{V} = 0 \tag{2.72}$$

expressed in the Cartesian coordinate system for a two-dimensional, steady, incompressible flow gives

$$\frac{\partial v}{\partial x} - \frac{\partial u}{\partial y} = 0 \tag{2.73}$$

When the stream function is introduced, the result is

$$\nabla^2 \psi = 0 \tag{2.74}$$

Thus, the stream function satisfies Laplace's equation for the steady, irrotational flow of an incompressible fluid.

Also, because of Eq. (2.72), the velocity $\mathbf{V}$ can be expressed as the gradient of a single-valued point function ϕ, the velocity potential. Requiring that the continuity equation be satisfied for the steady flow of an incompressible fluid gives

$$\nabla \cdot \mathbf{V} = \nabla \cdot (\nabla \phi) = \nabla^2 \phi = 0 \tag{2.75}$$

which is the Laplace equation for the velocity potential. It should be noted that Eq. (2.75) is valid regardless of the dimension of the flow.

The implications of Eqs. (2.74) and (2.75) are quite far reaching. At first glance it may appear that these equations only enforce conservation of mass for an irrotational flow. However, as a consequence of the irrotationality of the flow, a solution to these equations which satisfies the Bernoulli equation anywhere (such as on the boundaries) will satisfy the Bernoulli equation having the same constant throughout the flow domain. Thus, if the flow can be assumed to be incompressible, steady, and irrotational, a solution which satisfies the continuity and Euler momentum equations can be obtained by solving the Laplace equation for stream function or velocity potential, subject to the correct boundary conditions. The correct boundary conditions may vary from problem to problem, but the conditions imposed must be consistent with a flow which satisfies the continuity and Euler momentum equations. Many analytical and numerical procedures are available for solving Laplace's equation. Thus, it is only a modest task to obtain at least an approximate solution for steady, incompressible, inviscid flows.

If the viscous and heat conduction terms are dropped from the energy equation (2.21), one obtains

$$\frac{\partial(\rho H)}{\partial t} + \frac{\partial}{\partial x_j}(\rho u_j H) = \frac{\partial p}{\partial t} \tag{2.76}$$

Using the continuity equation, Eq. (2.76) can be written as

$$\rho \frac{DH}{Dt} = \frac{\partial p}{\partial t} \tag{2.77}$$

which, for a steady flow, can be written as

$$\mathbf{V} \cdot \nabla H = 0 \tag{2.78}$$

This equation can be integrated along a streamline to give

$$H = i + \frac{V^2}{2} = \text{constant} \tag{2.79}$$

It is observed that Eq. (2.79) will satisfy Eq. (2.78) exactly throughout the flow. This means that if a constant value of H is implied by the boundary conditions of a flow, i.e., H = constant or $\partial H/\partial n = 0$ at all boundaries, then H = constant is a solution to Eq. (2.78). Such an inviscid flow frequently occurs, and is known as an *isoenergetic* or *homoenergic* flow. It is interesting to note that Eq. (2.79) becomes identical to Eq. (2.71) for the isentropic flow of a perfect gas.

It is worth noting that for a perfect gas with constant specific heat, Eq. (2.79) can be written as

$$T + \frac{V^2}{2c_p} = \text{constant} \tag{2.80}$$

In this form, the constant can be identified as the stagnation temperature T_0. If a stagnation process is also isentropic, the stagnation pressure is given by

$$p_0 = p\left(\frac{T_0}{T}\right)^{\gamma/(\gamma-1)} \tag{2.81}$$

It is possible to derive additional relationships which prove to be quite useful in particular applications. Some of these are based on the first and second laws of thermodynamics, which result in the following relationship among properties:

$$T\,ds = di - \frac{dp}{\rho} \tag{2.82}$$

Upon combining this with the Euler momentum equation, it is possible to show that

$$\frac{\partial V}{\partial t} - \mathbf{V} \times \boldsymbol{\omega} = T\,\nabla s - \nabla i - \nabla\left(\frac{V^2}{2}\right) \tag{2.83}$$

where $\boldsymbol{\omega}$ is the vorticity vector,

$$\boldsymbol{\omega} = \nabla \times \mathbf{V}$$

Equation (2.83) is known as Crocco's equation. This equation provides a relationship between vorticity and entropy. Using Eq. (2.83) along with Eq. (2.78), it can be shown that the entropy remains constant along a streamline for a steady, inviscid, adiabatic flow. Furthermore, for an irrotational, isoenergic flow, Crocco's equation indicates that the entropy remains constant everywhere.

The Mach number, an important dimensionless parameter, is defined as the ratio of the local speed of the fluid to the speed of sound, a, defined by

$$a = \left(\frac{\partial p}{\partial \rho}\right)_s^{1/2} \tag{2.84}$$

where the subscript s indicates an isentropic process. For a perfect gas, p/ρ^{γ} = constant and Eq. (2.84) becomes

$$a = (\gamma R T)^{1/2} \tag{2.85}$$

where R is the gas constant from the perfect-gas equation of state.

The velocity potential can be used in a restricted class of compressible flows to obtain a simple equation governing the inviscid flow. This equation is not as simple as the Laplace equation obtained in the incompressible case, but it is nevertheless useful in applications. The flow is assumed to be steady, irrotational, and isentropic. The velocity components are defined as the gradient of the potential ϕ, just as in the incompressible case. In Cartesian coordinates this gives

$$u = \frac{\partial \phi}{\partial x} \qquad v = \frac{\partial \phi}{\partial y} \qquad w = \frac{\partial \phi}{\partial z} \tag{2.86}$$

These velocity components are substituted into the continuity equation to obtain

$$\frac{\partial}{\partial x}(\rho \phi_x) + \frac{\partial}{\partial y}(\rho \phi_y) + \frac{\partial}{\partial z}(\rho \phi_z) = 0 \tag{2.87}$$

The Euler momentum and energy equations reduce to Eq. (2.69) with the assumptions of steady, irrotational, and isentropic flow. This provides a means of evaluating the derivatives of density appearing in Eq. (2.87) in terms of the velocity potential. After simplifying, Eq. (2.87) becomes

$$\left(1 - \frac{\phi_x^2}{a^2}\right)\phi_{xx} + \left(1 - \frac{\phi_y^2}{a^2}\right)\phi_{yy} + \left(1 - \frac{\phi_z^2}{a^2}\right)\phi_{zz} - \frac{2\phi_x\phi_y}{a^2}\phi_{xy} - \frac{2\phi_x\phi_z}{a^2}\phi_{xz} - \frac{2\phi_y\phi_z}{a^2}\phi_{yz} = 0 \tag{2.88}$$

where the subscripts denote differentiation. Equation (2.88) is known as the *velocity potential equation*. For an incompressible flow, $a \to \infty$ and Eq. (2.88) reduces to Laplace's equation.

Reduced forms of Eq. (2.88), such as the transonic small-disturbance equation [4] and the Prandtl-Glauert equation [4], are sometimes useful in applications. Because further simplifications of Eq. (2.88) are common in the literature, Eq. (2.88) is sometimes referred to as the *full* potential equation.

The velocity potential equation can be used for both supersonic and subsonic flows. The main restriction on its use in applications arises from the assumption of isentropic, irrotational flow. The entropy (and often the vorticity) of the flow changes across a shock wave. Numerical solutions to Eq. (2.88) are capable of capturing a shock in the solution domain, but the solutions show no change in entropy or rotation. When the normal Mach number upstream of a shock is less than about 1.3, the errors introduced by the assumptions of isentropic irrotational flow are usually very small.

Viscous Flow. In this subsection the boundary-layer form of the governing equations are presented. These equations are obtained from the full Navier-Stokes equations by dropping those terms which are small when the effects of viscosity and heat conduction are limited to a region which is thin relative to the characteristic length of the object immersed in the flow.

The concept of a boundary layer originated with Ludwig Prandtl in 1904. Prandtl reasoned from experimental evidence that for sufficiently large Reynolds numbers ($\mathrm{Re} = \rho UL/\mu$), a thin region existed near a solid boundary where viscous effects were at least as important as inertia effects no matter how small the viscosity of the fluid might be. Prandtl used an order-of-magnitude analysis as the basis for eliminating terms from the governing equations. His conclusions were that second derivatives of the velocity components in the streamwise direction were negligible compared to corresponding derivatives transverse to the main flow direction and that the entire momentum equation for the transverse direction could be neglected. Applying a similar order-of-magnitude analysis to the energy equation indicates that for sufficiently large values of the Péclet number ($\mathrm{Pe} = \mathrm{Re}\,\mathrm{Pr}$), conduction effects in the streamwise direction can be neglected and the dissipation function can be greatly simplified.

The perceptive work of Prandtl in 1904 set the stage for the analysis of a wide range of important problems in convective heat transfer. Although the boundary-layer equations were challenging in view of the analytical techniques available in 1904, they were nevertheless just within reach of existing and emerging methods. It was not until the modern digital computer became widely used in the 1960s and 1970s that numerical solution of more complete mathematical formulations became commonplace.

The Euler and boundary-layer equations are not the only "reduced" forms of the conservation equations in common use today. Other forms intermediate in complexity between the Navier-Stokes and boundary-layer equations have been found useful in numerical solutions to particular classes of problems. Generally, these reduced forms allow a few terms to be neglected in the Navier-Stokes equations, but do not permit the neglect of an entire momentum equation as in the case of the boundary-layer equations. Consequently, these reduced forms require an order of magnitude greater effort for solution than do the boundary-layer equations. A discussion of these forms can be found in texts which emphasize the computational aspects of convective problems [4]. The boundary-layer forms of the conservation equations will be given below because of their wide range of applicability in convective problems. Other reduced forms of the Navier-Stokes equations are somewhat less important for the understanding of convection processes and will not be covered in detail in this chapter.

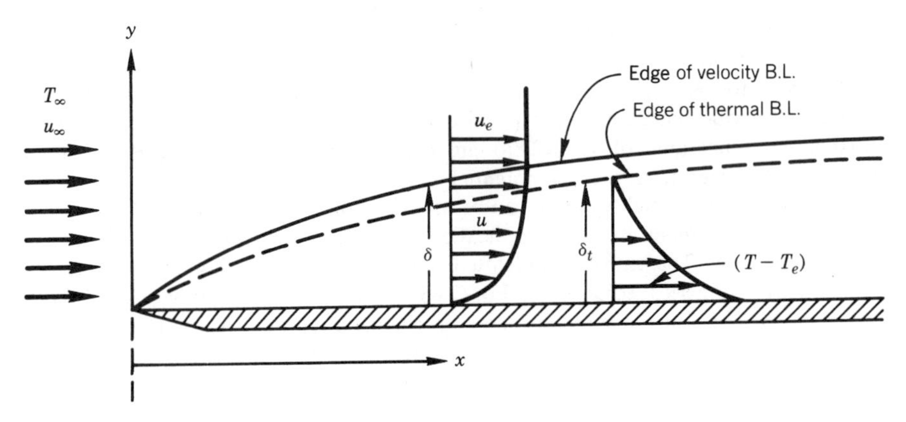

Figure 2.2. Boundary-layer configuration and coordinate system. (Adapted from [4] by permission.)

Historically, the boundary-layer equations were first developed for steady, two-dimensional, incompressible, constant-property flow along an isothermal surface. The thickness of the viscous and thermal boundary layers are assumed to be small relative to a characteristic length in the primary flow direction. That is, $\delta/L \ll 1$ and $\delta_t/L \ll 1$. See Fig. 2.2 for the flow configuration and coordinate system being used for the boundary layer. Details of the order-of-magnitude analysis can be found in most textbooks covering convective heat transfer [2, 4] and will not be repeated. The Reynolds number and Péclet numbers were assumed to be of the same order of magnitude in the analysis. The results for steady, two-dimensional, incompressible, constant-property flow can be written as follows:

Continuity:

$$\frac{\partial u}{\partial x} + \frac{\partial v}{\partial y} = 0 \tag{2.89}$$

Momentum:

$$u\frac{\partial u}{\partial x} + v\frac{\partial u}{\partial y} = -\frac{1}{\rho}\frac{dp}{dx} + \nu\frac{\partial^2 u}{\partial y^2} \tag{2.90}$$

Energy:

$$u\frac{\partial T}{\partial x} + v\frac{\partial T}{\partial y} = \alpha\frac{\partial^2 T}{\partial y^2} + \frac{\beta T u}{\rho c_p}\frac{dp}{dx} + \frac{\mu}{\rho c_p}\left(\frac{\partial u}{\partial y}\right)^2 \tag{2.91}$$

where ν is the kinematic viscosity μ/ρ, and α is the thermal diffusivity $k/\rho c_p$. It should be pointed out that the last two terms in Eq. (2.91) were retained from the order-of-magnitude analysis on the basis that the Eckert number $\mathrm{Ec} = 2(T_0 - T_\infty)/(T_w - T_\infty)$ was of the order of 1. Should Ec become of the order 0.1 or smaller for a particular flow, neglecting those terms should be permissible.

To complete the mathematical formulation, initial and boundary conditions must be specified. The steady boundary-layer momentum and energy equations are mathematically parabolic with the streamwise direction being the marching direction. Initial distributions of u and T must be provided. The usual boundary conditions are

$$u(x,0) = v(x,0) = 0$$

$$T(x,0) = T_w(x) \quad \text{or} \quad \left.\frac{\partial T}{\partial y}\right|_{y=0} = \frac{q(x)}{k}$$

$$\lim_{y\to\infty} u(x,y) = u_e(x), \qquad \lim_{y\to\infty} T(x,y) = T_e(x) \tag{2.92}$$

where the subscript e refers to conditions at the edge of the boundary layer. The pressure gradient term in Eqs. (2.90) and (2.91) is to be evaluated from the given boundary information. With $u_e(x)$ specified, dp/dx can be evaluated from an application of the equations which govern the inviscid outer flow (Euler's equations), giving $dp/dx = -\rho u_e \, du_e/dx$.

Next, the boundary-layer approximation is extended to an incompressible, constant-property two-dimensional turbulent flow. Under the incompressible assumption, $\rho' = 0$ and the Reynolds equations simplify considerably. As before, it is assumed that $\delta/L \ll 1$, $\delta_t/L \ll 1$, but experimental evidence must be used for guidance in establishing the magnitude estimates for the Reynolds stress and heat-flux terms. Experiments indicate that the Reynolds stresses can be at least as large as the laminar counterparts, and that $\overline{u'^2}, \overline{u'v'}, \overline{v'^2}$, while differing in magnitudes and distribution somewhat, are nevertheless of the same order of magnitude in the boundary layer. In the energy equation, $\overline{T'v'}$ and $\overline{T'u'}$ are observed to be of the same order of magnitude and at least as large as the laminar heat-flux term. Triple correlations such as $\overline{u'u'u'}$ are expected to be an order of magnitude smaller than double correlations. Again it is assumed that the Prandtl and Eckert numbers are near 1 in order of magnitude. With these assumptions, the boundary-layer form of the equations for two-dimensional incompressible turbulent flow are

Continuity:

$$\frac{\partial \bar{u}}{\partial x} + \frac{\partial \bar{v}}{\partial y} = 0 \tag{2.93}$$

Momentum:

$$\rho \bar{u} \frac{\partial \bar{u}}{\partial x} + \rho \bar{v} \frac{\partial \bar{u}}{\partial y} = -\frac{dp}{dx} + \mu \frac{\partial^2 \bar{u}}{\partial y^2} - \rho \frac{\partial}{\partial y}\left(\overline{u'v'}\right) \tag{2.94}$$

Energy:

$$\rho c_p \bar{u} \frac{\partial \bar{T}}{\partial x} + \rho c_p \bar{v} \frac{\partial \bar{T}}{\partial y} = \beta \bar{T} \bar{u} \frac{\partial p}{\partial x} + k \frac{\partial^2 \bar{T}}{\partial y^2} - \rho c_p \frac{\partial}{\partial y}\left(\overline{v'T'}\right)$$

$$+ \mu \left(\frac{\partial \bar{u}}{\partial y}\right)^2 - \rho \overline{v'u'} \frac{\partial \bar{u}}{\partial y} \tag{2.95}$$

It should be noted that only one Reynolds stress term and one Reynolds heat-flux remain in the governing equations after the boundary-layer approximation is invoked. The first and last two terms on the right-hand side of Eq. (2.95) can be neglected in some applications. However, it is not correct to neglect these terms categorically for incompressible flows. Equation (2.95) can be easily solved by numerical methods in its entirety, so further reductions should not be made unless it is very clear that the terms neglected will indeed be negligible. The boundary conditions remain unchanged for turbulent flow. For completeness it should be mentioned that terms do appear in the y-momentum equation for turbulent flow which are of the same order of magnitude as those included in Eqs. (2.93)–(2.95), namely,

$$\frac{1}{\rho}\frac{\partial \bar{p}}{\partial y} = -\frac{\partial}{\partial y}\left(\overline{v'^2}\right).$$

These terms have not be listed above with the boundary-layer equations, because they contribute no information about the mean velocities or temperature.

The order-of-magnitude reduction of the Reynolds equations to boundary-layer form is a lengthier process for compressible flow. Only the results will be presented here. Details of the arguments for elimination of terms are given by Schubauer and Tchen [7], van Driest [8], and Cebeci and Smith [9]. As was the case for incompressible flow, guidance must be obtained from experimental observations in assessing the magnitudes of turbulence quantities. An estimate must be made for $\rho'/\bar{\rho}$ for compressible flows.

Measurements in gases for Mach numbers less than about 5 indicate that temperature fluctuations are nearly isobaric for adiabatic flows. This suggests that $T'/\overline{T} \approx -\rho'/\bar{\rho}$. However, there is evidence that appreciable pressure fluctuations exist (8 to 10% of the mean wall static pressure) at $M_e = 5$, and it is speculated that $p'/\bar{p}$ increases with increasing Mach number. In the absence of specific experimental evidence to the contrary, it is common to base the order-of-magnitude estimates of fluctuating terms on the assumption that the pressure fluctuations are small. This appears to be a safe assumption for $M_e \leq 5$, and good predictions based on this assumption have been noted for Mach numbers as high as 7.5. The isobaric assumption will be adopted here. It is primarily the correlation terms involving the density fluctuations which may increase in magnitude with increasing Mach number above $M_e \approx 5$.

The difference between $\tilde{u}$ and $\bar{u}$ vanishes under the boundary-layer approximation. This follows because $\overline{\rho' u'}$ is expected to be small compared to $\bar{\rho}\bar{u}$ and can be neglected in the momentum equation. Also, $\overline{T} = \tilde{T}$ and $\overline{H} = \tilde{H}$ are observed to be consistent with the boundary-layer approximation. On the other hand, $\overline{\rho' v'}$ and $\bar{\rho}\bar{v}$ are both of about the same order of magnitude in a thin shear layer. Thus, $\bar{v} \neq \tilde{v}$.

Below, the unsteady boundary-layer equations for a compressible fluid are written in a form applicable to both two-dimensional and axisymmetric turbulent flow. For convenience the use of bars over time-mean quantities will be dropped, and the quantity $\tilde{v} = (\bar{\rho}\bar{v} + \overline{\rho' v'})/\bar{\rho}$ will be utilized. The equations are also valid for laminar flow when the terms involving fluctuating quantities are set equal to zero. The coordinate system is indicated in Fig. 2.3.

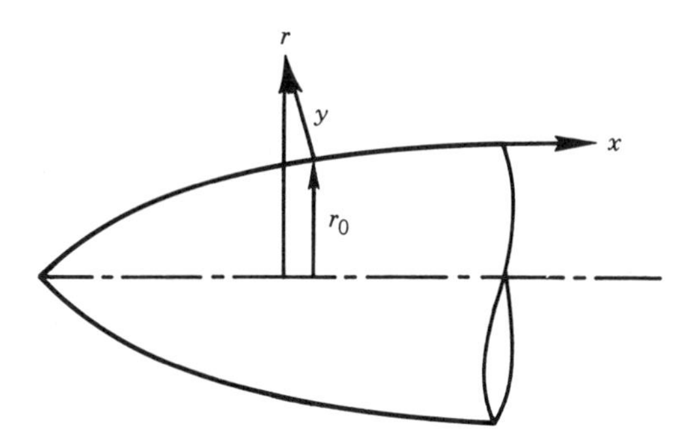

Figure 2.3. Notation and coordinate system for boundary-layer flow over an axisymmetric body. (Adapted from [4] by permission.)

Continuity:

$$\frac{\partial \rho}{\partial t} + \frac{\partial}{\partial x}(r^m \rho u) + \frac{\partial}{\partial y}(r^m \rho \tilde{v}) = 0 \tag{2.96}$$

Momentum:

$$\rho\frac{\partial u}{\partial t} + \rho u\frac{\partial u}{\partial x} + \rho\tilde{v}\frac{\partial u}{\partial y} = -\frac{dp}{dx} + \frac{1}{r^m}\frac{\partial}{\partial y}\left[r^m\left(\mu\frac{\partial u}{\partial y} - \rho\overline{u'v'}\right)\right] \tag{2.97}$$

Energy:

$$\rho\frac{\partial H}{\partial t} + \rho u\frac{\partial H}{\partial x} + \rho\tilde{v}\frac{\partial H}{\partial y} = \frac{\partial p}{\partial t} + \frac{1}{r^m}\frac{\partial}{\partial y}\left[r^m\left\{\frac{\mu}{\Pr}\frac{\partial H}{\partial y} - \rho c_p\overline{v'T'} + u\left[\left(1 - \frac{1}{\Pr}\right)\mu\frac{\partial u}{\partial y} - \rho\overline{v'u'}\right]\right\}\right] \tag{2.98}$$

State:

$$\rho = \rho(p, T) \tag{2.99}$$

In the above, m is a flow index equal to unity for axisymmetric flow ($r^m = r$) and equal to zero for two-dimensional flow ($r^m = 1$). The energy equation (2.98) has been written in terms of the total enthalpy. This is the most common choice for compressible flows, especially when numerical methods of solution are to be utilized, because the percentage variation of H across the flow is nearly always less than the variation in T. Thus, for the same grid, it is expected that the H distribution can be determined more accurately than the T distribution from solving the energy equation. On the other hand, use of the static temperature as the primary thermal variable is possible, and numerous examples of this can be found in the literature.

It should be noted that the boundary-layer equations for compressible flow are not significantly more complex than for incompressible flow. Only one Reynolds stress and one heat-flux term appear, regardless of whether the flow is compressible or incompressible. As for purely laminar flows, the main difference is in the property variations of μ, k, and ρ for the compressible case, which nearly always requires that a solution be obtained for some form of the energy equation, even when heat transfer results are not of primary interest.

The boundary-layer approximation remains valid for a flow in which the turning of the mainstream results in a three-dimensional flow as long as velocity derivatives with respect to only one coordinate direction are large. That is, the three-dimensional boundary layer remains "thin" with respect to only one coordinate direction. The three-dimensional unsteady boundary-layer equations in Cartesian coordinates, applicable to a compressible turbulent flow, are given below (the y direction is normal to the wall):

Continuity:

$$\frac{\partial \rho}{\partial t} + \frac{\partial \rho u}{\partial x} + \frac{\partial \rho\tilde{v}}{\partial y} + \frac{\partial \rho w}{\partial z} = 0 \tag{2.100}$$

x momentum:

$$\frac{\partial u}{\partial t} + \rho u \frac{\partial u}{\partial x} + \rho v \frac{\partial u}{\partial y} + \rho w \frac{\partial u}{\partial z} = -\frac{\partial p}{\partial x} + \frac{\partial}{\partial y}\left(\mu \frac{\partial u}{\partial y} - \overline{\rho u'v'}\right) \quad (2.101)$$

z momentum:

$$\frac{\partial w}{\partial t} + \rho u \frac{\partial w}{\partial x} + \rho \tilde{v} \frac{\partial w}{\partial y} + \rho w \frac{\partial w}{\partial z} = -\frac{\partial p}{\partial z} + \frac{\partial}{\partial y}\left(\mu \frac{\partial w}{\partial y} - \rho\overline{w'v'}\right) \quad (2.102)$$

Energy:

$$\frac{\partial H}{\partial t} + \rho u \frac{\partial H}{\partial x} + \rho \tilde{v} \frac{\partial H}{\partial y} + \rho w \frac{\partial H}{\partial z} = \frac{\partial p}{\partial t}$$

$$+ \frac{\partial}{\partial y}\left\{\frac{\mu}{\mathrm{Pr}}\frac{\partial H}{\partial y} - \rho c_p \overline{v'T'} + \mu\left(1 - \frac{1}{\mathrm{Pr}}\right)\left[u\frac{\partial u}{\partial y} + w\frac{\partial w}{\partial y}\right] - \rho u \overline{v'u'} - \rho w \overline{v'w'}\right\}$$

$$(2.103)$$

For a three-dimensional flow, the boundary-layer approximation permits H to be written as

$$H = c_p T + \frac{u^2}{2} + \frac{w^2}{2}$$

For external flows, the pressure gradient terms can be evaluated from a solution to the inviscid flow (Euler) equations.

It is common to employ body-intrinsic curvilinear coordinates to compute the three-dimensional boundary layers occurring on wings and other shapes of practical interest. Often, this curvilinear coordinate system is nonorthogonal. An example of this can be found in Cebeci et al. [10]. The orthogonal system is somewhat more common (see, for example, Blottner and Ellis [11]). One coordinate, x_2, is almost always taken to be orthogonal to the body surface. This convention will be followed here.

Below we record the three-dimensional boundary-layer equations in the orthogonal curvilinear coordinate system described previously. Typically, x_1 will be directed roughly in the primary flow direction and x_3 will be in the crossflow direction. The metric coefficients (h_1, h_2, h_3) are as defined previously; however, h_2 will be taken as unity as a result of the boundary-layer approximation. In addition, use will be made of the geodesic curvatures of the surface coordinate lines,

$$K_1 = \frac{1}{h_1 h_3}\frac{\partial h_1}{\partial x_3} \quad \text{and} \quad K_3 = \frac{1}{h_1 h_3}\frac{\partial h_3}{\partial x_1} \quad (2.104)$$

With this notation, the boundary-layer form of the conservation equations for a compressible, turbulent flow can be written:

Continuity:

$$\frac{\partial}{\partial x_1}(\rho h_3 u_1) + \frac{\partial}{\partial x_2}(h_1 h_3 \rho \tilde{u}_2) + \frac{\partial}{\partial x_3}(\rho h_1 u_3) = 0 \quad (2.105)$$

x_1 momentum:

$$\frac{\rho u_1}{h_1}\frac{\partial u_1}{\partial x_1} + \rho\tilde{u}_2\frac{\partial u_1}{\partial x_2} + \frac{\rho u_3}{h_3}\frac{\partial u_1}{\partial x_3} + \rho u_1 u_3 K_1 - \rho u_3^2 K_3$$

$$= -\frac{1}{h_1}\frac{\partial p}{\partial x_1} + \frac{\partial}{\partial x_2}\left(\mu\frac{\partial u_1}{\partial x_2} - \rho\overline{u_1' u_2'}\right) \tag{2.106}$$

x_3 momentum:

$$\frac{\rho u_1}{h_1}\frac{\partial u_3}{\partial x_1} + \rho\tilde{u}_2\frac{\partial u_3}{\partial x_2} + \frac{\rho u_3}{h_3}\frac{\partial u_3}{\partial x_3} + \rho u_1 u_3 K_3 - \rho u_1^2 K_1$$

$$= -\frac{1}{h_3}\frac{\partial p}{\partial x_3} + \frac{\partial}{\partial x_2}\left(\mu\frac{\partial u_3}{\partial x_2} - \rho\overline{u_3' u_2'}\right) \tag{2.107}$$

Energy:

$$\frac{\rho u_1}{h_1}\frac{\partial H}{\partial x_1} + \rho\tilde{u}_2\frac{\partial H}{\partial x_2} + \frac{\rho u_3}{h_3}\frac{\partial H}{\partial x_3}$$

$$= \frac{\partial}{\partial x_2}\left\{\frac{\mu}{\mathrm{Pr}}\frac{\partial H}{\partial x_2} - \rho c_p\overline{u_2' T'}\right.$$

$$\left. + \mu\left(1 - \frac{1}{\mathrm{Pr}}\right)\left(u_1\frac{\partial u_1}{\partial x_2} + u_3\frac{\partial u_3}{\partial x_2}\right) - \rho u_1\overline{u_2' u_1'} - \rho u_3\overline{u_2' u_3'}\right\} \tag{2.108}$$

As always, an equation of state, $\rho = \rho(p, T)$, is needed to close the system of equations for a compressible flow. The above equations remain valid for a laminar flow when the fluctuating quantities are set equal to zero.

2.2.5 Dimensionless Parameters for External Forced Convection

Dimensionless parameters play an important role in the analysis of convective flows. It is well known that the number of variables influencing friction, heat transfer, and other quantities of engineering interest can be reduced by arranging the variables into dimensionless groups. Dimensionless groups also frequently appear as parameters in nondimensional forms of the governing conservation equations. A few of these have already been introduced. The major nondimensional groups for external forced convection are defined in this section.

The Mach number ($M = V/a$) is the ratio of the local speed of the fluid to the local speed of sound. The speed of sound, a, is defined by Eq. (2.84). The Mach number provides one measure of the importance of compressibility effects in the flow.

The pressure coefficient [$C_p = (p_w - p_\infty)/\frac{1}{2}\rho_\infty V_\infty^2$] is the ratio of the difference between the surface and free-stream pressures to the free-stream dynamic pressure. The pressure coefficient is also known as the Euler number.

The Reynolds number ($\mathrm{Re}_L = \rho V L/\mu$) is the ratio of inertial to viscous forces. Various reference conditions may be used to define ρ, V, and μ. In particular, the quantity L is a characteristic length which may be defined in several ways. For example, it may be the distance along the body in the main flow direction measured from the stagnation point to the local point x of interest; it may be the characteristic dimension of the body immersed in the flow; or it may be the momentum thickness θ or the displacement thickness δ^* of the viscous flow.

The Prandtl number ($\mathrm{Pr} = \mu c_p/k$) is the ratio of the momentum diffusivity to the thermal diffusivity.

The Péclet number ($\mathrm{Pe} = \mathrm{Re\,Pr}$) is the product of the Reynolds and Prandtl numbers. It can be thought of as the ratio of transport by convection to transport by thermal diffusion, or as a Reynolds number in which the diffusivity of momentum is replaced by the thermal diffusivity.

The skin-friction coefficient ($c_f = 2\tau_w/\rho_e u_e^2$) is the wall shear stress divided by the dynamic pressure of the flow at the outer edge of the boundary layer. Sometimes, in pressure gradient flows in which the edge values are changing, the dynamic pressure of the undisturbed flow upstream of the object ($\rho_\infty V_\infty^2/2$) is used in the denominator. The average skin friction coefficient over an object of area A is defined as

$$c_f = \frac{\int_A \tau_w \, dA}{A \rho_\infty V_\infty^2/2}.$$

The Eckert number $[\mathrm{E_c} = u_e^2/c_p(T_w - T_e) = 2(T_{e,o} - T_e)/(T_w - T_e)]$ is proportional to the ratio of the temperature rise of the fluid in an adiabatic compression to the temperature difference between the wall and the fluid at the edge of the boundary layer. The Eckert number can be expressed in terms of the Mach number for a perfect gas:

$$\mathrm{Ec} = \frac{(\gamma - 1) M_e^2 T_e}{T_w - T_e}$$

The Nusselt number $[\mathrm{Nu} = q_w L/k(T_w - T_{\mathrm{ref}})]$ is one of the two major nondimensional parameters containing information on the wall heat flux. It can be thought of as the ratio of the actual wall heat flux to that which would occur by conduction alone across a layer of thickness L. In the above, q_w is the heat flux at the wall; L is a characteristic length which is usually the distance from the stagnation point to the point of interest, x; and T_{ref} is a reference temperature for the fluid. In low-speed flows, T_{ref} is usually taken as T_e, the static temperature at the edge of the thermal boundary layer. However, this is not correct for high-speed flows, as will be discussed below. The group $q_w/(T_w - T_{\mathrm{ref}})$ is also defined as the local heat transfer coefficient h through Newton's law of cooling. Thus, the Nusselt number can be written as $\mathrm{Nu} = hL/k$. It is logical to require that the temperature difference $T_w - T_{\mathrm{ref}}$ represent a "potential" for heat transfer. That is, when $T_w = T_{\mathrm{ref}}$ we should expect no heat transfer to occur. This will only be the case if T_{ref} is the *adiabatic wall temperature*, T_{aw}, of the fluid. For a simple adiabatic flow to stagnation, T_{aw} is expected to be equal to the stagnation temperature of the fluid, T_o. For flow *along* adiabatic surfaces, T_{aw} is found to be mostly a function of the Prandtl number of the fluid. The adiabatic wall temperature is related to the total and static temperatures through the recovery factor r as follows:

$$T_{aw} = T_e + r(T_{e,o} - T_e) \tag{2.109}$$

where $T_{e,o}$ is the stagnation temperature of the flow at the outer edge of the thermal boundary layer. The recovery factor can be well approximated as

$$r = \begin{cases} \mathrm{Pr}^{1/2} & \text{for laminar flow} \\ \mathrm{Pr}^{1/3} & \text{for turbulent flow} \end{cases}$$

Thus, the correct fluid temperature to use in Newton's law of cooling is T_{aw}, not T_e, but the difference is so small for low-speed flows ($M \leq 0.3$) that T_e is normally used. It is essential, however, that T_{aw} be used in high-speed flows. The average Nusselt number for a two-dimensional flow is defined as

$$\overline{\mathrm{Nu}} = \frac{1}{L}\int_0^L \frac{q_w \, dx}{k(T_w - T_{\mathrm{ref}})}$$

Wall heat flux is also nondimensionalized as the Stanton number [$\mathrm{St} = q_w/\rho_e u_e c_p(T_w - T_{\mathrm{ref}}) = \mathrm{Nu}/\mathrm{Re}\,\mathrm{Pr}$]. As with the Nusselt number, T_{ref} should be taken as the adiabatic wall temperature of the fluid. If the specific heat varies appreciably, it is more appropriate to use the static enthalpy in the definition of the recovery factor, in Newton's law of cooling, and in the Stanton number. Accordingly, we find

$$i_{aw} = i_e + r(i_{e,o} - i_e) \tag{2.110}$$

and

$$\mathrm{St} = \frac{q_w}{\rho_e u_e (i_w - i_{aw})}$$

The average Stanton number is defined as

$$\overline{\mathrm{St}} = \frac{1}{A}\int_A \mathrm{St}\, dA$$

2.3 TURBULENCE MODELS

2.3.1 Background

The need for turbulence modeling was pointed out in Sec. 2.2.3. In order to predict turbulent flows from solutions to the Reynolds equations, it becomes necessary to make closing assumptions about the apparent turbulent stress and heat-flux quantities. It is the implementation of the closing assumptions in order to evaluate the apparent turbulent stresses and heat fluxes appearing in the Reynolds equations that characterizes turbulence modeling.

All presently known turbulence models have limitations. The ultimate model has yet to be developed. On the other hand, it has generally been possible to develop models with reasonable accuracy only over a limited range of flow conditions. All proposed models should be verified by comparisons with experimental data, and care should be taken in interpreting predictions of models outside of the range of conditions over which they have been verified. Because of the need to rely heavily on experimental data in establishing turbulence models, numerical solutions to the Reynolds equations have been sometimes regarded as mere correlations of experimental data. This is perhaps the correct way to view such numerical simulations at the present time. An advantage of the computer simulations is that they usually provide considerable detailed information about the flow, much more than is obtained from a simple conventional correlation for the Nusselt number.

The literature on turbulence models is extensive. Scores of models have been suggested. None of them are both general and accurate. Only a few of the most commonly used models will be described in this section. The rationale which has guided the development of turbulence models will be briefly described.

2.3.2 Modeling Terminology

Boussinesq [12] suggested more than one hundred years ago that the apparent turbulent shearing stresses might be related to the rate of mean strain through an apparent scalar turbulent or "eddy" viscosity. For the general Reynolds stress tensor, the Boussinesq assumption gives

$$\overline{\rho u_i' u_j'} = \mu_T \left(\frac{\partial u_i}{\partial x_j} + \frac{\partial u_j}{\partial x_i} \right) - \tfrac{2}{3}\delta_{ij} \left(\mu_T \frac{\partial u_k}{\partial x_k} + \rho \bar{k} \right) \tag{2.111}$$

where μ_T is the turbulent viscosity and $\bar{k}$ is the kinetic energy of turbulence, $\bar{k} = \overline{u_i' u_i'}/2$. Following the convention introduced in Sec. 5.3.2, bars are being omitted over the time-mean variables.

By analogy with kinetic theory, by which the molecular viscosity for gases can be evaluated with reasonable accuracy, it might be expected that the turbulent viscosity could be modeled as

$$\mu_T = \rho v_T l \tag{2.112}$$

where v_T and l are characteristic velocity and length scales of the turbulence, respectively. The problem, of course, is to find suitable means for evaluating v_T and l.

Turbulence models to close the Reynolds equations can be divided into two categories according to whether or not the Boussinesq assumption is used. Models using the Boussinesq assumption will be referred to as *turbulent-viscosity* models. Most models currently employed in engineering calculations are of this type. Experimental evidence indicates that the turbulent-viscosity hypothesis is a valid one in many flow circumstances. There are exceptions, however, and there is no physical requirement that it hold. Models which effect closure of the Reynolds equations without this assumption include those known as *Reynolds-stress* or *stress-equation* models.

The other common classification of models is according to the number of supplementary partial differential equations which must be solved in order to supply the modeling parameters. This number ranges from zero for the simplest algebraic models to twelve for the most complex of the Reynolds stress models [13]. Reference is also sometimes made to the "order" of the closure. According to this terminology, a first-order closure evaluates the Reynolds stresses through functions of the mean velocity and geometry alone. A second-order closure employs a solution to a modeled form of a transport partial differential equation for one or more of the characteristics of turbulence.

A third category of turbulence models includes all of those that are not based entirely on the Reynolds equations. A promising computational approach known as "large-eddy simulation" falls into this category. In this approach [14], an attempt is made to resolve the large-scale turbulent motion from first principles by numerically solving a "filtered" set of equations governing this large-scale three-dimensional time-dependent motion. Turbulence modeling is employed to approximate the effects of the "subgrid" scale turbulence. Such calculations have shown much promise, but the technique is much too costly at present to be considered as an engineering tool.

2.3.3 Summary of Common Models

Much of the early work on turbulence modeling was done for flows in which the boundary-layer form of the conservation equations was adequate. For such flows, the modeling task reduces to finding expressions for $-\rho\overline{v'u'}$ and $-\rho c_p\overline{v'T'}$. A large fraction of external convective problems can still be solved through the use of the boundary-layer equations. Thus, the highest priority will be given to discussing ways in which $-\rho\overline{v'u'}$ and $-\rho c_p\overline{v'T'}$ can be evaluated.

Simple Algebraic or Zero-Equation Models. Algebraic turbulence models invariably utilize the Bossinesq assumption. One of the most successful models of this type was suggested by Prandtl in the 1920s:

$$\mu_T = \rho l^2 \left| \frac{\partial u}{\partial y} \right| \tag{2.113}$$

where l, a "mixing length," can be thought of as a transverse distance over which particles maintain their original momentum, somewhat on the order of a mean free path for the collision or mixing of globules of fluid. The product $l|\partial u/\partial y|$ can be interpreted as the characteristic velocity of turbulence, v_T. In Eq. (2.113), u is the component of velocity in the primary flow direction and y is the coordinate transverse to the primary flow direction.

For three-dimensional thin shear layers, Prandtl's formula can be interpreted as

$$\mu_T = \rho l^2 \left[\left(\frac{\partial u}{\partial y} \right)^2 + \left(\frac{\partial w}{\partial y} \right)^2 \right]^{1/2} \tag{2.114}$$

This formula treats the turbulent viscosity as a scalar and gives qualitatively correct trends, especially near the wall. There is increasing experimental evidence, however, that in the outer layer, the turbulent viscosity should be treated as a tensor (i.e., dependent upon the direction of strain) in order to provide the best agreement with measurements. For flows in corners or in other geometries where a single "transverse" direction is not clearly defined, Prandtl's formula must be modified further (see, for example, Patankar et al. [15]).

The evaluation of l in the mixing length varies with the type of flow being considered, wall boundary layer, jet, wake, etc. For flow along a solid surface (internal or external flow), good results are observed by evaluating l according to

$$l_i = \kappa y \left(1 - e^{-y^+/A^+}\right) \tag{2.115}$$

in the inner region closest to the solid boundaries and switching to

$$l_0 = C_1 \delta \tag{2.116}$$

when l_i predicted by Eq. (2.115) first exceeds l_0. The constant C_1 in Eq. (2.116) is usually assigned a value of 0.089, and δ is the velocity boundary-layer thickness.

In Eq. (2.115), κ is the von Kármán constant, usually taken as 0.41, and A^+ is the damping constant, most commonly evaluated as 26. The quantity in parentheses is the van Driest damping function [16] and is the most common expression used to bridge the gap between the fully turbulent region where $l = \kappa y$ and the viscous sublayer where $l \to 0$. The parameter y^+ is defined as

$$y^+ = \frac{y(|\tau_w|/\rho_w)^{1/2}}{\nu_w}$$

Numerous variations on the exponential function of Eq. (2.115) have been utilized in order to take account of effects of property variations, pressure gradients, blowing, and surface roughness. A discussion of modifications to account for several of these effects can be found in Cebeci and Smith [9]. It appears reasonably clear from comparisons in the literature, however, that the inner-layer model as stated [Eq. (2.115)] requires no modification to accurately predict the variable-property flows of gases with moderate pressure gradients on smooth surfaces.

The expression for l_i, Eq. (2.115), is responsible for producing the inner, "law-of-the-wall" region of the turbulent flow, and l_0 [Eq. (2.116)] produces the outer "wakelike" region. These two zones are indicated in Fig. 2.4, which depicts a typical velocity distribution for an incompressible turbulent boundary layer on a smooth impermeable plate using "law-of-the-wall" coordinates. $\mathrm{Re}_\theta = \rho_e u_e \theta / \mu_e$ is the Reynolds number based on the momentum thickness θ, which for two-dimensional flow is

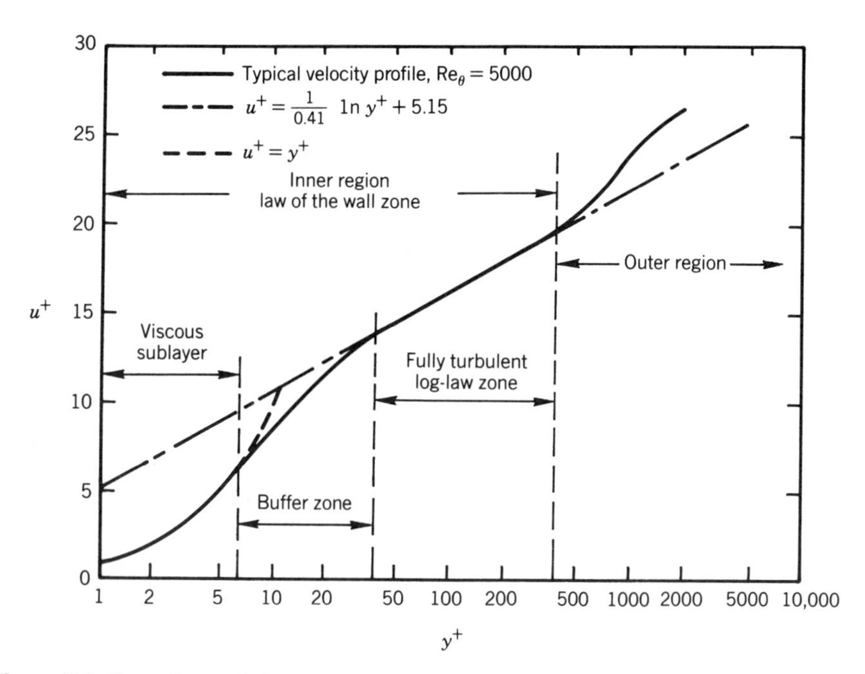

Figure 2.4. Zones in a turbulent boundary for a typical incompressible flow over a smooth flat plate. (Adapted from [4] by permission.)

defined as

$$\theta = \int_0^\infty \frac{\rho u}{\rho_e u_e}\left(1 - \frac{u}{u_e}\right) dy$$

The nondimensional velocity u^+ is defined as $u^+ = u/(|\tau_w|/\rho_w)^{1/2}$. The inner and outer regions are indicated in the figure. Under normal conditions, the inner law-of-the-wall zone only includes about 20% of the boundary layer. The log-linear zone is the characteristic "signature" of a turbulent wall boundary layer, although the law-of-the-wall plot changes somewhat in general appearance as the Reynolds and Mach numbers are varied.

It is worth noting that at low momentum-thickness Reynolds numbers (i.e., near the origin of the turbulent boundary layer), both inner and outer regions tend toward zero and problems might be expected with the two-region turbulence model employing Eqs. (2.115) and (2.116). The difficulty occurs because the smaller δ's occurring near the origin of the turbulent boundary layer are causing the switch to the outer model to occur before the wall damping effect has permitted the fully turbulent law-of-the-wall zone to develop. This causes the finite difference scheme using such a model to underpredict the wall shear stress. The discrepancy is nearly negligible for incompressible flow, but the effect is more serious for compressible flows, persisting at higher and higher Reynolds numbers as the Mach number increases, due to the relative thickening of the viscous sublayer from thermal effects [17]. Naturally, details of the effects are influenced by the extent of wall cooling in the compressible flow.

Predictions can be brought into good agreement with measurements at low Reynolds numbers by simply delaying the switch from the inner model, Eq. (2.115), to the outer model, Eq. (2.116) until $y^+ \geq 50$. If, at $y^+ = 50$ in the flow, $l/\delta \leq 0.089$, then no adjustment is necessary. On the other hand, if Eq. (2.115) predicts $l/\delta > 0.089$, then the mixing length becomes constant in the outer region at the value computed at $y^+ = 50$ by Eq. (2.115). This simple adjustment ensures the existence of the log-linear region in the flow, which is in agreement with the preponderance of measurements.

An alternative treatment to Eq. (2.116) is often used to evaluate the turbulent viscosity in the outer region [9]. This follows the Clauser formulation,

$$\mu_{T(\text{outer})} = \alpha \rho u_e |\delta_k^*| \tag{2.117}$$

where α brings in the low Reynolds number effects. Cebeci and Smith [9] recommend

$$\alpha = 0.0168 \frac{1.55}{1 + \pi} \tag{2.118}$$

where $\pi = 0.55[1 - \exp(-0.243 z^{1/2} - 0.298 z)]$ and $z = (\text{Re}_\theta/425 - 1)$. For Re_θ greater than 5000, $\alpha \approx 0.0168$. The parameter δ_k^* is the kinematic displacement thickness defined as

$$\delta_k^* = \int_0^\infty \left(1 - \frac{u}{u_e}\right) dy$$

Other modeling procedures have been used successfully for the inner and outer regions. The use of wall functions based on a Couette flow assumption (Patankar and Spalding [18]) in the near-wall region is advocated by some. This approach probably has not been quite as well refined to include variable properties, transpiration, and other near-wall effects as the van Driest function.

Closure for the Reynolds heat-flux term, $\rho c_p \overline{v'T'}$, is usually handled in algebraic models by a form of the Reynolds analogy which is based on the similarity between the transport of heat and momentum. The Reynolds analogy is applied to the apparent turbulent conductivity in the assumed Boussinesq form, $\rho c_p \overline{v'T'} = -k_T \, \partial T/\partial y$. In turbulent flow, this additional transport of heat is caused by the turbulent motion. Experiments confirm that the ratio of the diffusivities for the turbulent transport of heat and momentum, called the turbulent Prandtl number, $\mathrm{Pr}_T = \mu_T c_p / k_T$, is a well-behaved function across the flow. Most algebraic turbulence models do well by letting the turbulent Prandtl number be a constant near 1; most commonly, $\mathrm{Pr}_T = 0.9$. Experiments indicate the for wall shear flows Pr_T varies somewhat, from between 0.6 and 0.7 at the outer edge of the boundary layer to about 1.5 near the wall, although the evidence is not conclusive. Several semiempirical distributions for Pr_T have been proposed [9, 19, 20]. Using the turbulent Prandtl number, the apparent turbulent heat flux is related to the turbulent viscosity and mean flow variables as

$$-\rho c_p \overline{v'T'} = \frac{c_p \mu_T}{\mathrm{Pr}_T} \frac{\partial T}{\partial y} \tag{2.119}$$

and closure has been completed.

For other than boundary-layer flows, it may be necessary to model other Reynolds heat-flux terms. To do so, the turbulent conductivity $k_T = c_p \mu_T / \mathrm{Pr}_T$ is normally considered as a scalar and the Boussinesq-type approximation is extended to other components of the temperature gradient. As an example, we would evaluate $-\rho c_p \overline{u'T'}$ as

$$-\rho c_p \overline{u'T'} = \frac{c_p \mu_T}{\mathrm{Pr}_T} \frac{\partial T}{\partial x}$$

To summarize, a recommended base-line algebraic model for wall boundary layers consists of evaluating the turbulent viscosity by Prandtl's mixing-length formula, Eq. (2.113), where l is given by Eq. (2.115) for the inner region, and then using Eq. (2.116) with Eq. (2.113) for the outer region. Alternatively, the Clauser formulation, Eq. (2.117), can be used in the outer region. The apparent turbulent heat flux can be evaluated through Eq. (2.119) using a turbulent Prandtl number of 0.9. This simplest form of modeling has employed four empirical adjustable constants: κ, A^+, C_1 or α, and Pr_T.

Algebraic models have accummulated an impressive record of good performance for simple viscous flows but need to be modified in order to accurately predict flows with "complicating" features. It should be noted that compressible flows do not represent a "complication" in general. The turbulence structure of the flow appears to remain essentially unchanged for Mach numbers up through at least 5. Naturally, the variation of density and other properties must be taken account of in the form of the conservation equations used with the turbulence model. Wall roughness, transpiration, and strong pressure gradients are examples of complicating features which require that adjustments or extensions be made to the simplest form of algebraic turbulence models in order to obtain predictions in agreement with experimental measurements. Such adjustments will not be given in detail here. Recommended adjustments to account for the effects of surface roughness can be found in [9, 21–25]. Recommended modifications to account for wall blowing or suction can be found in [9, 21, 26–28]. Modifications to account for strong pressure gradients are discussed in [9, 21, 24, 26–28].

TABLE 2.1 Some One-Half-Equation Models

Model	Transport Equation Used as Basis for ODE	Model Parameter Determined by ODE Solution	References
A	Turbulence kinetic energy	l_∞	McDonald and Camerata [29] Kreskovsky et al. [30] McDonald and Kreskovsky [31]
B	Turbulence kinetic energy	l_∞	Chan [32]
C	Turbulence kinetic energy	l_∞	Adams and Hodge [24]
D	Empirical ODE for $\mu_{T(\text{outer})}$	$\mu_{T(\text{outer})}$	Shang and Hankey [33]
E	Empirical ODE for $\mu_{T(\text{outer})}$	$\mu_{T(\text{outer})}$	Reyhner [34]
F	Empirical ODE for l_∞	l_∞	Malik and Pletcher [35] Pletcher [36]
G	Turbulence kinetic energy	$\tau_{\max}$	Johnson and King [37]

Philosophically, the strongest motivation for turning to more complex models is the observation that the algebraic model evaluates the turbulent viscosity only in terms of *local* flow parameters, yet it would seem that a turbulence model ought to provide a mechanism by which effects upstream can influence the turbulence structure (and viscosity) downstream. Further, with the simplest models, ad hoc additions and corrections are frequently required to handle specific effects, and constants need to be changed to handle different classes of shear flows.

If the general form for the turbulent viscosity is accepted as $\mu_T = \rho v_T l$, then a logical way to extend the generality of turbulent viscosity models is to permit v_T or l or both to be more complex (and thus more general) functions of the flow capable of being influenced by upstream (historic) effects. This rationale serves to motivate several of the more complex turbulence models.

One-Half Equation Models. A one-half equation model is defined as one in which the value of the model parameter (v_T, l, or μ_T itself) is permitted to vary with the primary flow direction in a manner determined by the solution to an *ordinary* differential equation (ODE). The ODE usually results for either neglecting or assuming the variation of the model parameter with one coordinate direction. Extended mixing-length models and relaxation models fall into this category. A one-equation model is one in which an additional *partial* differential equation is solved for a model parameter. The main features of several one-half equation models are tabulated in Table 2.1.

The first three models in Table 2.1 differ in detail, although all three utilize an integral form of a transport equation for turbulence kinetic energy as a basis for letting the flow history influence the turbulent viscosity. Models of this type have been refined to allow prediction of transition, roughness effects, transpiration, pressure gradients, and qualitative features of relaminarization. Most of the test cases reported for the models have involved external rather than channel flows.

Although models D, E, and F appear to be purely empirical relaxation or lag models, Birch [38] shows that models of this type are actually equivalent to one-dimensional versions of transport partial differential equations for the quantities concerned except that these transport equations are not generally derivable from the Navier-Stokes equations. This is no serious drawback, since transport equations cannot be solved without considerable empirical simplification and modeling of terms; so that in the end, these transport equations tend to have a similar form characterized by generation, dissipation, diffusion, and source terms, regardless of the origin of the equation. Models F and G have been used with fairly good success in predicting boundary-layer flows with separation [36, 37].

One-Equation Models. The obvious shortcoming of algebraic viscosity models which normally evaluate v_T in the expression $\mu_T = \rho v_T l$ by $v_T = l|\partial u/\partial y|$ is that $\mu_T = k_T = 0$ whenever $\partial u/\partial y = 0$. This would suggest that μ_T and k_T would be zero at the centerline of a pipe, in regions near the mixing of a wall jet with a mainstream and in flow through an annulus or between parallel plates where one wall is heated and the other cooled. Measurements (and common sense) indicate that μ_T and k_T are *not* zero under all conditions whenever $\partial u/\partial y = 0$. The mixing-length models can be modified to overcome this deficiency, but this conceptual shortcoming provides motivation for considering other interpretations for μ_T and k_T. In fairness to the algebraic models, it should be mentioned that this defect is not always crucial, because Reynolds stresses and heat fluxes are frequently small when $\partial u/\partial y = 0$. Some examples illustrating this point are given in [39].

It was the suggestion of Prandtl and Komogorov in the 1940s to let v_T in $\mu_T = \rho v_T l$ be proportional to the square root of the kinetic energy of turbulence, $\bar{k} = \frac{1}{2}\overline{u_i' u_i'}$. Thus the turbulent viscosity can be evaluated as

$$\mu_T = C_k \rho l \bar{k}^{1/2} \tag{2.120}$$

and μ_T no longer becomes equal to zero when $\partial u/\partial y = 0$. The kinetic energy of turbulence is a measurable quantity and is easily interpreted physically. Now a means for predicting $\bar{k}$ will be considered.

A transport partial differential equation can be developed for $\bar{k}$ from the Navier-Stokes equations. For incompressible two-dimensional boundary-layer flows, the equation takes the form

$$\rho \frac{D\bar{k}}{Dt} = \mu \frac{\partial^2 \bar{k}}{\partial y^2} - \frac{\partial}{\partial y}\left(\rho \overline{v'k'} + \overline{v'p'}\right) - \rho \overline{v'u'} \frac{\partial u}{\partial y} - \mu \left[\overline{\left(\frac{\partial u'}{\partial y}\right)^2 + \left(\frac{\partial v'}{\partial y}\right)^2 + \left(\frac{\partial w'}{\partial y}\right)^2} \right] \tag{2.121}$$

which is commonly modeled as

$$\underbrace{\rho \frac{D\bar{k}}{Dt}}_{\substack{\text{Particle} \\ \text{rate} \\ \text{increase} \\ \text{of } \bar{k}}} = \underbrace{\frac{\partial}{\partial y}\left[\left(\mu + \frac{\mu_T}{\mathrm{Pr}_k}\right)\frac{\partial \bar{k}}{\partial y}\right]}_{\substack{\text{Diffusion} \\ \text{rate for } \bar{k}}} + \underbrace{\mu_T \left(\frac{\partial u}{\partial y}\right)^2}_{\substack{\text{Generation} \\ \text{rate for } \bar{k}}} - \underbrace{\frac{C_D \rho (\bar{k})^{3/2}}{l}}_{\substack{\text{Dissipation} \\ \text{rate for } \bar{k}}} \tag{2.122}$$

The physical interpretation of the various terms is indicated for Eq. (2.122). This modeled transport equation is then added to the system of PDEs to be solved for the problem at hand. Note that a length parameter l needs to be specified algebraically. In the above, Pr_k is a Prandtl number for turbulence kinetic energy (~ 1), $C_k = 0.548$, and $C_D \approx 0.164$ if l is taken as the ordinary mixing length.

The above modeling for the $\bar{k}$ transport equation is only valid in the fully turbulent regime, i.e., away from any wall damping effects. For typical wall flows, this means y^+ greater than about 30. Inner boundary conditions for the $\bar{k}$ equation are often supplied through the use of wall functions [40]. Another way of treating the inner boundary condition for k is to make use of the experimental observation that very near the wall convection and diffusion of $\bar{k}$ are usually negligible. Thus, generation and dissipation of $\bar{k}$ are in balance, and it can be shown that the turbulence kinetic energy model reduces to Prandtl's mixing-length formulation, Eq. (2.113), under these conditions. At the location where the diffusion and convection are first neglected, an inner boundary condition can be established for $\bar{k}$ as

$$\bar{k}(x, y_c) = \frac{\tau(y_c)}{\rho C_D^{2/3}} \tag{2.123}$$

where y_c is a point within the region where the logarithmic law of the wall is expected to be valid. For $y < y_c$ the Prandtl-type algebraic inner-region model [Eqs. (2.113) and (2.115)] can be used.

The one-equation model has been extended to compressible flows by Rubesin [41], and the results appear encouraging. Apparently for flows containing shock-wave interactions which greatly affect the stream turbulence level, the predictions of Rubesin's one-equation model provide a definite improvement over those from algebraic models. On the whole, however, the performance of most one-equation models (for both incompressible and compressible flows) has been disappointing in that relatively few cases have been observed in which these models offer an improvement over the predictions of the algebraic models. In fact, several flows can be predicted more accurately by the one-half-equation models than by the representative one-equation model of the Prandtl-Kolmogorov type, which merely alters the velocity of turbulence used in the viscosity expression. The reason for this may be that in most flows, an improvement in the specification of a characteristic length scale l will have more effect than a change in the velocity of turbulence, v_T, and many of the one-half equation models listed in Table 2.1 offer an improvement in this length scale.

Other one-equation models have been suggested which deviate somewhat from the Prandtl-Kolmogorov pattern. The most notable of these is by Bradshaw et al. [42]. The turbulence energy equation is used in the Bradshaw model, but the modeling is different both in the momentum equation, where the turbulent shearing stress is assumed proportional to $\bar{k}$, and in the turbulence energy equation. The details will not be given here, but an interesting feature of the Bradshaw method is that as a consequence of the form of modeling used for the turbulent transport terms, the system of equations becomes hyperbolic and can be solved by a procedure similar to the method of characteristics. The Bradshaw method has enjoyed good success in the prediction of the wall boundary layers. Even so, the predictions have not been notably superior to those of the algebraic models, one-half equation models, or other one-equation models.

One and One-Half-and Two-Equation Models. One conceptual advance made by moving from a purely algebraic mixing length model to a one-equation model was that the latter permitted one model parameter to vary throughout the flow, being governed by a PDE of its own. In the one-equation models, a length parameter still appears which is generally evaluated by an algebraic expression dependent upon only *local* flow parameters. It is reasonable to expect that the length scale in turbulence models should also depend on the upstream "history" of the flow and not just local flow conditions. An obvious way to provide more complex dependence of l on the flow is to derive a transport equation for the variation of l. If the equation for l added to the system is an ordinary differential equation, the resulting model is logically termed a one and one-half equation model. Applications of a one and one-half equation models can be found in [36, 39].

Frequently, the equation from which the length scale is obtained is a partial differential equation, and the model is then referred to as a two-equation turbulence model.

Although a transport PDE can be developed for a length scale, the terms of this equation are not easily modeled and some workers have experienced better success by solving a transport equation for a length-scale-related parameter rather than the length scale itself. This point is discussed by Launder and Spalding [40].

One of the most frequently used two-equation models is the $\bar{k} - \epsilon$ model first proposed by Harlow and Nakayama [43]. The description here follows the papers of Jones and Launder [44] and Launder and Spalding [40]. The parameter ϵ is a turbulence dissipation rate and is assumed to be related to other model parameters through $\epsilon = C_D \bar{k}^{3/2}/l$. The turbulent viscosity is related to ϵ through

$$\mu_T = C_k \rho l \bar{k}^{1/2} = \frac{C_\mu \rho \bar{k}^2}{\epsilon} \tag{2.124}$$

where $C_\mu = C_D C_k = 0.09$.

In the $\bar{k}$-ϵ model the turbulence kinetic energy is obtained by solving Eq. (2.122), but the last term on the right is recognized as $\rho\epsilon$. A parabolic transport equation for ϵ is added to close the system. For two-dimensional incompressible boundary-layer flow the equation takes the form

$$\rho \frac{D\epsilon}{Dt} = \frac{\partial}{\partial y}\left(\frac{\mu_T}{\mathrm{Pr}_\epsilon}\frac{\partial \epsilon}{\partial y}\right) + \frac{C_2 \mu_T \epsilon}{\bar{k}}\left(\frac{\partial u}{\partial y}\right)^2 - \frac{C_3 \rho \epsilon^2}{\bar{k}} \tag{2.125}$$

The terms on the right-hand side of Eq. (2.125) from left to right can be interpreted as the diffusion, generation, and dissipation rates of ϵ. Typical values of the model constants are tabulated in Table 2.2.

The above form of the transport equation for ϵ is not appropriate for the near-wall region, i.e., the viscous sublayer. This is just as noted for the turbulence kinetic energy equation (2.122) presented earlier. Inner boundary conditions for ϵ can be provided at

TABLE 2.2 Model Constants for $\bar{k}$-ϵ Two-Equation Model

C_μ	C_2	C_3	Pr_k	Pr_ϵ	Pr_T
0.09	1.44	1.92	1.0	1.3	0.9

the same point y_c used for imposing boundary conditions on $\bar{k}$ [see Eq. (2.123)]. At the point y_c, Prandtl's mixing-length formulation is assumed to be valid and

$$\epsilon = \frac{C_D \bar{k}^{3/2}}{l} = \frac{C_D[\bar{k}(y_c)]^{3/2}}{\kappa y}$$

The quantity $\bar{k}(y_c)$ can be evaluated as indicated in Eq. (2.123).

Many applications of the $\bar{k}$-ϵ model have made use of wall functions [40] to treat the near-wall region. Alternatively, additional terms have been added to the $\bar{k}$ and ϵ equations to extend their applicability to the viscous sublayer by Jones and Launder [44], Chien [95], and others. In this connection, the viscous sublayer is often referred to as the region of low turbulence Reynolds number $(\bar{k}^{1/2}l/\nu)$. This inner modeling is crucial for complex turbulent flows, as for example those containing separated regions or severe property variations. The uncertainty of such inner-region modeling for complex flows appears to limit the range of applicability of the $\bar{k}$-ϵ model (and nearly all other models) at the present time. The $\bar{k}$-ϵ models that have been modified so that they are applicable in the viscous sublayer [44, 45] are known as *low Reynolds number* $\bar{k}$-ϵ *models*. Details of such models will not be given here. Modifications to the $\bar{k}$-ϵ model to include the effects of buoyancy and streamline curvature on the turbulence structure have also been proposed. The most common $\bar{k}$-ϵ closure for the Reynolds heat-flux terms utilizes the same turbulent Prandtl number formulation as used with algebraic models [Eq. (2.119)].

Numerous other two-equation models have been suggested, the most frequently used being the Ng-Spalding [46] model and the Wilcox-Traci model [47], the latter being a modification to the earlier Saffman-Wilcox model [48]. All of these models employ a modeled form of the turbulence kinetic energy equation, but the modeling for the gradient diffusion term is different. The most striking difference, however, is in the choice of dependent variable for the second model transport equation from which the length scale is determined.

Reynolds Stress Models. Reynolds stress models (sometimes called stress-equation models) are those models that *do not* assume that the turbulent shearing stress is proportional to the rate of mean strain. That is, for a two-dimensional incompressible flow

$$-\rho\overline{u'v'} \neq \mu_T\left(\frac{\partial u}{\partial y} + \frac{\partial v}{\partial x}\right)$$

These models have been used to date largely as tools or subjects in turbulence research rather than to solve engineering problems. Thus, details will not be given here. Exact transport equations can be derived for the Reynolds stresses. However, these equations contain terms which must be modeled. Such modeling, which generally follows the pioneering work of Rotta [49], requires the solution of at least three additional transport PDEs. For a flow in which normal stresses are important, five additional equations are usually required. The most widely used Reynolds stress models at the present time are those of Hanjalic and Launder [50], Launder et al. [51], and Donaldson [52].

Reynolds stress models are not restricted by the Boussinesq approximation relating turbulent stresses to rates of mean strain and contain the greatest number of model PDEs and constants of the models discussed. Thus, it would seem that these models ought to have the best chance of emerging as "ultimate" turbulence models if success is to be achieved at all through the time-averaged Navier-Stokes equations. Nevertheless, these models still must utilize approximations and assumptions in modeling terms which presently cannot be measured. These Reynolds stress models are perhaps still in their infancy, and it may be some time yet before they have been tested and refined to the point that they become commonplace in engineering calculations. Since simpler models perform adequately for many flows, the expectation is that the Reynolds stress models may only be used in engineering predictions where the flow complexity demands it. At the present time, the Reynolds stress models have not even been tested for many types of complex flows.

A simplification of the Reynolds stress modeling known as an *algebraic* stress or flux model is gaining in popularity. This approach is discussed in detail in [53]. In the algebraic stress modeling, it is generally assumed that the transport of the Reynolds stresses is proportional to the transport of turbulence kinetic energy. For boundary-layer flows without buoyancy effects, the algebraic Reynolds stress model results in

$$-\overline{u'v'} = C_\mu \frac{\bar{k}^2}{\epsilon}\frac{\partial \bar{u}}{\partial y}$$

which is identical to the results obtained from the $\bar{k}$-ϵ model. However, in the algebraic Reynolds stress model, C_μ becomes a function of the ratio of production to dissipation of turbulence kinetic energy rather than a constant. These models show considerable promise as useful extensions of the $\bar{k}$-ϵ modeling approach.

2.4 FLOW OVER A FLAT PLATE

2.4.1 Incompressible Flow

A *flat plate* is a surface at constant pressure with a sharp leading edge. The boundary-layer equations in the form of Eqs. (2.89)–(2.91) are valid in this case. If the properties are assumed to be constant, the boundary-layer momentum and continuity equations can be solved independently of the energy equation. Historically, this was first done using the similarity variables proposed by Blasius [54], which permits the governing equations to be combined into a single third-order ODE. The solution of this equation is very well known, and tabulated results can be found many places, including [1]. The local skin friction coefficient is given by

$$c_f = 0.332\,\mathrm{Re}_x^{-1/2} \tag{2.126}$$

Pohlhausen [55] utilized the Blasius velocity solution to solve the energy equation under constant-property assumptions for low-speed flow where viscous dissipation can be neglected. The solution obtained by Pohlhausen for the local Nusselt number is well represented by

$$\mathrm{Nu} = \frac{hx}{k} = 0.332\,\mathrm{Re}_x^{1/2}\,\mathrm{Pr}^{1/3} \qquad \text{for} \quad 0.6 \le \mathrm{Pr} \le 10 \tag{2.127}$$

When written in terms of the Stanton number

$$\mathrm{St} = \frac{c_f}{2}\,\mathrm{Pr}^{-2/3} \tag{2.128}$$

a close relationship between skin friction and heat transfer is observed. When Pr = 1, the Stanton number is identically equal to $c_f/2$. The function of Prandtl number which serves as the proportionality factor between St and $c_f/2$ is sometimes referred to as the Reynolds analogy factor. Within this same range of Prandtl numbers, $\delta/\delta_T = \mathrm{Pr}^{1/3}$ for a constant-temperature plate. For Pr > 10, the following expression is recommended:

$$\mathrm{Nu} = 0.339\,\mathrm{Re}_x^{1/2}\,\mathrm{Pr}^{1/3} \tag{2.129}$$

For fluids of small Prandtl number, such as liquid metals, the local Nusselt number can be evaluated from

$$\mathrm{Nu} = 0.564(\mathrm{Re}_x\mathrm{Pr})^{1/2} \tag{2.130}$$

Integral methods [56] have been used to show that for constant-property laminar flow under uniform heat flux conditions, the local Nusselt number can be obtained from

$$\mathrm{Nu} = 0.453\,\mathrm{Re}_x^{1/2}\,\mathrm{Pr}^{1/3} \tag{2.131}$$

This equation is expected to be valid for moderate Prandtl number range 0.6 to 10. The effects of viscous dissipation and variable fluid properties are not included in the above expressions for the Nusselt number.

For constant fluid properties, the effect of viscous dissipation on laminar forced convection over a flat plate can be obtained by superimposing two solutions obtained for the energy equation by Pohlhausen [55]. The first solution is for the simple flat-plate problem without dissipation, resulting in the solution of Eq. (2.127) for Prandtl numbers near unity. The second is the "thermometer" problem of determining the temperature of an insulated plate when viscous dissipation is important. In that case the heat transfer is to be computed using the adiabatic wall temperature T_{aw} according to

$$q_w = h(T_w - T_{aw}) \tag{2.132}$$

as was indicated in Sec. 2.2.5. The adiabatic wall temperature is computed using the recovery factor according to

$$T_{aw} = T_e + r\frac{u_e^2}{2c_p}$$

where r is a function of Prandtl number for the laminar flat-plate flow.

The solution for r for the constant-property laminar boundary layer is shown in Fig. 2.5. Approximations to the solution are

$$r = \mathrm{Pr}^{1/2}, \qquad 0.5 \le \mathrm{Pr} < 47 \tag{2.133}$$

$$r = 1.9\,\mathrm{Pr}^{1/3}, \qquad \mathrm{Pr} > 47 \tag{2.134}$$

The heat transfer coefficient is still given by Eq. (2.127). The effects of viscous dissipation appear in the calculation of heat transfer for this case with constant properties only through the adiabatic wall temperature used in Newton's law of cooling. Further details on this solution are discussed in [1,2]. Average Nusselt numbers over a plate of length L can be easily obtained by integrating the local values

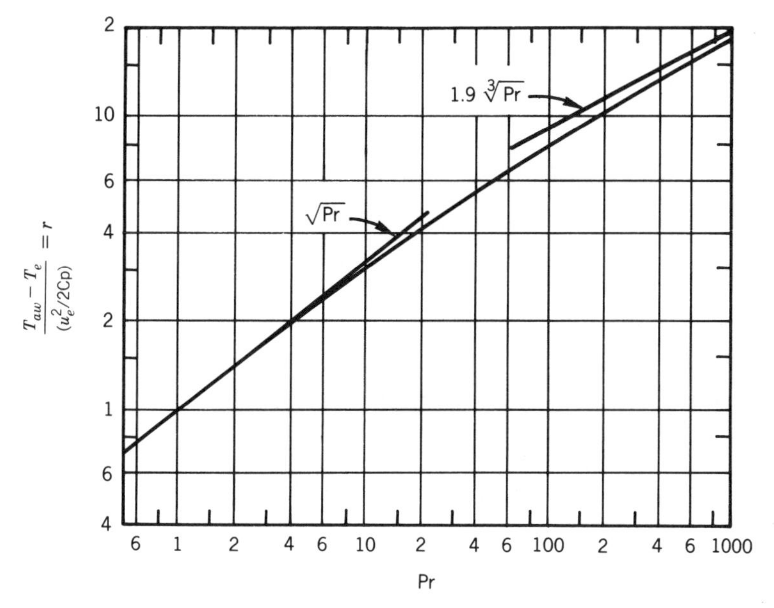

Figure 2.5. Dependence of adiabatic wall temperature on Prandtl number for a laminar boundary layer on a flat plate (adapted from [1] by permission).

over the length according to

$$\overline{\mathrm{Nu}} = \frac{\bar{h}L}{k} = \frac{1}{L}\int_0^L \frac{hx}{k}\,dx.$$

For turbulent flow over a flat plate under constant-property conditions, the local skin friction coefficient is given by

$$c_f = 0.0592\,\mathrm{Re}_x^{-1/5} \qquad \text{for} \quad 5 \times 10^5 \leq \mathrm{Re}_x < 10^7 \tag{2.135}$$

An expression valid for much higher Reynolds numbers was given by Schultz-Grunow [57]:

$$c_f = 0.185(\log_{10}\mathrm{Re}_x)^{-1.584} \tag{2.136}$$

It is interesting that the Stanton number for turbulent flow over an isothermal plate is well correlated by Eq. (2.128) when the skin friction coefficient is evaluated by Eq. (2.135). That is, the same Reynolds analogy factor, $\mathrm{Pr}^{-2/3}$, works for both laminar and turbulent boundary-layer flow in a limited range of Prandtl numbers. In terms of the Nusselt number, this gives

$$\mathrm{Nu} = 0.0296\,\mathrm{Re}_x^{4/5}\mathrm{Pr}^{1/3} \qquad \text{for} \quad 0.6 \leq \mathrm{Pr} \leq 60 \tag{2.137}$$

In a typical application, laminar-turbulent transition will occur on the flat plate. In order to obtain the average Nusselt number for the plate, both the laminar and turbulent portions of the flow need to be taken into account. Thus

$$\bar{h} = \frac{1}{L}\int_0^{x_c} h_{\text{lam}}\, dx + \frac{1}{L}\int_{x_c}^{L} h_{\text{turb}}\, dx$$

where it is assumed that transition occurs at $x = x_c$. Using Eqs. (2.127) and (2.134) gives

$$\overline{\text{Nu}} = \left[0.664\,\text{Re}_{x,c}^{1/2} + 0.037\left(\text{Re}_L^{4/5} - \text{Re}_{x,c}^{4/5}\right)\right]\text{Pr}^{1/3} \tag{2.138}$$

If the typical transition Reynolds number of $\text{Re}_{x,c} = 5 \times 10^5$ is assumed, Eq. (2.138) reduces to

$$\overline{\text{Nu}} = \left(0.037\,\text{Re}_L^{4/5} - 871\right)\text{Pr}^{1/3} \tag{2.139}$$

for $0.6 \leq \text{Pr} < 60$ and $5 \times 10^5 < \text{Re}_L \leq 10^8$. When $L \gg x_c$ or the boundary layer is tripped at the leading edge, the laminar portion of the flow can be neglected and Eqs. (2.138) and (2.139) reduce to

$$\overline{\text{Nu}} = 0.037\,\text{Re}_L^{4/5}\text{Pr}^{1/3} \tag{2.140}$$

For uniform heat flux, integral methods have been used [56] to obtain the following expression for the local Nusselt number:

$$\text{Nu} = 0.0308\,\text{Re}_x^{4/5}\text{Pr}^{1/3} \tag{2.141}$$

All of the above expressions for the Nusselt number are restricted to situations where the wall boundary conditions are imposed all along the plate starting from the leading edge. A common exception occurs when an *unheated starting length* ($T_w = T_\infty$) exists upstream of a heated section ($T_w \neq T_\infty$). For laminar flow, integral methods [56] have been used to obtain an approximate solution for this case in the form

$$\text{Nu} = \frac{\text{Nu}|_{\xi=0}}{\left[1 - (\xi/x)^{3/4}\right]^{1/3}} \tag{2.142}$$

where ξ is the value of x at which heating (or cooling) starts and $\text{Nu}|_{\xi=0}$ is the value of the local Nusselt number given by Eq. (2.127). For turbulent flow the corresponding expression for the local Nusselt number is

$$\text{Nu} = \frac{\text{Nu}|_{\xi=0}}{\left[1 - (\xi/x)^{9/10}\right]^{1/9}} \tag{2.143}$$

A wide range of wall boundary conditions can be handled if the convection solution is obtained by numerical methods. The wall temperature or wall heat flux can be varied in an arbitrary manner. Such an approach typically employs a finite difference procedure to solve the boundary-layer form of the momentum, energy, and continuity equations for a free stream of constant velocity and temperature. For flat-plate flows, the algebraic turbulence models discussed in Sec. 2.3.3 work well. Examples of such numerical procedures and computed results can be found in [4, 58].

2.4.2 Compressible Flow; Effects of Property Variations

All of the results for heat transfer presented in the previous subsection were obtained under the assumption that the fluid properties remained constant throughout the flow. Clearly, this is an idealization, since the properties of most fluids vary with temperature, and thus will vary throughout the thermal boundary layer. In this section, ways of including the effects of property variations will be presented.

For gases, the specific heat and Prandtl number do not vary significantly over a fairly wide range of temperatures and are nearly independent of pressure. For air near standard conditions in the stagnation state, the assumption of constant specific heat and Prandtl number is a reasonable one for flows at Mach numbers up through at least 5. The viscosity and thermal conductivity of gases increase with temperature, and the variation should be taken into account. Density variations for gases can be significant but can usually be determined as a function of temperature and pressure from the ideal-gas equation of state.

For most liquids, the specific heat and thermal conductivity are relatively independent of temperature, but the viscosity decreases significantly as the temperature increases. The density of liquids varies only a little with temperature. The Prandtl number for liquids varies significantly with temperature, due primarily to the strong temperature dependence of the viscosity.

When numerical methods are used to obtain solutions to the governing equations, it is a fairly easy task to let the properties vary in the solution procedure. Very little complication is added to the solution procedure in doing so. However, it has also been observed that many of the constant-property solutions and correlations can be "corrected" in a simple manner to take account of property variations. This is frequently the case for the flow of liquids in general and gases at low speeds where viscous dissipation is not significant.

For very small temperature differences (perhaps $< 5°C$ for liquids and $< 50°C$ for gases) between the free stream and wall, reasonably accurate predictions can be obtained by using the equations of this section with all properties evaluated at the *film temperature* defined as

$$T_f = \frac{T_w + T_e}{2} \tag{2.144}$$

Use of this reference temperature for property evaluation in gases is only recommended for low-speed ($M_e < 0.3$) flows. For greater temperature differences with liquids, there is relatively little specific information available. It is tentatively recommended that the correction factors provided elsewhere in this handbook for the flow of liquids in tubes be used.

For gases, as the velocity increases, the conversion of mechanical energy to thermal energy due to the effects of viscosity becomes increasingly important. This effect clearly alters the temperature distribution in the fluid. This was discussed earlier under the assumption of constant fluid properties. In general, the temperature variations throughout the fluid will influence fluid properties, which in turn will alter the skin friction and surface heat transfer. Extremely high velocities, generally associated with hypersonic flow (Mach numbers greater than about 7) in air lead to very high temperatures, dissociation and chemical reactions. These conditions will not be considered in this chapter.

According to Eq. (2.109), the adiabatic wall temperature for flat-plate flow can be related to the free-stream static and stagnation temperatures by

$$T_{aw} = T_\infty + r(T_0 - T_\infty) \tag{2.145}$$

For an ideal gas with constant specific heat, T_{aw} can be expressed in terms of free-stream Mach number:

$$T_{aw} = T_\infty\left(1 + \frac{\gamma - 1}{2} r M_\infty^2\right) \tag{2.146}$$

where γ is the ratio of specific heats c_p/c_v. From Eq. (2.146) it is easy to estimate the flight Mach number at which active cooling will be required in order to keep the surface temperature of a structure below a specified level.

Eckert [59] made the remarkable observation that if the specific heat can be treated as constant and all fluid properties are evaluated at an appropriate reference temperature T^*, the low-speed constant-property correlating equations for Nu can be used for air for Mach numbers up to 20, the errors being less than a few percent. The Eckert reference temperature is given by

$$T^* = 0.5(T_w + T_e) + 0.22(T_{aw} - T_e) \tag{2.147}$$

If the assumption of constant specific heat is not valid due to very large temperature differences, the total enthalpy can conveniently be used as the dependent variable in the energy equation. In this case, all properties should be evaluated at the temperature corresponding to the reference enthalpy i^*:

$$i^* = 0.5(i_w + i_e) + 0.22(i_{aw} - i_e) \tag{2.148}$$

Following the Eckert reference-property method, Eq. (2.127) can be used to compute the Nusselt number for laminar flow on a flat plate. The fluid properties in the Reynolds number, the Prandtl number, and the Nusselt number should be evaluated at the value obtained from Eqs. (2.147) or (2.148). The recovery factor is taken as $r = (\mathrm{Pr}^*)^{1/2}$, where the asterisk denotes that the Prandtl number should be evaluated at the reference temperature. The wall heat flux can then be computed from Eq. (2.132) using T_{aw} computed from Eq. (2.145) or Eq. (2.146). For turbulent flow on a flat plate, a similar procedure is followed. Equation (2.137) can be used to compute the Nusselt number, but the properties in this equation should be evaluated at the Eckert reference temperature. The recovery factor can be taken as $r = (\mathrm{Pr}^*)^{1/3}$ for turbulent flow. The wall heat flux is computed using Eq. (2.132) as for laminar flow.

Van Driest [60] used a method developed by Crocco [61] to obtain solutions to the compressible laminar boundary-layer equations for air at a constant Prandtl number of 0.75. The specific heat was assumed to be constant, and the viscosity was evaluated as a function of temperature through the use of the Sutherland equation

$$\mu = \frac{C_1 T^{3/2}}{T + C_2} \tag{2.149}$$

where C_1 and C_2 are constants for a given gas. For air, $C_1 = 1.458 \times 10^{-6}$ kg/(m · s · K$^{1/2}$) and $C_2 = 110.4$ K. Rather detailed results are given in [60] for skin friction coefficient, Stanton number, and velocity and temperature profiles. Van Driest's

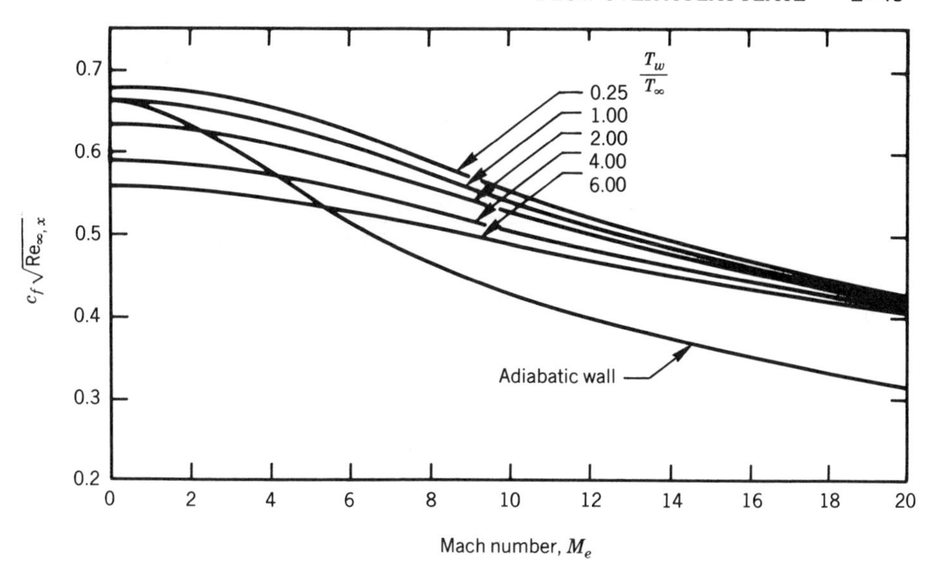

Figure 2.6. Local skin friction coefficient for compressible laminar boundary-layer flow along a flat plate, Pr = 0.75, according to van Driest [60].

solution for the skin friction coefficient is shown in Fig. 2.6 as a function of wall-to-stream temperature ratio and Mach number. For fixed temperature conditions, the skin friction coefficient is seen to decrease with increasing Mach number. For fixed free-stream conditions, lowering the wall temperature increases the skin friction coefficient. Similar trends are observed for the Stanton number, as can be seen in Fig. 2.7. Van Driest observed for Pr = 0.75 that the Stanton number from his solution was within 1% of that predicted by using Eq. (2.128), i.e., the Reynolds analogy factor is almost exactly $\mathrm{Pr}^{-2/3}$.

Thus, for laminar flat-plate flow in the compressible regime, the practicing engineer can choose from among several sources for information on heat transfer. For some purposes, the graphical solutions given in Fig. 2.7 or in [60] will suffice. The Eckert reference-property method can also be used along with the incompressible expression for Nusselt number. Finite difference solutions to the boundary-layer equations give very accurate results for this type of flow, and many engineers will have access to the required computer programs. Of these three sources, it is only the last which can be used for flows in which the wall temperature or heat flux varies in an arbitrary manner. With numerical methods it is also possible to solve combined conduction and convection problems.

For compressible turbulent flow over a flat plate, finite difference methods can be used to solve the boundary-layer equations with appropriate turbulence modeling. The algebraic turbulence models discussed in Sec. 2.3.3 are adequate for flat-plate flows, but as the Mach number increases, the low Reynolds number effects mentioned in that section become increasingly important. The algebraic turbulence model is easily modified to account for this [9, 17].

Several empirical formulas have been proposed for calculating the skin friction and heat transfer coefficients for compressible turbulent boundary layers on flat plates. Those developed by Spalding and Chi [62] and van Driest [63] have higher accuracy

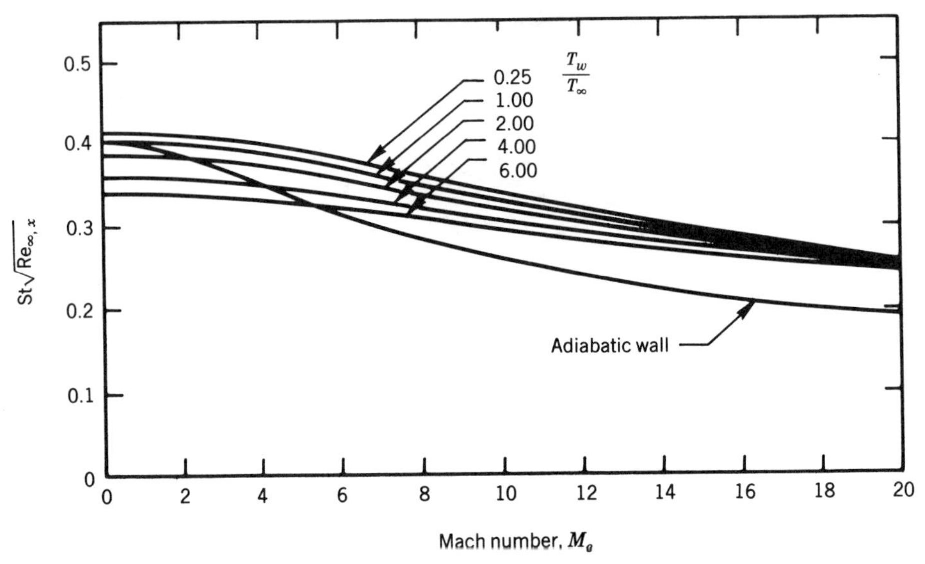

Figure 2.7. Local Stanton number for compressible laminar boundary-layer flow along a flat plate, Pr = 0.75, according to van Driest [60].

than the rest, according to studies by Hopkins and Keener [64] and Cary and Bertrum [65].

Both methods define compressibility factors by the following relationships:

$$c_{f,i} = c_f F_c \tag{2.150}$$

$$\mathrm{Re}_{\theta,i} = F_{\mathrm{Re}_\theta} \mathrm{Re}_\theta \tag{2.151}$$

$$\mathrm{Re}_{x,i} = \int_0^{\mathrm{Re}_x} \frac{F_{\mathrm{Re}_\theta}}{F_c}\, d\,\mathrm{Re}_x = F_{\mathrm{Re}_x} \mathrm{Re}_x \tag{2.152}$$

The subscript i denotes incompressible values, and the factors F_c, F_{Re_θ} and F_{Re_x} are functions of Mach number, ratio of wall temperature to total temperature, and recovery factor. Spalding and Chi assumed that a unique relationship exists between $c_f F_c$ and $F_{\mathrm{Re}_x} \mathrm{Re}_x$. The quantity F_c is obtained by means of a mixing-length theory, and F_{Re_x} is obtained semiempirically. According to Spalding and Chi [62],

$$F_c = \frac{T_{aw}/T_e - 1}{\left(\sin^{-1}\alpha + \sin^{-1}\beta\right)^2} \tag{2.153}$$

$$F_{\mathrm{Re}_\theta} = \left(\frac{T_{aw}}{T_e}\right)^{0.772} \left(\frac{T_w}{T_e}\right)^{-1.474} \tag{2.154}$$

$$\alpha = \frac{T_{aw}/T_e + T_w/T_e - 2}{\left[\left(T_{aw}/T_e + T_w/T_e\right)^2 - 4\left(T_w/T_e\right)\right]^{1/2}} \tag{2.155}$$

$$\beta = \frac{T_{aw}/T_e - T_w/T_e}{\left[\left(T_{aw}/T_e + T_w/T_e\right)^2 - 4\left(T_w/T_e\right)\right]^{1/2}} \tag{2.156}$$

In the van Driest method, F_c is evaluated as indicated above, but F_{Re_θ} is given by

$$F_{\mathrm{Re}_\theta} = \frac{\mu_e}{\mu_w} \tag{2.157}$$

A power-law formula is used to relate viscosity to temperature, $\mu \propto T^\omega$.

Van Driest actually proposed two expressions for the skin friction coefficient. One was based entirely on Prandtl's mixing-length expression, $l = \kappa y$, and the other on the von Kármán similarity law

$$l = \kappa \left| \frac{\partial u/\partial y}{\partial^2 u/\partial y^2} \right| \tag{2.158}$$

The expression based on the von Kármán similarity law, known as van Driest II, is in better agreement with experimental data than the expression based only on the mixing-length formula. Defining

$$G = \left[\left(\frac{\gamma - 1}{2} \right) \frac{M_e^2 T_e}{T_w} \right]^{1/2} \tag{2.159}$$

the van Driest II formula can be written as

$$\frac{0.242(\sin^{-1}\alpha + \sin^{-1}\beta)}{G\sqrt{c_f(T_w/T_e)}} = 0.41 + \log_{10}(\mathrm{Re}_x c_f) - \omega \log_{10}(T_w/T_e) \tag{2.160}$$

where x is the distance measured from the effective origin of the turbulent flow. According to van Driest, the average skin friction coefficient can be obtained from

$$\frac{0.242(\sin^{-1}\alpha + \sin^{-1}\beta)}{G\sqrt{\bar{c}_f(T_w/T_e)}} = \log_{10}(\mathrm{Re}_x \bar{c}_f) - \omega \log_{10}(T_w/T_e) \tag{2.161}$$

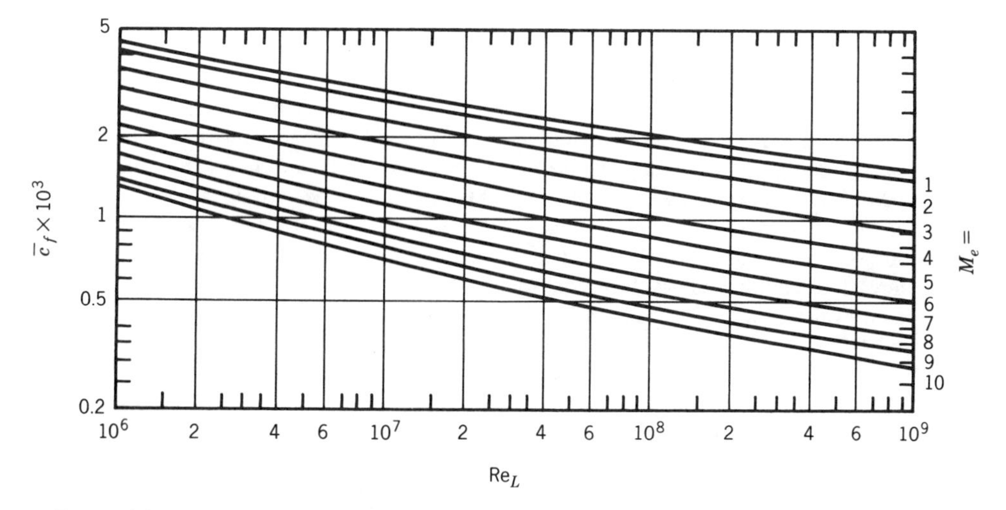

Figure 2.8. Average skin friction coefficient for the compressible turbulent boundary-layer flow of air along a flat plate according to van Driest II [63].

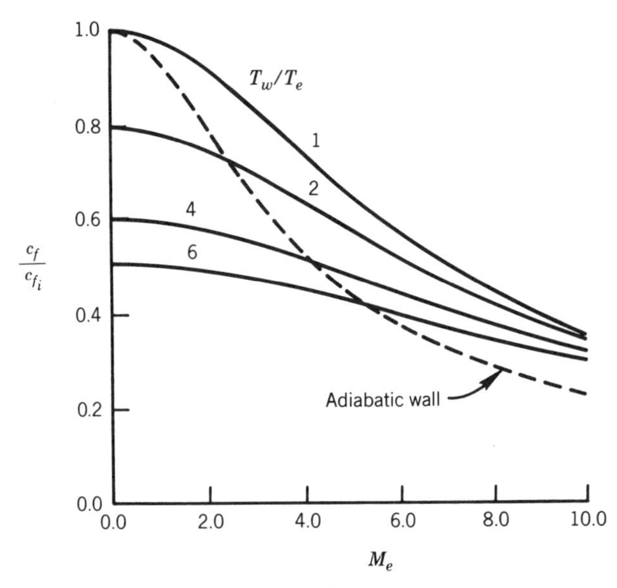

Figure 2.9. Effect of compressibility on local skin-friction coefficient for the turbulent flow of air along a flat plate according to van Driest II [63].

The variation of the average skin friction coefficient with Mach and Reynolds numbers can be seen in Fig. 2.8 for an adiabatic plate according to van Driest II. The recovery factor was assumed to be 0.89. The effects of compressibility on the local skin friction coefficient can be seen in Fig. 2.9 for $\mathrm{Re}_x = 10^7$. The effect of compressibility on the average skin friction is nearly identical.

Predictions of heat transfer are obtained for compressible turbulent flow over a flat plate by using a Reynolds analogy factor along with a prediction for skin friction. For Mach numbers less than about 5 and near-adiabatic wall conditions,

$$\mathrm{St} = 1.16 \frac{c_f}{2} \tag{2.162}$$

adequately represents the available experimental data. Thus, the skin friction coefficient should first be obtained from the van Driest II or Spalding-Chi prediction, and then the Stanton number can be computed by Eq. (2.162). More uncertainty exists about the appropriate Reynolds analogy factor for highly cooled walls at any Mach number and at Mach numbers greater than 5 at any wall temperature. Recent data [65] indicate that for Mach numbers greater than 6 and T_w/T_0 less than about 0.3, the Reynolds analogy factor is more nearly 1.0. The results presented by Cary [66] for a Mach number of 11.3 showed Reynolds analogy factors scattering between 0.8 and 1.4, with no particular trend evident for T_w/T_0. Studies by Hopkins et al. [67] suggest that for $5 < M_e < 7.5$ and $0.1 < T_w/T_{aw} < 0.6$, predictions of heat transfer will be within 10% using the van Driest II theory for skin friction and a Reynolds analogy factor of 1.0.

2.4.3 Effects of Blowing and Suction

One way to provide cooling for a surface exposed to a hot external stream is to inject a coolant into the hot boundary layer formed on the surface. This can be done through a porous section of the surface, through discrete holes or through slots. Cooling provided through a porous material which distributes the coolant fairly uniformly and in a direction normal to the surface is known as transpiration cooling. In this mode, the fluid at the surface possesses an effective component of velocity in the normal direction, v_w. When the fluid is injected through slots or large discrete holes, the cooling mode is usually referred to as film cooling. Often the slots or holes are not normal to the surface, so that the coolant emerges with a tangential component of velocity. Film or transpiration cooling is quite commonly used to cool turbine blades. The coolant may be a gas or a liquid which changes phase prior to or even after ejection from the surface.

Systems can be designed to operate more or less continuously with film or transpiration cooling. Another type of closely related cooling system is known as ablation. An ablation cooling system is one in which the outer solid material is designed to sublime or chemically react with the convecting fluid, absorbing heat in the process. This protects an underlying structure. The advantage of this method is its relative simplicity. It requires no active components such as pumps, ducting, or storage vessels. The main disadvantage is that once the ablating material is consumed, the cooling ends. That is, it is not suited for systems which must operate continuously, or for long periods of time.

The analysis of flows in which cooling is enhanced by transpiration, film cooling, or ablation is generally not easy. Only the simplest cases will be considered here.

Transpiration is the easiest mode to analyze. When the boundary-layer and coolant gases are the same and the blowing rates are moderate (so that the boundary-layer theory is applicable), the boundary-layer equations are applicable. The injection at the surface only changes the boundary conditions. Instead of being equal to zero, the normal component of velocity becomes equal to the injection velocity. There exist, in fact, similarity solutions for laminar boundary layers with blowing and suction. The similarity condition requires that the blowing rate v_w/u_e be proportional to $x^{-1/2}$. For this solution, the wall temperature is constant.

It is not easy in practice to adjust the blowing rates to be proportional to $x^{-1/2}$. With modern numerical methods, it is relatively easy to solve the boundary-layer equations without the similarity constraint to provide heat transfer predictions for arbitrary distributions of the blowing parameter. During the 1950s, however, before numerical techniques for solving the boundary-layer equations for nonsimilar flows were well developed, the similarity solutions provided considerable guidance on the general capabilities of transpiration cooling systems. Examples of these solutions can be found in the work of Emmons and Leigh [68], Hartnett and Eckert [69], and Low [70]. Numerical results for laminar boundary layers with uniform injection can be found in the work of Libby and Chen [71].

In many practical applications of film or transpiration cooling, the boundary-layer flow is turbulent. A considerable amount of experimental data has been accumulated for the transpiration of air into a turbulent low-speed air stream [72]. The experiments included flows with both blowing and suction over a range of pressure gradients. Property variations were small. The results can be well correlated making use of the parameters

$$B_f = \frac{2\rho_w v_w}{c_f \rho_e u_e} \tag{2.163}$$

and

$$B_h = \frac{\rho_w v_w}{\mathrm{St}\, \rho_e u_e} \tag{2.164}$$

For flows with zero pressure gradient, a simple analysis based on a Couette-flow approximation to account for the effects of transpiration leads to the correlation

$$\frac{c_f}{2} = 0.0287 \frac{\ln(1 + B_f)}{B_f} \mathrm{Re}_x^{-1/5} \tag{2.165}$$

or, with the understanding that c_{f0} is the skin friction coefficient without blowing to be evaluated at the same x Reynolds number as the flow with blowing,

$$\frac{c_f}{c_{f0}} = \frac{\ln(1 + B_f)}{B_f} \tag{2.166}$$

This result is in excellent agreement with the available experimental data. For the same condition of zero pressure gradient, the Stanton number can be obtained from

$$\mathrm{St}\,\mathrm{Pr}^{0.4} = 0.0287 \frac{\ln(1 + B_h)}{B_h} \mathrm{Re}_x^{-1/5} \tag{2.167}$$

The results with transpiration can be most easily generalized to include flows with pressure gradients by using the Reynolds number based on momentum thickness when dealing with the skin friction coefficient, and using the Reynolds number based on the enthalpy thickness Δ defined as

$$\Delta = \int_0^\infty \frac{\rho u (H - H_e)}{\rho_e u_e (H_w - H_e)}\, dy$$

when dealing with the Stanton number. For a wide range of pressure gradients, the skin friction coefficient for turbulent flows with transpiration can be obtained from

$$\frac{c_f}{2} = 0.0125 \left[\frac{\ln(1 + B_f)}{B_f} \right]^{1.25} (1 + B_f)^{1/4} \mathrm{Re}_\theta^{-1/4} \tag{2.168}$$

which can be developed from Eq. (2.165) making use of the momentum integral equation [56]. The Stanton number can be obtained from

$$\mathrm{St}\,\mathrm{Pr}^{0.4} = 0.0125 \left[\frac{\ln(1 + B_h)}{B_h} \right]^{1.25} (1 + B_h)^{1/4} \mathrm{Re}_\Delta^{-1/4} \tag{2.169}$$

The constants and Reynolds number functions in Eqs. (2.168) and (2.169) are the same as observed for flows without transpiration [56]. Thus, the results implied in Eqs. (2.168) and (2.169) can be written in the form

$$\left.\frac{c_f}{c_{f0}}\right|_\theta = \left.\frac{\mathrm{St}}{\mathrm{St}_0}\right|_\Delta = \left[\frac{\ln(1 + B)}{B} \right]^{1.25} (1 + B)^{1/4} \tag{2.170}$$

In Eq. (2.170), $B = B_f$ when c_f is to be determined, and $B = B_h$ when St is to be determined. When c_f is being evaluated, c_{f0} in Eq. (2.170) is taken as the skin friction coefficient for a flow without transpiration, but at the same value of Re_θ as the

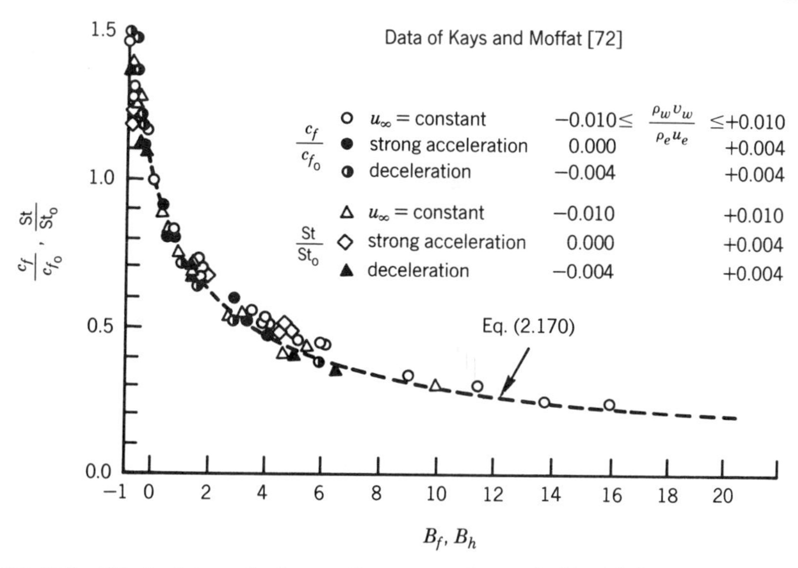

Figure 2.10. Effect of transpiration on Stanton number and skin friction coefficient for an incompressible turbulent boundary layer. Data of Kays and Moffat [72]. (Adapted from [56] by permission.)

transpired flow. Similarly, when St is being evaluated, St_0 is taken as the Stanton number for a flow without transpiration, but at the same Re_Δ as the transpired flow.

The range of applicability of Eq. (2.170) appears to be quite broad. Experimental results for St and c_f are shown in Fig. 2.10 along with Eq. (2.170). The data were obtained for flows with blowing and suction over a range of both favorable and adverse pressure gradients. Older reviews of the status of research on turbulent flows with transpiration can be found in [73, 74].

In general, blowing reduces both the local skin friction coefficient and the local Stanton number. The effect of suction is just the opposite. Qualitatively, the effect of blowing on the skin friction coefficient and the Stanton number is similar to that of imposing an adverse pressure gradient, whereas the effect of suction is similar to imposing a favorable pressure gradient. Suction is primarily used to control boundary-layer growth, especially to delay or eliminate separation by removing low-momentum fluid from the boundary layer. Suction can also be used to delay laminar-turbulent transition.

Predictions based on boundary-layer theory with mass injection can give some guidance in the design of ablation systems. However, the solution of the complete ablation problem with phase changes and chemical reactions is difficult and requires considerations which are beyond the scope of this chapter. Likewise, film cooling from slots or discrete holes is difficult to analyze, and few general results are available. Such flows can be dealt with through numerical methods, however.

The experimental results available for compressible turbulent flows with transpiration are limited in number and cover a relatively narrow range of flow conditions. For the most part, the data are for zero pressure gradient flows. Representative experimental data for compressible flows can be found in [75–80]. The experimental uncertainty is greater for the compressible flows, and no simple correlation can be recommended with confidence. The data trends are discussed in [81].

In recent years, finite difference solutions to the boundary-layer equations have been reasonably well established as being reliable for flows with transpiration. For laminar flows, the calculation proceeds in a very straightforward manner, the only change being a modification to the wall boundary condition. For the turbulent case, most investigators have reported a need to modify the turbulence model in order to accurately predict flows with all but the smallest blowing and suction rates. It appears that the modification can be implemented in algebraic models by making adjustments to the van Driest wall damping function. Several schemes for these modifications have been proposed. Descriptions of two of these can be found in [9, 26]. The effect of these modifications is to decrease the magnitude of the damping constant A^+ [see Eq. (2.115)] for flows with blowing and to increase its magnitude for flows with suction. Several comparisons of finite difference predictions with experimental measurements, including some at supersonic speeds, can be found in [9, 26].

2.5 FLOWS WITH PRESSURE GRADIENTS

2.5.1 Similarity Solutions for Laminar Flow

In a previous section, the similarity solution for the boundary layer on a flat plate was discussed. In 1931, Falkner and Skan [82] discovered the similarity transformation appropriate for two-dimensional incompressible wedge flow for which the inviscid solution is given by

$$u_e = Cx^m \tag{2.171}$$

where m is related to the wedge angle $\beta\pi$ by

$$\beta = \frac{2m}{1+m} \tag{2.172}$$

The inviscid solution, of course, provides the boundary condition needed at the outer edge of the boundary layer. Included in the wedge-flow family of boundary-layer solutions is the important special case of two-dimensional stagnation flow (see Fig. 2.11). For the stagnation flow indicated in Fig. 2.11, $\beta = m = 1$ and $u_e = Cx$. It has been shown that in the local neighborhood of a stagnation point in any symmetric incompressible two-dimensional flow, the solution to the Falkner-Skan equation for $\beta = m = 1$ is valid. This is because near the stagnation point, the velocity at the outer edge of the boundary layer varies linearly with distance. For example, for flow over a

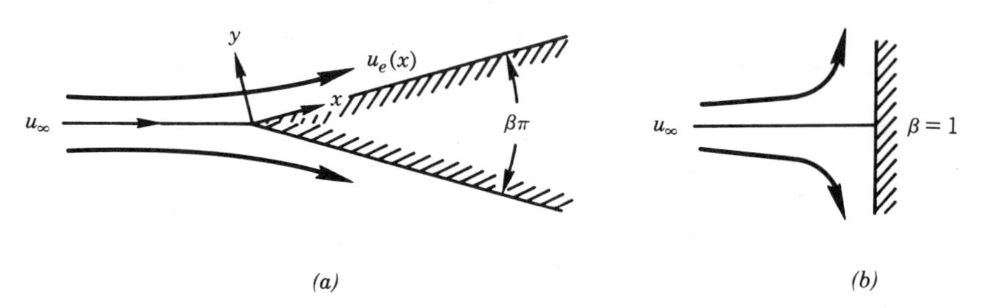

Figure 2.11. Wedge flow: (*a*) general configuration; (*b*) two-dimensional stagnation flow.

TABLE 2.3 Laminar Wedge Flow Results

β	m	$c_f \mathrm{Re}_x^{1/2}/2$	Case
1.6	5	2.6344	
1.0	1	1.2326	2D stagnation
0.5	$\frac{1}{3}$	0.75746	
0	0	0.33206	Flat plate
−0.14	−0.06542	0.16372	
−0.1988	−0.09041	0	Separation

TABLE 2.4 $\mathrm{Nu}_x / \mathrm{Re}_x^{1/2}$ for Laminar Wedge Flow

	$\mathrm{Nu}_x/\mathrm{Re}_x^{1/2}$				
m	Pr = 0.7	0.8	1.0	5.0	10.0
− 0.0753	0.242	2.53	0.272	0.457	0.570
0	0.292	0.307	0.332	0.585	0.730
0.111	0.331	0.348	0.378	0.669	0.851
0.333	0.384	0.403	0.440	0.792	1.013
1.0	0.496	0.523	0.570	1.043	1.344
4.0	0.813	0.858	0.938	1.736	2.236

circular cylinder placed with its axis normal to the flow, it is well known that the inviscid solution is given by $u_e = 2u_\infty \sin(x/R)$, where x is the distance along the surface measured from the stagnation point and R is the radius of the cylinder. Clearly, very near the stagnation point, $u_e \approx 2u_\infty x/R$.

The skin friction parameter from the solution to the Falker-Skan equation for some typical flows [83] is given in Table 2.3. The stagnation flow ($m = 1$) and the flat plate flow ($m = 0$) are the two wedge-flow cases which commonly occur in practice. Assuming constant properties and neglecting viscous dissipation, the energy equation can be readily solved once the solution for the velocity field has been obtained. Results for constant wall and free-stream temperatures were obtained by Eckert [84] and, more recently, by Evans [85]. A few of these heat transfer results are presented in terms of $\mathrm{Nu}_x/\mathrm{Re}_x^{1/2}$ for several values of Pr in Table 2.4. The heat transfer solution for the stagnation point in any symmetric two-dimensional flow (such as flow about a circular cylinder) is given by the similarity solution for $m = 1$. Near the stagnation point, u_e used in Re_x is proportional to x, so that the x cancels out of the expression $\mathrm{Nu}_x/\mathrm{Re}_x^{1/2}$ at the stagnation point. Thus, a finite, nonsingular solution is obtained for the heat transfer coefficient.

Similarity solutions also exist for certain axisymmetric flows. The relationship between similarity solutions for two-dimensional and axisymmetric flows is conveniently developed by using the Mangler [86] transformation. Details of solutions for axisymmetric flows can be found in [2, 58]. The Mangler transformation indicates that the stagnation-point solution for the flow about an axisymmetric body with a blunt nose (such as a sphere) can be obtained from the two-dimensional Falkner-Skan solution with $m = \frac{1}{3}$. The heat transfer results for an axisymmetric stagnation flow can be expressed as

$$\frac{\mathrm{Nu}_x}{\mathrm{Pr}^{0.4}\mathrm{Re}_x^{1/2}} = 0.76 \tag{2.173}$$

Near the stagnation point, u_e used in Re_x is again proportional to x, so that the x cancels out of Eq. (2.173). For the sphere, the limiting expression for u_e becomes $u_e = \frac{3}{2} u_\infty x/R$ as $x \to 0$, where R is the radius of the sphere. Axisymmetric bodies with sharp conical noses also admit similarity solutions near the nose. The results for the cone in incompressible flow are discussed in [58].

Flows very near stagnation regions are invariably laminar. Thus, the results given above are adequate for design purposes when the properties can be assumed to be constant. There is experimental evidence, however, that the heat transfer rate in stagnation regions is influenced by the turbulence level of the free stream. The mechanism responsible for this behavior is not well understood, and there are insufficient data to establish a reliable and accurate correlation to account for the effect.

Similarity solutions are also available for the compressible laminar boundary-layer equations. These are perhaps easiest to obtain if the compressible boundary-layer equations are first reduced to almost the same form as for incompressible flow through the use of the Illingworth-Stewartson transformation [87, 88]. The development of this transformation and the procedure for obtaining similarity solutions for compressible laminar flows are outlined in [1]. Sample results are also presented in [1]. As for the incompressible case, it is mainly the similarity flows for stagnation points and the flat plate that are of interest in applications. However, all similarity solutions are intrinsically important because of the great accuracy with which the numerical solutions to the resulting ordinary differential equations can be obtained. These solutions can provide benchmarks against which the accuracy of more approximate methods can be judged.

The heat transfer results for the stagnation line on a uniform-temperature circular cylinder in cross flow were obtained by a similarity solution to the boundary-layer equations by Cohen [89] for a compressible flow. For $u_\infty < 8840$ m/s the solution is well represented by

$$\frac{\mathrm{Nu}_w}{\mathrm{Pr}_w^{0.4}\mathrm{Re}_w^{1/2}} = 0.57\left(\frac{\mu_e \rho_e}{\mu_w \rho_w}\right)^{0.45} \tag{2.174}$$

where the properties in the expression on the left-hand side are to be evaluated at the wall temperature. Here

$$\mathrm{Nu}_w = \frac{q_w c_{pw} L}{(H_w - H_e) k_w} \quad \text{and} \quad \mathrm{Re}_w = \frac{(du_e/dx) L^2 \rho_w}{\mu_w}$$

where L is a reference length which cancels from the correlation. For incompressible flow, Eq. (2.174) gives results identical to that obtained from Table 2.4 for $m = 1$. The velocity gradient in the Reynolds number is to be obtained from the inviscid flow solution, which for a cylinder is well approximated for $4 \le M_\infty \le 10$ by modified Newtonian theory as

$$\frac{r_n}{u_\infty}\frac{du_e}{dx} = \frac{1.43}{u_\infty}\sqrt{\frac{p_{\mathrm{st}} - p_\infty}{\rho_{\mathrm{st}}}} \tag{2.175}$$

where r_n is the radius of the cylinder, and p_{st} and ρ_{st} are the inviscid flow conditions on the stagnation line of the cylinder.

A similar relationship was obtained by Cohen [89] for an axisymmetric stagnation point for speeds less than 9144 m/s, and for 10^{-4} atm $< p_{\mathrm{st}} < 10^2$ atm, 300 K $< T_w$

< 1722 K,

$$\frac{\mathrm{Nu}_w}{\mathrm{Pr}_w^{0.4}\mathrm{Re}_w^{1/2}} = 0.767\left(\frac{\mu_e \rho_e}{\mu_w \rho_w}\right)^{0.43} \tag{2.176}$$

The parameters Nu_w and Re_w are defined as for Eq. (2.174), and the velocity gradient in the Reynolds number is well approximated for a sphere for $3 \leq M_\infty \leq 10$ by modified Newtonian flow theory as

$$\frac{r_n}{u_\infty}\frac{du_e}{dx} = \frac{1.54}{u_\infty}\sqrt{\frac{p_{\mathrm{st}} - p_\infty}{\rho_{\mathrm{st}}}} \tag{2.177}$$

where r_n is the radius of the sphere, and p_{st} and ρ_{st} are the inviscid flow conditions at the stagnation point.

As with all heat transfer calculations for compressible flows, the adiabatic wall temperature (or enthalpy) is used as the fluid reference temperature in Newton's law of cooling to compute the heat flux at a stagnation point. However, it should be noted that the adiabatic wall temperature and enthalpy are identical to the stagnation (or total) values at a stagnation point.

2.5.2 General Laminar and Turbulent Flows

Finite difference solutions to the boundary-layer equations can be easily obtained for attached laminar flows in pressure gradients. In such cases, the pressure gradient can be obtained from numerical results for the inviscid flow. When the flow separates so that regions of recirculation are present, the solution procedure becomes more involved. An iterative viscous-inviscid interaction calculation procedure must be used, or, as an alternative, the full Navier-Stokes equations can be solved numerically throughout the flow. The situation is much the same for turbulent flow, except that turbulence modeling must be included in the computational approach. For mild to moderate pressure gradients, algebraic turbulence models work well with little or no adjustment. As was mentioned in Sec. 2.3.3, several investigators have suggested modifications to algebraic turbulence models to include the effects of pressure gradients. Turbulence modeling in flows with separation remains somewhat more uncertain, however.

Although certain classes of flows can be solved numerically in almost a routine fashion, limitations do exist. As a flow becomes turbulent, three-dimensional, or time-dependent, the difficulties (and generally, the cost) associated with obtaining reliable numerical predictions increase. Many flows over bodies tend to become unsteady as the Reynolds number increases, due to vortex shedding. These are especially difficult to predict with accuracy using current methods.

Relatively few extensive studies, either computational or experimental, have been reported for three-dimensional convective flows. For the most part, incompressible three-dimensional flows over bodies have been computed by solving the three-dimensional boundary-layer equations along with the governing equations for the inviscid flow. Often, results have been limited to stagnation regions and planes of symmetry. Reviews citing results obtained from numerical solutions to the three-dimensional boundary-layer equations have been prepared by Blottner [90] and Bushnell et al. [21].

In the compressible flow regime, the accurate prediction of the aerodynamic heating associated with the three-dimensional flow over bodies is of considerable interest. Solutions for the heat transfer rates for laminar boundary-layer flow at a three-dimensional stagnation point have been reported in [91, 92]. For flows over bodies at supersonic speeds, it is becoming increasingly common to obtain numerical solutions by solving a reduced (or "parabolized") form of the Navier-Stokes equations. These equations are solved in both viscous and inviscid regions and differ from the three-dimensional boundary-layer equations primarily through the inclusion of a momentum equation in all three coordinate directions. An early example of this approach was reported by Lubard and Helliwell [93]. More recently, Venkatapathy et al. [94] used this strategy to obtain numerical solutions, including heat transfer rates, for the three-dimensional flow about the Space Shuttle.

Heat transfer from several simple shapes which induce pressure gradients has been studied experimentally. These include the sphere and cylinders of various cross sections (circular, square, hexagonal) in cross flow, which are considered in detail in Chap. 6.

2.6 CONCLUDING REMARKS

The literature related to external flow forced convection is very extensive. Consequently, it has not been possible to include all of the available useful information in this chapter. It is intended, however, that the combination of the specific information provided plus the references cited will cover the most important aspects of the subject.

An attempt has been made to include the fundamental equations and concepts. This should serve the needs of those who will use computational methods to solve problems in convective heat transfer. This approach is becoming increasingly common. General formulas and data correlations for the configurations that occur most frequently in applications have also been included.

NOMENCLATURE

A^+	van Driest damping constant, Eq. (2.115)
a	speed of sound, Eq. (2.84), m/s, ft/s
B_f	blowing function, Eq. (2.163)
B_h	blowing function, Eq. (2.164)
C_1	constant, Eq. (2.116)
C_2, C_3	constants, Table 2.2, Eq. (2.125)
C_D	constant in Eq. (2.122)
C_k	constant, Eq. (2.120)
C_μ	constant, Table 2.2, Eq. (2.124)
C_p	pressure coefficient, $(p_w - p_\infty)/(\frac{1}{2}\rho_\infty V_\infty^2)$
c_f	local skin friction coefficient, $2\tau_w/\rho_e u_e^2$
c_p	specific heat at constant pressure, J/(kg · K), Btu/(lb$_m$ · °F)
c_v	specific heat at constant volume, J/(kg · K), Btu/(lb$_m$ · °F)
Ec	Eckert number, $u_e^2/[c_p(T_w - T_e)]$
F_c	turbulent boundary-layer transformation function, Eq. (2.153)
F_{Re_x}	turbulent boundary-layer transformation function, Eq. (2.152)
F_{Re_θ}	turbulent boundary-layer transformation function, Eq. (2.154), Eq. (2.157)

H	total enthalpy per unit mass, J/kg, Btu/lb_m
h	heat transfer coefficient, $W/(m^2 \cdot K)$, $Btu/(hr \cdot ft^2 \cdot °F)$
h_1, h_2, h_3	metric coefficients
i	enthalpy per unit mass, J/kg, Btu/lb_m
$\hat{\mathbf{i}}_1, \hat{\mathbf{i}}_2, \hat{\mathbf{i}}_3$	unit vectors in a generalized curvilinear coordinate system
K_1, K_3	geodesic curvatures, Eq. (2.104), m^{-1}, ft^{-1}
k	thermal conductivity, $W/(m \cdot K)$, $Btu/(hr \cdot ft \cdot °F)$
$\bar{k}$	kinetic energy of turbulence per unit mass, J/kg, Btu/lb_m
L	reference length, m, ft
l	mixing length, m, ft
M	Mach number = V/a
m	flow index = 0 (two-dimensional), = 1 (axisymmetric)
Nu	Nusselt number = hx/k
Pe	Péclet number = Re Pr
Pr	Prandtl number = $c_p \mu / k$
Pr_k	turbulent Prandtl number for turbulent kinetic energy; see Eq. (2.122)
Pr_ϵ	turbulent Prandtl number for dissipation rate of turbulent kinetic energy; see Eq. (2.125)
Pr_T	turbulent Prandtl number = $\mu_T c_p / k_T$
p	pressure, Pa, lb_f/ft^2
q	heat flux, W/m^2, $Btu/(hr \cdot ft^2)$
R	gas constant, $J/(kg \cdot K)$, $ft \cdot lb_f/(lb_m \cdot °R)$
Re_L	Reynolds number based on length L, $= \rho u_e L/\mu$
Re_x	Reynolds number based on length x
Re_θ	Reynolds number based on momentum thickness
Re_Δ	Reynolds number based on enthalpy thickness
r	radius, radial distance, m, ft
St	Stanton number = Nu/(Re Pr)
s	entropy per unit mass, $J/(kg \cdot K)$, $Btu/(lb_m \cdot °R)$
T	temperature, K, °R
t	time, s
u	velocity component in x direction, m/s, ft/s
u_1, u_2, u_3	velocity components in a generalized coordinate system, m/s, ft/s
u_r	velocity component in r direction, m/s, ft/s
u_θ	velocity component in θ direction, m/s, ft/s
u_z	velocity component in z direction, m/s, ft/s
u^+	nondimensional velocity = $u/(\tau_w/\rho_w)^{1/2}$
$\mathbf{V}$	velocity vector, m/s, ft/s
V	magnitude of velocity vector, m/s, ft/s
v	velocity component in y direction, m/s, ft/s
w	velocity component in z direction, m/s, ft/s
x	rectangular coordinate, m, ft
y	rectangular coordinate normal to surface, m, ft

y^+	nondimensional distance from surface $= (\tau_w/\rho_w)^{1/2} y/\nu_w$
z	rectangular coordinate, m, ft

Greek Symbols

α	turbulence model parameter, Eq. (2.118)
α	van Driest function, Eq. (2.155)
α	thermal diffusivity $= k/\rho c_p$, m^2/s, ft^2/s
β	wedge angle parameter, Eq. (2.172)
β	coefficient of thermal expansion $= -(1/\rho)(\partial \rho/\partial T)_p$
β	van Driest function, Eq. (2.156)
γ	ratio of specific heats $= c_p/c_v$
Δ	enthalpy thickness $= \int_0^\infty [\rho u(H - H_e)/\rho_e u_e(H_w - H_e)]\, dy$, m, ft
δ	boundary-layer thickness, m, ft
δ_k^*	kinematic displacement thickness, $\int_0^\infty (1 - u/u_e)\, dy$, m, ft
δ_t	thermal boundary-layer thickness, m, ft
δ_{ij}	Kronecker delta
ϵ	dissipation rate of turbulent kinetic energy, $J/(kg \cdot s)$, $Btu/(lb_m \cdot s)$
θ	momentum thickness, $\int_0^\infty (\rho u/\rho_e u_e)(1 - u/u_e)\, dy$, m, ft
κ	von Kármán mixing-length constant
μ	dynamic viscosity, $Pa \cdot s$, $lb_m/(ft \cdot s)$
μ'	second coefficient of viscosity, $Pa \cdot s$, $lb_m/(ft \cdot s)$
ν	kinematic viscosity, m^2/s, ft^2/s
Π	stress tensor
π	turbulence model parameter, Eq. (2.118)
ρ	density, kg/m^3, lb_m/ft^3
τ	viscous stress tensor
τ	shear stress, Pa, lb_f/ft^2
ϕ	dissipation function, $N/(m^2 \cdot s)$, $lb_f/(ft^2 \cdot s)$
ϕ	velocity potential function
ψ	stream function
ω	vorticity, s^{-1}

Subscripts

aw	adiabatic wall conditions
e	evaluated at boundary-layer edge
f	film temperature, Eq. (2.144)
i	incompressible value
max	maximum value
0	total value
st	stagnation conditions
T	turbulent quantity
w	wall conditions
∞	free-stream conditions

Superscripts and Accents

$\bar{}$	mean value
$'$	random fluctuating value
$*$	properties to be evaluated at reference enthalpy or temperature condition
$\tilde{}$	mass-weighted average
ω	power in viscosity-temperature relation

REFERENCES

1. H. Schlichting, *Boundary-Layer Theory*, 7th ed., translated by J. Kestin, McGraw-Hill, New York, 1979.
2. L. C. Burmeister, *Convective Heat Transfer*, Wiley-Interscience, New York, 1983.
3. J. F. Thompson, Z. U. A. Warsi, and C. W. Mastin, *Numerical Grid Generation*, Elsevier, New York, 1985.
4. D. A. Anderson, J. C. Tannehill and R. H. Pletcher, *Computational Fluid Mechanics and Heat Transfer*, Hemisphere, New York, 1984.
5. A. Favre, Equations des Gas Turbulents Compressibles: 1. Formes Générales, *J. Méc.*, Vol. 4, pp. 361–390, 1965.
6. J. A. Owczarek, *Fundamentals of Gas Dynamics*, International Textbook Co., Scranton, Pa., 1964.
7. G. B. Schubauer and C. M. Tchen, *Turbulent Flows and Heat Transfer*, High Speed Aerodynamics and Jet Propulsion, Vol. 5, Princeton U.P., Princeton, N.J., Sec. B, 1959.
8. E. R. van Driest, Turbulent Boundary Layer in Compressible Fluids, *J. Aerosp. Sci.*, Vol. 18, pp. 145–160, 1951.
9. T. Cebeci and A. M. O. Smith, *Analysis of Turbulent Boundary Layers*, Academic, New York, 1974.
10. T. Cebeci, K. Kaups, and J. A. Ramsey, A General Method for Calculating Three-Dimensional Compressible Laminar and Turbulent Boundary Layers on Arbitrary Wings, NASA CR-2777.
11. F. G. Blottner and M. A. Ellis, Finite-Difference Solution of the Incompressible Three-Dimensional Boundary Layer Equations for a Blunt Body, *Comput. Fluids*, Vol. 1, Pergamon, Oxford, pp. 133–158, 1973.
12. J. Boussinesq, Essai sur la Théorie des Equaux Courantes, *Mem. Présentés Acad. Sci.*, Paris, Vol. 23, p. 46, 1877.
13. C. duP. Donaldson and H. Rosenbaum, Calculation of Turbulent Shear Flows through Closure of the Reynolds Equations by Invariant Modeling, Aeronaut. Res. Assoc. of Princeton Report 127, 1968.
14. J. W. Deardorff, A Numerical Study of Three-Dimensional Turbulent Channel Flow at Large Reynolds Numbers, *J. Fluid Mech.*, Vol. 41, pp. 453–480, 1970.
15. S. V. Patankar, M. Ivanovic, and E. M. Sparrow, Analysis of Turbulent Flow and Heat Transfer in Internally Finned Tubes and Annuli, *J. Heat Transfer*, Vol. 101, pp. 29–37, 1979.
16. E. R. van Driest, On Turbulent Flow Near a Wall, *J. Aerosp. Sci.*, Vol. 23, pp. 1007–1011, 1956.
17. R. H. Pletcher, Prediction of Turbulent Boundary Layers at Low Reynolds Numbers, *AIAA J.*, Vol. 14, pp. 696–698, 1976.
18. S. V. Patankar and D. B. Spalding, *Heat and Mass Transfer in Boundary Layers*, 2nd ed., Intertext Books, London, 1970.

19. W. M. Kays, Heat Transfer to the Transpired Turbulent Boundary Layer, *Int. J. Heat Mass Transfer*, Vol. 15, pp. 1023–1044, 1972.

20. A. J. Reynolds, The Prediction of Turbulent Prandtl and Schmidt Numbers, *Int. J. Heat Mass Transfer*, Vol. 18, pp. 1055–1069, 1975.

21. D. M. Bushnell, A. M. Cary, Jr., and J. E. Harris, Calculation Methods for Compressible Turbulent Boundary Layers, von Kármán Institute for Fluid Dynamics, *Lecture Series 86 on Compressible Turbulent Boundary Layers*, Vol. 2, Rhode Saint Genese, Belgium, 1976.

22. H. McDonald and R. W. Fish, Practical Calculations of Transitional Boundary Layers, *Int. J. Heat Mass Transfer*, Vol. 16, pp. 1729–1744, 1973.

23. J. M. Healzer, R. J. Moffat, and W. M. Kays, The Turbulent Boundary Layer on a Rough Porous Plate: Experimental Heat Transfer with Uniform Blowing, Thermosciences Division, Report No. HMT-18, Dept. of Mech. Eng., Stanford Univ., Stanford, Calif., 1974.

24. J. C. Adams, Jr. and B. K. Hodge, The Calculation of Compressible, Transitional, Turbulent, and Relaminarizational Boundary Layers over Smooth and Rough Surfaces Using an Extended Mixing Length Hypothesis, AIAA Paper 77-682, Albuquerque, N. Mex., 1977.

25. G. H. Christoph and R. H. Pletcher, Prediction of Rough-Wall Skin-Friction and Heat Transfer, *AIAA J.*, Vol. 21, No. 4, pp. 509–515, 1983.

26. R. H. Pletcher, Prediction of Transpired Turbulent Boundary Layers, *J. Heat Transfer*, Vol. 96, pp. 89–94, 1974.

27. R. J. Baker and B. E. Launder, The Turbulent Boundary Layer with Foreign Gas Injection: II—Predictions and Measurements in Severe Streamwise Pressure Gradients, *Int. J. Heat Mass Transfer*, Vol. 17, pp. 293–306, 1974.

28. W. M. Kays and R. J. Moffat, The Behavior of Transpired Turbulent Boundary Layers, *Studies in Convection: Theory Measurement, and Applications*, Vol. 1, Academic, New York, pp. 223–319, 1975.

29. H. McDonald and F. J. Camerata, An Extended Mixing Length Approach for Computing the Turbulent Boundary Layer Development, *Proc. Computation of Turbulent Boundary Layers—1968 AFOSR-IFP-Stanford Conference*, Vol. 1, Stanford Univ., Stanford, Calif., pp. 83–98, 1968.

30. J. P. Kreskovsky, S. J. Shamroth, and H. McDonald, Parametric Study of Relaminarization of Turbulent Boundary Layers on Nozzle Walls, NASA CR-2370, 1974.

31. H. McDonald and J. P. Kreskovsky, Effect of Free Stream Turbulence on the Turbulent Boundary Layers, *Int. J. Heat Mass Transfer*, Vol. 17, pp. 705–716, 1974.

32. Y. Y. Chan, Compressible Turbulent Boundary Layer Computations Based on an Extended Mixing Length Approach, *Canad. Aeronaut. and Space Inst. Trans.*, Vol. 5, pp. 21–27, 1972.

33. J. S. Shang and W. L. Hankey, Jr., Supersonic Turbulent Separated Flows Utilizing the Navier-Stokes Equations, *Flow Separation*, AGARD-CCP-168, 1975.

34. T. A. Reyhner, Finite-Difference Solution of the Compressible Turbulent Boundary Layer Equations, *Proc. Computation of Turbulent Boundary Layers—1968 AFOSR-IFP-Stanford Conference*, Vol. 1, Stanford Univ., Stanford, Calif., pp. 375–383, 1968.

35. M. R. Malik and R. H. Pletcher, Computation of Annular Turbulent Flows with Heat Transfer and Property Variations, *Heat Transfer 1978, Proc. Sixth Int. Heat Transfer Conference*, Vol. 2, Hemisphere, Washington, pp. 537–542, 1978.

36. R. H. Pletcher, Prediction of Incompressible Turbulent Separating Flow, *J. Fluids Eng.*, Vol. 100, pp. 427–433, 1978.

37. D. A. Johnson and L. S. King, A Mathematically Simple Turbulence Closure Model for Attached and Separated Turbulent Boundary Layers, *AIAA J.*, Vol. 23, No. 11, pp. 1684–1692, 1985.

38. S. F. Birch, A Critical Reynolds Number Hypothesis and Its Relation to Phenomenological Turbulence Models, *Proc. 1976 Heat Transfer and Fluid Mechanics Inst.*, Stanford Univ., Stanford, Calif., 1976.

39. M. R. Malik and R. H. Pletcher, A Study of Some Turbulence Models for Flow and Heat Transfer in Ducts of Annular Cross-Section, *J. Heat Transfer*, Vol. 103, pp. 146–152, 1981.
40. B. E. Launder and D. B. Spalding, The Numerical Computation of Turbulent Flows, *Comput. Methods Appl. Mech. Eng.*, Vol. 3, pp. 269–289, 1974.
41. M. W. Rubesin, A One-Equation Model of Turbulence for Use with the Compressible Navier-Stokes Equations, NASA TM X-73-128, 1976.
42. P. Bradshaw, D. H. Ferriss, and N. D. Altwell, Calculation of Boundary Layer Development Using the Turbulent Energy Equation, *J. Fluid Mech.*, Vol. 28, pp. 593–616, 1967.
43. F. H. Harlow and P. I. Nakayama, Transport of Turbulence Energy Decay Rate, Los Alamos Scientific Laboratory Report LA-3584, Los Alamos, N. Mex., 1968.
44. W. P. Jones and B. E. Launder, The Prediction of Laminarization with a Two-Equation Model of Turbulence, *Int. J. Heat Mass Transfer*, Vol. 15, pp. 301–314, 1972.
45. W. Rodi, *Turbulence Models and their Application in Hydraulics*, Monograph, International Association of Hydraulic Research, Delft, the Netherlands, 1980.
46. K. H. Ng and D. B. Spalding, Turbulence Model for Boundary Layers Near Walls, *Phys. Fluids*, Vol. 15, pp. 20–30, 1972.
47. D. C. Wilcox and R. M. Traci, A Complete Model of Turbulence, AIAA Paper 76-351, San Diego, Calif., 1976.
48. P. G. Saffman and D. C. Wilcox, Turbulence Model Prediction for Turbulent Boundary Layers, *AIAA J.*, Vol. 12, pp. 541–546, 1974.
49. J. Rotta, Statistische Theorie nichthomogener Turbulenz, *Z. Phys.*, Vol. 129, pp. 547–572, 1951.
50. K. Hanjalic and B. E. Launder, A Reynolds Stress Model of Turbulence and Its Application to Asymmetric Shear Flows, *J. Fluid Mech.*, Vol. 52, pp. 609–638, 1972.
51. B. E. Launder, G. J. Reece, and W. Rodi, Progress in the Development of a Reynolds Stress Turbulence Closure, *J. Fluid Mech.*, Vol. 68, pp. 537–566, 1975.
52. C. duP. Donaldson, Calculation of Turbulent Shear Flows for Atmospheric and Vortex Motions, *AIAA J.*, Vol. 10, pp. 4–12, 1972.
53. W. Rodi, Progress in Turbulence Modeling for Incompressible Flows, AIAA Paper No. 81-0045, 1981.
54. H. Blasius, Grenzschichten in Flüssigkeiten mit Kleiner Reibung, *Z. Math. Phys.*, Vol. 56, pp. 1–37, 1908.
55. E. Pohlhausen, Der Wärmeaustausch zwischen festen Körpern und Flüssigkeiten mit kleiner Reibung und kleiner Wärmeleitung, *Z. Angew. Math. Mech.*, Vol. 1, pp. 115–121, 1921.
56. W. M. Kays and M. E. Crawford, *Convective Heat and Mass Transfer*, 2nd ed., McGraw-Hill, New York, 1980.
57. F. Schultz-Grunow, New Frictional Resistance Law for Smooth Plates, NACA Tech. Memo. 986, 1941, transl. of Neues Widerstandsgesetz für glatte Platten, *Luftfahrforschung*, Vol. 17, pp. 239–246, 1940.
58. T. Cebeci and P. Bradshaw, *Physical and Computational Aspects of Convective Heat Transfer*, Springer, New York, 1984.
59. E. R. G. Eckert, Engineering Relations for Heat Transfer and Friction in High-Velocity Laminar and Turbulent Boundary-Layer Flow over Surfaces with Constant Pressure and Temperature, *Trans. ASME*, Vol. 78, No. 6, pp. 1273–1283, 1956.
60. E. R. van Driest, Investigation of Laminar Boundary Layer in Compressible Fluids Using the Crocco Method, NACA Tech. Note 2597, 1952.
61. L. Crocco, Lo strato limite laminare nei gas, Monografie Sci. di Aeronaut. 3, Rome, 1946, transl. as North American Aviation Aerophys. Lab. Rep. APL/NAA/CF-1038, 1948.
62. D. B. Spalding and S. W. Chi, The Drag of a Compressible Turbulent Boundary Layer on a Smooth Flat Plate with and without Heat Transfer, *J. Fluid Mech.*, Vol. 18, pp. 117–143, 1964.

63. E. R. van Driest, The Problem of Aerodynamic Heating, *Aeronaut. Eng. Rev.*, Vol. 15, pp. 26–41, 1956.

64. E. J. Hopkins and E. R. Keener, Pressure Gradient Effects on Hypersonic Skin Friction and Boundary-Layer Profiles, *AIAA J.*, Vol. 10, p. 1141, 1972.

65. A. M. Cary and M. H. Bertram, Engineering Prediction of Turbulent Skin Friction and Heat Transfer in High Speed Flow, NASA TN D-7507, 1974.

66. A. M. Cary, Summary of Available Information on Reynolds Analogy for Zero-Pressure Gradient, Compressible Turbulent-Boundary-Layer Flow, NASA TN D-5560, 1970.

67. E. J. Hopkins, M. W. Rubesin, M. Inouye, E. R. Keener, G. G. Mateer, and T. E. Polek, Summary and Correlation of Skin-Friction and Heat-Transfer Data for a Hypersonic Turbulent Boundary Layer on Simple Shapes, NASA Tech. Note D-5089, 1969.

68. H. W. Emmons and D. Leigh, Tabulation of the Blasius Function with Blowing and Suction, Harvard Univ. Combust. Aerodyn. Lab. Tech. Rep. 9, Cambridge, Mass., 1953.

69. J. P. Hartnett and E. R. G. Eckert, Mass Transfer Cooling in the Laminar Boundary Layer with Constant Fluid Properties, *Trans. ASME*, Vol. 79, pp. 247–254, 1957.

70. G. M. Low, The Compressible Laminar Boundary Layer with Fluid Injection, NACA Tech. Note 3404, 1955.

71. P. A. Libby and K. Chen, Laminar Boundary Layer with Uniform Injection, *Phys. Fluids*, Vol. 8, pp. 568–574, 1965.

72. W. M. Kays and R. J. Moffat, The Behavior of Transpired Turbulent Boundary Layers, Stanford Univ. Dept. Mech. Eng. Rep. HMT-20, Stanford, Calif., 1975.

73. L. O. F. Jeromin, The Status of Research in Turbulent Boundary Layers with Fluid Injection, *Progress in Aeronautical Sciences*, Vol. 10, pp. 55–189, 1970.

74. D. Coles, A Survey of Data for Turbulent Boundary Layers with Mass Transfer, AGARD-CP-93, Turbulent Shear Flows, 1972.

75. R. L. P. Voisinet, Influence of Roughness and Blowing on Compressible Turbulent Boundary Layer Flow, Naval Surface Weapons Center TR 79-153, Silver Spring, Md., June 1979.

76. T. Tendeland and A. F. Okuno, The Effect of Fluid Injection on the Compressible Turbulent Boundary Layer—The Effect on Skin Friction of Air Injected into the Boundary Layer of a Cone at $M = 2.7$, NACA Res. Memo. A56D05, 1956.

77. C. C. Pappas and A. F. Okuno, Measurements of Heat Transfer and Recovery Factor of a Compressible Turbulent Boundary Layer on a Sharp Cone with Foreign Gas Injection, NASA Tech. Note D-2230, 1964.

78. E. R. Bartle and B. M. Leadon, The Effectiveness as a Universal Measure of Mass Transfer Cooling for a Turbulent Boundary Layer, *Proc. 1962 Heat Transfer and Fluid Mech. Inst.*, Stanford U.P., Stanford, Calif., pp. 27–41, 1962.

79. J. E. Danberg, Characteristics of the Turbulent Boundary Layer with Heat and Mass Transfer at Mach Number 6.7, *Proc. 5th U.S. Navy Symp. Aeroballistics*, U.S. Naval Ordnance Lab., 1961.

80. L. C. Squire, *Further Experimental Investigations of Compressible Turbulent Boundary Layers with Air Injection*, Aeronaut. Res. Council (Great Britain) Rep. and Memo. No. 3627, 1970.

81. J. P. Hartnett, Mass Transfer Cooling, *Handbook of Heat Transfer Applications*, ed. W. M. Rohsenow, J. P. Hartnett, and E. N. Ganic, Chap. 1, McGraw-Hill, New York, 1985.

82. V. M. Falkner and S. W. Skan, Some Approximate Solutions of the Boundary-Layer Equations, *Philos. Mag.*, Vol. 12, pp. 865–896, 1931; see also *Rep. Memorial Aeronaut. Res. Committee*, London, No. 1, p. 314, 1930.

83. D. R. Hartree, On an Equation Occurring in Falkner and Skan's Approximate Treatment of the Equation of the Boundary Layer, *Proc. Cambridge Philos. Soc.*, Vol. 33, pp. 223–239, 1937.

84. E. Eckert, Die Berechnung des Wärmeüberganges in der Laminaren Grenzschicht um strömter Körper, *VDI-Forschungsheft*, No. 416, Berlin, 1942.

85. H. L. Evans, Mass Transfer through Laminar Boundary Layers. Further Similar Solutions to the B-equation for $B = 0$, *Int. J. Heat Mass Transfer*, Vol. 5, pp. 35–37, 1962.

86. W. Mangler, Zusammenhang zwischen ebenen und rotationssymmetrichen Grenzschichten in kompressiblen Flüssigkeiten, *Z. Angew. Math. Mech.*, Vol. 28, pp. 97–103 1948.

87. C. R. Illingworth, Steady Flow in the Laminar Boundary Layer of a Gas, *Proc. Roy. Soc. London*, Vol. A199, p. 533, 1949.

88. K. Stewartson, Correlated Compressible and Incompressible Boundary Layers, *Proc. Roy. Soc. London*, Vol. A200, p. 84, 1949.

89. N. B. Cohen, Boundary-Layer Similar Solutions and Correlation Equations for Laminar Heat Transfer Distribution in Equilibrium Air at Velocities up to 41,000 feet per second, NASA Tech. Rep. R-118, 1961.

90. F. G. Blottner, Computational Techniques for Boundary Layers, *AGARD Lecture Series No. 73 on Computational Methods for Inviscid an Viscous Two- and Three-dimensional Flowfields*, pp. 3-1–3-51, 1975.

91. P. A. Libby, Heat and Mass Transfer at a General Three-dimensional Stagnation Point, *AIAA J.*, Vol. 5, pp. 507–517, 1967.

92. E. Reshotko, Heat Transfer to a General Three-dimensional Stagnation Point, *Jet Propulsion*, Vol. 26, pp. 58–60, 1958.

93. S. C. Lubard and W. S. Helliwell, Calculation of the Flow on a Cone at High Angle of Attack, *AIAA J.*, Vol. 12, pp. 965–974.

94. E. Venkatapathy, J. Rakich, and J. C. Tannehill, Numerical Solution of Space Shuttle Orbiter Flowfield, *J. Spacecraft and Rockets*, Vol. 21, No. 1, pp. 9–15, 1984.

95. K. Y. Chien, Prediction of Channel of Boundary-Layer Flows with a Low Reynolds Number Turbulence Model, *AIAA J.*, Vol. 20, No. 1, pp. 33–38, 1982.

3

LAMINAR CONVECTIVE HEAT TRANSFER IN DUCTS

R. K. Shah

Harrison Radiator Division, GM
Lockport, New York

M. S. Bhatti

Harrison Radiator Division, GM
Lockport, New York

3.1 INTRODUCTION

This chapter deals with laminar fluid flow and forced convection heat transfer characteristics of a variety of ducts which are of interest in a wide variety of heating and cooling devices used in aerospace, electronics, nuclear engineering, instrumentation, biomechanics, and slurry, oil, and water transport as well as in food, glass, polymer, and metal processing.

The results presented here are applicable to straight ducts with axially unchanging cross sections. Also, the duct walls in all cases are considered as smooth, nonporous, rigid, and stationary. Furthermore, the duct walls are assumed to be uniformly thin, so that the temperature distribution within the solid wall has negligible influence on the convective heat transfer in the flowing fluid.

The scope of the chapter is restricted to steady, incompressible and laminar flow of constant-property Newtonian fluids only. All forms of body forces are neglected. Also omitted are the effects of natural convection, phase change, mass transfer, and chemical reactions. The effects of thermal energy soures, viscous dissipation (i.e., internal friction), flow work (i.e., work done by pressure forces), and fluid axial conduction are included at appropriate places.

In the ensuing sections of this chapter, a uniform format is adopted to present information for the most important singly and doubly connected ducts covering the four types of flows to be described shortly. A limited amount of information is also presented for some unusual singly connected ducts which may be of interest in special situations. The singly connected ducts, such as circular, rectangular, triangular, elliptical, and the like, are characterized by the fact that the periphery of their bounding surface can be represented by a single closed curve which can be continuously contracted into a point without leaving the free-flow area of the duct cross section. The doubly connected ducts, such as annular concentric and eccentric geometries, are characterized by the fact that the peripheries of their bounding surfaces can be represented by a set of two closed curves which cannot be continuously contracted into a point without leaving the free-flow area of the duct cross section.

The most important fluid flow and heat transfer results are presented here in terms of mathematical expressions as well as in graphical form to display the general trends. Due to vastness of the subject and the space limitations, all the computed results are not presented in tabular form. However, frequent references are made to Shah and London's monograph [1], which contains an extensive tabulation of the computed results collected from worldwide sources.

3.1.1 Types of Laminar Duct Flows

There are four types of laminar duct flows, namely, fully developed, hydrodynamically developing, thermally developing, and simultaneously developing.† A brief description of these flows is given here with an aid of Fig. 3.1 which depicts a fluid with uniform velocity u_m and temperature T_e entering a duct of arbitrary cross section at $x = 0$.

Referring to Fig. 3.1*a*, suppose that the temperature of the duct wall is held at the entering fluid temperature ($T_w = T_e$) and there is no generation or dissipation of heat within the fluid. In this case, the fluid experiences no gain or loss of heat. In such an isothermal flow, the effect of viscosity gradually spreads across the duct cross section commencing at $x = 0$. The extent to which the viscous effects diffuse normally from the duct wall is represented by the hydrodynamic boundary-layer thickness δ, which varies with the axial coordinate x. In accordance with Prandtl's boundary-layer theory, the hydrodynamic boundary-layer thickness divides the flow field into two regions: a viscous region near the duct wall and an essentially inviscid region around the duct axis.

At $x = L_{hy}$, the viscous effects have completely spread across the duct cross section. The region $0 \leq x \leq L_{hy}$ is called the *hydrodynamic entrance region*, and the fluid flow in this region is called the *hydrodynamically developing flow*. As shown in Fig. 3.1*a*, the axial velocity profile in the hydrodynamic entrance region varies with all three space coordinates, i.e., $u = u(x, y, z)$. For hydrodynamically developed flow, the axial velocity profile becomes independent of the axial coordinate and varies with the transverse coordinates alone, i.e., $u = u(y, z)$.

After the flow becomes hydrodynamically developed ($x > L_{hy}$, Fig. 3.1*a*), suppose that the duct wall temperature is raised above the entering fluid temperature, i.e., $T_w > T_e$. In this case, the thermal effects diffuse gradually from the duct wall, commencing at $x = L_{hy}$. The extent of diffusion of the thermal effects is denoted by the thermal boundary-layer thickness δ_t, which also varies with the axial coordinate x. According to Prandtl's boundary-layer theory, the thermal boundary-layer thickness divides the flow field into two regions: a heat-affected region near the duct wall and an

† Throughout this chapter, thermally developing and hydrodynamically developed flow is simply referred to as "thermally developing flow"; thermally and hydrodynamically developing flow is referred to as "simultaneously developing flow."

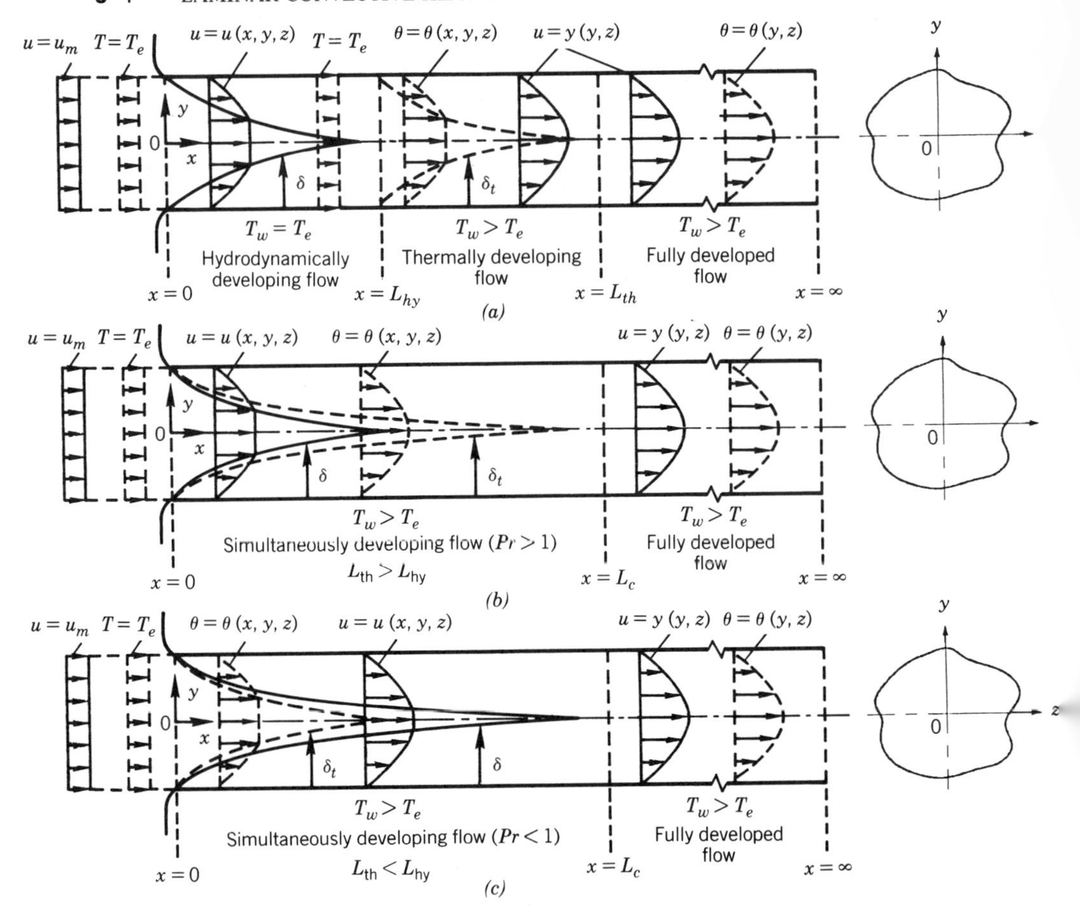

Figure 3.1. Types of laminar duct flows for constant wall temperature boundary condition: (a) hydrodynamically developing flow followed by thermally developing and hydrodynamically developed flow; (b) simultaneously developing flow, Pr > 1; (c) simultaneously developing flow Pr < 1. Solid lines denote the velocity profiles, the dashed lines the temperature profiles.

essentially unaffected region around the duct axis. At $x = L_{th}$, the thermal effects have completely spread across the duct cross section. The region $L_{hy} \le x \le L_{th}$ is termed the *thermal entrance region*, and the fluid flow in this region is called the *thermally developing flow*. It may be emphasized that the thermally developing flow is already hydrodynamically developed in Fig. 3.1a. As shown in this figure, in the thermal entrance region the local dimensionless fluid temperature $\theta = (T_w - T)/(T_w - T_m)$, where T_m is the fluid bulk mean temperature, varies with all three space coordinates, i.e., $\theta = \theta(x, y, z)$.

For $L_{th} \le x < \infty$ in Fig. 3.1a, the viscous and thermal effects have completely diffused across the duct cross section. This region is referred to as the *fully developed region*. The fluid flow in this region is termed the *fully developed flow*. In the fully developed region, the dimensionless temperature θ varies with the transverse coordinates alone although the local fluid temperature T varies with all three space coordinates, and fluid bulk mean temperature T_m varies with the axial coordinate alone.

The fourth type of flow, called *simultaneously developing flow*, is illustrated in Figs. 3.1b and c. In this case, the viscous and thermal effects diffuse simultaneously from the duct wall, commencing at $x = 0$. Depending on the value of the Prandtl number Pr, the two effects diffuse at different rates. In Fig. 3.1*b*, $\mathrm{Pr} > 1$ and $\delta > \delta_t$, whereas in Fig. 3.1*c*, $\mathrm{Pr} < 1$ and $\delta < \delta_t$. This relationship among Pr, δ, and δ_t is easy to infer from the definition of the Prandtl number, which for this purpose can be expressed as $\mathrm{Pr} = \nu/\alpha$, a ratio of kinematic viscosity to thermal diffusivity. The kinematic viscosity is the diffusion rate for momentum or for velocity in the same sense that the thermal diffusivity is the diffusion rate for heat or for temperature. When $\mathrm{Pr} = 1$, the viscous and thermal effects diffuse through the fluid at the same rate. This equality of diffusion rates does not guarantee that the hydrodynamic and thermal boundary layers in internal duct flows will be of the same thickness at a given axial location. The reason for this apparent paradox lies in the fact that with $\mathrm{Pr} = 1$, the applicable momentum and energy differential equations do not become analogous. In external laminar flow over a flat plate, on the other hand, the energy and momentum equations do become analogous when $\mathrm{Pr} = 1$, and for the case when the boundary conditions for the momentum and thermal problems are also analogous, we get $\delta = \delta_t$ for all values of x.

As depicted in Fig. 3.1*b* and *c*, within the region $0 \le x \le L_c$, the viscous and thermal effects diffuse simultaneously across the duct cross section. Accordingly, this region is referred to as the *combined entrance region*. It is apparent that the length L_c of the combined entrance region is dependent on Pr. For $\mathrm{Pr} > 1$, $L_c \approx L_{\mathrm{th}}$ and for $\mathrm{Pr} < 1$, $L_c \approx L_{\mathrm{hy}}$. It may also be noted that in the combined entrance region both the axial velocity and the dimensionless temperature vary with all three space coordinates, i.e., $u = u(x, y, z)$ and $\theta = \theta(x, y, z)$. The region $L_c \le x < \infty$ is the fully developed region, similar to the one depicted in Fig. 3.1*a* with axially invariant $u(y, z)$ and $\theta(y, z)$.

3.1.2 Fluid Flow Parameters

The fluid flow characteristics of all the ducts are expressed in terms of certain hydrodynamic parameters which are defined here. The various symbols are listed in the Nomenclature section at the end of the chapter.

For the hydrodynamically developing flow, the dimensionless axial distance x^+ is defined as

$$x^+ = \frac{x/D_h}{\mathrm{Re}} \tag{3.1}$$

where the hydraulic diameter D_h equals 4 times the duct cross-section area A_c divided by the wetted perimeter P. The hydraulic diameter is consistently used as the characteristic dimension in all the hydrodynamic and thermal parameters such as the Reynolds number $\mathrm{Re} = u_m D_h/\nu$ and the Nusselt number $\mathrm{Nu} = hD_h/k$.

The hydrodynamic entrance length L_{hy} is defined as the axial distance required to attain 99% of the ultimate fully developed maximum velocity when the entering flow is uniform. The dimensionless hydrodynamic entrance length is expressed as $L_{\mathrm{hy}}^+ = L_{\mathrm{hy}}/(D_h \mathrm{Re})$.

The *Fanning friction factor* f is defined as the ratio of the wall shear stress τ_w to the flow kinetic energy per unit volume, $\rho u_m^2/2g_c$:

$$f = \frac{\tau_w}{\rho u_m^2/2g_c} \tag{3.2}$$

In the hydrodynamic entrance region, the apparent Fanning friction factor f_{app} incorporates the combined effect of wall shear and the change in momentum flow rate due to the developing velocity profile. It is based on the total pressure drop from the duct inlet ($x = 0$) to the point of interest and is defined as

$$\Delta p^* \equiv \frac{p_0 - p}{\rho u_m^2/2g_c} = f_{\mathrm{app}} \frac{x}{r_h} \tag{3.3}$$

The incremental pressure drop number $K(x)$ in the hydrodynamic entrance region is defined as

$$K(x) = \left(f_{\mathrm{app}} - f_{\mathrm{fd}}\right)\frac{x}{r_h} \tag{3.4}$$

where f_{fd} is the Fanning friction factor for fully developed laminar flow. $K(x)$ is sometimes referred to as the incremental pressure defect. It increases monotonically from a value of zero at $x = 0$ to a constant value in the hydrodynamically developed region at $x > L_{\mathrm{hy}}$. This constant value is designated as $K(\infty)$, and in the viscometry literature it is referred to as *Hagenbach's factor*.

The Fanning friction factor, axial pressure drop, and incremental pressure drop number are related as

$$\Delta p^* = \left(f_{\mathrm{app}}\mathrm{Re}\right)(4x^+) = K(x) + (f\ \mathrm{Re})(4x^+) \tag{3.5}$$

In practical calculations, the following dimensional form of Eq. (3.5) is found to be useful:

$$\Delta p = \frac{4\left(f_{\mathrm{app}}\mathrm{Re}\right)\mu u_m x}{2g_c D_h^2} = \frac{4(f\ \mathrm{Re})\mu u_m x}{2g_c D_h^2} + \frac{K(x)\rho u_m^2}{2g_c} \tag{3.6}$$

3.1.3 Heat Transfer Parameters

The fluid bulk mean temperature, also referred to as the "mixing cup" or "flow average" temperature T_m, is defined as

$$T_m = \frac{1}{A_c u_m}\int_{A_c} uT\,dA_c \tag{3.7}$$

The circumferentially averaged but axially local heat transfer coefficient h_x is defined by

$$q_x'' = h_x(T_{w,m} - T_m) \tag{3.8}$$

where $T_{w,m}$ is the wall mean temperature and T_m is the fluid bulk mean temperature given by Eq. (3.7). In Eq. (3.8), the heat flux q_w'' and the temperature difference $T_{w,m} - T_m$ are vector quantities. In this equation, the direction of heat transfer is from the wall to the fluid, and consistently the temperature drop is from the wall to the fluid. In contrast, if q_w'' represents the heat flux from the fluid to the wall, the temperature difference entering Eq. (3.8) will be $T_m - T_{w,m}$.

The flow-length average heat transfer coefficient h_m is the integrated average of h_x from $x = 0$ to x, being given by

$$h_m = \frac{1}{x}\int_0^x h_x\,dx \tag{3.9}$$

The ratio of the convective conductance h to the pure molecular conductance k/D_h is defined as a Nusselt number Nu. The circumferentially averaged but axially local Nusselt number is defined as

$$\mathrm{Nu}_x = \frac{h_x D_h}{k} = \frac{q''_x D_h}{k(T_{w,m} - T_m)} \tag{3.10}$$

The mean Nusselt number based on h_m in the thermal entrance region is defined as

$$\mathrm{Nu}_m = \frac{1}{x}\int_0^x \mathrm{Nu}_x \, dx = \frac{h_m D_h}{k} = \frac{q''_m D_h}{k(\Delta T)_m} \tag{3.11}$$

The expressions for $(\Delta T)_m$ could be complicated and dependent upon thermal boundary conditions. Refer to [1] for specific formulas for the Ⓣ and (H1) boundary conditions.

The dimensionless axial distance x^* is defined as

$$x^* = \frac{x/D_h}{\mathrm{Pe}} = \frac{x/D_h}{\mathrm{Re}\,\mathrm{Pr}} \tag{3.12}$$

where $\mathrm{Pe} = u_m D_h/\alpha$ is the Péclet number and $\mathrm{Pr} = \nu/\alpha$ is the Prandtl number. It may be noted that x^* is related to x^+ simply as $x^* = x^+/\mathrm{Pr}$. The Graetz number is related to x^* as $\mathrm{Gz} = Wc_p/kL = P/(4D_h x^*)$.

The thermal entrance length L_{th} is defined as the axial distance required to achieve a value of the local Nusselt number Nu_x, which is 1.05 times the fully developed Nusselt number value. The dimensionless thermal entrance length is expressed as $L^*_{\mathrm{th}} = L_{\mathrm{th}}/(D_h \mathrm{Pe})$.

3.1.4 Thermal Boundary Conditions

In order to accurately interpret the highly sophisticated heat transfer results presented in the ensuing sections of this chapter, a clear understanding of the thermal boundary conditions imposed on the duct wall(s) is absolutely essential. A systematic exposition of the boundary conditions is provided by Shah and London [1]. We give here only a brief description of the specific boundary conditions that are pertinent to this chapter.

Table 3.1 contains a description of the eight most important thermal boundary conditions that are treated in the context of the singly connected ducts in Secs. 3.2 to 3.7. All boundary conditions of Table 3.1 are also applicable to doubly connected ducts. Since a doubly connected duct possesses two walls, each boundary condition of Table 3.1 can be independently applied to each wall, resulting in numerous possible combinations of the boundary conditions. Four types of fundamental boundary conditions analyzed in the context of doubly connected ducts are of special importance, as suitable combinations of them via superposition techniques lead to solutions of all problems of practical interest involving thermally developing flows. These boundary conditions are described near the beginning of Sec. 3.3.1 on p. 3 · 30. Table 3.2 contains a description of the three important thermal boundary conditions treated in the context of the doubly connected ducts covered in Secs. 3.8 to 3.10. The applications of the doubly connected ducts with Ⓣ, (H1) and (H2) boundary conditions are the same as those of the singly connected ducts with the corresponding boundary conditions. Hence, they are not included in Table 3.2.

It may be noted that (H1) to (H4) thermal boundary conditions for the symmetrically heated ducts with no sharp corners (e.g., circular, flat and concentric annular ducts) are identical. Hence, they will be designated simply as Ⓗ for these geometries.

TABLE 3.1. Thermal Boundary Conditions for Singly Connected Ducts

Designation	Description	Mathematical Formulation[a]	Applications
(T)	Uniform wall temperature, with circumferentially and axially constant wall temperature	$T_w = \text{constant}$	Condensers, evaporators, automotive radiators (at high liquid flow rates), with negligible wall thermal resistance
(T3)	Convective, with axially constant wall temperature and finite thermal resistance normal to the wall	$T_{w0} = T_{w0}(y, z)$ $T_w = T_w(x, y, z)$ $k\left(\frac{\partial T}{\partial n}\right)_w = h_0(T_{w0} - T_w)$ $\frac{1}{h_0} = \frac{\delta_w}{k_w} + \frac{1}{h_e}$	Same as for (T) except that the wall thermal resistance is finite in these applications
(T4)	Radiative, with axially constant environment temperature and wall heat flux proportional to the fourth power of the absolute wall temperature	$T_a = T_a(y, z)$ $T_w = T_w(x, y, z)$ $-k\left(\frac{\partial T}{\partial n}\right)_w = \epsilon_w \sigma (T_w^4 - T_a^4)$	High-temperature systems such as space radiators, liquid-metal exchangers, and exchangers involving heat-radiating gases
(H1)	Constant wall heat flux, with circumferentially constant wall temperature and axially constant wall heat flux	$q''_w = q''_w(y, z)$ $T_w = T_w(x)$	Electric resistance heating, nuclear heating, counterflow heat exchangers having nearly identical fluid capacity rates ($\rho u_m A_c c_p$), all involving highly conductive wall materials

(H2)	Uniform wall heat flux, axially and circumferentially	$q''_w = \text{constant}$	Same as (H1) except that the thermal conductivity of the wall material is low (e.g., glass-ceramic, Teflon) and the wall thickness is uniform
(H3)	Convective, with axially constant wall heat flux and finite thermal resistance normal to the wall	$q''_w = q''_w(y, z)$ $k\left(\frac{\partial T}{\partial n}\right)_w = h_0(T_{w0} - T_w)$	Same as (H1) except that the thermal resistance normal to the wall is finite and there is negligible heat conduction along the duct circumference
(H4)	Conductive, with axially constant wall heat flux and finite heat conduction along the wall circumference	$q''_w = q''_w(y, z)$ $\frac{q''_w}{k} - \left(\frac{\partial T}{\partial n}\right)_w + \frac{k_w \delta_w}{k}\left(\frac{\partial^2 T}{\partial s^2}\right)_w = 0$	Same as (H1) except that the circumferential heat conduction is finite
(H5)	Exponential wall heat flux, with circumferentially constant wall temperature and exponentially varying wall heat flux along the duct axis	$T_w = T_w(x)$ $q''_w = q''_0 \exp(mx^*)$	Parallel and counter flow heat exchangers with appropriate values of m

[a] See the Nomenclature section for meaning of the symbols. In particular, w denotes the interior and $w0$ the exterior of the duct wall; n and s designate normal and circumferential directions.

TABLE 3.2. Thermal Boundary Conditions for Doubly Connected Ducts

Designation	Description	Mathematical Formulation[a]
Ⓣ	Uniform wall temperature, with circumferentially and axially constant and equal temperatures at both walls	$T_{w1} = T_{w2} = \text{constant}$
(H1)	Unequal wall heat flux, with two heat fluxes of such magnitude that they give rise to circumferentially uniform and equal wall temperatures along the peripheries of the two walls	$q''_{w1} = q''_{w1}(y, z)$ $q''_{w2} = q''_{w2}(y, z)$ $T_{w1}(x) = T_{w2}(x)$
(H2)	Unequal wall heat flux, with two unequal and uniform heat fluxes of such magnitude that they give rise to equal mean wall temperatures along the peripheries of the two walls	$q''_{w1} = q''_{w1}(y, z)$ $q''_{w2} = q''_{w2}(y, z)$ $T_{w1,m}(x) = T_{w2,m}(x)$

[a] The subscripts 1 and 2 refer to the two walls of the doubly connected duct. If preferred, these subscripts may be replaced by i and o respectively denoting inside and outside duct walls.

3.2 CIRCULAR DUCT

The circular duct is the most widely used geometry in fluid flow and heat transfer devices. Accordingly, its fluid flow and heat transfer characteristics have been analyzed in great detail for various boundary conditions. Also available in the literature is a great deal of information on the effects of viscous dissipation, fluid axial conduction, thermal energy soures and axial momentum diffusion. Several results of practical interest are now outlined.

3.2.1 Fully Developed Flow

The fully developed velocity distribution for the circular duct with origin at the duct axis is given by

$$\frac{u}{u_m} = 2\left[1 - \left(\frac{r}{a}\right)^2\right] \tag{3.13}$$

which is widely known as the *Hagen-Poiseuille parabolic profile*. The mean velocity u_m and the fully developed friction factor corresponding to Eq. (3.13) are

$$u_m = -\frac{1}{8\mu}\left(\frac{dp}{dx}\right)a^2, \qquad f\,\text{Re} = 16 \tag{3.14}$$

The heat transfer results for the fully developed flow will now be presented for various thermal boundary conditions imposed on the duct wall. A brief description of the boundary conditions together with their alphanumeric designation is provided in Table 3.1.

Uniform Wall Temperature, Ⓣ. The fully developed Ⓣ temperature distribution in circular ducts for nondissipative flow in the absence of flow work, thermal energy sources, and fluid axial conduction is given exactly by the following rapidly convergent infinite series [2]:

$$\frac{T_w - T}{T_w - T_m} = \sum_{n=0}^{\infty} C_{2n}\left(\frac{r}{a}\right)^{2n} \tag{3.15}$$

where the coefficients C_{2n} are given by

$$C_0 = 1, \qquad C_2 = -\frac{\lambda_0^2}{2^2} = -1.828397$$

$$C_{2n} = \frac{\lambda_0^2}{(2n)^2}(C_{2n-4} - C_{2n-2}) \tag{3.16}$$

$$\lambda_0 = 2.7043644199 \tag{3.17}$$

An important requirement to be satisfied by the coefficients of Eq. (3.16) is that their sum must vanish. This constraint, conjoined with the bounds on the values of r/a (namely, $0 \le r/a \le 1$), guarantees the rapid convergence of the infinite series in Eq. (3.15). For $n = 0, 2, 4, 6, 8$, and 10, the sum of the coefficients in Eq. (3.16) is $1, 0.464460, 0.050564, 0.002441, 0.000066$, and 0, respectively. Thus for all practical purposes, the series in Eq. (3.15) may be considered fully convergent for $n = 10$.

The fluid bulk mean temperature in Eq. (3.15) is given by the following asymptotic formula applicable for $x^* > 0.0335$ [2]:

$$\frac{T_w - T_m}{T_w - T_e} = 0.819048 \exp(-2\lambda_0^2 x^*) \tag{3.18}$$

Referring to Eq. (3.15), it may be noted that although the local temperature T is a function of both the radial and axial coordinates and the fluid bulk mean temperature T_m is a function of the axial coordinate, the dimensionless temperature $(T_w - T)/(T_w - T_m)$ is a function of the radial coordinate alone. Such a temperature profile is referred to as hydrodynamically and thermally developed, or briefly as a fully developed temperature profile.

The fully developed Nusselt number corresponding to the temperature distribution of Eq. (3.15) can be shown to be

$$\mathrm{Nu_T} = \frac{\lambda_0^2}{2} = 3.6567935 \tag{3.19}$$

The influence of fluid axial conduction for the fully developed flow with the Ⓣ boundary condition is negligible when the Péclet number $\mathrm{Pe}(= u_m D_m/\alpha) > 10$. For lower Péclet numbers, the following asymptotic formulas presented by Michelsen and Villadsen [3] are recommended:

$$\mathrm{Nu_T} = \begin{cases} 4.180654 - 0.183460\,\mathrm{Pe} & \text{for } \mathrm{Pe} < 1.5 \\ 3.656794 + 4.487/\mathrm{Pe}^2 & \text{for } \mathrm{Pe} > 5 \end{cases} \tag{3.20}$$

The effect of viscous dissipation, i.e., heating or cooling of the fluid due to internal friction, is extremely important in situations such as flow of heavy oils in long pipelines, flow of lubricant between fast moving parts, flow of plastics through dies in high-speed extrusion, and flow of air around an earth satellite or a rocket. It is taken into account via a dimensionless parameter called the Brinkman number $\mathrm{Br} = \mu u_m^2/[Jg_c k(T_{w,m} - T_e)]$. Ou and Cheng [4] showed that when viscous dissipation is taken into consideration, the fully developed Nusselt number attains a value of $\frac{48}{5}$ independent of Br. This value is 2.6 times higher than the $\mathrm{Nu_T}$ value of Eq. (3.19).

Uniform Wall Heat Flux, Ⓗ. The fully developed temperature distribution and Nusselt number in a circular duct with the Ⓗ boundary condition for nondissipative flow in the absence of flow work, but with thermal energy sources, finite viscous dissipation, and finite fluid axial conduction, is given by the following set of equations derived by the present authors by recasting the results of Tyagi [5]:

$$\frac{T_w - T}{T_w - T_m} = 6\left[1 - \left(\frac{r}{a}\right)^2\right] \frac{(12 + \gamma) - (4 + \gamma)(r/a)^2}{44 + 3\gamma} \tag{3.21}$$

$$\frac{T_w - T_m}{q_w'' D_h/k} = \frac{44 + 3\gamma}{192} = \frac{1}{\mathrm{Nu_H}} \tag{3.22}$$

$$\gamma = S^* + 64\,\mathrm{Br}' \tag{3.23}$$

Here S^* is the dimensionless thermal energy source number $S^* = SD_h/q_w''$, and Br′ is the dimensionless Brinkman number $\mathrm{Br}' = \mu u_m^2/q_w'' D_h J g_c$ for the Ⓗ boundary condition.

For the case of negligible viscous dissipation and no thermal energy sources we have $\gamma = 0$, and from Eq. (3.22) $\mathrm{Nu_H} = \frac{48}{11} = 4.363636$, a value 19% higher than $\mathrm{Nu_T}$ of Eq. (3.19).

Since the fluid axial conduction is constant for the Ⓗ case, $\mathrm{Nu_H}$ of Eq. (3.22) is also valid for finite fluid axial conduction; the additional parameter Pe does not come into the picture for this case.

The effect of circumferential heat-flux variation could be important in practical applications involving radiantly heated tubes of low thermal conductivity materials like ceramics or for tubes with internal thermal energy generation. For the cosine heat-flux variation represented by $q_w''(\theta) = q_m''(1 + b\cos\theta)$, Reynolds [6] presented the following solutions for local peripheral variations for the case of $\gamma = 0$:

$$\mathrm{Nu_H}(\theta) = \frac{1 + b\cos\theta}{\frac{11}{48} + (b/2)\cos\theta} \tag{3.24}$$

$$\frac{T_w - T_m}{q_m'' D_h/k} = \frac{11}{48} + \frac{b}{2}\cos\theta \tag{3.25}$$

Convectively Heated or Cooled Duct Wall, Ⓣ3. For the Ⓣ3 boundary condition, the wall temperature is constant axially and the duct has a finite thermal resistance normal to the wall. This thermal resistance is incorporated in an external convective heat transfer coefficient h_e, which in turn is incorporated in the dimensionless Biot number defined as $\mathrm{Bi} = h_e D_h/k$. The Biot number could also include the effect of the wall thermal resistance, if any. In this case $\mathrm{Bi} = 1/R_w$, where $R_w = k\delta_w/k_w D_h + k/h_e D_h$. The Ⓗ and Ⓣ boundary conditions are the limiting cases of the (T3) boundary condition corresponding to Bi = 0 and ∞, respectively.

For constant h_e, Hickman [7] developed the following asymptotic formula:

$$\mathrm{Nu_{T3}} = \frac{\frac{48}{11} + \mathrm{Bi}}{1 + \frac{59}{220}\mathrm{Bi}} \tag{3.26}$$

The overall mean Nusselt number $\mathrm{Nu}_{o,m} = h_o D_h/k = q''_w D_h/[k(T_a - T_m)]$ is related to $\mathrm{Nu_{T3}}$ and Bi as [1]

$$\frac{1}{\mathrm{Nu}_{o,m}} = \frac{1}{\mathrm{Nu_{T3}}} + \frac{1}{\mathrm{Bi}} \tag{3.27}$$

Sparrow et al. [8] found that $\mathrm{Nu}_{o,m}$ is quite insensitive to the circumferential variation of Bi and hence h_e.

Radiantly Heated or Cooled Duct Wall, (T4). The (T4) boundary condition was investigated by Kadaner et al. [9]. They provided the following simple formula for the fully developed Nusselt number, reproducing their solution within 0.5%:

$$\mathrm{Nu_{T4}} = \frac{8.728 + 3.66\,\mathrm{Sk}(T_a/T_e)^3}{2 + \mathrm{Sk}(T_a/T_e)^3} \tag{3.28}$$

Here Sk is the dimensionless Stark number defined as $\mathrm{Sk} = \varepsilon_w \sigma T_e^3 D_h/k$, and T_a and T_e are the *absolute* temperatures respectively of the external environment and of the internal fluid at the point of impingement of the radiation flux.

$\mathrm{Nu_{T4}}$ of Eq. (3.28) reduces to $\mathrm{Nu_H}$ and $\mathrm{Nu_T}$ respectively for Sk = 0 and ∞. This indicates that the Ⓗ and Ⓣ boundary conditions are the limiting cases of the (T4) boundary condition.

Exponentially Varying Wall Heat Flux, (H5). The exponentially varying wall heat flux can be represented by $q''_w = q''_0 \exp(mx^*)$, where the exponent m can assume both positive and negative values corresponding to exponential growth or decay of the wall heat flux. The restricted range of the exponent, $-4\mathrm{Nu_T} \le m \le 0$ (where $\mathrm{Nu_T} = 3.6568$), is of special interest, as in this range the fluid in the duct can be convectively heated or cooled by an external environment with uniform temperature and uniform heat transfer coefficient. The case of $m = 0$ corresponds to the Ⓗ boundary condition, and $m = -4\mathrm{Nu_T}$ corresponds to the Ⓣ boundary condition. Since the Ⓣ and Ⓗ boundary conditions are the limiting cases of the (T3) and (T4) boundary conditions, it follows that the Ⓣ, Ⓗ, (T3), and (T4) boundary conditions constitute a subclass of the (H5) boundary condition. This relationship among the boundary conditions was clearly expounded by Sparrow and Patankar [10].

The Nusselt number $\mathrm{Nu_{H5}}$ determined by Shah and London [1] for $-51.36 \le m \le 100$ from the theoretical formula can be represented by the following correlation with a maximum error of 3%:

$$\mathrm{Nu}_{\mathrm{H5}} = 4.3573 + 0.0424m - 2.8368 \times 10^{-4}m^2 + 3.6250 \times 10^{-6}m^3$$

$$-7.6497 \times 10^{-8}m^4 + 9.1222 \times 10^{-10}m^5 - 3.8446 \times 10^{-12}m^6 \quad (3.29)^\dagger$$

3.2.2 Hydrodynamically Developing Flow

The problem of hydrodynamic flow development in a circular duct has stimulated numerous investigations; attempts to describe the flow theoretically extend back to 1891. Depending on the value of the Reynolds number, the various solutions can be categorized as (1) solutions involving boundary-layer theory simplifications valid for large Reynolds numbers, $\mathrm{Re} \to \infty$, (2) solutions involving Navier-Stokes equations with low Reynolds numbers, $\mathrm{Re} < 400$, and (3) creeping flow solutions with $\mathrm{Re} \to 0$. The results pertaining to each solution category are presented separately.

Solutions Involving Boundary-Layer Theory Simplifications. A variety of analytical and numerical techniques with a varying degree of approximation have been employed in developing the solutions in this category. In accordance with Prandtl's boundary-layer theory simplifications, these solutions neglect the effects of axial momentum diffusion and radial pressure variation. It may also be noted that all these solutions employ the simplest entry condition of uniform velocity at the duct inlet.

The various solutions in this category have been reviewed and classified in [1]. Among them, the numerical solution by Hornbeck [11] is believed to be the most accurate. The dimensionless axial velocity and pressure drop values computed by Hornbeck are presented in Table 3.3. According to these results,

$$L_{\mathrm{hy}}^+ = 0.0565, \qquad K(\infty) = 1.28 \qquad (3.30)$$

For practical computations, the dimensionless axial pressure drop in the hydrodynamic entrance region can be computed from the following correlation proposed by Shah as reported in [1]:

$$\Delta p^* = 13.74(x^+)^{1/2} + \frac{1.25 + 64x^+ - 13.74(x^+)^{1/2}}{1 + 0.00021(x^+)^{-2}} \qquad (3.31)$$

The Δp^* values computed from Eq. (3.31) are in excellent accord with the values in Table 3.3.

Solutions Involving the Navier-Stokes Equations. The solutions based on the boundary-layer theory simplifications are not very accurate in the neighborhood of the duct inlet, approximately one diameter both upstream and downstream of the inlet section at $x = 0$. In this region, the neglected effects of the axial momentum diffusion and radial pressure variation are of importance. They are taken into account by the solutions in this category. The proper accounting of the effects introduces the Reynolds number as a parameter in the solution and also requires careful specification of the inlet velocity profile.

† Note that $\mathrm{Nu}_{\mathrm{H5}} = 0$ for $m < -51.36$.

TABLE 3.3. Axial Velocity and Pressure Distribution in the Hydrodynamic Entrance Region of a Circular Duct [11]

x^+	Axial Velocity u/u_m											Pressure
	$r/a = 0$	0.1	0.2	0.3	0.4	0.5	0.6	0.7	0.8	0.9	1.0	Drop Δp^*
0.00000	1.0000	1.0000	1.0000	1.0000	1.0000	1.0000	1.0000	1.0000	1.0000	1.0000	0	0.0000
0.00050	1.1503	1.1503	1.1503	1.1503	1.1503	1.1503	1.1502	1.1485	1.1293	0.8434	0	0.3220
0.00125	1.2269	1.2269	1.2269	1.2269	1.2268	1.2264	1.2230	1.2016	1.0950	0.6893	0	0.5034
0.00250	1.3126	1.3126	1.3125	1.3124	1.3115	1.3068	1.2867	1.2144	1.0098	0.5908	0	0.7204
0.00375	1.3782	1.3781	1.3779	1.3770	1.3733	1.3596	1.3160	1.2000	0.9511	0.5417	0	0.8960
0.00500	1.4332	1.4331	1.4324	1.4299	1.4214	1.3959	1.3292	1.1814	0.9107	0.5102	0	1.0506
0.00750	1.5239	1.5232	1.5204	1.5120	1.4902	1.4395	1.3349	1.1476	0.8585	0.4720	0	1.3212
0.01000	1.5977	1.5960	1.5893	1.5727	1.5358	1.4623	1.3308	1.1218	0.8261	0.4496	0	1.5610
0.01250	1.6595	1.6562	1.6448	1.6188	1.5675	1.4751	1.3245	1.1023	0.8040	0.4346	0	1.7822
0.01750	1.7555	1.7488	1.7269	1.6831	1.6073	1.4874	1.3125	1.0757	0.7756	0.4159	0	2.1900
0.02250	1.8240	1.8142	1.7829	1.7244	1.6306	1.4927	1.3034	1.0588	0.7584	0.4047	0	2.5692
0.03000	1.8920	1.8785	1.8366	1.7626	1.6509	1.4962	1.2943	1.0433	0.7429	0.3947	0	3.1064
0.04000	1.9431	1.9266	1.8763	1.7901	1.6650	1.4981	1.2875	1.0321	0.7319	0.3877	0	3.7894
0.05000	1.9698	1.9517	1.8969	1.8042	1.6721	1.4990	1.2840	1.0264	0.7263	0.3840	0	4.4520
0.06250	1.9863	1.9672	1.9095	1.8128	1.6764	1.4996	1.2818	1.0229	0.7229	0.3818	0	5.2688
∞	2.0000	1.9800	1.9200	1.8200	1.6800	1.5000	1.2800	1.0200	0.7200	0.3800	0	—

When the effects of axial momentum diffusion and radial pressure variation are taken into account by the use of the Navier-Stokes equations, the velocity profiles display peculiar behavior in the neighborhood of the duct inlet ($x^+ < 0.005$) for $\mathrm{Re} \leq 400$. They exhibit a local minimum at the duct centerline and symmetrically placed maxima near the duct wall, unlike the convex velocity profile found by the boundary-layer analysis. This phenomenon is referred to as the velocity overshoot, and its physical existence has been confirmed by experimental measurements [1].

The details of various solutions in this category are available in [1]. They indicate that the effects of the axial momentum diffusion and radial pressure variation are important in the neighborhood of the duct inlet ($x^+ < 0.005$) for $\mathrm{Re} \leq 400$ only. For higher values of x^+, boundary-layer type analyses, such as the one by Hornbeck [11], are quite satisfactory.

Based on the analysis by Chen [12], it is found that the Reynolds number dependence of L_{hy}^+ and $K(\infty)$ is given by

$$L_{\mathrm{hy}}^+ = 0.056 + \frac{0.60}{\mathrm{Re}(1 + 0.035\,\mathrm{Re})} \tag{3.32}$$

$$K(\infty) = 1.20 + \frac{38}{\mathrm{Re}} \tag{3.33}$$

Creeping Flow Solutions. Creeping flow is the generic name given to flows with vanishingly small Reynolds numbers. It represents an asymptotic limit of laminar flow and occurs when the viscous forces completely overwhelm the inertia forces. The creeping flow problem is also referred to as the *Stokes flow problem*. The practical applications of this flow occur in processes involving highly viscous liquids at low velocities. Some notable examples are lubrication, viscometry, glass processing, and polymer processing.

The creeping flow problem in the entrance region of a circular duct has been solved by a number of investigators as summarized in [1]. Based on these investigations, it is observed that the hyrodynamic entrance length (L_{hy}/D_h) approaches the value 0.60 as $\mathrm{Re} \to 0$ with the inlet condition of uniform flow.

Weissberg [13] proposed the following expression for the pressure drop in the entrance region of a circular tube with creeping flow:

$$\Delta p^* = 64\left(x^+ + \frac{3\pi}{16\,\mathrm{Re}}\right) \tag{3.34}$$

Linehan and Hirsch [14] experimentally verified Eq. (3.34) within 6%. Note that while Eq. (3.34) is valid for $\mathrm{Re} \to 0$ (i.e., $\mathrm{Re} \leq 1$), Eq (3.31) is valid for high Reynolds number flow ($\mathrm{Re} \geq 400$). For $1 < \mathrm{Re} < 400$, no analytical expression for Δp^* is available. In this case, the numerical results based on a full set of Navier-Stokes equations must be used.

3.2.3 Thermally Developing Flow

This section deals with the laminar convection heat transfer problem in which the velocity profile is specified to be the fully developed parabolic distribution given by Eq. (3.13) but the temperature profile is allowed to develop under a variety of thermal boundary conditions specified at the duct wall. The results are presented for four boundary conditions.

Specified Axial Wall Temperature Distribution. Consider the case when the duct wall is maintained at a constant temperature different from the uniform temperature of the fluid at the entrance, and the fluid axial conduction, viscous dissipation, flow work, and thermal energy sources are negligible. This problem was first solved by Graetz [15] in 1885. Subsequently, Nusselt [16] solved the same problem independently in 1910. This celebrated problem is now widely known *as the Graetz problem or the Graetz-Nusselt problem*. Its solution is given by the following set of equations:

$$\theta = \frac{T_w - T}{T_w - T_e} = \sum_{n=0}^{\infty} C_n R_n\left(\frac{r}{a}\right) \exp(-2\lambda_n^2 x^*) \tag{3.35}$$

$$\theta_m = \frac{T_w - T_m}{T_w - T_e} = 8 \sum_{n=0}^{\infty} \frac{G_n}{\lambda_n^2} \exp(-2\lambda_n^2 x^*) \tag{3.36}$$

$$\mathrm{Nu}_{x,\mathrm{T}} = \frac{\sum_{n=0}^{\infty} G_n \exp(-2\lambda_n^2 x^*)}{2 \sum_{n=0}^{\infty} (G_n/\lambda_n^2) \exp(-2\lambda_n^2 x^*)} \tag{3.37}$$

$$\mathrm{Nu}_{m,\mathrm{T}} = -\frac{\ln \theta_m}{4x^*} \tag{3.38}$$

Here λ_n, $R_n(r/a)$, and C_n are the eigenvalues, eigenfunctions, and constants, respectively, and $G_n = -(C_n/2)R_n'(1)$, where $R_n'(1)$ is the derivative of $R_n(r/a)$ evaluated at $r/a = 1$. The values of λ_n, C_n, and G_n are furnished in Table 3.4, and those of $R_n(r/a)$ in Table 3.5.

Table 3.4 lists the first 11 eigenvalues and constants of the Graetz problem. The higher values can be computed from the following formulas developed by Newman

TABLE 3.4. Eigenvalues and Constants of the Graetz Problem [17]

n	λ_n	C_n	G_n
0	2.70436 44199	1.47643 54070	0.74877 4555
1	6.67903 14493	−0.80612 38956	0.54382 7956
2	10.67337 95381	0.58876 21541	0.46286 1060
3	14.67107 84627	−0.47585 04282	0.41541 8455
4	18.66987 18645	0.40502 18107	0.38291 9188
5	22.66914 33588	−0.35575 65063	0.35868 5566
6	26.66866 19960	0.31916 90532	0.33962 2164
7	30.66832 33409	−0.29073 58292	0.32406 2211
8	34.66807 38224	0.26789 11826	0.31101 4074
9	38.66788 33469	−0.24906 25329	0.29984 4038
10	42.66773 38055	0.23322 77932	0.29012 4676

TABLE 3.5. Eigenfunctions $R_n(r/a)$ of the Graetz Problem [17, 18][a]

	$R_n(r/a)$								
n	$r/a = 0.1$	0.2	0.3	0.4	0.5	0.6	0.7	0.8	0.9
0	0.98184	0.92889	0.84547	0.73809	0.61460	0.48310	0.35101	0.22426	0.10674
1	0.89181	0.60470	0.23386	−0.10959	−0.34214	−0.43218	−0.39763	−0.28449	−0.14113
2	0.73545	0.15247	−0.31521	−0.39208	−0.14234	0.16968	0.33149	0.30272	0.16262
3	0.53108	−0.23303	−0.35914	0.06793	0.31507	0.11417	−0.19604	−0.29224	−0.17762
4	0.30229	−0.40260	0.00054	0.29907	−0.07973	−0.25523	0.03610	0.25918	−0.18817
5	0.07488	−0.32121	0.28982	−0.04766	−0.20532	0.19750	0.10372	−0.20893	−0.19522
6	−0.12642	−0.07613	0.20122	−0.25168	0.19395	−0.01391	−0.18883	0.14716	0.19927
7	−0.28107	0.17716	−0.10751	0.03452	0.05514	−0.15368	0.20290	−0.07985	−0.20068
8	−0.37523	0.29974	−0.25305	0.22174	−0.20502	0.19303	−0.15099	0.01298	0.19967
9	−0.40326	0.23915	−0.08558	−0.02483	0.08126	−0.09176	0.05652	0.04787	−0.19645
10	−0.36817	0.04829	0.16645	−0.20058	0.13289	−0.06474	0.04681	−0.09797	0.19120
11	−0.28088	−0.15310	0.19847	0.01714	−0.15931	0.16099	−0.12577	0.13375	−0.18409
12	−0.15836	−0.24999	−0.00845	0.18456	−0.01927	−0.13393	0.15742	−0.15311	0.17527
13	−0.02118	−0.19545	−0.18955	−0.01074	0.15967	0.01258	−0.13539	0.15549	−0.16491
14	0.10953	−0.03182	−0.13083	−0.17183	−0.08560	0.10927	0.07069	−0.14189	0.15319

[a] For all values of n, $R_n(0) = 1$ and $R_n(1) = 0$.

[19]:

$$\lambda_n = \lambda + S_1\lambda^{-4/3} + S_2\lambda^{-8/3} + S_3\lambda^{-10/3} + S_4\lambda^{-11/3} + O(\lambda^{-14/3}) \tag{3.39}$$

$$G_n = \frac{C}{\lambda^{1/3}}\left[1 + \frac{L_1}{\lambda^{4/3}} + \frac{L_2}{\lambda^{6/3}} + \frac{L_3}{\lambda^{7/3}} + \frac{L_4}{\lambda^{10/3}} + \frac{L_5}{\lambda^{11/3}} + O(\lambda^{-4})\right] \tag{3.40}$$

where

$$\lambda = 4n + \tfrac{8}{3}, \qquad n = 0, 1, 2, \ldots \tag{3.41}$$

$$S_1 = 0.159152288, \qquad S_2 = 0.0114856354, \qquad S_3 = -0.224731440$$

$$S_4 = -0.033772601, \qquad C = 1.012787288 \tag{3.42}$$

and

$$L_1 = 0.144335160, \qquad L_2 = 0.115555556, \qquad L_3 = -0.21220305$$

$$L_4 = -0.187130142, \qquad L_5 = 0.0918850832 \tag{3.43}$$

The temperature distribution for the Graetz problem is displayed in Fig. 3.2 as presented by Grigull and Tratz [20]. The local and mean Nusselt numbers computed from Eqs. (3.37) to (3.43) and certain asymptotic formulas to be presented shortly are displayed in Fig. 3.3. From the tabulated values of the Nusselt number available in [1], the thermal entrance length for the Graetz problem as defined in Sec. 3.1.3 is found to be

$$L^*_{\text{th,T}} = 0.0334654 \tag{3.44}$$

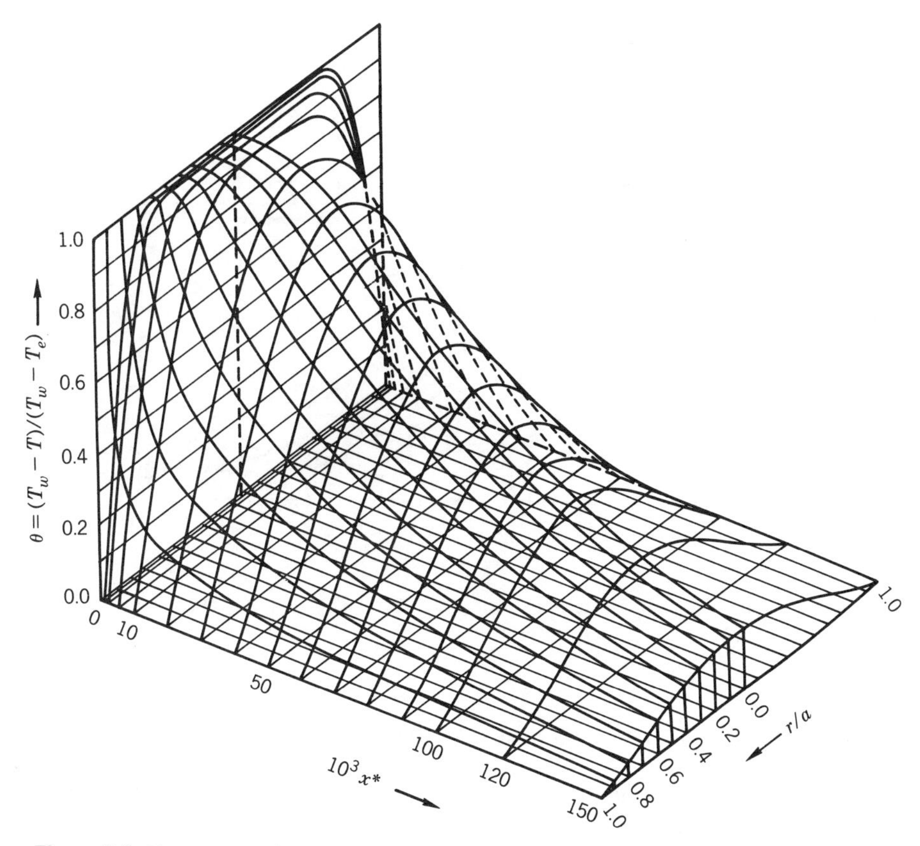

Figure 3.2. Temperature distribution in the thermal entrance region of a circular duct with the Ⓣ boundary condition (Graetz solution) [20].

The infinite series of Eq. (3.37) converges uniformly for all nonzero values of x^*, but the convergence is extremely slow as x^* approaches zero. Therefore, for $x^* < 10^{-4}$, Lévêque's asymptotic solution [21] is employed, which becomes increasingly accurate as $x^* \rightarrow 0$. The temperature distribution and the Nusselt number due to Lévêque are given by

$$\theta = \frac{T_w - T}{T_w - T_e} = \frac{1}{\Gamma(\frac{4}{3})}\left[\int_0^{\chi} e^{-\chi^3}\, d\chi\right] \tag{3.45}$$

where

$$\chi = \frac{1 - (r/a)}{(9x^*)^{1/3}} \tag{3.46}$$

$$\mathrm{Nu}_{x,\mathrm{T}} = \frac{2}{\Gamma(\frac{4}{3})(9x^*)^{1/3}} = \tfrac{2}{3}\mathrm{Nu}_{m,\mathrm{T}} \tag{3.47}$$

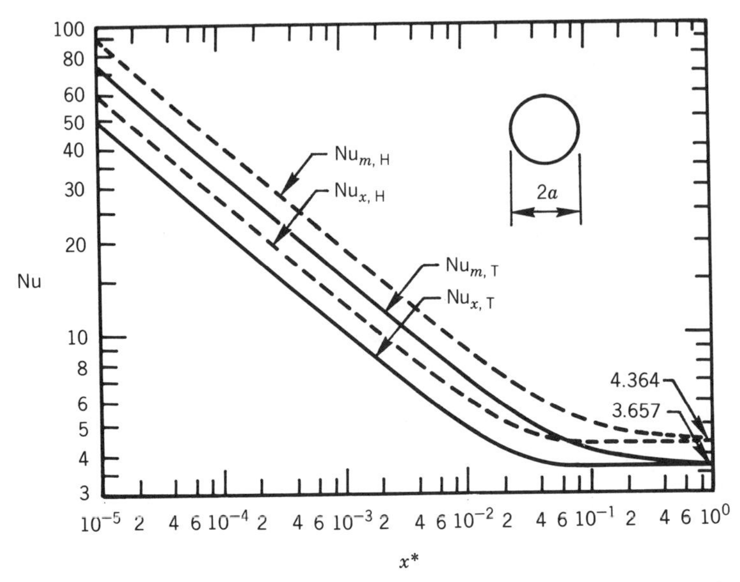

Figure 3.3. Local and mean Nusselt numbers in the thermal entrance region of a circular duct with the Ⓣ and Ⓗ boundary conditions [1].

The integral appearing in Eq. (3.45) has been tabulated by Abramowitz [22], and $\Gamma(\frac{4}{3})$ has the value 0.8929795116.

Based on the Graetz and extended Lévêque solutions, the local and mean Nusselt numbers can be computed from the following simple formulas [1]:

$$\mathrm{Nu}_{x,\mathrm{T}} = \begin{cases} 1.077x^{*-1/3} - 0.7 & \text{for} \quad x^* \le 0.01 \\ 3.657 + 6.874(10^3 x^*)^{-0.488} e^{-57.2x^*} & \text{for} \quad x^* > 0.01 \end{cases} \tag{3.48}$$

$$\mathrm{Nu}_{m,\mathrm{T}} = \begin{cases} 1.615x^{*-1/3} - 0.7 & \text{for} \quad x^* \le 0.005 \\ 1.615x^{*-1/3} - 0.2 & \text{for} \quad 0.005 < x^* < 0.03 \\ 3.657 + (0.0499/x^*) & \text{for} \quad x^* \ge 0.03 \end{cases} \tag{3.49}$$

The values calculated from the expressions in Eqs. (3.48) and (3.49) are higher (+) or lower (−) than those tabulated in [1] by ±0.5, ±0.2, ±2.2, ±3.0, and +2.1%, respectively. Except for the end points of the range, the error is much less than specified.

Hausen [23] presented the following correlation for the mean Nusselt numbers of Graetz's solution for the entire range of x^*:

$$\mathrm{Nu}_{m,\mathrm{T}} = 3.66 + \frac{0.0668}{x^{*1/3}(0.04 + x^{*2/3})} \tag{3.50}$$

whence by differentiating with respect to x^*, the present authors arrived at the following correlation for the local Nusselt numbers of Graetz's solution:

$$\mathrm{Nu}_{x,\mathrm{T}} = 3.66 + \frac{0.0018}{x^{*1/3}(0.04 + x^{*2/3})^2} \tag{3.51}$$

The predictions of Eqs. (3.50) and (3.51) are higher than the tabulated values in [1] by amounts ranging from 14% for $x^* < 0.0001$ to 0% for $x^* \to \infty$. For $0.0001 \le x^* \le 0.001$, they are 14 to 7% higher; for $0.001 \le x^* \le 0.01$, 7 to 3% higher, and for $0.01 \le x^* \le \infty$, 3 to 0% higher.

The effect of fluid axial conduction on the Graetz solution has been investigated quite extensively, as summarized in [1]. It is determined from the analysis of Hennecke [24]. It shows that except in the neighborhood of the duct inlet ($x^* < 10^{-2}$), the effect of fluid axial conduction is negligible for Pe > 50. Hennecke's analysis further shows that the thermal entrance length $L^*_{\mathrm{th,T}}$ increases from 0.033 of Eq. (3.44) for Pe = ∞ to 0.5 for Pe = 1.

The effect of viscous dissipation on the solution of the Graetz problem was first studied by Brinkman [25], after whom the viscous dissipation parameter Br is named. He treated the fluid viscosity as constant and considered the duct wall temperature to be the same as the entering fluid temperature. Ou and Cheng [4] studied the viscous dissipation effect in greater detail with constant fluid viscosity for the case where the duct wall temperature is different from the entering fluid temperature. The resulting Nusselt numbers exhibit peculiar behavior which is explained in [1].

Several other additional effects have also been investigated in the context of the Graetz problem. They include the effects of nonuniform inlet temperature profiles, axial variation of the wall temperature, circumferential variation of the wall temperature, internal thermal energy generation, nonparabolic velocity profiles, polynomial wall temperature variation, and sinusoidal wall temperature variation. For details refer to [1].

Convectively Heated or Cooled Duct Wall, Ⓣ3. The problem of heat transfer with the convective boundary condition Ⓣ3 in the absence of viscous dissipation, fluid axial conduction, flow work, and internal heat sources has been investigated quite extensively as described in [1]. The local Nusselt numbers $\mathrm{Nu}_{x,\mathrm{T3}}$ of Hsu [26] are presented in Fig. 3.4 with the Biot number Bi as the parameter. The curves corresponding to Bi = 0 and ∞ are respectively identical with the $\mathrm{Nu}_{x,\mathrm{H}}$ and $\mathrm{Nu}_{x,\mathrm{T}}$ curves of Fig. 3.3 since the Ⓗ and Ⓣ boundary conditions are the limiting cases of the Ⓣ3 boundary condition.

Radiatively Heated or Cooled Duct Wall, Ⓣ4. The problem of heat transfer with the radiative boundary condition Ⓣ4 in the absence of viscous dissipation, fluid axial conduction, flow work, and internal heat sources has been investigated in some depth, as summarized in [1]. The local Nusselt numbers normalized with respect to $\mathrm{Nu}_{x,\mathrm{H}}$ are given by the following formula due to Kadaner et al. [9]:

$$\frac{\mathrm{Nu}_{x,\mathrm{T4}}}{\mathrm{Nu}_{x,\mathrm{H}}} = 0.94 - \frac{0.0061 - 0.0053 \ln x^*}{1 + 0.0242 \ln x^*} \ln\left(\frac{\mathrm{Sk}}{2}\right) \tag{3.52}$$

It represents their analytical solution within 2% in the ranges $0.001 < x^* < 0.2$ and $0.2 < \mathrm{Sk} < 100$ for zero ambient temperature. $\mathrm{Nu}_{x,\mathrm{H}}$ of Eq. (3.52) is available from

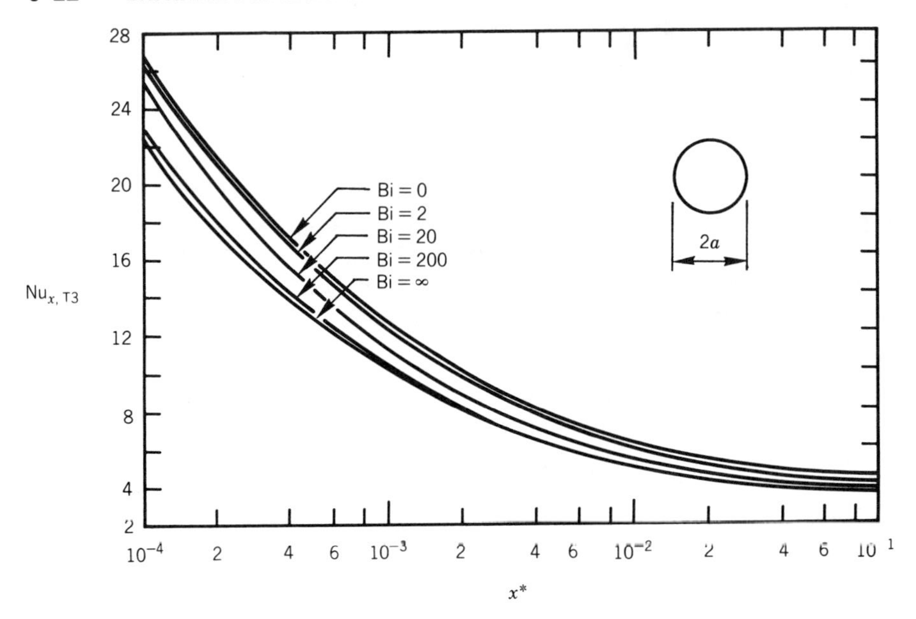

Figure 3.4. Local and mean Nusselt numbers in the thermal entrance region of a circular duct with the Ⓣ3 boundary condition [26].

Fig. 3.3. $\mathrm{Nu}_{x,\mathrm{T4}}$ with Sk = 0 and ∞ respectively correspond to $\mathrm{Nu}_{x,\mathrm{H}}$ and $\mathrm{Nu}_{x,\mathrm{T}}$ of Fig. 3.3, since the Ⓗ and Ⓣ boundary conditions are the limiting cases of the Ⓣ4 boundary condition.

Specified Heat Flux Distribution. Similar to the Graetz problem with the Ⓣ boundary condition, the Ⓗ thermal entry length problem is of great practical interest. It is identical to the Graetz problem except for the thermal boundary condition. According to Siegel et al. [27], the Ⓗ temperature distribution and the local Nusselt numbers are given by

$$\Theta = \frac{T - T_e}{q_w'' D_h / k} = 4x^* + \frac{1}{2}\left(\frac{r}{a}\right)^2 - \frac{1}{8}\left(\frac{r}{a}\right)^4 - \frac{7}{48} + \frac{1}{2}\sum_{n=1}^{\infty} C_n R_n\left(\frac{r}{a}\right)\exp\left(-2\beta_n^2 x^*\right) \tag{3.53}$$

$$\Theta_m = \frac{T_m - T_e}{q_w'' D_h / k} = 4x^* \tag{3.54}$$

$$\mathrm{Nu}_{x,\mathrm{H}} = \left(\frac{11}{48} + \frac{1}{2}\sum_{n=1}^{\infty} C_n R_n(1)\exp\left(-2\beta_n^2 x^*\right)\right)^{-1} \tag{3.55}$$

$$\mathrm{Nu}_{m,\mathrm{H}} = \left(\frac{11}{48} + \frac{1}{2}\sum_{n=1}^{\infty} C_n R_n(1)\frac{1 - \exp\left(-2\beta_n^2 x^*\right)}{2\beta_n^2 x^*}\right)^{-1} \tag{3.56}$$

TABLE 3.6. Eigenvalues and Constants for the Thermal Entry Length Solution for a Circular Duct with the Ⓗ Boundary Condition [28]

n	β_n^2	$R_n(1)$	C_n
1	25.679611	−0.49251658	0.40348318
2	83.861753	0.39550848	−0.17510993
3	174.16674	−0.34587367	0.10559168
4	296.53630	0.31404646	−0.073282370
5	450.94720	−0.29125144	0.055036482
6	637.38735	0.27380691	−0.043484355
7	855.849532	−0.25985296	0.035595085
8	1106.329035	0.24833186	−0.029908452
9	1388.822594	−0.23859024	0.025640098
10	1703.3278521	0.23019903	−0.022333685
11	2049.843045	−0.22286280	0.019706916
12	2438.366825	0.21637034	−0.017576456
13	2838.898142	−0.21056596	0.015818436
14	3281.436173	0.20533190	−0.014346369
15	3755.980271	−0.20057716	0.013098171
16	4262.529926	0.19623013	−0.012028202
17	4801.084747	−0.19223350	0.011102223
18	5371.644444	0.18854081	−0.010294071
19	5974.208812	−0.18511389	0.0095834495
20	6608.777727	0.18192104	−0.0089543767

Here β_n, $R_n(r/a)$, and C_n are eigenvalues, eigenfunctions, and constants, respectively. The first seven of these quantities were determined by Siegel et al. [27]. Hsu [28] extended their work and reported the first 20 values for each of β_n^2, $R_n(1)$, and C_n, which are listed in Table 3.6. In addition, Hsu [28] graphically presented the first ten eigenfunctions R_n for the radius range $0 \le r/a \le 1$. He also presented approximate formulas for higher eigenvalues and constants. Of particular interest are

$$\beta_n = 4n + \tfrac{4}{3} \tag{3.57}$$

$$R_n(1) = (-1)^n \, 0.774759003\beta_n^{-1/3} \tag{3.58}$$

$$C_n = (-1)^n \, 3.099036005\beta_n^{-4/3} \tag{3.59}$$

The Ⓗ temperature distribution as presented by Grigull and Tratz [20] is depicted in Fig. 3.5. The local Nusselt numbers $\mathrm{Nu}_{x,\mathrm{H}}$ for $x^* > 10^{-4}$ are presented in Fig. 3.3. From these results, the thermal entrance length is found to be

$$L^*_{\mathrm{th,H}} = 0.0430527 \tag{3.60}$$

The infinite series of Eqs. (3.55) and (3.56) converge uniformly for all nonzero value of x^*, but the convergence becomes extremely slow as x^* approaches zero. The

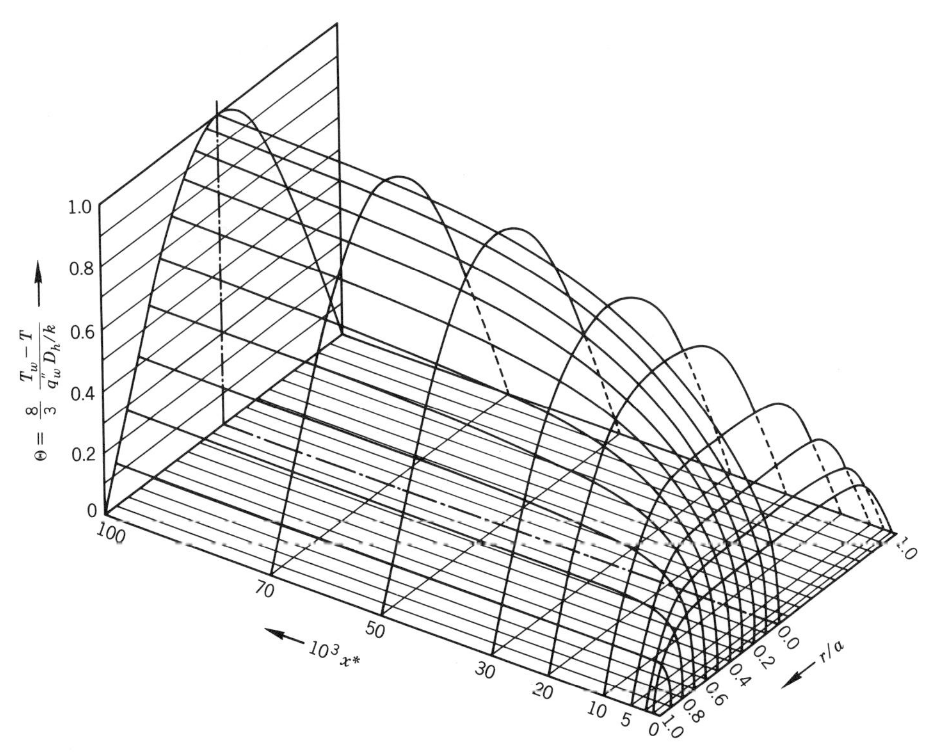

Figure 3.5. Temperature distribution in the thermal entrance region of a circular duct with the Ⓗ boundary condition [20].

Lévêque-type asymptotic formulas applicable in the neighborhood of the duct inlet are obtained by Bird et al. [29] as

$$\Theta = \frac{T - T_e}{q''_w D_h/k} = \frac{(9x^*)^{1/3}}{2}\left\{\frac{e^{-\chi^3}}{\Gamma(\frac{2}{3})} - \chi\left[1 - \frac{\Gamma(\frac{2}{3}, \chi^3)}{\Gamma(\frac{2}{3})}\right]\right\} \tag{3.61}$$

$$\mathrm{Nu}_{x,\mathrm{H}} = \frac{2\Gamma(\frac{2}{3})}{(9x^*)^{1/3}} = \tfrac{2}{3}\mathrm{Nu}_{m,\mathrm{H}} \tag{3.62}$$

Here χ is defined in Eq. (3.46), $\Gamma(\frac{2}{3}) = 1.3541179394$, and $\Gamma(\frac{2}{3}, \chi^3)$ is an incomplete gamma function.

Based on the Graetz-type and the Lévêque-type solutions for the Ⓗ thermal entrance length problem, the local and mean Nusselt numbers can be computed from [1]

$$\mathrm{Nu}_{x,\mathrm{H}} = \begin{cases} 1.302x^{*-1/3} - 1 & \text{for} \quad x^* \le 0.00005 \\ 1.302x^{*-1/3} - 0.5 & \text{for} \quad 0.00005 \le x^* \le 0.0015 \\ 4.364 + 8.68(10^3 x^*)^{-0.506} e^{-41x^*} & \text{for} \quad x^* \ge 0.0015 \end{cases} \tag{3.63}$$

$$\mathrm{Nu}_{m,\mathrm{H}} = \begin{cases} 1.953x^{*-1/3} & \text{for} \quad x^* \le 0.03 \\ 4.364 + 0.0722/x^* & \text{for} \quad x^* > 0.03 \end{cases} \tag{3.64}$$

These empirical equations yield $\mathrm{Nu}_{x,\mathrm{H}}$ and $\mathrm{Nu}_{m,\mathrm{H}}$ within ± 1 and $\pm 3\%$ respectively of the values from the exact solution.

The effect of fluid axial conduction on the Ⓗ thermal entrance length problem has been investigated quite extensively, as reviewed in [1]. From the analysis of Hennecke [24], it is concluded that the effect of fluid axial conduction is negligible for $\mathrm{Pe} > 10$ when $x^* \geq 0.005$.

The effect of viscous dissipation on the solution of the Ⓗ problem was first studied by Brinkman [25], who considered the duct wall to be adiabatic, i.e., $q''_w = 0$. He found the fluid temperature to rise rapidly near the wall for increasing x^*. Ou and Cheng [30] studied the viscous dissipation effect in greater detail for a nonadiabatic wall.

The effect of exponential variation of the wall heat flux as represented by $q''_w = q''_0 \exp(mx^*)$ can be readily taken into account by the superposition method. Based on the analysis by Siegel et al. [27] for the Ⓗ problem, the local Nusselt number for the (H5) problem can be determined from the following formula:

$$\mathrm{Nu}_{x,\mathrm{H5}} = \left(\sum_{n=1}^{\infty} \frac{-C_n R_n(1) \beta_n^2}{m + 2\beta_n^2} \left\{ 1 - \exp\left[-\left(m + 2\beta_n^2 \right) x^* \right] \right\} \right)^{-1} \tag{3.65}$$

The constants C_n, $R_n(1)$, and β_n^2 of Eq. (3.65) are available in Table 3.6 for n up to 20, and for higher values of n they can be determined from Eqs. (3.57) to (3.59).

Several other additional effects have also been investigated in the context of the Ⓗ thermal entrance length problem. They include the effects of nonuniform inlet temperature distribution, internal thermal energy generation, sinusoidal variation of wall heat flux, arbitrary wall heat-flux variation, and peripherally variable but axially constant wall heat flux. For details, refer to [1].

One of the principal idealizations involved in all the solutions presented so far has been that the duct has a uniform wall thickness and there does not exist simultaneous heat conduction in the wall in the axial, normal, and circumferential directions. In reality, this may not be true for thick-walled ducts, and the heat transfer problem for the solid wall must be analyzed simultaneously with the heat transfer problem for the fluid. This combined problem is referred to as a *conjugate problem*, and its solution entails several additional parameters. The conjugate problem for the circular duct has been analyzed extensively, as summarized by Barozzi and Pagliarini [31].

3.2.4 Simultaneously Developing Flow

All the heat transfer results presented so far have been based on the idealization of the fully developed velocity profile at the point in the duct where heating or cooling begins. For high Prandtl number fluids, such as viscous liquids, this idealization does not seriously restrict the usefulness of the results, because the velocity profile develops much more rapidly than the temperature profile. However, for the Prandtl number range near unity, which includes gases, the velocity and the temperature profiles develop at similar rates. Consequently, the idealization of a fully developed velocity profile at the duct inlet may lead to a considerable error in the predicted performance. In such cases, the results pertaining to simultaneously developing velocity and temperature fields should be applied. Such results for a variety of thermal boundary conditions are now summarized.

***Uniform Wall Temperature,* Ⓣ.** The Ⓣ simultaneously developing flow has been extensively analyzed in the literature, as reviewed in [1]. The most accurate $\mathrm{Nu}_{x,\mathrm{T}}$ and $\mathrm{Nu}_{m,\mathrm{T}}$ are presented in Figs. 3.6 and 3.7; the tabulated results are available in [1].

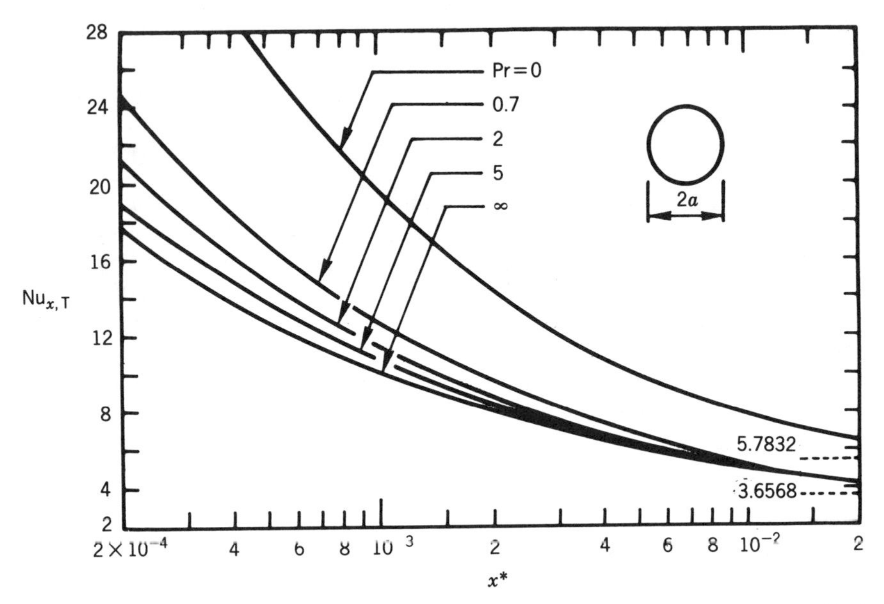

Figure 3.6. Local Nusselt numbers for simultaneously developing flow in a circular duct with the Ⓣ boundary condition [1].

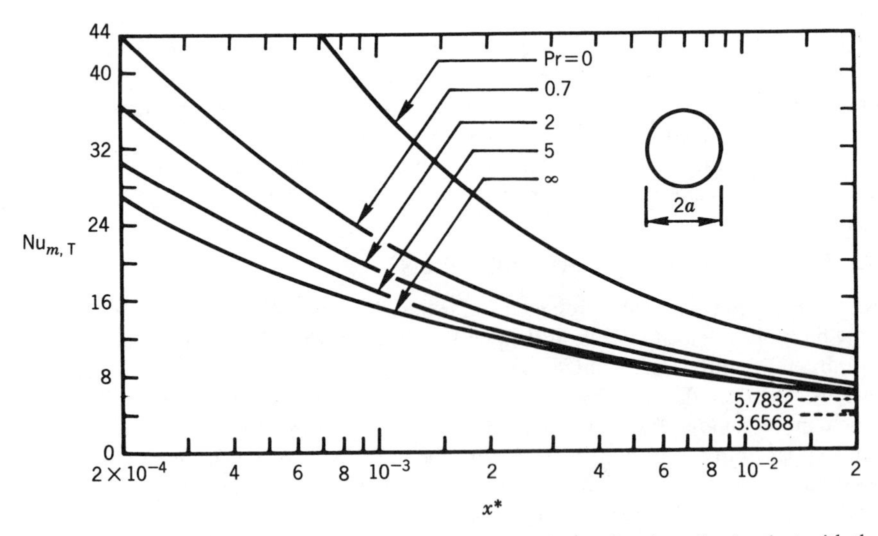

Figure 3.7. Mean Nusselt numbers for simultaneously developing flow in a circular duct with the Ⓣ boundary condition [1].

For the fluid with Pr = ∞, the velocity profile has already developed before the temperature profile starts developing. So this limiting case corresponds to the previously discussed case of thermally developing but hydrodynamically developed flow. Accordingly, the curves corresponding to Pr = ∞ in Figs. 3.6 and 3.7 are identical with the curves of Fig. 3.3. It must be emphasized that when the velocity profile is fully developed for any fluid (i.e., any Pr), the applicable Nusselt numbers are those corresponding to Pr = ∞ in Figs. 3.6 and 3.7.

For a fluid with Pr = 0, the temperature profile develops very much faster than the velocity profile. So in this limiting case, while the temperature profile develops, the velocity profile remains perfectly uniform (i.e., slug flow). This solution was first obtained by Graetz [15] in 1883. Graetz's solution was rediscovered by Lévêque [21] in 1928. It can be expressed by the following set of equations:

$$\theta = \frac{T_w - T}{T_w - T_e} = 2\sum_{n=1}^{\infty} \frac{\exp(-4\gamma_n^2 x^*)\, J_0[\gamma_n(r/a)]}{\gamma_n J_1(\gamma_n)} \tag{3.66}$$

$$\theta_m = \frac{T_w - T}{T_w - T_e} = 4\sum_{n=1}^{\infty} \frac{\exp(-4\gamma_n^2 x^*)}{\gamma_n^2} \tag{3.67}$$

$$\mathrm{Nu}_{x,T} = \frac{\sum_{n=1}^{\infty} \exp(-4\gamma_n^2 x^*)}{\sum_{n=1}^{\infty} \frac{\exp(-4\gamma_n^2 x^*)}{\gamma_n^2}} \tag{3.68}$$

$$\mathrm{Nu}_{m,T} = -\frac{\ln \theta_m}{4x^*} \tag{3.69}$$

Here $J_0[\gamma_n(r/a)]$ and $J_1(\gamma_n)$ are Bessel functions of the first kind and order 0 and 1,

TABLE 3.7. Eigenvalues for the Graetz Solution for Slug Flow in a Circular Duct with the Ⓣ Boundary Condition[a]

n	γ_n	n	γ_n
1	2.40482 55577	11	33.77582 02136
2	5.52007 81103	12	36.91709 83537
3	8.65372 97129	13	40.05842 57646
4	11.79153 44391	14	43.19979 17132
5	14.93091 77086	15	46.34118 83717
6	18.07106 39679	16	49.48260 98974
7	21.21163 66299	17	52.62405 18411
8	24.35242 15308	18	55.76551 07550
9	27.49347 91320	19	58.90698 39261
10	30.63460 64684	20	62.04846 91902

[a] The higher eigenvalues of the problem can be computed to a high degree of accuracy from the asymptotic formula

$$\gamma_n = \frac{\pi}{4}(4n-1) + \frac{1}{2\pi(4n-1)} - \frac{31}{6\pi^3(4n-1)^3} + \frac{3779}{15\pi^5(4n-1)^5} - \cdots$$

respectively. γ_n are the eigenvalues, being the zeros of $J_0(\gamma) = 0$. The first 20 eigenvalues are listed in Table 3.7.

The local and mean Nusselt numbers computed from Eqs. (3.68) and (3.69) are displayed in Figs. 3.6 and 3.7 corresponding to Pr = 0. Note that the curves corresponding to Pr = 0 approach the asymptotic value of 5.7832, whereas the remainder of the curves approach the asymptotic value of 3.6568.

The thermal entrance lengths for the simultaneously developing flow with the Ⓣ boundary condition are found to be [1, 32]

$$L^*_{\text{th, T}} = \begin{cases} 0.028 & \text{for} \quad \text{Pr} = 0 \\ 0.037 & \text{for} \quad \text{Pr} = 0.7 \\ 0.033 & \text{for} \quad \text{Pr} = \infty \end{cases} \tag{3.70}$$

Uniform Wall Heat Flux, Ⓗ. The simultaneously developing flow in a circular duct with the Ⓗ boundary condition has been extensively analyzed in the literature. The most accurate results are presented in Fig. 3.8, and tabulated values for Pr = 0.7, 2, 5, and ∞ are available in [1]. The results for Pr = 0 are obtained from the following set of exact equations derived from the transient temperature distribution of the analogous heat conduction problem [32]:

$$\Theta = \frac{T - T_e}{q''_w D_h / k} = 4x^* + \frac{1}{4}\left(\frac{r}{a}\right)^2 - \frac{1}{8} - \sum_{n=1}^{\infty} \frac{\exp(-4\lambda_n^2 x^*)\, J_0[\lambda_n (r/a)]}{\lambda_n^2 J_0(\lambda_n)} \tag{3.71}$$

$$\Theta_m = \frac{T_m - T_e}{q''_w D_h / k} = 4x^* \tag{3.72}$$

$$\text{Nu}_{x,\text{H}} = \left(\frac{1}{8} - \sum_{n=1}^{\infty} \frac{\exp(-4\lambda_n^2 x^*)}{\lambda_n^2} \right)^{-1} \tag{3.73}$$

$$\text{Nu}_{m,\text{H}} = \left(\frac{1}{8} - \sum_{n=1}^{\infty} \frac{1 - \exp(-4\lambda_n^2 x^*)}{4\lambda_n^2 x^*} \right)^{-1} \tag{3.74}$$

Here λ_n are the eigenvalues, being the zeros of $J_1(\lambda) = 0$, where $J_1(\lambda)$ is Bessel function of the first kind and first order. The first 20 eigenvalues are listed in Table 3.8.

The local Nusselt numbers computed from Eq. (3.73) are displayed in Fig. 3.8, corresponding to Pr = 0. The curve corresponding to Pr = ∞ in Fig. 3.8 is identical with the $\text{Nu}_{x,\text{H}}$ curve of Fig. 3.3. Also note that the curve corresponding to Pr = 0 approaches the asymptotic value of 8, whereas remainder of the curves approach the asymptotic value of 4.3636.

The thermal entrance lengths for the simultaneously developing Ⓗ problem are given by [1, 32]

$$L^*_{\text{th, H}} = \begin{cases} 0.042 & \text{for} \quad \text{Pr} = 0 \\ 0.053 & \text{for} \quad \text{Pr} = 0.7 \\ 0.043 & \text{for} \quad \text{Pr} = \infty \end{cases} \tag{3.75}$$

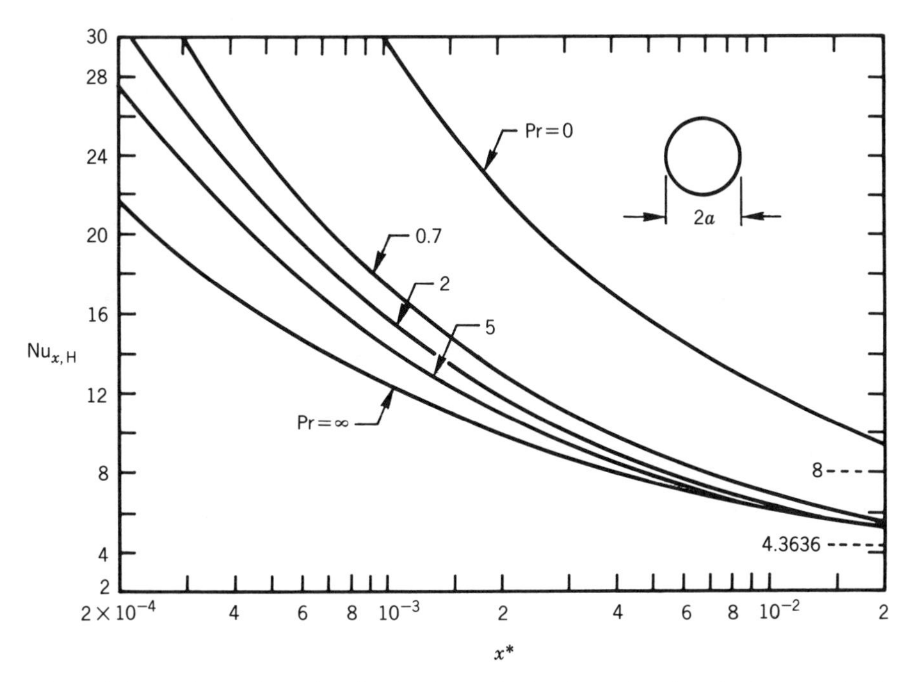

Figure 3.8. Local Nusselt numbers for simultaneously developing flow in a circular duct with the Ⓗ boundary condition [1, 32].

TABLE 3.8 Eigenvalues for the Ⓗ Problem Solution with Pr = 0 (Slug Flow) in a Circular Duct[a]

n	λ_n	n	λ_n
1	3.83170 59702	11	35.33230 75501
2	7.01558 66698	12	38.47476 62348
3	10.17346 81351	13	41.61709 42128
4	13.32369 19363	14	44.75931 89977
5	16.47063 00509	15	47.90146 08872
6	19.61585 85105	16	51.04353 51836
7	22.76008 43806	17	54.18555 36411
8	25.90367 20876	18	57.32752 54379
9	29.04682 85340	19	60.46945 78453
10	32.18967 99110	20	63.61135 66985

[a] The higher eigenvalues of the problem can be computed to a high degree of accuracy from the asymptotic formula

$$\lambda_n = \frac{\pi}{4}(4n+1) - \frac{3}{2\pi(4n+1)} + \frac{3}{2\pi^3(4n+1)^3} - \frac{1179}{5\pi^5(4n+1)^5} + \cdots$$

Convective Boundary Condition, Ⓣ3**.** The simultaneously developing flow with the convective boundary condition Ⓣ3 has been analyzed in great detail by Javeri [33]. His $\mathrm{Nu}_{x,\mathrm{T3}}$ prediction for a full range of Pr and $4 \leq \mathrm{Bi} \leq 2000$ are available in [1]. For $\mathrm{Bi} = \infty$, the $\mathrm{Nu}_{x,\mathrm{T3}}$ are identical to $\mathrm{Nu}_{x,\mathrm{T}}$ of Fig. 3.6, and for $\mathrm{Bi} = 0$ they are identical to $\mathrm{Nu}_{x,\mathrm{H}}$ of Fig. 3.8.

3.3 FLAT DUCT

Due to its simplicity, the problem of the flat duct (i.e., parallel plate channel), has attracted more investigation than perhaps is warranted by its practical utility. However, since this duct is the limiting geometry for the family of rectangular and concentric annular ducts, the results presented here are of considerable interest.

3.3.1 Fully Developed Flow

The fully developed velocity distribution and Fanning friction factor for a flat duct with hydraulic diameter $D_h = 4b$ (b being the half spacing between the plates) and origin at the duct axis are given by

$$\frac{u}{u_m} = \frac{3}{2}\left[1 - \left(\frac{y}{b}\right)^2\right] \tag{3.76}$$

$$u_m = -\frac{1}{3\mu}\left(\frac{dp}{dx}\right)b^2, \qquad f\,\mathrm{Re} = 24 \tag{3.77}$$

The heat transfer results for various thermal boundary conditions are presented next.

Uniform Wall Temperature and Wall Heat Flux, Ⓣ ***and*** Ⓗ**.** The two walls of the flat duct present the choice of imposing different boundary conditions on each wall, resulting in numerous combinations of the Ⓣ and Ⓗ boundary conditions. It can be shown that once the solutions to the following four fundamental boundary conditions (see Fig. 3.9) are obtained, then by simple superposition any combination of the Ⓣ and Ⓗ boundary conditions can be handled:

1. *Fundamental boundary condition of the first kind:* specification of the uniform temperature (different from the entering fluid temperature) at one wall with the other wall at the uniform entering fluid temperature.
2. *Fundamental boundary condition of the second kind:* specification of the uniform wall heat flux at one wall with the other wall insulated (i.e., adiabatic with zero heat flux).
3. *Fundamental boundary condition of the third kind:* specification of the uniform temperature (different from the entering fluid temperature) at one wall with the other wall insulated.
4. *Fundamental boundary condition of the fourth kind:* specification of the uniform wall heat flux at one wall with the other wall maintained at the entering fluid temperature.

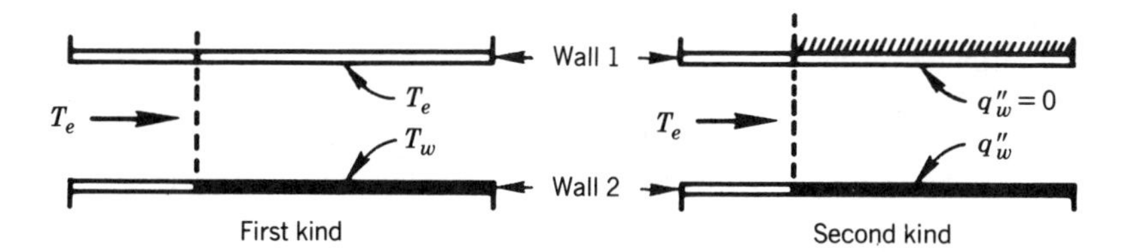

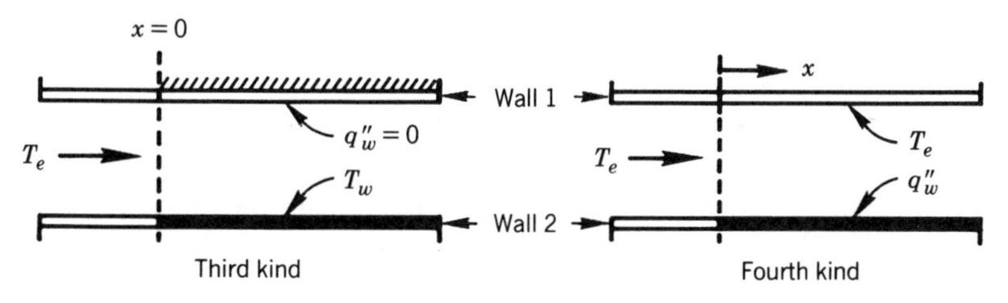

Figure 3.9. Four kinds of fundamental boundary conditions for a flat duct.

With the subscripts 1 and 2 referring to the walls labeled 1 and 2 in Fig. 3.9, the fully developed Nusselt numbers for the four solutions are [1]

First kind: $$\mathrm{Nu}_1 = \mathrm{Nu}_2 = 4 \tag{3.78}$$

Second kind: $$\mathrm{Nu}_1 = 0, \qquad \mathrm{Nu}_2 = 5.385 \tag{3.79}$$

Third kind: $$\mathrm{Nu}_1 = 0, \qquad \mathrm{Nu}_2 = 4.861 \tag{3.80}$$

Fourth kind: $$\mathrm{Nu}_1 = \mathrm{Nu}_2 = 4 \tag{3.81}$$

Of the numerous combinations of the Ⓣ and Ⓗ boundary conditions, the three combinations illustrated in Fig. 3.10 are of particular interest. The Nusselt numbers for these special cases are outlined below. They are defined by

$$\mathrm{Nu}_j = \frac{q''_{wj} D_h}{k(T_j - T_m)} \tag{3.82}$$

where j denotes wall 1 or 2 and T_j is the temperature of the jth wall.

Uniform Temperature at Each Wall. When $T_{w1} = T_{w2}$, we have $\mathrm{Nu}_1 = \mathrm{Nu}_2 = \mathrm{Nu}_\mathrm{T}$; the value of Nu_T is given by

$$\mathrm{Nu}_\mathrm{T} = 7.54070087 \tag{3.83}$$

When $T_{w1} \neq T_{w2}$, we have $\mathrm{Nu}_1 = \mathrm{Nu}_2 = 4$ as in Eq. (3.78).

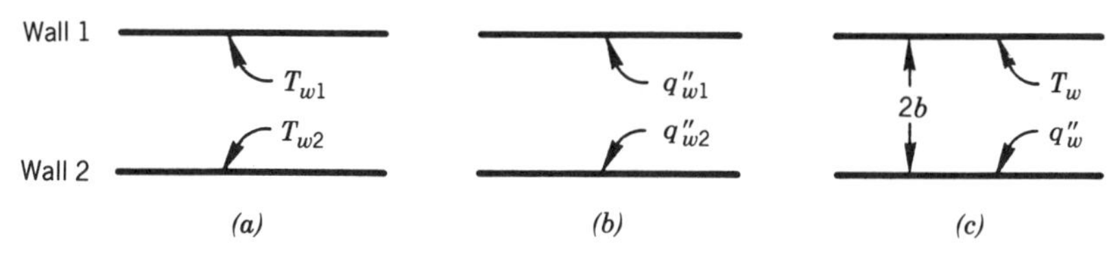

Figure 3.10. Thermal boundary conditions for three flat duct problems.

Cheng and Wu [34] considered the effect of viscous dissipation and developed the following formulas for the case $T_{w1} > T_{w2}$.

$$\mathrm{Nu}_1 = \frac{4(1 - 6\,\mathrm{Br})}{1 - \frac{48}{35}\mathrm{Br}}, \qquad \mathrm{Nu}_2 = \frac{4(1 + 6\,\mathrm{Br})}{1 + \frac{48}{35}\mathrm{Br}} \tag{3.84}$$

Pahor and Strand [35] and Grosjean et al. [36] considered the effect of fluid axial conduction for the case $T_{w1} = T_{w2}$ and presented the following asymptotic formulas:

$$\mathrm{Nu_T} = \begin{cases} 7.540(1 + 3.79/\mathrm{Pe}^2 + \cdots) & \text{for} \quad \mathrm{Pe} \gg 1 \\ 8.11742(1 - 0.030859\,\mathrm{Pe} + 0.0069436\,\mathrm{Pe}^2 - \cdots) & \text{for} \quad \mathrm{Pe} \ll 1 \end{cases} \tag{3.85}$$

When the effects of viscous dissipation and flow work are considered together for the case of $T_{w1} = T_{w2}$, Ou and Cheng [37] showed that for the fully developed flow $\mathrm{Nu_T} = 0$ and

$$\frac{T_w - T}{T_w - T_e} = \frac{9}{8}\mathrm{Br}\left[1 - \left(\frac{y}{b}\right)^2\right]^2, \qquad \frac{T_w - T_m}{T_w - T_e} = \frac{27}{35}\mathrm{Br} \tag{3.86}$$

Uniform Heat Flux at Each Wall. The fully developed temperature distribution for the case $q''_{w1} = q''_{w2}$, i.e., the Ⓗ problem, is given by

$$\frac{T_w - T_\mathrm{H}}{T_w - T_{m,\mathrm{H}}} = \frac{35}{136}\left[5 - 6\left(\frac{y}{b}\right)^2 + \left(\frac{y}{b}\right)^4\right], \qquad \frac{T_w - T_{m,\mathrm{H}}}{q''_w D_h/k} = \frac{17}{140} \tag{3.87}$$

When $q''_{w1} = q''_{w2}$, we have $\mathrm{Nu}_1 = \mathrm{Nu}_2 = \mathrm{Nu_H}$; the value of $\mathrm{Nu_H}$ is given by

$$\mathrm{Nu_H} = \frac{140}{17} = 8.2352941 \tag{3.88}$$

When $q''_{w1} \neq q''_{w2}$,

$$\mathrm{Nu}_1 = \frac{140}{26 - 9(q''_{w2}/q''_{w1})}, \qquad \mathrm{Nu}_2 = \frac{140}{26 - 9(q''_{w1}/q''_{w2})} \tag{3.89}$$

Tao [38] considered the effect of internal thermal energy generation on the Ⓗ problem, and Tyagi [5] extended his analysis by incorporating the effect of viscous dissipation. Based on Tyagi's analysis [5], the Nusselt number in terms of S^* and Br' is given by

$$\mathrm{Nu_H} = \frac{140}{17 + \frac{3}{4}S^* + 108\,\mathrm{Br}'} \tag{3.90}$$

Uniform Temperature at One Wall and Heat Flux at the Other. For the case of Fig. 3.10c,

$$\mathrm{Nu}_1 = 4.8608125, \quad \mathrm{Nu}_2 = 0 \qquad \text{for} \quad q''_w = 0 \tag{3.91}$$

$$\mathrm{Nu}_1 = \mathrm{Nu}_2 = 4 \qquad \text{for} \quad q''_w \neq 0 \tag{3.92}$$

Convective Boundary Condition, Ⓣ3. Based on the analysis by Hickman [7], the fully developed Nusselt number with the convective boundary condition at both the walls can be computed from

$$\mathrm{Nu_{T3}} = \frac{4620 + 561\,\mathrm{Bi}}{561 + 74\,\mathrm{Bi}} \tag{3.93}$$

Exponential Wall Heat-Flux Boundary Condition, Ⓗ5. The fully developed Nusselt numbers $\mathrm{Nu_{H5}}$ for the exponential heat flux of $q''_w = q''_0 \exp(mx^*)$ imposed on both the duct walls can be computed from the following equation developed by the present authors utilizing the Ⓗ5 thermal entrance length solution for the flat duct:

$$\mathrm{Nu_{H5}} = 8.2400 + 2.1611 \times 10^{-3} m - 4.4397 \times 10^{-5} m^2$$
$$+1.2856 \times 10^{-7} m^3 - 2.7035 \times 10^{-10} m^4 \tag{3.94}^{\dagger}$$

Equation (3.94) reproduces the tabulated results in [1] for $-80 \le m \le 100$ with a maximum deviation of 6%. Note that with $m = -30.16$, the Nusselt number $\mathrm{Nu_T}$ of Eq. (3.83) is obtained [1]. Likewise with $m = 0$, the Nusselt number $\mathrm{Nu_H}$ of Eq. (3.88) is recovered.

3.3.2 Hydrodynamically Developing Flow

The problem of laminar flow development in a flat duct has been solved in considerable detail by employing both boundary-layer theory idealizations and a complete set of Navier-Stokes equations. In addition, some solutions have also been developed for creeping flow. All the solutions up to 1975 are summarized in [1].

Among the boundary-layer type of solutions, the numerical solution by Bodoia and Osterle [39] is regarded as the most accurate. The dimensionless axial velocity and pressure drop computed by them are presented in Table 3.9. According to these results

$$L^+_{\mathrm{hy}} = 0.0110, \qquad K(\infty) = 0.6760 \tag{3.95}$$

Bhatti and Savery [40] provided a closed-form analytical solution to the hydrodynamic entrance length problem for the flat duct. The solution is quite accurate for engineering calculations. According to this analysis, the flow field is idealized as consisting of a viscous boundary layer along the duct wall $1 - \delta/b \le y/b \le 1$ and an essentially inviscid fluid core around the duct axis $0 \le y/b \le 1 - \delta/b$, where δ is the hydrodynamic boundary-layer thickness. The velocity components within the fluid core are given by

$$\frac{u}{u_m} = U(x) \tag{3.96}$$

$$\frac{v}{u_m} = \frac{280U^2(3 - 2U)}{\mathrm{Re}(U-1)^2(513 - 297U)}\left(\frac{y}{b}\right) \tag{3.97}$$

† Note that $\mathrm{Nu_{H5}} = 0$ for $m < -196.06$.

TABLE 3.9 Axial Velocity and Pressure Distribution in the Hydrodynamic Entrance Region of a Flat Duct [39]

	Axial Velocity u/u_m											Pressure
10^3x^+	$y/b = 0$	0.1	0.2	0.3	0.4	0.5	0.6	0.7	0.8	0.9	1.0	Drop Δp^*
0.0625	1.0615	1.0615	1.0615	1.0615	1.0615	1.0615	1.0615	1.0612	1.0551	0.9587	0	0.12420
0.125	1.0751	1.0751	1.0751	1.0751	1.0751	1.0751	1.0750	1.0725	1.0485	0.8655	0	0.15328
0.250	1.1013	1.1013	1.1013	1.1013	1.1012	1.1010	1.0993	1.0863	1.0132	0.7194	0	0.21006
0.375	1.1244	1.1244	1.1244	1.1244	1.1243	1.1234	1.1176	1.0874	0.9665	0.6204	0	0.26150
0.500	1.1443	1.1443	1.1443	1.1442	1.1438	1.1414	1.1290	1.0788	0.9204	0.5567	0	0.30648
0.625	1.1615	1.1615	1.1615	1.1613	1.1604	1.1555	1.1351	1.0655	0.8798	0.5136	0	0.34612
0.750	1.1767	1.1767	1.1766	1.1763	1.1745	1.1665	1.1373	1.0501	0.8455	0.4832	0	0.38164
1.000	1.2031	1.2030	1.2028	1.2017	1.1972	1.1813	1.1339	1.0185	0.7922	0.4427	0	0.44436
1.250	1.2259	1.2258	1.2252	1.2228	1.2144	1.1893	1.1253	0.9896	0.7535	0.4162	0	0.49984
1.500	1.2463	1.2460	1.2448	1.2406	1.2275	1.1928	1.1144	0.9644	0.7241	0.3971	0	0.55044
1.750	1.2648	1.2643	1.2623	1.2556	1.2373	1.1935	1.1030	0.9429	0.7011	0.3825	0	0.59746
2.000	1.2818	1.2811	1.2778	1.2684	1.2447	1.1923	1.0918	0.9246	0.6825	0.3708	0	0.64170
2.500	1.3121	1.3105	1.3043	1.2887	1.2542	1.1871	1.0715	0.8950	0.6541	0.3534	0	0.72384
3.125	1.3441	1.3412	1.3306	1.3067	1.2601	1.1786	1.0504	0.8677	0.6291	0.3383	0	0.81784
3.750	1.3707	1.3663	1.3511	1.3195	1.2626	1.1703	1.0337	0.8475	0.6112	0.3275	0	0.90498
5.000	1.4111	1.4039	1.3803	1.3357	1.2635	1.1565	1.0095	0.8917	0.5870	0.3132	0	1.06516
6.250	1.4388	1.4292	1.3993	1.3454	1.2628	1.1467	0.9938	0.8022	0.5720	0.3042	0	1.21262
9.375	1.4758	1.4629	1.4239	1.3573	1.2611	1.1336	0.9733	0.7796	0.5526	0.2926	0	1.55014
12.5	1.4903	1.4762	1.4336	1.3619	1.2604	1.1284	0.9653	0.7708	0.5451	0.2880	0	1.86538
62.5	1.4999	1.4850	1.4400	1.3640	1.2600	1.1249	0.9599	0.7650	0.5400	0.2500	0	6.67604
∞	1.5000	1.4850	1.4400	1.3650	1.2600	1.1250	0.9600	0.7650	0.5400	0.2850	0	∞

and those within the boundary layer by

$$\frac{u}{u_m} = \frac{U^3}{9(U-1)^2}\left[1 - \left(\frac{y}{b}\right)\right]\left[2\left(\frac{\delta}{b}\right) + \left(\frac{y}{b}\right) - 1\right] \tag{3.98}$$

$$\frac{v}{u_m} = \frac{280U^3(3-2U)}{9\,\mathrm{Re}\,(U-1)^4(513-297U)}$$

$$\times\left\{\frac{U(3-U)}{(U-1)}\left[\left(\frac{\delta}{b}\right) - \frac{1}{3}\left(1 - \frac{y}{b}\right)\right] - 3\right\}\left[1 - \left(\frac{y}{b}\right)\right]^2 \tag{3.99}$$

where U is implicitly expressed as a function of x^+ by

$$x^+ = \frac{594U^2 + 90U - 684 - 15U\ln(3-2U) - 1308U\ln U}{4480U} \tag{3.100}$$

The apparent friction factor, local pressure drop, and boundary-layer thickness are also expressible in terms of U, which in essence is treated as the independent variable:

$$f_{\mathrm{app}}\,\mathrm{Re} = \frac{\Delta p^*}{4x^+}, \qquad \Delta p^* = \tfrac{3}{140}\left[22U^2 - 10U - 12 - 15\ln(3-2U)\right] \tag{3.101}$$

$$\frac{\delta}{b} = \frac{3(U-1)}{U} \tag{3.102}$$

In the limit when the flow becomes hydrodynamically developed, U attains the value of $\frac{3}{2}$ and all the momentum transfer quantities asymptotically reduce to the known exact fully developed values.

The following correlation [1] predicts the apparent friction factors in excellent accord with the aforementioned numerical and analytical results:

$$f_{\mathrm{app}}\mathrm{Re} = \frac{3.44}{(x^+)^{1/2}} + \frac{24 + 0.674/(4x^+) - 3.44/(x^+)^{1/2}}{1 + 0.000029(x^+)^{-2}} \tag{3.103}$$

Chen [12] developed the following equation for $K(\infty)$ as a function of Re:

$$K(\infty) = 0.64 + \frac{38}{\mathrm{Re}} \tag{3.104}$$

Chen [12] also presented the following equation for the hydrodynamic entrance length:

$$\frac{L_{\mathrm{hy}}}{D_h} = 0.011\,\mathrm{Re} + \frac{0.315}{1 + 0.0175\,\mathrm{Re}} \tag{3.105}$$

3.3.3 Thermally Developing Flow

The thermal entrance solutions for a flat duct are presented for four types of thermal boundary conditions.

TABLE 3.10 Eigenvalues and Constants of the Flat Duct Thermal Entrance Length Problem with the Ⓣ Boundary Condition [17]

n	λ_n	C_n	G_n
0	1.68159 53222	1.20083 0379	0.85808 6674
1	5.66985 73459	−0.29916 0685	0.56946 2850
2	9.66842 24625	0.16082 6463	0.47606 5463
3	13.66766 14426	−0.10743 6641	0.42397 3730
4	17.66737 35653	0.07964 6080	0.38910 8706
5	21.66720 53243	−0.06277 5656	0.36346 5044
6	25.66709 64863	0.05151 9218	0.34347 5506
7	29.66702 10447	−0.04351 0736	0.32726 5745
8	33.66696 60687	0.03754 1808	0.31373 9318
9	37.66692 44563	−0.03293 3278	0.30220 4200

Specified Wall Temperature Distribution. The solution with equal and uniform temperatures at both duct walls is of special importance, as it constitutes the limiting case of rectangular and concentric annular ducts with the Ⓣ boundary condition. This solution was first developed by Nusselt [41] in 1923 and is expressible in terms of the following set of equations:

$$\theta = \frac{T_w - T}{T_w - T_e} = \sum_{n=0}^{\infty} C_n Y_n\left(\frac{y}{b}\right) \exp\left(-\tfrac{32}{3}\lambda_n^2 x^*\right) \tag{3.106}$$

$$\theta_m = \frac{T_w - T_m}{T_w - T_e} = 3 \sum_{n=0}^{\infty} \frac{G_n}{\lambda_n^2} \exp\left(-\tfrac{32}{3}\lambda_n^2 x^*\right) \tag{3.107}$$

$$\mathrm{Nu}_{x,\mathrm{T}} = \frac{8}{3} \frac{\sum_{n=0}^{\infty} G_n \exp\left(-\tfrac{32}{3}\lambda_n^2 x^*\right)}{\sum_{n=0}^{\infty} (G_n/\lambda_n^2) \exp\left(-\tfrac{32}{3}\lambda_n^2 x^*\right)} \tag{3.108}$$

$$\mathrm{Nu}_{m,\mathrm{T}} = -\frac{\ln \theta_m}{4x^*} \tag{3.109}$$

Here λ_n, $Y_n(y/b)$, and C_n are the eigenvalues, eigenfunctions, and constants, respectively, and $G_n = -(C_n/2)Y_n'(1)$, where $Y_n'(1)$ is the derivative of $Y_n(y/b)$ evaluated at $y/b = 1$. λ_n, C_n, and G_n values are listed in Table 3.10, and $Y_n(y/b)$ values in Table 3.11. Table 3.10 lists the first ten eigenvalues and constants. The higher values can be computed from the following formulas due to Sellars et al. [42]:

$$\lambda_n = 4n + \tfrac{5}{3} \tag{3.110}$$

$$C_n = (-1)^n 2.271141411 \lambda_n^{-7/6} \tag{3.111}$$

$$G_n = 1.012787291 \lambda_n^{-1/3} \tag{3.112}$$

TABLE 3.11 Eigenfunctions $Y_n(y/b)$ of the Flat Duct Thermal Entrance Length Problem with the Ⓣ Boundary Condition at Both Walls [17, 18][a]

	$Y_n(y/b)$								
n	$y/b = 0.1$	0.2	0.3	0.4	0.5	0.6	0.7	0.8	0.9
0	0.9859	0.9443	0.8772	0.7876	0.6793	0.5566	0.4238	0.2848	0.1429
1	0.8438	0.4262	−0.1205	−0.6345	−0.9832	−1.1013	−0.9973	−0.7311	−0.3787
2	0.5685	−0.3513	−0.9843	−0.8414	−0.0750	0.7540	1.1669	1.0499	0.5831
3	0.2036	−0.9213	−0.6343	0.6016	1.0350	0.1743	−0.9142	−1.2329	−0.7656
4	−0.1937	−0.9414	0.5015	0.8632	−0.6121	−0.9713	0.3409	1.2703	0.9285
5	−0.5606	−0.3986	1.0161	−0.5737	−0.6272	1.0360	0.3505	−1.1652	−1.0720
6	−0.8396	0.3828	0.2708	−0.8814	1.0315	0.3214	−0.9217	0.9305	1.1956
7	−0.9865	0.9352	−0.8106	0.5467	−0.0632	−0.6347	1.1775	−0.5942	−1.2988
8	−0.9783	0.9279	−0.8863	0.8980	−0.9893	1.1146	−1.0311	0.1934	1.3808
9	−0.8162	0.3651	0.1377	−0.5198	0.7263	−0.7588	0.5331	0.2283	−1.4411
10	−0.5257	−0.4163	0.9909	−0.9136	0.5023	−0.1659	0.1465	−0.6254	1.4792
11	−0.1526	−0.9485	0.6144	0.4927	−1.0632	0.9663	−0.7763	0.9556	−1.4951
12	0.2446	−0.9127	−0.5246	0.9282	0.2110	1.0428	1.1420	−1.1835	1.4888
13	0.6034	−0.3304	−1.0125	0.4653	0.9217	0.3382	−1.1194	1.2849	−1.4607
14	0.8672	0.4497	−0.2437	−0.9419	−0.8295	0.6197	0.7162	−1.2492	1.4113

[a] For all values of n, $Y_n(0) = 1$ and $Y_n(1) = 0$.

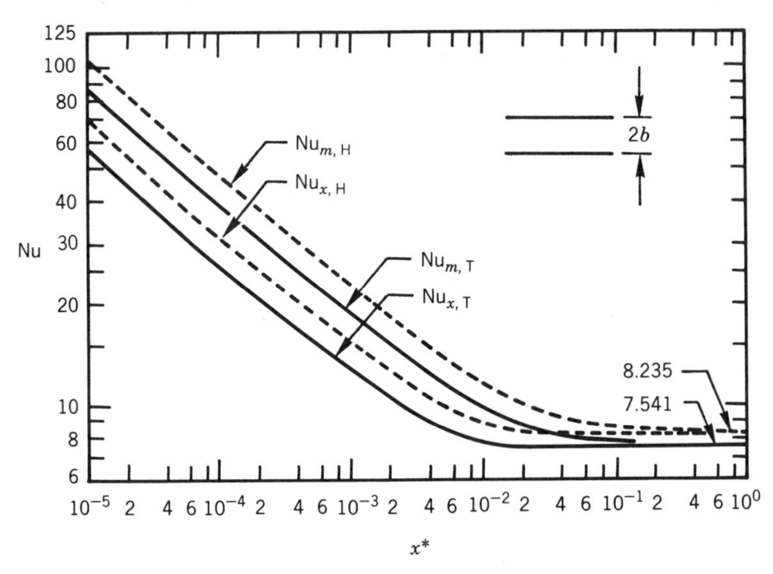

Figure 3.11. Local and mean Nusselt numbers in the thermal entrance region of a flat duct with the Ⓣ and Ⓗ boundary conditions [1].

The local and mean Nusselt numbers computed from the foregoing equations are displayed in Fig. 3.11. Tabulated values for Fig. 3.11 are available in [1]. From these results, the dimensionless thermal entrance length is found to be

$$L^*_{\text{th, T}} = 0.00797350 \tag{3.113}$$

Following Lévêque's analysis [21] for a circular duct, the asymptotic solution for a flat duct is found to be

$$\theta = \frac{T_w - T}{T_w - T_e} = \frac{1}{\Gamma(\frac{4}{3})}\left[\int_0^{\chi} e^{-\chi^3}\, d\chi\right] \tag{3.114}$$

where

$$\chi = \frac{1 - (y/b)}{2(6x^*)^{1/3}} \tag{3.115}$$

$$\text{Nu}_{x,\text{T}} = \frac{2}{\Gamma(\frac{4}{3})(6x^*)^{1/3}} = \tfrac{2}{3}\text{Nu}_{m,\text{T}} \tag{3.116}$$

Strictly speaking, this solution is applicable for very low values of x^*. However, the local and mean Nusselt numbers computed from Eq. (3.116) are found to be fairly accurate down to $x^* = 10^{-3}$, being up to 4% higher than the results in Fig. 3.11.

The following set of empirical equations is found to yield results within $\pm 3\%$ of those obtained from the foregoing theoretical results [1]:

$$\mathrm{Nu}_{x,\mathrm{T}} = \begin{cases} 1.233x^{*-1/3} + 0.4 & \text{for} \quad x^* < 0.001 \\ 7.541 + 6.874(10^3 x^*)^{-0.488} e^{-245x^*} & \text{for} \quad x^* > 0.001 \end{cases} \tag{3.117}$$

$$\mathrm{Nu}_{m,\mathrm{T}} = \begin{cases} 1.849x^{*-1/3} & \text{for} \quad x^* \le 0.0005 \\ 1.849x^{*-1/3} + 0.6 & \text{for} \quad 0.0005 < x^* \le 0.006 \\ 7.541 + 0.0235/x^* & \text{for} \quad x^* > 0.006 \end{cases} \tag{3.118}$$

The effect of fluid axial conduction on the flat duct solution with uniform and equal temperatures at both duct walls has been investigated for two types of initial conditions: uniform fluid temperature at the duct inlet $x = 0$, and uniform fluid temperature at $x = -\infty$ with the duct walls in the region $-\infty < x \le 0$ maintained at the entering fluid temperature [1]. The results of these analyses show that except in the neighborhood of the duct inlet ($x^* < 10^{-2}$), the effect of fluid axial conduction is negligible for $\mathrm{Pe} > 50$.

The effect of viscous dissipation and flow work on the flat duct solution with uniform and equal temperatures at both walls has been investigated by Ou and Cheng [37]. The peculiar behavior of their Nusselt numbers is explained in [1].

The effects of internal heat generation, fluid axial conduction, variation in inlet fluid temperature, sinusoidal variation of temperature along the duct axis, and internal thermal energy sources on the flat duct solution with equal temperatures at both duct walls has been investigated quite extensively, as reviewed in [1].

Specified Wall Heat-Flux Distribution. The thermal entrance length problem for a flat duct with uniform and equal heat fluxes at both walls was first investigated by Cess and Shaffer [43] and later in more detail by Sparrow et al. [44]. Their solution is expressed by the following set of equations:

$$\Theta = \frac{T - T_e}{q''_w D_h/k} = 4x^* + \frac{3}{16}\left(\frac{y}{b}\right)^2 - \frac{1}{32}\left(\frac{y}{b}\right)^4 - \frac{39}{1120} + \frac{1}{4}\sum_{n=1}^{\infty} C_n Y_n\left(\frac{y}{b}\right)\exp\left(-\tfrac{32}{3}\beta_n^2 x^*\right) \tag{3.119}$$

$$\Theta_m = \frac{T_m - T_e}{q''_w D_h/k} = 4x^* \tag{3.120}$$

$$\mathrm{Nu}_{x,\mathrm{H}} = \left[\frac{17}{140} + \frac{1}{4}\sum_{n=1}^{\infty} C_n Y_n(1)\exp\left(-\tfrac{32}{3}\beta_n^2 x^*\right)\right]^{-1} \tag{3.121}$$

$$\mathrm{Nu}_{m,\mathrm{H}} = \left[\frac{17}{140} + \frac{1}{4}\sum_{n=1}^{\infty} C_n Y_n(1)\frac{1 - \exp\left(-\frac{32}{3}\beta_n^2 x^*\right)}{\frac{32}{3}\beta_n^2 x^*}\right]^{-1} \tag{3.122}$$

Here β_n, $Y_n(y/b)$, and C_n are eigenvalues, eigenfunctions, and constants, respectively. The tabulated values of $Y_n(y/b)$ are not reported either in [43] or in [44]. However, β_n, $Y_n(1)$, and C_n values are tabulated in [43] for n up to 3 and in [44] for n up to 10. They

TABLE 3.12 Eigenvalues and Constants of the Flat Duct Thermal Entrance Length Problem with the Ⓗ Boundary Condition [43, 44]

n	β_n	$-C_n Y_n(1)$
1	4.287224	0.2222280
2	8.30372	0.0725316
3	12.3106	0.0373691
4	16.3145	0.0232829
5	20.3171	0.0161112
6	24.3189	0.0119190
7	28.3203	0.0092342
8	32.3214	0.0074013
9	36.3223	0.0060881
10	40.3231	0.0051116

are reproduced in Table 3.12. For higher values of n, Cess and Shaffer [43] provided the following asymptotic formulas:

$$\beta_n = 4n + \tfrac{1}{3} \tag{3.123}$$

$$C_n Y_n(1) = -2.401006045\beta_n^{-5/3} \tag{3.124}$$

Following Lévêque's analysis for the circular duct, the asymptotic solution for the flat duct is found to be

$$\frac{T - T_e}{q_w'' D_h/k} = \frac{(6x^*)^{1/3}}{2}\left[\frac{e^{-\chi^3}}{\Gamma(\frac{2}{3})} - \chi\left(1 - \frac{\Gamma(\frac{2}{3}, \chi^3)}{\Gamma(\frac{2}{3})}\right)\right] \tag{3.125}$$

$$\mathrm{Nu}_{x,\mathrm{H}} = \frac{2\Gamma(\frac{2}{3})}{(6x^*)^{1/3}} = \tfrac{2}{3}\mathrm{Nu}_{m,\mathrm{H}} \tag{3.126}$$

Here χ is defined by Eq. (3.115) and $\Gamma(\frac{2}{3}) = 1.3541179394$. Strictly speaking, Eq. (3.126) is applicable only for very low values of x^*. However, it is found to yield fairly accurate results down to $x^* = 10^{-3}$, being up to 4% higher than the exact results shown in Fig. 3.11 [1]. The dimensionless thermal entrance length for the problem is found to be [1]

$$L^*_{\mathrm{th,H}} = 0.0115439 \tag{3.127}$$

The following computationally expedient empirical equations are found to yield accurate results to within $\pm 3\%$ of the analytical results [1]:

$$\mathrm{Nu}_{x,\mathrm{H}} = \begin{cases} 1.490x^{*-1/3} & \text{for } x^* \le 0.0002 \\ 1.490x^{*-1/3} - 0.4 & \text{for } 0.0002 < x^* \le 0.001 \\ 8.235 + 8.68(10^3 x^*)^{-0.506} e^{-164x^*} & \text{for } x^* > 0.001 \end{cases} \tag{3.128}$$

$$\mathrm{Nu}_{m,\mathrm{H}} = \begin{cases} 2.236x^{*-1/3} & \text{for } x^* \le 0.001 \\ 2.236x^{*-1/3} + 0.9 & \text{for } 0.001 < x^* \le 0.01 \\ 8.235 + 0.0364/x^* & \text{for } x^* \ge 0.01 \end{cases} \tag{3.129}$$

The effect of fluid axial conduction for this problem was investigated by Jones [45] for the initial condition of uniform fluid temperature at $x = -\infty$ with the duct walls in the extended region $-\infty < x \leq 0$ maintained at the entering fluid temperature. Hsu [46] investigated the effect of fluid axial conduction for the initial condition of uniform fluid temperature at $x = -\infty$ with the duct walls in the extended region $-\infty < x \leq 0$ insulated. The results show that except in the neighborhood of the duct inlet ($x^* < 10^{-2}$), the effect of fluid axial conduction is negligible for Pe > 45.

Specified Convective Boundary Condition, Ⓣ3. The problem of heat transfer with the convective boundary condition Ⓣ3 at both duct walls in the absence of fluid axial conduction, flow work, and internal heat sources has been investigated extensively [1]. In addition, the problem with one wall insulated (Bi = 0) and the other subjected to convective heating or cooling has been solved [1].

Expotential Axial Wall Heat Flux, Ⓗ5. The local Nusselt numbers with the Ⓗ5 boundary condition at both duct walls can be determined from

$$\mathrm{Nu}_{x,\mathrm{H5}} = \frac{3}{8}\left(\sum_{n=1}^{\infty} \frac{-C_n Y_n(1)\beta_n^2}{\frac{32}{3}\beta_n^2 + m} \left\{1 - \exp\left[-\left(\tfrac{32}{3}\beta_n^2 + m\right)x^*\right]\right\}\right)^{-1} \tag{3.130}$$

where the constants β_n and $-C_n Y_n(1)$ are available in Table 3.12 for n up to 10, and those for $n > 10$ can be determined from Eqs. (3.123) and (3.124). The constant m is the coefficient in $q''_w = q''_0 \exp(mx^*)$.

All the solutions presented in this section so far ignore heat conduction in the solid duct walls. When heat conduction in the solid walls is taken into account concurrently with heat transfer in the fluid, the problem is referred to as the *conjugate problem*. For flat duct, the conjugate problem has been investigated quite extensively, as reviewed in [47].

3.3.4 Simultaneously Developing Flow

For the simultaneously developing flow in a flat duct, the results are available for (1) uniform temperature at one or both walls, (2) uniform heat flux at one or both walls, and (3) convective boundary conditions at both walls. These results are described in the following text.

Prescribed Uniform Wall Temperature. The problem of simultaneously developing flow with equal and uniform temperatures at both walls, i.e., the Ⓣ problem, has been studied by a number of investigators [1]. The most accurate results are those of Hwang and Fan [48], who performed an all-numerical analysis. They presented the mean Nusselt numbers $\mathrm{Nu}_{m,\mathrm{T}}$ for Pr = 0.1, 0.72, 10, 50. Their numerical results are well represented by the following correlation (due to Stephan [49]) valid for 0.1 < Pr < 1000:

$$\mathrm{Nu}_{m,\mathrm{T}} = 7.55 + \frac{0.024 x^{*-1.14}}{1 + 0.0358\,\mathrm{Pr}^{0.17} x^{*-0.64}} \tag{3.131}$$

The maximum deviation of Hwang and Fan's results from the predictions of Eq. (3.131) is 3%. Because of its simplicity and excellent agreement with the numerical results, the use of Eq. (3.131) is recommended for engineering calculations. Neither Hwang and Fan [48] nor Stephan [49] provided a simple formula for the local Nusselt numbers $Nu_{x,T}$. By differentiating Eq. (3.131) with respect to x^*, the present authors developed the following formula for the local Nusselt numbers applicable in the Prandtl number range of $0.1 < Pr < 1000$:

$$Nu_{x,T} = 7.55 + \frac{0.024x^{*-1.14}[0.0179\,Pr^{0.17}x^{*-0.64} - 0.14]}{[1 + 0.0358\,Pr^{0.17}x^{*-0.64}]^2} \tag{3.132}$$

For $Pr = \infty$, the results of Eqs. (3.108) and (3.109) for the fully developed velocity profile and developing temperature profile are applicable. For $Pr = 0$, the velocity profile never develops, i.e., the flow remains slug flow while the temperature profile develops. The exact solution for $Pr = 0$ was developed by Lévêque in 1928 [21]. It is represented by the following set of equations:

$$\theta = \frac{T_w - T}{T_w - T_e} = \frac{4}{\pi}\sum_{n=0}^{\infty}\frac{(-1)^n}{2n+1}\exp\left[-4\pi^2(2n+1)^2x^*\right]\cos\left[(2n+1)\frac{\pi}{2}\left(\frac{y}{b}\right)\right] \tag{3.133}$$

$$\theta_m = \frac{T_w - T_m}{T_w - T_e} = \frac{8}{\pi^2}\sum_{n=0}^{\infty}\frac{1}{(2n+1)^2}\exp\left[-4\pi^2(2n+1)^2x^*\right] \tag{3.134}$$

$$Nu_{x,T} = \frac{\pi^2\sum_{n=0}^{\infty}\exp\left[-4\pi^2(2n+1)^2x^*\right]}{\sum_{n=0}^{\infty}\frac{\exp\left[-4\pi^2(2n+1)^2x^*\right]}{(2n+1)^2}} \tag{3.135}$$

$$Nu_{m,T} = -\frac{\ln\theta_m}{4x^*} \tag{3.136}$$

The qualitative trends of $Nu_{x,T}$ and $Nu_{m,T}$ for a flat duct are similar to those for a circular duct displayed in Figs. 3.6 and 3.7. To conserve space they will not be displayed. The plotted $Nu_{m,T}$ results for $Pr = 0.1, 0.72, 10, 50$, and ∞ are available in [1]. The $Nu_{x,T}$ and $Nu_{m,T}$ values for $Pr = 0$ computed from Eqs. (3.135) and (3.136) approach an asymptotic limit of $\pi^2 \approx 9.8696$, whereas for the remaining values of Pr an asymptotic limit of 7.5407 is approached.

A lower bound for the thermal entrance $L^*_{th,T}$ for simultaneously developing flow in a flat duct was provided by Bhatti and Savery [50] as 0.0064 for $0.01 < Pr < 10{,}000$. This value was confirmed in a separate investigation by Das and Mohanty [51]. A more accurate value based on Eq. (3.132) is 0.01. From Eq. (3.135), the $L^*_{th,T}$ value for $Pr = 0$ is found to be 0.0091. This value corresponds to $Nu_{x,T} = 10.3631$.

The solution to the problem of simultaneously developing flow with one duct wall insulated and the other at a uniform temperature, i.e., the fundamental solution of the third kind (see Fig. 3.9), was obtained among others by Mercer et al. [52], who correlated their results by the following equation valid for $0.1 \le Pr \le 10$:

$$Nu_{m,T} = 4.86 + \frac{0.0606x^{*-1.2}}{1 + 0.0909\,Pr^{0.17}x^{*-0.7}} \tag{3.137}$$

The present authors arrived at the following formula for the local Nusselt numbers by differentiating Eq. (3.137) with respect to x^*:

$$\mathrm{Nu}_{x,\mathrm{T}} = 4.86 + \frac{0.0606 x^{*-1.2}[0.0455\,\mathrm{Pr}^{0.17} x^{*-0.7} - 0.2]}{[1 + 0.0909\,\mathrm{Pr}^{0.17} x^{*-0.7}]^2} \tag{3.138}$$

When Pr = 0, the exact solution to the fundamental problem of the third kind for simultaneously developing flow is given by the following set of equations obtained from the solution to the analogous transient heat conduction problem [32]:

$$\theta = \frac{T_w - T}{T_w - T_e} = 4 \sum_{n=0}^{\infty} \frac{(-1)^n}{(2n+1)\pi} \exp\left[-(2n+1)^2 \pi^2 x^*\right] \times \cos\left[\frac{(2n+1)\pi}{4}\left(1 + \frac{y}{b}\right)\right] \tag{3.139}$$

$$\theta_m = \frac{T_w - T_m}{T_w - T_e} = 8 \sum_{n=0}^{\infty} \frac{\exp\left[-(2n+1)^2 \pi^2 x^*\right]}{(2n+1)^2 \pi^2} \tag{3.140}$$

$$\mathrm{Nu}_{x,\mathrm{T}} = \frac{\sum_{n=0}^{\infty} \exp\left[-(2n+1)^2 \pi^2 x^*\right]}{2 \sum_{n=0}^{\infty} \frac{\exp\left[-(2n+1)^2 \pi^2 x^*\right]}{(2n+1)^2 \pi^2}} \tag{3.141}$$

$$\mathrm{Nu}_{m,\mathrm{T}} = -\frac{\ln \theta_m}{2x^*} \tag{3.142}$$

Prescribed Uniform Wall Heat Flux. The problem of simultaneously developing flow with *equal uniform heat fluxes at both walls*, i.e., the Ⓗ problem, has been solved by several investigators [1]. The most accurate results are those of an all-numerical analysis performed by Hwang and Fan [48]. The tabular $\mathrm{Nu}_{x,\mathrm{H}}$ results of Hwang and Fan for Pr = 0.01, 0.7, 1, and 10 are available in [1]. The results for Pr = ∞ are identical to those of Eqs. (3.119) to (3.122). The results for Pr = 0 can be calculated from the following set of exact equations derived from the transient temperature distribution of the analogous heat conduction problem [32]:

$$\Theta = \frac{T - T_e}{q_w'' D_h / k} = 4x^* + \frac{1}{8}\left(\frac{y}{b}\right)^2 - \frac{1}{24} - \frac{1}{2} \sum_{n=1}^{\infty} \frac{(-1)^n}{n^2 \pi^2} \exp(-16 n^2 \pi^2 x^*) \cos n\pi\left(\frac{y}{b}\right) \tag{3.143}$$

$$\Theta_m = \frac{T_m - T_e}{q_w'' D_h / k} = 4x^* \tag{3.144}$$

$$\mathrm{Nu}_{x,\mathrm{H}} = \left(\frac{1}{12} - \frac{1}{2} \sum_{n=1}^{\infty} \frac{\exp(-16 n^2 \pi^2 x^*)}{n^2 \pi^2}\right)^{-1} \tag{3.145}$$

$$\mathrm{Nu}_{m,\mathrm{H}} = \left(\frac{1}{12} - \frac{1}{32} \sum_{n=1}^{\infty} \frac{1 - \exp(-16 n^2 \pi^2 x^*)}{n^4 \pi^4 x^*}\right)^{-1} \tag{3.146}$$

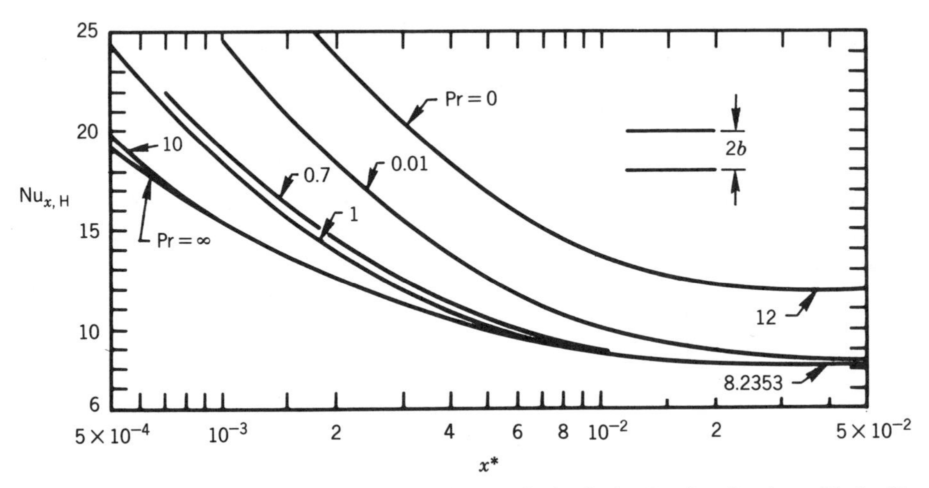

Figure 3.12. Local Nusselt numbers for simultaneously developing flow in a flat duct with the Ⓗ boundary condition [32, 48].

The foregoing $\mathrm{Nu}_{x,\mathrm{H}}$ results are displayed in Fig. 3.12. Note that the curve corresponding to $\mathrm{Pr} = 0$ approaches the asymptotic value of 12, whereas the remainder of the curves approach the asymptotic value of 8.2353.

The thermal entrance lengths based on the results of Fig. 3.12 are found to be 0.016, 0.030, 0.017, 0.014, 0.012, and 0.0115, respectively, for $\mathrm{Pr} = 0$, 0.01, 0.7, 1, 10, and ∞.

The solution to the problem of simultaneous development of velocity and temperature fields in a flat duct with a *constant (nonzero) heat flux at one wall with the other wall insulated (zero flux)*, i.e., the fundamental solution of the second kind (see Fig. 3.9), was obtained by Heaton et al. [53]. Their local Nusselt numbers for $\mathrm{Pr} = 0.01$, 0.7, and 10 evaluated at the heated wall are displayed in Fig. 3.13. The tabular results for these three Pr values are available in [1]. Included in Fig. 3.13 are the results for $\mathrm{Pr} = \infty$, which are taken from Table 3.38 (Sec. 3.8.3), corresponding to $r^* = 1$. The results for $\mathrm{Pr} = 0$ are calculated from the following set of exact equations derived from the transient temperature distribution of the analogous heat conduction problem [32]:

$$\Theta = \frac{T - T_e}{q_w'' D_h / k} = 2x^* + \frac{1}{16}\left(\frac{y}{b}\right)^2 + \frac{1}{8}\left(\frac{y}{b}\right) - \frac{1}{48} - \sum_{n=1}^{\infty} \frac{(-1)^n}{n^2\pi^2} \exp(-4n^2\pi^2 x^*) \cos\frac{n\pi}{2}\left(1 + \frac{y}{b}\right) \tag{3.147}$$

$$\Theta_m = \frac{T_m - T_e}{q_w'' D_h / k} = 2x^* \tag{3.148}$$

$$\mathrm{Nu}_{x,\mathrm{H}} = \left(\frac{1}{6} - \sum_{n=1}^{\infty} \frac{\exp(-4n^2\pi^2 x^*)}{n^2\pi^2}\right)^{-1} \tag{3.149}$$

$$\mathrm{Nu}_{m,\mathrm{H}} = \left(\frac{1}{6} - \sum_{n=1}^{\infty} \frac{1 - \exp(-4n^2\pi^2 x^*)}{4n^4\pi^4 x^*}\right)^{-1} \tag{3.150}$$

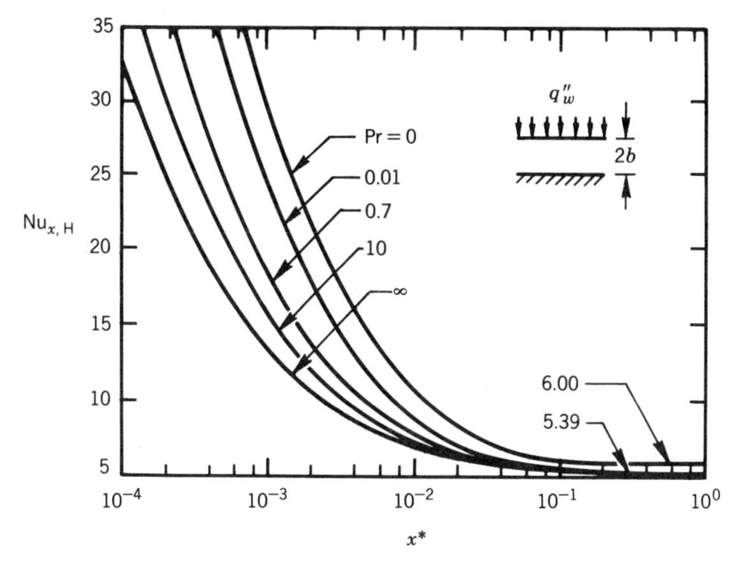

Figure 3.13. Local Nusselt numbers for simultaneously developing flow in a flat duct with uniform heat flux at one wall and the other wall insulated [32, 52].

Convective Heating or Cooling at Both Walls, (T3). The simultaneously developing flow with the (T3) boundary condition, i.e., convective heating or cooling at both walls, was analyzed by Javeri [54]. For $\mathrm{Bi} = \infty$, the $\mathrm{Nu}_{x,\mathrm{T3}}$ are identical to the $\mathrm{Nu}_{x,\mathrm{T}}$ of Eq. (3.132), and for $\mathrm{Bi} = 0$ they are identical to $\mathrm{Nu}_{x,\mathrm{H}}$ of Fig. 3.12.

3.4 RECTANGULAR DUCTS

The mathematical complexities inherent in a two-dimensional analysis have prevented an in-depth investigation of fluid flow and heat transfer characteristics of rectangular ducts. However, sufficient information of direct use to design engineers is available, and it is presented next.

3.4.1 Fully Developed Flow

The fully developed velocity profile for rectangular ducts has been determined exactly [55] using an analogy with the stress function of the theory of elasticity. Referring to the left inset to Fig. 3.14 for the coordinate system, the fully developed velocity profile for rectangular ducts is expressed as

$$u = -\frac{16}{\pi^3}\left(\frac{dp}{dx}\right)\frac{a^2}{\mu}\sum_{n=1,3,\ldots}^{\infty}\frac{(-1)^{(n-1)/2}}{n^3}\left(1-\frac{\cosh(n\pi y/2a)}{\cosh(n\pi b/2a)}\right)\cos\left(\frac{n\pi z}{2a}\right) \tag{3.151}$$

where the pressure gradient dp/dx is given in terms of u_m by

$$u_m = -\frac{1}{3}\left(\frac{dp}{dx}\right)\frac{a^2}{\mu}\left[1-\frac{192}{\pi^5}\left(\frac{a}{b}\right)\sum_{n=1,3,\ldots}^{\infty}\frac{1}{n^5}\tanh\left(\frac{n\pi b}{2a}\right)\right] \tag{3.152}$$

Equation (3.151) entails considerable computational complexity. To circumvent this difficulty, Purday [56] proposed a simple approximation in the following form:

$$\frac{u}{u_{max}} = \left[1 - \left(\frac{y}{b}\right)^n\right]\left[1 - \left(\frac{z}{a}\right)^m\right] \tag{3.153}$$

$$\frac{u_{max}}{u_m} = \left(\frac{m+1}{m}\right)\left(\frac{n+1}{n}\right) \tag{3.154}$$

where $n = 2$ and $m = 2.37$, 3.78, 5.19, 6.60, 13.6, and ∞ for $\alpha^* = 0.5$, $\frac{1}{3}$, 0.25, 0.20, 0.1, and 0, respectively. Natarajan and Lakshmanan [57] provided the following relations for the values of m and n:

$$m = 1.7 + 0.5\alpha^{*-1.4} \tag{3.155}$$

$$n = \begin{cases} 2 & \text{for} \quad \alpha^* \le \frac{1}{3} \\ 2 + 0.3\left(\alpha^* - \frac{1}{3}\right) & \text{for} \quad \alpha^* \ge \frac{1}{3} \end{cases} \tag{3.156}$$

The values of m and n from Eqs. (3.155) and (3.156) yield velocity profiles that are within 1% of those computed from the exact relations of Eqs. (3.151) and (3.152).

The fully developed incremental pressure drop number $K(\infty)$ for rectangular ducts has been determined by a number of investigators [1]. The analytical values of Miller and Han [58] are found to be in the closest agreement with the experimental values [1]. They are 1.433, 1.281, 0.931, and 0.658 for $\alpha^* = 1$, 0.50, 0.20, and 0, respectively. The predictions of the hydrodynamic entrance length L_{hy}^+ given by Wiginton and Dalton [59], believed to be the most accurate, are 0.090, 0.085, 0.075, and 0.080 for $\alpha^* = 1$, 0.50, 0.25, and 0.20, respectively.

The exact expression for the fully developed Fanning friction factor is

$$f\,\text{Re} = \frac{24}{\left(1 + \dfrac{1}{\alpha^*}\right)^2\left(1 - \dfrac{192}{\pi^5\alpha^*}\displaystyle\sum_{n=1,3,\ldots}^{\infty}\dfrac{\tanh(n\pi\alpha^*/2)}{n^5}\right)} \tag{3.157}$$

which is closely approximated (within +0.05%) by the following empirical equation [1]:

$$f\,\text{Re} = 24(1 - 1.3553\alpha^* + 1.9467\alpha^{*2} - 1.7012\alpha^{*3} + 0.9564\alpha^{*4} - 0.2537\alpha^{*5}) \tag{3.158}$$

The fully developed Nusselt numbers Nu_T for the case of prescribed uniform temperature at four walls are available in [1]. They are approximated within $\pm 0.1\%$ by the formula

$$\text{Nu}_T = 7.541(1 - 2.610\alpha^* + 4.970\alpha^{*2} - 5.119\alpha^{*3} + 2.702\alpha^{*4} - 0.548\alpha^{*5}) \tag{3.159}$$

Schmidt and Newell [60] considered the case of one or more walls being heated to a uniform temperature with the other wall(s) insulated. Their Nusselt numbers, as modified by Shah and London [1] by replacing the heated perimeter with the wetted perimeter, are displayed in Fig. 3.14.

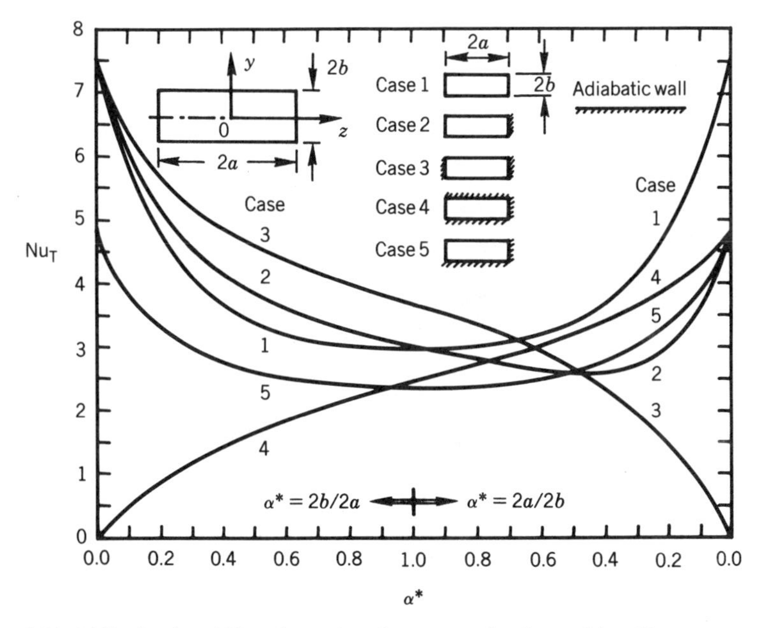

Figure 3.14. Fully developed Nusselt numbers for rectangular ducts with uniform temperature at one or more walls [1].

The fully developed Nusselt numbers with a prescribed wall heat flux are available for three types of thermal boundary conditions: (H1), (H2), and (H4). Marco and Han [55] invoked the analogy of small deflection of a thin plate simply supported along four edges and subjected to a uniform lateral load to arrive at the result

$$\mathrm{Nu}_{\mathrm{H1}} = \frac{64}{(1+\alpha^*)^2\pi^2} \frac{\left(\displaystyle\sum_{m=1,3,\ldots}^{\infty} \sum_{n=1,3,\ldots}^{\infty} \frac{1}{m^2 n^2 (m^2 + n^2\alpha^{*2})}\right)^2}{\displaystyle\sum_{m=1,3,\ldots}^{\infty} \sum_{n=1,3,\ldots}^{\infty} \frac{1}{m^2 n^2 (m^2 + n^2\alpha^{*2})^3}} \tag{3.160}$$

which can be approximated within $\pm 0.03\%$ by the formula

$$\mathrm{Nu}_{\mathrm{H1}} = 8.235(1 - 2.0421\alpha^* + 3.0853\alpha^{*2} - 2.4765\alpha^{*3} + 1.0578\alpha^{*4} - 0.1861\alpha^{*5}) \tag{3.161}$$

Savino and Siegel [61] investigated the effect of unequal heat fluxes. They found that poor convection due to low velocities in the corners and along the narrow wall causes peak temperatures to occur at the corners. Also, lower peak temperatures occur when heating takes place at the broad sides only.

Schmidt and Newell [60] determined the Nusselt numbers with one or more walls subjected to the (H1) boundary condition with the other wall(s) insulated. Their results, with appropriate modification by Shah and London [1], are displayed in Fig. 3.15.

The (H2) problem with heating at four walls was solved by several investigators [1]. Their results show that $\mathrm{Nu}_{\mathrm{H2}}$ is quite insensitive to change in α^*. For example, for

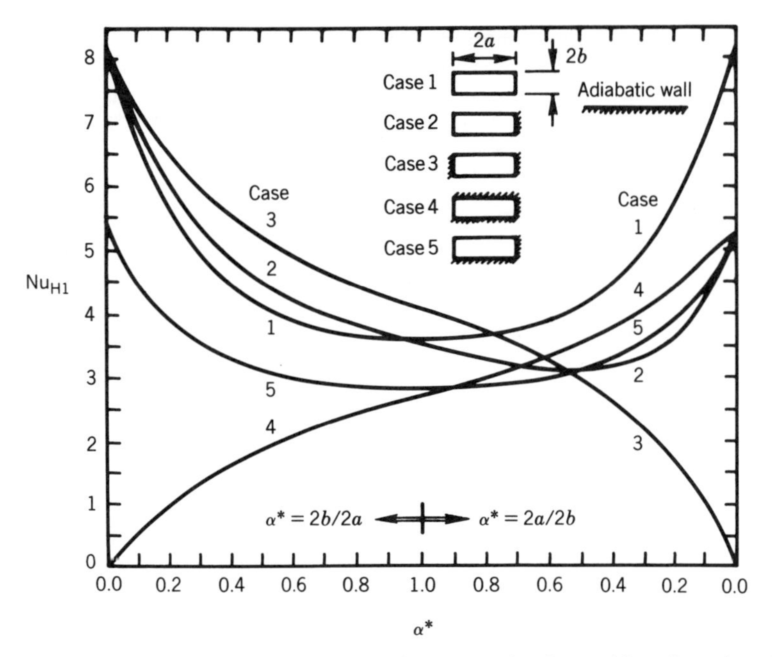

Figure 3.15. Fully developed Nusselt numbers for rectangular ducts with uniform heat flux at one or more walls [1].

$0.1 \leq \alpha^* \leq 1$, $\mathrm{Nu_{H2}}$ varies in the range $2.95 \leq \mathrm{Nu_{H2}} \leq 3.09$. Also, it is noted that as $\alpha^* \to 0$, $\mathrm{Nu_{H2}}$ does not approach 8.235, the value for the flat duct. This is because the imposed heat flux on the short sides continues to influence $\mathrm{Nu_{H2}}$ even as $\alpha^* \to 0$.

Han [62] treated the problem of a rectangular duct with a pair of narrow thin walls subjected to the (H1) boundary condition and the opposite pair of broad thick walls as extended surfaces or fins subjected to the (H4) boundary condition. The fully developed Nusselt numbers for this set of boundary condition are expressible as

$$\mathrm{Nu} = F(K, \alpha^*)\mathrm{Nu_{H1}} \tag{3.162}$$

where $\mathrm{Nu_{H1}}$ is given by Eq. (3.160) and F is the correction factor listed in Table 3.13

TABLE 3.13 Correction Factors $F(K, \alpha^*)$ to be Used in Conjunction with Eq. (3.162) [62]

	$F(K, \alpha^*)$			
K	$\alpha^* = 0.20$	0.25	0.50	1.00
0	0.552	0.828	2.348	3.872
4	2.580	2.840	3.568	3.972
8	3.108	3.292	3.752	3.984
16	3.492	3.592	3.864	3.992
40	3.776	3.828	3.944	3.996

as a function of the dimensionless fin parameter $K = k_w\delta_w/ka$, where δ_w is the thickness of the broad wall. The aspect ratio is $\alpha^* = 2b/2a$.

Siegel and Savino [63] considered the effect of peripheral wall heat conduction in the broad walls with nonconducting insulated narrow walls. They also extended their analysis to include different boundary conditions at the corners [64]. They found that the circumferential wall conduction substantially lowers the peak temperatures at the corners and the temperature gradients along the broad walls.

Lyczkowski et al. [65] and Iqbal et al. [66] determined the fully developed Nusselt numbers $\mathrm{Nu}_{\mathrm{H4}}$ for the (H4) boundary condition on all four walls of the square duct ($\alpha^* = 1$). With $K_p = k_w\delta_w/kD_h$ as a parameter, they found the $\mathrm{Nu}_{\mathrm{H4}}$ values to be 3.08, 3.41, 3.50, 3.52, 3.606, and 3.608 for $K_p = 0, 0.5, 1, 2, 100$, and ∞, respectively. In addition, Iqbal et al. [66] determined the $\mathrm{Nu}_{\mathrm{H4}}$ values for $\alpha^* = \frac{1}{2}$ and $\frac{1}{3}$. For $\alpha^* = \frac{1}{2}$, they reported $\mathrm{Nu}_{\mathrm{H4}} = 3.04$, 3.87, and 4.12 corresponding to $K_p = 0$, 1, and ∞, respectively. For the same three values of K_p, they found $\mathrm{Nu}_{\mathrm{H4}} = 2.95$, 4.24, and 4.79, respectively, for $\alpha^* = \frac{1}{3}$.

3.4.2 Hydrodynamically Developing Flow

There are several analytical and experimental investigations of the problem of laminar flow development in rectangular ducts. They are summarized in [1]. Based on comparisons with the experimental measurements, it is determined that the numerical results of Curr et al. [67] and the theoretical results of Tachibana and Iemoto [68] are in closest agreement with the measurements. The apparent Fanning friction factors of [67] are presented in Fig. 3.16.

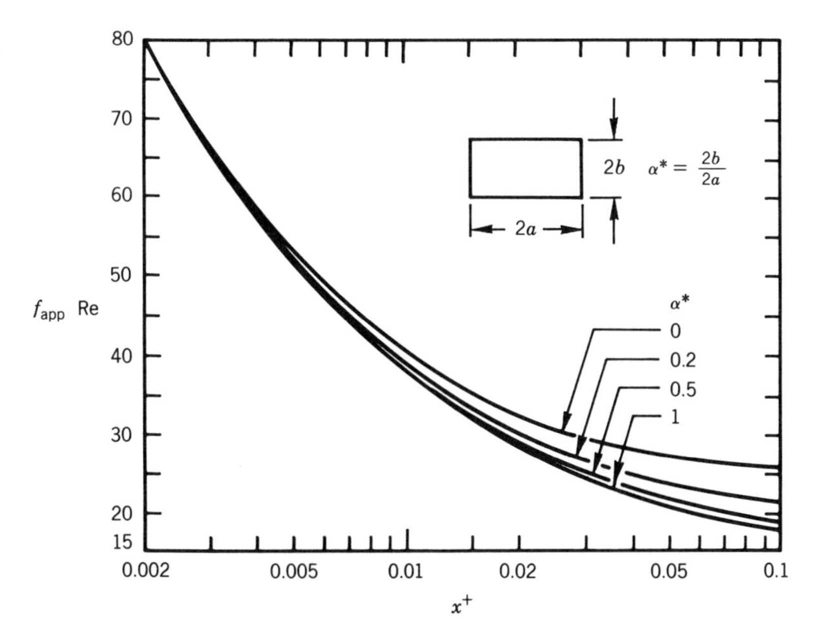

Figure 3.16. Apparent Fanning friction factors for hydrodynamically developing flow in rectangular ducts [67].

TABLE 3.14 Local and Mean Nusselt Numbers in the Thermal Entrance Region of Rectangular Ducts with the Ⓣ Boundary Condition [69]

$\frac{1}{x^*}$	$Nu_{x,T}$						$Nu_{m,T}$					
	$\alpha^* = 1.0$	0.5	$\frac{1}{3}$	0.25	0.2	$\frac{1}{6}$	1.0	0.5	$\frac{1}{3}$	0.25	0.2	$\frac{1}{6}$
0	2.65	3.39	3.96	4.51	4.92	5.22	2.65	3.39	3.96	4.51	4.92	5.22
10	2.86	3.43	4.02	4.53	4.94	5.24	3.50	3.95	4.54	5.00	5.36	5.66
20	3.08	3.54	4.17	4.65	5.04	5.34	4.03	4.46	5.00	5.44	5.77	6.04
30	3.24	3.70	4.29	4.76	5.31	5.41	4.47	4.86	5.39	5.81	6.13	6.37
40	3.43	3.85	4.42	4.87	5.22	5.48	4.85	5.24	5.74	6.16	6.45	6.70
60	3.78	4.16	4.67	5.08	5.40	5.64	5.50	5.85	6.35	6.73	7.03	7.26
80	4.10	4.46	4.94	5.32	5.62	5.86	6.03	6.37	6.89	7.24	7.53	7.77
100	4.35	4.72	5.17	5.55	5.83	6.07	6.46	6.84	7.33	7.71	7.99	8.17
120	4.62	4.93	5.42	5.77	6.06	6.27	6.86	7.24	7.74	8.13	8.39	8.63
140	4.85	5.15	5.62	5.98	6.26	6.47	7.22	7.62	8.11	8.50	8.77	9.00
160	5.03	5.34	5.80	6.18	6.45	6.66	7.56	7.97	8.45	8.86	9.14	9.35
180	5.24	5.54	5.99	6.37	6.63	6.86	7.87	8.29	8.77	9.17	9.46	9.67
200	5.41	5.72	6.18	6.57	6.80	7.02	8.15	8.58	9.07	9.47	9.79	10.01

3.4.3 Thermally Developing Flow

Wibulswas [69] solved the thermal entrance length problem for rectangular ducts with the Ⓣ boundary condition neglecting the effects of viscous dissipation, fluid axial conduction, and thermal energy sources in the fluid. His local and mean Nusselt numbers are presented in Table 3.14. Chandrupatla and Sastri [70] analyzed the Ⓣ, (H1), and (H2) thermal entrance length problem for $\alpha^* = 1$. Their results, presented in Table 3.15, are more accurate than those of [69] and hence are recommended for $\alpha^* = 1$.

TABLE 3.15 Local and Mean Nusselt Numbers in the Thermal Entrance Region of a Square Duct ($\alpha^* = 1$) with the Ⓣ, (H1), and (H2) Boundary Conditions [70]

$\frac{1}{x^*}$	$Nu_{x,T}$	$Nu_{m,T}$	$Nu_{x,H1}$	$Nu_{m,H1}$	$Nu_{x,H2}$	$Nu_{m,H2}$
0	2.975	2.975	3.612	3.612	3.095	3.095
10	2.976	3.514	3.686	4.549	3.160	3.915
20	3.074	4.024	3.907	5.301	3.359	4.602
25	3.157	4.253	4.048	5.633	3.481	4.898
40	3.432	4.841	4.465	6.476	3.843	5.656
50	3.611	5.173	4.720	6.949	4.067	6.083
80	4.084	5.989	5.387	8.111	4.654	7.138
100	4.357	6.435	5.769	8.747	4.993	7.719
133.3	4.755	7.068	6.331	9.653	5.492	8.551
160	—	—	6.730	10.279	5.848	9.128
200	5.412	8.084	7.269	11.103	6.330	9.891

TABLE 3.16 Local and Mean Nusselt Numbers in the Thermal Entrance Region of Rectangular Ducts with the (H1) Boundary Condition [69]

	$Nu_{x,H1}$				$Nu_{m,H1}$			
$\frac{1}{x^*}$	$\alpha^* = 1.0$	0.5	$\frac{1}{3}$	0.25	1.0	0.5	$\frac{1}{3}$	0.25
0	3.60	4.11	4.77	5.35	3.60	4.11	4.77	5.35
10	3.71	4.22	4.85	5.45	4.48	4.94	5.45	6.03
20	3.91	4.38	5.00	5.62	5.19	5.60	6.06	6.57
30	4.18	4.61	5.17	5.77	5.76	6.16	6.60	7.07
40	4.45	4.84	5.39	5.87	6.24	6.64	7.09	7.51
60	4.91	5.28	5.82	6.26	7.02	7.45	7.85	8.25
80	5.33	5.70	6.21	6.63	7.66	8.10	8.48	8.87
100	5.69	6.05	6.57	7.00	8.22	8.66	9.02	9.39
120	6.02	6.37	6.92	7.32	8.69	9.13	9.52	9.83
140	6.32	6.68	7.22	7.63	9.09	9.57	9.93	10.24
160	6.60	6.96	7.50	7.92	9.50	9.96	10.31	10.61
180	6.86	7.23	7.76	8.18	9.85	10.31	10.67	10.92
200	7.10	7.46	8.02	8.44	10.18	10.64	10.97	11.23

The thermal entrance lengths $L^*_{th,T}$ for rectangular ducts with fully developed laminar velocity profile are determined to be 0.008, 0.054, 0.049, and 0.041 for $\alpha^* = 0$, 0.25, 0.5, and 1, respectively [1].

Wibulswas [69] solved the thermal entrance length problem for the (H1) boundary condition. His results for negligible fluid axial conduction, viscous dissipation, and thermal energy sources are presented in Table 3.16. Perkins et al. [71] experimentally determined $Nu_{x,H1}$ for a square duct. Their experimental results are in excellent accord with the theoretical results in Table 3.16. The following correlation is provided by Perkins et al. [71] for their measurements:

$$Nu_{x,H1} = [0.277 - 0.152\exp(-38.6x^*)]^{-1} \tag{3.163}$$

The results of Chandrupatla and Sastri [70] for the square duct (H1) thermal entrance problem presented in Table 3.15 are also in excellent agreement with the results in Table 3.16.

The thermal entrance lengths $L^*_{th,H1}$ for rectangular ducts with fully developed laminar velocity profile are determined to be 0.0115, 0.042, 0.048, 0.057, and 0.066 for $\alpha^* = 0, 0.25, \frac{1}{3}, 0.5$, and 1, respectively [1].

Chandrupatla and Sastri [70] analyzed the (H2) thermal entrance length problem for a square duct. Their results are presented in Table 3.15.

In addition to the thermal entrance length results presented above for the (T), (H1), and (H2) boundary conditions, certain thermal entry length results involving a combination of boundary conditions on four walls are available, as are the results pertaining to a circumferentially uniform wall temperature which varies axially in a linear fashion [1].

TABLE 3.17 Local and Mean Nusselt Numbers for Simultaneously Developing Flow in Rectangular Ducts with the (T) and (H1) Boundary Conditions [69]

	$Nu_{x,H1}$				$Nu_{m,H1}$				$Nu_{m,T}$				
$\frac{1}{x^*}$	$\alpha^* = 1.0$	0.5	$\frac{1}{3}$	0.25	1.0	0.5	$\frac{1}{3}$	0.25	1.0	0.5	$\frac{1}{3}$	0.25	$\frac{1}{6}$
5	—	—	—	—	4.60	5.00	5.57	6.06	—	—	—	—	—
10	4.18	4.60	5.18	5.66	5.43	5.77	6.27	6.65	3.75	4.20	4.67	5.11	5.72
20	4.66	5.01	5.50	5.92	6.60	6.94	7.31	7.58	4.39	4.79	5.17	5.56	6.13
30	5.07	5.40	5.82	6.17	7.52	7.83	8.13	8.37	4.88	5.23	5.60	5.93	6.47
40	5.47	5.75	6.13	6.43	8.25	8.54	8.85	9.07	5.27	5.61	5.96	6.27	6.78
50	5.83	6.09	6.44	6.70	8.90	9.17	9.48	9.70	5.63	5.95	6.28	6.61	7.07
60	6.14	6.42	6.74	7.00	9.49	9.77	10.07	10.32	5.95	6.27	6.60	6.90	7.35
80	6.80	7.02	7.32	7.55	10.53	10.83	11.13	11.35	6.57	6.88	7.17	7.47	7.90
100	7.38	7.59	7.86	8.08	11.43	11.70	12.00	12.23	7.10	7.42	7.70	7.98	8.38
120	7.90	8.11	8.37	8.58	12.19	12.48	12.78	13.03	7.61	7.91	8.18	8.48	8.85
140	8.38	8.61	8.84	9.05	12.87	13.15	13.47	13.73	8.06	8.37	8.66	8.93	9.28
160	8.84	9.05	9.38	9.59	13.50	13.79	14.10	14.48	8.50	8.80	9.10	9.36	9.72
180	9.28	9.47	9.70	9.87	14.05	14.35	14.70	14.95	8.91	9.20	9.50	9.77	10.12
200	9.69	9.88	10.06	10.24	14.55	14.88	15.21	15.49	9.30	9.60	9.91	10.18	10.51
220	—	—	—	—	15.03	15.36	15.83	16.02	9.70	10.00	10.30	10.58	10.90

3.4.4 Simultaneously Developing Flow

Wibulswas [69] analyzed the simultaneously developing flow in rectangular ducts, neglecting the transverse velocity components. His results for the (T) and (H1) boundary conditions for air (Pr = 0.72) are presented in Table 3.17. Recently, Chandrupatla and Sastri [72] reported a more refined analysis for a square duct with the (H1) boundary condition without the neglect of the transverse velocity components. A comparison of their results for Pr = 0.72 with those of Table 3.17 revealed that neglect of the transverse velocity components could cause $Nu_{x,H1}$ to be lowered by nearly 13% at $x^* = 0.1$ in a square duct ($\alpha^* = 1$). This behavior is opposite to that of the Nusselt numbers in a circular duct. As discussed in [1], neglect of the radial velocity in a circular duct overestimates the Nusselt numbers. The reason for this difference is not apparent to the authors.

$Nu_{x,H1}$ and $Nu_{m,H1}$ results of Chandrupatla and Sastri [72] are presented in Table 3.18. They also serve to illustrate the effect of the Prandtl number on $Nu_{x,H1}$ and $Nu_{m,H1}$ for $\alpha^* = 1$. It may be noted that Pr = 0 in Table 3.18 corresponds to the case of slug flow throughout the duct. Likewise Pr = ∞ corresponds to the case of hydrodynamically developed flow throughout the duct.

Recently, Neti and Eichhorn [73] solved the problem of simultaneously developing flow in a square duct with the (T) boundary condition. They graphically presented extensive heat transfer results for a fluid with Pr = 6.

3.5 TRIANGULAR DUCTS

In view of the mathematical complexities inherent in a two-dimensional analysis, the fluid flow and heat transfer characteristics of triangular ducts have not been investigated in great detail. Although considerable information is available for fully developed

TABLE 3.18 Local and Mean Nusselt Numbers for Simultaneously Developing Flow in a Square Duct ($\alpha^* = 1$) with the (H1) Boundary Condition [72]

$\frac{1}{x^*}$	$\mathrm{Nu}_{x,\mathrm{H1}}$					$\mathrm{Nu}_{m,\mathrm{H1}}$				
	Pr = 0.0	0.1	1.0	10.0	∞	0.0	0.1	1.0	10.0	∞
200	14.653	11.659	8.373	7.329	7.269	21.986	17.823	13.390	11.200	11.103
133.3	12.545	9.597	7.122	6.381	6.331	19.095	15.391	11.489	9.737	9.653
100	11.297	8.391	6.379	5.816	5.769	17.290	13.781	10.297	8.823	8.747
80	10.459	7.615	5.877	5.480	5.387	16.003	12.620	9.461	8.181	8.111
50	9.031	6.353	5.011	4.759	4.720	13.622	10.475	7.934	7.010	6.949
40	8.500	5.883	4.683	4.502	4.465	12.647	9.601	7.315	6.533	6.476
25	7.675	5.108	4.152	4.080	4.048	10.913	8.043	6.214	5.682	5.633
20	7.415	4.826	3.973	3.939	3.907	10.237	7.426	5.782	5.347	5.301
10	7.051	4.243	3.687	3.686	3.686	8.701	5.948	4.783	4.580	4.549
0	7.013	3.612	3.612	3.612	3.612	7.013	3.612	3.612	3.612	3.612

flow involving equilateral triangular, isosceles triangular, right triangular, and even arbitrary triangular ducts, the information is quite scarce for hydrodynamically, thermally, and simultaneously developing flows.

3.5.1 Fully Developed Flow

Equilateral Triangular Duct. The fully developed velocity distribution and friction factor for the equilateral triangular duct of Fig 3.17*a* with the hydraulic diameter $D_h = 4b/3$ are given by [55]

$$\frac{u}{u_m} = \frac{15}{8}\left[\left(\frac{y}{b}\right)^3 - 3\left(\frac{y}{b}\right)\left(\frac{z}{b}\right)^2 - 2\left(\frac{y}{b}\right)^2 - 2\left(\frac{z}{b}\right)^2 + \frac{32}{27}\right] \tag{3.164}$$

$$u_m = -\frac{b^2}{15\mu}\left(\frac{dp}{dx}\right), \qquad f\,\mathrm{Re} = \tfrac{40}{3} = 13.333 \tag{3.165}$$

The Nusselt number $\mathrm{Nu_T}$ with uniform temperature at the three duct walls has been obtained by several investigators [1]. The recommended value is 2.49 [76]. The Nusselt number $\mathrm{Nu_{H1}}$ has also been obtained by several investigators. According to Tyagi [5]

$$\mathrm{Nu_{H1}} = \frac{\frac{28}{9}}{1 + \frac{1}{12}S^* + \frac{40}{11}\mathrm{Br}'} \tag{3.166}$$

The Nusselt number $\mathrm{Nu_{H2}}$ for the equilateral triangular duct, determined by Cheng [74], is 1.892.

The effect of duct corner rounding on fluid flow and heat transfer is important because the duct corners are rarely sharp, due to manufacturing processes. For the equilateral triangular duct of Fig. 3.17*b*, this effect was investigated by Shah [75]. His results are presented in Table 3.19. Note that $\bar{y}$ and $\bar{y}_{max}$ in Table 3.19 refer to the distances measured from the duct base to the centroid and to the point of maximum fluid velocity, respectively. Also, $T^*_{w,max}$ and $T^*_{w,min}$ included in Table 3.18 are the

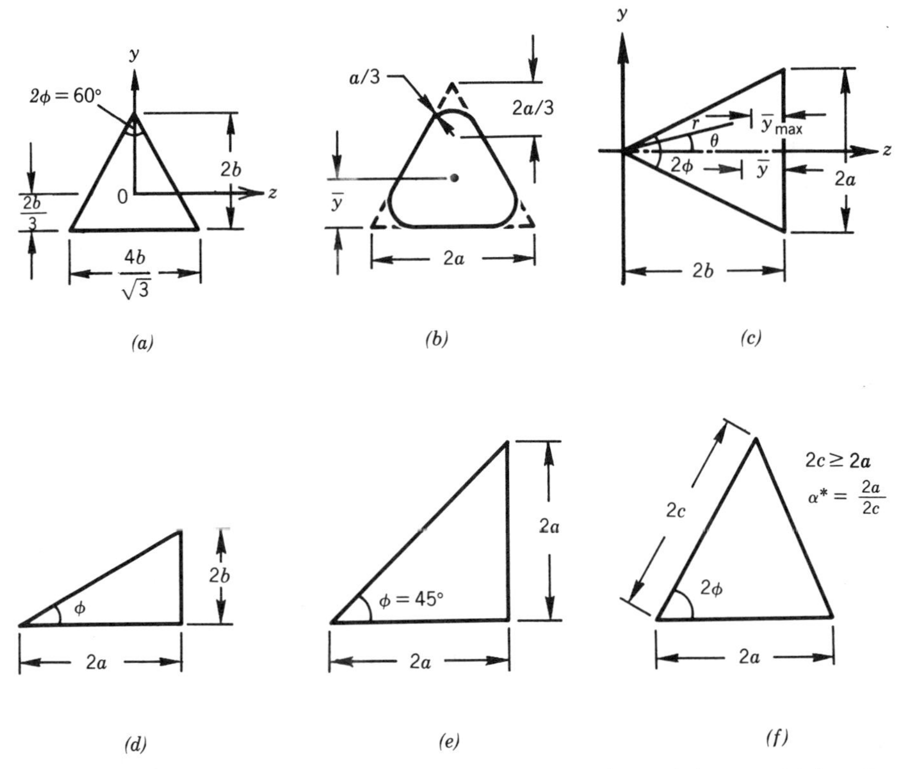

Figure 3.17. Triangular ducts: (a) equilateral, (b) equilateral with rounded corners, (c) isosceles, (d) right, (e) right-angled isosceles, (f) arbitrary.

TABLE 3.19 Fully Developed Fluid Flow and Heat Transfer Characteristics of Equilateral Triangular Ducts with Rounded Corners [1]

	No Rounded Corners	One Rounded Corner	Two Rounded Corners	Three Rounded Corners
$P/2a$	3.00000	2.77172	2.54343	2.31515
$A_c/(2a)^2$	0.43301	0.41399	0.39497	0.37594
$D_h/2a$	0.57735	0.59745	0.62115	0.64953
$\bar{y}/2a$	0.28868	0.26778	0.30957	0.28868
$\bar{y}_{max}/2a$	0.28868	0.28627	0.29117	0.28868
u_{max}/u_m	2.222	2.172	2.115	2.064
$K(\infty)$	1.818	1.698	1.567	1.441
L^+_{hy}	0.0398	0.0359	0.0319	0.0284
f Re	13.333	14.057	14.899	15.993
Nu_{H1}	3.111	3.401	3.756	4.205
Nu_{H2}	1.892	2.196	2.715	3.780
$T^*_{w,max}$	1.79	2.03	2.42	1.22
$T^*_{w,min}$	0.515	0.512	0.550	0.757

normalized maximum and minimum temperatures at the duct wall, defined as

$$T^*_{w,\max} = \frac{T_{w,\max} - T_c}{T_{w,m} - T_c}, \qquad T^*_{w,\min} = \frac{T_{w,\min} - T_c}{T_{w,m} - T_c} \tag{3.167}$$

where T_c denotes the fluid temperature at the duct centroid.

Isosceles Triangular Ducts. The fully developed flow and heat transfer characteristics of the isosceles triangular duct of Fig. 3.17*c* have been investigated rather extensively both theoretically and experimentally [1]. Figures 3.18 and 3.19 display the fluid flow and heat transfer results for these ducts. The tabular results are available in [1]. Recently, Haji-Sheikh et al. [76] reported more accurate values of $\mathrm{Nu_T}$ than those listed in [1] for $7.15° < 2\phi < 151.93°$.

The fully developed velocity distribution and friction factors for the isosceles triangular duct of Fig. 3.17*c* can be expressed by the following set of equations due to Migay [77], who solved the flow problem by invoking the analogy of torsion of a prismatic bar:

$$u = -\frac{1}{2\mu}\left(\frac{dp}{dx}\right)\frac{y^2 - z^2\tan^2\phi}{1 - \tan^2\phi}\left[\left(\frac{z}{2b}\right)^{B-2} - 1\right] \tag{3.168}$$

$$u_m = -\frac{2b^2}{3\mu}\left(\frac{dp}{dx}\right)\frac{(B-2)\tan^2\phi}{(B+2)(1-\tan^2\phi)} \tag{3.169}$$

$$f\,\mathrm{Re} = \frac{12(B+2)(1-\tan^2\phi)}{(B-2)\left[\tan\phi + (1+\tan^2\phi)^{1/2}\right]^2} \tag{3.170}$$

$$B = \left[4 + \tfrac{5}{2}(\cot^2\phi - 1)\right]^{1/2} \tag{3.171}$$

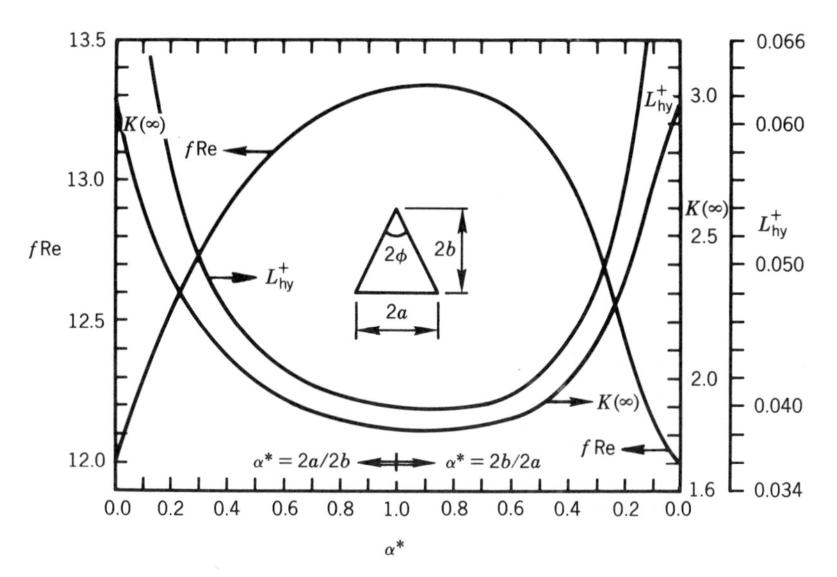

Figure 3.18. Fully developed fluid flow characteristics of isosceles triangular ducts [1].

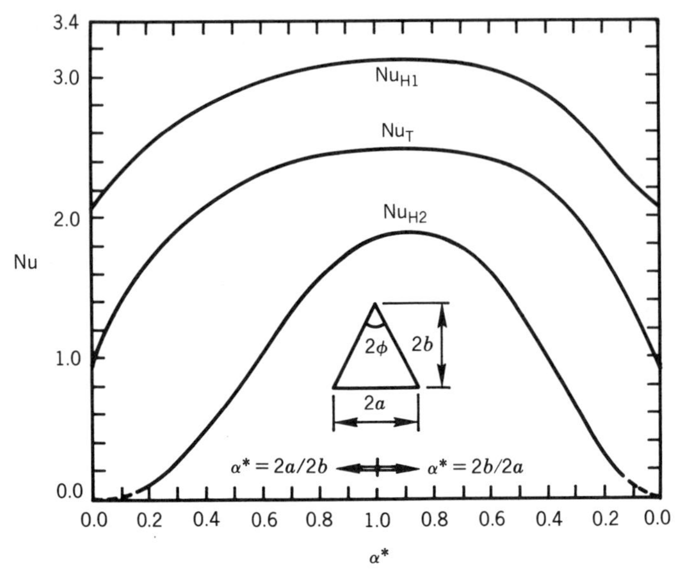

Figure 3.19. Fully developed Nusselt numbers for isosceles triangular ducts [1].

It is apparent from Eq. (3.170) that when $2\phi = 90°$, f Re is indeterminate. By the application of l'Hospital's rule, Migay [77] showed that for $2\phi = 90°$

$$f\,\mathrm{Re} = \frac{12(B+2)(1-3\tan^2\phi)}{\left\{\frac{3}{2}\tan\phi\left[4\tan^2\phi + \frac{5}{2}(1-\tan^2\phi)\right]^{-1/2} - 2\right\}\left[\tan\phi + (1+\tan^2\phi)^{1/2}\right]^2} \tag{3.172}$$

The f Re values computed from Eq. (3.170) and (3.172) are higher by about 1% than the values of Fig. 3.18 except for the limiting case of $2\phi = 180°$.

Schmidt and Newell [60] reported Nu_{H1} and Nu_T results for isosceles triangular ducts with one or more walls insulated. Their results are displayed in Figs. 3.20 and 3.21.

Right Triangular Ducts. The fully developed fluid flow and heat transfer characteristics of the right triangular duct of Fig. 3.17*d* have been investigated quite extensively, as summarized in [1]. The principal quantities of practical interest are displayed in Fig. 3.22. The Nu_T values in Fig. 3.22 are taken from Haji-Sheikh et al. [76]; Nu_{H1}, f Re, and $K(\infty)$ from Sparrow and Haji-Sheikh [78]; and Nu_{H2} from Iqbal et al. [79].

The fully developed velocity profiles and friction factors of the right-angled isosceles triangular duct of Fig. 3.17*e* are expressible in closed-form formulas which are given in [1].

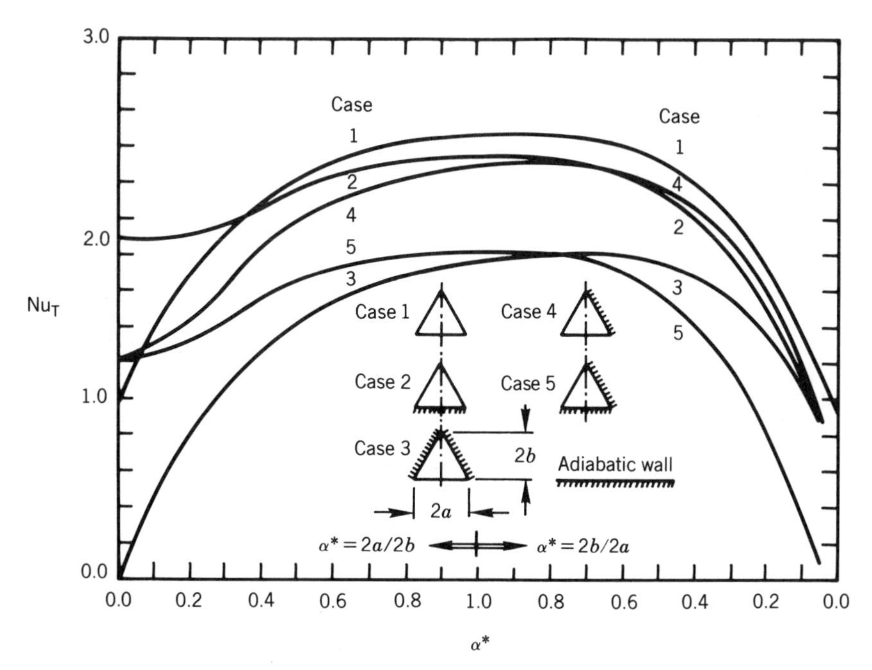

Figure 3.20. Fully developed Nusselt numbers for isosceles triangular ducts with one or more walls at uniform temperature [60].

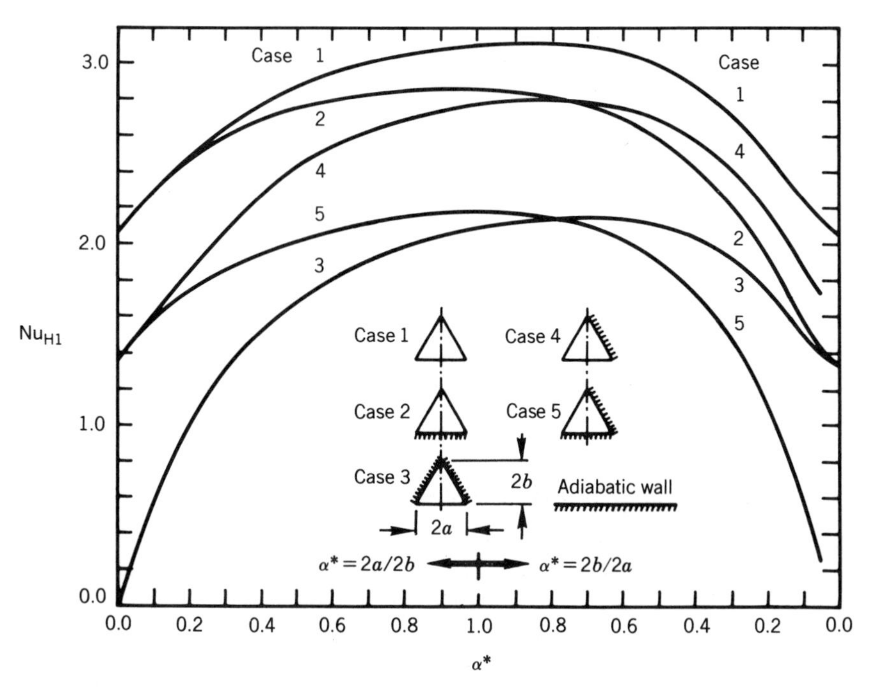

Figure 3.21. Fully developed Nusselt numbers for isosceles triangular ducts with uniform heat flux at one or more walls [60].

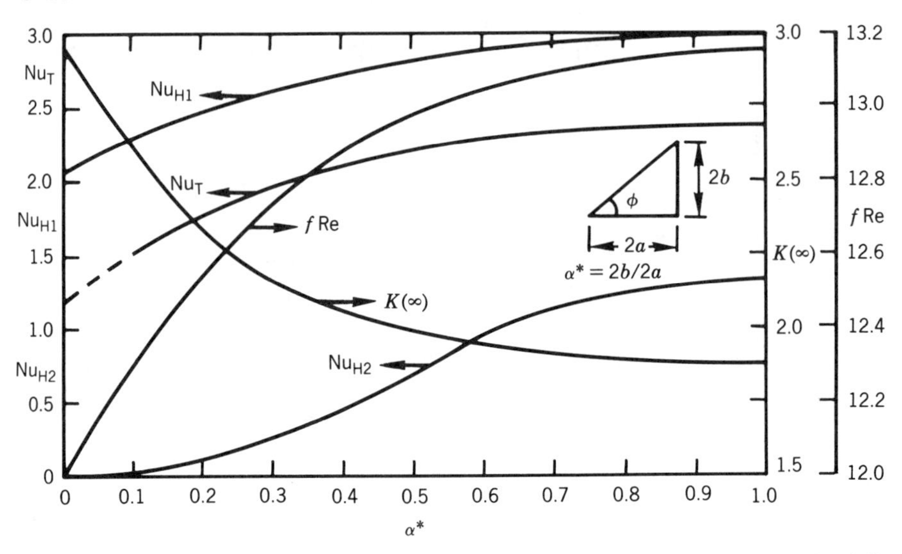

Figure 3.22. Thermal and hydrodynamic characteristics of right triangular ducts for fully developed flow [76, 78, 79].

Arbitrary Triangular Ducts. Nakamura et al. [80] analyzed arbitrary triangular ducts by a finite difference method. Their results for the arbitrary triangular duct of Fig. 3.17*f* are presented in Figs. 3.23 and 3.24 for $0.9 \leq \alpha^* \leq 0.2$. For $\alpha^* = 1$, the more accurate results for isosceles triangular ducts presented in [1] are displayed.

Semilet [81] proposed the following correlations for gas flow in arbitrary triangular ducts with $L/D_h > 50$ but with unspecified thermal boundary conditions:

$$f\,\text{Re} = 12.5 + 0.007\,\text{Re} \qquad \text{for} \quad 50 < \text{Re} < 1000 \tag{3.173}$$

$$\text{Nu} = 2.15 + 0.00245\,\text{Re} \qquad \text{for} \quad 50 < \text{Re} < 2000 \tag{3.174}$$

$$\text{Nu} = 2.15 + 2.31 \times 10^{-3}\text{Re} + 1.25 \times 10^{-7}\text{Re}^2 - 9.6 \times 10^{-2}\text{Re}^3 \qquad \text{for} \quad 50 < \text{Re} < 8000 \tag{3.175}$$

These correlations agree with the experimental and theoretical results from various sources within $\pm 15\%$.

3.5.2 Hydrodynamically Developing Flow

The hydrodynamic entrance length problem for the equilateral triangular duct shown in Fig. 3.17*a* has been solved by several investigators, as summarized in [1]. In addition, Fleming and Sparrow [82] solved the problem for the isosceles triangular duct of Fig. 3.17*c* with $2\phi = 30°$, and Aggarwala and Gangal [83] for the right-angled isosceles triangular duct of Fig. 3.17*e*. The results of these two investigations for $2\phi = 30°, 60°$, and $90°$ are believed to be the most accurate. They are displayed in Fig. 3.25 along with the results of Miller and Han [58].

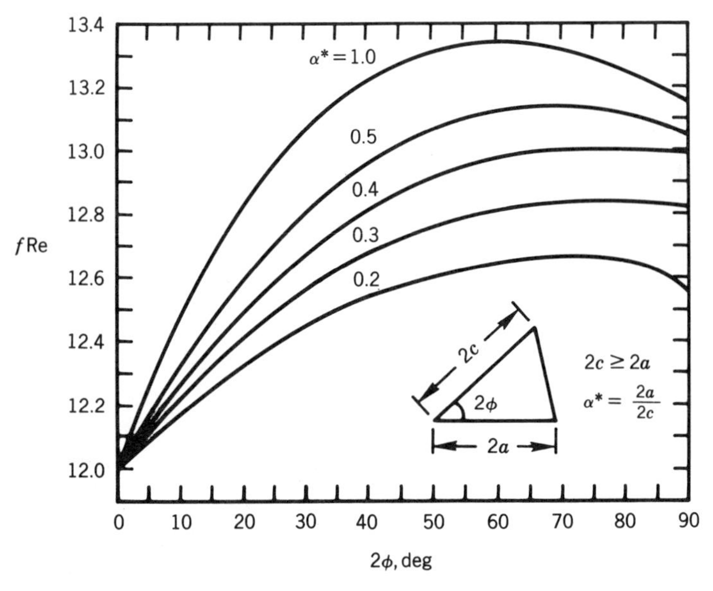

Figure 3.23. Fully developed friction factors for arbitrary triangular ducts [1, 80].

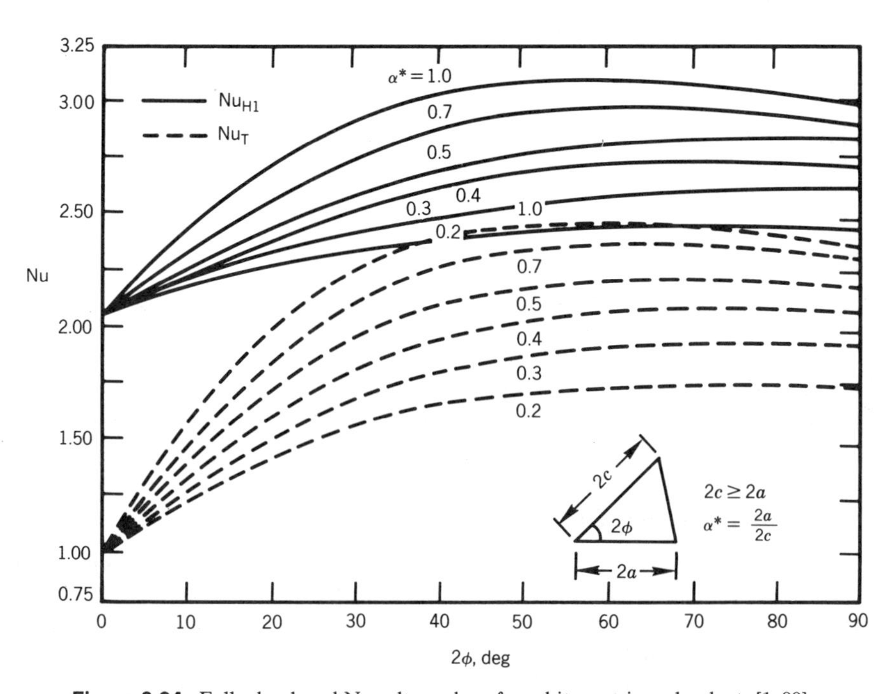

Figure 3.24. Fully developed Nusselt numbers for arbitrary triangular ducts [1, 80].

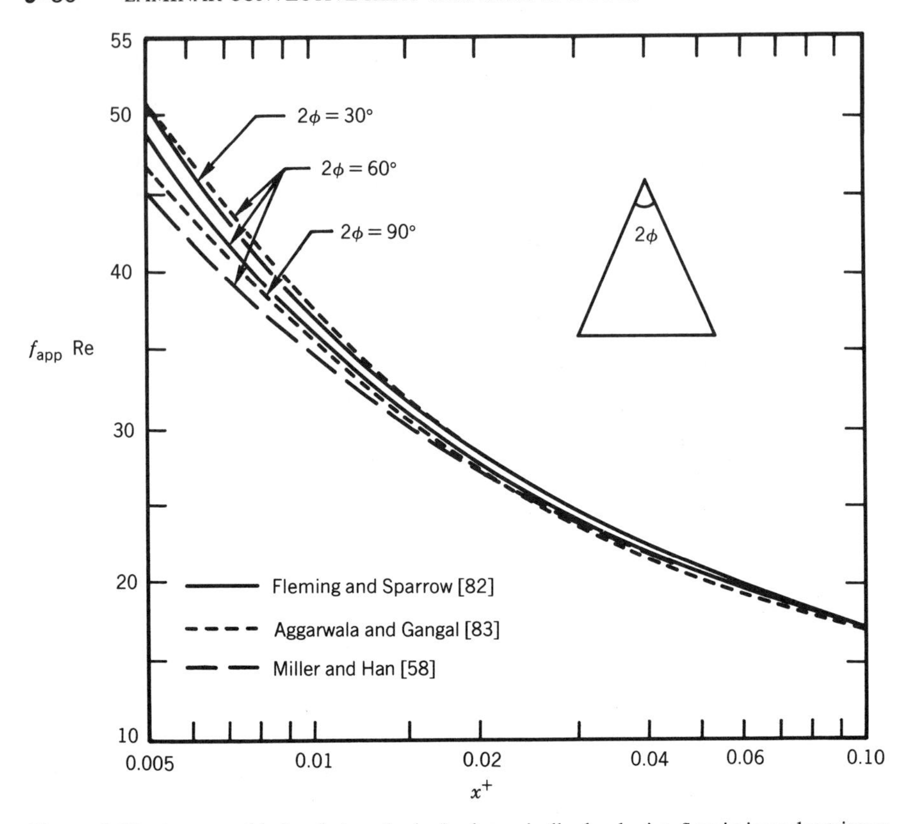

Figure 3.25. Apparent friction factors for hydrodynamically developing flow in isosceles triangular ducts [58, 82, 83].

3.5.3 Thermally Developing Flow

The thermal entrance length problem has been solved by Wibulswas [69] for equilateral (Fig. 3.17*a*) and right-angled isosceles triangular (Fig. 3.17*e*) ducts. His results for Pr = 0, 0.72, and ∞ are presented in Tables 3.20 and 3.21. These results are for simultaneously developing flow. However, the results for Pr = ∞ apply to all fluids in the thermal entrance region as Pr = ∞ implies that the flow is hydrodynamically developed.

3.5.4 Simultaneously Developing Flow

The problem of simultaneously developing flow for equilateral triangular and right-angled isosceles triangular ducts was solved numerically by Wibulswas [69] for a single fluid with Pr = 0.72. His results for the Ⓣ and Ⓗ1 boundary conditions are presented in Tables 3.20 and 3.21.

In an unpublished communication, Wibulswas and Tangsirimonkol [84] compared the numerical prediction of [69] with the experimental measurementals for the Ⓣ boundary condition. For the equilateral triangular duct, they correlated the results of Table 3.20 within +0.2% and −3% by

$$\mathrm{Nu}_{m,\mathrm{T}} = 1.594 x^{*\,-0.331} \tag{3.176}$$

TABLE 3.20 Local and Mean Nusselt Numbers for Thermally and Simultaneously Developing Flows in an Equilateral Triangular Duct [69]

$\frac{1}{x^*}$	$Nu_{x,T}$			$Nu_{m,T}$			$Nu_{x,H1}$			$Nu_{m,H1}$		
	Pr = ∞	0.72	0	∞	0.72	0	∞	0.72	0	∞	0.72	0
10	2.57	2.80	3.27	3.10	3.52	4.65	3.27	3.58	4.34	4.02	4.76	6.67
20	2.73	3.11	3.93	3.66	4.27	5.79	3.48	4.01	5.35	4.76	5.87	8.04
30	2.90	3.40	4.46	4.07	4.88	6.64	3.74	4.41	6.14	5.32	6.80	9.08
40	3.08	3.67	4.89	4.43	5.35	7.32	4.00	4.80	6.77	5.82	7.57	9.96
50	3.26	3.93	5.25	4.75	5.73	7.89	4.26	5.13	7.27	6.25	8.20	10.65
60	3.44	4.15	5.56	5.02	6.08	8.36	4.49	5.43	7.66	6.63	8.75	11.27
80	3.73	4.50	6.10	5.49	6.68	9.23	4.85	6.03	8.26	7.27	9.73	12.35
100	4.00	4.76	6.60	5.93	7.21	9.98	5.20	6.56	8.81	7.87	10.60	13.15
120	4.24	4.98	7.03	6.29	7.68	10.59	5.50	7.04	9.30	8.38	11.38	13.82
140	4.47	5.20	7.47	6.61	8.09	11.14	5.77	7.50	9.74	8.84	12.05	14.46
160	4.67	5.40	7.88	6.92	8.50	11.66	6.01	7.93	10.17	9.25	12.68	15.02
180	4.85	5.60	8.20	7.18	8.88	12.10	6.22	8.33	10.53	9.63	13.27	15.50
200	5.03	5.80	8.54	7.42	9.21	12.50	6.45	8.71	10.87	10.02	13.80	16.00

TABLE 3.21 Local and Mean Nusselt Numbers for Thermally and Simultaneously Developing Flows in Right-Angled Isosceles Triangular Ducts [69]

$\frac{1}{x^*}$	$Nu_{x,T}$			$Nu_{m,T}$			$Nu_{x,H1}$			$Nu_{m,H1}$		
	$Pr = \infty$	0.72	0	∞	0.72	0	∞	0.72	0	∞	0.72	0
10	2.40	2.52	3.75	2.87	3.12	4.31	3.29	4.00	5.31	4.22	5.36	6.86
20	2.53	2.76	4.41	3.33	3.73	5.35	3.58	4.73	6.27	4.98	6.51	7.97
30	2.70	2.98	4.82	3.70	4.20	6.48	3.84	5.23	6.85	5.50	7.32	8.68
40	2.90	3.18	5.17	4.01	4.58	6.97	4.07	5.63	7.23	5.91	7.95	9.20
50	3.05	3.37	5.48	4.28	4.90	7.38	4.28	5.97	7.55	6.25	8.50	9.67
60	3.20	3.54	5.77	4.52	5.17	7.73	4.47	6.30	7.85	6.57	8.99	10.07
80	3.50	3.85	6.30	4.91	5.69	8.31	4.84	6.92	8.37	7.14	9.80	10.75
100	3.77	4.15	6.75	5.23	6.10	8.80	5.17	7.45	8.85	7.60	10.42	11.32
120	4.01	4.43	7.13	5.52	6.50	9.18	5.46	7.95	9.22	8.03	10.90	11.77
140	4.21	4.70	7.51	5.78	6.82	9.47	5.71	8.39	9.58	8.40	11.31	12.14
160	4.40	4.96	7.84	6.00	7.10	9.70	5.95	8.80	9.90	8.73	11.67	12.47
180	4.57	5.22	8.10	6.17	7.33	9.94	6.16	9.14	10.17	9.04	12.00	12.75
200	4.74	5.49	8.38	6.33	7.57	10.13	6.36	9.50	10.43	9.33	12.29	13.04

They also correlated within +0.8% and −4% the results of Table 3.21 for the right-angled isosceles triangular duct by

$$\mathrm{Nu}_{m,\mathrm{T}} = 1.470x^{*-0.309} \tag{3.177}$$

On comparing the predictions of Eqs. (3.176) and (3.177) with the experimental measurements, Wibulswas and Tangsirimonkol [84] found that the numerical predictions are appreciably lower than the experimental values, especially in the neighborhood of the duct inlet. This is apparently due to neglect of the transverse velocity components in the numerical analysis. As in a square duct, so in triangular ducts, neglect of the transverse velocity components underestimates the Nusselt numbers. However, in a circular duct neglect of the radial velocity component overestimates the Nusselt numbers. The reason for this difference in circular and noncircular ducts is not clear to the authors, although it appears that the corner effect in noncircular ducts may be a contributing factor.

Wibulswas and Tangsirimonkol [84] correlated their experimental measurements for the equilateral triangular and right-angled isosceles triangular ducts by

$$\mathrm{Nu}_{m,\mathrm{T}} = 0.44x^{*-0.66} \tag{3.178}$$

and recommended that Eq. (3.178) be used for $1.82 \times 10^{-3} < x^* < 3.33 \times 10^{-2}$, and Eq. (3.176) be used for $x^* > 3.33 \times 10^{-2}$, for both cross sections.

3.6 ELLIPTICAL DUCTS

Elliptical geometry constitutes a useful family of ducts, ranging from a narrow lenticular passage to a circular one. A flat duct does not constitute a limiting case of elliptical geometry. Fully developed laminar fluid flow and heat transfer characteristics of elliptical ducts have been analyzed. Also, the hydrodynamic entrance length and the thermal entrance length problems have been solved. However, the problem of simultaneously developing flow has not received attention in the literature.

3.6.1 Fully Developed Flow

The fully developed velocity profile and friction factors for the elliptical duct shown as an inset to Fig. 3.26 are given by [1]

$$\frac{u}{u_m} = 2\left[1 - \left(\frac{z}{a}\right)^2 - \left(\frac{y}{b}\right)^2\right] \tag{3.179}$$

$$u_m = -\frac{1}{4\mu}\left(\frac{dp}{dx}\right)\left(\frac{b^2}{1+\alpha^{*2}}\right), \qquad f\,\mathrm{Re} = 2(1+\alpha^{*2})\left(\frac{\pi}{E(m)}\right)^2 \tag{3.180}$$

In the above equations, $m = 1 - \alpha^{*2}$ and $E(m)$ is the complete elliptic integral of the second kind. The hydraulic diameter of the elliptical duct is $D_h = \pi b/E(m)$, and the cross-sectional area $A_c = \pi ab$.

The f Re factors computed from Eq. (3.180) are displayed in Fig. 3.26, which also includes a curve for the hydrodynamic entrance length L_{hy}^+ determined from the

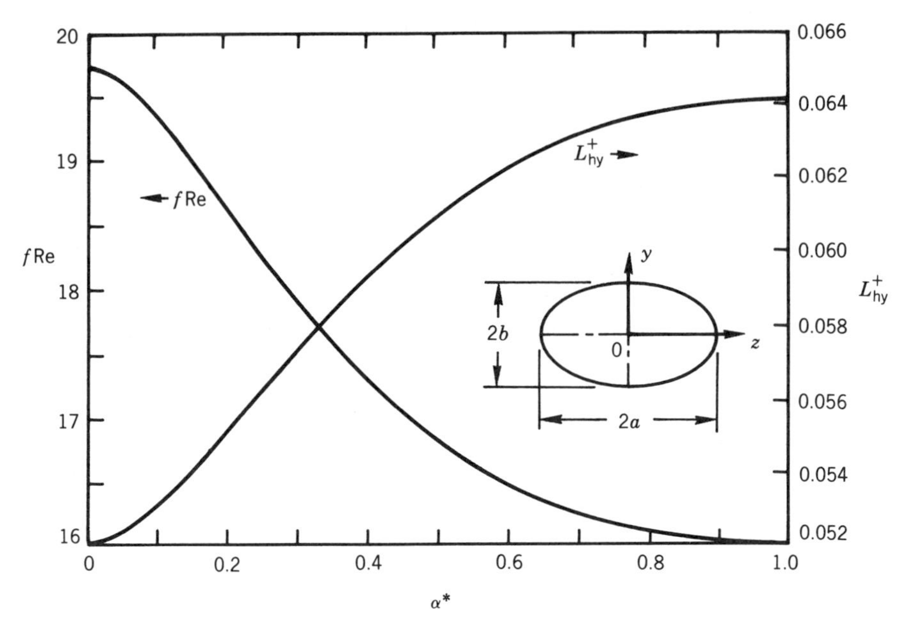

Figure 3.26. Fully developed friction factors and hydrodynamic entrance lengths for elliptical ducts [85].

following formula developed by Bhatti [85]:

$$L^+_{hy} = \frac{0.5132}{1 + \alpha^{*2}} \left(\frac{E(m)}{\pi} \right)^2 \tag{3.181}$$

The values of L^+_{hy} from this equation represent a substantial improvement over very approximate values reported in [1].

The fully developed incremental pressure drop number $K(\infty)$ for elliptical ducts is independent of the duct aspect ratio α^*, as elaborated by Bhatti [85]. Lundgren et al. [86] determined $K(\infty)$ as $\frac{4}{3}$ without solving the hydrodynamic entrance length problem, whereas the hydrodynamic entrance length analysis of Bhatti [85] led to a value of $\frac{7}{6}$. The experimental values of $K(\infty)$ for the circular duct, which are applicable to elliptical ducts also, range between 1.12 and 1.35. In view of such a wide variation in the measured values, it is recommended that the mean experimental value of 1.26 be used for practical computations.

Dunwoody [87] determined the fully developed Nusselt numbers $\mathrm{Nu_T}$ for elliptical ducts with $\alpha^* = \frac{1}{16}, \frac{1}{8}, \frac{1}{4}, \frac{1}{2}$, and $\frac{4}{5}$. His results are displayed in Fig. 3.27, including the limiting cases of $\mathrm{Nu_T}$ equal to 3.658 and 3.488 for α^* equal to 1 and 0, respectively.

The fully developed (H1) heat transfer problem for elliptical ducts with internal thermal energy sources was first investigated by Tao [38]. Tyagi [5] extended Tao's work by including the effect of viscous dissipation. The closed-form formulas of these investigators for various momentum and heat transfer quantities are rather complex. Recently, Bhatti [88] developed a simpler closed-form solution for the (H1) problem

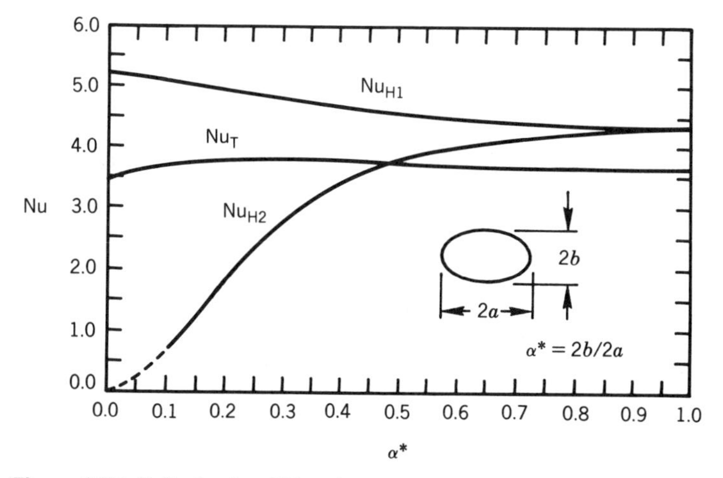

Figure 3.27. Fully developed Nusselt numbers for elliptical ducts [5, 38, 79, 87].

using the integral method. According to this analysis

$$\frac{T_w - T}{T_w - T_m} = \frac{6}{11}\left\{3 - \left[\left(\frac{z}{a}\right)^2 + \left(\frac{y}{b}\right)^2\right]\left[4 - \left(\frac{z}{a}\right)^2 - \left(\frac{y}{b}\right)^2\right]\right\} \tag{3.182}$$

$$\frac{T_w - T_m}{q_w'' D_h/k} = \frac{11}{6(1 + \alpha^{*2})}\left(\frac{E(m)}{\pi}\right)^2 \tag{3.183}$$

$$\mathrm{Nu}_{\mathrm{H1}} = \frac{6}{11}\left(\frac{\pi}{E(m)}\right)^2(1 + \alpha^{*2}) \tag{3.184}$$

For the limiting case of the circular duct, Eqs. (3.182) to (3.184) reduce to Eqs. (3.21) and (3.22) with $\gamma = 0$. The $\mathrm{Nu}_{\mathrm{H1}}$ values computed from Eq. (3.184) agree within $\pm 3\%$ with the values computed from the following more accurate formula due to Tao [38] and Tyagi [5]:

$$\mathrm{Nu}_{\mathrm{H1}} = 9\left(\frac{\pi}{E(m)}\right)^2\left(\frac{1 + 6\alpha^{*2} + \alpha^{*4}}{17 + 98\alpha^{*2} + 17\alpha^{*4}}\right)(1 + \alpha^{*2}) \tag{3.185}$$

These latter values are displayed in Fig. 3.27. Iqbal et al. [79] analyzed the (H2) problem for elliptical ducts. Their $\mathrm{Nu}_{\mathrm{H2}}$ results are also included in Fig. 3.27.

3.6.2 Hydrodynamically Developing Flow

The hydrodynamic entrance length problem for elliptical ducts has received scant attention. The only available theoretical solution is an integral solution developed by Bhatti [85]. Recently, Abdel-Wahed et al. [89] reported an experimental investigation of the problem for an elliptical duct with $\alpha^* = 0.5$.

According to [85], the velocity distribution in the hydrodynamic region of elliptical ducts is given by

$$\frac{u}{u_m} = \frac{2}{1+\eta} \qquad \text{for} \quad 0 \le \left(\frac{z}{a}\right)^2 + \left(\frac{y}{b}\right)^2 \le \eta \tag{3.186}$$

$$\frac{u}{u_m} = \frac{2\left[1 - (z/a)^2 - (y/b)^2\right]}{1-\eta^2} \qquad \text{for} \quad \eta \le \left(\frac{z}{a}\right)^2 + \left(\frac{y}{b}\right)^2 \le 1 \tag{3.187}$$

where η is a boundary-layer parameter representing the fraction of the duct cross section carrying inviscid flow. It is implicitly given as a function of x^+ by

$$16(1 + \alpha^{*2})\left(\frac{\pi}{E(m)}\right)^2 x^+ = \eta^2 - 1 - \ln \eta^2 \tag{3.188}$$

The apparent friction factors and incremental pressure drop numbers are expressed in terms of η, which in essence is treated as an independent variable:

$$f_{\text{app}}\text{Re} = \frac{\Delta p^*}{4x^+}, \qquad \Delta p^* = \frac{2(1-\eta)(1+3\eta) - (1+\eta)^2 \ln \eta^3}{3(1+\eta)^2} \tag{3.189}$$

$$K(x) = \frac{(3\eta^3 + 9\eta^2 + 21\eta + 7)(1-\eta)}{6(1+\eta)^2} \tag{3.190}$$

In the limit when the flow becomes hydrodynamically developed, η attains the value of zero and all the momentum transfer quantities asymptotically reduce to the known exact fully developed values. The apparent friction factors, incremental pressure drop numbers, and axial pressure drop computed from Eqs. (3.188) to (3.190) are displayed in Fig. 3.28. Note that the geometric factor $(1 + \alpha^{*2})[\pi/E(m)]^2$ appears in the abscissa as well as in the left-hand ordinate corresponding to the $f_{\text{app}}\text{Re}$ curve of Fig. 3.28.

The axial pressure drop measurements of Abdel-Wahed et al. [89] for an elliptical duct of aspect ratio $\alpha^* = 0.5$ are in excellent accord with the predictions of Eq. (3.189). However, their axial velocity measurements along the semimajor and semiminor axes exhibit peculiar wiggles in the neighborhood of the duct inlet ($x^+ < 0.02$) which are not predicted by the theoretical analysis. Farther downstream ($x^+ > 0.02$), the agreement between the experimental measurements and the analytical predictions is within $\pm 2\%$.

3.6.3 Thermally Developing Flow

The problem of thermally developing flow with the Ⓣ boundary condition has been solved by several investigators [1]. The most accurate results are those of Dunwoody [87]. His local Nusselt numbers $\text{Nu}_{x,\text{T}}$ are expressed in terms of a double infinite series. However, his mean Nusselt numbers $\text{Nu}_{m,\text{T}}$ are expressible in terms of the following

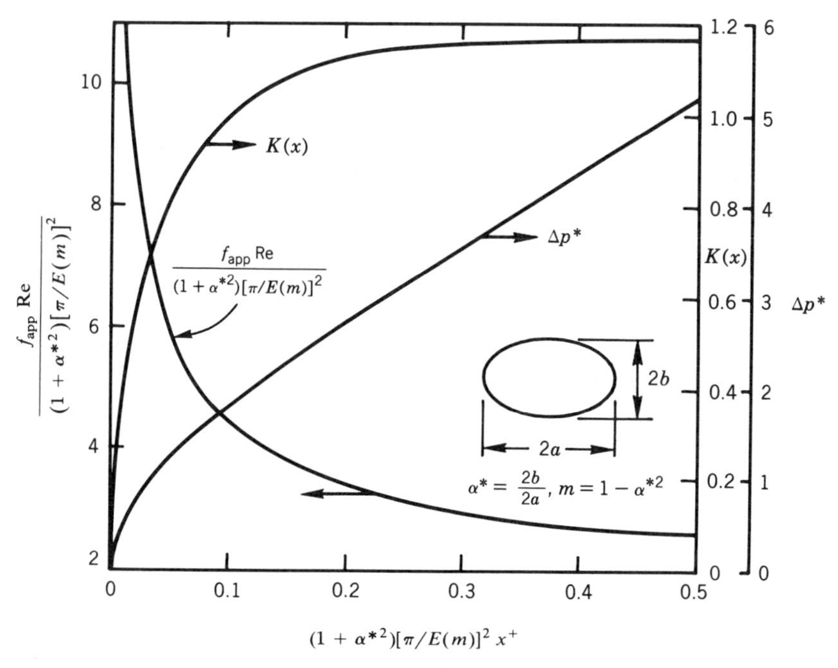

Figure 3.28. Fluid flow parameters in the hydrodynamic entrance region of elliptical ducts [85].

simple formula:

$$\mathrm{Nu}_{m,\mathrm{T}} = \frac{\lambda}{4}\left(1 + \frac{C}{x^*}\right) \tag{3.191}$$

The λ and C values for Eq. (3.191) are presented in Table 3.22 as functions of α^*.

Richardson [90] among others solved the Ⓣ problem in the thermal entrance region by invoking the simplifications offered by Lévêque's theory [21] and presented the following formula:

$$\mathrm{Nu}_{m,\mathrm{T}} = \frac{3}{\Gamma(\frac{4}{3})(9x^*)^{1/3}}\left(1 + \frac{(1-\alpha^*)^2 + (1-\alpha^*)^3}{36}\right) \tag{3.192}$$

TABLE 3.22 The Constants of Eq. (3.191) Representing Mean Nusselt Numbers in the Thermal Entrance Region of Elliptical Ducts with the Ⓣ Boundary Condition [87]

α^*	λ	C
4/5	14.67	0.0138
1/2	14.97	0.0158
1/4	15.17	0.0239
1/8	14.90	0.0388
1/16	14.59	0.0578

When $\alpha^* = 1$, Eq. (3.192) reduces to Lévêque's solution for a circular duct given by Eq. (3.47). The predictions of Eq. (3.192) agree with those of Eq. (3.191) within -7% down to $x^* = 0.05$. Strictly speaking, Eq. (3.192) is more accurate for low values of x^* such as $x^* < 0.005$, and Eq. (3.191) for high values such as $x^* > 0.005$.

Someswara Rao et al. [91] developed a Lévêque-type thermal entrance length solution for elliptical ducts with the (H1) boundary condition. Their mean Nusselt numbers can be represented by the following equation:

$$\mathrm{Nu}_{m,\mathrm{H1}} = \frac{2.61F}{x^{*1/3}} \tag{3.193}$$

where the factor F is a function of α^*. The numerical values of F furnished in tabular form in [91] can be represented with a maximum error of $\pm 3\%$ by

$$F = 1.2089 - 0.7951\alpha^* - 4.3011\alpha^{*2} + 23.8465\alpha^{*3} - 44.7053\alpha^{*4} + 37.0874\alpha^{*5} - 11.4809\alpha^{*6} \tag{3.194}$$

It may be noted that for $\alpha^* = 1$, Eq. (3.193) does not reduce to Eq. (3.62) derived from Lévêque's analysis for a circular duct.

3.7 ADDITIONAL SINGLY CONNECTED DUCTS

In addition to the singly connected ducts covered in the preceding sections, a number of other singly connected ducts have been analyzed. However, their fluid flow and heat transfer characteristics have not been studied in detail. The available information on the remaining singly connected ducts is presented in this section.

3.7.1 Sine Ducts

The fully developed fluid flow and heat transfer characteristics of a sine duct shown in Fig. 3.29*a* are presented in Tables 3.23 and 3.24 together with some geometrical characteristics. These results are based on the analyses by Shah [75] and Sherony and Solbrig [92]. $\bar{y}$ and $\bar{y}_{\max}$ in Table 3.23 denote the distances measured from the duct base to the centroid and to the point of the maximum velocity respectively. $T^*_{w,\max}$ and $T^*_{w,\min}$ in Table 3.24 are defined by Eq. (3.167).

3.7.2 Trapezoidal Ducts

Shah [75] analyzed the fully developed fluid flow and heat transfer characteristics of the trapezoidal duct shown in Fig. 3.29*b*. His results are presented in Figs. 3.30 and 3.31. Tabulated results are available in [1]. When $a \to 0$ (or equivalently $b/a \to \infty$), the trapezoidal duct reduces to an isosceles triangular duct; when $\phi = 90°$, it reduces to a rectangular duct.

Lawal and Mujumdar [93] numerically analyzed the problem of laminar flow development in a trapezoidal duct with $\phi = 72°$ and $2b/2a = 1.1902$. Their fluid flow results are presented in Table 3.25.

Lawal and Mujumdar [93] also analyzed simultaneously developing flow in a trapezoidal duct with $\phi = 72°$ and $2b/2a = 1.1902$. Their local Nusselt numbers $\mathrm{Nu}_{x,\mathrm{T}}$ for a fluid with $\mathrm{Pr} = 0.1$ are presented in Table 3.25.

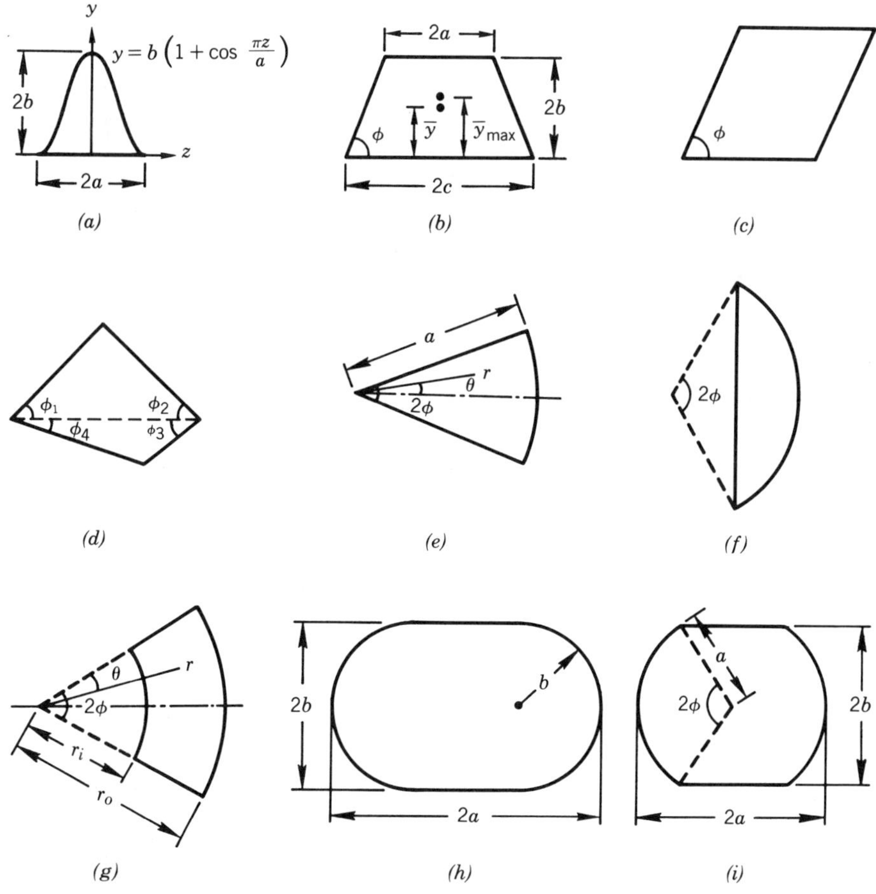

Figure 3.29. Some singly connected ducts: (a) sine, (b) trapezoidal, (c) rhombic, (d) quadrilateral, (e) circular sector, (f) circular segment, (g) annular sector, (h) stadium-shaped, (i) modified stadium-shaped.

TABLE 3.23 Geometrical and Fully Developed Flow Characteristics of Sine Ducts [75]

$\frac{2b}{2a}$	$\frac{P}{2a}$	$\frac{D_h}{2a}$	$\frac{\bar{y}}{2a}$	$\frac{\bar{y}_{max}}{2a}$	$\frac{u_{max}}{u_m}$
∞	—	—	—	—	3.825
2	5.1898	0.77074	0.75000	0.46494	2.288
$\frac{3}{2}$	4.2315	0.70897	0.56250	0.40964	2.239
1	3.3049	0.60516	0.37500	0.33390	2.197
$\sqrt{3}/2$	3.0667	0.56479	0.32476	0.30773	2.191
$\frac{3}{4}$	2.8663	0.52332	0.28125	0.28205	2.190
$\frac{1}{2}$	2.4637	0.40589	0.18750	0.21347	2.211
$\frac{1}{4}$	2.1398	0.23366	0.09375	0.11926	2.291
$\frac{1}{8}$	2.0375	0.12270	0.04688	0.06173	2.357
0	2.0000	0.00000	0.00000	0.00000	2.400

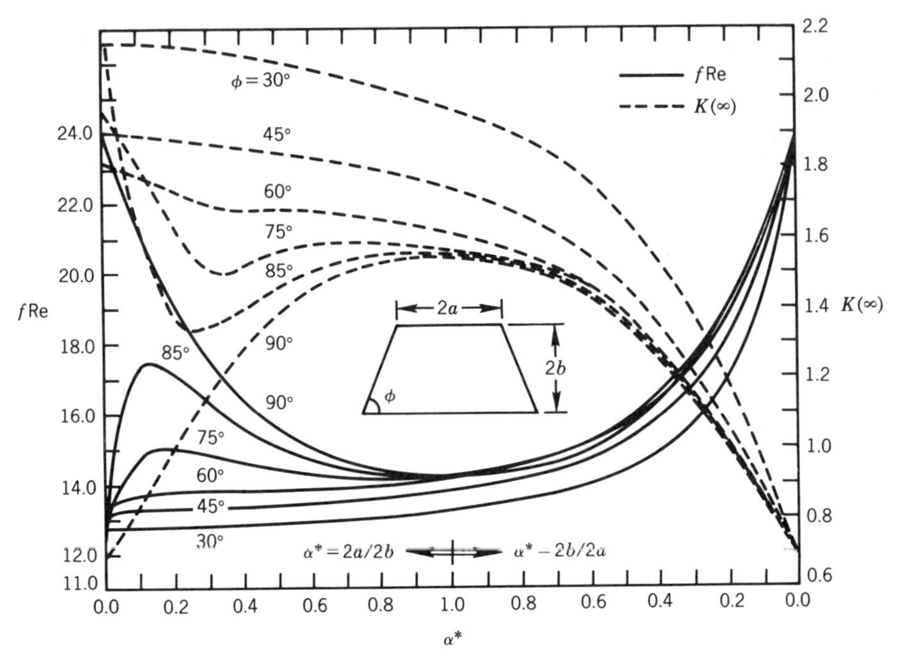

Figure 3.30. Fully developed fluid flow characteristics of trapezoidal ducts [75].

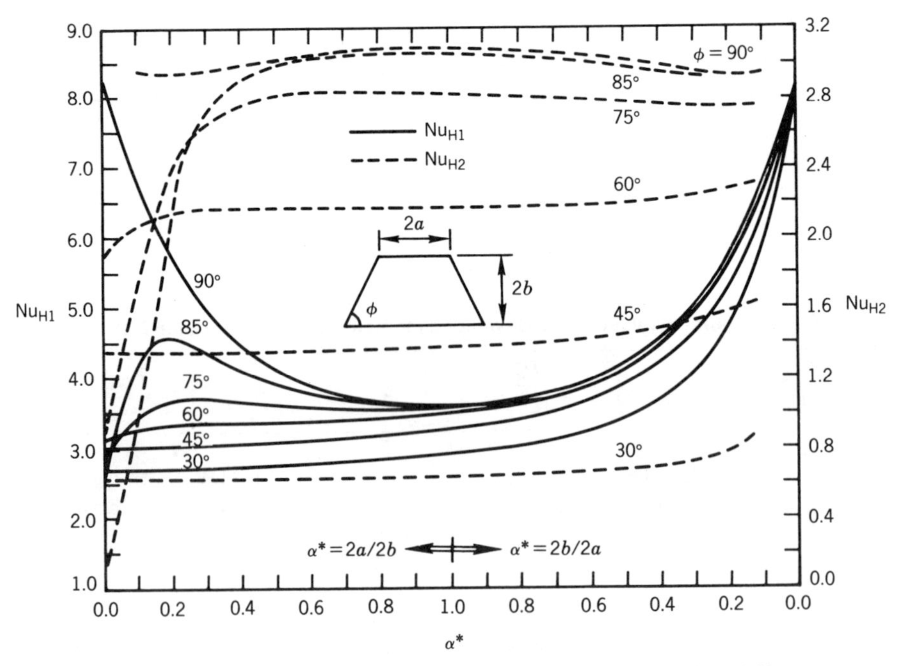

Figure 3.31. Fully developed Nusselt numbers for trapezoidal ducts [75].

TABLE 3.24 Fully Developed Fluid Flow and Heat Transfer Characteristics of Sine Ducts [75, 92]

$\frac{2b}{2a}$	$K(\infty)$	L_{hy}^+	f Re	Nu_T	Nu_{H1}	Nu_{H2}	$T_{w,max}^*$	$T_{w,min}^*$
∞	3.218	0.1701	15.303	0.739	2.521	0	—	—
2	1.884	0.0403	14.553	—	3.311	0.95	2.92	0.002
$\frac{3}{2}$	1.806	0.0394	14.022	2.60	3.267	1.38	2.93	0.257
1	1.744	0.0400	13.023	2.45	3.102	1.55	2.17	0.398
$\sqrt{3}/2$	1.739	0.0408	12.630	—	3.014	1.47	2.58	0.396
$\frac{3}{4}$	1.744	0.0419	12.234	2.33	2.916	1.34	2.93	0.379
$\frac{1}{2}$	1.810	0.0464	11.207	2.12	2.617	0.90	3.65	0.266
$\frac{1}{4}$	2.013	0.0553	10.123	1.80	2.213	0.33	4.16	0.099
$\frac{1}{8}$	2.173	0.0612	9.743	—	2.017	0.095	4.31	0.030
0	2.271	0.0648	9.600	1.178	1.920	0	—	—

TABLE 3.25 Hydrodynamic Entrance Region Fluid Flow Characteristics and Local Nusselt Numbers for Simultaneously Developing Flow for Pr = 0.1 in a Trapezoidal Duct (ϕ = 72°, $2b/2a$ = 1.1902) [93]

x^+	f_{app} Re	u_{max}/u_m	x^*	$Nu_{x,T}$
0.00075	140.6000	1.2006	0.00175	16.7062
0.00150	96.9333	1.2656	0.00250	13.7624
0.00750	46.5800	1.5573	0.00750	7.6264
0.03000	26.1900	1.9466	0.01500	5.1608
0.04875	22.1733	2.0556	0.07500	2.9888
0.08625	18.9333	2.1197	0.15000	2.5573
0.1	18.3125	2.1263		

3.7.3 Rhombic Ducts

The fully developed fluid flow and heat transfer characteristics of rhombic ducts shown in Fig. 3.29*c* were analyzed by Shah [75]. His results are presented in Figs. 3.32 and 3.33. Tabulated results are available in [1].

3.7.4 Quadrilateral Ducts

Nakamura et al. [94] analyzed the fully developed fluid flow and heat transfer characteristics of arbitrary polygonal ducts. Their numerical results for some quadrilateral ducts, shown in Fig. 3.29*d*, are presented in Table 3.26.

3.7.5 Regular Polygonal Ducts

The fully developed fluid flow and heat transfer characteristics of a regular polygonal duct with n equal sides each subtending an angle of $360°/n$ at the duct center have been analyzed by several investigators, as summarized in [1]. The fully developed friction factors and Nusselt numbers for these ducts are presented in Fig. 3.34.

Schenkel [95] presented the following formula for the fully developed friction factors for regular polygonal ducts:

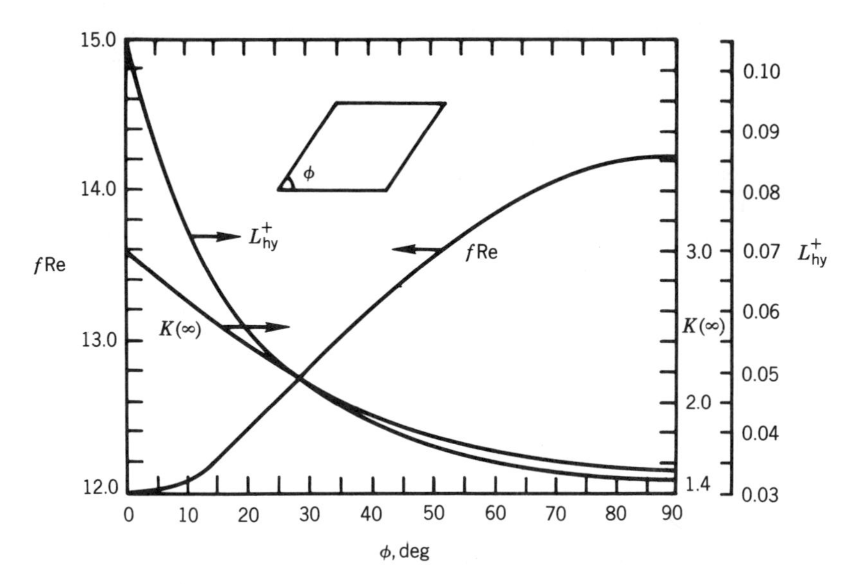

Figure 3.32. Fully developed fluid flow characteristics of rhombic ducts [75].

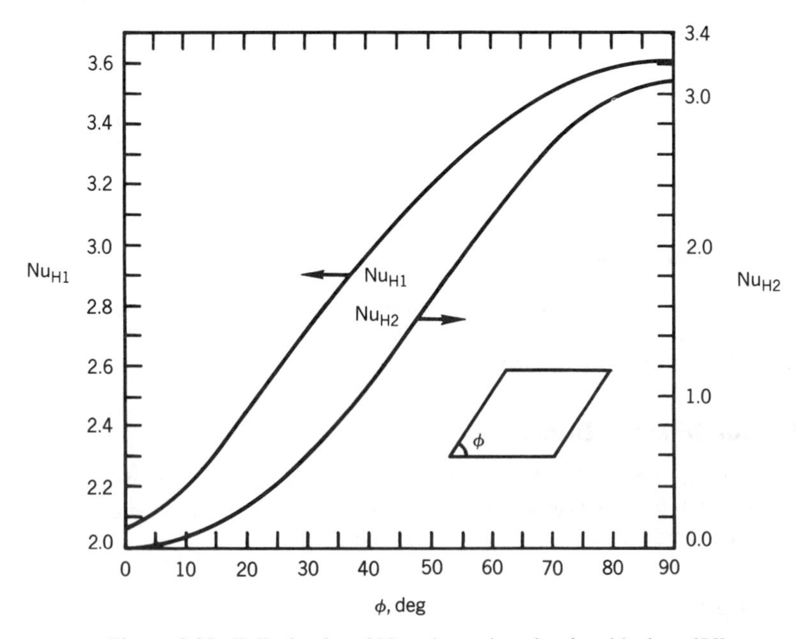

Figure 3.33. Fully developed Nusselt numbers for rhombic ducts [75].

TABLE 3.26 Fully Developed Friction Factors, Incremental Pressure Drop Numbers, and Nusselt Numbers for Some Quadrilateral Ducts [94]

ϕ_1 (deg)	ϕ_2 (deg)	ϕ_3 (deg)	ϕ_4 (deg)	f Re	$K(\infty)$	Nu_{H1}	Nu_{H2}
60	70	45	32.23	14.16	1.654	3.45	2.80
50	60	30	21.67	14.36	1.612	3.55	2.90
60	30	45	71.57	14.69	1.522	3.72	3.05
60	30	60	79.11	14.01	1.707	3.35	2.68

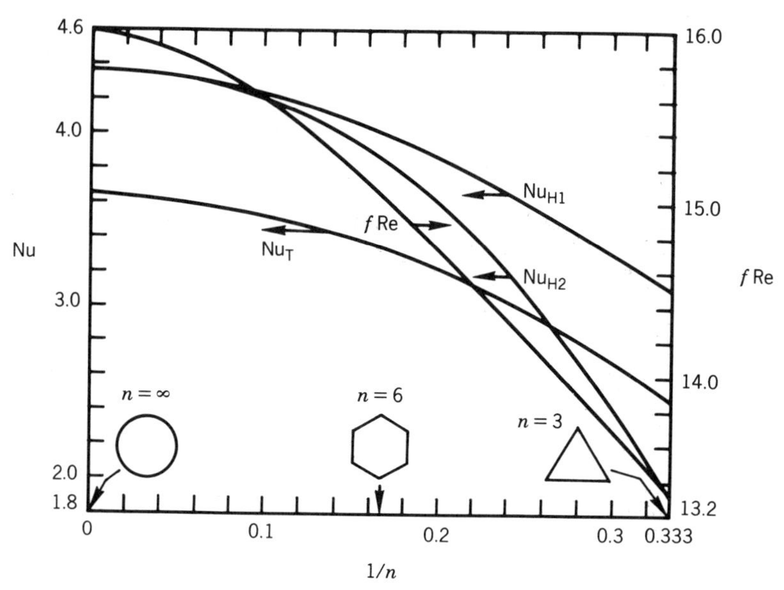

Figure 3.34. Fully developed friction factors and Nusselt numbers for regular polygonal ducts [1].

$$f\,\mathrm{Re} = 16\left(\frac{n^2}{0.44 + n^2}\right)^4 \tag{3.195}$$

The predictions of Eq. (3.195) are within ±1% of the tabulated values in [1], which are shown in Fig. 3.34.

Lawal and Mujumdar [93] numerically analyzed the problem of laminar flow development in a pentagonal duct ($n = 5$). Their fluid flow results are presented in Table 3.27.

Lawal and Mujumdar [93] also analyzed simultaneously developing flow in a pentagonal duct for a Newtonian fluid with Pr = 0.1. Their local Nusselt numbers for the Ⓣ boundary condition are presented in Table 3.27.

3.7.6 Circular Sector Ducts

Eckert and Irvine [96] analyzed the fully developed laminar flow through the circular sector duct (also referred to as a wedge-shaped duct) shown in Fig. 3.29*e* by invoking

TABLE 3.27 Hydrodynamic Entrance Region Fluid Flow Characteristics and Local Nusselt Numbers for Simultaneously Developing Flow for Pr = 0.1 in a Pentagonal Duct [93]

x^+	f_{app} Re	u_{max}/u_m	x^*	$\mathrm{Nu}_{x,T}$
0.00075	145.8667	1.2069	0.00175	17.0033
0.00150	101.200	1.2755	0.00250	13.5163
0.00750	46.3267	1.5541	0.00750	7.4950
0.03000	26.1517	1.9283	0.01500	5.4270
0.04875	22.2749	2.0210	0.07500	3.7295
0.08625	19.2023	2.0672	0.30000	3.3404
0.1	18.6230	2.0710		

the analogy of the torsion of a bar. Their formulas for the velocity profile, mean velocity, and friction factors are presented in [1].

Sparrow and Haji-Sheikh [78] extended the results of [96] to cover a wider range of 2ϕ. Their f Re and $K(\infty)$ results are presented in Fig. 3.35. Tabulated results are available in [1].

Schenkel [95] presented the following formula for the fully developed friction factors for circular sector ducts:

$$f\,\mathrm{Re} = 14.4\left[1 - \frac{1}{6}\left(1 - \frac{2\phi}{70}\right)^{1.371}\right] \quad \text{for } 2\phi \le 70° \tag{3.196}$$

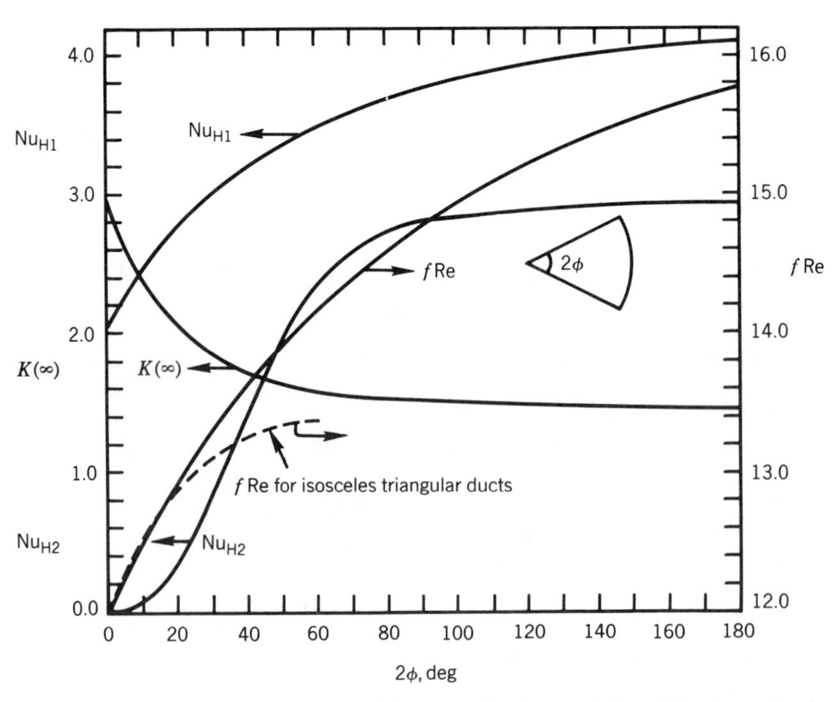

Figure 3.35. Fully developed fluid flow and heat transfer characteristics of circular sector ducts [78, 98].

TABLE 3.28 Flow Parameters for Hydrodynamically Developing Flow in Circular Sector Ducts [99]

x^+/L^+_{hy}	$2\phi = 11.25°$, $L^+_{hy} = 0.235$		$2\phi = 22.5°$, $L^+_{hy} = 0.144$		$2\phi = 45°$, $L^+_{hy} = 0.108$		$2\phi = 90°$, $L^+_{hy} = 0.0786$	
	f Re	$K(x)$	f Re	$K(x)$	f Re	$K(x)$	f Re	$K(x)$
0.001	109.3	0.207	147.9	0.177	180.4	0.171	226.0	0.154
0.003	66.09	0.335	81.62	0.281	98.27	0.266	115.7	0.241
0.006	48.27	0.456	60.18	0.377	63.69	0.352	78.29	0.314
0.010	38.52	0.568	48.30	0.469	53.91	0.432	60.26	0.380
0.020	28.93	0.758	35.64	0.628	39.59	0.567	43.93	0.492
0.030	25.09	0.887	30.20	0.739	33.47	0.662	37.37	0.571
0.040	22.70	0.991	27.02	0.830	29.89	0.738	33.50	0.635
0.050	21.18	1.076	24.92	0.901	27.51	0.801	30.88	0.689
0.070	19.90	1.211	22.20	1.019	24.41	0.903	27.44	0.778
0.100	17.34	1.364	19.91	1.153	21.78	1.021	24.40	0.881
0.150	15.82	1.540	17.81	1.313	19.40	1.165	21.58	1.007
0.200	14.95	1.664	16.65	1.429	18.05	1.269	19.95	1.100
0.250	14.47	1.756	15.89	1.517	17.09	1.350	18.87	1.171
0.300	14.07	1.828	15.35	1.588	16.54	1.412	18.07	1.229
0.350	13.83	1.885	14.95	1.646	16.06	1.464	17.51	1.275
0.400	13.59	1.930	14.65	1.695	15.66	1.510	17.05	1.315
0.450	13.51	1.970	14.41	1.735	15.34	1.546	16.70	1.347
0.500	13.35	1.997	14.22	1.770	15.11	1.577	16.41	1.375
0.600	13.19	2.051	13.95	1.825	14.79	1.625	15.96	1.418
0.700	13.03	2.091	13.76	1.868	14.55	1.661	15.67	1.451
0.800	12.95	2.118	13.63	1.902	14.39	1.689	15.46	1.475
0.900	12.90	2.139	13.52	1.930	14.23	1.710	15.31	1.494
1.000	12.87	2.156	13.47	1.951	14.15	1.728	15.19	1.509

where the angle 2ϕ is in degrees. The f Re values from Eq. (3.196) agree with the tabulated values of [1] presented in Fig. 3.35 within $\pm 1\%$ for $0 \leq 2\phi \leq 70°$.

Based on the analogy with the deflection of a uniformly loaded thin plate, simply supported around the rim, Eckert et al. [97] analyzed the (H1) fully developed temperature problem. Their temperature distribution is presented in [1].

Sparrow and Haji-Sheikh [78] extended the results of Eckert et al. [97] to cover a wider range of 2ϕ. Their $\mathrm{Nu}_{\mathrm{H1}}$ results are presented in Fig. 3.35.

Hu and Chang [98] analyzed the fully developed laminar problem for internally finned circular ducts with the (H2) boundary condition. A circular duct with n longitudinal full fins each of length a comprises n circular sector passages, each with an apex angle $2\phi = 360°/n$. The $\mathrm{Nu}_{\mathrm{H2}}$ results of Hu and Chang [98] are presented in Fig. 3.35.

A comparison of the results in Figs. 3.18, 3.19, and 3.35 shows that for $2\phi < 20°$ the results for the isosceles triangular duct agree rather closely with those for the circular sector duct.

Soliman et al. [99] numerically analyzed the problem of laminar flow development in circular sector ducts. Their $f_{\mathrm{app}}\mathrm{Re}$, $K(x)$, and L_{hy}^{+} results for $2\phi = 11.25$, 22.5, 45, and 90° are presented in Table 3.28.

3.7.7 Circular Segment Ducts

Sparrow and Haji-Sheikh [100] analyzed the fully developed laminar flow through the circular segment duct shown in Fig. 3.29f. Their f Re, $K(\infty)$, $\mathrm{Nu}_{\mathrm{H1}}$, and $\mathrm{Nu}_{\mathrm{H2}}$ results are presented in Fig. 3.36.

Hong and Bergles [101] obtained the thermal entrance length solution for a circular segment duct with $2\phi = 180°$, i.e., the semicircular duct. They presented the results for two thermal boundary conditions: (1) a constant wall heat flux along the axial flow direction and a constant wall temperature along the duct circumference, i.e., the (H1) boundary condition, (2) a constant wall heat flux along the axial flow direction and a constant wall temperature along the semicircular arc, with zero heat flux along the diameter. The local Nusselt numbers for these two boundary conditions are presented in Table 3.29.

3.7.8 Annular Sector Ducts

Sparrow et al. [102] analyzed the fully developed laminar flow through the annular sector duct (also referred to as a truncated sectorial duct) pictured in Fig. 3.29g. Their formulas for the velocity profile, mean velocity, and friction factors are presented in [1].

Recently, Niida [103] also obtained an analytical solution for the velocity distribution in an annular sector duct and expressed his solution in terms of an equivalent diameter. In addition, he provided an experimental verification of his pressure drop predictions.

Shah and London [1] computed f Re values to a high degree of accuracy using the analytical solution of Sparrow et al. [102]. The computed results are displayed in Fig. 3.37. Tabulated values are available in [1].

Based on the analogy of the torsion of a prismatic bar, Schenkel [95] derived the following approximate equation for f Re in an annular sector duct:

$$f\,\mathrm{Re} = \frac{24}{\left[1 - \dfrac{0.63}{\phi}\left(\dfrac{1 - r^*}{1 + r^*}\right)\right]\left[1 + \dfrac{1}{\phi}\left(\dfrac{1 - r^*}{1 + r^*}\right)\right]^2} \tag{3.197}$$

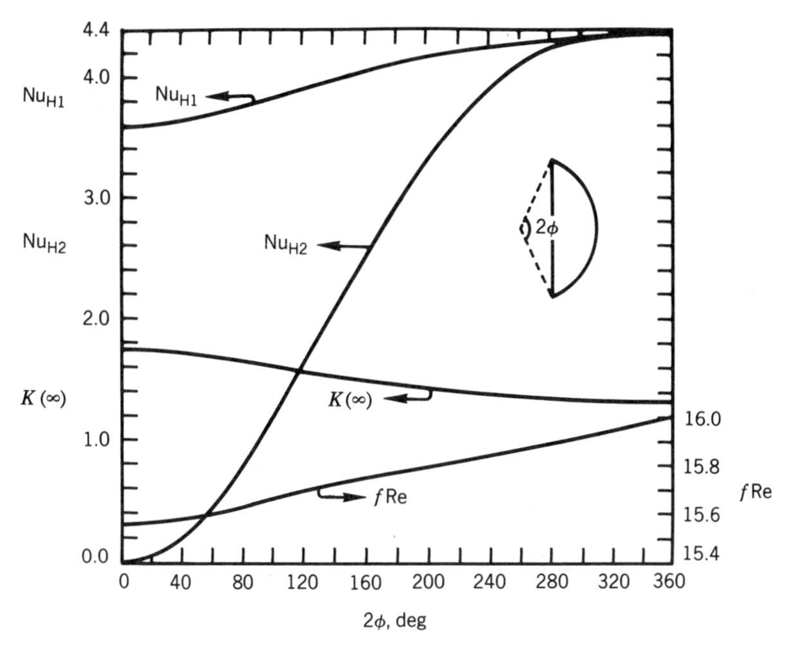

Figure 3.36. Fully developed fluid flow and heat transfer characteristics of circular segment ducts [100].

TABLE 3.29 Local Nusselt Numbers in the Thermal Entrance Region of Semicircular Duct [101]

x^*	$Nu_{x,H1}$ (D)	$Nu_{x,H1}$ (D, flat wall hatched)	x^*	$Nu_{x,H1}$ (D)	$Nu_{x,H1}$ (D, flat wall hatched)
0.000458	17.71	17.43	0.0279	4.767	4.339
0.000954	13.72	13.41	0.0351	4.562	4.037
0.00149	11.80	11.37	0.0442	4.429	3.830
0.00208	10.55	10.08	0.0552	4.276	3.686
0.00271	9.605	9.141	0.0686	4.217	3.543
0.00375	8.475	8.127	0.0849	4.156	3.425
0.00493	7.723	7.375	0.105	4.124	3.330
0.00627	7.137	6.788	0.130	4.118	3.265
0.00777	6.556	6.312	0.159	4.108	3.208
0.00946	6.300	5.912	0.196	—	3.171
0.0128	5.821	5.368	0.241	—	3.161
0.0168	5.396	4.935	0.261	—	3.160
0.0217	5.077	4.579	∞	4.089	3.160

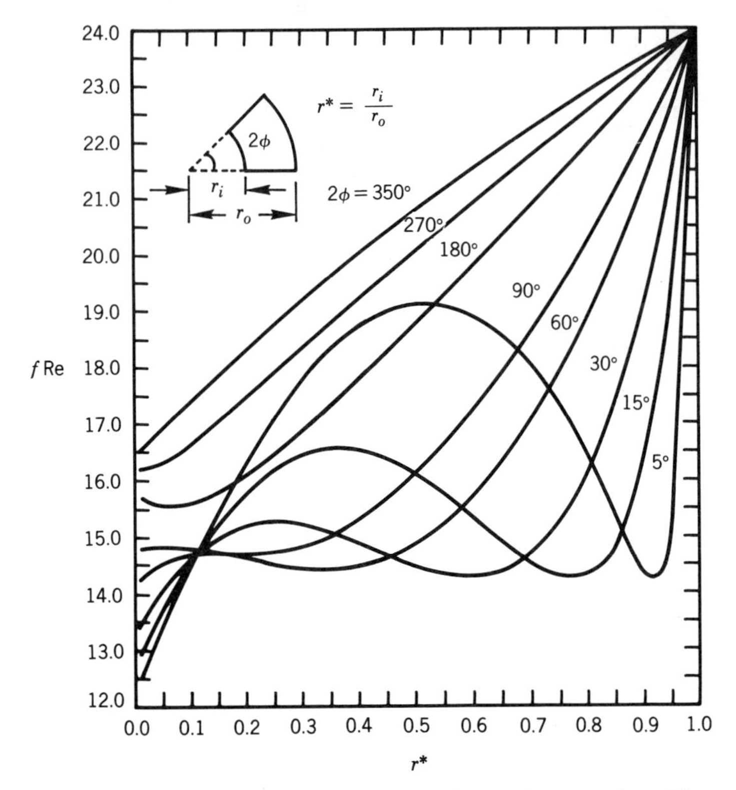

Figure 3.37. Fully developed friction factors for annular sector ducts [1].

This equation is applicable for $\phi \geq \phi_{\min}(r^*)$. The values of $\phi_{\min}$ for $r^* = 0, 0.1, 0.2, 0.3, 0.4, 0.5, 0.6, 0.7, 0.8$, and 0.9 are determined to be 60, 50, 42, 35, 28.5, 22.5, 17.5, 13, 8.5, and 4°, respectively. For $\phi \geq \phi_{\min}(r^*)$, the predictions of Eq. (3.197) are in excellent accord with the results tabulated in [1] and presented in Fig. 3.37.

Soliman [104] investigated the fully developed (H1) and (H2) problems for annular sector ducts. His Nu_{H1} and Nu_{H2} results are presented in Table 3.30.

Renzoni and Prakash [105] analyzed the simultaneously developing flow in annular sector ducts for air (Pr = 0.707). They treated the outer curved wall as adiabatic while imposing the (H1) boundary condition on the inner curved wall as well as on the two straight walls of the sector. Their fully developed friction factors, incremental pressure drop numbers, and hydrodynamic entrance lengths are presented in Table 3.31. It may be noted that L_{hy}^{+} values reported in Table 3.31 do not correspond to the definition given in Section 3.1.2. They are defined as the dimensionless axial distance corresponding to $f_{\text{app}}\text{Re} = 1.05 f\,\text{Re}$. This definition is similar to the definition of L_{th}^{*} given in Section 3.1.3. The fully developed Nusselt numbers (denoted simply as Nu_{fd}, since the thermal boundary condition employed in [105] does not correspond to any defined in Table 3.1) and the thermal entrance lengths determined by Renzoni and Prakash [105] are also presented in Table 3.31. The local Nusselt numbers and friction factors in the combined entrance region are available in [105].

TABLE 3.30 Fully Developed Nusselt Numbers for Annular Sector Ducts [104][a]

	$2\phi = 10°$		$15°$		$20°$		$30°$		$40°$		$60°$		$120°$	
r^*	Nu_{H1}	Nu_{H2}	Nu_{H1}	Nu_{H2}	Nu_{H1}	Nu_{H2}	Nu_{H1}	Nu_{H2}	Nu_{H1}	Nu_{H2}	Nu_{H1}	Nu_{H2}	Nu_{H1}	Nu_{H2}
0.05	2.668	0.0941	2.825	0.2288	2.961	0.4295	3.181	1.013	3.349	1.671	3.581	2.629	3.887	3.032
0.10	2.896	0.1097	3.041	0.2678	3.163	0.5059	3.354	1.187	3.490	1.939	3.660	2.818	3.863	3.020
0.15	3.137	0.1293	3.264	0.3203	3.367	0.6062	3.516	1.460	3.611	2.277	3.704	2.946	3.861	3.007
0.20	3.387	0.1552	3.488	0.3889	3.564	0.7516	3.656	1.839	3.697	2.583	3.715	3.022	3.893	2.996
0.25	3.643	0.1900	3.706	0.4877	3.742	0.9824	3.761	2.253	3.742	2.803	3.702	3.059	3.964	2.986
0.30	3.893	0.2393	3.903	0.6414	3.888	1.379	3.821	2.587	3.746	2.941	3.685	3.073	4.072	2.977
0.35	4.126	0.3152	4.065	0.9186	3.988	1.937	3.832	2.809	3.721	3.021	3.679	3.073	4.215	2.967
0.40	4.326	0.4414	4.177	1.436	4.031	2.427	3.802	2.945	3.683	3.061	3.697	3.066	4.392	2.956
0.45	4.474	0.6839	4.226	2.111	4.016	2.721	3.747	3.022	3.650	3.078	3.750	3.054	4.600	2.944
0.50	4.554	1.187	4.207	2.580	3.951	2.893	3.686	3.062	3.640	3.080	3.844	3.039	4.836	2.915
0.55	4.555	1.999	4.126	2.820	3.854	2.993	3.639	3.079	3.666	3.074	3.986	3.020	5.098	2.794
0.60	4.471	2.602	3.997	2.950	3.747	3.047	3.627	3.081	3.743	3.060	4.183	2.997	5.383	2.670
0.65	4.312	2.851	3.848	3.024	3.658	3.074	3.668	3.073	3.884	3.038	4.440	2.970	5.687	2.581
0.70	4.098	2.967	3.709	3.063	3.616	3.082	3.785	3.052	4.106	3.008	4.763	2.956	6.010	—
0.75	3.867	3.032	3.621	3.079	3.658	3.072	4.006	3.018	4.429	2.968	5.158	—	6.348	—
0.80	3.675	3.066	3.641	3.069	3.834	3.036	4.371	2.967	4.875	—	5.624	—	6.700	—
0.85	3.616	3.065	3.861	3.023	4.230	2.963	4.928	—	5.463	—	6.163	—	7.065	—
0.90	3.889	2.979	4.462	2.870	4.979	—	5.726	—	6.208	—	6.775	—	7.443	—
0.95	5.028	4.852	5.773	—	6.250	—	6.809	—	7.123	—	7.464	—	7.832	—

[a]Additional Nu_{H1} results for 2ϕ = 150, 180, 210, 240, 270, 300, 330, and 350° are available in [104].

TABLE 3.31 Fully Developed Fluid Flow and Heat Transfer Characteristics of Annular Sector Ducts [105]

2ϕ	r^*	$f\,\mathrm{Re}$	$K(\infty)$	L_{hy}^+	Nu_{fd}	L_{th}^*
15°	0.2	15.65	1.77	0.0775	3.433	0.1530
	0.5	16.01	1.32	0.0500	4.372	0.0924
	0.8	14.21	1.42	0.0529	3.340	0.0898
22.5°	0.2	15.35	1.64	0.0703	3.493	0.1320
	0.5	14.90	1.37	0.0516	3.933	0.0838
	0.8	15.03	1.33	0.0476	3.113	0.1090
45°	0.2	14.73	1.46	0.0574	3.461	0.1070
	0.5	14.29	1.42	0.0529	3.235	0.0972
	0.8	17.58	1.07	0.0303	3.327	0.1230

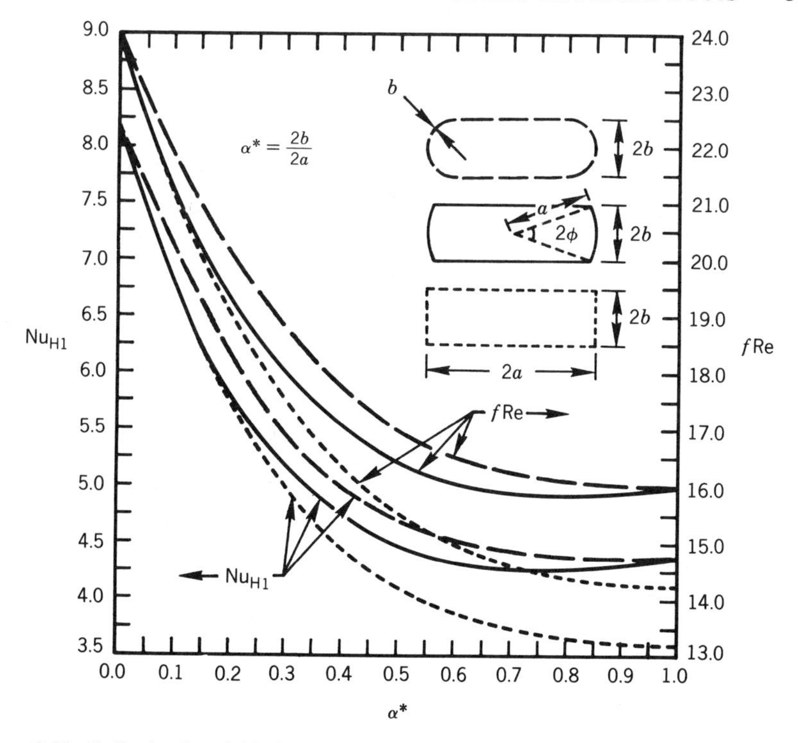

Figure 3.38. Fully developed friction factors and Nusselt numbers for stadium-shaped, modified stadium-shaped, and rectangular ducts [1].

3.7.9 Stadium-Shaped and Modified Stadium-Shaped Ducts

The stadium-shaped duct, shown in Fig. 3.29*h*, comprises two semicircular short sides of radius b connected by two long straight sides each of length $2(a - b)$. Zarling [106] determined the fully developed $f\,\mathrm{Re}$ and $\mathrm{Nu}_{\mathrm{H1}}$ values for these ducts. They are displayed in Fig. 3.38. Schenkel [95] also determined the fully developed $f\,\mathrm{Re}$ values for the stadium-shaped ducts. His graphical results are represented by the following expression valid for $0 \le \alpha^* \le 1$:

$$f\,\mathrm{Re} = 24(1 - 0.8765\alpha^* + 1.2753\alpha^{*2} - 1.3086\alpha^{*3} + 0.5765\alpha^{*4}) \quad (3.198)$$

where $\alpha^* = 2b/2a$. The predictions of Eq. (3.198) agree with the tabular values of [1] presented in Fig. 3.38 within $\pm 3\%$.

The modified stadium-shaped duct, shown in Fig. 3.29*i*, comprises two circular arcs subtended by an angle $2\phi < 180^\circ$ and two straight connecting sides. Cheng and Jamil [107] determined the fully developed $f\,\mathrm{Re}$ and $\mathrm{Nu}_{\mathrm{H1}}$ values for modified stadium-shaped ducts. Their results are presented in Fig. 3.38 which also includes the results for stadium-shaped and rectangular ducts for comparison.

3.7.10 Circumferentially Corrugated Circular Ducts

Hu and Chang [98] analyzed the fully developed fluid flow and heat transfer characteristics of circumferentially corrugated circular ducts with n sinusoidal corrugations over

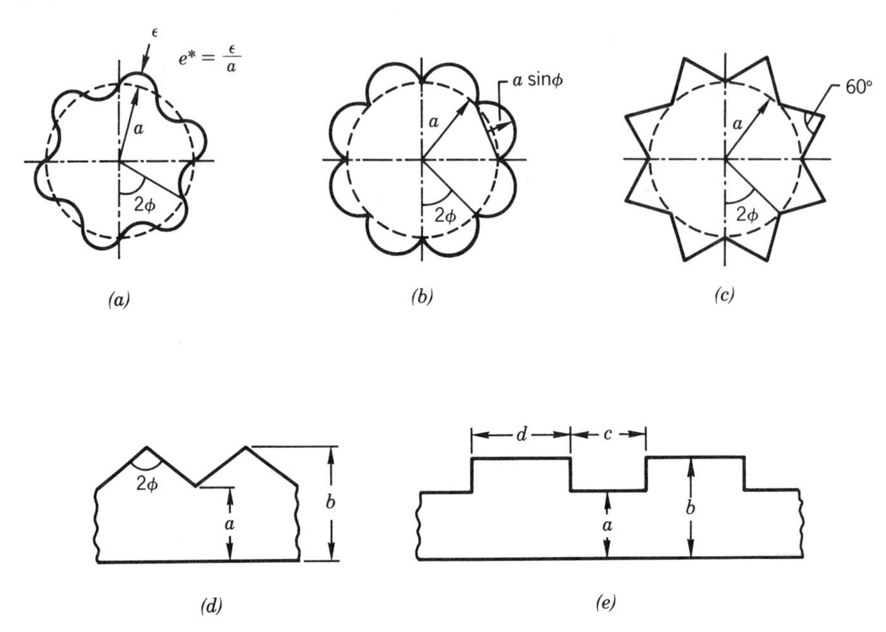

Figure 3.39. Some corrugated singly connected ducts: (a) circular with sinusoidal corrugations, (b) circular with semicircular corrugations (c) circular with triangular corrugations, (d) flat with spanwise-periodic triangular corrugations, (e) flat with spanwise-periodic rectangular corrugations.

the circumference as shown in Fig. 3.39a. The perimeter and hydraulic diameter of these ducts need to be evaluated numerically. However, their free flow area A_c (in the nomenclature of Fig. 3.39a) is given by $A_c = \pi a^2(1 + 0.5\epsilon^2)$.

The f Re, $\mathrm{Nu}_{\mathrm{H1}}$, and $\mathrm{Nu}_{\mathrm{H2}}$ values determined by Hu and Chang [98] are presented in Table 3.32 as functions of n and e^* defined in Fig. 3.39a. Their f Re values for $e^* = 0.06$ are included in Fig. 3.40 for comparison with the other two ducts. Note that the angle 2ϕ of Fig. 3.39a is related to n simply as $2\phi = 360°/n$.

Schenkel [95] determined the fully developed friction factors, displayed in Fig. 3.40, for the circular duct with semicircular corrugations as shown in Fig. 3.39b. For this duct

$$A_c = \pi a^2 \frac{\sin\phi}{\phi}\left[\frac{\pi}{2}\sin\phi + \cos\phi\right], \qquad P = \pi^2 a \frac{\sin\phi}{\phi} \tag{3.199}$$

Note that the radius of the semicircular corrugation is $a \sin\phi$.

The f Re values determined by Schenkel [95] for the duct with semicircular corrugations can be represented by the following expression with a maximum deviation of $\pm 2\%$:

$$f\,\mathrm{Re} = 6.4537 + 0.8350\phi - 3.6909 \times 10^{-2}\phi^2 + 8.6674 \times 10^{-4}\phi^3$$
$$-1.0588 \times 10^{-5}\phi^4 + 6.2094 \times 10^{-8}\phi^5 - 1.3261 \times 10^{-10}\phi^6 \tag{3.200}$$

where ϕ is in degrees. Equation (3.200) is valid for $0 \le 2\phi \le 180°$. When $2\phi = 180°$,

TABLE 3.32 Fully Developed Friction Factors and Nusselt Numbers for Circumferentially Corrugated Circular Ducts with Sinusoidal Corrugations [98]

n	e^*	f Re	$\mathrm{Nu}_{\mathrm{H1}}$	$\mathrm{Nu}_{\mathrm{H2}}$	$D_h/2a$
8	0.02	15.990	4.356	4.357	0.9986
	0.04	15.962	4.334	4.335	0.9944
	0.06	15.915	4.297	4.299	0.9874
	0.08	15.850	4.244	4.246	0.9776
	0.10	15.765	4.176	4.177	0.9650
	0.12	15.678	4.090	4.089	0.9501
12	0.02	15.952	4.340	4.340	0.9966
	0.04	15.806	4.267	4.267	0.9863
	0.06	15.559	4.142	4.140	0.9689
	0.08	15.200	3.962	3.956	0.9439
	0.10	14.711	3.723	3.709	0.9107
16	0.02	15.887	4.316	4.316	0.9938
	0.04	15.542	4.168	4.167	0.9747
	0.06	14.943	3.912	3.906	0.9418
	0.08	14.051	3.540	3.527	0.8934
24	0.02	15.679	4.245	4.245	0.9856
	0.04	14.671	3.875	3.870	0.9402
	0.06	12.872	3.231	3.219	0.8583

this geometry reduces to a circular duct for which Eq. (3.200) correctly predicts $f\,\mathrm{Re} = 16$.

Schenkel [95] also determined the fully developed friction factors, displayed in Fig. 3.40, for the circular duct having triangular corrugations with included angle 60° as shown in Fig. 3.39*c*. For this duct,

$$A_c = \pi a^2 \frac{\cos\phi + \sqrt{3}\sin\phi}{\phi}, \qquad P = 4\pi a \frac{\sin\phi}{\phi} \tag{3.201}$$

The f Re values determined by Schenkel [95] for the duct with triangular corrugation can be represented by the following expression with a maximum error of $\pm 1\%$:

$$f\,\mathrm{Re} = 3.8952 + 0.3692\phi - 3.2483 \times 10^{-3}\phi^2$$
$$-3.3187 \times 10^{-5}\phi^3 + 4.5962 \times 10^{-7}\phi^4 \tag{3.202}$$

where ϕ is in degrees. Equation (3.202) is valid for $0 \le 2\phi \le 120°$.

3.7.11 Flat Duct With Spanwise-Periodic Triangular Corrugations at One Wall

Sparrow and Charmchi [108] analyzed the fully developed fluid flow and heat transfer characteristics of a flat duct with a spanwise-periodic corrugated wall as pictured in Fig. 3.39*d*. The flow in the duct is in the direction perpendicular to the plane of the paper. The duct has an infinite extent in the spanwise direction, so that the end effects due to the short bounding walls are ignored as in a flat duct. The (H1) boundary

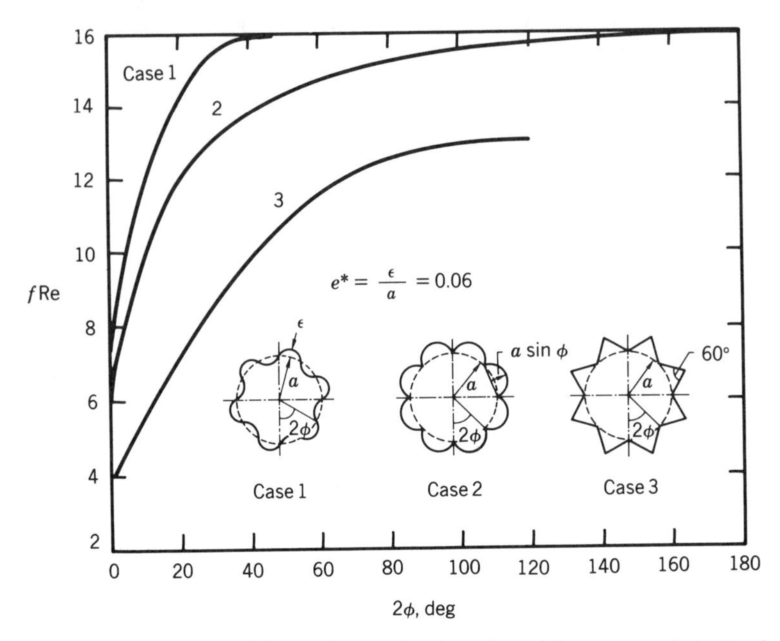

Figure 3.40. Fully developed friction factors for circumferentially corrugated circular ducts.

condition was imposed on the corrugated wall, while the flat wall was treated as adiabatic, i.e., insulated. This duct geometry together with the just-described thermal boundary conditions is encountered in present-day air-operated flat plate solar collectors. In the nomenclature of Fig. 3.39*d*, the cross-sectional area and perimeter of the duct are given by

$$A_c = n(b^2 - a^2)\tan\phi, \qquad P = 2n(b-a)\frac{1+\sin\phi}{1+\cos\phi} \tag{3.203}$$

where n is the number of triangular corrugations, each with included angle 2ϕ.

The f Re and $\mathrm{Nu_{H1}}$ values determined by Sparrow and Charmchi [108] are shown in Fig. 3.41. These values are replotted using the original graphs of [108], which had $a/(b-a)$ as the abscissa. This was done to provide a common basis of comparison for the results in Figs. 3.41 and 3.42 for the flat duct with triangular and rectangular corrugations at one wall.

When $a/b = 0$, the geometry of Fig. 3.39*d* reduces to an array of isosceles triangles. For this limiting case, f Re values of Fig. 3.41 are 12.750, 13.250, 13.478, and 13.125 for $2\phi = 20$, 40, 60, and 90°, respectively. These values agree with those in Fig. 3.18 within ±1%. Likewise, the $\mathrm{Nu_{H1}}$ values of Fig. 3.41 are within 1% of the $\mathrm{Nu_{H1}}$ values of Case 2 of Fig. 3.21. When $a/b = 1$, the geometry of Fig. 3.39*d* reduces to a flat duct. In this limiting case, $f\,\mathrm{Re} \to 24$ and $\mathrm{Nu_{H1}} \to 5.385$ [see Eqs. (3.77) and (3.89)].

3.7.12 Flat Duct with Spanwise-Periodic Rectangular Corrugations at One Wall

Sparrow and Chukaev [109] analyzed the fully developed fluid flow and heat transfer characteristics of a flat duct with spanwise-periodic rectangular corrugations at one

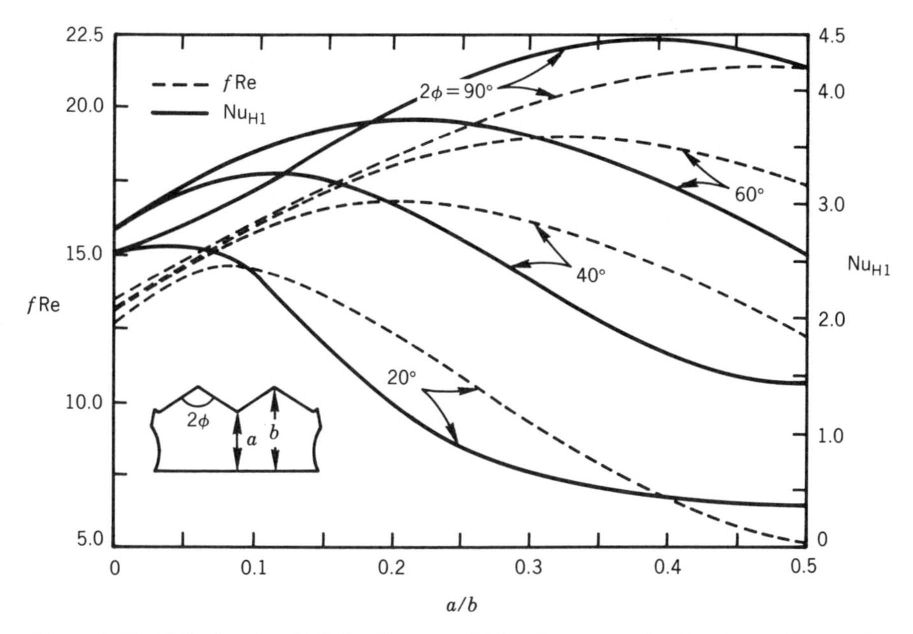

Figure 3.41. Fully developed friction factors and Nusselt numbers for flat ducts with spanwise-periodic triangular corrugations at one wall [108].

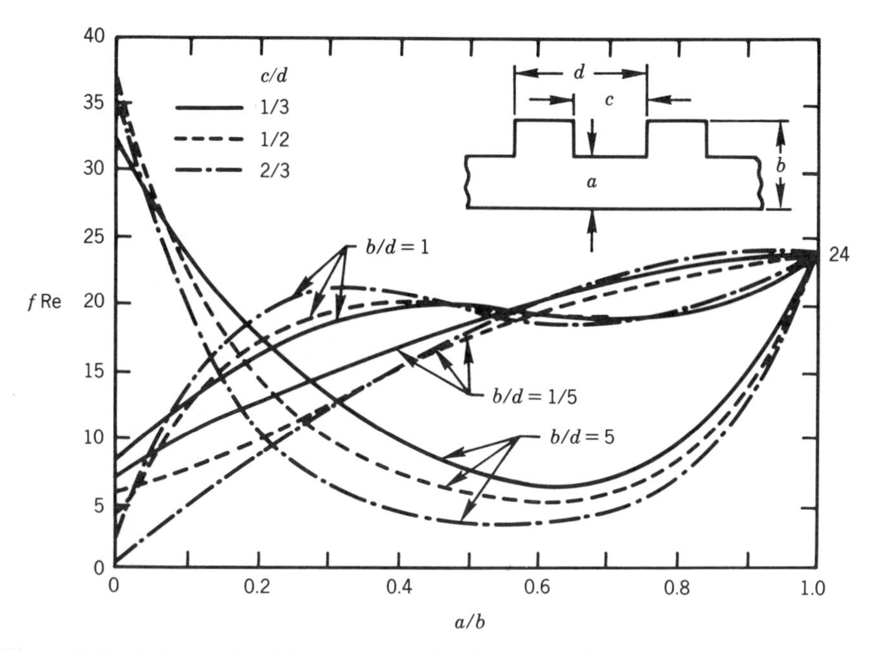

Figure 3.42. Fully developed friction factors for flat ducts with spanwise-periodic rectangular corrugations at one wall [109].

TABLE 3.33 Geometrical Characteristics and Fully Developed Friction Factors for Some Unusual Singly Connected Ducts

Geometry	Cross-sectional area, perimeter, and fully developed friction factor	Geometry	Cross-sectional area, perimeter, and fully developed friction factor
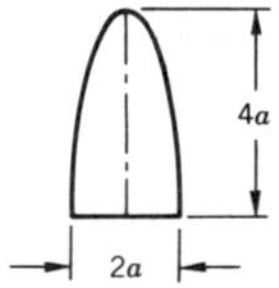 Parabolic duct	$A_c = \frac{16}{3} a^2$ $P = \left[2 + \sqrt{65} + \frac{1}{8}\ln(8 + \sqrt{65})\right] a$ $f\,\mathrm{Re} = 15.3261$	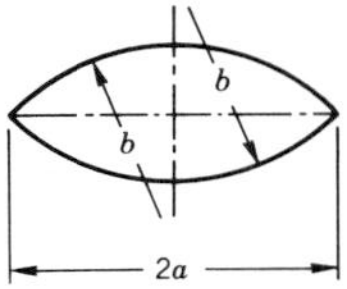 Football-shaped duct	$A_c = 2\left[\sin^{-1}\left(\frac{a}{b}\right) - \sqrt{\left(\frac{b}{a}\right)^2 - 1}\right] b^2$ $P = 4b \sin^{-1}\left(\frac{a}{b}\right)$ $f\,\mathrm{Re} = \begin{cases} 16.2541 & \text{for } b/a = 1.0417 \\ 17.0714 & \text{for } b/a = 1.250 \\ 18.6527 & \text{for } b/a = 2.125 \end{cases}$
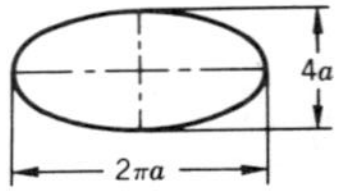 Cycloidal duct†	$A_c = 6\pi a^2$ $P = 16\pi a$ $f\,\mathrm{Re} = 16.9881$	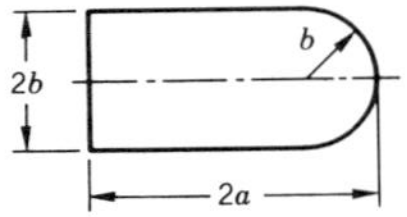 Semistadium-shaped or horseshoe-shaped duct	$A_c = 2b(2a - b)$ $P = (2a - b) + \pi b,\ \alpha^* = 2b/2a$ $f\,\mathrm{Re} = 24(1 - 0.4967\alpha^* - 0.8910\alpha^{*2} + 0.8655\alpha^{*3} + 0.3818\alpha^{*4} - 0.5693\alpha^{*5} + 0.1471\alpha^{*6})$ for $0 \le \alpha^* \le 2$
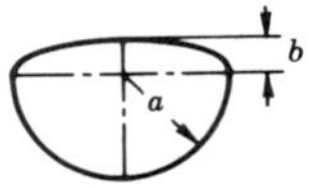 Elliptic-cum-circular duct	$A_c = (\pi/2)(a + b)a$ $P = a[\pi + 2E(m)]$, where $E(m)$ is the complete elliptic integral of the second kind with $m = 1 - \alpha^{*2}$, $\alpha^* = b/a$ $f\,\mathrm{Re} = 16.0196$ for $\alpha^* = \frac{1}{3}$ and $\frac{2}{3}$	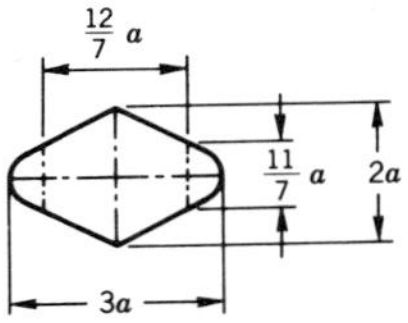 Radiator-shaped duct‡	$A_c = \frac{3a^2}{196}(200 + 33\pi)$ $P = \frac{a}{14}(63 + 4\sqrt{153})$ $f\,\mathrm{Re} = 15.7847$
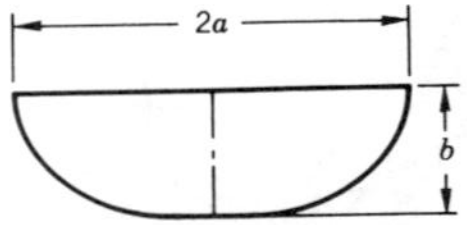 Semielliptical duct	$A_c = \frac{\pi}{2} ab$ $P = 2a[1 + E(m)]$, where $E(m)$ is the complete elliptic integral of the second kind with $m = 1 - \alpha^{*2}$, $\alpha^* = b/a$ $f\,\mathrm{Re} = \frac{3\pi^2(1.75 + 0.4875\alpha^{*2})}{2[1 + E(m)]^2}$	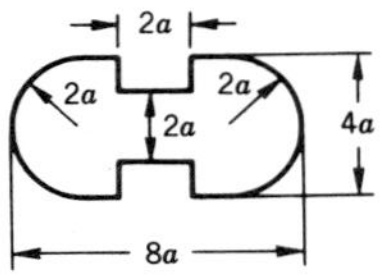 Anchor-shaped duct	$A_c = 4(\pi + 12)a^2$ $P = 4(\pi + 2)a$ $f\,\mathrm{Re} = 16.8356$

† The two branches of the cycloidal duct, symmetric about the horizontal axis, are the loci of a point on a circle of radius a as it rolls on the horizontal axis.
‡ The radiator-shaped duct comprises two central trapezoidal sections together with the semielliptic ends.

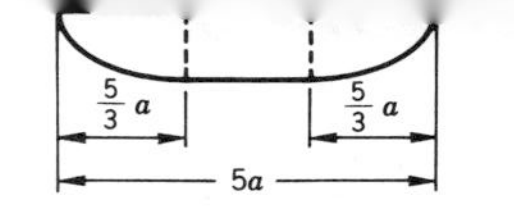

Rectangular duct with unilateral elliptical ends

$A_c = \ldots$

$P = 10.9212a$

$f\,\mathrm{Re} = 19.2522$

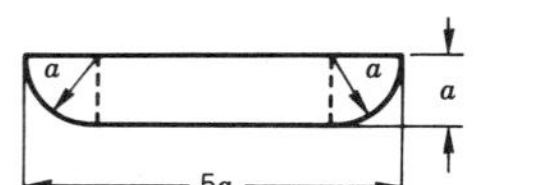

Rectangular duct with unilateral circular ends

$A_c = \left(3 + \frac{\pi}{2}\right)a^2$

$P = (8 + \pi)a$

$f\,\mathrm{Re} = 20.0728$

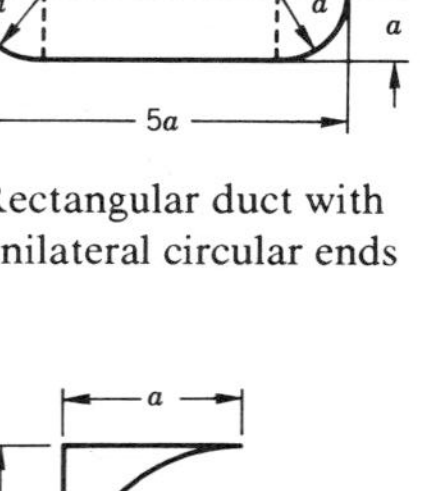

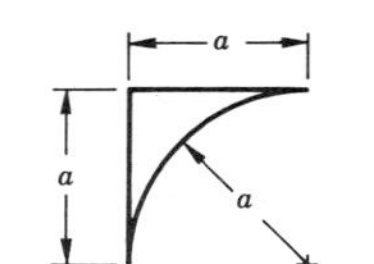

Duct with two straight sides and a quarter circular arc

$A_c = \left(1 - \frac{\pi}{4}\right)a^2$

$P = \left(2 + \frac{\pi}{2}\right)a$

$f\,\mathrm{Re} = 7.06$

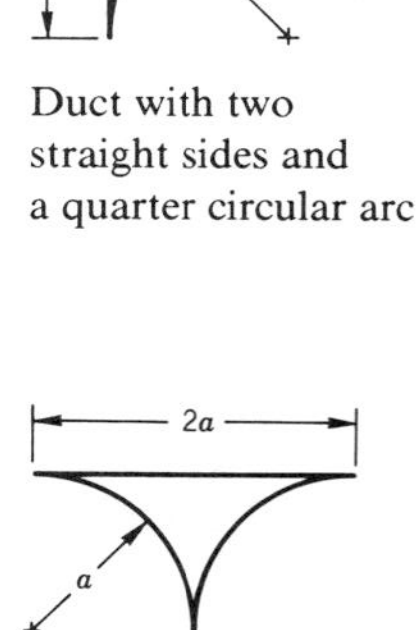

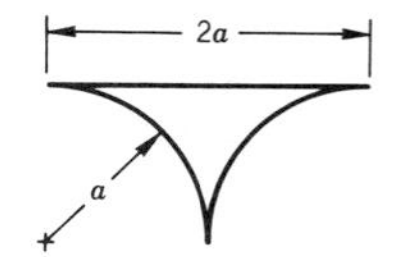

Duct with a straight side and two quarter circular arcs

$A_c = \left(4 - \frac{\pi}{2}\right)a^2$

$P = (2 + \pi)a$

$f\,\mathrm{Re} = 6.50$

Anchor-shaped duct

$A_c = \ldots 4$

$P = (\pi + \frac{5}{2})a$

$f\,\mathrm{Re} = 16.5152$

Waspwaist-shaped duct

$A_c = 2a^2[\pi - \phi + \sin\phi\,(2 - \cos\phi)]$

$P = 4a(\pi - \phi + 1 - \cos\phi)$

$$f\,\mathrm{Re} = \frac{32}{\pi}\,\frac{[\pi - \phi + \sin\phi\,(2 - \cos\phi)]^3}{F^4(\pi - \phi + 1 - \cos\phi)^2}$$

where for $0 \le 2\phi \le 180°$

$F = 0.5091 + 0.1851\phi + 0.0980\phi^2 - 0.0169\phi^3 - 0.1814\phi^4 + 0.0723\phi^5 + 0.0092\phi^6$

where ϕ is in radians.

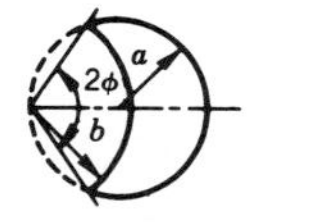

Star-shaped duct

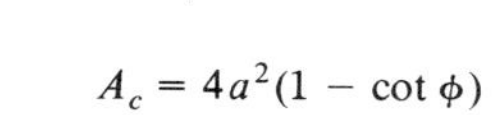

$A_c = 4a^2(1 - \cot\phi)$

$$P = \frac{8a}{\sin\phi}$$

$$f\,\mathrm{Re} = \frac{4}{\pi}\,\frac{(\sin\phi - \cos\phi)^3}{F^4 \sin\phi}$$

where for $45° \le \phi \le 90°$

$F = 0.7217\phi^3 - 3.1197\phi^2 + 4.9152\phi - 2.2731$

where ϕ is in radians.

Moon-shaped or Sickle-shaped duct

$A_c = (2a^2 - b^2)\phi + a^2 \sin 2\phi$

$P = 2(2a + b)\phi,\ \alpha^* = b/a$

$$f\,\mathrm{Re} = \frac{96[(2 - \alpha^{*2})\phi + \sin 2\phi]^3}{[(2 + \alpha^*)\phi]^2 F}$$

where

$F = \sin 4\phi - 4(3\alpha^{*2} - 2)\sin 2\phi + 32\alpha^{*3}\sin\phi - 6(\alpha^{*4} + 4\alpha^{*2} - 2)\phi$

TABLE 3.33 Continued

Geometry	Cross-sectional area, perimeter, and fully developed friction factor	Geometry	Cross-sectional area, perimeter, and fully developed friction factor
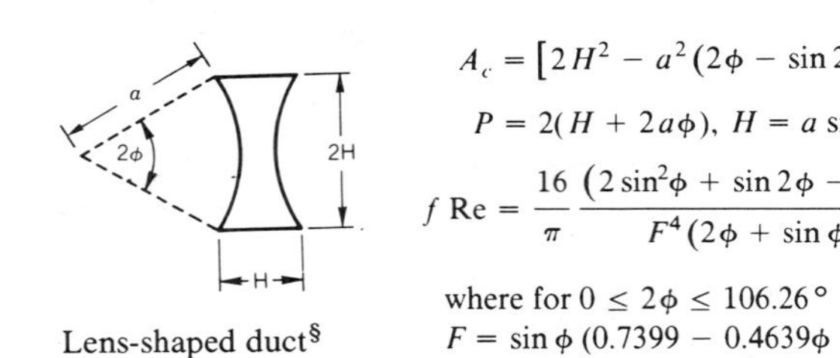 Lens-shaped duct[§]	$A_c = \left[2H^2 - a^2(2\phi - \sin 2\phi)\right]$ $P = 2(H + 2a\phi),\ H = a\sin\phi$ $f\,\mathrm{Re} = \dfrac{16}{\pi}\dfrac{(2\sin^2\phi + \sin 2\phi - 2\phi)^3}{F^4(2\phi + \sin\phi)}$ where for $0 \le 2\phi \le 106.26°$ $F = \sin\phi\,(0.7399 - 0.4639\phi$ $+ 0.6568\phi^2 - 2.0720\phi^3$ $+ 2.5873\phi^4 - 1.1896\phi^5)$ where ϕ is in radians	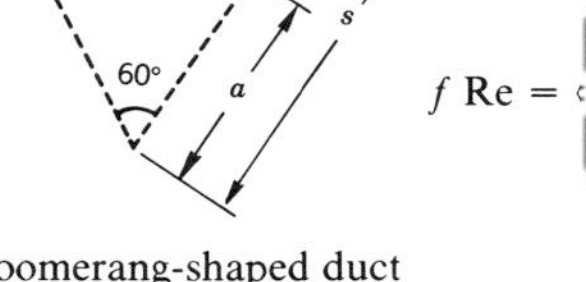 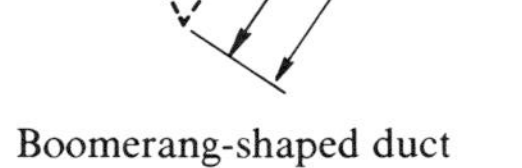 Boomerang-shaped duct	$A_c = \dfrac{s^2}{\sqrt{3}} - \dfrac{\pi}{6}a^2$ $P = 2\left(1 + \dfrac{1}{\sqrt{3}}\right)s - \left(2 - \dfrac{\pi}{3}\right)a$ $f\,\mathrm{Re} = \begin{cases} 14.2400 & \text{for } \dfrac{s}{a} = 1.732 \\ 13.5360 & \text{for } \dfrac{s}{a} = 1.299 \end{cases}$
H-shaped duct	$A_c = 7a^2$ $P = 16a$ $f\,\mathrm{Re} = 15.6060$	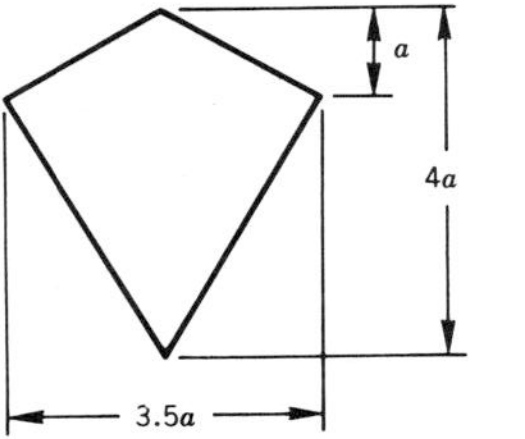 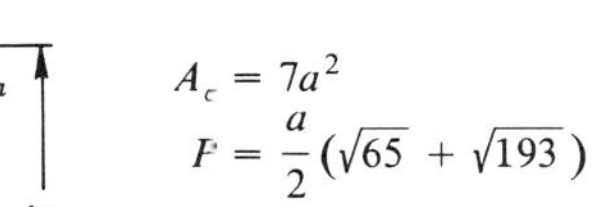 Kite-shaped duct	$A_c = 7a^2$ $P = \dfrac{a}{2}(\sqrt{65} + \sqrt{193})$ $f\,\mathrm{Re} = 14.0320$
I-shaped duct	$A_c = 120a^2$ $P = 52a$ $f\,\mathrm{Re} = 15.7728$	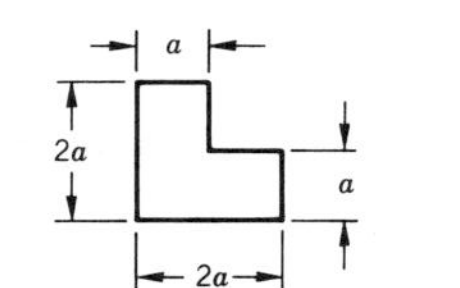 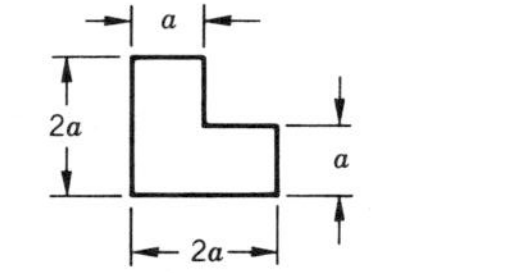 Stairstep-shaped or L-shaped duct	$A_c = 3a^2$ $P = 8a$ $f\,\mathrm{Re} = 14.3965$

[§]The angle $2\phi = 106.26°$ corresponds to the case when two arcs of the lens-shaped duct just contact each other.

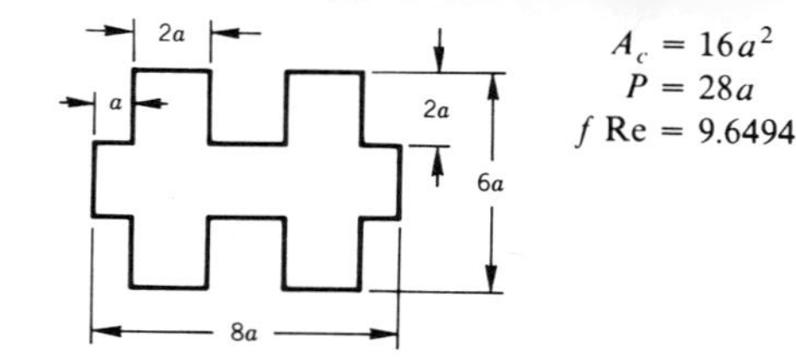

Dumbbell-shaped duct

$$A_c = 16a^2$$
$$P = 28a$$
$$f\,\mathrm{Re} = 9.6494$$

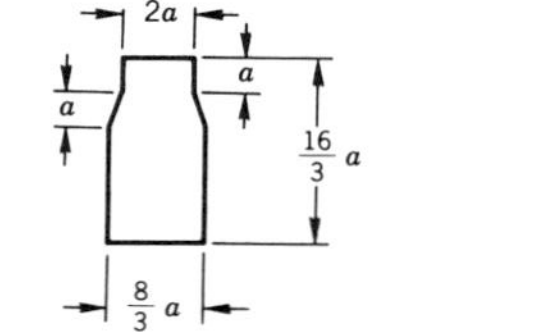

Milkcan-shaped duct

$$A_c = \tfrac{119}{9}a^2$$
$$P = \frac{a}{3}(40 + 2\sqrt{10})$$
$$f\,\mathrm{Re} = 16.0824$$

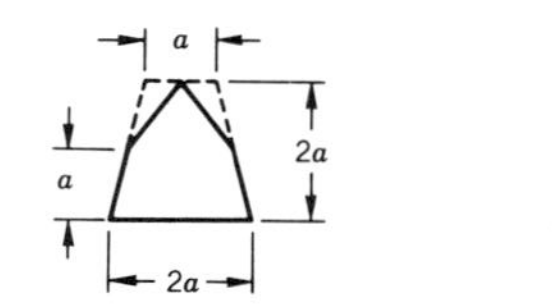

Atomic-bunker-shaped duct

$$A_c = \tfrac{5}{2}a^2$$
$$P = \frac{a}{2}(9 + \sqrt{17})$$
$$f\,\mathrm{Re} = 14.4324$$

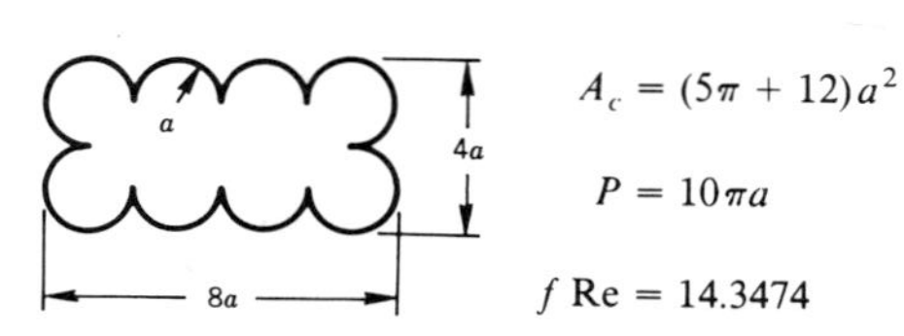

Tire-shaped duct

$$A_c = (5\pi + 12)a^2$$
$$P = 10\pi a$$
$$f\,\mathrm{Re} = 14.3474$$

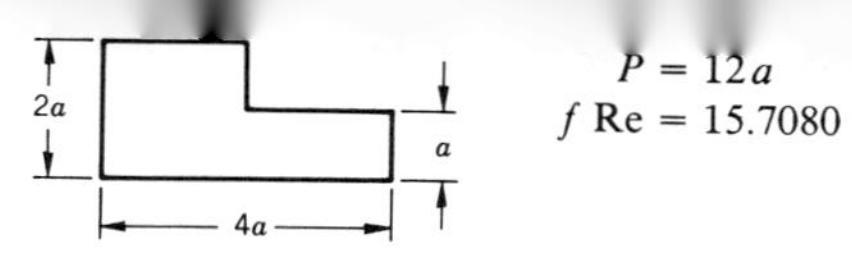

Stairstep-shaped or L-shaped duct

$$P = 12a$$
$$f\,\mathrm{Re} = 15.7080$$

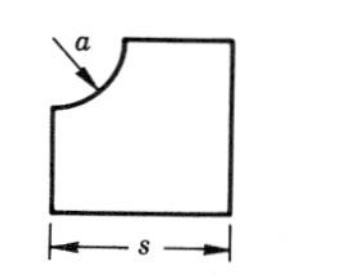

Square duct with one indented corner

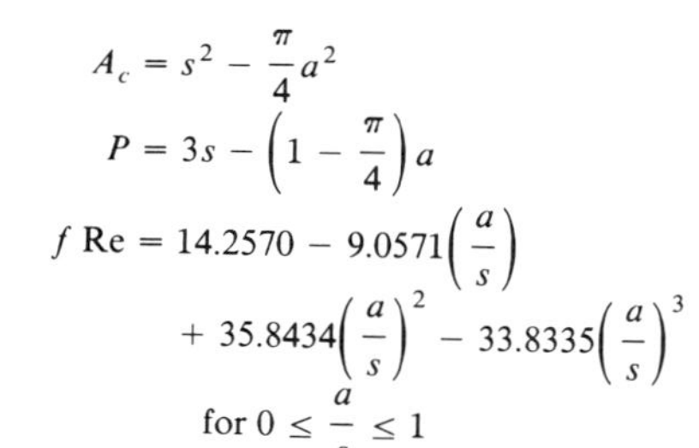

$$A_c = s^2 - \frac{\pi}{4}a^2$$
$$P = 3s - \left(1 - \frac{\pi}{4}\right)a$$
$$f\,\mathrm{Re} = 14.2570 - 9.0571\left(\frac{a}{s}\right) + 35.8434\left(\frac{a}{s}\right)^2 - 33.8335\left(\frac{a}{s}\right)^3$$
$$\text{for } 0 \le \frac{a}{s} \le 1$$

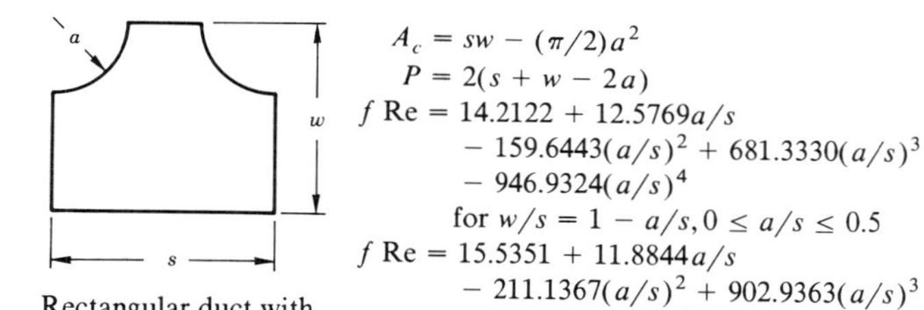

Rectangular duct with two indented corners

$$A_c = sw - (\pi/2)a^2$$
$$P = 2(s + w - 2a)$$
$$f\,\mathrm{Re} = 14.2122 + 12.5769a/s - 159.6443(a/s)^2 + 681.3330(a/s)^3 - 946.9324(a/s)^4$$
$$\text{for } w/s = 1 - a/s, 0 \le a/s \le 0.5$$
$$f\,\mathrm{Re} = 15.5351 + 11.8844a/s - 211.1367(a/s)^2 + 902.9363(a/s)^3 - 1199.708(a/s)^4$$
$$\text{for } w/s = 2 - 3(a/s), 0 \le a/s \le 0.5$$
$$f\,\mathrm{Re} = 15.5315 + 8.5289(a/s) - 185.2351(a/s)^2 + 807.7854(a/s)^3 - 1086.055(a/s)^4$$
$$\text{for } w/s = \tfrac{1}{4}[2 - 3(a/s)], 0 \le a/s \le 0.5$$

TABLE 3.33 Continued

Geometry	Cross-sectional area, perimeter, and fully developed friction factor	Geometry	Cross-sectional area, perimeter, and fully developed friction factor
2a, a, 4a, 8a Tire-shaped duct	$A_c = 22a^2$ $P = 20\sqrt{2}\,a$ $f\,\mathrm{Re} = 13.9422$	a; $n=3$, $n=4$, $n=5$, $n=6$, $n=7$, $n=8$ Cusped ducts	$A_c = na^2\left(\sin\frac{\pi}{n}\cos\frac{\pi}{n} + \cot\frac{\pi}{n}\cos\frac{\pi}{n} + \frac{\pi}{n} - \frac{\pi}{2}\right)$ $P = n\pi a\left(1 - \frac{2}{n}\right)$ where n is the number of concave circular arcs each of radius a forming the duct $f\,\mathrm{Re} = 5.5667 + 0.5253n - 0.0841n^2 + 0.0044n^3$ for $3 \le n \le 8$.
a, s Equilateral triangular duct with indented corners	$A_c = \frac{\sqrt{3}}{4}s^2 - \frac{\pi}{2}a^2$ $P = 3s - (6 - \pi)a$ $f\,\mathrm{Re} = \begin{cases} 13.7760 & \text{for } s/a = 6 \\ 13.6480 & \text{for } s/a = 3 \end{cases}$		
a, s, s Square duct with four indented corners	$A_c = s^2 - \pi a^2$ $P = 4s - (8 - \pi)a$ $f\,\mathrm{Re} = 14.2445 - 43.2622a/s + 277.0151(a/s)^2 - 439.7446(a/s)^3$ for $0 \le \frac{a}{s} \le 0.5$	a, b; $a<b$, $a=b$, $a>b$ Pascal's limaçon ducts¶	$A_c = \frac{\pi}{2}(a^2 + 2b^2)$ $P = 4(a+b)E(m)$ where $E(m)$ is the complete elliptic integral of the second kind with $m = \frac{2\alpha^{*3/2}}{1+\alpha^*},\ \alpha^* = \frac{a}{b}$ $f\,\mathrm{Re} = \frac{4[\pi/E(m)]^2(2+\alpha^{*2})^3}{(1+\alpha^*)^2(8+8\alpha^{*2}+\alpha^{*4})}$

¶Pascal's limaçon is described by a point on the periphery of a circle of diameter b as it rolls on the periphery of a fixed circle of diameter a. When $b = a$, Pascal's limaçon is called a cardioid, and when $b < a$ it is called a trisectrix. For a cardioid duct, the Nusselt numbers Nu_{H1} and Nu_{H2} have the values 4.2079 and 4.0966, respectively [1].

wall as shown in Fig. 3.39*e*. This duct geometry could simulate the channel-like cooling passages of electronic devices where the corrugations could be viewed as representing heat-generating components of the printed circuit boards.

The duct of Fig. 3.39*e* has an infinite extent in the spanwise direction so that the end effects due to the short bounding walls are ignored as in a flat duct. Two thermal problems were solved numerically by successively applying the (H1) boundary condition at each wall while keeping the other wall adiabatic. The heat transfer results were then obtained by superposition technique for arbitrary heating at both walls. Extensive heat transfer results are available in [109] for the three geometric parameters $b/d = \frac{1}{5}, 1, 5$; $c/d = \frac{1}{3}, \frac{1}{2}, \frac{2}{3}$; and $a/b = 0.1, 0.2, \ldots, 0.9$. The f Re results are displayed in Fig. 3.42 for the indicated values of the three parameters. As $a/b \to 1$ (vanishing corrugations), all of the f Re curves in Fig. 3.42 tend toward the value 24, which corresponds to that for the flat duct. Note that the plotted values in [109] are for $0.1 \le a/b \le 0.9$. The values shown in Fig. 3.42 beyond this range were obtained by the present authors using a polynomial regression.

3.7.13 Miscellaneous Singly Connected Ducts

The fully developed friction factors for some unusual singly connected ducts are presented in Table 3.33. The results for the moon-shaped or sickle-shaped ducts, cusped ducts, ducts with two straight sides and a quarter circular arc, and ducts with one straight side and two quarter circular arcs are based on the analytical solutions presented in [1]. The results for Pascal's limaçon ducts are due to Tao [110], and those for the remaining ducts were obtained by the present authors using a graphical method developed by Schenkel [95]. The method is called the 3*R* method because it entails the determination of three radii: hydraulic, equal area, and effective for the duct in question. The method also relies for interpolation on a knowledge of the friction factors for at least two ducts similar in shape to the duct in question.

3.8 CONCENTRIC ANNULAR DUCTS

The concentric annular duct pictured in the inset to Fig. 3.43 is of great technical importance, as it is used in numerous fluid flow and heat transfer devices involving two fluids. One fluid flows through the inside tube while the other flows through the annular passage between the two tubes forming the annular duct. One limiting case of the annular duct ($r^* = 1$) is the flat duct, while the other ($r^* = 0$) is a circular duct with an infinitesimal core at the center. Despite the presence of the infinitesimal core at the center, most of the fluid flow and heat transfer results for $r^* = 0$ turn out to be identical to those for a true circular duct without the infinitesimal core.

3.8.1 Fully Developed Flow

The fully developed velocity profile and friction factors for concentric annular ducts are given by [1]

$$u = -\frac{1}{4\mu}\left(\frac{dp}{dx}\right)r_o^2\left[1 - \left(\frac{r}{r_o}\right)^2 + 2r_m^{*2}\ln\left(\frac{r}{r_o}\right)\right] \tag{3.204}$$

$$u_m = -\frac{1}{8\mu}\left(\frac{dp}{dx}\right) r_o^2 \left(1 + r^{*2} - 2r_m^{*2}\right) \tag{3.205}$$

$$\frac{u_{\max}}{u_m} = \frac{2\left(1 - r_m^{*2} + 2r_m^{*2} \ln r_m^*\right)}{1 + r^{*2} - 2r_m^{*2}} \tag{3.206}$$

$$f_i \mathrm{Re} = -\frac{1}{\mu}\left(\frac{dp}{dx}\right)\frac{D_h}{u_m}\left(\frac{r_m^2 - r_i^2}{r_i}\right), \qquad D_h = 2(r_o - r_i) \tag{3.207}$$

$$f_o \mathrm{Re} = -\frac{1}{\mu}\left(\frac{dp}{dx}\right)\frac{D_h}{u_m}\left(\frac{r_o^2 - r_m^2}{r_o}\right) \tag{3.208}$$

$$f\,\mathrm{Re} = \frac{16(1 - r^*)^2}{1 + r^{*2} - 2r_m^{*2}} \tag{3.209}$$

Here r_m designates the radius where the maximum velocity occurs ($\partial u/\partial r = 0$). It is given by

$$r_m^* = \frac{r_m}{r_o} = \left(\frac{1 - r^{*2}}{2\ln(1/r^*)}\right)^{1/2} \tag{3.210}$$

In Eqs. (3.207) and (3.208), f_i and f_o respectively designate the fully developed friction factor at the inner and the outer walls. In Eq. (3.209), f stands for the circumferentially averaged fully developed friction factor, which is related to f_i and f_o by

$$f = \frac{f_i r_i + f_o r_o}{r_i + r_o} \tag{3.211}$$

The fully developed friction factors f Re computed from Eq. (3.209) are displayed in Fig. 3.43, which also includes circumferentially averaged fully developed Nusselt numbers $\mathrm{Nu_T}$ and $\mathrm{Nu_H}$ to be presented shortly.

The hydrodynamic entrance lengths L_{hy}^+ for concentric annular ducts with $r^* = 0$, 0.05, 0.10, 0.50, 0.75, and 1.00 are 0.0541, 0.0206, 0.0175, 0.0116, 0.0109, and 0.0108, respectively [1]. The fully developed incremental pressure drop numbers $K(\infty)$ for the aforementioned values of r^* are 1.25, 0.830, 0.784, 0.688, 0.678, and 0.674, respectively [1].

Natarajan and Lakshmanan [111] presented the following simple formula for f Re, which agrees with the values in Fig. 3.43 within $\pm 2\%$ for $r^* \geq 0.005$:

$$f\,\mathrm{Re} = 24 r^{*0.035} \tag{3.212}$$

Since one or both annulus surfaces can be heated independently, numerous thermal boundary conditions are possible for heat transfer to a flowing fluid in the annulus. Depending on the temperature or heat flux specified at the inner or the outer surface, four fundamental solutions have been developed for concentric annular ducts [112]. By using superposition techniques, it is possible to combine the four fundamental solutions to obtain a solution for any desired boundary condition.

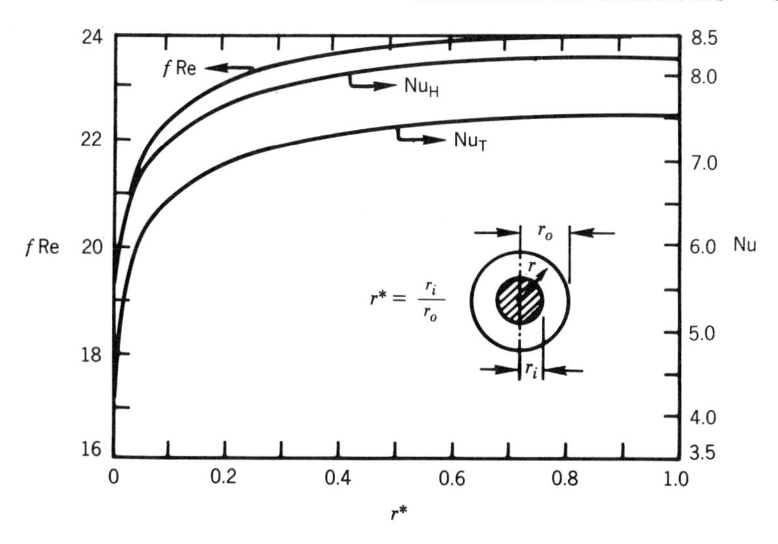

Figure 3.43. Fully developed friction factors and Nusselt numbers for concentric annular ducts [1].

The boundary conditions for the four fundamental problems alluded to above have already been described in Sec. 3.3.1 in the context of the flat duct. The nomenclature used in describing the corresponding solutions can be best explained with reference to the specific heat transfer parameters $\theta_{lj}^{(k)}$ and $\theta_{mj}^{(k)}$, which are the dimensionless duct wall and fluid bulk mean temperatures, respectively. The superscript k denotes the type of the fundamental solution according to the four types of the boundary conditions described in Sec. 3.3.1. Thus, $k = 1, 2, 3$, or 4. The first subscript l in $\theta_{lj}^{(k)}$ refers to the particular wall at which the temperature is evaluated. Thus, depending on whether the temperature is evaluated at the inner or the outer wall, $l = i$ or o. The second subscript j in $\theta_{lj}^{(k)}$ refers to the wall at which $T \neq T_e$ or $q_w'' \neq 0$, i.e., the active wall with a nonzero boundary condition. The significance of k and j in $\theta_{mj}^{(k)}$ is the same as in $\theta_{lj}^{(k)}$, while m denotes the mean, which in this instance refers to the fluid bulk mean temperature.

The fundamental solutions of the first, second, and third kinds for fully developed flow in concentric annular ducts are presented in Table 3.34 [1]. The additional heat transfer results pertaining to the four fundamental solutions can be obtained from the

TABLE 3.34 Fundamental Solutions of the First, Second, and Third Kinds for Fully Developed Flow in Concentric Annular Ducts [1]

r^*	$\Phi_{ii}^{(1)}$	$\mathrm{Nu}_{ii}^{(1)}$	$\mathrm{Nu}_{oo}^{(1)}$	$\theta_{mo}^{(2)} - \theta_{io}^{(2)}$	$\mathrm{Nu}_{ii}^{(2)}$	$\mathrm{Nu}_{oo}^{(2)}$	$\mathrm{Nu}_{ii}^{(3)}$	$\mathrm{Nu}_{oo}^{(3)}$
0	∞	∞	2.66667	0.145833	∞	4.36364	∞	3.6568
0.02	25.05098	30.17942	2.94836	0.127945	32.70512	4.73424	32.337	3.9934
0.05	12.68471	16.05843	3.01887	0.122568	17.81128	4.79198	17.460	4.0565
0.10	7.81730	10.45870	3.09528	0.116214	11.90578	4.83421	11.560	4.1135
0.25	4.32809	6.47139	3.26700	0.102207	7.75347	4.90475	7.3708	4.2321
0.50	2.88539	4.88896	3.52035	0.085513	6.18102	5.03653	5.7382	4.4293
1.00	2.00000	4.00000	4.00000	0.064286	5.38462	5.38462	4.8608	4.8608

following set of equations in conjunction with Table 3.34:

$$\Phi_{oo}^{(1)} = -\Phi_{oi}^{(1)} = r^*\Phi_{ii}^{(1)} = -r^*\Phi_{io}^{(1)} \tag{3.213}$$

$$\theta_{mi}^{(1)} = 1 - \theta_{mo}^{(1)} \tag{3.214}$$

$$\mathrm{Nu}_{io}^{(1)} = \mathrm{Nu}_{ii}^{(1)} = \frac{\Phi_{ii}^{(1)}}{1-\theta_{mi}^{(1)}} = \frac{\Phi_{ii}^{(1)}}{\theta_{mo}^{(1)}} \tag{3.215}$$

$$\mathrm{Nu}_{oi}^{(1)} = \mathrm{Nu}_{oo}^{(1)} = \frac{\Phi_{oo}^{(1)}}{1-\theta_{mo}^{(1)}} = \frac{r^*\Phi_{ii}^{(1)}}{\theta_{mi}^{(1)}} \tag{3.216}$$

$$\theta_{oi}^{(2)} - \theta_{mi}^{(2)} = r^*\left[\theta_{io}^{(2)} - \theta_{mo}^{(2)}\right] \tag{3.217}$$

$$\mathrm{Nu}_{ii}^{(2)} = \frac{1}{\theta_{ii}^{(2)} - \theta_{mi}^{(2)}} \tag{3.218}$$

$$\mathrm{Nu}_{oo}^{(2)} = \frac{1}{\theta_{oo}^{(2)} - \theta_{mo}^{(2)}} \tag{3.219}$$

$$\mathrm{Nu}_{oi}^{(2)} = \mathrm{Nu}_{io}^{(2)} = 0 \tag{3.220}$$

$$\Phi_{ii}^{(3)} = \Phi_{oo}^{(3)} = 0 \tag{3.221}$$

$$\theta_{oi}^{(3)} = \theta_{io}^{(3)} = \theta_{mi}^{(3)} = \theta_{mo}^{(3)} = 1 \tag{3.222}$$

$$\Phi_{oi}^{(4)} = \frac{1}{\Phi_{io}^{(4)}} = -r^* \tag{3.223}$$

$$\theta_{ii}^{(4)} = r^*\theta_{oo}^{(4)} = \frac{1}{\mathrm{Nu}_{ii}^{(1)}} + \frac{r^*}{\mathrm{Nu}_{oo}^{(1)}} \tag{3.224}$$

$$\mathrm{Nu}_{ii}^{(4)} = \mathrm{Nu}_{io}^{(4)} = \frac{1}{\theta_{ii}^{(4)} - \theta_{mi}^{(4)}} = \frac{1}{r^*\theta_{mo}^{(4)}} = \mathrm{Nu}_{ii}^{(1)} \tag{3.225}$$

$$\mathrm{Nu}_{oo}^{(4)} = \mathrm{Nu}_{oi}^{(4)} = \frac{1}{\theta_{oo}^{(4)} - \theta_{mo}^{(4)}} = \frac{r^*}{\theta_{mi}^{(4)}} = \mathrm{Nu}_{oo}^{(1)} \tag{3.226}$$

The direct practical applications of the four fundamental solutions presented above are rather limited. Their real utility lies in developing solutions to problems of practical interest. Three such problems are of great interest, and their solutions are presented next.

Constant Temperatures at Both Walls. For this problem,

$$T = \begin{cases} T_i & \text{at} \quad r = r_i, \quad x \geq 0 \\ T_o & \text{at} \quad r = r_o, \quad x \geq 0 \\ T_e & \text{at} \quad x \leq 0, \quad r_i < r < r_o \end{cases}$$

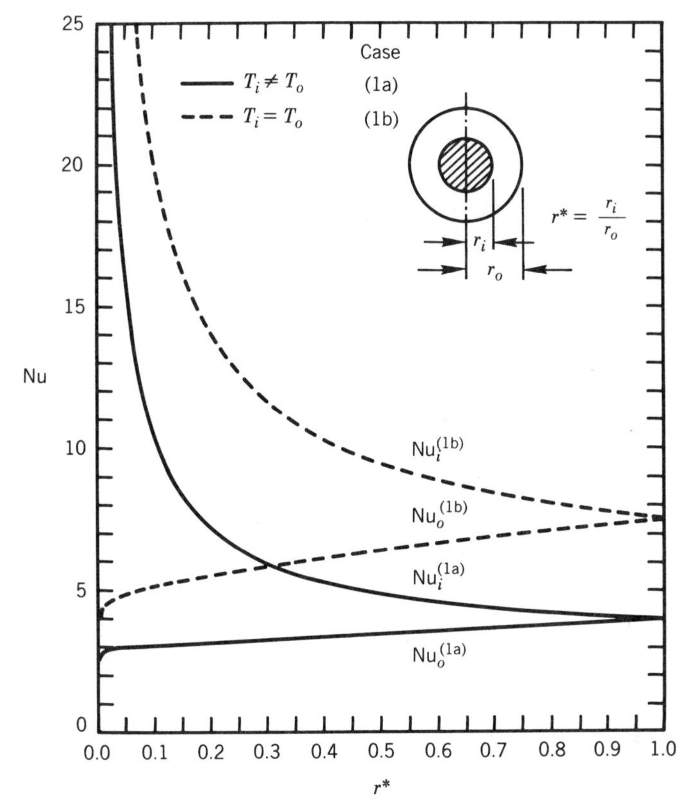

Figure 3.44. Fully developed Nusselt numbers for constant temperatures at both walls in concentric annular ducts [1].

When $T_i \neq T_0$, the problem is designated as 1a and the fully developed Nusselt numbers at the two walls are designated as $Nu_i^{(1a)}$ and $Nu_o^{(1a)}$. They are presented in Fig. 3.44. Tabulated values for this and the subsequent solutions are available in [1].

When $T_i = T_o$, the problem is designated as 1b and the fully developed Nusselt numbers at the two walls are designated as $Nu_i^{(1b)}$ and $Nu_o^{(1b)}$. They are presented in Fig. 3.44. In this case, it is useful to obtain a circumferentially averaged Nusselt number, designated as Nu_T. It can be obtained from $Nu_i^{(1b)}$ and $Nu_o^{(1b)}$ via the following relation:

$$Nu_T = \frac{Nu_o^{(1b)} + r^* Nu_i^{(1b)}}{1 + r^*} \tag{3.227}$$

The Nu_T values from Eq. (3.227) are presented in Fig. 3.43.

Constant Heat Fluxes at Both Walls. For this problem,

$$q_w'' = \begin{cases} q_i'' & \text{at} \quad r = r_i, \quad x \geq 0 \\ q_0'' & \text{at} \quad r = r_o, \quad x \geq 0 \end{cases}$$

$$T = T_e \quad \text{at} \quad x \leq 0, \quad r_i < r < r_o$$

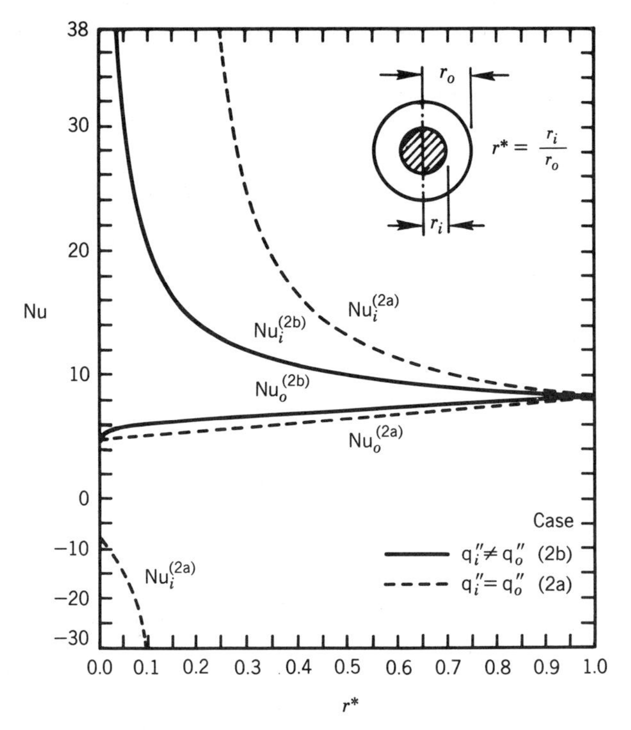

Figure 3.45. Fully developed Nusselt numbers for constant axial heat fluxes at both walls in concentric annular ducts [1].

When $q_i'' = q_o''$, the problem is designated as 2a and the fully developed Nusselt numbers at the two walls are designated as $\mathrm{Nu}_i^{(2a)}$ and $\mathrm{Nu}_o^{(2a)}$. They are presented in Fig. 3.45.

When $q_i'' \neq q_o''$, the problem is designated as 2b and the fully developed Nusselt numbers at the two walls are denoted as $\mathrm{Nu}_i^{(2b)}$ and $\mathrm{Nu}_o^{(2b)}$. They are presented in Fig. 3.45. A circumferentially averaged Nusselt number $\mathrm{Nu_H}$ can be obtained from $\mathrm{Nu}_i^{(2b)}$ and $\mathrm{Nu}_o^{(2b)}$ via a relation similar to Eq. (3.227). $\mathrm{Nu_H}$ thus calculated are shown in Fig. 3.43.

It may be noted that the heat flux is considered as positive if the heat transfer is from the wall to the fluid. Consequently, a negative Nusselt number as in Fig. 3.45 means that heat transfer takes place from the fluid to the wall. In aforementioned both cases, $T_w - T_m$ is positive. An infinite Nusselt number in Figs. 3.44 and 3.45 implies that $T_w = T_m$ and not an infinite heat flux.

Constant Temperature at One Wall with Constant Heat Flux at the Other. For this problem,

$$T = T_1 \quad \text{at} \quad r = r_1, \quad x \geq 0$$

$$q_w'' = q_2'' \quad \text{at} \quad r = r_2, \quad x \geq 0$$

$$T = T_e \quad \text{at} \quad x \leq 0, \quad r_i < r < r_o$$

Here the subscripts 1 and 2 can refer to either the inside or the outside wall. When $T_1 \neq T_2$, the problem is referred to as 4a, and when $T_1 = T_2$, it is referred to as 4b. It can be shown that

$$\mathrm{Nu}_i^{(4a)} = \mathrm{Nu}_i^{(4b)} = \mathrm{Nu}_i^{(1a)} \tag{3.228}$$

$$\mathrm{Nu}_o^{(4a)} = \mathrm{Nu}_o^{(4b)} = \mathrm{Nu}_o^{(1a)} \tag{3.229}$$

3.8.2 Hydrodynamically Developing Flow

There are several analytical investigations of the problem of laminar flow development in concentric annuli. They are summarized in [1]. Presented in Table 3.35 are the axial velocity and pressure distribution in the hydrodynamic entrance region for $r^* = 0.1$, 0.25, and 0.5 as determined by Kakaç and Yücel [113]. The results of Table 3.35 are in excellent accord with those of other analyses cited in [1]. The results for $r^* = 0$ are available in Table 3.3, and those for $r^* = 1$ in Table 3.9. It may be noted that for $r^* = 1$ the transverse coordinates of Tables 3.9 and 3.35 are related as $y/b = 2(r - r_i)/(r_o - r_i) - 1$ with $2b = r_o - r_i$.

The apparent Fanning friction factor in the hydrodynamic entrance region of concentric annuli can be determined from the following correlation developed by Shah [114]:

$$f_{\mathrm{app}}\mathrm{Re} = 3.44(x^+)^{-0.5} + \frac{K(\infty)/(4x^+) + f\,\mathrm{Re} - 3.44(x^+)^{-0.5}}{1 + C(x^+)^{-2}} \tag{3.230}$$

The values of $K(\infty)$, $f\,\mathrm{Re}$, and C entering Eq. (3.230) are given in Table 3.36. The results computed from Eq. (3.230) agree with various analytical predictions within $\pm 3\%$.

3.8.3 Thermally Developing Flow

The thermal entrance length solutions for concentric annular ducts are divided into five categories, which are discussed next.

Fundamental Solution of the First Kind. As indicated in Sec. 3.3.1, this solution applies to concentric annular ducts with one wall at a uniform temperature and the other at a uniform temperature different from the entering fluid temperature. The solution is presented in Table 3.37 as a function of x^* and r^*. The additional quantities of practical interest can be determined from the following relations in conjunction with Table 3.37:

$$\Phi_{x,oi}^{(1)} = -\theta_{x,mi}^{(1)}\mathrm{Nu}_{x,oi}^{(1)}, \qquad \theta_{x,ii}^{(1)} = 1, \quad \theta_{x,oi}^{(1)} = 0 \tag{3.231}$$

$$\Phi_{x,io}^{(1)} = -\theta_{x,mo}^{(1)}\mathrm{Nu}_{x,io}^{(1)}, \qquad \theta_{x,oo}^{(1)} = 1, \quad \theta_{x,io}^{(1)} = 0 \tag{3.232}$$

Fundamental Solution of the Second Kind. As indicated in Sec. 3.3.1, this solution applies to concentric annular ducts with one wall insulated and the other subject to uniform heat flux. The solution is presented in Table 3.38 as a function of x^* and r^*. The additional quantities of practical interest can be determined from the following

TABLE 3.35 Axial Velocity and Pressure Distribution in the Hydrodynamic Entrance Region of Concentric Annular Ducts [113]

	Axial Velocity, u/u_m					Pressure Drop
	$(r - r_i)/(r_o - r_i)$					
x^+	0.05	0.25	0.50	0.75	0.95	Δp^*
			$r^* = 0.10$			
0.00005	0.959	1.0540	1.0540	1.0540	0.9130	0.04041
0.0001	0.895	1.0710	1.0710	1.0710	0.8130	0.07753
0.0005	0.652	1.1530	1.1540	1.1480	0.4980	0.2608
0.001	0.580	1.2020	1.2090	1.1770	0.4060	0.3921
0.005	0.499	1.3090	1.4150	1.0940	0.2780	1.0099
0.01	0.498	1.3620	1.4870	1.0370	0.2500	1.5560
0.04	0.509	1.4149	1.5187	1.0004	0.2355	4.3013
∞	0.5128	1.4176	1.5187	0.9992	0.2351	∞
			$r^* = 0.25$			
0.00005	0.9470	1.0540	1.0540	1.0540	0.9100	0.04042
0.0001	0.8720	1.0710	1.0710	1.0710	0.8100	0.07748
0.0005	0.5800	1.1520	1.1530	1.1480	0.5015	0.26032
0.001	0.4980	1.1960	1.2090	1.1780	0.4100	0.39142
0.005	0.3970	1.2490	1.4140	1.1060	0.2840	1.00816
0.01	0.3840	1.2700	1.4840	1.0580	0.2580	1.55613
0.04	0.3846	1.2916	1.5095	1.0329	0.2473	4.38902
∞	0.3850	1.2922	1.5096	1.0327	0.2472	∞
			$r^* = 0.50$			
0.00005	0.9420	1.0540	1.0540	1.0540	0.9060	0.04050
0.0001	0.8610	1.0710	1.0710	1.0710	0.8040	0.07765
0.0005	0.5440	1.1510	1.1530	1.1490	0.5070	0.26037
0.001	0.4570	1.1900	1.2085	1.1800	0.4170	0.39128
0.005	0.3440	1.1990	1.4130	1.1250	0.2940	1.00697
0.01	0.3270	1.1950	1.4820	1.0860	0.2700	1.55491
0.03	0.3234	1.2002	1.5027	1.0705	0.2618	3.48400
∞	0.3235	1.2004	1.5028	1.0703	0.2617	∞

relations:

$$\theta^{(2)}_{x,mi} = \left(\frac{4r^*}{1 + r^*}\right)x^*, \qquad \Phi^{(2)}_{x,ii} = 1, \quad \Phi^{(2)}_{x,oi} = 0 \tag{3.233}$$

$$\theta^{(2)}_{x,mo} = \left(\frac{4}{1 + r^*}\right)x^*, \qquad \Phi^{(2)}_{x,oo} = 1, \quad \Phi^{(2)}_{x,io} = 0 \tag{3.234}$$

TABLE 3.36 Flow Parameters and Constants to be Used in Conjunction with Eq. (3.230) for Concentric Annular Ducts [114]

r^*	$K(\infty)$	f Re	C
0	1.250	16.000	0.000212
0.05	0.830	21.567	0.000050
0.10	0.784	22.343	0.000043
0.50	0.688	23.813	0.000032
0.75	0.678	23.967	0.000030
1.00	0.674	24.000	0.000029

Fundamental Solution of the Third Kind. As indicated in Sec. 3.3.1, this solution applies to concentric annular ducts with one wall insulated and the other at a uniform temperature different from the entering fluid temperature. The solution is presented in Table 3.39 as a function of x^* and r^*. The additional quantities of practical interest can be determined from the following relations:

$$\Phi_{x,oi}^{(3)} = 0, \qquad \theta_{x,ii}^{(3)} = 1, \qquad \text{Nu}_{x,oi}^{(3)} = 0 \tag{3.235}$$

$$\Phi_{x,io}^{(3)} = 0, \qquad \theta_{x,oo}^{(3)} = 1, \qquad \text{Nu}_{x,io}^{(3)} = 0 \tag{3.236}$$

Fundamental Solution of the Fourth Kind. As indicated in Sec. 3.3.1, this solution applies to concentric annular ducts with one wall at a uniform temperature equal to the entering fluid temperature and the other wall subjected to uniform heat flux. The solution is presented in Table 3.40 as a function of x^* and r^*. The additional quantities of practical interest can be determined from the following relations in conjunction with Table 3.40:

$$\Phi_{x,oi}^{(4)} = -\theta_{x,mi}^{(4)} \text{Nu}_{x,oi}^{(4)}, \qquad \Phi_{x,ii}^{(4)} = 1, \qquad \theta_{x,oi}^{(4)} = 0 \tag{3.237}$$

$$\Phi_{x,io}^{(4)} = -\theta_{x,mo}^{(4)} \text{Nu}_{x,io}, \qquad \Phi_{x,oo}^{(4)} = 1, \qquad \theta_{x,io}^{(4)} = 0 \tag{3.238}$$

The thermal entrance lengths for the four fundamental solutions are presented in Table 3.41. The thermal entrance length $L_{\text{th},j}^{*(k)}$ for the kth fundamental solution and jth heated wall is the value of x^* at which $\text{Nu}_{x,jj}^{(k)} = 1.05\,\text{Nu}_{jj}^{(k)}$.

Prescribed Wall Temperature and Heat-Flux Distributions. The utility of the four fundamental solutions presented above lies in synthesizing solutions to the problems of practical interest. The results pertaining to three such problems are presented below. They are obtained by superposing the fundamental solutions of the first, second, and fourth kinds. In these solutions, the Nusselt numbers for the inner and the outer walls are defined as

$$\text{Nu}_i = \frac{h_i D_h}{k} \qquad \text{where} \quad h_i = \frac{q_i''}{T_i - T_m} \tag{3.239}$$

$$\text{Nu}_o = \frac{h_o D_h}{k} \qquad \text{where} \quad h_o = \frac{q_o''}{T_o - T_m} \tag{3.240}$$

TABLE 3.37 Fundamental Solutions of the First Kind for Thermally Developing Flow in Concentric Annular Ducts [1]

x^*	$\Phi_{x,ii}^{(1)}$	$\theta_{x,mi}^{(1)}$	$\mathrm{Nu}_{x,ii}^{(1)}$	$\mathrm{Nu}_{x,oi}^{(1)}$	$\Phi_{x,oo}^{(1)}$	$\theta_{x,mo}^{(1)}$	$\mathrm{Nu}_{x,oo}^{(1)}$	$\mathrm{Nu}_{x,io}^{(1)}$
				$r^* = 0.02$				
0.00001	—	—	—	—	51.081	0.00303	51.236	—
0.0001	78.5	0.0011	78.5	—	23.033	0.01380	23.355	—
0.001	50.87	0.00519	51.14	—	9.993	0.06134	10.646	—
0.01	35.475	0.03328	36.697	0.043	3.881	0.25664	5.220	0.567
0.1	26.124	0.15146	30.787	2.748	0.835	0.75734	3.440	27.500
∞	25.051	0.16993	30.179	2.948	0.501	0.83006	2.948	30.179
				$r^* = 0.05$				
0.00001	—	—	—	—	51.627	0.00297	51.781	—
0.0001	52.0	0.0014	52.1	—	23.296	0.01355	23.616	—
0.001	30.43	0.00759	30.67	—	10.125	0.06031	10.774	—
0.01	19.397	0.04606	20.334	0.054	3.951	0.25247	5.286	0.166
0.1	13.269	0.19140	16.409	2.841	0.915	0.73191	3.413	14.856
∞	12.685	0.21009	16.058	3.019	0.634	0.78991	3.019	16.058
				$r^* = 0.10$				
0.00001	80.290	0.00043	80.324	—	52.186	0.00287	52.336	—
0.0001	40.682	0.00210	40.767	—	23.576	0.01308	23.888	—
0.001	21.949	0.01094	22.192	—	10.276	0.05832	10.912	—
0.01	12.918	0.06131	13.762	0.064	4.044	0.24530	5.359	0.155
0.1	8.199	0.23388	10.702	2.933	1.022	0.70058	3.413	9.792
∞	7.817	0.25256	10.459	3.095	0.782	0.74744	3.095	10.459
				$r^* = 0.25$				
0.00001	66.502	0.00079	66.555	—	53.276	0.00257	53.414	—
0.0001	31.947	0.00375	32.067	—	24.150	0.01176	24.438	—
0.001	15.843	0.01826	16.138	—	10.613	0.05273	11.204	—
0.01	8.236	0.09229	9.073	0.083	4.277	0.22473	5.517	0.130
0.1	4.567	0.31231	6.641	3.120	1.276	0.63474	3.494	6.141
∞	4.328	0.33120	6.471	3.267	1.082	0.66880	3.267	6.471
				$r^* = 0.50$				
0.00001	60.470	0.00121	60.543	—	54.613	0.00220	54.733	—
0.0001	28.295	0.00563	28.455	—	24.889	0.01007	25.142	—
0.001	13.339	0.02642	13.701	—	11.077	0.04549	11.605	—
0.01	6.341	0.12488	7.246	0.092	4.622	0.19773	5.761	0.116
0.1	3.073	0.38995	5.037	3.374	1.615	0.56325	3.698	4.671
∞	2.885	0.40982	4.889	3.520	1.443	0.59018	3.520	4.889
				$r^* = 1.0$				
0.00001	56.804	0.00171	56.901	—	56.804	0.00171	56.901	—
0.0001	26.141	0.00788	26.349	—	26.141	0.00788	26.349	—
0.001	11.895	0.03613	12.341	—	11.895	0.03613	12.341	—
0.01	5.235	0.16249	6.251	0.064	5.235	0.16249	6.251	0.064
0.1	2.168	0.47770	4.151	3.835	2.168	0.47770	4.151	3.835
∞	2.000	0.50000	4.000	4.000	2.000	0.50000	4.000	4.000

TABLE 3.38 Fundamental Solution of the Second Kind for Thermally Developing Flow in Concentric Annular Ducts [1]

x^*	$\theta^{(2)}_{x,ii}$	$\theta^{(2)}_{x,oi}$	$\mathrm{Nu}^{(2)}_{x,ii}$	$\theta^{(2)}_{x,oo}$	$\theta^{(2)}_{x,io}$	$\mathrm{Nu}^{(2)}_{x,oo}$
			$r^* = 0.02$			
0.0001	0.0115	—	86.9	0.035376	—	28.584
0.001	0.01859	—	54.01	0.079885	—	13.164
0.01	0.027036	0.000002	38.093	0.193018	0.000097	6.502
0.1	0.38397	0.005296	32.729	0.603071	0.264808	4.741
1.0	0.109008	0.075872	32.705	4.132796	3.793624	4.734
∞	∞	∞	32.705	∞	∞	4.734
			$r^* = 0.05$			
0.0001	0.01725	—	58.0	0.034990	—	28.894
0.001	0.03034	—	33.17	0.078920	—	13.314
0.01	0.048370	0.000005	21.521	0.190125	0.000108	6.578
0.1	0.075142	0.012947	17.827	0.589334	0.258931	4.799
1.0	0.246620	0.184348	17.811	4.018208	3.686958	4.792
∞	∞	∞	17.811	∞	∞	4.792
			$r^* = 0.10$			
0.0001	0.021194	—	47.265	0.034596	—	29.212
0.001	0.04043	—	24.96	0.077880	—	13.469
0.01	0.070738	0.000012	14.903	0.186700	0.000116	6.652
0.1	0.120267	0.024794	11.918	0.570190	0.247944	4.841
1.0	0.447629	0.352015	11.906	3.843224	3.520150	4.834
∞	∞	∞	11.906	∞	∞	4.834
			$r^* = 0.25$			
0.0001	0.026327	—	38.099	0.033832	—	29.840
0.001	0.053885	—	18.838	0.075739	—	13.786
0.01	0.105856	0.000030	10.219	0.179005	0.000118	6.803
0.1	0.208792	0.054572	7.764	0.523546	0.218290	4.913
1.0	0.928974	0.774448	7.753	3.403883	3.097793	4.905
∞	∞	∞	7.753	∞	∞	4.905
			$r^* = 0.50$			
0.0001	0.029345	—	34.233	0.032919	—	30.626
0.001	0.062474	—	16.356	0.073054	—	14.207
0.01	0.131911	0.000052	8.433	0.168984	0.000105	7.027
0.1	0.294840	0.090599	6.192	0.464851	0.181605	5.046
1.0	1.495118	1.290576	6.181	2.865214	2.581152	5.037
∞	∞	∞	6.181	∞	∞	5.037
			$r^* = 1.0$			
0.0001	0.031498	—	31.950	0.031498	—	31.950
0.001	0.068821	—	14.965	0.068821	—	14.965
0.01	0.153517	0.000080	7.490	0.153517	0.000080	7.490
0.1	0.385362	0.136066	5.395	0.385362	0.136066	5.395
0.5	1.185714	0.935714	5.385	1.185714	0.935714	5.385
∞	∞	∞	5.385	∞	∞	5.385

TABLE 3.39 Fundamental Solution of the Third Kind for Thermally Developing Flow in Concentric Annular Ducts [1]

x^*	$\Phi_{x,ii}^{(3)}$	$\theta_{x,oi}^{(3)}$	$\theta_{x,mi}^{(3)}$	$\mathrm{Nu}_{x,ii}^{(3)}$	$\Phi_{x,oo}^{(3)}$	$\theta_{x,io}^{(3)}$	$\theta_{x,mo}^{(3)}$	$\mathrm{Nu}_{x,oo}^{(3)}$
				$r^* = 0.02$				
0.01	35.394	0.00012	0.0331	36.775	3.8810	0.00118	0.25616	5.217
0.05	28.207	0.05729	0.13407	32.574	1.5241	0.34276	0.62191	4.031
0.1	24.666	0.16699	0.23740	32.345	0.68894	0.69722	0.82751	3.994
0.5	8.9413	0.69793	0.72350	32.337	0.00131	0.99942	0.99967	3.993
∞	0	1	1	32.337	0	1	1	3.993
				$r^* = 0.05$				
0.01	19.405	0.00020	0.04562	20.332	3.9517	0.00140	0.25254	5.287
0.05	14.605	0.07801	0.16980	17.592	1.5705	0.34310	0.61631	4.093
0.1	12.273	0.21677	0.29722	17.464	0.71766	0.69444	0.82311	4.057
0.5	3.2443	0.79290	0.81419	17.460	0.00148	0.99937	0.99963	4.057
∞	0	1	1	17.460	0	1	1	4.057
				$r^* = 0.10$				
0.01	12.920	0.00032	0.061221	13.762	4.0442	0.00153	0.24535	5.359
0.05	9.1601	0.10117	0.21362	11.648	1.6427	0.33688	0.60416	4.150
0.1	7.3655	0.26918	0.36295	11.562	0.76967	0.68400	0.81292	4.114
0.5	1.3705	0.86398	0.88144	11.560	0.00194	0.99920	0.99953	4.114
∞	0	1	1	11.560	0	1	1	4.114
				$r^* = 0.25$				
0.01	8.2382	0.00058	0.09229	9.076	4.2773	0.00159	0.22480	5.518
0.05	5.2488	0.14493	0.29346	7.429	1.8451	0.31296	0.56785	4.269
0.1	3.8763	0.36106	0.47417	7.372	0.92804	0.64888	0.78074	4.233
0.5	0.3664	0.93959	0.95028	7.371	0.00412	0.99844	0.99903	4.232
∞	0	1	1	7.371	0	1	1	4.232
				$r^* = 0.50$				
0.01	6.3404	0.00087	0.1250	7.246	4.6214	0.00147	0.19789	5.762
0.05	3.6492	0.18808	0.3692	5.785	2.1535	0.27967	0.51803	4.468
0.1	2.4675	0.44405	0.5700	5.739	1.1814	0.59887	0.73330	4.430
0.5	0.1156	0.97394	0.9798	5.738	0.01048	0.99644	0.99763	4.429
∞	0	1	1	5.738	0	1	1	4.429
				$r^* = 1.0$				
0.01	5.2421	0 00119	0.16254	6.260	5.2421	0.00119	0.16254	6.260
0.05	2.7028	0.23561	0.44867	4.902	2.7028	0.23561	0.44867	4.902
0.1	1.6468	0.52780	0.66124	4.861	1.6468	0.52780	0.66124	4.861
0.5	0.0337	0.99033	0.99306	4.861	0.0337	0.99033	0.99306	4.861
∞	0	1	1	4.861	0	1	1	4.861

TABLE 3.40 Fundamental Solution of the Fourth Kind for Thermally Developing Flow in Concentric Annular Ducts [1]

x^*	$\theta^{(4)}_{x,mi}$	$\mathrm{Nu}^{(4)}_{x,ii}$	$\mathrm{Nu}^{(4)}_{x,oi}$	$\theta^{(4)}_{x,mo}$	$\mathrm{Nu}^{(4)}_{x,oo}$	$\mathrm{Nu}^{(4)}_{x,io}$
			$r^* = 0.02$			
0.01	0.0007837	38.093	0.030	0.039215	6.809	—
0.05	0.0034689	32.689	1.976	0.196078	4.938	15.128
0.1	0.0052665	31.260	2.648	0.37264	4.529	22.761
0.5	0.0067801	30.181	2.948	1.1911	3.375	29.338
∞	0.0067834	30.179	2.948	1.6568	2.948	30.179
			$r^* = 0.05$			
0.01	0.0019050	21.522	0.037	0.037508	6.552	—
0.05	0.0084154	17.776	2.038	0.18544	4.807	8.355
0.1	0.012796	16.798	2.713	0.34771	4.379	12.442
0.5	0.016555	16.060	3.018	1.0081	3.289	15.728
∞	0.016563	16.058	3.019	1.2455	3.019	16.058
			$r^* = 0.10$			
0.01	0.0036361	14.902	0.042	0.036254	6.646	0.034
0.05	0.016093	11.864	2.093	0.17599	4.836	5.739
0.1	0.024625	11.072	2.778	0.32377	4.350	8.309
0.5	0.032288	10.460	3.095	0.83845	3.271	10.304
∞	0.032307	10.459	3.095	0.95614	3.095	10.459
			$r^* = 0.25$			
0.01	0.0079989	10.225	0.050	0.031985	6.802	0.066
0.05	0.035729	7.710	2.190	0.15253	4.880	3.800
0.1	0.055780	7.040	2.908	0.27133	4.321	5.322
0.5	0.076431	6.474	3.266	0.58532	3.344	6.421
∞	0.076523	6.471	3.267	0.61810	3.267	6.471
			$r^* = 0.50$			
0.01	0.013332	8.433	0.055	0.026664	7.026	0.064
0.05	0.060343	6.137	2.317	0.12508	4.998	3.007
0.1	0.096747	5.503	3.095	0.21530	4.400	4.125
0.5	0.14163	4.894	3.518	0.40000	3.554	4.870
∞	0.14203	4.889	3.520	0.40908	3.520	4.889
			$r^* = 1.0$			
0.01	0.02009	7.495	0.254	0.02009	7.495	0.254
0.05	0.09211	5.341	2.559	0.09211	5.341	2.559
0.1	0.15285	4.723	3.453	0.15285	4.723	3.453
0.5	0.24801	4.013	3.993	0.24801	4.013	3.993
∞	0.25000	4.000	4.000	0.25000	4.000	4.000

TABLE 3.41 Thermal Entrance Lengths for Thermally Developing Flows in Concentric Annular Ducts [1]

r^*	$L^{*(1)}_{\text{th},i}$	$L^{*(1)}_{\text{th},o}$	$L^{*(2)}_{\text{th},i}$	$L^{*(2)}_{\text{th},o}$
0.02	0.05840	0.1650	0.02699	0.03901
0.05	0.06488	0.1458	0.03043	0.03886
0.10	0.06953	0.1311	0.03334	0.03911
0.25	0.07621	0.1126	0.03726	0.04006
0.50	0.08237	0.1003	0.03975	0.04090
1.00	0.09023	0.09023	0.04101	0.04101
r^*	$L^{*(3)}_{\text{th},i}$	$L^{*(3)}_{\text{th},o}$	$L^{*(4)}_{\text{th},i}$	$L^{*(4)}_{\text{th},o}$
0.02	0.02252	0.03001	0.07962	0.04241
0.05	0.02429	0.02970	0.09493	0.6638
0.10	0.02558	0.02960	0.1101	0.5284
0.25	0.02720	0.02964	0.1309	0.3770
0.50	0.02829	0.02956	0.1721	0.2875
1.00	0.02913	0.02913	0.2201	0.2201

(i) Constant Temperatures at Both Walls. For this problem,

$$T = \begin{cases} T_i & \text{at} \quad r = r_i, \quad x \geq x_e \\ T_o & \text{at} \quad r = r_o, \quad x \geq x_e \\ T_e & \text{at} \quad x \leq x_e, \quad r_i < r < r_o \end{cases}$$

When $T_i \neq T_o$, the problem is designated as 1a and its solution is expressible in terms of the following set of equations:

$$T^{(1a)}_{x,m} = T_e + (T_i - T_e)\theta^{(1)}_{x,mi} + (T_o - T_e)\theta^{(1)}_{x,mo} \tag{3.241}$$

$$q''^{(1a)}_{x,i} = \frac{k}{D_h}\left[(T_i - T_e)\Phi^{(1)}_{x,ii} + (T_o - T_e)\Phi^{(1)}_{x,io}\right] \tag{3.242}$$

$$q''^{(1a)}_{x,o} = \frac{k}{D_h}\left[(T_i - T_e)\Phi^{(1)}_{x,oi} + (T_o - T_e)\Phi^{(1)}_{x,oo}\right] \tag{3.243}$$

$$\text{Nu}^{(1a)}_{x,i} = \frac{(T_i - T_e)\Phi^{(1)}_{x,ii} + (T_o - T_e)\Phi^{(1)}_{x,io}}{(T_i - T_e)(1 - \theta^{(1)}_{x,mi}) - (T_o - T_e)\theta^{(1)}_{x,mo}} \tag{3.244}$$

$$\text{Nu}^{(1a)}_{x,o} = \frac{(T_o - T_e)\Phi^{(1)}_{x,oo} + (T_i - T_e)\Phi^{(1)}_{x,oi}}{(T_o - T_e)(1 - \theta^{(1)}_{x,mo}) - (T_i - T_e)\theta^{(1)}_{x,mi}} \tag{3.245}$$

When $T_i = T_o$, the problem is designated as 1b. The circumferentially averaged Nusselt number for this problem can be determined from

$$\text{Nu}_{x,\text{T}} = \frac{\text{Nu}^{(1b)}_{x,o} + r^*\,\text{Nu}^{(1b)}_{x,i}}{1 + r^*} \tag{3.246}$$

(ii) Constant Heat Fluxes at Both Walls. For this problem,

$$q_w'' = \begin{cases} q_i'' & \text{at} \quad r = r_i, \quad x \geq x_e \\ q_o'' & \text{at} \quad r = r_o, \quad x \geq x_e \end{cases}$$

$$T = T_e \qquad \text{at} \quad x = x_e, \quad r_i < r < r_o$$

When $q_i'' \neq q_o''$, the problem is designated as 2a and its solution is expressible in terms of the following set of equations:

$$T_{x,i}^{(2a)} = T_e + \frac{D_h}{k}\left[q_i''\theta_{x,ii}^{(2)} + q_o''\theta_{x,io}^{(2)}\right] \tag{3.247}$$

$$T_{x,o}^{(2a)} = T_e + \frac{D_h}{k}\left[q_i''\theta_{x,oi}^{(2)} + q_o''\theta_{x,oo}^{(2)}\right] \tag{3.248}$$

$$T_{x,m}^{(2a)} = T_e + \frac{D_h}{k}\left[q_i''\theta_{x,mi}^{(2)} + q_o''\theta_{x,mo}^{(2)}\right] \tag{3.249}$$

$$\mathrm{Nu}_{x,i}^{(2a)} = \frac{q_i''}{q_i''\left[\theta_{x,ii}^{(2)} - \theta_{x,mi}^{(2)}\right] - q_o''\left[\theta_{x,mo}^{(2)} - \theta_{x,io}^{(2)}\right]} \tag{3.250}$$

$$\mathrm{Nu}_{x,o}^{(2a)} = \frac{q_o''}{q_o''\left[\theta_{x,oo}^{(2)} - \theta_{x,mo}^{(2)}\right] - q_i''\left[\theta_{x,mi}^{(2)} - \theta_{x,oi}^{(2)}\right]} \tag{3.251}$$

(iii) Constant Temperature at One Wall With Constant Heat Flux at the Other. For this problem,

$$T = T_1 \quad \text{at} \quad r = r_1, \quad x \geq x_e$$

$$q_w'' = q_2'' \quad \text{at} \quad r = r_2, \quad x \geq x_e$$

$$T = T_e \quad \text{at} \quad x \leq x_e, \quad r_i < r < r_o$$

Here the subscripts 1 and 2 can refer to either the inside or the outside wall. The thermal entrance length solution for this problem is given by

$$T_{x,2} = T_e + (T_1 - T_e)\theta_{x,21}^{(3)} + \frac{D_h}{k}q_2''\theta_{x,22}^{(4)} \tag{3.252}$$

$$T_{x,m} = T_e + (T_1 - T_e)\theta_{x,m1}^{(3)} + \frac{D_h}{k}q_2''\theta_{x,m2}^{(4)} \tag{3.253}$$

$$q_{w,1}'' = \frac{k}{D_h}(T_1 - T_e)\Phi_{x,11}^{(3)} + q_2''\Phi_{x,12}^{(4)} \tag{3.254}$$

$$\mathrm{Nu}_{x,1} = \frac{(T_1 - T_e)\Phi_{x,11}^{(3)} + (q_2''D_h/k)\Phi_{x,12}^{(4)}}{(T_1 - T_e)(1 - \theta_{x,m1}^{(3)}) - (q_2''D_h/k)\theta_{x,m2}^{(4)}} \tag{3.255}$$

$$\mathrm{Nu}_{x,2} = \frac{1}{(\theta_{x,22}^{(4)} - \theta_{x,m2}^{(4)}) + [(T_1 - T_e)k/q_2''D_h](\theta_{x,21}^{(3)} - \theta_{x,m1}^{(3)})} \tag{3.256}$$

In addition to the practical solutions presented above, several other solutions for circumferentially and axially varying temperatures and heat fluxes as well as for the convective boundary condition (T3) are available. They are summarized in [1].

3.8.4 Simultaneously Developing Flow

Kakaç and Yücel [113] solved the problem of simultaneously developing velocity and temperature fields in concentric annuli with $r^* = 0.1, 0.25, 0.5, 1$ and $\mathrm{Pr} = 0.01, 0.7, 10$, covering all four fundamental boundary conditions described in Sec. 3.3.1. Their results for $\mathrm{Pr} = 0.7$ are presented in Tables 3.42 to 3.44. Tabulated results for $\mathrm{Pr} = 0.01, 10$ and $r^* = 0.25, 5, 1$ are available in [113]. A graphical representation of the results of Tables 3.42 to 3.44 is available in [115]. The results of Tables 3.42 to 3.44 compare extremely well with the results cited in [1] for the restricted ranges of r^* and Pr.

It may be noted that the four fundamental solutions for the simultaneously developing flow cannot be superposed in the same fashion as the thermally developing flow solutions to synthesize solutions for any prescribed variation of axial wall temperature or heat flux. The dependence of the hydrodynamically developing velocity profile on the axial coordinate makes it impractical to exploit the superposition techniques for simultaneously developing flow. However, by the use of certain influence coefficients in conjunction with the four fundamental solutions, the local Nusselt numbers for several problems of practical interest can be easily evaluated. The influence coefficients θ_1^* through θ_{12}^* determined by Kakaç and Yücel [113] are presented in Tables 3.45 and 3.46. The use of these influence coefficients in determining the Nusselt numbers for four problems of practical interest is explained below.

The fundamental solution of the first kind (Table 3.42) is valid when one of the duct walls is at the entering fluid temperature. When this restriction is removed and the duct walls are allowed to attain uniform and equal or unequal temperatures T_i and T_o, then the local Nusselt numbers $\mathrm{Nu}_{x,i}$ and $\mathrm{Nu}_{x,o}$ at the two walls can be determined from

$$\frac{\mathrm{Nu}_{x,i}}{\mathrm{Nu}_{x,ii}^{(1)}} = \frac{1 - [(T_o - T_e)/(T_i - T_e)]\theta_1^*}{1 - [(T_o - T_e)/(T_i - T_e)]\theta_2^*} \tag{3.257}$$

$$\frac{\mathrm{Nu}_{x,o}}{\mathrm{Nu}_{x,oo}^{(1)}} = \frac{1 - [(T_i - T_e)/(T_o - T_e)]\theta_3^*}{1 - [(T_i - T_e)/(T_o - T_e)]\theta_4^*} \tag{3.258}$$

where $\mathrm{Nu}_{x,ii}^{(1)}$ and $\mathrm{Nu}_{x,oo}^{(1)}$ are available from Table 3.42 and $\theta_1^*, \theta_2^*, \theta_3^*, \theta_4^*$ are listed in Table 3.45.

The fundamental solution of the second kind (Table 3.43) is valid when one of the duct walls is adiabatic, i.e., either $q_i'' = 0$ or $q_o'' = 0$. When this restriction is removed and the duct walls are subject to uniform and equal or unequal wall fluxes q_i'' and q_o'', then the local Nusselt numbers $\mathrm{Nu}_{x,i}$ and $\mathrm{Nu}_{x,o}$ at the two walls can be determined from

$$\frac{\mathrm{Nu}_{x,i}}{\mathrm{Nu}_{x,ii}^{(2)}} = \frac{1}{1 - (q_o''/q_i'')\theta_5^*} \tag{3.259}$$

$$\frac{\mathrm{Nu}_{x,o}}{\mathrm{Nu}_{x,oo}^{(2)}} = \frac{1}{1 - (q_i''/q_o'')\theta_6^*} \tag{3.260}$$

TABLE 3.42 Fundamental Solution of the First Kind for Simultaneously Developing Flow in Concentric Annular Ducts for Pr = 0.7 [113]

r^*	x^*	$Nu^{(1)}_{x,ii}$	$Nu^{(1)}_{x,oi}$	$Nu^{(1)}_{x,oo}$	$Nu^{(1)}_{x,io}$
0.10	0.00005	68.030	—	57.450	—
	0.0001	46.990	—	36.860	—
	0.0005	26.960	—	17.480	—
	0.001	22.020	—	12.910	—
	0.005	15.030	—	6.690	—
	0.01	13.330	0.0649	5.310	0.2473
	0.05	11.162	2.3436	3.856	7.6064
	0.1	10.567	2.9307	3.353	9.7125
	∞	10.450	3.0970	3.095	10.4603
0.25	0.00005	63.500	—	56.990	—
	0.0001	41.700	—	36.690	—
	0.0005	21.310	—	17.620	—
	0.0010	16.660	—	13.000	—
	0.0025	12.460	—	8.930	—
	0.01	8.870	0.0843	5.400	0.1772
	0.05	7.099	2.5334	3.859	4.9398
	0.1	6.626	3.1222	3.394	6.1506
	∞	6.471	3.2669	3.267	6.4713
0.50	0.00005	61.930	—	56.310	—
	0.0001	39.820	—	36.560	—
	0.0005	19.170	—	17.700	—
	0.0010	14.600	—	13.110	—
	0.0025	10.510	—	9.060	—
	0.01	7.085	0.1066	5.640	0.1485
	0.05	5.465	2.7629	4.108	3.8121
	0.1	5.030	3.3749	3.658	4.6824
	∞	4.892	3.5228	3.518	4.8881
1.00	0.00005	58.850	—	58.850	—
	0.0001	37.615	—	37.615	—
	0.0005	18.140	—	18.140	—
	0.001	13.440	—	13.440	—
	0.005	7.484	—	7.484	—
	0.01	6.126	0.1152	6.126	0.1152
	0.5	4.576	3.1321	4.576	3.1321
	0.1	4.140	3.8392	4.140	3.8392
	∞	4.000	4.0000	4.000	4.0000

where $Nu^{(2)}_{x,ii}$ and $Nu^{(2)}_{x,oo}$ are available from Table 3.43 and θ_5^* and θ_6^* are listed in Table 3.45.

The fundamental solution of the third kind (Table 3.43) is valid when one of the walls is at a uniform temperature (different from the entering fluid temperature) and the other wall is adiabatic (i.e., zero heat flux). The fundamental solution of the fourth kind (Table 3.44) is valid when one of the walls is at a uniform temperature (different from the entering fluid temperature) and the other is at the entering fluid temperature. Suppose that the restrictions of the adiabatic wall and the wall being at the entering

TABLE 3.43 Fundamental Solutions of the Second and Third Kinds for Simultaneously Developing Flow in Concentric Annular Ducts for Pr = 0.7 [113]

r^*	x^*	$\mathrm{Nu}_{x,ii}^{(2)}$	$\mathrm{Nu}_{x,oo}^{(2)}$	$\mathrm{Nu}_{x,ii}^{(3)}$	$\mathrm{Nu}_{x,oo}^{(3)}$
0.10	0.00005	91.410	82.510	68.030	57.450
	0.0001	64.670	55.520	46.990	36.860
	0.0005	33.240	24.300	26.960	17.480
	0.001	26.350	17.660	22.020	12.910
	0.005	16.890	9.014	15.030	6.690
	0.01	14.630	7.044	13.330	5.313
	0.05	12.043	4.969	11.500	4.099
	0.1	11.840	4.841	11.416	4.045
	∞	11.900	4.834	11.560	4.113
0.25	0.00005	87.590	82.050	63.500	56.990
	0.0001	60.170	55.240	41.700	36.690
	0.0005	27.870	24.370	21.310	17.620
	0.0010	21.160	17.720	16.660	13.000
	0.0025	15.220	11.940	12.460	8.930
	0.01	10.190	7.100	8.870	5.397
	0.05	7.931	5.046	7.412	4.142
	0.1	7.759	4.915	7.357	4.084
	∞	7.735	4.904	7.370	4.232
0.50	0.00005	83.340	81.370	61.930	56.310
	0.0001	58.640	54.870	39.820	36.560
	0.0005	25.900	24.490	19.170	17.700
	0.0010	19.240	17.860	14.600	13.110
	0.0025	13.395	12.090	10.510	9.060
	0.10	8.500	7.250	7.085	5.639
	0.05	6.351	5.188	5.777	4.405
	0.1	6.190	5.044	5.734	4.378
	∞	6.181	5.036	5.738	4.429
1.00	0.00005	83.620	83.620	58.850	58.850
	0.0001	56.220	56.220	37.615	37.615
	0.0005	24.880	24.880	18.140	18.140
	0.001	18.270	18.270	13.440	13.440
	0.005	9.601	9.601	7.484	7.484
	0.01	7.631	7.631	6.126	6.126
	0.05	5.542	5.542	4.890	4.890
	0.1	5.387	5.387	4.847	4.847
	∞	5.384	5.384	4.860	4.860

fluid temperature are removed. Let $q_w'' = q_o''$ at $r = r_o$ and $T_w = T_i$ at $r = r_i$. In this case, the local Nusselt numbers $\mathrm{Nu}_{x,i}$ and $\mathrm{Nu}_{x,o}$ at the two walls are given by

$$\frac{\mathrm{Nu}_{x,i}}{\mathrm{Nu}_{x,ii}^{(3)}} = \frac{1 - [(q_o''D_h/k)/(T_i - T_e)]\theta_7^*}{1 - [(q_o''D_h/k)/(T_i - T_e)]\theta_8^*} \tag{3.261}$$

$$\frac{\mathrm{Nu}_{x,o}}{\mathrm{Nu}_{x,oo}^{(4)}} = \frac{1}{1 - [(T_i - T_e)/(q_o''D_h/k)]\theta_{12}^*} \tag{3.262}$$

where $\mathrm{Nu}_{x,ii}^{(3)}$ and $\mathrm{Nu}_{x,oo}^{(4)}$ are available from Tables 3.43 and 3.44, and θ_7^*, θ_8^*, θ_{12}^* are listed in Table 3.46.

TABLE 3.44 Fundamental Solution of the Fourth Kind for Simultaneously Developing Flow in Concentric Annular Ducts for Pr = 0.7 [113]

r^*	x^*	$\mathrm{Nu}_{x,ii}^{(4)}$	$\mathrm{Nu}_{x,oi}^{(4)}$	$\mathrm{Nu}_{x,oo}^{(4)}$	$\mathrm{Nu}_{x,io}^{(4)}$
0.10	0.00005	91.410	—	82.510	—
	0.0001	64.670	—	55.520	—
	0.0005	33.240	—	24.300	—
	0.001	26.350	—	17.660	—
	0.005	16.890	—	9.014	—
	0.01	14.626	—	7.044	—
	0.05	11.780	−2.0910	4.854	5.6765
	0.1	10.997	−2.7783	4.352	8.2064
	∞	10.450	−3.0960	3.111	10.4592
0.25	0.00005	87.590	—	87.590	—
	0.0001	60.170	—	60.170	—
	0.0005	27.870	—	24.370	—
	0.0010	21.160	—	17.720	—
	0.0025	15.220	—	11.940	—
	0.01	10.190	—	7.100	0.0939
	0.05	7.703	2.1797	4.884	3.7974
	0.1	7.034	2.9006	4.321	5.3139
	∞	6.471	3.2671	3.267	6.4714
0.50	0.00005	86.340	—	81.370	—
	0.0001	58.640	—	54.870	—
	0.0005	25.900	—	24.490	—
	0.0010	19.240	—	17.860	—
	0.0025	13.395	—	12.090	—
	0.01	8.497	0.0752	7.249	0.0752
	0.05	6.136	2.3150	5.000	3.0098
	0.1	5.502	3.0995	4.399	4.1245
	∞	4.890	3.5211	3.518	4.8912
1.00	0.00005	83.620	—	83.620	—
	0.0001	56.220	—	56.220	—
	0.0005	24.880	—	24.880	—
	0.001	18.270	—	18.270	—
	0.005	9.601	—	9.601	—
	0.01	7.631	0.0503	7.631	0.0503
	0.05	5.339	2.5453	5.339	2.5453
	0.1	4.719	3.4571	4.719	3.4571
	∞	4.000	4.0000	4.000	4.0000

If $T_w = T_o$ at $r = r_o$ and $q_w'' = q_i''$ at $r = r_i$, the local Nusselt numbers $\mathrm{Nu}_{x,i}$ and $\mathrm{Nu}_{x,o}$ at the two walls are given by

$$\frac{\mathrm{Nu}_{x,i}}{\mathrm{Nu}_{x,ii}^{(4)}} = \frac{1}{1 - [(T_o - T_e)/(q_i'' D_h/k)]\,\theta_{11}^*} \tag{3.263}$$

$$\frac{\mathrm{Nu}_{x,o}}{\mathrm{Nu}_{x,oo}^{(3)}} = \frac{1 - [(q_i'' D_h/k)/(T_o - T_e)]\,\theta_9^*}{1 - [(q_i'' D_h/k)/(T_o - T_e)]\,\theta_{10}^*} \tag{3.264}$$

TABLE 3.45 Influence Coefficients from Fundamental Solution of the First and Second Kinds for Simultaneously Developing Flow in Concentric Annular Ducts for Pr = 0.7 [113]

r^*	x^*	θ_1^*	θ_2^*	θ_3^*	θ_4^*	θ_5^*	θ_6^*
0.10	0.00005	—	0.0191	—	0.0022	0.0160	0.0015
	0.0001	—	0.0273	—	0.0033	0.0228	0.0020
	0.0005	—	0.0595	—	0.0086	0.0587	0.0044
	0.001	—	0.0891	—	0.0136	0.0944	0.0064
	0.005	—	0.1996	—	0.0460	0.3060	0.0165
	0.01	0.0054	0.2930	0.0011	0.0849	0.5289	0.0256
	0.05	0.4885	0.7167	0.2698	0.4434	1.2855	0.0538
	0.1	0.8442	0.9184	0.6908	0.7892	1.3705	0.0565
	∞	1.0000	1.0000	1.0000	1.0000	1.3835	0.0562
0.25	0.00005	—	0.0169	—	0.0044	0.0136	0.0031
	0.0001	—	0.0241	—	0.0066	0.0187	0.0043
	0.0005	—	0.0530	—	0.0159	0.0433	0.0096
	0.0010	—	0.0762	—	0.0242	0.0667	0.0140
	0.0025	—	0.1246	—	0.0440	0.1209	0.0238
	0.01	0.0056	0.2744	0.0020	0.1263	0.3242	0.0565
	0.05	0.4996	0.7182	0.3588	0.5474	0.7443	0.1189
	0.1	0.8596	0.9262	0.7909	0.8604	0.7897	0.1249
	∞	1.0000	1.0000	1.0000	1.0000	0.7932	0.1250
0.50	0.00005	—	0.0142	—	0.0072	0.0113	0.0051
	0.0001	—	0.0202	—	0.0106	0.0153	0.0070
	0.0005	—	0.0449	—	0.0245	0.0336	0.0160
	0.0010	—	0.0652	—	0.0365	0.0506	0.0235
	0.0025	—	0.1081	—	0.0637	0.0887	0.0400
	0.01	0.0051	0.2481	0.0030	0.1674	0.2252	0.0960
	0.05	0.4922	0.7058	0.4922	0.7058	0.4979	0.2037
	0.1	0.8624	0.9265	0.8280	0.8971	0.5270	0.2147
	∞	1.0000	1.0000	1.0000	1.0000	0.5288	0.2160
1.00	0.00005	—	0.0087	—	0.0087	0.0064	0.0064
	0.0001	—	0.0138	—	0.0138	0.0095	0.0095
	0.0005	—	0.0339	—	0.0339	0.0234	0.0234
	0.001	—	0.0503	—	0.0503	0.0354	0.0354
	0.005	—	0.1321	—	0.1321	0.0951	0.0951
	0.01	0.0041	0.2101	0.0041	0.2101	0.1512	0.1512
	0.05	0.4635	0.4745	0.4635	0.4745	0.3257	0.3257
	0.1	0.8512	0.9165	0.8512	0.9165	0.3427	0.3427
	∞	1.0000	1.0000	1.0000	1.0000	0.3460	0.3460

where $\mathrm{Nu}_{x,ii}^{(4)}$ and $\mathrm{Nu}_{x,oo}^{(3)}$ are available from Tables 3.43 and 3.44, and θ_9^*, θ_{10}^*, θ_{11}^* are listed in Table 3.46.

3.9 ECCENTRIC ANNULAR DUCTS

Eccentric annular ducts are encountered in practice quite frequently, as the manufacturing tolerances and the imposed service conditions tend to introduce eccentricities in nominally concentric annuli. Even moderate values of the eccentricity exert a dramatic influence on the flow rate through an annular duct with a large value of $r^* = r_i/r_o$ ($r^* \to 1$). This fact has long been recognized in a number of engineering applications such as journal bearings. This underscores the importance of the results pertaining to eccentric annular ducts. Although not investigated as extensively as the concentric annuli, a fair amount of information of practical interest is available for these ducts. Several useful fluid flow and heat transfer results are presented next. Note that $D_h = 2(r_o - r_i)$ for eccentric annular ducts.

TABLE 3.46 Influence Coefficients from Fundamental Solutions of the Third and Fourth Kinds for Simultaneously Developing Flow in Concentric Annular Ducts for Pr = 0.7 [113]

r^*	x^*	θ_7^*	θ_8^*	θ_9^*	θ_{10}^*	θ_{11}^*	θ_{12}^*
0.10	0.00005	—	0.0002	—	—	1.7416	0.1754
	0.0001	—	0.0004	—	0.0001	1.7578	0.1782
	0.0005	—	0.0018	—	0.0002	1.9608	0.1956
	0.001	—	0.0037	—	0.0004	2.2146	0.2198
	0.005	—	0.0188	—	0.0023	3.2452	0.3348
	0.01	0.0003	0.0387	0.0001	0.0051	3.9806	0.4318
	0.05	0.1102	0.2233	0.0217	0.0425	3.0630	0.5460
	0.1	0.3652	0.5080	0.0944	0.1374	1.3722	0.4086
	∞	∞	∞	∞	∞	0.0000	0.0000
0.25	0.00005	—	0.0002	—	0.0001	1.4746	0.3572
	0.0001	—	0.0003	—	0.0001	1.4424	0.3562
	0.0005	—	0.0016	—	0.0004	1.4536	0.3668
	0.0010	—	0.0033	—	0.0009	1.5770	0.3966
	0.0025	—	0.0083	—	0.0023	1.8228	0.4630
	0.01	0.0003	0.0353	0.0001	0.0106	2.5066	0.6700
	0.05	0.1105	0.2159	0.0448	0.0849	1.9258	0.7256
	0.1	0.3732	0.5168	0.1860	0.2612	0.9058	0.4882
	∞	∞	∞	∞	∞	0.0000	0.0000
0.50	0.00005	—	0.0002	—	0.0001	1.2192	0.5752
	0.0001	—	0.0003	—	0.0002	1.1744	0.5700
	0.0005	—	0.0014	—	0.0014	1.1354	0.5746
	0.0010	—	0.0027	—	0.0027	1.2106	0.6104
	0.0025	—	0.0070	—	0.0070	1.3650	0.6920
	0.01	0.0003	0.0307	0.0002	0.0170	1.8152	0.9452
	0.05	0.1036	0.1991	0.0813	0.1456	1.4446	0.9036
	0.1	0.3620	0.5034	0.2612	0.4486	0.7276	0.5518
	∞	∞	∞	∞	∞	0.0000	0.0000
1.00	0.00005	—	0.0001	—	0.0001	0.7224	0.7224
	0.0001	—	0.0002	—	0.0002	0.7670	0.7670
	0.0005	—	0.0010	—	0.0010	0.8152	0.8152
	0.001	—	0.0021	—	0.0021	0.8750	0.8750
	0.005	—	0.0112	—	0.0112	1.1202	1.1202
	0.01	0.0003	0.0241	0.0003	0.0241	1.3146	1.3146
	0.05	0.0881	0.1683	0.0881	0.1683	1.1274	1.1274
	0.1	0.3243	0.4548	0.3243	0.4548	0.6208	0.6208
	∞	∞	∞	∞	∞	0.0000	0.0000

3.9.1 Fully Developed Flow

Based on the analysis by Piercy et al. [116], the fully developed velocity distribution in the eccentric annulus (Fig. 3.46) can be expressed by the following set of equations:

$$u = -\frac{1}{\mu}\left(\frac{dp}{dx}\right) r_0^2 S^2 \left(C + A\eta + B - \frac{\cosh \eta - \cos \xi}{4(\cosh \eta + \cos \xi)}\right) \qquad (3.265)$$

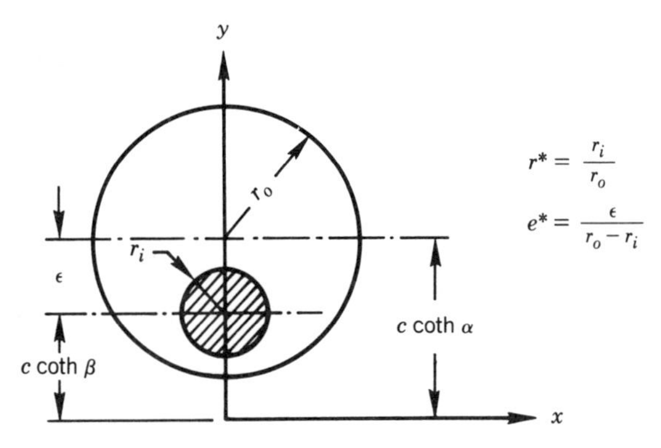

Figure 3.46. An eccentric annular duct.

where

$$C = \sum_{n=1}^{\infty} \frac{(-1)^n \cos n\xi}{\sinh[n(\beta - \alpha)]} \{ e^{-n\beta} \coth \beta \sinh[n(\eta - \alpha)]$$

$$- e^{-n\alpha} \coth \alpha \sinh[n(\eta - \beta)] \} \tag{3.266}$$

$$A = \frac{\coth \alpha - \coth \beta}{2(\alpha - \beta)}, \qquad B = \frac{\beta(1 - 2\coth \alpha) - \alpha(1 - 2\coth \beta)}{4(\alpha - \beta)} \tag{3.267}$$

$$S = \frac{1 - r^*}{2e^*}(1 - e^{*2})^{1/2} \left[\left(\frac{1 + r^*}{1 - r^*} \right)^2 - e^{*2} \right]^{1/2} \tag{3.268}$$

$$\alpha = \sinh^{-1} S, \qquad \beta = \sinh^{-1}(S/r^*) \tag{3.269}$$

The bipolar coordinates ξ and η are related to the Cartesian coordinates x and y of Fig. 3.46 by

$$\tan \xi = \frac{2cx}{x^2 + y^2 - c^2}, \qquad e^{-2\eta} = \frac{x^2 + (y - c)^2}{x^2 + (y + c)^2}, \qquad c = r_0 S \tag{3.270}$$

Based on Eq. (3.265), the friction factor for the eccentric annular ducts is determined to be [1]

$$f\,\mathrm{Re} = 16(1 - r^{*2})(1 - r^*)^2$$

$$\times \left(1 - r^{*4} + Z - 8e^{*2}(1 - r^*)^2 S^2 \sum_{n=1}^{\infty} \frac{n \exp[-n(\alpha + \beta)]}{\sinh[n(\beta - \alpha)]} \right)^{-1} \tag{3.271}$$

where S, α, and β are given by Eqs. (3.268) and (3.269), and

$$Z = \frac{4e^{*2}(1 - r^*)^2}{\alpha - \beta} S^2 \tag{3.272}$$

The foregoing formulas are applicable for $0 < e^* < 1$ and $0 \le r^* < 1$. The two limiting cases excluded in the above formulation are: (1) $e^* = 1, 0 \le r^* < 1$, and (2) $0 \le e^* \le 1$, $r^* = 1$. The applicable formulas for these cases are presented next. The third limiting case of $e^* = 0, 0 \le r^* \le 1$, corresponds to the concentric annular duct for which f Re is given by Eq. (3.209).

Based on the exact solution for torsion in a hollow shaft of unit eccentricity by Stevenson [117], the factor f Re for $e^* = 1, 0 \le r^* < 1$ is given by Tiedt [118] as

$$f \text{ Re} = \frac{16(1 - r^{*2})(1 - r^*)^2}{1 - r^{*4} - 4r^{*2}\psi'[1/(1 - r^*)]} \tag{3.273}$$

where ψ' is the so-called trigamma function with the argument $1/(1 - r^*)$. It is given

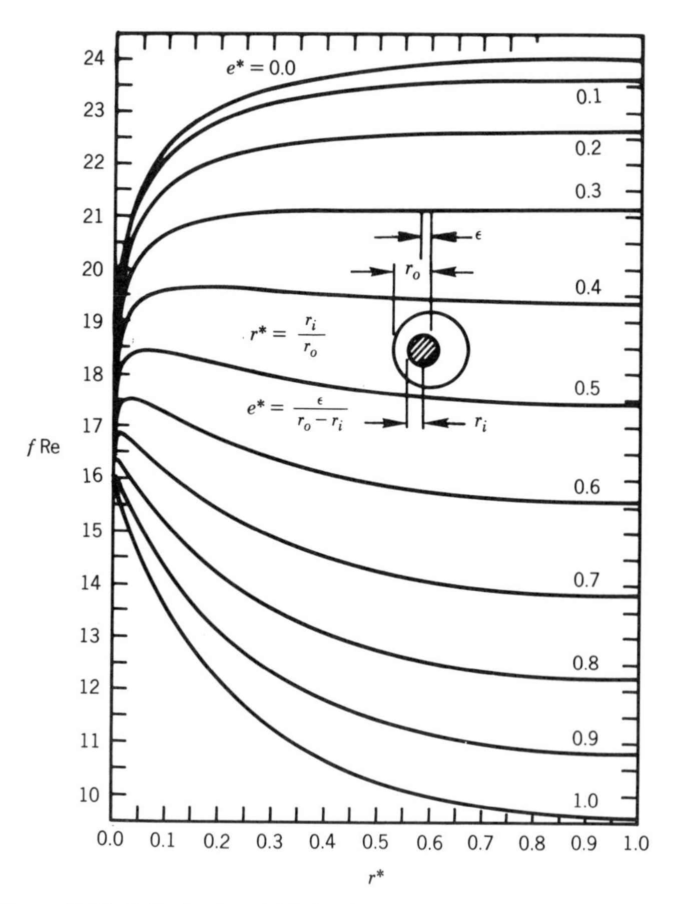

Figure 3.47. Fully developed friction factors for eccentric annular ducts [1].

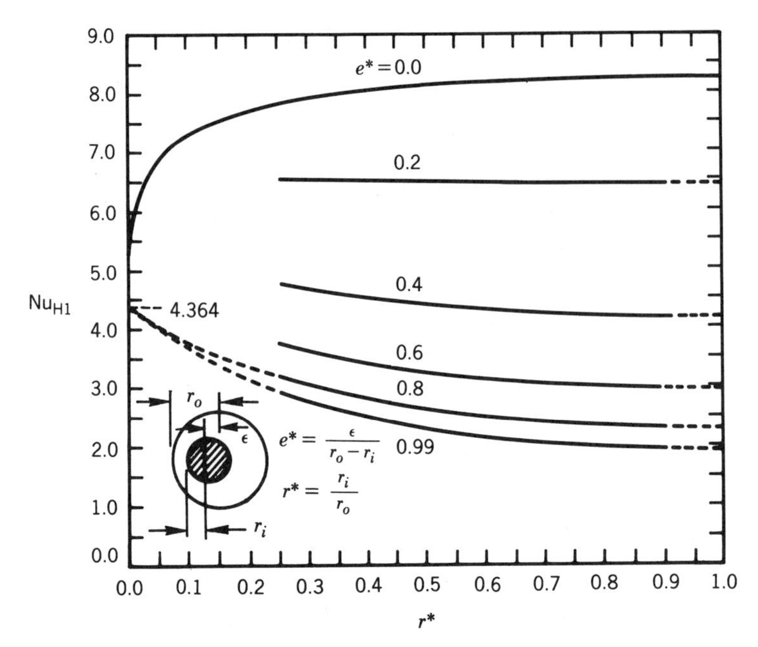

Figure 3.48. Fully developed Nusselt numbers for eccentric annular ducts [120].

by

$$\psi'\left(\frac{1}{1-r^*}\right) = \sum_{n=0}^{\infty}\left(\frac{1}{n + [1/(1-r^*)]}\right)^2 \tag{3.274}$$

Becker [119] derived the expression for the volumetric flow rate through an eccentric annulus for $0 \le e^* \le 1$, $r^* \to 1$. Based on these results, Tiedt [118] showed that

$$f\,\mathrm{Re} = \frac{24}{1 + 1.5e^{*2}} \tag{3.275}$$

The f Re factors computed from Eqs. (3.271), (3.273), and (3.275) are presented in Fig. 3.47. Tabulated results are available in [1].

Cheng and Hwang [120] analyzed the (H1) problem for eccentric annuli. Their results are presented in Fig. 3.48.

Trombetta [121] conducted a detailed study of the fundamental problems of the first, second, and fourth kinds for eccentric annuli. His tabular heat transfer results are available in [1].

The concentric annular duct relationships for the fundamental solutions of the first and second kinds [Eqs. (3.213) to (3.220)] are also applicable to eccentric annular ducts.

The concentric annular duct relationships (3.224) to (3.226), pertaining to the fundamental solutions of the fourth kind, are not valid for eccentric annular ducts. The applicable relationships are

$$\mathrm{Nu}_{ii}^{(4)} = \frac{1}{\theta_{ii}^{(4)} - \theta_{mi}^{(4)}}, \qquad \mathrm{Nu}_{oo}^{(4)} = \frac{1}{\theta_{oo}^{(4)} - \theta_{mo}^{(4)}} \tag{3.276}$$

$$\mathrm{Nu}_{oi}^{(4)} = \frac{r^*}{\theta_{mi}^{(4)}}, \qquad \mathrm{Nu}_{io}^{(4)} = \frac{1}{r^*\theta_{mo}^{(4)}} \tag{3.277}$$

The relationship (3.223) for the concentric annular duct also applies to the eccentric annular duct.

Similar to concentric annuli, the superposition technique can be used for eccentric annuli to synthesize solutions of practical interest utilizing the fundamental solutions of the first, second, and fourth kinds. However, for the case when equal and constant temperatures are prescribed at the two walls, the superposition technique does not yield the desired results. In this case, the solution may be derived from the asymptotic thermal entrance length solution.

To obtain $\mathrm{Nu}_{\mathrm{H2}}$ for eccentric annuli, first the ratio q_i''/q_o'' is obtained from Eqs. (3.247) and (3.248) with $x = \infty$ as

$$\frac{q_i''}{q_o''} = \frac{\theta_{oo}^{(2)} - \theta_{io}^{(2)}}{\theta_{ii}^{(2)} - \theta_{oi}^{(2)}} = \frac{\left(\theta_{oo}^{(2)} - \theta_{mo}^{(2)}\right) - \left(\theta_{io}^{(2)} - \theta_{mo}^{(2)}\right)}{\left(\theta_{ii}^{(2)} - \theta_{mi}^{(2)}\right) - \left(\theta_{oi}^{(2)} - \theta_{mi}^{(2)}\right)} \tag{3.278}$$

The Nusselt numbers at the two walls are then determined from the following relations:

$$\mathrm{Nu}_i^{(2b)} = \left(\left(\theta_{ii}^{(2)} - \theta_{mi}^{(2)}\right) + \frac{q_o''}{q_i''}\left(\theta_{io}^{(2)} - \theta_{mo}^{(2)}\right)\right)^{-1} \tag{3.279}$$

$$\mathrm{Nu}_o^{(2b)} = \left(\left(\theta_{oo}^{(2)} - \theta_{mo}^{(2)}\right) + \frac{q_i''}{q_o''}\left(\theta_{oi}^{(2)} - \theta_{mi}^{(2)}\right)\right)^{-1} \tag{3.280}$$

Finally, $\mathrm{Nu}_{\mathrm{H2}}$ is computed from

$$\mathrm{Nu}_{\mathrm{H2}} = \frac{\mathrm{Nu}_o^{(2b)} + r^*\,\mathrm{Nu}_i^{(2b)}}{1 + r^*} \tag{3.281}$$

The $\mathrm{Nu}_{\mathrm{H2}}$ values computed from the foregoing relations are presented in Fig. 3.49.

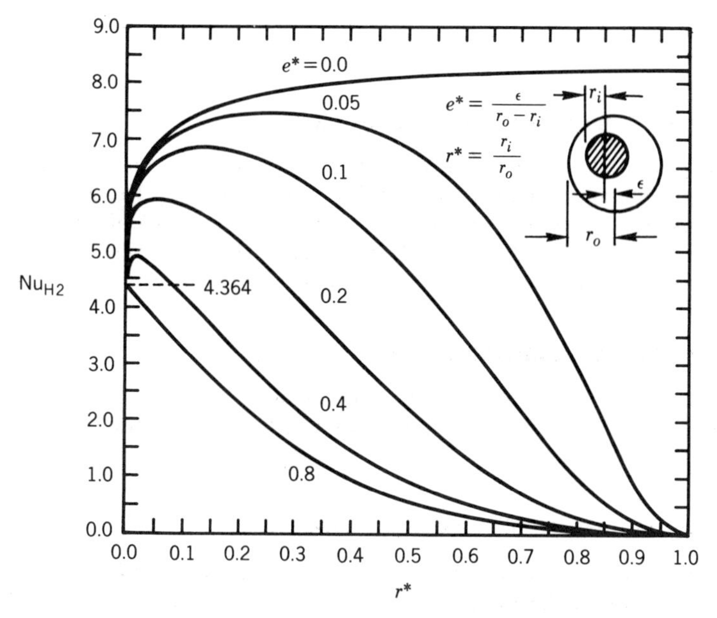

Figure 3.49. Fully developed Nusselt numbers for eccentric annular ducts [1].

TABLE 3.47 Flow Parameters in the Hydrodynamic Entrance Region of Eccentric Annuli [122]

x^+	$e^* = 0.5$ $r^* = 0.5$	$e^* = 0.5$ $r^* = 0.1$	$e^* = 0.7$ $r^* = 0.3$	$e^* = 0.9$ $r^* = 0.1$	$e^* = 0.9$ $r^* = 0.5$
			$f_{app}\mathrm{Re}$		
0.001	113.65	109.58	114.75	116.10	120.16
0.002	82.03	79.10	83.05	81.72	83.47
0.010	42.35	41.05	42.39	38.51	37.46
0.020	—	—	32.18	29.23	27.70
0.050	26.32	25.52	23.45	21.54	19.66
0.100	—	22.16	19.57	—	16.09
0.150	—	20.91	—	—	14.68
0.200	—	—	17.29	—	13.92
0.304	19.42	—	—	15.63	—
0.424	18.93	—	—	15.26	—
∞	17.67	18.35	14.86	14.33	11.42
			$K(\infty)$		
	2.143	1.535	1.959	1.571	2.060
			L_{hy}^+		
	0.254	0.0897	0.156	0.106	0.313
			u_{max}/u_m		
	2.373	2.149	2.277	2.163	2.324

3.9.2 Hydrodynamically Developing Flow

Feldman et al. [122] analyzed the hydrodynamically developing flow through eccentric annuli employing an idealized transverse flow model. Their $f_{app}\mathrm{Re}$, $K(\infty)$, L_{hy}^+, and u_{max}/u_m results are presented in Table 3.47. For the limiting cases of concentric annuli and circular ducts, the predictions of Feldman et al. [122] are in excellent agreement with the well-established results reported earlier.

3.9.3 Thermally Developing Flow

Feldman et al. [123] obtained the four fundamental solutions for an eccentric annular duct with $e^* = 0.5$ and $r^* = 0.5$. A partial set of their tabular results is available in [1].

3.9.4 Simultaneously Developing Flow

Feldman et al. [123] obtained a combined entrance length solution for the fundamental problem of the first kind, with the inner wall heated, for an eccentric annular duct with

$e^* = 0.5$ and $r^* = 0.5$. They reported the heat transfer results for fluids with Pr = 0.01, 1.0 and ∞. Their results are tabulated in [1].

3.10 ADDITIONAL DOUBLY CONNECTED DUCTS

Limited available information on fully developed flow in some additional doubly connected ducts is presented in this section.

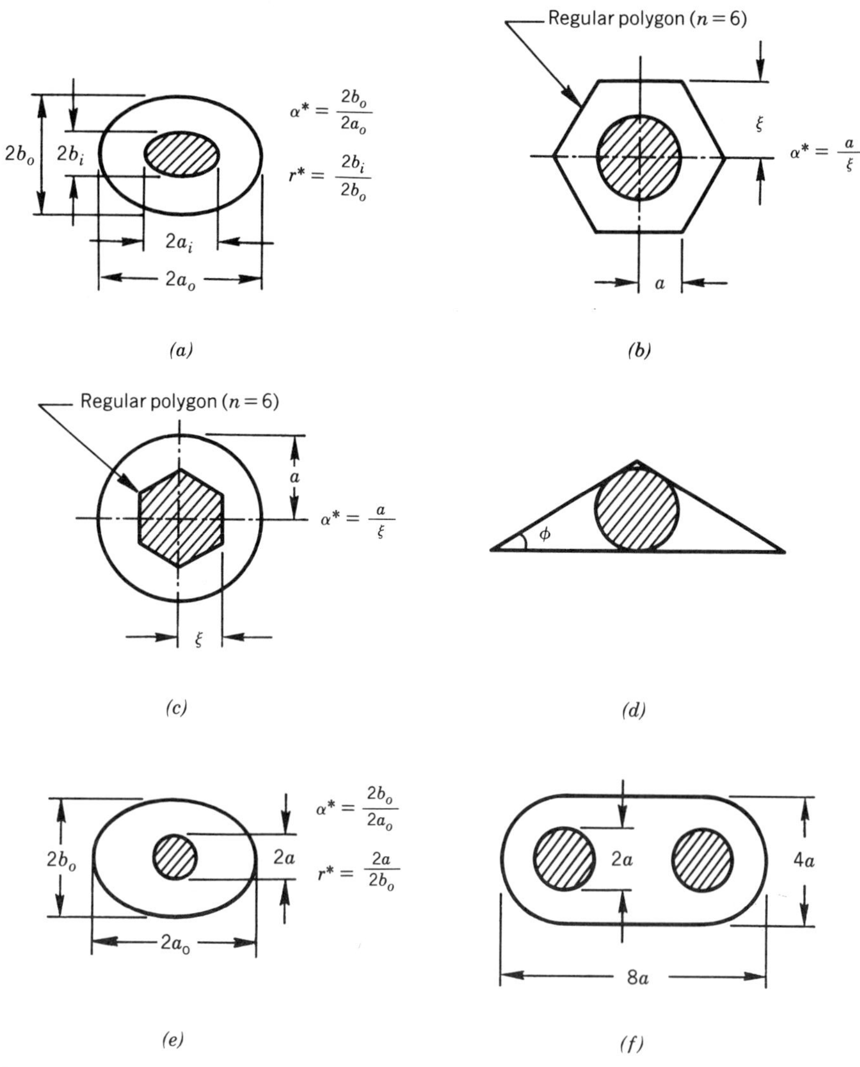

Figure 3.50. Some doubly connected ducts: (*a*) confocal elliptical, (*b*) regular polygonal with centered circular core, (*c*) circular with centered regular polygonal core, (*d*) isosceles triangular with inscribed centered circular core, (*e*) elliptical with centered circular core, (*f*) stadium-shaped with twin circular cores.

3.10.1 Confocal Elliptical Ducts

A confocal elliptical duct shown in Fig. 3.50a comprises two concentric elliptical walls. The major and the minor axes of the walls are related as $a_o^2 - b_o^2 = a_i^2 - b_i^2$.

Based on the analysis by Topakoglu and Arnas [124], the fully developed friction factor for confocal elliptical ducts is given by

$$f\,\mathrm{Re} = \frac{256 A_c^3}{\pi I_{oo} P^2 (a_o + b_o)^4} \tag{3.282}$$

where

$$I_{oo} = \tfrac{1}{4}(1 - \omega^4)\left(1 + \frac{m^8}{\omega^4}\right) - 2m^4 \frac{1 - \omega^2}{1 + \omega^2} + \frac{1}{4 \ln \omega}(1 - \omega^2)^2 \left(1 - \frac{m^4}{\omega^2}\right)^2 \tag{3.283}$$

$$\frac{A_c}{(a_o + b_o)^2} = \frac{\pi}{4}(1 - \omega^2)\left(1 + \frac{m^4}{\omega^2}\right) \tag{3.284}$$

$$\frac{P}{a_o + b_o} = 2\left[(1 + m^2) E_1 + \left(1 + \frac{m^2}{\omega^2}\right) \omega E_\omega\right] \tag{3.285}$$

$$\omega = \frac{a_i + b_i}{a_o + b_o} = \frac{\alpha^* r^* + \left[1 - \alpha^{*2}(1 - r^*)^2\right]^{1/2}}{1 + \alpha^*} \tag{3.286}$$

$$m = \left(\frac{1 - \alpha^*}{1 + \alpha^*}\right)^{1/2}, \qquad \alpha^* = \frac{b_o}{a_o}, \qquad r^* = \frac{b_i}{b_o} \tag{3.287}$$

E_1 and E_ω are the complete elliptical integrals of the second kind which are evaluated for the arguments $1 - b_o^2/a_o^2$ and $1 - b_i^2/a_i^2$, respectively. It may be noted that b_i/a_i is expressible in terms of ω and m by the relation

$$\frac{b_i}{a_i} = \frac{1 - (m^2/\omega^2)}{1 + (m^2/\omega^2)} \tag{3.288}$$

The f Re factors calculated from the foregoing equations are displayed in Fig. 3.51. The fully developed Nusselt numbers $\mathrm{Nu}_{\mathrm{H1}}$ determined from the analysis of Topakoglu and Arnas [124] are also displayed in Fig. 3.51.

3.10.2 Regular Polygonal Ducts with Centered Circular Cores

The fully developed friction factors for regular polygonal ducts with centered circular cores (Fig. 3.50b) are presented in Fig. 3.52. The results are based on the analysis by Ratkowsky and Epstein [125]. As $n \to \infty$, $f\,\mathrm{Re} \to 6.222$ for $\alpha^* = 1$, but $f\,\mathrm{Re} \to 16$ for $\alpha^* = 0$. The fully developed Nusselt numbers $\mathrm{Nu}_{\mathrm{H1}}$ for these ducts were determined by Cheng and Jamil [126] and are presented in Fig. 3.53.

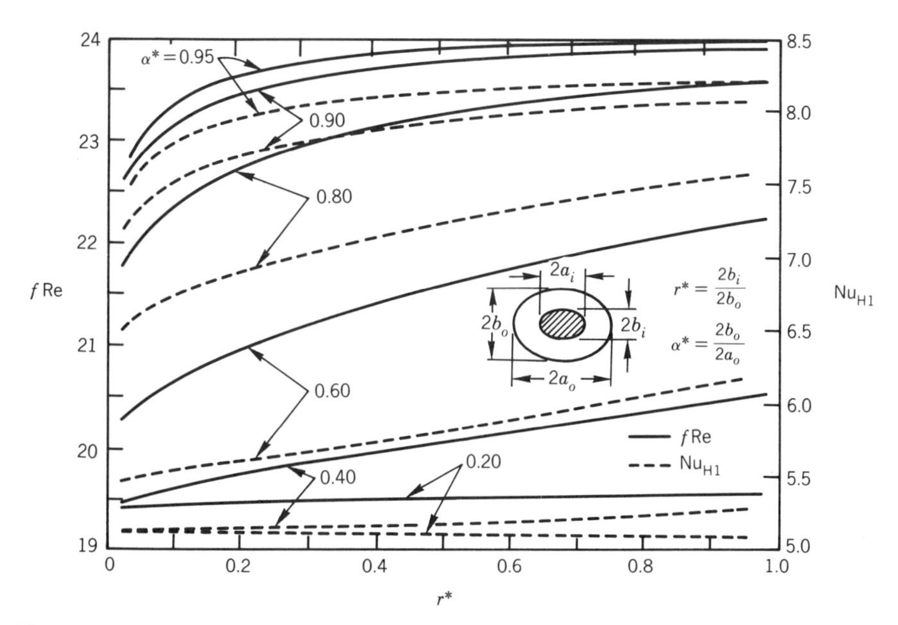

Figure 3.51. Fully developed friction factors and Nusselt numbers for confocal elliptical ducts [124].

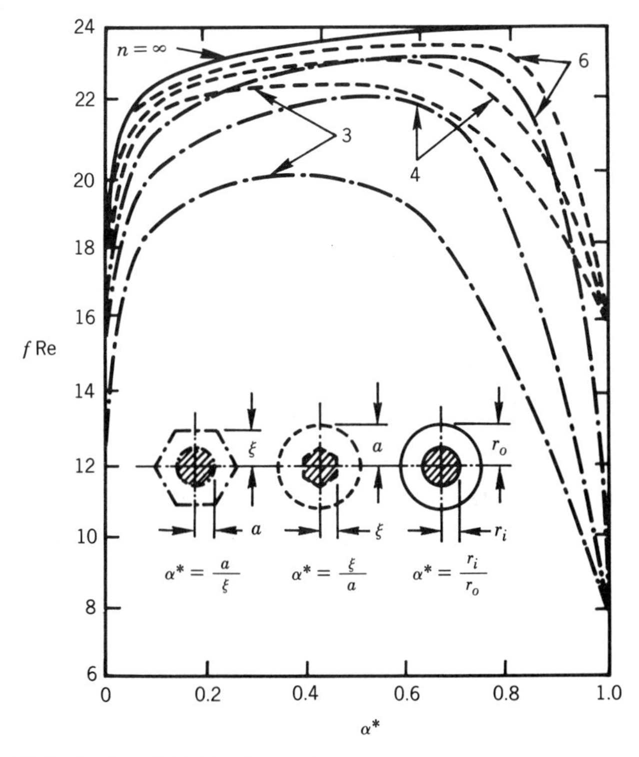

Figure 3.52. Fully developed friction factors for regular polygonal ducts with centered circular cores and circular ducts with centered regular polygonal cores [125, 127].

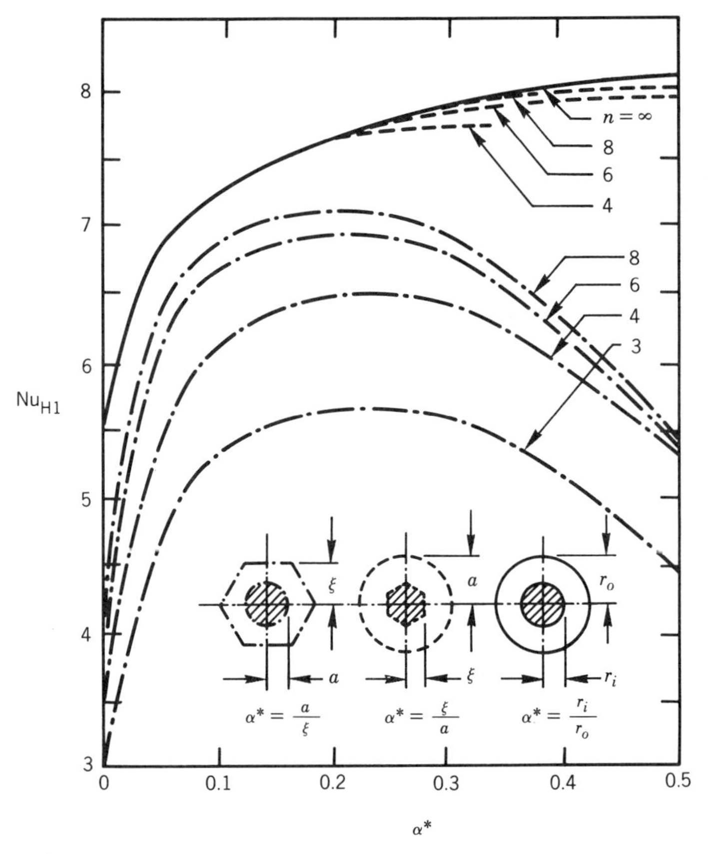

Figure 3.53. Fully developed Nusselt numbers for regular polygonal ducts with centered circular cores and circular ducts with centered regular polygonal cores [107, 126].

3.10.3 Circular Ducts with Centered Regular Polygonal Cores

The fully developed friction factors for circular ducts with centered regular polygonal cores (Fig. 3.50*c*) are presented in Fig. 3.52. They are based on the analysis by Hagen and Ratkowsky [127]. The fully developed Nusselt numbers Nu_{H1} for these ducts were determined by Cheng and Jamil [107]. They are presented in Fig. 3.53.

3.10.4 Isosceles Triangular Ducts with Inscribed Circular Cores

The fully developed friction factors for isosceles triangular ducts with inscribed circular cores (Fig. 3.50*d*) were determined by Bowen [128]. They can be represented by the following expression:

$$f\,\mathrm{Re} = 12.0000 - 0.1605\phi + 4.2883 \times 10^{-3}\phi^2$$
$$-1.0566 \times 10^{-4}\phi^3 + 1.6251 \times 10^{-6}\phi^4 - 1.04821 \times 10^{-8}\phi^5 \quad (3.289)$$

where ϕ is in degrees. $f\,\mathrm{Re}$ from this equation agrees with the tabular values in [1] within $\pm 2\%$.

TABLE 3.48 Fully Developed Friction Factors for Elliptical Ducts with Centered Circular Cores [129]

α^*	r^*	f Re
0.9	0.5	23.519
	0.6	23.435
	0.7	23.159
	0.95	16.816
0.7	0.5	21.694
	0.7	19.402
0.5	0.5	19.321

3.10.5 Elliptical Ducts with Centered Circular Cores

The fully developed friction factors for elliptical ducts with centered circular cores (Fig. 3.50*e*) are presented in Table 3.48. They are based on the analysis by Shivakumar [129].

3.10.6 Stadium-Shaped Ducts with Twin Circular Cores

Schenkel [95] determined the fully developed friction factor f Re as 17.68 for the stadium-shaped duct with twin circular cores (Fig. 3.50*f*) using the 3*R* graphical method mentioned in Sec. 3.7.13.

3.11 CLOSURE

A wide variety of ducts can be fabricated for a specific technical application requiring a preferred flow-passage configuration peculiar to the application. In this regard, Table 3.33 is particularly instructive, as it illustrates a great variety of duct shapes. In the preceding sections of the chapter, we have presented the most useful laminar fluid flow and heat transfer information for 62 singly connected and 8 doubly connected ducts covering a variety of thermal boundary conditions described in Tables 3.1 and 3.2.

Of the various thermal boundary conditions in Tables 3.1 and 3.2, the constant wall temperature Ⓣ and the constant heat flux (H1) and (H2) boundary conditions are of particular importance, as they constitute the usual extremes met in heat exchanger design. It is therefore deemed useful to bring together the key fluid flow and heat transfer characteristics of the most important ducts with the Ⓣ, (H1), and (H2) boundary conditions. They are summarized in Table 3.49, where several of the results are presented as computationally expedient correlations which reproduce the tabular results of [1] to a high degree of accuracy, indicated in parentheses at the end of the appropriate equations. The equations in Table 3.49 with no indicated accuracy level represent exact results.

The detailed information pertaining to each duct geometry can be easily located by referring to the Table of Contents of this chapter. A review of this information brings out the fact that there are several gaps in our knowledge of fluid flow and heat transfer characteristics of even the most commonly encountered duct geometries. As regards the less common ducts covered in Sections 3.7 and 3.10, the information is particularly

TABLE 3.49 Fully Developed Friction Factors and Nusselt Numbers for Technically Important Duct Geometries

Duct Geometry	D_h, α^*, r^*	$f\,\mathrm{Re}$, $\mathrm{Nu_T}$, $\mathrm{Nu_{H1}}$, $\mathrm{Nu_{H2}}$	
Circular (2a)	$D_h = 2a$	$f\,\mathrm{Re} = 16$ $\mathrm{Nu_T} = 3.6568$ $\mathrm{Nu_H} = \frac{48}{11} = 4.3636$ †	
Flat (2b)	$D_h = 4b$	$f\,\mathrm{Re} = 24$ $\mathrm{Nu_T} = 7.5407$ $\mathrm{Nu_H} = 140/17 = 8.2353$	
Rectangular (2a, 2b)	$D_h = \frac{4ab}{a+b} = \frac{4b}{1+\alpha^*}$ $\alpha^* = \frac{2b}{2a}$	For $0 \le \alpha^* \le 1$, $f\,\mathrm{Re} = 24(1 - 1.3553\alpha^* + 1.9467\alpha^{*2} - 1.7012\alpha^{*3} + 0.9564\alpha^{*4} - 0.2537\alpha^{*5})$	(0.05%)
		$\mathrm{Nu_T} = 7.541(1 - 2.610\alpha^* + 4.970\alpha^{*2} - 5.119\alpha^{*3} + 2.702\alpha^{*4} - 0.548\alpha^{*5})$	(0.1%)
		$\mathrm{Nu_{H1}} = 8.235(1 - 2.0421\alpha^* + 3.0853\alpha^{*2} - 2.4765\alpha^{*3} + 1.0578\alpha^{*4} - 0.1861\alpha^{*5})$	(0.03%)
		$\mathrm{Nu_{H2}} = 8.235(1 - 10.6044\alpha^* + 61.1755\alpha^{*2} - 155.1803\alpha^{*3} + 176.9203\alpha^{*4} - 72.9236\alpha^{*5})$	(7.0%)
Equilateral triangular (2a, 2b)	$D_h = \frac{4b}{3}$ $2a = \frac{4b}{\sqrt{3}}$	$f\,\mathrm{Re} = \frac{40}{3} = 13.3333$ $\mathrm{Nu_T} = 2.49$ $\mathrm{Nu_{H1}} = \frac{28}{9} = 3.1111$ $\mathrm{Nu_{H2}} = 1.892$	

†As explained in Section 3.1.4, $\mathrm{Nu_{H1}}$ and $\mathrm{Nu_{H2}}$ are denoted simply as $\mathrm{Nu_H}$ for circular, flat, and concentric annular ducts.

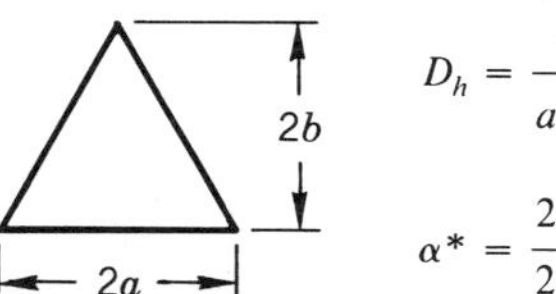

Isosceles triangular

$$D_h = \frac{4ab}{a + \sqrt{a^2 + b^2}}$$

$$\alpha^* = \frac{2b}{2a}$$

For $0 \le \alpha^* \le 1$,

$$f\,\text{Re} = 12(1 - 0.0115\alpha^* + 1.7099\alpha^{*2} - 4.3394\alpha^{*3} + 4.2732\alpha^{*4} - 1.5817\alpha^{*5} + 0.0599\alpha^{*6}) \quad (0.1\%)$$

$$\text{Nu}_\text{T} = 0.943(1 + 4.8340\alpha^* - 2.1738\alpha^{*2} - 4.0797\alpha^{*3} - 2.1220\alpha^{*4} + 11.3589\alpha^{*5} - 6.2052\alpha^{*6}) \quad (0.4\%)$$

$$\text{Nu}_\text{H1} = 2.059(1 + 0.7139\alpha^* + 2.9540\alpha^{*2} - 7.8785\alpha^{*3} + 5.6450\alpha^{*4} + 0.2144\alpha^{*5} - 1.1387\alpha^{*6}) \quad (0.3\%)$$

$$\text{Nu}_\text{H2} = \begin{cases} 1.088\alpha^* & \text{for } \alpha^* \le 0.125 \quad (0.0\%) \\ -0.2113(1 - 10.9962\alpha^* - 15.1301\alpha^{*2} + 16.5921\alpha^{*3} & \text{for } 0.124 < \alpha^* \le 1 \quad (5.0\%) \end{cases}$$

For $1 \le \alpha^* \le \infty$,

$$f\,\text{Re} = 12(\alpha^{*3} + 0.2595\alpha^{*2} - 0.2046\alpha^* + 0.0552)/\alpha^{*3} \quad (0.04\%)$$

$$\text{Nu}_\text{T} = 0.943(\alpha^{*5} + 5.3586\alpha^{*4} - 9.2517\alpha^{*3} + 11.9314\alpha^{*2} - 9.8035\alpha^* + 3.3754)/\alpha^{*5} \quad (0.4\%)$$

$$\text{Nu}_\text{H1} = 2.059(\alpha^{*5} + 1.2489\alpha^{*4} - 1.0559\alpha^{*3} + 0.2515\alpha^{*2} + 0.1520\alpha^* - 0.0901)/\alpha^{*5} \quad (0.04\%)$$

$$\text{Nu}_\text{H2} = \begin{cases} 0.912(\alpha^{*3} - 13.3739\alpha^{*2} + 78.9211\alpha^* - 46.6239)/\alpha^{*3} & \text{for } 1 \le \alpha^* < 8 \quad (5.7\%) \\ 0.312/\alpha^* & \text{for } 8 \le \alpha^* \le \infty \quad (0.0\%) \end{cases}$$

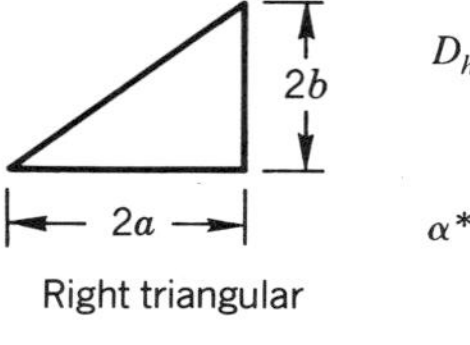

Right triangular

$$D_h = \frac{4ab}{a + b + \sqrt{a^2 + b^2}}$$

$$\alpha^* = \frac{2b}{2a}$$

For $0 \le \alpha^* \le 1$,

$$f\,\text{Re} = 12(1 + 0.27956\alpha^* - 0.2756\alpha^{*2} + 0.0591\alpha^{*3} + 0.0622\alpha^{*4} - 0.0290\alpha^{*5}) \quad (0.04\%)$$

$$\text{Nu}_\text{T} = 1.1731(1 + 3.1312\alpha^* - 3.5919\alpha^{*2} + 1.7893\alpha^{*3} - 0.3189\alpha^{*4}) \quad (0.2\%)$$

$$\text{Nu}_\text{H1} = 2.0581(1 + 1.2981\alpha^* - 2.1837\alpha^{*2} + 4.3496\alpha^{*3} - 6.2381\alpha^{*4} + 4.3140\alpha^{*5} - 1.0911\alpha^{*6}) \quad (0.1\%)$$

TABLE 3.49 Continued

Duct Geometry	D_h, α^*, r^*	$f\,\mathrm{Re}, \mathrm{Nu_T}, \mathrm{Nu_{H1}}, \mathrm{Nu_{H2}}$
		$\mathrm{Nu_{H2}} = \begin{cases} 0.2299\alpha^* & \text{for } 0 \le \alpha^* \le 0.125 \quad (0.0\%) \\ -0.4402\alpha^*(1 - 6.8176\alpha^* + 53.2849\alpha^* - 77.9848\alpha^{*3} + 33.5641\alpha^{*4}) & \text{for } 0.125 < \alpha^* \le 1 \quad (3.8\%) \end{cases}$
2b 2a Elliptical	$D_h = \dfrac{\pi b}{E(m)}$, $\alpha^* = \dfrac{2b}{2a}$, $m = 1 - \alpha^{*2}$	For $0 \le \alpha^* \le 1$, $f\,\mathrm{Re} = 2(1 + \alpha^{*2})\left(\dfrac{\pi}{E(m)}\right)^2$ $\mathrm{Nu_T} = 0.3536(1 + 0.9864\alpha^* - 0.7189\alpha^{*2} + 3.3364\alpha^{*3} - 3.0307\alpha^{*4} + 1.0130\alpha^{*5})[\pi/E(m)]^2$ (0.2%) $\mathrm{Nu_{H1}} = 9(1 + \alpha^{*2})\dfrac{\alpha^{*4} + 6\alpha^{*2} + 1}{17\alpha^{*4} + 98\alpha^{*2} + 17}\left(\dfrac{\pi}{E(m)}\right)^2$ $\mathrm{Nu_{H2}} = 0.3258\alpha^*(1 + 15.6397\alpha^* - 29.5117\alpha^{*2} + 16.2250\alpha^{*3})\left(\dfrac{\pi}{E(m)}\right)^2$ (4.3%)
2b 2a Sine	D_h is a function of α^* ‡ $\alpha^* = \dfrac{2b}{2a}$	For $0 \le \alpha^* \le 2$, $f\,\mathrm{Re} = 9.5687(1 + 0.0772\alpha^* + 0.8619\alpha^{*2} - 0.8314\alpha^{*3} + 0.2907\alpha^{*4} - 0.0338\alpha^{*5})$ (0.3%) $\mathrm{Nu_T} = 1.1791(1 + 2.7701\alpha^* - 3.1901\alpha^{*2} - 1.9975\alpha^{*3} - 0.4966\alpha^{*4})$ (0.5%) $\mathrm{Nu_{H1}} = 1.9030(1 + 0.4556\alpha^* + 1.2111\alpha^{*2} - 1.6805\alpha^{*3} + 0.7724\alpha^{*4} - 0.1228\alpha^{*5})$ (1.2%)

‡Hydraulic diameter of the sine duct is determined numerically and can be represented within ±1% of the numerical values reported in [1] by $D_h/2a = (1.0542 - 0.4670\alpha^* - 0.1180\alpha^{*2} + 0.1794\alpha^{*3} - 0.0436\alpha^{*4})\alpha^*$.

Geometry	D_h	Correlations
		$Nu_{H2} = \begin{cases} 0.76\alpha^* & \text{for } 0 \le \alpha^* \le 0.125 \quad (0.0\%) \\ -0.0202(1 - 32.0594\alpha^* - 216.1635\alpha^{*2} \\ +244.3812\alpha^{*3} - 82.4951\alpha^{*4} + 7.6733\alpha^{*5}) & \text{for } 0.125 < \alpha^* \le 2 \quad (4.2\%) \end{cases}$
Rhombic	$D_h = a \tan \dfrac{\phi}{2}$	For $0 \le \phi \le \pi/2$, $f\,Re = 12(1 - 0.0231\phi + 0.4994\phi^2 - 0.5002\phi^3 + 0.2054\phi^4 - 0.3356\phi^5)$ [§] (0.2%) $Nu_{H1} = 2.0564(1 + 0.3105\phi + 1.0330\phi^2 - 1.0572\phi^3 + 0.3867\phi^4 - 0.0563\phi^5)$ (0.2%) $Nu_{H2} = 0.0447\alpha^*(1 + 40.1477\alpha^* + 34.0984\alpha^{*2} - 35.8881\alpha^{*3} + 5.7293\alpha^{*4})$ (0.6%)
Regular polygonal ($n = 6$)	$D_h = 2a\cos\left(\dfrac{\pi}{n}\right)$ where a is the radius of the circumscribed circle	For $3 \le n \le 20$, $f\,Re = 8.3880(1 + 0.3015n - 0.0404n^2 + 0.0024n^3 - 0.00005n^4)$ (0.3%) $Nu_{H1} = 0.1908(1 + 7.9489n - 1.1383n^2 + 0.0712n^3 - 0.0016n^4)$ (0.9%) $Nu_{H2} = -2.2578(1 - 0.8051n + 0.0586n^2 - 0.0007n^3 - 0.0002n^4 + 0.000003n^5)$ (6.9%)
Circular sector	$D_h = \dfrac{2a\phi}{1 + \phi}$	For $0 \le 2\phi \le \pi$, $f\,Re = 12(1 + 0.5059\phi - 0.3948\phi^2 + 0.1875\phi^3 - 0.0385\phi^4)$ (0.1%) $Nu_{H1} = 2.0705(1 + 2.2916\phi - 2.5682\phi^2 + 1.4815\phi^3 - 0.3338\phi^4)$ (0.6%) $Nu_{H2} = \begin{cases} 0.1144\phi(28.7972\phi - 1) & \text{for } 0 \le \phi \le \pi/9 \quad (1.2\%) \\ 0.22691(1.7021\phi^4 - 9.9127\phi^3 \\ +14.3914\phi^2 + 4.2653\phi - 1) & \text{for } \pi/9 < \phi \le \pi \quad (5.2\%) \end{cases}$

[§] In this and in the subsequent expressions in this table, the numerical values of ϕ are to be used in radians.

TABLE 3.49 Continued

Duct Geometry	D_h, α^*, r^*	$f\,\mathrm{Re}, \mathrm{Nu_T}, \mathrm{Nu_{H1}}, \mathrm{Nu_{H2}}$	
Circular segment	$D_h = \dfrac{a(2\phi - \sin^2\phi)}{\phi + \sin\phi}$	For $0 \le 2\phi \le 2\pi$,	
		$f\,\mathrm{Re} = 15.556(1 - 0.0015\phi + 0.0185\phi^2 - 0.0123\phi^3 + 0.0035\phi^4 - 0.0004\phi^5)$	(0.01%)
		$\mathrm{Nu_{H1}} = 3.5878(1 + 0.0408\phi - 0.0157\phi^2 + 0.1483\phi^3 - 0.1312\phi^4 + 0.0434\phi^5 - 0.0050\phi^6)$	(0.2%)
		$\mathrm{Nu_{H2}} = 0.00085\phi(499.2830\phi + 1)$ for $0 \le \phi \le \pi/9$	(0.0%)
		$\mathrm{Nu_{H2}} = 0.0056(0.2321\phi^5 - 2\phi^4 - 8.0357\phi^3 + 93.7679\phi^2 - 10.4107\phi + 1)$ for $\pi/9 < \phi \le 2\pi$	(1.9%)
Concentric annular	$D_h = 2(r_o - r_i)$	For $0 \le r^* \le 1$,	
	$r^* = \dfrac{r_i}{r_o}$	$f\,\mathrm{Re} = \dfrac{16(1 - r^*)^2}{1 + r^{*2} - 2r_m^{*2}}, \quad r_m = \left(\dfrac{1 - r^{*2}}{2\ln(1/r^*)}\right)^{1/2}$	
		For $0 \le r^* \le 0.02$,	
		$\mathrm{Nu_T} = 3.657 + 98.95r^*$	(0.01%)
		$\mathrm{Nu_H} = 4.364 + 100.95r^*$	(0.01%)
		For $0.02 \le r^* \le 1$,	
		$\mathrm{Nu_T} = 5.3302(1 + 3.2904r^* - 12.0075r^{*2} + 18.8298r^{*3} - 9.6980r^{*4})$	(0.6%)
		$\mathrm{Nu_H} = 6.2066(1 + 2.3108r^* - 7.7553r^{*2} + 13.2851r^{*3} - 10.5987r^{*4} + 2.6178r^{*5} + 0.4680r^{*6})$	(1.4%)

lacking for flows other than fully developed flow. We hope that the present review of the subject will engender some interest among researchers to generate additional solutions to fill the gaps in our knowledge. We also hope that the fluid flow and heat transfer characteristics of the less common ducts described in Secs. 3.7 and 3.10 will be explored in greater detail to meet the demands of the emerging technical applications requiring newer and more complex flow passages.

NOMENCLATURE

A_c	flow cross-sectional area, m^2, ft^2
a	radius of a circular duct; also half width of a noncircular duct, m, ft
Bi	Biot number $= h_e D_h/k$
Br	Brinkman number for the boundary condition of constant axial wall temperature, $= \mu u_m^2/[Jg_c k(T_{w,m} - T_e)]$
Br′	Brinkman number for the boundary condition of constant axial wall heat flux, $= \mu u_m^2/(Jg_c q'' D_h)$
b	half spacing or half height of a duct, m, ft
c	half spacing or width of a duct; also a parameter for the eccentric annular duct defined by Eq. (3.270), m, ft
c_p	specific heat of the fluid at constant pressure, J/(kg · K), Btu/(lb_m ·° F)
D_h	hydraulic diameter of the duct $= 4A_c/P$, m, ft
$E(m)$	complete elliptic integral of the second kind with the argument m defined at appropriate places in the text
e^*	eccentricity $\epsilon/(r_0 - r_i)$ of the eccentric annular duct or amplitude ϵ/a of the circular duct with sinusoidal corrugation
F	a multiplicative factor entering various expressions
f	circumferentially averaged fully developed Fanning friction factor $= \tau_w/(\rho u_m^2/2g_c)$
f_{app}	apparent Fanning Friction factor $= \Delta p^*/(x/r_h)$
g_c	proportionality constant in Newton's second law of motion, $= 1$ (dimensionless) in SI units; 32.174 lb_m · ft/(lb_f · s^2)
h	convective heat transfer coefficient for fully developed flow, W/(m^2 · K), Btu/(hr · ft^2 · °F)
h_e	convective heat transfer coefficient for the duct exterior, W/(m^2 · K), Btu/(hr · ft^2 · °F)
J	mechanical-to-thermal energy conversion factor, $= 1$ (dimensionless) in SI units; $= 778.163$ lb_f · ft/Btu
$J_i(\ )$	Bessel functions of the first kind and orders 0 or 1 corresponding to $i = 0$ or 1
$K(x)$	incremental pressure drop number, defined by Eq. (3.4)
k	thermal conductivity of the fluid if either no subscript or the subscript f is used; thermal conductivity of the duct wall material if the subscript w is used, W/(m · K), Btu/(hr · ft · °F)
L	length of the duct, m, ft
L_{hy}	hydrodynamic entrance length, m, ft
L_{hy}^+	dimensionless hydrodynamic entrance length $= L_{hy}/D_h\mathrm{Re}$

L_{th} thermal entrance length, m, ft

L_{th}^* dimensionless thermal entrance length $= L_{th}/D_h \mathrm{Pe}$

m argument of complete integral of the second kind; also an exponent or a constant

Nu_{bc} circumferentially averaged Nusselt number for fully developed flow for the thermal boundary condition of Table 3.1 or 3.2

$\mathrm{Nu}_{x,bc}$ circumferentially averaged but axially local Nusselt number for the thermal entrance region for the specified thermal boundary condition defined by Eq. (3.10)

$\mathrm{Nu}_{m,bc}$ mean Nusselt number for the thermal entrance region for the specified thermal boundary condition, defined by Eq. (3.11)

$\mathrm{Nu}_{x,lj}^{(k)}$ local Nusselt number for a doubly connected duct $= \Phi_{lj}^{(k)}/(\theta_{lj}^{(k)} - \theta_{mj}^{(k)})$

$\mathrm{Nu}_{x,i}^{(1a)}$ local Nusselt number at inner wall of a doubly connected duct for specific thermal boundary condition 1a; Nusselt numbers for the other specific thermal boundary conditions are defined similarly

Nu_o overall Nusselt number associated with the (T3) boundary condition, defined by Eq. (3.27)

n number of sides of a duct; also a dimensionless constant

P wetted perimeter of the duct, m, ft

Pe Péclet number $= u_m D_h/\alpha = \mathrm{RePr}$

Pr Prandtl number $= \nu/\alpha$

p fluid static pressure, Pa, $\mathrm{lb}_f/\mathrm{ft}^2$

Δp fluid static pressure drop in the flow direction between two cross sections of interest, Pa, $\mathrm{lb}_f/\mathrm{ft}^2$

Δp^* dimensionless fluid static pressure drop $= \Delta p/(\rho u_m^2/2g_c)$

q_w'' wall heat flux, heat transfer rate per unit heat transfer area of the duct (average value with respect to perimeter), $\mathrm{W/m^2}$, $\mathrm{Btu/(hr \cdot ft^2)}$

Re Reynolds number $= u_m D_h/\nu$

r radial coordinate in the cylindrical coordinate system, m, ft

r_h hydraulic radius of the duct $= A_c/P$, m, ft

r^* aspect ratio for doubly connected ducts, explicitly defined for specific geometry in the text

s duct dimension, m, ft

S dimensionless parameter for eccentric annular duct defined by Eq. (3.268); also thermal energy source function, rate of thermal energy generated per unit volume of the fluid, $\mathrm{W/m^3}$, $\mathrm{Btu/(hr \cdot ft^3)}$

S^* thermal energy source number, $= SD_h^2/k(T_{w,m} - T_e)$ for constant wall temperature boundary condition, $= SD_h/q_w''$ for axially constant wall heat flux

Sk Stark number $= \epsilon_w \sigma T_e^3 D_h/k$

T fluid temperature, °C, K, °F, °R

T_a ambient fluid temperature °C, K, °F, °R

T_m fluid bulk mean temperature, defined by Eq. (3.7), °C, K, °F, °R

T_w wall temperature at the inside duct periphery, °C, K, °F, °R

$T_{w,m}$ circumferentially averaged wall temperature, °C, K, °F, °R

$T^*_{w,\max}$	dimensionless maximum wall temperature, defined by Eq. (3.167)
$T^*_{w,\min}$	dimensionless minimum wall temperature, defined by Eq. (3.167)
u	fluid axial velocity, fluid velocity component in x direction, m/s, ft/s
u_m	fluid mean axial velocity, m/s, ft/s
$u_{\max}$	fluid maximum axial velocity across the duct cross section for fully developed flow, m/s, ft/s
v	fluid velocity component in y or r direction, m/s, ft/s
W	fluid mass flow rate through the duct, $= \rho u_m A_c$, kg/s, lb_m/hr
w	fluid velocity component in z or θ direction, m/s, ft/s
x	axial (streamwise) coordinate in Cartesian or cylindrical coordinate system, m, ft
x^+	dimensionless axial coordinate for the hydrodynamic entrance region, $= x/D_h\text{Re}$
x^*	dimensionless axial coordinate for the thermal entrance region, $= x/D_h\text{Pe}$
y, z	Cartesian coordinates across the flow cross section, m, ft
$\bar{y}$	distance of centroid of the duct cross section measured from the base, m, ft
$\bar{y}_{\max}$	normal distance from the base to a point where $u_{\max}$ occurs in the duct cross section, m, ft

Greek symbols

α	fluid thermal diffusivity $= k/\rho c_p$, m²/s, ft²/s
α^*	duct aspect ratio, explicitly defined for specific geometry in the text
$\Gamma(\ )$	gamma function
γ	dimensionless parameter defined by Eq. (3.23)
δ	hydrodynamic boundary-layer thickness, m, ft
δ_t	thermal boundary-layer thickness, m, ft
δ_w	duct wall thickness, m, ft
ϵ	eccentricity of an eccentric annular duct (see Fig. 3.46) or amplitude of a circular duct with sinusoidal corrugations (see Fig. 3.39*a*), m, ft
ϵ_w	emissivity of the duct wall material
η	dimensionless hydrodynamic boundary-layer parameter for elliptical duct [see Eq. (3.188)]; also a bipolar coordinate [see Eq. (3.270)], m, ft
Θ	dimensionless fluid temperature for boundary condition of axially constant wall heat flux, $= (T - T_e)/(q''_w D_h/k)$
θ	angular coordinate in the cylindrical coordinate system, rad, deg; also dimensionless fluid temperature for boundary condition of axially constant wall temperature, $= (T - T_w)/(T_e - T_w)$
θ_m	dimensionless fluid bulk mean temperature $= (T_m - T_w)/(T_e - T_w)$
$\theta_j^{(k)}$	dimensionless fluid temperature for a doubly connected duct, defined in [1]
$\theta_{lj}^{(k)}$	dimensionless circumferentially averaged temperature of wall ($l = i$ for inner wall, $l = o$ for outer wall) for the fundamental boundary condition of kind k when inner or outer wall ($j = i$ or o) is heated or cooled; dimensionless fluid bulk mean temperature if $l = m$
θ_i^*	influence coefficients derived from the fundamental solutions

μ fluid dynamic viscosity coefficient, Pa · s, $lb_m/(hr \cdot ft)$

ν fluid kinematic viscosity coefficient $= \mu/\rho$, m^2/s, ft^2/s

ξ a duct dimension (see Fig. 3.50); also a bipolar coordinate [see Eq. (3.270)], m, ft

ρ fluid density, kg/m^3, lb_m/ft^3

σ Stefan-Boltzmann constant $= 5.6697 \times 10^{-8}$ $W/(m^2 \cdot K^4) = 0.1713 \times 10^{-8}$ $Btu/(hr \cdot ft^2 \cdot {}^\circ R^4)$

τ_w wall shear stress due to skin friction, Pa, lb_f/ft^2

$\Phi_j^{(k)}$ dimensionless heat flux at a point in the flow field for the jth wall of a doubly connected duct, defined in [1]

$\Phi_{lj}^{(k)}$ dimensionless wall heat flux defined in a manner similar to $\theta_{lj}^{(k)}$; $= q_l'' D_h / k(T_j - T_e)$ for $k = 1, 3$; $= q''/q_{lj}''$ for $k = 2, 4$

$\Phi_{m,\mathrm{T}}$ dimensionless mean wall heat flux for boundary condition of axially constant wall temperature, $= q_m'' D_h / k(T_w - T_e)$

$\Phi_{x,\mathrm{T}}$ dimensionless local wall heat flux for boundary condition of axially constant wall temperature, $= q_x'' D_h / k(T_w - T_e)$

ϕ apex angle or half-apex angle of a duct, rad, deg

χ similarity variable defined by Eqs. (3.46) and (3.115)

Subscripts

bc thermal boundary condition (refer to Tables 3.1 and 3.2 for the alphanumeric designation and meaning of various thermal boundary conditions)

c center or centroid; also combined entrance length

e initial value at the entrance of the duct or where the heat transfer starts

f fluid

fd fully developed flow

H Ⓗ boundary condition (see Tables 3.1 and 3.2 for a description of Ⓗ, (H1), (H2), (H3), (H4), and (H5) boundary conditions)

hy hydrodynamic

i inner surface of a doubly connected duct

j heated wall of a doubly connected duct, $= i$ or o

m mean

max maximum

min minimum

o outer surface of a doubly connected duct

T Ⓣ boundary condition (see Tables 3.1 and 3.2 for a description of Ⓣ, (T3), and (T4) boundary conditions)

th thermal

x denoting arbitrary section along the duct length, a local value as opposed to a mean value

w wall or fluid at the wall

∞ fully developed value at $x = \infty$

REFERENCES

1. R. K. Shah and A. L. London, *Laminar Flow Forced Convection in Ducts*, Supplement 1 to *Advances in Heat Transfer*, Academic, New York, 1978.
2. M. S. Bhatti, Fully Developed Temperature Distribution in a Circular Tube With Uniform Wall Temperature, unpublished paper, Owens-Corning Fiberglas Corporation, Granville, Ohio, 1985.
3. M. L. Michelsen and J. Villadsen, The Graetz Problem with Axial Heat Conduction, *Int. J. Heat Mass Transfer*, Vol. 17, pp. 1391–1402, 1974.
4. J. W. Ou and K. C. Cheng, Viscous Dissipation Effects on Thermal Entrance Heat Transfer in Laminar and Turbulent Pipe Flows With Uniform Wall Temperature, AIAA Paper No. 74-743 or ASME Paper No. 74-HT-50, 1974.
5. V. P. Tyagi, Laminar Forced Convection of a Dissipative Fluid in a Channel, *J. Heat Transfer*, Vol. 88, pp. 161–169, 1966.
6. W. C. Reynolds, Heat Transfer to Fully Developed Laminar Flow in a Circular Tube with Arbitrary Circumferential Heat Flux, *J. Heat Transfer*, Vol. 82, pp. 108–112, 1960.
7. H. J. Hickman, An Asymptotic Study of the Nusselt-Graetz Problem, Part 1: Large x Behavior, *J. Heat Transfer*, Vol. 96, pp. 354–358, 1974.
8. E. M. Sparrow, S. V. Patankar, and H. Shahrestani, Laminar Heat Transfer in a Pipe Subjected to Circumferentially Varying External Heat Transfer Coefficients, *Numer. Heat Transfer*, Vol. 1, pp. 117–127, 1978.
9. Y. S. Kadaner, Y. P. Rassadkin, and E. L. Spektor, Heat Transfer in Laminar Liquid Flow through a Pipe Cooled by Radiation, *Heat Transfer—Sov. Res.*, Vol. 3, No. 5, pp. 182–188, 1971.
10. E. M. Sparrow and S. V. Patankar, Relationships among Boundary Conditions and Nusselt Numbers for Thermally Developed Duct Flows, *J. Heat Transfer*, Vol. 99, pp. 483–485, 1977.
11. R. W. Hornbeck, Laminar Flow in the Entrance Region of a Pipe, *Appl. Sci. Res.*, Vol. A13, pp. 224–232, 1964.
12. R. Y. Chen, Flow in the Entrance Region at Low Reynolds Numbers, *J. Fluids Eng.*, Vol. 95, pp. 153–158, 1973.
13. H. L. Weissberg, End Correction for Slow Viscous Flow through Long Tubes, *Phys. Fluids*, Vol. 5, pp. 1033–1036, 1962.
14. J. H. Linehan and S. R. Hirsch, Entrance Correction for Creeping Flow in Short Tubes, *J. Fluids Eng.*, Vol. 99, pp. 778–779, 1977.
15. L. Graetz, Über die Wärmeleitungs fähigkeit von Flüssigkeiten (On the Thermal Conductivity of Liquids). Part 1, *Ann. Phys. Chem.*, Vol. 18, pp. 79–94, 1883; Part 2, *Ann. Phys. Chem.*, Vol. 25, pp. 337–357, 1885.
16. W. Nusselt, Die Abhängigkeit der Wärmeübergangszahl von der Rohrlänge (The Dependence of the Heat-Transfer Coefficient on the Tube Length), *VDI Z.*, Vol. 54, pp. 1154–1158, 1910.
17. G. M. Brown, Heat or Mass Transfer in a Fluid in Laminar Flow in a Circular or Flat Conduit, *AIChE J.*, Vol. 6, pp. 179–183, 1960.
18. B. K. Larkin, High–Order Eigenfunctions of the Graetz Problem, *AIChE J.*, Vol. 7, p. 530, 1961.
19. J. Newman, The Graetz Problem, *The Fundamental Principles of Current Distribution and Mass Transport in Electrochemical Cells*, ed. A. J. Bard, Vol. 6, pp. 187–352, Dekker, New York, 1973.
20. U. Grigull and H. Tratz, Thermischer einlauf in ausgebildeter laminarer Rohrströmung, *Int. J. Heat Mass Transfer*, Vol. 8, pp. 669–678, 1965.

21. M. A. Lévêque, Les lois de la transmission de chaleur par convection, *Ann. Mines, Mem.*, Ser. 12, Vol. 13, pp. 201–299, 305–362, 381–415, 1928.

22. M. Abramowitz, Table of the Integral $\int_0^x e^{-u^3}\,du$, *J. Math. Phys.*, Vol. 30, p. 162, 1951.

23. H. Hausen, Darstellung des Wärmeübergangs in Rohren durch verallgemeinerte Potenzbeziehungen, *VDI Z.*, Suppl. "Verfahrenstechnik," No. 4, pp. 91–98, 1943.

24. D. K. Hennecke, Heat Transfer by Hagen-Poiseuille Flow in the Thermal Development Region with Axial Conduction, *Wärme-Stoffübertrag.*, Vol. 1, pp. 177–184, 1968.

25. H. C. Brinkman, Heat Effects in Capillary Flow, *Appl. Sci. Res.*, Vol. A2, pp. 120–124, 1951.

26. C. J. Hsu, Exact Solution to Entry-Region Laminar Heat Transfer with Axial Conduction and the Boundary Conditions of the Third Kind, *Chem. Eng. Sci.*, Vol. 23, pp. 457–468, 1968.

27. R. Siegel, E. M. Sparrow, and T. M. Hallman, Steady Laminar Heat Transfer in a Circular Tube with Prescribed Wall Heat Flux, *Appl. Sci. Res.*,Vol. A7, pp. 386–392, 1958.

28. C. J. Hsu, Heat Transfer in a Round Tube with Sinusoidal Wall Heat Flux Distribution, *AIChE J.*, Vol. 11, pp. 690–695, 1965.

29. R. B. Bird, W. E. Stewart, and E. N. Lightfoot, *Transport Phenomena*, Wiley, New York, 1960.

30. J. W. Ou and K. C. Cheng, Viscous Dissipation Effects on Thermal Entrance Region Heat Transfer in Pipes with Uniform Wall Heat Flux, *Appl. Sci. Res.*, Vol. 28, pp. 289–301, 1973.

31. G. S. Barozzi and G. Pagliarini, A Method to Solve Conjugate Heat Transfer Problems: The Case of Fully Developed Laminar Flow in a Pipe, *J. Heat Transfer*, Vol. 107, pp. 77–83, 1985.

32. M. S. Bhatti, Limiting Laminar Heat Transfer in Circular and Flat Ducts by Analogy with Transient Heat Conduction Problems, unpublished paper, Owens-Corning Fiberglas Corporation, Granville, Ohio, 1985.

33. V. Javeri, Simultaneous Development of the Laminar Velocity and Temperature Fields in a Circular Duct for the Temperature Boundary Condition of the Third Kind, *Int. J. Heat Mass Transfer*, Vol. 19, pp. 943–949, 1976.

34. K. C. Cheng and R. S. Wu, Viscous Dissipation Effects on Convective Instability and Heat Transfer in Plane Poiseuille Flow Heated from Below, *Appl. Sci. Res.*, Vol. 32, pp. 327–346, 1976.

35. S. Pahor and J. Strand. A Note on Heat Transfer in Laminar Flow through a Gap, *Appl. Sci. Res.*, Vol. A10, pp. 81–84, 1961.

36. C. C. Grosjean, S. Pahor, and J. Strand, Heat Transfer in Laminar Flow through a Gap, *Appl. Sci. Res.*, Vol. A11, pp. 292–294, 1963.

37. J. W. Ou and K. C. Cheng, Effects of Pressure Work and Viscous Dissipation on Graetz Problem for Gas Flow in Parallel-Plate Channels, *Wärme-Stoffübertrag.*, Vol. 6, pp. 191–198, 1973.

38. L. N. Tao, On Some Laminar Forced-Convection Problems, *J. Heat Transfer*, Vol. 83, pp. 466–472, 1961.

39. J. R. Bodoia and J. F. Osterle, Finite Difference Analysis of Plane Poiseuille and Couette Flow Developments, *Appl. Sci. Res.*, Vol. A10, pp. 265–276, 1961.

40. M. S. Bhatti and C. W. Savery, Heat Transfer in the Entrance Region of a Straight Channel: Laminar Flow With Uniform Wall Heat Flux, *J. Heat Transfer*, Vol. 99, pp. 142–144, 1977.

41. W. Nusselt, Der Wärmeaustausch am Berieselungskühler, *VDI Z.*, Vol. 67, pp. 206–210, 1923.

42. J. R. Sellars, M. Tribus, and J. S. Klein, Heat Transfer to Laminar Flow in a Round Tube or Flat Conduit—the Graetz Problem Extended, *Trans. ASME*, Vol. 78, pp. 441–448, 1956.

43. R. D. Cess and E. C. Shaffer, Heat Transfer to Laminar Flow Between Parallel Plates with a Prescribed Wall Heat Flux, *Appl. Sci. Res.*, Vol. A8, pp. 339–344, 1959.

44. E. M. Sparrow, J. L. Novotny, and S. H. Lin, Laminar Flow of a Heat-Generating Fluid in a Parallel-Plate Channel, *AIChE J.*, Vol. 9, pp. 797–804, 1963.
45. A. S. Jones, Two-Dimensional Adiabatic Forced Convection at Low Péclet Number, *Appl. Sci. Res.*, Vol. 25, pp. 337–348, 1972.
46. C. J. Hsu, An Exact Analysis of Low Péclet Number Thermal Entry Region Heat Transfer in Transversely Nonuniform Velocity Fields, *AIChE J.*, Vol. 17, pp. 732–740, 1971.
47. S. Mori, T. Shinke, M. Sakakibara, and A. Tanimoto, Steady Heat Transfer to Laminar Flow Between Parallel Plates with Conduction in Wall, *Heat Transfer—Jpn. Res.*, Vol. 5, No. 4, pp. 17–25, 1976.
48. C. L. Hwang and L. T. Fan, Finite Difference Analysis of Forced Convection Heat Transfer in Entrance Region of a Flat Rectangular Duct, *Appl. Sci. Res.*, Vol. A13, pp. 401–422, 1964.
49. K. Stephan, Wärmeübergang und druckabfall bei nicht ausgebildeter Laminarströmung in Rohren und in ebenen Spalten, *Chem.-Ing.-Tech.*, Vol. 31, pp. 773–778, 1959.
50. M. S. Bhatti and C. W. Savery, Heat Transfer in the Entrance Region of a Straight Channel: Laminar Flow With Uniform Wall Temperature, *J. Heat Transfer*, Vol. 100, pp. 539–542, 1978.
51. R. Das and A. K. Mohanty, Forced Convection Heat Transfer in the Entrance Region of a Parallel Plate Channel, *Int. J. Heat Mass Transfer*, Vol. 26, pp. 1403–1405, 1983.
52. W. E. Mercer, W. W. Pearce, and J. E. Hitchcock, Laminar Forced Convection in the Entrance Region between Parallel Flat Plates, *J. Heat Transfer*, Vol. 89, pp. 251–257, 1967.
53. H. S. Heaton, W. C. Reynolds, and W. M. Kays, Heat Transfer in Annular Passages: Simultaneous Development of Velocity and Temperature Fields in Laminar Flow, *Int. J. Heat Mass Transfer*, Vol. 7, pp. 763–781, 1964.
54. V. Javeri, Heat Transfer in Laminar Entrance Region of a Flat Channel for the Temperature Boundary Condition of the Third Kind, *Wärme–Stoffübertrag.*, Vol. 10, pp. 137–144, 1977.
55. S. M. Marco and L. S. Han, A Note on Limiting Laminar Nusselt Number in Ducts with Constant Temperature Gradient by Analogy to Thin-Plate Theory, *Trans. ASME*, Vol. 77, pp. 625–630, 1955.
56. H. F. P. Purday, *Streamline Flow*, Constable, London, 1949; same as *An Introduction to the Mechanics of Viscous Flow*, Dover, New York, 1949.
57. N. M. Natarajan and S. M. Lakshmanan, Laminar Flow in Rectangular Ducts: Prediction of Velocity Profiles and Friction Factor, *Indian J. Technol.*, Vol. 10, pp. 435–438, 1972.
58. R. W. Miller and L. S. Han, Pressure Losses for Laminar Flow in the Entrance Region of Ducts of Rectangular and Equilateral Triangular Cross Sections, *J. Appl. Mech.*, Vol. 38, pp. 1083–1087, 1971.
59. C. L. Wiginton and C. Dalton, Incompressible Laminar Flow in the Entrance Region of a Rectangular Duct, *J. Appl. Mech.*, Vol. 37, pp. 854–856, 1970.
60. F. W. Schmidt and M. E. Newell, Heat Transfer in Fully Developed Laminar Flow through Rectangular and Isosceles Triangular Ducts, *Int. J. Heat Mass Transfer*, Vol. 10, pp. 1121–1123, 1967.
61. J. M. Savino and R. Siegel, Laminar Forced Convection in Rectangular Channels with Unequal Heat Addition on Adjacent Sides, *Int. J. Heat Mass Transfer*, Vol. 7, pp. 733–741, 1964.
62. L. S. Han, Laminar Heat Transfer in Rectangular Channels, *J. Heat Transfer*, Vol. 81, pp. 121–128, 1959.
63. R. Siegel and J. M. Savino, An Analytical Solution of the Effect of Peripheral Wall Conduction on Laminar Forced Convection in Rectangular Channels, *J. Heat Transfer*, Vol. 87, pp. 59–66, 1965.
64. J. M. Savino and R. Siegel, Extension of an Analysis of Peripheral Wall Conduction Effects for Laminar Forced Convection in Thin-Walled Rectangular Channels, NASA Tech. Note TN D-2860, 1965.

65. R. W. Lyczkowski, C. W. Solbrig, and D. Gidaspow, Forced Convective Heat Transfer in Rectangular Ducts—General Case of Wall Resistance and Peripheral Conduction, *Nucl. Eng. Design*, Vol. 67, pp. 357–378, 1981.

66. M. Iqbal, B. D. Aggarwala, and A. K. Khatry, On the Conjugate Problem of Laminar Combined Free and Forced Convection Through Vertical Non-Circular Ducts, *J. Heat Transfer*, Vol. 94, pp. 52–56, 1972.

67. R. M. Curr, D. Sharma, and D. G. Tatchell, Numerical Predictions of Some Three-Dimensional Boundary Layers in Ducts, *Comput. Methods. Appl. Mech. Eng.*, Vol. 1, pp. 143–158, 1972.

68. M. Tachibana and Y. Iemoto, Steady Laminar Flow in the Inlet Region of Rectangular Ducts, *Bull. JSME*, Vol. 24, No. 193, pp. 1151–1158, 1981.

69. P. Wibulswas, Laminar Flow Heat Transfer in Non-Circular Ducts, Ph.D. Thesis, London Univ., London, 1966.

70. A. R. Chandrupatla and V. M. K. Sastri, Laminar Forced Convection Heat Transfer of a Non-Newtonian Fluid in a Square Duct, *Int. J. Heat Mass Transfer*, Vol. 20, pp. 1315–1324, 1977.

71. K. R. Perkins, K. W. Shade, and D. M. McEligot, Heated Laminarizing Gas Flow in a Square Duct, *Int. J. Heat Mass Transfer*, Vol. 16, pp. 897–916, 1973.

72. A. R. Chandrupatla and V. M. K. Sastri, Laminar Flow and Heat Transfer to a Non Newtonian Fluid in an Entrance Region of a Square Duct With Prescribed Constant Axial Wall Heat Flux, *Numer. Heat Transfer*, Vol. 1, pp. 243–254, 1978.

73. S. Neti and R. Eichhorn, Combined Hydrodynamic and Thermal Development in a Square Duct, *Numer. Heat Transfer*, Vol. 6, pp. 497–510, 1983.

74. K. C. Cheng, Laminar Forced Convection in Regular Polygonal Ducts with Uniform Peripheral Heat Flux, *J. Heat Transfer*, Vol. 91, pp. 156–157, 1969.

75. R. K. Shah, Laminar Flow Friction and Forced Convection Heat Transfer in Ducts of Arbitrary Geometry, *Int. J. Heat Mass Transfer*, Vol. 18, pp. 849–862, 1975.

76. A. Haji-Sheikh, M. Mashena, and M. J. Haji-Sheikh, Heat Transfer Coefficient in Ducts with Constant Wall Temperature, *J. Heat Transfer*, Vol. 105, pp. 878–883, 1983.

77. V. K. Migay, Hydraulic Resistance of Triangular Channels in Laminar Flow (in Russian), *Izv. Vyssh. Uchebn. Zaved. Energ.*, Vol. 6, No. 5, pp. 122–124, 1963.

78. E. M. Sparrow and A. Haji-Sheikh, Laminar Heat Transfer and Pressure Drop in Isosceles Triangular, Right Triangular, and Circular Sector Ducts, *J. Heat Transfer*, Vol. 87, pp. 426–427, 1965.

79. M. Iqbal, A. K. Khatry, and B. D. Aggarwala, On the Second Fundamental Problem of Combined Free and Forced Convection through Vertical Non-Circular Ducts, *Appl. Sci. Res.*, Vol. 26, pp. 183–208, 1972.

80. H. Nakamura, S. Hiraoka, and I. Yamada, Laminar Forced Convection Flow and Heat Transfer in Arbitrary Triangular Ducts, *Heat Transfer—Jpn. Res.*, Vol. 2, No. 14, pp. 56–63, 1972.

81. Z. V. Semilet, Laminar Heat Transfer and Pressure Drop for Gas Flow in Triangular Ducts. *Heat Transfer—Sov. Res.*, Vol. 2, No. 1, pp. 100–105, 1970.

82. D. P. Fleming and E. M. Sparrow, Flow in the Hydrodynamic Entrance Region of Ducts of Arbitrary Cross Section, *J. Heat Transfer*, Vol. 91, pp. 345–354, 1969.

83. B. D. Aggarwala and M. K. Gangal, Laminar Flow Development in Triangular Ducts, *Trans. Can. Soc. Mech. Eng.*, Vol. 3, pp. 231–233, 1975.

84. P. Wibulswas and P. Tangsirimonkol, Laminar and Transition Forced Convection in Triangular Ducts With Constant Wall Temperature, Unpublished Paper, London Univ., London, 1978.

85. M. S. Bhatti, Laminar Flow in the Entrance Region of Elliptical Ducts, *J. Fluids Eng.*, Vol. 105, pp. 290–296, 1983.

86. T. S. Lundgren, E. M. Sparrow, and J. B. Starr, Pressure Drop Due to the Entrance Region in Ducts of Arbitrary Cross Section, *J. Basic Eng.*, Vol. 86, pp. 620–626, 1964.

87. N. T. Dunwoody, Thermal Results for Forced Convection through Elliptical Ducts, *J. Appl. Mech.*, Vol. 29, pp. 165–170, 1962.

88. M. S. Bhatti, Heat Transfer in the Fully Developed Region of Elliptical Ducts with Uniform Wall Heat Flux, *J. Heat Transfer*, Vol. 106, pp. 895–898, 1984.

89. R. M. Abdel-Wahed, A. E. Attia, and M. A. Hifni, Experiments on Laminar Flow and Heat Transfer in an Elliptical Duct, *Int. J. Heat Mass Transfer*, Vol. 27, pp. 2397–2413, 1984.

90. S. M. Richardson, Lévêque Solution for Flow in an Elliptical Duct, *Lett. Heat Mass Transfer*, Vol. 7, pp. 353–362, 1980.

91. S. Someswara Rao, N. C. Pattabhi Ramacharyulu, and V. V. G. Krishnamurty, Laminar Forced Convection in Elliptical Ducts, *Appl. Sci. Res.*, Vol. 21, pp. 185–193, 1969.

92. D. F. Sherony and C. W. Solbrig, Analytical Investigation of Heat or Mass Transfer and Friction Factors in a Corrugated Duct Heat or Mass Exchanger, *Int. J. Heat Mass Transfer*, Vol. 13, pp. 145–159, 1970.

93. A. Lawal and A. S. Mujumdar, Forced Convection Heat Transfer to a Power Law Fluid in Arbitrary Cross-Section Ducts, *Can. J. Chem. Eng.*, Vol. 62, pp. 326–333, 1984.

94. H. Nakamura, S. Hiraoka, and I. Yamada, Flow and Heat Transfer of Laminar Forced Convection in Arbitrary Polygonal Ducts, *Heat Transfer—Jpn. Res.*, Vol. 2, No. 4, pp. 56–63, 1974.

95. G. Schenkel, *Laminar Durchströmte Profilkanäle: Ersatzradien und Widerstandsbeiwerte*, Fortschritt-Berichte der VDI Zeitschriften, Reihe: Stromungstechnik, Vol. 7(62), 1981. (English translation available from the authors of this chapter.)

96. E. R. G. Eckert and T. F. Irvine, Jr., Flow in Corners of Passages with Noncircular Cross Sections, *Trans. ASME*, Vol. 78, pp. 709–718, 1956.

97. E. R. G. Eckert, T. F. Irvine, Jr., and J. T. Yen, Local Laminar Heat Transfer in Wedge-Shaped Passages, *Trans. ASME*, Vol. 80, pp. 1433–1438, 1958.

98. M. H. Hu and Y. P. Chang, Optimization of Finned Tubes for Heat Transfer in Laminar Flow, *J. Heat Transfer*, Vol. 95, pp. 332–338, 1973; for numerical results, see M. H. Hu, Flow and Thermal Analysis for Mechanically Enhanced Heat Transfer Tubes, Ph.D. Thesis, Dept. Mech. Eng., State Univ. of New York at Buffalo, 1973.

99. H. M. Soliman, A. A. Munis, and A. C. Trupp, Laminar Flow in the Entrance Region of Circular Sector Ducts, *J. Appl. Mech.*, Vol. 49, pp. 640–642, 1982.

100. E. M. Sparrow and A. Haji-Sheikh, Flow and Heat Transfer in Ducts of Arbitrary Shape with Arbitrary Thermal Boundary Conditions, *J. Heat Transfer*, Vol. 88, pp. 351–358, 1966; Discussion by C. F. Neville, *J. Heat Transfer*, Vol. 91, pp. 588–589, 1969.

101. S. W. Hong and A. E. Bergles, Augmentation of Laminar Flow Heat Transfer in Tubes by Means of Twisted-Tape Inserts, Tech. Rep. HTL-5, ISU-ERI-Ames-75011, Eng. Res. Inst., Iowa State Univ., Ames, 1974.

102. E. M. Sparrow, T. S. Chen, and V. K. Jonsson, Laminar Flow and Pressure Drop in Internally Finned Annular Ducts, *Int. J. Heat Mass Transfer*, Vol. 7, pp. 583–585, 1964.

103. T. Niida, Analytical Solution for the Velocity Distribution in Laminar Flow in an Annular-Sector Duct, *Int. Chem. Eng.*, Vol. 20, No. 2, pp. 258–265, 1980.

104. H. M. Soliman, Laminar Heat Transfer in Annular Sector Ducts, *J. Heat Transfer*, Vol. 109, pp. 247–249, 1987.

105. P. Renzoni and C. Prakash, Analysis of Laminar Flow and Heat Transfer in the Entrance Region of an Internally Finned Concentric Circular Annular Duct, *J. Heat Transfer*, Vol. 109, pp. 532–538, 1987.

106. J. P. Zarling, Application of Schwarz-Neumann Technique to Fully Developed Laminar Heat Transfer in Noncircular Ducts, *J. Heat Transfer*, Vol. 99, pp. 332–335, 1977.

107. K. C. Cheng and M. Jamil, Laminar Flow and Heat Transfer in Circular Ducts With Diametrically Opposite Flat Sides and Ducts of Multiply Connected Cross Sections, *Can. J. Chem. Eng.*, Vol. 48, pp. 333–334, 1970.

108. E. M. Sparrow and M. Charmchi, Heat Transfer and Fluid Flow Characteristics of Spanwise-Periodic Corrugated Ducts, *Int. J. Heat Mass Transfer*, Vol. 23, pp. 471–481, 1980.

109. E. M. Sparrow and A. Chukaev, Forced-Convection Heat Transfer in a Duct Having Spanwise-Periodic Rectangular Protuberances, *Numer. Heat Transfer*, Vol. 3, pp. 149–167, 1980.

110. L. N. Tao, Heat Transfer of Laminar Forced Convection in Indented Pipes, *Development in Mechanics*, ed. J. E. Lay and L. E. Malvern, Vol. 1, pp. 511–525, Plenum, New York, 1961.

111. N. M. Natarajan and S. M. Lakshmanan, Laminar Flow through Annuli: Analytical Method for Calculation of Pressure Drop, *Indian Chem. Eng.*, Vol. 5, No. 3, pp. 50–53, 1973.

112. R. E. Lundberg, W. C. Reynolds, and W. M. Kays, Heat Transfer with Laminar Flow in Concentric Annuli with Constant and Variable Wall Temperature and Heat Flux, NASA Technical Note TN D-1972, 1963.

113. S. Kakaç and O. Yücel, Laminar Flow Heat Transfer in an Annulus with Simultaneous Development of Velocity and Temperature Fields, Technical and Scientific Council of Turkey, TUBITAK, ISITEK No. 19, Ankara, Turkey, 1974.

114. R. K. Shah, A Correlation for Laminar Hydrodynamic Entry Length Solutions for Circular and Noncircular Ducts, *J. Fluids Eng.*, Vol. 100, pp. 177–179, 1978.

115. S. Kakaç and Y. Yener, *Convective Heat Transfer*, Publication No. 65, Middle East Technical Univ., Ankara, Turkey, distributed by Hemisphere Publishing Corp., New York, 1980.

116. N. A. V. Piercy, M. S. Hooper, and H. F. Winny, Viscous Flow through Pipes With Cores, *London Edinburgh Dublin Philos. Mag. J. Sci.*, Vol. 15, pp. 647–676, 1933.

117. A. C. Stevenson, The Centre of Flexure of a Hollow Shaft, *Proc. London Math. Soc.*, Ser. 2, Vol. 50, p. 536, 1949.

118. W. Tiedt, Berechnung des laminaren und turbulenten Reibungswiderstandes konzentrischer und exzentrischer Ringspalte. Part I, *Chem.–Ztg. Chem. Appar.*, Vol. 90, pp. 813–821, 1966; Part II, *Chem.-Ztg. Chem. Appar.*, Vol. 91, pp. 17–25, 1967; also as Tech. Ber. 4. Inst. Hydraul. Hydrol., Technische Hochschule, Darmstadt, 1968; English translation, Transl. Bur. No. 0151, 248 pp., Trans. Dev. Agency Libr., Montreal, 1971.

119. E. Becker, Strömungsvorgänge in ringförmigen Spalten und ihre Beziehung zum Poiseuilleschen Gesetz, *Forsch. Geb. Ingenieurwes., VDI*, Vol. 48, 1907.

120. K. C. Cheng and G. J. Hwang, Laminar Forced Convection in Eccentric Annuli, *AIChE J.*, Vol. 14, pp. 510–512, 1968.

121. M. L. Trombetta, Laminar Forced Convection in Eccentric Annuli, *Int. J. Heat Mass Transfer*, Vol. 14, pp. 1161–1173, 1971.

122. E. E. Feldman, R. W. Hornbeck, and J. F. Osterle, A Numerical Solution of Laminar Developing Flow in Eccentric Annular Ducts, *Int. J. Heat Mass Transfer*, Vol. 25, pp. 231–241, 1982.

123. E. E. Feldman, R. W. Hornbeck, and J. F. Osterle, A Numerical Solution of Developing Temperature for Laminar Developing Flow in Eccentric Annular Ducts, *Int. J. Heat Mass Transfer*, Vol. 25, pp. 243–253, 1982.

124. H. C. Topakoglu and O. A. Arnas, Convective Heat Transfer for Steady Laminar Flow between Two Confocal Elliptical Pipes with Longitudinal Uniform Wall Temperature Gradient, *Int. J. Heat Mass Transfer*, Vol. 17, pp. 1487–1498, 1974.

125. D. A. Ratkowsky and N. Epstein, Laminar Flow in Regular Polygonal Ducts with Circular Centered Cores, *Can. J. Chem. Eng.*, Vol. 46, pp. 22–26, 1968.

126. K. C. Cheng and M. Jamil, Laminar Flow and Heat Transfer in Ducts of Multiply Connected Cross Sections, ASME Paper No. 67-HT-6, 1967.

127. S. L. Hagen and D. A. Ratkowsky, Laminar Flow in Cylindrical Ducts Having Regular Polygonal Shaped Cores, *Can. J. Chem. Eng.*, Vol. 46, pp. 387–388, 1968.

128. B. D. Bowen, Laminar Flow in Unusual-Shaped Ducts, B. A. Sc. Thesis, Univ. of British Columbia, Vancouver, 1967.

129. P. N. Shivakumar, Viscous Flow in Pipes Whose Cross-Sections are Doubly Connected Regions, *Appl. Sci. Res.*, Vol. 27, pp. 355–365, 1973.

4

TURBULENT AND TRANSITION FLOW CONVECTIVE HEAT TRANSFER IN DUCTS

M. S. Bhatti

Harrison Radiator Division, GM
Lockport, New York

R. K. Shah

Harrison Radiator Division, GM
Lockport, New York

4.1 INTRODUCTION

Turbulent duct flows are of immense technological importance, as they occur frequently under normal operating conditions for a variety of heating and cooling devices in such diverse fields as aerospace, naval, nuclear, materials, mechanical, and chemical engineering. The main advantage of turbulent over laminar flows is that they are capable of providing vastly enhanced heat and mass transfer rates. However, this is at the expense of the increased friction losses accompanying the turbulent flows.

In this chapter, we shall present turbulent fluid flow and heat transfer results of practical interest for a variety of ducts, including circular, flat, rectangular, triangular, elliptical, trapezoidal, concentric annular, and eccentric annular. No results will be presented for turbulent flow over rod bundles, since these results are available in Chap. 7 of this Handbook.

A word of caution is in order in applying the results of this chapter to extremely small tubes with hydraulic diameters smaller than about 2 mm (0.1 in.). In such tubes, the turbulent eddy mechanism for fluid flow and heat transfer is suppressed by the physical size of the tube cross section resulting in lower friction factors and heat transfer coefficients. Unfortunately, definitive information on turbulent flow friction factors and heat transfer coefficients in small diameter tubes is not available for inclusion in the chapter.

The duct walls treated in the chapter are considered to be uniformly thin, straight in the flow direction with axially unchanging cross sections, nonporous, rigid, and stationary. The effect of duct wall roughness is considered wherever possible, since, in contrast with laminar flows, the wall roughness exerts a strong influence in enhancing both friction and heat transfer coefficients for turbulent flows. The scope of the chapter is restricted to steady, incompressible, and constant-property Newtonian fluids only. All forms of body forces are neglected, as are the effects of natural convection, phase change, mass transfer, chemical reactions, thermal energy sources, viscous dissipation, and fluid axial conduction.

4.1.1 Fluid Flow and Heat Transfer Parameters

The fluid flow and heat transfer characteristics of all ducts are described in terms of certain hydrodynamic and thermal parameters. Some of them have already been introduced in Secs. 3.1.2 and 3.1.3 in the context of the laminar flows in Chap. 3 of this Handbook. They include the Fanning friction factor f for fully developed flow [Eq. (3.2)], the apparent Fanning friction factor for hydrodynamically developing flow [Eq. (3.3)], the bulk mean, mixing-cup, or flow-average temperature T_m [Eq. (3.7)], the local heat transfer coefficient h_x [Eq. (3.8)], the flow-length average heat transfer coefficient h_m [Eq. (3.9)], the local Nusselt number Nu_x [Eq. (3.10)], and the mean Nusselt number Nu_m [Eq. (3.11)]. At this stage, additional parameters that are peculiar to turbulent duct flows will be defined.

Flow is considered turbulent when the fluid particles do not travel in a well-ordered pattern. In turbulent flow, fluid particles possess velocities with macroscopic fluctuations at any point in the flow field. Even in steady turbulent flow, the local velocity components transverse to the main flow direction change in magnitude with respect to time. In describing turbulent flow, it is convenient to separate it into a *mean motion* and a *fluctuating or eddying motion*. Denoting the time-average velocity of the axial flow by $\bar{u}$ and its axial fluctuating component by u', the instantaneous velocity u_i is presented as

$$u_i = \bar{u} + u' \tag{4.1}$$

with similar expressions for the instantaneous transverse velocity components v_i and w_i as well as the pressure p_i.

When the turbulent stream involves heat transfer, its instantaneous temperature T_i can be expressed in terms of time-average and fluctuating components $\overline{T}$ and T' as

$$T_i = \overline{T} + T' \tag{4.2}$$

The time-average temperature component $\overline{T}$ must not be confused with the bulk mean (mixing-cup) temperature T_m defined by Eq. (3.7). The former is a local quantity at a point, whereas the latter is a flow-average quantity at a cross section.

In order to simplify writing of the time-average equations, the conventional bar over the symbols for time-average turbulent quantities such as $\bar{u}$ and $\overline{T}$ will be omitted in the remainder of this chapter.

In analogy with Newton's law of friction for the laminar shear stress τ_l, a law of friction for the apparent turbulent stress τ_t was introduced by Boussinesq [1]. Thus

$$\tau_l = -\mu\frac{\partial u}{\partial y}, \qquad \tau_t = -\mu_t\frac{\partial u}{\partial y} \tag{4.3}$$

where μ_t is an apparent viscosity also referred to as *virtual* or *eddy viscosity*. Note that μ_t is not a property of the fluid like the dynamic viscosity μ, but depends on the time-average flow velocity.

In analogy with the molecular momentum diffusivity ν, eddy diffusivity for momentum ϵ_m is often required to characterize turbulent flow. ν and ϵ_m are defined as

$$\nu = \frac{\mu}{\rho}, \qquad \epsilon_m = \frac{\mu_t}{\rho} \tag{4.4}$$

In analogy with Fourier's law of heat conduction for laminar flow, a law of heat conduction for turbulent flow is introduced. These laws are

$$q_l = -k\frac{\partial T}{\partial y}, \qquad q_t = -k_t\frac{\partial T}{\partial y} \tag{4.5}$$

where k_t is an apparent (virtual or eddy) conductivity which is not a property of the fluid like the molecular thermal conductivity k, but depends on the time-average flow velocity as well as on the time-average bulk mean temperature.

In analogy with the thermal diffusivity α, an eddy diffusivity for heat transfer, ϵ_h, is defined:

$$\alpha = \frac{k}{\rho c_p}, \qquad \epsilon_h = \frac{k_t}{\rho c_p} \tag{4.6}$$

Furthermore, in analogy with the molecular Prandtl number Pr, a turbulent Prandtl number Pr_t is introduced:

$$\mathrm{Pr} = \frac{\nu}{\alpha}, \qquad \mathrm{Pr}_t = \frac{\epsilon_m}{\epsilon_h} \tag{4.7}$$

Since neither ϵ_m nor ϵ_h is a fluid property, Pr_t is also not a fluid property. Like ϵ_m and ϵ_h, it depends on the time-average flow velocity and the time-average bulk mean temperature.

Next, we introduce the turbulent friction or shear velocity u_t. From the definition of the friction coefficient $f = \tau_w/(\rho u_m^2/2g_c)$ [Eq. (3.2)], it is noted that $\tau_w g_c/\rho = \frac{1}{2}fu_m^2$ has the dimensions of velocity squared. Accordingly, a turbulent friction or shear velocity u_t is defined by the first inequality of the equation

$$u_t = \sqrt{\frac{\tau_w g_c}{\rho}} = u_m\sqrt{\frac{f}{2}} \tag{4.8}$$

Next, the dimensionless velocity u^+ and wall distance y^+ are defined in terms of u_t as

$$u^+ = \frac{u}{u_t} = \frac{u}{\sqrt{\tau_w g_c/\rho}} \tag{4.9}$$

$$y^+ = \frac{yu_t}{\nu} = \frac{y\sqrt{\tau_w g_c/\rho}}{\nu} \tag{4.10}$$

The set of dimensionless group u^+ and y^+ is frequently referred to as the *wall coordinates* as the wall shear stress τ_w enters their defining relations.

Finally, we note that in duct flows, the use of the Stanton number St is sometimes preferred to the use of the Nusselt number Nu, since in the fully developed laminar flow the expressions for the Stanton number turn out to be similar in form to those for the fully developed Fanning friction factor. Moreover, no characteristic duct dimension enters the definition of St [see Eq. (4.11)] below. The Stanton number is related to the Nusselt number via the Reynolds and Prandtl numbers simply as

$$\mathrm{St} = \frac{h}{\rho c_p u_m} = \frac{\mathrm{Nu}}{\mathrm{Re}\,\mathrm{Pr}}. \tag{4.11}$$

Throughout Chap. 4, however, the heat transfer coefficients for various ducts will be presented in terms of Nu rather than St to facilitate comparisons with the laminar flow heat transfer coefficients, which are presented in Chap. 3 in terms of Nu.

4.1.2 Characterization of Turbulent Duct Flows

A turbulent boundary layer in a duct may be visualized as comprising three distinct regions as depicted in Fig. 4.1. They are a *laminar sublayer* in the immediate vicinity of the duct wall, a *buffer layer*, and a prominent *turbulent core*. In the laminar sublayer, the fluid particles move in an orderly streamline pattern parallel to the duct wall. In the turbulent core, on the other hand, chunks of fluid move in a totally chaotic pattern, causing an intense mixing of the fluid. The fluid motion in the turbulent core is termed *eddying motion*. The fluid motion in the intervening buffer layer exhibits behavior that is intermediate between that of the fluid in the laminar sublayer and in the turbulent core.

The laminar sublayer plays a decisive role in controlling the rates of heat, mass, and momentum transfer as the major temperature, concentration, and velocity changes occur across it. It is important to know its thickness δ_l, which is given by

$$\delta_l = \frac{5\nu}{u_t} \tag{4.12}$$

where u_t denotes the friction velocity given by Eq. (4.8). This expression for δ_l was deduced analytically by Goldstein [2] from the criterion that at a Reynolds number $u_t\delta_l/\nu = 5$, a von Kármán vortex street just forms a single protrusion signaling the onset of eddying motion. Equation (4.12) is experimentally verified by several investigators. It is applicable to internal as well as external flow.

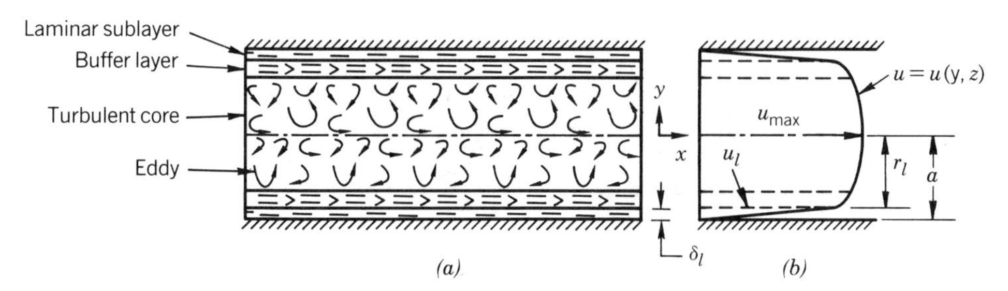

Figure 4.1. Fully developed turbulent flow in a smooth duct: (*a*) three sublayers of a turbulent boundary layer, (*b*) velocity profile.

The turbulent-core and buffer-layer portions of the duct flow are very difficult to characterize by purely analytical means. A significant step in this direction was taken by Reynolds [3], who recognized that a knowledge of the fluctuations occurring during turbulent flow would add materially to an understanding of turbulence. Accordingly, he modified the Navier-Stokes equations to include fluctuations of the velocity components. This led to the introduction of additional apparent turbulent stresses, now called *Reynolds stresses*, in the modified Navier-Stokes equations. Unfortunately, determination of the Reynolds stresses by purely analytical means is an impossible task, and this has created what is called the *turbulence closure problem* (see Chap. 2, Sec. 2.3).

Several attempts have been made to solve the turbulence closure problem by the introduction of a succession of *turbulence models*. The simplest one of them is Prandtl's celebrated mixing-length model, inspired by the kinetic theory of gases [4]. This model and its variations at the hands of Taylor [5], von Kármán [6], and Van Driest [7] have proven reasonably adequate for plain two-dimensional flows. For the general case of three-dimensional flows, higher-order turbulence models are required. They utilize one or more partial differential equations derived from the modified Navier-Stokes equations for quantities like the kinetic energy κ, the kinetic-energy dissipation ϵ, and components of the turbulent stress tensor τ_{ij}. The turbulence model employing the single partial differential equation for the turbulent kinetic energy in conjunction with the algebraic expression for the turbulence length scale (e.g., Prandtl's mixing length l) is referred to as the one-equation (κ-l) model. Another model, employing the partial differential equations for the turbulent kinetic energy and its dissipation, is called the two-equation (κ-ϵ) model. The more complicated multiequation models involve solution of the partial differential equations for all components of the turbulent stress tensor and are referred to as stress-equation models. A brief description of the turbulence models together with pertinent references is available in Refs. 8, 9 as well as in Sec. 2.3 of Chap. 2 in this Handbook. Suffice it to say here that as of this writing there is no consensus as to the ultimate turbulence model to solve the turbulence closure problem. The various models have been quite successful in providing useful solutions to many duct flows of practical interest. Several of the results obtained from them will be presented in the ensuing sections of this chapter.

As regards the turbulent heat transfer problem, the energy equation has been modified in a manner analogous to the modification of Navier-Stokes velocity equations. This has led to the introduction of *Reynolds heat fluxes*, which one seeks to determine via semiempirical algebraic, one-equation, and two-equation models. For a review of these models, Ref. 10 may be consulted. Many useful turbulent heat transfer results have been obtained by assuming an analogy between the processes of heat and momentum transfer. The original analogy is due to Reynolds [11], and improvements in it have been made by Prandtl [12], Taylor [13], von Kármán [14], Martinelli [15], and others.†

In addition to the semiempirical theory of turbulence mentioned above, a *statistical theory of turbulence* is being developed. The notable contributors to this effort have been Taylor, von Kármán, Kolmogoroff, Burgers, Townsend, Dryden, Lin, and Chandrasekhar. The accounts of the development of the statistical theory may be found in Refs. 16, 17, 18. For design purposes, it may be stated that notwithstanding some impressive advances made possible by the advent of high-speed digital computers in recent years, the statistical theory is far from predicting friction and heat transfer coefficients for turbulent duct flows.

† The mathematical expressions of these and other analogies are presented in Sec. 4.2.2.

4.1.3 Laminar-to-Turbulent Transition in Duct Flows

In laminar flow through a straight duct of uniform cross section, every fluid particle moves with a uniform velocity along a straight path with particles near the wall moving at a lower velocity than those near the duct axis. The observations show that this orderly laminar pattern transforms to a chaotic turbulent pattern when the Reynolds number $\text{Re} = u_m D_h/\nu$ exceeds a certain critical value called the *critical Reynolds number* Re_{crit}. This phenomenon was first studied by Reynolds [19] in 1883. Careful observations by Reynolds [19], Lindgren [20], and several other investigators reveal that laminar-to-turbulent transition is not a sudden phenomenon but occurs over a range of Re. Furthermore, it is observed that the transition starts in the duct core region rather than at the duct wall.

The numerical value of Re_{crit} depends strongly on the duct inlet conditions as well as on conditions at the duct interior such as surface roughness and flow pulsation. Disturbances such as noise and vibration on the exterior of the duct wall also influence Re_{crit}. The *upper limit* to which Re_{crit} can be driven with extreme precautions to eliminate various sources of disturbances is not known at present. However, there appears to be a certain *lower limit* below which the flow remains laminar even in the presence of strong disturbances. In the case of fully developed flow in circular duct, the lower limit for Re_{crit} is accepted to be 2300, whereas the highest value of the upper limit attained by Pfenninger [21] is 1.001×10^5.

The aforementioned lower value of Re = 2300 for a circular duct reduces to 153 when the hydraulic diameter in Re_{crit} is replaced by the momentum thickness δ_2 defined as

$$\delta_2 = \int_0^a \frac{u}{u_{\max}}\left(1 - \frac{u}{u_{\max}}\right) dr \tag{4.13}$$

The velocity distribution entering this definition for a circular duct is given by Eq. (3.13) with $u_{\max} = 2u_m$, yielding $\delta_2/D_h = \frac{1}{15}$. Remarkably enough, the lower value of Re_{crit} for a flat plate at zero incidence based on the momentum thickness evaluated from Blasius's solution is found to be 162, nearly the same as the value for a circular duct [22]. This suggests that $\text{Re}_{\text{crit}} = 162$ based on the momentum thickness may be taken as a general criterion for the onset of transition in duct flows with negative pressure gradients as well as for flow along a flat plate with zero pressure gradient.

Careful investigation of the process of laminar-to-turbulent transition shows that in a certain range of Re around the lower value of Re_{crit}, the flow exhibits intermittent behavior, i.e., it alternates in time between laminar and turbulent behavior. The physical nature of the intermittent flow can best be described in terms of an *intermittency factor* $\tilde{\gamma}$, which is defined as a fraction of time during which the flow exhibits turbulent characteristics. Thus $\tilde{\gamma} = 1$ signifies continuously turbulent flow while $\tilde{\gamma} = 0$ denotes continuously laminar flow. Figure 4.2 shows a plot of the intermittency factor $\tilde{\gamma}$ for a circular duct as measured by Rotta [23] at various axial locations with $\text{Re} = u_m D_h/\nu$ as a parameter.

The pertinent experimental information relating to Re_{crit} as well as friction and heat transfer coefficients in the transition region for various ducts will be provided in the sections below. Some analytical results based on the *hydrodynamic stability theory* developed by Tollmien [24] and Schlichting [25] will also be presented. This theory is restricted to disturbances so small as to be irrelevant to turbulent duct flows. However, it is useful in separating the stable and unstable flow regimes, thereby providing some information pertaining to transition duct flows.

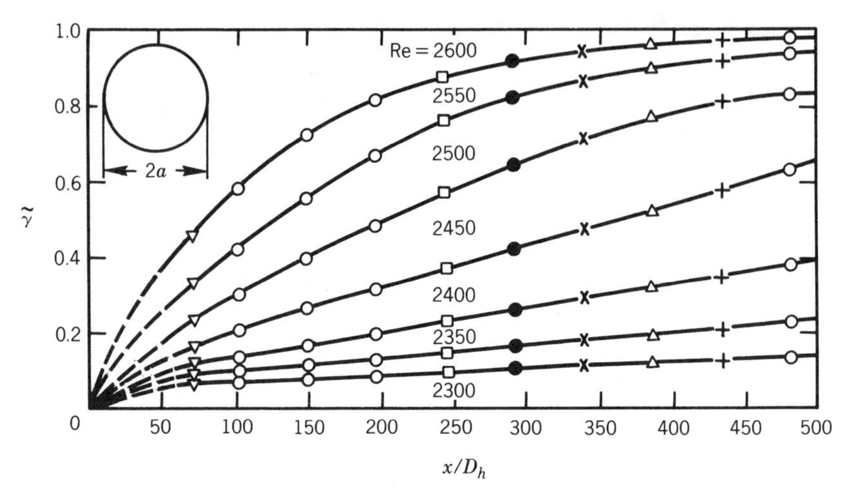

Figure 4.2. Intermittency factor $\tilde{\gamma}$ for a circular duct with symbols denoting the experimental points [23].

4.1.4 Turbulent-to-Laminar Transition in Duct Flows

Under certain conditions, a turbulent duct flow may revert to laminar flow. This process is known as *reverse transition* or *laminarization*. It is not fully explained yet, although the conditions under which it is likely to occur have been established. According to Patel and Head [26], the onset of turbulent-to-laminar transition can be expressed in terms of the following parameter, called the shear stress gradient parameter:

$$\Delta = \frac{g_c \nu m}{\rho u_t^3} \tag{4.14}$$

where m is the shear stress gradient, representing the variation of the local shear stress τ near the wall in the expression $\tau = \tau_w + my$, y being the distance from the wall. The shear velocity u_t in Eq. (4.14) is defined in Eq. (4.8). The critical value of Δ for turbulent-to-laminar transition is -0.009. For $\Delta < -0.009$, the turbulent flow in a duct as well as over a flat plate reverts to laminar flow.

Bankstone [27] has shown that for a heated turbulent gas flow in a duct, there is an additional mechanism responsible for turbulent-to-laminar transition. It has to do with the gas viscosity, which in general increases with an increase in the temperature. The increase in viscosity may lower Re to the upper $\mathrm{Re}_{\mathrm{crit}}$, triggering onset of the reverse transition.

4.1.5 Types of Turbulent Duct Flows

The turbulent duct flows can be divided into four categories: fully developed, hydrodynamically developing, thermally developing, and simultaneously developing. This division is identical to the one adopted for laminar duct flows in Chap. 3 of this Handbook. It may be useful at this point to review the four types of duct flows described in Sec. 3.1.1. In particular note that thermally developing flow is flow that is already hydrodynamically developed, whereas simultaneously developing flow is flow for which the hydrodynamic development occurs simultaneously with the thermal development. It is also useful to note at this point that for turbulent duct flow, the hydrodynamic entrance length and the thermal entrance length are characteristically much shorter than the corresponding lengths in laminar duct flow. Consequently, results on fully developed turbulent fluid flow and heat transfer are frequently used in design calculations without reference to the hydrodynamic and thermal entrance regions. However, caution must be exercised in using the fully developed results for the low Prandtl number liquid metals, since the entrance region effects are quite pronounced for such fluids even in turbulent duct flows.

In the ensuing sections of this chapter, the most useful and available fluid flow and heat transfer results will be presented for the aforementioned four types of flows in various ducts. In addition, the results will be provided for the transition flow described in Secs. 4.1.3 and 4.1.4 for each duct geometry.

4.1.6 Hydraulic and Equivalent Diameter Concepts

The hydraulic radius r_h of a duct cross section is defined as $r_h = A_c/P$, where A_c is the flow cross-sectional area and P is the wetted perimeter of the duct. The hydraulic diameter D_h of the duct is then defined as $D_h = 4r_h$. According to these definitions, the hydraulic and physical diameters of a circular duct are identical; however, the hydraulic radius is half the physical radius.

The hydraulic diameter is a convenient substitute for the characteristic physical dimension of a noncircular duct, and it leads to fairly good correlations between turbulent fluid flow and heat transfer characteristics of circular and noncircular ducts. The hydraulic diameter is also used for ducts involving laminar flow to provide a consistent basis of comparison with the turbulent flow results. However, for laminar flow itself this quantity does not lead to satisfactory correlations between circular and noncircular ducts.

The use of hydraulic diameter for ducts with very sharp corners (e.g., triangular and cusped ducts) leads to unacceptably large errors, of the order of 35%, in turbulent flow friction and heat transfer coefficients determined from the circular duct correlations; the errors may not be that large for other noncircular ducts. With the objective of improving turbulent flow predictions, several other linear dimensions have been proposed as substitutes for the hydraulic diameter. We present below expressions for several such characteristic dimensions. These substitute dimensions provide improved friction and heat transfer coefficients for specific ducts only.

Jones [28] introduced the laminar equivalent diameter D_l for rectangular ducts with sides $2a$ and $2b$ ($2a > 2b$), given by

$$\frac{D_l}{D_h} = \tfrac{2}{3}(1 + \alpha^*)^2\left(1 - \frac{192\alpha^*}{\pi^5}\sum_{n=0}^{\infty}\frac{1}{(2n+1)^5}\tanh\frac{(2n+1)\pi\alpha^*}{2}\right) \tag{4.15}$$

where the hydraulic diameter $D_h = 4b/(1 + \alpha^*)$ and the duct aspect ratio $\alpha^* = 2b/2a$. An approximate expression for D_l is

$$\frac{D_l}{D_h} = \tfrac{2}{3} + \tfrac{11}{24}\alpha^*(2 - \alpha^*) \tag{4.16}$$

which yields D_l values within $\pm 2\%$ of those given by Eq. (4.15).

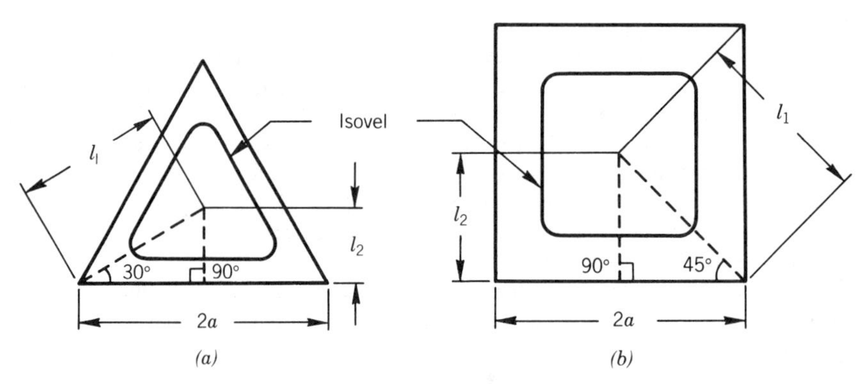

Figure 4.3. The characteristic lengths for (*a*) an equilateral triangular and (*b*) a square duct.

Extensive comparisons presented in Ref. 28 show that substitution of D_l from Eq. (4.15) for D_h in the circular duct correlation reduces the scatter of rectangular duct turbulent flow experimental friction-factor data points about the predicted values from about $\pm 20\%$ to $\pm 5\%$.

Ahmed and Brundrett [29] introduced a characteristic length D_l suitable for equilateral triangular and square ducts. This length is merely the sum of two lengths l_1 and l_2 indicated in Fig. 4.3. Physically, l_1 is the distance between the duct wall and the axis measured at the point of highest curvature of an isovel (curve of constant axial velocity). Likewise, l_2 is the distance between the duct wall and the axis measured at the point of lowest curvature of an isovel. A schematic representation of the isovels for equilateral triangular and square ducts is given in Fig. 4.3. It is seen from the figure that the two sets of isovels possess the highest curvature in the corner regions along the corner bisectors and the lowest curvature parallel to the duct sides along the side bisectors. With the aid of Fig. 4.3, it can be shown that for an equilateral triangular duct

$$D_l = l_1 + l_2 = \sqrt{3}\, a \tag{4.17}$$

and for a square duct

$$D_l = l_1 + l_2 = (1 + \sqrt{2})\, a \tag{4.18}$$

The fully developed turbulent flow friction factors and Nusselt numbers computed from the circular duct correlations with the use of the D_l values from Eqs. (4.17) and (4.18) are compared in Ref. 29 with the available experimental measurements. The agreement is found to be excellent within about $\pm 3\%$ for both equilateral triangular and square ducts.

Hodge [30] introduced the concept of a hydraulically effective zone for computing the fully developed Fanning friction factors for isosceles triangular ducts. According to his analysis, the perimeter P_z and the cross-sectional area A_z of the hydraulically effective zone are related to the corresponding quantities P and A of the actual duct by

$$\frac{P_z}{P} = \frac{(1 - \beta) + (1 + \beta)\sin\phi}{1 + \sin\phi} \tag{4.19}$$

$$\frac{A_z}{A} = 1 - \beta^2 \tag{4.20}$$

where 2ϕ is the apex angle and β is a scaling factor given by

$$\beta = \frac{5}{24}\frac{1 - \sin\phi}{1 + \sin\phi} \tag{4.21}$$

The fully developed turbulent flow Fanning friction factor f for an isosceles triangular duct is given in terms of the perimeter and area ratios of Eqs. (4.19) and (4.20) by

$$\frac{f}{f_c} = \left(\frac{P_z}{P}\right)^{1.25}\left(\frac{A_z}{A}\right)^{3} \tag{4.22}$$

where f_c is the Fanning friction factor for the same duct calculated from the circular duct relation via the hydraulic diameter.

The f values calculated from Eqs. (4.19) to (4.22) for $4000 \le \mathrm{Re} \le 10^4$ and $2\phi = 4.01°$, $7.96°$, $12.0°$, $22.3°$, $38.8°$, and $60°$ are 7% to 29% lower than the experimental values reported by Carlson and Irvine [31]. The use of D_h in the circular duct correlation, on the other hand, yields f values that are 7% to 25% higher than the experimental values.

Bandopadhayay and Ambrose [32] introduced a generalized length dimension which may be viewed as an average distance of the duct boundary from the point of the maximum axial velocity in the duct. For an isosceles triangular duct with an apex angle of 2ϕ, they presented the following expression for the generalized length D_g:

$$\frac{D_g}{D_h} = \frac{1}{2\pi}\left[3\ln\cot\frac{\theta}{2} - 2\ln\tan\frac{\phi}{2} - \ln\tan\frac{\theta}{2}\right] \tag{4.23}$$

where $\theta = (90° - \phi)/2$.

Substitution of D_g from Eq. (4.23) for D_h in the circular duct correlation yields the fully developed Fanning friction factor values for $4000 \le \mathrm{Re} \le 10^4$ and $2\phi = 4.01°$, $7.96°$, $12.0°$, $22.3°$, and $38.8°$ that are 0% to 6% higher than the experimental values reported by Carlson and Irvine [31]. The use of D_h in the circular duct correlation, on the other hand, yields f values that are 7% to 25% higher than the experimental values.

Jones and Leung [33] applied the concept of laminar equivalent diameter to concentric annular ducts and presented the following expression for D_l:

$$\frac{D_l}{D_h} = \frac{1 + r^{*2} + (1 - r^{*2})/\ln r^*}{(1 - r^*)^2} \tag{4.24}$$

where the duct aspect ratio $\alpha^* = r_i/r_o$ and the hydraulic diameter $D_h = 2(r_o - r_i)$, r_i and r_o being the radii of the inner and outer tubes, respectively, forming the duct. Substitution of D_l from Eq. (4.24) for D_h in the circular duct correlation reduces the scatter of the experimental fully developed friction factors from various sources to within about $\pm 5\%$ of the predicted values, compared to $\pm 20\%$ scatter with the use of D_h.

No general characteristic dimension has been identified that provides satisfactory correlation for all noncircular ducts. Various characteristic dimensions presented above give improved friction and heat transfer coefficients for the specfic noncircular ducts only.

4.1.7 Influence of Duct Surface Roughness

The duct wall roughness has little effect on laminar flow. However, it exerts a strong influence on turbulent flow. If the surface-roughness height is of the same order of magnitude as the laminar-sublayer thickness δ_l, it tends to break up the laminar sublayer, thus increasing the wall shear stress. In fact, if the surface is sufficiently rough, no laminar sublayer can exist. In that case, the apparent turbulent shear stresses are transmitted directly to the wall in the form of a profile drag. From a physical point of view, it is apparent that the ratio of the surface-roughness element height ε to the laminar-sublayer thickness δ_l must be a determining factor for the effect of roughness. It is seen from Eq. (4.12) that δ_l is proportional to ν/u_t, where u_t is the friction velocity. It follows from this relation that the ratio ε/δ_l must be proportional to $\varepsilon u_t/\nu$, which has the significance of a Reynolds number and is designated as a *roughness Reynolds number* Re_ε:

$$\mathrm{Re}_\varepsilon = \frac{\varepsilon u_t}{\nu} \tag{4.25}$$

Re_ε is found to be the most convenient parameter to identify various flow regimes from the standpoint of the roughness influence.

Nikuradse [34] performed systematic experiments with sand grains glued onto the interior of circular ducts. Based on the results of these experiments, three flow regimes were identified, depending on the manner in which the friction factor f varied with the relative roughness ε/a and the Reynolds number $\mathrm{Re} = 2u_m a/\nu$, where a is the radius of the circular duct and the height ε of the roughness element is in essence the size of the sieve used by Nikuradse to sift the sand. With the roughness Reynolds number Re_ε [Eq. (4.25)] as the parameter, Nikuradse identified the following three flow regimes depending on the variation of f with Re_ε and ε/a:

1. Hydraulically smooth regime, $0 \le \mathrm{Re}_\varepsilon \le 5$: $f = f(\mathrm{Re})$
2. Transition regime, $5 \le \mathrm{Re}_\varepsilon \le 70$: $f = f(\varepsilon/a, \mathrm{Re})$
3. Completely rough regime, $\mathrm{Re}_\varepsilon > 70$: $f = f(\varepsilon/a)$

In the hydraulically smooth regime, ε is so small that the sand grains are contained within the laminar sublayer. Hence f is not affected by ε; in other words, $f = f(\mathrm{Re})$. In the transition regime, the sand grains extend partly outside the laminar sublayer, exerting an additional resistance to flow, in the nature of a profile drag. This causes the friction coefficient to depend on ε/a as well as on Re, i.e., for the transition regime $f = f(\varepsilon/a, \mathrm{Re})$. Finally, in the completely rough regime, all sand grains reach outside the laminar sublayer, disrupting it completely. For this situation, the friction coefficient must depend on the size of the sand grains alone, i.e., $f = f(\varepsilon/a)$.

The roughness used by Nikuradse in his experiments does not represent the type of roughness encountered on the commercial duct surfaces. To circumvent this difficulty, Schlichting [35] introduced the concept of equivalent sand-grain roughness for roughness elements such as spheres, spherical segments, cones, and short triangles. Moody [36] determined the equivalent sand-grain roughness for eight types of commercially available duct surfaces. His results, presented in Fig. 4.4, are very useful in practical applications. Recently, Musker [37] proposed a relatively crude model for predicting roughness functions for naturally occurring surfaces from a knowledge of the surface geometry. His results are tentative.

We now turn to the effect of roughness on the heat transfer rate. Two distinct influences of the roughness elements are recognized. First, they increase the duct

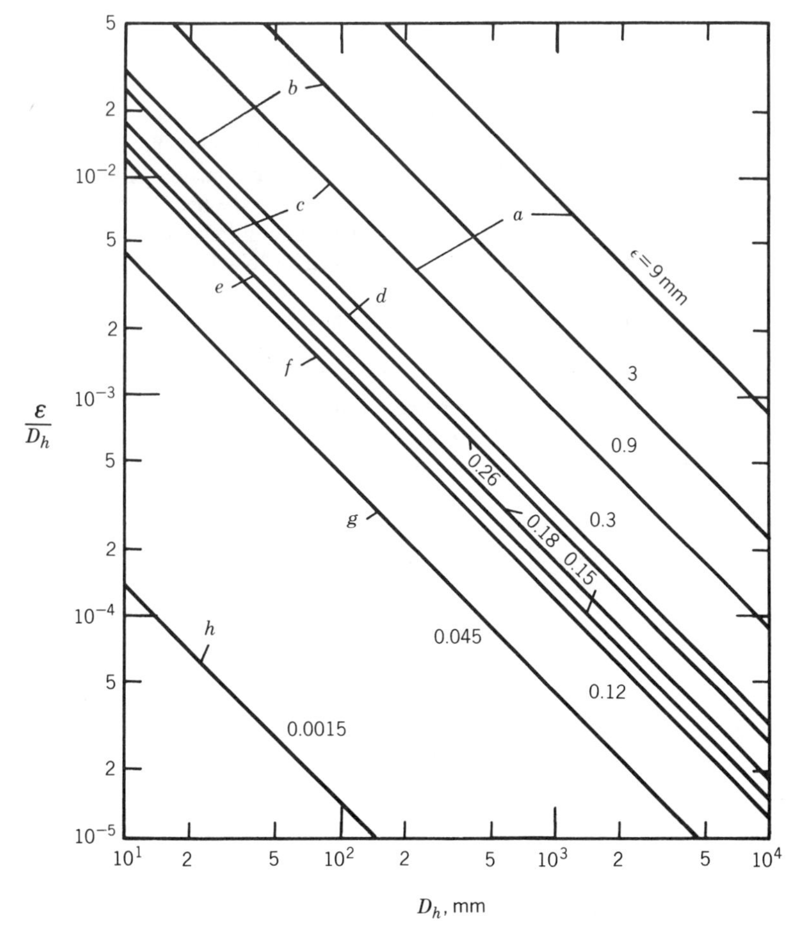

Figure 4.4. Equivalent relative sand roughness for commercial pipe surfaces: (*a*) riveted steel, (*b*) reinforced concrete, (*c*) wood, (*d*) cast iron, (*e*) galvanized steel, (*f*) bitumen-coated steel, (*g*) structural and forged steel, and (*h*) drawn pipe [36].

surface area, and second, they increase the heat transfer coefficient. This latter effect is brought about by the change in turbulence pattern close to the wall. Unfortunately, the increase in the heat transfer coefficient is accompanied by a proportionately larger increase in the friction coefficient. As a consequence, the heat transfer per unit of power consumption is usually lower for a hydrodynamically rough than for a smooth duct. The following simple empirical correlation, suggested by Norris [38], expresses the effect of roughness in turbulent duct flows:

$$\frac{\mathrm{Nu}}{\mathrm{Nu}_s} = \left(\frac{f}{f_s}\right)^n \quad \text{for} \quad \frac{f}{f_s} \le 4 \tag{4.26}$$

where $n = 0.68\,\mathrm{Pr}^{0.215}$ for $1 < \mathrm{Pr} < 6$. For $f/f_s > 4$, Norris observed that the Nusselt number $\mathrm{Nu} = hD_h/k$ no longer increases. For $\mathrm{Pr} > 6$, the value of $n = 1$ provides a conservative estimate of the effect of roughness on turbulent flow.

Equation (4.26) shows that the effect of roughness is more pronounced for high-Pr fluids than for low-Pr fluids since the exponent n has a lower value for the latter fluids. The physical explanation of this behavior is that for high-Pr fluids, the thermal resistance is concentrated very close to the wall because the thermal boundary layer is thin compared to the hydrodynamic boundary layer. For low-Pr fluids, on the other hand, the thermal resistance is distributed over a larger portion of the duct cross section because the thermal boundary layer is thicker than the hydrodynamic boundary layer. Since the roughness elements markedly affect the wall region by destabilizing the laminar sublayer, they are more effective in increasing Nu for high-Pr fluids, which have thinner laminar sublayers.

Another effect shown by Eq. (4.26) is that the heat transfer coefficient is not affected as strongly as the friction coefficient. The physical explanation of this behavior is as follows. The friction coefficient $f = \tau_w/(\rho u_m^2/2g_c)$ is a measure of the apparent wall shear stress τ_w, which is augmented markedly by the profile drag developed by the roughness element. The profile drag is the pressure or dynamic force generated by the faces of the roughness elements normal to the mean flow direction. As regards the heat transfer coefficient $h = q_w''/(T_w - T_m)$, there is no mechanism comparable to the profile drag to generate additional heat flux q_w''. Consequently, the heat transfer coefficient is affected less markedly than the friction coefficient.

Finally, when the roughness effect on friction coefficient becomes very large ($f/f_s >$ 4), no further increase in the heat transfer coefficient is possible. This is because the thermal resistance becomes essentially a conduction resistance at the surface.

4.1.8 Thermal Boundary Conditions

Thermal boundary conditions for laminar duct flows, described in Tables 3.1 and 3.2 of Chap. 3, are also applicable to turbulent duct flows treated in this chapter. However, there is generally no need of analyzing these boundary conditions for turbulent flows for fluids with Pr $\geq$ 0.5. The reason is as follows: At high Pr (e.g., air, water, and oil), the thermal resistance is primarily very close to the wall, yielding a temperature profile that is essentially flat over most of the cross section, regardless of the thermal boundary condition. For low-Pr fluids (e.g., liquid metals), the thermal resistance is distributed over the entire flow cross section, resulting in rounded temperature profiles similar to those for laminar flow of high- and low-Pr fluids. Such rounded temperature profiles are markedly influenced by thermal boundary conditions. Hence, the influence of the thermal boundary condition in turbulent flows is important for low-Pr fluids only.

4.2 CIRCULAR DUCT

Turbulent fluid flow and heat transfer characteristics of a circular duct have been explored in great detail, as this geometry finds widespread use in practical applications. An additional reason is that various flow friction and heat transfer correlations for a circular duct are found to apply to noncircular ducts with reasonable accuracy provided that the circular duct diameter in these correlations is replaced by the hydraulic diameter of the noncircular duct. Several results of practical interest for a circular duct, pertaining to transition, fully developed, hydrodynamically developing, thermally developing, and simultaneously developing flows, are now outlined.

4.2.1 Transition Flow

The first systematic experiments on the laminar-to-turbulent transition in a circular duct were performed by Reynolds [19] in 1883. By dimensional reasoning, Reynolds concluded that the dimensionless parameter $u_m D_h/\nu$ best characterizes the laminar and turbulent flow regimes in a circular duct. We now recognize this parameter as the Reynolds number Re.

Reynolds obtained two different values of $\mathrm{Re}_{\mathrm{crit}}$, viz., 3800 and 1.2×10^4, where the laminar flow changed to turbulent. Several investigators tried to verify Reynolds's results, particularly with regard to the values of $\mathrm{Re}_{\mathrm{crit}}$. However, the effort only led to confusion as the values of $\mathrm{Re}_{\mathrm{crit}}$ came to cover a range of 200 to 5.1×10^4. This latter value is due to Ekman [39], who—experimenting with Reynolds's original apparatus—concluded that there is no upper $\mathrm{Re}_{\mathrm{crit}}$ value if all sources of disturbances on the flow are carefully eliminated. The correctness of Ekman's finding was confirmed by Pfenninger [21], who, with extreme precautions to eliminate sources of disturbances, obtained a $\mathrm{Re}_{\mathrm{crit}}$ value of 1.001×10^5. This is recognized as the highest $\mathrm{Re}_{\mathrm{crit}}$ value as yet attained.

In 1921, Schiller [40] performed systematic experiments and provided a plausible explanation of the discrepancies in earlier work. He accepted Ekman's idea of no upper limit for $\mathrm{Re}_{\mathrm{crit}}$. However, he introduced a lower limit for $\mathrm{Re}_{\mathrm{crit}}$, defined as the value of Re at which a laminar flow remains laminar no matter how large a magnitude the disturbances may attain at the duct inlet. Schiller's careful experiments led to the sharply defined value of 2320 for the lower limit of $\mathrm{Re}_{\mathrm{crit}}$. This value, rounded to 2300, is now widely accepted as the lower limit of $\mathrm{Re}_{\mathrm{crit}}$ for a smooth circular duct. Recently, Šimonek [41] theoretically obtained a value of 2295 as the lower limit for $\mathrm{Re}_{\mathrm{crit}}$, in substantial agreement with the measurements of Schiller [40] and several other investigators.

Prengle and Rothfus [42] conducted careful experiments in the Re range of 1225 to 2.5×10^4 covering the transition flow, and presented the following expression for the fluid axial velocity u_l at the edge of the laminar sublayer (see Fig. 4.1):

$$u_l = \frac{2450\nu}{D_h} \tag{4.27}$$

The corresponding radial distance r_l from the center of the duct to the edge of the laminar sublayer is given by

$$\frac{r_l}{a} = \left(1 - \frac{1225}{\mathrm{Re}}\right)^{1/2} \tag{4.28}$$

where a is the duct radius. Note that the laminar sublayer thickness δ_l shown in Fig. 4.1 is related to r_l simply as $\delta_l = a - r_l$. Also note that according to Eq. (4.28), the laminar sublayer vanishes as $\mathrm{Re} \to \infty$.

Although the upper limit of $\mathrm{Re}_{\mathrm{crit}}$ is undefined, for most practical purposes the flow in the range $2300 \le \mathrm{Re} < 10^4$ may be regarded as transition flow. There have been some attempts to develop a single equation to calculate the friction factors for $2300 < \mathrm{Re} < \infty$ spanning the laminar, transition, and turbulent flow regimes. Wilson and Azad [43] presented a numerical method utilizing a single set of equations to predict mean flow characteristics in the range $100 < \mathrm{Re} < 5 \times 10^5$. Barr [44] developed a numerical method for predicting friction factors for the transition flow using Re as a linking parameter to combine laminar and turbulent flow friction-factor expressions. Neither of these analyses resulted in simple analytical expressions suitable for engineering computations. Churchill [45] constructed the following correlation for engineering calculations covering laminar, transition, and turbulent flow regimes:

$$\frac{2}{f} = \left\{ \frac{1}{\left[(8/\mathrm{Re})^{10} + (\mathrm{Re}/36{,}500)^{20} \right]^{1/2}} + \left[2.21 \ln\left(\frac{\mathrm{Re}}{7} \right) \right]^{10} \right\}^{1/5} \tag{4.29}$$

For laminar flow ($\mathrm{Re} < 2100$) as well as for transition cum turbulent flow ($\mathrm{Re} > 4000$), the predictions of Eq. (4.29) are in exact agreement with the well-established results namely Eq. (3.14) for laminar flow and the Colebrook equation for turbulent flow (see Table 4.2). For $2100 < \mathrm{Re} < 4000$, Eq. (4.29) is in fair to excellent agreement with the available experimental data; there is considerable uncertaintly about the exactitude of the available data from various sources.

Hrycak and Andrushkiw [47] presented the following interpolation formula for the transition flow covering $2100 \le \mathrm{Re} \le 4500$:

$$f = -3.10 \times 10^{-3} + 7.125 \times 10^{-6}\mathrm{Re} - 9.70 \times 10^{-10}\mathrm{Re}^2 \tag{4.30}$$

In its range of applicability, the predictions of Eq. (4.30) are within +3% and −9% of those of Eq. (4.29).

The present authors developed the following formula applicable to laminar, transition, and turbulent flow regimes:

$$f = A + \frac{B}{\mathrm{Re}^{1/m}} \tag{4.31}$$

For laminar flow ($\mathrm{Re} < 2100$), $A = 0$, $B = 16$, $m = 1$; for transition flow ($2100 < \mathrm{Re} \le 4000$), $A = 0.0054$, $B = 2.3 \times 10^{-8}$, $m = -\frac{2}{3}$, and for transition cum turbulent flow ($\mathrm{Re} > 4000$) $A = 1.28 \times 10^{-3}$, $B = 0.1143$, $m = 3.2154$. The accuracy of Eq. (4.31) is on par with that of Eq. (4.29). The main appeal of Eq. (4.31) is that it is computationally more expedient than Eq. (4.29); its drawback is that the constants A, B, m possess different values for the three flow regimes.

The following classical formula presented by Blasius [48] in 1913 is applicable for $4000 \le \mathrm{Re} \le 10^5$ covering a portion of the transition flow regime:

$$f = \frac{0.079}{\mathrm{Re}^{1/4}} \tag{4.32}$$

Its predictions agree with those of the most accurate implicit formula (see Table 4.2) within +2.6% and −1.3%.

The heat transfer results for transition flow are rather uncertain in view of a large number of parameters required to characterize the heat-affected transition flow. The following correlation developed by Churchill [45] for $0 < \mathrm{Pr} < \infty$ and $2100 \le \mathrm{Re} \le 10^6$, spanning laminar, transition, and turbulent flow regimes, is recommended for calculating transition flow Nusselt numbers:

$$\mathrm{Nu}^{10} = \mathrm{Nu}_l^{10} + \left\{ \frac{\exp[(2200 - \mathrm{Re})/365]}{\mathrm{Nu}_l^2} + \frac{1}{\mathrm{Nu}_t^2} \right\}^{-5} \tag{4.33}$$

where

$$\mathrm{Nu}_l = \begin{cases} 3.657 & \text{for (T) boundary condition} \\ 4.364 & \text{for (H) boundary condition} \end{cases} \tag{4.34}$$

$$\mathrm{Nu}_t = \mathrm{Nu}_0 + \frac{0.079(f/2)^{1/2}\mathrm{Re\,Pr}}{(1+\mathrm{Pr}^{4/5})^{5/6}} \tag{4.35}$$

$$\mathrm{Nu}_0 = \begin{cases} 4.8 & \text{for (T) boundary condition} \\ 6.3 & \text{for (H) boundary condition} \end{cases} \tag{4.36}$$

For $\mathrm{Re} \leq 2100$, Eq. (4.33) in conjunction with Eqs. (4.34) to (4.36) yields laminar flow Nu values of 3.657 and 4.364 corresponding to the (T) and (H) boundary conditions, respectively. For $2100 \leq \mathrm{Re} \leq 10^4$, it gives Nu values in agreement with the limited experimental results for transition flow cited in [45]. For $\mathrm{Re} > 10^4$, its predictions are in good agreement with the most accurate turbulent flow results. For detailed comparison of the turbulent flow results, refer to Table 4.4 in Sec. 4.2.2 below.

Patel and Head [26] showed that the shear stress gradient parameter Δ [Eq. (4.14)] for turbulent-to-laminar transition in a circular duct can be expressed as

$$\Delta = -\frac{2\sqrt{2}}{\mathrm{Re}\sqrt{f}} \tag{4.37}$$

Equation (4.37) in conjunction with the critical value of $\Delta = -0.009$ [see discussion after Eq. (4.14)] shows that turbulent-to-laminar transition in a circular duct can be expected when

$$\mathrm{Re}\sqrt{f} \geq 314 \tag{4.38}$$

For the nonisothermal flow, Bankstone [27] showed that turbulent-to-laminar transition is likely when

$$\frac{q''_w}{\rho c_p u_m T_m} \geq 1.05 \times 10^{-6}\mathrm{Re}^{0.8}\mathrm{Pr}^{-0.6} \tag{4.39}$$

where q''_w is the uniform wall heat flux. Furthermore, this reverse transition is essentially complete at

$$\frac{x}{D} = 8 \times 10^{-5}\frac{\mathrm{Re}}{q''_w/(\rho c_p u_m T_e)} \tag{4.40}$$

4.2.2 Fully Developed Flow

Fluid Flow. Using Blasius's formula for friction factor [Eq. (4.32)], Prandtl [49] derived the following *power law of velocity* distribution for fully developed turbulent flow in a smooth circular duct:

$$\frac{u}{u_{\max}} = \left(\frac{y}{a}\right)^{1/n}, \qquad \frac{u_m}{u_{\max}} = \frac{2n^2}{(n+1)(2n+1)} \tag{4.41}$$

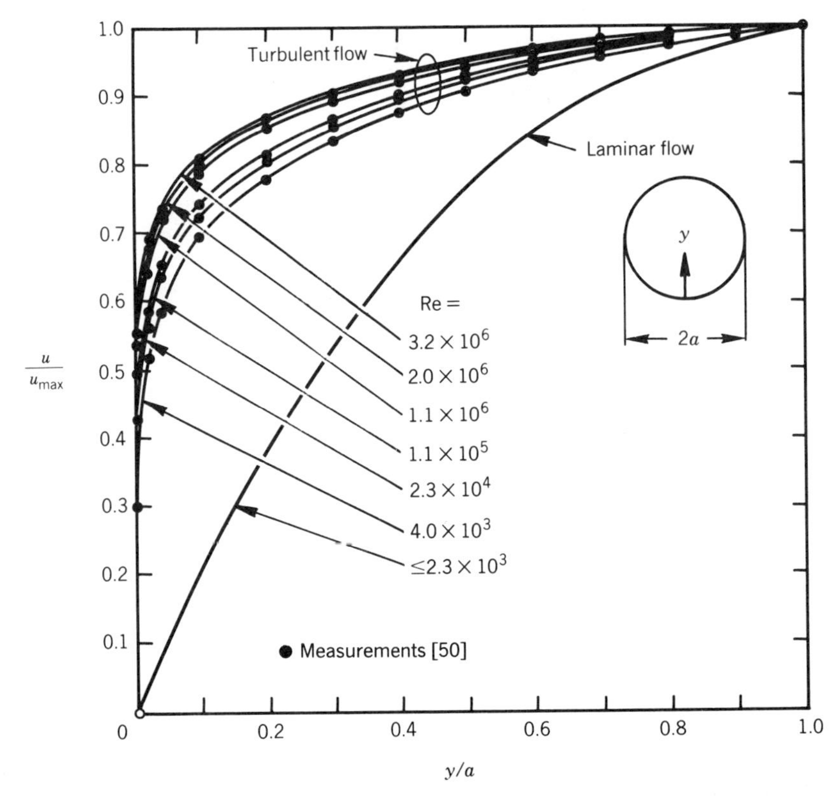

Figure 4.5. Fully developed turbulent velocity distribution in a smooth circular duct (power-law representation).

where $y = a - r$ is the radial distance measured from the duct wall. The exponent n in Eq. (4.41) varies slightly with Re. According to Nikuradse's measurements [50] at Re = 4000, 2.3×10^4, 1.1×10^5, 1.1×10^6, 2×10^6, 3.2×10^6, the values of the exponent n are 6, 6.6, 7, 8.8, 10, and 10, respectively. The main appeal of Eq. (4.41) is its simplicity; its major drawback is that the exponent n varies with Re.

The velocity profiles computed from Eq. (4.41) for the aforementioned values of Re are plotted in Fig. 4.5 together with Nikuradse's data points. Included in Fig. 4.5 for comparison is the Hagen-Poiseuille parabolic velocity profile [Eq. (3.13)] applicable to laminar flow (Re $\leq$ 2300). It is seen that as Re increases, the velocity profile becomes flatter over most of the duct cross section. Also, near the duct wall all the turbulent velocity profiles are significantly steeper then the laminar velocity profile.

The fact that the exponent $1/n$ of the power law of the velocity distribution [Eq. (4.41)] decreases with increasing Re suggests that the power-law expression must asymptotically approach some expression (valid for very high Re) which contains the logarithm of the independent variable y. This is because a logarithmic expression is the limit of a polynomial expression for very small values of the exponent. This consideration led Prandtl [51] to develop the following form of the velocity distribution:

$$\frac{u_{\max} - u}{u_t} = 2.5 \ln \frac{a}{y} \tag{4.42}$$

where u_t is the friction velocity defined by Eq. (4.8). This form is known as the *velocity-defect law*, because the left-hand side represents a dimensionless velocity difference. Equation (4.42) is applicable only in the turbulent core away from the wall. For this reason, it is often referred to as the *outer law* of velocity distribution. Its great advantage over Eq. (4.41) is that it can be extrapolated to arbitrarily large values of Re beyond the range covered by the experiments. For this reason, it is also referred to as the *universal velocity-defect law*.

von Kármán [6] obtained the following different form of the universal velocity-defect law:

$$\frac{u_{max} - u}{u_t} = -2.5\left[\ln\left(1 - \sqrt{1 - \frac{y}{a}}\right) + \sqrt{1 - \frac{y}{a}}\right] \tag{4.43}$$

A third form was developed by Wang [52] as

$$\frac{u_{max} - u}{u_t} = 2.5\left[\ln\frac{1 + \sqrt{1 - y/a}}{1 - \sqrt{1 - y/a}} - 2\tan^{-1}\sqrt{1 - \frac{y}{a}}\right.$$
$$-0.572\ln\frac{2.53 - y/a + 1.75\sqrt{1 - y/a}}{2.53 - y/a - 1.75\sqrt{1 - y/a}}$$
$$\left.+1.143\tan^{-1}\frac{1.75\sqrt{1 - y/a}}{0.53 + y/a}\right] \tag{4.44}$$

In 1855, Darcy [53] performed very careful measurements of velocity distribution in a smooth circular duct. The empirical formula based on his measurements can be written as

$$\frac{u_{max} - u}{u_t} = 5.08\left(1 - \frac{y}{a}\right)^{3/2} \tag{4.45}$$

Darcy's formula is in excellent accord with all the measurements in the range $0.25 < y/a < 1$.

The velocity distributions computed from Eqs. (4.42) to (4.45) are plotted in Fig. 4.6 and compared with the experimental data of Nikuradse [50]. It is seen that Eq. (4.44) is in overall best accord with the data. However, in view of its complexity, it is less useful than either Eq. (4.42) or Eq. (4.43). Darcy's formula, Eq. (4.45), is seen to be in excellent agreement with all points except those near the wall ($y/a < 0.25$).

It is now customary to represent the fully developed velocity in terms of the wall coordinates u^+ and y^+ defined by Eqs. (4.9) and (4.10). A number of analytical expressions have been developed for this purpose. They are summarized in Table 4.1

The formulas attributed to Prandtl [54] and Taylor [55] in Table 4.1 were not actually derived by these investigators in 1910 and 1916, respectively. However, they did introduce the idea of a sharp division between a laminar sublayer and a fully turbulent core (see Fig. 4.1). This idea, when applied to the experimental data, directly leads to the formulas attributed to Prandtl [54] and Taylor [55]. These formulas are frequently referred to as the universal velocity distribution law. von Kármán [14] was

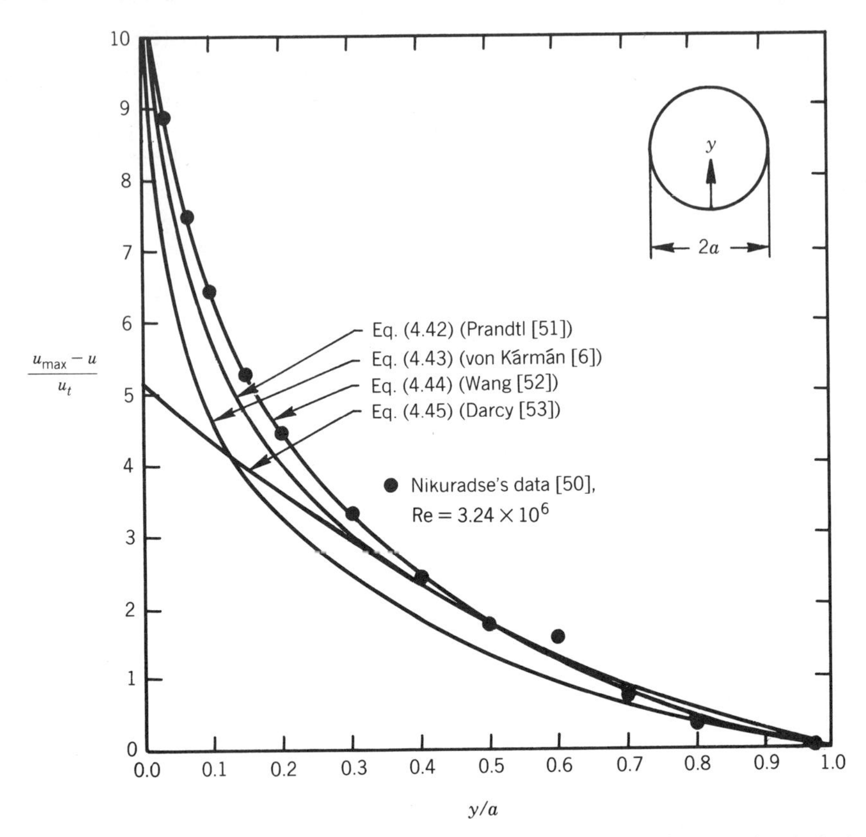

Figure 4.6. Fully developed turbulent velocity distribution in a smooth circular duct (velocity-defect representation).

the first investigator to divide the fully developed flow field into three layers by introducing the idea of an intervening buffer layer in addition to the laminar sublayer and the fully turbulent core introduced by Prandtl [54] and Taylor [55]. As seen from Table 4.1, all authors except Reichardt [56], Van Driest [7], and Spalding [59] found it necessary to employ at least two expressions, valid for different ranges of y^+, to describe the velocity profile adequately.

Figure 4.7 provides a comparison among the different expressions for u^+ listed in Table 4.1. The discrepancies among various expressions are of the same order of magnitude as the scatter in the available experimental measurements. Consequently, it is impossible to discriminate among various expressions. If the final choice is to be made on the basis of convenience, then the preferred expressions are those of Reichardt [56] and Spalding [59].

The fully developed velocity distribution in a rough circular duct can be described by the power-law formula [Eq. (4.41)] with the exponent n ranging between 4 and 5 [60]. Such a velocity distribution is displayed in Fig. 4.8, which also shows experimental points of Nikuradse [34] for various values of the relative roughness ε/D_h at $\mathrm{Re} = 10^6$.

The velocity-defect law for a smooth circular duct [Eq. (4.42)] applies unchanged to a rough circular duct, underscoring the fact that the turbulence mechanism in the turbulent core is independent of the conditions at the duct wall.

TABLE 4.1. Formulas for the Fully Developed Turbulent Velocity Profile in a Smooth Circular Duct

Investigators	Formulas for $u^+(y^+)$	Range of Validity
Prandtl [54], Taylor [55]	$u^+ = y^+$ $u^+ = 2.5 \ln y^+ + 5.5$	$0 \le y^+ \le 11.5$ $y^+ > 11.5$
von Kármán [14]	$u^+ = y^+$ $u^+ = 5 \ln y^+ - 3.05$ $u^+ = 2.5 \ln y^+ + 5.5$	$0 \le y^+ \le 5$ $5 \le y^+ \le 30$ $y^+ > 30$
Reichardt [56]	$u^+ = 2.5 \ln(1 + 0.4y^+) + 7.8\,[1 - \exp(-y^+/11) - (y^+/11)\exp(-0.33y^+)]$	All y^+
Deissler [57]	$u^+ = \int_0^{y^+} \frac{dy^+}{1 + n^2u^+y^+\left[1 - \exp(-n^2u^+y^+)\right]}$, $n = 0.124$ $u^+ = 2.78 \ln y^+ + 3.8$	$0 \le y^+ \le 26$ $y^+ \ge 26$
Van Driest [7]	$u^+ = \int_0^{y^+} \frac{2\,dy^+}{1 + \left\{1 + 0.64y^{+2}[1 - \exp(-y^+/26)]^2\right\}^{1/2}}$	All y^+
Rannie [58]	$u^+ = 14.53 \tanh(y^+/14.53)$ $u^+ = 2.5 \ln y^+ + 5.5$	$0 \le y^+ \le 27.5$ $y^+ > 27.5$
Spalding [59]	$y^+ = u^+ + 0.1108\,[\exp(0.4u^+) - 1 - 0.4u^+ - (0.4u^+)^2/2! - (0.4u^+)^3/3! - (0.4u^+)^4/4!]$	All y^+

In terms of the wall coordinates u^+ and y^+, the fully developed velocity distribution in a completely rough circular duct can be expressed as [60]

$$u^+ = 2.5 \ln \frac{y^+}{\mathrm{Re}_\varepsilon} + 8.5 \tag{4.46}$$

where the roughness Reynolds number Re_ε is defined by Eq. (4.25).

The velocity distribution given by Eq. (4.46) is plotted in Fig. 4.9, which consists of a family of parallel straight lines with Re_ε playing the role of a parameter. As discussed

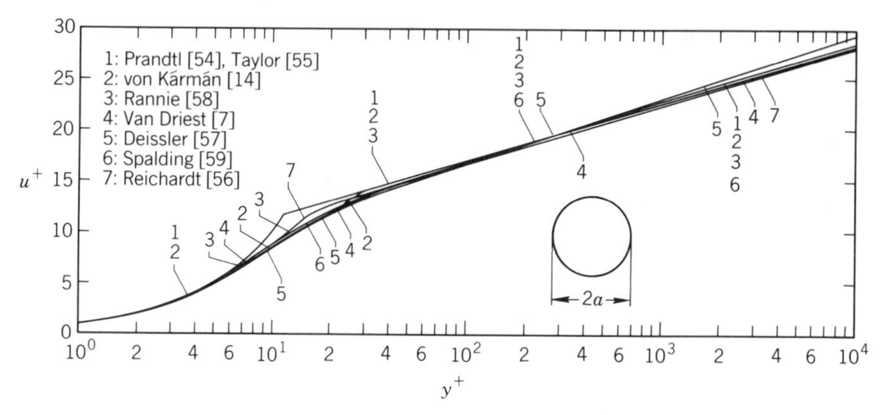

Figure 4.7. Fully developed turbulent velocity distribution in a smooth circular duct (wall-coordinate representation).

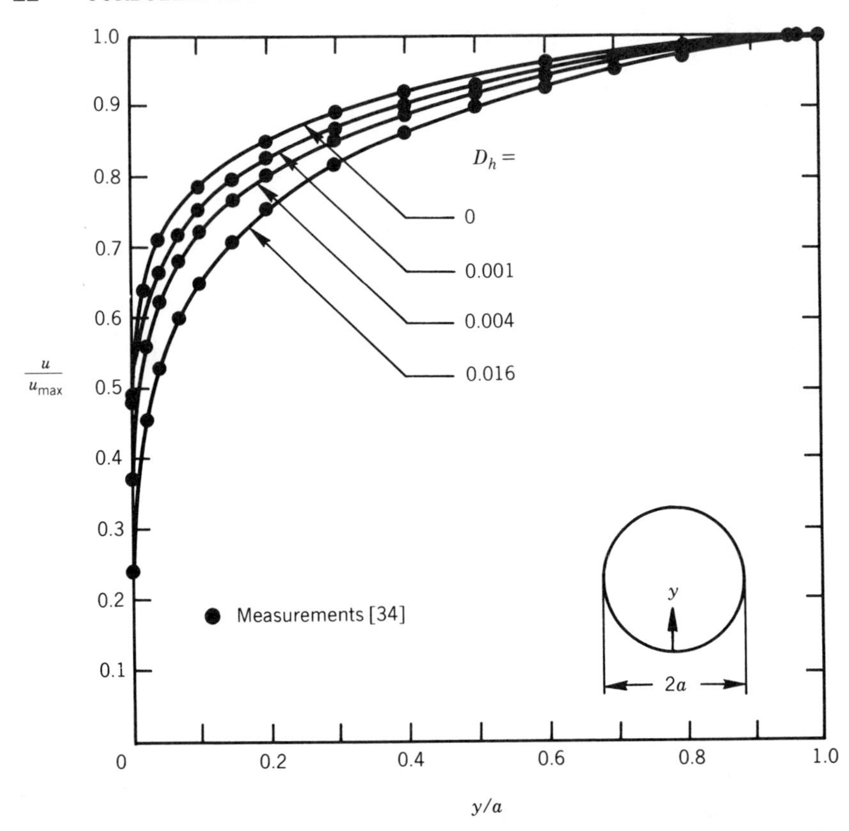

Figure 4.8. Fully developed turbulent velocity distribution in a rough circular duct at $\mathrm{Re} = 10^6$ (power-law representation).

in Sec. 4.1.7, the value $\mathrm{Re}_\varepsilon \leq 5$ corresponds to the hydraulically smooth regime, $5 \leq \mathrm{Re}_\varepsilon \leq 70$ corresponds to the transition from the hydraulically smooth to the completely rough regime, and $\mathrm{Re}_\varepsilon > 70$ corresponds to the completely rough regime. Included in Fig. 4.9 are the curves determined from Prandtl's [54] and Taylor's [55] formulas (see Table 4.1) for the smooth circular duct.

Referring to the abscissa of Fig. 4.9, it may be noted that $y^+ < 5$ is the laminar sublayer region, whereas $y^+ > 70$ is the fully turbulent region. The intervening region $(5 < y^+ < 70)$ is the transition region. The broken vertical lines in Fig. 4.9 demarcate the three regions.

It can be shown that the power-law velocity distribution [Eq. (4.41)] leads to the following expression for the friction factor:

$$f = \frac{C}{\mathrm{Re}^{1/m}} \tag{4.47}$$

where C is an experimentally determined constant and the exponent m is related to the exponent n of Eq. (4.41) by

$$m = \frac{n+1}{2} \tag{4.48}$$

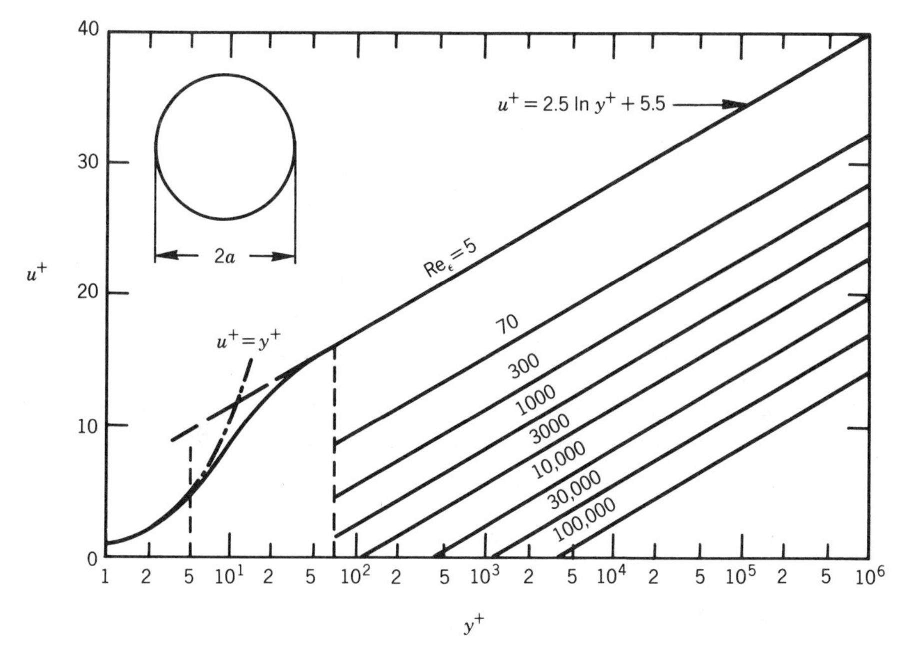

Figure 4.9. Fully developed turbulent velocity distribution in a rough circular duct (wall-coordinate representation).

According to Nikuradse's measurements [50], for Re = 4000, 2.3×10^4, 1.1×10^5, 1.1×10^6, 2×10^6, 3.2×10^6 the exponent $n = 6, 6.6, 7, 8.8, 10, 10$. Equation (4.48) then yields $m = 3.5, 3.8, 4, 4.9, 5.5, 5.5$ for the aforementioned values of Re. The corresponding values of C determined by the present authors from Nikuradse's measurements [50] are 0.1064, 0.0880, 0.0804, 0.0490, 0.0363, 0.0366. Recalling Eq. (4.32), it is seen that Blasius's celebrated formula corresponds to $n = 7$, $m = 4$, and $C = 0.079$; this constant C is within 2% of 0.0804 determined by the present authors.

Several friction factor correlations for fully developed turbulent flow in a smooth circular duct are presented in Table 4.2. All of them are based on the highly accurate experimental results reported by various investigators. The various experimental results agree quite well among themselves, and consequently the various correlations are also in excellent accord among themselves for the restricted Re ranges. The Prandtl-Kármán-Nikuradse (PKN) correlation is regarded as the most accurate. It is based on the universal velocity distribution law with the coefficients slightly modified to best fit the highly accurate experimental data of Nikuradse [50]. Frequently, it is referred to as the *universal law of friction*. The major drawback of the PKN formula is that it does not give f values explicitly, since f occurs on both sides of the formula. This being the case, use of the explicit formulas by Colebrook [65], Filonenko [66], or Techo et al. [67] is preferred. As noted in Table 4.2, these formulas are extremely close approximations to the PKN formula.

For the case of a rough circular duct, von Kármán [6] derived the following theoretical formula for the friction factor for fully rough regime:

$$\frac{1}{\sqrt{f}} = 1.763 \ln \frac{a}{\varepsilon} + 3.36 \tag{4.49}$$

TABLE 4.2 Fully Developed Turbulent Flow Friction Factor Correlations for a Smooth Circular Duct

Investigators	Correlation	Recommended Re Range	Remarks
Blasius [48]	$f = 0.0791\,\mathrm{Re}^{-0.25}$	4×10^3 to 10^5	Within +2.6% and −1.3% of PKN (see below)
McAdams [64]	$f = 0.046\,\mathrm{Re}^{-0.2}$	3×10^4 to 10^6	Within +2.6% and −0.4% of PKN
Present authors	$f = 0.0366\,\mathrm{Re}^{-0.1818}$	4×10^4 to 10^7	Within +2.4% and −3% of PKN
Nikuradse [50]	$f = 0.0008 + 0.0553\,\mathrm{Re}^{-0.237}$	10^5 to 10^7	Within −2% of PKN
Drew et al. [61]	$f = 0.0014 + 0.125\,\mathrm{Re}^{-0.32}$	4×10^3 to 5×10^6	Within +3% of PKN
Present authors	$f = 0.00128 + 0.1143\,\mathrm{Re}^{-0.311}$	4×10^3 to 10^7	Within +1.2% and −2% of PKN
Prandtl [62], Kármán [63], Nikuradse [50]	$\frac{1}{\sqrt{f}} = 1.7372 \ln(\mathrm{Re}\sqrt{f}) - 0.3946$	4×10^3 to 10^7	Classical correlation, here called PKN, has a theoretical basis and is valid for arbitrarily large Re. Its predictions agree with the extensive experimental measurements within ±2%.
Colebrook [65]	$\frac{1}{\sqrt{f}} = 1.5635 \ln\left(\frac{\mathrm{Re}}{7}\right)$	4×10^3 to 10^7	Mathematical approximation to PKN, yielding numerical values within ±1% of PKN
Filonenko [66]	$\frac{1}{\sqrt{f}} = 1.58 \ln \mathrm{Re} - 3.28$	10^4 to 10^7	Within ±1.8% of PKN
Techo et al. [67]	$\frac{1}{\sqrt{f}} = 1.7372 \ln \frac{\mathrm{Re}}{1.964 \ln \mathrm{Re} - 3.8215}$	10^4 to 10^7	Explicit form of PKN; agrees within ±0.1%

A nearly identical formula which Nikuradse [34] obtained experimentally is

$$\frac{1}{\sqrt{f}} = 1.737 \ln \frac{a}{\varepsilon} + 3.48 \tag{4.50}$$

An empirical formula correlating the entire transition regime ($5 < \mathrm{Re}_\varepsilon < 70$) was established by Colebrook and White and is reported in [65]. It is given by

$$\frac{1}{\sqrt{f}} = 3.48 - 1.7372 \ln \left(\frac{\varepsilon}{a} + \frac{9.35}{\mathrm{Re}\sqrt{f}} \right) \tag{4.51}$$

For $\varepsilon \to 0$, this formula transforms to the PKN correlation (see Table 4.2) valid for a hydraulically smooth duct. For $\mathrm{Re} \to \infty$, it transforms to Eq. (4.50) for the completely rough flow regime.

As a matter of historical interest, it may be noted that White's name does not appear as a coauthor of the paper [65] in which Eq. (4.51) was first reported. However, Colebrook made a special point of acknowledging White's contribution to the development of Eq. (4.51). Accordingly, in the literature Eq. (4.51) is referred to as the Colebrook-White correlation.

A drawback of Eq. (4.51) is that it is not an explicit equation for the computation of f, as f appears on both sides of the equation. A number of explicit equations for calculating f in a rough circular duct have been developed. They are presented in Table 4.3. Gregory and Fogarasi [77] made detailed comparisons of the predictions of the explicit equations in Table 4.3 with the predictions of the Colebrook-White implicit equation [Eq. (4.51)] in the ranges $4000 \leq \mathrm{Re} \leq 10^8$ and $2 \times 10^{-8} \leq \varepsilon/a \leq 0.1$. Based on simplicity and close agreement with the predictions of Eq. (4.51), they concluded that the explicit equation due to Chen [72] is the best available correlation for practical friction factor computations in a rough circular duct.

Moody's friction factor plot is shown in Fig. 4.10 in terms of the Fanning friction factor for laminar and turbulent flows in smooth and rough circular ducts. The relative roughness ε/D_h appears as a parameter for the turbulent flow curves. The dashed line demarcating the fully turbulent flow and the transition flow is given by $\sqrt{f} = 100/[\mathrm{Re}(\varepsilon/D_h)]$ [36].

The horizontal portions of the curves to the right of the dashed line are represented by Eq. (4.50). The downward-sloping portions of the curves to the left of the dashed line are represented by Eq. (4.51). The lowermost curve for the smooth turbulent flow is represented by the PKN correlation of Table 4.2. The downward-sloping line for the laminar flow is represented by Eq. (3.14). The relative roughness ε/D_h which appears as a parameter in Moody's friction-factor plot can be determined from Fig. 4.4 for a variety of commercially available pipes.

Heat Transfer. The fully developed temperature distribution in a smooth circular duct with uniform wall temperature, i.e., the Ⓣ boundary condition (see Table 3.1) can be represented by the following power law, analogous to the one for the velocity distribution [Eq. (4.41)]:

$$\frac{T_w - T}{T_w - T_c} = \left(\frac{y}{a} \right)^{1/n}, \qquad \frac{T_m - T_w}{T_c - T_w} = \frac{2n + 1}{2(n + 2)} \tag{4.52}$$

where T_c is the temperature at the duct axis. The expression for $(T_m - T_w)/(T_c - T_w)$

TABLE 4.3. Fully Developed Turbulent Flow Friction Factor Correlations for a Rough Circular Duct

Investigators	Correlation	Remarks
von Kármán [6]	$\frac{1}{\sqrt{f}} = 3.36 - 1.763 \ln \frac{\varepsilon}{a}$	This explicit theoretical formula is applicable for $\mathrm{Re}_\epsilon > 70$.
Nikuradse [34]	$\frac{1}{\sqrt{f}} = 3.48 - 1.737 \ln \frac{\varepsilon}{a}$	This experimentaly derived formula gives very nearly the same results as the foregoing formula, also for $\mathrm{Re}_\epsilon > 70$.
Colebrook and White [65[	$\frac{1}{\sqrt{f}} = 3.48 - 1.7372 \ln\left(\frac{\varepsilon}{a} + \frac{9.35}{\mathrm{Re}\sqrt{f}}\right)$	This implicit formula is applicable for $5 \le \mathrm{Re}_\varepsilon \le 70$ spanning the transition, hydraulically smooth, and completely rough flow regimes.
Moody [36]	$f = 1.375 \times 10^{-3}\left[1 + 21.544\left(\frac{\varepsilon}{a} + \frac{100}{\mathrm{Re}}\right)^{1/3}\right]$	Shows a maximum deviation of -15.78% from the Colebrook-White equation for $4000 \le \mathrm{Re} \le 10^8$ and $2 \times 10^{-8} \le \varepsilon/a \le 0.1$.
Wood [68]	$f = 0.08\left(\frac{\varepsilon}{a}\right)^{0.225} + 0.265\left(\frac{\varepsilon}{a}\right) + 66.69\left(\frac{\varepsilon}{a}\right)^{0.4} \mathrm{Re}^{-n}$ where $n = 1.778\left(\frac{\varepsilon}{a}\right)^{0.134}$	Applicable only for $\epsilon/a \ge 2 \times 10^{-5}$; shows a maximum deviation of 6.16% from the Colebrook-White equation for $4000 \le \mathrm{Re} \le 10^8$ and $2 \times 10^{-8} \le \varepsilon/a \le 0.1$.
Swamee and Jain [69]	$\frac{1}{\sqrt{f}} = 3.4769 - 1.7372 \ln\left[\frac{\varepsilon}{a} + \frac{42.48}{\mathrm{Re}^{0.9}}\right]$	Shows a maximum deviation of 3.19% from the Colebrook-White equation for $4000 \le \mathrm{Re} \le 10^8$ and $2 \times 10^{-8} \le \varepsilon/a \le 0.1$.

Jain [70]	$\frac{1}{\sqrt{f}} = 3.4841 - 1.7372 \ln\left[\frac{\varepsilon}{a} + \frac{42.5}{\mathrm{Re}^{0.9}}\right]$	Gives results comparable with those obtained from the preceding equation.
Churchill [71]	$f = 2\left[\left(\frac{8}{\mathrm{Re}}\right)^{12} + \frac{1}{(A_1 + B_1)^{3/2}}\right]^{1/12}$ where $A_1 = \left\{2.2088 + 2.457 \ln\left[\frac{\varepsilon}{a} + \frac{42.683}{\mathrm{Re}^{0.9}}\right]\right\}^{16}$ $B_1 = \left(\frac{37{,}530}{\mathrm{Re}}\right)^{16}$	Unlike other equations in the table, this equation applies to all three flow regimes: laminar, transition, and turbulent. Its predictions for laminar flow are in agreement with $f = 16/\mathrm{Re}$. The predictions for transition flow are subject to some uncertainty. However, the predictions for turbulent flow are comparable with those of the preceding two equations.
Chen [72]	$\frac{1}{\sqrt{f}} = 3.48 - 1.7372 \ln\left[\frac{\varepsilon}{a} - \frac{16.2426}{\mathrm{Re}} \ln A_2\right]$ where $A_2 = \frac{(\varepsilon/a)^{1.1098}}{6.0983} + \left(\frac{7.149}{\mathrm{Re}}\right)^{0.8981}$	This explicit equation is consistently in good agreement with the Colebrook-White equation for $4000 \le \mathrm{Re} \le 10^8$ and $2 \times 10^{-8} \le \varepsilon/a \le 0.1$, the maximum deviation being -0.39%.
Round [73]	$\frac{1}{\sqrt{f}} = 4.2146 - 1.5635 \ln\left[\frac{\varepsilon}{a} + \frac{96.2963}{\mathrm{Re}}\right]$	Comparable with Moody's equation.
Zigrang and Sylvester [74]	$\frac{1}{\sqrt{f}} = 3.4769 - 1.7372 \ln\left[\frac{\varepsilon}{a} - \frac{16.1332}{\mathrm{Re}} \ln A_3\right]$ where $A_3 = \frac{\varepsilon/a}{7.4} + \frac{13}{\mathrm{Re}}$	Shows a maximum deviation of $+0.96\%$ from the Colebrook-White equation for $4000 \le \mathrm{Re} \le 10^8$ and $2 \times 10^{-8} \le \varepsilon/a \le 0.1$

Investigators	Correlation	Remarks
Zigrang and Sylvester [74]	$\dfrac{1}{\sqrt{f}} = 3.4769 - 1.7372\ln\left[\dfrac{\varepsilon}{a} - 16.1332\ln A_4\right]$ where $A_4 = \dfrac{\varepsilon/a}{7.4} - 2.1802\ln\left[\dfrac{\varepsilon/a}{7.4} + \dfrac{13}{\text{Re}}\right]$	Predictions not distinguishably different from those of the preceding equation.
Haaland [75]	$\dfrac{1}{\sqrt{f}} = 3.4735 - 1.5635\ln\left[\left(\dfrac{\varepsilon}{a}\right)^{1.11} + \dfrac{63.6350}{\text{Re}}\right]$	Shows a maximum deviation of +1.21% from the Colebrook-White equation for $4000 \le \text{Re} \le 10^8$ and $2 \times 10^{-8} \le \varepsilon/a \le 0.1$.
Serghides [76]	$\dfrac{1}{\sqrt{f}} = A_5 - \dfrac{(A_5 - B_2)^2}{A_5 - 2B_2 + C_1}$ where $A_5 = -0.8686\ln\left(\dfrac{\varepsilon/a}{7.4} + \dfrac{12}{\text{Re}}\right)$, $B_2 = -0.8686\ln\left(\dfrac{\varepsilon/a}{7.4} + \dfrac{2.51A_5}{\text{Re}}\right)$, $C_1 = -0.8686\ln\left(\dfrac{\varepsilon/a}{7.4} + \dfrac{2.51B_2}{\text{Re}}\right)$	Shows a maximum deviation of +0.14% from the Colebrook-White equation for $4000 \le \text{Re} \le 10^8$ and $2 \times 10^{-8} \le \varepsilon/a \le 0.1$.
Serghides [76]	$\dfrac{1}{\sqrt{f}} = 4.781 - \dfrac{(A_5 - 4.781)^2}{(4.781 - 2A_5 + B_2)}$ where A_5 and B_2 are as defined with the preceding equation.	Shows a maximum deviation of −0.45% from the Colebrook-White equation for $4000 \le \text{Re} \le 10^8$ and $2 \times 10^{-8} \le \varepsilon/a \le 0.1$.

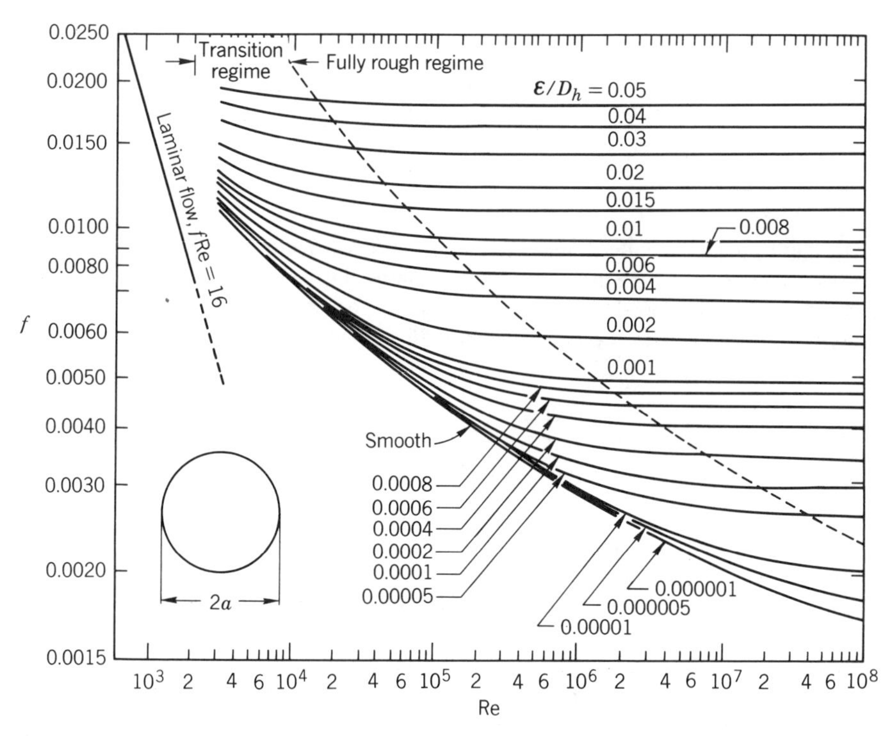

Figure 4.10. Moody's friction factor diagram for fully developed flow in a rough circular duct.

in Eq. (4.52) is obtained by substituting the values of u and T from Eqs. (4.41) and (4.52), respectively, in Eq. (3.7) and evaluating the resulting integral.

The value of the exponent n can vary from 6 to 10, depending on Re, as noted in the discussion pertaining to Eq. (4.41). Another point to be noted about Eq. (4.52) is that it is valid only for fluids with Pr not too far removed from unity.

Following von Kármán's proposal [14], Martinelli [15] analyzed the fully developed temperature distribution in a smooth circular duct and obtained the following expressions for $\mathrm{Pr} \gtrsim 0.1$:

For the laminar sublayer ($0 < y^+ < 5$):

$$\frac{T_w - T}{T_w - T_c} = \frac{\dfrac{\mathrm{Pr}}{\mathrm{Pr}_t}\left(\dfrac{y}{a}\right)\mathrm{Re}\sqrt{f}}{10\sqrt{2}\left[\dfrac{\mathrm{Pr}}{\mathrm{Pr}_t} + \ln\left(1 + 5\dfrac{\mathrm{Pr}}{\mathrm{Pr}_t}\right) + 0.5\ln\left(\dfrac{\mathrm{Re}\sqrt{f}}{60\sqrt{2}}\right)\right]} \tag{4.53}$$

For the buffer layer ($5 \le y^+ < 30$):

$$\frac{T_w - T}{T_w - T_c} = \frac{\dfrac{\mathrm{Pr}}{\mathrm{Pr}_t} + \ln\left\{1 + \dfrac{\mathrm{Pr}}{\mathrm{Pr}_t}\left[\left(\dfrac{y}{a}\right)\dfrac{\mathrm{Re}\sqrt{f}}{10\sqrt{2}} - 1\right]\right\}}{\dfrac{\mathrm{Pr}}{\mathrm{Pr}_t} + \ln\left(1 + 5\dfrac{\mathrm{Pr}}{\mathrm{Pr}_t}\right) + 0.5\ln\left(\dfrac{\mathrm{Re}\sqrt{f}}{60\sqrt{2}}\right)} \tag{4.54}$$

For the turbulent core ($y^+ \geq 30$):

$$\frac{T_w - T}{T_w - T_c} = \frac{\dfrac{\mathrm{Pr}}{\mathrm{Pr}_t} + \ln\left(1 + 5\dfrac{\mathrm{Pr}}{\mathrm{Pr}_t}\right) + 0.5\ln\left(\dfrac{\mathrm{Re}\sqrt{f}}{60\sqrt{2}}\right)\left(\dfrac{y}{a}\right)}{\dfrac{\mathrm{Pr}}{\mathrm{Pr}_t} + \ln\left(1 + 5\dfrac{\mathrm{Pr}}{\mathrm{Pr}_t}\right) + 0.5\ln\left(\dfrac{\mathrm{Re}\sqrt{f}}{60\sqrt{2}}\right)} \tag{4.55}$$

In addition to the fully developed Fanning friction factor f, Eqs. (4.53) to (4.55) contain as a parameter the turbulent Prandtl number Pr_t defined by Eq. (4.7). Malhotra and Kang [78] developed the following relation for the variation of the turbulent Prandtl number Pr_t with the molecular Prandtl number Pr in a circular duct covering the range $10^4 < \mathrm{Re} < 10^6$:

$$\mathrm{Pr}_t = \begin{cases} 1.01 - 0.09\,\mathrm{Pr}^{0.36} & \text{for} \quad 1 \leq \mathrm{Pr} \leq 145 \\ 1.01 - 0.11\ln \mathrm{Pr} & \text{for} \quad 145 < \mathrm{Pr} \leq 1800 \\ 0.99 - 0.29\,(\ln \mathrm{Pr})^{1/2} & \text{for} \quad 1800 < \mathrm{Pr} \leq 12500 \end{cases} \tag{4.56}$$

The technique employed in establishing the above relationship consisted in introducing an established value of the Nusselt number as a function of Re and Pr in the conservation equations of momentum and energy. The resulting equations were then solved iteratively to find the value of Pr_t. The correlations of Eq. (4.56) agree within $\pm 4\%$ with the numerical values reported in [78].

The radial temperature distribution calculated from Eqs. (4.53) to (4.55) is displayed in Fig. 4.11 for $\mathrm{Pr}_t = 1$ and $\mathrm{Re} = 10^4$. It may be noted that these equations are not applicable to the liquid metals with $\mathrm{Pr} \leq 0.03$.

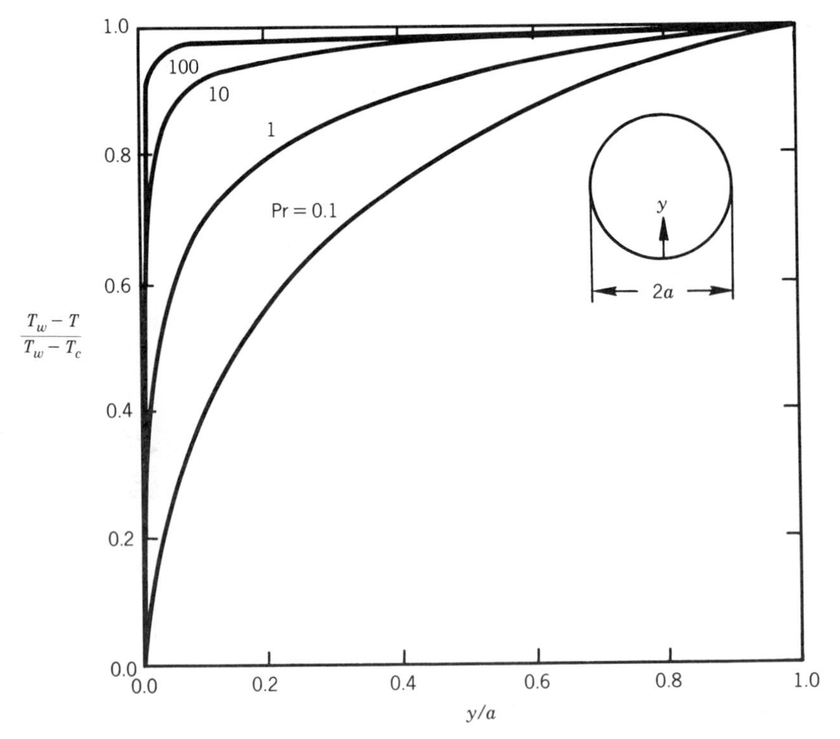

Figure 4.11. Fully developed turbulent temperature distribution in a smooth circular duct with the Ⓣ and Ⓗ boundary conditions at $\mathrm{Re} = 10^4$ and $\mathrm{Pr}_t = 1$.

Kays [79] developed the following expressions with $\mathrm{Pr}_t = 1$ for the fully developed temperature distribution in a circular duct with uniform wall heat flux, i.e., the Ⓗ boundary condition:

For the laminar sublayer ($0 < y^+ < 5$):

$$\frac{(T - T_w)\,u_t}{q''_w/\rho c_p} = \mathrm{Pr}\, y^+ \tag{4.57}$$

For the buffer layer ($5 \le y^+ < 30$):

$$\frac{(T - T_w)\,u_t}{q''_w/\rho c_p} = 5\left[\mathrm{Pr} + \ln\left(\frac{\mathrm{Pr}\, y^+}{5} - \mathrm{Pr} + 1\right)\right] \tag{4.58}$$

For the turbulent core ($y^+ \ge 30$):

$$\frac{(T - T_w)\,u_t}{q''_w/\rho c_p} = 2.5 \ln\frac{y^+}{30} + 5[\mathrm{Pr} + \ln(1 + 5\,\mathrm{Pr})] \tag{4.59}$$

The radial temperature distribution for $\mathrm{Re} = 10^4$ was calculated from Eqs. (4.57) to (4.59) and is also plotted in Fig. 4.11. It was found to be indistinguishable from the temperature distribution for the Ⓣ problem. It may be noted that Eqs. (4.57) to (4.59) are not applicable to low-Pr fluids such as the liquid metals ($\mathrm{Pr} \le 0.03$).

Kays and Crawford [22] developed the following single expression for the temperature distribution in a smooth circular duct with the Ⓗ boundary condition:

$$\frac{(T_w - T)\,u_t}{q''_w/\rho c_p} = 2.25\,\mathrm{Pr}_t \ln\left(y^+ \frac{1.5(1 + r/a)}{1 + 2(r/a)^2}\right) + 13.2\,\mathrm{Pr} - 5.8 \tag{4.60}$$

This equation is applicable in the Pr range of 0.5 to 5 and yields results in agreement with those obtained from Eqs. (4.57) to (4.59).

For very high- as well as very low-Pr fluids, no closed-form formulas are available for the temperature distribution for either the Ⓣ or Ⓗ boundary condition.

A large number of correlations, both theoretical and empirical, have been developed for the fully developed Nusselt number in a smooth circular duct. Shah and Johnson [80] brought these correlations together in tabular form. Tables 4.4 and 4.5 based partly on their tabulation, contain a summary of the correlations for $\mathrm{Pr} > 0.5$ (gases and liquids) and $\mathrm{Pr} < 0.1$ (liquid metals), respectively. The correlations in Table 4.4 are valid for both the Ⓣ and Ⓗ boundary conditions. However, as indicated in Table 4.5 and in Sec. 4.1.8, separate correlations for the Ⓣ and Ⓗ boundary conditions are required for $\mathrm{Pr} < 0.1$. The recommended correlations are the Gnielinski correlation [94] for $\mathrm{Pr} > 0.5$ and the Notter-Sleicher correlations [104] for $\mathrm{Pr} < 0.1$.

The Gnielinski correlation is based on comparisons with extensive experimental measurements by various investigators. Of nearly 800 measurements compared, 720 fall within $\pm 20\%$ of the predictions of the correlation. Given the uncertainties of various investigations stemming from fluid property variation and unaccounted effects like natural convection, this level of agreement is considered very satisfactory. In the judgment of the present authors, the level of agreement will improve to $\pm 10\%$ if the comparison is restricted to the most reliable experimental data.

The fully developed Nusselt numbers $\mathrm{Nu_H}$ computed from the aforementioned Gnielinski and Notter-Sleicher correlations over a wide range of Re and Pr are presented in Table 4.6 as well as in Fig. 4.12. For laminar flow ($\mathrm{Re} \le 2300$), $\mathrm{Nu_H}$ is independent of Pr and has a value of $\frac{48}{11}$ given by Eq. (3.22) with $\gamma = 0$.

TABLE 4.4. Fully Developed Turbulent Flow Nusselt Numbers Nu_T and Nu_H[†] in a Smooth Circular Duct for Gases and Liquids ($Pr > 0.5$)[‡]

Investigators	Correlations	Remarks
Reynolds [11]	$Nu = (f/2) Re\, Pr$	This equation, based on a single-layer model, is inferred from [11]. It is sometimes referred to as the "Reynolds analogy" and is theoretically valid for $Pr = 1$.
Nusselt [81]	$Nu = 0.024\, Re^{0.786} Pr^{0.45}$	This correlation, originated by Nusselt, has been modified by a number of investigators. For $Pr < 1$ and $10^3 \le Re \le 10^6$, its predictions are within $+4.4\%$ and -6.3% of the Gnielinski correlation (see below).
Prandtl [54], Taylor [55]	$Nu = \dfrac{(f/2) Re\, Pr}{1 + 5(f/2)^{1/2}(Pr - 1)}$	This equation is based on a two-layer model (laminar sublayer and turbulent core). It was derived independently by Prandtl in 1910 and by Taylor in 1916. For $Pr \le 10$ and $5 \times 10^3 \le Re \le 5 \times 10^6$, its predictions are within $+14.9\%$ and -11.1% of the Gnielinski correlation.
Dittus and Boelter [82]	$Nu = \begin{cases} 0.024\, Re^{0.8} Pr^{0.4} & \text{for heating} \\ 0.026\, Re^{0.8} Pr^{0.3} & \text{for cooling} \end{cases}$	These classical correlations were developed for $0.7 \le Pr \le 120$ and $2500 \le Re \le 1.24 \times 10^5$. The objective of providing different correlations for heating and cooling was to account for variation of the fluid properties with temperature. Compared to the Gnielinski correlation, predictions of the heating correlation for $10^4 \le Re \le 1.24 \times 10^5$ are: (1) 13.5% to 17% higher for air ($Pr = 0.7$), (2) 15% lower to 7% higher for water ($3 \le Pr \le 10$), and (3) 10% lower to 21% higher for oil ($Pr = 120$). Predictions of the cooling correlation for $10^4 \le Re \le 1.24 \times 10^5$ are: (1) 29% to 33% higher for air ($Pr = 0.7$), (2) 26% lower to 3% higher for water ($3 \le Pr \le 10$), and (3) 39% to 18% lower for oil ($Pr = 120$). For $Re = 2500$ and $0.7 \le Pr \le 120$, predictions of the two correlations are unacceptable being up to 94% higher than those of the Gnielinski correlation.
Kraussold [83]	$Nu = 0.024\, Re^{0.8} Pr^{1/3}$	For $0.7 \le Pr \le 5$ and $10^4 \le Re \le 10^5$, the pre-

Colburn [84]	$\mathrm{Nu} = (f/2)\mathrm{Re}\,\mathrm{Pr}^{1/3}$ $\mathrm{Nu} = 0.023\,\mathrm{Re}^{0.8}\mathrm{Pr}^{1/3}$	For $0.5 \le \mathrm{Pr} \le 3$ and $10^4 \le \mathrm{Re} \le 10^5$, both correlations predict Nu within +27.6% and −19.8% of the Gnielinski correlation.
von Kármán [14]	$\mathrm{Nu} = \dfrac{(f/2)\mathrm{Re}\,\mathrm{Pr}}{1 + 5(f/2)^{1/2}\left[\mathrm{Pr} - 1 + \ln\left(\dfrac{5\,\mathrm{Pr} + 1}{6}\right)\right]}$	This equation is based on a three-layer model (laminar sublayer, buffer layer, and turbulent core as shown in Fig. 4.1). For $0.5 \le \mathrm{Pr} \le 10$ and $10^4 \le \mathrm{Re} \le 5 \times 10^6$, its predictions are within +16.2% and −11% of the Gnielinski correlation.
Hausen [85]	$\mathrm{Nu} = 0.116(\mathrm{Re}^{2/3} - 125)\mathrm{Pr}^{1/3}[1 + (x/D)^{-2/3}]$	This correlation takes into account the thermal entrance length effect embodied in the last, bracketed factor. With this factor ignored, i.e., set equal to 1, its predictions for $0.7 \le \mathrm{Pr} \le 3$ and $10^4 \le \mathrm{Re} \le 10^5$ are within +22.1% and −16.2% of the Gnielinski correlation.
Prandtl [62]	$\mathrm{Nu} = \dfrac{(f/2)\mathrm{Re}\,\mathrm{Pr}}{1 + 8.7(f/2)^{1/2}(\mathrm{Pr} - 1)}$	This equation is based on the three-layer model (laminar sublayer, buffer layer, and turbulent core as shown in Fig.4.1). For $0.5 \le \mathrm{Pr} \le 5$ and $10^4 \le \mathrm{Re} \le 5 \times 10^6$, its predictions are within +10.5% and −17.7% of the Gnielinski correlation.
Drexel and McAdams [86]	$\mathrm{Nu} = 0.021\,\mathrm{Re}^{0.8}\mathrm{Pr}^{0.4}$	For $\mathrm{Pr} \le 0.7$ and $10^4 \le \mathrm{Re} \le 5 \times 10^6$, the predictions are within +13.7% and −3.6% of the Gnielinski correlation.
Bernado and Eian [87]	$\mathrm{Nu} = 0.048\,\mathrm{Re}^{0.73}\mathrm{Pr}^{0.4}$	For $0.5 \le \mathrm{Pr} \le 2000$ and $10^4 \le \mathrm{Re} \le 10^5$, the predictions are within +23.4% and −22.8% of the Gnielinski correlation.
Deissler [57]	Graphical results for $0.73 \le \mathrm{Pr} \le 3000$ and $5 \times 10^3 < \mathrm{Re} < 3 \times 10^5$; for $\mathrm{Pr} > 200$, the asymptotic formula: $\mathrm{Nu} = 0.0789\,\mathrm{Re}\sqrt{f}\,\mathrm{Pr}^{1/4}$	For $5 \le \mathrm{Pr} \le 2000$ and $10^4 \le \mathrm{Re} \le 10^5$, the graphical results are within +20% of Gnielinski. For $\mathrm{Pr} = 0.73$ and $10^4 \le \mathrm{Re} \le 10^5$, the results are 54% higher than those of Gnielinski.
Friend and Metzner [88]	$\mathrm{Nu} = \dfrac{(f/2)\mathrm{Re}\,\mathrm{Pr}}{1.2 + 11.8(f/2)^{1/2}(\mathrm{Pr} - 1)\mathrm{Pr}^{-1/3}}$	This equation is based on the experimental data for $50 < \mathrm{Pr} < 600$. For $50 \le \mathrm{Pr} \le 600$ and $5 \times 10^4 \le \mathrm{Re} \le 5 \times 10^6$, its predictions are within +7.8% and −1.7% of the Gnielinski correlation.

TABLE 4.4. Continued

Investigators	Correlations	Remarks
Petukhov, Kirillov, and Popov [89]	$\mathrm{Nu} = \dfrac{(f/2)\mathrm{Re}\,\mathrm{Pr}}{C + 12.7(f/2)^{1/2}(\mathrm{Pr}^{2/3} - 1)}$ where $C = 1.07 + 900/\mathrm{Re} - [0.63/(1 + 10\,\mathrm{Pr})]$ $\mathrm{Nu} = \dfrac{(f/2)\mathrm{Re}\,\mathrm{Pr}}{1.07 + 12.7(f/2)^{1/2}(\mathrm{Pr}^{2/3} - 1)}$	The first Petukhov et al. correlation agrees with the most reliable experimental data on heat and mass transfer to an accuracy of $\pm 5\%$. It is valid for $0.5 \le \mathrm{Pr} \le 10^6$ and $4000 \le \mathrm{Re} \le 5 \times 10^6$. The second is a simplified version of the first and is modified by Gnielinski (see below) to arrive at his correlation.
Hausen [90]	$\mathrm{Nu} = 0.037(\mathrm{Re}^{0.75} - 180)\mathrm{Pr}^{0.42}[1 + (x/D)^{-2/3}]$	This correlation takes into account the thermal entrance effect embodied in the last, bracketed factor. With this factor ignored, i.e., set equal to 1, its predictions for $0.7 \le \mathrm{Pr} \le 3$ and $10^4 \le \mathrm{Re} \le 10^5$ are within $+3.8\%$ and -21.1% of the Gnielinski correlation.
Mills [91]	$\mathrm{Nu} = 0.0397\,\mathrm{Re}^{0.73}\mathrm{Pr}^{1/3}$	For $\mathrm{Pr} \approx 0.7$ and $10^4 \le \mathrm{Re} \le 10^5$, the predictions are within $+5\%$ and -13.5% of the Gnielinski correlation.
Webb [92]	$\mathrm{Nu} = \dfrac{(f/2)\mathrm{Re}\,\mathrm{Pr}}{1.07 + 9(f/2)^{1/2}(\mathrm{Pr} - 1)\mathrm{Pr}^{1/4}}$	For $0.5 \le \mathrm{Pr} \le 100$ and $10^4 \le \mathrm{Re} \le 5 \times 10^6$, the predictions are within $+10.4\%$ and -7.3% of the Gnielinski correlation.
Sleicher and Rouse [93]	$\mathrm{Nu} = 5 + 0.015\,\mathrm{Re}^m\mathrm{Pr}^n$ $m = 0.88 - 0.24/(4 + \mathrm{Pr})$ $n = \frac{1}{3} + 0.5\exp(-0.6\,\mathrm{Pr})$ $\mathrm{Nu} = 5 + 0.012\,\mathrm{Re}^{0.83}(\mathrm{Pr} + 0.29)$	The predictions of the first correlation for $0.5 \le \mathrm{Pr} \le 2000$ and $5 \times 10^4 \le \mathrm{Re} \le 5 \times 10^6$ are within $+43\%$ and -10.3% of the Gnielinski correlation. The second correlation is applicable to gases only. For $\mathrm{Pr} \approx 0.7$ and $10^4 \le \mathrm{Re} \le 5 \times 10^6$ its predictions are within $+4\%$ and -12.5% of the Gnielinski correlation.
Gnielinski [94]	$\mathrm{Nu} = \dfrac{(f/2)(\mathrm{Re} - 1000)\mathrm{Pr}}{1 + 12.7(f/2)^{1/2}(\mathrm{Pr}^{2/3} - 1)}$ $\mathrm{Nu} = 0.0214(\mathrm{Re}^{0.8} - 100)\mathrm{Pr}^{0.4}$ $\mathrm{Nu} = 0.012(\mathrm{Re}^{0.87} - 280)\mathrm{Pr}^{0.4}$	The first Gnielinski correlation is a modified version of the second Petukhov et al. [89] correlation (see above) extending it to the $2300 \le \mathrm{Re} \le 5 \times 10^4$ range. For $0.5 \le \mathrm{Pr} \le 2000$ and $2300 \le \mathrm{Re} \le 5 \times 10^6$, it is in overall best accord with the experimental data; it agrees with the Petukhov et al. correlation within -2% and $+7.8\%$. Hence it is selected as the common basis of comparison for all the correlations in this

		table. The second correlation is for $0.5 \le \Pr \le 1.5$ and $10^4 \le \text{Re} \le 5 \times 10^6$, where it agrees with the first within +4% and −6%. The third correlation is for $1.5 \le \Pr \le 500$ and $3 \times 10^3 \le \text{Re} \le 10^6$, where it agrees with the first within −10%.
Churchill [45]	$(\text{Nu})^{10} = (\text{Nu}_l)^{10} + \left\{ \frac{e^{(2200-\text{Re})/365}}{(\text{Nu}_l)^2} + \frac{1}{(\text{Nu}_t)^2} \right\}^{-5}$ where $\text{Nu}_t = \text{Nu}_0 + \frac{0.079(f/2)^{1/2}\text{Re}\Pr}{(1+\Pr^{4/5})^{5/6}}$; $\text{Nu}_0 = 4.8$ for Ⓣ and 6.3 for Ⓗ; $\text{Nu}_l = 3.657$ for Ⓣ and 4.364 for Ⓗ	This general correlation was constructed for $0 < \Pr < 10^6$ and $10 < \text{Re} \le 10^6$, spanning the laminar, transition, and turbulent flow regimes. For $\text{Re} \le 2100$, it yields the laminar flow Nu values of 3.657 and 4.364 corresponding to the Ⓣ and Ⓗ boundary conditions, respectively, For $2100 \le \text{Re} \le 10^4$, it gives Nu values in agreement with the experimental results for the transition flow. For $0.5 \le \Pr \le 2000$ and $10^4 \le \text{Re} \le 10^6$, its predictions with the Ⓣ boundary conditions are within +17.1% and −11.9% of the Gnielinski correlation; with the Ⓗ boundary conditions, within +13.7% and −10.5%.
Polley [95]	$\text{Nu} = \exp[-3.796 - 0.205 \ln \text{Re} - 0.505 \ln \Pr - 0.0225(\ln \Pr)^2] \text{Re} \Pr$	For $\Pr \approx 0.7$ and $10^4 \le \text{Re} \le 5 \times 10^5$, the predictions are within +5.5% and −8% of the Gnielinski correlation.
Kays and Crawford [22]	$\text{Nu} = \frac{\sqrt{f/2}\,\text{Re}\Pr}{0.833\left[2.25 \ln(0.75\,\text{Re}\sqrt{f/2}) + 13.2\Pr - 5.8\right]}$; $\text{Nu} = 0.022\,\text{Re}^{0.8}\Pr^{0.5}$	The first equation is valid for $0.5 \le \Pr \le 5$, and the second for $0.5 \le \Pr \le 1$. For $10^4 \le \text{Re} \le 5 \times 10^6$, the predictions of the first equation are within +8.2% and −30.2% of the Gnielinski correlation; of the second, within +11.1% and −2.5%.
Sandall et al. [96]	$\text{Nu} = \frac{\sqrt{f/2}\,\text{Re}\Pr}{12.48\Pr^{2/3} - 7.853\Pr^{1/3} + 3.613 \ln \Pr + 5.8 + C}$ where $C = 2.78 \ln(\sqrt{f/8}\,\text{Re}/45)$	For $0.5 \le \Pr < 2000$ and $10^4 \le \text{Re} \le 5 \times 10^6$, the predictions are within +6.6% and −4% of the Gnielinski correlation.

†All the formulas in this table apply to the Ⓣ or Ⓗ boundary condition; hence there is no need to distinguish Nu_T and Nu_H in the table.
‡The friction factor f needed in some of the formulas may be calculated from the PKN, Colebrook, Filonenko, or Techo et al. correlation given in Table 4.2.
For the comparisons presented in Table 4.4, the Techo et al. correlation was employed in the computations.

TABLE 4.5. Fully Developed Turbulent Flow Nusselt Numbers Nu_T and Nu_H† in a Smooth Circular Duct for Liquid Metals ($Pr < 0.1$)

Investigators	Correlations	Remarks
Lyon [97]	$Nu_H = 7.0 + 0.025\,Pe^{0.8}$	For $0 \le Pr \le 0.1$ and $10^4 \le Re \le 5 \times 10^6$, the predictions are within $+32.8\%$ and -6.5% of the Notter-Sleicher correlation (see below).
Seban and Shimazaki [98]	$Nu_T = 5.0 + 0.025\,Pe^{0.8}$	For $0 \le Pr \le 0.1$ and $10^4 \le Re \le 5 \times 10^6$, the predictions are within $+39.9\%$ of the Notter-Sleicher correlation (see below).
Lubarsky and Kaufman [99]	$Nu_H = 0.625\,Pe^{0.4}$	For $0 \le Pr \le 0.1$ and $10^4 \le Re \le 10^5$, the predictions are within -42.7% of the Notter-Sleicher correlation.
Hartnett and Irvine [216]	$Nu = \frac{2}{3} Nu_{slug} + 0.015\,Pe^{0.8}$ ‡ where $Nu_{slug} = 5.78$ for Ⓣ $Nu_{slug} = 8.00$ for Ⓗ	For $0 \le Pr \le 0.1$ and $10^4 \le Re \le 5 \times 10^6$, the predictions of the Nu_T correlation are within -39.9% of the Notter-Sleicher correlation; the predictions of the Nu_H correlation are within -43.5% of the Notter-Sleicher correlation.
Sleicher and Tribus [100]	$Nu_T = 4.8 + 0.015\,Re^{0.91} Pr^{1.21}$ $Nu_H = 6.3 + 0.016\,Re^{0.91} Pr^{1.21}$	For $0 \le Pr \le 0.1$ and $10^4 \le Re \le 5 \times 10^6$, the predictions of the Nu_T correlation are within $+19.5\%$ and -33.4% of the Notter-Sleicher correlation; the predictions of the Nu_H correlation are within $+26.3\%$ and -32.5% of the Notter-Sleicher correlation.
Azer and Chao [101]	$Nu_T = 5 + 0.05\,Re^{0.77} Pr^{1.02}$	For $0 \le Pr \le 0.1$ and $10^4 \le Re \le 5 \times 10^5$, the predictions are within $+14.2\%$ and -18.6% of the Notter-Sleicher correlation.

Dwyer [102]	$\mathrm{Nu_H} = 7 + 0.025\left[\mathrm{Re\,Pr} - \frac{1.82\,\mathrm{Re}}{(\epsilon_m/\nu)_{max}^{0.14}}\right]^{0.8}$ where $(\epsilon_m/\nu)_{max} = 0.037\,\mathrm{Re}\sqrt{f}$[§]	For $0 \le \mathrm{Pr} \le 0.1$ and $10^4 \le \mathrm{Re} \le 5 \times 10^6$, the predictions are within +31.4% and −6.5% of the Notter-Sleicher correlation.
Skupinski et al. [103]	$\mathrm{Nu_H} = 4.82 + 0.0185\,\mathrm{Pe}^{0.827}$	For $0 \le \mathrm{Pr} \le 0.1$ and $10^4 \le \mathrm{Re} \le 5 \times 10^6$, the predictions are within +22.3% and −17.8% of the Notter-Sleicher correlation.
Notter and Sleicher [104]	$\mathrm{Nu_T} = 4.8 + 0.0156\,\mathrm{Re}^{0.85}\mathrm{Pr}^{0.93}$ $\mathrm{Nu_H} = 6.3 + 0.0167\,\mathrm{Re}^{0.85}\mathrm{Pr}^{0.93}$	These correlations are valid for $0.004 < \mathrm{Pr} < 0.1$ and $10^4 < \mathrm{Re} < 10^6$. They are based on numerical analysis [104] coupled with experimental verification [105].
Chen and Chiou [106]	$\mathrm{Nu_T} = 4.5 + 0.0156\,\mathrm{Re}^{0.85}\mathrm{Pr}^{0.86}$ $\mathrm{Nu_H} = 5.6 + 0.0165\,\mathrm{Re}^{0.85}\mathrm{Pr}^{0.86}$	For $0 \le \mathrm{Pr} \le 0.1$ and $10^4 \le \mathrm{Re} \le 5 \times 10^6$, the predictions of the $\mathrm{Nu_T}$ correlation are within +36.1% and −1.8% of the Notter-Sleicher correlation; the predictions of the $\mathrm{Nu_H}$ correlation are within +33.9% and −7.1% of the Notter-Sleicher correlation.
Lee [107]	$\mathrm{Nu_H} = 3.01\,\mathrm{Re}^{0.0833}$	This correlation is valid for $0.001 \le \mathrm{Pr} \le 0.02$ and $5 \times 10^3 \le \mathrm{Re} \le 10^5$, where its predictions are within +24.7% and −44.3% of the Notter-Sleicher correlation.

[†] For $\mathrm{Pr} < 0.1$, different formulas are required corresponding to the (T) and (H) boundary conditions; hence $\mathrm{Nu_T}$ and $\mathrm{Nu_H}$ are distinguished in this table.

[‡] This equation is of general applicability, requiring the knowledge of the slug Nusselt number $\mathrm{Nu_{slug}}$ corresponding to $\mathrm{Pr} = 0$ for the specific duct geometry and the pertinent boundary condition.

[§] This relation was derived by the present authors. Its predictions are in excellent accord with the values of $(\epsilon_m/\nu)_{max}$ reported graphically in [102]. The value of f in this relation may be computed from the Techo et al. correlation presented in Table 4.2.

TABLE 4.6. Fully Developed Turbulent Flow Nusselt Numbers Nu_H in a Smooth Circular Duct with Uniform Wall Heat Flux

	Nu_H				
Pr	$Re = 10^4$	5×10^4	10^5	5×10^5	10^6
0.0	6.30	6.30	6.30	6.30	6.30
0.001	6.37	6.57	6.78	8.19	9.71
0.003	6.49	7.04	7.64	11.56	15.77
0.01	6.88	8.57	10.4	22.4	35.3
0.03	7.91	12.6	17.7	51.0	86.9
0.06	9.4	18.3	28.0	91.5	159.9
0.09	10.8	23.9	37.9	130.6	230.3
0.5	24.7	84.8	144.2	510.3	891.4
0.7	29.4	104.5	179.8	648.8	1141
1.0	34.9	128.7	224.0	825.5	1462
3.0	56.5	226.8	406.6	1590	2877
10	90.0	381.9	700.0	2865	5280
30	133.5	582.1	1079	4526	8427
100	202.4	895.5	1670	7108	13,313
1000	440.1	1967	3683	15,829	29,774

The fully developed Nu_T for fluids with $Pr \geq 0.7$ are practically identical with the corresponding Nu_H shown in Fig. 4.12. This can be seen more clearly from the results in Table 4.7, based on the theoretical calculations by Siegel and Sparrow [108]. The fully developed Nu_T for low-Pr liquid metals, on the other hand, are significantly lower than the corresponding Nu_H shown in Fig. 4.12. This can be seen more clearly from the results in Table 4.8, based on the Notter-Sleicher correlations given in Table 4.5.

All the preceding heat transfer results pertain to uniform heat flux or uniform wall temperature around the duct circumference. The technically important problem of circumferentially varying but axially constant wall heat flux has been solved by Reynolds [109], Sparrow and Lin [110], and Gartner et al. [111]. In addition, the problem of circumferentially varying but axially constant wall temperature has been solved in Refs. 109 and 110. The only available experimental measurements for the aforementioned two circumferentially varying thermal boundary conditions are those of Black and Sparrow [112].

Based on the analysis by Gartner et al. [111], for the cosine heat flux variation

$$q''_w(\theta) = q''_m(1 + b\cos\theta) \tag{4.61}$$

at a given cross section of the duct, the local Nusselt number around the circumference can be determined from

$$Nu_{\theta,H} = \frac{1 + b\cos\theta}{1/Nu_H + (G_1 b/2)\cos\theta} \tag{4.62}$$

where Nu_H is the value for a uniform wall heat flux and b is the specified constant.

The circumferential heat flux function G_1 appearing in Eq. (4.62) is presented in Table 4.9 for wide ranges of Re and Pr. For laminar flow ($Re \leq 2300$), the value of G_1 is unity for all values of Pr. With $G_1 = 1$, Eq. (4.62) becomes identical with Eq. (3.24) for the laminar flow case.

The effect of surface roughness on the fully developed Nusselt number can be taken into account by the simple empirical correlation given by Eq. (4.26). Several other correlations have been proposed for determining the fully developed Nusselt numbers in the completely rough flow regime in a circular duct. They are summarized in Table 4.10. The friction factor f in this table is for fully rough flow and is given by the

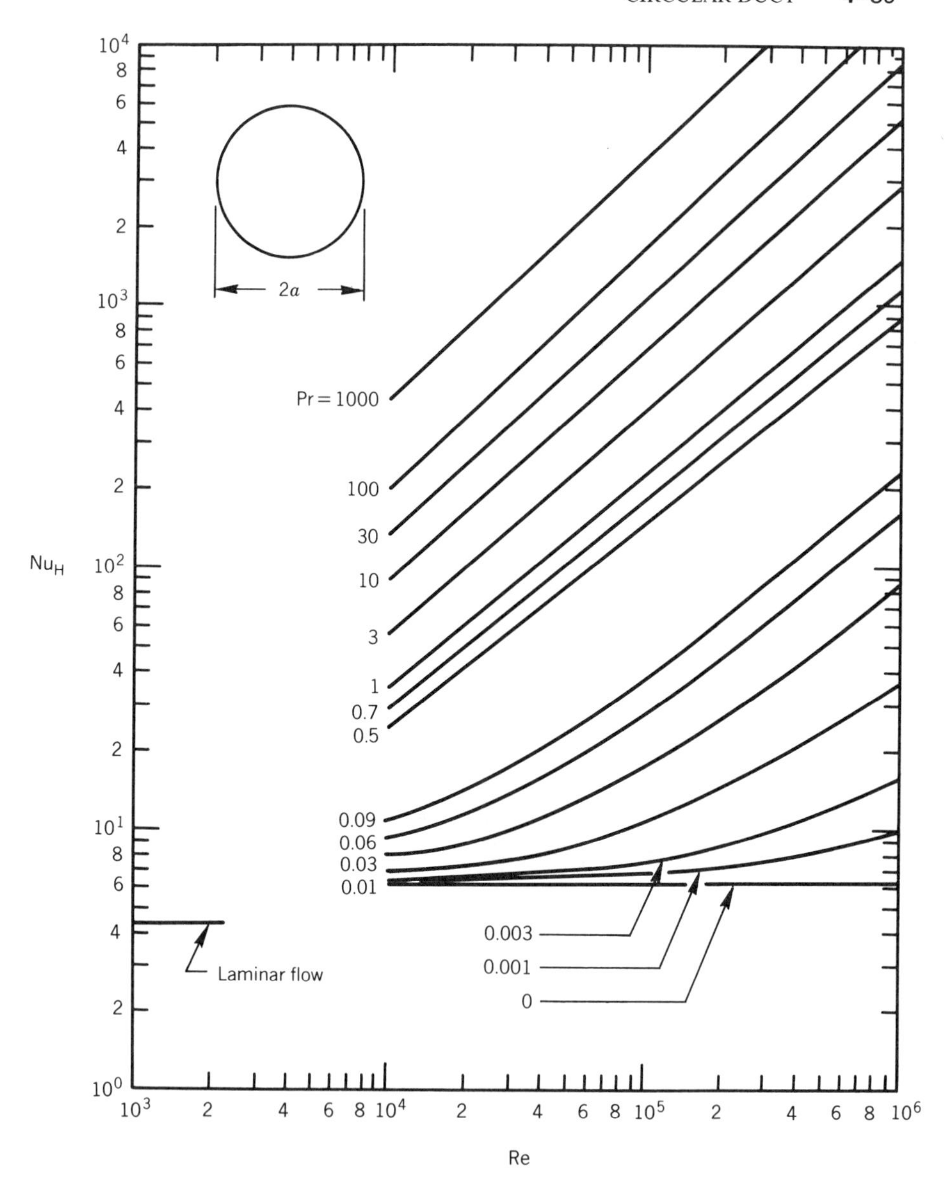

Figure 4.12. Nusselt numbers Nu_H for fully developed turbulent flow in a smooth circular duct.

Nikuradse correlation presented in Table 4.3. The recommended equations for practical calculations are the two correlations by the present authors given in Table 4.10.

4.2.3 Hydrodynamically Developing Flow

Several attempts have been made to solve the problem of turbulent flow development in a smooth circular duct starting with a uniform velocity at the duct inlet. This latter velocity can be obtained by providing a boundary-layer tripping device (e.g., a tripping ring, strip, or sandpaper) just upstream of the duct inlet. All the attempted solutions fall into the category of the integral method. Latzko [118], Ross [119], and Zhi-qing [120] solved the problem analytically by assuming a power–law velocity profile within

TABLE 4.7. Comparison of Fully Developed Turbulent Flow Nusselt Numbers Nu_H and Nu_T for a Smooth Circular Duct and High Prandtl Number ($Pr \geq 0.7$) Fluids [108]

Pr	Nu_H	Nu_T	Nu_H/Nu_T
	$Re = 5 \times 10^4$		
0.7	101.8	98.95	1.029
10	381.0	379.70	1.003
100	836.6	834.90	1.002
	$Re = 10^5$		
0.7	171.4	167.7	1.022
10	683.9	683.9	1.000
100	1529.0	1532.0	0.998
	$Re = 5 \times 10^5$		
0.7	599.7	592.9	1.011
10	2722.0	2730.0	0.997
100	6352.0	6386.0	0.995

TABLE 4.8. Comparison of Fully Developed Turbulent Flow Nusselt Numbers Nu_H and Nu_T for a Smooth Circular Duct and Low Prandtl Number ($Pr < 0.1$) Fluids

Pr	Nu_H	Nu_T	Nu_H/Nu_T
	$Re = 10^4$		
0.1	11.23	9.40	1.194
0.01	6.88	5.34	1.288
0.001	6.37	4.86	1.309
0.000	6.30	4.80	1.313
	$Re = 10^5$		
0.1	41.19	37.39	1.102
0.01	10.40	8.63	1.205
0.001	6.78	5.25	1.292
0.000	6.30	4.80	1.313
	$Re = 10^6$		
0.1	253.31	235.54	1.075
0.01	35.32	31.91	1.107
0.001	9.71	7.99	1.216
0.000	6.30	4.80	1.313

TABLE 4.9. Circumferential Heat Flux Function G_1 for Use in Conjunction with Eq. (4.62) [111]

Pr	G_1				
	Re = 10^4	3×10^4	10^5	3×10^5	10^6
0	1.000	1.000	1.000	1.000	1.000
0.001	0.9989	0.9937	0.9561	0.8059	0.4853
0.003	0.9929	0.9613	0.8005	0.5042	0.2300
0.01	0.9499	0.7915	0.4567	0.2185	0.0867
0.03	0.7794	0.4705	0.2055	0.0888	0.0336
0.7	0.1116	0.0442	0.0161	0.00644	0.00232
3	0.0440	0.0168	0.00594	0.00234	0.000824
10	0.0233	0.00879	0.00305	0.00120	0.000415
30	0.0145	0.00544	0.00184	0.000744	0.000250
100	0.00941	0.00354	0.00110	0.000484	0.000149

the boundary layer. Bowlus and Brighton [121] and Na and Lu [122] solved it numerically, by assuming a power-law velocity profile within the boundary layer. Holdhusen [123], Deissler [124], and Filippov [125] employed a logarithmic velocity profile within the boundary layer.

The closed-form analytical solution of Zhi-qing [120] is particularly suitable for engineering calculations and so is presented below. According to this analysis, the velocity distribution in the hydrodynamic entrance region is given by

$$\frac{u}{u_{\max}} = \begin{cases} (y/\delta)^{1/7} & \text{for } 0 \le y \le \delta \\ 1 & \text{for } \delta \le y \le a \end{cases} \tag{4.63}$$

$$\frac{u_m}{u_{\max}} = 1 - \frac{1}{4}\left(\frac{\delta}{a}\right) + \frac{1}{15}\left(\frac{\delta}{a}\right)^2 \tag{4.64}$$

where δ is the hydrodynamic boundary-layer thickness, which varies with the axial coordinate x in accordance with the relation

$$\frac{x/D_h}{\mathrm{Re}^{1/4}} = 1.4039\left(\frac{\delta}{a}\right)^{5/4}\left[1 + 0.1577\left(\frac{\delta}{a}\right) - 0.1793\left(\frac{\delta}{a}\right)^2 - 0.0168\left(\frac{\delta}{a}\right)^3 + 0.0064\left(\frac{\delta}{a}\right)^4\right] \tag{4.65}$$

The axial pressure drop Δp^*, the incremental pressure drop number $K(x)$, and the apparent Fanning friction factor f_{app} are given as

$$\Delta p^* = \left(\frac{u_{\max}}{u_m}\right)^2 - 1 \tag{4.66}$$

$$K(x) = \Delta p^* - 0.316\frac{x/D_h}{\mathrm{Re}^{1/4}} \tag{4.67}$$

$$f_{\text{app}}\,\mathrm{Re}^{1/4} = \frac{\Delta p^*}{4x/(D_h\,\mathrm{Re}^{0.25})} \tag{4.68}$$

TABLE 4.10. Nusselt Numbers for Fully Developed Turbulent Flow in the Fully Rough Flow Regime of a Circular Duct†

Investigators	Corrrelations	Remarks
Martinelli [15]	$\mathrm{Nu} = \dfrac{\mathrm{Re}\,\mathrm{Pr}\sqrt{f/2}}{5\left[\mathrm{Pr} + \ln(1 + 5\,\mathrm{Pr}) + 0.5\ln\left(\mathrm{Re}\sqrt{f/2}\,/60\right)\right]}$	This equation differs from that derived by Martinelli [15] for a smooth duct by the omission of the temperature ratio $(T_w - T_c)/(T_w - T_m)$.
Nunner [113]	$\mathrm{Nu} = \dfrac{\mathrm{Re}\,\mathrm{Pr}(f/2)}{1 + 1.5\,\mathrm{Re}^{-1/8}\mathrm{Pr}^{-1/6}\left[\mathrm{Pr}(f/f_s) - 1\right]}$	This correlation is valid for $\mathrm{Pr} \approx 0.7$; it does not give satisfactory results for $\mathrm{Pr} > 1$.
Dipprey and Sabersky [114]	$\mathrm{Nu} = \dfrac{\mathrm{Re}\,\mathrm{Pr}(f/2)}{1 + \sqrt{f/2}\left[5.19\,\mathrm{Re}_\varepsilon^{0.2}\mathrm{Pr}^{0.44} - 8.48\right]}$	This correlation is valid for $0.0024 \le \varepsilon/D_h \le 0.049$, $1.2 \le \mathrm{Pr} \le 5.94$, and $1.4 \times 10^4 \le \mathrm{Re} \le 5 \times 10^5$.
Owen and Thomson [60]	$\mathrm{Nu} = \dfrac{\mathrm{Re}\,\mathrm{Pr}(f/2)}{1 + 0.52\,\mathrm{Re}_\varepsilon^{0.45}\mathrm{Pr}^{0.8}\sqrt{f/2}}$	This equation correlates experimental results from various sources including those from Refs. [113] and [114].
Gowen and Smith [115]	$\mathrm{Nu} = \dfrac{\mathrm{Re}\,\mathrm{Pr}\sqrt{f/2}}{4.5 + \left[0.155\left(\mathrm{Re}\sqrt{f/2}\right)^{0.54} + \sqrt{2/f}\right]\sqrt{\mathrm{Pr}}}$	This correlation is valid for $0.021 \le \varepsilon/D_h \le 0.095$, $0.7 \le \mathrm{Pr} \le 14.3$, and $10^4 \le \mathrm{Re} \le 5 \times 10^4$.
Kawase and Ulbrecht [116]	$\mathrm{Nu} = 0.0523\,\mathrm{Re}\sqrt{\mathrm{Pr}}\,\sqrt{f}$	The predictions of this correlation are somewhat lower than those of the following correlation.
Kawase and De [117]	$\mathrm{Nu} = 0.0471\,\mathrm{Re}\sqrt{\mathrm{Pr}}\,\sqrt{f}(1.11 + 0.44\,\mathrm{Pr}^{-1/3} - 0.7\,\mathrm{Pr}^{-1/6})$	The predictions of this correlation are in reasonable agreement with the experimental data for $0.0024 < \varepsilon/D_h < 0.165$, $5.1 < \mathrm{Pr} < 390$, and $5000 < \mathrm{Re} < 5 \times 10^5$.
Present authors	$\mathrm{Nu} = \dfrac{(\mathrm{Re}\,\mathrm{Pr}(f/2))}{1 + \sqrt{f/2}\left(4.5\,\mathrm{Re}_\varepsilon^{0.2}\mathrm{Pr}^{0.5} - 8.48\right)}$	This correlation is valid for $0.5 < \mathrm{Pr} < 10$, $0.002 < \varepsilon/D_h < 0.05$, and $\mathrm{Re} > 10^4$. Its predictions are within $\pm 5\%$ of the available measurements.
Present authors	$\mathrm{Nu} = \dfrac{(\mathrm{Re} - 1000)\,\mathrm{Pr}(f/2)}{1 + \sqrt{f/2}\left[\left(17.42 - 13.77\mathrm{Pr}_t^{0.8}\right)\mathrm{Re}_\varepsilon^{0.5} - 8.48\right]}$ where Pr_t is given by Eq. (4.56).	This correlation is valid for $0.5 < \mathrm{Pr} < 5000$, $0.001 < \varepsilon/d_h < 0.05$, and $\mathrm{Re} > 2300$. Its predictions are within $\pm 15\%$ of the available measurements.

† The friction factor f needed to evaluate Nu may be calculated from Nikuradse's correlation in Table 4.3 for fully rough flow regime.

TABLE 4.11. Turbulent Momentum Transfer Results in the Hydrodynamic Entrance Region of a Smooth Circular Duct

$\frac{x/D_h}{\mathrm{Re}^{1/4}}$	$\frac{u_{max}}{u_m}$	Δp^*	$K(x)$	$f_{app}\,\mathrm{Re}^{1/4}$	$\frac{\delta}{a}$
0.0000	1.0000	0.0000	0.0000	∞	0.0
0.0800	1.0249	0.0505	0.0252	0.1578	0.1
0.1923	1.0497	0.1018	0.0410	0.1325	0.2
0.3213	1.0741	0.1537	0.0522	0.1196	0.3
0.4615	1.0981	0.2058	0.0600	0.1115	0.4
0.6093	1.1215	0.2577	0.0652	0.1057	0.5
0.7616	1.1442	0.3091	0.0684	0.1015	0.6
0.9153	1.1659	0.3594	0.0702	0.0982	0.7
1.0679	1.1867	0.4082	0.0700	0.0956	0.8
1.2167	1.2063	0.4551	0.0700	0.0935	0.9
1.3590	1.2245	0.4994	0.0700	0.0919	1.0

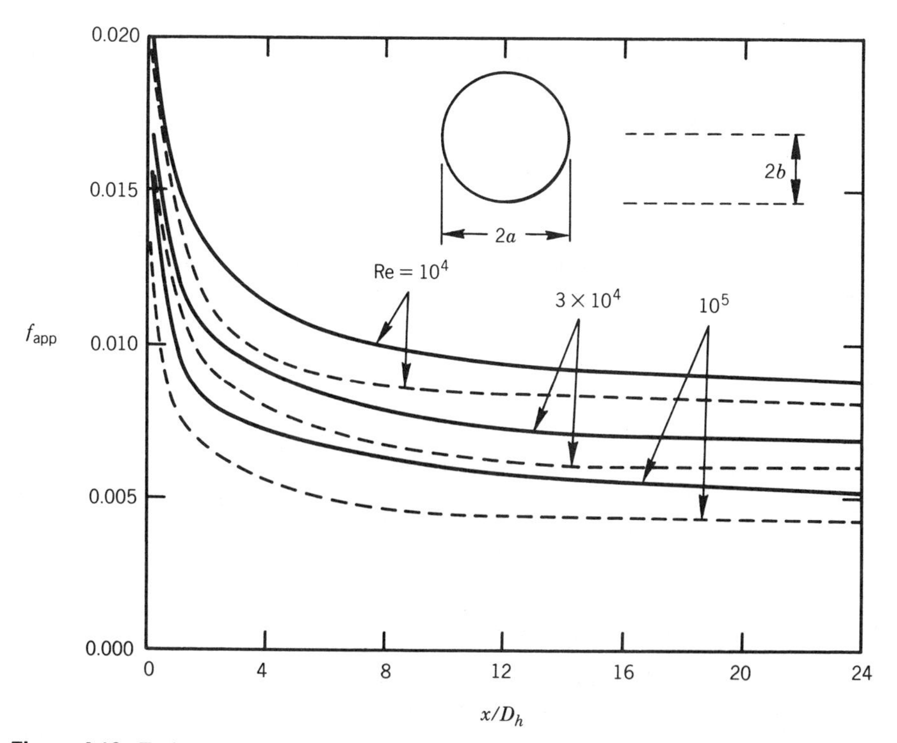

Figure 4.13. Turbulent flow apparent friction factors in the hydrodynamic entrance region of a smooth circular and a flat duct with uniform inlet velocity.

TABLE 4.12. Turbulent Flow Hydrodynamic Entrance Length Predictions for a Smooth Circular Duct

Investigators	Formula for L_{hy}/D_h	L_{hy}/D_h at $Re = 3.88 \times 10^5$
Latzko [118]	$0.625\,Re^{1/4}$	15.6
Zhi-qing [120]	$1.3590\,Re^{1/4}$	33.9
Bowlus and Brighton [121]	$6.1887 \ln Re - 46$	33.6
Na and Lu [122]	—	28.5
Holdhusen [123]	$2.0846 \ln Re - 5.6$	21.2
Filippov [125]	$2.44/\sqrt{f}$	41.6†
Barbin and Jones [126]	—	28.0‡

†Calculated using Techo et al. [67] formula for f from Table 4.2.
‡Experimental value.

The various momentum-transfer quantities calculated from Eqs. (4.64) to (4.68) by the present authors are presented in Table 4.11. A comparison of u_{max}/u_m values in Table 4.11 with the accurate experimental measurements of Barbin and Jones [126] for $Re = 3.88 \times 10^5$ shows agreement within $\pm 5\%$. The f_{app} values from Table 4.11 are also plotted in Fig. 4.13 for $Re = 10^4$, 3×10^4, and 10^5.

For the present purpose, the hydrodynamic entrance length L_{hy} is defined as the axial distance at which the hydrodynamic boundary layer growing from the duct wall reaches the duct axis. The L_{hy} predictions of various investigators in conformity with this definition are presented in Table 4.12. It is seen that the L_{hy}/D_h value at $Re = 3.88 \times 10^5$ due to Na and Lu [122] is in almost exact agreement with the experimental value of Barbin and Jones [126]. However, since Na and Lu [122] provide no general formula for L_{hy}/D_h, the use of Zhi-qing's [120] or Bowlus and Brighton's [121] formula is recommended for practical calculations; these formulas provide fairly good agreement with the experimental value.

Wang and Tullis [127] mathematically analyzed the turbulent flow in the hydrodynamic entrance region of a rough circular duct, employing a logarithmic velocity profile within the boundary layer. In addition, they reported experimental measurements of the pressure drop and velocity distribution for $0.00038 < \varepsilon/D_h < 0.0002$ and $7 \times 10^5 < Re < 3.7 \times 10^6$. The predictions of the boundary-layer growth, axial velocity, and pressure drop agree reasonably well with the measurements up to $x/D_h = 12$ but not for $x/D_h > 12$. Based on their measurements, Wang and Tullis [127] concluded that f_{app}/f for a rough circular duct is essentially independent of ε/D_h in the ranges $1.2 \times 10^6 < Re < 3.7 \times 10^6$ and $0.00038 < \varepsilon/D_h < 0.0002$.

4.2.4 Thermally Developing Flow

The problem of the fully developed turbulent velocity profile and developing temperature profile with uniform wall temperature in a smooth circular duct has been studied analytically quite extensively. This Ⓣ problem is often referred to as the *turbulent Graetz problem*.

Latzko [118] appears to be the first investigator to attack the turbulent Graetz problem, in 1921. Sleicher and Tribus [128] developed a solution primarily for $Pr < 1$. Becker [129] developed a solution for $0.1 < Pr < 100$. Notter and Sleicher [104] provided solutions for numerous Pr values in the range $0 < Pr < 10^4$. All these solutions are of infinite-series type, similar to the Graetz-problem solution for laminar

TABLE 4.13. Eigenvalues and Constants of the Turbulent Graetz Problem for $0 \leq Pr \leq 0.06$ [104]

Re	λ_0^2	λ_1^2	λ_2^2	λ_3^2	C_0	$-C_1$	C_2	$-C_3$	G_0	G_1	G_2	G_3
						Pr = 0						
10,000	9.87	54.27	135.1	252.4	1.559	0.973	0.750	0.623	0.910	0.824	0.769	0.726
20,000	10.03	54.95	136.4	254.5	1.564	0.986	0.769	0.647	0.918	0.849	0.812	0.786
50,000	10.16	55.48	137.5	256.4	1.568	0.996	0.782	0.661	0.926	0.867	0.839	0.822
100,000	10.25	55.86	138.4	257.8	1.571	1.001	0.787	0.668	0.931	0.876	0.852	0.837
200,000	10.35	56.27	139.3	259.4	1.573	1.006	0.793	0.673	0.935	0.886	0.865	0.853
500,000	10.48	56.82	140.5	261.7	1.575	1.011	0.798	0.678	0.940	0.895	0.876	0.865
						Pr = 0.002						
50,000	10.61	58.19	144.6	269.7	1.564	0.986	0.770	0.650	0.975	0.897	0.862	0.841
100,000	11.23	61.74	153.6	286.7	1.561	0.980	0.763	0.644	1.036	0.943	0.903	0.880
200,000	12.44	68.90	172.0	321.4	1.554	0.963	0.745	0.627	1.159	1.027	0.972	0.942
500,000	15.91	90.19	227.4	426.8	1.531	0.913	0.691	0.576	1.527	1.246	1.137	1.080
						Pr = 0.004						
50,000	11.23	61.94	154.3	288.1	1.558	0.972	0.755	0.636	1.041	0.939	0.894	0.867
100,000	12.52	69.57	173.9	325.3	1.550	0.955	0.735	0.618	1.175	1.028	0.967	0.932
200,000	15.03	84.91	213.8	400.9	1.534	0.920	0.698	0.582	1.441	1.191	1.092	1.040
500,000	21.96	129.6	331.9	626.8	1.497	0.840	0.616	0.506	2.203	1.589	1.376	1.277
1,000,000	32.57	202.6	527.7	1003.0	1.457	0.760	0.540	0.439	3.410	2.130	1.750	1.594

TABLE 4.13 Continued

Re	λ_0^2	λ_1^2	λ_2^2	λ_3^2	C_0	$-C_1$	C_2	$-C_3$	G_0	G_1	G_2	G_3
						Pr = 0.01						
10,000	10.41	57.61	143.8	269.2	1.554	0.957	0.732	0.606	0.970	0.857	0.788	0.738
20,000	11.24	62.33	155.6	290.9	1.552	0.958	0.738	0.616	1.050	0.928	0.868	0.829
50,000	13.39	75.23	188.9	353.9	1.540	0.932	0.710	0.593	1.277	1.079	0.996	0.949
100,000	16.75	96.15	243.8	458.5	1.520	0.888	0.665	0.550	1.642	1.290	1.154	1.085
200,000	23.00	137.2	352.4	666.3	1.490	0.824	0.600	0.491	2.34	1.644	1.406	1.296
500,000	39.31	252.9	665.6	1271.0	1.436	0.717	0.500	0.405	4.20	2.42	1.93	1.752
1,000,000	63.20	438.6	1176	2260	1.390	0.633	0.428	0.345	7.01	3.40	2.59	2.34
						Pr = 0.015						
50,000	15.33	87.37	220.9	415.0	1.526	0.901	0.677	0.561	1.491	1.198	1.079	1.015
100,000	20.37	119.9	306.7	579.0	1.499	0.844	0.620	0.509	2.050	1.495	1.295	1.198
200,000	29.56	182.5	474.3	901.3	1.463	0.771	0.548	0.446	3.09	1.972	1.625	1.478
500,000	53.03	357.8	953.5	1829.0	1.406	0.660	0.451	—	5.81	2.98	2.30	—
1,000,000	86.99	638.0	1731	3339	1.362	0.581	0.386	0.310	9.84	4.26	3.16	2.85
						Pr = 0.02						
10,000	11.18	62.33	156.4	293.4	1.544	0.937	0.709	0.583	1.055	0.901	0.814	0.754
20,000	12.86	72.30	181.9	341.0	1.537	0.925	0.701	0.580	1.228	1.029	0.937	0.879
50,000	17.31	100.1	254.6	479.3	1.513	0.873	0.648	0.534	1.713	1.314	1.158	1.077
100,000	23.97	144.2	371.9	704.3	1.483	0.809	0.584	0.477	2.46	1.685	1.422	1.302
200,000	35.92	228.5	599.5	1143.0	1.443	0.730	0.511	0.414	3.83	2.27	1.819	1.643
500,000	66.17	463.8	1247.0	2399.0	1.384	0.622	0.419	0.337	7.36	3.49	2.64	2.37
1,000,000	109.8	838.7	2293	4437	1.342	0.546	0.359	0.288	12.6	5.03	3.66	3.30

Pr = 0.03												
10,000	12.03	67.76	170.8	321.1	1.534	0.915	0.685	0.560	1.152	0.949	0.840	0.769
20,000	14.60	83.37	211.0	396.8	1.523	0.894	0.666	0.547	1.423	1.132	1.002	0.924
50,000	21.28	126.7	325.2	614.8	1.491	0.826	0.600	0.489	2.17	1.532	1.320	1.189
100,000	31.06	194.2	506.9	964.8	1.454	0.753	0.531	0.430	3.28	2.03	1.649	1.488
200,000	48.32	322.3	856.3	1641.0	1.412	0.673	0.461	0.371	5.29	2.79	2.16	1.938
500,000	91.50	678.1	1844	3562	1.356	0.571	0.377	0.303	10.38	4.39	3.22	2.90
1,000,000	154.0	1244	3435	6667	1.318	0.503	0.325	0.260	17.9	6.42	4.57	4.12
Pr = 0.04												
10,000	12.94	73.62	186.5	351.3	1.524	0.893	0.660	0.537	1.255	0.996	0.863	0.782
20,000	16.38	95.06	242.1	456.4	1.510	0.864	0.635	0.517	1.627	1.230	1.062	0.964
50,000	25.27	154.0	398.7	756.7	1.472	0.787	0.562	0.454	2.63	1.731	1.428	1.287
100,000	38.01	245.2	646.2	1234	1.434	0.712	0.494	0.397	4.10	2.34	1.847	1.651
200,000	60.29	417.5	1120	2153	1.391	0.633	0.427	0.342	6.71	3.26	2.46	2.20
500,000	115.8	894.5	2452	4750	1.337	0.536	0.350	0.281	13.30	5.19	3.74	3.37
1,000,000	196.5	1652	4593	8928	1.301	0.475	0.304	0.242	23.0	7.67	5.38	4.81
Pr = 0.06												
10,000	14.80	86.22	220.5	417.3	1.505	0.850	0.615	0.496	1.473	1.086	0.905	0.806
20,000	20.00	119.6	308.0	583.9	1.485	0.812	0.581	0.466	2.05	1.411	1.162	1.028
50,000	33.02	210.2	552.3	1054.0	1.442	0.726	0.504	0.404	3.54	2.08	1.642	1.451
100,000	51.57	349.7	933.7	1793	1.402	0.652	0.441	0.352	5.71	2.88	2.18	1.932
200,000	83.45	611.3	1660	3205	1.361	0.579	0.382	0.305	9.48	4.09	2.99	2.65
500,000	163.4	1334	3695	7173	1.313	0.493	0.318	0.253	19.0	6.66	4.70	4.18
1,000,000	279.3	2475	6947	13540	1.281	0.438	0.278	0.221	33.0	9.93	6.88	6.12

TABLE 4.14. Eigenvalues and Constants of the Turbulent Graetz Problem for 0.1 ≤ Pr ≤ 1 [104]

Re	λ_0^2	λ_1^2	λ_2^2	C_0	$-C_1$	C_2	G_0	G_1	G_2
				Pr = 0.1					
10,000	18.66	113.6	296.0	1.468	0.774	0.540	1.928	1.235	0.965
20,000	27.12	171.6	450.7	1.444	0.728	0.499	2.89	1.701	1.304
50,000	48.05	327.5	876.1	1.398	0.644	0.431	5.34	2.65	1.959
100,000	77.13	564.7	1,534	1.361	0.577	0.378	8.79	3.77	2.71
200,000	127.4	1,007	2,777	1.325	0.515	0.332	14.79	5.46	3.84
500,000	253.6	2,226	6,239	1.284	0.444	0.280	29.9	9.16	6.27
1,000,000	437.3	4,150	11,750	1.257	0.400	0.249	52.3	14.05	9.45
				Pr = 0.2					
10,000	27.92	190.4	514.3	1.395	0.634	0.417	3.06	1.472	1.049
20,000	43.78	312.5	849.1	1.371	0.591	0.381	4.93	2.17	1.493
50,000	82.45	636.8	1,754	1.334	0.527	0.334	9.54	3.62	2.45
100,000	136.3	1,126	3,134	1.306	0.478	0.301	16.02	5.35	3.60
200,000	229.3	2,033	5,715	1.279	0.435	0.270	27.3	8.09	5.38
500,000	462.6	4,515	12,840	1.249	0.385	0.236	55.5	14.21	9.32
1,000,000	804.6	8,416	24,090	1.229	0.353	0.214	97.2	22.5	14.58
				Pr = 0.4					
10,000	42.69	346.9	972.2	1.316	0.488	0.308	4.907	1.669	1.127
20,000	70.25	594.2	1,672	1.295	0.458	0.280	8.215	2.58	1.641
50,000	138.0	1,249	3,537	1.273	0.416	0.252	16.48	4.58	2.90
100,000	232.9	2,234	6,363	1.252	0.387	0.233	28.0	7.17	4.52
200,000	398.1	4,057	11,620	1.235	0.360	0.216	48.2	11.32	7.13
500,000	817.9	9,031	26,040	1.213	0.327	0.195	99.2	20.9	13.07
				Pr = 0.72					
10,000	64.38	646.8	1,870	1.239	0.369	0.227	7.596	1.829	1.217
20,000	109.0	1,119	3,240	1.231	0.352	0.208	13.06	2.95	1.784
50,000	219.0	2,350	6,808	1.220	0.333	0.193	26.6	5.63	3.32
100,000	375.9	4,183	12,130	1.210	0.319	0.185	45.8	9.25	5.48
200,000	651.2	7,539	21,940	1.200	0.302	0.177	79.6	15.05	9.10
500,000	1,357	16,630	48,540	1.190	0.282	0.165	166.0	28.9	17.5
				Pr = 1.0					
10,000	81.45	940.4	2,758	1.200	0.311	0.191	9.69	1.915	1.28
20,000	139.9	1,624	4,768	1.199	0.301	0.175	16.89	3.16	1.87
50,000	285.2	3,385	9,923	1.194	0.291	0.166	34.8	6.24	3.57
100,000	493.1	5,987	17,540	1.189	0.283	0.161	60.4	10.54	6.04
200,000	859.4	10,730	31,430	1.183	—	0.155	105.5	—	10.03
500,000	1,802	23,480	68,980	1.174	0.258	0.148	221.0	34.3	30.2

TABLE 4.15. Eigenvalues and Constants of the Turbulent Graetz Problem for $8 \le \mathrm{Pr} \le 10{,}000$ [104]

Re	λ_0^2	C_0	G_0
	Pr = 8		
10,000	176.6	1.056	21.6
20,000	313.5	1.056	38.7
50,000	685.6	1.054	85.4
100,000	1,232	1.054	154.0
200,000	2,271	1.054	284.0
500,000	5,020	1.052	625.0
1,000,000	9.369	1.052	1,170
	Pr = 20		
10,000	247.9	1.033	30.3
20,000	448.2	1.033	55.4
50,000	990.6	1.032	124.0
100,000	1,799	1.032	225.0
200,000	3,346	1.032	418.0
500,000	7,509	1.031	936.0
1,000,000	14,090	1.031	1,760
	Pr = 50		
10,000	348.0	1.019	42.6
20,000	631.1	1.019	78.1
50,000	1,393	1.018	174.0
100,000	2,570	1.018	321.0
200,000	4,778	1.018	598.0
500,000	10,800	1.018	1,350
1,000,000	20,420	1.018	2,550
	Pr = 100		
10,000	444.6	1.012	54.5
20,000	811.1	1.012	100.0
50,000	1,788	1.012	223.0
100,000	3,317	1.012	415.0
200,000	6,129	1.012	766.0
500,000	14,040	1.012	1,750
1,000,000	26,220	1.012	3,270
	Pr = 1000		
10,000	988.4	1.003	121.0
20,000	1,794	1.003	222.0
50,000	4,012	1.003	501.0
100,000	7,383	1.003	923.0
200,000	13,820	1.003	1,730
500,000	31,380	1.003	3,910
	Pr = 10,000		
10,000	2,132	1.001	261.0
20,000	3,907	1.001	484.0
50,000	8.695	1.001	1,090
100,000	16,140	1.001	2,020
200,000	29,840	1.001	3,732

flow reported in Sec. 3.2.3. In fact, Eqs. (3.35) to (3.38) apply unchanged to the solution of the turbulent Graetz problem.

The eigenvalues λ_n and constants G_n to be used in conjunction with Eqs. (3.36) to (3.38) for the turbulent Graetz problem are reported in Tables 4.13 to 4.15, based on the analysis of Notter and Sleicher [104]. Note that in contrast to the laminar flow solution, both λ_n and G_n are functions of Re and Pr. Also, the series for the turbulent flow solution converge much more rapidly, and so the few terms given in Tables 4.13 to 4.15 are quite adequate for most practical computations.

The thermal entrance lengths $L_{\text{th,T}}$ for the complete range of Pr for the turbulent Graetz problem are shown in Fig. 4.14. Here $L_{\text{th,T}}$ is defined as the axial distance x at which $\text{Nu}_{x,\text{T}} = 1.05\,\text{Nu}_\text{T}$, where $\text{Nu}_{x,\text{T}}$ and Nu_T are respectively the local and the fully developed Nusselt numbers with the Ⓣ boundary condition.

The thermal entrance Ⓗ problem is similar to the thermal entrance Ⓣ (i.e., the turbulent Graetz) problem discussed above except for the uniform wall heat flux boundary condition. An integral solution to the Ⓗ problem was developed by Deissler [124] for Pr = 0.01 and 0.73. Sparrow et al. [130] developed an accurate solution for Pr = 0.7, 10, and 100. This solution was later extended to Pr = 1, 3, 8, 25, 50, 75, 100, 125, 150 by Malina and Sparrow [131]. Notter and Sleicher [104] also provided a solution for $0 < \text{Pr} < 8$. These latter solutions are similar to the corresponding Ⓗ problem solution reported in Sec. 3.2.3. Equations (3.55) and (3.56) of that solution apply unchanged to the present turbulent flow solution.

The eigenvalues β_n^2 and constants $C_n R_n(1)$ to be employed in Eqs. (3.55) and (3.56) for the turbulent Ⓗ problem are given in Tables 4.16 and 4.17, based on the analysis by Notter and Sleicher [104]. Note that A_n in Tables 4.16 and 4.17 stands for $C_n R_n(1)$ in Eqs. (3.55) and (3.56).

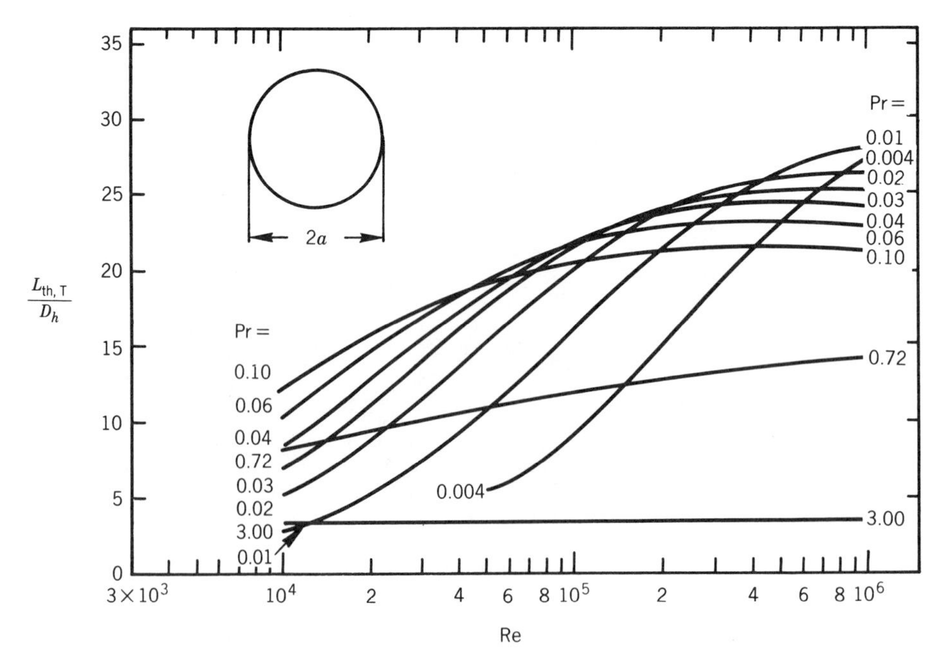

Figure 4.14. Thermal entrance lengths for the turbulent Graetz problem for a smooth circular duct [104].

TABLE 4.16. Eigenvalues and Constants of the Turbulent Thermal Entrance Length Solution for a Smooth Circular Duct with the Ⓗ Boundary Condition for $0 \leq Pr \leq 0.04$ [104]

Re	β_1^2	β_2^2	β_3^2	β_4^2	$-A_1$	$-A_2$	$-A_3$	$-A_4$
				Pr = 0				
10,000	28.31	94.26	197.5	337.9	0.156	0.0409	0.0243	0.0147
50,000	28.04	93.34	195.8	335.3	0.153	0.0474	0.0229	0.0136
100,000	28.09	93.45	196.1	335.7	0.152	0.0469	0.0227	0.0134
500,000	28.40	94.43	198.3	339.3	0.150	0.0459	0.0221	0.0130
				Pr = 0.002				
50,000	29.66	98.51	206.5	353.4	0.147	0.0457	0.0223	0.0132
100,000	31.61	104.7	219.2	374.9	0.138	0.0437	0.0213	0.0127
500,000	48.83	159.0	331.0	564.8	0.096	0.0329	0.0167	0.0102
				Pr = 0.004				
50,000	31.93	105.7	221.4	378.7	0.138	0.0438	0.0214	0.0128
100,000	36.37	119.7	250.3	427.5	0.123	0.0400	0.0198	0.0119
500,000	74.41	239.1	494.5	839.8	0.067	0.0249	0.0133	0.0084
				Pr = 0.010				
10,000	30.45	101.1	211.6	361.6	0.147	0.0471	0.0235	0.0143
50,000	40.12	131.8	275.0	469.3	0.144	0.0379	0.0190	0.0115
100,000	53.11	172.7	358.8	—	0.090	0.0314	0.0162	—
500,000	159.8	504.6	1032	1737	0.034	0.0144	0.0084	0.0056
				Pr = 0.015				
50,000	47.84	156.2	325.1	554.0	0.0982	0.0337	0.0173	0.0106
100,000	68.58	221.3	458.1	778.6	0.0718	0.0264	0.0140	0.0088
500,000	235.9	740.5	1506	2528	0.0237	0.0106	0.0065	0.0044
				Pr = 0.02				
10,000	33.55	111.0	231.8	395.8	0.136	0.0445	0.0225	0.0139
50,000	56.08	182.3	378.4	643.8	0.0857	0.0303	0.0158	0.0098
100,000	84.92	272.5	562.3	953.6	0.0595	0.0227	0.0124	0.0079
500,000	314.7	984.5	1997	3342	0.0181	0.0084	0.0053	0.0037
				Pr = 0.03				
10,000	37.13	122.4	255.2	435.1	0.125	0.0419	0.0215	0.0134
50,000	73.68	237.8	491.7	835.3	0.0676	0.0251	0.0135	0.0086
100,000	119.4	380.5	781.4	1320	0.0439	0.0177	0.0101	0.0066
500,000	477.7	1490	3011	5027	0.0121	0.0059	0.0038	0.0027
				Pr = 0.04				
10,000	41.06	134.9	280.8	478.0	0.115	0.0394	0.0206	0.0130
50,000	92.42	296.7	611.7	1037	0.0553	0.0213	0.0118	0.0077
100,000	155.8	494.1	1011	1704	0.0345	0.0145	0.0085	0.0057
500,000	646.0	2011	4059	6765	0.0091	0.0045	0.0030	0.0021

TABLE 4.17. Eigenvalues and Constants of the Turbulent Thermal Entrance Solution for a Smooth Circular Duct with the Ⓗ Boundary Condition for 0.06 ≤ Pr ≤ 0.72 [104]

Re	β_1^2	β_2^2	β_3^2	$-A_1$	$-A_2$	$-A_3$
			Pr = 0.06			
10,000	49.74	162.4	336.8	0.0979	0.0351	0.0190
50,000	132.4	422.3	866.7	0.0400	0.0163	0.00940
100,000	232.4	733.3	1494	0.0238	0.0105	0.00642
500,000	993.6	3090	6225	0.00594	0.00300	0.00204
			Pr = 0.10			
10,000	69.52	224.9	463.0	0.0737	0.0286	0.0165
50,000	219.6	695.9	1421	0.0250	0.0109	0.00667
100,000	396.9	1247	2531	0.0143	0.00663	0.00427
500,000	1718	5341	10750	0.00344	0.00176	0.00122
			Pr = 0.20			
10,000	128.7	410.4	833.1	0.0433	0.0194	0.0128
50,000	465.1	1467	2980	0.0122	0.00570	0.00377
100,000	849.7	2664	5392	0.00679	0.00329	0.00222
500,000	3633	11310	22780	0.00162	0.00830	0.00058
			Pr = 0.40			
10,000	258.1	812.8	1623	0.0234	0.0123	0.00946
50,000	982.3	3093	6268	0.00589	0.00288	0.00202
100,000	1793	5621	11370	0.00323	0.00160	0.00111
500,000	7551	23530	47450	0.00078	0.00039	0.00027
			Pr = 0.72			
10,000	519.5	1624	3202	0.0123	0.00738	0.00653
50,000	1952	6154	12480	0.00296	0.00147	0.00106
100,000	3510	11030	22340	0.00164	0.00081	0.00056
500,000	14310	44690	89830	0.000405	0.00020	—

The local Nusselt numbers $Nu_{x,T}$ and $Nu_{x,H}$ computed from Eqs. (3.37) and (3.55) respectively for the turbulent thermally developing flow are nearly identical for $Pr \geq 0.7$. In view of this fact, there is no need to provide the eigenvalues and constants for $Pr > 0.7$ in Table 4.17. For $Pr > 0.7$, use Tables 4.14 and 4.15 in conjunction with Eqs. (3.36) to (3.38). The equality of $Nu_{x,T}$ and $Nu_{x,H}$ for $Pr \geq 0.7$ can be clearly seen from Table 4.18 based on the computations by Siegel and Sparrow [108].

The thermal entrance lengths $L_{th,H}$ for a complete Pr range for the turbulent Ⓗ problem are plotted in Fig. 4.15. The $L_{th,H}$ values in Fig. 4.15 correspond to

TABLE 4.18. Comparison of Turbulent Flow Nusselt Numbers $Nu_{x,H}$ and $Nu_{x,T}$ at Re = 10^5 in the Thermal Entrance Region of a Smooth Circular Duct [108]

	Pr = 0.7			Pr = 10			Pr = 100		
$\frac{x}{D_h}$	$Nu_{x,H}$	$Nu_{x,T}$	$\frac{Nu_{x,H}}{Nu_{x,T}}$	$Nu_{x,H}$	$Nu_{x,T}$	$\frac{Nu_{x,H}}{Nu_{x,T}}$	$Nu_{x,H}$	$Nu_{x,T}$	$\frac{Nu_{x,H}}{Nu_{x,T}}$
2	220.2	211.6	1.041	733.6	732.0	1.002	1554	1556	0.999
5	196.4	190.2	1.033	711.3	710.4	1.001	1543	1545	0.999
10	183.6	178.4	1.029	697.7	697.2	1.001	1536	1539	0.998
20	175.2	170.8	1.026	688.5	688.3	1.000	1531	1534	0.998
30	172.8	168.8	1.024	685.5	685.4	1.000	1530	1532	0.999
∞	171.4	167.7	1.022	683.9	683.9	1.000	1529	1532	0.998

$Nu_{x,H} = 1.05\,Nu_H$, where $Nu_{x,H}$ and Nu_H are respectively the local and fully developed Nusselt numbers with the Ⓗ boundary condition. Comparison of the results in Figs. 4.14 and 4.15 shows that $L_{th,H}$ values are higher than $L_{th,T}$ values for $0.01 < Pr < 0.72$.

Certain correlations for the Nusselt numbers in the thermal entrance region of a smooth circular duct have been developed by various investigators. They are applicable for restricted ranges of Re, Pr, or Pe = Re Pr. Reynolds et al. [132] performed an

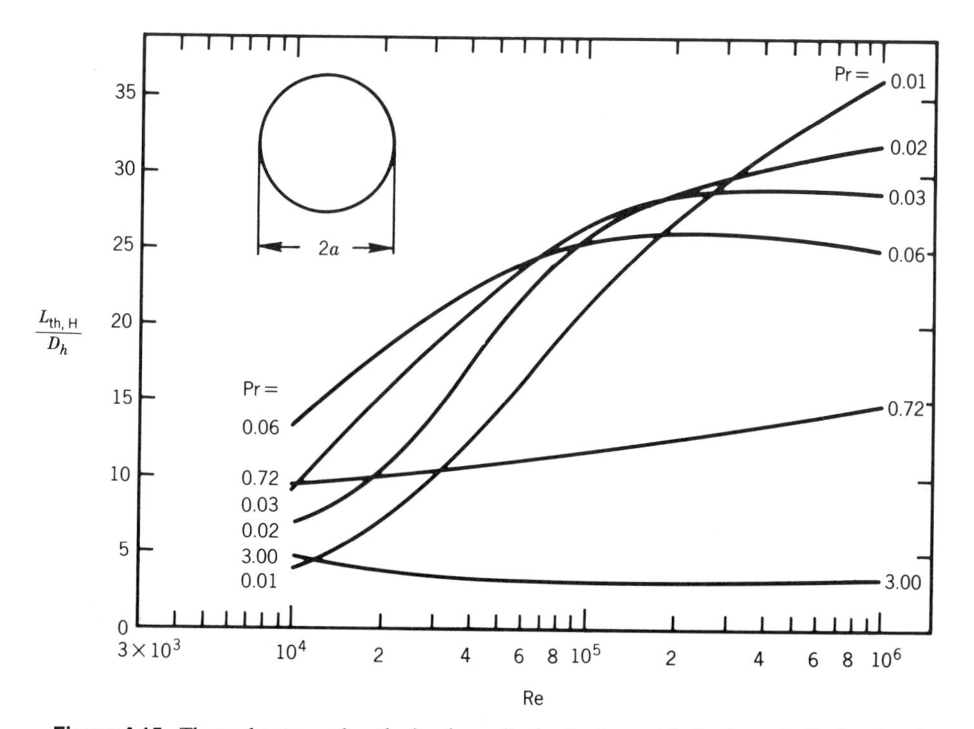

Figure 4.15. Thermal entrance lengths for thermally developing and hydrodynamically developed turbulent flow (Ⓗ boundary condition) in a smooth circular duct [104].

analysis similar to that of Sparrow et al. [130] for Pr = 0.71. Based on their analysis, Reynolds et al. [132] proposed the following correlation:

$$\frac{\mathrm{Nu}_{x,\mathrm{H}}}{\mathrm{Nu}_{\mathrm{H}}} = 1 + \frac{0.8(1 + 70{,}000\,\mathrm{Re}^{-3/2})}{x/D_h} \tag{4.69}$$

valid for $3000 < \mathrm{Re} < 5 \times 10^4$ and $\mathrm{Pr} = 0.71$. The predictions of this correlation agree with those of their analysis within ±5% for $x/D_h \geq 2$. The predictions of their analysis in turn agree with those of Sparrow et al. [130] within ±1.1%. For the fully developed Nusselt number Nu_H in Eq. (4.69), Reynolds et al. [132] found that their Nu_H predictions agree with those of the Dittus-Boelter correlation in Table 4.3 within ±4%.

Al-Arabi [133] developed the following correlation for the mean Nusselt number Nu_m for the thermally developing flow with the Ⓣ or Ⓗ boundary condition:

$$\frac{\mathrm{Nu}_m}{\mathrm{Nu}_\infty} = 1 + \frac{C}{x/D_h} \tag{4.70}$$

where Nu_∞ stands for the fully developed Nusselt number Nu_T or Nu_H and

$$C = \frac{(x/D_h)^{0.1}}{\mathrm{Pr}^{1/6}}\left(0.68 + \frac{3000}{\mathrm{Re}^{0.81}}\right) \tag{4.71}$$

This correlation is valid for $x/D_n > 3$, $500 < \mathrm{Re} < 10^5$, and $0.7 < \mathrm{Pr} < 75$. Its predictions agree within ±12% with the experimental measurements for Pr = 0.7.

Chen and Chiou [106] performed an analysis in the thermal entrance region for liquid metals (Pr < 0.03) and proposed the following correlations for the local and mean Nusselt numbers valid for $x/D_h > 2$ and Pe > 500:

$$\frac{\mathrm{Nu}_{x,\mathrm{T}}}{\mathrm{Nu}_\mathrm{T}} = 1 + \frac{2.4}{x/D_h} - \frac{1}{(x/D_h)^2} \tag{4.72}$$

$$\frac{\mathrm{Nu}_{m,\mathrm{T}}}{\mathrm{Nu}_\mathrm{T}} = 1 + \frac{7}{x/D_h} + \frac{2.8}{x/D_h}\ln\left(\frac{x/D_h}{10}\right) \tag{4.73}$$

$$\mathrm{Nu}_\mathrm{T} = 4.5 + 0.0156\,\mathrm{Re}^{0.85}\mathrm{Pr}^{0.86} \tag{4.74}$$

For the Ⓗ problem, Eqs. (4.72) and (4.73) are applicable with $\mathrm{Nu}_{x,\mathrm{H}}$, $\mathrm{Nu}_{m,\mathrm{H}}$, and Nu_H respectively replacing $\mathrm{Nu}_{x,\mathrm{T}}$, $\mathrm{Nu}_{m,\mathrm{T}}$ and Nu_T. The recommended expression for Nu_H is

$$\mathrm{Nu}_\mathrm{H} = 5.6 + 0.0165\,\mathrm{Re}^{0.85}\mathrm{Pr}^{0.86} \tag{4.75}$$

Based on their experimental study, Genin et al. [134] presented the following correlation for liquid metals (Pr < 0.03) valid for 190 < Pe < 1800:

$$\mathrm{Nu}_{x,\mathrm{H}} = \mathrm{Nu}_\mathrm{H} + 0.006\left(\frac{x/D_h}{\mathrm{Pe}}\right)^{-1.2} \tag{4.76}$$

reproducing their experimental results within ±9%. For improved agreement to within

±6%, they proposed

$$\mathrm{Nu}_{x,\mathrm{H}} = \mathrm{Nu}_\mathrm{H} + 0.006(x/D_h)^{-n}\mathrm{Pe}^{1.2}, \qquad n = (x/D_h)^{0.12} \tag{4.77}$$

For the fully developed Nusselt number Nu_H in Eqs. (4.76) and (4.77), Lyon's formula given in Table 4.5 was recommended, as it was found to be in satisfactory agreement with the measurements of [134].

The thermal entrance length $L_{\mathrm{th,H}}$ corresponding to $\mathrm{Nu}_{x,\mathrm{H}} = 1.05\,\mathrm{Nu}_\mathrm{H}$ was found to be given by [134]

$$\frac{L_{\mathrm{th,H}}}{D_h} = \frac{0.04\,\mathrm{Pe}}{1 + 0.002\,\mathrm{Pe}} \tag{4.78}$$

The predictions of Eq. (4.78) for $190 < \mathrm{Pe} < 1800$ are in good agreement with Fig. 4.15.

Siegel and Sparrow [135] took into account the effect of internal heat generation in the solution for the thermal entrance Ⓗ problem. Knowles and Sparrow [136] studied the effect of nonuniform heating at the duct wall on the same problem. Kays and Nicoll [137] measured the effect on the local Nusselt numbers of linearly varying wall heat flux along the duct length. Hall and Price [138] investigated the effect on the local Nusselt numbers of exponentially and sinusoidally varying wall heat fluxes along the duct length. Lee [107] explored the effect of axial fluid conduction on the solution of the Ⓗ problem for liquid metals.

4.2.5 Simultaneously Developing Flow

Deissler [124] theoretically solved the problem of simultaneously developing velocity and temperature fields in a smooth circular duct for $\mathrm{Pr} = 0.73$, starting with uniform velocity and temperature profiles at the duct inlet. In order to realize the uniform velocity profile at the duct inlet, a boundary-layer tripping device such as a ring or strip has to be placed just upstream of the duct inlet to induce turbulence. The local Nusselt numbers computed by Deissler [124] for the Ⓣ and Ⓗ boundary conditions are displayed in Fig. 4.16. The two sets of results corresponding to the Ⓣ and Ⓗ boundary conditions are practically indistinguishable for $x/D_h > 8$.

The simultaneously developing turbulent flow in a smooth duct is affected appreciably by the type of duct entrance configuration. Boelter et al. [139] and Mills [91] carried out extensive experimental investigations to study this effect in a smooth circular duct using air as the working fluid. Boelter et al. [139] employed the Ⓣ boundary condition, whereas Mills [91] employed the Ⓗ boundary condition. A summary of the local Nusselt numbers determined by Boelter et al. [139] for Re close to 5×10^4 is presented in Fig. 4.17 for five different entrance configurations. The lowest curve in Fig. 4.17 is for the case of the thermally developing flow for which the velocity profile is fully developed, as indicated by the long calming section preceding the heated duct section. All other curves lie substantially above this curve, underscoring the decisive role played by the duct entrance configuration in determining the Nusselt number. The peculiar behavior of the square-inlet curve for $x/D_h < 2$ is apparently caused by the flow contraction followed by reexpansion in the vicinity of the duct inlet.

In practical applications, the mean Nusselt number Nu_m is more useful than the local Nusselt number Nu_x. Accordingly, the present authors have expressed Nu_m for

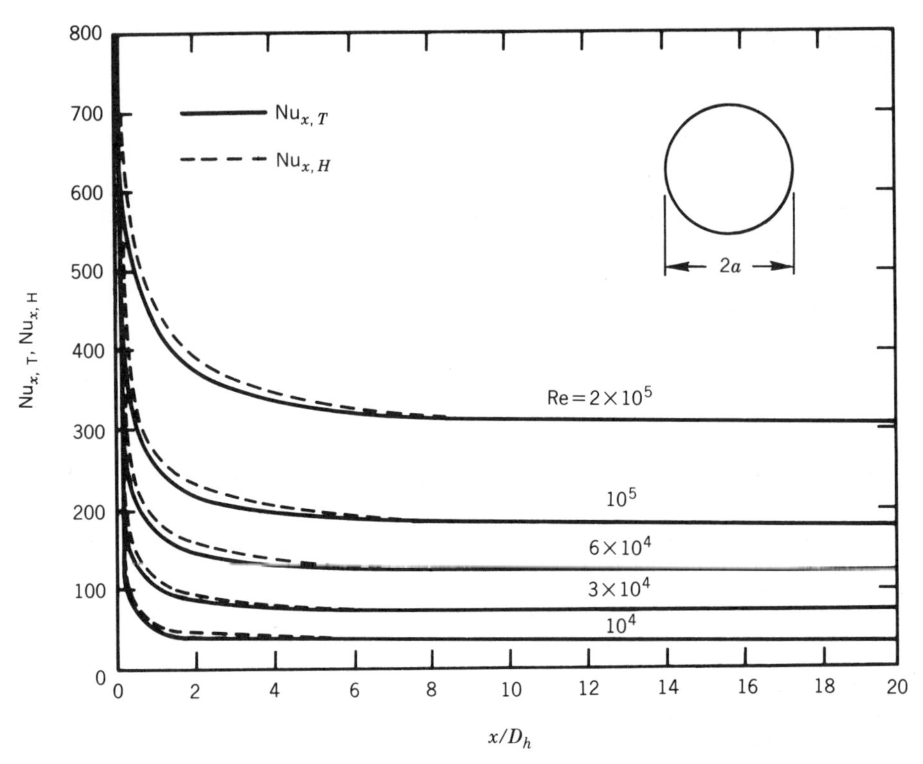

Figure 4.16. Local Nusselt numbers $Nu_{x,T}$ and $Nu_{x,H}$ for simultaneously developing turbulent flow in a smooth circular duct for Pr = 0.73 [124].

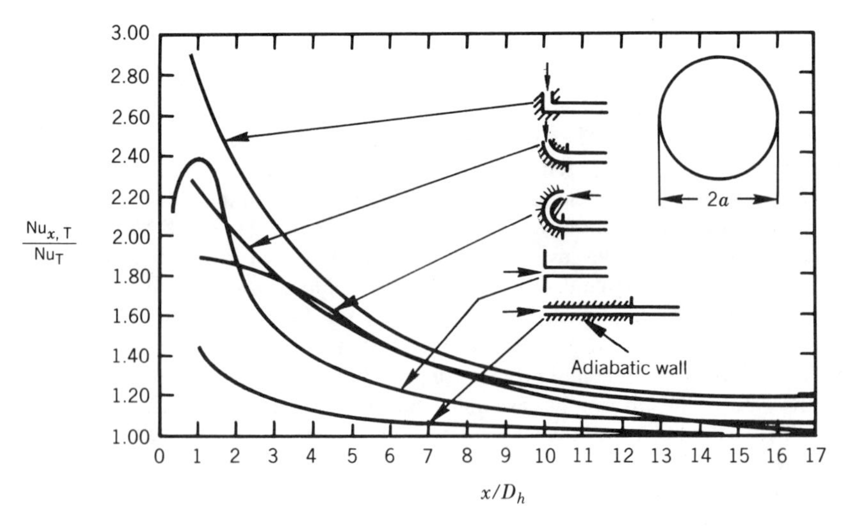

Figure 4.17. Normalized local Nusselt numbers for turbulent flow in the entrance region of a smooth circular duct with various entrance configurations for Pr = 0.7 and Re ≈ 5 × 10[4] [139].

the five entrance configurations shown in Fig. 4.17 by formulas of the type

$$\frac{\mathrm{Nu}_m}{\mathrm{Nu}_\infty} = 1 + \frac{C}{(x/D_h)^n} \tag{4.79}$$

where Nu_∞ stands for the fully developed Nusselt number $\mathrm{Nu_H}$ or $\mathrm{Nu_T}$. The magnitudes of the coefficient C and the exponent n for various entrance configurations were determined from the $\mathrm{Nu}_{m,\mathrm{H}}$ measurements of Mills [91] for air (Pr = 0.7). Table 4.19 contains a listing of the resulting formulas. In general, these formulas are valid for $x/D_h > 3$, and their predictions agree with the experimental values of Mills [91] within $\pm 3\%$. Although the values of C and n were determined for the Ⓗ boundary condition, the formulas in Table 4.19 may be applied to both the Ⓗ and Ⓣ boundary conditions, especially for high Re values. Hall and Khan [140] experimentally studied the effect of the Ⓣ and Ⓗ boundary conditions on the local Nusselt numbers for simultaneously developing flow in a circular duct for airflow (Pr = 0.7). Their results show that significant differences between $\mathrm{Nu}_{x,\mathrm{T}}$ and $\mathrm{Nu}_{x,\mathrm{H}}$ are observed for $\mathrm{Re} < 3 \times 10^4$; for higher Re values the differences diminish rapidly.

Molki and Sparrow [141] expressed the mean Nusselt numbers for simultaneously developing flow in a circular duct with a square entrance and the Ⓣ boundary condition for Pr = 2.5 by the empirical formula given by Eq. (4.79). The values of C and n for the range $9000 \le \mathrm{Re} \le 8.8 \times 10^4$ determined by them are

$$C = 23.99\,\mathrm{Re}^{-0.230}, \qquad n = 0.815 - 2.08 \times 10^{-6}\mathrm{Re} \tag{4.80}$$

The predictions of Eq. (4.79) with the C and n values of Eq. (4.80) agree within $\pm 5\%$ with the measurements of Sparrow and Molki [142] for Pr = 2.5 and $x/D_h > 2$.

In a series of investigations, Sparrow and coworkers [143–147] studied turbulent fluid flow and heat transfer in circular ducts with inlet disturbances induced by a variety of blockages. In another series of studies, Sparrow and coworkers [148–150] investigated the influence of plenum-related losses on turbulent fluid flow and heat transfer in circular ducts. The aforementioned investigations employed the oil-lamp-black technique for flow visualization and the naphathalene-sublimation technique to determine the heat transfer coefficients.

Chen and Chiou [106] presented the correlations for the simultaneously developing flow of liquid metals (Pr < 0.03) in a smooth circular duct with uniform velocity profile at the inlet. These correlations, valid for $2 \le x/D_h < 35$ and Pe > 500, are given by

$$\frac{\mathrm{Nu}_x}{\mathrm{Nu}_\infty} = 0.88 + \frac{2.4}{x/D_h} - \frac{1.25}{(x/D_h)^2} - A \tag{4.81}$$

$$\frac{\mathrm{Nu}_x}{\mathrm{Nu}_\infty} = 1 + \frac{5}{x/D_h} + \frac{1.86}{x/D_h}\ln\left(\frac{x/D_h}{10}\right) - B \tag{4.82}$$

where for the Ⓣ boundary condition

$$A = \frac{40 - x/D_h}{190}, \qquad B = 0.09 \tag{4.83}$$

and for the Ⓗ boundary condition,

$$A = B = 0 \tag{4.84}$$

TABLE 4.19. Ratio of Mean to Fully Developed Turbulent Flow Nusselt Number in the Entrance Region of a Smooth Circular Duct with Various Entrance Configurations for Pr = 0.7

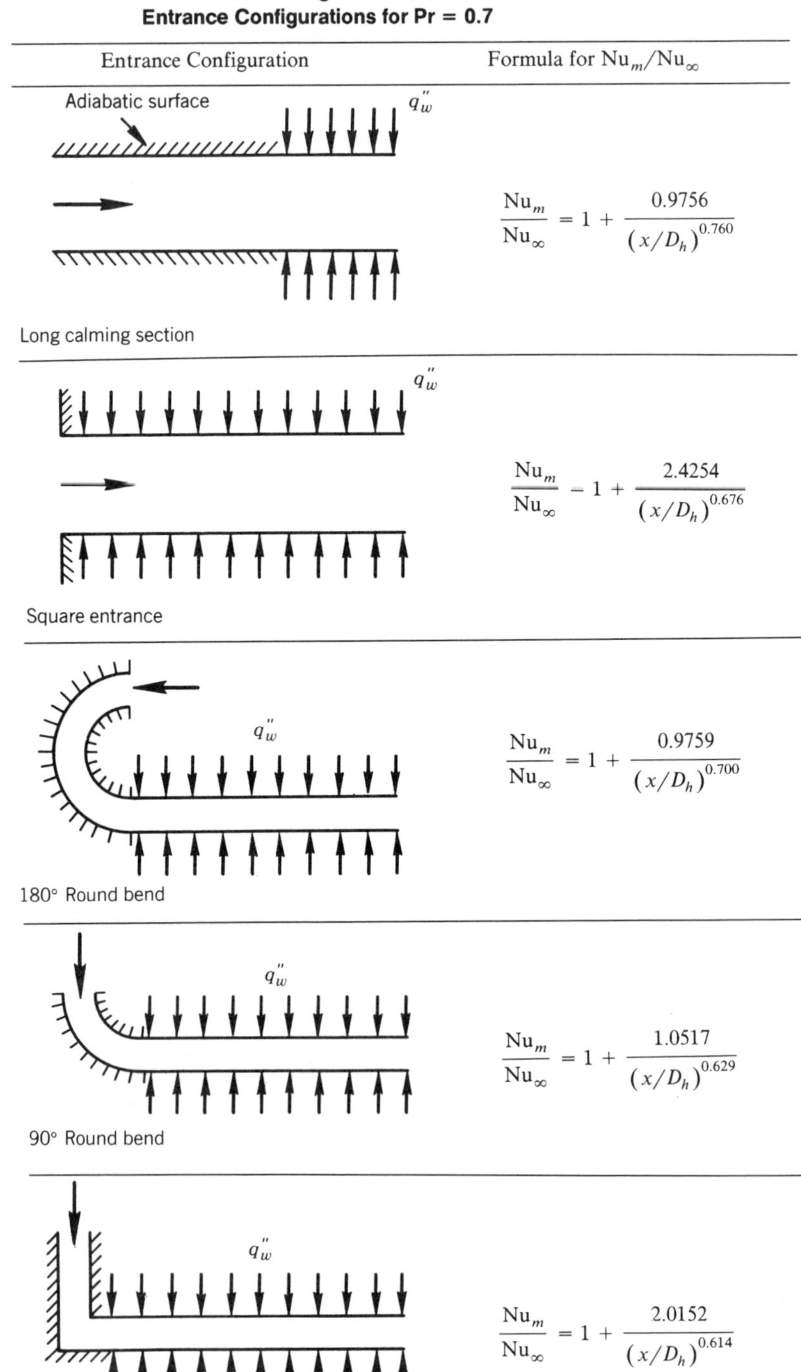

Entrance Configuration	Formula for Nu_m/Nu_∞
Long calming section	$\frac{Nu_m}{Nu_\infty} = 1 + \frac{0.9756}{(x/D_h)^{0.760}}$
Square entrance	$\frac{Nu_m}{Nu_\infty} = 1 + \frac{2.4254}{(x/D_h)^{0.676}}$
180° Round bend	$\frac{Nu_m}{Nu_\infty} = 1 + \frac{0.9759}{(x/D_h)^{0.700}}$
90° Round bend	$\frac{Nu_m}{Nu_\infty} = 1 + \frac{1.0517}{(x/D_h)^{0.629}}$
90° Elbow	$\frac{Nu_m}{Nu_\infty} = 1 + \frac{2.0152}{(x/D_h)^{0.614}}$

Here Nu_x stands for the local Nusselt number $Nu_{x,T}$ or $Nu_{x,H}$, and correspondingly Nu_∞ stands for the fully developed Nusselt number Nu_T or Nu_H.

4.3 FLAT DUCT

Turbulent and transition fluid flow and heat transfer characteristics of a flat duct (i.e., parallel-plate channel) have been studied quite extensively, as this duct constitutes the limiting geometry for the family of rectangular and concentric annular ducts. Several results of practical interest pertaining to transition, fully developed, hydrodynamically developing, thermally developing, and simultaneously developing flows are presented next.

4.3.1 Transition Flow

The lower limit of the critical Reynolds number Re_{crit} for a flat duct has not been determined as precisely as that for a circular duct. The reported experimental values of $Re_{crit} = u_m D_h/\nu$, where the hydraulic diameter $D_h = 4b$, b being the half spacing between the flat duct walls, range between 2200 and 3400. Beavers et al. [153], experimenting with a variety of entrance configurations, determined $Re_{crit} = 3400$ for almost a flat duct ($\alpha^* = 0.0145$) with a symmetric rounded entrance as shown in Table 4.20, which also includes the values for three other entrance configurations. The results in Table 4.20 underscore the importance of the entrance configuration in determining Re_{crit}.

Beavers et al. [153] experimentally studied the effect of the sources of disturbance on the lower limit of Re_{crit} and concluded that the stationary sources of disturbance such as an undulating wire, a notched rod, or an ensemble of rods situated in the upstream plenum chamber connected to the flat duct yield a value of 3400, identical to the one for the symmetric rounded entrance. However, a nonstationary source such as a flat-bladed stirrer located in the upstream plenum lowered Re_{crit} to 2600. Furthermore, Beavers et al. [153] noted that when the flat duct with the symmetric square entrance is provided with a source of disturbance (such as a metal strip) situated within the duct downstream of the entrance, then Re_{crit} attains its lowest value, 2200. For a completely disturbed flow at the inlet, Hanks [154] arrived at a theoretical Re_{crit} value of 2288, in agreement with the experimental value of 2285 determined by Davies and White [155] for a rectangular duct with $\alpha^* = 2b/2a = 0.006$.

Chen and Sparrow [156] theoretically studied the laminar-to-turbulent transition phenomenon in the entrance region of a flat duct using the hydrodynamic stability theory mentioned in Sec. 4.1.3. They found that Re_{crit} decreases monotonically with increasing distance from the duct inlet, approaching a fully developed value of 1.5384×10^4. This is significantly higher than the experimentally observed value of 3400, underscoring the limitations of the linear hydrodynamic stability theory. Gupta and Garg [157] refined the analysis by Chen and Sparrow [156] and predicted lower Re_{crit} values near the duct inlet. However, their prediction of the fully developed Re_{crit} value is identical with that of Chen and Sparrow [156].

The following friction factor interpolation formula developed by Hrycak and Andrushkiw [47] for the transition flow is recommended in the range $2200 \le Re \le 4000$:

$$f = -2.56 \times 10^{-3} + 4.085 \times 10^{-6}Re - 5.5 \times 10^{-10}Re^2 \tag{4.85}$$

TABLE 4.20. Critical Reynolds Numbers for a Smooth Flat Duct† with Various Entrance Configurations [153]

Entrance Configuration	Re_{crit}
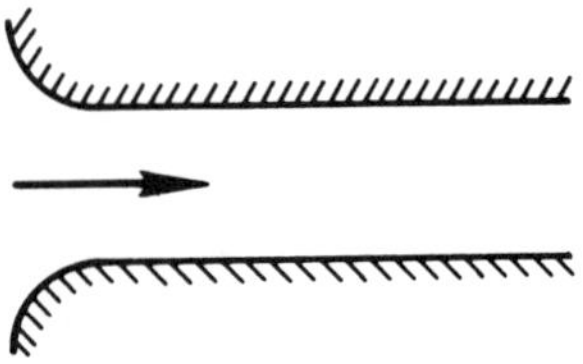 Symmetric rounded entrance	3400
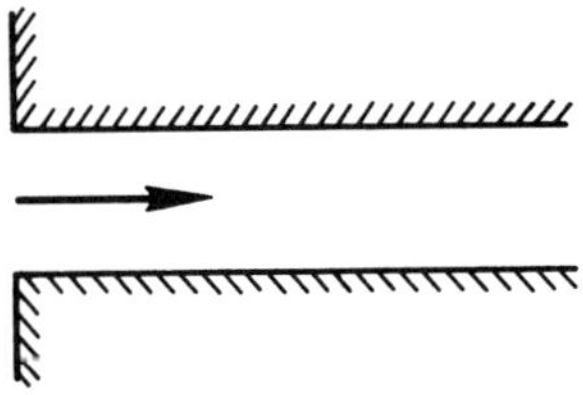 Symmetric square entrance	3100
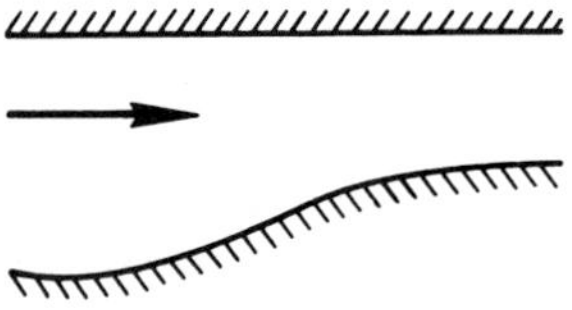 Asymmetric curved entrance	2700
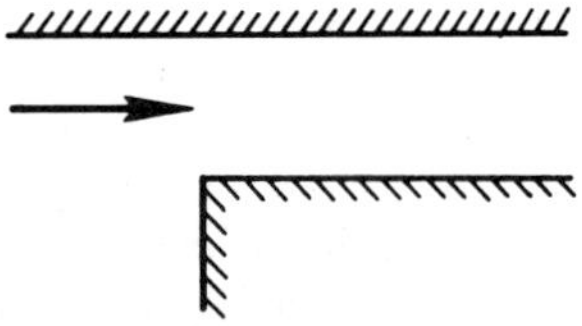 Asymmetric square entrance	2600

†The duct in question is a rectangular duct with $\alpha^* = 2b/2a = 0.0145$, $2a$ and $2b$ being the sides of the duct with $2a > 2b$.

The mean Nusselt number in the thermal entrance region of a flat duct with uniform wall temperature at both walls in the range $2300 < Re < 6000$ is given by [85]:

$$\mathrm{Nu}_{m,\mathrm{T}} = 0.116(\mathrm{Re}^{2/3} - 160)\mathrm{Pr}^{1/3}\left[1 + \left(\frac{x}{D_h}\right)^{-2/3}\right] \tag{4.86}$$

Patel and Head [26] showed that the shear stress gradient parameter Δ [Eq. (4.14)] for turbulent-to-laminar transition in a flat duct can be expressed as

$$\Delta = -\frac{4\sqrt{2}}{\mathrm{Re}\sqrt{f}} \tag{4.87}$$

Equation (4.87) in conjunction with the critical value of $\Delta = -0.0094$ [26] shows that turbulent-to-laminar transition in a flat duct is possible when

$$\mathrm{Re}\sqrt{f} = 602 \tag{4.88}$$

f in Eqs. (4.87) and (4.88) is for fully developed turbulent flow (see Table 4.2).

4.3.2 Fully Developed Flow

Fluid Flow. Pai [158] developed the following polynomial form of the velocity profile for a smooth-walled flat duct with spacing $2b$ between the plates:

$$\frac{u}{u_{\max}} = 1 - \frac{n-s}{n-1}\left(\frac{y}{b}\right)^2 - \frac{s-1}{n-1}\left(\frac{y}{b}\right)^{2n} \tag{4.89}$$

$$\frac{u_m}{u_{\max}} = 1 - \frac{n-s}{3(n-1)} - \frac{s-1}{(n-1)(2n+1)} \tag{4.90}$$

where y is the transverse distance measured from the duct axis, and s and n are functions of Re for turbulent flow. With $s = 1$, Eqs. (4.89) and (4.90) reduce to Eq. (3.76) which represents the fully developed laminar flow in a flat duct.

From the experimental measurements of Laufer [159] at $\mathrm{Re} = 4.28 \times 10^4$, Pai [158] estimated the values of s and n as $s = 11.06$ and $n = 16$. The present authors developed the following general relationships for s and n applicable in the range $4000 \le \mathrm{Re} \le 10^5$:

$$s = 0.004\,\mathrm{Re}^{3/4}, \qquad n = 0.00625\,\mathrm{Re}^{3/4} - 2.0625 \tag{4.91}$$

These expressions are based on the circular duct experimental data of Nikuradse [50]. For $\mathrm{Re} = 4.28 \times 10^4$, Eq. (4.91) predicts $s = 11.9$ and $n = 16.5$ which are in satisfactory agreement with the values estimated by Pai [158].

Figures 4.18 provides a comparison between the predictions of Eq. (4.89) with $s = 11.06$ and $n = 16$ and the measurements of Laufer [159]. The agreement is seen to be excellent. Also included in Fig. 4.18 for comparison is the laminar flow velocity profile determined from Eq. (3.76) or Eq. (4.89) with $s = 1$.

Goldstein [160] developed the following velocity-defect form of the velocity distribution in a smooth-walled flat duct:

$$\frac{u_{\max} - u}{u_t} = -3.39\left[\ln\left(1 - \sqrt{\frac{y}{b}}\right) + \sqrt{\frac{y}{b}}\right] - 0.172 \tag{4.92}$$

where y is the normal distance measured from the duct axis. The hypotheses of the theory upon which Eq. (4.92) is based break down in the middle of the duct. Therefore, Eq. (4.92) does not apply at the duct axis.

The predictions of Eq. (4.92) are compared in [160] with the experimental measurements of Dönch [161] and Nikuradse [162]. The agreement is found to be quite good especially with the former measurements.

Rothfus et al. [163] and Dwyer [164] showed that the fully developed velocity distribution in a smooth flat duct can be represented in terms of the wall coordinates (u^+, y^+) by the smooth-duct formulas for a circular duct given in Table 4.1, provided

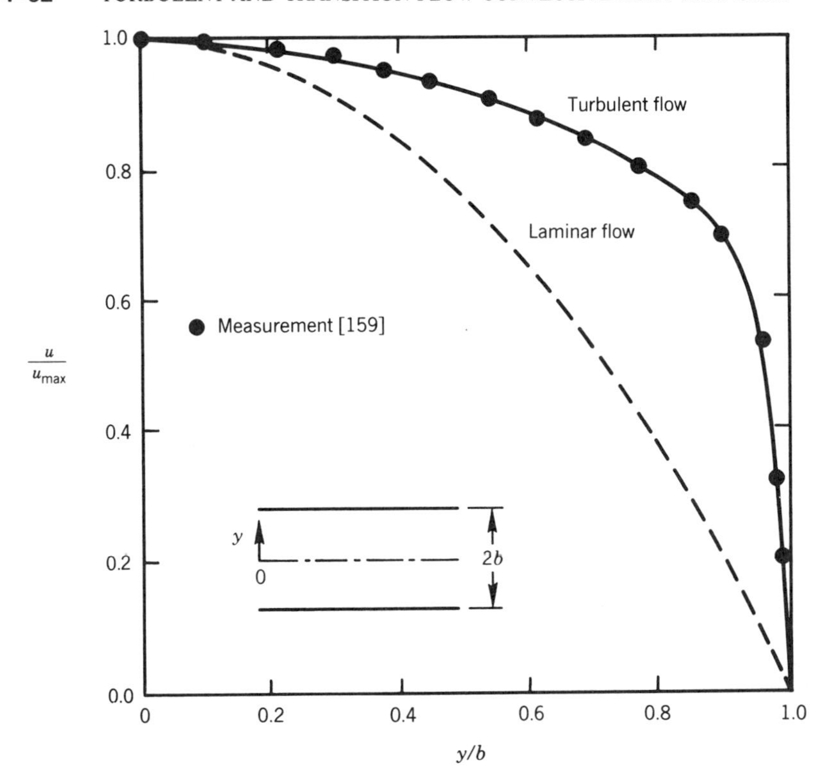

Figure 4.18. Fully developed turbulent velocity distribution in a smooth flat duct at Re = 4.28×10^4.

that the wall coordinates are modified as follows

$$y^+ = \frac{yu_t/\nu}{(u_m/u_{\max})_c} \tag{4.93}$$

$$u^+ = \frac{u}{u_t}\left(\frac{u_m}{u_{\max}}\right)_c \tag{4.94}$$

where $(u_m/u_{\max})_c$ is the ratio of the mean to the maximum velocity and u_t is the friction velocity defined by Eq. (4.8). The agreement of the velocity distribution in terms of u^+ and y^+ of Eqs. (4.93) and (4.94) with the experimental measurements presented in [163] is excellent.

Based on very accurate experimental measurements, Beavers et al. [165] presented the following Fanning friction-factor formula for $5000 < \text{Re} < 3 \times 10^4$:

$$f = \frac{0.1268}{\text{Re}^{0.3}} \tag{4.95}$$

The f factors of Eq. (4.95) are a maximum of 5% higher than those of the Blasius formula [Eq. (4.32)], and this maximum difference occurs at the lower end of the range (Re = 5000); the difference is less than 1% at Re = 3×10^4.

TABLE 4.21. Comparison of Fully Developed Turbulent Flow Friction Factors for a Smooth Flat Duct

			f			
Re	Re_c, Eq. (4.97)[†]	f_c, Eqs. (4.31), (4.32)[‡]	Eq. (4.32)	Eqs. (4.95), (4.96)[§]	Eq. (4.97)	Sage et al. [167]
6,960	3,267	0.0098	0.0087	0.0089	0.0087	0.0077
9,110	4,276	0.0096	0.0081	0.0082	0.0085	0.0078
17,500	8,215	0.0083	0.0069	0.0068	0.0073	0.0072
36,400	17,086	0.0069	0.0057	0.0063	0.0061	0.0063
53,200	24,972	0.0063	0.0052	0.0057	0.0055	0.0056

† $(u_m/u_{\max})_c$ in Eq. (4.97) was calculated from Eq. (4.41) as 0.8167, corresponding to $n = 7$. Also, $(u_m/u_{\max})$ in Eq. (4.97) was calculated from Eq. (4.90) as 0.8699.

‡ The first value in the column was calculated from Eq. (4.31), and the rest from Eq. (4.32) using the Re_c values in column 2 of this table.

§ The first three values in this column were calculated from Eq. (4.95), and the rest from Eq. (4.96).

Based on a comprehensive survey of the available data, Dean [166] developed the following formula for $1.2 \times 10^4 < \text{Re} < 1.2 \times 10^6$:

$$f = \frac{0.0868}{\text{Re}^{1/4}} \tag{4.96}$$

The predictions of Eq. (4.96) at Re = 5000 and 3×10^4 are respectively 9.5% and 14.6% higher than those of Eq. (4.95). In the range $5000 \le \text{Re} \le 3 \times 10^4$, use of Eq. (4.95) is recommended.

Rothfus and Monrad [167] expressed f for a flat duct in terms of f for a circular duct via the relation

$$\sqrt{\frac{f}{f_c}} = \frac{(u_m/u_{\max})_c}{u_m/u_{\max}} = \frac{2\,\text{Re}_c}{\text{Re}} \tag{4.97}$$

where the subscript c refers to the circular duct.

The predictions of Eqs. (4.95), (4.96), and (4.97) for the applicable Re ranges are compared in Table 4.21 with the experimental measurements of Sage et al. for $6960 \le \text{Re} \le 5.32 \times 10^4$. These latter measurements are reported in [167]. Also included in Table 4.21 are the predictions of the Blasius formula [Eq. (4.32)]. It is seen that for Re > 9110, the predictions of Eqs. (4.95), (4.96), and (4.97) are only in slightly better accord with the measurements than the predictions of Eq. (4.32). It may therefore be concluded that use of the hydraulic diameter [employed in Eq. (4.32)] is reasonably effective in predicting the friction factors for a flat duct from the circular duct correlations.

Heat Transer. Kays and Leung [168] presented comprehensive turbulent heat transfer results for arbitrarily prescribed heat fluxes at the two duct walls. Based on their analysis, the fully developed Nusselt number Nu_H can be determined from

$$\text{Nu}_\text{H} = \frac{\text{Nu}}{1 - \gamma\theta^*} \tag{4.98}$$

where γ is the ratio of the prescribed heat fluxes at the two duct walls. The Nusselt number Nu and the influence coefficient θ^* entering Eq. (4.98) are presented in Table 4.22 as functions of Re and Pr. To illustrate use of the information in Table 4.22 in conjunction with Eq. (4.98), consider the case of turbulent airflow (Pr = 0.7) at Re = 10^4 in a flat duct with uniform heat fluxes of arbitrary magnitude ($-1 \leq \gamma \leq 1$) prescribed at the two walls. From Table 4.22, corresponding to Pr = 0.7 and Re = 10^4, we have Nu = 27.8 and $\theta^* = 0.22$. First, suppose that the ratio of the heat fluxes $\gamma = 0$. Then from Eq. (4.98), $\mathrm{Nu_H} = 27.8$, which is the Nusselt number for only one wall heated and the other insulated. Next, suppose that $\gamma = 1$. Then Eq. (4.98) yields $\mathrm{Nu_H} = 35.6$, which is the Nusselt number for uniform heat fluxes of equal magnitudes at both walls. Finally, suppose that $\gamma = -1$. Then according to Eq. (4.98), $\mathrm{Nu_H} = 22.9$, which is the Nusselt number for heat transfer into one wall and out of the other.

In order to verify whether $\mathrm{Nu_H}$ values for $\gamma = 1$ can be determined from the circular duct correlations, the predictions of Eq. (4.98) were compared with those of the Gnielinski correlation (Table 4.4) for $0.5 \leq \mathrm{Pr} \leq 2000$ and the Notter-Sleicher correlation (Table 4.5) for Pr < 0.1. The comparisons are shown in the following tabulation.

Pr Range	Re Range	$\mathrm{Nu_{flat\ duct}}/\mathrm{Nu_{cir}}$
$0.5 \leq \mathrm{Pr} \leq 100$	$10^4 \leq \mathrm{Re} \leq 3 \times 10^4$	up to +1.23
$0.5 \leq \mathrm{Pr} \leq 100$	$3 \times 10^4 \leq \mathrm{Re} \leq 10^6$	−1.075 to −1.089
Pr = 1000	$10^4 \leq \mathrm{Re} \leq 10^6$	up to −1.23
$0 < \mathrm{Pr} \leq 0.003$	$10^4 \leq \mathrm{Re} \leq 10^6$	+1.57 to −1.055

From these comparisons, it is concluded that the flat duct $\mathrm{Nu_H}$ can be determined within ±9% using the circular duct correlations for $0.5 \leq \mathrm{Pr} \leq 100$ and $3 \times 10^4 \leq \mathrm{Re} \leq 10^6$.

Sparrow and Lin [169] performed a theoretical analysis for the case $\gamma = 1$ analogous to the analysis reported in Ref. 130 for a circular duct. They found that for $0.7 \leq \mathrm{Pr} \leq 100$ and $10^4 \leq \mathrm{Re} \leq 5 \times 10^5$, $\mathrm{Nu_H}$ for a flat duct can be determined quite accurately from the circular duct correlation using the hydraulic diameter. This finding is consistent with the aforementioned conclusion reached from comparison with the Gnielinski correlation.

Similar to the results for a circular duct, it is found that the fully developed turbulent Nusselt numbers with the Ⓣ and Ⓗ boundary conditions in a flat duct are nearly identical for Pr > 0.7 and Re > 10^5. Kakaç and Paykoç [170] numerically analyzed the fully developed and thermally developing problems in a flat duct with the Ⓣ and Ⓗ boundary conditions. Their results for Pr = 0.73 are available in Ref. 171. They show that for both fully developed and thermally developing flows the Nusselt numbers with the two boundary conditions are very close.

Certain empirical correlations are developed to calculate $\mathrm{Nu_H}$ for liquid metals (Pr < 0.03) flowing turbulently in a smooth flat duct. For $\gamma = 0$, three such correlations–presented by Buleev [172], Dwyer [173], and Duchatelle and Vautrey [174], respectively—are

$$\mathrm{Nu_H} = 5.1 + 0.02\,\mathrm{Pe}^{0.8} \tag{4.99}$$

$$\mathrm{Nu_H} = 5.6 + 0.01905\,\mathrm{Pe}^{0.8} \tag{4.100}$$

$$\mathrm{Nu_H} = 5.14 + 0.0127\,\mathrm{Pe}^{0.8} \tag{4.101}$$

TABLE 4.22. Nusselt Numbers and Influence Coefficients for Fully Developed Turbulent Flow in a Smooth Flat Duct with Uniform Heat Flux at One Wall and the Other Wall Insulated [168]†

	$Re = 10^4$		$Re = 3 \times 10^4$		$Re = 10^5$		$Re = 3 \times 10^5$		$Re = 10^6$	
Pr	Nu	θ^*	Nu	θ^*	Nu	θ^*	Nu	θ^*	Nu	θ^*
0.0	5.70	0.428	5.78	0.445	5.80	0.456	5.80	0.460	5.80	0.468
0.001	5.70	0.428	5.78	0.445	5.80	0.456	5.88	0.460	6.23	0.460
0.003	5.70	0.428	5.80	0.445	5.90	0.450	6.32	0.450	8.62	0.422
0.01	5.80	0.428	5.92	0.445	6.70	0.440	9.80	0.407	21.5	0.333
0.03	6.10	0.428	6.90	0.428	11.0	0.390	23.0	0.330	61.2	0.255
0.5	22.5	0.256	47.8	0.222	120	0.193	290	0.174	780	0.157
0.7	27.8	0.220	61.2	0.192	155	0.170	378	0.156	1,030	0.142
1.0	35.0	0.182	76.8	0.162	197	0.148	486	0.138	1,340	0.128
3.0	60.8	0.095	142	0.092	380	0.089	966	0.087	2,700	0.084
10.0	101	0.045	214	0.045	680	0.045	1,760	0.045	5,080	0.046
30.0	147	0.021	367	0.022	1,030	0.022	2,720	0.023	8,000	0.024
100.0	210	0.009	514	0.009	1,520	0.010	4,030	0.010	12,000	0.011
1,000.0	390	0.002	997	0.002	2,880	0.002	7,650	0.002	23,000	0.002

† For laminar flow ($Re < 2300$), $Nu = 5.385$ and $\theta^* = 0.346$ for all values of Pr.

For $0 \leq \Pr \leq 0.04$ and $10^4 \leq \mathrm{Re} \leq 10^5$, the predictions of Eq. (4.99) are within +56.3% and −12% of the results in Table 4.22. The predictions of Eq. (4.100) for the same Pr and Re ranges are within +55.7% and −3.4% of the results in Table 4.22. Finally, for the same Pr and Re ranges, the predictions of Eq. (4.101) are within +24.3% and −11.4% of the results in Table 4.22. Based on this comparison, use of Eq. (4.101) is recommended for low-Pr fluids ($\Pr < 0.03$).

Dwyer [173] also presented the following empirical correlation for $\gamma = 1$:

$$\mathrm{Nu_H} = 9.49 + 0.00596\,\mathrm{Pe}^{0.688} \tag{4.102}$$

For $0 \leq \Pr \leq 0.03$ and $10^4 \leq \mathrm{Re} \leq 10^5$, the predictions of Eq. (4.102) are within +37% and −11% of the results computed from Eq. (4.98) in conjunction with Table 4.22.

For $0.01 < \Pr < 1$ and $10 < \mathrm{Pe} < 10^5$, Seban [176] presented the following correlation for a flat duct with uniform temperature at one wall and the other wall insulated:

$$\mathrm{Nu_T} = 5.8 + 0.02\,\mathrm{Pe}^{0.8} \tag{4.103}$$

Kakaç and Price [177] theoretically investigated the fully developed turbulent flow in a flat duct with exponentially varying wall heat flux along the duct walls.

The information on the effect of duct wall roughness is quite sparse for a flat duct. For the time being, the fully developed fluid flow and heat transfer results for a rough circular duct given in Sec. 4.2.2 are recommended for a flat duct, with the use of the hydraulic diameter.

4.3.3 Hydrodynamically Developing Flow

The apparent friction factors f_{app} in the hydrodynamic entrance region of a smooth flat duct with uniform velocity at the duct inlet have been determined by Deissler [124] by an integral method. They are presented in Fig. 4.13 along with the results for a circular duct.

Na and Lu [122] also performed an integral analysis of the hydrodynamic entrance problem for a smooth flat duct and concluded that the hydrodynamic entrance length L_{hy}, judged by the merging of the hydrodynamic boundary layers growing from the two walls commencing at the duct inlet, is equal to $13.75D_h$ at $\mathrm{Re} = 2.21 \times 10^5$. The experimental measurements by Byrne et al. [178] of the hydrodynamic parameters of the momentum and displacement thickness are in excellent accord with the predictions of Na and Lu [122].

Shcherbinin and Shklyar [179] have discussed the application of various turbulence models to the analysis of the hydrodynamic entrance region for a smooth flat duct. Kobata et al. [180] have explored the effect of various types of inlet disturbances on the turbulent flow development in a smooth flat duct.

4.3.4 Thermally Developing Flow

The thermally developing turbulent flow in a flat duct with uniform and equal temperatures at the two walls (i.e., the Ⓣ boundary condition) has been solved by Sakakibara and Endo [181] and by Shibani and Özişik [175]. The solution by Sakakibara and Endo [181] is readily expressible in terms of Eqs. (3.106) to (3.109) presented for laminar flow.

TABLE 4.23. Eigenvalues and Constants for the Turbulent Thermal Entrance Length Solution for a Smooth Flat Duct with the Ⓣ Boundary Condition (Symmetric Heating) [181]†

n	Pr = 0.01		Pr = 0.1		Pr = 0.7		Pr = 10	
	λ_n	G_n	λ_n	G_n	λ_n	G_n	λ_n	G_n
				Re = 10^4				
0	1.827497	0.960790	2.346001	1.669267	3.537461	4.056980	5.702075	10.802955
1	5.692406	0.835600	7.608357	1.080093	14.362314	1.125396	46.184547	0.784344
2	9.554691	0.756055	13.120667	0.792200	26.311537	0.772568	85.737071	1.355758
3	13.436811	0.684588	18.696392	0.701383	37.673305	0.977139	114.119186	2.646274
4	17.334806	0.625434	24.201008	0.695717	48.045119	1.224250	145.165458	1.808341
5	21.242563	0.577446	29.630975	0.695640	58.347764	1.108809	178.899218	2.025040
6	25.156713	0.539413	35.043809	0.675657	69.106903	0.931110	209.443743	2.162400
7	29.075272	0.508446	40.460120	0.644934	80.037567	0.868743	241.765985	1.782791
8	32.996097	0.483637	45.879494	0.603474	90.895604	0.903912	274.816338	1.850689
9	36.742346	0.463312	51.328292	0.555912	101.489525	0.939343	306.211937	1.903596
				Re = 5×10^4				
0	1.912004	1.035521	3.527715	3.827682	6.460615	13.601305	11.409717	43.202670
1	5.932245	0.926977	8.623147	2.204825	26.780945	3.328302	90.445510	2.289603
2	9.947142	0.873805	20.351132	1.564034	49.187683	1.887531	172.661348	1.332138
3	13.971657	0.837950	29.014831	1.418690	70.804246	1.660553	249.391491	1.396456
4	17.995642	0.816434	37.473274	1.367410	91.863581	1.466552	324.535346	1.607540
5	22.013858	0.799183	45.903775	1.236662	113.357272	1.300462	399.658134	2.399156
6	26.030544	0.779091	54.454974	1.116344	134.946554	1.343659	469.116784	4.169534
7	30.051246	0.757480	63.022707	1.073363	156.049031	1.495806	534.173088	5.131340
8	34.075648	0.738703	71.513251	1.049389	176.976859	1.621927	601.154079	4.599756
9	38.099639	0.723218	80.001866	0.996582	197.923670	1.818685	670.834714	3.686269

TABLE 4.23. Continued

n	Pr = 0.01 λ_n	Pr = 0.01 G_n	Pr = 0.1 λ_n	Pr = 0.1 G_n	Pr = 0.7 λ_n	Pr = 0.7 G_n	Pr = 10 λ_n	Pr = 10 G_n
				Re = 10^5				
0	1.990190	1.123400	4.375285	5.942880	8.470391	23.372875	15.430524	78.803350
1	6.175360	0.992972	14.788268	3.094337	35.828748	5.381725	122.364719	4.049093
2	10.370246	0.922960	25.973789	2.139305	65.941439	3.061719	233.424860	2.165841
3	14.580808	0.885030	37.073823	1.990705	94.764022	2.683785	337.044890	1.909180
4	18.785272	0.872191	47.840923	1.939199	122.844238	2.261690	439.802209	1.662779
5	22.975895	0.863827	58.610288	1.742381	151.640439	1.880653	545.076992	1.710149
6	27.163225	0.847313	69.550567	1.594006	180.548215	1.790845	647.917288	2.191135
7	31.358092	0.827645	80.450000	1.560083	208.933514	1.767783	748.047835	2.913008
8	35.558171	0.814226	91.233572	1.507466	237.429236	1.689111	847.402202	4.157575
9	39.754790	0.806821	102.073116	1.402132	266.194134	1.718213	942.247363	6.060920
				Re = 5×10^5				
0	2.536235	1.881412	7.714196	18.589975	16.119222	85.688550	31.116542	323.270700
1	7.998224	1.429666	27.952842	7.240240	72.067816	16.797820	250.662777	15.761505
2	13.631325	1.187660	50.111539	4.781244	133.439628	9.613410	477.682956	8.166020
3	19.299358	1.141790	71.590134	4.655515	191.457385	8.522315	688.203127	6.585335
4	24.879737	1.182594	92.278310	4.424427	248.137230	7.001135	898.347608	4.903591
5	30.393820	1.183590	113.313549	3.886888	306.443721	5.786745	1114.31912	4.081324
6	35.937719	1.133990	134.575334	3.741947	364.439371	5.434265	1326.17457	3.843262
7	41.524240	1.108558	155.454295	3.744517	421.432873	4.988365	1536.58206	3.454989
8	47.093094	1.122997	176.294676	3.509672	479.186024	4.410366	1750.71768	3.223083
9	52.625266	1.128003	197.417361	3.331574	537.246585	4.187492	1963.19010	3.289714

† The eigenvalues λ_n and constants G_n in the table are related to the quantities λ_m, C_m, $Y_m'(0)$ of [181] by $\lambda_n = \sqrt{3/32}\,\lambda_m$ and $G_n = C_m Y_m'(0)/2$.

TABLE 4.24. Eigenvalues and Constants for Turbulent Thermal Entrance Length Solution for a Smooth Flat Duct with Uniform Temperature at One Wall and the Other Wall Insulated [181]†

	Pr = 0.01		Pr = 0.1		Pr = 0.7		Pr = 10	
n	λ_n	G_n	λ_n	G_n	λ_n	G_n	λ_n	G_n
				$Re = 10^4$				
0	0.968458	0.531344	1.308335	1.011678	2.205480	3.107447	3.920320	10.211430
1	2.923466	0.457068	4.010807	0.747596	7.621578	1.171898	24.206960	0.788558
2	4.875412	0.432468	6.747202	0.591440	13.484355	0.606671	45.245995	0.391225
3	6.829239	0.411184	9.512223	0.479151	19.397228	0.419345	65.235148	0.368511
4	8.785355	0.390378	12.283628	0.407800	25.116321	0.362503	81.249605	0.446745
5	10.743336	0.370221	15.043211	0.366680	30.513698	0.365989	94.550641	0.835026
6	12.703336	0.351551	17.783345	0.347442	35.676726	0.428095	108.795040	1.471252
7	14.664262	0.334557	20.506406	0.341876	40.818818	0.529056	124.089740	1.300522
8	16.625948	0.319415	23.219705	0.343990	46.016512	0.620729	140.779890	0.923790
9	18.588112	0.305945	25.930335	0.347894	51.276761	0.636564	157.260790	0.811979
				$Re = 5 \times 10^4$				
0	1.002373	0.563600	2.002667	2.400574	4.066854	10.605850	7.849310	40.846480
1	3.020234	0.493450	6.192429	1.653793	14.269258	3.713388	47.494718	3.031625
2	5.034749	0.473168	10.476835	1.216118	25.399310	1.808061	89.265897	1.168320
3	7.049987	0.457370	14.817110	0.945009	36.692407	1.180802	131.116409	0.765203
4	9.065908	0.443580	19.152291	0.796458	47.796434	0.944262	171.253626	0.655889
5	11.082104	0.431858	23.445561	0.726463	58.583068	0.860229	209.875710	0.664361
6	13.098193	0.422772	27.692657	0.701659	69.100167	0.820202	247.000912	0.669594
7	15.113976	0.415992	31.911983	0.699416	79.516928	0.795569	283.744238	0.737863
8	17.129453	0.410987	36.127402	0.692082	91.222213	0.735588	320.263129	0.738082
9	19.144783	0.406955	40.357242	0.672078	100.595277	0.689600	356.141418	0.873307

TABLE 4.24. Continued

n	Pr = 0.01		Pr = 0.1		Pr = 0.7		Pr = 10	
	λ_n	G_n	λ_n	G_n	λ_n	G_n	λ_n	G_n
				Re = 10^5				
0	1.044401	0.612353	2.519408	3.834979	5.362251	18.448815	10.617552	74.527050
1	3.147239	0.535528	7.858294	2.474531	19.065763	6.117300	64.246380	5.484345
2	5.247462	0.509005	13.369376	1.711920	34.081925	2.912350	120.838350	2.067949
3	7.348892	0.486864	18.961708	1.288604	49.286918	1.900553	177.500805	1.311530
4	9.451019	0.468166	24.528884	1.079291	64.191634	1.530374	231.820199	1.072304
5	11.552982	0.453706	30.015398	0.994397	78.628376	1.403342	284.205477	1.015756
6	13.654203	0.444265	35.423694	0.976925	92.688998	1.334024	334.910466	0.930933
7	15.754567	0.439037	40.792179	0.989007	106.640404	1.274243	385.723599	0.918377
8	17.854306	0.436785	46.162410	0.983390	120.731063	1.144552	437.185848	0.794651
9	19.953801	0.435823	51.561790	0.951585	135.036083	1.043057	489.418461	0.844775
				Re = 5×10^5				
0	1.367311	1.073120	4.615004	12.989110	10.426118	67.935950	21.866644	303.459700
1	4.143042	0.880886	14.891367	6.762215	38.278605	19.884170	131.579975	22.480190
2	6.932025	0.760654	25.804529	3.988501	69.035225	9.024950	247.839127	8.328860
3	9.732016	0.663989	36.882212	2.825447	100.141050	5.895590	363.958658	5.240980
4	12.529569	0.599636	47.787524	2.364957	130.430914	4.808732	474.839098	4.152434
5	15.315741	0.566130	58.392651	2.245881	159.546290	4.500556	581.485289	3.843706
6	18.090866	0.558606	68.763546	2.271366	187.868760	4.274631	684.839671	3.249870
7	20.860663	0.568553	79.064262	3.327053	216.041811	4.073542	788.969623	3.034212
8	23.630019	0.585438	89.431146	2.259735	244.615966	3.572037	894.985377	2.301723
9	26.401115	0.598425	99.917932	2.126929	273.670956	3.262675	1002.99624	2.309664

† The eigenvalues λ_n and constants G_n are related to the quantities λ_m, C_m, $Y_m'(0)$ of [181] by $\lambda_n = \sqrt{\frac{3}{32}}\,\lambda_m$ and $G_n = C_m Y_m'(0)/2$.

TABLE 4.25. Nusselt Numbers and Influence Coefficients for Thermally Developing Turbulent Flow in a Smooth Flat Duct with Uniform Heat Flux at One Wall and the Other Wall Insulated [22]

Pr = 0.01

	Re = 7104		73,712		495,164	
x/D_h	Nu	θ^*	Nu	θ^*	Nu	θ^*
1	8.33	0.233	23.5	0.076	60.2	0.058
3	6.52	0.378	16.1	0.133	45.1	0.063
10	6.11	0.417	11.3	0.284	32.0	0.131
30	6.10	0.417	9.36	0.399	24.8	0.265
100	6.10	0.417	9.13	0.414	21.9	0.349
300	6.10	0.417	9.13	0.414	21.8	0.353

Pr = 0.1

	Re = 7096		73,612		494,576	
x/D_h	Nu	θ^*	Nu	θ^*	Nu	θ^*
1	19.7	0.056	75.2	0.018	241	0.005
3	14.3	0.122	56.2	0.016	194	0.023
10	10.7	0.267	42.4	0.115	155	0.062
30	9.44	0.352	34.8	0.233	132	0.147
100	9.34	0.359	32.1	0.290	120	0.219
300	9.34	0.359	32.1	0.291	120	0.219

Pr = 1.0

	Re = 7096		73,612		494,576	
x/D_h	Nu	θ^*	Nu	θ^*	Nu	θ^*
1	47.3	0.013	234	0.005	940	0.000
3	37.9	0.033	203	0.018	851	0.009
10	31.5	0.089	177	0.049	761	0.030
30	28.0	0.173	160	0.114	697	0.077
100	27.1	0.200	152	0.155	661	0.123

Pr = 10.0

	Re = 7096		73,612		494,576	
x/D_h	Nu	θ^*	Nu	θ^*	Nu	θ^*
1	102	0.004	602	0.004	2925	0.000
3	88.6	0.012	575	0.008	2829	0.003
10	81.9	0.027	550	0.018	2724	0.010
30	78.6	0.057	532	0.041	2640	0.027
100	77.5	0.070	522	0.057	2590	0.045

The eigenfunctions $Y_n(y/b)$ for turbulent flow required in Eq. (3.106) are not presented in Ref. 181. However, the eigenvalues λ_n and the constants G_n for turbulent flow required in Eqs. (3.107) to (3.109) are available in Ref. 181. These quantities recomputed in the notation of Eqs. (3.107) and (3.109) by the present authors are presented in Table 4.23. With the help of this tabular information, $\mathrm{Nu}_{x,\mathrm{T}}$ and $\mathrm{Nu}_{m,\mathrm{T}}$ can be computed from Eqs. (3.107) to (3.109) as functions of Re and Pr. Some computed $\mathrm{Nu}_{x,\mathrm{T}}$ results are presented in [181] in a graphical form.

Hatton and Quarmby [182] as well as Sakakibara and Endo [181] solved the thermally developing turbulent flow problem in a flat duct with one wall at uniform temperature and the other insulated. As described in Sec. 3.3.1, this solution is referred to as the fundamental solution of the third kind. The local and mean Nusselt numbers for this problem can be expressed by the following set of equations:

$$\mathrm{Nu}_{x,\mathrm{T}} = \frac{16}{3} \frac{\sum_{n=0}^{\infty} G_n \exp\left(-\frac{32}{3}\lambda_n^2 x^*\right)}{\sum_{n=0}^{\infty}\left(G_n/\lambda_n^2\right)\exp\left(-\frac{32}{3}\lambda_n^2 x^*\right)} \tag{4.104}$$

$$\mathrm{Nu}_{m,T} = -\frac{\ln \theta_m}{2x^*} \tag{4.105}$$

where

$$\theta_m \equiv \frac{T_w - T_m}{T_w - T_e} = \frac{3}{2}\sum_{n=0}^{\infty}\frac{G_n}{\lambda_n^2}\exp\left(-\frac{32}{3}\lambda_n^2 x^*\right), \qquad x^* = \frac{x/D_h}{\mathrm{Re\,Pr}} \tag{4.106}$$

The eigenvalues λ_n and the constants G_n to be used in conjunction with Eqs. (4.104) to (4.106) are presented in Table 4.24. Based on the results of Sakakibara and Endo [181], the values in Table 4.24 were recomputed by the present authors in the notation of Eq. (4.104).

Hatton and Quarmby [182], Hatton et al. [183], and Sakakibara [184] solved the problem of thermally developing turbulent flow in a flat duct with uniform heat flux at one wall and the other wall insulated. As mentioned in Sec. 3.3.1, this solution is referred to as the fundamental solution of the second kind. It can be employed to solve for any combination of heat fluxes at the two walls via Eq. (4.98). The Nusselt numbers Nu and the influence coefficients θ^* needed for this purpose are presented in Table 4.25. They were computed by Kays and Crawford [22], based on the analysis of Hatton and Quarmby [182].

Faggiani and Gori [185] studied the effect of axial fluid conduction on thermally developing flow in a flat duct. For $7060 \le \mathrm{Re} \le 7.362 \times 10^4$ and $0.001 < \mathrm{Pr} < 0.1$, they concluded that the influence of the axial fluid conduction is to significantly decrease the Nusselt numbers in the thermal entrance region.

4.3.5 Simultaneously Developing Flow

The information on simultaneously developing flow in a flat duct is extremely sparse. Duchatelle and Vautrey [174] experimentally determined the local Nusselt numbers $\mathrm{Nu}_{x,\mathrm{H}}$ for simultaneously developing flow in a flat duct with uniform heat flux at one wall and the other wall insulated. Their results for NaK with $\mathrm{Pr} = 0.02$ are shown in Fig. 4.19. Included in the figure is a reference line about which the data points are evenly scattered. It is seen that the ratio $\mathrm{Nu}_{x,\mathrm{H}}/\mathrm{Nu}_{\mathrm{H}}$, where Nu_{H} is the fully developed Nusselt number [given by Eq. (4.101)], attains the value of unity at

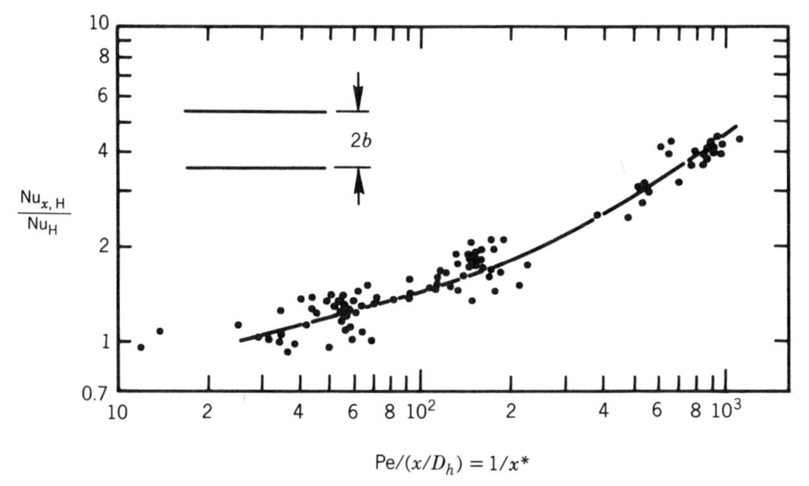

Figure 4.19. Normalized local Nusselt numbers for simultaneously developing turbulent flow in a smooth flat duct for Pr = 0.02 with one wall heated and the other wall insulated [174].

$\mathrm{Pe}/(x/D_h) = 25$. Thus, for low-Pr fluids such as NaK, the entrance length L_{th} for the simultaneously developing flow is estimated to be $L_{\mathrm{th}}/D_h = \mathrm{Pe}/25$. For the thermally developing but hydrodynamically developed flows with the same fluids, Duchatelle and Vautrey [174] estimate the entrance length as $L_{\mathrm{th}}/D_h = \mathrm{Pe}/80$.

4.4 RECTANGULAR DUCTS

Turbulent fluid flow and heat transfer characteristics of rectangular ducts are not explored as extensively as those of circular ducts. The available information indicates that the circular duct results can be applied fairly accurately to rectangular ducts by the use of the hydraulic diameter D_h given by

$$D_h = \frac{4ab}{a + b} = \frac{4b}{1 + \alpha^*} \tag{4.107}$$

where $2a$ and $2b$ are the lengths of the two sides of a rectangular duct with $2a > 2b$, and the duct aspect ratio $\alpha^* = 2b/2a$.

Several useful results pertaining to rectangular ducts with arbitrary values of α^* in the range $0 < \alpha^* \leq 1$ are presented next. The results corresponding to $\alpha^* = 0$ have already been presented in the preceding Sec. 4.3.

4.4.1 Transition Flow

The lower limit of the critical Reynolds number $\mathrm{Re}_{\mathrm{crit}}$ for rectangular ducts has been established by several experimental investigations. The entrance configuration exerts a rather marked influence on $\mathrm{Re}_{\mathrm{crit}}$. This is brought out in Table 4.26, which contains the experimental measurements of various investigations for the two types of entrance configurations shown in the table. Included in this table are the results for $\alpha^* = 0$, i.e., a flat duct taken from Table 4.20. The results for $\alpha^* = 0.1$, 0.2, 0.3333, and 1 are due

TABLE 4.26. Lower Limits of Critical Reynolds Numbers for Smooth Rectangular Ducts

α^*	Re_{crit}	α^*	Re_{crit}
0	3400	0	3100
0.1	4400	0.01	2920
0.2	7000	0.2	2500
0.3333	6000	0.2555	2400
1.0	4300	0.3425	2360
		1.0	2200

to Hartnett et al. [186]. The rest of the results for $\alpha^* = 0.01$, 0.2555, and 0.3425 are due to Davies and White [155], Allen and Grunberg [187], and Cornish [188], respectively.

It was pointed out by Allen and Grunberg [187] that for a rectangular duct with an abrupt entrance, Re_{crit} is inversely proportional to the ratio $(u_{max}/u_m)_l$ for laminar flow in a rectangular duct. This ratio can be determined from Eqs. (3.154) to (3.156) if one knows the duct aspect ratio α^*. Evaluating the proportionality constant from the condition that for laminar flow in a flat duct ($\alpha^* = 0$) one has $u_{max}/u_m = \frac{3}{2}$ and $\mathrm{Re}_{crit} = 3100$ (Table 4.26), the present authors arrived at the formula

$$\mathrm{Re}_{crit} = \frac{4650}{(u_{max}/u_m)_l} \tag{4.108}$$

for rectangular ducts with $0 \le \alpha^* \le 1$. The values of Re_{crit} calculated from Eq. (4.108) for various values of α^* are at the most 8% higher than the experimental measurements for an abrupt entrance presented in Table 4.26.

To the knowledge of the authors, there are no reliable formulas for the friction and heat transfer coefficients developed specifically for transition flow in rectangular ducts of various aspect ratios. However, it has been observed by a number of investigators that the circular duct formulas for fully developed turbulent flows apply quite well to the rectangular ducts with the use of the hydraulic diameter. Accordingly, the formulas in Sec. 4.2.1 may be applied to rectangular ducts with the hydraulic diameter $D_h = 4ab/(a+b)$ replacing the circular duct diameter $2a$. A better choice for substitution in the circular duct formulas appears to be the laminar equivalent diameter D_l given by Eq. (4.15) or (4.16). D_l provides $\pm 5\%$ agreement with the experimentally determined fully developed friction coefficients determined by various investigators for rectangular ducts, compared to $\pm 20\%$ agreement provided by D_h [28].

4.4.2 Fully Developed Flow

Fluid Flow. The fully developed laminar flow through a straight circular or noncircular duct is unidirectional, and its axial velocity depends only on the cross-sectional coordinates. The fully developed turbulent flow through a straight circular duct is similar to its laminar counterpart. However, the fully developed turbulent flow through straight noncircular ducts with sharp corners (e.g., rectangular, triangular, and trapezoidal) possesses a nonzero flow normal to the duct axis. In turbulent duct flows, it is common practice to refer to the axial flow as the *primary flow* and the normal flow as the *secondary flow*, as originally suggested by Prandtl [189]. The fully developed secondary flow, like the fully developed primary flow, depends only on the cross-sectional coordinates of the duct. It is quite small (approximately 1% of the magnitude of the axial mean velocity). However, it exerts a measurable effect in distorting the axial velocity profile and in increasing the friction coefficients in the ducts by approximately 10%.

In 1926, Nikuradse [190] was the first investigator to experimentally detect distortion of the axial velocity profiles in fully developed turbulent flows in rectangular and triangular ducts. Prandtl [189] explained this distortion in terms of the secondary flow. During a subsequent experimental investigation in 1930, Nikuradse [191] obtained photographic evidence of the existence of secondary flow in noncircular ducts by squirting a milky fluid into the flow field and tracing the lines of constant brightness. More than 30 years elapsed after Nikuradse's initial observation of the secondary flow before detailed measurements of the turbulence structure were made by Hoagland [192], Leutheusser [193], and Brundrett and Baines [194]. Subsequently, Gessner and Jones [195] made more extensive measurements of turbulent flow in a rectangular duct and from a simplified analysis provided a clear explanation of the existence of the secondary flow. They showed that the secondary flow is the result of small differences in magnitude of the opposing forces exerted by the Reynolds stresses and static pressure gradients in a plane normal to the duct axis.

In recent years, a significant effort has been directed toward the computation of turbulent secondary flows in rectangular and triangular ducts. The earlier works in this direction were heavily empirical [196–197]. However, the more recent work [198–203] has minimized dependence on empiricism by utilizing the Reynolds stress transport equations. In two recent papers, Speziale [204–205] has examined the usefulness of various turbulence models in predicting the secondary flows in noncircular ducts.

Figure 4.20*a* shows the curves of constant axial velocity (isovels) exprerimentally determined by Nikuradse [190] in a rectangular duct with $\alpha^* = \frac{2}{7}$. It is clear from the figure that the velocities at the corners are comparatively large, resulting in bulging of the isovels in the corners. As mentioned above, this distortion of the isovels is due to the secondary flow, which moves the fluid toward the corner along the bisector of the angle and then outward in both directions. A schematic diagram of the secondary flow pattern in a rectangular duct is presented in Fig. 4.20*b*.

The fully developed turbulent friction factors for rectangular ducts of various aspect ratios can be determined from the circular duct formulas given in Sec. 4.2.2 by substituting the hydraulic diameter $D_h = 4ab/(a + b)$ for the circular duct diameter $D_h = 2a$. The applicability of the hydraulic diameter to the turbulent flow in a rectangular duct ($\alpha^* = \frac{2}{7}$) is illustrated in Fig. 4.21. The figure also contains plots of f against Re based on D_h for a square and a trapezoidal duct. For turbulent flow (Re > 2000), the curves are calculated from the Blasius formula [Eq. (4.32)] derived for a circular duct. They agree very well with the experimental measurements of Nikuradse [191] and Schiller [206]. For laminar flow (Re < 2000), the experimental data points for

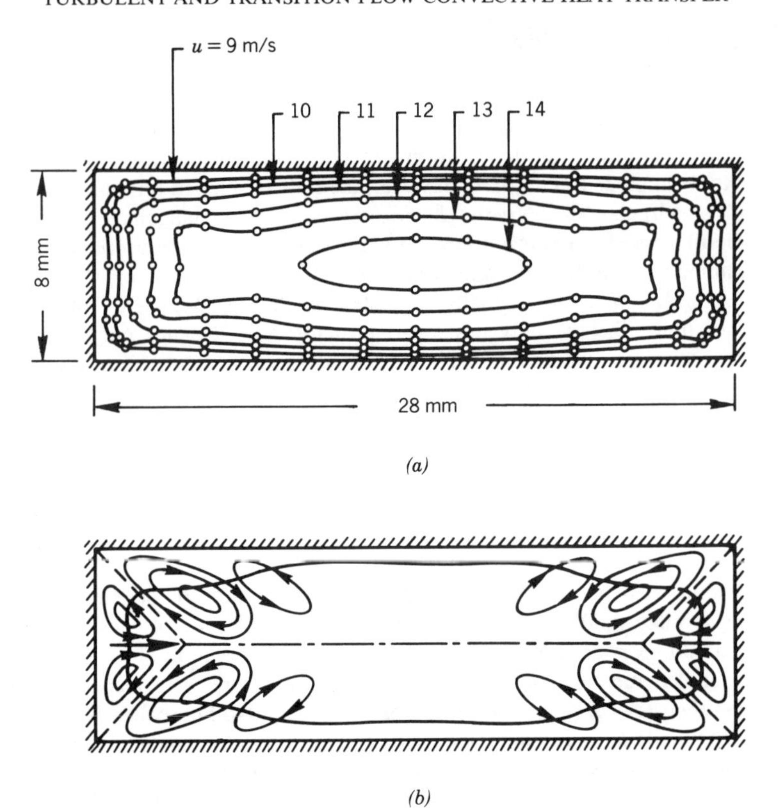

Figure 4.20. Fully developed turbulent velocity distribution at $\text{Re} = 6 \times 10^5$ in a rectangular duct with $\alpha^* = \frac{2}{7}$: (a) primary-flow isovels, (b) secondary-flow pattern [190].

various noncircular ducts do not fall on the laminar flow curve for a circular duct given by Eq. (3.14), i.e., $f\,\text{Re} = 16$. The continuous curves for laminar flow in various ducts are represented by $f\,\text{Re} = C$. The theoretical values of C for square ($\alpha^* = 1$), rectangular ($\alpha^* = \frac{2}{7}$) and trapezoidal ducts are 14.25, 17.75, and 14.15, respectively. The values of C for the square and rectangular ducts were computed from Eq. (3.157), whereas the value for the trapezoidal duct was obtained from Fig. 3.30.

Jones [28] introduced the laminar equivalent diameter D_l for rectangular ducts given by Eq. (4.15) or (4.16). Use of D_l in place of D_h in the circular duct correlations such as the Blasius formula (see Table 4.2) reduces the scatter of the experimental measurements from $\pm 20\%$ to $\pm 5\%$. Hence, substitution of D_l for D_h is recommended for calculating f values for rectangular ducts from the circular duct correlations of Table 4.2.

The present authors performed calculations for f values using the Techo et al. correlation of Table 4.2. On comparing these results with the experimental measurements for rectangular ducts ($0 \le \alpha^* \le 1$) in the range $5000 \le \text{Re} \le 10^7$, they arrived at the following correlation for rectangular ducts:

$$f = (1.0875 - 0.1125\alpha^*) f_c \tag{4.108a}$$

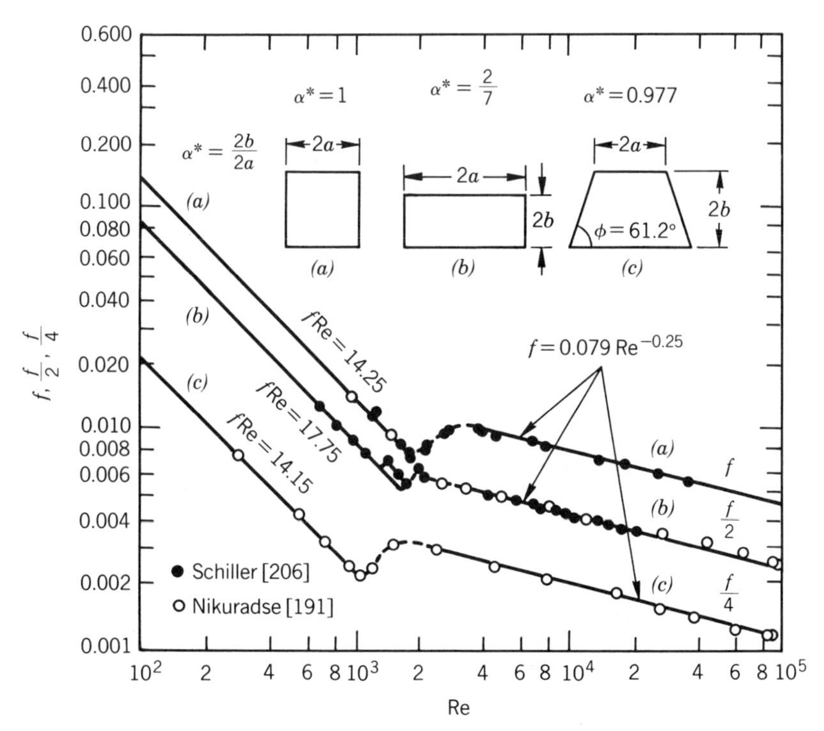

Figure 4.21. Friction factors for fully developed laminar, transition, and turbulent flows in smooth square, rectangular, and trapezoidal ducts [190].

where f_c is the friction factor for circular duct given by the Techo et al. correlation (Table 4.2).

The predictions of Eq. (4.108a) are on par with those determined by substituting D_l of Eq. (4.16) in the Techo et al. correlation.

The effect of rib-roughened walls on the fully developed turbulent friction factors in rectangular ducts has been explored by Wilkie et al. [207] and Han [208].

Heat Transfer. For most practical computations of the fully developed Nusselt numbers in rectangular ducts, the circular duct correlations are sufficiently accurate if the hydraulic diameter or the laminar equivalent diameter [Eq. (4.15) or (4.16)] replaces the circular duct diameter in the Reynolds and Nusselt numbers.

Recently, there have been several numerical studies to predict the fully developed Nusselt numbers in rectangular ducts using a variety of turbulence closure models. Patankar and Acharya [209] developed a simple mixing-length turbulence model which neglects the secondary flow. The Nusselt number predictions of [209] for $\alpha^* = 0.1, 0.2, 1$ are in excellent accord with the measurements of Brundrett and Burroughs [210]. The predictions of Launder and Ying [211] for $\alpha^* = 1$ are in slightly better accord with the measurements of Ref. 210, as the turbulence model employed by them takes into account the secondary flow.

Novotny et al. [212] experimentally determined the Nusselt numbers for turbulent flow of air (Pr = 0.7) in rectangular ducts ($\alpha^* = 0.1, 0.2, 1$) with the two shorter walls insulated and the two longer walls subject to uniform heat fluxes of equal magnitude.

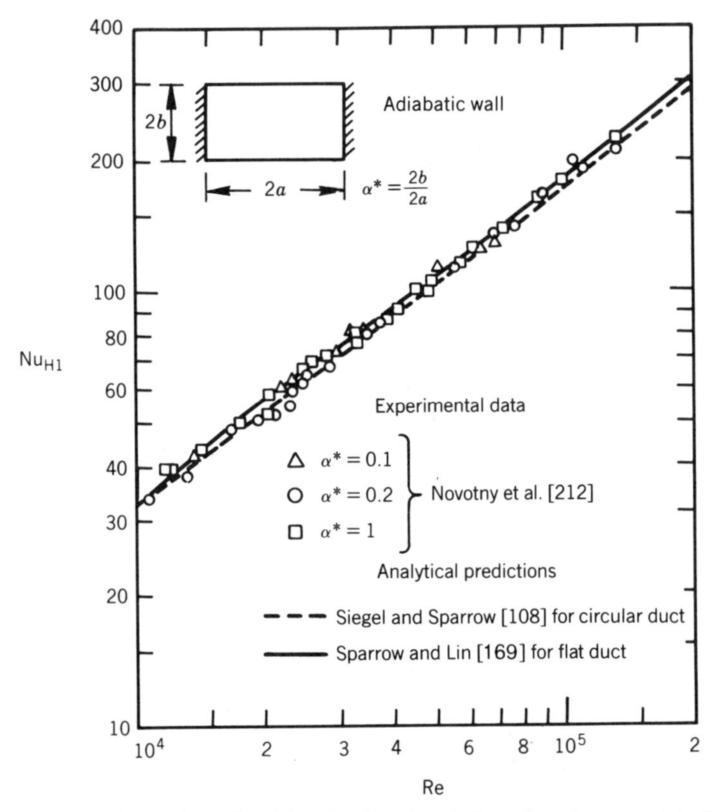

Figure 4.22. Nusselt numbers for fully developed turbulent flow in symmetrically heated rectangular ducts with shorter walls insulated for Pr = 0.7 [212].

The boundary condition on the longer walls corresponds to the (H1) boundary condition. Their experimental Nu_{H1} results in the range $10^4 < Re < 10^5$ are bracketed by the Nu_H analytical predictions of Siegel and Sparrow [108] for a circular duct and those of Sparrow and Lin [169] for a flat duct. These two sets of predictions are in close agreement among themselves, the flat duct predictions being 5% to 10% higher than the circular duct predictions in the range $10^4 < Re < 10^5$. Figure 4.22 shows the experimentally determined Nu_{H1} and their comparison with the predictions of Refs. 108, 169. An inspection of the experimental data points in Fig. 4.22 for $\alpha^* = 0.1, 0.2, 1$ and their close agreement with the flat duct ($\alpha^* = 0$) predictions shows that in a symmetrically heated rectangular duct, (i.e., with the same heat flux at the opposing walls), the Nu_{H1} are quite insensitive to α^*.

Sparrow et al. [213] experimentally investigated the effect of asymmetrical heating on Nu_{H1} for turbulent flow of air (Pr = 0.7) in a rectangular duct ($\alpha^* = 0.2$), employing the same test setup and apparatus as used by Novotny et al. [212]. They studied two cases of asymmetrical heating: (1) uniform heat flux at one long wall with the remaining three walls insulated; (2) uniform heat fluxes of unequal magnitude at the two long walls, the flux at one wall being twice that at the other, and the short walls insulated. Their experimental data are presented in Fig. 4.23, which also includes the data points from Fig. 4.22 for comparison with the case of symmetrical heating.

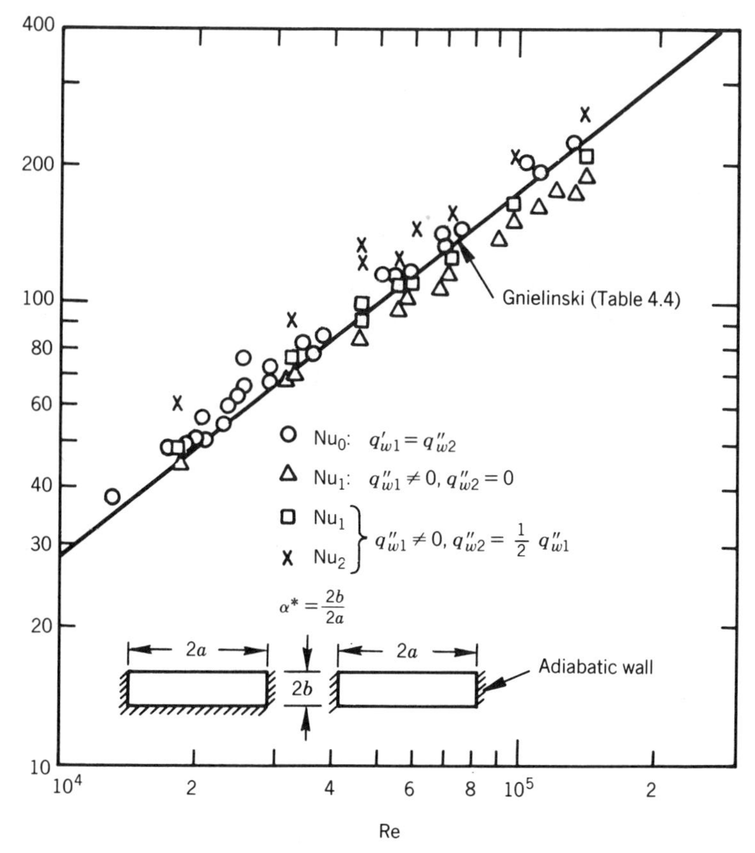

Figure 4.23. Nusselt numbers for fully developed turbulent flow in a symmetrically and an asymmetrically heated smooth rectangular duct ($\alpha^* = 0.2$) for Pr = 0.7 [213].

Included in Fig. 4.23 is a reference line representing the Gnielinski correlation for $0.5 < \text{Pr} < 2000$. This correlation is valid for uniform heating at four walls, i.e., the (H1) boundary condition.

The notation associated with the data points in Fig. 4.23 needs an explanation. The more strongly heated long wall is denoted as wall 1 (heat flux q''_{w1}), and the less strongly heated wall is denoted as wall 2 (heat flux q''_{w2}). The Nusselt numbers Nu_{H1} at the two walls are denoted simply as Nu_1 and Nu_2. For the case of symmetrical heating ($q''_{w1} = q''_{w2}$), $\text{Nu}_1 = \text{Nu}_2 = \text{Nu}_0$. The open circles represent the data for symmetrical heating, and the triangles represent the data for the heating only at one long wall. For the case with $q''_{w2} = q''_{w1}/2$, the squares represent the Nusselt number at the more strongly heated wall 1, and the crosses represent the Nusselt number at the less-heated wall 2. It is noted that all the data points for the asymmetrical heating lie within $\pm 20\%$ of those for the symmetrical heating.

On the basis of analytical considerations, Madsen [214] proposed a generalized heat transfer coefficient to describe the overall heat transfer characteristics of an asymmetrically heated flat duct. The purpose of introducing this coefficient was to render the fully developed Nusselt numbers independent of the asymmetry of the heating. For a rectangular duct, the redefinition of Madsen's overall coefficient $\bar{h}$ given by Sparrow et al. [213] is

$$\bar{h} = \frac{q''_{w1} + q''_{w2}}{(T_{w1} - T_m)_1 + (T_{w2} - T_m)_2} \tag{4.109}$$

The overall Nusselt numbers computed via Eq. (4.109) with the use of the hydraulic diameter $D_h = 4ab/(a + b)$ bring all of the data points of Fig. 4.23 for asymmetrical heating into virtual coincidence with the results for symmetrical heating, shown separately in Fig. 4.22.

Sparrow and Cur [215] reported extremely accurate experimental values of the fully developed Nusselt numbers for a fluid with Pr = 2.5 flowing turbulently in a rectangular duct with $\alpha^* = \frac{1}{18}$. For the case of symmetric heating (i.e., equal and uniform temperatures at the long walls with the short walls insulated), their results are represented by

$$\mathrm{Nu_T} = 0.0500\,\mathrm{Re}^{0.76} \tag{4.110}$$

For the case of asymmetric heating (i.e., one long wall at a uniform temperature and the remaining three insulated), their results are represented by

$$\mathrm{Nu_T} = 0.0464\,\mathrm{Re}^{0.76} \tag{4.111}$$

It should be emphasized that both Eqs. (4.110) and (4.111) are valid for Pr = 2.5 and $10^4 \leq \mathrm{Re} \leq 4.5 \times 10^4$. Also Eqs. (4.110) and (4.111) show that for the same Re, the asymmetric heating reduces the Nusselt numbers by about 8% for the specific case.

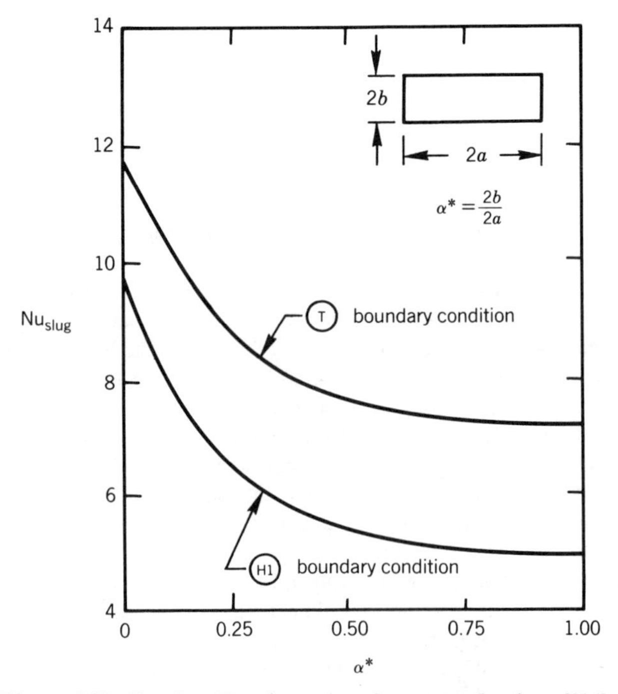

Figure 4.24. Slug flow Nusselt numbers for rectangular ducts [216].

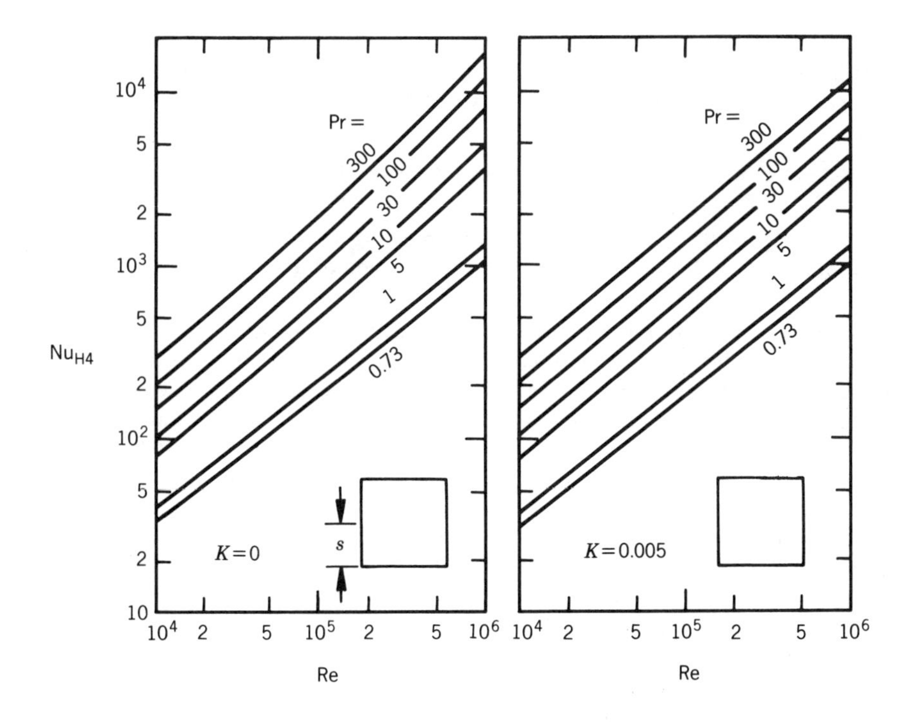

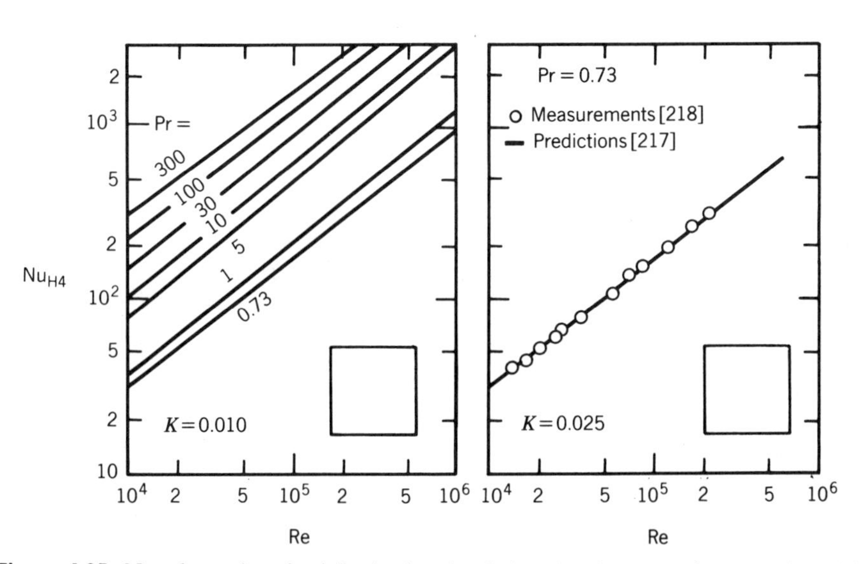

Figure 4.25. Nusselt numbers for fully developed turbulent flow in a smooth square duct with the (H4) boundary condition [217].

A simple correlation is available for estimating fully developed Nusselt numbers for turbulent flow of liquid metals in rectangular ducts with the Ⓣ and (H1) boundary conditions. This was derived by Hartnett and Irvine [216] for a uniform velocity distribution (slug flow) and a pure molecular-conduction heat transfer mechanism. This is a good approximation for liquid metals with $\Pr \to 0$. This correlation is given by

$$\mathrm{Nu} = \tfrac{2}{3}\mathrm{Nu}_{\mathrm{slug}} + 0.015\,\mathrm{Pe}^{0.8} \tag{4.112}$$

Here $\mathrm{Nu}_{\mathrm{slug}}$ is the Nusselt number corresponding to slug flow ($\Pr = 0$) through rectangular ducts. It is given in Fig. 4.24 as a function of α^* for rectangular ducts for the Ⓣ and (H1) boundary conditions.

Deissler and Taylor [217] developed an analytical solution for a square duct ($\alpha^* = 1$) with the (H4) boundary condition, i.e., axially constant wall heat flux and finite heat conduction along the wall circumference (see Table 3.1). Their predictions are presented in Fig. 4.25 for four values of the wall conduction parameter $K = ks/k_w\delta_w$, where k is the fluid conductivity, k_w is the thermal conductivity of the wall material, s is the distance between the corner and midpoint of the wall, and δ_w is the wall thickness. The value $K = 0$ corresponds to uniform wall temperature along the circumference, i.e., the (H1) boundary condition (see Table 3.1). Figure 4.25 provides a comparison between the prediction of Ref. 217 and the measurements of Lowdermilk et al. [218] for the case of $\Pr = 0.73$ and $K = 0.025$.

Based on the foregoing results for $\Pr > 0.5$, we can draw several useful conclusions pertaining to the relative magnitudes of the Nusselt numbers with heating at one, two, and four walls vis-a-vis circular duct correlation. First, on comparing the square duct ($\alpha^* = 1$) results of Fig. 4.25, for $K = 0$, with the predictions of the Gnielinski correlation (Table 4.4) for circular duct, it is noted that for $0.73 \le \Pr \le 300$ and $10^4 \le \mathrm{Re} \le 10^6$, the square duct results are approximately 6% lower than those predicted by the Gnielinski correlation. It is already pointed out in Sec. 4.3.2 on flat duct ($\alpha^* = 0$) that for $0.5 \le \Pr \le 100$ and $3 \times 10^4 \le \mathrm{Re} \le 10^6$, the Nusselt numbers are within +8% and −9% of those predicted by the Gnielinski correlation. Thus, it is reasonable to conclude that for rectangular ducts ($0 \le \alpha^* \le 1$), with heating at four walls, the Nusselt numbers can be determined within ±9% by the Gnielinski correlation for circular duct in the ranges $0.5 \le \Pr \le 100$ and $10^4 < \mathrm{Re} \le 10^6$.

Referring to Fig. 4.23, it is seen that with equal heating at two long walls (data points with circles), the Nusselt numbers for $\Pr = 0.7$ and $10^4 < \mathrm{Re} < 10^5$ are about 10% higher than those with equal heating at four walls represented by the Gnielinski correlation. As pointed out in Fig. 4.22 and the associated discussion, these results are quite insensitive to the variation in α^*. Next, referring to Eq. (4.110), it is found that with heating at two long walls for $\Pr = 2.5$ and $10^4 \le \mathrm{Re} \le 4.5 \times 10^4$, the Nusselt numbers are 10% lower than those predicted by the Gnielinski correlation. Thus, it is concluded that for $0.7 \le \Pr \le 2.5$ and $10^4 < \mathrm{Re} < 10^5$, the Nusselt numbers with equal heating at two long walls for $0 \le \alpha^* \le 1$ can be determined within ±10% from the circular duct correlation.

For heating at one long wall (data points with triangles in Fig. 4.23), it is seen that for $\Pr = 0.7$ and $10^4 < \mathrm{Re} < 10^5$, the Nusselt numbers are about 20% lower than those with equal heating at four walls represented by the Gnielinski correlation. Also, Eq. (4.111) shows that with heating at one long wall for $\Pr = 2.5$ and $10^4 \le \mathrm{Re} \le 4.5 \times 10^4$, the Nusselt numbers are 19% lower than those predicted by the Gnielinski correlation. Thus, it is concluded that for $0.7 \le \Pr \le 2.5$ and $10^4 < \mathrm{Re} < 10^5$, the Nusselt numbers for heating at one long wall can be estimated within +20% from the circular duct correlation. This is true in all probability for $0 \leq \alpha^* \leq 1$ although the results from which this inference is drawn are for $\alpha^* = 0.06$ and 0.2.

4.4.3 Hydrodynamically Developing Flow

Hartnett et al. [186] studied hydrodynamically developing flow in rectangular ducts with $\alpha^* = 0.1, 0.2, 1$. For a smooth entrance configuration, shown in Table 4.26, they found that the flow in the entrance region can remain laminar for Re considerably greater than 2200, thereby causing the hydrodynamic entrance length to be considerably longer than that for an abrupt entrance configuration, also shown in Table 4.26. For Re = 3000, they estimated the hydrodynamic entrance length to be $L_{hy}/D_h = 40$ with an abrupt entrance for $\alpha^* = 0.1, 0.2, 1$. For Re > 4000, the estimated hydrodynamic entrance length was $L_{hy}/D_h < 20$ with an abrupt entrance for $\alpha^* = 0.1, 0.2, 1$.

The analysis by Hartnett et al. [186] did not take into account the effect of the secondary flow on the hydrodynamically developing flow. A number of turbulence models of various degrees of sophistication are being developed to take account of the secondary flow in rectangular ducts. A succinct review of these models is given in Refs. 219–220.

The axial variation of the duct centerline velocity and the apparent Fanning friction factor in the hydrodynamic entrance region of a square duct are presented in Fig. 4.26. These results are predicted by Emery et al. [219]. The fully developed velocity distribution for a square duct predicted by them at $x/D_h = 96$ and $\text{Re} = 7.5 \times 10^4$ is in excellent accord with the measurements of Alexopoulos [221] at $x/D_h = 94$ and $\text{Re} = 7.7 \times 10^4$.

The local peaking of the axial centerline velocity distribution in Fig. 4.26 occurs at x/D_h, where the turbulent hydrodynamic boundary layers growing along the four walls merge; it is apparently due to some shear-layer interaction. The local peaking and leftward shift of the peaks with decrease in Re is observed in other numerical and experimental studies mentioned in Ref. 219. The initial decrease in f_{app} in Fig. 4.26 is probably also attributable to the shear-layer interaction. This effect is observed experimentally in low-α^* rectangular ducts, also mentioned in Ref. 219.

4.4.4 Thermally Developing Flow

No analytical or experimental results for thermally developing turbulent flow in rectangular ducts are available. However, the results for a circular duct, given in Sec. 4.2.4, may be used with the hydraulic diameter D_h or the laminar equivalent diameter D_l [Eq. (4.15) or (4.16)] replacing the circular duct diameter. While using the circular duct results, it should be borne in mind that the secondary flow in rectangular ducts leads to the formation of temperature hot spots in the corner region. The circular duct results omit the effect of the hot spots and therefore they must be viewed only as an approximation. The results presented in Sec. 4.3.4 for thermally developing flow in a flat duct ($\alpha^* = 0$) may also provide a reasonably good estimate for a rectangular duct with α^* not too large.

4.4.5 Simultaneously Developing Flow

Sparrow and Cur [215] conducted a careful investigation of the problem of simultaneously developing flow in a rectangular duct, invoking the analogy of heat and mass transfer processes. They reported experimental Nusselt numbers for a fluid with

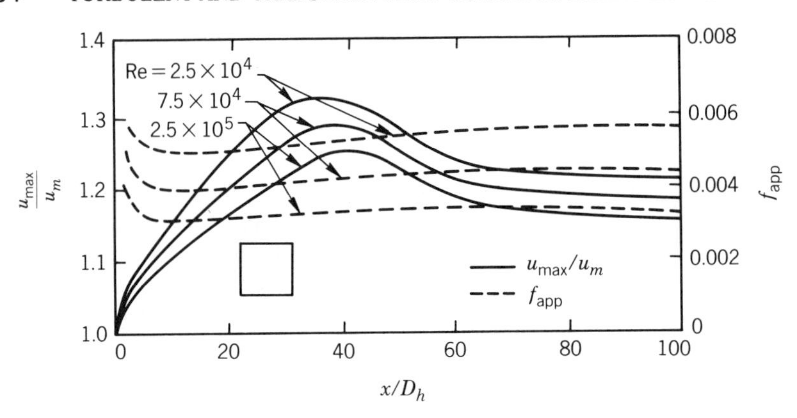

Figure 4.26. Turbulent flow centerline velocity and apparent friction factors in the hydrodynamic entrance region of a smooth square duct [219].

Pr = 2.5 flowing turbulently in the entrance region of a rectangular duct with $\alpha^* = \frac{1}{18}$. The reported results are for symmetric heating as well as asymmetric heating. For the symmetric heating the two long walls were isothermal (i.e., had the Ⓣ boundary condition) while the two short walls were adiabatic (i.e., insulated). For the asymmetric heating, one of the long walls was isothermal while the remaining three walls were adiabatic. The inlet configuration employed was the square inlet, i.e., the sharp-edged

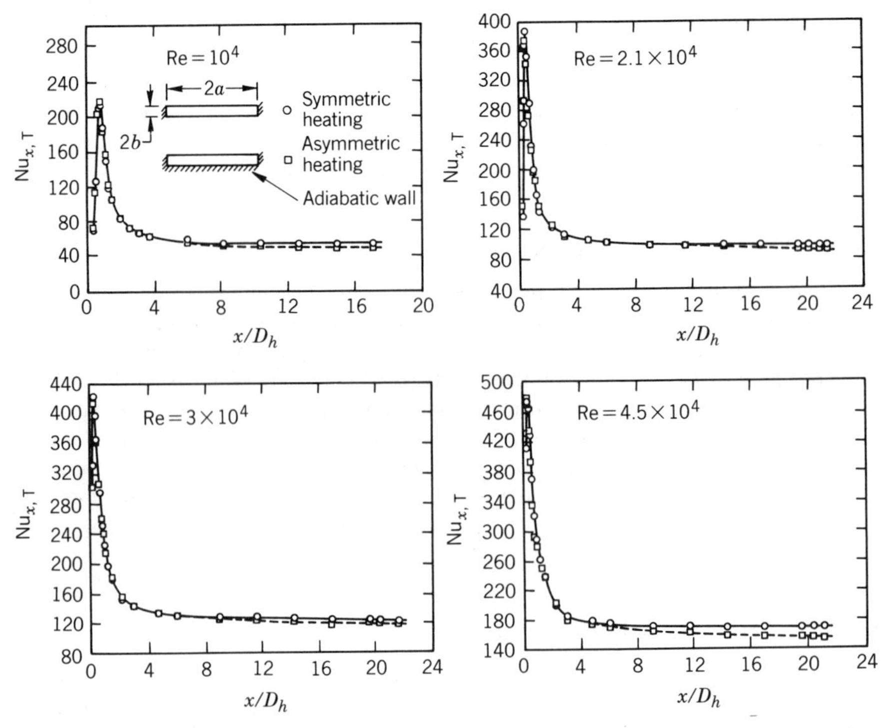

Figure 4.27. Local Nusselt numbers $Nu_{x,T}$ for simultaneously developing turbulent flow in a smooth rectangular duct with $\alpha^* = 2b/2a = \frac{1}{18}$ for Pr = 2.5 [215].

abrupt entrance shown in Table 4.26. For this entrance configuration, it was found that flow separation and reattachment occurred near the duct inlet, and this flow characteristic played a decisive role in shaping the axial distribution of the local heat transfer coefficient in the entrance region.

Figure 4.27 shows the variation of the local Nusselt numbers $\mathrm{Nu}_{x,\mathrm{T}}$ along the length of the duct for four values of Re. The novel features of the $\mathrm{Nu}_{x,\mathrm{T}}$ vs. x/D_h distribution shown in Fig. 4.27 are the initial rapid rise of $\mathrm{Nu}_{x,\mathrm{T}}$ and attainment of the peak. These features had not been reported in the literature before. The information available in the literature shows a steady decrease of $\mathrm{Nu}_{x,\mathrm{T}}$ with increasing x/D_h. The initial moderate values of the $\mathrm{Nu}_{x,\mathrm{T}}$, the rapid rise, the sharp peak, and the subsequent rapid decline correspond to the successive processes of flow separation, reattachment, and redevelopment. It is seen from Fig. 4.27 that the differences between the $\mathrm{Nu}_{x,\mathrm{T}}$ for the cases of symmetric and asymmetric heating are modest (about 7% in the fully developed region). The fully developed Nusselt numbers Nu_T for the two cases are given by Eqs. (4.110) and (4.111).

If the thermal entrance L_{th} is defined as corresponding to $\mathrm{Nu}_{x,\mathrm{T}} = 1.05\ \mathrm{Nu}_\mathrm{T}$, then for $\mathrm{Pr} = 2.5$ the entrance length for symmetric heating was found to lie in the range $5 \le L_{\mathrm{th}}/D_h \le 7$, while for asymmetric heating it was found to lie in the range $10 \le L_{\mathrm{th}}/D_h \le 13$ [215].

Sukomel et al. [222] also reported local Nusselt numbers for simultaneously developing flow in a rectangular duct with a smooth rounded entrance configuration of the inlet, shown in Table 4.26.

4.5 TRIANGULAR DUCTS

The triangular duct geometry has been analyzed in considerable detail for fully developed turbulent flow in sharp-cornered equilateral, isosceles, and right triangular ducts. A limited amount of information is also available for fully developed turbulent flow in a scalene triangular duct with two rounded corners. The other types of turbulent flows in triangular ducts have not been studied in any appreciable detail. A narrow-angle isosceles triangular duct involving hydrodynamically developing, thermally developing, and simultaneously developing turbulent flows has been investigated to a limited extent. In addition, hydrodynamically developing turbulent flow in a scalene triangular duct with two rounded corners and thermally developing flow in a sharp-cornered and a rounded-corner equilateral triangular duct has been investigated in limited detail.

4.5.1 Transition Flow

The experience with wide-angle triangular ducts with apex angles of the order of 30° is similar to that discussed in Sec. 4.4.1. With a square inlet configuration, a lower limit of $\mathrm{Re}_{\mathrm{crit}}$ is consistently observed at about 2000 in triangular ducts. As regards narrow-apex-angle isosceles triangular ducts with apex angles of the order of 10°, there is considerable uncertainty about the lower limit of $\mathrm{Re}_{\mathrm{crit}}$. This is attributed to the coexistence of laminar and turbulent flows in the corner region even for moderately high Re values.

4.5.2 Fully Developed Flow

Fully developed fluid flow and heat transfer results for various triangular ducts are presented below under separate captions.

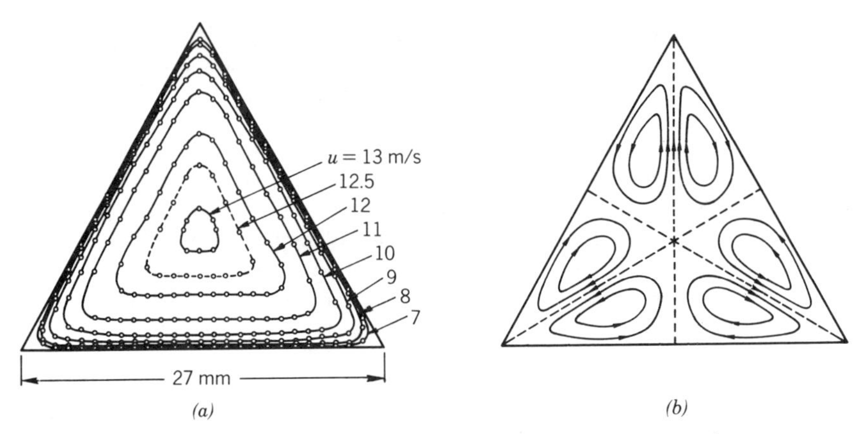

Figure 4.28. Fully developed turbulent velocity distribution at Re = 2.28×10^5 in an equilateral triangular duct: (*a*) primary-flow isovels, (*b*) secondary-flow pattern [191].

Equilateral Triangular Duct. Nikuradse [191] was the first investigator to deal with fully developed turbulent flow in an equilateral triangular duct. The classical velocity measurements of Nikuradse [191] are shown in Fig. 4.28*a*. Nikuradse [191] was also the first investigator to confirm the existence of the secondary flow in noncircular ducts (see Sec. 4.4.2) through a flow visualization study. The secondary-flow pattern in an equilateral triangular duct is portrayed in Fig. 4.28*b*. It consists of six counterrotating cells bounded by the corner bisectors. For each cell, the circulation is from the high-momentum central core region to the corner region via the corner bisector, with return along the wall and midwall bisector. Aly et al. [223] measured the maximum secondary-flow velocity in an equilateral triangular duct as about 1.5% of the primary-flow mean velocity u_m in the range $5.3 \times 10^4 < \text{Re} < 1.07 \times 10^5$. Aly et al. [223] also presented the fully developed velocity distribution in an equilateral triangular duct by the following wall coordinate (u^+, y^+) representation:

$$u^+ = 2.47 \ln y^+ + 5.08 \tag{4.113}$$

This equation is a least-square fit to 44 experimental measurements in the range $5.3 \times 10^4 < \text{Re} < 1.07 \times 10^5$. It is quite similar to von Kármán's equation in Table 4.1 for $y^+ > 30$. In the range $30 \le y^+ \le 3000$, Eq. (4.113) yields results that are within $\pm 3\%$ of those predicted by von Kármán's equation.

The fully developed friction factors for an equilateral triangular duct were measured by Altemani and Sparrow [224] under isothermal as well as nonisothermal conditions over the range $4000 < \text{Re} < 8 \times 10^4$. Their data are well represented by the following correlation due to Malák et al. [225]:

$$f = \frac{0.0425}{\text{Re}^{0.2}} \tag{4.114}$$

In the range $4000 < \text{Re} < 8 \times 10^4$, the predictions of Eq. (4.114) are 5.4% to 22.8% lower than those of the Blasius formula given by Eq. (4.32). The earlier experimental measurements by Nikuradse [191] and Schiller [206] for an equilateral triangular duct are presented in Fig. 4.29, where they are also compared with the predictions of the

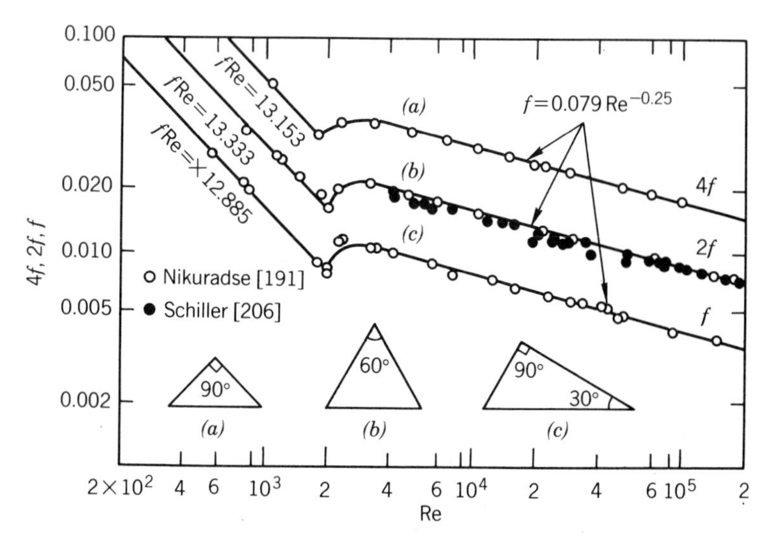

Figure 4.29. Friction factors for fully developed turbulent flow in smooth right-angled isosceles, equilateral, and right-angled scalene triangular ducts [190].

Blasius formula. The experimental measurements by Nikuradse [191] appear to be quite closely represented by the Blasius formula, whereas the measurements by Schiller [206] appear to be overestimated. In view of these comparisons, Eq. (4.114) is recommended for an equilateral triangular duct.

Aly et al. [223] developed a computational model for fully developed turbulent flow in an equilateral triangular duct. The f predictions of this model are in almost exact agreement with their measurements [223], which in turn are in excellent agreement with the predictions of Eq. (4.114).

It is apparent from the foregoing comparisons that the turbulent flow friction factors for equilateral triangular duct cannot be accurately determined from the circular duct correlation such as the Blasius formula. However, it is found that in the circular duct correlation if the hydraulic diameter $D_h = 2\sqrt{3}\,a$, $2a$ being the length of the equilateral duct side, is replaced by the equivalent diameter $D_l = \sqrt{3}\,a$, then the friction factors can be determined quite accurately. The equivalent diameter D_l was proposed by Ahmed and Brundrett [29]; refer to Sec. 4.1.6 for the physical significance of D_l. The calculations performed by the present authors show that for $4000 \le \text{Re} \le 8 \times 10^4$, the substitution of D_l in the Blasius formula yields f values that are within $+3\%$ and -11% of the measurements represented by Eq. (4.114); without the use of D_l, the Blasius formula yields f values that are up to $+22.8\%$ of the experimental measurements.

The fully developed Nusselt numbers for air (Pr = 0.7) in an equilateral triangular duct with the (H1) boundary condition on two walls and the third insulated are given by

$$\text{Nu}_{\text{H1}} = 0.019\,\text{Re}^{0.781} \tag{4.115}$$

This correlation is due to Altemani and Sparrow [224], and it represents their experimental measurements over the range $4000 < \text{Re} < 6 \times 10^4$ within $\pm 4\%$.

For $\text{Pr} = 0.7$ and $4000 \le \text{Re} \le 6 \times 10^4$, the predictions of the Gnielinski correlation (Table 4.4) using the Blasius formula [Eq. (4.32)] for f are within $+18\%$ of those of Eq. (4.115). When D_h in the Blasius formula is replaced by D_l given by Eq. (4.17), the Gnielinski correlation predicts $\mathrm{Nu}_{\mathrm{H1}}$ up to 9% higher than the experimental results.

Deissler and Taylor [217] developed an analytical solution for an equilateral triangular duct with the (H4) boundary condition, i.e., axially constant wall heat flux and finite heat conduction along the wall circumference. Their predictions are presented in Fig. 4.30 for four values of the wall conduction parameter $K = ks/k_w\,\delta_w$,

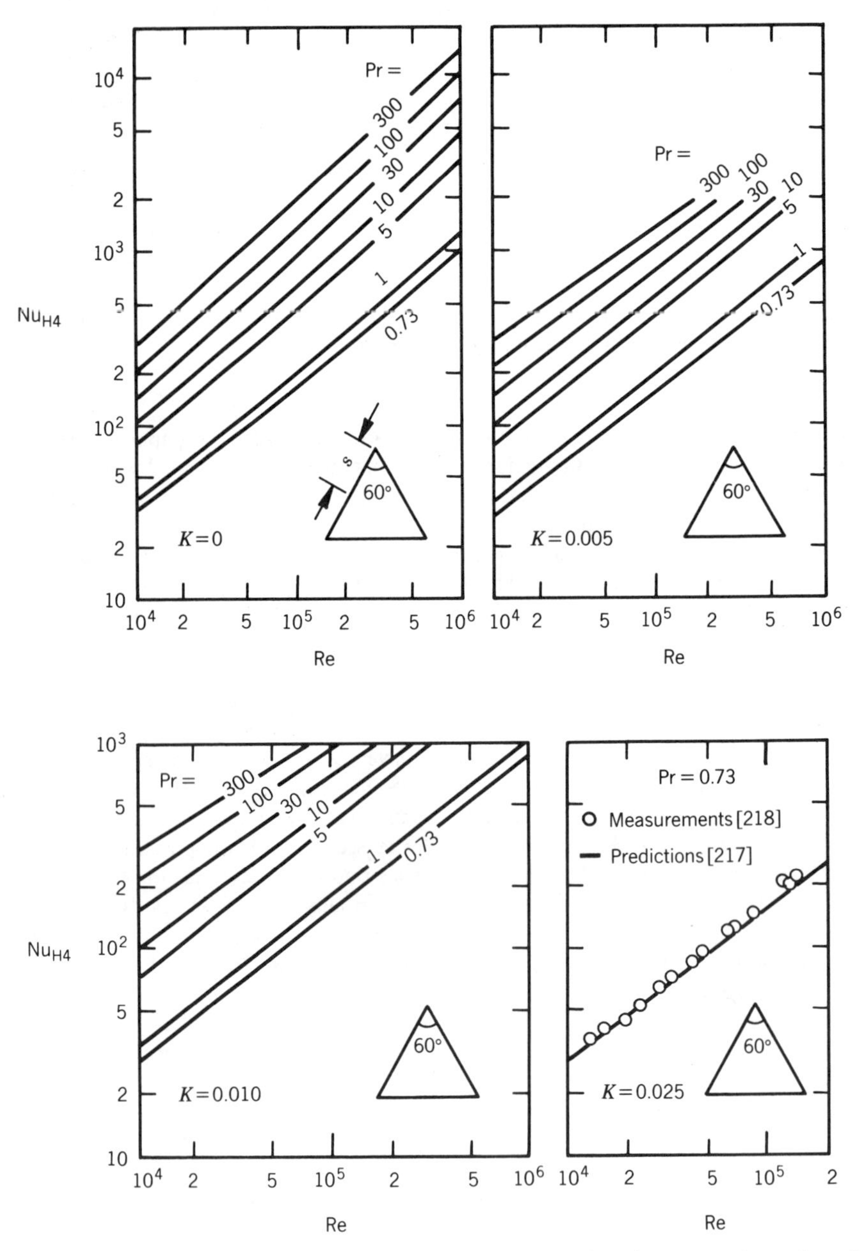

Figure 4.30. Nusselt numbers for fully developed turbulent flow in a smooth equilateral triangular duct with the (H4) boundary condition [217].

where k is the fluid thermal conductivity, k_w is the thermal conductivity of the wall material, s is the distance between the corner and midpoint of the wall (see inset in Fig. 4.30), and δ_w is the wall thickness. The value $K = 0$ corresponds to the (H1) boundary condition. Figure 4.30 also provides a comparison between the predictions of Ref. 217 and the measurements of Lowdermilk et al. [218] for Pr = 0.73 and $K = 0.025$.

On comparing the equilateral triangular duct Nusselt numbers given in Fig. 4.30 with the predictions of the Gnielinski correlation (Table 4.4), it is found that for $0.73 \le \mathrm{Pr} \le 300$ and $10^4 \le \mathrm{Re} \le 10^6$, the results of Fig. 4.30 for $K = 0$ are approximately 6% lower than those predicted by the Gnielinski correlation using the Techo et al. correlation (Table 4.2) for f. When D_h in the Techo et al. correlation is replaced by D_l given by Eq. (4.17), the agreement improves to 2%.

Isosceles Triangular Ducts. Among the family of triangular ducts, a narrow-angle isosceles triangular duct has received the most attention in the literature on turbulent fluid flow and heat transfer. A peculiarity of turbulent flow in this duct is that the viscous layer in the corner region can become large relative to the distance between the adjoining wall surfaces. This leads to the coexistence of laminar and turbulent flows in the corner region. As a result, the friction factors and heat transfer coefficients in a narrow-angle isosceles triangular duct tend to be low. As Re increases, the apparent laminar region decreases, and it is only at very high values of Re that fully turbulent behavior is observed.

The possible coexistence of laminar and turbulent flows in a narrow-angle triangular duct had been a point of controversy among various investigators. Eckert and Irvine [226] were the first to demonstrate such coexistence in the corner region, based on their flow measurements and flow visualization experiments in isosceles triangular ducts with apex angles $2\phi = 11.5°$ and $24.8°$. Figure 4.31 represents results of the measurements performed by Eckert and Irvine [226] in the isosceles triangular duct with $2\phi = 11.5°$. These results represent the laminar region thickness δ_l as a function of the duct height c measured from the apex for airflow. δ_l was determined visually by the use of smoke injection. The figure shows that at Re = 2000 the flow is laminar over 40% of the duct height, whereas at Re = 10^4 it is laminar over 10% of the duct height.

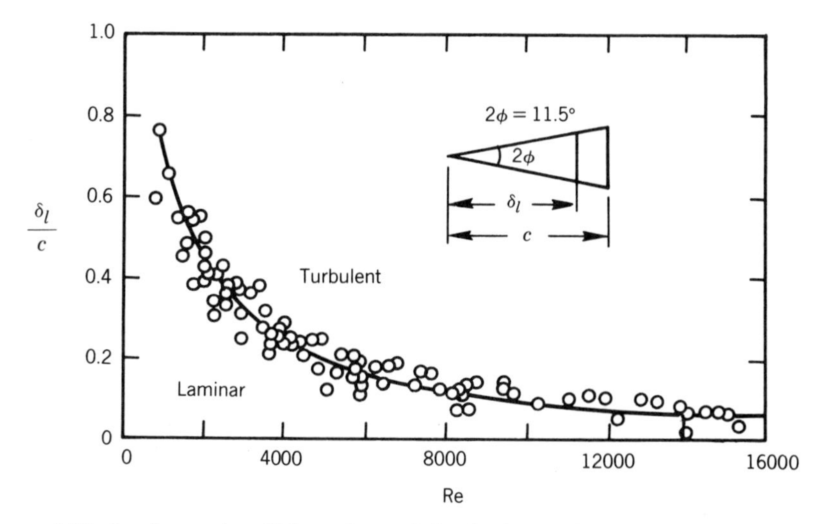

Figure 4.31. Laminar-region thickness in coexisting laminar and turbulent flows in a smooth isosceles triangular duct with apex angle $2\phi = 11.5°$ [226].

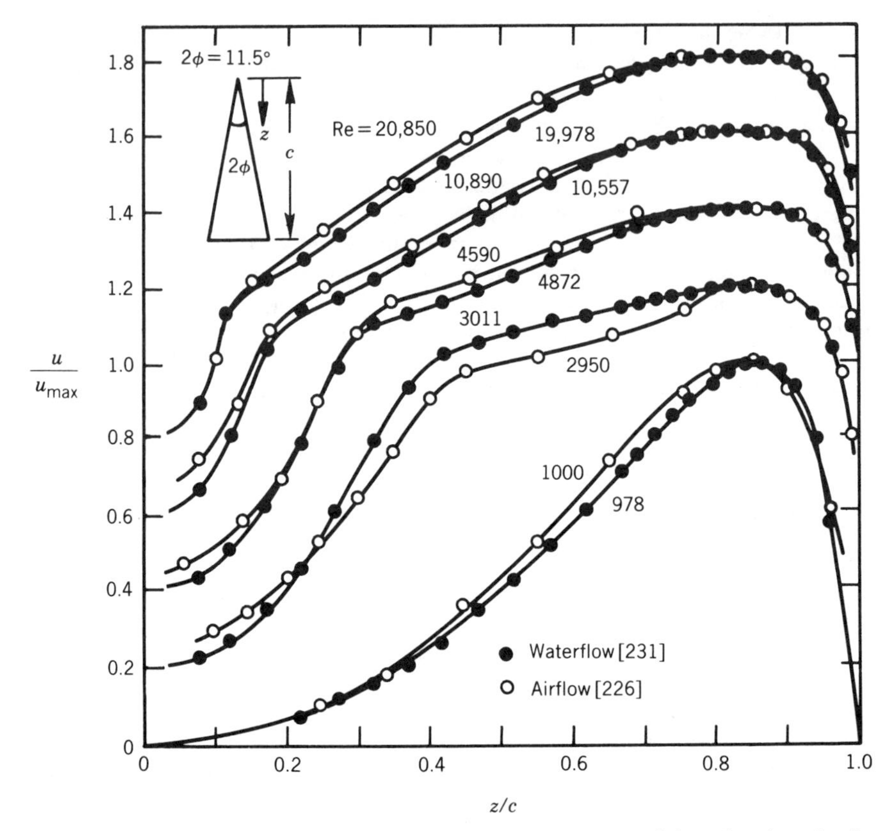

Figure 4.32. Fully developed turbulent velocity distribution in a smooth isosceles triangular duct with apex angle $2\phi = 11.5°$.

Hanks and coworkers [227–229] criticized the flow visualization techniques employed by Eckert and Irvine [226] and other investigators, casting doubt on the coexistence of laminar and turbulent flows in narrow-angle triangular ducts. More recent experimental investigations reported by Bandopadhayay and Hinwood [230] and Tung and Irvine [231] have confirmed the results of Eckert and Irvine [226] and conclusively established the coexistence of laminar and turbulent flows in such ducts. Tung and Irvine [231] provide the following empirical relation for the laminar-region thickness δ_l for $2\phi = 11.5°$ (see Fig. 4.31):

$$\frac{\delta_l}{c} = \frac{41.071}{\mathrm{Re}^{7/12}} \tag{4.116}$$

where c is defined in the inset of Fig. 4.31. Equation (4.116) is applicable to waterflow and shows that $\delta_l \to 0$ as $\mathrm{Re} \to \infty$.

Tung and Irvine [231] measured the fully developed point velocities for waterflow in an isosceles triangular duct with the apex angle $2\phi = 11.5°$. Their velocity measurements along the midplane through the apex are presented in Fig. 4.32, which also includes the measurements made earlier by Eckert and Irvine [226] for airflow in an isosceles triangular duct with the same apex angle. Rapley and Gosman [232] numerically predicted the fully developed turbulent velocity distribution in an isosceles triangular duct with $2\phi = 11.5°$, in good agreement with the measurements of Eckert and Irvine [226].

Usui et al. [233] numerically predicted the secondary-flow pattern in isosceles triangular ducts with apex angles $5.7° < 2\phi < 60°$. The maximum secondary-flow velocity in the range $5000 < \text{Re} < 5 \times 10^4$ was predicted to be approximately 2% of the mean axial velocity u_m for the primary flow.

Carlson and Irvine [234] measured fully developed friction factors in isosceles triangular ducts with $2\phi = 4.01°$, 7.96°, 12.0°, 22.3°, and 38.8°, covering the laminar, transition, and turbulent flow regimes. Their fully developed friction factors in the range $5000 < \text{Re} < 10^4$ can be represented by

$$f = \frac{C}{\text{Re}^{0.25}} \tag{4.117}$$

where C is a function of the apex angle 2ϕ given by

$$C = 0.060759 + 0.078631\phi - 0.078093\phi^2 - 0.202421\phi^3 + 0.282280\phi^4 \tag{4.118}$$

where ϕ is in radians. Equation (4.118) was developed by the present authors to fit the experimental data of Carlson and Irvine [234] for $4.01° < \phi < 38.8°$. Equations (4.117) and (4.118) reproduce the aforementioned experimental measurements within $\pm 0.1\%$. In the range $5000 \le \text{Re} \le 10^4$, the predictions of Eqs. (4.117) and (4.118) for $2\phi = 4°$ are 16.7% lower than those of the Blasius formula [Eq. (4.32)] using the hydraulic diameter. For $2\phi = 38°$ and the same Re range, the predictions of Eqs. (4.117) and (4.118) are 5.2% lower than those of the Blasius formula. This shows that for narrow-angle triangular ducts the influence of the coexisting laminar flow is reflected in the reduced friction factors calculated from the Blasius formula. Rapley and Gosman [232] numerically predicted fully developed friction factors for an isosceles triangular duct with $2\phi = 11.5°$ in good agreement with the measurements of Carlson and Irvine [234]. Tung and Irvine [231] reported measurements of the fully developed friction factor for waterflow in an isosceles triangular duct with $2\phi = 11.5°$. These measurements are in good agreement with the airflow measurements reported by Carlson and Irvine [234].

It is apparent from the foregoing comparisons that the turbulent flow friction factors for narrow apex angle isosceles triangular ducts cannot be accurately determined from the circular duct correlation such as the Blasius formula. However, it is found that if the hydraulic diameter $D_h = 4ab/(a + \sqrt{a^2 + b^2})$, $2a$ and $2b$ being the lengths of the base and height respectively of isosceles triangular ducts, is replaced by the so-called generalized length D_g given by Eq. (4.23), then the friction factors can be predicted quite accurately. D_g was proposed by Bandopadhayay and Ambrose [32]; refer to Sec. 4.1.6 for the physical significance of D_g. The calculations performed by the present authors show that for $5000 \le \text{Re} \le 10^4$, the substitution of D_g in the Blasius formula yields f values for isosceles triangular ducts with the apex angle $4° \le 2\phi \le 39°$ that are 0 to 6% higher than the experimental measurements represented by Eq. (4.117). Without the use of D_g in the Blasius formula, the f values are 5 to 17% higher. For $2\phi = 60°$, i.e., the equilateral triangular duct, the f values with the use of D_g are 10% to 15% higher than the experimental measurements represented by Eq. (4.114). Without the use of D_g, the Blasius formula predicts f values that are 10 to 22% higher than the measurements represented by Eq. (4.114). For $2\phi = 60°$, it was found that the use D_l [given by Eq. (4.17)] in the Blasius formula gives f values that are within 2% to 4% of the measurements represented by Eq. (4.114). Finally, for $2\phi = 90°$, it was found that the direct use of the Blasius formula with D_h yields the f values that are in excellent agreement with the experimental data of Nikuradse [191] as shown in Fig. 4.29.

In the view of the foregoing comparisons, the following recommendations are made to determine f factors for isosceles triangular ducts. (1) for $0 < 2\phi < 60°$, use the circular duct correlations from Table 4.2 with D_h replaced by D_g of Eq. (4.23). (2) For $2\phi = 60°$, use the circular duct correlations with D_h replaced by D_l of Eq. (4.17). (3) For $2\phi = 90°$, use the circular duct correlations with the characteristic dimension as D_h. (4) For $60° < 2\phi < 90°$, the use of circular duct correlations with D_h is probably accurate enough. (5) For $2\phi > 90°$, no definite recommendation could be made at this time due to lack of the experimental data.

Eckert and Irvine [235] measured fully developed Nusselt numbers for air (Pr = 0.7) in an isosceles triangular duct ($2\phi = 11.5°$) with the (H4) boundary condition covering the range $4300 < \mathrm{Re} < 2.4 \times 10^4$. For the heat conduction parameter $K = kD_h/k_w\,\delta_w = 0.0417$, their experimental measurements are correlated by

$$\mathrm{Nu}_{\mathrm{H4}} = 0.0325\,\mathrm{Re}^{0.66} \tag{4.119}$$

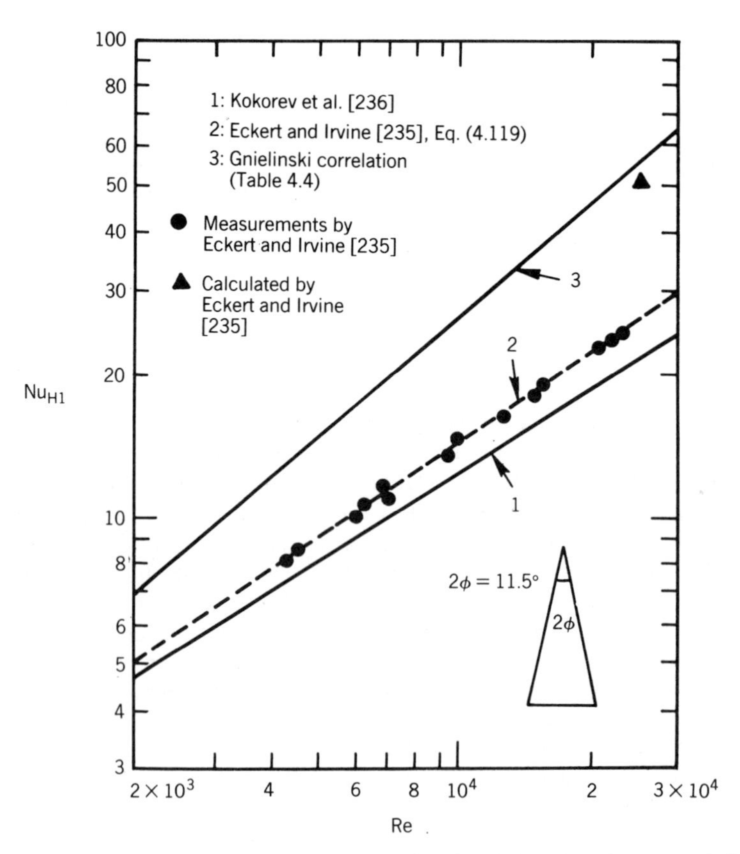

Figure 4.33. Nusselt numbers $\mathrm{Nu}_{\mathrm{H1}}$ for fully developed turbulent flow in a smooth isosceles triangular duct with apex angle $2\phi = 11.5°$ for airflow (Pr = 0.7).

The wall temperature entering the determination of Nu_{H4} in Eq. (4.119) is the circumferentially averaged value.

Kokorev et al. [236] developed an integral solution to the problem of fully developed turbulent flow in an isosceles triangular duct with the (H1) boundary condition. Their Nusselt number predictions for air (Pr = 0.7) are presented as curve 1 in Fig. 4.33, which also contains the predictions of Eq. (4.119) as curve 2. The experimental

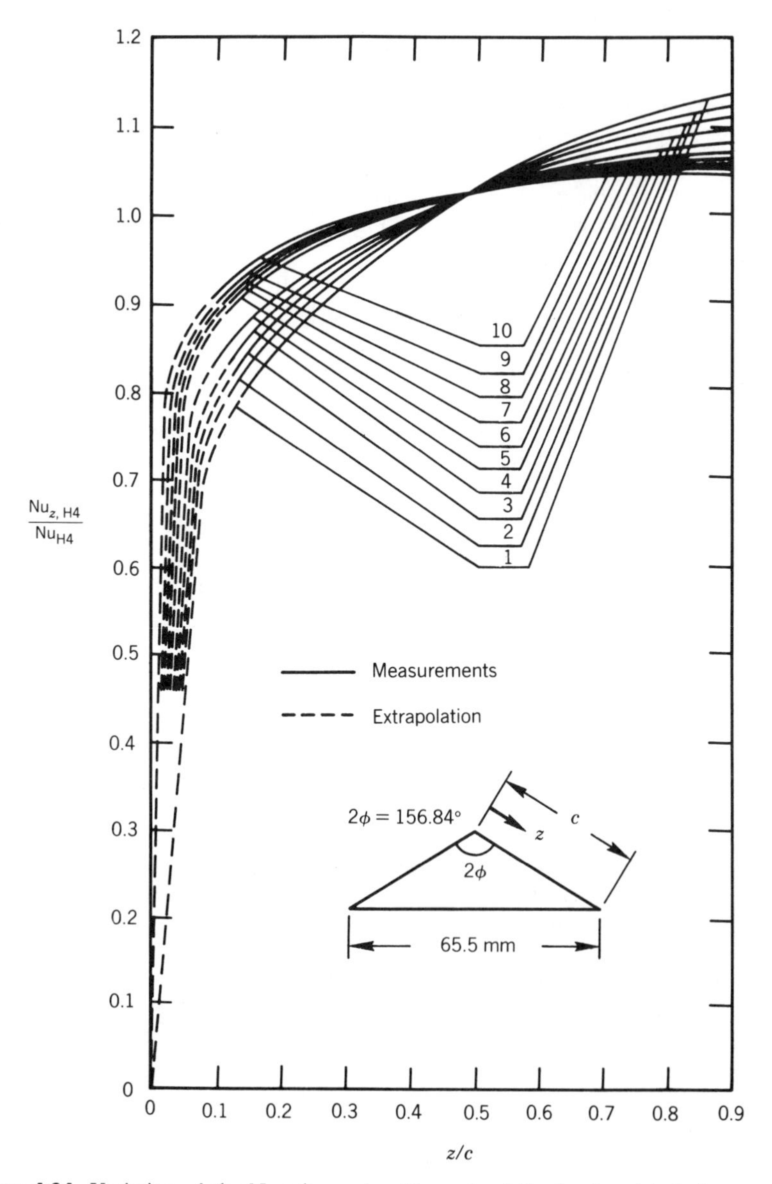

Figure 4.34. Variation of the Nusselt number Nu_{H4} for fully developed turbulent waterflow along the side of an isosceles triangular duct with apex angle $2\phi = 156.84°$ and wall conduction parameter $K = 1.0526$ [237]. See the text for Re values for curves 1 through 10.

measurements of Eckert and Irvine [235] are shown by the solid circles in Fig. 4.33, which also contains an additional single point, shown by a solid triangle, that was calculated by Eckert and Irvine [235] by the method of Deissler and Taylor [217]. Finally, the topmost curve, labeled 3, in Fig. 4.33 represents the Gnielinski correlation for a circular duct (see Table 4.4). It is apparent that the measured $\mathrm{Nu}_{\mathrm{H1}}$ values are approximately 40% lower than the values calculated from the circular duct correlation with the use of the hydraulic diameter. Hence the hydraulic diameter is not suitable for calculating the Nusselt numbers in narrow-angle triangular ducts. The method of Deissler and Taylor [217], which does not take into account secondary flows and the presence of exaggerated laminar sublayers in corner regions, is also not suited for this purpose. The predictions of Kokorev et al. [236] are in satisfactory agreement with the measurements, being about 8% lower.

Tokarev [237] performed experimental measurements in an isosceles triangular duct with apex angle $2\phi = 156.84°$ for waterflow ($2.71 < \mathrm{Pr} < 2.98$), covering the range $10^4 < \mathrm{Re} < 10^5$. For the (H4) boundary condition with the heat conduction parameter $K = ks/k_w\delta_w = 2.7815$ where $s = c/2$ (see Fig. 4.34), their circumferentially averaged Nusselt numbers $\mathrm{Nu}_{\mathrm{H4}}$ are given by

$$\mathrm{Nu}_{\mathrm{H4}} = 0.047\,\mathrm{Re}^{0.64} \tag{4.120}$$

Tokarev [237] also reported the local Nusselt numbers $\mathrm{Nu}_{\mathrm{H4}}(z)$ expressed as a function of the distance z from the vertex of the apex angle $2\phi = 156.84°$. Note that z is a cross-sectional spanwise coordinate as shown in the inset of Fig. 4.34, which shows the local Nusselt numbers $\mathrm{Nu}_{\mathrm{H4}}(z)$ measured by Tokarev [237]. The local Nusselt numbers are normalized by the circumferentially averaged Nusselt numbers $\mathrm{Nu}_{\mathrm{H4}}$ given by Eq. (4.120). The curves labeled 1 through 10 in Fig. 4.34 correspond to $\mathrm{Re} = 1.19 \times 10^4, 1.47 \times 10^4, 2.55 \times 10^4, 3.35 \times 10^4, 4.46 \times 10^4, 5.06 \times 10^4, 6.62 \times 10^4, 7.01 \times 10^4, 8.10 \times 10^4$, and 9.03×10^4, respectively.

The local Nusselt numbers of Fig. 4.34 are correlated by [237]

$$\mathrm{Nu}_{\mathrm{H4}}(z) = A\,\mathrm{Re}^n \tag{4.121}$$

where

$$n = \frac{0.0428}{[(z/c) + 0.3]^{1.5}} + 0.57 \tag{4.122}$$

The values of A in Eq. (4.121) are dependent on z/c. $A = 0$, 0.0136, 0.0228, 0.0332, 0.0400, 0.0490, 0.0613, 0.0692, 0.0772, and 0.0830, respectively for $z/c = 0$, 0.1, 0.2, 0.3, 0.4, 0.5, 0.6, 0.7, 0.8, and 0.9.

The fully developed turbulent flow Nusselt numbers $\mathrm{Nu}_{\mathrm{H1}}$ for an isosceles triangular duct with liquid metal flow ($\mathrm{Pr} < 0.03$) can be determined from Eq. (4.112). The slug flow Nusselt number $\mathrm{Nu}_{\mathrm{slug}}$ in this equation is given in Fig. 4.35. For narrow-apex-angle isosceles triangular ducts ($2\phi < 30°$), $\mathrm{Nu}_{\mathrm{slug}}$ may be taken from the curve for a circular-sector duct given in Fig. 4.35.

Usui et al. [238] employed an electrochemical technique to measure the fully developed Nusselt numbers Nu_{T} for $\mathrm{Pr} = 1630$ in an isosceles triangular duct with apex angle $2\phi = 11.4°$, covering the range $10^4 < \mathrm{Re} < 3 \times 10^4$. Their Nusselt numbers Nu_{T}, based on the mean wall temperature around the circumference, are presented in Fig. 4.36. These Nusselt numbers are nearly 50% lower than those computed from the circular duct correlation due to Gnielinski [94] (given in Table 4.4) with the

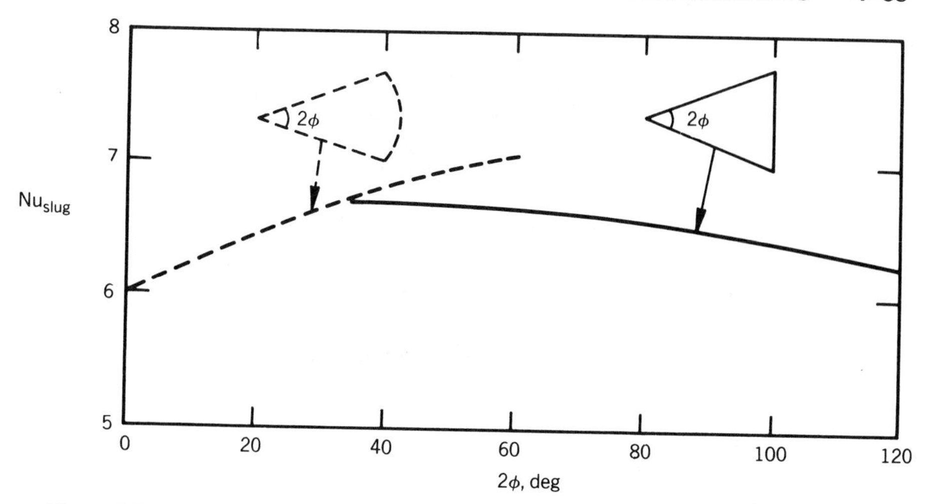

Figure 4.35. Slug flow Nusselt numbers for isosceles triangular and circular-sector ducts [216].

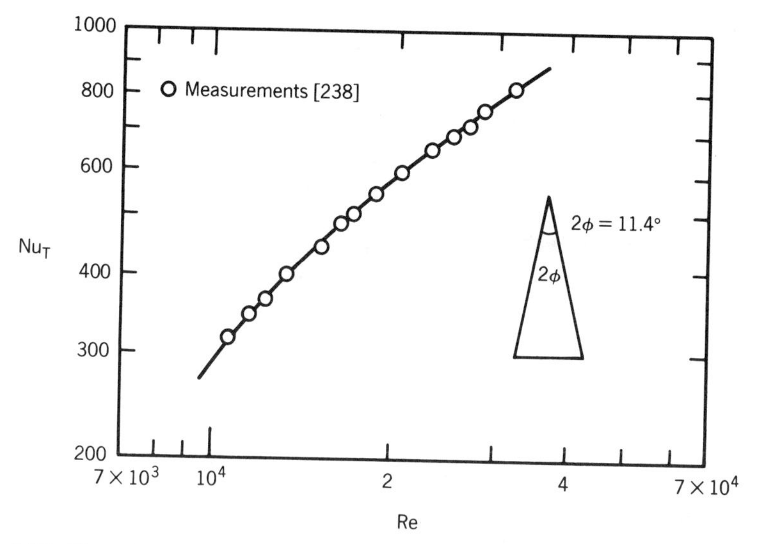

Figure 4.36. Circumferentially averaged experimental Nusselt numbers Nu_T for fully developed turbulent flow in a smooth isosceles triangular duct with apex angle $2\phi = 11.4°$ for $Pr = 1630$ [238].

use of the hydraulic diameter. Thus once again it is noted that the hydraulic-diameter concept breaks down in the case of narrow-apex-angle triangular ducts.

Right-Triangular Ducts. Nikuradse [191] measured fully developed velocity profiles in a right-angled isosceles triangular duct. The velocity profiles for this duct are presented in Fig. 4.37, which includes the primary-flow isovels and the secondary-flow pattern.

The fully developed turbulent friction factors for two right-angled isosceles triangular ducts including right-angled isosceles triangular duct are presented in Fig. 4.29.

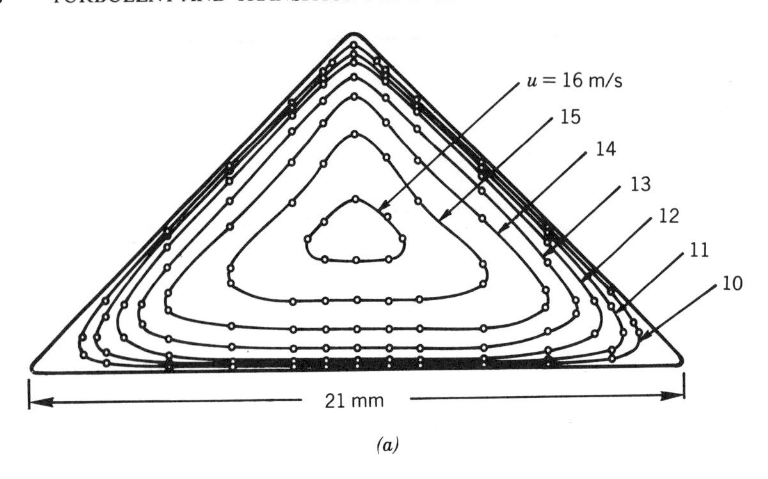

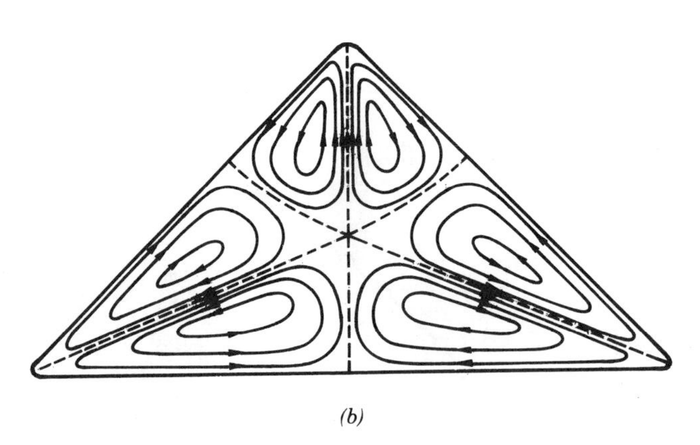

Figure 4.37. Fully developed turbulent velocity distribution at $Re = 8.1 \times 10^4$ in a right-angled isosceles triangular duct: (*a*) primary-flow isovels, (*b*) secondary-flow pattern [191].

These results are in excellent agreement with the predictions of the Blasius formula [Eq. (4.32)], affirming the applicability of the hydraulic-diameter concept to these ducts.

Scalene Triangular Duct with Two Rounded Corners. Obot and Adu-Wusu [239] reported measured fully developed Fanning friction factors in a smooth scalene triangular duct with two rounded corners, as shown in the inset to Fig. 4.38. The measured values are displayed in Fig. 4.38 together with the values for an isosceles triangular duct with the apex angle $2\phi = 38.8°$ for comparison purposes. These latter values were measured by Carlson and Irvine [234]. Figure 4.38 shows that for $Re \leq 1700$ the scalene triangular duct data are adequately represented by the circular duct line $f\,Re = 16$ whereas the isosceles triangular duct is not represented by this line. This difference appears to be due to the rounded corners of the scalene triangular duct. Figure 4.38 shows that for this latter duct transition to turbulent flow occurs over the range $1700 < Re < 1900$. Beyond $Re = 2000$, the scalene triangular duct data are about 17% lower than the Blasius relation $f = 0.079\,Re^{-0.25}$ [Eq. (4.32)] predictions.

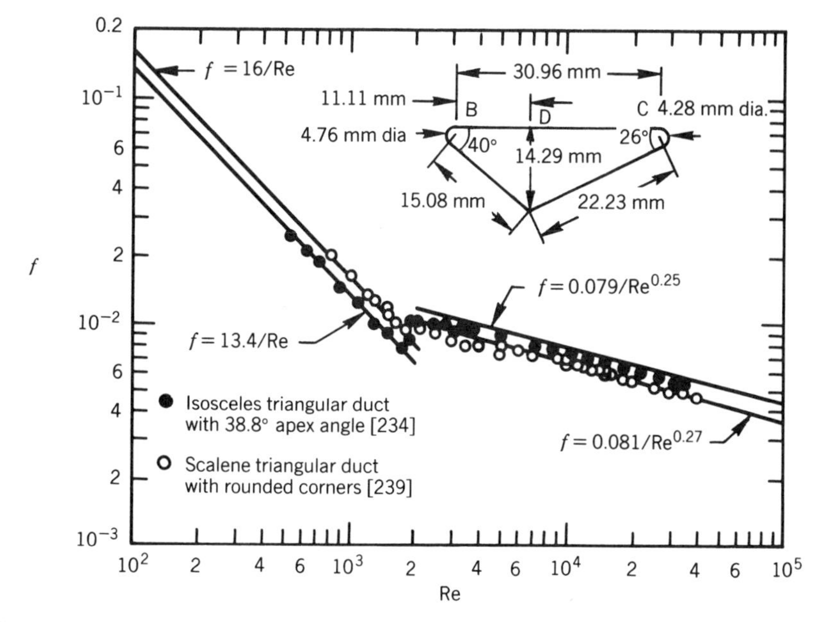

Figure 4.38. Friction factors for fully developed turbulent flow in a smooth scalene triangular duct with two rounded corners shown in the inset and an isosceles triangular duct with 38.8° apex angle [239].

The former data are representable by the relation $f = 0.081\ \mathrm{Re}^{-0.27}$. For laminar flow, the isosceles triangular duct data are represented by $f\ \mathrm{Re} = 13.4$, which yields about 14% lower values than the scalene triangular duct values. For turbulent flow, the isosceles triangular duct ($2\phi = 38.8°$) data are represented by the Blasius formula.

Obot [240] measured fully developed Nusselt numbers for turbulent airflow (Pr = 0.7) in the scalene triangular duct with two rounded corners shown in the inset of Fig. 4.38. The experiments employed a square inlet and the (H1) boundary condition. The measured $\mathrm{Nu}_{\mathrm{H1}}$ values are presented in Fig. 4.39 for axial locations $x/D_h = 55$ and 100, where fully developed conditions prevailed. The correlating relations for peripherally local $\mathrm{Nu}_{\mathrm{H1}}$ at the three duct walls are included. The circumferentially averaged $\mathrm{Nu}_{\mathrm{H1}}$ values for the scalene triangular duct are representable by the second Colburn correlation given in Table 4.4.

Obot et al. [241] conducted an experimental investigation of friction in a scalene triangular duct having two or three rib-roughened sides and two rounded corners. The shape of the duct is the same as shown in the inset of Fig. 4.38. The parametric study covered the ranges $0.13 < \varepsilon/a < 0.298$ and $2000 < \mathrm{Re} < 3 \times 10^4$. The study concluded that with three roughened sides, good estimates of friction factors can be obtained using the generalized friction relations for a circular duct discussed in Chap. 17.

4.5.3 Hydrodynamically Developing Flow

Obot and Adu-Wusu [239] measured the turbulent pressure drop and point velocities in the hydrodynamic entrance region of a scalene triangular duct with a square inlet for airflow. The duct cross section is shown as an inset to Fig. 4.38. The hydrodynamic

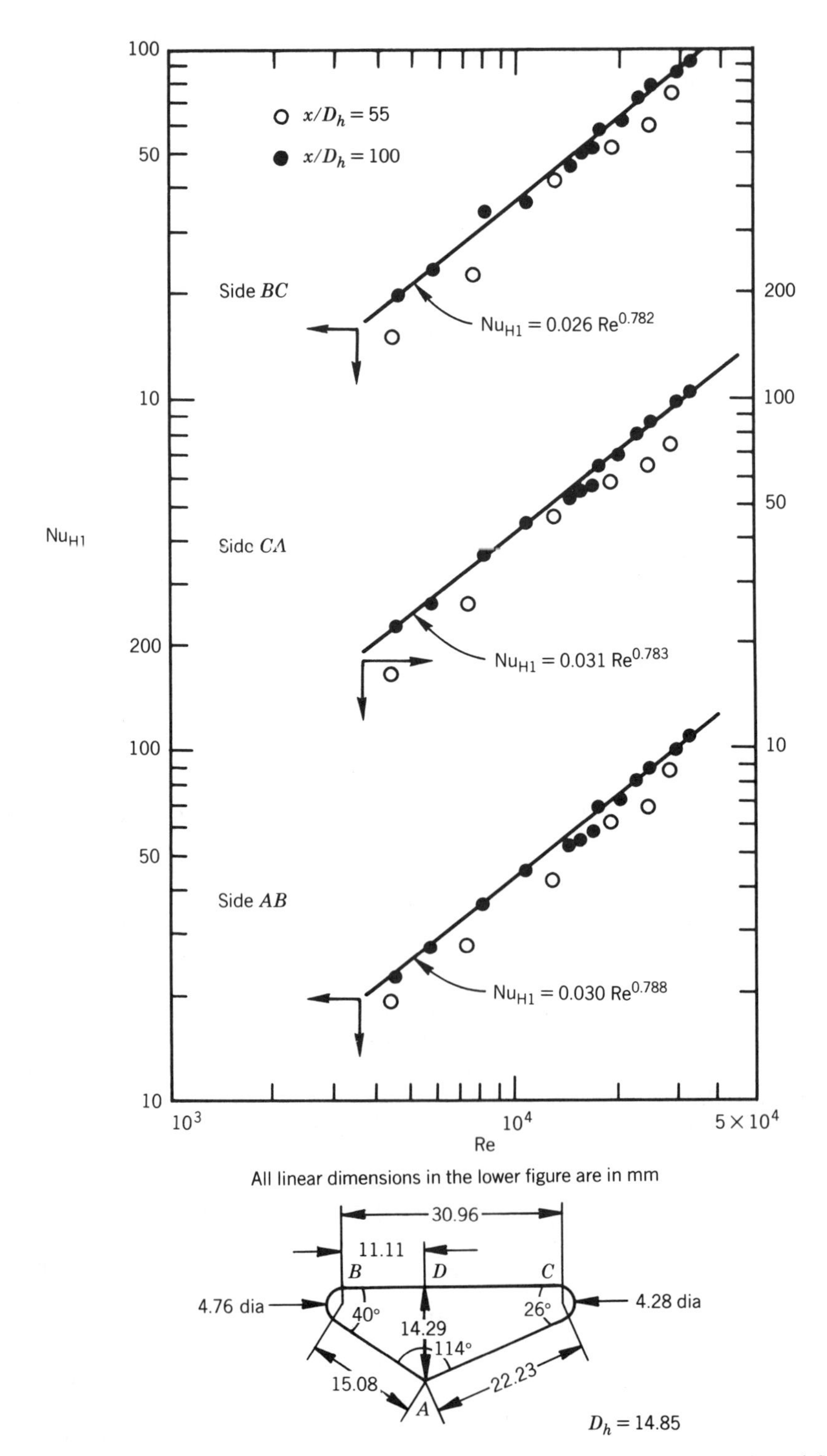

Figure 4.39. Peripherally local Nusselt numbers Nu_{H1} for fully developed turbulent airflow ($Pr = 0.7$) in a smooth scalene triangular duct with two rounded corners [240].

entrance length for this scalene triangular duct for $5400 < \mathrm{Re} < 3.17 \times 10^4$ is estimated to be $L_{hy}/D_h > 25$.

4.5.4 Thermally Developing Flow

Altemani and Sparrow [224] performed experimental measurements in the thermal entrance region of an equilateral triangular duct for airflow (Pr = 0.7) with the (H1) boundary condition on the two walls and the third wall insulated. Their local Nusselt numbers $\mathrm{Nu}_{x,\mathrm{H1}}$ and the thermal entrance lengths are presented in Figs. 4.40 and 4.41, respectively. The L_{th}/D_h values in Fig. 4.41 are based on 5% approach to the fully developed $\mathrm{Nu}_{\mathrm{H1}}$.

The flow in the corners of an equilateral triangular duct with rounded corners is not as stagnant as that in a sharp-cornered equilateral triangular duct. Consequently, the hydraulic-diameter concept works well with the former duct. This means that the

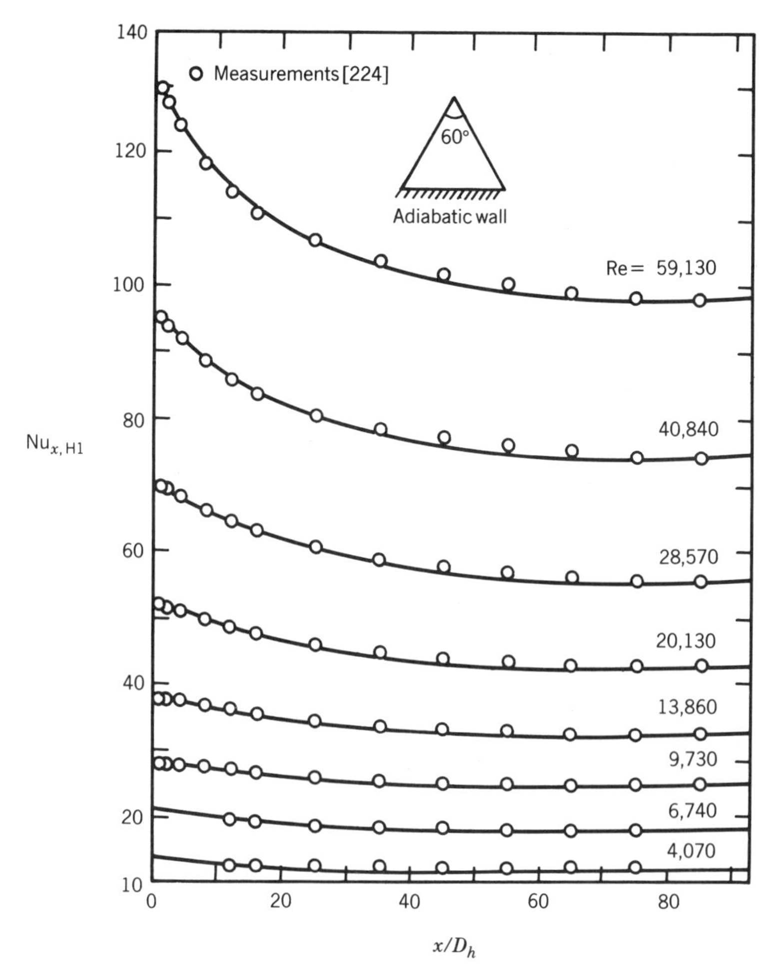

Figure 4.40. Local Nusselt numbers $\mathrm{Nu}_{x,\mathrm{H1}}$ for thermally developing and hydrodynamically developed turbulent airflow (Pr = 0.7) in a smooth equilateral triangular duct [224].

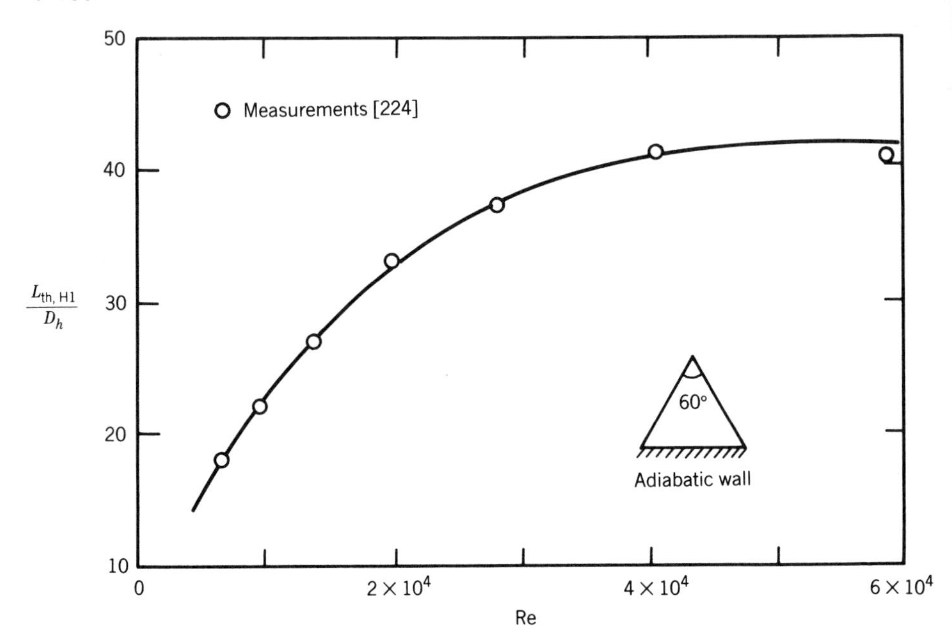

Figure 4.41. Thermal entrance lengths for thermally developing and hydrodynamically developed turbulent airflow (Pr = 0.7) in a smooth equilateral triangular duct [224].

circular duct correlations can be applied with good assurance to an equilateral triangular duct with rounded corners.

Campbell and Perkins [242] reported experimental data over the range $6000 < \mathrm{Re} < 4 \times 10^4$ for local friction and heat transfer coefficients with the (H1) boundary condition on all three walls of an equilateral triangular duct having rounded corners with a ratio of the corner radius of curvature to the hydraulic diameter of 0.15. The reported results are for hydrodynamically developed flow in the thermal entrance region with local wall (T_w) to fluid bulk mean (T_m) temperature ratio in the range $1.1 < T_w/T_m < 2.11$. The measured friction factors are correlated by

$$\frac{f}{f_{\mathrm{iso}}} = \left(\frac{T_w}{T_m}\right)^{-0.40+(x/D_h)^{-0.67}} \tag{4.123}$$

where f_{iso} is the isothermal friction factor, which can be obtained either from the Blasius formula [Eq. (4.32)] or the PKN formula (Table 4.3). The kinematic viscosity ν entering $\mathrm{Re} = u_m D_h/\nu$ in these calculation needs to be taken at the duct wall temperature T_w. Equation (4.123) is valid for $14.5 < x/D_h < 72$ and $1.10 < T_w/T_m < 2.11$.

The measured Nusselt numbers of Campbell and Perkins [242] are correlated by

$$\mathrm{Nu}_{x,\mathrm{H1}} = 0.021\,\mathrm{Re}^{0.8}\mathrm{Pr}^{0.4}\left(\frac{T_w}{T_m}\right)^{0.7}\Phi \tag{4.124}$$

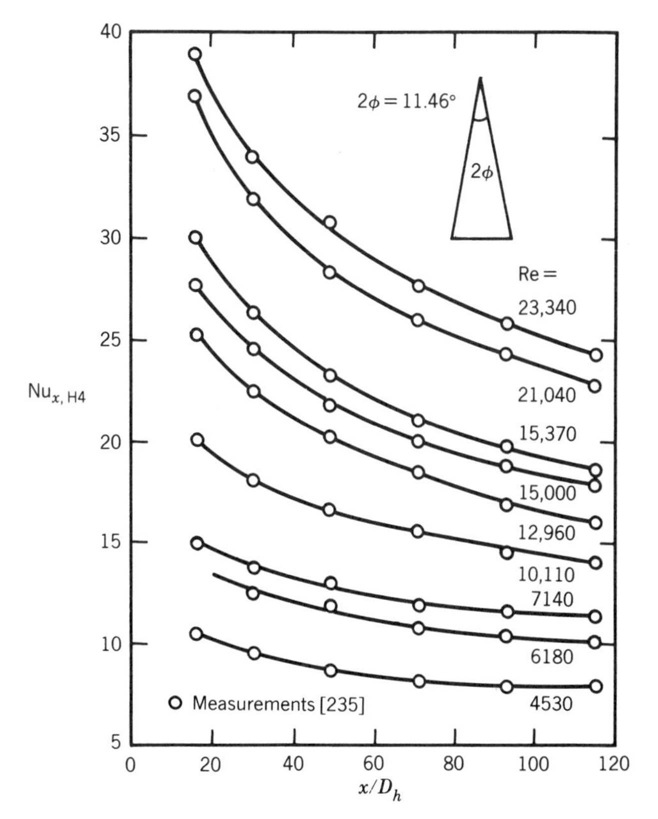

Figure 4.42. Circumferentially averaged local Nusselt numbers $\mathrm{Nu}_{x,\mathrm{H4}}$ for thermally developing and hydrodynamically developed airflow (Pr = 0.7) in a smooth triangular duct with apex angle $2\phi = 11.5°$ and wall conduction parameter $K = 0.0417$ [235].

For $6 < x/D_h \leq 50$, the correction factor Φ is given by

$$\Phi = \left[1 + \left(\frac{x}{D_h}\right)^{-0.7}\left(\frac{T_w}{T_m}\right)^{0.7}\right] \tag{4.125}$$

and for $x/D_h > 50$, $\Phi = 1$. Equation (4.124) is valid for $6 < x/D_h < 123$ and $1.10 < T_w/T_m < 2.11$. The fluid properties ν, α, and k entering Eq. (4.124) need to be evaluated at the fluid bulk mean temperature T_m.

Eckert and Irvine [235] reported circumferentially averaged local Nusselt numbers $\mathrm{Nu}_{x,\mathrm{H4}}$ in the thermal entrance region of an isosceles triangular duct with the apex angle $2\phi = 11.5°$ for airflow (Pr = 0.7). These results are presented in Fig. 4.42 for the heat conduction parameter $K = kD_h/k_w\delta_w = 0.0417$. It is seen that the thermal entrance length $L_{\mathrm{th}}/D_h > 100$. This value is at least an order of magnitude higher than that obtained for a circular duct. This suggests that the flow development in a narrow-angle triangular duct is significantly slower than that in a circular duct.

4.5.5 Simultaneously Developing Flow

Eckert and Irvine [235] reported circumferentially averaged local Nusselt numbers $\mathrm{Nu}_{x,\mathrm{H4}}$ for simultaneously developing airflow (Pr = 0.7) in an isosceles triangular duct with an apex angle $2\phi = 11.5°$ and with the heat conduction parameter $K = kD_h/k_w\delta_w$

= 0.0417 covering the range $4800 < \mathrm{Re} < 2.3 \times 10^4$. These results are found to be nearly identical to those for thermally developing flow shown in Fig. 4.42. It should be noted that these values were obtained with a square inlet.

4.6 ADDITIONAL SINGLY CONNECTED DUCTS

The remaining singly connected duct geometries analyzed in the literature are covered in this section. The available information for these ducts pertains mainly to the fully developed flows.

4.6.1 Elliptical Ducts

Cain and Duffy [243] measured fully developed velocity distributions in two elliptical ducts with $\alpha^* = 0.5$ and 0.6667 covering the range $2 \times 10^4 < \mathrm{Re} < 1.3 \times 10^5$. The measured point velocities for $\alpha^* = 0.5$ along the major and minor axes are presented in Fig. 4.43. The wall coordinates u^+ and y^+ appearing in this figure are defined in Eqs. (4.9) and (4.10), respectively.

The solid lines in Fig. 4.43 represent the u^+—y^+ correlation of von Kármán for a circular duct (see Table 4.1). The predictions of the von Kármán correlation are within +4.6% and −2.2% of the measurements along the major axis, and are within 3% of the measurements along the minor axis.

Cain et al. [244] presented velocity-distribution measurements in an elliptical duct ($\alpha^* = 0.5$) at $\mathrm{Re} = 1.2 \times 10^5$. Their measurements, shown in Fig. 4.44, point to the existence of the secondary flow in elliptical ducts, as evidenced by distortion of the isovels. Based on the shape of the isovels in Fig. 4.44a, the present authors have hypothesized the secondary flow pattern in an elliptical duct. This flow pattern, shown

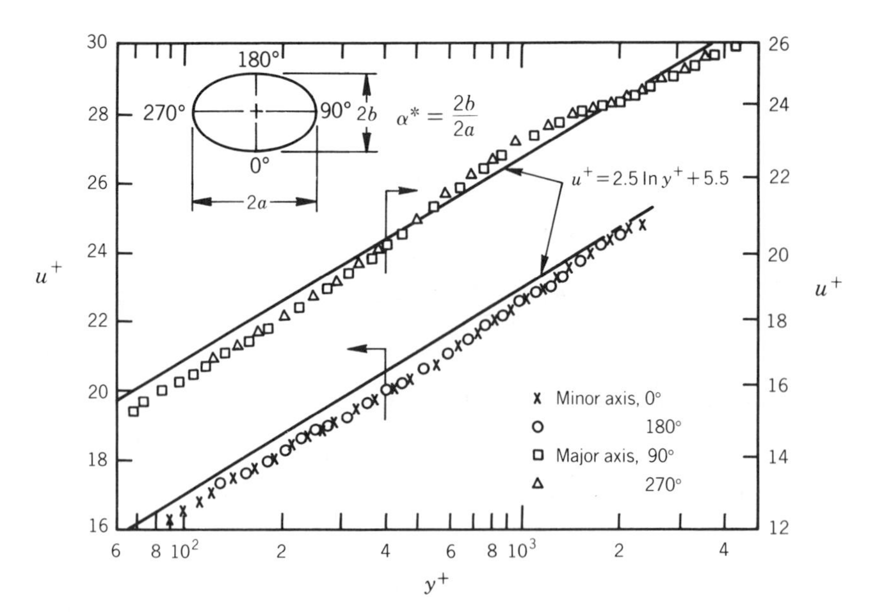

Figure 4.43. Fully developed turbulent velocity distribution along the major and minor axes of a smooth elliptical duct with $\alpha^* = 0.5$ (wall-coordinate representation) [244].

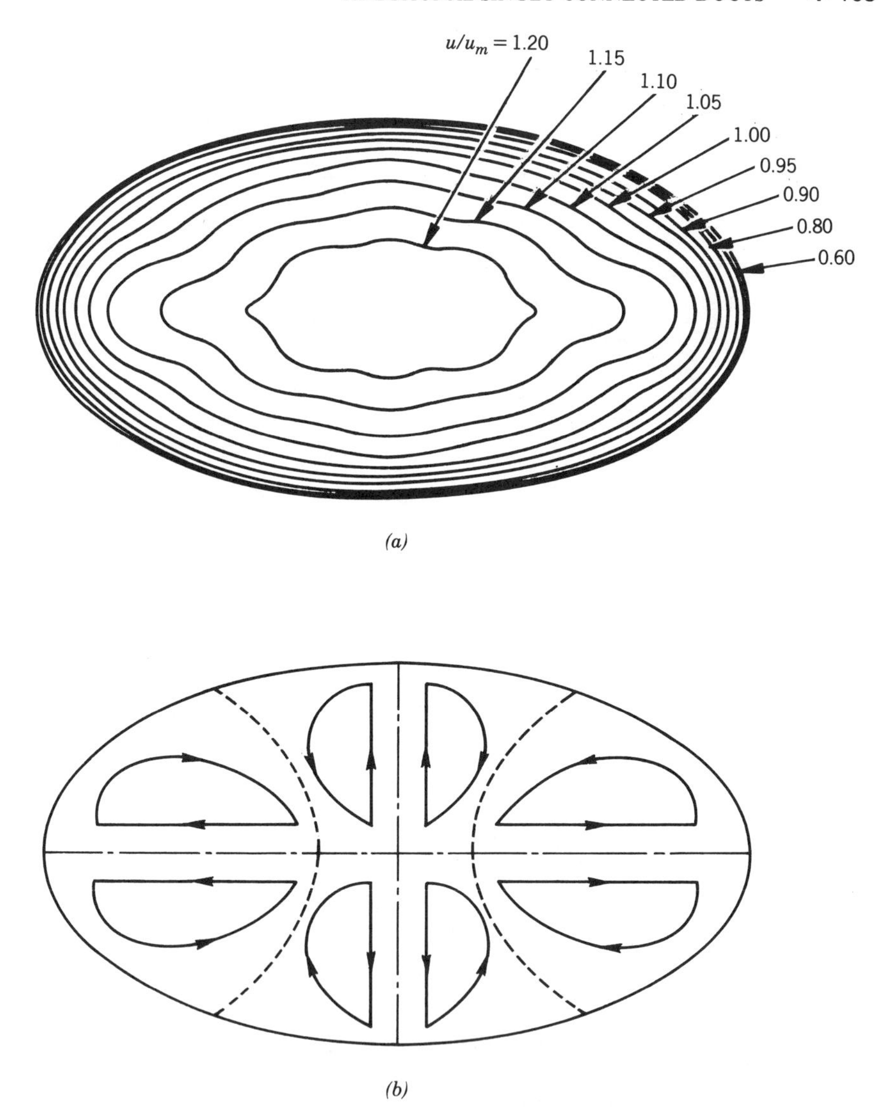

Figure 4.44. Fully developed turbulent velocity distribution at $\mathrm{Re} = 1.2 \times 10^5$ in an elliptical duct with $\alpha^* = 0.5$: (a) primary-flow isovels, (b) secondary-flow pattern.

in Fig. 4.44b, consists of eight counterrotating cells bounded by the major and minor axes. Despite the continuous change of curvature in elliptical geometry, the secondary flow appears to be as pronounced as that in sharp-cornered ducts such as rectangular (Fig. 4.20) and triangular (Figs. 4.28, 4.37) ones.

Measurements of the fully developed Fanning friction factor in elliptical ducts with $\alpha^* = 0.316$ and 0.415 were reported by Barrow and Roberts [245] for waterflow in the range $10^3 < \mathrm{Re} < 3 \times 10^5$. Cain and Duffy [243] reported additional measurements for elliptical ducts with $\alpha^* = 0.5$ and 0.6667 in the range $2 \times 10^4 < \mathrm{Re} < 1.3 \times 10^5$. Based on these two sets of measurements, the present authors developed the following

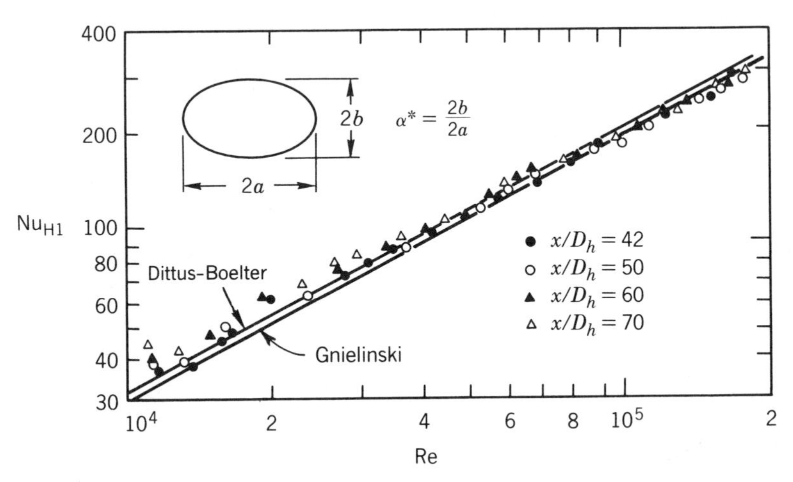

Figure 4.45. Nusselt numbers Nu_{H1} for fully developed turbulent airflow (Pr = 0.7) in a smooth elliptical duct with $\alpha^* = 0.5$ [246].

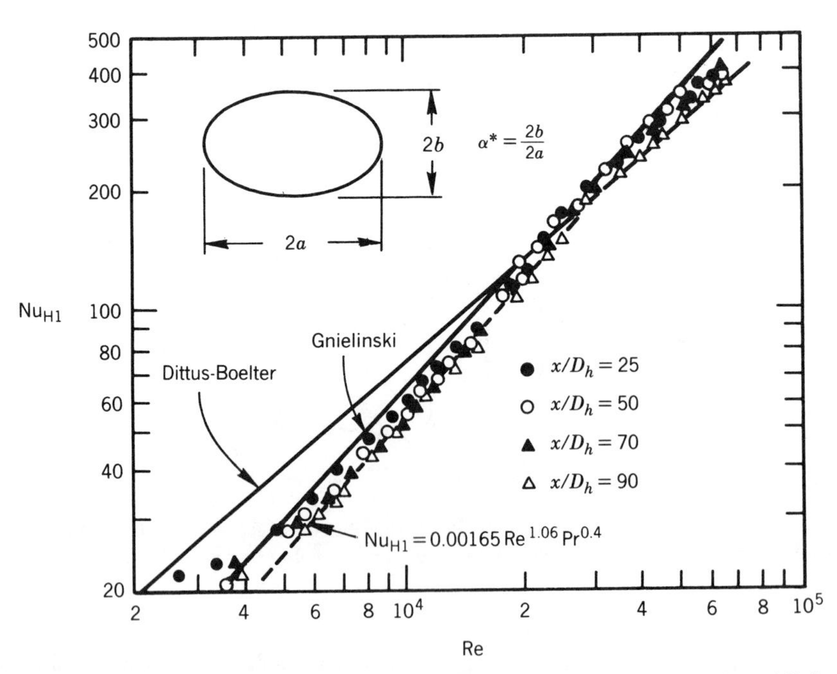

Figure 4.46. Nusselt numbers Nu_{H1} for fully developed turbulent waterflow (Pr = 6.5) in a smooth elliptical duct with $\alpha^* = 0.3415$ [246].

Fanning friction factor correlation valid for $0.316 < \alpha^* < 1$:

$$\frac{f}{f_c} = 0.4443 + 2.2168\alpha^* - 2.0431\alpha^{*2} + 0.3821\alpha^{*3} \tag{4.126}$$

where f_c is the friction factor for a circular duct ($\alpha^* = 1$), given as a function of Re by the Blasius formula [Eq. (4.32)].

Equation (4.126) shows that for $\alpha^* = 0.6667$, 0.5, and 0.415, the f values are respectively 13%, 8%, and 5% higher than the f_c values. For $\alpha^* = 0.316$, the f value is 5% lower than the f_c value. These departures from f_c are as reported in Refs. 243, 245.

Cain et al. [246] measured fully developed Nusselt numbers in elliptical ducts with $\alpha^* = 0.5$ and 0.666 for turbulent airflow (Pr = 0.7) with the (H1) boundary condition, covering the range $10^4 < \text{Re} < 2 \times 10^5$. Their results for $\alpha^* = 0.5$ are presented in Fig. 4.45. Close agreement is found between these results and the predictions of the two circular duct correlations Dittus-Boelter and Gnielinski given in Table 4.4. Similar agreement is found for $\alpha^* = 0.666$. This indicates that in the range $0.5 \leq \alpha^* \leq 1$ the Nusselt numbers can be determined quite accurately from the circular duct correlations.

Cain et al. [246] also reported fully developed Nusselt numbers in elliptical ducts with $\alpha^* = 0.3415$ and 0.3750 for turbulent waterflow (Pr = 6.5) with the (H1) boundary condition in the range $2.5 \times 10^3 < \text{Re} < 8 \times 10^4$. Their results for $\alpha^* = 0.3415$ are presented in Fig. 4.46. These measurements, as well as those for $\alpha^* = 0.3750$ for $\text{Re} < 2.5 \times 10^4$, are not in accord with the Dittus-Boelter correlation. However, the Gnielinski correlation accurately represents the data for the entire Re range. In the

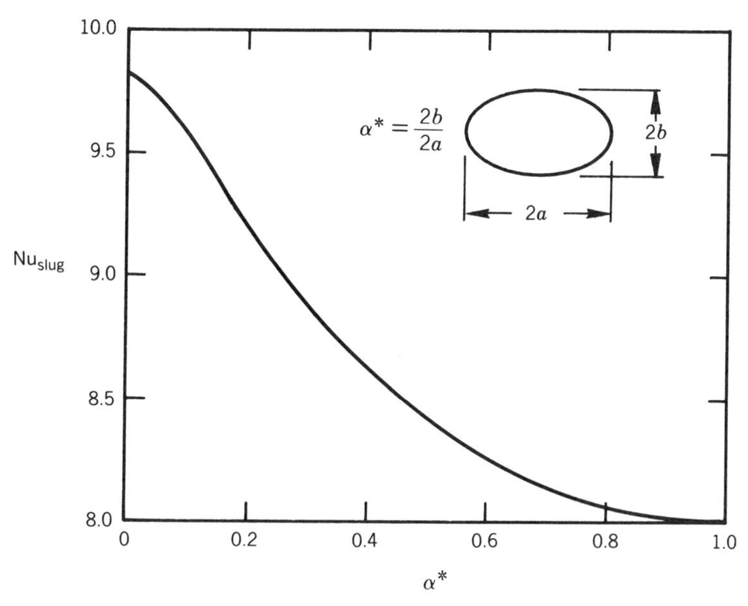

Figure 4.47. Slug flow Nusselt numbers for elliptical ducts with the (H1) boundary condition [216].

range $4000 < \mathrm{Re} < 2 \times 10^4$, the data of Cain et al. [246] are correlated by

$$\mathrm{Nu}_{\mathrm{H1}} = 0.00165\,\mathrm{Re}^{1.06}\mathrm{Pr}^{0.4} \tag{4.127}$$

which was presented earlier by Barrow and Roberts [245] to correlate their data on waterflow in an elliptical duct with $\alpha^* = 0.284$ and $4000 < \mathrm{Re} < 2 \times 10^4$. Thus Eq. (4.127) is the best fit to experimental data for the low aspect ratio ($0.28 \le \alpha^* \le 0.38$) elliptical ducts for $4000 < \mathrm{Re} < 2 \times 10^4$ and it aggrees with the Gnielinski correlation within -5%.

Ryadno and Kochubei [247] developed an approximate analytical solution to the unsteady convective heat transfer problem for elliptical ducts. Their steady-state results for airflow in a circular duct in the range $10^4 < \mathrm{Re} < 5 \times 10^4$ are in good agreement with the predictions of the Petukhov-Kirillov-Popov and Gnielinski correlations given in Table 4.4 with the f value obtained from the Filonenko correlation given in Table 4.3.

Based on the foregoing comparisons, it is concluded that the Gnielinski correlation for circular duct can be confidently employed to determine the fully developed Nusselt numbers for elliptical ducts for $\mathrm{Pr} \ge 0.5$.

The fully developed Nusselt numbers for liquid metals can be computed from Eq. (4.112), which requires a knowledge of the slug Nusselt number $\mathrm{Nu}_{\mathrm{slug}}$ corresponding to $\mathrm{Pr} = 0$. For elliptical ducts with the (H1) boundary condition, values of $\mathrm{Nu}_{\mathrm{slug}}$ are given in Fig. 4.47, based on the computations by Hartnett and Irvine [216].

4.6.2 Trapezoidal Ducts

The classical fully developed velocity profiles measured by Nikuradse [191] in a trapezoidal duct are presented in Fig. 4.48, which also includes the schematic secondary-flow pattern for a trapezoidal duct.

The fully developed Fanning friction factors for the trapezoidal duct of Fig. 4.48*a* are presented in Fig. 4.21. The results in this figure show that the turbulent flow f values for the duct in question can be calculated from the Blasius formula [Eq. (4.32)], affirming the applicability of the hydraulic-diameter concept to the trapezoidal duct.

Rodet [248] conducted an experimental study to measure the fully developed turbulent velocity distribution in a trapezoidal duct with the corner angle $\phi = 75°$. Nakayama et al. [203] developed a numerical solution for the fully developed turbulent flow in the trapezoidal duct with the corner angle $\phi = 75°$. The agreement between the predictions of Ref. 203 and the measurements of Ref. 248 is satisfactory. The predicted friction factors of Ref. 203 agree with those of the Blasius formula [Eq. (4.32)].

4.6.3 Circular Ducts With Rectangular Indentations

The classical paper of Nikuradse [191] included information on two rather unusual ducts: a circular duct with a single rectangular indentation and a circular duct with twin rectangular indentations. The velocity distributions in these two ducts are shown in Figs. 4.49 and 4.50, respectively. These figures also include the schematic secondary-flow patterns.

The fully developed friction factors for the two ducts are presented in Fig. 4.51, which also includes the friction factors for a rectangular duct with $\alpha^* = \frac{2}{7}$ for comparison. The turbulent friction factors for all three ducts can be calculated from the Blasius formula [Eq. (4.32)], attesting to the applicability of the hydraulic-diameter concept to the ducts in question.

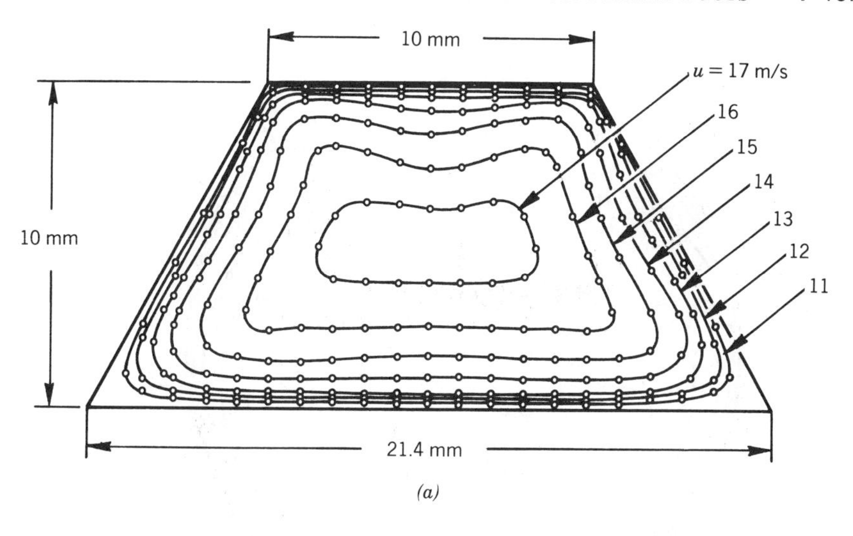

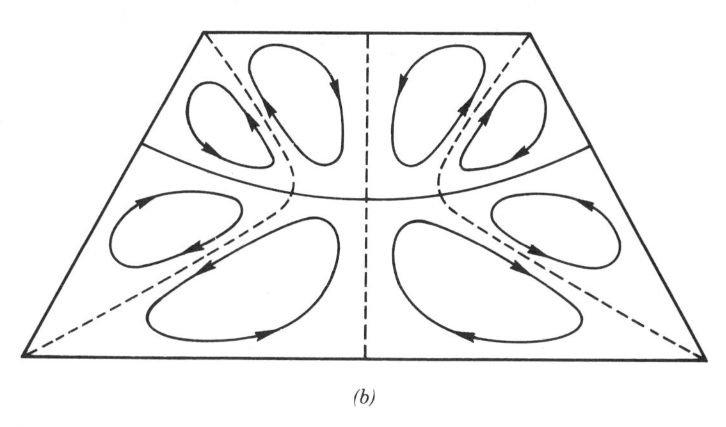

Figure 4.48. Fully developed turbulent velocity distribution at Re = 1.19×10^5 in a trapezoidal duct: (*a*) primary-flow isovels, (*b*) secondary-flow pattern [191].

4.6.4 Ducts Formed by Intersection of Circular Rods with Flat Plates

Gun and Darling [249] measured fully developed Fanning friction factors for waterflow in three types of ducts formed by intersection of circular rods with flat plates, shown as insets to Fig. 4.52. The experiments, spanning the range $200 < \mathrm{Re} < 10^5$, covered laminar, transition, and turbulent flows. The lower limits of $\mathrm{Re}_{\mathrm{crit}}$ for the three types of ducts (labeled center, side, and corner sections in Fig. 4.52) are 900, 1000, and 1100, respectively.

Figure 4.52 shows that for laminar flow, there is good agreement between the measurements and predictions of the theoretical formula $f\,\mathrm{Re} = C$, where the values of C for the three types of ducts are 6.50, 6.50, and 7.06, respectively [249]. For the turbulent flow, comparisons are made with the predictions of the Blasius formula [Eq. (4.32)]. The measured values are found to be lower than the predictions of the Blasius

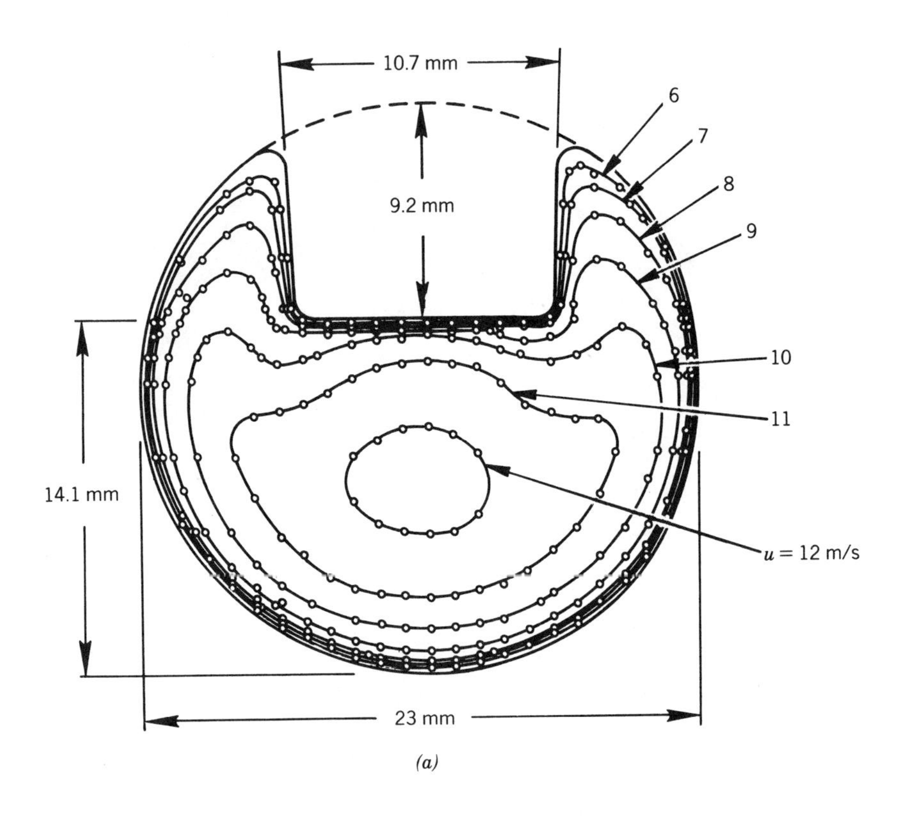

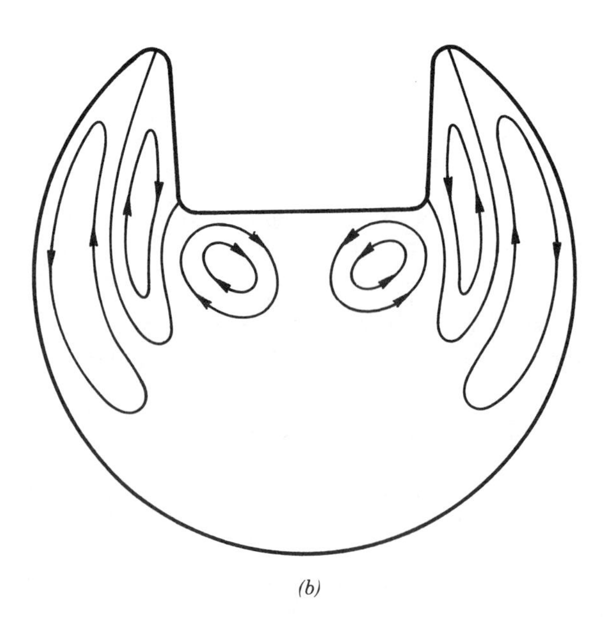

Figure 4.49. Fully developed turbulent velocity distribution at $Re = 1.09 \times 10^5$ in a smooth circular duct with a single rectangular indentation: (a) primary-flow isovels, (b) secondary-flow pattern [191].

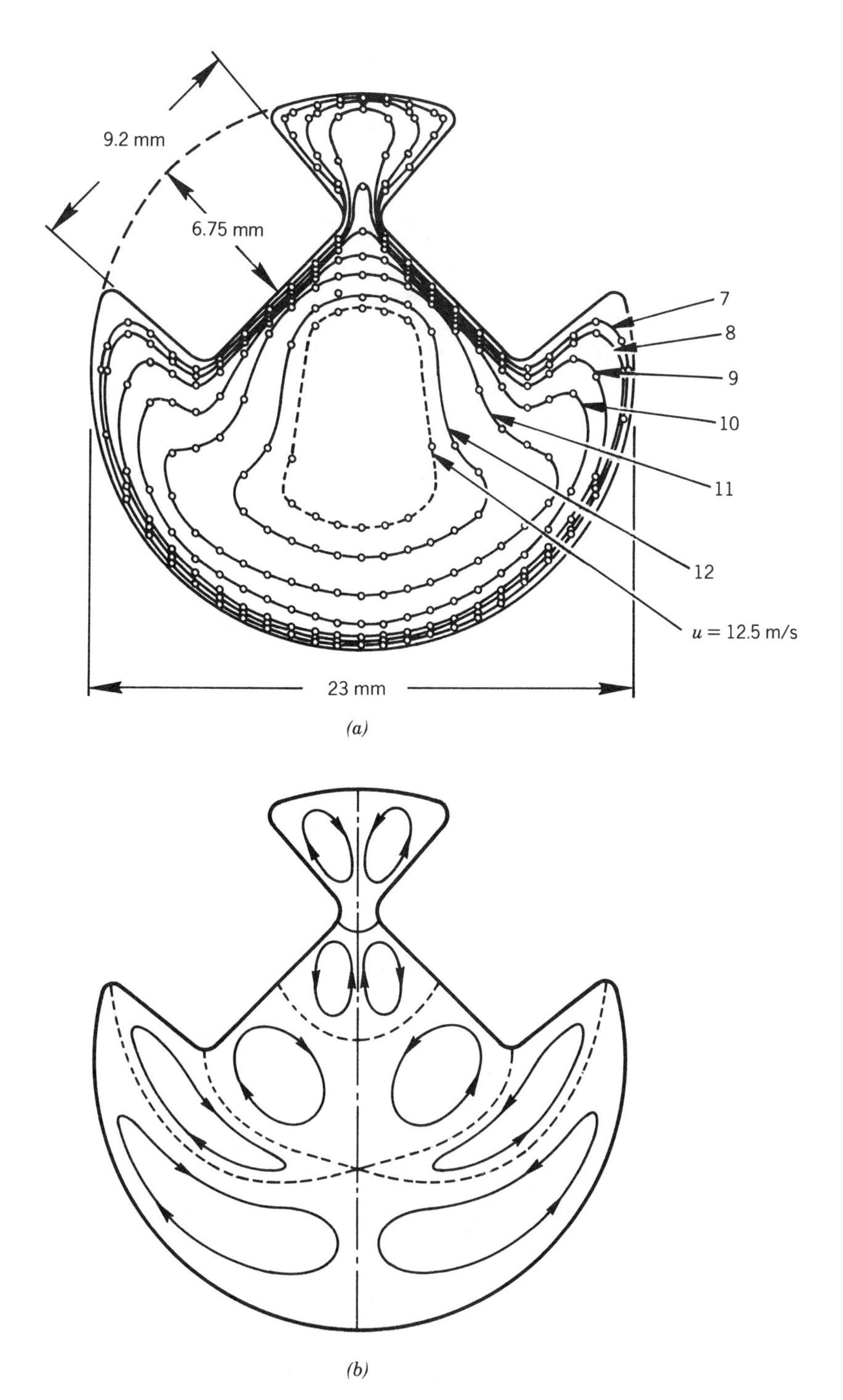

Figure 4.50. Fully developed turbulent velocity distribution at $\mathrm{Re} = 7.7 \times 10^4$ in a smooth circular duct with twin rectangular indentations: (a) primary-flow isovels, (b) secondary-flow pattern [191].

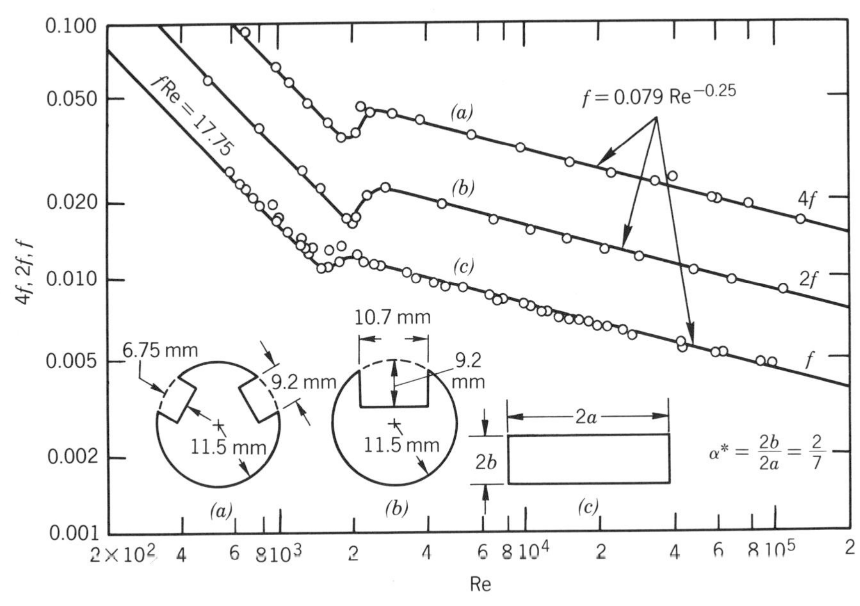

Figure 4.51. Friction factors for fully developed turbulent flow in smooth circular ducts with rectangular indentations and in a rectangular duct [191].

formula for all three types of the ducts by about 26%. This indicates that the hydraulic-diameter concept does not apply to turbulent flow in these three sharp-cornered ducts.

Gun and Darling [249] successfully correlated the fully developed turbulent flow friction factors of Fig. 4.52 within $\pm 5\%$ of the measurements by the formula

$$\frac{f}{f_c} = \left(\frac{C}{C_c}\right)^{0.45} \exp\left(-\frac{\mathrm{Re} - 3000}{10^6}\right) \tag{4.128}$$

where f_c is the turbulent flow friction factor for the circular duct, which can be determined from the PKN formula of Table 4.2 or from the Blasius formula [Eq. (4.32)]. The values of C (the constant in the laminar flow relation $f\,\mathrm{Re} = C$) for the three types of ducts (labeled center, side, and corner in Fig. 4.52) are 6.50, 6.50, and 7.06, respectively. The value of C_c is 16 for fully developed laminar flow in a circular duct.

Barrow et al. [250] also reported data on the fully developed turbulent friction factor for the corner section in Fig. 4.52 over the range $10^4 < \mathrm{Re} < 3 \times 10^4$. These measurements are in excellent accord with the measurements due to Gun and Darling [249] shown in Fig. 4.52.

Mohandes and Knudsen [251] presented fully developed Fanning friction factors f for the duct formed from two flat plates and two circular rods as shown in Table 4.27. This duct geometry is characterized by the rod diameter d and the spacing s between the two rod centers. The duct geometry shown as side section in the inset to Fig. 4.52 is a special case of the geometry studied by Mohandes and Knudsen [251]; it corresponds to $s/d = 1$.

Using the values of C in the laminar flow relation $f\,\mathrm{Re} = C$, the fully developed turbulent flow Fanning friction factors f for the duct of Table 4.27 can be calculated from Eq. (4.128). Here experimentally measured values of C, covering the range

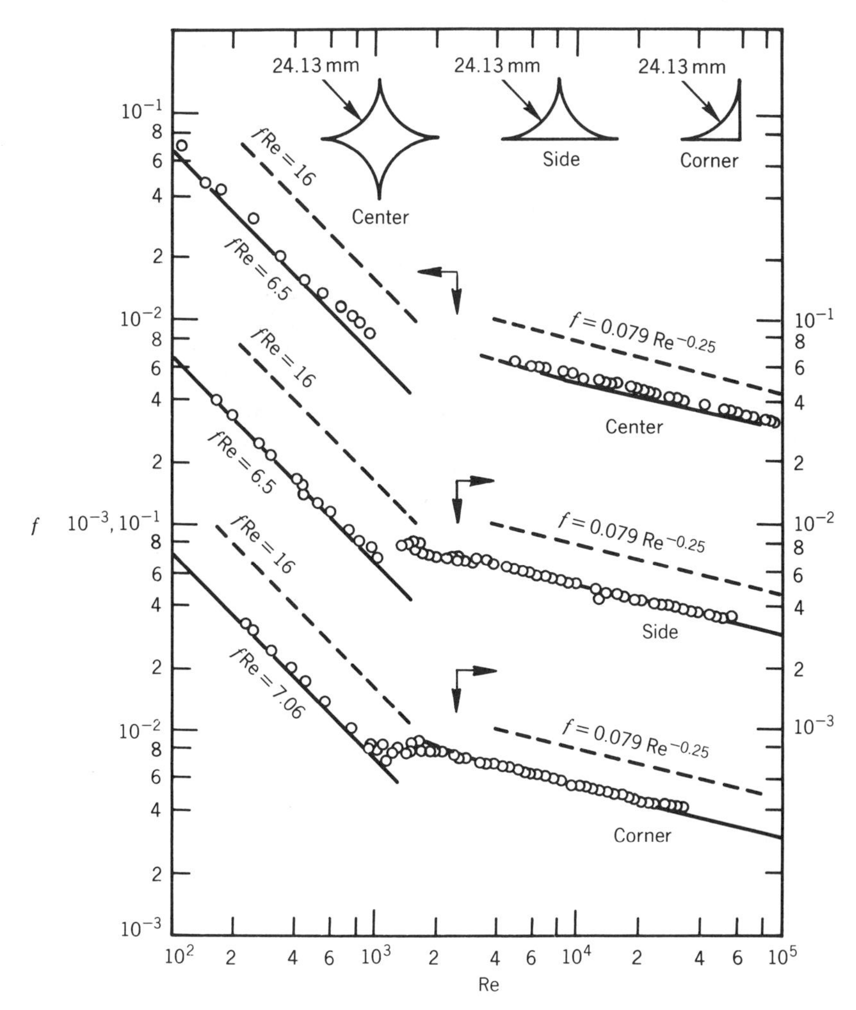

Figure 4.52. Friction factors for fully developed turbulent flow in a smooth duct formed by intersection of flat plates and circular rods.

$1 \le s/d \le 1.64$, are provided by Mohandes and Knudsen [251] as presented in Table 4.27.

For $s/d = 1$, Table 4.27 shows three values for C, viz., 5.482, 5.186, and 6.526, corresponding to $d = 12.70$, 25.40, and 38.10 mm, respectively. The first two values appear low when compared with the value 6.50 measured by Gun and Darling [249] for $s/d = 1$ and $d = 48.26$ mm. For $s/d = 1$, the duct shapes for all values of d are geometrically similar, and it is expected that the C values will not vary with d. Consequently, for this case it is recommended that in Table 4.27 the C values of 5.482 and 5.186 be ignored in preference to 6.526, which is in close agreement with the value provided by Gun and Darling [249].

Barrow et al. [250] measured fully developed Nusselt numbers in the corner section shown in Fig. 4.53 for the (H1) boundary condition on the curved wall with the two straight walls insulated. The experiments were for airflow (Pr = 0.7), covering the range $2 \times 10^4 < \text{Re} < 5 \times 10^4$. In addition to the experimental data, the predictions of the Dittus-Boelter (curve 1) and Petukhov-Kirillov-Popov (curves 2, 3) correlations of

TABLE 4.27. Experimentally Determined Values of f Re = C for Fully Developed Laminar Flow in Smooth Ducts Formed from Circular Rods and Flat Plates [251]

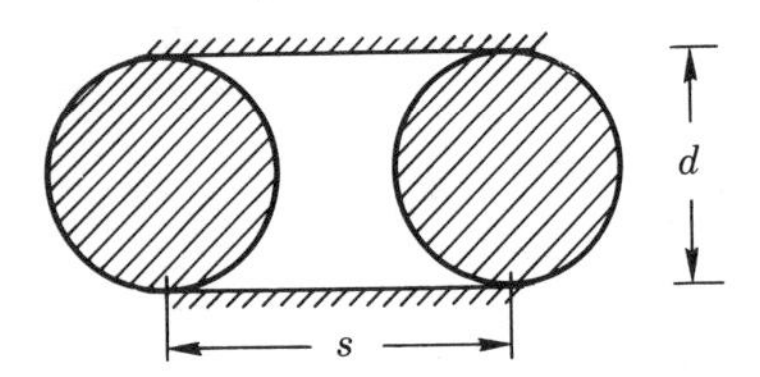

Rod Diameter d, mm	Rod Pitch s, mm	s/d	Hydraulic Diameter D_h, mm	C
12.70	12.70	1.000	21.20	5.482
	12.95	1.020	23.00	5.482
	13.21	1.040	24.77	7.515
	13.72	1.080	28.23	7.111
	14.73	1.160	34.84	7.140
	16.76	1.320	46.97	6.636
	20.83	1.640	67.61	6.941
25.40	25.40	1.000	42.41	5.186
	25.65	1.010	44.21	6.019
	25.91	1.020	46.00	7.369
	26.42	1.040	49.54	6.857
	27.43	1.080	56.46	7.225
	29.46	1.160	69.69	6.257
38.10	38.10	1.000	63.61	6.526
	38.35	1.0067	65.42	6.746
	38.61	1.0133	67.21	7.574
	39.12	1.0267	70.78	7.780
	40.13	1.0533	77.80	7.870
	42.16	1.1067	91.43	6.941
	42.77	1.1227	95.42	7.960

Table 4.4 are presented in Fig. 4.53. For curve 2, the friction factor entering the Petukhov-Kirillov-Popov correlation was computed from the Filonenko correlation of Table 4.2. Following the proposal by Altemani and Sparrow [224], the friction factors experimentally determined by Barrow et al. [250] were also employed in predicting the Nusselt numbers (curve 3) from the Petukhov-Kirillov-Popov correlation. None of the curves shows satisfactory agreement with the measured values. The differences between the measured and predicted values are attributable to the possible coexistence of the laminar and turbulent flows in the corner regions, as with the narrow-apex-angle isosceles triangular duct discussed in Sec. 4.5.2.

4.6.5 Some Unusual Ducts

Table 3.33 in Chap. 3 provides formulas for fully developed laminar friction factors for numerous singly connected ducts. All formulas are in the form f Re = C. Knowing the value of C for a specific duct from Table 3.33, the friction factor for fully developed

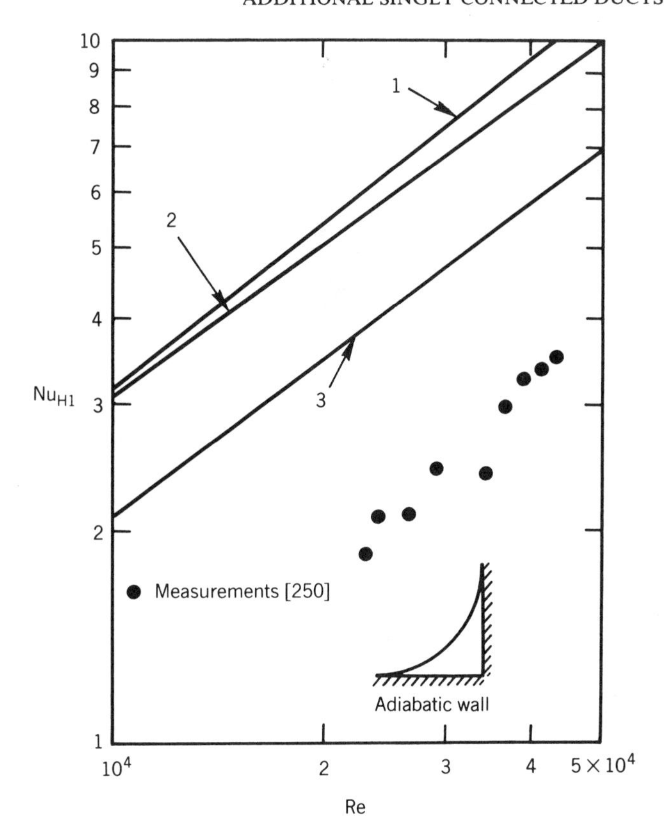

Figure 4.53. Nusselt numbers for fully developed turbulent airflow (Pr = 0.7) in a smooth corner duct with two flat sides insulated [250]; see the test for curves 1, 2 and 3.

turbulent flow in the duct can be determined from Eq. (4.128). In the absence of turbulent friction factor measurements for all duct geometries listed in Table 3.33, it is not possible to assert how accurate the calculated friction factors will be for a specific duct. However, since Eq. (4.128) correlates within $\pm 5\%$ the measured turbulent flow friction factors for the ducts shown in Fig. 4.52 as well as rectangular and concentric annular ducts [249], it appears that Eq. (4.128) in conjunction with Table 3.33 may be applied with reasonable assurance to calculate fully developed turbulent flow friction factors for various noncircular ducts.

In addition to Eq. (4.128), a few more formulas have been proposed in the literature to calculate turbulent flow friction factors using laminar flow solutions. Rehme [252] derived the following relation for turbulent flow friction factor f for a noncircular duct:

$$\sqrt{\frac{2}{f}} = A\left(5.5 + 2.5 \ln \mathrm{Re}\sqrt{\frac{2}{f}}\right) - G^* \tag{4.129}$$

where A and G^* are purely geometrical factors derived from analytical considerations. These factors are presented in Ref. 252 as functions of $4C$ where $C = f\,\mathrm{Re}$ is the constant for fully developed laminar flow through the duct in question. Rehme [252] compared the prediction of Eq. (4.129) with the available turbulent friction factors for isosceles triangular ducts, eccentric annular ducts, longitudinal flow over rod bundles arranged in triangular and square arrays, and three- and four-sided cusped ducts, i.e., ducts formed by tangent circular rods. Good agreement, within about $\pm 5\%$, was found between the predictions and measurements.

Malák et al. [225] presented an alternative formula for determining the fully developed turbulent friction factors for noncircular ducts from a knowledge of fully developed laminar friction factors for noncircular ducts. According to [225], the turbulent flow friction factor f can be determined from

$$\sqrt{\frac{K_t}{f}} = 1.7372 \ln \frac{\text{Re}\sqrt{f}}{K_t^{3/2}} - 0.4041 \tag{4.130}$$

where

$$K_t = \frac{3C + 16}{64} \tag{4.131}$$

Here $C = f\,\text{Re}$ is the constant for fully developed laminar flow through the duct in question.

Malák et al. [225] compared the predictions of Eq. (4.130) with the available turbulent flow measurements for longitudinal flow over rod bundles arranged in triangular and square arrays and for ducts formed from flat plates and circular rods. The turbulent flow friction factors computed by Malák et al. [225] as well as by Rehme [252] are within about $\pm 5\%$ of the measurements.

Knowing the fully developed turbulent friction factor f for a noncircular duct from any of the formulas presented above, the fully developed turbulent Nusselt number for the same duct may be calculated from the Gnielinski correlation for a circular duct given in Table 4.4. This recommendation is based on the results obtained by Altemani and Sparrow [224] using experimentally determined turbulent friction factors for an unsymmetrically heated triangular duct.

Altemani and Sparrow [224] found that when the experimentally determined friction factors for the equilateral triangular duct are substituted in the circular duct Nusselt number correlation by Petukhov-Kirillov-Popov, the resulting Nusselt numbers agreed quite closely with the experimental values. Use of the friction factors calculated from the circular duct correlations (e.g., the P-K-N correlation of Table 4.2), did not give satisfactory Nusselt numbers for the equilateral triangular duct. Thus it appears that once the duct-specific friction factors for the noncircular ducts are available, circular duct correlations for the Nusselt numbers may be successfully employed to determine the Nusselt numbers for the noncircular ducts. We recommend the Gnielinski correlation, instead of Petukhov-Kirillov-Popov correlation, for the circular duct since it is applicable to Re as low as 2300.

4.7 CONCENTRIC ANNULAR DUCTS

Concentric annular ducts are of great technological importance, as they are frequently used in a variety of fluid flow and heat transfer devices. These ducts are geometrically characterized by the radius ratio $r^* = r_i/r_o$, where r_i and r_o are the inside and outside radii of the two walls forming the concentric annular duct. From an analytical point of view, the two limiting cases of a concentric annular duct are a circular duct ($r^* = 0$) and a flat duct ($r^* = 1$). These two geometries are important in their own right and

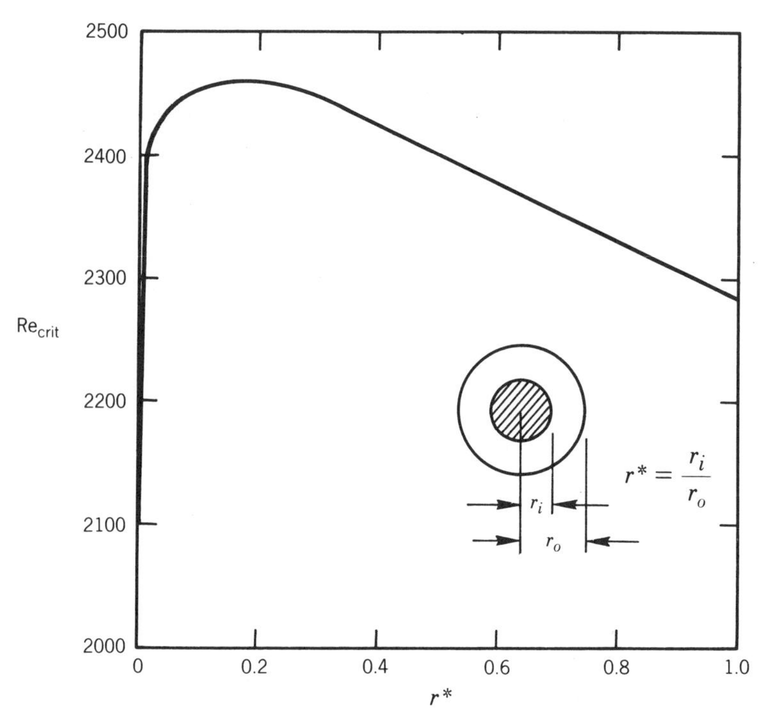

Figure 4.54. Lower limits of critical Reynolds numbers for concentric annular ducts with uniform velocity at the inlet [154].

have already been treated at some length in Secs. 4.2 and 4.3. In the present section are presented several results of practical interest for concentric annular ducts, mainly with intermediate values of the radius ratio ($0 < r^* < 1$).

4.7.1 Transition Flow

Hanks [154] determined the lower limit of $\mathrm{Re}_{\mathrm{crit}}$ for concentric annular ducts from a theoretical consideration for the case of a uniform flow at the duct inlet. His results, presented in Fig. 4.54, are within $\pm 3\%$ of the selected measurements by various investigators with air and water as test fluids [154]. A striking feature of the curve in Fig. 4.54 is that it exhibits a maximum $\mathrm{Re}_{\mathrm{crit}}$ value of 2462 at $r^* = 0.15$. The $\mathrm{Re}_{\mathrm{crit}}$ values at $r^* = 0$ and 1 are 2100 and 2285, respectively. These two values are in reasonably good agreement with the widely accepted values of 2300 and 2200 for $r^* = 0$ and 1, respectively.

Walker and Rothfus [253] reported axial velocity measurements for the transition flow of water in a smooth concentric annular duct with $r^* = 0.331$. Walker et al. [254] reported extensive measurements of the Fanning friction factor f_o at the outer duct wall, covering laminar, transition, and turbulent flows in eight smooth concentric annuli with $r^* = 0$, 0.0260, 0.0667, 0.1251, 0.1653, 0.3312, 0.4987, and 1. The f_o values are reported as a function of the Reynolds number based on $2(r_o^2 - r_m^2)/r_o$ as the characteristic dimension, where r_m is the radius of maximum velocity, which is assumed to be coincident with the radius of zero shear stress.

4.7.2 Fully Developed Flow

Fluid Flow. Knudsen and Katz [255] presented the following power–law velocity distribution valid for $\mathrm{Re} > 10^4$ for fully developed turbulent flow in smooth concentric annular ducts for $0 \le r^* \le 1$:

$$\frac{u}{u_{\max}} = \left(\frac{r_o - r}{r_o - r_m}\right)^{0.142} \qquad \text{for} \quad r_m \le r \le r_o \tag{4.132}$$

$$\frac{u}{u_{\max}} = \left(\frac{r - r_i}{r_m - r_i}\right)^{0.102} \qquad \text{for} \quad r_i \le r \le r_m \tag{4.133}$$

$$\frac{u_m}{u_{\max}} = 0.876 \tag{4.134}$$

The radius of maximum velocity (zero stress), r_m, entering Eqs. (4.132) and (4.133) can be determined from the following relation presented by Kays and Leung [168]:

$$r_m^* = \frac{r_m}{r_o} = r^{*0.343}(1 + r^{*0.657} - r^*) \tag{4.135}$$

A noteworthy point about Eq. (4.135) is that the r_m values predicted by it are independent of Re. This finding is in accord with the r_m measurements by Brighton and Jones [256], Ivey [257], and Jonsson and Sparrow [258].

The comparisons between the $u/u_{\max}$ predictions of Eqs. (4.132) and (4.133) over the range $10^4 < \mathrm{Re} < 7 \times 10^4$ and the experimental measurements for the isothermal as well as nonisothermal flow presented in [255] are quite good.

Bailey [259] developed the following form of the velocity-defect law for fully developed turbulent flow in smooth concentric annular ducts:

$$\frac{u_{\max} - u}{u_{t,o}} = -2.5 \ln\left[1 - \frac{r^2 - r_m^2}{r_o^2 - r_m^2}\left(\frac{r_o}{r}\right)\right] \qquad \text{for} \quad r_m \le r \le r_o \tag{4.136}$$

$$\frac{u_{\max} - u}{u_{t,i}} = -2.5 \ln\left[1 - \frac{r_m^2 - r^2}{r_m^2 - r_i^2}\left(\frac{r_i}{r}\right)\right] \qquad \text{for} \quad 0 \le r \le r_m \tag{4.137}$$

The friction velocities $u_{t,o}$ and $u_{t,i}$ entering Eqs. (4.136) and (4.137) are given by

$$u_{t,o} = \sqrt{\frac{\tau_o g_c}{\rho}}, \qquad u_{t,i} = \sqrt{\frac{\tau_i g_c}{\rho}} \tag{4.138}$$

where τ_o and τ_i are the wall shear stresses due to skin friction at the outer and inner walls, respectively. The radius of maximum velocity r_m can again be determined from Eq. (4.135).

The predictions of Eq. (4.136) and (4.137) have been compared with the experimental measurements for isothermal as well as nonisothermal flow over the range $10^4 < \mathrm{Re} < 7 \times 10^4$ and are found to be quite good [255].

Knudsen and Katz [255] developed the following velocity distributions in terms of the wall coordinates u^+ and y^+:

$$u_o^+ = 3.0 + 2.6492 \ln y_o^+ \qquad \text{for} \quad r_m \le r \le r_o \tag{4.139}$$

$$u_i^+ = 6.2 + 1.9109 \ln y_i^+ \qquad \text{for} \quad r_i \le r \le r_m \tag{4.140}$$

where

$$u_o^+ = \frac{u}{u_{t,o}}, \qquad u_i^+ = \frac{u}{u_{t,i}} \tag{4.141}$$

$$y_o^+ = \frac{u_{t,o}(r_o - r)}{\nu}, \qquad y_i^+ = \frac{u_{t,i}(r - r_i)}{\nu} \tag{4.142}$$

The friction velocities $u_{t,o}$ and $u_{t,i}$ entering Eqs. (4.141) and (4.142) are defined in Eq. (4.138).

Comparisons between the predictions of Eq. (4.139) and (4.140) and the experimental measurements for the isothermal as well as nonisothermal flow over the range $4000 < \text{Re} < 7 \times 10^4$ show that Eqs. (4.139) and (4.140) are in satisfactory agreement with the measurements for $y_o^+ \geq 30$ and $y_i^+ \geq 40$, respectively [255].

Extensive friction-factor data exist for fully developed turbulent flow in smooth concentric annular ducts. A critical review of these data is presented by Jones and Leung [33]. Based on this review, it is recommended that the fully developed friction factor formulas for a smooth circular duct given in Table 4.2 be used for calculating the friction factors for concentric annular ducts. In using these formulas, the circular duct diameter 2a needs to be replaced by the laminar equivalent diameter for concentric annular ducts. This latter diameter is given by Eq. (4.24). Jones and Leung [33] have shown that with the use of D_l in the circular duct correlations, the calculated results for concentric annular ducts agree with the experimental measurements from various sources within $\pm 5\%$ compared to $\pm 20\%$ agreement with the use of D_h.

The present authors performed calculations for f values using the Techo et al. correlation of Table 4.2. On comparing these results with the experimental measurements for concentric annular ducts ($0 \leq r^* \leq 1$) in the range $5000 \leq \text{Re} \leq 10^7$, they arrived at the following correlation for concentric annular ducts:

$$f = (1 + 0.0925 r^*) f_c \tag{4.142a}$$

where f_c is the friction factor for circular duct given by the Techo et al. correlation (Table 4.2).

The predictions of Eq. (4.142a) are on par with those determined by substituting D_l of Eq. (4.24) in the Techo et al. correlation.

Olson and Sparrow [260] measured fully developed incremental pressure drop numbers $K(\infty)$, defined by Eq. (3.4) in Chap. 3, in concentric annuli with $r^* = 0$, 0.3125, and 0.5 and three types of inlet configurations shown in Table 4.28. The tests covered the range $1.6 \times 10^4 < \text{Re} < 7 \times 10^4$.

The values of $K(\infty)$ in Table 4.28 for the square entrance are seen to be substantially larger than those for the rounded entrance without a trip. This suggests that the losses due to flow separation that occur with the square entrance are dominant. Also, the values of $K(\infty)$ for the rounded entrance with a trip are higher than those without a trip, suggesting once again that the flow separation induced by the trip enhances the entrance-related losses reflected in $K(\infty)$. The negative value of $K(\infty)$ for $r^* = 0.5$ with the rounded entrance indicates that the actual pressure drop in the hydrodynamic entrance region is less than the fully developed pressure drop. The reason appears to be the absence of a boundary-layer tripping device such as sharp corners or a tripping strip.

The laminar flow $K(\infty)$ values for $r^* = 0$ and 0.5, presented on p. 3.93, are 1.25 and 0.688, respectively. These values with the rounded entrance are substantially higher than the corresponding turbulent flow values in Table 4.28 for rounded entrances with

TABLE 4.28. Turbulent Flow Fully Developed Incremental Pressure Drop Numbers $K(\infty)$ for Concentric Annular Ducts [260]

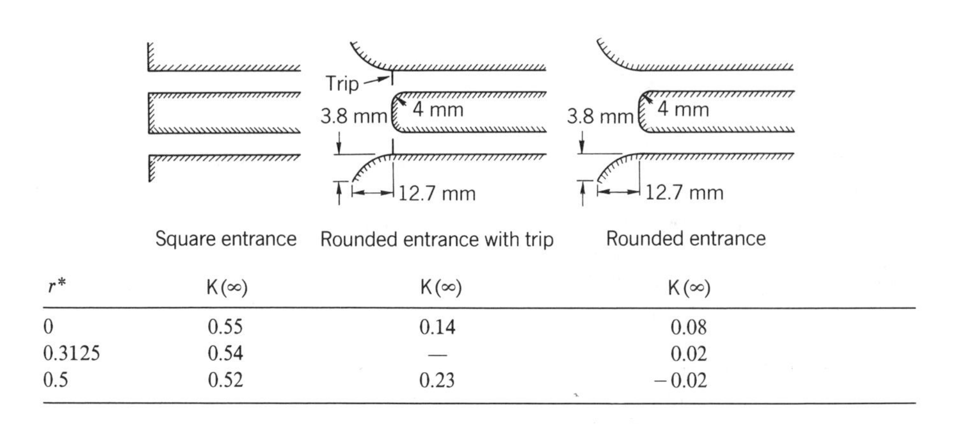

r^*	Square entrance $K(\infty)$	Rounded entrance with trip $K(\infty)$	Rounded entrance $K(\infty)$
0	0.55	0.14	0.08
0.3125	0.54	—	0.02
0.5	0.52	0.23	−0.02

and without trip. Since $K(\infty)$ is a measure of all the entrance-region-related effects sustained by the fluid, it follows that in laminar flow the entrance-region-related effects are substantially more dominant than in turbulent flow. The primary reason for this dominance is the significantly longer hydrodynamic entrance lengths for laminar flow than for turbulent flow.

Heat Transfer. The fully developed Nusselt numbers Nu_o and Nu_i at the outer and inner surfaces of a smooth concentric annular duct can be determined from the following relations for arbitrarily prescribed values of uniform heat fluxes q_o'' and q_i'' at the outer and inner surfaces:

$$\mathrm{Nu}_o = \frac{h_o D_h}{k} = \frac{\mathrm{Nu}_{oo}}{1 - (q_i''/q_o'')\theta_o^*} \tag{4.143}$$

$$\mathrm{Nu}_i = \frac{h_i D_h}{k} = \frac{\mathrm{Nu}_{ii}}{1 - (q_o''/q_i'')\theta_i^*} \tag{4.144}$$

where

$$q_o'' = h_o(T_o - T_m), \qquad q_i'' = h_i(T_i - T_m) \tag{4.145}$$

T_o and T_i in Eq. (4.145) denote the duct wall temperatures at the outer and inner surfaces. The temperature difference $T_o - T_i$ is given by

$$T_o - T_i = \frac{D_h}{k}\left[q_o''\left(\frac{1}{\mathrm{Nu}_{oo}} + \frac{\theta_i^*}{\mathrm{Nu}_{ii}}\right) - q_i''\left(\frac{1}{\mathrm{Nu}_{ii}} + \frac{\theta_o^*}{\mathrm{Nu}_{oo}}\right)\right] \tag{4.146}$$

The Nusselt numbers Nu_{oo} and Nu_{ii} as well as the influence coefficients θ_o^* and θ_i^* entering Eqs. (4.143), (4.144), and (4.146) have been determined by Kays and Leung [168]. They are given in Tables 4.29 to 4.32 for wide ranges of Re and Pr and for $r^* = 0.1$, 0.2, 0.5, and 0.8. For $r^* = 1$, the applicable results are given in Table 4.22.

TABLE 4.29. Nusselt Numbers and Influence Coefficients for Fully Developed Turbulent Flow in a Concentric Annular Duct ($r^* = 0.10$) with Uniform Heat Flux at One Wall and the Other Wall Insulated [168]

HEATING FROM OUTER TUBE WITH INNER TUBE INSULATED

	Re = 10^4		3×10^4		10^5		3×10^5		10^4	
Pr	Nu_{oo}	θ_o^*	Nu_{oo}	θ_o^*	Nu_{oo}	θ_o^*	Nu_{oo}	θ_o^*	Nu_{oo}	θ_o^*
0	6.00	0.077	6.12	0.079	6.32	0.081	6.50	0.084	6.68	0.085
0.001	6.00	0.077	6.12	0.079	6.40	0.082	6.60	0.082	7.20	0.082
0.003	6.00	0.077	6.24	0.081	6.55	0.083	7.34	0.082	10.8	0.071
0.01	6.13	0.076	6.50	0.081	7.80	0.077	12.1	0.067	26.4	0.052
0.03	6.45	0.076	7.95	0.075	13.7	0.065	28.2	0.051	71.8	0.036
0.5	24.8	0.039	53.4	0.032	134	0.028	320	0.025	860	0.022
0.7	29.8	0.032	66.0	0.028	167	0.024	409	0.022	1,100	0.020
1.0	36.5	0.026	81.8	0.023	212	0.021	520	0.019	1,430	0.017
3	61.5	0.013	147	0.013	395	0.012	1,000	0.012	2,830	0.011
10	99.2	0.006	246	0.006	685	0.006	1,780	0.006	5,200	0.006
30	143	0.003	360	0.003	1,030	0.003	2,720	0.003	8,030	0.003
100	205	0.002	525	0.002	1,500	0.002	4,030	0.002	12,100	0.002
1000	378		980		2,850		7,600		23,000	

HEATING FROM INNER TUBE WITH OUTER TUBE INSULATED

	Re = 10^4		3×10^4		10^5		3×10^5		10^6	
Pr	Nu_{ii}	θ_i^*	Nu_{ii}	θ_i^*	Nu_{ii}	θ_i^*	Nu_{ii}	θ_i^*	Nu_{ii}	θ_i^*
0	11.5	1.475	11.5	1.502	11.5	1.500	11.5	1.460	11.6	1.477
0.001	11.5	1.475	11.5	1.502	11.5	1.480	11.7	1.462	12.3	1.410
0.003	11.5	1.475	11.5	1.475	11.7	1.473	12.6	1.391	17.0	1.124
0.01	11.8	1.482	11.8	1.442	13.5	1.323	19.4	1.090	39.0	0.760
0.03	12.5	1.472	14.1	1.330	21.8	1.027	42.0	0.760	103	0.526
0.5	40.8	0.632	81.0	0.486	191	0.394	443	0.339	1,160	0.294
0.7	48.5	0.512	98.0	0.407	235	0.338	550	0.292	1,510	0.269
1.0	58.5	0.412	120	0.338	292	0.286	700	0.256	1,910	0.232
3	93.5	0.202	206	0.175	535	0.162	1,300	0.152	3,720	0.148
10	140	0.089	328	0.081	890	0.078	2,300	0.078	6,700	0.077
30	195	0.041	478	0.039	1,320	0.038	3,470	0.038	10,300	0.040
100	272	0.017	673	0.015	1,910	0.015	5,030	0.016	15,200	0.018
1000	486	0.004	1,240	0.003	3,600	0.003	9,600	0.004	28,700	0.004

TABLE 4.30. Nusselt Numbers and Influence Coefficients for Fully Developed Turbulent Flow in a Concentric Annular Duct ($r^* = 0.20$) with Uniform Heat Flux at One Wall and the Other Wall Insulated [168]

HEATING FROM OUTER TUBE WITH INNER TUBE INSULATED

	Re = 10^4		3×10^4		10^5		3×10^5		10^6	
Pr	Nu_{oo}	θ_o^*	Nu_{oo}	θ_o^*	Nu_{oo}	θ_o^*	Nu_{oo}	θ_o^*	Nu_{oo}	θ_o^*
0	5.83	0.140	5.92	0.145	6.10	0.151	6.16	0.152	6.35	0.157
0.001	5.83	0.140	5.92	0.144	6.10	0.151	6.30	0.154	6.92	0.153
0.003	5.83	0.140	6.00	0.146	6.22	0.150	6.90	0.150	10.2	0.136
0.01	5.95	0.140	6.20	0.146	7.40	0.144	11.4	0.131	24.6	0.102
0.03	6.22	0.140	7.55	0.140	12.7	0.125	26.3	0.098	80.0	0.074
0.5	22.5	0.071	51.5	0.064	130	0.055	310	0.049	823	0.044
0.7	29.4	0.063	64.3	0.055	165	0.049	397	0.044	1,070	0.040
1.0	35.5	0.051	80.0	0.046	206	0.042	504	0.039	1,390	0.035
3	60.0	0.026	145	0.026	390	0.024	980	0.024	2,760	0.023
10	98.0	0.013	243	0.013	680	0.012	1,750	0.012	4,980	0.012
30	142	0.004	360	0.006	1,030	0.006	2,700	0.006	7,850	0.006
100	205	0.003	520	0.003	1,500	0.003	4,000	0.003	12,000	0.003
1,000	380	0.001	980	0.001	2,830	0.001	7,500	0.001	22,500	0.001

HEATING FROM INNER TUBE WITH OUTER TUBE INSULATED

	Re = 10^4		3×10^4		10^5		3×10^5		10^6	
Pr	Nu_{ii}	θ_i^*	Nu_{ii}	θ_i^*	Nu_{ii}	θ_i^*	Nu_{ii}	θ_i^*	Nu_{ii}	θ_i^*
0	8.40	1.009	8.30	1.028	8.30	1.020	8.30	1.038	8.30	1.020
0.001	8.40	1.009	8.40	1.040	8.30	1.020	8.40	1.014	8.90	0.976
0.003	8.40	1.009	8.40	1.027	8.50	1.025	9.05	0.980	12.5	0.834
0.01	8.50	1.000	8.60	1.018	9.70	0.944	14.0	0.796	33.6	0.748
0.03	9.00	1.012	10.1	0.943	15.8	0.771	31.7	0.600	81.0	0.374
0.5	31.2	0.520	64.0	0.398	157	0.333	370	0.295	980	0.262
0.7	38.6	0.412	79.8	0.338	196	0.286	473	0.260	1,270	0.235
1.0	46.8	0.339	99.0	0.284	247	0.248	600	0.229	1,640	0.209
3	77.4	0.172	175	0.151	465	0.143	1,150	0.137	3,250	0.135
10	120	0.120	290	0.074	800	0.072	2,050	0.073	6,000	0.077
30	172	0.036	428	0.034	1,210	0.035	3,150	0.036	9,300	0.038
100	243	0.014	617	0.014	1,760	0.015	4,630	0.016	13,800	0.016
1,000	448	0.004	1,400	0.002	3,280	0.002	8,800	0.004	26,000	0.003

TABLE 4.31. Nusselt Numbers and Influence Coefficients for Fully Developed Turbulent Flow in a Concentric Annular Duct ($r^* = 0.50$) with Uniform Heat Flux at One Wall and the Other Wall Insulated [168]

	HEATING FROM OUTER TUBE WITH INNER TUBE INSULATED									
	Re = 10^4		3×10^4		10^5		3×10^5		10^6	
Pr	Nu_{oo}	θ_o^*	Nu_{oo}	θ_o^*	Nu_{oo}	θ_o^*	Nu_{oo}	θ_o^*	Nu_{oo}	θ_o^*
0	5.66	0.281	5.78	0.294	5.80	0.296	5.83	0.302	5.95	0.310
0.001	5.66	0.281	5.78	0.294	5.80	0.296	5.92	0.302	6.40	0.304
0.003	5.66	0.281	5.78	0.294	5.85	0.294	6.45	0.301	9.00	0.278
0.01	5.73	0.281	5.88	0.289	6.80	0.289	10.3	0.264	22.6	0.217
0.03	6.03	0.279	7.05	0.284	11.6	0.258	24.4	0.214	64.0	0.163
0.5	22.6	0.162	49.8	0.142	125	0.123	298	0.111	795	0.098
0.7	28.3	0.137	62.0	0.119	158	0.107	380	0.097	1,040	0.090
1.0	34.8	0.111	78.0	0.101	200	0.092	490	0.085	1,340	0.078
3	60.5	0.059	144	0.058	384	0.055	960	0.054	2,730	0.052
10	100	0.028	246	0.028	680	0.028	1,750	0.028	5,030	0.028
30	143	0.013	365	0.013	1,030	0.014	2,700	0.014	8,000	0.015
100	207	0.006	530	0.006	1,500	0.006	4,000	0.006	12,000	0.006
1,000	387	0.001	990	0.001	2,830	0.001	7,600	0.001	23,000	0.001

	HEATING FROM INNER TUBE WITH OUTER TUBE INSULATED									
	Re = 10^4		3×10^4		10^5		3×10^5		10^6	
Pr	Nu_{ii}	θ_i^*	Nu_{ii}	θ_i^*	Nu_{ii}	θ_i^*	Nu_{ii}	θ_i^*	Nu_{ii}	θ_i^*
0	6.28	0.620	6.30	0.632	6.30	0.651	6.30	0.659	6.30	0.654
0.001	6.28	0.620	6.30	0.632	6.30	0.651	6.40	0.659	6.75	0.644
0.003	6.28	0.620	6.30	0.632	6.40	0.656	6.85	0.637	9.40	0.585
0.01	6.37	0.622	6.45	0.636	7.30	0.623	10.8	0.540	23.2	0.427
0.03	6.75	0.627	7.53	0.598	12.0	0.533	24.8	0.430	65.5	0.333
0.5	24.6	0.343	52.0	0.292	130	0.253	310	0.229	835	0.208
0.7	30.9	0.300	66.0	0.258	166	0.225	400	0.206	1,080	0.185
1.0	38.2	0.247	83.5	0.218	212	0.208	520	0.183	1,420	0.170
3	66.8	0.129	152	0.121	402	0.115	1,010	0.114	2,870	0.111
10	106	0.059	260	0.059	715	0.059	1,850	0.059	5,400	0.061
30	153	0.028	386	0.027	1,080	0.028	2,850	0.031	8,400	0.032
100	220	0.006	558	0.006	1,600	0.006	4,250	0.007	12,600	0.007
1,000	408	0.002	1,040	0.002	3,000	0.002	8,000	0.002	24,000	0.002

TABLE 4.32 Nusselt Numbers and Influence Coefficients for Fully Developed Turbulent Flow in a Concentric Annular Duct ($r^* = 0.80$) with Uniform Heat Flux at One Wall and the Other Wall Insulated [168]

HEATING FROM OUTER TUBE WITH INNER TUBE INSULATED

Pr	Re = 10^4		3×10^4		10^5		3×10^5		10^6	
	Nu_{oo}	θ_o^*	Nu_{oo}	θ_o^*	Nu_{oo}	θ_o^*	Nu_{oo}	θ_o^*	Nu_{oo}	θ_o^*
0	5.65	0.379	5.70	0.386	5.75	0.398	5.80	0.407	5.85	0.409
0.001	5.65	0.379	5.70	0.386	5.75	0.398	5.88	0.406	6.25	0.407
0.003	5.65	0.379	5.70	0.386	5.84	0.397	6.35	0.407	8.80	0.374
0.01	5.75	0.381	5.85	0.386	6.72	0.390	9.95	0.361	21.0	0.286
0.03	6.10	0.388	6.90	0.380	11.1	0.339	23.2	0.290	62.0	0.216
0.5	22.4	0.225	48.0	0.191	121	0.169	292	0.153	790	0.136
0.7	28.0	0.192	61.0	0.166	156	0.150	378	0.136	1,020	0.122
1.0	34.8	0.159	76.5	0.141	197	0.129	483	0.120	1,330	0.111
3	61.3	0.083	142	0.079	382	0.078	960	0.076	2,730	0.073
10	100	0.039	243	0.039	670	0.039	1,740	0.040	5,050	0.040
30	146	0.019	365	0.019	1,040	0.020	2,720	0.021	8,000	0.022
100	209	0.008	533	0.008	1,500	0.009	4,000	0.009	12,000	0.010
1,000	385	0.002	1,000	0.002	2,870	0.002	7,720	0.002	23,000	0.002

HEATING FROM INNER TUBE WITH OUTER TUBE INSULATED

Pr	Re = 10^4		3×10^4		10^5		3×10^5		10^6	
	Nu_{ii}	θ_i^*	Nu_{ii}	θ_i^*	Nu_{ii}	θ_i^*	Nu_{ii}	θ_i^*	Nu_{ii}	θ_i^*
0	5.87	0.489	5.90	0.505	5.92	0.515	5.95	0.525	5.97	0.528
0.001	5.87	0.489	5.90	0.505	5.92	0.515	6.00	0.518	6.33	0.516
0.003	5.87	0.489	5.90	0.505	6.03	0.485	6.40	0.504	8.80	0.468
0.01	5.95	0.485	6.07	0.506	6.80	0.493	10.0	0.452	21.7	0.382
0.03	6.20	0.478	7.05	0.485	11.4	0.445	23.0	0.357	61.0	0.276
0.5	22.9	0.268	49.5	0.250	123	0.214	296	0.193	800	0.174
0.7	28.5	0.244	62.3	0.212	157	0.186	384	0.172	1,050	0.160
1.0	35.5	0.200	78.3	0.181	202	0.166	492	0.154	1,350	0.140
3	63.0	0.108	145	0.102	386	0.097	973	0.096	2,750	0.093
10	102	0.051	248	0.051	693	0.052	1,790	0.051	5,150	0.051
30	147	0.027	370	0.027	1,050	0.028	2,750	0.029	8,100	0.030
100	215	0.010	540	0.010	1,540	0.010	4,050	0.011	12,100	0.012
1,000	393	0.002	1,000	0.002	2,890	0.002	7,700	0.002	23,000	0.002

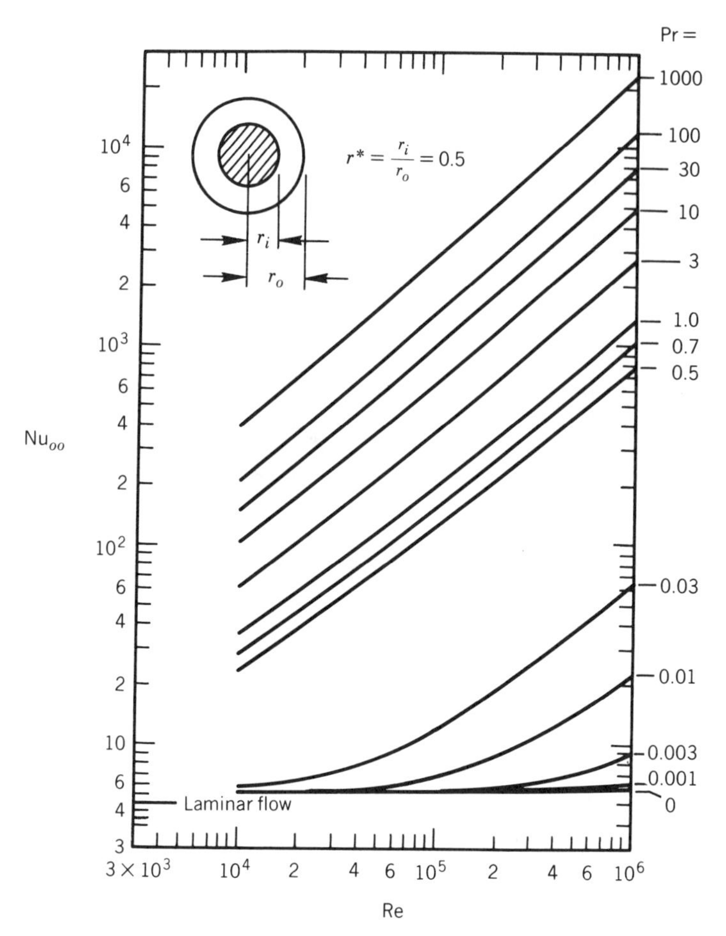

Figure 4.55. Theoretical Nusselt numbers Nu_{oo} for fully developed turbulent flow in a smooth concentric annular duct with $r^* = 0.5$ [168].

In using Eqs. (4.143), (4.144), and (4.146) for any combination of the heat fluxes q_o'' and q_i'', the following sign convention should be used: the heat flux is positive whenever there is heat transfer to the fluid and negative whenever there is heat transfer from the fluid. Thus, for example, when the heat is supplied to the fluid at the inner wall and removed at the outer wall, q_i'' is positive and q_o'' is negative. When the heat is supplied to the fluid at the inner as well as the outer wall, both q_i'' and q_o'' are positive.

It is apparent from Eq. (4.143) that $\mathrm{Nu}_o = \mathrm{Nu}_{oo}$ when $q_i'' = 0$. That means Nu_{oo} represents the Nusselt number at the outer wall when the outer wall alone is heated with the inner wall insulated. A similar interpretation applies to Nu_{ii} of Eq. (4.144). The theoretical Nusselt numbers Nu_{oo} and Nu_{ii} determined by Kays and Leung [168] are presented in Figs. 4.55 and 4.56 as functions of Re for a single value of $r^* = 0.5$ and a wide range of Pr. In Fig. 4.57, on the other hand, Nu_{oo} and Nu_{ii} are presented as functions of r^* and Re for a single value of Pr = 0.7. The interplay among Re, Pr, and r^* in shaping the Nu_{oo} and Nu_{ii} values is easy to discern from Figs. 4.55 to 4.57.

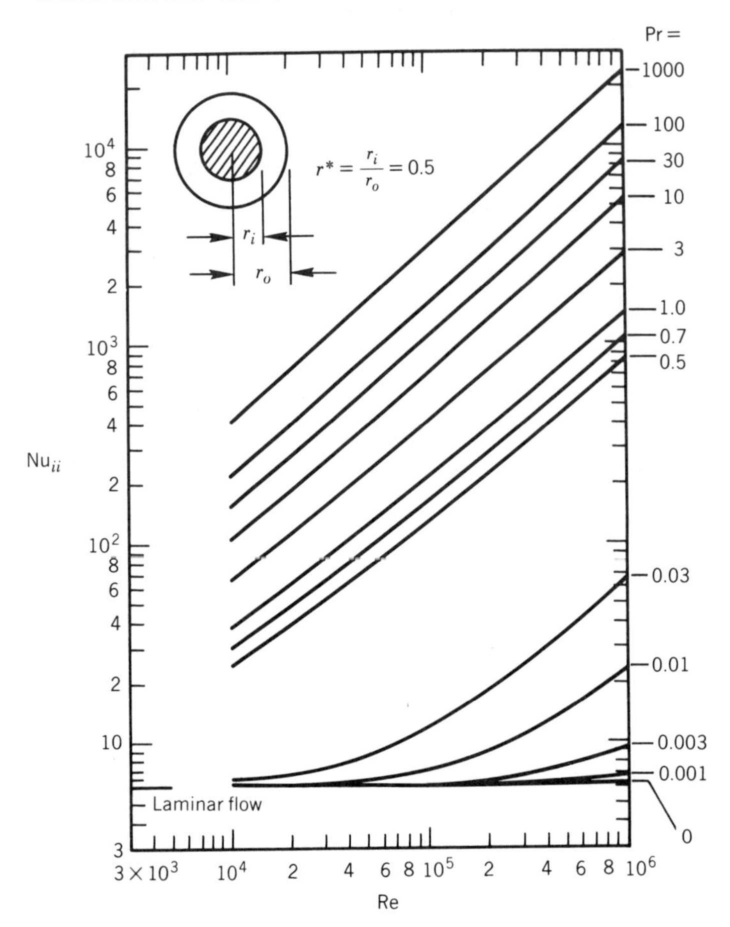

Figure 4.56. Theoretical Nusselt numbers Nu_{ii} for fully developed turbulent flow in a smooth concentric annular duct with $r^* = 0.5$ [168].

As described in Sec. 3.3.1. of Chap. 3, the solutions presented in Tables 4.29 to 4.32 are the fundamental solutions of the second kind. In accordance with the notation adopted in Sec. 3.8.1, Nu_{oo}, Nu_{ii}, θ_o^*, and θ_i^* should be denoted as $\mathrm{Nu}_{oo}^{(2)}$, $\mathrm{Nu}_{ii}^{(2)}$, $\theta_o^{*(2)}$, and $\theta_i^{*(2)}$ to emphasize the fact that these values pertain to the fundamental solution of the second kind. However, the superscript (2) is dropped throughout Sec. 4.7, as no information is presented relating to the remaining three fundamental solutions. The fundamental solutions of the first, third, and fourth kinds for turbulent flow are not reported in the literature. Two additional fundamental solutions of the second kind have been presented by Barrow [261] and Wilson and Medwell [262]. However, the solution by Kays and Leung [168] presented here is by far the most comprehensive.

Nusselt Numbers for Gases and Liquids, Pr > 0.5. Because of the possibilities of applying different heat fluxes at the two walls of concentric annuli with $r^* > 0$, the

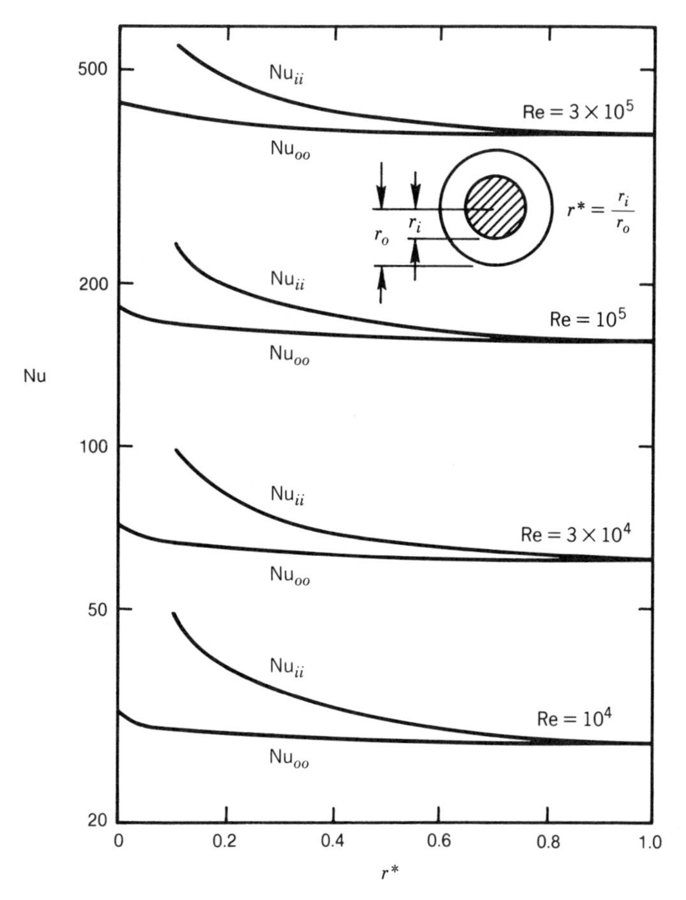

Figure 4.57. Theoretical Nusselt numbers Nu_{oo} and Nu_{ii} for fully developed turbulent flow in smooth concentric annular ducts for Pr = 0.7 [168].

Nusselt numbers Nu_o and Nu_i for these ducts cannot be determined from the circular duct correlations with a high level of accuracy. In order to ascertain how closely Nu_o and Nu_i compare with the circular duct Nusselt number Nu_c calculated from the Gnielinski correlation, the present authors performed detailed calculations and obtained the ratios $\mathrm{Nu}_o/\mathrm{Nu}_c$ and $\mathrm{Nu}_i/\mathrm{Nu}_c$ with the aid of Tables 4.22 and 4.29 to 4.32 in conjunction with Eqs. (4.143) and (4.144). The calculations covered $0.5 \leq \mathrm{Pr} \leq 100$, $10^4 \leq \mathrm{Re} \leq 10^6$, $0 \leq r^* \leq 1$, and $-1 \leq q_i''/q_o'' \leq 1$. In all calculations, the hydraulic diameter D_h based on the wetted perimeter rather than the heated perimeter was employed in the definitions of the Nusselt and Reynolds numbers. Some useful results of the calculations are summarized below.

Heating at Outer Wall with Inner Wall Adiabatic

$$\mathrm{Nu}_o = \mathrm{Nu}_{oo}, \qquad \mathrm{Nu}_i = 0$$

$$0.87 \leq \mathrm{Nu}_o/\mathrm{Nu}_c \leq 1.13$$

Heating at Inner Wall with Outer Wall Adiabatic

$$\mathrm{Nu}_o = 0, \qquad \mathrm{Nu}_i = \mathrm{Nu}_{ii}$$

$$0.90 \le \mathrm{Nu}_i/\mathrm{Nu}_c \le 1.66$$

Equal Heating at Two Walls ($q_i''\,/\,q_o'' = 1$)

$$0.90 \le \mathrm{Nu}_o/\mathrm{Nu}_c \le 1.23$$

$$0.91 \le \mathrm{Nu}_i/\mathrm{Nu}_c \le 4.52$$

Heating at One Wall and Equal Cooling at Other Wall ($q_i''\,/\,q_o'' = -1$)

$$0.68 \le \mathrm{Nu}_o/\mathrm{Nu}_c \le 1.10$$

$$0.68 \le \mathrm{Nu}_i/\mathrm{Nu}_c \le 1.43$$

Unequal Heating at Two Walls

$$0.90 \le \mathrm{Nu}_o/\mathrm{Nu}_c \le 1.14 \quad \text{for} \quad q_i''/q_o'' = 0.25$$

$$0.90 \le \mathrm{Nu}_o/\mathrm{Nu}_c \le 1.15 \quad \text{for} \quad q_i''/q_o'' = 0.50$$

$$0.90 \le \mathrm{Nu}_o/\mathrm{Nu}_c \le 1.17 \quad \text{for} \quad q_i''/q_o'' = 0.75$$

$\mathrm{Nu}_i/\mathrm{Nu}_c$ values for $q_i''/q_o'' = 0.25$, 0.5, and 0.75 were found to span much wider range including negative values which occur when $(q_o''/q_i'')\theta_i^* > 1$ [see Eq. (4.144)].

Heating at One Wall and Unequal Cooling at Other Wall

$$0.79 \le \mathrm{Nu}_o/\mathrm{Nu}_c \le 1.11 \quad \text{for} \quad q_i''/q_o'' = -0.25$$

$$0.75 \le \mathrm{Nu}_o/\mathrm{Nu}_c \le 1.10 \quad \text{for} \quad q_i''/q_o'' = -0.50$$

$$0.73 \le \mathrm{Nu}_o/\mathrm{Nu}_c \le 1.10 \quad \text{for} \quad q_i''/q_o'' = -0.75$$

$$0.41 \le \mathrm{Nu}_i/\mathrm{Nu}_c \le 1.26 \quad \text{for} \quad q_i''/q_o'' = -0.25$$

$$0.58 \le \mathrm{Nu}_i/\mathrm{Nu}_c \le 1.35 \quad \text{for} \quad q_i''/q_o'' = -0.50$$

$$0.64 \le \mathrm{Nu}_i/\mathrm{Nu}_c \le 1.40 \quad \text{for} \quad q_i''/q_o'' = -0.75$$

It is apparent from the foregoing results that Nu_o and Nu_i for $r^* > 0$ cannot be determined accurately for different q_i''/q_o'' using the circular duct correlation. This being the case, recourse has to be taken to Eq. (4.143) and (4.144) to calculate Nu_o and Nu_i for q_i''/q_o'' of interest.

Nusselt Numbers for Liquid Metals, $Pr < 0.1$. Rensen [263] measured fully developed Nusselt numbers in a concentric annulus ($r^* = 0.5409$) with the inner wall

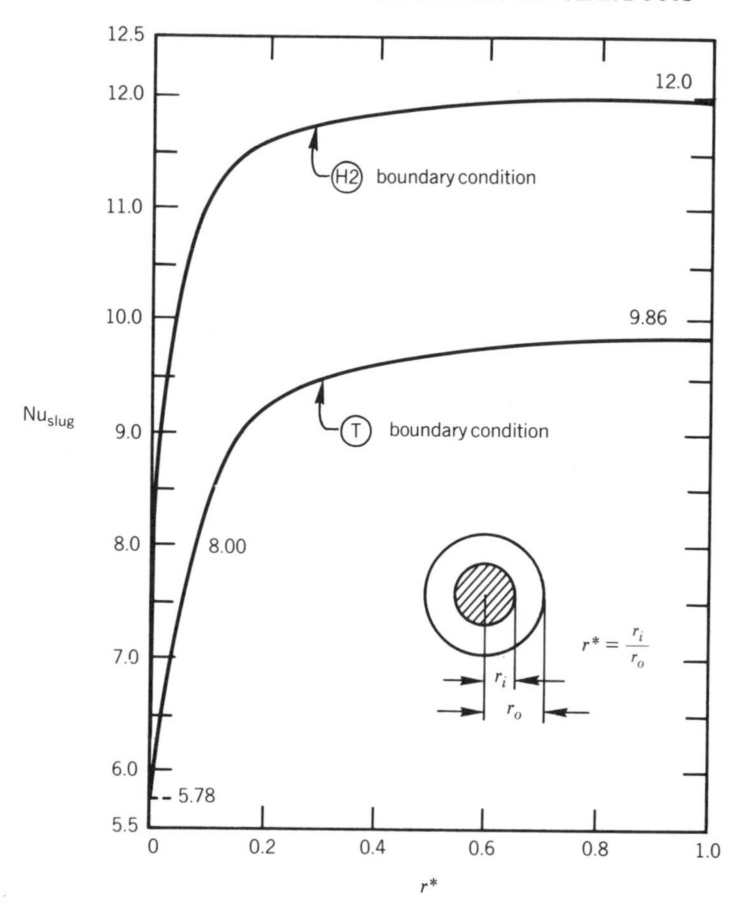

Figure 4.58. Slug flow Nusselt numbers for smooth concentric annular ducts with the Ⓣ and (H2) boundary conditions [216].

subjected to a uniform heat flux and the outer wall insulated. The experiments with liquid sodium as the test fluid covered the ranges $0.0047 \leq \mathrm{Pr} \leq 0.0059$ and $6000 < \mathrm{Re} < 6 \times 10^4$. Under these conditions, Rensen [263] correlated his fully developed Nusselt numbers at the inner wall within $\pm 5\%$ by

$$\mathrm{Nu}_{ii} = 5.75 + 0.022\ \mathrm{Pe}^{0.8} \tag{4.147}$$

Hartnett and Irvine [216] presented an analytic solution for liquid metals with $\mathrm{Pr} \to 0$ as given by Eq. (4.112). It requires a knowledge of the slug Nusselt numbers, which for the Ⓣ and (H2) boundary conditions are given in Fig. 4.58. For the fundamental solution of the second kind, i.e., with one wall subject to uniform heat flux and the other wall insulated, the slug Nusselt numbers are given in Fig. 4.59.

The prediction of Eq. (4.112) for $r^* = 0.5409$ in conjunction with Fig. 4.59 for the case of the inner wall heated and the outer wall insulated are 32% lower than those of Eq. (4.147). This latter equation is valid in the range $28 \leq \mathrm{Pe} \leq 354$.

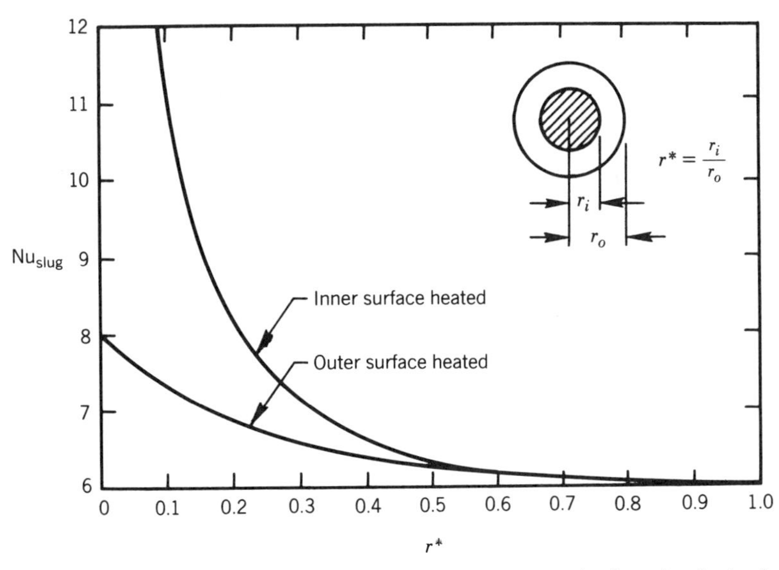

Figure 4.59. Slug flow Nusselt numbers for smooth concentric annular ducts for the fundamental solution of the second kind [216].

A comparison between the predictions of Eq. (4.112) for $r^* = 1$ in conjunction with Fig. 4.59 and Eq. (4.101), which is valid for $r^* = 1$, shows that in the range $0 \le \mathrm{Pe} \le 10^5$, the predictions of Eq. (4.112) are within $\pm 17\%$ of those of Eq. (4.101). Additional comparisons of the predictions of Eq. (4.112) for $r^* = 0$ with the Ⓣ and Ⓗ boundary conditions are presented in Table 4.5 on p. 4.36.

Dwyer [102] developed semiempirical equations for liquid metal flow (Pr < 0.03) in concentric annuli ($0 < r^* \le 1$) with one wall subjected to uniform heat flux and the other wall insulated. For the case of the outer wall heated, the semiempirical equations are

$$\mathrm{Nu}_{oo} = A_o + B_o(\beta \mathrm{Pe})^{n_o} \tag{4.148}$$

where

$$A_o = 5.26 + \frac{0.05}{r^*} \tag{4.149}$$

$$B_o = 0.01848 + \frac{0.003154}{r^*} - \frac{0.0001333}{r^{*2}} \tag{4.150}$$

$$n_o = 0.78 - \frac{0.01333}{r^*} + \frac{0.000833}{r^{*2}} \tag{4.151}$$

$$\beta = 1 - \frac{1.82}{\mathrm{Pr}(\epsilon_m/\nu)^{1.4}_{\max}} \tag{4.152}$$

in which $(\epsilon_m/\nu)_{max}$ can be calculated from the relation

$$\left(\frac{\epsilon_m}{\nu}\right)_{max} = \frac{1}{2}\left(\frac{\epsilon_m}{\nu}\right)_{max,\,c} \tag{4.153}$$

An expression for $(\epsilon_m/\nu)_{max,\,c}$ applicable to a circular duct ($r^* = 0$) was developed by the present authors and is given by

$$\left(\frac{\epsilon_m}{\nu}\right)_{max,\,c} = 0.037\ \mathrm{Re}\sqrt{f} \tag{4.154}$$

in which the Fanning friction factor f can be calculated from the explicit formula due to Techo et al. [67] given in Table 4.2. The predictions of Eqs. (4.153) and (4.154) for $0 \le r^* \le 1$ are in excellent agreement with the results for $(\epsilon_m/\nu)_{max}$ given graphically in [102].

For the case of the inner wall heated, the semiempirical equations of Dwyer [102] are

$$\mathrm{Nu}_{ii} = A_i + B_i(\beta\ \mathrm{Pe})^{n_i} \tag{4.155}$$

where

$$A_i = 4.63 + \frac{0.686}{r^*} \tag{4.156}$$

$$B_i = 0.02154 - \frac{0.000043}{r^*} \tag{4.157}$$

$$n_i = 0.752 + \frac{0.01657}{r^*} - \frac{0.000883}{r^{*2}} \tag{4.158}$$

The values of β for this case can also be calculated from Eqs. (4.152) to (4.154). Both Eqs. (4.148) and (4.155) are valid for Pe values above the critical values. For $\mathrm{Pe} < \mathrm{Pe}_{crit}$, the sole mode of heat transfer is molecular conduction for liquid metals. For $\mathrm{Pr} = 0.005$, 0.01, 0.02, and 0.03, the critical Pe values are 270, 300, 330, and 345, respectively [102].

The foregoing values of Nu_{oo} and Nu_{ii} are in best agreement with the most recent experimental data for liquid metals obtained under conditions of continuous purification. For $300 \le \mathrm{Pe} \le 10^5$, the predictions of Eq. (4.148) for $r^* = 1$ are within +20.2% and −13.3% of the predictions of Eq. (4.101), representing the experimental data of Duchatelle and Vautrey [174] obtained under conditions of continuous purification for flow of NaK ($0.001 < \mathrm{Pr} < 0.021$) in the range $200 < \mathrm{Pe} < 1200$. In the restricted range of $300 \le \mathrm{Pe} \le 1000$, the predictions of Eq. (4.148) with $r^* = 1$ are within +6.9% and −13.3% of the predictions of Eq. (4.101). A comparison between the predictions of Eq. (4.155) with $r^* = 0.5409$ and Eq. (4.147), representing the measurements of Rensen [263] for flow of liquid sodium ($0.0047 \le \mathrm{Pr} \le 0.0059$), showed that in the range $300 \le \mathrm{Pe} \le 10^5$, the predictions of Eq. (4.155) are within +26% of the predictions of Eq. (4.147).

Nusselt Numbers for Circumferentially Varying Heat Fluxes. Sutherland and Kays [264] solved the problem of fully developed turbulent flow in concentric annular ducts with circumferentially varying heat fluxes on the two walls. Let the heat fluxes at the

TABLE 4.33. Circumferential Heat Flux Functions for Fully Developed Turbulent Flow in Smooth Concentric Annular Ducts for Use with Eqs. (4.161) and (4.162) [266]

$r^* = r_i/r_o$	Pr	Re	R_{ii}	R_{io}	R_{oo}	R_{oi}
0.20	Laminar		0.135	0.260	0.677	0.0521
	0	10^4	0.135	0.260	0.677	0.0521
		10^5	0.135	0.260	0.677	0.0521
		10^6	0.135	0.260	0.677	0.0521
	0.01	10^4	0.133	0.256	0.688	0.0513
		10^5	0.119	0.220	0.553	0.0440
		10^6	0.0404	0.0544	0.134	0.0109
	1.00	10^4	0.0244	0.0207	0.0616	0.00413
		10^5	0.00436	0.00267	0.00928	0.000535
		10^6	0.000662	0.0003398	0.00127	0.000675
	10.00	10^4	0.00839	0.00172	0.0129	0.000341
		10^5	0.00127	0.000235	0.00185	0.0000471
		10^6	0.000173	0.0000320	0.000244	0.0000064
0.50	Laminar		0.833	1.33	1.67	0.667
	0	10^4	0.833	1.33	1.67	0.667
		10^5	0.833	1.33	1.67	0.667
		10^6	0.833	1.33	1.67	0.667
	0.01	10^4	0.821	1.32	1.64	0.657
		10^5	0.710	1.13	1.41	0.565
		10^6	0.183	0.272	0.344	0.136
	1.00	10^4	0.0764	0.0995	0.137	0.0484
		10^5	0.0114	0.0131	0.0193	0.00655
		10^6	0.00162	0.00166	0.00252	0.00119
	10.00	10^4	0.00135	0.00774	0.0186	0.00425
		10^5	0.00197	0.00115	0.00272	0.000575
		10^6	0.000270	0.000158	0.000360	0.0000794
0.80	Laminar		9.11	11.1	11.4	8.89
	0	10^4	9.11	11.1	11.4	8.89
		10^5	9.11	11.1	11.4	8.89
		10^6	9.11	11.1	11.4	8.89
	0.01	10^4	8.99	11.0	11.2	8.78
		10^5	7.75	10.0	10.3	7.56
		10^6	1.89	2.29	2.36	1.84
	1.00	10^4	0.681	0.830	0.865	0.648
		10^5	0.0934	0.110	0.116	0.0880
		10^6	0.0120	0.0140	0.0148	0.0112
	10.00	10^4	0.0656	0.0643	0.0749	0.0559
		10^5	0.00917	0.00965	0.0111	0.00770
		10^6	0.00127	0.00132	0.00152	0.00105
0.90	Laminar		42.9	47.4	47.6	42.6
	0	10^4	42.9	47.4	47.6	42.6
		10^5	42.9	47.4	47.6	42.6
		10^6	42.9	47.4	47.6	42.6
	0.01	10^4	42.3	46.8	47.0	42.1
		10^5	36.5	40.3	40.5	36.3
		10^6	8.90	9.82	9.89	8.84
	1.00	10^4	3.16	3.54	3.57	3.12
		10^5	0.429	0.471	0.477	0.424
		10^6	0.0546	0.0597	0.0606	0.0538
	10.00	10^4	0.281	0.280	0.290	0.271
		10^5	0.0386	0.0413	0.0428	0.0371
		10^6	0.00531	0.00564	0.00584	0.00509

outer and inner walls be expressed as

$$q_o''(\theta) = q_{o,m}''(1 + A_o \cos\theta) \tag{4.159}$$

$$q_i''(\theta) = q_{i,m}''(1 + A_i \cos\theta) \tag{4.160}$$

where θ is the angular coordinate, $q_{o,m}''$ and $q_{i,m}''$ are the mean values of $q_o''(\theta)$ and $q_i''(\theta)$ around the circumferences, and A_o and A_i are constants.

The local Nusselt numbers $\mathrm{Nu}_o(\theta)$ and $\mathrm{Nu}_i(\theta)$ on the outer and inner surfaces, respectively, for any angular position θ can then be evaluated from

$$\mathrm{Nu}_o(\theta) = \frac{h_o(\theta) D_h}{k} = \frac{\mathrm{Nu}_{oo}(1 + A_o \cos\theta)}{(1 + \mathrm{Nu}_{oo} A_o R_{oo} \cos\theta) - (q_i''/q_o'')(\theta_o^* - \mathrm{Nu}_{oo} A_i R_{oi} \cos\theta)} \tag{4.161}$$

$$\mathrm{Nu}_i(\theta) = \frac{h_i(\theta) D_h}{k} = \frac{\mathrm{Nu}_{ii}(1 + A_i \cos\theta)}{(1 + \mathrm{Nu}_{ii} A_i R_{ii} \cos\theta) - (q_o''/q_i'')(\theta_i^* - \mathrm{Nu}_{ii} A_o R_{io} \cos\theta)} \tag{4.162}$$

where

$$h_o(\theta) = \frac{q_o''(\theta)}{T_o(\theta) - T_m}, \qquad h_i(\theta) = \frac{q_i''(\theta)}{T_i(\theta) - T_m} \tag{4.163}$$

The Nusselt numbers Nu_{oo} and Nu_{ii} and the influence coefficients θ_o^* and θ_i^* are available from Tables 4.29 to 4.32. The values of R_{ii}, R_{oo}, R_{io}, and R_{oi} are furnished in Table 4.33 as functions of Re, Pr, and r^*.

4.7.3 Hydrodynamically Developing Flow

The only studies of the hydrodynamically developing turbulent flow in concentric annular ducts have been those of Rothfus et al. [265], Olson and Sparrow [260], and Okiishi and Serovy [266]. Rothfus et al. [265] measured the apparent Fanning friction factors $f_{\mathrm{app},i}$ at the inner surfaces of two concentric annuli ($r^* = 0.3367$ and 0.5618) with a square entrance. The experiments covered the range $970 < \mathrm{Re} < 4.85 \times 10^4$. The resulting apparent friction factors for $r^* = 0.5618$ are presented in Fig. 4.60 as the ratio $f_{\mathrm{app},i}/f_i$, where f_i is the fully developed friction factor at the inner surface. For $r^* = 0.5618$, the measured values of f_i are 0.01, 0.008, and 0.0066 for Re = 6000, 1.5×10^4, and 3×10^4, respectively [265].

Knowing $f_{\mathrm{app},i}$ from Fig. 4.60, the apparent friction factor $f_{\mathrm{app},o}$ at the outer wall can be determined from

$$\frac{f_{\mathrm{app},o}}{f_{\mathrm{app},i}} = \frac{r^*(1 - r_m^{*2})}{r_m^{*2} - r^{*2}} \tag{4.164}$$

where r_m^* is given by Eq. (4.135). Knowing both $f_{\mathrm{app},o}$ and $f_{\mathrm{app},i}$, the perimeter-average friction factor f_{app} can be determined from

$$f_{\mathrm{app}} = \frac{f_{\mathrm{app},o} r_o + f_{\mathrm{app},i} r_i}{r_o + r_i} \tag{4.165}$$

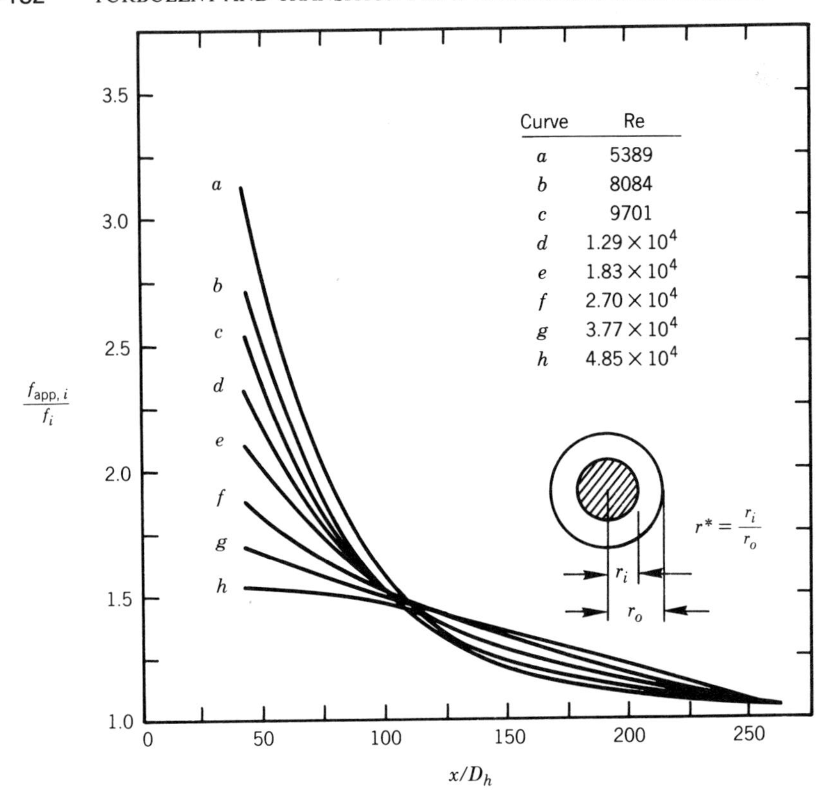

Figure 4.60. Normalized apparent friction factors for turbulent flow in the hydrodynamic entrance region of a smooth concentric annular duct ($r^* = 0.5618$) [265].

Olson and Sparrow [260] performed static pressure gradient measurements in concentric annuli ($r^* = 0, 0.3125$, and 0.5) with a square as well as a rounded entrance as shown in Table 4.28. The tests covered the range $1.6 \times 10^4 < \mathrm{Re} < 7 \times 10^4$. Based on the criterion of 5% approach to the fully developed pressure gradient, Olson and Sparrow [260] estimated the hydrodynamic entrance length as $20 < L_{hy}/D_h < 25$ for ducts with the square as well as the rounded entrance without a trip. For the rounded entrance with a trip, L_{hy}/D_h was estimated to be 15.

Okiishi and Serovy [266] performed static pressure drop and axial velocity measurements in concentric annuli ($r^* = 0.344$ and 0.531) with a square as well as a rounded entrance without a trip. The experiments covered the range $7 \times 10^4 < \mathrm{Re} < 1.6 \times 10^5$.

4.7.4 Thermally Developing Flow

Kays and Leung [168] presented an experimentally obtained solution to the problem of thermally developing flow in four concentric annular ducts ($r^* = 0.192, 0.255, 0.376$, and 0.500). The results are for the case of one wall at uniform heat flux and the other insulated, i.e., the fundamental solution of the second kind. According to this solution, the local Nusselt numbers $\mathrm{Nu}_{x,o}$ and $\mathrm{Nu}_{x,i}$ at the outer and inner surfaces are given by

$$\mathrm{Nu}_{x,o} = \frac{\mathrm{Nu}_{x,oo}}{1 - \theta^*_{x,o} q''_i / q''_o}, \qquad \mathrm{Nu}_{x,i} = \frac{\mathrm{Nu}_{x,ii}}{1 - \theta^*_{x,i} q''_o / q''_i} \tag{4.166}$$

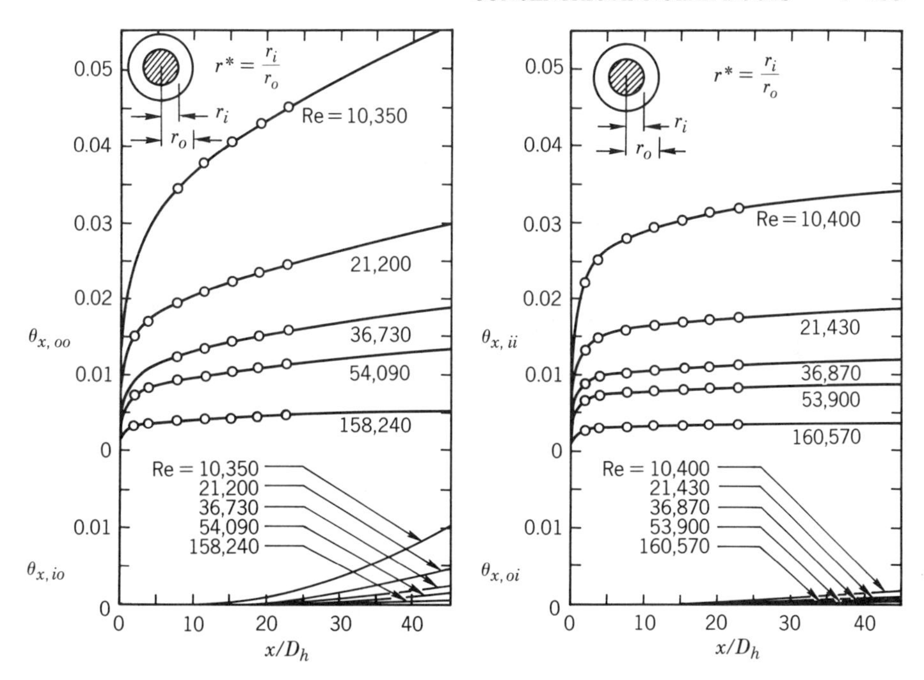

Figure 4.61. Experimental results for the fundamental solution of the second kind for thermally developing and hydrodynamically developed turbulent flow in a smooth concentric annular duct with $r^* = 0.192$ for Pr = 0.7 [168].

where q_o'' and q_i'' are the uniform heat fluxes at the outer and inner walls. Both q_o'' and q_i'' are taken as positive whenever there is heat transfer to the fluid, and negative whenever there is heat transfer out of the fluid. The Nusselt numbers $\mathrm{Nu}_{x,oo}$ and $\mathrm{Nu}_{x,ii}$ and the influence coefficients $\theta_{x,o}^*$ and $\theta_{x,i}^*$ are given by

$$\mathrm{Nu}_{x,oo} = \frac{1}{\theta_{x,oo} - \theta_{x,mo}}, \qquad \mathrm{Nu}_{x,ii} = \frac{1}{\theta_{x,ii} - \theta_{x,mi}} \tag{4.167}$$

$$\theta_{x,o}^* = \frac{\theta_{x,mi} - \theta_{x,oi}}{\theta_{x,oo} - \theta_{x,mo}}, \qquad \theta_{x,i}^* = \frac{\theta_{x,mo} - \theta_{x,io}}{\theta_{x,ii} - \theta_{x,mi}} \tag{4.168}$$

The nondimensional temperatures $\theta_{x,oo}$, $\theta_{x,ii}$, $\theta_{x,oi}$, and $\theta_{x,io}$ for $r^* = 0.192$ and 0.5 are presented in Figs. 4.61 and 4.62. Additional graphical results for $r^* = 0.255$ and 0.376 are available in [168]. The nondimensional temperatures $\theta_{x,mo}$ and $\theta_{x,mi}$ required in Eqs. (4.167) and (4.168) can be computed from

$$\theta_{x,mo} = \frac{4(x/D_h)}{\mathrm{Re}\,\mathrm{Pr}(1 + r^*)}, \qquad \theta_{x,mi} = \frac{4r^*(x/D_h)}{\mathrm{Re}\,\mathrm{Pr}(1 + r^*)} \tag{4.169}$$

The foregoing solution is limited to a fluid with Pr = 0.7, $10^4 < \mathrm{Re} < 1.61 \times 10^5$, and $0.192 < r^* < 0.5$. However, cross-plotting and interpolation could be employed to increase the generality of the results in terms of Re and r^*.

Quarmby and Anand [267] provided an eigenvalue solution to the fundamental problem of the second kind for four concentric annuli ($r^* = 0.02$, 0.1067, 0.1778, and

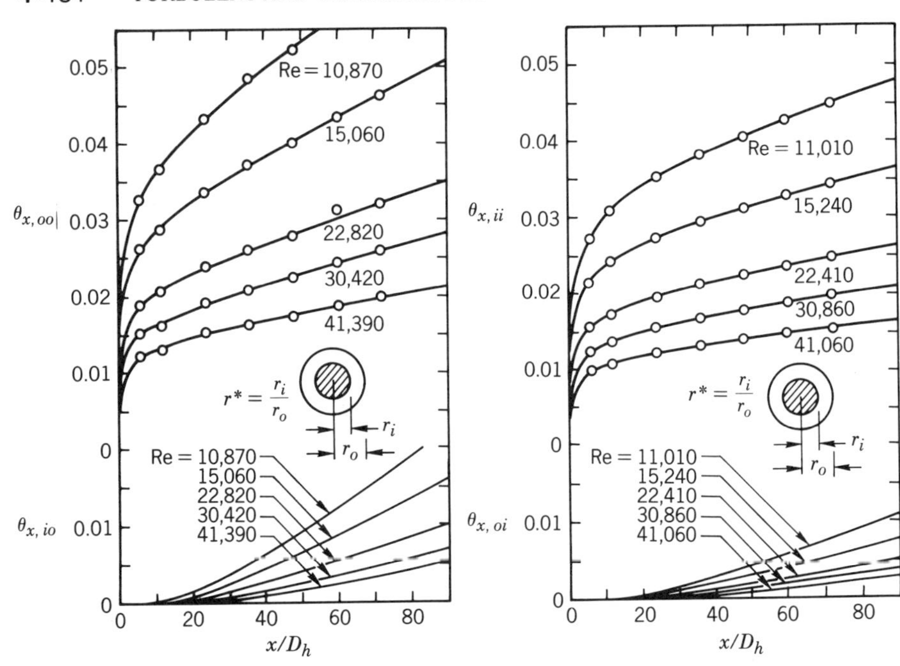

Figure 4.62. Experimental results for the fundamental solution of the second kind for thermally developing and hydrodynamically developed turbulent flow in a smooth concentric annular duct with $r^* = 0.5$ for Pr = 0.7 [168].

0.3472) for three fluids with Pr = 0.01, 0.7, and 1000 covering the range $2 \times 10^4 <$ Re $< 2.4 \times 10^5$. The solution shows good agreement with the experimental results for the four annuli for Pr = 0.7. Some results for $r^* = 0.02$ compare well with the results for a circular duct for all three Pr.

Rensen [263] measured the local Nusselt numbers $\mathrm{Nu}_{x,ii}$ in the thermal entrance region of a concentric annular duct ($r^* = 0.5409$) with the inner wall subjected to the uniform heat flux and the outer wall insulated. The experiments, performed with liquid sodium as the test fluid, covered the ranges 6000 < Re < 6×10^4 and $0.0047 \leq \mathrm{Pr} \leq 0.0059$. Figure 4.63 shows the variation of the local Nusselt number $\mathrm{Nu}_{x,ii}$ with the axial distance x/D_h for various Re and Pr = 0.0054.

Rensen [263] also reported the thermal entrance lengths L_{th}/D_h based on 5% and 1% approach of the local Nusselt number $\mathrm{Nu}_{x,ii}$ to the fully developed value Nu_{ii}. These thermal entrance lengths for Pr = 0.0054 and various Re values are presented in Table 4.34.

4.7.5 Simultaneously Developing Flow

There appears to be little information available on simultaneously developing flow in concentric annuli. An integral analysis was presented by Roberts and Barrow [268], who also made heat transfer measurements on simultaneously developing flow in concentric annuli. They obtained good agreement between the predictions and measurements. Their conclusion was that the Nusselt numbers for simultaneously develop-

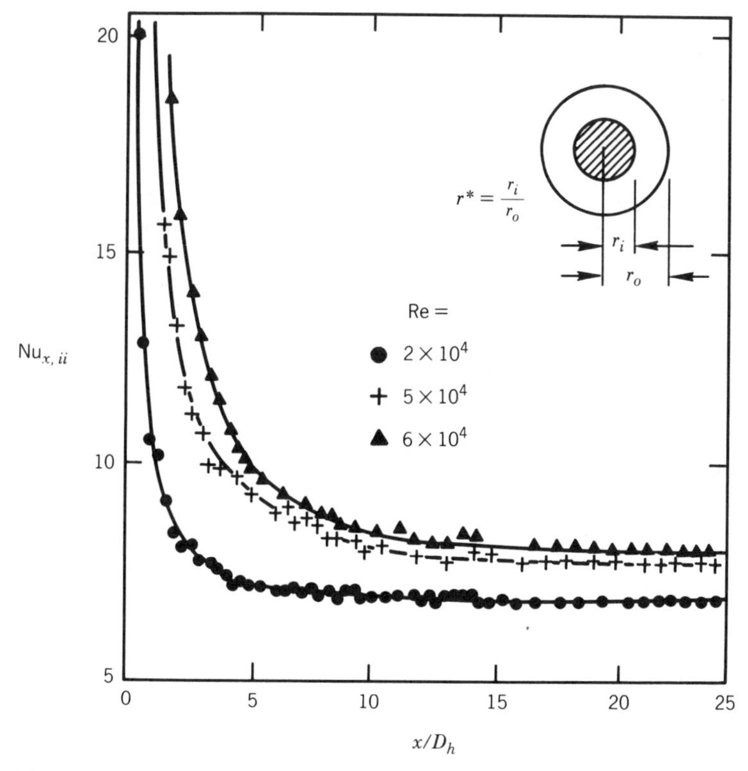

Figure 4.63. Local Nusselt numbers $\mathrm{Nu}_{x,ii}$ for thermally developing and hydrodynamically developed turbulent flow in a smooth concentric annular duct ($r^* = 0.5409$) for Pr = 0.0054 with the inner wall subjected to uniform heat flux and the outer wall insulated [263].

TABLE 4.34. Thermal Entrance Lengths for Thermally Developing Liquid Metal Flow (Pr = 0.0054) in a Smooth Concentric Annular Duct (r^* = 0.5409) with Inner Wall Subjected to Uniform Heat Flux and Outer Wall Insulated [263]

Re	L_{th}/D_h	
	$\mathrm{Nu}_{x,ii}/\mathrm{Nu}_{ii} = 1.05$	$\mathrm{Nu}_{x,ii}/\mathrm{Nu}_{ii} = 1.01$
6,000	2†	3
8,000	2.7	4
10,000	3.2	5
20,000	6	12
50,000	13	25
60,000	13.5	25

†For comparison, it may be noted that for laminar flow, L_{th}/D_h corresponding to $\mathrm{Nu}_{x,ii}/\mathrm{Nu}_{ii} = 1.05$ in a concentric annular duct ($r^* = 0.5$) has a value of about 80 at Re = 2000 (see Table 3.41 of Chap. 3).

ing flow were not significantly different from those for the thermally developing flow with a hydrodynamically developed velocity profile.

4.8 ECCENTRIC ANNULAR DUCTS

Turbulent fluid flow and heat transfer characteristics of eccentric annular ducts have received only sparse attention in the literature. The available information pertains only to the fully developed and hydrodynamically developing flows.

4.8.1 Transition Flow

To the authors' knowledge, the transition flow in eccentric annular ducts has not been studied either experimentally or analytically. Intuitively, it is expected that the duct dimensionless eccentricity $e^* = \epsilon/(r_o - r_i)$, where ϵ is the distance between the centers of the two circular walls, must play a decisive role in shaping the $\mathrm{Re}_{\mathrm{crit}}$ value for an eccentric annulus. For $e^* \to 0$, the $\mathrm{Re}_{\mathrm{crit}}$ values for an eccentric annulus can be taken from Fig. 4.54, since $e^* = 0$ corresponds to the concentric annular duct. For $e^* \to 1$, the inner duct wall tends to contact the outer wall. This could lead to coexistence of laminar and turbulent flows in the vicinity of the contact point between the inner and outer walls. In a duct with coexisting laminar and turbulent flows, transition does not occur simultaneously over the entire cross section. Such a state of affairs in an eccentric annulus with $e^* \to 1$ may push the upper limit of $\mathrm{Re}_{\mathrm{crit}}$ to higher values than for $e^* \to 0$.

4.8.2 Fully Developed Flow

Fluid Flow. Jonsson and Sparrow [269] conducted a careful experimental investigation of fully developed turbulent flow in smooth eccentric annular ducts with $r^* = 0.281, 0.561, 0.750$, $0 < e^* < 1$, and $1.8 \times 10^4 < \mathrm{Re} < 1.8 \times 10^5$.

In terms of the wall coordinates u^+ and y^+, the velocity distribution under the aforementioned conditions is representable as [269]

$$u_o^+ = 2.56 \ln y_o^+ + 4.9 \tag{4.170}$$

$$u_i^+ = 2.44 \ln y_i^+ + 4.9 \tag{4.171}$$

where

$$u_o^+ = u/u_{t,o}, \qquad u_i^+ = u/u_{t,i} \tag{4.172}$$

$$y_o^+ = y_o u_{t,o}/\nu, \qquad y_i^+ = y_i u_{t,i}/\nu \tag{4.173}$$

with the subscripts o and i referring to the outer and inner walls of the annulus. The friction velocity u_t in Eqs. (4.172) and (4.173) is defined by Eq. (4.8), and y is the radial distance measured from the wall. Equations (4.170) and (4.171) are valid for $y_o^+ > 30$ and $y_i^+ > 30$.

Jonsson and Sparrow [269] provided the velocity measurements graphically in terms of the wall coordinates (u^+, y^+) as well as the velocity-defect representation. In addition, they presented isovels for all three eccentric annuli with $r^* = 0.281$, 0.561, and 0.750. These isovel maps are quite instructive in understanding the roles played by r^*, e^*, and Re in shaping the velocity field in eccentric annuli. Accordingly, the isovel

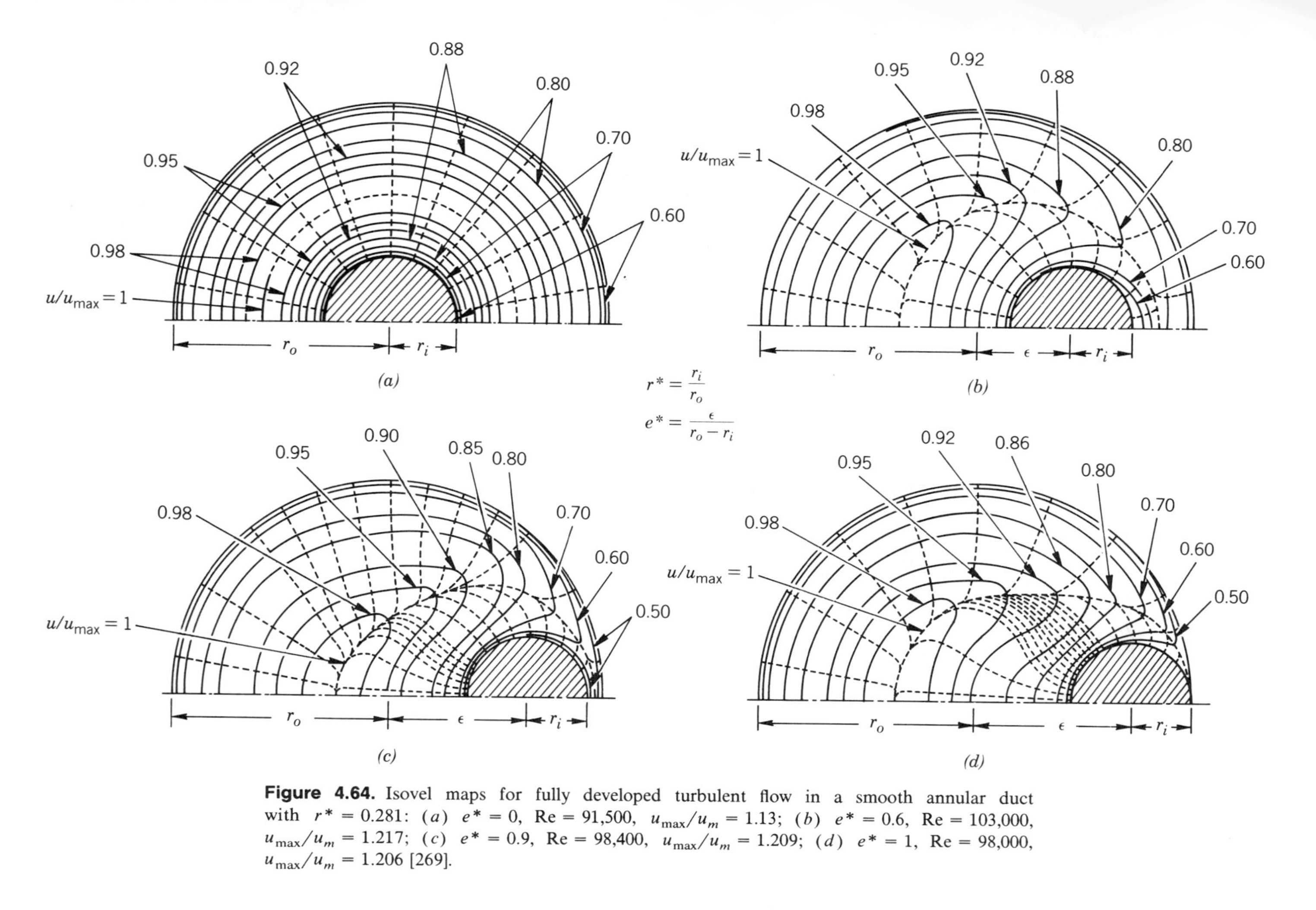

Figure 4.64. Isovel maps for fully developed turbulent flow in a smooth annular duct with $r^* = 0.281$: (*a*) $e^* = 0$, Re = 91,500, $u_{max}/u_m = 1.13$; (*b*) $e^* = 0.6$, Re = 103,000, $u_{max}/u_m = 1.217$; (*c*) $e^* = 0.9$, Re = 98,400, $u_{max}/u_m = 1.209$; (*d*) $e^* = 1$, Re = 98,000, $u_{max}/u_m = 1.206$ [269].

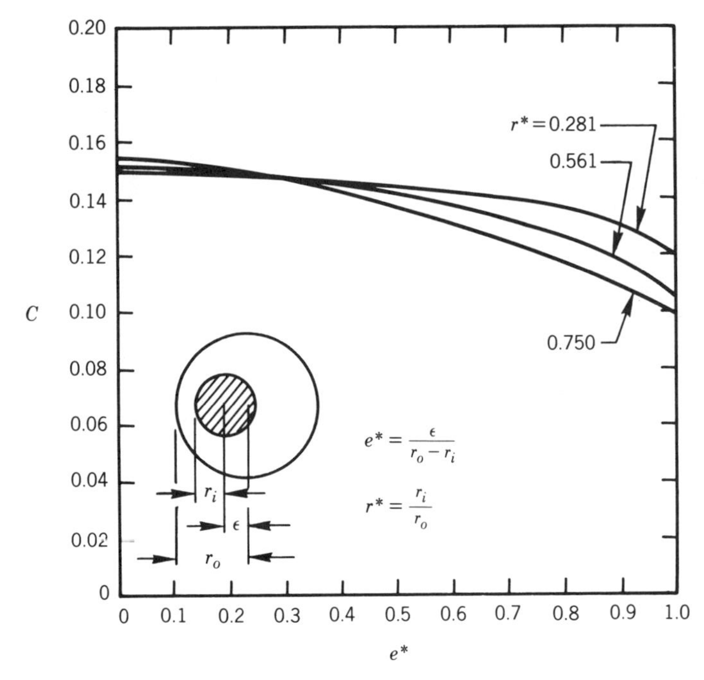

Figure 4.65. Empirical constant C of Eq. (4.174) as a function of e^* and r^* of eccentric annuli [269].

maps for $r^* = 0.281$ are presented in Fig. 4.64. Additional isovel maps for $r^* = 0.561$, and 0.750 are available in [269]. The dashed lines in Fig. 4.64 are the so-called gradient lines, which for fully developed flow coincide with the lines of zero shear. Also, it may be noted that the gradient lines are normal to the isovels, shown by the continuous lines.

Jonsson and Sparrow [269] correlated their circumferentially averaged fully developed Fanning friction factors for smooth eccentric annuli by a power-law relationship of the type

$$f = \frac{C}{\mathrm{Re}^n} \tag{4.174}$$

where C is a strong function of e^* and a relatively weak function of r^*; it is independent of Re. Figure 4.65 shows the variation of C with e^* and r^*. As regards the exponent n, Jonsson and Sparrow [269] found a single value $n = 0.18$ to provide the most satisfactory correlation for all r^*, e^*, and Re.

Eccentricity in an annular duct causes the local shear stresses and hence the local friction factors f_i and f_o to vary around the circumference of each of the two surfaces. This circumferential variation of f_i and f_o for four eccentric annuli with $r^* = 0.90$, 0.75, 0.50, 0.25 and $e^* = 0.01$, 0.2, 0.4, 0.6, 0.8, 0.99 is presented in Fig. 4.66, based on the analysis by Jonsson and Sparrow [270]. The continuous curves represent the results of an analysis for fully developed laminar flow in eccentric annuli developed by Jonsson and Sparrow [270]. The data points shown along the laminar flow curves are for the aforementioned fully developed turbulent flow in eccentric annuli with $r^* =$

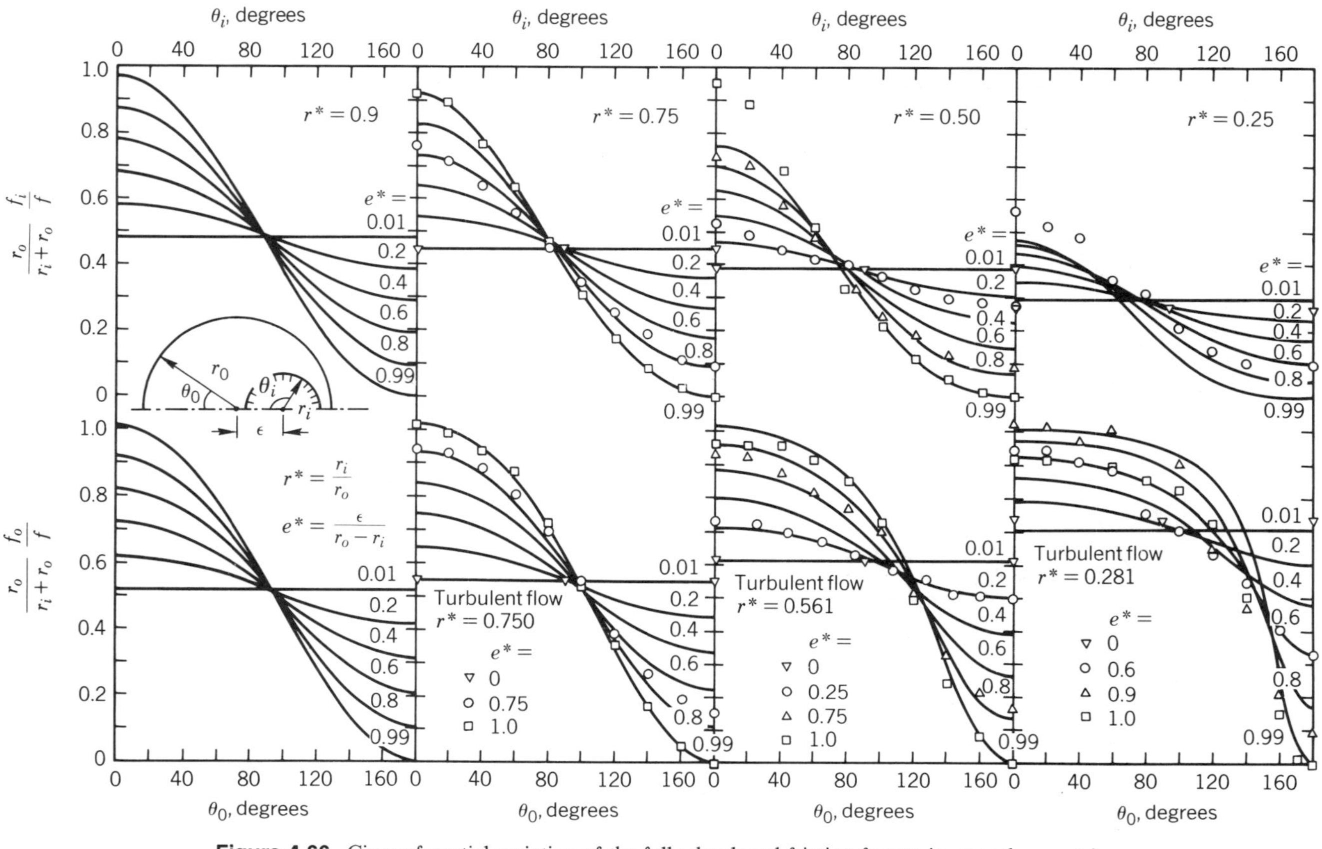

Figure 4.66. Circumferential variation of the fully developed friction factors in smooth eccentric annular ducts [270].

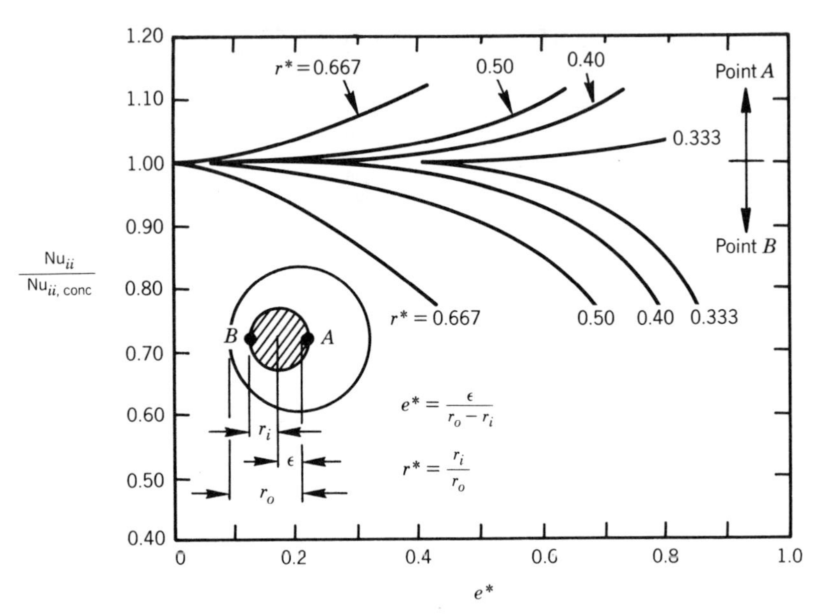

Figure 4.67. Nusselt numbers Nu_{ii} for fully developed turbulent waterflow (Pr = 6.5) in smooth eccentric annular ducts with the inner wall subjected to a uniform heat flux and the outer wall insulated [275].

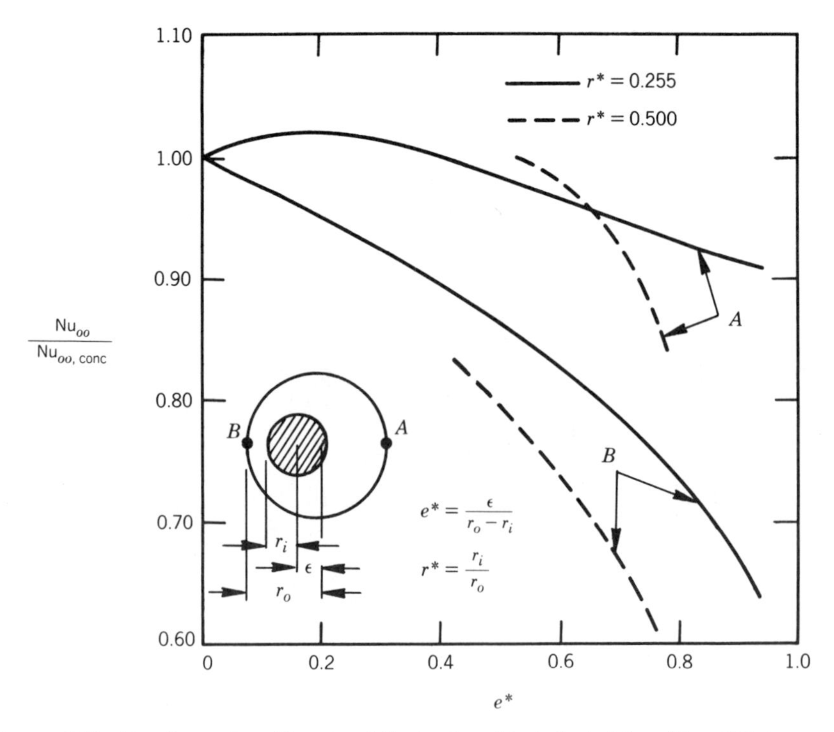

Figure 4.68. Nusselt numbers Nu_{oo} for fully developed turbulent airflow (Pr = 0.7) in smooth eccentric annular ducts with the outer wall subjected to a uniform heat flux and the inner wall insulated [276].

0.750, 0.561, and 0.281 [269]. At the higher r^* values of 0.75 and 0.50, the agreement between the laminar theory and the turbulent data is remarkably good. For the lowest r^* value of 0.25, the agreement is not quite so good. The local friction coefficients f_i and f_o in Fig. 4.66 are normalized by the overall friction coefficient f, which can be obtained from Fig. 3.47 in Chap. 3 on p. 3.113.

Lee and Barrow [271] also conducted an experimental study of the fully developed turbulent flow in three eccentric annuli with $r^* = 0.2581, 0.3872, 0.6128$, $0 < e^* < 1$, and $10^4 < \text{Re} < 5 \times 10^4$. They presented the velocity distribution and friction-factor data in a graphical form.

Some analytical predictions of the fully developed velocity distribution in eccentric annuli have been made by Deissler and Taylor [272], Yu and Dwyer [273], and Ricker et al. [274].

Heat Transfer. Judd and Wade [275] experimentally determined fully developed Nusselt numbers Nu_{ii} for waterflow (Pr = 6.5) in four eccentric annuli ($r^* = 0.3333$, 0.40, 0.50, and 0.6667) for varying eccentricity ($0 < e^* < 1$). The thermal boundary condition employed was uniform heat flux at the inner wall (i.e., the (H1) boundary condition) with the outer wall insulated. The measurements were made at the two extremes of the wall spacing: at the points of the maximum and minimum separation between the walls. These points are labeled A and B in the inset of Fig. 4.67, which presents the results obtained by Judd and Wade [275]. The Nusselt numbers Nu_{ii} for

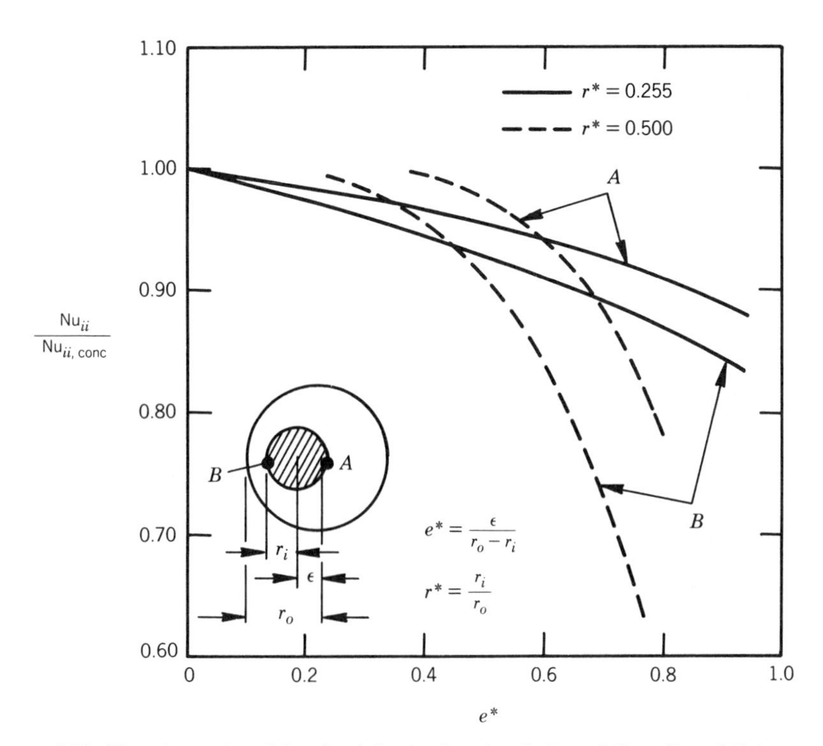

Figure 4.69. Nusselt numbers Nu_{ii} for fully developed turbulent airflow (Pr = 0.7) in smooth eccentric annular ducts with the inner wall subjected to a uniform heat flux and the outer wall insulated [276].

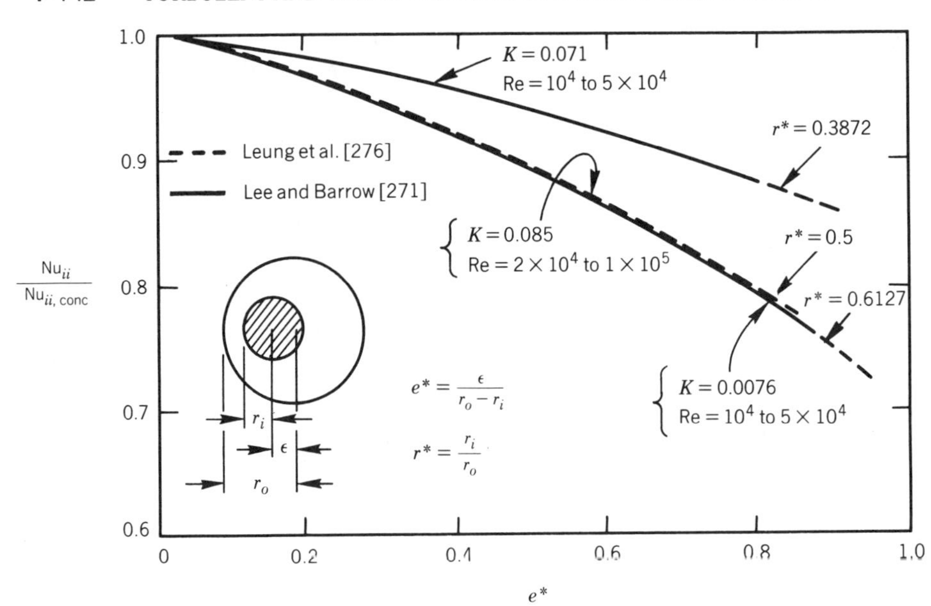

Figure 4.70. Nusselt numbers Nu_{ii} for fully developed turbulent airflow ($Pr = 0.7$) in smooth eccentric annular ducts with the inner wall subjected to the (H4) boundary condition and the outer wall insulated [271].

the eccentric annuli in Fig. 4.67 have been normalized by the Nusselt numbers $Nu_{ii,\text{conc}}$ of the concentric annuli for the same r^* values as the eccentric annuli.

Leung et al. [276] measured the fully developed Nusselt numbers Nu_{ii} and Nu_{oo} for airflow ($Pr = 0.7$) in two eccentric annuli ($r^* = 0.255$ and 0.500) for varying eccentricity ($0 < e^* < 1$). The thermal boundary conditions employed were (H1) at the inner wall with the outer wall insulated and (H1) at the outer wall with the inner wall

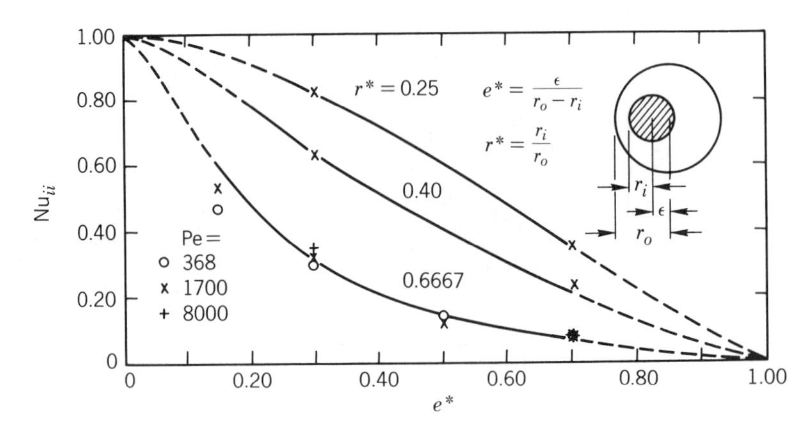

Figure 4.71. Nusselt numbers Nu_{ii} for fully developed turbulent liquid metal flow ($0.005 \leq Pr \leq 0.03$) in smooth eccentric annular ducts with uniform heat flux at the inner wall and the outer wall insulated [273].

insulated. The results of Leung et al. [276] are presented in Figs. 4.68 and 4.69. These results are in the same form as those of Judd and Wade [275] in Fig. 4.67.

Lee and Barrow [271] studied the fully developed heat transfer characteristics of eccentric annuli for airflow (Pr = 0.7) with the outer wall insulated and the inner subjected to the (H4) boundary condition. The experiments covered the range $10^4 <$ Re $< 5 \times 10^4$. Their experimentally measured Nusselt numbers Nu_{ii}, normalized by the corresponding Nusselt numbers $\mathrm{Nu}_{ii,\mathrm{conc}}$ for the concentric annuli ($e^* = 0$), are presented in Fig. 4.70, which also includes some results presented by Leung et al. [276]. Of particular interest is the effect of the wall conduction parameter $K = kr_i/k_w\delta_w$ in shaping the Nusselt number Nu_{ii} in Fig. 4.70.

Yu and Dwyer [273] presented an analytical solution for fully developed turbulent flow of liquid metals (0.005 < Pr < 0.03) in eccentric annuli. Their calculated Nusselt numbers Nu_{ii} for eccentric annuli ($r^* = 0.25$, 0.40, and 0.6667) covering the ranges $0 < e^* < 1$ and $368 \leq \mathrm{Pe} \leq 8000$ are presented in Fig. 4.71. The thermal boundary condition employed in these calculations is uniform heat flux at the inner wall (i.e., the (H1) boundary condition) with the outer wall insulated.

4.8.3 Hydrodynamically Developing Flow

The information on hydrodynamically developing flows is extremely sparse. Jonsson [277] obtained experimental information on the pressure gradient for the hydrodynamically developing flow in three eccentric annuli ($r^* = 0.281$, 0.561, and 0.750) covering the range $1.8 \times 10^4 <$ Re $< 1.8 \times 10^5$ for four e^* values of 0, 0.5, 0.9, and 1.0. He presented his results in a graphical form by normalizing the local pressure gradients by the fully developed pressure gradients and plotting them against x/D_h.

The hydrodynamic entrance lengths L_{hy}/D_h for the eccentric annuli ($r^* = 0.281$, 0.561, and 0.750) are presented in Table 4.35. These results are deduced by Jonsson [277] from his pressure gradient measurements. The entrance lengths in Table 4.35 correspond to the approach of the pressure gradient to within 2% of the fully developed value.

4.8.4 Thermally Developing Flow

No results are available for thermally developing turbulent flow with hydrodynamically developed velocity profiles in eccentric annuli. Judging from the fully developed turbulent flow results in Figs. 4.67 to 4.71, it is inferred that for $e^* < 0.5$, the local Nusselt numbers for the thermally developing flow in eccentric annuli may be estimated to within $\pm 15\%$ from the corresponding results for concentric annuli ($e^* = 0$) presented in Sec. 4.7.4.

TABLE 4.35. Turbulent Flow Hydrodynamic Entrance Lengths for Smooth Eccentric Annular Ducts [277]

	L_{hy}/D_h			
r^*	$e^* = 0$	0.5	0.9	1.0
0.281	29	32	38	38
0.561	26	38	59	78
0.750	28	50	69	91

4.8.5 Simultaneously Developing Flow

No results are available for simultaneously developing turbulent flow in eccentric annuli. To a crude approximation, for $e^* < 0.5$, the local Nusselt numbers for the simultaneously developing turbulent flow in eccentric annuli may be estimated from the corresponding results for concentric annuli ($e^* = 0$) presented in Sec. 4.7.5. As mentioned there, the results for the simultaneously developing flow are expected to be close to those for the thermally developing flow.

4.9 CLOSURE

In this chapter, we have presented the most important turbulent fluid flow and heat transfer results pertaining to a variety of ducts, paralleling the treatment in Chap. 3 on laminar flow. The turbulent duct flows have not been investigated as exhaustively as their laminar counterparts. The main reason is that the transport equations for turbulent flow are more complex, requiring a turbulence model for the Reynolds stresses to solve the fluid flow problem and an additional model for the Reynolds fluxes to solve the heat transfer problem. No universal turbulence models are available that can be applied to all duct geometries. Within the scope of the chapter, it has not been possible to provide a systematic exposition of the available turbulence models with a view to appraising the merits of each model. The main emphasis of the chapter has been to present the most important fluid flow and heat transfer results of direct interest to design engineers. This includes analytical, numerical, and experimental results, and empirical correlations.

A limitation of the results in this chapter is that they may not give accurate friction factors and heat transfer coefficients when applied to extremely small tubes with hydraulic diameters smaller than about 2 mm (0.1 in.). In such tubes, lower friction factors and heat transfer coefficients generally prevail than those predicted by the formulas in the chapter. The reason is that the turbulent eddy mechanism for fluid flow and heat transfer is suppressed by the physical size of the tube cross section. Unfortunately, definitive information on turbulent flow friction factors and heat transfer coefficients is not available in the literature.

Unlike the laminar flow results, the turbulent flow results for $\mathrm{Pr} > 0.5$ which covers the common fluids like air, water, and various oils (but not liquid metals) are rather insensitive to the effects of the thermal boundary conditions as well as to the entrance effects associated with the hydrodynamically, thermally, and simultaneously developing flows. Consequently, these flows with different thermal boundary conditions have not been analyzed in great detail in the literature. The results for fully developed turbulent flow with uniform wall temperature or uniform wall heat flux provide fairly accurate estimates for the aforementioned turbulent flows. Table 4.36 serves as a ready reference for the recommended fully developed turbulent flow correlations for the most useful smooth-walled duct geometries for $\mathrm{Pr} > 0.5$.

Another notable difference between laminar and turbulent flows is that the duct wall roughness has little effect on the results for the former flow, whereas it exerts a strong influence on the results for the latter flow. Therefore, results on the influence of the duct wall roughness are provided wherever possible.

As evidenced by the amount of space allocated to it, circular duct geometry has been investigated most exhaustively. Several of the circular duct results can be utilized to make fairly good estimates of the friction and heat transfer coefficients for noncircular ducts through the use of the hydraulic diameter. This concept yields results for

TABLE 4.36. Fully Developed Turbulent Flow Friction Factors and Nusselt number (Pr > 0.5) for Technically Important Smooth-Walled Ducts

Duct Geometry	Characteristic Dimension	Recommended Correlations†
Circular, 2a *(a)*	$D_h = 2a$	Friction factor correlation by the present authors for $2300 \le \text{Re} \le 10^7$: ‡ $f = A + \dfrac{B}{\text{Re}^{1/m}}$ where $A = 0.0054$, $B = 2.3 \times 10^{-8}$, $m = -\frac{2}{3}$ for $2300 \le \text{Re} \le 4000$ and $A = 1.28 \times 10^{-3}$, $B = 0.1143$, $m = 3.2154$ for $4000 \le \text{Re} \le 10^7$. Nusselt number correlation by Gnielinski for $2300 \le \text{Re} \le 5 \times 10^6$: § $\text{Nu} = \dfrac{(f/2)(\text{Re} - 1000)\text{Pr}}{1 + 12.7(f/2)^{1/2}(\text{Pr}^{2/3} - 1)}$
Flat, 2b *(b)*	$D_h = 4b$	Use circular duct f and Nu correlations. Predicted f are up to 12.5% lower and predicted Nu are within ±9% of the most reliable experimental results.

TABLE 4.36. Continued

Duct Geometry	Characteristic Dimension	Recommended Correlations†
2b, 2a Rectangular *(c)*	$D_h = \dfrac{4ab}{a+b}, \quad \alpha^* = \dfrac{2b}{2a}$ $\dfrac{D_l}{D_n} = \dfrac{2}{3} + \dfrac{11}{24}\alpha^*(2-\alpha^*)$	f Factors: (1) Substitute D_l for D_h in the circular duct correlation, and calculate f from the resulting equation. (2) Alternatively, calculate f from $f = (1.0875 - 0.1125\alpha^*)f_c$ where f_c is the friction factor for the circular duct using D_h. In both cases, predicted f factors are within $\pm 5\%$ of the experimental results. Nusselt Numbers: (1) With uniform heating at four walls, use circular duct Nu correlation for an accuracy of $\pm 9\%$ for $0.5 \le \mathrm{Pr} \le 100$ and $10^4 \le \mathrm{Re} \le 10^6$. (2) With equal heating at two long walls, use circular duct correlation for an accuracy of $\pm 10\%$ for $0.5 < \mathrm{Pr} < 10$ and $10^4 \le \mathrm{Re} \le 10^5$. (3) With heating at one long wall only, use circular duct correlation to get approximate Nu values for $0.5 < \mathrm{Pr} < 10$ and $10^4 \le \mathrm{Re} \le 10^6$. These calculated values may be up to 20% higher than the actual experimental values.
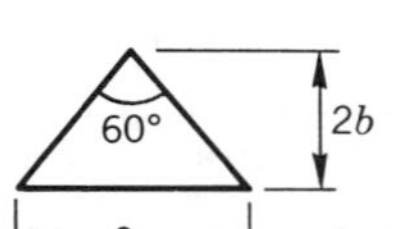 Equilateral triangular *(a)*	$D_h = 2\sqrt{3}\,a = \dfrac{4b}{3}$ $D_l = \sqrt{3}\,a = \dfrac{2b}{3\sqrt{3}}$	Use circular duct f and Nu correlations with D_h replaced by D_l. Predicted f are within $+3\%$ and -11% and predicted Nu within $+9\%$ of the experimental values.

Geometry	Hydraulic diameter	Recommendation
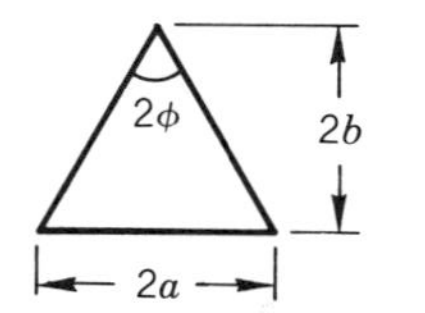 Isosceles triangular	$D_h = \dfrac{4ab}{a + \sqrt{a^2 + b^2}}$ $\dfrac{D_g}{D_h} = \dfrac{1}{2\pi}\left[3 \ln \cot\dfrac{\theta}{2} + 2 \ln \tan\dfrac{\phi}{2} - \ln \tan\dfrac{\theta}{2}\right]$ where $\theta = (90° - \phi)/2$.	For $0 < 2\phi < 60°$, use circular duct f and Nu correlations with D_h replaced by D_g; for $2\phi = 60°$, replace D_h by D_l (see above) and for $60° < 2\phi \leq 90°$ use circular duct correlations directly with D_h. Predicted f and Nu are within +9% and −11% of the experimental values. No recommendations can be made for $2\phi > 90°$ due to lack of the experimental data.
(b)		
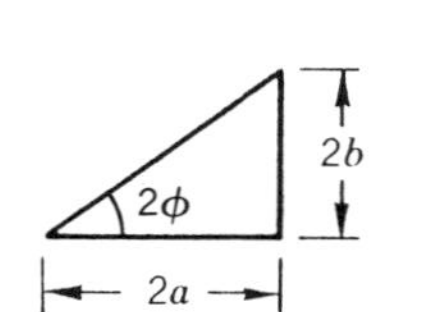 Right triangular	$D_h = \dfrac{4ab}{a + b + \sqrt{a^2 + b^2}}$	For $30° \leq 2\phi \leq 45°$, use circular duct f and Nu correlations. Predicted f and Nu are within ±5% of the experimental measurements (Fig. 4.29). No recommendations can be made for 2ϕ values outside the range $30° \leq 2\phi \leq 45°$ due to lack of the experimental results.
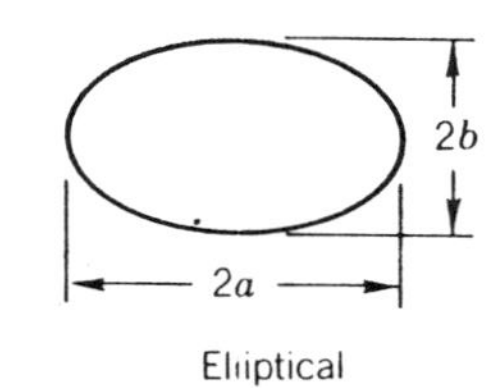 Elliptical	$D_h = \dfrac{\pi b}{E(m)}, \quad \alpha^* = \dfrac{2b}{2a}$ $m = 1 - \alpha^{*2}$, $E(m)$ is the complete elliptic integral of the second kind.	Use circular duct f and Nu correlations. Predicted f are within +13% and −5% and predicted Nu within ±10% of the experimental results.

TABLE 4.36. Continued

Duct Geometry	Characteristic Dimension	Recommended Correlations†
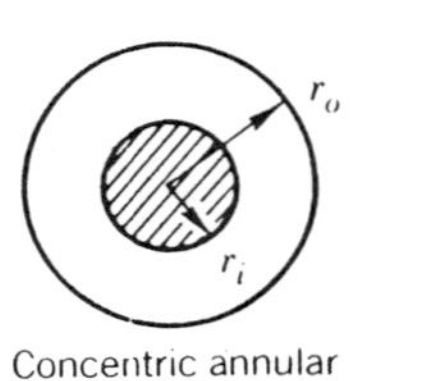 Concentric annular	$D_h = 2(r_o - r_i)$, $r^* = \frac{r_i}{r_o}$ $\frac{D_l}{D_n} = \frac{1 + r^{*2} + (1 - r^{*2})/\ln r^*}{(1 - r^*)^2}$	f Factors: (1) Substitute D_l for D_h in the circular duct correlation, and calculate f from the resulting equation. (2) Alternatively, calculate f from $f = (1 + 0.0925r^*)f_c$ where f_c is the friction factor for the circular duct using D_h. In both cases, predicted f factors are within $\pm 5\%$ of the experimental results. Nusselt Numbers: In all the following recommendations, use D_h with *wetted* perimenter in Nu and Re: (1) Nu at the outer wall *can* be determined from the circular duct correlation within the accuracy of about $\pm 10\%$ regardless of the heating/cooling condition at the inner wall. (2) Nu at the inner wall *cannot* be determined accurately regardless of the heating/cooling condition at the outer wall. (3) As summarized in Sec. 4.7.2, use Eqs. (4.143) and (4.144) in conjunction with Tables 4.22 and 4.29 to 4.32 to calculate Nu_o and Nu_i for the Ⓗ boundary condition on each wall for arbitrary q_i''/q_o''. (4) In the absence of experimental results for the Ⓣ boundary condition, compute Nu_o and Nu_i from Eqs. (4.143) and (4.144) as an approximation.

†The friction factor and Nusselt number correlations for the circular duct are the most reliable and agree with a large amount of the experimental data within $\pm 2\%$ and $\pm 10\%$, respectively. The correlations for all other duct geometries are not as good as those for the circular duct on an absolute basis.

‡This correlation is presented as Eq. (4.31) in the text. In contrast with other correlations of Table 4.2, it spans the transition flow regime ($2300 \le Re \le 4000$). If desired, the other well-known correlations of Table 4.2 may be used for $Re \ge 4000$.

§Predictions of the Gnielinski correlation are compared in Table 4.4 with those of other correlations reported in the literature.

non-sharp-cornered noncircular ducts that are generally within ±15% of the experimental values. Several other characteristic dimensions have been proposed as substitutes for the hydraulic diameter for specific duct geometries, yielding results within ±5%. Refer to Table 4.36 for the most useful of these dimensions. Unfortunately, none of these characteristic dimensions appears to be as broadly and simply applicable as the hydraulic diameter. There is a need to discover a universal characteristic dimension applicable to all duct geometries. The usefulness of the hydraulic diameter appears to break down for noncircular ducts with acute-angled corners (around 20°). The calculated friction factors and Nusselt numbers with the hydraulic diameter are about 35% lower than the experimental results for such ducts. The coexistence of laminar and turbulent flows in the corner regions of such ducts seems to contribute to this breakdown, since the laminar flow results for noncircular ducts do not correlate well via the hydraulic diameter. One consequence of the coexistence of laminar and turbulent flows is the reduction of the friction and heat transfer coefficients, as these coefficients possess lower values for laminar flows.

A unique feature of the fully developed turbulent flow in noncircular ducts is the presence of secondary flow. Though small in magnitude (approximately 1 to 2% of the axial mean velocity), the secondary flow is found to exert a significant effect on the turbulent fluid flow and heat transfer characteristics of noncircular ducts. Both friction and heat transfer coefficients tend to be approximately 10% higher in the region of the duct cross section dominated by the secondary flow than in the region uninfluenced by it.

The secondary-flow patterns in noncircular ducts can also provide valuable clues to the existence of hot or cold spots in turbulent duct flows. With this objective in mind, secondary-flow patterns have been provided for several noncircular ducts. These patterns can be inferred from the isovel maps of the primary flow through the ducts in question.

The transition flow regime in ducts represents a rather gray area between the laminar and turbulent flow regimes. It is quite desirable to learn about the transition flow in noncircular ducts, as certain compact heat exchangers employing noncircular ducts operate in this regime. Unfortunately, this subject has not been explored systematically. Moreover, determination of the transition flow friction and heat transfer coefficients is not a matter of straightforward interpolation between the laminar and turbulent flow results.

We would like to close this chapter with the hope that the information contained in it will prove useful to designers and researchers. Furthermore, we hope that the state of the art presented will engender some interest among researchers in finding new solutions that will fill the gaps in our understanding of transition and turbulent duct flows.

NOMENCLATURE

A_c	flow cross-sectional area, m^2, ft^2
a	radius of a circular duct; also half width of a noncircular duct, m, ft
b	half spacing or half height of a duct, m, ft
C	proportionality constant for the variation of f with Re or Re^{-n}
c	half spacing or width of a duct, m, ft

c_p	specific heat of fluid at constant pressure, $J/(kg \cdot K)$, $Btu/(lb_m \cdot °F)$
D_h	hydraulic diameter of duct, $= 4A_c/P$, m, ft
D_l	laminar equivalent diameter [see Eqs. (4.15), (4.16), and (4.24)], m, ft
e^*	eccentricity of an eccentric annular duct, $= \epsilon/(r_o - r_i)$
f	circumferentially averaged fully developed Fanning friction factor $= \tau_w/(\rho u_m^2/2g_c)$
f_{app}	apparent Fanning friction factor in the hydrodynamic entrance region, $= \Delta p^*/(x/r_h)$
g_c	proportionality constant in Newton's second law of motion, $= 1$ and dimensionless in SI units; $= 32.174\ lb_m \cdot ft/(lb_f \cdot s^2)$
(H), (H1), (H2), (H4)	constant wall heat flux boundary conditions (refer to Table 3.1 in Chap. 3)
h	convective heat transfer coefficient for fully developed flow, $W/(m^2 \cdot K)$, $Btu/(hr \cdot ft^2 \cdot °F)$
J	mechanical-to-thermal energy conversion factor, $= 1$ and dimensionless in SI units; $= 778.163\ lb_f \cdot ft/Btu$
$K(x)$	incremental pressure drop number, defined by Eq. (3.4) in Chap. 3
K	wall conductivity parameter $= ks/k_w\delta_w$, where s is a duct-specific dimension appropriately defined in the text
k, k_f	thermal conductivity of fluid, $W/(m \cdot K)$, $Btu/(hr \cdot ft \cdot °F)$
k_t	apparent, virtual, or eddy conductivity for turbulent flow [see Eq. (4.5)], $W/(m \cdot K)$, $Btu/(hr \cdot ft \cdot °F)$
k_w	thermal conductivity of duct wall material, $W/(m \cdot K)$, $Btu/(hr \cdot ft \cdot °F)$
L_{hy}	hydrodynamic entrance length, m, ft
L_{th}	thermal entrance length, m, ft
Nu_{bc}	circumferentially averaged Nusselt number for fully developed flow for the thermal boundary condition of Table 3.1 or 3.2 in Chap. 3
$Nu_{x,bc}$	circumferentially averaged but axially local Nusselt number for the thermal entrance region for the specified thermal boundary condition, defined by Eq. (3.10) in Chap. 3
$Nu_{m,bc}$	mean Nusselt number for the thermal entrance region for the specified thermal boundary condition, defined by Eq (3.11) in Chap. 3
$Nu_{x,lj}^{(k)}$	local Nusselt number for a concentric or eccentric annular duct, $= \Phi_{lj}^{(k)}/[\theta_{lj}^{(k)} - \theta_{mj}^{(k)}]$
$Nu_{x,i}^{(k)}, Nu_{x,o}^{(k)}$	local Nusselt number at inner and outer walls of a concentric or eccentric annular duct
P	wetted perimeter of duct, m, ft
Pe	Péclet number $= u_m D_h/\alpha = Re\,Pr$
Pr	Prandtl number $= \nu/\alpha$

Pr_t	turbulent Prandtl number $= \epsilon_m/\epsilon_h$
p	fluid mean static pressure, Pa, $\mathrm{lb}_f/\mathrm{ft}^2$
Δp	fluid static pressure drop in the flow direction between two cross sections of interest, Pa, $\mathrm{lb}_f/\mathrm{ft}^2$
Δp^*	fluid static pressure drop $= \Delta p/(\rho u_m^2/2g_c)$
q''_w	wall heat flux, i.e., heat transfer rate per unit heat transfer area of the duct (average value with respect to perimeter), $\mathrm{W/m^2}$, $\mathrm{Btu/(hr \cdot ft^2)}$
Re	Reynolds number $= u_m D_h/\nu$
Re_ϵ	roughness Reynolds number $= \epsilon u_t/\nu$
$\mathrm{Re}_{\mathrm{crit}}$	critical Reynolds number (see Sec. 4.2.1)
r	radial coordinate in the cylindrical coordinate system, m, ft
r_i, r_o	inner and outer tube radii of a concentric or eccentric annular duct, m, ft
r_h	hydraulic radius of the duct, $= A_c/P$, m, ft
r^*	aspect ratio r_i/r_o of a concentric or eccentric annular duct
s	duct dimension, m, ft
T, $\overline{T}$	fluid time-average temperature [see Eq. (4.2)], °C, °F
T_i	fluid instantaneous temperature [see Eq. (4.2)], °C, °F
T'	fluid fluctuating temperature [see Eq. (4.2)], °C, °F
T_m	fluid bulk mean temperature, defined by Eq. (3.7) in Chap. 3, °C, °F
T_w	wall temperature at the inside duct periphery, °C, °F
$T_{w,m}$	circumferentially averaged wall temperature, °C, °F
Ⓣ	constant wall temperature boundary conditions (refer to Table 3.1 in Chap. 3)
u, $\bar{u}$	fluid time-average axial velocity, fluid time-average velocity component in x direction [see Eq. (4.1)], m/s, ft/s
u_i	fluid instantaneous axial velocity component in x direction [see Eq. (4.1)], m/s, ft/s
u'	fluid fluctuating axial velocity component in x direction [see Eq. (4.1)], m/s, ft/s
u_m	bulk mean axial velocity averaged over the duct cross section, m/s, ft/s
$u_{\max}$	maximum value of the fluid time-average axial velocity across the duct cross section for fully developed turbulent flow, m/s, ft/s
u_t	turbulent friction or shear velocity $= \sqrt{\tau_w g_c/\rho}$, m/s, ft/s
u^+	wall coordinate $= u/u_t$, dimensionless
v, $\bar{v}$	fluid time-average velocity component in y or r direction, m/s, ft/s
w, $\overline{w}$	fluid time-average velocity component in z or θ direction, m/s, ft/s
x	axial (streamwise) coordinate in the Cartesian or cylindrical coordinate system, m, ft

x^+	axial coordinate for the hydrodynamic entrance region, $= x/D_h\text{Re}$
x^*	axial coordinate for the thermal entrance region, $= x/D_h\text{Pe}$
y	Cartesian coordinate across the flow cross section; distance measured from the duct wall, m, ft
y^+	wall coordinate $= yu_t/\nu$
z	Cartesian coordinate across the flow cross section; also distance from the apex of a triangle, m, ft

Greek Symbols

α	fluid thermal diffusivity $= k/\rho c_p$, m^2/s, ft^2/s
α^*	duct aspect ratio, explicitly defined for specific geometry in the text
γ	ratio of heat fluxes at two walls of a flat duct
$\tilde{\gamma}$	intermittency factor characterizing turbulent flow (see Sec. 4.1.3)
Δ	shear stress gradient parameter [see Eq. (4.14)], dimensionless
δ	hydrodynamic boundary layer thickness (see Fig. 3.1), m, ft
δ_l	laminar-sublayer thickness (see Fig. 4.1), m, ft
δ_t	thermal boundary layer thickness (see Fig. 3.1), m, ft
δ_w	duct wall thickness, m, ft
ϵ	distance between centers of two circles of an eccentric annular duct (see Fig. 4.64), m, ft
ε	height of surface roughness element, m,
ϵ_h	apparent, virtual, or eddy thermal diffusivity for turbulent flow [see Eq. (4.6)], $= k_t/\rho c_p$, m^2/s, ft^2/s
ϵ_m	apparent, virtual, or eddy kinematic viscosity coefficient for turbulent flow, $= \mu_t/\rho$ [see Eq. (4.4)], m^2/s, ft^2/s
Θ	dimensionless temperature for axially constant wall heat flux boundary condition, $= (T - T_e)/(q''_w D_h/k)$
θ	angular coordinate in the cylindrical coordinate system, rad, deg
θ	dimensionless fluid temperature for axially constant wall temperature boundary condition, $= (T - T_w)/(T_e - T_w)$
θ_m	fluid bulk mean temperature $= (T_m - T_w)/(T_e - T_w)$
$\theta_j^{(k)}$	fluid temperature for a doubly connected duct
$\theta_{lj}^{(k)}$	circumferentially averaged temperature of wall ($l = i$ for inner wall, $l = o$ for outer wall) for the fundamental boundary condition of kind k when inner wall ($j = i$) or outer wall ($j = o$) is heated or cooled
$\theta_{mj}^{(k)}$	fluid bulk mean temperature for fundamental boundary condition of kind k when inner wall ($j = i$) or outer wall ($j = o$) is heated or cooled.
θ_i^*	influence coefficients derived from fundamental solutions of the second kind, $= (\theta_{mo}^{(2)} - \theta_{io}^{(2)})/(\theta_{ii}^{(2)} - \theta_{mi}^{(2)})$

θ_o^*	influence coefficients derived from fundamental solutions of the second kind, $= (\theta_{mi}^{(2)} - \theta_{oi}^{(2)})/(\theta_{oo}^{(2)} - \theta_{mo}^{(2)})$
μ	fluid dynamic viscosity coefficient, Pa · s, $lb_m/(hr \cdot ft)$
μ_t	apparent, virtual, or eddy viscosity coefficient for turbulent flow [see Eq. (4.3)], Pa · s, $lb_m/(hr \cdot ft)$
ν	kinematic fluid viscosity coefficient $= \mu/\rho$, m^2/s, ft^2/s
ρ	fluid density, kg/m^3, lb_m/ft^3
τ	shear stress, Pa, lb_f/ft^2
τ_l	shear stress due to laminar flow, Pa, lb_f/ft^2
τ_t	apparent shear stress due to turbulent flow, Pa, lb_f/ft^2
τ_w	wall shear stress due to skin friction, Pa, lb_f/ft^2
$\Phi_i^{(k)}, \Phi_o^{(k)}$	dimensionless heat flux at a point in the flow field for the inner or outer wall of a concentric or eccentric annular duct
$\Phi_{lj}^{(k)}$	dimensionless wall heat flux defined in a manner similar to $\theta_{lj}^{(k)} = q_l''/k(T_j - T_e)$ for $k = 1, 3$, and as q_l''/q_{lj}'' for $k = 2, 4$
ϕ	apex angle of a duct, rad, deg

Subscripts

bc	thermal boundary condition (refer to Tables 3.1 and 3.2 in Chap. 3 for the alphanumeric designation and meaning of various thermal boundary conditions)
c	center or centroid; also circular duct
e	initial value at the entrance of the duct or where the heat transfer starts
f	fluid
fd	fully developed
H	Ⓗ boundary condition
hy	hydrodynamic
i	inner surface of a doubly connected duct
j	heated wall of a doubly connected duct, $j = i$ or o
l	laminar
m	mean, bulk mean
max	maximum
min	minimum
o	outer surface of a doubly connected duct
s	smooth
T	Ⓣ boundary condition
t	turbulent
th	thermal
x	arbitrary section along the duct length, a local value as opposed to a mean value
w	wall or fluid at the wall
∞	fully developed value at $x = \infty$

REFERENCES

1. J. Boussinesq, Théorie de l'écoulement tourbillant, *Mem. Pres. Acad. Sci.*, Paris, Vol. XXIII, p. 46, 1877.
2. S. Goldstein, A Note on Roughness, Aeronautical Research Council, London, RM 1763, 1936.
3. O. Reynolds, On the Dynamic Theory of Incompressible Viscous Fluids and the Determination of the Criterion, *Philos. Trans. Roy. Soc.*, Vol. T186A, pp. 123–164, 1894.
4. L. Prandtl, Bemerkung über den Wärmeübergang im Rohr, *Phys. Z.*, Vol. 29, pp. 487–489, 1928.
5. G. I. Taylor, The Transport of Vorticity and Heat Through Fluids in Turbulent Motion, *Philos. Trans. Roy. Soc.*, Vol. A215, pp. 1–26, 1915.
6. T. von Kármán, Mechanische Ähnlichkeit und Turbulenz, *Nachr. Ges. Wiss. Göttingen, Math. Phys. Klasse*, No. 5, pp. 58–76, 1930; English transl., NACA TM 611, 1931.
7. E. R. Van Driest, On Turbulent Flow Near a Wall, *J. Aerosp. Sci.*, Vol. 23, pp. 1007–1011, 1956.
8. W. C. Reynolds, Computation of Turbulent Flows, *Ann. Rev. Fluid Mech.*, Vol. 8, pp. 183–208, 1976.
9. V. S. Arpaci and P. S. Larsen, *Convective Heat Transfer*, Prentice-Hall, Englewood Cliffs, N.J., 1984.
10. P. Bradshaw, *Turbulence—Topics in Applied Physics*, Springer, New York, 1976.
11. O. Reynolds, On the Extent and Action of the Heating Surface for Steam Boilers, *Proc. Manchester Lit. Philos. Soc.*, Vol. 14, pp. 7–12, 1874.
12. L. Prandtl, Über die ausgebildete Turbulenz, *Z. Angew. Math. Mech.*, Vol. 5, pp. 136–139, 1925; English transl., NACA TM 1231, 1949.
13. G. I. Taylor, Conditions at the Surface of a Hot Body Exposed to the Wind, Tech. Report, *British Advisory Committee for Aeronautics*, Vol. 2, No. 272, pp. 423–429, May 1916.
14. T. von Kármán, The Analogy between Fluid Friction and Heat Transfer, *Trans. ASME*, Vol. 61, pp. 705–710, 1939.
15. R. C. Martinelli, Heat Transfer to Molten Metals, *Trans. ASME*, Vol. 69, pp. 947–959, 1947.
16. T. von Kármán, The Fundamentals of the Statistical Theory of Turbulence, *J. Aerosp. Sci.*, Vol. 4, pp. 131–138, 1937.
17. G. K. Batchelor, *The Theory of Homogeneous Turbulence*, Cambridge U.P., New York, 1967.
18. A. A. Townsend, *The Structure of Turbulent Shear Flow*, Cambridge U.P., New York, 1976.
19. O. Reynolds, An Experimental Investigation of the Circumstances which Determine Whether the Motion of Water Shall be Direct or Sinuous and of the Law of Resistance in Parallel Channels, *Philos. Trans. Roy. Soc.*, Vol. 174, pp. 935–982, 1883.
20. E. R. Lindgren, Some Aspects of Change between Laminar and Turbulent Flow of Liquids in Cylindrical Tubes, *Ark. Phys.*, Vol. 7, No. 23, pp. 293–308, 1953.
21. W. Pfenniger, Experiments with Laminar Flow in the Inlet Length of a Tube at High Reynolds Numbers With and Without Boundary Layer Suction, Technical Report, Northrop Aircraft Inc., Hawthorne, Calif., May 1952.
22. W. M. Kays and M. E. Crawford, *Convective Heat and Mass Transfer*, 2nd ed., McGraw-Hill, New York, pp. 161–163, 1980.
23. J. Rotta, Experimenteller Beitrag zur Entstehung turbulenter Strömung im Rohr, *Ing. Arch.*, Vol. 24, pp. 258–281, 1956.
24. W. Tollmien, Über die Entstehung der Turbulenz, I, Mitteilung, *Nachr. Ges. Wiss. Göttingen, Math. Phys. Klasse*, pp. 21–44, 1929; English transl., NACA TM 609, 1931.

25. H. Schlichting, Amplitudenverteilung und Energiebilanz der kleinen Störungen bei der Plattenströmung, *Nachr. Ges. Wiss. Göttingen, Math. Phys. Klasse*, Fachgruppe I, pp. 47–78, 1935; English transl., NACA TM 1265.

26. V. C. Patel and M. R. Head, Reversion of Turbulent to Laminar Flow, *J. Fluid Mech.*, Vol. 34, pp. 371–392, 1968.

27. C. A. Bankstone, The Transition from Turbulent to Laminar Gas Flow in a Heated Pipe, *J. Heat Transfer*, Vol. 92, pp. 569–579, 1970.

28. O. C. Jones, Jr., An Improvement in the Calculation of Turbulent Friction in Rectangular Ducts, *J. Fluids Eng.*, Vol. 98, pp. 173–181, 1976.

29. S. Ahmed and E. Brundrett, Characteristic Lengths for Non-Circular Ducts, *Int. J. Heat Mass Transfer*, Vol. 14, pp. 157–159, 1971.

30. R. I. Hodge, Frictional Pressure Drop in Noncircular Ducts, *J. Heat Transfer*, Vol. 83, pp. 384–385, 1961.

31. L. W. Carlson and T. F. Irvine, Jr., Fully Developed Pressure Drop in Triangular Shaped Ducts, *J. Heat Transfer*, Vol. 83, pp. 441–444, 1961.

32. P. C. Bandopadhayay and C. W. Ambrose, A Generalized Length Dimension for Noncircular Ducts, *Lett. Heat Mass Transfer*, Vol. 7, pp. 323–328, 1980.

33. O. C. Jones, Jr., and J. C. M. Leung, An Improvement in the Calculation of Turbulent Friction in Smooth Concentric Annuli, *J. Fluids Eng.*, Vol. 103, pp. 615–623, 1981.

34. J. Nikuradse, Strömungsgesetze in rauhen Rohren, *Forsch. Arb. Ing.-Wes.*, No. 361, 1933; English transl., NACA TM 1292.

35. H. Schlichting, Experimentelle Untersuchungen zum Rauhigkeitsproblem, *Ing.-Arch.*, Vol. 7, pp. 1–34, 1936; English transl., *Proc. ASME*, 1936.

36. L. F. Moody, Friction Factors for Pipe Flow, *Trans. ASME*, Vol. 66, pp. 671–684, 1944.

37. A. J. Musker, Universal Roughness Functions for Naturally-Occurring Surfaces, *Trans. CSME*, Vol. 6, No. 1, pp. 1–6, 1980–81.

38. R. H. Norris, Some Simple Approximate Heat Transfer Correlations for Turbulent Flow in Ducts with Rough Surfaces, *Augmentation of Convective Heat and Mass Transfer*, ASME, New York, Dec. 1970.

39. V. W. Ekman, On the Change from Steady to Turbulent Motion of Liquids, *Ark. Mat. Astron. OCH Fys.*, Vol. 6, No. 12, pp. 1–16, 1910.

40. L. Schiller, Experimentelle Untersuchungen zum Turbulenzproblem, *Z. Angew. Math. Mech.*, Vol. 1, pp. 436–441, 1921.

41. J. Šimonek, Turbulent Transport in the Transition Flow Region, *Int. J. Heat Mass Transfer*, Vol. 27, pp. 2415–2420, 1984.

42. R. S. Prengle and R. R. Rothfus, Transition Phenomena in Pipes and Annular Cross Sections, *Ind. Eng. Chem.*, Vol. 47, pp. 379–386, 1955.

43. N. W. Wilson and R. S. Azad, A Continuous Prediction Method for Fully Developed Laminar, Transition, and Turbulent Flows in Pipes, *J. Appl. Mech.*, Vol. 42, pp. 51–54, 1975.

44. D. I. H. Barr, The Transition From Laminar to Turbulent Flow, *Proc. Inst. Civ. Engrs.*, Part 2, Vol. 69, pp. 555–562, 1980.

45. S. W. Churchill, Comprehensive Correlating Equations for Heat, Mass and Momentum Transfer in Fully Developed Flow in Smooth Tubes, *Ind. Eng. Chem. Fundam.*, Vol. 16, No. 1, pp. 109–116, 1977.

46. V. C. Patel and M. R. Head, Some Observations on Skin Friction and Velocity Profiles in Fully Developed Pipe and Channel Flows, *J. Fluid Eng.*, Vol. 38, pp. 181, 1969.

47. P. Hrycak and R. Andrushkiw, Calculation of Critical Reynolds Number in Round Pipes and Infinite Channels and Heat Transfer in Transition Regions, *Heat Transfer 1974*, Vol. II, pp. 183–187, 1974.

48. H. Blasius, Das Ähnlichkeitsgesetz bei Reibungsvorgängen in Flüssigkeiten, *Forschg. Arb. Ing.-Wes.*, No. 131, Berlin, 1913.

49. L. Prandtl, Über den Reibungswiderstand Strömender Luft. *Ergebnisse der Aerodynamics Versuchstalt zu Göttingen*, Edition 3, pp. 1–5, 1927.

50. J. Nikuradse, Gesetzmäßigkeiten der turbulenten Strömung in glatten Rohren, *Forsch. Arb. Ing.-Wes.*, No. 356, 1932; English transl., NASA TT F-10, 359, 1966.

51. L. Prandtl, Neuere Ergebnisse der Turbulenzforschung, *VDI Z.*, Vol. 77, No. 5, pp. 105–114, 1933; English transl., NACA TM 720, 1933.

52. C. Wang, On the Velocity Distribution of Turbulent Flow in Pipes and Channels of Constant Cross Section, *J. Appl. Mech.*, Vol. 68, pp. A85–A90, 1946.

53. H. Darcy, Recherches expérimentales relatives aux mouvements de l'eau dans tuyauz, *Mem. Prés. Acad. des Sci. Inst. France*, Vol. 15, p. 141, 1858.

54. L. Prandtl, Eine Beziehung zwischen Wärmeaustausch und Strömungswiderstand der Flüssigkeit, *Z. Physik*, Vol. 11, pp. 1072–1078, 1910.

55. G. I. Taylor, Conditions at the Surface of a Hot Body Exposed to the Wind, *British Advisory Committee for Aeronautics*, Vol. 2, R & M No. 272, pp. 423–429, 1916.

56. H. Reichardt, Vollständige Darstellung der turbulenten Geschwindigkeitsverteilung in glatten Leitungen, *Z. Angew. Math. Mech.*, Vol. 31, pp. 208–219, 1951.

57. R. G. Deissler, Analysis of Turbulent Heat Transfer, Mass Transfer and Friction in Smooth Tubes at High Prandtl and Schmidt Numbers, NACA Tech. Rep. 1210, 1955 (supersedes NACA Tech. Note 3145, 1954).

58. W. D. Rannie, Heat Transfer in Turbulent Shear Flow, *J. Aeronaut. Sci.*, Vol. 23, p. 485, 1956.

59. D. B. Spalding, A Single Formula for the "Law of the Wall," *J. Appl. Mech.*, Vol. 28, pp. 455–458, 1961.

60. H. Schlichting, *Boundary Layer Theory*, 7th ed., McGraw-Hill, 1979.

61. T. B. Drew, E. C. Koo, and W. H. McAdams, The Friction Factor for Clean Round Pipes, *Trans. AIChE*, Vol. 28, pp. 56–72, 1932.

62. L. Prandtl, *Führrer durch die Stömungslehre*, Vieweg, Braunschweig, p. 359, 1944; English transl., Blackie, London, 1952.

63. T. von Kármán, Turbulence and Skin Friction, *J. Aerosp. Sci.*, Vol. 7, pp. 1–20, 1934.

64. W. H. McAdams, *Heat Transmission*, 3rd ed., McGraw-Hill, 1954.

65. C. F. Colebrook, Turbulent Flow in Pipes with Particular Reference to the Transition Region between the Smooth and Rough Pipes Laws, *J. Inst. Civ. Eng.*, Vol. 11, pp. 133–156, 1939.

66. G. K. Filonenko, Hydraulic Resistance in Pipes (in Russian), *Teploenergetika*, Vol. 1, No. 4, pp. 40–44, 1954.

67. R. Techo, R. R. Tickner, and R. E. James, An Accurate Equation for the Computation of the Friction Factor for Smooth Pipes from the Reynolds-Number, *J. Appl. Mech.*, Vol. 32, p. 443, 1965.

68. D. J. Wood, An Explicit Friction Factor Relationship, *Civ. Eng.*, pp. 60–61, 1966.

69. P. K. Swamee and A. K. Jain, Explicit Equations for Pipe-Flow Problems, *J. Hydraulic Div. ASCE*, Vol. 102, pp. 657–664, 1976.

70. A. K. Jain, Accurate Explicit Equation for Friction Factor, *J. Hydraulic Div. ASCE*, Vol. 102, pp. 674–677, 1976.

71. S. W. Churchill, Friction-Factor Equation Spans all Fluid Flow Regimes, *Chem. Eng.*, pp. 91–92, 1977.

72. N. H. Chen, An Explicit Equation for Friction Factor in Pipe, *Ind. Eng. Chem. Fund.*, Vol. 18, pp. 296–297, 1979.

73. G. F. Round, An Explicit Approximation for the Friction Factor–Reynolds Number Relation for Rough and Smooth Pipes, *Can. J. Chem. Eng.*, Vol. 58, pp. 122–123, 1980.

74. D. J. Zigrang and N. D. Sylvester, Explicit Approximations to the Solution of Colebrook's Friction Factor Equation, *AIChE J.*, Vol. 28, pp. 514–515, 1982.

75. S. E. Haaland, Simple and Explicit Formulas for the Friction Factor in Turbulent Pipe Flow, *J. Fluids Eng.*, Vol. 105, pp. 89–90, 1983.
76. T. K. Serghides, Estimate Friction Factor Accurately, *Chem. Eng.*, pp. 63–64, 1984.
77. G. A. Gregory and M. Fogarasi, Alternate to Standard Friction Factor Equation, *Oil and Gas J.*, pp. 120–127, Apr. 1985.
78. A. Malhotra and S. S. Kang, Turbulent Prandtl Number in Circular Pipes, *Int. J. Heat Mass Transfer*, Vol. 27, pp. 2158–2161, 1984.
79. W. M. Kays, *Convective Heat and Mass Transfer*, 1st ed., McGraw-Hill, 1966.
80. R. K. Shah and R. S. Johnson, Correlations for Fully Developed Turbulent Flow Through Circular and Noncircular Channels, *Proc. Sixth National Heat and Mass Transfer Conf., Indian Inst. of Technology, Madras, India*, pp. D-75–D-95, 1981.
81. W. Nusselt, Wärmeübergang in Rohrleitungen, *Forsch.-Arb. Ing.-Wes.*, No. 89, Berlin, 1910.
82. P. W. Dittus and L. M. K. Boelter, Heat Transfer in Automobile Radiators of the Tubular Type, *Univ. Calif. Pub. Eng.*, Vol. 2, No. 13, pp. 443–461, Oct. 17, 1930; reprinted in *Int. Comm. Heat Mass Transfer*, Vol. 12, pp. 3–22, 1985.
83. H. Kraussold, Heat Transfer to Fluids in Tubes in Conditions of Turbulent Flow (in German), *Forsch. Ing.-Wes.*, Vol. 4, pp. 39–44, 1933.
84. A. P. Colburn, A Method of Correlating Forced Convection Heat Transfer Data and a Comparison With Fluid Friction, *Trans. AIChE*, Vol. 19, pp. 174–210, 1933; reprinted in *Int. J. Heat Mass Transfer*, Vol. 7, pp. 1359–1384, 1964.
85. H. Hausen, Darstellung des Wärmeüberganges in Rohren durch Verallgeneinerte Potenzbeziehungen, *VDI Z.*, No. 4, pp. 91–98, 1943.
86. R. E. Drexel and W. H. McAdams, Heat Transfer Coefficients for Air Flowing in Round Tubes, and around Finned Cylinders, NACA ARR No. 4F28; also Wartime Report W-108, 1945.
87. E. Bernardo and C. S. Eian, Heat Transfer Tests of Aqueous Glycol Solutions in an Electrically Heated Tube, NACA Wartime Report E-316, 1945 superseding NACA ARR E5F07.
88. W. L. Friend and A. B. Metzner, Turbulent Heat Transfer inside Tubes and the Analogy among Heat, Mass, and Momentum Transfer, *AIChE J.*, Vol. 4, pp. 393–402, 1958.
89. B. S. Petukhov and V. V. Kirillov, The Problem of Heat Exchange in the Turbulent Flow of Liquids in Tubes (in Russian), *Teploenergetika*, Vol. 4, No. 4, pp. 63–68, 1958; see also B. S. Petukhov and V. N. Popov, Theoretical Calculation of Heat Exchange in Turbulent Flow in Tubes of an Incompressible Fluid with Variable Physical Properties, *High Temp.*, Vol. 1, No. 1, pp. 69–83, 1963.
90. H. Hausen, Neue Gleichungen für die Wärmeübertragung bei freier oder erzwungener Stromüng, *Allg. Warmetchn.*, Vol. 9, No. 4/5, pp. 75–79, 1959.
91. A. F. Mills, Experimental Investigation of Turbulent Heat Transfer in the Entrance Region of a Circular Conduit, *J. Mech. Eng. Sci.*, Vol. 4, pp. 63–77, 1962.
92. R. L. Webb, A Critical Evaluation of Analytical Solutions and Reynolds Analogy Equations for Turbulent Heat and Mass Transfer in Smooth Tubes, *Wärme- und Stoffübertrag.*, Vol. 4, pp. 197–204, 1971.
93. C. A. Sleicher and M. W. Rouse, A Convenient Correlation for Heat Transfer to Constant and Variable Property Fluids in Turbulent Pipe Flow, *Int. J. Heat Mass Transfer*, Vol. 18, pp. 677–683, 1975.
94. V. Gnielinski, New Equations for Heat and Mass Transfer in Turbulent Pipe and Channel Flow, *Int. Chem. Eng.*, Vol. 16, pp. 359–368, 1976.
95. G. T. Polley, Correlations for Forced Convection Heat Transfer, *Chem. Eng.*, London, pp. 233–234, 1979.
96. O. C. Sandall, O. T. Hanna, and P. R. Mazet, A New Theoretical Formula for Turbulent Heat and Mass Transfer with Gases or Liquids in Tube Flow, *Can. J. Chem. Eng.*, Vol. 58, pp. 443–447, 1980.

97. R. N. Lyon, Liquid Metal Heat Transfer Coefficients, *Chem. Eng. Progr.*, Vol. 47, No. 2, pp. 75–79, 1951.

98. R. A. Seban and T. T. Shimazaki, Heat Transfer to a Fluid Flowing Turbulently in a Smooth Pipe with Walls at Constant Temperature, *Trans. ASME*, Vol. 73, pp. 803–809, 1951.

99. B. Lubarsky and S. J. Kaufman, Review of Experimental Investigations of Liquid-Metal Heat Transfer, NACA TN 3336, 1955.

100. C. A. Sleicher, Jr. and M. Tribus, Heat Transfer in a Pipe with Turbulent Flow and Arbitrary Wall-Temperature Distribution, *Trans. ASME*, Vol. 79, pp. 789–797, 1957.

101. N. T. Azer and B. T. Chao, Turbulent Heat Transfer in Liquid Metals—Fully Developed Pipe Flow with Constant Wall Temperature, *Int. J. Heat Mass Transfer*, Vol. 3, pp. 77–83, 1961.

102. O. E. Dwyer, Eddy Transport in Liquid Metal Heat Transfer, *AIChE J.*, Vol. 9, pp. 261–268, 1963.

103. E. Skupinski, J. Tortel, and L. Vautrey, Determination des coefficients de convection d'un allage sodium-potassium dans un tube circulaire, *Int. J. Heat Mass Transfer*, Vol. 8, pp. 937–951, 1965.

104. R. H. Notter and C. A. Sleicher, A Solution to the Turbulent Graetz Problem III. Fully Developed and Entry Region Heat Transfer Rates, *Chem. Eng. Sci.*, Vol. 27, pp. 2073–2093, 1972.

105. C. A. Sleicher, A. S. Awad, and R. H. Notter, Temperature and Eddy Diffusivity Profiles in NaK, *Int. J. Heat Mass Transfer*, Vol. 16, pp. 1565–1575, 1973.

106. C. J. Chen and J. S. Chiou, Laminar and Turbulent Heat Transfer in the Pipe Entrance Region for Liquid Metals, *Int. J. Heat Mass Transfer*, Vol. 24, pp. 1179–1189, 1981.

107. S. L. Lee, Liquid Metal Heat Transfer in Turbulent Pipe Flow with Uniform Wall Flux, *Int. J. Heat Mass Transfer*, Vol. 26, pp. 349–356, 1983.

108. R. Siegel and E. M. Sparrow, Comparison of Turbulent Heat-Transfer Results for Uniform Wall Heat Flux and Uniform Wall Temperature, *J. Heat Transfer*, Vol. 82, pp. 152–153, 1960.

109. W. C. Reynolds, Turbulent Heat Transfer in a Circular Tube with Variable Circumferential Heat Flux, *Int. J. Heat Mass Transfer*, Vol. 6, pp. 445–454, 1963.

110. E. M. Sparrow and S. H. Lin, Turbulent Heat Transfer in a Tube With Circumferentially-Varying Temperature or Heat Flux, *Int. J. Heat Mass Transfer*, Vol. 6, pp. 866–867, 1963.

111. D. Gartner, K. Johannsen, and H. Ramm, Turbulent Heat Transfer in a Circular Tube with Circumferentially Varying Thermal Boundary Conditions, *Int. J. Heat Mass Transfer*, Vol. 17, pp. 1003–1018, 1974.

112. A. W. Black and E. M. Sparrow, Experiments on Turbulent Heat Transfer in a Tube with Circumferentially Varying Thermal Boundary Conditions, *J. Heat Transfer*, Vol. 89, pp. 258–268, 1967.

113. W. Nunner, Wärmeübergang und Druckabfall in Rauhen Rohren, *VDI-Forschungsheft 445*, Ser. B, Vol. 22, pp. 5–39, 1956; English transl., No. 786, Atomic Energy Research Establishment, Harwell, U.K.

114. D. F. Dipprey and R. H. Sabersky, Heat and Momentum Transfer in Smooth and Rough Tubes at Various Prandtl Numbers, *Int. J. Heat Mass Transfer*, Vol. 6, pp. 329–353, 1963.

115. R. A. Gowen and J. W. Smith, Turbulent Heat Transfer from Smooth and Rough Surfaces, *Int. J. Heat Mass Transfer*, Vol. 11, pp. 1657–1673, 1968.

116. Y. Kawase and J. J. Ulbrecht, Turbulent Heat and Mass Transfer in Dilute Polymer Solutions, *Chem. Eng. Sci.*, Vol. 37, pp. 1039–1046, 1982.

117. Y. Kawase and A. De, Turbulent Heat and Mass Transfer in Newtonian and Dilute Polymer Solutions Flowing through Rough Tubes, *Int. J. Heat Mass Transfer*, Vol. 27, pp. 140–142, 1984.

118. H. Latzko, Der Wärmeübergang and einen turbulenten Flüssigkeitsoder Gasstrom, *Z. Angew. Math. Mech.*, Vol. 1, No. 4, pp. 268–290, 1921; English transl., NACA TM 1068, 1944.

119. D. Ross, Turbulent Flow in the Entrance Region of a Pipe, *Trans. ASME*, Vol. 78, pp. 915–923, 1956.

120. W. Zhi-qing, Study on Correction Coefficients of Laminar and Turbulent Entrance Region Effect in Round Pipe, *Appl. Math. Mech.*, Vol. 3, No. 3, pp. 433–446, 1982.

121. D. A. Bowlus and J. A. Brighton, Incompressible Turbulent Flow in the Inlet Region of a Pipe, *J. Basic Eng.*, Vol. 90, pp. 431–433, 1968.

122. T. Y. Na and Y. P. Lu, Turbulent Flow Development Characteristics in Channel Inlets, *Appl. Sci. Res.*, Vol. 27, pp. 425–439, 1973.

123. J. S. Holdhusen, The Turbulent Boundary Layer in the Inlet Region of a Smooth Pipe, Ph.D. Thesis, Univ. of Minnesota, Minneapolis, 1952.

124. R. G. Deissler, Analysis of Turbulent Heat Transfer and Flow in the Entrance Regions of Smooth Passages, NACA TN 3016, 1953.

125. G. V. Filippov, On Turbulent Flow in the Entrance Length of a Straight Tube of Circular Cross-Section, *Sov. Phys.—Techn. Phys.*, Vol. 32, No. 8, pp. 1681–1686, 1958.

126. A. R. Barbin and J. B. Jones, Turbulent Flow in the Inlet Region of a Smooth Pipe, *J. Basic Eng.*, Vol. 85, pp. 29–34, 1963.

127. J. Wang and J. P. Tullis, Turbulent Flow in the Entry Region of a Rough Pipe, *J. Fluids Eng.*, Vol. 96, pp. 62–68, 1974.

128. C. A. Sleicher and M. Tribus, Heat Transfer in a Pipe with Turbulent Flow and Arbitrary Wall Temperature Distribution, *1956 Heat Transfer and Fluid Mechanics Institute*, Stanford Univ. Press, Stanford, Calif., pp. 59–78, 1956.

129. H. L. Becker, Heat Transfer in Turbulent Tube Flow, *Appl. Sci. Res.*, Ser. A, Vol. 6, pp. 147–191, 1956.

130. E. M. Sparrow, T. M. Hallman, and R. Siegel, Turbulent Heat Transfer in the Thermal Entrance Region of a Pipe with Uniform Heat Flux, *Appl. Sci. Res.*, Ser. A, Vol. 7, pp. 37–52, 1957.

131. J. A. Malina and E. M. Sparrow, Variable-Property, Constant-Property, and Entrance Region Heat Transfer Results for Turbulent Flow of Water and Oil in a Circular Tube, *Chem. Eng. Sci.*, Vol. 19, pp. 953–962, 1964.

132. H. C. Reynolds, T. B. Swearingen, and D. M. McEligot, Thermal Entry for Low Reynolds Number Turbulent Flow, *J. Basic Eng.*, Vol. 91, pp. 87–94, 1969.

133. M. Al-Arabi, Turbulent Heat Transfer in the Entrance Region of a Tube, *Heat Transfer Eng.*, Vol. 3, pp. 76–83, 1982.

134. L. G. Genin, E. V. Kudryavtseva, Yu. A. Pakhotin, and V. G. Sviridov, Temperature Fields and Heat Transfer for a Turbulent Flow of Liquid Metal on an Initial Thermal Section, *Teplofiz. Vysokikh Temp.*, Vol. 16, No. 6, pp. 1243–1249, 1978.

135. R. Siegel and E. M. Sparrow, Turbulent Flow in a Circular Tube with Arbitrary Internal Heat Sources and Wall Heat Transfer, *J. Heat Transfer*, Vol. 81, pp. 280–290, 1959.

136. G. R. Knowles and E. M. Sparrow, Local and Average Heat Transfer Characteristics for Turbulent Airflow in an Asymmetrically Heated Tube, *J. Heat Transfer*, Vol. 101, pp. 635–641, 1979.

137. W. M. Kays and W. B. Nicoll, The Influence of Nonuniform Heat Flux on the Convection Conductances in a Nuclear Reactor, TR No. 33, Dept. Mech. Eng., Stanford Univ., 1957.

138. W. B. Hall and P. H. Price, The Effect of a Longitudinally Varying Wall Heat Flux on the Heat Transfer Coefficient for Turbulent Flow in a Pipe, *Int. Dev. in Heat Transfer*, ASME/Inst. Mech. Engrs., Vol. III, 1961.

139. L. M. K. Boelter, G. Young, and H. W. Iverson, An Investigation of Aircraft Heaters XXVII —Distribution of Heat Transfer Rate in the Entrance Section of a Circular Tube, NACA TN 1451, 1948.

140. W. B. Hall and S. A. Khan, Experimental Investigation into the Effect of the Thermal Boundary Condition on Heat Transfer in the Entrance Region of a Pipe, *J. Mech. Eng. Sci.*, Vol. 6, pp. 250–256, 1964.

141. M. Molki and E. M. Sparrow, An Empirical Correlation for the Average Heat Transfer Coefficient in Circular Tubes, *J. Heat Transfer*, Vol. 108, pp. 482–484, 1986.

142. E. M. Sparrow and M. Molki, Turbulent Heat Transfer Coefficients in an Isothermal-Walled Tube for Either a Built-in or Free Inlet, *Int. J. Heat Mass Transfer*, Vol. 27, pp. 669–675, 1984.

143. E. M. Sparrow, K. K. Koram, and M. Charmchi, Heat Transfer and Pressure Drop Characteristics Induced by a Slat Blockage in a Circular Tube, *J. Heat Transfer*, Vol. 102, pp. 64–70, 1980.

144. E. M. Sparrow and J. E. O'Brien, Heat Transfer Coefficients on the Downstream Face of an Abrupt Enlargement or Inlet Constriction in a Pipe, *J. Heat Transfer*, Vol. 102, pp. 408–414, 1980.

145. E. M. Sparrow and U. Gurdal, Heat Transfer at an Upstream Facing Surface Washed by Fluid en Route to an Aperture in the Surface, *Int. J. Heat Mass Transfer*, Vol. 24, pp. 851–857, 1981.

146. E. M. Sparrow, M. Molki, and S. R. Chastain, Turbulent Heat Transfer Coefficients and Fluid Flow Patterns on the Face of a Centrally Positioned Blockage in a Duct, *Int. J. Heat Mass Transfer*, Vol. 24, pp. 507–519, 1981.

147. E. M. Sparrow and A. Chaboki, Swirl-Affected Turbulent Fluid Flow and Heat Transfer in a Circular Tube, *J. Heat Transfer*, Vol. 106, pp. 766–773, 1984.

148. S. C. Lau, E. M. Sparrow, and J. W. Ramsey, Effect of Plenum Length and Diameter on Turbulent Heat Transfer in a Downstream Tube and on Plenum-Related Pressure Losses, *J. Heat Transfer*, Vol. 103, pp. 415–422, 1981.

149. E. M. Sparrow and L. D. Bosmans, Heat Transfer and Fluid Flow Experiments with a Tube Fed by a Plenum Having Nonaligned Inlet and Exit, *J. Heat Transfer*, Vol. 105, pp. 56–63, 1983.

150. M. Molki and E. M. Sparrow, In-Tube Heat Transfer for Skewed Inlet Flow Caused by Competition among Tubes Fed by the Same Plenum, *J. Heat Transfer*, Vol. 105, pp. 870–877, 1983.

151. D. A. Wesley and E. M. Sparrow, Circumferentially Local and Average Turbulent Heat-Transfer Coefficients in a Tube Downstream of a Tee, *Int. J. Heat Mass Transfer*, Vol. 19, pp. 1205–1214, 1976.

152. P. Souza Mendes and E. M. Sparrow, Periodically Converging-Diverging Tubes and Their Turbulent Heat Transfer, Pressure Drop, Fluid Flow, and Enhancement Characteristics, *J. Heat Transfer*, Vol. 106, pp. 55–63, 1984.

153. G. S. Beavers, E. M. Sparrow, and R. A. Magnuson, Experiments on the Breakdown of Laminar Flow in a Parallel-Plate Channel, *Int. J. Heat Mass Transfer*, Vol. 13, pp. 809–815, 1970.

154. R. W. Hanks, The Laminar-Turbulent Transition for Flow in Pipes, Concentric Annuli, and Parallel Plates, *AIChE J.*, Vol. 9, pp. 45–48, 1963.

155. S. J. Davies and C. M. White, An Experimental Study of the Flow of Water in Pipes of Rectangular Section, *Proc. Roy. Soc.*, Vol. A119, pp. 92–107, 1928.

156. T. S. Chen and E. M. Sparrow, Stability of the Developing laminar Flow in a Parallel-Plate Channel, *J. Fluid Mech.*, Vol. 30, pp. 209–224, 1967.

157. S. C. Gupta and V. K. Garg, Effect of Velocity Distribution on the Stability of Developing Flow in a Channel, *J. Phys. Soc. Japan*, Vol. 50, pp. 673–680, 1981.

158. S. I. Pai, On Turbulent Flow between Parallel Plates, *J. Appl. Mech.*, Vol. 20, pp. 109–114, 1953.

159. J. Laufer, Some Recent Measurements in a Two-Dimensional Turbulent Channel, *J. Aerosp. Sci.*, Vol. 17, pp. 277–287, 1950.

160. S. Goldstein, The Similarity Theory of Turbulence and Flow between Parallel Planes and through Pipes, *Proc. Roy. Soc.*, Vol. 159A, pp. 473–496, 1937.

161. F. Dönch, Divergente und konvergente Strömungen mit kleinen Öffnungswinkeln, Diss. Göttingen, 1925; VDI-Forschungsheft, No. 292, 1926.

162. J. Nikuradse, Untersuchungen über die Strömungen des Wassers in Konvergenten und Dirergenten Kanälen, *Forschungsarbeiten auf dem Gebiete des Ingenieurwesens*, Heft 289, Berlin, 1929.

163. R. R. Rothfus, J. E. Walker, and G. A. Whan, Correlation of Local Velocities in Tubes, Annuli, and Parallel Plates, *AIChE J.*, Vol. 4, pp. 240–245, 1958.

164. O. E. Dwyer, Heat Transfer to Liquid Metals Flowing Turbulently between Parallel Plates, *Nucl. Sci. Eng.*, Vol. 21, pp. 79–89, 1965.

165. G. S. Beavers, E. M. Sparrow, and J. R. Lloyd, Low Reynolds Number Flow in Large Aspect Ratio Rectangular Ducts, *J. Basic Eng.*, Vol. 93, pp. 296–299, 1971.

166. R. B. Dean, Reynolds Number Dependence of Skin Friction and Other Bulk Flow Variables in Two-Dimensional Rectangular Duct Flow, *J. Fluids Eng.*, Vol. 100, pp. 215–223, 1978.

167. R. R. Rothfus and C. C. Monrad, Correlation of Turbulent Velocities for Tubes and Parallel Plates, *Ind. Eng. Chem.*, Vol. 47, pp. 1144–1149, 1955.

168. W. M. Kays and E. Y. Leung, Heat Transfer in Annular Passages: Hydrodynamically Developed Turbulent Flow with Arbitrarily Prescribed Heat Flux, *Int. J. Heat Mass Transfer*, Vol. 6, pp. 537–557, 1963.

169. E. M. Sparrow and S. H. Lin, Turbulent Heat Transfer in a Parallel-Plate Channel, *Int. J. Heat Mass Transfer*, Vol. 6, pp. 248–249, 1963.

170. S. Kakaç and S. Paykoç, Analysis of Turbulent Forced Convection Heat Transfer between Parallel Plates, *Middle East Tech. Univ., J. Pure Appl. Sci.*, Vol. 1, No. 1, pp. 27–47, 1968.

171. S. Kakaç and Y. Yener, *Convective Heat Transfer*, Publication No. 65, Middle East Tech Univ., Ankara, Turkey, distributed by Hemisphere Publishing Corp., New York, 1980.

172. N. I. Buleev, *Problems of Heat Transfer*, Publishing House of the U.S.S.R. Acad. Sci., Moscow, 1959.

173. O. E. Dwyer, Heat Transfer to Liquid Metals Flowing Turbulently between Parallel Plates, *Nucl. Sci. Eng.*, Vol. 21, pp. 79–89, 1965.

174. L. Duchatelle and L. Vautrey, Determination des coefficients de convection d'un alliage NaK en ecoulement turbulent entre plaques planes paralleles, *Int. J. Heat Mass Transfer*, Vol. 7, pp. 1017–1031, 1964.

175. A. A. Shibani and M. N. Özişik, A Solution to Heat Transfer in Turbulent Flow between Parallel Plates, *Int. J. Heat Mass Transfer*, Vol. 20, pp. 565–573, 1977.

176. R. A. Seban, Heat Transfer to a Fluid Flowing Turbulently between Parallel Walls with Asymmetric Wall Temperatures, *Trans. ASME*, Vol. 72, pp. 789–795, 1950.

177. S. Kakaç and P. H. Price, On the Constant Heat Transfer Boundary Condition on Forced Convection Heat Transfer in a Channel, *Middle East Tech. Univ., J. Pure Appl. Sci.*, Vol. 2, No. 3, pp. 239–262, 1969.

178. J. Byrne, A. P. Hatton, and P. G. Marriott, Turbulent Flow and Heat Transfer in the Entrance Region of a Parallel Wall Passage, *Proc. Inst. Mech. Engrs.*, Vol. 184, Pt. 1, No. 39, 1969–70.

179. V. I. Shcherbinin and F. R. Shklyar, Analysis of the Hydrodynamic Entrance Region of a Plane Channel with the Application of Various Turbulence Models, *Teplofiz. Vysokikh Temp.*, Vol. 18, pp. 1026–1031, 1980.

180. T. Kobata, Y. Yoshino, S. Masuda, and I. Ariga, Turbulent Flow in Entrance Region of Two-Dimensional Straight Channel with Inlet Disturbance, *Bull. JSME*, Vol. 26, No. 220, pp. 1711–1718, 1983.

181. M. Sakakibara and K. Endo, Analysis of Heat Transfer for Turbulent Flow between Parallel Plates, *Int. Chem. Eng.*, Vol. 18, pp. 728–733, 1976.

182. A. P. Hatton and A. Quarmby, The Effect of Axially Varying and Unsymmetrical Boundary Conditions on Heat Transfer with Turbulent Flow between Parallel Plates, *Int. J. Heat Mass Transfer*, Vol. 6, pp. 903–914, 1963.

183. A. P. Hatton, A. Quarmby, and I. Grundy, Further Calculations on the Heat Transfer with Turbulent Flow between Parallel Plates, *Int. J. Heat Mass Transfer*, Vol. 7, pp. 817–823, 1964.

184. M. Sakakibara, Analysis of Heat Transfer in the Entrance Region with Fully Developed Turbulent Flow between Parallel Plates—the Case of Uniform Wall Heat Flux, *Mem. Fac. of Eng. Fukui Univ.*, Vol. 30, No. 2, pp. 107–120, 1982.

185. S. Faggiani and F. Gori, Influence of Streamwise Molecular Heat Conduction of the Heat Transfer Coefficient for Liquid Metals in Turbulent Flow between Parallel Planes, *J. Heat Transfer*, Vol. 102, pp. 292–296, 1980.

186. J. P. Hartnett, J. C. Y. Koh, and S. T. McComas, A Comparison of Predicted and Measured Friction Factors for Turbulent Flow through Rectangular Ducts, *J. Heat Transfer*, Vol. 84, pp. 82–88, 1962.

187. J. Allen and N. D. Grunberg, The Resistance to the Flow of Water along Smooth Rectangular Passages and the Effect of a Slight Convergence or Divergence of the Boundaries, *Philos. Mag.*, Ser. 7, pp. 490–502, 1937.

188. R. J. Cornish, Flow in a Pipe of Rectangular Cross-Section, *Proc. Roy. Soc. London*, Vol. A120, pp. 691–700, 1928.

189. L. Prandtl, Über den Reibungswiderstand Strömender Luft, Ergeb. Aerodyn. Versuch., Göttingen, III ser., 1927; English transl., NACA TM 435, 1927.

190. J. Nikuradse, Untersuchungen über die Geschwindigkeitsverteilung in turbulenten Strömungen, *VDI-Forschungsheft*, No. 281, 1926.

191. J. Nikuradse, Untersuchungen über turbulent Strömung in nicht kreisförmigen Rohren, *Ing.-Arch.*, No. 1, pp. 306–332, 1930.

192. L. C. Hoagland, Fully Developed Turbulent Flow in Straight Rectangular Ducts—Secondary Flow, its Causes and Effects on the Primary Flow, Ph.D. Thesis, Mass. Inst. of Tech., Cambridge, Mass., 1960.

193. H. J. Leutheusser, Turbulent Flow in Rectangular Ducts, *Proc. ASCE, J. Hydraul. Div.*, Vol. 89, pp. 1–19, 1963.

194. E. Brundrett and W. D. Baines, The Prediction and Diffusion of Vorticity in Duct Flow, *J. Fluid Mech.*, Vol. 19, pp. 375–394, 1964.

195. F. B. Gessner and J. B. Jones, On Some Aspects of Fully Developed Turbulent Flow in Rectangular Channels, *J. Fluid Mech.*, Vol. 23, pp. 689–713, 1965.

196. B. Krajewski, Determination of Turbulent Velocity Field in Rectangular Duct with Non-Circular Cross-Section, *Int. J. Heat Mass Transfer*, Vol. 13, pp. 1819–1824, 1970.

197. J. A. Ligget, C. L. Chiu, and L. S. Miao, Secondary Currents in a Corner, *J. Heat Transfer*, Vol. 95, pp. 453–457, 1973.

198. B. E. Launder and W. M. Ying, Secondary Flows in Ducts of Square Cross-Section, *J. Fluid Mech.*, Vol. 54, pp. 289–295, 1972.

199. D. Naot, A. Shavit, and M. Wolfshtein, Numerical Calculation of Reynolds Stresses in a Square Duct with Secondary Flow, *Wärme–und Stoffübertragung*, Vol. 7, pp. 171–161, 1974.

200. F. B. Gessner and A. F. Emery, A Reynolds Stress Model for Turbulent Corner Flows—Part I: Development of the Model, *J. Fluids Eng.*, Vol. 96, pp. 261–268, 1976.

201. F. B. Gessner and J. K. Po, A Reynolds Stress Model for Turbulent Corner Flows—Part II: Comparisons between Theory and Experiment, *J. Fluids Eng.*, Vol. 99, pp. 296–277, 1977.

202. A. DeMuren and W. Rodi, Calculation of Turbulence-Driven Secondary Motion in Noncircular Duct, *J. Fluid Mech.*, Vol. 140, pp. 189–222, 1984.

203. A. Nakayama, W. L. Chow, and D. Sharma, Calculation of Fully Turbulent Flows in Ducts of Arbitrary Cross-Section, *J. Fluid Mech.*, Vol. 128, pp. 199–217, 1983.

204. C. G. Speziale, On Turbulent Secondary Flows in Pipes of Noncircular Cross-Section, *Int. J. Eng. Sci.*, Vol. 20, pp. 863–872, 1982.

205. C. G. Speziale, The Dissipation Rate Correlation and Turbulent Secondary Flows in Noncircular Ducts, *J. Fluids Eng.*, Vol. 108, pp. 118–120, 1986.

206. L. Schiller, Über den Strömungswiderstand von Rohren verschiedenen Querschnitts und Rauhigkeitsgrades, *Z. Angew. Math. Mech.*, Vol. 3, pp. 2–13, 1923.

207. D. Wilkie, M. Cowin, P. Burnett, and T. Burgoyne, Friction Factor Measurements in a Rectangular Channel with Walls of Identical and Non-Identical Roughness, *Int. J. Heat Mass Transfer*, Vol. 10, pp. 611–621, 1967.

208. J. C. Han, Heat Transfer and Friction in Channels with Two Opposite Rib-Roughened Walls, *J. Heat Transfer*, Vol. 106, pp. 774–781, 1984.

209. S. V. Patankar and S. Acharya, Development of Turbulence Model for Rectangular Passages, *Trans. CSME*, Vol. 8, pp. 146–149, 1984.

210. E. Brundrett and P. R. Burroughs, The Temperature Inner Law and Heat Transfer for Turbulent Air Flow in a Vertical Square Duct, *Int. J. Heat Mass Transfer*, Vol. 10, pp. 1133–1142, 1967.

211. B. E. Launder and W. M. Ying, Prediction of Flow and Heat Transfer in Ducts of Square Cross-Section, *Proc. Inst. Mech. Engrs.*, Vol. 187, 37/33, pp. 455–461, 1973.

212. J. L. Novotny, S. T. McComas, E. M. Sparrow, and E. R. G. Eckert, Heat Transfer for Turbulent Flow in Rectangular Ducts with Two Heated and Two Unheated Walls, *AIChE J.*, Vol. 10, pp. 466–470, 1964.

213. E. M. Sparrow, J. R. Lloyd, and C. W. Hixon, Experiments on Turbulent Heat Transfer in an Asymmetrically Heated Rectangular Duct, *J. Heat Transfer*, Vol. 88, pp. 170–174, 1966.

214. N. Madsen, Comments on the Effect of Axially Varying and Unsymmetrical Boundary Condition on Heat Transfer with Turbulent Flow between Parallel Plates, *Int. J. Heat Mass Transfer*, Vol. 7, pp. 1143–1144, 1964.

215. E. M. Sparrow and N. Cur, Turbulent Heat Transfer in a Symmetrically or Asymmetrically Heated Flat Rectangular Duct with Flow Separation at Inlet, *J. Heat Transfer*, Vol. 104, pp. 82–89, 1982.

216. J. P. Hartnett and T. F. Irvine, Jr., Nusselt Values for Estimating Liquid Metal Heat Transfer in Noncircular Ducts, *AIChE J.*, Vol. 3, pp. 313–317, 1957.

217. R. C. Deissler and M. F. Taylor, Analysis of Turbulent Flow and Heat Transfer in Noncircular Passages, NASA TR R-31, 1959.

218. W. H. Lowdermilk, W. F. Wieland, and J. N. B. Livingood, Measurements of Heat Transfer and Friction Coefficients for Flow of Air in Noncircular Ducts at High Surface Temperatures, NACA RN E53J07, 1954.

219. A. F. Emery, P. K. Neighbors, and F. B. Gessner, The Numerical Prediction of Developing Turbulent Flow and Heat Transfer in a Square Duct, *J. Heat Transfer*, Vol. 102, pp. 51–57, 1980.

220. F. B. Gessner and A. F. Emery, The Numerical Prediction of Developing Turbulent Flow in Rectangular Ducts, *J. Fluids Eng.*, Vol. 103, pp. 445–455, 1981.

221. C. C. Alexopoulos, Temperature and Velocity Distribution and Heat Transfer for Turbulent Flow in a Square Duct, M. A. Sc. Thesis, Dept. Mech. Eng., Univ. Toronto, 1964.

222. A. S. Sukomel, V. I. Velichko, Yu. G. Abrosimov, and D. F. Gustev, An Investigation of Heat Transfer in the Entry Section of a Rectangular Duct, *Teploenergetika*, Vol. 22, No. 3, pp. 81–83, 1975.

223. A. M. M. Aly, A. C. Trupp, and A. D. Gerrard, Measurements and Predictions of Fully Developed Turbulent Flow in an Equilateral Triangular Duct, *J. Fluid Mech.*, Vol. 85, pp. 57–83, 1978.

224. C. A. C. Altemani and E. M. Sparrow, Turbulent Heat Transfer and Fluid Flow in an Unsymmetrically Heated Triangular Duct, *J. Heat Transfer*, Vol. 102, pp. 590–597, 1980.

225. J. Malák, J. Hejna, and J. Schmid, Pressure Losses and Heat Transfer in Non-Circular Channels with Hydraulically Smooth Walls, *Int. J. Heat Mass Transfer*, Vol. 18, pp. 139–149, 1975.

226. E. R. G. Eckert and T. F. Irvine, Jr., Flow in Corners of Passages with Noncircular Cross Section, *Trans. ASME*, Vol. 78, pp. 709–718, 1956.

227. R. W. Hanks and J. C. Brooks, Birefringent Flow Visualization of Transitional Flow in Isosceles Triangular Duct, *AIChE J.*, Vol. 16, pp. 483–489, 1970.

228. R. W. Hanks and R. C. Cope, Laminar-Turbulent Transitional Flow Phenomena in Isosceles Triangular Cross-Section Ducts, *AIChE J.*, Vol. 16, pp. 528–535, 1970.

229. R. C. Cope and R. W. Hanks, Transitional Flow in Isosceles Triangular Ducts, *Ind. Eng. Chem. Fundam.*, Vol. 11, pp. 106–117, 1972.

230. P. C. Bandopadhayay and J. B. Hinwood, On the Coexistence of Laminar and Turbulent Flow in Narrow Triangular Duct, *J. Fluid Mech.*, Vol. 59, pp. 775–783, 1973.

231. S. S. Tung and T. F. Irvine, Jr., Experimental Study of the Flow Viscoelastic Fluid in a Narrow Isosceles Triangular Duct, *Studies in Heat Transfer*, ed. J. P. Hartnett, T. F. Irvine Jr., E. Pfender, and E. M. Sparrow, Hemisphere, New York, pp. 309–329, 1978.

232. C. W. Rapley and A. D. Gosman, The Prediction of Turbulent Flow and Heat Transfer in a Narrow Isosceles Triangular Duct, *Int. J. Heat Mass Transfer*, Vol. 27, pp. 253–262, 1984.

233. H. Usui, H. Fukuma, and Y. Sano, Fully Developed Turbulent Flow in Isosceles Triangular Ducts, *J. Chem. Eng. Jpn.*, Vol. 16, pp. 13–18, 1983.

234. L. W. Carlson and T. F. Irvine, Jr., Fully Developed Pressure Drop in Triangular Shaped Corners, *J. Heat Transfer*, Vol. 83, pp. 441–444, 1961.

235. E. R. G. Eckert and T. F. Irvine, Jr., Pressure Drop and Heat Transfer in a Duct with Triangular Cross-Section, *J. Heat Transfer*, Vol. 82, pp. 125–138, 1960.

236. L. S. Kokorev, A. S. Korsun, B. N. Kostyunin, and V. I. Petrovichev, Hydraulic Drag and Heat Transfer in Turbulent Flow of Liquids in Triangular Channels, *Heat Transfer—Sov. Res.*, Vol. 3, pp. 56–65, 1971.

237. Y. I. Tokarev, Experimental Investigation of Heat Transfer in a Channel Triangular in Cross-Section, *Thermal Eng.*, Vol. 25, pp. 605–606, 1979.

238. H. Usui, Y. Sano, and H. Fukuma, Turbulence Measurements and Mass Transfer in Fully Developed Flow in a Triangular Duct with a Narrow Apex Angle, *Int. J. Heat Mass Transfer*, Vol. 25, pp. 615–624, 1982.

239. N. T. Obot and K. Adu-Wusu, The Flow Pattern in a Scalene Triangular Duct Having Two Rounded Corners, *J. Fluids Eng.*, Vol. 107, pp. 455–459, 1985.

240. N. T. Obot, Heat Transfer in a Smooth Scalene Triangular Duct With Two Rounded Corners, *Int. Comm. Heat Mass Transfer*, Vol. 12, pp. 251–258, 1985.

241. N. T. Obot, E. B. Esen and K. Adu-Wusu, Pressure Drop for Rib-Roughened Scalene Triangular Duct Having Two Rounded Corners, *Int. Comm. Heat Mass Transfer*, Vol. 14, pp. 11–20, 1987.

242. D. A. Campbell and H. C. Perkins, Variable Property Turbulent Heat and Momentum Transfer for Air in a Vertical Rounded Corner Triangular Duct, *Int. J. Heat Mass Transfer*, Vol. 11, pp. 1003–1012, 1968.

243. D. Cain and J. Duffy, An Experimental Investigation of Turbulent Flow in Elliptical Ducts, *Int. J. Mech. Sci.*, Vol. 13, pp. 451–459, 1971.

244. D. Cain, A. Roberts, and H. Barrow, A Theoretical Study of Fully Developed Flow and Heat Transfer in Elliptical Ducts, *Recent Developments in Compact High Duty Heat Exchangers*, Inst. Mech. Engrs., London, pp. 39–46, Oct. 1972.

245. H. Barrow and A. Roberts, Flow and Heat Transfer in Elliptic Ducts, *Heat Transfer 1970*, Paper No. FC 4.1, Versailles, Sept. 1970.

246. D. Cain, A. Roberts, and H. Barrow, An Experimental Investigation of Turbulent Flow and Heat Transfer in Elliptical Ducts, *Wärme- und Stoffübertragung*, Vol. 2, pp. 101–107, 1973.

247. A. A. Ryadno and A. A. Kochubei, Calculation of Nonsteady Convective Heat Transfer for Turbulent Viscous Incompressible Flow in a Tube of Elliptical Cross Section, *Inzh.-Fiz. Zh.*, Vol. 15, pp. 912–915, 1978.

248. E. Rodet, Etude de l'écoulement d'un fluide dans un tunnel prismatique de section trapezoïdale, *Publ. Sci. et Tech. Min. Air*, No. 369, 1960.

249. D. J. Gunn and C. W. W. Darling, Fluid Flow and Energy Losses in Non-Circular Conduits, *Trans. Inst. Chem. Engrs.*, Vol. 41, pp. 163–173, 1963.

250. H. Barrow, A. K. A. Hassan, and C. Avgerinos, Peripheral Temperature Variation in the Wall of a Non-Circular Duct—an Experimental Investigation, *Int. J. Heat Mass Transfer*, Vol. 77, pp. 1031–1032, 1984.

251. A. Mohandes and J. G. Knudsen, Friction Factors in Noncircular Ducts With Sharp Corners, *Can. J. Chem. Eng.*, Vol. 57, pp. 109–111, 1979.

252. K. Rehme, Simple Method of Predicting Friction Factors of Turbulent Flow in Non-circular Channels, *Int. J. Heat Mass Transfer*, Vol. 16, pp. 933–950, 1973.

253. J. E. Walker and R. R. Rothfus, Transitional Velocity Patterns in a Smooth Concentric Annulus, *AIChE J.*, Vol. 5, pp. 51–54, 1959.

254. J. E. Walker, G. A. Whan, and R. R. Rothfus, Fluid Friction in Noncircular Ducts, *AIChE J.*, Vol. 3, pp. 484–489, 1957.

255. J. G. Knudsen and D. L. Katz, *Fluid Dynamics and Heat Transfer*, Robert E. Krieger, Huntingdon, N.Y., pp. 191–193, 1979.

256. J. A. Brighton and J. B. Jones, Fully Developed Turbulent Flow in Annuli, *J. Basic Eng.*, Vol. 86, pp. 835–844, 1964.

257. C. M. Ivey, The Position of Maximum Velocity in Annular Flow, M.Sc. Thesis, Univ. of Windsor, Canada, 1965.

258. V. K. Jonsson and E. M. Sparrow, Turbulent Diffusivity for Momentum Transfer in Concentric Annuli, *J. Basic Eng.*, Vol. 88, pp. 550–552, 1966.

259. R. V. Bailey, Heat Transfer to Liquid Metals in Concentric Annuli, ORNL521, Oak Ridge Natl. Lab., Tech. Div. Eng. Res. Sec., 1950.

260. R. M. Olson and E. M. Sparrow, Measurements of Turbulent Flow Development in Tubes and Annuli with Square or Rounded Entrances, *AIChE J.*, Vol. 9, pp. 766–770, 1963.

261. H. Barrow, Fluid Flow and Heat Transfer in an Annulus with a Heated Core Tube, *Proc. Inst. Mech. Engrs.*, London, Vol. 169, p. 1113, 1955.

262. N. W. Wilson and J. O. Medwell, An Analysis of Heat Transfer for Fully Developed Turbulent Flow in Concentric Annuli, *J. Heat Transfer*, Vol. 90, pp. 43–50, 1968.

263. Q. Rensen, Experimental Investigation of Turbulent Heat Transfer to Liquid Sodium in the Thermal Entrance Region of an Annulus, *Nucl. Eng. Design*, Vol. 68, pp. 397–404, 1981.

264. W. A. Sutherland and W. M. Kays, Heat Transfer in an Annulus with Variable Circumferential Heat Flux, *Int. J. Heat Mass Transfer*, Vol. 7, pp. 1187–1194, 1964.

265. R. R. Rothfus, C. C. Monrad, K. G. Sikchi, and W. J. Heideger, Isothermal Skin Friction in Flow through Annular Sections, *Ind. Eng. Chem.*, Vol. 47, pp. 913–918, 1955.

266. T. H. Okiishi and G. K. Serovy, An Experimental Study of the Turbulent Flow Boundary Layer Development in Smooth Annuli, *J. Basic Eng.*, Vol. 89, pp. 823–836, 1967.

267. A. Quarmby and R. K. Anand, Turbulent Heat Transfer in the Thermal Entrance Region of Concentric Annuli with Uniform Wall Heat Flux, *Int. J. Heat Mass Transfer*, Vol. 13, pp. 395–411, 1970.

268. A. Roberts and H. Barrow, Turbulent Heat Transfer to Air in the Vicinity of the Entry of an Internally Heated Annulus, *Proc. Inst. Mech. Engrs.*, Vol. 182, Pt. 3H, pp. 268–276, 1967.

269. V. K. Jonsson and E. M. Sparrow, Experiments on Turbulent Flow Phenomena in Eccentric Annular Ducts, *J. Fluid Mech.*, Vol. 25, pp. 65–68, 1966.

270. V. K. Jonsson and E. M. Sparrow, Results of Laminar Flow Analysis and Turbulent Flow Experiments for Eccentric Annular Ducts, *AIChE J.*, Vol. 11, pp. 1143–1145, 1965.

271. Y. Lee and H. Barrow, Turbulent Flow and Heat Transfer in Concentric and Eccentric Annuli, *Proc. Thermodyn. and Fluid Mech. Convention*, Inst. Mech. Engrs., London, Paper No. 12, Apr. 1964.

272. R. G. Deissler and M. F. Taylor, Analysis of Fully Developed Turbulent Heat Transfer in an Annulus of Various Eccentricities, NACA TN 3451, 1955.

273. W. S. Yu and O. E. Dwyer, Heat Transfer to Liquid Metals Flowing Turbulently in Eccentric Annuli—I, *Nucl. Sci. Eng.*, Vol. 24, pp. 105–117, 1966.

274. T. W. Ricker, J. H. T. Wade, and N. W. Wilson, On the Velocity Fields in Eccentric Annuli, ASME Paper 68-WA/FE-35, 1968.

275. R. L. Judd and J. H. T. Wade, Forced Convection Heat Transfer in Eccentric Annular Passages, *Heat Transfer and Fluid Mechanics Institute*, pp. 272–288, Stanford Univ., Stanford, Calif., 1963.

276. E. Y. Leung, W. M. Kays, and W. C. Reynolds, Heat Transfer with Turbulent Flow in Concentric and Eccentric Annuli With Constant and Variable Heat Flux, TR AHT-4, Mech. Eng. Dept., Stanford Univ., Stanford, Calif., 1962.

277. V. K. Jonsson, Experimental Studies of Turbulent Flow Phenomena in Eccentric Annuli, Ph.D. Thesis, Univ. Minnesota, Minneapolis, Minn. 1965.

5

CONVECTIVE HEAT TRANSFER IN CURVED DUCTS

R. K. Shah

Harrison Radiator Division, GM
Lockport, New York

S. D. Joshi

Phillips Petroleum Company
Bartlesville, Oklahoma

5.1 INTRODUCTION

Curved tubes may be classified as helical coils, spirals, and bends. The design information and discussion in this chapter is restricted to helical coils and spirals shown in Fig. 5.1. Results for bends and fittings are presented in Chapter 10. Curved tubes are used in chemical reactors, agitated vessels, storage tanks, and heat recovery systems. Industries such as dairy and food processing, refrigeration, and hydrocarbon processing extensively use curved tube heat exchangers. An extensive use is also found in medical equipment such as kidney dialysis machines. In oil fields, the use of curved tubes as an inline viscometer is quite common.

Nomenclature for helical and spiral coils is illustrated in Fig. 5.1. A helical coil of circular cross section is characterized by tube radius a, coiled tube curvature radius R, and coil pitch b, as shown in Fig. 5.1a. The figure also shows the vertical and horizontal reference planes along which velocity and temperature profiles are normally measured. The helical coil in Fig. 5.1a will be referred to as a "horizontal" coil throughout this chapter, since the tube in each turn is approximately horizontal. If this coil had been turned 90°, it would be referred to as a "vertical" coil. The radius of curvature R of the coiled tube is constant for a helical coil, while it continuously increases for a spiral coil. A "simple" or Archimedean spiral coil of circular cross

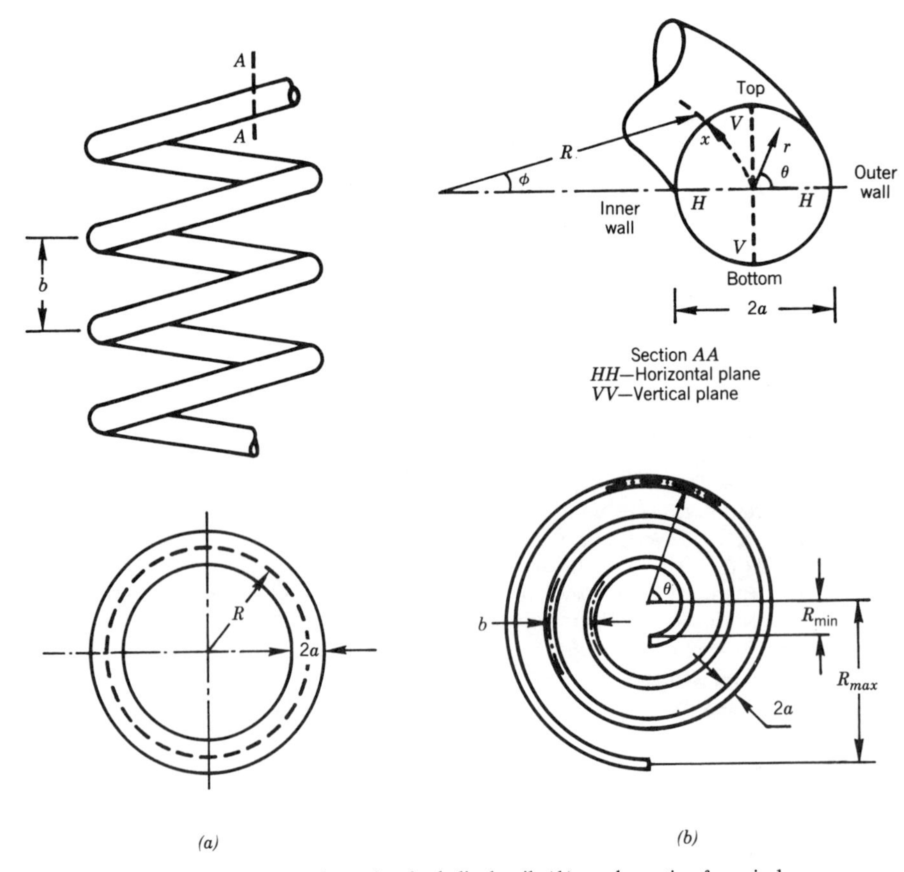

Figure 5.1. (a) A schematic of a helical coil, (b) a schematic of a spiral.

section is characterized by the tube radius a, the *constant* pitch b, and the minimum and maximum radii of curvature (R_{min} and R_{max}) of the beginning and the end of the spiral. Unless specified otherwise, results for only simple spirals are covered in this chapter.

As compared to straight tubes, curved tubes are compact and yield higher heat-transfer coefficients and friction factors; however, mechanical cleaning of curved tubes is difficult. As fluid flows within a curved tube, it experiences a centrifugal force along with the axial pressure gradient. The centrifugal force causes secondary flow velocities, resulting in an increased main flow (axial) velocity near the tube outer wall, and decreased axial velocity near the tube inner wall. At the tube outer wall, the higher velocity decreases the thermal resistance considerably, resulting in high heat transfer coefficients. However, this is also accompanied by a higher friction factor and possibly higher pressure drop than that in an equivalent straight tube.

In a curved tube, the heat transfer rate and pressure drop are dependent upon the following parameters: Reynolds number, Prandtl number, Newtonian or non-Newtonian fluid, wall thermal boundary condition, coil-to-tube radius ratio, tube cross section, length-to-diameter ratio, and coil pitch. The influence of these parameters on heat transfer and flow friction of curved tubes have been described in many papers [1–85], including some review articles [1–4]. The purpose of this chapter is to summarize experimental and theoretical results for laminar, transitional, and turbulent flow, and recommend correlations for curved-tube heat transfer and flow friction. The scope of this chapter is restricted to a steady-state single-phase flow in stationary ducts. Most of the recommended correlations are for helical coils, but wherever available, spiral-coil results are also presented. Because of time and space limitations, the subject of centrifugal instability in curved channels is not treated here, but the reader may refer to Akiyama et al. [86, 87] and Komiyama et al. [88].

5.2. PROBLEM FORMULATION

In order to determine the heat transfer and pressure drop characteristics of curved coils theoretically, the continuity, momentum, and energy equations need to be solved. At any cross section in a curved tube, a curvature-induced centrifugal force generates a secondary flow over the normal axial flow. This results in the presence of all three velocity components, even in the case of fully developed flow.

Generally, the theoretical or numerical solutions are obtained by solving the appropriate differential equations and boundary conditions. The momentum equations for laminar flow are presented by Sankariah and Rao [5] in the toroidal coordinate system, and by Patankar et al. [6] in the curvilinear cylindrical coordinate system. The momentum equations for turbulent flow are presented by Patankar et al. [7] in the curvilinear cylindrical coordinate system. The energy equation for laminar flow is presented by Tyagi and Sharma [8] in the toroidal coordinate system, and the time-average energy equation is presented for turbulent flow by Kreith [9] in the curvilinear cylindrical coordinate system. Because of space limitation, these equations will not be duplicated here.

While the boundary condition for the velocity problem is clear—no slip (zero velocity components) at the wall—a large variety of thermal boundary conditions can be specified for the temperature problem; many such boundary conditions are summarized in Table 3.1 of Chapter 3. The most important and also limiting boundary conditions are the Ⓣ, (H1), and (H2) boundary conditions. Ⓣ refers to axially and peripherally constant wall temperature. (H1) refers to axially constant heat flux at the wall with peripherally constant wall temperature. (H2) refers to axially and peripherally constant heat flux at the wall. For more details, refer to Chapter 3. It is important to note that in experimental coil heat transfer studies with the constant wall heat flux boundary condition, it is difficult to achieve exact (H1) or (H2) boundary conditions. The difficulty arises because tube bending operations distort the tube wall thickness, and secondary flow alters the fluid temperature profile.

Most of the dimensionless groups associated with the flow in a straight duct are presented in [10]. Two additional dimensionless groups for curved tubes are the Dean number and helical coil number. Dean [11, 12] pointed out the existence of a secondary flow in a curved tube and introduced a parameter to take account of it. This parameter is now called the Dean number and is defined as

$$\mathrm{De} = \mathrm{Re}\left(\frac{a}{R}\right)^{1/2} \tag{5.1}$$

In a helical coil, the effective radius of curvature R_c of each turn is influenced by the coil pitch b, and is given by [82]

$$R_c = R\left[1 + \left(\frac{b}{2\pi R}\right)^2\right] \tag{5.2}$$

Use of R_c instead of R in the Dean number definition results in a new number, referred to as the helical coil number He, defined as

$$\mathrm{He} = \mathrm{Re}\left(\frac{a}{R_c}\right)^{1/2} = \mathrm{De}\left[1 + \left(\frac{b}{2\pi R}\right)^2\right]^{-1/2} \tag{5.3}$$

The definitions of the peripherally and axially local and mean heat transfer coefficients and Nusselt numbers are the same as those by Shah and London [10]. They are restated here for completeness. The peripherally local heat transfer coefficient h_p is defined by

$$q_p'' = h_p(T_w - T_m) \tag{5.4}$$

where t_w is the local temperature on the duct periphery, t_m is the fluid bulk mean temperature at the cross section, and q_p'' is the heat flux at the point of concern on the duct periphery. The peripherally average, but axially local, heat transfer coefficient h_x is defined by

$$q_x'' = h_x(T_{w,m} - T_m) \tag{5.5}$$

where $t_{w,m}$ is the peripheral mean wall temperature (peripheral integrated average of t_w). Note that h_x may or may not be the peripheral integrated average of h_p, because of $t_{w,m}$. The flow-length average heat transfer coefficient h_m is the integrated average of h_x from $x = 0$ to x:

$$h_m = \frac{1}{x}\int_0^x h_x\,dx \tag{5.6}$$

Correspondingly, the Nusselt numbers $\mathrm{Nu}_{p,\mathrm{bc}}$, $\mathrm{Nu}_{x,\mathrm{bc}}$, $\mathrm{Nu}_{m,\mathrm{bc}}$, and $\mathrm{Nu}_{\mathrm{bc}}$ are defined below, where bc denotes a specific thermal boundary condition such as Ⓣ, Ⓗ1 and Ⓗ2. A local peripheral Nusselt number is defined as

$$\mathrm{Nu}_{p,\mathrm{bc}} = \frac{h_p D_h}{k} = \frac{q_p'' D_h}{k(T_w - T_m)} \tag{5.7}$$

The peripheral average but axially local Nusselt number is defined as

$$\mathrm{Nu}_{x,\mathrm{bc}} = \frac{h_x D_h}{k} = \frac{q_x'' D_h}{k(T_{w,m} - T_m)} \tag{5.8}$$

The mean (flow-length average) Nusselt number in the thermal entrance region is defined as

$$\mathrm{Nu}_{m,\mathrm{bc}} = \frac{h_m D_h}{k} = \frac{1}{x}\int_0^x \mathrm{Nu}_{x,\mathrm{bc}}\, dx \tag{5.9}$$

The peripheral average Nusselt number in the hydrodynamically and thermally fully developed region is simply defined without the first subscript p, x, or m as follows:

$$\mathrm{Nu}_{\mathrm{bc}} = \frac{hD_h}{k} \tag{5.10}$$

In the fully developed region, $\mathrm{Nu}_{m,\mathrm{bc}}$ approaches $\mathrm{Nu}_{x,\mathrm{bc}}$ and both approach $\mathrm{Nu}_{\mathrm{bc}}$.

5.3 LAMINAR FLOW THROUGH COILS OF CIRCULAR CROSS SECTION

The laminar flow results are described in four subsections: fully developed flow, hydrodynamically developing flow, thermally developing and hydrodynamically developed flow, and simultaneously developing flow. For convenience, wherever feasible, each subsection is further divided into two sub-subsections: fluid flow and heat transfer.

5.3.1 Fully Developed Flow

Fluid Flow. In fluid flow, the quantities of design interest are velocity profiles and friction factors, and are summarized below.

Velocity Profiles. In a flow through a coil, the centrifugal force strongly influences the velocity profile. In the case of the helical coil flow, the constant radius of curvature generates a constant centrifugal force resulting in the establishment of "fully developed flow." However, for a flow through a spiral, the radius of curvature is continuously varying, resulting in a continuous varying centrifugal force; the Dean number is continuously varying as well. Therefore, "fully developed flow" is established in a spiral coil only as a limiting condition when R/a or De becomes large. In the case of a helical coil, starting with Dean [11, 12], several theoretical and experimental results are available [13, 14]. However, velocity profile results are not available for a spiral coil.

Dean's [11, 12] velocity profiles based on perturbation analysis are applicable only for De < 20. Mori and Nakayama [14] have obtained the following fully developed velocity profile using boundary-layer idealizations (De > 100) for a coil with $R \gg a$:

In the core region,

$$u_r = \frac{\nu D}{a}\cos\theta, \qquad u_\theta = -\frac{\nu D}{a}\sin\theta, \qquad u_x = \left(\frac{B\nu}{a} + \frac{Cr\nu}{Da^2}\cos\theta\right) \tag{5.11}$$

In the boundary layer,

$$u_r = 0 \tag{5.12}$$

$$u_\theta = \frac{D\nu\sin\theta}{a}\left(\frac{12a - 6\delta}{\delta}\right)\delta^* + \left(\frac{9\delta - 24a}{\delta}\right)\delta^{*2} + \left(\frac{12a - 4\delta}{\delta}\right)\delta^{*3} \tag{5.13}$$

and

$$u_x = \frac{\nu}{a}\left[\left(B + \frac{C}{D}\frac{(a-\delta)\cos\theta}{a}\right)(2\delta^* - \delta^{*2})\right] + \frac{\nu C\delta\cos\theta}{Da^2}(\delta^* - \delta^{*2}) \tag{5.14}$$

where

$$B = \frac{0.5\mathrm{Re}}{1 - \dfrac{2\delta}{3a} + \dfrac{\delta^2}{6a^2}}, \qquad C = \frac{\dfrac{2a\,\mathrm{Re}}{\delta}}{1 - \dfrac{2\delta}{3a} + \dfrac{\delta^2}{6a^2}} \tag{5.15a}$$

and

$$D = 0.9656\,\mathrm{De}^{1/2} + 1.65, \qquad \delta = \frac{4.63a}{\mathrm{De}^{1/2}} - \frac{0.766a}{\mathrm{De}} \tag{5.15b}$$

The above velocity profiles show good agreement with experimental data [14, 15] and numerical predictions [6] for De > 100. For De < 20, the velocity profile is not significantly different from that in a straight tube flow. The velocity profiles for intermediate values (20 < De < 100) are not available. Figure 5.2 shows a typical variation in the velocity profile with the Dean number, based on the analysis of Patankar et al. [6]. The velocity peak moves toward the outer wall as the Dean number increases.

Friction Factors— Helical Coils. The results of several theoretical and experimental friction factor studies demonstrate that coiled-tube friction factors are higher than those in a straight tube for a given Reynolds number. As shown in Fig. 5.3, friction factors at the outer wall in a helical coil are substantially higher than those obtained in a straight tube. Conversely, friction factors at the inner wall are almost the same as those obtained in a straight tube [16]. The overall effect is an increase in friction factors.

The major difference between various experimental and theoretical correlations is the way they account for an increase in the friction factor due to the coil curvature. Many investigators, such as Srinivasan et al. [17], use only the Dean number in their correlation, claiming that it alone is sufficient to account for an increase in the friction

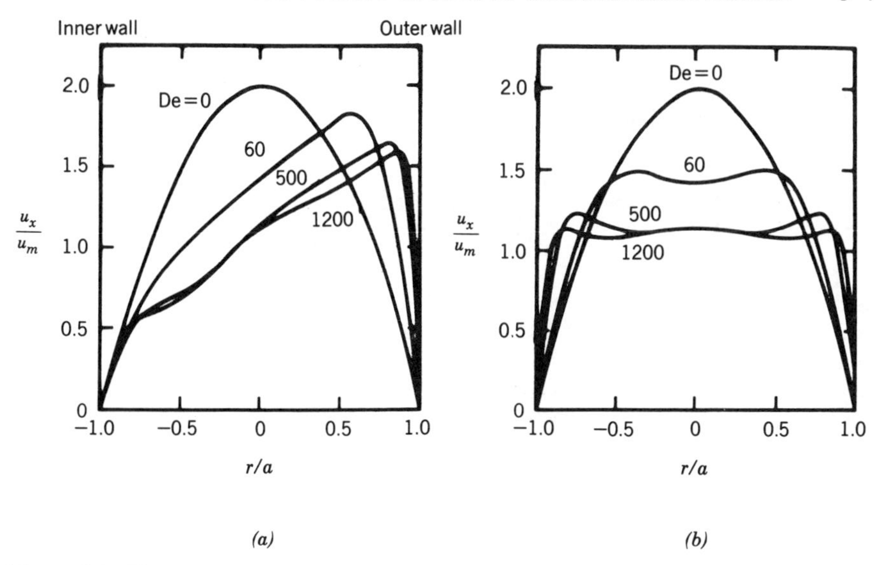

Figure 5.2. The influence of the Dean number on axial velocity profiles in a horizontal curved tube: (a) horizontal plane, (b) vertical plane.

factor due to the coil curvature. They proposed the following correlation for their experimental data with several coils ($7 < R/a < 104$):

$$\frac{f_c}{f_s} = \begin{cases} 1 & \text{for} \quad \text{De} < 30 \\ 0.419\,\text{De}^{0.275} & \text{for} \quad 30 < \text{De} < 300 \\ 0.1125\,\text{De}^{0.5} & \text{for} \quad \text{De} > 300 \end{cases} \tag{5.16}$$

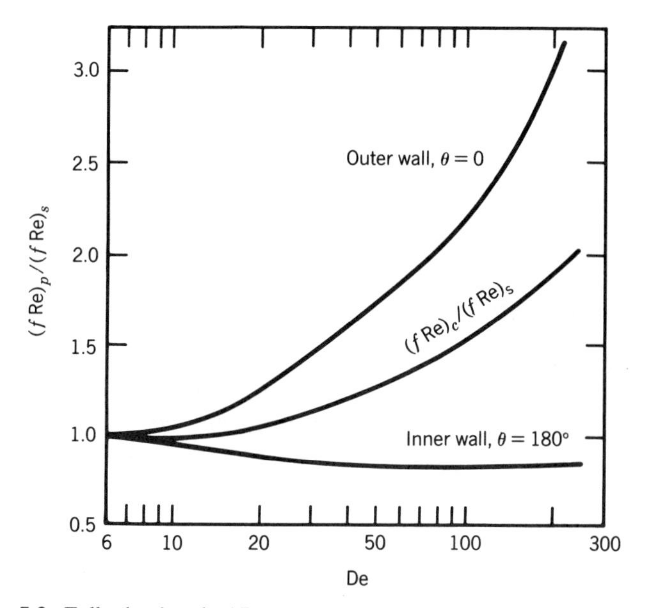

Figure 5.3. Fully developed $(f\,\text{Re})_p/(f\,\text{Re})_s$ as a function of the Dean number [16].

On the other hand, some correlations include a separate R/a term in addition to a De term to account for the coil-curvature effect. For example, Manlapaz and Churchill [18] reviewed available experimental data and theoretical predictions and recommended the following correlation using a regression analysis:

$$\frac{f_c}{f_s} = \left[\left(1.0 - \frac{0.18}{\left[1 + (35/\mathrm{De})^2\right]^{0.5}}\right)^m + \left(1.0 + \frac{a/R}{3}\right)^2\left(\frac{\mathrm{De}}{88.33}\right)\right]^{0.5} \tag{5.17}$$

where $m = 2$ for $\mathrm{De} < 20$, $m = 1$ for $20 < \mathrm{De} < 40$, and $m = 0$ for $\mathrm{De} > 40$. The friction factor ratios calculated using the above two correlations show excellent agreement with each other (deviation of less than 9%), as long as $R/a > 7$. Even at $R/a = 3$, the f_c/f_s value is almost the same as that for $R/a = 7$. This demonstrates that De alone is sufficient to account for the increase in the friction factor for coils with $R/a > 3$. Hence, for coils with $R/a > 3$, we recommend the use of either Eq. (5.16) or Eq. (5.17). For coils with $R/a < 3$, we recommend Eq. (5.17).

Manlapaz and Churchill [18] suggested using the helical coil number He instead of the Dean number De in Eq. (5.17) to account for changes in the friction factor due to the coil pitch. However, their own theoretical predictions, other predictions [19], and experimental data [20] demonstrate that the influence of the coil pitch on the friction factors is very small.

Kubair and Kuloor [21] measured nonisothermal friction factors and proposed the following correlation, which accounts for the temperature-dependent viscosity:

$$f_{c,\,\mathrm{nonisothermal}} = 0.91 f_{c,\,\mathrm{isothermal}}\left(\frac{\mu_w}{\mu_m}\right)^{0.25} \tag{5.18}$$

Friction Factors— Spiral Coils. Kubair and Kuloor [22–24] have measured friction factors in three spirals, and Srinivasan et al. [17] have measured friction factors in five spirals. We recommend the following correlation for design purposes, since it includes the influence of spiral length and pitch [17]:

$$f_c = \frac{0.62\left(n_2^{0.7} - n_1^{0.7}\right)^2}{\mathrm{Re}^{0.6}(b/a)^{0.3}} \tag{5.19}$$

where n_1 and n_2 are the numbers of turns from the origin to the start and the end of a spiral. This equation is valid for $500 < \mathrm{Re}(b/a)^{0.5} < 20{,}000$ and $7.3 < b/a < 15.5$.

Critical Reynolds Number. A transition from laminar to turbulent flow is identified by a critical Reynolds number, Re_{crit}. In a curved duct flow, it is difficult to identify Re_{crit} by a change in the slope of the curve for the friction factor vs. Reynolds number because of the gradual change. In contrast, in a straight-duct flow, the curve of friction factor vs. Reynolds number shows a distinct discontinuity, facilitating the identification of Re_{crit}. After reviewing a large amount of data reported in the literature, the following correlation is recommended for design purposes [17].

$$\mathrm{Re}_{crit} = 2100\left[1 + 12\left(\frac{R}{a}\right)^{-0.5}\right] \tag{5.20}$$

This correlation is correct in the limiting case of straight-tube flow; i.e., as $R/a \to \infty$, the equation reduces to the straight-tube $Re_{crit} = 2100$. In addition, it correlates most of the published data within $\pm 15\%$ deviation.

In spiral flow, the radius of curvature varies along the spiral, and therefore, the flow does not have a single critical Reynolds number. Kubair and Kuloor [22] have suggested the use of the arithmetic-average radius of curvature of the spiral in their helical-coil correlation. On the other hand, Srinivasan et al. [17] have proposed correlations for minimum and maximum Re_{crit} that would occur at the R_{max} and R_{min} locations in the spiral. Their equations, an extension of the above helical-coil correlation, correlate their data for spirals very well:

$$(Re_{crit})_{min} = 2100\left[1 + 12\left(\frac{R_{max}}{a}\right)^{-0.5}\right] \tag{5.21}$$

$$(Re_{crit})_{max} = 2100\left[1 + 12\left(\frac{R_{min}}{a}\right)^{-0.5}\right] \tag{5.22}$$

Heat Transfer. The quantities of design interest are temperature profiles and Nusselt numbers. For the fully developed heat transfer, the quantities of design interest are described on the basis of boundary conditions: uniform wall temperature Ⓣ, and uniform wall heat flux (H1) and (H2).

Fully Developed Temperature Profiles. In coils, as with the velocity profiles, the secondary flow distorts the temperature profiles, pushing the temperature peak toward the tube outer wall; this results in a higher heat transfer rate at the coil outer wall than at the inner wall. Figure 5.4 shows that either increasing De or increasing Pr causes a higher distortion of the temperature profile. Increasing De augments secondary flow, while increasing Pr augments thermal convection.

The temperature-profile equation given in the next two subsections are for helical-coil flow. These profiles could be also used for spiral-coil flow by using R_{ave} instead of R in the calculation of the Dean number. This would give a profile at the average radius of curvature. Similarly, a profile at the spiral coil exit could be calculated by the use of R_{max}. However, the following equations should not be used to calculate a developing profile at the coil entrance.

Ⓣ *Temperature Profiles.* Several investigators have reported theoretical temperature profiles [25–27]. They show good agreement with each other. However, no experimental data are available to confirm them. Mori and Nakayama [25] presented the following temperature profile using a boundary-layer type solution (De > 100) for $Pr \le 1$ and $R/a > 1$:

In the core region,

$$\frac{T_w - T}{T_w - T_m} = F = N\left[1 + \frac{EB}{D}\left(\frac{r}{a}\cos\theta\right) + \frac{E}{2D^2}(C + EB^2)\left(\frac{r}{a}\cos\theta\right)^2 + \frac{EB}{D}\left(\frac{EC}{2D^2} + \frac{E^2B^2}{6D^2}\right)\left(\frac{r}{a}\cos\theta\right)^3\right] \tag{5.23}$$

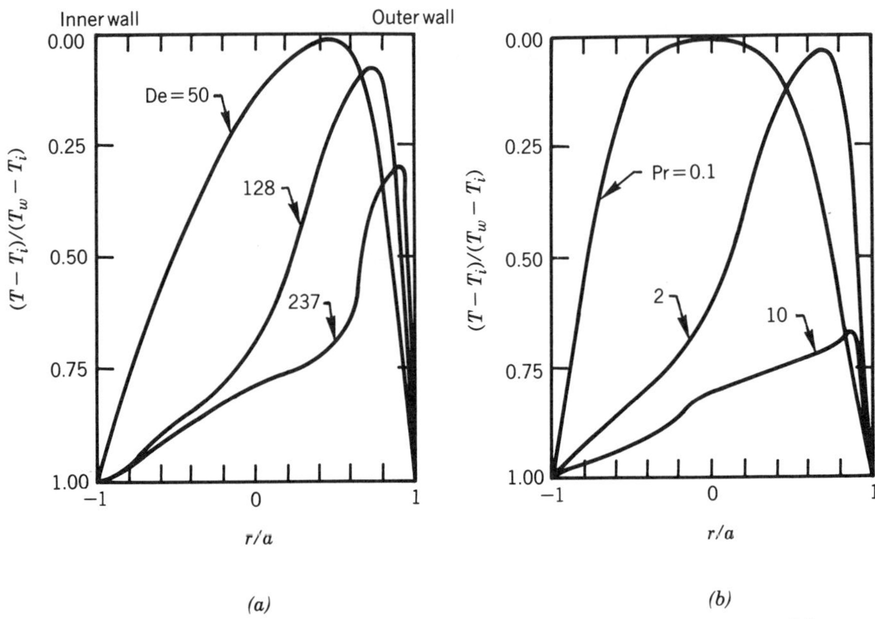

Figure 5.4. The dimensionless horizontal temperature profile for the Ⓣ boundary condition as a function of (a) Dean number and (b) Prandtl number for a curved tube [26].

where

$$N = \left(1 + \frac{3}{8}\frac{EC}{D^2} + \frac{1}{8}\frac{E^2B^2}{D^2}\right)^{-1}, \qquad E = \frac{8a}{\delta_t \mathrm{Re}\,\mathrm{Pr}} \tag{5.24}$$

$$\delta_t = \begin{cases} 0.2\delta\left[2 + (10/\mathrm{Pr}^2 - 1)^{1/2}\right] & \text{for} \quad \mathrm{Pr} \le 1 \\ 0.189\delta\left[1 + (1 + 19.5/\mathrm{Pr}^2)^{1/2}\right] & \text{for} \quad \mathrm{Pr} \ge 1 \end{cases} \tag{5.25}$$

where B, C, D, and δ are given by Eqs. (5.15a) and (5.15b).

In the boundary layer,

$$\frac{T_w - T}{T_w - T_m} = F_1\left[\frac{2(a - r)}{\delta_t} - \left(\frac{a - r}{\delta_t}\right)^2\right] \tag{5.26}$$

Here F_1 represents the core-region dimensionless temperature profile at the edge of a thermal boundary layer. F_1 is the modified F, calculated by substituting $r = a - \delta_t$ in Eq. (5.23).

The above equations are valid for $\mathrm{Pr} \le 1$, $\mathrm{De} > 100$, and $R/a > 4$. At present, no solutions are available for $\mathrm{De} < 100$, but a straight-tube temperature profile may be used at low Dean numbers.

To obtain a simple δ_t expression for fluids with $\mathrm{Pr} > 1$, Mori and Nakayama [25] evaluated energy integrals of the thermal boundary layer over the hydrodynamic boundary layer rather than over the thermal boundary layer. The authors explain that boundary-layer profiles are approximate, and therefore changing the integration range should not cause a significant error in the computed temperature profile [89]. Thus, Eqs. (5.23) through (5.26) could also be used to calculate temperatures profiles for fluids with Pr > 1.

(H1) *Temperature Profiles.* In 1932, Hawes made the first attempt to measure the temperature profile [28]. Since then, several theoretical and a few experimental studies have reported temperature-profile data [14, 16, 30]. Mori and Nakayama [14] used a boundary-layer type solution (De > 100) for Pr ≤ 1 and $R/a > 1$, and reported the following temperature profile:

In the core region,

$$\frac{T_w - T}{a(dT/dx)} = g = B' + \frac{C(r/a)^2}{2D^2} + \frac{B(r/a)\cos\theta}{D} \tag{5.27}$$

where

$$B' = \frac{\delta_t \operatorname{Re}\operatorname{Pr}}{8a} - \frac{a \operatorname{Re}}{D^2 \delta} \tag{5.27a}$$

In the boundary layer,

$$\frac{T_w - T}{a(dT/dx)} = g_{\delta t}\left(\frac{2(a-r)}{\delta_t} - \frac{(a-r)^2}{\delta_t^2}\right)$$
$$+ \frac{\delta_t}{a}\left(\frac{C}{aD^2}(a-\delta_t) + \frac{B}{D}\cos\theta\right)\left[\left(\frac{a-r}{\delta_t}\right)^2 - \left(\frac{a-r}{\delta_t}\right)^3\right] \tag{5.28}$$

where $g_{\delta t}$ represents the core-region dimensionless temperature profile at the edge of a thermal boundary layer, which is calculated by substituting $r = a - \delta_t$ into Eq. (5.27).

The parameters B, C, D, and δ are calculated using Eqs. (5.15a) and (5.15b), and δ_t from Eq. (5.25).

The temperature profiles calculated using the above equations show excellent agreement with air data [14]. Their experimental data for the vertical temperature profile show an excellent agreement with the predictions of Patankar et al. [6]. However, for a horizontal temperature profile, a small discrepancy is observed between Mori and Nakayama's experimental data [14] and Patankar et al.'s theoretical predictions. The discrepancy is seen only for the results from the inner coil wall to the coil center. As discussed earlier in connection with (T) profiles, Mori and Nakayama [14, 89] calculated an approximate temperature profile for fluids with Pr > 1. Therefore, Eqs. (5.27) and (5.28) could also be used to calculate the (H1) temperature profile for fluids with Pr > 1.

(T) *Nusselt Numbers— Helical Coils.* Several theoretical and experimental studies have reported Nusselt numbers for flow through a helical coil subjected to the (T) boundary condition [25–27, 31, 32]. In Fig. 5.5, experimental and theoretical results are

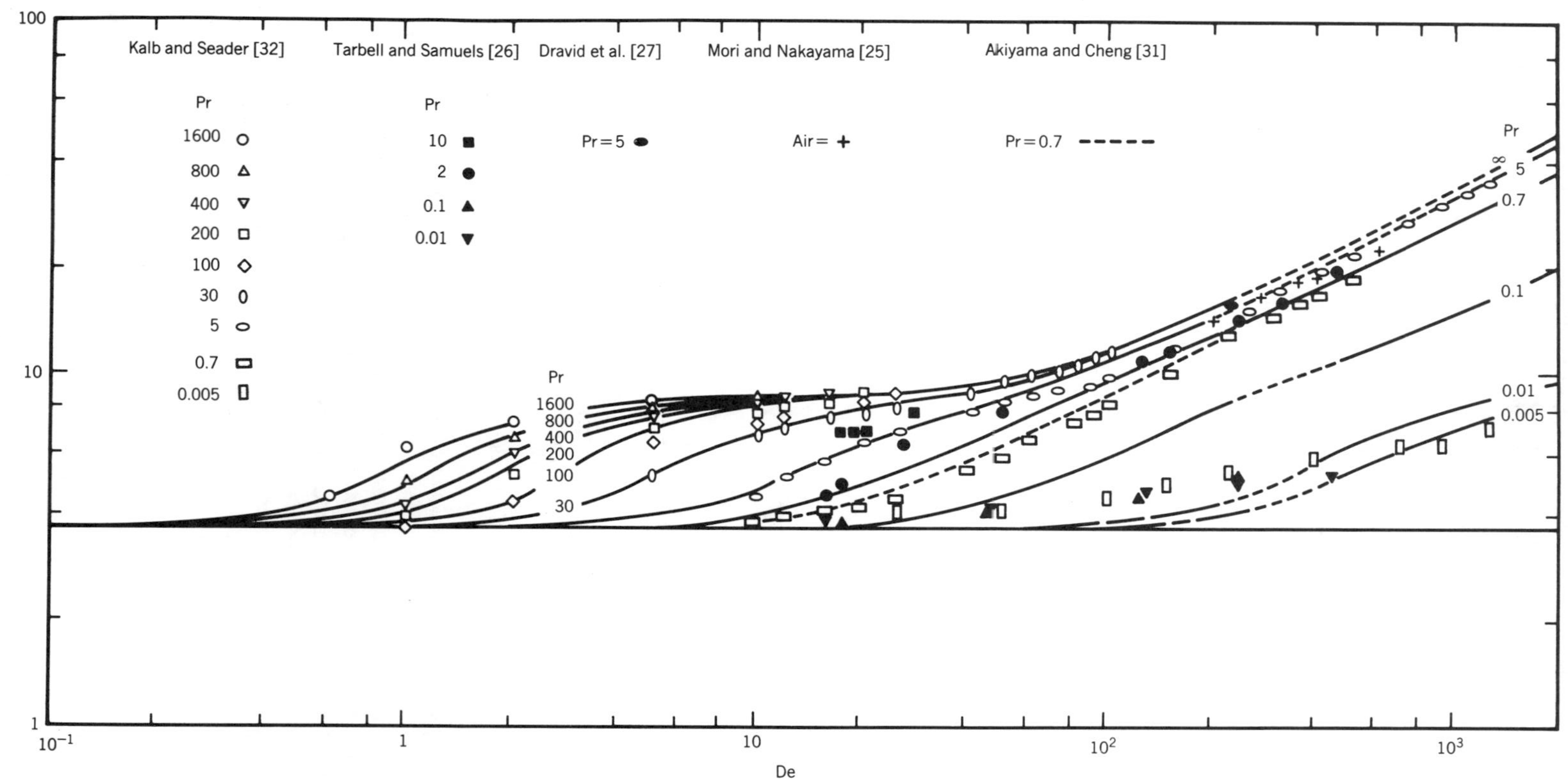

Figure 5.5. A comparison of the recommended design correlation [Eq. (5.29), drawn as solid lines] with theoretical and experimental Nusselt numbers for the Ⓣ boundary condition [19].

compared with the following Manlapaz-Churchill correlation [19] based on a regression analysis of the available data:

$$\mathrm{Nu}_T = \left[\left(3.657 + \frac{4.343}{x_1}\right)^3 + 1.158\left(\frac{\mathrm{De}}{x_2}\right)^{3/2}\right]^{1/3} \tag{5.29}$$

where

$$x_1 = \left(1.0 + \frac{957}{\mathrm{De}^2 \mathrm{Pr}}\right)^2, \qquad x_2 = 1.0 + \frac{0.477}{\mathrm{Pr}} \tag{5.30}$$

Although Nusselt numbers calculated using Eq. (5.29) are somewhat higher for $\mathrm{Pr} = 0.1$ and somewhat lower for $\mathrm{Pr} = 0.01$ at intermediate De values, the agreement between the data and the correlation is fairly good, as shown in Fig. 5.5.

Ⓣ *Nusselt Numbers— Spiral Coils.* Kubair and Kuloor [24, 33] obtained the Nusselt numbers for two spirals that were enclosed in a steam chamber, using glycerol solutions. They suggested the following relation, which uses fluid properties calculated at an arithmetic-mean temperature:

$$\mathrm{Nu}_T = \left(1.98 + 1.8\frac{a}{R_{\mathrm{ave}}}\right)\mathrm{Gz}^{0.7} \tag{5.31}$$

The correlation application range is $9 \le \mathrm{Gz} < 1000$, $80 < \mathrm{Re} < 6000$, and $20 < \mathrm{Pr} < 100$. Although the correlation appears to be for the thermal entrance length, the fully developed Nusselt number may be calculated by substituting $\mathrm{Gz} = 20$. The correlation does not include the influence of geometric parameters such as coil pitch, number of turns, etc. This indicates a need for more work in this area to obtain a more general correlation.

(H1) *Nusselt Numbers— Helical Coils.* Manlapaz and Churchill [19] derived Eq. (5.32) by performing a regression analysis on the available Nusselt number results:

$$\mathrm{Nu}_{\mathrm{H1}} = \left[\left(4.364 + \frac{4.636}{x_3}\right)^3 + 1.816\left(\frac{\mathrm{De}}{x_4}\right)^{3/2}\right]^{1/3} \tag{5.32}$$

where

$$x_3 = \left(1.0 + \frac{1342}{\mathrm{De}^2 \mathrm{Pr}}\right)^2, \qquad x_4 = 1.0 + \frac{1.15}{\mathrm{Pr}} \tag{5.33}$$

Figure 5.6 compares Eq. (5.32) with some of the available theoretical predictions [30, 34] and the experimental Nusselt number data [14, 27]. The figure shows a fairly good agreement between the correlation and most of the available data.

(H2) *Nusselt Numbers— Helical Coils.* Tyagi and Sharma [8] and Manlapaz and Churchill [19] calculated Nusselt numbers for the (H2) boundary condition. Tyagi and Sharma's perturbation-analysis results demonstrate the negligible influence of viscous dissipation on Nusselt numbers for flows with $\mathrm{De} < 30$. However, their results are limited, since the data are restricted to flows with $\mathrm{De} < 30$. Numerical results of

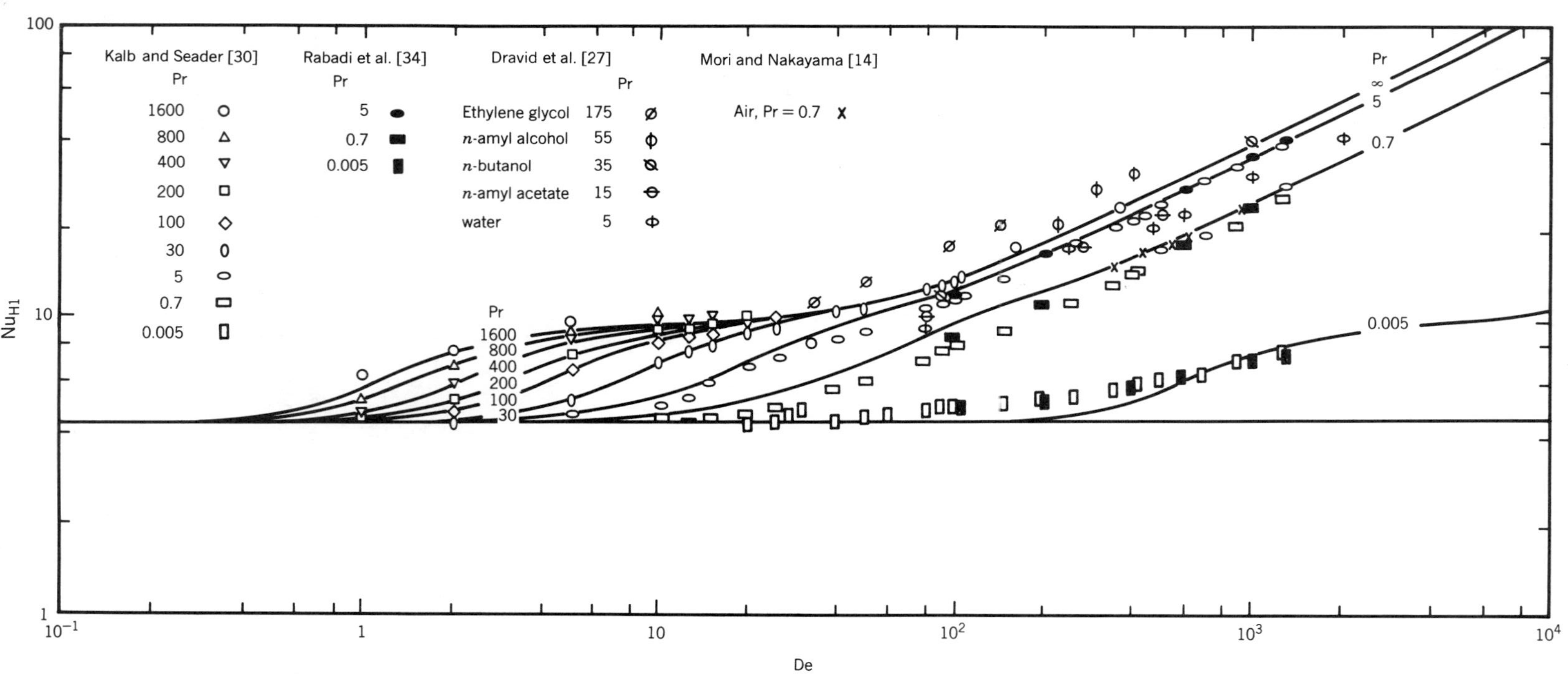

Figure 5.6. A comparison of the recommended design correlation [Eq. (5.32), drawn as solid lines] with theoretical and experimental Nusselt numbers for the (H1) boundary condition [19].

TABLE 5.1 Numerically Calculated Nu_{H2} for Helical Coils of Circular Cross Section [19]

					$Nu_{p,H2}$			
R/a	b/R	Re	De	He	Pr = 0.1	0.3162	1.0	10.0
5.0	0.0	9.196	4.113	4.113	4.642	4.639	4.633	4.620
	0.5	9.197	4.113	4.100	4.462	4.639	4.633	4.620
	1.0	9.194	4.112	4.061	4.462	4.640	4.634	4.621
5.0	0.0	46.70	20.88	20.88	4.769	4.759	4.936	8.447
	0.5	47.72	20.89	20.83	4.768	4.758	4.934	8.438
	1.0	46.79	20.93	20.67	4.765	4.755	4.929	8.414
10.0	0.0	392.6	124.14	124.14	5.604	7.541		
	0.5	393.0	124.29	123.90	5.602	7.535		
	1.0	394.4	124.72	123.17	5.596	7.518		
5.0	0.0	402.5	180.01	180.01	6.058	9.312		
	0.5	403.1	180.28	179.71	6.078	9.307		
	1.0	404.9	181.07	178.82	6.071	9.292		
10.0	0.0	1008	318.8	318.8	7.120	14.30		
	0.5	1009	319.1	318.1	7.114	14.27		
	1.0	1013	320.5	316.5	7.103	14.23		
5.0	0.0	1043	466.6	466.6	9.680			
	0.5	1045	467.4	465.9	9.600			
	1.0	1051	469.8	464.0	9.588			

Manlapaz and Churchill are presented in Table 5.1 which indicate almost no influence of the coil pitch on the Nusselt number. The (H2) results show qualitative agreement with the (H1) correlation, Eq. (5.32). Certainly, more results are required for the (H2) boundary condition.

Peripheral Variation in Nusselt Numbers. Several investigators have noted substantial peripheral variation in the fully developed Nusselt numbers for the (T) and (H1) boundary conditions. As shown in Fig. 5.7 for De = 898, $Nu_{p,H1}$ at the outer wall is about 11 times more than that for the straight-tube asymptotic $Nu_{H1} = 4.36$. At the inner wall, $Nu_{p,H1}$ is only about 2, half the straight-tube asymptotic value. The region of about $\pm 50°$ from the inner wall shows $Nu_{p,H1}$ values less than 4.36, while the rest of the 260° shows $Nu_{p,H1}$ values substantially higher than 4.36. This clearly shows why curved coils provide higher heat transfer rates than those for straight tubes.

Influence of Coil Pitch and Curvature Radius on Nusselt Numbers. To account for the coil-pitch effect on Nusselt numbers, Manlapaz and Churchill [19] recommend using the helical coil number He instead of the Dean number De in the correlations of Eqs. (5.29) and (5.32). However, their own predictions for the (H2) boundary condition indicate that the influence of the coil pitch ($b/R = 0$ to 1) on heat transfer is very small. Until more results are available to prove otherwise, the effect of the coil pitch on Nusselt numbers may be neglected.

For the (T) boundary condition, the nondependence of Nu_T on R/a has been shown for coils with $R/a > 10$ [32]. We compared Tarbell and Samuels's [26] Nu_T predictions for $R/a = 3$, 10, and 30 with the correlation of Eq. (5.29), which includes a

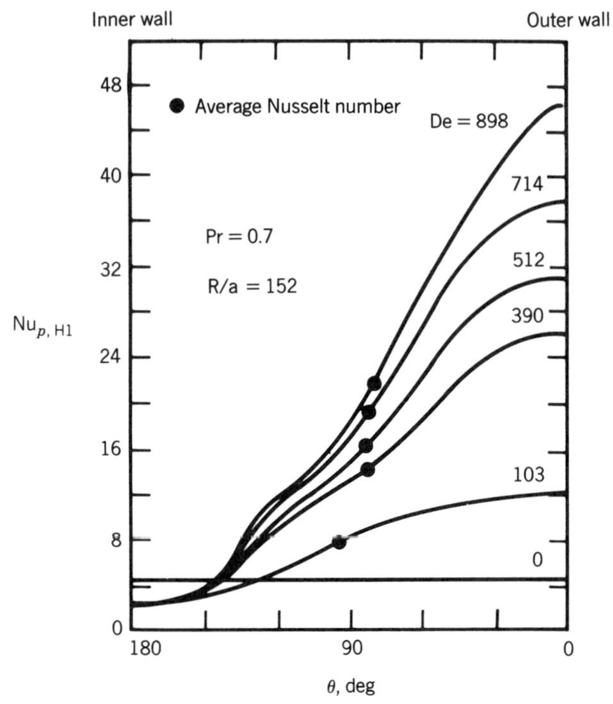

Figure 5.7. Peripheral variation of fully developed Nusselt number for the (H1) boundary condition for a curved circular tube [30].

De term, but not a separate R/a term. In spite of the absence of a separate R/a term, Eq. (5.29) correlated data of Tarbell and Samuels [26] within $\pm 12\%$, indicating that the De term alone is sufficient to account for the curvature effect as long as $R/a \geq 3$. Presently, results are not available to assess the influence of $R/a < 3$ on Nusselt numbers.

Similarly, for the (H1) boundary condition, the influence of R/a on $\mathrm{Nu}_{\mathrm{H1}}$ was found to be very small for coils with $R/a > 10$ [30]. Based on the influence for the (T) boundary condition, the influence of R/a on the (H1) Nusselt numbers is probably negligible for coils with $R/a \geq 3$. Therefore, the design correlation (5.32) may be used to predict $\mathrm{Nu}_{\mathrm{H1}}$ for coils with $R/a \geq 3$.

Influence of Boundary Conditions on Nusselt Numbers. Some theoretical calculations and experimental data show almost the same Nusselt numbers for the (H1) and (T) boundary conditions [25, 35]. As noted earlier, in a curved-tube flow, an increase in De enhances secondary flow, increases fluid mixing, and augments heat transfer. Similarly, an increase in Pr enhances thermal convection, increases fluid mixing, and augments heat transfer. This results in Nusselt numbers almost independent of the wall thermal boundary condition.

Mori and Nakayama's [14, 25] first theoretical approximation showed identical Nusselt numbers for (H1) and (T) boundary conditions. Janssen and Hoogendoorn [35] measured the same Nusselt numbers for (H1) and (T) boundary conditions using a glycerol-water mixture (Pr > 1). In fact, they recommend the same Nusselt number

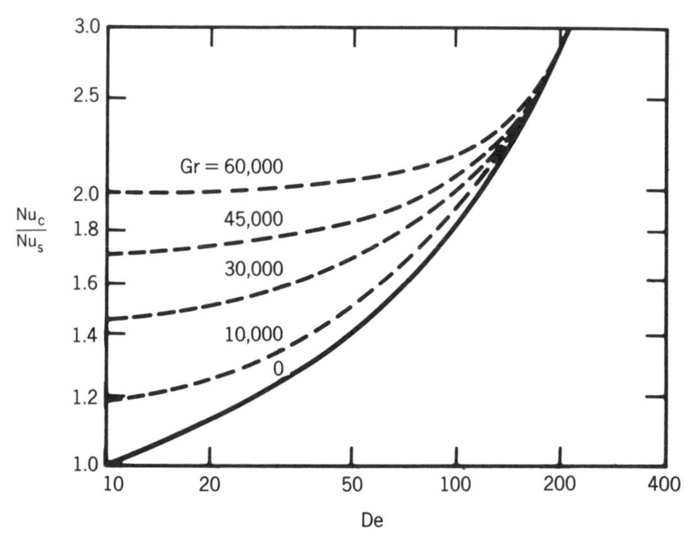

Figure 5.8. The ratio of the Nusselt number for a horizontal curved circular tube to that for a straight tube as a function of the Grashof and Dean numbers [39, 40].

correlations for either the Ⓣ or (H1) boundary condition. Similarly, Kalb and Seader's [30, 32] Nusselt numbers for (H1) and Ⓣ boundary conditions show less than 15% deviation from each other. In addition, the (H1) and Ⓣ Nusselt numbers calculated using the recommended design correlations, Eqs. (5.29) and (5.32), show less than $\pm 15\%$ deviation from each other for fluids with Pr = 0.01 to 100 and De = 100 to 1500. This demonstrates a very small influence of the wall thermal boundary condition on the curved-tube Nusselt numbers. However, no experimental data exist for liquid-metal flow in curved tubes.

The Influence of Variable Viscosity and Density. The results discussed so far have assumed constant fluid properties. The variation in temperature-dependent properties, especially viscosity and density, affects the heat transfer rates. The conventional Sieder-Tate [90] viscosity ratio, $(\mu_m/\mu_w)^{0.14}$, is used to take account of the change in heat transfer due to the viscosity variation [36, 37]. Since no specific data are available, for design purpose it is recommended to use Eqs. (5.29) and (5.32) with their right-hand sides multiplied by $(\mu_m/\mu_w)^{0.14}$.

The density variation results in a free convection superimposed on a forced convection, normally resulting in an augmented heat transfer rate. Theoretical studies have shown that the superimposed free convection influences coil heat transfer only if De < 150 [38–41]. The free convection in a horizontal coil shifts the point of maximum Nusselt number away from the coil outer wall toward the bottom. Similarly, in a vertical coil, the point of maximum Nusselt number shifts to the coil bottom [41]. Figure 5.8 shows the calculated increase in Nusselt numbers due to buoyancy at various Dean numbers [39, 40].

A more general experimental correlation is given by Abul-Hamayel and Bell [37] that accounts for the density and viscosity variations in coil-tube heat transfer. Water, ethylene glycol, and *n*-butyl alcohol in a coiled tube with the (H1) boundary condition were used to obtain the following correlation:

$$\mathrm{Nu_{H1}} = \left[4.36 + 2.84\left(\frac{\mathrm{Gr}'}{\mathrm{Re}^2}\right)^{3.94}\right]\left[1 + 0.9348\left(\frac{\mathrm{Gr}'}{\mathrm{De}^2}\right)^{2.78} x_5\right]$$
$$\times\left[1 + 0.0276\,\mathrm{De}^{0.75}\mathrm{Pr}^{0.197}\left(\frac{\mu_m}{\mu_w}\right)^{0.14}\right] \qquad (5.34)$$

where

$$x_5 = \exp\left(-\frac{1.33\,\mathrm{Gr}'}{\mathrm{De}^2}\right) \qquad (5.35)$$

This correlation is valid for $92 < \mathrm{Re} < 5500$, $2.2 < \mathrm{Pr} < 101$, and $760 < \mathrm{Gr}' < 10^6$. It is correct insofar as it reduces to the constant-property coil-tube correlation on neglecting viscosity and density variations. In addition, it reduces to the straight-tube forced convection Nusselt number value of 4.36 on neglecting the coil effect ($\mathrm{De} \to 0$). Equation (5.34) is recommended for design purposes when fluid properties are highly temperature-dependent.

5.3.2 Hydrodynamically Developing Flow

Austin and Seader [42] and others [43, 44] experimentally studied flow development in a curved channel. Only Austin and Seader [42] correlated their entry lengths in terms of the angle of tube curvature to attain "fully developed" velocity profiles for four coils ($R/a = 6.9$, 9.1, 14.4, and 24.1):

$$\phi = 49\left(\mathrm{De}\frac{a}{R}\right)^{0.33} \qquad \text{for} \quad 190 \le \mathrm{De} \le 950. \qquad (5.36)$$

The velocity profile was considered parabolic at the coil entry. In most cases, they found ϕ to be between 90 and 200°, indicating a very short entrance length.

Yao and Berger [45] calculated the curved-duct entrance length using the boundary-layer theory. The results from their proposed equation differ significantly from the experimental data [42, 43]. Experimental data of Agrawal et al. [43] show good agreement with Eq. (5.36), even though they employed an uniform coil inlet velocity profile. This indicates that the hydrodynamic entry length is probably unaffected by the coil inlet velocity profile.

5.3.3 Thermally Developing and Hydrodynamically Developed Flow

Ⓣ ***Nusselt Numbers — Helical Coils.*** The numerical calculations [26, 27, 35, 46] as well as experimental data [35] indicate the coil thermal entrance length to be 20% to 50% shorter than that observed in a straight tube. In addition, all investigators observed significant Nusselt number oscillations in the entrance region. As seen in Fig. 5.9, for fluids with $\mathrm{Pr} = 0.1$, the entrance region shows a typical Graetz solution similar to the one obtained in a straight tube [26]. However, at higher Pr, large oscillations in the Nusselt numbers are observed in the entrance region. In addition, as Pr increases, oscillations start at higher Gz values (i.e., lower values of x^*). The secondary flow causes these oscillations by exposing the tube wall alternately to the hot and cold fluids, resulting in the oscillatory entrance region Nusselt numbers [27].

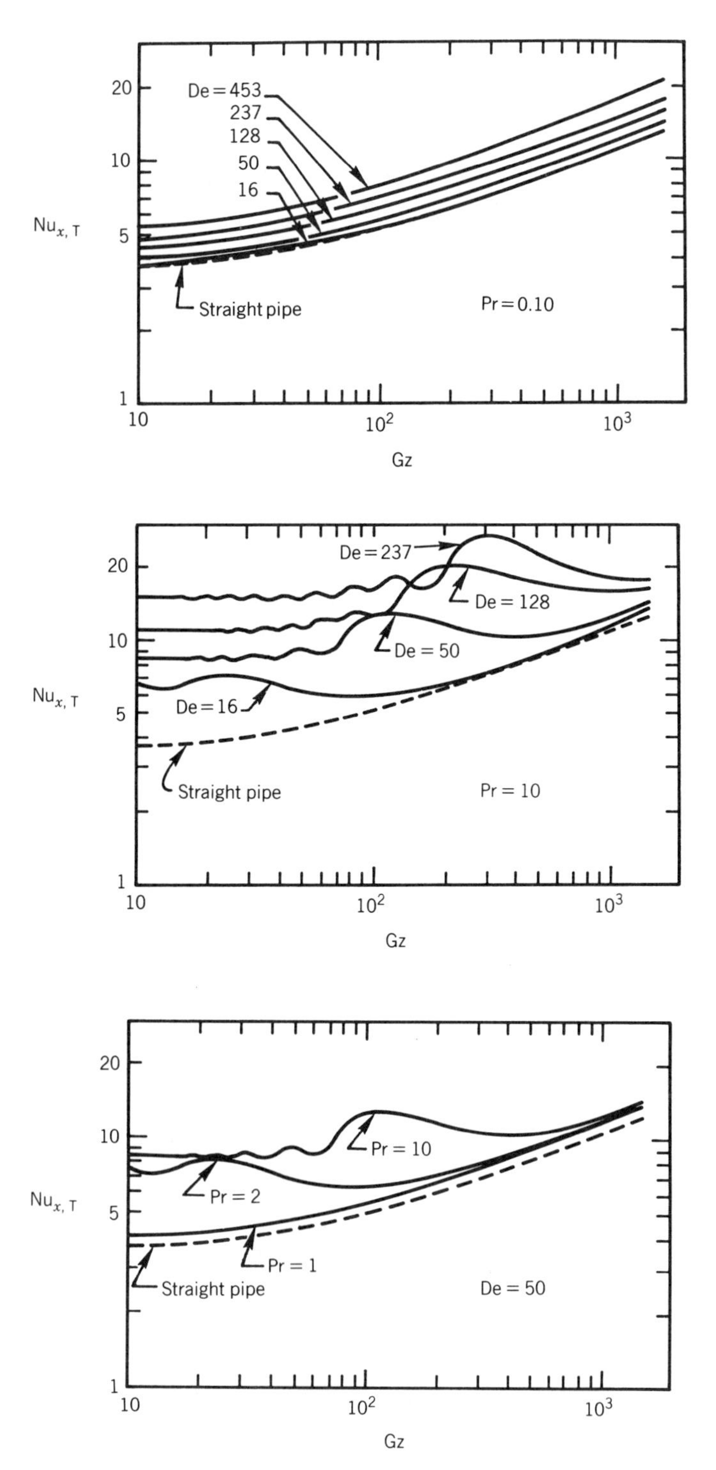

Figure 5.9. The influence of Prandtl number on the Nusselt numbers in the entrance region of a curved circular tube for the Ⓣ boundary condition [26].

Janssen and Hoogendoorn [35], using their numerical solution, proposed the following formula to calculate peripherally average thermal entrance region Nusselt numbers for $20 < \mathrm{Pr} < 450$ and $\mathrm{Re} < \mathrm{Re}_{\mathrm{crit}}$:

$$\mathrm{Nu}_{x,T} = \left(\frac{0.32 + 3(a/R)}{0.86 - 0.8(a/R)}\right)\mathrm{Re}^{0.5}\mathrm{Pr}^{0.33}\left(\frac{2a}{x}\right)^{x_6}\left(\frac{\mu_m}{\mu_w}\right)^{0.14} \tag{5.37}$$

where $x_6 = 0.14 + 0.8(a/R)$. The results of the above equation show fairly good agreement with experimental data [35]. It is important to note that Eq. (5.37) yields average values and does not account for the observed oscillations.

Ⓣ ***Nusselt Numbers — Spiral Coils.*** Kubair and Kuloor [24] conducted experiments with glycerol-water solutions and proposed Eq. (5.31) to calculate entrance region data by using R_{ave} instead of R in the Dean number. Their own data fit their correlation very well, but it may be coil-specific. More data are needed in this area.

(H1) ***Nusselt Numbers — Helical Coils.*** Several investigators have reported experimental data and theoretical predictions [27, 35, 47]. Similarly to the Ⓣ boundary condition, the entrance region shows oscillatory Nusselt numbers, and the oscillation intensity increases with Pr values. Janssen and Hoogendoorn [35] also proposed the following equation to calculate peripherally average thermal entrance region Nusselt numbers:

$$\mathrm{Nu}_{x,\mathrm{H1}} = \left(0.32 + 3\frac{a}{R}\right)\mathrm{Re}^{0.5}\mathrm{Pr}^{0.33}\left(\frac{2a}{x}\right)^{x_6} \tag{5.38}$$

where x_6 was defined just after Eq. (5.37). The above equation does not account for oscillations, but gives an average value. Janssen and Hoogendoorn [35] provided the following equation for the thermal entry length:

$$L_{\mathrm{th}}^* = \frac{L_{\mathrm{th}}}{D_h \mathrm{Re}\,\mathrm{Pr}} \le \frac{15.7\mathrm{Pr}^{-0.8}}{\mathrm{De}} \tag{5.39}$$

The above equation indicates that the thermal entry length is mainly determined by the secondary flow, rather than by the thermal diffusivity as in the straight tubes.

(H2) ***Nusselt Numbers — Helical Coils.*** Dravid et al. [27] calculated the thermal entrance length with the (H2) boundary condition. They also proposed an equation to determine the oscillation wavelength for the (H2) case. However, no specific $\mathrm{Nu}_{x,\mathrm{H2}}$ results are given.

5.3.4 Thermally and Hydrodynamically Developing Flow

Only a very limited number of results ($\mathrm{De} < 30$) are available for this case [48]. The lack of results is probably because of the very short entry lengths observed at high De and Pr values.

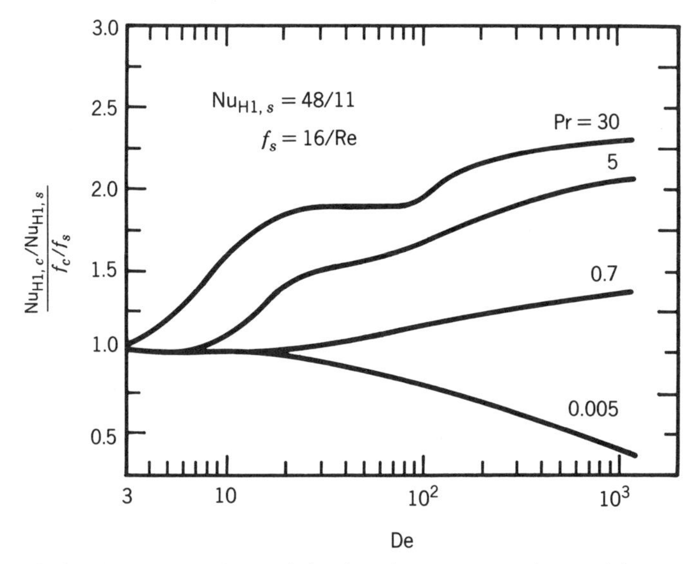

Figure 5.10. Effectiveness of curved circular tubes as compared to straight tubes [30].

5.3.5 Assessment for the Use of Coiled Tubes in Laminar Flow

The entrance region of a helical coil is about 20 to 50% shorter than that of a straight tube. Therefore, for most engineering applications, especially with De $\geq$ 200, the design can be based on fully developed values without significant errors.

The relative performance of a curved tube to a straight tube is shown in Fig. 5.10 in which the ratio $(\mathrm{Nu}_{H1,c}/\mathrm{Nu}_{H1,s})/(f_c/f_s)$ is plotted against the Dean number. A value of this ratio greater than 1 indicates better performance of coiled tubes than straight tubes. The figure shows that coiled tubes perform better than straight tubes for fluids with Pr $\geq$ 0.7; the performance of coiled tubes improves with increasing Pr. Thus, to augment tube-side heat transfer and to save space, a coiled tube is preferable. If tube-side fouling is significant, it may be necessary to devise a chemical cleaning method, since mechanical cleaning of a coilded tube is difficult.

5.4 TURBULENT FLOW THROUGH COILS OF CIRCULAR CROSS SECTION

Most of the turbulent fluid flow and heat transfer analyses are limited to fully developed flow. Limited data on turbulent developing flow indicate flow becoming fully developed within the first half turn of the coil [49], and probably even sooner, since the entrance length for developing turbulent flow is usually shorter than that for the laminar flow.

5.4.1 Fluid Flow

Velocity Profiles. The shape of the fully developed turbulent velocity profile is similar to that of a fully developed laminar flow, with the point of maximum velocity shifted toward the outer wall. Mori and Nakayama [50] calculated the velocity profile by employing a mixing-length concept and using the $\frac{1}{7}$-power-law velocity distribution corrected for the peripheral angle. Their calculated velocity profiles using the boundary-layer theory for coils with $R/a > 1$ are given as follows:

In the core region,

$$u_r = \frac{G\nu}{a}\cos\theta, \qquad u_\theta = -\frac{G\nu}{a}\sin\theta \tag{5.40}$$

$$u_x = \frac{H\nu}{a}\left(1 + \frac{0.4r\cos\theta}{a}\right) \tag{5.41}$$

In the boundary-layer region,

$$u_r = 0 \tag{5.42}$$

$$u_\theta = -\frac{G\nu}{a}\sin\theta\left[\frac{1}{3}\left(\frac{8}{\delta/a} - 1\right)\left(\frac{1-(r/a)}{\delta/a}\right) - \frac{4}{3}\left(\frac{2}{\delta/a} - 1\right)\left(\frac{1-(r/a)}{\delta/a}\right)^{1/7}\right] \tag{5.43}$$

$$u_x = \frac{\nu u_\delta}{a}\left(\frac{1-(r/a)}{\delta/a}\right)^{1/7} \tag{5.44}$$

where

$$G = 0.0855\mathrm{Re}^{0.8}\left(\frac{R}{a}\right)^{-0.1}, \qquad \frac{\delta}{a} = 0.5545\mathrm{Re}^{-0.2}\left(\frac{R}{a}\right)^{0.4} \tag{5.45}$$

$$u_\delta = H(1 + 0.7\cos\theta + 0.1\cos^2\theta + \cdots)^{4/7} \tag{5.46}$$

$$H = \frac{\mathrm{Re}}{2.0 - 0.5(\delta/a) + 0.133(\delta/a)^2} \tag{5.47}$$

The axial velocity profiles calculated from the above equations are in good agreement with Mori and Nakayama's own experimental measurements and other numerical results [7].

Friction Factors — Helical Coils. Several experimental and limited theoretical studies have proposed correlations to calculate friction factors for turbulent flow in a helical coil. Ito [29] proposed the following correlation:

$$f_c\left(\frac{R}{a}\right)^{0.5} = 0.00725 + 0.076\left[\mathrm{Re}\left(\frac{R}{a}\right)^{-2}\right]^{-0.25} \qquad \text{for} \quad 0.034 < \mathrm{Re}\left(\frac{R}{a}\right)^{-2} < 300 \tag{5.48}$$

Srinivasan et al. [17] obtained extensive friction factor data and proposed the following correlation:

$$f_c\left(\frac{R}{a}\right)^{0.5} = 0.084\left[\mathrm{Re}\left(\frac{R}{a}\right)^{-2}\right]^{-0.2} \quad \text{for} \quad \mathrm{Re}\left(\frac{R}{a}\right)^{-2} < 700 \text{ and } 7 < \frac{R}{a} < 104 \tag{5.49}$$

These correlations show fairly good agreement with each other. Hence, either could be used for design purposes. They also show good agreement (within $\pm 10\%$) with experimental data for air [51] and water [52], and with numerical predictions by Patankar et al. [7].

In contrast with laminar flow, the peripheral variation in the friction factor is not significant for turbulent flow. For Re = 25,000 to 135,000, the local friction factor is about 1.5 times more and 0.5 times less than that in a straight tube at the coil outer and inner wall, respectively [49]. Similarly to the laminar flow case, the coil pitch ($b/a = 0$ to 25.4) does not affect turbulent friction factors [20].

The correlations discussed so far are for a turbulent flow through a smooth tube. In practice, tube roughness may cause higher friction factors than those predicted using the smooth-tube correlations [53]. Presently, no explicit correlations are available to account for tube-roughness effects.

Rogers and Mayhew [52] measured nonisothermal friction factors for flow of water through the three helical coils, and recommended a nonisothermal friction factor correlation as

$$f_{\text{nonisothermal}} = f_{\text{isothermal}}\left(\frac{\mathrm{Pr}_m}{\mathrm{Pr}_w}\right)^{-0.33} \tag{5.50}$$

where $f_{\text{isothermal}}$ is calculated using Ito's correlation, Eq. (5.48).

Friction Factors— Spiral Coils. Kubair and Kuloor [22] measured friction factors in three spiral coils using water. Srinivasan et al. [17] also measured friction factors in five sprials for water and fuel-oil flow. Their experimental correlation is

$$f_c = \frac{0.0074\left(n_2^{0.9} - n_1^{0.9}\right)^{1.5}}{\left[\mathrm{Re}(b/a)^{0.5}\right]^{0.2}} \tag{5.51}$$

The above correlation is valid for $40{,}000 < \mathrm{Re}(b/a)^{0.5} < 150{,}000$ and $7.3 < b/a < 15.5$.

5.4.2 Heat Transfer

Aspects of heat transfer are further discussed in two parts: (1) temperature profiles, and (2) Nusselt numbers. It is important to recall here that in turbulent flow heat transfer, temperature profiles and Nusselt numbers are independent of the thermal boundary condition for $\mathrm{Pr} \geq 0.7$.

Temperature Profiles. Mori and Nakayama [50] and Hogg [49] reported theoretically calculated and experimentally measured temperature profiles. Mori and Nakayama employed boundary-layer idealizations for the calculations. Their calculated profiles, given below, were confirmed by their own experimental data for air flow through a curved tube subjected to the (H1) boundary condition:

In the core region,

$$\frac{T_w - T_m}{a(dT/dx)} = H\left(\frac{H'}{H} + \frac{0.2}{G}\left(\frac{r}{a}\right)^2 + \frac{r}{Ga}\cos\theta\right) \tag{5.52}$$

In the boundary layer,

$$\frac{T_w - T_m}{a(dT/dx)} = g_{1\delta}\left(\frac{1-(r/a)}{\delta/a}\right)^{1/7} \tag{5.53}$$

where

$$g_{1\delta} = H\left[\frac{H'}{H} + \frac{0.2}{G}\left(1-\frac{\delta}{a}\right)^2 + \frac{\cos\theta}{G}\left(1-\frac{\delta}{a}\right)\right] \tag{5.54}$$

$$H' = \frac{(\mathrm{Pr}^{0.67} - 0.16724)(10.64\mathrm{Re}^{0.2})(R/a)^{0.1}}{1 + 0.1875(\delta/a)} \tag{5.55}$$

and G, H, and δ/a are obtained from Eqs. (5.45) and (5.47). The above equations are applicable for gases with $\mathrm{Pr} \leq 1$. The validity of these equations for $\mathrm{Pr} > 1$ is not known.

Nusselt Numbers — Helical Coils. The measured local Nusselt numbers along the coil outer wall are about 1.5 times higher than those obtained in straight-tube flow. Conversely, at the coil inner wall, the local Nusselt numbers are about half those obtained in straight-tube flow [49]. Thus, in a helical-coil turbulent flow the mean Nusselt number is probably only 20 to 30% higher than in a straight tube.

About fifteen experimental and two theoretical correlations are available to calculate mean Nusselt numbers. We have compared Nusselt numbers calculated from these correlations for three values of Re ($\mathrm{Re} = 10^4$, 5.5×10^4, and 10^5) and four values of R/a ($R/a = 10$, 20, 50, and 100). The calculations were restricted to the range of application of each correlation. The calculated values show the curved-tube Nusselt numbers to be 10 to 30% higher than those observed in a straight tube. The results indicate some discrepancy in calculated Nusselt numbers for $\mathrm{Re} = 10^4$. However, for $\mathrm{Re} > 10^4$, the calculated values from various correlations exhibit only ± 10 percent deviation from each other. Thus, any correlation could be used for $\mathrm{Re} > 10^4$. Schmidt's correlation [54] has the largest application range and is as follows:

$$\frac{\mathrm{Nu}_c}{\mathrm{Nu}_s} = 1.0 + 3.6\left[1 - \left(\frac{a}{R}\right)\right]\left(\frac{a}{R}\right)^{0.8} \tag{5.56}$$

It is valid for $2 \times 10^4 < \mathrm{Re} < 1.5 \times 10^5$ and $5 < R/a < 84$. This correlation was developed using air and water in coils subjected to the (H1) boundary condition.

For low Reynolds numbers, Pratt's correlation [55] is recommended:

$$\frac{\mathrm{Nu}_c}{\mathrm{Nu}_s} = 1 + 3.4\frac{a}{R} \qquad \text{for} \quad 1.5 \times 10^3 < \mathrm{Re} < 2 \times 10^4 \tag{5.57}$$

This correlation is based on water and isopropyl alcohol. To include the influence of temperature-dependent properties, especially viscosity for liquids, Orlov and Tselishchev [56] recommend the following correlation due to Mikheev:

$$\frac{\mathrm{Nu}_c}{\mathrm{Nu}_s} = \left(1 + 3.54\frac{a}{R}\right)\left(\frac{\mathrm{Pr}_m}{\mathrm{Pr}_w}\right)^{0.25} \quad \text{for} \quad \frac{R}{a} > 6 \tag{5.58}$$

Note that the discrepancy introduced by the different constants (3.4 vs. 3.54) of Eqs. (5.57) and (5.58) is negligible and within the uncertainty of the correlations.

Nusselt Numbers— Spiral Coils. The only experimental data for spiral coils are reported by Orlov and Tselishchev [56]. They report that Mikheev's correlation, Eq. (5.58), represents their data within $\pm 15\%$ deviation when an average radius of curvature of a spiral was used in the correlation. This indicates that most helical coil correlations can be used for spiral coils if the average radius of curvature of the spiral, R_{ave}, is used in the correlations.

5.4.3 Assessment for the Use of Coiled Tubes in Turbulent Flow

A large number of experimental and theoretical correlations are available to predict friction factors and Nusselt numbers for a turbulent flow through a helical coil. The turbulent flow friction factors in coiled tubes are about 30 to 40% higher than those in straight tubes. For example, at $\mathrm{Re}(R/a)^{-2} = 10$ and 200, the turbulent flow friction factors in coiled tubes are about 15% and 28% higher than those in straight tubes, respectively. Calculated resutls also indicate that the increase in heat transfer over the straight-tube value due to the coil curvature is less than 30%; moreover, for $R/a > 20$, the heat transfer increase is less than 10%. Thus, other than space saving, a coiled tube does not offer any significant advantages over a straight tube for turbulent flow.

5.5 NON-NEWTONIAN FLUID FLOW THROUGH COILS OF CIRCULAR CROSS SECTION

Non-Newtonian fluids may be broadly classified as purely viscous fluids (inelastic) and viscoelastic fluids. Purely viscous fluids may be further categorized as time-dependent and time-independent. The time-independent purely viscous non-Newtonian fluids are further classified as those with and without a yield stress.

The power-law fluids are the most common non-Newtonian fluids without a yield stress. They exhibit the following power-law relationship between shear stress and shear rate:

$$\tau = -K'\left(\frac{\partial u}{\partial y}\right)^n \tag{5.59}$$

If $n < 1$, the fluid is called pseudoplastic; if $n > 1$, it is called dilatant; and if $n = 1$, it is the well-known Newtonian fluid. For further details on non-Newtonian fluids, refer to Chapter 20.

For a fully developed flow through a straight tube, the following relationships for the wall shear stress ($\tau_w = 2a\,\Delta P/4L$), wall shear rate, and shear-dependent viscosity

μ_n, are well known [57]:

$$\tau_w = K''\left(\frac{4u_m}{a}\right)^n, \qquad K'' = K'\left(\frac{3n+1}{4n}\right)^n \tag{5.60}$$

$$-\left(\frac{\partial u}{\partial y}\right)_w = \frac{3n+1}{4n}\left(\frac{4u_m}{a}\right) \tag{5.61}$$

$$\mu_n = \frac{\tau_w}{4u_m/a} = K''\left(\frac{4u_m}{a}\right)^{n-1} \tag{5.62}$$

Using the following generalized definition of the Reynolds number, it can be shown that the power-law fluids follow the well-known Fanning friction factor equation for the fully developed laminar flow of a Newtonian fluid in a circular tube:

$$\mathrm{Re}_{\mathrm{gen}} = \frac{2\rho a u_m}{\mu_n} = \frac{2^n \rho u_m^{2-n} a^n}{8^{n-1} K''} \tag{5.63}$$

$$f = \frac{16}{\mathrm{Re}_{\mathrm{gen}}} \tag{5.64}$$

The above results are for flow through a straight circular tube. In a curved circular tube, due to the secondary flow, the wall shear stress is expected to be higher. Similarly, at a given wall shear stress, the average fluid velocity in a curved tube is lower than that in a straight tube. Hence, the following equations are used to calculate an effective viscosity based on the curved-tube shear stress and corresponding generalized Reynolds number, for non-Newtonian fluid through a curved tube [58]:

$$\mu_{n,c} = \frac{\tau_{w,c}}{4u_{m,c}/a} = K''\left(\frac{4u_{m,c}}{a}\right)^{n-1} = K''\left(\frac{\tau_{w,c}}{K''}\right)^{(n-1)/n} \tag{5.65}$$

$$\mathrm{Re}_{\mathrm{gen},c} = \frac{2\rho a u_{m,c}}{\mu_{n,c}} \tag{5.66}$$

5.5.1 Fluid Flow

Velocity Profiles. The fully developed power-law velocity profile calculated using either a boundary-layer solution [59] or a numerical method [60] shows that decreasing n tends to flatten the velocity profile and dampen the secondary flow. This is consistent with the observation for the flow of a power-law fluid through a straight tube. For $n = 1.25$ to $n = 0.5$, the influence of n on the velocity profile is not significant, and the velocity profile can be represented by the Newtonian profile ($n = 1$) given in Eqs. (5.11) to (5.14).

Friction Factors — Laminar Flow. The friction factor for the flow of power-law fluids through a curved tube has been investigated for purely viscous (inelastic) and viscoelastic fluids. Some viscoelastic fluids reduce the drag, and consequently reduce the friction factor and pressure drop.

Helical Coils, Inelastic Power-Law Fluid — Friction Factors. Mashelkar and Devarajan [59] used a boundary-layer method to derive the following theoretical correlation:

$$f_c = (9.069 - 9.438n + 4.374n^2)\left(\frac{R}{a}\right)^{-0.5}(\mathrm{De_{gen}})^{0.122n-0.768} \quad (5.67)$$

This equation is valid for $70 \le \mathrm{De_{gen}} \le 400$, $10 \le R/a \le 100$, and $0.5 \le n \le 1$. Here $\mathrm{De_{gen}} = \mathrm{Re_{gen}}\,(R/a)^{-1/2}$. This correlation shows good agreement with Mashelkar and Devarajan's own experimental data [85] and other theoretical predictions [60, 61]. The only drawback of the correlation is that it does not reduce to the correlation for Newtonian fluid flow in a curved tube when n is set equal to unity.

Helical Coils, Viscoelastic Power-Law Fluid— Friction Factors. As noted earlier, for power-law fluids Eq. (5.59) describes a general relationship between shear stress and shear rate. Viscoelastic fluids exhibit an additional normal stress component described as

$$\tau_{11} - \tau_{22} = D_1\left(\frac{\partial u}{\partial y}\right)^{x_7} \quad (5.68)$$

where D_1 and x_7 are constants dependent upon the liquid, and τ_{11} and τ_{22} are normal stress components. Various polymer solutions such as polyacrylamide (PAA—Dow Chemical), and polyethelene oxide (PEO-WSR—Union Carbide) exhibit viscoelastic behavior.

Mashelkar and Devarajan [85] observed a limited drag reduction (about 5 to 10%) in the laminar region in their extensive experiments using eight different viscoelastic liquids in four different coils with $R/a = 7.4$, 22, 83, and 103. They arrived at the following correlation for viscoelastic fluids:

$$f_{c,\,\mathrm{viscoelastic}} = f_{c,\,\mathrm{inelastic}}(1 - 0.03923\mathrm{Wi}^{0.2488}) \quad (5.69)$$

where $f_{c,\,\mathrm{inelastic}}$ is obtained from Eq. (5.67), and the Wissenberg number Wi, defined as the ratio of elastic stress to viscous stress evaluated at the edge of the velocity boundary layer, is given by

$$\mathrm{Wi} = \left(\frac{\tau_{11} - \tau_{22}}{\tau_{12}}\right)_\delta \quad (5.70)$$

$$\mathrm{Wi} = \frac{D_1}{K'}\left(\frac{u_m}{a}\right)^{x_7-n}(\mathrm{De_{gen}})^{-(x_7-n)/(n+1)} \quad (5.71)$$

Spiral Coils, Inelastic Power-Law Fluid— Friction Factors. Rajasekharan et al. [62, 63] reported limited data on three spiral coils using carboxymethylcellulose (CMC), CPM, and sodium silicate solutions with n varying from 0.47 to 2.13. They [63] correlated their data on 0.3% CMC solution ($n = 0.9$) as

$$f_c = 0.02936\exp\left(-123.1\frac{a}{R_{\mathrm{ave}}}\right)(\mathrm{Re_{gen}})^{-0.008431(R_{\mathrm{ave}}/a)^{1.419}} \quad (5.72)$$

This correlation does not include the influence of the power-law index, coil length, coil pitch, and number of turns. Therefore, it should be used with caution.

Critical Reynolds Number— Helical coil. Using data of Mujawar and Raja Rao [64], the critical Reynolds number for power-law fluids can be correlated as

$$\mathrm{Re}_{\mathrm{gen,crit}} \approx \left[2100 D_2 \left(\frac{R}{a}\right)^{-x_8}\right]^{1/(1-x_9)} \qquad \text{for} \quad 0.7 < n < 1 \tag{5.73}$$

where $D_2 = 47.969 - 153.8n + 166.22n^2 - 60.132n^3$, and $x_8 = x_9/2 = 0.4375n - 0.2575$.

Fully Developed Turbulent Flow. Turbulent flow data for power-law fluid flow through a helical coil are reported by Mishra and Gupta [58] and by Rajasekharan et al. [62, 65]. The correlations proposed by these authors are:

Mishra and Gupta [58]:

$$f_c = \frac{0.079}{(\mathrm{Re}_{\mathrm{gen},c})^{0.25}} + 0.0075\left(\frac{R}{a}\right)^{-0.5} \tag{5.74}$$

valid for $0.02 < [0.079/(\mathrm{Re}_{\mathrm{gen},c})^{0.25}](R/a)^{0.5} < 0.04$ and $9 < R/a < 25$.

Rajasekharan et al. [65]:

$$f_c = \frac{0.079}{n^5(\mathrm{Re}_{\mathrm{gen}})^{2.63/(10.5)^n}} + 0.012\left(\frac{R}{a}\right)^{-0.5} \tag{5.75}$$

valid for $6000 < \mathrm{Re}_{\mathrm{gen}} < 30{,}000$ and $10 \le R/a \le 27$.

If $n = 1$, Eqs. (5.74) and (5.75) are almost identical, reducing to a turbulent friction factor for a coil [see Eqs. (5.48) and (5.49)]. The difference between the two correlations is mainly due to different definitions for the viscosity used in the calculation of $\mathrm{Re}_{\mathrm{gen}}$ and $\mathrm{Re}_{\mathrm{gen},c}$. Therefore, either equation could be used for design purposes.

5.5.2 Heat Transfer

Only three heat transfer studies, one numerical [60] and two experimental [62, 66], are reported for laminar flow of a non-Newtonian, power-law fluid through curved ducts.

Fully Developed Flow. Hsu and Patankar [60, 61] have numerically computed fully developed Nusselt numbers for the (H1) boundary condition for a power-law fluid. Their results are presented in Table 5.2 for various power-law indices and Prandtl numbers. In general, the power-law fluids with $n > 1$ exhibit higher Nusselt numbers than the Newtonian fluids. Conversely, power-law fluids with $n < 1$ exhibit lower Nusselt numbers than the Newtonian fluids. Presently, no fully developed results are available for the (T) boundary condition. However, as pointed out in Sec. 5.3.1, the wall thermal boundary condition does not significantly affect Nusselt numbers. Therefore, (H1) results may be used for the (T) boundary condition without introducing significant errors.

Thermally Developing and Hydrodynamically Developed Flow. Oliver and Asghar [66] have correlated their (T) boundary-condition experimental data with various

TABLE 5.2 Ratio of (H1) Nusselt Numbers for a Helical Coil to That for a Straight Tube for a Non-Newtonian Power-Law Fluid [6][a]

n	Pr	$De_{gen} = 50$	100	200	400	700	1000
0.5	1	1.127	1.378	1.741	2.250	2.803	3.203
	5	1.604	2.150	2.970	4.341	5.984	7.459
	10	1.886	2.529	3.561	5.268	7.459	9.546
	20	2.170	3.013	4.320	6.490	9.377	12.24
	50	2.760	3.962	5.795	8.913	13.32	17.09
	100	3.414	4.994	7.417	11.74	17.38	21.28
	200	4.287	6.385	9.672	15.64	21.49	24.44
	500	5.900	8.955	14.20	21.07	25.29	27.18
	1000	7.544	11.84	18.56	24.44	27.18	28.24
0.75	1	1.279	1.650	2.204	3.040	3.980	4.731
	5	1.864	2.466	3.488	5.043	6.821	8.420
	10	2.139	2.910	4.177	6.098	8.398	10.50
	20	2.533	3.555	5.132	7.576	10.62	13.53
	50	3.333	4.755	6.976	10.53	15.22	19.17
	100	4.021	6.043	8.954	13.93	19.75	23.32
	200	5.162	7.732	11.73	18.37	24.00	26.66
	500	7.310	10.93	17.17	23.99	27.55	29.10
	1000	9.100	14.62	21.77	27.11	29.33	29.99
1	1	1.453	1.920	2.635	3.712	4.835	5.752
	5	2.062	2.773	3.964	5.660	7.562	9.166
	10	2.429	3.346	4.812	6.943	9.418	11.60
	20	2.956	4.148	6.004	8.753	12.10	15.17
	50	3.941	5.637	8.226	12.31	17.48	21.33
	100	4.995	7.172	10.61	16.36	22.23	25.44
	200	6.257	9.189	14.02	21.15	26.12	28.41
	500	8.753	13.18	20.21	26.35	29.33	30.48
	1000	11.34	17.64	24.75	28.87	30.71	31.16
1.25	1	1.658	2.154	2.971	4.187	5.403	6.362
	5	2.295	3.134	4.468	6.362	8.467	10.20
	10	2.760	3.859	5.544	7.953	10.69	13.12
	20	3.438	4.842	6.970	10.10	13.96	17.33
	50	4.655	6.596	9.614	14.43	20.09	23.62
	100	5.894	8.421	12.51	19.13	24.79	27.37
	200	7.387	10.88	16.72	23.86	28.07	29.94
	500	10.41	15.86	23.23	28.54	30.64	31.58
	1000	13.71	20.89	27.13	30.41	31.81	32.05

[a] $Nu_{H1,s}$ = 4.746, 4.501, 4.364, and 4.275 for n = 0.5, 0.75, 1, and 1.25, respectively; n = power-law index.

power-law, viscoelastic polyacrylamide solutions ($0.4 < n < 1$) in different coils (R/a = 12.7, 16.8, 26.7, 32, and 43) as

$$Nu_{x,T} = 1.75\left(\frac{3n+1}{4n}\right)^{0.33} Gz^{0.33}\left(1 + 0.36\, De_{gen}^{0.25}\right)\left(\frac{K'_m}{K'_w}\right)^{0.14} \quad \text{for} \quad 4 < De_{gen} \leq 60 \qquad (5.76)$$

and

$$\mathrm{Nu}_{x,T} = 1.75\left(\frac{3n+1}{4n}\right)^{0.33} \mathrm{Gz}^{0.33}\left(1 + 0.118\mathrm{De}_{\mathrm{gen}}^{0.5}\right)\left(\frac{K'_m}{K'_w}\right)^{0.14}$$

$$\text{for} \quad 60 \leq \mathrm{De}_{\mathrm{gen}} < 2000 \qquad (5.77)$$

Equations (5.76) and (5.77) take account of the change in the consistency index K' due to the temperature variation. In addition, they have correct form in that as $\mathrm{De}_{\mathrm{gen}} \to 0$, they reduce to the classical constant-property, non-Newtonian Pigford solution for the thermal entrance region in a straight tube [91]. Moreover, they further reduce to the classical Lévêque-type constant-property Newtonian thermal entrance length solution if $n = 1$.

Equations (5.76) and (5.77) are for power-law viscoelastic fluids. As noted earlier, the friction factors for a viscoelastic fluid in laminar flow are not significantly different from those for inelastic fluids. Similarly, it is conjectured that laminar flow Nusselt numbers for power-law inelastic or viscoelastic fluids may be the same. Therefore, the correlation of Eqs. (5.76) and (5.77) may be used as design equations for laminar flow of power-law fluids until specific experimental data/correlations are available.

5.6 FLOW THROUGH COILS OF NONCIRCULAR CROSS SECTION

Analyses and experiments are reported in the literature for curved coils having square, rectangular, elliptical [82], and concentric annular [83, 84, 92] cross sections. However, most of the usable data for a practicing engineer are restricted to curved helical coils with square cross section. Results are summarized here for square, rectangular, and parallel-plate cross sections. All ducts are curved coils unless noted otherwise.

5.6.1 Fully Developed Laminar Flow

Velocity Profiles— Helical Coils. Available results for elliptical curved coils and annuli are limited to De < 30, and hence are not reproduced here. In the case of square cross-section coils, various investigators note that up to De = 100, the secondary flow velocity profile shows two vortices similar to those observed in a circular cross-section helical coil. However, for De > 100, four vortices appear in a square cross-section coil, and these additional vortices reduce the sharpness of the velocity profile and also reduce the peak velocities. These extra vortices vanish at about De = 500. The exact value of De for their appearances and disappearances depends on the channel aspect ratio [67].

Mori et al. [68] have obtained a boundary-layer solution (De > 100) for flow through a curved square channel. Their calculated velocity profiles show a good agreement with their own experimental data for air. The general pattern of their velocity profile is very much similar to that observed in a circular cross-section coil. The effect of additional vortices between De = 100 and 500 is not significantly noticeable in the velocity profile of Mori et al. To calculate the velocity profile, they divided the channel boundary layer into three regions along three walls as shown in Fig. 5.11. Their velocity profile is given next. The second subscript (i, ii, or iii) for the velocity components indicates the boundary-layer regions shown in Fig. 5.11:

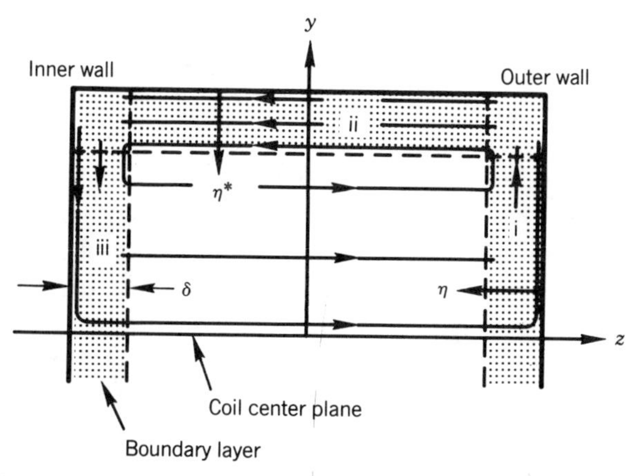

Figure 5.11. Boundary-layer regions in a curved square duct [68].

In a core region,

$$u_x = u_m\left[1.0 + \frac{\lambda}{D_3}\left(0.5\frac{z}{a}\right)\right] \tag{5.78}$$

$$u_z = u_m C_1, \qquad u_y = 0 \tag{5.79}$$

In the boundary layer,

$$\frac{u_{x,\,\mathrm{i}}}{u_m} = \left[\frac{4a}{\delta}\left(1.0 + \frac{\lambda}{2D_3}\right) - \frac{\lambda}{D_3}\right]\eta - \left(1.0 + \frac{\lambda}{2D_3}\right)\frac{4a^2\eta^2}{\delta^2} \tag{5.80}$$

$$\frac{u_{z,\,\mathrm{i}}}{u_m} = \left(\frac{\eta^2}{\delta} - \frac{8a\eta^3}{3\delta^2} + \frac{2a^2\eta^4}{\delta^3}\right)\frac{24a^2D_3}{\delta} \tag{5.81}$$

$$\frac{u_{y,\,\mathrm{i}}}{u_m} = \left(\frac{\eta}{\delta} - \frac{4a\eta^2}{\delta^2} + \frac{4a^2\eta^3}{\delta^3}\right)\frac{24ayD_3}{\delta} \tag{5.82}$$

$$\frac{u_{x,\,\mathrm{ii}}}{u_m} = \left(1.0 + \frac{\lambda z}{2D_3 a}\right)\left[\frac{4a\eta^*}{\delta} - \left(\frac{2a\eta^*}{\delta}\right)^2\right] \tag{5.83}$$

$$\frac{u_{z,\,\mathrm{ii}}}{u_m} = \frac{4a^2D_3}{\delta}\left(\frac{n^*}{\delta}\right)\left(\frac{3\delta}{a} - 6\right) + \frac{8a^3D_3}{\delta}\left(\frac{\eta^*}{\delta}\right)^2\left(12 - \frac{9\delta}{2a}\right)$$
$$+ \frac{16a^4D_3}{\delta}\left(\frac{2\delta}{a} - 6\right)\left(\frac{\eta^*}{\delta}\right)^3 \tag{5.84}$$

$$u_{y,\,\mathrm{ii}} = 0 \tag{5.85}$$

$$\frac{u_{x,\,\mathrm{iii}}}{u_m} = \left[\frac{\lambda}{D_3} + \frac{4a}{\delta}\left(1.0 - \frac{\lambda}{2D_3}\right)\right]\eta - \left(1.0 - \frac{\lambda}{2D_3}\right)\frac{4a^2\eta^2}{\delta^2} \tag{5.86}$$

$$u_{z,\,\mathrm{iii}} = u_{z,\,\mathrm{i}} \quad \text{and} \quad u_{y,\,\mathrm{iii}} = -u_{y,\,\mathrm{i}} \tag{5.87}$$

where

$$D_3 = \frac{1.541}{\mathrm{Re}}(1.41\mathrm{De})^{0.5} \qquad \lambda = \frac{2.668}{\mathrm{Re}}(1.41\mathrm{De})^{0.5}$$

$$\delta = 5.99a(1.41\mathrm{De})^{-0.5}$$

$$\eta = 0.5 - 0.5\frac{|z|}{a} \qquad \eta^* = 0.5 - 0.5\frac{|y|}{a} \tag{5.88}$$

Equations (5.78) to (5.88) are recommended for the calculation of velocity profiles in a curved square channel for fully developed laminar Newtonian fluid flow with De > 100. At present, no simple correlations are available to calculate the influence of the duct aspect ratio on velocity profiles.

Friction Factors — Square Cross-Section Helical Coils. Based on the comparison of various theoretical [67–69] and experimental [70] results, the following correlations, Eqs. (5.89)–(5.91), seem to give the best fit of the available experimental data and theoretical predictions. These equations were obtained from Refs. 67, 69, and 68, respectively; however, their application range has been modified to obtain the best fit of the available data:

$$\frac{(f\,\mathrm{Re})_c}{(f\,\mathrm{Re})_s} = 0.1520\mathrm{De}^{0.5}(1.0 - 0.216\mathrm{De}^{0.5} + 0.473\mathrm{De}^{-1} + 111.6\mathrm{De}^{-1.5} - 256.1\mathrm{De}^{-2}) \quad \text{for} \quad \mathrm{De} < 100 \tag{5.89}$$

$$\frac{(f\,\mathrm{Re})_c}{(f\,\mathrm{Re})_s} = 0.2576\mathrm{De}^{0.39} \quad \text{for} \quad 100 < \mathrm{De} < 1500 \tag{5.90}$$

$$\frac{(f\,\mathrm{Re})_c}{(f\,\mathrm{Re})_s} = 0.1115\mathrm{De}^{0.5} \quad \text{for} \quad \mathrm{De} > 1500 \tag{5.91}$$

The influence of the coil pitch on the friction factor has been found to be negligible [71, 72]. As expected, at high pitch values ($b/a = 200$ to 600), the coil pitch shows a little influence on friction factors [72].

Friction Factors — Rectangular Cross-Section Helical Coils. Cheng et al. [67] have correlated their numerical results for the aspect ratio $\alpha^* = 0.5$, 1, 2, and 5 as

$$\frac{f_c}{f_s} = C_0\mathrm{De}^{*0.5}\left(1.0 + \frac{C_1}{\mathrm{De}^{*0.5}} + \frac{C_2}{\mathrm{De}^*} + \frac{C_3}{\mathrm{De}^{*1.5}} + \frac{C_4}{\mathrm{De}^{*2}}\right) \quad \text{for} \quad \mathrm{De}^* \le 700 \text{ or } \mathrm{De} \le 435 \tag{5.92}$$

where C_0, C_1, C_2, C_3, and C_4 are constants listed in Table 5.3. Here $\mathrm{De}^* = \mathrm{Re}(D_h/R)^{1/2}$. The results of this equation are in fairly good agreement with the other numerical results [69].

Temperature Profiles — Helical Coils. Mori et al. [68] have obtained temperature profiles in a square cross-sectional channel using boundary-layer approximations. Their

TABLE 5.3 Laminar Flow in a Helical Coil of Rectangular Cross Section: Constants for Eq. (5.92) [67]

α^*	C_0	C_1	C_2	C_3	C_4
0.5	0.0974	4.366	−13.56	131.8	−182.6
1.0	0.1278	−0.257	0.699	187.7	−512.2
2.0	0.2736	−24.79	325.2	−1591.0	2728.0
5.0	0.0805	−5.218	104.4	−202.8	0.0

solution for the (H1) boundary condition and $\Pr < 1$ is given below. The subscripts i, ii, and iii in the solutions correspond to the regions depicted in Fig. 5.11.

In a core region,

$$T_m = D_4 + \frac{1}{D_3}\left(\frac{0.5z}{a}\right) + \frac{0.5\lambda}{D_3^2}\left(\frac{0.5z}{a}\right)^2 \tag{5.93}$$

In the boundary layer,

$$\begin{aligned} T_{\mathrm{i}} &= 0.5T_m\left(1.0 - \frac{\delta_t}{a}\right)\left[\frac{4a\eta}{\delta_t} - \left(\frac{2a\eta}{\delta_t}\right)^2\right] \\ &\quad + \frac{0.5\delta_t}{aD_3}\left[1.0 + \frac{0.5\lambda}{D_3}\left(1.0 - \frac{\delta_t}{a}\right)\right]\left[\left(\frac{2a\eta}{\delta_t}\right)^2 - \left(\frac{2a\eta}{\delta_t}\right)^3\right] \end{aligned} \tag{5.94}$$

$$T_{\mathrm{ii}} = T_m\left[\frac{4a\eta^*}{\delta_t} - \left(\frac{2a\eta^*}{\delta_t}\right)^2\right] \tag{5.95}$$

$$\begin{aligned} T_{\mathrm{iii}} &= 0.5T_m\left(\frac{\delta_t}{a} - 1.0\right)\left[\frac{4a\eta}{\delta_t} - \left(\frac{2a\eta}{\delta_t}\right)^2\right] \\ &\quad - \frac{0.5\delta_t}{a}\left[\frac{1}{D_3} - \frac{0.5\lambda}{D_3^2}\left(1.0 - \frac{\delta_t}{a}\right)\right]\left[\left(\frac{2a\eta}{\delta_t}\right)^2 - \left(\frac{2a\eta}{\delta_t}\right)^3\right] \end{aligned} \tag{5.96}$$

where

$$D_4 = 0.225\,\mathrm{Re}\,(1.414\,\mathrm{De})^{-0.5} \quad \text{and} \quad \delta_t = 0.851\frac{\delta}{\Pr} \quad \text{for } \Pr < 1 \tag{5.97a}$$

$$D_4 = [(0.375\Pr\delta_t/\delta) - 0.0937]\mathrm{Re}(1.414\,\mathrm{De})^{-0.5} \quad \text{for } \Pr > 1 \tag{5.97b}$$

and D_3, λ, δ, η, and η^* are defined in Eq. (5.88).

The above equations are in fairly good agreement with the experimental data for air [68]. Therefore, Eqs. (5.93) through (5.97a) are recommended for the (H1) boundary condition when $\mathrm{De} > 100$, $\Pr < 1$, and $R/a > 1$. As discussed earlier, it is important to note that Eqs. (5.93)–(5.96) and Eq. (5.97b) give an approximate temperature profile for fluids with $\Pr > 1$ [89]. In Eq. (5.97b) δ_t/δ values are 1.0, 0.739, 0.441, 0.262, 0.253, and 0.250 for $\Pr = 0.851$, 1, 3, 10, 30, and ∞.

Nusselt Numbers — Square Cross-Section Helical Coils. Several theoretical correlations are available to calculate Nusselt numbers in square cross-section coils. The following correlation by Cheng et al. [73] is particularly recommended:

$$\mathrm{Nu}_{\mathrm{H2}} = \mathrm{Nu}_{\mathrm{T}} = 0.152 + 0.627(1.414\,\mathrm{De})^{0.5}\,\mathrm{Pr}^{0.25} \tag{5.98}$$

which is valid for $0.7 \le \mathrm{Pr} \le 5$ and $20 \le \mathrm{De} \le 705$. This correlation shows a good agreement with the experimental data for air for the (H1) boundary condition [68]. In addition, it also represents the (T) boundary condition predictions very well [74].

As noted earlier, the influence of the wall thermal boundary condition on the coil Nusselt number is not significant for $\mathrm{Pr} \ge 0.7$. Therefore, the above equations could be used for (H1), (H2), and (T) boundary conditions. Beyond the range of the parameters for the above correlation, the use of the appropriate correlation for circular cross-section coiled tubes is recommended with the substitution of the appropriate hydraulic diameter for $2a$.

The influence of free convection on heat transfer and friction factor in a square channel with the (H2) boundary condition is reported by Akiyama et al. [75]. They recommend the following correlation to take account of natural convection combined with forced convection:

$$\mathrm{Nu}_{\mathrm{H2}} = (1.0 + 0.275\,\mathrm{De}^*\,\mathrm{Gr}^{-0.5})\mathrm{Nu}_s \tag{5.99}$$

where Nu_s is for natural convection in a straight horizontal duct with a square cross section [75]:

$$\mathrm{Nu}_s = 0.525(16\,\mathrm{Gr})^{0.175} \tag{5.100}$$

Nusselt Numbers — RectangularCross – Section Helical Coils. Butuzov et al. [76] determined experimentally the Nusselt numbers for laminar flow of water through three curved rectangular coils ($\alpha^* = 0.45, 0.67$, and 1)† subjected to the (H1) boundary condition. Their Nusselt numbers for De = 100 to 1000 exhibit a very little influence of the duct aspect ratio. They found that their experimental Nusselt numbers for rectangular cross-section coils were 17% lower than those for circular cross-section coils of the same hydraulic diameter. This contradicts the Nusselt number values obtained in straight ducts of circular and rectangular cross sections. Thus, it appears that the correlations for circular tubes with the appropriate hydraulic diameter may be used in laminar flow for rectangular ducts without introducing significant errors until more definite results are available.

5.6.2 Hydrodynamically Developing Laminar Flow

The only solution available is for the parallel-plate curved channel. So [77] formulated correlations for $R/a > 2.5$ for the entry lengths and friction factors based on his

† Butuzov et al. [76] did not clearly define the duct aspect ratio in relation to the coil curvature.

numerical results; here $2a$ is the distance between parallel plates:

$$\left(\frac{L_{\text{hy}}}{a}\right)_{\text{inner wall}} = 0.026\left(1 + 1.153\frac{a}{R}\right)\text{Re} \tag{5.101}$$

$$\left(\frac{L_{\text{hy}}}{a}\right)_{\text{outer wall}} = 0.026\left(1 + 3.153\frac{a}{R}\right)\text{Re} \tag{5.102}$$

$$\left(\frac{f_c}{f_s}\right)_{\text{inner wall}} = 1.0 + \frac{a}{R} + \frac{0.626a\,\text{Re}}{24L}\left(1.0 - \frac{1.51a}{R}\right) \tag{5.103}$$

$$\left(\frac{f_c}{f_s}\right)_{\text{outer wall}} = 1.0 + \frac{a}{R} + \frac{0.626a\,\text{Re}}{L}\left(1.0 - \frac{2.785a}{R}\right) \tag{5.104}$$

5.6.3 Thermally Developing and Hydrodynamically Developed Laminar Flow

Numerical solutions for thermally developing flow in a curved square duct and rectangular ducts are given by several authors [73–75]. The results of Cheng et al. [73] for the Ⓣ boundary condition are plotted in Fig. 5.12. The figure demonstrates the oscillatory Nusselt number behavior in the entrance region, very similar that found in a circular curved tube. Similar oscillatory trends are also observed for the (H1) boundary condition. The results indicate that the thermal entrance length $L^*_{\text{th}} = 0.02$ which is about 2 to 3 times shorter than that for a straight duct.

5.6.4 Thermally and Hydrodynamically Developing Laminar Flow

Helical Coils. The experimental data for thermally and hydrodynamically developing flow are reported by Shchukin and Filin [78] for a short square duct with the (H1) boundary condition and with water as the test fluid. Their correlations of data seem to assume free convection alone controlling coiled-tube heat transfer up to De = 800. This is contrary to most of the other results. Thus no reliable test data are available for simultaneously developing laminar flow in a curved square coil or a curved coil of any other noncircular cross section.

Spiral Plate Heat Exchangers. Normally, a spiral plate heat exchanger consists of two spiral channels with rectangular cross section. Two different fluids flow through the two channels. In laminar flow, Coons et al. [79] found the measured heat transfer coefficients to be only 35% greater than those in a straight duct. They found such limited improvement because they used a wrong equation to calculate straight-duct Nusselt numbers. Using a correct straight-tube equation, the calculated enhancements over straight-tube Nusselt numbers turn out to be of the order of 60 to 70%. Coon et al. proposed the following equation for Nusselt numbers, based on the arithmetic mean temperature difference, to correlate their viscous-oil heating data:

$$\text{Nu}_{am} = 8.4\text{Gz}^{0.2} \tag{5.105}$$

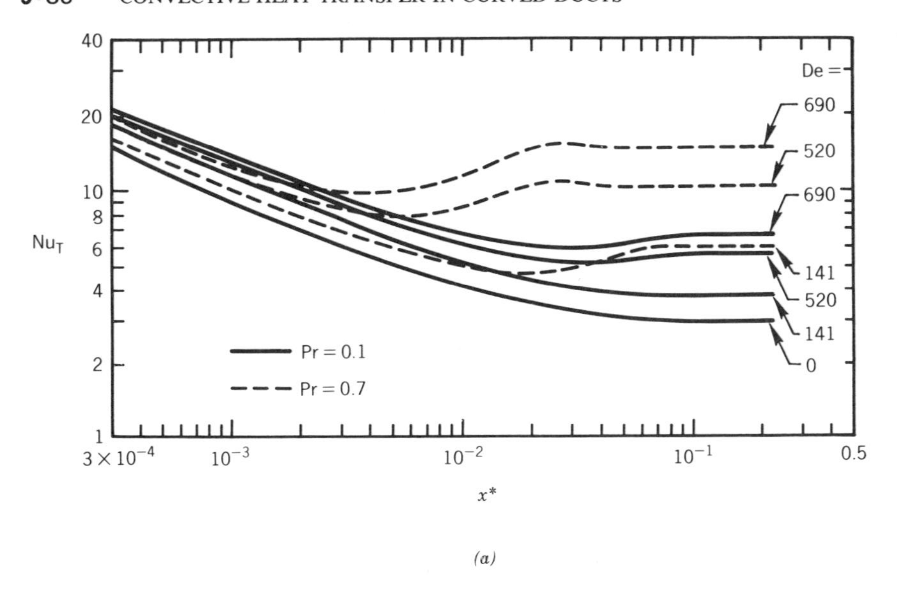

(a)

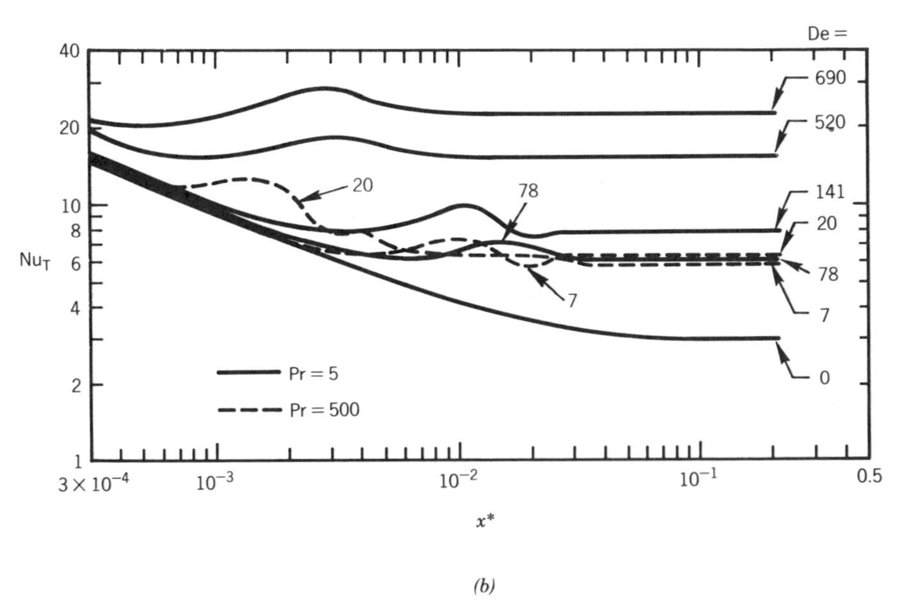

(b)

Figure 5.12. The influence of Dean number on the local Nusselt numbers for the Ⓣ boundary condition for a curved square tube [73].

5.6.5 Turbulent Flow

Friction Factors. Kadambi's [80] air friction factor data for Re > 8000 for two curved rectangular ducts are well predicted by a circular-tube correlation [such as Eq. (5.48)] when the hydraulic diameter of the rectangular tube is used. However, for Re < 8000 the friction factors for a curved rectangular duct were higher than those for the curved circular tube. Higher friction factors were also observed by Butuzov et al.

[76]. Their experiments included two rectangular ducts and a square duct with water and Freon as working fluids. They have correlated their extensive test results as [80]

$$\frac{f_c}{f_s} = 0.435 \times 10^{-3} \mathrm{Re}^{*0.96} \left(\frac{R}{d^*} \right)^{0.22} \tag{5.106}$$

where d^* represents the short channel side and is used as a characteristic dimension in Re^*. In the above equation, f_s represents the friction factor in a straight duct with the same aspect ratio as that of a curved coil. The application range for the correlation is given as $450 \leq \mathrm{Re}^*(d^*/R)^{0.5} \leq 7500$ and $25 \leq R/d^* \leq 164$. Thus, Eq. (5.106) may be used for curved rectangular ducts for $\mathrm{Re}^* < 8000$, and Eq. (5.48) or (5.49) for $\mathrm{Re}^* \geq 8000$, with a replaced by $0.5D_h$, where D_h is the hydraulic diameter of the rectangular duct.

Helical Coils, Rectangular Cross Section. Experimental Nusselt numbers for turbulent flow through a curved rectangular duct are reported by Butuzov et al. [76] and Kadambi [80]. Butuzov et al. [76] have correlated their experimental data on water and Freon in three helical coils (two rectangular and one square cross section) as

$$\frac{\mathrm{Nu}_c}{\mathrm{Nu}_s} = 0.117 \times 10^{-2} \mathrm{Re}^{*0.93} \left(\frac{R}{d^*} \right)^{0.24} \tag{5.107}$$

which is valid for $450 \leq \mathrm{Re}^*(R/d^*)^{0.5} \leq 7500$ and $25 \leq R/d^* \leq 164$. Nu_s is an asymptotic Nusselt number for a straight duct of the same aspect ratio as that of a curved duct under consideration, and d^* is a shorter channel side.

Helical Coils, Parallel-Plate Cross Section. Kreith [9] has obtained integronumerical solution for the Nusselt number for flow through a curved parallel-plate channel subjected to the (H1) boundary condition. Figure 5.13 shows a plot of Kreith's numerical results. As noted in all other cases, an increase in Pr is similar to an increase in De in augmenting Nusselt numbers. He also noted that the Nusselt numbers at the outer wall were higher than those at the inner wall.

Spiral Plate Heat Exchanger. For a spiral plate heat exchanger, the flow passage cross section is rectangular with a small aspect ratio. Tagri and Jayaraman [81] have reported experimental water-to-water heat transfer data for such an exchanger. The authors [81] have correlated their results for $\mathrm{Re} > 6000$ as

$$\frac{\mathrm{Nu}_c}{\mathrm{Nu}_s} = \left(1.0 + 1.77 \frac{D_h}{R_{\mathrm{ave}}} \right) \tag{5.108}$$

where Nu_s is evaluated using the straight-tube Dittus-Boelter correlation.

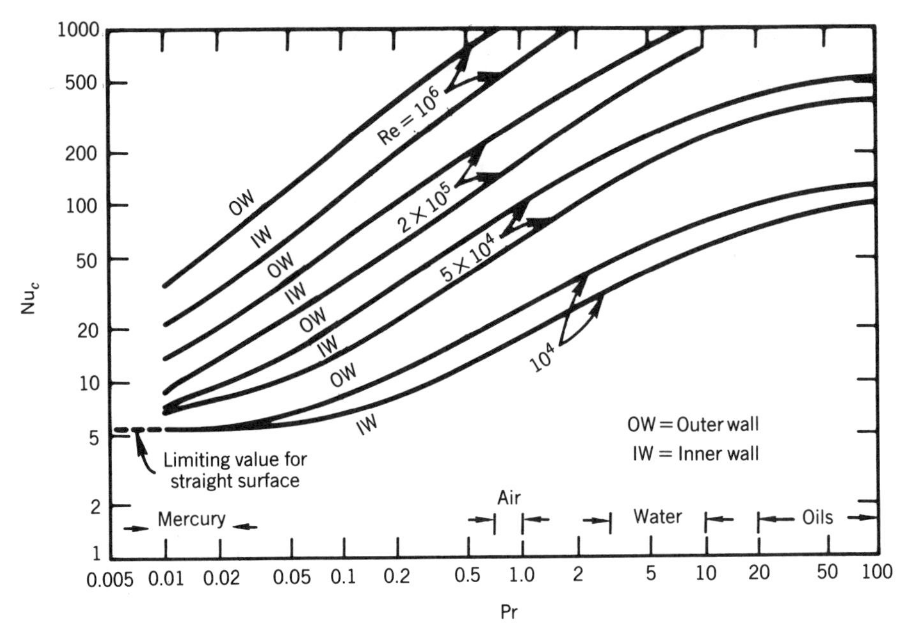

Figure 5.13. Turbulent flow Nusselt numbers in curved ducts of parallel-plate cross section [9].

5.7 CONCLUDING REMARKS

About 200 papers on curved ducts were reviewed and the results (correlations) were compared. Based on this extensive information, the present condensed article has been prepared. The results are summarized for laminar and turbulent flow of Newtonian and non-Newtonian fluids in coils of circular and noncircular cross section, and correlating equations for pertinent results are identified in Table 5.4 for coils of circular cross section. The results of each section and subsection indicate the present state of the art and should provide guidelines for future research.

ACKNOWLEDGMENTS

The authors are grateful to Professor A. E. Bergles of Rensselaer Polytechnic Institute, Dr. W. Nakayama of Hitachi, Ltd., and Professor K. C. Cheng of the University of Alberta for providing constructive suggestions on this article.

NOMENCLATURE

a	tube inside radius or half duct width for a noncircular cross-section channel, m, ft
A	flow cross-sectional area, m^2, ft^2
b	coil pitch (see Fig. 5.1), m, ft
c	half channel height of a rectangular duct in the direction perpendicular to the radius of curvature, m, ft

TABLE 5.4 Summary of Major Correlations for Curved Ducts of Circular Cross Section

Type of Fluid	Specific Quantities	Pertinent Solutions
	HELICAL COILS	
Newtonian fluid—laminar flow	Fully developed flow:	
	Velocity profiles	Eqs. (5.11)–(5.14)
	Friction factors—constant properties	Eqs. (5.16), (5.17)
	Friction factors—variable properties	Eq. (5.18)
	(T) temperature profiles	Eqs. (5.23), (5.26)
	(H1) temperature profiles	Eqs. (5.27), (5.28)
	(T) Nusselt numbers	Eq. (5.29)
	(H1) Nusselt numbers	Eq. (5.32)
	(H2) Nusselt numbers	Table 5.1
	Hydrodynamically developing flow:	
	Entrance length	Eq. (5.36)
	Thermally developing and hydrodynamically developed flow:	
	(T) Nusselt numbers	Eq. (5.37)
	(H1) Nusselt numbers	Eq. (5.38)
Newtonian fluid—turbulent flow	Velocity profiles	Eqs. (5.40)–(5.44)
	Friction factors—constant properties	Eqs. (5.48), (5.49)
	Friction factors—variable properties	Eq. (5.50)
	Temperature profiles	Eqs. (5.52), (5.53)
	Nusselt numbers—constant properties	Eqs. (5.56), (5.57)
	Nusselt numbers—variable properties	Eq. (5.58)
Non-Newtonian fluids	Fully developed laminar flow:	
	Friction factors, power-law inelastic fluids	Eq. (5.67)
	Friction factors, power-law viscoelastic fluids	Eq. (5.69)
	(H1) Nusselt numbers	Table 5.2
	Thermally developing and hydrodynamically developed laminar flow:	
	(T) Nusselt numbers	Eqs. (5.76), (5.77)
	Fully developed turbulent flow:	
	Friction factors, power-law inelastic fluids	Eqs. (5.74), (5.75)
	SPIRAL COILS	
Newtonian fluids	Laminar flow friction factors	Eq. (5.19)
	Laminar flow (T) Nusselt numbers	Eq. (5.31)
	Turbulent flow friction factors	Eq. (5.51)
	Turbulent flow Nusselt numbers	Eq. (5.58) with R_{ave}
Non-Newtonian fluids	Laminar friction factors, power-law, inelastic fluid	Eq. (5.72)

c_p	specific heat of fluid at constant pressure, J/(kg · K), Btu/(1b$_m$ · °F)
D_h	hydraulic diameter of the duct = 4 A/p = inside diameter for a circular tube, m, ft
De	Dean number = $\mathrm{Re}\sqrt{a/R} = \mathrm{Re}\sqrt{0.5\ D_h/R}$
De*	modified Dean number = $\mathrm{Re}\sqrt{D_h/R}$
$\mathrm{De}_{\mathrm{gen}}$	generalized Dean number = $\mathrm{Re}_{\mathrm{gen}}\sqrt{a/R}$
f_c	curved-tube Fanning friction factor = $\tau_w/(\rho u_m^2/2)$
f_s	straight-tube Fanning friction factor for fully developed laminar flow
Gr	modified Grashof number = $\beta g a^4 q''/k\nu^2$
Gr′	Grashof number = $8\beta g a^3\ \Delta T/\nu^2$
Gz	Graetz number = $\dot{m}c_p/kL = p/(4D_h x^*)$
h	heat transfer coefficient, W/(m^2 · K), Btu/(hr · ft^2 · °F)
(H1)	thermal boundary condition referring to constant axial wall heat flux with constant peripheral wall temperature
(H2)	thermal boundary condition referring to axially and circumferentially uniform and constant wall heat flux
He	helical coil number [see Eq. (5.2)]
k	fluid thermal conductivity, W/(m · K), Btu/(hr · ft · °F)
K'	consistency index for non-Newtonian fluids, Pa · s^n, lb$_f$ · s^n/ft^2
K''	modified consistency index for non-Newtonian fluids, Pa · s^n, lb$_f$ · s^n/ft^2
L	tube length, m, ft
L_{hy}	hydrodynamic entrance length, m, ft
L_{th}	thermal entrance length, m, ft
L_{th}^*	$= L_{\mathrm{th}}/(D_h \mathrm{RePr})$
$\dot{m}$	mass flow rate, kg/hr, lb$_m$/hr
n	non-Newtonian fluid power-law index [see Eq. (5.59)]
n_1	number of coil turns at the beginning of the spiral = $L/(2\pi bN) - N/2$
n_2	number of coil turns at the end of the spiral = $L/(2\pi bN) + N/2$
N	number of spiral coil turns = $n_2 - n_1$
Nu	Nusselt number = $h\ D_h/k$
$\mathrm{Nu}_{\mathrm{bc}}$	peripherally averaged mean Nusselt number in the fully developed region for a given boundary condition bc
$\mathrm{Nu}_{m,\mathrm{bc}}$	mean Nusselt number in the thermal entrance length for a given boundary condition bc
$\mathrm{Nu}_{p,\mathrm{bc}}$	peripherally local Nusselt number in the fully developed region for a given boundary condition bc
$\mathrm{Nu}_{x,\mathrm{bc}}$	peripherally averaged axially local Nusselt number in the thermal entrance region for a given boundary condition bc
p	duct wetted perimeter, m, ft
P	fluid static pressure, Pa, lb$_f$/ft^2
Pr	Prandtl number = $\mu c_p/k$
q	heat transfer rate, W, Btu/hr
q'	heat transfer rate per unit length, W/m, Btu/(hr · ft)

q''	heat flux, W/m^2, $Btu/(hr \cdot ft^2)$
r	radial distance, m, ft
R	radius of curvature (see Fig. 5.1), m, ft
Re	Reynolds number $= \rho u_m D_h/\mu$
Re_{gen}	generalized Reynolds number [see Eq. (5.63)], $Re_{gen,c}$ defined by Eq. (5.66)
R_{ave}	mean radius of curvature for a spiral, m, ft
T	temperature, K, °C, °R, °F
Ⓣ	thermal boundary condition referring to axially and peripherally constant wall temperature
u_m	mean axial velocity, m/s, ft/s
u_x	axial velocity along the curvilinear axial coordinate, m/s, ft/s
u_r	radial velocity, m/s, ft/s
u_θ	angular velocity, m/s, ft/s
x	axial distance along the axis of the curved tube, m, ft
x^*	dimensionless length $= x/D_h \mathrm{Re\,Pr}$
y, z	Cartesian coordinates across the flow cross section, m, ft

Greek Symbols

α	fluid thermal diffusivity, m^2/s, ft^2/s
α^*	aspect ratio of a rectangular channel, $= 2c/2a$
β	coefficient of thermal expansion, 1/K, 1/°R
δ	momentum boundary-layer thickness, m, ft
δ^*	dimensionless radial location within the boundary layer, $= (a - r)/\delta$
δ_t	thermal boundary-layer thickness [see Eq. (5.25)], m, ft
Δ	prefix denoting a difference or change
ϵ_H	eddy diffusivity for heat transfer, m^2/s, ft^2/s
ϵ_M	eddy diffusivity for momentum, m^2/s, ft^2/s
θ	angular coordinate in the cylindrical coordinate system (see Fig. 5.1), rad, deg
μ	dynamic viscosity, Pa · s, $lb_m/(hr \cdot ft)$
μ_{eff}	effective dynamic viscosity, Pa · s, $lb_m/(hr \cdot ft)$
μ_n	effective dynamic viscosity of non-Newtonian fluid, Pa · s, $lb_m/(hr \cdot ft)$
μ_t	turbulent viscosity, Pa · s, $lb_m/(hr \cdot ft)$
ρ	fluid density, kg/m^3, lb_m/ft^3
τ	wall shear stress, Pa, lb_f/ft^2
ϕ	angle of tube curvature, rad, deg

Subscripts

bc	boundary condition
c	curved coil or duct
H1	(H1) boundary condition

H2	(H2) boundary condition
in	inlet
m	bulk mean value
p	peripheral value
s	straight duct
T	(T) boundary condition
w	wall
x	axial
δ	at the edge of velocity boundary layer
0	tube center

REFERENCES

1. P. S. Srinivasan, S. S. Nandapurkar, and F. A. Holland, Pressure Drop and Heat Transfer in Coils, *Chem. Eng.*, pp. CE113–CE119, May 1968.
2. S. A. Berger, L. Talbot, and L. S. Yao, Flow in Curved Pipes, *Ann. Rev. Fluid Mech.*, Vol. 15, pp. 461–512, 1983.
3. V. Gnielinski, Correlations for the Pressure Drop in Helically Coiled Tubes, *Int. Chem. Eng.*, Vol. 26, No. 1, pp. 36–44, 1986.
4. U. Baurmeister and H. Brauer, Laminar Flow and Heat Transfer in Helically and Spirally Coiled Tubes, *VDI Forschungsheft*, No. 593, pp. 2–48, 1979.
5. M. Sankariah and Y. V. N. Rao, Analysis of Steady Laminar Flow of an Incompressible Newtonian Fluid through Curved Pipes of Small Curvature, ASME Paper No. 72-WA/FE-19, 1972.
6. S. V. Patankar, V. S. Pratap, and D. B. Spalding, Prediction of Laminar Flow and Heat Transfer in Helically Coiled Pipes, *J. Fluid Mech.*, Vol. 62, Part 3, pp. 539–551, 1974.
7. S. V. Patankar, V. S. Pratap, and D. B. Spalding, Prediction of Turbulent Flow in Curved Pipes, *J. Fluid Mech.*, Vol. 67, Part 3, pp. 583–595, 1975.
8. V. P. Tyagi and V. K. Sharma, An Analysis of Steady Fully Developed Heat Transfer in Laminar Flow with Viscous Dissipation in a Curved Circular Duct, *Int. J. Heat Mass Transfer*, Vol. 18, pp. 69–78, 1975.
9. F. Kreith, The Influence of Curvature on Heat Transfer to Incompressible Fluids, *Trans. ASME*, Vol. 77, pp. 1247–1256, 1955.
10. R. K. Shah and A. L. London, *Laminar Flow Forced Convection in Ducts*, Supplement 1 to *Advances in Heat Transfer*, Academic, New York, 1978.
11. W. R. Dean, Note on the Motion of Fluid in a Curved Pipe, *Philos. Mag.*, Ser. 7, Vol. 4, pp. 208–223, 1927.
12. W. R. Dean, The Stream-line Motion of Fluid in a Curved Pipe, *Philos. Mag.*, Ser. 7, Vol. 5, No. 30, pp. 673–695, 1928.
13. L. R. Austin and J. D. Seader, Fully Developed Viscous Flow in Coiled Circular Pipes, *AIChE J.*, Vol. 19, pp. 85–94, 1973.
14. Y. Mori and W. Nakayama, Study on Forced Convective Heat Transfer in Curved Pipes (1st Report, Laminar Region), *Int. J. Heat Mass Transfer*, Vol. 8, pp. 67–82, 1965.
15. M. Adler, Flow in Curved Tube, *Z. Angew. Math. Mech.*, Vol. 14, pp. 257–275, 1934.
16. M. Akiyama and K. C. Cheng, Boundary Vorticity Method for Laminar Forced Convection Heat Transfer in Curved Pipes, *Int. J. Heat Mass Transfer*, Vol. 14, pp. 1659–1675, 1971.
17. P. S. Srinivasan, S. S. Nandapurkar, and F. A. Holland, Friction Factors for Coils, *Trans. Inst. Chem. Eng.*, Vol. 48, pp. T156–T161, 1970.

18. R. L. Manlapaz and S. W. Churchill, Fully Developed Laminar Flow in a Helically Coiled Tube of Finite Pitch, *Chem. Eng. Commun.*, Vol. 7, pp. 57–78, 1980.

19. R. L. Manlapaz and S. W. Churchill, Fully Developed Laminar Convection from a Helical Coil, *Chem. Eng. Commun.*, Vol. 9, pp. 185–200, 1981.

20. P. Mishra and S. N. Gupta, Momentum Transfer in Curved Pipes. I. Newtonian Fluids, *Ind. Eng. Chem. Process Des. Dev.*, Vol. 18, pp. 130–137, 1979.

21. V. Kubair and N. R. Kuloor, Non-isothermal Pressure Drop Data for Liquid Flow in Helical Coils, *Indian J. Technol.*, Vol. 3, pp. 5–7, 1965.

22. V. Kubair and N. R. Kuloor, Flow of Newtonian Fluids in Archimedian Spiral Tube Coils: Correlation of the Laminar, Transition and Turbulent Flows, *Indian J. Technol.*, Vol. 4, pp. 3–8, 1966.

23. V. Kubair and N. R. Kuloor, Non-isothermal Pressure Drop Data for Spiral Tube Coils, *Indian J. Technol.*, Vol. 3, pp. 382–383, 1965.

24. V. Kubair and N. R. Kuloor, Heat Transfer to Newtonian Fluids in Coiled Pipes in Laminar Flow, *Int. J. Heat Mass Transfer*, Vol. 9, pp. 63–75, 1966.

25. Y. Mori and W. Nakayama, Study on Forced Convective Heat Transfer in Curved Pipes (3rd Report, Theoretical Analysis under the Condition of Uniform Wall Temperature and Practical Formulae), *Int. J. Heat Mass Transfer*, Vol. 10, pp. 681–695, 1967.

26. J. M. Tarbell and M. R. Samuels, Momentum and Heat Transfer in Helical Coils, *Chem. Eng. J.—Lausanne (Netherlands)*, Vol. 5, pp. 117–127, 1973.

27. A. N. Dravid, K. A. Smith, E. W. Merrill, and P. L. T. Brian, Effect of Secondary Fluid Motion on Laminar Flow Heat Transfer in Helically Coiled Tubes, *AIChE J.*, Vol. 17, pp. 1114–1122, 1971.

28. W. B. Hawes, Some Sidelights on the Heat Transfer Problem, *Trans. Inst. Chem. Eng.*, Vol. 10, pp. 161–167, 1932.

29. H. Ito, Friction Factors for Turbulent Flow in Curved Pipes, *J. Basic Eng.*, Vol. 81, pp. 123–134, 1959.

30. C. E. Kalb and J. D. Seader, Heat and Mass Transfer Phenomena for Viscous Flow in Curved Circular Tubes, *Int. J. Heat Mass Transfer*, Vol. 15, pp. 801–817, 1972.

31. M. Akiyama and K. C. Cheng, Laminar Forced Convection Heat Transfer in Curved Pipes with Uniform Wall Temperature, *Int. J. Heat Mass Transfer*, Vol. 15, pp. 1426–1431, 1972.

32. C. E. Kalb and J. D. Seader, Fully Developed Viscous-Flow Heat Transfer in Curved Circular Tubes with Uniform Wall Temperature, *AIChE J.*, Vol. 20, pp. 340–346, 1974.

33. V. Kubair and N. R. Kuloor, Heat Transfer to Newtonian Fluids in Spiral Coils at Constant Tube Wall Temperature in Laminar Flow, *Indian J. Technol.*, Vol. 3, pp. 144–146, 1965.

34. N. J. Rabadi, J. C. F. Chow, and H. A. Simon, An Efficient Numerical Procedure for the Solution of Laminar Flow and Heat Transfer in Coiled Tubes, *Numer. Heat Transfer*, Vol. 2, pp. 279–289, 1979.

35. L. A. M. Janssen and C. J. Hoogendoorn, Laminar Convective Heat Transfer in Helical Coiled Tubes, *Int. J. Heat Mass Transfer*, Vol. 21, pp. 1197–1206, 1978.

36. S. Manafzadeh, J. C. F. Chow, and H. A. Simon, Heat Transfer in Curved Tubes, *Proc.* 1983 *ASME-JSME Thermal Eng. Conf.*, Vol. 8, pp. 21–26, ASME, New York, 1983.

37. M. A. Abul-Hamayel and K. J. Bell, Heat Transfer in Helically Coiled Tubes with Laminar Flow, ASME Paper No. 79-WA/HT-11, 1979.

38. J. Prusa and L. S. Yao, Heat Transfer of Fully Developed Flow in Curved Tubes, ASME Paper No. 81-HT-39, 1981.

39. J. B. Lee, H. A. Simon, and J. C. F. Chow, Buoyancy in Laminar Curved Tube Flows, *Proc.* 1983 *ASME-JSME Thermal Eng. Conf.*, Vol. 8, pp. 133–139, ASME, New York, 1983.

40. J. B. Lee, H. A. Simon, and J. C. F. Chow, Buoyancy in Developed Laminar Curved Tube Flows, *Int. J. Heat Mass Transfer*, Vol. 28, pp. 631–640, 1985.

41. L. S. Yao and S. A. Berger, Flow in Heated Curved Pipes, *J. Fluid Mech.*, Vol. 88, Part 2, pp. 339–354, 1978.

42. L. R. Austin and J. D. Seader, Entry Region for Steady Viscous Flow in Coiled Circular Pipes, *AIChE J.*, Vol. 20, pp. 820–822, 1974.

43. Y. Agrawal, L. Talbot, and K. Gong, Laser Anemometer Study of Flow Development in Curved Circular Pipes, *J. Fluid Mech.*, Vol. 85, Part 3, pp. 497–518, 1978.

44. G. H. Keulegan and K. H. Beij, Pressure Losses for Fluid Flow in Curved Pipes, *J. Res. Natl. Bur. Standards*, Vol. 18, pp. 89–114, 1937.

45. L. S. Yao and S. A. Berger, Entry Flow in a Curved Pipe, *J. Fluid Mech.*, Vol. 67, Part 1, pp. 177–196, 1975.

46. M. Akiyama and K.C. Cheng, Laminar Forced Convection in the Thermal Entrance Region of Curved Pipes with Uniform Wall Temperature, *Can. J. Chem. Eng.*, Vol. 52, pp. 234–240, 1974.

47. M. Akiyama and K. C. Cheng, Graetz Problem in Curved Pipes with Uniform Wall Heat Flux, *Appl. Sci. Res.*, Vol. 29, pp. 401–418, 1974.

48. L. S. Yao, Heat Convection in a Horizontal Curved Pipe, *J. Heat Transfer*, Vol. 106, pp. 71–77, 1984.

49. G. W. Hogg, The Effect of Secondary Flow on Point Heat Transfer Coefficients for Turbulent Flow Inside Curved Tubes, Ph.D. Thesis, Univ. of Idaho, 1968.

50. Y. Mori and W. Nakayama, Study on Forced Convective Heat Transfer in Curved Pipes (2nd Report, Turbulent Region), *Int. J. Heat Mass Transfer*, Vol. 10, pp. 37–59, 1967.

51. B. E. Boyce, J. G. Collier, and J. Levy, Hold-up and Pressure Drop Measurements in the Two-Phase Flow of Air-Water Mixtures in Helical Coils, *Co-current Gas Liquid Flow*, Plenum Press, U.K., pp. 203–231, 1969.

52. G. F. C. Rogers and Y. R. Mayhew, Heat Transfer and Pressure Loss in Helically Coiled Tubes with Turbulent Flow, *Int. J. Heat Mass Transfer*, Vol. 7, pp. 1207–1216, 1964.

53. A. E. Ruffell, The Application of Heat Transfer and Pressure Drop Data to the Design of Helical Coil Once Through Boilers, Symp. Multiphase Flow Systems, Univ. of Strathclyde, Paper 15, published in *Inst. Chem. Eng. Symp. Ser.*, No. 38, pp. 15–21, 1974.

54. E. F. Schmidt, Wärmeübergang und Druckverlust in Rohrschlangen, *Chem. Ing. Tech.*, Vol. 39, pp. 781–789, 1967.

55. N. H. Pratt, The Heat Transfer in a Reaction Tank Cooled by Means of a Coil, *Trans. Inst. Chem. Eng.*, Vol. 25, pp. 163–180, 1947.

56. V. K. Orlov and P. A. Tselishchev, Heat Exchange in a Spiral Coil with Turbulent Flow of Water, *Thermal Eng.* (translated from *Teploenergetika*), Vol. 11, No. 12, pp. 97–99, 1964.

57. A. H. P. Skelland, *Non-Newtonian Flow and Heat Transfer*, Wiley, New York, 1967.

58. P. Mishra and S. N. Gupta, Momentum Transfer in Curved Pipes. 2. Non-Newtonian Fluids, *Ind. Eng. Chem. Process Des. Dev.*, Vol. 18, pp. 137–142, 1979.

59. R. A. Mashelkar and G. V. Devarajan, Secondary Flows of Non-Newtonian Fluids: Part I—Laminar Boundary Layer Flow of a Generalised Non-Newtonian Fluid in a Coiled Tube, *Trans. Inst. Chem. Eng.*, Vol. 54, pp. 100–107, 1976.

60. C. F. Hau and S. V. Patankar, Analysis of Laminar Non-Newtonian Flow and Heat Transfer in Curved Tubes, *AIChE J.*, Vol. 28, pp. 610–616, 1982.

61. C. F. Hsu, Personal Communication, Shell Development Co., Houston, Texas, Apr. 1985.

62. S. Rajasekharan, V. G. Kubair, and N. R. Kuloor, Heat Transfer to Non-Newtonian Fluids in Coiled Pipes in Laminar Flow, *Int. J. Heat Mass Transfer*, Vol. 13, pp. 1583–1594, 1970.

63. S. Rajasekharan, V. G. Kubair, and N. R. Kuloor, Isothermal Laminar Flow of Non-Newtonian Fluids in Spiral Tube Coils, *Indian J. Technol.*, Vol. 2, pp. 149–152, 1964.

64. B. A. Mujawar and M. Raja Rao, Flow of Non-Newtonian Fluids through Helical Coils, *Ind. Eng. Chem. Process Des. Dev.*, Vol. 17, pp. 22–27, 1978.

65. S. Rajasekharan, V. G. Kubair, and N. R. Kuloor, Flow of Non-Newtonian Fluids through Helical Coils, *Indian J. Technol.*, Vol. 8, pp. 391–397, 1970.

66. D. R. Oliver and S. M. Asghar, Heat Transfer to Newtonian and Viscoelastic Liquids during Laminar Flow in Helical Coils, *Trans. Inst. Chem. Eng.*, Vol. 54, pp. 218–224, 1976.

67. K. C. Cheng, R. C. Lin, and J. W. Ou, Fully Developed Laminar Flow in Curved Rectangular Channels, *J. Fluids Eng.*, Vol. 98, pp. 41–48, 1976.

68. Y. Mori, Y. Uchida, and T. Ukon, Forced Convective Heat Transfer in a Curved Channel with a Square Cross Section, *Int. J. Heat Mass Transfer*, Vol. 14, pp. 1787–1805, 1976.

69. K. C. Cheng and M. Akiyama, Laminar Forced Convection Heat Transfer in Curved Rectangular Channels, *Int. J. Heat Mass Transfer*, Vol. 13, pp. 471–490, 1970.

70. J. A. Baylis, Experiments on Laminar Flow in Curved Channels of Square Section, *J. Fluid Mech.*, Vol. 48, Part 3, pp. 417–422, 1971.

71. B. Joseph, E. P. Smith, and R. J. Adler, Numerical Treatment of Laminar Flow in Helically Coiled Tubes of Square Cross Section: Part 1–Stationary Helically Coiled Tubes, *AIChE J.*, Vol. 21, pp. 965–979, 1975.

72. J. H. Masliyah and K. Nandakumar, Fully Developed Laminar Flow in a Helical Tube of Finite Pitch, *Chem. Eng. Commun.*, Vol. 29, pp. 125–138, 1984.

73. K. C. Cheng, R. C. Lin, and J. W. Ou, Graetz Problem in Curved Square Channels, *J. Heat Transfer*, Vol. 97, pp. 244–248, 1975.

74. K. C. Cheng, R. C. Lin, and J. W. Ou, Graetz Problem in Curved Rectangular Channels with Convective Boundary Condition—the Effect of Secondary Flow on Liquid Solidification—Free Zone, *Int. J. Heat Mass Transfer*, Vol. 18, pp. 996–999, 1975.

75. M. Akiyama, K. Kikuchi, K. C. Cheng, M. Suzuki, and I. Nishiwaki, Mixed Laminar Convection of the Thermal Entry Region in Curved Rectangular Channels, *Proc.* 1983 *ASME-JSME Thermal Eng. Conf.*, Vol. 3, pp. 27–33, ASME, New York, 1983.

76. A. I. Butuzov, M. K. Bezrodnyy, and M. M. Pustovit, Hydraulic Resistance and Heat Transfer in Forced Flow in Rectangular Coiled Tubes, *Heat Transfer—Sov. Res.*, Vol. 7, No. 4, pp. 84–88, 1975.

77. R. M. C. So, Entry Flow in Curved Channels, ASME Paper No. 76-FE-G, 1976.

78. V. K. Shchukin and V. A. Filin, Convective Heat Transfer in Short Curved Channels, *J. Eng. Phys.*, Vol. 12, No. 2, pp. 78–82, 1967.

79. K. W. Coons, A. W. Hargis, P. Q. Hewes, and F. T. Weems, Spiral Heat Exchanger Heat-Transfer Characteristics, *Chem. Eng. Prog.*, Vol. 43, No. 8, pp. 405–414, 1947.

80. V. Kadambi, Heat Transfer and Pressure Drop in a Helically Coiled Rectangular Duct, ASME Paper No. 83-WA/HT-1, 1983.

81. N. N. Tagri and R. Jayaraman, Heat Transfer Studies on a Spiral Plate Heat Exchanger, *Trans. Inst. Chem. Eng.*, Vol. 40, pp. 161–168, 1962.

82. L. C. Truesdell, Jr. and R. J. Adler, Numerical Treatment of Fully Developed Laminar Flow in Helically Coiled Tubes, *AIChE J.*, Vol. 16, pp. 1010–1015, 1970.

83. J. A. C. Humphrey, Numerical Calculations of Developing Laminar Flow in Pipes of Arbitrary Curvature Radius, *Can. J. Chem. Eng.*, Vol. 56, pp. 151–164, 1978.

84. J. N. Kapur, V. P. Tyagi, and R. C. Srivastava, Streamline Flow through a Curved Annulus, *Appl. Sci. Res., Sect. A.*, Vol. 14, pp. 253–267, 1964.

85. R. A. Mashelkar and G. V. Devarajan, Secondary Flow of Non-Newtonian Fluids: Part II. Frictional Losses in Laminar Flow of Purely Viscous and Viscoelastic Fluids Through Coiled Tubes, *Trans. Inst. Chem. Eng.*, Vol. 54, pp. 108–114, 1976.

86. M. Akiyama, K. Kikuchi, M. Suzuki, I. Nishiaki, K. C. Cheng, and J. Nakayama, Numerical Analysis and Flow Visualization on the Hydrodynamic Entrance Region of Laminar Flow in Curved Square Channels, *Trans. JSME*, Vol. 47, No. 422, pp. 1960–1970, Oct. 1981.

87. M. Akiyama, K. Kikuchi, J. Nakayama, M. Suzuki, I. Nishiwaki, and K. C. Cheng, Two Stage Developments of Entry Flow with an Interaction of Boundary-Wall and Dean's Instability Type Secondary Flows, *Trans. JSME*, Vol. 47, No. 421, pp. 1705–1714, Sept. 1981.

88. Y. Komiyama, F. Mikami, K. Okui, and T. Hori, Laminar Forced Convection Heat Transfer in Curved Channels of Rectangular Cross Section, *Heat Transfer–Jpn. Res.*, Vol. 13, No. 3, pp. 68–91, 1984.

89. W. Nakayama, Personal Communication, Hitachi Mechanical Engineering Research Lab., Tsuchiura-Shi, Ibaraki, Japan, 1985.

90. E. N. Sieder and G. E. Tate, Heat Transfer and Pressure Drops of Liquid in Tubes, *Ind. Eng. Chem.*, Vol. 28, pp. 1429–1435, 1936.

91. R. L. Pigford, Nonisothermal Flow and Heat Transfer inside Vertical Tubes, *Chem. Eng. Progress Symp. Ser.* No. 17, Vol. 51, pp. 79–92, 1955.

92. S. Garimella, R. N. Christensen, and D. E. Richards, Experimental Investigation of Heat Transfer in Curved Annular Ducts, ASME Paper No. 84-WA/HT-28, 1984.

6

CONVECTIVE HEAT TRANSFER IN CROSS FLOW

A. Žukauskas

Academy of Sciences of the Lithuanian SSR,
Vilnius, USSR

6.1 SPECIFIC FEATURES OF CROSS FLOW FLUID DYNAMICS AND HEAT TRANSFER

The fundamentals of heat transfer by forced convection are presented in Chaps. 1 and 2. We deal here with heat transfer from tubes of different cross sections and bodies of other geometries, which are applied in modern heat exchangers. Heat transfer is closely related to fluid dynamics. That is why heat transfer is considered simultaneously with fluid dynamics.

Fluid dynamics and heat transfer around curvilinear bodies are complex processes, mainly dependent on the fluid type and the Reynolds number

$$\mathrm{Re} = \frac{ud}{\nu} \tag{6.1}$$

where u and d are the reference (or characteristic) velocity and diameters, respectively, and ν the kinematic viscosity.

In real fluids, due to their viscosity, a laminar boundary layer is formed on the front part of a body, its thickness increasing downstream. It also involves a longitudinal pressure gradient caused by the curved surface. From a two-dimensional equation of momentum (see Chap. 1)

$$u\frac{\partial u}{\partial x} + v\frac{\partial u}{\partial y} = -\frac{1}{\rho}\frac{dP}{dx} + \frac{1}{\rho}\frac{\partial \tau}{\partial y} \tag{6.2}$$

with the boundary condition $u = v = 0$ when $y = 0$. A relation between shear stress derivative at $y = 0$ and the pressure gradient is

$$\left(\frac{\partial \tau}{\partial y}\right)_{y=0} = \frac{dP}{dx} \tag{6.3}$$

The shear stress at the wall for a laminar boundary layer is determined by the near-wall velocity gradient

$$\tau_w = \mu\left(\frac{\partial u}{\partial y}\right)_{y=0} \tag{6.4}$$

A decrease or a zero value of the shear stress τ_w is reflected by a corresponding change in the velocity gradient $\partial u/\partial y$. In the classical theory of flow separation, the boundary-layer separation point is assumed to be at $(\partial u/\partial y)_{y=0} = 0$. From Eq. (6.4), we see that this is the point where the shear stress acquires a zero value, $\tau_w = 0$, so that the boundary layer can separate from the surface (see Fig. 6.1). The separation point is followed by an inverse flow, where the velocity vectors of near-wall fluid masses are in opposite directions. The inverse layer contacts the boundary layer and curls up in a vortex, which begins to rotate.

Each of the above phenomena of fluid dynamics is reflected in the local heat transfer. For a heat flux to appear, there must be a temperature gradient in the flow. The local heat transfer coefficient is defined by

$$h = -\frac{k}{T_w - T_\infty}\left(\frac{\partial T}{\partial y}\right)_{y=0} \tag{6.5}$$

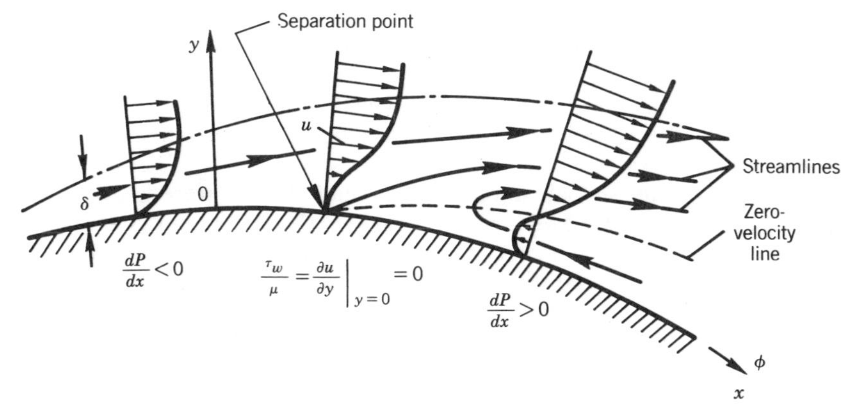

Figure 6.1. Velocity distribution on a curvilinear surface near the separation point.

With a laminar boundary layer developing on the front part of a curvilinear body in cross flow, the heat transfer coefficient decreases with increasing thickness, and is lower in fluids of lower heat conduction. Heat transfer increases in the rear recirculation region, downstream from the boundary-layer separation.

Heat transfer from tubes and other bodies in cross flow is determined by the stream velocity, turbulence level, physical properties of the fluid, thermal load, heat flux direction, geometry of the body, and some other factors. The most general dimensionless description is

$$\mathrm{Nu} = f\left(\mathrm{Re}, \mathrm{Pr}, \mathrm{Tu}, \frac{\mu}{\mu_w}, \frac{k}{k_w}, \frac{c_p}{c_{p,w}}, \frac{\rho}{\rho_w}\right) \tag{6.6}$$

To predict heat transfer by Eq. (6.6), it is usually written in the following way:

$$\mathrm{Nu} = c\,\mathrm{Re}^m \mathrm{Pr}^n \tag{6.7}$$

where the Nusselt number $\mathrm{Nu} = hd_o/k$, the Prandtl number $\mathrm{Pr} = \mu c_p/k$, and the Reynolds number $\mathrm{Re} = u_0 d_o/\nu$.

Fluid physical properties in the heat transfer equation are described by the Prandtl number, which is about 0.7 for most gases. Therefore, from experimental data on heat transfer in air or in other gases, we have

$$\mathrm{Nu} = c\,\mathrm{Re}^m \tag{6.8}$$

The process of heat transfer involves a change of temperature, and consequently, variable fluid physical properties. Thus to account for the effect of fluid physical properties in heat transfer implies the prediction of the effect of their change on a temperature variation across the boundary layer. We encounter a problem of choosing the so-called characteristic temperature, or reference temperature, for the physical properties.

Different approaches to evaluating temperature dependence of fluid physical properties are currently in use. We suggest the bulk mean temperature of the fluid as the reference temperature for moderate temperature gradients. This approach is simple and sufficiently accurate for practical purposes.

A marked influence of fluid physical properties on the heat transfer in viscous fluids is related to their dependence on the heat flux direction and on the temperature gradient. Experimental data, with reference to the bulk mean temperature, show that heat transfer coefficients are higher for wall-to-fluid heat transfer (fluid heating) than for fluid-to-wall heat transfer (fluid cooling). The difference increases with the temperature gradient.

To account for the heat flux direction, when fluid physical properties refer to the fluid bulk mean temperature, we introduce ratio $\mathrm{Pr}/\mathrm{Pr}_w$ with a proper power index p, where Pr_w stands for Pr evaluated at the wall temperature. Thus we arrive at the following relation for the heat transfer from bodies in cross flow of viscous fluids:

$$\mathrm{Nu} = c\,\mathrm{Re}^m \mathrm{Pr}^n \left(\frac{\mathrm{Pr}}{\mathrm{Pr}_w} \right)^p \tag{6.9}$$

where fluid properties refer to the bulk mean temperature. For gases, Pr is constant and $\mathrm{Pr}/\mathrm{Pr}_w \approx 1$.

Both the fluid dynamics and heat transfer are also influenced by free-stream turbulence, geometry, surface roughness, etc.

A large number of heat exchangers employ tube bundles or other arrays. Both the fluid dynamics and heat transfer over bundles are again different from those over single tubes, because of the additional influence of the neighboring tubes. Still other types of flow are induced by the application of smooth or rough-surface or finned tubes.

6.2 HEAT TRANSFER FOR SINGLE TUBES AND BODIES

6.2.1 Fluid Dynamics over a Single Tube and a Sphere

The velocity distribution over a circular cylinder in cross flow of an ideal fluid (no boundary layer) is described by

$$u = u_\infty \sin\phi \left[1 + \left(\frac{r_o}{r_1} \right)^2 \right] \tag{6.10}$$

Here r_o and r_1 are outside radius of the cylinder and the radial distance from its axis, respectively. The velocity is larger at smaller distances, and on the surface itself it is

$$u = 2u_\infty \sin\phi \tag{6.11}$$

We now relate Eq. (6.11) to the Bernoulli equation

$$P + \tfrac{1}{2}\rho u_\infty^2 = \text{constant} \tag{6.12}$$

and find a functional relation between the velocity of the flow and the pressure coefficient

$$C_P = \frac{2(P - P_\infty)}{\rho u_\infty^2} = 1 - 4\sin^2\phi \tag{6.13}$$

Thus from Fig. 6.2, the pressure coefficient for potential flow has two maxima at $\phi = 0° = 360°$ and $\phi = 180°$, and two minima in the medial cross section. The pressure coefficient is distributed symmetrically in ideal fluids.

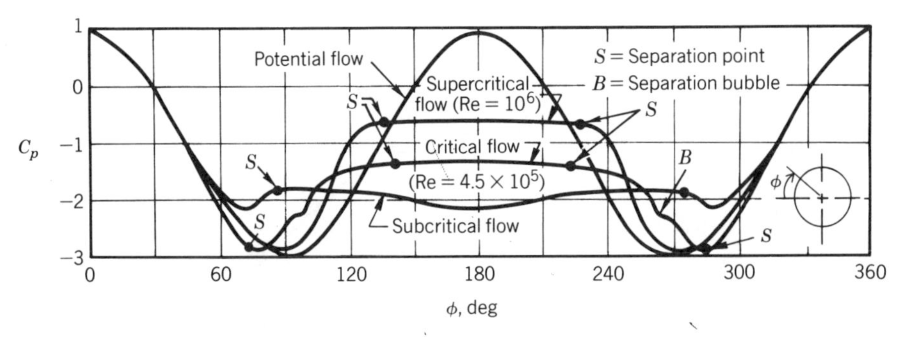

Figure 6.2. Circumferential distribution of the pressure coefficient over a cylinder in cross flow.

The distribution of the pressure coefficient is a reflection of the flow phenomena. On a cylinder in a real fluid, their interaction with the viscous force gives rise to a laminar boundary layer, which is formed on the front part and whose thickness increases downstream. The main determining parameters of this layer are the Reynolds number Re and the turbulence level Tu. In Fig. 6.2, the curve for the subcritical flow represents the circumferential distribution of the pressure coefficient on a cylinder in the medium range of Re ($Re < 2 \times 10^5$). The kink in the curve at a point $\phi \approx 80°$ corresponds to the boundary-layer separation and to the formation of a complex vortical flow in the rear.

With an increase of Re ($Re > 2 \times 10^5$), the flow enters the critical regime, and the corresponding distribution of C_P is shown in Fig. 6.2. The laminar boundary layer separates on the front part point S, forms a separation bubble, and later reattaches at point B. Reattachment is followed by a turbulent boundary layer, which withstands the increased pressure gradient and finally separates at $\phi \approx 140°$.

In the supercritical regime, at $Re > 0.6 \times 10^6$, the laminar-turbulent transition occurs in a nonseparated boundary layer, and the transition point is shifted upstream as shown in Fig. 6.2. The separation of the turbulent boundary layer occurs at ϕ between 120° and 140°.

The location of a laminar-turbulent transition in the boundary layer depends both on the Reynolds number and on the turbulence level.

Figure 6.3 shows the dynamic behavior of the laminar-turbulent transition as a function of Re and Tu, according to measurements of fluid dynamic and thermal parameters [1, 2]. For example, the transition from the laminar boundary layer initiates at $\phi = 80°$ for $Re = 4.3 \times 10^5$ and Tu = 7%.

On a sphere, the distribution of pressure and velocity is analogous to that on a circular cylinder [3].

6.2.2 Drag on a Cylinder and a Sphere

The total drag is generated by the friction force F_f and pressure force F_w acting on a body in cross flow. At very low Re, a cylinder and a sphere are streamlined, and their drag consists mainly of friction. Figure 6.4 presents the local skin friction coefficient $\tilde{c}_f$ on a circular cylinder for various Re [2, 4]. We see that $\tilde{c}_f$ increases from zero at $\phi = 0°$ to a maximum value at $\phi = 60°$, and again diminishes. The point which indicate zero friction coefficient represents the boundary-layer separation. With a further increase of Re, the contribution of inertial forces begins to grow, so that in a highly vortical flow, skin friction drag constitutes just a few percent of the total drag.

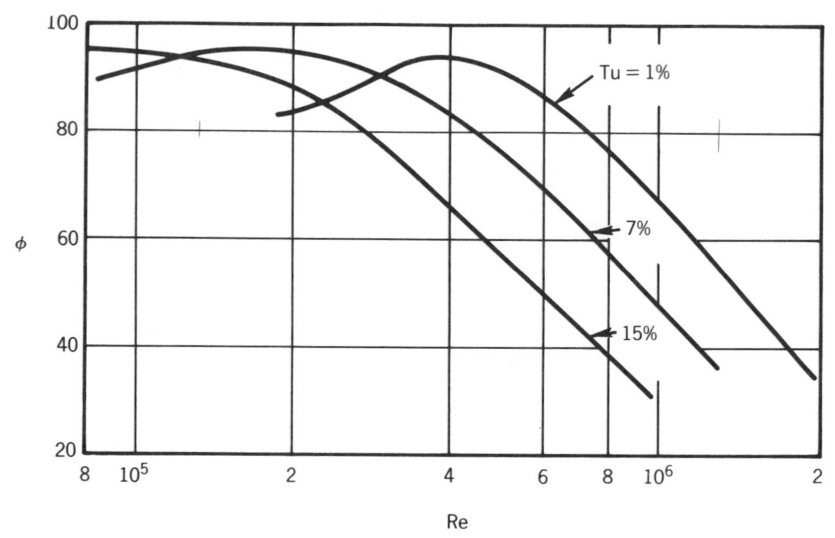

Figure 6.3. Locations of the separation point of the laminar boundary layer for $\mathrm{Re} < 2 \times 10^5$, and of the laminar-turbulent boundary-layer transition point for $\mathrm{Re} \geq 2 \times 10^5$, as functions of Re and Tu.

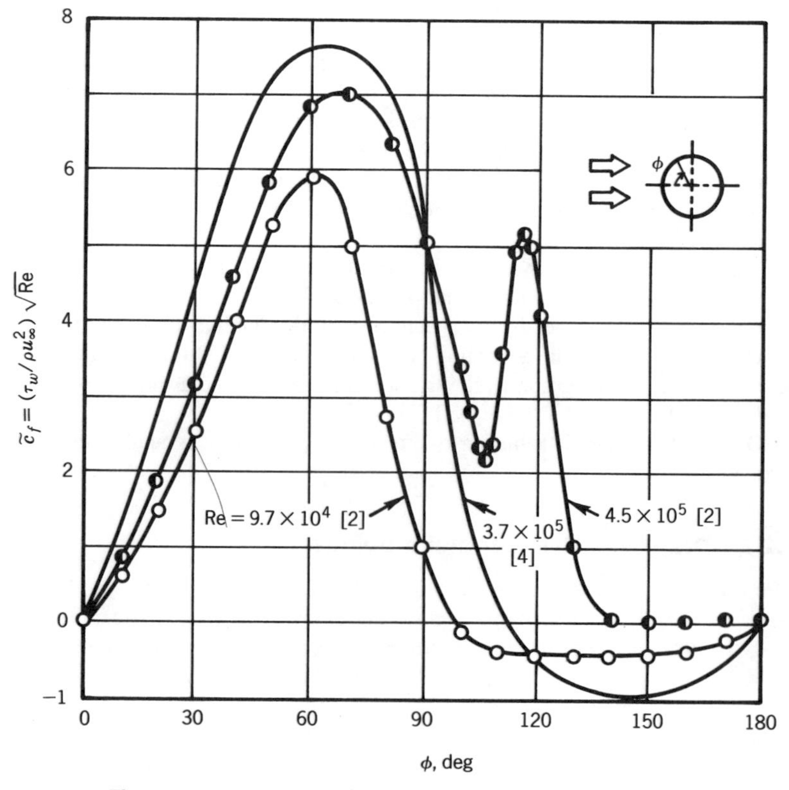

Figure 6.4. Friction coefficient of a cylinder for variable Re.

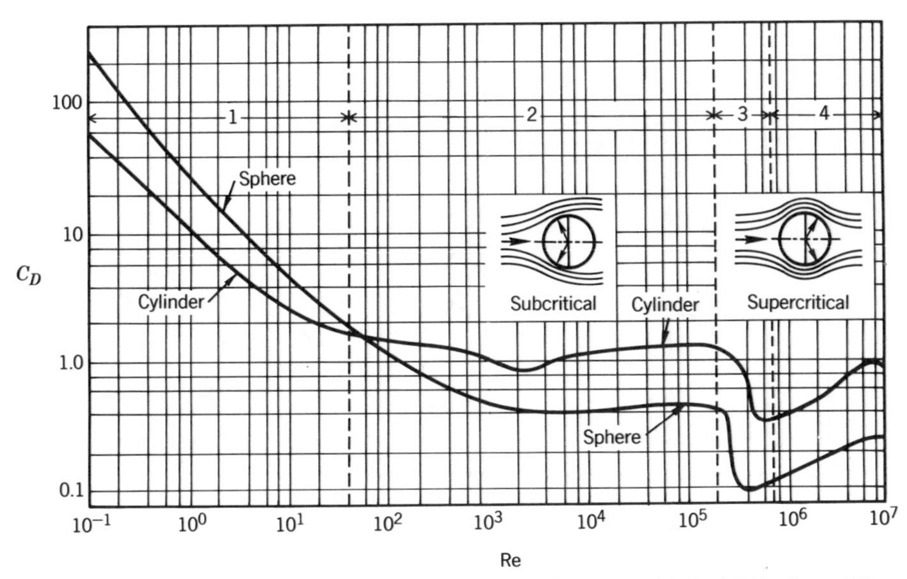

Figure 6.5. Total drag coefficient of a cylinder and a sphere for variable Re [1]. Regions of Re: (1) laminar, (2) subcritical, (3) critical and (4) supercritical.

A dimensionless expression for the total drag coefficient is

$$C_D = \frac{F_f + F_w}{\frac{1}{2}\rho u_\infty^2 d_o L} \tag{6.14}$$

The drag coefficient for a circular cylinder and a sphere is shown in Fig. 6.5 as a function of Re. The drag coefficient decreases significantly with increasing Reynolds number in the low range of Re (region 1), due mostly to the contribution of the skin friction. In the subcritical flow regime (region 2), C_D changes insignificantly with Re. In the critical flow regime (region 3), the total drag coefficient sharply decreases with Re. This is due to a much narrower wake caused by the turbulent boundary layer and a downstream shift of its separation. In the supercritical flow regime (region 4), the total drag increases again because the streamlines are displaced by the thicker turbulent boundary layer.

Figure 6.5 includes a curve of the drag coefficient of a sphere, which is similar to that of a circular cylinder.

An increase in the turbulence level leads to an earlier onset of the critical flow regime with corresponding changes of the total drag coefficient as shown in Fig. 6.6.

Figure 6.7 shows the effect of surface roughness on the drag coefficient of a circular cylinder in the critical flow regime [2, 4]. Here the surface roughness causes a higher drag coefficient and an earlier onset of the critical flow regime.

6.2.3 Local Heat Transfer for a Cylinder and a Sphere

The variable fluid flow over a cylinder in cross flow gives rise to similar variations of the local heat transfer. On the front part, up to the separation of a laminar boundary layer, the heat transfer can be determined by either approximate or exact analytical

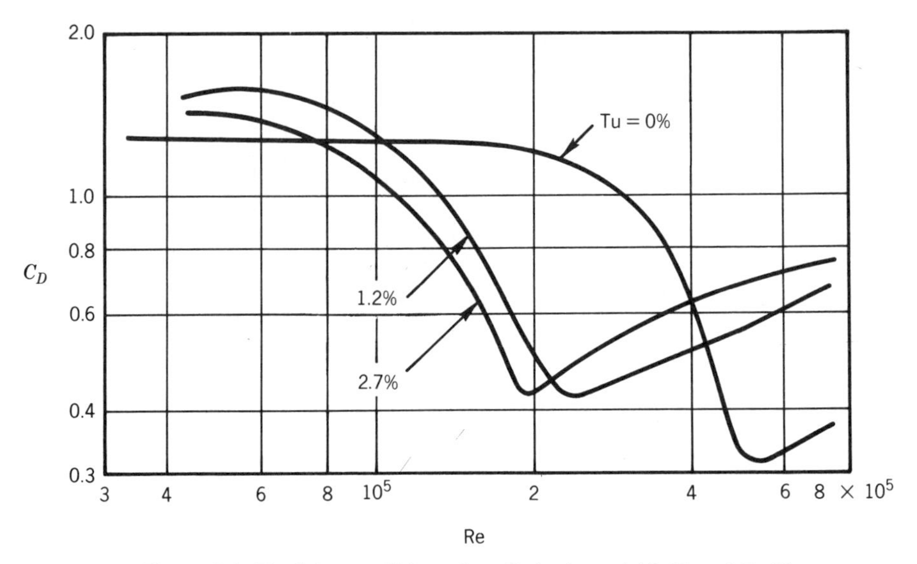

Figure 6.6. Total drag coefficient of a cylinder for variable Tu and Re [1].

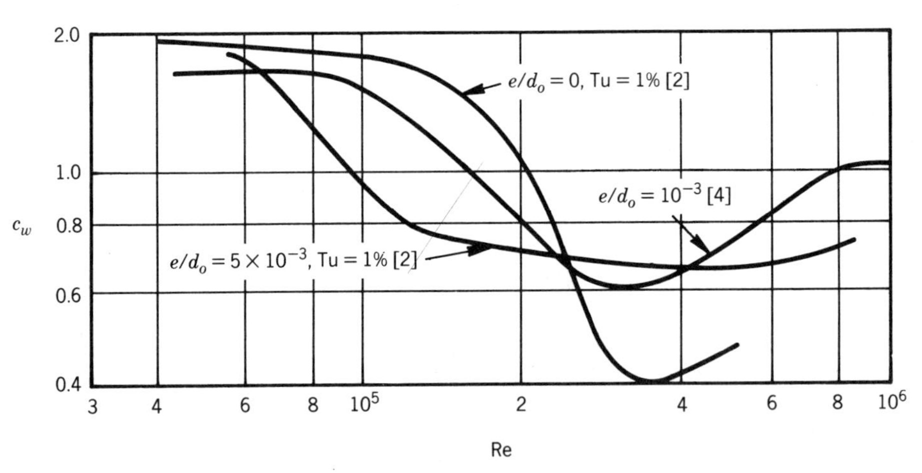

Figure 6.7. The effect of surface roughness of a cylinder on its pressure drag coefficient.

techniques [5, 6]. In the lower range of Re (Fig. 6.8, Re = 500), the heat transfer on the front part of a cylinder is at its maximum. It gradually decreases with the development of a laminar boundary layer. At higher values of Re (Fig. 6.8, Re = 10^4), heat transfer gradually increases downstream of the laminar boundary-layer separation and is mainly determined experimentally [1].

Two heat transfer minima (Fig. 6.8, Re = 2×10^5) are observed in the critical flow regime. The first is at the separation of a laminar boundary layer (ϕ about 84°), where a separation bubble appears. The second minimum is at the separation of a turbulent boundary layer.

In the supercritical flow regime, the first heat transfer minimum (Fig. 6.8, Re = 2×10^6) corresponds to the laminar-turbulent transition in the boundary layer ($\phi \approx 30°$).

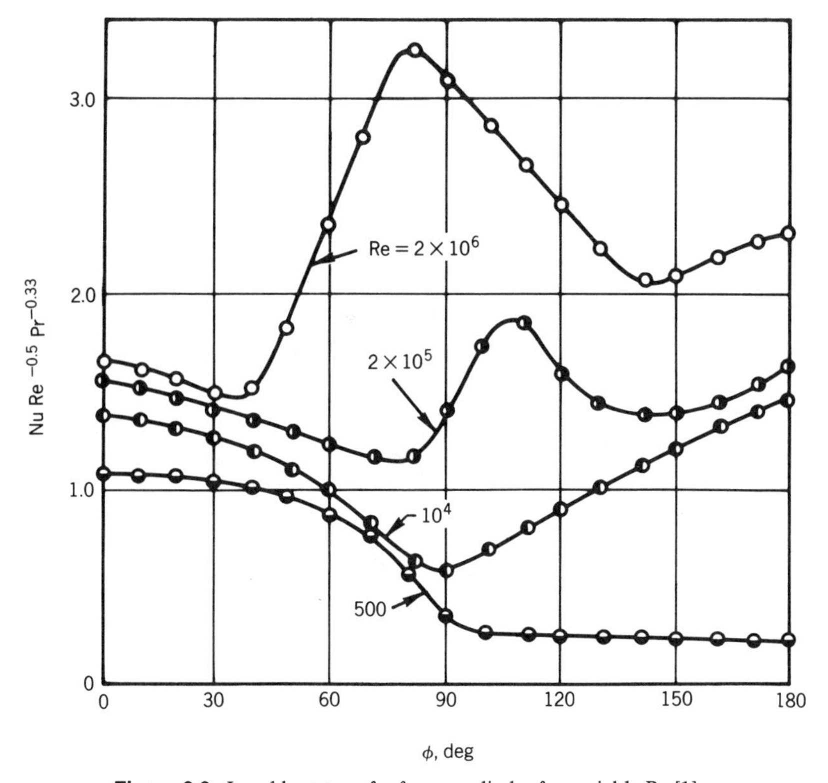

Figure 6.8. Local heat transfer from a cylinder for variable Re [1].

The second minimum is at the separation of a turbulent boundary layer. Although the results of Fig. 6.8 are for Tu = 1%, we note again that the location of the laminar-turbulent transition depends on Re and Tu.

In the supercritical flow regime, an increase of turbulence level causes an upstream shift of the first heat transfer minimum, so that the region of the laminar boundary layer is narrowed. As a result, total heat transfer from the front part is considerably augmented. Effects of higher turbulence levels are in general insignificant in the rear part in the subcritical and critical (not shown in Fig. 6.9), or supercritical flow regime (shown in Fig. 6.9). A higher turbulence level causes heat transfer augmentation on the front part of a cylinder [2].

Heat transfer can be augmented by different influences on the laminar boundary layer, where the thermal resistance is the highest. A higher level of turbulence augments the heat transfer through an external influence on the boundary layer. Internal influences are equally effective. Thus surface elements on a rough heat transfer surface turbulize the laminar boundary layer or even destroy it. Special studies suggest that a 60 to 80% augmentation of the heat transfer can be achieved with an optimal surface roughness [2, 7]. With an increase of Pr, the thermal resistance concentrates in the viscous sublayer, so that heat transfer augmentation can be achieved by lower surface elements. The additional turbulization of the boundary layer by surface elements is similar to the effects of higher turbulence levels. With higher surface elements, heat transfer augmentation is accompanied by an earlier onset of the critical flow regime.

Local heat transfer on a sphere is similar to that on a circular cylinder [8].

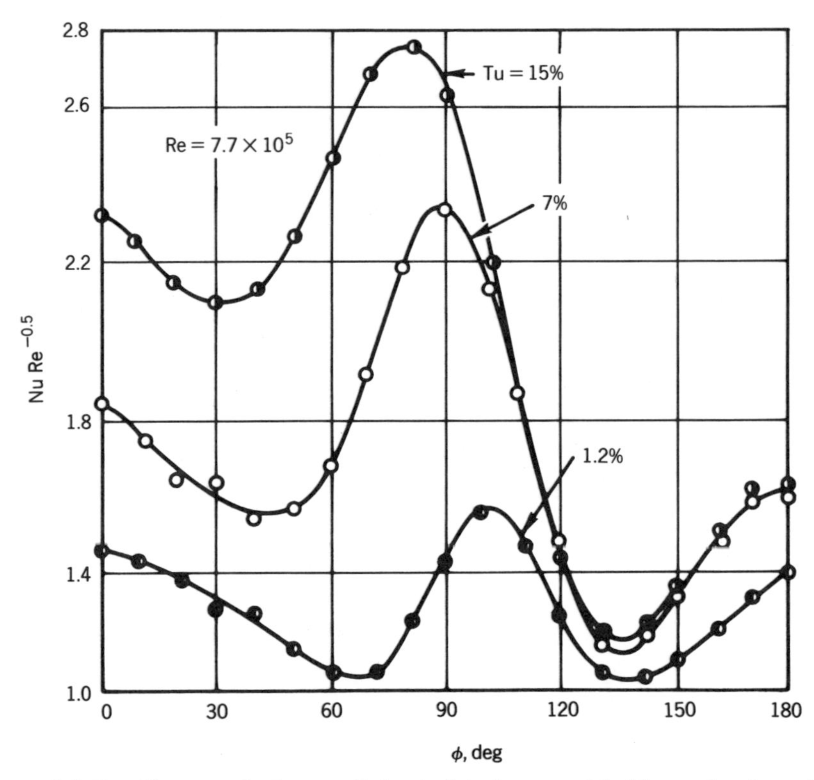

Figure 6.9. Local heat transfer from a cylinder to air in the supercritical flow regime for variable Tu [1].

6.2.4 Average Heat Transfer for a Cylinder and a Sphere

The average Nusselt number for cross flow around a cylinder depends upon the Reynolds and Prandtl numbers. It is shown in Fig. 6.10 for air and water heating with constant fluid properties. From Eq. (6.9),†

$$K = \frac{\mathrm{Nu}}{\mathrm{Pr}^n (\mathrm{Pr}/\mathrm{Pr}_w)^p} = c\,\mathrm{Re}^m \propto \mathrm{Re}^m \tag{6.15}$$

At low values of Re, the slope of the curve, which corresponds to the exponent m of Re, varies from 0.4 to 0.5. In the higher subcritical range of Re, it increases to 0.6. In the critical flow regime ($\mathrm{Re} = 2 \times 10^5$), the behavior of the heat transfer is difficult to define, but in the supercritical flow regime the exponent of Re increases to 0.8. Thus the exponent m of Re varies from 0.4 to 0.8 [1].

The exponent n of Pr is equal to 0.37 in the subcritical flow regime, and 0.4 in the supercritical flow regime (0.37 is shown in Fig. 6.10, but 0.4 is more accurate).

With physical properties evaluated at the bulk mean temperature, their effect on the heat transfer for heating and cooling is satisfactorily approximated by $(\mathrm{Pr}/\mathrm{Pr}_w)^p$. For

† For constant fluid properties, $(\mathrm{Pr}/\mathrm{Pr}_w)^p = 1$ and $n = 0.37$ for the ordinate of Fig. 6.10.

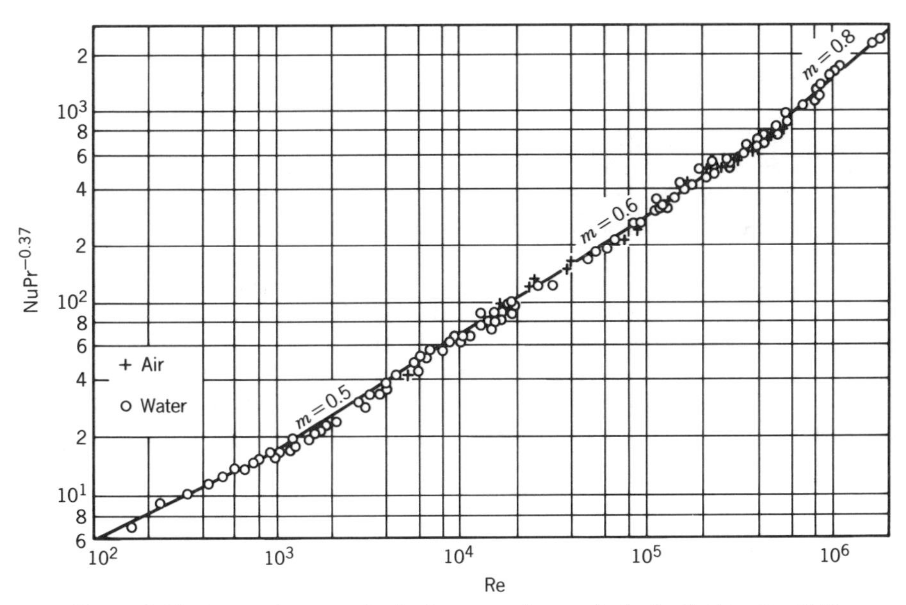

Figure 6.10. Average heat transfer for flow normal to a cylinder for fluid heating [1, 2].

wall-to-fluid heat transfer, the value of p is about 0.25; for fluid-to-wall heat transfer, it is 0.20. A mean value of $p \approx 0.25$ can be assumed for both heat flux directions at moderate temperature differences. The data in Fig. 6.10 are satisfactorily described by Eq. (6.9). For the heat transfer from a cylinder in cross flow [1, 2], we recommend several formulas for different ranges of Re as shown in Table 6.1. Here $\mathrm{Re} = u_0 d_o/\nu$, and the fluid properties are evaluated at the bulk mean temperature.

The heat transfer to and from a sphere is similar to that for a cylinder and is determined mainly by Re. Heat transfer for a sphere is described by the following equations for $\mathrm{Re} < 7 \times 10^4$:

$$\mathrm{Nu} = 2 + (0.4\,\mathrm{Re}^{0.5} + 0.06\,\mathrm{Re}^{0.7})\mathrm{Pr}^{0.4}\left(\frac{\mathrm{Pr}}{\mathrm{Pr}_w}\right)^{0.25} \tag{6.16a}$$

For $4 \times 10^5 < \mathrm{Re} < 5 \times 10^6$ and $\mathrm{Pr} = 0.71$,

$$\mathrm{Nu} = (495.9 + a\,\mathrm{Re} + b\,\mathrm{Re}^2 + c\,\mathrm{Re}^3)\mathrm{Pr}^{0.4} \tag{6.16b}$$

where $a = 5.767 \times 10^{-4}$, $b = 0.288 \times 10^{-9}$, and $c = -3.58 \times 10^{-17}$.

TABLE 6.1. Heat Transfer Correlations for a Cylinder in Cross Flow

Recommended Correlation[a]	Range of Re
$\mathrm{Nu} = 0.76\,\mathrm{Re}^{0.4}\mathrm{Pr}^{0.37}(\mathrm{Pr}/\mathrm{Pr}_w)^p$	10^0–4×10^1
$\mathrm{Nu} = 0.52\,\mathrm{Re}^{0.5}\mathrm{Pr}^{0.37}(\mathrm{Pr}/\mathrm{Pr}_w)^p$	4×10^1–10^3
$\mathrm{Nu} = 0.26\,\mathrm{Re}^{0.6}\mathrm{Pr}^{0.37}(\mathrm{Pr}/\mathrm{Pr}_w)^p$	10^3–2×10^5
$\mathrm{Nu} = 0.023\,\mathrm{Re}^{0.8}\mathrm{Pr}^{0.4}(\mathrm{Pr}/\mathrm{Pr}_w)^p$	2×10^5–10^7

[a] $p = 0.25$ for fluid heating, and $p = 0.20$ for fluid cooling.

6.2.5 Factors Influencing Heat Transfer

Correlations for the heat transfer from cylinders and spheres, presented in the previous section, have been determined at moderate levels of turbulence $\mathrm{Tu} < 1\%$, and hence should be applied preferably in the same conditions. A number of studies [1, 2, 9, 10] suggest augmentation of both average and local heat transfer from cylinders and spheres in cross flow by higher levels of turbulence. In the predictions of the heat transfer, several variables should be considered, but the turbulence level is the most important of them.

An increase of Tu from 1.2 to 15% yields a 50% augmentation of the average heat transfer. An analysis of experimental data [1] also suggests that with $\mathrm{Tu} > 1\%$, the critical flow regime is reached at $\mathrm{Re}\,\mathrm{Tu} > 150{,}000$ where Tu is the turbulence level in percent. Thus to predict the average heat transfer by equations from Table 6.1 at $\mathrm{Tu} \geq 1\%$, an additional factor of $\mathrm{Tu}^{0.15}$ should be introduced, and the general formula becomes

$$\mathrm{Nu} = c\,\mathrm{Re}^m \mathrm{Pr}^n \left(\frac{\mathrm{Pr}}{\mathrm{Pr}_w} \right)^p \mathrm{Tu}^{0.15} \tag{6.17}$$

At low values of Re, heat transfer from a cylinder in cross flow is influenced by free convection, and the influence must be duly evaluated. Free convective motion in the medium is caused by a buoyancy force, which is described by the Grashof number $\mathrm{Gr} = g\beta\,\Delta T\,L^3/\nu^2$.

In the case of mixed convection, the relation becomes

$$\mathrm{Nu} = f(\mathrm{Re}, \mathrm{Gr}, \mathrm{Pr}, \Phi) \tag{6.18}$$

where Φ is the angle between forced convection motion and buoyancy force.

There exist several techniques for evaluating free convection in the case of mixed convection [1, 11, 12]. The concept of an *effective Reynolds number* is suggested by Hatton et al. [11]; its value depends on the angle Φ. The technique of vector summation was applied by Van der Hegge Zijnen [12], so that heat transfer in mixed convection could be determined as a sum of free-convection heat transfer and forced-convection heat transfer:

$$(\mathrm{Nu} - 0.35)^2 = (0.24\,\mathrm{Gr}^{1/8} + 0.41\,\mathrm{Gr}^{1/4})^2 + (0.5\,\mathrm{Re}^{0.5})^2 \tag{6.19}$$

With an increase of Re, heat transfer approaches asymptotically the forced convection value, and Eq. (6.19) for air flow reduces to [12]

$$\mathrm{Nu} = 0.35 + 0.5\,\mathrm{Re}^{0.5} \tag{6.20}$$

Heat transfer from rough-surface cylinders has been studied by Achenbach [4, 7] in air, and by Žukauskas et al. [2] in water (Fig. 6.11). In air, the average heat transfer from cylinders could be augmented with an increase of relative roughness e/d_o, where e is the roughness height. Similar to the effect of turbulence level, a higher surface roughness causes the onset of the critical flow regime at lower Re. In viscous liquids, thermal resistance is concentrated in the viscous sublayer. For efficient heat transfer in such flows, lower surface elements are preferable.

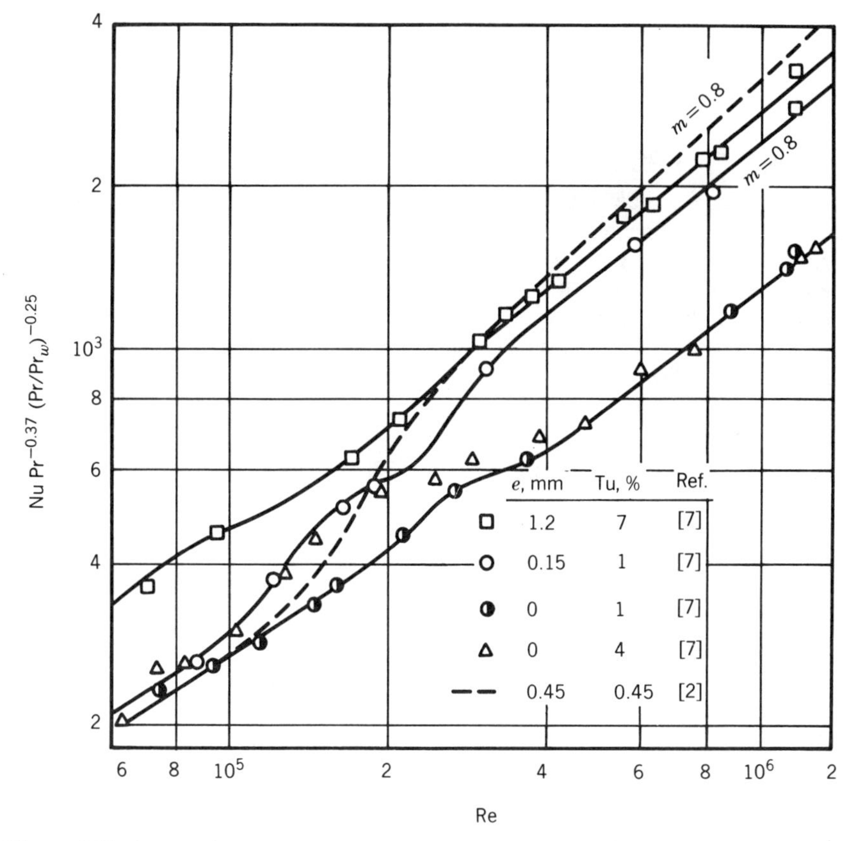

Figure 6.11. Average heat transfer from a rough surface cylinder for different Tu. The data from Ref. 7 are for air; those from Ref. 2 are for water.

6.2.6 Heat Transfer from Tubes of Different Shapes

Cross flow heat exchangers employ tubes of circular, elliptical, and other cross-section geometries. Different geometries introduce additional complications in the heat transfer prediction techniques. Žukauskas and Žiugžda [2] proposed the following relation for average heat transfer from elliptical and circular tubes:

$$\mathrm{Nu}_{d_1} = 0.27\,\mathrm{Re}_{d_1}^{0.6}\,\mathrm{Pr}^{0.37}\left(\frac{\mathrm{Pr}}{\mathrm{Pr}_w}\right)^{0.25} \tag{6.21}$$

with d_1, the flow-parallel elliptic axis, used as the characteristic dimension in Nu and Re.

Knowledge of the heat transfer behavior in different parts over bodies wetted by either laminar or turbulent boundary layers, enables one to predict heat transfer by the following equation [13, 14, 37]:

$$\mathrm{Nu}_{L_0} = c + \sqrt{\mathrm{Nu}_{L,\mathrm{lam}}^2 + \mathrm{Nu}_{L,\mathrm{turb}}^2} \tag{6.22}$$

where L_0 is the stream length, or length along the surface, e.g. $L_0 = \pi d_0/4$ for $\phi = 0$

to 90° along the circular-cylinder surface. According to Gnielinski [14], $c = 0.3$. The value of $\mathrm{Nu}_{L,\mathrm{lam}}$ has to be calculated from equations for a laminar boundary layer, and $\mathrm{Nu}_{L,\mathrm{turb}}$ from turbulent boundary-layer correlations on smooth flat plates.

6.2.7 Heat Transfer From Tubes in Narrow Channels

Heat transfer from a tube bounded by two flat walls depends on the ratio of the tube diameter (d_0) to the channel height or spacing between the two plates (e_1).

Heat transfer studies [1, 2] at Re from 10^3 to 10^6 showed that at constant Re and Pr, heat transfer increases with $\psi = d_0/e_1$ and depends on the blockage factor. The experimental results are satisfactorily approximated by the relations in Table 6.1, but the reference mean velocity u_0 in the minimum space in Re must be replaced by a reduced velocity value $u_* = (1 - \psi^2)u_0$. Then for the subcritical flow regime, the Nu correlation from Table 6.1 is modified as

$$\mathrm{Nu} = 0.26\left[(1 - \psi^2)\mathrm{Re}\right]^{0.6}\mathrm{Pr}^{0.37}\left(\frac{\mathrm{Pr}}{\mathrm{Pr}_w}\right)^{0.25} \tag{6.23}$$

and for the supercritical flow regime,

$$\mathrm{Nu} = 0.023\left[(1 - \psi^2)\mathrm{Re}\right]^{0.8}\mathrm{Pr}^{0.4}\left(\frac{\mathrm{Pr}}{\mathrm{Pr}_w}\right)^{0.25} \tag{6.24}$$

If the free-stream velocity u_∞ is known, the reference velocity u_{**} in Re, according to Perkins and Leppert [13], can be found from the following equation:

$$u_{**} = u_\infty\left(1 - \frac{\pi}{4}\frac{d_0}{e_1}\right)^{-1} \tag{6.25}$$

6.2.8 The Effect of Yaw Angle on the Heat Transfer from a Cylinder

The average heat transfer from a single tube decreases with a decrease in the yaw angle, and is particularly low at small yaw angles. The effect is similar to tube bundles as shown in Fig. 6.22 (Sec. 6.3.3). A vast body of results on the heat transfer from yawed cylinders has been generalized by Morgan [5].

6.3 HEAT TRANSFER AND PRESSURE DROP FOR SMOOTH-TUBE BUNDLES

6.3.1 Fluid Dynamics in Smooth-Tube Bundles

A circular tube bundle is one of the most common heat transfer surfaces, particularly in shell-and-tube exchangers. Fluid flow is ideally normal to the tubes ($\beta = 90°$), but some tubes operate at different yaw angles β to the flows.

The most common tube arrays are staggered and inline (Fig. 6.12), although other arrangements are possible. Bundles are described by the ratio of the transverse ($X_t^* = X_t/d_o$), longitudinal ($X_l^* = X_l/d_o$), or diagonal ($X_d^* = X_d/d_o$) pitch to the tube diameter.

Inside a bundle, the flow converges in the intertube spaces and forms a highly turbulent flow over the inner tubes. The recirculation region in the rear of an inner tube

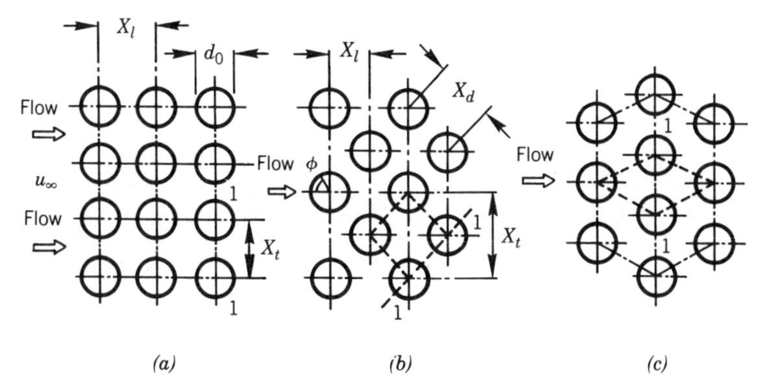

Figure 6.12. Most common types of tube bundle arrangements: (a) inline, (b)–(c) staggered. Minimum intertube spacing at section 1-1 between two tubes.

is smaller than in a single tube. The situation is governed by the relative pitches and the bundle geometry. The more compact a bundle is, the larger is the difference from the single-tube situation. Some differences depend on the number of longitudinal rows because of the inlet-outlet effects.

Fluid flow inside a staggered bundle may be compared to a periodically narrowing and widening channel.

Flow inside inline bundles approaches that in straight channels, and the mean velocity distribution in the minimum intertube space of a transverse row is highly influenced by the relative pitches.

The leading tubes induce a vortical flow and a variable velocity distribution around the inner tubes. At low Re, the inside flow is predominantly laminar with large vortices in the recirculation regions. Their effect on the front parts of inner tubes is eliminated by viscous forces and by negative pressure gradients. Laminar boundary layers are still formed on the inner tubes which separate and form recirculation region in the rear. This pattern may be called a predominantly laminar flow. It is observed at Re < 1000.

Significant changes are introduced at higher values of Re. The intertube flow becomes vortical and highly turbulent. On inner tubes, in spite of high turbulence, laminar boundary layers are still observed.

A negative pressure gradient on the front part of an inner tube causes an acceleration of the flow. The boundary layer is thin and changes but little with the distance from the front stagnation point.

Both the intensity of turbulence and its generation in the intertube spaces are governed by the bundle geometry and Re. With shorter transverse pitches, the velocity fluctuations become more intensive.

The turbulence level of the main flow can influence fluid dynamics only over the first and second rows. A tube bundle acts as a turbulizing grid, and establishes a specific level of turbulence. Highly turbulent transient flows are observed on inner tubes in the intertube spaces.

In most bundles, steady-state flow begins on the third row. On an inner tube, the distribution of pressure and velocity is very different from that on a single tube. In Fig. 6.13, we see circumferential distribution of the pressure coefficient over the leading rows and over the fourth row of an inline and a staggered bundle with $X_t^* = 2.0$ and $X_l^* = 2.0$ at Re $= 10{,}800$.

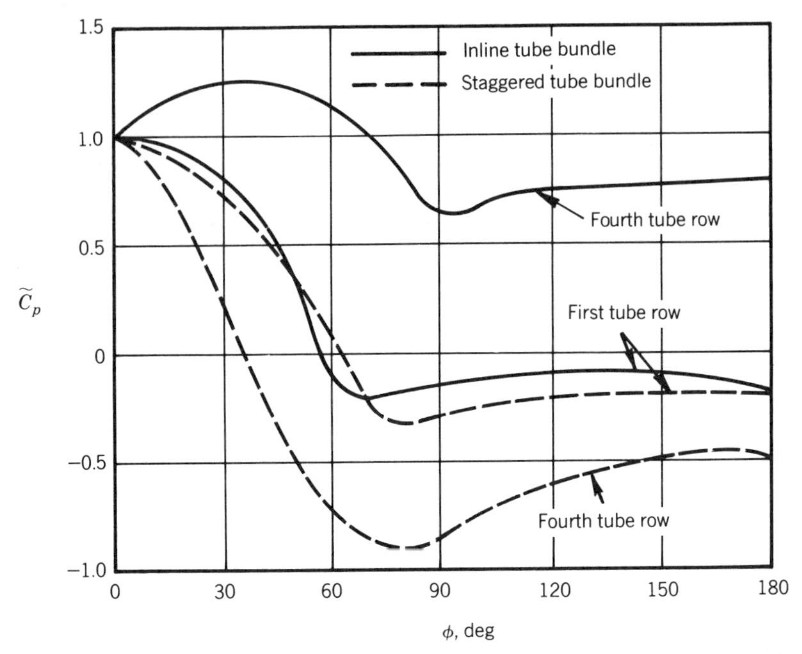

Figure 6.13. Pressure coefficient on a tube in the first and fourth tube rows of an inline bundle and a staggered bundle; air as the test fluid [1].

The pressure coefficient of a tube in a bundle is given by

$$\tilde{C}_P = 1 - \frac{P_{\phi=0} - P_\phi}{\frac{1}{2}\rho u_0^2} \tag{6.26}$$

where u_0 is the mean velocity in the minimum intertube space. It is obvious from Fig. 6.13 that in the subcritical flow regime,† fluid flow phenomenon over tubes of the leading row is very similar to that over a single tube. But on an inner tube of a staggered bundle, higher pressure coefficients than on a single tube precede the separation. On an inner tube of an inline bundle, a maximum pressure coefficient is observed at $\phi = 40°$, at the point of attack. This means that there are two points of attack and two pressure maxima on a tube inside an inline bundle.

On a tube inside a staggered bundle, in a way similar to a single tube, the stream is split at the front stagnation point, and a laminar boundary layer begins to develop. A certain influence of Re is also evident. But unlike a single tube, based on heat transfer results, the boundary layer on a tube inside a staggered bundle separates at $\phi = 150°$. The separation is preceded by a laminar-turbulent transition, where the transition point is dependent on Re. At high values of Re, the boundary-layer separation fluctuates in the range of $\phi = (150 \pm 5)°$ [1].

In symmetric staggered bundles of increased compactness (such as $X_t^* = X_l^* = 1.25$ in Fig. 6.14), $\tilde{C}_P$ becomes a function of Re as early as $\phi \approx 20°$ for a tube inside the

† The subcritical, critical, and supercritical regimes for a tube bundle are defined the same way as for a single tube, as shown in Fig. 6.5.

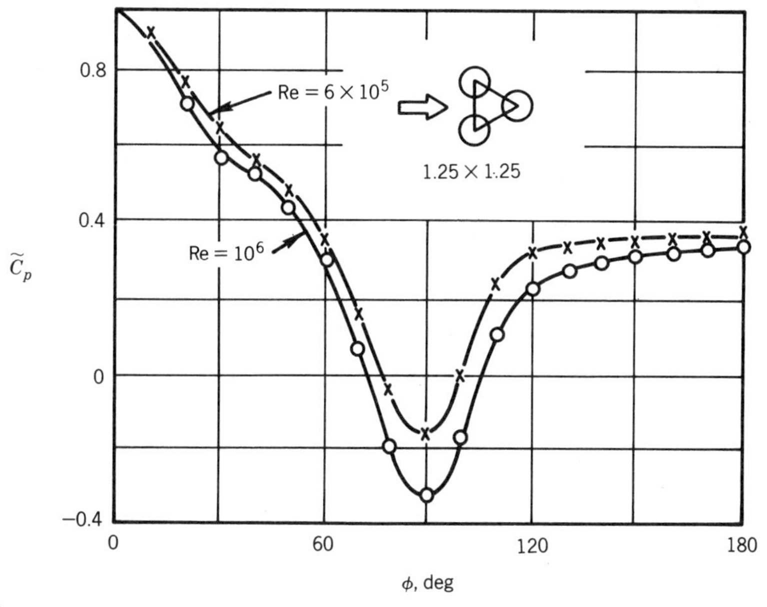

Figure 6.14. Pressure coefficient on a tube in the inner rows of a staggered bundle in air [1].

tube bundle. For ϕ from 30° to 55°, the curves have a slight kink due to the flow separated from the preceding tubes. The distribution of $\tilde{C}_P$ on the rest part of a tube is the same as in an asymmetric staggered bundle. In symmetric inline bundles (such as $X_t^* \times X_l^* = 1.25 \times 1.25$) in the subcritical range of Re, a maximum value of $\tilde{C}_P$ is observed at $\phi = 40°$ as shown in Fig. 6.15. Its minimum value lies at ϕ from 93 to 97°. In the rear part of a tube, $\tilde{C}_P$ increases, but is nearly constant after the separation. The curves of $\tilde{C}_P$ for inner rows of inline bundles become similar to those for staggered bundles, when the supercritical flow regime is established.

In studies of the velocity distribution in tube bundles, the effect of transverse pitch should never be ignored. The rate of velocity growth in intertube spaces is mainly determined by the transverse pitch, and it increases drastically at narrower pitches.

The mean velocity in the minimum intertube space of a transverse tube row for a variable transverse pitch and a constant flow rate is given by

$$u_0 = u_\infty \frac{X_t^*}{X_t^* - 1} \tag{6.27}$$

and the mean velocity in a free flow area at an angle ϕ from the front stagnation point (see Fig. 6.12) is given by

$$u_\phi = u_\infty \frac{X_t^*}{X_t^* - \sin \phi} \tag{6.28}$$

The velocity distributions in the intertube spaces are notably different in staggered and inline bundles.

Over inner tubes of staggered bundles, the point of attack is located at the stagnation point, and the maximum velocity occurs in the minimum intertube spaces.

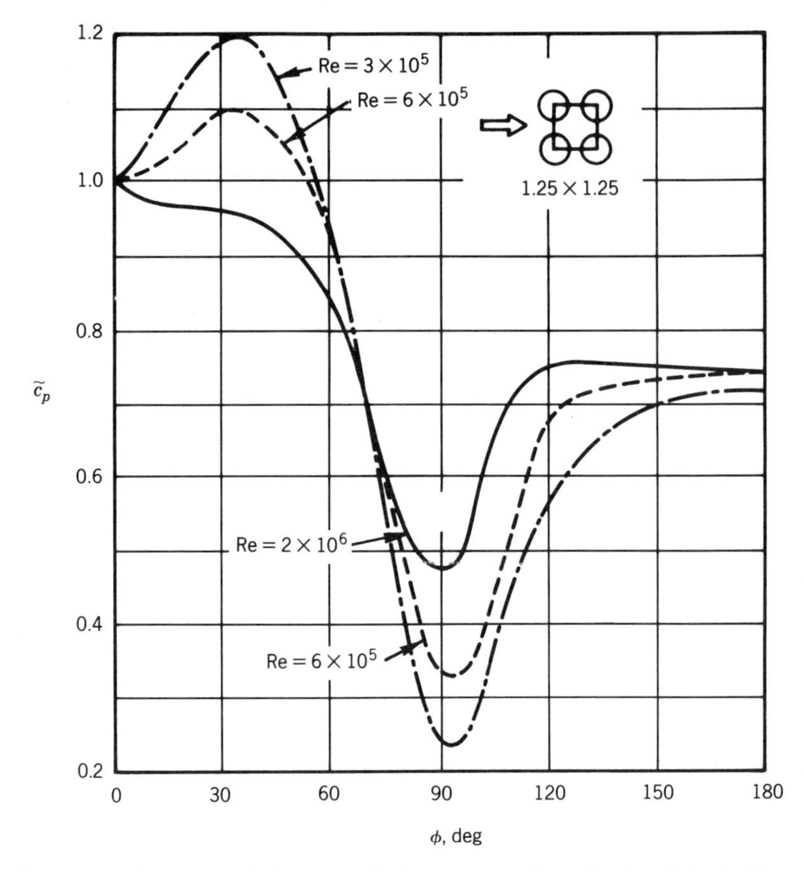

Figure 6.15. Pressure coefficient on a tube in inner rows of an inline bundle in air [1].

The fluid dynamics is mainly determined by the location of the minimum free flow area at either $\phi = 90$ or $\phi \gtrless 90°$. In the rear part of an inner tube, the velocity sharply decreases, and even acquires a negative value in the recirculation region as shown in Fig. 6.16.

Over inner tubes of inline bundles, the point of attack lies at $\phi \approx 40°$; the velocity is at its minimum there and increases downstream. Further circumferential variations of local velocities are similar to those in staggered bundles.

6.3.2 Drag on Smooth-Tube Bundles

In tube bundles in cross flow, the total drag also consists of friction and pressure (or profile or form) drag. The drag on a tube in a bundle is described by its skin friction coefficient c_f and pressure drag coefficient c_w. The skin friction coefficient is defined as

$$c_f = \frac{2\tau_w}{\rho u_0^2} \tag{6.29}$$

where τ_w is the shear stress at the wall of a tube.

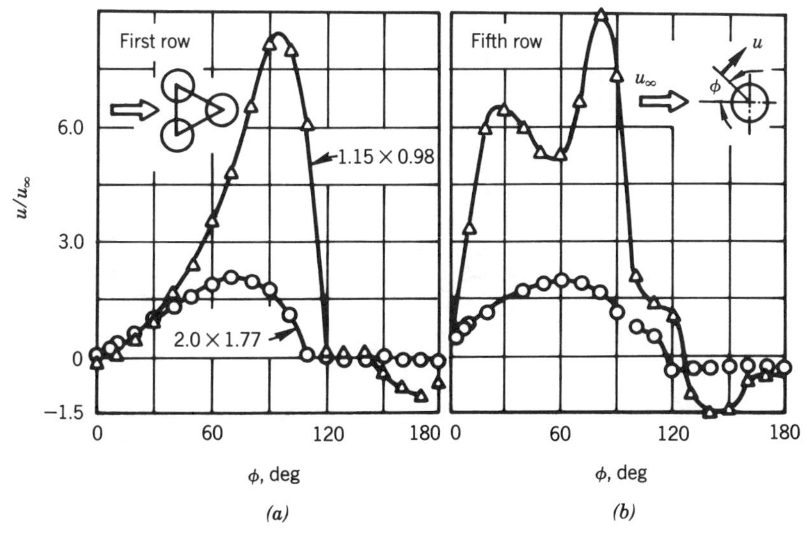

Figure 6.16. Variation of local velocity over a tube in (*a*) the leading row and (*b*) the fifth row of a staggered bundle at the distance $y/d_o = 0.019$ [1].

The local shear stress τ_w is determined as a product of fluid viscosity and velocity gradient at the wall. Thus for known velocity distributions in the boundary layer, the values of $\tau_w(\phi)$ may be determined for a region from the front stagnation point to either the rear stagnation point or the separation point. The total friction drag is evaluated by integrating $\tau_w(\phi)$ over the circumference.

The pressure drag coefficient is defined as

$$c_w = \frac{2F_w}{\rho u_0^2 A} \tag{6.30}$$

where

$$F_w = \Delta A \sum_{i=1}^{n} P_i \cos \phi_i \tag{6.31}$$

$$\Delta A = \frac{\pi d_o L}{j} \tag{6.32}$$

F_w is the longitudinal component of the pressure force, A is the tube cross section perpendicular to the flow, L is the tube length, ΔA is the tube surface element between two generating lines on the tube, drawn through the centers of two measured areas, j is the number of measured areas, P_i is the pressure on the ith measured area over the circumference, and ϕ_i is the angle measured from the stagnation point to the center of the area ΔA around the circumference.

The local skin friction coefficients of three staggered bundles in the lower, medium, and higher ranges of Re are presented in Fig. 6.17. As already mentioned, $c_f = 0$ at the points of attack and of separation. With an increase of ϕ, the value of c_f increases and reaches a maximum at ϕ from 50 to 80°, depending upon Re and the bundle geometry. After the separation, the value of c_f becomes negative, though it never reaches large absolute values.

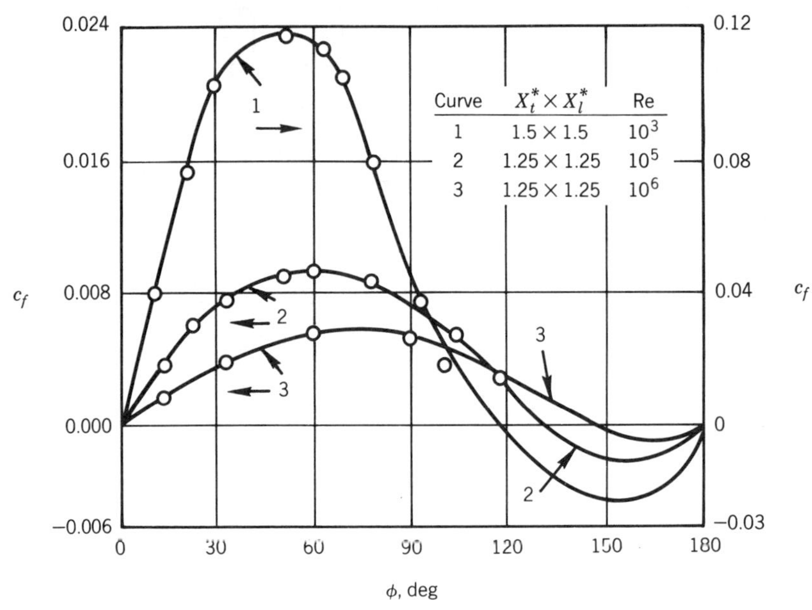

Figure 6.17. Friction coefficient over a tube in a staggered bundle for variable $X_t^* \times X_l^*$ and Re; aviation oil as the test fluid [1].

Over inner tubes of inline and staggered 2.0 × 2.0 bundles, the pressure drag coefficient begins to decrease sharply from Re = 10^2 (Fig. 6.18) and becomes stable at Re from 3×10^3 to 10^4 as well as at Re from 10^5 to 10^6.

At low Re, the values of c_f and c_w are of the same order. At higher Re, the value of c_w exceeds significantly that of c_f. For example, at Re = 3×10^5, the ratio $c_f/c_w = 0.01$ for staggered bundles, and becomes even less at higher Re.

The combined friction and pressure drag over tube bundles in cross flow constitute their hydraulic drag. Consequently, the hydraulic drag depends on the configuration of a bundle. The hydraulic drag is also proportional to the number of longitudinal rows n;

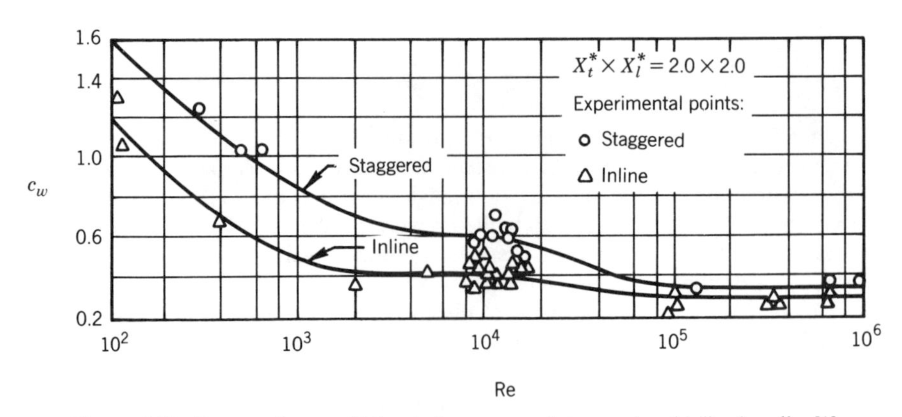

Figure 6.18. Pressure drag coefficient in inner rows of staggered and inline bundles [1].

and for small numbers of tube rows the contribution of the input-output losses of kinetic energy is significant. They may be significant in a short bundle (a small number of tube rows). The total drag in a heat exchanger also depends on the physical properties of the fluid.

On the basis of the above considerations, the pressure drop over a bundle is implicitly given by

$$\Delta P = f(u_0, X_t, X_l, d_o, n, \mu, \rho), \tag{6.33}$$

or in a dimensionless form by

$$\text{Eu} = f\left(\text{Re}, \frac{X_t}{d_o}, \frac{X_l}{d_o}\right) \tag{6.34}$$

where the Euler number $\text{Eu} = 2\,\Delta P/(\rho u_0^2 n)$ and is defined per tube row. Here u_0 is the mean velocity in the minimum intertube spacing.

The results are usually correlated in the following form for engineering use:

$$\text{Eu} = c\,\text{Re}^r \tag{6.35}$$

The design data for the hydraulic drag over banks of tubes are useful for engineering. Such experimental results are presented in Figs. 6.19 and 6.20 from Žukauskas et al. [16] and from recent experiments with different fluids carried out at the Institute of Physical and Technical Problems of Energetics, Academy of Sciences of the Lithuanian SSR. Closed-form equations for the results of Figs. 6.19 and 6.20 are presented in [36]. At high values of Re, the data coincide with the results of numerous measurements by Hammeke et al. [17] and Niggeschmidt [18], and at lower values of Re with the results of Bergelin [19].

The charts of average Eu per tube row for multirow inline bundles are presented in Fig. 6.19 as a function of Re and X_l^* $(= X_t^*)$. For other tube pitches $(X_l^* \neq X_t^*)$, a correction factor χ is first determined from the inset of Fig. 6.19, and then Eu/χ from the main figure to find Eu for a specified inline tube bundle.

The charts of average Eu per tube row for multirow staggered tube bundles with 30° tube layout $[X_t = X_d$ and $X_l = (\sqrt{3}/2)X_t]$ are presented in Fig. 6.20 as functions of Re and X_t^*. For other tube pitches $[X_l \neq (\sqrt{3}/2)X_t]$, the correction factor χ is first obtained from the inset of the figure, and used in Eu/χ obtained from the main figure for a given X_t to find Eu for a specified staggered tube bundle. Eu values determined from Figs. 6.19 and 6.20 have an uncertainty within $\pm 10\%$.

The pressure drop of a multirow bundle is then given by

$$\Delta P = \left(\frac{\text{Eu}}{\chi}\right)\chi \cdot \tfrac{1}{2}\rho u_0^2 \cdot n = \xi \frac{\rho u_0^2}{2} n, \tag{6.36}$$

Here ξ is the hydraulic drag coefficient.

The value of Eu for the whole bundle increases (at a decreasing rate) with the number of tube rows. An average value of Eu for one row is found by dividing the total

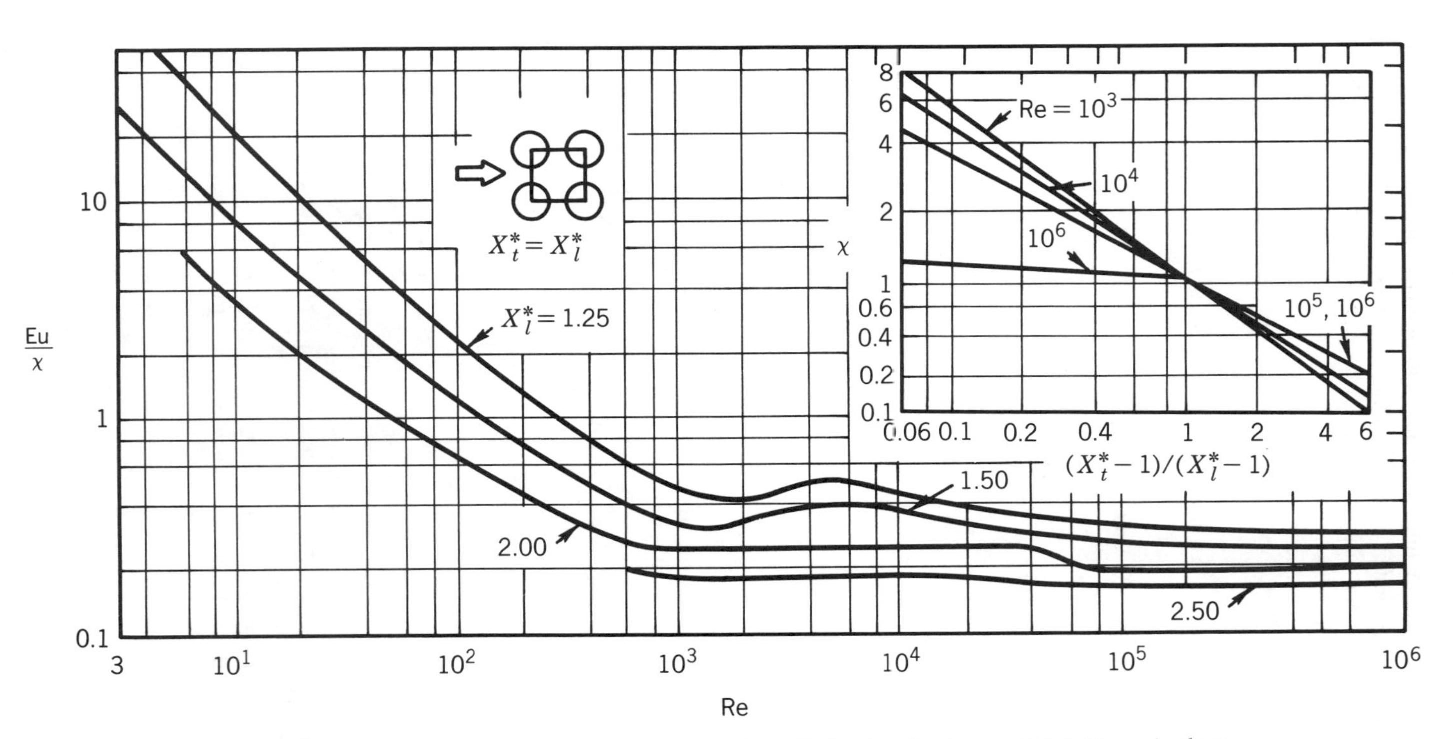

Figure 6.19. A chart for the hydraulic drag coefficient of inline bundles for $n > 9$ [16]. The main chart has the longitudinal pitch as a parameter. The test fluids are air, water, and various oils.

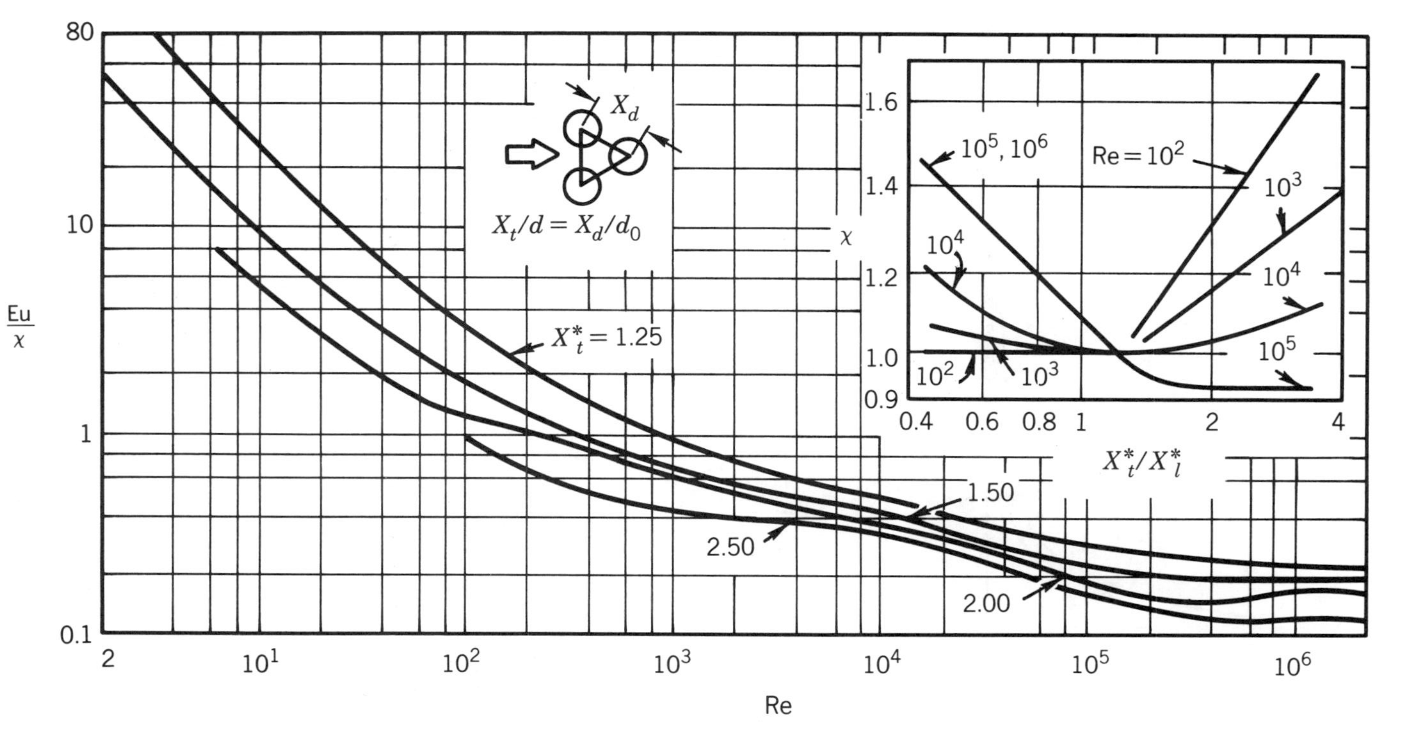

Figure 6.20. A chart for the hydraulic drag coefficient of staggered bundles for $n > 9$ [16]. The main chart has the transverse pitch as a parameter. The test fluids are air, water, and various oils.

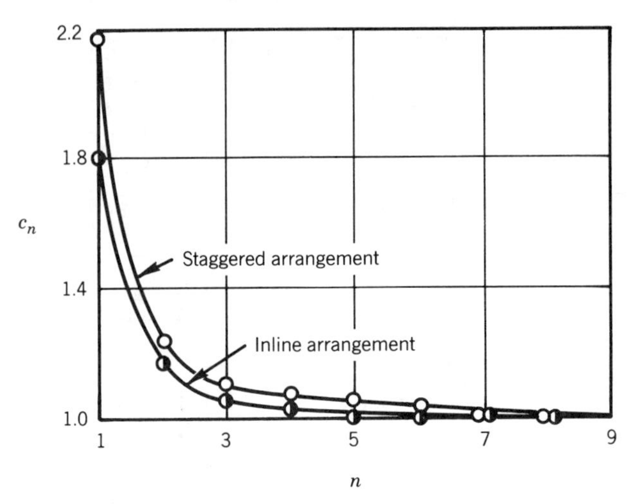

Figure 6.21. Correction factor c_n for the hydraulic drag of short staggered and inline bundles [1].

Eu value by the number of rows. Experimental results indicate that the actual drag on the leading rows is significantly higher than this average value. Therefore, to determine hydraulic drag of short bundles or of leading rows, use Fig. 6.21 for c_n, defined as

$$c_n = \frac{\mathrm{Eu}_n}{\mathrm{Eu}} \tag{6.37}$$

Here Eu_n is the average Euler number per tube row for the n-row bundle with $n \leq 9$, and Eu is the average Euler number per tube row for the n-row bundle with $n > 9$ and is obtained from Fig. 6.19 or 6.20.

The charts in Figs. 6.19 and 6.20 refer to isothermal conditions. They also apply to nonisothermal flows if the fluid physical properties are evaluated at the bulk mean temperature and a correction is applied to account for variable fluid properties for liquids as follows:

$$\mathrm{Eu}_b = \mathrm{Eu}\left(\frac{\mu_w}{\mu}\right)^p \tag{6.38}$$

Here Eu_b is the Euler number for both heating and cooling, Eu the Euler number for isothermal conditions, and μ_w and μ are the liquid dynamic viscosities at the wall temperature and at the bulk mean temperature, respectively. For $\mathrm{Re} > 10^3$, $p \approx 0$. For $\mathrm{Re} \lesssim 10^3$,

$$p = -0.0018\,\mathrm{Re} + 0.28 \qquad \text{for liquid heating} \tag{6.39}$$

$$p = -0.0026\,\mathrm{Re} + 0.43 \qquad \text{for liquid cooling} \tag{6.40}$$

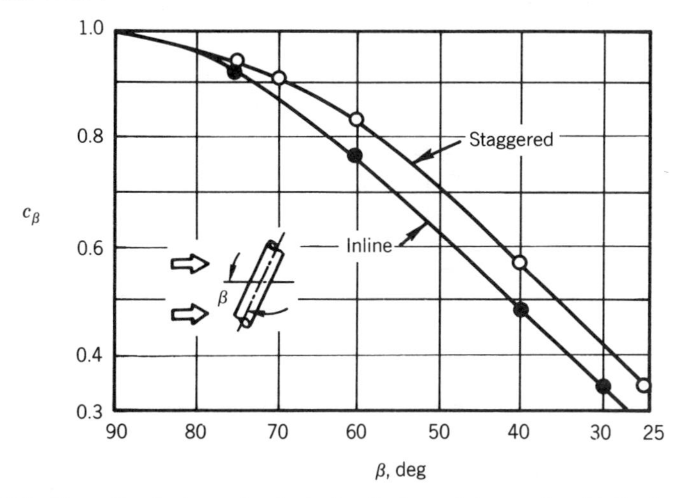

Figure 6.22. Correction factor c_β for the hydraulic drag of staggered and inline bundles [1].

Many other investigations have been reported worldwide, presenting Eu or its variants for various tube bundles. The most notable is by Gaddis and Gnielinski [20], who suggested curves and formulas for the pressure drop for inline and staggered tube bundles.

6.3.3 Drag on Yawed-Tube Bundles

The drag on a yawed-tube bundle depends, as in crossflow cases, on the bundle geometry and pitches, as well as on the Reynolds number, the yaw angle, and some other factors. Figures 6.19 and 6.20 apply to obtain their drag, but with a correction factor for the effect of the yaw angle β.

Based on the experimental results, the drag decreases with decreasing β at constant flow rates. The decrease of drag for $\beta < 90°$ is related to an altered flow in the bundle. The hydraulic drag on a tube consists of the friction drag and pressure drag, and any change of β involves a change of the ratio between the two components, as well as of the flow phenomena.

To determine the hydraulic drag on a yawed-tube bundle ($\beta < 90°$), a correction factor

$$c_\beta = \frac{\mathrm{Eu}_\beta}{\mathrm{Eu}_{\beta=90°}} \tag{6.41}$$

must be introduced for the yaw-angle effect. It is a ratio of Euler numbers defined in Eq. (6.41) for a constant flow rate and is presented in Fig. 6.22.

6.3.4 Local Heat Transfer from a Tube in a Bundle

The general laws governing the distribution of local heat transfer from a tube in a bundle are the same as those for a single tube. The circumferential distribution of the heat transfer is governed by flow phenomena, which in turn depend on the bundle geometry; e. g., two points of attack, and consequently two heat transfer maxima, exist

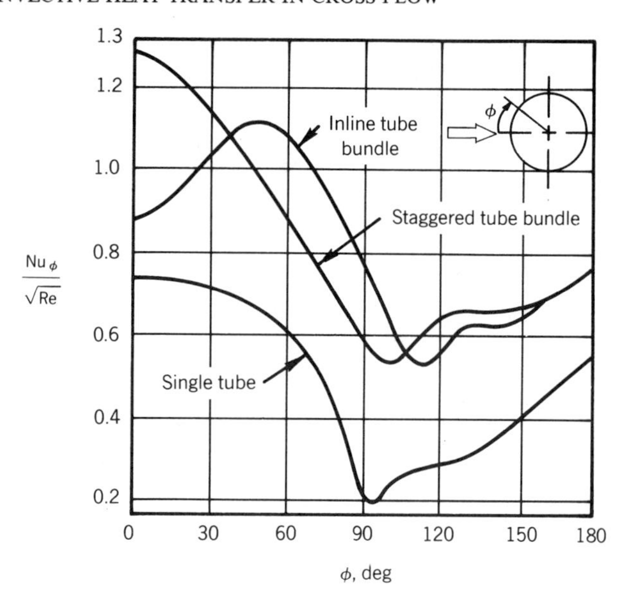

Figure 6.23. Local heat transfer from a single tube, a tube in an inner row of a staggered bundle, and a tube in an inner row of an inline bundle for $X_t^* \times X_l^* = 2 \times 2$ and $\mathrm{Re} = 1.4 \times 10^4$ [16]. Air is the test fluid.

on a tube inside an inline bundle. The flow over a tube inside a staggered bundle is somewhat similar to that in a single tube, but with higher turbulence in the surrounding fluid. A laminar boundary layer on the front part, formed by a split stream for an inline bundle, exists only at the lower values of Re.

Local heat transfer from a tube inside an inline bundle and a staggered bundle is compared with that from a single tube in Fig. 6.23. Inside a staggered bundle, heat transfer from the front part is higher than that on a single tube, because of the impingement of the two upstream fluid streams on the front part and because of higher turbulence. On the remaining part, heat transfer is higher because of higher turbulence. Inside an inline bundle, maximum heat transfer is observed at $\phi \approx 50°$, which is the point of attack of a stream coming from the preceding intertube space. Because of the lateral point of attack, which is reflected in a sharp heat transfer maximum, a laminar boundary layer begins its development not at $\phi = 0°$, but at ϕ from 30 to 50°. At higher angular distances, the heat transfer decreases with growth of the laminar boundary layer thickness.

Figure 6.24 shows peripheral local heat transfer coefficients for tubes in an inline bundle. Tubes of the second row and onwards are under the influence of the leading row; therefore their heat transfer is different. A steady state in the inline bundles is established from the fourth row onwards. Local heat transfer is similar for inline bundles of different relative pitches ($1.25 \le X_t^* \le 2.0$), with the boundary-layer separation at $\phi \approx 120°$.

Now we consider peripheral local heat transfer coefficients for tubes in a staggered bundle. As shown in Fig. 6.25, the trend is similar to the single-tube case for the leading row. A tube in the second row is under a stream which comes from the leading row, so that the heat transfer from its front part is strongly increased.

The flow is turbulized in the two leading rows, and this is immediately reflected in the heat transfer. The circumferential heat transfer from a tube in the third row is

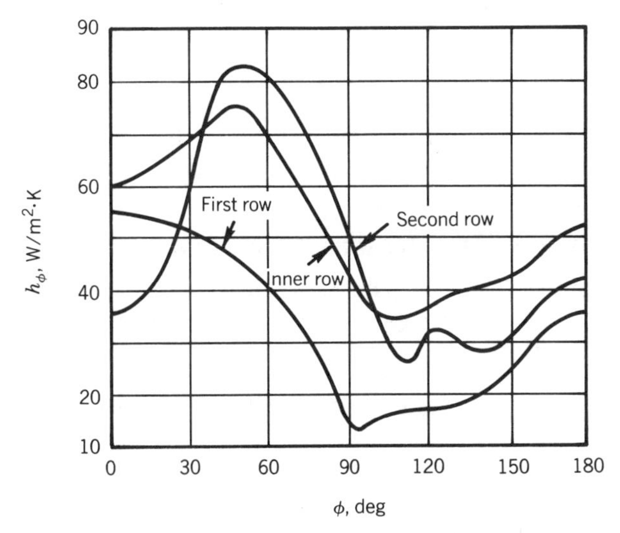

Figure 6.24. Local heat transfer coefficients from a tube in the leading row, the second row, and an inner row of an inline $X_t^* \times X_l^* = 2.0 \times 2.0$ bundle at Re $= 1.4 \times 10^4$ [1, 16]. Air is the test fluid.

higher than that from a tube in the second row. Over the subsequent rows, the steady-state behavior is the same as over the third row. Heat transfer from the rear parts of inner tubes is always higher than that from the first-row tubes, because of higher turbulence.

In the critical flow regime, heat transfer increases at $\phi \approx 120°$, the point of the laminar-turbulent transition. At $\phi \approx 150°$, boundary-layer separation is observed in

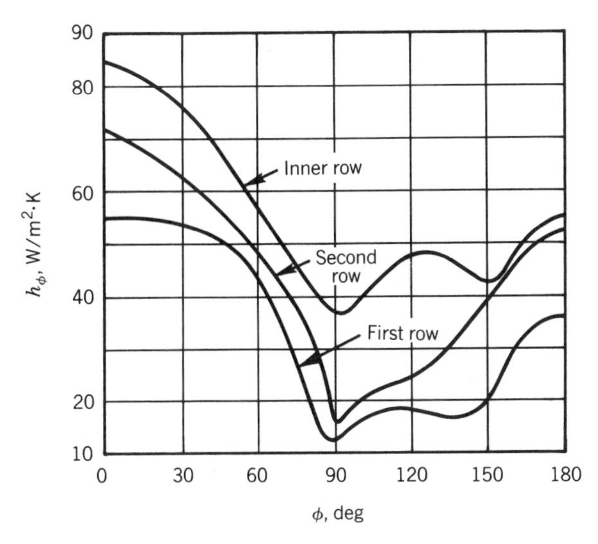

Figure 6.25. Local heat transfer coefficients from a tube in the leading row, the second row, and an inner row ($n \geq 4$) of a staggered $X_t^* \times X_l^* = 2.0 \times 2.0$ bundle at Re $= 1.4 \times 10^4$ [1, 16]. Air is the test fluid.

most configurations. In the supercritical flow regime (for $Re > 2 \times 10^5$) in staggered bundles, the first heat transfer minimum corresponds to the laminar-turbulent transition, which begins to shift upstream and reaches $\phi \approx 25°$ at $Re = 10^6$. The second heat transfer minimum, which corresponds to the separation of a turbulent boundary layer, is practically stable in staggered bundles, and remains at $\phi \approx 150°$ even at higher Re.

6.3.5 Average Heat Transfer from a Tube in a Bundle

The average heat transfer from bundles of smooth tubes is generally determined by Eq. (6.9). Studies [1, 16, 21] in the ranges of Pr from 1 to 10,000 and Re from 1 to 2×10^6 suggest a constant exponent of Pr, which is $n = 0.36$ for the subcritical flow regime and $n = 0.40$ for the supercritical flow regime at low Prandtl numbers ($0.2 \leq Pr \leq 1$). The exponent of the ratio Pr/Pr_w can safely be approximated with a constant value of $p = 0.25$.

The effect of the bundle arrangement on the average heat transfer varies with Re. In the low range of Re ($\lesssim 50$), the heat transfer from a tube in the leading row actually coincides with the single-tube and inner-tube values. In the higher range of Re (> 50), heat transfer rates from inner tubes are higher than those from the leading row because of increased intertube turbulence due to the leading rows acting as turbulizers. In most bundles, steady-state heat transfer is established from the third or the fourth row on. A comparison with heat transfer coefficients from the leading rows and from the inner rows illustrates heat transfer augmentation in bundles due to higher turbulence.

As a rule, heat transfer from inner tubes increases with decreasing X_l. This is in agreement with observations on single tubes mounted at different distances from a turbulizer grid. An exception is presented by inline bundles of short X_l, where heat transfer may be reduced even more with further reduction in X_l. This phenomenon is related to a less intensive recirculation in the rear, especially for low or medium values of X_l.

Because of the higher turbulence, under the influence of a variable tube longitudinal pitch, heat transfer from the inner tube may be from 30 to 100% higher than from the leading row. In most cases, heat transfer from the second row is lower than from the further rows, except for inline bundles at low Re.

Figure 6.26 shows the heat transfer from a staggered and an inline 2.0×2.0 bundle. At $Re < 10^3$, the heat transfer from a tube in the leading row of the inline bundle is about 25% higher than that from the inner tubes. Any inner tube is in the wake of a preceding one, and its heat transfer is decreased accordingly by a lower velocity in the wake—the so-called "shadow" effect.

In contrast to the inline bundle, heat transfer in the staggered case is lower in the leading row than in the inner rows. For a 2.0×2.0 bundle at $Re \approx 30$, the difference reaches 7%, but increases to 35% at higher Re. This is due to different flow phenomena in the two configurations.

In the range of predominantly laminar fluid flow, the exponent of the Reynolds number varies widely with Re. At $Re < 10^2$, we have $m = 0.33$ to 0.4, but $m = 0.5$ at higher Re ($\lesssim 10^3$), and reaches 0.6 or even 0.63 with the increase of Re for the predominantly turbulent fluid flow in the subcritical zone ($Re < 2 \times 10^5$).

For inline bundles at higher Re, the heat transfer from inner rows exceeds that of the first row, as for staggered bundles.

Figure 6.27 presents the average heat transfer from the first and inner rows of a 1.5×1.25 inline bundle, and from the inner rows of 2.0×1.25 inline bundle. At $Re < 2 \times 10^5$, the exponent m of Re is between 0.6 and 0.65, but at higher Re it

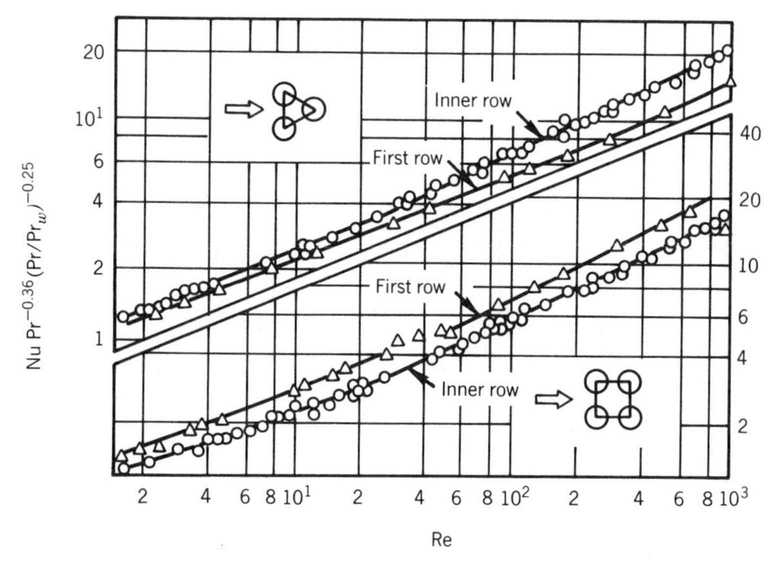

Figure 6.26. Average heat transfer in different rows of a staggered and an inline bundle of $X_t^* \times X_l^* = 2.0 \times 2.0$ at low Re for an inner row and the leading row [1,16]. Aviation oil is the test fluid.

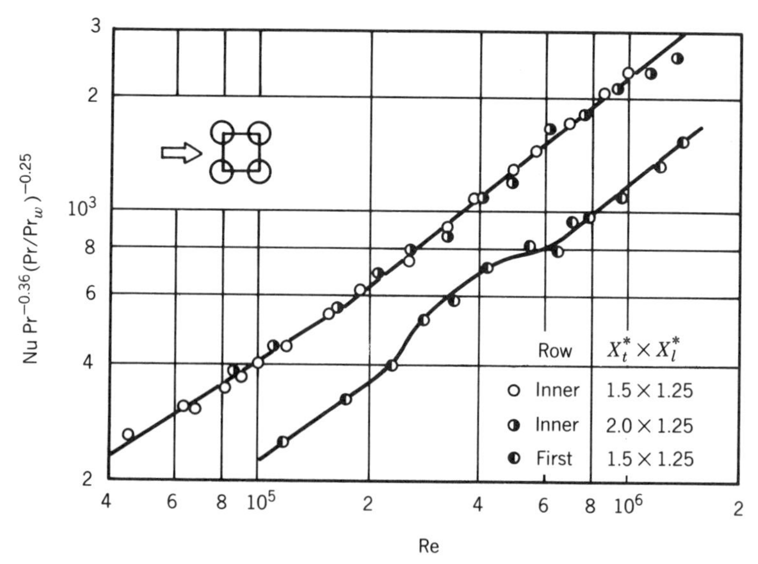

Figure 6.27. Average heat transfer over a tube in an inline bundle for variable $X_t^* \times X_l^*$ [1,16].

increases to a value between 0.76 and 0.8. This growth of m is related to the upstream movement of the laminar-turbulent transition in the boundary layers, as explained in the analysis of the local heat transfer. As a result, with an increase of Re, a larger circumferential part of any inner tube is covered by a turbulent boundary layer. Experiments suggest that the transition to an augmented heat transfer occurs in a

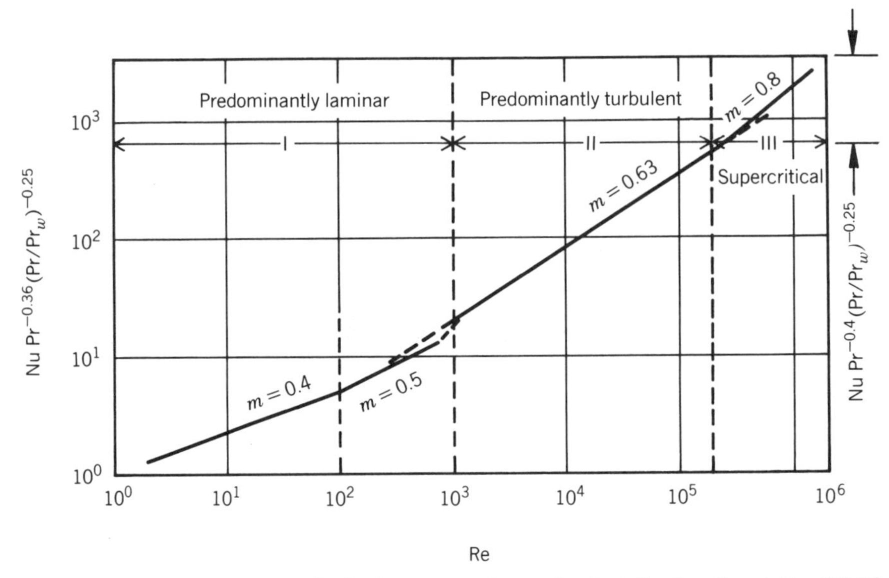

Figure 6.28. Average heat transfer for inner rows of smooth-tube inline bundles at $\beta = 90°$ [1]. The Pr range covered is from 0.7 to 5000.

different way on the first row and on an inner row. For the first row, the heat transfer function K clearly deviates from the exponent law at Re from 2.5×10^5 to 7×10^5, which is similar to a single tube in low turbulence. We already know that this deviation is due to the separation bubble and an altered flow in the rear, and we have already noted the absence of such deviation over a single cylinder in a highly turbulent flow. A similar situation occurs in inner rows of both staggered and inline bundles. For staggered bundles, the value of m is somewhat higher at $\text{Re} > 2 \times 10^5$, and equal to 0.8.

The studies [1, 16] of numerous bundles of different geometries in the subcritical flow regime suggest that in staggered arrangements, heat transfer increases with a decreasing longitudinal pitch, and to a lesser extent, with a decreasing transverse pitch. The variation of c in Eq. (6.9) may be represented by a geometrical parameter X_t^*/X_l^* with an exponent of 0.2 for $X_t^*/X_l^* < 2$. For $X_t^*/X_l^* > 2$, a constant value $c = 0.40$ may be assumed. For inline bundles, the effect of change in either longitudinal or transverse pitch is not so evident, and $c = 0.27$ may be assumed, with certain reservations, for the whole subcritical regime ($10^3 < \text{Re} < 2 \times 10^5$).

The average heat transfer from inline bundles in cross flow ($\beta = 90°$) is shown in Fig. 6.28, and correlating equations are presented in Table 6.2. The uncertainty of these results is within $\pm 15\%$.

TABLE 6.2. Heat Transfer Correlations for Inline Tube Bundles for $n > 16$ [1]

Recommended Correlations	Range of Re
$\text{Nu} = 0.9\,\text{Re}^{0.4}\text{Pr}^{0.36}(\text{Pr}/\text{Pr}_w)^{0.25}$	10^0–10^2
$\text{Nu} = 0.52\,\text{Re}^{0.5}\text{Pr}^{0.36}(\text{Pr}/\text{Pr}_w)^{0.25}$	10^2–10^3
$\text{Nu} = 0.27\,\text{Re}^{0.63}\text{Pr}^{0.36}(\text{Pr}/\text{Pr}_w)^{0.25}$	10^3–2×10^5
$\text{Nu} = 0.033\,\text{Re}^{0.8}\text{Pr}^{0.4}(\text{Pr}/\text{Pr}_w)^{0.25}$	2×10^5–2×10^6

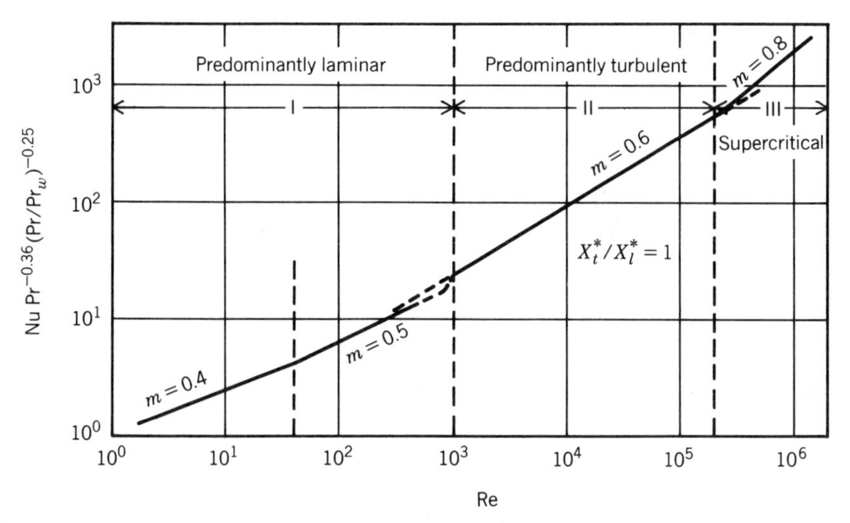

Figure 6.29. Average heat transfer for inner rows of smooth-tube staggered bundles at $\beta = 90°$ [1]. For low Re, $0.7 \leq \text{Pr} \leq 5000$. For high Re, $0.7 \leq \text{Pr} \leq 5$.

TABLE 6.3. Heat Transfer Correlations for Staggered Tube Bundles for $n > 16$ [1]

Recommended Correlations	Range of Re
$\text{Nu} = 1.04\,\text{Re}^{0.4}\text{Pr}^{0.36}(\text{Pr}/\text{Pr}_w)^{0.25}$	10^0–5×10^2
$\text{Nu} = 0.71\,\text{Re}^{0.5}\text{Pr}^{0.36}(\text{Pr}/\text{Pr}_w)^{0.25}$	5×10^2–10^3
$\text{Nu} = 0.35(X_t^*/X_l^*)^{0.2}\text{Re}^{0.6}\text{Pr}^{0.36}(\text{Pr}/\text{Pr}_w)^{0.25}$	10^3–2×10^5
$\text{Nu} = 0.031(X_t^*/X_l^*)^{0.2}\text{Re}^{0.8}\text{Pr}^{0.36}(\text{Pr}/\text{Pr}_w)^{0.25}$	2×10^5–2×10^6

The average heat transfer from staggered bundles in cross flow ($\beta = 90°$) is presented in Fig. 6.29, and correlating equations are presented in Table 6.3. The uncertainty of these results is within $\pm 15\%$.

For a general correlation, a correction $\tilde{c}_n$ for the number of tube rows should be introduced because the shorter the bundle, the lower its average heat transfer. The influence of the number of tube rows becomes negligible only for $n > 16$, as found from Fig. 6.30.

6.3.6 The Effect of Yaw Angle on the Heat Transfer

Heat transfer surfaces can be mounted for crossflow operation at $\beta = 90°$, or at other yaw angles $\beta < 90°$. To avoid overheating, the circumferential distribution of heat transfer and heat flux densities should be known for arbitrary yaw angles.

Based on experiments, Žukauskas [1] and Kazakevich [22] have determined the effect of yaw angle on the local and mean Nusselt numbers with different fluids. Their results are summarized in terms of a correction factor $\tilde{c}_\beta$ as shown in Fig. 6.31 and defined as

$$\tilde{c}_\beta = \frac{\text{Nu}_\beta}{\text{Nu}_{\beta=90°}} \tag{6.42}$$

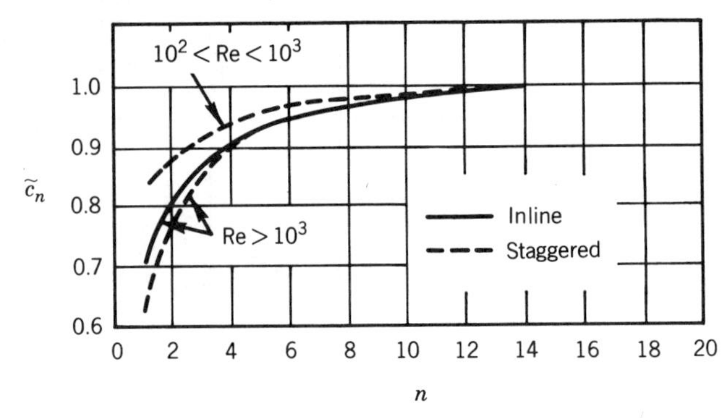

Figure 6.30. Correction factor for the number of rows for the average heat transfer from tube bundles [1, 16].

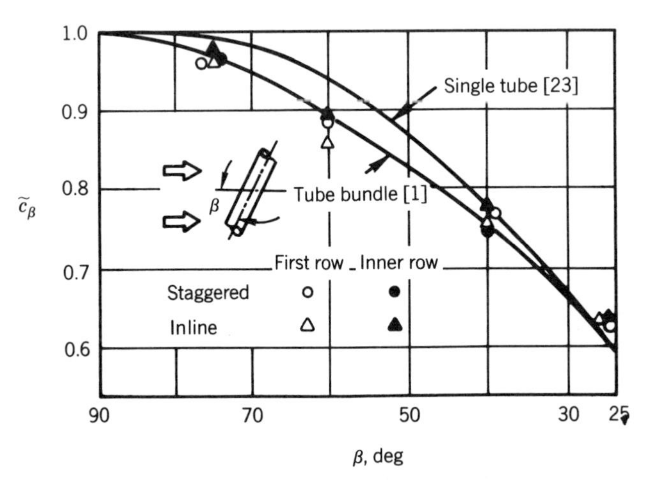

Figure 6.31. Correction factor for the yaw angle for heat transfer from bundles.

Here both Nu_β and $Nu_{\beta=90°}$ are evaluated at the same Re. The data for staggered and inline configurations are approximated by a single curve, and $\tilde{c}_\beta$ decreases with decreasing β.

The plot also includes data on the effect of yaw angle on the heat transfer from a single tube, from Mikheyev [23]. A similar dependence is evident.

6.4 HEAT TRANSFER AND PRESSURE DROP FOR ROUGH-TUBE BUNDLES

6.4.1 Fluid Dynamics in Rough-Tube Bundles

We consider rough surfaces with two-dimensional and three-dimensional protrusions. Of course, they display different flow phenomena over different surface elements and in their wakes. Such isolated surface roughness elements act as local turbulizers.

The flow over surface roughness elements also depends on their intervening spaces. For sparsely distributed surface roughness elements, in-between recirculation regions are possible. If the spaces are large enough, reattachment and boundary-layer formation can be observed. According to the available data, boundary-layer reattachment occurs at a distance of 5 to 8 times the roughness height.

To describe a rough surface, we introduce the notion of relative roughness e/δ or e/d_o, where e is the roughness height, δ is the velocity boundary-layer thickness, and d_o is the outside diameter of a tube. For turbulent boundary layers, we prefer a dimensionless value of the surface roughness height $e^+ = u^*e/\nu$, where $u^* = (\tau_w/\rho)^{1/2}$ is the friction velocity. Let us distinguish three different types of flow over rough surfaces:

Nonseparated flow, when surface roughness elements are submerged in the viscous sublayer ($e^+ < 5$), and the drag coefficient is close to that of a smooth tube. In this type of flow, the drag coefficient is solely a function of the Reynolds number.

Partial effect of roughness ($5 < e^+ < 70$), when surface roughness elements are higher than the viscous sublayer thickness. Their tips protrude from the viscous sublayer, so that local separations and vorticities are formed. The drag coefficient is influenced both by the Reynolds number and by the height of surface roughness elements.

Complete effect of surface roughness ($e^+ > 70$), when surface roughness elements significantly protrude from the viscous sublayer, each of them forming a separated flow. The drag coefficient is significantly increased.

According to experiments in staggered and inline bundles [24, 25], surface roughness changes the pressure coefficient from that with smooth tubes. With an increase of the surface roughness and the Reynolds number, the separation point of the turbulent boundary layer moves from $\phi = 150$ to 120°.

Results are summarized next for the drag and heat transfer for staggered bundles of $X_l^* \times X_t^* = 1.25 \times 0.935, 1.25 \times 1.25, 2.0 \times 2.0, 2.06 \times 1.37$; Re from 10^3 to 2×10^6; and Pr from 0.7 to 220 [26–28].

6.4.2 Drag on Rough-Tube Bundles

In the subcritical and critical flow regimes, under the influence of surface roughness, c_f increases over a large part of the circumference (ϕ from 0 to 120°) up to the separation point. In the supercritical flow regime for the same surface roughness, c_f shows a small decrease. With higher e/d_o in the supercritical flow, the value of c_f increases sharply on that part of the circumference where the velocity u has large gradients. All this illustrates a large effect of surface roughness on the flow in the vicinity of the wall. The hydraulic drag of rough-surface bundles is larger than that of smooth bundles for $e^+ > 5$, because of the additional energy losses in the wake of each surface roughness element. Eu increases with increasing e/d_o.

Figure 6.32 presents the measured hydraulic drag on smooth and rough staggered bundles at Re up to 2×10^6. In the higher range of Re, the hydraulic drag coefficient reaches an asymptotic value, dependent upon e/d_o.

The effect of the height of surface elements on the hydraulic drag over a 2×2 staggered bundle is shown in Fig. 6.33,[†] which gives a curve $c_{\text{Eu}} = f(\text{Re})$, where

$$c_{\text{Eu}} = \frac{\text{Eu}_r}{\text{Eu}} \tag{6.43}$$

[†] The results of Fig. 6.33 are presented in closed-form Correlations in Ref. 36.

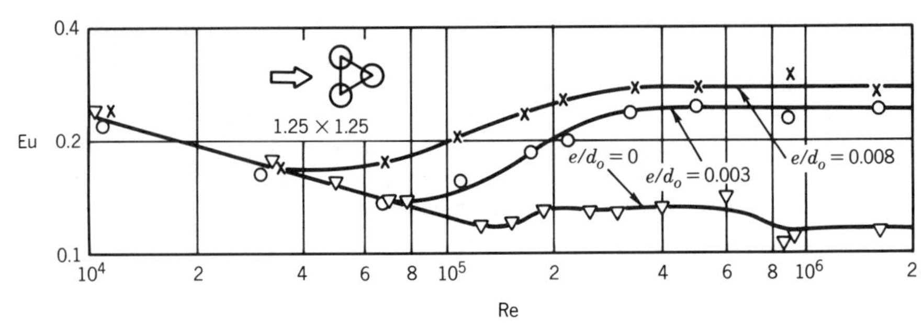

Figure 6.32. Hydraulic drag of a rough-tube staggered bundle of $X_t^* \times X_l^* = 1.25 \times 1.25$ for variable e/d_o [1].

Eu_r is the Euler number for rough tubes and Eu is the value for smooth tubes, both on a per-tube-row basis. It is quite evident that for lower surface roughness elements ($e/d_o = 0.001$), the value of c_{Eu} begins to increase at $Re = 2 \times 10^5$. With larger e/d_o, the increase in c_{Eu} commences at a significantly lower Re [1]

6.4.3 Heat Transfer for Rough-Tube Bundles

Rough surfaces are introduced as a means of augmenting heat transfer by specific fluid dynamic changes in the boundary layer. For similar conditions, the laminar-turbulent transition in the boundary layer occurs at lower Reynolds numbers on a rough surface than on a smooth one. Surface roughness generates local vorticities and enhances heat transport from the wall. The intensity of such heat transport is governed by the surface roughness height e and the velocity boundary layer thickness δ. For e significantly lower than δ, vorticities generated by surface roughness elements do not exert any noticeable influence on the heat transfer. Heat transfer augmentation in gases is observed when the values of e and δ are about the same.

The effect of surface roughness on heat transfer depends also on the Prandtl number of the fluid. For higher Pr, the effect is more pronounced because of a larger concentration of thermal resistance in the vicinity of the wall. In viscous liquids, lower surface roughness elements are necessary in order to turbulize the viscous layer.

Figure 6.34 presents average heat transfer of rough staggered tube bundles in the range of Re from 4×10^2 to 2×10^6, X_t^* from 1.25 to 2.0, and X_l^* from 0.935 to 2.0 [26–28]. At $Re \lesssim 2000$, the effect of surface roughness on heat transfer is observed only

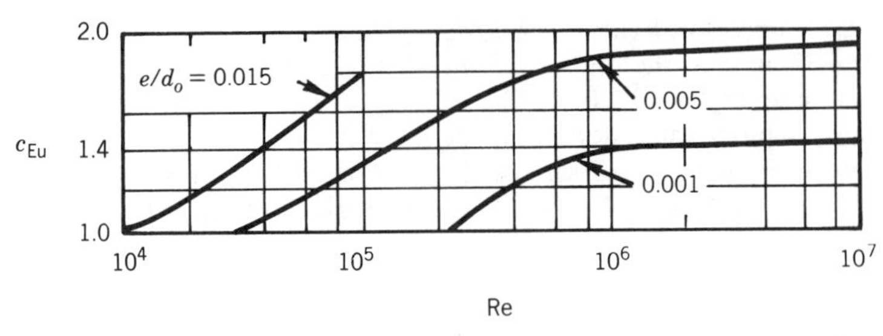

Figure 6.33. Correction factor for the effect of surface roughness in the prediction of hydraulic drag [1]. The results are for a staggered bundle ($X_t^* = X_l^* = 2.0$) using water as the test fluid.

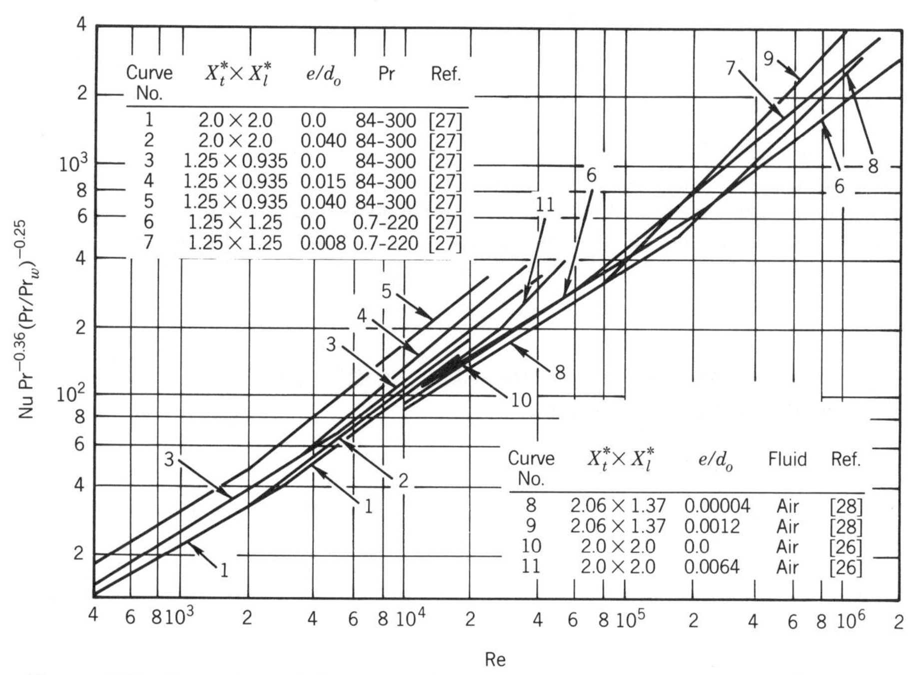

Figure 6.34. Comparison of the average heat transfer from staggered bundles for various $X_t^* \times X_l^*$ and e/d_o.

in viscous liquids, but in air it is observed at Re from 10^4 to 7×10^4. Thus in air, inside a compact 1.25×1.25 bundle, the heat transfer augmentation is 14% at Re $= 10^5$, but reaches 75% at Re $= 10^6$.

The effect of surface roughness height on the average heat transfer in air and in a liquid with Pr = 84 is shown in Fig. 6.35†, in terms of c_{Nu} vs. Re for staggered bundles, where

$$c_{\text{Nu}} = \frac{\text{Nu}_r}{\text{Nu}} \tag{6.44}$$

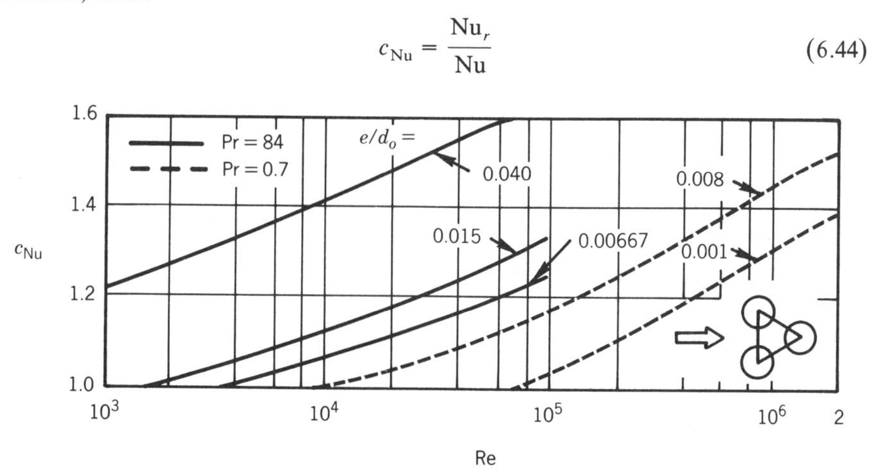

Figure 6.35. Correction factor for the effect of surface roughness in the prediction of average heat transfer for a 1.25×0.935 staggered tube bundle at various Pr and e/d_o [1].

† The results are accurate within $\pm 15\%$.

Nu_r is the Nusselt number for rough tubes, and Nu is the Nusselt number for smooth tubes. c_{Nu} represents the heat transfer augmentation due to surface roughness on staggered bundles [27].

Figure 6.35 reveals that on rough surfaces, heat transfer augmentation commences at lower Re in viscous liquids than in air, and the rate of augmentation c_{Nu} is higher.

Heat transfer augmentation by surface roughness is accompanied by an increase of the hydraulic drag coefficient because the flow energy is consumed in the generation of local vorticities. Thus to augment heat transfer in compact bundles, surface elements of e from 0.4 to 0.8 mm are recommended for liquids, and of 0.8 to 2.0 mm for air.

6.5 HEAT TRANSFER AND PRESSURE DROP FOR FINNED-TUBE BUNDLES

6.5.1 Finned-Tube Bundles

Fins and other extended surfaces are more and more commonly used in heat exchangers today. Higher heat transport from finned tubes is achieved both through the higher heat transfer coefficients and through the extended surfaces.

Many different fin geometries are currently in use, though the most common are annular and continuous fins. Design data are available in the open literature for fabrication, and experiments have been performed on many finned surfaces, such as individually finned tubes and flat, wavy, or louvered continuous fins. Finned surfaces are common in recuperative heat exchangers with gas flows in order to improve their performance. Because of high heat transfer coefficients associated with liquid flows, only fins of low heights (having reasonably high fin efficiency) are used.

No universal equivalent diameter has been found to correlate the design data on finned tubes. Many different expressions are available in the literature for the equivalent diameter. We choose the outside tube diameter of the base tube in correlating the experimental data and in all of the correlations presented in this section.

High fins introduce changes in the interfin flow phenomena, and smaller interfin spaces give large increase in flow resistance of the bundles. Hence, an important geometrical variable for finned tubes is the fin factor F, which is defined as the finned-tube outside total area (primary plus secondary) divided by the bare base-tube surface area [29]. For tubes with annular fins, it is given by

$$F = \frac{l}{S}\left[\frac{2e_f}{l}\left(1 + \frac{e_f}{d_o}\right) + \frac{\delta_f}{l}\left(1 + \frac{2e_f}{d_o}\right) + 1\right] \tag{6.45}$$

where $l = (S - \delta_f)$ is the interfin spacing (Fig. 6.36), δ_f is the fin thickness, and d_o is the tube diameter at the fin base.

Designers of finned tubes in different industrial applications operate with e_f/l as an important design variable.

6.5.2 Fluid Dynamics in Finned-Tube Bundles

An inner tube contains a flow of increased turbulence; therefore its separation point is farther downstream, and the recirculation region in its rear is narrower, than for a tube in the first row.

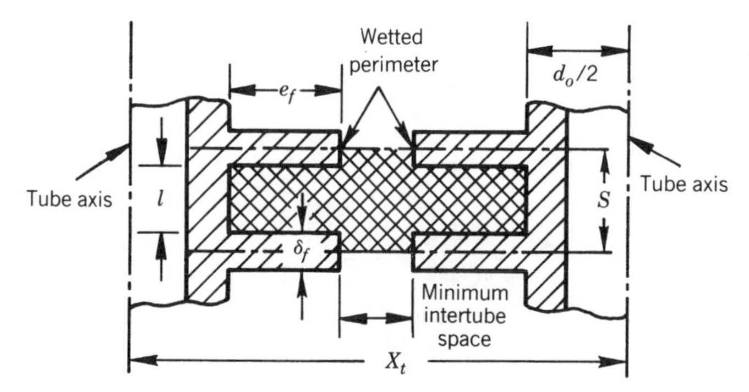

Figure 6.36. Nomenclature for flow over a bank of annular finned tubes. Fluid flow direction is perpendicular to the plane of the paper.

Deviations in flow phenomena from the smooth-tube case are negligible for low fins and large interfin spacing. With an increase in the fin height and a decrease in the interfin spacing, the flow approaches slot-flow behavior. The pressure drop in slot flow may be predicted by the general correlations for laminar and turbulent flows with a properly evaluated longitudinal pressure gradient.

Staggered bundles of finned tubes have higher heat transfer coefficients than inline bundles, but with higher hydraulic drag.

The velocity distribution in the outer boundary layer is derived from the distribution of surface pressure. Measurements of the pressure coefficient distribution revealed highly complicated behavior on fin tips, as compared to fin roots. At fin roots, the value of $\tilde{C}_p$ has a minimum at ϕ between 75 and 90°, followed by an increase and steady-state behavior at $\phi = 100°$. With an increase of Re, the extreme points are shifted upstream [29].

6.5.3 Drag on Finned-Tube Bundles

The transition to a flat f-vs.-Re curve is more complex in bundles of finned tubes than in smooth ones because of the large number of variables. There is a general tendency to earlier transition and a long transition region with Re from 6×10^4 to 1.9×10^5. An analysis of experimental results suggests earlier transitions to the flat behavior in more compact bundle arrangements because of generally higher turbulence [29].

Shorter interfin spaces cause splitting of the flow into separate streams. As a result, vorticities are extinguished and the transition to a predominantly turbulent flow is delayed.

Figure 6.37 presents generalized results by Stasiulevičius and Skrinska [29] measured on multirow staggered bundles of tubes with helical fins in terms of Eu′ vs. Re. Here

$$\mathrm{Eu}' = \mathrm{Eu}\, X_t^{*\,0.55} X_l^{*\,0.5} F^{-0.5} \tag{6.46}$$

where $X_t^* = X_t/d_o$, $X_l^* = X_l/d_o$, d_o is the tube diameter at the fin base, and F is defined in Eq. (6.45).

Numerous studies of tube bundles with helical fins were performed at the Institute of Physical and Technical Problems of Energetics, Academy of Sciences of the Lithuanian SSR. They covered a wide range of the Reynolds number and geometrical parameters of helical and annular fins. The studies resulted in the following correla-

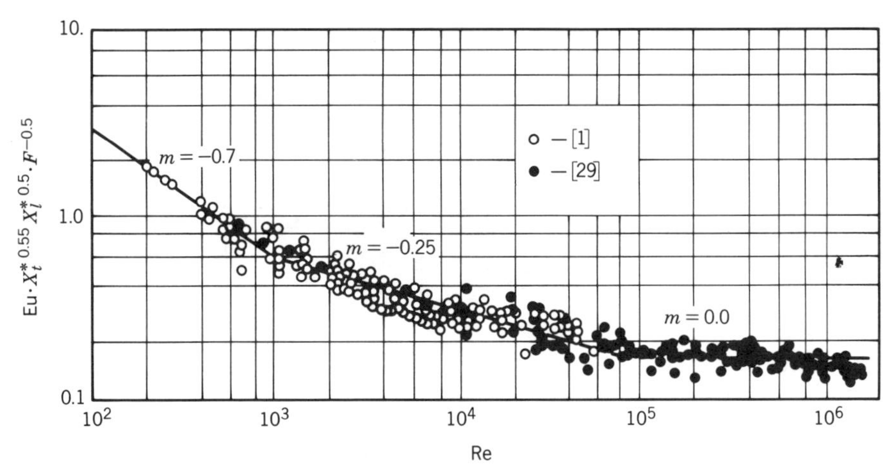

Figure 6.37. Hydraulic drag of different staggered bundles. m is the slope of the curves; see Eqs. (6.47)–(6.49).

tions for hydraulic drag on staggered bundles of finned tubes:

$$\mathrm{Eu} = 67.6\,\mathrm{Re}^{-0.7} X_t^{*\,-0.55} X_l^{*\,-0.5} F^{0.5} \qquad \text{for} \quad 10^2 < \mathrm{Re} \le 10^3 \tag{6.47}$$

$$\mathrm{Eu} = 3.2\,\mathrm{Re}^{-0.25} X_t^{*\,-0.55} X_l^{*\,-0.5} F^{0.5} \qquad \text{for} \quad 10^3 < \mathrm{Re} \le 10^5 \tag{6.48}$$

$$\mathrm{Eu} = 0.18 X_t^{*\,-0.55} X_l^{*\,-0.5} F^{0.5} \qquad \text{for} \quad 10^5 \le \mathrm{Re} \le 1.4 \times 10^6 \tag{6.49}$$

Equation (6.47) is valid for $1.5 \le F \le 16.0$, $1.13 \le X_t^* \le 2.0$, and $1.06 \le X_l^* \le 2.00$. Equations (6.48) and (6.49) are valid for $1.9 \le F \le 16$, $1.6 \le X_t^* \le 4.13$, and $1.2 \le X_l^* \le 2.35$. Eu predicted from Eqs. (6.47) to (6.49) are accurate within $\pm 15\%$.

For inline bundles with helical and annular fins, Lokshin and Fomina [30] suggested a relation

$$\mathrm{Eu} = 0.068 F^{0.5} \eta^{-0.4} \tag{6.50}$$

where $\eta = (X_t^* - 1)/(X_l^* - 1)$, for Re from 10^3 to 10^5, F from 1.8 to 16.3, X_t^* from 2.38 to 3.13, and X_l^* from 1.2 to 2.35.

Because of simple fabrication techniques, low fouling rates, and high heat transfer coefficients, cylindrical, conical, or parabolic pins and hemispherical protrusions are also used on the tubes. For the hydraulic drag of staggered tube bundles with cylindrical pins, Lyshevskii et al. [31] suggested the following equation for Re from 2×10^3 to 10^4, F from 2.88 to 4.95, X_t^* from 1.65 to 4.35, and X_l^* from 1.27 to 1.79:

$$\mathrm{Eu} = 5.62 \psi^{-2.1} F^{0.6} \tag{6.51}$$

where $\psi = [(X_t/d_o - 1)/(X_d/d_o - 1)]$. For inline bundles having $X_t^* = 2.2$ and X_l^* from 1.47 to 2.05, the following correlation is applicable [31]:

$$\mathrm{Eu} = 0.69 F^{0.15} X_l^* \tag{6.52}$$

More details on the hydraulic drag of finned-tube bundles are presented by Žukauskas [1], Schmidt [32], and Yudin [33].

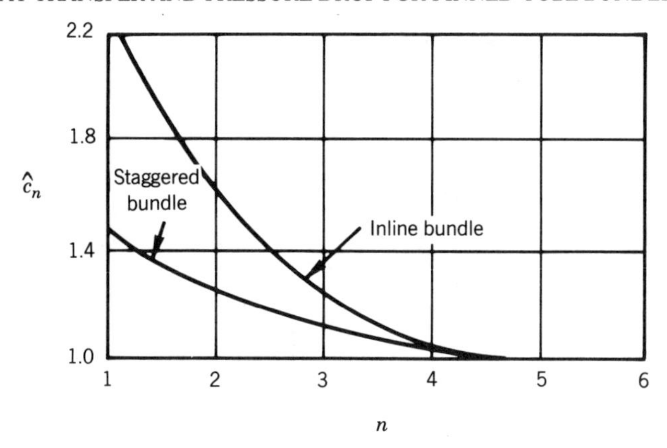

Figure 6.38. Correction factor for the drag of short staggered and inline bundles as a function of the number of tube rows, valid only for $\text{Re} \geq 10^4$ [1].

It has been determined experimentally that the Euler number for short ($n \leq 5$) finned bundles may be significantly higher than the experimental values of Eu for deep bundles. For a short bundle, a correction factor $\hat{c}_n$ is presented in Fig. 6.38 [33]. It is known to be valid only for $\text{Re} \geq 10^4$, since no data are available for $\text{Re} < 10^4$. However, it is believed that the results of Fig. 6.38 should also be applicable for $\text{Re} < 10^4$.

Thus the pressure drop across a finned-tube bundle is given by

$$\Delta P = \text{Eu}\left(\tfrac{1}{2}\rho u_0^2\right) n\hat{c}_n \tag{6.53}$$

where Eu is the drag coefficient of an inner row in a long bundle, and $\hat{c}_n$ is the correction factor for a short bundle from Fig. 6.38. Use $\hat{c}_n = 1$ for $n \geq 6$.

6.5.4 Heat Transfer for Finned-Tube Bundles

Higher heat transfer rates are found on the fin tips, based on the local heat transfer coefficients over the tubes and on the fins as measured by Stasiulevičius and Skrinska [29], and Neal and Hitchock [34]. The increasing thickness of the boundary layers leads to lower heat transfer coefficients on both the base tube and the fins. They increase again in the rear regions. Heat transfer is usually higher on fin tips, because the hydrodynamic boundary layers are thinner. Local heat transfer coefficients are higher on fin tips than on fin roots, but their circumferential distribution is not uniform. A general correlation is possible only on the basis of the average heat transfer coefficient h based on the average temperature of the whole extended surface (primary plus secondary surface).

Figure 6.39 presents a curve for the heat transfer behavior of staggered tube bundles with helical fins [29, 33]. Correlations for mean Nusselt numbers for staggered finned-tube bundles are presented in Table 6.4, based on the results of [29, 33]. The bulk mean temperature, average extended surface temperature, mean velocity in the minimum intertube space, and base-tube diameter are used in evaluating dimensionless groups of these correlations.

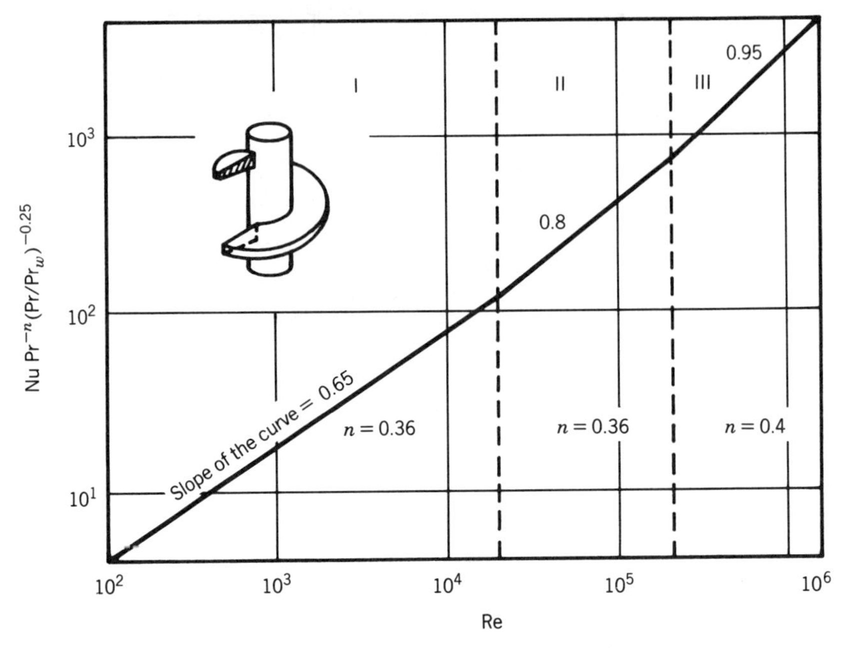

Figure 6.39. Correlation of the heat transfer from staggered bundles of finned tubes [1]; specific values of the exponent n to Pr are shown in the figure for three ranges of Re.

The average Nusselt numbers for inline bundles are correlated by Schmidt [32] as

$$\text{Nu} = 0.303\,\text{Re}^{0.625} F^{-0.375}\,\text{Pr}^{0.36}\left(\frac{\text{Pr}}{\text{Pr}_w}\right)^{0.25} \tag{6.54}$$

for Re from 5×10^3 to 10^5, X_t^* from 1.72 to 3.0, X_l^* from 1.8 to 4.0, and F from 5 to 12.

Heat transfer from bundles of tubes with low fins ($e_f = 1.4$, $S = 1.25$ mm) was determined by Groehn [35]. His results were approximated by

$$\text{Nu} = 0.072\,\text{Re}^{0.74}\text{Pr}^{0.36} \qquad \text{for} \quad 5 \times 10^3 < \text{Re} < 3.5 \times 10^4 \tag{6.55}$$

$$\text{Nu} = 0.137\,\text{Re}^{0.68}\text{Pr}^{0.36} \qquad \text{for} \quad 3.5 \times 10^4 < \text{Re} < 2.3 \times 10^5 \tag{6.56}$$

$$\text{Nu} = 0.051\,\text{Re}^{0.76}\text{Pr}^{0.36} \qquad \text{for} \quad 2.3 \times 10^5 < \text{Re} < 10^6 \tag{6.57}$$

We also present prediction equations for the heat transfer from other extended surfaces. Average Nusselt numbers for staggered tube bundles with cylindrical pins are presented by Lyshevskii et al. [31] as

$$\text{Nu} = 0.108 F^{0.55}\psi^{1.1}\text{Re}^{0.7}\text{Pr}^{0.36}\left(\frac{\text{Pr}}{\text{Pr}_w}\right)^{0.25} \tag{6.58}$$

for Re from 2×10^3 to 10^4, F from 2.88 to 4.95, X_t^* from 1.65 to 4.35, and X_l^* from 1.27 to 1.79.

TABLE 6.4. Heat Transfer Correlations for Staggered Tube Bundles with Helical Fins [29, 33]

Recommended Correlations	Range of Re
$\mathrm{Nu} = 0.192(X_t^*/X_l^*)^{0.2}(S/d_o)^{0.18}(e_f/d_o)^{-0.14} \times \mathrm{Re}^{0.65}\mathrm{Pr}^{0.36}(\mathrm{Pr}/\mathrm{Pr}_w)^{0.25}$	10^2–2×10^4
$\mathrm{Nu} = 0.0507(X_t^*/X_l^*)^{0.2}(S/d_o)^{0.18}(e_f/d_o)^{-0.14} \times \mathrm{Re}^{0.8}\mathrm{Pr}^{0.36}(\mathrm{Pr}/\mathrm{Pr}_w)^{0.25}$, $1.1 < X_t^* < 4.0$, $1.03 < X_l^* < 2.5$, $0.06 < S/d_o < 0.36$, $0.07 < e_f/d_o < 0.715$	2×10^4–2×10^5
$\mathrm{Nu} = 0.0081(X_t^*/X_l^*)^{0.2}(S/d_o)^{0.18}(e_f/d_o)^{-0.14} \times \mathrm{Re}^{0.95}\mathrm{Pr}^{0.4}(\mathrm{Pr}/\mathrm{Pr}_w)^{0.25}$, $2.2 < X_t^* < 4.2$, $1.27 < X_l^* < 2.2$, $0.125 < S/d_o < 0.28$, $0.125 < e_f/d_o < 0.6$	2×10^5–2×10^6

Average Nusselt numbers for inline tube bundles with cylindrical pins for X_l^* from 1.47 to 2.05 and $X_t^* = 2.2$ are given by [31]

$$\mathrm{Nu} = 0.428F^{0.33}X_l^{*0.44}\mathrm{Re}^{0.54}\mathrm{Pr}^{0.36}\left(\frac{\mathrm{Pr}}{\mathrm{Pr}_w}\right)^{0.25} \tag{6.59}$$

In the lower range of Re, bundles of pinned tubes are more efficient than bundles of helical finned tubes. The correlations of Eqs. (6.54)–(6.59) are based on the test data for air and water.

The average heat transfer from leading rows of finned tubes in bundles of different arrangements may be either higher or lower than from the inner tubes, depending on the arrangement. Nu_n for short bundles (for $n \leq 6$) is determined by

$$\mathrm{Nu}_n = \check{c}_n \mathrm{Nu} \tag{6.60}$$

where Nu is the average Nusselt number for a large number of tube rows ($n > 6$), and $\check{c}_n$ is the correction factor from Fig. 6.40 depending on the number of tube rows. A

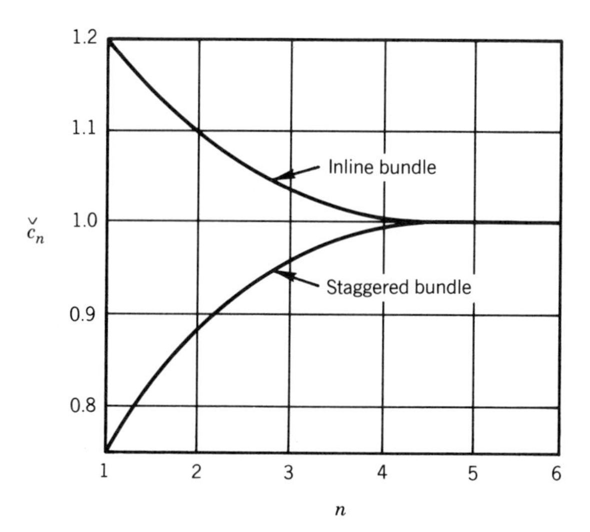

Figure 6.40. Correction factor for the heat transfer from short staggered and inline bundles of $X_t^* = X_l^* = 2$ with helical fins [1].

certain effect of Re is also observed. The correction factor for short bundles is $\check{c}_n = 1$ from the fourth row on.

The different behavior of curves for $\check{c}_n$ for inline and staggered bundles may be explained as follows. Because of the turbulizing action of the leading rows in staggered arrangements, the heat transfer from inner rows is higher than that from the leading rows. In contrast, due to the "shadow" effect in inline arrangements, the heat transfer from inner rows is lower than that from the leading rows.

NOMENCLATURE

A	minimum intertube space, m^2, ft^2
C_D	drag coefficient for flow normal to a single tube or a sphere, defined by Eq. (6.14)
C_P	pressure coefficient for flow normal to a single tube, defined by Eq. (6.13)
$\tilde{C}_p$	pressure coefficient for cross flow to a tube in a bundle, defined by Eq. (6.26)
c	constant
c_f	skin friction coefficient $= 2\tau_w/\rho u_0^2$
$\tilde{c}_f$	skin friction coefficient for a single cylinder, $= (\tau_w/\rho u_\infty^2)\sqrt{\mathrm{Re}}$
c_p	specific heat at constant pressure, J/(kg · K), Btu/(lb_m · °F)
c_w	pressure drag coefficient $= F_w/(\frac{1}{2}\rho u_\infty^2 d_o L)$
c_β	correction factor for drag in a yawed-tube bundle, defined by Eq. (6.41)
d_o	outside diameter of a circular tube, m, ft
Eu	n-row average Euler number $= 2\,\Delta p/\rho u_0^2 n$
Eu_n	n-row average Euler number for $n < 9$
e	roughness height, m, ft
e_1	channel height, m, ft
e_f	fin height, m, ft
e^+	dimensionless height of surface elements, $= eu^*/\nu$
F	ratio of extended surface area (primary plus secondary) to the bare base-tube area of a finned-tube bundle
f	function of
Gr	Grashof number $= g\tilde{\beta}\,\Delta T\, L^3/\nu^2$
g	gravitational acceleration, m/s^2, ft/s^2
h	average heat transfer coefficient, W/(m^2 · K), Btu/(hr · ft^2 · °F)
h_ϕ	local heat transfer coefficient, W/(m^2 · K), Btu/(hr · ft^2 · °F)
k	thermal conductivity, W/(m · K), Btu/(hr · ft · °F)
k_w	thermal conductivity of the wall material, W/(m · K), Btu/(hr · ft · °F)
K	complex similarity number $= \mathrm{Nu}\,\mathrm{Pr}^{-n}(\mathrm{Pr}/\mathrm{Pr}_w)^{-0.25}$
L	length, m, ft
l	width, m, ft
m	power index of Re
Nu	average Nusselt number $= hd_o/k$
Nu_ϕ	local Nusselt number $= h_\phi d_o/k$
n	number of tube rows in the flow direction
n	exponent of Pr

P	pressure, Pa, lb_f/ft^2
Pr	Prandtl number $= \mu c_p/k$
ΔP	total pressure drop for n tube rows in a tube bundle, Pa, lb_f/ft^2
p	exponent of Pr/Pr_w
Re	Reynolds number $= u_0 d_o/\nu$
r	exponent; see Eq. (6.35)
T	temperature, °C, K, °F, °R
Tu	turbulence intensity
ΔT	temperature difference, °C, °F
u	velocity component in the x direction, m/s, ft/s
u_0	mean velocity in the minimum free-flow area or intertube spacing, m/s, ft/s
u^*	friction velocity $= (\tau_w/\rho)^{1/2}$
u_{**}	reference velocity for restricted channels, Eq. (6.25)
X_d	tube diagonal pitch (see Fig. 6.12), m, ft
X_d^*	ratio of diagonal pitch to tube diameter for cross flow to a tube bundle, $= X_d/d_o$
X_l	tube longitudinal pitch (see Fig. 6.12), m, ft
X_t	tube transverse pitch (see Fig. 6.12), m, ft
X_l^*	ratio of longitudinal pitch to tube diameter for cross flow to a tube bundle, $= X_l/d_o$
X_t^*	ratio of transverse pitch to tube diameter for cross flow to a tube bundle, $= X_t/d_o$
x	distance, m, ft
y	distance normal to surface measured from the tube surface, m, ft

Greek Symbols

β	yaw angle, defined in a sketch in Fig. 6.22, rad, deg
$\tilde{\beta}$	coefficient of thermal expansion, K^{-1}, $°R^{-1}$
δ	hydrodynamic boundary-layer thickness, m, ft
δ_f	fin thickness, m, ft
μ	dynamic viscosity, Pa · s, $lb_m/(hr \cdot ft)$
ν	kinematic viscosity, m^2/s, ft^2/s
ξ	hydraulic drag coefficient = Eu
ρ	density, kg/m^3, lb_m/ft^3
τ	shear stress between fluid layers, Pa, lb_f/ft^2
τ_w	shear stress at wall, Pa, lb_f/ft^2
Φ	angle between forced convection motion and buoyancy force, rad, deg
ϕ	angle measured along the tube perimeter from the front stagnation point, rad, deg
Ψ	$(X_t^* - 1)/(X_d^* - 1)$
ψ	d_o/e_1

Subscripts

lam	laminar
r	rough surface

turb turbulent
w wall condition
x local value at distance x
ϕ local value at angle ϕ

Superscripts

$+$ dimensionless
∞ free-stream conditions

REFERENCES

1. A. A. Žukauskas, *Konvektivnyi Perenos v Teploobmennikakh* (*Convective Transfer in Heat Exchangers*), Nauka, Moscow, 472, pp., 1982.
2. A. Žukauskas and J. Žiugžda, *Heat Transfer of a Cylinder in Cross Flow*, Hemisphere, Washington, 208 pp., 1985.
3. J. Knudsen and D. Katz, *Fluid Dynamics and Heat Transfer*, McGraw-Hill, New York, 576 pp., 1958.
4. E. Achenbach, Influence of Surface Roughness on the Cross Flow around a Circular Cylinder, *J. Fluid Mech.*, Vol. 46, Pt. 2, pp. 321–335, 1971.
5. S. Kakaç and Y. Yener, *Convective Heat Transfer*, Middle East Technical Univ., Ankara, Turkey, Distributed by Hemisphere, New York, 512 pp., 1980.
6. N. Frössling, Verdunstung, Wärmeübergang und Geschwindigkeitsverteilung bei zweidimensionaler und rotationsimmetrischer Grenzschichtströmung, Lunds. Univ. Arsskr., Avd. 2, Vol. 36, No. 4, pp. 25–35, 1940.
7. E. Achenbach, The Effect of Surface Roughness on the Heat Transfer from a Circular Cylinder to the Cross Flow of Air, *Int. J. Heat Mass Transfer*, Vol. 20, pp. 359–369, 1977.
8. E. Achenbach, Heat Transfer from Spheres up to Re = $6 \cdot 10^6$, *Heat Transfer* 1978, Vol. 5, pp. 341–346, 1978.
9. E. P. Dyban, E. Ya. Epick, and L. G. Kozlova, Combined Influence of Turbulence Intensity and Longitudinal Scale and Air Flow Acceleration on Heat Transfer of Circular Cylinder, *Heat Transfer* 1974, Vol. II, pp. 310–319, 1974.
10. I. Kestin, The Effect of Free-Stream Turbulence on Heat Transfer Rates, *Adv. Heat Transfer*, Vol. 3, pp. 1–32, 1966.
11. A. P. Hatton, D. D. James, and H. W. Swise, Combined Forced and Natural Convection with Low-Speed Air Flow over Horizontal Cylinders, *J. Fluid Mech.*, Vol. 42, pp. 15–31, 1970.
12. B. G. Van der Hegge Zijnen, Modified Correlation Formulae for the Heat Transfer by Forced Convection from Horizontal Cylinders, *Appl. Sci. Res.*, Ser. A, Vol. 6, pp. 129–140, 1956.
13. H. Perkins and G. Leppert, Local Heat Transfer Coefficients on a Uniformly Heated Cylinder, *Int. J. Heat Mass Transfer*, Vol. 7, No. 2, pp. 143–156, 1964.
14. V. Gnielinski, Berechnung mittlerer Wärme- und Stoffübergangskoeffizienten an laminar und turbulent überströmten Einzelkörpern mit Hilfe einer einheitlichen Gleichung, *Forsch. Ingenieurw.*, Vol. 41, No. 5, pp. 145–153, 1975.
15. V. T. Morgan, Heat Transfer from Cylinders, *Adv. Heat Transfer*, Vol. 11, pp. 199–264, 1975.
16. A. A. Žukauskas, V. I. Makarevičius, and A. A. Šlančiauskas, *Teplootdacha Puchkov Trub v Poperechnom Potoke Zhidkosti* (*Heat Transfer in Banks of Tubes in Cross Flow of Fluid*), Mintis, Vilnius, 192 pp., 1968.
17. K. Hammeke, E. Heinecke, and F. Scholz, Wärmeübergangs und Drückverlustmessungen an querangeströmten Glattrohrbündeln, insbesondere bei Hohen Reynoldszahlen, *Int. J. Heat Mass Transfer*, Vol. 10, pp. 427–446, 1967.

18. W. Niggeschmidt, Drüskverlust und Wärmeübergang bei fluchtenden, versetzten und teilversetzten querangeströmten Rohrbündeln, Dissertation, Univ. of Darmstadt, 1975.

19. O. P. Bergelin, G. A. Brown, and S. C. Doberstein, Heat Transfer and Fluid Friction during Flow Across Banks of Tubes, *Trans. ASME*, Vol. 74, No. 6, pp. 953–960, 1952.

20. E. S. Gaddis and V. Gnielinski, Pressure Drop in Cross Flow across Tube Bundles, *Int. J. Chem. Eng.*,Vol. 25, No. 1, pp. 1–15, 1985.

21. A. A. Žukauskas, Heat transfer from Tubes in Cross-Flow, *Adv. Heat Transfer*, Vol. 8, pp. 93–160, 1972.

22. F. P. Kazakevich, Issledovaniye Teplootdachi Puchkov Trub pri Razlichnykh Uglakh Ataki Glavnogo Potoka (Heat Transfer from Tube Bundles for Different Yaw Angles), *Teploenergetika*, No. 8, pp. 22–28, 1958.

23. M. A. Mikheyev, *Fundamentals of Heat Transfer*, Peace Publishers, Moscow, 376 pp., 1957.

24. E. Achenbach, Influence of Surface Roughness on the Flow through a Staggered Tube Bank, *Wärme- und Stoffübertragung*, Vol. 4, pp. 120–126, 1971.

25. E. Achenbach, On the Cross Flow through In-Line Tube Banks with Regard to the Effect of Surface Roughness, *Wärme- und Stoffübertragung*, Vol. 4, pp. 152–155, 1971.

26. P. Puchkov, Vliyaniye Sherokhovatosti na Teplootdachu Puchkov Trub v Poperechnom Potoke (The Effect of Roughness on the Heat Transfer of Tube Banks in Cross Flow), *Kotlostroyenye*, No. 4, pp. 5–6, 1948.

27. A. A. Žukauskas, R. V. Ulinskas, and P. I. Daunoras, Teploobmen i Gidravlicheskoe Soprotivleniye Puchkov Sherokhovatykh Trub v Shirokom Intervale Pr (Heat Transfer and Hydraulic Drag of Rough Tube Bundles in a Wide Range of Pr), *Lietuvos TSR Mokslu Akad. Darbai*, Ser. B, No. 3, 1984.

28. H. G. Groehn and F. Scholz, Änderung von Wärmeübergang und Strömungswiderstand in querangeströmten Rohrbündeln unter dem Einfluss verschiedener Ruhigkeiten sowie Anmerkungen zur Wahl der Stoffwertbazungstemperaturen, *Heat Transfer 1970*, Paper FC.7.10, Vol. 3, pp. 1–11, 1970.

29. J. Stasiulevičius and A. Skrinska, *Teplootdacha Poperechno Obtekaemykh Puchkov Rebristykh Trub (Heat Transfer in Banks of Finned Tubes in Cross Flow)*, Mintis, Vilnius, 243 pp., 1974.

30. V. A. Lokshin and V. N. Fomina, Obobschenye Materialov po Eksperimentalinomu Issledovaniyu Rebristykh Puchkov Trub (Interpretation of Experimental Results on the Heat Transfer of Finned Tube Bundles), *Teploenergetika*, No. 6, pp. 36–39, 1978.

31. A. S. Lyshevskii, V. G. Sokolov, V. M. Sychev, and A. A. Kutukov, Issledovaniye Teplootdachi i Soprotivleniya Puchkov Trub s Tsilindricheskoi Oshipovkoi pri Malykh Chislakh Reynoldsa Poperechno Obtekayuschego Glavnogo Potoka (A Study of the Heat Transfer and Drag of Bundles of Tubes with Cylindrical Pins in Cross Flow at Low Reynolds Numbers), *Teplosnabzheneye i Teplomassoperenos*, Rostova-na-Donu, pp. 63–71, 1977.

32. Th. E. Schmidt, Der Wärmeübergang an Rippenrohre und die Berechnung von Rohrbündel-Wärmeaustauschern, *Kältetechn.*, Vol. 15, No. 4, pp. 98–102; No. 12, pp. 370–378, 1963.

33. F. V. Yudin, *Teploobmen Poperechnoorebrennykh Trub (Heat Transfer From Tubes with Transverse Fins)*. Mashynostroenye, Leningrad, 189 pp., 1982.

34. S. B. H. C. Neal and J. A. Hitchcock, A Study of the Heat Transfer Processes in Banks of Finned Tubes in Cross Flow, using a Large Scale Model Technique, *Proc. 3rd Int. Heat Transfer Conf.*, Chicago, Vol. III, pp. 290–298, 1966.

35. H. G. Groehn, Wärme und Stromungstechnische Untersuchungen an einem querdurchstromten Rohrbündel-Wärmeaustauscher mit niedrig berippten Rohren bei grossen Reynolds-Zahlen, Jülich, 1962, 1977.

36. A. Žukauskas and R. Ulinskas, Banks of Plain and Finned Tubes, *Heat Exchanger Design Handbook*, Vol. 2, Section 2.2.4, Hemisphere, New York, 1983.

37. V. Gnielinski, Forced Convection around Immersed Bodies, *Heat Exchanger Design Handbook*, Vol. 2, Sec. 2.5.2, Hemisphere, New York, 1983.

7

CONVECTIVE HEAT TRANSFER OVER ROD BUNDLES

Klaus Rehme

Kernforschungszentrum Karlsruhe
Karlsruhe, F.R.G.

7.1 INTRODUCTION

Longitudinal flow over tube or rod bundles is common in most fuel elements of nuclear power reactors. The principal heat-exchanging equipment, the core, is composed of a large number of parallel fuel rods. Generally, the core is subdivided into subassemblies, the fuel elements, which consist of a certain number of fuel rods arranged in a regular array. To allow adequate and regularly distributed space for the coolant flowing parallel to the axes of the fuel rods, suitable spacing devices are used, which also provide rigidity to the fuel element [1].

Applications of this geometry are also encountered in shell-and-tube heat exchangers, boilers, condensers, etc.

7.1.1 Geometrical Considerations

The possible geometrical arrangements of rods inside rod bundles are infinite. Three principal types of arrangements will be distinguished under the assumption of equal rod diameters in the following: (1) a regular triangular array of rods contained in a hexagonal channel (Fig. 7.1*a*), (2) a regular square array of rods contained in a square channel (Fig. 7.1*b*), and (3) a circular array of rods contained in a circular channel (Fig. 7.1*c*). All these rod-bundle arrangements can be subdivided into three types of subchannels: central, wall, and corner subchannels.

The main geometrical parameters of rod bundles are the rod outside diameter D; the distance between the rod centers (pitch) P; the distance W from the wall, defined as rod diameter plus the shortest distance between a rod and the channel wall; and the number n_R of rods in a rod bundle. Table 7.1 lists the number of subchannels, their flow areas, and their wetted perimeters for regular triangular and square arrays. The number of rods, n_R, can be calculated by the equation of Table 7.1 from the number N of rings around the central rod for the regular triangular and the number N of rods in one row for the regular square array (see Fig. 7.1). Similar concise formulas are not possible for circular arrays, since the radii of the different rings (r_1, r_2 in Fig. 7.1*c*) are not the same. The hydraulic diameter for each subchannel of Table 7.1 can be determined by $D_{h,i} = 4A_i/P_{W,i}$.

7.1.2 Fundamentals

Single-phase fluid flow and heat transfer in rod bundles with constant flow cross sections in the axial direction is a three-dimensional problem. Therefore, the three-dimensional conservation equations for momentum, energy, and mass have to be solved, taking into account the relevant boundary conditions to obtain the velocity and temperature distributions.

For laminar flow and heat transfer, analytical and numerical solutions of the conservation equations are possible without additional idealizations. The situation is different for turbulent flow and heat transfer. Basically, two different approaches exist to obtain solutions for velocity and temperature distributions in turbulent flow.

The standard approach is subchannel analysis, by which mean velocities and fluid temperatures averaged over the subchannels are computed. Surface temperatures, peripherally averaged in the subchannels, are then obtained through Nusselt number correlations. A detailed survey of subchannel methods has been published by Sha [2]. Subchannel analysis requires Nusselt numbers and friction factors, which will be discussed in the following.

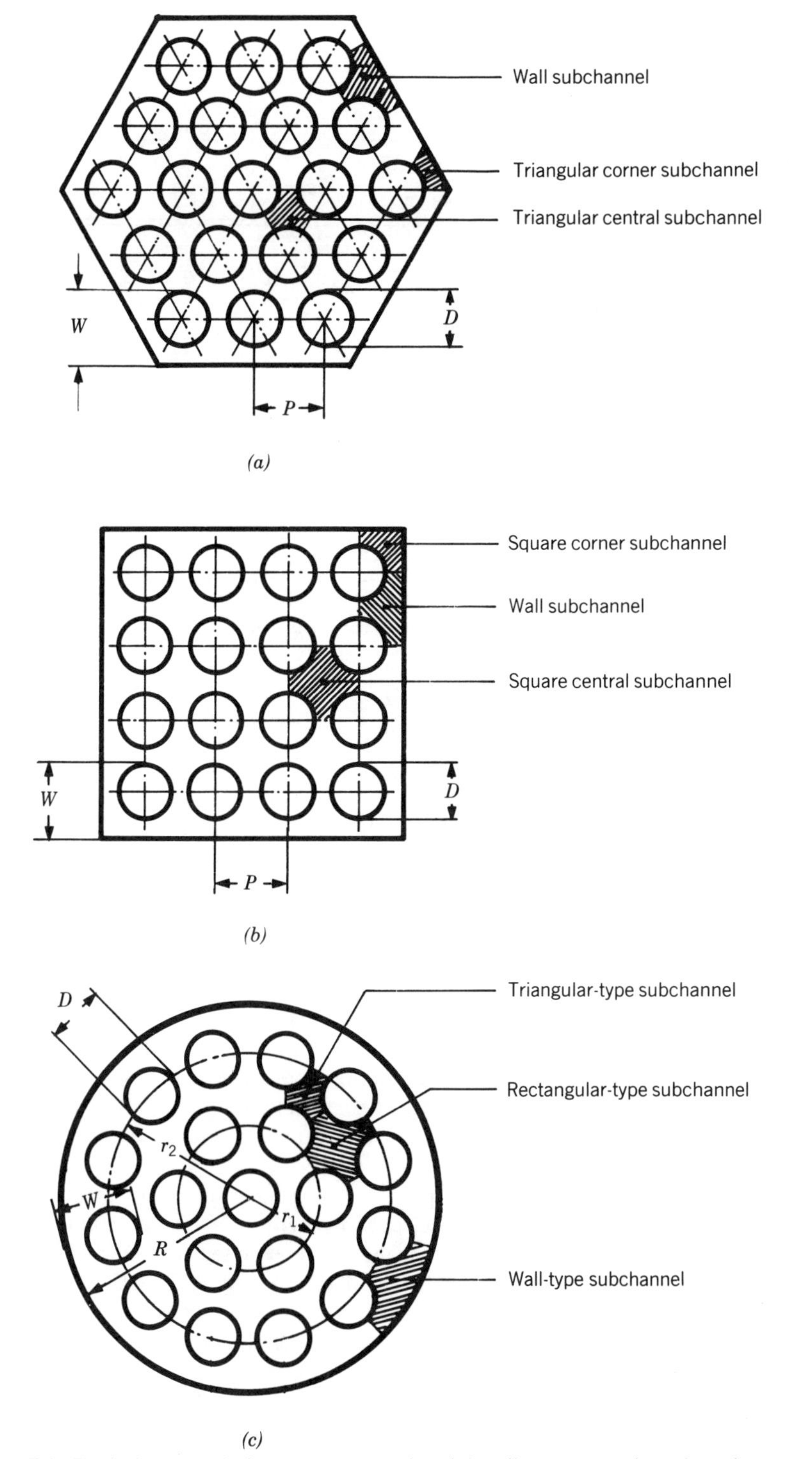

Figure 7.1. Typical geometrical arrangements of rod bundles: (*a*) regular triangular array, -hexagonal channel ($N = 2$); (*b*) regular square array, -square channel ($N = 4$); (*c*) circular array, -circular channel.

TABLE 7.1 Geometrical Parameters of Triangular and Square Array Rod Bundles

Subchannel Type i	Number n_i	Flow Area A_i	Wetted Perimeter P_{wi}
		REGULAR TRIANGULAR ARRAY	
		Total number of rods, $n_R = 1 + 3N(N+1)$	
Central	$6N^2$	$\frac{\sqrt{3}}{4}P^2 - \frac{\pi}{8}D^2$	$\frac{\pi}{2}D$
Wall	$6N$	$\left(W - \frac{D}{2}\right)P - \frac{\pi}{8}D^2$	$\frac{\pi}{2}D + P$
Corner	6	$\frac{1}{\sqrt{3}}\left(W - \frac{D}{2}\right)^2 - \frac{\pi}{24}D^2$	$\frac{\pi}{6}D + \frac{2}{\sqrt{3}}\left(W - \frac{D}{2}\right)$
		REGULAR SQUARE ARRAY	
		Total number of rods, $n_R = N^2$	
Central	$(N+1)^2$	$P^2 - \frac{\pi}{4}D^2$	πD
Wall	$4(N-1)$	$\left(W - \frac{D}{2}\right)P - \frac{\pi}{8}D^2$	$\frac{\pi}{2}D + P$
Corner	4	$\left(W - \frac{D}{2}\right)^2 - \frac{\pi}{16}D^2$	$\frac{\pi}{4}D + 2\left(W - \frac{D}{2}\right)$

To obtain detailed velocity and temperature distributions in the flow cross section of rod bundles, so-called local analyses are performed. Finite difference or finite element methods are used to discretize the conservation equations and to compute solutions under simplifying idealizations for the transport of momentum and heat by turbulence. Examples of local analyses in turbulent flow are reported by Slagter [3], Chen et al. [4], Ramachandra [5], Shimizu [6], and others. Local-analysis codes do not require Nusselt numbers and friction factors, but require empirical correlations of turbulent transport properties.

It is beyond the scope of this chapter to discuss subchannel and local analyses in more detail. However, the basic theoretical and experimental results from the literature that are necessary to apply subchannel analysis will be presented. These are the friction factors and Nusselt numbers for laminar and turbulent flow through subchannels of rod bundles. First, the solutions for laminar flow will be discussed. These solutions are valid for ordinary fluids, e.g. water and gases, as well as for liquid metals as working fluids. However, free convection becomes important at low Reynolds numbers in liquid metals and usually influences heat transfer.† For turbulent flow, the Nusselt numbers

†Chapter 8 of this handbook should be consulted for criteria regarding free convection effects.

are presented for ordinary fluids and liquid metals in different sections of this chapter. Finally, the effects of spacers on pressure drop and heat transfer are presented.

7.2 LAMINAR FLOW

Longitudinal laminar flow through rod bundles has been analyzed by many investigators. One important result was obtained by Schmid [7], who investigated the influence of the channel wall in a semi-infinite rod bundle arranged in a square array and bounded by a fixed wall on one side. Schmid found that only the flow rates through the first and second rows of subchannels are affected by the wall. This influence exceeds 1% for $P/D > 2.5$ and $\xi > 1.3$ or for $P/D > 2.5$ and $\xi < 0.6$. Here ξ is the ratio of the hydraulic diameter of a wall subchannel to that of a central subchannel. Schmid's results thus indicated that the treatment of flow through rod bundles by individual subchannels is quite accurate for the indicated range of P/D and ξ. By this treatment, it is implicitly assumed that no momentum transport occurs across the boundary between neighboring subchannels. This idealization of isolated subchannels is definitely valid for laminar flow through rod bundles, at least in the range of parameters given by Schmid. In regular and infinite arrays of rods (central subchannels only) with axisymmetric and equal heat generation, this assumption is also valid for laminar flow heat transfer. However, when the rod array is bounded by a fixed duct wall, heat generation by rods varies across the rod bundle, large rod spacings exist, and entrance effects are present, then subchannel analysis can only be considered an approximation which will result in conservative solutions in most cases for the subchannel of maximum geometric irregularity or maximum heat generation [8]. In such cases, a multiregion, multicell analysis has to be performed. Theoretical tools for such an analysis are available [2, 3, 4] but have not been applied to the problem of laminar flow heat transfer.

Therefore, only subchannel solutions will be discussed in the following. All subchannel solutions are obtained by solving the conservation equations. The methods used to obtain solutions have been discussed in detail by Johannsen [1, 8] and by Shah and London [9].

For laminar flow through rod bundles, almost all knowledge of friction factors and Nusselt numbers stems from theoretical analysis. Only very limited experimental data are found in the literature. This is mainly because pressure drops in laminar flow are small and therefore difficult to measure. On the other hand, heat transfer in laminar flow is often augmented by natural convection.

7.2.1 Central Subchannel — Regular Triangular Array

***Fully Developed Flow: f* Re.** Many solutions have been obtained for fully developed laminar flow through a central subchannel in regular triangular array. Solutions for f Re are reported by Rosenberg [10], Sparrow and Loeffler [11], Axford [12, 13], Shih [14], Rehme [15, 16], Oberjohn as reported by Johannsen [1], Malák et al. [17], Meyder [18], Ramachandra [5], Mikhailov [19], and Subbotin et al. [20]. The tabulated results of the different authors agree within 2%, and are represented by a smooth curve in Fig. 7.2.

The results by Sparrow and Loeffler [11] and Ramachandra [5], presented graphically, agree with the results of the other authors. Dwyer and Berry [21] tabulated the dimensionless "pressure-drop–flow parameter", from which f Re can easily be computed by multiplying it by 0.5 × (flow area) × (hydraulic diameter)^2. Sholokov et al.

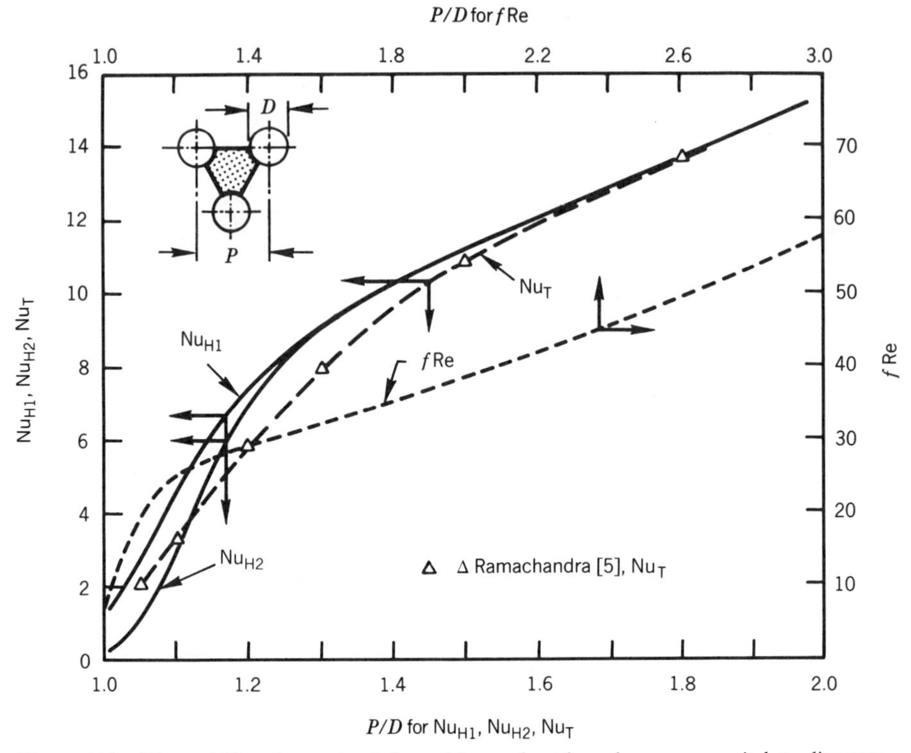

Figure 7.2. f Re and Nu of central subchannel in regular triangular array vs. pitch-to-diameter ratio for fully developed laminar flow.

[22] presented f Re factors for $P/D = 1.0$ to 1.5. Their results of an electric simulation agree satisfactorily with the data of Fig. 7.2; however, their numerical values do not.

The f Re data can be approximated by the following equations:

$$f\,\mathrm{Re} = \begin{cases} 51.777(P/D - 1)^{0.404} & \text{for } 1.02 \le P/D < 1.12 \quad (7.1) \\ 36.713(P/D - 1)^{0.24} & \text{for } 1.12 \le P/D < 1.6 \quad (7.2) \\ 38.947(P/D - 1)^{0.372} & \text{for } 1.6 \le P/D < 2.0 \quad (7.3) \\ \dfrac{16(r_*^2 - 1)^3}{4r_*^2 \ln r_* - 3r_*^4 + 4r_*^2 - 1} & \text{for } P/D > 2.1 \quad (7.4) \end{cases}$$

where

$$r_* = \frac{P}{D}\sqrt{\frac{2\sqrt{3}}{\pi}} \approx 1.05\frac{P}{D} \tag{7.5}$$

These equations agree within 1.5% with the data of Axford [12, 13] and Dwyer and Berry [21].

The equation for $P/D > 2.1$ is a theoretical solution for f Re in the inner zone of a concentric annulus (zone between the surface of the inner rod of an annulus and the radius of zero shear stress), denoted by $(f\,\mathrm{Re})_{\mathrm{eaz}}$ [23]. Martelli [24] has provided an alternative expression for $f\,\mathrm{Re}$ for $1.2 \le P/D \le 1.5$ which is accurate to $\pm 1.5\%$ of [15]. Morosova and Nomofilov [25] have reported a formula for $f\,\mathrm{Re}$ for $P/D = 1.05$ to 1.8 which is up to 13% higher than Eqs. (7.1) to (7.3). Cheng and Todreas [26] developed formulas for $f\,\mathrm{Re}$ for $1.0 \le P/D \le 1.5$ which agree with Eqs. (7.1) and (7.2) within 1.5%.

Sparrow and Loeffler [11] and Ramachandra [5] also presented graphs of the variation in wall shear stresses along the periphery of the rods.

***Fully Developed Heat Transfer:* $\mathrm{Nu}_{\mathrm{H1}}$.** The (H1) problem† for fully developed laminar-flow heat transfer through a central subchannel in regular triangular array was analyzed by Sparrow et al. [27], Dwyer and Berry [21], Hsu [28], and Ramachandra [5]. The results for $\mathrm{Nu}_{\mathrm{H1}}$ of these investigators agree with each other within 1%, and are represented by smoothed curves in Fig. 7.2.

The correlation by Subbotin et al. [29] [limiting case for $\epsilon_K \to \infty$ of Eq. (7.10)] agrees with the results of Dwyer and Berry [21] within 2.8% in the range $1.0 \le P/D \le 2.0$:

$$\mathrm{Nu}_{\mathrm{H1}} = 7.55\frac{P}{D} - \frac{6.3}{(P/D)^{17(P/D)(P/D-0.81)}} \tag{7.6}$$

Subbotin [30] provided an alternative expression for $\mathrm{Nu}_{\mathrm{H1}}$ for $1.3 \le P/D \le 2.0$ which agrees with the results of [21] within 1%. Based on the results by Axford [31], Martelli [24] developed a $\mathrm{Nu}_{\mathrm{H1}}$ correlation for $1.1 \le P/D \le 1.7$ which is accurate to 1.8% of [21].

Sparrow et al. [27], Dwyer and Berry [21], and Ramachandra [5] also graphically presented heat-flux variations along the periphery of the rod for a range of P/D ratios.

***Fully Developed Heat Transfer:* $\mathrm{Nu}_{\mathrm{H2}}$.** The (H2) problem for fully developed laminar heat transfer through a central subchannel in regular triangular array has been analyzed by Dwyer and Berry [21], Hsu [28], and Ramachandra [5]. The results of these authors are in excellent agreement (within $\pm 0.5\%$) and are shown in Fig. 7.2. $\mathrm{Nu}_{\mathrm{H2}}$ is less than $\mathrm{Nu}_{\mathrm{H1}}$ for $P/D < 1.4$. For $P/D > 1.4$, the distribution of the peripheral surface temperature is uniform with good approximation; therefore, $\mathrm{Nu}_{\mathrm{H2}}$ is equal to $\mathrm{Nu}_{\mathrm{H1}}$.

For $P/D > 1.4$, $\mathrm{Nu}_{\mathrm{H2}}$ can be calculated by the formula of Subbotin [30] [Eq. (7.6) above].

The correlation by Subbotin et al. [29] [limiting case $\epsilon_K = 0.01$ of Eq. (7.10)] agrees with the results of Dwyer et al. [21] within 5% for $P/D > 1.02$:

$$\mathrm{Nu}_{\mathrm{H2}} = \left[7.55\frac{P}{D} - \frac{6.3}{(P/D)^{17(P/D)(P/D-0.81)}}\right] \times \left[1 - \frac{3.6P/D}{1.048(P/D)^{20} + 3.2}\right] \tag{7.7}$$

valid in the range $1.0 \le P/D \le 2.0$.

† The (H1), (H2), (H5), and (T) boundary conditions are defined in Chapter 3.

Dwyer and Berry [21] and Ramachandra [5] also presented graphs of the temperature variation around the rod periphery.

Fully Developed Heat Transfer: Nu_T. The only available analysis of the Ⓣ problem has been performed by Ramachandra [5]. The Nu_T values computed for the P/D range between 1.05 and 1.8 are shown in Fig. 7.2. For the whole range of P/D ratios investigated, Nu_T is less than Nu_{H1}. For $P/D = 1.8$, Nu_T approaches the Nu_{H1} values. Nu_T is higher than Nu_{H2} for $P/D < 1.125$, which is due to the strong peripheral variation in the surface temperature for the (H2) problem.

Developing Flow. Ramachandra [5] analyzed the development of the pressure drop for three P/D ratios: 1.05, 1.1, and 1.5. It should be noted that for fully developed flow and $P/D = 1.5$, the pressure gradient determined by Ramachandra [5] is about 7% lower than the pressure gradient calculated from the f Re data of Fig. 7.2 and Eq. (7.2).

Martelli [24] developed an approximate correlation for the incremental pressure drop in the hydrodynamic entrance region based on the data for concentric annuli by Sparrow and Lin [32].

Developing Heat Transfer. The problem of developing heat transfer has been analyzed by Ramachandra [5] for the (H2) boundary condition only. At the beginning of the temperature development, a fully developed velocity profile at constant fluid temperature and a uniform wall temperature were prescribed. The values of Nu_{H2} computed for three P/D ratios (1.05, 1.1, and 1.2) show that the thermal entrance length decreases with increasing P/D.

Based on the data by Lundberg et al. [33] for concentric annuli, Martelli [24] developed approximate correlations for multipliers to the fully developed Nusselt number, which he applied to central subchannels of rod bundles arranged in a triangular array.

7.2.2 Central Subchannel — Regular Square Array

Fully Developed Flow: f Re. Solutions for f Re for fully developed laminar flow through central subchannels in a regular square array have been obtained by Sparrow and Loeffler [11], Shih [14], Rehme [15, 16], Oberjohn as reported by Johannsen [1], Malák et al. [17], Meyder [18], Kim [34], Ramachandra [5], and Ohnemus [35]. The f Re of these authors (except for Ohnemus) agree within 1% and are shown in Fig. 7.3. The graphical results presented by Sparrow and Loeffler [11] and Ramachandra [5] are also in excellent agreement with those of Fig. 7.3. Gunn and Darling [36] calculated f Re = 6.5 for $P/D = 1.0$, which agrees well with the value of 6.60 of Rehme [15, 16]. fRe for the central subchannel in a regular square array can be approximated by the equation

$$f\,\mathrm{Re} = 40.70\left(\frac{P}{D} - 1\right)^{0.435} \tag{7.8}$$

in the range $1.05 \le P/D \le 2.0$ with a maximum error of $\pm 2\%$ [35].

For $P/D > 2.8$, f Re can be calculated from Eq. (7.4), the solution for the annular zone [11]. For the square array, the equivalent annular zone parameter r_* [to be used in

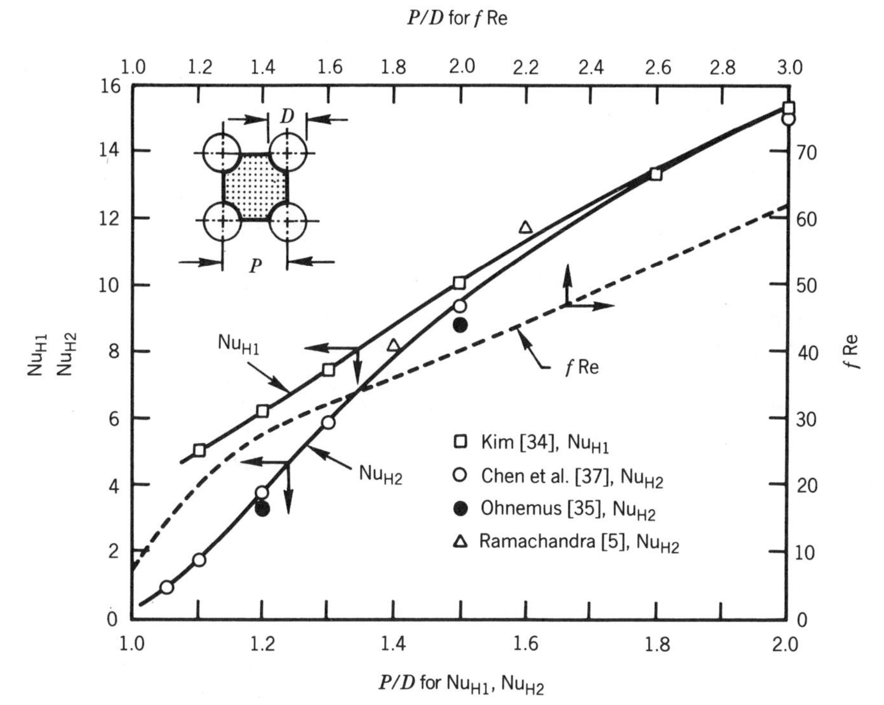

Figure 7.3. f Re and Nu of central subchannels in a regular square array vs. pitch-to-diameter ratio for fully developed laminar flow.

Eq. (7.4)] is

$$r_* = \sqrt{\frac{4}{\pi}}\frac{P}{D} \approx 1.128\frac{P}{D}. \tag{7.9}$$

For the central subchannel, it is interesting to note that f Re for a triangular array is higher than that for a square array if $P/D < 1.8$, but lower if $P/D > 1.8$. This can be seen also from the annular zone solution, valid for $P/D > 2.8$, because the annular-zone parameter, $r_* = r_{\tau=0}/r_i$, with $r_{\tau=0}$ as the radius of the zero-shear position and r_i as the radius of the inner tube of an annulus, is higher for a square array than for a triangular array at the same P/D [compare Eqs. (7.9) and (7.5)].

The analysis of Ohnemus [35] with gases as coolants shows that f Re, either for fully developed conditions or for simultaneously developing flow and heat transfer, increases with increasing heat flux if temperature-dependent properties are taken into account.

Sparrow and Loeffler [11] and Ramachandra [5] also presented graphs of the wall shear-stress variation around the rod periphery.

***Fully Developed Heat Transfer:* $\mathrm{Nu_{H1}}$.** The (H1) problem for central subchannels in a regular square array has been analyzed by Kim [34] in the range of $P/D = 1.1$ to 4.0. His data are shown in Fig. 7.3. A comparison of Figs. 7.2 and 7.3 reveals that the $\mathrm{Nu_{H1}}$ values for a square array are lower than those for a triangular array in the range of

$1.15 < P/D < 1.95$. Nu_{H1} of a square array is higher than Nu_{H1} of a triangular array for $P/D > 1.95$. This trend is in agreement with the f Re data, because the annular-zone parameter r_* is higher for a square array than for a triangular array at the same P/D. Kim [34] also graphically presented the variation in heat flux along the rod periphery.

***Fully Developed Heat Transfer:* Nu_{H2}.** The (H2) problem of central subchannels in a regular square array has been analyzed by Oberjohn as reported in [1], Ramachandra [5], Ohnemus [35], Chen et al. [37], and Kim et al. [38]. The Nu_{H2} data obtained for a range of $P/D = 1.05$ to 4.0 are displayed in Fig. 7.3 for P/D ratios between 1.0 and 2.0. As expected, Nu_{H2} is lower than Nu_{H1}; the Nusselt numbers for the (H1) and (H2) problems are coincident for $P/D > 1.8$ because the temperature distribution along the rod periphery then becomes uniform. All data computed by different methods agree satisfactorily for $P/D < 1.2$. For $P/D \geq 1.2$, however, the data by Ohnemus [35] and Chen et al. [37], both computed by the BODYFIT code [39], are lower than the data obtained by Oberjohn as reported in [1] and Kim et al. [38]. This is due to the mesh size used for the calculations. The values of Nu_{H2} computed by Zachmann [40] using the BODYFIT code clearly show that a reduced mesh size results in higher Nu_{H2}, especially for $P/D \geq 1.2$.

The results by Ramachandra [5], presented graphically, are higher than those of Oberjohn as reported in [1] and Kim et al. [38] for $P/D > 1.2$. For $P/D > 1.5$, the Nu_{H2} data by Ramachandra [5] exceed the Nu_{H1} data of Kim [34]. Therefore, it is concluded that Ramachandra's Nu_{H2} data are too high for $P/D > 1.5$.

The effect of temperature-dependent properties on Nu_{H2} has been investigated by Ohnemus [35]. His results for CO_2, He, and N_2 and for different heat fluxes show that the fully developed Nu_{H2} are not affected by temperature-dependent properties.

TABLE 7.2 f_xRe of Developing Flow in Central Subchannels Arranged in Square Arrays for Isothermal Flow

	f_x Re				
10^3x^+	$P/D = 1.05$ [35]	1.1 [35]	1.2 [40]	1.5 [40]	2.0 [35]
1.0	—	—	—	—	84.05
1.5	—	—	—	—	69.16
2	—	—	—	—	60.03
3	—	—	—	—	51.42
4	—	—	—	—	48.09
5	—	—	—	45.78	46.38
8	—	—	38.91	38.06	43.96
10	—	36.21	35.79	36.56	43.15
15	—	27.71	31.3	34.09	42.12
20	—	24.44	28.37	32.86	41.75
30	19.88	20.04	24.87	31.30	—
40	16.29	18.25	23.05	30.64	—
50	14.31	16.98	21.98	30.38	—
75	12.30	15.63	20.84	30.15	—
100	11.57	15.05	20.46	30.20	—
150	10.98	14.65	20.29	—	—
200	10.76	14.56	20.27	—	—

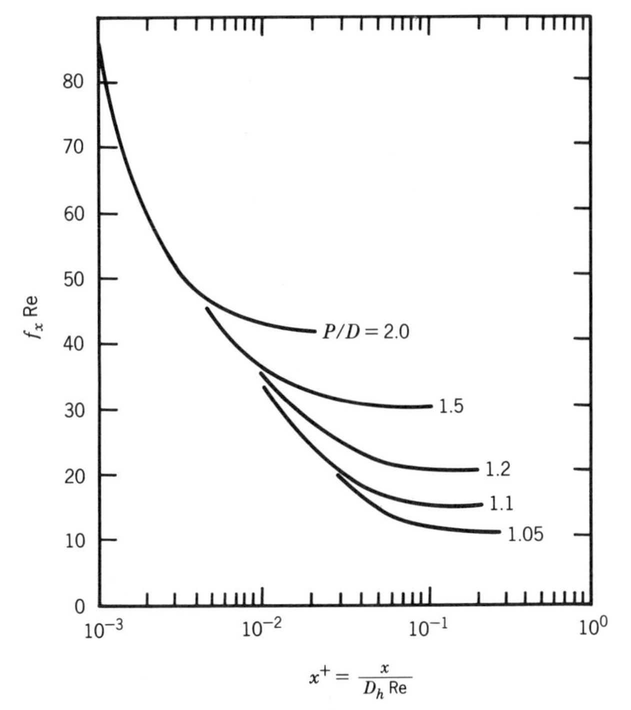

Figure 7.4. f Re of central subchannels in regular square array vs. the nondimensional axial coordinate for hydrodynamically developing laminar flow [35, 40].

Developing Flow. Developing flow through a central subchannel in a regular square array has been analyzed by Ohnemus [35] and Zachmann [40]. The results of their calculations for isothermal flow are presented in Table 7.2 and in Fig. 7.4 as f_xRe versus the nondimensional hydrodynamic entrance length $x^+ = x/(D_h\text{Re})$. Here, f_x is the local friction factor at the axial distance x from the entrance. It is obvious from Fig. 7.4 that the entrance length to arrive at the fully developed f Re increases with decreasing P/D ratio.

Based on the analysis and results of Ohnemus [35], it is concluded that f_xRe increases with increasing heat flux for simultaneous development of flow and heat transfer when the coolant is a gas (CO_2, He, N_2), and when the influence of temperature-dependent fluid properties is included. The nondimensional entrance length to reach fully developed flow conditions also increases with increasing applied heat flux.

Developing Heat Transfer. The (H2) Nusselt numbers in the thermal entrance region have been computed by Ohnemus [35], Zachmann [40], and Chen et al. [37] for the case of simultaneous development of flow and heat transfer, and are shown in Fig. 7.5. The entrance length increases with decreasing P/D. The Nusselt numbers strongly increase near the entrance (as x^* decreases) as expected. The increase in Nu_{H2} in the entrance region is stronger (Table 7.3) for lower P/D ratios. Ohnemus [35] and Zachmann [40] obtained $\text{Nu}_{x,\text{H2}}$ for three different gases as coolant (CO_2, He, and N_2) and considering temperature-dependent properties. Chen et al. [37] obtained $\text{Nu}_{x,\text{H2}}$ for water as a coolant. In the entrance region, $\text{Nu}_{x,\text{H2}}$ according to Zachmann and Ohnemus (Pr =

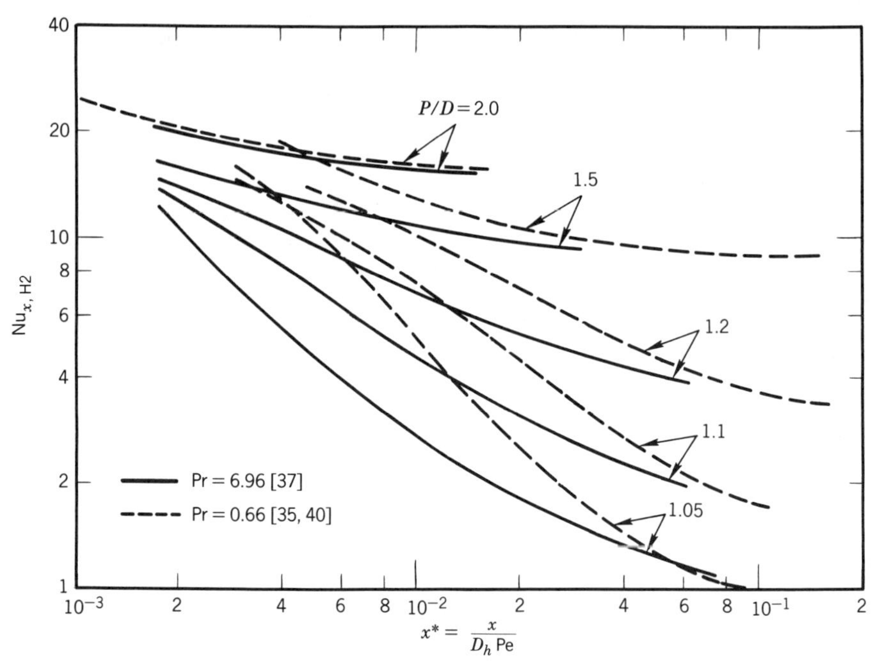

Figure 7.5. Nusselt numbers for thermally developing laminar flow in central subchannels of a regular square array for the (H2) boundary condition.

TABLE 7.3 Nusselt Numbers for Simultaneous Development of Flow and Heat Transfer in Central Subchannels Arranged in a Square Array[a]

	$Nu_{x,H2}$				
$10^3 x^*$	$P/D = 1.05$ [35]	1.1 [35]	1.2 [40]	1.5 [40]	2.0 [35]
1	—	—	—	—	25.21
1.5	—	—	—	—	22.05
2	—	—	—	—	20.33
3	15.74	14.43	—	—	18.51
4	12.88	12.31	—	18.52	—
5	10.29	11.02	13.91	16.96	—
8	6.77	8.52	11.03	13.46	—
10	5.29	7.44	9.93	12.46	—
15	3.33	5.56	8.20	11.22	—
20	2.44	4.42	7.12	10.54	—
30	1.71	3.20	5.75	9.75	—
40	1.41	2.61	4.95	9.33	—
50	1.25	2.27	4.45	9.09	—
75	1.05	1.87	3.81	8.85	—
100	0.97	1.70	3.55	8.79	—
150	0.90	—	3.37	8.77	—

[a]According to Ohnemus [35] and Zachmann [40] for $q'' = 10^3$ W/m^2 and helium as coolant (Pr = 0.66).

0.66) are higher than those according to Chen et al. (Pr = 6.96). This is to be expected, because $\mathrm{Nu}_{x,\mathrm{H2}}$ increases with decreasing Prandtl number for a given x^* [9]. Ohnemus [35] reports that $\mathrm{Nu}_{x,\mathrm{H2}}$ is unaffected by the temperature-dependent properties for gases and by minor variations in Pr for three gases he considered. At higher values of $x^* = x/(D_h \mathrm{Pe})$, when fully developed conditions are approached, the data by Zachmann [40], Ohnemus [35], and Chen et al. [37] are in excellent agreement, as seen in Fig. 7.5.

7.2.3 Central Subchannel: Additional Thermal Boundary Conditions

Gräber [41] analyzed the (H5) heat transfer problem for a central subchannel in a regular triangular array by applying the model of an equivalent annular zone. He presented $\mathrm{Nu}_{\mathrm{H5}}/\mathrm{Nu}_{\mathrm{H1}}$ graphically as a function of P/D and the following parameter: the ratio of the temperature gradient along the wall transferring heat to the temperature gradient in the fluid at a point where the temperature gradient normal to the wall vanishes.

Dwyer and Berry [42] studied a more general boundary condition of constant heat flux in the axial direction, prescribed at the inner surface of a tube, and took into account axial and peripheral heat conduction within the tube wall (multiregion analysis). This condition is more relevant to nuclear fuel elements, where the heat is generated inside the tube (in the fuel) and conducted through the cladding (tube) to the coolant. As expected, the Nusselt numbers calculated are between the limiting cases of the (H1) and (H2) boundary conditions, since the circumferential variations of the cladding surface temperature are diminished by peripheral conduction within the cladding. For a central subchannel in a regular triangular array, Dwyer and Berry [42] tabulated Nusselt numbers and circumferential temperature and heat-flux variations as a function of the P/D ratio, the ratio k_2/k_3 of the thermal conductivities of the cladding (k_2) and the fluid (k_3), and the ratio r_1/r_2 of the inner and outer radii of the cladding.

The conductivity effect in the cladding (k_2) *and* the fuel (k_1) was taken into account by Trombetta [43] for distinct sets of k_1/k_2, k_2/k_3, and r_1/r_2 for a triangular array and P/D ratios between 1.0 and 2.0. Subbotin [30] considered these effects in a more general way and presented the following correlation for Nu for a central subchannel in a regular triangular array with a volumetric uniform heat source in the fuel:

$$\mathrm{Nu} = \left(7.55\frac{P}{D} - \frac{6.3}{(P/D)^{17(P/D)(P/D-0.81)}}\right) \times \left(1 - \frac{3.6P/D}{(P/D)^{20}\left(1 + 2.5\epsilon_K^{0.86}\right) + 3.2}\right) \tag{7.10}$$

where the *thermal modeling parameter* ϵ_K takes into account the effects of heat conduction in the fuel, cladding and fluid. This parameter of thermal similarity is defined as

$$\epsilon_K = \frac{k_2}{k_3}\frac{1 - \Lambda_0(r_1/r_2)^{12}}{1 + \Lambda_0(r_1/r_2)^{12}} \tag{7.11}$$

with

$$\Lambda_0 = \frac{k_2 - k_1}{k_2 + k_1} \tag{7.12}$$

The limiting cases of ϵ_k are

$$\epsilon_K = \begin{cases} 0.01 & \text{for peripherally constant heat flux, } \mathrm{Nu_{H2}} \\ \infty & \text{for peripherally constant temperature, } \mathrm{Nu_{H1}} \end{cases}$$

The limiting case $\epsilon_K = 0.01$ is approached for a thermal conductivity of the cladding (k_2) approaching zero. For infinite conductivity of the cladding material, the limiting case of $\epsilon_K \to \infty$ is obtained. The results of Eq. (7.10) are presented graphically by Subbotin et al. [44]. The dimensionless peripheral variation of surface temperatures, defined by

$$\Delta T^{\max} = \frac{t_w^{\max} - t_w^{\min}}{q''D} 2k_3 \tag{7.13}$$

is correlated by Subbotin et al. [45] for laminar flow as

$$\Delta T_{\mathrm{lam}}^{\max} = \frac{0.022}{(P/D)^{3(P/D-1)^{0.4}} - 0.99}\left\{1 - \tanh\left[\frac{1.2e^{-26.4(P/D-1)} + \ln \epsilon_K}{0.84 + 0.2\left(\dfrac{P/D - 1.06}{0.06}\right)^2}\right]\right\} \tag{7.14}$$

Hsu [28] analyzed the same problem and tabulated Nu for the case of a volumetric uniform heat source and $k_2/k_3 = 0.3$, $k_1/k_2 = 0.1$, $r_1/r_2 = 0.875$, typical values for a sodium-cooled nuclear reactor with oxide fuel and stainless-steel cladding, for a P/D range of 1.05 to 2.0.

The same problem, including conduction effects in the fuel and cladding, was analyzed by Cieszko and Kolodziej [46] for a central subchannel in both regular triangular and square arrays. They presented the results graphically for a wide range of P/D, k_1/k_2, k_2/k_3, and r_1/r_2. While their calculated Nusselt numbers for a square array are in between $\mathrm{Nu_{H1}}$ and $\mathrm{Nu_{H2}}$ of Fig. 7.3 as expected, the results for the triangular array are higher than $\mathrm{Nu_{H1}}$ of Fig. 7.2, which is clearly not correct.

Hsu [47] also considered the effect of lateral rod displacement on heat transfer in rod bundles of a triangular array by a multiregion analysis. Hsu tabulated rod-averaged Nusselt numbers in the range of $P/D = 1.1$ to 1.2 for uniform heat generation in the fuel and for various sets of k_1/k_2, k_2/k_3, and r_1/r_2. Hsu [47] analyzed heat transfer in longitudinal laminar flow with one displaced rod in a triangular array for the (H1) and (H2) problems. Hsu [48] graphically presented $\mathrm{Nu_{H1}}$ and $\mathrm{Nu_{H2}}$ for $P/D = 1.0$ to 2.0 and displacements of up to 80% of the maximum possible displacement. His results for the symmetric case (i.e., without a displaced rod) are in excellent agreement with those of Dwyer and Berry [21]. Hsu's results show that the rod-average Nusselt numbers of the displaced rod decrease from the symmetric case with increasing displacement. The drop below the $\mathrm{Nu_{H1}}$ and $\mathrm{Nu_{H2}}$ of the symmetric case increases with increasing P/D ratio.

7.2.4 Wall Subchannel

The wall subchannel shape and size (flow area and wetted perimeter) are identical for regular triangular and square arrays for the same W, P, and D. Hence, no distinction has been made between the wall subchannels of different arrays in the following.

***Fully Developed Flow: f* Re.** The pressure drop in wall subchannels of rod bundles has been analyzed by Rehme [15, 16, 49] for a wide range of P/D and W/D ratios. The f Re factors are presented in Table 7.4. Robinson [50] studied the pressure drop in wall subchannels for $P/D = 1.225$ and W/D in the range between 1.0206 and 1.1738. His f Re factors are in excellent agreement with those of Table 7.4. Herzog [52] investigated f Re for $P/D = W/D = 1.2, 1.4, 1.6$ and for $P/D = 1.4$, $W/D = 1.2, 1.6$. The f Re factors from this study are also in good agreement with Table 7.4: within 0.6% for a fine mesh grid used for the calculations, and within 4% for a coarse mesh grid. Gunn and Darling [36] calculated $f\,\mathrm{Re} = 6.50$ for $P/D = W/D = 1.0$, which agrees within 1% with that of Table 7.4.

Mohanty and Sahoo [254] computed f Re of twelve wall subchannels for P/D in the range between 1.2 and 2.5 and W/D in the range between 1.15 and 2.0. Their results are in close agreement with those of Table 7.4.

Figure 7.6 shows f Re versus P/D for three characteristic W/D: (1) $W/D = P/D$ (gap between rods equal to gap between rod and channel) (2) $W/D = 0.5(P/D - 1) + 1$ (gap between rod and channel = 0.5 × gap between rods), and (3) $W/D = 2(P/D - 1) + 1$ (gap between rod and channel = 2 × gap between rods).

For the special case of $P/D = W/D$, Cheng and Todreas [26] developed the following correlations:

$$f\,\mathrm{Re} = \begin{cases} 6.545 + 138.63(P/D - 1) - 370(P/D - 1)^2 & \text{for } 1.0 \le P/D \le 1.1 \\ 11.1 + 64.175(P/D - 1) - 66.9(P/D - 1)^2 & \text{for } 1.1 < P/D \le 1.5 \end{cases} \tag{7.15}$$

which agree with the results of Table 7.4 within 1.0%.

TABLE 7.4 *f* Re of Wall Subchannels for Fully Developed Laminar Flow [15, 16]

	f Re										
P/D	$W/D = 1.0$	1.02	1.05	1.1	1.2	1.25	1.35	1.5	1.75	2.0	3.0
1.0	6.56	8.46	10.98	13.99	16.68	—	—	17.27	—	16.46	15.46
1.02	7.22	9.15	11.70	14.75	—	17.93	—	17.91	—	17.20	14.42
1.05	7.98	9.93	12.52	15.70	—	19.18	—	19.16	—	18.00	16.98
1.1	8.74	10.68	13.31	16.69	—	20.69	—	—	19.94	19.26	17.98
1.2	8.88	10.67	13.24	16.89	21.39	—	23.56	23.63	22.98	22.18	20.24
1.35	—	—	—	—	20.54	—	24.72	—	—	—	—
1.5	7.35	8.54	10.35	13.35	18.63	—	—	26.56	27.72	27.68	26.21
1.75	—	—	9.17	11.49	15.94	—	—	25.10	28.35	29.55	31.94
2.0	6.99	7.70	8.84	10.72	—	16.11	19.35	23.24	27.43	29.94	33.11
3	8.66	9.14	9.85	—	13.15	—	—	18.88	22.80	25.96	—
4	—	—	—	12.38	—	14.73	—	18.01	—	23.45	28.76
5	12.12	—	12.93	—	14.98	—	—	—	—	—	—
7	—	—	—	—	—	—	—	18.90	—	21.70	—
10	16.67	16.88	17.16	17.60	—	18.66	—	19.94	—	21.79	24.77

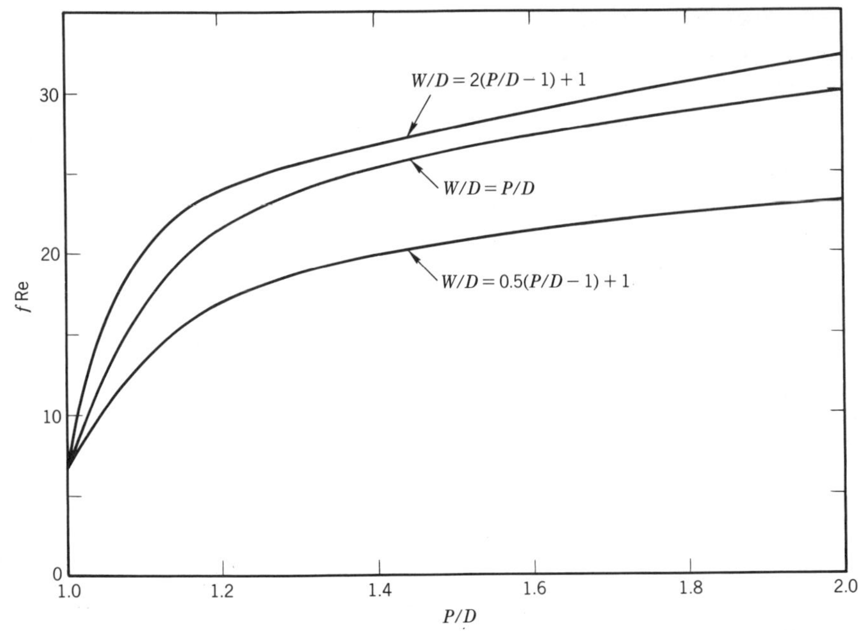

Figure 7.6. f Re of wall subchannels versus pitch-to-diameter ratio for fully developed laminar flow.

When the rod is heated, the fully developed f Re increases with increasing heat flux, when the calculations are made with helium as the working fluid and with temperature-dependent fluid properties [52].

***Fully Developed Heat Transfer:* $\mathbf{Nu_{H2}}$.** Herzog [52] analyzed the (H2) problem for five geometries and for two different heat fluxes at the rod wall, taking into account temperature-dependent fluid properties with helium as a working fluid. The Nu_{H2} for the higher heat flux were slightly lower (up to 6%) than those for the lower heat flux. The difference may be due to the coarse mesh grid used.

Mohanty and Sahoo [254] computed Nu_{H2} of twelve wall subchannels for P/D in the range between 1.2 and 2.5 and W/D in the range between 1.5 and 2.0, for constant fluid properties.

TABLE 7.5 $\mathbf{Nu_{H2}}$ for Fully Developed Laminar Flow in Wall Subchannels

	Nu_{H2}				
P/D	$W/D = 1.15$	1.2	1.4	1.5	1.6
1.2	3.07	3.45	3.85	3.87	—
1.4	—	4.49[a]	5.53[a]	—	5.00[a]
1.5	—	—	6.86	—	—
1.6	—	—	—	—	7.14[a]
1.75	—	—	—	7.77	8.13

[a] From Herzog [52]; other data from Mohanty and Sahoo [254].

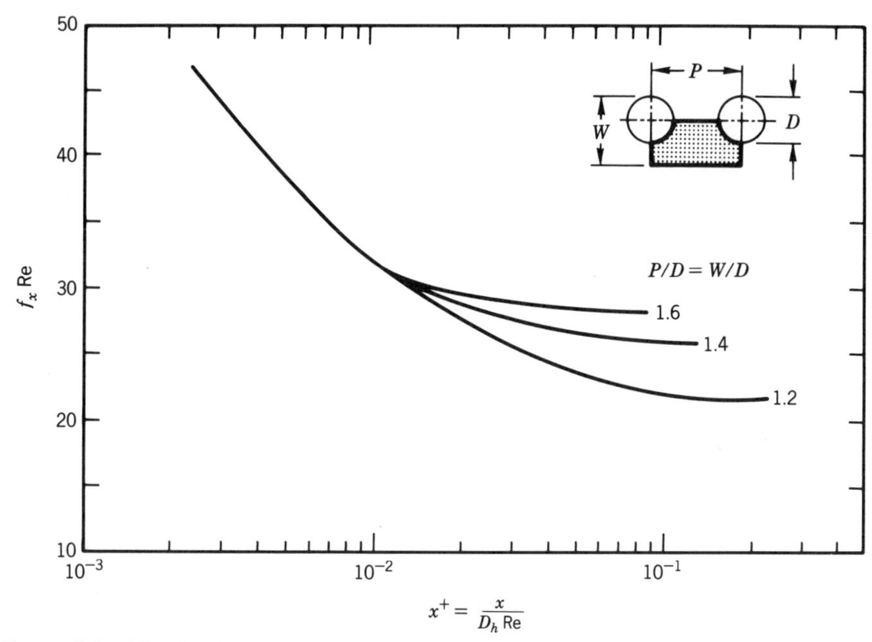

Figure 7.7. f Re for hydrodynamically developing laminar flow in wall subchannels with $P/D = W/D$, vs. the nondimensional axial coordinate [52].

Table 7.5 shows some of the results of both investigations. The Nu_{H2} data of Herzog [52] are for the lower heat flux. The tabulated Nu_{H2} for $P/D = W/D = 1.2$ by Mohanty and Sahoo agrees within 5% with $\text{Nu}_{\text{H2}} = 3.30$ due to Herzog.

***Developing Flow: f* Re.** The only data for f_xRe in the hydrodynamic entrance region were computed by Herzog [52] and are shown in Fig. 7.7 for $P/D = W/D$. As expected, the f_xRe factors are higher in the entrance region than those for fully developed flow. The entrance length increases with decreasing $P/D = W/D$. From the tabulated f_xRe by Herzog [52], it is obvious that f_xRe for heated rods and simultaneously developing flow and heat transfer, calculated for helium and taking temperature-dependent properties into account, is higher than for isothermal flow.

***Developing Heat Transfer:* Nu_{H2}.** Herzog [52] also analyzed the thermal entrance region in wall subchannels for the case of simultaneous development of flow and heat transfer with helium (Pr = 0.66) as coolant for the geometries of Table 7.5. His results of $\text{Nu}_{x,\text{H2}}$ show that the thermal entrance length decreases with increasing $P/D = W/D$. The results computed for $q'' = 10^2$ and 10^3 W/m² are almost identical.

7.2.5 Corner Subchannel — Triangular Array

***Fully Developed Flow: f* Re.** Laminar flow through corner subchannels of a regular triangular array has been analyzed by Ratkowsky as reported in [9], Cheng and Jamil [53], and Rehme [15, 16]. Their f Re factors are presented in Fig. 7.8. Agreement among the three data sets is better than 1%, except for the range of $1.03 < W/D < 1.15$, where Rehme's data seem to be too high. The f Re factors increase strongly between

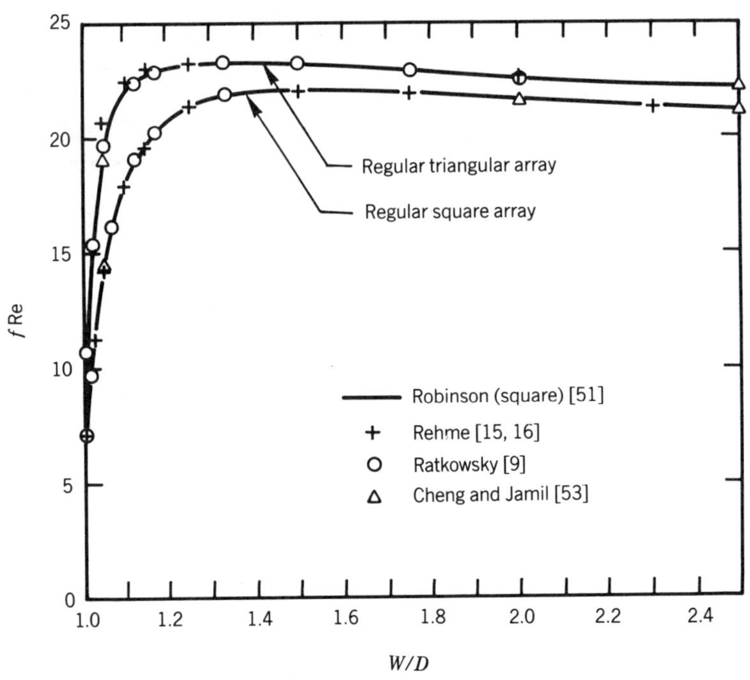

Figure 7.8. f Re vs. wall-to-diameter ratio for fully developed laminar flow in corner subchannels.

$W/D = 1.0$ and 1.1; for higher W/D ratios, f Re reaches a maximum at values close to the f Re of annuli with high radius ratios ($f\,\mathrm{Re} = 23$ to 24) before decreasing slightly. The recent results by Mohanty and Sahoo [254] are in close agreement with Rehme's [15, 16] data.

Cheng and Todreas [26] developed the following correlation for f Re in corner subchannels of a regular triangular array:

$$f\,\mathrm{Re} = \begin{cases} 6.745 + 409(W/D - 1) - 2513(W/D - 1)^2, & 1.0 \le W/D \le 1.1 \\ 21.815 + 9.6475(W/D - 1) - 13.78(W/D - 1)^2, & 1.1 < W/D \le 1.5 \end{cases} \tag{7.16}$$

This equation fits the data within $\pm 1.1\%$.

Fully Developed Heat Transfer: **$\mathrm{Nu_{H1}}$.** The (H1) heat transfer problem for a corner subchannel in a regular triangular array was analyzed by Cheng and Jamil [53]. They reported $\mathrm{Nu_{H1}}$ for the range of $W/D = 1.5$ to ∞, which is outside the range of most practical applications.

Fully Developed Heat Transfer: **$\mathrm{Nu_{H2}}$.** Mohanty and Sahoo [254] investigated the (H2) heat transfer problem for a corner subchannel in a regular triangular array in the range of $W/D = 1.05$ to 2.0. Their results are shown in Fig. 7.9.

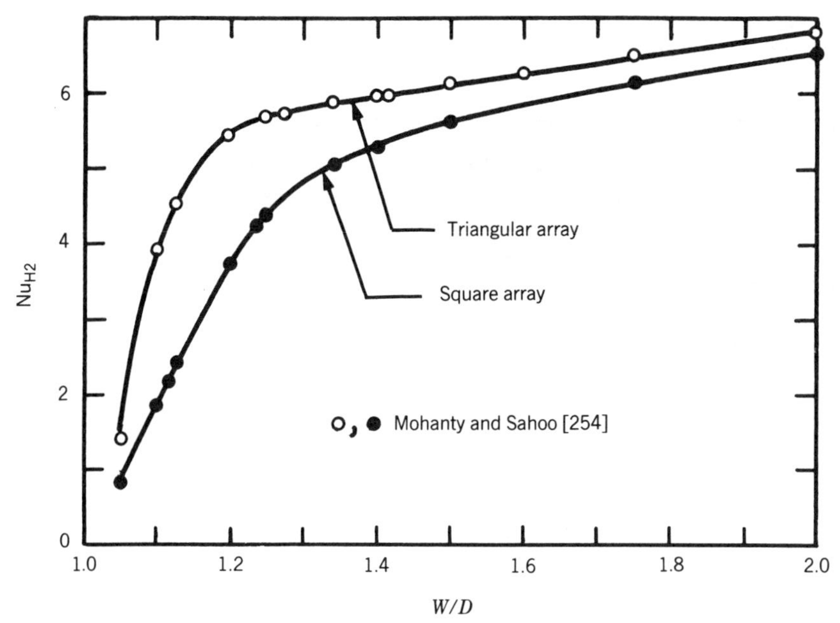

Figure 7.9. Nu_{H2} vs. pitch-to-diameter ratio for fully developed laminar flow in corner subchannels: (*a*) regular triangular array, (*b*) regular square array.

7.2.6 Corner Subchannel — Square Array

Fully Developed Flow: f Re. Fully developed laminar flow through corner subchannels in a regular square array has been analyzed by Rehme [15,16], Ratkowsky as reported in [9], Cheng and Jamil [53], and Robinson [50]. The f Re factors of Rehme [15], Cheng and Jamil [53], and Ratkowsky (reported in [9]) are presented in Fig. 7.8. These results are also in excellent agreement with the graphical f Re of Robinson [51] as shown in Fig. 7.8. It should be noted that the statement "excellent agreement of Robinson's data with Rehme's result is not confirmed" made by Johannsen [1] is incorrect, because of a misinterpretation of the W/D ratio. In the present nomenclature, W/D of Robinson [50] is $0.5(P/D + 1)$. Gunn and Darling [36] calculated $f\,\mathrm{Re} = 7.06$ for $W/D = 1.0$, which is in good agreement with that of Fig. 7.8. The recent results by Mohanty and Sahoo [254] for the range of $W/D = 1.05$ to 2.0 agree closely with the results of Fig. 7.8.

The f Re factors for a regular square array are lower than those of a regular triangular array for the same W/D.

***Fully Developed Heat Transfer:* $\mathrm{Nu}_{\mathrm{H1}}$.** As for the triangular array, the (H1) heat transfer problem for a corner subchannel in a regular square array was analyzed by Cheng and Jamil [53]. They reported $\mathrm{Nu}_{\mathrm{H1}}$ for the range of $W/D = 1.5$ to ∞, which is outside the range of most practical applications. In the range considered, $\mathrm{Nu}_{\mathrm{H1}}$ for a triangular array are higher than those for a square array.

***Fully Developed Heat Transfer:* $\mathrm{Nu}_{\mathrm{H2}}$.** Mohanty and Sahoo [254] published $\mathrm{Nu}_{\mathrm{H2}}$ computed for a corner subchannel in a regular square array in the range $W/D = 1.05$ to 2.0. Their results are shown in Fig. 7.9. The values of $\mathrm{Nu}_{\mathrm{H2}}$ for a corner subchannel in a triangular array are higher than those in a square array, which is in agreement with the trends observed for f Re in corner subchannels.

7.2.7 Finite Rod Bundles in Regular Triangular and Square Arrays

***Fully Developed Laminar Flow: f* Re.** The f Re factors for fully developed laminar flow through rod bundles of any size can be computed from the known f Re factors of the individual subchannels by a method proposed by Rehme [15]. Under the idealizations of (1) a constant pressure drop in all subchannels and (2) a total flow rate through the rod bundle under consideration which is equal to the sum of the flow rates through the n individual subchannels, the f Re factor of a rod bundle can be obtained by the following equation [15]:

$$\frac{1}{f\,\mathrm{Re}} = \sum_{i=1}^{n} \frac{1}{(f\,\mathrm{Re})_i} \left(\frac{P_W}{P_{W_i}} \right)^2 \left(\frac{A_i}{A} \right)^3 \tag{7.17}$$

With the indices c for central, w for wall, and co for corner subchannels and the respective n_i, A_i and P_{W_i} from Table 7.1, Eq. (7.17) can be rewritten as

$$\frac{1}{f\,\mathrm{Re}} = \frac{n_\mathrm{c}}{(f\,\mathrm{Re})_\mathrm{c}} \left(\frac{P_W}{P_{W,\mathrm{c}}} \right)^2 \left(\frac{A_\mathrm{c}}{A} \right)^3 + \frac{n_\mathrm{w}}{(f\,\mathrm{Re})_\mathrm{w}} \left(\frac{P_W}{P_{W,\mathrm{w}}} \right)^2 \left(\frac{A_\mathrm{w}}{A} \right)^3$$

$$+ \frac{n_\mathrm{co}}{(f\,\mathrm{Re})_\mathrm{co}} \left(\frac{P_W}{P_{W,\mathrm{co}}} \right)^2 \left(\frac{A_\mathrm{co}}{A} \right)^3 \tag{7.18}$$

where the total flow cross section A and the total wetted perimeter P_W of the rod bundle are

$$A = n_\mathrm{c} A_\mathrm{c} + n_\mathrm{w} A_\mathrm{w} + n_\mathrm{co} A_\mathrm{co} \tag{7.19}$$

$$P_W = n_\mathrm{c} P_{W,\mathrm{c}} + n_\mathrm{w} P_{W,\mathrm{w}} + n_\mathrm{co} P_{W,\mathrm{co}}. \tag{7.20}$$

EXAMPLE. Consider a rod bundle of 19 rods in a regular triangular array ($N = 2$) with $P/D = 1.266$ and $W/D = 1.04$.

From Table 7.1 we have $n_\mathrm{c} = 24$, $n_\mathrm{w} = 12$, $n_\mathrm{co} = 6$. From Fig. 7.2, $(f\,\mathrm{Re})_\mathrm{c} = 26.85$; from Table 7.4 (by interpolation) or from [49], $(f\,\mathrm{Re})_\mathrm{w} = 11.95$; and from Fig. 7.8, $(f\,\mathrm{Re})_\mathrm{co} = 18.95$. With A_i and $P_{W,i}$ from Table 7.1 and A and P_W from Eqs. (7.19) and (7.20),

$$f\,\mathrm{Re} = 16.33$$

is calculated from Eq. (7.18). It agrees with the numerical value f Re = 16.21 calculated by Ullrich [54], within 0.8%.

The validity of Eq. (7.18) to calculate f Re of rod bundles using the subchannel f Re factors has been proven for $1.1 \le P/D \le 2.3$ and $1.04 \le W/D \le 2.5$. A comparison with numerically computed f Re factors for 20 rod bundles [15] and 11 rod bundles [54], and experimental f Re factors for 8 rod bundles [55] and 4 rod bundles [36] indicates that the subchannel analysis predicts the f Re factors accurately to within 5%, and the agreement is within 3% in most cases. In these comparisons, the maximum

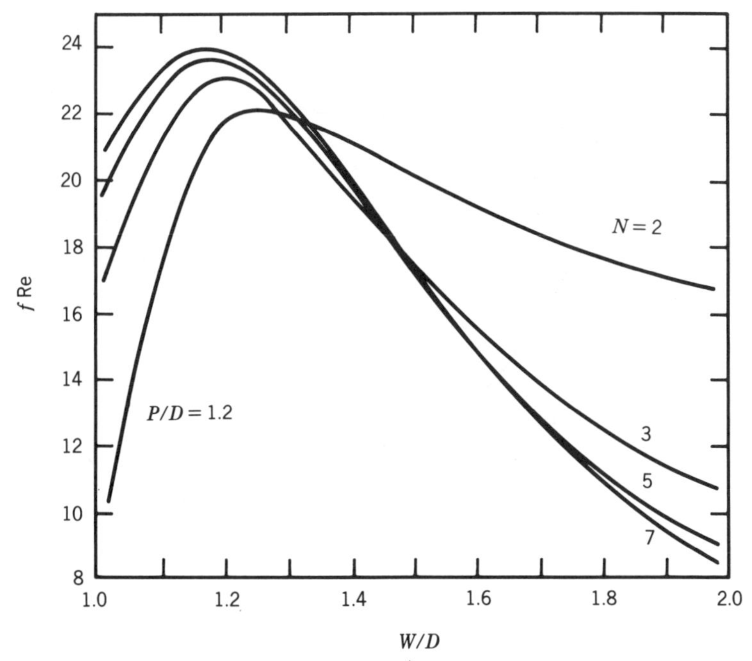

Figure 7.10. f Re for fully developed laminar flow in rod bundles in a regular triangular array at $P/D = 1.2$: effect of W/D ratio and the bundle size [1].

number of rods employed in a bundle was 37. As the number of rods increases, the accuracy of the subchannel analysis increases due to a somewhat reduced "wall effect". Since the subchannel analysis is an accurate tool, the f Re factors for an important family of rod bundles have been computed by Johannsen [1] using this tool and are presented in Figs. 7.10 to 7.12.

Figure 7.10 shows the dependence of f Re on variables W/D and bundle size N for $P/D = 1.2$. f Re initially increases with W/D, reaches a maximum in the neighborhood of $P/D = W/D$, and then decreases continuously for the ranges of W/D and N shown. The maximum f Re value increases and is shifted toward a smaller W/D with increasing bundle size; the change from $N = 2$ to $N = 7$ corresponds to an increase in the number of rods from 19 to 169. This behavior of f Re, which is similar when P/D is varied and W/D kept constant, may be explained by the fact that f Re reaches a maximum for a uniform velocity distribution among the different subchannels. This occurs in rod bundles with nearly uniform hydraulic diameters of the subchannels, which condition is met most closely in the neighborhood of $P/D = W/D$.

It is often assumed that the influence of the channel wall may be neglected for rod bundles with a large enough number of rods ($N > 5$), and f Re of the central subchannel may be applied to the total rod bundle. This is not quite true. In Fig. 7.11 the deviation of f Re between rod bundles and central subchannels is plotted against the number of rings, N, for a constant $P/D = 1.1$ and different W/D ratios. It clearly shows that, even when $P/D = W/D$, the f Re factor of rod bundles is considerably smaller than f Re of the central subchannel. With increasing W/D ratio, the convergence of f Re toward the central subchannel value becomes extremely slow [1].

Even for bundle sizes of practical application and $P/D = W/D$, the ratio of the f Re of the bundle to that of the central subchannel does not approach unity, as shown

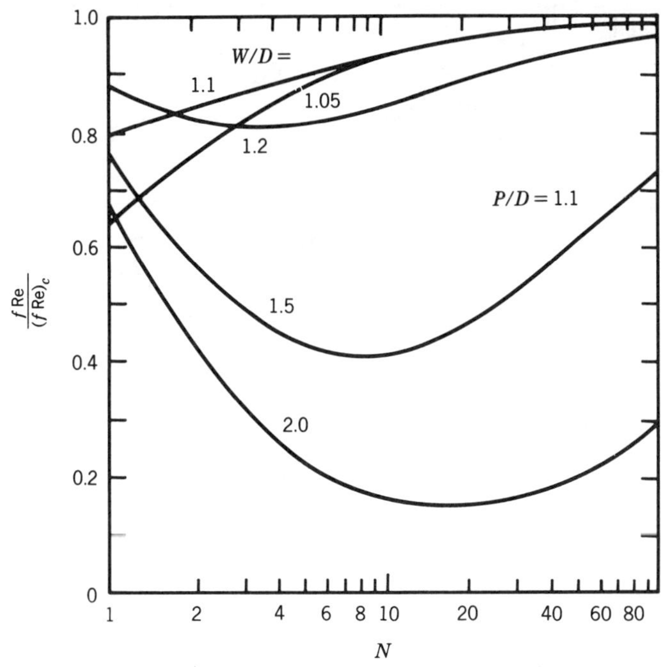

Figure 7.11. Ratio of f Re for fully developed laminar flow in rod bundles in a regular triangular array to f Re for a central subchannel with $P/D = 1.1$: effect of W/D ratio and bundle size [1].

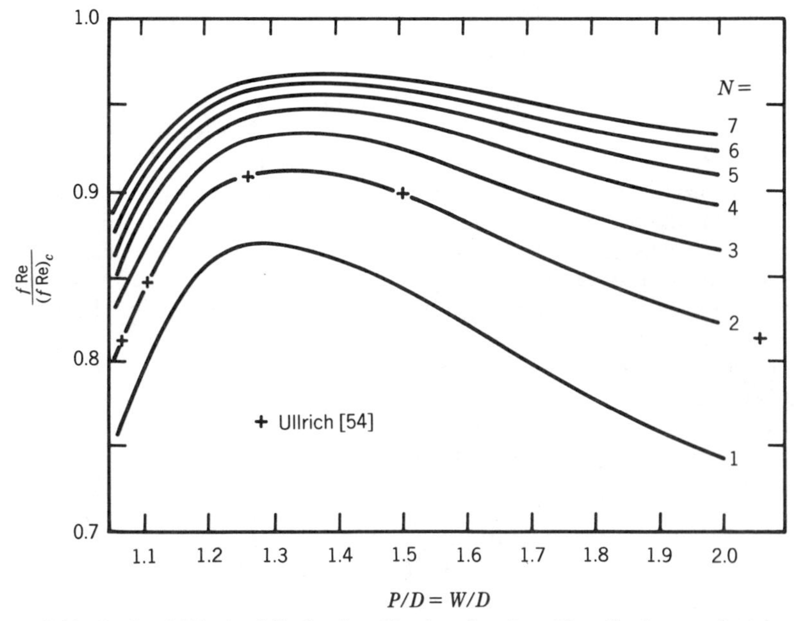

Figure 7.12. Ratio of f Re for fully developed laminar flow in rod bundles in a regular triangular array to f Re for a central subchannel with $P/D = W/D$: effect of bundle size [1].

in Fig. 7.12. Some f Re factors, calculated numerically by Ullrich [54] for bundles of 19 rods, are included in Fig. 7.12, demonstrating the excellent accuracy obtained by the subchannel method of Rehme [15].

Fully Developed Heat Transfer. In contrast to f Re, there is no simple rule to calculate the temperature distribution for fully developed heat transfer in laminar flow through rod bundles. Subchannel solutions for Nusselt numbers may be adequate to study the average laminar heat transfer in finite rod bundles, but the analysis based on subchannel solutions is inappropriate when investigating local effects, such as thermal stresses, local hot spots, and maximum or minimum surface temperatures along the periphery of the rods. The peripheral variations of surface temperatures and heat fluxes in rod bundles are due to internal heat generation and asymmetric heat dissipation as a result of nonidentical flow passage geometries.

At high P/D ratios, e.g. $P/D > 1.5$, peripheral wall temperature variations in all types of subchannels are small enough to be neglected, provided $P/D \approx W/D$. Hence, the designer of such heat transfer equipment may be satisfied with knowing the subchannel Nusselt numbers to estimate wall surface temperatures and heat transfer rates [8]. At low P/D ratios, such as those in nuclear-reactor fuel assemblies, peripheral variations of the surface temperature may be much larger than the average temperature drop between the wall and fluid bulk temperatures. In this case, single-region solutions will give wrong answers to the problem of the maximum temperature in a rod bundle. Application of multiregion solutions with specified heat flux at the inner surface of the cladding and with heat conduction in the cladding taken into account will result in smaller errors, but not in correct answers [8].

The peripheral variations of temperatures and heat fluxes depend on (1) the geometrical parameters of the rod bundle (P/D, W/D, number of rods, triangular or square arrays), (2) the thermal properties of the fluid as well as of the structures, and (3) the thermal boundary conditions. Thus the number of independent and dependent variables is very large, which makes it completely impracticable to both compute and compile results for all possible relevant combinations of independent parameters [8]. Computer programs are necessary to enable the designer to evaluate all necessary data for any set of parameters of interest.

The problem of laminar flow heat transfer through rod bundles is discussed in depth by Johannsen [8].

7.2.8 Cell Solutions

To overcome the difficulty of the heat transfer problem for laminar flow in rod bundles to a certain extent, Hsu [28] calculated average Nusselt numbers for wall and corner rods in a regular triangular array, i.e., for the regions with the highest asymmetries in the flow cross section. These cell solutions include an analysis of a wall cell (the wall subchannel together with the neighboring central subchannel) and a corner cell (the corner subchannel together with the neighboring central subchannel) as shown in Fig. 7.13.

Hsu [28] tabulated $\mathrm{Nu}_{\mathrm{H1}}$ and $\mathrm{Nu}_{\mathrm{H2}}$ for wall and corner cells in the range of $P/D = W/D = 1.05$ to 1.8. He used an equivalent diameter (D_e) for the Nusselt number, which did not include the channel perimeter. Therefore, Hsu's Nusselt numbers have been recalculated using the hydraulic diameter, including the channel

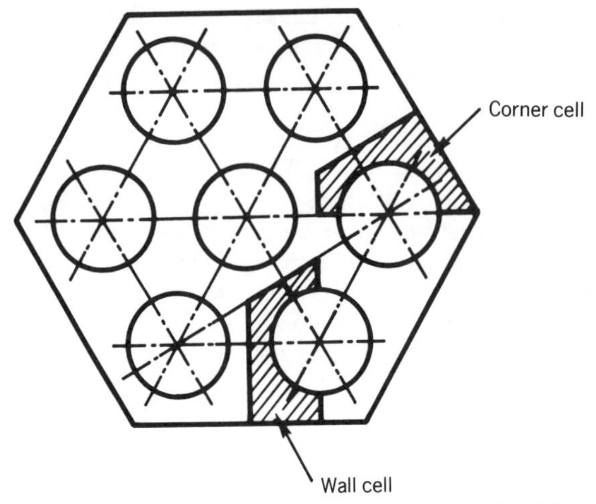

Figure 7.13. Definition of flow domain of corner and wall cells of rod bundles arranged in a regular triangular array.

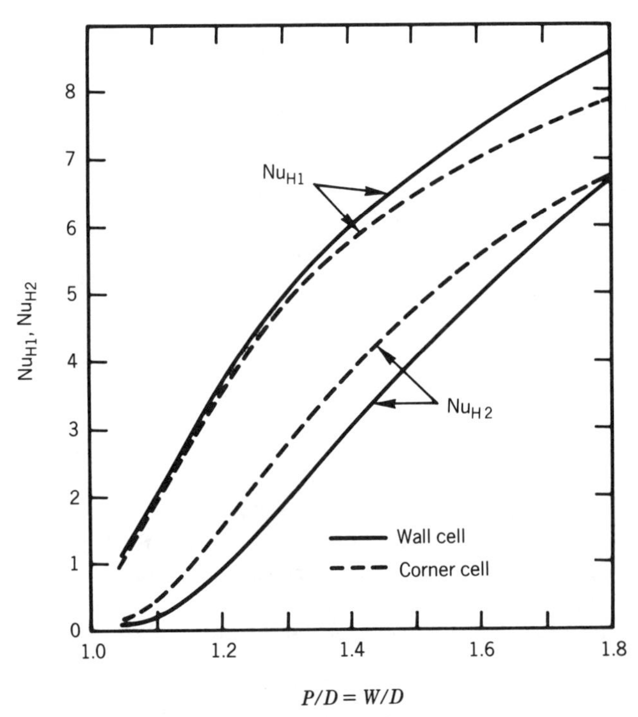

Figure 7.14. Nusselt numbers for fully developed laminar flow in wall and corner cells for $P/D = W/D$, vs. pitch-to-diameter ratio [28].

perimeter. It can easily be shown that, for the geometries investigated ($P/D = W/D$),

$$\frac{D_h}{D_e} = \begin{cases} \dfrac{\pi\sqrt{3}}{\pi\sqrt{3} - 1 + (P/D)(2 + \sqrt{3})} & \text{for a corner cell} \quad (7.21) \\[2ex] \dfrac{\pi}{\pi + P/D} & \text{for a wall cell} \quad (7.22) \end{cases}$$

The $\mathrm{Nu_{H1}}$ and $\mathrm{Nu_{H2}}$ data for the wall cell and corner cell are shown in Fig. 7.14. As expected, $\mathrm{Nu_{H1}}$ is higher than $\mathrm{Nu_{H2}}$, and both Nusselt numbers increase with increasing $P/D = W/D$. $\mathrm{Nu_{H1}}$ of a corner cell is slightly lower than $\mathrm{Nu_{H1}}$ of a wall cell. However, $\mathrm{Nu_{H2}}$ of a corner cell is higher than $\mathrm{Nu_{H2}}$ of a wall cell at the same $P/D = W/D$.

Hsu [28] also calculated Nusselt numbers for wall and corner cells applying a constant volumetric heat source within the rods and taking into account heat conduction in the fuel and the cladding (multiregion analysis). He analyzed the problem of a typical sodium-cooled nuclear reactor with fuel rods consisting of oxide fuel core and stainless-steel cladding ($k_1/k_2 = 0.1$ and $r_1/r_2 = 0.875$). His computed Nusselt numbers, for the range $P/D = W/D = 1.05$ to 2.0 for wall and corner cells, are very close to the $\mathrm{Nu_{H2}}$ results of Fig. 7.14. Hsu [28] also presented graphs of the peripheral variations of inner and outer cladding wall temperatures and wall heat fluxes for typical geometrical cases.

Multicell solutions for wall and central subchannels have been obtained by Milbauer [57], who computed variations of wall shear stresses and of surface temperatures for rod displacements under the assumption of a constant heat flux at the inner surface of the tube. He presented the results graphically.

7.2.9 Concentric Rod Bundles

A concise presentation of results for concentric rod bundles, similar to those for bundles arranged in a regular triangular or square array, is impossible because of the larger number of geometric parameters needed to characterize the geometry.

The f Re of concentric rod bundles can be computed by Eq. (7.18), provided the f Re of the wall subchannels with curved boundary are known. Mohanty and Sahoo [254] computed f Re for fifteen different geometries of wall subchannels with curved boundary in the range of $P/D = 1.083$ to 1.822 and $W/D = 1.125$ to 1.82, and for three different included angles of 120, 105, and 100°, which correspond to clusters of 7, 19, and 37 rods, respectively. Mohanty and Sahoo showed that the superposition [Eq. (7.18)] of the subchannel f Re agrees within 2.6% with the f Re from the literature for five rod clusters. The literature f Re were obtained by analysis of a sector of the rod cluster.

Mohanty and Sahoo [254] also presented $\mathrm{Nu_{H2}}$ for fifteen different wall subchannels with a curved boundary. However, superposition of the subchannel $\mathrm{Nu_{H2}}$ to obtain $\mathrm{Nu_{H2}}$ of the cluster was not successful (see Sec. 7.2.7).

Concentric rod bundles have been used as fuel elements in some nuclear reactors. Therefore, the problem of laminar flow and heat transfer for concentric rod bundles has been studied in some detail. Axford [58] analyzed the velocity distribution for $n = 3, 4, 5, 6$, and 7 tubes in two arrangements, but reported no f Re factors. Courtaud et al. [59] experimentally investigated a rod bundle of six rods surrounding the central rod for different radii r_1 of the ring of six rods between two limits: (1) the six rods touching the central rod, and (2) the six rods touching the outer circular tube. Courtaud et al. [59] also calculated f Re factors numerically.

Min et al. [60, 61] analyzed rod bundles with one ring of different numbers of rods around the central rod and reported f Re factors. Mottaghian and Wolf [62–64] considered tube bundles with more than one ring around the central tube. They developed an analytical method able to handle an arbitrary number of rods with different radii, placed on concentric rings around the central rod. Flow through rod bundles with one ring of six rods around the central rod was analyzed by Chen [65, 66]. Chen also investigated the effect of displacing one rod of the outer ring. Chen's results for the symmetric case are in very good agreement with those by Axford [58] and Min et al. [61]. For a cluster with a displaced rod, Chen [66] found that the dimensionless pressure gradient, which is proportional to f Re, decreases as the displacement increases. This means that, for a fixed pressure drop, the flow rate increases with displacement. Zarling [67, 68] analyzed laminar flow and heat transfer in concentric rod bundles with 5, 7, 9 and 11 rods. His f Re factors are tabulated in [9]. He considered two boundary conditions for the heat transfer problem [67]: (1) the rods were maintained at one constant temperature while the shell was maintained at another constant temperature, and (2) the rods were held at a uniform heat flux while the shell was held at a constant temperature. Zarling and Min [67, 69, 70] also analyzed the (T) and (H1) thermal entrance length problems for symmetrical one-ring rod bundles.

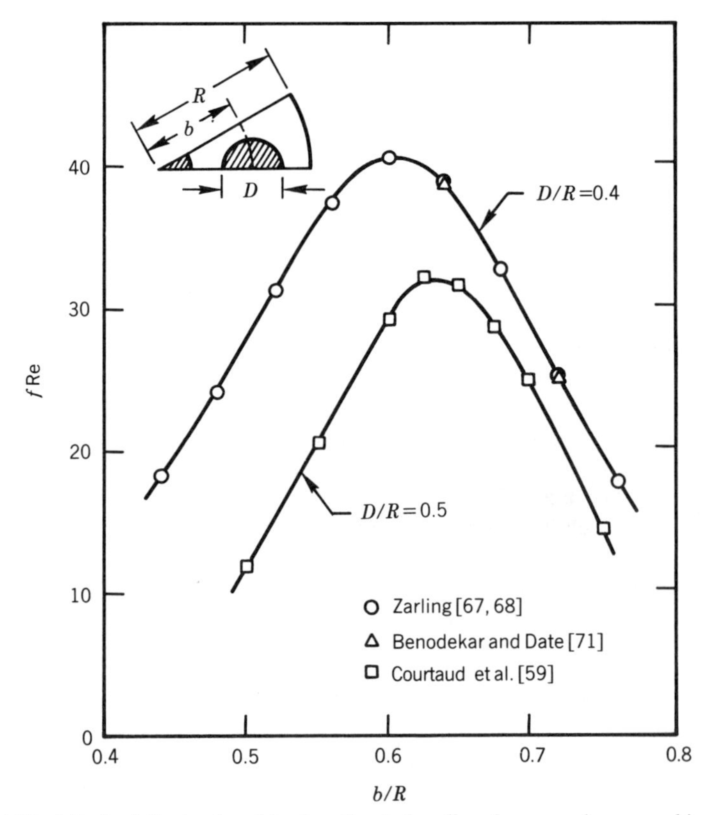

Figure 7.15. f Re for fully developed laminar flow in bundles of seven rods arranged in a circular array, vs. the radius ratio.

Benodekar and Date [71] analyzed laminar flow and the (H2) problem in circular rod bundles with one ring of 4, 6, 8, and 10 rods around the central rod, with two rings of different rod numbers, and with 18 rods placed in two rings around the central rod. They tabulated f Re and $\mathrm{Nu}_{\mathrm{H1}}$ for 47 different rod bundles [71]. Benodekar and Date [72] also investigated the thermal entrance length problem for both the (H2) boundary condition and sinusoidal heat flux in the axial direction with peripherally constant heat flux, for the case of fully developed velocity profile. Nusselt numbers and dimensionless temperatures and temperature differences are presented for clusters of 7 and 19 rods.

Mohanty and Ray [73] investigated laminar flow through circular tube bundles with one ring of different numbers of rods around the central rod. However, their f Re factors do not agree with those independently obtained by various researchers. The special case of symmetrical and displaced clusters of three rods inside a circular tube, which occurs in underground, pipe-type electrical cable systems, was analyzed by Chern and Chato [74]. The authors treat thermally developing heat transfer for hydrodynamically developed flow under the condition of constant heat flux in the axial

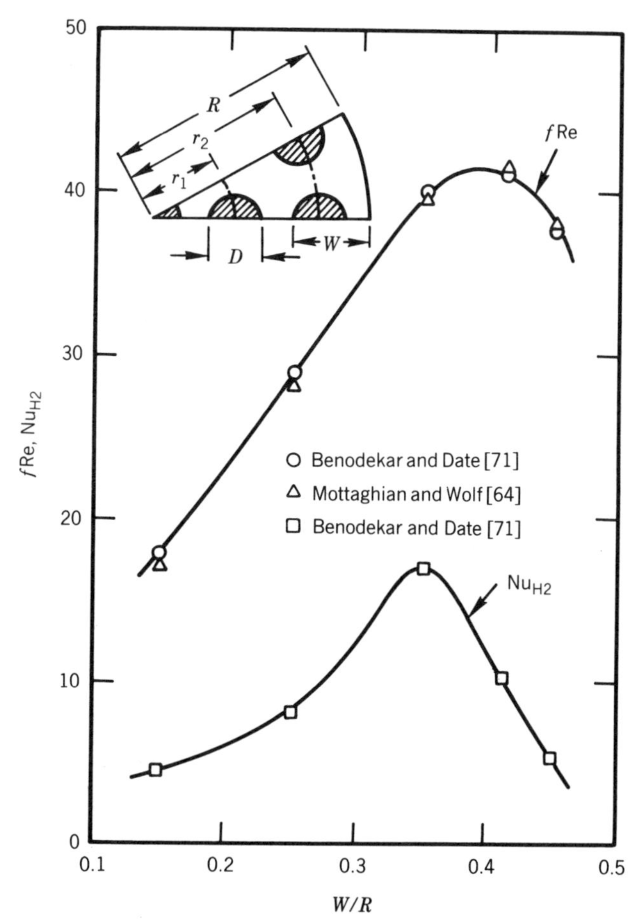

Figure 7.16. f Re and Nusselt numbers for fully developed laminar flow in bundles of 19 rods in a circular array, vs. the nondimensional distance of the outer row of rods from the shroud.

direction for a high Prandtl number fluid. They graphically present f Re factors and Nusselt numbers for three different geometrical configurations.

To illustrate some features of circular rod bundles, Fig. 7.15 displays f Re of clusters of seven rods (six rods symmetrically placed on one ring around the central rod) vs. the ratio b/R and for two ratios of rod diameter to shell radius. As expected, the same trends are exhibited as are found for rod bundles in a regular triangular array. f Re reaches a maximum for a b/R ratio at which the velocity distribution among the different subchannels is uniform. The data of all authors agree closely. Figure 7.16 shows f Re for clusters of 19 rods placed on two concentric rings around the central rod vs. the dimensionless wall spacing W/R for the case of the distance between the outer and inner rings of rods $(r_2 - r_1)$ kept constant. Again, the f Re factors increase with increasing wall spacing to a maximum value, and then decrease with further increase in the wall spacing. The results of Benodekar and Date [71] and of Mottaghian and Wolf [63] are found to be in good agreement. Nu_{H2}, also plotted in Fig. 7.16, varies in the same fashion, but its behavior is more peaked than that of f Re and its maximum occurs at a W/R ratio lower than that for the maximum f Re.

7.3 TURBULENT FLOW

In contrast to laminar flow through rod bundles, the pressure drop and heat transfer of turbulent flow through rod bundles has mainly been investigated experimentally. Many more experiments have been performed on pressure drop in rod bundles than on heat transfer, because heat transfer test sections are very expensive.

Quite a number of attempts have been reported in the literature to treat the flow and heat transfer problem in rod bundles theoretically. However, the validity of the theoretical work, especially the needed idealizations and modeling of turbulence, can only be assessed by comparison with the experimental data.

Subchannel analysis is a common practice in computing temperature distributions (subchannel averaged) for turbulent flow in rod bundles used as nuclear fuel elements. Measuring subchannel friction factors and heat transfer coefficients is difficult, since real rod bundles are enclosed in channel walls (shroud or wrapper tube). These walls affect the pressure drop and, above all, the mass flow distribution among the subchannels. Therefore, subchannel friction factors can be established only by detailed measurements of the distributions of wall shear stresses and velocities, which are very cumbersome to perform. Many attempts have been made to simulate infinite arrays of rod bundles (central subchannels); but, except for $P/D = 1$ (touching rods), walls always have to be used to close the channels.

7.3.1 Fully Developed Flow: Friction Factors — Triangular Array

The experimental results reported in the literature will be considered first. Rod bundles with 3 to 217 rods were used for the experiments with P/D ratios from 1.0 to 2.37 and W/D ratios from 1.0 to 2.44 [49, 50, 55, 56, 59, 75–117]. The Reynolds numbers ranged up to 7×10^5. Many different shapes were used for the channels surrounding the rod bundles in the triangular arrays: (1) circular, (2) hexagonal, (3) rectangular, (4) rhombic, (5) triangular, (6) scalloped ducts, and (7) quasi-infinite (formed by a wall in the gap between the rods).

Most of the investigations were performed on a single geometry. Systematic experiments were performed by Galloway and Epstein [55, 56] for 19-rod bundles in hexagonal ducts, by Eifler and Nijsing [86] for quasi-infinite arrays, by Courtaud et al. [59]

with seven rods in the same circular channel but for different geometrical arrangements. Ibragimov et al. [90] examined three rods in different arrangements in a triangular channel. Presser [92] measured the pressure drop of rod bundles in circular tubes and in quasi-infinite arrays for a wide range of P/D ratios. Subbotin et al. [100] performed experiments with 19-rod bundles of the same geometry but with different channels, and Rehme [103] measured the pressure drop of 25 different rod bundles surrounded by hexagonal ducts.

The experimental data up to 1970 have been reviewed by Rehme [103]. The ratio of the friction factor measured in rod bundles to that in a smooth circular tube is shown in Fig. 7.17 for two Reynolds numbers: (a) Re = 10^4 and (b) Re = 10^5. This figure is an update of [103] including all data published since 1970. The results by Salikov et al. [75] have been reevaluated with the channel dimensions not given in [75], but reported by Inayatov [118]. The friction factor was calculated as $f = 2\,\mathrm{Eu}\, D_h/D_{h,\infty}$ and the Reynolds number as $\mathrm{Re} = \mathrm{Re}_s D_h/D$, using values of Eu and Re_s tabulated by Salikov [75]. Here $D_{h,\infty}$ is the hydraulic diameter of an infinite array. Some early data, which are obviously in error [77, 81], have been omitted. The friction factors for the circular tube used are $f = 0.0079$ for Re = 10^4 and $f = 0.00455$ for Re = 10^5, from the equation by Maubach [119].

There is considerable scatter in experimental data. This is mainly due to the channels and the distances between channel and outer row of rods used. As has been discussed for laminar flow, the distance between rods and channel wall, or the W/D ratio, affects the overall friction factor. A maximum friction factor is obtained for cases in which the average velocity in the different subchannels is the same. To a first approximation, this condition is met for equal hydraulic diameters in the different subchannels. If the average velocities in the different subchannels are different, the friction factor will always be lower. For turbulent flow, this effect is not as pronounced as for laminar flow. Systematic experiments by Courtaud et al. [59], Ibragimov et al. [90], and Presser [92] clearly show these trends.

Figure 7.17 includes the friction factor of turbulent flow in the inner zone of an annulus [119, 120], which should be the upper limit for the friction factor in rod bundles, as is the case for laminar flow.

Before conclusions are drawn, theoretical work on friction factors and the correlations developed from experimental evidence will be discussed.

The fundamental theoretical work on velocity distribution and pressure drop in rod bundles was presented by Deissler and Taylor [121]. It was followed by theoretical work of Russian authors: Osmachkin [122], Buleev et al. [123], Kokorev et al. [124], and Ibragimov et al. [125, 126]. Results on friction factors were also presented by Nijsing et al. [127], Gräber [41], Vonka [128], Aranovich [129], Eifler and Nijsing [130, 131], Subbotin et al. [29, 132], Ramm and Johannsen [133], Ushakov [134, 135], Meyder [136, 137], Gosse and Schiestel [138], and Ramachandra [5]. All theoretical investigations considered the central subchannel of rod bundles in a regular triangular array (infinite array). The results differ widely.

Most theoretical results fall in one of two categories according as the velocity profiles were obtained from (1) the law of the wall for circular tubes applied to rod bundles, or (2) measurements in concentric annuli by Brighton and Jones [139], especially for the radius ratio 0.0625. These latter results were misinterpreted in that the coincidence of the positions of zero shear stress and maximum velocity profile was assumed, which is not true for strongly asymmetric velocity profiles, as the experimental data by Rehme [140] for radius ratios as small as 0.02 definitely show. For very low radius ratios, there is in fact a slight deviation from the law of the wall in the nondimensional velocity profiles of the inner zone of an annulus [140]. However, this is

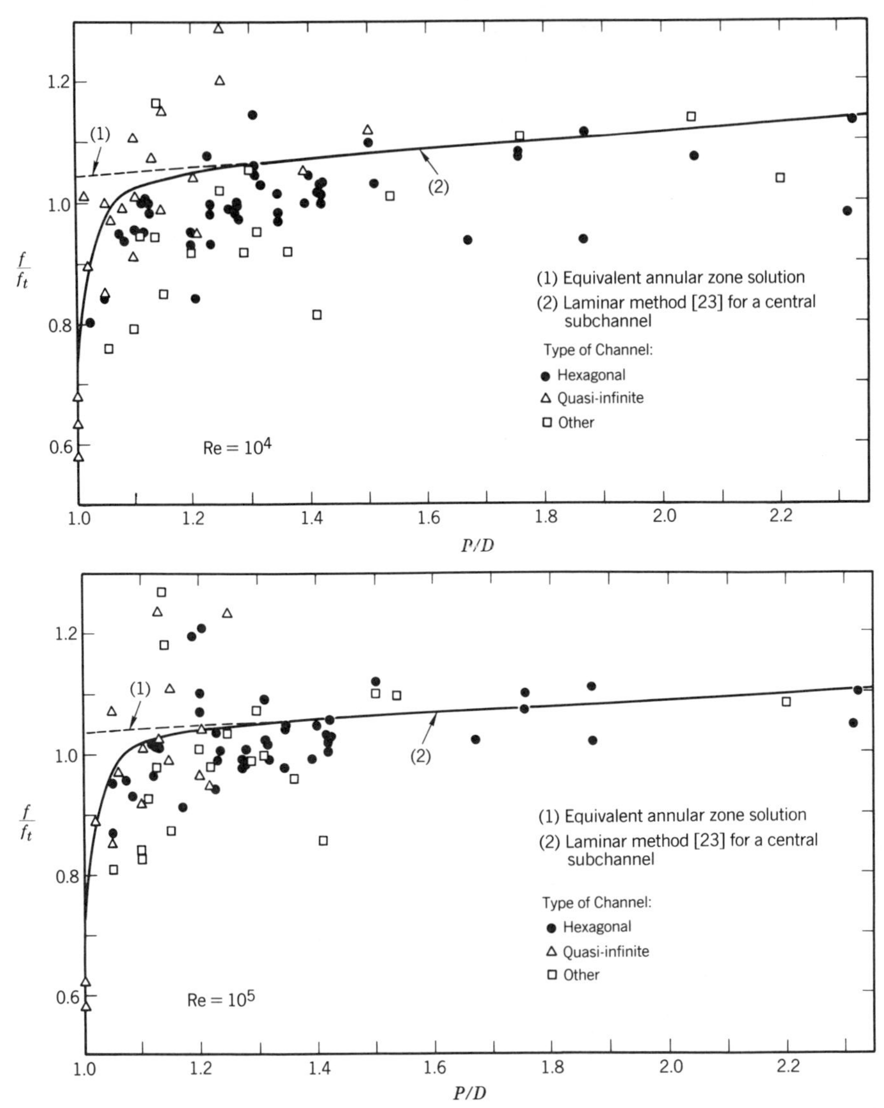

Figure 7.17. Experimental results on fully developed turbulent flow: friction factors of rod bundles arranged in a triangular array normalized with respect to the friction factor of circular tubes, vs. pitch-to-diameter ratio: (*a*) Re = 10^4, (*b*) Re = 10^5.

caused by the strong asymmetry of the velocity profile, which is not present in rod bundles. Moreover, all measurements of velocity profiles in rod bundles show that the law of the wall for circular tubes is valid also in rod bundles [86, 91, 99, 107, 112, 114, 141] to a good approximation.

Therefore, it is concluded that the upper limit for the friction factor in rod bundles of the triangular array is the friction factor of the inner annular zone based on the law of the wall for the velocity profile. This solution is shown in Fig. 7.17. Theoretical work in agreement with this solution has been done by Deissler and Taylor [121], Buleev et al. [123] for $P/D \leq 1.2$, Kokorev et al. [124], Nijsing et al. [127] for $P/D \leq 1.15$, Vonka [128], Aranovich [129], Ramm and Johannsen [133], Meyder [136], and Ramachandra [5] (maximum $P/D = 1.35$). The prediction of Rapley [142,143] for $P/D = 1$ agrees closely with the experimental friction factors.

Most of the experimental results shown in Fig. 7.17 agree within 5% with the upper limit of the friction factor predicted from the annular-zone solution. Some experimental data, mostly from early investigations, are higher than the annular-zone solution. Experimental error is probable, due to the influence of surface roughness (especially for $\mathrm{Re} = 10^5$), inaccurate knowledge of the flow cross section and wetted perimeter, possible effects of spacers on the pressure drop, and the fact that some of the pressure drop results were obtained as by-products of heat transfer measurements.

The empirical correlations for the friction factor in triangular array rod bundles by Mikhaylov et al. [85], Bogdanov [144], and Inayatov [118] are too simple and based on limited data. The correlations by Ushakov [134,135], Subbotin and Ushakov [132], and Morosova and Nomofilov [25] result in friction factors which are too high for $P/D > 1.2$. The empirical correlations by Presser [92],

$$f = \frac{A_1}{4}\mathrm{Re}^{-0.25} \qquad \text{for} \quad 10^4 \leq \mathrm{Re} \leq 5 \times 10^4 \tag{7.23}$$

$$f = \frac{A_1}{4}\mathrm{Re}^{-0.2} \qquad \text{for} \quad 5 \times 10^4 \leq \mathrm{Re} \leq 2 \times 10^5 \tag{7.24}$$

with

$$A_1 = 0.171 + 0.012\frac{P}{D} - 0.07e^{-50(P/D-1)} \tag{7.25}$$

are recommended for an infinite triangular array and $1 < P/D < 2$. Equation (7.25), due to Presser [92], agrees within 2% with the annular-zone solution for $P/D > 1.2$. The annular-zone solution, shown in Figure 7.17, can be correlated by

$$\frac{f}{f_t} = 1.045 + 0.071\left(\frac{P}{D} - 1\right) \qquad \text{for} \quad \mathrm{Re} = 10^4 \tag{7.26}$$

$$\frac{f}{f_t} = 1.036 + 0.054\left(\frac{P}{D} - 1\right) \qquad \text{for} \quad \mathrm{Re} = 10^5 \tag{7.27}$$

recommended by Rehme [103] for $P/D > 1.2$. For $P/D < 1.2$, Presser's correlation agrees within 2% with the solution obtained by the laminar method (see Sec. 7.3.3), which is also presented in Fig. 7.17.

7.3.2 Fully Developed Flow: Friction Factors — Square Array

For square arrays, fewer experimental investigations exist than that for triangular arrays. Rod bundles with 4 to 100 rods have been used in the range of P/D ratios between 1.0 and 1.67 and W/D ratios between 1.041 and 1.67, respectively [36, 55, 56, 76, 78, 79, 92, 114, 141, 145–152]. No systematic experiments, varying all relevant geometrical parameters have been performed.

The experimental data up to 1972 have been reviewed by Marek et al. [149]. The ratio of the friction factors measured in rod bundles to the friction factor of a smooth

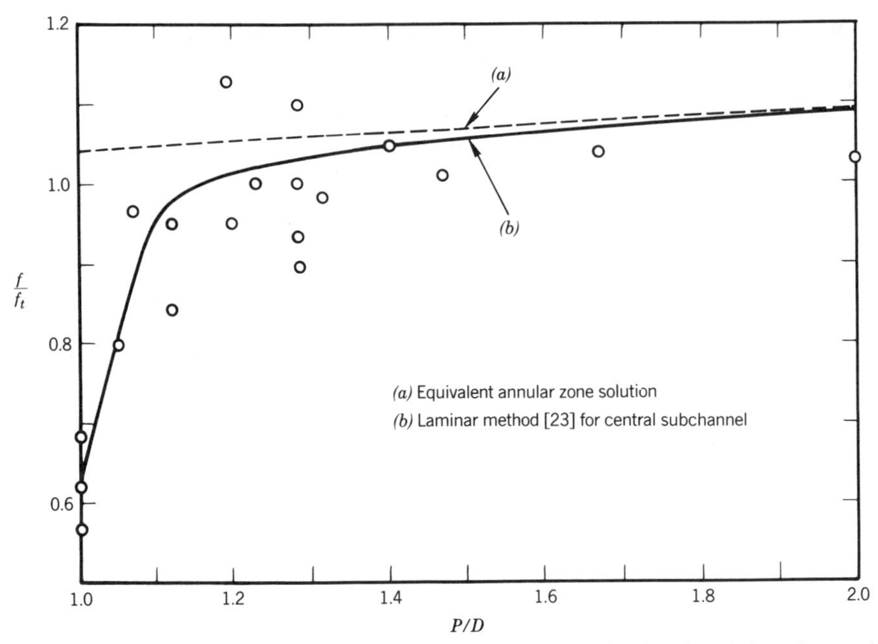

Figure 7.18. Experimental results of friction factors of fully developed turbulent flow rod bundles arranged in a square array, normalized with respect to the friction factor of circular tubes, vs. pitch-to-diameter ratio.

circular tube is shown in Fig. 7.18 for Re = 10^5. Figure 7.18 includes the friction factor for turbulent flow in the inner zone of an annulus [119, 120], which should be the upper limit for the friction factor.

Theoretical solutions for the friction factor in rod bundles arranged in square arrays were obtained by most of the authors who reported solutions for triangular arrays: Deissler and Taylor [121], Osmachkin [122], Buleev et al. [123], Kokorev et al. [124], Ibragimov et al. [125, 126], Gräber [41], Eifler and Nijsing [130], Subbotin et al. [29, 132], Ramm and Johannsen [133], Meyder [136, 137], and Gosse and Schiestel [138]. As already discussed for rod bundles in triangular arrays, the theoretical work based on, or fitted to, velocity profiles measured in annuli by Brighton and Jones [139] does not agree with the experimental observations.

The empirical correlations by Presser [92]

$$f = \frac{A_1}{4}\mathrm{Re}^{-0.25} \qquad \text{for} \quad 10^4 \le \mathrm{Re} \le 5 \times 10^4 \tag{7.28}$$

$$f = \frac{A_1}{4}\mathrm{Re}^{-0.2} \qquad \text{for} \quad 5 \times 10^4 \le \mathrm{Re} \le 2 \times 10^5 \tag{7.29}$$

with

$$A_1 = 0.181 + 0.0108\frac{P}{D} - 0.132e^{-20(P/D-1)} \tag{7.30}$$

are recommended for an infinite square array and $1 < P/D < 2$. Friction factors from Eq. (7.30) are in close agreement with the annular-zone friction factor for $P/D > 1.5$.

7.3.3 Fully Developed Flow: Subchannel Friction Factors via Laminar Solutions

As mentioned before, subchannel friction factors for turbulent flow can only be determined experimentally by detailed measurements of the distributions of wall shear stress and velocity, because the mass flow rate or average velocity in a subchannel can be determined precisely only by integrating the measured velocity distribution. Only few experimental results are reported, except for the limiting case of touching rods and/or channel walls ($P/D = 1$, $W/D = 1$), of friction factors determined by pressure drop measurement.

Subbotin et al. [99] presented friction factors for central subchannels in triangular arrays for $P/D = 1.05$, 1.1, 1.2; Kjellström [107] for $P/D = 1.22$; and Hejna et al. [112, 153] for $P/D = 1.17$. The experimental results for $P/D = 1$ are presented in Fig. 7.17 for triangular arrays and Fig. 7.18 for square arrays. Mohandes and Knudsen [101, 102] reported friction factors for wall subchannels with $W/D = 1$ and 14 different P/D ratios between 1.0 and 1.64. Rehme determined friction factors of wall subchannels in the ranges of P/D between 1.036 and 1.4 and W/D between 1.026 and 1.4 in 15 different geometries; some of his results are presented in [49], and the references for the other test sections may be found in [154]. Friction factors for a corner subchannel of a square array and $W/D = 1$ are reported by Gunn and Darling [36] and Barrow et al. [155].

There have been several attempts to apply the knowledge of $f\,\mathrm{Re}$ in laminar flow to predict friction factors in turbulent flow. The empirical correlation by Gunn and Darling [36] was based on limited data and overpredicts the turbulent friction factor [50, 59]. The procedure by Malák et al. [17] results in friction factors higher than experimental data [26, 102, 151]. The relationship between laminar and turbulent flow friction factors developed by Rehme [23] seems to work well. This method is based on the law of the wall for the velocity profile. The equation for the turbulent friction factor can be written as

$$\sqrt{\frac{2}{f}} = A_2\left[2.5 \ln \mathrm{Re}\sqrt{\frac{f}{2}} + 5.5\right] - G^* \tag{7.31}$$

with two geometry parameters, A_2 and G^*, which depend on $f\,\mathrm{Re}$ for laminar flow:

$$A_2 = \begin{cases} 1 & \text{for } f\,\mathrm{Re} \ge 16 \\ 1 + 0.554 \log_{10}\left(\dfrac{16}{f\,\mathrm{Re}}\right) & \text{for } f\,\mathrm{Re} < 16 \end{cases} \tag{7.32}$$

and $G^* = f(f\,\mathrm{Re})$ can be determined from correlations developed by Cheng and Todreas [26]:

$$G^* = \begin{cases} 2.553 + 3.872 \log_{10}(f\,\mathrm{Re}) - 1.042(\log_{10}(f\,\mathrm{Re}))^2 & \text{for } 6 < f\,\mathrm{Re} \le 16 \\ 6.615 - 3.376 \log_{10}(f\,\mathrm{Re}) + 2.159(\log_{10}(f\,\mathrm{Re}))^2 & \text{for } 16 < f\,\mathrm{Re} < 31.25 \\ 1.663 + 3.151 \log_{10}(f\,\mathrm{Re}) & \text{for } 31.25 \le f\,\mathrm{Re} < 250 \end{cases} \tag{7.33}$$

The equations for G^* from Cheng and Todreas [26] have been modified, because the authors used the Darcy friction factor ($4f$).

The value of A_2 from Eq. (7.32) for f Re < 16 is slightly higher than in the original work [23], due to better agreement with experimental data. The friction law [Eq. (7.31)] was checked against experimental data from many noncircular channels, including rod bundles [23]. The experimental subchannel friction factors mentioned above are predicted to within 6%. The prediction for most of the channels is much closer.

EXAMPLE. Consider the central subchannel of a triangular array with $P/D = 1.0$. From Fig. 7.2, $f \text{Re} = 6.5$. From Eq. (7.32), $A_2 = 1.217$. From Eq. (7.33), $G^* = 5.012$. Then

$$\sqrt{\frac{2}{f}} = 1.217\left[2.5 \ln \text{Re}\sqrt{\frac{f}{2}} + 5.5\right] - 5.012$$

For $\text{Re} = 5 \times 10^4$ this equation yields $f = 0.003244$. With $f_t = 0.005297$ we have $f/f_t = 0.61$. This compares with the experimental $f/f_t = 0.57$ of Eifler and Nijsing [86], $f/f_t = 0.6$ of Levchenko et al. [91], and $f/f_t = 0.547$ of Krett et al. [114]. Thus the predicted value is in satisfactory agreement with the experimental results for this extreme case of a noncircular channel.

7.3.4 Fully Developed Heat Transfer: Ordinary Fluids — Triangular Array

Comparing the experimental data by different authors [75–78, 81, 82, 84, 87, 89, 92, 93, 117, 156–166] measured in different test sections is difficult. Most of the data were obtained for constant heat flux in the axial direction with finite peripheral wall heat conduction ((H4) boundary condition). For $P/D > 1.2$, the peripheral variations of heat flux and surface temperature can be neglected for an infinite triangular array; for this case, the equivalent-annulus solution for Nu is approached.

Therefore, all experimental data are compared with the equivalent-annulus solution. The most reliable correlation for the equivalent annulus was developed by Petukhov and Roizen [167] and is based on numerous experimental data. All experimental data are based on the hydraulic diameter of an infinite array,

$$\frac{D_{h,\infty}}{D} = \frac{2\sqrt{3}}{\pi}\left(\frac{P}{D}\right)^2 - 1 \tag{7.34}$$

Some original data based on other definitions have been recalculated. The experimental Nusselt numbers Nu, shown in Fig. 7.19 as Nu/Nu_t, were calculated from correlations presented by the authors or taken from graphs presented, for one Reynolds number within the range of the experiments: 10^4, 5×10^4, or 10^5 was chosen. The Nusselt numbers are related to the Nusselt number Nu_t of a circular tube at the Reynolds number chosen and to the Prandtl number of the experiment. Nu_t is calculated from the correlation by Petukhov and Roizen [167]:

$$\text{Nu}_t = \frac{(f/2)\text{Re Pr}}{h_1 + 12.7\sqrt{f/2}\,(\text{Pr}^{2/3} - 1)} \tag{7.35}$$

where

$$h_1 = 1.07 + \frac{900}{\text{Re}} - \frac{0.63}{1 + 10\,\text{Pr}} \tag{7.36}$$

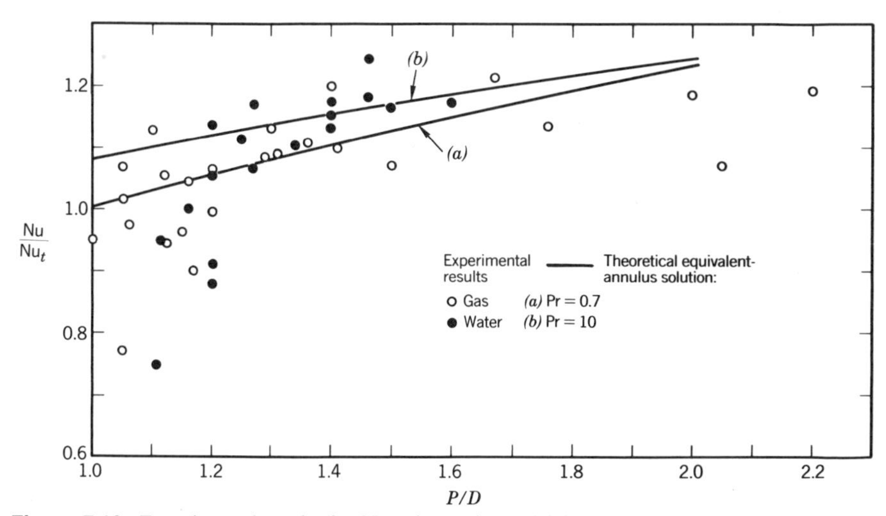

Figure 7.19. Experimental results for Nusselt numbers of fully developed turbulent flow in rod bundles arranged in a triangular array, normalized with respect to the Nusselt number of circular tubes, as a function of the pitch-to-diameter ratio.

and

$$f = (3.64 \log_{10} \text{Re} - 3.28)^{-2} \tag{7.37}$$

is the friction factor of a circular tube. Equation (7.35) is valid in the ranges $4 \times 10^3 \leq \text{Re} \leq 6 \times 10^5$ and $0.7 \leq \text{Pr} \leq 5 \times 10^5$.

The equivalent-annulus solution, included in Fig. 7.19 for Pr = 0.7 and Pr = 10, is calculated from the correlation by Petukhov and Roizen [167]:

$$\frac{\text{Nu}}{\text{Nu}_t} = [1 - \phi(\text{Pr})]\left(\frac{1}{r^*}\right)^{n(\text{Pr})} \tag{7.38}$$

where

$$\phi(\text{Pr}) = \frac{0.45}{2.4 + \text{Pr}} \tag{7.39}$$

$$n(\text{Pr}) = 0.16\,\text{Pr}^{-0.15} \tag{7.40}$$

valid for $0.2 \leq r^* \leq 1$, $10^4 \leq \text{Re} \leq 10^6$, and $0.7 \leq \text{Pr} \leq 100$.

The radius ratio r^* of an annulus with the same flow cross section as a triangular array is given by

$$\frac{1}{r^*} = \sqrt{\frac{2\sqrt{3}}{\pi}\frac{P}{D}} \tag{7.41}$$

The hydraulic diameters of an annulus and an infinite triangular array are different. To take this into account, both in the Nusselt number and in the Reynolds number,

Nu_t of Eq. (7.35) has been approximated by

$$\mathrm{Nu}_t = \begin{cases} 0.02087\,\mathrm{Re}^{0.7878} & \text{for} \quad \mathrm{Pr} = 0.7 \qquad (7.42) \\ 0.0393\,\mathrm{Re}^{0.848} & \text{for} \quad \mathrm{Pr} = 10 \qquad (7.43) \end{cases}$$

which is within $\pm 1\%$ of Eq. (7.35) for $5 \times 10^3 \leq \mathrm{Re} \leq 10^5$.

The hydraulic diameter of an equivalent annulus can be expressed by

$$D_{h,e} = D\left(\sqrt{\frac{2\sqrt{3}}{\pi}\frac{P}{D}} - 1\right) \tag{7.44}$$

In terms of the hydraulic diameter [Eq. (7.34)] of a triangular array, the ratio of the hydraulic diameters of a triangular array and of an equivalent annulus can be written as

$$\frac{D_h}{D_{h,e}} = \frac{(2\sqrt{3}/\pi)(P/D)^2 - 1}{\sqrt{2\sqrt{3}/\pi\, P/D} - 1} = \sqrt{\frac{2\sqrt{3}}{\pi}\frac{P}{D}} + 1 \tag{7.45}$$

Introducing Eq. (7.45) in Eq. (7.38) yields

$$\frac{\mathrm{Nu}}{\mathrm{Nu}_t} = 0.855\left(\sqrt{\frac{2\sqrt{3}}{\pi}\frac{P}{D}}\right)^{0.1688}\left(\sqrt{\frac{2\sqrt{3}}{\pi}\frac{P}{D}} + 1\right)^{0.2122} \quad \text{for} \quad \mathrm{Pr} = 0.7 \tag{7.46}$$

$$\frac{\mathrm{Nu}}{\mathrm{Nu}_t} = 0.9637\left(\sqrt{\frac{2\sqrt{3}}{\pi}\frac{P}{D}}\right)^{0.1133}\left(\sqrt{\frac{2\sqrt{3}}{\pi}\frac{P}{D}} + 1\right)^{0.152} \quad \text{for} \quad \mathrm{Pr} = 10. \tag{7.47}$$

These correlations are valid for $P/D > 1.2$ and represent an upper limit for P/D ratios from 1.0 to 4.0.

The systematic measurements by Presser [92] showed that maximum Nusselt numbers are obtained for geometrical arrangements in which the flow distribution is uniform, similar to the friction factor. For nonuniform flow distribution in the different subchannels, the Nusselt numbers are always less than the maximum values, which, for $P/D > 1.2$, should agree with the annulus correlations. Experimental data much higher than the equivalent-annulus solution are suspected to be erroneous and hence are not included in Fig. 7.19. The maximum values in Fig. 7.19 agree satisfactorily with the equivalent-annulus correlations for $P/D > 1.2$, considering the experimental inaccuracies. For $P/D < 1.2$, the experimental Nusselt numbers are lower than the annulus solution.

The limiting case of $P/D = 1$ was investigated by Bobkov et al. [165]. The results show that Nu strongly depends on the thermal boundary condition. For a case approaching the (H2) boundary condition, Nu is only 25% of Nu_t. With increasing peripheral heat conduction, Nu increases.

Theoretical solutions for Nusselt numbers in rod bundles depend on the turbulent transport properties assumed for momentum and heat. These properties are not well known. Therefore, most of the Nusselt numbers obtained theoretically do not agree with the equivalent-annulus solution based on experimental data.

The Nusselt numbers due to Nijsing et al. [127] for the range of $P/D = 1.05$ to 1.15 seem to be reasonable. The equivalent-annulus solution by Gräber [41] results in

Nusselt numbers considerably higher than those of Petukhov and Roizen [167]. Meyder [136, 137] and Taylor et al. [168] do not present Nusselt numbers. Gosse and Schiestel [138] computed Nusselt numbers considerably higher than the correlation by Petukhov and Roizen for $P/D > 1.3$ and Pr = 0.7 and 10. Their correlation of $\mathrm{Nu}/\mathrm{Nu}_t$ does not agree with the results presented graphically. Ramachandra [5] presented Nusselt numbers for two P/D ratios which are in good agreement with those of Petukhov and Roizen.

The correlation of Nu as a function of P/D by Inayatov [169, 170] does not agree with the equivalent-annulus solution; those by Subbotin [30] and Markóczy [164] may be used for $P/D < 1.2$. Rieger's correlation [161] for $P/D \geq 1.25$ and $\mathrm{Pr} \geq 1$ and the correlation by Borishanskiy et al. [163] for $P/D > 1.2$ agree with the equivalent-annulus data. The correlation established by Presser [92] is in agreement with the equivalent-annulus results, but may be used for $P/D > 1.2$ only.

7.3.5 Fully Developed Heat Transfer: Ordinary Fluids — Square Array

The situation for rod bundles arranged in square arrays is similar to that encountered in triangular arrays; however, fewer experimental investigations exist [76, 78, 92, 149, 152, 155, 171]. The experimental data on Nu, divided by Nu_t of circular tubes, are shown in Fig. 7.20 and compared with the equivalent-annulus solution, as developed from the correlations by Petukhov and Roizen [167]:

$$\frac{\mathrm{Nu}}{\mathrm{Nu}_t} = 0.855\left(\sqrt{\frac{4}{\pi}\frac{P}{D}}\right)^{0.1688}\left(\sqrt{\frac{4}{\pi}\frac{P}{D}}+1\right)^{0.2122} \quad \text{for} \quad \mathrm{Pr} = 0.7 \tag{7.48}$$

$$\frac{\mathrm{Nu}}{\mathrm{Nu}_t} = 0.9637\left(\sqrt{\frac{4}{\pi}\frac{P}{D}}\right)^{0.1133}\left(\sqrt{\frac{4}{\pi}\frac{P}{D}}+1\right)^{0.152} \quad \text{for} \quad \mathrm{Pr} = 10 \tag{7.49}$$

Except for one result by Dingee and Chastain [78] for $P/D = 1.27$ (omitted), the

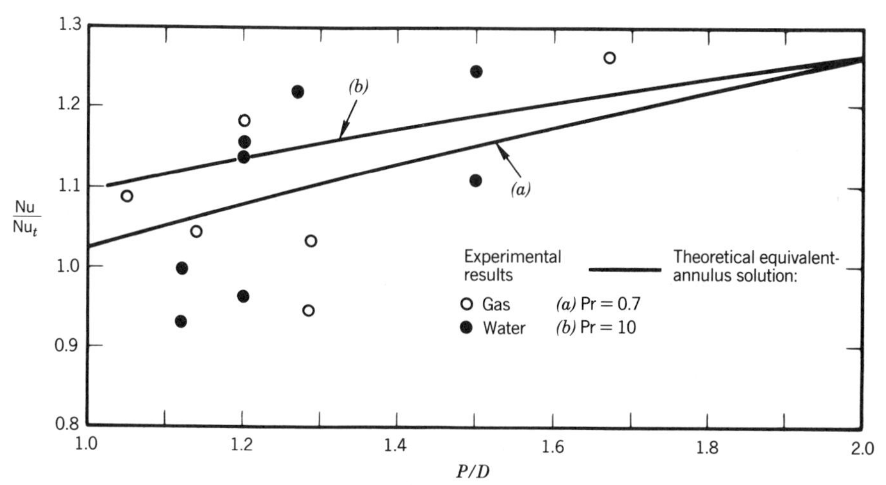

Figure 7.20. Experimental results for Nusselt numbers of fully developed turbulent flow in rod bundles arranged in a square array, normalized with respect to the Nusselt number of circular tubes, as a function of the pitch-to-diameter ratio.

experimental data are in reasonable agreement with the equivalent-annulus solution as the upper limit for Nu.

Most theoretical and empirical correlations result in Nusselt numbers higher than those due to Petukhov and Roizen [167] for $P/D > 1.3$. The exception is the theoretical result by Kokorev et al. [172], which is considerably lower in the range of $P/D = 1.1$ to 1.5. For $P/D < 1.3$, the correlation by Gosse and Schiestel [138] may be used; however, Nu_t of circular tubes given by Eq. (7.35) should be used in the correlation.

Kokorev et al. [172] calculated Nusselt numbers for the (H1) and (H2) boundary conditions and found that $\mathrm{Nu}_{\mathrm{H1}} = \mathrm{Nu}_{\mathrm{H2}}$ for $P/D > 1.25$. For $P/D < 1.25$, $\mathrm{Nu}_{\mathrm{H1}} > \mathrm{Nu}_{\mathrm{H2}}$, as expected.

The experimental investigation by Barrow et al. [155] for a two-cusp duct (corner subchannel at $W/D = 1.0$) resulted in $\mathrm{Nu}/\mathrm{Nu}_t = 0.38$.

7.3.6 Developing Flow and Heat Transfer: Ordinary Fluids

Measurements by Presser [92] show that the additional pressure drop in the hydraulic entrance length is small, about 2 to 3% of the average velocity head for 14 test sections of 7 to 61 rods arranged both in triangular and square array and having L/D_h from 36 to 135 and air as the working fluid. However, in rod bundles, an additional pressure drop occurs due to redistribution of mass flow rates among the subchannels. For uniform average velocity distribution in the subchannels of the rod bundle, this redistribution pressure drop can be neglected, as a first approximation, when the hydraulic diameters in the different subchannels are equal [92]. For very nonuniform velocity distributions in rod bundles, Presser [92] estimated the redistribution pressure drop to be of the order of 0.1 to 0.3 times the average velocity head.

The hydraulic entrance lengths in rod bundles also depend on the average flow distribution among the subchannels. Presser [92] found $x/D_h = 20$ for a uniform flow distribution, and $x/D_h = 40$ for very nonuniform velocity distributions, using air as the fluid.

As far as the Nusselt numbers in the thermal entrance length are concerned, Presser [92] correlated as follows his experimental data from six test sections of 7 to 61 rods arranged in a triangular array and from one test section in a square array with air ($\mathrm{Pr} = 0.7$):

$$\frac{\mathrm{Nu}_x}{\mathrm{Nu}_\infty} = 1 + \frac{0.7}{x/D_h} \tag{7.50}$$

for $1.2 \le P/D \le 2.2$, $\mathrm{Re} \ge 10^5$, and $x/D_h > 3$. For $\mathrm{Re} < 10^5$, the Nusselt numbers in the thermal entrance are higher than given by Eq. (7.50). From Eq. (7.50), the thermal entrance length, defined as the duct length required to achieve $\mathrm{Nu}_x/\mathrm{Nu}_\infty = 1.05$ [9], is $x/D_h = 14$. Hoffman et al. [157] reported the thermal entrance length $x/D_h = 20$ for $P/D = 1.71$. The experiments by Lel'chuk et al. [166] with a bundle of seven rods arranged in a triangular array and $P/D = 1.17$ were correlated by

$$\mathrm{Nu}_x = 0.015(1 + 0.9413e^{-0.0424x/D_h})\mathrm{Re}^{0.8}\mathrm{Pr}^{0.4} \tag{7.51}$$

for $x/D_h > 5$. Equation (7.51) yields a thermal entrance length of $x/D_h = 69$ for air ($\mathrm{Pr} = 0.7$), considerably higher than given by Eq. (7.50).

From the calculations by Vonka and Boonstra [173] for two connected central subchannels of a triangular array with $P/D = 1.3$ and $\mathrm{Pr} = 0.7$ at $\mathrm{Re} = 10^5$, it can be concluded that the Nusselt numbers reach fully developed turbulent flow conditions within $x/D_h = 30$ to 40, but the heat transport among central subchannels is not fully developed at $x/D_h = 290$.

7.3.7 Fully Developed Heat Transfer: Liquid Metals

Since 1960, the practice of cooling the fuel elements of fast breeder reactors by liquid metals has greatly stimulated research on liquid metal heat transfer. As far as heat transfer is concerned, the most important property of liquid metals is their excellent thermal conductivity compared to ordinary fluids. Prandtl numbers of liquid metals are of the order of magnitude 10^{-3} to 10^{-2}.

Many experimental investigations have been performed to study heat transfer to liquid metals in turbulent flow through rod bundles [174–196]. These investigations will not be discussed in detail, because empirical correlations have been developed which are valid in a wide range of geometrical, thermal, and flow boundary conditions. An overview of the experimental research and the heat transfer correlations developed up to 1973 was presented by Weinberg [197].

Experiments on heat transfer to liquid metals are difficult because of possible depositions of impurities on the heat transfer surfaces (contact thermal resistance) because of nonwetting of the surfaces. Most important, however, are the small temperature differences between surface and fluid due to the high thermal conductivity of liquid metals. These small temperature differences are difficult to measure precisely. On the other hand, it is not required that heat transfer correlations be very precise, for surface temperatures calculated by means of correlations are not strongly affected by uncertainties in the correlations, due to the small temperature differences between surface and fluid.

The experimental investigations [174–196] have been performed on rod bundles arranged in a triangular array, the geometry used in fuel elements of fast reactors, in the ranges of $P/D = 1.0$ to 1.95 and $\mathrm{Pe} = \mathrm{Re}\,\mathrm{Pr} = 2$ to 4500 with 7 to 37 rods. Only one experiment was performed with a rod cluster arranged in a square array, with 9 heated rods in a cluster of 25, for the limiting case of $P/D = 1.0$, by Ushakov et al. [146].

Many theoretical attempts have been made to analyze the heat transfer problem in liquid metals. The main difficulty in theoretical treatment is that the turbulent transport properties, especially the turbulent Prandtl number, are not known with sufficient precision. Different correlations for the turbulent Prandtl number have been applied by Friedland and Bonilla [198], Osmachkin [122], Buleev et al. [123, 199], and Ramm and Johannsen [200]. In some investigations turbulent transport of heat was neglected, e.g., by Nijsing and Eifler [201], Pfann [202], and Wolf and Johannsen [203]. The problem of the turbulent Prandtl number is also discussed by Rust [204]. The correlation by Bobkov et al. [205–207] for the turbulent Prandtl number was used by Nijsing and Eifler [208, 209], Bobkov et al. [210, 211] and Pfann [212, 213].

Based on the analysis of heat transfer to liquid metals for the limiting case of vanishing Prandtl number by Zhukov et al. [214] and Ushakov et al. [215, 216], a general correlation of heat transfer to liquid metal in turbulent flow through rod bundles was developed at the Institute of Physics and Energy (FEI), Obninsk. This correlation was first published by Subbotin [30] and slightly modified by Subbotin et al. [29]. The final equations were reported by Ushakov et al. [135, 217, 218]. This general correlation was checked against numerous results, both experimental data and theoretical correlations, and can be recommended for rod bundles arranged in the

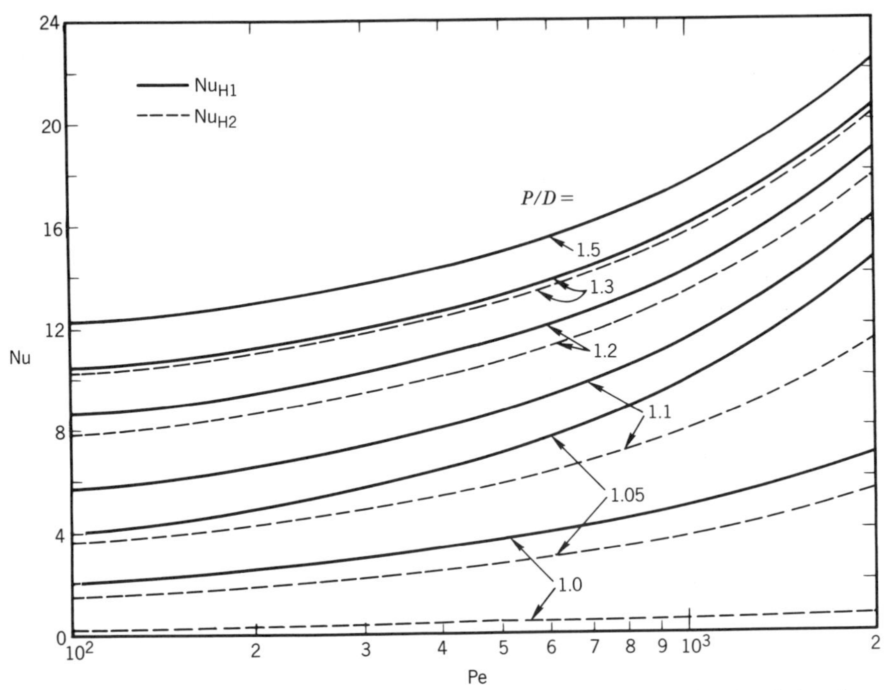

Figure 7.21. Nusselt numbers for heat transfer to liquid metals in fully developed turbulent flow in rod bundles arranged in a triangular array, vs. the Péclet number of the two limiting cases: (H1) boundary condition (full lines) and (H2) boundary condition (dashed lines) [135].

triangular array:

$$\mathrm{Nu} \approx \mathrm{Nu}_{\mathrm{lam}} + \frac{3.67}{90(P/D)^2}\left[1 - \frac{1}{\frac{1}{6}\left[(P/D)^{30} - 1\right] + \sqrt{1.24\epsilon_K + 1.15}}\right]\mathrm{Pe}^{m_1} \quad (7.52)$$

$$m_1 = 0.56 + 0.19\frac{P}{D} - 0.1\left(\frac{P}{D}\right)^{-80} \quad (7.53)$$

with $\mathrm{Nu}_{\mathrm{lam}}$ the Nusselt number for laminar flow given by Eq. (7.10), and ϵ_K defined by Eq. (7.11). Equation (7.52) is valid in the ranges $1.0 \le P/D \le 2.0$, $1 \le \mathrm{Pe} \le 4000$, and $0.01 \le \epsilon_K < \infty$.

Figure 7.21 shows $\mathrm{Nu}_{\mathrm{H1}}$ and $\mathrm{Nu}_{\mathrm{H2}}$ versus the Péclet number with P/D as a parameter. For $P/D > 1.3$, the effect of ϵ_K diminishes and Eq. (7.52) reduces to

$$\mathrm{Nu} = 7.55\frac{P}{D} - 20\left(\frac{P}{D}\right)^{-13} + \frac{3.67}{90(P/D)^2}\mathrm{Pe}^{0.19(P/D)+0.56} \quad (7.54)$$

valid in the ranges $1.3 < P/D \le 2.0$ and $1 < \mathrm{Pe} \le 4000$.

Thermal-hydraulic design of rod bundles is usually performed by subchannel codes which compute fluid and surface temperatures averaged over the subchannel. Therefore, it is important to know the peripheral variations of rod surface temperatures superimposed on the average temperatures.

The dimensionless peripheral variation of surface temperatures, defined by Eq. (7.13), is correlated by Subbotin et al. [29, 30] and Ushakov [135, 218] as

$$\Delta T^{\max} = \frac{\Delta T_{\mathrm{lam}}^{\max}}{1 + \gamma \, \mathrm{Pe}^{\beta}} \tag{7.55}$$

$$\gamma = 0.008(1 + 0.03\epsilon_K) \tag{7.56}$$

$$\beta = 0.65 + \frac{51 \log_{10}(P/D)}{(P/D)^{20}} \tag{7.57}$$

valid in the ranges $1 \leq P/D \leq 1.15$, $1 < \mathrm{Pe} < 2000$, and $\epsilon_K > 0.2$. Here $\Delta T_{\mathrm{lam}}^{\max}$ is the maximum temperature variation for laminar flow, given by Eq. (7.14).

The agreement between Eq. (7.55) and most of the experimental data presented in [29] is better than 10%.

Correlations of Nusselt numbers for heat transfer to liquid metals for design purposes, e.g. by Kazimi and Carelli [219] and Tang et al. [220], are only approximate and in some ranges very pessimistic, and therefore should not be used.

7.3.8 Developing Heat Transfer: Liquid Metals

Knowledge of the thermal entrance length in liquid metal cooled rod bundles is poor. Subbotin et al. [29] found that fully developed thermal conditions are not reached within $L/D_h < 200$ for $\mathrm{Pe} > 100$ in the peripheral subchannels of rod bundles (wall and corner subchannels). Their experimental data were obtained in a rod bundle with $P/D = 1.15$ and $W/D = 1.075$. For lower P/D ratios, the thermal entrance length is even longer because the heat transfer between subchannels is reduced. Subbotin et al. also found that increasing the W/D ratio for a fixed P/D ratio increases the thermal entrance length. This is in agreement with the results by Möller and Tschöke [195, 221] obtained in a 19-rod bundle with $P/D = 1.31$ and $W/D = 1.19$. They observed even higher thermal entrance lengths for $\mathrm{Pe} > 350$. For a corner subchannel and $\mathrm{Pe} = 150$, the thermal entrance length was measured to be $L/D_h \approx 70$.

The results of both investigations show an almost linear increase in the peripheral temperature variations around rods in corner and wall subchannels for the full axial heated length, which is $x/D_h = 100$ in [195] and $x/D_h = 200$ in [29]. These long thermal entrance lengths in rod bundles are caused by the heat transport between subchannels, as clearly demonstrated by calculations of Vonka and Boonstra [173]. Subbotin et al. [29] mention that the use of helical fins as spacers reduces the thermal entrance length.

7.4 TRANSITION FLOW

There is no critical Reynolds number at which transition from laminar to turbulent flow occurs in rod bundles. Figure 7.22 shows all experimental data on the pressure drop for the onset and completion of the transition. The onset of the transition is defined by the first deviation from the laminar $f\,\mathrm{Re}$, and its completion is defined by the curve of the friction factor vs. Re becoming parallel to that in fully developed turbulent flow.

The data scatter widely, especially for the onset of the transition. This may be due not only to the experimental conditions, to a certain extent, but also to the fact that the

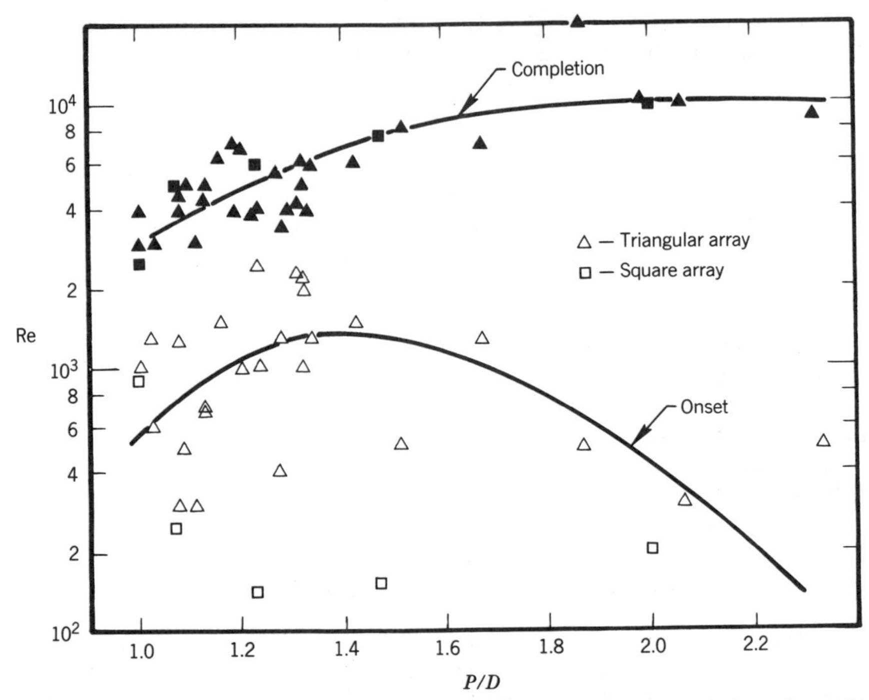

Figure 7.22. Experimental data of Reynolds numbers of the onset (open symbols) and completion (full symbols) of the transition from laminar to turbulent flow.

data shown have been obtained on test sections over a wide range of geometrical parameters such as the number of rods, triangular or square arrays, and W/D ratios. There is some indication that the onset of the transition in rod bundles arranged in square arrays starts at lower Reynolds numbers than in triangular arrays. For low P/D ratios, the transition to turbulent flow begins at Reynolds numbers below 10^3. To some extent this is affected by the hydraulic diameter. The Reynolds number for the onset of the transition to turbulent flow seems to increase with increasing P/D ratio up to $P/D \approx 1.5$; for higher P/D ratios, the Reynolds number for the onset decreases.

The scatter of the data for the Reynolds number at which the transition to turbulent flow is completed is much smaller, and there is a clear trend, showing that the completion Reynolds number increases from 3000 for $P/D = 1$ to 10^4 for $P/D > 1.8$. The wide range of Reynolds numbers for transition from laminar to turbulent flow at high P/D ratios may be affected by test sections being too short to produce fully developed conditions.

Friction factors for the transition regime are reported by Morosova and Nomofilov [25]. However, as stated before, the correlation presented by the authors results in friction factors for turbulent flow which are higher than the annular-zone solutions. The procedure of Morosova and Nomofilov might be useful for constants fitted against reliable data.

As far as heat transfer is concerned, no correlations or experimental data exist for ordinary fluids. For liquid metals, however, the correlations by Subbotin [30] and Ushakov [135, 218] are valid in the full range from laminar to turbulent flow.

7.5 EFFECTS OF SPACERS

Spacers are used to fix the rods in the rod bundle. Two basically different types of spacers are used in rod bundles:

Spacer grids defining the distance between the rods and relative to the wrapper tube. These are arranged at fixed planes along the rod bundles.

Spacers connected with the rods and extending over the entire length of the rods, such as wire wraps (helical wires) or helical fins.

Wire wraps or helical fins are of special interest for small P/D ratios, whereas spacer grids are commonly used for greater clearances between the rods, say for $P/D > 1.15$.

7.5.1 Pressure Drop — Spacer Grids

The pressure drop caused by spacer grids has been investigated experimentally for numerous designs. Experimental results have been reported by Le Tourneau et al. [79], Rehme [97, 222], Grillo et al. [147, 148], Marek et al. [149], Voj and Scholven [105], Rehme and Trippe [111, 49], Korotaev et al. [223], and Ito and Mawatari [224]. The pressure drop strongly depends not only on the blockage of the flow cross section by the spacer grid, but also on the axial length of the spacer. The pressure drop of spacer grids can be drastically reduced by smoothing the leading edge of the grids [49]. The pressure drop, of course, depends on the design of the grid; therefore, precise data on the pressure drop of spacer grids can be determined only by the measurement of the pressure drop itself for the spacer grid under consideration. As the pressure drop depends on many parameters, attempts to develop correlations of general validity have been unsuccessful.

7.5.2 Heat Transfer — Spacer Grids

Spacer grids also affect heat transfer. Due to the blockage of the flow cross section downstream of a spacer grid, the velocity and temperature distributions redevelop, and for ordinary fluids heat transfer is enhanced in the region of heat transfer development. Results on heat transfer improvement downstream of spacer grids have been presented by Hoffmann et al. [88], Vlček and Weber [225], Hudina and Nöthiger [226], Krett and Majer [227], and Marek and Rehme [228, 229]. Detailed experimental investigations by Hassan and Rehme [230, 231] show that the results can be expressed in terms of a ratio of the Nusselt number affected by the spacer grid, Nu_x, to the undisturbed Nusselt number in the rod bundle, Nu_0:

$$\mathrm{Nu}^* = \frac{\mathrm{Nu}_x}{\mathrm{Nu}_0}. \tag{7.58}$$

There is a typical shape of Nu^* in the flow direction which can be approximated as follows: upstream of the spacer, the Nusselt number rises linearly up to a maximum; there is a constant maximum below the spacer and a gradual drop to the undisturbed value downstream of the spacer grid. This drop in Nusselt numbers shows some similarity to an entrance effect.

The following correlations are recommended in the light of the experimental results:

Upstream of the spacer grid,

$$\mathrm{Nu}^* = 1 + \frac{\mathrm{Nu}^*_{\max} - 1}{1 + A_3} \qquad \text{for} \quad -1 \le \frac{x}{D_h} \le A_3 \tag{7.59}$$

$$A_3 = \frac{L_A}{2D_h} \qquad \text{for} \quad \mathrm{Re} < 3000 \tag{7.60}$$

$$A_3 = \frac{L_A}{D_h} \qquad \text{for} \quad \mathrm{Re} > 3000 \tag{7.61}$$

L_A is the axial length of the spacer grid, x the length in the axial direction measured downstream of the leading edge of the spacer grid, and D_h the hydraulic diameter outside the spacer grid in the rod bundle. The maximum relative Nusselt number is given as

$$\mathrm{Nu}^*_{\max} = \min(\mathrm{Nu}^*_{\max,1}, \mathrm{Nu}^*_{\max,2})$$

where

$$\mathrm{Nu}^*_{\max,1} = 1 + 0.174\,\mathrm{Re}^{0.5}\epsilon^2 \qquad \text{for} \quad \mathrm{Re} < 3000 \tag{7.62}$$

$$\mathrm{Nu}^*_{\max,2} = 1 + \epsilon^{2.4}(6.38 + 4550\,\mathrm{Re}^{-0.8}) \qquad \text{for} \quad \mathrm{Re} > 3000 \tag{7.63}$$

Here ϵ is the blockage ratio, defined as the cross section of the spacer grid projected in the axial direction divided by the flow cross section undisturbed by the spacer grid.

$$\mathrm{Nu}^* = K(x^*)^m \tag{7.64}$$

Here $x^* = x/(D_h \mathrm{Pe})$ is the dimensionless axial coordinate in the thermal entrance length, and the expressions for K and m are given below:

$$K = 4.42 - 1.05\log_{10}\mathrm{Re} - 2.25\epsilon \qquad \text{for} \quad \mathrm{Re} < 3000 \tag{7.65}$$

$$K = 0.426 + 0.113\log_{10}\mathrm{Re} - 2.25\epsilon \qquad \text{for} \quad \mathrm{Re} > 3000 \tag{7.66}$$

provided $K \ge 0.895 - 2.25\epsilon$, which is the minimum value for all other cases. The exponent m is given as

$$m = 1.855 \times 10^{-3}\epsilon^2\mathrm{Re} \qquad \text{for} \quad 600 < \mathrm{Re} < 3000 \tag{7.67}$$

$$m = 30.34\epsilon^2\mathrm{Re}^{-0.253} \qquad \text{for} \quad \mathrm{Re} > 3000 \tag{7.68}$$

provided $m \le 4\epsilon^2$, which is the maximum value for all other cases.

Equations (7.59) to (7.68) are valid in the ranges $600 \le \mathrm{Re} \le 2 \times 10^5$, $0.25 \le \epsilon \le 0.35$, and $-10 \le x/D_h \le 33$.

The data reported in the literature for $\epsilon < 0.25$ are described satisfactorily by the correlation due to Hassan [230].

According to the experimental data of Engel and Bishop [232], heat transfer is improved by spacer grids also for liquid metal flow. Möller and Tschöke [221], however, found slightly elevated temperatures directly below the spacer grid.

7.5.3 Pressure Drop — Wire Wraps and Helical Fins

Pressure Drop. The experimental pressure drops in rod bundles with wire wraps obtained before 1967 have been reviewed by Rehme [233, 234]. Based on a systematic investigation of 75 different geometries, Rehme [233, 234, 222] developed a general correlation of the friction factor in wire-wrapped rod bundles. This correlation, which is presented below, was confirmed by the experimental investigations of Hamid and Quaijum [235], Wakasugi and Kakehi [236], Hoffmann [237], Cornet and Lamotte [238], Sarno et al. [239], and Tirelli [240], and by the results from seven of twelve test sections studied by Reihman [116] in the range of validity. Reihman's other data are 12 to 23% higher than Rehme's correlation, probably because of the rather large tolerances of the test sections [26]. The data of Reihman, therefore, show some inconsistencies. The validity of Rehme's correlation is also confirmed by McAreavy and Betts [241] and by Cheng and Todreas [26].

The correlation by Rehme [222],

$$f = \left[\frac{16}{\mathrm{Re}\sqrt{F}} + \frac{0.0204}{(\mathrm{Re}\sqrt{F})^{0.133}} \right] F \frac{P_{\mathrm{B}}}{P_{\mathrm{tot}}} \tag{7.69}$$

is based on an effective velocity in the wire-wrapped rod bundle with the ratio of the effective to average velocity $u_{eff}/u_m = \sqrt{F}$, where

$$F = \left(\frac{P}{D} \right)^{0.5} + \left[7.6 \frac{d_m}{H} \left(\frac{P}{D} \right)^2 \right]^{2.16} \tag{7.70}$$

The hydraulic diameter in the Reynolds number of Eq. (7.69) and for pressure drop evaluation ($\Delta p = 2fL\rho u_m^2/D_h$) includes the cross section and the wetted perimeter of the wires, taking into account that the cross section of the wire normal to the rod bundle axis is an ellipse. d_m is the mean diameter of the wire wraps, which is $d_m = P$ for contact between rods and wires and $d_m = D + h$ for contact among fins, with h the height of the fins. The correlation has been found to be adequate for rods with three or six helical fins with fin-to-fin contact by Tschöke [242] and Hoffmann [237]. H is the pitch of the wire wraps, and the ratio $P_{\mathrm{B}}/P_{\mathrm{tot}}$ takes into account the size of the rod bundle. P_{B} is the wetted perimeter of rods and wires, and $P_{\mathrm{tot}} = P_{\mathrm{B}} + P_{\mathrm{Ch}}$ is the total wetted perimeter of the rod bundle, including the wetted perimeter P_{Ch} of the channel walls.

The ranges of validity of Eq. (7.69), which have been extended by Cheng and Todreas [26] beyond the original ranges [233], are $2 \times 10^3 < \mathrm{Re} < 3 \times 10^5$, $8 < H/d_m < 50$, $1.1 < P/D < 1.42$, and $7 < n_R < 217$. The experimental data by Sheynina [94] and Subbotin et al. [100] cannot be compared with Eq. (7.69), as some geometrical parameters are missing or the geometrical data given are inconsistent.

TABLE 7.6 Coefficients Q_{ij} of Equation (7.71) [247]

	Q_{ij}				
j	$i = 1$	2	3	4	5
1	−7.204995	24.93619	−21.43300	3.218632	1.762412
2	−20113.61	68163.57	−86357.49	48474.87	−10172.29
2	195675.5	−676074.3	871670.6	−497416.4	106161.8

For $P/D < 1.1$, experimental data are reported by Sheynina [94], Chiu et al. [243], Engel et al. [244], Cheng and Todreas [245], and from a systematic study by Marten et al. [246]. On the basis of the new data, Marten et al. [247] developed a correlation for the factor F of Eq. (7.70), which is valid in the ranges $1.04 < P/D < 1.42$ and $8.3 < H/d_m < 16.7$:

$$F = \sum_{j=1}^{3} \sum_{i=1}^{5} Q_{j,i} \left(\frac{P}{D} \right)^{i-1} \left(\frac{H}{d_m} \right)^{1-j} \tag{7.71}$$

where the coefficients $Q_{j,i}$, obtained by a regression analysis, are presented in Table 7.6.

Novendstern [248] developed a semiempirical model to predict pressure losses in wire-wrapped rod bundles, which is based on an empirical friction-factor multiplier for the friction factor of circular tubes. The multiplier was correlated on the basis of selected data due to Reihman [249], Baumann et al. [95], and Rehme [233, 250]. The preliminary data of Rehme [250] were in error; the corrected data from [233] supersede them. Novendstern's correlation is widely used and has been recommended by Ushakov [135]. Carajilescov and Fernandez [251] presented a semiempirical model for friction factors in wire-wrapped rod bundles on a limited data base [26], which is not able to predict overall friction factors with sufficient accuracy.

Very few results have been published on the pressure drop at laminar flow conditions. The results by Subbotin et al. [100], Chiu et al. [243], Engel et al. [244], Spencer and Markley [252], Marten et al. [247], and Efthimiadis [253] are not conclusive. Therefore, no general correlation is available for the pressure drop of laminar flow through wire-wrapped rod bundles. A correlation is presented by Engel et al. [244] on a limited data base with dimensional parameters.

Subchannel friction factors for laminar, transition, and turbulent flow through wire-wrapped rod bundles, i.e., "flow split parameters," can be calculated from the model developed by Cheng and Todreas [26]. For reasons of space, the model and correlations for a wide range of parameters are not outlined here; the reader is referred to the fundamental and excellent work by Cheng and Todreas, developed on the most complete data base currently existing.

Transition from laminar to turbulent flow occurs over a wide range of Reynolds numbers. Cheng and Todreas [26] presented simple correlations for the onset and completion of transition flow over wire-wrapped rod bundles in triangular arrays as

$$\mathrm{Re}_{\mathrm{crit}} = 300 \times 10^{1.7(P/D-1)} \qquad \text{for onset of transition} \tag{7.72}$$

$$\mathrm{Re}_{\mathrm{crit}} = 1000 \times 10^{0.7(P/D-1)} \qquad \text{for completion of transition} \tag{7.73}$$

7.5.4 Heat Transfer — Wire Wraps and Helical Fins

Experimental data on Nusselt numbers in wire-wrapped rod bundles have not been published. Subbotin et al. [29] and Ushakov [135] mention that liquid metal heat transfer in rod bundles is only slightly affected by helical fins for $P/D > 1.1$, when the hydraulic diameter in Nu and Pe is evaluated considering the cross sections and wetted perimeters of the fins; however, mixing between subchannels is enhanced by wire wraps or helical fins. For $P/D < 1.1$, Ushakov [135] presents correlations for the maximum peripheral temperature variation in wall subchannels, both for bare rod bundles and for clusters of rods with helical fins. Moreover, correlations are presented for all subchannels, with and without fillers in the wall subchannels, which are used to obtain a more uniform temperature distribution across a rod bundle.

NOMENCLATURE

A	free flow area, m^2, ft^2
A_1	dimensionless constant [see Eq. (7.25)]
A_2	geometry coefficient [see Eq. (7.32)]
A_3	ratio of lengths [see Eq. (7.60)]
b	difference of radii in circular arrangements, m, ft
c_p	specific heat at constant pressure, J(kg · K), Btu/(lb$_m$ · °F)
D	rod outside diameter, m, ft
D_h	hydraulic diameter, m, ft
$D_{h,\infty}$	hydraulic diameter of infinite array of rod bundles, m, ft
d_m	mean diameter of wire wraps, m, ft
Eu	Euler number = $\Delta p/\rho u_m^2$
e	base of natural logarithms
F	dimensionless factor [see Eq. (7.70)]
f	Fanning friction factor = $\tau_w/(\rho u_m^2/2) = [\Delta p/(\rho u_m^2/2)]\,(D_h/4L)$
G^*	geometry coefficient [see Eq. (7.33)]
H	pitch of wire wraps, m, ft
(H1)	boundary condition of constant axial wall heat flux with constant peripheral wall temperature
(H2)	boundary condition of constant axial wall heat flux with constant peripheral wall heat flux
(H4)	boundary condition of constant axial wall heat flux with finite normal wall thermal resistance
(H5)	boundary condition of exponential wall heat flux with constant peripheral wall temperature
h	heat transfer coefficient, W/(m^2 · K), Btu/(hr · ft^2 · °F)
h_1	dimensionless constant [see Eq. (7.36)]
K	dimensionless coefficient [see Eq. (7.65)]
k	thermal conductivity, W/(m · K), Btu/(hr · ft · °F)
L_A	length of spacer grid in axial direction, m, ft
$\log_{10}$	logarithm with base 10
ln	natural logarithm

m	exponent [see Eq. (7.67)]
m_1	exponent [see Eq. (7.53)]
N	positive integer
Nu	Nusselt number $= hD_h/k$
Nu*	Nusselt number ratio [see Eq. (7.58)]
n	positive integer
n_i	number of individual subchannels
n_R	number of rods
$n(\mathrm{Pr})$	constant [see Eq. (7.40)]
P	pitch or distance between rod centers, m, ft
Pe	Péclet number = Re Pr
Pr	Prandtl number $= \mu c_p/k$
P_W	wetted perimeter, m, ft
Δp	pressure drop, Pa, $\mathrm{lb}_f/\mathrm{ft}^2$
Q	dimensionless coefficient [see Eq. (7.71)]
q''	heat flux, $\mathrm{W/m^2}$, $\mathrm{Btu/(hr \cdot ft^2)}$
R	radius of shroud for circular arrays, m, ft
Re	Reynolds number $= \rho u_m D_h/\mu$
r	radial coordinate, m, ft
r_1	inner radius of cladding, m, ft
r_2	outer radius of cladding, m, ft
r_i	inner radius of annulus, m, ft
r_o	outer radius of annulus, m, ft
r_*	dimensionless radial coordinate $= r_{\tau=0}/r_i$
r^*	dimensionless radial coordinate $= r_i/r_o$
$r_{\tau=0}$	radius of zero shear stress in annulus, m, ft
T	temperature, °C, K, °F, °R
Ⓣ	constant wall temperature boundary condition
ΔT	temperature difference, °C, K, °F, °R
$\Delta T^{\max}$	dimensionless peripheral variation of surface temperature [see Eq. (7.13)]
tanh	hyperbolic tangent function
u_m	mean axial velocity, m/s, ft/s
W	wall distance (see Fig. 7.1), m, ft
x	axial coordinate or distance, m, ft
x^*	dimensionless axial coordinate for the thermal entrance region, $= x/D_h \mathrm{Pe}$
x^+	dimensionless axial coordinate for the hydrodynamic entrance region, $= x/D_h \mathrm{Re}$

Greek symbols

β	exponent [see Eq. (7.57)]
γ	coefficient [see Eq. (7.56)]
ϵ	relative blockage due to spacer grid [see Eq. (7.62)]
ϵ_K	thermal modeling parameter [see Eq. (7.11)]

μ	dynamic viscosity, Pa · s, $lb_m/(hr \cdot ft)$
Λ_o	ratio of conductivities [see Eq. (7.12)]
ξ	ratio of hydraulic diameter of wall to that of the central subchannel
ρ	density, kg/m^3, lb_m/ft^3
τ	shear stress, Pa, lb_f/ft^2
π	transcendental number = 3.14159...
$\phi(Pr)$	constant [see Eq. (7.39)]

Subscripts

B	bundle
b	bulk fluid condition
c	central subchannel
Ch	channel
co	corner subchannel
e	equivalent
eaz	equivalent annular zone
H1	(H1) boundary condition
H2	(H2) boundary condition
H5	(H5) boundary condition
i	individual
lam	laminar
m	average value
max	maximum
min	minimum
T	(T) boundary condition
t	circular tube
tot	total
w	wall or wall subchannel
x	local
1	fuel
2	cladding
3	fluid
∞	fully developed conditions

Superscripts

max	maximum
min	minimum

REFERENCES

1. K. Johannsen Longitudinal Flow over Tube Bundles, *Low Reynolds Number Flow Heat Exchangers*, ed. S. Kakaç, R. K. Shah, and A. E. Bergles, Hemisphere, New York, pp. 229–273, 1983.

2. W. T. Sha, An Overview on Rod-Bundle Thermal-Hydraulic Analysis, *Nucl. Eng. Design*, Vol. 62, pp. 1–24, 1980.

3. W. Slagter, Finite Element Solution of Axial Turbulent Flow in a Bare Rod Bundle Using a One-Equation Turbulence Model, *Nucl. Sci. Eng.*, Vol. 82, pp. 243–259, 1982.

4. B. C.-J. Chen, S. P. Vanka, and W. T. Sha, Some Recent Computations of Rod Bundle Thermal Hydraulics Using Boundary Fitted Coordinates, *Nucl. Eng. Design*, Vol. 62, pp. 123–135, 1980.

5. V. Ramachandra, The Numerical Prediction of Flow and Heat Transfer in Rod-Bundle Geometries, Ph.D. Thesis, Mech. Eng. Dept., Imperial College of Science and Technology, London, 1979.

6. T. Shimizu, Two-Dimensional Steady-State Thermal and Hydraulic Analysis Code for Prediction of Detailed Temperature Fields around Distorted Fuel Pin in LMFBR Assembly: SPOTBOW1, *Thermal Hydraulics of Nuclear Reactors*, ed. M. Merilo, Am. Nucl. Soc. Publ., Santa Barbara, Calif., Vol. 2, pp. 1355–1364, 1983.

7. J. Schmid, Longitudinal Laminar Flow in an Array of Circular Cylinders, *Int. J. Heat Mass Transfer*, Vol. 9, pp. 925–937, 1966.

8. K. Johannsen, Heat Exchangers Having Longitudinal Flow over Tubes or Rods, *Low Reynolds Number Flow Heat Exchangers*, ed. S. Kakaç, R. K. Shah, and A. E. Bergles, Hemisphere, New York, pp. 275–297, 1983.

9. R. K. Shah and A. L. London, *Laminar Flow Forced Convection in Ducts*, *A Source Book for Compact Heat Exchanger Analytical Data*, Advances in Heat Transfer, Supplement 1, Academic, New York, 1978.

10. H. Rosenberg, Numerical Solution of the Velocity Profile in Axial Laminar Flow through a Bank of Touching Rods in a Triangular Array, *Trans. Am. Nucl. Soc.*, Vol. 1, pp. 55–57, 1958.

11. E. M. Sparrow and A. L. Loeffler, Jr., Longitudinal Laminar Flow between Cylinders Arranged in Regular Array, *AIChE J.*, Vol. 5, pp. 325–330, 1959.

12. R. A. Axford, Multiregion Analysis of Temperature Fields and Heat Fluxes in Tube Bundles with Internal, Solid, Nuclear Heat Sources, LA-3167, Los Alamos Scientific Lab., Los Alamos, N. Mex., 1964.

13. R. A. Axford, Two-Dimensional, Multiregion Analysis of Temperature Fields in Reactor Tube Bundles, *Nucl. Eng. Design*, Vol. 6, pp. 25–42, 1967.

14. F. S. Shih, Laminar Flow in Axisymmetric Conduits by a Rational Approach, *Can. J. Chem. Eng.*, Vol. 45, pp. 285–294, 1967.

15. K. Rehme, Laminarströmung in Stabbündeln, *Chemie-Ingenieur-Technik*, Vol. 43, pp. 962–966, 1971.

16. K. Rehme, Laminarströmung in Stabbündeln, *Reaktortagung 1971*, Deutsches Atomforum, Bonn, Germany, pp. 130–133, 1971.

17. J. Malák, J. Hejna, and J. Schmid, Pressure Losses and Heat Transfer in Non-circular Channels with Hydraulically Smooth Walls, *Int. J. Heat Mass Transfer*, Vol. 18, pp. 139–149, 1975.

18. R. Meyder, Solving the Conservation Equations in Fuel Rod Bundles Exposed to Parallel Flow by Means of Curvilinear-Orthogonal Coordinates, *J. Comp. Phys.*, Vol. 17, pp. 53–67, 1975.

19. M. D. Mikhailov, Finite Element Analysis of Turbulent Heat Transfer in Rod Bundles, *Turbulent Forced Convection in Channels and Bundles*, ed. S. Kakaç and D. B. Spalding, Vol. 1, pp. 259–277, 1979.

20. V. I. Subbotin, M. Kh. Ibragimov, et al., *Hyrdrodynamics and Heat Transfer in Nuclear Power Systems*, Atomizdat, Moscow, 1975.

21. O. E. Dwyer and H. C. Berry, Laminar-Flow Heat Transfer for In-Line Flow through Unbaffled Rod Bundles, *Nucl. Sci. Eng.*, Vol. 42, pp. 81–88, 1970.

22. A. A. Sholokov, N. I. Buleev, Yu. I. Gribanov, and V. E. Minashin, Laminar Fluid Flow in a Bundle of Rods (in Russian), *Inzh. Fiz. Zh.*, Vol. 14, pp. 389–394, 1968; English translation cited in [1].

23. K. Rehme, Simple Method of Predicting Friction Factors of Turbulent Flow in Non-circular Channels, *Int. J. Heat Mass Transfer*, Vol. 10, pp. 933–950, 1973.

24. A. Martelli, Thermo- und fluiddynamische Analyse von gasgekühlten Brennelementbüdneln, KfK 2436, EUR 5508d, Kernforschungszentrum Karlsruhe, F.R.G., 1977.

25. S. I. Morosova and E. B. Nomofilov, Hydraulic Resistance Coefficients in the Transition Range for Fluid Flow through Rod Bundles in Triangular Array (in Russian), *Heat Transfer and Hydrodynamics of Single-Phase Flow in Rod Bundles*, collected papers, ed. V. M. Borishanskij and P. A. Ushakov, Izd. Nauka, Leningrad, 1979.

26. S.-K. Cheng and N. E. Todreas, Constitutive Correlations for Wire-Wrapped Subchannel Analysis under Forced and Mixed Convection Conditions, DOE/ET/37240-108 TR, Mass. Inst. of Technol., Cambridge, Mass., 1984.

27. E. M. Sparrow, A. L. Loeffler, Jr., and H. A. Hubbard, Heat Transfer to Longitudinal Laminar Flow between Cylinders, *Trans. ASME, J. Heat Transfer*, Vol. 83, pp. 415–422, 1961.

28. C.-J. Hsu, Laminar- and Slug-Flow Heat Transfer Characteristics of Fuel Rods Adjacent to Fuel Subassembly Walls, *Nucl. Sci. Eng.*, Vol. 49, pp. 398–404, 1972.

29. V. I. Subbotin, P. A. Ushakov, A. V. Zhukov, N. M. Matyukhin, Yu. S. Jur'ev, and L. K. Kudryatseva, *Heat Transfer in Cores and Blankets of Fast Breeder Reactors*, collection of reports, Vol. 2, Symp. of CMEA Countries: present and future work on creating AES with fast reactors, Obninsk, 1975.

30. V. I. Subbotin, Heat Exchange and Hydrodynamics in Channels of Complex Geometry, *Heat Transfer 1974*, Proc. 5th Int. Heat Transfer Conf., Tokyo, Vol. VI, pp. 89–104, 1974.

31. R. A. Axford, Two-Dimensional Multiregion Analysis of Temperature Fields in Reactor Tube Bundles, *Nucl. Eng. Design*, Vol. 6, pp. 25–42, 1967.

32. E. M. Sparrow and S. H. Lin, The Developing Laminar Flow and Pressure Drop in the Entrance Region of Annular Ducts, *J. Basic Eng.*, Vol. 86, pp. 827–834, 1964.

33. R. E. Lundberg, P. A. McCuen, and C. W. Reynolds, Heat Transfer in Annular Passages, Hydrodynamically Developed Laminar Flow with Arbitrarily Prescribed Wall Temperatures or Heat Fluxes, *Int. J. Heat Mass Transfer*, Vol. 6, pp. 495–529, 1963.

34. J. H. Kim, Heat transfer in longitudinal laminar flow along circular cylinders in square array, *Fluid Flow and Heat Transfer over Rod or Tube Bundles*, ed. S. C. Yao and P. A. Pfund, ASME, New York, pp. 155–161, 1979.

35. J. Ohnemus, Wärmeübergang und Druckverlust in einem Zentralkanal eines Stabbündels in quadratischer Anordnung, Diplomarbeit, Inst. für Neutronenphysik und Reaktortechnik, Kernforschungszentrum Karlsruhe, 1982.

36. D. J. Gunn and C. W. W. Darling, Fluid Flow and Energy Losses in Non-circular Conduits, *Trans. Inst. Chem. Eng.*, Vol. 41, pp. 163–173, 1963.

37. B. C.-J. Chen, T. H. Chien, W. T. Sha, and J. H. Kim, 3-D Solution of Flow in an Infinite Square Array of Circular Tubes by Using Boundary-Fitted Coordinate System, *Numerical Grid Generation*, ed. J. F. Thompson, Elsevier, pp. 619–632, 1982.

38. J. H. Kim, B. C.-J. Chen, T. H. Chien, and W. T. Sha, Heat Transfer in Longitudinal Laminar Flow between Cylinders Arranged in Regular Array, to be published; cited in [37].

39. B. C.-J. Chen, W. T. Sha, M. L. Doria, R. C. Schmitt, and J. F. Thompson, BODYFIT-1FE: *A Computer Code for Three-Dimensional Steady State/Transient Single-Phase Rod-Bundle Thermal Hydraulic Analysis*, NUREG/CR-1874, ANL-80-127, Argonne Nat. Lab., 1980.

40. H. Zachmann, Wärmeübergang und Druckverlust in einem Zentralkanal eines Stabbündels in quadratischer Anordnung, Studienarbeit, Inst. für Neutronenphysik und Reaktortechnik, Kernforschungszentrum Karlsruhe, 1982.

41. H. Gräber, Der Wärmeübergang in glatten Rohren, zwischen parallelen Platten, in Ringspalten und längs Rohrbündeln bei exponentieller Wärmeflußverteilung in erzwungener laminarer oder turbulenter Strömung, *Int. J. Heat Mass Transfer*, Vol. 13, pp. 1645–1703, 1970.

42. O. E. Dwyer and H. C. Berry, Effects of Cladding Thickness and Thermal Conductivity on Heat Transfer for Laminar In-Line Flow through Rod Bundles, *Nucl. Sci. Eng.*, Vol. 42, pp. 69–80, 1970.

43. M. L. Trombetta, Multiregional Analysis of Temperature Fields in Reactor Tube Bundles, *Nucl. Eng. Design*, Vol. 11, pp. 132–136, 1969.

44. V. I. Subbotin, P. A. Ushakov, A. V. Zhukov, and N. M. Matyukhin, Calculation of the Heat Exchange in Core Flow and of Hydrodynamics in Laminar Flow of Coolants in Regular Fuel Element Lattices, *Sov. At. Energy*, Vol. 33, pp. 959–960, 1973.

45. V. I. Subbotin, M. Kh. Ibragimov, P. A. Ushakov, V. P. Bobkov, A. V. Zhukov, and Yu. S. Yurev, *Hydrodynamic and Heat Transfer in Nuclear Power Stations* (*Fundamentals of Calculation*) (in Russian), Atomizdat, Moscow, 1975.

46. M. Cieszko and J. A. Kolodziej, Określenie wymiany ciepla pomiedzy wiazka regularnych pretów i plynem podczas podluźnego laminarnego przeplywu metoda kollokacji brzegowej, *Mech. Teor. i Stosowana*, Vol. 4, pp. 605–619, 1981.

47. C.-J. Hsu, Multiregion Analysis of the Effect of Lateral Rod Displacement on Heat Transfer in Laminar Flow through Closely Packed Reactor Fuel Rods, *Progress in Heat and Mass Transfer*, ed. O. E. Dwyer, Vol. 7, Heat Transfer in Liquid Metals, Pergamon, Oxford, U.K., pp. 219–237, 1973.

48. C.-J. Hsu, The Effect of Lateral Rod Displacement on Laminar-Flow Transfer, *J. Heat Transfer*, Vol. 94, pp. 169–173, 1972.

49. K. Rehme and G. Trippe, Pressure Drop and Velocity Distribution in Rod Bundles with Spacer Grids, *Nucl. Eng. Design*, Vol. 62, pp. 349–359, 1980.

50. D. P. Robinson, Subchannel Friction Factor for Rod Bundles: Laminar Flow Predictions and Their Application to Turbulent Flows, AAEW-M-1656, UKAEA, Winfrith Atomic Energy Establishment, U.K., 1979.

51. D. P. Robinson, personal communication, British-American Tobacco Comp. Ltd, Regent's Park Road, Southampton, England, 1985.

52. H.-J. Herzog, *Wärmeübergang und Druckverlust in einem Wandkanal von Stabbündeln*, Diplomarbeit, Inst. für Neutronenphysik und Reaktortechnik, Kernforschungszentrum Karlsruhe, 1983.

53. K. C. Cheng and M. Jamil, Laminar Flow and Heat Transfer in Ducts of Multiply Connected Cross Sections, ASME Paper 67-HT-6, 1967.

54. R. Ullrich, Analyse der ausgebildeten Laminarströmung in längsangeströmten, endlichen, hexagonalen Stabbündeln, Dr.-Ing. Dissertation, TUBIK 36, Inst. für Kerntechnik, Techn. Univ. Berlin, Berlin, F.R.G., 1974.

55. L. R. Galloway and N. Epstein, Longitudinal Flow between Cylinders in Square and Triangular Arrays and in a Tube with Square-Edged Entrance, *Am. Inst. Chem. Eng. Symp. Ser.*, No. 6, pp. 3–15, 1965.

56. L. R. Galloway, Longitudinal Flow between Cylinders in Square and Triangular Arrays, Ph.D. Thesis, Univ. of British Columbia, Canada, 1964.

57. P. Milbauer, Vliyanie geometricheskikh povrezhdenii na skorostnye i temperaturnye polya v periferiinoi oblasti toplivnoi kassety bystrogo reaktora pri laminarnom techenii teplonositelya (otchet), ÚJV 4576-T, Řež, Č.S.S.R., 1978; German translation, KfK-tr-643, Kernforschungszentrum Karlsruhe, 1980.

58. R. A. Axford, Longitudinal Laminar Flow of an Incompressible Fluid in Finite Tube Bundles with $m + 1$ Tubes, LA-3418, Los Alamos Sci. Lab., Univ. of California, Los Alamos, N. Mex., 1965.

59. M. Courtaud, R. Ricque, and B. Martinet, Etude des pertes de charge dans des conduites circulaires contenant un faisceau de barreaux, *Chem. Eng. Sci.*, Vol. 21, pp. 881–893, 1966.
60. T. C. Min, A Two-Dimensional Analysis of Heat Transfer and Fluid Flow in a Rod Cluster with Special Attention to Asymmetry, Ph.D. Thesis, Univ. of Tennessee, Knoxville, Tenn., 1969.
61. T. C. Min, H. W. Hoffman, T. C. Tucker, and F. N. Peebles, An Analysis of Axial Flow through a Circular Channel Containing Rod Clusters, *Developments in Theoretical and Applied Mechanics*, ed. W. A. Shaw, Pergamon, New York, Vol. 3, pp. 667–690, 1967.
62. R. Mottaghian, Analytische Lösungen für die laminaren Geschwindigkeitsfelder in längsangeströmten, endlichen Stabbündeln mit beliebiger Anordnung, Dr.-Ing. Dissertation, Inst. für Kerntechnik, Tech. Univ. Berlin, Report TUBIK 25, 1973.
63. R. Mottaghian and L. Wolf, Fully Developed Laminar Flow in Finite Rod Bundles of Arbitrary Arrangements, *Trans. ANS*, Vol. 15, p. 876, 1972.
64. R. Mottaghian and L. Wolf, A Two-Dimensional Analysis of Laminar Flow in Rod Bundles of Arbitrary Arrangement, *Int. J. Heat Mass Transfer*, Vol. 17, pp. 1121–1128, 1974.
65. K. Chen, A Two-Dimensional Analysis of Heat Transfer and Fluid Flow in a Rod Cluster with Special Attention to Asymmetry, Ph.D. Thesis, Purdue Univ., Lafayette, Ind., 1970.
66. K. Chen, Longitudinal Laminar Flow in Asymmetrical Finite Bundles of Rods, *Nucl. Eng. Design*, Vol. 25, pp. 207–216, 1973.
67. J. P. Zarling, Analytical Investigation of Laminar Forced Convective Heat Transfer in a Finite Rod Bundle, Ph.D. Thesis, Michigan Technological Univ., Houghton, Mich., 1971.
68. J. P. Zarling, Laminar-Flow Pressure Drop in Symmetrical Finite Rod Bundles, *Nucl. Sci. Eng.*, Vol. 61, pp. 282–285, 1976.
69. J. P. Zarling and T. C. Min, An Analysis of Heat Transfer in the Thermal Entrance Region to Fluids with High Péclet Moduli for Axial Flow through a Circular Shell Containing Tube Banks, ASME Paper No. 73-HT-29, 1973.
70. J. P. Zarling and T. C. Min, Forced Convective Heat Transfer in the Thermal Entrance Region of a Finite Uniform-Heat-Flux Rod Bundle, *Heat Transfer 1974, Tokyo*, AIChE, New York, Vol. 2, pp. 203–207, 1974.
71. R. W. Benodekar and A. W. Date, Numerical Prediction of Heat-Transfer Characteristics of Fully Developed Laminar Flow through a Circular Channel Containing Rod Clusters, *Int. J. Heat Mass Transfer*, Vol. 21, pp. 935–945, 1978.
72. R. W. Benodekar and A. W. Date, Prediction of Heat Transfer Characteristics of Thermally Developing Laminar Flow in Finite Nuclear-Rod Clusters, *J. Thermal Eng.*, Vol. 2, pp. 73–84, 1982.
73. A. K. Mohanty and D. K. Ray, Fluid Flow Through a Circular Tube Containing Rod Clusters, *Fluid Flow and Heat Transfer over Rod or Tube Bundles*, ed. S. C. Yao and P. A. Pfund, Am. Soc. Mech. Eng., New York, pp. 121–128, 1979.
74. S. Y. Chern and J. C. Chato, A Finite-Element Technique to Determine the Friction Factor and Heat Transfer for Laminar Flow in a Pipe with Irregular Cross Section, *Numer. Heat Transfer*, Vol. 1, pp. 453–470, 1978.
75. A. P. Salikov, Ya. L. Polynovskij, and K. I. Belyakov, Investigation of Heat Transfer and Pressure Drop of Axial Flow through Smooth Rod Bundles (in Russian), *Teploenergetica*, Vol. 1, pp. 13–17, 1967.
76. J. L. Wantland, Compact Tubular Heat Exchangers, *Reactor Heat Transfer Conference of 1956*, New York, TID-7529, (Pt. 2), Book 2, pp. 525–548, 1956.
77. P. Miller, J. J. Byrnes, and D. M. Benforado, Heat Transfer to Water Flowing Parallel to a Rod Bundle, *AIChE J.*, Vol. 2, pp. 226–234, 1956.
78. D. A. Dingee and J. W. Chastain, Heat Transfer from Parallel Rods in Axial Flow, *Reactor Heat Transfer Conference of 1956*, New York, TID-7529 (Pt. 2), Book 2, pp. 462–501, 1956.
79. B. W. Le Tourneau, R. E. Grimble, and J. E. Zerbe, Pressure Drop of Parallel Flow through Rod Bundles, *Trans. Am. Soc. Mech. Eng.*, Vol. 79, pp. 1751–1758, 1957.

80. V. I. Subbotin, P. A. Ushakov, and B. N. Gabrianovich, Hydraulischer Widerstand bei durch Flüssigkeiten längsumströmten Stabbündeln, *Kernenergie*, Vol. 4, pp. 658–660, 1961; cf. *At. Energiya*, Vol. p. 308, 1960.

81. A Draycott and K. R. Lawther, Improvement of Fuel Element Heat Transfer by Use of Roughened Surfaces and the Application to a 7-Rod Cluster, *Int. Dev. in Heat Transfer*, ASME, New York, pp. 543–552, 1961/2.

82. L. D. Palmer and L. L. Swanson, Measurements of Heat-Transfer Coefficients, Friction Factors, and Velocity Profiles for Air Flowing Parallel to Closely Spaced Rods, *Int. Dev. Heat Transfer*, ASME, New York, pp. 535–542, 1961/2.

83. E. D. Waters, Effect of Wire Wraps on Pressure Drop for Axial Turbulent Flow through Rod Bundles, HW-65173, Hanford Atomic Products Operation, Richland, Wash., 1963.

84. E. V. Firsova, Investigation of Heat Transfer and Hydraulic Resistance in a Longitudinal Flow of Water past a Bundle of Tubes (in Russian), *Inzh.-Fiz. Zh.*, Vol. 6, pp. 17–22, 1963.

85. A. I. Mikhaylov, E. K. Kalinin, and G. A. Dreytser, Investigation of Hydraulic Resistance to the Longitudinal Flow of Air over a Staggered Tube Bundle (in Russian), *Inzh. Fiz. Zh.*, Vol. 7, pp. 42–46, 1964.

86. W. Eifler and R. Nijsing, Fundamental Studies of Fluid Flow and Heat Transfer in Fuel Element Geometries. Pt. II. Experimental Investigation of Velocity Distribution and Flow Resistance in a Triangular Array of Parallel Rods, EUR 2193e, European Atomic Energy Community, 1965; cf. *Nucl. Eng. Design*, Vol. 5, pp. 22–42, 1967.

87. W. A. Sutherland and W. M. Kays, Heat Transfer in Parallel Rod Arrays, GEAP-4637, General Electric Co., San Jose, Calif., 1965.

88. H. W. Hoffmann, C. W. Miller, G. L. Sozzi, and W. A. Sutherland, Heat Transfer in Seven-Rod Clusters, Influence of Liner and Spacer Geometry on Superheat Fuel Performance, GEAP-5289, General Electric Co., San Jose, Calif., 1966.

89. J. Šimonek, Přispěvec k problematice přestupu tepla v proutkovém palivovém článku, *Jaderná Energie*, Vol. 12, pp. 246–249, 1966.

90. M. Kh. Ibragimov, I. A. Isupov, and V. I. Subbotin, Calculation and Experimental Study of Velocity Fields in a Complicated Channel (in Russian), *Liquid Metals*, ed. P. L. Kirillov, V. I. Subbotin, and P. A. Ushakov, Atomizdat, Moscow, pp. 234–250, 1967.

91. Yu. D. Levchenko, V. I. Subbotin, P. A. Ushakov, and A. V. Sheynina, Velocity Distribution in a Tightly Packed Rod Bundle (in Russian), *Liquid Metals*, ed. P. L. Kirillov, V. I. Subbotin, and P. A. Ushakov, Atomizdat, Moscow, pp. 223–234, 1967.

92. K. H. Presser, Wärmeübergang und Druckverlust an Reaktorbrennelementen in Form längsdurchströmter Rundstabbündel, JÜL-486-RB, Kernforschungsanlage Jülich, F.R.G., 1967.

93. K. H. Presser, Stoffübergang und Druckverlust an parallel angeströmten Stabbündeln in einem großen Bereich von Reynolds-Zahlen und Teilungsverhältnissen, *Int. J. Heat Mass Transfer*, Vol. 14, pp. 1235–1259, 1971.

94. A. V. Sheynina, Pressure Drop of Axial Flow Through Rod Bundles (in Russian), *Liquid Metals*, ed. P. L. Kirillov, V. I. Subbotin, and P. A. Ushakov, Atomizdat, Moscow, pp. 210–223, 1967.

95. W. Baumann, V. Casal, H. Hoffmann, R. Möller, and K. Rust, Brennelemente mit wendelförmigen Abstandshaltern für Schnelle Brutreaktoren, KfK 768, EUR 3694d, Kernforschungszentrum Karlsruhe, Karlsruhe, F.R.G., 1968.

96. W. Eifler, Über die turbulente Geschwindigkeitsverteilung und Wandreibung in Strömungskanälen verschiedener Querschnitte, Dr.-Ing. Dissertation, TH Darmstadt, F.R.G., 1968.

97. K. Rehme, Widerstandsbeiwerte von Gitterabstandshaltern für Reaktorbrennelemente, *Atomkernenergie*, Vol. 15, pp. 127–30, 1970.

98. B. Kjellström and A. Stenbäck, Pressure Drop Velocity Distributions and Turbulence Distributions for Flow in Rod Bundles, AE-RV-145, Aktiebolaget Atomenergi, Studsvik, Sweden, 1970.

99. V. I. Subbotin, P. A. Ushakov, Yu. D. Levchenko, and A. M. Aleksandrov, Velocity Fields in Turbulent Flow past Rod Bundles, *Heat Transfer—Sov. Res.*, Vol. 3, pp. 9–35, 1971.

100. V. I. Subbotin, B. N. Gabrianovich, and A. V. Sheynina, Hydraulic Resistance with Longitudinal Streamline Flow for Bundles of Plain and Finned Rods, *Sov. At. Energy*, Vol. 33, pp. 1031–1034, 1973.

101. A. Mohandes, Friction Losses for Water Flowing in Noncircular Ducts, M.Sc. Thesis, Oregon State Univ., Corvallis, Oregon, 1972.

102. A. Mohandes and J. G. Knudsen, Friction Factors in Noncircular Ducts with Sharp Corners, *Can. J. Chem. Eng.*, Vol. 57, pp. 109–111, 1979.

103. K. Rehme, Pressure Drop Performance of Rod Bundles in Hexagonal Arrangements, *Int. J. Heat Mass Transfer*, Vol. 15, pp. 2499–2517, 1972.

104. P. Voj, D. Markfort, and E. Ruppert, A Thermal-Hydraulic Analysis for Fuel Elements with Liquid-Metal Cooling, *Progress in Heat and Mass Transfer, Vol. 7, Heat Transfer in Liquid Metals*, ed. O. E. Dwyer, Pergamon, Oxford, pp. 179–193, 1973.

105. P. Voj and K. Scholven, Druckverlustmessungen an Abstandshaltergittern für SNR-300-Brennelment, ITB 74.34, INTERATOM, Bensberg, F.R.G., 1974.

106. A. A. Volobik, A. M. Krapivin, and P. I. Bystrov, Longitudinal Hydrodynamic Characteristics of Triangular Rod Bundles with Flat or Corrugated Casing, *High Temperature*, Vol. 14, pp. 209–211, 1976.

107. B. Kjellström, Studies of Turbulent Flow Parallel to a Rod Bundle of Triangular Array, AE-487, Aktiebolaget Atomenergi, Studsvik, Sweden, 1974.

108. P. Voj and K. Scholven, Widerstandsbeiwerte von funkenerodierten und punktgeschweißten Wabengittern, ITB, 75.75, INTERATOM, Bensberg, F.R.G., 1975.

109. A. C. Trupp, The Structure of Turbulent Flow in Triangular Array Rod Bundles, Ph.D. Thesis, Univ. of Manitoba, Winnipeg, Canada, 1973.

110. A. C. Trupp and R. S. Azad, The Structure of Turbulent Flow in Triangular Array Rod Bundles, *Nucl. Eng. Design*, Vol. 32, pp. 47–84, 1975.

111. K. Rehme, Pressure Drop of Spacer Grids in Smooth and Roughened Rod Bundles, *Nucl. Technol.*, Vol. 33, pp. 314–317, 1977.

112. J. Hejna, J. Červenka, and F. Mantlík, Rezul'taty izmerenii lokal'nykh gidrodinamicheskikh kharakteristik v puchke sterzhnei s geometricheskim povreszhdeniem. I. Otchet, ÚJV 4156T, Řež, Č.S.S.R., 1977.

113. J. Hejna and F. Mantlík, Turbulent Flow in Rod Bundles with Geometrical Disturbances, *Nucl. Technol.*, Vol. 59, pp. 509–525, 1982.

114. V. Krett, I. Vlček, J. Majer, and J. Smolik, Experimental Investigation of Fluiddynamic Characteristics of Rod Bundles with $s/d = 1$ Arranged in Triangular and Square Array (in Russian), *COMECON-Symposium Teplofizika i gidrodynamika aktivnoi zony i parogeneratorov dlya bystrykh reaktorov*, Marianské Laznenĕ, Č.S.S.R., Prague, Vol. 1, paper ML 78/06, pp. 90–103, 1978.

115. M. Fakory and N. Todreas, Experimental Investigation of Flow Resistance and Wall Shear Stress in the Interior Subchannel of a Triangular Array of Parallel Rods, *J. Fluids Eng.*, Vol. 101, pp. 429–435, 1979.

116. T. C. Reihman, Experimental Study of Pressure Drop in Wire-Wrapped Rod Bundles, *Tube Bundle Thermal-Hydraulics*, ed. P. A. Pfund, S. C. Yao, C. D. Morgan, and S. Cho, Am. Soc. Mech. Eng. Publ., New York, pp. 19–25, 1982.

117. O. S. Vasilevich, L. I. Kolykhan, and A. V. Sheynina, Experimental Study of Heat Transfer and Hydraulic Resistance in Rod Bundles, Akad. Nauk Belorusskoj SSR, Vesti Akad. Navuk BSSSR, Ser. Fiz-energet. Navuk, Minsk, No. 1, pp. 107–111, 1983.

118. A. Ya. Inayatov, A New Correlation of the Average Surface Resistance Coefficient for Axial Fluid Flow through Rod Bundles (in Russian), *Izv. AN Usb. SSR*, Ser. Tekhn. Nauk, No. 2, pp. 59–62, 1968.

119. K. Maubach, Reibungsgesetze turbulenter Strömungen, *Chem. Ing. Techn.*, Vol. 42, pp. 995–1004, 1970.

120. K. Maubach, Reibungsgesetze turbulenter Strömungen in geschlossenen, glatten und rauhen Kanälen von beliebigem Querschnitt, Dr.-Ing. Dissertation, Univ. of Karlsruhe, F.R.G., 1969.

121. R. G. Deissler and M. F. Taylor, Analysis of Axial Turbulent Flow and Heat Transfer through Banks of Rods or Tubes, *Reactor Heat Transfer Conference of 1956*, TID-7529, Pt. 1, Book 2, pp. 416–461, 1957.

122. V. S. Osmachkin, Some Problems of Heat-Transfer in Liquid-Cooled Reactors, *Third United Nations Int. Conf. Peaceful Uses of Atomic Energy*, A/Conf.28/P/326 USSR, 1964.

123. N.I. Buleev, K. N. Polosukhina, and V. K. Pyshin, Hydraulic Resistance and Heat Transfer in a Turbulent Liquid Stream in a Lattice of Rods, *High Temp.*, Vol. 2, pp. 673–681, 1964.

124. L. S. Kokorev, A. S. Korsun, and V. I. Petrovichev, Méthodes approchées de calcul de la distribution des vitesses dans le écoulements turbulents des canaux à section non circulaire, Rapport Inzherno-Fizicheskii Inst., Moscow, 1966; cf. CEA-tr-R-1922, 1968.

125. M. Kh. Ibragimov, I. A. Isupov, L. L. Kobzar, and V. I. Subbotin, Calculation of the Tangential Stresses at the Wall of a Channel and the Velocity Distribution in a Turbulent Flow of Liquid, *At. Energiya*, Vol. 21, pp. 101–107, 1966.

126. M. Kh. Ibragimov, I. A. Isupov, L. L. Kobzar, and V. I. Subbotin, Calculation of Hydraulic Resistivity Coefficients for Turbulent Fluid Flow in Channels of Noncircular Cross Section, *At. Energiya*, Vol. 23, pp. 300–305, 1967.

127. R. Nijsing, I. Gargantini, and W. Eifler, Analysis of Fluid Flow and Heat Transfer in a Triangular Array of Parallel Heat Generating Rods, *Nucl. Eng. Design*, Vol. 4, pp. 375–398, 1966.

128. V. Vonka, Numerical Calculation of Velocity and Temperature Profiles of an Incompressible Fluid Flowing In-Line through a Triangular Rod Bundle under Fully Developed Heat Transfer Conditions, *Zürich Club*, *Gas Cooled Fast Reactor Heat Transfer Specialists Meeting*, Würenlingen, 1970.

129. E. Aranovich, A Method for the Determination of the Local Turbulent Friction Factor and Heat Transfer Coefficient in Generalized Geometries, *J. Heat Transfer*, Vol. 93, pp. 61–68, 1971.

130. W. Eifler and R. Nijsing, Calculation of Turbulent Velocity Distribution and Wall Shear Stress for Axial Turbulent Flow in Infinite Arrays of Parallel Rods, *Wärme- und Stoffübertragung*, Vol. 2, pp. 246–256, 1969.

131. W. Eifler and R. Nijsing, Berechnung der turbulenten Geschwindigkeitsverteilung und Wandreibung in asymmetrischen Stabbündeln, *Atomkernenergie*, Vol. 18, pp. 189–197, 1971.

132. V. I. Subbotin and P. A. Ushakov, Calculation of the Hydrodynamic Characteristics of Rod Clusters, *Heat Transfer—Sov. Res.*, Vol. 3, pp. 157–176, 1971.

133. H. Ramm and K. Johannsen, Hydrodynamics and Heat Transfer in Regular Arrays of Circular Tubes, *1972 Int. Seminar on Recent Developments in Heat Exchangers,* Trogir, Yugoslavia, 1972.

134. P. A. Ushakov, Calculation of the Hydrodynamic Characteristics of the Longitudinal Flow of a Liquid around Regular Arrays of Fuel Element Rods, *High Temp.*, Vol. 12, pp. 89–95, 1974.

135. P. A. Ushakov, Problems of Fluid Dynamic and Heat Transfer in Cores of Fast Reactors, *COMECON-Symposium Teplofizika i gidrodinamika aktivnoi zony i parogeneratorov dlya bystrykh reaktorov*, Marianské Lazně, Č.S.S.R., Vol. 1, paper ML 78/01, pp. 14–35, Praha, 1978.

136. R. Meyder, Bestimmung des turbulenten Geschwindigkeits- und Temperaturfeldes in Stabbündeln mit Hilfe von krummlinig orthogonalen Koordinaten, Dr.-Ing. Dissertation, Univ. of Karlsruhe, F.R.G., 1974.

137. R. Meyder, Turbulent Velocity and Temperature Distribution in the Central Subchannel of Rod Bundles, *Nucl. Eng. Design*, Vol. 35, pp. 181–189, 1975.

138. J. Gosse and R. Schiestel, Convection forcée turbulente dans les faisceaux de tubes en attaque longitudinale, *Rev. Gén. Therm.*, No. 206, pp. 75–86, 1979.

139. J. A. Brighton and J. B. Jones, Fully Developed Turbulent Flow in Annuli, *J. Basic Eng.*, Vol. 86, pp. 835–844, 1964.

140. K. Rehme, Turbulent Flow in Smooth Concentric Annuli with Small Radius Ratios, *J. Fluid Mech.*, Vol. 64, Pt. 2, pp. 263–287, 1974.

141. J. D. Hooper, Developed Single Phase Turbulent Flow through a Square-Pitched Rod Cluster, *Nucl. Eng. Design*, Vol. 60, pp. 365–379, 1980.

142. C. W. Rapley, The Limiting Case of Rods Touching in Turbulent Flow through Rod Bundles, *Proc. 3rd. Int. Conf. Num. Meth. Laminar and Turbulent Flow*, Univ. of Washington, Seattle, Pineridge Press, Swansea, pp. 93–103, U.K., 1983.

143. C. W. Rapley, Turbulent Flow in a Duct with Cusped Corners, *Int. J. Numer. Methods Fluids*, Vol. 5, pp. 155–167, 1985.

144. F. F. Bogdanov, Friction Factors for Longitudinal Streamline Flow through Tube Bundles, *At. Energiya*, Vol. 23, pp. 46–47, 1967.

145. C. W. W. Darling, Fluid Flow and Energy Losses in Irregular Conduits of Constant Flow Area, *M.Sc. Thesis, Dept. of Chemical Eng., Queen's Univ., Kingston, Ontario*, 1961.

146. P. A. Ushakov, V. I. Subbotin, B. N. Gabrianovich, V. D. Talanov, and I. P. Sviridenko, Heat Transfer and Hydraulic Resistance in Tightly Packed Corridor Bundle of Rods, *Sov. At. Energy*, Vol. 13, pp. 761–768, 1963.

147. P. Grillo and V. Marinelli, Single- and Two-Phase Pressure Drops on a 16-Rod Bundle, *Nucl. Appl. Technol.*, Vol. 9, pp. 682–693, 1970.

148. P. Grillo and G. Mazzone, Single- and Two-Phase Pressure Drops on a 6 × 6 Rod Bundle at 70 atm, *Nucl. Technol.*, Vol. 15, pp. 25–35, 1972.

149. J. Marek, K. Maubach, and K. Rehme, Heat Transfer and Pressure Drop Performance of Rod Bundles Arranged in Square Arrays, *Int. J. Heat Mass Transfer*, Vol. 16, pp. 2215–2228, 1973.

150. R. Gerard and W. B. Baines, Turbulent Flow in Very Noncircular Conduit, *J. Hydraul. Div., Proc. Am. Soc. Civil Eng.*, Vol. 103, No. HY8, pp. 829–842, 1977.

151. R. Eichhorn, H. C. Kao, and S. Neti, Measurements of Shear Stress in a Square Array Rod Bundle, *Nucl. Eng. Design*, Vol. 56, pp. 385–391, 1980.

152. M. Khattab, A. Mariy, and M. Habib, Experimental Heat Transfer in Tube Bundle (Part II), *Atomkernenergie-Kerntechnik*, Vol. 45, pp. 93–97, 1984.

153. J. Hejna, J. Červenka, and F. Mantlik, Rezul'taty izmerenii lokal'nykh gidrodinamicheskikh kharakteristik v puchke sterzhnei s geometricheskim povrezhdeniem. II. Otchet, ÚJV 4175-T, Řež, Č.S.S.R., 1978.

154. K. Rehme, The Structure of Turbulent Flow through Rod Bundles, *Proc. 3rd. Int. Top. Mtg. Nuclear Reactor Thermal Hydraulics,* ed. C. Chiu and G. Brown, Vol. 2, paper 16.A Newport, R.I., 1985.

155. H. Barrow, A. K. A. Hassan, and C. Avgerinos, Peripheral Temperature Variation in the Wall of a Noncircular Duct—an Experimental Investigation, *Int. J. Heat Mass Transfer*, Vol. 27, pp. 1031–1037, 1984.

156. A. J. Inajatov and M. A. Michiejew, Heat Transfer in Longitudinal Flow through Rod Bundles (in Russian), *Teploenergetika*, Vol. 3, pp. 48–50, 1957.

157. H. W. Hoffman, J. L. Wantland, and W. J. Stelzman, Heat Transfer with Axial Flow in Rod Clusters, *Int. Dev. in Heat Transfer*, ASME, New York, pp. 553–560, 1961/62.

158. K. Koziol, Heat Transfer in Turbulent Flow on the Shell Side of Nonbaffled Exchangers with Standard and Squeezed Tubes (in Polish), *Chemija stosowana*, Vol. 4B, pp. 359–392, 1965.

159. J. D. Redman, G. McKee, and I. C. Rule, The Influence of Surface Heat Flux Distribution and Surface Temperature Distribution on Turbulent Forced Convective Heat Transfer in Clusters of Tubes in Which the Flow of Coolant is Parallel to the Axes of the Tubes, *Proc. 3rd. Int. Heat Transfer Conference,* Vol. I, pp. 186–198, 1966.

160. E. K. Kalinin, G. A. Dreytser, and A. K. Kozlov, Heat Transfer in Longitudinal Flow through Rod Bundles in Triangular Array with Different Pitch-to-Diameter Ratios (in Russian), *Inzh. Fiz. Zh.*, Vol. 16, pp. 47–53, 1969.

161. M. Rieger, Experimentelle Untersuchung des Wärmeübergangs in parallel durchströmten Rohrbündeln bei konstanter Wärmestromdichte im Bereich mittlerer Prandtl-Zahlen, *Int. J. Heat Mass Transfer*, Vol. 12, pp. 1421–1447, 1969.

162. F. Tachibana, A. Oyama, M. Akiyama, and S. Kondo, Measurement of Heat Transfer Coefficients for Axial Air Flow through Eccentric Annulus and Seven-Rod Cluster, *J. Nucl. Sci. Technol.*, Vol. 6, pp. 207–214, 1969.

163. V. M. Borishanskiy, M. A. Gotovskiy, and E. V. Firsova, The Effect of Pitch on Heat Transfer from Rod Clusters in Longitudinal Turbulent Coolant Flow ($\Pr \geq 1$), *Heat Transfer—Sov. Res.*, Vol. 3, pp. 91–99, 1971.

164. G. Markóczy, Konvektive Wärmeübertragung in längsangeströmten Stabbündeln bei turbulenter Strömung, *Wärme- und Stoffübertragung*, Vol. 5, pp. 204–212, 1972.

165. V. P. Bobkov, M. Kh. Ibragimov, V. F. Sinyavskii, and N. A. Tychinskii, Heat Exchange for Flow of Water in a Densely Packed Triangular Bundle of Rods, *Sov. At. Energy*, Vol. 37, pp. 823–827, 1974.

166. V. L. Lel'chuk, K. F. Shuyskaya, A. G. Sorokin, and O. N. Bragina, Heat Transfer in the Inlet Length of Fuel-Elements Modeling Rod Bundle in Longitudinal Air Flow, *Heat Transfer—Sov. Res.*, Vol. 9, pp. 100–104, 1977.

167. B. S. Petukhov and L. I. Roizen, Generalized Dependences for Heat Transfer in Tubes of Annular Cross Section, *Teplofiz. Vysokikh Temp.*, Vol. 12, pp. 565–569, 1974.

168. C. Taylor, C. E. Thomas, and K. Morgan, Turbulent Heat Transfer via the F.E.M.: Heat Transfer in Rod Bundles, *Int. J. Numer. Meth. Fluids*, Vol. 3, pp. 363–375, 1983.

169. A. Ya. Inayatov, Variation of Heat Transfer along a Bundle of Rods, with Streamline Coolant Flow, *Izv. Akad. Nauk UZSSR Ser. Tekh. Nauk*, No. 4, pp. 50–54, 1972.

170. A. Ya. Inayatov, Correlation of Data on Heat Transfer. Flow Parallel to Tube Bundles at Relative Tube Pitches of $1.1 < s/d < 1.6$, *Heat Transfer—Sov. Res.*, Vol. 7, pp. 84–88, 1975.

171. J. R. Parette and R. E. Grimble, Average and Local Heat Transfer Coefficients for Parallel Flow through a Rod Bundle, WAPD-TH 180, Westinghouse Electric Corp., Pittsburgh, Pa., 1956.

172. L. S. Kokorev, A. S. Korsun, and V. I. Petrovichev, Calculation of Stable Heat Transfer from Rod Bundles in Longitudinal Flow, *Heat Transfer—Sov. Res.*, Vol. 3, pp. 44–55, 1971.

173. V. Vonka and B. H. Boonstra, Calculated Heat Transfer Development in Bundles, *Nucl. Eng. Design*, Vol. 31, pp. 337–345, 1974.

174. V. I. Subbotin, P. A. Ushakov, B. N. Gabrianovich, and A. V. Zhukov, Heat Exchange during the Flow of Mercury and Water in a Tightly Packed Rod Pile, *Sov. At. Energy*, Vol. 9, pp. 1001–1009, 1961.

175. A. J. Friedland, O. E. Dwyer, M. W. Maresca, and C. F. Bonilla, Heat Transfer to Mercury in Parallel Flow through Bundles of Circular Rods, *Int. Dev. in Heat Transfer*, Am. Soc. Mech. Eng., New York, Part III, pp. 526–534, 1961.

176. V. M. Borishanskii and E. V. Firsova, Heat Exchange in the Longitudinal Flow of Metallic Sodium past a Tube Bank, *At. Energiya*, Vol. 14, pp. 584–585, 1963.

177. V. M. Borishanskii and E. V. Firsova, Heat Exchange in Separated Bundles of Rods with Metallic Sodium Flowing Longitudinally, *At. Energiya*, Vol. 16, pp. 457–458, 1964.

178. V. I. Subbotin, P. A. Ushakov, et al., Heat Removal of Reactor Fuel Elements Cooled by Liquid Metals, *Proc. 3rd Int. Conf. Peaceful Uses At. Energy,* Geneva, Vol. 8, United Nations, New York, pp. 192–203, 1964.

179. M. W. Maresca and O. E. Dwyer, Heat Transfer to Mercury Flowing In-Line through a Bundle of Circular Tubes, *J. Heat Transfer*, Vol. 86, pp. 180–196, 1964.

180. B. Nimmo and O. E. Dwyer, Heat Transfer to Mercury Flowing In-Line through a Rod Bundle, *J. Heat Transfer*, Vol. 87, pp. 312–313, 1965.

181. S. Kalish and O. E. Dwyer, Heat Transfer to NaK Flowing through Unbaffled Rod Bundles, *Int. J. Heat Mass Transfer*, Vol. 10, pp. 1533–1558, 1967.

182. V. M. Borishanskii, M. A. Gotovskiy, and E. V. Firsova, Heat Exchange When a Liquid Flows Longitudinally through a Bundle of Rods Arranged in a Triangular Lattice, *At. Energiya*, Vol. 22, pp. 318–320, 1967.

183. M. Pashek, Study of Local Heat Transfer and Thermal Fields in a Seven-Rod Bundle with Axial Flow of Sodium, UJV-1815, Nuclear Research Inst., Řež, Č.S.S.R., 1967.

184. V. I. Subbotin, P. A. Ushakov, A. V. Zhukov, and V. D. Talanov, The Temperature Distribution in Liquid Metal Cooled Fuel Elements, *At. Energiya*, Vol. 22, pp. 372–378, 1967.

185. V. M. Borishanskii, M. A. Gotovskiy, and E. V. Firsova, Heat Transfer in Liquid Metal Flowing Longitudinally in Wetted Rod Bundles, *At. Energiya*, Vol. 27, pp. 549–552, 1969.

186. P. J. Hlavac, O. E. Dwyer, and M. A. Helfant, Heat Transfer to Mercury Flowing In-Line through an Unbaffled Rod Bundle: Experimental Study of the Effect of Rod Displacement on Rod-Average Heat Transfer Coefficients, *J. Heat Transfer*, Vol. 91, pp. 568–580, 1969.

187. A. V. Zhukov, V. I. Subbotin, and P. A. Ushakov, Heat Transfer from Loosely Spaced Rod Clusters to Liquid Metal Flowing in the Axial Direction, *Liquid Metals*, English Transl., NASA Report NASA-TT-F522, p. 149, 1969.

188. A. V. Zhukov, L. K. Kudryatseva, E. Ya. Sviridenko, V. I. Subbotin, D. V. Talanov, and P. A. Ushakov, Experimental Study of Temperature Fields of Fuel Elements, Using Models, *Liquid Metals*, English Transl., NASA Report, NASA-TT-F522, p. 170, 1969.

189. V. I. Subbotin, P. A. Ushakov, A. V. Zhukov, and E. Ya. Sviridenko, Temperature Fields of Fuel Elements in the BOR Reactor Core, *Sov. At. Energy*, Vol. 28, pp. 620–621, 1970.

190. H. Gräber and M. Rieger, Experimental Study of Heat Transfer to Liquid Metals Flowing In-Line through Tube Bundles, *Progress in Heat and Mass Transfer*, ed. O. E. Dwyer, Vol. 7, Heat Transfer in Liquid Metals, Pergamon, Oxford, pp. 151–166, 1973.

191. H. Gräber and M. Rieger, Experimentelle Untersuchung des Wärmeübergangs an Flüssigmetalle (NaK) in parallel durchströmten Rohrbündeln bei konstanter und exponentieller Wärmeflußdichteverteilung, *Atomkernenergie*, Vol. 19, pp. 23–40, 1972.

192. A. R. Marchese, Experimental Study of Heat Transfer to NaK Flowing In-Line through a Tightly Packed Rod Bundle, *13th Nat. Heat Transfer Conf.*, Denver, Colo., AIChE Paper No. 36, Am. Inst. Chem. Eng., 1972.

193. A. V. Zhukov, N. M. Matyukhin, E. V. Nomofilov, A. P. Sorokin, and V. S. Yurev, Temperature Distributions in Non-standard and Deformed Fuel Pin Grids of Fast Reactors, *COMECON-Symposium Teplofizika i gidrodinamika aktivnoi zony i parogeneratorov dlya bystrykh reaktorov*, Marianské Lazně, Č.S.S.R., paper ML-78/11, 1978; English transl.: Risley Transl. 4235, Risley, Warrington, U.K.

194. A. V. Zhukov, N. M. Matyukhin, and E. Ya. Sviridenko, Einfluß von Geometriestörungen auf die Temperaturfelder und den Wärmeübergang in den Unterkanälen von Brennelementbündeln Schneller Reaktoren, FEI–979, Fiziko-energeticheskii Institut, Obninsk, 1980; German transl., KfK-tr-675, 1982.

195. R. Möller and H. Tschöke, Experimental Determination of Cladding Temperature Fields in the Critical Regions of Rod Bundles with Turbulent Sodium Flow and Comparison with Calculations, *Heat Transfer 1978*, Vol. 5, pp. 29–34, 1978.

196. R. Möller and H. Tschöke, Local Temperature Distributions in the Critical Duct Wall Zones of LMFBR Core Elements. Status of Knowledge, Unsettled Problems and Possible Solutions, *Proc. ANS/ASME Int. Top. Mtg. Nuclear Reactor Thermal Hydraulics*, Saratoga, N.Y., Vol. 3, pp. 1871–1881, 1980.

197. D. Weinberg, Temperaturfelder in Bündeln mit Na-Kühlung, *Thermo- und fluiddynamische Unterkanalanalyse der Schnellbrüter-Brennelemente und ihre Relation zur Brennstabmechanik*, ed. P. Voj, KfK-2232, Kernforschungszentrum Karlsruhe, F.R.G., 1975.

198. A. J. Friedland and C. F. Bonilla, Analytical Study of Heat Transfer Rates for Parallel Flow of Liquid Metals through Tube Bundles: II, *AIChE J.*, Vol. 7, pp. 107–112, 1961.

199. N. I. Buleev and R. Ya. Mironovich, Heat Transfer in Turbulent Fluid Flow in a Triangular Array of Rods, *High Temp.*, Vol. 10, pp. 925–931, 1972.

200. H. Ramm and K. Johannsen, A Phenomenological Turbulence Model and its Applications to Heat Transport in Infinite Rod Arrays with Axial Turbulent Flow, *J. Heat Transfer*, Vol. 97, pp. 231–237, 1975.

201. R. Nijsing and W. Eifler, Analysis of Liquid Metal Heat Transfer in Assemblies of Closely Spaced Fuel Rods, *Nucl. Eng. Design*, Vol. 10, pp. 21–54, 1969.

202. J. Pfann, Heat Transfer in Turbulent Longitudinal Flow through Unbaffled Assemblies of Fuel Rods, *Nucl. Eng. Design*, Vol. 25, pp. 217–247, 1973.

203. L. Wolf and K. Johannsen, Two-Dimensional Multiregion Analysis of Temperature Fields in Finite Rod Bundles Cooled by Liquid Metals, *Proc. 1st. Int. Conf. Structural Mechanics in Reactor Technology,* Berlin, Vol. 6, Part L, Paper 2/6, 1971

204. J. H. Rust, A Parametric Analysis of Lyon's Integral Equation for Liquid-Metal Heat Transfer Coefficients, *Nucl. Eng. Design*, Vol. 16, pp. 223–236, 1971.

205. V. P. Bobkov, M. Kh. Ibragimov, and V. I. Subbotin, Calculating the Coefficient of Turbulent Heat Transfer for a Liquid Flowing in a Tube, *Sov. At. Energy*, Vol. 24, pp. 545–550. 1968.

206. V. P. Bobkov, M. Kh. Ibragimov, and G. I. Sabelev, The Calculation of the Coefficient of Turbulent Heat Diffusion in Channels of Noncircular Cross Section, *High Temp.*, Vol. 6, pp. 645–651, 1968.

207. V. P. Bobkov and M. Kh. Ibragimov, Diffusion of Heat with the Turbulent Flow of Liquids with Different Prandtl Numbers, *High Temp.*, Vol. 8, pp. 97–101, 1970.

208. R. Nijsing and W. Eifler, Temperature Fields in Liquid-Metal-Cooled Rod Assemblies, *Progress in Heat and Mass Transfer*, *Vol. 7*, *Heat Transfer in Liquid Metals*, ed. O. E. Dwyer, Pergamon, Oxford, pp. 115–149, 1973.

209. R. Nijsing and W. Eifler, Axially Varying Heat Flux Effects in Tubes, Flat Ducts and Widely Spaced Rod Bundles Cooled by a Turbulent Flow of Liquid Metal, *Nucl. Eng. Design*, Vol. 23, pp. 331–346, 1972.

210. V. P. Bobkov, M. Kh. Ibragimov, and V. I. Subbotin, Generalized Relationships for Heat Transfer in the Fuel Assemblies of Nuclear Reactors with Liquid Metal Cooling, *Teplofiz. Vys. Temp.*, Vol. 10, pp. 795–803, 1972.

211. V. P. Bobkov and N. K. Savanin, Calculation of Heat Transfer for Turbulent Flow of Liquid Metals through Channels of Cylindrical Rods Arranged in Infinite Square Array, FEI-644, Inst. of Physics and Energy, Obninsk, 1975; German transl., KfK-tr-634, 1979.

212. J. Pfann, Turbulent Heat Transfer to Longitudinal Flow through a Triangular Array of Circular Rods, *Nucl. Eng. Design*, Vol. 34, pp. 203–219, 1975.

213. J. Pfann, A New Description of Liquid Metal Heat Transfer in Closed Conduits, *Nucl. Eng. Design*, Vol. 41, pp. 149–163, 1977.

214. A. V. Zhukov, N. M. Matyukhin, A. B. Muzhanov, Ye. Ya. Sviridenko, and P. A. Ushakov, Methods of Calculating Heat Exchange of Liquid Metals in Regular Lattices of Fuel Elements and Some New Experimental Data, FEI-404, Inst. of Physics and Energy, Obninsk; Transl., AEC TR-7549, 1973.

215. P. A. Ushakov, Calculation of the Temperature Fields of Bundles of Fuel Elements in Axial Turbulent Flows of Heat-Transfer Media with Vanishing Small Prandtl Numbers, *High Temp.*, Vol. 12, pp. 677–685, 1974.

216. P. A. Ushakov, A. V. Zhukov, and N. M. Matyukhin, Thermal Fields of Rod-Type Fuel Elements Situated in Regular Lattices Streamlined by a Heat Carrier, *High Temp.*, Vol. 14, pp. 482–488, 1976.

217. P. A. Ushakov, A. V. Zhukov, and N. M. Matyukhin, Heat Transfer to Liquid Metals in Regular Arrays of Fuel Elements, *High Temp.*, Vol. 15, pp. 868–873, 1977.

218. P. A. Ushakov, Problems of Heat Transfer in Cores of Fast Breeder Reactors, *Heat Transfer and Hydrodynamics of Single-Phase Flow in Rod Bundles* (in Russian), collected papers, ed. V. M. Borishanskij and P. A. Ushakov, Izd. Nauka, Leningrad, 1979.

219. M. S. Kazimi and M. D. Carelli, Heat Transfer Correlations for Analysis of CRBRP Assemblies, CRBRP-ARD-0034, Westinghouse Electric Corp., Madison, Penn., 1976.

220. Y. S. Tang, R. D. Coeffield, Jr., and R. A. Markley, *Thermal Analysis of Liquid-Metal Fast Breeder Reactors*, Am. Nucl. Soc., 1978.

221. R. Möller and H. Tschöke, Steady-State, Local Temperature Fields with Turbulent Liquid Sodium Flow in Nominal and Disturbed Bundle Geometries with Spacer Grids, *Nucl. Eng. Design*, Vol. 62, pp. 69–80, 1980.

222. K. Rehme, Pressure Drop Correlations for Fuel Element Spacers, *Nucl. Technol.*, Vol. 17, pp. 15–23, 1973.

223. O. I. Korotaev, P. I. Puchkov, and E. D. Fedorovich, Hydraulic Resistance of Spacer Elements Made from Sleeves for Bundles of Rod-Type Fuel Elements, *Thermal Eng.*, Vol. 27, pp. 683–686, 1980.

224. A Ito and K. Mawatari, Ring Grid Spacer Pressure Loss, Experimental Study and Evaluation Method, *Thermal Hydraulics of Nuclear Reactors*, ed. M. Merilo, Vol. 2, pp. 972–978, Am. Nucl. Soc., Santa Barbara, Calif., 1983.

225. J. Vlček and P. Weber, The Experimental Investigation of Film Heat-Transfer Coefficient in the Fuel Element Spacers Area, ZJE-66, Škoda Works, Nucl. Power Plants Div., Information Centre Plzen, Č.S.S.R., 1970.

226. M. Hudina and N. Nöthiger, Experimental Study of Local Heat Transfer Under and Near Grid Spacers Developed for GCFR, TM-IN-526, Swiss Federal Inst. for Reactor Res., Würenlingen, Switzerland, 1973.

227. V. Krett and J. Majer, Temperature Field Measurement in the Region of Spacing Elements, ZJE-114, Škoda Works, Nucl. Power Construction Dept., Information Centre Plzen, Č.S.S.R., 1971.

228. J. Marek and K. Rehme, Experimentelle Untersuchung der Temperaturverteilung unter Abstandshaltern in glatten und rauhen Stabbündeln, KfK 2128, Kernforschungszentrum Karlsruhe, F.R.G., 1975.

229. J. Marek and K. Rehme, Heat Transfer in Smooth and Roughened Rod Bundles Near Spacer Grids, *Fluid Flow and Heat Transfer over Rod or Tube Bundles*, ed. S. C. Yao and P. A. Pfund, Am. Soc. Mech. Eng., New York, pp. 163–170, 1979.

230. M. A. Hassan, Wärmeübergang im Abstandshalterbereich gasgekühlter Stabbündel, Dr.-Ing. Thesis, Univ. of Karlsruhe, F.R.G., 1980; cf. KfK 2954, Kernforschungszentrum Karlsruhe, F.R.G., 1980.

231. M. A. Hassan and K. Rehme, Heat Transfer Near Spacer Grids in Gas-Cooled Rod Bundles, *Nucl. Technol.*, Vol. 52, pp. 401–414, 1981.

232. F. C. Engel and A. A. Bishop, Experimental Determination of Local Heat Transfer Coefficients in an Annular Sodium Flow Passage At and Near a Flow Obstruction, *Liquid-Metal Heat Transfer and Fluid Dynamics*, ed. J. C. Chen and A. A. Bishop, Am. Soc. Mech. Eng., New York, pp. 50–57, 1970.

233. K. Rehme, Systematische experimentelle Untersuchung der Abhängigkeit des Druckverlustes von der geometrischen Anordnung für längsdurchströmte Stabbündel mit Spiraldrahtabstandshaltern, Dr.-Ing. Thesis, Univ. of Karlsruhe, 1967; cf. External Report INR-4/68-16, Kernforschungszentrum Karlsruhe, 1968.

234. K. Rehme, Druckverlust in Stabbündeln mit Spiraldraht-Abstandshaltern, *Forsch. Ing.-Wes.*, Vol. 35, pp. 107–112, 1969.

235. M. A. Hamid and M. A. Quaijum, Pressure Measurements in Model Fuel Bundles for Fast Breeder Reactor, *The Nucleus*, Vol. 6, pp. 77–81, 1969.

236. K. Wakasugi and I. Kakehi, Pressure Drop Performance of Fuel Pin Bundle with Spiral Wire Spacer, *J. Nucl. Sci. Technol.*, Vol. 8, pp. 167–172, 1971.

237. H. Hoffmann, Experimentelle Untersuchungen zur Kühlmittelquervermischung und zum Druckabfall in Stabbündeln mit wendelförmigen Abstandshaltern. Einfluß der Ergebnisse auf die Auslegung von Brennelementen Schneller Natrium-gekühlter Reaktoren, Dr.-Ing. Thesis, Univ. Karlsruhe, F.R.G., 1973; cf. KfK 1843, Kernforschungszentrum Karlsruhe, F.R.G., 1973.

238. G. Cornet and G. Lamotte, Pressure Measurements in a SNR–Mark II Fuel Element with Spiral Wires, Test Results and Interpretation, BN-7707-02, Belgonucleaire, Bruxelles, Belgium, 1977.

239. A. Sarno, P. Gori, and G. Andalo, Local Pressure and Velocity Measurements in a Water 19-Rod Bundle Using a Wire Wrap Spacer System, RT/ING (79)36, Com. Naz. Energia Nucl., Bologna, Italy, 1979.

240. D. Tirelli, Pressure Drop Characteristics of PEC Fuel Element, *2nd IAHR Int. Spec. Mtg. on Thermal-Hydraulics in LMFBR Rod Bundles*, Roma, 1982.

241. C. G. McAreavy and C. Bettes, A Review of Theoretical and Experimental Studies Underlying the Thermal-Hydraulic Design of Fast Reactor Fuel Elements, *Spec. Mtg. on Thermodynamics of FBR Fuel Subassemblies under Nominal and Non-nominal Operating Conditions*, IWGFR/29, International Working Group on Fast Reactors, IAEA, Karlsruhe, pp. 7–22, 1979.

242. H. Tschöke, Experimentelle Bestimmung des Druckverlustes an einem 37-Stabbündel aus Rohren mit 6 integralen Wendelrippen pro Stab als Abstandshalter, KfK 1038, Kernforschungszentrum Karlsruhe, F.R.G., 1970.

243. C. Chiu, N. E. Todreas, and W. M. Rohsenow, Pressure Drop Measurements in LMFBR Wire-Wrapped Blanket Bundles, *Trans. Am. Nucl. Soc.*, Vol. 22, pp. 541–543, 1979.

244. F. C. Engel, R. A. Markley, and A. A. Bishop, Laminar Transition and Turbulent Parallel Flow Pressure Drop across Wire-Wrap-Spaced Rod Bundles, *Nucl. Sci. Eng.*, Vol. 69, pp. 290–296, 1979.

245. S.-K. Cheng and N. E. Todreas, Fluid Mixing Studies in a Hexagonal 37 Pin Wire-Wrapped Rod Bundle, DOE/ET/37240-96TR, M.I.T., Cambridge, Mass., 1982.

246. K. Marten, S. Yonekawa, and H. Hoffmann, Experimental Investigations on Pressure Drop in Tightly Packed Bundles with Wire Wrapped Rods, *2nd IAHR Int. Spec. Mtg. on Thermal-Hydraulics in LMFBR Rod Bundles,* Roma, 1982.

247. K. Marten, S. Yonekawa, H. Hoffmann, and W. Hame, Experimentelle Bestimmung von Druckverlustbeiwerten in eng gepackten hexagonalen Stabbündeln mit Wendeldraht-Abstandshaltern. Empirische Korrelation zur Berechnung der Druckverlustbeiwerte für enge und weite Stabpackungen, KfK 4038, Kernforschungszentrum Karlsruhe, F.R.G., 1986.

248. E. H. Novendstern, Turbulent Flow Pressure Drop Model for Fuel Rod Assemblies Utilizing a Helical Wire-Wrap Spacer System, *Nucl. Eng. Design*, Vol. 22, pp. 19–27, 1972.

249. T. C. Reihman, *An Experimental Study of Pressure Drop in Wire-Wrapped FFTF Fuel Assemblies*, BNWL-1207, Battelle Northwest Lab., Richland, Wash., 1969.

250. K. Rehme, Messung von Reibungsbeiwerten an längs angeströmten Stabbündeln mit verschiedenen Abstandshaltern, durchgeführt am Wasserversuchsstand des INR, unpublished internal document, EURFNR-14P, 1965.

251. P. Carajilescov and E. Fernandez y Fernandez, Semi-empirical Model for Friction Factors in LMFBR Wire-Wrapped Rod Bundles, *Thermal Hydraulics of Nuclear Reactors*, ed. M. Merilo, Am. Nucl. Soc., Santa Barbara, Calif., Vol. 2, pp. 1318–1325, 1983.

252. D. R. Spencer and R. A. Markley, Friction Factor Correlation for 217-Pin Wire-Wrap Spaced LMFBR Fuel Assemblies, *Trans. Am. Nucl. Soc.*, Vol. 39, pp. 1014–1015, 1980.

253. A. Efthimiadis, Mixed Convection and Hydrodynamic Modeling of Flows in Rod Bundles, Ph.D. Thesis, Nucl. Eng. Dept., M.I.T., Cambridge, Mass., 1984.

254. A. K. Mohanty and K. M. Sahoo, Laminar Convection in Wall Subchannels and Transport Rates for Finite Rod-Bundle Assemblies by Superposition, *Nucl. Eng. Design*, Vol. 92, pp. 169–180, 1986.

8

CONVECTIVE HEAT TRANSFER IN LIQUID METALS

C. B. Reed

Engineering Division
Argonne National Laboratory
Argonne, Illinois

8.1 INTRODUCTION

Liquid metals have significantly higher thermal conductivity and lower specific heat than ordinary liquids and gases. In convective heat transfer, this is reflected by the low Prandtl number (Pr) of liquid metals, compared to ordinary fluids. Typically, for liquid metals $0.003 \lesssim \mathrm{Pr} \lesssim 0.06$, while for other liquids and gases $\mathrm{Pr} \gtrsim 0.2$. Figure 8.1 shows relative magnitudes of Prandtl number for several fluids. In forced convection systems under laminar flow conditions, molecular conduction of heat controls the heat transfer process, irrespective of whether the coolant is a liquid metal or an ordinary fluid. Hence, there is no fundamental difference between the thermal behavior of the two types of fluids under these conditions, and accordingly nondimensional correlations developed to describe the heat transfer performance of ordinary fluids can be applied equally well to liquid metals in spite of their low Prandtl numbers.

Under turbulent flow conditions, however, eddy conduction of heat becomes important and the process of heat transfer is determined by both molecular and eddy

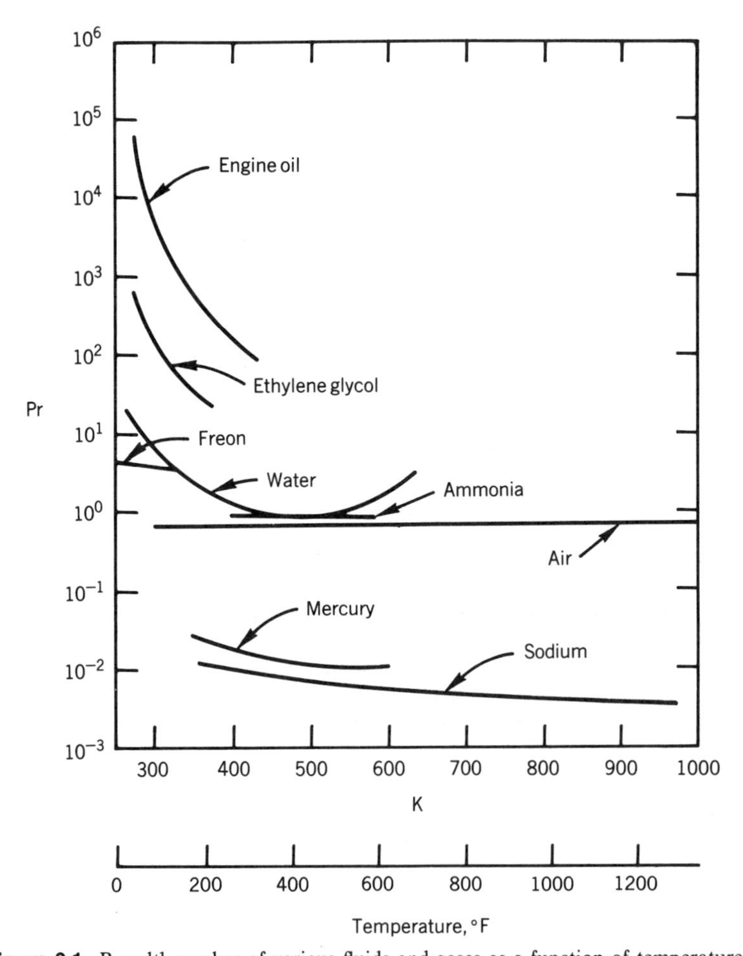

Figure 8.1. Prandtl number of various fluids and gases as a function of temperature.

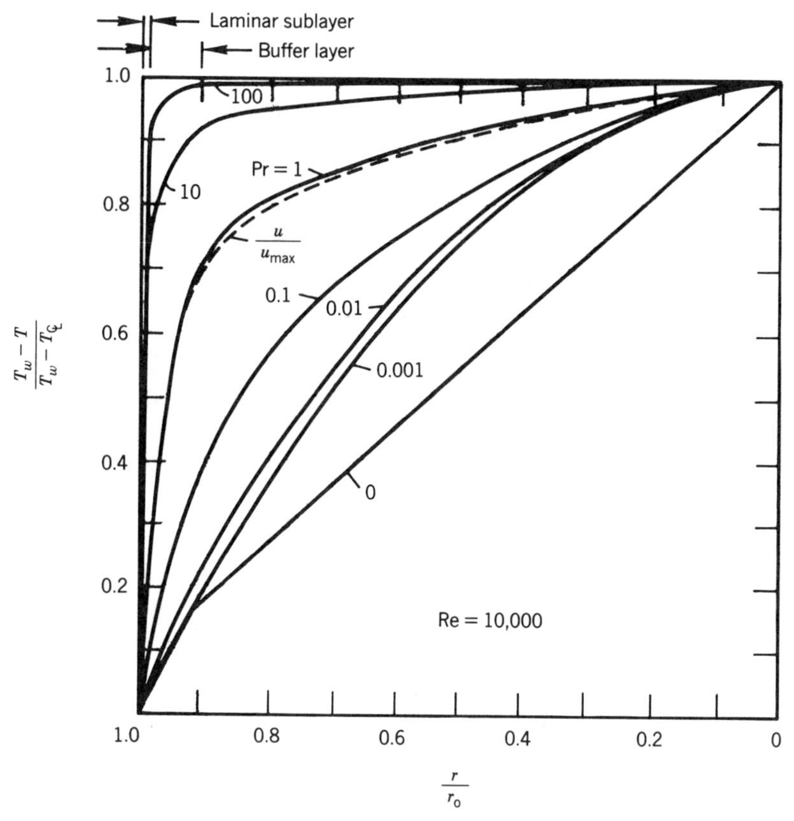

Figure 8.2. Effect of Prandtl number on the temperature profile for turbulent flow in a long pipe [41].

conduction over the various flow regions in the fluid stream. Whereas in ordinary fluids molecular conduction is only significant near the wall (in the laminar sublayer), in a liquid metal the magnitude of the molecular conductivity is of the same order as that of the eddy conductivity, and accordingly, the effects of molecular conduction are not felt only in the boundary layers, but extend well into the turbulent core of the fluid. (This Prandtl number effect on the fluid temperature profiles is shown in Fig. 8.2. In this context, the Prandtl number can be thought of as expressing the ratio of the viscous boundary-layer thickness to the thermal boundary-layer thickness.) Therefore the fundamental details of the heat transfer mechanism in liquid metals differ significantly from those observed in ordinary fluids, and as a result, relationships developed to calculate heat transfer coefficients for turbulent flows of ordinary fluids cannot be used.

A further consequence of the greater importance of molecular conduction of heat in turbulent liquid metal flow is that the concept of hydraulic diameter cannot be used so freely to correlate heat transfer data from systems which differ in configuration but retain a similar basic flow pattern. For example, in ordinary fluids, basic heat transfer data for flow through circular pipes can be used to predict Nusselt numbers (Nu) for flow parallel to a rod bundle by evaluating the hydraulic diameter for the latter and using this in the nondimensional correlations for the circular tube. Such methods are found to be invalid for liquid metal systems, and accordingly theoretical or experimental heat transfer relationships must be developed to deal with each specific configuration [1].

Available liquid metal heat transfer data show quite a bit of scatter, even more than ordinary convective heat transfer data. The earlier heavy-metal data (Hg and Pb-Bi) show the most variation. Several phenomena have been proposed to explain the scatter and lack of correlation with theoretical predictions. They include: nonwetting or partial wetting, gas entrainment, the possibility of oxides or other surface contaminants, and mixed convection effects. Following an extensive amount of experimental investigation on the effects of wetting on liquid metal heat transfer, a general consensus has been reached on the subject. This is that wetting or lack of wetting, in and of itself, does not significantly affect liquid metal heat transfer. However, nonwetting combinations of liquid metals and solid surfaces can suffer more readily from gas-entrainment problems at the solid-liquid interface; impurities and particles can more easily become trapped at a nonwetting solid-liquid interface, thus reducing heat transfer. Hence, care should be taken to avoid these problems in system designs. Also, recent, more detailed experiments show that mixed convection effects are an additional source of the differences in Nusselt number measurements. This topic is covered more thoroughly in Sec. 8.3.1.

Finally, in liquid metal systems, uniform wall temperature boundary conditions yield lower Nusselt numbers than do uniform wall heat flux boundary conditions for the same Péclet number (Pe). This is in contrast to ordinary fluid systems, having relatively high Pr, in which the two boundary conditions make little difference on Nu.

8.2 LIQUID METAL HEAT TRANSFER IN FULLY DEVELOPED LAMINAR FLOWS

As mentioned in Sec. 8.1, liquid metals behave as ordinary fluids in laminar heat transfer. Hence, the reader is referred to other chapters of this handbook which cover laminar heat transfer for the appropriate geometry. However, free convection becomes important at low Reynolds numbers (Re) in liquid metals and usually influences the heat transfer. Section 8.3.1 should be consulted for criteria regarding free-convection effects.

8.3 LIQUID METAL HEAT TRANSFER IN FULLY DEVELOPED TURBULENT FLOWS

8.3.1 Free Convection Distortion in Liquid Metal Heat Transfer

Buoyancy forces can become important in forced convection liquid metal heat transfer, especially for heavy metals such as mercury. Free convection distortion of temperature profiles in NaK have been observed by Schrock [2] at Re of 12,000 to 16,000, and in mercury at Re as high as 315,000 by Gardner [3], both in horizontal pipe flow. Figure 8.3 shows, as an example, some temperature profiles in horizontal mercury flow in which free convection distortion is present. Free convection effects were also observed by Kowalski [4] in vertical pipe flows of mercury at Re as high as 90,000. Figure 8.4 shows velocity profiles in vertical upward flow of mercury distorted by free convection effects at Re = 60,000. Experimental results show that the velocity profile distorts rapidly as the heat flux is increased; at high heat flux (i.e., high Z since $Z \propto q$), a

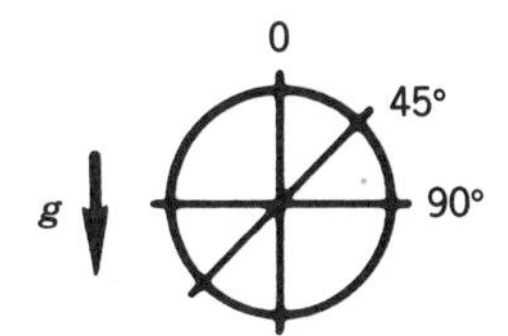

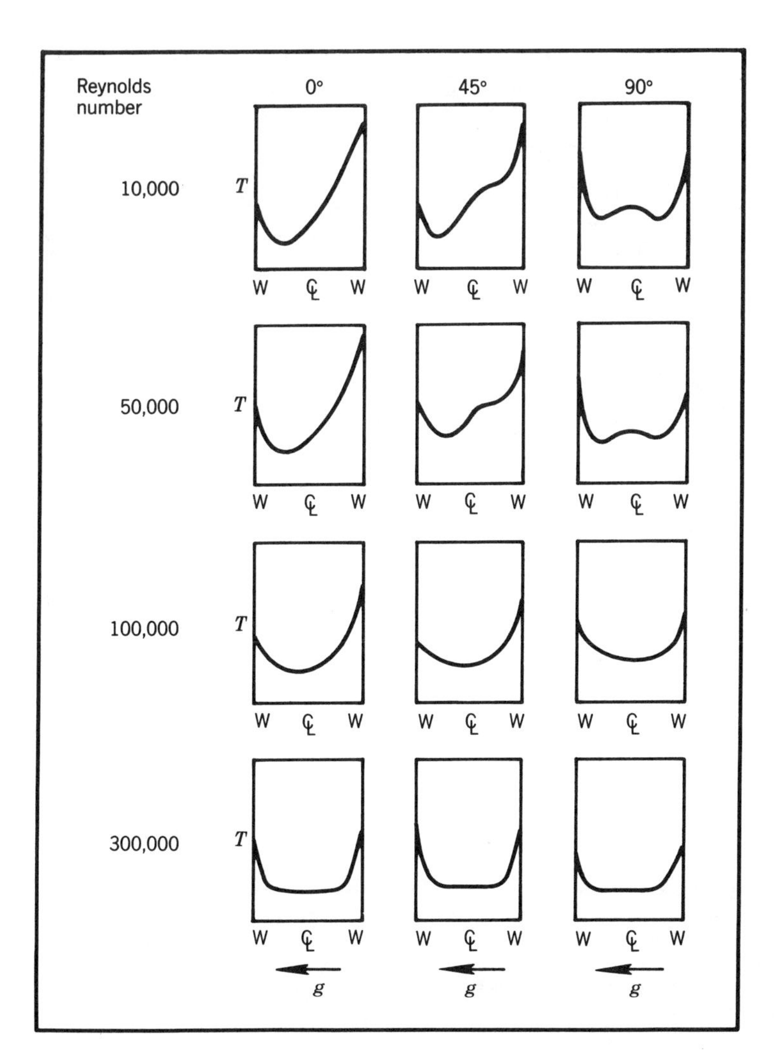

Figure 8.3. Temperature profiles in horizontal pipe flow of mercury, showing free convection distortion effects [3].

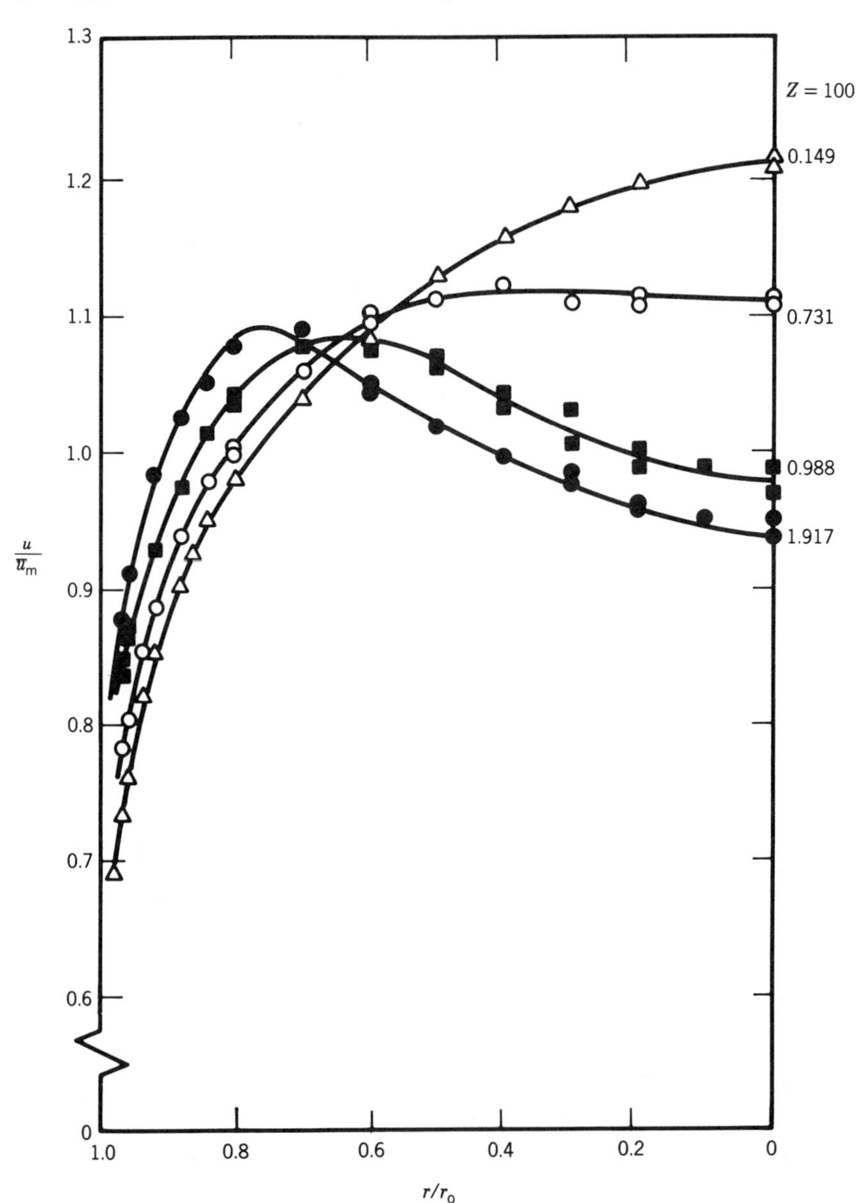

Figure 8.4. Velocity profiles in vertical pipe flow of mercury, showing free convection distortion effects [43].

limiting profile shape is approached, with the centerline velocity well below the mean. Hence, Nu in pipe flow depends on several variables, such as whether the pipe is vertical or horizontal, whether the flow is upward or downward, whether the fluid is heated or cooled, and the value of the Grashof number or Rayleigh number. Figure 8.5 shows variations in the local Nusselt number around the perimeter of a horizontal pipe having free-convection distortion.

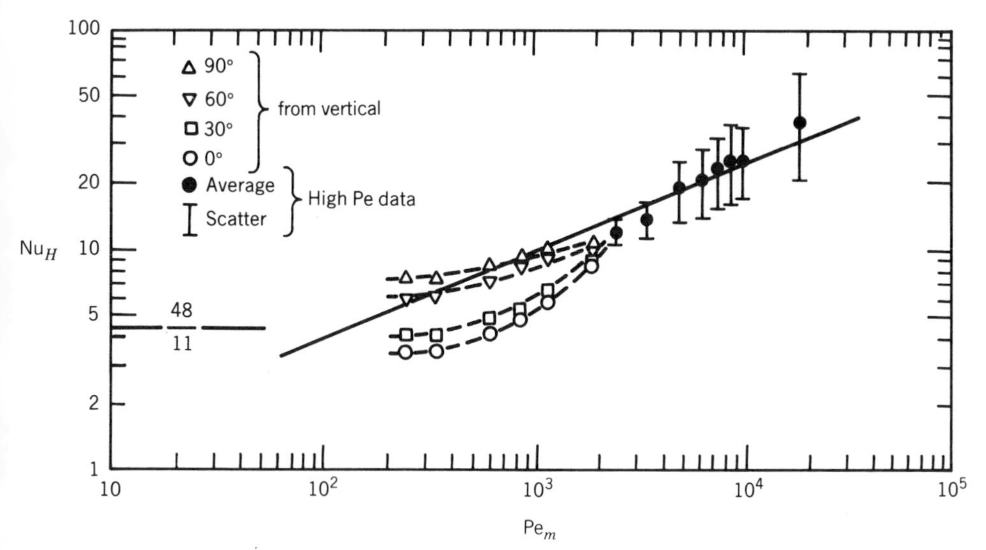

Figure 8.5. Local Nusselt number around the perimeter of a horizontal pipe flow of mercury, showing free convection distortion effects on Nu [3].

The following criterion, due to Buhr [5], can be used to estimate whether free convection will influence the Nusselt number in pipe flow:

$Z < 20 \times 10^{-4}$: insignificant free convection.
$Z > 20 \times 10^{-4}$: free convection affects forced convection Nu.

where

$$Z = \frac{\mathrm{Ra}'_m}{\mathrm{Re}_m} \frac{D_h}{L} \quad \text{and} \quad \mathrm{Ra}'_m = \mathrm{Gr}'_m \, \mathrm{Pr}_m$$

The prime is used to distinguish the Grashof number using the axial temperature difference from the usual one using the radial temperature difference:

$$\mathrm{Gr}'_m = \frac{D_h^3 \beta g \, \Delta T}{\nu^2} \qquad \text{where} \quad \Delta T = \frac{dT}{dx} D_h .$$

This criterion is roughly valid for both vertical and horizontal flows. The fluid properties should be evaluated at the bulk mean temperature $T_m = (T_{\text{in}} + T_{\text{out}})/2$.

When the flow is dominated by forced convection, the following qualitative picture of the effects of free convection on the forced convection heat transfer is observed. Buoyancy forces enhance turbulent heat transfer to liquid metals for downward flow and retard heat transfer for upward flow. Buoyancy forces influence the convective heat transfer indirectly through their effect on the shear stress distribution in the liquid metal. In vertically downward flow, the buoyancy forces work against the mean-flow direction and increase the isothermal shear stress near the heated surface. For vertically upward flow, the shear stress is reduced by the buoyancy forces. The increase in shear stress in the downward-flowing case leads to an increase in turbulence production,

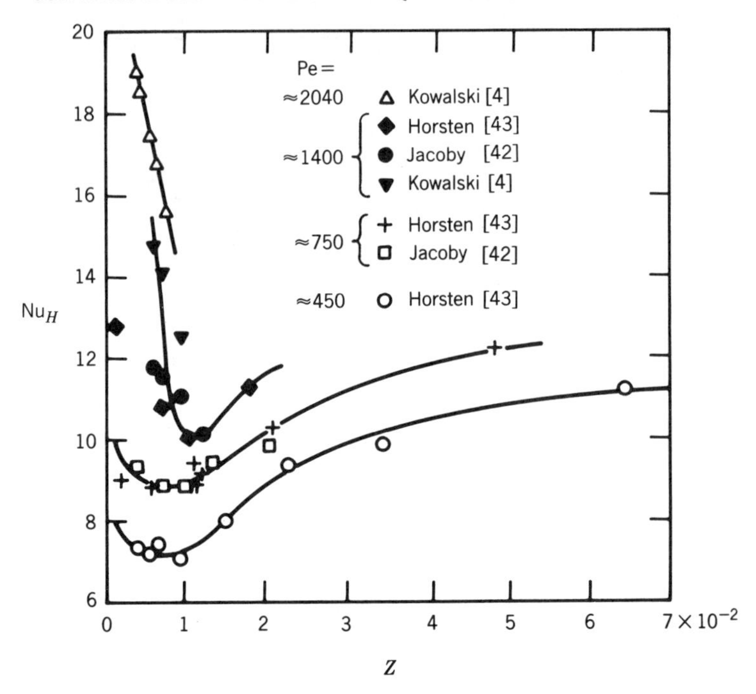

Figure 8.6. Dependence of Nusselt number on parameter Z, showing free convection effects on Nu in vertically upward pipe flow [4, 42, 43].

which will enhance the heat transfer. The opposite is true for upward flow. For conditions of strong influence of buoyancy, enhancement of heat transfer occurs for both upward and downward flow, the mechanism being buoyancy-induced production of turbulence [6]. These effects of buoyancy on heat transfer to liquid metals are less marked than for ordinary fluids because of the reduced importance of turbulent eddy conduction at low Pe.

This qualitative picture is supported by the data in Fig. 8.6, which show that the Nusselt number is a strong function of the parameter Z—initially decreasing, then increasing as Z is increased. The parameter Z is proportional to the heat flux. As the buoyancy forces begin to become comparable to the inertia forces of the flow, Nu reaches a minimum, begins to increase, and finally appears to reach a limiting value.

It can be seen that the effects of free convection distortion can affect the measured value of Nu and could, therefore, explain some of the well-known scatter in the Nu data for liquid-metal heat transfer. Needless to say, the effects of free convection distortion should be considered when dealing with liquid metal heat transfer systems.

8.3.2 Pipe Flow

For flows in which free convection effects are not important (see Sec. 8.3.1), and which are free from surface contamination and gas entrainment, the following relation is recommended for uniform-heat-flux conditions:

$$\mathrm{Nu}_H = 5.0 + 0.025\,\mathrm{Pe}_m^{0.8} \tag{8.1}$$

This equation is based on the well-known Lyon-Martinelli [7] equation with the leading coefficient adjusted to best fit the very large body of experimental data. Equation (8.1) was first proposed by Subbotin et al. [8]. It is valid for $\mathrm{Pe}_m > 100$ and $L/D_h > 60$; the subscript m indicates that the fluid properties are evaluated at the mean bulk temperature $T_m = (T_{\mathrm{in}} + T_{\mathrm{out}})/2$.

There is little experimental information for the case of uniform wall temperature; only Sleicher et al. [9] and Gilliland et al. [10] (taken from Azer and Chao [11]) are judged reliable. However, there is ample analytical evidence that Nusselt numbers for uniform wall temperature are lower than for uniform heat flux (including the theoretical laminar asymptotes: Nu = 4.36 for uniform heat flux and Nu = 3.66 for uniform wall temperature). Hence, the following equation for the uniform-wall-temperature boundary condition is recommended under the same other conditions as for Eq. (8.1):

$$\mathrm{Nu}_T = 3.3 + 0.02\,\mathrm{Pe}_m^{0.8} \tag{8.2}$$

This equation was developed by the author to be recommended here because it represents a good fit to the data of Sleicher et al. [9] and Gilliland et al. [10] (see Fig. 8.7) and it retains the simple, classical dependence on $\mathrm{Pe}_m^{0.8}$. There is an analytical basis for an explicit Pr dependence in Nu correlations for liquid metals; however, there is little quantitative experimental support to justify including Pr in correlations presented here. The reader is referred to Azer and Chao [11], Leslie [12], Sleicher et al. [9], and El Hadidy et al. [13] for examples of analytical treatments producing explicit Nu dependences on Pr, velocity-profile, and temperature-profile assumptions. Equation (8.2) is recommended for $\mathrm{Pe}_m > 100$ and $L/D_h > 60$.

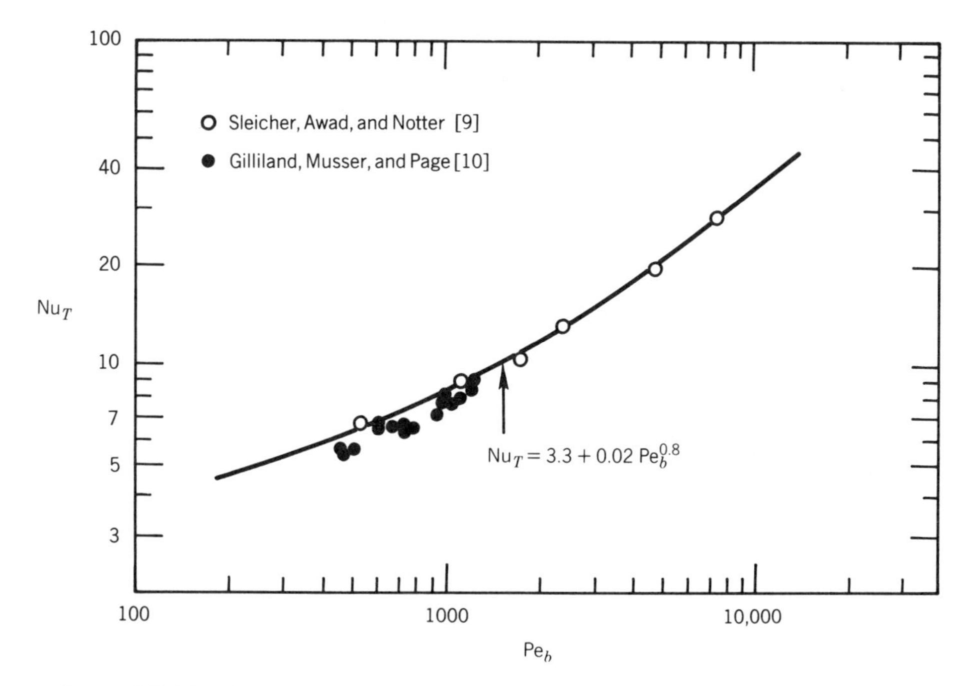

Figure 8.7. Nusselt number data and correlation for uniform wall temperature boundary condition in horizontal [9] and vertical [10] pipe flow.

8.3.3 Flow in Annuli

According to Kottowski [14], for $r_2/r_1 < 1.4$ the following modified Lyon equation fits experimental data on heat flow across either the inner wall (r_1) or the outer wall (r_2) in annular flow:

$$\mathrm{Nu} = 0.75\left(\frac{r_2}{r_1}\right)^{0.3}\left(7.0 + 0.025\,\mathrm{Pe}_m^{0.8}\right) \tag{8.3}$$

where the hydraulic diameter of the annulus must be used ($D_h = D_2 - D_1$). For $1.4 < r_2/r_1 < 10$, this equation is expected to be valid as well, though for heat transfer across the outer wall only, the inner wall being adiabatic.

More complex expressions for liquid metal heat transfer in annuli are given by Dwyer [1] and compared extensively with the data there.

8.3.4 Flow between Parallel Plates

Duchatelle and Vautrey [15] present the only experimental results for liquid metal heat transfer between parallel plates. The experimental conditions represented unilateral heat transfer (heat flux through one wall, adiabatic at the other wall) under conditions of uniform heat flux. Those results are well represented by the empirical expression

$$\mathrm{Nu}_H = 5.85 + 0.000341\,\mathrm{Pe}_m^{1.29} \tag{8.4}$$

For bilateral heat transfer (heat flux through both walls) under uniform and equal heat-flux conditions, the following equation, proposed by Dwyer [16], is recommended:

$$\mathrm{Nu}_H = 9.49 + 0.0596\,\mathrm{Pe}_m^{0.688} \tag{8.5}$$

More extensive relationships for cases where the heat fluxes are unequal can be found in Dwyer [1].

8.3.5 Single Cylinder in 90° Cross Flow

For a cylinder with constant heat flux, the local heat transfer conditions vary around the cylinder from the forward to the rear stagnation point, and consequently the local Nusselt number also varies. The Nusselt number is highest at the forward stagnation point and reaches a minimum at or near the rear stagnation point, as shown in Fig. 8.8. This is because the thermal boundary layer increases in thickness from the forward to the rear stagnation point. Figure 8.8 also shows the local Nusselt number for air flow. The fundamental influence of the high thermal conductivity of the liquid metal is reflected in the difference in shape between the Nu_ϕ profile for the liquid metal and the one for air. It can be seen that the fluid mechanics of the flow around the cylinder has a much greater influence on the Nu_ϕ for the air than for the liquid metal. An average Nusselt number is defined, based on the heat flux, the average wall temperature $\overline{T}_w$ of the cylinder, the bulk mean temperature T_m of the fluid, and the outside diameter of the cylinder. The fluid properties should be evaluated at the film temperature, defined as $T_f = (\overline{T}_w + T_m)/2$. Use of the film temperature in this way minimizes the effect of variable fluid properties. Variables so evaluated have a subscript f, e.g., Pe_f. The data of Andreevskii [17] for this case are well correlated by the following equation:

$$\mathrm{Nu}_f = 0.65\,\mathrm{Pe}_f^{0.5} \tag{8.6}$$

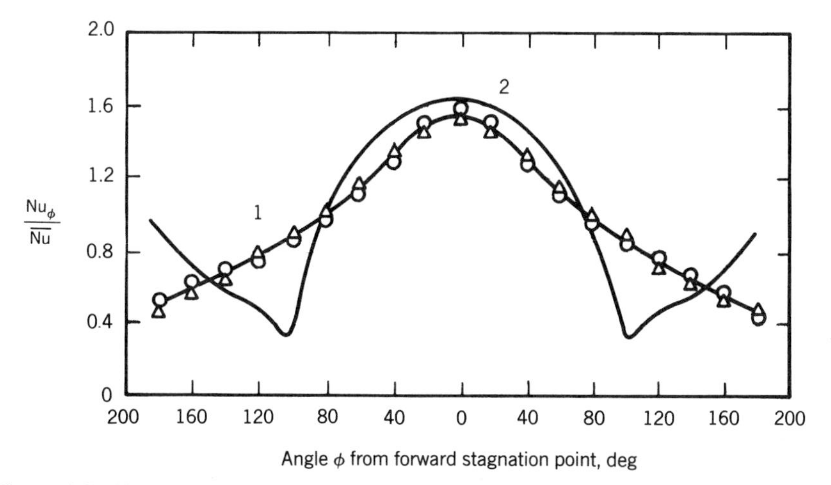

Figure 8.8. Change of relative Nusselt number around the perimeter of a single cylinder in cross flow with uniform heat flux. Curve 1: sodium (O: Pe = 0.25, Re = 4000; △: Pe = 125, Re = 20,000). Curve 2: air (Re = 16,000) [17].

The empirical coefficient 0.65 in the above equation is below the theoretical values—1.015 for uniform wall temperature and 1.145 for uniform wall heat flux—given by Hsu [18]. In both experimental work and commercial practice, however, the thermal boundary condition is neither, but something in between. The coefficient 0.65 should represent all practical situations; hence, the above equation is recommended for engineering applications.

8.3.6 Flow in Tube Bundles

Flow across a Staggered Tube Bank. Results for mercury and for sodium can be correlated to within ±12% over the Péclet number range 50 to 4000 by the following empirical equation given by Kottowski [14]:

$$\mathrm{Nu}_f = \mathrm{Pe}_f^{0.5} \tag{8.7}$$

which is recommended for use in such conditions. Nu_f is based on a mean heat transfer coefficient obtained by dividing the mean heat flux from the tube by the circumferential average of the temperature differences between tube and bulk fluid at points equispaced around the periphery of the tube. For each tube, the bulk temperature of the flowing stream is evaluated at the location in the tube bank corresponding to the axis of the tube in question. The cross-flow velocity in Pe_f is based on the minimum flow area.

Flow Parallel to Tube Bundles (In-Line Flow). Theoretical expressions for calculating the rate of heat transfer to a liquid metal in parallel flow through a bundle of rods have been developed. These analyses follow the general method of Lyon [7], assuming an annular model in which the boundary of zero shear (hexagonal for rods on triangular pitch, square for square pitch) surrounding a typical rod is replaced by a circular boundary enclosing the same area. This circular boundary is then treated as the circle of zero shear within a larger annulus, thus enabling the flow distribution within

TABLE 8.1 Slug Flow Nusselt Numbers for Flow through Tubes of Simple Geometrical Shape[a]

Geometry	Boundary Condition[b]	$Nu_{s,m}$
Circle	(a)	5.80[c]
	(b)	8.00[c]
Square	(a)	4.93
	(b)	7.03
Equilateral triangle	(b)	6.67
Infinite slot	(a)	9.87
	(b)	12.00[d]
Infinite slot with	(a)	4.93
one wall insulated	(b)	6.00[d]
90° isosceles triangle	(b)	6.55

[a] From Hartnett and Irvine [19].
[b] Boundary condition (a): constant wall temperature. Boundary condition (b): Heat input per unit length constant and wall temperature constant around the periphery of the duct at a given axial position.
[c] These values are included for completeness of the table as taken from [19]. When substituted into Eq. (8.8), the resulting expressions do not agree with Eqs. (8.1) and (8.2). Equations (8.1) and (8.2) are recommended because they are based on a fit to the experimental data.
[d] These values are included for completeness of the table as taken from [19]. Equations (8.4) and (8.5) are recommended because they are based on a fit to the experimental data.

the circle to be determined from annulus data. The other principal assumptions are constant heat flux and fully developed turbulent flow.

Chapter 7 deals extensively with heat transfer over tube bundles, and Sec. 7.3.7 covers fully developed heat transfer to liquid metals in turbulent flow through tube bundles specifically. The reader is referred to that section for the detailed correlations to be used for flow parallel to tube bundles. Section 7.3.8 should be consulted regarding the thermal entrance length in liquid metal-cooled tube bundles.

8.3.7 Other Channel Shapes

Average Nusselt numbers for flow in channels of noncircular shape can be estimated using an equation initially proposed by Hartnett and Irvine [19] but later modified by Kottowski [14]:

$$Nu = \tfrac{2}{3} Nu_{s,m} + 0.025\, Pe_m^{0.8} \tag{8.8}$$

where $Nu_{s,m}$ is the Nusselt number for slug flow. Values of $Nu_{s,m}$ are given in Table 8.1. This equation is valid when free convection effects are negligible (see Sec. 8.3.1) and when heat transfer surfaces are clean and there is no gas entrainment. The fluid properties are evaluated at the bulk mean temperature, as described in Sec. 8.3.2.

8.4 THERMAL ENTRANCE LENGTHS

Chen and Chiou [20] have presented a rather extensive set of analytical results for thermal entrance effects in both laminar and turbulent pipe flow of liquid metals. Both

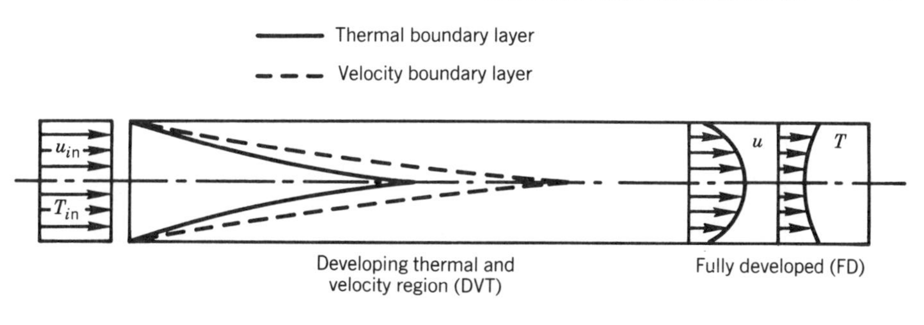

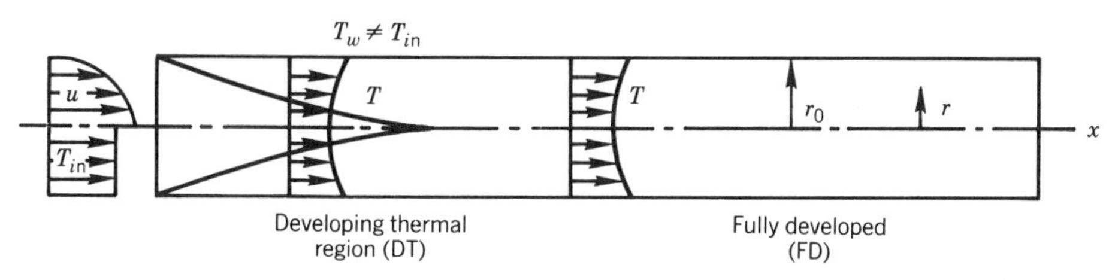

Figure 8.9. Regions for thermal entrance length analysis in pipe flow [20].

cases of constant heat flux and constant wall temperature were analyzed. A modified van Driest–Cebeci mixing-length model was employed for the analysis of turbulent flow. Three flow regions, shown in Fig. 8.9, were considered: fully developed, developing thermal, and developing thermal and velocity. In the fully developed region (FD), both velocity and temperature profiles are developed. In the developing thermal region (DT), the velocity profile is fully developed but the temperature profile is developing. In the developing thermal and velocity region (DTV), both the velocity and temperature profiles are developing.

8.4.1 Laminar Entrance Lengths

Uniform Wall Temperature. Figure 8.10 shows the predictions of Chen and Chiou [20] for the DT region in the form of $\mathrm{Nu}_{x,T}$ vs. $(x/D_h)/\mathrm{Pe}_m$ (the inverse Graetz number) for both air and NaK. These predictions agree well with previous predictions of Kays [21] for air; however, there are no data available for liquid metals with which to compare the predictions.

Uniform Wall Heat Flux. Figure 8.11 is a plot from Chen and Chiou [20] for the DTV region. These predictions are for $\mathrm{Pr}_m = 0.02$ and agree well with previous predictions of McMordie and Emery [22]; however, no data are available. In the DTV region, $\mathrm{Nu}_{x,H}$ is a function of both the Graetz number $[\mathrm{Pe}_m/(x/D_h)]$ and Pr_m.

8.4.2 Turbulent Entrance Lengths

Uniform Wall Temperature. Figure 8.12 presents predictions from Chen and Chiou [20] of $\mathrm{Nu}_{x,T}/\mathrm{Nu}_T$ vs. x/D_h for the DT region for two values of Re. Data from Awad [23] compare well with the predictions. It can be seen in this case that the entrance

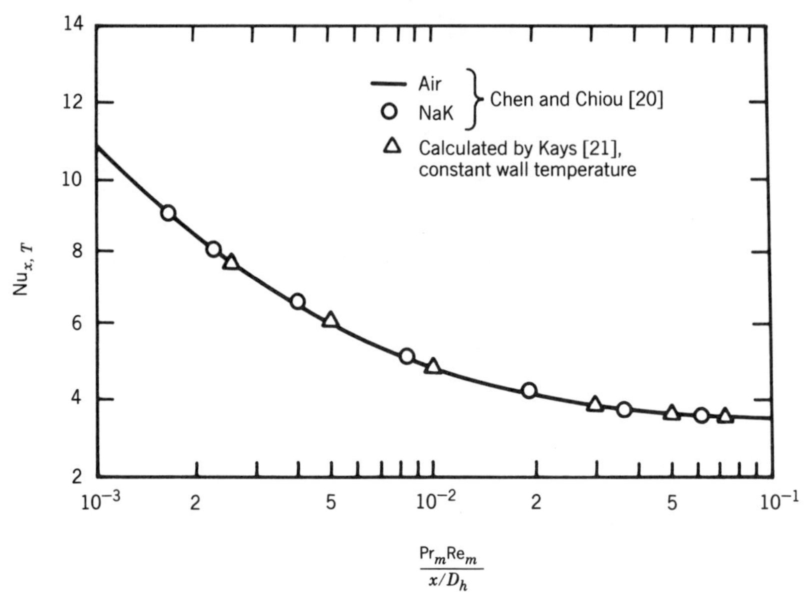

Figure 8.10. Laminar pipe flow heat transfer in the developing thermal (DT) region with isothermal wall boundary condition [20].

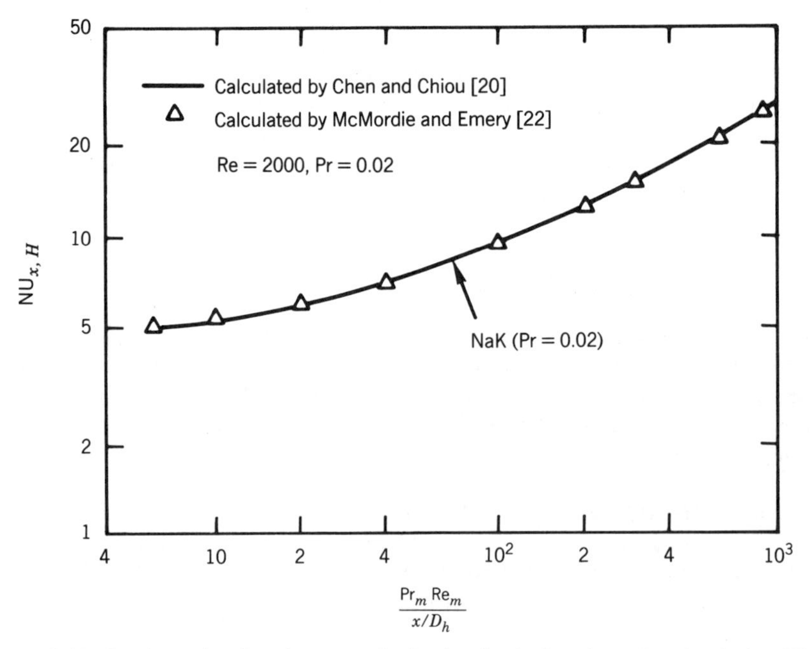

Figure 8.11. Laminar pipe flow heat transfer in the developing thermal and velocity (DTV) region with constant wall heat flux [20].

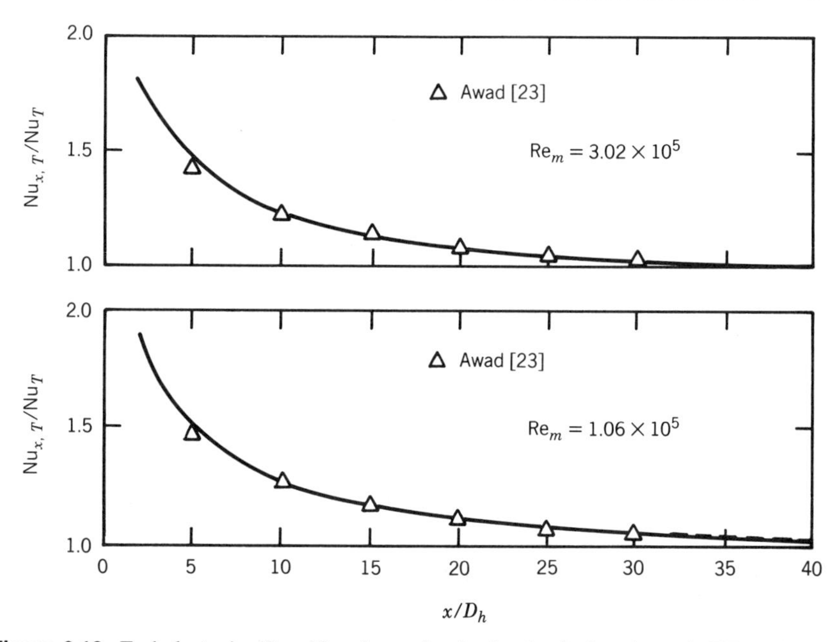

Figure 8.12. Turbulent pipe flow Nusselt number in the developing thermal (DT) region with constant wall temperature [20].

effects last for only 30 to 40 diameters. In the DT region, Chen and Chiou [20] give the following expressions for the local and average Nusselt numbers, respectively:

$$\frac{Nu_{x,T}}{Nu_T} = 1 + \frac{2.4}{x/D_h} - \frac{1}{(x/D_h)^2} \qquad \text{for} \quad x/D_h > 2 \text{ and } Pe_m > 500 \tag{8.9}$$

and

$$\frac{Nu_{m,T}}{Nu_T} = 1 + \frac{7}{L/D_h} + \frac{2.8}{L/D_h} \ln\left(\frac{L/D_h}{10}\right) \qquad \text{for} \frac{L}{D_h} > 2 \text{ and } Pe_m > 500. \tag{8.10}$$

In the DTV region, the local Nusselt number is given by the following expression:

$$\frac{Nu_{x,T}}{Nu_T} = 0.88 + \frac{2.4}{x/D_h} - \frac{1.25}{(x/D_h)^2} - \frac{40 - (x/D_h)}{190} \qquad \text{for} \quad 2 \le \frac{x}{D_h} < 35, \tag{8.11}$$

and the average Nusselt number is given by

$$\frac{Nu_{m,T}}{Nu_T} = 0.91 + \frac{5}{L/D_h} + \frac{1.86}{L/D_h} \ln\left(\frac{L/D_h}{10}\right) \qquad \text{for } 2 \le \frac{L}{D_h} \le 35 \tag{8.12}$$

Uniform Wall Heat Flux. In the DT region, Chen and Chiou [20] recommend using Eqs. (8.9) and (8.10) for the local and average Nusselt numbers, respectively, after substituting the uniform wall heat flux subscript for the uniform wall temperature subscripts.

In the DTV region, Chen and Chiou [20] give the following expressions:

$$\frac{\mathrm{Nu}_{x,H}}{\mathrm{Nu}_H} = 0.88 + \frac{2.4}{x/D_h} - \frac{1.25}{(x/D_h)^2} \qquad \text{for} \quad 2 \leq \frac{x}{D_h} < 35 \tag{8.13}$$

and

$$\frac{\mathrm{Nu}_{m,H}}{\mathrm{Nu}_H} = 1 + \frac{5}{L/D_h} + \frac{1.86}{L/D_h}\ln\left(\frac{L/D_h}{10}\right) \qquad \text{for } 2 \leq \frac{L}{D_h} \leq 35 \tag{8.14}$$

The analysis of Chen and Chiou [20] leading to Eqs. (8.9) through (8.14) was carried out for $0.004 \lesssim \mathrm{Pr}_m \lesssim 0.1$.

8.4.3 Influence of Axial Conduction in the Thermal Entrance Region

Recent numerical work by Lee [24, 25] found that below $\mathrm{Pe}_m \simeq 100$, axial heat conduction becomes important for both uniform wall heat flux and uniform wall temperature in the thermal entrance region. Above $\mathrm{Pe}_m \simeq 100$ it is negligible in the thermal entrance region, and by definition it is negligible for all Pe_m in the thermally fully developed region.

8.5 EFFECT OF VARIABLE PROPERTIES

When large temperature differences and high heat fluxes are encountered, the effect of physical property variation on the Nusselt number can become important. Chen and Chiou [20] considered this issue analytically for the case of fully developed pipe flow. Their results for the Nusselt number were presented in the following way:

$$\mathrm{Nu} = \mathrm{Nu}_0 \left(\frac{T_b}{T_{\mathrm{in}}}\right)^n \tag{8.15}$$

where the subscript zero refers to a Nusselt number calculated assuming constant physical properties evaluated at T_b. The exponent n is a function of the working fluid, the temperature range, and whether heating or cooling is being considered. Table 8.2 summarizes the resulting expressions for n obtained by Chen and Chiou [20] for two specific liquid metals: Na and NaK eutectic. These results should be used with considerable caution, since no data are available for comparison.

8.6 NATURAL CONVECTION HEAT TRANSFER IN LIQUID METALS

The very high thermal conductivity of liquid metals has a major effect on natural convection heat transfer as well. Thermal boundary-layer thicknesses in liquid metals are many times greater than in ordinary fluids, as can be seen from Fig. 8.13, where the dimensionless thermal boundary-layer thickness η is roughly 11 times greater for $\mathrm{Pr} = 0.01$ than for $\mathrm{Pr} = 10$. The influence of the Prandtl number on velocity profiles in

TABLE 8.2 Predicted Values of *n* for Use in Eq. (8.15)[a]

n	T_b Range, K	For
	SODIUM	
	CONSTANT WALL HEAT FLUX	
$\exp(5.9 \times 10^{-3}\ T_b - 6.9)$	600 to 1000	Heating
0	370 to 600	Heating
0.25	370 to 1000	Cooling
	CONSTANT WALL TEMPERATURE	
$0.08 + 2.2 \times 10^{-4}\ T_b$	600 to 1000	Heating
0.08	370 to 600	Heating
0.16	370 to 1000	Cooling
	NaK EUTECTIC	
	CONSTANT WALL HEAT FLUX	
−0.24	400 to 1000	Heating
−0.15	400 to 1000	Cooling
	CONSTANT WALL TEMPERATURE	
−0.2	400 to 1000	Heating and cooling

[a] Chen and Chiou [20].

natural convection is also marked, as shown in Fig. 8.14. It can be seen by looking at the velocity profiles in Fig. 8.14 that the larger thermal boundary-layer thickness for liquid metals results in a much thicker velocity profile as well. This is because, of course, the buoyancy forces are the driving body forces on the fluid motion. And in liquid metals, the buoyancy forces extend much farther away from the heated surface than they do in ordinary fluids, thus creating a wider velocity profile.

Whereas in ordinary fluids the Nusselt number in natural convection flows is a function of $Gr_f Pr_f$ ($= Ra_f$, the Rayleigh number), in liquid metal flows (where $Pr_f \ll 1$) the Nusselt number becomes a function of $Gr_f Pr_f^2$ (often referred to as the Boussinesq number).

The reader is cautioned in this Sec. 8.6 to ascertain carefully whether it is the average Nusselt number or the local Nusselt number that is being considered and to use the recommended correlations accordingly. Also, in this section, all fluid properties should be evaluated at the film temperature (the average of the surface temperature and the bulk mean fluid temperature). Use of the film temperature in this way minimizes the effect of variable fluid properties according to Sparrow and Gregg [26]. Variables so evaluated will bear a subscript f, e.g., $Nu_{x,H,f}$.

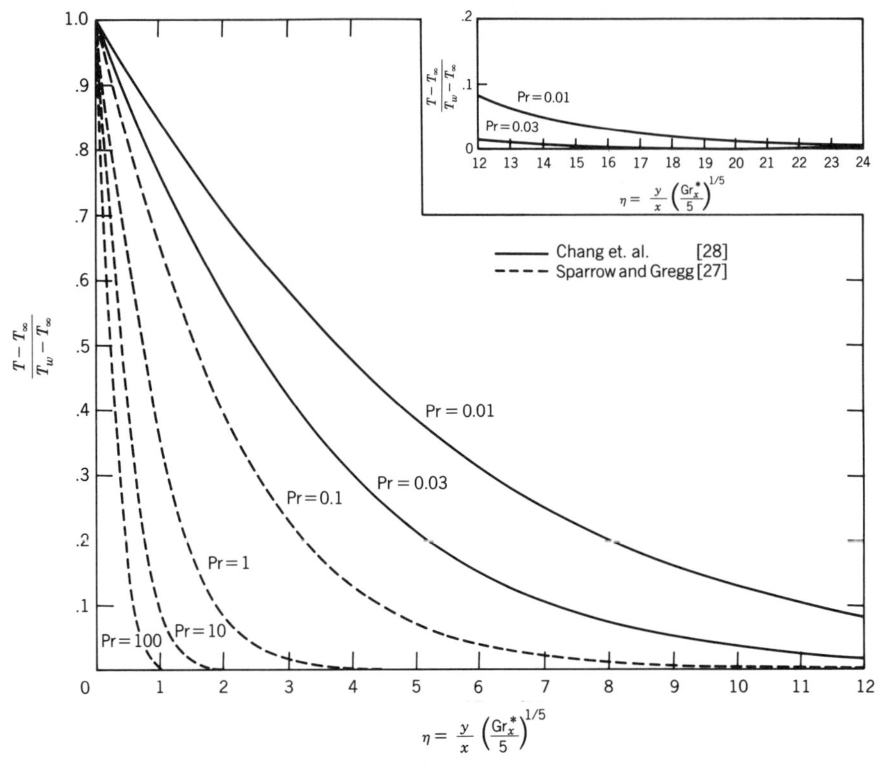

Figure 8.13. Dimensionless temperature distributions for various Prandtl numbers in natural convection from a heated vertical surface with constant heat flux [28].

8.6.1 Vertical Plates in Laminar Flow

Uniform Heat Flux. Numerical work by Sparrow and Gregg [27] and Chang et al. [28] can be cast into the following equation for the local Nusselt number:

$$\mathrm{Nu}_{x,H,f} = \left(\frac{\mathrm{Gr}^*_{x,f}}{5}\right)^{1/5} \frac{1}{\Theta(0)} \tag{8.16}$$

where $\mathrm{Gr}^*_{x,f}$ is a modified Grashof number defined as

$$\mathrm{Gr}^*_{x,f} = \frac{g\beta q x^4}{k\nu^2} \tag{8.17}$$

(which is more convenient for results obtained at uniform heat flux). $\Theta(0)$ is a dimensionless temperature difference evaluated at the wall. Values of $\Theta(0)$ as a function of Pr_f are listed in Table 8.3. If one fits a relationship of the form

$$\Theta(0) = A\,\mathrm{Pr}_f^{-2/5} \tag{8.18}$$

to the Table 8.3 values and substitutes the resulting expression into Eq. (8.16), the

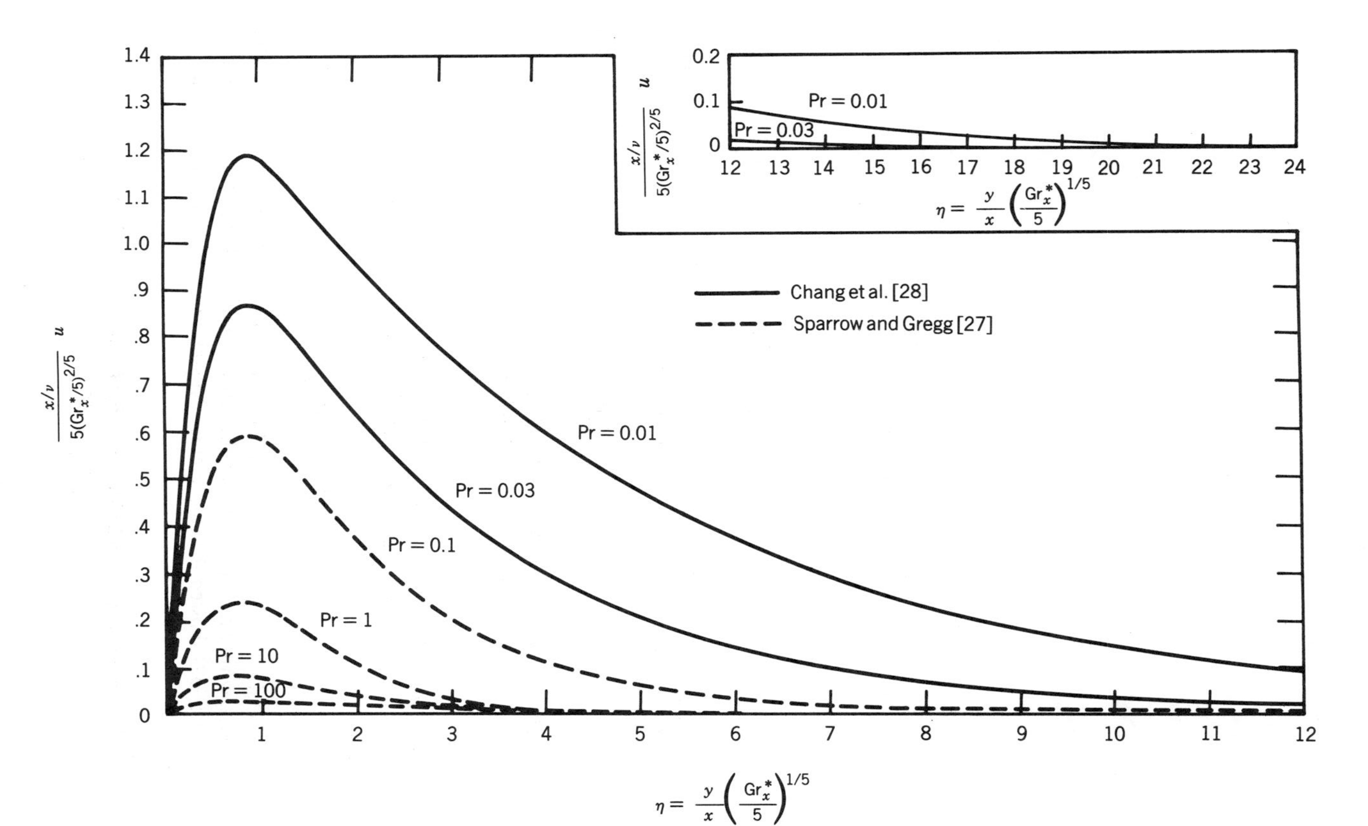

Figure 8.14. Dimensionless velocity distributions for various Prandtl numbers in natural convection from a heated vertical surface with constant heat flux [28].

TABLE 8.3 Predicted Values of Θ(0) for Use in Eq. (8.16)

Pr_f	$\Theta(0)$	Ref.
0.01	6.304	[28]
0.03	4.198	[28]
0.10	2.751	[27, 28]

result is as follows:

$$\mathrm{Nu}_{x,H,f} = \left(\frac{\mathrm{Gr}^*_{x,f}\mathrm{Pr}_f^2}{5}\right)^{1/5}\frac{1}{A} \tag{8.19}$$

which follows the classical dependence on the Boussinesq number mentioned earlier. The value of A is found to be unity within a few percent. Sheriff and Davies [29] used this method to arrive at the following relation, which fits most of the available data within $\pm 7\%$ and is therefore recommended:

$$\mathrm{Nu}_{x,H,f} = 0.732\left(\mathrm{Gr}^*_{x,f}\mathrm{Pr}_f^2\right)^{1/5} \tag{8.20}$$

Sparrow and Gregg [27] showed that the average Nusselt number for a plate of length L is related to the local Nusselt number at $x = L$ by

$$\mathrm{Nu}_{m,H,f} = \tfrac{6}{5}\mathrm{Nu}_{L,H,f} \tag{8.21}$$

in the case of uniform heat flux. Here the average heat transfer coefficient $\bar{h}$ used in $\mathrm{Nu}_{L,H,f}$ is based on the average temperature difference $\overline{T_w - T_\infty}$ along the plate length L.

Uniform Wall Temperature. Sparrow and Gregg [30], LeFevre [31], and Ostrach [32] obtained solutions of the following form for local Nusselt number in this case:

$$\mathrm{Nu}_{x,T,f} = f(\mathrm{Pr}_f)\left(\mathrm{Gr}_{x,f}\mathrm{Pr}_f^2\right)^{1/4} \tag{8.22}$$

where the values of $f(\mathrm{Pr}_f)$ are given in Table 8.4. Although there are no available experimental data with which to compare this correlation, it should be as reliable as Eq. (8.21). Ostrach [32] showed that the average Nusselt number for a plate of length L

TABLE 8.4 Predicted Values of $f(\mathrm{Pr}_f)$ for Use in Eq. (8.22)

Pr_f	$f(\mathrm{Pr}_f)$	Ref.
0	0.6004	[31]
0.003	0.5827	[30]
0.008	0.5729	[30]
0.01	0.5715	[31, 32]
0.02	0.5582	[30]
0.03	0.5497	[30]
0.1	0.5160	[31]

is

$$\mathrm{Nu}_{m,T,f} = \tfrac{4}{3}\mathrm{Nu}_{L,T,f} \tag{8.23}$$

in the case of uniform wall temperature. Here the average heat transfer coefficient $\bar{h}$ in $\mathrm{Nu}_{L,T,f}$ is based on the average heat flux $\bar{q}$ along the plate length L.

8.6.2 Heated Downward-Facing (Cooled Upward-Facing) Plates in Laminar Flow

Uniform Heat Flux. The experimental results of Sheriff and Davies [29] for the average Nusselt number at $\mathrm{Gr}^*_{a,f} \sim 10^{10}$ were observed to be roughly 15% higher than the approximate integral predictions of Fujii et al. [33], given as follows:

$$\mathrm{Nu}_{m,H,f} = 0.522\left(\mathrm{Gr}^*_{a,f}\mathrm{Pr}_f^2\right)^{1/6} \tag{8.24}$$

where $\mathrm{Gr}^*_{a,f}$ is the modified Grashof number of Eq. (8.17) based on the half width a of an infinite strip. Analytical results are given for other geometries by Fujii et al. [33].

Uniform Wall Temperature. Clifton and Chapman [34] obtained the following average Nusselt number correlation for low Prandtl number fluids:

$$\mathrm{Nu}_{m,T,f} = 0.5212\left(\mathrm{Gr}_{a,f}\mathrm{Pr}_f^2\right)^{1/5} \tag{8.25}$$

There are no known experimental data with which to compare this result.

8.6.3 Heated Upward-Facing (Cooled Downward-Facing) Plates

Laminar Flow, Uniform Wall Temperature. The following theoretical expression for the local Nusselt number under isothermal conditions was extrapolated from the work of Pera and Gebhart [35] by Sheriff and Davies [29]:

$$\mathrm{Nu}_{x,T,f} = 0.48\left(\mathrm{Gr}_{x,f}\mathrm{Pr}_f^2\right)^{1/5} \tag{8.26}$$

where x is the distance from the leading edge of the plate. Experimental data of Kudryavtsev et al. [36] in the range of $\mathrm{Gr}_{x,f}$ from 10^6 to 10^8 are 0 to 25% above this prediction and are questionable because of possible edge effects due to the small size of the apparatus. Based on these considerations, Eq. (8.26) is cautiously recommended.

Turbulent Flow, Uniform Wall Temperature. The experimental results of McDonald and Connolly [37] for the average Nusselt number are correlated by the following equation in the Grashof number range $6 \times 10^8 < \mathrm{Gr}_{D,T} < 5 \times 10^9$:

$$\mathrm{Nu}_{m,T,f} = 0.262\left(\mathrm{Gr}_{D,T}\mathrm{Pr}_f^2\right)^{0.35} \tag{8.27}$$

The exponent near $\frac{1}{3}$ is indicative of the turbulent heat transfer regime. The Grashof number is based on the diameter D of the horizontal disk used in the experiment.

8.6.4 Inclined Plates in Laminar Flow, Uniform Heat Flux

A first-order estimate of the local Nusselt number for heated downward-facing surfaces inclined at an angle γ from the vertical can be determined by replacing g with $g \cos \gamma$ in the Grashof number in Eq. (8.20). Experimental data cited by Sheriff and Davies [29] for $\gamma \approx 75°$ were roughly 10% lower than the following relation (which they suggested on the basis of the above argument):

$$\mathrm{Nu}_{x,H,f}(\gamma) = 0.732\left(\mathrm{Gr}^*_{x,f} \cos \gamma \, \mathrm{Pr}_f^2\right)^{1/5} \tag{8.28}$$

The data were in the range $10^5 < \mathrm{Gr}^*_{x,f} < 10^{11}$. Better agreement is expected at lower values of γ, in part because at such large γ, parts of the thermal boundary layer may actually be below the leading edge of the plate.

8.6.5 Heated Horizontal Cylinders in Laminar Flow

The average Nusselt number data of Hyman et al. [38] for natural convection from horizontal cylinders is well correlated by the following equation:

$$\mathrm{Nu}_f = 0.53\left(\mathrm{Gr}_{d,f} \mathrm{Pr}_f^2\right)^{1/4} \tag{8.29}$$

where the Grashof number is based on the diameter of the horizontal cylinder, and the fluid properties are evaluated at the average of the surface and bulk mean temperatures. Although no detailed comparison has been made, the data of Michiyoshi et al. [39] and the theory of Levy [40] on circumferentially local $\mathrm{Nu}_{\phi,f}$ in mercury both fall below Eq. (8.29) by roughly 10%. It should be mentioned that Levy's theory was developed for both laminar and turbulent flow, assuming uniform wall temperature, while the data of Hyman et al. [38] and Michiyoshi et al. [39] were collected at approximately uniform wall heat flux.

8.6.6 Transition from Laminar to Turbulent Natural Convection

The transition from laminar to turbulent boundary-layer flow occurs in the range $10^8 < \mathrm{Gr}_{x,f} < 10^{10}$. However, on log-log plots of $\mathrm{Nu}_{x,f}$ vs. $\mathrm{Gr}_{x,f}$, the breakpoint is less sharp than for ordinary fluids. This is due mainly to the great influence of molecular thermal conductivity in the case of liquid metals and also to the generally larger scatter found in liquid metal heat transfer data.

8.7 SUMMARY

Table 8.5 summarizes all the correlations presented in this chapter on convective heat transfer in liquid metals.

TABLE 8.5 Correlations for Liquid-Metal Convective Heat Transfer

Reference	Eq.	Correlation	Limitations	Limitations—Flow	Comments
Buhr [5]		$Z = \frac{Ra'_m}{Re_m} \frac{D_h}{L}$	$Z > 20 \times 10^{-4}$	Free convection important	$\mathrm{Ra}'_m = \mathrm{Gr}'_m \mathrm{Pr}_m$
Subbottin [8]	(8.1)	$\mathrm{Nu}_H = 5.0 + 0.025\,\mathrm{Pe}_m^{0.8}$	$\mathrm{Pe}_m > 100$ $L/D_h > 60$	Turbulent pipe flow	Uniform heat flux
Reed (present work)	(8.2)	$\mathrm{Nu}_T = 3.3 + 0.02\,\mathrm{Pe}_m^{0.8}$	$\mathrm{Pe}_m > 100$ $L/D_h > 60$	Turbulent pipe flow	Uniform wall temperature
Kottowski [14]	(8.3)	$\mathrm{Nu} = 0.75\left(\frac{r_2}{r_1}\right)^{0.3} \times (7.0 + 0.025\,\mathrm{Pe}_m)^{0.8}$	$r_2/r_1 < 1.4$ or $1.4 < r_2/r_1 < 10$	Annular flow	
Duchatrelle and Vautrey [15]	(8.4)	$\mathrm{Nu}_H = 5.85 + 0.000341\,\mathrm{Pe}_m^{1.29}$		Flow between parallel plates	Unilateral heat transfer
Dwyer [16]	(8.5)	$\mathrm{Nu} = 9.49 + 0.0596\,\mathrm{Pe}_m^{0.688}$		Flow between parallel plates	Bilateral heat transfer
Andreevskii [17]	(8.6)	$\mathrm{Nu}_f = 0.65\,\mathrm{Pe}_f^{0.5}$		Single cylinder in 90° cross flow	
Kottowski [14]	(8.7)	$\mathrm{Nu}_f = \mathrm{Pe}_f^{0.5}$	$50 < \mathrm{Pe}_f < 4000$	Flow across a staggered tube bank	*Hg* and *Na* data
Hartnett and Irvine [19]	(8.8)	$\mathrm{Nu} = \frac{2}{3}\mathrm{Nu}_{s,m} + 0.025\,\mathrm{Pe}_m^{0.8}$		Various channel shapes	See Table 8.1 for $\mathrm{Nu}_{s,m}$

TABLE 8.5 Continued

Reference	Eq.	Correlation	Limitations	Limitations—Flow	Comments
Chen and Chiou [20]	(8.9)	$\frac{Nu_{x,bc}}{Nu_{bc}} = 1 + \frac{2.4}{x/D_h} - \frac{1}{(x/D_h)^2}$	$2 < x/D_h$ $500 < \mathrm{Pe}_m$ $0.004 \lesssim \mathrm{Pr}_m \lesssim 0.1$	Thermal entrance flow in a round pipe	Uniform wall temperature, Uniform heat flux
Chen and Chiou [20]	(8.10)	$\frac{Nu_{m,bc}}{Nu_{bc}} = 1 + \frac{7}{L/D_h} + \frac{2.8}{L/D_h}\ln\left(\frac{L/D_h}{10}\right)$	$2 < L/D_h$ $500 < \mathrm{Pe}_m$ $0.004 \lesssim \mathrm{Pr}_m \lesssim 0.1$	Thermal entrance flow in a round pipe	Uniform wall temperature, Uniform heat flux
Chen and Chiou [20]	(8.11)	$\frac{Nu_{x,H}}{Nu_H} = 0.88 + \frac{2.4}{x/D_h} - \frac{1.25}{(x/D_h)^2} - \frac{40 - (x/D_h)}{190}$	$2 \le x/D_h < 35$ $0.004 \lesssim \mathrm{Pr}_m \lesssim 0.1$	Thermal entrance flow in a round pipe	Uniform heat flux
Chen and Chiou [20]	(8.12)	$\frac{Nu_{m,H}}{Nu_H} = 0.91 + \frac{5}{L/D_h} + \frac{1.86}{L/D_h}\ln\left(\frac{L/D_h}{10}\right)$	$2 \le L/D_h \le 35$ $0.004 \lesssim \mathrm{Pr}_m \lesssim 0.1$	Thermal entrance flow in a round pipe	Uniform heat flux

Chen and Chiou [20]	(8.13)	$\frac{Nu_{x,H}}{Nu_H} = 0.88 + \frac{2.4}{x/D_h} - \frac{1.25}{(x/D_h)^2}$	$2 \leq x/D_h < 35$ $0.004 \lesssim Pr_m \lesssim 0.1$	Thermal entrance flow in a round pipe	Uniform heat flux
Chen and Chiou [20]	(8.14)	$\frac{Nu_{m,H}}{Nu_H} = 1 + \frac{5}{L/D_h} + \frac{1.86}{L/D_h}\ln\left(\frac{L/D_h}{10}\right)$	$2 \leq L/D_h \leq 35$ $0.004 \lesssim Pr_m \lesssim 0.1$	Thermal entrance flow in a round pipe	Uniform heat flux
Chen and Chiou [20]	(8.15)	$Nu = Nu_0\left(\frac{T_b}{T_{in}}\right)^n$	$0.004 \lesssim Pr_m \lesssim 0.1$	Fully developed pipe flow Variable fluid properties	See Table 8.2
Chang et al. [28]	(8.16)	$Nu_{x,H,f} = \left(\frac{Gr^*_{x,f}}{5}\right)^{1/5}\frac{1}{\Theta(0)}$		Vertical plates in laminar flow	Uniform heat flux $Gr^*_{x,f} = g\beta qx^4/k\nu^2$ For $\Theta(0)$ see Table 8.3
Sheriff and Davies [29]	(8.20)	$Nu_{x,H,f} = 0.732(Gr^*_{x,f}Pr_f^2)^{1/5}$		Vertical plates in laminar flow	Uniform heat flux Eq. (8.16) modified
Sparrow and Gregg [27]	(8.21)	$Nu_{m,H,f} = \frac{6}{5}Nu_{L,H,f}$			Uniform heat flux
Sparrow and Gregg [30], LeFevre [31], Ostrach [32]	(8.22)	$Nu_{x,T,f} = f(Pr_f)(Gr_{x,f}Pr_f^2)^{1/4}$		Vertical plates in laminar flow	Uniform wall temperature Values of $f(Pr_f)$ In Table 8.4 No experimental data

TABLE 8.5 Continued

Reference	Eq.	Correlation	Limitations	Limitations—Flow	Comments
Ostrach [32]	(8.23)	$\mathrm{Nu}_{m,T,f} = \frac{4}{3}\mathrm{Nu}_{L,T,f}$		Heated downward-facing plates in laminar flow in	Uniform wall temperature
Fujii [33]	(8.24)	$\mathrm{Nu}_{m,H,f} = 0.522(\mathrm{Gr}^*_{a,f}\mathrm{Pr}_f^2)^{1/6}$			Uniform heat flux $\mathrm{Gr}^*_{a,f}$ = modified Grashof no. Data are 15% higher
Clifton and Chapman [34]	(8.25)	$\mathrm{Nu}_{m,T,f} = 0.5212(\mathrm{Gr}_{a,f}\mathrm{Pr}_f^2)^{1/5}$		Heated downward-facing plates in laminar flow	Uniform wall temperature No experimental data
Pera and Gebhart [35]	(8.26)	$\mathrm{Nu}_{x,T,f} = 0.48(\mathrm{Gr}_{x,f}\mathrm{Pr}_f^2)^{1/5}$		Heated upward-facing plates (cooled downward), laminar flow	Uniform wall temperature Expression is 0–25% below data
McDonald and Connolly [37]	(8.27)	$\mathrm{Nu}_{m,T,f} = 0.262(\mathrm{Gr}_{D,T}\mathrm{Pr}_T^2)^{0.35}$	$6 \times 10^8 < \mathrm{Gr}_{D,T} < 5 \times 10^9$	Heated upward-facing plates (cooled downward), turbulent flow	Uniform wall temperature
Sheriff and Davies [29]	(8.28)	$\mathrm{Nu}_{x,H,f}(\gamma) = 0.732(\mathrm{Gr}^*_{x,f}\cos\gamma\ \mathrm{Pr}_f^2)^{1/5}$	$1 \times 10^5 < \mathrm{Gr}^*_{x,f} < 1 \times 10^{11}$	Inclined plates in laminar flow	Eq. (8.20) modified
Hyman et al [38]	(8.29)	$\mathrm{Nu}_f = 0.53(\mathrm{Gr}_{d,f}\mathrm{Pr}_f^2)^{1/4}$		Natural convection from horizontal cylinders in laminar flow	Data collected at uniform heat flux

NOMENCLATURE

A	constant in Eq. (8.19)
c_p	specific heat at constant pressure, J/(kg · K), Btu/(lb_m · °F)
D_h	hydraulic diameter = 4 (minimum free flow area)/(wetted perimeter), m, ft
D_1	inner diameter of annulus, m, ft
D_2	outer diameter of annulus, m, ft
$\partial/\partial T$	partial derivative with respect to T
d/dx	total derivative with respect to x
f	function of
Gr	Grashof number = $g\beta \Delta T L^3/\nu^2$
Gr*	modified Grashof number = $g\beta\, qx^4/k\nu^2$
Gr′	Grashof number based on axial temperature difference = $(D_h^3 \beta g \Delta T)/\nu^2$ where $\Delta T = (dT/dx)/D_h$
g	gravitational acceleration, m/s^2, ft/s^2
h	heat transfer coefficient, W/(m^2 · K), Btu/(hr · ft^2 · °F)
k	thermal conductivity, W/(m · K), Btu/(hr · ft · °F)
L	length, m, ft
ln	natural logarithm
Nu	Nusselt number = $h D_h/k$
Nu_{bc}	circumferentially averaged Nusselt number for fully developed flow, thermal boundary condition
$Nu_{x,bc}$	circumferentially averaged but axially local Nusselt number for the thermal entrance region for the specified thermal boundary condition
$Nu_{m,bc}$	mean Nusselt number for the thermal entrance region for the specified thermal boundary condition
n	exponent in Eq. (8.15)
Pe	Péclet number = Re Pr
Pr	Prandtl number = $\mu c_p/k$
q	heat flux, W/m^2, Btu/(hr · ft^2)
Ra	Rayleigh number = Gr Pr
Re	Reynolds number = uD_h/ν
r	radial coordinate, m, ft
r_0	inside radius of a pipe, m, ft
r_1	inner radius of annulus, m, ft
r_2	outer radius of annulus, m, ft
T	temperature, °C, K, °F, °R
u	velocity, m/s, ft/s
x	axial coordinate, m, ft
y	transverse coordinate, m, ft
Z	free convection distortion parameter = $(Ra'_m/Re_m)\, D_h/L$

Greek Symbols

β	coefficient of thermal expansion = $-(1/\rho)(\partial\rho/\partial T)_p$, K^{-1}, $°R^{-1}$
Δ	finite difference
γ	angle of inclination from vertical, rad, deg
η	dimensionless distance = $(y/x)(Gr_x^*/5)^{1/5}$
$\Theta(0)$	dimensionless temperature function
μ	dynamic viscosity, Pa · s, $lb_m/(hr \cdot ft)$
ν	kinematic viscosity, m^2/s, ft^2/s
ρ	density, kg/m^3, lb_m/ft^3

Subscripts

a	based on half-width dimension a
D	based on disk diameter D
d	based on cylinder diameter d
f	fluid properties evaluated at the film temperature
H	constant wall heat flux boundary condition
h	hydraulic parameter (see D_h)
in	inlet bulk fluid condition
L	based on length L
m	fluid properties evaluated at bulk mean temperature
out	outlet bulk fluid condition
p	evaluated at constant pressure
s	slug flow
T	constant wall temperature boundary condition
w	evaluated at the wall
x	based on length x
ϕ	evaluated at angle ϕ from forward stagnation point
∞	evaluated at ∞ (fully developed)
1	inner diameter of annulus
2	outer diameter of annulus

Superscripts and accents

$\bar{}$	average value
*	modified number

REFERENCES

1. O. E. Dwyer, Liquid-Metal Heat Transfer, *Sodium-NaK Engineering Handbook*, Vol. 2, pp. 73–191, 1976.
2. S. L. Schrock, Eddy Diffusivity Ratios in Liquid Metals, Ph.D. Thesis, Purdue Univ., West Lafayette, Ind., 1964.
3. R. A. Gardner, Magneto-Fluid-Mechanic Pipe Flow in a Transverse Magnetic Field With and Without Heat Transfer, Ph.D. Thesis, Purdue Univ., West Lafayette, Ind., 1969.

4. D. J. Kowalski, Free Convection Distortion in Turbulent Mercury Pipe Flow, M.S. Thesis, Purdue Univ., West Lafayette, Ind., 1974.
5. H. O. Buhr, Heat Transfer to Liquid Metals, with Observations on the Effect of Superimposed Free Convection in Turbulent Flow, Ph.D. Thesis, Univ. of Cape Town, Cape Town, South Africa, 1967.
6. J. D. Jackson, Turbulent Mixed Convection Heat Transfer to Liquid Sodium, *Int. J. Heat Fluid Flow*, Vol. 4, pp. 107–111, 1983.
7. R. N. Lyon, Forced-Convection Heat-Transfer Theory and Experiments with Liquid Metals, USAEC Report ORNL-361, Oak Ridge National Laboratory, 1949.
8. V. I. Subbotin, A. K. Papovyants, P. L. Kirillov, and N. N. Ivanovskii, A Study of Heat Transfer to Molten Sodium in Tubes, *At. Energiya (USSR)*, Vol. 13, pp. 380–382, 1962.
9. C. A. Sleicher, A. S. Awad, and R. H. Notter, Temperature and Eddy Diffusivity Profiles in NaK, *Int. J. Heat Mass Transfer*, Vol. 16, pp. 1565–1575, 1973.
10. E. R. Gilliland, R. J. Musser, and W. R. Page, Heat Transfer to Mercury, *Gen. Disc. on Heat Transfer, Inst. Mech. Eng. and ASME*, London, pp. 402–404, 1951.
11. N. Z. Azer and B. T. Chao, Turbulent Heat Transfer in Liquid Metals—Fully Developed Pipe Flow with Constant Wall Temperature, *Int. J. Heat Mass Transfer*, Vol. 3, pp. 77–83, 1961.
12. D. C. Leslie, A Recalculation of Turbulent Heat Transfer to Liquid Metals, *Lett. Heat Mass Transfer*, Vol. 4, pp. 25–33, 1977.
13. M. A. El Hadidy, F. Gori, and D. B. Spalding, Further Results on the Heat Transfer to Low-Prandtl-Number Fluids in Pipes, *Numer. Heat Transfer*, Vol. 5, pp. 107–117, 1982.
14. H. M. Kottowski, Thermohydraulics of Liquid Metals, *Lecture Series* 1983-07, von Karman Institute for Fluid Dynamics, pp. 1–47, 1983.
15. L. Duchatelle and L. Vautrey, Determination des Coefficients de Convection d'un Alliage NaK en Ecoulement Turbulent entre Plaques Planes Paralleles (in French), *Int. J. Heat Mass Transfer*, Vol. 7, pp. 1017–1031, 1964.
16. O. E. Dwyer, Heat Transfer to Liquid Metals Flowing Turbulently between Parallel Plates, *Nucl. Sci. Eng.*, Vol. 21, pp. 79–89, 1965.
17. A. A. Andreevskii, Heat Transfer in Transverse Flow of Molten Sodium around a Single Cylinder, *Sov. J. Atomic Energy*, Vol. 7, No. 3, pp. 745–747, 1961.
18. C. J. Hsu, Analytical Study of Heat Transfer to Liquid Metals in Cross-Flow through Rod Bundles, *Int. J. Heat Mass Transfer*, Vol. 7, pp. 431–446, 1964.
19. J. P. Hartnett and T. F. Irvine, Jr., Nusselt Values for Estimating Turbulent Liquid Metal Heat Transfer in Noncircular Ducts, *AIChE J.*, Vol. 3, pp. 313–317, 1957.
20. C-J. Chen and J. S. Chiou, Laminar and Turbulent Heat Transfer in the Pipe Entrance Region for Liquid Metals, *Int. J. Heat Mass Transfer*, Vol. 24, pp. 1179–1189, 1981.
21. W. M. Kays, Numerical Solutions for Laminar-Flow Heat Transfer in Circular Tubes, *Trans. ASME*, Vol. 77, pp. 1265–1274, 1955.
22. R. K. McMordie and A. F. Emery, A Numerical Solution for Laminar-Flow Heat Transfer in Circular Tubes with Axial Conduction and Developing Thermal and Velocity Fields, *J. Heat Transfer*, Vol. 89, pp. 11–16, 1967.
23. A. S. Awad, Heat Transfer and Eddy Diffusivity in NaK in a Pipe at Uniform Wall Temperature, Ph.D. Thesis, Univ. of Washington, Seattle, 1965.
24. S-L. Lee, Forced Convection Heat Transfer in Low Prandtl Number Turbulent Flows: Influence of Axial Conduction, *Can. J. Chem. Eng.*, Vol. 60, pp. 482–486, 1982.
25. S-L. Lee, Liquid Metal Heat Transfer in Turbulent Pipe Flow with Uniform Wall Flux, *Int. J. Heat Mass Transfer*, Vol. 26, pp. 349–356, 1983.
26. E. M. Sparrow and J. L. Gregg, The Variable Fluid-Property Problem in Free Convection, *Trans. ASME*, Vol. 80, pp. 879–886, 1958.
27. E. M. Sparrow and J. L. Gregg, Laminar Free Convection from a Vertical Plate with Uniform Surface Heat Flux, *Trans. ASME*, Vol. 78, pp. 435–440, 1956.

28. K. S. Chang, R. G. Akins, L. Burris, Jr., and S. G. Bankoff, Free Convection of a Low Prandtl Number Fluid in Contact with a Uniformly Heated Vertical Plate, USAEC Report ANL-6835, Argonne Nat. Lab., 1964.

29. N. Sheriff and N. W. Davies, Liquid Metal Natural Convection from Plane Surfaces: A Review Including Recent Sodium Measurements, *Int. J. Heat Fluid Flow*, Vol. 1, pp. 149–154, 1979.

30. E. M. Sparrow and J. L. Gregg, Details of Exact Low Prandtl Number Boundary-Layer Solutions for Forced and Free Convection, *US NASA Memo* 2.27.59*E*, 1959.

31. E. J. LeFevre, Laminar Free Convection from a Vertical Plane Surface, *Ninth Int. Congr. Appl. Mech.*, pp. 168–174, 1956.

32. S. Ostrach, An Analysis of Laminar Free-Convection Flow and Heat Transfer about a Flat Plate Parallel to the Direction of the Generating Body Force, NASA Report 1111, 1953.

33. T. Fujii, H. Honda, and I. Morioka, A Theoretical Study of Natural Convection Heat Transfer from Downward-Facing Horizontal Surfaces with Uniform Heat Flux, *Int. J. Heat Mass Transfer*, Vol. 16, pp. 611–627, 1973.

34. J. V. Clifton and A. J. Chapman, Natural-Convection on a Finite-Size Horizontal Plate, *Int. J. Heat Mass Transfer*, Vol. 12, pp. 1573–1584, 1969.

35. L. Pera and B. Gebhart, Natural Convection Boundary Layer Flow over Horizontal and Slightly Inclined Surfaces, *Int. J. Heat Mass Transfer*, Vol. 16, pp. 1131–1146, 1973.

36. A. P. Kudryavtsev, D. M. Ovechkin, D. N. Sorokin, V. I. Subbotin, and A. A. Tsyganak, Experimental Study of Heat Transfer from Horizontal Flat Surface to Sodium under Free Convection, USAEC Report BNL-tr-167 (transl. by S. J. Amoretty from *Zhidkie Metally*, Atomizdat, Moscow, 1967, pp. 131–136), Brookhaven Nat. Lab., 1967.

37. J. S. McDonald and T. J. Connolly, Investigation of Natural Convection Heat Transfer in Liquid Sodium, *Nucl. Sci. Eng.*, Vol. 8, pp. 369–377, 1960.

38. S. C. Hyman, C. F. Bonilla, and S. W. Ehrlich, Natural-Convection Transfer Processes: I. Heat Transfer to Liquid Metals and Nonmetals at Horizontal Cylinders, *Chem. Eng. Progr., Symp. Ser.* No. 5, Vol. 49, pp. 21–31, 1953.

39. I. Michiyoshi, O. Takahashi, and A. Serizawa, Natural Convection Heat Transfer from a Horizontal Cylinder to Mercury under Magnetic Field, *Int. J. Heat Mass Transfer*, Vol. 19, pp. 1021–1029, 1976.

40. S. Levy, Integral Methods in Natural-Convection Flow, *J. Appl. Mech.*, Vol. 22, pp. 515–522, 1955.

41. R. C. Martinelli, Heat Transfer to Molten Metals, *Trans. ASME*, Vol. 69, pp. 947–959, 1947.

42. J. K. Jacoby, Free Convection Distortion and Eddy Diffusivity Effects in Turbulent Mercury Heat Transfer, M.S. Thesis, Purdue Univ., West Lafayette, Ind., 1972.

43. E. A. Horsten, Combined Free and Forced Convection in Turbulent Flow of Mercury, Ph. D. Thesis, Univ. of Cape Town, Cape Town, South Africa, 1971.

9

CONVECTIVE HEAT TRANSFER WITH ELECTRIC AND MAGNETIC FIELDS

J. H. Davidson and F. A. Kulacki

College of Engineering
Colorado State University
Fort Collins, Colorado

P. F. Dunn

University of Notre Dame
Notre Dame, Indiana

9.1 INTRODUCTION

Electrohydrodynamics (EHD) refers to the coupling of an electric field and a velocity field in a dielectric fluid continuum, whereas magnetohydrodynamics (MHD) encompasses the phenomena arising when an electromagnetic field is applied to an electrically conducting fluid. Applications of EHD and MHD in convective heat transfer are many and varied. This chapter is restricted to EHD of polar gases and MHD of electrically conductive fluids.

Electric field effects on heat transfer in polar gases generally take place via a modification of the gas velocity and temperature boundary layers. The use of a nonuniform electric field and the electric or corona wind has been most frequently studied and used in thermal systems applications. Only MHD as applied to heat transfer in liquid metals is addressed in this chapter. For areas not covered here, the reader is referred to review articles by Romig [105, 106] (which include MHD effects in conducting gases), the annual symposium proceedings on Engineering Aspects of Magnetohydrodynamics, and two recent status reports [21, 93]. Liquid metal heat transfer in the presence of magnetic fields in fusion-reactor coolant blankets is discussed in Ref. 23.

For the user of the information presented in this chapter, it is important to know that no generalized correlations of heat transfer, such as those for forced and free convection without electric and magnetic fields, are available. Much of information available for electric fields today is restricted to specialized studies, and detailed knowledge of experimental apparatus or numerical schemes is needed to correctly interpret results. Also, no complete analytical or numerical solution of the coupled electrohydrodynamic and fluid-dynamic equations has been obtained. Thus, it is extremely important to consider both experimental and mathematical limitations of the existing literature in new applications.

9.2 BASIC CONCEPTS IN ELECTROHYDRODYNAMICS (EHD)

9.2.1 Governing Equations

The governing equations of EHD are derived from Maxwell's macroscopic equations of electrodynamics, Ohm's law, and the classical fluid-dynamic conservation equations of mass, momentum, and energy. The most general form of the equations of fluid dynamics and electrodynamics are well established, and the reader is referred to Refs. 2, 39, 38, 54. Electrohydrodynamics is restricted to those problems for which there is no externally applied magnetic field and any induced magnetic field is negligible. The equations are based on the concept of a single-species fluid continuum and only apply to the region outside an active corona discharge. The discharge zone is, however, extremely small relative to the interelectrode space, and the theory applies in a practical sense to the entire region between electrodes. In this case, for an isotropic fluid which is free of surface forces, the reduced Maxwell equations are

$$\mathbf{E} = -\Delta\phi \tag{9.1}$$

$$\nabla \cdot (\epsilon \mathbf{E}) = \rho_c \tag{9.2}$$

$$\nabla \cdot \mathbf{J} = \frac{\partial \rho_c}{\partial t} \tag{9.3}$$

These equations can be further reduced for the steady state, where

$$\nabla \cdot \mathbf{J} = 0 \tag{9.4}$$

This simplification is reasonable when the electric field is either completely steady, as in the case of a direct current (DC) field, or at most has low-frequency components.

The current density is given by Ohm's law as

$$\mathbf{J} = K|\rho_c|\mathbf{E} + \rho_c \mathbf{u} \tag{9.5}$$

Thus the total current is composed of that due to the local body force, $K|\rho_c|\mathbf{E}$, and that due to the convective transport of charge, $\rho_c \mathbf{u}$. When the magnitude of the bulk fluid velocity U is much less than the ion drift velocity $V_i = KE$, the convection term may be dropped from Ohm's law. Typically, as shown in Table 9.1, ionic mobilities for gases commonly used in engineering applications are greater than 10^{-4} m^2/(V · s). The maximum spark breakdown potential in gases is approximately 15×10^5 V/m for a variety of electrode configurations. Considering a self-sustained corona discharge operating well below sparkover with a field strength of 50×10^4 V/m, values of V_i on the order of 50 m/s are possible. Thus the neglect of the convective current for this situation is reasonable when the mean velocity is on the order of 5 m/s.

The basic equations for an incompressible fluid under the influence of an electric field are conservation of mass,

$$\nabla \cdot \mathbf{u} = 0 \tag{9.6}$$

TABLE 9.1 Ionic Mobilities at 0° C and 760 Torr [15]

	K^-, 10^{-4} m^2/(V · s)	K^+, 10^{-4} m^2/(V · s)
Air (dry)	2.1	1.36
Air (very pure)	2.5	1.8
Ar	1.70	1.37
Ar (very pure)	206	1.31
Cl_2	0.74	0.74
CCl_4	0.31	0.30
C_2H_2	0.83	0.78
C_2H_5Cl	0.38	0.36
C_2H_5OH	0.37	0.36
CO	1.14	1.10
CO_2 (dry)	0.98	0.84
H_2	8.15	5.9
H_2 (very pure)	7900	
HCl	0.62	0.53
H_2O (at 100° C)	0.95	1.1
H_2S	0.56	0.62
He	6.3	5.09
He (very pure)	500	5.09
N_2	1.84	1.27
N_2 (very pure)	145	1.28
NH_3	0.66	0.56
N_2O	0.90	0.82
Ne	9.9	
O_2	1.8	1.31
SO_2	0.41	0.41

and the incompressible Navier-Stokes equations with an electric body force term $\mathbf{F}_e$,

$$\rho\left[\frac{\partial \mathbf{u}}{\partial t} + (\mathbf{u} \cdot \nabla)\mathbf{u}\right] = -\nabla p + \mu\nabla^2\mathbf{u} + \rho\mathbf{g} + \mathbf{F}_e \tag{9.7}$$

This body force, which results from collisions of ions and neutral molecules, is given by

$$\mathbf{F}_e = \rho_c\mathbf{E} - \tfrac{1}{2}E^2\,\nabla\epsilon + \tfrac{1}{2}\,\nabla\left[E^2\rho\left(\frac{\partial \epsilon}{\partial \rho}\right)_T\right] \tag{9.8}$$

The first term results from the presence of free charge and is the Coulombic force. The second term results from the force of a nonuniform electric field on the dielectric permittivity. This force is normally much weaker than the Coulombic force except at the interface between two fluids where the permittivity changes abruptly [71]. The final contribution to the body force is the electrostrictive force. When considering gases, this term can be simplified by using the linear relationship [54]

$$\rho\frac{\partial \epsilon}{\partial \rho} = \epsilon - \epsilon_0 \tag{9.9}$$

Kronig and Schwartz [47] and Lykoudis and Yu [67] investigated the role of electrostrictive forces in natural convection. In both liquids and gases, the electrostriction term can be lumped into a modified pressure in the momentum equation given by

$$P = p - \rho\frac{E^2}{2}\left(\frac{\partial \epsilon}{\partial \rho}\right)_T \tag{9.10}$$

Thus in the absence of a free surface, the electrostrictive contribution to the body force has no effect on the velocity field, but merely affects the pressure distribution in the fluid. Only the Coulombic force and the force resulting from permittivity gradients in the fluid contribute to modifications in the fluid velocity field that exists in the absence of the electric field. Generally, for gases in which a corona discharge (and thus space charge) exists, nonuniform permittivity effects and electrostriction are negligible compared to charge-density effects. In this case, the electric body force term reduces to

$$\mathbf{F}_e = \rho_c\mathbf{E} \tag{9.11}$$

Using Ohm's law, Eq. (9.5), this body force may be represented as $\mathbf{J}/K$ for a positive corona discharge and as $-\mathbf{J}/K$ for a negative discharge.

The energy equation for an incompressible fluid subject to the EHD idealizations is

$$\frac{\partial T}{\partial t} + \mathbf{u} \cdot \nabla T = \nabla \cdot (\alpha\nabla T) + \frac{\nu}{c_p}\Phi + \frac{K|\rho_c|E^2}{\rho c_p} \tag{9.12}$$

where the existence of a space charge in the fluid contributes the Joule heating term $K|\rho_c|E^2$. Joule heating need only be considered when extraordinarily high current densities are present.

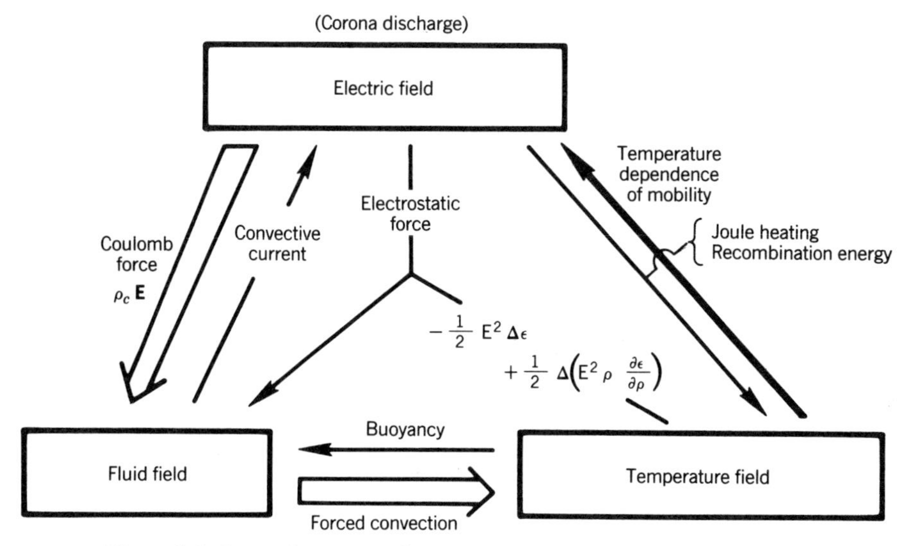

Figure 9.1. Interactions among the electric, fluid, and temperature fields [138].

The couplings between the electrodynamic, momentum, and energy equations are dependent on the electrical and thermophysical properties of the fluid. An electric field may affect convective heat transfer through four major mechanisms: the Coulombic force, electrostriction, dielectric gradients, and Joule heating. The Coulombic and permittivity gradient terms in the electric body force expression of Eq. (9.8) couple the fluid-dynamic equations to Maxwell's equations. The electrodynamic equations are in turn coupled to the momentum equation through the convective current term $\rho_c\mathbf{u}$ in Ohm's law. Joule heating directly couples the electrical equations to the energy equation. If bouyancy is effected, the fluid and electrical equations are coupled. A fifth mechanism involves the temperature dependence of ionic mobility. This mechanism has recently begun to receive attention [34, 121, 122]; however, the voltages at which it is important are much less than those associated with space-charge effects. The interaction among the electric, fluid, and temperature fields is illustrated in Fig. 9.1.

In the formulation of a specific heat transfer problem, one or more of the available coupling mechanisms may be insignificant compared to the others, so that the problem can be greatly simplified [24, 57, 81]. In the case of a corona discharge in a gas, the major interactions are those indicated in Fig. 9.1 by broad arrows. In most practical situations, Joule heating is negligible and thus the temperature field is influenced only indirectly through modification of the flow field by the electric body force $\mathbf{F}_e$. When the convective current in Ohm's law is negligible, the electrical equations may be solved independently of the hydrodynamic equations.

9.2.2 Dimensionless Groups

Nondimensionalization of the EHD equations yields the standard groups for correlating convective heat transfer data. These include the Reynolds, Prandtl, Grashoff, and Nusselt numbers. In addition to these familiar parameters, the electric field effect introduces new dimensionless parameters represented by the ratio of the electric body force and the fluid inertial or viscous forces. The formulation chosen for these ratios varies in the literature.

Using a geometric length scale L, a velocity scale U_0, and a current scale I_0, the dimensionless momentum equation is

$$\frac{\partial \mathbf{u}}{\partial t} + (\mathbf{u} \cdot \nabla)\mathbf{u} = -\nabla p + \frac{1}{\text{Re}}\nabla^2\mathbf{u} + \frac{\text{Gr}}{\text{Re}^2}\theta + \text{Ne}^2\mathbf{J} \tag{9.13}$$

where the electric body force F_e is represented as J/K, p is the static pressure nondimensionalized by ρU_0^2, and θ is given by $(T - T_1)/(T_1 - T_2)$. The dimensionless electric number Ne given by,

$$\text{Ne} = \left(\frac{I_0/KL}{\rho U_0^2}\right)^{1/2} \tag{9.14}$$

is the ratio of the electric body force to the inertial force. There is unfortunately no standard nondimensionalization in the literature. Numerous authors present this ratio in terms of a reference space-charge density ρ_{c0} rather than a current density I_0/L^2. A general representation of the space-charge dimensionless parameter is

$$\chi = \left(\frac{\rho_{c0}\phi_0}{\rho U_0^2}\right)^{1/2} \tag{9.15}$$

where ϕ_0 is a reference electric potential, usually chosen to be the operating voltage at the corona discharge electrode. Using this scheme, a characteristic electric wind velocity may be defined as

$$U_e = \left(\frac{\rho_{c0}\phi_0}{\rho}\right)^{1/2} \tag{9.16}$$

from which an electric Reynolds number is defined as

$$\text{Re}_{\text{EHD}} = \frac{U_e L}{\nu} \tag{9.17}$$

This common approach to nondimensionalization is awkward, because the space-charge scale is not physically measurable. The total current I_0 is measurable in all cases.

The dimensionless incompressible energy equation is given by

$$\frac{\partial \theta}{\partial t} + \mathbf{u} \cdot \nabla\theta = \frac{1}{\text{Pr Re}}\nabla \cdot \mathbf{q}'' + \frac{\text{Ec}}{\text{Re}}\Phi + \frac{\text{Ec}}{\text{Re}}\text{Ne}^2\left(\frac{K\,\Delta\phi}{\nu}\right)\frac{J^2}{\rho_c} \tag{9.18}$$

where the ratio K $\Delta\phi/\nu$ represents the ratio of the electric body force to viscous forces. The space-charge density is nondimensionalized by the quantity $K\,\Delta\phi\,L/I_0$. Since Joule heating is normally assumed negligible in corona discharges, correlations for the

Nusselt number generally include the Reynolds number, Prandtl number, and either form of the force ratio given in Eqs. (9.14) and (9.15).

9.2.3 Basic Physics of the Corona Discharge

The term "corona discharge" refers to a self-sustaining electrical discharge produced in a gas. A stable discharge requires the presence of two electrodes, one with a much smaller radius of curvature than the other. Additionally, the gap between the two electrodes must be large compared to the radius of the smaller electrode. As the electrical potential is raised between the two electrodes, the gas in the immediate vicinity of the electrode surface which has the highest degree of curvature is ionized. These discharges are termed positive or negative according to the polarity of the electrically stressed electrode.

The idealized situation of a small-diameter wire on the axis of a long cylinder is used to illustrate the corona discharge process. In the case of a positive discharge wire, free electrons are drawn from the interelectrode space in toward the wire. There, in the intense electrical field, they collide with neutral molecules and produce numerous electron–positive-ion pairs. The newly created electrons are in turn accelerated and produce further ionization. This cumulative process, referred to as the Townsend electron avalanche, is responsible for the sustained corona discharge. The electron avalanches move toward the wire discharge electrode. Positive ions formed by the electron avalanches are accelerated away from the ionization region near the wire and collide with neutral molecules as they move in the direction of decreasing field strength toward the outer cylinder. These ions do not have sufficient energy to cause emission of electrons by positive ion bombardment at the cylinder cathode or to significantly ionize the gas in the region between the two electrodes. Positive ions fill most of the gas volume and are responsible for all of the current outside the active ionization region near the wire. The effect of positive space charge is to stabilize the discharge process. The discharge appears in air as a bluish glowing region of gas, ideally extending uniformly over the entire surface of the wire.

Negative corona discharges are quite different in their charge-generation process. In this case, the electrons formed by chance ionizing events gain energy from the electric field and produce positive ions and other electrons by collision. The positive ions formed in this avalanche process are accelerated toward the wire. Additional electrons are generated by subsequent electron ejection from the wire surface, which results from bombardment of the wire by the positive ions. As the electrons move into the weaker electrical field away from the wire, they collide with neutral gas molecules and form negative ions. All the current in the interelectrode space outside the discharge region is carried by these negative ions. Because the discharge is dependent on the electrode surface for electrons, the discharge is characterized not by a uniform sheath of glowing gas surrounding the wire, but rather by discrete, intensely bright, glowing tufts of ionized gas distributed nonuniformly along the wire. If the discharge electrode surface is oxidized or marred by other surface irregularities, positive discharges also display inhomogenities in the discharge structure.

The electrode potential at which current begins to flow is termed the "breakdown" or "threshold" potential ϕ_0. The exact value of ϕ_0 depends on the geometry of the electrodes and the gas composition. The gas composition also determines the type of ions that are formed in the corona discharge process. The species with the lowest ionization potential will normally appear in the greatest concentration. Table 9.2 contains the ionization potentials in several polyatomic molecules. Once the breakdown

TABLE 9.2 First Ionization Potentials of Polyatomic Molecules [15]

Gas	Ionization Potential, eV	Probable Ion
H_2	15.37	H_2^+
N_2	15.57	N_2^+
O_2	12.5	O_2^+
CO	14.1	CO^+
CO_2	14.0	CO_2^+
NO	9.5	NO^+
NO_2	11.0	NO_2^+
N_2O	12.9	N_2O^+
H_2O	12.59	H_2O^+

potential is exceeded, the current increases almost parabolically with voltage until the sparkover potential is reached. This is illustrated in Fig. 9.2, where data for a wire-cylinder system are shown.

Determination of the spatial values of the electric field intensity and current for various geometries in the presence of a corona discharge is complicated by the presence of space charge in the gas. The simplest geometry for the purposes of analysis is the coaxial wire-cylinder electrode system just described. Prior to the initiation of a corona, the electric field is given by

$$E(r) = \frac{\phi(R_2) - \phi(R_1)}{r \ln(R_2/R_1)} \tag{9.19}$$

Above the corona threshold potential, the introduction of a nonzero space-charge

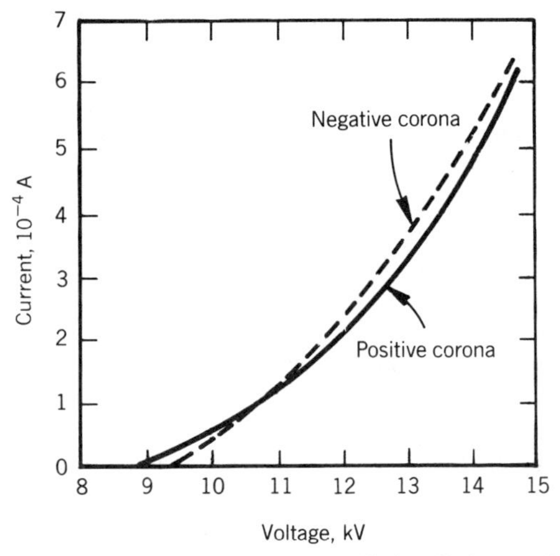

Figure 9.2. Current-voltage relation [15] for a wire-cylinder discharge. (Wire diameter = 0.41 mm, cylinder length = 25 cm, cylinder diameter = 4.45 cm, $T = 25°$ C, relative humidity = 29.2%, $P = 746.8$ Torr.)

density results in an electric field given as

$$E(r) = \left[\frac{IL}{2\pi\epsilon K} + \left(\frac{R_1}{r}\right)\left(E_0^2 - \frac{IL}{2\pi\epsilon K}\right)\right]^{1/2} \tag{9.20}$$

The threshold electric field E_0 at the wire surface is given for air by the semiempirical formula [89]

$$E_0 = 31.0\left(1 + \frac{0.308}{R_1^{1/2}}\right) \tag{9.21}$$

where the wire radius is in centimeters and E_0 is in kV/cm. Neglecting higher order terms, the corona current-voltage relation is given by

$$\frac{\phi(R_1) - \phi_0}{\phi_0}\ln\left(\frac{R_2}{R_1}\right) = (1 + \xi)^{1/2} - 1 - \ln\left(\frac{1 + (1 + \xi)^{1/2}}{2}\right) \tag{9.22}$$

where ϕ_0 is related to E_0 by

$$\phi_0 = E_0 R_1 \ln\left(\frac{R_1}{R_2}\right) \tag{9.23}$$

The quantity ξ is defined as

$$\xi = \left(\frac{R_2}{E_0 R_1}\right)^2\left(\frac{I}{2\pi\epsilon K}\right) \tag{9.24}$$

For small currents near the corona threshold, the total current is given by

$$I = \frac{8\pi\epsilon LK\phi(R_2)[\phi(R_2) - \phi_0]}{R_2^2 \ln(R_2/R_1)} \tag{9.25}$$

For a wire-plate geometry, Cooperman [16] and Lagarias [52] pioneered experimental techniques for the measurement of electric fields, and Penny and Matick [92] used a Langmuir probe to measure potentials. Cooperman [17] argued that the electric potential could be represented as the sum of the electrostatic potential and that due to a uniformly distributed space charge. He used a conformal mapping and the method of images to obtain an approximate expression for the potential, from which he derived an expression for the current-voltage relationship for the wire-plate electrodes as

$$I = \frac{LK\phi(\phi - \phi_0)}{d^2 \ln(4d/\pi R_1)} \tag{9.26}$$

for a wire spacing less than $1.2d$. This expression is subject to the assumption of uniform space charge and is only valid at low current levels.

White [135] summarized the early work on the electrical properties of corona discharges in both the wire-cylinder and the wire-plate geometry. More recent numerical studies [35, 55, 56, 60, 139] compare well with experimental measurements. Hay's

[33] recent application of a perturbation method to Cooperman's analytical work provides an approximate analytical expression for the current density distribution and thus the body force in a wire-plate geometry.

9.2.4 Basic Fluid Mechanics of the Corona Wind

In the study of electrical coronas, the corona wind, discovered by Hauksbee [32] more than 275 years ago when he felt a weak blowing sensation from a high voltage tube held to his face, emerged as a novelty and remained a curiosity for many years. The first quantitative analysis of the phenomena is Chattock's [9–11] study of ion mobilities in the needle-to-plate geometry. He developed the relationship between electric wind pressure and corona current for a continuous discharge from a point to a plane. Much later, Löb [61] extended Chattock's pressure vs. current relationship to other geometries. Steutzer [116–118] developed a theory describing ion-drag pressure generation in gases and liquids. A comprehensive survey of the early theoretical and experimental work on the corona wind is given by Robinson [104].

The corona wind is characteristic of asymmetric electric field corona discharges, such as are found in the point-to-plane and wire-to-plate electrode arrangements, or any geometry in which the discharge exhibits a spatial nonuniformity. Modification of the gas flow field by the corona discharge depends on the electrode geometry and the structure of the discharge. Situations in which there is no induced gas motion are characterized by electrode geometries in which the current-density vector is irrotational. Flippen [24] has recently shown that in only three geometries (concentric spheres, concentric cylinders, and parallel plates) is it possible to have an irrotational electric body force in the presence of space charge. For all other geometries, the presence of a discharge results in some modification to the flow field. The body force may induce a secondary flow or create a flow where none would otherwise exist, and in some instances generate turbulence.

The corona wind resulting from the usual spotty discharge, whether located at a point or along a wire, may be pictured as a jet of gas originating at the discharge and streaming toward the grounded electrode. Depending on the electrode geometry and the direction and velocity of any primary gas flow, this jet flow may result in large-scale circulating motions in the interelectrode space. Additionally, it is well established that in channel flows the corona discharge generates turbulence. Heat transfer augmentation is due to both impingement of the electric wind on the grounded electrode surface and the interaction of the jet with the free-stream flow. As expected from the nondimensionalization of the governing equations the effect of the corona discharge on the flow field, and as a result on the heat transfer coefficients, diminishes with increasing free-stream velocity.

9.3 EHD IN EXTERNAL BOUNDARY LAYERS

9.3.1 Impingement by Single Corona Discharges

Point-plane (Fig. 9.3) and wire-plane electrode systems (Fig. 9.4) can be used to create an impingement flow to a heated surface in both free and forced convection. Effective limits for heat transfer augmentation in each case are determined by the electrode geometry, general flow features, and ranges of Reynolds and Grashof numbers. In free convection from external surfaces, there is usually no limit to the increase in Nusselt

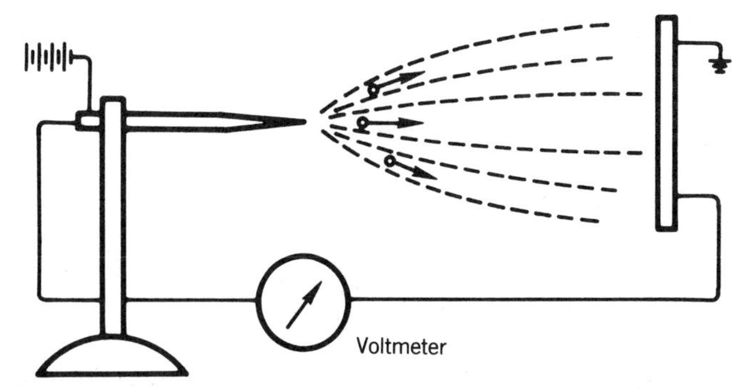

Figure 9.3. Schematic of induced flow field for point-plane electrode system [129].

number with electric-wind impingement (at least, to the point of spark breakdown). For forced convection flows, however, there is an upper limit on Reynolds numbers beyond which an augmentation due to electric-wind impingement does not exist.

Point-Plane and Wire-Plane Electrodes. Several studies [69, 75, 137] of electric-wind impingement on flat surfaces in free convection have been made for a single wire, or point, discharge source and flat plane electrodes. These types of electrode systems form the basis for the fundamental relation between increased heat transfer coefficient and discharge current.

Assuming zero charge density in the interelectrode region, and a discharge electrode diameter much smaller than the spacing (i.e., $R_1/d \ll 1$), the vertical component of the

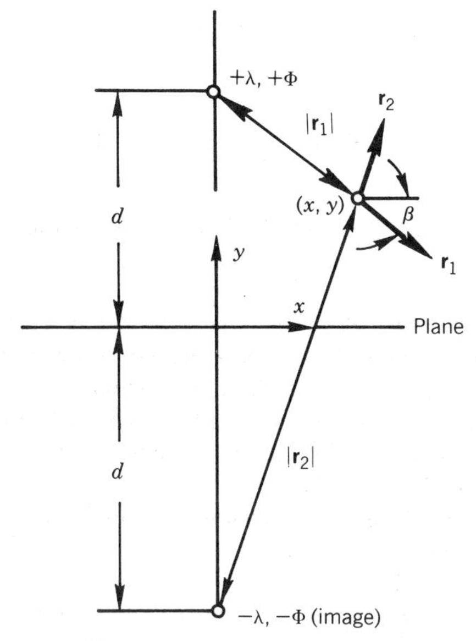

Figure 9.4. Fundamental wire-and-plane electrode system. The wire is treated as a line source of charge at a distance d above the plane. The image point is at a distance $-d$ from the plane.

field at the plate ($y = 0$ and $E_x = 0$) is

$$E_y = \frac{2(\phi/d)}{\ln(2d/R_1)}\left(\frac{1}{1+X^2}\right) \tag{9.27}$$

where $X = x/d$, and the potential is given by

$$\phi = \frac{\lambda}{2\pi\epsilon}\ln\left(\frac{2d}{R_1}\right) \tag{9.28}$$

The pressure rise above the ambient pressure under the corona discharge is

$$p - p_{\text{atm}} = \text{constant} \cdot \frac{\epsilon E_{y,\,\text{max}}^2}{2} \tag{9.29}$$

This relation is an approximate one that neglects space-charge effects but agrees fairly well with experimental information.

By combining Eqs. (9.28) and (9.29), one has

$$\frac{p - p_{\text{atm}}}{p_0 - p_{\text{atm}}} = \left(\frac{1}{1+X^2}\right)^2 \tag{9.30}$$

where the pressure at the origin is

$$p_0 - p_{\text{atm}} = \text{constant} \cdot 2\epsilon\left(\frac{\phi/d}{\ln(2d/R_1)}\right)^2 \tag{9.31}$$

For a single point electrode over a flat plate, the field at $y = 0$ for constant space charge is

$$E_y = -\frac{\lambda}{2\pi\epsilon d}\left(\frac{1}{1+X^2}\right)^{3/2} \tag{9.32}$$

Expressions similar to Eqs. (9.30) to (9.32) can be derived to approximate the pressure distribution.

Free Convection Systems. Based on data from corona wind impingement and visual observation of the thinning of the thermal boundary layer in the stagnation region, the flow from a wire electrode can be modeled approximately as a two-dimensional plane jet [69]. Its impingement on the surface can be considered to interact with the velocity and thermal boundary layers with the free-stream velocity following the pressure-vs.-distance relation given by Eq. (9.30).

A von Kármán–Pohlhausan boundary-layer analysis was done by Marco and Velkoff and compared with experimental data for a flat plate [69]. First, the assumption was made that the space charge ρ_c is a scalar constant, so that the flow field outside the momentum boundary layer could be considered irrotational. This is not strictly true, and numerical solutions should be sought when possible. For thin boundary layers, the velocity at the edge of the boundary layer is obtained from Bernouli's equation and the pressure-distance relation of Eq. (9.30), and is given by

$$\frac{\rho U_\delta^2}{2} = p_0 \left[1 - \left(\frac{1}{1 + X^2}\right)^2\right] \tag{9.33}$$

For a parabolic velocity and temperature distribution in the velocity and temperature boundary layers, i.e.,

$$\frac{u}{u_\delta} = \frac{2y}{\delta} - \frac{y^2}{\delta^2} \quad \text{and} \quad \theta = \frac{2y}{\delta_T} - \frac{y^2}{\delta_T^2},$$

the local heat transfer coefficient near the origin for air is given by

$$h_0 = 0.756k(\rho C^* I)^{1/4}(d\mu)^{-1/2} \tag{9.34}$$

and, generally, for large x/d,

$$h_x \approx 0.43 \frac{k}{\delta_*}\left[\frac{(\rho C^* I)^{1/4}}{(d\mu)^{1/2}}\left(\frac{\Lambda^{9/2}}{F(X)}\right)^{1/2}\right] \tag{9.35}$$

where

$$\delta_*^2 = \left(\frac{\delta_T}{\delta}\right)^2 = \frac{-4}{\Pr \delta_*(\delta_* - 5)}$$

$$\Lambda = \left[1 - \left(\frac{1}{1 + X^2}\right)^2\right]$$

$$F(x) = \int_0^X \Lambda^4 \, dx.$$

In Eq. (9.35), the constant C^* depends on the electrode and surface geometry and is the slope of the curve of stagnation pressure vs. total current divided by the total current.

If linear approximations to the velocity and temperature distributions in the boundary layers are used, the heat transfer coefficient near the origin is given by

$$h_0 \approx 0.67k(\rho C^* I)^{1/4}(d\mu)^{-1/2} \tag{9.36}$$

and generally, for large x/d,

$$h_x \approx 0.34 \frac{k}{\delta_*}\left[\frac{(I\rho C^* I)^{1/4}}{(d\mu)^{1/2}}\left(\frac{\Lambda^5}{F_1(X)}\right)^{1/2}\right] \tag{9.37}$$

where

$$F_1(X) = \int_0^X \Lambda^{9/2} \, dX$$

While these results are approximate, their comparison with experimental data is quite good. Figure 9.5 presents results for a horizontal, downward-facing heated plate

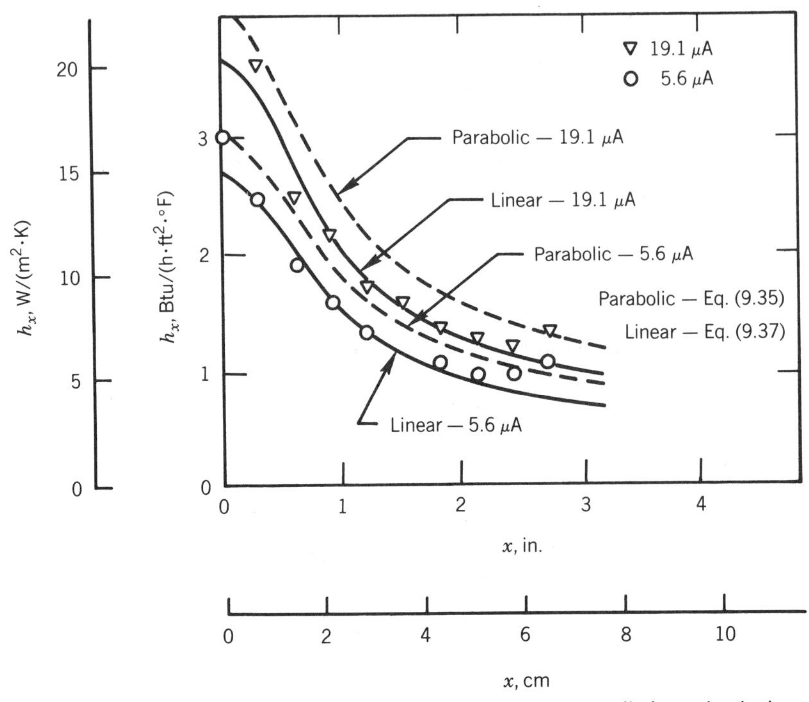

Figure 9.5. Convective heat transfer coefficient under a single corona discharge impinging on a downward-facing heated flat plate [69]. The electrode is a wire running the entire width of the plate.

with corona-wind impingement from below. It is seen that the approximate solutions bracket the experimental data quite well. Both the linear and parabolic solutions predict that local and average heat transfer coefficients will be proportional to $I^{1/4}$, and this is borne out in Fig. 9.6. It may be noted that the $I^{1/4}$ dependence in the heat transfer coefficient can be predicted by an approximate analysis via the Euler equation in the interelectrode space [43]. Generally, this current dependence is valid only when the ionic mobility is independent of the electrode field strength.

A numerical analysis of the impingement flow and heat transfer problem for a single wire electrode underneath a horizontal heated plate has been recently done in conjunction with a fundamental experimental study [137, 138]. The numerical calculations were carried out over a domain bounded by rigid walls at $y = 3d$ and $x = 3d$, whereas the experimental domain was open in the horizontal plane and above the wire (Fig. 9.7). The advantage of the numerical calculations is that one can include charge-density variation and its coupling to the electric field and velocity distributions.

The method of solution involves decoupling the electric field problem first and then using the resulting charge-induced body force in the momentum equation to obtain the velocity field. The heat transfer problem is then solved with the velocity field determined. The equations can be written in stream-function–vorticity form for a strictly laminar flow. Two dimensionless parameters characterize the solution. These are

$$\Gamma = \frac{d^3 J_y = 0}{\epsilon K \phi_{y=0}^2}\left(\frac{\rho_{c,\,y=0}}{\epsilon d}\right)\left(\frac{d}{\phi_{y=0}}\right) \tag{9.38}$$

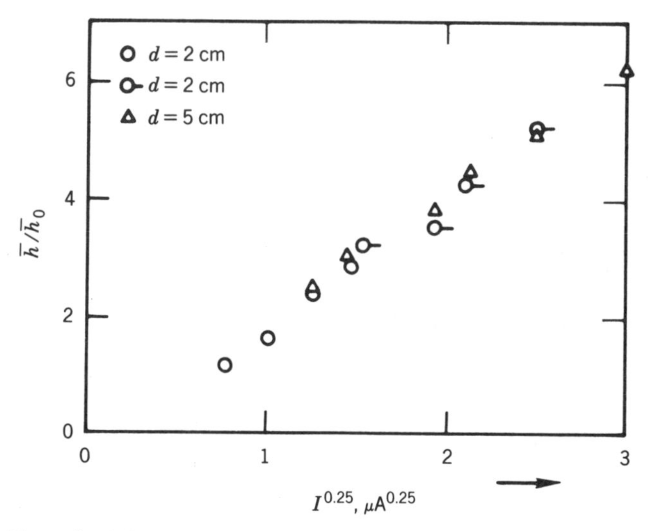

Figure 9.6. Normalized free convective heat transfer coefficients under a single corona-wind impingement on a downward-facing heated flat plate [69].

which represents the ratio of the electric field produced by space charge to the average field along the axis between the wire and the wall, and

$$\mathrm{Re}_{\mathrm{EHD}} = \frac{d}{\nu}\left(\frac{J_{y=0}d}{\rho K}\right)^{1/2} \tag{9.39}$$

which is the electric Reynolds number.

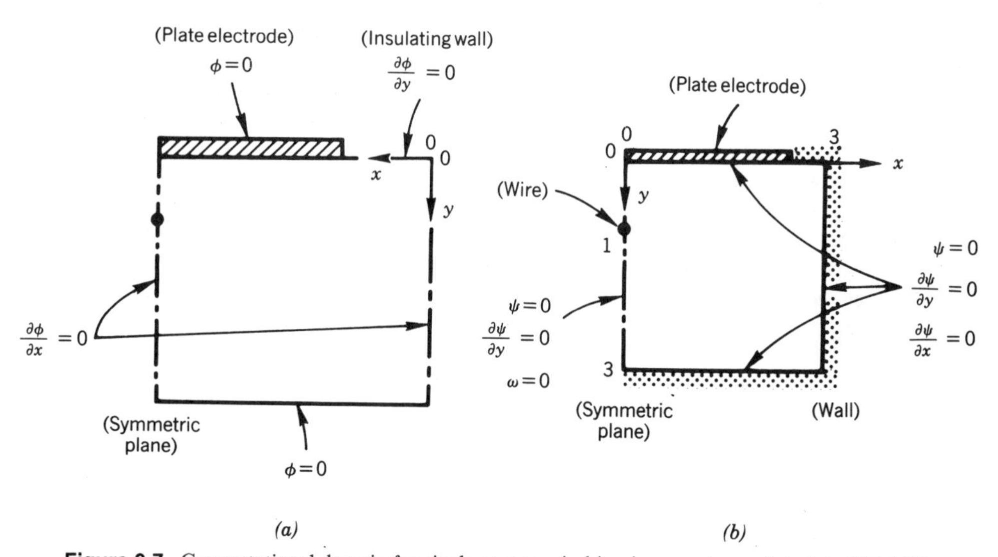

Figure 9.7. Computational domain for single corona-wind impingement on a flat plate [137, 138].

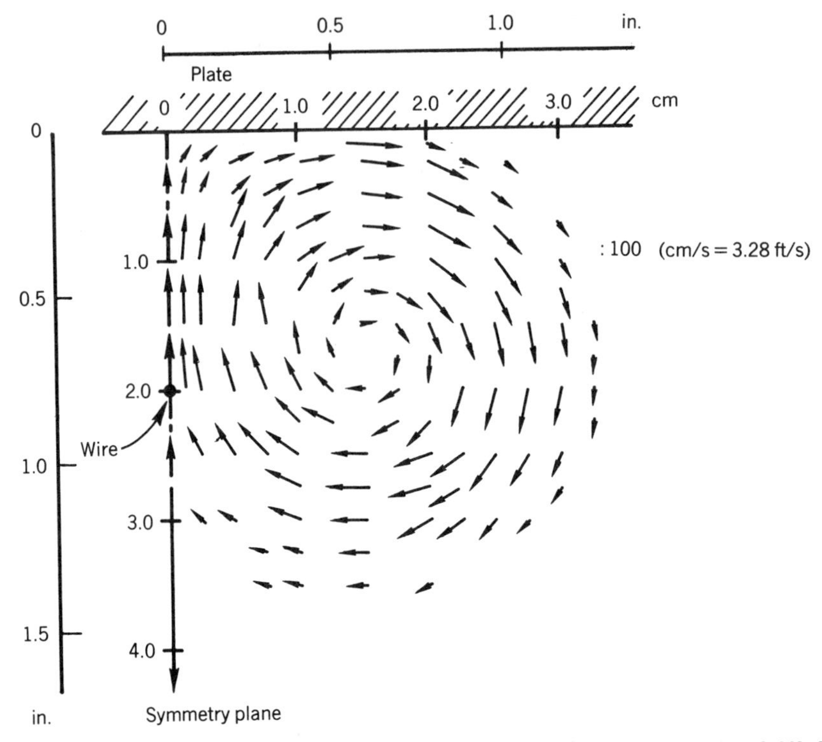

Figure 9.8. Velocity field produced by a single corona discharge over a flat plate [138]. The solution is for laminar flow. Flow distortions due to the presence of the wire are not taken into account.

Numerical solutions are available for $\mathrm{Re}_{\mathrm{EHD}} = 1290$, $\Gamma = 0.89$, and other conditions comparable to experiments conducted by Yabe et al. [137]. The velocity field is shown in Fig. 9.8 and does not take into account the distortion of the flow field by the corona wire and the inevitable turbulence of the electric wind. The heat transfer results are dependent on the electric Péclet number, $\mathrm{Pe}_E = \mathrm{Re}_{\mathrm{EHD}}\mathrm{Pr}$, because of the neglect of free-convection effects. Local and stagnation point heat transfer coefficients agree well with experimental results (Figs. 9.9 and 9.10). For local heat transfer coefficients, the agreement between theory and experiments is best near the stagnation point.

A wholly experimental study for a horizontal heated plate underneath a single corona discharge has been reported by Mitchell and Williams [75]. Their apparatus was an electrically heated foil, 22.9 cm square, backed by an epoxy–fiber-glass board. Average heat transfer coefficients over the length x have been correlated by the following expression:

$$\overline{\mathrm{Nu}_x} = 3.82\,\mathrm{Pr}^{1/3}\left(\frac{x_n}{d}\right)^{0.496}\left(\frac{\rho\epsilon(\phi - \phi_0)^2}{\mu^2}\right)^{0.216} \tag{9.40}$$

where x_n is the lateral distance from the wire location over which the average Nusselt number is desired.

For a vertical plate in free convection, impingement by a corona discharge from a single wire has not been investigated as extensively as for the horizontal plate. Some

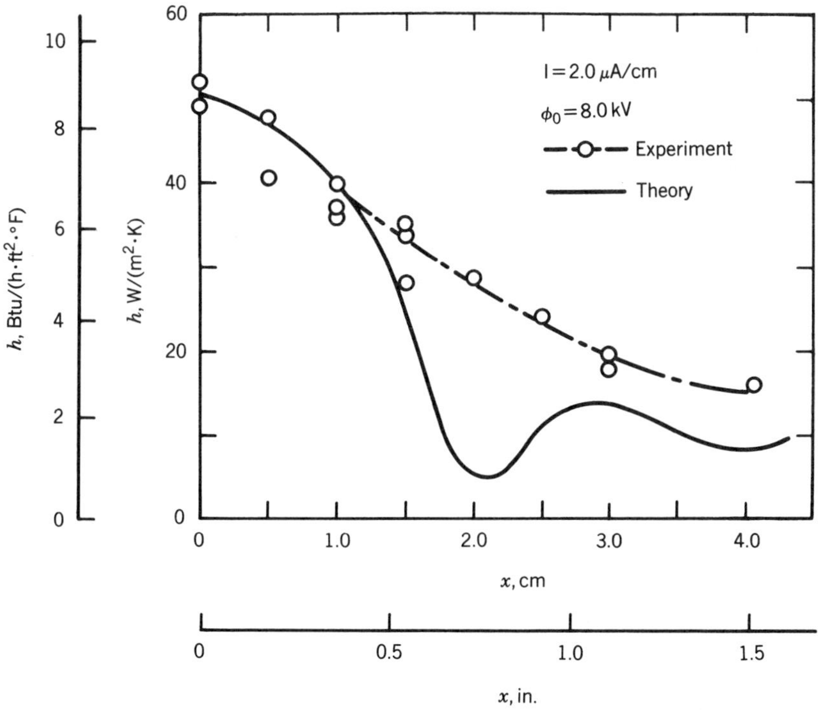

Figure 9.9. Heat transfer coefficients in the vicinity of the stagnation region of a wire-plane electrode system for a downward-facing heated plate [137]. Differences between theory and experiment are possibly due to secondary flow produced by a restricted computational domain.

data [84] are presented in Fig. 9.11. Note here that the discharge current must be above a certain level to fully eliminate secondary flows.

The use of a single point discharge electrode has been considered in a few experimental studies [43, 75]. No analytical or numerical studies of the heat transfer and fluid flow problems have been attempted. Figure 9.12 presents augmented heat transfer coefficients for various gases. Mitchell and Williams [75] have obtained laboratory data for a point-plane electrode system similar to the wire-plate electrode pair described above. The radial average heat transfer coefficients can be represented by

$$\overline{\mathrm{Nu}}_r = 0.295\,\mathrm{Pr}^{1/3}\left(\frac{r}{D}\right)^{0.206}\left(\frac{\rho\epsilon(\phi-\phi_0)^2}{\mu^2}\right)^{0.393} \tag{9.41}$$

where r is the radial distance from the discharge point over which the heat transfer coefficient is desired.

9.3.2 Impingement by Multiple Corona Discharges

Free-Convection Systems. Multiple corona discharges to an external boundary-layer flow involve corona-wind interaction in both the discharge region and the boundary layers on the surface. Electrode geometry, electrode distance from the plate, distance

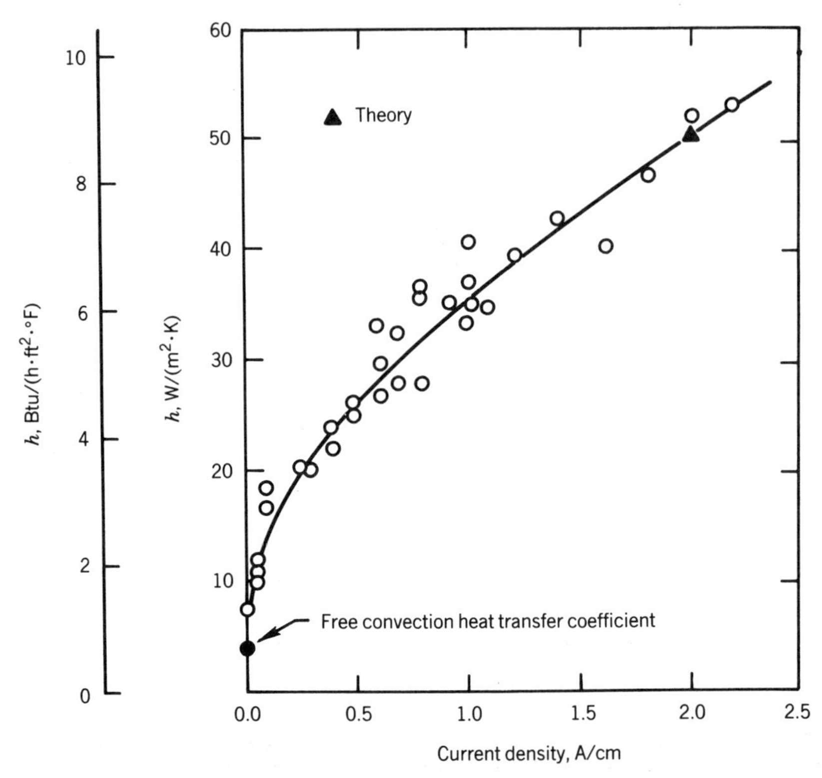

Figure 9.10. Stagnation-point heat transfer coefficients for a wire-plane system [137].

between electrodes, and type of gas will determine the increase in the heat transfer coefficient above free convection values. A fundamental numerical and experimental study to predict flow-field interaction and heat transfer coefficients have been done by Yabe et al. [137] for a downward-facing flat plate, and experimental work has been done by Franke [25] for a vertical plate.

Yabe et al. [137] solved the problem numerically on a finite difference domain similar to that in Fig. 9.7. No generalized correlations for the Nusselt number were presented, however. Franke's experimental study [25] of a multiple-wire electrode system was conducted with alternate electrodes at opposite polarity. The electrical ground plane was some distance away from the plates, and the discharge electrodes were at the plate surface. Heat transfer coefficients were augmented by the secondary flow. Figure 9.13 shows average augmented heat transfer coefficients as a function of the electric field power.

Forced-Convection Systems. Multiple-electrode coronas in forced convection have somewhat a greater practical value than a single discharge. Despite this, only a few fundamental studies have appeared from which broadly applicable data can be obtained. Sadek et al. [107] presented experimental measurements of augmented mass transfer coefficients and used the analogy between heat and mass transfer to generalize the results. The boundary layer on the plate was subjected to impingement flow from an array of wires or points (common household pins). The wet surface was maintained

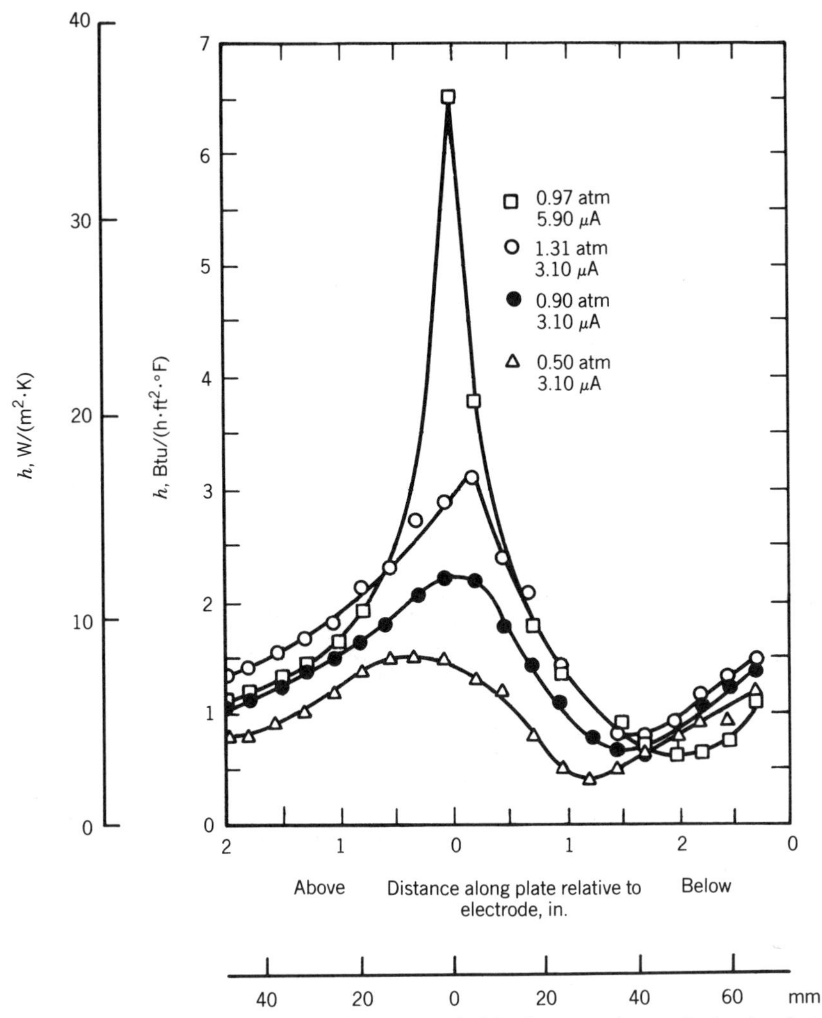

Figure 9.11. Heat transfer coefficients for corona-wind impingement from a single wire electrode to a heated vertical plate [84] in air. Wire diameter is 0.1 mm, and $d = 20$ mm.

at ground potential. Drying rates were determined by mass lost over a given time, and average mass transfer coefficients were determined from the usual definition using the partial pressure at the sponge surface. With corona discharge present, mass transfer rates were found to increase directly with corona voltage, or the square root of total corona current. Their data are correlated by

$$\mathrm{Nu} = \mathrm{Nu}_0(1 + 1.85\chi) \tag{9.42}$$

where χ is the ratio of ion drag forces to momentum forces in the free stream and is given by

$$\chi = \text{constant} \cdot \left(\frac{\epsilon}{\rho}\right)^{1/2}\left(\frac{\phi - \phi_0}{dU}\right) \tag{9.43}$$

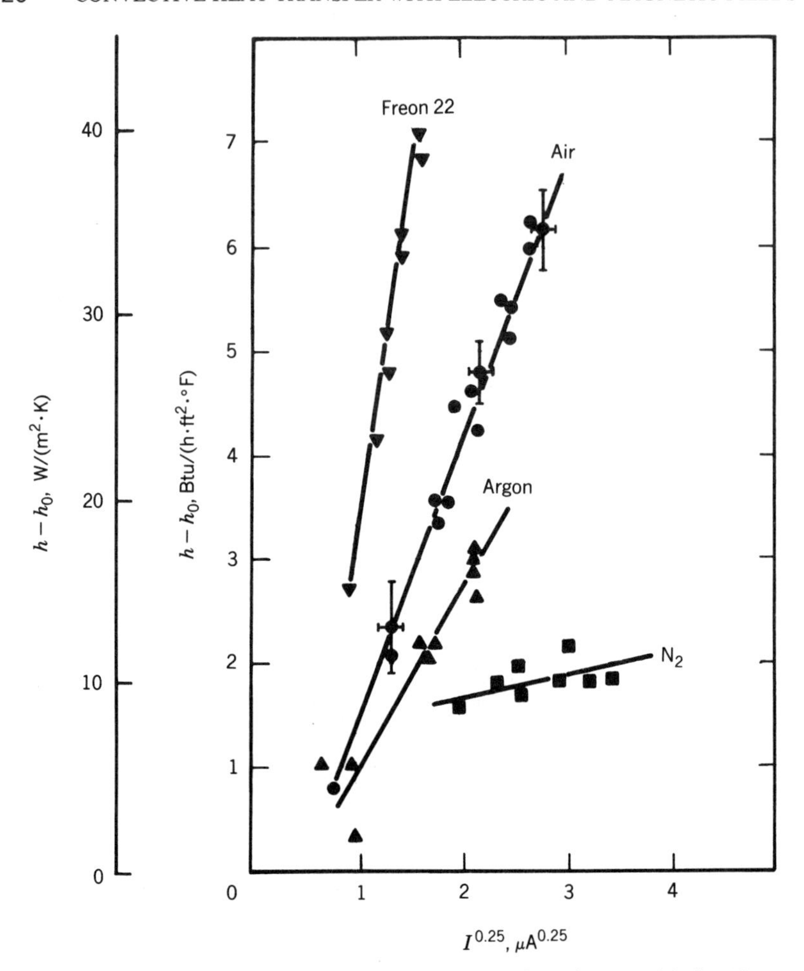

Figure 9.12. Increase in convective heat transfer coefficients for a downward-facing 10-cm square plate at 212° F (100° C) [43]. A point electrode is 5 cm from the plate.

or equivalently,

$$\chi = \text{constant} \cdot \left(\frac{I}{K \rho U^2} \right)^{1/2} \tag{9.44}$$

and $0.1 < \chi < 10$. This relation represents the data well with the constant set equal to unity for either point or wire electrodes over a range of Reynolds number based on the length of the mass transfer surface of 1000 to 10,000. Since the distance of the wet surface from the leading edge of the plate was not reported, the range of local Reynolds numbers was actually larger than reported.

A study somewhat similar to that of Sadek et al. was reported by Velkoff and Godfrey [127] for wire electrodes parallel to the plate in the direction of the flow and for a uniformly heated plate. Horizontally averaged temperature distributions were obtained using a Mach-Zhender interferometer with $d = 6.35$ mm and a wire spacing of 12.7 mm. Free-stream speeds varied from 9.8 to 1.9 m/s, and local Nusselt numbers

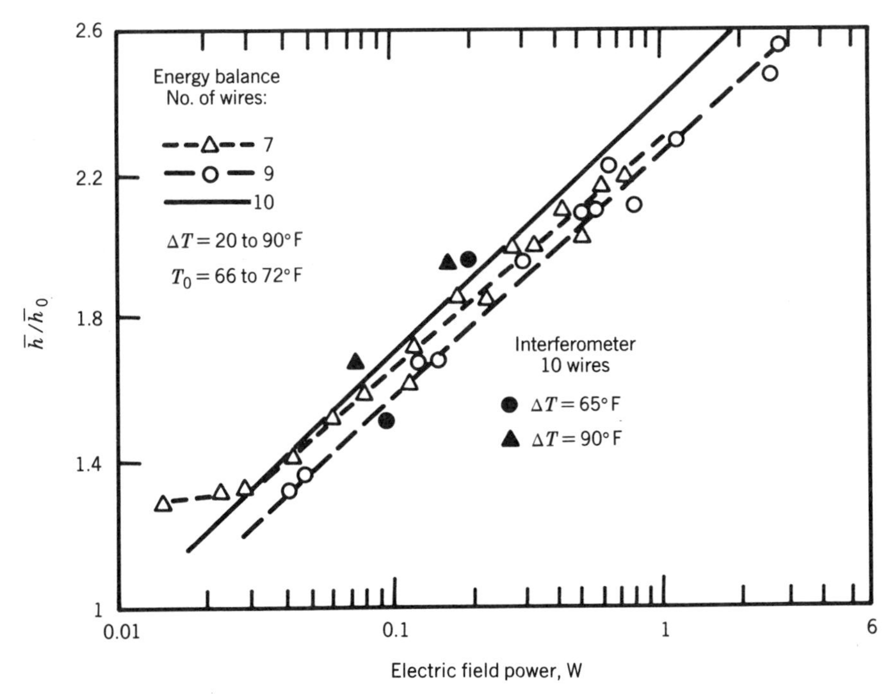

Figure 9.13. Augmented heat transfer coefficients for a heated plane with multiple electrodes [26].

were measured at two longitudinal locations from the leading edge (178 and 203 mm). The impingement of the corona wind produced marked increases in average Nusselt number for low free-stream velocities. For $U/U_{f,\max} > 30$, where $U_{f,\max}$ is the maximum velocity for free convection, no augmention was measured. This result was independent of corona current over the range $200 < I < 1000$ μA. Experimental data are shown in Fig. 9.14.

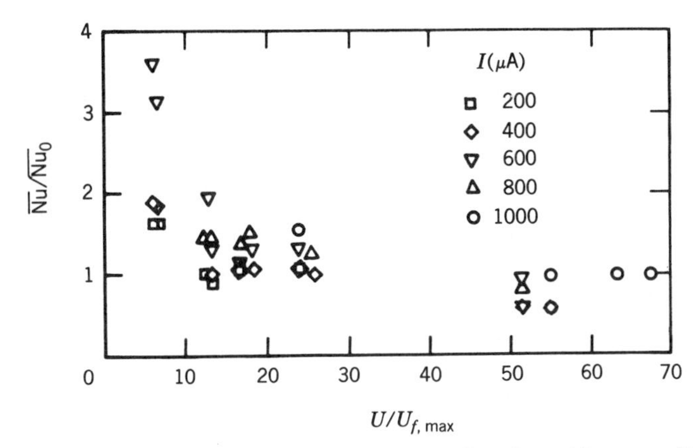

Figure 9.14. Nusselt numbers in forced convection over a flat plate with corona-discharge wire running parallel to direction of flow [127].

9.4 EHD IN CONFINED FLOWS

9.4.1 Free Convection Systems

Free convection augmented by corona discharge in enclosures has received very little attention. This is, perhaps, a result of the inherent difficulty in making accurate measurements in free convection and the complicated recirculating flow that can result from the induced corona wind. The previously discussed study by Yabe et al. [137] of a single-wire discharge above a flat plate actually treated recirculating flow in a rectangular enclosure with a partly heated upper wall. A specialized enclosure flow has been investigated by Franke and Hutson [26]. Wire electrodes were mounted near the inner surface of a vertical cylinder and alternately held at opposite polarities. Recirculating flows tended to break up the boundary layer on the wall and produce an increase in the average heat transfer coefficient. No correlations of results were given.

9.4.2 Tube Flows

In tubes, a corona discharge introduced on a small-diameter wire located at the centerline has been found to increase friction factors and heat transfer rates for low Reynolds number (usually laminar) flows. The augmentation of convective heat transfer in this geometry is attributed both to corona-wind impingement on the flow boundary and to the interaction of the wind with the bulk gas flow. The extent to which the discharge affects heat transfer is dependent upon both the discharge structure and characteristics of the primary bulk gas flow. At large Reynolds numbers, bulk convection dominates and the corona discharge process has negligible effect on wall transport rates.

For tube flows with a corona wire along the axis, theory indicates that a uniform discharge along the central wire does not result in the formation of a corona wind. The nonperturbing nature of a uniform corona discharge in this geometry is verified by the close agreement found by Weaver [134] between measurements of the pressure differential in a horizontal cylinder and the analytical solution for the radial pressure distribution,

$$\Delta p(r) = \frac{I}{2\pi LK} \ln\left(\frac{R_2}{r}\right) \tag{9.45}$$

Earlier investigations of the effects of ionization on laminar flow and heat transfer in this geometry [49, 77, 109, 124–126] indicate that the effects are not restricted to modifying the radial pressure distribution, but that the discharge is responsible for a jetlike corona wind which creates circulating gas patterns within the tube. These apparent discrepancies in data are explained by differences in the structure of the discharge along the wire. The discharge is normally characterized by discrete discharge spots along the wire. This localized discharge structure provides an explanation for the measured increases in pressure drop and heat transfer rates. The interaction of the jetlike corona wind originating at each discharge spot with the bulk gas flow, and the possible impingement of the jets at the flow boundary, are the key elements in the heat transfer augmentation.

Moss and Grey [80] explained the hydrodynamics of EHD heat transfer augmentation in terms of the physical model depicted in Fig. 9.15. At the threshold voltage, a few discrete discharges appear along the central wire. These discharges set up the corona-wind jet depicted, which produces a region of impingement. As the voltage is

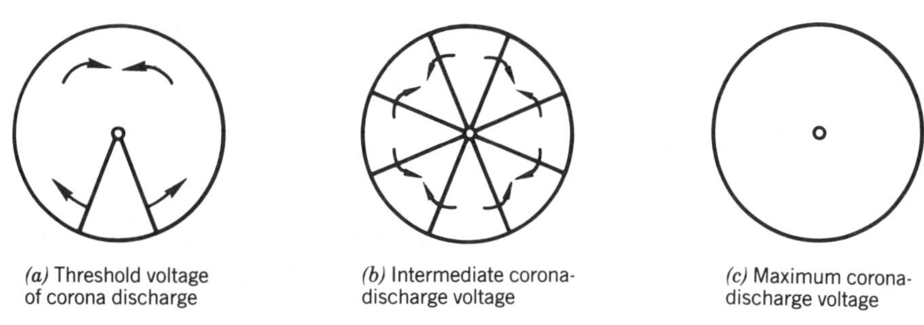

Figure 9.15. Hydrodynamics of EHD-augmentation in tube flow according to Moss and Grey [80].

increased, more discharge spots appear and numerous impingement flows are formed. As the discharge spots become more numerous, recirculating regions of flow are produced. At some voltage level, the discharge spots are so close together that the discharge can be considered nearly uniform. At this point, no corona wind exists and heat transfer rates revert back to their original values at a given Reynolds number. It should be noted that this sequence of events is only possible at low Reynolds numbers. Secondary flows generated by the corona wind are effective only when the bulk momentum forces are not so strong that the radially directed corona jets are swept downstream.

Velkoff [124–126] investigated the effects of positive ionization on laminar flow and heat transfer in a tube of 32-mm (1.25-in.) diameter with a corona wire of 0.1-mm (0.004-in.) diameter. Data on the average heat transfer coefficient obtained in this geometry for CO_2 and a starting length of 13 diameters are shown in Fig. 9.16. Heat

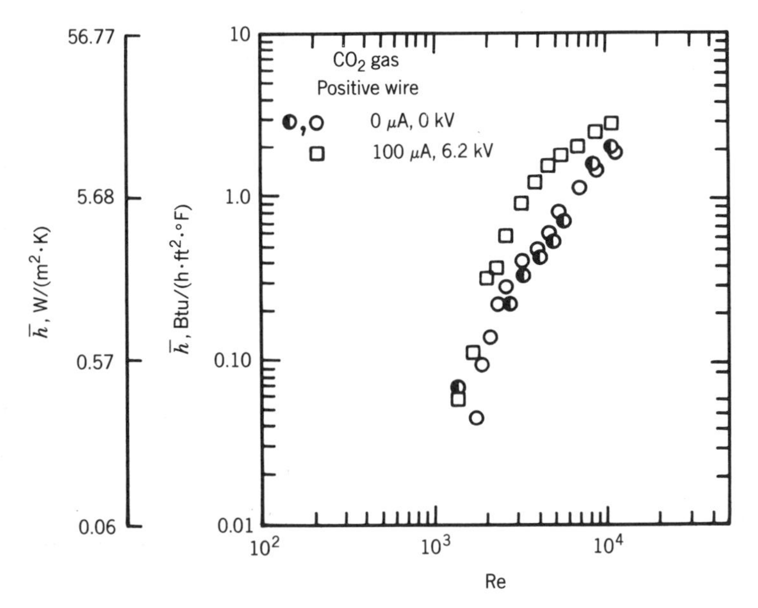

Figure 9.16. Augmented heat transfer coefficients [124, 126] for CO_2 flowing in a constant wall temperature tube with a positively charged corona wire on the centerline. The tube length-to-diameter ratio is 51 : 1 [124, 126].

transfer coefficients double from the laminar to the turbulent flow regime; but at a Reynolds number of approximately 10,000, the augmentation tends to decrease. A subsequent study [109] also established a peaking trend in augmentation of the Nusselt number with increasing Reynolds number. Unfortunately the early data are too sparse to permit a correlation of heat transfer coefficients.

Moss and Grey [80] conducted a study similar to that of Velkoff in which nitrogen was used as the test fluid. Typical data are shown in Fig. 9.17, where the augmented heat transfer coefficients are represented in terms of the corona power normalized by the total heat transfer rate without the corona discharge. The trend in the data is explained by their physical model discussed earlier. A method proposed to calculate the magnitude of the corona wind velocity resulting from the discharge at the central wire is based on relating the pressure drop as expressed in Eq. (9.45) to the velocity through Bernoulli's equation. An estimate of corona-wind velocity obtained in this manner is given by

$$U_e = \left[\frac{1}{2\pi LK}\ln\left(\frac{R_2}{R_1}\right)\right]^{1/2} \tag{9.46}$$

This result neglects viscosity effects and is not accurate at large values of corona currents.

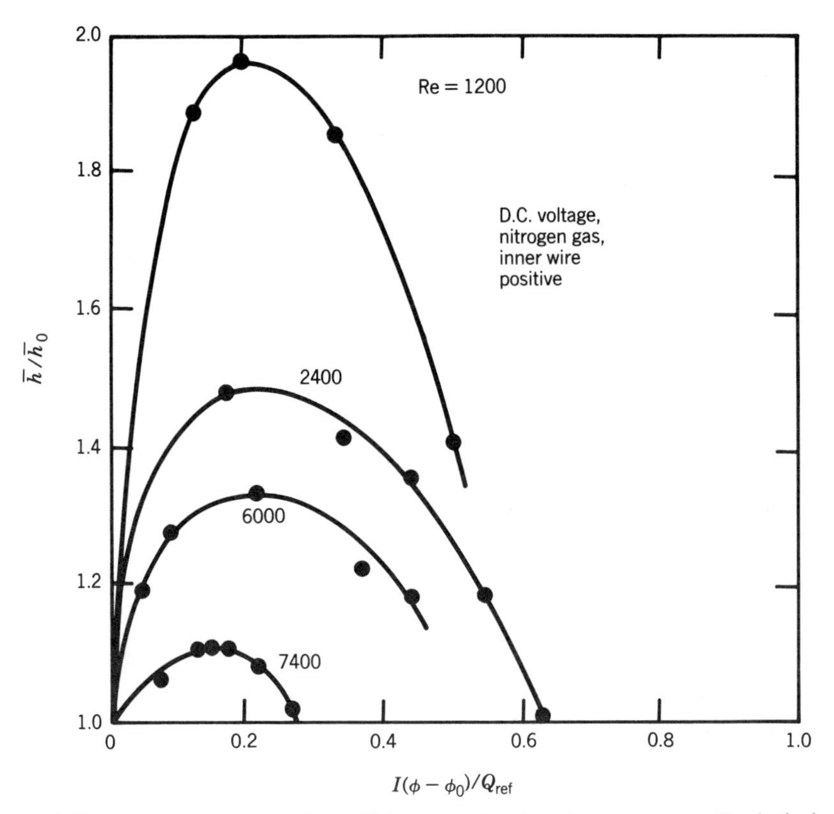

Figure 9.17. Augmented heat transfer coefficients as a function of corona power. Q_{ref} is the heat transfer rate to the gas when no electrical power is applied to the corona wire [80].

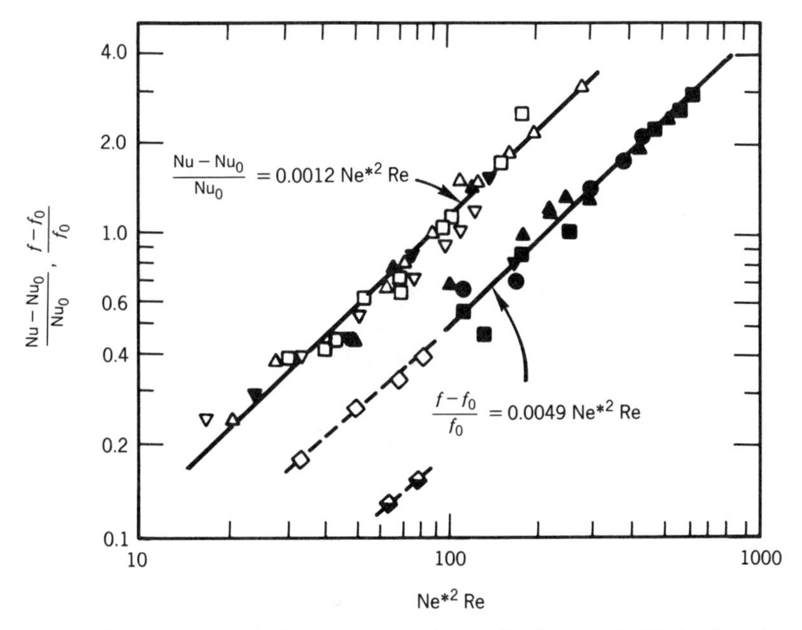

Figure 9.18. Heat transfer and friction-factor results of Mizushina et al. [77] for forced convection in tubes with a negatively charged corona wire along the centerline.

More recently, Mizushina et al. [77] conducted an extensive study of EHD-augmented forced convection in an annulus with a corona-discharge wire along the inside-tube centerline. The annulus was operated as a counterflow heat exchanger with hot air flowing in the inside tube and cooling water flowing in the outer jacket. Correlations of their data for Nusselt numbers and friction factors are presented in Fig. 9.18 in terms of the dimensionless group

$$\mathrm{Ne}^* = \frac{R_2 I}{\pi K R_1^2 \rho U^2} \tag{9.47}$$

When the Reynolds number is well below the value corresponding to transition to turbulent flow, significant levels of augmentation are seen and the heat transfer data are correlated by

$$\frac{\Delta\,\mathrm{Nu}}{\mathrm{Nu}_0} = 0.012\mathrm{Ne}^{*2}\mathrm{Re} \tag{9.48}$$

Similarly the friction factor is correlated by

$$\frac{\Delta f}{f_0} = 0.0049\mathrm{Ne}^{*2}\mathrm{Re} \tag{9.49}$$

At higher Reynolds numbers where the usual turbulent heat transfer correlations hold, no correlation including the EHD parameter was found.

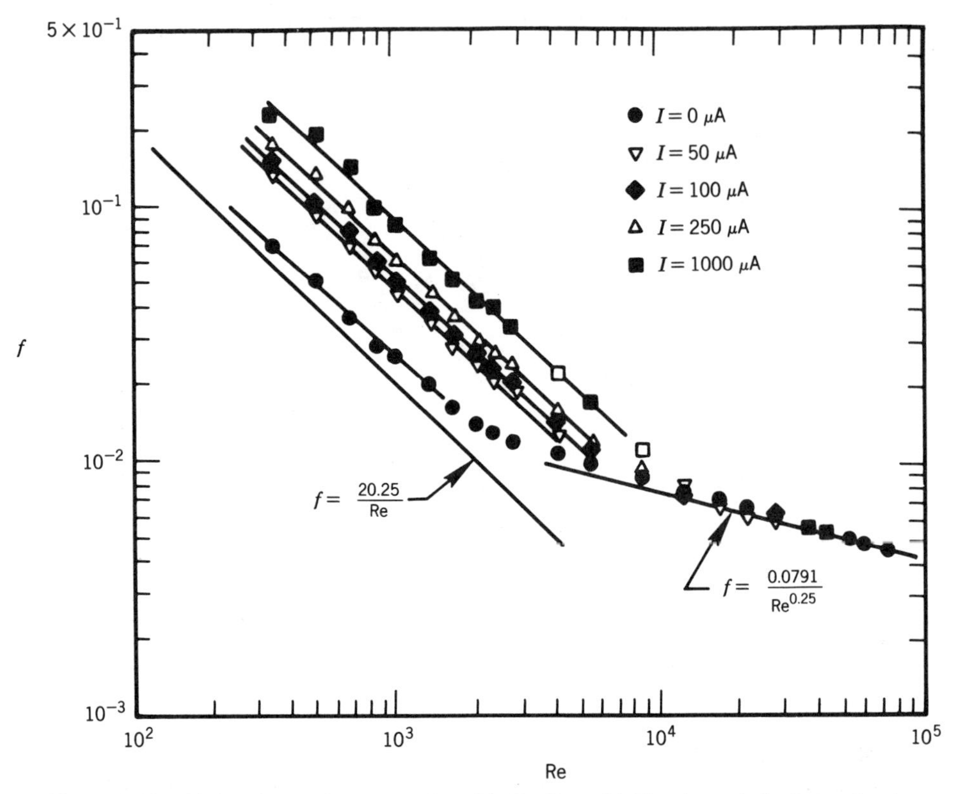

Figure 9.19. Friction factors for an annulus with $R_2/R_1 = 74$. The theoretical relation for the friction factor in fully developed laminar flow is also shown [49].

Recent experimental data obtained by Kulacki et al. [49], in a study of EHD effects on catalytic combustion of hydrogen in air, show how friction factors for an annulus approach the Blasius relation for turbulent flow in a smooth-wall tube at large Reynolds numbers. An experimental reactor comprising a tube and positive corona central wire with a radius ratio of approximately 74 was used. Friction factors for the annulus are shown in Fig. 9.19 along with the theoretical relation for friction factor in fully developed laminar flow.

9.4.3 Channel Flows

In flat rectangular channels, corona discharges from longitudinal or transverse wire electrodes are spatially nonuniform due to the geometry of the electrode configuration. Thus, unlike tube flows, even a uniform discharge along the wire results in an electrically induced flow modification. Past experimental and theoretical studies in this geometry have centered on the hydrodynamic process which results from the presence of the discharge in an otherwise well-understood flow field. The corona-discharge interaction with the bulk gas flow can result in circulation zones in the vicinity of the discharge wires and substantial increases in free-stream turbulence and diffusivity levels. Unfortunately, an extension of these results to EHD augmentation of convective heat transfer has not yet appeared in the literature.

The early study of a uniform wire discharge in a laminar flow by Ramadan and Soo [98] and the more recent work by Yamamoto and Velkoff [139] provide an analytical basis for the early hypotheses of the existence of circulating secondary flow patterns in channel flows. Yamamoto and Velkoff show how cross flow affects the corona wind from either single or multiple wires at the centerline of a two-dimensional channel. The streamline contours indicate that, at a constant discharge current, regions of recirculating laminar flow between the wire and the channel walls are moved downstream as the Reynolds number is increased. Above a Reynolds number of 5000, secondary flows are inconsequential in comparison with the bulk momentum transport. Recent experimental [19, 20, 59] studies with a negative discharge indicate that, in the presence of a localized spot discharge, the interaction of the jetlike corona wind with low-speed cross flows (gas speeds under approximately 3 m/s) causes as much as a threefold increase in turbulence levels.

Friction factors measured in a $\frac{5}{8}$-in. × 5-in. rectangular duct in which ten equally spaced positive corona discharge wires operated at equal current levels are shown in Fig. 9.20 [90]. The data are presented in terms of both the total current and the

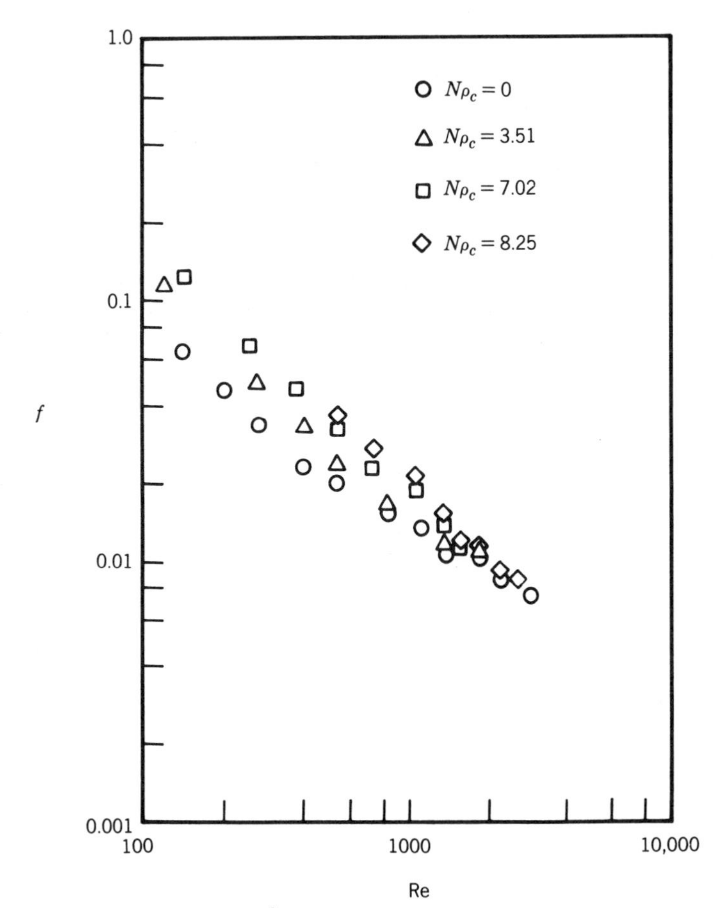

Figure 9.20. Friction factors for a $\frac{5}{8}$-in. × 5-in. rectangular duct with ten positive corona-discharge wires [90].

dimensionless parameter

$$N\rho_c = \left(\frac{2I\epsilon}{A_w dK} \right)^{1/2} \frac{d^2}{\mu K} \tag{9.50}$$

As reported for tube flows, friction factors were found to increase at low Reynolds numbers with increasing corona current. No appreciable EHD effects were noted at Reynolds numbers above 1000.

9.5 BASIC CONCEPTS OF MAGNETOHYDRODYNAMICS (MHD)

9.5.1 Governing Equations

In classical MHD theory, the fluid is considered to be a continuum. The transport coefficients, e.g., electrical conductivity, are assumed to be isotropic and the fluid to be electrically neutral. It is assumed further that the dielectric constant ϵ and the permeability μ_e are scalars, the net space-charge density ρ_c is neglected, and the convection (displacement) and polarization currents are ignored.

When an electromagnetic field is applied to an electrically conducting fluid at rest, four forces can arise [85]. These are electrostatic (forces applied on particles with free electric charges), ponderomotive (the macroscopic summation of the elementary Lorentz forces applied on charged particles), electrostrictive (forces resulting from variations in the dielectric constant with the mass density of the fluid), and magnetostrictive (forces arising from variations in the magnetic permeability with the mass density of the fluid). Typically, in MHD the ponderomotive force is the only one of the above forces that is comparable to other hydrodynamic forces. (There are exceptions–for example, in electrostrictive natural convection, in which ϵ is a function of the mass density of the fluid [47, 67].) Maxwell's equations for the fixed (laboratory) reference frame [14], written for the rationalized MKS (m-kg-s) system or SI units [39], subject to the aforementioned idealizations, are

$$\nabla \cdot \mathbf{B} = 0 \tag{9.51}$$

$$\nabla \cdot \mathbf{D} = 0 \tag{9.52}$$

$$\nabla \times \mathbf{H} = \mathbf{J} \tag{9.53}$$

$$\nabla \times \mathbf{E} = -\frac{\partial \mathbf{B}}{\partial t} \tag{9.54}$$

where

$$\mathbf{B} = \mu_e \mathbf{H} \quad \text{and} \quad \mathbf{D} = \epsilon \mathbf{E} \tag{9.55}$$

Ohm's law for this case is

$$\mathbf{J} = \sigma(\mathbf{E} + \mathbf{u} \times \mathbf{B}) \tag{9.56}$$

in which $\mathbf{E}$ is measured in the laboratory reference frame. As shown by this equation, the motion of a conducting fluid through an applied magnetic field contributes to the current density **J**. This, by virtue of Eqs. (9.53) and (9.55), implies that the applied field, in turn, will be altered. The resultant, or total, field is described by the magnetic induction equation,

$$\frac{\partial \mathbf{B}}{\partial t} = \frac{1}{\sigma \mu_e} \nabla^2 \mathbf{B} + \nabla \times (\mathbf{u} \times \mathbf{B}) \tag{9.57}$$

This equation is derived from Eqs. (9.56), (9.53), and (9.55) and illustrates the coupling of the electromagnetic and hydrodynamic fields. The electrical conductivity and density are assumed constant.

The governing hydrodynamic equations are the equations of conservation of mass, momentum, and energy. The momentum equation is

$$\rho \left[\frac{\partial \mathbf{u}}{\partial t} + (\mathbf{u} \cdot \nabla) \mathbf{u} \right] = -\nabla p + \mu \nabla^2 \mathbf{u} + \rho \mathbf{g} + \mathbf{J} \times \mathbf{B} \tag{9.58}$$

Magnetic field interaction with the flow occurs through the ponderomotive force, $\mathbf{J} \times \mathbf{B}$.

The energy equation is

$$\rho c_p \frac{DT}{Dt} = -\nabla \cdot \mathbf{q}'' + \Phi + q''' + \frac{J^2}{\sigma} \tag{9.59}$$

The term J^2/σ represents the dissipative energy resulting from Joule heating of the conducting fluid. Thus, the effect of an applied magnetic field enters the energy equation explicitly through Joule heating and implicitly through the viscous-dissipation and convective terms.

9.5.2 Dimensionless Groups

The dimensionless momentum equation is

$$\rho \frac{D\mathbf{u}}{Dt} = -\nabla p + \frac{\mathrm{Gr}}{\mathrm{Re}^2} \theta + \frac{1}{\mathrm{Re}} \nabla^2 \mathbf{u}$$

$$+ \frac{M^2}{\mathrm{Re}} (\kappa \mathbf{E} \times \mathbf{B}) + \frac{M^2}{Re} (\mathbf{u} \times \mathbf{B} \times \mathbf{B}) \tag{9.60}$$

where all vector operators are dimensionless. Also,

$$M^2 = \frac{B_0^2 L^2 \sigma}{\mu} \propto \frac{\text{ponderomotive force}}{\text{viscous force}} \tag{9.61}$$

and

$$\kappa = \frac{E_0}{U_0 B_0} \propto \frac{\text{applied electric field}}{\text{induced electric field}} \tag{9.62}$$

The parameter M^2/Re in Eq. (9.60) represents the ratio of ponderomotive to inertia forces. This ratio is referrred to as the magnetic interaction parameter N. The direction of the applied electric field is specified by the sign of κ.

The dimensionless energy equation is

$$\rho\frac{D\theta}{Dt} = -\frac{1}{\mathrm{Pr\,Re}}\nabla\cdot\mathbf{q}'' + \frac{\mathrm{Ec}}{\mathrm{Re}}\Phi + \mathrm{Ec}\,q''' + \frac{\mathrm{Ec}\,M^2}{\mathrm{Re}}(\mathbf{E}+\mathbf{u}\times\mathbf{B})^2 \qquad (9.63)$$

in which the scaling for the energy source is $L/\rho u_0^3$.

The dimensionless magnetic induction equation is

$$\frac{\partial\mathbf{B}}{\partial t} = \frac{1}{\mathrm{Re}_m}\nabla^2\mathbf{B} + \nabla\times(\mathbf{u}\times\mathbf{B}) \qquad (9.64)$$

For small values of Re_m, the applied field is altered solely by diffusion. For large values of Re_m, it is altered solely by convection. When $\mathrm{Re}_m = 0$, the electromagnetic and hydrodynamic equations are decoupled. In most terrestrial applications, $\mathrm{Re}_m \ll 1$ and weak interaction is assumed. Re_m, however, can approach a value of unity in large sodium electromagnetic flowmeters.

9.5.3 Basic Physics of Magnetic Field Effects in Electrically Conducting Liquids

As shown by the governing equations of laminar MHD flow, an applied magnetic field can affect the temperature of a liquid metal directly through Joule heating, and indirectly through altering the liquid metal velocity distribution and thereby convection and viscous dissipation. Joule heat generation is significant primarily in situations in which an electric field is applied externally or in which current flows through an external circuit or conducting duct walls. Usually viscous heat dissipation is negligible except in situations with very high velocity gradients, in which viscous and Joule heatings can be of the same order. In most situations concerning heat transfer between the liquid metal and a physical boundary, the heat transfer is affected by the alternation of the velocity gradient at the wall by the magnetic field. However, because liquid metals have low Prandtl numbers, the heat transfer is governed mostly by conduction. In turbulent MHD flow, heat transfer is affected primarily by the magnetic damping of turbulence.

Noticeable interaction of the magnetic field with the flow field occurs if the magnetic interaction parameter, $N = M^2/\mathrm{Re}$, is on the order of 1 or greater. The resulting ponderomotive force, $\mathbf{J}\times\mathbf{B}$, has two components that interact with the flow. The first component, $\sigma(\mathbf{u}\times\mathbf{B})\times\mathbf{B}$, always acts to decelerate the flow. The second component, $\sigma\mathbf{E}\times\mathbf{B}$, can act either to accelerate or to decelerate the flow, depending upon the direction of $\mathbf{E}$. It will accelerate the flow if $\mathbf{E}$ is opposite in direction to $\mathbf{u}\times\mathbf{B}$. When $\mathbf{E} = -\mathbf{u}\times\mathbf{B}$, the current density $\mathbf{J}$, and hence Joule dissipation, become zero. In this case, a deceleration of the flow occurs and the heat transfer is decreased. In general, it is not straightforward to determine the change in heat transfer as the result of an applied magnetic field. In some flow situations, the application of a magnetic field can increase the heat transfer rate; in others, it can decrease the rate.

9.6 MHD IN CONFINED FLOWS

9.6.1 Channel Flow

The most basic channel-flow problem is the Hartmann problem [105]. This is a one-dimensional incompressible laminar flow problem in which the spacing between the wall electrodes is large compared to that between the insulating side walls. The flow is fully developed, and no axial currents exist. The magnetic field is applied normal to the side walls. An electric field is established externally between the electrodes.

Romig [105] examined specialized cases of the Hartmann problem in which the mass flow is held constant as M and κ are varied, and in which κ is held constant as M is varied. As M is increased, convection near the wall increases and the temperature becomes more uniform. Internal Joule and viscous-dissipation heating also increase with increasing M. Viscous dissipation is maximum at the wall. The magnitude of Joule heating also depends on κ. When $\kappa = -1$ (the electrically insulated case), internal heating occurs very close to the walls. When $\kappa = 0$ (the open-circuit case) most of the Joule heating occurs near the center of the channel.

Blum et al. [3, 5] theoretically investigated the case of heat transfer for developed Hartmann flow through a channel with electrically insulating walls in a transverse magnetic field. For cases of either constant wall temperature or constant wall heat flux, the application of a magnetic field increases the heat transfer by approximately 30 and 45%, respectively. Most of this initial rise in heat transfer occurs when $M \lesssim 100$. This is indicative of the Hartmann effect, i.e., the flattening of the velocity profile with a concomitant increase in the velocity gradient at the wall.

Various aspects of heat transfer in a spatially developing laminar flow between parallel conducting walls with various applied magnetic field orientations have been considered theoretically by Rajaram and Yu [97]. Similar theoretical entrance studies with an applied transverse magnetic field have been conducted by Perlmutter and Siegal [94], Hsia [37], and others [97]. The entrance length for flow development was found to depend upon M, Re_m, $\hat{\phi}$, and the direction of the applied magnetic field. The entrance length for velocity decreased with increasing strength of an applied transverse field, whereas with a strong field applied parallel to the flow, it was found to increase even beyond the length for the ordinary hydrodynamic case. The inclination of the magnetic field was determined to have no effect on the heat transfer for low Pr with constant wall temperature. For low Pr with constant wall heat flux, the heat transfer depended weakly on the field inclination, increasing slightly as the inclination increased toward the parallel-field case.

The change in heat transfer for a rectangular channel that occurs in the transition region from laminar to turbulent flow has been measured by Kovner et al. [44]. The data (Fig. 9.21) reveal that the maximum reduction in heat transfer occurs at a Reynolds number equal to approximately twice the critical Reynolds number at a given M. The critical Reynolds number is that at which the flow becomes fully developed turbulent. The values of Re_{crit} are given in [8]. The magnitude of this reduction in heat transfer is proportional to M and decreases as M^2/Re decreases.

The case of turbulent flow between two parallel walls with constant heat flux in a transverse magnetic field was considered analytically by Krasil'nikov [46]. He obtained an expression for the fluid velocity from the semiempirical theory of Kovner [45]. Heat transfer results were approximated by the semiempirical formula [46]

$$\text{Nu} = 10.0 + 0.025\left(\frac{\text{Pe}}{1 + (236M^2/\text{Re})}\right)^{0.8} \tag{9.65}$$

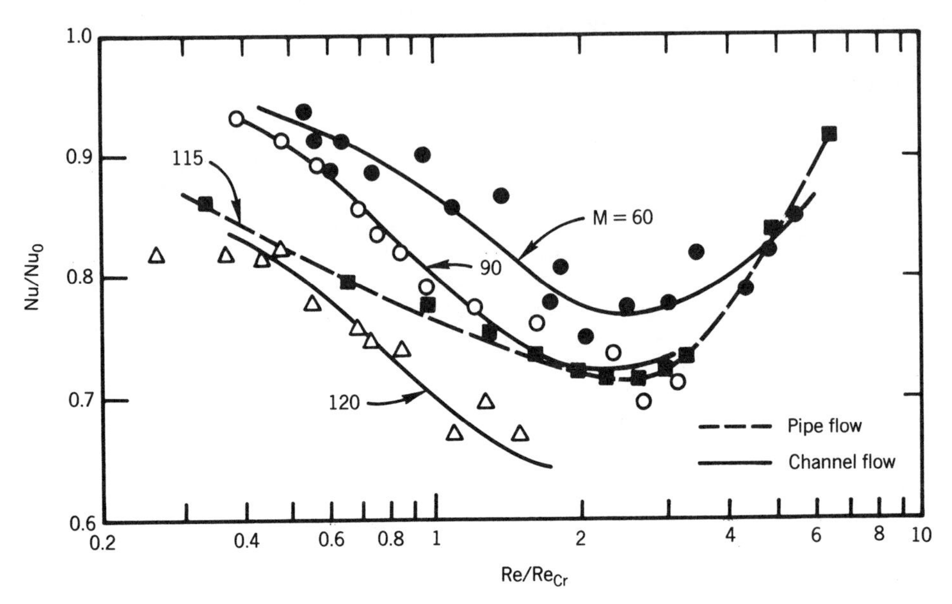

Figure 9.21. Heat Transfer in a transverse magnetic field during laminar-to-turbulent flow transition (adapted from Ref. 8).

in which the characteristic length is the channel width, and the heat transfer coefficient in Nu is based upon the difference between the wall and bulk mean fluid temperatures.

Branover [8] cited experiments performed by Krasil'nikov to study the effect of a longitudinal magnetic field on heat transfer in turbulent rectangular plane-parallel channel flow using gallium (Pr = 0.019). The data obtained for both the ordinary hydrodynamic ($M = 0$) and MHD ($M = 120$) cases are shown in Fig. 9.22. For a constant Pe, application of the longitudinal field was found to reduce the heat transfer. This reduction was caused by the suppression of turbulence and the average velocity gradient near the wall by the magnetic field. Over the range of gathered data

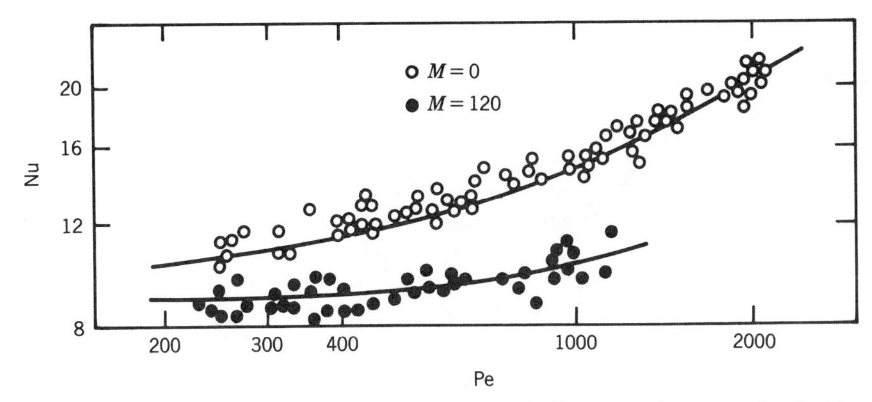

Figure 9.22. Heat transfer data for turbulent flow in a high-aspect ratio rectangular duct in a transverse magnetic field (adapted from Ref. 8).

($8 \lesssim \text{Nu} \lesssim 12$, $M = 120$, and $200 \lesssim \text{Pe} \lesssim 1200$), the MHD data were best approximated by

$$\text{Nu} = 9.0 + \frac{0.006\,\text{Pe}}{1 + \left(14.8M^2/\text{Re}\right)} \tag{9.66}$$

in which the characteristic length is the channel width. The heat transfer coefficient was determined from the heat flux through the wall, the difference between the wall inside surface temperature and the mean mixed temperature of the liquid. The characteristic temperature for the thermal conductivity was the arithmetic mean of the test-section inlet and outlet liquid temperatures.

Heat transfer experiments on combined free- and forced-convection flow through a vertical channel with conducting walls under the influence of a transverse magnetic field were conducted by Yang and Yu [140]. Application of the field to turbulent flow was found initially to reduce the heat transfer by suppressing free convection and turbulence. At higher field strengths, the flow became laminarized and the heat transfer increased because of the Hartmann effect. The point of minimum heat transfer was found to decrease linearly with increasing Gr/Re^2, i.e., from $\text{Re}/M = 217$ at $\text{Gr}/\text{Re}^2 = 0$ to $\text{Re}/M = 80$ at $\text{Gr}/\text{Re}^2 = 0.4$.

9.6.2. Pipe Flow

As in the case of channel flow, the heat transfer for laminar flow through a pipe with electrically insulating walls increases in the presence of an applied transverse magnetic field. Mittal [76] examined the intermediate Hartmann number cases of $M = 0.8$, 2.0, 2.8, and 4.0, and found that the temperature profile and heat flux at the wall acquire an angular dependence because of the symmetry of the applied transverse field. Increases in the local Nusselt number as much as 100% occurred when $M = 4.0$.

The analytical results of Gardner [27] for a constant wall heat flux show that the average Nusselt number increases approximately 60% as M increases from 0 to 500. (Here the characteristic length for Nu and M is the pipe diameter.) Most of the increase occurs from $1 < M < 100$. The Hartmann number range over which this increase occurs is the same as that predicted by Blum et al. [3, 5] for the analogous channel-flow case.

Experimental heat transfer studies for transition and moderate turbulent flow in an electrically insulated pipe were conducted by Gardner et al. [29] using mercury ($\text{Pr} = 0.023$). These results are shown in Fig. 9.23, in which the characteristic length chosen was the pipe diameter. The data for $\text{Re} < 10^4$ show no effect on heat transfer, because heat is transferred primarily by conduction. As Re is increased, a decrease in heat transfer then occurs because of the damping of turbulence by the magnetic field. For higher Re, Nu increases because the inertial force becomes much larger than the ponderomotive force and turbulent mixing dominates.

In the turbulent flow regime, experiments were carried out by Gardner and Lykoudis [28]. The effect of a transverse magnetic field on local and average heat transfer was measured for flow through an electrically insulated pipe with constant wall heat flux. For $\text{Re} < 50{,}000$, the local heat transfer coefficient depends upon angular orientation with respect to gravity and the applied field. This free-convection effect exists up to $\text{Re} = 315{,}000$. As the strength of the magnetic field increased, the centerline temperature of the fluid was lowered, and the temperature near the wall increased. These results demonstrate that the overall influence of the applied field is to inhibit convective heat transfer.

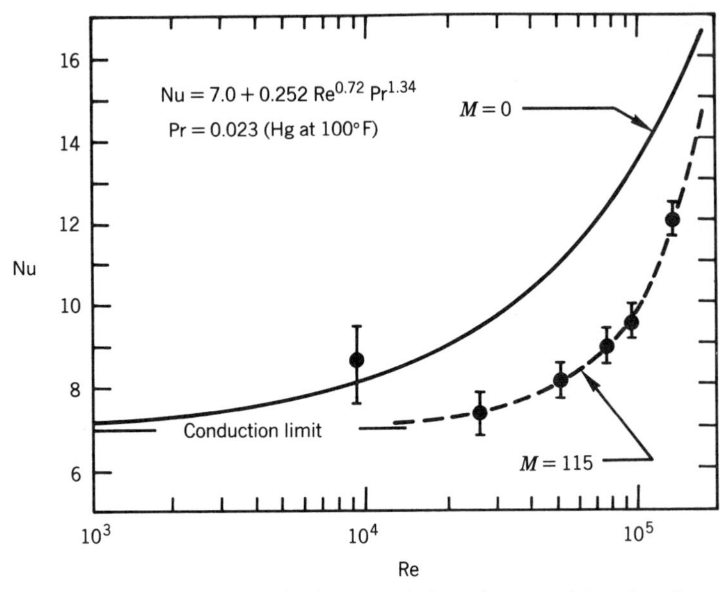

Figure 9.23. Heat transfer during laminar-to-turbulent flow transition for pipe flow in a transverse magnetic field [29].

The effect of a longitudinal magnetic field on heat transfer for turbulent flow of gallium (Pr = 0.019) in an electrically insulated pipe was measured by Kovner et al. [44]. Their results are shown in Fig. 9.24. The overall effect of the longitudinal field is to decrease the heat transfer rate. For low values of Pe ($\lesssim$ 200), the magnitude of this effect decreases. At Pe $\approx$ 700, suppression of the heat transfer rate is greatest. For high values of Pe ($\gtrsim$ 2000), the applied field has little effect, primarily because the inertial force becomes much greater than the ponderomotive force, i.e., M^2/Re becomes low.

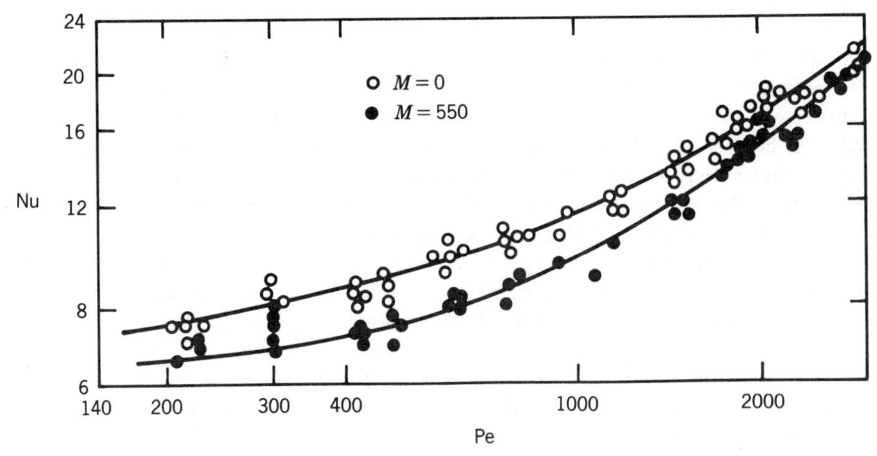

Figure 9.24. Heat transfer data for turbulent pipe flow in a longitudinal magnetic field (adapted from Ref. 8).

The experimental heat transfer results are best described by the expression

$$\mathrm{Nu} = 6.5 + \frac{0.005\,\mathrm{Pe}}{1 + 1890(M/\mathrm{Re})^{1.7}} \tag{9.67}$$

in which the characteristic length is the pipe diameter. (Here the Nusselt number is for fully developed turbulent flow based upon the difference between the inside wall surface temperature fluid bulk mean temperature.) The heat transfer coefficient is determined in the same manner as that for Eq. (9.65) and the thermal conducitivity is based upon the bulk mean temperature. The mean temperature ranged from 22 to 33 K, and the temperature difference from 1.4 to 2.5 K. A theoretical expression similar to the above was developed by Lykoudis [65], based upon his theory of turbulence damping due to the presence of a magnetic field [143, 144]:

$$\mathrm{Nu} = \mathrm{Nu}_c + \frac{\mathrm{Nu}_0 - \mathrm{Nu}_c}{1 + \left(250M^2/\mathrm{Re}^{1.75}\right)} \tag{9.68}$$

in which the characteristic length is the pipe diameter, Nu_c is the Nusselt number value for pure conduction ($\mathrm{Nu}_c \approx 7.0$), and Nu_0 is that for the ordinary hydrodynamic case. Here, Nu is based upon a heat transfer coefficient defined in terms of the difference between the inside wall temperature and the bulk mean temperature, and the thermal conductivity is based upon the bulk mean temperature. This expression also agrees well with the data of Kovner et al. [44].

9.7 MHD IN EXTERNAL FLOWS AND IN NATURAL CONVECTION

The effect of an applied magnetic field on heat transfer in external flows has been investigated mainly for the cases of flat-plate boundary-layer and blunt-body stagnation-point flows. The works published in these areas are theoretical and appeared in the late 1950s and early 1960s with application to space-vehicle surface heating upon reentry. Because the air in this situation was ionized and therefore conducting, it was envisioned that the application of a transverse magnetic field could be utilized to reduce the local velocity and skin friction drag and thereby the heat transfer to the vehicle's surface.

The classical works in these areas have been reviewed thoroughly by Romig [105]. In particular, the reader is referred to papers by Rossow [145], Bush [146], and Lykoudis [147, 148]. These include the theoretical treatments of MHD heat transfer of flow over a flat plate for the incompressible case, assuming a constant magnetic flux density and either a constant or a variable electrical conductivity [145], and for the compressible case, assuming variable electrical conductivity and variable magnetic flux density [146], constant electrical conductivity, and either constant or variable magnetic flux density [147], or variable electrical conductivity and constant magnetic flux density [148]. These studies show in general that as the boundary layer develops the heat transfer is reduced, and more specifically that the heat transfer is affected by variations of electrical conductivity with temperature, of temperature with velocity, and of magnetic flux density with distance, as well as by the temperature of the surface.

In MHD free convection, the application of a magnetic field reduces the magnitude of heat transfer because ponderomotive forces retard the motion induced by buoyancy. The reader is referred to a recent review article by Lykoudis [142], which covers the

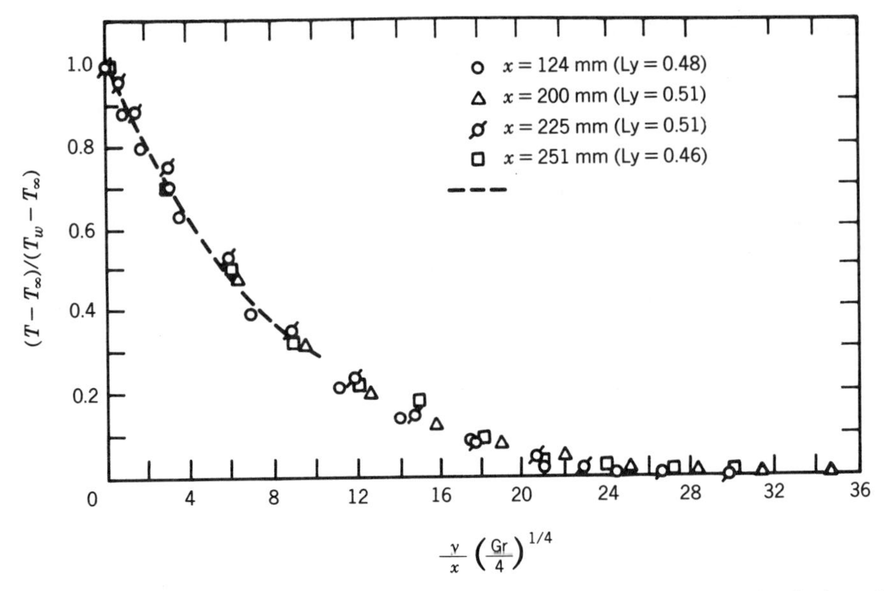

Figure 9.25. Temperature profiles for a heated vertical plate in a spatially varying horizontal magnetic field [87].

same subject material. Free convection heat transfer from a heated vertical plate to a liquid metal has been studied both theoretically and experimentally.

Sparrow and Cess [115] examined the case of laminar free convection from a heated vertical plate with a constant magnetic field normal to the plate. Their results were recast [105] in terms of a parameter equal to the mean Nusselt number divided by $\mathrm{Gr}^{1/4}$. This parameter decreases from its hydrodynamic value in direct proportion to the Lykoudis number, which represents the ratio of the ponderomotive force to the square root of the product of the buoyancy and inertia forces. Previously in the literature (e.g., [105, 110, 141]), the Lykoudis number was defined as $2M^2/\sqrt{\mathrm{Gr}}$. Recently, however, it has been defined as $\mathrm{Ly} = M^2/\sqrt{\mathrm{Gr}}$ [133, 142]. The more recent definition is used herein.

Similarity solutions were obtained [30, 64] for conditions like the aforementioned case but with the applied magnetic field varying as $x^{-1/4}$ in the vertical direction, x being the distance from the leading edge. Experimental confirmation of these solutions was obtained by using mercury as the working fluid [87]. Experimental results for the case of $\mathrm{Ly} \approx 0.5$ are compared with the theory in Fig. 9.25. At a given value of the similarity coordinate, the dimensionless temperature was found to decrease with increasing magnetic flux density (not shown in Fig. 9.25). For values of Ly up to 1.2, similarity appeared to be maintained, although no exact theoretical solution was available for comparison.

Romig [105] has compared the theoretical predictions of the mean heat transfer parameter for the $x^{-1.4}$ similarity case with that of a constant applied magnetic field. She found that for liquid metals when $\mathrm{Ly} < 0.5$, the mean heat transfer is not reduced as effectively as when the field is variable.

Seki et al. [110] conducted both experimental and numerical studies on the heat transfer from a vertical plate with uniform heat flux for the case in which the magnetic field was applied parallel to gravity. The magnetic field increases the surface temperature, and thereby increases Gr and decreases Nu. Data for $M/\mathrm{Gr} < 6 \times 10^{-6}$ ($0 \lesssim M \lesssim 400$, $2 \times 10^7 < \mathrm{Gr} < 5 \times 10^8$) were best approximated by

$$\frac{\mathrm{Nu}}{\mathrm{Nu}_0} = 1 - 1.3 \times 10^5 \frac{M}{\mathrm{Gr}} + 7.5 \times 10^9 \left(\frac{M}{\mathrm{Gr}} \right)^2 \tag{9.69}$$

in which the characteristic length is the heating-surface half height, and the characteristic temperature difference used to determine the heat transfer coefficient is that between the surface temperature at half height and the cold wall, which ranged from 0 to 60 K. These results may be limited to cases having a similar ratio (= 0.4) of heated section height to spacing between the hot and cold walls, because of possible thermal interference by the cold wall. Compared to the theoretical predictions of Sparrow and Cess [115] for the case of a field applied normal to the wall, the overall reduction in heat transfer for the parallel field case is less at a given value of Ly.

Papailiou and Lykoudis [88] conducted experiments on a free-convection turbulent boundary layer along a vertical wall with constant heat flux subjected to an applied, horizontal magnetic field. The magnetic field reduced convective heat transfer along the plate. The thermal boundary-layer temperature and thickness increased with the strength of the applied field. Heat transfer coefficients were expressed in terms of total Nusselt numbers based upon the length of the heated wall and the average temperature difference between the wall and free stream along the boundary layer. These data are shown in Fig. 9.26. As Ly increases, a reduction in the overall heat transfer coefficient occurs. The change of slope in the curve at Ly ≈ 0.33 corresponds to laminarization of the turbulent flow. Based upon measured mean temperature profiles, turbulence intensity distributions, and temperature spectra along the wall, the transition from turbulent to laminar flow occurs at a constant value of the ratio of $\mathrm{Gr\,Pr}/M$. For six experimental cases, $\mathrm{Gr\,Pr}/M = 1.2 \times 10^9$. Further analysis [86] showed that below this value a rapid drop in turbulence intensity occurs, as well as marked changes in the turbulence structure.

A relation between the overall heat transfer and Ly has been found also for the case of the natural convection of mercury in a vertical cylindrical container with a heated bottom surface. In experiments by Wagner [133], data were obtained at various saturated pressures in the presence of an applied horizontal magnetic field (Fig. 9.27).

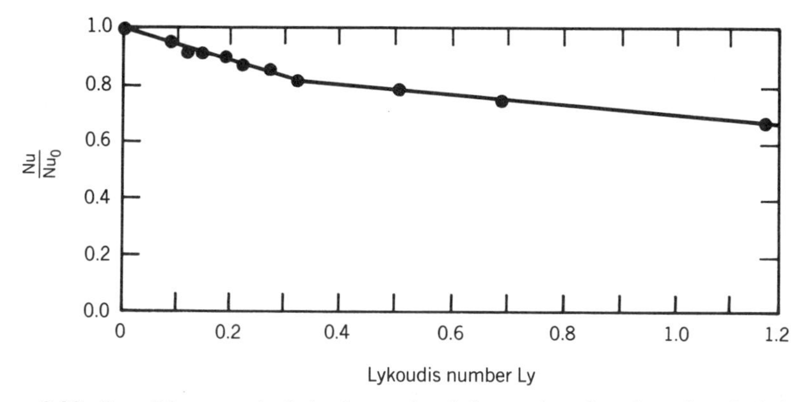

Figure 9.26. Overall heat transfer in laminar and turbulent regimes for a heated vertical plate in a transverse magnetic field [88].

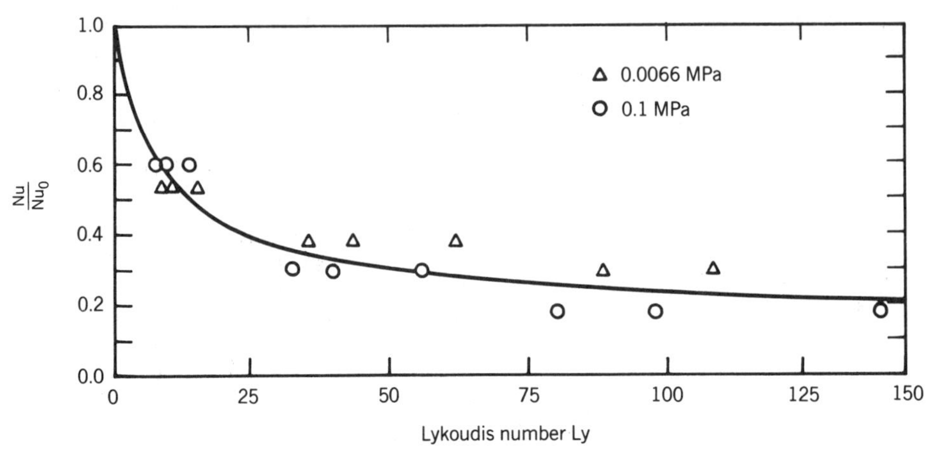

Figure 9.27. Overall heat transfer for natural convection from a heated horizontal surface inside a vertical cylinder in a transverse magnetic field (adapted from Ref. 133).

Reductions of up to 80% in the Nusselt number were measured. The correlation best describing the data is [142]

$$\frac{\mathrm{Nu}}{\mathrm{Nu}_0} = \frac{1}{(1 + 0.15\,\mathrm{Ly})^{0.5}} \tag{9.70}$$

Here, the heat transfer coefficient in Nu is based upon the difference between the temperature of heat transfer surface and liquid bulk mean temperature.

Recent experiments on natural convection heat transfer from finite horizontal cylinders treat magnetic field orientations in all three directions normal to the axis of the cylinder. Measurements in mercury with the magnetic field oriented along the axis were reported by Lykoudis and Dunn [66]. Detailed local heat transfer measurements in mercury for the other two field orientations have been presented by Michiyoshi et al. [72]. Blum and Kronkalns [4] reported data on free-convection heat transfer between a horizontal cylinder and a ferroliquid with the magnetic field normal to the axis of the cylinder. Similar experiments were reported by Kronkalns and Blum [48] for a high-Pr lithium-ammonia solution with the magnetic field normal to the axis and parallel to gravity. For all these cases, the application of the magnetic field reduced natural convection heat transfer from the cylinder.

In the experiments of Lykoudis and Dunn [66], the magnetic field suppressed free convection to the conduction limit. In the experiments of Michiyoshi et al. [72], for a fixed Gr, the Nusselt number (determined from temperature measurements around the cylinder circumference) decreased with increasing M. This decrease gradually levelled off at high M. Dunn [22] derived a semiempirical expression that correlated the data from both these experiments:

$$\frac{\mathrm{Nu} - \mathrm{Nu}_c}{\mathrm{Nu}_0 - \mathrm{Nu}_c} = \left[\frac{\mathrm{Ly}}{\sqrt{C_6}} + \left(1 + \frac{\mathrm{Ly}^2}{C_6}\right)^{1/2}\right]^{-1/2} \tag{9.71}$$

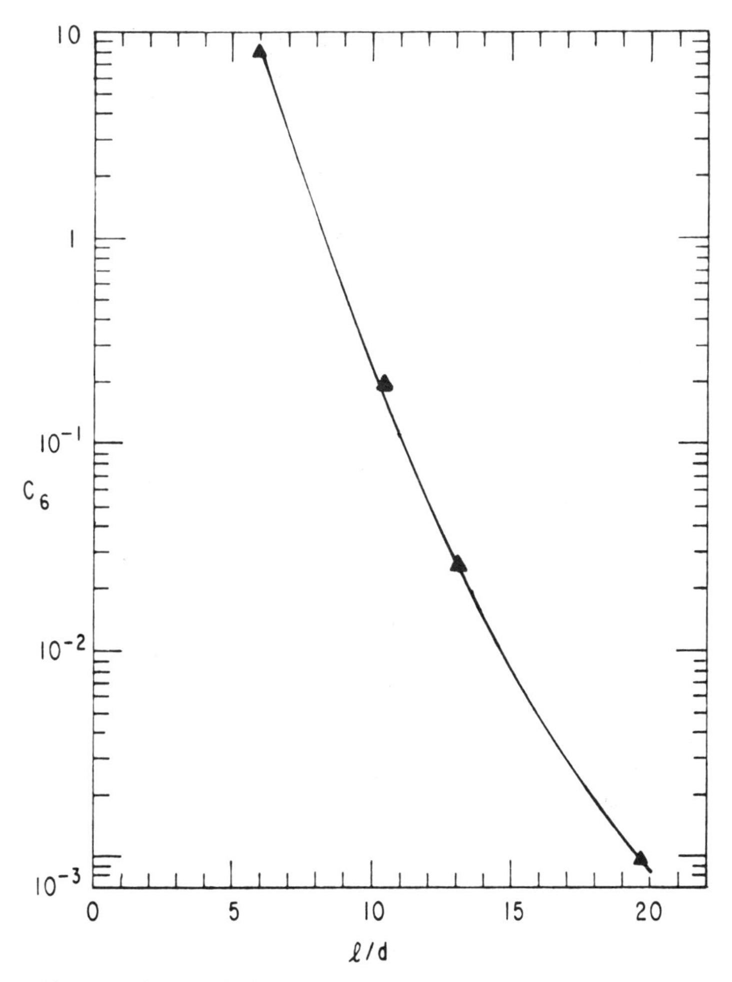

Figure 9.28. Empirical constant C_6 versus cylinder aspect ratio l/d [22].

in which C_6 is a function of the cylinder aspect ratio as shown in Fig. 9.28 [22] and the characteristic length is the cylinder diameter. This expression is compared with the experimental data in Fig. 9.29, in which cases 1 and 2 are from Ref. 72 and case 3 from Ref. 66. For cases 1 and 2, Nu represents a local Nusselt number whose heat transfer coefficient is based upon the temperature difference between the cylinder surface and the ambient mercury. This difference ranged from 0 to 80 K. For case 3, Nu represents an average Nusselt number based upon the temperature difference between the probe's surface (as determined from the probe's overheat ratio) and the ambient mercury. This difference varied from 12 to 27 K. The semiempirical expression predicts the reduction in both local heat transfer and overall heat transfer. Close agreement between theory and experiment is obtained in all three cases for $(\mathrm{Nu} - \mathrm{Nu}_c)/(\mathrm{Nu}_0 - \mathrm{Nu}_c) \gtrsim 0.2$. For lower values, theory and experiment do not compare well because conduction has become the dominant mode of heat transfer.

The effect of a magnetic field on forced convection from a horizontal cylinder (a hot-film probe) was measured also in the experiments of Lykoudis and Dunn [66]. In

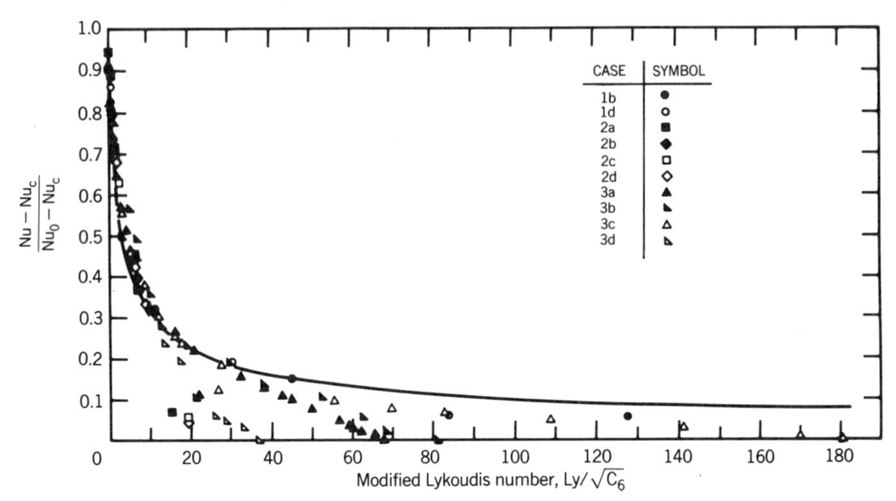

Figure 9.29. Heat transfer for natural convection from heated horizontal cylinders in magnetic fields of different orientations (adapted from Ref. 22).

their experiments, the value of the interaction parameter N was of order 1. Heat transfer data were gathered over the ranges $0 < \text{Re} < 130$ and $0 < M < 4.7$. Their results are shown in Fig. 9.30. For a given value of Re, the heat transfer from the probe decreased with increasing M. Decreases of up to 50% in the Nusselt number were measured at $\text{Re} \simeq 100$. Similar experiments were conducted at low values (< 0.005) of the interaction parameter N by Platnieks [96], and no effect was found.

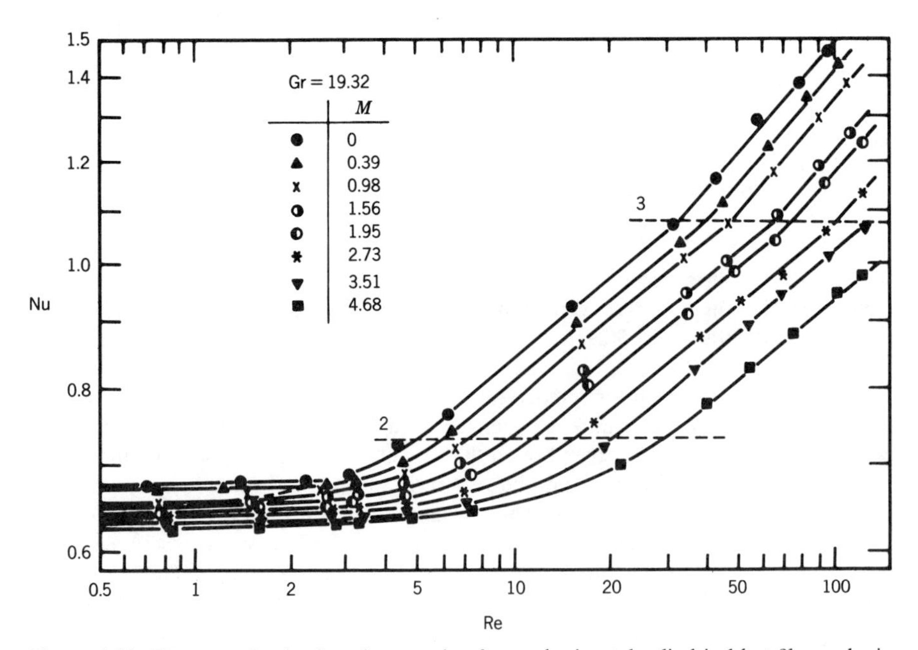

Figure 9.30. Heat transfer for forced convection from a horizontal cylindrical hot-film probe in an axially aligned magnetic field [66].

NOMENCLATURE

A_W	wall area, m^2, ft^2
B	magnetic flux density, T, Wb/m^2, Wb/ft^2
C, C^*	constants depending on geometry
C_6	empirical constant
c_p	specific heat, J/(kg · K), Btu/(lb$_m$ · °F)
D	dielectric displacement field, C/m^2, C/ft^2
d	spacing between electrode and corona discharge source, m, ft
$\mathbf{E}$	electric field vector (E_x, E_y, E_z), V/m, V/ft
E	average electric field, V/m, V/ft
E_0	threshold electric field, V/m, V/ft
Ec	Eckert number $= U^2/c_p\ (T_2 - T_1)$
F	load factor
$\mathbf{F}_e$	electric body force, C/(m^2 · s), C/ft^2 · s)
f	Fanning friction factor $= \tau_\omega/(\rho U^2/2)$
f_0	Fanning friction with no electric or magnetic field
g	gravitational acceleration, m/s^2, ft/s^2
Gr	Grashof number $= \rho^2 g\beta\,\Delta T\,L^3/\mu^2$
H	magnetic field intensity, A/m, A/ft
h	heat transfer coefficient, W/(m^2 · K), Btu/(h · ft^2 ·° F)
h_0	stagnation heat transfer coefficient, W/(m^2 · K), Btu/(h · ft^2 · °F)
h_x	local heat transfer coefficient, W/(m^2 · K), Btu/(h · ft^2 · °F)
I	electric current, A
I_0	electric current scale, A
i	enthalpy per unit mass, J/kg, Btu/lb$_m$
$\mathbf{J}$	current density (J_x, J_y, J_z), A/m^2, A/ft^2
J	magnitude of current density, A/m^2, A/ft^2
K	ion mobility, m^2/(V · s), ft^2/(V · s)
k	thermal conductivity, W/(m · K), Btu/(h · ft · °F)
L	characteristic length, m, ft
Ly	Lykoudis number $= M^2/\sqrt{\mathrm{Gr}}$
L_0	length scale, m, ft
M	Hartmann number $= BL\sqrt{\sigma/\mu}$
N	Interaction parameter $= M^2/\mathrm{Re}$
Ne	dimensionless electric number, Eq. (9.15)
Ne*	dimensionless number, Eq. (9.48)
Nu	Nusselt number $= hd/k$
Nu$_0$	Nusselt number with no electric or magnetic field
$N\rho_c$	dimensionless charge number, Eq. (9.50)
P	modified pressure, Eq. (9.10), Pa, lb$_f$/ft^2
p	pressure, Pa, lb$_f$/ft^2
Pe	Péclet number = Pr Re

Pr	Prandtl number = ν/α
q'''	volumetric heat generation, W/m³, Btu/(h · ft²)
q''	heat flux, W/m², Btu/(h · ft)
q''_w	magnitude of wall heat flux, W/m², Btu/(h · ft³)
Re	Reynolds number = UL/ν
Re_{cr}	critical Reynolds number
Re_{EHD}	electrical Reynolds number, Eq. (9.17)
Re_m	magnetic Reynolds number, $\sigma\mu_e UL$
R_1	inner radius, m, ft
R_2	outer radius, m, ft
r	radius, m, ft
T	temperature, °C, K, °F, °R
T_w	wall temperature, °C, K, °F, °R
T_1, T_2	reference temperature, °C, K, °F, °R
T_∞	free-stream temperature, °C, K, °F, °R
t	time, s
U	bulk velocity, m/s, ft/s
U_e	characteristic corona wind velocity, m/s, ft/s
U_0	velocity scale, m/s, ft/s
$\mathbf{u}$	velocity vector (u, v, w), m/s, ft/s
V	voltage, V
V_i	ion drift velocity, m/s, ft/s
X	dimensionless length, Eq. (9.27)
x	Cartesian coordinate, m, ft
y	Cartesian coordinate, m, ft

Greek Symbols

α	thermal diffusivity, m²/s, ft²/s
β	coefficient of thermal expansion, K^{-1}, $°R^{-1}$
δ	boundary-layer thickness, m, ft
δ_T	thermal boundary-layer thickness, m, ft
ϵ	dielectric constant, gas permittivity, C/(m · V), C/(ft · V)
ϵ_0	permittivity of free space, C/(m · V), C/(ft · V)
θ	dimensionless temperature, Eq. (9.13)
κ	(applied electric field)/(induced electric field)
λ	linear charge density, C/m, C/ft, or point charge density, C/m², C/ft²
μ	dynamic viscosity, Pa · s, lb_m/(h · ft)
μ_e	magnetic permeability, H/m, H/ft
ν	kinematic viscosity = μ/ρ
ρ	density, kg/m³, lb_m/ft^3
ρ_c	space-charge density, C/m³, C/ft³
ρ_{c0}	reference space-charge density, C/m³, C/ft³

σ	electrical conductivity, S/m, S/ft
τ_ω	fluid shear stress at wall, Pa, lb_f/ft^2
χ	dimensionless space-charge number, Eq. (9.15)
ϕ	electric potential, V
ϕ_0	reference or threshold electric potential, V
$\hat{\phi}$	conductance ratio

REFERENCES

1. P. H. G. Allen, Electric Stress and Heat Transfer, *Brit. J. Appl. Phys.*, Vol. 10, pp. 347–351, 1959.
2. G. K. Batchelor, *An Introduction to Fluid Mechanics*, Cambridge U.P., New York, 1967.
3. E. Ya. Blum, The Influence of Magnetic Fields on Heat Exchange in Electroconductive-Fluid Duct Flows, Ph.D. Thesis, Riga, U.S.S.R., 1967.
4. E. Ya. Blum and G. E. Kronkalns, Free Convective Heat Transfer on a Magnetic Cylinder in a Uniform Magnetic Field, *Magnetohydrodynamics*, Vol. 15, No. 3, pp. 264–269, 1979.
5. E. Ya. Blum, M. V. Zake, U. I. Ivanov, and Yu. A. Mikhajlov, *Heat and Mass Transfer in an Electromagnetic Field*, Zinaitne, Riga, U.S.S.R., 1967.
6. E. Bonjour, J. Mercier, and L. Weil, Electro-convection Effects on Heat Transfer, *Chem. Eng. Prog.*, Vol. 58, No. 7, pp. 63–66, 1962.
7. G. R. Bopp, The Role of Electrohydrodynamics in Transport Phenomena, *Chem. Eng. Prog.*, Vol. 63, p. 74, 1967.
8. H. Branover, *Magnetohydrodynamic Flow in Ducts*, Wiley, New York, 1978.
9. A. P. Chattock, On the Velocity and Mass of Ions in the Electric Wind in Air, *Philos. Mag. J. Sci.*, Vol. 48, pp. 401–420, 1899.
10. A. P. Chattock, On the Pressure of the Electric Wind in Hydrogen Containing Traces of Oxygen, *Philos. Mag.*, S. 6, Vol. 19, No. 112, pp. 449–460, 1910.
11. A. P. Chattock, On the Specific Velocities of Ions in the Discharge from Points, *Philos. Mag.*, S. 6, Vol. 1, No. 1, pp. 79–96, 1901.
12. T. Chaung and H. R. Velkoff, Analytical Studies of the Effects of Ionization on Fluid Flows, Tech. Report No. 6, RF Project 1864, Ohio State Univ. Res. Foundation, June 1967.
13. T. H. Chaung and H. R. Velkoff, Frost Formation on a Non-uniform Electric Field, Paper, 48a, *Symposium on Heat Transfer with Change of Phase*, *Part I*, AIChE Sixty-Third Annual Meeting, 1970.
14. B.-T. Chu, Thermodynamics of Electrically Conducting Fluids, *Phys. Fluids*, Vol. 2, No. 5, pp. 473–484, 1959.
15. J. D. Cobine, *Gaseous Conductors*, McGraw-Hill, New York, 1941.
16. P. Cooperman, A New Technique for the Measurement of Corona Field Strength and Current Density in Electrical Precipatation, *Trans. AIEE*, Vol. 75, pp. 64–67, 1956.
17. P. Cooperman, A Theory for Space-Charge-Limited Currents with Application to Electrical Precipitation, *Trans. AIEE*, Vol. 79, pp. 47–50, 1960.
18. T. G. Cowling, *Magnetohydrodynamics*, Wiley, New York, 1957.
19. J. H. Davidson and E. J. Shaughnessy, Turbulence Generation by Electric Body Forces, *Experiments in Fluids*, Vol. 4, pp. 17–26, 1986.
20. J. H. Davidson and E. J. Shaughnessy, Mean Velocity and Turbulent Intensity Profiles in a Large Scale Laboratory Precipitator, ASME Paper No. 84-JPGC-APC-1, 1984.
21. E. D. Doss, T. R. Johnson, M. Petrick, and W. C. Redman, Major Remaining Technical Issues in Coal-Fired MHD Technology, *22nd Symp. Eng. Aspects MHD,* Mississippi State Univ., Miss., 1984.

22. P. F. Dunn, Magnetohydrodynamic Natural Convection Heat Transfer from Horizontal Cylinders, *Int. J. Heat Mass Transfer*, Vol. 26, pp. 1413–1416, 1983.

23. M. A. Abdou, FINESSE: A Study of the Issues, Experiments and Facilities for Fusion Nuclear Technology Research and Development, Interim Report, No. UCLA-ENG-84-30, Center for Plasma Phys. and Fusion Eng., Univ. of California, Los Angeles, 1984.

24. L. D.Flippen, *Electrohydrodynamics*, Doctoral Dissertation, Duke Univ., 1982.

25. M. E. Franke, Effects of Vortices Induced by Corona Discharge on Free Convection Heat Transfer from a Vertical Plate, *ASME J. Heat Transfer*, Vol. 91, pp. 427–433, 1969.

26. M. E. Franke and K. E. Hutson, Effects of Corona Discharge on the Free Convection Heat Transfer inside a Vertical Hollow Cylinder, ASME Paper No. 82-WA/HT-20, 1982.

27. R. A. Gardner, Laminar Pipe Flow in a Transverse Magnetic Field with Heat Transfer, *Int. J. Heat Mass Transfer*, Vol. 11, pp. 1076–1081, 1968.

28. R. A. Gardner and P. S. Lykoudis, Magneto-Fluid Mechanics Pipe Flow in a Transverse Magnetic Field, Part 2, Heat Transfer, *J. Fluid Mech.*, Vol. 48, pp. 129–141, 1971.

29. R. A. Gardner, K. L. Uherka, and P. S. Lykoudis, Influence of a Transverse Magnetic Field on Forced Convection Liquid Metal Heat Transfer, *AIAA J.*, Vol. 4, No. 5, pp. 848–852, 1966.

30. A. S. Gupta, Steady and Transient Free Convection of an Electrically Conducting Fluid from a Vertical Plate in the Presence of a Magnetic Field, *Appl. Sci. Res.*, Vol. A9(5), pp. 319–333, 1960.

31. D. S. Harney, Aerodynamic Study of the Electric Wind, Thesis, Calif. Inst. of Technol., 1957; also, ASTIA Doc. No. AD-134400, 1957.

32. F. Hauksbee, *Physico-Mechanical Experiments on Various Subjects*, 1st ed., London, 1709, pp. 46–47.

33. J. C. Hay, The Electric Field and Vorticity Production in a Wire-Plate Precipitator with Uniform Discharge, Master's Thesis, Duke Univ., 1984.

34. J. F. Hoburg, Temperature-Gradient-Driven Electrohydrodynamics Instability with Unipolar Injection in Air, *J. Fluid Mech.*, Vol. 132, pp. 231–245, 1983.

35. J. F. Hoburg and J. L. Davis, Wire-Duct Precipitator Field and Charge Computation Using Finite Element and Characteristics Methods, *J. Electrostatics*, Vol. 14, No. 2, pp. 187–199, 1983.

36. R. E. Holmes and S. J. Basham, A Dry Cooling System for Steam Power Plants, Paper No. 719158, *Proc. 1971 Inter-Society Energy Conversion Conf.*, Soc. Automotive Eng., New York, 1972.

37. E. S. Hsia, Effects of Wall Electrical Conductance and Induced Magnetic Field on MHD Channel Heat Transfer with Developing Thermal and Velocity Fields, ASME Paper No. 77-HT-63, Am. Soc. Mech. Eng., New York, 1977.

38. W. F. Hughes and F. J. Young, *The Electrodynamics of Fluids*, Wiley, New York, 1966.

39. J. D. Jackson, *Classical Electrodynamics*, Wiley, New York, 1962.

40. D. C. Jolly and J. R. Melcher, Electroconvective Instability in a Fluid Layer, *Proc. Int. Symp. Electrohydrodynamics*, Mass. Inst. Technol., pp. 110–111, 1969.

41. T. B. Jones, Electrohydrodynamically Enhanced Heat Transfer in Liquids—A Review, Report NSF/Eng74-24113/RR2/77, Dept. of Electrical Eng., Colorado State Univ., Mar. 1977.

42. L. E. Kalikhman, *Elements of Magnetogasdynamics*, Saunders, Philadelphia, 1967.

43. K. G. Kibler and H. G. Carter, Jr., Electro-cooling in Gases, *J. Appl. Phys.*, Vol. 2, No. 10, pp. 4436–4440, 1974.

44. D. S. Kovner, E. Yu. Krasil'nikov, and I. G. Panevin, Experimental Investigation of the Effect of a Longitudinal Magnetic Field on Convective Heat Exchange in a Turbulent Duct Flow of Conducting Liquid, *Magnetohydrodynamics*, Vol. 2, No. 4, pp. 60–63, 1966.

45. D. S. Kovner, On the Use of the Locality Hypothesis in Turbulent Flow of a Conducting Fluid in a Magnetic Field, *Magnetohydrodynamics*, Vol. 1, No. 2, pp. 7–12, 1965.

46. E. Yu. Krasil'nikov, The Effect of a Transverse Field on Convective Heat Transfer in Conductive-Fluid Duct Flow, *Magnetohydrodynamics*, Vol. 1, No. 3, pp. 26–28, 1965.

47. R. Kronig and N. Schwartz, On the Theory of Heat Transfer from a Wire in an Electric Field, *Appl. Sci. Res.*, Vol. A1, pp. 35–46, 1949.

48. G. E. Kronkalns and E. Ya. Blum, Natural MHD Convection in a Horizontal Cylinder in Metal-Ammonia Solutions, *Magnetohydrodynamics*, Vol. 12, No. 3, pp. 294–298, 1976.

49. F. A. Kulacki, S. Boriah, and S. A. Martin, Corona Discharge Augmentation of the Catalytic Combustion of Hydrogen in the Diffusion Controlled Regime, *Int. J. Hydrogen Energy*, Vol. 6, pp. 73–95, 1981.

50. F. A. Kulacki and J. A. Daubenmier, A Preliminary Study of Electrohydrodynamically Augmented Baking, *J. Electrostatics*, Vol. 5, pp. 325–336, 1978.

51. F. A. Kulacki and J. M. Kevra, Corona Wind Augmented Baking, *Drying '82,* ed. A. S. Majumdar, Hemisphere, New York, pp. 183–195, 1982.

52. J. S. Lagarias, Field-Strength Measurements in Parallel-Plate Precipitators, *Trans. AIEE*, Vol. 78, p. 427, 1957.

53. E. W. Laing, *Plasma Physics*, Sussex U.P., England, 1976.

54. L. D. Landau and E. M. Lifshitz, *Electrodynamics of Continuous Media*, Pergamon, New York, 1960.

55. P. A. Lawless and L. E. Sparks, A Mathematical Model for Calculating Effects of Back Corona in Wire-Duct Electrostatic Precipitators, *J. Appl. Phys.*, Vol. 51(1), p. 242, 1980.

56. P. A. Lawless and L. E. Sparks, Prediction of Voltage-Current Curves for Novel Electrodes: Arbitrary Wire Electrodes on Axis, *Fourth Symposium on the Transfer and Utilization of Particulate Control Technology*, 1984.

57. B. R. Lazaranko, F. P. Grosu, and M. K. Bologa, Convective Heat Transfer Enhancement by Electric Fields, *Int. J. Heat Mass Transfer*, Vol. 18, pp. 1433–1441, 1975.

58. C.-O. Lee, Electrohydrodynamic Cellular Bulk Convection Induced by a Temperature Gradient, *Phys. Fluids*, Vol. 13, pp. 789–795, 1972.

59. G. L. Leonard, M. Mitchner, and S. A. Self, An Experimental Study of the Electrohydrodynamic Flow in Electrostatic Precipitators, *J. Fluid Mech.*, Vol. 127, pp. 123–140, 1983.

60. G. Leutert and B. Bohlen, The Spatial Trend of Electric Field Strength and Space Charge Density in Plate-Type Electrostatic Precipitators, *Staub-Reinhalt. Luft*, Vol. 32, No. 7, pp. 27–33, 1972.

61. E. Löb, Beitrag über die Druckwirkugen von Ionenstromen in atmosphärische Licht bei verschiedenen Entladungsanordnungen, *Arch. Elek. Übertrag*, Vol. 8, pp. 85–90, 1854.

62. L. B. Loeb, *Fundamental Processes of Electrical Discharge in Gases*, Wiley, New York, 1939.

63. L. B. Loeb, *Electrical Coronas. Their Basic Physical Mechanisms*, Univ. of Calif. Press, 1965.

64. P. S. Lykoudis, Natural Convection of an Electrically Conducting Fluid in the Presence of a Magnetic Field, *Int. J. Heat Mass Transfer*, Vol. 15, pp. 25–34, 1962.

65. P. S. Lykoudis, Short Description of Current Work in the MFM Laboratory of Purdue University, *MHD—Flows and Turbulence*, ed. H. Branover, Wiley, New York, pp. 103–118, 1976.

66. P. S. Lykoudis and P. F. Dunn, Magneto-Fluid Mechanics Heat Transfer from Hot Film Probes, *Int. J. Heat Mass Transfer*, Vol. 16, pp. 1439–1452, 1973.

67. P. S. Lykoudis and C. P. Yu, The Influence of Electrostrictive Forces in Natural Thermal Convection *Int. J. Heat Mass Transfer*, Vol. 6, pp. 853–862, 1963.

68. M. R. Malik, L. M. Weinstein, and M. Y. Hussanini, Ion Wind Drag Reduction, Paper AIAA-83-0231, Am. Inst. Aeronautics and Astronautics, New York, 1983.

69. S. M. Marco and H. R. Velkoff, Effect of Electrostatic Fields on Free Convection Heat Transfer from Flat Plates, ASME Paper No. 63-HT-9, 1963.

70. E. W. McDaniel and E. A. Mason, *The Mobility and Diffusion of Ions in Gases*, Wiley, New York, 1973.

71. J. R. Melcher, *Field-Coupled Surface Waves: A Comparative Study of Surface Coupled Electrohydrodynamic and Magnetohydrodynamic Systems*, MIT Press, Cambridge, Mass., 1963.

72. I. Michiyoshi, O. Takahasi, and A. Serizawa, Natural Convection Heat Transfer from a Horizontal Cylinder to Mercury under Magnetic Field, *Int. J. Heat Mass Transfer*, Vol. 19, pp. 1021–1029, 1976.

73. A. S. Mitchell, Heat Transfer by a Corona Wind Heat Exchanger, Paper No. 78-WA/HT-43, Am. Soc. Mech. Eng., New York, 1978.

74. A. S. Mitchell and L. E. Williams, A Study of the Thermal Performance of a Pin Fin for Corona Wind Cold Plate for Use in Avionics, Final Report, Contract No. N00019-77-C-0191, Naval Air System Command (AIR-52022), 1978.

75. A. S. Mitchell and L. E. Williams, Heat Transfer by the Corona Wind Impinging on a Flat Surface, *J. Electrostatics*, Vol. 5, pp. 309–324, 1978.

76. M. L. Mittal, Heat Transfer by Laminar Flow in a Circular Pipe under Transverse Magnetic Field, *Int. J. Heat Mass Transfer*, Vol. 7, pp. 239–246, 1964.

77. T. Mizushina, H. Ueda, T. Matsumota, and K. Waga, Effect of Electrically Induced Convection on Heat Transfer of Air Flow in an Annulus, *J. Chem. Eng. Japan*, Vol. 9, No. 2, pp. 97–102, 1975.

78. A. D. Moore, Electrostatics, *Sci. Am.*, Vol. 47, pp. 47–58, 1972.

79. A. D. Moore, ed., *Electrostatics and Its Applications*, Wiley, New York, 1973.

80. R. A. Moss and J. Grey, Heat Transfer Augmentation by Steady and Alternating Electric Fields, *Proc. 1966 Heat Transfer and Fluid Mech. Inst.,* ed. W. H. Giedt and S. Levy, Stanford U.P., 1966, pp. 210–235.

81. D. A. Nelson and E. J. Shaughnessy, Electric Field Effects on Natural Convection in Enclosures, *Heat Transfer in Enclosures*, ed. R. D. Douglass and A. F. Emery, HTD Vol. 39, Am. Soc. Mech. Eng., New York, pp. 13–20, 1984.

82. I. Newton, *Optics*, 2nd ed., London, pp. 315–316, 1718.

83. K. J. Nygaard, Electric Wind Gas Discharge Anemometer, *Rev. Sci. Inst.*, Vol. 36, No. 9, pp. 1320–1323, 1960.

84. R. J. O'Brien and A. J. Shine, Some Effects of an Electric Field on Heat Transfer from a Vertical Plate in Free Convection, *ASME J. Heat Transfer*, Vol. 89, pp. 114–116, 1967.

85. W. Panofsky and M. Phillips, *Classical Electricity and Magnetism*, Addison-Wesley, Reading, Mass., 1956.

86. D. D. Papailiou, Statistical Characteristics of a Turbulent Free-Convection Flow in the Absence and Presence of a Magnetic Field, *Int. J. Heat Mass Transfer*, Vol. 23, pp. 889–895, 1980.

87. D. D. Papailiou and P. S. Lykoudis, Magnetic-Fluid-Mechanic Laminar Natural Convection —An Experiment, *Int. J. Heat Mass Transfer*, Vol. 11, pp. 1385–1391, 1968.

88. D. D. Papailiou and P. S. Lykoudis, Magneto-Fluid Mechanics Free Convection Turbulent Flow, *Int. J. Heat Mass Transfer*, Vol. 17, pp. 1181–1189, 1974.

89. F. W. Peek, Jr., *Dielectric Phenomena in High Voltage Engineering*, 3rd ed., McGraw-Hill, New York, 1929.

90. E. J. Pejack and H. R. Velkoff, The Effect of Transverse Ion Current on the Flow of Air in a Flat Duct, Tech. Report, Contract No. DA-31-124-ARO-D-246, Ohio State Univ. Res. Foundation, Feb. 1967.

91. F. M. Penning, *Electrical Discharges in Gases*, Macmillan, New York, 1957.

92. G. W. Penny and R. E. Matick, Potential in a DC Corona Field, *Trans. AIEE*, Vol. 79, pp. 91–99, 1960.

93. M. Petrick and B. Ya. Shumyatsky, eds., *Open Cycle Magnetohydrodynamic Electrical Power Generation*, A Joint U.S./U.S.S.R. publication, Argonne Nat. Lab., 1978.

94. M. Perlmutter and R. Siegal, Heat Transfer to an Electrically Conducting Fluid Flowing in a Channel with a Transverse Magnetic Field, NASA IN-D-875, Nat. Aeronaut. and Space Admin., 1961.

95. D. L. Phon and J.-L. LaForte, The Influence of Electro-Freezing on Ice Formation on High-Voltage DC Transmission Lines, *Cold Region Sci. Tech.*, Vol. 4, pp. 15–25, 1981.

96. I. A. Platnieks, Comparison of the Hot Wire Anemometer and Conduction Methods for Mercury Measurements, *Magnetohydrodynamics*, Vol. 7. No. 3, pp. 140–142, 1971.

97. S. Rajaram and C. P. Yu, Heat Transfer in Developing MHD Channel Flows in an Inclined Magnetic Field, ASME Paper No. 80-WA/HT-12, 1980.

98. O. E. Ramadan and S. L. Soo, Electrohydrodynamic Secondary Flow, *Phys. Fluids*, Vol. 12, pp. 1943–1945, 1969.

99. B. L. Reynolds and R. E. Holmes, Heat Transfer in a Corona Discharge, *Mech. Eng.*, pp. 44–49, Oct. 1976.

100. O. Rho, C.-J. Lee, and L. Trefethen, Effects of Radial Electrostatic Fields on Natural Convection and Forced Convection Heat Transfer in Annuli, *Proc. Int. Symp. Electrohydrodynamics*, Mass. Inst. of Technol., pp. 209–212, 1969.

101. M. Robinson, Movement of Air in the Electric Wind of the Corona Discharge, *Trans. AIEE*, Vol. 80, pp. 143–150, 1961.

102. M. Robinson, Convective Heat Transfer at the Surface of a Corona Electrode, *Int. J. Heat Mass. Transfer*, Vol. 13, pp. 263–274, 1970.

103. M. Robinson, Electrostatic Precipitation, *Electrostatics and Its Applications*, ed. A. E. Moore, Wiley, New York, 1973, pp. 180–249.

104. M. Robinson, Effects of Corona Discharge on Electric Wind Convection and Eddy Diffusion in an Electrostatic Precipitator, Report No. HASL-301, Health and Safety Lab., U.S. Energy Research and Development Admin., New York, Feb. 1976.

105. M. F. Romig, The Influence of Electric and Magnetic Fields on Heat Transfer to Electrically Conducting Fluids, *Adv. Heat Transfer*, Vol. 1, pp. 267–354, 1964.

106. M. F. Romig, Electric and Magnetic Fields, *Handbook of Heat Transfer*, ed. W. M. Rohsenow and J. P. Hartnett, McGraw-Hill, New York, Chap. 11, pp. 1–29, 1973.

107. S. E. Sadek, R. G. Fax, and M. Hurwitz, The Influence of Electric Fields on Convective Heat and Mass Transfer from a Horizontal Surface under Forced Convection, *J. Heat Transfer*, Vol. 94, pp. 144–148, 1972.

108. S. D. Savkar, Dielectrophoretic Effects in Laminar Forced Convection between Two Parallel Plates, *Phys. Fluids*, Vol. 14, No. 12, pp. 2670–2679, 1971.

109. N. M. Schnurr, The Effect of a Radial Electric Field on Heat Transfer to Air Flowing Through a Circular Duct., Ph.D. Diss., Ohio State Univ., 1965.

110. M. Seki, H. Kawamura, and K. Sanokawa, Natural Convection of Mercury in a Magnetic Field Parallel to the Gravity, *J. Heat Transfer*, Vol. 101, pp. 227–232, 1979.

111. H. Senftleben and W. Braun, Der Einfluss Elektrischer Felder auf den Wärmestrom in Gasen, *Z. Phys.*, Vol. 102, pp. 480–506, 1936.

112. L. Sharpe, Jr, and F. A. Morrison, Jr., Numerical Analysis of Heat and Mass Transfer from Fluid Spheres in an Electric Field, ASME Paper No. 83-WA/HT-29, 1983.

113. A. F. Smelewicz, M. P. Majchar, J. B. Nystrom, and W. W. Durgin, Augmentation of Free Surface Heat and Mass Transfer Due to Electrostatic Fields, ASME Paper No. 83-WA/HT-97, 1983.

114. C. Sozou, On Fluid Motions Induced by an Electric Current Source, *J. Fluid Mech.*, Vol. 48, pp. 25–32, 1971.

115. E. M. Sparrow and R. D. Cess, The Effect of a Magnetic Field on Free Convection Heat Transfer, *Int. J. Heat Mass Transfer*, Vol. 3, pp. 267–274, 1961.

116. O. M. Steutzer, Ion Drag Pressure Generation, *J. Appl. Phys.*, Vol. 30, pp. 984–994, 1959.
117. O. M. Steutzer, Ion Drag Pumps, *J. Appl. Phys.*, Vol. 31, pp. 132–146, 1960.
118. O. M. Steutzer, Instability of Certain Electrodynamic Systems, *Phys. Fluids*, Vol. 2, pp. 642–648, 1960.
119. J. J. Thompson and G. P. Thompson, *Conduction of Electricity through Gases*, 3rd ed., Cambridge U.P., 1928.
120. J. S. Townsend, *Electricity in Gases*, Oxford U.P., Oxford, 1915.
121. R. J. Turnbull, Effect of a Non-uniform Alternating Electric Field on the Thermal Boundary Layer Near a Heated Vertical Plate, *J. Fluid Mech.*, Vol. 49, pp. 693–703, 1971.
122. R. J. Turnbull, Instability of a Thermal Boundary Layer in a Constant Electric Field, *J. Fluid Mech.*, Vol. 47, pp. 231–239, 1971.
123. A. B. Vatazhin and V. I. Grabowski, On Two-Dimensional Electro-Gas Dynamic Flows with Allowance for the Inertia of Charged Particles, *P.P.M.*, Vol. 40, pp. 65–73, 1976.
124. H. R. Velkoff, An Analysis of the Effect of Ionization on the Laminar Flow of a Dense Gas in a Channel, Report RID-TDR-63-4009, Air Force Aero-Propulsion Lab., Wright Patterson AFB, Ohio, 1963.
125. H. R. Velkoff, An Exploratory Investigation of the Effects of Ionization on the Flow and Heat Transfer with a Dense Gas, Tech. Doc. ASD-TDR-63-642, Wright-Patterson AFB, Ohio, Nov. 1963.
126. H. R. Velkoff, The Effects of Ionization on the Flow and Heat Transfer of a Dense Gas in a Transverse Electrical Field, *Proc. 1964 Heat Transfer and Fluid Mech. Inst.,* ed. W. H. Giedt and S. Levy, Stanford U.P., Palo Alto, Calif., pp. 260–275, 1964.
127. H. R. Velkoff and R. D. Godfrey, Low-Velocity Heat Transfer to a Flat Plate in the Presence of a Corona Discharge in Air, *J. Heat Transfer*, Vol. 101, pp. 157–165, 1979.
128. H. R. Velkoff and J. Ketchem, Effect of an Electrostatic Field on Boundary Layer Transition, *AIAA J.*, Vol. 6, pp. 1381–1383, 1966.
129. H. R. Velkoff, Evaluating the Interaction of Electrostatic Fields with Fluid Flows, ASME Paper No. 71-DE-41, 1971.
130. H. R. Velkoff, E. J. Pejack, and T. H. Chaung, Electrostatically Induced Secondary Flows in a Channel, *Electric Field Effects on Unit Operations Symposium*, AIChE 69th National Meeting, May 1971.
131. H. R. Velkoff and F. A. Kulacki, Electrostatic Cooling, ASME Paper No. 77-DE-36, New York, 1977.
132. H. R. Velkoff, Electrofluidmechanics: Investigation of the Effects of Electrostatic Fields on Heat Transfer and Boundary Layers, Tech. Doc. ASD-TDR-62-650, Propulsion Lab., Aeronaut. Systems Div., Wright-Patterson AFB, Ohio, 1962.
133. L. Y. Wagner, Single and Two-Phase Liquid Metal Heat Transfer under the Influence of a Magnetic Field, Ph.D. Diss., Purdue Univ., 1981.
134. M. K. Weaver, The Pressure Distribution in a Gas in Electrohydrostatic Equilibrium, Master's Thesis, Duke Univ., 1984.
135. H. J. White, *Electrostatic Precipitation*, Pergamon, Oxford, 1963.
136. H. Windishmann, Investigation of Corona Discharge Cooling (CDC) of a Horizontal Plate Under Free Convection, Preprint No. 21, 14th National Heat Transfer Conference, Am. Inst. Chem. Eng., New York, 1973.
137. A. Yabe, Y. Mori, and K. Hijikata, Heat Transfer Augmentation on a Downward-Facing Flat Plate by Non-uniform Electric Fields, *Proc. Fifth Int. Heat Transfer Conf.*, Vol. 3, Hemisphere, New York, pp. 171–176, 1978.
138. A. Yabe, Y. Mori, and K. Hijikata, EHD Study of the Corona Wind between Wire and Plate Electrodes, *AIAA J.*, Vol. 16, pp. 340–345, 1978.
139. T. Yamamoto and H. R. Velkoff, Electrohydrodynamics in an Electrostatic Precipitator, *J. Fluid Mech.*, Vol. 108, pp. 1–18, 1981.

140. H. K. Yang and C. P. Yu, Experimental Study of Mixed Convection Heat Transfer in an MHD Channel, *AIAA J.*, Vol. 12, pp. 1740–1743, 1974.

141. D. D. Gray, The Laminar Plume above a Line Heat Source in a Transverse Magnetic Field, *Appl. Sci. Res.*, Vol. 33, pp. 437–457, 1977.

142. P. S. Lykoudis, Natural Convection of Electrically Conducting Fluids in the Presence of Magnetic Fields, *Natural Convection, Fundamentals and Applications*, eds. S. Kakaç, W. Aung, and R. Viskanta, Hemisphere, New York, pp. 1100–1117, 1985.

143. P. S. Lykoudis and E. C. Brouillette, Magneto-Fluid-Mechanic Channel Flow. II. Theory, *Phys. Fluids*, Vol. 10, pp. 1002–1007, 1967.

144. P. S. Lykoudis, Damping of Shear Turbulence in the Presence of Magnetic Fields: A Semi-empirical Approach, *MHD—Flows and Turbulence II*, eds. H. Branover and A. Yakhot, Israel Universities Press, Jerusalem, pp. 271–277, 1980.

145. V. J. Rossow, On the Flow of Electrically Conducting Fluids Over a Flat Plate in the Presence of a Transverse Magnetic Field, NACA TN 3971, May 1957.

146. W. Bush, On the Laminar Compressible Boundary Layer in the Presence of an Applied Magnetic Field, Phys. Res. Lab., Space Technol. Lab., Inc., STL/TR-59-0000-00668, Oct. 1959.

147. P. S. Lykoudis, On a Class of Compressible Laminar Boundary Layers with Pressure Gradient for an Electrically Conducting Fluid in the Presence of a Magnetic Field, *Proc. 9th Ann. Congress Int. Astronaut. Fed.,* Amsterdam, Holland, 1958, Springer-Verlag, Austria, 1959.

148. P. S. Lykoudis, Velocity Overshoots in Magnetic Boundary Layers, *J. Aerospace Sci.*, Vol. 28, pp. 896–897, 1961.

10

CONVECTIVE HEAT TRANSFER IN BENDS AND FITTINGS

S. D. Joshi

Phillips Petroleum Company
Bartlesville, Oklahoma

R. K. Shah

Harrison Radiator Division, GM
Lockport, New York

10.1 INTRODUCTION

Bends and fittings are commonly used in pipelines and flow lines. In tubular heat exchangers, bends are used as return lines. In some applications, bends are heated. However, in other applications, pipes leading to and from a bend are heated while the bend itself is not heated.

As shown in Fig. 10.1, bends are described by a bend angle ϕ in degrees and a bend curvature ratio R/a. Pipe and tube fittings such as 90° or 180° elbows are a special kind of pipe bend with a threaded inlet and outlet. In addition, fittings exhibit variations in pipe cross-sectional area.

Available friction factor and Nusselt number data for flow through bends and fittings [1–56] are summarized in this chapter. Presently available Nusselt number data are limited to bends. The scope of this chapter is restricted to fittings that geometrically resemble bends. These fittings include bends with various turning radii and turning angles. As an engineering approximation, the heat transfer results for bends may be used for geometrically similar smooth fittings. The following discussion is only for smooth bends unless noted otherwise.

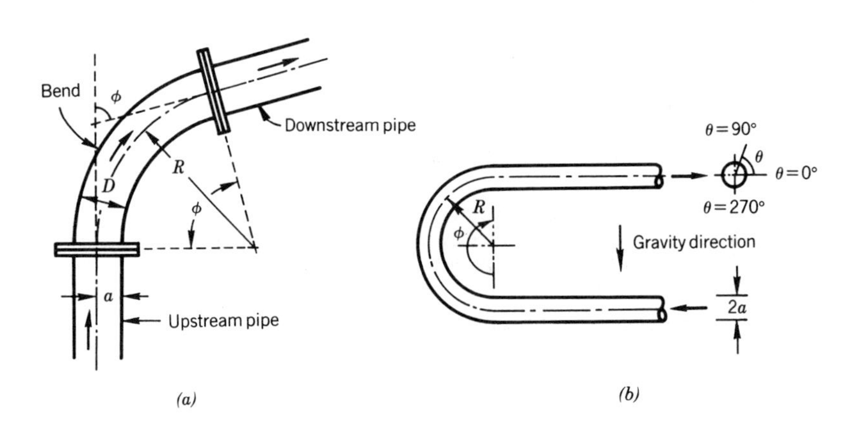

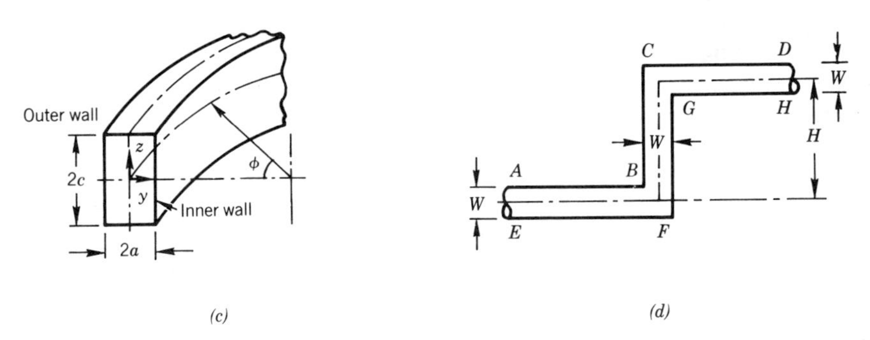

Figure 10.1. Schematic diagrams of various bend shapes: (a) a bend with $\phi < 90°$, (b) 180° bend, (c) rectangular cross-section bend, and (d) two 90° bends in series.

Geometrically, a bend represents a curved duct of short length. Therefore, fluid flow and heat transfer in a bend resembles that observed in an entrance length of a curved duct (see Section 5.1). As noted in Chapter 5, the fluid experiences a centrifugal force due to the bend curvature. This force, superimposed on the primary force due to the axial pressure gradient, generates a secondary motion transverse to primary axial flow. This results in a distortion of the fluid velocity and temperature profiles at any cross section in the bend. As shown in Figs. 5.2 and 5.4, the distorted profiles display a shift in peak velocity and temperature away from the tube center and toward the tube outer wall. As depicted in Figs. 5.3 and 5.7, the profile distortion results in higher heat transfer rates and friction factors at the tube outer wall than those at the inner wall. In flow lines with bends, the profile distortions persist to about 10 to 30 diameters in the straight pipe downstream of a bend. The exact length of straight pipe affected depends on the bend geometry, Reynolds number, and Dean number. Thus, an upstream bend may influence heat transfer in a downstream straight pipe.

Secondary flows, which can be relatively strong in bends, diminish the influence of wall thermal boundary conditions on the Nusselt number. Therefore, the Nusselt number correlations provided in this chapter could probably be used for any wall thermal boundary condition for gases and liquids (excluding liquid metals). In addition, a bending process distorts the wall thickness of a bend. Hence, the temperature or heat flux at any given cross section in a bend will not be uniform. Therefore, the boundary condition does not correspond to the (H1) or (H2) as described in Chapter 3. We shall simply refer to it as the boundary condition of axially constant heat flux.

Most of the available data are restricted to 90° and 180° bends of circular and rectangular cross sections. Therefore, the following discussion is divided into two major parts based upon the flow cross-section shape.

10.2 BENDS WITH A CIRCULAR CROSS SECTION

This section is divided into the following five subsections: fluid flow, heat transfer in 90° bends, heat transfer in pipes downstream of a 90° bend, heat transfer in 180° bends, and heat transfer in pipes downstream of a 180° bend.

10.2.1 Fluid Flow

Quantities of design interest are velocity profiles and fluid friction factors. Due to space limitation, details of velocity profiles are not included here. For laminar flow in a 180° bend, Humphrey et al. [1] have reported numerical predictions for developing and developed velocity profiles. Olson and Snyder [2] experimentally measured laminar velocity profiles in a 300° bend. Turbulent velocity profiles in a 180° bend are reported by Azzola and Humphrey [3,4]. Rowe [5] and Weske [6] have reported theoretical solutions for the velocity profiles in pipes with different bend curvatures and bend angles. Azzola and Humphrey [3,4] have noted that in turbulent flow, the flow distortion in a bend decays significantly over about 5 diameters into the downstream straight pipe, but persists considerably further downstream.

In the case of bends, bend loss coefficients are normally given instead of friction factors. The total pressure drop in a bend is the sum of the following three components: frictional head loss due to the length of the bend, head loss due to curvature, and head loss due to excess pressure drop in a downstream pipe because of the profile distortion. The pressure drops, other than frictional loss, are combined and expressed as a loss coefficient K^*. Thus, the total pressure drop is represented as

$$\Delta P = \left(\frac{4fL}{D_h} + K^* \right) \frac{\rho u_m^2}{2} \tag{10.1}$$

Ito [7] combined the terms in the first factor and defined a total loss coefficient K as

$$\Delta P = K \frac{\rho u_m^2}{2} \tag{10.2a}$$

and thus

$$K = \frac{4f_c L}{D_h} = \frac{4fL}{D_h} + K^* = 2f\left(\frac{\pi}{180} \phi \right) \frac{R}{a} + K^* \tag{10.2b}$$

where f_c represents a bend friction factor, f is the friction factor for a straight pipe at the Reynolds number in the bend, and ϕ is the bend angle in degrees. Here f is given by

$$f = \begin{cases} 0.0791 \text{ Re}^{-0.25} & \text{for } 3 \times 10^4 < \text{Re} \le 10^5 \\ 0.0008 + 0.005525 \text{ Re}^{-0.237} & \text{for } 10^5 < \text{Re} \le 10^7 \end{cases} \tag{10.2c}$$

In most fittings, the ratio L/D_h is very small, and frictional losses due to bend and fitting lengths are very small. As a result, in many engineering applications, $K^* \approx K$ in Eq. (10.2b).

Friction Factors In Smooth Bends. Powle [8] has reported laminar friction factors for flow through smooth 90° bends. Her graphical correlation, presented in Fig. 10.2, was obtained by correlating and extrapolating the experimental results of Kittredge and Rowley [17]. Koli and Powle [9] also reported extensive laminar friction factors for flow through 30°, 45°, 60°, 75° and 90° bends. However, they inform us these results [9] are in error and should not be used for design purpose.

Idelchik [10] reported the following equation to calculate laminar friction factors in smooth bends of any angle $\phi < 360°$:

$$f_c = 5 \text{ Re}^{-0.65} (R/a)^{-0.175} \qquad \text{for} \quad 50 < \text{De} \le 600 \tag{10.3a}$$

$$f_c = 2.6 \text{ Re}^{-0.55} (R/a)^{-0.225} \qquad \text{for} \quad 600 < \text{De} \le 1400 \tag{10.3b}$$

$$f_c = 1.25 \text{ Re}^{-0.45} (R/a)^{-0.275} \qquad \text{for} \quad 1400 < \text{De} < 5000 \tag{10.3c}$$

These correlations were developed using hydrodynamic entrance length data for smooth coiled tubes. Equation (10.3c) also includes limited turbulent flow data. The above equations indicate that in laminar flow, the friction factor depends only upon Re and R/a and is independent of the bend angle ϕ. This finding is quite contrary to that observed for turbulent flow (as presented in the next paragraph), where the friction factor is found to be strongly dependent upon the bend angle ϕ. Until more experimental data are available, it is difficult to assess the validity of Eqs. (10.3).

For turbulent flow, Ito [7] obtained extensive experimental data for 45°, 90°, and 180° bends with R/a varying from 2 to 15. He correlated his friction factor data for $2 \times 10^4 < \text{Re} < 4 \times 10^5$ as

$$K = \begin{cases} 0.00873\, B\phi f_c R/a & \text{for} \quad \text{Re}(R/a)^{-2} < 91 \\ 0.00241\, B\phi \text{ Re}^{-0.17} (R/a)^{0.84} & \text{for} \quad \text{Re}(R/a)^{-2} > 91 \end{cases} \tag{10.4}$$

where ϕ is a bend angle in degrees and f_c is the curved tube friction factor obtained

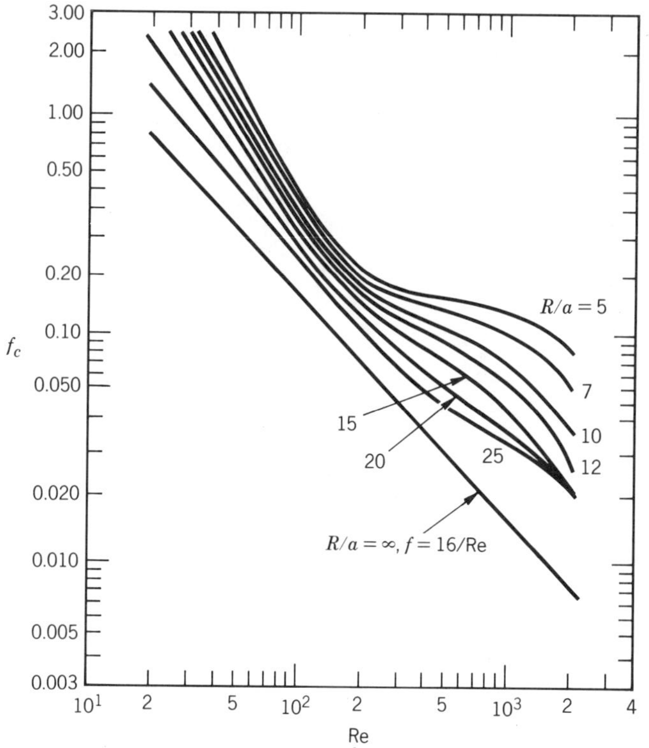

Figure 10.2. Laminar friction factors for flow through 90° bends of circular cross section.

from Eq. (5.48).† B is a numerical constant whose value is determined using the following correlations:

For $\phi = 45°$,‡

$$B = 1 + 14.2(R/a)^{-1.47} \tag{10.5}$$

For $\phi = 90°$,

$$B = \begin{cases} 0.95 + 17.2(R/a)^{-1.96} & \text{for} \quad R/a < 19.7 \\ 1 & \text{for} \quad R/a > 19.7 \end{cases} \tag{10.6}$$

For $\phi = 180°$,

$$B = 1 + 116(R/a)^{-4.52} \tag{10.7}$$

Figure 10.3a and b depict typical K values at $\text{Re} = 2 \times 10^5$. It is important to note that in the limiting case as $R \to \infty$, the bend friction loss coefficient should reduce to the straight-pipe value. Ito has noted that for $R/a > 100$, K values for bends and straight tubes are almost identical.

For turbulent flow, Powle [8] also recommends to use the following correlation for K^* in Eq. (10.2b); it is based on an empirical equation that appeared in the Soviet

† $f_c(R/a)^{0.5} = 0.00725 + 0.076[\text{Re}(a/R)^2]^{-0.25}$ for $0.034 < \text{Re}(a/R)^2 < 300$ [Eq. (5.48)].
‡ No R/a range is indicated in [7].

TABLE 10.1 Values of the Constant C_ϕ in Eq. (10.8b)

Bend Turning Angle ϕ, deg	C_ϕ
20	0.29
40	0.56
60	0.77
80	0.93
100	1.06
120	1.16
140	1.25
160	1.32
180	1.38

literature.

$$K^* = B(\phi)\left[0.051 + 0.38(R/a)^{-1}\right] \tag{10.8}$$

where

$$B(\phi) = \begin{cases} 1 & \text{for } \phi = 90^\circ \\ 0.9\sin\phi & \text{for } \phi \le 70^\circ \\ 0.7 + 0.35(\phi^\circ/90^\circ) & \text{for } \phi \ge 100^\circ \end{cases} \tag{10.8a}$$

K values calculated from Eq. (10.2b) using K^* of Eq. (10.8) agree with K values of Ito from Eq. (10.4) within $\pm 20\%$ for $\phi = 45^\circ$, 90°, and 180°, $2 \times 10^4 \le \mathrm{Re} \le 2 \times 10^5$ and $3 \le R/a \le 15$, except for the extreme combinations of Re and R/a. Thus, Ito's correlation may be used for $\phi = 45^\circ$, 90°, and 180° while Powle's correlation may be used for other values of ϕ.

In turbulent flow, Benedict [11] presented a correlation for the loss coefficient for a bend with any turning angle $\phi < 360^\circ$ as follows.

$$K_\phi = C_\phi K_{90} \tag{10.8b}$$

where K_{90} represents the loss coefficient for a 90° bend. The values for multiplying factor C_ϕ are listed in Table 10.1. The K_{90} values could be obtained using Eq. (10.4) with $\phi = 90^\circ$. It is important to note that C_ϕ of Eq. (10.8b) should be dependent upon R/a and Re as indicated through Eqs. (10.4)–(10.8). Single values of C_ϕ noted for various angles in Table 10.1 are approximate and are applicable only for standard elbows with $R/a = 2$ or 3.

In turbulent flow, the pipe roughness significantly influences bend pressure loss coefficients and friction factors. Extensive pressure drop data for various rough bends are given in Refs. 10, 12. In addition, Refs. 10, 12 also include data on bends with square, elliptical, and rectangular cross sections.

Friction Factors in Fittings. As shown in Fig. 10.4, several types of fittings are available [12–16]. The curvature ratio of standard fittings is $R/a = 2$, while that of long-radius fittings is normally $R/a = 3$. At a given Reynolds number, the pressure drop in a fitting is larger than that obtained in a corresponding smooth bend. Some

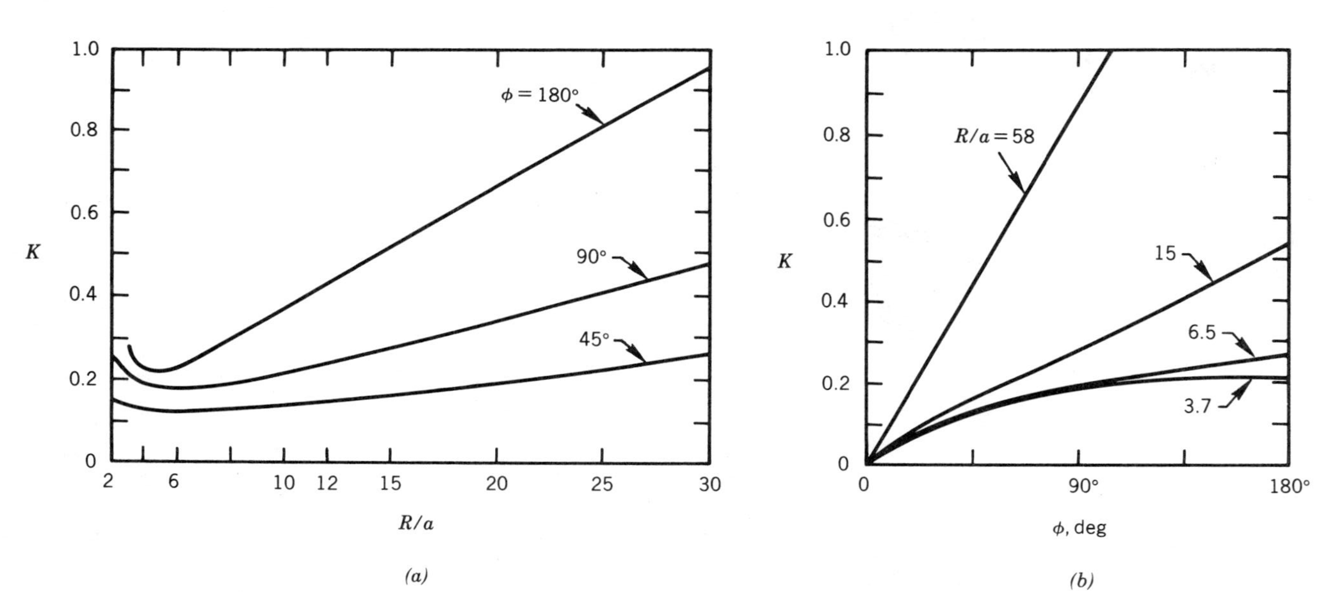

Figure 10.3. Pressure loss coefficient K as a function of the bend angle ϕ and curvature ratio R/a for $\mathrm{Re} = 2 \times 10^5$ [7].

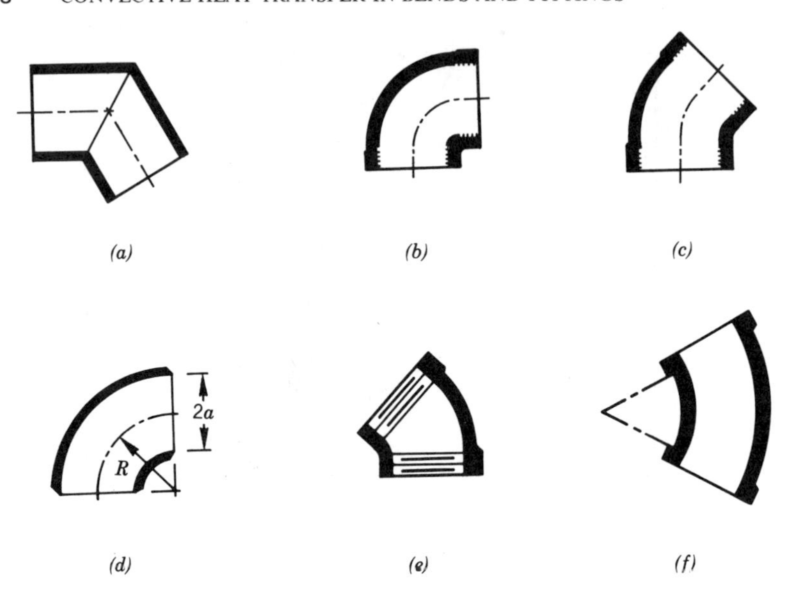

Figure 10.4. Various fittings: (*a*) miter bend, (*b*) 90° standard elbow, (*c*) 45° standard elbow, (*d*) 90° flanged or butt-welded elbow, (*e*) regular screwed 45° elbow, and (*f*) long-radius flanged 45° elbow.

fitting shapes add contraction and expansion losses due to changes in fluid flow area along the flow length, resulting in an additional pressure drop [12–16].

Hooper [16] compiled available laminar and turbulent flow pressure drop data for standard and long radius elbows (45°, 90°, 180°), miter bends, tees, and valves. The compiled data were presented as the following single correlation:

$$K^* = K_1 \mathrm{Re}^{-1} + K_\infty (1 + 0.5a^*) \tag{10.9}$$

Here, a^* is a fitting radius in inches, K_1 is K^* at Re = 1, and K_∞ is K^* as $\mathrm{Re} \to \infty$. The values of K_1 and K_∞ for various fittings are presented in Table 10.2. Figure 10.5 shows excellent agreement between the above correlation and experimental data for a standard $\frac{1}{2}$-in. elbow [17].

10.2.2 Heat Transfer in 90° Bends

Very limited experimental data are available for a 90° bend [18, 19]. The experimental data indicate that a bend has only a small influence on the heat transfer in a pipe upstream of a bend. However, it does exert a significant influence downstream. Only limited theoretical results are available to determine Nusselt numbers for air heating in a 90° bend [20].

Presently, no data or correlations are available to predict laminar flow Nusselt numbers in a 90° bend. Due to the lack of data, as an engineering approximation, we suggest using helical coil thermal entrance length Nusselt number correlations, Eqs. (5.37) and (5.38). As noted next, these equations may result in overpredicted values of Nusselt numbers.

TABLE 10.2 Constants K_1 and K_∞ of Eq. (10.9) for Fittings [16]

		Fitting Type		K_1	K_∞
Elbows	90°	Standard (R/a = 2), screwed		800	0.40
		Standard (R/a = 2), flanged/welded		800	0.25
		Long-radius (R/a = 3), all types		800	0.20
		Mitered Elbows (R/a = 3)	1 weld (90° angle)	1000	1.15
			2 weld (45° angles)	800	0.35
			3 weld (30° angles)	800	0.30
			4 weld ($22\frac{1}{2}$° angles)	800	0.27
			5 weld (18° angles)	800	0.25
	45°	Standard (R/a = 2), all types		500	0.20
		Long-radius (R/a = 3), all types		500	0.15
		Mitered, 1 weld, 45° angle		500	0.25
		Mitered, 2 weld, $22\frac{1}{2}$° angle		500	0.15
	180°	Standard (R/a = 2), screwed		1000	0.60
		Standard (R/a = 2), flanged/welded		1000	0.35
		Long-radius (R/a = 3), all types		1000	0.30
Tees	Used as elbow	Standard, screwed		500	0.70
		Long-radius, screwed		800	0.40
		Standard, flanged or welded		800	0.80
		Stub-in-type branch		1000	1.00
	Run-through tee	Screwed		200	0.10
		Flanged or welded		150	0.50
		Stub-in-type branch		100	0.00
Valves	Gate, ball, plug	Full line size, β' = 1.0		300	0.10
		Reduced trim, β' = 0.9		500	0.15
		Reduced trim, β' = 0.8		1000	0.25
		Globe, standard		1500	4.00
		Globe, angle or Y-type		1000	2.00
		Diaphragm, dam type		1000	2.00
		Butterfly		800	0.25
	Check	Lift		2000	10.00
		Swing		1500	1.50
		Tilting-disk		1000	0.50

Ede's [18] results for water heating in turbulent flow primarily dealt with heat transfer in a pipe downstream of a bend. Tailby and Staddon [19] measured Nusselt numbers for turbulent cooling of air in a 90° bend. The bend was immersed in water, thus simulating a constant wall temperature boundary condition. The authors have reported an increase in Nusselt numbers to be about 20 to 30% above the straight-tube values. They also noticed higher Nusselt numbers at the outer wall than those at the inner wall.

Tailby and Staddon [19] reported only peak Nusselt numbers at the outer wall, which are presented in Table 10.3. Also compared in the table are the values of the peak Nusselt numbers with those calculated using fully developed Nusselt numbers for a helical coil from Eq. (5.56). The comparison indicates that peak Nusselt numbers for air cooling in a 90° bend are smaller than the average increment expected in a fully developed flow. In contrast, for air heating Iacovides and Launder [20] report the ratio

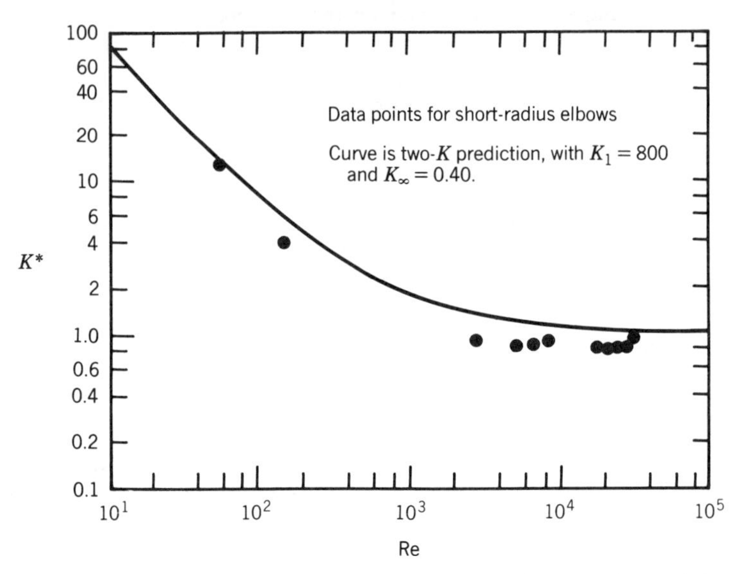

Figure 10.5. A comparison of experimental data and Eq. (10.9) for K^* values of $\frac{1}{2}$-in. standard elbow [16].

of outside- and inside-wall Nusselt numbers to be nearly 5 : 1 at a flow length of 75° from the bend inlet (Re = 4300, De = 1800, and Pr = 0.7). Tailby and Staddon [19] also noted small Nusselt number increments for air cooling in 180° bends. They explained that in a bend secondary flow pushes heavier fluid particles toward the outer wall and lighter ones toward the inner wall. The bend heating augments the secondary flow resulting in significantly higher heat transfer coefficients at the outer wall. Although bend cooling reduces the secondary flow, it still gives high heat transfer coefficients at the outer wall. However, the increment in heat transfer above the straight-tube values is not as great as that observed during fluid heating.

Tailby and Staddon [19] proposed the following correlation for turbulent air cooling in a 90° bend:

$$\mathrm{Nu}_x = 0.0336\,\mathrm{Re}^{0.81}\mathrm{Pr}^{0.4}\left(\frac{R}{a}\right)^{-0.06}\left(\frac{x}{D}\right)^{-0.06} \tag{10.10}$$

TABLE 10.3 Ratio of Peak Nusselt Numbers at the Bend Outer Wall to Nu for a Straight Pipe for Turbulent Flow Air Cooling through a 90° Bend [19]

R/a	Distance to Peak Nu in Pipe Diameters	$\mathrm{Nu}_{\mathrm{peak}}/\mathrm{Nu}_s$	$\mathrm{Nu}_c/\mathrm{Nu}_s$ [Eq. (5.56)]
Miter bend	—	2.0	—
2.5	2.5	1.42	—
4.0	4.0	1.36	—
6.0	5.0	1.29	1.59
14.0	7.0	1.22	1.25

The application range of the above correlation is $2.3 \leq R/a \leq 14$, $7 \leq x/D \leq 30$, and $10^4 \leq \text{Re} \leq 5 \times 10^4$. The correlation suggests that for the turbulent flow fluid cooling in a 90° bend, the average heat transfer increase in a bend is only about 20 to 30% above the straight-tube values. It is important to note that the above correlation is valid only for fluid cooling. Moreover, the available numerical predictions for heating are also limited [20]. Hence, for the fluid-heating case, probably the fully developed turbulent flow heat transfer correlation, Eq. (5.56), may be used.

10.2.3 Heat Transfer in a Pipe Downstream of a 90° Bend

In laminar flow, Ede's [18] water-heating data, plotted in Fig. 10.6, show that the bend affects heat transfer in the downstream straight pipe for a distance of about 10 diameters. In addition, the upstream heated length shows an influence on heat transfer in the downstream pipe. The longer the length of heated upstream pipe, the higher is the downstream pipe Nusselt number. The plot for an unheated upstream length in Fig. 10.6 shows Nusselt numbers of the same order of magnitude as those observed in a thermally developing flow in a straight tube. No simple correlation is available to predict the incremental heat transfer. The limited laminar results shown in Figs. 10.6 and 10.7 could be used to obtain Nusselt numbers in a heated pipe located downstream of a 90° bend. The results presented in Fig. 10.7 are for the case of a heated bend and heated downstream pipe.

In turbulent flow, heat transfer in the downstream pipe was found to be independent of upstream heating [18, 19]. Here also water data depicted in Fig. 10.7 could be used when the fluid is being heated. The following correlation could be used to predict downstream pipe heat transfer during fluid cooling [19]:

$$\text{Nu}_x = 0.0366\,\text{Re}^{0.8}\text{Pr}^{0.4}F_1 \qquad \text{for} \quad 0 \leq x/D \leq 30 \tag{10.11}$$

where F_1, plotted in Fig. 10.8, depends upon the ratio x/D. In general, fluid heating in a bend yields higher heat transfer coefficients than fluid cooling.

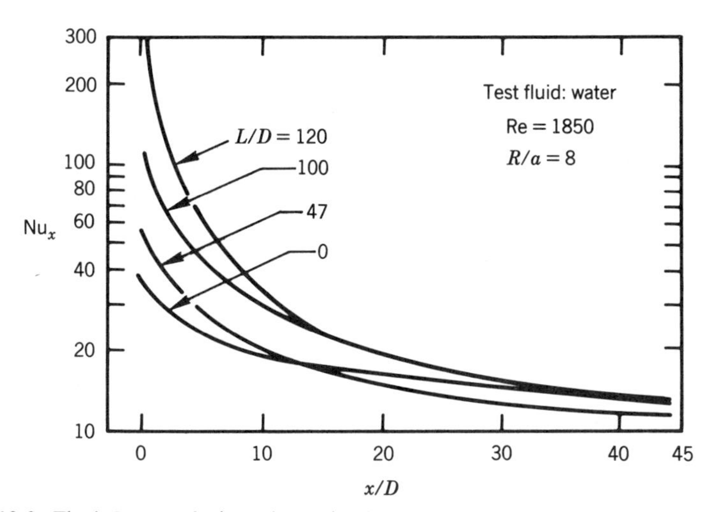

Figure 10.6. The influence of a heated straight pipe upstream of a bend on the constant heat flux Nusselt numbers in a straight pipe downstream of a 90° bend [18]. L/D values indicate upstream-pipe heated length.

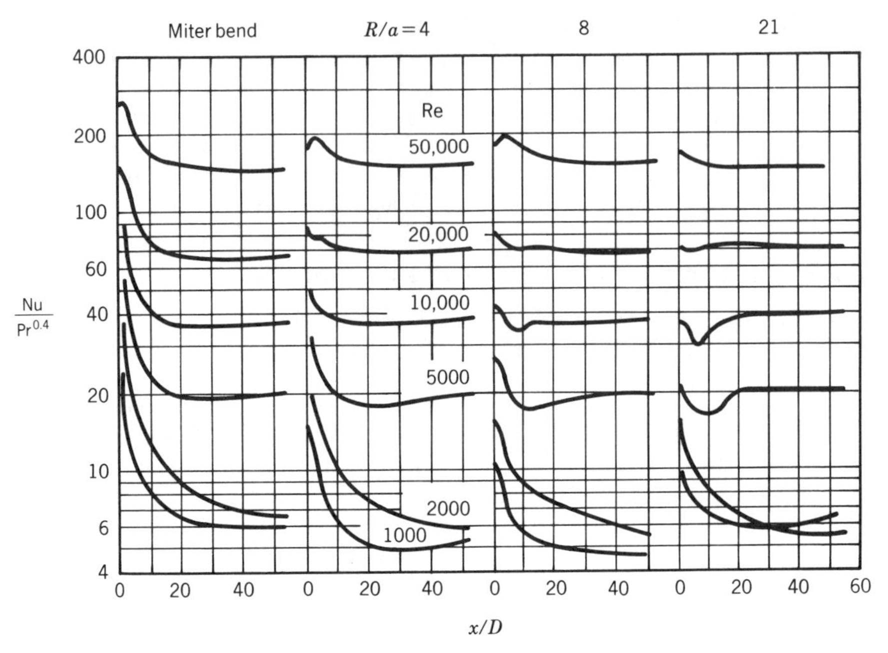

Figure 10.7. Constant heat flux Nusselt numbers in a straight pipe located downstream of a 90° bend [18]. The bend as well as the downstream pipe is heated. The upstream pipe is unheated.

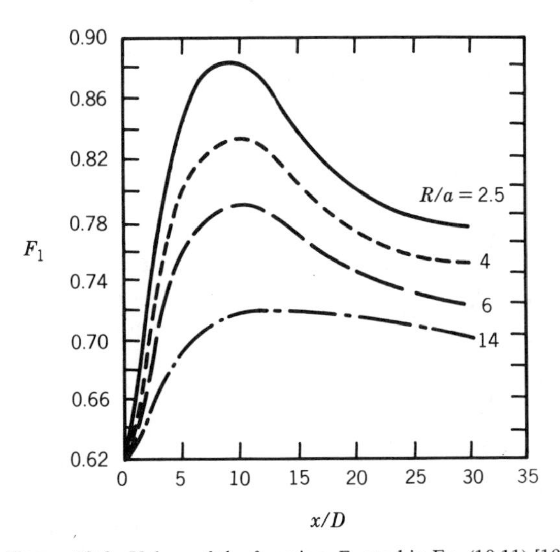

Figure 10.8. Values of the function F_1 used in Eq. (10.11) [19].

In summary, for turbulent flow through a 90° bend, the overall effect of incremental heat transfer in a bend and its downstream pipe is equivalent to increasing the length of a heated pipe by about eight diameters for an elbow, and three diameters for bends with $R/a = 8$ and 4 [18, 19].

10.2.4 Heat Transfer in 180° Bends

Eight experimental studies [21–28] have reported heat transfer results for flow through a 180° bend. For laminar flow, the most extensive results are reported by Moshfeghian and Bell [24] for four 180° bends ($R/a = 4.84$, 7.66, 12.32, and 25.62). Data were obtained by placing the 180° bend in a vertical plane as shown in Fig. 10.1*b* with the bend outlet pipe above the bend inlet pipe. The bend inlet and outlet pipes, along with the bend, were heated electrically, thus approximating the boundary condition of axially constant heat flux.† Water, Dowtherm G®, and ethylene glycol were used as working fluids. They observed higher heat transfer coefficients in the bend as well as downstream of the bend. Their experimental data on laminar flow showed no consistent trends, and they were unable to obtain a satisfactory correlation. As noted earlier, as with a 90° bend, the helical-coil entrance region correlations, Eqs. (5.37) and (5.38), may be used to estimate laminar flow Nusselt numbers in a 180° bend. However, the predicted values may be higher than the experimental values.

For turbulent flow, Baughn et al. [25] measured temperature profiles for air flow through a bend with $R/a = 6.75$. The heated bend wall was maintained at a constant wall temperature. The temperature profile exhibited profile distortion with a peak temperature shifting toward the tube outer wall. Iacovides and Launder [26] have reported a numerical solution for a temperature profile which shows excellent agreement with the experimental data of Baughn et al. [25].

For turbulent flow, experimental data [24, 25, 28] and theoretical predictions [26] are available to calculate variations in peripheral local Nusselt numbers. The theoretical predictions show fairly good agreement with the experimental data near the bend inner wall. However, theoretical predictions underestimate the Nusselt numbers by about 50 to 80% at the bend outer wall. Figure 10.9 shows a typical circumferential and axial variation in local Nusselt numbers for air heating in a bend subjected to the Ⓣ boundary condition [25]. Substantial circumferential variations in Nusselt numbers occur even at 15° from the bend inlet. The peak circumferential variation occurs at 90°, showing about a 3 : 1 ratio of Nusselt numbers at the duct outer and inner walls. The variation in peripheral Nusselt numbers is seen even at a distance of 6 diameters in a downstream straight pipe.

Moshfeghian and Bell [24] measured turbulent flow Nusselt numbers in a heated bend using water, Dowtherm®, and ethylene glycol as working fluids. Along with the bend, the upstream and downstream straight pipes were also heated. The authors [24] presented the following correlation of their extensive experimental data for the turbulent flow Nusselt numbers:

$$\mathrm{Nu}_x = 0.0285\,\mathrm{Re}^{0.81}\mathrm{Pr}^{0.4}\left(\frac{x}{D}\right)^{0.046}\left(\frac{R}{a}\right)^{-0.133}\left(\frac{\mu_m}{\mu_w}\right)^{0.14} \tag{10.12}$$

The correlation is valid for $4.8 \le R/a \le 26$, $10^4 \le \mathrm{Re} \le 3 \times 10^4$, and $0 < x/D \le \pi R/(2a)$. It shows less than 8% deviation from the experimental data.

† This boundary condition does not fit the designation Ⓗ1 or Ⓗ2 used in Chapter 3, since the temperature or heat flux across a cross section in a bend is *not* uniform.

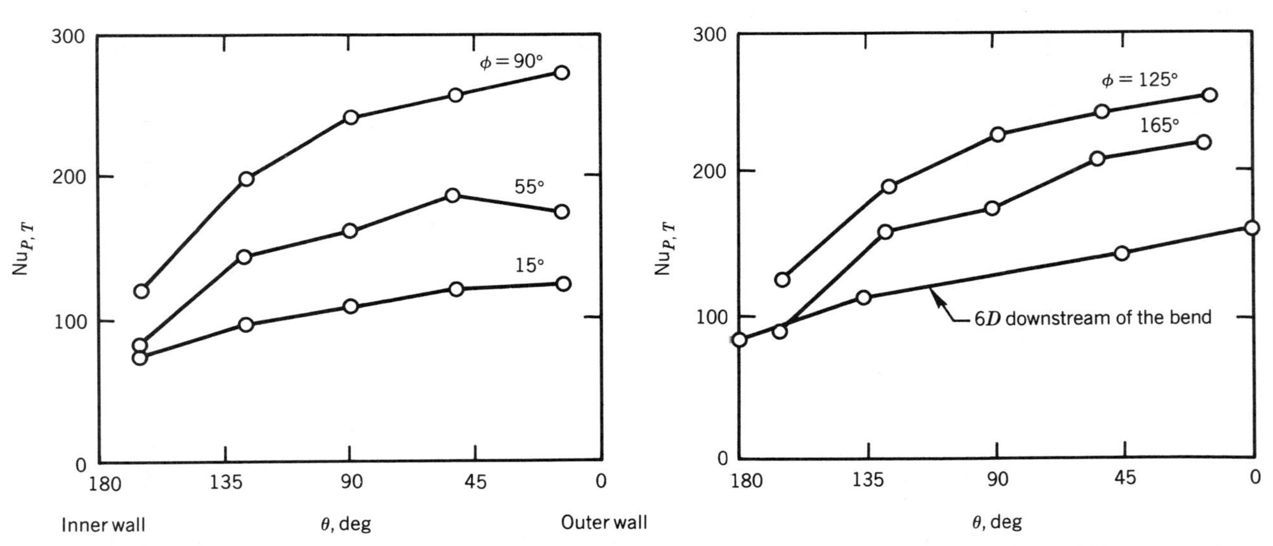

Figure 10.9. Experimental Ⓣ circumferential Nusselt number variation for air heating for different ϕ along a 180° bend and its downstream pipe. $Re = 6 \times 10^4$, $De = 2.31 \times 10^4$, $R/a = 6.75$ [25]. Here $Nu_{p,T}$ is based on the pipe wall temperature minus pipe center line temperature difference.

10.2.5 Heat Transfer in a Pipe Downstream of a 180° Bend

Mehta and Bell [27] measured Nusselt numbers in a heated pipe ($D = 15.75$ mm) downstream of an unheated 180° bend. Their laminar flow experimental results for Re = 80 to 600, using the three liquids of [24], show that the unheated bend ($R/a = 7.7$) has no influence on the downstream pipe heat transfer. At low Reynolds numbers, natural convection strongly influences heat transfer in a downstream pipe. Combined free- and forced-convection Nusselt number correlations for a straight pipe, such as those proposed by Marcos and Bergles [57], are useful in predicting laminar flow heat transfer. However, when a 180° bend as well as the downstream pipe is heated, laminar flow Nusselt numbers in the entry portion of a downstream pipe seems to be influenced by the upstream bend. In such cases, a downstream pipe shows an influence of both the natural convection and decaying swirl flow. Moshfeghian and Bell [24] have correlated their laminar flow results for four 180° bends ($R/a = 4.84$, 7.66, 12.32, and 25.62) for $\mathrm{Re} < 2100$ and $\pi R/2a < x/D < 160$ as shown below:

$$\mathrm{Nu}_x = 0.00275 C_1 C_2 [\mathrm{Re}\exp(C_3)] \mathrm{Pr}^{0.4} \left(\frac{\mu_m}{\mu_w}\right)^{0.14} \tag{10.13}$$

where

$$C_1 = 1.0 + 8.5\left(\frac{\mathrm{Gr}}{\mathrm{Re}^2}\right)^{0.429} \tag{10.14}$$

$$C_2 = 1.0 + 4.79\exp\left[-2.11\left(\frac{x}{D}\right)^{-0.237}\right] \tag{10.15}$$

$$C_3 = 0.733 + 14.33\left(\frac{R}{a}\right)^{-0.593}\left(\frac{x}{D}\right)^{-1.619} \tag{10.16}$$

The correlation exhibits less than ±16% deviation with the experimental data.

For turbulent liquid flow, Moshfeghian and Bell [24] have proposed the following correlation to calculate the local Nusselt numbers in a downstream pipe when the downstream pipe as well as the upstream bend is heated:

$$\mathrm{Nu}_x = 0.031\,\mathrm{Re}^{0.825}\mathrm{Pr}^{0.4}\left(\frac{\mu_m}{\mu_w}\right)^{0.14}\left(\frac{x}{D}\right)^{-0.116}\left(\frac{R}{a}\right)^{-0.048} \tag{10.17}$$

The natural convection effects are negligible in turbulent flow and hence are not included here. The above correlation is valid for $10^4 \le \mathrm{Re} \le 3 \times 10^4$, $\pi R/2a < x/D < 160$, and $4.8 \le R/a \le 26$. The correlation predicts Nu_x within ±6% deviation from the experimental data.

At present, turbulent flow Nusselt numbers in a downstream pipe have not been reported in the literature when the upstream bend is unheated. However, as noted earlier in the case of a 90° bend, the turbulent flow downstream-pipe heat transfer is independent of the heating condition of the upstream bend. A similar conclusion may also be valid for a 180° bend. Therefore, until experimental data are available, Eq. (10.17) may be used to calculate turbulent liquid flow Nusselt numbers in a downstream pipe regardless of the heating boundary condition of an upstream bend.

10.3 BENDS WITH RECTANGULAR CROSS SECTIONS

Limited results are available on fluid flow and heat transfer through a bend having a rectangular cross section. However, most of the available results are for the special case of a square cross section. As shown in Fig. 10.1*c*, for a rectangular bend, $2a$ represents the channel depth in the same plane as that for a radius of curvature.

10.3.1 Fluid Flow

The developing *laminar* velocity profiles for a flow through a 90° bend of square cross section are reported by Humphrey et al. [29] and others [30, 31]. Humphrey et al. [29] obtained theoretical solutions and experimental measurements of velocity profiles for water flowing through a bend of square cross section with $R/a = 4.6$. The experimental data show good agreement with the theoretical predictions. Yamashita et al. [32] have reported theoretical velocity profiles for laminar flow through a 90° miter bend. Similarly, for *turbulent* flow through a 90° bend of square cross section, Taylor et al. [30] and Humphrey et al. [33] have reported velocity profiles. Taylor et al. [30] measured velocity profiles for turbulent air flow in 90° bends with $R/a = 4.6$ and 14. Humphrey et al. [33] have reported experimental data and theoretical predictions for turbulent flow through a 90° bend with $R/a = 4.6$. Their experimental data show good agreement with the theoretical predictions.

For a 180° bend with a square cross section, Chang et al. [34, 35] and others [36, 38] have reported theoretical as well as experimental turbulent velocity profiles. In addition, Chang et al. [34, 35] and Johnson and Launder [36, 37] have reported velocity profiles in a straight pipe downstream of a 180° bend. In general, the theoretical predictions show fairly good agreement with experimental data, although some of the detailed variations of the velocity profiles in the bend are not reproduced by the calculations.

10.3.2 Friction Factors

Shiragami and Inoue [39] measured friction factors for water flow in four 90° bends with square cross section, and having the curvature ratios $R/a = 3.4$, 6.9, 13.7, and 27.4. For laminar flow, they proposed the following correlation for De = 100 to 400:

$$\frac{f_c}{f_s} = 0.291\left[\mathrm{Re}\left(\frac{R}{a}\right)^{-0.5}\right]^{0.35} \qquad \text{for} \quad 3.4 \le \frac{R}{a} \le 27.4 \tag{10.18}$$

It may be noted that Eqs. (10.18) and (10.3a) are quite close since $f_s = 14.23/\mathrm{Re}$.

For turbulent flow through a 90° square cross section bend, Shiragami and Inoue [39] proposed the following correlation:

$$\frac{f_c}{f_s} = 0.0040\left[\mathrm{Re}\left(\frac{R}{a}\right)^{-0.2}\right]^{0.87} \qquad \text{for} \quad \frac{R}{a} \ge 13.7 \tag{10.19}$$

Their measured friction factors in turbulent flow for bends with $R/a = 3.4$ and 6.9 were significantly higher than those determined from Eq. (10.19). The authors observed flow separation for $R/a = 3.4$ and 6.9, which may have resulted in higher friction factor measurements [39].

Presently, no friction factor correlation is available for 180° bends.

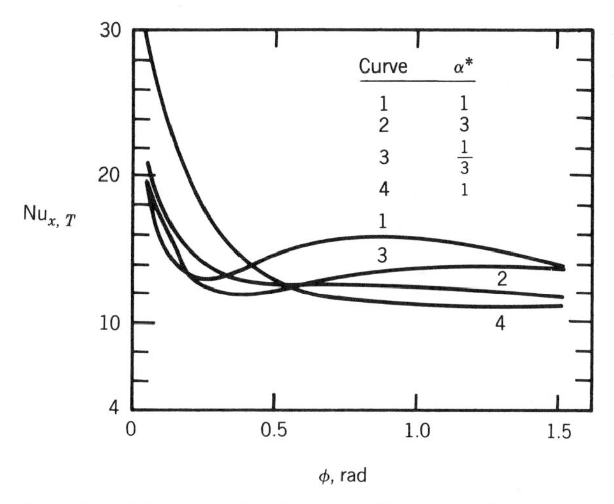

Figure 10.10. Ⓣ peripherially averaged, axially local Nusselt-number variation along the length of 90° bends of rectangular cross section. In curves 1, 2, and 3, the fluid flow is fully developed at the heated bend inlet. Curve 4 is for uniform velocity profile at the entrance. De = 368, R/D_h = 2.3 [40].

10.3.3 Heat Transfer in 90° Bends

Yamashita et al. [32] have reported temperature profiles in a miter bend of square cross section.

Limited data are available for the calculation of temperature profiles and Nusselt numbers in a 90° bend with rectangular cross section. Yee et al. [40] theoretically calculated thermally developing Nusselt numbers for air heating in bends subjected to the Ⓣ boundary condition. The results are reported only for De = 368 for a laminar flow through rectangular bends ($\alpha^* = \frac{1}{3}$, 1, and 3) having $R/a = 4.6$. As shown in Fig. 10.10, the duct inlet velocity profile has a strong influence on Nusselt numbers, while the duct aspect ratio has a lesser influence. Metzner and Larsen [41] experimentally measured Nusselt numbers for turbulent air flow in 38.1 mm × 12.7 mm rectangular cross-sectional duct ($\alpha^* = 3$). The ratios of outer and inner bend wall radii were 2 and 3. The authors found that the average heat transfer coefficients were about 20 to 30% more than those obtained in straight ducts.

In laminar flow at low Reynolds numbers, natural convection can influence the heat transfer in a 90° bend. Chilukuri and Humphrey [42] investigated the influence of free convection on heat transfer in a bend with a square cross section. The numerical results were obtained for air heating in a bend oriented in a vertical plane and subjected to the Ⓣ boundary condition. Their limited predictions for a thermally developing flow for Re = 797, De = 367, and $R/a = 4.6$ are shown in Fig. 10.11. As expected, natural convection augments the forced convection Nusselt numbers for vertical upflow and reduces them for vertical downflow.

10.3.4 Heat Transfer in 180° Bends

Data are available to determine Nusselt numbers and temperature profiles in square bends [35–37]. Nusselt numbers are also available for rectangular ducts [43–45]. All the available data are for turbulent flow.

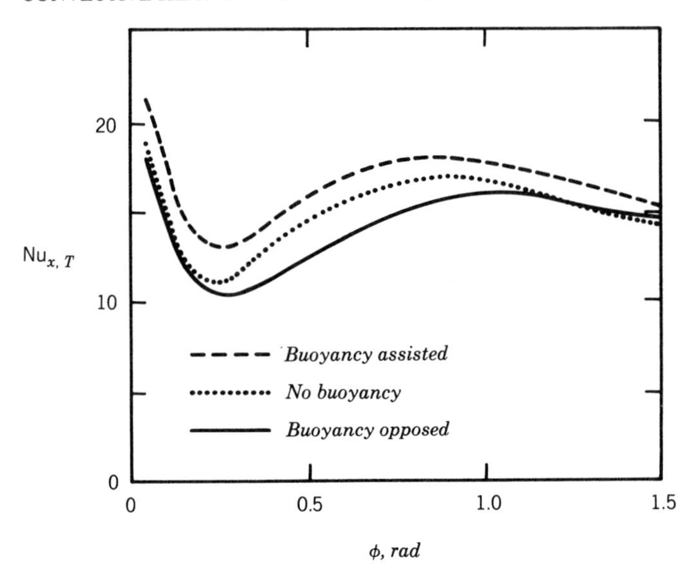

Figure 10.11. Ⓣ peripherally averaged, axially local Nusselt number variation along the length of a 90° vertical bend of square cross section. Re = 747, De = 367, $R/a = 4.6$, Pr = 0.7, $Gr^+ = 3.14 \times 10^5$ [42].

Chang et al. [35] have reported theoretical temperature profiles and Nusselt numbers for Re $= 5.67 \times 10^4$ for a square cross-section bend. Their data for a bend with $R/a = 6.70$ show about a 2:1 ratio of outer- to inner-wall Nusselt numbers. In general, the peripheral mean Nusselt numbers are about 30% higher than those in a straight duct. The 30% Nusselt number increment over the straight-tube values is similar to the 20 to 30% increment reported by Metzger and Larson [41] for 90° bends. Chang et al. [35] also observed that a strong nonuniformity in Nusselt numbers persists for about 10 diameters of pipe length in a downstream straight duct. Johnson and Launder [36, 37] obtained similar results for a bend with a square cross section ($R/a = 6.70$). They measured Nusselt numbers for axially constant heat flux boundary condition for turbulent air flow through the bend (Re $= 9.9 \times 10^3$ to 9.2×10^4). Figure 10.12 depicts typical results, showing higher Nusselt numbers at the bend outer wall than those at the bend inner wall. Johnson and Launder [36] found that the intensity of the secondary velocity in a turbulent inlet velocity profile has no significant influence on Nusselt numbers in 180° bends (not shown in Fig. 10.12).

Yang and Liao [43] reported experimental data for three rectangular cross-section bends ($R/a = 12.95$, 18.4, and 39). All bends had an aspect ratio of 10. As shown in Fig. 10.13, their experimental data for Re = 7000 to 26,500 with air heating show good agreement with their theoretical analysis. The bends were preceded by a long heated straight duct, resulting in fully developed flow at the bend entrance. It is important to note that in their experiments, only the inner and outer walls of the bend and entrance duct were heated electrically, simulating the boundary condition of axially constant heat flux. However, the side walls were unheated and insulated. Their data clearly indicate that the outer-wall Nusselt numbers are substantially higher than those at the inner wall.

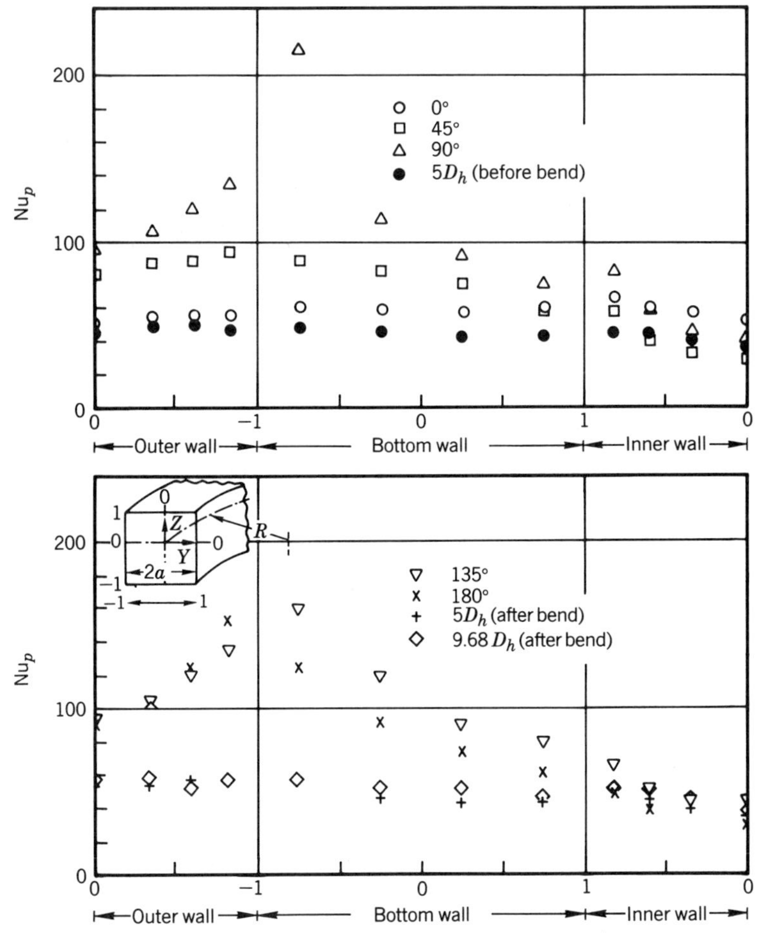

Figure 10.12. Development of the peripheral local Nusselt number in a 180° bend of square cross section. Fully developed flow at the bend entrance (entrance length = $72\,D_h$, heated entrance length = $57D_h$). Air heating in a bend subjected to the boundary condition of axially constant heat flux, Re = 9880, R/a = 6.70 [36].

Metzger and Sahm [44, 45] obtained Nusselt numbers for a sharp 180° turn. Such turns are usually encountered in cooling passages of high-temperature gas turbines. Figure 10.14 shows a schematic diagram of their experimental apparatus. As shown in the figure, the heater locations are the top surface (segments 1 to 5), bottom surface (segments 13 to 17), inlet side wall (segments 6 to 8), exit side wall (segments 10 to 12), and end turn (segment 9). Thus, all walls, except the dividing center wall, are heated and are maintained at the same constant wall temperature. Figure 10.15 shows a typical variation in the Nusselt numbers, which are highest on the wall directly downstream of a 180° bend. The authors have correlated their results with air heating for $10^4 \le \text{Re} \le 6 \times 10^4$ as

$$\text{Nu}_{m,j} = B_1 \text{Re}^{B_2} \tag{10.20}$$

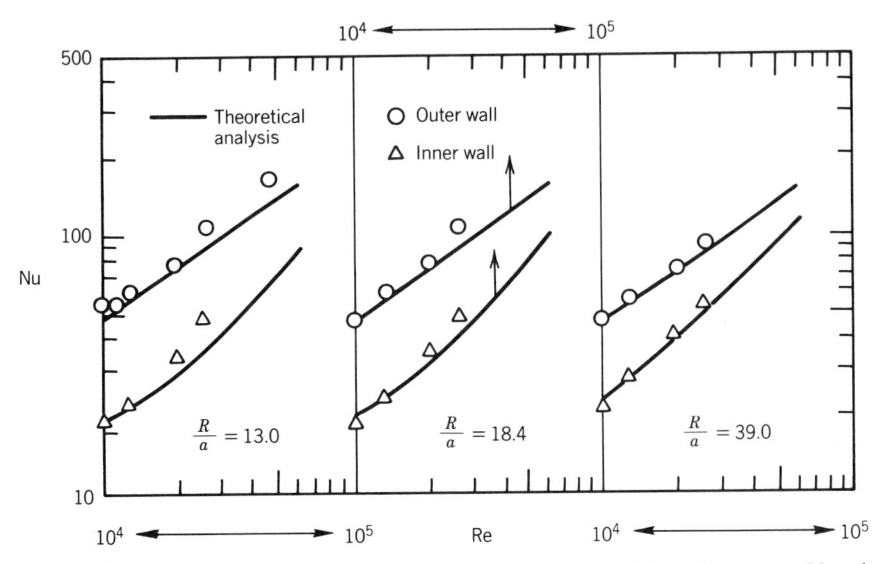

Figure 10.13. A comparison of experimental and theoretical turbulent flow mean Nusselt numbers for air flow through rectangular cross-section 180° bends. All bends have the same aspect ratio, $2c/2a = 10$. Nusselt numbers are averaged over respective outer and inner wall areas [43].

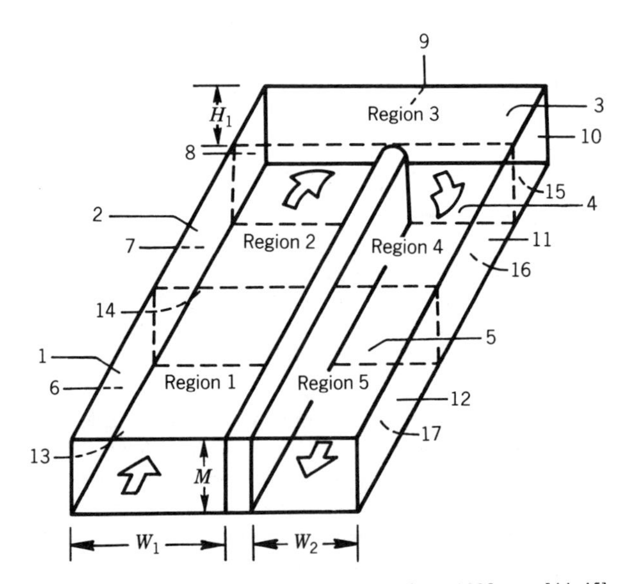

Figure 10.14. Schematic geometry of a sharp 180° turn [44, 45].

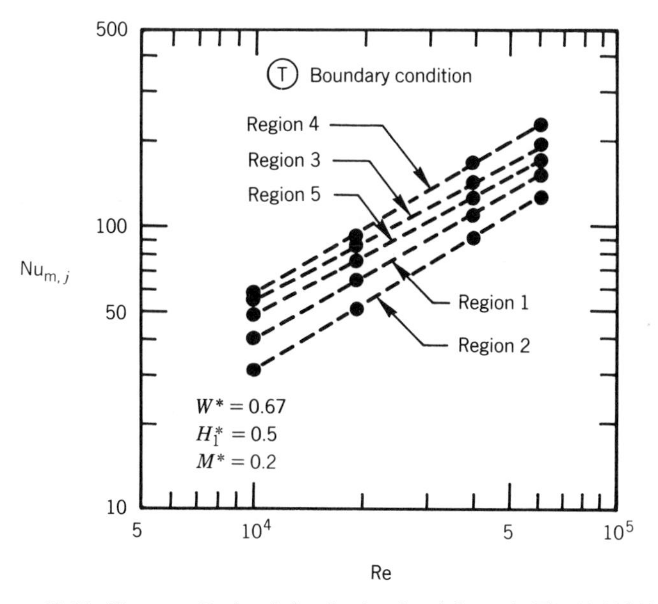

Figure 10.15. $Nu_{m,j}$ vs. Re for air heating in a bend shown in Fig. 10.14 [44, 45].

where subscript j indicates the region shown in Fig. 10.14, and Re is defined on the basis of the hydraulic diameter of the inlet channel (dimensions $M \times W_1$). The constants B_1 and B_2, presented in Table 10.4, depend upon the flow region, wall spacing, and end clearance.

10.3.5 Heat Transfer in Bends with Bend Angle Less Than 250°

Shchukin and Filin [46] have obtained experimental Nusselt number data for water heating in a square cross-section bend subjected to the boundary condition of axially constant heat flux. The experimental channel had $R/a = 6.12$ and had about a 250° bend with large inlet- and exit-flow mixing chambers. The authors measured Nusselt numbers at 13 locations along the bend. In their experiment, free convection was quite significant ($Gr = 4.5 \times 10^6$), and hence they divided their results into three regions: the free convection region ($De < 690$), mixed convection region ($690 < De < 2400$), and forced convection region ($De > 2400$). Shchukin and Filin correlated their peripheral average Nusselt numbers for the free- and forced-convection regions for water heating in a square bend ($R/a = 6.12$) as shown below:

$$\mathrm{Nu}_x = \left[0.91 + 2.95\left(\frac{x}{D_h}\right)^{-0.83}\right] B_3 \qquad \text{for} \quad \mathrm{De} < 690 \tag{10.21}$$

$$\mathrm{Nu}_x = \left[0.96 + 2.41\left(\frac{x}{D_h}\right)^{-1.025}\right] B_4 \qquad \text{for} \quad \mathrm{De} > 2400 \tag{10.22}$$

TABLE 10.4 Constants for Eq. (10.20) for the Bend Geometry of Fig. 10.14 [44, 45]

W^*	H_1^*	Region	B_1	B_2
0.67	0.4	1	0.04360	0.7420
		2	0.02495	0.7757
		3	0.08364	0.7105
		4	0.03619	0.7992
		5	0.09037	0.6940
	0.5	1	0.04022	0.7493
		2	0.02275	0.7839
		3	0.09320	0.6926
		4	0.04128	0.7845
		5	0.08221	0.6943
	0.6	1	0.04883	0.7324
		2	0.03025	0.7577
		3	0.06622	0.7265
		4	0.05907	0.7469
		5	0.09201	0.6786
1.0	0.4	1	0.04371	0.7450
		2	0.02396	0.7847
		3	0.12350	0.6717
		4	0.07339	0.7546
		5	0.12109	0.6784
	0.5	1	0.03699	0.7613
		2	0.02421	0.7849
		3	0.05790	0.7375
		4	0.06487	0.7651
		5	0.07621	0.7164
	0.6	1	0.03571	0.7633
		2	0.02603	0.7763
		3	0.07289	0.7059
		4	0.07246	0.7536
		5	0.07939	0.7092
1.5	0.4	1	0.03429	0.7718
		2	0.02408	0.7857
		3	0.14491	0.6542
		4	0.12281	0.7210
		5	0.08377	0.7334
	0.5	1	0.04513	0.7449
		2	0.02729	0.7731
		3	0.12321	0.6610
		4	0.10515	0.7350
		5	0.06917	0.7465
	0.6	1	0.01988	0.8188
		2	0.00843	0.8827
		3	0.01630	0.8581
		4	0.01830	0.9033
		5	0.01710	0.8720

where

$$B_3 = 0.51(\mathrm{Gr\,Pr})^{0.25}\left(\frac{\mathrm{Pr}}{\mathrm{Pr}_w}\right)^{0.25} \tag{10.23}$$

$$B_4 = 0.0575\,\mathrm{Re}^{0.33}\mathrm{De}^{0.42}\mathrm{Pr}^{0.43}\left(\frac{\mathrm{Pr}}{\mathrm{Pr}_w}\right)^{0.25} \tag{10.24}$$

In the above correlation, all properties except for Pr_w are based on the fluid bulk mean temperature. A correlation is not available for the intermediate Dean numbers ($690 < \mathrm{De} < 2400$).

10.3.6 Bends with Either Outer or Inner Wall Heating

Few papers [47–52] report Nusselt number data for 180° rectangular cross-section bends having either an inner or outer bend wall heated and the remaining three walls unheated. Friction factor data for these cases are not available. An engineering approximation of the turbulent flow friction factors for a rectangular duct may be made by replacing a in Eq. (10.4) with $D_h/2$.

Experimental heat transfer correlations are given in Refs. 47–49, and the fundamental heat transfer and fluid mechanics studies are reported in Refs. 50–52. Seki et al. [47] measured Nusselt numbers for water heating in three 180° bends. A boundary condition of constant wall heat flux was imposed only on the outer wall of the bend. The Nusselt numbers were measured in three rectangular cross-section bends. In all bends, the channel height was 400 mm and the curvature radius of an inner wall was 121 mm. The bend channel widths were 9, 34, and 55 mm, giving aspect ratios 44.44, 11.76, and 7.27 and curvature ratios $R/a = 27.88$, 8.18, and 5.40, respectively. The bends were preceded by an unheated entrance length. The authors have correlated their turbulent flow mean Nusselt numbers for the outer wall of a 180° bend as

$$\mathrm{Nu}_m^* = 0.0255\,\mathrm{Re}^{*0.443}\mathrm{De}^{*0.443}\mathrm{Pr}^{0.415} \tag{10.25}$$

The above correlation is valid for $5 \times 10^3 \le \mathrm{Re}^* \le 8 \times 10^4$, $4 \le \mathrm{Pr} \le 13$, and $5.4 \le R/a \le 27.9$. The fluid properties in the above correlation as well as in other correlations in this section [Eqs. (10.26)–(10.33)] were evaluated at the bend-inlet fluid temperature. Note that Nu_m^*, Re^*, and De^* in Eq. (10.25) are based on $4a$ as the characteristic length and *not* D_h. Also note that T_i (and *not* T_m) is used in the definition of Nu_m^*.

Seki et al. proposed the following relation to calculate local Nusselt numbers in the turbulent flow [47]:

$$\mathrm{Nu}_x^* = 447.745\,\mathrm{Re}_x^{*1.497}\mathrm{De}_x^{*-1.596}\mathrm{Pr}^{0.412}B_5^{0.960} \quad \text{for} \quad 4 \le \mathrm{Pr} \le 13, \quad 0 \le x/R \le 1 \tag{10.26}$$

where

$$B_5 = 2\frac{u_m}{R}\frac{\partial u}{\partial y} \tag{10.27}$$

and

$$u\sqrt{\frac{\rho}{\tau_w}} = 5.6\log_{10}\left(\frac{y}{\nu}\sqrt{\frac{\tau_w}{\rho}}\right) + 4.9 \tag{10.28}$$

The authors do not clearly state the location at which $\partial u/\partial y$ is to be evaluated, but it might be at the tube wall.

Seki et al. [48, 49] have also reported only Nusselt numbers for four 180° bends of rectangular cross section. Only the inner bend wall was heated; the other three were unheated. In all bends, the channel height was 400 mm and curvature radius of an inner wall was 121 mm. The bend channel widths were 80, 60, 40, and 15 mm, giving aspect ratios $\alpha^* = 5.0$, 6.67, 10.0, and 26.7 and bend curvature ratios $R/a = 4.03$, 5.03, 7.05, and 17.13, respectively. The bends were preceded by an unheated entrance length. The Nusselt numbers at the inner wall, subjected to the boundary condition of constant wall heat flux, were lower than those observed in a straight tube flow. The authors correlated their water-flow mean Nusselt numbers for $8 \times 10^3 \le \mathrm{Re}^* < 8 \times 10^4$, $6.5 \le \mathrm{Pr} \le 8.5$, and $4.0 \le R/a \le 17.5$ as

$$\mathrm{Nu}_m^* = 0.0208\,\mathrm{Re}^{*0.997}\mathrm{De}^{*-0.116}\mathrm{Pr}^{0.39} \qquad \text{for} \quad 0 \le \phi \le 90° \tag{10.29}$$

$$\mathrm{Nu}_m^* = 0.0196\,\mathrm{Re}^{*1.161}\mathrm{De}^{*-0.360}\mathrm{Pr}^{0.376} \qquad \text{for} \quad 90 \le \phi \le 180° \tag{10.30}$$

The above correlations agree within ±10% deviation with the experimental data. The authors [48, 49] also recommended the following two alternate correlations for the mean Nusselt numbers over the entire length of a 180° bend, i.e., for $0 \le \phi \le 180°$.

$$\mathrm{Nu}_m^* = 0.0129\,\mathrm{Re}^{*0.807}\left(\frac{\mathrm{Re}^*}{\mathrm{De}^*}\right)^{0.161}\mathrm{Pr}^{0.379}\left[1.0 + \left(\frac{\mathrm{Re}^*}{\mathrm{De}^*}\right)^{0.25}\right]^{0.807} \tag{10.31}$$

$$\mathrm{Nu}_m^* = 0.0318\,\mathrm{Re}^{*0.971}\mathrm{De}^{*-0.194}\mathrm{Pr}^{0.389} \tag{10.32}$$

The preceding correlations are valid for $8 \times 10^3 < \mathrm{Re}^* < 8 \times 10^4$, $6.5 < \mathrm{Pr} < 8.5$, and $4.0 \le R/a \le 17.5$ [48, 49]: The correlations (10.29) through (10.32) are based on the same data. Therefore, either one of them could be used for design purposes. The local Nusselt number is calculated as [49]

$$\mathrm{Nu}_x^* = 763.834\,\mathrm{Re}_x^{*2.546}\mathrm{De}_x^{*-2.814}\mathrm{Pr}^{0.402}B_5^{1.072} \tag{10.33}$$

where B_5 is defined in Eq. (10.27).

10.4 TWO 90° MITER BENDS IN SERIES

Figure 10.1*d* shows a typical diagram of two 90° miter bends in series and associated nomenclature. Izumi et al. [53] have reported experimental Nusselt number data, and

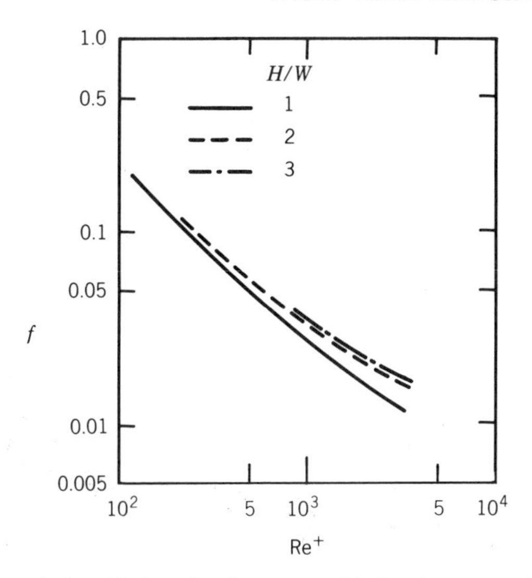

Figure 10.16. Theoretical predictions for the average friction factor as a function of Reynolds number for two 90° bends in series [54].

Amano [54–56] has reported theoretical Nusselt numbers. Results in these references are described next.

10.4.1 Fluid Flow

Amano [54–56] has reported theoretical velocity profiles for laminar and turbulent flow through a two-dimensional (small aspect ratio rectangular) bend as in Fig. 10.1*d*, having $H/W = 1$, 2, and 3. Both in laminar and in turbulent flows, flow separation occurs at point B, with subsequent reattachment on either the wall CD if $H/W = 1$ or on the wall BC if $H/W > 1$ (see Fig. 10.1*d*).

Figure 10.16 shows the average friction factor vs. Reynolds number characteristics for two bends in series. The dependence of the friction factor on the Reynolds number changes slightly from $H/W = 1$ to 2, but almost no change is observed between $H/W = 2$ and 3. This indicates a change in flow pattern from the case of $H/W = 1$ to 2, but almost no such change between $H/W = 2$ and 3.

10.4.2 Heat Transfer

Figure 10.17 shows a comparison of experimental data on air heating [53] and theoretical predictions [54–56] of laminar and turbulent flow Nusselt numbers for the boundary condition of axially constant heat flux. The figure exhibits about 20–30% discrepancy between experimental data and theoretical predictions. The average Nusselt numbers seem to be independent of the ratio H/W. Izumi et al. [53] also reported local Nusselt numbers slightly dependent upon H/W. Based on his theoretical predictions, Amano [54] proposed the following laminar flow correlation for average Nusselt

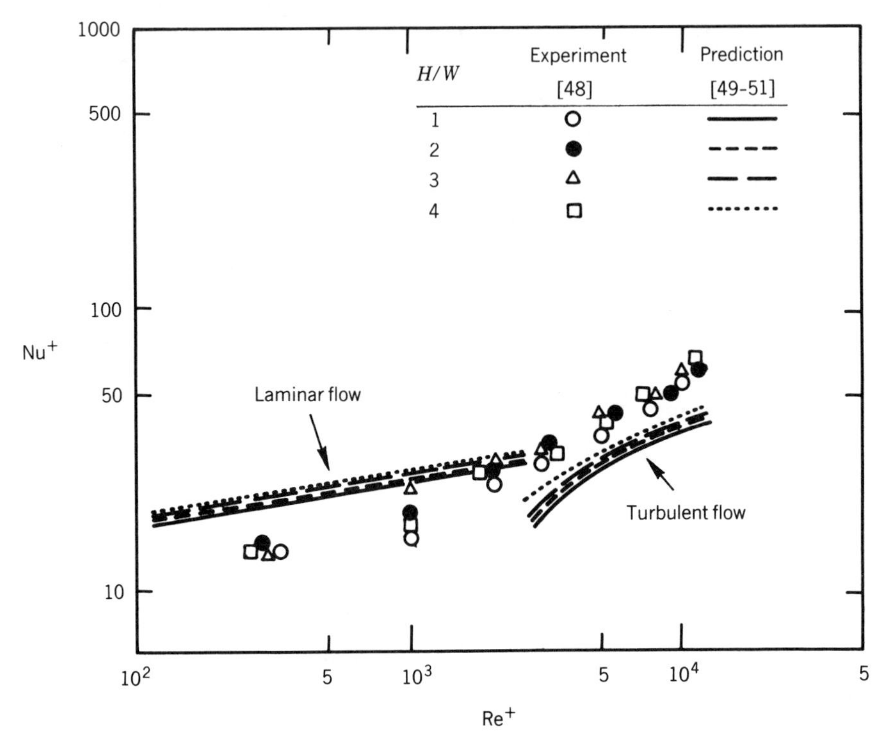

Figure 10.17. A comparison of experimental data and theoretical predictions of average Nusselt numbers for air flow (Pr = 0.7) through two 90° bends in series [56].

numbers:

$$\mathrm{Nu}^+ = \frac{(q''_w/k)2W}{\overline{T}_w - T_m} = 7.31(\mathrm{Re}^+)^{0.18} \qquad \text{for} \quad 300 < \mathrm{Re}^+ < 2000 \text{ and } \mathrm{Pr} = 0.7 \tag{10.34}$$

where $\mathrm{Re}^+ = 2W\rho u_m/\mu$. For turbulent flow, Izumi et al. [53] proposed the following experimental correlation:

$$\mathrm{Nu}^+ = 0.177(B_6\mathrm{Re}^+)^{0.5} \tag{10.35}$$

where

$$B_6 = 12.3 - \left(\frac{H}{W} - 2.8\right)^2$$

The above correlation is valid for $1 \le H/W \le 4$, $8 \times 10^3 \le B_6\mathrm{Re} \le 1.5 \times 10^3$, and $\mathrm{Pr} = 0.7$.

10.5 CONCLUDING REMARKS

An attempt has been made to present the pressure drop and heat transfer in bends and fittings of circular and rectangular cross sections. Many handbooks and a few papers (such as Refs. 10, 12) present extensively the pressure loss coefficients for bends, fittings, valves, tees, etc., and hence no attempt is made here to compile this information; primary emphasis is on heat transfer results in the bends. Most information available is for 90° bends. The experimental and theoretical results presented in the text for laminar and turbulent flow friction factors and Nusselt numbers indicate the present state of the art; the gaps in the information should provide indications for future research. Whenever correlations are not available for a specific bend, the appropriate helical-coil correlation, presented in Chapter 5, may be used for engineering estimates.

NOMENCLATURE

a — tube inside radius or half duct width for a noncircular cross-section channel (see Fig. 10.1), m, ft

A — flow cross-sectional area, m^2, ft^2

c — half channel height of a rectangular duct in the direction perpendicular to the radius of curvature (see Fig. 10.1c), m, ft

c_p — specific heat of fluid at constant pressure, J/(kg · K), Btu/(lb_m · °F)

D — pipe inside diameter, m, ft

D_h — hydraulic diameter of the duct $= 4A/p$ = inside diameter for a circular tube, m, ft

De — Dean number $= \mathrm{Re}\sqrt{0.5D_h/R}$

De* — $\mathrm{Re}^*\sqrt{2a/R}$ for rectangular channels

De_x^* — $\mathrm{Re}_x\sqrt{2a/R}$ for rectangular channels

f — Fanning friction factor $= \tau_w/(\rho u_m^2/2)$

f_c — curved-bend Fanning friction factor

f_s — straight-tube Fanning friction factor for fully developed laminar flow

Gr — Grashof number $= g\beta\rho^2 D_h^3(\overline{T}_w - T_m)/\mu^2$

Gr^+ — modified Grashof number $= g\beta\rho^2 D_h^3(T_w - T_i)/\mu^2$

h — heat transfer coefficient, W/(m^2 · K), Btu/(hr · ft^2 · °F)

h_m — average heat transfer coefficient $= (1/L)\int_0^L h_x\,dx$, W/(m^2 · K), Btu/(hr · ft^2 · °F)

h_m^* — average heat transfer coefficient $= (1/L)\int_0^L h_x^*\,dx$, W/($m^2$ · K), Btu/(hr · ft^2 · °F)

h_x — peripheral average axially local heat transfer coefficient $= (0.5/\pi)\int_0^{2\pi} h_\theta\,d\theta$, W/($m^2$ · K), Btu/(hr · ft^2 · °F)

h_x^* — peripheral average axially local heat transfer coefficient $= (0.5/\pi)\int_0^{2\pi} h_\theta^*\,d\theta$, W/($m^2$ · K), Btu/(hr · ft^2 · °F)

h_θ — local heat transfer coefficient $= q_\theta''/(T_{w,\theta} - T_m)$, W/($m^2$ · K), Btu/(hr · ft^2 · °F)

h_θ^*	modified local heat transfer coefficient $= q_\theta''/(T_{w,\theta} - T_i)$, W/(m^2 · K), Btu/(hr · ft^2 · °F)
H	bend pitch (see Fig. 10.1*d*), m, ft
H_1	wall clearance (see Fig. 10.14), m, ft
H_1^*	$H_1/(W_1 + W_2)$ (see Fig. 10.14)
k	fluid thermal conductivity, W/(m · K), Btu/(hr · ft · °F)
K	modified pressure loss coefficient $= 4f_c L/D_h$
K^*	pressure loss coefficient, defined in Eq. (10.1)
L	tube length or bend length $= R\phi$, m, ft
M	channel depth (see Fig. 10.14), m, ft
M^*	$M/(W_1 + W_2)$ (see Fig. 10.14)
Nu	Nusselt number $= hD_h/k$
Nu^+	average Nusselt number, defined in Eq. (10.34)
Nu_m	average Nusselt number $= h_m D_h/k$
Nu_m^*	average Nusselt number for rectangular channels, $= 4ah_m^*/k$
Nu_p	peripheral local Nusselt number $= h_\theta D_h/k$
Nu_x	peripherally average, axially local Nusselt number $= h_x D_h/k$
Nu_x^*	peripherally average, axially local Nusselt number for rectangular channels, $= 4ah_x^*/k$
p	duct wetted perimeter, m, ft
P	fluid static pressure, Pa, lb$_f$/ft^2
Pr	Prandtl number $= \mu c_p/k$
q	heat transfer rate, W, Btu/hr
q''	heat flux, W/m^2, Btu/hr · ft^2
r	radial distance, m, ft
R	radius of curvature of the bend centerline (see Fig. 10.1), m, ft
R_0	radius of curvature of the bend outer wall, m, ft
Re	Reynolds number $= \rho u_m D_h/\mu$
Re*	modified Reynolds number for rectangular channels, $= 4a\rho u_m/\mu$
Re^+	Reynolds number $= 2W\rho u_m/\mu$
Re_x	Reynolds number $= \rho u_m x/\mu$
T	temperature, K, °C, °F, °R
$\overline{T}$	average temperature, K, °C, °F, °R
Ⓣ	thermal boundary condition referring to axially and peripherally constant wall temperature
u, u_x	axial velocity, m/s, ft/s
u_m	mean axial velocity, m/s, ft/s
W	duct width (see Fig. 10.1*d*), m, ft
W_1, W_2	duct widths (see Fig. 10.14), m, ft
W^*	W_1/W_2
x	axial distance along the axis of a bend measured from the bend inlet, m, ft
y	distance measured from a bend wall in a direction perpendicular to the bend wall, m, ft

Y	coordinate for rectangular and square channels, $= (\|y - a\|/a)$ (see Fig. 10.1c)
Z	coordinate for rectangular and square channels, $= (\|z - a\|/a)$ (see Fig. 10.1c)

Greek Symbols

α^*	aspect ratio of a rectangular cross-section bend, $= 2c/2a$
β	coefficient of fluid thermal expansion, K^{-1}, $°R^{-1}$
β'	ratio of orifice to pipe inside diameter
Δ	prefix denoting a difference or change
θ	angular coordinate in the cylindrical coordinate system (see Fig. 10.1b), rad, deg
μ	fluid density, kg/m^3, lb_m/ft^3
τ	shear stress, Pa, lb_f/ft^2
ϕ	bend angle, rad, deg

Subscripts

c	curved coil or duct
i	inlet
m	bulk mean value
p	peripheral value
s	straight-duct value
w	wall condition
x	axial value
θ	at angle θ

REFERENCES

1. J. A. C. Humphrey, H. Iacovides, and B. E. Launder, Some Numerical Experiments on Developing Laminar Flow in Circular-Sectioned Bends, *J. Fluid Mech.*, Vol. 154, pp. 357–375, 1985.
2. D. E. Olson and B. Snyder, The Growth of Swirl in Curved Circular Pipes, *Phys. Fluids*, Vol. 26, No. 2, pp. 347–349, 1983.
3. J. Azzola and J. A. C. Humphrey, Developing Turbulent Flow in a 180° Curved Pipe and Its Downstream Tangent, Presented at Second International Symposium on Application of Laser Anemometry to Fluid Mechanics, Lisbon, Portugal, 2–4 July, 1986. See Report No. LBL-17681, Univ. of Calif., Lawrence Berkeley Lab., 1984.
4. J. Azzola, J. A. C. Humphrey, H. Iacovides, and B. E. Launder, Developing Turbulent Flow in a U-Bend of Circular Cross-Section: Measurement and Computation, *J. Fluids Eng.*, Vol. 108, pp. 214–221, 1986.
5. M. Rowe, Measurements and Computations of Flow in Pipe Bends, *J. Fluid Mech.*, Vol. 43, Pt. 4, pp. 771–783, 1970.
6. J. R. Weske, Investigations of the Flow in Curved Ducts at Large Reynolds Numbers, *J. Appl. Mech.*, Vol. 15, pp. 344–348, 1948.
7. H. Ito, Pressure Losses in Smooth Pipe Bends, *J. Basic Eng.*, Vol. 82, pp. 131–143, 1960.

8. U. S. Powle, Energy Losses in Smooth Bends, *Mech. Eng. Bull.* (India), Vol. 12, No. 4, pp. 104–109, 1981.
9. B. W. Koli and U. S. Powle, Energy Losses for Laminar Flow Through Pipe Bends, *Proc. 13th National Conf. Fluid Mech. and Fluid Power*, Tiruchirapalli, India, pp. 107–112, 1984.
10. I. E. Idelchik, *Handbook of Hydraulic Resistance*, 2nd ed., Hemisphere, New York, 1986.*
11. R. P. Benedict, *Fundamentals of Pipe Flow*, Wiley, New York, 1980.
12. D. S. Miller, *Internal Flow Systems*, Vol. 5 in Fluid Engineering Series, Br. Hydromech. Res. Assoc., 1978.
13. *Flow of Fluids Through Valves, Fittings, and Pipes*, Technical Paper No. 410, Crane Co., New York, 1982.
14. L. L. Simpson and M. L. Weirick, Designing Plant Piping, *Chemical Engineering/Deskbook Issue*, pp. 35–48, 3 Apr. 1978.
15. A. Vazsonyi, Pressure Loss in Elbows and Duct Branches, *Trans. ASME*, Vol. 66, pp. 177–183, Apr. 1944.
16. W. B. Hooper, The Two-K Method Predicts Head Losses in Pipe Fittings, *Chem. Eng.*, pp. 96–100, 24 Aug. 1981.
17. C. P. Kittredge and D. S. Rowley, Resistance Coefficients for Laminar and Turbulent Flow Through $\frac{1}{2}$ Inch Valves and Fittings, *Trans. ASME*, Vol. 79, pp. 1759–1766, 1957.
18. A. J. Ede, The Effect of a Right-Angled Bend on Heat Transfer in a Pipe, *Int. Dev. Heat Transfer*, ASME, New York, pp. 634–642, 1962.
19. S. R. Tailby and P. W. Staddon, The Influence of 90° and 180° Pipe Bends on Heat Transfer from an Internally Flowing Gas Stream, *Heat Transfer 1970*, Vol. 2, Paper No. FC 4.5, 1970.
20. H. Iacovides and B. E. Launder, The Computation of Momentum and Heat Transport in Turbulent Flow around Pipe Bends, *Inst. Chem. Engrs.* Symp. Ser. No. 86, pp. 1097–1114, 1984.
21. A. J. Ede, The Effect of a 180° Bend on Heat Transfer to Water in a Tube, *3rd Int. Heat Transfer Conf.*, Vol. 1, pp. 99–103, 110, 1966.
22. J. Lis and M. J. Thelwell, Experimental Investigation of Turbulent Heat Transfer in a Pipe Preceded by a 180° Bend, *Proc. Inst. Mech. Eng.*, Vol. 178, Pt. 31(iv), pp. 17–28, 1963–64.
23. B. M. Kavalenko, Heat Exchange with Viscous Flow of Liquid in U-Tubes, *Thermal Eng.*, Vol. 17, No. 11, pp. 113–116, 1970.
24. M. Moshfeghian and K. J. Bell, Local Heat Transfer Measurements in and Downstream From a U-Bend, ASME Paper No. 79-HT-82, 1979.
25. J. W. Baughn, H. Iacovides, D. C. Jackson, and B. E. Launder, Local Heat Transfer Measurements in Turbulent Flow around a 180° Pipe Bend, *J. Heat Transfer*, Vol. 109, pp. 43–48, 1987.
26. H. Iacovides and B. E. Launder, ASM Predictions of Turbulent Momentum and Heat Transfer in Coils and U-Bends, *Proc. 4th Int. Conf. on Numer. Methods in Laminar and Turbulent Flow*, pp. 1023–1045, Swansea, U.K., 9–12 July 1985.
27. N. D. Mehta and K. J. Bell, Laminar Flow Heat Transfer in a Tube Preceded by a 180° Bend, *Heat Transfer—Sov. Res.*, Vol. 13, No. 6, pp. 71–80, Nov.–Dec. 1981.
28. S. Kakaç, Y. Göğüş, and M. R. Özgü, Investigation of the Effect of Turns on Turbulent Forced Convection Heat Transfer in Pipes, *Heat Exchangers: Design and Theory Sourcebook*, ed. N. H. Afgan and E. U. Schlünder, McGraw-Hill, New York, pp. 637–662, 1974.
29. J. A. C. Humphrey, A. M. K. Taylor, and J. H. Whitelaw, Laminar Flow in a Square Duct of Strong Curvature, *J. Fluid Mech.*, Vol. 83, Pt. 3, pp. 509–527, 1977.

* This English edition came to authors' attention after this chapter was completed.

30. A. M. K. P. Taylor, J. H. Whitelaw, and M. Yianneskis, Curved Ducts with Strong Secondary Motion: Velocity Measurements of Developing Laminar and Turbulent Flow, *J. Fluids Eng.*, Vol. 104, pp. 350–358, 1982.

31. V. I. Grabovskii and G. V. Zhestkov, Calculation of Laminar Flow of Compressible Gas in the Presence of Heat Transfer in Two-Dimensional Curvilinear Channel, *Fluid Dyn.*, Vol. 18, No. 2, pp. 180–187, Mar.–Apr. 1983.

32. H. Yamashita, G. Kushida, and R. Izumi, Study on Three-Dimensional Flow and Heat Transfer in Miter-Bend, *Bull. JSME*, Vol. 27, No. 231, pp. 1905–1912, Sept. 1984.

33. J. A. C. Humphrey, J. H. Whitelaw, and G. Yee, Turbulent Flow in a Square Duct with Strong Curvature, *J. Fluid Mech.*, Vol. 103, pp. 443–463, 1981.

34. S. M. Chang, J. A. C. Humphrey, and A. Modavi, Turbulent Flow in a Strongly Curved U-Bend and Downstream Tangent of Square Cross-Sections, *PCH PhysicoChem. Hydrodyn.*, Vol. 4, No. 3, pp. 243–269, 1983.

35. S. M. Chang, J. A. C. Humphrey, R. W. Johnson, and B. E. Launder, Turbulent Momentum and Heat Transport in Flow through a 180° Bend of Square Cross-Section, *Proc. 4th Symp. Turbulent Shear Flow*, Univ. of Karlsruhe, pp. 6.20–6.25, Sept. 1983.

36. R. W. Johnson and B. E. Launder, Local Nusselt Number and Temperature Field in Turbulent Flow through a Heated Square Sectioned U-bend, *Int. J. Heat and Fluid Flow*, Vol. 6, No. 3, pp. 171–180, Sept. 1985.

37. R. W. Johnson and B. E. Launder, Local Heat Transfer Behavior in Turbulent Flow Around a 180 deg Bend of Square Cross Section, ASME Paper No. 85-GT-68, 1985.

38. J. A. C. Humphrey, S. M. Chang, and A. Modavi, Developing Turbulent Flow in 180° Bend and Downstream Tangent of Square Cross-Sections, Report No. LBL-14844, Lawrence Berkeley Lab., Berkeley, Calif., 1982.

39. N. Shiragami and I. Inoue, Pressure Losses in Square Section Bends, *J. Chem. Eng. Japan*, Vol. 14, No. 3, pp. 173–177, 1981.

40. G. Yee, R. Chilukuri, and J. A. C. Hymphrey, Developing Flow and Heat Transfer in Strongly Curved Ducts of Rectangular Cross Section, *J. Heat Transfer*, Vol. 102, pp. 285–291, 1980.

41. D. E. Metzger and D. E. Larson, Use of Melting Point Surface Coating for Local Convection Heat Transfer Measurements in Rectangular Channel Flow with 90-deg Turns, *J. Heat Transfer*, Vol. 108, pp. 48–54, 1986.

42. R. Chilukuri and J. A. C. Humphrey, Numerical Computation of Buoyancy-Induced Recirculation in Curved Square Duct Laminar Flow, *Int. J. Heat Mass Transfer*, Vol. 24, pp. 305–314, 1981.

43. J. W. Yang and N. Liao, Turbulent Heat Transfer in Rectangular Ducts with 180° Bend, *Heat Transfer 1974*, Vol. 2, pp. 169–172, 1974.

44. D. E. Metzger and M. K. Sahm, Heat Transfer around Sharp 180 Degree Turns in Smooth Rectangular Channels, ASME Paper No. 85-GT-122, 1985.

45. M. K. Sahm, Heat Transfer in Smooth 180° Sharp Turns, M.S. Thesis, Mech. Eng. Dept., Arizona State Univ., 1983.

46. V. K. Shchukin and V. A. Filin, Convective Heat Transfer in Short Curved Channels, *J. Eng. Phys.*, Vol. 12, No. 2, pp. 78–82, 1967.

47. N. Seki, S. Fukusako, and M. Yoneta, Heat Transfer from the Heated Concave Wall of a Return Bend with Rectangular Cross Section, *Wärme- und Stoffübertragung*, Vol. 17, No. 1, pp. 17–26, 1982.

48. N. Seki, S. Fukusako, and M. Yoneta, Turbulent Heat-Transfer Characteristics along the Heated Convex Wall of a Rectangular Cross-Sectional Return Bend, *Wärme- und Stoffübertragung*, Vol. 17, No. 2, pp. 85–92, 1983.

49. N. Seki, S. Fukusako, and M. Yoneta, Heat Transfer from the Heated Convex Wall of a Return Bend with Rectangular Cross Section, *J. Heat Transfer*, Vol. 105, pp. 64–69, 1983.

50. T. Wang and T. W. Simon, Heat Transfer and Fluid Mechanics Measurements in Transitional Boundary Layers on Convex-Curved Surfaces, ASME Paper No. 85-HT-60, 1985.
51. R. E. Mayle, M. F. Blair, and F. C. Kopper, Turbulent Boundary Layer Heat Transfer on Curved Surfaces, *J. Heat Transfer*, Vol. 101, pp. 521–525, 1979.
52. T. W. Simon and R. J. Moffat, Convex Curvature Effects on the Heated Turbulent Boundary Layer, *Heat Transfer 1982*, Vol. 3, pp. 295–300, 1982.
53. R. Izumi, K. Oyakawa, S. Kaga, and H. Yamashita, Fluid Flow and Heat Transfer in Corrugated Wall Channels, *JSME J.*, Vol. 47, No. 416, pp. 657–665, 1981.
54. R. A. Amano, Laminar Heat Transfer in a Channel with Two Right-Angled Bends, *J. Heat Transfer*, Vol. 106, pp. 591–596, 1984.
55. R. S. Amano, A Numerical Study of Turbulent Heat Transfer in a Channel with Bends, ASME Paper No. 85-HT-20, 1985.
56. R. S. Amano, Turbulent Heat Transfer in a Channel with Two Right-Angle Bends, *Int. J. Heat Mass Transfer*, Vol. 28, pp. 2177–2179, 1985.
57. S. M. Marcos and A. E. Bergles, Experimental Investigation of Combined Forced and Free Laminar Convection in Horizontal Tubes, *J. Heat Transfer*, Vol. 97, pp. 212–219, 1975.

11

TRANSIENT FORCED CONVECTION IN DUCTS

Y. Yener

Northeastern University
Boston, Massachusetts

S. Kakaç

University of Miami
Coral Gables, Florida

11.1 INTRODUCTION

Steady and unsteady duct flows with transient forced convection have recently received greater attention in connection with the increasingly greater use of automatic control devices for the accurate control of fluid flow in high-performance heat transfer systems. The accurate regulation of fluid flow is especially important, for example, when the positive control of industrial heat exchangers must be assured, which requires better understanding and more precise evaluation of thermal transients.

Thermal transients can be due to startup, shutdown, power surge, pump failure, etc., during an operation, or due to such operating conditions as time-varying inlet temperatures and/or flow rates. Thermal transients in ducts may also arise because of time-dependent wall heat flux, wall temperature, or internal heat generation, as in the flow channels of nuclear reactors. Accurate prediction of the transient response of thermal systems is highly important, not only to provide for an effective control system, but also for the understanding of adverse effects such as reduced thermal performance and severe thermal stresses that they can produce, with eventual mechanical failure.

Transient response of thermal systems can be classified as (1) *step response*, (2) *frequency response*, and (3) *impulse response*. The frequency response is the long-time behavior of a thermal system subjected to a periodically varying operating condition. The response then also varies periodically with time. The step response characterizes the behavior following a sudden change in the operating conditions. The step response asymptotically approaches the steady-state behavior of the new operating conditions. An impulse response is the behavior when the disturbance has a large amplitude over an infinitesimal time duration.

The literature on transient forced convection in ducts is small but growing. Some of the important contributions are listed in the Refs. 1, 2, 4, 6–46.

This chapter is mainly concerned with the study of transient thermal response of duct flows. The parallel-plate channel and the circular tube, which are the two commonly encountered geometries in practice, are considered with both laminar and turbulent flows.

11.2 TRANSIENT LAMINAR FORCED CONVECTION IN DUCTS

In this section, first a short review of the literature on transient laminar forced convection in parallel-plate channels and circular tubes is given, and then some fundamental problems are discussed.

One of the earliest works on transient laminar forced convection was the calculation of transient temperatures in pipes and heat exchangers by Dusinberre [1], who presented explicit iteration formulas and numerical computation guides. The case of a compressible fluid flowing through an insulated tube with an exponential or step-function fluid inlet temperature was analyzed by Rizika [2]. The transient conditions for a compressible fluid flowing through a heat exchanger were also partially analyzed in [2]. Rizika extended this work in [4] to obtain the transient solutions due to thermal lags in flowing incompressible fluid systems. Both the case of an incompressible fluid with a step-function temperature input flowing through a circular tube, and the case of an incompressible fluid flowing in a simple heat exchanger, were examined in detail. An example which demonstrates the transient condition at the exit of a simple (condensing-steam–water) heat exchanger was also presented.

Sparrow and Siegel [6] made an analysis of transient laminar forced convection in the thermal entrance region of circular tubes. They first determined the thermal responses to step changes both in wall temperature and in wall heat flux, using an integral formulation of the energy equation in connection with the method of characteristics. Then, using the linearity of the energy equation, they generalized the step-function results for arbitrary time-dependent boundary conditions, and expressed their results in the form of integrals which can easily be evaluated for particular applications. Siegel and Sparrow in [10] also made a similar analysis in the thermal entrance region of flat ducts.

Siegel in [9] investigated laminar slug flow in a circular tube and a parallel-plate channel where the walls were given a step change in heat flux or, alternately, a step change in temperature. Siegel [13] also investigated laminar forced convection both in a circular tube and in a parallel-plate channel with arbitrary time variations in wall temperature. The velocity distributions in both cases were assumed to be steady and fully developed. First the analyses were done for a step change in wall temperature, and then the results were generalized for arbitrary time variations. The method used was not an exact solution, but involved an approximation. The validity of the approximation was tested by comparing the results with the exact ones available for part of the solution, and good agreement was obtained. It was also demonstrated that the slug flow assumption does, in fact, lead to the essential physical behavior of the systems considered, although the numerical results were somewhat in error.

Perlmutter and Siegel [16] studied transient heat transfer with unsteady laminar flow between two parallel plates. The transients were caused by simultaneously changing the fluid pumping pressure and either the wall temperature or the wall heat flux. During the solution, the time-dependent slug flow simplification was made. Within this limitation, exact solutions for the fluid temperature distribution were obtained for a step change in wall temperature or wall heat flux with a simultaneous step change in the pumping pressure. Then, using superposition, solutions for more involved situations were developed. Perlmutter and Siegel in [17] analyzed transient heat transfer in a two-dimensional unsteady incompressible laminar duct flow between two parallel plates with a step change in wall temperature. Some results were also presented for the case where the transient heat conduction through the bounding walls was taken into consideration. Siegel and Perlmutter [19] also made an analysis of unsteady incompressible laminar forced convection between two parallel plates with the wall heat flux varying with both time and axial position. The flow velocity was assumed constant over the channel cross section (slug flow assumption), but allowed to vary with time. General analytical expressions were derived that could be used for computing the transient heat transfer in a channel with the wall heat flux varying with time and axial position.

Siegel [18] analyzed laminar forced convection between two parallel plates for slug flow conditions by including the heat capacity of the walls and with the wall heating assumed variable with axial position and time. The walls were assumed to be sufficiently thin or highly conducting so that the temperature variations across the wall thicknesses could be neglected. The method used involved coupling the heat transfer behavior within the fluid to that in the wall and solving the resultant equations together with the energy equation. After the introduction of a general method of solution, some illustrative examples were considered. These included uniform wall heating varying sinusoidally in time, and heating varying sinusoidally with axial distance and exponentially in time.

References 7, 8, 11, 15 are a series of papers on the dynamic response of heat exchangers having time-varying internal heat sources. In these papers, theoretical results are also compared with experiments. In addition, Kardas [20] studied the heat transfer from parallel flat plates to fluids flowing between them with an inlet temperature varying with time. He presented an analytical solution of the unidirectional regenerator problem.

Namatame [24] presented a modified quasi-steady solution for the transient temperature response of an annular slug flow in which the dependence of the surface temperatures upon axial position was taken into consideration.

Campo and Yoshimura [32] made a theoretical study to describe the influence of randomly varying ambient temperatures on the heat transfer performance of a fully developed flow through a parallel-plate channel. In the analysis, the energy equation was simplified by the use of a lumped formulation in the transverse direction of the channel, and the time-dependent fluid temperature in the axial direction was obtained.

Lin and Shih [38] studied the unsteady thermal-entrance heat transfer for fully developed laminar flow of power-law non-Newtonian fluids in pipes and plate slits with step change in surface temperature, by a method called the instant-local similarity method which uses the concept of the extended Lévêque method by restricting the solution to large Graetz numbers and converting the energy equation to a boundary-layer-type equation. The effects of the flow index, viscous dissipation, and Graetz number on the heat transfer rate were demonstrated with numerical solutions.

Sucec [39] presented an improved quasi-steady approach, which took into approximate account both the effect of thermal history and of the thermal-energy storage capacity of the fluid in problems of transient conjugated laminar forced convection. The method was applied to two problems in a parallel-plate channel in which the finite-thermal-capacity walls and fluid were both at constant temperature initially, when the transient was initiated by either a step change in fluid inlet temperature or a sinusoidal variation with time. Exact solutions were given for slug flow and for a linear velocity profile.

By the use of finite difference numerical schemes, Lin et al. [42, 43] solved the transient two-dimensional energy equation for various flow conditions. The first paper studied the thermal-entrance heat transfer in laminar pipe flows subjected to a step change in ambient temperature. The transient thermal-entrance laminar flow heat transfer resulting from a step change in both the pressure gradient and the inlet temperature was studied in the second paper. Lin et al. [44] also studied heat transfer in the thermal entrance region of laminar pipe flows resulting from a step change in the inlet temperature, when coupled with the unsteady temperature variations in the surrounding enclosure, representing a refrigerator cabinet. The Nusselt number, fluid bulk mean temperature, and pipe wall temperature were presented over wide ranges of the parameters involved.

Recently, Chen et al. [45] gave a direct numerical solution for transient laminar heat transfer inside a circular duct subjected to a step change in either wall temperature or heat flux.

An extensive review of the work on transient laminar forced convection in ducts has been given recently by Kakaç and Yener [41].

11.2.1 Transient Laminar Forced Convection in Circular Tubes with Step Change in Wall Temperature

Consider a circular tube as shown in Fig. 11.1. A steady and fully developed laminar flow passes through the tube in the x direction. The tube wall and fluid are initially isothermal at temperature T_i. Let the temperature of the tube wall for $x \geq 0$ be instantaneously changed (say at time $t = 0$) to a new value T_w and be maintained at this value for all times thereafter.

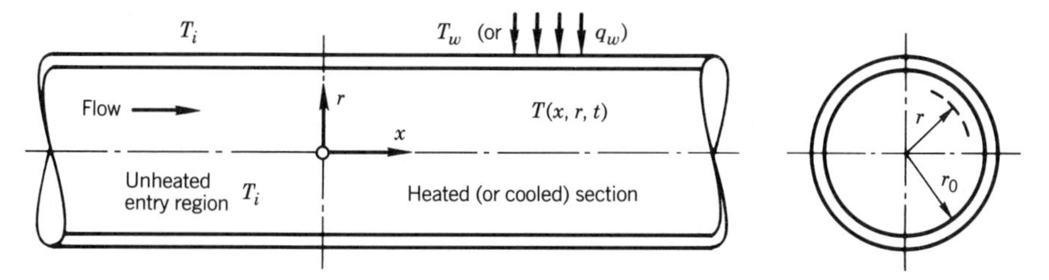

Figure 11.1. Coordinate system for circular tube geometry.

The transient temperature distribution $T(x, r, t)$ in the tube for $x \geq 0$ and times $t > 0$ will satisfy the unsteady energy equation for fully developed laminar flow in a circular tube; that is,

$$\frac{\partial T}{\partial t} + u\frac{\partial T}{\partial x} = \alpha\frac{1}{r}\frac{\partial}{\partial r}\left(r\frac{\partial T}{\partial r}\right) \tag{11.1}$$

where the fluid properties have been assumed to be constant, and viscous dissipation and axial heat conduction have been neglected. The initial, inlet, and boundary conditions are given by

$$T(x, r, 0) = T_i, \qquad T(0, r, t) = T_i \tag{11.2a, b}$$

$$\left(\frac{\partial T}{\partial r}\right)_{r=0} = 0, \qquad T(x, r_0, t) = T_w \tag{11.2c, d}$$

In the following subsections, two solutions of the foregoing problem are presented for slug flow and parabolic velocity distribution in the tube.

Solution for Slug Flow. If the velocity distribution u in Eq. (11.1) is assumed to be uniform over the tube cross section and equal to the mean flow velocity u_m, then the formulation of the problem can be rewritten in the following dimensionless form:

$$\frac{\partial \theta}{\partial \tau} + \frac{1}{2}\frac{\partial \theta}{\partial \xi} = \frac{1}{\eta}\frac{\partial}{\partial \eta}\left(\eta\frac{\partial \theta}{\partial \eta}\right) \tag{11.3}$$

$$\theta(\xi, \eta, 0) = 1, \qquad \theta(0, \eta, \tau) = 1 \tag{11.4a, b}$$

$$\left(\frac{\partial \theta}{\partial \eta}\right)_{\eta=0} = 0, \qquad \theta(0, 1, \tau) = 0 \tag{11.4c, d}$$

where

$$\theta = \frac{T - T_w}{T_i - T_w}, \qquad \eta = \frac{r}{r_0}, \qquad \xi = \frac{2x/D}{\mathrm{Re}\,\mathrm{Pr}}, \qquad \tau = \frac{\alpha t}{r_0^2} \tag{11.4e, f, g, h}$$

with

$$\mathrm{Re} = \frac{u_m D}{\nu}, \qquad \mathrm{Pr} = \frac{\nu}{\alpha}, \qquad D = 2r_0 \tag{11.4i, j, k}$$

The method of solution as described in [9] is as follows: The fluid which was at $x = 0$ when the transient began does not reach a position x until a time x/u_m has elapsed. Thus, in the region where $x/u_m \geq t$, the heat transfer process will not be influenced by the inlet condition. Consequently, in this region the fluid at any cross section undergoes the same transient heating process as that at any other cross section, with the effect of heat convection being zero. Therefore, when $t \leq x/u_m$ (i.e., $\tau \leq 2\xi$), the solution at x is governed by

$$\frac{\partial \theta}{\partial \tau} = \frac{1}{\eta} \frac{\partial}{\partial \eta}\left(\eta \frac{\partial \theta}{\partial \eta}\right) \tag{11.5}$$

with

$$\theta(\eta, \tau = 0) = 1 \tag{11.6a}$$

$$\left(\frac{\partial \theta}{\partial \eta}\right)_{\eta=0} = 0, \qquad \theta(\eta = 1, \tau) = 0 \tag{11.6b, c}$$

The solution of this problem can readily be found to be [47]

$$\theta = 2 \sum_{n=1}^{\infty} e^{-\lambda_n^2 \tau} \frac{J_0(\lambda_n \eta)}{\lambda_n J_1(\lambda_n)}, \qquad \tau \leq 2\xi \tag{11.7}$$

where λ_n are the positive zeros of $J_0(\lambda) = 0$, and J_0 and J_1 are Bessel functions of the first kind and of zero and first orders, respectively.

Now consider the steady-state solution, which is governed by

$$\frac{1}{2} \frac{\partial \theta}{\partial \xi} = \frac{1}{\eta} \frac{\partial}{\partial \eta}\left(\eta \frac{\partial \theta}{\partial \eta}\right) \tag{11.8}$$

with

$$\theta(\xi = 0, \eta) = 1 \tag{11.9a}$$

$$\left(\frac{\partial \theta}{\partial \eta}\right)_{\eta=0} = 0, \qquad \theta(\xi, \eta = 1) = 0 \tag{11.9b, c}$$

Since the differential equations (11.5) and (11.8) and the conditions (11.6a, b, c) and (11.9a, b, c) are the same in τ and 2ξ, the steady-state solution will have the same form as Eq. (11.7); that is,

$$\theta = 2 \sum_{n=1}^{\infty} e^{-2\lambda_n^2 \xi} \frac{J_0(\lambda_n \eta)}{\lambda_n J_1(\lambda_n)} \tag{11.10}$$

This result matches the initial transient solution, Eq. (11.7), when $\xi = \frac{1}{2}\tau$, and therefore it must be the solution for $\tau \geq 2\xi$. Thus, Eqs. (11.7) and (11.10) together represent the complete solution of the problem for all times $t > 0$. Furthermore, these results show that the steady-state temperature distribution at a particular cross section is reached over the time period that it takes for the fluid to flow from the entrance to that particular section, i.e., x/u_m.

Solution for Parabolic Velocity Distribution. For fully developed steady laminar flow in a tube of radius r_0, the velocity distribution is given by

$$\frac{u}{u_m} = 2\left[1 - \left(\frac{r}{r_0}\right)^2\right] \tag{11.11}$$

where u_m is the mean flow velocity. With this parabolic profile, the energy equation (11.1) can be rewritten in dimensionless form as

$$\frac{\partial \theta}{\partial \tau} + (1 - \eta^2)\frac{\partial \theta}{\partial \xi} = \frac{1}{\eta}\frac{\partial}{\partial \eta}\left(\eta \frac{\partial \theta}{\partial \eta}\right) \tag{11.12}$$

where the definitions of various dimensionless quantities are the same as in the previous case. Furthermore, the same dimensionless initial, inlet, and the boundary conditions, i.e., Eq. (11.4), also apply to this case.

To develop the general solution, first consider the steady-state solution.

Steady-State Solution. Under steady-state conditions, the energy equation (11.12) reduces to

$$(1 - \eta^2)\frac{\partial \theta_s}{\partial \xi} = \frac{1}{\eta}\frac{\partial}{\partial \eta}\left(\eta \frac{\partial \theta_s}{\partial \eta}\right) \tag{11.13}$$

with the following inlet and boundary conditions:

$$\theta_s(0, \eta) = 1 \tag{11.14a}$$

$$\left(\frac{\partial \theta_s}{\partial \eta}\right)_{\eta=0} = 0, \qquad \theta_s(\xi, 1) = 0 \tag{11.14b, c}$$

The problem consisting of the differential equation (11.13) and the conditions (11.14) is the classical Graetz problem, and its solution is given by [35]

$$\theta_s(\xi, \eta) = \sum_{n=0}^{\infty} C_n e^{-\lambda_n^2 \xi} R_n(\eta) \tag{11.15}$$

where λ_n are the eigenvalues and $R_n(\eta)$ are the corresponding eigenfunctions of the following eigenvalue problem:

$$\frac{d^2 R_n}{d\eta^2} + \frac{1}{\eta}\frac{dR_n}{d\eta} + \lambda_n^2(1 - \eta^2)R_n = 0 \tag{11.16a}$$

$$\frac{dR_n(0)}{d\eta} = 0, \qquad R_n(1) = 0 \tag{11.16b, c}$$

and the constants C_n are given by

$$C_n = -\frac{2}{\lambda_n(\partial R_n/\partial \lambda_n)_{\eta=1}} \tag{11.17}$$

Transient Solution. Let the turbulent solution be given by a series expansion about the steady-state solution as

$$\theta(\xi, \eta, \tau) = \sum_{n=0}^{\infty} C_n F_n(\xi, \tau) R_n(\eta) \tag{11.18}$$

For large times the functions F_n should, by comparison with Eq. (11.15), converge to

$$F_n = e^{-\lambda_n^2 \xi} \tag{11.19}$$

As suggested by Siegel [13], the following approximation is now introduced: Multiply Eq. (11.12) by η and then integrate from 0 to 1 to obtain

$$\int_0^1 \eta \frac{\partial \theta}{\partial \tau}\, d\eta + \int_0^1 \eta(1-\eta^2) \frac{\partial \theta}{\partial \xi}\, d\eta = \left(\frac{\partial \theta}{\partial \eta} \right)_{\eta=1} \tag{11.20}$$

Substitution of Eq. (11.18) into Eq. (11.20) yields

$$\frac{\partial F_n}{\partial \tau} \int_0^1 \eta R_n\, d\eta - \frac{1}{\lambda_n^2} \frac{\partial F_n}{\partial \xi} \frac{dR_n(1)}{d\eta} - F_n \frac{dR_n(1)}{d\eta} = 0, \qquad n = 0, 1, 2, \ldots \tag{11.21}$$

where the following relation has also been used [35]:

$$\int_0^1 \eta(1-\eta^2) R_n(\eta)\, d\eta = -\frac{1}{\lambda_n^2} \frac{dR_n(1)}{d\eta} \tag{11.22}$$

The differential equation (11.21) can be solved by the *method of characteristics*, and the result is [13]

$$\theta(\xi, \eta, \tau) = \frac{T - T_w}{T_i - T_w} = \sum_{n=0}^{\infty} C_n R_n(\eta) \begin{Bmatrix} e^{-\beta_n \tau}, & \tau \le a_n \xi \\ e^{-\lambda_n^2 \xi}, & \tau \ge a_n \xi \end{Bmatrix} \tag{11.23}$$

where $a_n = \lambda_n^2 / \beta_n$ and

$$\beta_n = -\frac{dR_n(1)/d\eta}{\int_0^1 \eta R_n\, d\eta} \tag{11.24}$$

The solution (11.23) satisfies all the required conditions of the problem, converges exactly to the steady-state solution for large times, and is approximate to the extent that it satisfies an integrated form of the energy equation, i.e., Eq. (11.20).

The wall heat flux can be evaluated from Fourier's law,

$$q''_w = k \left(\frac{\partial T}{\partial r} \right)_{r=r_0} \tag{11.25}$$

which yields

$$\frac{q''_w r_0}{k(T_w - T_i)} = -\sum_{n=0}^{\infty} C_n \frac{dR_n(1)}{d\eta} \begin{Bmatrix} e^{-\beta_n \tau}, & \tau \le a_n \xi \\ e^{-\lambda_n^2 \xi}, & \tau \ge a_n \xi \end{Bmatrix} \tag{11.26}$$

where q''_w is defined as the rate of heat transfer per unit surface area to the fluid at the wall. Figure 11.2 gives, together with the results of Chen et al. [45] by finite differences,

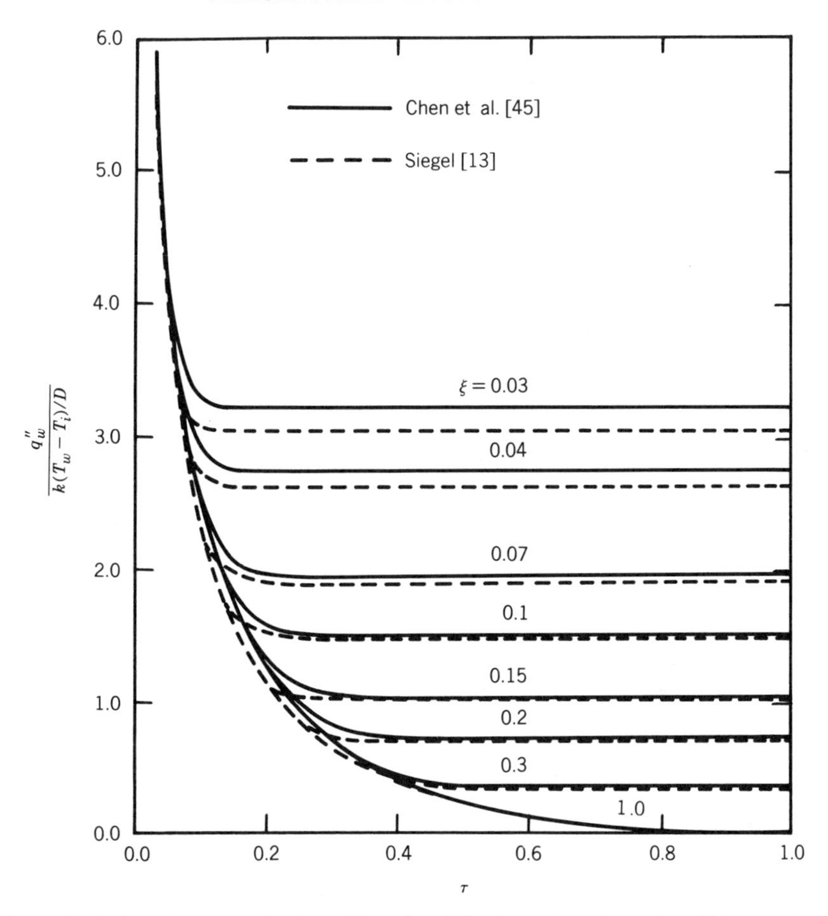

Figure 11.2. Transient variation in wall heat flux following a step change in wall temperature for fully developed steady laminar flow in a circular tube.

the dimensionless wall heat flux from Eq. (11.26) vs. τ for various values of ξ as computed by Siegel [13].

A quantity of practical importance is the time required to reach the steady-state conditions at any location x. The steady-state time, $\tau_s = \alpha\, t_s/r_0^2$, is given in Fig. 11.3 as a function of ξ, where τ is defined as the time required for the local heat flux to approach within 5% of the value reached for infinite time. Two lines are also drawn in Fig. 11.3. The line $\tau_s = \xi$ represents a lower bound on the steady-state time and is determined by the fact that the heat transfer process cannot be stabilized at a location x until a time of at least $x/u_{\max}$ has elapsed. The upper line is obtained from the slug flow solution, which gives a steady-state time of $t_s = x/u_m$ or $\tau_s = 2\xi$. As Fig. 11.3 shows, the slug flow solution underestimates the steady-state times for small values of ξ, and overestimates them for large ξ. Physically, for small values of ξ the establishment of a steady state depends on the convection process in the thin thermal boundary layers near the wall, where the velocities are smaller than indicated by the slug flow approximation, and accordingly the slug flow solution yields lower steady-state times. On the other hand, for large values of ξ, heat has already penetrated all the way across the tube and the fluid temperature near the wall is already close to the temperature of

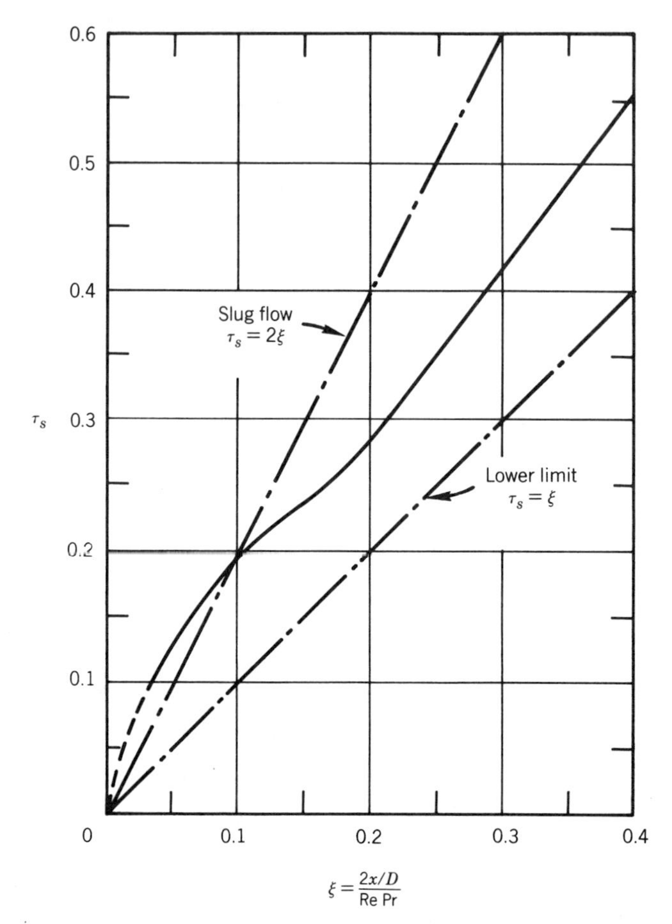

Figure 11.3. Time to reach steady state after a step change in wall temperature for fully developed steady laminar flow in a circular tube [13].

the tube wall. In this region, the establishment of a steady state is evidently more dependent on the velocities in the central portion of the tube cross section, which are, in fact, higher than the slug flow velocity. The steady-state times are therefore overestimated by the slug flow solution.

The transient heat transfer problem considered here has been solved in [6] using an integral method. However, in this reference only the thermal entrance region is considered and the results do not extend far down the tube. On the other hand, although the present series solutions, Eqs. (11.23) and (11.26), are valid for the entire length of the tube, many terms are required in the calculations for regions very close to the tube entrance. Thus, the results presented here and those of [6] can be used simultaneously to obtain information for all positions along the tube length.

11.2.2 Transient Laminar Forced Convection in Circular Tubes with Arbitrary Time Variations in Wall Temperature

In the previous section, results describing the transient behavior following a step change in wall temperature have been presented. Since the energy equation (11.1) is

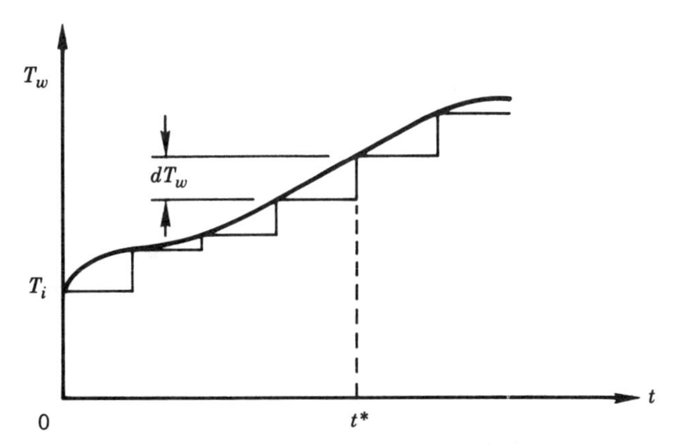

Figure 11.4. Representation of an arbitrary time-dependent wall temperature.

linear, by using a superposition technique these results can be generalized to apply for arbitrary time variations in wall temperature.

Let $T_w(t)$ represent the wall temperature as some arbitrary function of time. At any instant, T_w is spatially uniform over the walls. As illustrated in Fig. 11.4, this arbitrary wall temperature variation can be visualized as a series of differential steps. The effects of those steps can then be superposed to determine the response for an arbitrary variation in T_w.

First consider a system isothermal at T_i, and let the tube wall be given a step change dT_w in temperature at t^*. Then, from the result (11.26), the wall heat-flux response to this differential step will be given by

$$dq_w'' = -\frac{k}{r_0}\sum_{n=0}^{\infty} C_n \frac{dR_n(1)}{d\eta}\begin{Bmatrix} e^{-\beta_n(\tau-\tau^*)}, & 0 < (\tau-\tau^*) \le a_n\xi \\ e^{-\lambda_n^2\xi}, & \tau-\tau^* \ge a_n\xi \end{Bmatrix} dT_w \qquad (11.27)$$

where $\tau^* = t^*/r_0^2$ and all other parameters are as defined in the previous section. As explained in [13], when this result is integrated over an arbitrary wall temperature variation, the variation in wall heat flux q_w'' is obtained as

$$\frac{q_w''(\xi,\tau)r_0}{k} = -\sum_{n=0}^{N-1} C_n \frac{dR_n(1)}{d\eta}\left[e^{-\lambda_n^2\xi}\{T_w(\tau-a_n\xi)-T_i\} - \beta_n\int_{\tau-a_n\xi}^{\tau} e^{-\beta_n(\tau-\tau^*)}\{T_w(\tau^*)-T_i\}\,d\tau^*\right]$$
$$+\sum_{n=N}^{\infty}\beta_n C_n\frac{dR_n(1)}{d\eta}\int_0^{\tau} e^{-\beta_n(\tau-\tau^*)}\{T_w(\tau^*)-T_i\}\,d\tau^* \qquad (11.28)$$

where, for a given τ, the value of N is found from the relation

$$a_{N-1}\xi < \tau \le a_N\xi \qquad (11.29)$$

and for $N = 0$ the first summation is defined as

$$\sum_{n=0}^{-1} \equiv 0 \qquad (11.30)$$

and also $a_{-1} \equiv 0$. In evaluating the heat flux at a particular location $\xi = \xi_i$, for early times τ will be less than all of the $a_n \xi_i$, and therefore only the second summation from $n = 0$ to $n = \infty$ is needed. For later times, more and more terms are needed from the first summation. For very large times, however, only the first summation needs to be considered.

If the transient starts from an already established initial steady-state heat transfer situation in which the wall is at a uniform temperature T_w different from the inlet fluid temperature T_i, then the heat transfer behavior can be determined from the above results by first letting the system go through an initial transient process with a step change in wall temperature from T_i to T_w, and keeping the wall temperature at T_w until the steady state is reached. Then the specified transient is initiated, and the results for this part of the computation yield the desired response from the initial steady-state heat transfer condition.

11.2.3 Transient Laminar Forced Convection in Circular Tubes with Step Change in Wall Heat Flux

Attention is again directed to a hydrodynamically fully developed steady laminar flow through a circular tube of radius r_0 (see Fig. 11.1) where the tube wall and the fluid are initially isothermal at T_i, but the tube wall is subjected to a constant heat flux q''_w for $x \geq 0$ and for times $t > 0$. In this case the energy equation in dimensionless form is also given by Eq. (11.12); however, the dimensionless temperature $\theta(\xi, \eta, \tau)$ is defined as

$$\theta = \frac{T - T_i}{q''_w r_0 / k} \tag{11.31}$$

whereas the definitions of the other dimensionless quantities are as before. The inlet and the boundary conditions are then given accordingly by

$$\theta(\xi, \eta, 0) = 0, \qquad \theta(0, \eta, \tau) = 0 \tag{11.32a, b}$$

$$\left(\frac{\partial \theta}{\partial \eta}\right)_{\eta=0} = 0, \qquad \left(\frac{\partial \theta}{\partial \eta}\right)_{\eta=1} = 1 \tag{11.32c, d}$$

Chen et al. [45] solved this problem numerically by finite differences, and Fig. 11.5 shows their results for the transient wall temperature distribution vs. the dimensionless time τ for various values of the dimensionless axial distance ξ.

11.2.4 Transient Laminar Forced Convection in a Parallel-Plate Channel with Step Change in Wall Temperature

Consideration is now given to a parallel-plate channel as shown in Fig. 11.6. A steady and fully-developed laminar flow passes through the channel in the x direction. The channel walls and the fluid are initially isothermal at temperature T_i. The temperature of the channel walls is suddenly changed at $t = 0$ to a new value T_w and maintained at this value for all times thereafter.

The transient temperature distribution $T(x, r, t)$ in the channel for $x \geq 0$ and $t > 0$ will satisfy

$$\frac{\partial T}{\partial t} + u \frac{\partial T}{\partial x} = \alpha \frac{\partial^2 T}{\partial y^2} \tag{11.33}$$

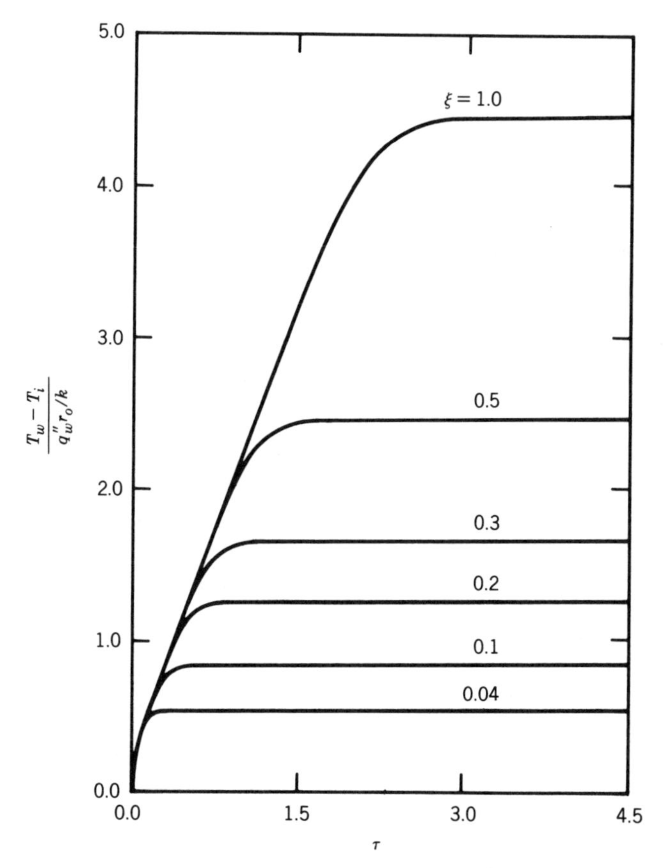

Figure 11.5. Transient variation in the wall temperature following a step change in the wall heat flux for fully developed steady laminar flow through a circular tube of radius r_0 [45].

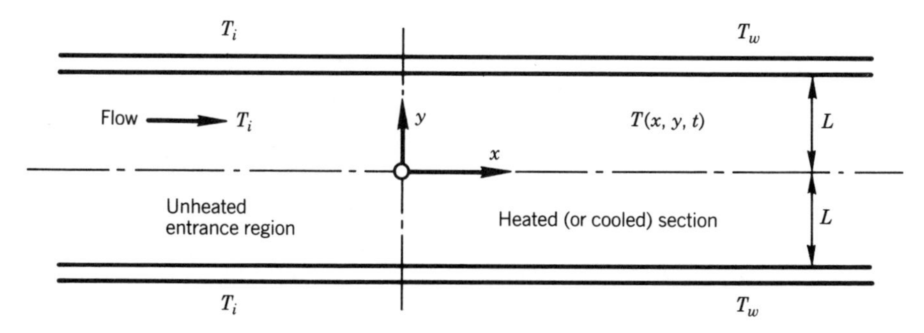

Figure 11.6. Coordinate system for parallel-plate channel geometry.

with the following initial, inlet, and boundary conditions:

$$T(x, y, 0) = T_i, \qquad T(0, y, t) = T_i \tag{11.34a, b}$$

$$\left(\frac{\partial T}{\partial y}\right)_{y=0} = 0, \qquad T(x, L, t) = T_w \tag{11.34c, d}$$

where the fluid properties have been considered constant, and viscous dissipation and axial heat conduction have been neglected. In addition, the velocity distribution is given by

$$\frac{u}{u_m} = \frac{3}{2}\left[1 - \left(\frac{y}{L}\right)^2\right] \tag{11.35}$$

where u_m is the mean flow velocity in the channel.

This problem was first solved by Siegel and Sparrow [10], who obtained an approximate solution in the thermal entrance region using an integral formulation of the energy equation (11.33), together with the method of characteristics. Later, Siegel

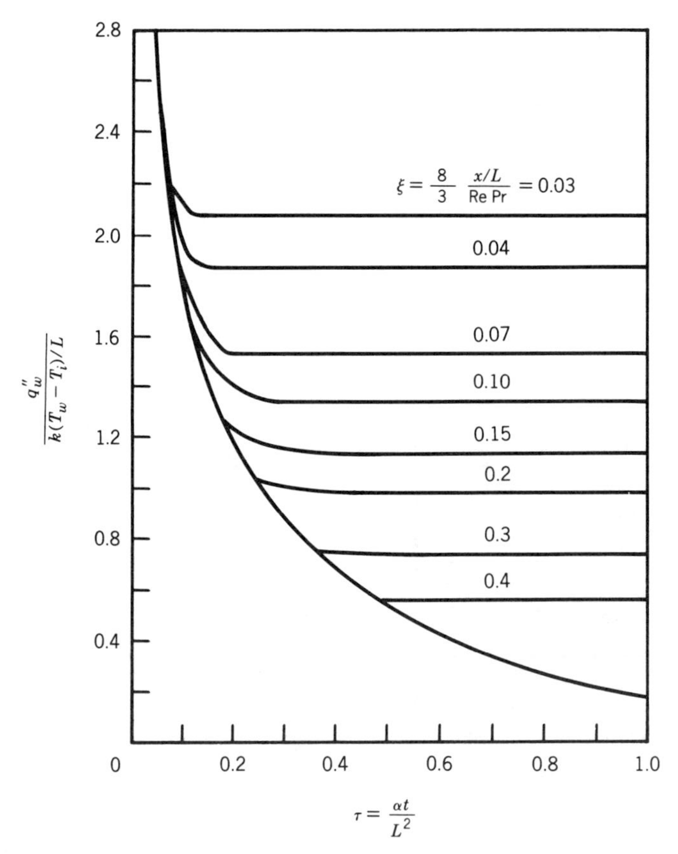

Figure 11.7. Transient variation in wall heat flux following a step change in wall temperature for fully developed steady laminar flow in a parallel-plate channel [13].

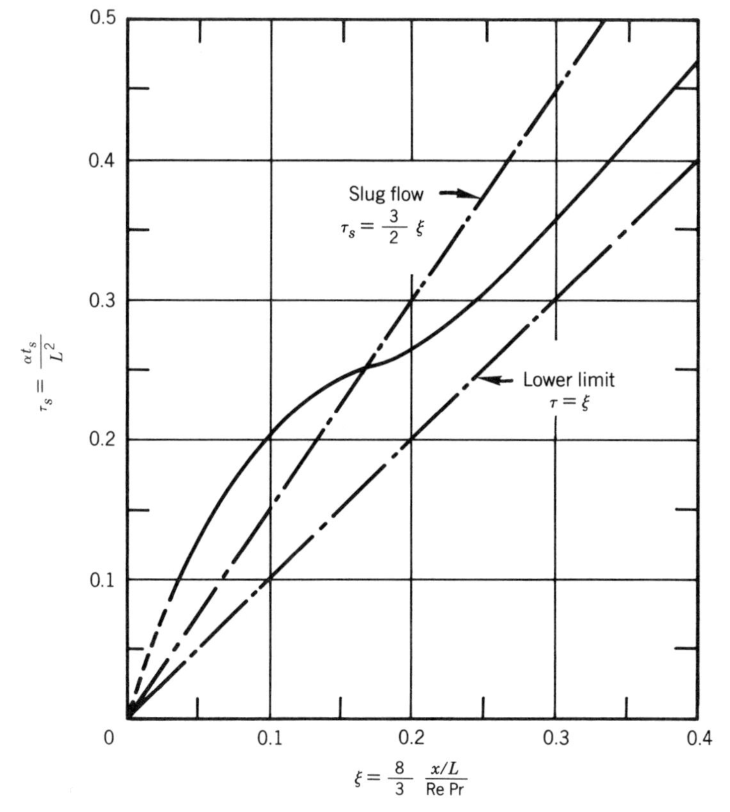

Figure 11.8. Time to reach steady state after a step change in wall temperature for fully developed steady laminar flow in a parallel-plate channel [13].

[13] developed a solution by a method similar to the one discussed in Sec. 11.2.1 for laminar flow through a circular tube with the parabolic velocity distribution [Eq. (11.11)]. However, instead of using the eigenfunctions of the corresponding Graetz problem in his expressions of the steady-state and transient temperature distributions, he employed the eigenfunctions that would result in the solution of the same problem with the slug flow assumption. Figures 11.7 and 11.8 show his results for the local transient heat flux and the steady-state times $\tau_s = \alpha t_s/L^2$ as a function of distance along the channel length, respectively. In these two figures the Reynolds number Re is defined in terms of the hydraulic diameter $D_h = 4L$, where L is the half distance between the plates.

11.2.5 Transient Laminar Forced Convection in a Parallel-Plate Channel with Unsteady Flow

Consider again the same parallel-plate channel shown in Fig. 11.6. In this section, the transient heat transfer phenomena that occur in this channel when there are simultaneous changes in fluid pumping pressure and wall heating conditions are discussed. Let there be a hydrodynamic entrance region, so that the flow is always fully developed for $x > 0$. Therefore, in the fully developed region, the velocity distribution, although time-dependent, does not vary with the axial position along the channel. Furthermore, if the fluid is assumed incompressible, then the velocity distribution for $x > 0$ will be governed by

$$\frac{\partial u}{\partial t} = -\frac{1}{\rho}\frac{\partial p}{\partial x} + \nu\frac{\partial^2 u}{\partial y^2} \tag{11.36}$$

where $\partial p/\partial x$ is a function of time only.

Step Change in Both Wall Temperature and Pressure Gradient from an Unheated Initial Condition. Initially, let the flow through the channel be steady with a mean velocity u_1 and isothermal at temperature T_i. At time $t = 0$, the pressure gradient is abruptly changed so that the fluid velocity undergoes a transient to a new mean value u_2. At $t = 0$, when the pressure gradient is changed, the temperature of the bounding walls is also changed to a new value T_w and maintained at this value for $t > 0$.

If the fluid properties are assumed constant, and viscous dissipation and axial conduction are neglected, then the unsteady temperature distribution $T(x, y, t)$ in the channel for $x \geq 0$ and $t > 0$ will satisfy Eq. (11.33), together with the conditions of

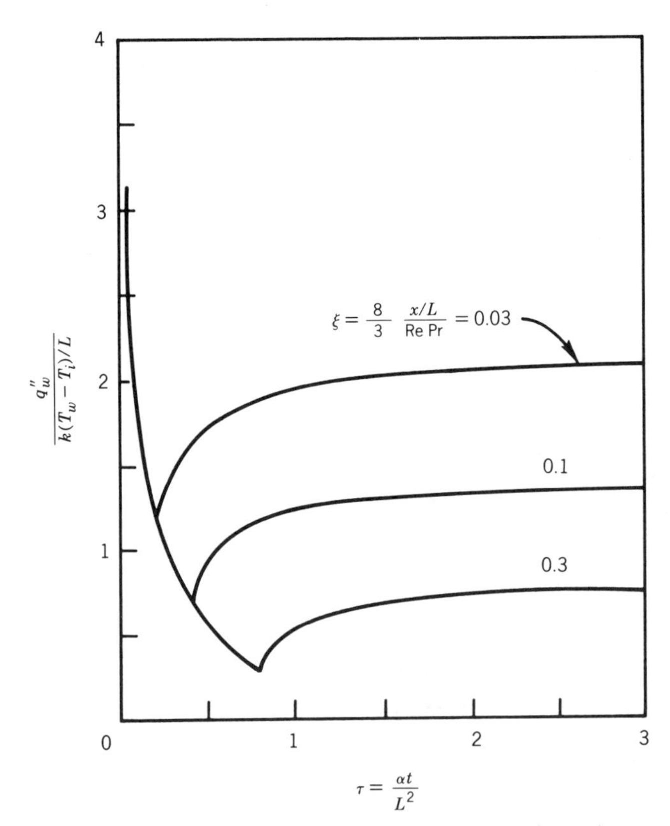

Figure 11.9. Transient variation in wall heat flux following a step change in pressure gradient and wall temperature. Pr = 0.7; $u_1 = 0$ [17].

Eqs. (11.34a, b, c, d). However, the velocity distribution u in Eq. (11.33) will be given by the solution of Eq. (11.36).

Perlmutter and Siegel [17] obtained an analytical solution of this problem by expanding the transient temperature distribution in a series in the same form as the steady-state solution of the problem. They evaluated the expansion coefficients in their time and axial-coordinate dependence by restricting the expansion to satisfy an integrated form of the energy equation (11.33) and then solving resulting partial differential equation by the method of characteristics. Once they obtained the transient temperature distribution, they calculated the variation of the heat flux q''_w to the fluid at the channel walls from Fourier's law. Figure 11.9 gives their calculations for the variation of the wall heat flux with time for a fluid with Pr = 0.7 and for the special case where $u_1 = 0$, i.e., initially there is no flow and both the channel and the fluid are isothermal at T_i. In this figure, the Reynolds number Re is defined in terms of the hydraulic diameter $D_h = 4L$ and the mean velocity u_2. As seen in Fig. 11.9, at each axial location the wall heat flux goes through a minimum and then increases toward the constant steady-state value. The reason for the initial decrease in q''_w is that after the initiation of the transient, the heat transfer at any location proceeds as if the channel were of infinite extent until the fluid particles that were at $x = 0$ at the beginning of the

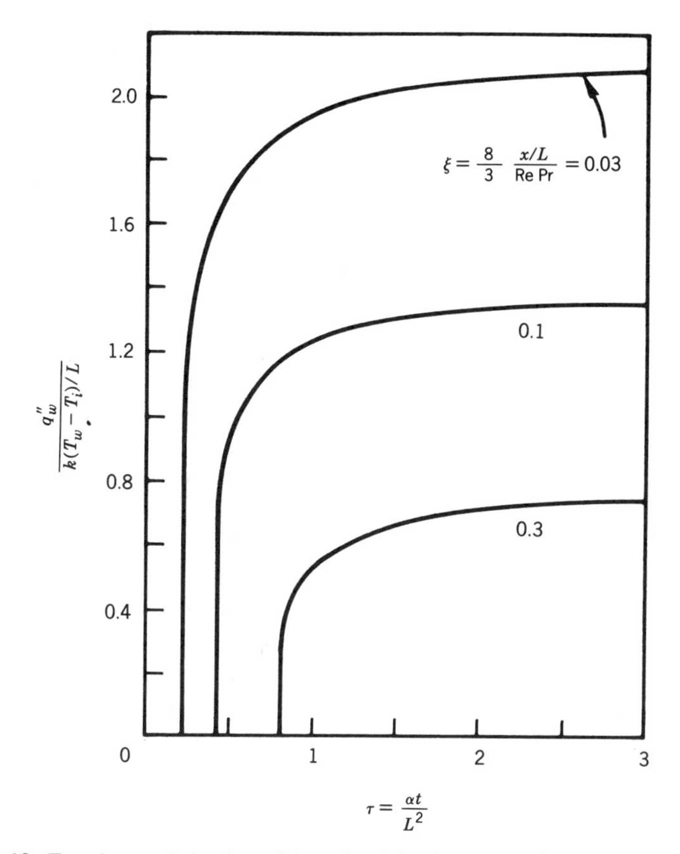

Figure 11.10. Transient variation in wall heat flux following a step change in pressure gradient with initial steady heating. Pr = 0.7; $u_1 = 0$ [17].

transient reach that location. For this early part of the process, the convective term drops out of the energy equation and the transient temperature distribution becomes independent of the axial direction.

Step Change in Pressure Gradient Only, with Initial Steady Heating. Now consider the situation where the transient is caused by a sudden change in the pumping pressure when there is a steady-state heat transfer process in the channel with the inlet temperature T_i and the temperature of the walls T_w. By following an analysis similar to the one in the previous case, Perlmutter and Siegel [17] also developed a solution to this problem, and Fig. 11.10 gives their results for the variation of the wall heat flux q''_w for a fluid with Pr = 0.7 and for the special case where $u_1 = 0$. As Fig. 11.10 shows, since there is no flow and the axial conduction was neglected, there is no heat transfer during the early transient period. After the fluid particles with the mean velocity u_2 reach a specific location, then heat transfer begins at that location and the wall heat flux rises toward its steady-state value.

Step Change in Both Pressure Gradient and Wall Temperature with Initial Steady Heating. Consider a more general case where the pressure gradient and the wall temperature are suddenly changed to new values when initially there is a steady-state heat transfer process in the channel with nonzero flow velocity. The resulting transient can be evaluated by a superposition of the solutions to the previous two simpler cases. The details of this superposition are explained in [17], and they will not be repeated here.

11.3 TRANSIENT TURBULENT FORCED CONVECTION IN DUCTS

The literature on transient turbulent forced convection is sparse. Abbrecht and Churchill [12] presented the results of an experimental investigation of heat transfer in the thermal entrance region following a step change in wall temperature in fully developed turbulent flow in a tube. Radial and longitudinal temperature gradients, radial heat fluxes, and eddy diffusivities for heat and momentum were computed from the measurements.

Sparrow and Siegel [14] investigated transient turbulent heat transfer in the thermal entrance region of a tube whose wall temperature varies arbitrarily with time. As a first step, the heat transfer response to a step jump in wall temperature was analyzed, and then this was generalized by a superposition technique to apply to arbitrary time variations. Use of the generalized results was illustrated by application to the case where the wall temperature variation was linear with time. The method used permitted the heat transfer coefficient to vary with time and position in accordance with the energy conservation principle.

Kakaç [21] analyzed transient heat transfer in incompressible turbulent flow between two parallel plates for a step jump in wall heat flux or wall temperature. The variations of the fluid velocity and effective diffusivity over the channel cross section were taken into account. It was assumed that the velocity profile was fully developed throughout the length of the channel. The thermal response of the system was obtained by solving the energy equation for air on a digital computer. The Nusselt number was presented, in the forms of graphs, as a function of time and space. A method was also discussed to obtain the velocity distribution from the distribution of the turbulent eddy diffusivity of momentum.

Kakaç in [29] presented a general closed-form solution to the transient energy equation under boundary conditions of zero wall temperature or zero heat flux for the decay of the inlet and initial temperature distributions in an incompressible turbulent flow between two parallel plates. However, the eigenfunctions and the corresponding eigenvalues were left to be determined to complete the solution.

Gärtner [30] analyzed the unsteady convective heat transfer in a hydrodynamically stabilized steady turbulent flow of a viscous incompressible fluid in a concentric annulus with the wall heat flux varying with time. The formulation permitted the heat transfer coefficient to vary with time and position. The energy equation was solved using the method of superposition and separating variables by finite integral transforms. The use of the generalized results was illustrated by application to the case where the wall heat flux varies exponentially with time.

Kawamura in [31] examined the variation of the heat transfer coefficient experimentally in a steady and turbulent flow through a circular tube cooled by water and heated stepwise with time. In addition, a numerical analysis was made for the same configuration, the results of which agreed well with the experimental results. Furthermore, an analytical expression for the variation of heat transfer coefficient was obtained. The time required for the heat transfer coefficient to reach its steady-state value was also studied.

Other important contributions in the field of transient turbulent forced convection are given in the references of the papers cited herein.

11.3.1 Transient Turbulent Forced Convection in Circular Tubes

Consider the circular tube shown in Fig. 11.1. Let there be a steady and fully developed turbulent flow through this tube, and the tube wall and fluid be initially isothermal at T_i. Assume that the tube wall is given an instantaneous step in temperature (say at time $t = 0$) to reach a new value T_w, and maintained at T_w for all times thereafter. The starting point in the analysis is the unsteady energy equation for fully developed turbulent flow in a circular tube:

$$\frac{\partial T}{\partial t} + u\frac{\partial T}{\partial x} = \frac{1}{r}\frac{\partial}{\partial r}\left[r(\alpha + \epsilon_h)\frac{\partial T}{\partial r}\right] \tag{11.37}$$

The initial, inlet, and boundary conditions are

$$T(x, r, 0) = T_i, \qquad T(0, r, t) = T_i \tag{11.38a,b}$$

$$\left(\frac{\partial T}{\partial r}\right)_{r=0} = 0, \qquad T(x, r_0, t) = T_w \tag{11.38c,d}$$

Following an approach similar to the one discussed in Sec. 11.2.1, Sparrow and Siegel [14] first obtained the steady-state solution of this problem and then expanded the transient solution about the steady-state conditions, which was only required to satisfy the integrated form of the energy equation. The following is their result for the transient temperature distribution:

$$\frac{T - T_w}{T_i - T_w} = \sum_{n=1}^{\infty} C_n F_n(x^+, t^+) R_n(r^+) \tag{11.39}$$

where

$$F_n = \begin{cases} \exp\left[\dfrac{(r_0^+)^3 \, dR_n(r_0^+)/dr^+}{\Pr \int_0^{r_0^+} r^+ R_n \, dr^+} t^+ \right], & t^+ \le a_n x^+ \\ \exp\left[-\dfrac{4\beta_n^2}{\text{Re}} x^+ \right], & t^+ \ge a_n x^+ \end{cases} \tag{11.40}$$

Here, β_n and $R_n(r^+)$ are the eigenvalues and eigenfunctions, respectively, of the following eigenvalue problem:

$$\frac{d}{dr^+}\left(r^+ \gamma \frac{dR_n}{dr^+} \right) + \left(\frac{2\beta_n^2}{\text{Re}} \frac{r^+}{r_0^+} u^+ \right) R_n = 0 \tag{11.41a}$$

$$\frac{dR_n(0)}{dr^+} = 0, \qquad R_n(r_0^+) = 0 \tag{11.41b,c}$$

and

$$C_n = \frac{\int_0^{r_0^+} r^+ u^+ R_n \, dr^+}{\int_0^{r_0^+} r^+ u^+ R_n^2 \, dr^+} \tag{11.42}$$

Various quantities in the above equations are defined as follows:

$$x^+ = \frac{x}{2r_0}, \qquad r^+ = r\frac{\sqrt{\tau_w/\rho}}{\nu}, \qquad r_0^+ = r_0\frac{\sqrt{\tau_w/\rho}}{\nu}$$

$$u^+ = \frac{u}{\sqrt{\tau_w/\rho}}, \qquad t^+ = \frac{\nu t}{r_0^2}$$

where τ_w is the wall shear stress and

$$\gamma = \frac{\alpha + \epsilon_h}{\nu}$$

Once the temperature distribution (11.39) is available, the wall heat-flux variation for the entire transient period can be calculated from Fourier's law, and the result is

$$\frac{q_w'' r_0}{(T_w - T_i)k} = -r_0^+ \sum_{n=0}^{\infty} C_n F_n(x^+, t^+) \frac{dR_n(r_0^+)}{dr^+} \tag{11.43}$$

Sparrow and Siegel [14] evaluated Eq. (11.43) for several combinations of Reynolds number and Prandtl number. Figure 11.11 is a representative result from [14], where the heat transfer responses at various positions ranging from $x/D = 2$ to $x/D = 100$ are given. At any position along the tube, initially the heat transfer is only by pure diffusion and follows the envelope curve, decreasing with increasing time. Then, at a certain time, for example at $t^+ = 0.00078$ for $x/D = 20$, convection begins to act and the curve breaks away from the pure-diffusion envelope, with the heat transfer continuing to decrease until the horizontal steady state is reached. Sparrow and Siegel

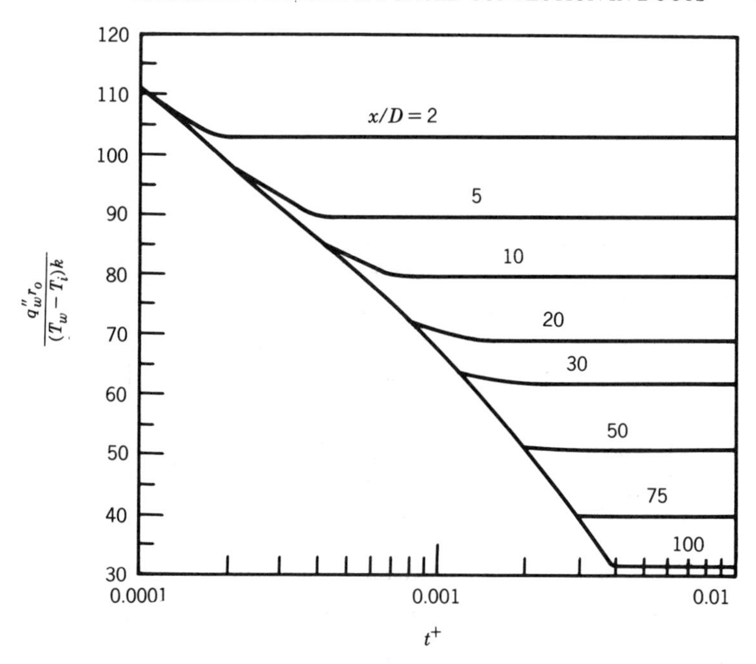

Figure 11.11. Wall heat-flux response to a step jump in wall temperature for fully developed turbulent flow through a circular pipe [14].

[14] in their numerical evaluations used the following correlations given in [3]:

$$\frac{du^+}{dy^+} = \left\{1 + (0.124)^2 u^+ y^+ \left[1 - e^{-(0.124)^2 u^+ y^+}\right]\right\}^{-1}, \qquad 0 \le y^+ \le 26 \quad (11.44)$$

$$u^+ = \frac{1}{0.36} \ln\left(\frac{y^+}{26}\right) + 12.8493, \qquad y^+ \ge 26 \quad (11.45)$$

where $y^+ = r_0^+ - r^+$. The total diffusivity was evaluated from [5]:

$$\gamma = \frac{1}{\text{Pr}} + (0.124)^2 u^+ y^+ \left[1 - e^{-(0.124)^2 u^+ y^+}\right], \qquad 0 \le y^+ < 26 \quad (11.46)$$

$$\gamma = \frac{1}{\text{Pr}} + 0.36 y^+ \left(1 - \frac{y^+}{r_0^+}\right) - 1, \qquad y^+ > 26 \quad (11.47)$$

The value of γ at $y^+ = 26$ was taken as the average of Eqs. (11.46) and (11.47). The -1 appearing on the right-hand side of Eq. (11.47) was retained for $26 < y^+ < r_0^+/2$ and deleted for larger values of y^+.

The steady-state times t_s, defined as the time period required for the heat transfer to come to within 5% of the steady-state value, were also calculated as a function of position by Sparrow and Siegel [14], and their results are given here in Fig. 11.12. As this figure shows, the steady-state time decreases as the Prandtl number increases, but by no more than a factor of 3 for this Prandtl number range. The Reynolds number

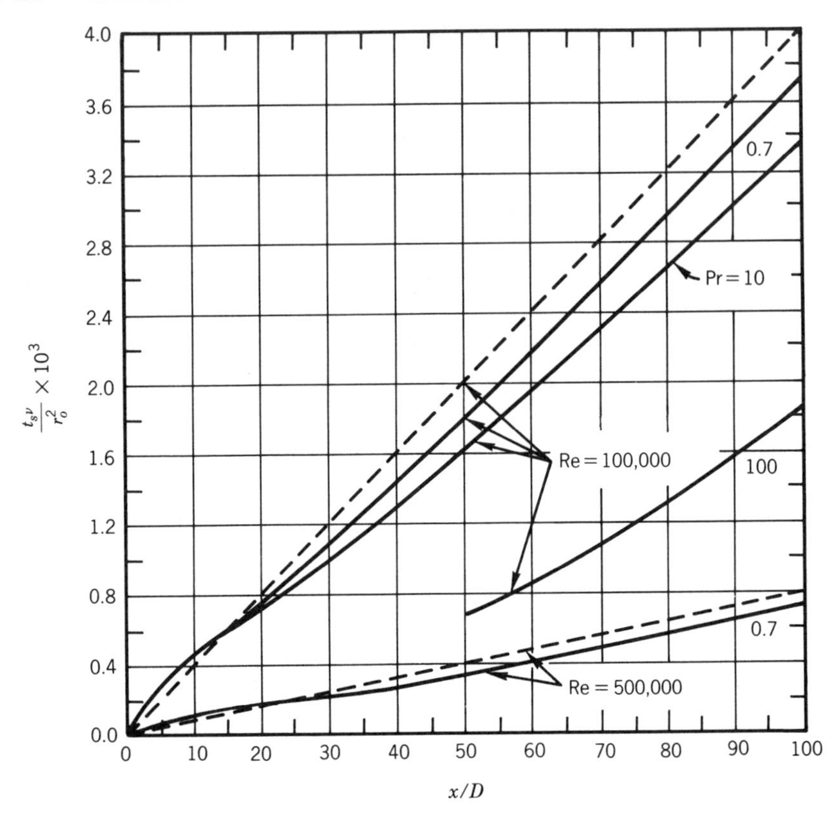

Figure 11.12. Steady-state times after a step jump in wall temperature for fully developed turbulent flow through a circular pipe [14].

also has a significant effect on the steady-state times, which are approximately in inverse proportion to the Reynolds number. Also appearing in this figure are two straight dashed lines corresponding to the time x/u_m, which approximately represents the time at which the heat transfer process at any position x begins to be influenced by the convection of fluid from the tube entrance. This simple relation $t_s = x/u_m$, which is $\tau_s = 4(x/D)/\text{Re}$ in dimensionless form, gives a fairly good estimate for a Prandtl number around unity, but tends to overestimate the steady-state times as the Prandtl number increases. However, for the purpose of providing an order-of-magnitude estimate, x/u_m appears to be useful.

11.3.2 Transient Turbulent Forced Convection in a Parallel-Plate Channel

Kakaç [21] made a numerical analysis by finite differences of transient forced convection for a hydrodynamically fully developed incompressible steady turbulent flow in a parallel-plate channel when there is a step change in wall heat flux or wall temperature. He used experimentally determined values for the eddy diffusivities of momentum and of heat in his calculations, and presented the variation of the Nusselt number as a function of time and axial position along the channel. Figures 11.13 and 11.14 show

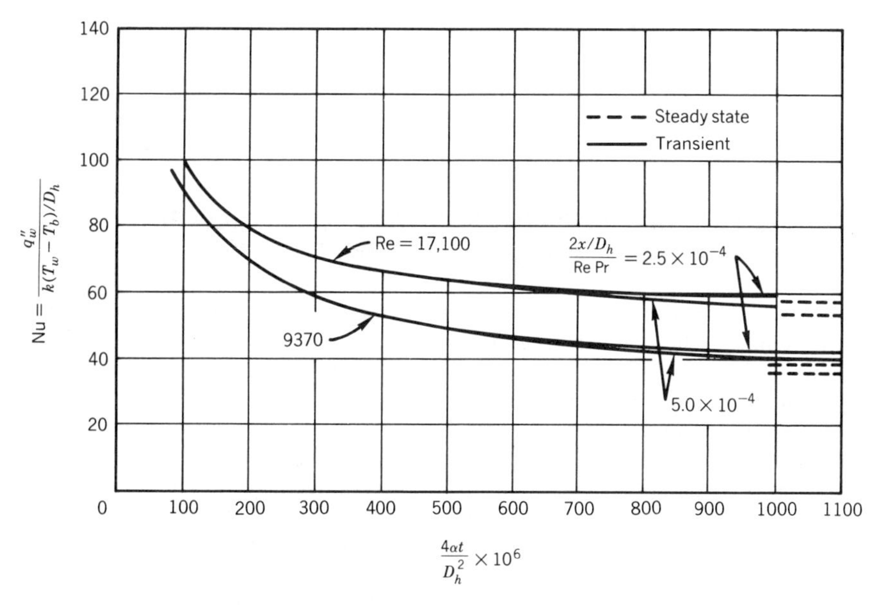

Figure 11.13. Transient Nusselt numbers for a step change in wall temperature for fully developed turbulent flow in a parallel-plate channel. Pr = 0.73 [21].

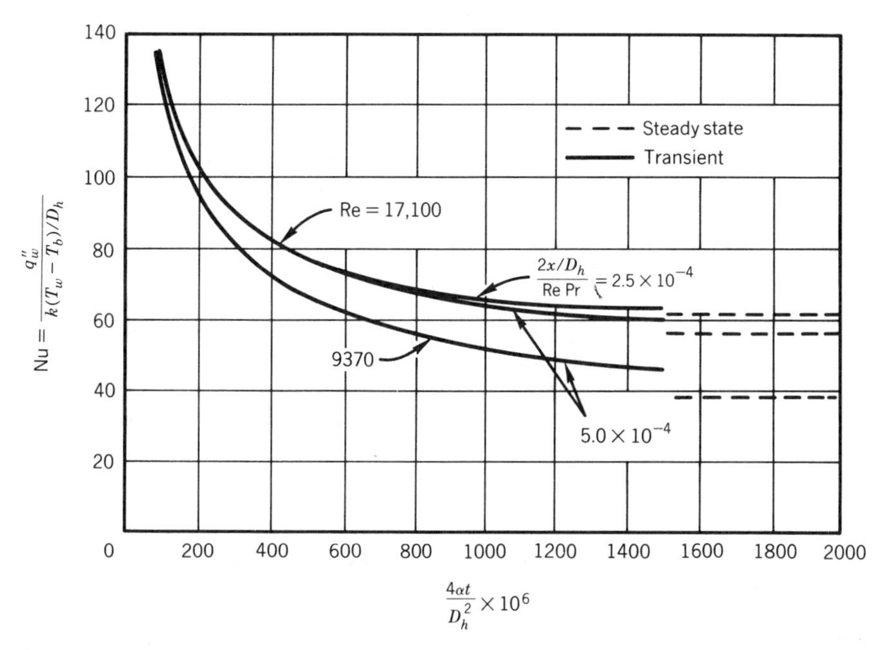

Figure 11.14. Transient Nusselt numbers for a step change in wall heat flux for fully developed turbulent flow in a parallel-plate channel. Pr = 0.73 [21].

two of his calculations for the local Nusselt number for air (Pr = 0.73) at two locations along the channel and for two Reynolds numbers, for step changes in wall temperature and in wall heat flux, respectively. In these figures q_w'' is the wall heat flux to the fluid, T_m is the fluid bulk mean temperature, and $D_h = 4L$, with L the half distance between the channel walls.

11.4 ANALYSIS OF TRANSIENT FORCED CONVECTION FOR TIMEWISE VARIATION OF INLET TEMPERATURE

The periodic thermal response of duct flows to imposed cyclic variations in thermal conditions has also been investigated. Sparrow and De Farias [22] made an analysis of unsteady laminar heat transfer in a parallel-plate channel with periodically varying inlet temperature. The midplane of each wall was considered insulated, and the wall temperature was dynamically determined by a balance of heat transfer and energy storage. In the analytical formulation, the commonly used quasi-steady assumption was lifted in favor of the local application of the energy equation, the solution of which involved an eigenvalue problem with complex eigenvalues and eigenfunctions. Numerical evaluation of the analytical results provided the time and space dependence of the wall and bulk temperatures and of the Nusselt number. In addition, results for the overall performance of the channel as a heat exchanger were presented in terms of the energy carried across the exit cross section relative to that carried across the entrance section. For comparison purposes, results for the overall performance were also derived using the quasi-steady model. It was found that for a range of operating conditions the quasi-steady model was able to give accurate performance predictions, especially when it was used in conjunction with spatially varying heat transfer coefficients.

Kakaç and Yener [26] obtained an exact solution to the transient energy equation for laminar slug flow of an incompressible fluid in a parallel-plate channel with time-varying inlet temperature. The results were confirmed experimentally by the frequency response method for a limited range of Reynolds number.

Acker and Fourcher [37] studied the laminar flow in a storage unit in thermal periodic regime. The energy equations were solved simultaneously both for the wall and for the fluid flow between two parallel plates by Laplace transforms with the slug flow assumption and when there is a sinusoidal variation in the inlet fluid temperature.

Sucec and Sawant [46] made a study of the unsteady, conjugated laminar forced convection in a parallel-plate channel with periodically varying inlet fluid temperature. They obtained the wall and the fluid bulk temperatures as a function of distance along the channel and of time for a sinusoidal inlet temperature variation, the channel walls being adiabatic on their outside surfaces and communicating thermally with the fluid across their inside surfaces.

11.4.1 Heat Transfer in Laminar Slug Flow through a Parallel-Plate Channel with Time-Varying Inlet Temperature

The parallel-plate channel under consideration is shown in Fig. 11.15. The fluid entering the heated section has a temperature which is spatially uniform across the entrance section but varies sinusoidally with time as

$$T(0, y, t) = T_0 + (\Delta T)_0 \sin \beta t \tag{11.48}$$

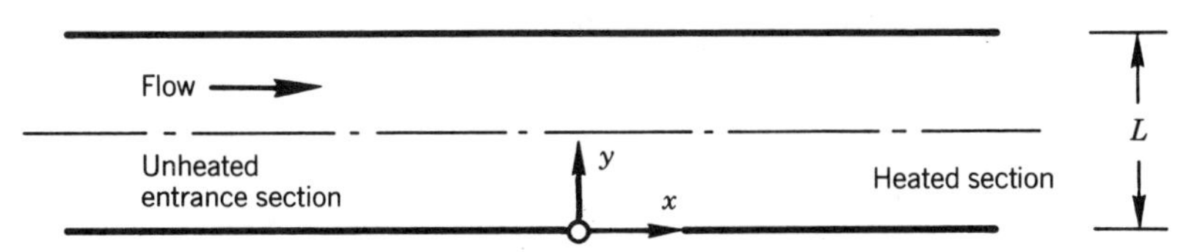

Figure 11.15. Coordinate system for parallel-plate channel.

where T_0 is the cycle mean temperature, $(\Delta T)_0$ is the amplitude, and β is the inlet frequency. The following idealizations are made in the analysis:

Flow between the plates is steady, fully developed, and laminar.
Viscous dissipation is negligible.
Axial conduction is negligible with respect to bulk transport in the x direction. This is a reasonable assumption when the Péclet number exceeds 100.
Fluid properties are constant.
Thermal resistance of the channel walls is negligible.

The starting point of the analysis is again the unsteady energy equation for a fully developed laminar flow in a parallel-plate channel, which can be written as

$$\frac{\partial \theta}{\partial t} + u\frac{\partial \theta}{\partial x} = \alpha\frac{\partial^2 \theta}{\partial y^2} \tag{11.49}$$

where

$$\theta(x, y, t) = \frac{T(x, y, t) - T_0}{(\Delta T)_0} \tag{11.50}$$

with the following inlet and boundary conditions:

$$\theta(0, y, t) = \sin \beta t \tag{11.51a}$$

$$\left(\frac{\partial \theta}{\partial y}\right)_{y=0} = 0, \qquad \left(k\frac{\partial \theta}{\partial y} + h\theta\right)_{y=L} = f(x) \tag{11.51b, c}$$

where the function $f(x)$ is given in Table 11.1 for various boundary conditions at $y = L$. One obtains the temperature boundary condition at $y = L$ by setting $k = 0$ and $h = 1$, and the heat-flux boundary condition by setting $h = 0$. When h and k are finite, Eq. (11.51c) means that the boundary at $y = L$ is losing heat by convection to the environment at temperature $T_\infty(x)$.

The foregoing problem can be separated into two as follows:

$$\theta(x, y, t) = \theta_1(x, y) + \theta_2(x, y, t) \tag{11.52}$$

where $\theta_1(x, y)$ and $\theta_2(x, y, t)$ are solutions of the following problems:

$$u\frac{\partial \theta_1}{\partial x} = \alpha\frac{\partial^2 \theta_1}{\partial y^2} \tag{11.53a}$$

TABLE 11.1. The Function $f(x)$

Boundary Condition at $y = L$	Function $f(x)$
First kind ($k = 0$, $h = 1$)	$\dfrac{T_w(x) - T_0}{(\Delta T)_0}$
Second kind ($h = 0$)	$\dfrac{q_w(x)}{(\Delta T)_0}$
Third kind (k and h finite)	$h\dfrac{T_\infty(x) - T_0}{(\Delta T)_0}$

with

$$\theta_1(0, y) = 0 \tag{11.53b}$$

$$\left(\frac{\partial \theta_1}{\partial y}\right)_{y=0} = 0, \qquad \left(k\frac{\partial \theta_1}{\partial y} + h\theta_1\right)_{y=L} = f(x) \tag{11.53c, d}$$

and

$$\frac{\partial \theta_2}{\partial t} + u\frac{\partial \theta_2}{\partial x} = \alpha\frac{\partial^2 \theta_2}{\partial y^2} \tag{11.54a}$$

with

$$\theta_2(0, y, t) = \sin \beta t \tag{11.54b}$$

$$\left(\frac{\partial \theta_2}{\partial y}\right)_{y=0} = 0, \qquad \left(k\frac{\partial \theta_2}{\partial y} + h\theta_2\right)_{y=L} = 0 \tag{11.54c, d}$$

To simplify the method of analysis, the velocity profile u across the entire flow area of the channel will be taken constant (i.e., slug flow idealization).

Solution for $\theta_1(x, y)$. The solution of the problem given by Eqs. (11.53) can be written as [26]

$$\theta_1(x, y) = \sum_{n=1}^{\infty} \frac{\cos \lambda_n y}{N_n} \int_0^x e^{-(\alpha\lambda_n^2/u)(x-x')} A_n(x')\, dx' \tag{11.55}$$

where λ_n and $A_n(x)$ are given in Tables 11.2 and 11.3, and N_n is defined by

$$N_n = \frac{L}{2} + \frac{1}{4\lambda_n}\sin 2L\lambda_n \tag{11.56}$$

Solution for $\theta_2(x, y, t)$. The solution of the problem given by Eqs. (11.54) can be written as [26]

$$\theta_2(x, y, t) = \begin{cases} \sin\left[\beta\left(t - \dfrac{x}{u}\right)\right] \displaystyle\sum_{n=1}^{\infty} \frac{\cos \lambda_n y \sin \lambda_n L}{\lambda_n N_n} e^{-(\alpha\lambda_n^2/u)x}, & h \neq 0 \\ \sin\left[\beta\left(t - \dfrac{x}{u}\right)\right], & h = 0 \end{cases} \tag{11.57}$$

TABLE 11.2. Eigenvalues

Boundary Condition at $y = L$	λ_n
First kind ($k = 0$, $h = 1$)	$\lambda_n = \frac{2n-1}{L}\frac{\pi}{2}$, $n = 1, 2, \ldots$
Second kind ($h = 0$)	$\lambda_n = \frac{n-1}{L}\pi$, $n = 1, 2, \ldots$
Third kind (k and h are finite)	Positive roots of $\lambda_n \tan \lambda_n L = h/k$

TABLE 11.3. The function $A_n(x)$ in Eq. (11.55)

Boundary Condition at $y = L$	$A_n(x)$
First kind	$\frac{\alpha\lambda_n}{u} f(x) \sin \lambda_n$
Second and third kinds	$\frac{\alpha}{uk} f(x) \cos \lambda_n$

When the boundary condition for $\theta(x, y, t)$ at $y = L$ is homogeneous, i.e., when $f(x) = 0$, then $\theta_1(x, y)$ becomes identically zero, and in that case

$$\theta(x, y, t) = \theta_2(x, y, t)$$

When the temperature or the convection boundary condition is homogeneous, the walls lose heat in such a way that each mode of $\theta_2(x, y, t)$ decays exponentially along the duct and this decay is inversely proportional to the velocity u. Therefore, as the velocity u is increased, the rate of decay decreases. It is also seen that the phase lag along the tube is linear with the slope β/u, and as the velocity u is increased, this slope decreases.

When the heat-flux boundary condition is homogeneous, there will be no heat conduction in the y direction. Since the axial diffusion of heat has already been neglected, the amplitude of $\theta_2(x, r, t)$ remains constant. The phase lag, however, is the same as in the other two cases because of the convention in the x direction.

11.4.2 A General Solution to the Transient Forced Convection Energy Equation for Timewise Variation of the Inlet Temperature

In this section, following the work of Kakaç [29], formal solutions for the decay of a periodically varying inlet temperature in a fully developed turbulent flow between two parallel plates with linear and homogeneous boundary conditions are given.

Consider a steady, fully developed turbulent flow through a parallel-plate channel whose walls are separated by a distance L as shown in Fig. 11.15. Neglecting axial diffusion and viscous dissipation, and assuming constant fluid properties, the energy equation governing the conduction (in the y direction) and the convection (in the x

direction) can be written as

$$\frac{\partial T}{\partial t} + u(y)\frac{\partial T}{\partial x} = \frac{\partial}{\partial y}\left(D(y)\frac{\partial T}{\partial y}\right) \tag{11.58}$$

where $u(y)$ is the fully developed velocity profile and $D(y)$ is the effective diffusivity, which is assumed to be a function of y only. Suppose that the system satisfying Eq. (11.58) is subject to a periodic inlet condition of the form

$$T(0, y, t) = e^{i\beta t}, \qquad i = \sqrt{-1} \tag{11.59}$$

and two linear homogeneous boundary conditions of the following general forms:

$$\left(a_1 T + b_1 \frac{\partial T}{\partial y}\right)_{y=0} = 0 \tag{11.60a}$$

$$\left(a_2 T + b_2 \frac{\partial T}{\partial y}\right)_{y=L} = 0 \tag{11.60b}$$

where a_i and b_i, $i = 1, 2$, are given real constants. If a periodic solution of the form

$$T(x, y, t) = e^{i\beta t} X(x) Y(y)$$

for the decay of the inlet condition of Eq. (11.59) is assumed, then it can be shown that the solution for $T(x, y, t)$ is given by [29]

$$T(x, y, t) = \sum_{n=1}^{\infty} c_n e^{-\alpha_n x}[P_n(y)\cos(\beta t - \delta_n x) - Q_n(y)\sin(\beta t - \delta_n x)]$$

$$+ i\sum_{n=1}^{\infty} c_n e^{-\alpha_n x}[P_n(y)\sin(\beta t - \delta_n x) + Q_n(y)\cos(\beta t - \delta_n x)] \tag{11.61}$$

where $P_n(y)$ and $Q_n(y)$ are the eigenfunctions and α_n are the eigenvalues of the following coupled eigenvalue problem:

$$\frac{d}{dy}\left(D\frac{dP_n}{dy}\right) = -\alpha_n u P_n + (\delta_n u - \beta) Q_n \tag{11.62a}$$

$$\frac{d}{dy}\left(D\frac{dQ_n}{dy}\right) = -\alpha_n u Q_n - (\delta_n u - \beta) P_n \tag{11.62b}$$

with

$$a_1 P_n(0) + b_1 \frac{dP_n(0)}{dy} = 0, \qquad a_1 Q_n(0) + b_1 \frac{dQ_n(0)}{dy} = 0 \tag{11.63a, b}$$

$$a_2 P_n(L) + b_2 \frac{dP_n(L)}{dy} = 0, \qquad a_2 Q_n(L) + b_2 \frac{dQ_n(L)}{dy} = 0 \tag{11.63c, d}$$

and

$$\delta_n = \beta \frac{\int_0^L (P_n^2 + Q_n^2)\, dy}{\int_0^L u(P_n^2 + Q_n^2)\, dy} \tag{11.64}$$

As an example, let the system satisfying Eq. (11.58) be subjected to a periodic inlet condition given by

$$T(0, y, t) = T_m + (\Delta T)_0 \sin \beta t \tag{11.65}$$

and two linear homogeneous boundary conditions of the following forms:

$$\left(\frac{\partial T}{\partial y}\right)_{y=0} = 0 \quad \text{and} \quad T(x, L) = T_w \tag{11.66a, b}$$

The solution will then be given by

$$\frac{T - T_m}{(\Delta T)_0} = \theta_1(x, y) + \sum_{n=1}^{\infty} c_n e^{-\alpha_n x} \sqrt{P_n^2(y) + Q_n^2(y)}\, \sin(\beta t - \delta_n x + \epsilon_n) \tag{11.67}$$

where

$$\epsilon_n = \tan^{-1} \frac{Q_n(y)}{P_n(y)} \tag{11.68}$$

and $\theta_1(x, y)$ satisfies the following problem:

$$u \frac{\partial \theta_1}{\partial x} = \frac{\partial}{\partial y}\left(D \frac{\partial \theta_1}{\partial y}\right) \tag{11.69}$$

$$\theta_1(0, y) = 0 \tag{11.70a}$$

$$\left(\frac{\partial \theta_1}{\partial y}\right)_{y=0} = 0, \qquad \theta_1(x, L) = \frac{T_w - T_m}{(\Delta T)_0} \tag{11.70b, c}$$

In regions away from the inlet, only the first term in the series in Eq. (11.67) needs to be considered. Hence, the asymptotic solution, deleting the subscript 1, becomes

$$\frac{T - T_m}{(\Delta T)_0} = \theta_1(x, y) + c e^{-\alpha x} \sqrt{P^2(y) + Q^2(y)}\, \sin(\beta t - \delta x + \epsilon) \tag{11.71}$$

It is to be noted that solutions developed so far are also valid for laminar flow.

The form of Eq. (11.71) suggests that the results can best be confirmed experimentally by the frequency response method, and the first values of the eigenvalues α_n and δ_n can be determined for various values of the inlet frequency for a wide range of the Reynolds number.

An experimental setup can be designed and used to study the decay of sinusoidal inlet conditions for turbulent forced convection in various channel geometries and to obtain experimentally the first eigenvalue and other parameters appearing in the general solutions by the frequency analysis.

At a fixed Reynolds number, the changes of amplitudes and phases of temperature waves along the channel can be measured. Phase data can be taken with respect to the thermocouple nearest to the inlet heater. An experiment can be carried out for the different frequency values of the sinusoidal variation of heat input to the inlet heater.

From the series of such measurements, the coefficients α and δ can be measured as a function of the Reynolds number, inlet frequency, and distance along the channel. The results of the general solution given in this section were confirmed experimentally by the frequency response method for a limited range of Reynolds number [34].

11.5 CONCLUDING REMARKS

A state-of-the-art review of transient forced convection in ducts has been given. Basic solution methods, together with some important solutions, have also been introduced for two geometries, namely parallel plates and circular tubes. These are the two geometries most commonly used in fluid flow and heat transfer devices. For further solutions and applications the reader is referred to the references cited at the end of this chapter.

NOMENCLATURE

c_p	specific heat at constant pressure, J/(kg · K), Btu/(lb$_m$ · °F)
D	diameter = $2r_0$, m, ft
D_h	hydraulic diameter = $4L$, m, ft
h	heat transfer coefficient, W/(m^2 · K), Btu/(hr · ft^2 · °F)
J_0, J_1	Bessel functions of the first kind and of zero and first orders
k	thermal conductivity, W/(m · K), Btu/(hr · ft · °F)
L	half the distance between parallel plates; distance between parallel plates, m, ft
P_n	eigenfunction
Pr	Prandtl number = $\mu c_p/k = \nu/\alpha$
p	pressure, Pa, lb$_f$/ft^2
Q_n	eigenfunction
q''	heat flux, W/m^2, Btu/(hr · ft^2)
R_n	eigenfunction
Re	Reynolds number = $\rho u D_h/\mu$
r	radial coordinate, m, ft
r_0	tube radius, m, ft
r_0^+	dimensionless radius = $r_0\sqrt{\tau_w/\rho}/\nu$
T	temperature, °C, K, °F, °R
T_0	cycle mean temperature, °C, K, °F, °R
t	time, s
t^+	dimensionless time = $\nu t/r_0^2$
t_s	steady-state time, s
u_m	mean flow velocity, m/s, ft/s

u	velocity component in x direction, m/s, ft/s
u^+	dimensionless velocity $= u/\sqrt{\tau_w/\rho}$
x	distance parallel to flow direction along the ducts, m, ft
x^+	dimensionless x coordinate $= x/D$
y	transverse distance in parallel-plate channels, m, ft
y^+	dimensionless variable $= r_0^+ - r^+$

Greek Symbols

α	thermal diffusivity $= k/\rho c_p$, m^2/s, ft^2/s
α_n	eigenvalues
β	inlet frequency, 1/s
β_n	eigenvalues
ϵ_h	eddy diffusivity of heat, m^2/s, ft^2/s
ϵ_m	eddy diffusivity of momentum, m^2/s, ft^2/s
η	dimensionless r coordinate $= r/r_0$
θ	dimensionless temperature, defined by Eq. (11.4e) or (11.31)
λ	eigenvalues
μ	dynamic viscosity, Pa · s, lbm/(hr · ft)
ν	kinematic viscosity, m^2/s, ft^2/s
ξ	dimensionless x coordinate $= (2x/D)/(\mathrm{Re\,Pr})$
ρ	density, kg/m^3, lb_m/ft^3
τ	dimensionless time $= \alpha t/r_0^2$
τ	shear stress, Pa^2, lb_f/ft^2
τ_s	dimensionless steady-state times $= \nu t_s/r_0^2$, $\alpha t_s/L^2$
$(\Delta T)_0$	amplitude of inlet temperature variation, °C, K, °F, °R

Subscripts

i	inlet condition
m	bulk mean condition
s	steady-state conditions
w	wall condition
∞	environment condition

REFERENCES

1. G. M. Dusinberre, Calculation of Transient Temperatures in Pipes and Heat Exchangers by Numerical Methods, *Trans. ASME*, Vol. 76, pp. 421–426, 1954.
2. J. W. Rizika, Thermal Lags in Flowing Incompressible Fluid Systems Containing Heat Capacitors, *Trans. ASME*, Vol. 78, pp. 411–420, 1954.
3. R. G. Deissler, Analysis of Turbulent Heat Transfer, Mass Transfer, and Friction in Smooth Tubes at High Prandtl and Schmidt Numbers, NACA Report 1210, 1955.
4. J. W. Rizika, Thermal Lags in Flowing Incompressible Fluid Systems Containing Heat Capacitors, *Trans. ASME*, Vol. 78, pp. 1407–1413, 1956.

5. E. M. Sparrow, T. M. Hallman, and R. Siegel, Turbulent Heat Transfer in the Thermal Entrance Region of a Pipe with Uniform Heat Flux, *Appl. Sci. Res.*, Vol. 7, Sec. A, pp. 37–52, 1957.
6. E. M. Sparrow and R. Siegel, Thermal Entrance Region of a Circular Tube under Transient Heating Conditions, *Proc. Third U.S. Nat. Congr. Appl. Mech.*, pp. 817–826, 1958.
7. J. A. Clark, V. S. Arpaci, and K. M. Treadwell, Dynamic Response of Heat Exchangers Having Internal Heat Sources—Part I, *Trans. ASME*, Vol. 80, pp. 612–624, 1958.
8. V. S. Arpaci and J. A. Clark, Dynamic Response of Heat Exchangers Having Internal Heat Sources—Part II, *Trans. ASME*, Vol. 80, pp. 625–634, 1958.
9. R. Siegel, Transient Heat Transfer for Laminar Slug Flow in Ducts, *Trans. ASME, J. Appl. Mech.*, Vol. 81E, pp. 140–142, 1959.
10. R. Siegel and E. M. Sparrow, Transient Heat Transfer for Laminar Forced Convection in the Thermal Entrance Region of Flat Ducts, *Trans. ASME, J. Heat Transfer*, Vol. 81C, pp. 29–36, 1959.
11. V. S. Arpaci and J. A. Clark, Dynamic Response of Heat Exchangers Having Internal Heat Sources—Part III, *Trans. ASME, J. Heat Transfer*, Vol. 81C, pp. 253–266, 1959.
12. P. H. Abbrecht and S. W. Churchill, The Thermal Entrance Region in Fully Developed Turbulent Flow, *AIChE J.*, Vol. 6, No. 2, p. 268, 1960.
13. R. Siegel, Heat Transfer for Laminar Flow in Ducts with Arbitrary Time Variation in Wall Temperature, *Trans. ASME, J. Appl. Mech.*, Vol. 82E, pp. 241–249, 1960.
14. E. M. Sparrow and R. Siegel, Unsteady Turbulent Heat Transfer in Tubes, *Trans. ASME, J. Heat Transfer*, Vol. 82C, pp. 170–180, 1960.
15. J. W. Yang, J. A. Clark, and V. S. Arpaci, Dynamic Response of Heat Exchangers Having Internal Heat Sources—Part IV, *Trans. ASME, J. Heat Transfer*, Vol. 83C, pp. 321–388, 1961.
16. M. Perlmutter and R. Siegel, Unsteady Laminar Flow in a Duct with Unsteady Heat Addition, *Trans. ASME, J. Heat Transfer*, Vol. 83, pp. 432–440, 1961.
17. M. Perlmutter and R. Siegel, Two-Dimensional Unsteady Incompressible Laminar Duct Flow with a Step Change in Wall Temperature, *Int. J. Heat Mass Transfer*, Vol. 3, pp. 94–107, 1961.
18. R. Siegel, Forced Convection in a Channel with Wall Heat Capacity and with Wall Heating Variable with Axial Position and Time, *Int. J. Heat Mass Transfer*, Vol. 6, pp. 607–620, 1963.
19. R. Siegel and M. Perlmutter, Laminar Heat Transfer in a Channel with Unsteady Flow and Wall Heating Varying with Position and Time, *Trans. ASME, J. Heat Transfer*, Vol. 85, pp. 358–365, 1963.
20. A. Kardas, On a Problem in the Theory of the Unidirectional Regenerators, *Int. J. Heat Mass Transfer*, Vol. 9, p. 567, 1966.
21. S. Kakaç, Transient Turbulent Flow in Ducts, *Wärme- und Stoffübertragung*, Vol. 1, pp. 169–176, 1968.
22. E. M. Sparrow and F. N. De Farias, Unsteady Heat Transfer in Ducts with Time Varying Inlet Temperature and Participating Walls, *Int. J. Heat Mass Transfer*, Vol. 11, pp. 837–853, 1968.
23. C. A. Chase, Jr., D. Gidaspow, and R. E. Pech, A Generator-Prediction of Nusselt Number, *Int. J. Heat Mass Transfer*, Vol. 12, pp. 727–736, 1969.
24. K. Namatame, Transient Temperature Response of an Annular Flow with Step Change in Heat Generating Rod, *J. Nucl. Sci. Technol.*, Vol. 6, pp. 291–600, 1969.
25. E. K. Kalinin and G. A. Dreitser, Unsteady Convective Heat Transfer and Hydrodynamics in Channels, *Adv. Heat Transfer*, Vol. 8, pp. 367–502, 1970.
26. S. Kakaç and Y. Yener, Exact Solution of the Transient Forced Convection Energy Equation for Timewise Variation of Inlet Temperature, *Int. J. Heat Mass Transfer*, Vol. 16, pp. 2205–2214, 1973.

27. T. W. Schanatz, E. P. Russo, and O. Tanner, Transient Temperature Distribution for Fully Developed Laminar Flow in a Tube, *Heat Transfer 1974*, Vol. 5, pp. 160–164, 1974.

28. H. Kawamura, Analysis of Transient Turbulent Heat Transfer in an Annulus: Part I: Heating Element with a Finite (Nonzero) Heat Capacity and no Thermal Resistance, *Heat Transfer—Japan Res.*, *Vol.* 3, No. 1, pp. 45–68, 1974.

29. S. Kakaç, A General Analytical Solution to the Equation of Transient Forced Convection with Fully Developed Flow, *Int. J. Heat Mass Transfer*, Vol. 18, pp. 1449–1453, 1975.

30. D. Gärtner, Instationärer Wärmeübergan bei Turbulenter Ringspaltströmung, *Wärme- und Stoffübertragung*, Vol. 9, pp. 179–191, 1976.

31. H. Kawamura, Experimental and Analytical Study of Transient Heat Transfer for Turbulent Flow in a Circular Tube, *Int. J. Heat Mass Transfer*, Vol. 20, pp. 443–450, 1977.

32. A. Campo and T. Yoshimura, Random Heat Transfer in Flat Channels with Timewise Variation of Ambient Temperature, *Int. J. Heat Mass Transfer*, Vol. 22, pp. 5–12, 1979.

33. S. Kakaç, Transient Heat Transfer by Forced Convection in Channels, *Turbulent Forced Convection in Channels and Bundles*, ed. S. Kakaç and D. B. Spalding, Hemisphere, New York, Vol. 2, pp. 865–880, 1979.

34. S. Kakaç and Y. Yener, Frequency Response Analysis of Transient Turbulent Forced Convection for Timewise Variation of Inlet Temperature, *Turbulent Forced Convection in Channels and Bundles*, ed. S. Kakaç and D. B. Spalding, Hemisphere, New York, Vol. 2, pp. 865–880, 1979.

35. S. Kakaç and Y. Yener, *Convective Heat Transfer*, METU, Ankara, Turkey, 1980.

36. R. K. Shah, The Transient Response of Heat Exchangers, *Heat Exchangers: Thermal-Hydraulic Fundamentals and Design*, ed., S. Kakaç, A. E. Bergles, and F. Mayinger, pp. 915–953, Hemisphere, New York, 1981.

37. M. T. Acker and B. Fourcher, Analyse in Regime Thermique Périodique du Cauplage Conduction-Convection entre un Fluide en Ecoulement Laminaire et une Paroi de Stokage, *Int. J. Heat Mass Transfer*, Vol. 24, pp. 1201–1210, 1981.

38. H. T. Lin and Y. P. Shih, Unsteady Thermal Entrance Heat Transfer of Power-Law Fluids in Pipes and Plate Slits, *Int. J. Heat Mass Transfer*, Vol. 24, pp. 1531–1539, 1981.

39. J. Sucec, An Improved Quasi-steady Approach for Transient Conjugated Forced Convection Problems, *Int. J. Heat Mass Transfer*, Vol. 24, pp. 1711–1722, 1981.

40. M. D. Mikhailov, Mathematical Modelling of Heat Transfer in Single Duct and Double-Pipe Exchangers, *Low Reynolds Number Flow Heat Exchangers*, ed. S. Kakaç, R. K. Shah, and A. E. Bergles, pp. 137–164, Hemisphere, New York, 1983.

41. S. Kakaç and Y. Yener, Transient Laminar Forced Convection in Ducts, *Low Reynolds Number Flow Heat Exchangers*, ed., S. Kakaç, R. K. Shah, and A. E. Bergles, pp. 205–227, Hemisphere, New York, 1983.

42. T. F. Lin, K. H. Hawks, and W. Leidenfrost, Unsteady Thermal Entrance Heat Transfer in Laminar Pipe Flows with a Step Change in Ambient Temperature, *Wärme- und Stöffubertragung*, Vol. 17, pp. 125–132, 1983.

43. T. F. Lin, K. H. Hawks, and W. Leidenfrost, Transient Thermal Entrance Heat Transfer in Laminar Flows with a Step Change in Pumping Pressure, *Wärme- und Stoffübertragung*, Vol. 17, pp. 201–209, 1983.

44. T. F. Lin, K. H. Hawks, and W. Leidenfrost, Transient Conjugated Heat Transfer between a Cooling Coil and Its Surrounding Enclosure, *Int. J. Heat Mass Transfer*, Vol. 16, pp. 1661–1667, 1983.

45. S. C. Chen, N. K. Anand, and D. R. Tree, Analysis of Transient Laminar Convective Heat Transfer inside a Circular Duct, *J. Heat Transfer*, Vol. 105, pp. 922–924, 1983.

46. J. Sucec and A. M. Sawant, Unsteady Conjugated Forced Convection Heat Transfer in a Parallel Plate Duct, *Int. J. Heat Mass Transfer*, Vol. 27, pp. 95–101, 1984.

47. S. Kakaç and Y. Yener, *Heat Conduction*, 2nd ed., Hemisphere, New York and Springer, Berlin, 1985.

12

BASICS OF NATURAL CONVECTION

Yogesh Jaluria

Rutgers University
New Brunswick, New Jersey

12.1 INTRODUCTION

The convective mode of heat transfer is generally divided into two basic processes. If the motion of the fluid arises from an external agent—for instance, a fan, a blower, the wind, or the motion of the heated object itself—then the process is termed *forced convection*. If, on the other hand, no such externally induced flow is provided and the flow arises "naturally" from the effect of a density difference, resulting from a temperature or concentration difference, in a body force field such as the gravitational field, then the process is termed *natural convection*. The density difference gives rise to buoyancy forces due to which the flow is generated. A heated body cooling in ambient air generates such a flow in the region surrounding it. The buoyant flow arising from heat rejection to the atmosphere, heating of rooms, fires, and many other such heat transfer processes, both natural and artificial, are other examples of natural convection. Several recent books and reviews may be consulted for detailed presentations of this subject. See, for instance, the books by Turner [1], Jaluria [2], Kakaç et al. [3] and Gebhart et al. [4].

The main difference between natural and forced convection lies in the nature of the fluid flow generation. In forced convection, the externally imposed flow is generally known, whereas in natural convection it results from an interaction of the density difference with the gravitational (or some other body force) field, and is therefore invariably linked with and dependent on the temperature and concentration fields. Thus the motion that arises is not known at the onset and has to be determined from a consideration of the heat and mass transfer processes coupled with fluid flow mechanisms. Also, in practice, velocities in natural convection are usually much smaller than those in forced convection.

The above differences between natural and forced convection make the analysis of, as well as experimentation on, processes involving natural convection much more complicated than those involving forced convection. Special techniques and methods have therefore been devised to study the former process with a view to providing information on the flow and on the heat and mass transfer rates.

In order to understand the physical nature of natural convection transport, let us consider natural convection heat transfer from a heated vertical surface placed in an extensive quiescent medium at a uniform temperature, as shown in Fig. 12.1. If the plate surface temperature T_w is greater than the ambient temperature T_∞, the fluid adjacent to the vertical surface gets heated, becomes lighter (assuming it expands on heating), and rises. Fluid from the neighboring areas rushes in to take the place of this rising fluid. If the vertical surface is initially at temperature T_∞ and then, at a given instant, heat is turned on, say through an electric current, in order to heat it to a temperature T_w, the flow undergoes a transient before the flow shown is achieved. It is the analysis and study of this steady flow that yields the desired information on heat transfer rates, flow rates, temperature field, etc. The flow for a cooled surface is downward, as shown in Fig. 12.1*b*.

The heat transfer from the vertical surface may be expressed in terms of the usual relationship between the heat transfer rate q and the temperature difference between the surface and the ambient, given as

$$q = h_m A(T_w - T_\infty), \tag{12.1}$$

where h_m is termed the convective heat transfer coefficient and A is the total area of the vertical surface. The coefficient h_m depends on the flow configuration, fluid properties, dimensions of the heated body, and generally also on the temperature

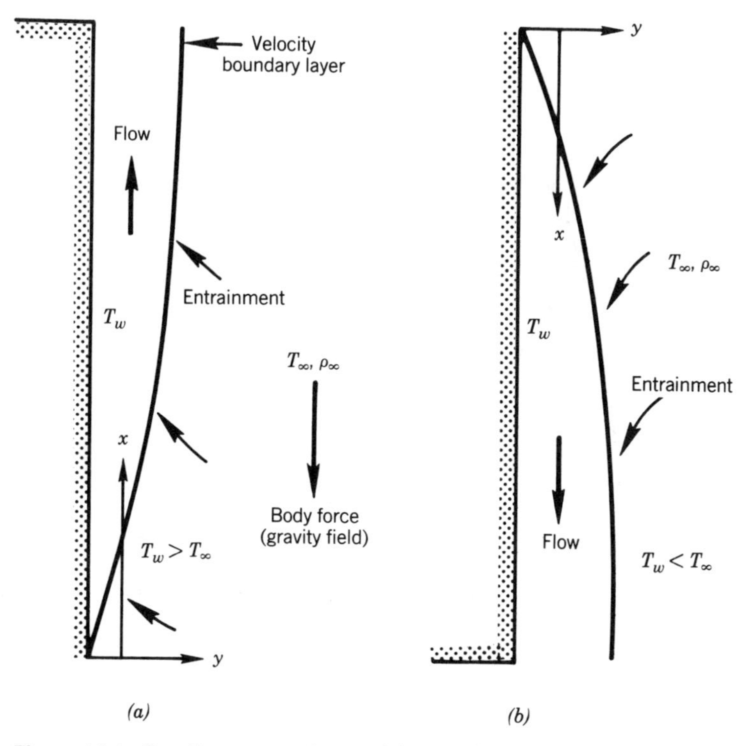

Figure 12.1. Coordinate system for natural-convection flow over a vertical surface.

difference, because of which the dependence of q on $T_w - T_\infty$ is not linear. Since the fluid motion becomes zero at the surface due to the no-slip condition, the heat transfer from the heated surface to the fluid in its immediate vicinity is by conduction. It is therefore given by Fourier's law as

$$q = -kA\left(\frac{\partial T}{\partial y}\right)_0 \tag{12.2}$$

where the gradient is evaluated at the surface, $y = 0$, in the fluid, k being the thermal conductivity of the fluid. From this equation, it is obvious that the natural convection flow largely affects the temperature gradient at the surface, since the remaining parameters would remain essentially unaltered. The purpose of an analysis is, therefore, largely to determine this gradient, which in turn depends on the flow, temperature field, and fluid properties.

The heat transfer coefficient h_m represents an integrated value for the heat transfer from the entire surface, since, in general, the local value h_x would vary with the vertical distance from the leading edge ($x = 0$) of the vertical surface. The local heat transfer coefficient h_x is defined by the equation

$$q_x'' = h_x(T_w - T_\infty) \tag{12.3}$$

Here q_x'' is the heat transfer per unit area per unit time at a location x, where the surface temperature difference is $T_w - T_\infty$, which may itself be a function of x. The

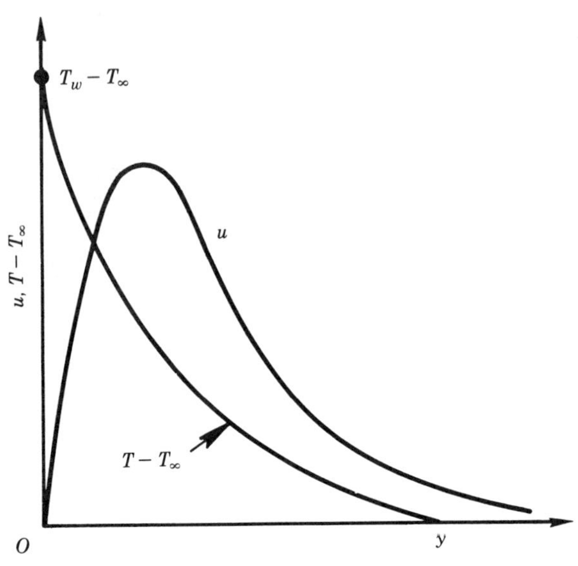

Figure 12.2. Sketch of the velocity and temperature distributions in natural-convection flow over a vertical surface.

average heat transfer coefficient h_m, defined in Eq. (12.1), is obtained from the above relationship through integration over the entire area. Both h_m and h_x are generally given in terms of the nondimensional parameter called the Nusselt number Nu. Again, an overall (or average) value, Nu_m, and a local value, Nu_x, may be defined as

$$\mathrm{Nu}_x = \frac{h_x x}{k}, \qquad \mathrm{Nu}_m = \frac{h_m L}{k} \tag{12.4}$$

where L is the height of the vertical surface and thus represents a characteristic dimension.

The fluid far from the vertical surface is stagnant, this being an extensive medium. The fluid next to the surface is also stationary due to the no-slip condition. The result is that flow exists in a layer adjacent to the surface, with zero velocity on either side, as shown in Fig. 12.2. The temperature varies from T_w to T_∞. Therefore, the maximum velocity occurs at some distance away from the vertical surface. Its exact location and magnitude has to be determined through analysis or experimentation. The flow near the bottom or the leading edge ($x = 0$) is laminar, being a well-ordered and well-layered flow. However, as the flow proceeds vertically upward or downward, the flow gets more and more disorderly and disturbed, eventually becoming completely disorderly and random, a condition termed turbulent flow. The flow region between the laminar and the turbulent flow is termed the transition region; its location and spread depend on several variables, such as the temperature of the surface and the fluid, and the nature and magnitude of external disturbances in the vicinity of the flow. Most of the processes encountered in nature are generally turbulent, though in industry, flows are often in the laminar or transition regime. A determination of the regime of flow and its consequent effect on the flow parameters and heat transfer rates is therefore important.

Natural convection flow may also arise in enclosed areas; see Chapter 13. This flow, generally termed internal natural convection, is very different in many ways from the external convection considered above for a vertical heated surface; the latter surface was considered to be immersed in an extensive, quiescent, isothermal medium. In this chapter, we shall discuss only external natural convection, internal flows being considered in the next chapter. Let us now proceed to the governing equations in natural convection.

12.2 BASIC MECHANISMS AND GOVERNING EQUATIONS

12.2.1 Governing Equations

The governing equations for a convective heat transfer process are obtained by considerations of mass and energy conservation and of the balance between the rate of momentum change and applied forces. These may be written, for constant viscosity μ and zero bulk viscosity, as

$$\frac{D\rho}{Dt} = \frac{\partial \rho}{\partial t} + \mathbf{V} \cdot \nabla \rho = -\rho \nabla \cdot \mathbf{V} \tag{12.5}$$

$$\rho \frac{D\mathbf{V}}{Dt} = \rho\left(\frac{\partial \mathbf{V}}{\partial t} + \mathbf{V} \cdot \nabla \mathbf{V}\right) = \mathbf{F} - \nabla p + \mu \nabla^2 \mathbf{V} + \frac{\mu}{3} \nabla(\nabla \cdot \mathbf{V}) \tag{12.6}$$

$$\rho c_p \frac{DT}{Dt} = \rho\left(\frac{\partial T}{\partial t} + \mathbf{V} \cdot \nabla T\right) = \nabla \cdot (k \nabla T) + q''' + \beta T \frac{Dp}{Dt} + \mu \Phi_v \tag{12.7}$$

where $\mathbf{V}$ is the velocity vector, T the local temperature, $\mathbf{F}$ the body force per unit volume, p the static pressure, t the time, ρ the fluid density, c_p the specific heat at constant pressure, β the coefficient of thermal expansion of the fluid, Φ_v the viscous dissipation (which is the irreversible part of the energy transfer due to viscous forces), and q''' the energy generation per unit volume. The total, or particle, derivative D/Dt may be expressed in terms of local derivative as $(\partial/\partial t + \mathbf{V} \cdot \nabla)$.

In natural convection flows, the basic driving force arises from the temperature field. The temperature variation causes a difference in density, which then results in a buoyancy force due to the presence of the body force field. For a gravitational field, the body force $\mathbf{F} = \rho \mathbf{g}$, where $\mathbf{g}$ is the gravitational acceleration. It is the variation of ρ that gives rise to the flow. The temperature field is linked with the flow, and all the above equations are coupled through the variation of the density ρ. Therefore, these equations have to be solved simultaneously to give the distributions, in space and time, of the velocity, pressure, and temperature fields. Due to this added complexity in the analysis of the flow, several simplifying assumptions and approximations are generally made in natural convection.

In the momentum equation, the local static pressure p may be broken down into two terms: one, p_a, due to the hydrostatic pressure, and the other, p_d, due to the motion of the fluid. The former, coupled with the body force acting on the fluid, constitutes the driving mechanism for the flow. Thus, $p = p_a + p_d$, and if ρ_∞ is the density in the ambient medium, we write

$$\begin{aligned} \mathbf{F} - \nabla p &= (\rho \mathbf{g} - \nabla p_a) - \nabla p_d \\ &= (\rho \mathbf{g} - \rho_\infty \mathbf{g}) - \nabla p_d = (\rho - \rho_\infty)\mathbf{g} - \nabla p_d \end{aligned}$$

If $\mathbf{g}$ is downward and the x direction upward ($\mathbf{g} = -\mathbf{i}g$), as is generally the case for vertical buoyant flows, then

$$\mathbf{F} - \nabla p = (\rho_\infty - \rho) g \mathbf{i} - \nabla p_d,$$

where $\mathbf{i}$ is the unit vector in the x direction and g is the magnitude of the gravitational acceleration. Therefore, the resulting governing equations for natural convection are the continuity equation (12.5), the energy equation (12.7), and the momentum equation, which becomes

$$\rho \frac{D\mathbf{V}}{Dt} = (\rho_\infty - \rho)\mathbf{g} - \nabla p_d + \mu \nabla^2 \mathbf{V} + \frac{\mu}{3} \nabla(\nabla \cdot \mathbf{V}) \tag{12.8}$$

12.2.2 Approximations in Natural Convection

The governing equations for natural convection flow are coupled elliptic partial differential equations, and are therefore of considerable complexity. The major problems in obtaining a solution to these equations lie in the inevitable variation of the density ρ with temperature or concentration, and in their partial elliptic nature. Several approximations are generally made to considerably simplify these equations. Two of the most important among these are the Boussinesq and the boundary-layer approximations.

The Boussinesq approximations involve two aspects. First, the density variation in the continuity equation is neglected. Thus, the continuity equation (12.5) becomes $\nabla \cdot \mathbf{V} = 0$. Second, the density difference, which causes the flow, is approximated as a pure temperature effect, i.e., the effect of pressure on the density is neglected. In fact, the density difference is estimated as

$$\rho_\infty - \rho = \rho\beta(T - T_\infty) \tag{12.9}$$

where T_∞ is the ambient temperature. These approximations are very extensively employed for a very wide range of problems in natural convection. An important condition for the validity of these approximations is that $\beta(T - T_\infty) \ll 1$ [2]. Therefore, the approximations are valid for small temperature differences. However, they are not valid near the density extremum of water, where a linear dependence of ρ on T may not be assumed [5].

Another approximation made in the governing equations pertains to the extensively employed boundary-layer assumption. The basic concepts involved in employing the boundary-layer approximation in natural convection flows are very similar to those in forced flow. The main difference lies in the fact that the pressure in the region beyond the boundary layer is hydrostatic, instead of being imposed by an external flow, and that the velocity outside the layer is zero. However, the basic treatment and analysis remain the same. It is assumed that the flow and the energy, or mass, transfer, from which it arises, are predominantly restricted to a thin region close to the surface. Beyond this region, the fluid is stationary. Several experimental studies have corroborated this assumption. As a consequence, the gradients along the surface are assumed much smaller than those normal to it.

The main consequences of the boundary-layer approximations are that the axial diffusion terms in the momentum and energy equations are neglected in comparison with the transverse diffusion terms. The transverse momentum balance is neglected, since it is found to be of negligible importance compared to the axial balance. Also, the velocity and thermal boundary layer thickness, δ and δ_T respectively, are found to be of orders of magnitude given by

$$\delta \sim \frac{L}{\mathrm{Gr}^{1/4}} \tag{12.10}$$

$$\frac{\delta_T}{\delta} \sim \frac{1}{\mathrm{Pr}^{1/2}} \tag{12.11}$$

where Gr is the Grashof number based on a characteristic length L, and Pr is the Prandtl number. These are defined as

$$\mathrm{Gr} = \frac{g\beta L^3(T_w - T_\infty)}{\nu^2}, \qquad \mathrm{Pr} = \frac{\mu c_p}{k} = \frac{\nu}{\alpha} \tag{12.12}$$

where ν is the kinematic viscosity and α the thermal diffusivity of the fluid.

The resulting boundary-layer equations for a two-dimensional variable-fluid-property flow are obtained as

$$\frac{\partial u}{\partial x} + \frac{\partial v}{\partial y} = 0 \tag{12.13}$$

$$u\frac{\partial u}{\partial x} + v\frac{\partial u}{\partial y} = g\beta(T - T_\infty) + \frac{1}{\rho}\frac{\partial}{\partial y}\left(\mu\frac{\partial u}{\partial y}\right) \tag{12.14}$$

$$\rho c_p\left(u\frac{\partial T}{\partial x} + v\frac{\partial T}{\partial y}\right) = \frac{\partial}{\partial y}\left(k\frac{\partial T}{\partial y}\right) + q''' + \beta Tu\frac{\partial p_a}{\partial x} + \mu\left(\frac{\partial u}{\partial y}\right)^2 \tag{12.15}$$

where the last two terms in the energy equation are the dominant terms from pressure work and viscous dissipation effects. Here, u and v are the velocity components in the x and y directions, respectively. Though these equations are written for a vertical, two-dimensional flow, similar approximations can be carried out for several other flow circumstances, such as axisymmetric flow over a vertical cylinder and the wake above a concentrated heat source.

There are several other approximations that are commonly employed in the analysis of natural convection flows. The fluid properties—except the density, for which the Boussinesq approximations are generally employed—are often taken as constant. The viscous dissipation and pressure work terms are generally small and are neglected. However, the importance of various terms can be best considered by nondimensionalizing the governing equations and boundary conditions, as outlined next.

12.2.3 Dimensionless Parameters

In an attempt to characterize the natural convection transport processes, a study of the basic nondimensional parameters must be carried out. These parameters are of considerable importance not only in simplifying the governing equations, but also in guiding the experiments that have to be carried out to obtain the relevant data for the process, and in the presentation of the data for subsequent design of equipment.

In natural convection, there is no free-stream velocity, and a convection velocity V_c is employed for the nondimensionalization of the velocity $\mathbf{V}$, where V_c is given by

$$V_c = [g\beta L(T_w - T_\infty)]^{1/2} \tag{12.16}$$

The governing equations may be nondimensionalized by employing the following dimensionless variables (indicated by primes in this subsection only), together with the volumetric heating rate q''':

$$\mathbf{V}' = \frac{\mathbf{V}}{V_c}, \qquad p' = \frac{p}{\rho V_c^2}, \qquad \theta' = \frac{T - T_\infty}{T_w - T_\infty}$$

$$\Phi_v' = \Phi_v \frac{L^2}{V_c^2}, \qquad t' = \frac{dt}{t_c}, \qquad \nabla' = L\nabla, \qquad (\nabla')^2 = L^2\nabla^2 \tag{12.17}$$

where t_c is a characteristic time scale. If the governing equations are nondimensionalized with the above transformations, one obtains

$$\nabla' \cdot \mathbf{V}' = 0 \tag{12.18}$$

$$\mathrm{Sr}\frac{\partial \mathbf{V}'}{\partial t'} + \mathbf{V}' \cdot \nabla'\mathbf{V}' = -\mathbf{e}\theta' - \nabla' p_d' + \frac{1}{\sqrt{\mathrm{Gr}}}\nabla^2\mathbf{V}' \tag{12.19}$$

$$\mathrm{Sr}\frac{\partial \theta'}{\partial t'} + \mathbf{V}' \cdot \nabla'\theta' = \frac{1}{\mathrm{Pr}\sqrt{\mathrm{Gr}}}\nabla'^2\theta' + q''' + \beta T\left[\frac{g\beta L}{c_p}\mathrm{Sr}\frac{\partial p'}{\partial t'} + \frac{g\beta L}{c_p}\mathbf{V}' \cdot \nabla' p'\right] + \frac{g\beta L}{c_p}\frac{1}{\sqrt{\mathrm{Gr}}}\Phi_v' \tag{12.20}$$

where $\boldsymbol{e}$ is the unit vector in the direction of the gravitational force.

Here, $\mathrm{Sr} = L/(V_c t_c)$ is the Strouhal number, and q''' is nondimensionalized with $\rho c_p(T_w - T_\infty)V_c/L$. It is clear from the above equations that $\sqrt{\mathrm{Gr}}$ replaces Re, which arises in forced convection. Similarly, the Eckert number is replaced by $g\beta L/c_p$, which now determines the importance of the pressure and viscous-dissipation terms. The Grashof number indicates the relative importance of the buoyancy term as compared to the viscous term. A large value of Gr, therefore, indicates small viscous effects in the momentum equation, similar to the physical significance of Re in forced flow. The Prandtl number Pr gives a comparison between momentum and thermal diffusion. Thus, the Nusselt number may be expressed as a function of Gr and Pr for steady flows, if pressure work and viscous dissipation are neglected.

12.3 LAMINAR NATURAL CONVECTION FLOW OVER FLAT SURFACES

12.3.1 Vertical Surfaces

The classical problem of natural-convection heat transfer from an isothermal heated vertical surface, shown in Fig. 12.1, with the flow assumed to be steady and laminar and the fluid properties (except density) taken as constant, has been of interest to investigators for a very long time. Viscous-dissipation effects are neglected, and no heat source is considered within the flow. The problem is therefore considerably simplified, though the complexities due to the coupled partial differential equations remain. The governing differential equations may be obtained from Eqs. (12.13)–(12.15).

An important method for finding the boundary-layer flow over a heated vertical plate is the similarity variable method. A stream function $\psi(x, y)$ is first defined so that it satisfies the continuity equation. Thus,

$$u = \frac{\partial \psi}{\partial y}, \qquad v = -\frac{\partial \psi}{\partial x} \tag{12.21}$$

Then, the similarity variable η, and dimensionless stream function f, and the temperature θ are defined as follows so as to convert the governing partial differential equations into ordinary differential equations. For flow over a vertical isothermal surface, these are

$$\eta = \frac{y}{x}\left(\frac{\mathrm{Gr}_x}{4}\right)^{1/4}, \qquad \psi = 4\nu f(\eta)\left(\frac{\mathrm{Gr}_x}{4}\right)^{1/4}, \qquad \theta = \frac{T - T_\infty}{T_w - T_\infty} \tag{12.22}$$

where

$$\mathrm{Gr}_x = \frac{g\beta(T_w - T_\infty)x^3}{\nu^2} \tag{12.22a}$$

The boundary conditions are

$$\text{at } y = 0: \quad u = v = 0, \ T = T_w; \qquad \text{as } y \to \infty; \quad u \to 0, \ T \to T_\infty$$

These may also be written in terms of the similarity variables.

The governing equations are obtained from the above similarity transformation as

$$f''' + 3ff'' - 2(f')^2 + \theta = 0 \tag{12.23}$$

$$\frac{\theta''}{\mathrm{Pr}} + 3f\theta' = 0 \tag{12.24}$$

with boundary conditions

$$\text{at } \eta = 0: \quad f = f' = 1 - \theta = 0; \qquad \text{as } \eta \to \infty: \quad f' \to 0, \theta \to 0$$

or

$$f(0) = f'(0) = 1 - \theta(0) = f'(\infty) = \theta(\infty) = 0 \tag{12.25}$$

where the primes indicate differentiation with respect to the similarity variable η.

These equations have been considered by several investigators. Schuh [6] gave solutions for various values of the Prandtl number, employing approximate methods. Ostrach [7] numerically obtained the solution for the Pr range of 0.01 to 1000. The velocity and temperature profiles thus obtained are shown in Figs. 12.3 and 12.4. An increase in Pr is found to cause a decrease in thermal boundary-layer thickness and an increase in the absolute value of the temperature gradient at the surface. The dimensionless maximum velocity is also found to decrease and the velocity gradient at the surface to decrease with increasing Pr, indicating the effect of greater viscous forces. The location of this maximum value is found to shift to higher η as Pr is decreased.

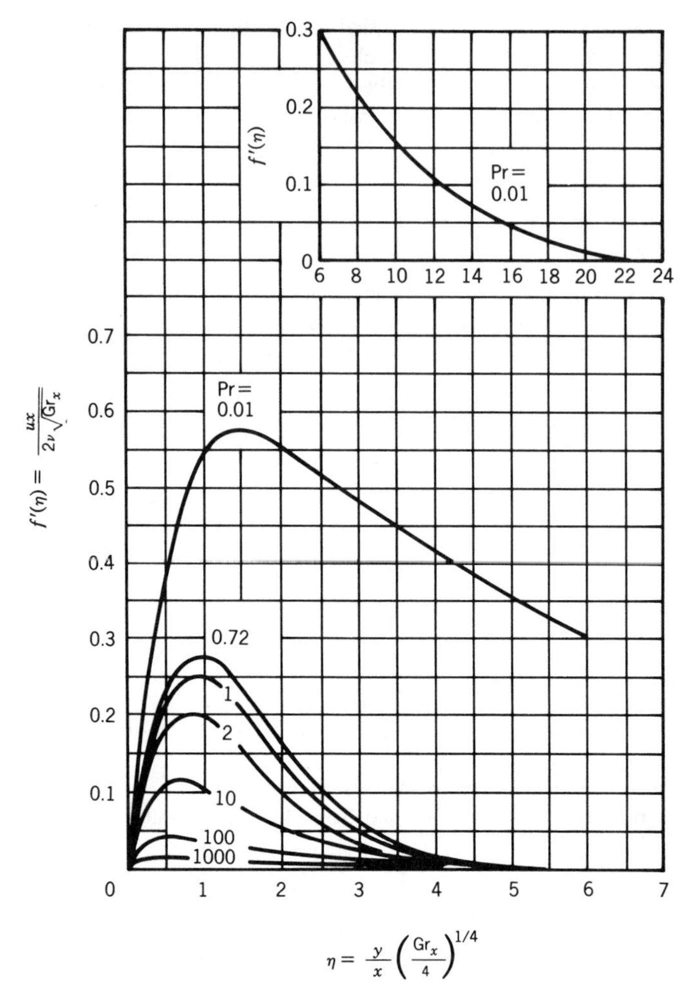

Figure 12.3. Velocity variation across the boundary layer for flow over an isothermal vertical surface [7].

The velocity boundary-layer thickness is also found to increase as Pr is decreased to low values.

Now, the heat transfer from a heated surface may be obtained as

$$q_x'' = -k\left(\frac{\partial T}{\partial y}\right)_0 = -k(T_w - T_\infty)\frac{1}{x}\left(\frac{\mathrm{Gr}_x}{4}\right)^{1/4}\left(\frac{d\theta}{d\eta}\right)_0$$

$$= [-\theta'(0)]\frac{k(T_w - T_\infty)}{x}\left(\frac{\mathrm{Gr}_x}{4}\right)^{1/4}$$

Since

$$\mathrm{Nu}_x = \frac{h_x x}{k} = \frac{q_x''}{T_w - T_\infty}\frac{x}{k}$$

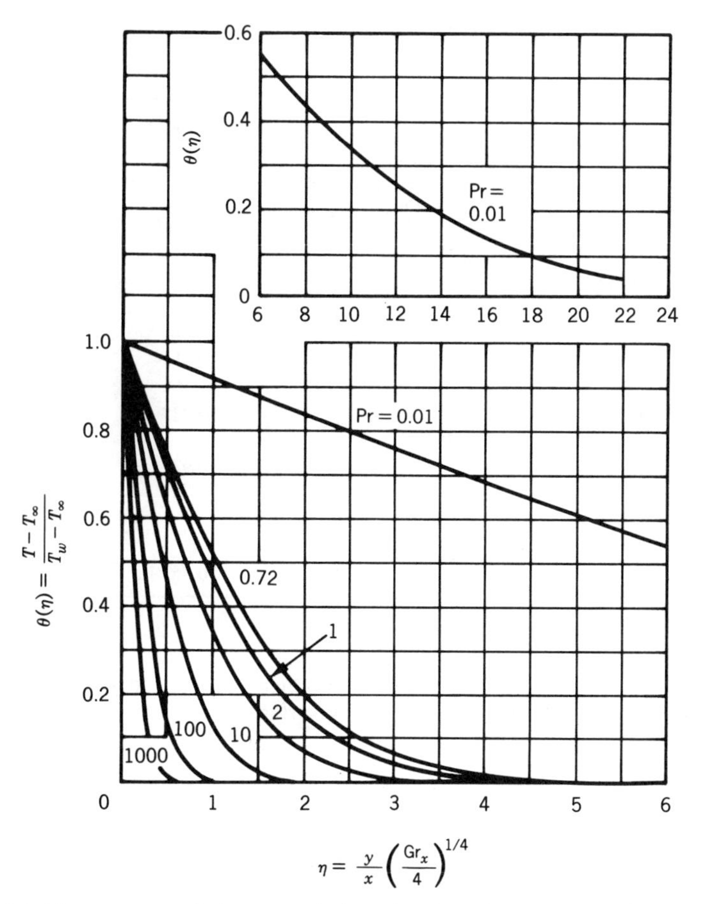

Figure 12.4. Temperature variation across the boundary layer for flow over an isothermal vertical surface [7].

we have for an isothermal surface (denoted by subscript T)

$$\mathrm{Nu}_{x,T} = [-\theta'(0)]\left(\frac{\mathrm{Gr}_x}{4}\right)^{1/4}$$

$$= \frac{-\theta'(0)}{\sqrt{2}}\mathrm{Gr}_x^{1/4} = \phi(\mathrm{Pr})\mathrm{Gr}_x^{1/4}$$

where $\phi(\mathrm{Pr}) = [-\theta'(0)]/\sqrt{2}$. Therefore, the local surface heat transfer coefficient $h(x)$ varies as

$$h_{x,T} = Bx^{-1/4}, \qquad \text{where} \quad B = \frac{k[-\theta'(0)]}{\sqrt{2}}\left(\frac{g\beta(T_w - T_\infty)}{\nu^2}\right)^{1/2}$$

TABLE 12.1 Computed Values of the Parameter $\phi(\text{Pr})$ for a Vertical Heated Surface [9]

Pr	$\phi(\text{Pr})$ (Isothermal)	$\phi(\text{Pr}, \frac{1}{5})$ (Uniform Heat Flux) $n = 1/5$
0	$0.600\ \text{Pr}^{1/2}$	$0.711\ \text{Pr}^{1/2}$
0.01	0.0570	0.0669
0.72	0.357	
0.733		0.410
1.0	0.401	
2.0	0.507	
2.5		0.616
5.0	0.675	
6.7		0.829
7.0	0.754	
10	0.826	0.931
10^2	1.55	1.74
10^3	2.80	
10^4	5.01	
∞	$0.503\ \text{Pr}^{1/4}$	$0.563\ \text{Pr}^{1/4}$

The average value of the heat transfer coefficient, $h_{m,T}$, may be obtained by averaging over the entire length of the vertical surface:

$$h_{m,T} = \frac{1}{L}\int_0^L h_{x,T}\, dx = \frac{B}{L}\cdot\frac{4}{3}\cdot L^{3/4} = \tfrac{4}{3}h_L$$

and

$$\text{Nu}_{m,T} = \frac{4}{3}\cdot\frac{-\theta'(0)}{\sqrt{2}}\cdot \text{Gr}^{1/4} = \tfrac{4}{3}\phi(\text{Pr})\text{Gr}^{1/4} \tag{12.26}$$

The values of $\phi(\text{Pr})$ can be obtained from a numerical solution of the governing differential equations. Values obtained at various Pr are listed in Table 12.1.

In several problems of practical interest, the surface from which heat transfer occurs is nonisothermal. The two families of surface temperature variation which give rise to similarity in the governing laminar boundary-layer equations have been shown by Sparrow and Gregg [8] to be the power-law and exponential distributions:

$$T_w - T_\infty = Nx^n \quad \text{and} \quad T_w - T_\infty = Me^{mx} \tag{12.27}$$

where N, M, n and m are constants. The power-law distribution is of particular interest, since it represents many practical circumstances. The isothermal surface is obtained for $n = 0$. A uniform heat flux condition, $q_x'' =$ constant, arises for $n = \frac{1}{5}$. A thermal plume due to a line heat source at $x = 0$ is obtained for $n = -\frac{3}{5}$. It can also be shown that physically realistic solutions are obtained for $1 > n \geq -\frac{3}{5}$ [2, 8].

The local Nusselt number Nu_x is obtained as

$$\frac{\text{Nu}_x}{\text{Gr}_x^{1/4}} = \frac{-\theta'(0)}{\sqrt{2}} = \phi(\text{Pr}, n) \tag{12.28}$$

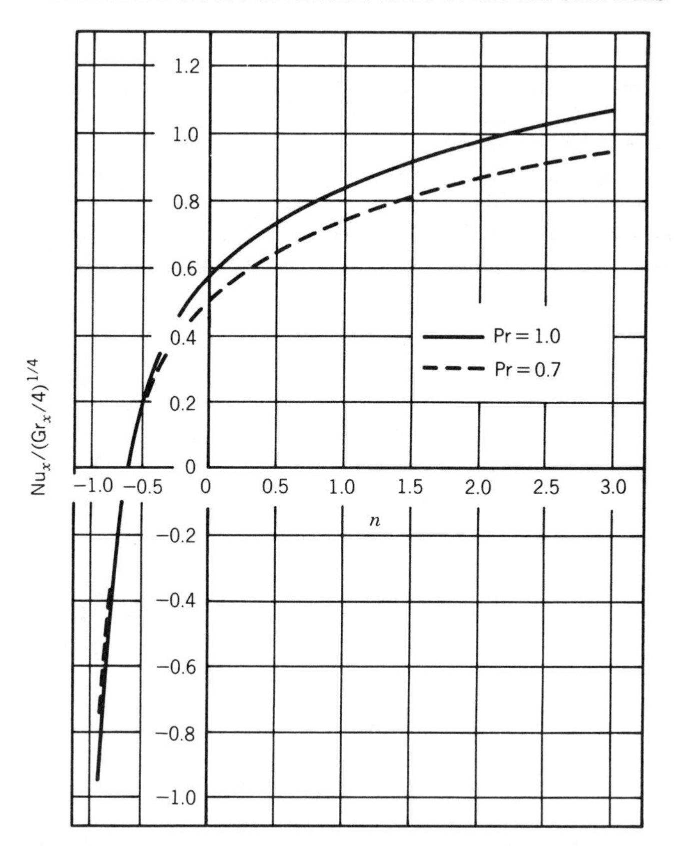

Figure 12.5. Dependence of the local Nusselt number on the value of n for a power-law surface temperature distribution [8].

The function $\mathrm{Nu}_x/\mathrm{Gr}_x^{1/4}$ is plotted against n in Fig. 12.5. For $n < -\frac{3}{5}$, the function is found to be negative, indicating the physically unrealistic circumstance of heat transfer to the surface for $T_w > T_\infty$. The surface is adiabatic for $n = -\frac{3}{5}$.

For the case of uniform heat flux, $n = \frac{1}{5}$ and $q''_x = q''$, a constant. Therefore,

$$q'' = k[-\theta'(0)]N\left(\frac{g\beta N}{4\nu^2}\right)^{1/4}$$

which gives

$$N = \left(\frac{q''}{k[-\theta'(0)]}\right)^{4/5}\left(\frac{4\nu^2}{g\beta}\right)^{1/5} \tag{12.29}$$

For a given heat flux q'' at a vertical surface, which may be known, for example from the electrical input into the surface, the temperature of the surface varies as $x^{1/5}$, and its value may be determined as a function of the heat flux and fluid properties from the above relationship. The parameter $-\theta'(0)$ is obtained from a numerical solution of the governing equations for $n = 0.2$, at the given value of Pr. Some results obtained from Gebhart [9] are shown in Table 12.1.

12.3.2 Inclined and Horizontal Surfaces

In many interesting and important cases of natural convection, the flow is generated as a consequence of the thermal input from a surface which is itself curved or inclined with respect to the direction of the gravity field. Consider, first, a flat surface at a small inclination γ from the vertical. Boundary-layer approximations, similar to those for a vertical surface, may be made for this flow. It can be shown that if x is taken along the surface and y normal to it, the continuity and energy equations (12.13) and (12.15) remain unchanged and the x-direction momentum equation becomes

$$u\frac{\partial u}{\partial x} + v\frac{\partial u}{\partial y} = g\beta(T - T_\infty)\cos\gamma + \frac{1}{\rho}\frac{\partial}{\partial y}\left(\mu\frac{\partial u}{\partial y}\right) \tag{12.30}$$

The problem is identical to that for flow over a vertical surface except that g is replaced by $g\cos\gamma$. Therefore, a replacement of g by $g\cos\gamma$ in all the relationships derived earlier would give the results for an inclined surface. This implies using $\mathrm{Gr}_x\cos\gamma$ for Gr_x. However, this also assumes equal rates of heat transfer on the two sides of the surface.

The above procedure for obtaining the heat transfer rate from an inclined surface was first suggested theoretically by Rich [10], and his data are in general agreement with the anticipated values. The data obtained by Vliet [11] for a uniform-flux, heated surface in air and in water indicate the validity of the above procedure up to inclination angles as large as 60°. Therefore, the replacement of g by $g\cos\gamma$ in the Grashof number is appropriate for inclination angles up to around 45° and, to a close approximation, up to a maximum angle of 60°. Detailed experimental results on this were obtained by Fujii and Imura [12]. They also discuss the separation of the boundary layer for the inclined surface facing upward.

Natural convection over horizontal surfaces is a problem of considerable importance and interest in technology and in nature. Rotem and Claassen [13] found solutions for the boundary-layer equations, for flow over a semi-infinite isothermal horizontal surface. Various values of Pr, including the extreme cases, were treated. Experimental results indicated the existence of a boundary layer near the leading edge on the upper side of a heated horizontal surface. Equations were presented for the power-law case, $T_w - T_\infty = Nx^n$, and solved for $n = 0$. Pera and Gebhart [14] have considered flow over surfaces slightly inclined from the horizontal.

For a semi-infinite horizontal surface with a single leading edge, as shown in Fig. 12.6, the governing equations are the continuity and energy equations, as given in Eqs. (12.13) and (12.15), and the momentum equations of the form given below:

$$u\frac{\partial u}{\partial x} + v\frac{\partial u}{\partial y} = \frac{1}{\rho}\frac{\partial}{\partial y}\left(\mu\frac{\partial u}{\partial y}\right) - \frac{1}{\rho}\frac{\partial p_d}{\partial x} \tag{12.31a}$$

$$g\beta(T - T_\infty) = \frac{1}{\rho}\frac{\partial p_d}{\partial y} \tag{12.31b}$$

Therefore, the dynamic or motion pressure p_d drives the flow. Physically, the upper side of a heated surface heats up the fluid adjacent to it, which being lighter tends to rise. This results in a negative pressure gradient, which causes a boundary-layer flow over the surface. Similar considerations apply for the lower side of a cooled surface.

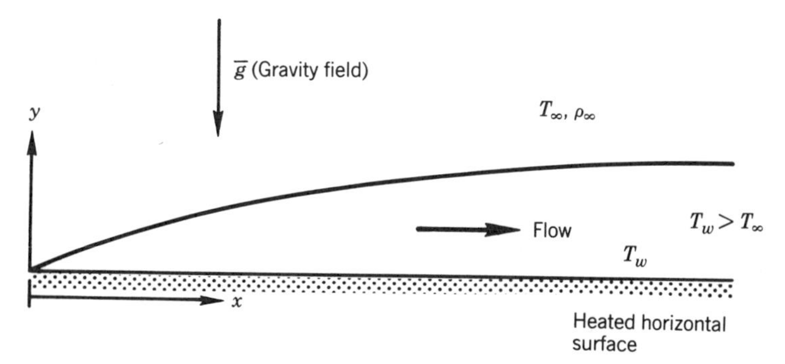

Figure 12.6. Coordinate system for the natural-convection boundary-layer flow over a semi-infinite horizontal surface.

This problem may be solved by similarity analysis, discussed above for vertical surfaces. The similarity variables, given by Pera and Gebhart [14], are

$$\eta = \frac{y}{x}\left(\frac{\mathrm{Gr}_x}{5}\right)^{1/5}, \qquad \psi = 5\nu f(\eta)\left(\frac{\mathrm{Gr}_x}{5}\right)^{1/5} \tag{12.32}$$

Figure 12.7 shows the computed velocity and temperature profiles for flow over a heated horizontal surface facing upward.

The local Nusselt number for horizontal surfaces is given by Pera and Gebhart [14] for both the uniform-temperature and the uniform-heat-flux surface conditions. The Nusselt number was found to be approximately proportional to $\mathrm{Pr}^{1/4}$ over the range

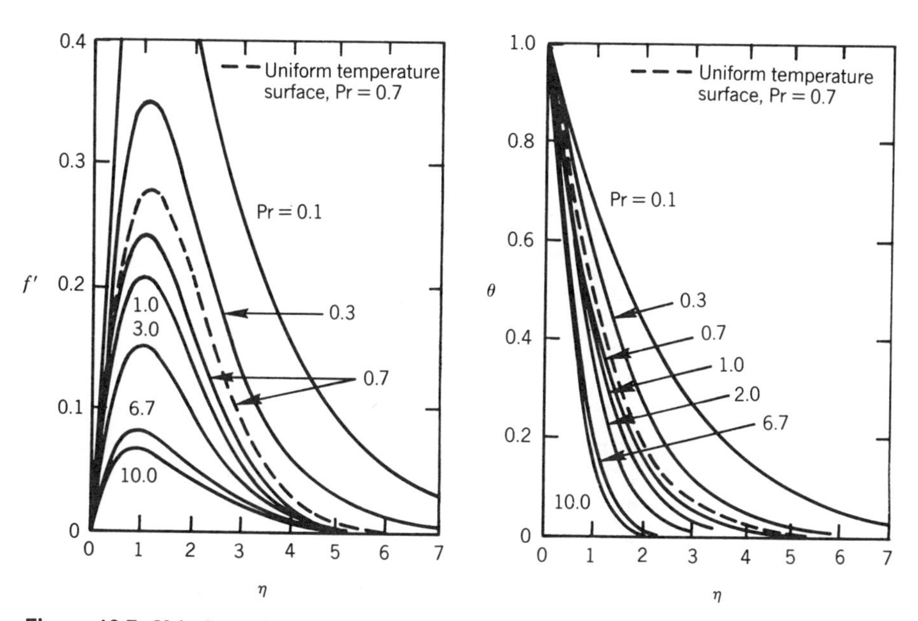

Figure 12.7. Velocity and temperature profiles for natural convection over a horizontal surface, with a uniform heat flux [14].

0.1 to 100, and the expressions obtained are

$$\mathrm{Nu}_{x,T} = \frac{h_x x}{x} = 0.394\,\mathrm{Gr}_x^{1/5}\,\mathrm{Pr}^{1/4} \tag{12.33a}$$

for a uniform-temperature surface and

$$\mathrm{Nu}_{x,H} = 0.5013\,\mathrm{Gr}_x^{1/5}\,\mathrm{Pr}^{1/4} \tag{12.33b}$$

for a uniform-flux surface. For the isothermal surface, the average Nusselt number would be $\frac{5}{3}$ times the value of the local Nusselt number at $x = L$.

In conclusion, the problem of natural convection from inclined surfaces can be treated in terms of small inclinations from the vertical and horizontal positions, detailed results on which are available. For intermediate values of γ, an interpolation between the above two regimes may be carried out to determine the resulting heat transfer. This regime has not received much attention, though some experimental results are available such as those of Fujii and Imura [12].

12.4 LAMINAR NATURAL CONVECTION FLOW IN OTHER CONFIGURATIONS

12.4.1 Horizontal Cylinder and Sphere

Much of the information on natural convection over heated surfaces, discussed earlier, has been obtained through similarity methods. However, neither the horizontal cylindrical nor the spherical configuration gives similarity, and for these cases several other methods have been employed for obtaining a solution to the governing equations. Among the earliest detailed considerations of these flows was that by Merk and Prins [15], who employed integral methods, taking the velocity and thermal boundary-layer thicknesses to be equal and denoted by δ. The variation of the local Nusselt number with ϕ, the angular position from the lower stagnation point $\phi = 0°$, is shown in Fig. 12.8 for a horizontal cylinder and for a sphere. The peripheral local Nusselt number Nu_ϕ decreases downstream due to the increase in the boundary-layer thickness, which is theoretically predicted to be infinite at $\phi = 180°$, resulting in a zero value for Nu_ϕ. Merk and Prins have indicated the inapplicability of the analysis for $\phi \geq 165°$ due to boundary-layer separation and realignment into a plume flow.

The mean value of the Nusselt number, $\mathrm{Nu}_{m,T}$ is given by Merk and Prins [15] for a horizontal, isothermal cylinder as

$$\mathrm{Nu}_{m,T} = \frac{h_m D}{k} = C(\mathrm{Pr})(\mathrm{Gr}\,\mathrm{Pr})^{1/4} \tag{12.34}$$

where Gr is also based on the diameter D, and $C(\mathrm{Pr})$ was obtained as 0.436, 0.456, 0.520, 0.523, and 0.523 for Pr values of 0.7, 1.0, 10.0, 100.0, and ∞, respectively. The same expression is given for spheres, with $C(\mathrm{Pr})$ given for the above Pr values as 0.474, 0.497, 0.576, 0.592, and 0.595, respectively.

There are many other analytical and experimental studies of natural convection over spheres. Being of interest in chemical processes, this configuration has been studied in detail for mass transfer also. Chiang et al. [16] solved the governing equations, using a series method, and presented heat transfer results. Trends similar to those discussed

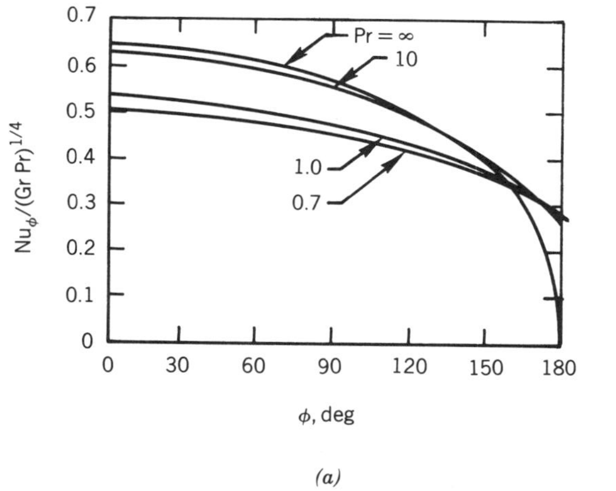

(a)

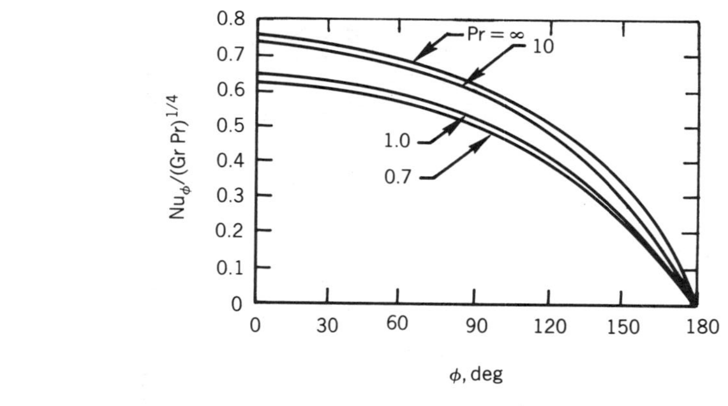

(b)

Figure 12.8. The variation of the local Nusselt number with downstream angular position ϕ for (a) a horizontal cylinder and (b) a sphere [15].

above were obtained. A considerable amount of experimental work has been done on the heat transfer from spheres. Amato and Tien [17] have discussed such studies and have given the heat transfer correlation as

$$\mathrm{Nu}_{m,T} = 2 + 0.5(\mathrm{Gr\,Pr})^{1/4} \tag{12.35}$$

12.4.2 Vertical Cylinder

Natural convection over vertical cylinders is also a very important problem, being relevant to many applications, such as flow over tubes (as in nuclear reactors), over cylindrical heating elements, and over various closed bodies (including the human body) that can be approximated as a vertical cylinder. For large values of D/L, where D is the diameter of the cylinder and L its length, the flow is close to that over a flat plate, since the boundary-layer thickness is small compared to the diameter of the cylinder. As a result, the governing equations become the same as those for a flat plate. However, since this comparison is really based on the boundary-layer thickness, which in turn depends on the Grashof number, the deviation of the results obtained for a vertical cylinder from those for a flat plate must be given in terms of D/L and the Grashof number. By studying this deviation, Sparrow and Gregg [18] obtained the following criterion for Pr values of 0.72 and 1.0 for a difference in heat transfer of less than 5% from the flat plate solution:

$$\frac{D}{L} \geq \frac{35}{\mathrm{Gr}^{1/4}} \tag{12.36}$$

where Gr is the Grashof number based on L.

When D/L is not large enough to ignore the effect of curvature, the relevant governing equations must be solved. Sparrow and Gregg [18] employed similarity methods for obtaining a solution to these equations. Minkowycz and Sparrow [19] obtained the solution using the local nonsimilarity method. LeFevre and Ede [20] employed an integral method to solve the governing equations and gave the following expression for the Nusselt number $\mathrm{Nu}_{m,T}$ based on the height L of the cylinder:

$$\mathrm{Nu}_{m,T} = \frac{h_m L}{k} = \frac{4}{3}\left[\frac{7\,\mathrm{Gr\,Pr}^2}{5(20 + 21\,\mathrm{Pr})}\right]^{1/4} + \frac{4(272 + 315\,\mathrm{Pr})\,L}{35(64 + 63\,\mathrm{Pr})\,D} \tag{12.37}$$

where Gr is also based on L.

12.4.3 Transients

We have so far considered steady natural-convection flows, in which the velocity and temperature fields do not vary with time. However, transient effects are important in many practical circumstances. The change in the thermal condition causing the natural-convection flow could be a sudden or a periodic one, leading to a variation in the flow. The startup and shutdown of systems, such as furnaces and nuclear reactors, involves considerations of transient natural convection, if an externally induced flow is not present.

If the heat input at a surface is suddenly changed from zero to a specific value, the steady natural-convection flow is eventually obtained, following a transient flow which occurs for a certain period of time. The moment the heat is turned on, the surface gets heated, this change being essentially a step variation if the thermal capacity of the body is very small. In response to this sudden change, the fluid adjacent to the surface gets heated and rises. However, it is initially unaffected by flow at other portions of the surface. This implies that the fluid element behaves essentially as an isolated one, and the heat transfer mechanisms are therefore largely unaffected by the fluid motion. Consequently, the initial transport mechanism is predominantly conduction and can be approximated as a one-dimensional conduction problem till the leading-edge effect, which propagates downstream along the flow, is felt at a given location x. The heat transfer rates due to pure conduction being much smaller than those due to convection, it is to be expected that, for a step change in the heat-flux input, there may initially be an overshoot in the temperature, above the steady-state value. Similarly, for a step change in temperature, a lower heat flux is expected initially, ultimately approaching the steady-state value, as the flow itself progresses through a transient regime to the steady flow.

At the initial stages of the transient, the solution for a step change in the surface temperature, or in the heat flux, is independent of the vertical location and is of the form obtained for semi-infinite conduction solutions. Employing Laplace transforms for a step change in the heat flux, the solution is obtained as

$$\tilde{\theta} = \frac{2q''\sqrt{\alpha t}}{k}\left[\frac{e^{-\eta^2}}{\sqrt{\pi}} - \eta \operatorname{erfc} \eta\right] \tag{12.38}$$

where $\eta = y/2\sqrt{\alpha t}$, α being the thermal diffusivitycient of thermal of the fluid: erfc η is the conjugate of the error function; and q'' is the constant heat-flux input imposed at time $t = 0$, starting from a no-flow condition. The temperature $\tilde{\theta}$ is simply the excess over the initial temperature T_∞. The heat transfer coefficient is obtained from the above as:

$$h = \frac{q''}{[\tilde{\theta}]_0} = \frac{k}{2}\sqrt{\frac{\pi}{\alpha t}} \tag{12.39}$$

Similarly, for a step change in the surface temperature, the solution is

$$\frac{T - T_\infty}{T_w - T_\infty} = \theta = \operatorname{erfc} \eta \tag{12.40}$$

The velocity profile is obtained by substituting the above temperature solution into the momentum equation and solving the resulting equation by Laplace transforms to obtain $u(y)$.

Numerical solutions of the governing boundary-layer equations have been obtained by Hellums and Churchill [21] for a vertical surface subjected to a step change in the surface temperature. The results converge to the steady-state solution at large time and show a minimum in the local Nusselt number during the transient; see Fig. 12.9. An integral method of analysis for transient natural convection has also been developed for a time-dependent heat input and for a finite thermal capacity of the surface element. This work has been summarized by Gebhart [9] and is based on the analytical and experimental work of Gebhart and coworkers, as referenced in the above paper. This analysis is particularly suited to practical problems, since it considers the element thermal capacity and determines the temperature variation with time over the entire transient regime.

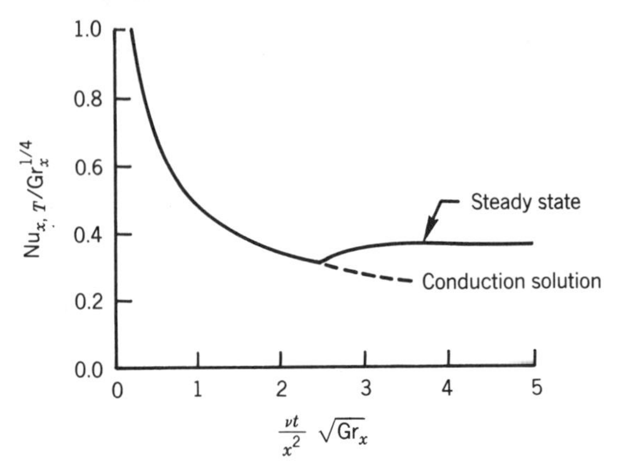

Figure 12.9. Transient variation of the heat transfer rate for a step change in the surface temperature of a vertical plate [21].

Churchill [22] has given a correlation for the transient natural convection from a heated vertical plate, subjected to a step change in the heat flux. The thermal capacity of the plate was taken as negligible, and the local Nusselt number is given as

$$[\mathrm{Nu}_{x,H}]^n = \left(\frac{\pi x^2}{4\alpha t}\right)^{n/2} + \left\{\frac{\mathrm{Ra}_x/10}{\left[1+(0.437/\mathrm{Pr})^{9/16}\right]^{16/9}}\right\}^{n/4},$$

where

$$\mathrm{Ra}_x = \frac{g\beta(T_w - T_\infty)x^3}{\nu\alpha} \tag{12.41}$$

Employing the available experimental information, the chosen value of n is 6, and with this value the above correlation was found to give Nusselt number values quite close to the measured ones. No temperature overshoot is considered, since the experimental studies of Gebhart [9] showed no significant overshoot. For a step change in surface temperature, Churchill and Usagi [23] have also obtained an empirical correlation approximating the entire time span.

12.5 TURBULENT FLOW

12.5.1 Transition from Laminar to Turbulent Flow

In natural convection, as in forced convection, one of the most important questions is whether the flow is laminar or turbulent, since the transport processes depend strongly on the flow regime. Near the leading edge, or end, of a body, the flow is well ordered and well layered. The fluctuations and disturbances, if any, are small in magnitude compared to the mean flow, and the processes can be defined in terms of the laminar governing equations and mechanisms, as discussed in the preceding sections. However, as the flow proceeds downstream from the leading edge, it undergoes transition to turbulent flow, which is characterized by random disturbances of large magnitude. The flow is then a combination of a mean and a fluctuating component, due to these large disturbances, which being random can be described by statistical methods. In several natural-convection flows of interest, the flow lies in the unstable or in the transition regime. It is therefore important to study these regimes and the basic processes underlying the transition to turbulence.

In a study of the transition of a laminar flow to turbulence, the conditions under which a disturbance in the flow amplifies as it proceeds downstream form a very important consideration. This refers to the stability of the flow, an unstable one leading to a growth in disturbances. These disturbances can enter the flow from various sources, such as building vibrations, fluctuations in heat input to the heated surface, vibrations in equipment, etc., and depending on the conditions (frequency, location, etc.), they may grow in amplitude due to a balance of buoyancy, pressure, and viscous forces. This form of instability is termed hydrodynamic stability and is of particular relevance to disturbance growth, leading to turbulence.

The disturbances gradually amplify to large enough magnitudes to cause distortion in the mean velocity and temperature profiles due to secondary mean flows. This leads to the formation of a shear layer, which fosters further amplification of the disturbances, and concentrated turbulent bursts results. These bursts then increase in magnitude, and the fraction of time they occur increases, eventually crowding out the remaining laminar flow and giving rise to a completely turbulent flow. The general mechanisms underlying transition are shown in Fig. 12.10 from the work of Jaluria and Gebhart [24].

12.5.2 Turbulence

Most natural-convection flows of interest, in nature and in technology, are turbulent. The velocity, pressure, and temperature at a given point do not remain constant with time, but vary irregularly at high frequency. There is a considerable amount of mixing, with fluid packets moving around irregularly, giving rise to the observed fluctuations in the velocity and temperature fields, rather than the well-ordered and well-layered characteristics of laminar flow. Due to the importance of turbulent natural-convection flows, a considerable amount of effort, experimental and analytical, has been directed at understanding and determining the transport mechanisms and the rates of energy transfer. The work done in forced flows has been even more extensive, and in fact much of our understanding of turbulent flows in natural convection is derived from this work. Transport mechanisms in turbulent flow are obviously very different from those in laminar flow, and some of the basic considerations are given below.

In describing a turbulent flow, the fluctuating or eddying motion is superimposed on a mean motion. The flow may therefore be described in terms of the time-averaged values of the velocity components (denoted as $\overline{u}$, $\overline{v}$, and $\overline{w}$) and the disturbance or fluctuating quantities (u', v', and w'). The instantaneous value of each of the velocity components is then given as

$$\begin{aligned} u &= \overline{u} + u' \\ v &= \overline{v} + v' \\ w &= \overline{w} + w' \end{aligned} \tag{12.42}$$

Similarly, pressure and temperature in the flow may be written as

$$\begin{aligned} p &= \overline{p} + p' \\ T &= \overline{T} + T' \end{aligned} \tag{12.43}$$

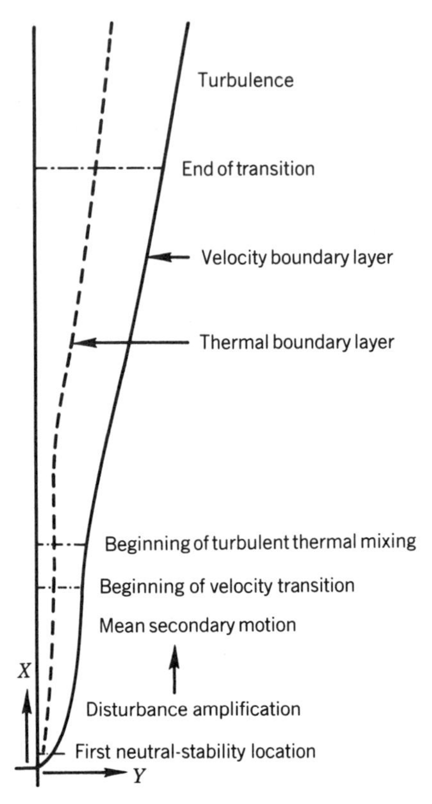

Figure 12.10. Growth of the boundary layer and the sequence of events during transition in water, Pr = 6.7 [24].

The time averages are found by integrating the local instantaneous value of the particular quantity at a given point over a sufficiently long time interval—long compared to the time period of the fluctuations. For steady turbulence, the time-averaged quantities do not vary with time, and by the definition of the averaging process, the time averages of the fluctuating quantities are zero. For unsteady turbulence, the time-averaged quantities themselves vary with time. Here we shall consider only the case of steady turbulence, so that the average quantities are independent of time and allow a representation of the flow and the transport processes in terms of time-independent variables.

If the above instantaneous quantities are inserted into the governing continuity, momentum, and energy equations and a time average taken, additional transport terms due to the turbulent eddies arise. An important concept employed for treating these additional transport components is that of eddy viscosity ϵ_M and diffusivity ϵ_H. Momentum and heat transfer processes consist of a molecular component and an eddy component. This may be expressed as

$$\frac{\tau}{\rho} = (\nu + \epsilon_M)\frac{d\bar{u}}{dy} \tag{12.44}$$

$$\frac{q_x''}{\rho c_p} = (\alpha + \epsilon_H)\frac{d\bar{T}}{dy} \tag{12.45}$$

where τ is the total shear stress and q_x'' the heat flux. For isotropic turbulence, ϵ_M and ϵ_H are independent of direction and are of the form $-\overline{u'v'}/(\partial\bar{u}/\partial y)$ and $\overline{v'T'}/(\partial\bar{T}/\partial y)$, respectively.

If the above relationships for τ and q'' are introduced into the governing equations for mean flow, obtained by time-averaging the equations written for the total instantaneous flow in the boundary-layer form, one obtains

$$\frac{\partial\bar{u}}{\partial x} + \frac{\partial\bar{v}}{\partial y} = 0 \tag{12.46}$$

$$\bar{u}\frac{\partial\bar{u}}{\partial x} + \bar{v}\frac{\partial\bar{u}}{\partial y} = g\beta(\bar{T} - T_\infty) + \frac{\partial}{\partial y}\left[(\nu + \epsilon_M)\frac{\partial\bar{u}}{\partial y}\right] \tag{12.47}$$

$$\bar{u}\frac{\partial\bar{T}}{\partial x} + \bar{v}\frac{\partial\bar{T}}{\partial y} = \frac{\partial}{\partial y}\left[(\alpha + \epsilon_H)\frac{\partial\bar{T}}{\partial y}\right] \tag{12.48}$$

where the viscous dissipation and energy source terms have not been included. Therefore, a replacement of ν and α by $\nu + \epsilon_M$ and $\alpha + \epsilon_H$, respectively, gives the governing equations for turbulent flow. However, ϵ_M and ϵ_H are functions of the flow, so experimental results or various turbulence models are used for approximating them.

Several turbulence models have been developed in recent years and employed for solving various turbulent flows of practical interest. Some of these are summarized in Chap. 2. Among these, the k-ϵ model, where k is the turbulence kinetic energy and ϵ the rate of dissipation of turbulence energy, has been employed most extensively. Both these quantities are calculated from their governing differential equations for the mean flow. Launder and Spalding [25] discuss this and other turbulence models in detail and also give various relevant constants and functions of the k-ϵ model.

Various researchers have employed different turbulence models to simulate complex turbulent flows of interest in industry and in nature. The algebraic eddy-viscosity model has been a popular choice because of its simplicity and because the empirical constants in higher-order models may not be available for a given flow circumstance. A considerable amount of work has been done on recirculating turbulent flows in enclosures, such as those due to fire in a room. See, for instance, the review paper on such flows by Yang and Lloyd [26].

Experimental work on turbulent natural convection has been done to somewhat larger extent than the analytical work. Still, the data available for the various flow configurations and conditions encountered are few. Cheesewright [27] measured velocity and temperature profiles and provided heat transfer data in air. In turbulent flow, the generalized temperature profiles were found not to change significantly downstream, a fact which was employed by Cheesewright to determine the end of the transition regime. The efforts of Vliet and Liu [28] and of Vliet [29] were directed at vertical and inclined uniform flux surfaces. The turbulent heat transfer data for a vertical surface were correlated by the expression

$$\mathrm{Nu}_{x,H} = 0.568(\mathrm{Gr}_x^*\,\mathrm{Pr})^{0.22} \tag{12.49a}$$

where

$$\mathrm{Gr}_x^* = \frac{g\beta q'' x^4}{k\nu^2} \tag{12.49b}$$

over a $Gr_x^* \, Pr$ range of 2×10^{13} to 10^{16}. Here, Gr_x^* is a Grashof number based on the heat flux. Turbulence levels $u'/\bar{u}$ as high as 30% were observed, and the spreading out of the velocity profile with a decrease in the nondimensional velocity, as observed by Jaluria and Gebhart [24], was obtained. For inclined surfaces, Vliet [29] found the exponent to vary from 0.22 for vertical surfaces to 0.25 for horizontal. Both water and air were employed. Vliet and Ross [30] considered an inclined surface with constant heat flux and found their data to correlate well with the expression

$$Nu_{x,H} = 0.17(Gr_x^* \, Pr)^{0.25} \tag{12.50}$$

and therefore a weak dependence of h_x on x is obtained. In the Grashof number Gr_x^* above, g was replaced by $g \cos^2 \gamma$, where γ is the angle at which the surface is inclined with the vertical. Several other correlations are given in the following section.

12.6 EMPIRICAL CORRELATIONS

In several problems of practical interest, the heat transfer and flow processes are so complicated that the analytical methods discussed earlier cannot be employed easily and one has to depend on experimental data. Over the years, a considerable amount of heat transfer information for various flow configurations and conditions has been gathered. Some of this information has already been presented in Secs. 12.3.1, 12.3.2, 12.4.1, and 12.4.2 The present section gives some of the typical results for several important cases. The results included here are only a small fraction of what is available in the literature, and the attempt is only to present useful results in a few frequently encountered problems and to indicate the general features of the empirical relationships. Unless mentioned otherwise, all fluid properties are to be evaluated at the *film* temperature $T_f = (T_w + T_\infty)/2$.

12.6.1 Vertical Flat Surfaces

To cover the range of Rayleigh number, Ra = Gr Pr, from laminar to turbulent flow, the following correlations have been given and employed over the years for isothermal surfaces, i.e., with constant wall temperature boundary condition (McAdams [31], Warner and Arpaci [32]):

$$Nu_{m,T} = \begin{cases} 0.59\, Ra^{1/4} & \text{for} \quad 10^4 < Ra < 10^9 \quad \text{(laminar)} \\ 0.10\, Ra^{1/3} & \text{for} \quad 10^9 < Ra < 10^{13} \quad \text{(turbulent)} \end{cases} \tag{12.51}$$

The above equations are applicable to Pr values not too different from 1.0. Churchill and Chu [33] have recommended the following correlation, which may be applied over a wide range of Ra:

$$Nu_{m,T} = \left\{ 0.825 + \frac{0.387\, Ra^{1/6}}{\left[1 + (0.492/Pr)^{9/16}\right]^{8/27}} \right\}^2 \quad \text{for} \quad 10^{-1} < Ra < 10^{12} \tag{12.52}$$

For laminar flow, slightly greater accuracy is obtained with the correlation (Churchill and Chu [33])

$$\mathrm{Nu}_{m,T} = 0.68 + \frac{0.67\,\mathrm{Ra}^{1/4}}{\left[1 + (0.492/\mathrm{Pr})^{9/16}\right]^{4/9}} \qquad \text{for} \quad 0 < \mathrm{Ra} < 10^9 \qquad (12.53)$$

Here, $\mathrm{Nu}_{m,T}$ and Ra are based on the plate height L. The above correlations, Eqs. (12.52) and (12.53), are to be preferred over the others, since they have the best agreement with experimental data. The local Nusselt numbers $\mathrm{Nu}_{x,T}$ can be obtained from the results of [31, 32, 33].

For the uniform heat flux case, the heat transfer results given by Vliet and Liu [28] in water indicate the following relationships.

For laminar flow,

$$\mathrm{Nu}_{x,H} = 0.60(\mathrm{Gr}_x^* \,\mathrm{Pr})^{1/5} \qquad \text{for} \quad 10^5 < \mathrm{Gr}_x^* \,\mathrm{Pr} < 10^{13} \qquad (12.54\text{a})$$

$$\mathrm{Nu}_{m,H} = 1.25\,\mathrm{Nu}_{L,H} \qquad \text{for} \quad 10^5 < \mathrm{Gr}^* \,\mathrm{Pr} < 10^{11} \qquad (12.54\text{b})$$

For turbulent flow,

$$\mathrm{Nu}_{x,H} = 0.568(\mathrm{Gr}_x^* \,\mathrm{Pr})^{0.22} \qquad \text{for} \quad 10^{13} < \mathrm{Gr}_x^* \,\mathrm{Pr} < 10^{16} \qquad (12.55\text{a})$$

$$\mathrm{Nu}_{m,H} = 1.136\,\mathrm{Nu}_{L,H} \qquad \text{for} \quad 2 \times 10^{13} < \mathrm{Gr}^* \,\mathrm{Pr} < 10^{16} \qquad (12.55\text{b})$$

where $\mathrm{Nu}_{L,H}$ represents the Nusselt number at $x = L$ and $\mathrm{Gr}^* = g\beta q'' L^4/k\nu^2$. In a later study, Vliet and Ross [30] obtained a closer corroboration for data in air with the following relationships:

$$\mathrm{Nu}_{x,H} = 0.55(\mathrm{Gr}_x^* \,\mathrm{Pr})^{0.2} \qquad \text{for laminar flow} \qquad (12.56\text{a})$$

and

$$\mathrm{Nu}_{x,H} = 0.17(\mathrm{Gr}_x^* \,\mathrm{Pr})^{0.25} \qquad \text{for turbulent flow} \qquad (12.56\text{b})$$

$\mathrm{Nu}_{m,H}$ can be obtained by computing the mean temperature difference and using the overall heat transfer rate provided in [30].

12.6.2 Inclined and Horizontal Flat Surfaces

As discussed earlier, the results obtained for vertical surfaces may be employed for surfaces inclined at an angle γ up to about 45° with the vertical, by replacing g with $g\cos\gamma$ in the Grashof number. For inclined surfaces with constant heat flux, Vliet and Ross [30] have suggested the use of Eq. (12.54a) for laminar flow, with the replacement of Gr_x^* by $\mathrm{Gr}_x^* \cos\gamma$ for both upward- and downward-facing heated inclined surfaces. In the turbulent region also, Eq. (12.55a) is suggested, with Gr_x^* the same as that for a vertical surface for an upward-facing heated surface, and with Gr_x^* replaced by $\mathrm{Gr}_x^* \cos^2\gamma$ for a downward-facing surface.

Several correlations for inclined surfaces, under various conditions, were given by Fujii and Imura [12]. For an inclined plate with heated surface facing upward with approximately constant heat flux, the correlation obtained is of the form

$$\mathrm{Nu}_{m,H} = 0.14\left[(\mathrm{Gr\,Pr})^{1/3} - (\mathrm{Gr}_{cr}\,\mathrm{Pr})^{1/3}\right] + 0.56(\mathrm{Gr}_{cr}\,\mathrm{Pr}\cos\gamma)^{1/4}$$

$$\text{for} \quad 10^5 < \mathrm{Gr\,Pr}\cos\gamma < 10^{11} \text{ and } 15° < \gamma < 75° \quad (12.57)$$

where Gr_{cr} is the critical Grashof number at which the Nusselt number starts deviating from the laminar relationship, which is the second expression on the right-hand side of the above equation. The above correlation applies for $\mathrm{Gr} > \mathrm{Gr}_{cr}$. For various inclination angles, the value of Gr_{cr} is also given by the authors. For $\gamma = 15, 30, 60$, and $70°$, Gr_{cr} is given as 5×10^9, 2×10^9, 10^8, and 10^6, respectively. For inclined heated surfaces facing downward, the expression given is

$$\mathrm{Nu}_{m,T} = 0.56(\mathrm{Gr\,Pr}\cos\gamma)^{1/4}$$

$$\text{for} \quad 10^5 < \mathrm{Gr\,Pr}\cos\gamma < 10^{11}, \quad \gamma < 88° \quad (12.58)$$

The fluid properties are evaluated at $T_w - 0.25\ (T_w - T_\infty)$, and β at $T_\infty + 0.25\ (T_w - T_\infty)$.

For horizontal surfaces, several classical expressions exist. For heated isothermal surfaces facing downward, or cooled ones facing upward, the correlation given by McAdams [31] is

$$\mathrm{Nu}_{m,T} = 0.27\,\mathrm{Ra}^{1/4} \qquad \text{for} \quad 3 \times 10^5 \lesssim \mathrm{Ra} \lesssim 3 \times 10^{10} \qquad (12.59)$$

Fujii and Imura [12] give the corresponding correlation as

$$\mathrm{Nu}_{m,T} = 0.58\,\mathrm{Ra}^{1/5} \qquad \text{for} \quad 10^6 < \mathrm{Ra} < 10^{11} \qquad (12.60)$$

Over the overlapping range of the two studies [12, 13], the agreement between the two $\mathrm{Nu}_{m,T}$'s is very good.

For the heated isothermal horizontal surface facing upward, and cold surface facing downward, the correlations for heat transfer are given by McAdams [31] as

$$\mathrm{Nu}_{m,T} = 0.54\,\mathrm{Ra}^{1/4} \qquad \text{for} \quad 10^5 \lesssim \mathrm{Ra} \lesssim 10^7 \qquad (12.61a)$$

and

$$\mathrm{Nu}_{m,T} = 0.15\,\mathrm{Ra}^{1/3} \qquad \text{for} \quad 10^7 \lesssim \mathrm{Ra} \lesssim 10^{10} \qquad (12.61b)$$

The corresponding correlation given by Fujii and Imura [12] for an approximately uniform heat flux condition is

$$\mathrm{Nu}_{m,H} = 0.14\,\mathrm{Ra}^{1/3} \qquad \text{for} \quad \mathrm{Ra} > 2 \times 10^8 \qquad (12.62)$$

12.6.3 Cylinders and Spheres

A considerable amount of information exists on natural-convection heat transfer from a cylinder. For vertical cylinders of large diameter, ascertained from Eq. (12.36), the relationships for vertical flat plates may be employed. For cylinders of small diameter correlations for Nu are suggested in terms of the Rayleigh number Ra, where Ra and Nu are based on the diameter D of the cylinder.

The horizontal cylinder has been of interest to several investigators. McAdams [31] gives the correlation for isothermal cylinders as

$$\mathrm{Nu}_{m,T} = 0.53\,\mathrm{Ra}^{1/4} \qquad \text{for} \quad 10^4 < \mathrm{Ra} < 10^9 \tag{12.63a}$$

and

$$\mathrm{Nu}_{m,T} = 0.13\,\mathrm{Ra}^{1/3} \qquad \text{for} \quad 10^9 < \mathrm{Ra} < 10^{12} \tag{12.63b}$$

The Rayleigh number Ra is based on the cylinder diameter D. For smaller values of Ra, graphs are presented by McAdams [31]. A general expression of the form $\mathrm{Nu}_{m,T} = C\,\mathrm{Ra}^n$, with a tabulation of C and n, is given by Morgan [34]. Recently, Churchill and Chu [35] have given a correlation covering a very wide range of Ra for isothermal cylinders:

$$\mathrm{Nu}_{m,T} = \left[0.60 + 0.387\left\{\frac{\mathrm{Ra}}{\left[1 + (0.559/\mathrm{Pr})^{9/16}\right]^{16/9}}\right\}^{1/6}\right]^2,$$

$$\text{for} \quad 10^{-5} \lesssim \mathrm{Ra} \lesssim 10^{12} \tag{12.64}$$

Again, $\mathrm{Nu}_{m,T}$ and Ra are based on the cylinder diameter D. This correlation is convenient to use and agrees closely with experimental results. Thus, it is recommended for horizontal cylinders.

For natural convection from spheres too, several experimental studies have provided heat transfer correlations. Amato and Tien [17] have listed the correlations for $\mathrm{Nu}_{m,T}$ obtained from various investigations of heat and mass transfer. In a review paper, Yuge [36] suggested the following correlation for heat transfer from isothermal spheres in air

TABLE 12.2 Summary of Natural Convection Correlations for External Flows over Isothermal Surfaces

Geometry	Recommended Correlation[a]	Range	Ref.
1. Vertical flat surfaces	$\mathrm{Nu}_{m,T} = \left\{0.825 + \frac{0.387\,\mathrm{Ra}^{1/6}}{\left[1 + (0.492/\mathrm{Pr})^{9/16}\right]^{8/27}}\right\}^2$	$10^{-1} < \mathrm{Ra} < 10^{12}$	33
2. Inclined flat surfaces	Above equation with g replaced with $g\cos\gamma$	$\gamma \lesssim 60°$	
3. Horizontal flat surfaces (a) Heated, facing upward	$\mathrm{Nu}_{m,T} = 0.54\,\mathrm{Ra}^{1/4}$ $\mathrm{Nu}_{m,T} = 0.15\,\mathrm{Ra}^{1/3}$	$10^5 \lesssim \mathrm{Ra} \lesssim 10^7$ $10^7 \lesssim \mathrm{Ra} \lesssim 10^{10}$	31
(b) Heated, facing downward	$\mathrm{Nu}_{m,T} = 0.27\,\mathrm{Ra}^{1/4}$	$3 \times 10^5 \lesssim \mathrm{Ra} \lesssim 3 \times 10^{10}$	31
4. Horizontal cylinders	$\mathrm{Nu}_{m,T} = \left\{0.6 + \frac{0.387\,\mathrm{Ra}^{1/6}}{\left[1 + (0.559/\mathrm{Pr})^{9/16}\right]^{8/27}}\right\}^2$	$10^{-5} \lesssim \mathrm{Ra} \lesssim 10^{12}$	35
5. Spheres	$\mathrm{Nu}_{m,T} = 2 + 0.43\,\mathrm{Ra}^{1/4}$	$\mathrm{Pr} \approx 1$ $1 \lesssim \mathrm{Ra} \lesssim 10^5$	36

[a] $\mathrm{Nu}_{m,T}$ and Ra are based on height L for the vertical plate, length L for inclined and horizontal surfaces, and diameter D for horizontal cylinders and spheres. All fluid properties are evaluated at the film temperature $T_f = (T_w + T_\infty)/2$.

and gases over a Grashof number range $1 < \text{Gr} < 10^5$, where Gr and $\text{Nu}_{m,T}$ are based on the diameter D:

$$\text{Nu}_{m,T} = 2 + 0.43\,\text{Ra}^{1/4} \qquad \text{for} \quad \text{Pr} \approx 1 \text{ and } 1 < \text{Ra} < 10^5 \tag{12.65}$$

For heat transfer in water, Amato and Tien [17] obtained the correlation for isothermal spheres as

$$\text{Nu}_{m,T} = 2 + C\,\text{Ra}^{1/4} \qquad \text{for} \quad 3 \times 10^5 \leq \text{Ra} \leq 8 \times 10^8 \tag{12.66}$$

where $C = 0.500 \pm 0.009$, which gave a mean deviation of less than 11%. Several of the important correlations presented earlier are summarized in Table 12.2.

12.7 SUMMARY

This chapter discusses the basic considerations relevant to natural convection flows. External buoyancy-induced flows are considered, and the governing equations are obtained. The approximations generally employed in the analysis of these flows are outlined. The important dimensionless parameters are derived in order to discuss the importance of the basic processes that govern these flows. Laminar flows over various kinds of surfaces are discussed, and the solutions obtained are presented, particularly those derived from similarity analysis. The heat transfer results and the characteristics of the resulting velocity and temperature fields are discussed. Also considered are transient and turbulent flows. The governing equations for turbulent flow are given, and experimental results for various flow configurations are presented. The frequently employed empirical correlations for heat transfer by natural convection from various kinds of surfaces and bodies are also included. Thus, this chapter presents the basic aspects that underlie natural convection and also the heat transfer correlations that may be employed for practical applications.

NOMENCLATURE

c_p	specific heat at constant pressure, J/(kg · K), Btu/(lb$_m$ ·° F)
D	diameter of cylinder or sphere, m, ft
f	dimensionless stream function, defined in Eq. (12.22)
$\mathbf{F}$	body force per unit volume, N/m^3, lb$_f$/ft^3
$\mathbf{g}$	gravitational acceleration, m/s^2, ft/s^2
Gr_x	local Grashof number $= g\beta\,\Delta T x^3/\nu^2$
Gr	Grashof number $= g\beta\,\Delta T L^3/\nu^2$
Gr*	heat-flux Grashof number $= g\beta L^4 q''/k\nu^2$
h_x	local heat transfer coefficient at a cross section, W/(m^2 · K), Btu/(hr · ft^2 °F)
h_m	average heat transfer coefficient, W/(m^2 · K), Btu/(hr · ft^2 °F)
h_ϕ	local Nusselt number at an angular position ϕ, W/(m^2 · K), Btu/(hr · ft^2 ·° F)
k	thermal conductivity, W/m · K, Btu/(hr · ft ·° F)
L	characteristic length, height of vertical plate, m, ft

m, n	exponents in exponential and power-law distributions
M, N	constants employed for exponential and power-law distributions of surface temperature
Nu_x	local Nusselt number = $h_x x/k$
Nu_m	average Nusselt number = $h_m L/k$
Nu_ϕ	peripheral local Nusselt number = $h_\phi D/k$
p	pressure, Pa, $\mathrm{lb}_f/\mathrm{ft}^2$
Pr	Prandtl number = $c_p \mu/k$
q	total heat transfer, W, Btu/hr
q_x''	local heat flux, $\mathrm{W/m^2}$, $\mathrm{Btu/(hr \cdot ft^2)}$
q''	constant surface heat flux, $\mathrm{W/m^2}$, $\mathrm{Btu/(hr \cdot ft^2)}$
q'''	volumetric heat source, $\mathrm{W/m^3}$, $\mathrm{Btu/(hr \cdot ft^3)}$
Ra	Rayleigh number = Gr Pr
Ra_x	local Rayleigh number = $\mathrm{Gr}_x \mathrm{Pr}$
Sr	Strouhal number = $L/V_c t_c$
t	time, s
t_c	characteristic time, s
ΔT	temperature difference = $T_w - T_\infty$, °C, K, °F, °R
T	local temperature, °C, K, °F, °R
T_w	wall temperature, °C, K, °F, °R
T_∞	ambient temperature, °C, K, °F, °R
u, v, w	velocity components in x, y, z directions, respectively, m/s, ft/s
$\mathbf{V}$	velocity vector, m/s, ft/s
V_c	convection velocity = $(g\beta L \Delta T)^{1/2}$, m/s, ft/s
x, y, z	coordinate distances, m, ft

Greek Symbols

α	thermal diffusivity, $\mathrm{m^2/s}$, $\mathrm{ft^2/s}$
β	coefficient of thermal expansion = $-(1/\rho)(\partial\rho/\partial T)_p$, $\mathrm{K^{-1}}$, $°\mathrm{R}^{-1}$
γ	inclination with the vertical
δ	boundary-layer thickness, m, ft
δ_T	thermal boundary-layer thickness, m, ft
ϵ_M, ϵ_H	eddy viscosity and eddy diffusivity, respectively, $\mathrm{m^2/s}$, $\mathrm{ft^2/s}$
η	similarity variable = $(y/x)(\mathrm{Gr}_x/4)^{1/4}$
μ	dynamic viscosity, Pa · s, $\mathrm{lb}_m/(\mathrm{s \cdot ft})$
ν	kinematic viscosity, $\mathrm{m^2/s}$, $\mathrm{ft^2/s}$
Φ_v	viscous dissipation, $\mathrm{s^{-2}}$
ψ	stream function, $\mathrm{m^2/s}$, $\mathrm{ft^2/s}$
θ	dimensionless temperature = $(T - T_\infty)/(T_w - T_\infty)$

Subscripts

H	constant wall heat flux boundary condition
T	constant wall temperature boundary condition

REFERENCES

1. J. S. Turner, *Buoyancy Effects in Fluids*, Cambridge U.P., Cambridge, U.K., 1973.
2. Y. Jaluria, *Natural Convection Heat and Mass Transfer*, Pergamon, Oxford, U.K., 1980.
3. S. Kakaç, W. Aung, and R. Viskanta, eds., *Natural Convection: Fundamentals and Applications*, Hemisphere, New York, 1985.
4. B. Gebhart, Y. Jaluria, R. L. Mahajan, and B. Sammakia, *Buoyancy Induced Flows and Transport*, Hemisphere, New York, 1987, to appear.
5. B. Gebhart, Buoyancy Induced Fluid Motions Characteristic of Applications in Technology, *J. Fluids Eng.*, Vol. 101, pp. 5–28, 1979.
6. H. Schuh, Boundary Layers of Temperature, *Boundary Layers*, by W. Tollmien, British Ministry of Supply, German Doc. Center, Ref. 3220T, Sec. B.6, 1948.
7. S. Ostrach, An Analysis of Laminar Free Convection Flow and Heat Transfer about a Flat Plate parallel to the Direction of the Generating Body Force, NACA Tech. Rep. 1111, 1953.
8. E. M. Sparrow and J. L. Gregg, Similar Solutions for Free Convection from a Non-isothermal Vertical Plate, *J. Heat Transfer*, Vol. 80, pp. 379–386, 1958.
9. B. Gebhart, Natural Convection Flows and Stability, *Adv. Heat Transfer*, Vol. 9, pp. 273–348, 1973.
10. B. R. Rich, An Investigation of Heat Transfer from an Inclined Flat Plate in Free Convection, *Trans. ASME*, Vol. 75, pp. 489–499, 1953.
11. G. C. Vliet, Natural Convection Local Heat Transfer on Constant Heat Flux Inclined Surfaces, *J. Heat Transfer*, Vol. 9, pp. 511–516, 1969.
12. T. Fujii and H. Imura, Natural Convection from a plate with Arbitrary Inclination, *Int. J. Heat Mass Transfer*, Vol. 15, pp. 755–767, 1972.
13. Z. Rotem and L. Claassen, Natural Convection above unconfined Horizontal Surfaces, *J. Fluid Mech.*, Vol. 39, pp. 173–192, 1969.
14. L. Pera and B. Gebhart, Natural Convection Boundary Layer Flow over Horizontal and Slightly Inclined Surfaces, *Int. J. Heat Mass Transfer*, Vol. 16, pp. 1131–1146, 1972.
15. H. J. Merk and J. A. Prins, Thermal Convection in Laminar Boundary Layers I, II, and III, *Appl. Sci. Res.*, Vol. A4, pp. 11–24, 195–206, 207–221, 1953–1954.
16. T. Chiang, A. Ossin, and C. L. Tien, Laminar Free Convection from a Sphere, *J. Heat Transfer*, Vol. 86, pp. 537–542, 1964.
17. W. S. Amato and C. Tien, Free Convection Heat Transfer from Isothermal Spheres in Water, *Int. J. Heat Mass Transfer*, Vol. 15, pp. 327–339, 1972.
18. E. M. Sparrow and J. L. Gregg, Laminar Free Convection Heat Transfer from the Outer Surface of a Vertical Circular Cylinder, *Trans. ASME*, Vol. 79, pp. 1823–1829, 1956.
19. W. J. Minkowycz and E. M. Sparrow, Local Nonsimilar Solutions for Natural Convection on a Vertical Cylinder, *J. Heat Transfer*, Vol. 96, pp. 178–183, 1974.
20. E. J. LeFevre and A. J. Ede, Laminar Free Convection from the Outer Surface of a Vertical Circular Cylinder, *Proc. 9th Int. Congr. Appl. Mech.*, Brussels, Vol. 4, pp. 175–183, 1956.
21. J. D. Hellums and S. W. Churchill, Transient and Steady State, Free and Natural Convection, Numerical Solutions: Part I, The Isothermal, Vertical Plate, *AIChE J.*, Vol. 8, pp. 690–692, 1962.
22. S. W. Churchill, Transient Laminar Free Convection from a Uniformly Heated Vertical Plate, *Lett. Heat Mass Transfer*, Vol. 2, pp. 311–317, 1975.
23. S. W. Churchill and R. Usagi, A Standardized Procedure for the Production of Correlations in the Form of a Common Empirical Equation, *Indust. Eng. Chem. Fund.*, Vol. 13, pp. 39–46, 1974.
24. Y. Jaluria and B. Gebhart, On Transition Mechanisms in Vertical Natural Convection Flow, *J. Fluid Mech.*, Vol. 66, pp. 309–337, 1974.

25. B. E. Launder and D. B. Spalding, *Mathematical Models of Turbulence*, Academic, London, 1972.

26. K. T. Yang and J. R. Lloyd, Turbulent Buoyant Flow in Vented Simple and Complex Enclosures, *Natural Convection: Fundamentals and Applications*, ed. S. Kakaç, W. Aung, and R. Viskanta, pp. 303–329, Hemisphere, New York, 1985.

27. R. Cheesewright, Turbulent Natural Convection from a Vertical Plane Surface, *J. Heat Transfer*, Vol. 90, pp. 1–8, 1968.

28. G. C. Vliet and C. K. Liu, An Experimental Study of Turbulent Natural Convection Boundary Layers, *J. Heat Transfer*, Vol. 91, pp. 517–531, 1969.

29. G. C. Vliet, Natural Convection Local Heat Transfer on Constant Heat Flux Inclined Surfaces, *J. Heat Transfer*, Vol. 91, pp. 511–516, 1969.

30. G. C. Vliet and D. C. Ross, Turbulent, Natural Convection on Upward and Downward Facing Inclined Heat Flux Surfaces, *J. Heat Transfer*, Vol. 97, pp. 549–555, 1975.

31. W. H. McAdams, *Heat Transmission*, 3rd ed., McGraw-Hill, New York, 1954.

32. C. Y. Warner and V. S. Arpaci, An Experimental Investigation of Turbulent Natural Convection in Air at Low Pressure along a Vertical Heated Flat Plate, *Int. J. Heat Mass Transfer*, Vol. 11, pp. 397–406, 1968.

33. S. W. Churchill and H. H. S. Chu, Correlating Equations for Laminar and Turbulent Free Convection from a Vertical Plate, *Int. J. Heat Mass Transfer*, Vol. 18, pp. 1323–1329, 1975.

34. V. T. Morgan, The Overall Convective Heat Transfer from Smooth Circular Cylinders, *Adv. Heat Transfer*, Academic, New York, Vol. 11, pp. 199–264, 1975.

35. S. W. Churchill and H. H. S. Chu, Correlating Equations for Laminar and Turbulent Free Convection from a Horizontal Cylinder, *Int. J. Heat Mass Transfer*, Vol. 18, pp. 1049–1053, 1975.

36. T. Yuge, Experiments on Heat Transfer from Spheres Including Combined Natural and Forced Convection, *J. Heat Transfer*, Vol. 82, pp. 214–220, 1960.

13

NATURAL CONVECTION IN ENCLOSURES

K. T. Yang

University of Notre Dame
Notre Dame, Indiana

13.1 INTRODUCTION

This chapter deals with single-phase natural-convection phenomena in enclosures. By definition, enclosures are finite spaces bounded by walls and filled with fluid media. Also included in this definition are cases where internal partitions or obstacles may be present. Natural convection in such enclosures is induced by buoyancy caused by a body force field such as gravity together with density variations within the fluid. Such density variations may be due to external heating or cooling through the bounding walls, to the presence of internal heat sources or sinks, to concentration changes in the fluid as a result of mass transfer, or to any combination of these processes.

The significance of the enclosure natural-convection phenomena can best be appreciated by noting several important application areas. The proper design of furnaces must necessarily take into account the contribution of turbulent buoyant flows of the flue gases arising from the fuel bed. In the operation of solar collectors, natural-convection effects, which contribute to energy losses to the environment, must be minimized to increase the collector efficiency. Similarly, heat losses though double windows are also affected by natural convection between the window panes. Natural convection also plays a dominant role in energy transfer in other energy storage systems, ranging from hot- or chilled-water storage tanks to large solar ponds where stratification conditions must be properly maintained to facilitate energy storage and removal. Also, in thermal storage systems utilizing phase-change materials, natural convection significantly affects the energy transfer process at the liquid-solid interface. In the cooling of electronic equipment and devices such as circuit boards and chips, natural convection may be the only permissible mode of cooling and determines the operational limits. In growing high-purity crystals, understanding of natural convection in the enclosure enables the designer to minimize contamination due to convection effects in the melt. In such important appliances as kitchen ovens and household water heaters, the dominant mode of energy transfer is again natural convection, which must be properly analyzed in the design of such appliances. Another practically important class of natural-convection phenomena in enclosures concerns the spread of fire and smoke in rooms, corridors, and other confined spaces. A proper understanding of the associated turbulent buoyant flows is essential in the development of countermeasures against the hazards of unwanted fires.

Such applications have contributed much to the recent interest among heat transfer specialists in the study of enclosure natural-convection phenomena. Perhaps a lesser but equally significant reason for this recent interest is that significant advances have been made recently in our understanding of other related physical processes such as turbulence, combustion, and radiative transfer, as well as in the development of mathematical tools such as numerical solution of nonlinear elliptical partial differential equations. As a result, many of the important enclosure natural-convection problems can now be properly formulated and successfully analyzed.

Natural convection flows in enclosures are also known as buoyancy-driven enclosure flows. The importance of buoyancy is quite obvious here. One quantitative characteristic of such flows is that the velocity components are all of essentially the same order of magnitude, except in some subregions of the flow where one component may be more important than the others. This basic characteristic of the enclosure flows becomes clear in looking at several specific examples of enclosures in accordance with the general definition given at the beginning of this chapter. These are illustrated in Fig. 13.1.

Two-dimensional enclosures

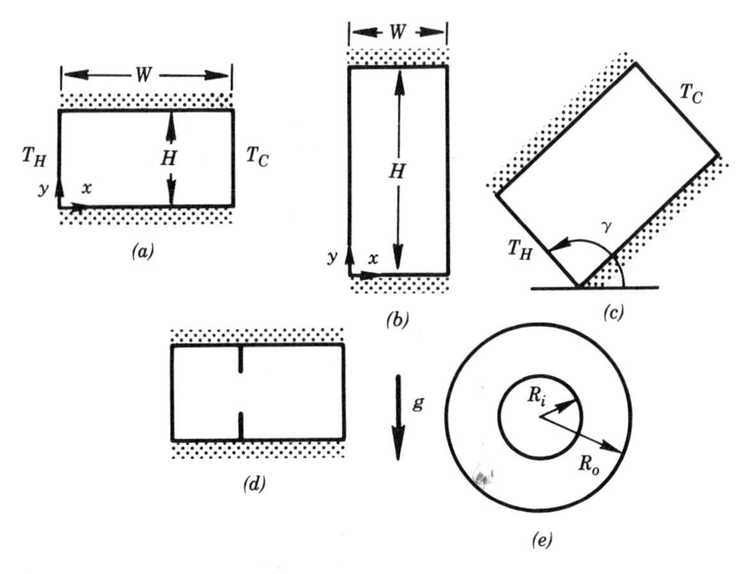

Three-dimensional enclosures

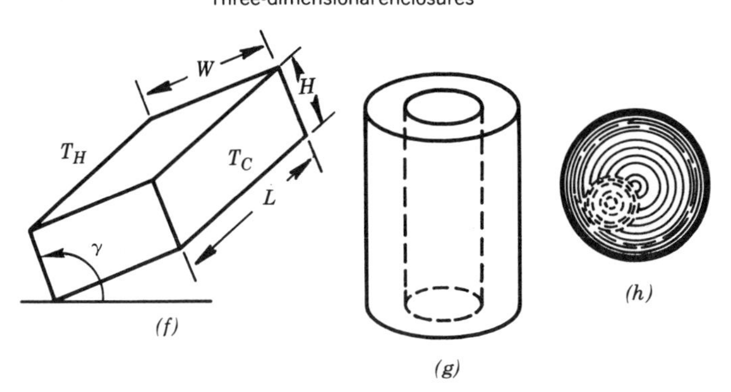

Figure 13.1. Examples of two-dimensional and three-dimensional enclosures.

Figure 13.1*a* is a two-dimensional rectangular enclosure heated differentially at the two ends ($T_H > T_C$) and thermally insulated on the two sides. This enclosure has two characteristic dimensions: the height H and the width W. Usually W is taken to be the dimension separating the surfaces where temperatures are specified. It is clear that the buoyancy-driven flow throughout this enclosure is in general of the recirculating type, even though boundary layers may exist close to the walls. When the aspect ratio α^* defined by H/W is less than unity, the enclosure is known as a shallow enclosure, or shallow cavity.

Figure 13.1*b* shows a vertical enclosure with $\alpha^* > 1$, and here again boundary layers may exist close to the hot and cold walls, but the recirculating flow in the core region definitely cannot be described as that of boundary or shear layers.

A more general rectangular enclosure is shown in Fig. 13.1*c*, where the enclosure is tilted relative to the gravity field. For zero tilt angle γ, we have an enclosure heated at the top. Figure 13.1*a* and *b* are special cases of Fig. 13.1*c* with $\gamma = 90°$. When $\gamma = 180°$, we have the heated-from-below situation.

A more complex situation is shown in Fig. 13.1*d*, where the rectangular enclosure also contains a pair of vertical partitions, and the buoyancy-driven flow tends to be slower than without the partitions. This shows a simple way of suppressing natural-convection flow.

Natural-convection flows are not limited to rectangular configurations. Figure 13.1*e* illustrates an annular enclosure bounded by two concentric cylinders which are differentially heated. The overall flow pattern is in the form of two symmetrical kidney-shaped streamlines. Here again it is expected that the predominant flow will be recirculating flow. A more general class of two-dimensional annular enclosures is similar to that in Fig. 13.1*e* except that the two cylinders are eccentric to each other.

All real enclosures are three-dimensional, and three realistic examples are shown in Fig. 13.1*f*, *g*, and *h*. Figure 13.1*f* represents an enclosure in the form of a parallelepiped, sometimes known as a box enclosure. The enclosure can be tilted relative to gravity in two different planes, and a great variety of thermal boundary conditions can be imposed. Figure 13.1*g* shows a truncated annular enclosure, which can also be tilted, and Fig. 13.1*h* shows a three-dimensional enclosure formed by two eccentric spheres. The flow patterns in such three-dimensional enclosures are also characterized by complex recirculating flows.

While the determination of the overall heat transfer rates across such enclosures does not necessarily depend on a prior knowledge of the flows, it is essential that the details of the flow field also be obtained. Only then can we gain physical insight into the interactions of the various mechanisms involved in natural convection, such as is important in many applications where the flow must be controlled. Consequently, it is not surprising that much of the emphasis in enclosure natural-convection studies in the recent literature is placed not only on the overall heat transfer rates, but also on the flow field and its related phenomena such as stability and transition.

The primary objective of this chapter is to provide some basic information on the physics of the enclosure natural-convection phenomena, the mathematical formulation of the natural-convection problem, solution techniques that are available, some significant results in the field including both theoretical and experimental data, and a brief description of recent studies of interaction of the basic enclosure natural-convection phenomena with other heat transfer processes. The contents of this chapter complement several well known and excellent reviews that have appeared in the recent literature [1–7]. In order to focus on natural convection in enclosures, horizontal fluid layers will not be covered in this chapter; they now constitute an extensive field, especially in view of recent advances. Readers interested in this field are referred to the classical texts of Chandrasekhar [8] and Turner [9], and a recent review by Catton [10].

In this chapter, the mathematical formulation of the enclosure natural-convection problem and the associated dimensionless parameters are given in Sec. 13.2, and solution techniques for the governing differential equations are described in Sec. 13.3. Section 13.4 presents some theoretical and experimental results for laminar natural convection in two-dimensional and three-dimensional enclosures, and the corresponding results for turbulent flow are described in Sec. 13.5. In Sec. 13.6, recent advances in the study complex phenomena dealing with the interaction of natural convection with other physical processes in enclosures are briefly reviewed. Finally, in Sec. 13.7, several useful heat transfer correlation equations for buoyant enclosure flows are given.

13.2 MATHEMATICAL FORMULATION AND DIMENSIONLESS PARAMETERS

13.2.1 Mathematical Formulation for Three-Dimensional Enclosure Flows

The governing differential equations for natural-convection enclosure flows are the conservation equations of mass, momentum, and energy. For a compressible variable-property fluid, these equations may be written in vector form [11] as follows:

$$\frac{\partial \rho}{\partial t} + \nabla \cdot (\rho \mathbf{V}) = 0 \tag{13.1}$$

$$\rho \frac{D\mathbf{V}}{Dt} = \nabla \cdot \boldsymbol{\tau} - \nabla p + \rho \mathbf{B} \tag{13.2}$$

$$\rho c_p \frac{DT}{Dt} = \nabla \cdot \mathbf{q}_c + S \tag{13.3}$$

where ρ is the fluid density, t is the time variable, $\mathbf{V}$ is the velocity vector, $\boldsymbol{\tau}$ is the stress tensor, p is the static pressure, $\mathbf{B}$ is the body force vector, c_p is the fluid specific heat at constant pressure, T is the temperature, $\mathbf{q}_c$ is the conduction flux, and S is the source term. Also, ∇ denotes the gradient operator, $\nabla \cdot$ the divergence operator, and D/Dt the substantial derivative $(\partial/\partial t + \mathbf{V} \cdot \nabla)$. To complete these equations for the unknowns $\mathbf{V}$, p, and T, we also need to add constitutive relations for $\boldsymbol{\tau}$ and $\mathbf{q}_c$, an equation of state, and specific information on transport property variations and the nature of $\mathbf{B}$ and S. It should be noted that the energy equation (13.3) does not contain the dissipation and pressure work terms, since both are small and can thus be neglected for most natural-convection phenomena [12]. For laminar flows of a Newtonian Fourier fluid, the constitutive relations for $\boldsymbol{\tau}$ and $\mathbf{q}_c$ are given by

$$\boldsymbol{\tau} = \mu \nabla \mathbf{V} + \mu (\nabla \mathbf{V})^T, \qquad \mathbf{q}_c = k \nabla T \tag{13.4}$$

where μ is the dynamic viscosity, k is the thermal conductivity, and $(\ )^T$ represents the transpose.

For enclosure natural-convection phenomena, the driving force is provided by the body force vector $\mathbf{B}$ in the momentum equation (13.2), which arises from an imposed field such as gravity, a centrifugal force, an electrostatic field, or the like. The predominant body force field in enclosure natural-convection studies is gravity, in which case the body-force vector $\mathbf{B}$ becomes the gravitational acceleration vector $\mathbf{g}$. Similarly, the volumetric heat source term S in the energy equation (13.3) can arise from various sources: a heat-generating or a reacting fluid, or one that absorbs or emits radiation.

Even for complex physical situations, it is seldom necessary to deal with the complete equations given in (13.1), (13.2), and (13.3); some simplifications are usually possible. One common simplification is to treat the flow as being incompressible except for the slight variation in the density that gives rise to the buoyancy force. This refers to the now well-known Boussinesq approximation:

$$\frac{\rho}{\rho_0} = 1 - \beta (T - T_0) \tag{13.5}$$

where $\beta = -(1/\rho)(\partial\rho/\partial T)_p$, and the subscript 0 refers to a reference condition. The Boussinesq approximation can be shown to be valid in many applications as long as the temperature variations are not large [13, 14]. If in addition the properties can be taken as constant under this condition, the governing equations for the laminar flow case can be simplified to

$$\nabla \cdot \mathbf{V} = 0 \tag{13.6}$$

$$\rho \frac{D\mathbf{V}}{Dt} = \mu \nabla^2 \mathbf{V} - \nabla(p - p_0) - \rho\beta(T - T_0)\mathbf{g} \tag{13.7}$$

$$\rho c_p \frac{DT}{Dt} = k \nabla^2 T + S \tag{13.8}$$

where p_0 and T_0 both now refer to the hydrostatic conditions. These equations form the basis for many enclosure natural-convection analyses in recent years. However, where temperature variations in the enclosure are substantial, use must be made of the more complete equations (13.1), (13.2), and (13.3). Examples are the spread of fire in compartments [15] and interactions between natural convection and gas radiation in enclosures [16, 17].

For illustration purposes only, we present here the governing equations in a Cartesian coordinate system for a three-dimensional tilted-box enclosure as shown in Fig. 13.1f filled with a Newtonian fluid in laminar motion with variable properties corresponding to equations given in (13.1), (13.2), and (13.3), respectively:

$$\frac{\partial \rho}{\partial t} + \frac{\partial(\rho u)}{\partial x} + \frac{\partial(\rho v)}{\partial y} + \frac{\partial(\rho w)}{\partial z} = 0 \tag{13.9}$$

$$\begin{aligned}
\rho\left(\frac{\partial u}{\partial t} + u\frac{\partial u}{\partial x} + v\frac{\partial u}{\partial y} + w\frac{\partial u}{\partial z}\right)
&= -\frac{\partial(p - p_0)}{\partial x} + \frac{\partial}{\partial x}\left[\tfrac{4}{3}\mu\frac{\partial u}{\partial x} - \tfrac{2}{3}\mu\left(\frac{\partial v}{\partial y} + \frac{\partial w}{\partial z}\right)\right] \\
&= \frac{\partial}{\partial y}\left[\mu\left(\frac{\partial u}{\partial y} + \frac{\partial v}{\partial x}\right)\right] + \frac{\partial}{\partial y}\left[\mu\left(\frac{\partial w}{\partial x} + \frac{\partial u}{\partial z}\right)\right] \\
&\quad + \rho g_x \beta(T - T_0)
\end{aligned} \tag{13.10}$$

$$\begin{aligned}
\rho\left(\frac{\partial v}{\partial t} + u\frac{\partial v}{\partial x} + v\frac{\partial v}{\partial y} + w\frac{\partial v}{\partial z}\right)
&= -\frac{\partial(p - p_0)}{\partial y} + \frac{\partial}{\partial x}\left[\mu\left(\frac{\partial u}{\partial y} + \frac{\partial v}{\partial x}\right)\right] \\
&\quad + \frac{\partial}{\partial y}\left[\tfrac{4}{3}\mu\frac{\partial v}{\partial y} - \tfrac{2}{3}\mu\left(\frac{\partial u}{\partial x} + \frac{\partial w}{\partial z}\right)\right] \\
&\quad + \frac{\partial}{\partial z}\left[\mu\left(\frac{\partial w}{\partial y} + \frac{\partial v}{\partial z}\right)\right] + \rho g_y \beta(T - T_0)
\end{aligned} \tag{13.11}$$

$$\rho\left(\frac{\partial w}{\partial t}+u\frac{\partial w}{\partial x}+v\frac{\partial w}{\partial y}+w\frac{\partial w}{\partial z}\right)$$

$$=-\frac{\partial(p-p_0)}{\partial z}+\frac{\partial}{\partial x}\left[\mu\left(\frac{\partial w}{\partial x}+\frac{\partial u}{\partial z}\right)\right]+\frac{\partial}{\partial y}\left[\mu\left(\frac{\partial v}{\partial z}+\frac{\partial w}{\partial y}\right)\right]$$

$$+\frac{\partial}{\partial z}\left[\tfrac{4}{3}\mu\frac{\partial w}{\partial z}-\tfrac{2}{3}\mu\left(\frac{\partial u}{\partial x}+\frac{\partial v}{\partial y}\right)\right]+\rho g_z\beta(T-T_0) \qquad (13.12)$$

$$\rho c_p\left(\frac{\partial T}{\partial t}+u\frac{\partial T}{\partial x}+v\frac{\partial T}{\partial y}+w\frac{\partial T}{\partial z}\right)$$

$$=\frac{\partial}{\partial x}\left(k\frac{\partial T}{\partial x}\right)+\frac{\partial}{\partial y}\left(k\frac{\partial T}{\partial y}\right)+\frac{\partial}{\partial z}\left(k\frac{\partial T}{\partial z}\right)+S \qquad (13.13)$$

where (x, y, z) are the coordinates, (u, v, w) are the velocity components of the velocity vector **V**, and (g_x, g_y, g_z) are the components of the gravitational acceleration vector **g**. Furthermore, $c_p = c_p(p, T)$, $\mu = \mu(p, T)$, and $k = k(p, T)$ need to be specified. If the enclosure medium is a perfect gas, then β is the reciprocal of the absolute temperature T_0. All these equations must be solved simultaneously for the five unknowns u, v, w, p and T.

For other enclosure configurations, it may be desirable to write the governing equations in other coordinate systems. For cylindrical and spherical geometries, the corresponding equations are also very well known and can be found in the recent text by Arpaci and Larsen [18]. In another interesting case dealing with laminar natural convection between two eccentric horizontal cylinders, governing equations in a cylindrical bipolar coordinate system can be used [19].

To complete the mathematical formulation of the enclosure natural-convection problem, boundary and initial conditions for the governing differential equations need to be specified. For the momentum field, the no-slip conditions for the velocity components at the bounding walls and a specified hydrostatic pressure field are all that is necessary. For the temperature field, the situation is slightly more complex, due to the great variety of thermal conditions that can be specified. The simplest are the usual imposed wall temperatures or wall heat fluxes, which also include the insulated wall as a special case. Other possibilities are walls with heat capacities and walls where convective conditions may be specified. Also, it is possible to have a combination of above conditions even on a single wall.

There is also a class of enclosure natural-convection problems where there are physical openings in the bounding walls. In the extreme case where a whole wall is missing, the enclosure is known as an open enclosure or open cavity. When the openings are small they are referred to as vents. Open cavities are found in devices for the cooling of electronic components [20]; vents include windows and doorways for rooms and compartments [15].

Obviously, boundary conditions also need to be specified at these openings before solutions to the governing equations can be attempted. A difficulty arises in that some of the flow conditions are not known physically at the openings. On the other hand, the elliptic nature of the momentum and energy equations dictates that solutions are to be sought for the regions both inside and outside the enclosure so that they can be joined at the opening to yield the physical conditions there [20, 21]. One common practice is to use what is now known as the natural conditions [15], as follows: Figure 13.2 shows a

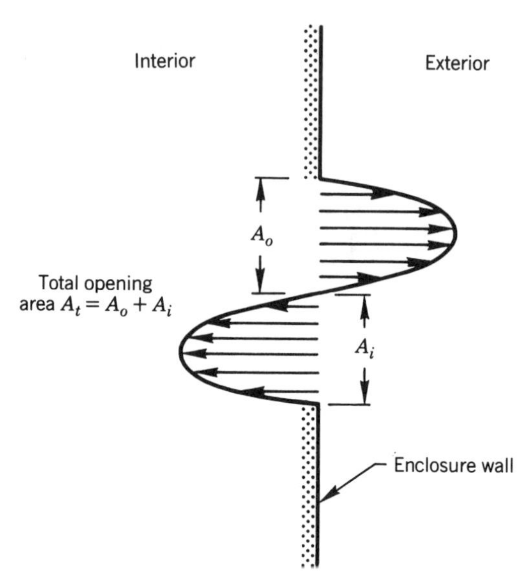

Figure 13.2. Typical enclosure opening.

typical opening which physically connects the enclosure interior to the outside. The natural conditions can then be written as follows:

$$\begin{aligned} &\text{for } A_o \qquad p = p_0, \quad \frac{\partial \mathbf{V}}{\partial n} = 0, \quad \frac{\partial T}{\partial n} = 0 \\ &\text{for } A_i \qquad p = p_0, \quad \frac{\partial \mathbf{V}}{\partial n} = 0, \quad T = T_0 \end{aligned} \tag{13.14}$$

where n is coordinate normal to the plane of the opening, T_0 is the ambient temperature, and p_0 is the ambient pressure. The physical implication of these conditions is that the flow field persists as it approaches the opening from both sides and that the fluid there leaves the enclosure at the same interior temperatures, but carries with it the ambient temperature when it enters the enclosure at the opening. It is clear that these conditions are artificial, and at best only represent approximations to the real conditions. However, in studies dealing with shallow cavities [20] and enclosures with doorways [21], it has been found that these natural conditions do represent reasonable approximations.

For a transient or unsteady problem, the proper initial conditions are prescribed initial velocity and temperature fields, which also include the special cases where the initial field may either be that corresponding to a steady-state solution or one that has no motion and is maintained at a uniform temperature.

13.2.2 Formulations for Turbulent Flow and Turbulence Modeling

The field of turbulent natural convection in enclosures is now also in active development because of several important areas of application. Examples are heat transfer and fluid flow in furnaces [22] and fire and smoke spread in vented enclosures [15]. In these applications, the very high Grashof numbers reached are caused either by large temperature differences resulting from intense heating, by the large size of the enclosures, or by both. The conservation equations (13.1), (13.2), and (13.3) can also be applied to turbulent enclosure natural-convection problems, except that the dependent variables are all interpreted as mean quantities. In addition, the stress tensor τ must now also include the Reynolds stress tensor τ_t, and similarly turbulent heat fluxes must also be added to the mean energy equation. In order to gain closure, these second-order correlations must be modeled before the turbulent flow problem can be solved. The development of these closure models represent one of the most active areas of research in fluid mechanics and heat transfer in recent years [24, 25].

13.2.3 Nondimensionalization and Scaling

It is a common practice in enclosure natural-convection analysis to first nondimensionalize the governing equations and the boundary and initial conditions before a solution is attempted. In the nondimensionalization process, a series of normalizing factors or scales must be introduced. For an enclosure natural-convection problem, such scales include a characteristic length L_c, a characteristic velocity U_c, a characteristic temperature T_c, and a characteristic temperature difference ΔT_c. While the choice of these characteristic quantities is arbitrary, the physical implication of a given choice and its effect on the resulting nondimensional equations are not well appreciated and understood, as pointed out recently by Ostrach [4]. The major difficulty lies in the fact that enclosure natural-convection phenomena are multiple-scale phenomena with different scales operating in different regimes of the flow, and also possibly at different times. If exact solutions were obtainable, then any nondimensionalization would be suitable. Unfortunately, no such exact solutions exist, even for the simpler two-dimensional enclosure problems, and only approximate solutions can be attempted. If such attempts do not take into account of the different scales involved, significant physics may be lost. Unfortunately, since in the usual practice only a single set of scales is utilized, this may indeed happen in many of the enclosure natural-convection solutions obtained so far.

The difficulty can best be illustrated by considering the following great variety of characteristic velocities that can be used for the two-dimensional enclosure problem [4]:

$$U_c = \begin{cases} \nu/L_c & \text{(viscous vs. inertia)} \\ \alpha/L_c & \text{(convection vs. conduction)} \\ (\alpha\nu)^{1/2}/L_c & \text{(combination)} \\ g\beta\,\Delta T_c\,L_c^2/\nu & \text{(buoyancy vs. viscous)} \\ (g\beta\,\Delta T_c\,L_c)^{1/2} & \text{(buoyancy vs. inertia)} \end{cases} \tag{13.15}$$

and so on, where ν is the kinetic viscosity, α is the thermal diffusivity, and L_c can be taken as H (or W). It is thus seen that each choice represents a specific balance between a pair of mechanisms, and it is clear that there is no reason to expect that any of these balances will be operative throughout the flow in the enclosure.

The resolution of this scaling difficulty in a given enclosure natural-convection problem is not simple [26, 27]. For the time being, it is important to realize that nondimensionalization of the governing equations by using a single set of characteristic or normalizing quantities must be carried out with caution until rational procedures for the enclosure problem are developed for general use.

13.2.4 Dimensionless Parameters

Dimensionless parameters which govern the enclosure natural-convection phenomena can normally be obtained by a standard dimensional analysis. It is generally known that for any enclosure flow problem, the Nusselt number can be written functionally as

$$\mathrm{Nu}_H = \mathrm{Nu}_H(\mathrm{Ra}_H, \mathrm{Pr}, \alpha^*) \tag{13.16}$$

where Nu_H is the average Nusselt number hH/k (with h the average coefficient of heat transfer), Ra_H is the Rayleigh number $g\beta(T_H - T_C)H^3/\alpha\nu$, and α^* is representative of one or more dimensionless geometrical ratios.

There are additional parameters that may be equally important, depending on the specific problems under consideration. Most important of these is the temperature-difference ratio $\theta_0 = (T_H - T_C)/T_C$, which determines whether the fluid in the enclosure is a Boussinesq fluid, and becomes a separate parameter, affecting both the flow and temperature fields, whenever it exceeds about 0.1 [14]. It has a special significance when thermal radiation effects become important [28].

Another dimensionless parameter which appears often in turbulent enclosure natural-convection problems is the gradient Richardson number Ri, defined by $g\beta(\partial T/\partial y)/(\partial u/\partial y)^2$, which is seen to be a local parameter denoting the ratio of buoyancy to inertia forces. Its physical significance lies in the effect of buoyancy on the turbulence field.

Another commonly encountered parameter for turbulent enclosure flow is the turbulent Prandtl number $\mathrm{Pr}_t = \epsilon_M/\epsilon_H$, where ϵ_M and ϵ_H are the turbulent momentum diffusivity and turbulent thermal diffusivity, respectively. As is now generally known, a quantitative knowledge of the turbulent Prandtl number eliminates the need for a seperate turbulent model for the turbulent heat fluxes [29].

Finally, it may be mentioned that other parameters may appear in an enclosure natural-convection problem in which additional modes of heat transfer occur.

13.3 MATHEMATICAL AND NUMERICAL ANALYSES

13.3.1 Classification of Methods of Analysis

Obtaining a solution to the nonlinear Navier-Stokes equations of motion has always been a challenge to fluid mechanicians and applied mathematicians alike. Even though exact solutions do exist [30], they are restricted to very specialized cases and hence are not very useful. Much of the effort has therefore been devoted to developing approximate solution techniques which can be applied to a variety of problems; this represents one of most fruitful areas of research in viscous flow theory in recent years.

For enclosure natural-convection phenomena as described by Eqs. (13.1), (13.2), and (13.3), the momentum and energy equations are coupled through the body force term, which depends on the temperature field. As a result, these equations must be solved simultaneously and hence represent an additional level of complexity in obtaining the solutions. As a matter of fact, simple enclosure natural-convection problems have often beeen used as a vehicle for the development of new analyses and new numerical algorithms [2, 4, 31].

There appears to be no exact solution available for enclosure natural-convection problems, with the exception of limiting cases as the Rayleigh number vanishes. All existing methods of analysis are approximate in nature, but are applicable to a variety of enclosure-flow problems. They can be broadly classified into three categories as follows:

1. Asymptotic analysis (analytical) [1, 4, 26, 32, 33]
2. Discretization methods (numerical) [31, 34, 35]
   Finite difference methods
   Finite element methods
3. Hybrid and other numerical methods

Detailed descriptions of these methods are beyond the scope of this Chapter, and readers are directed to the references above. In the following sections emphasis will be placed on the advantages and disadvantages of each of the methods as well as on the recent developments in each area.

For purposes of illustration only, attention in the following sections is directed to the simple enclosure natural-convection problem associated with the two-dimensional enclosures given in Fig. 13.1(a) and (b). Under the usual conditions of an incompressible Boussinesq fluid with constant properties, the governing differential equations, in accordance with Eqs. (13.9) to (13.13), can be written as

$$\frac{\partial u}{\partial x} + \frac{\partial v}{\partial y} = 0 \tag{13.17}$$

$$u\frac{\partial u}{\partial x} + v\frac{\partial u}{\partial y} = -\frac{1}{\rho}\frac{\partial(p - p_0)}{\partial x} + \nu\nabla^2 u \tag{13.18}$$

$$u\frac{\partial v}{\partial x} + v\frac{\partial v}{\partial y} = -\frac{1}{\rho}\frac{\partial(p - p_0)}{\partial y} + g\beta(T - T_0) + \nu\nabla^2 v \tag{13.19}$$

$$u\frac{\partial T}{\partial x} + v\frac{\partial T}{\partial y} = \alpha\nabla^2 T \tag{13.20}$$

where the origin of the coordinates (x, y) is placed at the lower left corner of the enclosure and gravity is in the $-y$ direction. Also, T_0 can simply be taken as the cold wall temperature T_C, and p_0 is the hydrostatic pressure in the enclosure. The corresponding boundary conditions are simply

$$\begin{aligned} &x = 0, && 0 \le y \le H, && u = v = 0; && T = T_H \\ &x = W, && 0 \le y \le H, && u = v = 0; && T = T_C \\ &y = 0, H, && 0 < x < W, && u = v = 0; && \frac{\partial T}{\partial y} = 0 \end{aligned} \tag{13.21}$$

13.3.2 Asymptotic Analysis

In an asymptotic analysis, no attempt is made to obtain uniformly valid solutions throughout the flow and temperature fields, even though this is the ultimate goal [32]. Rather, limiting solutions for subregions of the fields are first obtained by letting one of the dimensionless parameters become either very large or vanishingly small. Attempt is then made to obtain asymptotic expansions around these limiting solutions, which are then matched asymptotically for the intermediate regions.

Unfortunately, the procedure for obtaining the terms in the asymptotic expansions becomes increasingly complex when higher order terms are included, and there is always the question of the radii of convergence of these expansions. For the enclosure natural-convection problem under consideration here, the solutions from the asymptotic analysis depend on which limiting solutions are utilized. Furthermore, sometimes there are nonuniformities in the expansions and the solutions can only be constructed by introducing additional ad hoc conditions or idealizations.

For the problem formulated in Eqs. (13.17) to (13.21), it can easily be shown that the three dimensionless parameters governing the phenomenon are the same ones given in Eq. (13.16). Even for a given Prandtl number, there are several possible limiting cases [1, 5] as follows:

Ra_H	α^*	Limit
$\to 0$	Arbitrary	Conduction
$\to \infty$	Arbitrary	Boundary layer
Fixed	$\to \infty$	Tall enclosure
Fixed	$\to 0$	Shallow enclosure

Other limiting cases can also be identified based on $Pr \to 0$ and $Pr \to \infty$ [4, 26]. While each of these limits has its own unique characteristics and is valid only in its own region, asymptotic expansions around these limits can be constructed and asymptotically matched to yield uniformly valid solutions in the overlapped regions. Unfortunately, not all such expansions are known at the present time.

The case associated with the conduction limit is the easiest to understand physically. For vanishing Ra_H, the circulating convective motion is very slow, and heat transfer across the enclosure is dominated by conduction [36]. Asymptotic solutions can be constructed for both the temperature and stream function which characterize the flow field in terms of ascending powers of Ra_H. The leading term for the temperature expansion is simply the linear temperature variation, and that for the stream function is of the order of Ra_H.

For the case of the boundary-layer limit with Ra_H approaching infinity, it is expected that thin boundary layers will form along both the hot and cold walls. The core region is relatively stagnant and stably stratified. How the wall boundary layers can be properly linked to the flow in the core has become a central, but yet unsettled issue in constructing the asymptotic solutions for this problem [4]. Batchelor [36] was the first to set the stage for the development of such a solution. While the boundary layers along the isothermal walls could be scaled in the usual manner in accordance with the boundary-layer theory, Batchelor chose for the core region dominant behaviors given by an isothermal core with a constant vorticity. These were subsequently found to be incorrect.

A more realistic asymptotic analysis was made by Gill [37] and recently clarified by Blythe, Daniels, and Simpkins [38]. Since this analysis is representative of all asymptotic analyses for the enclosure problem, it is instructive here to describe several key steps in the analysis.

Since the boundary-layer thickness along the hot wall is of the order of $Ra_H^{-1/4}$, the boundary-layer region there can be scaled according to

$$\bar{x} = \frac{x}{H} Ra_H^{1/4}, \qquad \bar{y} = \frac{y}{H} \tag{13.22}$$

and the corresponding asymptotic expansions can be constructed [39] as

$$\psi(x^*, y^*) = \mathrm{Ra}_H^{1/4} \sum_{r=0}^{\infty} \psi_r(\bar{x}, \bar{y})\, \mathrm{Ra}_H^{-r/4}$$

$$\theta(x^*, y^*) = \sum_{r=0}^{\infty} \theta_r(\bar{x}, \bar{y})\, \mathrm{Ra}_H^{-r/4} \tag{13.23}$$

where ψ is a dimensionless stream function obtained by normalizing the usual stream function with the thermal diffusivity α, and θ is a dimensionless temperature variable defined as $(T - T_C)/(T_H - T_C)$. It is noted that coefficient functions ψ_r and θ_r are also dependent on Pr. The coordinates x^* and y^* are x and y normalized by means of the height H. In the core region, no scaling is needed, and the corresponding expansions can therefore be written as

$$\psi(x^*, y^*) = \mathrm{Ra}_H^{1/4} \sum_{r=0}^{\infty} \psi_{cr}(x^*, y^*)\, \mathrm{Ra}_H^{-r/4}$$

$$\theta(x^*, y^*) = \sum_{r=0}^{\infty} \theta_{cr}(x^*, y^*)\, \mathrm{Ra}_H^{-r/4} \tag{13.24}$$

Here the coefficient functions are also functions of Pr. The governing differential equations (13.17) to (13.20) can now be recast into those based on the new dependent variables ψ and θ by eliminating the continuity equation and also eliminating the pressure terms in the momentum equations by cross differentiations, resulting in

$$\frac{\partial \psi}{\partial x^*}\frac{\partial \nabla^2 \psi}{\partial y^*} - \frac{\partial \psi}{\partial y^*}\frac{\partial^2 \Delta \psi}{\partial x} = \mathrm{Pr}\left(\nabla^4 \psi - \mathrm{Ra}_H \frac{\partial \theta}{\partial x^*}\right) \tag{13.25}$$

$$\frac{\partial \psi}{\partial x^*}\frac{\partial \theta}{\partial y^*} - \frac{\partial \psi}{\partial y^*}\frac{\partial \theta}{\partial x^*} = \nabla^2 \theta \tag{13.26}$$

When the core expansions of Eq. (13.24) are substituted into Eqs. (13.25) and (13.26) and we let $\mathrm{Ra}_H \to \infty$, the following is obtained:

$$\theta_{c0} = \theta_{c0}(y^*)$$

$$\psi_{c0} = \psi_{c0}(y^*) \tag{13.27}$$

indicating that the core region is vertically stratified. These functions, however, are not known and can only be determined by matching asymptotically with the solutions for the wall boundary-layer regions, including those along the adiabatic surfaces. For instance, as the boundary layer along the hot wall extends into the core, the matching requires

$$\theta_{c0}(y^*) = \lim_{\bar{x} \to \infty} \theta_0(\bar{x}, \bar{y}) = \theta_{0,\infty}(\bar{y})$$

$$\psi_{c0}(y^*) = \lim_{\bar{x} \to \infty} \psi_0(\bar{x}, \bar{y}) = \psi_{0,\infty}(\bar{y}) \tag{13.28}$$

The asymptotic solution given by Gill [37] deals with the specific case of large Pr ($\mathrm{Pr} \to \infty$). It is also to be noted that the present case is for $\mathrm{Ra}_H \to \infty$ with a fixed aspect ratio α^*. When Eqs. (13.25) and (13.26) are recast in terms of the boundary-layer variables $\bar{x}$ and $\bar{y}$ and these limits applied, the boundary-layer equations become

$$\frac{\partial^3 \psi_0}{\partial \bar{x}^3} = \theta_0 - \theta_{0,\infty}(\bar{y}) \tag{13.29}$$

$$\frac{\partial \psi_0}{\partial \bar{x}} \frac{\partial \theta_0}{\partial \bar{y}} - \frac{\partial \psi_0}{\partial \bar{y}} \frac{\partial \theta_0}{\partial \bar{x}} = \frac{\partial^2 \theta_0}{\partial \bar{x}^2} \tag{13.30}$$

The boundary conditions at the wall $\bar{x} = 0$ are simply

$$\psi_0 = \frac{\partial \psi_0}{\partial x} = 0, \qquad \theta_0 = 1 \tag{13.31}$$

and as $\bar{x} \to \infty$

$$\psi_0 \to \psi_{0,\infty}(\bar{y}), \qquad \theta_0 \to \theta_{0,\infty}(\bar{y}) \tag{13.32}$$

Additional conditions must be imposed on $\psi_{0,\infty}$ and $\theta_{0,\infty}$ before solution of Eqs. (13.29) and (13.30) can be attempted. These are known as the centrosymmetry conditions and are based on symmetries from both the hot and cold walls:

$$\psi_{0,\infty}(\bar{y}) = \psi_{0,\infty}(1 - \bar{y}), \qquad \theta_{0,\infty}(\bar{y}) = 1 - \theta_{0,\infty}(1 - \bar{y}) \tag{13.33}$$

In Gill's treatment [37], the coupled equations (13.29) and (13.30) are solved by the standard Oseen linearization technique. The final equations for $\psi_{0,\infty}(\bar{y})$ and $\theta_{0,\infty}(\bar{y})$ include two arbitrary constants which are determined by the ad hoc mass-flux hypothesis that

$$\psi_{0,\infty}(0) = \psi_{0,\infty}(1) = 0 \tag{13.34}$$

The physical interpretation of this hypothesis is that the top and bottom corners of the enclosure are impermeable, so that the mass from the boundary layer empties completely into the core. The mathematical implication here is that the boundary layers along the insulated walls are sufficiently thin to be completely neglected. Therefore, the asymptotic matching between the boundary layer and the core need only be carried out in the x direction. It is also to be noted that in the limit as $\mathrm{Ra}_H \to \infty$, the vertical boundary-layer thickness also becomes vanishingly small, and then Gill's solution is uniformly valid for $\mathrm{Pr} \to \infty$.

Attempts have been made to improve Gill's solution so that the results can be extended to cases of large but finite Ra_H. One such attempt has been given by Bejan [40], in which the mass-flux hypothesis is replaced by a more general energy-flux hypothesis which allows for both impermeable and adiabatic horizontal surfaces. A similar analysis for $\mathrm{Pr} < 1$ has more recently been given by Graebel [41].

A proper treatment must necessarily include the boundary layers on the horizontal surfaces. Unfortunately, such a treatment is not known to have been attempted for arbitrary Pr, nor for $\mathrm{Pr} \to \infty$.

This example of asymptotic analysis illustrates one of the common shortcomings of this approach. It is in general very difficult to construct properly matched asymptotic expansions that are uniformly valid. Any compromise in this regard often requires the introduction of ad hoc conditions and consequently is not very satisfying as a formal analysis. On the other hand, despite these shortcomings, the asymptotic analysis does provide considerable physical insight into the flow and temperature fields in the neighborhood of the limiting conditions considered.

In addition to the conduction limit and boundary-layer limit just described, there are also cases in which the aspect ratio is allowed to approach its limits for fixed Rayleigh number. In this class, there are the tall-enclosure limit as $\alpha^* \to \infty$ and the shallow-enclosure limit as $\alpha^* \to 0$. In the asymptotic sense, the tall-enclosure limit is essentially that of one-dimensional conduction. For high Rayleigh numbers, where convection becomes important, the flow patterns and heat transfer characteristics are well predicted by the boundary-layer limit solutions described previously. For the shallow-enclosure limit, a great deal has been learned from asymptotic analyses over the years. The special case of $\mathrm{Ra} \sim 1$ and $\alpha^* \ll 1$, known as the Hadley limit, has received special attention due to its possible application to shallow solar ponds, reactors, crystal growth, and geophysical phenomena, as well as to the fact that the thermal diffusion is significantly different from that of the boundary-layer limit situations. Asymptotic analyses for the Hadley limit have been carried out by Hart [42–44] and Cormack et al [45], and have recently been reiterated and clarified by Simpkins and Chen [33].

Another limiting condition represents a cross between the boundary-layer limit and the shallow-enclosure limit with small α^* and fixed and variable Ra_H. Bejan and Tien [46] analyzed the heat transfer by means of asymptotic analyses for the region of vanishing Rayleigh number ($\mathrm{Ra}_H \to 0$), the intermediate region ($\mathrm{Ra}_H \sim 1$), and the boundary-layer region ($\mathrm{Ra}_H \to \infty$). Even though the condition that the core flow is parallel to the insulated horizontal surfaces is only correct for the Hadley and boundry-layer limits, this condition has also been applied to other cases with aspect ratios less than unity. This is somewhat unfortunate, since it is now known that the condition of parallel core flow is essentially valid only for aspect ratios less than 0.1 [47]. This instance illustrates yet another shortcoming of the asymptotic analysis: that it is in general very difficult to determine the region of validity of the asymptotic solutions by obtaining the solutions themselves.

13.3.3 Numerical Methods

Numerical methods represent a useful alternative to asymptotic analyses in treating enclosure natural-convection problems. In view of the general accessibility of mainframe computers, such methods have proven to be increasingly popular. The entire temperature and flow fields can be calculated for a given set of Ra_H, Pr, and α^*, and numerical solutions have often been used to ascertain the validity of the asymptotic solutions. On the other hand, there are also shortcomings in the numerical methods which cast some degree of uncertainty on the calculated results, as will be discussed later in this section.

In the following subsections on the various numerical methods, the specific purpose is not to describe in detail the procedure in each of the methods, but rather to address the critical issues involved. An understanding of these issues is important in carrying out the numerical solution to any enclosure natural-convection problem. For readers interested in learning the details of the methods, pertinent references are cited in the following.

13.3.4 Finite Difference Methods

Among the many numerical methods, finite difference methods have been the most popular for the analysis of buoyant enclosure flow problems [48, 31, 34, 35]. There are also more or less standard computer codes available for recirculating flows with or without buoyancy. Good examples are TEACH [49], SIMPLE [34, 50], COMIX [51], and UNDSAFE [52]. While these finite difference methods are indeed versatile and produce good results in many instances, they are not without pitfalls. The primary purpose of this section is therefore to identify and discuss several critical issues in the use of finite difference methods for buoyant enclosure flow problems. These issues must be borne in mind for reasonable assurance that the results in a given numerical study are correct.

One critical issue relative to the finite difference methods is that of false or artificial diffusion. It is now well known [34] that upwind differencing for the convection terms produces numerical false diffusion which may overwhelm the physical diffusion when convection effects are dominant. As a result, calculations based on upwind differencing may lead to gross errors [53], particularly in buoyancy-driven enclosure flows at high Rayleigh numbers. For instance, in the experimental study of Bajorek and Lloyd [54], oscillations have been observed in the corner regions of a square enclosure at Rayleigh numbers as low as 10^6, while the numerical solutions based on upwind differences [55] are completely stable and steady for the flow field.

Attempts have been made to remedy the false-diffusion problem by using higher-order differencing schemes for the convective terms [53, 56, 57]. For the two-dimensional flow problem, these higher order schemes all utilize more cells surrounding the base cell which is being calculated. The most promising is the QUICK (quadratic upstream interpolation for convective kinematics) scheme developed by Leonard [53], which has been shown to reduce false diffusion significantly. Its extension to three-dimensional natural-convection flows in box enclosures has also met with some success recently [58]. Also, in a recent study of the basic square-enclosure problem for turbulent flows at very large Rayleigh numbers (up to $\mathrm{Ra} = 10^{16}$) by Markatos and Pericleous [50], the upwind-differencing scheme is used, but the grid is successively refined until the solution becomes grid-independent. The rationale here is that since false diffusion depends on the cell size, it is expected that different false-diffusion levels would lead to different solutions. If the solution becomes grid-independent, then false diffusion must not be an important factor. Whether this rationale is true in general remains to be seen.

The choice of cell sizes in finite difference calculations is evidently an important issue. To obtain high resolution in the computed results and ensure good accuracy without the contamination of false diffusion, cell sizes should be as small as possible, though not so small as to overtax the available computing facilities. Numerical stability requirements must also be adhered to [48]. Extrapolation formulas are available to extrapolate finite difference calculations to zero cell size [59], but their generality is not certain, especially where nonlinear effects are large. At the present time, the goals of numerical computations with finite differences should be to obtain results that are grid-independent.

A more critical issue, however, is the controversy between uniform and nonuniform grid systems. It can be argued that for the same number of calculation cells, a nonuniform grid allows for better resolution in regions where large changes in the physical behavior occur. On the other hand, a case can be made for the uniform grid system where the physics is not sufficiently clear to allow for a predetermined nonuniform grid. Further controversy arises because of the lack of very accurate bench-mark cases with which calculation results based on uniform and nonuniform grid systems can be compared [60, 61]. At the present time, the use of nonuniform grid systems also suffers the lack of rigorous criteria as to what degree of nonuniformity of cell sizes should be used in a given problem. Research is critically needed here to develop rational criteria, which somehow should be tied to the physical phenomena.

A related issue for finite difference calculations is the use of body-fitted coordinate systems. One major deficiency of the finite difference method is the difficulty in fitting a finite difference grid to the bounding surfaces of the enclosure, which can be quite complex. The use of body-fitted coordinates eliminates the problem by transforming the physical enclosure domain to a simple domain, in which the finite difference calculations are carried out. The difficulty, however, is that the body-fitted coordinates must be numerically generated in accordance with the numerical solution to a separate elliptic differential equation. This represents a major area of research in recent years [62].

Also, for buoyant flow in a closed enclosure, any transient or unsteady heating or cooling from the boundaries affects the average pressure in the enclosure. Consequently, in the numerical calculations the average pressure must be allowed to vary in accordance with the thermodynamic conditions [63]. If the numerical solution is based on primitive variables, then a second global pressure correction must be applied to account for this constant-volume process [64].

13.3.5 Finite Element Methods

The development of finite element methods for fluid flow and heat transfer problems has been more recent [31, 35, 65–67] than that of the finite difference methods, even though both methods had their origins in pure conduction analyses. The intent of this section is to address several aspects of the general methods and also to briefly delineate the relative merits and defects of the two numerical methods.

As in their applications to conduction problems, the inherent advantage of the finite element methods is that complex boundaries can be accommodated rather readily and therefore they are particularly suited for realistic enclosures where complex shapes are encountered. On the other hand, the finite element methods do contain two approximations to the governing equations, rather than one as in the finite difference methods. In the finite element methods, (1) the governing equations are only satisfied by an average relative to certain weighting functions, and (2) the calculation domain must be discretized and corresponding interpolation formulas evaluated for each resulting element. Only the latter approximation is utilized in the finite difference methods.

Another inherent difference between the two methods is that it is convenient in the finite difference methods to carry out the calculations by marching in time, while the finite element methods involve inherently time-independent calculations. In the latter case, final solutions for time-dependent cases may be obtained by solving simultaneous ordinary differential equations in time or by replacing the time derivatives with backward differences. Also it maybe pointed out that calculations of the finite element methods in general require the simultaneous solutions of very large set of algebraic equations, which may overtax the computing facilities. On the other hand, because of the use of nonlinear interpolation formulas or basis functions, less nodes are required to yield the same degree of accuracy.

There are also several critical issues that need to be addressed. At the present time, the weighted-residual integral formulation is the dominant finite element formulation used in problems of fluid flow and heat transfer; of the methods based on it, the Galerkin method is the most popular. There are several other integral formulations which can be used in finite element calculations, notably the simple central integration method, the method of least squares, the penalty-functional method, and the collocation method [31]. There is a critical need to examine closely which of these methods produce the most satisfactory results in terms of computational efficiency and accuracy for buoyant enclosure flow problems. At the present time, very little is known about this question.

As in the finite difference methods, the choice of elements in the finite element methods and their size distribution also represent an important issue. For fluid flow problems, the most popular elements in two-dimensional situations are triangles and rectangles. The nonuniform element-size distributon does not create any difficulty and can be generated to place more elements in regions with large flow and temperature variations. Body-fitted coordinates and various schemes of grid generation can also be used in finite element methods [62]. The critical problem here is still the lack of rational criteria for setting up optimal finite element grid in a given problem for most efficient solution.

Numerical stability and convergence criteria for finite element calculations are very difficult to determine for viscous enclosure flow problems. The usual practice is to obtain them by numerical experimentation with various time steps and element sizes. Improvement of numerical stability by means of upwind schemes in the interpolation formulas has been achieved recently, but the solution of the associated false-diffusion problem is still not completely at hand [68]. There is every indication that in the foreseeable future the finite element methods will become a strong competitor to finite difference methods, particularly in view of their capability of dealing with problems involving complex boundaries.

Finally, there is the important issue for both finite difference and finite element calculations concerning the computer on which the calculations are carried out. The existing mainframe computers are rather inadequate for three-dimensional calculations. Supercomputers that are based on vector and parallel processing have been increasingly accessible to the scientific research community, and represent a powerful tool to deal with many unresolved issues of natural convection in enclosures. Unfortunately, to use supercomputers efficiently requires a very different coding strategy and may even indicate a change in the basic algorithms in the numerical methods, and more research is critically needed in this regard [69].

13.3.6 Hybrid and Other Numerical Methods

While the finite difference and finite element methods are by far the dominant numerical methods for studies in buoyant enclosure flow problems, other methods are available which are semianalytical or hybrid and based on rather different approaches [31, 70]. Good examples are the spectral methods [71, 72], the vortex methods [73], and the more recently developed finite analytic method [74, 75]. These methods have just begun to be applied to buoyant enclosure problems, and their potential here remains to be demonstrated.

13.4 LAMINAR NATURAL CONVECTION IN ENCLOSURES

13.4.1 Two-Dimensional Buoyant Enclosure Flow: Analyses

Laminar buoyant flows in two-dimensional enclosures are the best studied of all buoyant enclosure flows. The two-dimensionality greatly simplifies the analyses, and yet most of the physics involved in general buoyant enclosure flows is still retained. Experimental measurements are simpler to carry out than three-dimensional experiments, and the results can be used directly for comparison with those from analyses.

The purpose of this section is to summarize the more important recent results from two-dimensional analyses relative to the steady flow and temperature fields.

For rectangular enclosures with $\alpha^* > 1$ and differentially heated vertical side walls, asymptotic solutions for the boundary-layer regime at high Rayleigh numbers have been given by Gill [37], Bejan [40], and Graebel [41], as already discussed. Many more solutions, dealing with both insulated and perfectly conducting end walls for the vertical slot problem, have been obtained by numerical methods. Earlier studies have been reviewed by Catton [2], and more recent studies are given by Korpela et al. [76] and by Schinkel et al. [77]. The general flow behavior and heat transfer characteristics are now fairly well established; they depend on the Rayleigh and Prandtl numbers and the aspect ratio. In accordance with the classification proposed by Eckert and Carlson [78], the flow behavior at a given Rayleigh number undergoes a change from a boundary-layer type, through a transition regime, and finally to one dominated by conduction, as the aspect ratio increases from slightly greater than unity to large values.

More recent findings suggest even more complex flow behavior. In the boundary-layer flow regime, a unicell flow structure exists and the core region is stratified. In the transition regime, the core stratification persists, while flows start to appear in the core region. Typical examples of the isotherms and streamlines in this regime are shown in Fig. 13.3 for various Ra_H and α^* and end boundary conditions, in accordance with the numerical results of Schinkel et al. [77]. When α^* becomes very large, the isotherms all become essentially vertical and the flow reverts to a unicell structure. The heat transfer is dominated by conduction, and increases as α^* increases.

However, as pointed out by Bergholz [79] and Korpela et al. [76], for Rayleigh numbers Ra_H greater than about 2.5×10^7 there is another distinct flow regime, lying between the transition regime and the conduction regime for a given aspect ratio, and characterized by regularly spaced multicellular flow in the core region. These cells do not affect the overall heat transfer much, since most of the energy is transferred in the end regions away from the cells. From the results of the stability analysis of Bergholz [79], the heat transfer across the slot decreases as the aspect ratio is reduced in the conduction regimes, and achieves a minimum when multicellular flow first appears. A correlation can be determined to relate the aspect ratio to the Grashof number when such a minimum occurs, and is given by

$$\alpha^{*3} + 5\alpha^{*2} = 1.25 \times 10^{-4}\,\mathrm{Gr}_H \tag{13.35}$$

where $\mathrm{Gr}_H = \mathrm{Ra}_H/\mathrm{Pr}$. This equation is useful for determining the optimum spacing for minimum heat transfer through a double-pane window [76].

For laminar flow and heat transfer in two-dimensional rectangular shallow enclosures with $\alpha^* < 1$, many of the characteristics have been clarified by asymptotic analyses in the different limiting flow regimes as already discussed in Sec. 13.3.2. These characteristics have since been verified by numerical solutions to the governing equations [80–82], and typical examples are shown in Fig. 13.4 for isotherms and streamlines for two Rayleigh numbers and $\mathrm{Pr} = 1.0$ and $\alpha^* = 0.1$, based on the primitive-variable calculations of Tichy and Gadgil [82]. Since the calculations are based on upwind differencing for large cell Péclet numbers, there is a question of their accuracy because of false-diffusion effects, as noted previously. However these results are certainly expected to depict the global behavior of the flow and temperature fields adequately. In view of the high Rayleigh numbers, this behavior corresponds to the limits as $\mathrm{Ra}_H \to \infty$ for fixed small aspect ratios. As noted earlier, these limits are different from that for fixed Ra_H as $\alpha^* \to 0$ [45]. It is seen in Fig. 13.4 that the flow is characterized by thin layers lining all four walls and the high velocities in these layers

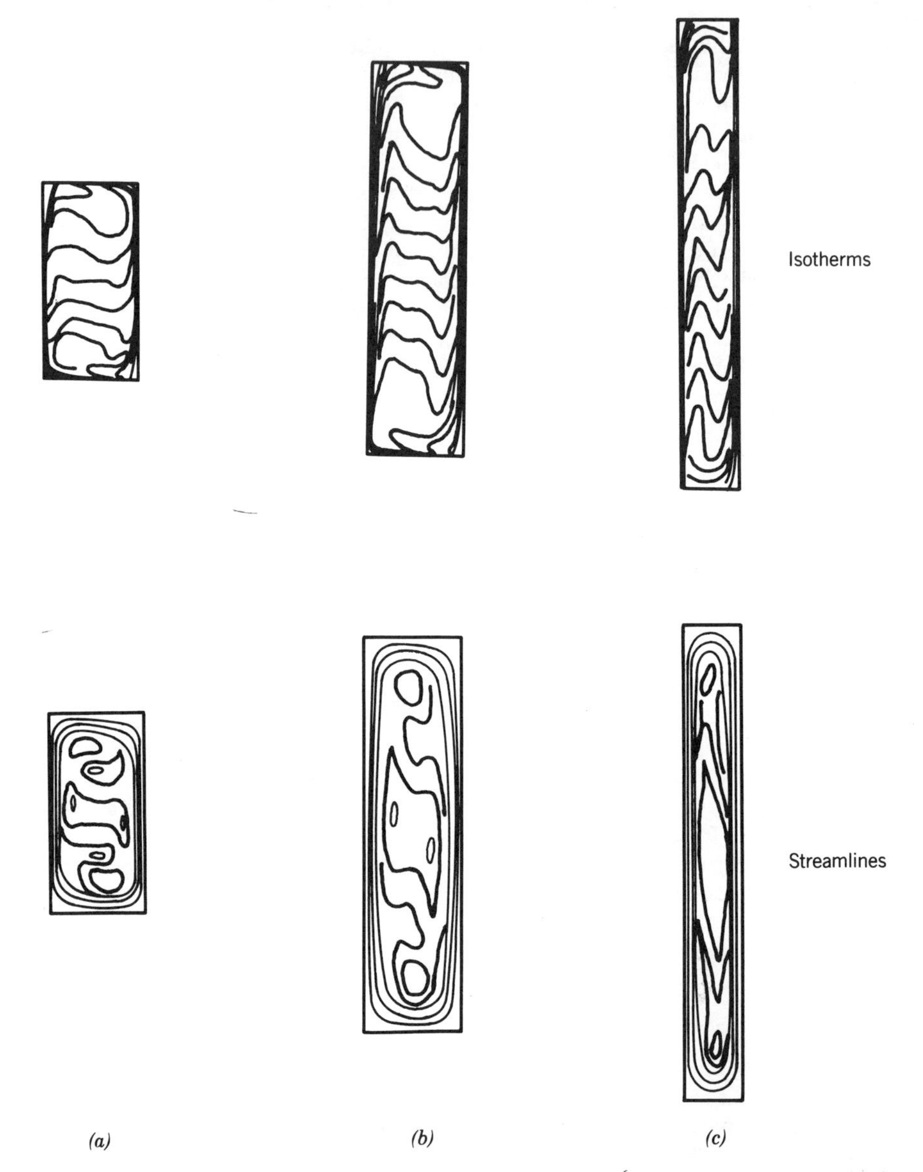

Figure 13.3. Isotherms and streamlines: (*a*) $Ra_H = 1.32 \times 10^6$, $\alpha^* = 2$, perfectly conducting walls; (*b*) $Ra_H = 1.56 \times 10^6$, $\alpha^* = 4$, perfectly conducting walls; (*c*) $Ra_H = 1.6 \times 10^6$, $\alpha^* = 8$, adiabatic walls [77].

drive the core flow. Almost all the temperature drops occur in the vertical end-wall regions, leaving the core region essentially stratified. The streamlines in the core are nearly parallel, and as the Rayleigh number increases, the core region extends almost the entire length of the enclosure.

Calculations for small Prandtl numbers down to $Pr = 0.01$ have also been made by Shiralkar and Tien [81], and it has been found that for $Pr < 1$ there exists a transition

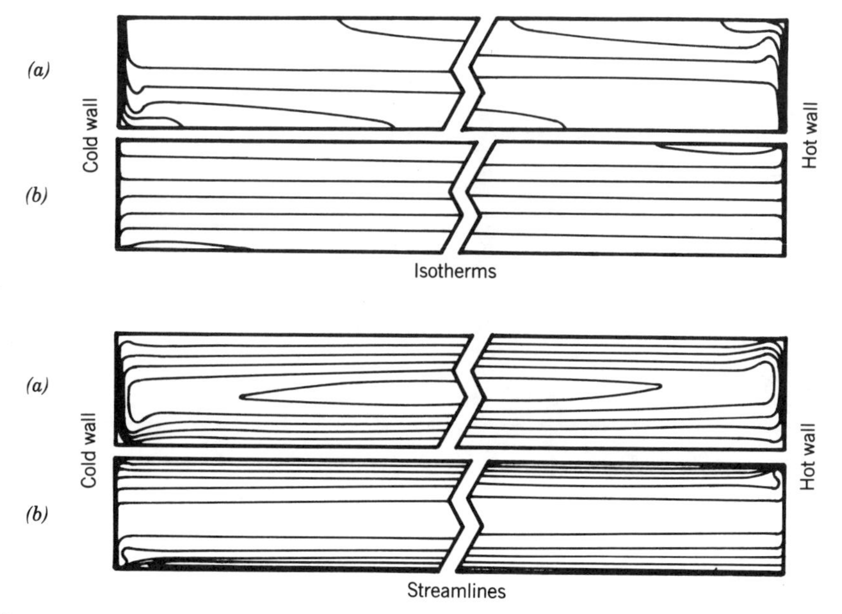

Figure 13.4. Isotherms and streamlines, Pr = 1.0, $\alpha^* = 0.1$: (a) Ra = 10^6, (b) Ra = 10^8 [82].

regime outside the end-wall boundary-layer region in which inertia effects dominate, resulting in a very large Prandtl number effect as Pr → 0.

While asymptotic analyses have provided much physical insight into buoyant rectangular enclosure flow for both $\alpha^* > 1$ and $\alpha^* < 1$, they are not quantitatively applicable to the corresponding square-enclosure problem with $\alpha^* = 1$, even though some of the basic physics is still retained. As a result, the analysis of the square-enclosure problem falls into the domain of numerical solutions; and in view of its geometrical simplicity and well-defined boundary conditions, is often used as a standard problem for developing or testing different numerical methods. Because no exact solutions exist for the natural-convection-dominated range of Rayleigh numbers, it s difficult to assess the relative accuracies of the results based on these numerical methods. A concerted effort, however, has been made to compare the various numerical results for this square enclosure flow problem with a Boussinesq fluid for Pr = 0.71 and Rayleigh numbers of 10^3, 10^4, 10^5, and 10^6 in an international "competition" [83] which has attracted 36 entries from nine different countries. The results are compared in terms of the maximum velocities and their positions at mid section, and in terms of the average Nu_H on the hot wall and the positions of its maxima and minima. Details of this comparison exercise are given in [60, 84]. The main conclusions are that there is a good consensus on the average Nu_H for the various Rayleigh numbers, but that other local characteristics from different methods show considerable scatter.

As later discussed by Quon [61], such comparisons of local characteristics are not very meaningful in view of the very different grid systems used in the entries. Quon [61] and Markatos and Pericleous [50] have subsequently made further calculations on this problem, emphasizing the effects of grid distribution and high Rayleigh numbers. One set of results for $\mathrm{Ra}_H = 10^7$ and Pr = 0.71 in terms of streamlines and isotherms is given in Fig. 13.5, in which the hot wall is on the right. The boundary-layer regimes can be clearly seen, the core region is again highly stratified, and there is also a clear

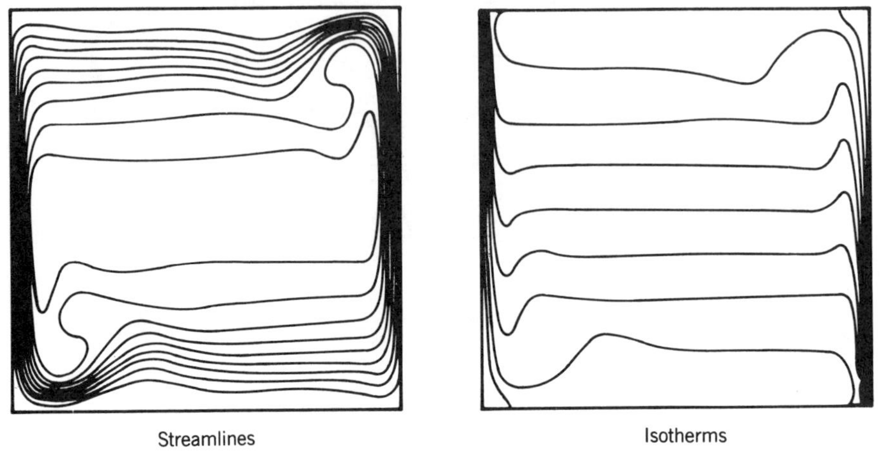

Figure 13.5. Streamlines and isotherms, Ra = 10^7, $\alpha^* = 1.0$, Pr = 0.71 [61].

indication of the considerable penetration of the wall layers near the horizontal wall into the core flow at this Rayleigh number. The vertical velocity profiles near the hot wall for $\mathrm{Ra}_H = 10^6$ and Pr = 0.71 calculated by Quon [61], Winters [85], and Gartling [60] are shown in Fig 13.6. This is one of the cases covered by the "competition" [83], and the data from Winters [85] and Gartling [60] were calculated by finite element methods. While the overall agreement appears to be good, considerable scatter does occur, especially near the velocity maxima.

Heat transfer results for vertical enclosures based on analytical analyses and numerical computations have been correlated in the recent literature. Typical examples will be given in Sec. 13.7.

While problems in vertical rectangular enclosures have received a great deal of attraction in recent years, special attention has also been directed to the corresponding problems in which the same enclosures are tilted relative to the direction of gravity. The design of tilted solar collectors represents an important application. The mechanisms of the flow behavior and its stability encountered in the vertical enclosures still play a role in the tilted enclosures. However, the unstable stratification and thermal instability effects usually associated with the heated-from-below phenomena are expected to provide additional physical mechanisms affecting the flow and heat transfer. The physical phenomena are much more complicated, and the tilt angle γ (Fig. 13.1*c*) is an additional parameter that must be accommodated in any analysis of the tilted-enclosure problem.

Earlier analyses were occupied with the interaction of hydrodynamic and thermal instabilities for γ close to 180°, as reviewed by Catton [2]. In addition, heat transfer results have also been obtained for a range of Rayleigh numbers and aspect ratios at various tilt angles. For fixed values of Ra_H and α^*, heat transfer is by conduction only for sufficiently small γ. As γ increases, convection becomes more important due to increased buoyancy along the isothermal walls and the effect of unstable stratification. At $\gamma = 90°$, the buoyancy effect reaches a maximum, while the increasing effect of unstable stratification persists, thus giving rise to a maximum heat transfer at $\gamma > 90°$. The heat transfer then decreases until a critical tilt angle γ_c is reached, after which the heat transfer rises until $\gamma = 180°$. Prior to $\gamma = \gamma_c$, the flow inside the enclosure has a two-dimensional unicell structure, as long as the combination of Rayleigh number and aspect ratio is such that the flow is hydrodynamically stable. At $\gamma = \gamma_c$ and beyond,

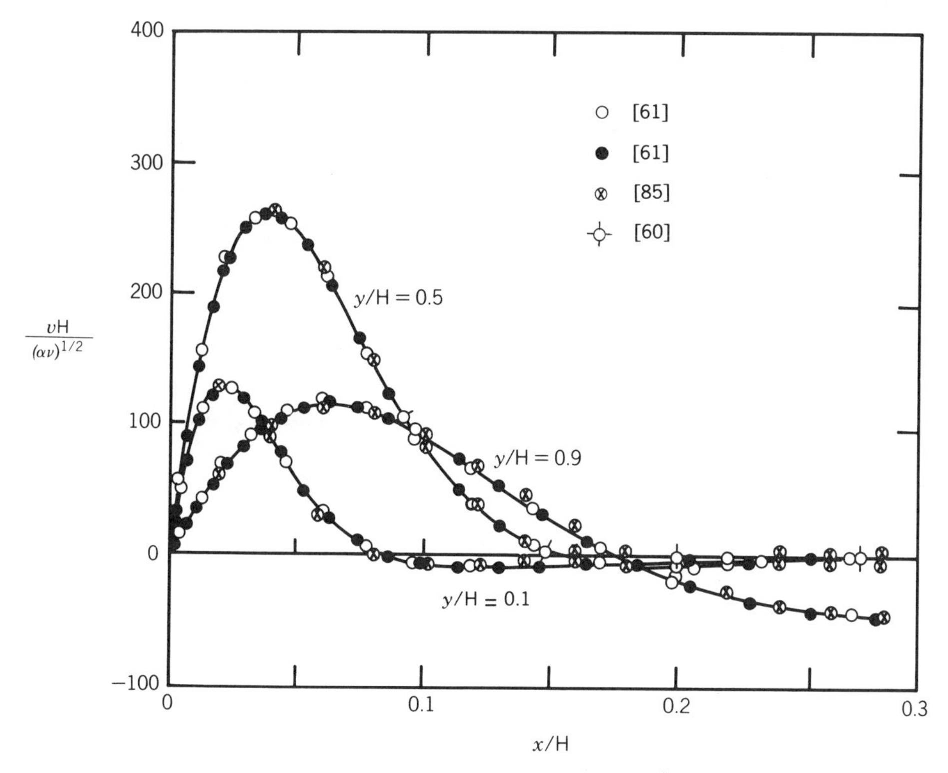

Figure 13.6. Vertical velocities near the hot wall, Ra = 10^6, $\alpha^* = 1.0$ [61].

both stability analysis and experiments have shown that the unicell structure becomes unstable and the flow undergoes a transition to a three-dimensional roll [2]. In this region, the flow behavior can no longer be predicted by two-dimensional calculations. The critical tilt angle γ_c has been experimentally determined by Arnold et al. [86] as a function of the aspect ratio, and does not appear to depend strongly on the Rayleigh number. Values of γ_c are given in Table 13.1. Attempts have been made recently to carry out three-dimensional numerical computations to predict this functional relationship [58, 87, 88].

Typical isotherms and streamlines for a tilted enclosure are shown in Fig. 13.7 for an air-filled enclosure for $Ra_H = 10^5$, $\alpha^* = 1.0$, and $\theta_0 = 0.5$ [89]. It is seen that even at $\gamma = 45°$, conduction still dominates. However, at $\gamma = 135°$ the flow becomes much

TABLE 13.1 Critical Tilt Angle γ_c

α^*	γ_c (deg)
1	165
3	127
6	120
12	113
> 12	110

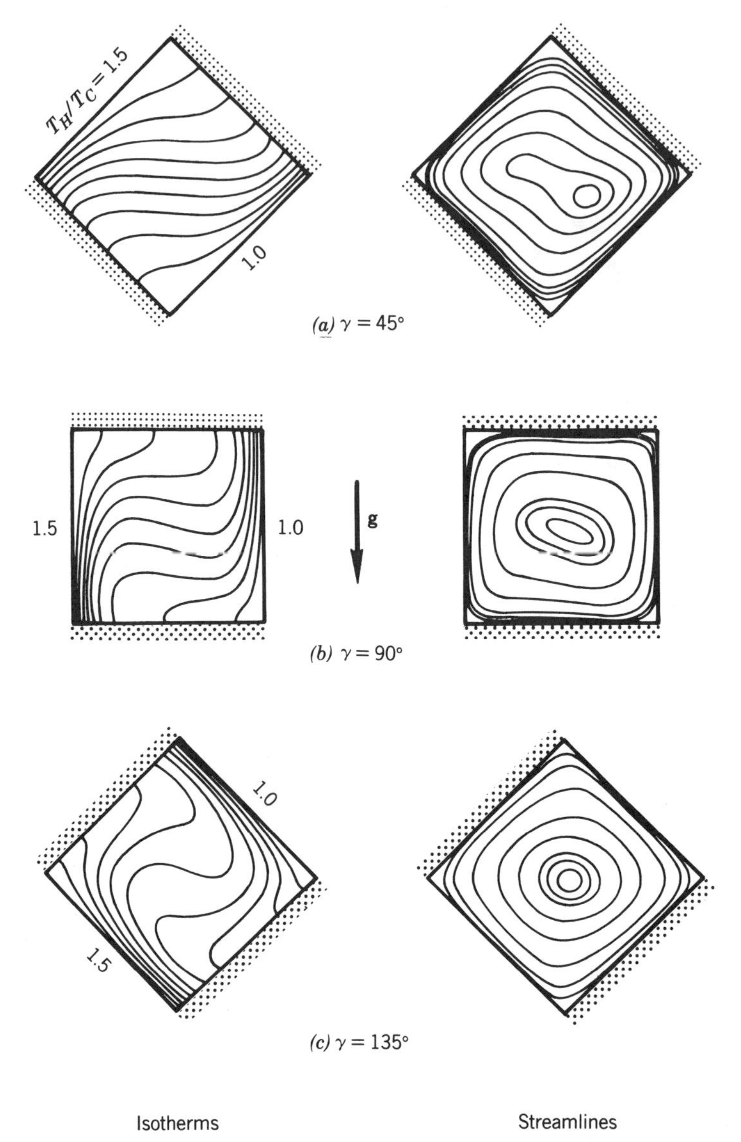

Figure 13.7. Isotherms and streamlines for tilted square enclosures, Ra = 10^5.

more vigorous and the isotherms already show some similarity to those for thermally unstable conditions.

From the results of two-dimensional numerical calculations, it is possible to correlate the heat transfer results as a function of Ra_H, α^*, and γ, as follows [2]:

$$\frac{\mathrm{Nu}_H(\gamma)}{\alpha^*} = \begin{cases} 1 + \left(\dfrac{\mathrm{Nu}_H(90°)}{\alpha^*} - 1\right)\sin\gamma & \text{for} \quad 0 < \gamma \leq 90° \\ \dfrac{\mathrm{Nu}_H(90°)}{\alpha^*}(\sin\gamma)^{1/4} & \text{for} \quad 90° \leq \gamma < \gamma_c \end{cases} \tag{13.36}$$

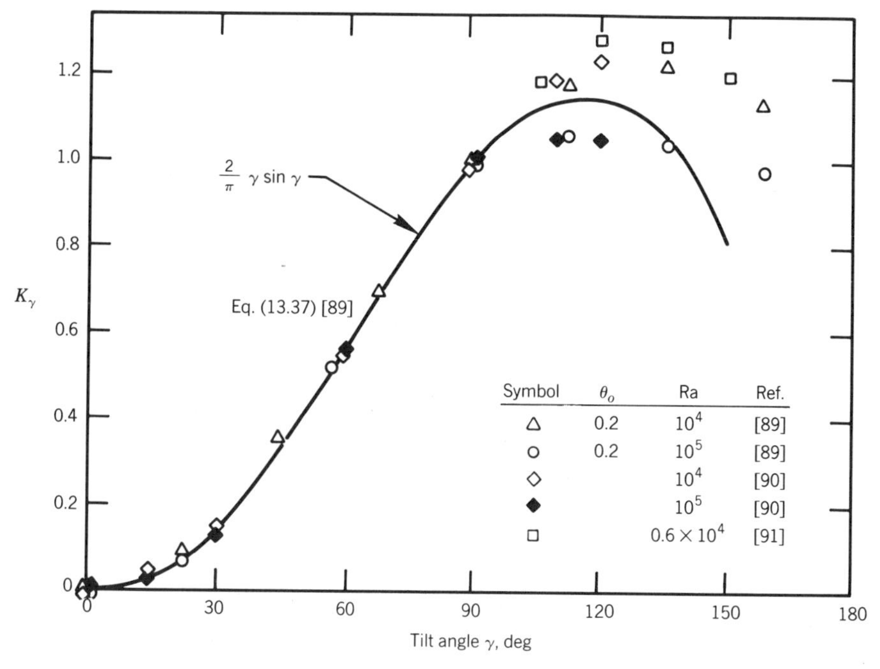

Figure 13.8. K_γ variations at different tilt angles γ [89].

It is particularly interesting to note that the Nusselt number can be directly scaled in terms of its corresponding value at $\gamma = 90°$.

In a more recent numerical study dealing with air-filled tilted enclosures with large temperature differences, Zhong et al. [89] have proposed a more general correlation for $\alpha^* = 1.0$ in the region $0 < \gamma \leq 90°$ as follows:

$$K_\gamma = \frac{\mathrm{Nu}_H(\gamma) - \mathrm{Nu}_H(0°)}{\mathrm{Nu}_H(90°) - \mathrm{Nu}_H(0°)} = \frac{2}{\pi}\gamma \sin \gamma \tag{13.37}$$

where $\mathrm{Nu}_H(0°)$ is that for pure conduction. The correlation parameter K_γ is shown in Fig. 13.8 together with data from the literature. It is also shown [89] that this correlation is valid for Ra_H up to 10^6 and θ_0 up to 2.0.

So far the discussion has been concentrated on empty rectangular enclosures. Other two-dimensonal enclosures such as those shown in Fig. 13.1*d* and *e* have also attracted much attention. However, because of their geometrical complexities, analyses for these enclosures are essentially limited to numerical solutions. Partitioned rectangular enclosures are important in ventilation studies in single and connecting rooms and also in solar collectors, where baffles are utilized to reduce convection effects. Though several experimental studies have been carried out (see [54]), there are very few numerical investigations on the same problem. Chang et al. [55] have treated natural convection in the standard square enclosure, fitted with symmetric adiabatic vertical partitions connected to the adiabatic horizontal walls. The height of each partition is less than half the height of the enclosure, so that there is an opening at mid height. The results show that the partitions essentially act as flow barriers to reduce the convective flows inside the enclosure, thus achieving lower heat transfer rates. Another pertinent study in this regard is the one carried out by Kaviany [92], and treats a square enclosure with a

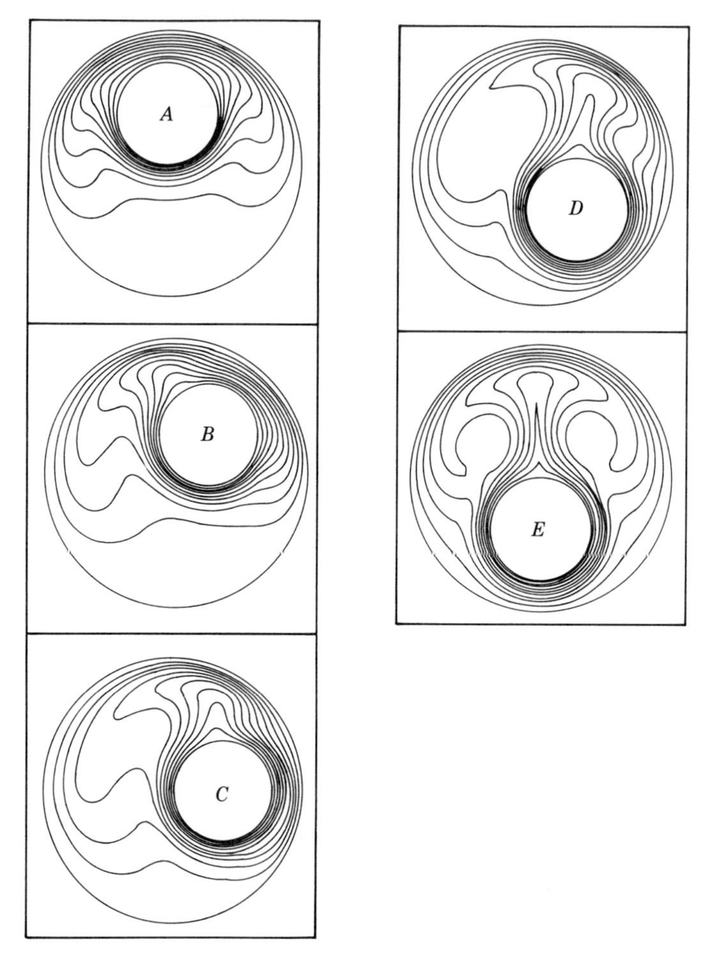

Figure 13.9. Isotherms and streamlines for annmuli of the same eccentricity, but in different angular positions; $\mathrm{Ra}_L = 10^4$, $\tilde{\epsilon} = 0.623$, $R_i R_o = 0.3846$ [19].

half-cylindrical protuberance located at center of the floor. The general conclusion is similar to that of [55]; in addition, it is found that the protuberance loses its effect as $\mathrm{Ra} \to \infty$.

Two-dimensional cylindrical annuli represent another geometry for which several numerical solutions have been obtained. For the concentric annulus, streamlines, isotherms, and average heat transfer data have been given recently by Cho et al. [19] and Farouk and Guceri [93]. The corresponding solutions for eccentric cylindrical annuli have been obtained by Cho et al. [19] and Prusa and Yao [94]. Since all these solutions compare well with the experimental data of Kuehn and Goldstein [95], the numerical approaches can be considered valid. It is of interest to note that a bipolar coordinate system of coordinates is used in [19], while a radial transformation technique is used in [94]. For illustration, isotherms and streamlines for eccentric cylindrical annuli with the same eccentricity $\tilde{\epsilon}$, but at different angular positions, are shown in Fig. 13.9 for $\mathrm{Ra}_L = 10^4$ and $R_i/R_o = 0.3846$, where $\tilde{\epsilon}$ is defined as the ratio of the distance of the centers to $R_o - R_i$, and Ra_L is the Rayleigh number based on a

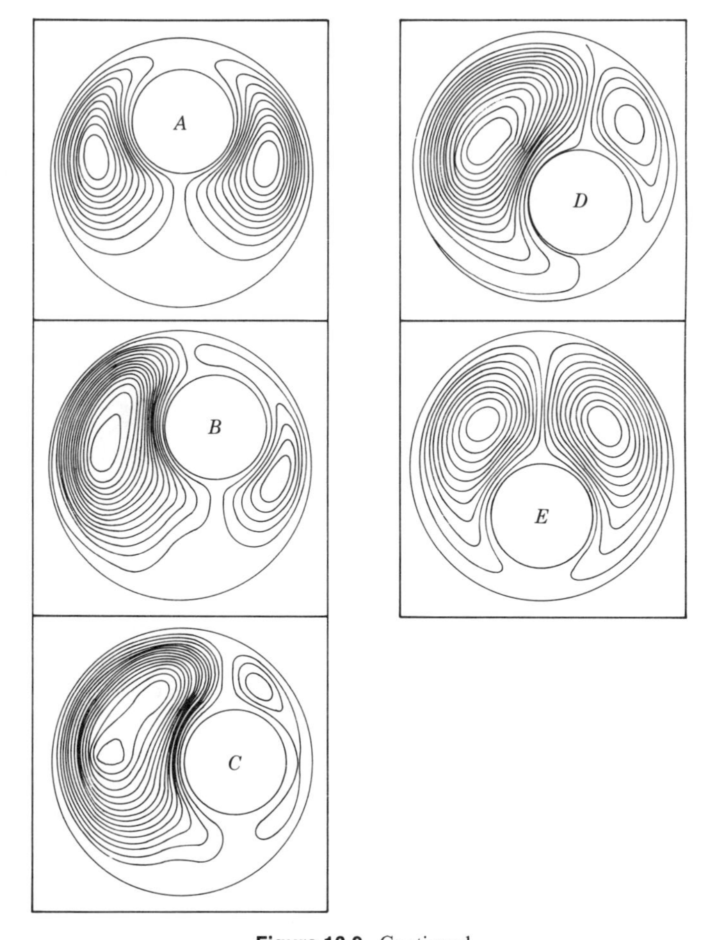

Figure 13.9. Continued.

characteristic length $R_o - R_i$. Here the inner cylinder is maintained at T_H. Numerical studies dealing with some variations of this problem may be found in [96, 97].

Two additional geometries of two-dimensional enclosures have been numerically treated. One deals with a square prism located concentrically in a horizontal circular cylinder [98], and the other with a right triangular enclosure depicting an attic space [99].

13.4.2 Two-Dimensonal Buoyant Enclosure Flow: Experimental Studies

Ever since the pioneering work of Elder [100, 101] on the details of the flow field in vertical rectangular enclosures, experimental studies on a variety of essentially two-dimesnional enclosures have received much attention. These studies serve two important functions. One is to provide heat transfer data for practically important enclosures and their correlations as functions of Rayleigh and Prandtl numbers and other geometrical parameters. The other is to serve as a complementary part of asymptotic, numerical, and stability analyses, either to provide a means to validate them or to furnish proper

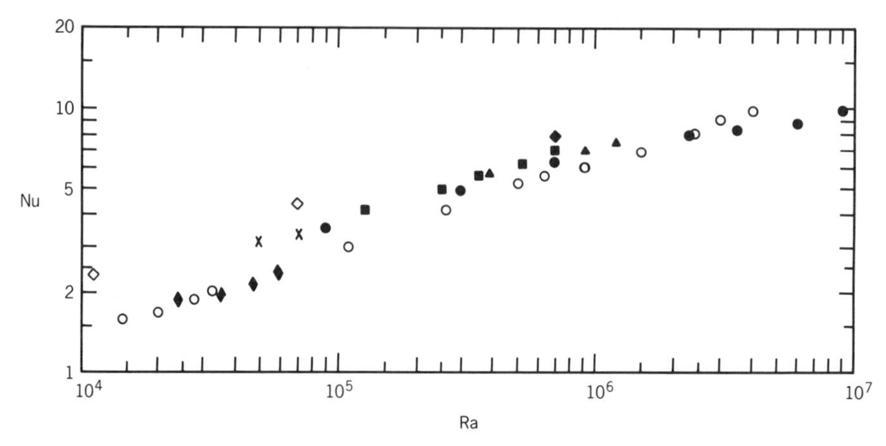

Figure 13.10. Experimental average Nusselt numbers for square enclosures [57]. Sources: ■, air [57]; ▲, CO_2 [54]; ○, [86]; ●, [109]; ◇, [110], × [111]; ◆, [112].

physical insight to guide them. It is clear that for the latter purpose, detailed velocity and temperature measurements are also needed.

Many experimental techniques are available for the study of buoyant enclosure flow. A recent review of these techniques has been given by Hoogendoorn [102]. Average and local Nusselt numbers can readily be measured by the standard caloric technique or the use of heat-flow meters. On the other hand, measurements of local temperatures and velocities are much more difficult. For the temperature field, the most popular technique is Mach-Zehnder interferometry [103], while holographic interferometry [102, 104] and differential interferometry [105] have also seen increased use in enclosure flow studies. For velocity-field measurements, flow visualization techniques, which give mostly qualitative information on the flow field, have been very popular, with the use of smoke, dye, or electrochemical tracers. For detailed quantitative velocity measurements, the laser Doppler velocimeter (LDV) is the most popular and perhaps the most accurate device at the present time [102, 106–108].

Known heat transfer data and their correlations for various two-dimensional enclosures, among others, have recently been compiled by Raithby and Hollands [6]. The extensiveness of this compilation attests to the recent interest in buoyant enclosure flow. There is, however, a critical issue relative to correlating the heat transfer data for enclosure flows and to comparing them with those from the analyses, analytical or numerical. Consider, as an example, the case of a differentially heated square enclosure with adiabatic horizontal walls. Several sets of data from Refs. 54, 86, 109–112 are shown in Fig. 13.10. It is seen that there exists considerable scatter among the data, and similar observations have also been made by Bejan [5]. This degree of scatter poses a problem in that any correlation is subject to uncertainties difficult to tolerate in design practice, and also in that these data cannot appropriately be used for validating the corresponding analytical and numerical solutions.

Most of this problem can be traced to the difficulty of controlling the thermal boundary conditions at the enclosure walls [89, 113]. In analyses, differentially heated rectangular enclosures are treated with either adiabatic or perfectly conducting horizontal walls. Unfortunately, these conditions are both very difficult to achieve in the laboratory, and the actual horizontal-wall conditions lie somewhere between adiabatic and perfectly conducting [54, 89]. To illustrate this difficulty, Zhong [114] has utilized

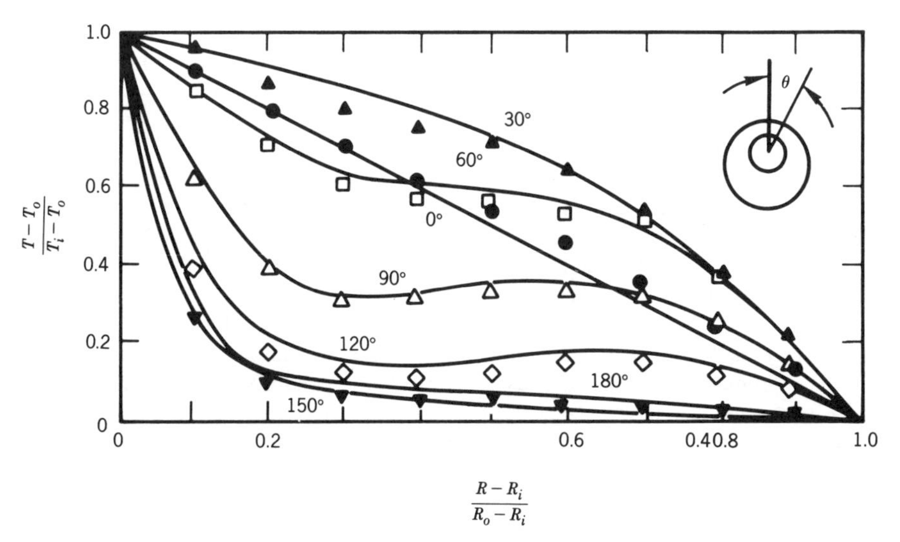

Figure 13.11. Radial temperature distributions for eccentric annulus [19]. Curves: calculations [19]; plotted points: experimental data [95].

the experimentally determined wall temperature variations along all four walls of a square enclosure [54] to obtain a numerical solution, and has shown that the results agree with the experimental data much more closely than do the numerical solutions for either adiabatic or perfectly conducting horizontal walls. It is thus quite possible that the scatter shown in Fig. 13.10 is primarily due to the different actual wall conditions in the various experiments.

Despite the aforementioned difficulty, experimental studies have continued in many cases to provide validation of theoretical analyses. A good example is shown in Fig. 13.11 for the radial temperature distributon in an eccentric annulus, where the numerical results [19] compare very well with the experimental data of Kuehn and Goldstein [95]. Other recent measurements on vertical, inclined, and parallelogrammatic enclosures in the range of $\alpha^* < 1$ can be found in Refs. 115 to 118, where the transition between unicell and multicell structures in the flow field as a function of the tilt angle is again clarified. A particular study worthy of note is that of Simpkins and Dudderar [119], who have made velocity measurements in shallow enclosures with $\alpha^* < 0.1$ and delineated the conditions under which the configuration of thin layers on the horizontal walls changes to one where those layers fill the entire depth of the enclosure.

13.4.3 Three-Dimensional Buoyant Enclosure Flow

Many physically realistic situations of buoyant enclosure flows are inherently three-dimensional. Both analyses and experiments for such three-dimensional phenomena are difficult to do. Theoretical studies are limited to perturbation analysis, which usually has a limited range of validity, and to numerical solutions, which may not be accurate in view of the necessity of using coarse grids. Experimentally, while caloric methods can still be utilized for overall heat transfer measurements, basic two-dimensional nonintrusive methods such as Mach-Zehnder interferometry cannot be used for three-dimensional studies, and even flow visualization techniques are much more difficult to carry out. However, in view of the increasing accessibility of large and fast computing facilities, numerical solutions with the desired field resolutions are now possible. Furthermore, the development of experimental techniques has also advanced to the state where three-dimensoinal experimental investigations are becoming more routine. A good example in this regard is the use of the laser Doppler velocimeter for three-dimensional flow measurements. Consequently, there are indications that three-dimensional buoyant enclosure flow will soon become an active field of research.

For the box geometry as shown in Fig. 13.1f, two aspect ratios are needed to define the geometry: $\alpha_x^* = H/W$ and $\alpha_z^* = L/W$, where the width W is the distance separating the isothermal surfaces, H is the height, and L is the depth. When $\alpha_z^* \rightarrow \infty$, α_x^* reduces to α^* for the two-dimensional case. If the box is arbitrarily tilted, then another angle is needed to specify the box orientation in addition to the tilt angle γ.

In early numerical studies, emphasis was placed on conditions under which three-dimensionality in the flow becomes important. Mallinson and de Vahl Davis [120] obtained numerical solutions to the vertical box problem with adiabatic horizontal and end vertical walls, and found that small depth, low Rayleigh number, and low Prandtl number all increase the three-dimensionality in the flow and that there is a spiraling motion from the side walls toward the center from both ends. This latter result agrees at least qualitatively with the velocity field measured by Morrison and Tran for a box with $\alpha_x^* = \alpha_z^* = 5$ and $\mathrm{Ra}_H = 6.25 \times 10^4$ by means of a laser Doppler velocimeter [121]. This experimental study also shows that when heat loss is permitted at the two vertical ends, the resulting three-dimensional flows destroy the centrosymmetry for two-dimensional rectangular enclosures, as has also been found in a recent LDV study by Bilski [122].

Buoyant flow in inclined box enclosures is also the subject of a series of numerical studies with complementary experimental observations by Ozoe et al. [87, 88, 106, 110, 123]. The emphasis in these studies is on determining the critical tilt angles at which the Nusselt number achieves a minimum. As already discussed, this minimum corresponds to the transition from unicell structure to that of longitudinal rolls. These numerical studies do produce such a transition, and from the resulting velocity fields it is possible to determine the transition process. One interesting aspect of the calculations done by Ozoe et al. [88] is that for rectangular enclosures of arbitrary aspect ratio, it is only necessary to carry out calculations for a single roll cell with either free-free boundaries (for cells away from the rigid walls) or rigid-free boundaries (for cells next to a wall). From the experimental observations, these cells all have aspect ratios close to unity at $\mathrm{Ra}_H = 4 \times 10^3$ and $\mathrm{Pr} = 10$, the conditions for the calculations. Results of these single-roll-cell calculations are shown in Fig. 13.12, which clearly illustrates the occurrence of the maximum Nusselt numbers. It has also been found that a slight increase (about 10%) in the cell width does not affect these results. In a recent study of the same problem, Yang et al. [58] have utilized an improved numerical procedure to compare the results based on air directly with the experimental data, and good agreement has been found. In addition, the transition process characterized by a change of the direction of the cell axis by 90° is further clarified.

Another three-dimensional geometry which has received much attention is a long horizontal circular cylinder with differentially heated end walls. This geometry is of some practical interest in dealing with stress concentrations in dead-leg pipes connected to hot main pipes and with crystal growth by physical or chemical vapor transport in a cylindrical ampul. Theoretical analyses have been carried out by Bejan and Tien [124], Kimura and Bejan [125], and Schiroky and Rosenberger [126]. Their results show that the warm fluid from the hot end wall runs toward the cold wall in the upper half of the pipe, while there is a counterflow of cold fluid in the lower half of the pipe. In addition, spiral secondary flows are superimposed on the longitudinal flows.

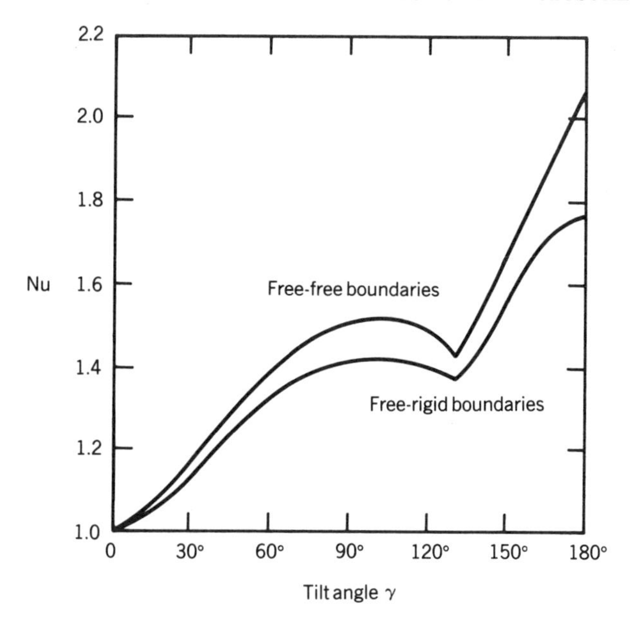

Figure 13.12. Effect of tilt angle on average Nusselt number of cell [88].

Recently, Schiroky and Rosenberger [126] have applied a boundary-layer analysis similar to that of two-dimensional rectangular enclosures and provided results very close to the experimental data of Kimura and Bejan [125] in the Rayleigh number range (based on the radius) between 10^7 and 10^9. Also recently, Schiroky and Rosenberger [108] have measured detailed three-dimensional velocity fields in a gas-filled horizontal pipe with a radius-to-length ratio of 0.1 and a wide range of Rayleigh numbers from 74 to 1.3×10^6, by means of a laser Doppler velocimeter. It has been found that there are pronounced three-dimensoinal effects in the end-wall regions and that part of the flow changes its direction before reaching the end-wall regions.

When a circular cylinder differentially heated at the end walls is tilted, the flow pattern is expected to be very complex. For a tilted cylindrical annulus, it is found that the flow structure is that of a coaxial double helix; and despite the complex flow patterns, the average Nusselt number only increases slightly when the tilt angle varies from 0° (horizontal) to 90° (vertical) [128, 129].

The recent literature is also rich in heat transfer data and their correlations for three-dimensional enclosures with enclosed heated bodies. Nusselt number correlations for concentric and eccentric spheres have been given by Raithby and Hollands [6]. A generalization to very general enclosures and inner bodies has been given recently by Warrington and Powe [130], and correlations are also available for enclosures bounded by concentric and eccentric vertical cylinders [131], enclosures with off-center inner bodies [132], and cylindrical tube bundles in a cubic enclosure [133].

13.5 TURBULENT NATURAL CONVECTION IN ENCLOSURES

13.5.1 Two-Dimensional Turbulent Buoyant Flow in Enclosures

When the Rayleigh number becomes sufficiently high, enclosure flow undergoes transition to turbulent flow. The transition process has been documented by the pioneering experimental study of Elder [100,101] for natural convection in vertical enclosures with $10 < \alpha_x^* < 30$, $1 < \alpha_z^* < 5$, and high Pr fluids (Pr = 1000). In his observations of the flow patterns in a vertical slot, at $Ra_H > 8 \times 10^8\, Pr^{12}$, traveling waves, similar to those found on isolated vertical plates, grow independently on both hot and cold walls. For $Ra_H \approx 10^{10}$, the traveling waves become highly irregular and there is intense interaction between the wall region and the core. When $Ra_H > 10^{10}$, the horizontal temperature gradient in the enclosure disappears and the mean velocity vanishes in the core.

The calculations of turbulent buoyant flow in an enclosure, as pointed out previously, require phenomenological turbulent models for closure. Various turbulence models† and their applications have been reviewed in Refs. 24, 25, and it is noted that such models are subject to less uncertainty for forced flows then for buoyant flows. These models range from simple zero-equation or algebraic models to high order turbulent stress models which include partial differential equations for all components of the turbulent stresses. In between lie the popular two-equation κ-ϵ models. The algebraic models, though easy to use, do suffer the shortcoming that no details of the turbulence field can be determined. On the other hand, the high-order stress models are difficult to use and are also subject to high degrees of uncertainty in view of the many constants and functionals that need to be introduced. Consequently, the κ-ϵ models represent a good compromise between ease of use and the physics involved.

For steady two-dimensoinal turbulent buoyant flows in enclosures, the governing differential equations for the turbulent kinetic energy κ and the kinematic rate of dissipation ϵ can be written respectively as [134]

$$\frac{\partial}{\partial x}(\rho u \kappa) + \frac{\partial}{\partial y}(\rho v \kappa) = \frac{\partial}{\partial x}\left(\Gamma_\kappa \frac{\partial \kappa}{\partial x}\right) + \frac{\partial}{\partial y}\left(\Gamma_\kappa \frac{\partial \kappa}{\partial y}\right) + G_\kappa - \rho\epsilon \tag{13.38}$$

$$\frac{\partial}{\partial x}(\rho u \epsilon) + \frac{\partial}{\partial y}(\rho v \epsilon) = \frac{\partial}{\partial x}\left(\Gamma_\epsilon \frac{\partial \epsilon}{\partial x}\right) + \frac{\partial}{\partial y}\left(\Gamma_\epsilon \frac{\partial \epsilon}{\partial y}\right) + C_1 \frac{\epsilon}{\kappa} G_\epsilon - \rho C_2 \frac{\epsilon^2}{\kappa} \tag{13.39}$$

where Γ_κ and Γ_ϵ are the exchange coefficients for the diffusive transport of κ and ϵ, respectively, and

$$\Gamma_\kappa = \frac{\mu_t}{\sigma_\kappa}, \qquad \Gamma_\epsilon = \frac{\mu_t}{\sigma_\epsilon}, \qquad \mu_t = \frac{C_\mu \rho \kappa^2}{\epsilon}$$

$$G_\kappa = \mu_t\left(\frac{\partial u}{\partial y} + \frac{\partial v}{\partial x}\right)^2 - \beta g \frac{\mu_t}{\sigma_\kappa}\frac{\partial T}{\partial y} \tag{13.40}$$

$$G_\epsilon = \mu_t\left(\frac{\partial u}{\partial y} + \frac{\partial v}{\partial x}\right)^2$$

Here μ_t is the turbulent viscosity, and σ_κ and σ_ϵ play a similar role of the turbulent Prandtl number. In addition.

$$C_1 = 1.44, \quad C_2 = 1.92, \quad C_\mu = 0.09, \quad \sigma_\kappa = 1.0, \quad \sigma_\epsilon = 1.314 \tag{13.41}$$

It is seen here that the generation term G_κ includes a buoyancy term which accounts for

†A brief review of turbulence models is also given in Chap. 2.

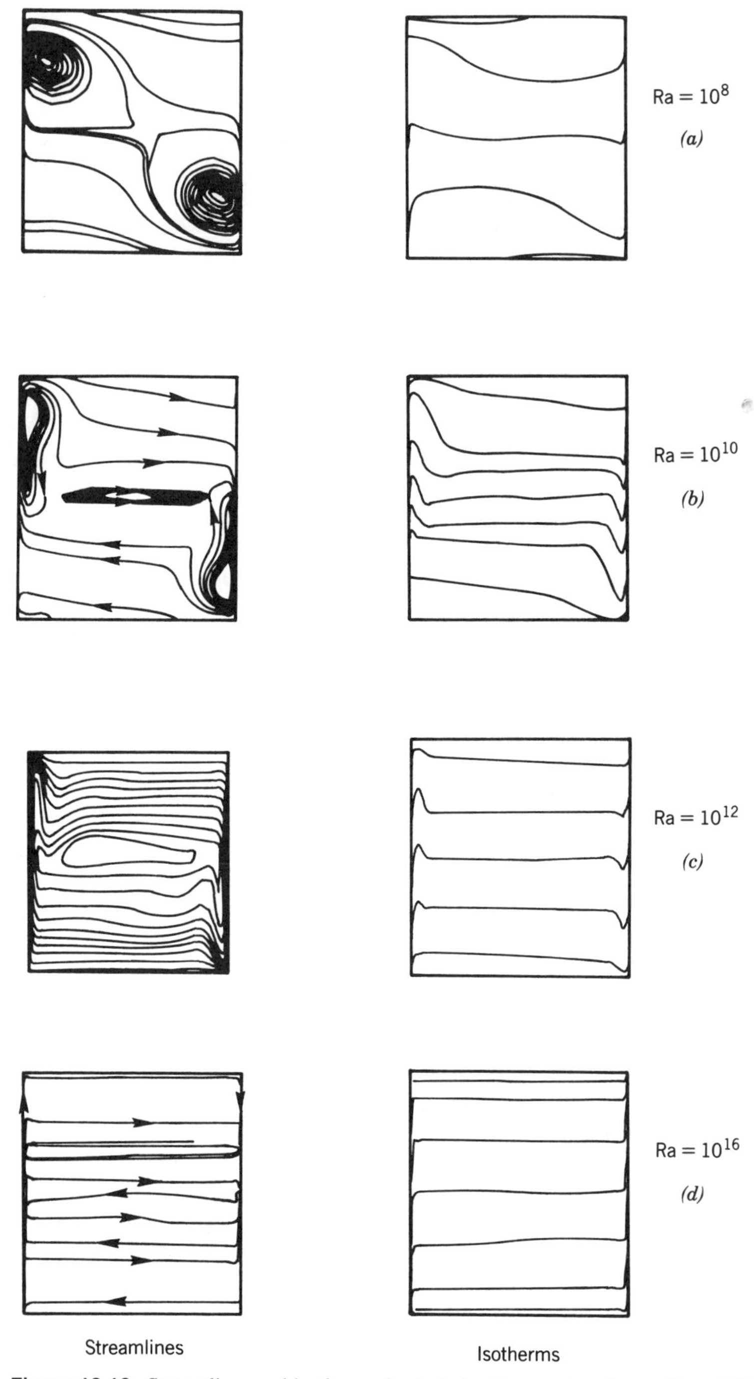

Figure 13.13. Streamlines and isotherms for turbulent bouyant enclosure flows [50].

the effects of stratification. This turbulence model involves constants and wall functions which are determined from experimental data for simple shear flows, and consequently there is a degree of uncertainty as to whether these experimentally determined quantities are still valid for buoyant flow in enclosures.

This model has been used in a numerical study of buoyancy-induced smoke flow in two-dimensional rectangular enclosures by Markatos et al. [135], and also in a numerical study for a square enclosed enclosure for Ra_H in the range between 10^8 to 10^{16} [50]. In the latter study, the enclosure is differentially heated with adiabatic horizontal walls, and the calculated streamlines and isotherms are shown in Fig. 13.13. At $Ra_H = 10^8$, the secondary vortices that existed in the lower Rayleigh number range are now connected toward the vertical walls. At $Ra_H = 10^{10}$, the center vortex reappears, while the secondary vortices start to interact strongly with the wall layers. These flow patterns persist at $Ra_H = 10^{12}$, even though the secondary vortices now completely disappear. At $Ra_H = 10^{16}$, the flow is essentially in stratified layers. For Ra_H at 10^{10} and 10^{12}, temperatures are stratified in the core, but close to the vertical walls the effect of the secondary vortices can still be seen. At $Ra_H = 10^{16}$, the temperature gradients in the x direction completely disappear away from the vertical walls. Almost all the flows are now confined in very thin layers next to all four walls. Experimental results are needed to compare with these numerical findings.

One important application for turbulent buoyant enclosure flows is the spread of fire and smoke in rooms with and without partitions due to fire sources located on the floors of the rooms. This type of enclosure also has openings on either the vertical walls (doorways or windows) or the horizontal walls (ceiling and floor vents). Based on an algebraic turbulence model, a series of numerical studies have been made for two-dimensoinal rectangular enclosures with vents and complex interior partitions, using natural boundary conditions as shown in Fig. 13.2. These studies have recently been summarized by Yang and Lloyd [15]. It suffices to mention here that the collective results show that the vents and internal partitions play very important roles in affecting the flow in such vented enclosures.

13.5.2 Three-Dimensional Turbulent Buoyant Flow in Enclosures

Three-dimensoinal enclosures are always more realistic than two-dimensional ones, but very few studies have been devoted to turbulent buoyant flow in three-dimensional enclosures. From the point of view of numerical solutions, the lack of an appropriate turbulence model for three-dimensional flows is always troublesome, even though some progress has been made in this regard recently [136]. For experimental studies, the same difficulties as those encountered in three-dimensional laminar flows are still present. Flow visualization is even more difficult because of the turbulent diffusion effects in the flow, and more development is needed to apply the LDV technique for turbulent flow measurements in such three-dimensoinal enclosures. Nansteel and Greif [137] has experimentally studied buoyant flows in water-filled enclosures with two and three-dimensonal partitions. The aspect ratios are $\alpha_x^* = 0.5$ and $\alpha_z^* = 2.688$, and the range of Rayleigh numbers Ra_H is 1.25×10^9 to 1.25×10^{10}. The three-dimensional vertical partition is placed midway along W and is fitted with a center opening of constant width and variable height in the form of a doorway. It has been found from flow visualization that the doorway induces much three-dimensionality in the flow and causes localized turbulence and flow instability.

Calculations for three-dimensonal turbulent buoyant flows in rectangular enclosures with vents have been attempted. Of particular interest is that in the numerical and experimental study of Satoh et al. [138] for a three-dimensional rectangular enclosure

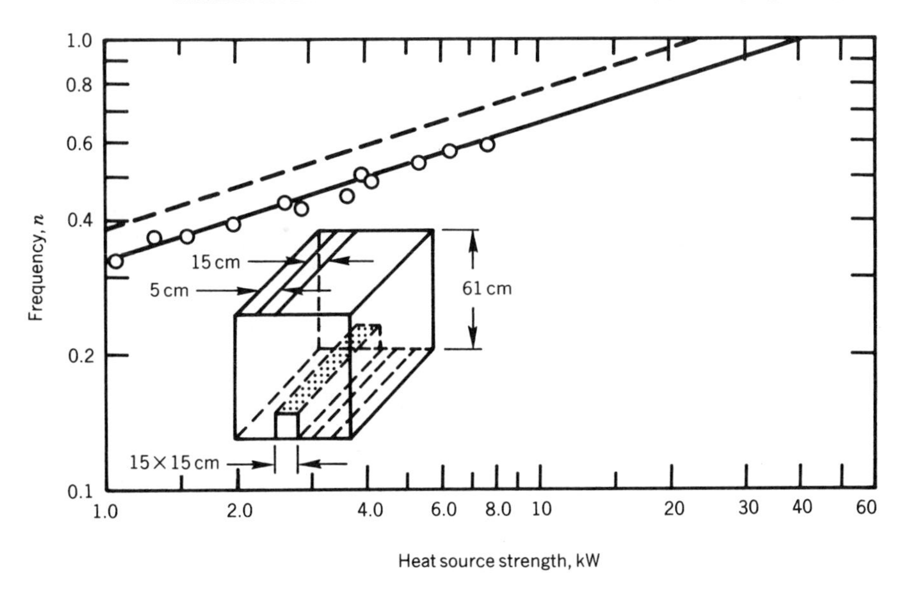

Figure 13.14. Variation of frequency of oscillations with heat source strength for a vertically vented enclosure [138].

with horizontal vents and a heat source located midway on the floor exit, the flow at the ceiling vent is oscillatory and the frequency of oscillation depends on the heat source strength, as shown in Fig. 13.14. Even though the two-dimensional calculations give the same slope, they overpredict the frequencies, while the three-dimensional calculation gives quite adequate results.

13.6 OTHER PHYSICAL EFFECTS IN BUOYANT ENCLOSURE FLOWS

In many physical processes involving buoyancy-driven flows in enclosures, the flows are subjected to additional physical effects which may greatly alter the flow and heat transfer in the enclosures. The purpose of this section is simply to give an overview, with an emphasis on the physical aspects of these combined phenomena.

13.6.1 Transient and Unsteady Phenomena

As is now clear, the internal dynamics of the flow field in an enclosure is sensitive to changes in the Rayleigh number. In a transient or unsteady mode—as represented, for instance, by the buoyant flow in a rectangular enclosure when the hot wall undergoes a step increase in temperature—the effective Rayleigh number during the transient changes from zero at first to the steady-state value in the long run. The added complexity in the development of the flow pattern may even lead to oscillatory behavior when the steady state is reached. Patterson and Imberger [139] and Patterson [140] have performed an ordering analysis for shallow enclosures to delineate the conditions under which the approach to steady state is oscillatory, and the results are substantiated by the experiments of Yewell et al. [141]. In a recent numerical and experimental study, Nicolette et al. [64] have treated transient laminar buoyant flow in a square enclosure filled with air or water, when one vertical wall undergoes a step decrease in temperature while the other three walls are maintained adiabatic. The numerical solution also incorporates a separate global pressure correction to satisfy the thermodynamic conditions. The enclosure both starts and ends with zero motion. During the transient, however, the Nusselt number at the cold wall approaches the final state through a series of abrupt changes due to the dynamics of primary and secondary vortices. Several numerical studies on the transient responses of two-dimensional rectangular and horizontal cylindrical annulus enclosures are also known, and have recently been reviewed by Nicolette [142].

13.6.2 Effects of Complex Boundary Conditions

The boundary conditions so far discussed include only isothermal, adiabatic, and perfectly conducting walls. Since it is known that buoyant enclosure flows are sensitive to the boundary conditions [4], numerical and experimental studies have also been carried out with other boundary conditions which are more complex. The case of uniform heat flux along the heated vertical wall of a rectangular enclosure has been treated numerically by Kimura and Bejan [143], and as expected, the resulting hot-wall temperature increases upward, and the interaction between the boundary layers and the core is similar to that for the case of a vertical isothermal hot wall. Another numerical study by Chao et al. [144] deals with laminar natural convection in a three-dimensional inclined rectangular box with the lower surface half heated and half insulated. Very different flow patterns can be expected in this case. Conjugate problems involving the interaction between the buoyant enclosure flow and conduction in the wall or convection outside the wall have also received some interest [145, 146], and the results show significant wall-temperature nonuniformity.

13.6.3 Interaction between Radiation and Natural Convection in Enclosures

When an enclosure is filled with a gas (with or without scattering particles), radiation effects may become important even for moderate temperature differences. This is indeed the case for such applications as window enclosures, cooling of electronic equipment, solar collectors, smoke and fire spread in rooms, and furnace flows. The radiation effects can be either passive for a nonparticipating medium or active for a participating medium.

The calculation of radiative transfer in multidimensional enclosures is not a simple task and is often complicated further by the use of realistic spectral models. A recent comprehensive review of the interaction between natural convection and a participating gas medium in enclosure flow has been given by Yang and Lloyd [28]. Even though several methods for the calculation of multidimensional radiative transfer in enclosures do accommodate particle scattering, there is hardly any work showing how particle scattering contributes to the heat transfer in the enclosures.

13.6.4 Other Interaction Effects

The problem of combined natural convection and forced convection is technically very important, and is covered in detail in Chap. 15 for internal flow.

Effects of the interaction of natural convection with other physical processes in enclosures are not limited to those already described above. Here we mention several such effects, just to illustrate the breadth of recent research activities which are either important in special applications or significant in the phenomena that they address.

One such interaction effect is melting and solidification in an enclosure, as recently reviewed by Viskanta [147]. When the Rayleigh number associated with the liquid phase is large, natural convection plays a significant role in the movement of the solid-liquid interface, and analysis based on pure conduction often leads to gross errors in determining the melting and solidification rates.

Another phenomenon which has received much attention in recent years concerns natural convection in enclosures filled with porous media [148]. These have important applications in geothermal systems and also in insulation. The more important current activities in this area address more realistic conditions [149] than the simple Darcy formulation. Furthermore, experimental investigations are difficult to perform and represent a critical area of research.

Another interesting phenomenon associated with buoyant enclosure flows is the effect of a density extremum—when the enclosure is filled, for instance, with water having temperature variations crossing the maximum density at 4°C. In such instances the buoyancy-force variations in the enclosure become very complex and significantly affect the heat transfer across the enclosure [150–152].

Finally, mention is here made of the studies dealing with heat transfer augmentation in buoyant enclosure flows by means of corona discharge [153] and by using roughness [154].

13.7 HEAT TRANSFER CORRELATION EQUATIONS FOR NATURAL CONVECTION IN ENCLOSURES

Heat transfer correlation equations for buoyant enclosure flows in terms of appropriate Nusselt numbers can be obtained both from the overall heat transfer measurements or from numerical simulations by integration of local heat transfer rates. In view of the scarcity of experimental data covering the full range of buoyant enclosure flow cases, a strong reliance on numerically generated results can be observed in published correlation in this area [6, 155]. To complement what has already been presented in this chapter, several useful correlations for buoyant enclosure flows are now presented in this final section. However, the emphasis again will be placed on vertical enclosures. More specifically, correlations will not be given for the heated-from-below phenomena. For such phenomena, the reader is referred to correlations given in Refs. 6, 155.

Many of the heat transfer results deal with limiting conditions such as for high and low Grashof numbers and high and low aspect ratios. These results can be made to collapse on single correlation curves by using the joining technique of Churchill and Usagi [156], as given by Churchill [155], and several correlations indicated below have been obtained by this means.

13.7.1 Two-Dimensional Vertical Rectangular Enclosures

In this section, correlation equations are given for the standard vertical rectangular enclosure with differentially heated vertical walls and insulated horizontal walls ($\gamma = 90°$), and the Nusselt number is expressed as a function of the parameters Ra_H, Pr, and α^*. All properties are evaluated at the average temperature of the hot and cold walls. Table 13.2 shows those correlations for the three distinct regions of the aspect ratio α^*—namely, large α^*, α^* near unity, and small α^*—which are based on both experimental and numerical results.

TABLE 13.2 Correlation Equations for Vertical Two-Dimensional Rectangular Enclosures (Fig. 13.1a, b)

Aspect Ratio α^*	Laminar	Turbulent
Large	$\mathrm{Nu}_H = \dfrac{0.364[\mathrm{Ra}_H f_1(\mathrm{Pr})]^{1/4}}{G(\mathrm{Ra}_H, \alpha^*)}$ [36, 40, 155] $f_1(\mathrm{Pr}) = \left[1 + \left(\dfrac{0.5}{\mathrm{Pr}}\right)^{9/16}\right]^{16/9}$ $G(\mathrm{Ra}_H, \alpha^*) = \left[1 + \dfrac{0.231}{(\mathrm{Ra}_H \alpha^{*4})^{1/4}}\right]^2$ $\alpha^* > 5,\ 10^2 < \mathrm{Ra}_H \alpha^{*-4} < 10^6$	$\mathrm{Nu}_H = 0.05[\mathrm{Ra}_H f_1(\mathrm{Pr})]^{1/3}$ [155] $\mathrm{Ra}_H \alpha^{*-4} > 10^6$
≈ 1	$\mathrm{Nu}_H = \alpha^* \left\{ \left[\left(1 + \dfrac{\mathrm{Ra}_H^2 \alpha^{*-7}}{20{,}000}\right)^{-8} + \left(\dfrac{\mathrm{Ra}_H \alpha^{*-2} f_1(\mathrm{Pr})}{57G}\right)^{-2} \right]^{-3/8} + \dfrac{\mathrm{Ra}_H \alpha^{*-3} f_1(\mathrm{Pr})}{8000} \right\}^{-1/3}$ [36, 40, 155] $\mathrm{Ra}_H < 10^9,\ \mathrm{Pr} \approx 0.7,\ \alpha^* \approx 1$	
	$\mathrm{Nu}_H = a\mathrm{Ra}_H^b$ [81] Pr / a / b: 0.01 / 0.1344 / 0.259 0.03 / 0.1521 / 0.266 0.06 / 0.1613 / 0.271 0.10 / 0.1605 / 0.277 $\mathrm{Ra}_H < 10^6,\ \alpha^* = 1$	$\mathrm{Nu}_H = \begin{cases} 0.082\ \mathrm{Ra}_H^{0.329} & \text{for } 10^6 < \mathrm{Ra}_H \le 10^{12} \\ 1.325\ \mathrm{Ra}_H^{0.245} & \text{for } 10^{12} < \mathrm{Ra}_H < 10^{16} \end{cases}$ [50]
Small	$\mathrm{Nu}_H = 1 + \left\{ \left(\dfrac{\mathrm{Ra}_H \alpha^*}{602.4}\right)^{-0.8} + \left[\left(\dfrac{\mathrm{Ra}_H f_1(\mathrm{Pr})}{10.66}\right)^{1/5} \alpha^{*-1} \right]^{-0.4} \right\}^{-2.5}$ [46, 155] $\alpha^* < 0.6,\ \mathrm{Pr} \ge 0.7,\ \mathrm{Ra}_H < 10^8$	

13.7.2 Two-Dimensional Horizontal Annuli

A correlation equation for differentially heated horizontal annuli formed by concentric and eccentric circular cylinders has been given by Kuehn and Goldstein [157], based on extensive experimental and numerical heat transfer results, for the conduction, laminar boundary-layer flow, and turbulent boundary-layer flow regimes with Rayleigh numbers (based on $L_c = R_o - R_i$ and the overall temperature difference) up to 10^{10}, a wide range of Prandtl numbers, and arbitrary R_o/R_i ratios greater than unity. One limitation is that the location of the inner cylinder must be such that the boundary layers at the two cylindrical surfaces do not overlap.

The correlation equation for the convection part may be written as follows:

$$\mathrm{Nu}_{D_i\mathrm{conv}} = \frac{hD_i}{k} = \frac{2}{\ln\left(\dfrac{1 + 2/\mathrm{Nu}_i}{1 - 2/\mathrm{Nu}_o}\right)} \tag{13.42}$$

where

$$\mathrm{Nu}_i = \left\{\left[0.518\,\mathrm{Ra}_{D_i}{}^{1/4} f_2(\mathrm{Pr})\right]^{15} + \left(0.1\,\mathrm{Ra}_{D_i}{}^{1/3}\right)^{15}\right\}^{1/15} \tag{13.43}$$

$$\mathrm{Nu}_o = \left(\left\{\left[\left(\frac{2}{1 - e^{-1/4}}\right)^{5/3} + \left[0.587 f_3(\mathrm{Pr})\,\mathrm{Ra}_{D_o}{}^{1/4}\right]^{5/3}\right]^{3/5}\right\}^{15} + \left(0.1\,\mathrm{Ra}_{D_o}{}^{1/3}\right)^{15}\right)^{1/15} \tag{13.44}$$

with

$$f_2(\mathrm{Pr}) = \left[1 + \left(\frac{0.559}{\mathrm{Pr}}\right)^{3/5}\right]^{-5/12}$$

$$f_3(\mathrm{Pr}) = \left[\left(1 + \frac{0.6}{\mathrm{Pr}^{0.7}}\right)^{-5} + \left(0.4 + 2.6\,\mathrm{Pr}^{0.7}\right)^{-5}\right]^{-1/5}$$

In Eq. (13.42), Ra_{D_i} and Ra_{D_o} are the Rayleigh numbers based on D_i and D_o as characteristic lengths, respectively, and h is an average coefficient of heat transfer based on the difference between the inner cylinder temperature and the fluid bulk mean temperature determined by equating the heat transfer at the two cylindrical surfaces [157]. For the conduction part, which prevails as the Rayleigh number approaches zero, the heat transfer by conduction between two circular cylinders is given by

$$\mathrm{Nu}_{D_i\mathrm{cond}} = \frac{2}{\cosh^{-1}\left\{\left[D_i^2 + D_o^2 - (D_o - D_i)^2\tilde{\epsilon}^2\right]/2D_iD_o\right\}} \tag{13.45}$$

where $\tilde{\epsilon}$ is the eccentricity, normalized with respect to $R_o - R_i$. Equations (13.42) and (13.45) can then be combined, resulting in a correlation equation for the overall Nusselt number valid for any Rayleigh number:

$$\mathrm{Nu}_{D_i} = \left[\left(\mathrm{Nu}_{D_i\mathrm{conv}}\right)^{15} + \left(\mathrm{Nu}_{D_i\mathrm{cond}}\right)^{15}\right]^{1/15} \tag{13.46}$$

It can also be shown that by taking the limit as $D_i \to 0$, Eq. (13.42) reduces to

$$\mathrm{Nu}_{D_o\mathrm{conv}} = \frac{2}{-\ln(1 - 2/\mathrm{Nu}_o)} \tag{13.47}$$

with Nu_0 given by Eq. (13.44), valid for quasi-steady natural convection in a horizontal cylinder. The corresponding overall Nusselt number can again be obtained by adding a conduction part similar to that in Eq. (13.46).

13.7.3 Concentric and Eccentric Spheres

A heat transfer correlation equation for concentric and eccentric spheres has been given by Raithby and Hollands [6] as follows:

$$\mathrm{Nu} = \frac{q(D_o - D_i)}{2\pi D_o D_i (T_i - T_o) k} = [\mathrm{Nu}_{\mathrm{cond}}, \mathrm{Nu}_{\mathrm{conv}}]_{\mathrm{max}} \tag{13.48}$$

where $[\ ]_{\mathrm{max}}$ indicates the higher value of the two Nusselt numbers, and

$$\mathrm{Nu}_{\mathrm{cond}} = \frac{D_o - D_i}{\tilde{\Phi}(\eta_i) D_o - \tilde{\Phi}(\eta_o) D_i} \tag{13.49}$$

$$\mathrm{Nu}_{\mathrm{conv}} = 1.16 f_4(\mathrm{Pr}) \left(\frac{D_o - D_i}{2D_i} \right)^{1/4} \frac{\mathrm{Ra}^{1/4}}{\left[(D_i/D_o)^{3/5} + (D_o/D_i)^{4/5} \right]^{5/4}} \tag{13.50}$$

where

$$\tilde{\Phi}(\eta) = \begin{cases} 0.659\eta^{0.42}, & \text{for } 0 \le \eta \le 1.2 \\ \dfrac{2\cosh\eta - 1}{2\cosh\eta} & \text{for } \eta > 1.2 \end{cases}$$

$$\eta_i = \cosh^{-1}\left[\frac{D_o^2 - D_i^2 - (D_o - D_i)^2 \tilde{\epsilon}^2}{2D_i(D_o - D_i)\tilde{\epsilon}} \right]^{1/2}$$

$$\eta_o = \cosh^{-1}\left\{ \frac{(D_o - D_i)\tilde{\epsilon}}{D_o} + \frac{D_i}{D_o}\left[\frac{D_o^2 - D_i^2 - (D_o - D_i)^2 \tilde{\epsilon}^2}{2D_i(D_o - D_i)\tilde{\epsilon}} \right]^{1/2} \right\}$$

$$f_4 = \frac{4}{3} \frac{0.503}{\left[1 + (0.492/\mathrm{Pr})^{9/16}\right]^{4/9}}$$

In Eq. (13.50), the Rayleigh number is based on the gap size $(R_o - R_i)$. As in the case of horizontal annuli, Eq. (13.48) is, strictly speaking, only valid for sufficiently high Rayleigh numbers so that the boundary layers do not overlap. When compared to available experimental data, however, it has been shown that this equation can be considered valid for $\mathrm{Ra} \le 6 \times 10^8$, $5 \le \mathrm{Pr} \le 400$, $1.25 \le D_o/D_i \le 2.5$, and $0 \le \tilde{\epsilon} \le 0.75$ for downward displacement. For upward displacement with $0.25 \le \tilde{\epsilon} \le 0.75$, Eq. (13.48) underpredicts the Nusselt number by about 10%.

13.7.4 Three-Dimensional Bodies and Their Enclosures

All experimental heat transfer data for natural convection between three-dimensional inner bodies and their enclosures have recently been compiled by Warrington and Powe [130], and a generalized correlation has been attempted. These data specifically cover concentrically located isothermal spherical, cylindrical, and cubical inner bodies and their isothermal cubical and spherical enclosures. The generalized correlation is given by

$$\mathrm{Nu}_b = 0.585\,\mathrm{Ra}_b^{*0.236} \tag{13.51}$$

with an average deviation of less than 15% from the respective experimental data. In the above equation, all properties are evaluated at the mean of the inner-body and enclosure-surface temperatures, and

$$\mathrm{Nu}_b = \frac{h\tilde{b}}{k}, \qquad \mathrm{Ra}_b^* = \mathrm{Ra}_b \frac{R_o - R_i}{R_i}, \qquad \mathrm{Ra}_b = \frac{\rho g \beta \tilde{b}^3 (T_i - T_o)}{\mu k}$$

where h is the average natural-convection coefficient of heat transfer, which specifically excludes conduction and radiation effects; $\tilde{b}$ is a boundary-layer length, occupied by the boundary layer on the inner body; R_i is the radius of a hypothetical inner sphere of volume equal to that of the inner body; and R_o is the corresponding hypothetical radius of an equivalent outer body. The range of validity of Eq. (13.51) is given by $0.7 < \mathrm{Pr} < 1.4 \times 10^4$, $4.6 \times 10^5 < \mathrm{Ra}_b < 4.0 \times 10^{10}$, and $(R_o - R_i)/R_i > 0.45$.

For $(R_o - R_i)/R_i < 0.45$, a different correlation equation is used:

$$\frac{q_{\mathrm{conv}}}{q_{\mathrm{cond}}} = 0.2\,\mathrm{Ra}_b^{0.239} \left(\frac{R_o - R_i}{\tilde{b}} \right)^{0.717} \tag{13.52}$$

with an average deviation of about 14% from the experimental data. The range of validity in Pr and Ra is the same as above.

Equations (13.51) and (13.52) can be used to predict natural-convection heat transfer between a three-dimensional inner body and its enclosure whenever correlations for specific geometries and fluid media are not available.

13.8 CONCLUDING REMARKS

It is clear that natural convection phenomena in enclosures represent one of the basic heat transfer disciplines that have seen many advances in recent years. However, much clarification and resolution are still needed in addressing several issues in the phenomena which notably include three-dimensional flow, stability and transition, turbulence, unsteady effects, and others. In addition, there is also a great need for careful experimental studies which really hold the key to significant further advances in this area.

NOMENCLATURE

A_i inflow area, m², ft² (Fig. 13.2)

A_o outflow area, m², ft² (Fig. 13.2)

A_t	total opening area, m², ft² (Fig. 13.2)
a, b	correlation constants
$\tilde{b}$	boundary-layer length, m, ft
$\mathbf{B}$	body force vector, m/s², ft/s²
C_1	constant in turbulence model, $C_1 = 1.44$
C_2	constant in turbulence model, $C_2 = 1.92$
C_μ	constant in turbulence model, $C_\mu = 0.09$
c_p	specific heat at constant pressure, J/(kg · K), Btu/(lb$_m$ · °F)
D_i	inner cylinder diameter, m, ft
D_o	outer cylinder diameter, m, ft
f_1–f_4	functions of Prandtl number
G	function defined in Table 13.2
Gr	Grashof number based on $L_c = g\beta \Delta T L_c^3/\nu^2$
G_ϵ	turbulence generation for the ϵ field, kg/(m · s³), lb$_m$/(ft · s³)
G_κ	turbulence generation for the κ field, kg/(m · s³), lb$_m$/(ft · s³)
g	gravitational acceleration, m/s², ft/s²
$\tilde{g}$	mean squared temperature fluctuations, °C, K, °F, R
$\mathbf{g}$	gravity vector, m/s², ft/s²
H	height of enclosure, m, ft
h	heat transfer coefficient, W/(m² · K), Btu/(hr · ft² · °F)
K_γ	correlation function for tilted enclosures
k	thermal conductivity, W/(m · K), Btu/(hr · ft · °F)
L	enclosure depth, m, ft
L_c	characteristic length, m, ft
Nu	Nusselt number $= hL_c/k$; $\mathrm{Nu}_b = h\tilde{b}/k$, $\mathrm{Nu}_{D_i} = hD_i/k$, $\mathrm{Nu}_{D_o} = hD_o/k$, $\mathrm{Nu}_H = hH/k$
n	normal coordinate, m, ft
$\tilde{n}$	oscillation frequency, s
Pr	Prandtl number $= c_p\mu/k$
Pr_t	turbulent Prandtl number $= \epsilon_M/\epsilon_H$
p	pressure, Pa, lb$_f$/ft²
p_o	hydrostatic pressure, Pa (N/m²), lb$_f$/ft²
q	heat transfer rate, W, Btu/hr
q_c	conduction flux, W/m², Btu/(hr · ft²)
r	expansion index
Ra	Rayleigh number = Gr Pr
Ri	gradient Richardson number $= g\beta(\partial T/\partial y)/(\partial u/\partial y)^2$
R_i	inner radius, m, ft
R_o	outer radius, m, ft
S	heat source, W/m³, Btu/(hr · ft³)
T	temperature, °C, K, °F, °R
T_c	characteristic temperature, °C, K, °F, °R
t	time, s

U_c	characteristic velocity defined in Eq. (13.15), m/s, ft/s
u, v, w	velocity components in x, y, z directions, respectively, m/s, ft/s
$\mathbf{V}$	velocity vector, m/s, ft/s
W	enclosure width, m, ft (Fig. 13.1)
x, y, z	rectangular coordinates, m, ft
$\bar{x}, \bar{y}$	coordinates defined in Eq. (13.22)
x^*, y^*	coordinates x/H, y/H, respectively

Greek Symbols

α	thermal diffusivity, m^2/s, ft^2/s
α^*	two-dimensional aspect ratio (Fig. 13.1); $\alpha^* = H/W$, $\alpha_x^* = H/W$, $\alpha_z^* = L/W$
β	coefficient of volumetric expansion, K^{-1}, R^{-1}
Γ_ϵ	exchange coefficient for the ϵ field, kg/(m · s), lb_m/(ft · s)
Γ_κ	exchange coefficient for the κ field, kg/(m · s), lb_m/(ft · s)
γ	tilt angle, deg (Fig. 13.1)
γ_c	critical tilt angle, deg
ΔT_c	characteristic temperature difference, °C, K, °F, °R
ϵ	kinematic rate of dissipation, m^2/s^3, ft^2/s^3
$\tilde{\epsilon}$	eccentricity
ϵ_H	turbulent (eddy) diffusivity for heat transfer, m^2/s, ft^2/s
ϵ_M	turbulent (eddy) diffusivity for momentum, m^2/s, ft^2/s
η	dimensionless geometric parameter
θ	temperature variable $= (T - T_C)/(T_H - T_C)$
θ_0	$(T_H - T_C)/T_C$
κ	turbulent kinetic energy, m^2/s^2, ft^2/s^2
μ	dynamic viscosity, Pa · s, lb_m/(ft · s)
μ_t	turbulent (eddy) viscosity, Pa · s, lb_m/(ft · s)
ν	kinematic viscosity, m^2/s, ft^2/s
ρ	density, kg/m^3, lb_m/ft^3
σ_ϵ	Prandtl number for the ϵ field
σ_κ	Prandtl number for the κ field
τ	stress tensor, Pa, lb_f/ft^2
τ_t	Reynolds stresses, Pa, lb_f/ft^2
$\tilde{\Phi}$	dimensionless function of η
ψ	dimensionless stream function
ω	vorticity, s^{-1}

Subscripts

C	cold wall
cond	conduction part
conv	convection part

H	hot wall
H	dimensionless numbers based on H
L	dimensionless numbers based on L
0	reference quantities
t	turbulent quantities

REFERENCES

1. S. Ostrach, Natural Convection in Enclosures, *Adv. Heat Transfer*, Vol. 8, pp. 161–227, 1972.
2. I. Catton, Natural Convection in Enclosures, *Heat Transfer 1978*, Vol. 6, pp. 13–43, 1978.
3. Y. Jaluria, *Natural Convection Heat and Mass Transfer*. Pergamon, Oxford, 1980.
4. S. Ostrach, Natural Convection Heat Transfer in Cavities and Cells, *Heat Transfer 1982*, Vol. 1, pp. 365–379, 1982.
5. A. Bejan, *Convection Heat Transfer*, Wiley-Interscience, New York, 1984.
6. G. D. Raithby and K. G. T. Hollands, Natural Convection, *Handbook of Heat Transfer Fundamentals*, ed. W. M. Rohsenow, J. P. Hartnett, and E. N. Ganić, 2nd ed., Chap. 6, McGraw-Hill, New York, 1985.
7. F. A. Kulacki and D. E. Richards, Natural Convection in Plane Layers and Cavities with Volumetric Energy Sources, *Natural Convection: Fundamentals and Applications*, ed. S. Kakaç, W. Aung, and R. Viskanta, pp. 179–255, Hemisphere, Washington, 1985.
8. S. Chandrasekhar, *Hydrodynamic and Hydromagnetic Stability*, Clarendon, Oxford, 1961.
9. J. S. Turner, *Buoyancy Effects in Fluids*, Cambridge U.P., Cambridge, U.K., 1973.
10. I. Catton, Natural Convection in Horizontally Unbounded Plane Layers, *Natural Convection: Fundamentals and Applications*, ed. S. Kakaç, W. Aung, and R. Viskanta, pp. 97–134, Hemisphere, Washington, 1985.
11. R. B. Bird, W. E. Stewart, and E. N. Lightfoot, *Transport Phenomena*, Wiley, New York, 1960.
12. B. Gebhart, *Heat Transfer*, 2nd ed., McGraw-Hill, New York, 1971.
13. D. D. Gray and A. Giorgini, The Validity of the Boussinesq Approximation for Liquids and Gases, *Int. J. Heat Mass Transfer*, Vol. 19, pp. 545–551, 1976.
14. Z. Y. Zhong, K. T. Yang, and J. R. Lloyd, Variable Property Effects in Laminar Natural Convection in a Square Enclosure, *J. Heat Transfer*, Vol. 107, pp. 133–138, 1985.
15. K. T. Yang and J. R. Lloyd, Turbulent Buoyant Flow in Vented Simple and Complex Enclosures, *Natural Convection: Fundamentals and Applications*, ed. S. Kakaç, W. Aung, and R. Viskanta, pp. 303–329, Hemisphere, Washington, 1985.
16. L. C. Chang, K. T. Yang, and J. R. Lloyd, Radiation–Natural Convection Interactions in Two-Dimensional Complex Enclosures, *J. Heat Transfer*, Vol. 105, pp. 89–95, 1983.
17. Z. Y. Zhong, K. T. Yang, and J. R. Lloyd, Variable-Property Natural Convection in Tilted Enclosures with Thermal Radiation, *Numerical Methods in Heat Transfer*, Vol. III, ed. W. R. Lewis and K. Morgan, Chap. 9, pp. 195–214, Wiley, New York, 1985.
18. V. S. Arpaci and P. S. Larsen, *Convection Heat Transfer*, Prentice-Hall, Englewood Cliffs, NJ, 1984.
19. C. H. Cho, K. S. Chang, and K. H. Park, Numerical Simulation of Natural Convection in Concentric and Eccentric Horizontal Cylindrical Annuli, *J. Heat Transfer*, Vol. 104, pp. 624–630, 1982.
20. Y. L. Chan and C. L. Tien, A Numerical Study of Two-Dimensional Laminar Natural Convection in Shallow Open Cavities, *Int. J. Heat Mass Transfer*, Vol. 28, pp. 603–612, 1985.

21. J. R. Lloyd, K. T. Yang, and V. K. Liu, A Numerical Study of One-Dimensional Surface, Gas, and Soot Radiation for Turbulent Buoyant Flows in Enclosures, *Proc. 1st Nat. Conf. Numerical Methods in Heat Transfer*, pp. 142–161, 1979.
22. E. E. Khalil, Numerical Computations of Turbulent Swirling Flames in Axisymmetric Combustors, *Flow Mixing and Heat Transfer in Furnaces*, ed. K. H. Khalil, F. M. El Mahallawy, and E. E. Khalil, pp. 231–246, Pergamon, Oxford, 1978.
23. B. E. Launder, Heat and Mass Transport, *Turbulence*, 2nd ed., ed. P. Bradshaw, Topics in Applied Physics, Vol. 12, pp. 232–287, Springer-Verlag, Berlin, 1978.
24. J. Mathieu and D. Jeandel, *Simulation of Turbulence Models and Their Applications*, Vol. 1, Editions Eyrolles, Paris, 1984.
25. B. E. Launder, W. C. Reynolds and W. Rodi, *Turbulence Models and Their Applications*, Vol. 2, Editions Eyrolles, Paris, 1984.
26. A. Nayfeh, *Perturbation Methods*, Wiley, New York, 1973.
27. J. Lee, Prediction of Natural Convection Flow Patterns in Low Aspect Ratio Enclosures, Ph.D. Thesis, Dept. of Mech. and Aerospace Eng., Case Western Reserve Univ., May 1982.
28. K. T. Yang and J. R. Lloyd, Natural Convection–Radiation Interaction in Enclosures, *Natural Convection: Fundamentals and Applications*, ed. S. Kakaç, W. Aung, and R. Viskanta, pp. 381–410, Hemisphere, Washington, 1985.
29. A. J. Reynolds, The Prediction of Turbulent Prandtl and Schmidt Numbers, *Int. J. Heat Mass Transfer*, Vol. 18, pp. 1055–1069, 1975.
30. H. Schlichting, *Boundary Layer Theory*, 6th ed., McGraw-Hill, New York, 1968.
31. T. M. Shih, *Numerical Heat Transfer*, Hemisphere, Washington, 1984.
32. M. Van Dyke, *Perturbation Methods in Fluid Mechanics*, Academic, New York, 1964.
33. P. G. Simpkins and K. S. Chen, Natural Convection in Horizontal Containers with Applications to Crystal Growth, *Natural Convection: Fundamentals and Applications*, ed. S. Kakaç, W. Aung, and R. Viskanta, pp. 1010–1032, Hemisphere, Washington, 1985.
34. S. V. Patankar, *Numerical Heat Transfer and Fluid Flow*, Hemisphere, Washington, 1980.
35. K. E. Torrance, Numerical Methods in Heat Transfer, *Handbook of Heat Transfer Fundamentals*, 2nd ed., ed. W. M. Rohsenow, J. P. Hartnett, and E. N. Ganić, pp. 5-1–5-85, McGraw-Hill, 1985.
36. G. K. Batchelor, Heat Transfer by Free Convection across a Closed Cavity between Vertical Boundaries at Different Temperatures, *Quart. Appl. Math.*, Vol. 12, pp. 209–233, 1954.
37. A. E. Gill, The Boundary Layer Regime for Convection in a Rectangular Cavity, *J. Fluid Mech.*, Vol. 26, pp. 515–536, 1966.
38. P. A. Blythe, P. G. Daniels, and P. G. Simpkins, Thermal Convection in a Cavity: The Core Structure Near the Horizontal Boundaries, *Proc. Roy. Soc. London*, Vol. A387, p. 367, 1983.
39. K. T. Yang and E. W. Jerger, First-Order Perturbations of Laminar Free-Convection Boundary Layers on a Vertical Plate, *J. Heat Transfer*, Vol. 86, pp. 107–115, 1964.
40. A. Bejan, Note on Gill's Solution for Free Convection in a Vertical Enclosure, *J. Fluid Mech.*, Vol. 90, pp. 561–568, 1979.
41. W. P. Graebel, The Influence of Prandtl Number on Free Convection in a Rectangular Cavity, *Int. J. Heat Mass Transfer*, Vol. 24, pp. 125–131, 1981.
42. J. E. Hart, Stability of Thin Nonrotating Hadley Circulations, *J. Atmos. Sci.*, Vol. 29, pp. 687–697, 1972.
43. J. E. Hart, Low Prandtl Number Convection Between Differentially Heated End Walls, *Int. J. Heat Mass Transfer*, Vol. 26, pp. 1069–1074, 1983.
44. J. E. Hart, A Note on the Stability of Low Prandtl Number Hadley Circulations, *J. Fluid Mech.*, Vol. 132, p. 271, 1983.
45. D. E. Cormack, L. G. Leal, and J. Imberger, Natural Convection in a Shallow Cavity with Differentially Heated End Walls. Pt. 1, Asymptotic Theory, *J. Fluid Mech.*, Vol. 65, pt. 2, pp. 209–229, 1974.

46. A. Bejan and C. L. Tien, Laminar Natural Convection Heat Transfer in a Horizontal Cavity with Different End Temperatures, *J. Heat Transfer*, Vol. 100, pp. 641–647, 1978.

47. S. Ostrach, R. R. Loka, and A. Kumar, National Convection in Low Aspect Ratio Rectangular Enclosures, *Natural Convection in Enclosures*, ed. K. Torrance and I. Catton, ASME HTD—Vol. 8, pp. 1–10, 1980.

48. P. J. Roache, *Computational Fluid Dynamics*, Hermosa, Albuquerque, NM, 1976.

49. A. D. Gosman, W. M. Pun, and D. B. Spalding, Lecture Notes for Course Entitled Calculations of Recirculating Flows, Mech. Eng. Dept., Imp. Coll. Sci. Tech., London, 1973.

50. N. C. Markatos and K. A. Pericleous, Laminar and Turbulent Natural Convection in an Enclosed Cavity, *Int. J. Heat Mass Transfer*, Vol. 27, pp. 755–772, 1984.

51. W. T. Sha, H. M. Domanus, R. C. Schmitt, J. J. Oras, and E. I. H. Liu, COMMIX-1: A Three Dimensional Transient Simple-Phase Component Computer Program for Thermal-Hydraulic Analysis, Argonne Nat. Lab., TR ANL-77-96, NUREG/CR-0785, 1978.

52. K. T. Yang and V. K. Liu, UNDSAFE II. A Computer Code for Buoyant Turbulent Flow in an Enclosure with Thermal Radiation, TR-7-9002-78-3, Dept. Aero. Mech. Eng., Univ. Notre Dame, 166 pp., 1978.

53. B. P. Leonard, A Convectively Stable, Third-Order Accurate Finite-Difference Method for Steady Two-Dimensional Flow and Heat Transfer, *Numerical Properties and Methodologies in Heat Transfer*, ed. T. M. Shih, pp. 211–226, Hemisphere, Washington, 1983.

54. S. M. Bajorek and J. R. Lloyd, Experimental Investigation of Natural Convection in Partitioned Enclosures, *J. Heat Transfer*, Vol. 104, pp. 527–532, 1982.

55. L. C. Chang, J. R. Lloyd, and K. T. Yang, A Finite Difference Study of Natural Convection in Complex Enclosures, *Heat Transfer 1982*, Vol. 2, pp. 183–188, 1982.

56. Y. Lecointe, J. Piquet, and M. Visonneau, "Mehrstellen" Technique for the Numerical Solution of Unsteady Incompressible Viscous Flow in Enclosures, *Numerical Properties and Methodologies in Heat Transfer*, ed. T. M. Shih, pp. 183–200, Hemisphere, Washington, 1983.

57. M. M. Gupta, R. P. Manohan, and J. W. Stephenson, A Fourth-Order, Cost Effective and Stable Finite-Difference Scheme for the Convection-Diffusion Equation, *Numerical Properties and Methodologies in Heat Transfer*, ed. T. M. Shih, pp. 201–209, Hemisphere, Washington, 1983.

58. H. Q. Yang, K. T. Yang, and J. R. Lloyd, Flow Transition in Laminar Buoyant Flow in a Three-Dimensional Tilted Rectangular Enclosure, *Heat Transfer 1986*, 1986.

59. S. W. Churchill, P. Chao, and H. Ozoe, Extrapolation of Finite-Difference Calculations of Laminar Natural Convection in Enclosures to Zero Grid Size, *Numer. Heat Transfer*, Vol. 4, pp. 39–51, 1981.

60. I. P. Jones and C. P. Thomson, eds., Numerical Solution for a Comparison Problem on Natural Convection in an Enclosed Cavity, AERE-R-9955, HMSO, 1981.

61. C. Quon, Effects of Grid Distribution on the Computation of High Rayleigh Number Convection in a Differentially Heated Cavity, *Numerical Properties and Methodologies in Heat Transfer*, ed. T. M. Shih, pp. 261–281, Hemisphere, Washington, 1983.

62. J. F. Thompson, ed., *Numerical Grid Generation*, Elsevier, New York, 1982.

63. E. Leonardi and J. A. Reizes, Convective Flows in Closed Cavities with Variable Fluid Properties, *Numerical Methods in Heat Transfer*, ed. R. W. Lewis, K. Morgan, and O. C. Zienbiewicz, pp. 387–412, Wiley, Chichester, England, 1981.

64. V. F. Nicolette, K. T. Yang, and J. R. Lloyd, Transient Cooling by Natural Convection in a Two-Dimensional Square Enclosure, *Int. J. Heat Mass Transfer*, Vol. 28, pp. 1721–1732, 1985.

65. S. F. Shen, Finite-Element Methods in Fluid Mechanics, *Annual Review of Fluid Mechanics*, Vol. 9, pp. 421–445, Annual Reviews, Palo Alto, CA, 1977.

66. T. J. Chung, *Finite Element Analysis in Fluid Dynamics*, McGraw-Hill, New York, 1981.

67. A. J. Baker, *Finite Element Computational Fluid Mechanics*, Hemisphere, Washington, 1983.

68. J. C. Heinrich, R. S. Huyakorn, O. C. Zienkiewicz, and A. R. Mitchell, An "Upwind" Finite Element Scheme for Two-Dimensional Convective Transport Equation, *Int. J. Numer. Meth. Eng.*, Vol. 11, pp. 131–143, 1977.

69. T. M. Shih, ed., *NSF and NASA Workshop on Parallel Computations in Heat Transfer and Fluid Flows*, Dept. Mech. Eng., Univ. of Maryland, College Park, MD, 1984.

70. T. M. Shih, A Literature Survey on Numerical Heat Transfer, *Numer. Heat Transfer*, Vol. 5, pp. 369–420, 1982.

71. S. A. Orszag, Spectral Methods for Problems in Complex Geometries, *J. Comput. Phys.*, Vol. 37, pp. 70–92, 1980.

72. T. D. Taylor and J. W. Murdock, Application of Spectral Methods to the Solution of Navier-Stokes Equations, *Comput. Fluids*, Vol. 9, pp. 255–263, 1981.

73. A. Leonard, Vortex Methods for Flow Simulation, *J. Comput. Phys.*, Vol. 37, pp. 287–335, 1980.

74. C. J. Chen and Y. H. Yoon, Finite Analytic Numerical Solutions of Axisymmetric Navier-Stokes and Energy Equations, *J. Heat Transfer*, Vol. 105, pp. 639–645, 1983.

75. V. Talaie and C. J. Chen, Finite Analytic Solutions of Steady and Transient Natural Convection in Two-Dimensional Rectangular Enclosures, unpublished paper, University of Iowa, 1985.

76. S. A. Korpela, Y. Lee, and J. E. Drummond, Heat Transfer through a Double Pane Window, *J. Heat Transfer*, Vol. 104, pp. 539–544, 1982.

77. W. M. M. Schinkel, S. J. M. Linhorst, and C. J. Hoogendoorn, The Stratification in Natural Convection in Vertical Enclosures, *J. Heat Transfer*, Vol. 105, pp. 267–272, 1983.

78. E. R. G. Eckert and W. O. Carlson, Natural Convection in an Air Layer Enclosed by Two Vertical Plates with Different Temperatures, *Int. J. Heat Mass Transfer*, Vol. 2, pp. 106–120, 1961.

79. R. F. Bergholz, Instability of Steady Natural Convection in a Vertical Fluid Layer, *J. Fluid Mech.*, Vol. 84, p. 743, 1978.

80. D. E. Cormack, L. G. Leal, and J. H. Seinfield, Natural Convection in a Shallow Cavity with Differentially Heated End Walls, Pt. 2, Numerical Solution, *J. Fluid Mech.*, Vol. 65, Pt. 2, pp. 231–246, 1974.

81. G. S. Shiralkar and C. L. Tien, A Numerical Study of Laminar Natural Convection in Shallow Cavities, *J. Heat Transfer*, Vol. 103, pp. 226–231, 1981.

82. J. Tichy and A. Gadgil, High Rayleigh Number Laminar Convection in Low Aspect Ratio Enclosures with Adiabatic Horizontal Walls and Differentially Heated Vertical Walls, *J. Heat Transfer*, Vol. 104, pp. 103–110, 1982.

83. G. de Vahl Davis, I. P. Jones, and P. J. Roache, Natural Convection in an Enclosed Cavity: A Comparison Problem, *J. Fluid Mech.*, Vol. 95, Pt. 4, inside back cover, 1979.

84. G. de Vahl Davis and I. P. Jones, Natural Convection in a Square Cavity—A Comparison Exercise, *Int. J. Numer. Meth. Fluids*, Vol. 3, pp. 227–248, 1983.

85. K. H. Winters, A Numerical Study of Natural Convection in a Square Cavity, AERE-R-9747, HMSO, 1980.

86. J. N. Arnold, I. Catton, and D. K. Edwards, Experimental Investigation of Natural Convection in Inclined Rectangular Regions of Differing Aspect Ratios, *J. Heat Transfer*, Vol. 98, pp. 67–71, 1976.

87. H. Ozoe, K. Yamamoto, and S. W. Churchill, Three-Dimensional Numerical Analysis of Natural Convection in an Inclined Channel with a Square Cross Section, *AIChE J.*, Vol. 25, pp. 709–716, 1979.

88. H. Ozoe, K. Fujii, N. Lior, and S. W. Churchill, Long Rolls Generated by Natural Convection in an Inclined, Rectangular Enclosure, *Int. J. Heat Mass Transfer*, Vol. 26, pp. 1427–1438, 1983.

89. Z. Y. Zhong, J. R. Lloyd, and K. T. Yang, Variable-Property Natural Convection in Tilted Square Cavities, *Numerical Methods in Thermal Problems*, Vol. III, ed. R. W. Lewis, J. A. Johnson, and W. R. Smith, pp. 968–979, Pineridge Press, Swansea, U.K., 1983.

90. I. Catton, P. S. Ayysswamy, and R. M. Clever, Natural Convection Flow in a Finite Rectangular Slot Arbitrarily Oriented with Respect to the Gravity Vector, *Int. J. Heat Mass Transfer*, Vol. 17, pp. 173–184, 1974.

91. A. Ozoe, S. W. Churchill, T. Okamoto, and H. Sayama, Natural Convection in Doubly Inclined Rectangular Boxes, *Heat Transfer 1978*, Vol. 2, pp. 293–298, 1978.

92. M. Kaviany, Effect of a Protuberance on Thermal Convection in a Square Cavity, *J. Heat Transfer*, Vol. 106, pp. 830–834, 1984.

93. B. Farouk and S. I. Guceri, Laminar and Turbulent Natural Convection in the Annulus between Horizontal Concentric Cylinders, *J. Heat Transfer*, Vol. 104, pp. 631–636, 1982.

94. J. Prusa and L. S. Yao, Natural Convection Heat Transfer Between Eccentric Horizontal Cylinders, *J. Heat Transfer*, Vol. 105, pp. 108–116, 1983.

95. T. H. Kuehn and R. J. Goldstein, An Experimental Study of Natural Convection Heat Transfer in Concentric and Eccentric Horizontal Cylindrical Annuli, *J. Heat Transfer*, Vol. 100, pp. 635–640, 1978.

96. J. H. Lee and T. S. Lee, Natural Convection in the Annuli between Horizontal Confocal Elliptic Cylinders, *Int. J. Heat Mass Transfer*, Vol. 24, pp. 1739–1742, 1981.

97. S. S. Kwon, T. H. Kuehn, and T. S. Lee, Natural Convection in the Annulus between Horizontal Circular Cylinders with Three Axial Spacers, *J. Heat Transfer*, Vol. 104, pp. 118–124, 1982.

98. K. S. Chang, Y. H. Won, and C. H. Cho, Patterns of Natural Convection around a Square Cylinder Placed Concentrically in a Horizontal Circular Cylinder, *J. Heat Transfer*, Vol. 105, pp. 273–280, 1983.

99. V. A. Akinsete and T. A. Coleman, Heat Transfer by Steady Laminar Free Convection in Triangular Enclosures, *Int. J. Heat Mass Transfer*, Vol. 25, pp. 991–998, 1982.

100. J. W. Elder, Laminar Free Convection in a Vertical Slot, *J. Fluid Mech.*, Vol. 23, pp. 77–98, 1965.

101. J. W. Elder, Turbulent Free Convection in a Vertical Slot, *J. Fluid Mech.*, Vol. 23, pp. 99–111, 1965.

102. C. J. Hoogendoorn, Experimental Methods in Natural Convection, *Natural Convection: Fundamentals and Applications*, ed. S. Kakaç, W. Aung, and R. Viskanta, pp. 674–696, Hemisphere, Washington, 1985.

103. R. J. Goldstein, Optical Techniques for Temperature Measurement, *Measurements in Heat Transfer*, ed. E. R. G. Eckert and R. J. Goldstein, 2nd ed., pp. 241–293, Hemisphere, Washington, 1976.

104. R. J. Collier, C. B. Burckhardt, and C. H. Lin, *Optical Holography*, Academic, New York, 1971.

105. H. H. Oertel, Three-Dimensional Convection within Rectangular Boxes, *Natural Convection in Enclosures*, ed. K. E. Torrance and I. Catton, ASME HTD-Vol. 8, pp. 11–16, 1980.

106. H. Ozoe, M. Ohmuro, A. Mouri, S. Wishiwa, H. Soyama, and S. W. Churchill, Laser-Doppler Measurements of the Velocity along a Heated Vertical Wall of a Rectangular Enclosure, *J. Heat Transfer*, Vol. 105, pp. 782–787, 1983.

107. S. J. M. Linthorst, W. M. M. Schinkel, and C. J. Hoogendoorn, Flow Structure with Natural Convection in Inclined Air-Filled Enclosures, *J. Heat Transfer*, Vol. 103, pp. 535–539, 1983.

108. G. H. Schiroky and F. Rosenberger, Free Convection of Gases in a Horizontal Cylinder with Differentially Heated End Walls, *Int. J. Heat Mass Transfer*, Vol. 27, pp. 587–598, 1984.

109. W. M. M. Schinkel and C. J. Hoogendoorn, An Interferometric Study of the Local Heat Transfer by Natural Convection in Inclined Airfilled Enclosures, *Heat Transfer 1978*, Vol. 6, pp. 287–292, 1978.

110. H. Ozoe, H. Sayama, and S. W. Churchill, Natural Convection in an Inclined Rectangular Channel at Various Aspect Ratios and Angles—Experimental Measurements, *Int. J. Heat Mass Transfer*, Vol. 18, pp. 1425–1431, 1975.

111. B. A. Meyer, J. W. Mitchell, and M. M. El-Wakil, Natural Convection Heat Transfer in Moderate Aspect Ratio Enclosures, *J. Heat Transfer*, Vol. 101, pp. 655–659, 1979.

112. J. G. A. DeGraaf and E. M. F. Van Der Held, The Relation between the Heat Transfer and Convective Phenomenon in Enclosed Plane Air Layers, *Appl. Sci. Res.* Vol. 3, p. 393, 1953.

113. S. M. ElSherbiny, K. G. T. Hollands, and G. D. Raithby, Effect of Thermal Boundary Conditions on Natural Convection in Vertical and Inclined Air Layers, *J. Heat Transfer*, Vol. 104, pp. 515–520, 1982.

114. Z. Y. Zhong, Variable Property Natural Convection with Thermal Radiation Interaction in Square Enclosures, Ph.D. Diss., Univ. of Notre Dame, 191 pp., Aug. 1983.

115. V. Sernas and E. I. Lee, Heat Transfer in Air Enclosures of Aspect Ratio Less than One, *J. Heat Transfer*, Vol. 103, pp. 617–622, 1981.

116. R. A. Wirtz, J. Righi, and F. Zirilli, Measurements of Natural Convection across Tilted Rectangular Enclosures of Aspect Ratio 0.1 and 0.2, *J. Heat Transfer*, Vol. 104, pp. 521–526, 1982.

117. J. G. Symons and M. K. Peck, Natural Convection Heat Transfer through Inclined Longitudinal Slots, *J. Heat Transfer*, Vol. 106, pp. 824–829, 1984.

118. N. Seki, S. Fukusako, and A. Yamaguchi, An Experimental Study of Free Convective Heat Transfer in a Parallelogrammic Enclosure, *J. Heat Transfer*, Vol. 105, pp. 433–439, 1983.

119. P. G. Simpkins and T. D. Dudderar, Convection in Rectangular Cavities with Differentially Heated End Walls, *J. Fluid Mechanics*, Vol. 110, pp. 433–456, 1981.

120. G. D. Mallinson and G. de Vahl Davis, Three-Dimensional Natural Convection in a Box: A Numerical Study, *J. Fluid Mech.*, Vol. 83, 1–31, 1977.

121. G. L. Morrison and V. Q. Tran, Laminar Flow Structure in Vertical Free Convective Cavities, *Int. J. Heat Mass Transfer*, Vol. 21, pp. 103–213, 1978.

122. S. M. Bilski, Experimental Investigation of the Natural Convection Velocity Field in Square and Complex Enclosures, M.S. Diss., Dept. of Mech. Eng., Univ. of Notre Dame, Feb. 1984.

123. H. Ozoe, S. W. Churchill, T. Okamoto, and H. Sayama, Three-Dimensional Natural Convection in Inclined Rectangular Enclosures, *Proc. PACHEC-II*, Vol. 1, pp. 24–31, AIChE, New York, 1977.

124. A. Bejan and C. L. Tien, Fully Developed Natural Counterflow in a Long Horizontal Pipe with Different End Temperatures, *Int. J. Heat Mass Transfer*, Vol. 21, pp. 701–708, 1978.

125. S. Kimura and A. Bejan, Numerical Study of Natural Circulation in a Horizontal Duct with Different End Temperatures, *Wärme- und Stoffübertragung*, Vol. 14, pp. 269–280, 1980.

126. G. H. Schiroky and F. Rosenberger, High Rayleigh Number Heat Transfer in a Horizontal Cylinder with Adiabatic Wall, *Int. J. Heat Mass Transfer*, Vol. 27, pp. 630–633, 1984.

127. S. Kimura and A. Bejan, Experimental Study of Natural Convection in a Horizontal Cylinder with Different End Temperatures, *Int. J. Heat Mass Transfer*, Vol. 23, pp. 1117–1126, 1980.

128. H. Ozoe and T. Shibata, Natural Convection in an Inclined Circular Cylinder Annulus Heated and Cooled on its End Plates, *Int. J. Heat Mass Transfer*, Vol. 24, pp. 727–737, 1981.

129. Y. Tabata, K. Iwashige, F. Fukada, and S. Hasegawa, Three-Dimensional Natural Convection in an Inclined Cylindrical Annulus, *Int. J. Heat Transfer*, Vol. 27, pp. 747–754, 1984.

130. R. O. Warrington, Jr. and R. E. Powe, The Transfer of Heat by Natural Convection between Bodies and Their Enclosures, *Int. J. Heat Mass Transfer*, Vol. 28, pp. 319–330, 1985.

131. E. M. Sparrow and M. Charmchi, Natural Convection Experiments in an Enclosure between Eccentric or Concentric Vertical Cylinders of Different Height and Diameter, *Int. J. Heat Mass Transfer*, Vol. 26, pp. 133–143, 1983.

132. E. M. Sparrow, P. C. Stryker, and M. A. Ansari, Natural Convection in Enclosures with Off-Center Innerbodies, *Int. J. Heat Mass Transfer*, Vol. 27, pp. 49–56, 1984.

133. R. O. Warrington, Jr. and G. Crupper, Natural Convection Heat Transfer between Cylindrical Tube Bundles and Cubic Enclosures, *J. Heat Transfer*, to be published.

134. B. E. Launder and D. B. Spalding, The Numerical Computation of Turbulent Flows, *Comp. Math. Appl. Mech. Eng.*, Vol. 3, pp. 269–289, 1974.

135. N. C. Markatos, M. R. Malin, and G. Cox, Mathematical Modeling of Buoyancy Induced Smoke Flow in Enclosures, *Int. J. Heat Mass Transfer*, Vol. 25, pp. 63–75, 1982.

136. W. T. Sha and B. E. Launder, A General Model for Turbulent Momentum and Heat Transfer in Liquid Metals, Argonne Nat. Lab. Rpt. ANL-77-78, Mar. 1979.

137. M. W. Nansteel and R. Greif, An Investigation of Natural Convection in Enclosures with Two- and Three-Dimensional Partitions, *Int. J. Heat Mass Transfer*, Vol. 27, pp. 561–571, 1984.

138. K. Satoh, K. T. Yang, and J. R. Lloyd, Flow and Temperature Oscillations of Fire in a Cubic Enclosure with Ceiling and Flow Vents. Part 3—Nondimensional Relationship between Oscillatory Frequency and Heat Source Strength, Rpt. No. 57, pp. 79–92, Fire Res. Inst., Japan, 1984.

139. J. C. Patterson and J. Imberger, Unsteady Natural Convection in a Rectangular Cavity, *J. Fluid Mech.*, Vol. 100, pp. 65–86, 1980.

140. J. C. Patterson, On the Existence of an Oscillatory Approach to Steady Natural Convection in Cavities, *J. Heat Transfer*, Vol. 106, pp. 104–108, 1984.

141. R. Yewell, D. Poulikakos, and A. Bejan, Transient Natural Convection Experiments in Shallow Enclosures, *J. Heat transfer*, Vol. 104, pp. 533–538, 1982.

142. V. F. Nicolette, Jr., Transient Natural Convection Inside Enclosures, Ph.D. Diss., Univ. of Notre Dame, 202 pp., May 1984.

143. S. Kimura and A. Bejan, The Boundary Layer Natural Convection Regime in a Rectangular Cavity with Uniform Heat Flux from the Side, *J. Heat Transfer*, Vo. 106, pp. 98–103, 1984.

144. P. K.-B. Chao, H. Ozoe, S. W. Churchill, and N. Lior, Laminar Natural Convection in an Inclined Rectangular Box with the Lower Surface Half-Heated and Half-Insulated, *J. Heat Transfer*, Vol. 105, pp. 425–432, 1983.

145. D. M. Kim and R. Viskanta, Effect of Wall Heat Conduction on Natural Convection Heat Transfer in a Square Enclosure, *J. Heat Transfer*, Vol. 107, pp. 139–146, 1985.

146. C. Prakash and D. Kaminski, Conjugate Natural Convection in a Square Enclosure: Effect of Conduction in One of the Vertical Walls, *Heat Transfer in Enclosures*, ed. R. W. Douglass and A. F. Emery, ASME HTD-Vol. 39, pp. 49–54, 1984.

147. R. Viskanta, Natural Convection in Melting and Solidification, *Natural Convection: Fundamentals and Applications*, ed. S. Kakaç, W. Aung, and R. Viskanta, pp. 845–877, Hemisphere, Washington, 1985.

148. P. A. Blythe, P. G. Daniels, and P. G. Simpkins, Limiting Behaviors in Porous Media Cavity Flows, *Natural Convection Fundamentals and Applications*, ed. S. Kakaç, W. Aung, and R. Viskanta, pp. 600–611, Hemisphere, Washington, 1985.

149. C. L. Tien and J. T. Hong, Natural Convection in Porous Media under Non-Darcian and Non-uniform Permeability Conditions, *Natural Convection: Fundamentals and Applications*, ed. S. Kakaç, W. Aung, and R. Viskanta, pp. 573–587, Hemisphere, Washington, 1985.

150. T. Hung Nguyen, P. Vasseur, and L. Robbilard, Natural Convection between Horizontal concentric Cylinders with Density Inversion of Water for Low Rayleigh Numbers, *Int. J. Heat Mass Transfer*, Vol. 25, pp. 1559–1568, 1982.

151. P. Vasseur, L. Robbilard, and B. Chandra Shekar, Natural Convection Heat Transfer of Water within a Horizontal Cylindrical Annulus with Density Inversion Effects, *J. Heat Transfer*, Vol. 105, pp. 117–123, 1983.

152. H. Inaba and T. Kukuda, An Experimental Study of Natural Convection in an Inclined Rectangular Cavity Filled with Water at its Density Extremum, *J. Heat Transfer*, Vol. 108, pp. 109–115, 1984.

153. M. E. Franke and K. E. Hutson, Effects of Corona Discharge on Free Convection Heat Transfer inside a Vertical Hollow Cylinder, *J. Heat Transfer*, Vol. 106, pp. 347–351, 1984.

154. R. Anderson and M. Bohn, Heat Transfer Enhancement in Natural Convection Enclosure Flow, *Heat Transfer in Enclosures*, ed. R. W. Douglass and A. F. Emery, ASME HTD-Vol. 39, pp. 29–38, 1984.

155. S. W. Churchill, Free Convection in Layers and Enclosures, *Heat Exchanger Design Handbook*, Section 2.5.8, Hemisphere, Washington, 1983.

156. S. W. Churchill and R. Usagi, A General Expression for the Correlation of Rates of Transfer and Other Phenomena, *AIChE J.*, Vol. 18, pp. 1121–1128, 1972.

157. T. H. Kuehn and R. J. Goldstein, Correlating Equation for Natural Heat Transfer between Horizontal Circular Cylinders, *Int. J. Heat Mass Transfer*, Vol. 19, pp. 1127–1134, 1976.

14

MIXED CONVECTION IN EXTERNAL FLOW

T. S. Chen and B. F. Armaly

University of Missouri — Rolla
Rolla, Missouri

14.1 INTRODUCTION

Mixed convection flows, or combined forced and free convection flows, arise in many transport processes in engineering devices and in nature. Such a process occurs when the effect of the buoyancy force in forced convection or the effect of forced flow in free convection becomes significant. The effect is especially pronounced in situations where the forced-flow velocity is low and the temperature difference is large. In mixed convection flows, the forced-convection effects and the free-convection effects are of comparable magnitude. Mixed convection processes may be divided into external flow over immersed bodies (such as flat plates, cylinders and wires, spheres, and moving surfaces), free-boundary flow (such as plumes, jets, and wakes), and internal flow in ducts (such as pipes, channels, and enclosures).

The domain of mixed convection regime is generally defined as the region $a \leq \mathrm{Gr}/\mathrm{Re}^n \leq b$, where a and b are, respectively, the lower and upper bounds of the domain. The buoyancy parameter, $\mathrm{Gr}/\mathrm{Re}^n$, provides a measure of the influence of free convection in comparison with that of forced convection on the flow. The power n depends on the flow configuration and the surface heating condition. Outside the mixed convection region, $a \leq \mathrm{Gr}/\mathrm{Re}^n \leq b$, either the pure forced-convection or the pure free-convection analysis can be used to describe accurately the flow or the temperature field. When the free convection is dominant over the forced convection, the relevant mixed-convection parameter is given by $\mathrm{Re}^n/\mathrm{Gr}$, the reciprocal of the buoyancy parameter. Buoyancy forces can enhance the surface heat transfer rate when they aid the forced flow, and vice versa. Buoyancy forces also play a significant role in the incipience of flow instabilities, and they can be responsible for either delaying or speeding the transition from laminar to turbulent flow. Forced convection is the dominant mode of transport when $\mathrm{Gr}/\mathrm{Re}^n \to 0$, whereas free convection is the dominant mode when $\mathrm{Gr}/\mathrm{Re}^n \to \infty$ or $\mathrm{Re}^n/\mathrm{Gr} \to 0$.

Available correlation equations for local and average Nusselt numbers in mixed convection will be presented in this chapter for external flow over immersed bodies of various configurations. Critical Reynolds and Grashof numbers for the incipience of the instability of laminar flow will also be presented for selected flow geometries. A brief description of the basic governing equations for laminar mixed convection boundary-layer flows is given below in order to identify the important mixed-convection parameters.

14.2 FUNDAMENTALS OF MIXED CONVECTION

Consider a forced flow aligned parallel to a semi-infinite flat plate that forms an acute angle γ with the vertical. The forced flow is above the plate when γ is measured in the clockwise direction and below the plate when it is measured in the counterclockwise direction. In the absence of heat generation and viscous dissipation, the boundary-layer equations under the Boussinesq approximation and for constant-property, steady-state, and steady flow conditions can be written as [1]

$$\frac{\partial u}{\partial x} + \frac{\partial v}{\partial y} = 0 \tag{14.1}$$

$$u\frac{\partial u}{\partial x} + v\frac{\partial u}{\partial y} = \pm g\beta \sin\gamma \frac{\partial}{\partial x}\int_y^\infty (T - T_\infty)\, dy \pm g\beta \cos\gamma\,(T - T_\infty) + \nu\frac{\partial^2 u}{\partial y^2} \tag{14.2}$$

$$u\frac{\partial T}{\partial x} + v\frac{\partial T}{\partial y} = \alpha\frac{\partial^2 T}{\partial y^2} \tag{14.3}$$

The first term on the right-hand side of Eq. (14.2) represents the buoyancy-induced streamwise pressure gradient, with the plus and minus signs pertaining, respectively, to flow above and flow below the plate. The second term represents the streamwise component of the buoyancy force, and its plus and minus signs refer, respectively, to upward and downward forced flows. The last term is the viscous force term.

In dimensionless form, Eqs. (14.1) to (14.3) become

$$\frac{\partial U}{\partial \chi} + \frac{\partial V}{\partial \eta} = 0 \tag{14.4}$$

$$U\frac{\partial U}{\partial \chi} + V\frac{\partial U}{\partial \eta} = \frac{1}{\mathrm{Re}}\frac{\partial^2 U}{\partial \eta^2} + \frac{\mathrm{Gr}}{\mathrm{Re}^2}\left[\cos\gamma\,\theta + \sin\gamma\,\frac{\partial}{\partial \chi}\int_{\eta}^{\infty}\theta\,d\eta\right] \tag{14.5}$$

$$U\frac{\partial \theta}{\partial \chi} + V\frac{\partial \theta}{\partial \eta} = \frac{1}{\mathrm{Re\,Pr}}\frac{\partial^2 \theta}{\partial \eta^2} \tag{14.6}$$

where

$$\chi = \frac{x}{l}, \qquad \eta = \frac{y}{l}, \qquad U = \frac{u}{u_\infty}, \qquad V = \frac{v}{u_\infty}, \qquad \theta = \frac{T - T_\infty}{T_w - T_\infty} \tag{14.7}$$

and

$$\mathrm{Re} = \frac{u_\infty l}{\nu}, \qquad \mathrm{Gr} = \frac{g\beta(T_w - T_\infty)l^3}{\nu^2}, \qquad \mathrm{Pr} = \frac{\nu}{\alpha} \tag{14.8}$$

In Eqs. (14.7) and (14.8), l is a characteristic length, u_∞ is a characteristic velocity (e.g. the free-stream velocity), $T_w - T_\infty$ is the temperature difference between the wall and the free stream, Re is the Reynolds number, Gr is the Grashof number, and Pr is the Prandtl number. From an energy balance at the wall, $q_w = -k(\partial T/\partial y)_{y=0} = h(T_w - T_\infty)$, the Nusselt number $\mathrm{Nu} = hl/k$ can be expressed by

$$\mathrm{Nu} = -\left(\frac{\partial \theta}{\partial \eta}\right)_{\eta=0} \tag{14.9}$$

Since $\partial\theta/\partial\eta$ depends on Re, Gr, and Pr, Eq. (14.9) provides a functional relationship between Nu and Re, Gr, and Pr:

$$\mathrm{Nu} = \mathrm{Nu}(\mathrm{Re}, \mathrm{Gr}, \mathrm{Pr}) \tag{14.10}$$

The precise expression for the Nusselt number as a function of Re, Gr, and Pr depends on the flow configuration and the surface heating condition.

14.3 CORRELATION EQUATIONS FOR NUSSELT NUMBERS

Mixed convection in laminar boundary-layer flow adjacent to vertical, inclined, and horizontal flat plates, as governed by Eqs. (14.1) to (14.3), has been extensively analyzed for the cases of uniform wall temperature (UWT) and uniform surface heat flux (UHF) as boundary conditions. Extensive analyses have also been conducted for other flow geometries, such as horizontal cylinders in cross flow, vertical cylinders, spheres, and moving sheets. Experimental studies on mixed convection have been reported for flows along vertical, inclined, and horizontal plates, vertical plates in cross flow, and flows across horizontal cylinders and spheres. Pertinent references for analytical and experimental studies will be mentioned in appropriate sections in this chapter.

Based on a critical comparison between numerical and experimental results, the local Nusselt number Nu for the mixed convection regime in laminar boundary-layer flow can be correlated very well by an equation of the form [2]

$$\mathrm{Nu}^n = \mathrm{Nu}_F^n \pm \mathrm{Nu}_N^n \tag{14.11}$$

in which Nu_F and Nu_N are, respectively, the Nusselt numbers for pure forced convection and pure free convection, and n is a constant exponent. The plus sign in Eq. (14.11) is for the buoyancy-assisting case, and the minus sign is for the buoyancy-opposing case. Equation (14.11) can be expressed in other forms as

$$\left(\frac{\mathrm{Nu}}{\mathrm{Nu}_F}\right)^n = 1 \pm \left(\frac{\mathrm{Nu}_N}{\mathrm{Nu}_F}\right)^n \tag{14.12}$$

or simply as

$$Y^n = 1 \pm X^n \tag{14.13}$$

where $Y = \mathrm{Nu}/\mathrm{Nu}_F$ and $X = \mathrm{Nu}_N/\mathrm{Nu}_F$. For laminar boundary-layer flows, the Nusselt number expressions for the pure forced convection and the pure free convection assume the respective forms

$$\mathrm{Nu}_F = A(\mathrm{Pr})\mathrm{Re}^{1/2}, \qquad \mathrm{Nu}_N = B(\mathrm{Pr})\mathrm{Gr}^m \tag{14.14}$$

where $A(\mathrm{Pr})$ and $B(\mathrm{Pr})$ are functions that depend on the Prandtl number and m is a constant exponent that depends on the flow configuration and the surface heating condition in pure free convection flow. Substituting Eq. (14.14) into Eq. (14.13) and solving for $\mathrm{Nu}/\mathrm{Nu}_F$ results in a correlation equation for the mixed-convection Nusselt number as

$$\frac{\mathrm{Nu}\,\mathrm{Re}^{-1/2}}{A(\mathrm{Pr})} = \left\{1 \pm \left[\frac{B(\mathrm{Pr})\,\xi^m}{A(\mathrm{Pr})}\right]^n\right\}^{1/n} \tag{14.15}$$

where $\xi = (\mathrm{Gr}/\mathrm{Re}^{1/2m})$ represents the buoyancy parameter. For a flat plate making an angle γ with the vertical, Eqs. (14.14) and (14.15) remain valid in the range of angles $0° \le \gamma < 85°$ if Gr is replaced with $\mathrm{Gr}\cos\gamma$. In addition, it is emphasized here that the correlation equations in the forms of Eqs. (14.14) and (14.15) are applicable to both local and average Nusselt numbers when appropriate characteristic lengths are used in Re, Gr, and ξ, and the Pr-dependent functions $A(\mathrm{Pr})$ and $B(\mathrm{Pr})$ are properly defined. This can be seen by examining Table 14.1, where correlation results are presented.

The correlation equations for the various flow configurations under different surface heating conditions (UWT and UHF) will be presented in the following sections. The

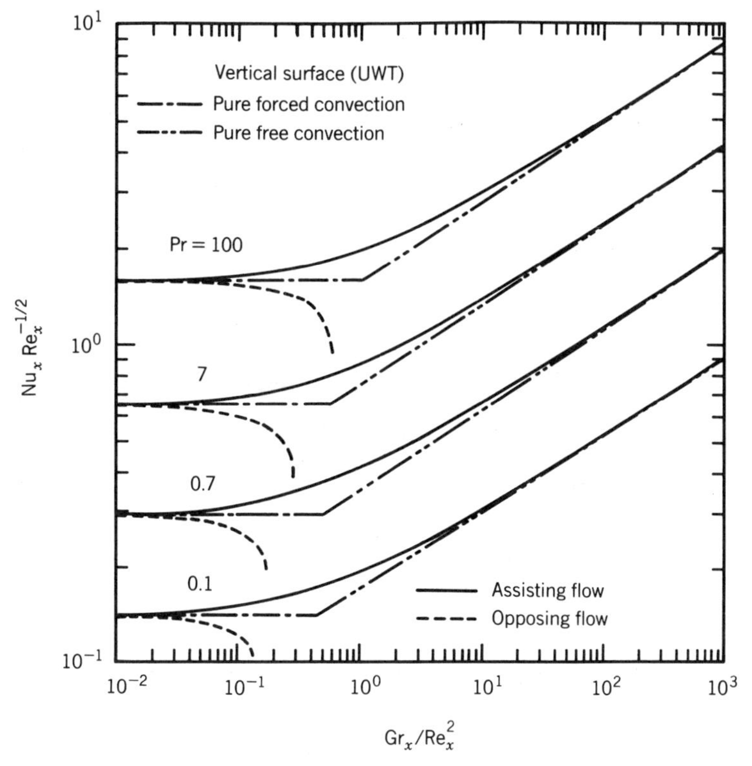

Figure 14.1. Calculated local Nusselt numbers for flow along a vertical flat plate with uniform wall temperature (UWT) [3].

validity of these correlations has been verified by comparisons with available numerical and experimental mixed-convection results.

14.4 CORRELATIONS FOR FLAT PLATES

The proposed form for correlating mixed-convection heat transfer results, Eq. (14.12) or (14.15), has been applied to laminar boundary-layer flow adjacent to horizontal, vertical, and inclined flat plates. Available mixed-convection results for these geometries can be found, for example, in Refs. 3 to 7, and some of the results on the local Nusselt number Nu_x are summarized in Figs. 14.1 through 14.10. The results which are presented in these figures cover a wide range of Prandtl numbers, $0.1 \leq Pr \leq 100$, for both the buoyancy-assisting and the buoyancy-opposing flow cases under the surface heating conditions of UWT and UHF. Experimental measurements, which compare very well with predictions, are also presented for the vertical plate (Fig. 14.5) and the 45° inclined flat plate (Fig. 14.6). Figures 14.1 through 14.6 illustrate the region where mixed-convection results deviate from the pure free-convection or the pure forced-convection values. It should be stated that in the buoyancy-opposing flow cases a flow reversal occurs at small values of buoyancy parameters as shown in the figures, thus

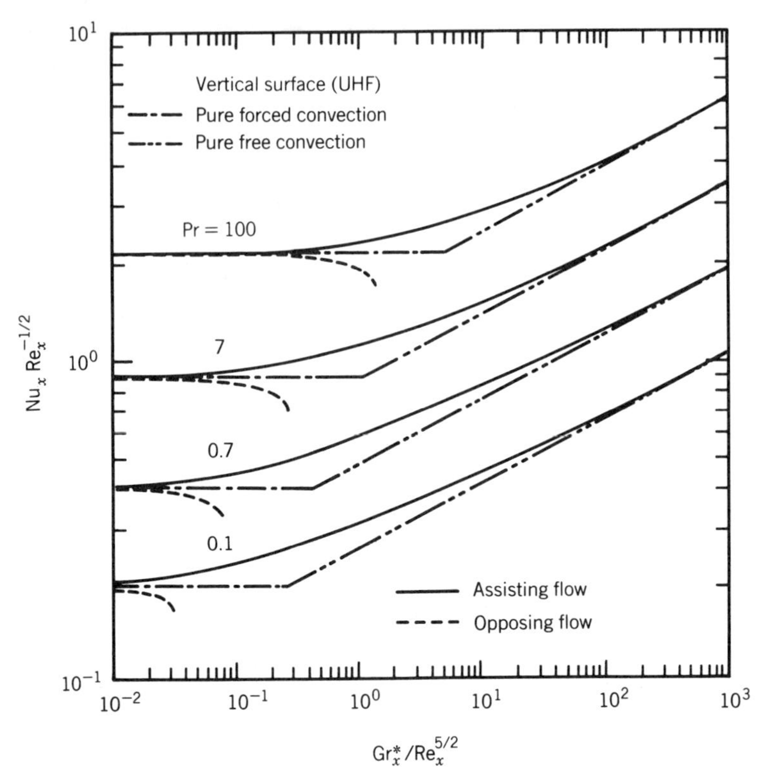

Figure 14.2. Calculated local Nusselt numbers for flow along a vertical flat plate with uniform surface heat flux (UHF) [5].

terminating the laminar boundary-layer flow regime. These predicted mixed-convection results from analysis as well as the experimental data are used to establish the range of applicability and the accuracy of the proposed simple mixed-convection correlations. For these flow geometries, the general form of the proposed mixed-convection correlation is expressed by Eq. (14.15).

The appropriate constants m and n, characteristic lengths x and L, and functions $A(\mathrm{Pr})$ and $B(\mathrm{Pr})$, as well as various forms of the buoyancy parameter ξ for the different geometries and boundary conditions, are defined in Table 14.1. The expressions for the various functions and dimensionless parameters are summarized in Table 14.2. The various forms of Reynolds and Grashof numbers that appear in Table 14.1 are defined in its footnote. The fluid properties are evaluated at the film temperature $T_f = (T_w + T_\infty)/2$.

Comparisons between the correlated results and the numerical values for the UWT case are presented in Figs. 14.7 through 14.10 in terms of Y-vs.-X plots, where $Y = \mathrm{Nu}\,\mathrm{Re}^{-1/2}/A(\mathrm{Pr})$ and $X = B(\mathrm{Pr})\xi^m/A(\mathrm{Pr})$. Excellent agreement between correlated and numerical results, with deviations of less than 5%, exists for buoyancy-assisting flow. The correlations for buoyancy-opposing flow start to fail near the flow separation region because boundary-layer assumptions are no longer valid in and near that region. They do, however, provide satisfactory agreement, with deviations smaller

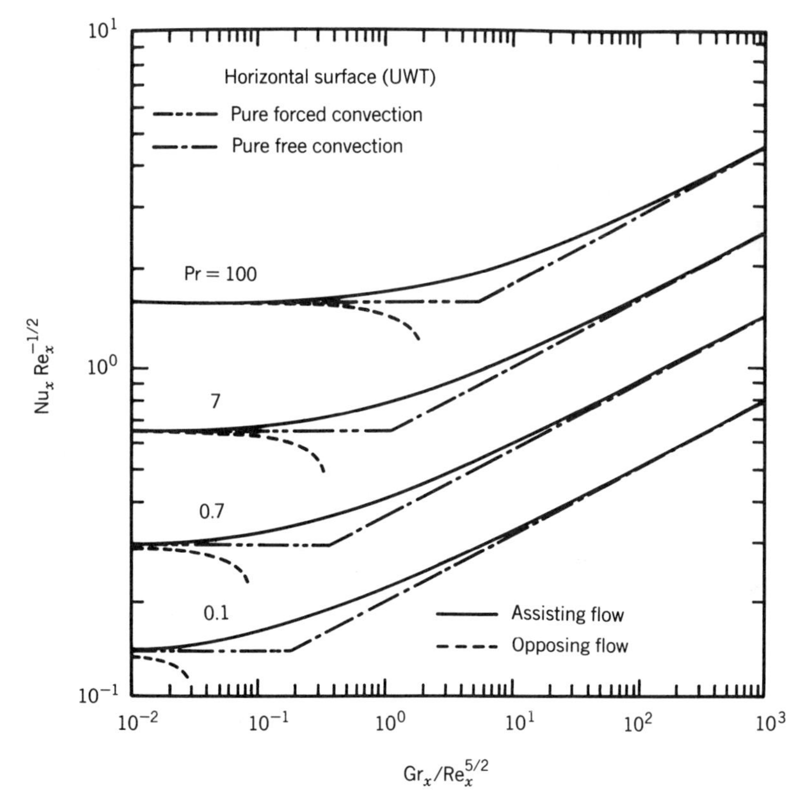

Figure 14.3. Calculated local Nusselt numbers for flow over a horizontal flat plate with uniform wall temperature (UWT) [3].

than 10%, in the region of nonseparating buoyancy-opposing flow. The results that are presented for comparisons in Figs. 14.7 through 14.9 are for the UWT case. However, similarly good correlations have been obtained for the UHF case [5]. A representative correlation that covers all inclination angles, $0° \leq \gamma \leq 90°$, for both UWT and UHF conditions for $Pr = 0.7$ is illustrated in Fig. 14.10.

It can be concluded that good agreement exists between correlated and predicted results for the vertical, inclined, and horizontal flat plates, as indicated by Figs. 14.1 through 14.10.

14.5 CORRELATIONS FOR CONTINUOUS MOVING SHEETS

The effects of buoyancy force on the laminar heat transfer from a moving continuous sheet in a quiescent fluid (equivalent to an extrusion process) have been analyzed for the cases when the sheet is moving in either the vertical, an inclined, or the horizontal direction [12–15]. Results are reported for the UWT and the UHF boundary condi-

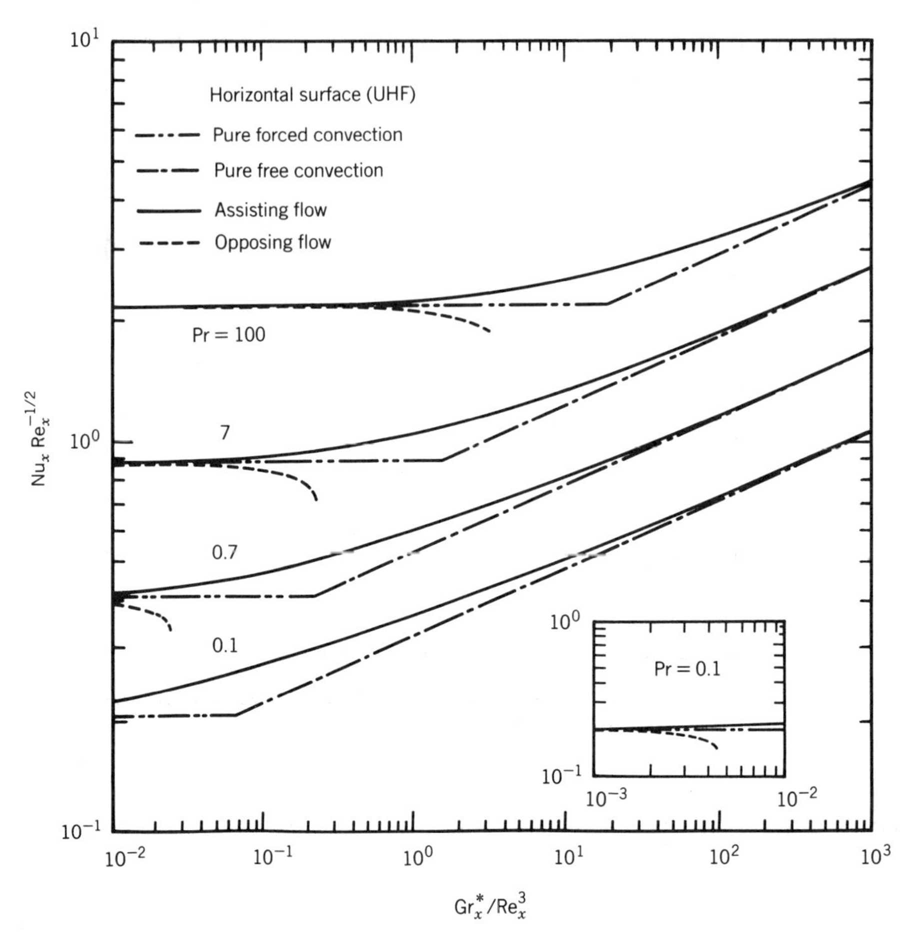

Figure 14.4. Calculated local Nusselt numbers for flow over a horizontal flat plate with uniform surface heat flux (UHF) [5].

tions in buoyancy-assisting and buoyancy-opposing situations. These results are used to develop and validate simple correlations for the Nusselt number in this flow regime. These correlations could be expressed in terms of the pure forced-convection and the pure free-convection results as described in Eq. (14.12) or in the $Y(X)$ form, Eq. (14.15), where $Y = \mathrm{Nu\,Re}^{-1/2}/A(\mathrm{Pr})$ and $X = B(\mathrm{Pr})\xi^m/A(\mathrm{Pr})$, with the appropriate functions $A(\mathrm{Pr})$ and $B(\mathrm{Pr})$ and constants m and n as given in Tables 14.1 and 14.2. The Reynolds number appearing in these correlations is based on the velocity of the moving sheet, u_0. Comparisons between the predicted and the correlated local Nusselt numbers are presented in Fig. 14.11 for the vertical and inclined cases and in Fig. 14.12 for the horizontal case. The results shown are for Pr = 0.7 and Pr = 7, for both the UWT and the UHF boundary conditions. The correlated local Nusselt numbers appear to agree very well with the predicted values, with a maximum deviation of 5%.

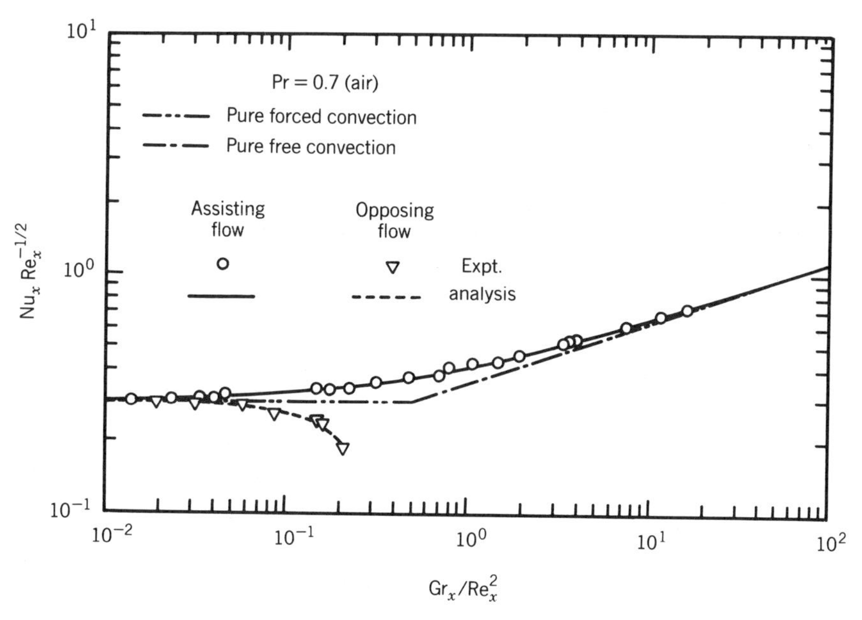

Figure 14.5. Measured and calculated local Nusselt numbers for air flow along an isothermal vertical flat plate [4].

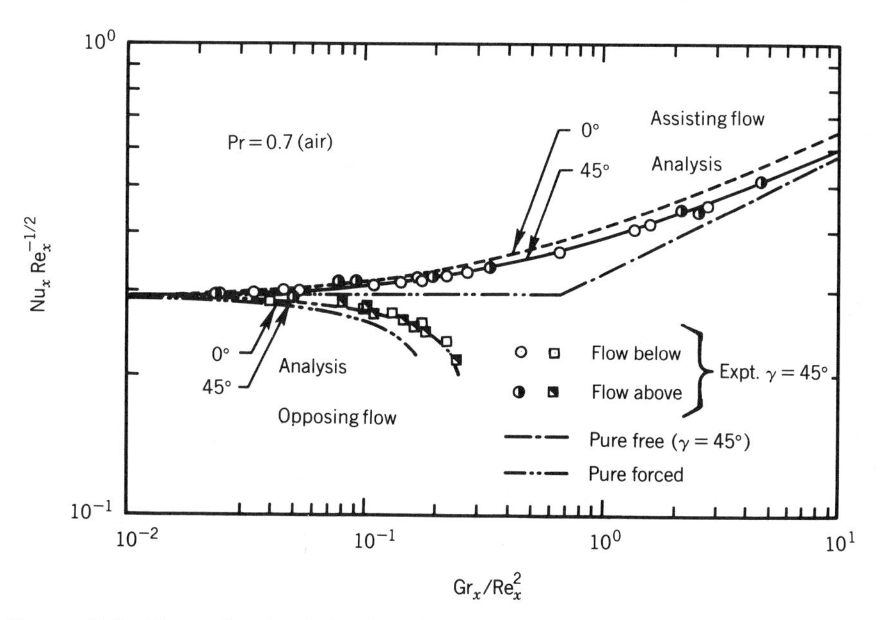

Figure 14.6. Measured and calculated local Nusselt numbers for air flow on an isothermal inclined flat plate [7].

TABLE 14.1 Recommended Constants, Functions, and Characteristic Length[a] for Mixed Convection, Eq. (14.15)

Geometry and Boundary Conditions	Charact. Length	$A(\mathrm{Pr})^b$	$B(\mathrm{Pr})^b$	ξ	m	n	Applicable Range
Vertical and inclined plate							Boundary-layer flow $0.1 \le \mathrm{Pr} \le 100$ $0 \le \gamma \le 85°$
Local							
UWT [3]	x	$F_1(\mathrm{Pr})$	$F_2(\mathrm{Pr})$	$(\mathrm{Gr}_x \cos\gamma)/\mathrm{Re}_x^2$	$\frac{1}{4}$	3	$10^3 \le \mathrm{Re}_x \le 10^5$, $\mathrm{Gr}_x < 10^9$
UHF [5]	x	$G_1(\mathrm{Pr})$	$G_2(\mathrm{Pr})$	$(\mathrm{Gr}_x^* \cos\gamma)/\mathrm{Re}_x^{5/2}$	$\frac{1}{5}$	3	$10^3 \le \mathrm{Re}_x \le 10^5$, $\mathrm{Gr}_x^* < 10^{11}$
Average							
UWT [3]	L	$2F_1(\mathrm{Pr})$	$4F_2(\mathrm{Pr})/3$	$(\mathrm{Gr}_L \cos\gamma)/\mathrm{Re}_L^2$	$\frac{1}{4}$	3	$10^3 \le \mathrm{Re}_L \le 10^5$, $\mathrm{Gr}_L < 10^9$
UHF [5]	L	$2G_1(\mathrm{Pr})$	$5G_2(\mathrm{Pr})/4$	$(\mathrm{Gr}_L^* \cos\gamma)/\mathrm{Re}_L^{5/2}$	$\frac{1}{5}$	3	$10^3 \le \mathrm{Re}_L \le 10^5$, $\mathrm{Gr}_L^* < 10^{11}$
Horizontal plate							Boundary-layer flow $0.1 \le \mathrm{Pr} \le 100$ $\gamma = 90°$
Local							
UWT [3]	x	$F_1(\mathrm{Pr})$	$F_3(\mathrm{Pr})$	$\mathrm{Gr}_x/\mathrm{Re}_x^{5/2}$	$\frac{1}{5}$	3	$10^3 < \mathrm{Re}_x < 10^5$, $\mathrm{Gr}_x < 10^7$
UHF [5]	x	$G_1(\mathrm{Pr})$	$G_3(\mathrm{Pr})$	$\mathrm{Gr}_x^*/\mathrm{Re}_x^3$	$\frac{1}{6}$	3	$10^3 < \mathrm{Re}_x < 10^5$, $\mathrm{Gr}_x^* < 10^8$
Average							
UWT [3]	L	$2F_1(\mathrm{Pr})$	$5F_3(\mathrm{Pr})/3$	$\mathrm{Gr}_L/\mathrm{Re}_L^{5/2}$	$\frac{1}{5}$	3	$10^3 < \mathrm{Re}_L < 10^5$, $\mathrm{Gr}_L < 10^7$
UHF [5]	L	$2G_1(\mathrm{Pr})$	$3G_3(\mathrm{Pr})/2$	$\mathrm{Gr}_L^*/\mathrm{Re}_L^3$	$\frac{1}{6}$	3	$10^3 < \mathrm{Re}_L < 10^5$, $\mathrm{Gr}_L^* < 10^8$
Continuous moving sheets							Boundary-layer flow $0.1 \le \mathrm{Pr} \le 100$
Vertical and Inclined							$0 \le \gamma \le 85°$
Local							
UWT [12, 14]	x	$F_4(\mathrm{Pr})$	$F_2(\mathrm{Pr})$	$(\mathrm{Gr}_x \cos\gamma)/\mathrm{Re}_x^2$	$\frac{1}{4}$	3	$\mathrm{Re}_x < 10^5$, $\mathrm{Gr}_x \le 10^9$
UHF [12, 14]	x	$G_4(\mathrm{Pr})$	$G_2(\mathrm{Pr})$	$(\mathrm{Gr}_x^* \cos\gamma)/\mathrm{Re}_x^{5/2}$	$\frac{1}{5}$	3	$\mathrm{Re}_x < 10^5$, $\mathrm{Gr}_x^* \le 10^{10}$

Average							
UWT [12, 14]	L	$2F_4(\mathrm{Pr})$	$4F_2(\mathrm{Pr})/3$	$(\mathrm{Gr}_L \cos\gamma)/\mathrm{Re}_L^2$	$\frac{1}{4}$	3	$\mathrm{Re}_L < 10^5$, $\mathrm{Gr}_L < 10^9$
UHF [12, 14]	L	$2G_4(\mathrm{Pr})$	$5G_2(\mathrm{Pr})/4$	$(\mathrm{Gr}_L^* \cos\gamma)/\mathrm{Re}_L^{5/2}$	$\frac{1}{5}$	3	$\mathrm{Re}_L < 10^5$, $\mathrm{Gr}_L^* < 10^{10}$
Horizontal							$\gamma = 90°$
Local							
UWT [13, 15]	x	$F_4(\mathrm{Pr})$	$F_3(\mathrm{Pr})$	$\mathrm{Gr}_x/\mathrm{Re}_x^{5/2}$	$\frac{1}{5}$	3	$\mathrm{Re}_x < 10^5$, $\mathrm{Gr}_x < 10^7$
UHF [13, 15]	x	$G_4(\mathrm{Pr})$	$G_3(\mathrm{Pr})$	$\mathrm{Gr}_x^*/\mathrm{Re}_x^3$	$\frac{1}{6}$	3	$\mathrm{Re}_x < 10^5$, $\mathrm{Gr}_x^* < 10^8$
Average							
UWT [13, 15]	L	$2F_4(\mathrm{Pr})$	$5F_3(\mathrm{Pr})/3$	$\mathrm{Gr}_L/\mathrm{Re}_L^{5/2}$	$\frac{1}{5}$	3	$\mathrm{Re}_L < 10^5$, $\mathrm{Gr}_L < 10^7$
UHF [13, 15]	L	$2G_4(\mathrm{Pr})$	$3G_3(\mathrm{Pr})/2$	$\mathrm{Gr}_L^*/\mathrm{Re}_L^3$	$\frac{1}{6}$	3	$\mathrm{Re}_L < 10^5$, $\mathrm{Gr}_L^* < 10^8$
Vertical cylinders in longitudinal flow							Boundary-layer flow $0 \le K_1 \le 4$, $\mathrm{Pr} = 0.7$
Local							
UWT [16]	x	$H_1(K_1)$	$H_2(K_2)$	$\mathrm{Gr}_x/\mathrm{Re}_x^2$	$\frac{1}{4}$	5	$\mathrm{Re}_x < 10^5$, $\mathrm{Gr}_x < 10^9$
UHF [17]	x	$I_1(K_1)$	$I_2(K_2^*)$	$\mathrm{Gr}_x^*/\mathrm{Re}_x^{5/2}$	$\frac{1}{5}$	5	$\mathrm{Re}_x < 10^5$, $\mathrm{Gr}_x^* < 10^{10}$
Horizontal cylinders in cross flow							Boundary-layer flow $\mathrm{Pr} = 0.7$
Local							
UWT [18]	d, ϕ	$0.992\,M_1(\phi)$	$0.440\,M_2(\phi)$	$\mathrm{Gr}_d/\mathrm{Re}_d^2$	$\frac{1}{4}$	3.5	$\mathrm{Re}_d < 10^5$, $\mathrm{Gr}_d < 10^9$
UHF [19]	d, ϕ	$0.992\,N_1(\phi)$	$0.519\,N_2(\phi)$	$\mathrm{Gr}_d^*/\mathrm{Re}_d^{5/2}$	$\frac{1}{5}$	4	$\mathrm{Re}_d < 10^5$, $\mathrm{Gr}_d^* < 10^{10}$
Spheres in cross flow							Boundary-layer flow $\mathrm{Pr} = 0.7$
Local							
UWT [28]	d, ϕ	$1.153\,M_3(\phi)$	$0.544\,M_4(\phi)$	$\mathrm{Gr}_d/\mathrm{Re}_d^2$	$\frac{1}{4}$	3.5	$\mathrm{Re}_d < 10^5$, $\mathrm{Gr}_d \le 10^9$
UHF [29]	d, ϕ	$1.153\,N_3(\phi)$	$0.615\,N_4(\phi)$	$\mathrm{Gr}_d^*/\mathrm{Re}_d^{5/2}$	$\frac{1}{5}$	4	$\mathrm{Re}_d < 10^5$, $\mathrm{Gr}_d^* < 10^{10}$

[a] $\mathrm{Re}_l = u_\infty l/\nu$, $\mathrm{Gr}_l = g\beta(T_w - T_\infty)l^3/\nu^2$, and $\mathrm{Gr}_l^* = g\beta q_w l^4/k\nu^2$ are based on a characteristic length l ($l = x$, L, or d as needed). Fluid properties are evaluated at the film temperature $T_f = (T_w + T_\infty)/2$.

[b] The functions in these formulas are defined in Table 14.2.

TABLE 14.2 Functions Used in Table 14.1

Function[a]	Ref.
$F_1(\mathrm{Pr}) = 0.339\,\mathrm{Pr}^{1/3}[1 + (0.0468/\mathrm{Pr})^{2/3}]^{-1/4}$	[8]
$F_2(\mathrm{Pr}) = 0.75\,\mathrm{Pr}^{1/2}[2.5(1 + 2\,\mathrm{Pr}^{1/2} + 2\,\mathrm{Pr})]^{-1/4}$	[9]
$F_3(\mathrm{Pr}) = (\mathrm{Pr}/5)^{1/5}\mathrm{Pr}^{1/2}[0.25 + 1.6\,\mathrm{Pr}^{1/2}]^{-1}$	[3]
$F_4(\mathrm{Pr}) = 1.886\,\mathrm{Pr}^{13/32} - 1.445\,\mathrm{Pr}^{1/3}$	[15]
$G_1(\mathrm{Pr}) = 0.464\,\mathrm{Pr}^{1/3}[1 + (0.0207/\mathrm{Pr})^{2/3}]^{-1/4}$	[10]
$G_2(\mathrm{Pr}) = \mathrm{Pr}^{2/5}[4 + 9\,\mathrm{Pr}^{1/2} + 10\,\mathrm{Pr}]^{-1/5}$	[11]
$G_3(\mathrm{Pr}) = (\mathrm{Pr}/6)^{1/6}\mathrm{Pr}^{1/2}[0.12 + 1.2\,\mathrm{Pr}^{1/2}]^{-1}$	[5]
$G_4(\mathrm{Pr}) = 2.845\,\mathrm{Pr}^{13/32} - 2.095\,\mathrm{Pr}^{1/3}$	[15]
$H_1(K_1) = 0.311 + 0.127K_1 - 0.0046K_1^2$ where $K_1 = 4(x/r_0)\mathrm{Re}_x^{-1/2}$	[16]
$H_2(K_2) = 0.353 + 0.155K_2 - 0.0105K_2^2$ where $K_2 = 2(x/r_0)(\mathrm{Gr}_x/4)^{-1/4} = (4\mathrm{Gr}_x/\mathrm{Re}_x^2)^{-1/4}K_1$	[16]
$I_1(K_1) = 0.420 + 0.123K_1 - 0.0041K_1^2$ where $K_1 = 4(x/r_0)\mathrm{Re}_x^{-1/2}$	[17]
$I_2(K_2^*) = 0.483 + 0.120K_2^* - 0.0044K_2^{*2}$ where $K_2^* = 2(x/r_0)(\mathrm{Gr}_x^*/5)^{-1/5} = (32\,\mathrm{Gr}_x^*/5\,\mathrm{Re}_x^{5/2})^{-1/5}K_1$	[17]
$M_1(\phi) = 1 - 0.018\phi - 0.105\phi^2 - 0.029\phi^3$	[18]
$M_2(\phi) = 1 - 0.0003\phi - 0.0424\phi^2 - 0.0010\phi^3$	[18]
$M_3(\phi) = 1 - 0.012\phi - 0.153\phi^2 - 0.016\phi^3$	[28]
$M_4(\phi) = 1 - 0.0020\phi - 0.0685\phi^2 - 0.0053\phi^3$	[28]
$N_1(\phi) = 1 - 0.013\phi - 0.054\phi^2 - 0.023\phi^3$	[19]
$N_2(\phi) = 1 - 0.0004\phi - 0.0201\phi^2 - 0.0011\phi^3$	[19]
$N_3(\phi) = 1 - 0.009\phi - 0.107\phi^2 - 0.012\phi^3$	[29]
$N_4(\phi) = 1 - 0.0007\phi - 0.0415\phi^2 - 0.0018\phi^3$	[29]

[a] ϕ is in radians.

14.6 CORRELATIONS FOR VERTICAL CYLINDERS IN LONGITUDINAL FLOW

Available results on laminar mixed convection for vertical cylinders in longitudinal flow are limited to fluids having a Prandtl number of 0.7 and to cylinders with fairly small values of the radius of curvature and buoyancy parameter. The results of Chen and Mucoglu [16, 17] for the UWT and UHF cases under buoyancy-assisting condition are used to develop and validate the proposed simple mixed-convection correlations for this geometry. These correlations can be expressed in the form of Eq. (14.15) with appropriate functions $A(\mathrm{Pr})$ and $B(\mathrm{Pr})$ and constants m and n as given in Tables 14.1 and 14.2. Comparisons between the predicted results and the correlations are shown in Fig. 14.13 as Y vs. X for both the UWT and the UHF cases, where

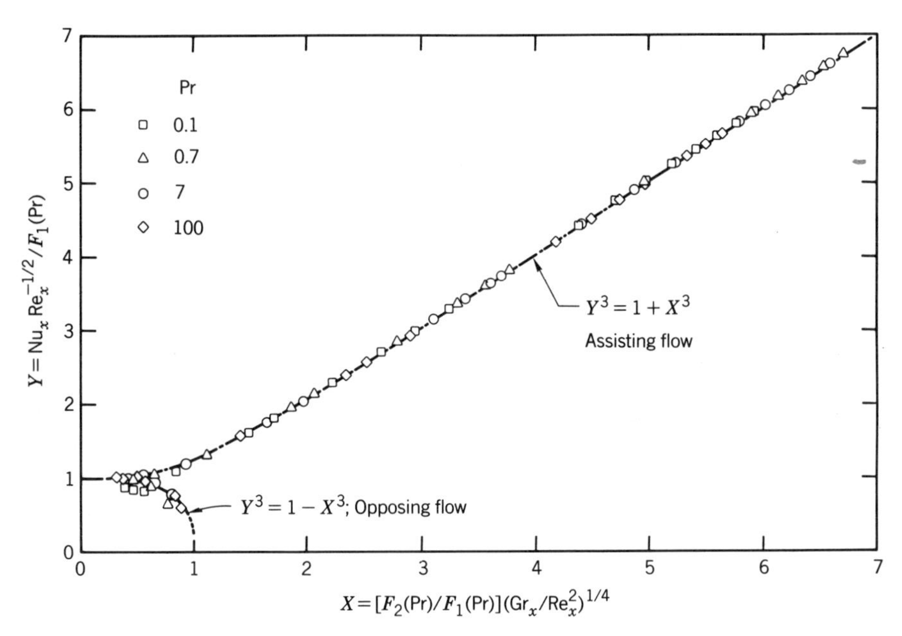

Figure 14.7. Comparison between predicted and correlated local Nusselt numbers: vertical flat plate, UWT [3].

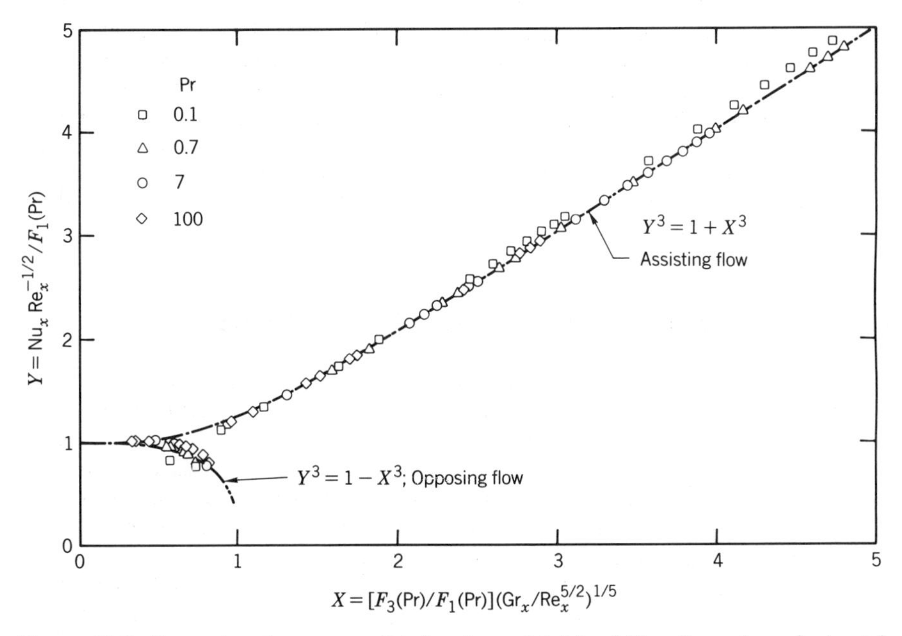

Figure 14.8. Comparison between predicted and correlated local Nusselt numbers: horizontal flat plate, UWT [3].

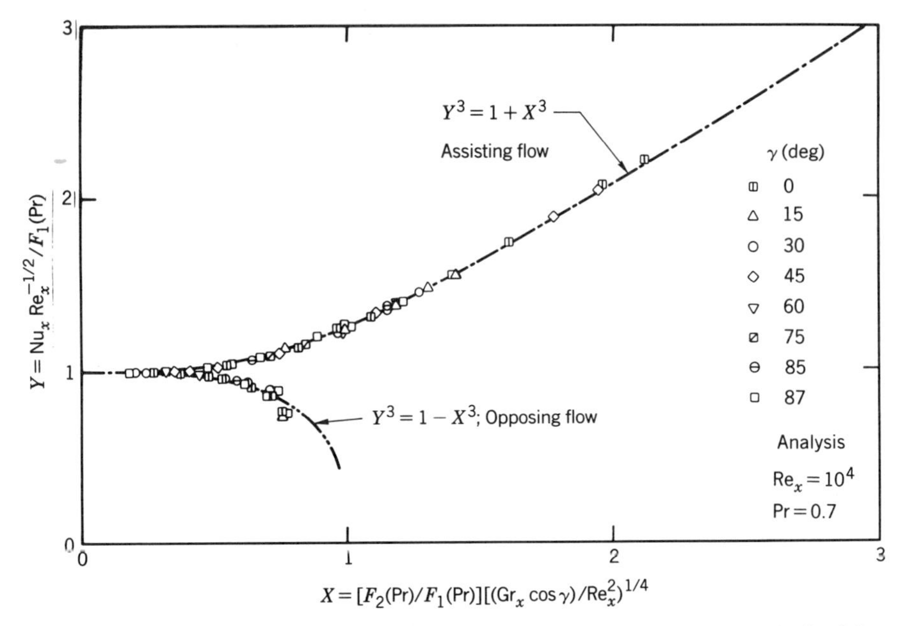

Figure 14.9. Comparison between predicted and correlated local Nusselt numbers: inclined flat plate, UWT [3].

$Y = \mathrm{Nu}\,\mathrm{Re}^{-1/2}/A(\mathrm{Pr})$ and $X = B(\mathrm{Pr})\xi^m/A(\mathrm{Pr})$. For the limited range of available numerical results, Pr = 0.7 and $1 \le K_1 \le 4$, where K_1 is the radius of curvature of the cylinder, the proposed mixed-convection correlation for this geometry has an exponent of $n = 5$ for both UWT and UHF instead of the 3 that is used for flat plates. The maximum deviation between the predicted and correlated results is 4% for the UWT case and 8% for the UHF case. Additional mixed-convection results are needed to extend the range of applicability of such a correlation to fluids having Prandtl numbers other than 0.7 and to cylinders having a larger radius of curvature and under a stronger buoyancy-force effect.

14.7 CORRELATIONS FOR HORIZONTAL CYLINDERS

The direction of the forced flow, relative to that of the resulting buoyancy force, in mixed convection flow across cylinders must be considered when calculating the heat transfer. Assisting, opposing, and cross flows are accepted terms for three forced-flow directions in this geometry. In assisting flow, the forced flow is in the same direction as the buoyancy force (vertically upward for a heated cylinder); in opposing flow it is exactly in the opposite direction (vertically downward for a heated cylinder). In cross flow, the forced flow is perpendicular to the buoyancy force.

The flow field across a cylinder will normally develop a complex separation region downstream from the stagnation point, and this feature has limited the boundary-layer type of calculations to the unseparated portion of the flow domain. Mucoglu and Chen [18, 19] reported on such a calculation for air flow, Pr = 0.7, for UWT and UHF boundary conditions. Their results can be used to develop and validate mixed-convection correlations for the local Nusselt number in the unseparated flow region around

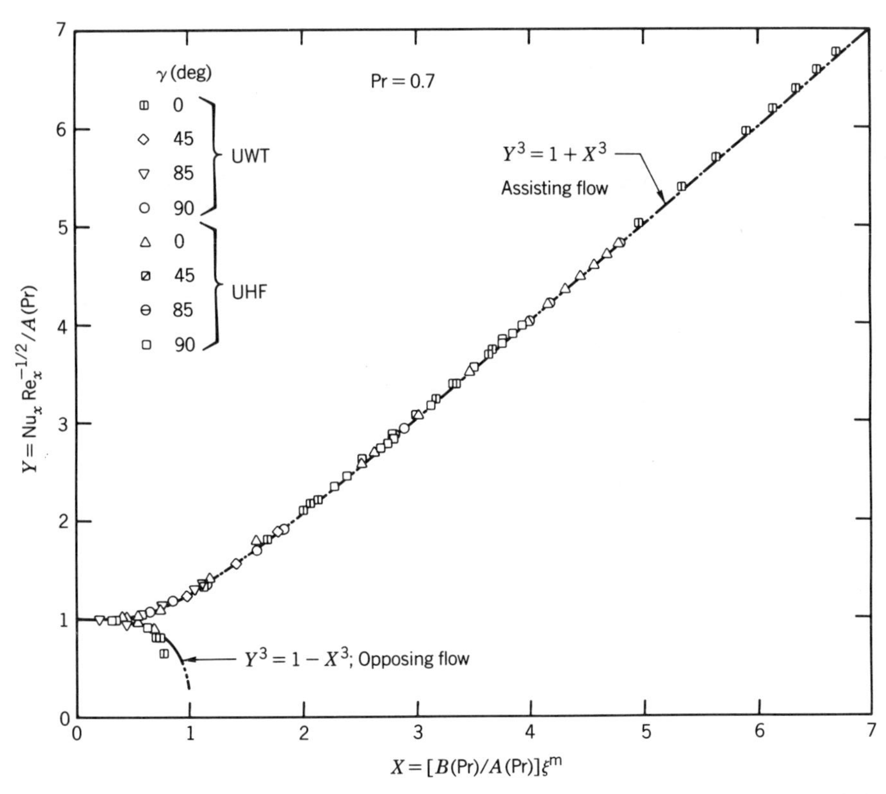

Figure 14.10. Comparison between predicted and correlated local Nusselt numbers: vertical, inclined, and horizontal flat plates, UWT and UHF, Pr = 0.7 [3, 5].

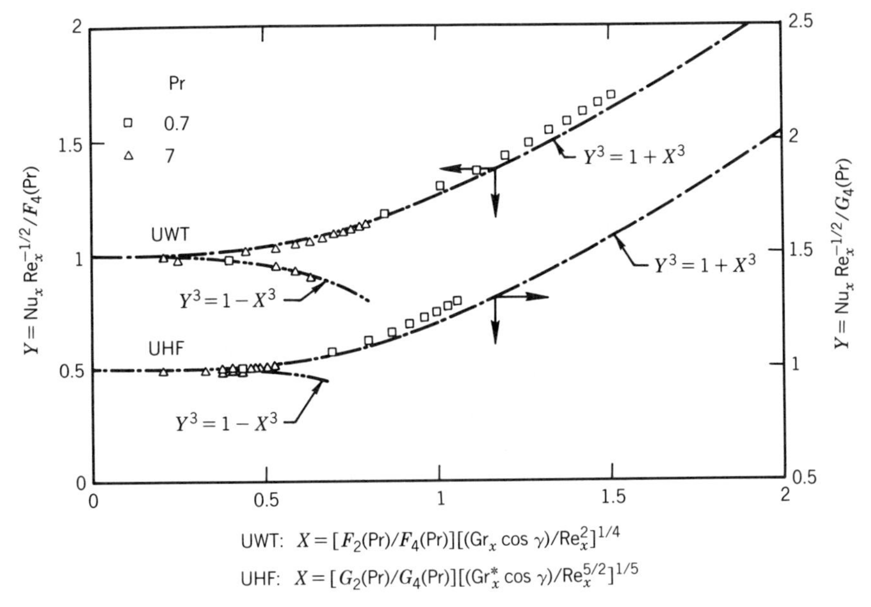

Figure 14.11. Comparison between predicted and correlated local Nusselt numbers: vertical and inclined moving sheets, UWT and UHF [12, 14].

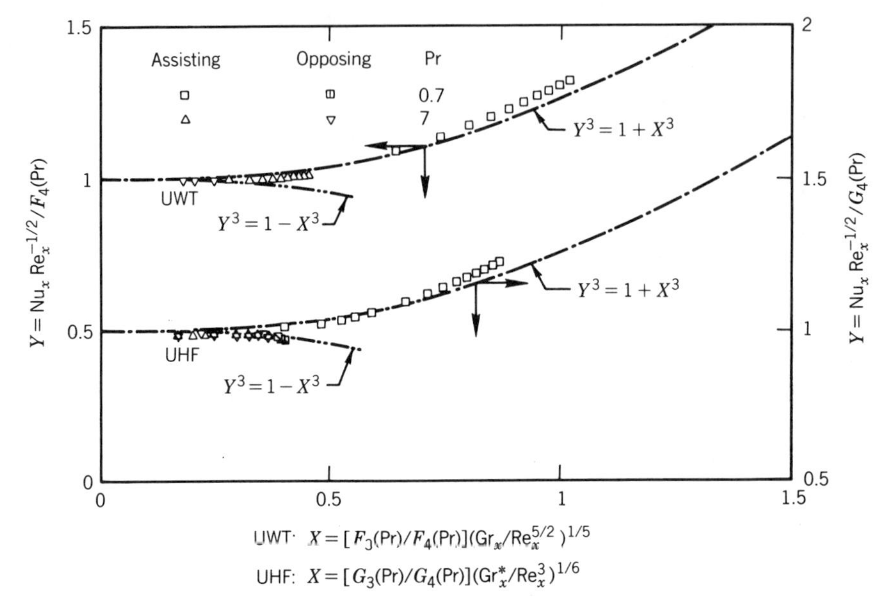

Figure 14.12. Comparison between predicted and correlated local Nusselt numbers: horizontal moving sheets, UWT and UHF [13, 15].

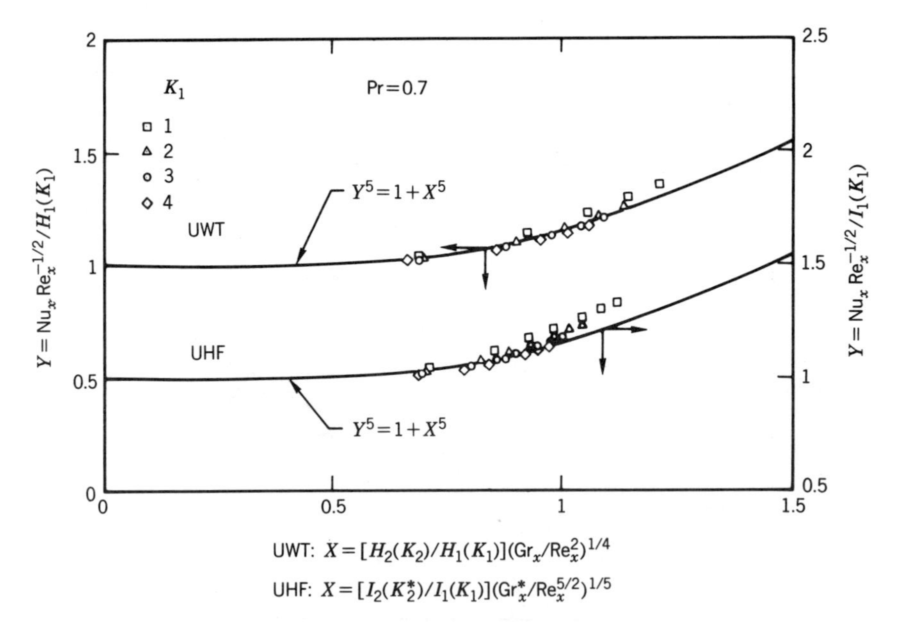

Figure 14.13. Comparison between predicted and correlated local Nusselt numbers: vertical cylinders in longitudinal flow, UWT and UHF, Pr = 0.7 [16, 17].

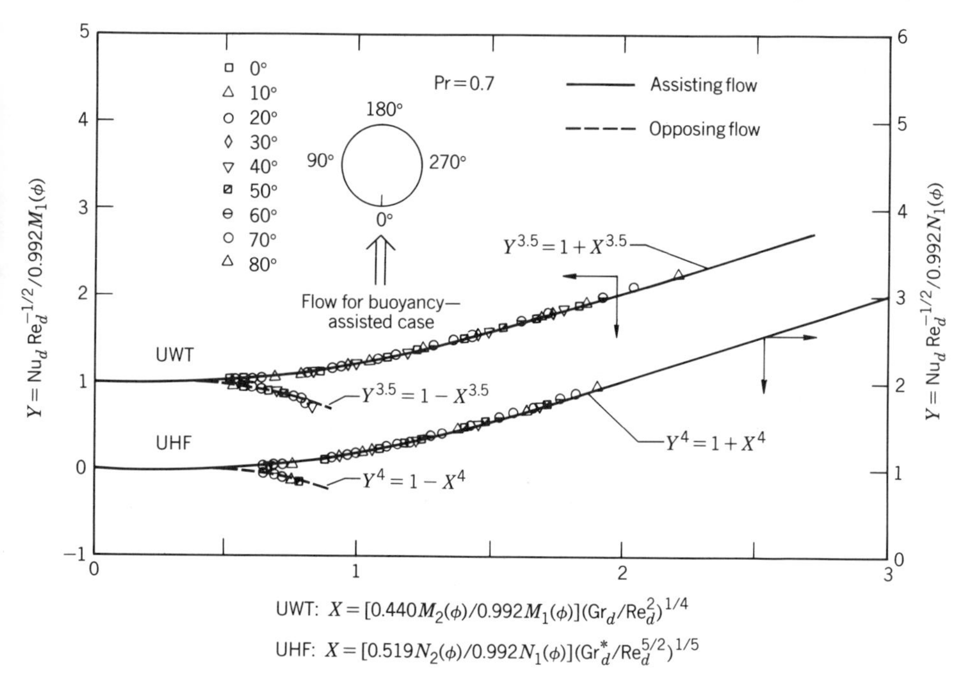

Figure 14.14. Comparison between predicted and correlated local Nusselt numbers: horizontal cylinders in cross flow, UWT and UHF, Pr = 0.7 [18, 19].

the circumference of the cylinders, i.e. from 0° (the stagnation point) to approximately 90°. The resulting correlations for this geometry can be expressed in a form equivalent to Eq. (14.15), and the appropriate functions and constants for these correlations are included in Tables 14.1 and 14.2. These correlations express the local mixed-convection Nusselt number, $\mathrm{Nu}_d(\phi)$, as a function of the angle ϕ, measured in radians, from the stagnation point. Comparisons between the predicted mixed convection results and the proposed simple correlation, Eq. (14.15), are presented in Fig. 14.14. The figure illustrates an excellent agreement between the correlated and the predicted results for all angles from the stagnation point to about 90° degrees for both assisting and opposing flows with UWT and UHF boundary conditions.

Measurements and predictions of the average mixed-convection Nusselt number for cylinders in air flow, Pr = 0.7, have been reported by Oosthuizen and Madan [20, 21], Badr [22, 23], Hatton et al. [24], and Nakai and Okazaki [25] for assisting, opposing, and cross flows, covering a wide range of Reynolds and Grashof numbers. The measurements by Oosthuizen and Madan [20, 21], which cover the flow regime of $10^2 < \mathrm{Re}_d < 3 \times 10^3$ and $2.5 \times 10^4 < \mathrm{Gr}_d < 3 \times 10^5$, can be correlated by the following relations [26]:

1. For assisting flow ($\phi = 0°$),

$$\left(\frac{\overline{\mathrm{Nu}}_d}{\overline{\mathrm{Nu}}_{dF}}\right)^{3.5} = 1 + 1.396\ \Omega^{3.5} \tag{14.16}$$

where $\overline{Nu}_{dF}$ is the average Nusselt number for the pure forced convection limit and is given by

$$\overline{\mathrm{Nu}}_{dF} = 0.464\,\mathrm{Re}_d^{0.5} + 0.0004\,\mathrm{Re}_d \tag{14.17}$$

and

$$\Omega = \left(\frac{\mathrm{Gr}_d}{\mathrm{Re}_d^2}\right)^{1/4} \tag{14.18}$$

2. For cross flow ($\phi = 90°$),

$$\left(\frac{\overline{\mathrm{Nu}}_d}{\overline{\mathrm{Nu}}_{dF}}\right)^7 = 1 + 6.275\,\Omega^7 \tag{14.19}$$

3. For opposing flow ($\phi = 180°$) in the region of $\Omega < 0.7$,

$$\left(\frac{\overline{\mathrm{Nu}}_d}{\overline{\mathrm{Nu}}_{dF}}\right)^{2.5} = 1 - \Omega^{2.5} \tag{14.20}$$

and in the region of $\Omega > 0.7$

$$\left(\frac{\overline{\mathrm{Nu}}_d}{\overline{\mathrm{Nu}}_{dF}}\right) = 1.2\,\Omega - 0.2 \tag{14.21}$$

Comparisons between measured and correlated results are shown in Fig. 14.15. All fluid properties are evaluated at the film temperature T_f.

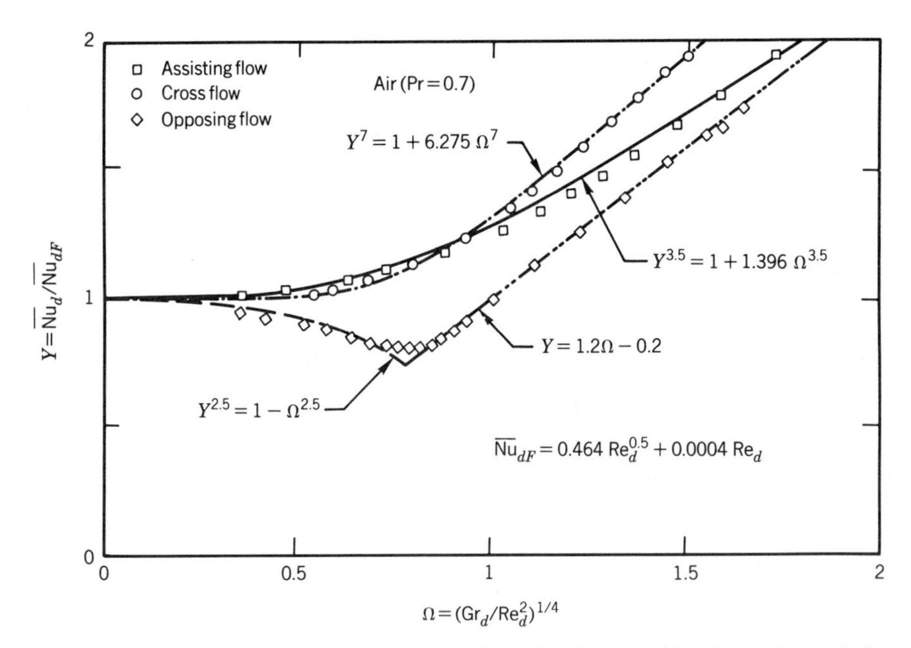

Figure 14.15. Comparison between measured and correlated average Nusselt numbers: air flow across isothermal horizontal cylinders, UWT, $10^2 < \mathrm{Re}_d < 3 \times 10^3$, $2.5 \times 10^4 < \mathrm{Gr}_d < 3 \times 10^5$ [26].

Analytical calculations of the average Nusselt number for mixed convection flow across horizontal cylinders have been reported by Badr [22, 23] for the flow domain of $1 < \mathrm{Re}_d < 60$ and $0 < \mathrm{Gr}_d < 7200$. His reported results can be correlated by the following relations [26]:

1. For assisting flow ($\phi = 0°$)

$$\frac{\overline{\mathrm{Nu}}_d}{\overline{\mathrm{Nu}}_{dF}} = 1 + 0.16\frac{\mathrm{Gr}_d}{\mathrm{Re}_d^2} - 0.015\left(\frac{\mathrm{Gr}_d}{\mathrm{Re}_d^2}\right)^2 \tag{14.22}$$

2. For cross flow ($\phi = 90°$)

$$\frac{\overline{\mathrm{Nu}}_d}{\overline{\mathrm{Nu}}_{dF}} = 1 + 0.05\frac{\mathrm{Gr}_d}{\mathrm{Re}_d^2} + 0.003\left(\frac{\mathrm{Gr}_d}{\mathrm{Re}_d^2}\right)^2 \tag{14.23}$$

3. For opposing flow ($\phi = 180°$)

$$\frac{\overline{\mathrm{Nu}}_d}{\overline{\mathrm{Nu}}_{dF}} = 1 - 0.37\frac{\mathrm{Gr}_d}{\mathrm{Re}_d^2} + 0.150\left(\frac{\mathrm{Gr}_d}{\mathrm{Re}_d^2}\right)^2 \tag{14.24}$$

where $\overline{\mathrm{Nu}}_{dF}$ is the averaged Nusselt number for the pure forced-convection limit, which can be correlated from predicted values in this flow regime by [26]

$$\overline{\mathrm{Nu}}_{dF} = 1.01 + 9.1 \times 10^{-2}\,\mathrm{Re}_d - 7.3 \times 10^{-4}\,\mathrm{Re}_d^2 \tag{14.25}$$

Comparisons between the above correlations and the predicted results are shown in Fig. 14.16.

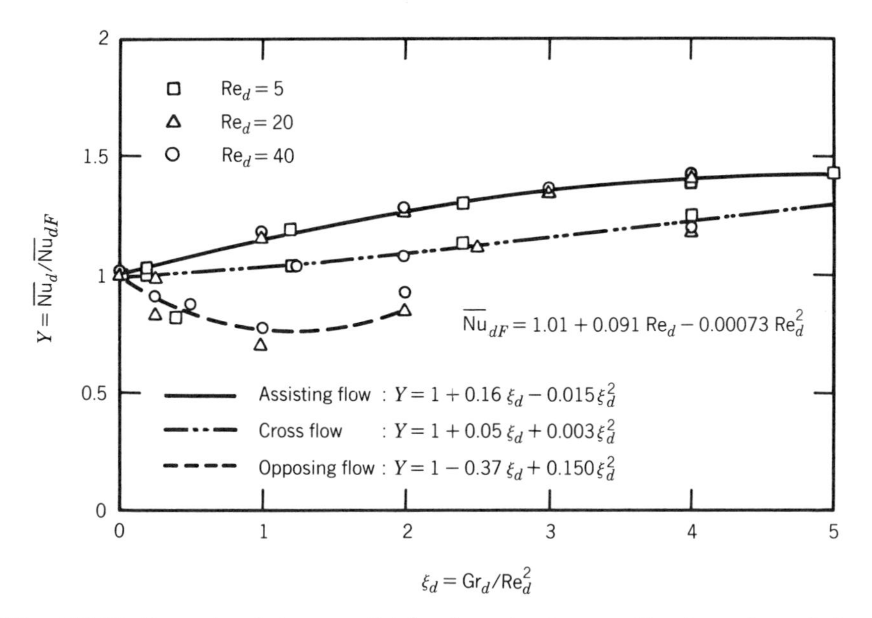

Figure 14.16. Comparison between predicted and correlated average Nusselt numbers: air flow across isothermal horizontal cylinders, $1 < \mathrm{Re}_d < 60$, $0 < \mathrm{Gr}_d < 7200$ [26].

Measurements of the average Nusselt number for this geometry by Hatton et al. [24] cover the flow domain of $10^{-2} < \text{Re}_d < 40$ and $10^{-3} < \text{Gr}_d < 10$. The results have been well correlated by the following relations:

$$\overline{\text{Nu}}_d \left(\frac{T_f}{T_\infty} \right)^{-0.154} = 0.384 + 0.581 R_f^{0.439} \tag{14.26}$$

where

$$R_f = \text{Re}_d \left(1 + 2.06 \frac{\text{Ra}^{0.418} \cos\phi}{\text{Re}_d} + 1.06 \frac{\text{Ra}^{0.836}}{\text{Re}_d^2} \right)^{1/2} \tag{14.27}$$

with Ra denoting the Rayleigh number ($\text{Ra} = \text{Gr}_d \text{Pr}$) and T_f the film temperature $[T_f = (T_w + T_\infty)/2]$. The angle ϕ is measured from the vertically upward direction of the forced flow (assisting $\phi = 0°$, opposing $\phi = 180°$, and cross $\phi = 90°$). The fluid properties are evaluated at the average film temperature T_f. Equation (14.26) correlates the measured mixed-convection data [24] very well for all forced-flow directions except for opposing flow in the region of $0.25 < \text{Ra}^{0.418}/\text{Re}_d < 2.5$, where the deviations between the measured and the correlated results exceed 10%.

Analysis and measurements of the average Nusselt number for mixed convection air flow across horizontal cylinders of length-to-diameter ratio $2 \times 10^4 \leq L/d \leq 2.5 \times 10^4$ have been reported by Nakai and Okazaki [25] for the flow domain of very low Reynolds and Grashof numbers, with $10^{-3} < \text{Re}_d < 10^{-1}$ and $10^{-6} < \text{Gr}_d < 6.5 \times 10^{-5}$ (i.e., creeping flow). Their measured results can be correlated by the following relations [26]:

1. For assisting flow ($\phi = 0°$)

$$\frac{\overline{\text{Nu}}_d}{\overline{\text{Nu}}_{dF}} = 1.295 \left(\frac{\text{Gr}_d}{\text{Re}_d^2} \right)^{0.0472} \tag{14.28}$$

2. For cross flow ($\phi = 90°$)

$$\frac{\overline{\text{Nu}}_d}{\overline{\text{Nu}}_{dF}} = 1.257 \left(\frac{\text{Gr}_d}{\text{Re}_d^2} \right)^{0.0613} \tag{14.29}$$

3. For opposing flow ($\phi = 180°$)

$$\frac{\overline{\text{Nu}}_d}{\overline{\text{Nu}}_{dF}} = 1.281 \left(\frac{\text{Gr}_d}{\text{Re}_d^2} \right)^{0.0795} \tag{14.30}$$

The Nusselt number for pure forced convection for this flow regime is given by analysis as [27]

$$\overline{\text{Nu}}_{dF} = \frac{2}{\ln[5.435/(\text{Pr}\,\text{Re}_d)]} \tag{14.31}$$

The above correlations provide a simple and accurate method for evaluating the average mixed-convection Nusselt number in this flow regime of very low Reynolds

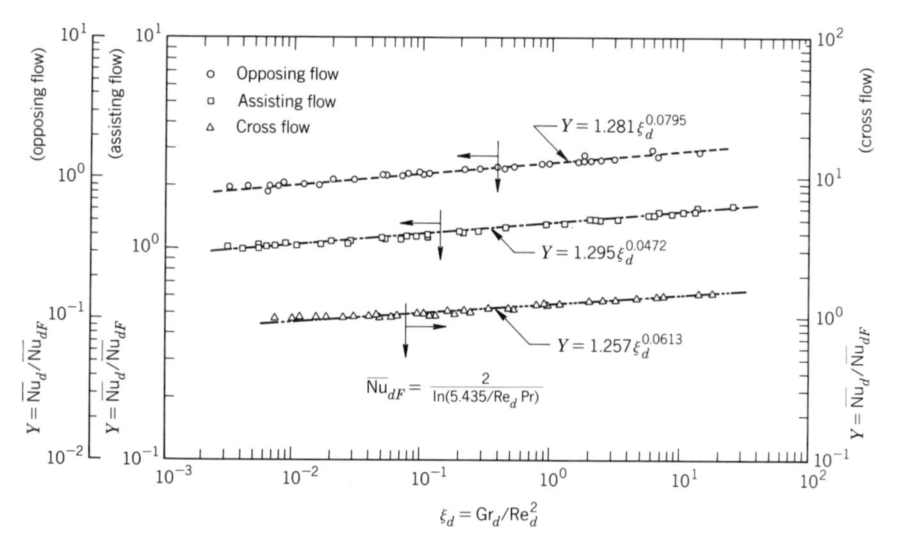

Figure 14.17. Comparison between measured and correlated average Nusselt numbers: air flow across isothermal horizontal cylinders, $10^{-3} < \mathrm{Re}_d < 10^{-1}$, $10^{-6} < \mathrm{Gr}_d < 6.5 \times 10^{-5}$ [26].

and Grashof numbers, as shown in Fig. 14.17, in which the correlated results and the measured data are in excellent agreement. It should be noted that in the correlations fluid properties are evaluated at the film temperature T_f except those in Gr_d, which are based on the free-stream temperature T_∞.

14.8 CORRELATIONS FOR SPHERES

Mixed convection flow across spheres exhibits similar features to those discussed for cylinders in Sec. 14.7. Boundary-layer calculations for the unseparated region of the flow domain have been reported by Chen and Mucoglu [28, 29] for air flow, $\mathrm{Pr} = 0.7$, with UWT and UHF boundary conditions. These results are used in a fashion similar to that used for cylinders to develop and validate mixed-convection correlations for the local mixed-convection Nusselt number in the unseparated region of the flow. The correlations can be expressed in terms of Eq. (14.15) with constants and functions as defined in Tables 14.1 and 14.2. The good agreement between the correlated and numerical results is shown in Fig. 14.18.

Measurements of the average Nusselt number for spheres in mixed convection air flow, $\mathrm{Pr} = 0.7$, have been reported by Yuge [30] for assisting, opposing, and cross flows in the region where $3.5 < \mathrm{Re}_d < 5.9 \times 10^3$ and $1 < \mathrm{Gr}_d < 10^5$. The measured mixed-convection results can be correlated well by the following relations [26]:

1. For assisting and cross flows ($\phi = 0°$ and $\phi = 90°$)

$$\frac{\overline{\mathrm{Nu}}_d - 2}{N_R} = \left[1 + \left(\frac{N_G}{N_R}\right)^{3.5}\right]^{1/3.5} \tag{14.32}$$

where

$$N_R = \overline{\mathrm{Nu}}_{dF} - 2 = 0.493\,\mathrm{Re}_d^{1/2} \tag{14.33}$$

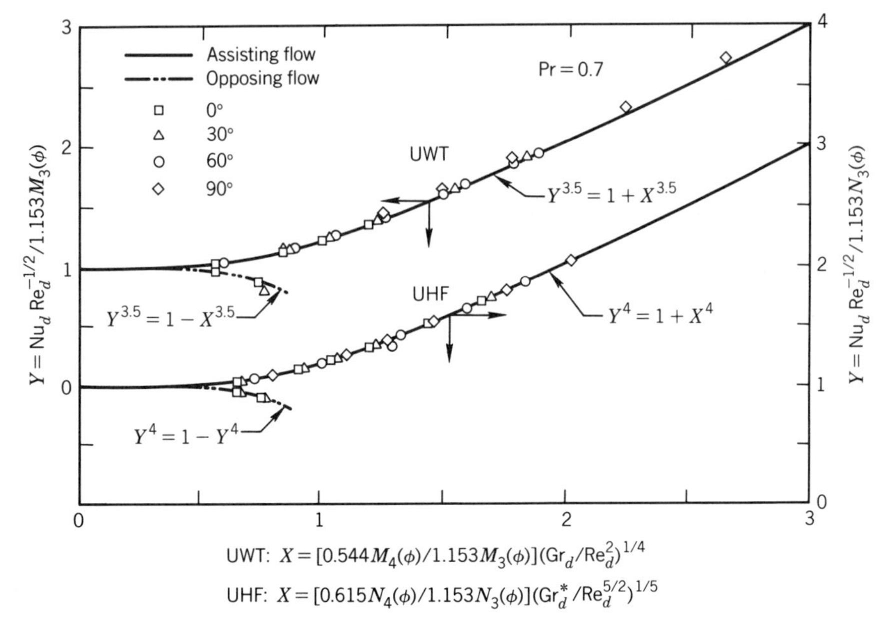

Figure 14.18. Comparison between predicted and correlated local Nusselt numbers: spheres in cross flow, UWT and UHF, Pr = 0.7 [28, 29].

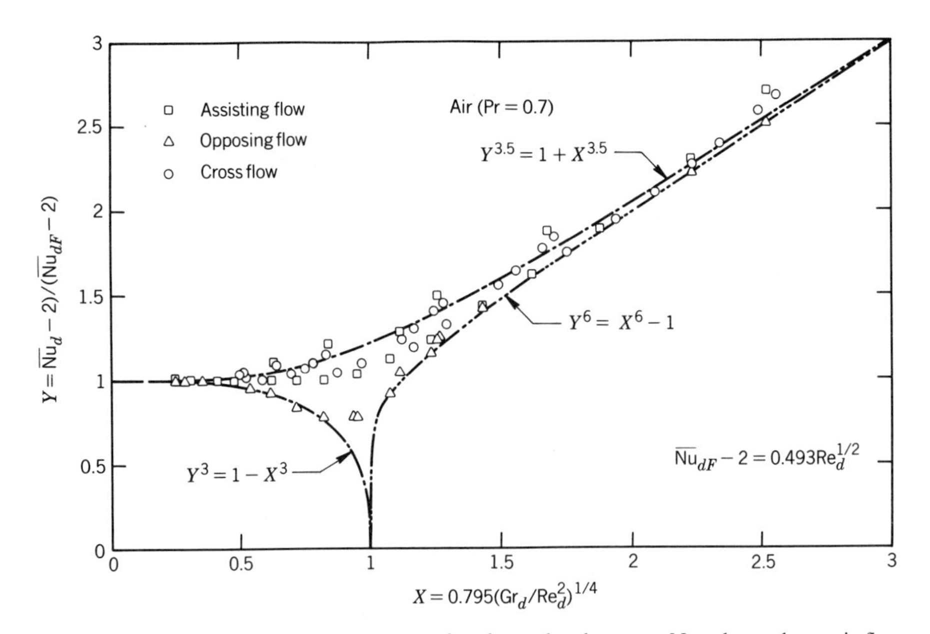

Figure 14.19. Comparison between measured and correlated average Nusselt numbers: air flow across isothermal spheres, $3.5 < \mathrm{Re}_d < 5.9 \times 10^3$, $1 < \mathrm{Gr}_d < 10^5$ [26].

and

$$N_G = \overline{\mathrm{Nu}}_{dN} - 2 = 0.392\,\mathrm{Gr}_d^{1/4} \tag{14.34}$$

such that $N_G/N_R = 0.795(\mathrm{Gr}_d/\mathrm{Re}_d^2)^{1/4}$, and $\overline{\mathrm{Nu}}_{dF}$ and $\overline{\mathrm{Nu}}_{dN}$ are, respectively, the average mixed-convection Nusselt numbers for the pure forced-convection and the pure free-convection limits.

2. For opposing flow ($\phi = 180°$), in the region where $N_G/N_R < 1$,

$$\frac{\overline{\mathrm{Nu}}_d - 2}{N_R} = \left[1 - \left(\frac{N_G}{N_R}\right)^3\right]^{1/3} \tag{14.35}$$

and in the region where $(N_G/N_R) > 1$,

$$\frac{\overline{\mathrm{Nu}}_d - 2}{N_R} = \left[\left(\frac{N_G}{N_R}\right)^6 - 1\right]^{1/6} \tag{14.36}$$

Comparisons of these mixed-convection correlations with measured results reveal reasonably good agreement, as shown in Fig. 14.19. In the correlation equations, all fluid properties are evaluated at the film temperature T_f.

14.9 CORRELATIONS FOR VERTICAL FLAT PLATES IN CROSS FLOW

Mixed convection adjacent to a vertical flat plate in horizontal cross flow parallel to the plate has been studied both analytically for UWT [31, 32] and experimentally for UHF [33]. For this flow configuration the buoyancy force is responsible for making the flow and thermal fields three-dimensional. The reported results on the average Nusselt number for laminar cross flow adjacent to a square plate ($L = H$) with UWT condition and Pr = 0.7 (air) are correlated by the equation [31]

$$\overline{\mathrm{Nu}}_L \mathrm{Re}_L^{-1/2} = 0.584\left[1 + 0.433\frac{\mathrm{Gr}_H}{\mathrm{Re}_L^2}\right]^{1/4} \tag{14.37}$$

in the range of buoyancy parameter $0 \le \mathrm{Gr}_L/\mathrm{Re}_L^2 \le 5$. Equation (14.37) corresponds to the correlation equation (14.15) with $n = 4$, $Y = \mathrm{Nu}\,\mathrm{Re}_L^{-1/2}/0.584$, and $X = (0.433\,\mathrm{Gr}_H/\mathrm{Re}_L^2)^{1/4}$.

For the case of laminar flow with the UHF condition, the experimental local Nusselt numbers in air flow (Pr = 0.7) have been correlated [33] as

$$\mathrm{Nu}_x \mathrm{Re}_x^{-1/2} = 0.402\left\{1 + \left[1.202\frac{x}{z}\left(\frac{\mathrm{Gr}_z^*}{\mathrm{Re}_x^{5/2}}\right)^{1/5}\right]^{3.2}\right\}^{1/3.2} \tag{14.38}$$

for $\mathrm{Re}_x < 10^5$ and $\mathrm{Gr}^*_z < 10^9$, where x is measured in the horizontal direction along the plate (in the forced-flow direction) and z is measured in the vertical direction along the plate (in the buoyancy-force direction). Again, Eq. (14.38) has the form of Eq. (14.15), with $n = 3.2$, $Y = \mathrm{Nu}_x\mathrm{Re}_x^{-1/2}/0.402$ and $X = 1.202\,(x/z)(\mathrm{Gr}^*_z/\mathrm{Re}_x^{5/2})^{1/5}$.

The corresponding correlation for the average Nusselt number can be expressed by

$$\overline{\mathrm{Nu}}_L \mathrm{Re}_L^{-1/2} = 0.804 \left\{ 1 + \left[0.752 \frac{L}{H} \left(\frac{\mathrm{Gr}^*_H}{\mathrm{Re}_L^{5/2}} \right)^{1/5} \right]^{3.2} \right\}^{1/3.2} \tag{14.39}$$

It should be emphasized here that the above correlations for vertical plates in cross flow are limited to fluids having $\mathrm{Pr} = 0.7$ and that fluid properties are evaluated at the free-stream temperature T_∞. Correlation equations for fluids with Prandtl numbers other than 0.7 can be developed only when the results from analyses and experiments become available.

14.10 TURBULENT HEAT TRANSFER CORRELATIONS FOR FLAT PLATES

Reported studies on turbulent heat transfer in mixed convection are lacking in the literature. Turbulent mixed convection on vertical and horizontal flat plates under the UWT condition has been analyzed by Chen et al. [34, 35] by employing a modified mixing-length model that accounts for the buoyancy-force effect. Their calculations for $\mathrm{Pr} = 0.7$ yield local Nusselt numbers that converge to the limit of pure forced convection, but underpredict by 20% the available results in the limit of pure free convection. The local Nusselt number for turbulent flow over a flat plate under UWT for $0.5 \le \mathrm{Pr} \le 1.0$ and $5 \times 10^5 \le \mathrm{Re}_x \le 5 \times 10^6$ is given by [36]

$$\mathrm{Nu}_x \mathrm{Re}_x^{-4/5} = 0.0287\, \mathrm{Pr}^{0.6} \tag{14.40}$$

The local Nusselt number for turbulent free convection along an isothermal vertical flat plate for all Prandtl numbers and for Gr_x to 10^{12} is correlated by Churchill and Chu [37] as

$$\mathrm{Nu}_x \mathrm{Gr}_x^{-1/3} = 0.15\, \mathrm{Pr}^{1/3} \left[1 + \left(\frac{0.492}{\mathrm{Pr}} \right)^{9/16} \right]^{-16/27} \tag{14.41}$$

For turbulent free convection over a heated horizontal flat plate facing upward, the local Nusselt number for $\mathrm{Gr}_x \mathrm{Pr} > 5 \times 10^8$ is given by Fujii and Imura [38] as

$$\mathrm{Nu}_x \mathrm{Gr}_x^{-1/3} = 0.13\, \mathrm{Pr}^{1/3} \tag{14.42}$$

Equation (14.42), obtained under neither the UWT nor the UHF condition, was also verified later by Imura et al. for an isothermal horizontal flat plate [39].

If Eq. (14.12) is employed to propose a correlation for the local Nusselt number in mixed convection, the resulting form is

$$\frac{\mathrm{Nu}_x \mathrm{Re}_x^{-4/5}}{F(\mathrm{Pr})} = \left\{ 1 + c \left[\frac{G(\mathrm{Pr})}{F(\mathrm{Pr})} \left(\frac{\mathrm{Gr}_x}{\mathrm{Re}_x^{12/5}} \right)^{1/3} \right]^n \right\}^{1/n} \tag{14.43}$$

where $F(\mathrm{Pr}) = 0.0287\, \mathrm{Pr}^{0.6}$ and $G(\mathrm{Pr}) = 0.15\, \mathrm{Pr}^{1/3} [1 + (0.492/\mathrm{Pr})^{9/16}]^{-16/27}$ for a vertical plate and $0.13\, \mathrm{Pr}^{1/3}$ for a horizontal flat plate. The analytical local Nusselt numbers of Chen et al. [34, 35] for $\mathrm{Pr} = 0.7$ agree fairly well with the proposed

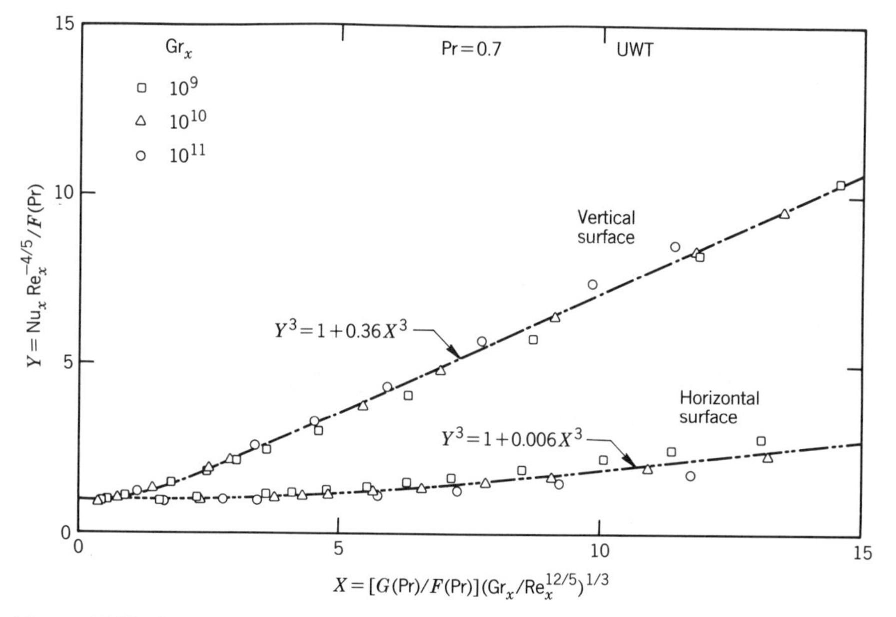

Figure 14.20. Comparison between predicted and correlated local Nusselt numbers: turbulent flow, vertical and horizontal flat plates, UWT, Pr = 0.7 [34, 35].

correlation equation (14.43) with $n = 3$ and with $c = 0.006$ and 0.36 for horizontal and vertical plates, respectively. This comparison is illustrated in Fig. 14.20. The corresponding average Nusselt number can be correlated by

$$\frac{\mathrm{Nu}\,\mathrm{Re}_L^{-4/5}}{1.25F(\mathrm{Pr})} = \left\{1 + c\left[\frac{G(\mathrm{Pr})}{1.25F(\mathrm{Pr})}\left(\frac{\mathrm{Gr}_L}{\mathrm{Re}_L^{12/5}}\right)^{1/3}\right]^n\right\}^{1/n} \tag{14.44}$$

with the same c and n values as in Eq. (14.43).

Turbulent mixed convection from an isothermal, heated, vertical flat plate in cross flow was analyzed by Plumb and Evans [32]. Their results for the average Nusselt number for a square plate ($H = L = 1$ m) in the ranges of $2 \times 10^5 \leq \mathrm{Re}_L \leq 1.3 \times 10^6$ and $0 \leq \mathrm{Gr}_H/\mathrm{Re}_L^2 \leq 3$ for Pr = 0.7 can be correlated by the equation

$$\frac{\mathrm{Nu}\,\mathrm{Re}_L^{-4/5}}{0.029} = \left\{1 + \left[3.225\left(\frac{\mathrm{Gr}_H}{\mathrm{Re}_L^{12/5}}\right)^{1/3}\right]^4\right\}^{1/4} \tag{14.45}$$

Equation (14.45) is similar in form to Eq. (14.43), but has an exponent of $n = 4$.

Measurements of turbulent mixed convection from a vertical flat plate ($L = 2.954$ m long and $H = 3.030$ m high) in air in cross flow under the UHF condition by Siebers et al. [33] have provided the local and average Nusselt numbers. Their proposed correlation equation for local Nusselt numbers for Pr = 0.7 can be written in the form as

$$\frac{\mathrm{Nu}_x\mathrm{Re}_x^{-4/5}}{0.025} = \left\{1 + \left[7.067\frac{x}{z}\left(\frac{\mathrm{Gr}^*_z}{\mathrm{Re}_x^{16/5}}\right)^{1/4}\left(\frac{T_w}{T_\infty}\right)^{0.295}\right]^{3.2}\right\}^{1/3.2} \tag{14.46}$$

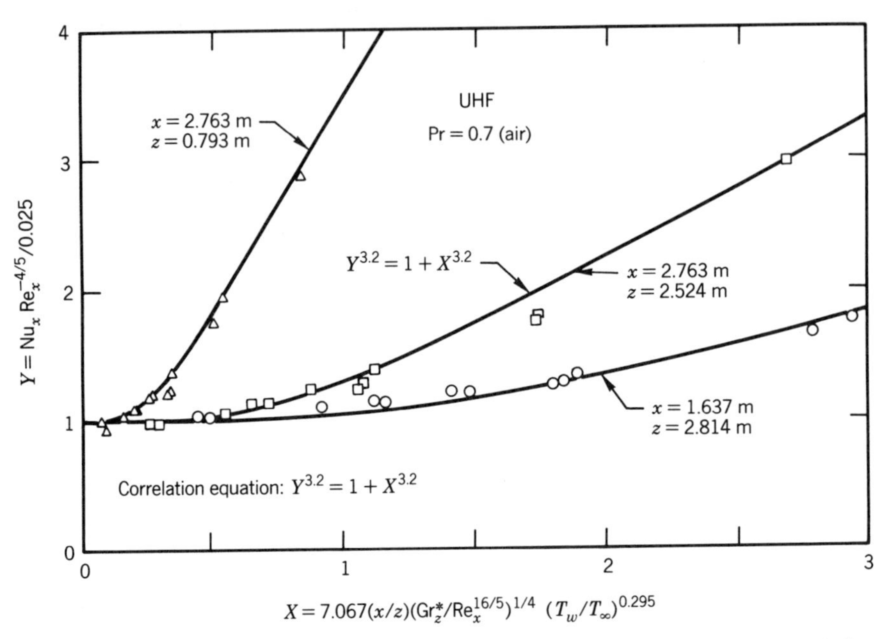

Figure 14.21. Comparison between measured and correlated local Nusselt numbers: turbulent air flow across a vertical flat plate, UHF [33].

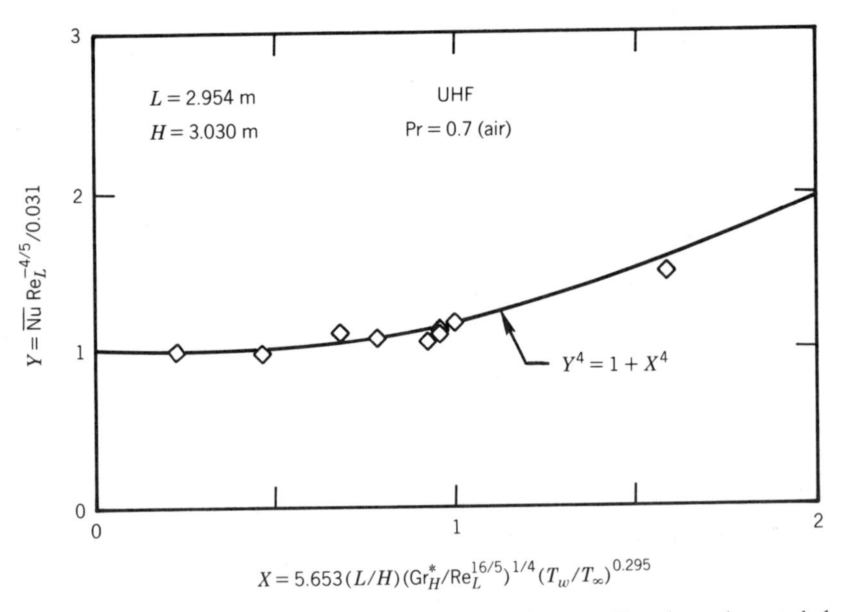

Figure 14.22. Comparison between measured and predicted average Nusselt numbers: turbulent air flow across a vertical flat plate, UHF [33].

in the range $0 \le Gr_H/Re_L^2 \le 30$, with $Re_x < 2 \times 10^6$, $Gr^*_z < 2 \times 10^{12}$, and $40 \le T_w \le 600°C$.

The corresponding expression for the average Nusselt number can be derived and expressed for Pr = 0.7 as

$$\frac{\overline{Nu}\, Re_L^{-4/5}}{0.031} = \left\{1 + \left[5.653 \frac{L}{H} \left(\frac{Gr_H^*}{Re_L^{16/5}}\right)^{1/4} \left(\frac{T_w}{T_\infty}\right)^{0.295}\right]^4\right\}^{1/4} \tag{14.47}$$

with $n = 4$ providing a better fit than $n = 3.2$.

Figure 14.21 illustrates a comparison between the measured local Nusselt number for a vertical flat plate ($L = 2.954$ m and $H = 3.030$ m) in cross flow of air [33] at three (x, z) locations, as indicated in the figure, and the proposed correlation equation (14.46). A comparison between the measured average Nusselt number over the entire plate and the correlation equation (14.47) is shown in Fig. 14.22. As can be seen from the two figures, the proposed correlations for both the local and average Nusselt numbers agree well with the experimental results. In Eqs. (14.46) and (14.47) all fluid properties are evaluated at the free-stream temperature T_∞.

14.11 INSTABILITY AND TRANSITION

Buoyancy forces play a significant role in affecting the laminar flow regime in mixed convection. Their presence may enhance or diminish the stability of laminar flow and hence alter the transport characteristics of the mixed convection regime. The instability of laminar mixed convection flow and its subsequent transition to turbulent flow can be induced by the wave mode of instability, by the vortex mode of instability, or by both modes. One of the major criteria for use in determining the incipience of the instability in laminar mixed convection flows is the relationship between the critical Reynolds number $Re_{x,c}$ and the critical Grashof number $Gr_{x,c}$, that is, the relationship between the minimum Reynolds and Grashof numbers that will cause the laminar flow to become unstable.

From the analyses of wave instability by the linear theory, it has been found that for a strong forced flow with weak buoyancy force, an aiding buoyancy force has a stabilizing effect on flow along a vertical plate [40], but a destabilizing effect on flow over a horizontal flat plate [41]. These trends are both reversed when the buoyancy force opposes the forced flow. For the laminar mixed convection flow adjacent to an inclined flat plate, an increase in the inclination angle from the vertical has a destabilizing effect for an aiding buoyancy force, but a stabilizing effect for an opposing buoyancy force [42]. For a strong free convection flow with very weak forced flow along a vertical flat plate, an aiding free stream has a stabilizing effect, whereas an opposing free stream tends to destabilize the flow [43].

The analysis of vortex instability by the linear theory for mixed convection flow over a heated, isothermal, horizontal flat plate has provided the relationships between the critical Reynolds number and the critical Grashof number as $Gr_{x,c}/Re_{x,c}^{3/2} = 0.447$ for Pr = 0.7 and $Gr_{x,c}/Re_{x,c}^{3/2} = 0.434$ for Pr = 7 in the Reynolds number range of $10^3 \le Re_{x,c} \le 10^7$ [44]. On the other hand, experiments on isothermal, heated, horizontal flat plates provide relationships for the onset of vortex instability as $Gr_{x,c}/Re_{x,c}^{3/2} = 192$ for air [45] and $Gr_{x,c}/Re_{x,c}^{3/2} = 46$ to 100, with an average value of 78, for water [46]. Thus, for a given Reynolds number, the linear theory predicts a much lower critical Grashof number than that observed in experiments. In another

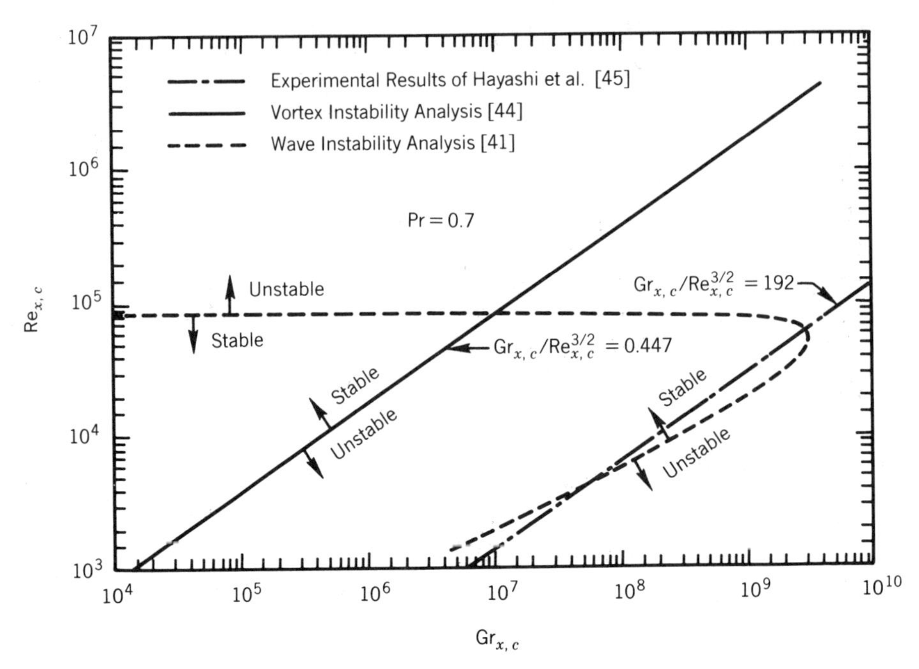

Figure 14.23. Instability and stability domains of mixed convection flow over an isothermal horizontal flat plate: wave and vortex modes, Pr = 0.7 [44].

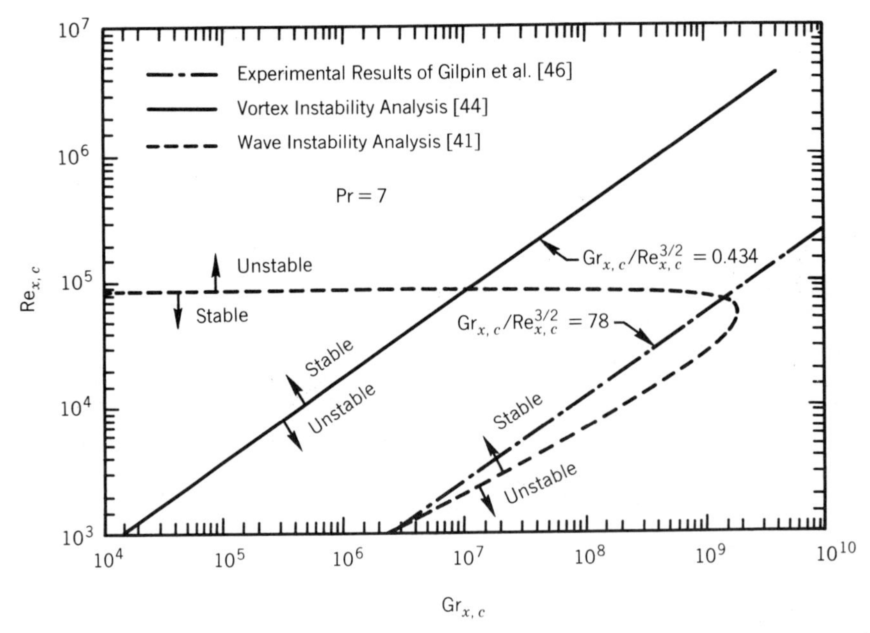

Figure 14.24. Instability and stability domains of mixed convection flow over an isothermal horizontal flat plate: wave and vortex modes, Pr = 7 [44].

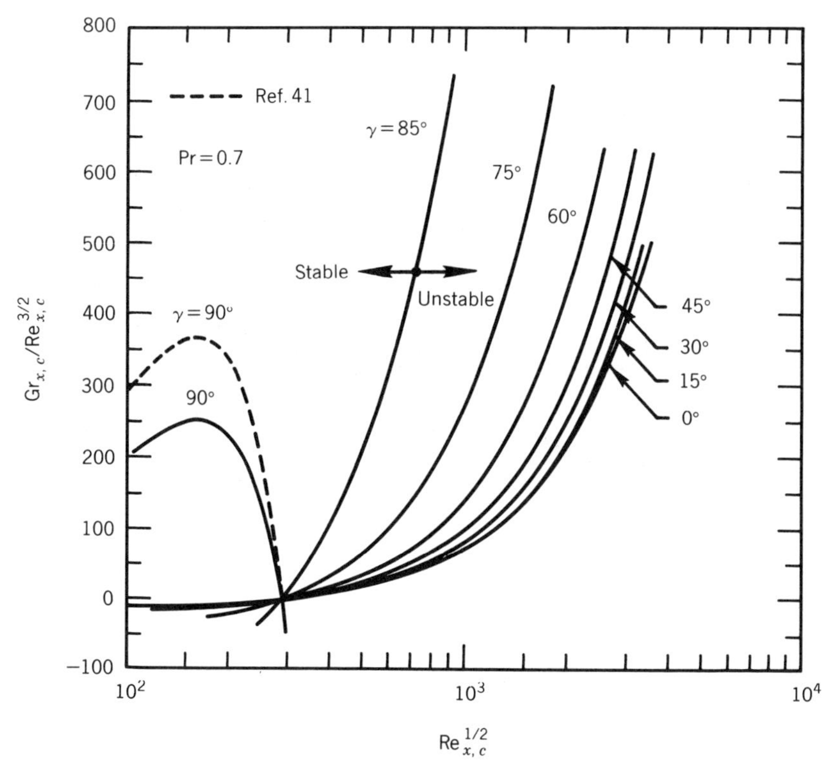

Figure 14.25. Instability and stability domains of mixed convection flow on isothermal inclined flat plates: wave mode, Pr = 0.7 [42].

experiment with water flow over an isothermal, heated, horizontal flat plate [39], it has been found that the transition from laminar forced convection to turbulent free convection in water is due to the incipience, growth, and subsequent breakdown of the longitudinal vortex rolls, and is characterized by the parameter $Gr_x/Re_x^{3/2}$ in the range of $100 \leq Gr_x/Re_x^{3/2} \leq 300$.

Figures 14.23 and 14.24 illustrate the relationship between the critical Reynolds number and the critical Grashof number, for Pr = 0.7 and 7 respectively, for both wave and vortex modes of instability in mixed convection flow over an isothermal, heated, horizontal flat plate [44]. The regions of stable and unstable flow are as marked in each figure. The region enclosed by the dashed line represents flow that is stable with respect to the wave mode of instability, while the region outside the line represents unstable flow. For the vortex mode of instability, the region below the solid line represents an unstable situation; the region above, a stable situation.

Figure 14.25 illustrates the wave instability characteristics of mixed convection flow adjacent to isothermal, heated, inclined flat plates [42] for $\gamma = 0°$ (vertical orientation) to 90° (horizontal orientation) for the forced-flow-dominated case. The results are for Pr = 0.7. Each curve in the plane of $Gr_{x,c}/Re_{x,c}^{3/2}$ vs. $Re_{x,c}^{1/2}$ separates the stable region to the left of the curve from the unstable region to the right for that particular inclination angle.

The wave and vortex instability characteristics of mixed convection flows on isothermal, heated, inclined flat plates for Pr = 0.7 are compared in Fig. 14.26, which is

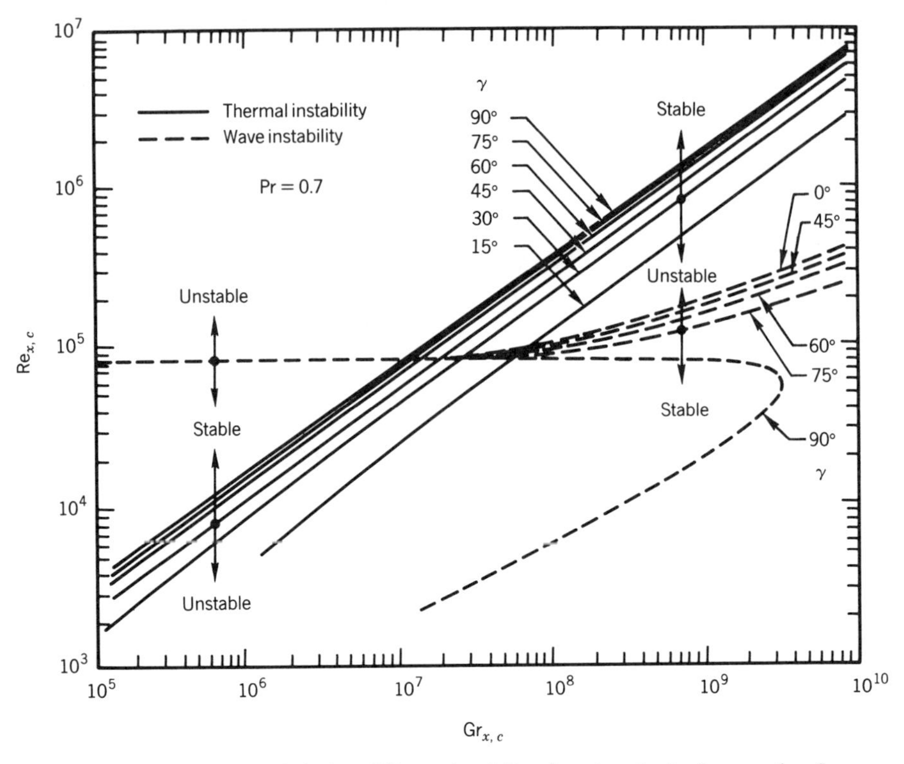

Figure 14.26. Comparison of the instability and stability domains of mixed convection flow on isothermal inclined flat plates between wave and vortex modes, Pr = 0.7 [47].

taken from [47]. For the wave mode of instability, at an inclination angle γ, any (Re_x, Gr_x) combination that lies above the dashed line represents an unstable flow situation, while any combination that lies below the line represents a stable flow situation. With regard to the vortex mode of instability, each solid line at a given angle separates the stable region above the line from the unstable region below the line. From a comparison between the wave and vortex instability results for Pr = 0.7 in mixed convection flows on isothermal, heated, inclined plates, it can be generally concluded that the flow is more susceptible to the vortex mode of instability than the wave mode for all angles of inclination when $Gr_x \leq 10^7$ and $Re_x \leq 8.4 \times 10^4$. On the other hand, the flow is more susceptible to the wave mode of instability than the vortex mode when $Gr_x \leq 10^7$ and $Re_x > 8.4 \times 10^4$. When $Gr_x > 10^7$, and at a given inclination angle, the first incipience of flow instability could be due to either the wave mode, the vortex mode, or a combined mode, depending on the values of Re_x and Gr_x. It should be noted, however, that the results shown in Fig. 14.26 are based on the linear theory, and the stated conclusions need to be further verified by experimental results, which are lacking for mixed convection flows adjacent to inclined flat plates.

14.12 CONCLUSIONS

In this chapter correlation equations for Nusselt numbers in external mixed convection flows over various flow configurations are summarized. The correlation equations presented are mostly for laminar flow. This is because studies on turbulent mixed convection flows are currently lacking. It has been demonstrated that the local and average Nusselt numbers for mixed convection in external flows can be correlated in a form given by Eq. (14.11) or (14.15) in terms of the respective Nusselt numbers for the limiting cases of pure forced convection and pure free convection. The constant exponent n in the equation can be determined if some analytical or experimental Nusselt number results for the mixed convection flow are available for a certain Prandtl number. The exponent n lies generally between 3 and 4 for laminar boundary-layer flows. Thus, an estimate of mixed convection Nusselt numbers in laminar boundary-layer flow may be made rather accurately for a certain geometry from Eq. (14.11) or (14.15) provided the correlation equations for the respective Nusselt numbers in pure forced convection and pure free convection are known. However, care must be exercised in such a practice.

ACKNOWLEDGMENT

The assistance of Mr. N. Ramachandran, a Ph.D. candidate in the Department of Mechanical and Aerospace Engineering, University of Missouri—Rolla, in the preparation of this chapter is sincerely appreciated.

NOMENCLATURE

$A(\text{Pr})$	function of Prandtl number [Eq. (14.14), Table 14.1]
$B(\text{Pr})$	function of Prandtl number [Eq. (14.14), Table 14.1]
d	diameter of cylinders or spheres, m, ft
$F(\text{Pr})$	function of Prandtl number [Eq. (14.43)]
$F_1(\text{Pr}),\cdots,F_4(\text{Pr})$	functions of Prandtl number (Table 14.2)
g	gravitational acceleration, m/s^2, ft/s^2
$G(\text{Pr})$	function of Prandtl number [Eq. (14.43)]
$G_1(\text{Pr}),\cdots,G_4(\text{Pr})$	functions of Prandtl number (Table 14.2)
Gr	Grashof number = $g\beta(T_w - T_\infty)l^3/v^2$
Gr_x	local Grashof number = $g\beta(T_w - T_\infty)x^3/v^2$
Gr_z	local Grashof number = $g\beta(T_w - T_\infty)z^3/v^2$
Gr_H	Grashof number = $g\beta(T_w - T_\infty)H^3/v^2$
Gr_L	Grashof number = $g\beta(T_w - T_\infty)L^3/v^2$
Gr_d	Grashof number = $g\beta(T_w - T_\infty)d^3/v^2$
Gr_x^*	modified local Grashof number = $g\beta q_w x^4/kv^2$
Gr_z^*	modified local Grashof number = $g\beta q_w z^4/kv^2$
Gr_H^*	modified Grashof number = $g\beta q_w H^4/kv^2$
Gr_L^*	modified Grashof number = $g\beta q_w L^4/kv^2$
Gr_d^*	modified Grashof number = $g\beta q_w d^4/kv^2$
$\text{Gr}_{x,c}$	local critical Grashof numbers = $g\beta(T_w - T_\infty)x_c^3/v^2$
h	local heat transfer coefficient, W/(m^2 · K), Btu/(hr · ft^2 · °F)
$\bar{h}$	average heat transfer coefficient, W/(m^2 · K), Btu/(hr · ft^2 · °F)
H	height of plate, m, ft
$H_1(K_1)$, $H_2(K_2)$	functions of curvature of cylinder (Table 14.2)

$I_1(K_1), I_2(K_2^*)$	functions of curvature of cylinder (Table 14.2)
k	thermal conductivity, W/m · K, Btu/(hr · ft · °F)
K_1, K_2, K_2^*	curvature parameters of cylinder (Table 14.2)
l	characteristic length, m, ft
L	length of plate, m, ft
m	constant exponent (Table 14.1)
$M_1(\phi), \cdots, M_4(\phi)$	functions of angle ϕ (Table 14.2)
n	constant exponent (Table 14.1)
$N_1(\phi), \cdots, N_4(\phi)$	functions of angle ϕ (Table 14.2)
Nu	Nusselt number $= hl/k$
Nu_x	local Nusselt number $= hx/k$
$\mathrm{Nu}_d(\phi)$	local Nusselt number $= hd/k$
$\overline{\mathrm{Nu}}_d$	average Nusselt number $= \bar{h}d/k$
$\overline{\mathrm{Nu}}_L$	average Nusselt number $= \bar{h}L/k$
Pr	Prandtl number $= \nu/\alpha$
q_w	local surface heat flux, W/m^2, Btu/(hr · ft^2)
r_0	radius of cylinder, m, ft
Ra	Rayleigh number $= \mathrm{Gr}_d \mathrm{Pr}$
Re	Reynolds number $= u_\infty l/k$
Re_x	local Reynolds number $= u_\infty x/\nu$
Re_L	Reynolds number $= u_\infty L/\nu$
Re_d	Reynolds number $= u_\infty d/\nu$
$\mathrm{Re}_{x,c}$	local critical Reynolds number $= u_\infty x_c/\nu$
T	fluid temperature, °C, K, °F, °R
T_f	film temperature $= (T_w + T_\infty)/2$, °C, K, °F, °R
u	axial velocity component, m/s, ft/s
v	normal velocity component, m/s, ft/s
U	dimensionless axial velocity component $= u/u_\infty$
V	dimensionless normal velocity component $= v/u_\infty$
x	axial coordinate, m, ft
y	normal coordinate, m, ft
z	spanwise coordinate, m, ft
X	ratio of Nusselt numbers $= \mathrm{Nu}_N/\mathrm{Nu}_F$ [Eq. (14.13)]
Y	ratio of Nusselt numbers $= \mathrm{Nu}/\mathrm{Nu}_F$ [Eq. (14.13)]

Greek Symbols

α	thermal diffusivity, m^2/s, ft^2/s
β	volumetric coefficient of thermal expansion, K^{-1}, °R^{-1}
γ	angle of inclination from the vertical, deg
η	dimensionless normal coordinate $= y/l$
θ	dimensionless temperature $= (T - T_\infty)/(T_w - T_\infty)$
ν	kinematic viscosity, m^2/s, ft^2/s

ξ	buoyancy parameter (Table 14.1)
ξ_d	buoyancy parameter = $\mathrm{Gr}_d/\mathrm{Re}_d^2$
ϕ	angle from stagnation point = x/r_0, radians
χ	dimensionless axial coordinate = x/l
Ω	buoyancy parameter = $(\mathrm{Gr}_d/\mathrm{Re}_d^2)^{1/4}$

Subscripts

c	critical condition
F	pure forced convection
N	pure free convection
w	condition at wall
∞	condition at free stream

REFERENCES

1. T. S. Chen, B. F. Armaly, and W. Aung, Mixed Convection in Laminar Boundary-Layer Flow, *Natural Convection: Fundamentals and Applications*, ed. S. Kakac, W. Aung, and R. Viskanta, pp. 699–725, Hemisphere, New York, 1985.
2. S. W. Churchill, A Comprehensive Correlating Equation for Laminar, Assisting, Forced and Free Convection, *AIChE J.*, Vol. 23, pp. 10–16, 1977.
3. T. S. Chen, B. F. Armaly, and N. Ramachandran, Correlations for Laminar Mixed Convection Flows on Vertical, Inclined, and Horizontal Flat Plates. *ASME J. Heat Transfer*, Vol. 108, pp. 835–840, 1986.
4. N. Ramachandran, B. F. Armaly, and T. S. Chen, Measurements and Predictions of Laminar Mixed Convection Flow Adjacent to a Vertical Surface, *ASME J. Heat Transfer*, Vol. 107, pp. 636–641, 1985.
5. B. F. Armaly, T. S. Chen, and N. Ramachandran, Correlations for Laminar Mixed Convection on Vertical, Inclined, and Horizontal Flat Plates with Uniform Surface Heat Flux. *Int. J. Heat Mass Transfer*, Vol. 30, pp. 405–408, 1987.
6. A. Mucoglu and T. S. Chen, Mixed Convection on Inclined Surfaces, *ASME J. Heat Transfer*, Vol. 101, pp. 422–426, 1979.
7. N. Ramachandran, B. F. Armaly, and T. S. Chen, Measurements of Laminar Mixed Convection from an Inclined Surface. *ASME J. Heat Transfer*, Vol. 109, pp. 146–150, 1987.
8. S. W. Churchill and H. Ozoe, Correlations for Laminar Forced Convection in Flow Over an Isothermal Flat Plate and in Developing and Fully Developed Flow in an Isothermal Tube, *ASME J. Heat Transfer*, Vol. 95, pp. 416–419, 1973.
9. A. J. Ede, Advances in Free Convection, *Advances in Heat Transfer*, Vol. 4, pp. 1–64, Academic Press, New York, 1967.
10. S. W. Churchill and H. Ozoe, Correlations for Laminar Forced Convection with Uniform Surface Heating in Flow over a Plate and in Developing and Fully Developed Flow in a Tube, *ASME J. Heat Transfer*, Vol. 95, pp. 78–84, 1973.
11. T. Fujii and M. Fujii, The Dependence of Local Nusselt Number on Prandtl Number in the Case of Free Convection along a Vertical Surface with Uniform Heat Flux, *Int. J. Heat Mass. Transfer*, Vol. 19, pp. 121–122, 1976.
12. A. Moutsoglou and T. S. Chen, Buoyancy Effects in Boundary Layers on Inclined, Continuous Moving Sheets, *ASME J. Heat Transfer*, Vol. 102, pp. 371–373, 1980.
13. T. S. Chen and F. A. Strobel, Buoyancy Effects in Boundary Layer Adjacent to a Continuous, Moving Horizontal Flat Plate, *ASME J. Heat Transfer*, Vol. 102, pp. 170–172, 1980.

14. N. Ramachandran, B. F. Armaly, and T. S. Chen, Correlations for Laminar Mixed Convection in Boundary Layers Adjacent to Inclined, Continuous Moving Sheets. To be published in *Int. J. Heat Mass Transfer*, 1987.
15. N. Ramachandran, B. F. Armaly, and T. S. Chen, Correlations for Laminar Mixed Convection in Boundary Layers Adjacent to Horizontal, Continuous Moving Sheets. To be published in *ASME J. Heat Transfer*, 1987.
16. T. S. Chen and A. Mucoglu, Buoyancy Effects on Forced Convection along a Vertical Cylinder, *ASME J. Heat Transfer*, Vol. 97, pp. 198–203, 1975.
17. A. Mucoglu and T. S. Chen, Buoyancy Effects on Forced Convection along a Vertical Cylinder with Uniform Surface Heat Flux, *ASME J. Heat Transfer*, Vol. 98, pp. 523–525, 1976.
18. A. Mucoglu and T. S. Chen, Analysis of Combined Forced and Free Convection across a Horizontal Cylinder, *Can. J. Chem. Eng.*, Vol. 55, pp. 265–271, 1977.
19. A. Mucoglu and T. S. Chen, Mixed Convection across a Horizontal Cylinder with Uniform Surface Heat Flux, *ASME J. Heat Transfer*, Vol. 99, pp. 679–682, 1977.
20. P. H. Oosthuizen and S. Madan, Combined Convective Heat Transfer from Horizontal Cylinders in Air, *ASME J. Heat Transfer*, Vol. 92, pp. 194–196, 1970.
21. P. H. Oosthuizen and S. Madan, The Effect of Flow Direction on Combined Convective Heat Transfer from Cylinders to Air, *ASME J. Heat Transfer*, Vol. 93, pp. 240–242, 1971.
22. H. M. Badr, Laminar Combined Convection from a Horizontal Cylinder—Parallel and Contra Flow Regimes, *Int. J. Heat Mass Transfer*, Vol. 27, pp. 15–27, 1984.
23. H. M. Badr, A Theoretical Study of Laminar Mixed Convection from a Horizontal Cylinder in a Cross Stream, *Int. J. Heat Mass Transfer*, Vol. 26, pp. 639–653, 1983.
24. A. P. Hatton, D. D. James, and H. W. Swire, Combined Forced and Natural Convection with Low-Speed Air Flow over Horizontal Cylinders, *J. Fluid Mech.*, Vol. 42, pp. 17–31, 1970.
25. S. Nakai and T. Okazaki, Heat Transfer from a Horizontal Circular Wire at Small Reynolds and Grashof Numbers—II. Mixed Convection, *Int. J. Heat Mass Transfer*, Vol. 18, pp. 397–413, 1975.
26. B. F. Armaly, T. S. Chen, and N. Ramachandran, Correlations for Mixed Convection Flows across Horizontal Cylinders and Spheres, submitted to *ASME J. Heat Transfer*.
27. S. Nakai and T. Okazaki, Heat Transfer from a Horizontal Circular Wire at Small Reynolds and Grashof Numbers—I. Pure Convection, *Int. J. Heat Mass Transfer*, Vol. 18, pp. 387–396, 1975.
28. T. S. Chen and A. Mucoglu, Analysis of Mixed Forced and Free Convection about a Sphere, *Int. J. Heat Mass Transfer*, Vol. 20, pp. 867–875, 1977.
29. A. Mucoglu and T. S. Chen, Mixed Convection about a Sphere with Uniform Surface Heat Flux, *ASME J. Heat Transfer*, Vol. 100, pp. 542–544, 1978.
30. Y. Yuge, Experiments on Heat Transfer from Spheres Including Combined Natural and Forced Convection, *ASME J. Heat Transfer*, Vol. 82, pp. 214–220, 1960.
31. G. H. Evans and O. A. Plumb, Laminar Mixed Convection from a Vertical Heated Surface in a Cross Flow, *ASME J. Heat Transfer*, Vol. 104, pp. 554–558, 1982.
32. O. A. Plumb and G. H. Evans, Turbulent Mixed Convection from a Vertical Heated Surface in a Cross Flow, *Proceedings of the ASME-JSME Thermal Engineering Joint Conference*, Vol. 3, pp. 47–53, 1983.
33. D. L. Siebers, R. G. Schwind, and R. J. Moffat, Experimental Mixed Convection Heat Transfer from a Large, Vertical Surface in a Horizontal Flow, Rep. No. HMT-36, Thermosciences Div., Dept. of Mech. Eng., Stanford Univ., February 1983.
34. T. S. Chen, B. F. Armaly, and M. M. Ali, Turbulent Mixed Convection along a Vertical Plate, *ASME J. Heat Transfer*, Vol. 109, pp. 251–253, 1987.
35. M. M. Ali, T. S. Chen, and B. F. Armaly, Mixed Convection in Turbulent Boundary Layer Flow over a Horizontal Plate, ASME Paper No. 83-WA/HT-5, 1983.

36. W. M. Kays and M. E. Crawford, *Convective Heat and Mass Transfer*, 2nd ed., Chapter 12, McGraw-Hill, New York, 1980.

37. S. W. Churchill and H. H. S. Chu, Correlating Equations for Laminar and Turbulent Free Convection from a Vertical Plate, *Int. J. Heat Mass Transfer*, Vol. 18, pp. 1323–1329, 1975.

38. T. Fujii and H. Imura, Natural Convection Heat Transfer from a Plate with Arbitrary Inclination, *Int. J. Heat Mass Transfer*, Vol. 15, pp. 755–767, 1972.

39. H. Imura, R. R. Gilpin, and K. C. Cheng, An Experimental Investigation of Heat Transfer and Buoyancy Induced Transition from Laminar Forced Convection to Turbulent Free Convection over a Horizontal Isothermally Heated Plate, *ASME J. Heat Transfer*, Vol. 100, pp. 429–434, 1978.

40. A. Mucoglu and T. S. Chen, Wave Instability of Mixed Convection Flow along a Vertical Flat Plate, *Numer. Heat Transfer*, Vol. 1, pp. 267–283, 1978.

41. T. S. Chen and A. Mucoglu, Wave Instability of Mixed Convection Flow over a Horizontal Flat Plate, *Int. J. Heat Mass Transfer*, Vol. 22, pp. 185–196, 1978.

42. T. S. Chen and A. Moutsoglou, Wave Instability of Mixed Convection Flow on Inclined Surfaces, *Numer. Heat Transfer*, Vol. 2, pp. 497–509, 1979.

43. V. P. Carey and B. Gebhart, The Stability and Disturbance-Amplification Characteristics of Vertical Mixed Convection Flow, *J. Fluid Mech.*, Vol. 127, pp. 185–201, 1983.

44. A. Moutsoglou, T. S. Chen, and K. C. Cheng, Vortex Instability of Mixed Convection Flow over a Horizontal Flat Plate, *ASME J. Heat Transfer*, Vol. 103, pp. 257–261, 1981.

45. Y. Hayashi, A. Takimoto, and K. Hori, Heat Transfer in Laminar Mixed Convection Flow over a Horizontal Flat Plate (in Japanese), *Proceedings of the 14th Japan Heat Transfer Symposium*, pp. 4–6, 1977.

46. R. R. Gilpin, H. Imura, and K. C. Cheng, Experiments on the Onset of Longitudinal Vortices in Horizontal Blasius Flow Heated from Below, *ASME J. Heat Transfer*, Vol. 100, pp. 71–77, 1978.

47. T. S. Chen, A. Moutsoglou, and B. F. Armaly, Thermal Instability of Mixed Convection Flow over Inclined Surfaces, *Numer. Heat Transfer*, Vol. 5, pp. 343–352, 1982.

15

MIXED CONVECTION IN INTERNAL FLOW

Win Aung

National Science Foundation
Washington, D.C.

15.1 INTRODUCTION

In previous chapters, pure forced convection and pure free (natural) convection have been discussed; combined free and forced convection (or mixed convection) has been treated in Chap. 14 for external flow. The present chapter concerns combined convection in internal flow situations.

Design information for mixed convection should reflect the interacting effects of free and forced convection. It is important to realize that heat transfer in mixed convection can be significantly different from its values in both pure free and pure forced convection. For example, in a vertical circular tube the laminar mixed convection heat transfer coefficient in buoyancy-assisted flow (usually upflow when the fluid is heated) can be as much as 5 times its value in pure forced convection [1]. On the other hand, in buoyancy-opposed flow (usually downflow when the fluid is heated), the laminar mixed convection heat transfer can be lower than that for pure forced flow. In turbulent flow, the heat transfer is often reduced in assisted flow and increased in opposed flow, compared with pure forced convection. Some of these effects are qualitatively displayed in Fig. 15.1 for a vertical circular tube. The curves are based on the analytical predictions for laminar flow by Martinelli and Boelter [2], and the turbulent flow data by Herbert and Sterns [3], Byrne and Ejiogu [4] and Eckert and Diaguila [5]. For a horizontal duct, temperature variations in the fluid lead to the possibility of counterrotating transverse vortices that are superimposed on the streamwise main flow. This so-called "secondary flow" can also increase the heat transfer significantly. Thus, buoyancy influences internal forced-convection heat transfer in ways that depend on whether the flow is laminar or turbulent, upflow or downflow, and on duct geometry as well as orientation. This makes it difficult and sometimes even dangerous to make *a priori* assumptions concerning buoyancy effects in internal flow.

With laminar, buoyancy-aided flow in a vertical parallel-plate channel (see Fig. 15.12 below), free convection causes the thermal development length to be shortened but the flow development length to be elongated considerably, and the streamwise velocity profile can be profoundly distorted also, leading even to the possibility of reversed flow [6, 7]. Experimental evidence of the profile deviations from the parabolic shape for constant-property fully developed laminar flow was reported by Watzinger and Johnson [8], among others. In general, increased free-convection effects near a heat transfer wall cause more cold fluid to be drawn to the wall, thereby increasing the heat transfer.

Inasmuch as the flow development length is a function of the buoyancy effects, the ratio of the length of a duct to its significant transverse dimension is important in mixed convection. For a duct at a uniform wall temperature (UWT), the fluid temperature tends toward the wall value, and the secondary flow introduced by buoyancy diminishes and becomes negligible at distances far (in comparison to its transverse dimension) from the entrance. For a duct heated at a uniform heat flux (UHF), the wall and fluid temperatures increase continuously along the duct, and the increase becomes linear with distance for a "long" duct. In both heating conditions, when the fluid properties are constant, a thermally "fully developed" situation is achieved at a large (in relation to the transverse dimension) distance from the duct entrance. An additional parameter that can have an effect on the heat transfer is the presence of an unheated section of the duct so that the flow is hydrodynamically fully developed prior to entering the heated section. In low Reynolds number laminar flows, the influence of the initial velocity profile can be quite significant; this effect, however, is not important for moderate to large Prandtl number fluids.

Relatively reliable correlations now exist for both upflow and downflow in vertical circular tubes involving various fluids. Augmenting the results of earlier investigators

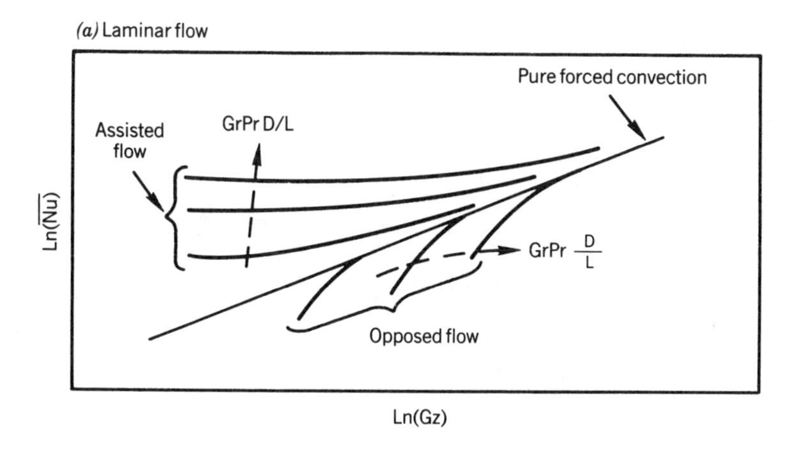

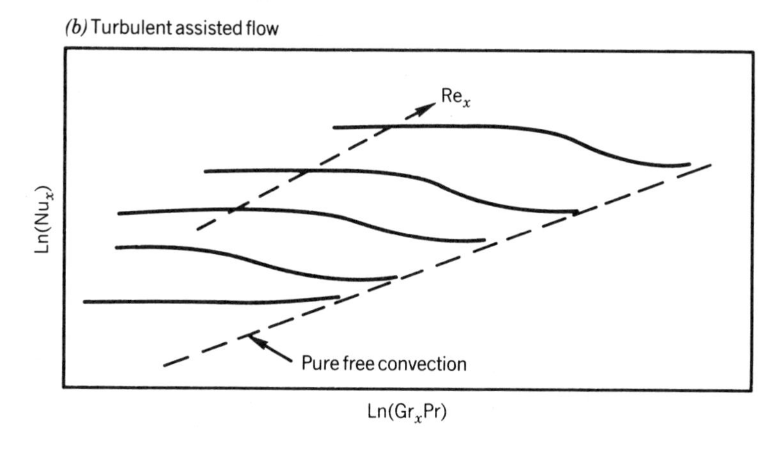

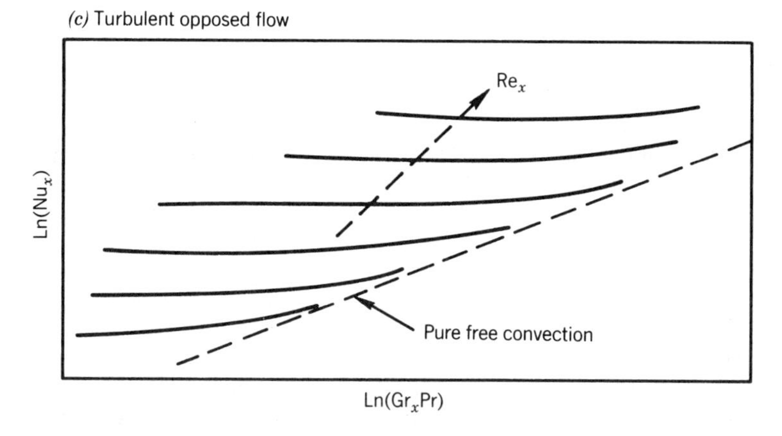

Figure 15.1. Nusselt numbers in laminar and turbulent mixed convection: (*a*) laminar flow, (*b*) buoyancy-assisted turbulent flow, and (*c*) buoyancy-opposed turbulent flow.

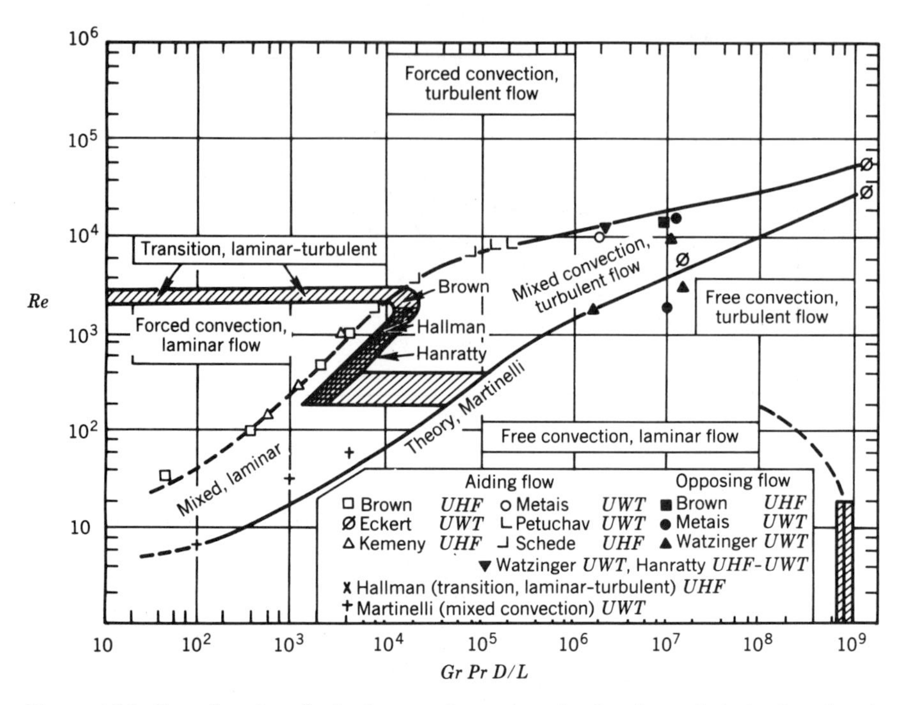

Figure 15.2. Free, forced, and mixed convection regimes for flow in vertical circular tubes for $10^{-2} < \mathrm{Pr}\, D/L < 1$ [9]. The results are valid for both upflow and downflow, and for UWT and UHF boundary conditions.

with their own, Metais and Eckert [9] have recommended the use of the flow regime maps of Figs. 15.2 and 15.3. In these figures, flow regime boundaries have been determined to be the locations where the mixed convection heat transfer does not deviate by more than 10% from pure forced convection or pure free convection. For forced flow represented by a given Reynolds number, the value of the parameter $\mathrm{Gr}\,\mathrm{Pr}\, D/L$ indicates whether it is necessary to consider buoyancy effects. Note that Fig. 15.2 applies for both upflow and downflow, and for both UWT and UHF conditions. Figure 15.3 applies for UWT, and indicates only the boundary between forced convection and mixed convection. In these figures, Gr_4 is based on the tube diameter and the difference between the wall and fluid bulk mean temperatures, and $\mathrm{Re} = u_m D_h/\nu$. All fluid properties should be determined at the film temperature.

The present chapter deals only with ducts in the vertical and horizontal orientations. Also of importance are inclined tubes, which are used in solar energy collection applications.

As in applying any heat transfer result, it is important to distinguish between UWT and UHF boundary conditions when using the correlations recommended in this chapter. The UWT condition approximates applications in condensers, evaporators, and any heat exchanger in which one fluid has a much larger heat capacity than the other; for these, the correlations for Nusselt numbers permit the evaluation of the heat transfer coefficients and hence the heat loads. In situations involving UHF,[†] such as in

[†] Because of the influence of the secondary flows, the peripheral wall temperature is not uniform. Hence, the UHF boundary condition may approximate the (H2) boundary condition described in Chapter 3.

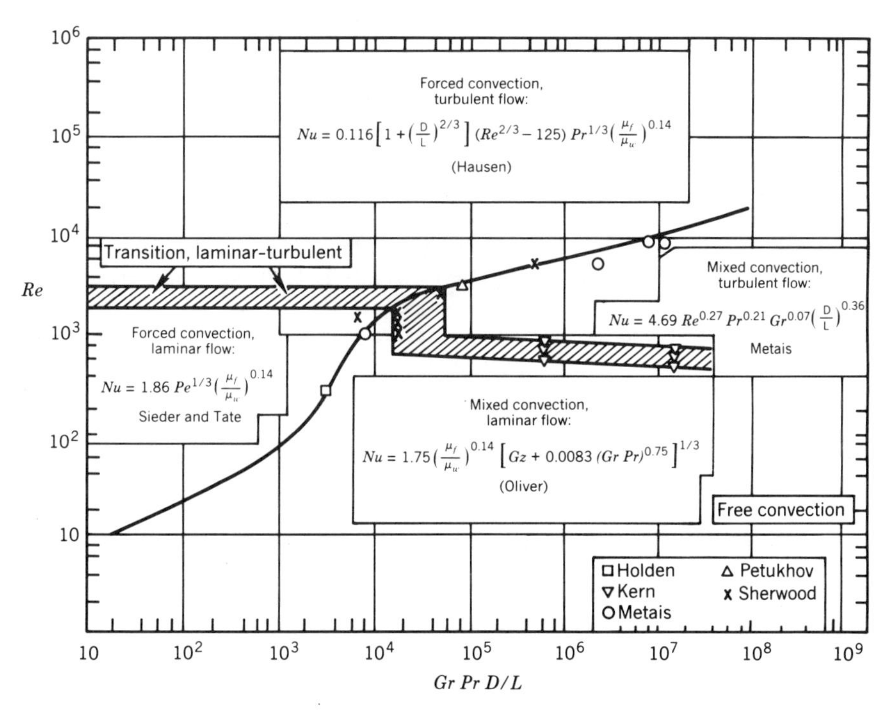

Figure 15.3. Free, forced and mixed convection regimes for flow in horizontal circular tubes for $10^{-2} < \mathrm{Pr}\, D/L < 1$ [9]. The results are valid for the UWT boundary condition. Refer to Chaps. 3 and 4 for more accurate correlations for laminar and turbulent forced convection duct flow.

certain solar collectors and electronic cooling applications, the heat fluxes are known and the Nusselt number can be used to obtain the wall temperatures.

Another effect to be considered in applying mixed convection correlations is peripheral heat-flux variation. The latter can be caused by nonuniform heating of the tube, which is inherent in heat transfer tubes having nonaxisymmetric outside flow streams. For tubes with thick enough walls and high enough thermal conductivities, peripheral variation of the heat flux is diminished by heat conduction; for tubes with thin walls, the problem can be severe.

Peripheral variations of heat transfer can also result from secondary flow. As will be noted, secondary flows are frequently present in mixed convection in horizontal tubes. For a horizontal circular tube at UWT, the circumferential Nusselt number can differ by as much as a factor of 4, with the maximum occurring at the lowest point and the minimum at the highest point on the circumference of the tube. This problem has been considered in some detail by Yousef and Tarasuk [10] for UWT and by Bergles and Simonds [11] for UHF.

For the sake of including the best available information, this chapter contains a number of graphs which are derived from theoretical investigations. Some of the results contained in these graphs have not been verified experimentally but they are useful as rough design guides. A clear distinction is made between tube orientations (vertical or horizontal), between boundary conditions (UHF or UWT), and between results for fully developed flow and those for developing flow. Thus, subsections dealing with simultaneous hydrodynamic and thermal development refer to cases where an unheated entrance length is absent, or to theoretical studies in which a flat velocity profile is assumed at the entrance to the heat transfer section. Thermally and hydrodynamically fully developed flow implies invariance of the results with streamwise distance, and hence $Nu_{fd} = Nu_m$. Note that thermally developing flow includes the hypothetical case where the flow is hydrodynamically fully developed throughout a duct, as is assumed in a number of theoretical investigations. Experiments, e.g. [12], have shown that this case is closely approached in practice in situations where the heat transfer section is preceded by a sufficiently long hydrodynamic development section, the implication here being that the distortion of the velocity profile in the heat transfer section does not cause the heat transfer results to deviate materially from those computed using the assumption of hydrodynamically fully developed flow.

Unless otherwise specified, thermophysical properties appearing in graphs and correlations contained in this chapter should be evaluated at the *film* temperature, obtained by the arithmetic average of the wall and bulk mean temperatures. The characteristic dimension in Re, and in various definitions of Gr and Ra, is consistently used as the hydraulic diameter for all duct geometries throughout this chapter.

15.2 GOVERNING EQUATIONS AND PARAMETERS

The general equations of continuity, momentum, and energy in various coordinate systems are given in Chap. 1. Consider now the special case of mixed convection in a vertical tube. Assume steady flow and heat transfer, no internal heat generation, negligible viscous dissipation, pressure change only in the streamwise direction, no axial diffusion of heat and momentum, and axially symmetric flow and heat transfer. The equations in cylindrical coordinates expressing the conservation of mass, momentum, and energy are, respectively,

$$\frac{1}{r}\frac{\partial}{\partial r}(\rho r v) + \frac{\partial}{\partial x}(\rho u) = 0 \tag{15.1}$$

$$\rho\left(v\frac{\partial u}{\partial r} + u\frac{\partial u}{\partial x}\right) = -\frac{dP}{dx} \mp \rho g + \frac{1}{r}\frac{\partial}{\partial r}\left(r\mu\frac{\partial u}{\partial r}\right) \tag{15.2}$$

$$\rho c_p\left[v\frac{\partial T}{\partial r} + u\frac{\partial T}{\partial x}\right] = \frac{1}{r}\frac{\partial}{\partial r}\left(rk\frac{\partial T}{\partial r}\right) \tag{15.3}$$

The left-hand side of Eq. (15.2) represents the inertia forces; the terms on the right-hand side denote the pressure gradient, buoyancy, and friction forces, respectively. With the axial coordinate x oriented vertically up, the negative sign in the buoyancy term (ρg) is used when the basic forced convection is directed upwards (upflow), while the positive sign is used for downflow. The terms on the left side of Eq. (15.3) represent the convective heat transport, and the right side is the heat conduction term.

In the absence of heat transfer, the temperature in the duct assumes a uniform value that is equal to that at the duct entrance. In addition, if there is no forced flow, the pressure gradient is that due to the hydrostatic pressure. Thus, Eq. (15.2) may be written for this special case as

$$0 = -\frac{dP_1}{dx} \mp \rho_1 g \tag{15.4}$$

In the above, P_1 is the hydrostatic pressure corresponding to ρ_1 and T_1. Subtracting Eq. (15.4) from Eq. (15.2), we have

$$\rho\left(v\frac{\partial u}{\partial r}+u\frac{\partial u}{\partial x}\right)=-\frac{d}{dx}(P-P_1)\mp(\rho-\rho_1)g+\frac{1}{r}\frac{\partial}{\partial r}\left(r\mu\frac{\partial u}{\partial r}\right) \tag{15.5}$$

The so-called Boussinesq approximation of the momentum equation, Eq. (15.5), is now introduced. First, for small temperature variations, the density may be expressed as

$$\rho=\rho_1[1-\beta(T-T_1)] \tag{15.6}$$

where subscript 1 indicates conditions at the tube entrance, and the thermal expansion coefficient is assumed constant and is defined as

$$\beta=-\frac{1}{\rho}\left(\frac{\partial\rho}{\partial T}\right)_P \tag{15.7}$$

Hence,

$$\rho-\rho_1=-\rho_1\beta(T-T_1)$$

Equation (15.5) becomes

$$\rho\left(v\frac{\partial u}{\partial r}+u\frac{\partial u}{\partial x}\right)=-\frac{d}{dx}(P-P_1)\pm\rho_1 g\beta(T-T_1)+\frac{1}{r}\frac{\partial}{\partial r}\left(r\mu\frac{\partial u}{\partial r}\right) \tag{15.8}$$

For constant properties [except for density in the buoyancy force term in Eq. (15.8)], the governing equations (15.1), (15.8), and (15.3) become, in nondimensional form,

$$\frac{\partial V}{\partial R}+\frac{V}{R}+\frac{\partial U}{\partial X}=0 \tag{15.9}$$

$$\frac{1}{Pr}\left(V\frac{\partial U}{\partial R}+U\frac{\partial U}{\partial X}\right)=-\frac{dP^*}{dX}+\frac{1}{R}\frac{\partial U}{\partial R}+\frac{\partial^2 U}{\partial R^2}\pm\frac{\mathrm{Gr}_1}{\mathrm{Re}}\theta \tag{15.10}$$

$$V\frac{\partial\theta}{\partial R}+U\frac{\partial\theta}{\partial X}=\frac{1}{R}\frac{\partial\theta}{\partial R}+\frac{\partial^2\theta}{\partial R^2} \tag{15.11}$$

where

$$U=\frac{u}{u_m},\qquad V=\frac{\mathrm{Re\,Pr}\,v}{u_m} \tag{15.11a}$$

$$R=\frac{r}{D/2},\qquad X=\frac{2x/D}{\mathrm{Re\,Pr}} \tag{15.11b}$$

$$P^*=\frac{P-P_1}{\mathrm{Pr}\,\rho u_m^2},\qquad \theta=\frac{T_w-T}{T_w-T_1} \tag{15.11c}$$

In the above, $\mathrm{Re}=\rho u_m D/\mu$. The plus and minus signs in Eqs. (15.8) and (15.10) apply respectively to upward and downward flows, when x is oriented vertically up.

In Eqs. (15.9)–(15.11), the dependent variables are U, V, P, and θ. Since there are only three equations for determining these variables, an additional relation must be obtained to complete the problem definition. This is furnished by the constraint relation, which requires that the mass flow at any axial location must remain constant. Thus,

$$\int_0^{D/2} 2\pi ur\,dr = \frac{\pi}{4} D^2 u_m \tag{15.12a}$$

or

$$\int_0^1 UR\,dR = \frac{1}{2} \tag{15.12b}$$

While this is simply an alternative statement of Eq. (15.9), it is adequate, since one of the variables, P^*, is not a complete field variable as are U, V, and θ (i.e., P^* is a function only of X).

Solution of Eqs. (15.9)–(15.12) yields U, V, and θ as functions of R, X, Pr, and $\mathrm{Gr}_1/\mathrm{Re}$, and P as a function of X, Pr, and $\mathrm{Gr}_1/\mathrm{Re}$. In addition, results for the heat transfer parameters may be derived. At the tube wall, an energy balance gives

$$q''_w = -k\left(\frac{\partial T}{\partial r}\right)_w = h_x(T_w - T_b)$$

where the bulk mean temperature is given by

$$T_b = \frac{\int_0^{D/2} \rho uTr\,dr}{\int_0^{D/2} \rho ur\,dr} \tag{15.13}$$

The local heat transfer coefficient at any cross section x is

$$h_x = -\frac{k}{T_w - T_b}\left(\frac{\partial T}{\partial r}\right)_w \tag{15.14}$$

For horizontal ducts, the heat transfer coefficient around the duct circumference varies in mixed convection. In that case, h_x represents the circumferential average value, with $(\partial T/\partial r)_w$ and T_w replaced by the respective circumferential average values at the location x. The Nusselt number defined using Eq. (15.14) and the tube hydraulic diameter as the characteristic length is

$$\mathrm{Nu}_x = \frac{h_x D_h}{k} = -\frac{2}{\theta_b}\left(\frac{\partial \theta}{\partial R}\right)_w = \frac{q''_x D_h}{k(\overline{T}_w - T_b)}$$

$$= \mathrm{Nu}\left(X, \mathrm{Pr}, \frac{\mathrm{Gr}_1}{\mathrm{Re}}\right) \tag{15.15}$$

The heat transfer coefficient can also be defied through the use of the arithmetic-mean temperature difference or the log-mean temperature difference,

$$\Delta T_{\mathrm{am}} = \frac{(T_w - T_1) + (T_w - T_b)}{2} = T_w - \overline{T}_b, \qquad \Delta\theta_{\mathrm{am}} = \frac{1 + \theta_b}{2} \tag{15.16a}$$

where $\overline{T}_b = (T_1 + T_b)/2$. In dimensionless form, ΔT_{am} can be represented as

$$\Delta T_{\text{lm}} = \frac{(T_w - T_b) - (T_w - T_1)}{\ln \dfrac{T_w - T_b}{T_w - T_1}}, \qquad \Delta\theta_{\text{lm}} = \frac{\theta_b - 1}{\ln \theta_b} \tag{15.16b}$$

where T_b and T_1 are the bulk mean temperatures at section x and inlet, respectively. Thus,

$$\bar{h}_{\text{am}} = -\frac{k}{\Delta T_{\text{am}}} \overline{\left(\frac{\partial T}{\partial r}\right)_w} \tag{15.17}$$

$$\overline{\text{Nu}}_{\text{am}} = \frac{\bar{h}_{\text{am}} D_h}{k} = -\frac{2}{\Delta\theta_{\text{am}}} \overline{\left(\frac{\partial \theta}{\partial R}\right)_w} \tag{15.18}$$

Similar relations may be derived for $\overline{\text{Nu}}_{\text{lm}}$. Note that

$$\text{Nu}_{\text{am}} = \frac{2(\theta_b - 1)}{(1 + \theta_b)\ln \theta_b} \tag{15.19}$$

Finally, an axial-mean Nusselt number may be defined as

$$\text{Nu}_{x,\,\text{lm}} = \frac{\bar{h} D_h}{k} = -\frac{2}{X}\int_0^X \left(\frac{\partial \theta}{\partial R}\right)_w dX \tag{15.20}$$

where

$$\bar{h} = \frac{1}{x}\int_0^x h\, dx \tag{15.20a}$$

The parameter Gr_1/Re appearing in Eq. (15.10) expresses the importance of buoyancy-induced flow relative to forced flow in internal mixed convection. This parameter is related to the Graetz number by

$$\frac{\text{Gr}}{\text{Re}} = \frac{\text{Gr}\,\text{Pr}\,D_h/L}{\text{Gz}} = \frac{\text{Gr}\,\text{Pr}\,D_h}{L} x^* \tag{15.21a}$$

$$= \frac{\text{Ra}\,D_h/L}{\text{Gz}} = \frac{\text{Ra}\,D_h}{L} x^* \tag{15.21b}$$

There are a number of different definitions used for Gr and hence Ra as follows; as a result, a distinction is made with different subscripts throughout this chapter. The Grashof numbers based on the temperature difference are defined as

$$\text{Gr}_1 = \frac{g\beta(\overline{T}_w - T_1) D_h^3}{\nu^2} \qquad \text{Gr}_2 = \frac{g\beta(T_h - T_1) D_h^3}{\nu^2}$$

$$\text{Gr}_3 = \frac{g\beta(\overline{T}_w - T_b) D_h^3}{\nu^2}, \qquad \text{Gr}_4 = \frac{g\beta(\overline{T}_w - \overline{T}_b) D_h^3}{\nu^2} \tag{15.22a}$$

$$\text{Gr}_5 = \frac{g\beta(T_w - T_{c1}) D_h^3}{\nu^2}, \qquad \text{Gr}_6 = \frac{g\beta\,\Delta T_{lm}\, D_h^3}{\nu^2}$$

The Grashof numbers based on the heat flux are defined as

$$\mathrm{Gr}^* = \frac{g\beta q'' D_h^4}{\nu^2 k}, \qquad \mathrm{Gr}_1^* = \frac{g\beta u_m (dT_w/dx) L^5}{\alpha \nu^2} \tag{15.22b}$$

The Rayleigh numbers based on the temperature difference are defined as

$$\mathrm{Ra}_1 = \mathrm{Gr}_1 \mathrm{Pr} = \frac{g\beta(\overline{T}_w - T_1) D_h^3}{\alpha\nu}, \qquad \mathrm{Ra}_2 = \mathrm{Gr}_2 \mathrm{Pr} = \frac{g\beta(T_h - T_1) D_h^3}{\alpha\nu}$$
$$\mathrm{Ra}_3 = \mathrm{Gr}_3 \mathrm{Pr} = \frac{g\beta(\overline{T}_w - T_b) D_h^3}{\alpha\nu} \tag{15.23a}$$

The Rayleigh numbers based on the heat flux are defined as

$$\mathrm{Ra}^* = \frac{g\beta(dT_w/dx) D_h^4}{\alpha\nu}, \qquad \mathrm{Ra}_1^* = \frac{g\beta(dT_w/dx) D_h^4 \rho_1^2 \mathrm{Pr}}{\mu^2}$$
$$\mathrm{Ra}_2^* = \frac{g\beta q_{bt}'' D_h^4 \mathrm{Pr}}{\nu^2 k}, \qquad \mathrm{Ra}_3^* = \frac{g\beta q_w'' D_h^4 \mathrm{Pr}}{\nu^2 k} \tag{15.23b}$$

Here

$\overline{T}_w$ = circumferential average wall temperature
T_1 = fluid inlet temperature
T_b = fluid bulk mean temperature at x
$\overline{T}_b$ = fluid bulk mean temperature from $x = 0$ to x
T_{cl} = fluid temperature at tube centerline

See the Nomenclature section for the other symbols. If the subscript b or w is used with any Gr or Ra, it means all fluid properties in that dimensionless group are evaluated at the fluid bulk mean temperature or the wall temperature respectively.

Equations (15.9)–(15.12) constitute a set of differential equations of the parabolic type. Since the equations are nonlinear, solutions of the full system are obtainable only by numerical analysis; see for example, the book by Hornbeck [13]. These equations or similar versions thereof have been most frequently used for mixed convection problems in internal flow. The system of governing equations for mass, momentum, and energy becomes of the elliptic type when the axial diffusion of heat and momentum is included. For this type of problem, a proper formulation requires that boundary conditions be specified outside the tube exit and entrance. In principle this creates no difficulty, but in implementation it becomes quite cumbersome. For this reason, theoretical results that are based on the full Navier-Stokes and energy equations are very scarce. For most practical situations, information is derived from either equations of the parabolic type or experimental measurements.

15.3 LAMINAR MIXED CONVECTION IN VERTICAL DUCTS

The available information for combined free and forced laminar convection in vertically oriented channels is presented in this section. The material is organized according to the duct geometry, and to whether the flow is fully developed or developing.

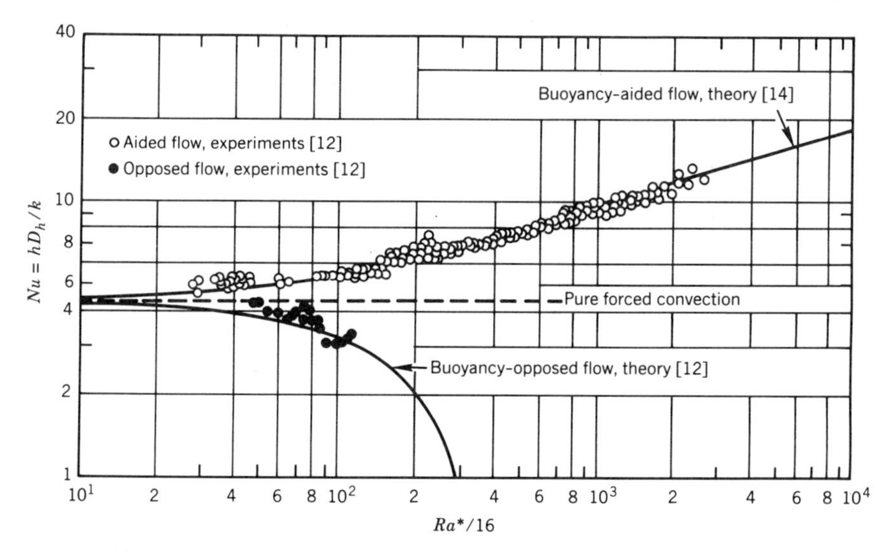

Figure 15.4. Effect of buoyancy on the fully developed Nusselt number for laminar assisted flow in uniformly heated (UHF) vertical circular tubes.

15.3.1 Hydrodynamically and Thermally Fully Developed Flow in Vertical Circular Tubes

For vertical circular tubes, solutions of simplified forms of Eqs. (15.2) and (15.3), under conditions of thermally and hydrodynamically fully developed flow and constant fluid properties [except for the density in Eq. (15.2)], may be obtained by analytical means [12, 14]. The results are applicable for uniformly heated (UHF) tubes with buoyancy-aided flow and buoyancy-opposed flow. The predicted variations of Nusselt number with Rayleigh number for both aided and opposed flows are indicated in Fig. 15.4. The Nusselt number is defined using h from Eq. (15.14) for fully developed flow. As the Rayleigh number increases, the Nusselt number increases above the pure forced-convection value of 4.36 when buoyancy force is oriented in the direction of forced flow; however, in buoyancy-opposed flow the heat transfer decreases with an increase in buoyancy. The quantitative predictions shown in Fig. 15.4 are in close agreement with the experimental measurements as shown in the figure. To use this figure, fluid properties should be evaluated at the *film* temperature. Note that in opposed flow, the absolute value of the Rayleigh number is to be used in Fig. 15.4.

15.3.2 Thermally Developing Flow in Vertical Circular Tubes

The heat transfer in the thermal entrance regions of vertical tubes is described by the coupled system of Eqs. (15.9) to (15.12). Assuming that the velocity profile is fully developed upstream of the heat transfer section, a numerical integration technique may be employed with the appropriate thermal boundary condition. As described previously, a marching scheme may be used for either the UWT or the UHF case. At the entrance to the heat transfer section, the axial velocity distribution is given by

$$U = 2(1 - R^2) \tag{15.24}$$

TABLE 15.1. Thermal Entrance Length for Vertical Circular Tube with Buoyancy-Assisted Laminar Flow at UHF and Fully Developed Velocity Profile at Entrance

Ra*	L^*_{th}
0	0.043
800	0.040
1600	0.034
3200	0.025
4800	0.022
6400	0.023
9600	0.032
12800	0.040
16000	0.048

The effects of buoyancy cause this profile to suffer a dramatic distortion as the fluid moves through the tube. Specifically, in buoyancy-assisted flow, the centerline velocity decreases, while the velocity near the wall, where the buoyancy force is dominant, increases [15]. After a minimum is reached, the centerline velocity starts increasing until, for constant fluid properties, at a large distance from the tube entrance, the profile resumes the fully developed shape given by Eq. (15.24). Experiments have shown that the thermal entrance length first decreases as the Rayleigh number increases, then increases at large Rayleigh numbers. Table 15.1 indicates the dimensionless thermal entrance length as a function of Rayleigh number for a buoyancy-assisted laminar flow in a UHF vertical tube where the flow is hydrodynamically fully developed at the entrance. The data are based on measurements taken by Hallman [12], and thermally fully developed flow is defined as the condition in which the Nusselt number reaches within 5% of the fully developed flow value. To use the table, physical properties should be evaluated at the *film* temperature. For buoyancy-opposed flow, the limited data obtained by Hallman indicate that the thermal entrance length may be estimated roughly by the equation

$$L^*_{\mathrm{th}} = \left(\frac{x}{D_h \operatorname{Re} \operatorname{Pr}} \right)_{\mathrm{fd}} = 0.05. \tag{15.25}$$

provided that $|\mathrm{Ra}^*| < 1920$.

At high heating rates, property variations are significant, and one must add terms dealing with viscous dissipation and pressure work to the energy equation (15.11). A number of approaches may be used to represent the property variations with temperature; for further details, see Chap. 18. For density changes, the relations used include, in addition to Eq. (15.6),

$$\rho = \rho_0 \left(c_1 + c_2 T - c_3 T^2 \right) \tag{15.26}$$

$$\rho = \frac{P}{\tilde{R} T} \tag{15.27}$$

Correspondingly, variations of the dynamic viscosity may be expressed by

$$\frac{1}{\mu} = \frac{1}{\mu_1}[1 + b(T - T_1)] \tag{15.28}$$

$$\mu = \mu_1 + (\mu_B - \mu_1)\exp[-c(T - T_1)] \tag{15.29}$$

$$\frac{\mu}{\mu_1} = \left(\frac{T}{T_1}\right)^a \tag{15.30}$$

Equations (15.6) and (15.28) were used by Pigford [16] in an approximate solution in which the inertia terms were neglected in the momentum equation. The problem considered was the vertical tube at UHF. In Eq. (15.28), b is positive for liquids and negative for gases. Graphical values of $1/\mu$ at different temperatures for water and machine oil are given by Pigford [16]. Worsøe-Schmidt and Leppert [17] applied an implicit finite difference scheme for the case of hydrodynamic fully developed flow, and employed Eqs. (15.27) and (15.30). The variations of the specific heat and thermal conductivity were represented by

$$\frac{c_p}{c_{p,1}} = \left(\frac{T}{T_1}\right)^{c'} \tag{15.31}$$

$$\frac{k}{k_1} = \left(\frac{T}{T_1}\right)^d \tag{15.32}$$

The exponents in Eqs. (15.30) to (15.32) were taken to be $a = 0.67$, $c' = 0.12$, and $d = 0.71$. The arithmetic-mean Nusselt number Nu_{am} [see Eq. (15.19)] obtained by Worsøe-Schmidt and Leppert for UHF and by Marner and McMillan [15] for UWT, and the axial-mean Nusselt number [see Eq. (15.20)] obtained by Martinelli and Boelter [2], are compared in [15]. The results of [2] are in close agreement with those of [15] at high Prandtl numbers; but at low values of Pr, the approximate theory of [2] predicts higher values of the heat transfer than the more complete theory of [15].

15.3.3 Hydrodynamically and Thermally Developing Flow in Vertical Circular Tubes

Temperature-dependent property variations of the forms indicated by Eqs. (15.26) and (15.29) may be utilized in a numerical solution for simultaneously developing velocity (i.e., the velocity profile is assumed to be flat at the entrance to the heat transfer section) and temperature profiles, for laminar upflow at UWT or UHF. For this type of problem, the pressure drop parameter may be defined as the difference between the buoyancy pressure and friction terms; the former is the difference between the static pressure prevailing with the fluid heated and the static pressure that would exist if the fluid in the tube remains at the temperature at the tube entrance. In dimensionless notation, the pressure drop parameter becomes

$$\Delta\bar{P} = \frac{\int_0^x g(\rho_m - \rho_1)\,dx - \Delta P_f}{\rho u_1^2} \tag{15.33}$$

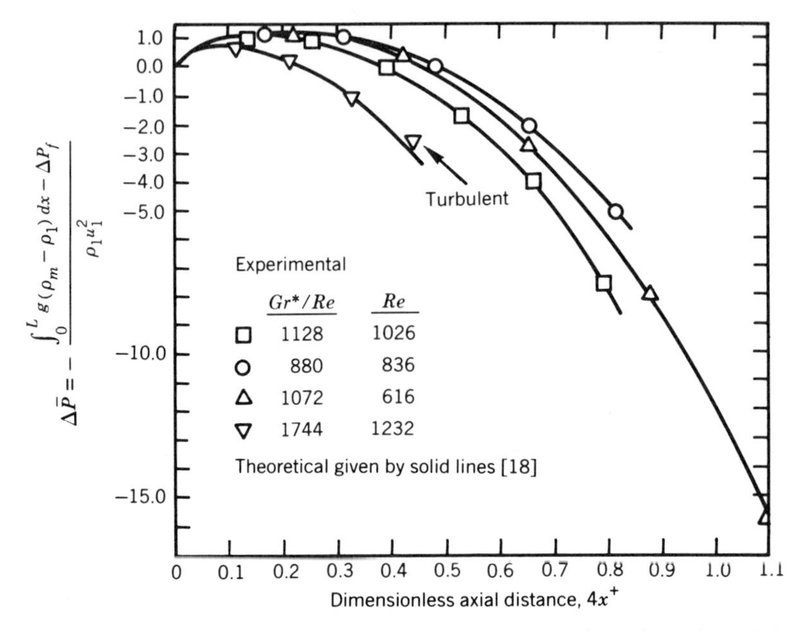

Figure 15.5. Axial distribution of the pressure gradient in the hydrodynamic and thermal development region of a vertical tube at UHF [18].

where ΔP_f is the pressure drop caused by friction, ρ_m is the cross-section-averaged fluid density at any given axial location, and x is the distance from the entrance. Defined in this fashion, $\Delta \bar{P}$ becomes negative when x is sufficiently large in UHF [18], as shown in Fig. 15.5.

The influence of a variable viscosity on the axial distribution of the local Nusselt nuber Nu_x [see. Eq. (15.15)] for UWT is given in [18] and shows that a larger temperature difference, and hence, a greater fluid viscosity variation leads to a higher local Nusselt nuber in the entrance region of the tube.

As discussed in Sec. 15.2, the inclusion of axial-diffusion effects changes the system of governing equations to an elliptic type. One of the resultant difficulties then is how to handle the entrance and exit conditions properly. The difficulty at the inlet stems from the migration of the thermal effects upstream of $x = 0$ at low Péclet numbers, which alters the inlet velocity and temperature profiles at $x = 0$, while at the outlet the conditions are dependent on the solutions inside the tube that are being sought. In the case of a constant-fluid-property flow in an infinitely long tube at UWT, the exit velocity and temperature profiles may be assumed to be those for fully developed flow. This problem may then be attacked using parabolic equations by first employing an axial coordinate transformation from the semi-infinite domain $0 \le X \le \infty, 0 \le R \le 1$ to the finite domain $0 \le \xi \le 1, 0 \le R \le 1$ by using the parameter

$$\xi = 1 - \frac{1}{1 + \tilde{c}X} \tag{15.34}$$

where $\tilde{c}$ is a transformation parameter. This approach was used by Zeldin and Schmidt [19] in a study of laminar upflow in a vertical UWT tube. The fluid properties are considered constant, except for the density in the buoyancy force term. The results

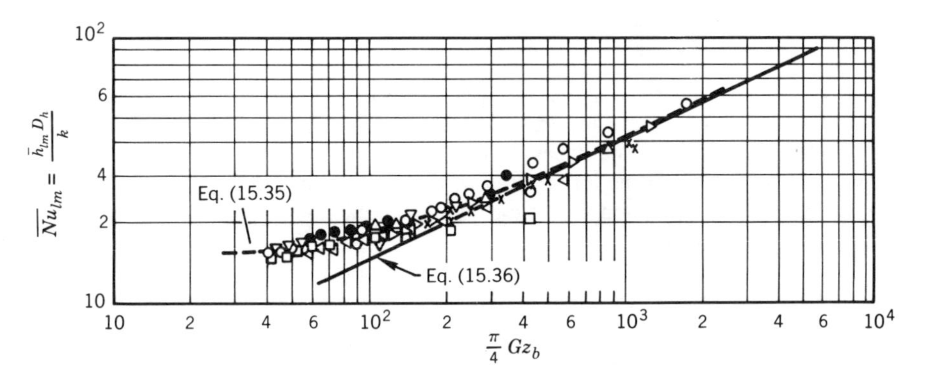

Figure 15.6. Average heat transfer for hydrodynamically and thermally developing assisted flow in vertical tubes at UWT [20].

show that flow reversal occurs for $\mathrm{Gr}_1/\mathrm{Re} > 89$ for a uniform entrance velocity profile, and for $\mathrm{Gr}_1/\mathrm{Re} > 97$ when the profile is fully developed at the entrance. When $\mathrm{Gr}_1/\mathrm{Re}$ is larger than these critical values, no convergent solutions could be obtained.

For a vertical tube with uniform wall temperature and forced upflow having a flat velocity profile at the entrance to the heated section, Jackson et al., [20] proposed the following more general correlation for the average Nusselt number for the heated section considered, which is of length L_x:†

$$\overline{\mathrm{Nu}}_{\mathrm{lm}} = 1.128\left\{\mathrm{Re}_b\,\mathrm{Pr}_b\,\frac{D_h}{L_x} + \left[3.02\left(\mathrm{Gr}_1\,\mathrm{Pr}_w\,\frac{D_h}{L}\right)_w \mathrm{Pr}_w\right]^{0.4}\right\}^{1/2} \tag{15.35}$$

The above correlation is valid in the following range:

$$40.2 \le \mathrm{Gz}_b = \frac{\mathrm{Re}_b\,\mathrm{Pr}_b\,D_h}{L} \le 1710$$

$$1.05 \times 10^5 \le \left(\mathrm{Gr}_1\,\mathrm{Pr}\,\frac{D_h}{L}\right)_w \le 1.30 \times 10^6$$

where Gz_b is based on properties evaluated at the bulk mean temperature of the fluid and $(\mathrm{Gr}_1\,\mathrm{Pr})_w$ is based on properties evaluated at the wall temperature. Equation (15.35) is compared with experimental data for air in Fig. 15.6. Also shown is the equation that applies when a uniform velocity profile exists throughout the test section [20]; that is,

$$\overline{\mathrm{Nu}}_{\mathrm{lm}} = 1.126(\mathrm{Gz}_b)^{1/2} \tag{15.36}$$

The experimental data in Fig. 15.6 are for $L = 1.52$ m, and $(\mathrm{Gr}_1\,\mathrm{Pr}\,D_h/L)_w$ varies from 1.05×10^5 to 1.30×10^5. The buoyancy parameter $\mathrm{Gr}_{1,w}/\mathrm{Re}_b$ can be evaluated by employing Eq. (15.22a). This parameter ranges from approximately 580 to 2220. In deriving this parameter, it has been assumed that, approximately, $\mu_w = \mu_b$, $c_{p,w} = c_{p,b}$, and $k_w = k_b$.

† The length of the entire heated section was L, and it was divided into three sections.

15.3.4 Hydrodynamically and Thermally Fully Developed Flow in Vertical Annuli

Mixed convection in vertical annuli is important in the design of coolant channels for power transformers, nuclear reactors, components for turbomachinery, double-pipe heat exchangers, and certain types of catalytic devices. While most equipment is designed for operation in the turbulent flow regime, laminar flow has to be considered for reduced power operation or during natural-circulation cooling following pump failure in a nuclear reactor.

Consider laminar convection in a circular, concentric vertical annulus. Under the conditions listed at the beginning of Sec. 15.2, the governing equations for mass, momentum, and energy, respectively, are given by Eqs. (15.1), (15.8) and (15.3). For fully developed flow, $v = 0$ and $u = u(r)$; hence, for constant-property flow and heat transfer, the terms in Eq. (15.1) are identically zero, and the convective terms on the left side of Eq. (15.8) are also zero. In addition, the first term on the left side of Eq. (15.3), the radial convection term, is zero. Consequently, the governing equations are the following, with the application of Eq. (15.4):

$$\mu\left(\frac{1}{r}\frac{\partial u}{\partial r} + \frac{\partial^2 u}{\partial r^2}\right) - \left(\frac{dP}{dx} \pm \rho_1 g\right) \pm (T - T_1)\rho_1 g\beta = 0 \tag{15.37}$$

$$u\frac{\partial T}{\partial x} = \frac{k}{\rho c_p}\left(\frac{1}{r}\frac{\partial T}{\partial r} + \frac{\partial^2 T}{\partial r^2}\right) \tag{15.38}$$

The global constraint condition expressed by Eq. (15.12a) now becomes

$$\int_{r_i}^{r_o} 2\pi u r\, dr = \frac{\pi}{4}\left(D_o^2 - D_i^2\right) u_m \tag{15.39}$$

Again, as in circular tube flow, the plus sign in Eq. (15.37) applies for upflow (buoyancy-aided flow) and the minus sign for downflow (buoyancy-opposed flow).

Sherwin [21] solved Eqs. (15.37)–(15.39) by first casting them in nondimensional forms. A fourth-order ordinary differential equation for the axial velocity then results, which he solved, obtaining a closed-form expression for the nondimensional velocity in terms of Bessel functions and modified Bessel functions of zero order. The solution to Eqs. (15.37) and (15.38) can be formulated also in terms of Kelvin functions for buoyancy-assisted (upward) flow in vertical annuli [22] with the boundary conditions

$$\text{at} \quad r = r_i: \quad U = 0, \qquad \tilde{\theta} = 1$$

$$\text{at} \quad r = r_o: \quad U = 0, \qquad \frac{d\tilde{\theta}}{dR^*} = 0$$

which correspond to uniform inner wall temperature and insulated outer wall. Here

$$R^* = \frac{r}{D_o - D_i} = \frac{r}{D_h}, \qquad \tilde{\theta} = \frac{T_w - T}{\tilde{P}\tau}$$

$$\tilde{P} = \frac{D_h^3}{\alpha\nu}\left(\frac{dP}{\rho_m\, dx} + g\right), \qquad \tau = D_h\frac{dT_w}{dx}$$

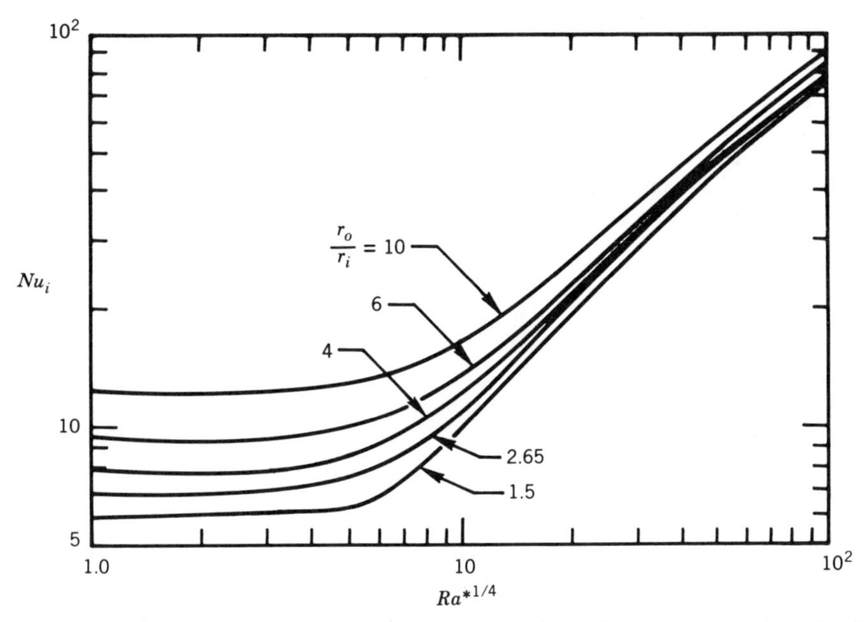

Figure 15.7. Nusselt number for fully developed, assisted flow in a vertical annulus with the inner wall maintained at UWT and the outer wall insulated [22]. [Nu_i and Ra* are defined in Eq. (15.40).]

The fully developed Nusselt number and Rayleigh number are defined as

$$\mathrm{Nu}_i = \frac{\int_{R_i^*}^{R_o^*} R^* U \, dR^*}{\int_{R_i^*}^{R_o^*} R^* U \tilde{\theta} \, dR^*} \left(\frac{d\tilde{\theta}}{dR^*} \right)_{r_i} = \frac{h_i D_h}{k}$$

$$\mathrm{Ra}^* = \frac{g\beta(dT_w/dx)\, D_h^4}{\alpha\nu} \tag{15.40}$$

Experimental measurements for $r_o/r_i = 2.5$ reported in [22] are on the average 45% higher than the predictions. Recent experimental data for $r_o/r_i = 1.17$ reported by Zaki et al. [23] indicate somewhat closer agreement with the theory of [22]. In general, the latter shows that in assisted, hydrodynamically and thermally fully developed flow, buoyancy effects are negligible for $\mathrm{Ra}^* < 10^3$. A step increase in Nu_i is evident beyond this value for all radius ratios, and the variation may be represented approximately as $\mathrm{Nu}_i \propto \mathrm{Ra}^{*1/4}$. The trends are indicated in Fig. 15.7. The validity of the theoretical predictions is subject to further verification; hence this figure should be used only as a guide.

For hydrodynamically and thermally fully developed buoyancy-assisted flow, the effects of viscous dissipation have been investigated by Rokerya and Iqbal [24]. Their theoretical calculations, employing Kelvin functions, show that the effect of viscous dissipation on the velocity and temperature field is very small. The Nusselt number, however, decreases as viscous dissipation increases.

15.3.5 Thermally Developing Flow in Vertical Annuli

The equations that describe mixed convection in the developing flow in a vertical, concentric annuli are given in Eqs. (15.1), (15.8), and (15.3), subject to the conditions stipulated immediately above these equations. For buoyancy-opposed (downward) flow with UHF where the velocity profile is assumed to be fully developed and fluid properties are constant, bouyancy effects are shown to introduce substantial radial velocities in the redeveloping momentum field [25]. We define the Grashof and Reynolds numbers as

$$\mathrm{Gr}_q = \frac{\beta g \rho^2 r_e^4 q'' r_i}{k \mu^2 (r_o + r_i)} = \frac{r_i}{r_o + r_i} \frac{\mathrm{Gr}^*}{16} \tag{15.41}$$

$$\mathrm{Re} = \frac{\rho u_m D_h}{\mu} \tag{15.42}$$

where

$$D_h = D_o - D_i \tag{15.43}$$

Criteria for the onset of flow reversal at fixed values of the nondimensional distance x^+ have been obtained in [26]. Results for $r_o/r_i = 3$ are shown in Fig. 15.8, indicating

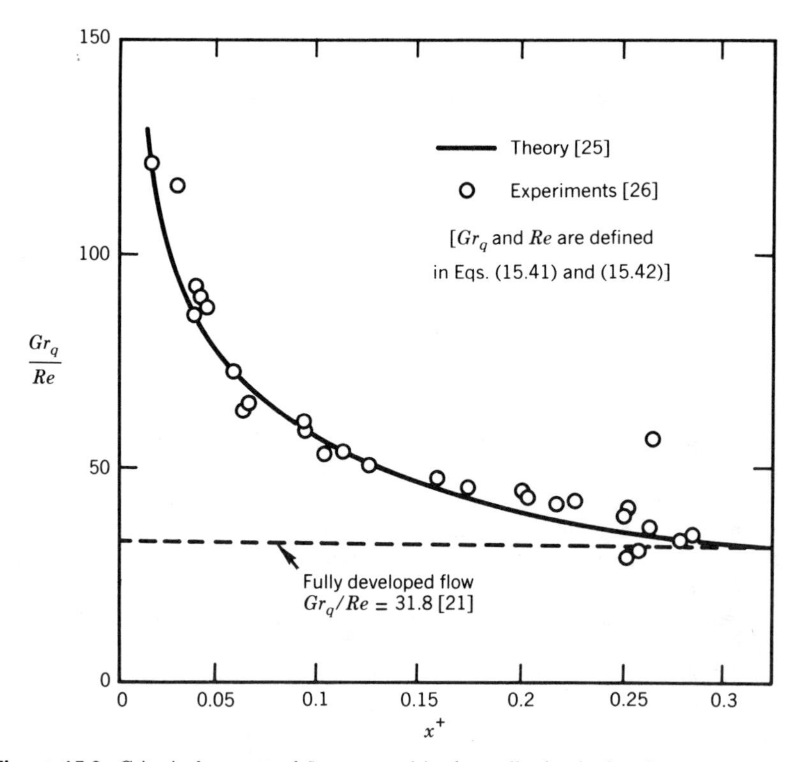

Figure 15.8. Criteria for onset of flow reversal in thermally developing, buoyancy-opposed flow in vertical annuli of $r_o/r_i = 3$ [26].

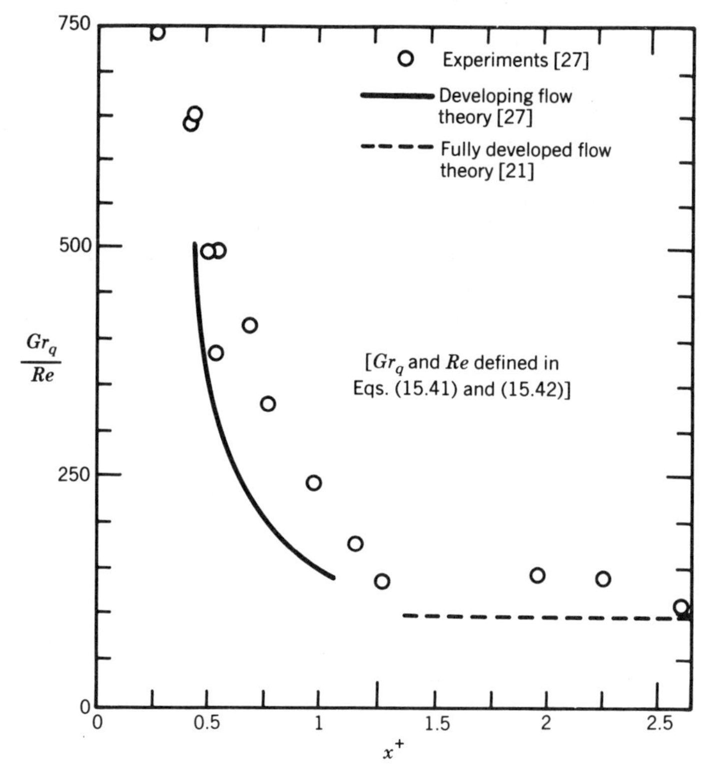

Figure 15.9. Criteria for onset of flow reversal in thermally developing, buoyancy-assisted flow in vertical annuli of $r_o/r_i = 3$ [27].

good agreement with the experimental data. In addition, at large values of x^+, the numerical results tend towards the value 31.8 which is predicted by the theory for thermally and hydrodynamically fully developed flow [21].

The numerical solution for the corresponding problem of upflow in a vertical concentric annulus has been obtained by Sherwin and Wallis [27]. The predicted values of $\mathrm{Gr}_q/\mathrm{Re}$ for the onset of flow reversal is plotted against the dimensionless axial distance in Fig. 15.9. In this case, the experimental measurements are underpredicted. The trends in both upflow and downflow are similar to that obtained by Lawrence and Chato [18] for developing, upflow mixed convection in a heated vertical circular tube.

The effects of temperature-dependent properties (including density) for laminar gas flows in vertical annuli have been considered [28] for upflow in an annulus with $r_o/r_i = 4$; however, the results do not include buoyancy effects. Property variations are represented by Eqs. (15.30)–(15.32), with $a = 0.67$, $c' = 0.095$, and $d = 0.805$. The axial velocity profile at the entrance of the annuli is assumed to be either fully developed or a flat profile. In either case, the effects of temperature-dependent properties on the Nusselt number are found to be slight. The dominant factors affecting the heat transfer are the variations of density and thermal conductivity, but these effects act in opposite directions, leading to little overall impact on the Nusselt number.

Only limited information is available on the distribution of the local Nusselt number for mixed convection in vertical annuli. Malik and Pletcher [29] reported their predicted Nusselt number for an annular duct with $r_o/r_i = 2.63$ for assisted flow of water. The inner wall was uniformly heated (UHF) and the outer wall insulated. The case considered is hydrodynamically fully developed but thermally developing flow, with variable fluid properties. By and large, however, no systematic study of this problem is available.

Developing mixed convection for laminar boundary-layer flow in a vertical annulus with a rotating inner cylinder has been investigated by El-Shaarawi and Sarhan [30]. Both aiding and opposed flows are investigated for a fluid with Pr = 0.7. The radius ratio considered is 1.11. The possibility of the nonexistence of a laminar solution is considered by the authors, and this condition is related to the vanishing of the velocity gradient normal to the wall—preceding, presumably, the onset of reversed flow. The hydrodynamic development length and the distance from the entrance of the annulus to the place where the axial velocity gradient at the wall becomes zero are both dependent on the heating condition and whether buoyancy is aiding or opposing the forced flow, in addition to the rotational speed of the inner cylinder. Thus, when buoyancy forces aid the forced flow, and the inner cylinder wall is isothermal while the outer stationary wall is adiabatic, an increase in the inner-cylinder rotational speed would move the location of zero wall gradient in the direction to decrease the hydrodynamic development length. At a fixed Gr_1/Re, the inner-cylinder rotation causes an increase in the local heat transfer coefficient and the bulk temperature if the inner-cylinder wall is heated, and vice versa if the outer-cylinder wall is heated.

15.3.6 Hydrodynamically and Thermally Fully Developed Flow between Vertical Parallel Plates

Constant-property fully developed mixed convection between vertical parallel plates has been of interest in research for many years. Early work includes studies by Ostrach [31] and Lietzke [32]. Studies have also been conducted by Cebeci et al. [33] and by Aung and Worku [6]. The work by these investigators has shown that mixed convection between parallel plates exhibits both similarities and contrasts with the flow in a vertical tube.

The vertical parallel-plate configuration is applicable in the design of cooling systems for electronic equipment and of finned cold plates in general. When the spacing between the plates is small relative to the height of the channel, the fully-developed-flow approximation can be invoked. In a constant-property, two-dimensional hydrodynamically fully developed flow, $v = 0$ and $\partial u/\partial x = 0$. It may be shown that the momentum and energy equations reduce, with the aid of Eq. (15.6), to

$$0 = -\frac{d}{dx}(P - P_1) \mp \rho_1 g\beta(T - T_1) + \mu\frac{d^2u}{dy^2} \tag{15.44}$$

$$\rho c_p u\frac{\partial T}{\partial x} = k\frac{\partial^2 T}{\partial y^2} \tag{15.45}$$

where, with the direction of x pointing vertically upward, the plus sign in Eq. (15.44) corresponds to buoyancy-assisted flow and the minus sign to opposed flow. For plate walls heated to constant but not necessarily equal temperatures, the fluid temperature in thermally fully developed flow with constant properties is at most a function of the transverse coordinate. Hence, $\partial T/\partial x = 0$, and Eq. (15.45) reduces to

$$\frac{d^2T}{dy^2} = 0 \tag{15.46}$$

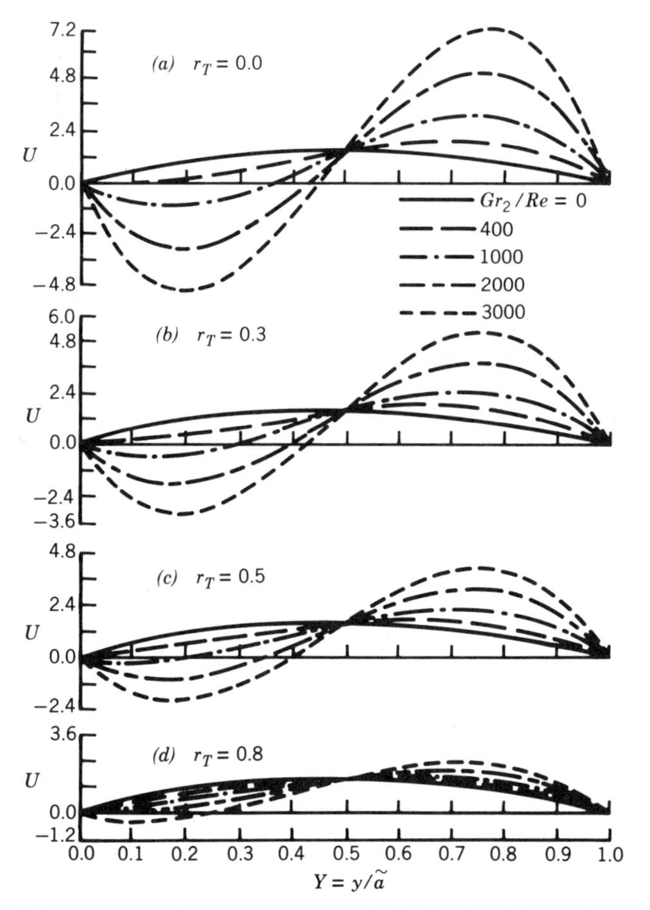

Figure 15.10. Velocity distributions at various r_T and Gr_2/Re for fully developed, buoyancy-assisted flow in a parallel-plate vertical channel at UWT [6].

Assuming that the axial pressure gradient is constant in fully developed flow, Eq. (15.44) becomes

$$\mu \frac{d^2 u}{dy^2} \pm \rho_1 g \beta (T - T_1) + \alpha' = 0 \tag{15.47}$$

where α' is a constant to be found. The boundary conditions are

$$\text{at} \quad y = 0: \quad u = 0, \qquad T = T_c$$

$$\text{at} \quad y = \tilde{a}: \quad u = 0, \qquad T = T_h$$

The above problem has been solved [6] using dimensionless parameters. For buoyancy-assisted flow, Fig. 15.10 displays the velocity distribution across the duct at various wall temperature difference ratios $r_T = (T_c - T_1)/(T_h - T_1)$ with Gr_2/Re as a parameter. Here Gr_2 is defined by Eq. (15.22a). The hydraulic diameter for parallel plates is

$$D_h = 2\tilde{a} \tag{15.48}$$

From the closed-form solution for U, the criterion for the existence of reversed flow may be deduced. The result is

$$(1 - r_T)\frac{\mathrm{Gr}_2}{\mathrm{Re}} > 288, \qquad r_T < 1 \tag{15.49}$$

Note that there is no reversed flow when the two walls are heated to the same temperature, i.e., $r_T = 1$.

The above discussion is based on no axial conduction within the fluid. The inclusion of the latter effect adds another term, $k\,\partial^2 T/\partial x^2$, to the right side of Eq. (15.45). For thermally fully developed flow in a channel with specified wall heat fluxes, the temperature must have the following general form ($\tilde{\tau}$ a constant, $\tilde{T}$ a temperature function of y):

$$T = \tilde{\tau}x + \tilde{T}(y)$$

which may be substituted into Eq. (15.45). The elimination of the velocity between Eqs. (15.44) and (15.45) then leads to a fourth-order ordinary differential equation which can be solved in a straightforward manner. Having obtained the temperature field, the velocity may be determined. This approach, which is also applied in [6], has been used by Rao and Morris [34] for both buoyancy-assisted and -opposed flows. These authors considered uniform heating on one wall while the other wall was thermally insulated. The relations between the Nusselt number and the Rayleigh number, and between the friction factor times the Reynolds number and the Rayleigh number, are shown in Fig. 15.11. Here the Nusselt number is as defined in Eq. (15.15) with the wall temperature gradient evaluated only at the heated wall. The Rayleigh number Ra_1^* is defined by Eq. (15.23b). For assisted flow, the Nusselt number increases with the Rayleigh number, while the opposite is true for opposed flow. The behavior of the product of friction

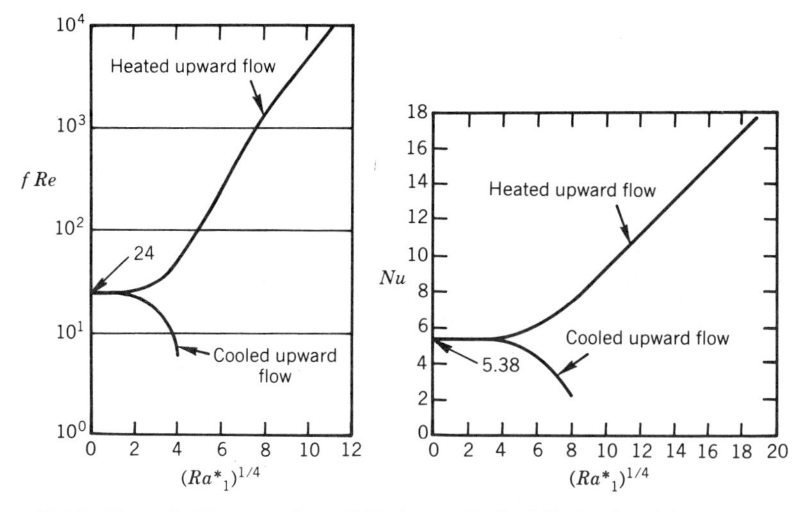

Figure 15.11. Theoretical heat transfer and friction results for fully developed, buoyancy-assisted and -opposed flows in a parallel-plate vertical channel with one wall at UHF and the other insulated [34].

factor and Reynolds number is similar to that of the Nusselt number. In general, these behaviors are similar to those for laminar flow in a uniformly heated vertical tube.

15.3.7 Hydrodynamically and Thermally Developing Flow between Vertical Parallel Plates

There is only meager information available concerning developing mixed convection between vertical parallel plates. The quantitative effects of buoyancy in this problem cannot be extrapolated from results available for circular tube flow.

One of the features that distinguishes the flow between parallel plates from tube flow is the possibility of asymmetric heating on the two walls in the former case. Under certain conditions, this gives rise to flow reversal, even in buoyancy-assisted flows. Sparrow et al. [35] provided insight into the phenomenon of reversed flow in a parallel plate channel for the limiting case of free convection. An analysis of the mixed convection in a channel with symmetric uniform temperature and symmetric uniform-flux heating has been presented by Yao [36]. His study provides information on the flow structure in the developing region and reveals the different length scales accompanying the different convective mechanisms operative in the developing flow region. While no quantitative information was presented by Yao, he conjectured that fully developed flow might consist of periodic reversed flow.

Quantitative information on the temperature and velocity fields has been provided in a numerical study reported by Aung and Worku [7]. These authors note that the hydrodynamic development distance is dramatically increased by buoyancy effects. With asymmetric heating, the bulk temperature is a function of $\mathrm{Gr}_z/\mathrm{Re}$ and r_T, and decreases as r_T is reduced. The axial variation of the bulk mean temperature for $r_T = 1$ (two walls at the same temperature) is plotted in Fig. 15.12. Buoyancy effects are noticeable through a large segment of the channel, but not near the channel entrance or far downstream from it. At large values of the dimensionless distance, all curves converge to the value of 1.

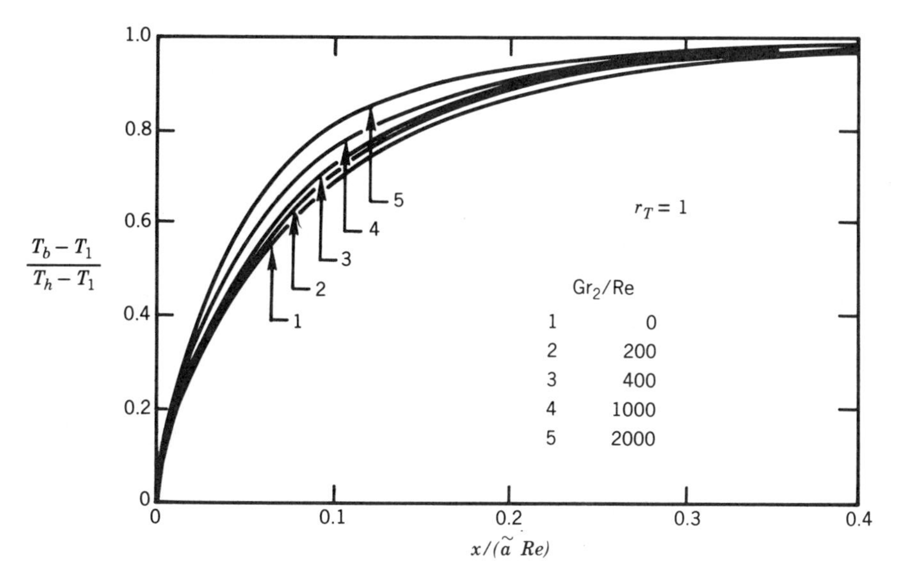

Figure 15.12. Effect of buoyancy on the bulk temperature of air in a parallel plate vertical channel with UWT and $r_T = 1$ [7].

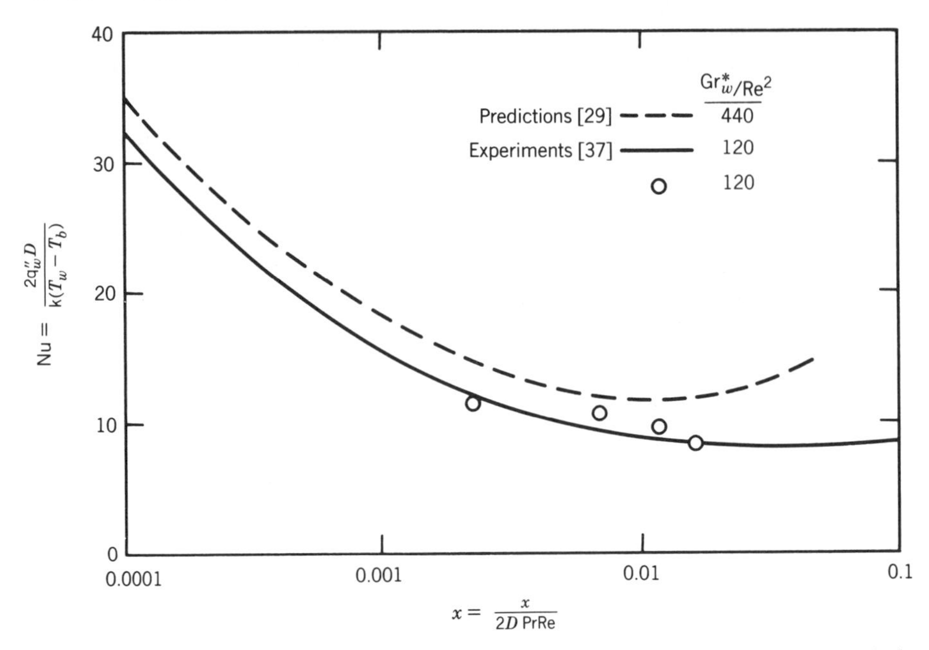

Figure 15.13. Local Nusselt numbers for upflow of ethylene glycol in a parallel plate vertical channel with one wall heated at UHF and the other wall insulated [29].

Figure 15.13 shows the distributions of the local Nusselt number along a channel with a fully developed flow velocity profile at the entrance, and with one plate heated at UHF while the other is insulated. The theoretical results, obtained by Malik and Pletcher [29], are for buoyancy-assisted laminar flow of ethylene glycol in a vertical concentric annulus having $r_o/r_i \approx 1$, including variable-fluid-property effects; thus the results are applicable for a parallel plate vertical duct, as is verified by the data of Joshi and Bergles [37] for a parallel plate duct. In Fig. 15.13, buoyancy effects are expressed in term of the parameter $\text{Gr}^*_w/\text{Re}^2$ where, $\text{Gr}^*_w = g\beta q''_w D_h^4/(\nu^2 k)$. Here, the subscript w refers to the heated wall. It may be noted that with a nearly 400% increase in $\text{Gr}^*_w/\text{Re}^2$, Nu_x is increased by approximately 25%. The tail-up behavior of Nu_x for $\text{Gr}^*_w/\text{Re}^2 = 440$ corresponds to the onset of flow reversal adjacent to the cool wall, as discussed in Sec. 15.3.6.

In summary, it may be noted that there is little quantitative information available on mixed convection in parallel plate vertical channels, even for laminar flow. This is perhaps one of those areas in heat transfer which would benefit from more research, especially experimental studies.

15.4 LAMINAR MIXED CONVECTION IN HORIZONTAL DUCTS

Mixed convection in horizontal ducts gives rise to secondary flows, which can cause increases in the friction factor and heat transfer and a decrease in the thermal entrance length, and so induce an early transition to turbulence. The flow and heat transfer become rotationally asymmetric, and are three-dimensional under conditions of developing flow. For steady flow with constant fluid properties with the conventional

Boussinesq approximation, the governing equations for flow in a horizontal tube are:

CONTINUITY EQUATION

$$\frac{\partial}{\partial r}(rv) + \frac{\partial w}{\partial \phi} + \frac{\partial (ru)}{\partial x} = 0 \tag{15.50}$$

r-MOMENTUM EQUATION

$$v\frac{\partial v}{\partial r} + \frac{w}{r}\frac{\partial v}{\partial \phi} + u\frac{\partial v}{\partial x} - \frac{w^2}{r}$$
$$= -\frac{1}{\rho}\frac{\partial P}{\partial r}$$
$$+\nu\left[\frac{\partial}{\partial r}\left(\frac{1}{r}\frac{\partial (rv)}{\partial r}\right) + \frac{1}{r^2}\frac{\partial^2 v}{\partial \phi^2} - \frac{2}{r^2}\frac{\partial w}{\partial \phi} + \frac{\partial^2 v}{\partial x^2}\right]$$
$$-g\beta(T_w - T)\cos\phi \tag{15.51}$$

ϕ-MOMENTUM EQUATION

$$v\frac{\partial w}{\partial r} + \frac{w}{r}\frac{\partial w}{\partial \phi} + u\frac{\partial w}{\partial x} + \frac{vw}{r}$$
$$= -\frac{1}{\rho r}\frac{\partial P}{\partial \phi}$$
$$+\nu\left[\frac{\partial}{\partial r}\left(\frac{1}{r}\frac{\partial (rw)}{\partial r}\right) + \frac{1}{r^2}\frac{\partial^2 w}{\partial \phi^2} + \frac{2}{r^2}\frac{\partial v}{\partial \phi} + \frac{\partial^2 w}{\partial x^2}\right]$$
$$+g\beta(T_w - T)\sin\phi \tag{15.52}$$

x-MOMENTUM EQUATION

$$v\frac{\partial u}{\partial r} + \frac{w}{r}\frac{\partial u}{\partial \phi} + u\frac{\partial u}{\partial x}$$
$$= -\frac{1}{\rho}\frac{\partial P}{\partial x}$$
$$+\nu\left[\frac{1}{r}\frac{\partial}{\partial r}\left(r\frac{\partial u}{\partial r}\right) + \frac{1}{r^2}\frac{\partial^2 u}{\partial \phi^2} + \frac{\partial^2 u}{\partial x^2}\right] \tag{15.53}$$

TABLE 15.2. Correlations for Mixed Convection in Horizontal Ducts

Geometry	Flow Conditions	Authors	Boundary Conditions	Properties Evaluated at
Circular tubes	Thermally and hydrodynamically fully developed	Mori and Futagami [41]	UHF	Film temperature
		Mori and Futagami [41]	UHF	Film temperature
		Morcos and Bergles [51]	UHF	Film temperature
	Thermally developing, but hydrodynamically fully developed (at entrance)	Depew and August [58]	UWT	Bulk mean temperature
	Simultaneously developing velocity and temperature profiles	Jackson, Spurlock, and Purdy [53]	UWT	Wall temperature
		Yousef and Tarasuk [54]	UWT	Bulk mean temperature
		Yousef and Tarasuk [54]	UWT	Bulk mean temperature
Circular concentric annuli	Hydrodynamically fully developed (at entrance)	Hattori [63]	Outer wall UWT, inner wall insulated	Bulk mean temperature
		Hattori [63]	Outer wall insulated, inner wall UWT	Bulk mean temperature
Parallel plates	Thermally developing	Osborne and Incropera [74]	UHF	Local bulk temperature
		Osborne and Incropera [74]	UHF	Local bulk temperature

Correlations	Eq.	Range of Applicability
$\dfrac{\text{Nu}}{\text{Nu}_F} = 0.04085(\text{Re Ra}^*)^{1/2}, \quad \text{Pr} = 0.72$	(15.55)	$1.3 \times 10^6 < \text{Re Ra}^* < 5.6 \times 10^6$
$\dfrac{\text{Nu}}{\text{Nu}_F} = 0.04823(\text{Re Ra}^*)^{1/2}, \quad \text{Pr} = 1.0$	(15.56)	
$\overline{\text{Nu}}_\text{F} = \left\{ (4.36)^2 + \left[0.145 \left(\dfrac{\text{Gr}_f^* \, \text{Pr}_f^{1.35}}{\text{Pw}_f^{*0.25}} \right)^{0.265} \right]^2 \right\}^{1/2}$		$3 \times 10^4 < \text{Ra}_{3,f} < 10^6$ $4 < \text{Pr} < 175$ $2 < \text{Pw} < 66$
$\overline{\text{Nu}}_\text{am} \left(\dfrac{\mu_w}{\mu_b} \right)^{0.14} = 1.75 \left[\text{Gz}_b^* + 0.12 \left(\text{Gz}_b^* \, \text{Gr}_{4,b}^{1/3} \text{Pr}_b^{0.36} \right)^{0.88} \right]^{1/3}$	(15.61)	$L/D = 28.4$ Fluid; Gz_b^*; $\text{Gr}_{4,b}$ Water; 25–338; 0.702×10^5–5.82×10^5 Ethyl alcohol; 36–712; 2.68×10^5–9.91×10^5 Glycerol water; 53.1–188.3; 5.10×10^2–8.99×10^2
$\overline{\text{Nu}}_\text{lm} = 2.67 \left[(\text{Gz}_w^*)^2 + (0.0087)^2 \text{Ra}_{1,w}^{1.5} \right]^{1/6}$	(15.62)	$60 \leq \text{Gz}_\text{w}^* \leq 1300$
$\overline{\text{Nu}}_\text{lm} (\dfrac{\mu_w}{\mu_b})^{0.14} = 1.75 \left[\text{Gz}_b^* + 0.245 \left(\text{Gz}_b^{*1.5} \, \text{Gr}_{6,b}^{1/3} \right)^{0.882} \right]^{1/3}$	(15.63)	$0.0073 < x^* < 0.040$ $20 < \text{Gz}_b^* < 110$ $1 \times 10^4 < \text{Gr}_{6,b} < 8.7 \times 10^4$
$\overline{\text{Nu}}_\text{lm} \left(\dfrac{\mu_w}{\mu_b} \right)^{0.14} = 0.969 (\text{Gz}_\text{b}^*)^{0.82}$	(15.64)	$0.040 < x^* < 0.25$ $3.2 < \text{Gz}_b^* < 20$ $0.8 \times 10^4 < \text{Gr}_{6,b} < 4 \times 10^4$
$\text{Nu}_o = 0.38 \, \text{Gr}_7^{0.20} \, \text{Pr}^{0.28}$	(15.65)	See Fig. 15.17
$\text{Nu}_i = 0.44 \, \text{Gr}_7^{0.20} \, \text{Pr}^{0.28} \left(\dfrac{D_o}{D_i} \right)^{0.35}$	(15.66)	See Fig. 15.17
$\text{Nu}_{x,t} = 1.490 \, \text{Gz}^{1/3}$	(15.67)	$0 \leq q'' \leq 6000 \text{ W/m}^2$ $0 \leq q_t''/q_{bt}'' \leq 2$ $\text{Re} < 2800$
$\dfrac{\text{Nu}_{x,bt}}{\text{Nu}_{x,F}} = \left(1 + 0.00365 \dfrac{\text{Ra}_2^{*3/4}}{\text{Gz}} \right)^{1/3}$	(15.69)	q_t''/q_{bt}''; Re, Ra_2^* ranges 0; $400 \leq \text{Re} \leq 2600$, $8.32 \times 10^7 \leq \text{Ra}_2^* \leq 4.16 \times 10^{10}$ 1; $400 \leq \text{Re} \leq 2600$, $4.16 \times 10^7 \leq \text{Ra}_2^* \leq 4.16 \times 10^{10}$ 2; $1300 \leq \text{Re} \leq 2600$, $8.32 \times 10^7 \leq \text{Ra}_2^* \leq 6.8 \times 10^9$

ENERGY EQUATION

$$v\frac{\partial T}{\partial r} + \frac{w}{r}\frac{\partial T}{\partial \phi} + u\frac{\partial T}{\partial x} = \frac{k}{\rho c_p}\left[\frac{1}{r}\frac{\partial}{\partial r}\left(r\frac{\partial T}{\partial r}\right) + \frac{1}{r^2}\frac{\partial^2 T}{\partial \phi^2} + \frac{\partial^2 T}{\partial x^2}\right] \quad (15.54)$$

In the following, the available information for mixed convection in horizontal ducts is presented in accordance with duct geometry and whether the flow is fully developed or developing. A summary of the important correlations is presented in Table 15.2.

15.4.1 Thermally and Hydrodynamically Fully Developed Flow in Horizontal Circular Tubes

A significant amount of information exists for mixed convection in horizontal circular tubes. Much of the quantitative understanding today is based on empirical information. Theoretical studies are limited to small ranges of Rayleigh number; numerical solutions are available, but only for large Prandtl number fluids.

When the flow is fully developed thermally and hydrodynamically, the problem represented in Eqs. (15.50) to (15.54) may be reduced to a two-dimensional one. Solutions can be obtained by perturbation techniques; the results apply only for extremely long tubes [38–40]. The velocity and temperature distributions may be subdivided into a core region and a boundary-layer region, and theoretical results may then be obtained by using an integral approach. For Pr ≈ 1 with UHF, the following correlations have been derived [41]:

$$\frac{\text{Nu}}{\text{Nu}_F} = 0.04085(\text{Re}\,\text{Ra}^*)^{1/2} \quad \text{for} \quad \text{Pr} = 0.72 \quad (15.55)$$

$$\frac{\text{Nu}}{\text{Nu}_F} = 0.04823(\text{Re}\,\text{Ra}^*)^{1/2} \quad \text{for} \quad \text{Pr} = 1.0 \quad (15.56)$$

In the above, Nu and Nu_F are defined as in Eq. (15.15) except that h is the circumferential average value; $\text{Nu}_F = 4.364$ is the pure forced-convection result. As a result of fully developed flow, both Nu and Nu_F are independent of the axial distance. Equation (15.55) has been verified against experimental measurements in the range $1.3 \times 10^6 < \text{Re}\,\text{Ra}^* < 5.6 \times 10^6$ [41]. The thermophysical properties should be evaluated at the film temperature $T_f = (T_w + T_b)/2$.

To obtain results that are valid over a wider range of the operating parameters, numerical methods must be used. In fully developed flow, the axial component of velocity is independent of axial distance. The pressure terms in the resulting simplified momentum equations in the r and ϕ directions may be eliminated by cross differentiation. A stream function can be defined such that

$$rv = \frac{\partial \psi}{\partial \phi}, \qquad w = -\frac{\partial \psi}{\partial r} \quad (15.57)$$

thereby satisfying the continuity equation. This reduces the number of equations to be solved from five to three, and the dependent variables are u, ψ, and T. Newell and Bergles [42] followed this method and extended the studies of [38–41] for UHF tubes with either infinite or very low tube thermal conductivity. An infinite-conductivity tube is shown to exhibit a much higher Nusselt number, and stems from a lesser degree of thermal stratification and therefore a larger driving force for secondary flow.

Experimental measurements dealing with hydrodynamically and thermally fully developed flow in horizontal tubes heated at UHF are rather abundant, and include the early data for air [43, 44], water [11, 45–47], and ethylene glycol [48, 49]. For ethylene glycol, studies in the ranges Re = 6 to 300, Gr_3 = 0 to 22400, Pr = 26 to 500, and Gz = 3 to 4800 show that the Nusselt number and the pressure gradient are functions only of the Graetz number and the wall-to-bulk viscosity ratio for both the hydrodynamically and thermally developing and fully developed flow regions [48]. With water in fully developed flow, the heat transfer coefficients can be 3 to 4 times the pure forced-flow values [11]. Large variations in the peripheral temperatures are present with gas flows when the tubes are uniformly heated, and the heat transfer is lower than the pure forced-flow value when the Reynolds number is small, and is higher than the pure forced-flow value when the Reynolds number is large [50].

For uniform heat flux, Morcos and Bergles [51] obtained empirical correlations for the Nusselt number. The flow was hydrodynamically and thermally fully developed in a circular tube. Their experimental study was conducted using electrically heated glass and stainless-steel tubes, with distilled water and ethylene glycol as working fluids. The circumferentially averaged Nusselt number, which is independent of axial distance, depends on the thermal boundary condition imposed and hence on the type of tube employed. Using the correlation technique recommended by Churchill and Usagi [52], Morcos and Bergles [51] gave the following correlation for the circumferentially averaged Nusselt number, $\overline{\text{Nu}}_f$, which is axially constant and which incorporates variable-fluid-property and tube-wall effects:

$$\overline{\text{Nu}}_f = \left\{ (4.36)^2 + \left(0.145 \left[\frac{\text{Gr}_f^* \, \text{Pr}_f^{1.35}}{(\text{Pw}^*)_f^{0.25}} \right]^{0.265} \right)^2 \right\}^{1/2}$$

In this equation, $\overline{\text{Nu}}_f$ is given explicitly, except for the iteration necessary for calculating the film temperature. Properties should be evaluated at this temperature where noted; otherwise they should be evaluated at the bulk temperature. Here, $\text{Gr}^* = g\beta q_w'' D^4/(k\nu^2)$, $\overline{\text{Nu}} = \bar{h}D_h/k$, and $\text{Pw}^* = kD_h/(k_w t)$. An alternate form of this correlation is indicated graphically and compared with experimental data in Fig.

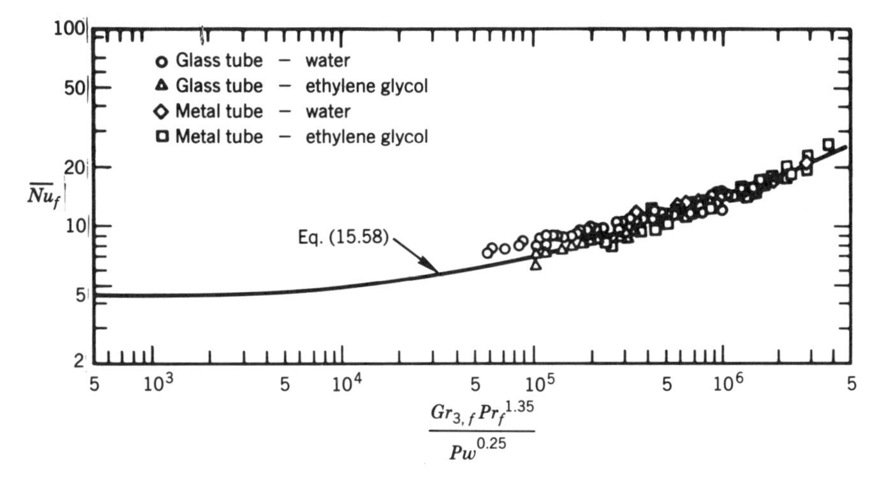

Figure 15.14. Nusselt numbers for hydrodynamically and thermally fully developed flow in a horizontal UHF tube, including the effects of property variations [51].

15.14. The equation represented by the curve in Fig. 15.14 is

$$\overline{\mathrm{Nu}}_f = \left\{ (4.36)^2 + \left[0.055 \left(\frac{\mathrm{Gr}_{3,f}\, \mathrm{Pr}_f^{1.35}}{\mathrm{Pw}^{0.25}} \right)^{0.4} \right]^2 \right\}^{1/2} \tag{15.58}$$

where $\mathrm{Gr}_{3,f} = g\beta(\overline{T}_w - T_b) D_h^3/\nu^2$ and $\mathrm{Pw} = \bar{h} D_h^2/(k_w t)$. These correlations are valid in the ranges $3 \times 10^4 < \mathrm{Ra}_{3,f} < 10^6$, $4 < \mathrm{Pr}_f < 175$, and $2 < \mathrm{Pw} < 66$.

15.4.2 Thermally Developing Flow in Horizontal Circular Tubes

Fundamental understanding of the heat transfer behavior of combined free and forced convection in the thermal entrance region of horizontal ducts is very limited. As may be expected, the early studies were based on experiments, but a number of theoretical investigations have been carried out using a variety of approximations. For large Prandtl number fluids, for example, the inertia terms in Eqs. (15.51) to (15.53), i.e., the left sides of these equations, may be neglected. The implication here is that the secondary flow is not significant in the momentum equations but is important in the energy equation.

Ou and Cheng [55] solved this problem by using the stream function defined as in Eq. (15.57); in addition, they also introduced the vorticity

$$\omega = \left(\frac{\partial^2}{\partial r^2} + \frac{1}{r}\frac{\partial}{\partial r} + \frac{1}{r^2}\frac{\partial^2}{\partial \phi^2} \right) \psi \tag{15.59}$$

The boundary condition used was axially uniform heat flux, but the wall temperature was assumed to be uniform circumferentially at any fixed axial location. Computed streamline patterns and isotherms from Ou and Cheng [55] for $\mathrm{Ra}_1 = 8 \times 10^5$ indicate that, along the axial direction, the secondary flow first grows and then decays.† Near the tube entrance, the isotherms are nearly concentric circles; but further downstream, the isotherms near the bottom of the tube become distorted as the local heat transfer deteriorates in the lower region. The variations in the distances between the isotherms around the tube indicate the variation of the heat transfer rates around the tube. A circumferentially averaged but axially local Nusselt number may be defined by extending the definition of Eq. (15.15). Thus,

$$\mathrm{Nu}_1 = -\frac{2}{\pi \theta_b} \int_0^{\pi} \left(\frac{\partial \theta}{\partial R} \right)_w d\phi$$

An axially averaged Nusselt number may then be defined as

$$\overline{\mathrm{Nu}}_1 = -\frac{4\int_0^X \int_0^{\pi} \left(\frac{\partial \theta}{\partial R} \right)_w d\phi\, dX}{\pi X (1 + \theta_b)}$$

† This feature distinguishes a tube with uniform wall temperature from one with uniform heat flux. In the latter, buoyancy effects persist throughout a tube even in thermally fully developed flow in which a fixed wall-minus-fluid temperature difference is established as both the wall and fluid temperatures increase linearly.

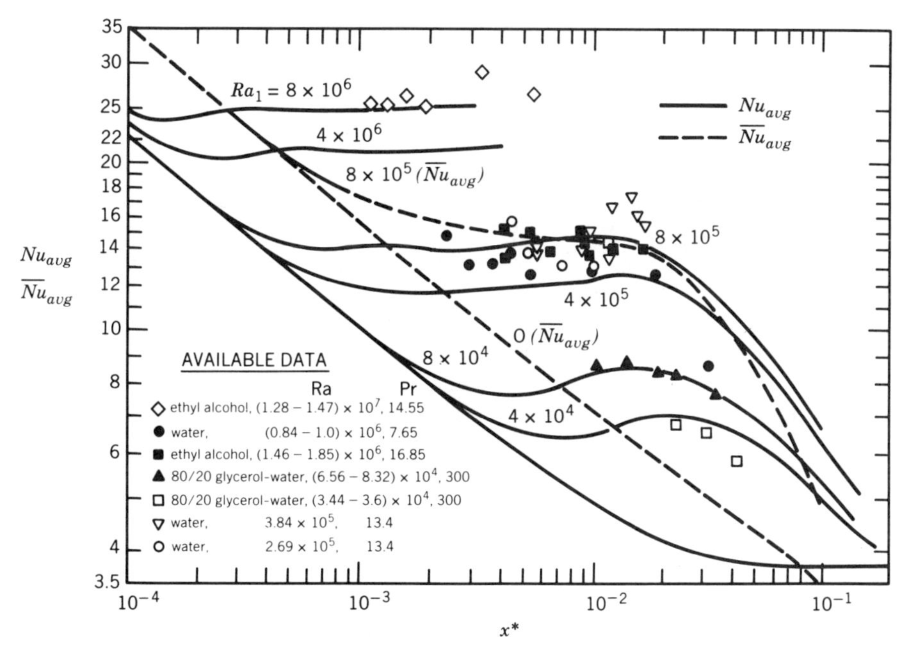

Figure 15.15. Effect of Rayleigh number on axial distribution of the local Nusselt number for a fluid of large Prandtl number in the thermal entrance region of a UWT horizontal tube [55].

The corresponding definitions using the axial temperature gradient are

$$\mathrm{Nu}_2 = -\frac{1}{2\pi\theta_b}\int_0^{\pi}\int_0^1 \frac{\partial\theta}{\partial Z}\left[2(1-R^2)\right] R\,dR\,d\phi$$

$$\overline{\mathrm{Nu}}_2 = -\int_0^{X}\int_0^{\pi}\int_0^1 \frac{\partial\theta}{\partial X}\left[2(1-R^2)\right] R\,dR\,d\phi\,dX \cdot \frac{1}{\pi X(1+\theta_b)}$$

The averages of the two definitions are then defined as

$$\mathrm{Nu}_{\mathrm{avg}} = \tfrac{1}{2}(\mathrm{Nu}_1 + \mathrm{Nu}_2) \qquad \text{(local Nusselt numbers at } x\text{)}$$

$$\overline{\mathrm{Nu}}_{\mathrm{avg}} = \tfrac{1}{2}(\overline{\mathrm{Nu}}_1 + \overline{\mathrm{Nu}}_2) \qquad \text{(mean Nusselt numbers from } x = 0 \text{ to } x\text{)}$$

Note that, in defining $\overline{\mathrm{Nu}}$, the average of the bulk temperatures at $x = 0$ and $x = x$ has been used; this temperature equals $(1 + \theta_b)/2$. Results for $\mathrm{Nu}_{\mathrm{avg}}$ and $\overline{\mathrm{Nu}}_{\mathrm{avg}}$ from Ou and Cheng [55] under UWT boundary conditions are shown at various values of Ra_l in Fig. 15.15. Also shown for comparison are experimental data from Oliver [56], Depew and Zenter [57], and Depew and August [58]. The complex behavior of the Nusselt number underscores the difficulty in obtaining a general correlation for the heat transfer.

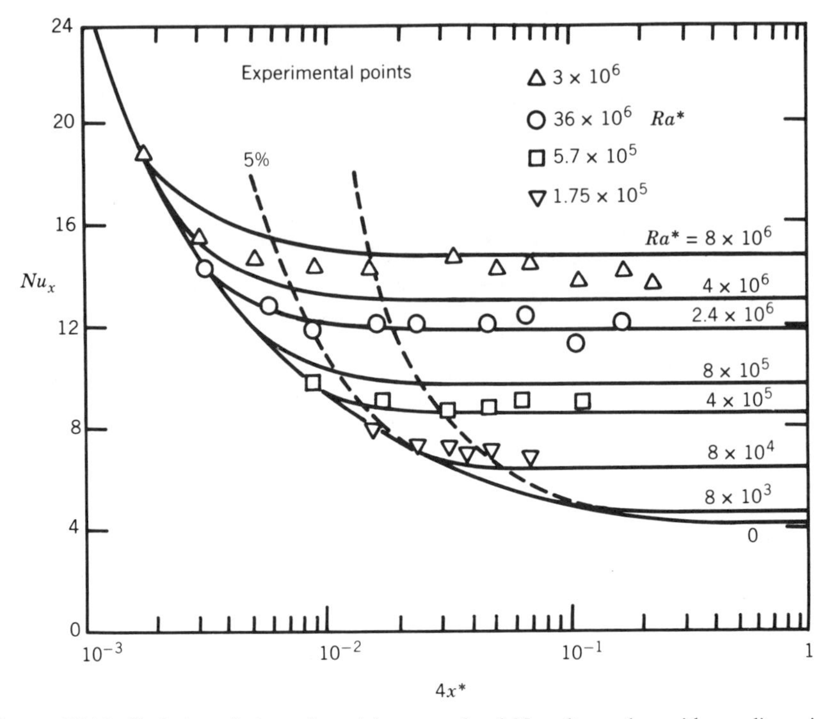

Figure 15.16. Variation of circumferential average local Nusselt number with nondimensional axial distance for laminar flow of a large Prandtl number fluid in the thermal entrance region of a UHF horizontal tube [60].

For uniformly heated tubes, numerical solutions based on the assumption of large Prandtl number show that the thermal entrance length is reduced to as low as 10% of the value for pure forced flow [59]. The circumferentially averaged local heat transfer coefficient is more than 300% the forced-flow prediction. The predicted results [60] are compared with available data in Fig. 15.16. For flows with temperature-dependent viscosity, the circumferentially averaged local Nusselt number may be correlated in terms of the viscosity parameter $\hat{\tau}$ [59]:

$$\hat{\tau}\Delta T = -\frac{1}{\mu}\left(\frac{d\mu}{dT}\right)\Delta T \tag{15.60}$$

where ΔT is the local average wall temperature minus the bulk mean temperature. The recommended correlations in [59] predict the existing experimental data well, but could lead to unrealistic results at very low Rayleigh numbers where both secondary flow and viscosity effects are negligible.

A number of empirical correlations exist for thermally developing mixed convection with hydrodynamically fully developed flow at entrance in horizontal heated tubes. For UWT and using water, ethyl alcohol, and a mixture of glycerol and water, Depew and August [58] generalized the earlier correlations of Oliver [56] and Brown and Thomas [61] to include short tubes. Their proposed correlation [58] for the average Nusselt number is

$$\overline{\mathrm{Nu}}_{\mathrm{am},\,b}\left(\frac{\mu_w}{\mu_b}\right)^{0.14} = 1.75\left[\mathrm{Gz}_b^* + 0.12\left(\mathrm{Gz}_b^*\mathrm{Gr}_{4,\,b}^{1/3}\mathrm{Pr}_b^{0.36}\right)^{0.88}\right]^{1/3} \tag{15.61}$$

This equation generally correlates to within ±40% the experimental data obtained by the aforementioned investigators and by Kern and Othmer [62]. In Eq. (15.64), all properties are evaluated at $\overline{T}_b = (T_1 + T_{b,\,L})/2$. The equation is based on data in the ranges $\mathrm{Gz}_b^* = 25$ to 700, $\mathrm{Re}_b < 1800$, and $\mathrm{Gr}_{4,\,b} = 500$ to 1×10^7.

The influence of free convection on forced laminar flow for simultaneously developing velocity and temperature profiles in UWT tubes has been investigated, and the correlations have been verified for air. For turbulent flow (Re > 7000), the effect of free convection is negligible and the Nusselt number is given by that for pure forced flow [53]. For Re < 3500, where the flow is laminar, mixed convection effects are important, and the recommended equation for the average Nusselt number for air is [53]

$$\overline{\mathrm{Nu}}_{\mathrm{lm}} = 2.67\left[(\mathrm{Gz}_w^*)^2 + (0.0087)^2\mathrm{Ra}_{1,\,w}^{1.5}\right]^{1/6} \qquad \text{for} \quad 60 \le \mathrm{Gz}_w^* \le 1300 \tag{15.62}$$

In the lower Graetz number regimes, Yousef and Tarasuk [10, 54] have examined the three-dimensional temperature distribution in the development region using an interferometric approach. They give the following correlations for laminar flow of air:

1. For

   $$0.0073 \times x^* < 0.040,\ 20 < \mathrm{Gz}_b^* < 110,\ \text{and}\ 1 \times 10^4 < \mathrm{Gr}_{6,\,b} < 8.7 \times 10^4$$

   one has†

   $$\overline{\mathrm{Nu}}_{\mathrm{lm}}\left(\frac{\mu_w}{\mu_b}\right)^{0.14} = 1.75\left[\mathrm{Gz}_b^* + 0.245\left(\mathrm{Gz}_b^{*1.5}\mathrm{Gr}_{6,\,b}^{1/3}\right)^{0.882}\right]^{1/3} \tag{15.63}$$

   Equation (15.63) correlates the available experimental data to within +3% and −30%.

2. For $0.040 < x^* < 0.25$, $3.2 < \mathrm{Gz}_b^* < 20$, and $0.8 \times 10^4 < \mathrm{Gr}_{6,\,b} < 4 \times 10^4$ one has†

   $$\overline{\mathrm{Nu}}_{\mathrm{lm}}\left(\frac{\mu_w}{\mu_b}\right)^{0.14} = 0.969(\mathrm{Gz}_b^*)^{0.82} \tag{15.64}$$

   Equation (15.64) gives a representation of the experimental data to within +7.8% and −6.1%.

15.4.3 Flow in Circular, Concentric Horizontal Annuli

Existing results on mixed convection in a horizontal circular, concentric annulus are derived mainly from theoretical studies. These include work by Hattori [63] and Nguyen et al. [64]. The general findings are that the secondary flows can strongly

†Although the viscosity variation correction factor $(\mu_w/\mu_b)^{0.14}$ normally is used only for liquids, Yousef and Tarasuk [10, 54] used it for air. Since this ratio varies only slightly for air, it should not make a significant difference in the results.

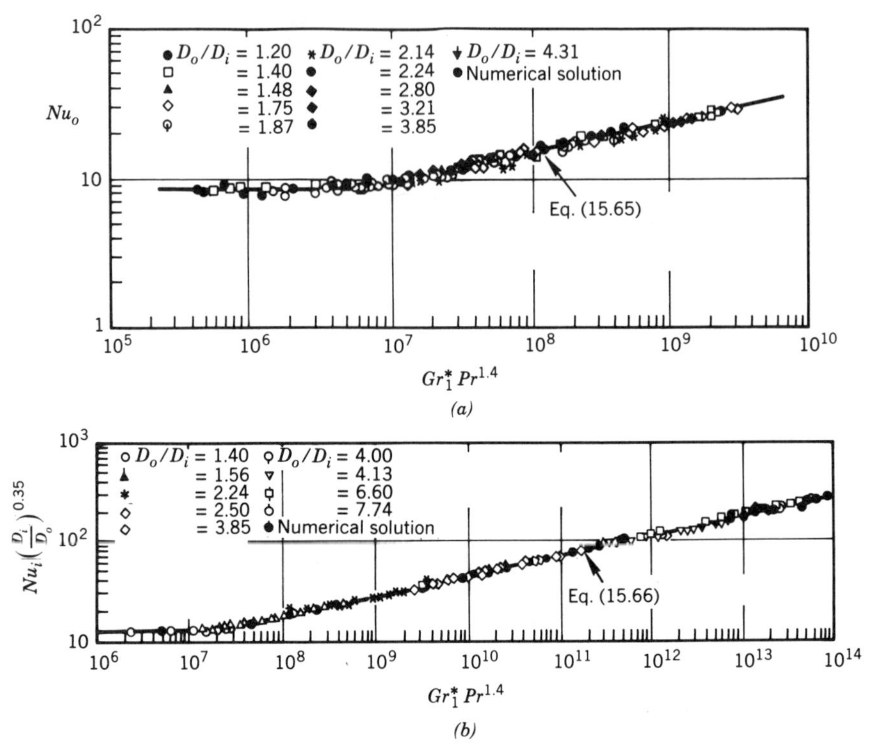

Figure 15.17. Local Nusselt numbers for horizontal circular, concentric annuli: (a) outer wall heated at UHF and inner wall insulated; (b) outer wall insulated and inner wall heated at UHF [63].

distort the velocity and temperature profiles and significantly increase the heat transfer. The circumferentially averaged Nusselt number depends on the heating condition. Figure 15.17a and b show numerical results obtained by Hattori [63] for two cases: outer cylinder heated at UHF and inner cylinder insulated, and outer cylinder insulated and inner cylinder heated at UHF. The flow is hydrodynamically fully developed. Good agreement is shown with the experimental data obtained in a complementary study by the author. In the figures, the Nusselt and Grashof numbers are defined as

$$\mathrm{Nu}_i = \frac{1}{2\pi}\int_0^{2\pi} \mathrm{Nu}_{\phi,i}\, d\phi, \qquad \mathrm{Nu}_o = \frac{1}{2\pi}\int_0^{2\pi} \mathrm{Nu}_{\phi,o}\, d\phi$$

$$\mathrm{Nu}_{\phi,i} = \frac{h_{\phi,i}(D_o - D_i)}{k}, \qquad \mathrm{Nu}_{\phi,0} = \frac{h_{\phi,o}(D_o - D_i)}{k}$$

$$h_{\phi,i} = -\frac{k}{(T_{w,i} - T_b)}\left(\frac{\partial T}{\partial r}\right)_{r=r_i}, \qquad h_{\phi,o} = -\frac{k}{(T_{w,o} - T_b)}\left(\frac{\partial T}{\partial r}\right)_{r=r_o}$$

$$\mathrm{Gr}_1^* = \frac{g\beta u_m (dT_w/dx)\, L^5}{\alpha \nu^2}$$

where $L = (D_o^2 - D_i^2)(q_o'' + q_i'')/(D_o q_o'' + D_i q_i'')$.

In Fig. 15.17*a* and *b*, the experimental and numerical results are shown to agree well with the respective heat transfer correlations, which are:

FOR OUTER WALL AT UWT AND INNER WALL INSULATED

$$\mathrm{Nu}_o = 0.38\,\mathrm{Gr}_1^{*0.20}\,\mathrm{Pr}^{0.28} \tag{15.65}$$

FOR OUTER WALL INSULATED AND INNER WALL AT UWT

$$\mathrm{Nu}_i = 0.44\,\mathrm{Gr}_1^{*0.20}\,\mathrm{Pr}^{0.28}\left(\frac{D_o}{D_i}\right)^{0.35} \tag{15.66}$$

15.4.4 Thermally Fully Developed Flow between Horizontal Parallel Plates

In the absence of free convection, the Nusselt number and the pressure drop parameter f Re both reach constant values when the flow is laminar and hydrodynamically and thermally fully developed and the heat fluxes on the horizontal plates are held constant. In this case, $\mathrm{Nu}_{fd} = 8.235$ and $f\,\mathrm{Re} = 24$ as presented in Chap. 3. When free convection effects are present, as measured by the product Re Ra, secondary flows set in as Re Ra reaches a critical value. For fully developed flow, the conditions marking the onset of this phenomenon have been studied theoretically by Nakayama et al. [65] and confirmed experimentally by Akiyama et al., [66] and by Ostrach and Kamotani [67], among others. Photographic evidence of the onset and subsequent formation of longitudinal vortices in the postcritical flow regime has been reported by Hwang and Cheng [68]. The photographs were taken near the exit of a horizontal, rectangular channel with a large width-to-height ratio so that a parallel-plate duct was approximated closely. Flow visualization was accomplished by injecting smoke into the main flow. The value of Re Ra* was increased from 4.7×10^5 to 1.49×10^6. The dimensionless wave number was shown to remain relatively constant.

The wave number after the onset of longitudinal vortex rolls may be assumed to be that given by the linear stability analysis. This approximation is justified, since the wave number is rather insensitive to the change of the product Re Ra*, at least in the immediate postcritical regime. The flow structure and heat transfer can be thus isolated for detailed analysis. For plane Poiseuille flow (with a fully developed temperature profile), Hwang and Cheng [68] obtained solutions for Pr Re Ra* values up to four times the critical value, thereby extending the applicability of earlier solutions acquired by means of a perturbation theory by Mori and Uchida [69]. For a given finite value of Pr Re Ra*, the effect of the Prandtl number on the friction parameter f Re is negligible when $\mathrm{Pr} \geq 10$. Hence for fully developed flow, the effect of secondary flow on the flow result is important only when the Prandtl number is small. For $\mathrm{Pr} \geq 2$, the Nusselt number exhibits asymptotic behavior.

15.4.5 Thermally Developing Flow between Horizontal Parallel Plates

Buoyant-force effects on laminar heat transfer in the hydrodynamic and thermal development region of a horizontal, parallel plate duct have been studied numerically by Naito [70]. The Prandtl number studied was 0.71, and the Reynolds number was less

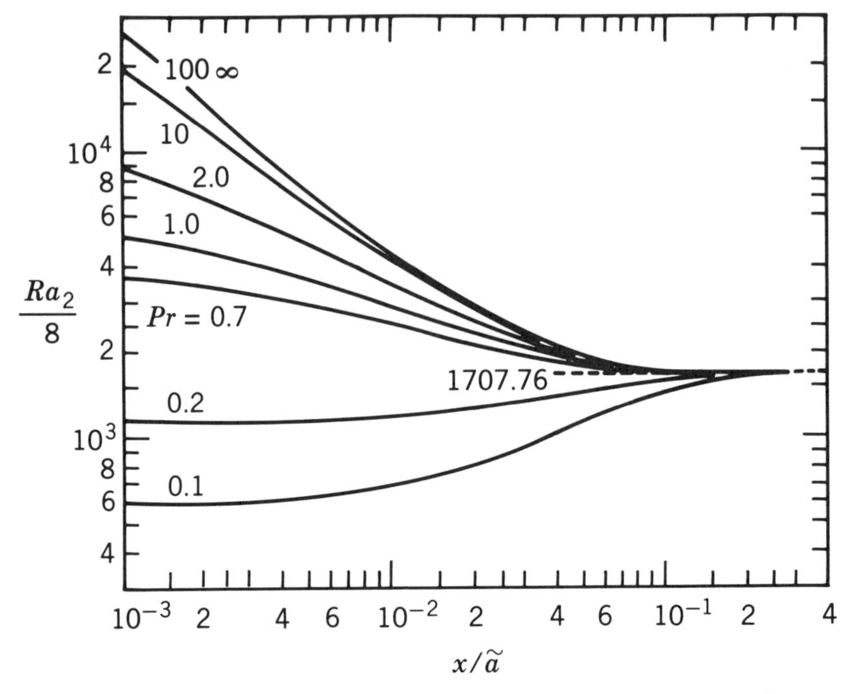

Figure 15.18. Effect of Prandtl number on critical Rayleigh number marking the onset of instability for Pe → ∞ [71].

than 300. Of concern was the subcritical range with the Rayleigh number less than 1800. For hydrodynamically fully developed but thermally developing laminar flow in a horizontal, parallel plate channel, the conditions giving rise to longitudinal vortex-type secondary flow have been determined theoretically by Hwang and Cheng [71]. They considered the case where the top plate is cooled and maintained at the same temperature as the fluid at the entrance (i.e. $T_c = T_1$), and the bottom plate is heated to a uniform temperature T_h. The problem is similar to the Bénard problem with a superimposed hydrodynamically fully developed laminar flow, and the vortex rolls with axes parallel to the basic flow have their counterparts in the Bénard cells, except that they are infinitely elongated. Unlike the case of thermally fully developed flow, the critical Rayleigh number is a function of the axial distance. For infinitely large Péclet numbers, the critical Rayleigh number decreases monotonically with increasing streamwise distance for large Prandtl number fluids, and increases monotonically with distance for small Prandtl number fluids. At the end of the thermal development region, the critical Rayleigh number Ra_2 is independent of the Prandtl number and assumes the value 13662.1 (= 8 × 1707.76). This behavior is depicted in Figure 15.18. This problem was later examined experimentally by Kamotani and Ostrach [72], who found that for air the critical Rayleigh number was much higher than that predicted in [71]. Heat transfer in the entrance region for thermally developing flow has been measured, in a horizontal, parallel-plate channel at UWT with the upper wall cooled and the bottom wall heated, by Kamotani et al. [73] and found to be several times the value in the absence of secondary flow, as shown in Fig. 15.19.

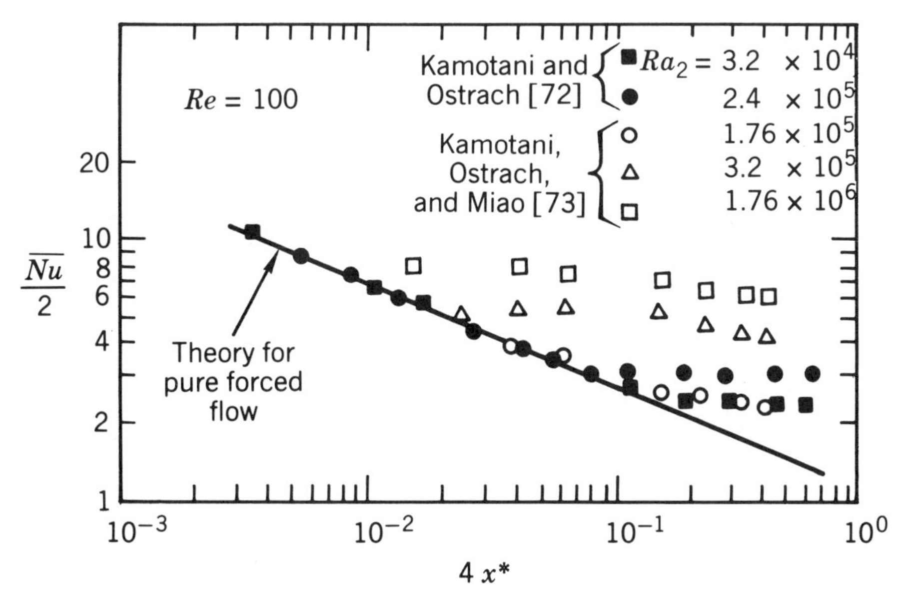

Figure 15.19. Enhanced heat transfer in the entrance region of a horizontal parallel plate channel having UWT with the upper wall cooled and the lower wall heated [73].

The effects of buoyancy on the heat transfer in a thermally developing horizontal parallel plate channel, where the velocity profile is fully developed at the entrance to the heated section, have been investigated experimentally by Osborne and Incropera [74] for the situation where both the top and bottom plates are heated at uniform heat fluxes that are not necessarily equal. The heating of the top plate causes the adjacent thermal boundary layer to be stratified so that there is negligible penetration by the buoyancy-induced secondary flow originating from the bottom plate. Forced convection therefore dominates the heat transfer from the top plate, according to the correlation

$$\mathrm{Nu}_{x,t} = 1.490\,\mathrm{Gz}^{1/3} \tag{15.67}$$

In Eq. (15.67), $\mathrm{Nu}_{x,t}$ is the local Nusselt number on the top plate, defined using the difference between the local plate temperature and the local bulk temperature; properties are determined at the local bulk mean temperature given by the energy balance:

$$T_{b,x} = T_1 + \frac{(q_t'' + q_{bt}'')\,x}{u_m \tilde{a} \rho c_p} \tag{15.68}$$

For the bottom plate, a correlation can be developed based on the scheme suggested by Churchill [75], namely,

$$\mathrm{Nu}_x^n = \mathrm{Nu}_{x,F}^n + \mathrm{Nu}_{x,N}^n$$

The resultant correlation due to Osborne and Incropera [74] is

$$\frac{\mathrm{Nu}_{x,bt}}{\mathrm{Nu}_{x,F}} = \left(1 + 0.00365\frac{\mathrm{Ra}_2^{*\,3/4}}{\mathrm{Gz}}\right)^{1/3} \tag{15.69}$$

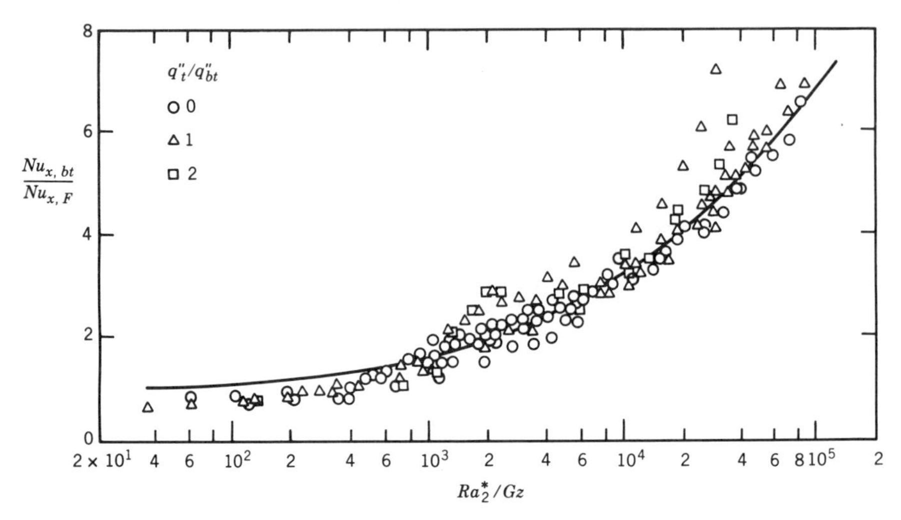

Figure 15.20. Nusselt number for the heated bottom plate of a parallel plate horizontal duct.

Here the thermally developing Nusselt number $\mathrm{Nu}_{x,F}$ for pure forced convection is given by [87]. This equation is compared satisfactorily with data for water in Fig. 15.20 for $q_t''/q_{bt}'' \leq 2$, where the subscripts t and bt stand for the top and bottom plate, respectively. For $q_t''/q_{bt}'' > 2$, the above correlation significantly underpredicts the experimental data. For the ranges of validity of Eqs. (15.67) and (15.69), see Table 15.2.

15.4.6 Flow in Rectangular Horizontal Channels

The effects of buoyancy-induced secondary flow on heat transfer in the thermal entrance region of horizontal rectangular ducts may be assessed using the same approach described in Sec. 15.4.2. Employing the assumption of large Prandtl number, Cheng et al. [76] investigated the influence of Rayleigh number on the Nusselt number at various aspect ratios γ [width divided by height (in gravity direction) of duct]. The duct walls are heated at uniform heat flux, and the flow is assumed hydrodynamically fully developed. The results are shown in Fig. 15.21 for $\gamma = 2$. The local Nusselt number is not affected by buoyancy up to a certain entrance length that depends on the magnitude of the Rayleigh number. The curve at a given Rayleigh number branches out from the pure forced-convection curve (Rayleigh number zero) and, after reaching a minimum, levels off to a constant value as the flow becomes thermally fully developed. The effect of buoyancy is to decrease the thermal entrance length.

An investigation of the buoyancy effects in the simultaneous hydrodynamic and thermal development region has been conducted by Abou-Ellail and Morcos [77]. For a duct with aspect ratio $\gamma = 1$, predicted local Nusselt number is given for Prandtl numbers ranging from 1 to 20. In the thermal entrance region, an increase in the Prandtl number causes a decrease in the Nusselt number. For Pr = 20, the results of [77] are higher than those of [76] in the thermal entrance region by a maximum of about 20%. An increase in the aspect ratio in general leads to an increase in the local Nusselt number and in the thermal entrance length. At a Rayleigh number of 10^5, the local Nusselt number in the thermally fully developed region is approximately 200% higher than the pure forced-convection value.

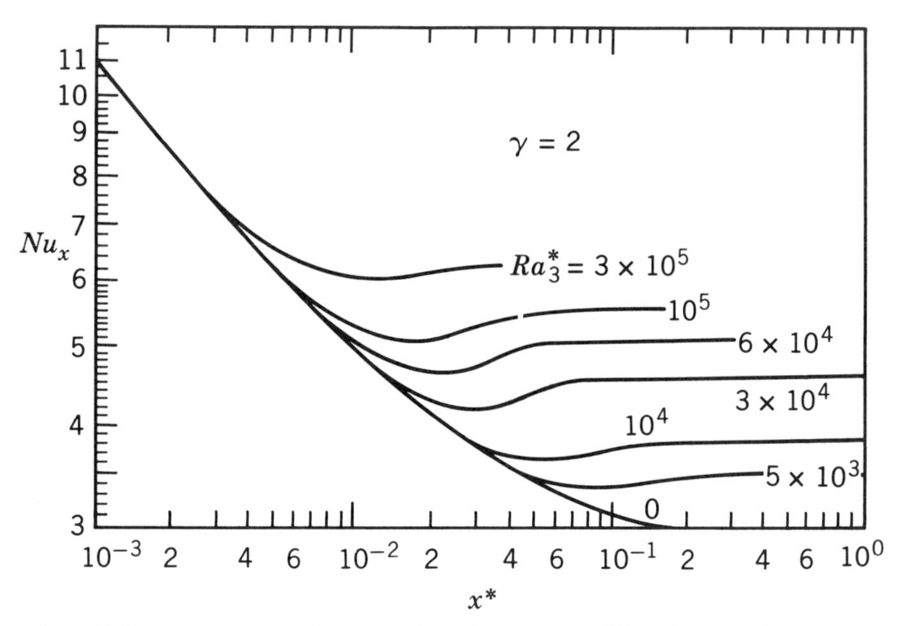

Figure 15.21. Axial variation of the circumferential-average local Nusselt number for a rectangular duct of aspect ratio 2 [76].

15.5 TRANSITION FROM LAMINAR TO TURBULENT FLOW

There is scant information available concerning mixed convection in the postlaminar regimes of transitional and turbulent flow. Much of the available data relates to the vertical circular tube. Even for this geometry, the current understanding of heat transfer in the laminar-to-turbulent flow transition regime is still limited; however, criteria for the onset of transition are available [19, 78, 79]. For the horizontal tube, some limited experimental data are also available [43].

15.5.1 Transitional Upward Flow in Vertical Circular Tubes

For hydrodynamically fully developed flow in a vertical tube, the transition from laminar to an unstable flow has been studied for water by Scheele et al. [79]. They observed the breakup of a thin stream of dye injected into the center of a tube upstream of the heat transfer section. For natural convection in the direction of forced flow (upflow) with uniform heating (UHF) of the tube, the first instability appears in the form of a sinuous motion in the dye filament. The amplitude of the disturbances increases, and eventually the dye filament breaks up, as the ratio Gr_5/Re is increased. The onset of the initial disturbance is preceded by a flatness in the velocity profile that appears at $Gr_5/Re = 88$; however, the critical value of Gr_5/Re depends on the ratio of the tube length to diameter. For downflow with uniform flux heating, the first instability consists of a slight asymmetry of the dye filament upon emergence from the heated section. An increase of Gr_5/Re triggers intermittent bursts of a highly disturbed flow. The critical value of Gr_5/Re is 252 and is independent of Re. For downflow with uniform wall temperature (UWT) heating, an asymmetric flow pattern develops that involves reversed flow on one side of the tube [79].

Indications are that for buoyancy-aided flow, the mechanism of transition is similar to that in boundary-layer flow over a flat plate. Here, transition consists of the appearance of regular oscillations which gradually grow in extent and amplitude until the disturbance breaks into fluctuating motion that is characteristic of turbulent flow. For buoyancy-opposed flow, transition consists of an asymmetric flow (resulting from an originally symmetric fully developed flow) which gives rise to reversed flow on one side of the tube. The extent of the reversed flow increases in size as Gr_5/Re increases, leading to an eddying flow. The transition to an eddying motion occurs suddenly.

The flow oscillations that accompany the transition process lead to fluctuations in wall temperatures, and these can in turn be used to indicate transition. Using this approach, Hallman [12] found that for buoyancy-assisted, hydrodynamically fully developed flow (at tube entrance) in a UHF vertical circular tube, the location of transition depends on the heating rate and the flow rate. At a constant flow rate, an increase in heating rate causes the point of transition to travel upstream (i.e., down the tube). If the heating rate is held fixed, increasing the flow rate moves the transition point downstream. Based on Hallman's data, the following correlation may be used to predict the location of transition:

$$Gr_3 \, Pr = 2664 \, Gz^{1.83} \tag{15.70a}$$

15.5.2 Transitional Flow in Horizontal Circular Tubes

In a horizontal circular tube, the transition to turbulent flow is strongly affected by the presence of secondary flow. With a high initial turbulence level at the entrance to the tube, the onset of secondary flow tends to suppress the turbulence, while at a low initial turbulence level, secondary flow tends to increase the turbulence. Consequently, when the initial turbulence level is low, the experimental critical Reynolds number Re_c, defined as the value where an intermittency begins to appear in the flow [43], decreases with $Re\,Ra^*$ but can be greater than 6000 at $Re\,Ra^* = 1.5 \times 10^4$. At a high level of initial turbulence such as that associated with a turbulence generator, Re_c increases with $Re\,Ra^*$ but is only about 2500 at $Re\,Ra^* = 1.5 \times 10^4$. For $Re\,Ra^* < 5 \times 10^5$ with UHF tubes, Mori et al. [43] recommend the following equation when the initial turbulence level is low:

$$Re_c = \frac{Re_{c0}}{1 + 0.14 \times 10^{-5} Re\,Ra^*} \tag{15.70b}$$

In the above, Re_{c0} is the critical Reynolds number without heating and can be more than 6 times larger than the high-initial-intensity critical Reynolds number of approximately 2000 [80]. Mori et al. [43] obtained a value of $Re_{c0} = 7700$. For high initial turbulence levels, Mori et al. [43] recommend, for UHF tubes,

$$Re_c = 128(Re\,Ra^*)^{1/4} \tag{15.70c}$$

15.6 TURBULENT MIXED CONVECTION IN DUCTS

For vertical upward flow in turbulent mixed convection in heated tubes, fairly well-established criteria for the onset of buoyancy-induced impairment of heat transfer are available. No satisfactory correlating equation, however, is available yet for the heat transfer. In downflow, a satisfactory correlation now exists. For horizontal tubes, experimental evidence indicates that the effect of buoyancy is negligible in turbulent flow.

15.6.1 Vertical Ducts

From the quantitative information presented in Secs. 15.3 and 15.4, it is clear that in laminar mixed convection in a vertical duct, the heat transfer is improved (by virtue of the increased velocity near the wall) for aided flow but is worsened in opposed flow (since near-wall velocities are reduced). The situation is quite different in turbulent flow, where the heat transfer is sometimes less than the pure forced-convection value when natural convection aids forced convection, and where in opposed flow the heat transfer is generally larger than the corresponding forced convection value.

The phenomenon of heat transfer impairment in vertical heated upflow can be explained by a two-layer model [81]. In this concept, the fluid in the layer close to the heated wall experiences a buoyancy force owing to the reduced density. Acting in the direction of motion, this force tends to decrease the shear stress in the layer away from the wall. Consequently, turbulence production is reduced across the tube, resulting in laminarization. A simple approximate analysis [82] leads to the following criterion for the onset of buoyancy-induced impairment of heat transfer:

$$\frac{\overline{\mathrm{Gr}}}{\mathrm{Re}^{2.7}} \leq 10^{-5} \tag{15.71}$$

where $\overline{\mathrm{Gr}} = g(\rho_b - \bar{\rho})D_h^3/(\bar{\rho}\nu^2)$. The integrated density $\bar{\rho}$ is defined as

$$\bar{\rho} = \frac{1}{T_w - T_b}\int_{T_b}^{T_w} \rho \, dT \tag{15.71a}$$

In tubes heated at specified fluxes (UHF), the heat transfer impairment leads to sharp peaks in the local wall temperatures, which have been observed by Ackerman [83], among others. The thermal impairment does not persist at higher buoyancy, however, since the shear stress changes sign and energy inputs to the turbulent motion start to increase, as does the thermal performance of the tube. As a result, in buoyancy-aided flow, the heat transfer from the tube is impaired in the low ranges of the Grashof number, but recovers and may even exceed the pure forced-convection value at high Grashof numbers. Quantitative evidence of this behavior has been reported originally by Fewster [84] for upflow of carbon dioxide and water at supercritical pressures. Figure 15.22 shows the situation in which the temperatures of the UHF wall are below the pseudocritical temperature, the temperature at which c_p becomes maximum. It may be noted that the criterion of Eq. (15.71) is supported by the data shown in Fig. 15.22. Included in Fig. 15.22 are the UWT upflow data of Herbert and Sterns [3], of which more will be said later in this section.

For turbulent mixed convection in buoyancy-opposed flow in vertical UHF tubes, the heat transfer is generally enhanced over that for pure forced convection. A semiempirical equation for this situation has been developed by Jackson and Hall [82]:

$$\frac{\mathrm{Nu}}{\mathrm{Nu}_F} = \left[1 + 2750\left(\frac{\overline{\mathrm{Gr}}}{\mathrm{Re}^{2.7}}\right)^{0.91}\right]^{1/3} \tag{15.72}$$

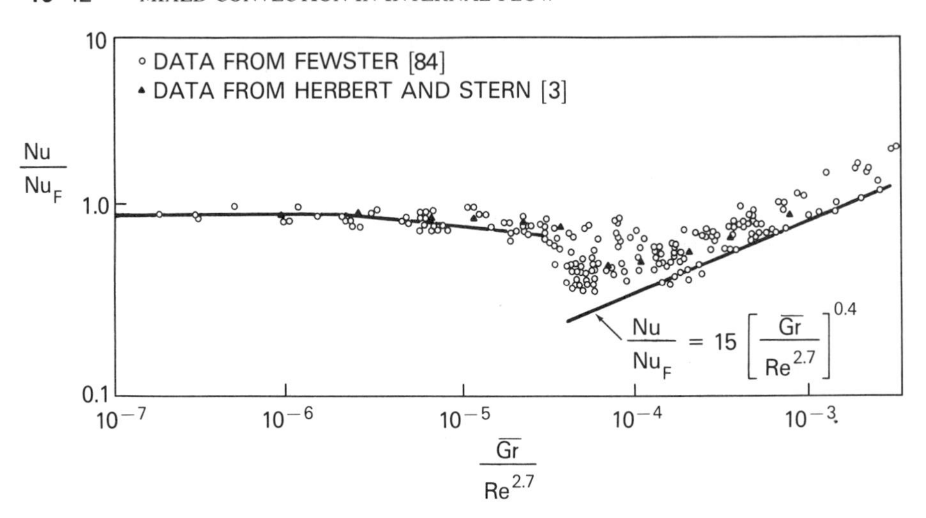

Figure 15.22. Average Nusselt number for buoyancy-assisted turbulent flow, normalized with Nusselt number for pure forced convection turbulent flow, in a vertical tube.

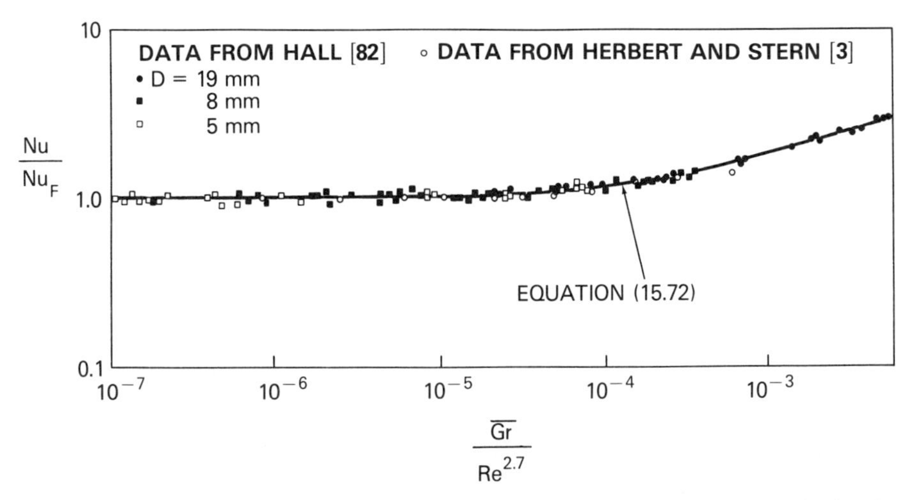

Figure 15.23. Average Nusselt number for buoyancy-opposed turbulent flow, normalized with Nusselt number for pure forced convection turbulent flow, in a vertical tube.

This equation is compared with experimental data for supercritical-pressure water in Fig. 15.23.

For vertical tubes at UWT, in the high Re range, Herbert and Sterns [3] have found that buoyancy effects on the heat transfer are negligible in aided turbulent mixed convection when Re exceeds a certain apparent critical value Re_{ac}. Their data were based on experiments with water, with Pr varying from 1.8 to 2.2 and Gr_4 from 2.0×10^7 to 2.6×10^7, approximately. The value of Re_{ac} may be calculated from the following equation:

$$Re_{ac} = 3000 + 0.00027\, Gr_4\, Pr \tag{15.73}$$

Thus, when Re is greater than Re_{ac}, the Nusselt number is given by the correlation for pure forced convection. When $Re < Re_{ac}$, Herbert and Sterns [3] suggest the following correlation for buoyancy-aided turbulent flow:

$$Nu = 8.5 \times 10^{-2}(Gr_4\, Pr)^{1/3} \tag{15.74}$$

This equation should be used in the ranges Re = 4500 to 15,000, $D = 0.0127$ to 0.0254 m, $L = 0.254$ to 3.30 m, Pr = 1.8 to 2.2; and $Gr_4 = 3 \times 10^6$ to 30×10^6. For buoyancy-opposed turbulent convection, the data of Herbert and Sterns [3] indicate that buoyancy effects may be neglected for Re > 15,000. For Re < 15,000, the data are correlated by the equation

$$Nu = 0.56\, Re^{0.47} Pr^{0.4} \tag{15.75}$$

which gives values higher than those for pure forced convection for UWT tubes. In Eqs. (15.73) to (15.75), all properties are evaluated at the film temperature. Using the raw experimental data given in [3], comparisons may be made between the Nu/Nu_F results of [3] with those of Fewster [84] and Jackson and Hall [82]. These comparisons are shown in Figs. 15.22 and 15.23. The agreement is very good. To effect conversion between $\overline{Gr}$ used in [82] and Gr_4 used in [3], the following approximations have been made:

$$\bar{\rho} \approx \rho_f$$

$$\rho_b - \bar{\rho} \approx -\rho_b \beta(T_w - \bar{T}_f)$$

$$\approx \rho_b \beta \frac{T_w - \bar{T}_b}{2}$$

$$\approx \rho_b \beta \frac{T_w - T_m}{2}$$

$$\overline{Gr} = \frac{g(\rho_b - \bar{\rho})\, D^3}{\bar{\rho} \nu^2}$$

$$\approx \frac{g\beta\rho^2 (T_w - T_m)\, D^3}{2\mu^2}$$

$$= \frac{Gr}{2}$$

In the above the overbar designates the arithmetic mean of the values at the inlet and outlet of the tube.

For vertical tubes at UWT in the low Re range, the turbulent mixed convection Nu is independent of Re in both aided and opposed flow. Testing with air in aided flow in the range Re = 385 to 4930, Brown and Gauvin [85] found that Nu may be predicted to ±7% in the range $Gr = 5 \times 10^6$ to 1×10^7 by the following equation for pure free convection [86]:

$$Nu = 0.13(Gr_5\, Pr)^{1/3} \tag{15.76}$$

For opposed flow in the range Re = 378 to 6900, Brown and Gauvin [85] show that Nu in the fully developed turbulent flow regime, which is independent of Re as noted above, is about 45% higher than the value given by Eq. (15.76). Here, Nu and Gr_5 are defined with the wall-to-center temperature difference, and properties are evaluated at the arithmetic mean between the wall and center temperatures.

15.6.2 Horizontal Tubes

Experimental data obtained by Mori et al. [43] show that for turbulent flow of air in a horizontal tube heated at UHF, the influence of buoyancy is completely overwhelmed by the turbulent motion. This is contrary to the case of laminar flow, which is discussed in Sec. 15.4.1. Thus, the Nusselt number for turbulent mixed convection in horizontal tubes may be computed using the standard correlation for pure forced-convection turbulent flow.

NOMENCLATURE

A	heat transfer surface area, m^2, ft^2
A_0	free flow area, m^2, ft^2
a	constant
$\tilde{a}$	spacing between parallel plates, m, ft
B	coefficient
b	constant
C	constant
c, c'	constants
c_p	specific heat at constant pressure, J/(kg · K), Btu/(lb$_m$ · °F)
D	tube inside diameter, m, ft
d	constant
D_h	hydraulic diameter $= 4A_0/S = 4$ (minimum free flow area)/(wetted perimeter), m, ft
D_i	inside diameter of annulus, m, ft
D_o	outside diameter of annulus, m, ft
Gr	Grashof number $= g\beta\,\Delta T\,D_h^3/\nu^2$, see Eq. (15.22a) for specific definitions
Gr*	heat-flux Grashof number $= g\beta q'' D_h^4/(16\nu^2 k)$, see Eq. (15.22b) for specific definitions
Gr_q	Grashof number; defined in Eq. (15.41)
$\overline{Gr}$	Grashof number $= g(\rho_b - \bar{\rho})D_h^3/(\bar{\rho}\nu^2)$
Gz	Graetz number = (Re Pr)(D_h/L)
Gz*	(π/4)Gz
f	Fanning friction factor $= -(D/\rho u_m^2)[\partial P/\partial x]$
g	gravitational acceleration, m/s^2, ft/s^2
h	heat transfer coefficient, W/(m^2 · K), Btu/(hr · ft^2 · °F)
h_x	local heat transfer coefficient at a section x in the thermal entrance region, W/(m^2 · K), Btu/(hr · ft^2 · °F)

$\bar{h}$	mean heat transfer coefficient from $x = 0$ to x in the thermal entrance region, $W/(m^2 \cdot K)$, $Btu/(hr \cdot ft^2 \cdot °F)$
k	thermal conductivity, $W/(m \cdot K)$, $Btu/hr \cdot ft \cdot °F)$
L_{th}	thermal entrance length, m, ft
L_{th}^*	$= L_{th}/(D_h \mathrm{Re\,Pr})$
Nu	Nusselt number for thermally and hydrodynamically fully developed flow
Nu_x	circumferential-average axially local Nusselt number $= h_x D_n/k$
$\overline{\mathrm{Nu}}$	axial-mean Nusselt number for the thermal entrance region defined by Eq. (15.18); $\overline{\mathrm{Nu}}$ is based on $\bar{h}$ defined by Eq. (15.20a)
$\overline{\mathrm{Nu}}_{am}$	arithmetic-average Nusselt number, defined in Eq. (15.18)
P	static pressure, Pa, lb_f/ft^2
P^*	nondimensional pressure, defined in Eq. (15.11c)
Pe	Péclet number = Re Pr
Pr	Prandtl number $= \mu c_p/k$
P_w	$hD_h^2/(k_w t)$
P_w^*	$kD_h/(k_w t)$
ΔP	pressure drop, Pa, lb_f/ft^2
$\Delta \bar{P}$	dimensionless pressure drop, defined by Eq. (15.33)
q	heat transfer rate, W, Btu/hr
q''	heat flux (heat transfer rate per unit area), W/m^2, $Btu/(hr \cdot ft^2)$
R	$r/(D/2)$ for a circular tube
R^*	$r/(D_o - D_i)$ for concentric annuli
$\tilde{R}$	gas constant, $kJ/(kg \cdot K)$, $lb_f \cdot ft/(lb_m \cdot °R)$
R	normalized radial distance in cylindrical coordinates, $= 2r/D$
r, ϕ, x	cylindrical coordinates (m, rad, m), (ft, deg, ft)
r_e	equivalent radius $= r_o - r_i$
r_i, r_o	inner and outer radius of a concentric annulus
Ra	Rayleigh number = Gr Pr
Re	Reynolds number $= \rho u_m D_h/\mu$
Re_c	critical Reynolds number indicating transition from laminar to turbulent flow
r_T	wall temperature difference parameter $= (T_c - T_1)/(T_h - T_1)$
S	perimeter of duct cross section, m, ft
St	Stanton number = Nu/Re Pr
T	temperature, °C, K, °F, °R
T_b	fluid bulk temperature at a section x, defined by Eq. (15.13), °C, K, °F, °R
$\bar{T}_b$	mean fluid bulk temperature from $x = 0$ to x, °C, K, °F, °R
T_w	wall temperature, local value at a peripheral point if the distinction is needed, °C, K, °F, °R
$\bar{T}_w$	circumferential-average wall temperature, °C, K, °F, °R
ΔT	characteristic temperature difference, °C, K, °F, °R

ΔT_{lm}	log-mean temperature difference, defined by Eq. (15.16b), °C, K, °F, °R
ΔT_{am}	arithmetic-mean temperature difference, defined by Eq. (15.16a), °C, K, °F, °R
t	tube wall thickness, m, ft
U	normalized velocity component in x direction $= u/u_m$
UHF	uniform heat flux boundary condition
UWT	uniform wall temperature boundary condition
u	velocity component in x direction, m/s, ft/s
V	normalized velocity component in R or Y direction
v	velocity component in r or y direction, m/s, ft/s
w	velocity component in ϕ direction, m/s, ft/s
X	$2x/(D_h \mathrm{Re}\,\mathrm{Pr})$
x	axial coordinate in Cartesian and radial coordinates, m, ft
x^*	$x/(D_h \mathrm{Re}\,\mathrm{Pr}) = X/2 = 1/\mathrm{Gz}$
x^+	$x/(D_h \mathrm{Re})$
y, z	transverse Cartesian coordinates, m, ft

Greek Symbols

α	fluid thermal diffusivity, m^2/s, ft^2/s
α'	constant
β	coefficient of thermal expansion defined in Eq. (15.7), K^{-1}, $°R^{-1}$
γ	aspect ratio (ratio of width to height) of rectangular duct
θ	dimensionless temperature $= (T_w - T)/(T_w - T_1)$
μ	dynamic viscosity, Pa · s, $lb_m/(hr \cdot ft)$
ν	kinematic viscosity, m^2/s, ft^2/s
ρ	density, kg/m^3, lb_m/ft^3
$\bar{\rho}$	integrated density across the flow cross section, see Eq. (15.71a), kg/m^3, lb_m/ft^3
ϕ	angular (azimuthal) coordinate, rad, deg
ψ	stream function
ω	vorticity, s^{-1}

Subscripts

am	arithmetic-mean value
avg	overall average value
b	bulk value
bt	bottom plate
c	pertaining to cold wall
F	pure forced convection

f	evaluated at fluid film temperature $(T_w + T_b)/2$
fd	fully developed flow value
h	pertaining to hot wall
i	value on inner wall
L	based on channel length
lm	logarithmic-mean value
m	mean value
N	pure natural convection
o	value on outer wall
t	top plate
w	value at wall, or properties evaluated at the wall temperature
z	based on z
ϕ	local value at an angle ϕ
1	value at inlet of duct

Diacritical

average value from $x = 0$ to x.

REFERENCES

1. R. C. Martinelli, C. J. Southwell, G. Alves, H. L. Craig, E. B. Weinberg, N. F. Lansing, and L. M. K. Boelter, Heat Transfer and Pressure Drop for a Fluid Flowing in the Viscous Region through a Vertical Pipe, *Trans. AIChE*, Vol. 38, pp. 493–530, 1942.
2. R. C. Martinelli and L. M. K. Boelter, The Analytical Prediction of Superposed Free and Forced Viscous Convection in a Vertical Pipe, *Univ. Calif. Publ. Eng.*, Vol. 5, No. 2, pp. 23–58, 1942.
3. L. S. Herbert and U. J. Sterns, Heat Transfer in Vertical Tubes—Interaction of Forced and Free Convection, *Chem. Eng. J.*, Vol. 4, pp. 46–52, 1972.
4. J. E. Byrne and E. Ejiogu, Combined Free and Forced Convection Heat Transfer in a Vertical Pipe, *Proc. Inst. Mech. Eng.*, Symposium on Heat and Mass Transfer by Combined Forced and Natural Convection, Paper C118/71, pp. 40–46, 1971.
5. E. R. G. Eckert and A. D. Diaguila, Convective Heat Transfer for Mixed, Free, and Forced Flow through Tubes, *Trans. ASME*, Vol. 76, pp. 497–504, 1954.
6. W. Aung and G. Worku, Theory of Fully Developed Combined Convection Including Flow Reversal, *J. Heat Transfer*, Vol. 108, pp. 485–488, 1986.
7. W. Aung and G. Worku, Developing Flow and Flow Reversal in a Vertical Channel with Asymmetric Wall Temperatures, *J. Heat Transfer*, Vol. 108, pp. 299–307, 1986.
8. A. Watzinger and D. H. Johnson, Wärmeübertragung von Wasser an Rohrwand bei senkrechter Strömung im Übergangsgebiet zwischen laminarer und turbulenter Strömung, *Forsch. Geb. Ingenieurwesens*, Vol. 10, pp. 182–196, 1939.
9. B. Metais and E. R. G. Eckert, Forced, Mixed, and Free Convection Regimes, *J. Heat Transfer*, Vol. 86, pp. 295–296, 1964.
10. W. W. Yousef and J. D. Tarasuk, An Interferometric Study of Combined Free and Forced Convection in a Horizontal Isothermal Tube, *J. Heat Transfer*, Vol. 103, pp. 249–256, 1981.

11. A. E. Bergles and R. R. Simonds, Combined Forced and Free Convection for Laminar Flow in Horizontal Tubes with Uniform Heat Flux, *Int. J. Heat Mass Transfer*, Vol. 14, pp. 1989–2000, 1971.

12. T. M. Hallman, Experimental Study of Combined Forced and Free Laminar Convection in a Vertical Tube, NASA TN D-1104, 1961.

13. R. W. Hornbeck, Numerical Marching Techniques for Fluid Flows with Heat Transfer, NASA SP-297, 1973.

14. T. M. Hallman, Combined Forced and Free-Laminar Heat Transfer in Vertical Tubes with Uniform Internal Heat Generation, *Trans. ASME*, Vol. 78, No. 8, pp. 1831–1841, 1956.

15. W. J. Marner and H. K. McMillan, Combined Free and Forced Laminar Convection in a Vertical Tube with Constant Wall Temperature, *J. Heat Transfer*, Vol. 92, pp. 559–562, 1970.

16. R. L. Pigford, Nonisothermal Flow and Heat Transfer Inside Vertical Tubes, *Chem. Eng. Prog. Symp. Ser.*, Vol. 51, pp. 79–92, 1955.

17. P. M. Worsøe-Schmidt and G. Leppert, Heat Transfer and Friction for Laminar Flow of Gas in a Circular Tube at High Heating Rate, *Int. J. Heat Mass Transfer*, Vol. 8, pp. 1281–1301, 1965.

18. W. T. Lawrence and J. C. Chato, Heat Transfer Effects on the Developing Laminar Flow inside Vertical Tubes, *J. Heat Transfer*, Vol. 88, pp. 215–222, 1966.

19. B. Zeldin and F. W. Schmidt, Developing Flow with Combined Forced-Free Convection in an Isothermal Vertical Tube, *J. Heat Transfer*, Vol. 94, pp. 211–223, 1972.

20. T. W. Jackson, W. B. Harrison, and W. C. Boteler, Combined Free and Forced Convection in a Constant-Temperature Vertical Tube, *Trans. ASME*, Vol. 80, pp. 739–745, 1958.

21. K. Sherwin, Laminar Convection in Uniformly Heated Vertical Concentric Annuli, *Br. Chem. Eng.*, Vol. 13, No. 11, pp. 1580–1585, 1968.

22. D. Maitra and K. Sabba Raju, Combined Free and Forced Convection Laminar Heat Transfer in a Vertical Annulus, *J. Heat Transfer*, Vol. 97, pp. 135–137, 1975.

23. G. M. Zaki, M. S. El-Genk, T. E. William, and J. S. Philbin, Experimental Heat Transfer Studies for Water in an Annulus at Low Reynolds Number, *Fundamentals of Forced and Mixed Convection*, ed. F. A. Kulacki and R. D. Boyd, HTD-Vol. 42, pp. 113–120, ASME, New York, 1985.

24. M. S. Rokerya and M. Iqbal, Effects of Viscous Dissipation on Combined Free and Forced Convection through Vertical Concentric Annuli, *Int. J. Heat Mass Transfer*, Vol. 14, pp. 491–494, 1971.

25. K. Sherwin and J. D. Wallis, A Theoretical Study of Combined Natural and Forced Laminar Convection for Developing Flow Down Vertical Annuli, *Heat Transfer 1970*, Vol. IV, Paper NC 3.9, 1970.

26. K. Sherwin and J. D. Wallis, A Study of Laminar Convection for Flow down Vertical Annuli, *Proc. Inst. Mech. Eng.*, Vol. 182, Pt. 3H, Paper 34, pp. 330–335, 1967–1968.

27. K. Sherwin and J. D. Wallis, Combined Natural and Forced Laminar Convection for Upflow through Heated Vertical Annuli, *Proc. Inst. Mech. Eng.*, Symp. Heat and Mass Transfer by Combined Forced and Natural Convection, Paper C112/71, 1971.

28. R. W. Shumway and D. M. McEligot, Heated Laminar Gas Flow in Annuli with Temperature-Dependent Transport Properties, *Nucl. Sci. and Eng.*, Vol. 46, pp. 394–407, 1971.

29. M. R. Malik and R. H. Pletcher, Calculation of Variable Property Heat Transfer in Ducts of Annular Cross Section, *Numer. Heat Transfer*, Vol. 3, pp. 241–257, 1980.

30. M. A. I. El-Shaarawi and A. Sarhan, Combined Forced-Free Laminar Convection in the Entry Region of a Vertical Annulus with a Rotating Inner Cylinder, *Int. J. Heat Mass Transfer*, Vol. 25, No. 2, pp. 175–186, 1982.

31. S. Ostrach, Combined Natural and Forced-Convection Laminar Flow Heat Transfer of Fluids with and without Heat Sources in Channels with Linearly Varying Wall Temperatures, NACA Tech. Note 3141, 1954.

32. A. F. Lietzke, Theoretical and Experimental Investigation of Heat Transfer by Laminar Natural Convection between Parallel Plates, NACA Rep. 1223, 1954.

33. T. Cebeci, A. A. Khattab, and R. LaMont, Combined Natural and Forced Convection in Vertical Ducts, *Heat Transfer 1982*, Vol. 2, pp. 419–424, 1982.

34. T. L. S. Rao and W. D. Morris, Superimposed Laminar Forced and Free Convection between Vertical Parallel Plates When One Plate Is Uniformly Heated and the Other Is Thermally Insulated, *Proc. Inst. Mech. Eng.*, Vol. 182, Pt. 3H, pp. 374–381, 1967–68.

35. E. M. Sparrow, G. M. Chrysler, and L. F. Azevedo, Observed Flow Reversals and Measured-Predicted Nusselt Numbers for Natural Convection in a One-Sided Heated Vertical Channel, *J. Heat Transfer*, Vol. 106, No. 2, pp. 325–332, 1984.

36. L. S. Yao, Free and Forced Convection in the Entry Region of a Heated Vertical Channel, *Int. J. Heat Mass Transfer*, Vol. 26, No. 1, pp. 65–72, 1983.

37. S. D. Joshi and A. E. Bergles, Heat Transfer in Cooling Channels of Power Transformers, Ann. Rep., ISU-ERI-AMES-78287, Affiliate Research Program in Electrical Power, Engineering Research Institute, Iowa State University, Ames, Iowa, pp. 7.1–7.15, 1978.

38. B. R. Morton, Laminar Convection in Uniformly Heated Horizontal Pipes at Low Rayleigh Numbers, *Quart. J. Mech. Appl. Math.*, Vol. 12, pp. 410–420, 1959.

39. M. Iqbal and J. W. Stachiewicz, Influence of Tube Orientation on Combined Free and Forced Laminar Convection Heat Transfer, *J. Heat Transfer*, Vol. 88, pp. 109–116, 1966.

40. G. N. Faris and R. Viskanta, An Analysis of Laminar Combined Forced and Free Convection Heat Transfer in a Horizontal Tube, *Int. J. Heat Mass Transfer*, Vol. 12, pp. 1295–1309, 1969.

41. Y. Mori and K. Futagami, Forced Convective Heat Transfer in Uniformly Heated Horizontal Tubes (2nd Report, Theoretical Study), *Int. J. Heat Mass Transfer*, Vol. 10, pp. 1801–1813, 1967.

42. P. H. Newell, Jr. and A. E. Bergles, Analysis of Combined Free and Forced Convection for Fully Developed Laminar Flow in Horizontal Tubes, *J. Heat Transfer*, Vol. 92, pp. 83–93, 1970.

43. Y. Mori, K. Futagami, S. Tokuda, and M. Nakamura, Forced Convective Heat Transfer in Uniformly Heated Horizontal Tubes, 1st Report—Experimental Study on the Effect of Buoyancy, *Int. J. Heat Mass Transfer*, Vol. 9, pp. 453–463, 1966.

44. S. T. McComas, Combined Free and Forced Convection in a Horizontal Circular Tube, *J. Heat Transfer*, Vol. 88, pp. 147–153, 1966.

45. B. S. Petukhov and A. F. Polyakov, Experimental Investigation of Viscogravitational Fluid Flow in a Horizontal Tube, *High Temp.*, Vol. 5, pp. 75–81, 1967.

46. B. S. Petukhov and A. F. Polyakov, Flow and Heat Transfer in Horizontal Tubes Under Combined Effect of Forced and Free Convection, *Heat Transfer 1970*, Vol. IV, Paper NC 3.7, 1970.

47. R. L. Shannon and C. A. Depew, Combined Free and Forced Laminar Convection in a Horizontal Tube with Uniform Heat Flux, *J. Heat Transfer*, Vol. 90, pp. 353–357, 1968.

48. R. L. Shannon and C. A. Depew, Forced Laminar Flow Convection in a Horizontal Tube with Variable Viscosity and Free Convection Effects, *J. Heat Transfer*, Vol. 91, pp. 251–258, 1969.

49. D. P. Siegworth, R. P. Mikesell, T. C. Readal, and T. J. Hanratty, Effect of Secondary Flow on the Temperature Field and Primary Flow in a Heated Horizontal Tube, *Int. J. Heat Mass Transfer*, Vol. 12, pp. 1535–1553, 1969.

50. N. A. Hussain and S. T. McComas, Experimental Investigation of Combined Convection in a Horizontal Circular Tube with Uniform Heat Flux, *Heat Transfer 1970*, Vol. 4, Paper No. NC 3.4, Elsevier, Amsterdam, 1970.

51. S. M. Morcos and A. E. Bergles, Experimental Investigation of Combined Forced and Free Laminar Convection in Horizontal Tubes, *J. Heat Transfer*, Vol. 97, pp. 212–219, 1975.

52. S. W. Churchill and R. Usagi, A General Expression for the Correlation of Rates of Transfer and Other Phenomena, *AIChE J.*, Vol. 18, pp. 1121–1128, 1972.

53. T. W. Jackson, J. M. Spurlock, and K. R. Purdy, Combined Free and Forced Convection in a Constant Temperature Horizontal Tube, *AIChE J.*, Vol. 7, pp. 38–45, 1961.

54. W. W. Yousef and J. D. Tarasuk, Free Convection Effects on Laminar Forced Convective Heat Transfer in a Horizontal Isothermal Tube, *J. Heat Transfer*, Vol. 104, pp. 145–152, 1982.

55. J-W. Ou and K. C. Cheng, Natural Convection Effects on Graetz Problem in Horizontal Isothermal Tubes, *Int. J. Heat Mass Transfer*, Vol. 20, pp. 953–960, 1977.

56. D. R. Oliver, The Effect of Natural Convection on Viscous-Flow Heat Transfer in Horizontal Tubes, *Chem. Eng. Sci.*, Vol. 17, pp. 335–350, 1962.

57. C. A. Depew and R. C. Zenter, Laminar Flow Heat Transfer and Pressure Drop with Freezing at the Wall, *Int. J. Heat Mass Transfer*, Vol. 12, pp. 1710–1714, 1969.

58. C. A. Depew and S. E. August, Heat Transfer Due to Combined Free and Forced Convection in a Horizontal and Isothermal Tube, *J. Heat Transfer*, Vol. 93, pp. 380–384, 1971.

59. S. W. Hong and A. E. Bergles, Theoretical Solutions for Combined Forced and Free Convection in Horizontal Tubes with Temperature-Dependent Viscosity, *J. Heat Transfer*, Vol. 98, pp. 459–465, 1976.

60. S. W. Hong, S. M. Morcos, and A. E. Bergles, Analytical and Experimental Results for Combined Forced and Free Laminar Convection in Horizontal Tubes, *Heat Transfer 1974*, Vol. III, pp. 154–158, Japan Soc. Mech. Eng., Tokyo, 1974.

61. A. R. Brown and M. A. Thomas, Combined Free and Forced Convection Heat Transfer for Laminar Flow in Horizontal Tubes, *J. Mech. Eng. Sci.*, Vol. 7, No. 4, pp. 440–448, 1965.

62. D. Q. Kern and D. F. Othmer, Effect of Free Convection on Viscous Heat Transfer in Horizontal Tubes, *Trans. AIChE*, Vol. 39, pp. 517–555, 1943.

63. N. Hattori, Combined Free and Forced-Convection Heat Transfer for Fully Developed Laminar Flow in Horizontal Concentric Annuli (Numerical Analysis), *JSME Trans.*, Vol. 45, pp. 227–239, 1979.

64. T. H. Nguyen, V. Vasseur, L. Robillard, and B. Chandra Shekar, Combined Free and Forced Convection of Water between Horizontal Concentric Cylinders, *J. Heat Transfer*, Vol. 105, pp. 498–504, 1983.

65. W. Nakayama, G. J. Hwang, and K. C. Cheng, Thermal Instability in Plane Poiseuille Flow, *J. Heat Transfer*, Vol. 92, pp. 61–68, 1970.

66. M. Akiyama, G. J. Hwang, and K. C. Cheng, Experiments on the Onset of Longitudinal Vortices in Laminar Forced Convection between Horizontal Plates, *J. Heat Transfer*, Vol. 93, pp. 335–341, 1971.

67. S. Ostrach and Y. Kamotani, Heat Transfer Augmentation in Laminar Fully Developed Channel Flow by Means of Heating from Below, *J. Heat Transfer*, Vol. 97, pp. 220–225, 1975.

68. G. J. Hwang and K. C. Cheng, Finite Amplitude Convection with Longitudinal Vortices in Plane Poiseuille Flow—The Effect of Uniform Axial Temperature Gradient, *Int. J. Heat Mass Transfer*, Vol. 15, pp. 789–800, 1972.

69. Y. Mori and Y. Uchida, Forced Convective Heat Transfer between Horizontal Flat Plates, *Int. J. Heat Mass Transfer*, Vol. 9, pp. 803–817, 1966.

70. E. Naito, Buoyant Force Effects on Laminar Flow and Heat Transfer in the Entrance Region between Horizontal Parallel Plates, *Heat Transfer—Jpn. Res.*, Vol. 13, pp. 80–95, 1984.

71. G. J. Hwang and K. C. Cheng, Convective Instability in the Thermal Entrance Region of a Horizontal Parallel-Plate Channel Heated from Below, *J. Heat Transfer*, Vol. 95, pp. 72–77, 1973.

72. Y. Kamotani and S. Ostrach, Effect of Thermal Instability on Thermally Developing Laminar Channel Flow, *J. Heat Transfer*, Vol. 98, pp. 62–66, 1976.

73. Y. Kamotani, S. Ostrach, and H. Miao, Convective Heat Transfer Augmentation in Thermal Entrance Region by Means of Thermal Instability, *J. Heat Transfer*, Vol. 101, pp. 222–226, 1979.

74. D. G. Osborne and F. P. Incropera, Laminar, Mixed Convection Heat Transfer for Flow between Horizontal Parallel Plates with Asymmetric Heating, *Int. J. Heat Mass Transfer*, Vol. 28, No. 1, pp. 207–217, 1985.

75. S. W. Churchill, A Comprehensive Correlating Equation for Laminar, Assisting, Forced and Free Convection, *AIChE J.*, Vol. 23, pp. 10–16, 1977.

76. K. C. Cheng, S. W. Hong, and G. J. Hwang, Buoyancy Effects on Laminar Heat Transfer in the Thermal Entrance Region of Horizontal Rectangular Channels with Uniform Wall Heat Flux for Large Prandtl Number Fluid, *Int. J. Heat Mass Transfer*, Vol. 15, pp. 1819–1836, 1972.

77. M. M. M. Abou-Ellail and S. M. Morcos, Buoyancy Effects in the Entrance Region of Horizontal Rectangular Channels, *J. Heat Transfer*, Vol. 105, pp. 924–928, 1983.

78. G. F. Scheele and T. J. Hanratty, Effect of Natural Convection Instabilities on Rates of Heat Transfer at Low Reynolds Numbers, *AIChE J.*, Vol. 9, pp. 183–185, 1963.

79. G. F. Scheele, E. M. Rosen, and T. J. Hanratty, Effect of Natural Convection on Transition to Turbulence in Vertical Pipes, *Can. J. Chem. Eng.*, Vol. 38, pp. 67–73, 1960.

80. J. Rotta, Experimenteller Beitrag zur Entstehung turbulenter Strömung im Rohr, *Ing.-Arch.*, Vol. 24, p. 258, 1956.

81. W. B. Hall, Heat Transfer Near the Critical Point, *Adv. in Heat Transfer*, Vol. 7, pp. 1–86, 1971.

82. J. D. Jackson and W. B. Hall, Influences of Buoyancy on Heat Transfer to Fluids Flowing in Vertical Tubes under Turbulent Conditions, *Turbulent Forced Convection in Channels and Bundles*, ed. S. Kakaç, and D. B. Spalding, Hemisphere, pp. 613–673, 1979.

83. J. W. Ackerman, Pseudoboiling Heat Transfer to Supercritical Pressure Water in Smooth and Ribbed Tubes, *J. Heat Transfer*, Vol. 92, pp. 490–498, 1970.

84. J. Fewster, Heat Transfer to Supercritical Pressure Fluids, Ph.D. Thesis, Univ. of Manchester, Manchester, U.K., 1975.

85. C. K. Brown and W. H. Gauvin, Combined Free-and-Forced Convection, *Can. J. Chem. Eng.*, Vol. 43, pp. 306–318, 1965.

86. W. H. McAdams, *Heat Transmission*, McGraw-Hill, New York, p. 172, 1954.

87. R. K. Shah and A. L. London, *Laminar Flow Forced Convection in Ducts*, Academic, New York, 1978.

16

CONVECTIVE HEAT TRANSFER IN POROUS MEDIA

Adrian Bejan

Duke University
Durham, North Carolina

16.1 INTRODUCTION

The subject of heat and mass transfer by convection through fluid saturated porous media represents an important development and an area of rapid growth in contemporary heat transfer research. Although the mechanics of flows in porous media has preoccupied engineers and scientists for more than a century, the phenomenon of convection heat transfer has achieved the status of a separate field of research only during the last two decades. For an introduction to the fluid mechanics of flows through porous media, the reader is directed to fluid mechanics monographs such as Muskat [1], Bear [2], Scheidegger [3], and Greenkorn [4]. Heat transfer reviews and monographs have so far been written only in conjunction with certain research aspects of convection through porous media, for example, applications to geothermal-reservoir engineering (Cheng [5, 6], O'Sullivan [7], McKibbin [8] and to thermal-insulation engineering (Bejan [9]); a review of the published work on natural convection from 1977 to 1984 is given by Nield [10]. The subject of forced and natural convection through porous media appeared for the first time in a heat-transfer textbook in 1984 (Bejan [11]). The objective of the present chapter is to provide the reader with an understanding of the breadth of research on convection in porous media, and with a collection of useful engineering results that covers the main developments in this field.

16.2 FUNDAMENTAL PRINCIPLES

The discussion of convection through a porous medium saturated with fluid (liquid or gas) is based on a series of special concepts that are not found in the pure-fluid convection chapters of this handbook. Examples of such concepts are the porosity and the permeability of the porous medium, and the volume-averaged properties of the fluid flowing through the porous medium. The object of this section is to define these special concepts of convection through porous media and to list the conservation laws that govern the convection phenomenon.

The *porosity* of the porous medium is defined as

$$\phi = \frac{\text{void volume contained in porous medium sample}}{\text{total volume of porous medium sample}} \tag{16.1}$$

The engineering heat transfer results assembled in this chapter refer primarily to fluid-saturated porous media that can be modeled as nondeformable, homogeneous, and isotropic. In such media, the volumetric porosity ϕ is the same as the area ratio (void area contained in sample cross section)/(total area of sample cross section).

The phenomenon of convection through the porous medium is described in terms of *volume-averaged* quantities such as temperature, pressure, concentration, and velocity components. Each volume-averaged quantity $\langle \psi \rangle$ is defined through the operation

$$\langle \psi \rangle = \frac{1}{V} \iiint_V \psi \, dV \tag{16.2}$$

where ψ is the actual value of the quantity at a point inside the sample volume V. Alternatively, the volume-average quantity equals the value of that quantity averaged over the total volume occupied by the porous medium.

16.2.1 Mass Conservation

The principle of mass conservation or mass continuity applied locally in a small region of the fluid saturated porous medium is

$$\frac{D\rho}{Dt} + \rho\nabla \cdot \boldsymbol{v} = 0 \tag{16.3}$$

where D/Dt is the material derivative operator

$$\frac{D}{Dt} = \frac{\partial}{\partial t} + u\frac{\partial}{\partial x} + v\frac{\partial}{\partial y} + w\frac{\partial}{\partial z} \tag{16.4}$$

and where $\boldsymbol{v}$ (u, v, w) is the volume-averaged velocity vector (Fig. 16.1*a*). For example, the volume-averaged velocity component u in the x direction is equal to ϕu_p, where u_p is the average velocity through the pores. In many single-phase flows through porous media, the density variations are small enough so that the $D\rho/Dt$ term may be neglected in Eq. (16.3). The incompressible flow model has been invoked in the development of the majority of the analytical and numerical results reviewed in this chapter. (Note: the incompressible flow model should not be confused with the incompressible-substance model encountered in thermodynamics [12].)

16.2.2 Momentum Conservation (the Darcy Flow Model and More General Models)

The most frequently used model for volume-averaged flow through a porous medium is the Darcy flow model [13]. According to this model, the volume-averaged velocity in a certain direction is directly proportional to the net pressure gradient in that direction, e.g.

$$u = \frac{K}{\mu}\left(-\frac{\partial P}{\partial x}\right) \tag{16.5}$$

In three dimensions, and in the presence of a body acceleration vector $\boldsymbol{g} = (g_x, g_y, g_z)$ (Fig. 16.1*a*), the Darcy flow model is

$$\boldsymbol{v} = \frac{K}{\mu}(-\nabla P + \rho \boldsymbol{g}). \tag{16.6}$$

The proportionality factor K in Darcy's model is the *permeability* of the porous medium. The units of K are m^2 or ft^2. In general, the permeability is an empirical constant that can be determined by measuring the pressure drop and the flow rate through a column-shaped sample of porous material, as suggested by Eq. (16.5). The permeability can also be estimated from simplified models of the labyrinth formed by the interconnected pores. Modeling the pores as a bundle of parallel capillary tubes of radius r_0 yields (Bear [2])

$$K = \frac{\pi r_0^4}{8}\frac{\tilde{N}}{A} \tag{16.7}$$

where $\tilde{N}$ is the number of tubes counted on a cross section of area A. Modeling the pores as a stack of parellel capillary fissures of width b and fissure-to-fissure spacing

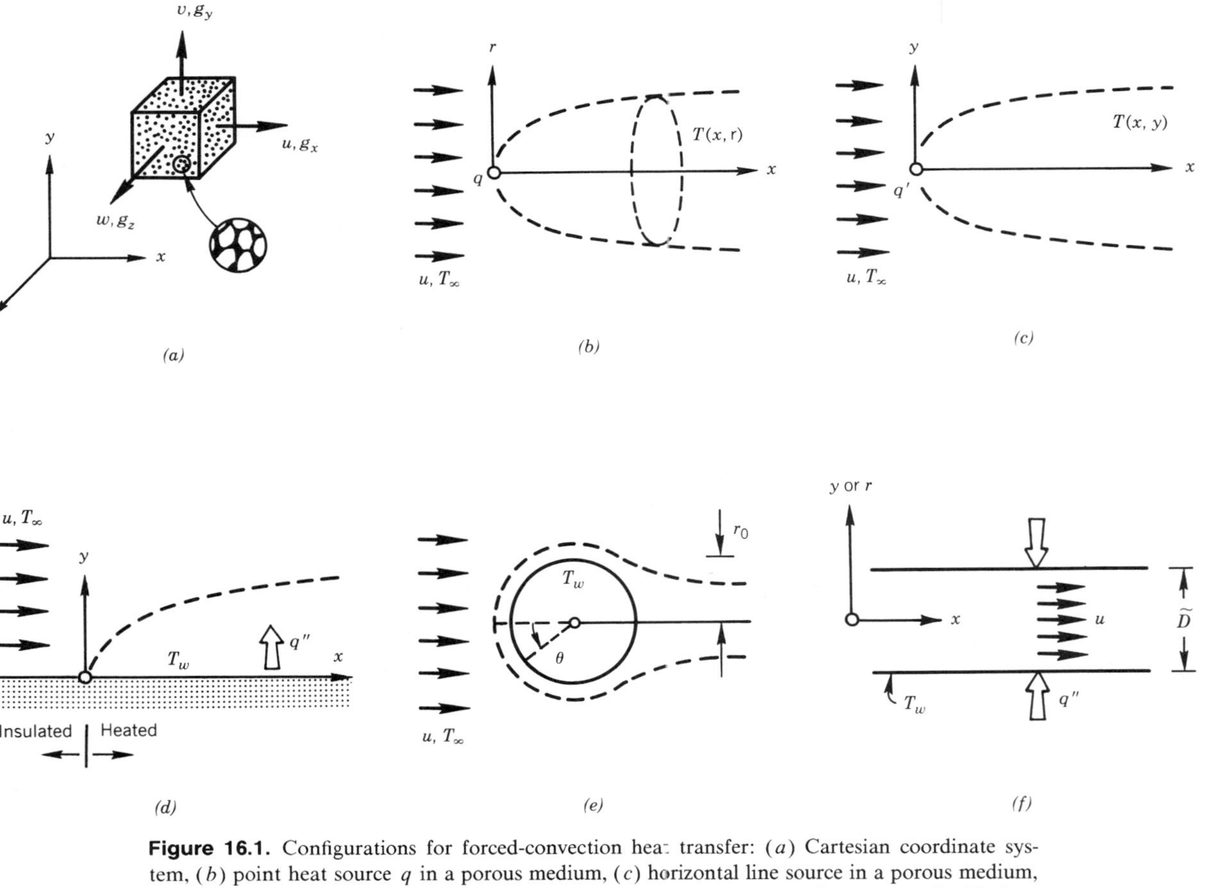

Figure 16.1. Configurations for forced-convection heat transfer: (a) Cartesian coordinate system, (b) point heat source q in a porous medium, (c) horizontal line source in a porous medium, (d) boundary-layer development over a flat surface in a porous medium, (e) cylindrical or spherical porous medium, (f) long duct filled with a porous medium.

$\tilde{a} + \tilde{b}$ yields the permeability formula (Bear [2])

$$K = \frac{\tilde{b}^3}{12(\tilde{a} + \tilde{b})} \tag{16.8}$$

Modeling the porous medium as a collection of solid spheres of diameter d, Kozeny [14] obtained the formula

$$K \sim \frac{d^2 \phi^3}{(1 - \phi)^2} \tag{16.9}$$

The Darcy flow model, Eq. (16.6), is valid in circumstances where the so-called "local pore Reynolds number" based on the local volume-averaged speed $(u^2 + v^2 + w^2)^{1/2}$ and $K^{1/2}$ is smaller than one (Ward [15]). At pore Reynolds numbers of order 1 and greater, the measured relationship between pressure gradient and volume-averaged velocity is correlated by Forschheimer's [16] modification of Darcy's model (16.5):

$$-\frac{\partial P}{\partial x} = \frac{\mu}{K} u + b \rho u^2 \tag{16.10}$$

In three dimensions and in the presence of body acceleration, the Forschheimer modification of the Darcy flow model is

$$\mathbf{v} + \frac{b \rho K}{\mu} |\mathbf{v}| \mathbf{v} = \frac{K}{\mu}(-\nabla P + \rho \mathbf{g}) \tag{16.11}$$

The experimental measurements published by Ward [15] suggest that as the local pore Reynolds number exceeds ~ 10, Forschheimer's constant b approaches asymptotically the value

$$b = 0.55 K^{-1/2} \tag{16.12}$$

Extensive measurements involving gas flow through columns of packed spheres, sand, and pulverized coal allowed Ergun [17] to propose the following correlations for K and b as

$$K = \frac{d^2 \phi^3}{150(1 - \phi)^2} \tag{16.13}$$

$$b = \frac{1.75(1 - \phi)}{\phi^3 d} \tag{16.14}$$

Another modification of the Darcy flow model was introduced by Brinkman [18] in order to account for the transition from Darcy flow to highly viscous flow (without porous matrix), in the limit of extremely high permeability:

$$\mathbf{v} = \frac{K}{\mu}(-\nabla \rho + \rho \mathbf{g}) + K \nabla^2 \mathbf{v} \tag{16.15}$$

The two modifications to the Darcy flow model discussed above, the Forschheimer model (16.11) and the Brinkman model (16.15), were used simultaneously by Vafai and Tien [19] in a study of forced-convection boundary-layer heat transfer. In the presence of gravitational acceleration, Vafai and Tien's momentum equations would read

$$\boldsymbol{v} + \frac{b\rho K}{\mu}|\boldsymbol{v}|\boldsymbol{v} = \frac{K}{\mu}(-\nabla P + \rho \boldsymbol{g}) + K\nabla^2 \boldsymbol{v} \tag{16.16}$$

It should be emphasized that none of the above models account adequately for the transition from porous medium flow to pure fluid flow as the permeability K increases. Note that in the high-K limit the terms that survive in Eq. (16.15) or Eq. (16.16) account for momentum conservation only in highly viscous flows in which the effect of fluid inertia is negligible relative to pressure and friction forces. A model that bridges the entire gap between the Darcy-Forschheimer model and the Navier-Stokes equations was developed by Vafai and Tien (for details see Ref. [19]):

$$\frac{\nu}{K}\boldsymbol{v} + b|\boldsymbol{v}|\boldsymbol{v} = -\frac{D\boldsymbol{v}}{Dt} - \frac{1}{\rho}\nabla P + \nu\nabla^2 \boldsymbol{v} + \boldsymbol{g} \tag{16.17}$$

As the permeability K increases, the left-hand side vanishes and gives way to the complete vectorial Navier-Stokes equation for Newtonian constant-property flow.

The vast majority of the heat transfer engineering results available in the literature and highlighted in this chapter are based on the original Darcy flow model, Eq. (16.6). Some of the modifications to Darcy flow outlined above have been used only sporadically; whenever available, the effect of the departure from Darcy flow conditions is discussed in conjunction with the engineering results of this chapter.

16.2.3 Energy Conservation

The first law of thermodynamics applied to a point in a homogeneous porous medium saturated with single-phase fluid yields (see, for example, Bejan [11, pp. 351–354]):

$$(\rho c_p)_f \left(\sigma \frac{\partial T}{\partial t} + \boldsymbol{v} \cdot \nabla T \right) = k \nabla^2 T + q''' + \frac{\mu}{K}(\boldsymbol{v})^2 \tag{16.18}$$

where σ is the heat-capacity ratio,

$$\sigma = \phi + (1 - \phi)\frac{(\rho c)_s}{(\rho c_p)_f} \tag{16.19}$$

and where $(\rho c)_s$ and $(\rho c_p)_f$ denote the heat capacity per unit volume of solid matrix material and fluid, respectively. In the derivation of Eq. (16.18), it was assumed that locally the fluid and the solid components are in thermal equilibrium. The first term on the right-hand side represents the effect of thermal diffusion, and k is the effective thermal conductivity of the porous medium with the fluid in it. The second term on the right-hand side, q''' (W/m^3), represents the effect of internal heat generation per unit volume of porous medium (solid plus fluid). The last term in Eq. (16.18) represents the internal heating rate per unit volume due to fluid friction or the extrusion of the fluid through the pores. This last term is based on the assumption that the Darcy flow model (16.6) is applicable.

Most of the engineering results assembled in this chapter refer to saturated porous media without internal heat generation and without internal heating due to fluid friction. The energy conservation statement for these cases is

$$\sigma \frac{\partial T}{\partial t} + \boldsymbol{v} \cdot \nabla T = \alpha \nabla^2 T \tag{16.20}$$

where α is the effective thermal diffusivity of the saturated porous medium. Note that α is calculated by dividing the effective thermal conductivity of the porous medium (with the fluid in it) by the heat capacity of the fluid phase alone.

16.2.4 Conservation of Chemical Species

In situations where the fluid that saturates the porous structure is a mixture of two or more chemical species, the equation that expresses the conservation of species i is (Bejan [11, pp. 332, 333])

$$\phi \frac{\partial C}{\partial t} + \boldsymbol{v} \cdot \nabla C = D \nabla^2 C + \dot{m}_i''' \tag{16.21}$$

In this equation C is the concentration of i, expressed in kilograms of i per unit volume of porous medium; D is the mass diffusivity of i through the porous medium with the fluid mixture in it, and $\dot{m}_i'''$ is the number of kilograms of i produced by a chemical reaction per unit time and per unit volume of porous medium. The porosity factor ϕ appearing in the first term of Eq. (16.21) is sometimes omitted in the literature (see, for example, Cheng [5, p. 26]).

16.3 FORCED CONVECTION

Most of the published work on heat transfer through porous media refers to natural or free convection. However, in an engineering overview such as the present chapter it is important to include, and in some cases derive, a number of pivotal results for forced convection. The heat transfer results listed next refer to a uniform unidirectional seepage flow u through a homogeneous and isotropic porous medium, as shown in the five geometric configurations sketched in Fig. 16.1*b*–*f*. The results are based on the idealization that the solid and fluid phases are locally in thermal equilibrium; this idealization breaks down in many chemical engineering applications of forced convection through porous beds, and in the functioning of periodic (regenerative) heat exchanges with porous matrices.

16.3.1 Point Source

The temperature field $T(x, r)$ downstream from a point heat source of strength q buried in a fluid saturated porous medium is (Bejan [11, pp. 301–303])

$$T - T_\infty = \frac{q}{4\pi k x} \exp\left(-\frac{u r^2}{4 \alpha x} \right) \tag{16.22}$$

This result (Fig. 16.1*b*) is valid in the limit where convection overwhelms diffusion as a longitudinal heat transfer mechanism, i.e., where $ux/\alpha \gg 1$.

16.3.2 Horizontal Line Source

The two-dimensional temperature field $T(x, y)$ (Fig. 16.1c) generated by a line heat source of steady strength q' is (Bejan [11, pp. 301–303])

$$T - T_\infty = \frac{q'}{(\rho c_p)_f (4\pi u \alpha x)^{1/2}} \exp\left(-\frac{uy^2}{4\alpha x}\right) \tag{16.23}$$

Equation (16.23) holds in the convection-dominated regime, $ux/\alpha \gg 1$. When this criterion fails, the T field is governed by pure diffusion and its analytical form may be derived by classical methods (e.g. Grigull and Sandner [20, pp. 125–129]).

16.3.3 Boundary Layers

Consider the uniform flow (u, T_∞) parallel to a solid wall heated to a constant temperature T_0 (Fig. 16.1d). The local Nusselt number is (Bejan [11, p. 358])

$$\mathrm{Nu}_x = \frac{q''(x)x}{(T_w - T_\infty)k} = 0.564\left(\frac{ux}{\alpha}\right)^{1/2} \tag{16.24}$$

where it is worth noting that the heat flux $q''(x)$ decreases as $x^{-1/2}$. In the case where the wall downstream from $x = 0$ is heated with uniform flux q'', the local Nusselt number is (Cheng [21]; also Bejan [11, p. 358])

$$\mathrm{Nu}_x = \frac{q''x}{[T_w(x) - T_\infty]k} = 0.886\left(\frac{ux}{\alpha}\right)^{1/2} \tag{16.25}$$

Equations (16.24) and (16.25) hold if $ux/\alpha \gg 1$, i.e., downstream from $x = 0$ so that the longitudinal heat transfer is dominated by convection. Mass transfer counterparts to the results of Eqs. (16.24) and (16.25) are obtained through the notation change $\mathrm{Nu}_x \to \mathrm{Sh}_x$, $q'' \to j''$, $T \to C$, $k \to D$, $\alpha \to D$, where Sh_x and j'' are the local Sherwood number and local mass flux. From a fluid mechanics standpoint, the results of Eqs. (16.24) and (16.25) are consistent with the Darcy-Forschheimer flow model of Eq. (16.11) for a homogeneous and isotropic porous medium. The special effect of the flow resistance provided by the solid wall is illustrated through numerical examples by Vafai and Tien [19].

16.3.4 Sphere and Cylinder

The boundary-layer heat transfer regime for a sphere or a cylinder of radius r_0 cooled or heated by forced convection in Darcy flow was documented by Cheng [22]. With reference to the angular coordinate θ drawn in the cross section shown in Fig. 16.1e, the peripheral local Nusselt numbers for a sphere and a cylinder are, in order,

$$\mathrm{Nu}_\theta = 0.564\left(\frac{ur_0\theta}{\alpha}\right)^{1/2} \left(\tfrac{3}{2}\theta\right)^{1/2} \sin^2\theta \left(\tfrac{1}{3}\cos^3\theta - \cos\theta + \tfrac{2}{3}\right)^{-1/2} \tag{16.26}$$

$$\mathrm{Nu}_\theta = 0.564\left(\frac{ur_0\theta}{\alpha}\right)^{1/2} (2\theta)^{1/2} \sin\theta\,(1 - \cos\theta)^{-1/2} \tag{16.27}$$

The peripheral local Nusselt number is defined as $\mathrm{Nu}_\theta = q'' r_0 \theta / k(T_w - T_\infty)$. The results (16.26) and (16.27) are valid in the boundary-layer regime, i.e., when $\mathrm{Nu}_\theta \gg 1$ (this criterion is the same as requiring that the boundary-layer length $r_0\theta$ be much greater than the thermal boundary-layer thickness).

16.3.5 Confined Flow

Consider a seepage flow of volume-averaged velocity u through a porous medium that fills a long duct whose wall is heated to a constant temperature T_0 (Fig. 16.1f). Since the velocity profile corresponds to the slug (uniform) flow in a porous medium, the fully developed heat transfer regime with slug flow in a tube of diameter $\tilde{D}$ is (Rohsenow and Choi [23])

$$\mathrm{Nu} = \frac{q''\tilde{D}}{(T_w - T_m)k} = 5.78 \tag{16.28}$$

The temperature T_m is the bulk mean temperature of the fluid in a cross section through the confined porous medium; $T_m = A^{-1}\iint_A T\,dA$ for slug flow. If the heating arrangement along the solid walls that confine the porous medium can be modeled as one of uniform heat flux, the fully developed Nusselt number is Nu = 8 in a circular duct of diameter $\tilde{D}$. The corresponding Nusselt numbers for the fully developed regime in a porous medium sandwiched between two parallel plates with plate-to-plate spacing $\tilde{D}$ are (Hwang and Fan [24])

$$\mathrm{Nu} = \begin{cases} 5.0 \text{ for constant wall temperature} \\ 6.0 \text{ for constant wall heat flux} \end{cases} \tag{16.29}$$

where the Nusselt number Nu is defined as in Eq. (16.28), i.e., it is based on $\tilde{D}$. The forced-convection results discussed in this section are valid at longitudinal locations situated sufficiently far from the entrance to the duct filled with porous medium. Since the thermal boundary-layer thickness in the entrance region scales as $(\alpha x/u)^{1/2}$ (Bejan [11, p. 357]), the criterion for the validity of the fully developed results (16.26) to (16.29) is $(\alpha x/u)^{1/2} \gg \tilde{D}$.

16.4 NATURAL CONVECTION

In addition to the features of the porous-medium model discussed already, most of the engineering results available for buoyancy-driven flows are based on the idealization that the Boussinesq approximation holds,

$$\rho \approx \rho_0[1 - \beta(T - T_0)] \tag{16.30a}$$

where β is the thermal expansion coefficient

$$\beta = -\frac{1}{\rho}\left(\frac{\partial \rho}{\partial T}\right)_P \tag{16.30b}$$

The main geometrical configurations into which Sec. 16.4 is divided are sketched in Figs. 16.2, 16.3, and 16.5. Note that the gravitational acceleration points in the negative y direction, in other words, the body acceleration vector $\mathbf{g}$ appearing in Eq. (16.6) and Fig. 16.1a has $(0, -g, 0)$ as components.

16.4.1 Point Source

The convection generated in a porous medium by a concentrated heat source has been studied in two limits: first, the *low Rayleigh number* regime where the temperature distribution is primarily due to thermal diffusion, and second, the *high Rayleigh number* regime where the flow driven by the source is a slender vertical plume (this second regime may be called the "boundary-layer regime"). Starting with the low Rayleigh number regime, the transient flow and temperature fields around a constant-strength heat source q that starts to generate heat at $t = 0$ is (Bejan [25])

$$\psi = \alpha K^{1/2}\mathrm{Ra}_q \frac{\tau^{1/2}}{8\pi}\sin^2\phi\left(2\eta\,\mathrm{erfc}\,\eta + \frac{1}{\eta}\,\mathrm{erf}\,\eta - \frac{2}{\pi^{1/2}}\exp(-\eta^2)\right) \tag{16.31}$$

$$\begin{aligned} T - T_\infty = \frac{q}{kK^{1/2}}\Bigg[&\frac{1}{4\pi R}\mathrm{erfc}\left(\frac{R}{2\tau^{1/2}}\right) + \mathrm{Ra}_q \frac{\cos\phi}{64\pi^2\tau^{1/2}} \\ &\times\left(\frac{1}{\eta} - \frac{4}{3\pi^{1/2}} + \frac{6}{5\pi^{1/2}}\eta^2 - \frac{16}{45\pi}\eta^3 - \frac{152}{315\pi^{1/2}}\eta^4\right. \\ &+ \frac{64}{315\pi}\eta^5 + \frac{517}{3780\pi^{1/2}}\eta^6 - \frac{992}{14175\pi}\eta^7 \\ &\left. - \frac{2039}{69300\pi^{1/2}}\eta^8 + \frac{2591}{155929\pi}\eta^9 + \cdots\right) \end{aligned} \tag{16.32}$$

The notation used in Eqs. (16.30) and (16.31) is

$$\eta = \frac{R}{2\tau^{1/2}}, \qquad R = \frac{r}{K^{1/2}}, \qquad \tau = \frac{\alpha t}{\sigma K} \tag{16.33a}$$

$$\mathrm{Ra}_q = \frac{Kg\beta q}{\alpha\nu k} = \text{Rayleigh number based on point-source strength} \tag{16.33b}$$

The stream function ψ is defined in the axisymmetric spherical coordinates of Fig. 16.2*a* via

$$v_r = \frac{1}{r^2\sin\phi}\frac{\partial\psi}{\partial\phi}, \qquad v_\phi = -\frac{1}{r\sin\phi}\frac{\partial\psi}{\partial r} \tag{16.33c}$$

The transient solution in Eqs. (16.31)–(16.33) is valid in the range $0 < \eta < 1$.

The steady-state flow and temperature fields around a constant point source q in the low Rayleigh number regime is (Bejan [25])

$$\psi = \frac{\alpha r}{8\pi}\left[\mathrm{Ra}_q\sin^2\phi + \frac{\mathrm{Ra}_q^2}{24\pi}\sin\phi\,\sin 2\phi - \frac{5\mathrm{Ra}_q^3}{18{,}432\pi^2}(8\cos^4\phi - 3) + \cdots\right] \tag{16.34}$$

$$T - T_\infty = \frac{q}{4\pi kr}\left[1 + \frac{\mathrm{Ra}_q}{8\pi}\cos\phi + \frac{5Ra_q^2}{768\pi^2}\cos 2\phi + \frac{\mathrm{Ra}_q^3}{55{,}296\pi^3}\cos\phi\,(47\cos^2\phi - 30) - \cdots\right] \tag{16.35}$$

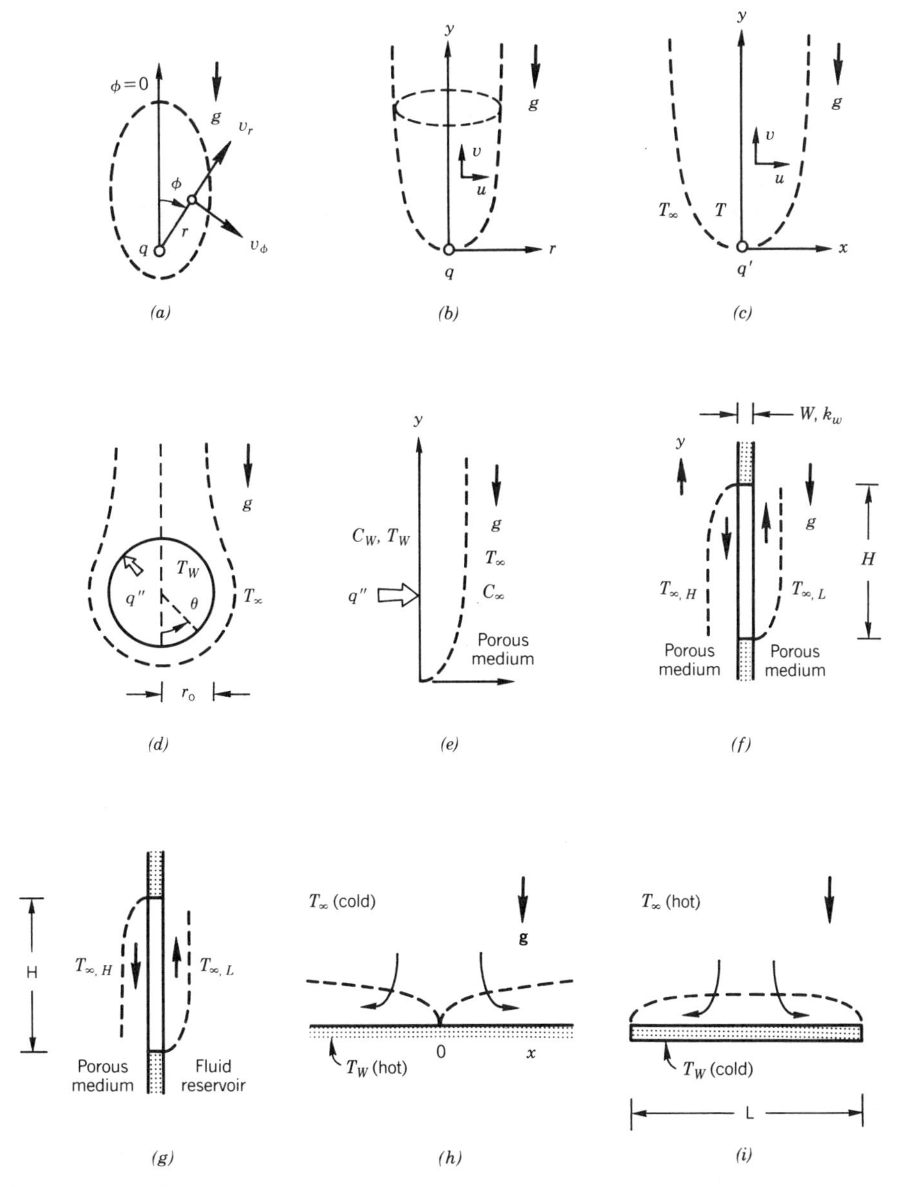

Figure 16.2. Configurations for natural-convection heat transfer in external flow: (a) point heat source, low Rayleigh number regime, (b) point heat source, high Rayleigh number regime, (c) horizontal line source, (d) impermeable sphere or horizontal cylinder imbedded in a porous medium, (e) impermeable vertical wall, (f) vertical partition imbedded in a porous medium, (g) vertical wall separating a porous medium and a fluid reservoir, (h) hot surface facing upward in a porous medium, (i) cold surface facing upward in a porous medium.

This steady-state Ra_q series solution is sufficiently accurate if Ra_q is of the order of 20 or less. Hickox and Watts [26] obtained numerical solutions for the point-source steady-state problem at Rayleigh numbers in the range 10^{-1} to 10^2. The same authors reported analytical results for the limit $\mathrm{Ra}_q \to 0$ and numerical results in the Ra_q range 10^{-1} to 10^2 for steady-state flow near a point source located at the base of a semi-infinite porous medium bounded from below by an impermeable and insulated surface. In a subsequent paper, Hickox [27] shows that the transient and steady-state solutions for the point source in the strict $\mathrm{Ra} \to 0$ limit can be superimposed in order to predict the flow and temperature fields around buried objects of more complicated geometries.

In the high Rayleigh number regime, the point source generates a vertical plume flow whose thermal boundary-layer thickness scales as $y\,\mathrm{Ra}_q^{-1/2}$, where y is the vertical position along the plume axis (Fig. 16.2b). The analytical solution for the flow and temperature field is constructed by Bejan [11, pp. 376–380], after a suggestion by Wooding [28], as

$$\frac{T - T_\infty}{q/(ky)} = \frac{v}{(\alpha/y)\mathrm{Ra}_q} = \frac{2C_1^2}{1 + (C_1\eta/2)^2} \tag{16.36}$$

where $\eta = (r/y)\mathrm{Ra}_q^{1/2}$ and $C_1 = 0.141$. The solution holds provided the plume region is slender, i.e., when $\mathrm{Ra}_q^{1/2} \gg 1$.

16.4.2 Horizontal Line Source

In the two-dimensional frame of Fig. 16.2c, the temperature and flow fields generated in a porous medium at high Rayleigh numbers by a horizontal line source of strength q' (W/m) are

$$\frac{T - T_\infty}{(q'/k)\mathrm{Ra}_{q'}^{-1/3}} = \frac{v}{(\alpha/y)\mathrm{Ra}_{q'}^{2/3}} = \frac{C_2^2/6}{\cosh^2(C_2\eta/6)} \tag{16.37}$$

where $C_2 = 1.651$, $\eta = (x/y)\mathrm{Ra}_{q'}^{1/3}$, and $\mathrm{Ra}_{q'} = Kg\beta y q'/(\alpha\nu k)$ is the Rayleigh number based on line source strength. The derivation of this solution can be found in Wooding [28], in Bejan [29, pp. 206–208], and as a special case of vertical boundary-layer convection in Cheng and Minkowycz [30]. The boundary-layer solution in Eq. (16.37) is valid at sufficiently high Rayleigh numbers, $\mathrm{Ra}_{q'}^{1/3} \gg 1$. The low Rayleigh number regime for convection near a horizontal line source (in an infinite medium or near a vertical insulated and impermeable surface) is described by Nield and White [31].

16.4.3 Sphere and Horizontal Cylinder

With reference to the coordinate system shown in the circular cross section sketched in Fig. 16.2d, the local Nusselt numbers for boundary-layer convection around a impermeable sphere or a horizontal cylinder imbedded in an infinite porous medium are, in order,

$$\mathrm{Nu}_\theta = 0.444\,\mathrm{Ra}_\theta^{1/2}\left(\tfrac{3}{2}\theta\right)^{1/2}\sin^2\theta\left(\tfrac{1}{3}\cos^3\theta - \cos\theta + \tfrac{2}{3}\right)^{-1/2} \tag{16.38}$$

$$\mathrm{Nu}_\theta = 0.444\,\mathrm{Ra}_\theta^{1/2}(2\theta)^{1/2}\sin\theta\,(1 - \cos\theta)^{-1/2} \tag{16.39}$$

where $\text{Nu}_\theta = q''r_0\theta/k(T_w - T_\infty)$ and $\text{Ra}_\theta = Kg\beta\theta r_0(T_w - T_\infty)/(\alpha\nu)$. These steady-state results have been reported by Cheng [22]; they are valid provided the boundary-layer region is slender enough, i.e., if $\text{Nu}_\theta \gg 1$. The transient flow and heat transfer near a horizontal cylinder are described by Ingham et al. [32].

16.4.4 Vertical Walls

The heat transfer between a fluid-saturated reservoir and a vertical impermeable wall of different temperature is effected by boundary-layer flow if the wall-reservoir temperature difference is large enough. Similarity solutions for the single-wall natural-convection boundary-layer problem have been reported independently by Avduyevskiy et al. [33] and by Cheng and Minkowycz [30]. The local Nusselt number for a vertical isothermal wall (Fig. 16.2*e*) is

$$\text{Nu}_y = 0.444\,\text{Ra}_y^{1/2} \tag{16.40}$$

where $\text{Nu}_y = q''y/(T_w - T_\infty)k$ and $\text{Ra}_y = Kg\beta y(T_w - T_\infty)/(\alpha\nu)$. Equation (16.40) is valid in the boundary-layer regime, $\text{Ra}_y^{1/2} \gg 1$. In cases where the vertical wall can be modeled as being heated with uniform heat flux, the local Nusselt number reported by Cheng and Minkowycz [30] can be written as (Bejan [11, p. 363])

$$\text{Nu}_y = 0.772\,\text{Ra}_y^{*1/3} \tag{16.41}$$

where $\text{Ra}_y^* = Kg\beta y^2q''/(\alpha\nu k)$. Equation (16.41) holds in the boundary-layer regime, $\text{Ra}_y^{*1/3} \gg 1$. The effect of linear thermal stratification (dT_∞/dy = constant) in the single-wall configuration of Fig. 16*e* is shown using integral analysis by Bejan [11, pp. 367–371]. The transient development of the boundary layer near a heated vertical wall has been considered by Cheng and Pop [34].

The heat transfer through a vertical impermeable wall that divides a porous medium into two semi-infinite reservoirs at different temperatures (Fig. 16.2*f*) was studied numerically by Bejan and Anderson [35]. The overall Nusselt number results for this configuration are correlated within 1% by the expression

$$\text{Nu} = 0.382(1 + 0.615\omega)^{-0.875}\text{Ra}_H^{1/2} \tag{16.42}$$

where $\text{Nu} = q''_{\text{avg}}H/(T_{\infty,H} - T_{\infty,L})k$, and where q''_{avg} is the heat flux averaged over the entire height H. In addition, $\text{Ra}_H = Kg\beta H(T_{\infty,H} - T_{\infty,L})/(\alpha\nu)$; the wall thickness parameter ω is defined as $\omega = (Wk/Hk_w)\text{Ra}_H^{1/2}$, where k_w is the thermal conductivity of wall material.

In thermal insulation and architectural applications, the porous media on both sides of the vertical partition of Fig. 16.2*f* may be thermally stratified. If the stratification on both sides is the same and linear, so that the vertical temperature gradient far enough from the wall is $dT/dy = b_1(T_{\infty,H} - T_{\infty,L})/H$, where b_1 is a constant, and if the partition is thin enough so that $\omega = 0$, then it is found that the overall Nusselt number increases substantially with the degree of stratification (Bejan and Anderson [35]): in the range $0 < b_1 < 1.5$, these findings can be summarized as

$$\text{Nu} = 0.382\left(1 + 0.662b_1 - 0.073b_1^2\right)\text{Ra}_H^{1/2} \tag{16.43}$$

Another configuration of engineering interest is sketched in Fig. 16.2*g*: a vertical impermeable surface separates a porous medium of temperature $T_{\infty,H}$ from a fluid reservoir of temperature $T_{\infty,L}$. When both sides of the interface are lined by boundary layers, the overall Nusselt number may be estimated as (Bejan and Anderson [36])

$$\mathrm{Nu} = \left[(0.638)^{-1} + (0.888B)^{-1}\right]^{-1}\mathrm{Ra}_{H,f}^{1/4} \tag{16.44}$$

where $\mathrm{Nu} = q''_{\mathrm{avg}}H/(T_{\infty,H} - T_{\infty,L})k$ and $B = k\,\mathrm{Ra}_H^{1/2}/(k_f \mathrm{Ra}_{H,f}^{1/4})$. The parameter k_f is the fluid-side thermal conductivity, and the fluid-side Rayleigh number $\mathrm{Ra}_{H,f} = g(\beta/\alpha\nu)_f H^3(T_{\infty,H} - T_{\infty,L})$ (see Chap. 20 in this handbook). Equation (16.44) is valid in the regime where both boundary layers are distinct, $\mathrm{Ra}_H^{1/2} \gg 1$ and $\mathrm{Ra}_{H,f}^{1/4} \gg 1$; it is also assumed that the fluid on the right side of the partition in Fig. 16.2g has a Prandtl number of order 1 or greater.

Referring again to the single-wall geometry of Fig. 16.2e, the corresponding problem in a porous medium saturated with water near the temperature of maximum density was solved by Ramilison and Gebhart [37]. In place of the Boussinesq model of Eq. (16.30), Ramilison and Gebhart used $\rho = \rho_m\,[1 - \alpha_m|T - T_m|^q]$, where ρ_m and T_m are the maximum density and the temperature of the state of maximum density. The parameters ρ_m, T_m, q, and α_m (not to be confused with the thermal diffusivity α) depend on the pressure and salinity, and are reported by Gebhart and Mollendorf [38]. Data on the local Nusselt number that correspond to Eq. (16.40) are reported by Ramilison and Gebhart [37] in graphical form; a closed-form analytical substitute for this graphical information is found by the present author for pure water at atmospheric pressure as,

$$\mathrm{Nu}_y \approx 0.42\left|0.35 - \frac{T_m - T_\infty}{T_w - T_\infty}\right|^{0.46}\left(\frac{2\alpha_m Kgy(T_w - T_\infty)}{\alpha\nu}\right)^{1/2} \tag{16.45}$$

Equation (16.45) is accurate within 1% in the range $-16 < (T_m - T_\infty)/(T_w - T_\infty) < 5$.

A geometric configuration that is related to that of Fig. 16.2e is the flow along the outer surface of a vertical cylinder of radius r_0 imbedded in a porous medium. The heat transfer rate to Darcy boundary-layer flow in this configuration was reported by Minkowycz and Cheng [39] for a variety of power-law distributions of wall temperature along the vertical. In the case of an isothermal cylinder (T_w), the expression for the local Nusselt number is

$$\mathrm{Nu}_y = 0.444\,\mathrm{Ra}_y^{1/2}\left(1 + 0.6\frac{y}{r_0}\mathrm{Ra}_y^{-1/2}\right) \tag{16.46}$$

The breakdown of the Darcy flow model in vertical boundary-layer natural convection is the subject of a number of recent studies (Plumb and Huenefeld [40], Bejan and Poulikakos [41], Nield and Joseph [42]). Assuming the Forschheimer modification of the Darcy flow model, Eq. (16.9), at local pore Reynolds numbers greater than ~ 10 the local Nusselt number for the vertical wall configuration of Fig. 16.2e approaches the following limits (Bejan and Poulikakos [41]):

$$\mathrm{Nu}_y = 0.494\,\mathrm{Ra}_{\infty,y}^{1/4} \qquad \text{for isothermal wall} \tag{16.47}$$

$$\mathrm{Nu}_y = 0.804\,\mathrm{Ra}_{\infty,y}^{*1/5} \qquad \text{for constant heat flux wall} \tag{16.48}$$

where $\mathrm{Ra}_{\infty,y} = g\beta y^2(T_w - T_\infty)/(b\alpha^2)$ and $\mathrm{Ra}_{\infty,y}^* = g\beta y^3 q''/(kb\alpha^2)$. Equations

(16.47) and (16.48) are valid provided $G \ll 1$, where $G = (\nu/K)[bg\beta(T_w - T_\infty)]^{-1/2}$. In the intermediate range between the Darcy limit and the inertia-dominated limit, i.e., in the range where G is of order one, the numerical results of Bejan and Poulikakos [41] for a vertical isothermal wall are correlated here within 2% by the closed-form expression

$$\mathrm{Nu}_y = \left[(0.494)^n + (0.444G^{-1/2})^n\right]^{1/n}\mathrm{Ra}_{\infty,y}^{1/4} \tag{16.49}$$

where $n = -3$. The form of Eq. (16.49) is based on the general correlation method proposed by Churchill and Usagi [43] (see also Churchill [44]).

The heat transfer results summarized in this section apply also to configurations where the vertical wall is inclined (slightly) to the vertical. In such cases, the gravitational acceleration that appears in the definition of the Rayleigh-type numbers in this section must be replaced by the gravitational acceleration component that acts along the nearly vertical wall. However, in the inclined-wall geometry the boundary-layer flow may be unstable to certain disturbances. This aspect of natural convection is documented by Hsu and Cheng [45–47] and Hsu et al. [48], for both inclined and horizontal walls.

16.4.5 Horizontal Walls

With reference to Fig. 16.2h, the boundary-layer flow in the vicinity of a heated horizontal surface that faces upward was studied by Cheng and Chang [49]. Measuring x horizontally away from the vertical plane of symmetry of the flow, the local Nusselt number for an isothermal wall is

$$\mathrm{Nu}_x = 0.42\,\mathrm{Ra}_x^{1/3} \tag{16.50}$$

where $\mathrm{Nu}_x = q''x/k(T_w - T_\infty)$ and $\mathrm{Ra}_x = Kg\beta x(T_w - T_\infty)/(\alpha\nu)$. The local Nusselt number for a horizontal wall heated with uniform flux is

$$\mathrm{Nu}_x = 0.859\,\mathrm{Ra}_x^{*1/4} \tag{16.51}$$

where $\mathrm{Ra}_x^* = Kg\beta x^2 q''/(k\alpha\nu)$. Equations (16.50) and (16.51) are valid in the boundary-layer regime, $\mathrm{Ra}_x^{1/3} \gg 1$ and $\mathrm{Ra}_x^{*1/4} \gg 1$, respectively. They also apply to porous media bounded from above by a cold surface; this new configuration is obtained by rotating Fig. 16.2h by 180°. The transient heat transfer associated with suddenly changing the temperature of the horizontal wall is documented by Pop and Cheng [50].

The other horizontal wall configuration, the upward-facing cold plate of Fig. 16.2i, was studied by Kimura et al. [51]. The overall Nusselt number in this configuration is

$$\mathrm{Nu} = 1.47\,\mathrm{Ra}_L^{1/3}, \tag{16.52}$$

where $\mathrm{Nu} = q'/k(T_\infty - T_w)$ and $\mathrm{Ra}_L = Kg\beta L(T_\infty - T_w)/(\alpha\nu)$, and where q' is the overall heat transfer rate through the upward-facing cold plate of length L. The result of Eq. (16.52) holds if $\mathrm{Ra}_L \gg 1$, and applies equally to hot horizontal plates facing downward in an isothermal porous medium.

16.4.6 Confined Layers Heated from the Side

The most basic geometric configuration of a porous layer heated in the horizontal direction is sketched in Fig. 16.3a. Provided the Darcy flow model is valid, the character of the heat and fluid flow driven by buoyancy depends on two parameters: the geometric aspect ratio H/L, and the Rayleigh number based on height, $\mathrm{Ra}_H = Kg\beta H(T_h - T_c)/\alpha\nu$. According to the scale analysis reported by Poulikakos and Bejan [52], there exist four heat transfer regimes, i.e., four ways to calculate the overall heat transfer rate $q' = \int_0^H q''\, dy$:

Regime I. The pure conduction regime, defined by $\mathrm{Ra}_H \ll 1$: in this regime q' is approximately equal to the pure conduction estimate $kH(T_h - T_c)/L$.

Regime II. The conduction dominated regime in tall layers, defined by $H/L \gg 1$ and $(L/H)\mathrm{Ra}_H^{1/2} \ll 1$: in this regime the heat transfer rate scales as $q' \gtrsim kH(T_h - T_c)/L$.

Regime III. The convection-dominated regime (or high Rayleigh number regime), defined by $\mathrm{Ra}_H^{-1/2} < H/L < \mathrm{Ra}_H^{1/2}$: in this regime q' scales as $k(T_h - T_c)\mathrm{Ra}_H^{1/2}$.

Regime IV. The convection-dominated regime in shallow layers, defined by $H/L \ll 1$ and $(H/L)\mathrm{Ra}_H^{1/2} \ll 1$: here the heat transfer rate scales as $q \lesssim k(T_h - T_c)\mathrm{Ra}_H^{1/2}$

Considerable analytical, numerical, and experimental work has been done to estimate more accurately the overall heat transfer rate q' or the overall Nusselt number

$$\mathrm{Nu} = \frac{q'}{kH(T_h - T_c)/L} \tag{16.53}$$

Note that unlike the single-wall configurations of Fig. 16.2e to i, in confined layers of thickness L the Nusselt number is defined as the ratio (actual heat transfer rate)/(pure conduction heat transfer rate). An analytical solution that covers smoothly the four heat transfer regimes was reported in parametric form by Bejan and Tien [53]:

$$\mathrm{Nu} = K_1 + \tfrac{1}{120}K_1^3\left(\mathrm{Ra}_H\frac{H}{L}\right)^2 \tag{16.54}$$

where $K_1(H/L, \mathrm{Ra}_H)$ is obtained by solving the system

$$\tfrac{1}{120}\delta_e \mathrm{Ra}_H^2 K_1^3\left(\frac{H}{L}\right)^3 = 1 - K_1 = \tfrac{1}{2}K_1\frac{H}{L}\left(\frac{1}{\delta_e} - \delta_e\right) \tag{16.55}$$

This result is displayed in chart form in Fig. 16.4 along with numerical results reported by Hickox and Gartling [54]. Two asymptotic values of this solution are of interest:

$$\mathrm{Nu} \sim 0.508\frac{L}{H}\mathrm{Ra}_H^{1/2} \qquad \text{as} \quad \mathrm{Ra}_H \to \infty \tag{16.56}$$

$$\mathrm{Nu} \sim 1 + \frac{1}{120}\left(\mathrm{Ra}_H\frac{H}{L}\right)^2 \qquad \text{as} \quad \frac{H}{L} \to 0 \tag{16.57}$$

The heat transfer in the convection-dominated regime III is represented well by Eq.

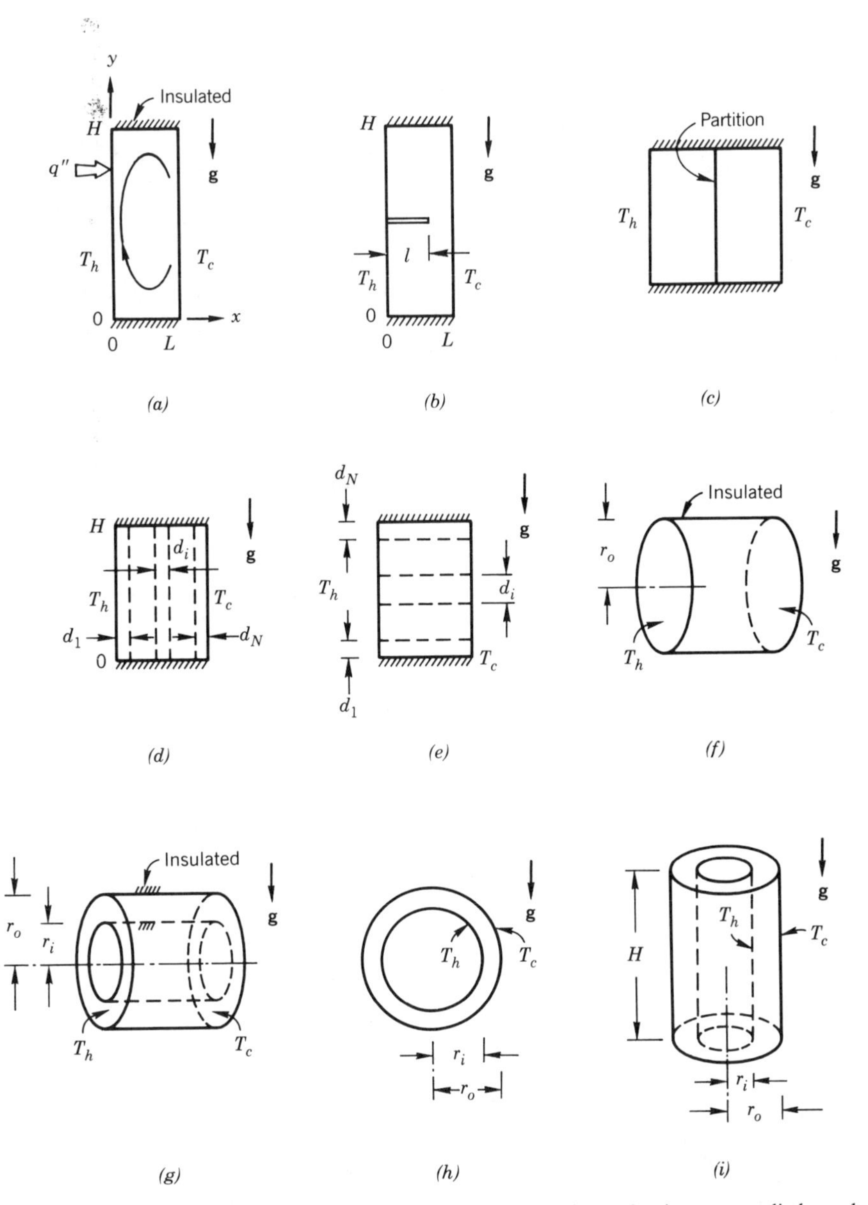

Figure 16.3. Configurations for natural-convection heat transfer in confined porous media heated from the side: (a) rectangular enclosure, (b) rectangular enclosure with a horizontal partial partition, (c) rectangular enclosure with a vertical full partition midway, (d) rectangular enclosure made up of N vertical sublayers of different K and α, (e) rectangular enclosure made up of N horizontal sublayers of different K and α, (f) horizontal cylindrical enclosure, (g) horizontal cylindrical annulus with axial heat flow, (h) horizontal cylindrical or spherical annulus with radial heat flow, (i) vertical cylindrical annulus with radial heat flow.

(16.56) or by alternative results reported specifically for the high Rayleigh number regime: Weber [55] obtained $\text{Nu} = 0.577\,(L/H)\,\text{Ra}_H^{1/2}$; this formula overestimates by roughly 14% experimental and numerical data from three sources (Bejan [11, p. 398]). More refined theories for regime III have been proposed by Bejan [56] and Simpkins and Blythe [57], where the constant that appears in $\text{Nu} \propto (L/H)\text{Ra}_H^{1/2}$ is replaced by a weak function of both H/L and Ra_H. For expedient engineering calculations involving heat transfer dominated by convection, Fig. 16.4 is recommended for shallow layers and Eq. (16.56) for regime III, $\text{Ra}_H^{-1/2} < H/L < \text{Ra}_H^{1/2}$.

In the field of thermal insulation engineering, a more appropriate model for heat transfer in the configuration of Fig. 16.3*a* is the case where the heat flux q'' is distributed uniformly along the two vertical sides of the porous layer. In the high Rayleigh number regime (regime III), the overall heat transfer rate is given by (Bejan [58])

$$\text{Nu} = \frac{1}{2}\left(\frac{L}{H}\right)^{4/5} \text{Ra}_H^{*2/5} \tag{16.58}$$

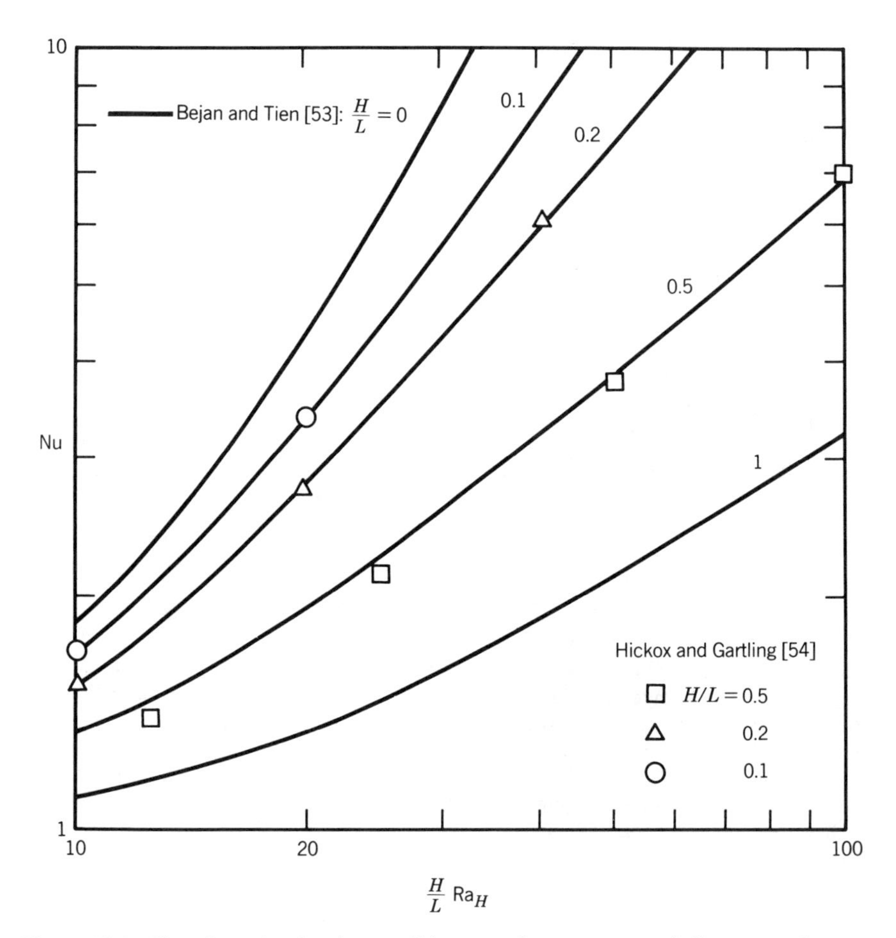

Figure 16.4. Chart for estimating the overall heat transfer rate across a shallow porous layer (see Fig. 16.3a with $H/L < 1$) heated in the end-to-end direction [11].

where $Ra_H^* = Kg\beta H^2 q''/(\alpha\nu k)$. The overall Nusselt number is defined as in Eq. (16.53), where $T_h - T_c$ is now the height-averaged temperature difference between the two sides of the rectangular cross section. The result (16.58) holds in the high Rayleigh number regime $Ra_H^{*\,-1/3} < H/L < Ra_H^{*\,1/3}$.

The presence of impermeable partitions (flow obstructions) in the confined porous medium can have a dramatic effect on the overall heat transfer rate across the enclosure (Bejan [59]). With reference to the two-dimensional geometry of Fig. 16.3*b*, in the convection-dominated regimes III and IV the overall heat transfer rate decreases steadily as the horizontal partition length l approaches L, i.e., as the partition divides the porous layer into two shorter layers [59]. The horizontal partition l has practically no effect in regimes I and II, where the overall heat transfer rate is dominated by conduction. If the partition is oriented vertically (Fig. 16.3*c*), then in the convection-dominated regime the overall heat transfer rate is approximately 40% of what it would have been in the same porous medium without the internal partition.

The heat transfer conclusions discussed so far, in connection with porous media confined in spaces heated from the side, apply to cases in which the medium can be modeled as homogeneous. It has been shown that the nonuniformity of permeability and thermal diffusivity can have a dominating effect on the overall heat transfer rate (Poulikakos and Bejan [60]). In cases where the properties vary so that the porous layer can be modeled as a sandwich of vertical sublayers of different permeability and diffusivity (Fig. 16.3*d*), an important parameter is the ratio of the peripheral sublayer thickness (d_1) to the thermal boundary-layer thickness ($\delta_{T,1}$) based on the properties of the d_1 sublayer (note that $\delta_{T,1}$ scales as $H\,Ra_{H,1}^{-1/2}$, where the Rayleigh number $Ra_{H,1} = K_1 g\beta H(T_h - T_c)/\alpha_1\nu$, where the subscript 1 represents the properties of the d_1 sublayer). If $d_1 > \delta_{T,1}$, then the heat transfer through the left side of the porous system of Fig. 16.3*d* is impeded by a thermal resistance of order $\delta_{T,1}/(k_1 H)$. If the sublayer situated next to the right wall (d_N) has exactly the same properties as the d_1 sublayer, and if $\delta_{T,1} < d_1, d_N$, then the overall heat transfer rate in the convection-dominated regime can be estimated using Eq. (16.56) in which both Nu and Ra_H are based on the properties of the peripheral layers.

In cases where the porous-medium inhomogeneity is such that the $H \times L$ system may be modeled as a sandwich of N horizontal sublayers (Fig. 16.3*e*), the scale of the overall Nusselt number in the convection-dominated regime can be evaluated as (Poulikakos and Bejan [60])

$$\mathrm{Nu} \sim 2^{-3/2} Ra_{H,1}^{1/2} \frac{L}{H} \sum_{i=1}^{N} \frac{k_i}{k_1} \left(\frac{K_i d_i \alpha_1}{K_1 d_1 \alpha_i} \right)^{1/2} \tag{16.59}$$

where both Nu and $Ra_{H,1}$ are based on the properties of the d_1 sublayer (Fig. 16.3*e*). The correlation in Eq. (16.59) was tested via numerical experiments in two-layer systems by Poulikakos and Bejan [60]. Although the applicability of Eq. (16.59) is suggested by scale analysis, this correlation remains to be verified experimentally in future studies of natural convection in horizontally layered porous media with $N > 2$.

Related to the two-dimensional convection driven by heating from the side in Fig. 16.3*a* is the convection heat transfer occurring in a porous medium confined to a horizontal cylindrical shape whose disk-shaped ends are at different temperatures (Fig. 16.3*f*). A parametric solution for the horizontal cylinder problem is reported in a paper by Bejan and Tien [53]. The corresponding phenomenon in a porous medium in the shape of a horizontal cylinder with annular cross section (Fig. 16.3*g*), is discussed by Bejan and Tien [61].

An important geometric configuration in thermal insulation engineering is a horizontal annular space filled with fibrous or granular insulation (Fig. 16.3*h*). Note that in this configuration the heat transfer is radial between the concentric cylindrical surfaces of radii r_i and r_o, unlike in the earlier sketch (Fig. 16.3*g*), where the cylindrical surfaces were insulated and the heat transfer was axial. Experimental measurements and numerical solutions for the overall heat transfer in the configuration of Fig. 16.3*h* have been reported by Caltagirone [62] and Burns and Tien [63]. These authors' results are reported either graphically or in tabular form for a discrete sequence of cases investigated in the laboratory or on the computer. A much-needed engineering correlation for the convection-dominated regime can be developed based on a scale analysis of the type described in Bejan [11, p. 194]. Using the data of Caltagirone [62] in the range $1.19 \leq r_o/r_i \leq 4$, the following correlation is obtained by the present author:

$$\mathrm{Nu} = \frac{q'_{\text{actual}}}{q'_{\text{conduction}}} \approx 0.44\,\mathrm{Ra}_{r_i}^{1/2}\,\frac{\ln(r_o/r_i)}{1 + 0.916(r_i/r_o)^{1/2}} \tag{16.60}$$

where $\mathrm{Ra}_{r_i} = Kg\beta r_i(T_h - T_c)/(\alpha\nu)$ and $q'_{\text{conduction}} = 2\pi k(T_h - T_c)/\ln(r_o/r_i)$. This correlation is valid in the convection-dominated limit, $\mathrm{Nu} \gg 1$.

Porous media confined to the space formed between two concentric spheres are also an important component in thermal insulation engineering. Figure 16.3*h* can be interpreted as a vertical cross section through the concentric-sphere arrangement. Numerical heat transfer solutions for discrete values of Rayleigh number and radius ratio are reported graphically in Burns and Tien [63]. Using the analytical method outlined in Bejan [11, p. 194], the data that correspond to the convection-dominated regime ($\mathrm{Nu} \gtrsim 1.5$) are correlated within 2% by the scaling-correct expression developed by the present author

$$\mathrm{Nu} = \frac{q_{\text{actual}}}{q_{\text{conduction}}} = 0.756\,\mathrm{Ra}_{r_i}^{1/2}\,\frac{1 - r_i/r_o}{1 + 1.422(r_i/r_o)^{3/2}} \tag{16.61}$$

where $\mathrm{Ra}_{r_i} = Kg\beta r_i(T_h - T_c)/(\alpha\nu)$ and $q_{\text{conduction}} = 4\pi k(T_h - T_c)/(r_i^{-1} - r_o^{-1})$. In terms of the Rayleigh number based on the insulation thickness, $\mathrm{Ra}_{r_o - r_i} = Kg\beta(r_o - r_i)(T_h - T_c)/(\alpha\nu)$, the correlation of Eq. (16.61) transforms to

$$\mathrm{Nu} = 0.756\,\mathrm{Ra}_{r_o - r_i}^{1/2}\,\frac{\left[r_i/r_o - (r_i/r_o)^2\right]^{1/2}}{1 + 1.422(r_i/r_o)^{3/2}} \tag{16.62}$$

In this form, the Nusselt number expression has a maximum in r_i/r_o (at $r_i/r_o = 0.301$); this maximum was noted empirically by Burns and Tien [63], and is now explained theoretically here by the scale analysis on which Eqs. (16.61) and (16.62) are based.

Heat-transfer by natural convection through an annular porous insulation oriented vertically (Fig. 16.3*i*) was investigated numerically by Havstad and Burns [64] and experimentally by Prasad et al. [65]. For systems where both vertical cylindrical surfaces may be modeled as isothermal (T_h and T_c), Havstad and Burns correlate their results with the five-constant empirical formula

$$\mathrm{Nu} = 1 + a_1\left[\frac{r_i}{r_o}\left(1 - \frac{r_i}{r_o}\right)\right]^{a_2} \mathrm{Ra}_{r_o}^{a_4}\left(\frac{H}{r_o}\right)^{a_5} \exp\left(-a_3\frac{r_i}{r_o}\right) \tag{16.63}$$

where $a_1 = 0.2196$, $a_2 = 1.334$, $a_3 = 3.702$, $a_4 = 0.9296$, and $a_5 = 1.168$, and where $\mathrm{Ra}_{r_o} = Kg\beta r_o(T_h - T_c)/(\alpha\nu)$. The Nusselt number is defined as $\mathrm{Nu} = q_{\mathrm{actual}}/q_{\mathrm{conduction}}$, where $q_{\mathrm{conduction}} = 2\pi kH(T_h - T_c)/\ln(r_o/r_i)$. The correlation of Eq. (16.63) fits the numerical data in the range $1 \leq H/r_o \leq 20$, $0 < \mathrm{Ra}_{r_o} < 150$, $0 < r_i/r_o \leq 1$, and $1 < \mathrm{Nu} < 3$. In the boundary-layer convection regime (at high Rayleigh and Nusselt numbers), the scale analysis of this two-boundary-layer problem suggests the following scaling law (Bejan [11, p. 194]):

$$\mathrm{Nu} = c_1 \frac{\ln(r_o/r_i)}{c_2 + r_o/r_i} \frac{r_o}{H} \mathrm{Ra}_H^{1/2}, \tag{16.64}$$

where $\mathrm{Ra}_H = Kg\beta H(T_h - T_c)/(\alpha\nu)$. Experimental data in the convection-dominated regime $\mathrm{Nu} \gg 1$ are needed in order to determine the constants c_1 and c_2 (note that Havstad and Burns's data are for moderate Nusselt numbers $1 < \mathrm{Nu} < 3$, i.e., for cases where pure conduction plays an important role).

The heat transfer results reviewed in this section for the geometries of Fig. 16.3 are all based on the idealization that the surface that surrounds the porous medium is impermeable. With reference to the two-dimensional geometry of Fig. 16.3*a*, the heat transfer through a shallow porous layer with one or both end surfaces permeable is anticipated theoretically by Bejan and Tien [53]. Subsequent laboratory measurements and numerical solutions for Ra_H values up to 120 validate the theory (Haajizadeh and Tien [66]).

Natural convection in cold water saturating the porous-medium configuration of Fig. 16.3*a* was considered by Poulikakos [67]. Instead of the linear approximation in Eq. (16.30), Poulikakos used the parabolic model

$$\rho_m - \rho = \gamma\rho_m(T - T_m)^2, \tag{16.64a}$$

where $\gamma \approx 8.0 \times 10^{-6}\ \mathrm{K}^{-2}$ and $T_m = 3.98°\mathrm{C}$ for pure water at atmospheric pressure. The parabolic density model is valid in the temperature range 0 to 10°C. In the convection-dominated regime $\mathrm{Nu} \gg 1$, the scale analysis illustrated in Bejan [11, p. 194] leads to a Nusselt number correlation of the following kind:

$$\mathrm{Nu} = c_3 \frac{L/H}{\mathrm{Ra}_{\gamma h}^{-1/2} + c_4 \mathrm{Ra}_{\gamma c}^{-1/2}} \tag{16.65}$$

where $\mathrm{Ra}_{\gamma h} = Kg\gamma H(T_h - T_m)^2/(\alpha\nu)$, $Ra_{\gamma c} = Kg\gamma H(T_m - T_c)^2/(\alpha\nu)$, and where the Nusselt number is defined in Eq. (16.53). For the convection-dominated regime, Poulikakos [67] tabulates numerical results primarily for the case $T_c = 0°\mathrm{C}$, $T_h = 7.96°\mathrm{C}$: using these data, for cases in which T_c and T_h are symmetrically positioned around T_m (*e*., when $\mathrm{Ra}_{\gamma h} = \mathrm{Ra}_{\gamma c}$), the scaling-correct correlation in Eq. (16.65) takes the form

$$\mathrm{Nu} \approx 0.26 \frac{L}{H} \mathrm{Ra}_{\gamma h}^{1/2} \tag{16.66}$$

In other words, the two constants that appear in Eq. (16.65) satisfy the relationship $c_3 \approx 0.26\ (1 + c_4)$. More experimental data for the high Rayleigh number regime in vertical layers with $\mathrm{Ra}_{\gamma h} \neq \mathrm{Ra}_{\gamma c}$ are needed in order to determine c_3 and c_4 uniquely.

16.4.7 Confined Layers Heated from Below

Assuming that the fluid that saturates the porous matrix expands upon heating ($\beta > 0$), the heat transfer mechanism in a porous layer heated from below (Fig. 16.5*a*) may be dominated by convection if the temperature difference $T_h - T_c$ exceeds a certain critical value. By analogy with the phenomenon of Bénard convection in a pure fluid, in the convection regime the flow consists of finite-size cells that multiply (become more slender) discretely as the destabilizing temperature difference $T_h - T_c$ increases. If $T_h - T_c$ does not exceed the critical value necessary for the onset of

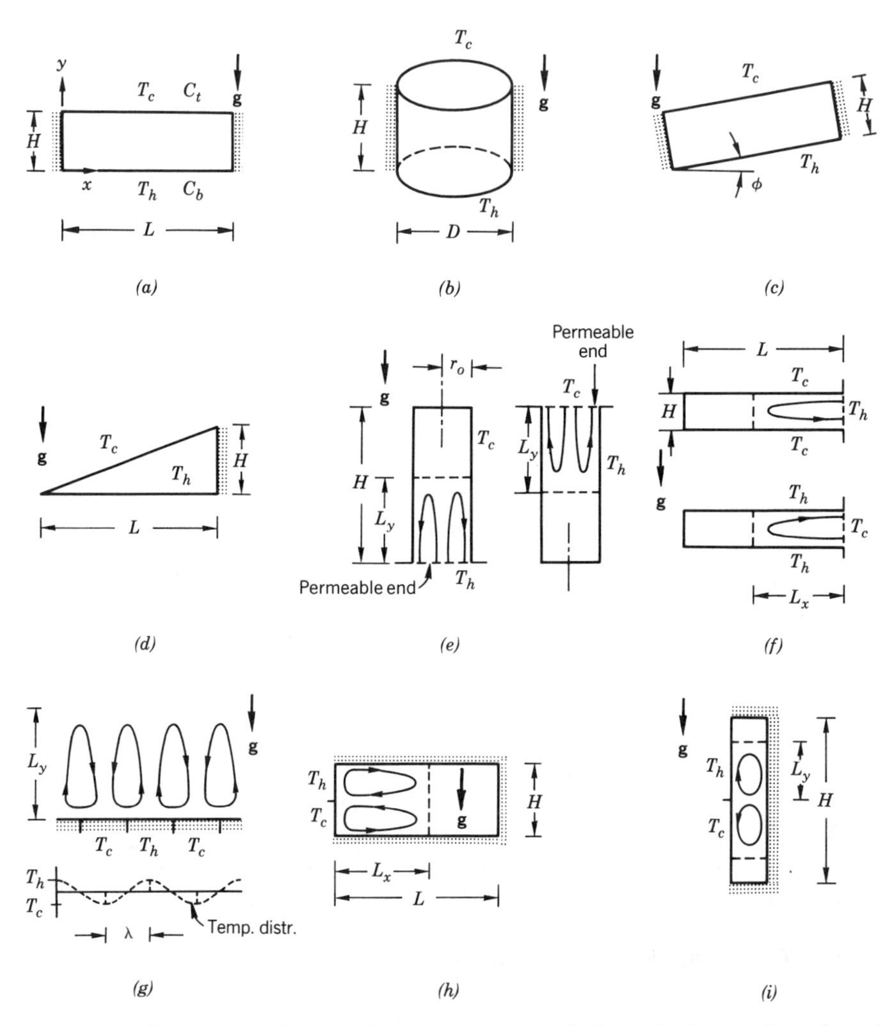

Figure 16.5. Configurations for natural-convection heat transfer in confined porous layers heated from below (*a*) to (*d*), and due to penetrative flows (*e*) to (*i*): (*a*) rectangular enclosure, (*b*) vertical cylindrical enclosure, (*c*) inclined rectangular enclosure, (*d*) wedge-shaped enclosure, (*e*) vertical cylindrical enclosure, (*f*) horizontal rectangular enclosure, (*g*) semi-infinite porous medium bounded by a horizontal surface with alternate zones of heating and cooling, (*h*) shallow rectangular enclosure heated and cooled from one vertical wall only, (*i*) slender rectangular enclosure heated and cooled from one vertical wall only.

convection, the heat transfer mechanism through the layer of thickness H is that of pure thermal conduction. If $\beta > 0$ and the porous layer is heated from above, i.e., if T_h and T_c change places in Fig. 16.5a, then the fluid remains stably stratified and the heat transfer is again due to pure thermal conduction: $q' = kL(T_h - T_c)/H$.

The onset of convection in an infinitely long porous layer heated from below was examined on the basis of linearized hydrodynamic stability analysis by Horton and Rogers [68] and Lapwood [69]. For fluid layers confined between impermeable and isothermal horizontal walls, these authors found that convection is possible if the Rayleigh number based on height, $\mathrm{Ra}_H = Kg\beta H(T_h - T_c)/(\alpha\nu)$, exceeds 39.48. For a history of the early theoretical and experimental work on the onset of Bénard convection in porous media, and for a rigorous generalization of the stability analysis to convection driven by combined buoyancy effects (Sec. 16.4.9), the reader is directed a seminal paper by Nield [70], where it is shown that the critical Rayleigh number for the onset of convection in infinitely shallow layers depends to a certain extent on the heat and fluid flow conditions imposed along the two horizontal boundaries.

Of practical interest in heat transfer engineering is the heat transfer rate at Rayleigh numbers that are higher than critical. There has been a considerable amount of analytical, numerical, and experimental work devoted to this issue. Reviews of these advances may be found in Cheng [5, 6], Nield [10], Combarnous and Bories [71], and Combarnous [72]. Figure 16.6 shows Cheng's [5] compilation of results from nine sources concerning the convection-dominated regime in a porous layer heated from below. Convection heat transfer occurs at Rayleigh numbers above approximately 40. The scale analysis of the convection regime (Bejan [11, pp. 412–414]) concludes that the Nusselt number should increase linearly with the Rayleigh number, whence the

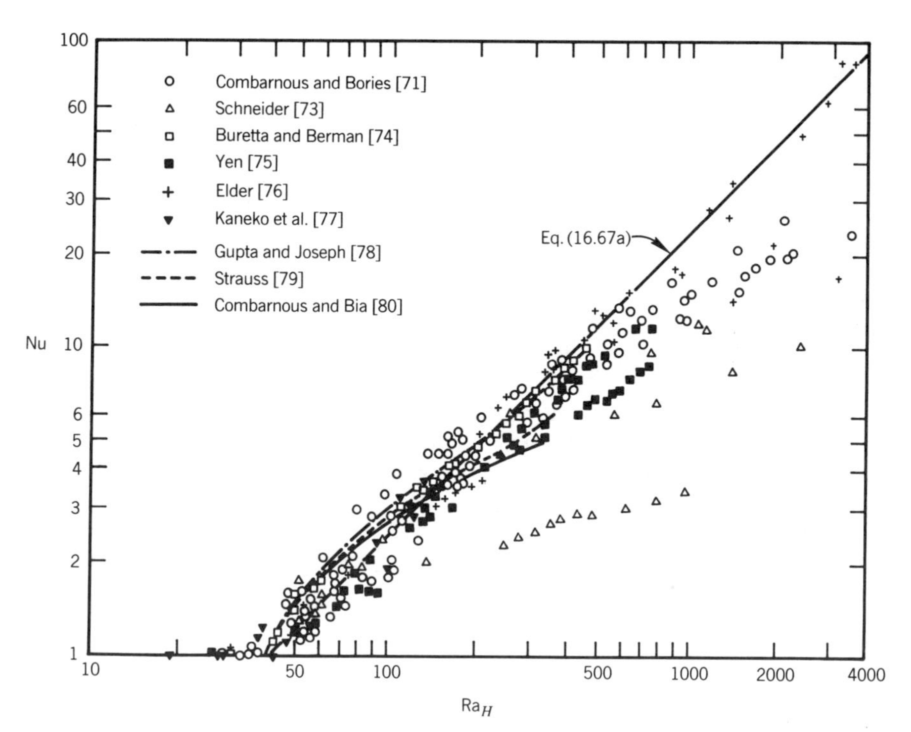

Figure 16.6. Summary of overall heat transfer through porous layers heated from below [5, 11].

formula

$$\mathrm{Nu} \approx \tfrac{1}{40}\mathrm{Ra}_H \qquad \text{for} \quad \mathrm{Ra}_H > 40 \tag{16.67a}$$

This linear relationship is confirmed by numerical heat transfer calculations at large Rayleigh numbers in a porous medium that obeys the Darcy flow model (Kimura [81]). However, as shown in Fig. 16.6, the theoretical scaling law (16.67) serves as an upper bound for some of the high-Ra_H experimental data available in the literature.

The deviation of some experimental high-Ra_H measurements from Eq. (16.67a) has attracted considerable attention. Using a boundary-layer analysis of Darcy flow, Robinson and O'Sullivan [82] have argued that Nu should increase as $\mathrm{Ra}_H^{2/3}$. Combarnous and Bories [71] link the same deviation to the failure of the homogeneous-porous-medium model, i.e., to the breakdown of the idealization of infinite heat transfer coefficient between the fluid and the solid matrix. The recent work of Georgiadis and Catton [83] shows that the discrepancy between the high-Ra_H measurements of Fig. 16.6 can be accounted for by discarding the Darcy flow model and using the more general Forschheimer-Brinkman model, Eq. (16.16). This direction of research appears to be promising: in fact, by redoing the scale analysis that led to Eq. (16.67a) using the Forschheimer model instead of the Darcy model, the present author found that Nu should increase as ([103]; note added in proof)

$$\mathrm{Nu} \sim \mathrm{Ra}_H^{1/2}\left(\frac{H}{bK}\frac{\nu}{\alpha}\right)^{1/2}. \tag{16.67b}$$

It seems that the Nusselt number increases as Ra_H^n, where the exponent n decreases from 1 to 1/2 as inertial effects become important. In addition, the proportionality constant in the $\mathrm{Nu} \propto \mathrm{Ra}_H^n$ relationship decreases as the medium Prandtl number ν/α decreases. The important group $(H\nu/bK\alpha)$ identified in eq. (16.67b) may be regarded as the "porous medium Prandtl number."

The heat transfer characteristics discussed so far in this section refer to layers whose length/height ratio is considerably greater than one. Natural-convection studies have also been reported for porous layers confined in rectangular parallelepipeds heated from below (Beck [84], Holst and Aziz [85]), horizontal circular cylinders (Zebib [86], Bories and Deltour [87], Bau and Torrance [88]), and horizontal annular cylinders (Bau and Torrance [89]). The general conclusion of these studies is that the lateral walls have a convection-suppression effect. For example, in a circular cylinder of diameter D and height H (Fig. 16.5b), in the limit $D \ll H$ the critical condition for the onset of convection is $\mathrm{Ra}_H = 13.56\,(H/D)^2$ (Bau and Torrance [88]). Regarding heat transfer rates in the convection-dominated regime, the information of Fig. 16.6 applies if the lateral dimension (perpendicular to gravity) of the confined system is greater than the horizontal length scale of a single convection cell, i.e., greater than $H\,\mathrm{Ra}_H^{-1/2}$ (Bejan [11, p. 413]).

In inclined porous layers that deviate from the horizontal position through an angle ϕ (Fig. 16.5c), convection sets in at Rayleigh numbers that satisfy the criterion (Combarnous and Bories [71])

$$\mathrm{Ra}_H > \frac{39.48}{\cos\phi}, \tag{16.68}$$

where it is assumed that the boundaries are isothermal and impermeable. The average heat transfer rate at high Rayleigh numbers can be estimated with the formula

$$\mathrm{Nu} = 1 + \sum_{s=1}^{\infty} k_s \left(1 - \frac{4\pi^2 s^2}{\mathrm{Ra}_H \cos\phi} \right) \tag{16.69}$$

where $k_s = 0$ if $\mathrm{Ra}_H \cos\phi < 4\pi^2 s^2$ and $k_s = 2$ if $\mathrm{Ra}_H \cos\phi \geq 4\pi^2 s^2$.

In a porous medium confined in a wedge-shaped (or attic-shaped) space cooled from above (Fig. 16.5*d*), convection consisting of a single counterclockwise cell exists even in the limit $\mathrm{Ra}_H \to 0$ (Bejan and Poulikakos [90]). Numerical solutions of transient high Rayleigh number convection in wedge-shaped layers show the presence of a Bénard-type instability at high enough Rayleigh numbers (Poulikakos and Bejan [91]). When $H/L = 0.2$, the instability occurs above $\mathrm{Ra}_H \approx 620$; this critical Rayleigh number was found to increase as H/L increases.

The onset of convection in the layer of Fig. 16.5*a* saturated with water near the state of maximum density has been studied using linear stability analysis (Sun et al. [92]) and time-dependent numerical solutions of the complete governing equations (Blake et al. [93]). In both studies, the condition for the onset of convection is reported graphically or numerically for a discrete series of cases. The numerical results of Blake et al. [93] for layers with $T_c = 0°\mathrm{C}$ and $5°\mathrm{C} \leq T_h \leq 8°\mathrm{C}$ can be used to derive

$$\frac{KgH}{\alpha\nu} > 1.25 \times 10^5 \exp[\exp(3.8 - 0.446 T_h)] \qquad \text{for } T_h \text{ in °C} \tag{16.70}$$

as an empirical dimensionless criterion for the onset of convection. Finite-amplitude heat and fluid flow results for Rayleigh numbers $Kg\gamma(T_h - T_c)^2 H/(\alpha\nu)$ of up to 10^4 (i.e., about 50 times greater than critical) are also reported in Blake et al. [93].

Nuclear-safety considerations have led to the study of natural convection in horizontal saturated porous layers heated volumetrically at a rate q'''. Boundary conditions and observations regarding the onset of convection and overall Nusselt numbers vary from one study to another, as indicated in one of the more recent of such studies (Kulacki and Freeman [94]). Generally, it is found that convection sets in at so-called internal Rayleigh numbers

$$\mathrm{Ra}_I = \left(\frac{k\beta}{\alpha\nu} \right)_f \frac{KgH^3 q'''}{2k} \tag{16.71}$$

in the range 33 to 46, where the subscript f indicates properties of the fluid alone. Top and bottom surface temperature measurements in the convection-dominated regime are adequately represented by (Buretta and Berman [74]; see also Rhee et al. [95])

$$\frac{q''' H^2}{2k(T_h - T_c)} \approx 0.116\, \mathrm{Ra}_I^{0.573} \tag{16.72}$$

where T_h and T_c are the resulting bottom and top temperatures if q''' is distributed throughout the layer of Fig. 16.5*a*. The empirical correlation in Eq. (16.72) is based on experiments that reach into the high $\mathrm{Ra_I}$ range of 10^3 to 10^4.

16.4.8 Penetrative Flows

There are situations where the confining geometry is such that the buoyancy-driven flow penetrates the porous medium only partially. This class of natural-convection phenomena may be categorized as *penetrative flows* (Bejan [11, pp. 403–407]). With reference to the two vertical cylindrical configurations sketched in Fig. 16.5*e*, if the cylindrical space is slender enough, the flow penetrates vertically to a distance (Bejan [96])

$$L_y = 0.085 r_o \mathrm{Ra}_{r_o} \tag{16.73}$$

where $\mathrm{Ra}_{r_o} = Kg\beta r_o(T_h - T_c)/(\alpha\nu)$ and $L_y < H$. The overall convection heat transfer rate through the permeable horizontal end is

$$q = 0.255 r_o k(T_h - T_c)\mathrm{Ra}_{r_o} \tag{16.74}$$

A similar partial penetration mechanism is encountered in the two horizontal geometries of Fig. 16.5*f*. The length of lateral penetration, L_x, and the convection heat transfer rate in the two-dimensional geometry, q' (W/m), are (Bejan [97])

$$L_x = 0.16 H\,\mathrm{Ra}_H^{1/2} \tag{16.75}$$

$$q' = 0.32 k(T_h - T_c)\mathrm{Ra}_H^{1/2} \tag{16.76}$$

where $\mathrm{Ra}_h = Kg\beta(T_h - T_c)H/(\alpha\nu)$ and $L_x < L$.

In a semi-infinite porous medium bounded from below or from above by a horizontal surface with alternating zones of heating and cooling (Fig. 16.5*g*), the buoyancy-driven flow penetrates vertically to a height or depth approximately equal to $\lambda\,\mathrm{Ra}_\lambda^{1/2}$, where $\mathrm{Ra}_\lambda = Kg\beta\lambda(T_h - T_c)/(\alpha\nu)$ and λ is the distance between a heated zone and an adjacent cooled zone (Poulikakos and Bejan [98]). Numerical heat transfer results are reported in the same paper for the Ra_λ range 1 to 100.

In a porous medium heated and cooled along the same vertical wall of height H, the incomplete penetration can be either horizontal (Fig. 16.5*h*) or vertical (Fig. 16.5*i*) (Poulikakos and Bejan [99]). In the case of incomplete horizontal penetration, the penetration length and convective heat transfer rate scale as

$$L_x \sim H\,\mathrm{Ra}_H^{1/2} \tag{16.77}$$

$$q' \sim k(T_h - T_c)\mathrm{Ra}_H^{1/2} \tag{16.78}$$

These order-of-magnitude results are valid if $\mathrm{Ra}_H^{1/2} < L/H$ and $\mathrm{Ra}_H \gtrsim 1$. The corresponding scales of incomplete vertical penetration (Fig. 16.5*i*) are

$$L_y \sim H\left(\frac{L}{H}\right)^{2/3} \mathrm{Ra}_H^{-1/3} \tag{16.79}$$

$$q' \sim k(T_h - T_c)\left(\frac{L}{H}\mathrm{Ra}_H\right)^{1/3} \tag{16.80}$$

and are valid if $\mathrm{Ra}_H^{1/2} > L/H$ and $\mathrm{Ra}_H^{1/2} > H/L$. The penetrative flows of Fig. 16.5*h* and *i* occur when the heated section T_h is situated above the cooled section T_c. When the positions of T_h and T_c are reversed, the buoyancy-driven flow fills the entire space $H \times L$ (Poulikakos and Bejan [99]).

16.4.9 Combined Heat and Mass Transfer

It is fitting to close this chapter with a look at an emerging subfield in porous-media research, namely natural convection due to combined buoyancy effects of temperature and concentration gradients. In geophysical fluid mechanics, this topic is referred to also as "double-diffusive" or "thermohaline" convection. The subfield has its origins in Nield's [70] pioneering paper on the stability of a horizontal porous layer subjected to heat and mass transfer in the vertical direction. With reference to Fig. 16.5*a*, Nield considered the additional effect of the concentration of constituent i maintained along the bottom wall (C_b) and the top wall (C_t). Instead of the density-temperature relation of Eq. (16.30a), he used

$$\rho \approx \rho_b[1 - \beta(T - T_h) - \beta_c(C - C_b)] \tag{16.81}$$

where $\beta_c = -(1/\rho)(\partial\rho/\partial C)_P$ is the concentration expansion coefficient. For saturated porous layers confined between impermeable walls with uniform T and C distributions, Nield found that convection is possible if

$$\mathrm{Ra}_H + \mathrm{Ra}_{D,H} > 39.48 \tag{16.82}$$

where $\mathrm{Ra}_H = Kg\beta H(T_h - T_c)/(\alpha\nu)$ and $\mathrm{Ra}_{D,H} = Kg\beta_c H(C_b - C_t)/(\nu D)$, with D the mass diffusivity of constituent i through the solution-saturated porous medium. Therefore, since β_c can be positive or negative, the effect of mass transfer from below can be respectively either to decrease or increase the critical Ra_H for the onset of convection. Alternatives to Eq. (16.82) for horizontal porous layers subjected to other boundary conditions are also presented in Nield's paper [70].

In the single-wall vertical flow configuration of Fig. 16.2*e*, the additional buoyancy effect caused by the imposed concentration difference $C_w - C_\infty$ can either aid or oppose the familiar flow due to $T_w - T_\infty$. Combined heat and mass transfer has only recently come under scrutiny (Bejan [11, pp. 335–338], Bejan [29, pp. 187–189], and Bejan and Khair [100]). An important role is played by the buoyancy ratio $N = \beta_c(C_w - C_\infty)/[\beta(T_w - T_\infty)]$. In heat-transfer-driven flows ($|N| \ll 1$) the heat transfer rate is given by Eq. (16.40), and the overall mass transfer rate can be estimated based on the scaling laws

$$\frac{j'}{D(C_w - C_\infty)} \sim \begin{cases} \mathrm{Ra}_H^{1/2}\mathrm{Le}^{1/2} & \text{for} \quad \mathrm{Le} \gg 1 \\ \mathrm{Ra}_H^{1/2}\mathrm{Le} & \text{for} \quad \mathrm{Le} \ll 1 \end{cases} \tag{16.83}$$

where j' [kg/(s · m)] is the overall mass transfer rate of constituent i, and Le is the Lewis number of the solution-saturated porous medium, α/D. On the other hand, in mass-transfer-driven situations ($|N| \gg 1$), the overall mass transfer rate is

$$\frac{j'}{D(C_w - C_\infty)} = 0.888(\mathrm{Ra}_H \mathrm{Le}|N|)^{1/2} \tag{16.84}$$

for all Lewis numbers, whereas the overall Nusselt number obeys the scaling laws

$$\frac{q'}{k(T_w - T_\infty)} \sim \begin{cases} (\mathrm{Ra}_H|N|)^{1/2} & \text{for} \quad \mathrm{Le} \ll 1 \\ \mathrm{Le}^{-1/2}(\mathrm{Ra}_H|N|)^{1/2} & \text{for} \quad \mathrm{Le} \gg 1 \end{cases} \tag{16.85}$$

The order-of-magnitude results of Eqs. (16.83) and (16.85) agree within 15% with overall heat and mass transfer calculations based on similarity solutions to the same problem (Bejan and Khair [100]). With reference to Fig. 16.3*a*, where the vertical walls are now maintained at different temperatures and concentrations, the heat and mass transfer due to convection driven by combined buoyancy effects was documented in terms of numerical experiments by Trevisan and Bejan [101, 102].

ACKNOWLEDGMENT

This chapter was prepared with the support of the National Science Foundation through Grant No. MEA-82-07779, of Duke University through a paid sabbatical leave, and of the University of Western Australia through the F. Mosey Visiting Scholarship to the Faculty of Engineering.

NOMENCLATURE

b	coefficient in Forschheimer's modification of Darcy's law, m^{-1}, ft^{-1}
C	constituent concentration, kg/m^3, lb_m/ft^3
d	sphere diameter, m, ft
D	mass diffusivity, m^2/s, ft^2/s
$\tilde{D}$	diameter of a circular tube, or distance between parallel plates m, ft
g	gravitational acceleration, m/s^2, ft/s^2
H	height, m, ft
j'	constituent mass flux, $kg/(m \cdot s)$, $lb_m/(ft \cdot s)$
K	permeability, defined by Eq. (16.5) or (16.6), m^2, ft^2
k	effective thermal conductivity of the porous medium with fluid inside, $W/(m \cdot K)$, $Btu/(hr \cdot ft \cdot °F)$
L	length, m, ft
L_x	penetration length, m, ft
L_y	penetration height, m, ft
Le	Lewis number $= \alpha/D$
N	buoyancy ratio $= \beta_c(C_w - C_\infty)/[\beta(T_w - T_\infty)]$
Nu	Nusselt number $= hD/k$, hL/k, or as indicated in the text
P	pressure, Pa, lb_f/in^2.
q	heat transfer rate, W, Btu/hr
q'	heat transfer rate per unit length, W/m, $Btu/(hr \cdot ft)$
q''	heat flux (heat transfer rate per unit area), W/m^2, $Btu(hr \cdot ft^2)$
q'''	volumetric heat generation rate, W/m^3, $Btu/(hr \cdot ft^3)$
r	radial coordinate, m, ft
r_i	inner radius, m, ft
r_o	outer radius, m, ft
Ra_I	internal Rayleigh number $= (k\beta/\alpha\nu)_f KgH^3q'''/2k$
Ra_y	Darcy-modified Rayleigh number $= Kg\beta y(T_w - T_\infty)/(\alpha\nu)$
Ra_y^*	Rayleigh number based on heat flux, $= Kg\beta y^2 q''/(\alpha\nu k)$
$Ra_{\infty,y}$	Rayleigh number for inertial flow, $= g\beta y^2(T_w - T_\infty)/(b\alpha^2)$

$Ra^*_{\infty, y}$ Rayleigh number for inertial flow, based on heat flux, $= g\beta y^3 q''/(kb\alpha^2)$

Ra_H Rayleigh number, $= Kg\beta H(T_h - T_c)/\alpha\nu$ for free convection in an enclosure, $= Kg\beta H(T_{\infty,H} - T_{\infty,L})/\alpha\nu$ for free convection over a vertical wall; Ra_{r_i}, Ra_{r_o}, $Ra_{r_o - r_i}$, and Ra_λ are defined in a manner similar to Ra_H for an enclosure with H replaced by r_i, r_o, $r_o - r_i$, and λ respectively.

Ra^*_H Rayleigh number based on heat flux, $= Kg\beta H^2 q''/\alpha\nu k$

$Ra_{\gamma c}$ Rayleigh number for the cold side of a porous medium saturated with fluid near the density maximum, $= Kg\gamma H(T_m - T_c)^2/\alpha\nu$

$Ra_{\gamma h}$ Rayleigh number for the hot side of a porous medium saturated with fluid near the density maximum, $= Kg\gamma H(T_h - T_m)^2/\alpha\nu$

t time, s, hr

T temperature, °C, K, °F, °R

T_c cold-side temperature, °C, K, °F, °R

T_h hot-side temperature, °C, K, °F, °R

T_m temperature of water at maximum density for pure water at atmospheric pressure = 3.98 °C = 39.16 °F

u velocity component in the x direction, m/s, ft/s

v velocity component in the y direction, m/s, ft/s

w velocity component in the z direction, m/s, ft/s

x, y, z Cartesian coordinates, m, ft

Greek Symbols

α thermal diffusivity, m^2/s, ft^2/s

β coefficient of thermal expansion, K^{-1}, $°R^{-1}$

β_C coefficient of concentration expansion, m^3/kg, ft^3/lb_m

μ viscosity, Pa · s, $lb_m/(ft \cdot s)$

ν kinematic viscosity, m^2/s, ft^2/s

ρ density, kg/m^3, lb_m/ft^3

η similarity variable

σ capacity ratio

τ dimensionless time, defined in Eq. (16.33a)

ψ stream function, m^2/s, ft^2/s

ϕ porosity

ω $= (Wk/Hk_w)Ra_H^{1/2}$

Subscripts

b bottom wall

c cold side

f fluid (liquid or gas) phase

h hot side

m bulk property

m property of the state of maximum density

s	solid phase
w	wall condition
x	local value, in the x direction
y	local value, in the y direction
θ	local value, in the θ direction
0	reference property
∞	condition sufficiently far from the wall

REFERENCES

1. M. Muskat, *The Flow of Homogeneous Fluids through Porous Media*, McGraw-Hill, New York, 1937; 2nd printing, Edwards, Ann Arbor, MI, 1946.
2. J. Bear, *Dynamics of Fluids in Porous Media*, American Elsevier, New York, 1972.
3. A. E. Scheidegger, *The Physics of Flow through Porous Media*, Macmillan, New York, 1957.
4. R. A. Greenkorn, *Flow Phenomena in Porous Media*, Dekker, New York, 1983.
5. P. Cheng, Heat Transfer in Geothermal Systems, *Adv. Heat Transfer*, Vol. 14, pp. 1–105, 1978.
6. P. Cheng, Geothermal Heat Transfer, *Handbook of Heat Transfer Applications*, ed. W. M. Rohsenow, J. P. Hartnett, and E. Ganić, 2nd ed., McGraw-Hill, New York, Chap. 11, 1985.
7. M. J. O'Sullivan, Convection with Boiling in a Porous Layer, *Convective Flows in Porous Media*, ed. R. A. Wooding and Z. White, DSIR Sci. Info. Publishing Centre, P. O. Box 9741, Wellington, New Zealand, 1985.
8. R. McKibbin, Thermal Convection in Layered and Anisotropic Porous Media: A Review, *Convective Flows in Porous Media*, ed. R. A. Wooding and Z. White, DSIR Sci. Info. Publishing Centre, P. O. Box 9741, Wellington, New Zealand, 1985.
9. A. Bejan, A. Synthesis of Analytical Results for Natural Convection Heat Transfer across Rectangular Enclosures, *Int. J. Heat Mass Transfer*, Vol. 23, pp. 723–726, 1980.
10. D. A. Nield, Recent Research on Convection in a Saturated Porous Medium, *Convective Flows in Porous Media*, ed. R. A. Wooding and Z. White, DSIR Sci. Info. Publishing Centre, P. O. Box 9741, Wellington, New Zealand, 1985.
11. A. Bejan, *Convection Heat Transfer*, Wiley, New York, 1984.
12. A. Bejan, Engineering Thermodynamics, *Mechanical Engineers' Handbook*, ed. M. Kutz, Wiley, New York, Chap. 54, 1986.
13. H. Darcy, *Les Fontaines Publiques de la Ville de Dijon*, Victor Dalmont, Paris, 1856.
14. J. Kozeny, Über kapillare Leitung des Wassers im Boden, *Sitzber. Akad. Wiss. Wien, Math-naturw. Kl.*, Vol. 136, Abt. IIa, p. 277, 1927.
15. J. C. Ward, Turbulent Flow in Porous Media, *J. Hydraul. Div. ASCE*, Vol. 90, No. HY5, pp. 1–12, 1964.
16. P. H. Forschheimer, *Z. Ver. Dtsch. Ing.*, Vol. 45, pp. 1782–1788, 1901.
17. S. Ergun, Fluid Flow through Packed Columns, *Chem. Eng. Progr.*, Vol. 48, No. 2, pp. 89–94, 1952.
18. H. C. Brinkman, A Calculation of the Viscous Force Extended by a Flowing Fluid on a Dense Swarm of Particles, *Appl. Sci. Res.*, Vol. A1, pp. 26–34, 1947.
19. K. Vafai and C. L. Tien, Boundary and Inertia Effects on Flow and Heat Transfer in Porous Media, *Int. J. Heat Mass Transfer*, Vol. 24, pp. 195–203, 1981.
20. U. Grigull and H. Sandner, *Heat Conduction*, transl. by J. Kestin, Hemisphere, Washington, 1984.

21. P. Cheng, Combined Free and Forced Convection Flow about Inclined Surfaces in Porous Media, *Int. J. Heat Mass Transfer*, Vol. 20, pp. 807–814, 1977.

22. P. Cheng, Mixed Convection about a Horizontal Cylinder and a Sphere in a Fluid-Saturated Porous Medium, *Int. J. Heat Mass Transfer*, Vol. 25, pp. 1245–1246, 1982.

23. W. M. Rohsenow and H. Y. Choi, *Heat, Mass and Momentum Transfer*, Prentice-Hall, Englewood Cliffs, NJ, 1961.

24. C. L. Hwang and L. T. Fan, Finite Difference Analysis of Forced Convection Heat Transfer in Entrance Region of a Flat Rectangular Duct, *Appl. Sci. Res.*, Vol. 13A, pp. 401–422, 1964.

25. A. Bejan, Natural Convection in an Infinite Porous Medium with a Concentrated Heat Source, *J. Fluid Mech.*, Vol. 89, pp. 97–107, 1978.

26. C. E. Hickox and H. A. Watts, Steady Thermal Convection from a Concentrated Source in a Porous Medium, *J. Heat Transfer*, Vol. 102, pp. 248–253, 1980.

27. C. E. Hickox, Thermal Convection at Low Rayleigh Number from Concentrated Sources in Porous Media, *J. Heat Transfer*, Vol. 103, pp. 232–236, 1981.

28. R. A. Wooding, Convection in a Saturated Porous Medium at Large Rayleigh or Péclet Number, *J. Fluid Mech.*, Vol. 15, pp. 527–544, 1963.

29. A. Bejan, *Solutions Manual for Convection Heat Transfer*, Wiley, New York, 1984.

30. P. Cheng and W. J. Minkowycz, Free Convection about a Vertical Plate Embedded in a Saturated Porous Medium with Application to Heat Transfer from a Dike, *J. Geophys. Res.*, Vol. 82, pp. 2040–2044, 1977.

31. D. A. Nield and S. P. White, Natural Convection in an Infinite Porous Medium Produced by a Line Heat Source, *Mathematical Models in Engineering Science*, ed. A. McNabb, R. A. Wooding, and M. Rosser, Dept. Sci. and Indust. Res., Wellington, New Zealand, 1982.

32. D. B. Ingham, J. H. Merkin, and I. Pop. The Collision of Free-Convective Boundary Layers on a Horizontal Cylinder Embedded in a Porous Medium, *Q. J. Mech. Appl. Math.*, Vol. 36, pp. 313–335, 1983.

33. V. S. Avduyevskiy, V. N. Kalashnik, and R. M. Kopyatkevich, Investigation of Free-Convection Heat Transfer in Gas-Filled Porous Media at High Pressures (English trans. of the Russian conference paper given in 1976 at the 5th All-Union Heat and Mass Transfer Conference, Minsk), *Heat Transfer—Sov. Res.*, Vol. 10, No. 5, pp. 136–144, 1978.

34. P. Cheng and I. Pop, Transient Free Convection about a Vertical Flat Plate Embedded in a Porous Medium, *Int. J. Eng. Sci.*, Vol. 22, pp. 253–264, 1984.

35. A. Bejan and R. Anderson, Heat Transfer across a Vertical Impermeable Partition Imbedded in a Porous Medium, *Int. J. Heat Mass Transfer*, Vol. 24, pp. 1237–1245, 1981.

36. A. Bejan and R. Anderson, Natural Convection at the Interface between a Vertical Porous Layer and an Open Space, *J. Heat Transfer*, Vol. 105, pp. 124–129, 1983.

37. J. M. Ramilison and B. Gebhart, Buoyancy Induced Transport in Porous Media Saturated with Pure or Saline Water at Low Temperatures, *Int. J. Heat Mass Transfer*, Vol. 23, pp. 1521–1530, 1980.

38. B. Gebhart and J. C. Mollendorf, A. New Density Relation for Pure and Saline Water, *Deep Sea Res.*, Vol. 24, pp. 831–841, 1977.

39. W. J. Minkowycz and P. Cheng, Free Convection about a Vertical Cylinder Embedded in a Porous Medium, *Int. J. Heat Mass Transfer*, Vol. 19, pp. 805–813, 1976.

40. O. A. Plumb and J. S. Huenefeld, Non-Darcy Natural Convection from Heated Surfaces in Saturated Porous Media, *Int. J. Heat Mass Transfer*, Vol. 24, pp. 765–768, 1981.

41. A. Bejan and D. Poulikakos, The Non-Darcy Regime for Vertical Boundary Layer Natural Convection in a Porous Medium, *Int. J. Heat Mass Transfer*, Vol. 27, pp. 717–722, 1984.

42. D. A. Nield and D. D. Joseph, Effects of Quadratic Drag on Convection in a Saturated Porous Medium, *Phys. Fluids*, Vol. 28, pp. 995–997, 1985.

43. S. W. Churchill and R. Usagi, A General Expression for the Correlation of Rates of Heat Transfer and Other Phenomena, *AIChE J.*, Vol. 18, pp. 1121–1128, 1972.

44. S. W. Churchill, The Development of Theoretically Based Correlations for Heat and Mass Transfer, *Lat. Am. J. Heat Mass Transfer*, Vol. 7, pp. 207–229, 1983.

45. C. T. Hsu and P. Cheng, Vortex Instability in Buoyancy-Induced Flow over Inclined Heated Surfaces in Porous Media, *J. Heat Transfer*, Vol. 101, pp. 660–665, 1979.

46. C. T. Hsu and P. Cheng, Vortex Instability of Mixed Convective Flow in a Semi-infinite Porous Medium Bounded by a Horizontal Surface, *Int. J. Heat Mass Transfer*, Vol. 23, pp. 789–798, 1980.

47. C. T. Hsu and P. Cheng, The Onset of Longitudinal Vortices in Mixed Convective Flow over an Inclined Surface in a Porous Medium, *J. Heat Transfer*, Vol. 102, pp. 544–549, 1980.

48. C. T. Hsu, P. Cheng, and G. M. Homsy, Instability of Free Convection Flow over a Horizontal Impermeable Surface in a Porous Medium, *Int. J. Heat Mass Transfer*, Vol. 21, pp. 1221–1228, 1978.

49. P. Cheng and I. D. Chang, Buoyancy Induced Flows in a Saturated Porous Medium Adjacent to Impermeable Horizontal Surfaces, *Int. J. Heat Mass Transfer*, Vol. 19, pp. 1267–1272, 1976.

50. I. Pop and P. Cheng, The Growth of a Thermal Layer in a Porous Medium Adjacent to a Suddenly Heated Semi-infinite Horizontal Surface, *Int. J. Heat Mass Transfer*, Vol. 26, pp. 1574–1576, 1983.

51. S. Kimura, A. Bejan, and I. Pop, Natural Convection Near a Cold Plate Facing Upward in a Porous Medium, *J. Heat Transfer*, Vol. 107, pp. 819–825, 1985.

52. D. Poulikakos and A. Bejan, Unsteady Natural Convection in a Porous Layer, *Phys. Fluids*, Vol. 26, pp. 1183–1191, 1983.

53. A. Bejan and C. L. Tien, Natural Convection in a Horizontal Porous Medium Subjected to an End-to-End Temperature Difference, *J. Heat Transfer*, Vol. 100, pp. 191–198, 1978; also errata: Vol. 105, pp. 683–684, 1983.

54. C. E. Hickox and D. K. Gartling, A Numerical Study of Natural Convection in a Horizontal Porous Layer Subjected to an End-to-End Temperature Difference, *J. Heat Transfer*, Vol. 103, pp. 797–802, 1981.

55. J. E. Weber, The Boundary Layer Regime for Convection in a Vertical Porous Layer, *Int. J. Heat Mass Transfer*, Vol. 18, pp. 569–573, 1975.

56. A. Bejan, On the Boundary Layer Regime in a Vertical Enclosure Filled with a Porous Medium, *Lett. Heat Mass Transfer*, Vol. 6, pp. 93–102, 1979.

57. P. G. Simpkins and P. A. Blythe, Convection in a Porous Layer, *Int. J. Heat Mass Transfer*, Vol. 23, pp. 881–887, 1980.

58. A. Bejan, The Boundary Layer Regime in a Porous Layer with Uniform Heat Flux from the Side, *Int. J. Heat Mass Transfer*, Vol. 26, pp. 1339–1346, 1983.

59. A. Bejan, Natural Convection Heat Transfer in a Porous Layer with Internal Flow Obstructions, *Int. J. Heat Mass Transfer*, Vol. 26, pp. 815–822, 1983.

60. D. Poulikakos and A. Bejan, Natural Convection in Vertically and Horizontally Layered Porous Media Heated from the Side, *Int. J. Heat Mass Transfer*, Vol. 26, pp. 1805–1814, 1983.

61. A. Bejan and C. L. Tien, Natural Convection in Horizontal Space Bounded by Two Concentric Cylinders with Different End Temperatures, *Int. J. Heat Mass Transfer*, Vol. 22, pp. 919–927, 1979.

62. J. P. Caltagirone, Thermoconvective Instabilities in a Porous Medium Bounded by Two Concentric Horizontal Cylinders, *J. Fluid Mech.*, Vol. 76, pp. 337–362, 1976.

63. P. J. Burns and C. L. Tien, Natural Convection in Porous Media Bounded by Concentric Spheres and Horizontal Cylinders, *Int. J. Heat Mass Transfer*, Vol. 22, pp. 929–939, 1979.

64. M. A. Havstad and P. J. Burns, Convective Heat Transfer in Vertical Cylindrical Annuli Filled with a Porous Medium, *Int. J. Heat Mass Transfer*, Vol. 25, pp. 1755–1766, 1982.

65. V. Prasad, F. A. Kulacki, and M. Keyhani, Natural Convection in Porous Media, *J. Fluid Mech.*, Vol. 150, pp. 89–119, 1985.

66. M. Haajizadeh and C. L. Tien, Natural Convection in a Rectangular Porous Cavity with One Permeable Endwall, *J. Heat Transfer*, Vol. 105, pp. 803–808, 1983.

67. D. Poulikakos, Maximum Density Effects on Natural Convection in a Porous Layer Differentially Heated in the Horizontal Direction, *Int. J. Heat Mass Transfer*, Vol. 27, pp. 2067–2075, 1984.

68. C. W. Horton and F. T. Rogers, Convection Currents in a Porous Medium, *J. Appl. Phys.*, Vol. 16, pp. 367–370, 1945.

69. E. R. Lapwood, Convection of a Fluid in a Porous Medium, *Proc. Camb. Phil. Soc.*, Vol. 44, pp. 508–521, 1948.

70. D. A. Nield, Onset of Thermohaline Convection in a Porous Medium, *Water Resources Res.*, Vol. 4, pp. 553–560, 1968.

71. M. A. Combarnous and S. A. Bories, Hydrothermal Convection in Saturated Porous Media, *Adv. Hydrosci.*, Vol. 10, pp. 231–307, 1975.

72. M. A. Combarnous, Natural Convection in Porous Media and Geothermal Systems, *Heat Transfer 1978*, Vol. 6, pp. 45–59, 1978.

73. K. J. Schneider, Investigation of the Influence of Free Thermal Convection in Heat Transfer Through Granular Material, Paper 11-4, *Proc. 11th Int. Congr. Refrig.*, pp. 247–254, 1963.

74. R. J. Buretta and A. S. Berman, Convective Heat Transfer in a Liquid Saturated Porous Layer, *J. Appl. Mech.*, Vol. 43, pp. 249–253, 1976.

75. Y. C. Yen, Effects of Density Inversion on Free Convective Heat Transfer in Porous Layer Heated from Below, *Int. J. Heat Mass Transfer*, Vol. 17, pp. 1349–1356, 1974.

76. J. W. Elder, Steady Free Convection in a Porous Layer Heated from Below, *J. Fluid Mech.*, Vol. 27, pp. 29–48, 1967.

77. T. Kaneko, M. F. Mohtadi, and K. Aziz, An Experimental Study of Natural Convection in Inclined Porous Media, *Int. J. Heat Mass Transfer*, Vol. 17, pp. 485–496, 1974.

78. V. P. Gupta and D. D. Joseph, Bounds for Heat Transport in a Porous Layer, *J. Fluid Mech.*, Vol. 57, pp. 491–514, 1973.

79. J. M. Strauss, Large Amplitude Convection in Porous Media, *J. Fluid Mech.*, Vol. 64, pp. 51–63, 1974.

80. M. A. Combarnous and P. Bia, Combined Free and Forced Convection in Porous Media, *Soc. Pet. Eng. J.*, Vol. 11, pp. 399–405, 1971.

81. S. Kimura, G. Schubert and J. M. Straus, Route to Chaos in Porous-Medium Thermal Convection, *J. Fluid Mech.*, Vol. 166, pp. 305–324, 1986.

82. J. L. Robinson and M. J. O'Sullivan, A Boundary Layer Model of Flow in a Porous Medium at High Rayleigh Number, *J. Fluid Mech.*, Vol. 75, pp. 459–467, 1976.

83. J. Georgiadis and I. Catton, Prandtl Number Effect on Bénard Convection in Porous Media, *J. Heat Transfer*, Vol. 108, pp. 284–290, 1986.

84. J. L. Beck, Convection in a Box of Porous Material Saturated with Fluid, *Phys. Fluids*, Vol. 15, pp. 1377–1383, 1972.

85. P. H. Holst and K. Aziz, Transient Three-Dimensional Natural Convection in Confined Porous Media, *Int. J. Heat Mass Transfer*, Vol. 15, pp. 73–90, 1972.

86. A. Zebib, Onset of Natural Convection in a Cylinder of Water Saturated Porous Media, *Phys. Fluids*, Vol. 21, pp. 699–700, 1978.

87. S. Bories and A. Deltour, Influence des Conditions aux limites sur la Convection Naturelle dans un Volume Poreux Cylindrique, *Int. J. Heat Mass Transfer*, Vol. 23, pp. 765–771, 1980.

88. H. H. Bau and K. E. Torrance, Low Rayleigh Number Thermal Convection in a Vertical Cylinder Filled with Porous Materials and Heated from Below, *J. Heat Transfer*, Vol. 104, pp. 166–172, 1982.

89. H. H. Bau and K. E. Torrance, Onset of Convection in a Permeable Medium between Vertical Coaxial Cylinders, *Phys. Fluids*, Vol. 24, pp. 382–385, 1981.

90. A. Bejan and D. Poulikakos, Natural Convection in an Attic-Shaped Space Filled with Porous Material, *J. Heat Transfer*, Vol. 104, pp. 241–247, 1982.

91. D. Poulikakos and A. Bejan, Numerical Study of Transient High Rayleigh Number Convection in an Attic-Shaped Porous Layer, *J. Heat Transfer*, Vol. 105, pp. 476–484, 1983.

92. Z. S. Sun, C. Tien, and Y. C. Yen, Onset of Convection in a Porous Medium Containing Liquid with a Density Maximum, *Heat Transfer 1970*, Vol. IV, Paper NC 2.11, 1970.

93. K. R. Blake, A. Bejan, and D. Poulikakos, Natural Convection Near 4°C in a Water Saturated Porous Layer Heated from Below, *Int. J. Heat Mass Transfer*, Vol. 27, pp. 2355–2364, 1984.

94. F. A. Kulacki and R. G. Freeman, A Note on Thermal Convection in a Saturated, Heat-Generating Porous Layer, *J. Heat Transfer*, Vol. 101, pp. 169–171, 1979.

95. S. J. Rhee, V. K. Dhir, and I. Catton, Natural Convection Heat Transfer in Beds of Inductively Heated Particles, *J. Heat Transfer*, Vol. 100, pp. 78–85, 1978.

96. A. Bejan, Natural Convection in a Vertical Cylindrical Well Filled with Porous Medium, *Int J. Heat Mass Transfer*, Vol. 23, pp. 726–729, 1980.

97. A. Bejan, Lateral Intrusion of Natural Convection into a Horizontal Porous Structure, *J. Heat Transfer*, Vol. 103, pp. 237–241, 1981.

98. D. Poulikakos and A. Bejan, Penetrative Convection in Porous Medium Bounded by a Horizontal Wall with Hot and Cold Spots, *Int. J. Heat Mass Transfer*, Vol. 27, pp. 1749–1758, 1984.

99. D. Poulikakos and A. Bejan, Natural Convection in a Porous Layer Heated and Cooled along One Vertical Side, *Int. J. Heat Mass Transfer*, Vol. 27, pp. 1879–1891, 1984.

100. A. Bejan and K. R. Khair, Heat and Mass Transfer by Natural Convection in a Porous Medium, *Int. J. Heat Mass Transfer*, Vol. 28, pp. 909–918, 1985.

101. O. V. Trevisan and A. Bejan, Natural Convection with Combined Heat and Mass Transfer Buoyancy Effects in a Porous Medium, *Int. J. Heat Mass Transfer*, Vol. 28, pp. 1597–1611, 1985.

102. O. V. Trevisan and A. Bejan, Mass and Heat Transfer by Natural Convection in a Vertical Slot Filled with Porous Medium, *Int. J. Heat Mass Transfer*, Vol. 29, pp. 403–415, 1986.

103. A. Bejan, The Basic Scales of Natural Convection Heat and Mass Transfer in Fluids and Fluid-Saturated Porous Media, *Int. Comm. Heat Mass Transfer*, Vol. 14, pp. 107–123, 1987.

17

ENHANCEMENT OF SINGLE-PHASE HEAT TRANSFER

Ralph L. Webb

Department of Mechanical Engineering
The Pennsylvania State University
University Park, Pennsylvania 16802

17.1 INTRODUCTION

The subject of enhanced heat transfer has developed to the stage that it is of serious interest for heat-exchanger application. Industry has used enhanced heat transfer surfaces for two purposes: (1) to obtain compact and less expensive heat exchangers, and (2) to increase the UA value of the heat exchanger. The higher UA may be exploited in either of two ways: (1) to obtain an increased heat exchange rate for fixed fluid inlet temperatures, or (2) to reduce the mean temperature difference for the heat exchange; this provides increased thermodynamic process efficiency and yields a saving of operating costs.

Table 17.1 lists 13 techniques to provide enhancement. These techniques are segregated into two groupings—passive and active. Passive techniques employ special surface geometries or fluid additives for enhancement. Active techniques require external power, such as electric or acoustic fields and surface vibration. Recently published bibliographies provide citations to the technical literature [1] and U.S. Patents [2] on the technology of enhanced heat transfer. Table 17.1 lists the number of citations given in Ref. [1] for enhancement of single-phase heat transfer.

The majority of commercially interesting enhancement techniques are passive ones. The active techniques have not found commercial interest because of the capital and operating cost of the enhancement device and problems associated with vibration or acoustic noise. This chapter will limit discussion to the passive techniques.

Special surface geometries provide enhancement by establishing a higher hA per unit base surface area. Three basic methods are employed to increase the hA:

1. Increase h without an appreciable area (A) increase.
2. Increase of A without appreciably changing h.
3. Increase of both h and A.

Examples of the above three methods are:

Method 1. Surface roughness or turbulence promoters

Method 2. Internally finned tubes or externally finned tubes with plain fins

Method 3. The louver fin geometry used in automotive radiators or the segmented finned-tube geometry used in finned-tube heat exchangers

This chapter will discuss applications of the passive enhancement techniques for laminar and turbulent flow to three basic flow geometries:

1. Internal flow in circular tubes and axial flow in rod bundles
2. Channel flow in closely spaced parallel plate channels, typical of plate-fin and plate-type heat exchangers
3. External flow normal to tubes and tube banks

For application to a two-fluid tubular heat exchanger, enhancement may be desired for the inner side, the outer side, or both sides of the tube. If enhancement is applied to the inner and outer tube surfaces, a doubly enhanced tube results. Applications for such doubly enhanced tubes include condenser and evaporator tubes. Depending on the design application, the two-phase heat transfer process may occur on the tube side or the shell side.

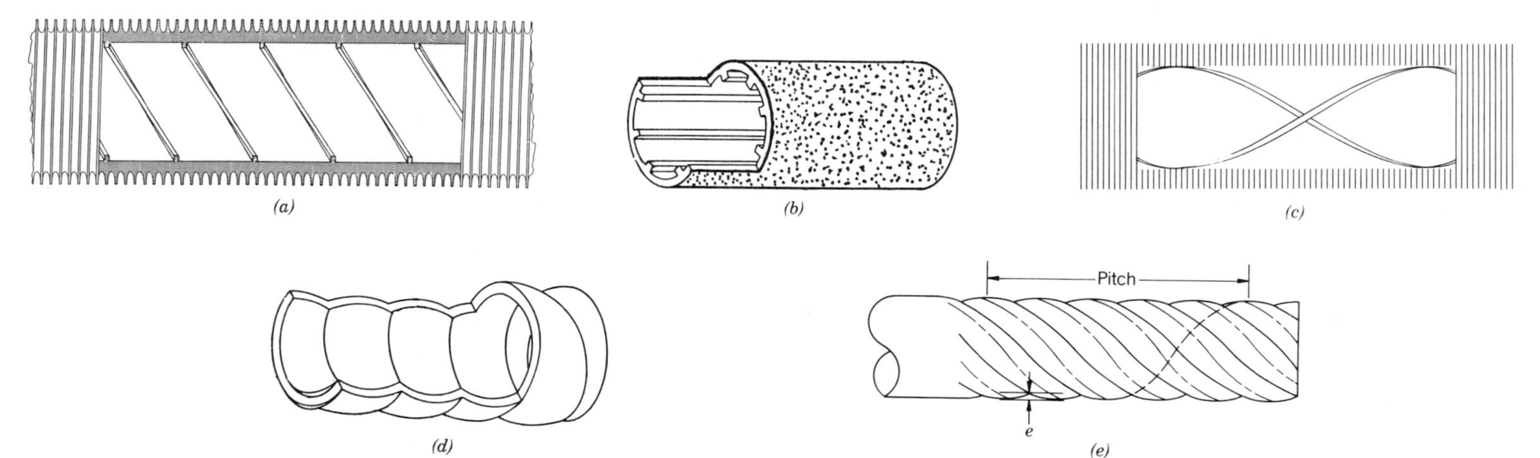

Figure 17.1. Five methods of making doubly enhanced tubes: (*a*) helical rib roughness on inner surface and integral fins on outer surface, (*b*) internal fins on inner surface and porous boiling coating on outer surface, (*c*) twisted tape insert on inner surface and integral fins on outer surface, (*d*) corrugated roughness on inner and outer surfaces, (*e*) corrugated strip rolled in a cylinder form and seam-welded.

Consider, for example, a shell-side condenser with cooling water on the tube side. The preferred enhancement geometry for the condensing side may be substantially different from that desired for the water side. Hence, one must be aware of possible manufacturing limitations or possibilities that affect independent formation of the enhancement geometry on the inner and outer surfaces of a tube. Figure 17.1 illustrates five basic approaches that may be employed to provide doubly enhanced tubes. Those shown in Fig. 17.1*a*, *b*, and *c* allow independent shell-side and tube-side enhancement geometries. However, the outside enhancement of the corrugated tube in Fig. 17.1*d* results from the forming process of the tube-side ridges. Such a tube may give good tube-side enhancement but poorer shell-side performance than would result from a different shell-side enhancement geometry. Webb [3] discusses possibilities for manufacturing doubly enhanced condenser tubes. Users should also be aware that it may not be possible to make a given enhancement geometry in all materials of interest. For example, the internally finned tube of Fig. 17.1*a* can easily be made as an aluminum

TABLE 17.1. Classification of Augmentation Techniques

Augmentation Technique	Single-Phase Natural Convection	Single-Phase Forced Convection
PASSIVE TECHNIQUES[a]		
Treated surfaces	—[b]	—
Rough surfaces	7	418
Extended surfaces	23	416
Displaced enhancement devices	—	59
Swirl flow devices	—	140
Coiled tubes	—	142
Surface tension devices	—	—
Additives for liquids	3	22
Additives for gases	—	211
ACTIVE TECHNIQUES[c]		
Mechanical aids	16	60
Surface vibration	52	30
Fluid vibration	44	127
Electric or magnetic fields	50	53
Injection or suction	6	25
Jet impingement	—	17
COMPOUND ENHANCEMENT[d]		
	2	50

[a] No external power required.
[b] Not applicable.
[c] External power required.
[d] Two or more techniques.

extrusion, but it would be extremely difficult to make from a very hard material, such as titanium. Webb [4] discusses particular enhancement geometries that have been commercially manufactured and their materials of manufacture. Table 17.1 shows that a variety of enhancement techniques are possible. For making decisions, the various enhancement techniques may be considered to be in competition with one another. The favored method will provide the highest performance per unit cost for the material of interest.

17.2 PERFORMANCE EVALUATION CRITERIA (PEC)

A quantitative method is required to evaluate the performance improvement provided by a given enhancement. The basic performance characteristic of an enhanced surface is defined by curves of the j and f vs. Reynolds number. One possibility for quantifying the performance improvement is to calculate the ratios j/j_p and f/f_p, where subscript p is for a plain surface at the same Re. However, this method is not recommended, because it does not define the actual performance improvement, subject to specific operating constraints.

The preferred evaluation method sets a performance objective (e.g., reduced surface area) and calculates the performance improvement relative to a reference design (e.g., plain tubes) for a given set of operating conditions and design constraints (e.g., constant pumping power). Hence, this method defines the improvement of the objective function (e.g., percent surface area decrease) relative to a heat exchanger without enhancement. Possible performance objectives of interest include:

1. Reduced heat transfer surface material for fixed heat duty and pressure drop
2. Increased UA, which may be exploited in two ways:
   a. To obtain increased heat duty
   b. To secure reduced LMTD for fixed heat duty
3. Reduced pumping power for fixed heat duty

Objective 1 allows a smaller heat-exchanger size and, hopefully, reduced capital cost. Objectives 2b and 3 offer reduced operating cost. The reduced LMTD of objective 2b will effect improved system thermodynamic efficiency, yielding lower system operating cost. Objectives 2b and 3 are particularly important if "life cycle" cost analysis is of interest. A more costly augmented surface will be justified if the operating-cost savings are sufficiently high.

The major operational variables include the heat transfer rate, pumping power (or pressure drop), heat exchanger flow rate, and fluid velocity, which affect the exchanger flow frontal area. A PEC is established by selecting one of the operational variables for the performance objective, subject to design constraints on the remaining variables.

The design constraints placed on the exchanger flow rate and velocity cause key differences among the possible PEC relations. The increased friction factor of augmented surfaces may require reduced velocity to satisfy a fixed pumping-power (or pressure drop) constraint. If the exchanger flow rate is held constant, it may be necessary to increase the flow frontal area to satisfy the pumping-power constraint. However, if the mass flow rate is reduced, it is possible to maintain constant flow frontal area at reduced velocity. When the exchanger flow rate is reduced, it may operate at higher thermal effectiveness to provide the required heat duty. This may significantly reduce the performance potential of the augmented surface if the design

TABLE 17.2. Performance Evaluation Criteria for $d_i / d_{ip} = 1$

	Fixed						Consequences						
Case	Geom	W	P	q	ΔT_i	Objective	$\frac{N}{N_p}$	$\frac{L}{L_p}$	$\frac{W}{W_p}$	$\frac{Re}{Re_p}$	$\frac{P}{P_p}$	$\frac{q}{q_p}$	$\frac{T_i}{T_{ip}}$
FG-1a	N, L	×			×	$\uparrow q$	1	1	1	1[b]	> 1	< 1	1
FG-1b	N, L	×		×		$\downarrow \Delta T_i$	1	1	1	1[b]	1	1	> 1
FG-2a	N, L		×		×	$\uparrow q$	1	1	< 1	< 1	1	> 1	1
FG-2b	N, L		×	×		$\downarrow \Delta T_i$	1	1	< 1	< 1	1	1	< 1
FG-3	N, L			×	×	$\downarrow P$	1	1	< 1	< 1	< 1	1	1
FN-1	N		×	×	×	$\downarrow L$	1	< 1	< 1	< 1	1	1	1
FN-2	N	×		×	×	$\downarrow L$	1	< 1	1	1[b]	< 1	1	1
FN-3	N	×		×	×	$\downarrow P$	1	< 1	1	1[b]	< 1	1	1
VG-1		×	×	×	×	$\downarrow NL$	> 1[c]	< 1	1	< 1[c]	1	1	1
VG-2a	NL[a]	×	×		×	$\uparrow q$	> 1[c]	< 1	1	< 1[c]	1	> 1	1
VG-2b	NL[a]	×	×	×		$\downarrow \Delta T_i$	> 1[c]	< 1	1	< 1[c]	1	1	< 1
VG-3	NL[a]	×		×	×	$\downarrow P$	< 1	< 1	1	< 1[c]	< 1	1	1

[a]The product of N and L is constant in cases VG-2 and VG-3.
[b]For internal roughness. For internal fins, $Re/Re_p = D_h S / d_i S_p$.
[c]Roughness with high-Pr fluids may not result in $N/N_p < 1$ (or $Re/Re_p < 1$).

effectiveness is sufficiently high. In many cases the heat exchanger flow rate is specified and thus flow-rate reduction is not permitted.

Table 17.2 defines PECs for 12 cases of interest with flow inside enhanced and smooth tubes of the same envelope diameter. The PECs are segregated by three different geometry constraints.

FG Criteria. The cross-sectional envelope area and tube length are held constant. The FG criteria may be thought of as a retrofit situation in which there is a one-for-one replacement of plain surfaces with enhanced surfaces of the same basic geometry (e.g., tube envelope diameter, tube length, and number of tubes for in-tube flow). The FG-2 criteria have the same objective functions as FG-1 but require the enhanced surface design to operate at the same pumping power as the reference smooth-tube design. In most cases, this will require the enhanced exchanger to operate at reduced flow rate. The FG-3 criterion seeks reduced pumping power for fixed heat duty.

FN Criteria. These criteria maintain fixed flow frontal area and allow the length of the heat exchanger to be a variable. These criteria seek reduced surface area (FN-1) or reduced pumping power (FN-2) for constant heat duty.

VG Criteria. In many cases, a heat exchanger is sized for a required thermal duty with a specified flow rate. In these situations, the FG and FN criteria are not applicable. Because the tube-side velocity must be reduced to accommodate the higher friction characteristics of the augmented surface, it is necessary to increase the flow area to maintain constant flow rate. This is accomplished by using a greater number of parallel flow circuits. Maintenance of a constant exchanger flow rate avoids the penalty encountered in the previous FG and FN cases of operating at higher thermal effectiveness.

Calculation of the performance improvement for any of the 12 cases in Table 17.2 requires algebraic relations which quantify the objective function and constraints. The detailed algebraic equations for the various cases are given in Ref. [5]. The concept of the PEC analysis will be explained for the case of a prescribed wall temperature or a prescribed heat flux. The heat transfer enhancement, $hA/h_pA_p \equiv K/K_p$, is given by

$$\frac{K}{K_p} = \frac{\mathrm{St}}{\mathrm{St}_p}\frac{A}{A_p}\frac{G}{G_p} \tag{17.1}$$

where subscript p refers to the reference, or smooth surface. The relative friction power is

$$\frac{P}{P_p} = \frac{f}{f_p}\cdot\frac{A}{A_p}\left(\frac{G}{G_p}\right)^3 \tag{17.2}$$

It is assumed that the smooth-tube operating conditions (G_p, f_p, and St_p) are known. The key to the use of Eqs. (17.1) and (17.2) is to determine the mass velocity G in the enhanced tube that satisfies the objective function and constraints. Consider, for example, case VG-1. This case sets $K/K_p = P/P_p = 1$. By eliminating A/A_p from Eqs. (17.1) and (17.2), one obtains

$$\frac{G}{G_p} = \left(\frac{\mathrm{St}/\mathrm{St}_p}{f/f_p}\right)^{1/2} \tag{17.3}$$

It is assumed that equations for St and f as a function of G are known. Hence, one iteratively solves Eq. (17.3) for G/G_p. Once G is known, the solution of Eq. (17.1) for A/A_p is easily performed.

The simple example described above does not provide for the most general conditions, for which one must take account of thermal resistances on the outer tube surface and the wall and fouling resistances. Webb [5] outlines the analysis method for this more general and more realistic situation.

The PEC defined in Table 17.2 may also be interpreted for flow normal to tube banks (bare or finned) and for plate-fin heat exchangers [6].

The FN and FG cases maintain constant cross-sectional flow area ($N/N_p = 1$). Cases FG-2 and FN-1 constrain $P/P_p = 1$, for which a reduced flow rate ($W/W_p < 1$) may be required to satisfy the constraint $P/P_p = 1$. It is possible that the potential benefits of an enhanced surface will be lost if the heat exchanger is required to operate with $W/W_p < 1$ [5]. When the flow rate is reduced, the LMTD is reduced (for constant ΔT_i), and additional surface is required to compensate for the reduced LMTD. The VG cases avoid this problem by increasing the flow cross-sectional area sufficiently to maintain $W/W_p = 1$; this is accomplished by adding tubes in parallel. Hence, when a pumping-power constraint is applied, the greatest performance benefit of an enhanced surface will be realized by use of the VG criteria. This implies that enhanced surfaces will yield smaller benefits in retrofit applications with fixed flow area than in new designs, for which the flow frontal area may be increased over that of the corresponding plain tube design.

The PEC defined in Table 17.2 may also be interpreted for flow normal to tube banks (bare or finned) and for plate-fin heat exchangers, which are discussed in Section 17.3 and illustrated by Fig. 17.2. Table 17.3 defines how the variables used in Table 17.2 (flow in tubes) should be interpreted for the Fig. 17.3 geometries.

TABLE 17.3. Interpretation of Table 17.2 PECs for Fig. 17.3 Geometries

Variable	Flow in Tubes	Fig. 2 Geometries
Flow area	SN	S_f
Mass flow rate W	SNG	$SG = S_f G_f$
Surface area	πdNL	βV
W/W_p	$\frac{N}{N_p}\frac{G}{G_p}$	$\frac{\sigma}{\sigma_p}\frac{S_f}{S_f p}\frac{G}{G_p}$

17.3 EXTENDED SURFACES FOR GASES

In forced-convection heat transfer between a gas and a liquid, the heat transfer coefficient of the gas may be 10 to 50 times smaller than that of the liquid. The use of extended surfaces will reduce the gas-side thermal resistance. However, the resulting gas-side resistance may still exceed that of the liquid. In this case, it will be advantageous to consider use of specially configured extended surfaces which provide increased heat transfer coefficients. Such special surface geometries may provide heat transfer coefficients 50 to 150% higher than those given by plain extended surfaces. For heat transfer between gases, such enhanced surfaces will provide a substantial heat-exchanger size reduction. The heat-exchanger (recuperator) geometries of interest are shown in Fig. 17.2.

This section contains a description of special extended surface geometries which provide increased heat transfer coefficients for gases. The use of these surface geometries in plate-fin and finned-tube heat exchangers will be discussed.

17.3.1 Plate-Fin Heat Exchangers

Figure 17.3 shows six enhanced surface geometries. Typical fin spacings are 300 to 800 fins/m. Due to the small hydraulic diameter and low density of gases, these surfaces are usually operated with $500 < \mathrm{Re} < 1500$ (hydraulic-diameter basis). To be effective, the enhancement technique must be applicable to low Reynolds number flows. The use of surface roughness will not provide appreciable enhancement for such flows. Two basic concepts have been extensively employed to provide enhancement:

1. Special channel shapes, such as the wavy channel, which provide mixing due to secondary flows or boundary-layer separation within the channel.
2. Repeated growth and wake destruction of boundary layers. This concept is employed in the offset strip fin, the louvered fin, and to some extent the perforated fin.

Offset Strip Fin. Figure 17.4 illustrates the enhancement principle of the Fig. 17.3*d* offset strip fin (OSF). A laminar boundary layer is developed on the short strip length, followed by its dissipation in the wake region between strips. Typical strip lengths are 3 to 6 mm, and the Reynolds number (based on strip length) is well within the laminar region. The enhancement provided by an OSF of $l/D_h = 1.88$ is shown by Fig. 17.5. The OSF and plain fins are taken from Figs. 10-58 and 10-27, respectively, of Kays and London [7]. The table in Fig. 17.3 compares the dimensions with the OSF scaled to the

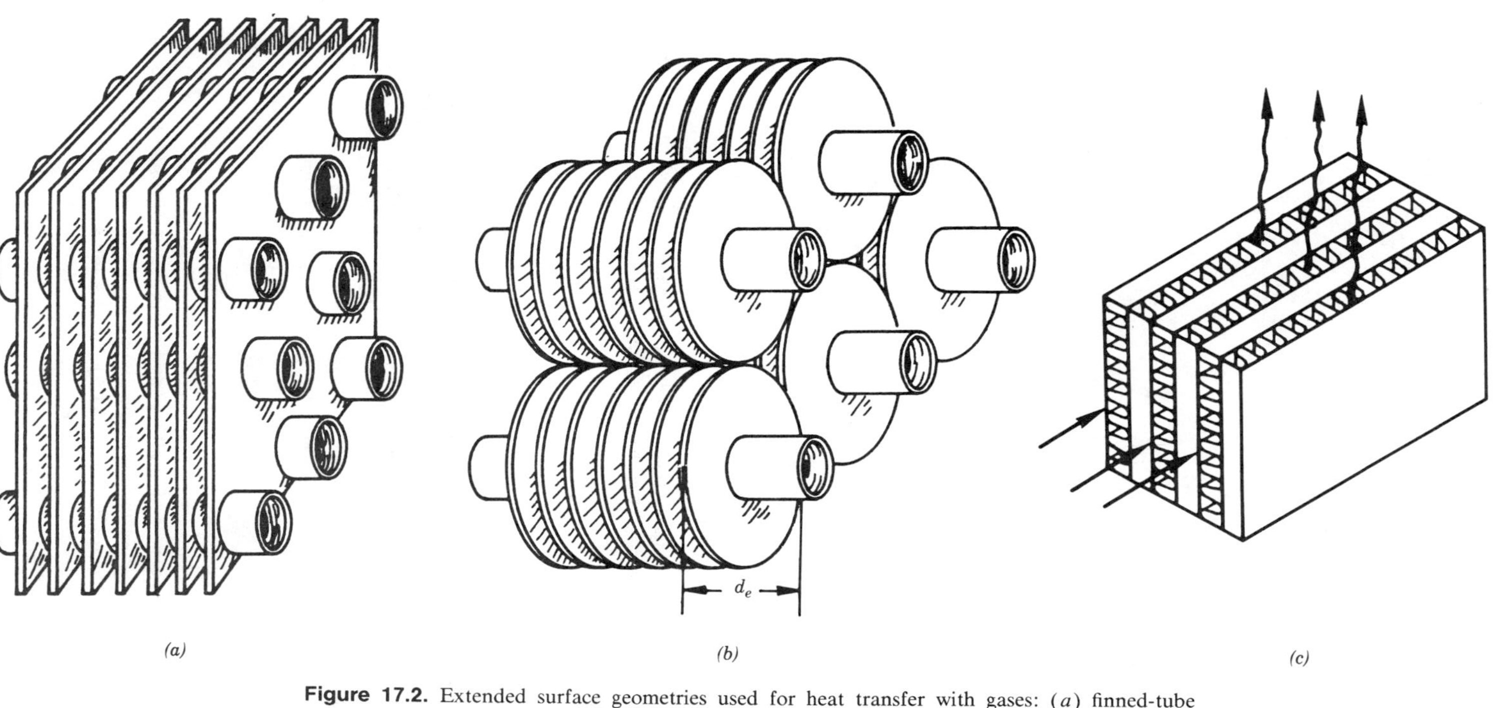

Figure 17.2. Extended surface geometries used for heat transfer with gases: (*a*) finned-tube heat-exchanger with flat fins, (*b*) individually finned tubes, (*c*) plate-fin exchanger.

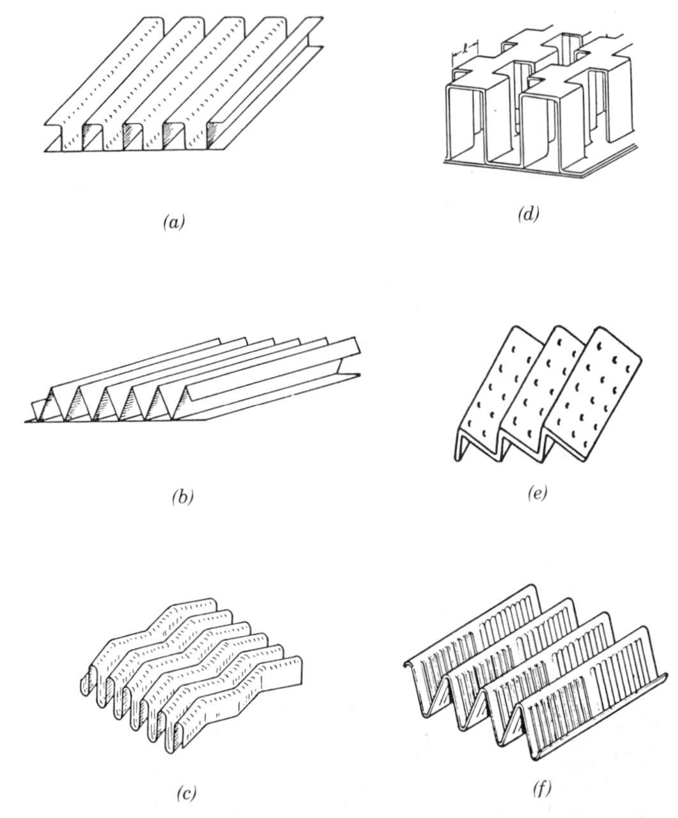

Figure 17.3. Plate-fin exchanger surface geometries: (a) plain rectangular fins, (b) plain triangular, (c) wavy, (d) offset strip, (e) perforated, (f) louvered.

same hydraulic diameter as that of the plain fin. The scaled dimensions (fin height and fin thickness) of the OSF are approximately equal to those of the plain fin surface. Therefore, the performance difference is due to the enhancement provided by the short strip lengths (6.6 mm) of the OSF. At Re = 1000, the j factor of the OSF is 2.5 times higher than for the plain fin, and the friction factor is 3.0 times higher. Considering the j/f ratio as an "efficiency index" ratio between heat transfer and friction, the OSF yields a 150%-increased heat transfer coefficient with a ratio of heat transfer to friction (j/f) 83% as high as that of the plain fin. Greater enhancement will be obtained by using shorter strip lengths.

PEC EXAMPLE 1

A more realistic comparison of the OSF and plain-fin performance is obtained using case VG-1 of Table 17.2. Assume that the plain fin (subscript p) operates at $Re_p = 834$. Equation (17.3), which applies to case VG-1, defines the value of G/G_p at which both surfaces satisfy $1 = K/K_p = P/P_p = W/W_p$. It is necessary to solve Eq. (17.3) iteratively for G/G_p, reading the j and f values for the OSF surface from Fig. 17.5. The

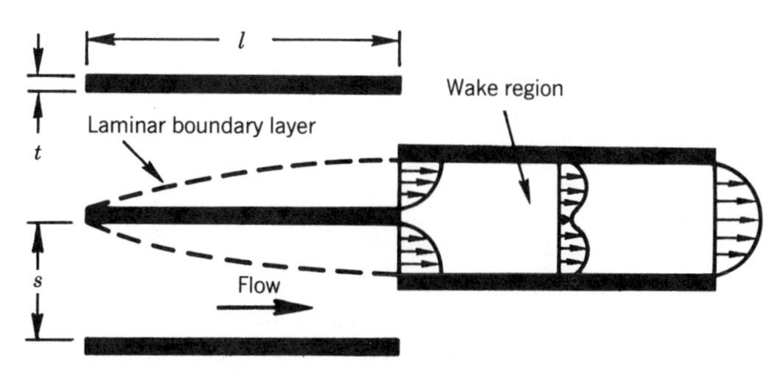

Figure 17.4. Boundary-layer and wake region of the offset strip fin [9].

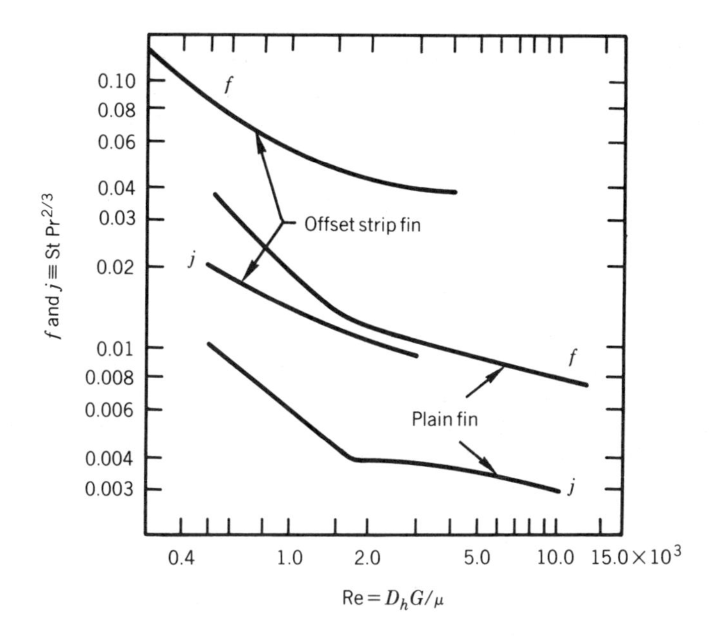

	Surface Geometry	
	Plain	Offset Strip
Surface designation	10–27	10–58[a]
Fins/m	437	437
Plate spacing (mm)	12.2	11.9
Hydraulic diam (mm)	3.51	3.51
Fin thickness (mm)	0.20	0.24
Offset-strip length in flow dir. (mm)	—	6.6

[a] Dimensions geometrically scaled to give same hydraulic diameter as plain fin.

Figure 17.5. Comparison of j and f for the offset-strip-fin and the plain-fin surface geometries [7].

final solution of Eq. (17.3) yields $G/G_p = 0.923$. So Re = 0.923 × 834 = 765. The j and f values of the OSF at Re = 765 are now known.

The surface area ratio A/A_s is obtained using Eq. (17.1):

$$\frac{K}{K_p} = 1 = \frac{j}{j_p}\frac{A}{A_p}\frac{G}{G_p}$$

Substituting the known values of j/j_p and G/G_p in the above equation, one calculates $A/A_p = 0.446$. The OSF geometry requires only 44.6% as much surface area to provide the same hA as the plain-fin geometry. One may show that the flow frontal area must be increased 10% to maintain $P/P_p = 1$.

Kays [8] proposed a simple, approximation model to predict the curves of j and f vs. Re for the OSF with the following idealizations: (1) laminar boundary layers on the fins, (2) that the boundary layers developed on the fin are totally dissipated in the wake region between fins. Using the equations for laminar flow on a flat plate in a free stream, Kays's analysis gives

$$j = 0.664\,\mathrm{Re}_l^{-0.5} \tag{17.4}$$

$$f = \frac{C_D t}{2l} + 1.328\,\mathrm{Re}_l^{-0.5} \tag{17.5}$$

The first term in Eq. (17.5) represents the form drag on the plate. This form drag contribution is proportional to the fin thickness t, and it has a negligible effect on the heat transfer coefficient. Kays suggested the use of $C_D = 0.88$, based on potential flow normal to a thin plate. Although this model is only approximate, it will allow the designer to predict the effect of strip length and thickness.

Joshi and Webb [9] have developed a more sophisticated theoretical model to predict the characteristics of j and f vs. Re for the OSF array. This model properly includes all geometric factors of the array (s/b, t/l, t/s) and heat transfer to the base surface area to which the fins are attached. The model employs the numerical solution of Sparrow and Liu [8] to calculate the j and f factors on the fin surface. Sparrow and Liu's analysis is for a zero-thickness fin and assumes laminar flow on the fin and wake regions. Joshi and Webb [9] show that the laminar model progressively fails when the Reynolds number exceeds a certain critical value. The j and f factors on the fin surface are predicted by a different equation when the transition Reynolds number is exceeded.

Multiple-regression power-law correlations have been developed by Joshi and Webb [9] and Wieting [10] to predict the characteristics of j and f vs. Re for the OSF array. The correlations of Refs. [9] and [10] were based on data from 21 and 22 OSF heat exchangers, respectively. Both references present separate correlations for the laminar and nonlaminar wake regions. The flow-visualization measurements of Joshi and Webb show that the Reynolds number ($= GD_h/\mu$) at which the wake departs from laminar flow is given by

$$\mathrm{Re}^* = 257.2\left(\frac{l}{s}\right)^{1.23}\left(\frac{t}{l}\right)^{0.58}\left[\frac{t}{D_h} + 1.328\left(\frac{l}{D_h}\right)^{0.5}\mathrm{Re}^{-0.5}\right] \tag{17.6}$$

The j and f correlations for Re < Re* are given by

$$f_L = 8.118\,\mathrm{Re}^{-0.742}\left(\frac{l}{D_h}\right)^{-0.411}\left(\frac{s}{b}\right)^{-0.160} \tag{17.7a}$$

$$j_L = 0.534\,\mathrm{Re}^{-0.560}\left(\frac{l}{D_h}\right)^{-0.147}\left(\frac{s}{b}\right)^{-0.137} \tag{17.7b}$$

By examining j and f curves of the 21 test geometries, a Reynolds number of $\mathrm{Re}_D = \mathrm{Re}_D^* + 1000$ was taken as the lower limit for the nonlaminar correlation. The following equations were developed for the nonlaminar region [9]:

$$f_T = 1.36\,\mathrm{Re}^{-0.198}\left(\frac{l}{D_h}\right)^{-0.781}\left(\frac{t}{D_h}\right)^{0.534} \tag{17.8a}$$

$$j_T = 0.242\,\mathrm{Re}^{-0.368}\left(\frac{l}{D_h}\right)^{-0.322}\left(\frac{t}{D_h}\right)^{0.089} \tag{17.8b}$$

The j factor in the transition region between Re* < Re < Re* + 1000 is a log-linear interpolation between j_L (at Re*) and j_T (at Re* + 1000). A similar equation is used for the friction factor.

The correlations predicted 82% of the f data and 91% of the j data within ±15%. The geometric parameters of the test arrays on which Eqs. (17.7) and (17.8) were based are $0.13 < s/b < 1.0$, $0.012 < t/l < 0.048$, and $0.04 < t/s < 0.20$.

The Wieting correlation [10] assumes Re* = 1000. Equation (17.6) is expected to provide a more accurate prediction of the transition Reynolds number.

Louvered Fin. The louvered-fin geometry of Fig. 17.3f bears a similarity to the OSF. Rather than offsetting the slit strips, the entire slit fin is rotated 20 to 60° relative to the air flow direction. The louvered surface is the standard geometry for automotive radiators. Current radiators use a louver strip width (in the air flow direction) of 1.0 to 1.25 mm. For the same strip width, the louvered-fin geometry provides heat transfer coefficients comparable to the OSF. Shah and Webb [11] provide a description of modified louvered-fin geometries and give references to several data sources.

Based on tests of 32 louvered-fin geometries, Davenport [12] developed multiple regression correlations for the j and f vs. Re for

$$j = 0.249\,\mathrm{Re}_l^{-0.42} l_h^{0.33} H^{0.26}\left(\frac{l_L}{H}\right)^{1.1} \qquad (300 < \mathrm{Re} < 4000) \tag{17.9a}$$

$$f = 5.47\,\mathrm{Re}_l^{-0.72} l_h^{0.37} l^{0.2} H^{0.23}\left(\frac{l_L}{H}\right)^{0.89} \qquad (70 < \mathrm{Re} < 1000) \tag{17.9b}$$

$$f = 0.494\,\mathrm{Re}_l^{-0.39} H^{0.46}\left(\frac{l_h}{l}\right)^{0.33}\left(\frac{l}{H}\right)^{1.1} \qquad (1000 < \mathrm{Re} < 4000) \tag{17.9c}$$

Equation (17.9) contains dimensional terms; the required dimensions are millimeters. Figure 17.6 defines the terms in the equations. The characteristic dimension in the Reynolds number limits is D_h, not l_L as used in the correlations. Note that the corrugation has a flat base rather than being open as shown in Fig. 17.3*f*. In the author's opinion, Eqs. (17.9) are nevertheless applicable to Fig. 17.3*f* corrugation geometry, which is used in automotive radiators. The author has used Eq. (17.9*a*) to predict the heat transfer performance of automotive radiators and has found agreement within ±15% with the test data. However, the pressure drop was underpredicted 30 to 40%.

Wavy Fin. Kays and London [7] provide curves of j and f vs. Re for two wavy-fin geometries (Fig. 17.3c). Their performance is competitive with that of the OSF. No specific correlations exist for the j and f characteristics of wavy or herringbone fins. Two studies [13, 14] of wavy channel geometries used in plate-type heat exchangers provide additional data for small aspect ratio channels. Goldstein and Sparrow [15] used a mass transfer technique to measure the local mass transfer coefficient distribution for a herringbone wave configuration. They propose that the enhancement results from Goertler vortices that form as the flow passes over the concave wave surfaces. These are counterrotating vortices which have a corkscrew-like flow pattern. Flow visualization studies by this author show local zones of flow separation and reattachment on the concave surfaces. The redeveloping boundary layer from the reattachment point also contributes to heat transfer enhancement.

Perforated Fin. This surface geometry (Fig. 17.3*e*) is made by punching a pattern of spaced holes in the fin material before it is folded to form the U-shaped flow channels. If the porosity of the resulting surface is sufficiently high, enhancement can occur due to boundary-layer dissipation in the wake region formed by the holes. However, Shah [16] has shown that little enhancement occurs for Re < 2000 if the heat transfer coefficient is based on the plate area before the holes were punched. Moderate enhancement may occur in the transition and turbulent flow regimes, Re > 2000, depending on the hole size and the plate porosity. Shah provides a very detailed evaluation of the perforated fin based on his study of test data on 68 perforated-fin geometries.

Plain Fin. If plain fins are used (Fig. 17.3*a* and *b*), the flow channel will have a rectangular or triangular cross section. If the flow is turbulent, standard equations for turbulent flow in circular tubes may be used to calculate j and f, provided Re is based on the hydraulic diameter D_h. If the Re based on hydraulic diameter is less than 2000, one may use theoretical laminar flow solutions for j and f. Values of j and f for developing and fully developed laminar flow are given in Chapter 3 for a variety of duct shapes. Criteria to establish the value of x/D_h at which the flow becomes fully developed are also given in Chapter 3.

Entrance Length Effects. Extended surface geometries for gases are typically designed for operation at Re < 2000. If the flow length is sufficiently short, it is possible that the average j and f over the flow length are higher than the fully developed values. Plain-fin geometries are more susceptible to entrance length effects than are the enhanced geometries of Fig. 17.3*c*, *d* and *f*. Because of the periodic flow interruptions, it is unlikely that entrance region effects would exist for interrupted-fin heat ex-

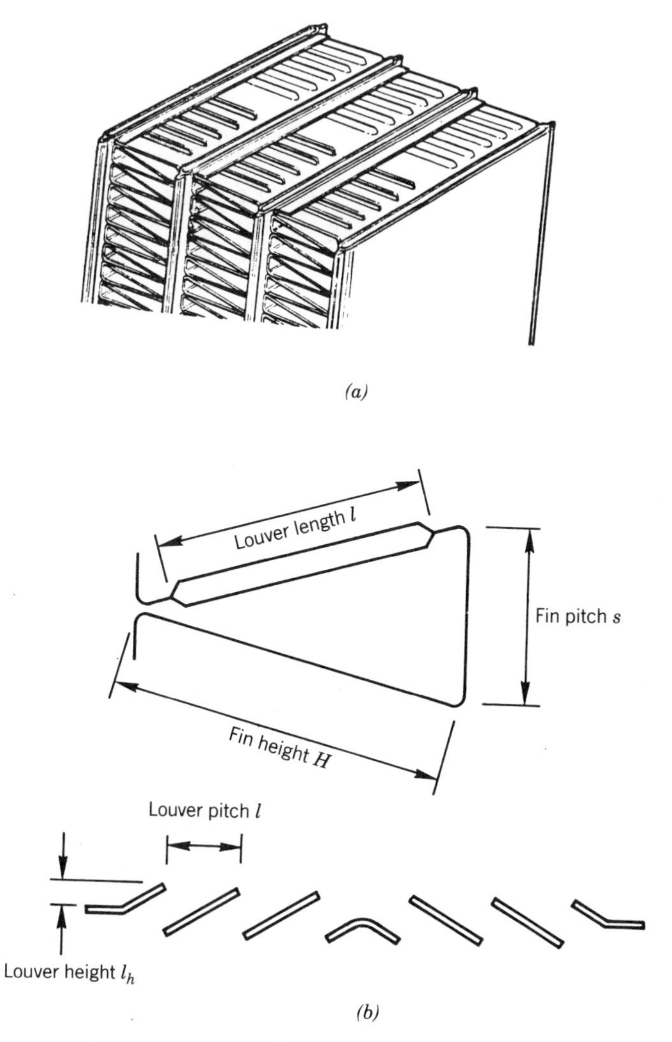

Figure 17.6. Louvered-fin geometry tested by Davenport [12]: (*a*) illustration of core geometry; (*b*) definition of parameters in Eqs. (17.9).

changers. However, this may not be a good assumption for plain-fin geometries. If developing flow exists over more than, say, 20% of the flow length, it is possible that the average Nu and f are moderately higher than the fully developed values. In this case, one should determine the average Nu and f over the air flow length.

The entrance region effect on heat transfer for several plain-fin channel geometries is illustrated in Fig. 17.7. This figure shows the ratio of the average Nu to the fully developed Nu for a constant wall temperature boundary condition with a developed velocity profile. Consider, for example, air flow (Pr = 0.7) in an equilateral triangular-shaped channel whose dimensionless flow length is $L^* = 0.10$. The mean Nu over the $L^* = 0.10$ flow length is 42% greater than the fully developed value.

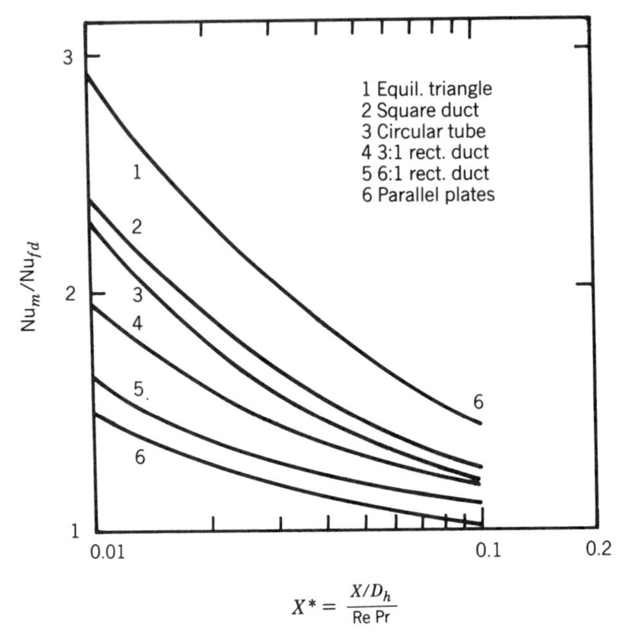

Figure 17.7. Nu_m/Nu_{fd} for laminar entrance region flow with developed velocity profile in different channel shapes for constant wall temperature boundary condition [11].

17.3.2 Finned-Tube Heat Exchangers

Figure 17.2*a* shows the finned-tube geometry with continuous flat plate fins on staggered tubes. An inline tube geometry is seldom used because it provides substantially lower performance than the staggered tube geometry [17].

Plain Fins. Correlations to predict the j and f factors vs. Reynolds number for plain fins on staggered tubes were developed by McQuiston [18] and by Gray and Webb [19]. The heat transfer correlation for four or more tube rows of a staggered tube geometry is [19]

$$j_4 = 0.14\,\text{Re}_d^{-0.328}\left(\frac{S_t}{S_l}\right)^{-0.50}\left(\frac{s}{d_o}\right)^{0.31} \tag{17.10}$$

Equation (17.10) implies that the heat transfer coefficient is stabilized by the fourth tube row; hence the j factor for more than four tube rows is the same as that for a four-row exchanger. The equation

$$\frac{j_N}{j_4} = 0.991\left[2.4\,\text{Re}_d^{-0.092}\left(\frac{N}{4}\right)^{-0.031}\right]^{0.607(4-N)} \tag{17.11}$$

predicts the j factor for one- to three-tube-row exchangers, relative to that of a four-row exchanger. Equations (17.10) and (17.11) correlated 89% of the data for 16 heat exchangers within 10%.

Gray and Webb [19] also developed a friction-factor correlation for plain flat fins, which is independent of the number of tube rows, and given by

$$f = f_{\mathrm{f}} \frac{A_{\mathrm{f}}}{A} + f_{\mathrm{tb}} \left(1 - \frac{A_{\mathrm{f}}}{A}\right)\left(1 - \frac{t}{s}\right) \tag{17.12}$$

This correlation assumes that the pressure drop is composed of two terms. The first term of Eq. (17.12) represents the drag force on the fins, and the second term represents the drag force on the tubes. The friction factor associated with the fins (f_{f}) is given by

$$f_{\mathrm{f}} = 0.508\,\mathrm{Re}_d^{-0.521} \left(\frac{S_t}{d_o}\right)^{1.318} \tag{17.13}$$

The friction factor associated with the tubes (f_{tb}) is obtained from a standard correlation for flow normal to a bank of bare tubes (see Chap. 6). The f_{tb} is calculated at the same mass velocity G that exists in the finned-tube exchanger. Equation (17.12) correlated 90% of the data for 19 heat exchangers within $\pm 20\%$.

The range of dimensionless variables used in the development of Equations (17.10) to (17.13) are $500 < \mathrm{Re}_d < 24{,}700$, $1.97 < S_t/d_o < 2.55$, $1.7 < S_l/d_o < 2.58$, and $0.08 < s/d_o < 0.64$.

Enhanced Fin Geometries. The wavy (or herringbone) fin and the offset strip fin (also referred to as parallel louver) are the major enhanced surface geometries used for flat fins on circular tubes. Figure 17.8 shows the wavy fin geometry applied to circular tubes. This figure shows the geometrical dimensions which influence the heat transfer and friction characteristics. The combination of tubes plus a special surface geometry establishes a very complex flow geometry. No correlations have been published in the open literature for prediction of the j and f factors. However, the air-conditioning industry uses wavy fins, and its product brochures contain rating data. The heat transfer coefficient of the wavy fin is 50 to 70% greater than that of a plain (flat) fin.

The OSF concept has been applied to finned-tube heat exchangers with flat fins for dry cooling towers and for refrigerant condensers. Figure 17.9 shows the heat transfer coefficients of the OSF and a plain fin used in a two-row staggered-tube coil having 966 fins/m on 10-mm-diameter tubes [20]. At 3 m/s air velocity, the OSF provides 78% higher heat transfer coefficient than the plain fin. The OSF of Fig. 17.5 provides 150% higher heat transfer coefficient than the plain fin at the same velocity. Further examination of Figs. 17.5 and 17.9 shows that the heat transfer coefficient of Fig. 17.9 is 90% greater than that of the Fig. 17.5 plain fin geometry. Thus, the flow acceleration and fluid mixing in the wake of the tube provides a substantial enhancement for plain fins on tubes.

Generalized empirical correlations for j and f vs. Re have not been developed for OSF finned tubes. However, Nakayama and Xu [20] propose an empirical correlation to define the enhancement level (h/h_p) of an OSF geometry having 2.0-mm strip width (in the flow direction) and 0.2 mm fin thickness.

17.3.3 Individually Finned Tubes

Helically wrapped or extruded fins on circular tubes as shown in Fig. 17.2b are frequently used in the process industries and in combustion heat-recovery equipment.

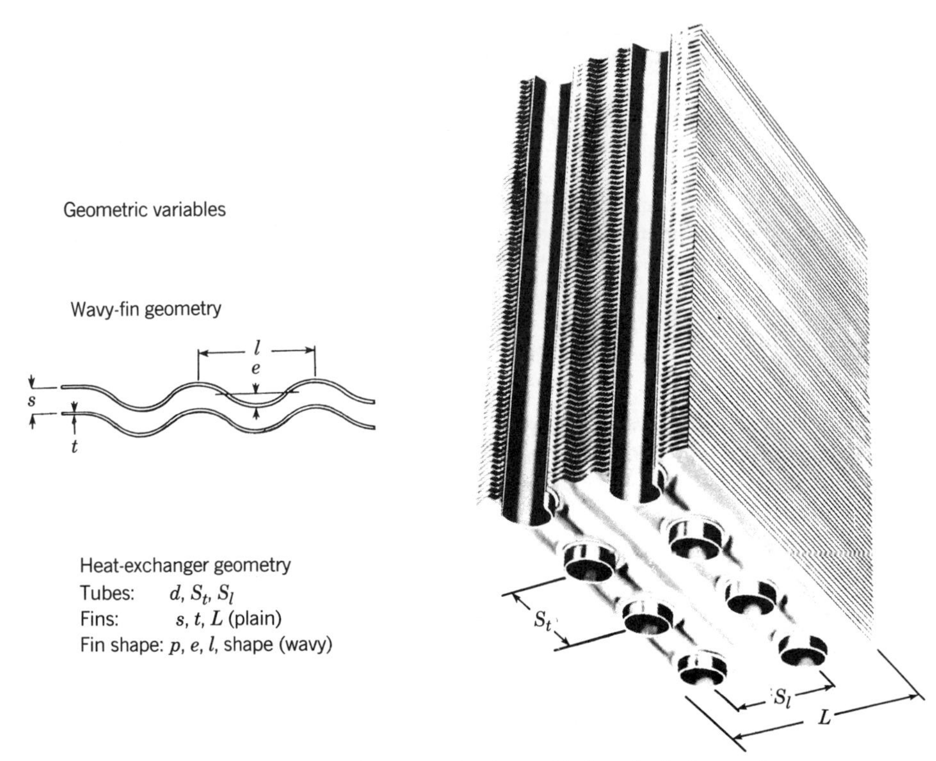

Figure 17.8. Geometric variables of the wavy flat fin used in finned-tube heat exchangers.

Both plain and enhanced fin geometries are used. A staggered tube layout is used, especially for high fins ($e/d_o > 0.2$).

Plain Fins. A substantial number of performance data have been published, and several heat transfer and pressure drop correlations have been proposed. The correlations must take account of the three tube-bank variables (d_o, S_t, and S_l) and the three fin geometry variables (t, e, s). Webb [21] provides a survey of the published data and correlations.

The recommended correlations for a staggered-tube layout are given by Briggs and Young [22] for heat transfer and Robinson and Briggs [23] for pressure drop. Both correlations are empirically based and are valid for four or more tube rows. The heat transfer correlation is

$$j = 0.134\,\mathrm{Re}_d^{-0.319}\left(\frac{s'}{e}\right)^{0.2}\left(\frac{s'}{t}\right)^{0.11} \tag{17.14}$$

Equation (17.14) is based on air flow over 14 equilateral triangular tube banks and covers the following ranges: $1100 < \mathrm{Re}_d < 18{,}000$, $0.13 < s'/e < 0.63$, $1.0 < s'/t <$

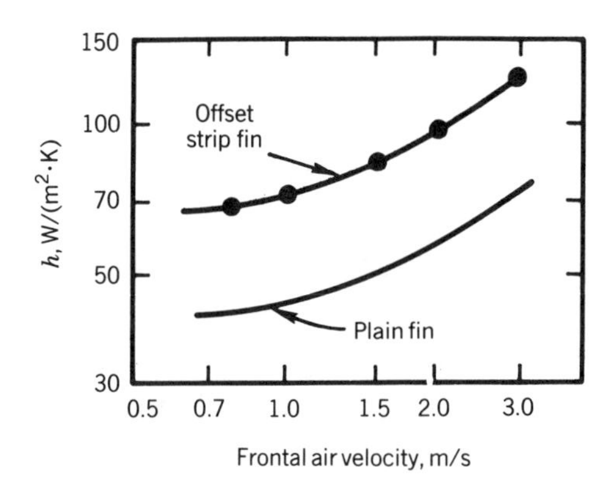

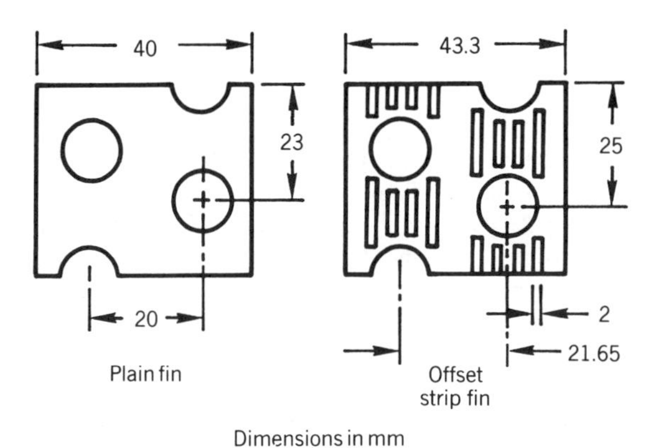

Figure 17.9. Comparison of the heat-transfer coefficient for the offset strip fin and the plain fin geometries for 9.5-mm-diameter tubes, 525 fins/m, and 0.2-mm fin thickness [20].

6.6, $0.09 < e/d_o < 0.69$, $0.01 < t/d_o < 0.15$, $1.5 < S_t/d_o < 8.2$, $11.1 \leq d_o \leq 40.9$ mm, and fin density 246 to 768 fins per meter. The standard deviation was 5.1%.

The isothermal friction correlation [23] written in terms of the tube-bank friction factor f_{tb} is

$$f_{tb} = 9.47 \, \mathrm{Re}_d^{-0.316} \left(\frac{S_t}{d_o} \right)^{-0.927} \left(\frac{S_t}{S_d} \right)^{0.515} \qquad (17.15)$$

Equation (17.15) is based on isothermal air flow data over 17 triangular pitch tube banks (15 equilateral and two isosceles). The following ranges were covered: $2000 < \mathrm{Re}_d < 50{,}000$, $0.15 < s'/e < 0.19$, $3.8 < s'/t < 6.0$, $0.35 < e/d_o < 0.56$, $0.01 < t/d_o < 0.03$, $1.9 < S_t/d_o < 4.6$. The standard deviation of the correlated data was 7.8%.

Equation (17.15) is recommended with reservations, because it does not contain any of the fin geometry variables (e, s', or t). Because only a small range of s'/e was covered in the tests, it is possible that the correlation may fail outside the range of the s'/e tested.

Although the data on which Equation (17.14) is based included low-fin data, e.g., $e/d_o \sim 0.1$, Rabas et al. [24] developed more accurate j and f correlations for low fin heights and fin spacings. The correlations are given below with the exponents rounded off to two significant digits:

$$j = 0.292\,\mathrm{Re}^{n}\left(\frac{s'}{d_o}\right)^{1.12}\left(\frac{s'}{e}\right)^{0.26}\left(\frac{t}{s'}\right)^{0.67}\left(\frac{d_e}{d_o}\right)^{0.47}\left(\frac{d_e}{t}\right)^{0.77} \tag{17.16}$$

where $n = -0.415 + 0.0346(d_e/s')$;

$$f = 3.805\,\mathrm{Re}^{-0.234}\left(\frac{s'}{d_e}\right)^{0.25}\left(\frac{e}{s'}\right)^{0.76}\left(\frac{d_o}{d_e}\right)^{0.73}\left(\frac{d_o}{S_t}\right)^{0.71}\left(\frac{S_t}{S_l}\right)^{0.38} \tag{17.17}$$

The equations are valid for staggered tubes with $N \geq 6$, $5000 < \mathrm{Re} < 25{,}000$, $1.3 < s'/e < 1.5$, $0.01 < s/t < 0.06$, $e/d_o \sim 0.10$, $0.01 < t/d_o < 0.02$, and $1.3 < S_t/d_o < 1.5$. The equations predicted 94% of the j data and 90% of the f data within $\pm 15\%$.

A staggered tube layout gives higher values for j at the same Re, especially for high fins ($e/d_o > 0.3$). Hence, the in line tube layout is not recommended for $e/d_o > 0.3$. Designers interested in in line finned tube banks should refer to Schmidt [25], who developed a heat-transfer correlation based on data from 11 sources.

Enhanced Fin Geometries. Figure 17.10 shows some of the enhanced fin geometries that have been used on circular tubes. In Webb's Table 2 [21], references are provided to information on performance of the Fig. 17.10 fin geometries. All of the geometries provide enhancement by the periodic development of thin boundary layers on small-diameter wires or flat strips, followed by their dissipation in the wake region between elements.

Perhaps the most popular enhancement geometry is the segmented fin (Fig. 17.10*d*), which is similar in concept to the offset strip fin shown in Fig. 17.9. The segmented fin is used in a wide range of applications, from air conditioning to boiler economizers. Figure 17.11 shows two versions of the segmented fin used in air-conditioning applications [26, 27]. In Fig. 17.11*b*, a 0.15-mm-thick aluminum fin strip is tension-wound on the tube and bonded to the tube with an epoxy resin. The fin segment width is 0.75 mm. Figure 17.12 shows the j and f performance of the Fig. 17.11*a* and *b* fin geometries and compares their performance with that of plain fins [28].

The data of Fig. 17.12 are for eight rows on a triangular pitch with $d_o = 12.7$ mm and $t = 0.51$ mm. Note that the Fig. 17.12 fins have a negative fin tip clearance. Figure 17.12 shows that the Fig. 17.11*a* and *b* geometries have approximately the same j and f for the same fin spacing. At $\mathrm{Re}_d = 4000$, we have $j_2/j_1 \approx 2.1$ and $f_2/f_1 \approx 4.2$, where subscript 1 refers to the plain fin. The substantial heat transfer enhancement is accompanied by a large friction increase. One is interested in increasing the value of hA/L by the use of an enhanced fin geometry. The Fig. 17.11*a* fin has only 43% as much surface area as a plain fin [28]. Hence, with $j_2/j_1 = 2.1$, one would obtain

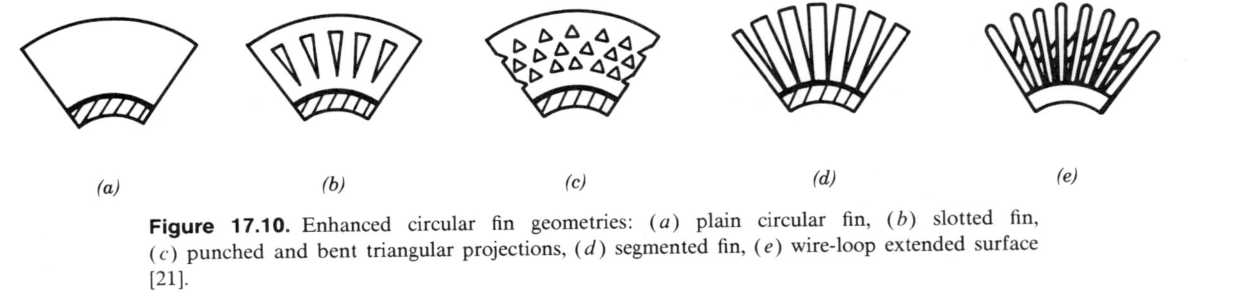

Figure 17.10. Enhanced circular fin geometries: (*a*) plain circular fin, (*b*) slotted fin, (*c*) punched and bent triangular projections, (*d*) segmented fin, (*e*) wire-loop extended surface [21].

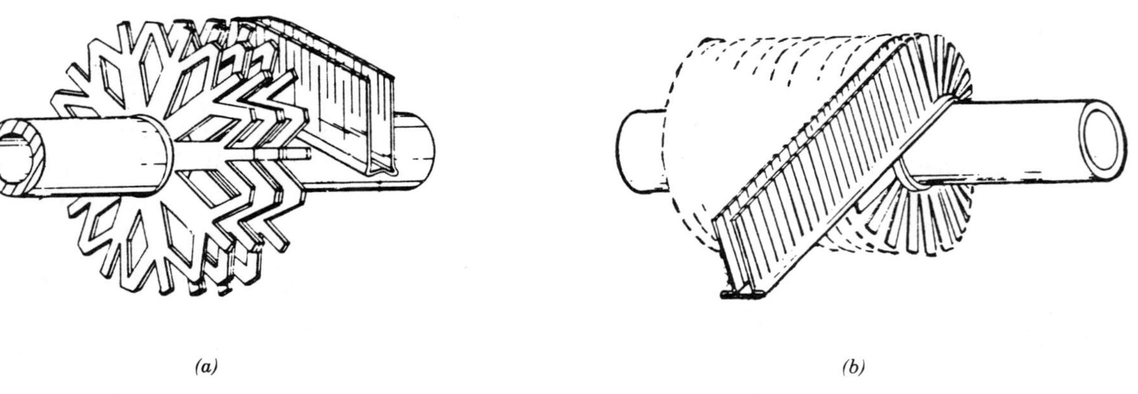

Figure 17.11. Segmented fins used in air-conditioning applications described by (*a*) LaPorte et al. [26] and (*b*) Abbott et al. [27], and tested by Eckels and Rabas [28].

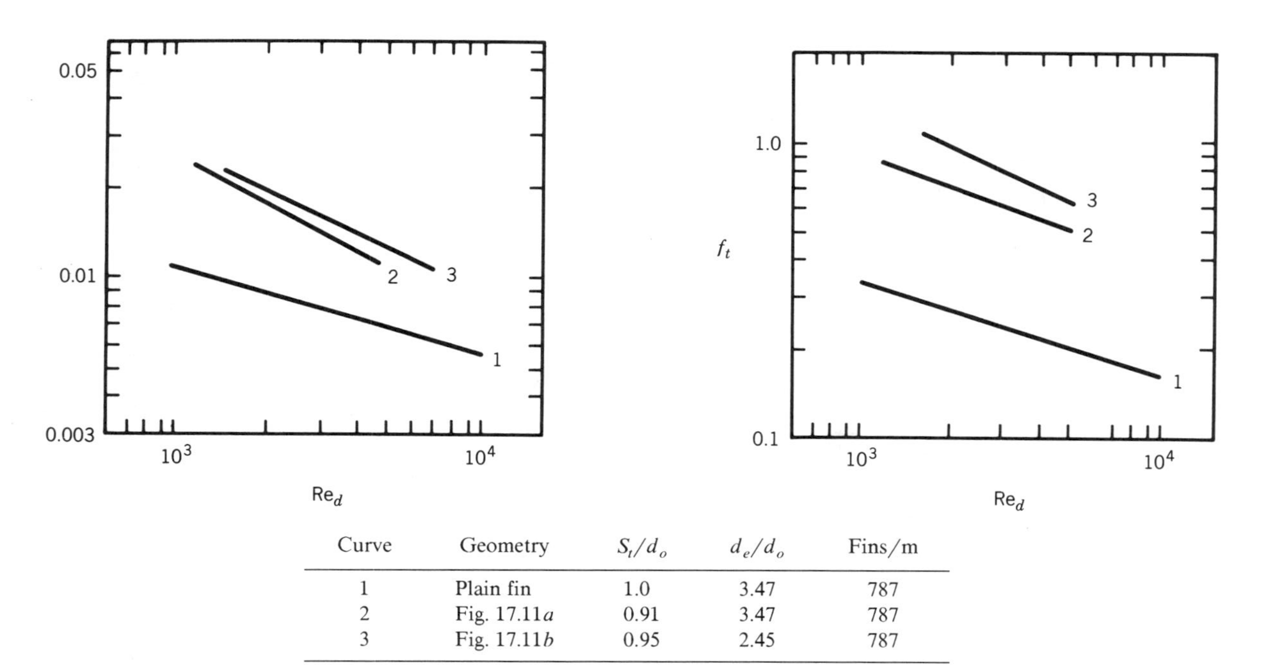

Curve	Geometry	S_t/d_o	d_e/d_o	Fins/m
1	Plain fin	1.0	3.47	787
2	Fig. 17.11*a*	0.91	3.47	787
3	Fig. 17.11*b*	0.95	2.45	787

Figure 17.12. Plots of j and f vs. Re_d of Fig. 17.11: enhanced fin geometries compared with plain circular finned-tube geometry [28].

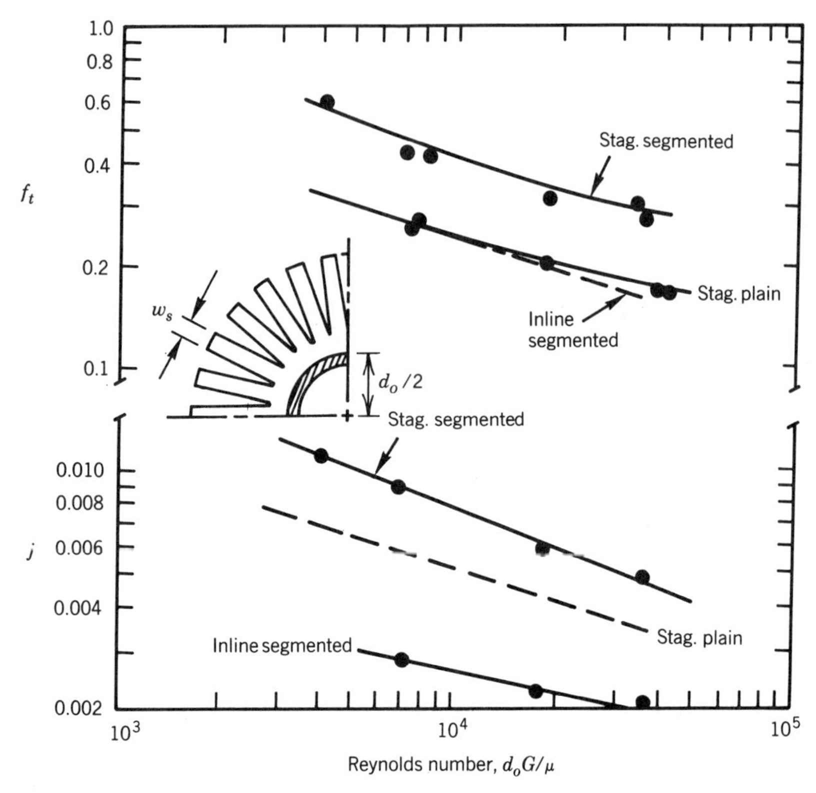

Figure 17.13. Comparison of segmented fins (staggered and inline) with plain staggered fin-tube geometry [29]. $S_t/d_o = 2.25$, $e/d_o = 0.51$, $s'/e = 0.12$, $s'/t = 2.51$, $w_s/e = 0.17$.

$h_1A_1/h_2A_2 \approx 0.90$. Thus, the Fig. 17.11*a* geometry would yield a 57% saving of fin material for the same *hA* and base tube length.

A steel segmented fin geometry has been used for boiler economizers and waste heat recovery boilers. Figure 17.13 shows the curves of j and f vs. Re_d for segmented-fin geometries with four staggered and seven inline tube rows [29]. Also shown are the j and f curves for a staggered plain fin geometry having the same geometrical parameters as the staggered segmented geometry. The plain-fin j and f values were calculated using Eqs. (17.14) and (17.15) respectively. For a staggered tube geometry, Fig. 17.13 shows that the j factor of the segmented fin is 40% greater than that of the plain fin geometry. The figure also shows that the heat transfer performance of the seven-row inline segmented fin geometry is poorer than that of the staggered plain fin geometry, while having approximately the same friction factor. Weierman [30] gives empirical design correlations for steel segmented and plain fin geometries. Rabas and Eckels [31] present additional data on steel segmented-fin tubes.

17.3.4 Oval and Flat Tube Geometries

Tube shapes of oval and flat cross section are also used for individually finned tubes. Figure 17.14 compares the performance of staggered banks of oval and circular finned

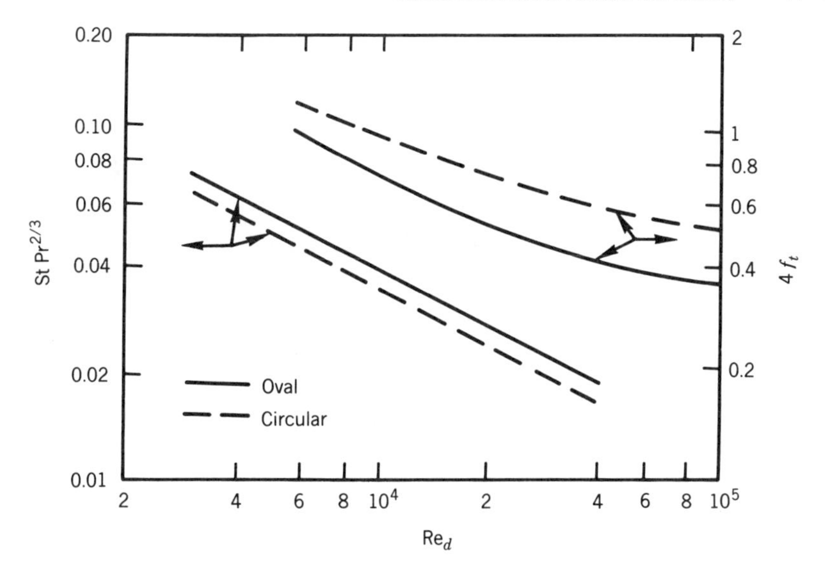

	Circular	Oval
Tube diam. d_o (mm)	2.9	19.9/35.2
Fin height l (mm)	9.8	10/9.3
Fin thickness t (mm)	0.4	0.4
Face pitch S_t/d_o (mm)	1.03	1.05
Row pitch S_l/d_o (mm)	1.15	1.04
Fins/m	312	312

Figure 17.14. Heat transfer and flow-friction characteristics of circular and oval finned tubes in a staggered arrangement [32].

tubes tested by Brauer [32]. Both banks have 312-fin/m, 10-mm-high fins on approximately the same transverse and longitudinal pitches. The oval tubes gave 15% higher heat transfer coefficient and 25% lower pressure drop than the circular tubes. The performance advantage of the oval tubes results from lower form drag on the tubes and a smaller wake region on the fin behind the tube. The use of oval tubes may not be practical unless the tube-side design pressure is sufficiently low.

Higher design pressures are possible using flattened aluminum tubes made by an extrusion process [33]. Such tubes can be made with internal membranes which strengthen the tube and allow for a high tube-side design pressure. A variety of fin and tube shapes may be made by aluminum extrusions of different shapes. Figure 17.15 shows a patented finned tube concept made from an aluminum extrusion [32]. The fins are formed from the thick wall using a modified high-speed punch press without creating scrap material. The punch press slits the thick aluminum wall and simultaneously bends the chip outward to form the fin. The process is applicable to virtually any cross-section geometry of the extrusion. Designs have been made with circular and flattened tube geometries. Haberski and Raco [34] show photographs of a number of geometries which have been fabricated. Cox [35] provides test data on a circular tube geometry, and Cox and Jallouk [36] on the Fig. 17.15 geometry.

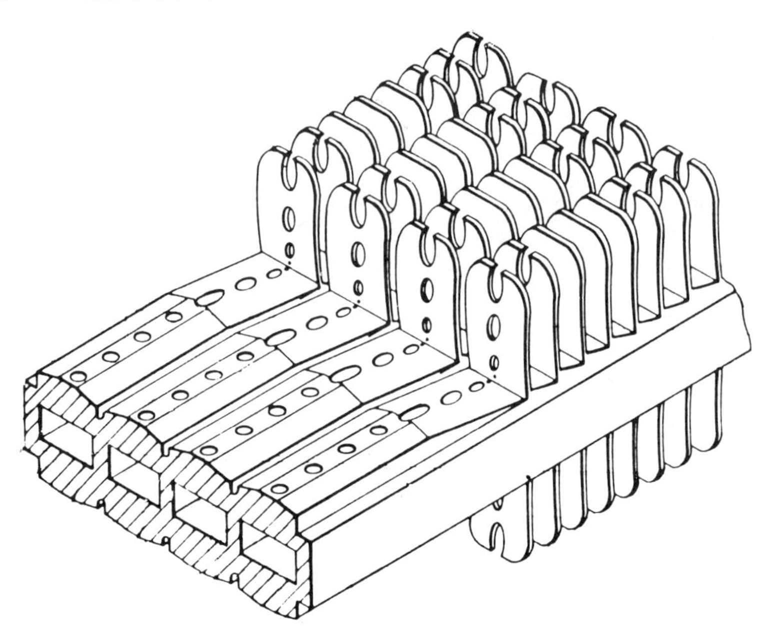

Figure 17.15. Finned tube made from thick-walled aluminum extrusion [33].

17.3.5 Row Effects — Staggered and Inline Layout

The published correlations are generally for deep tube banks and do not account for row effects. The heat transfer coefficient will decrease with rows in an inline bank due to the bypass effects, but the coefficient may increase with rows in a staggered bank. This is because the turbulent eddies shed from the tubes cause good mixing in the downstream fin region. As an approximate rule, one may assume that the heat transfer coefficient for a staggered tube bank of the Fig. 17.2a and b finned tubes has attained its asymptotic value at the fourth tube row [37].

Inline tube banks generally have a smaller heat transfer coefficient than staggered tube banks. At low Re_d ($\mathrm{Re}_d < 1000$) with deep tube banks ($N \geq 8$), the heat transfer coefficient may be as small as 60% of the staggered-tube value [38]. The heat transfer coefficient of the inline bank increases as Re_d is increased; at $\mathrm{Re}_d = 50{,}000$ with $N \geq 8$, the ratio of inline to staggered coefficients may approach 0.80.

There is a basic difference in the flow phenomena in staggered and inline finned-tube banks. Figure 17.16 shows the flow patterns through staggered and inline tube arrangements [32]. The streamlines are shown by the dashed lines. The low-velocity wake regions, or dead spaces, are shaded with fine dots. A much greater fraction of the fin surface area is contained in the low-velocity wake region for the inline arrangement than for the staggered arrangement. Consequently, the inline arrangement will have a lower surface-average heat transfer coefficient. Outside the shaded zone, particularly between the fin tips, a strong bypass stream exists. Because of poor mixing between the wake stream and bypass stream, the weaker wake stream is quickly heated; its mixed temperature is greater than that of the bypass stream. Thus, the actual temperature difference between the surface and the wake stream is much less than indicated by an

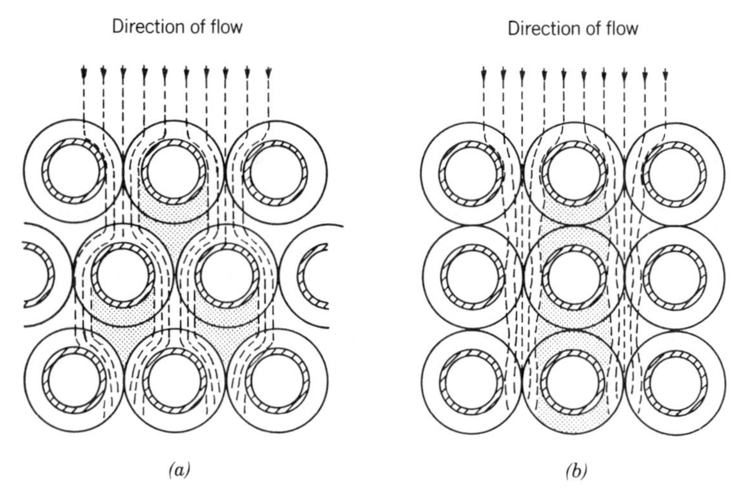

Figure 17.16. Flow patterns observed by Brauer [32] for: (a) staggered and (b) inline finned-tube banks.

overall LMTD based on the mixed outlet temperature. The staggered arrangement provides a good mixing of the wake and bypass streams after each tube row.

The superiority of the staggered fin geometry is shown in Fig. 17.13 [29] for $e/d_o = 0.51$ segmented fins. The staggered bank (st) has four rows on an equilateral triangular pitch ($S_t/d_o = 2.25$), and the seven-row inline (il) bank has $S_t/d_o = S_l/d_o = 2.25$. At $\mathrm{Re}_d = 10{,}000$, $j_{st}/j_{il} = 2.17$ and $f_{st}/f_{il} = 1.73$. Figure 17.17 shows the row effect of the inline segmented fin geometry. It shows that the bypass effect reduces the performance of the inline geometry as the number of rows increases. Rabas and Huber [38] have performed tests of the performance differences of staggered and inline banks of tubes having plain fins. Their work shows that the inline bank attains an asymptotic

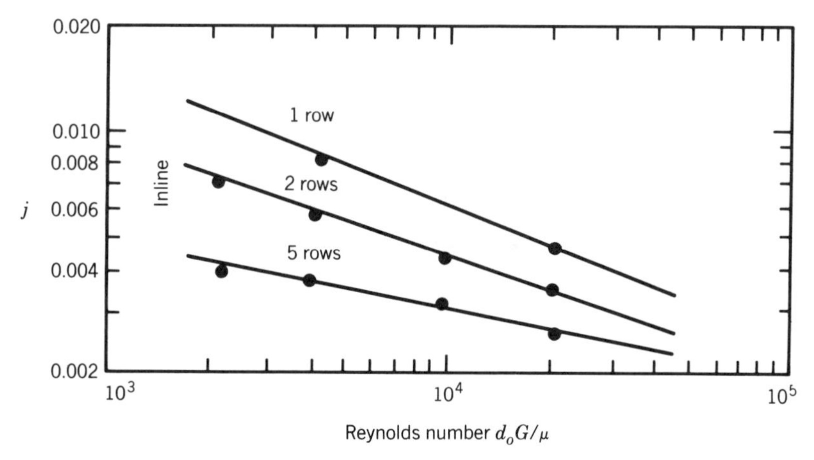

Figure 17.17. Row effect of in line tube banks [29] for the Fig. 17.13 in line and segmented fin tubes ($S_t/d_o = S_l/d_o = 2.25$).

value of j and f. In general, their work shows that plain, inline finned-tube banks yield the lowest performance for: (1) low Re_d, (2) large S_t/d_o, and (3) small s/d_o.

17.4 PACKINGS FOR GAS-GAS REGENERATORS

Regenerators, either rotary or valved [7], are commonly used to transfer heat from combustion products to inlet combustion air. The valved type has two identical packings which alternately serve the hot and cold streams and are switched by quick-operating valves. Any of the corrugated-plate fin geometries discussed in Sec. 17.3.1 may be applied to generators. Because the hot and cold streams are normally at different pressures, any packing geometry that allows significant transverse flow leakage will reduce the performance of a rotary regenerator. The louvered and offset-strip fin geometries are susceptible to this problem. However, one may design around it by

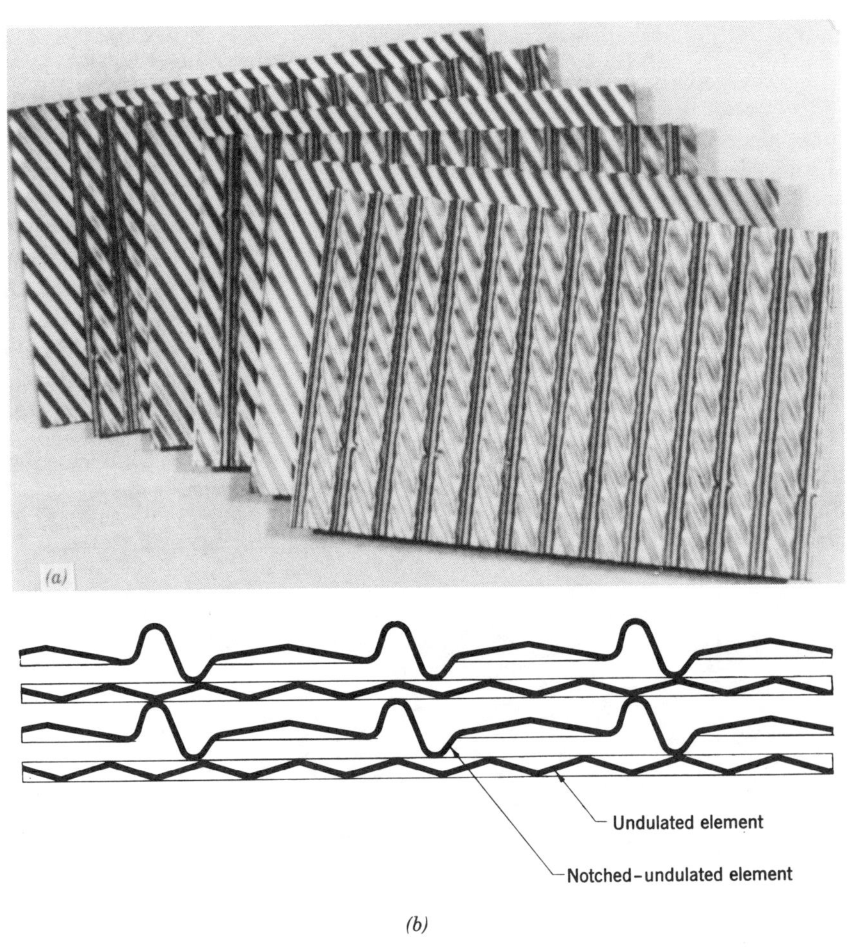

Figure 17.18. Matrix geometries for rotary regenerators: (a) illustration of plate stacking, (b) cross section of stacked plates. Courtesy of Combustion Engineering Air Preheater Division, Wellsville, NY.

including full-height continuous fins aligned with the flow at discrete spacings. The "brick checkers" geometry commonly used in the valved-type glass furnace regenerator is essentially an offset strip fin (Fig. 17.3d) having a large fin thickness.

Packings having small hydraulic diameter (small s) will provide the highest heat transfer coefficients; for laminar flow, $h \propto 1/D_h$. However, the fouling characteristics of the hot gas may limit the size of the flow passage. Considerably smaller passage size may be used in ventilation heat-recovery regenerators than in those used for heat recovery from coal-fired exhaust gases or glass furnace exhausts. Coal-fired electric utility plants frequently use corregated plate packings, such as that illustrated in Fig. 17.18. Conceivably, corrugated plate geometries similar to those used in plate-type heat exchangers may also be used.

17.5 EXTENDED SURFACES FOR LIQUIDS

Extended surfaces used with liquids may be on the inner or outer surface of the tubes, as shown in Fig. 17.1a. Because liquids have higher heat transfer coefficients than gases, fin efficiency considerations require shorter fins with liquids than with gases. The thermal conductance per unit tube length is $\eta_0 hA/L$. Typical finned surfaces provide an A/L in the range of 1.5 to 3 times that of a bare tube. Tube materials other than copper and aluminum may have relatively low thermal conductivity, which will also restrict the possible fin height if a moderately high fin efficiency is desired.

17.5.1 Externally Finned Tubes

Equations (17.16) and (17.17) are recommended for a staggered layout of low integral fins. Corresponding equations have not been developed for inline tube layouts. However, Brauer's j data [32] for an $e/d_o = 0.07$ inline tube bank was within 20% of that of a staggered bank having the same e/d_o, S_t/d_o, and S_l/d_o. The inline friction factor was approximately 35% smaller than that of the staggered bank. It appears that the performance decrement of inline banks having low fins (e.g., $e/d_o = 0.1$) is not nearly as severe as for high fins (e.g., $e/d_o = 0.4$).

17.5.2 Internally Finned Tubes

The flow in internally finned tubes (Fig. 17.19) may be either laminar or turbulent. When used to cool viscous fluids, such as oil, it is possible that the flow will be laminar.

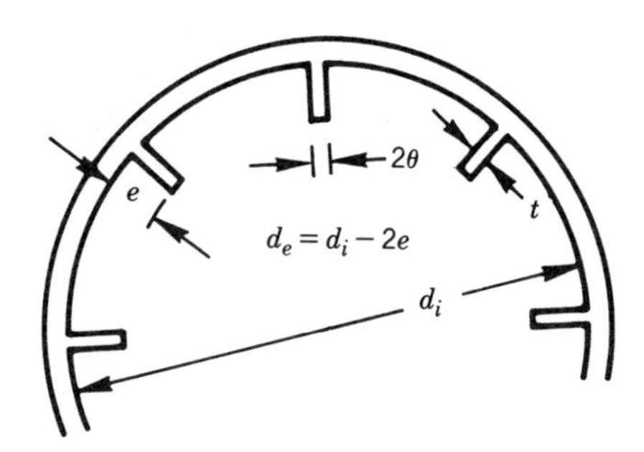

Figure 17.19. Cross section of an internally finned tube.

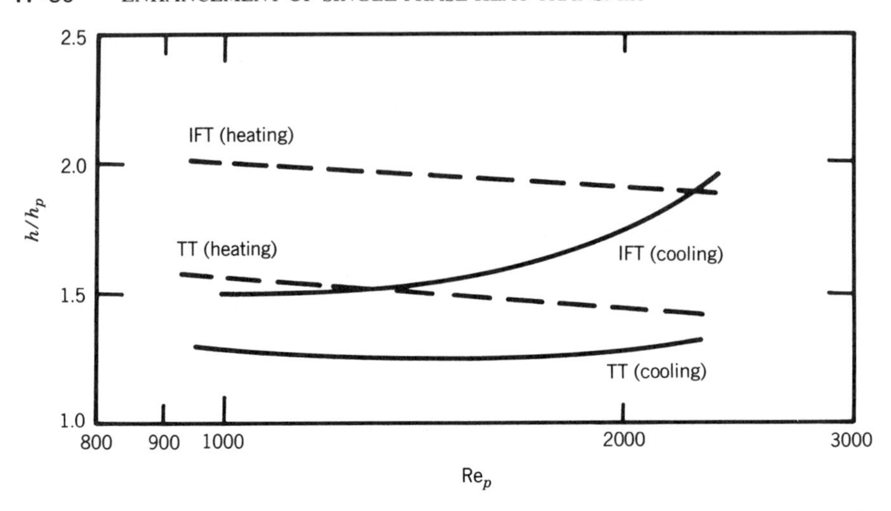

Figure 17.20. Performance comparison of internally finned tubes (IFT, $n_f = 16$, $e/d_i = 0.084$, $\alpha = 27°$) and twisted tape insert (TT, $\alpha = 30°$, $Y = 5.4$, $t/d_i = 0.053$) for laminar flow using case FG-2a of Table 17.2 [40].

Laminar Flow. Watkinson et al. [39] report Nu and f data ($50 < \text{Re} < 3000$) for steam heating of oil ($180 < \text{Pr} < 350$) in 18 different internally finned tubes. Marner and Bergles [40] report Nu and f for a tube with 16 axial fins ($e/d_i = 0.026$) for both heating and cooling conditions ($24 < \text{Pr} < 85$, $380 < \text{Re} < 3470$). Figure 17.20 shows h/h_p of their data for PEC FG-2a (Table 17.2). This figure also contains results for a twisted tape insert ($\alpha = 30°$) illustrated by Fig. 17.24. Figure 17.20 shows that: (1) the internal-fin performance is better in heating than in cooling, and (2) the internal-fin geometry is superior to the twisted tape.

Soliman et al. have numerically solved the energy equation for fully developed laminar flow in internally finned tubes and predicted the Nusselt number for constant wall temperature [41] and constant heat flux [42] boundary conditions. They assume constant fluid properties and no free-convection effects. The ranges of fin geometries analyzed are $0.1 < e/d_i < 0.4$, $4 < n_f < 32$, and $1.5 < \theta < 3°$, where n_f is the number of fins and 2θ is the fin included angle shown in Fig. 17.19. The analysis includes the effect of fin thermal conductivity as defined by the parameter $\theta^* \equiv \theta k_f/k$, where k and k_f are the thermal conductivities of the fluid and fin material respectively. Soliman and Feingold [43] have analytically solved the momentum equation for the friction factor. Table 17.4 shows the calculated Nu′ and f, relative to the plain-tube value (subscript p). The friction factor and Nu′ refer to the diameter and surface area of a plain tube. Hence, the ratio Nu'/Nu_p is the ratio of the conductance (hA/L) of the finned tube to that of the plain tube. Similarly, the pressure drop increase at fixed velocity is given by f/f_p. Table 17.4 also shows the area ratio (A/A_p). The table includes Nu'/Nu_p for $\theta k_f/k$ of 5 and ∞. Several important conclusions may be drawn from Table 17.4:

1. For $e/D < 0.3$, the maximum Nu'/Nu_p occurs with $n_f = 8$. However, f/f_p continues to increase for $n_f > 8$. Hence the use of $n_f > 8$ is of no value in providing increased hA/L.

TABLE 17.4. Enhancement Ratios Provided by Internally Finned Tubes for Fully Developed Laminar Flow, as Predicted by Soliman et al. [41–43][a]

				$(\mathrm{Nu}'/\mathrm{Nu}_p)_T$		$(\mathrm{Nu}'/\mathrm{Nu}_p)_{H1}$	
n_f	$\frac{e}{d_i}$	$\frac{A}{A_p}$	$\frac{f}{f_p}$	$\theta^* = 5$	$\theta^* = \infty$	$\theta^* = 5$	$\theta^* = \infty$
4	0.1	1.26	1.24	1.04	1.04	1.05	1.07
	0.2	1.51	1.91	1.28	1.30	1.38	1.45
	0.3	1.76	3.28	2.25	2.44	2.47	2.84
	0.4	2.02	4.80	3.73	4.40	3.61	4.52
8	0.1	1.51	1.57	1.06	1.06	1.08	1.10
	0.2	2.02	3.53	1.29	1.31	1.50	1.56
	0.3	2.58	8.67	2.41	2.50	3.79	4.84
	0.4	3.07	14.5	8.07	9.64	7.81	10.45
16	0.1	2.02	2.02	1.03	1.03	1.06	1.06
	0.2	3.04	5.93	1.09	1.10	1.18	1.21
	0.3	4.06	22.2	1.45	1.47	2.07	2.18
	0.4	5.07	60.8	8.59	8.66	17.4	24.4
24	0.1	2.53	2.26	1.01	1.01	1.02	1.02
	0.2	4.06	6.99	1.02	1.03	1.05	1.06
	0.3	5.58	31.7	1.13	1.13	1.29	1.32
	0.4	7.11	172.0	3.29	3.30	9.31	10.5
32	0.1	3.04	2.36	1.00	1.00	1.00	1.01
	0.2	5.08	6.99	1.00	1.00	1.01	1.02
	0.3	7.11	36.3	1.03	1.03	1.08	1.09
	0.4	9.15	355.0	1.66	1.66	2.81	2.91

[a] Subscript refers to plain tube, and $\theta^* = \theta k_m/k$.

2. $\mathrm{Nu}'/\mathrm{Nu}_p < A/A_p$, except for the highest fins at $4 < n_f < 16$. However, the friction-factor increase is greater than the area increase for all geometries.
3. The $\mathrm{Nu}'/\mathrm{Nu}_p$ for $\theta k_f/k = 5$ is within 10% of that for $\theta k_f/k = \infty$. The case $\theta k_f/k = \infty$ corresponds to 100% fin efficiency.

Table 17.5 shows values of $\theta k_m/k$ for different material-fluid combinations of interest. Examination of this table shows that the values listed in Table 17.4 for $\theta k_f/k = \infty$ are of primary interest. For $\theta k_f/k = \infty$, one may apply the conventionally used formulas for fin efficiency (η_f) to the Nu for $\theta k_f/k = \infty$ and obtain acceptable accuracy. Soliman et al. [41–43] have not attempted to compare their predicted Nu and

TABLE 17.5. Values of $\theta k_f/k$ for Different Fluid-Material Combinations at 25°C

	$\theta k_f/k$		
	Air	Water	Oil
Material	Pr = 0.7	6	1200
Copper	16000	670	2700
Aluminum	8000	330	1300
Steel	2000	80	400

f values for constant properties and fully developed flow with available data [39, 40]. It is likely that experimental values for heating will exceed the Table 17.4 Nusselt numbers. The experimental data are typically for a certain tube length, which may include entrance region and free-convection effects. If the fluid is heated, such experimental Nu values may be substantially greater than the values predicted by Soliman. However, cooling tends to reduce the Nu′ and counters the enhancement provided by free-convection and entrance region effects; this is seen in Fig. 17.20.

Bergles and Joshi [44] provide additional comparisons of the theoretical and experimental performance of internally finned tubes in laminar flow. Watkinson et al. [39] also present data on helical internally finned tubes. It appears that the fin helix angle adds to the enhancement.

Soliman's predictions [42] of Nu′ for constant heat flux are substantially below the experimental values of Marner and Bergles [40]. The data of Marner and Bergles also show a strong Prandtl number effect, while Soliman's results are independent of Pr. A partial explanation for the difference is that Marner and Bergles's data are not in the fully developed region.

Turbulent Flow. Carnavos [45] developed empirical correlations of Nu and f vs. Re for turbulent flow in internally finned tubes. His correlation was based on tests of 21 surface geometries, including helix angles up to 30°. The data are for heating of fluids ($6 < \Pr < 30$) and span $10{,}000 < \text{Re} < 60{,}000$. Carnavos attempted to correlate the data using the hydraulic diameter in standard St and f equations for turbulent flow in plain tubes. He found that this method underpredicted Nu and f for axial fins. He then developed geometry-dependent correction factors to correlate the data. Webb and Scott [46] restated the correlations in terms of the fundamental dimensional geometry parameters. The Carnavos correlations for straight and helical fins as stated by Webb and Scott [46] are

$$\frac{\text{Nu}}{\text{Nu}_p} = \frac{hD_h/k}{h_p d_i/k} = \left[\frac{d_i}{d_{im}}\left(1 - \frac{2e}{d_i}\right)\right]^{-0.2}\left(\frac{d_i D_h}{d_{im}^2}\right)^{0.5} \sec^3\alpha \tag{17.18}$$

$$\frac{f}{f_p} = \frac{d_{im}}{d_i}\sec^{0.75}\alpha \tag{17.19}$$

The quantity d_{im} is the tube inside diameter that would exist if the fins were melted and the material returned to the tube wall. The friction factor and Nusselt number in the above equations are defined in terms of the total surface area and hydraulic diameter, and $\text{Re} = D_h G/\mu$. Carnavos used the Dittus-Boelter and the Blasius equations to calculate Nu_p and f_p respectively . These equations are

$$\text{Nu}_p = \frac{h_p d_i}{k} = 0.023\,\text{Re}_p^{0.8}\Pr^{0.4} \tag{17.20}$$

$$f_p = 0.046\,\text{Re}_p^{-0.2} \tag{17.21}$$

Equations (17.18) and (17.19) correlated Carnavos's data on 21 tube geometries within $\pm 8\%$. The ranges of dimensionless geometric parameters covered by the correlated data are $0.03 < e/d_i < 0.24$, $1.4 < \pi d_i/n_f e < 7.3$, $0.1 < t/e < 0.3$, and

$0 < \alpha < 30°$. If one desires to calculate the enhancement ratio in terms of h' (based on $\pi d_i L$), i.e., $\mathrm{Nu}'/\mathrm{Nu}_p$, one writes

$$\frac{\mathrm{Nu}'}{\mathrm{Nu}_p} = \frac{\mathrm{Nu}}{\mathrm{Nu}_p} \cdot \frac{d_i}{D_h}\left(1 + \frac{2n_f e}{\pi d_i}\right) \tag{17.22}$$

Similarly, f'/f_p is given by

$$\frac{f'}{f_p} = \frac{f}{f_p} \cdot \frac{d_i}{D_h} \tag{17.23}$$

Webb and Scott [46] have applied cases VG-1, 2, and 3 of the Table 17.2 PEC to turbulent flow in internally finned tubes and determined their performance relative to plain tubes. A key purpose of this analysis was to determine preferred internal fin geometries. This analysis assumes all of the thermal resistance is on the tube side. The results for the VG-1 analysis are presented in PEC Example 2 below.

Yampolsky [47] has tested an interesting variant of an internally finned tube (Fig. 1*e*). This tube is formed by corrugating strip material and then rolling it in a circular form such that the corrugations are at a helix angle $\alpha = 30°$. Turbulent flow data for $\alpha = 30°$ show that $j/j_p \simeq 1$, where f, j, and Re are based on the hydraulic diameter. The tube tested, whose diameter was 31.8 mm (1.25 in.), provided a 60% internal area increase.

PEC EXAMPLE 2

Assume that a plain tube heat exchanger has been designed to provide heat duty q with specified flow rates and inlet temperatures. The tubes are 19.2 mm O.D. with 0.7 mm wall thickness. The design operates at Re_p with N_p tubes per pass and tube length L_p per pass. One desires to use internally finned tubes to reduce the total length of tubing (NL). The design constraints are those of case VG-1 of Table 17.2. Assuming the total thermal resistance is on the tube side, Eqs. (17.1) to (17.3) apply. Using Eqs. (17.22) and (17.23), one solves Eq. (17.3) for G/G_p. For constant inside diameter, $G/G_p = (\mathrm{Re}/\mathrm{Re}_p)(d_i/D_h)$.

The geometric parameters of the finned tube are e, t, n_f, and α. For a given tube geometry, one solves Eq. (17.3) iteratively for G/G_p and calculates $\mathrm{Re}/\mathrm{Re}_p$. The constraints on Eqs. (17.1) and (17.2) are $K/K_p = P/P_p = 1$. With the known Re, one calculates St and f of the finned tube and then solves for A/A_p using Eq. (17.1) or (17.2). This $A/A_p = NL/N_p L_p$. The ratio of tube material in the finned and plain tube exchangers is:

$$\frac{V_m}{V_{m,p}} = \frac{N}{N_p}\frac{L}{L_p}\frac{M}{M_p} \tag{17.24}$$

where M is the tube weight per unit length. Figure 17.21*a* shows the effect of varying the geometry parameters e/d_i and n_f with $e/t = 3.5$ and $\alpha = 0$. This figure shows that the smallest $V_m/V_{m,p}$ occurs with $e = 1$ to 1.5 mm and decreases with increasing n_f when $e \approx 1$. Figure 17.21*b* shows the effect of helix angle for $e/d_i = 0.084$ and $e/t = 3.5$. The $V_m/V_{m,p}$ may be reduced nearly 50% with $\alpha = 30°$. The $\alpha = 30°$ curve shows that n_f has little effect on $V_m/V_{m,p}$. However, as n_f increases, G_p/G must

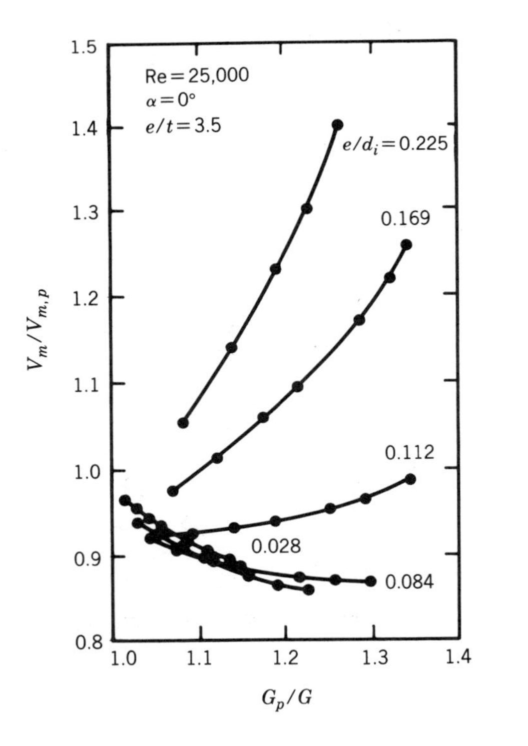

(a)

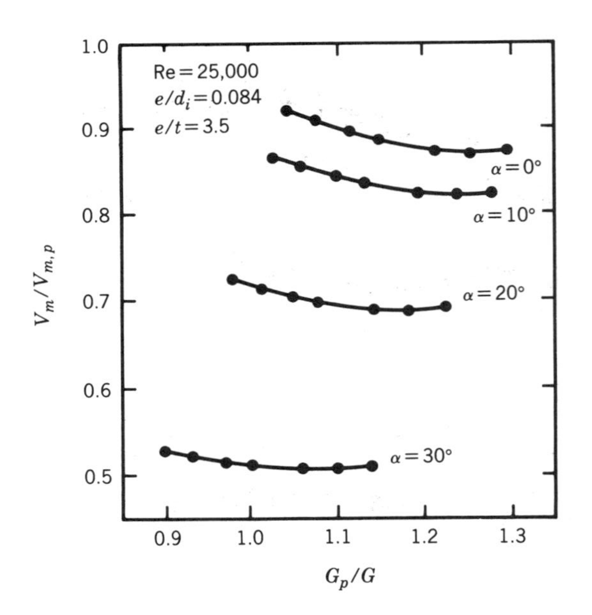

(b)

Figure 17.21. Performance comparison of internally finned tubes and plain tube ($d_i = 17.78$ mm) for case VG-2 of Table 17.2 [46]: (*a*) effect of e/d_i and n_f for $e/t = 3.5$ and $\alpha = 0$, (*b*) effect of n_f and α for $e/d_i = 0.084$ and $e/t = 3.5$. The points on each curve define n_f (from the left, $n_f = 5$, 8, 12, 16, 25, 32, and 40 fins).

increase to meet the constraint $P/P_p = 1$. Since heat exchanger cost is quite sensitive to shell diameter, $12 \leq n_f \leq 16$ appears to be a good choice. Because $S/S_p = G_p/G$, selection of the smaller G_p/G will reduce the shell diameter, and hence the shell cost. The preferred tube geometry ($e/d_i = 0.084$, $n_f \approx 12$, and $\alpha = 30°$) provides a 58% reduction of tubing length and a 48% reduction of tubing material (weight) relative to a plain tube design.

This example assumed no thermal resistance on the shell side. Webb [5] describes how the analysis may be extended to account for shell-side, fouling, and tube-wall resistances. The value of $V_m/V_{m,p}$ will decrease when these additional resistances are present.

17.6 INSERT DEVICES

This class refers to devices that are inserted inside a smooth tube. Several basic techniques have been investigated:

1. Devices that cause the flow to swirl along the flow length
2. An extended surface insert device that provides thermal contact with the tube wall (this may also swirl the flow)
3. A wall-attached insert device that mixes the fluid at the tube wall
4. A displaced insert device that is displaced from the tube wall and causes periodic mixing of the gross flow

The first three methods have found commercial use, but only specialized geometries of the fourth type have been used. The twisted-tape insert device is discussed in Sec. 17.7.

17.6.1 Displaced Enhancement Devices

Figure 17.22*a* and *b* illustrates two types of displaced insert devices tested by Koch [48] in laminar and turbulent flow. These devices periodically mix the gross flow structure and accelerate the local velocity near the wall. Koch found that these devices have substantially higher pressure drop than the type 1, 2, or 3 insert devices listed above. Theoretical reasoning suggests that, for turbulent flow, the fluid should be mixed in the viscous-dominated region near the wall, where the thermal resistance is large. The devices shown in Fig. 17.22*a* and *b* mix the flow in the core region and experience quite high profile drag forces, which substantially increase the pressure drop. Other displaced insert devices that have been employed include bristle brushes, static mixer devices (Fig. 17.22*c*), and flow-driven propellers, all of which promote mixing across the total flow cross section. Bergles and Joshi [44] compare the performance of such displaced insert devices with the swirl and wall-attached insert devices for laminar flow. Koch [48] provides a similar comparison for turbulent flow.

The Fig. 17.22*d* displaced wire-coil insert device is somewhat different from the other displaced insert devices. It causes mixing in a narrow region close to the tube wall. Thomas [49] tested this device for turbulent flow of water in an annulus. He concluded the most favorable St/f performance was obtained when a pair of wires were spaced approximately nine wire diameters, followed by a second pair separated approximately 75 wire diameters from the first pair. The enhancement was provided by

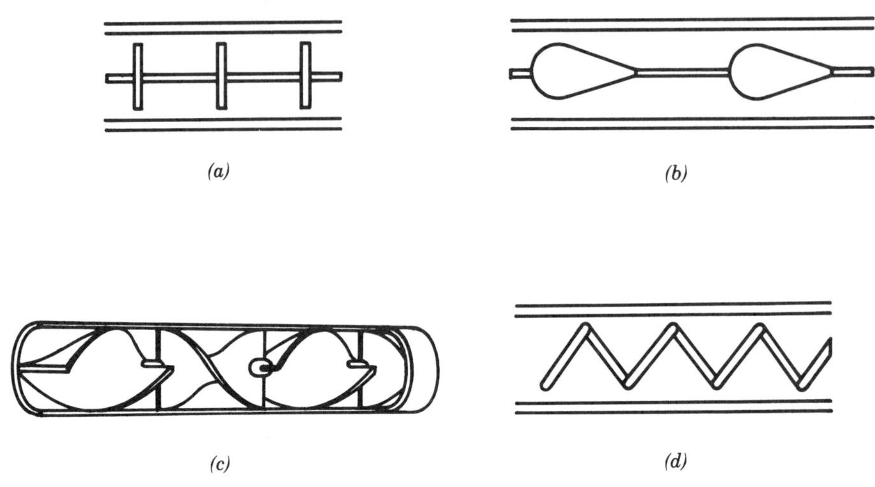

Figure 17.22. Displaced tube insert devices for flow in tubes: (*a*) streamline shape, (*b*) disks, (*c*) static mixer, (*d*) wire coil.

an increased velocity gradient at the surface and by interaction of the cylinder wake with the fluid in the boundary layer. Tests of this displaced wire coil in a circular tube have not been reported.

17.6.2 Wire-Coil Inserts

This enhancement is made by tightly wrapping a coil-spring wire of diameter e on a circular rod. The coil outside diameter, d_c, is made slightly larger than the tube inside diameter, d_i. When the coil spring is pulled through the tube, the wires form a roughness of height e at helix angle $\arcsin(d_i/d_c)$ and spacing $p = \pi d_c \cos\alpha$. This device appears similar to Fig. 17.22*d*, except the wires are in contact with the tube wall. It is necessary that the coil spring force the wire tightly against the tube wall to hold the wire in place and prevent tube-wall erosion. This requires a helix angle of 25° or more. The dimensionless geometric parameters that influence the heat transfer and friction characteristics are α, e/d_i, and p/e. For $\alpha = 25°$ and $e = 1.0$ mm in a 17.6-mm-diameter tube, the resulting values of e/d_i and p/e would be 0.057 and 119 respectively. This example is given to show that practical considerations limit the dimensions that influence the performance of the enhancement. A roughness of $e/d = 0.057$ results in quite high pressure drop compared to preferred roughness geometries that are formed integral to the tube wall.

Uttawar and Raja Rao [50] tested seven different wire-coil insert geometries in laminar flow ($30 < \mathrm{Re} < 675$) for an oil ($300 < \mathrm{Pr} < 675$) for heating of the oil. The ranges of insert geometries tested were $0.08 < e/d_i < 0.13$ and $32 < \alpha < 61°$. Because the heated tube length was only 60 diameters, the flow was not fully developed. The measured enhancement levels $\mathrm{Nu}/\mathrm{Nu}_p$ were between 1.5 and 4.0. The friction increases were considerably less than the Nusselt number increase. The heat transfer data were correlated by

$$\mathrm{Nu} = 1.65(\tan\alpha)\mathrm{Re}^m \mathrm{Pr}^{0.35}\left(\frac{\mu}{\mu_w}\right)^{0.14} \tag{17.25a}$$

where $m = 0.25(\tan\alpha)^{-0.38}$. The Nu, f, and Re are based on the volumetric hydraulic diameter D_v. A friction-factor correlation was not developed. However, when defined in terms of D_v, the friction factor was only 5 to 8% higher than the smooth-tube value for $\mathrm{Re}_v < 180$.

Kumar and Judd [51] developed an empirical heat transfer correlation for the Nusselt number based on their test data for $0.108 < e/d_i < 0.15$ and $8 < p/e < 47$. The relationship

$$\mathrm{Nu}' = 0.175\left(\frac{p}{d_i}\right)^{-0.35} \mathrm{Re}_d^{0.7}\mathrm{Pr}^{1/3} \tag{17.25b}$$

correlated their data with 7.5% rms deviation. Note that the correlation does not contain the parameter e/d_i. Apparently their e/d_i values were so large that they were in the "fully rough" regime. With Nu′ known, one calculates the friction factor by the Kumar-Judd correlation

$$f\,\mathrm{Re}_d = 24736(\mathrm{Nu}'\mathrm{Pr}^{-0.33})^{3.5} \tag{17.26}$$

17.6.3 Extended Surface Insert

Figure 17.23 shows this device. The insert device is formed as an aluminum extrusion. After inserting the extrusion in the tube, the tube is drawn to obtain a tight mechanical joint between it and the insert. The aluminum extrusion is normally formed with five legs, although the number of legs is a design choice. By twisting the extrusion before its insertion in the tube, one may also promote a swirling flow.

Hilding and Coogan [52] provide data on j and f vs. Re for a six-legged straight extrusion in turbulent flow. One may predict the turbulent flow j and f characteristics using an appropriate turbulent flow equation for smooth tubes with the tube diameter replaced by the hydraulic diameter D_h.

Trupp and Lau [53] have predicted the Nu and f for laminar flow in a tube having full-height fins of infinite thermal conductivity. The included angle between the fin legs was varied from 8 to 180°.

Analytical predictions for the Nusselt number of the device must take account of the fin efficiency and consider the possibility that a thermal contact resistance may exist between the aluminum and tube contact surfaces. The contact resistance may be negligible for flow of gases but may be appreciable for liquids. The author is unaware of test data for turbulent flow of liquids.

This insert device is not often used for liquids and gases. Its pressure drop and cost are believed to be unfavorable compared to other enhancement devices that may provide equal performance.

Figure 17.23. Extruded aluminum, star-shaped insert.

17.7 SWIRL-FLOW INSERT DEVICES

A variety of insert devices that cause a swirl flow in the tube have been investigated. The twisted tape insert shown in Fig. 17.24 has been extensively investigated. Variants of the twisted tape that have been evaluated include mechanically rotated twisted tapes, short sections of twisted tape at the tube inlet or periodically spaced along the tube length, and periodically spaced "propeller" inserts that are mechanically rotated by the flowing fluid. This discussion will present details only for the stationary twisted tape insert. Bergles and Joshi [44] present a survey of the performance of the different types of swirl-flow devices for laminar flow.

The insert shown in Fig. 17.24 consists of a thin strip that is twisted through 360° per axial distance p_t. Bergles and Joshi [44] describe their tapes by the twist ratio $y = p_t/2d_i$. The helix angle of the tape is given by

$$\tan \alpha = \frac{\pi d_i}{p_t} = \frac{\pi}{2y} \tag{17.27}$$

In order to allow easy insertion of the tape, there is usually a small clearance between the tape width and the tube inside diameter. This clearance results in poor contact between the tape and the tube wall. If the tape is made of low thermal conductivity material, and the fluid thermal conductivity is small, the heat transfer from the tape may be quite small. Heat transfer enhancement may occur for three reasons:

1. The tape reduces the hydraulic diameter D_h, which effects an increased heat transfer coefficient, even for zero tape twist.
2. The twist of the tape causes a tangential velocity component u_θ. Hence, the speed of the flow is increased—particularly near the wall. The heat transfer enhancement is a result of the increased shear stress at the wall. Thorsen and Landis [54] show that centrifugal forces caused by the tangential velocity component may contribute to the enhancement by mixing fluid from the core region with fluid in the wall region. However, this will occur only when the flowing fluid is being heated. The cold high-density core fluid is forced outward to mix with the warm low-density fluid near the wall. If the fluid is being cooled, the centrifugal force acts to maintain thermal stratification of the fluid.
3. There may be heat transfer from the tape, if good thermal contact with the wall exists.

17.7.1 Laminar Flow

Bergles and Joshi [44] provide an excellent summary of performance data on the twisted tape for constant wall temperature and constant heat flux boundary conditions. Figure 17.25 shows Nusselt number data for the two boundary conditions. The data are presented as a function of entrance region parameters because not all of the data are for the fully developed condition. The ethylene glycol data of Marner and Bergles [40] in Fig. 17.25*a* show an enhancement level about 300% over the smooth-tube value. Note that the data for heating are above those for cooling at the lower Gz values. Figure 17.26*a* shows the friction-factor data for the data of Fig. 17.25*a*. The friction data of Marner and Bergles [40] are shown by curve 3. Figure 17.20 shows h/h_p for their data. This figure shows that the performance of the twisted tape is better in

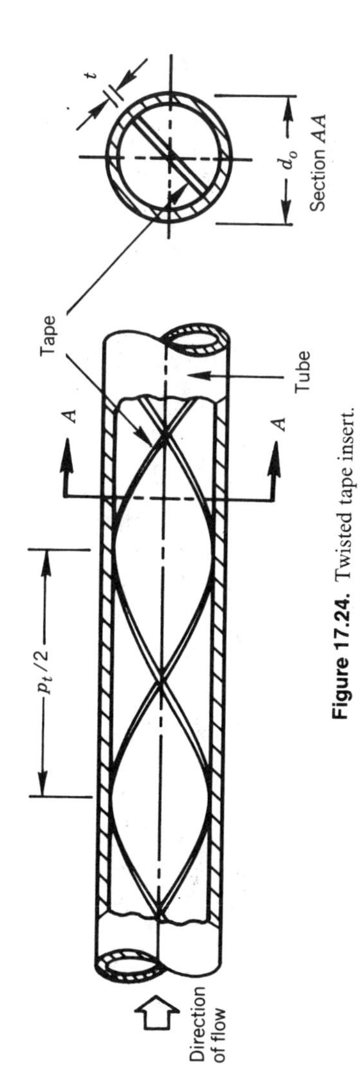

Figure 17.24. Twisted tape insert.

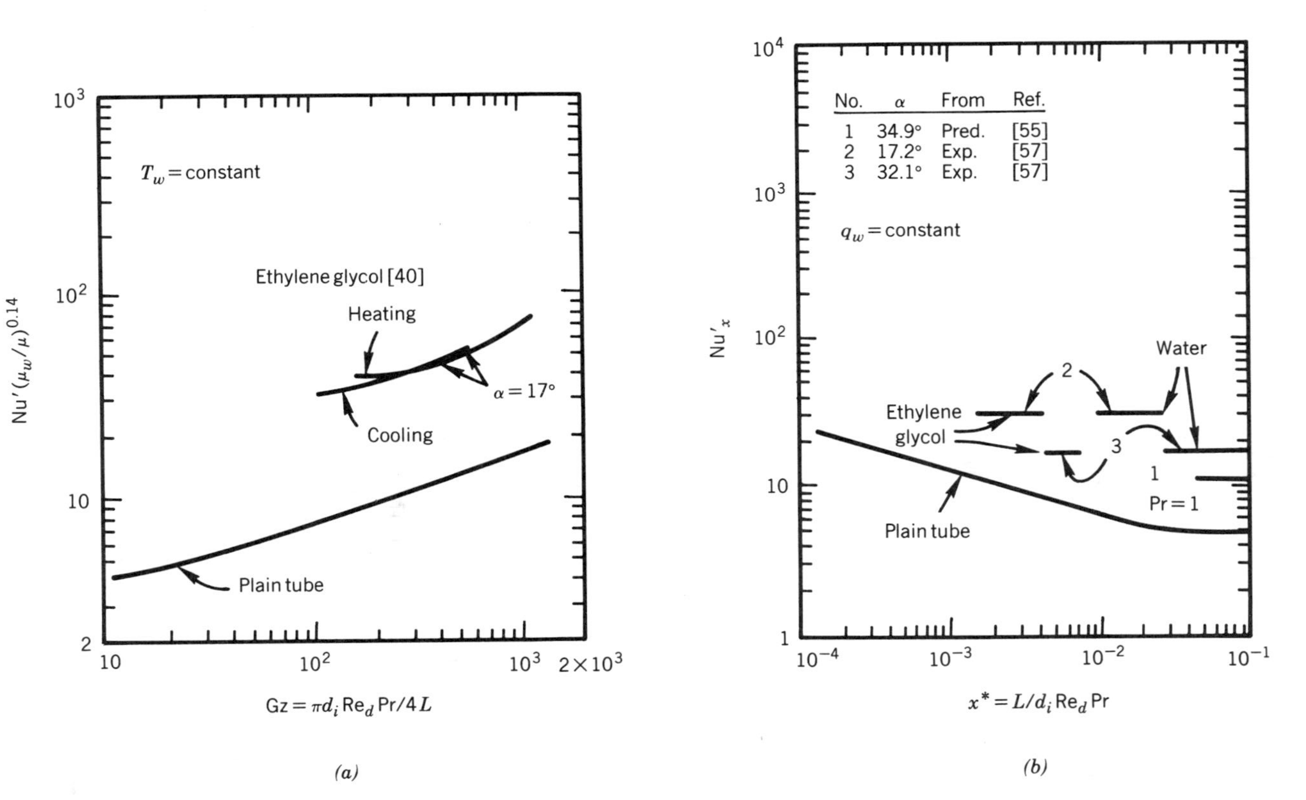

Figure 17.25. Nusselt number for laminar flow in tubes with twisted tape insert [44]: (*a*) constant wall temperature, (*b*) constant heat flux.

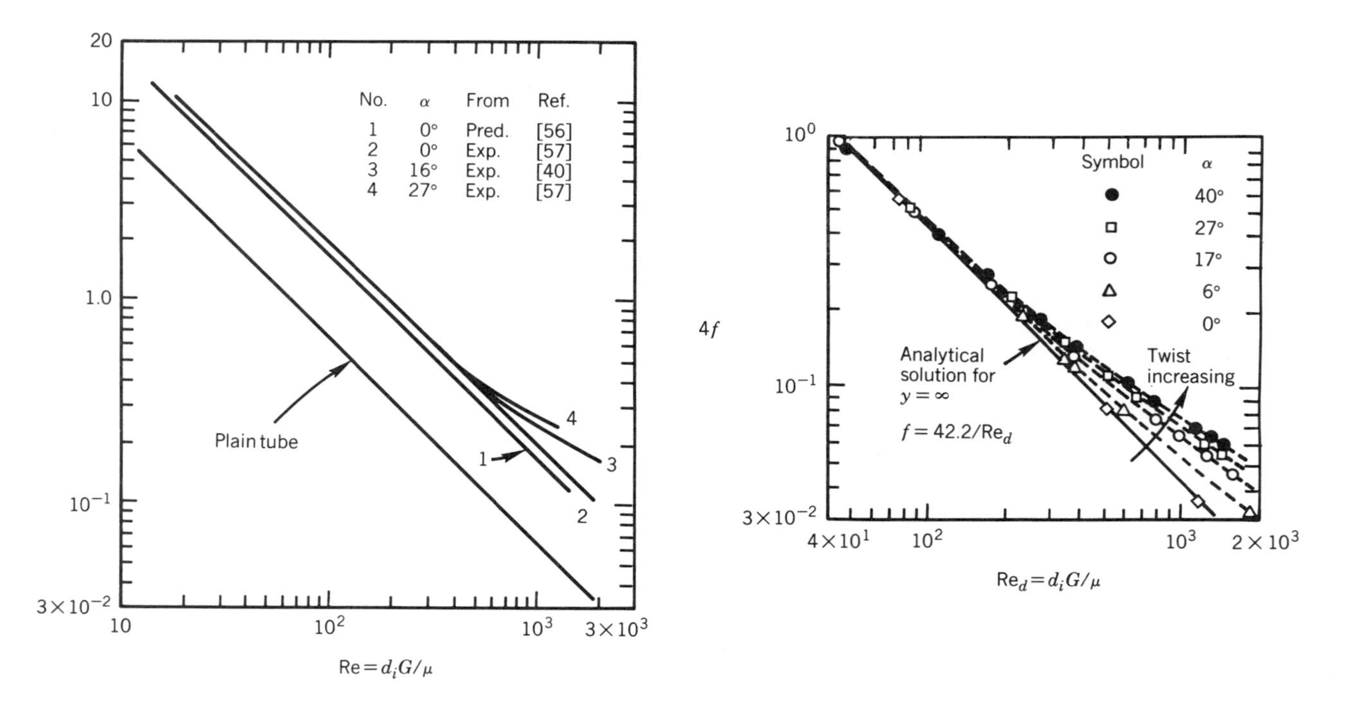

Figure 17.26. Friction factor in tubes with twisted tape insert for laminar flow: (*a*) experimental and predicted [44], (*b*) predicted [55].

heating than in cooling. It also shows that the performance of the internally finned tube is better than that of the twisted tape.

Date [55] has performed numerical predictions for the twisted tape in fully developed laminar flow for constant heat flux with constant properties. His analysis for a loosely fitting tape with Pr = 1 is shown by curve 1 in Fig. 17.25*b*. Date's analysis [55] also shows that the Nusselt number increases with increasing Prandtl and Reynolds numbers, contrary to laminar flow in smooth tubes, which is independent of Re and Pr. Note that the values of Nu as reported by Date [55] are 50% low, due to a computational error. This is noted by Hong and Bergles [57].

Hong and Bergles [57] have developed an empirical correlation for twisted-tape data taken using water and ethylene glycol with constant heat flux. Two twisted tapes were used, which had helix angles of $\alpha = 17$ and $32°$. Their heat transfer correlation is given by

$$\mathrm{Nu}' = 5.172\left[1 + 0.005484\left(\frac{\mathrm{Re}_d}{y}\right)^{1.25}\mathrm{Pr}^{0.7}\right]^{0.5} \tag{17.28}$$

where $y = \pi/(2\tan\alpha)$.

The author is unaware of correlations for Nu suitable for a constant wall temperature boundary condition. The Nusselt number should be smaller than for the constant heat flux boundary condition.

Date [55] has developed empirical correlations to fit his numerical predictions of the friction factor, which are shown in Fig. 17.26*b*. The equations, as corrected and modified by Shah and London [56, p. 580] for zero tape thickness, are

$$f\,\mathrm{Re}_d = 42.23 \qquad \text{for} \quad \frac{\mathrm{Re}_d}{y} < 6.7 \tag{17.29a}$$

$$f\,\mathrm{Re}_d = 38.4\left(\frac{\mathrm{Re}_d}{y}\right)^{0.05} \qquad \text{for} \quad 6.7 < \frac{\mathrm{Re}_d}{y} < 100 \tag{17.29b}$$

$$f\,\mathrm{Re}_d = C\left(\frac{\mathrm{Re}_d}{y}\right)^{0.3} \qquad \text{for} \quad \frac{\mathrm{Re}_d}{y} > 100 \tag{17.29c}$$

$$C = 8.82y - 2.12y^2 + 0.211y^3 - 0.0069y^4$$

As indicated in Fig. 17.26*b*, the $\alpha = 0$ curve is for a tape of zero thickness and no twist. Increasing the twist causes $f\,\mathrm{Re}_d$ to depart from the value 42.23. Shah and London [56] recommend that the right-hand sides of Eqs. (17.29) be multiplied by a correction factor ζ to account for the effect of finite tape thickness t. This factor is defined in the Nomenclature Section at the end of this chapter.

17.7.2 Turbulent Flow

Thorsen and Landis [54] recognized that buoyancy effects arising from density variations in the centrifugal field should have an effect on heat transfer. They showed that the swirl-flow-induced buoyancy effect should depend on the dimensionless group $\mathrm{Gr}/\mathrm{Re}^2$, which may be written as

$$\frac{\mathrm{Gr}}{\mathrm{Re}^2} = \frac{2D_h\beta_T\,\Delta T\tan\alpha}{d_i} \tag{17.30}$$

Thorsen and Landis measured the heat transfer coefficient for heating and cooling of water in tubes having tapes with three different helix angles: $\alpha = 11.1$, 16.9, and 26.5°. The heating data were correlated by

$$\mathrm{Nu} = 0.021F\left(1 + 0.25\sqrt{\frac{\mathrm{Gr}}{\mathrm{Re}}}\right)\mathrm{Re}^{0.8}\mathrm{Pr}^{0.4}\left(\frac{T_w}{T_b}\right)^{-0.32} \tag{17.31}$$

and the cooling data by

$$\mathrm{Nu} = 0.023F\left(1 + 0.25\sqrt{\frac{\mathrm{Gr}}{\mathrm{Re}}}\right)\mathrm{Re}^{0.8}\mathrm{Pr}^{0.3}\left(\frac{T_w}{T_b}\right)^{-0.1} \tag{17.32}$$

where

$$F = 1 + 0.004872\frac{\tan^2\alpha}{d_i(1 + \tan^2\alpha)}$$

and d_i is in meters.

Lopina and Bergles [58] attempted to include the increased speed of the flow (caused by the spiral flow) and the centrifugal buoyancy effect by using a simple superposition model. The model also includes the possibility that the tape may act as an extended surface. The model may be expressed as

$$q = q_{\mathrm{sc}} + q_{\mathrm{cc}} + q_{\mathrm{f}} \tag{17.33}$$

The term q_{sc} stands for swirl convection and is predicted using an appropriate equation for turbulent flow in plain tubes with the Reynolds number calculated in terms of D_h and a modified velocity u_{mod} to include the speed of the swirl flow at the wall. The heat transfer coefficient for the swirl convection term is given by

$$\frac{h_{\mathrm{sc}}D_h}{k} = 0.023\,\mathrm{Re}^{0.8}\,\mathrm{Pr}^{0.4} \tag{17.34a}$$

where

$$\mathrm{Re} = \frac{u_{\mathrm{mod}}D_h}{\nu} \qquad \text{with} \qquad u_{\mathrm{mod}} = u_m\sqrt{1 + \tan^2\alpha} \tag{17.34b}$$

The term q_{cc} in Eq. (17.33) represents the centrifugal convection effect identified by Thorsen and Landis. Lopina and Bergles use an equation for turbulent natural convection from a horizontal plate and replace the gravity force g by a radial acceleration g_r:

$$g_r = \frac{2u_\theta^2}{d_i} = \frac{2(u_{\mathrm{mod}}\tan\alpha)^2}{d_i} \tag{17.35}$$

The heat transfer coefficient for the centrifugal convection term is calculated by

$$\frac{h_{\mathrm{cc}}D_h}{k} = 0.12(\mathrm{Gr}'\mathrm{Pr})^{1/3} \tag{17.35a}$$

where

$$\mathrm{Gr}' = \frac{4.94\beta_T \, \Delta T \, D_h \mathrm{Re}^2}{d_i y^2} \tag{17.35b}$$

The term q_f in Eq. (17.33) takes account of heat transfer from the tape as an extended surface. Evaluation of this term requires knowledge of the contact resistance between the tube wall and the tape. For poor thermal contact, $q_f = 0$. See [58] for details on calculation of q_f.

Based on his evaluation of published data, Date [59] proposed an empirical correlation for the friction factor for the range $5000 < \mathrm{Re} < 70{,}000$ and $0 < y < 1.5$, or $0 < \alpha < 46°$. The correlation is

$$\frac{f}{f_p} = \left(\frac{y}{y-1}\right)^m \tag{17.36a}$$

where

$$m = 1.15 + \frac{0.15(70000 - \mathrm{Re})}{65000} \tag{17.36b}$$

and f_p is the friction factor in a plain tube, given by

$$f_p = 0.046 \, \mathrm{Re}_d^{-0.2} \tag{17.37}$$

17.8 ROUGHNESS

Considerable data exist for the single-phase forced-convection flow over rough surfaces. Data exist for six different flow geometries:

1. Flat plates
2. Circular tubes
3. Longitudinal flow in rod bundles
4. Annuli having roughness on the outer surface of the inner tube
5. Flow normal to circular tubes

Internally roughened channels are finding increasing practical use. The refrigeration industry uses roughness on the water side of water-chiller evaporators and condensers. Water side roughness also appears to offer economic benefits for use in electric utility steam condensers [60]. Many papers have been published on work related to the use of roughened fuel rods in gas-cooled nuclear reactors [61]. Work is in progress on water cooled gas turbine blades having an internal roughness [62].

There are many possible roughness geometries. Figure 17.27 is the author's attempt to catalog the possible geometries. This figure shows three basic roughness families. For any basic type, the key dimensionless variables are the relative roughness height (e/d), the relative roughness spacing (p/e), and the shape of the roughness element. The ridge-and-groove-type roughness may also be applied at a helix angle [63]; this is not shown in Fig. 17.27. For a specific roughness type, a family of geometrically similar roughness is possible simply by changing e/d, maintaining constant p/e and α. Thus,

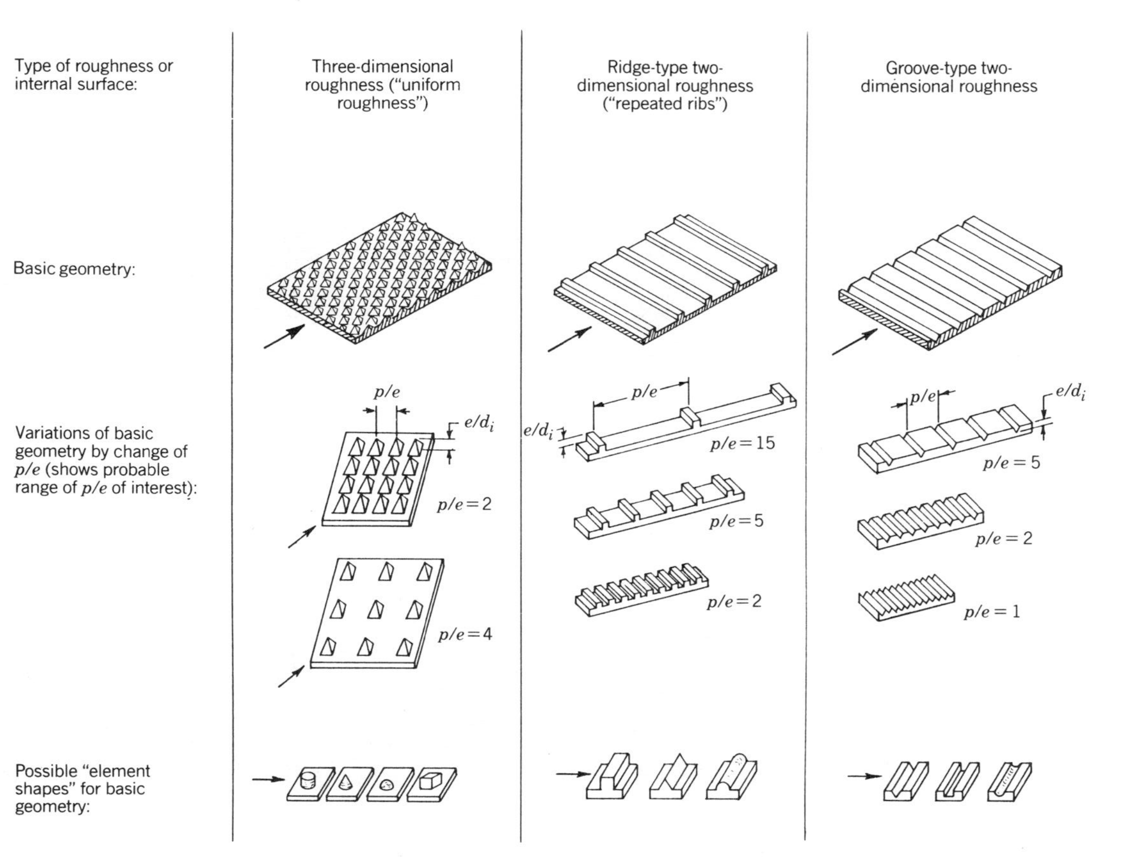

Figure 17.27. Catalog of roughness geometries [74].

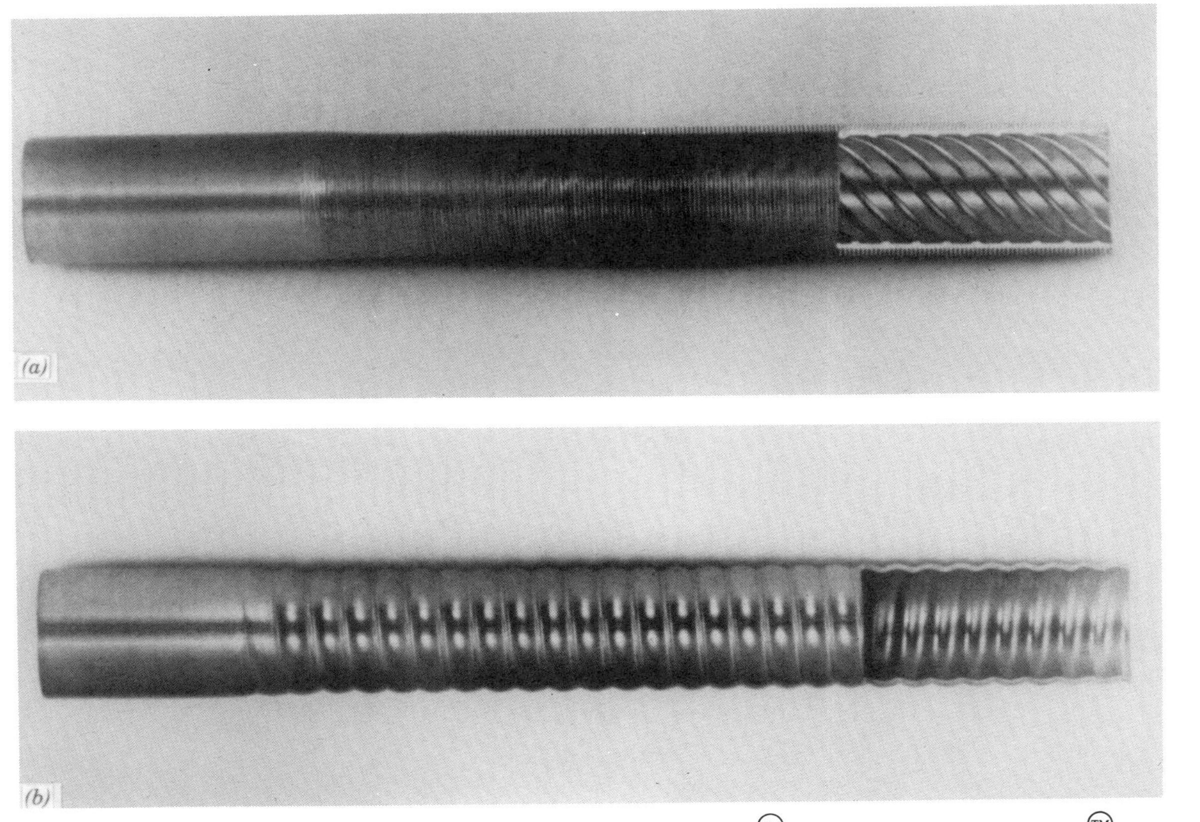

Figure 17.28. Illustration of commercially available enhanced tubes: (*a*) corrugated KORODENSE™ tube, (*b*) helical rib TURBOCHIL™ tube. Courtesy of Wolverine Tube Div., Decatur, AL.

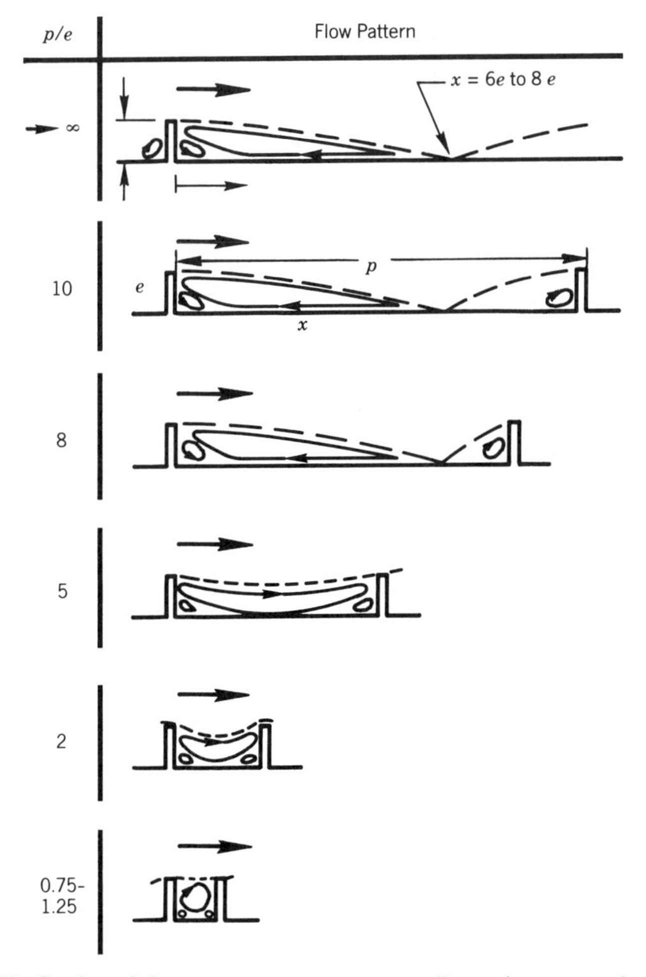

Figure 17.29. Catalog of flow patterns over transverse rib roughness as a function of rib spacing [64].

the designer is faced with choosing among thousands of possible specific roughness geometries and sizes.

Figure 17.28 shows two commercially used roughness geometries [70]. Their geometric parameters are e (rib height), p (rib spacing), α (helix angle), and the rib shape. The Fig. 17.28*a* tube has internal helical ribs and is normally made with low integral fins on the external surface. The internal ribs are made by cold deformation of the metal into a grooved internal mandrel to form the helical ridges. The Fig. 17.28*b* helically corrugated tube is made by rolling a sharp-edged wheel on the outer surface of the tube.

Figure 17.29 shows the flow pattern [64] in the vicinity of the ribs ($\alpha = 90°$) as a function of the rib spacing (p/e). The flow separates at the rib and reattaches 6 to 8 rib heights downstream from the rib. The heat transfer coefficient attains its maximum value at the reattachment point [65]. Figure 17.29 shows that reattachment does not occur for $p/e < 8$. The highest average coefficient occurs for $10 < p/e < 15$.

17.8.1 Heat Transfer and Friction Correlations

Consider a two-dimensional roughness, such as illustrated by Fig. 17.28*a*. A geometrically similar family of roughness geometries will exist if p/e, α, and the rib shape are held constant. Such a family of roughened tubes will differ in their e/d values.

The friction data for the different e/d tubes may be correlated using the friction similarity model developed by Nikuradse and described by Schlichting [66]. This model is based on the "law of the wall" velocity distribution for flow over rough surfaces [66]. In Nikuradse's model, the friction data for tubes with different e/d_i will fall on the same curve when plotted in the form $B(e^+)$ vs. e^+, where

$$B(e^+) = \sqrt{\frac{2}{f}} + 2.5 \ln \frac{2e}{d_i} + 3.75 \tag{17.38}$$

Nikuradse (as reported in [66]) correlated his friction-factor data for six geometrically similar sand-grain roughness geometries ($0.004 < e/d < 0.0679$) using Eq. (17.38). The variable e^+ is called the roughness Reynolds number and may be written in the following two equivalent forms:

$$e^+ = \frac{eu^*}{\nu} = \frac{e}{d_i} \mathrm{Re} \sqrt{\frac{f}{2}} \tag{17.39}$$

Dipprey and Sabersky [67] have developed a heat transfer correlation that is applicable to any type of geometrically similar surface roughness. Their model applies the heat–momentum-transfer analogy to flow over rough surfaces. The model proposes that the data for geometrically similar roughness (different e/d_i) will fall on the same curve when plotted in the form $\bar{g}(e^+)\,\mathrm{Pr}^n$ vs. e^+, where

$$\bar{g}(e^+)\,\mathrm{Pr}^n = \frac{f/(2\,\mathrm{St}) - 1}{\sqrt{f/2}} + B(e^+) \tag{17.40}$$

Equations (17.38) and (17.40) may be used to correlate the data for any family of geometrically similar roughnesses. However, the functions $B(e^+)$ and $g(e^+)$ may be different for different roughness families. One uses Eqs. (17.38) and (17.40) to calculate the friction factor and Stanton number as follows:

1. Select e^+ and calculate $B(e^+)$ and $\bar{g}(e^+)$ from the experimentally derived correlations for $B(e^+)$ and $\bar{g}(e^+)$.
2. Calculate the friction factor using Eq. (17.38).
3. Calculate the Stanton number with Eq. (17.40). Solving Eq. (17.40) for St yields

$$\mathrm{St} = \frac{f/2}{1 + \sqrt{f/2}\,[\bar{g}(e^+)\mathrm{Pr}^n - B(e^+)]} \tag{17.41}$$

Dipprey and Sabersky [67] found that the Prandtl number exponent was 0.44 for sand-grain roughness. However, Webb et al. [64] found $n = 0.57$ for transverse-rib roughness. Further commentary on the Prandtl number dependence will be given later.

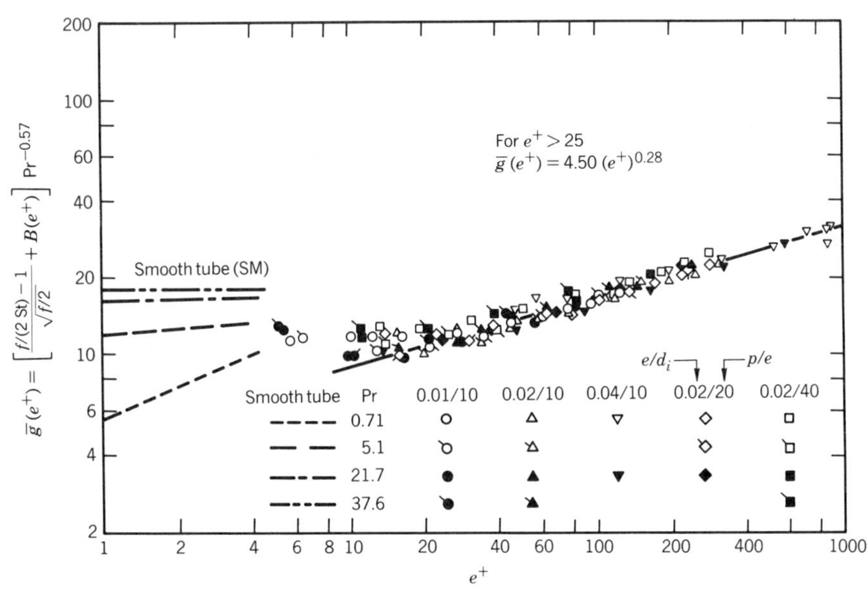

Figure 17.30. Heat transfer correlation for transverse-rib roughness ($\alpha = 90°$) by Webb et al. [64].

The literature contains correlations for $B(e^+)$ and $\bar{g}(e^+)$ applicable to the following goemetries.

1. Sand-grain roughness tested by Dipprey and Sabersky [67].
2. Transverse-rib roughness ($\alpha = 90°$) for $p/e = 10$, 20, and 40 as measured by Webb et al. [64].
3. Helical-rib roughness ($30 \leq \alpha \leq 70°$) with $p/e = 15$ reported by Gee and Webb [63]. Nakayama et al. [69] provide additional data for helical ribs, including $0 \leq \alpha \leq 80°$.

Figure 17.30 shows the function $\bar{g}(e^+)$ determined by Webb et al. for transverse-rib roughness [64]. Only the data for $p/e = 10$ are for geometrically similar roughness. Figure 17.30 shows that Eq. (17.40) does an excellent job of correlating the three e/d_i values (0.01, 0.02, and 0.04) for $p/e = 10$. The figure also shows that the nonsimilar $p/e = 20$ and 40 data also fall on the same correlating line as the $p/e = 10$ data. There is no reason to expect this, so it is regarded fortuitious. The data of Fig. 17.30 spanned $0.7 \leq \mathrm{Pr} \leq 37.6$. The data are correlated by a Prandtl number exponent of $n = 0.57$ [cf. Eq. (17.40)].

How does the helix angle α effect the St and f characteristics? Figure 17.31 shows the St and f data of Gee and Webb [63] for $p/e = 15$ and $e/d_i = 0.01$ as a function of helix angle. As α increases, the friction factor drops faster than does the Stanton number. Gee and Webb found that the Maximum St/f occurs at $\alpha \approx 45°$.

A commercial version of the Fig. 17.28a helical-rib tube is available and is known as the TURBOCHIL(TM) tube [70]. This tube has low integral fins on the outer tube surface. The TURBOCHIL(TM) tube has $\alpha = 47°$, $p/e = 11.1$, and $e/d_i = 0.0264$. The functions $\bar{g}(e^+)$ and $B(e^+)$ for this tube are given by the following equations, which were

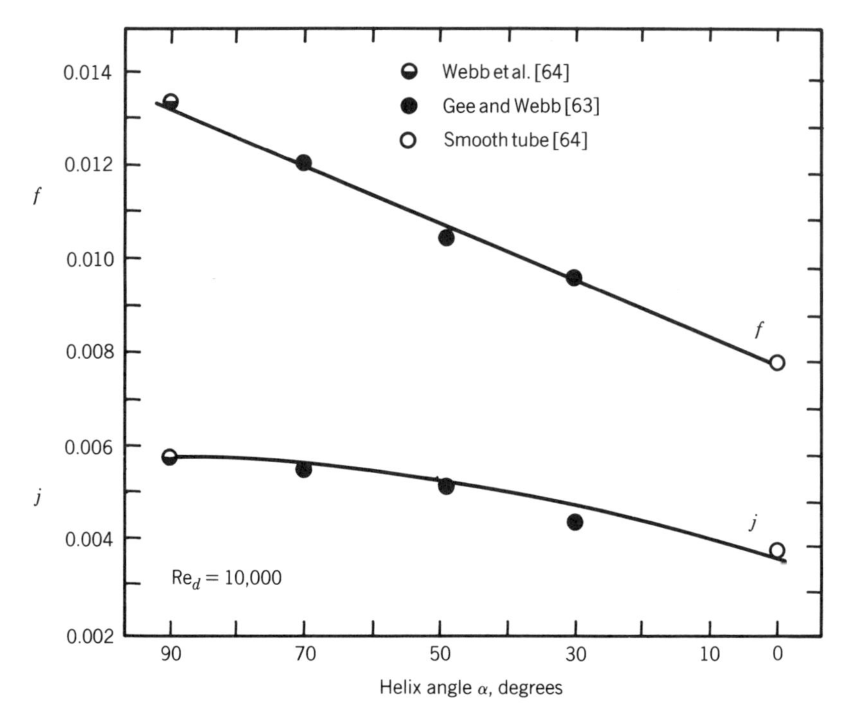

Figure 17.31. Effect of helix angle on f and St for $e/d_i = 0.01$, $p/e = 15$ [63].

developed by the present author and are valid for $e^+ > 25$:

$$\bar{g}(e^+) = 7.68(e^+)^{0.136} \tag{17.42}$$

$$B(e^+) = 1.7 + 2.06 \ln e^+ \tag{17.43}$$

Equations (17.42) and (17.43) may be used in Equation (17.41) to predict the Stanton number of the TURBOCHIL(TM) tube. The friction factor is calculated using Eq. (17.43) in Eq. (17.38). Equations (17.42) and (17.43) may also be used to calculate St and f for any value of e/d_i provided the value of p/e is maintained at 11.1. Withers [71] provides additional information on helical-rib roughness for different values of p/e and α.

The Fig. 17.28*b* helically corrugated tube has been tested by Sethumadhavan and Raja Rao [72]. They tested five different roughness geometries with water and a glycol-water solution ($5.2 < \Pr < 32$) and determined the correlating functions $B(e^+)$ and $\bar{g}(e^+)$. They recommended the following correlations for $\bar{g}(e^+)$ and $B(e^+)$ vs. e^+ for $10 < p/e < 40$:

$$\bar{g}(e^+) = 8.6\,(e^+)^{0.13} \qquad \text{for} \quad 3 < e^+ < 200 \tag{17.44}$$

$$B(e^+) = 0.4(e^+)^{0.164}\left(p\frac{d_i - e}{e^2}\right)^{1/3} \qquad \text{for} \quad 25 < e^+ < 180 \tag{17.45}$$

TABLE 17.6. Predicted h/h_p and f/f_p for Re = 25000, Pr = 10.4

Tube	e/d_i	p/e	h/h_p	f/f_p
KORODENSE	0.025	20.3	1.93	2.27
THERMOEXCEL-CC	0.025	46.7	1.59	1.90
TURBOCHIL	0.0264	11.1	1.99	1.83

The rms deviations of Eqs. (17.44) and (17.45) are 7.0% and 6.1%, respectively. The data on which the correlations are based span $0.0118 < e/d < 0.0303$ and $12.7 < p/e < 56.6$. The authors also show that the data of Withers [71] are in good agreement with Eq. (17.45).

Commercial versions of the Fig. 17.28*b* helically corrugated tube are available. The KORODENSE tube [68] has $e/d_i = 0.025$ and $p/e = 20.3$. The THERMOEXCEL-CC™ tube [71], which has an enhanced outer surface, has $e/d_i = 0.025$ and $p/e = 47$.

The present author has used the above correlation to predict the enhancement ratios h/h_p and f/f_p for the KORODENSE, TURBOCHIL, and THERMOEXCEL-CC tubes. Table 17.6 shows the enhancement ratios for water (Pr = 10.4) at Re = 25,000. The best performance is provided by the TURBOCHIL tube. It provides a 99% greater heat transfer coefficient than the smooth tube with a friction factor only 83% greater than the smooth tube.

17.8.2 Prandtl-Number Dependence

The correlations of $\bar{g}(e^+)$ and $B(e^+)$ vs. e^+ do not allow easy physical interpretation. The practitioner prefers to see curves of St and f vs. Re. One may use the functions $B(e^+)$ and $\bar{g}(e^+)$ in Eqs. (17.38) and (17.40) to generate curves of St and f vs. Re for different e/d_i values. Webb [74] shows such curves for $p/e = 10$, $\alpha = 90°$ rib roughness. Figure 17.32 shows curves of $\mathrm{St}/\mathrm{St}_p$ vs. e^+ for $p/e = 10$, $\alpha = 90°$ rib roughness. These curves show that $\mathrm{St}/\mathrm{St}_p$ increases with Pr. Figure 17.32 also shows that a maximum value of $\mathrm{St}/\mathrm{St}_p$ is attained at $e^+ = 20$ for Pr > 5.1. However, at large e^+, the curves tend to approach the same value asymptotically. The interpretation of these curves is that:

At high e^+ (high Re), the rough and smooth tubes have the same Prandtl number dependence.

At lower e^+, the rough surface has a stronger Prandtl number effect than a plain surface.

Figure 17.32 suggests that the rough surface will be of greater benefit for high Prandtl number fluids than for low Prandtl number fluids.

Additional information in Ref. 73 shows that $\mathrm{St}/\mathrm{St}_p > f/f_p$ for Pr > 21.7. This means that the increase in heat transfer coefficient is actually greater than the friction increase. Other experimental studies of helical and transverse rib roughness [72, 75] have confirmed that the Prandtl number exponent is $n = 0.57$. However, Dipprey and Sabersky [67] found $n = 0.44$ for sand-grain roughness. Their tests spanned a much smaller range of Pr (1.2 < Pr < 5.9) than that of Webb et al. [64]. For sand-grain roughness, the data [67] show that $\mathrm{St}/\mathrm{St}_p > f/f_p$ for Pr = 5.9. More work remains to be done to determine the degree to which the roughness type affects the Prandtl number dependence of rough surfaces.

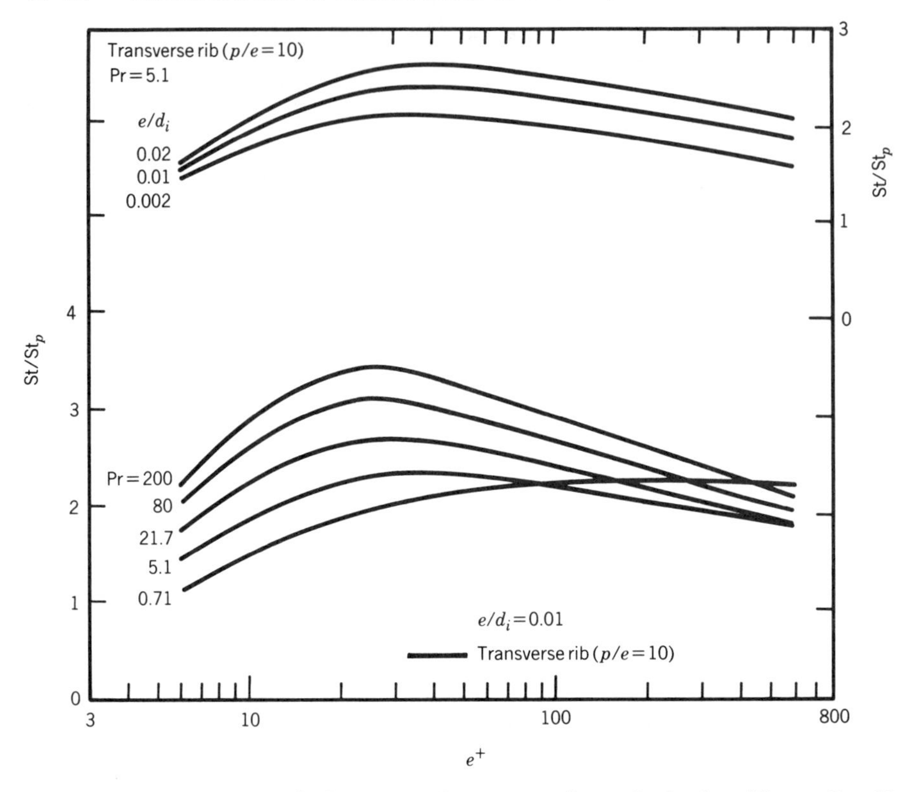

Figure 17.32. $\mathrm{St}/\mathrm{St}_s$ vs. e^+ for $p/e = 10$ transverse-rib roughness for different Prandtl numbers [74].

17.8.3 Heat-Transfer Design Methods

Since there are a large number of different basic roughness types and a range of possible e/d_i for each type, the designer is faced with a confusing array of choices. In order to approach the question, one must first define the design objective and establish design constraints. Possible design objectives are listed in Table 17.2. Assume that tube-side roughness will be used to reduce the size of a heat exchanger for case VG-1 of Table 17.2. Thus, one is interested in calculating A/A_p for constant heat duty, flow rate, and pumping power. One may place limits on the roughness choices by defining a hierarchy of questions:

1. What is the A/A_p for a given roughness type and e/d_i?
2. Will another e/d_i (for the same basic roughness type) give a smaller A/A_p?
3. Will another basic roughness type (and e/d_i) provide further reduction of A/A_p?

First, question 1 will be addressed for the VG-1 criterion. Assume, for example, that a heat exchanger has been designed that has a plain surface on the tube side. The tube-side mass flow rate and pressure drop and the UA value of the plain-tube heat exchanger are known. Will a tube-side roughness be of benefit? If the dominant thermal resistance is on the tube side, roughness is of potential benefit. Hence, the designer wants to calculate A/A_p for $P/P_p = UA/U_pA_p = 1$.

For simplicity, assume that the total thermal resistance is on the tube side. Then the constraints are satisfied by Eq. (17.3). The required velocity in the rough tube that will satisfy the pumping-power constraint [Eq. (17.2)] is unknown. Hence, one must use an iterative calculation procedure to solve Eqs. (17.2) and (17.3) for A/A_p and G/G_p (or $\mathrm{Re}_d/\mathrm{Re}_{d,p}$). The solution procedure is complicated by the fact that the equations for the rough-tube St and f are written in terms of the roughness Reynolds number e^+ rather than the pipe Reynolds number Re_d. Two solution procedures are possible and will be described.

Method 1. Assume that empirical curve fits of St and f vs. Re_d (or G) for the selected e/d_i have been developed. Using these equations, one solves Eq. (17.3) for G/G_p. With G/G_p known, the Re_d of the rough tube is calculated, and St and f are then readily calculated. Next, one calculates the U value for the heat exchanger, and the area ratio A/A_p is given by U_p/U. Note that the pumping-power constraint is inherently satisfied by Eq. (17.3).

Method 2. The second method avoids the need for the curve fits of St and f vs. Re_d and uses Eq. (17.41). Using Eq. (17.39) one may write

$$\frac{\mathrm{Re}_d}{\mathrm{Re}_{d,p}} = \frac{e^+}{\mathrm{Re}_{d,p}} \frac{d_i}{e} \sqrt{\frac{2}{f}} \tag{17.46}$$

If the rough and smooth tubes are of the same inside diameter, $\mathrm{Re}_d/\mathrm{Re}_{d,p} = G/G_p$. Using Eq. (17.41) with $n = 0.57$ and Eq. (17.46) in Eq. (17.3) gives

$$\frac{e^+}{\mathrm{Re}_p} \frac{d_i}{e} \sqrt{\frac{2}{f}} = \left(\frac{f_p/2\,\mathrm{St}_p}{1 + \sqrt{f/2}\left[\bar{g}(e^+)\mathrm{Pr}^{0.57} - B(e^+)\right]} \right)^{1/2} \tag{17.47}$$

The values of Re_p, St_p, and f_p for the smooth tube and e/d_i are known. Equation (17.47) is solved iteratively for e^+ using Eq. (17.38) to eliminate $\sqrt{2/f}$. One uses the $\bar{g}(e^+)$ and $B(e^+)$ equations for the roughness family of interest. The iterative solution gives the value of e^+ that satisfies Eq. (17.47). With e^+ known, one easily computes St, f, and Re of the rough tube and proceeds to calculate A/A_p as for method 1.

Using either method 1 or method 2, one obtains the predicted A/A_p and G/G_p. One desires $G/G_p \approx 1$. If the calculated $G/G_p < 1$, it will be necessary to increase the number of tubes in the heat exchanger. For $\mathrm{Pr} \geq 20$, one may find $G/G_p > 1$. Is it possible that a different e/d_i will provide a lower A/A_p or a more desirable G/G_p? This is question 2. One may repeat the analysis for different e/d_i to answer this question. Webb and Eckert [76] address this question for $p/e = 10$, $\alpha = 90°$ rib roughness, using PEC VG-1, -2, and -3 of Table 17.2. For this roughness type, they conclude that the criterion $e^+ = 20$ will provide the best value for the objective function. If one can specify the desired e^+, Eq. (17.47) may be solved iteratively for the required e/d_i. Such evaluations have not been done for different roughness types, e.g., $p/e = 15$, $\alpha = 45°$ helical rib roughness. Hence, one cannot state the preferred e^+ for different helix angles α.

TABLE 17.7. Comparison of Performance of Different Tube-Side Enhancement Geometries for Case VG-1 of Table 17.2

Geometry	α	e/d_i	N/N_p	L/L_p	$1 - V_m/V_{m,p}$
Helical fins ($n_f = 32$)	30	0.0420	1.01	0.37	0.42
Transverse ribs ($p/e = 10$)	90	0.0060	1.16	0.46	0.53
Axial fins ($n_f = 32$)	0	0.0420	1.04	0.49	0.54
Corrugated ($p/e = 19.5$)	82	0.0090	1.02	0.56	0.57
Helical ribs ($p/e = 15$)	45	0.0069	0.96	0.60	0.58

17.8.4 Preferred Roughness Type

What is the best roughness type (question 3)? There is no direct way to answer this question without performing a parametric study of candidate roughness types. Webb [77] uses a case-study method to compare the performance improvements of different roughness types. This analysis was performed for case VG-1 of Table 17.2. The design conditions of the reference plain-tube heat exchanger are $\mathrm{Re}_p = 46{,}200$, $h_o = 27{,}260$ W/(m · K), $h_i = 6814$ W/(m · K), and $R_{fi} = 0.000044$ m^2 · K/W (fouling resistance). Five basic types of tube-side enhancements were evaluated. The enhanced tubes are listed in order of decreasing material savings $V_m/V_{m,p}$, where V_m is the tube material volume in the heat exchanger. The ratio of the tube lengths in the enhanced and plain tube designs is L/L_p, and the relative number of tubes in the exchangers is N/N_p, which is equal to G_p/G for the roughness geometries. A summary of the key results of the study is shown in Table 17.7. The table provides several conclusions:

The greatest material savings are provided by the helical internal fins. The savings of the helical fins is 12% greater than for the axial internal fins. The helical fins can be operated at approximately the same velocity as the plain-tube design.

The smallest material saving (42%) is provided by the helical ribs. However, the helical ribs can be operated at a velocity only 4% lower than the plain-tube design. The material saving of the corrugated tube is approximately the same as for the helical ribs. However, the corrugated tube must be operated at a lower velocity.

The transverse ribs provide 5% greater material saving than the helical ribs; however, the velocity must be reduced to 84% of the plain tube value, which requires a larger heat exchanger diameter.

The heat exchanger using the helical internal fins is only 37% as long as in the plain-tube design.

Although the helical-internal-fin geometry appears to be superior to the roughness geometries, one must consider whether it can be made in the material of interest and define the relative cost of the enhanced tubes. Presently, it appears that the helical rib roughness is of significantly lower cost than an internally finned tube. Second, one must consider whether the design fouling factor R_{fi} can be maintained. There is essentially no published information on comparative fouling characteristics of the different enhanced tubes. Leitner [78] reports fouling test of an on-line brush-clean system used in the condenser tubes of a large refrigeration system. The brush system maintained the U value of the TURBOCHIL tube within 8% of the clean-tube value for cooling-tower water.

17.9 APPLICATION TO NATURAL CONVECTION

The previously discussed enhancement techniques may also be applied to natural-convection flows. A typical application involves the use of finned surfaces for cooling electronic equipment. Axial finned plates may be made by aluminum extrusion. Orienting the finned plates in the vertical direction, the flow geometry may be one of several forms: (1) a finned plate, (2) a finned plate with a "shroud" near the fin tips to channel the air in the inter-fin region, or (3) a parallel-plate channel formed by a pair of vertical finned plates.

Welling and Wooldridge [79] and Starner and McManus [80] provide data on heat transfer from a vertical finned plate. Jones and Smith [81] and Bar-Cohen [82] investigated the optimum fin spacing and fin thickness for the fin array. Sparrow et al. [83] theoretically evaluated the effect of an adiabatic shroud near the fin tips. Sparrow and Prakash [84] investigated the benefits of the offset strip fin geometry (Fig. 17.3*d*) relative to continuous axial fins.

Data on forced convection in finned channels may be applied to the case of natural convection in finned channels. The flow rate in the channel depends on the buoyancy force and friction factor of the channel. The buoyancy force depends on the heat transferred in the channel. Hence, one must iteratively solve the energy and momentum equations to determine the heat transfer and flow rate. Kraus and Bar-Cohen [85] illustrate this technique for the case of a plain, parallel-plate channel. The f and Nu vs. Re developed for forced convection in internally finned channels may be applied to the problem of natural convection in finned channels of the same geometry, provided the correlations are valid for the Re_d range of interest. For example, the friction-factor correlation developed by Webb and Scott [46] for forced convection in finned parallel plate channels may be applied to the natural-convection problem. If the flow is dominated by forced convection ($\mathrm{Gr}/\mathrm{Re}^2 \ll 1$), the forced-convection heat transfer correlation may be used for the heat transfer prediction. The problem of natural convection flow in a channel having wall roughness may be handled in the same way, using friction-factor and heat transfer correlations developed for forced convection.

Users are cautioned that the increased friction factor associated with an enhanced surface will lead to increased pressure drop for the FG and FN cases of Table 17.2. Hence, the flow rate will be reduced to satisfy the available buoyancy force. It is possible that no heat transfer improvement will result because of the reduced flow rate.

17.10 CLOSURE

The technology of enhanced heat transfer has been under serious development for approximately 20 years. The refrigeration and automotive industries are strong users of enhanced heat transfer surfaces. Other industries are rapidly becoming involved with the technology. This chapter discusses surface geometries for single-phase forced convection. Special surface geometries also exist for boiling and condensation. Tube geometries exist that are designed for boiling or condensation on one side and single-phase convection on the other side. References 1 and 2 provide citations on both single-phase and two-phase enhancement. Review articles that cover special surface geometries for condensation and boiling are Refs. 3 and 86 respectively.

NOMENCLATURE

A	total heat transfer surface area (both primary and secondary, if any) on one side of a direct-transfer exchanger; total heat transfer surface area of a regenerator, m², ft²
A_f	fin or extended surface area on one side of the exchanger, m², ft²
A_i	tube outside surface area, considering it as if there were no fins, m, ft
$B(e^+)$	correlating function for rough tubes [Eq. (17.38)]
b	distance between two plates (fin height) in a plate-fin exchanger, m², ft²
C_D	drag coefficient
c_p	specific heat of fluid at constant pressure, J/(kg · K), Btu/(1b$_m$ · F)
D_h	hydraulic diameter of flow passages, $= 4SL/A$, m, ft
D_v	volumetric hydraulic diameter = 4 × (void volume)/(total surface area)
d_e	diameter at tip of fins for a finned tube (Fig. 17.2*b*), m, ft
d_i	tube inside diameter; or diameter to the base of internal fins; roughness; m, ft
d_{im}	internal diameter if enhancement material is uniformly returned to tube wall thickness, m, ft
d_o	tube outside diameter; fin root diameter for a finned tube; m, ft
e	fin height, roughness height; m, ft
e^+	roughness Reynolds number $= eu^*/\nu$
f	Fanning friction factor $\Delta P \rho D_h / 2LG^2$
$f_{\rm f}$	friction factor of fins in Eq. (17.12), $= 2\rho \Delta P_f A_c / A_f G^2$
$f_{\rm tb}$	modified Fanning friction factor per tube row, $= \Delta P \rho / 2NG^2$
G	mass velocity based on the minimum flow area, kg/(m² · s), 1b$_m$/(ft² · s)
Gr	Grashof number $= g_r \beta \Delta T D_h^3 / \nu^2$
Gz	Graetz number $= \pi d_i \, \mathrm{Re}\,\mathrm{Pr}/4L$
g	acceleration due to gravity, $= 9.806$ m/s², $= 32.17$ ft/s²
g_r	radial acceleration defined by Eq. (17.35), m/s², ft/s²
H	louver fin height, defined in Fig. 17.6, m, ft
h	heat transfer coefficient based on A, W/(m² · K), Btu/(hr · ft² ·°F)
h'	heat transfer coefficient based on primary surface area (secondary surface excluded), W/(m² · K), Btu/(hr · ft² ·°F)
j	Colburn factor $= \mathrm{St}\,\mathrm{Pr}^{2/3}$
k	thermal conductance $= hA$, W/K, Btu/(hr ·°F)
k	thermal conductivity of fluid, W/(m · K), Btu/(hr · ft ·°F)
k_f	thermal conductivity of fin material, W/(m · K), Btu/(hr · ft ·°F)
L	fluid flow (core) length on one side of the exchanger, m, ft
LMTD	log-mean temperature difference, K, °F

l	strip flow length of OSF or louvre pitch of louvre fin, m, or ft
l_h	louver height shown on Fig. 17.6, m, ft
l_L	louver length shown on Fig. 17.6, m, ft
M	mass of tube material, kg, lb_m
N	number of tube rows in the flow direction; number of tubes in heat exchanger
Nu	Nusselt number $= hD_h/k$
Nu′	Nusselt number $= h'd/k$ for inside tube ($d = d_i$), outside tube ($d = d_0$)
NTU	number of heat transfer units, $= UA/(Wc_p)_{\min}$
n_f	number of fins in internally finned tube, dimensionless
P	fluid pumping power, W, hp
Pr	Prandtl number $= \mu c_p/k$
p	axial spacing between roughness elements, m, ft
p_t	pitch for 360° revolution of internal tape or fin, m, ft
Δp	fluid static pressure drop on one side of a heat-exchanger core, Pa, lb_f/ft^2
Δp_f	pressure drop assignable to fin area, Pa, lb_f/ft^2
q	heat transfer rate in exchanger, W, Btu/hr
R_{fi}	tube-side fouling resistance, $m^2 \cdot K/W$ or $ft^2 \cdot hr \cdot °F/Btu$
Re	Reynolds number based on the hydraulic diameter, $= GD_h/\mu$
Re*	defined by Eq. (17.6)
Re_d	Reynolds number based on the tube diameter, $= Gd/\mu$, $d = d_i$ for flow inside tube and $d = d_o$ for flow outside tube
Re_l	Reynolds number based on the interruption length, $= Gl/\mu$, GX_l/μ
Re_S	Reynolds number based on S_l
Re_v	Reynolds number based on D_v
S	flow cross-sectional area in minimum flow area, m^2, ft^2
S_d	diagonal pitch $= (S_t^2 + S_l^2)^{1/2}$
S_f	flow frontal area of heat exchanger, m^2, ft^2
S_l	longitudinal tube pitch, m, ft
S_t	transverse tube pitch, m, ft
St	Stanton number $= h/Gc_p$
s	fin pitch, center-to-center spacing, m, ft
s'	spacing between two fins, $= s - t$, m, ft
ΔT	temperature difference between hot and cold fluids, K or °F
t	thickness of fin or twisted tape, m, ft
ΔT_i	temperature difference between hot and cold inlet fluids, K, °F
U	overall heat transfer coefficient, $W/(m^2 \cdot K)$, $Btu/(hr \cdot ft^2 \cdot F)$
u_m	fluid mean axial velocity at the minimum free flow area, m/s, ft/s
u_{mod}	$= u_m(1 + \tan^2\alpha)^{1/2}$
u_θ	tangential velocity, m/s, ft/s
u^*	friction velocity $= (\tau_w/\rho)^{1/2}$, m/s, ft/s
V	heat exchanger total volume, m^3, ft^3
V_m	heat exchanger tube material volume, m^3, ft^3

W	fluid mass flow rate $= \rho u_m S$, kg/s, lb_m/s
x	Cartesian coordinate along the flow direction, m, ft
x^*	x/D_h Re Pr
y	twist ratio $= p_t/2d_i = \pi/(2 \tan \alpha)$

Greek Letters

α	helix angle relative to tube axis, $= \pi d_i/p_t$, rad
β_T	volume coefficient of thermal expansion, K^{-1}, $°R^{-1}$
γ	fin density, fins/m, fins/ft
ζ	$\left(\frac{\pi}{\pi + 2}\right)^2 \left(\frac{\pi + 2 - 2t/d_i}{\pi - 4t/d_i}\right)^2 \left(\frac{\pi}{\pi - 4t/d_i}\right)$
η_f	fin efficiency; temperature effectiveness of the fin
η_o	surface efficiency of finned surface, $= 1 - (1 - \eta_f)A_f/A$
2θ	included angle of fin cross section normal to flow, rad, deg
θ^*	$\theta k_m/k$
μ	fluid dynamic viscosity coefficient, Pa · s, lb_m/(s · ft)
ν	kinematic viscosity, m/s^2 or ft/s^2
ρ	fluid density, kg/m^3, lb_m/ft^3
τ_w	wall shear stress, Pa, lb_f/ft^2

Subscripts

fd	fully developed flow
m	average value over flow length
p	plain tube or surface
w	evaluated at wall temperature
x	local value

REFERENCES

1. A. E. Bergles, V. Nirmalan, G. H. Junkhan, and R. L. Webb, Bibliography on Augmentation of Convective Heat and Mass Transfer II, Heat Transfer Lab. Rep. HTL-31, ISU-ERI-Ames-84221, Iowa State Univ., Dec. 1983.
2. R. L. Webb, A. E. Bergles, and G. H. Junkhan, Bibliography of U.S. Patents of Augmentation of Convective Heat and Mass Transfer, Heat Transfer Lab. Rep. HTL-32, ISU-ERI-Ames-81070, Iowa State Univ., Dec. 1983.
3. R. L. Webb, The Use of Enhanced Surface Geometries in Condensers, *Power Condenser Heat Transfer Technology*, ed. P. J. Marto and R. H. Nunn, Hemisphere, Washington, pp. 287–324, 1981.
4. R. L. Webb, Special Surface Geometries for Heat Transfer Augmentation, *Developments in Heat Exchanger Technology-I*, ed. D. Chisholm, Applied Science Publishers, London, Chap. 7, 1980.
5. R. L. Webb, Performance Evaluation Criteria for Use of Enhanced Heat Transfer Surfaces in Heat Exchanger Design, *Int. J. Heat and Mass Transfer*, Vol. 24, pp. 715–726, 1981.
6. R. L. Webb, Performance Evaluation Criteria for Air-Cooled Finned Tube Heat Exchanger Surface Geometries, *Int. J. Heat and Mass Transfer*, Vol. 25, p. 1770, 1982.

7. W. M. Kays and A. L. London, *Compact Heat Exchangers*, 3rd ed., McGraw-Hill, New York, 1984.
8. W. M. Kays, Compact Heat Exchangers, AGARD Lecture Ser. No. 57 on Heat Exchangers, ed. J. J. Ginoux, AGARD-LS-57-72, Jan. 1972.
9. H. M. Joshi and R. L. Webb, Prediction of Heat Transfer and Friction in the Offset Strip Fin Array, *Int. J. Heat and Mass Transfer*, Vol. 30, No. 1, pp. 69–84, 1987.
10. A. R. Wieting, Empirical Correlations for Heat Transfer and Flow Friction Characteristics of Rectangular Offset Fin Heat Exchangers, *J. Heat Transfer*, Vol. 97, pp. 488–490, 1975.
11. R. K. Shah and R. L. Webb, Compact and Enhanced Heat Exchangers, *Heat Exchangers: Theory and Practice*, eds. J. Taborek, G. F. Hewitt, and N. H. Afgan, Hemisphere, New York, pp. 425–468, 1982.
12. C. J. Davenport, Correlations for Heat Transfer and Flow Friction Characteristics of Louvered Fin, *Heat Transfer—Seattle 1983*, AIChE Symp. Ser., No. 225, Vol. 79, pp. 19–27, 1984.
13. G. Rosenblad and A. Kullendorf, Estimating Heat Transfer Rates from Mass Transfer Studies on Plate Heat Exchanger Surfaces, *Wärme-Stoffübertragung*, Vol. 8, pp. 187–191, 1975.
14. K. Okada, M. Ono, T. Tomimura, T. Okuma, H. Konno, and S. Ohtani, Design and Heat Transfer Characteristics of New Plate Type Heat Exchanger, *Heat Transfer—Jpn. Res.*, Vol. 1, No. 1, pp. 90–95, 1972.
15. L. J. Goldstein and E. M. Sparrow, Heat/Mass Transfer Characteristics for Flow in a Corrugated Wall Channel, *J. Heat Transfer*, Vol. 99, pp. 187–195, 1977.
16. R. K. Shah, Perforated Heat Exchanger Surfaces: Part 2—Heat Transfer and Flow Friction Characteristics, ASME Paper 75-WA/HT-9, 1975.
17. R. L. Webb, Heat Transfer and Friction Characteristics for Finned Tubes Having Plain Fins, *Low Reynolds Number Flow Heat Exchangers*, ed. S. Kakaç, R. K. Shah, and A. E. Bergles, Hemisphere, Washington, pp. 431–450, 1983.
18. F. C. McQuiston, Correlation for Heat, Mass and Momentum Transport Coefficients for Plate-Fin-Tube Heat Transfer Surfaces with Staggered Tube, *ASHRAE Trans.*, Vol. 84, Pt. 1, pp. 294–309, 1978.
19. D. L. Gray and R. L. Webb, Heat Transfer and Friction Correlations for Plate Fin-and-Tube Heat Exchangers Having Plain Fins, *Heat Transfer 1986*, Vol. 6, pp. 2745–2750, 1986.
20. W. Nakayama and L. P. Xu, Enhanced Fins for Air-Cooled Heat Exchangers—Heat Transfer and Friction Factor Correlations, *Proc. 1983 ASME-JSME Thermal Eng. Conf.*, Vol. 1, pp. 495–502, 1983.
21. R. L. Webb, Air-Side Heat Transfer in Finned Tube Heat Exchangers, *Heat Transfer Eng.*, Vol. 1, No. 3, pp. 33–49, 1980.
22. D. E. Briggs and E. H. Young, Convection Heat Transfer and Pressure Drop of Air Flowing across Triangular Pitch Banks of Finned Tubes, *Chem. Eng. Prog. Symp. Ser.*, Vol. 59, No. 41, pp. 1–10, 1963.
23. K. K. Robinson and D. E. Briggs, Pressure Drop of Air Flowing Across Triangular Pitch Banks of Finned Tubes, *Chem. Eng. Prog. Symp. Ser.*, No. 64, No. 62, pp. 177–184, 1966.
24. T. J. Rabas, P. W. Eckles, and R. A. Sabatino, The Effect of Fin Density on the Heat Transfer and Pressure Drop Performance of Low Finned Tube Banks, ASME Paper No. 80-HT-97, 1980.
25. E. Schmidt, Heat Transfer at Finned Tubes and Computations of Tube Bank Heat Exchangers, *Kältetechnik*, Vol. 15, No. 4, p. 98, 1963; Vol. 15, No. 12, p. 370, 1963.
26. G. E. LaPorte, C. L. Osterkorn, and S. M. Marino, *Heat Transfer Fin Structure*, U.S. Patent 4,143,710, March 13, 1979.
27. R. W. Abbott, R. H. Norris, and W. A. Spofford, Compact Heat Exchangers in General Electric Products—Sixty Years of Advances in Design and Manufacturing Technologies, *Compact Heat Exchangers—History, Technology Advancement and Mechanical Design Problems*, ed. R. K. Shah, C. F. McDonald, and C. P. Howard, Book No. G00183, HTD—Vol. 10, ASME, New York, pp. 37–55, 1980.

28. P. W. Eckels and T. J. Rabas, Heat Transfer and Pressure Drop Performance of Finned Tube Bundles, *J. Heat Transfer*, Vol. 107, pp. 205–213, 1985.

29. C. Weierman, J. Taborek, and W. J. Marner, Comparison of the Performance of Inline and Staggered Banks of Tubes with Segmented Fins, 15th National Heat Transfer Conference, San Francisco, AIChE Paper 6, 1975.

30. C. Weierman, Correlations Ease the Selection of Finned Tubes, *Oil and Gas J.*, pp. 94–100, Sept. 6, 1976.

31. T. J. Rabas and P. W. Eckels, Heat Transfer and Pressure Drop Performance of Segmented Surface Tube Bundles, ASME Paper 75-HT-45, 1985.

32. H. Brauer, Compact Heat Exchangers, *Chem. Prog. Eng.*, London, Vol. 45, No. 8, pp. 451–460, Aug. 1964.

33. J. M. O'Connor and S. F. Pasternak, *Method of Making a Heat Exchanger*, U.S. Patent 3,947,941, Apr. 6, 1976.

34. R. J. Haberski and R. J. Raco, Engineering Analysis and Development of an Advanced Technology, Low Cost, Dry Cooling Tower Heat Transfer Surface, Curtis-Wright Corp. Rep. C00-2774-1, Mar. 1, 1976.

35. B. Cox, Heat Transfer and Pumping Power Performance in Tube Banks—Finned and Bare, ASME Paper 73-HT-27, 1973.

36. B. Cox and P. A. Jallouk, Methods for Evaluating the Performance of Compact Heat Transfer Surfaces, *J. Heat Transfer*, Vol. 95, pp. 464–469, 1973.

37. R. L. Webb, Heat Transfer and Friction Characteristics for Finned Tubes Having Plain Fins, *Low Reynolds Number Flow Heat Exchangers*, ed. S. Kakaç, R. K. Shah, and A. E. Bergles, Hemisphere, Washington, pp. 431–450, 1983.

38. T. J. Rabas and F. V. Huber, Row Number Effects on the Heat Transfer Performance of Inline Finned Tube Banks, ASME Paper 84-HT-97, 1984.

39. A. P. Watkinson, P. L. Miletti, and G. R. Kubanek, Heat Transfer and Pressure Drop of Internally Finned Tubes in Laminar Oil Flow, ASME Paper 75-HT-41, 1975.

40. W. J. Marner and A. E. Bergles, Augmentation of Tube-Side Laminar Flow Heat Transfer by Means of Twisted Tape Inserts, Static Mixer Inserts and Internally Finned Tubes, *Heat Transfer 1978*, Hemisphere, Washington, Vol. 2, pp. 583–588, 1978.

41. H. M. Soliman, T. S. Chau, and A. C. Trupp, Analysis of Laminar Heat Transfer in Internally Finned Tubes with Uniform Outside Wall Temperature, *J. Heat Transfer*, Vol. 102, pp. 598–604, 1980.

42. H. M. Soliman, The Effect of Fin Material on Laminar Heat Transfer Characteristics of Internally Finned Tubes, *Advances in Enhanced Heat Transfer*, ed. J. M. Chenoweth, J. Kaellis, J. W. Michel, and S. Shenkman, Book No. 100122, ASME, New York, pp. 95–102, 1979.

43. H. M. Soliman and A. Feingold, Analysis of Fully Developed Laminar Flow in Longitudinally Internally Finned Tubes, *Chem. Eng. J.*, Vol. 14, pp. 119–128, 1977.

44. A. E. Bergles and S. D. Joshi, Augmentation Techniques for Low Reynolds Number In-Tube Flow, *Low Reynolds Number Flow in Heat Exchangers*, S. Kakaç, R. K. Shah, and A. E. Bergles, Hemisphere, Washington, pp. 694–720, 1983.

45. T. C. Carnavos, Heat Transfer Performance of Internally Finned Tubes in Turbulent Flow, *Heat Transfer Eng.*, Vol. 4, No. 1, pp. 32–37, 1980.

46. R. L. Webb and M. J. Scott, Analytic Prediction of the Friction Factor for Turbulent Flow in Internally Finned Channels, *J. Heat Transfer*, Vol. 103, pp. 423–428, 1981.

47. Yampolsky, J. S., Spirally Fluted Tubing for Enhanced Heat Transfer, *Heat Exchangers—Theory and Practice*, ed. J. Taborek, G. F. Hewitt, and N. Afgan, Hemisphere, Washington, pp. 945–952, 1983.

48. R. Koch, Druckverlust und Wärmeübergang bei Verwirbeiter Strömung, *Verein Deutscher Ing.—Forschungsheft*, Ser. B, Vol. 24, No. 469, pp. 1–44, 1958.

49. D. G. Thomas, Enhancement of Forced Convection Mass Transfer Coefficient Using Detached Turbulence Promoters, *Ind. Eng. Chem. Process Des. Dev.*, Vol. 6, pp. 385–390, 1967.

50. S. B. Uttawar and M. Raja Rao, Augmentation of Laminar Flow Heat Transfer in Tubes by Means of Wire Coil Inserts, *J. Heat Transfer*, Vol. 107, pp. 930–935, 1985.

51. R. Kumar and R. L. Judd, Heat Transfer with Coiled Wire Turbulence Promoters, *Canad. J. Chem. Eng.*, Vol. 48, pp. 378–383, 1970.

52. W. E. Hilding and C. H. Coogan, Jr., Heat Transfer and Pressure Drop in Internally Finned Tubes, *ASME Symp. Air Cooled Heat Exchangers*, ASME, New York, pp. 57–84, 1964.

53. A. C. Trupp and A. C. Y. Lau, Fully Developed Laminar Heat Transfer in Circular Sector Ducts with Isothermal Walls, *J. Heat Transfer*, Vol. 106, pp. 467–469, 1984.

54. R. Thorsen and F. Landis, Friction and Heat Transfer Characteristics in Turbulent Swirl Flow Subjected to Large Transverse Temperature Gradients, *J. Heat Transfer*, Vol. 90, pp. 87–89, 1968.

55. A. W. Date, Prediction of Fully-Developed Flow in a Tube Containing a Twisted Tape, *Int. J. Heat Mass Transfer*, Vol. 17, pp. 845–859, 1974.

56. R. K. Shah and A. L. London, *Laminar Forced Convection in Ducts*, Academic, New York, 1978.

57. S. W. Hong and A. E. Bergles, Augmentation of Laminar Flow Heat Transfer in Tubes by Means of Twisted Tape Inserts, *J. Heat Transfer*, Vol. 98, pp. 251–256, 1976.

58. R. F. Lopina and A. E. Bergles, Heat Transfer and Pressure Drop in Tape-Generated Swirl Flow of Single-Phase Water, *J. Heat Transfer*, Vol. 91, pp. 434–442, 1969.

59. A. W. Date, Flow in Tubes Containing Twisted Tapes, *Heating and Ventilating Eng.*, Vol. 47, pp. 240–249, 1973.

60. R. L. Webb, T. S. Hui, and L. Haman, Enhanced Tubes in Electric Utility Steam Condensers, *Heat Transfer in Heat Rejection Systems*, ed. S. Sengupta and Y. F. Mussalli, Book No. G00265, HTD—Vol. 37, ASME, New York, pp. 17–26, 1984.

61. M. Dalle and L. Meyer, Turbulent Convection Heat Transfer from Rough Surfaces with Two-dimensional Ribs, *Int. J. Heat Mass Transfer*, Vol. 22, pp. 583–620, 1979.

62. D. E. Metzger, C. Z. Fan, and J. W. Pennington, Heat Transfer and Flow Friction Characteristics of Very Rough Transverse Ribbed Surfaces with and without Pin Fins, *Proc. 1983 ASME-JSME Thermal Eng. Conf.*, Vol. 1, pp. 429–436, 1983.

63. D. L. Gee and R. L. Webb, Forced Convection Heat Transfer in Helically Rib-Roughened Tubes, *Int. J. Heat Mass Transfer*, Vol. 23, pp. 1127–1136, 1980.

64. R. L. Webb, E. R. G. Eckert, and R. J. Goldstein, Heat Transfer and Friction in Tubes with Repeated-Rib Roughness, *Int. J. Heat Mass Transfer*, Vol. 14, pp. 601–617, 1971.

65. F. J. Edwards and N. Sheriff, The Heat Transfer and Friction Characteristics of Forced Convection Air Flow over a Particular Type of Rough Surface, *Int. Dev. Heat Transfer*, ASME, pp. 415–426, 1961.

66. H. Schlicting, *Boundary-Layer Theory*, 7th ed., McGraw-Hill, New York, pp. 600–620, 1968.

67. D. F. Dipprey and R. H. Sabersky, Heat and Momentum Transfer in Smooth and Rough Tubes at Various Prandtl Numbers, *Int. J. Heat Mass Transfer*, Vol. 6, pp. 329–353, 1963.

68. R. L. Webb, E. R. G. Eckert, and R. J. Goldstein, Generalized Heat Transfer and Friction Correlations for Tubes with Repeated-Rib Roughness, *Int. J. Heat Mass Transfer*, Vol. 15, pp. 180–184, 1972.

69. W. Nakayama, K. Takahashi, and T. Daikoku, Spiral Ribbing to Enhance Single-Phase Heat Transfer inside Tubes, *Proc. ASME-JSME Thermal Eng. Joint Conf.*, Honolulu, ASME, New York, Vol. 1, pp. 503–510, 1983.

70. K. J. Bell and A. C. Mueller, *Engineering Data Book II*, Wolverine Tube Div., Decatur, AL, 1984.

71. J. G. Withers, Tube-Side Heat Transfer and Pressure Drop for Tubes Having Helical Internal Ridging with Turbulent/Transitional Flow of Single-Phase Fluid, Part 2, Multiple-Helix Ridging, *Heat Transfer Eng.*, Vol. 2, No. 2, pp. 43–50, 1980.
72. R. Sethumadhavan and M. Raja Rao, Turbulent Flow Friction and Heat Transfer Characteristics of Single- and Multi-start spirally Enhanced Tubes, *J. Heat Transfer*, Vol. 108, pp. 55–61, 1986.
73. Anon., *Hitachi High-Performance Heat Transfer Tubes*, Cat. No. EA-500, Hitachi Cable, Tokyo, 1984.
74. R. L. Webb, Toward a Common Understanding of the Performance and Selection of Roughness for Forced Convection, *Studies in Heat Transfer: A Festschrift for E. R. G. Eckert*, ed. J. P. Hartnett et al., Hemisphere, Washington, pp. 257–272, 1979.
75. D. A. Dawson and O. Trass, Mass Transfer at Rough Surfaces, *Int. J. Heat Mass Transfer*, Vol. 15, pp. 1317–1336, 1972.
76. R. L. Webb and E. R. G. Eckert, Application of Rough Surfaces to Heat Exchanger Design, *Int. J. Heat Mass Transfer*, Vol. 15, pp. 1647–1658, 1972.
77. R. L. Webb, Performance, Cost Effectiveness and Water-Side Fouling Considerations of Enhanced Tube Heat Exchangers for Boiling Service with Tube-Side Water Flow, *Heat Transfer Eng.*, Vol. 3, No. 3, pp. 84–98, 1982.
78. G. F. Leitner, Controlling Chiller Tube Fouling, *ASHRAE J.*, Vol. 6, No. 2, pp. 40–43, 1980.
79. J. R. Welling and C. B. Wooldridge, Free Convection Heat Transfer Coefficients from Rectangular Fins, *J. Heat Transfer*, Vol. 87, pp. 439–444, 1965.
80. K. E. Starner and H. N. McManus, An Experimental Investigation of Free Convection Heat Transfer from Rectangular Fin Arrays, *J. Heat Transfer*, Vol. 85, pp. 273–278, 1963.
81. C. D. Jones and L. F. Smith, Optimum Arrangement of Rectangular Fins on Horizontal Surfaces for Free Convection Heat Transfer, *J. Heat Transfer*, Vol. 92, pp. 6–10, 1970.
82. A. Bar-Cohen, Fin Thickness for an Optimized Natural Convection Array of Rectangular Fins, *J. Heat Transfer*, Vol. 101, pp. 564–566, 1979.
83. E. M. Sparrow, R. B. Baliga, and S. V. Patankar, Forced Convection Heat Transfer from a Shrouded Fin Array with and without Tip Clearance, *J. Heat Transfer*, Vol. 100, pp. 572–579, 1978.
84. E. M. Sparrow and C. Prakash, Enhancement of Natural Convection Heat Transfer by a Staggered Array of Discrete Vertical Plates, *J. Heat Transfer*, Vol. 102, pp. 215–220, 1980.
85. A. D. Kraus and A. Bar-Cohen, *Thermal Analysis and Control of Electronic Equipment*, McGraw-Hill, New York, pp. 305–306, 1983.
86. R. L. Webb, The Evolution of Enhanced Surface Geometries for Nucleate Boiling, *Heat Transfer Eng.*, Vol. 2, pp. 46–69, 1981.
87. E. M. Sparrow and C. H. Liu, Heat-Transfer, Pressure Drop and Performance Relationship for In-Line, Staggered, and Continuous Plate Heat Exchangers, *Int. J. Heat Mass Transfer*, Vol. 22, pp. 1613–1625, 1979.

18

THE EFFECT OF TEMPERATURE-DEPENDENT FLUID PROPERTIES ON CONVECTIVE HEAT TRANSFER

S. Kakaç

University of Miami
Coral Gables, Florida

18.1 INTRODUCTION

In single-phase convective heat transfer solutions given in previous chapters, it has been generally assumed that the fluid properties are constant. When applied to practical heat transfer problems with large temperature differences between the surface and the fluid, the constant-property assumption could cause significant errors, since the transport properties of most fluids vary with temperature, which influences the variation of velocity and temperature through the boundary layer or over the flow cross section of a duct. In this chapter, internal flow forced convection and natural convection will be considered with temperature-dependent fluid properties. External flow forced convection is treated in Chapter 2.

For most liquids, the specific heat, thermal conductivity, and density are nearly independent of temperature, but the viscosity decreases markedly with increasing temperature. The Prandtl number of liquids varies with temperature in much the same manner as the viscosity.

In case of gases, convective heat transfer always causes considerable property variation. Density, thermal conductivity, and viscosity all vary at approximately the same rate as the absolute temperature. The variation with absolute temperature is approximately the same for both viscosity and thermal conductivity for most of the gases and their mixtures. If one represents this variation as a power-law dependence, the exponent to the temperature falls in the range from 0.7 to 0.8. The specific heat varies only slightly with temperature, and the Prandtl number does not vary significantly. As regards the density, most mixtures can be represented by the perfect-gas idealization.

Therefore, when the difference between the fluid bulk and the wall temperature is high, the effect of property variation is to alter the velocity and temperature profiles, resulting in different heat transfer and friction coefficients than would be obtained if properties were constant. The distorted laminar velocity profiles are shown in Fig. 18.1. When a liquid is heated, the fluid near the wall is less viscous than the fluid in the center. Consequently, the velocity of the heated liquid is larger than that of the unheated liquid at the same average velocity and temperature. The distortion of the parabolic velocity profile will be in opposite direction with cooling. For gases, the conditions are reversed because the viscosity increases with increasing temperature, but in addition to the viscosity, variations in density can also alter the profile. In a gas, acceleration induced by the change in density creates distortions in the opposite direction from those due to the viscosity variation, and the former tend to dominate.

The analytical investigation of the effect of property variations on heat transfer is a highly complicated task for various reasons. First of all, the variations of properties with temperature differ from one fluid to another. Sometimes it is impossible to express these variations in analytical form.

For practical applications, a reliable and appropriate correlation based on the constant-property assumption can be modified and/or corrected so that it may be used when the variable-property effect becomes important.

Two methods of correcting constant-property correlations for the variable-property effect have been employed: namely *the reference temperature* method and *the property ratio* method. In the former, a characteristic temperature is chosen at which the properties appearing in nondimensional groups are evaluated so that the constant-property results at that temperature may be used to consider the variable-property behavior; in the latter, all properties are taken at the bulk temperature or the free-stream temperature, and then all variable-property effects are lumped into a function of the ratio of one property evaluated at the surface temperature to that

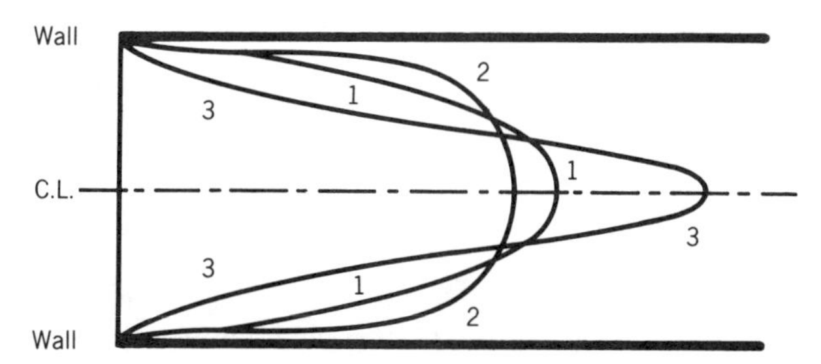

Figure 18.1. The effect of the property variations on velocity distribution in laminar flow. Curve 1: isothermal flow; curve 2: heating of liquid or cooling of gas; curve 3: cooling of liquid or heating of gas.

property evaluated at bulk temperature. Some correlations may involve a modification or combination of these two methods.

For *liquids*, the variation of viscosity is responsible for most of the property effects. Therefore, the variable-property Nusselt numbers and friction factors in the property ratio method for liquids are correlated by

$$\frac{\mathrm{Nu}}{\mathrm{Nu}_{\mathrm{cp}}} = \left(\frac{\mu_b}{\mu_w}\right)^n \tag{18.1a}$$

$$\frac{f}{f_{\mathrm{cp}}} = \left(\frac{\mu_b}{\mu_w}\right)^m \tag{18.1b}$$

where μ_b is the viscosity evaluated at the bulk mean temperature (or free-stream temperature in the case of external flows), μ_w is the viscosity evaluated at the wall temperature, and cp refers to the constant-property solution. The friction coefficient usually employed is the so-called Fanning friction factor based on the wall shear rather than the pressure drop.

For *gases*, the viscosity, thermal conductivity, and density vary with the absolute temperature. Therefore, in the property-ratio method, temperature corrections of the following forms are found to be adequate in practical applications for the temperature-dependent-property effects in gases:

$$\frac{\mathrm{Nu}}{\mathrm{Nu}_{\mathrm{cp}}} = \left(\frac{T_w}{T_b}\right)^n \tag{18.2a}$$

$$\frac{f}{f_{\mathrm{cp}}} = \left(\frac{T_w}{T_b}\right)^m \tag{18.2b}$$

where T_b and T_w are the absolute bulk mean (free-stream in the case of external flows) and wall temperatures, respectively.

It must be noted that the constant-property portion of the specific correlation is evaluated in terms of parameters and conditions defined by its author(s).

Extensive theoretical and experimental investigations with variable properties have been reported in the literature to obtain the values of the exponents n and m which will be cited in the following sections of this chapter.

18.2 FORCED CONVECTION IN DUCTS

Convective heat transfer in ducts is encountered in a wide variety of engineering situations. Different investigators performed extensive experimental and theoretical studies with various fluids. As a result, they formulated relations for the Nusselt number vs. the Reynolds and Prandtl numbers for a wide range of these dimensionless groups.

In heat transfer systems used in different fields of engineering, the heat exchange equipment is designed to operate in laminar and turbulent flow regimes. In the following sections, heat transfer and friction coefficients in laminar and turbulent forced convection in duct flow with temperature-dependent properties will be considered.

18.2.1 Laminar Flow in Ducts

To investigate theoretically the effect of variable fluid properties on heat transfer and friction coefficients in laminar flow, the velocity and temperature profiles should be calculated at a particular cross section employing certain idealizations. The steady-state two-dimensional momentum and energy equations with negligible heat dissipation in a circular duct can be written (see Tables 1.5 and 1.6 of Chapter 1)

$$\rho\left(v\frac{\partial u}{\partial r}+u\frac{\partial u}{\partial x}\right)=-\frac{\partial p}{\partial x}+\frac{1}{r}\frac{\partial}{\partial r}\left(\mu r\frac{\partial u}{\partial r}\right)+\frac{\partial}{\partial x}\left(\mu\frac{\partial u}{\partial x}\right) \tag{18.3}$$

$$\rho c_p\left(u\frac{\partial T}{\partial x}+v\frac{\partial T}{\partial r}\right)=\frac{\partial}{\partial r}\left(k\frac{\partial T}{\partial r}\right)+\frac{k}{r}\frac{\partial T}{\partial r}+\frac{\partial}{\partial x}\left(k\frac{\partial T}{\partial x}\right) \tag{18.4}$$

where x represent the distance along the channel.

For fully developed velocity and temperature profiles (fully developed conditions) under the constant heat flux boundary condition, Eqs. (18.3) and (18.4) reduce to

$$\frac{dp}{dx}=\frac{1}{r}\frac{\partial}{\partial r}\left(r\mu\frac{\partial u}{\partial r}\right) \tag{18.5}$$

$$\rho c_p u\frac{dT}{dx}=k\frac{d^2T}{dr^2}+\frac{dT}{dr}\left(\frac{dk}{dr}+\frac{k}{r}\right) \tag{18.6}$$

Since with variable fluid properties, dT/dx can still be assumed to be constant to a good approximation for constant heat flux boundary condition, $\partial k/\partial x$ and v in Eq. (18.4) are small and can be neglected.

For constant fluid properties (ρ, c_p, μ, k), Eqs. (18.5) and (18.6) give

$$u=u_{\max}\left(1-\frac{r^2}{r_i^2}\right) \tag{18.7}$$

$$T=T_w-\frac{c_p\rho u_{\max}}{16k}\frac{dT}{dx}\left(3r_i^2-4r^2+\frac{r^4}{r_i^2}\right) \tag{18.8}$$

For a given problem, the velocity and temperature profiles for constant fluid properties can be taken as a first approximation. By the use of an appropriate relationship of property variation with temperature, Eq. (18.5) can be integrated numerically to obtain the second approximation for the velocity profile. This velocity profile can be used to integrate Eq. (18.6) to obtain the second approximation of the temperature profile with the fluid properties based on the first approximation of the temperature profile. The number of approximations required for convergence will depend on the rate of heat input. The mixed mean fluid temperature, Nusselt number, and friction coefficient can be calculated as in the constant-property solutions.

The above procedure can also be applied to the flow between parallel plates. For constant fluid properties, the following equations can be obtained for fully developed velocity and temperature profiles for the flow between two parallel plates under the constant heat flux boundary condition:

$$u = u_{\max}\left(1 - \frac{y^2}{d^2}\right) \tag{18.9}$$

$$T = T_w - \frac{u_{\max}\rho c_p}{2k}\frac{dT}{dx}\left(\frac{5}{6}d^2 - y^2 + \frac{y^4}{6d^2}\right) \tag{18.10}$$

where y is measured from the centerline. With the velocity and temperature profiles for constant fluid properties as the first approximations, the method of iteration can again be applied as in the case of the circular tube. The Nusselt number and the friction factor can be found in an analogous manner. By the use of numerical methods, the two coupled equations of momentum and energy for the flow between parallel plates can also be solved directly to obtain velocity and temperature distributions.

Laminar Flow of Liquids. Deissler [1] carried out a numerical analysis as described above for laminar flow through a circular duct at constant heat flux boundary condition for liquid viscosity variation with temperature given by

$$\frac{\mu}{\mu_w} = \left(\frac{T}{T_w}\right)^{-1.6} \tag{18.11}$$

and obtained $n = 0.14$ to be used with Eq. (18.1a). This has been used widely to correlate experimental data for laminar flow for $\mathrm{Pr} > 0.6$.

Deissler [1] also obtained $m = -0.58$ for heating, and $m = -0.50$ for cooling of liquids to be used with Eq. (18.1b).

Yang [2] obtained the solution for both constant wall heat flux Ⓗ and constant wall temperature Ⓣ boundary conditions by assuming a viscosity dependence of a liquid on temperature as

$$\frac{\mu}{\mu_w} = \left[1 + A\left(\frac{T_w - T}{T_w - T_i}\right)\right]^{-1} \tag{18.12}$$

where A is a constant. His predictions for both Ⓗ and Ⓣ boundary conditions were correlated with $n = 0.11$ in Eq. (18.1a), and he concluded that the effect of thermal boundary conditions is small and the influence on the friction coefficient is very substantial. He also found that the correction for variable properties is the same for developing and developed regions.

Shannon and Depew [3] and Joshi and Bergles [4] carried out a similar analysis for the constant heat flux boundary condition. The μ-T relation they used in their calculation is

$$\mu = \mu_i \exp[-\gamma(T - T_i)] \tag{18.13}$$

where γ is a viscosity parameter defined as $\gamma = -(d\mu/dT)/\mu$ = constant. For $\mu/\mu_w < 5$, their fully developed prediction gives $n = 0.14$, which is essentially the same as that of Deissler [1]. Shannon and Depew [3] have done a similar analysis for the constant heat flux boundary condition, but their predictions differ from Yang's, particularly in the entrance region, for which they recommended $n = 0.3$.

A simple empirical correlation has been proposed by Sieder and Tate [5] to predict the mean Nusselt number for *laminar* flow in a circular duct at constant wall temperature

$$\mathrm{Nu}_b = 1.86\left(\mathrm{Re}_b \mathrm{Pr}_b \frac{d}{L}\right)^{1/3} \left(\frac{\mu_b}{\mu_w}\right)^{0.14} \tag{18.14}$$

which is valid for T_w = constant, smooth tubes, $0.48 < \mathrm{Pr}_b < 16{,}700$, and $0.0044 < \mu_b/\mu_w < 9.75$. This correlation has been recommended by Whitaker [6] for values of

$$\left[\mathrm{Re}_b\, \mathrm{Pr}_b \left(\frac{d}{L}\right)\right]^{1/3} \left(\frac{\mu_b}{\mu_w}\right)^{0.14} \gtrsim 2 \tag{18.15}$$

All physical properties are evaluated at the fluid bulk mean temperature except μ_w, which is evaluated at the wall temperature.

It is not surprising that alternative correlations have been proposed for specific fluids. Oskay and Kakaç [7] performed experimental studies with mineral oil in laminar flow through a circular duct under constant wall heat flux boundary condition in the range of $0.8 \times 10^3 < \mathrm{Re}_b < 1.8 \times 10^3$ and $1 < T_w/T_b < 3$, and suggested that the viscosity-ratio exponent for Nu should be increased to 0.152 for mineral oil.

Kuznetsova [8] made experiments with transformer oil and fuel oil in the range of $400 < \mathrm{Re}_b < 1900$ and $170 < \mathrm{Pr}_b < 640$, and recommended

$$\mathrm{Nu}_b = 1.23\left(\mathrm{Re}_b \mathrm{Pr}_b \frac{d}{L}\right)^{0.4} \left(\frac{\mu_w}{\mu_b}\right)^{-1/6} \tag{18.16}$$

Test [9] conducted an analytical and experimental study of heat transfer and fluid flow behavior for laminar flow in a circular duct for liquid with viscosity varying with temperature. The analytical approach is a numerical solution of continuity, momentum, and energy equations. The experimental approach involves the use of a hot-wire technique for determination of velocity profiles. He obtained the following correlation for the local Nusselt number:

$$\mathrm{Nu}_b = 1.4\left(\mathrm{Re}_b \mathrm{Pr}_b \frac{d}{L}\right)^{1/3} \left(\frac{\mu_b}{\mu_w}\right)^{n} \tag{18.17}$$

where

$$n = \begin{cases} 0.05 \text{ for heating liquids} \\ \frac{1}{3} \text{ for cooling liquids} \end{cases}$$

He also obtained the friction factor as

$$f = \frac{16}{\mathrm{Re}} \frac{1}{0.89} \left(\frac{\mu_b}{\mu_w} \right)^{0.2} \tag{18.18}$$

Equations (18.14) and (18.17) should not be applied to extremely long ducts.

Laminar Flow of Gases. Deissler [1] first studied the effects of variable properties in laminar flow of air through a circular duct. He made the following simplifying idealizations: (1) constant heat flux boundary condition, (2) negligible heat dissipation, (3) negligible entrance effects, and (4) negligible heat conduction in the direction of flow compared with the radial direction. He further assumed that the specific heat of air is constant, and that the viscosity and conductivity are both proportional to $T^{0.8}$. Deissler concluded from his results that the evaluation of the conductivity at $T_r = T_b - 0.27(T_w - T_b)$ for the Nusselt number, and the viscosity at $T_r = T_b + 0.58(T_w - T_b)$ for the friction factor, eliminates the effects of variable fluid properties on these parameters. The variations of the viscosity and conductivity of air with temperature are not identical as has been assumed by Deissler. On the other hand, because of density changes with temperature, there is always a radial component of the velocity, and under large temperature differences between the wall and the mixed mean temperature of fluid, it is not possible to have a fully developed velocity profile. Sze [10] investigated the effects of temperature-dependent properties on heat transfer and friction through a circular tube and parallel plates for fully developed laminar flow of air with a constant heat flux boundary condition. He used experimentally determined physical properties of air. Sze's method of solution is in general similar to Deissler's. Sze concluded that the effect of temperature-dependent fluid properties on the Nusselt number in laminar flow can be largely eliminated by evaluating the thermal conductivity at $T_r = T_b - 0.18(T_w - T_b)$ for circular tubes and at $T_r = T_b + 0.13(T_w - T_b)$ for parallel plates. Similarly, the effect of temperature-dependent fluid properties on the friction factor in laminar flow can be largely eliminated by evaluating the viscosity at the average of the mixed mean fluid temperature and wall temperature for both the circular tube and parallel plates.

The first reasonably complete solution for laminar heat transfer to a gas flowing in a tube with temperature-dependent property variation was developed by Worsøe-Schmidt [11]. He solved the governing equations with a finite difference technique for fully developed gas flow through a circular tube. Heating and cooling with a constant surface temperature and heating with constant heat rate are considered. In this solution, the radial velocity is included. He concluded that near the entrance, and also well downstream, the results can be satisfactorily correlated for heating ($1 < T_w/T_b < 3$) by $n = 0$, $m = 1.00$ and for cooling ($0.5 < T_w/T_b < 1$) by $n = 0.0$, $m = 0.81$.

Calculations by Sze [10] and Kays and Crawford [12] for flow between parallel plates, using the fully developed flow idealization and employing a procedure similar to that of Deissler [1], give the same results for the friction coefficient as were obtained for the circular tube under the same idealizations. But the results for the Nusselt number are substantially different.

Laminar forced convection and fluid flow in ducts have been studied extensively, and numerous results are available for circular and noncircular ducts under various boundary conditions. These results have been compiled by Shah and London [13] and in Chapter 3; they may be used for the constant-property portion of the correlations (18.1) and (18.2). The exponents of n and m are summarized in Table 18.1. For fully developed laminar flow, $n = 0.14$ is generally recommended for laminar heating of liquids.

TABLE 18.1. Exponents *n* and *m* Associated with Eqs. (18.1) and (18.2) for Laminar Forced Convection through Circular Ducts, Pr > 0.5

No. Reference	Fluid	Condition	n	m	Limitations
1 Deissler [1]	Liquid	Laminar heating	0.14	−0.58	Fully developed flow, q''_w = constant, Pr > 0.6, $\mu/\mu_w = (T/T_w)^{-1.6}$
	Liquid	Laminar cooling	0.14	−0.50	
2 Yang [2]	Liquid	Laminar heating	0.11	—	Developing and fully developed regions of a circular duct, T_w = constant, q''_w = constant
3 Shannon and Depew [3]	Liquid	Laminar heating	0.3	—	At entrance of tube, $\mu = \mu_i \exp[-\gamma(T - T_i)]$, $\gamma = -(1/\mu)d\mu/dT$ = constant, q''_w = constant
	Liquid	Laminar heating	0.14	—	Fully developed region
4 Joshi and Bergles [4]	Liquid	Laminar heating	0.14	—	$\mu(T)$ as in No. 3, $\mu_w/\mu_b < 5$, q''_w = constant
5 Worsøe-Schmidt [11]	Gas	Laminar heating	0	1.00	Developing and fully developed regions, q''_w = constant, T_w = constant, $1 < T_w/T_b < 3$
	Gas	Laminar cooling	0	0.81	T_w = constant, $0.5 < T_w/T_b < 1$

18.2.2 Turbulent Flow in Circular Ducts

The analysis of the effect of temperature-dependent fluid properties on heat transfer and friction in turbulent flow is much more complicated than that in laminar flow. In laminar flow, the shear stress for a certain velocity gradient depends solely on the viscosity, which is a property of the fluid and depends only on the temperature of the fluid. This is not the case in turbulent flow. The effect of eddy viscosity is more important over a major portion of the flow. Eddy viscosity is a function of the flow configuration which varies over the flow cross section and is indirectly dependent on temperature.

Hence, to obtain the variation of eddy viscosity with temperature, the change in flow configuration due to temperature changes has to be known. Similarly, heat transfer in turbulent flow depends not only on the thermal conductivity, which is a temperature-dependent physical property of the fluid, but also on eddy diffusivity, which is a function of the Reynolds number and Prandtl number.

For a steady, fully developed turbulent flow through a circular duct in the absence of axial diffusion, negligible viscous dissipation, low Mach number, and insignificant body forces such as gravity, the boundary-layer equations of continuity, momentum, and energy are given in Chapter 1 by Eqs. (1.96), (1.97), and (1.98):

$$\frac{\partial(\rho u)}{\partial x} + \frac{1}{r}\frac{\partial}{\partial r}(\rho v r) = 0 \tag{18.19}$$

$$\rho\left(u\frac{\partial u}{\partial x} + v\frac{\partial u}{\partial r}\right) = -\frac{\partial p}{\partial x} + \frac{1}{r}\left[\frac{\partial}{\partial r}r(\mu + \rho\epsilon_M)\frac{\partial u}{\partial r}\right] \tag{18.20}$$

$$\rho c_p\left(u\frac{\partial T}{\partial x} + v\frac{\partial T}{\partial r}\right) = \frac{1}{r}\left[\frac{\partial}{\partial r}r(k + \rho c_p\epsilon_H)\frac{\partial T}{\partial r}\right] \tag{18.21}$$

With temperature-dependent properties, the energy equation becomes nonlinear and the x-momentum equation becomes coupled to the thermal energy equation.

By the use of the expressions available in the literature for eddy diffusivity of momentum and heat (turbulence models), numerical solution of Eqs. (18.19), (18.20), and (18.21) can be carried out with property variations. Accurate predictions of heat transfer and friction coefficients depend on the use of accurate relationships for distributions of velocity and eddy diffusivities of heat and momentum, which require the application of numerical methods. ϵ_M and ϵ_H are largely governed by turbulence and therefore can be assumed to be independent of temperature; the effect of temperature-dependent fluid properties on ϵ_M and ϵ_H would only be indirect through their influence on flow configurations. In most of the analyses, it was assumed that the eddy diffusivity of momentum, ϵ_M, is equal to the eddy diffusivity of heat, ϵ_H, i.e., the turbulent Prandtl number $\mathrm{Pr}_t = \epsilon_M/\epsilon_H$ is unity. But the turbulent Prandtl number is not constant across the boundary layer; it varies continuously from the wall to the duct centerline. Very close to the wall, Pr_t is well above unity. In the turbulent core region, it is well above unity for liquid metals ($\mathrm{Pr} < 0.03$), it is slightly lower than unity for gases and it is 0.8 to 0.9 for high Prandtl number liquids [14].

Extensive efforts have been made to obtain empirical correlations that either represent a best fit curve to the experimental data or have the constants in the theoretical equations adjusted to best fit the experimental data. An example of the latter is the correlation given by Petukhov and Popov [15]. Their theoretical calculations for the case of fully developed turbulent flow with constant properties in a circular tube with constant heat flux boundary condition yielded the following correlation, which is based on the three-layer turbulent boundary-layer model with constants adjusted to match the experimental data:

$$\mathrm{Nu}_b = \frac{(f/2)\mathrm{Re}_b\mathrm{Pr}_b}{(1 + 13.6f) + \left(11.7 + 1.8\,\mathrm{Pr}_b^{-1/3}\right)(f/2)^{1/2}\left(\mathrm{Pr}_b^{2/3} - 1\right)}, \tag{18.22}$$

where

$$f = (3.64\log_{10}\mathrm{Re}_b - 3.28)^{-2} \tag{18.23}$$

and is defined as $f = \tau_w / \frac{1}{2}\rho u_m^2$.

Equation (18.22) is applicable for fully developed turbulent flow in the range $10^4 < \mathrm{Re}_b < 5 \times 10^5$ and $0.5 < \mathrm{Pr}_b < 2000$ with 1% error, and in the range $5 \times 10^5 < \mathrm{Re}_b < 5 \times 10^6$ and $200 < \mathrm{Pr}_b < 2000$ with 1 to 2% error. Equation (18.22) is also applicable to rough tubes. A simpler correlation has also been given by Petukhov and Kirillov as reported in [16] as

$$\mathrm{Nu}_b = \frac{(f/2)\mathrm{Re}_b\mathrm{Pr}_b}{1.07 + 12.7(f/2)^{1/2}\left(\mathrm{Pr}_b^{2/3} - 1\right)} \tag{18.24}$$

Equation (18.24) predicts the results in the range $10^4 < \mathrm{Re}_b < 5 \times 10^6$ and $0.5 < \mathrm{Pr}_b < 200$ with 5 to 6% error, and in the range $0.5 < \mathrm{Pr}_b < 2000$ with 10% error.

Webb [17] has examined a range of data for fully developed turbulent flow in smooth tubes; he concluded that the relation developed by Petukhov and Popov, given above, provides the best agreement with the measurements. Sleicher and Rouse [18] correlated analytical and experimental results for the range $0.1 < \mathrm{Pr}_b < 10^4$, $10^4 < \mathrm{Re}_b < 10^6$, obtaining

$$\mathrm{Nu}_b = 5 + 0.015\,\mathrm{Re}_b^m\mathrm{Pr}_b^n \tag{18.25}$$

with

$$m = 0.88 - \frac{0.24}{4 + \mathrm{Pr}_b}$$

$$n = \frac{1}{3} + 0.5 \exp(-0.6\, \mathrm{Pr}_b)$$

Gnielinski [19] further modified the Petukhov-Kirillov correlation by comparing it with experimental data so that the correlation covers a lower Reynolds number range ($2300 \leq \mathrm{Re} \leq 5 \times 10^6$). Gnielinski recommended the following correlation:

$$\mathrm{Nu}_b = \frac{(f/2)(\mathrm{Re}_b - 1000)\mathrm{Pr}_b}{1 + 12.7(f/2)^{1/2}\left(\mathrm{Pr}_b^{2/3} - 1\right)} \tag{18.26}$$

where

$$f = (1.58 \ln \mathrm{Re}_b - 3.28)^{-2} \tag{18.27}$$

A large number of heat transfer correlations have been established for fully developed flow in circular and noncircular channels because of their widespread application in heat transfer equipment. A compilation of such correlations for circular and noncircular channels has been summarized by Kays and Perkins [20], by Shah and Johnson [31], and in Chapter 4. Experimental data and correlations for laminar, transition, and turbulent flow through a circular duct have also been compiled by Rogers [22].

Some of the recommended constant-property, fully developed, turbulent flow heat transfer and friction-factor correlations are summarized in Table 18.2 and Table 18.3 respectively, which can be used for the constant-property portion of the correlations associated with Eqs. (18.1) and (18.2) (see also Tables 4.2 and 4.3 in Chap. 4). The comparison of some of the constant-property correlations for $\mathrm{Re} = 8 \times 10^4$ in the range of $0.5 < \mathrm{Pr} < 15$ are given in Fig. 18.2 [25].

Turbulent Liquid Flow in Ducts. Petukhov [16] reviewed the status of heat transfer and wall friction in fully developed turbulent pipe flow with both constant and variable physical properties.

Analyzing the experimental data, one can assume that the functional relation of Nu with Re and Pr and the relation between f and Re with variable physical properties are the same as in the case of constant properties. This is a good approximation and has been confirmed by the experimental analysis for liquids with variable viscosity and for gases with variable physical properties [16].

To choose the correct value of n in Eq. (18.1a), the heat transfer experimental data corresponding to heating and cooling for several liquids over a wide range of values μ_w/μ_b were collected by Petukhov [16]. He found that the data are well correlated by

$$\frac{\mu_w}{\mu_b} < 1, \quad n = 0.11 \qquad \text{for heating liquids} \tag{18.28}$$

$$\frac{\mu_w}{\mu_b} > 1, \quad n = 0.25 \qquad \text{for cooling liquids} \tag{18.29}$$

which are applicable for fully developed turbulent flow in the range of $10^4 < \mathrm{Re}_b <$

TABLE 18.2. Correlations for Fully Developed Turbulent Forced Convection through a Circular Duct with Constant Properties, Pr > 0.5

No.	Reference	Correlation	Remarks and Limitations
1	Prandtl (see[12, 14])	$\mathrm{Nu}_b = \dfrac{(f/2)\mathrm{Re}_b\mathrm{Pr}_b}{1 + 8.7(f/2)^{1/2}(\mathrm{Pr}_b - 1)}$	Based on three-layer turbulent boundary-layer model.
2	McAdams [23]	$\mathrm{Nu}_b = 0.021\,\mathrm{Re}_b^{0.8}\mathrm{Pr}_b^{0.4}$	Based on data for common gases; recommended for Prandtl numbers ≈ 0.7.
3	Petukhov and Kirillov [16]	$\mathrm{Nu}_b = \dfrac{(f/2)\mathrm{Re}_b\mathrm{Pr}_b}{1.07 + 12.7(f/2)^{1/2}(\mathrm{Pr}_b^{2/3} - 1)}$ $f = (3.64\log_{10}\mathrm{Re}_b - 3.28)^{-2}$	Based on three-layer model with constants adjusted to match experimental data.
4	Webb [17]	$\mathrm{Nu}_b = \dfrac{(f/2)\mathrm{Re}_b\mathrm{Pr}_b}{1.07 + 9(f/2)^{1/2}(\mathrm{Pr}_b - 1)\mathrm{Pr}^{-1/4}}$ $f = (1.58\ln\mathrm{Re}_b - 3.28)^{-2}$	Theoretically based. Webb found No. 3 better at high Pr and this one the same at other Pr.
5	Sleicher and Rouse [18]	$\mathrm{Nu}_b = 5 + 0.015\,\mathrm{Re}_b^m\mathrm{Pr}_b^n$ $m = 0.88 - 0.24/(4 + \mathrm{Pr}_b)$ $n = \dfrac{1}{3} + 0.5\exp(-0.6\,\mathrm{Pr}_b)$	Based on numerical results obtained for $0.1 < \mathrm{Pr}_b < 10^4$, $10^4 < \mathrm{Re}_b < 10^6$. Within 10% of No. 6 for $\mathrm{Re}_b > 10^4$.
		$\mathrm{Nu}_b = 5 + 0.012\,\mathrm{Re}_b^{0.83}(\mathrm{Pr}_b + 0.29)$	Simplified correlation for gases, $0.6 < \mathrm{Pr}_b < 0.9$.
6	Gnielinski [19]	$\mathrm{Nu}_b = \dfrac{(f/2)(\mathrm{Re}_b - 1000)\mathrm{Pr}_b}{1 + 12.7(f/2)^{1/2}(\mathrm{Pr}_b^{2/3} - 1)}$ $f = (1.58\ln\mathrm{Re}_b - 3.28)^{-2}$	Modification of No. 3 to fit experimental data at low Re ($2300 < \mathrm{Re}_b < 10^4$). Valid for $2300 < \mathrm{Re}_b < 5 \times 10^6$ and $0.5 < \mathrm{Pr}_b < 2000$.
		$\mathrm{Nu}_b = 0.0214(\mathrm{Re}_b^{0.8} - 100)\mathrm{Pr}_b^{0.4}$	Simplified correlation for $0.5 < \mathrm{Pr} < 1.5$. Agrees with No. 4 within -6% and $+4\%$.
		$\mathrm{Nu}_b = 0.012(\mathrm{Re}_b^{0.87} - 280)\mathrm{Pr}_b^{0.4}$	Simplified correlation for $1.5 < \mathrm{Pr} < 500$. Agrees with No. 4 within -10% and $+0\%$ for $3 \times 10^3 < \mathrm{Re}_b < 10^6$.
7	Kays and Crawford [12]	$\mathrm{Nu}_b = 0.022\,\mathrm{Re}_b^{0.8}\mathrm{Pr}_b^{0.5}$	Modified Dittus-Boelter correlation for gases ($\mathrm{Pr} \approx 0.5$ to 1.0). Agrees with No. 6 within 0 to 4% for $\mathrm{Re}_b \geq 5000$.

5×10^6, $2 < \mathrm{Pr}_b < 140$, and $0.08 < \mu_w/\mu_b < 40$. The value of Nu_{cp} in Eq. (18.1a) is calculated from Eq. (18.22) or Eq. (18.24).

Petukhov [16] collected data from various investigators for variable viscosity influence on friction in water for both heating and cooling, and suggested the following correlations for the friction factor:

$$\frac{\mu_w}{\mu_b} < 1, \quad \frac{f}{f_{cp}} = \frac{1}{6}\left(7 - \frac{\mu_b}{\mu_w}\right) \quad \text{for heating liquids} \tag{18.30}$$

$$\frac{\mu_w}{\mu_b} > 1, \quad \frac{f}{f_{cp}} = \left(\frac{\mu_w}{\mu_b}\right)^{0.24} \quad \text{for cooling liquids} \tag{18.31}$$

TABLE 18.3. Turbulent Flow Isothermal Fanning-Friction-Factor Correlations for Smooth Circular Ducts

No.	Reference[a]	Correlation[b]	Remarks and Limitations
1	Blasius	$f = \tau_w / \frac{1}{2}\rho \mu_m^2 = 0.0791\,\text{Re}^{-1/4}$	This approximate explicit equation agrees with No. 3 within ±2.5%. $4 \times 10^3 < \text{Re} < 10^5$.
2	Drew, Koo, and McAdams	$f = 0.00140 + 0.125\,\text{Re}^{-0.32}$	This correlation agrees with No. 3 within −0.5% and +3%. $4 \times 10^3 < \text{Re} < 5 \times 10^6$.
3	von Kármán and Nikuradse	$1/\sqrt{f} = 1.737 \ln(\text{Re}\sqrt{f}) - 0.4$ or $1/\sqrt{f} = 4 \log_{10}(\text{Re}\sqrt{f}) - 0.4$, approximated as $f = (3.64 \log_{10}\text{Re} - 3.28)^{-2}$	von Kármán's theoretical equation with the constants adjusted to best fit Nikuradse's experimental data. Also referred to as the Prandtl correlation. Should be valid for very high values of Re. $4 \times 10^3 < \text{Re} < 3 \times 10^6$.
		$f \approx 0.046\,\text{Re}^{-0.2}$	This approximate explicit equation agrees with the above within −0.4% and +2.2%. $3 \times 10^4 < \text{Re} < 10^6$.
4	Filonenko	$f = 1/(1.58 \ln \text{Re} - 3.28)^2$	Agrees with No. 3 within ±0.5% for $3 \times 10^4 < \text{Re} \le 10^7$ and within ±1.8% at $\text{Re} = 10^4$. $10^4 < \text{Re} < 5 \times 10^5$.
5	Techo, Tickner, and James	$\frac{1}{f} = \left(1.7372 \ln \frac{\text{Re}}{1.964 \ln \text{Re} - 3.8215}\right)^2$	An explicit form of No. 3; agrees with it within ±0.1%. $10^4 < \text{Re} < 2.5 \times 10^8$.

[a] Cited in Refs. 13, 14, 16, 21, 24.

[b] Properties are evaluated at bulk temperatures.

The friction factor for an isothermal flow, f_{cp}, can be calculated by the use of Table 18.3 or directly from Eq. (18.23) for the range of $0.35 < \mu_w/\mu_b < 2$, $10^4 < \text{Re}_b < 23 \times 10^4$, and $1.3 < \text{Pr}_b < 10$.

Turbulent Gas Flow in Ducts. Heat transfer and friction coefficients for turbulent fully developed gas flow in a circular duct were obtained theoretically by Petukhov and Popov [15] by assuming physical properties ρ, c_p, k, and μ as given functions of temperature. This analysis is valid only for small subsonic velocities, since the variations of density with pressure and heat dissipation in the flow were neglected. The eddy diffusivity of momentum was extended to the case of variable properties. The turbulent Prandtl number was taken to be unity (i.e., $\epsilon_H = \epsilon_M$). The analyses were carried out for hydrogen and air for the following range of parameters: $0.37 < T_w/T_b < 3.1$ and $10^4 < \text{Re}_b < 5.8 \times 10^6$ for air, and $0.37 < T_w/T_b < 3.7$ and $10^4 < \text{Re}_b < 5.8 \times 10^6$ for hydrogen. The analytical results are correlated by Eq. (18.2), where Nu_{cp} is given by Eq. (18.22) or Eq. (18.24), and the following values for n are obtained:

$$\frac{T_w}{T_b} < 1, \quad n = -0.36 \qquad \text{for cooling gases} \tag{18.32}$$

$$\frac{T_w}{T_b} > 1, \quad n = -\left[0.3 \log_{10}\left(\frac{T_w}{T_b}\right) + 0.36\right] \qquad \text{for heating gases} \tag{18.33}$$

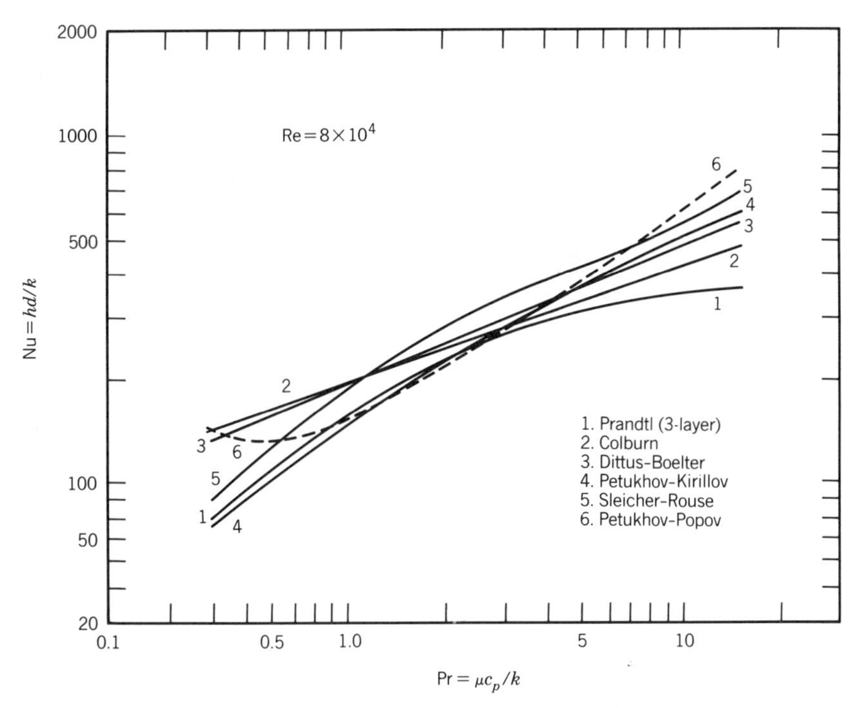

Figure 18.2. Comparison of some of the constant-property correlations for Re = 80×10^4 and $0.5 < \mathrm{Pr} < 15$ [25].

With these values for n, Eq. (18.2a) describes the solution for air and hydrogen with an accuracy of $\pm 4\%$. For simplicity, one can take n to be constant for heating as $n = -0.47$; then Eq. (18.2a) describes the solution for air and hydrogen within $\pm 6\%$. These results have also been confirmed experimentally and it can be used for practical calculations when $1 \leq T_w/T_b \leq 4$.

From the comparison of the analytical results with experimental data, for the prediction of friction factor for air and hydrogen for fully developed turbulent flow, the following values of m for Eq. (18.2b) are recommended [16]:

$$\frac{T_w}{T_b} > 1, \quad m = -0.6 + 5.6\,\mathrm{Re}_w^{-0.38} \qquad \text{for heating gases} \tag{18.34}$$

$$\frac{T_w}{T_b} < 1, \quad m = -0.6 + 0.79\,\mathrm{Re}_w^{-0.11} \qquad \text{for cooling gases} \tag{18.35}$$

where $\mathrm{Re}_w = u_m d\rho_w/\mu_w$. The above results are applicable in the range of $0.37 < T_w/T_b < 3.7$ and $14 \times 10^3 \leq \mathrm{Re}_w \leq 10^6$ with 2 to 3% error. If $m = -0.52$ for heating and $m = -0.38$ for cooling, Eq. (18.2b) describes the data to within 7% accuracy in the first case and 4% accuracy in the second. It has been confirmed theoretically and experimentally that n decreases with increasing T_w/T_b.

The friction coefficient in the circular duct flow of a gas with a large temperature difference between the wall and flow has also been studied [26–28]. The data obtained in these studies are not in agreement with each other.

TABLE 18.4. Exponents *n* and *m* Associated with Eqs. (18.1) and (18.2) for Turbulent Forced Convection through Circular Ducts

No.	Reference	Fluid	Condition	n	m	Limitations
1	Petukhov [16]	Liquid	Turbulent heating	0.11	—	$10^4 < \mathrm{Re}_b < 1.25 \times 10^5$, $2 < \mathrm{Pr}_b < 140$
		Liquid	Turbulent cooling	0.25	—	$0.08 < \mu_w/\mu_b < 1$
						$< \mu_w/\mu_b < 40$
		Liquid	Turbulent heating	—	Eq. (18.30)	$10^4 < \mathrm{Re}_b < 23 \times 10^4$, $1.3 < \mathrm{Pr}_b < 10^4$,
		Liquid	Turbulent cooling	—	-0.24	$0.35 < \mu_w/\mu_b < 1$
					or ≈ -0.25	$< \mu_w/\mu_b < 2$
2	Petukhov and Popov [15]	Gas	Turbulent heating	-0.47	—	$10^4 < \mathrm{Re}_b < 4.3 \times 10^6$, $1 < T_w/T_b < 3.7$
		Gas	Turbulent cooling	-0.36	—	$1 < T_w/T_b < 1$
		Gas	Turbulent heating	—	-0.52	$14 \times 10^3 < \mathrm{Re}_w^* \le 10^6$, $1 < T_w/T_b < 3.7$
		Gas	Turbulent cooling	—	-0.38	$0.37 < T_w/T_b < 1$
3	Worsøe-Schmidt [11]	Gas	Turbulent heating	—	-0.264	$1 \le T_w/T_b \le 4$
4	McElligot et al. [26]	Gas	Turbulent heating	—	-0.1	$1 < T_w/T_b < 2.4$

For fully developed velocity and temperature profiles ($L/d > 55$), Perkins and Worsøe-Schmidt [27] found that when $1 \leq T_w/T_b \leq 4$, $m = -0.264$.

McEligot et al. [26] recommend a value of -0.1 instead of -0.264 for $1.0 < T_w/T_b < 2.4$.

The exponents n and m of the correlations in Eqs. (18.1) and (18.2) for turbulent flow obtained by various investigators under various conditions are summarized in Table 18.4. In view of their experimental verification, items 1 and 2 are recommended.

A great number of experimental studies are available in the literature for the heat transfer between the tube wall and the gas flow with large temperature differences and temperature-dependent physical properties. The majority of the work deals with gas heating at constant wall temperature in a circular duct; experimental studies on gas cooling are limited.

The results of heat transfer measurements at large temperature differences between the wall and the gas flow are usually presented as

$$\mathrm{Nu}_b = C\,\mathrm{Re}_b^{0.8}\mathrm{Pr}_b^{0.4}\left(\frac{T_w}{T_b}\right)^n \tag{18.36}$$

For fully developed temperature and velocity profiles (i.e., $L/d > 60$), C becomes constant and n becomes independent of L/d.

A number of heat transfer correlations have been developed for variable-property fully developed turbulent liquid and gas flow in a circular duct, some of which are also summarized in Tables 18.5 and 18.6.

18.2.3 Turbulent Flow in Noncircular Ducts

Heat transfer and friction coefficients for turbulent flow in noncircular ducts are given in Chapter 4. For turbulent flow a common practice is to employ the hydraulic diameter in the circular duct correlations to predict Nu and f for noncircular ducts. For most noncircular smooth ducts, the accurate constant-property experimental friction factors (pressure drop) are within $\pm 10\%$ of those predicted using the smooth circular duct correlation with hydraulic diameter as a characteristic dimension. The constant-property experimental Nusselt numbers are also within ± 10 to 15% except for some sharp-cornered and narrow channels. This order of accuracy is adequate for most engineering calculations for overall heat transfer and pressure drop calculations.

Many attempts have been reported in the literature to arrive at a universal characteristic dimension in turbulent flows that will correlate the constant-property friction factors and Nusselt numbers for all noncircular ducts [46–48]. It must be emphasized that any improvement made by these attempts is only a few percent, and therefore the circular duct correlations may be adequate for many engineering applications.

In addition to the generalized attempts, correlations for specific noncircular ducts have been developed in the literature, and they are summarized below for rectangular, triangular, elliptical, and concentric annular ducts [21].

Rectangular Channels. Jones [49] obtained a correlation for turbulent friction factors in smooth rectangular ducts by modifying the definition of the Reynolds number

TABLE 18.5. Turbulent Forced Convection Correlations in Circular Ducts for Liquids with Variable Properties, Pr > 0.5

No.	Reference	Correlation	Comments and Limitations
1	Colburn [29]	$\mathrm{St}_b \mathrm{Pr}_f^{2/3} = 0.023\,\mathrm{Re}_f^{-0.2}$	$L/d > 60$, $\mathrm{Pr}_b > 0.6$, $T_f = (T_b + T_w)/2$; inadequate for large $T_w - T_b$.
2	Sieder and Tate [5]	$\mathrm{Nu}_b = 0.023\,\mathrm{Re}_b^{0.8}\mathrm{Pr}_b^{1/3}\left(\dfrac{\mu_b}{\mu_w}\right)^{0.14}$	$L/d > 60$, $\mathrm{Pr}_b > 0.6$, for moderate $T_w - T_b$.
3	Norris and Sims [30]	$\mathrm{St}_b = 0.006\,\mathrm{Pr}_b^{-0.8}\left(\dfrac{\mu_b}{\mu_w}\right)^{0.14}$	*Transitional*, $3500 \le \mathrm{Re}_b \le 11{,}000$, experiments on cooling of oils. *Undeveloped* velocity profile. $35 < \mathrm{Pr}_b < 140$.
4	Yakovlev (see [22])	$\mathrm{Nu}_b = 0.0277\,\mathrm{Re}_b^{0.8}\mathrm{Pr}_b^{0.36}\left(\dfrac{\mathrm{Pr}_b}{\mathrm{Pr}_w}\right)^{0.11}$	Use of Prandtl group was first suggested by the author in 1960.
5	Petukhov and Kirillov [16]	$\mathrm{Nu}_b = \dfrac{(f/8)\mathrm{Re}_b \mathrm{Pr}_b}{1.07 + 12.7\sqrt{f/8}\,(\mathrm{Pr}_b^{2/3} - 1)}\left(\dfrac{\mu_b}{\mu_w}\right)^n$	$L/d > 60$, $0.08 < \mu_w/\mu_b < 40$, $10^4 < \mathrm{Re}_b < 5 \times 10^6$, $2 < \mathrm{Pr}_b < 140$, $f = (1.82 \log \mathrm{Re}_b - 1.64)^{-2}$, $n = 0.11$ (heating), $n = 0.25$ (cooling).
6	Malina and Sparrow [31]	$\mathrm{Nu}_b = 0.023\,\mathrm{Re}_b^{0.8}\mathrm{Pr}_b^{1/3}\left(\dfrac{\mu_b}{\mu_w}\right)^{0.05}$	Water, oil, $3 < \mathrm{Pr}_b < 75$, $L/d = 30$.
7	Hufschmidt, Burck and Riebold [32]	$\mathrm{Nu}_b = \dfrac{(f/8)\,Re_b\,Pr_b}{1.07 + 12.7\sqrt{f/8}\,(Pr_b^{2/3} - 1)}\left(\dfrac{\mathrm{Pr}_b}{Pr_w}\right)^{0.11}$	Water, $2 \times 10^4 < \mathrm{Re}_b < 6.4 \times 10^5$, $0.1 < \mathrm{Pr}_b/\mathrm{Pr}_w < 10$, $f = (1.82 \log \mathrm{Re}_b - 1.64)^{-2}$.
8	Everett (see [22])	$\mathrm{Nu}_b = 0.0225\,\mathrm{Re}_b^{0.795}\mathrm{Pr}_b^{x}\left(\dfrac{\mu_b}{\mu_w}\right)^n$ $x = 0.495 - 0.225 \ln \mathrm{Pr}$ $n = \begin{cases} \left(\dfrac{\mathrm{Re}}{870{,}000}\right)^{0.84} & \text{for Re} < 62{,}500 \\ 0.11 & \text{for Re} > 62{,}500 \end{cases}$	$\mathrm{Re}_b > 4 \times 10^3$.

9 Hackl and Gröll [33]	$\frac{\mathrm{Nu}_b}{\mathrm{Nu}_{cp}} = 0.645\frac{\mu_w}{\mu_b} + 0.355$	Experimental results with oils, $4 \times 10^3 < \mathrm{Re}_b < 11 \times 10^3$.
10 Kuznetsova [8]	$\mathrm{Nu}_b = 0.013\,\mathrm{Re}_b^{0.8}\mathrm{Pr}_b^{0.4}\left(\frac{\mu_b}{\mu_w}\right)^{-0.11}$	Transformer oil, fuel oil, $2000 < \mathrm{Re}_b < 8000$, $70 < \mathrm{Pr}_b < 200$.
11 Oskay and Kakaç [7, 34]	$\mathrm{Nu} = 0.023\,\mathrm{Re}_b^{0.8}\mathrm{Pr}_b^{0.4}\left(\frac{\mu_b}{\mu_w}\right)^{0.262}$	Water, $L/d > 10$, $1.2 \times 10^4 < \mathrm{Re}_b < 4 \times 10^4$.
	$\mathrm{Nu} = 0.023\,\mathrm{Re}_b^{0.8}\mathrm{Pr}_b^{0.4}\left(\frac{\mu_b}{\mu_w}\right)^{0.487}$	30% glycerine-water mixture $L/d > 10$, $0.89 \times 10^4 < \mathrm{Re}_b < 2.0 \times 10^4$.
12 Hausen [35]	$\mathrm{Nu}_b = 0.0235(\mathrm{Re}_b^{0.8} - 230)(1.8\,\mathrm{Pr}_b^{0.3} - 0.8) \times \left[1 + \left(\frac{d}{L}\right)^{2/3}\right]\left(\frac{\mu}{\mu_w}\right)^{0.14}$	Altered form of equation presented in 1959 [22].
13 Sleicher and Rouse [18]	$\mathrm{Nu}_b = 5 + 0.015\,\mathrm{Re}_f^m\mathrm{Pr}_w^n$	$L/d > 60$, $0.1 < \mathrm{Pr}_b < 10^5$,
	$m = 0.88 - 0.24/(4 + \mathrm{Pr}_w)$ $n = \frac{1}{3} + 0.5\,e^{-0.6\,\mathrm{Pr}_w}$	$10^4 < \mathrm{Re}_b < 10^6$
	$\mathrm{Nu}_b = 0.015\,\mathrm{Re}_f^{0.88}\mathrm{Pr}_w^{1/3}$	$\mathrm{Pr}_b > 50$
	$\mathrm{Nu}_b = 4.8 + 0.015\,\mathrm{Re}_f^{0.85}\mathrm{Pr}_w^{0.93}$	$\mathrm{Pr}_b < 0.1$, uniform wall temperature
	$\mathrm{Nu}_b = 6.3 + 0.0167\,\mathrm{Re}_f^{0.85}\mathrm{Pr}_w^{0.93}$	$\mathrm{Pr}_b < 0.1$, uniform wall heat flux
14 Gregorig [36]	$\frac{\mathrm{Nu}}{\mathrm{Nu}_{cp}} = \left(\frac{\mathrm{Pr}_b}{\mathrm{Pr}_w}\right)^P$ $P = \frac{0.264\left(\frac{1}{2} - \chi\right)^{0.20} + 0.0757}{\mathrm{Pr}_b^{0.05}\mathrm{Re}_b^{0.02}\left(\frac{1}{2} - \chi\right)^{0.01}}$ with $\chi = \frac{\mathrm{Pr}_f - \mathrm{Pr}_w}{\mathrm{Pr}_b - \mathrm{Pr}_w} - \frac{1}{2}$	$L/d > 70$, cooling of liquids only, $T_f = (T_b + T_w)/2$. Correlation of others' results.

TABLE 18.6. Turbulent Forced Convection Correlations in Circular Ducts for Gases with Variable Physical Properties

No.	Reference	Correlation	Gas	Limitations
1	Humble, Lowdermilk, and Desmon [38]	$\mathrm{Nu}_b = 0.023\,\mathrm{Re}_b^{0.8}\mathrm{Pr}_b^{0.4}(T_w/T_b)^n$ $T_w/T_b < 1$, $n = 0$ (cooling) $T_w/T_b > 1$, $n = -0.55$ (heating)	Air	$30 < L/d < 120$, $7 \times 10^3 < \mathrm{Re}_b < 3 \times 10^5$, $0.46 < T_w/T_b < 3.5$, $L/d > 60$
2	Bialokoz and Saunders (see [16])	$\mathrm{Nu}_b = 0.022\,\mathrm{Re}_b^{0.8}\mathrm{Pr}_b^{0.4}(T_w/T_b)^{-0.5}$	Air	$29 < L/d < 72$, $1.24 \times 10^5 < \mathrm{Re}_b < 4.35 \times 10^5$, $1.1 < T_w/T_b < 1.73$
3	Taylor and Kirchgessner [39]	$\mathrm{Nu}_f = 0.021\,\mathrm{Re}_f^{0.8}\mathrm{Pr}_f^{0.4}$	He	$L/d = 60, 92$, $3.2 \times 10^3 < \mathrm{Re}_b < 60 \times 10^3$, $1.6 < T_w/T_b < 3.9$, $L/d > 60$
4	McCarthy and Wolf [40]	$\mathrm{Nu}_f = 0.045\,\mathrm{Re}_f^{0.8}\mathrm{Pr}_f^{0.4}(L/d)^{-0.15}(T_w/T_b)^{0.7}$	H_2, He	$21 < L/d < 67$, $5 \times 10^3 < \mathrm{Re}_b < 15 \times 10^5$, $1.5 < T_w/T_b < 9.9$
5	Barnes and Jackson [41]	$\mathrm{Nu}_b = 0.023\,\mathrm{Re}_b^{0.8}\mathrm{Pr}_b^{0.4}(T_w/T_b)^n$ $n = \begin{cases} -0.4 & \text{for air} \\ -0.185 & \text{for He} \\ -0.27 & \text{for } CO_2 \end{cases}$	Air, He, CO_2	$1.2 < T_w/T_b < 2.2$, $4 \times 10^3 < \mathrm{Re}_b < 6 \times 10^4$, $L/d > 60$
6	Wieland [42]	$\mathrm{Nu}_f = 0.021\,\mathrm{Re}_f^{0.8}\mathrm{Pr}_f^{0.4}$	He	$L/d = 250$, $T_w/T_b < 2.8$, far from entrance.

7 Taylor [43]	$Nu_f = 0.021\, Re_f^{0.8} Pr_f^{0.4}$	H_2, He	$L/d = 77$, $1.5 < T_w/T_b < 5.6$
8 McElligot et al. [26]	$Nu_b = 0.021\, Re_b^{0.8} Pr_b^{0.4} (T_w/T_b)^{-0.5}$	Air, He, N_2	$L/d > 30$, $1 < T_w/T_b < 2.5$, $1.5 \times 10^4 < Re_{i,b} < 2.33 \times 10^5$
	$Nu_b = 0.021\, Re_b^{0.8} Pr_b^{0.4} (T_w/T_b)^{-0.5} \times [1 + (L/d)^{-0.7}]$		$L/d > 5$, local values
9 Perkins and Worsøe-Schmidt [27]	$Nu_b = 0.024\, Re_b^{0.8} Pr_b^{0.4} (T_w/T_b)^{-0.7}$	N_2	$L/d > 40$, $1.24 < T_w/T_b < 7.54$, $18.3 \times 10^3 < Re_{i,b} < 2.8 \times 10^5$
	$Nu_w = 0.023\, Re_w^{0.8} Pr_w^{0.4}$		Properties evaluated at wall temperature, $L//d > 24$
	$Nu_b = 0.024\, Re_b^{0.8} Pr_b^{0.4} (T_w/T_b)^{-0.7} \times [1 + (L/d)^{-0.7} (T_w/T_b)^{0.7}]$		$1.2 \le L/d \le 144$
10 Petukov, Kirillov, and Maidanic (see [16])	$Nu_b = 0.021\, Re_b^{0.8} Pr_b^{0.4} (T_w/T_b)^n$, $n = -[0.9 \log(T_w/T_b) + 0.205]$	N_2	$80 < L/d < 100$, $13 \times 10^3 < Re_b < 3 \times 10^5$, $1 < T_w/T_b < 6$
11 Deissler and Prester [44]	$Nu_r = Re_r^{3/4}/31$	Air, He, Ar, H_2	$Re_b > 10^4$, properties evaluated at $T_r = T_b + 0.4(T_w - T_b)$, $L/d > 60$
12 Kutateladze and Leontiev (see [16])	$\dfrac{Nu_b}{Nu_{cp}} = \dfrac{f_b}{f_{cp}} = \left(\dfrac{2}{\sqrt{T_w/T_b} + 1} \right)^2$	gases	Fully developed flow
13 Sleicher and Rouse [18]	$Nu_b = 5 + 0.012\, Re_f^{0.83} (Pr_w + 0.29)$		$0.6 < Pr_b < 0.9$, fully developed flow
14 Gnielinski (see [45])	$Nu_b = 0.0214 (Re_b^{0.8} - 100) Pr_b^{0.4} (T_b/T_w)^{0.45} \times [1 + (d/L)^{2/3}]$	Air, He, CO_2	$0.5 < Pr_b < 1.5$, for heating of gases; from literature.
15 Dalle-Donne and Bowditch (see [20])	$Nu_b = 0.022\, Re_b^{0.8} Pr_b^{0.4} (T_w/T_b)^{-(0.29 + 0.0019x/d)}$	Air, He,	$10^4 < Re_b < 10^5$, $18 < L/d < 316$

so that it employs a *laminar equivalent diameter* D_e:

$$\mathrm{Re}^* = \frac{GD_e}{\mu}$$

where

$$D_e = \frac{16}{K_f} \cdot D_h = \Phi D_h \tag{18.37}$$

and K_f ($= f\,\mathrm{Re}$) is a constant for fully developed laminar flow for the rectangular duct in question. Since K_f is dependent upon the aspect ratio α^*, it is approximately given by

$$\Phi = \frac{16}{K_f} \approx \tfrac{2}{3} + \tfrac{11}{24}\alpha^*(2 - \alpha^*) \tag{18.38}$$

Once the Reynolds number is modified to Re*, Jones showed that the turbulent f factors for rectangular ducts can be predicted by the Kármán-Nikuradse equation with Re changed by Re*. Once the friction factor is determined by this procedure, the pressure drop ΔP is computed by the use of the hydraulic diameter D_h. Based on the Jones correlation, it can be shown that the f factors are lower by 3% for a square duct and are higher by 11% for infinite parallel plates than for a circular tube at the same Re.

Nusselt numbers for rectangular ducts can be computed by the use of Eq. (18.24) or Eq. (18.26) with f factors computed by using Re*, while Re is based on hydraulic diameter D_h. The Nusselt numbers for a square duct computed by this procedure are 2 to 4% lower than those for a circular tube at Pr = 0.7, and 1 to 2% lower at Pr = 10. For parallel plates, they are higher by 7 to 11% and by 5 to 7% than those for circular tubes at Pr = 0.7 and 10 respectively.

Triangular Ducts. The turbulent flow friction factors for isosceles triangular channels due to Carlson and Irvine [50] are correlated as

$$f = \left[0.06057 + 7.22 \times 10^{-4}(2\phi) - 9.413 \times 10^{-6}(2\phi)^2\right]\mathrm{Re}^{-0.25} \tag{18.39}$$

where 2ϕ is the apex angle of the triangle in degrees. The correlation is based on experimental data for $8° \le 2\phi \le 60°$. Note that the bracketed factor is 0.0791 for the circular tube (Table 18.3). Nusselt numbers can be calculated from Eq. (18.24) or Eq. (18.26) by the use of f computed from Eq. (18.39).

Elliptical Ducts. Fully developed turbulent flows have been investigated in [51, 52]. The friction factors for elliptical ducts with aspect ratios of 0.667 and 0.500 were consistently 8% and 13% higher than those by Kármán-Nikuradse correlation for the complete range of $2 \times 10^4 < \mathrm{Re} < 1.3 \times 10^5$ [32]. Jones's procedure of employing laminar equivalent diameter would correct these friction factors only by 0.5% and 1.3% respectively. Cain and Duffy [51] attributed this departure partially to the existence of the secondary flows.

Cain [52] obtained the heat transfer results for the aforementioned two elliptical ducts with air flow. His Nusselt numbers agree with the Dittus-Boelter correlation for

$10^4 < \text{Re} \leq 2 \times 10^5$. He also tested two more elliptical ducts (aspect ratio 0.375 and 0.341) with water flow over $2.5 \times 10^3 \leq \text{Re} \leq 8 \times 10^4$. For Re > 25,000, the Nusselt numbers agreed with those from the Dittus-Boelter correlation; however, for Re < 25,000 they were lower than the latter and were correlated by

$$\text{Nu}_b = 0.00165\,\text{Re}_b^{1.06}\,\text{Pr}_b^{0.4} \tag{18.40}$$

Cain [52] attributed this reduction in Nu to the presence of laminar or transition flow in the channel corners. Cain's Nusselt number results are in agreement with the test results of Barrow and Roberts [53] for the elliptical duct of aspect ratio 0.284 with water as a test fluid. The available data are insufficient to provide generalized correlations for elliptical ducts.

Concentric Annuli. Based on observed friction factors higher than those for the circular tube, Kays and Perkins [20] recommend using f factors for concentric annuli that are 1.10 times those for a circular tube. Recently, Jones and Leung [54] carefully scrutinized all experimental data and recommended the Kármán-Nikuradse correlation for concentric annuli with the Reynolds number $\text{Re}^* = GD_e/\mu$ based on the laminar equivalent diameter D_e defined as

$$D_e = \frac{16}{K_f} D_h = \Phi D_h, \qquad \text{where} \quad \Phi = \frac{1}{(1 - r^*)^2}\left(1 + r^{*2} - \frac{1 - r^{*2}}{\ln(1/r^*)}\right) \tag{18.41}$$

Here r^* is the ratio of inner radius to outer radius of the concentric annuli. The f factors computed by this procedure are higher than the circular tube correlation by 0 to 11% for r^* varying from 0 to 1 at the same Re based on D_h. Extensive Nusselt number tabulations for turbulent flow are available for concentric annuli depending on the inner and outer tube boundary conditions [14].

The Nusselt number and the friction coefficient with property variation in turbulent flow can be calculated for engineering applications by use of Eqs. (18.1) and (18.2) with Table 18.2, and Table 18.4 where Re is based on D_h.

Further correlations for laminar and turbulent forced convection in ducts and for flow normal to tube banks for constant and variable properties can be found in Ẑukauskas's work [55, 56] and in Chapter 6.

18.3 TURBULENT FORCED CONVECTION IN DUCTS AT SUPERCRITICAL PRESSURE

The special characteristic of fluids near the critical point is that the variation of their physical properties with temperature becomes extremely rapid. At the critical points, the isotherm has zero slope and saturated-liquid and saturated-vapor states are identical. In this section, the results on the heat transfer and friction coefficient in single-phase forced convection at supercritical pressure and at subcritical or pseudocritical temperatures (the critical temperature being that at which the specific heat peaks) are briefly summarized. Since the pressure variation across most thermal boundary layers is negligible, it is the variation of properties along a constant-pressure line that is important. The variation of ρ, c_p, μ, k, and i for CO_2 at 1.00×10^7 Pa (100 bar) is shown in Fig. 18.3. The temperature at which the specific heat reaches a peak is known as the pseudocritical temperature, T_{pc}. As the pressure is increased, this temperature

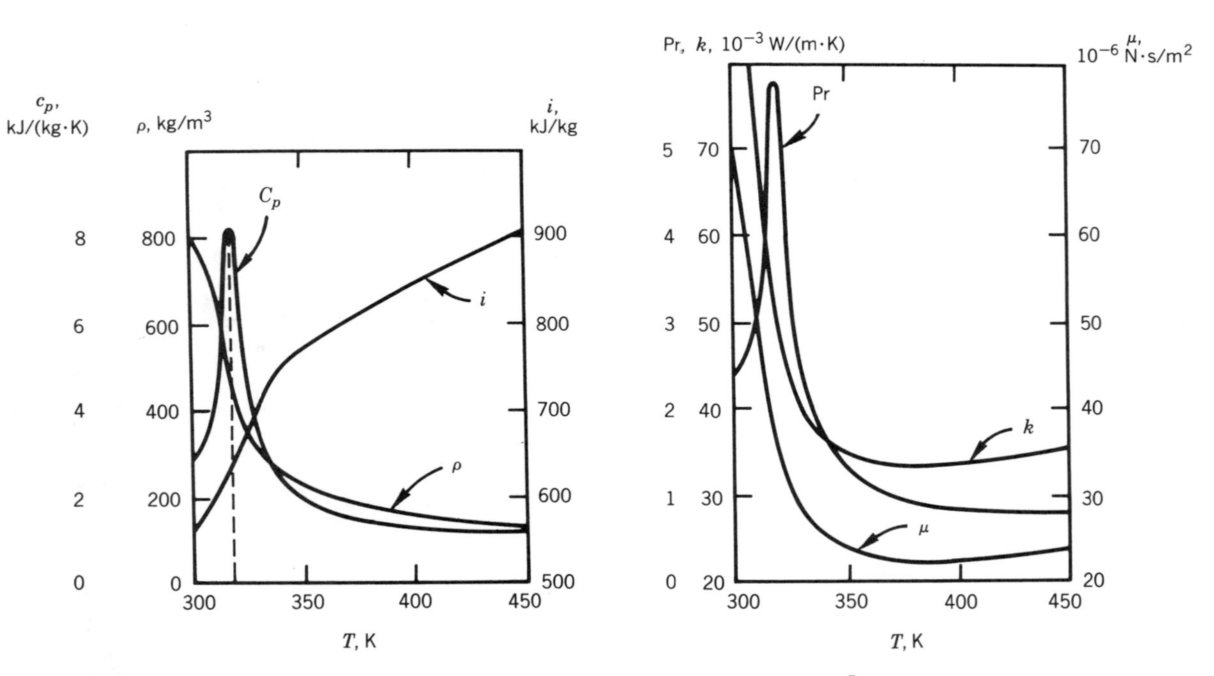

Figure 18.3. Physical properties of carbon dioxide at $P = 1.00 \times 10^7$ Pa = 100 bar [16].

increases, the maximum value of the specific heat falls, and the variation of the other properties becomes less severe. This behavior is followed by many other fluids. There is evidence for some fluids of a sharp peak in conductivity near the critical point. Unfortunately, at practical heat fluxes, the peaking of conductivity would occur in a very thin layer of fluid. In the case of heating, the density and viscosity increase rapidly from the wall to the duct axis.

18.3.1 Special Features of Heat Transfer at Supercritical Pressure

Forced convection measurements at supercritical pressures have been made using a wide range of fluids such as water, carbon dioxide, nitrogen, hydrogen, and helium. By far the most extensive collections of data are those for water and carbon dioxide, and fortunately the accuracy of physical properties for these fluids is good [57].

Review papers on supercritical-pressure heat transfer have been given by Petukhov [16, 58], Khan [59], Hall et al. [60], Hall [61], Hendricks et al. [62], and Jackson et al. [63].

As a result of the strong dependence of physical properties on temperature, processes involving heat transfer in supercritical pressure fluids are generally more complex than for ordinary fluids. The energy equation becomes nonlinear, with the result that the proportionality between the flux and temperature difference no longer exists. Density variations cause changes in the velocity, either by virtue of thermal expansion or because of buoyancy; these effects combined with locally high specific heat can be important influences on convection. The diffusion of heat through boundary layers can be strongly influenced by variations of thermal conductivity and specific heat.

At low heat fluxes, and therefore with small temperature variations across the flow, irrespective of the change of physical properties with temperature, the calculation of such a heat transfer process can be carried out as in the case of uniform properties. It should be noted that temperature changes do occur in the direction of flow even at low heat flux, so that the properties change in this respect. Therefore, for the limiting case ($T_w - T_b$ is small) of very low heat flux, the established relationship between Nu, Re, and Pr for constant properties can be used on a local basis. This relationship is often expressed in the form

$$\mathrm{Nu} = C\,\mathrm{Re}^m \mathrm{Pr}^n \qquad (18.42)$$

The behavior of the heat transfer coefficient in the vicinity of the pseudocritical temperature significantly depends on the specific heat variation, and one might expect the heat transfer coefficient h to vary in a similar manner. The data of Swenson et al. [64] illustrate the enhancement of heat transfer in the pseudocritical region for water at 2.33×10^7 and 3.10×10^7 Pa (Fig. 18.4). It will be noted that the influence of pressure in increasing T_{pc} and in reducing the peak value of c_p is clearly reflected in the variation of the heat transfer coefficient.

The explanation of the enhancement effect in the pseudocritical region becomes apparent when one considers the factors which influence the thermal resistance of the wall layer, the region through which the diffusion of heat takes place mainly by molecular action. This resistance depends both on the thermal conductivity of fluid and on the extent of the region. As temperature increases through the pseudocritical range, the thermal conductivity falls, but at the same time the extent of the high-resistance layer is reduced (turbulence is less damped as a result of the reduction in viscosity, and is more effective in diffusing heat as a result of high values of specific heat). The

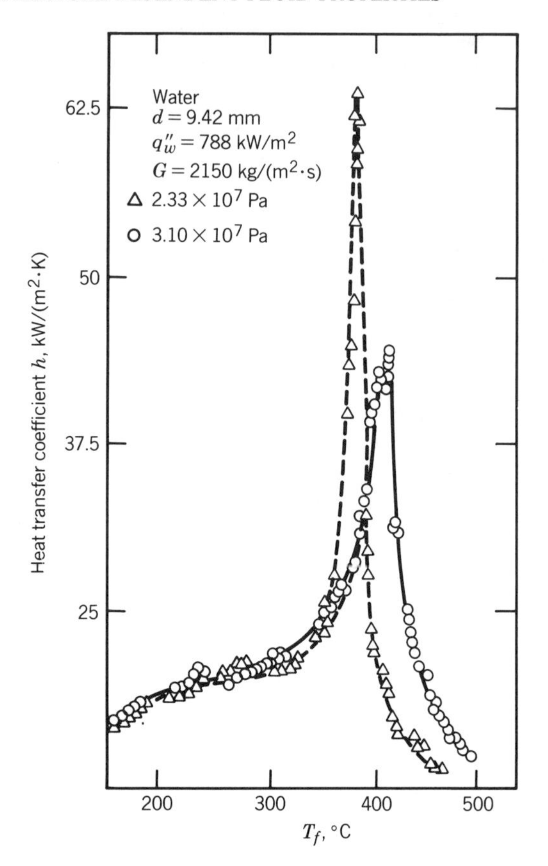

Figure 18.4. Enhancement of heat transfer in the pseudocritical region [64].

thinning of the thermal layer due to the latter mechanisms lowers the thermal resistance by an amount which far outweighs the opposing effect of the reduced thermal conductivity. The result is that the heat transfer coefficient is strongly enhanced. Beyond the pseudocritical temperature the specific heat falls to a low value, with the result that the effectiveness of turbulent diffusion of heat is reduced and the heat transfer coefficient is much lower. On the other hand, the minimum value of h in the case of cooling ($q''_w < 0$) is higher than that in the case of heating ($q''_w > 0$) and decreases when q''_w and T_w increase. For cooling, h decreases with increasing q''_w at a smaller rate than for heating.

An increase in heat flux has the effect of diminishing the enhancement [65, 66]. Whereas at low heat flux the high value of specific heat associated with temperatures near T_{pc} is felt across much of the boundary layer, with increase of heat flux the peak values become concentrated in one part of it. Thus the integrated effect of the peaking of c_p is reduced. Coupled with this, the influence of the low conductivity of the fluid adjacent to the wall is strengthened by the increased range of temperature covered.

With further increase in heat flux, the trends described above continue to the stage where the enhancement effect is inhibited completely. This is illustrated by the data of Kondratev [67] for water at 2.53×10^7 Pa (Fig. 18.5). At very high heat flux, severe impairment of heat transfer develops, followed further downstream by a recovery

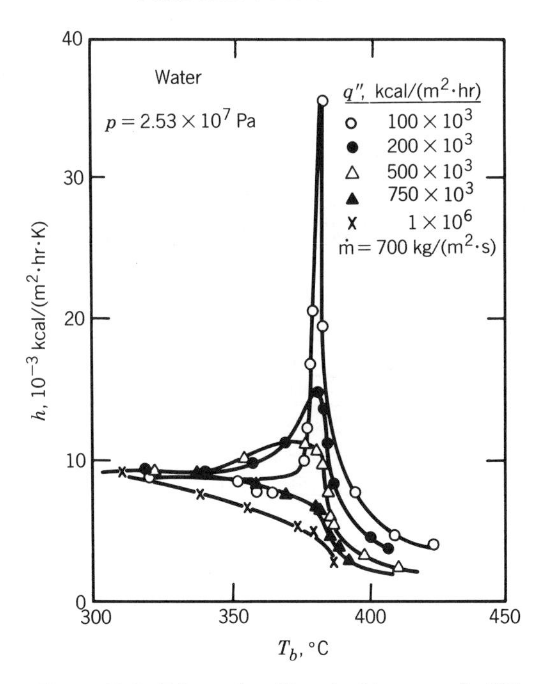

Figure 18.5. Enhanced and impaired heat transfer [67].

where the bulk temperature passes through T_{pc}. Such effects are clearly shown in the data of Domin [68] for water at 2.27×10^7 Pa, which are shown in Fig. 18.6, and of others [69, 70, 71]. The "broad" peaks on the wall temperature distributions of Fig. 18.6 are a characteristic of forced convection heat transfer to supercritical pressure fluids at high heat flux. They should not be confused with peaks due to influences of buoyancy under conditions of mixed convection with upward flow in heated tubes. The latter type are usually much more localized and are not restricted to the range of bulk temperature in the vicinity of T_{pc}.

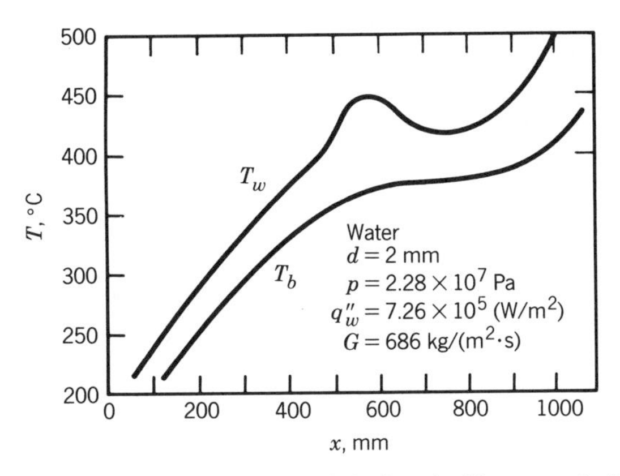

Figure 18.6. Wall temperature peak for impaired heat transfer [68].

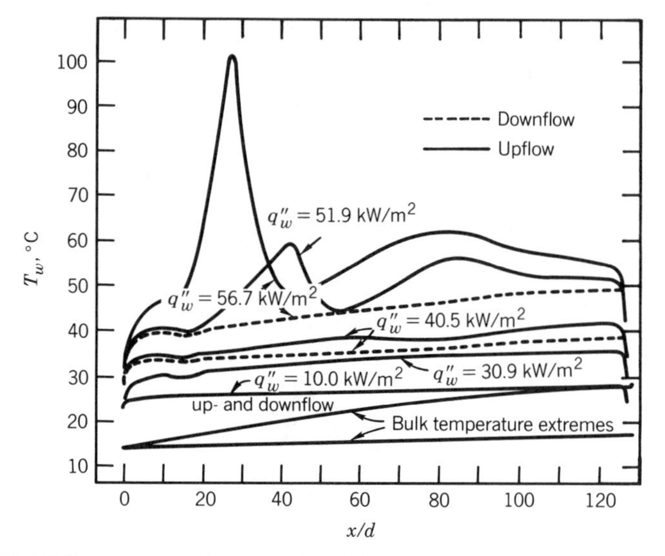

Figure 18.7. Wall temperature for upward and downward flow in a vertical tube—effects of buoyancy in carbon dioxide, $p = 7.59 \times 10^6$ Pa, $d = 19$ mm, $T_{b,\text{in}} = 14$°C, $\dot{m} = 16$ kg/s [75].

The changeover from enhanced to impaired forced convection is not just controlled by heat flux; mass velocity also turns out to be an important variable [72].

The conditions under which the impairment of heat transfer might become significant have been studied, and a criterion for the onset of impairment of forced convection at high heat flux has been suggested [72–74].

The examples discussed so far have all involved conditions where buoyancy effects have been negligible. The presence of significant influences of buoyancy can be detected in the case of vertical and inclined tubes because results for upward and downward flow are then found to differ. For horizontal and inclined tubes, buoyancy causes circumferential variations of heat transfer [65].

Under conditions of suitably low mass velocity, buoyancy effects develop with increase in heat flux, often when the flux reaches a level which causes the surface temperature to exceed T_{pc}; density variations sufficient to give buoyancy-induced flow then suddenly occur. This is evident in rather a dramatic form in the examples given in Figs. 18.7 and 18.8 for carbon dioxide [75, 76].

The localized impairment of heat transfer for upward flow in vertical tubes has been explained in terms of partial laminarization of the flow by Hall and Jackson [73], and the impaired heat transfer on the upper part of horizontal tubes has been attributed to stratification of the flow. It turns out (although it is not obvious from the examples cited) that the heat transfer is enhanced by buoyancy for downward flow in vertical tubes and also over the lower part of horizontal tubes. By comparing Figs. 18.7 and 18.8, it can be seen that impairment of heat transfer due to buoyancy occurs more readily in the case of the horizontal tube. A considerable amount of information has been produced in recent years on the influences of buoyancy on heat transfer to supercritical pressure fluids. Much of it has been collected and reviewed by Jackson et al. [63].

It is clearly important to be able to ascertain for a given set of operating conditions whether effects of buoyancy are likely to be significant or not. This matter has been

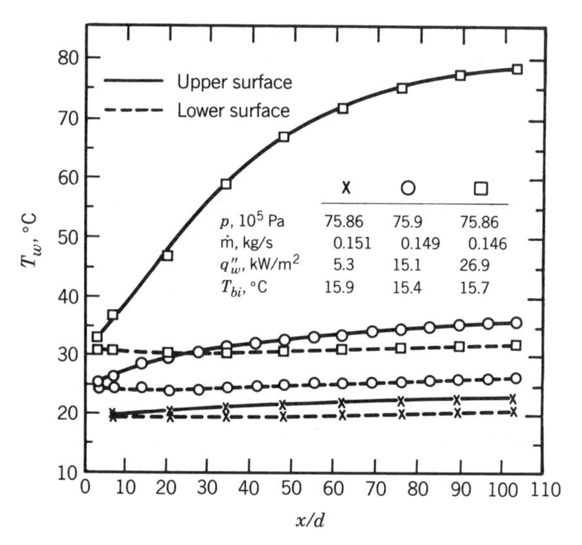

Figure 18.8. Temperature along the upper and lower surfaces of a horizontal tube, $d = 22.2$ mm, filled with carbon dioxide [76].

given detailed attention for both the vertical and horizontal cases in Ref. 63. Theoretical considerations lead to the following criteria for negligible buoyancy effects:

$$\frac{\overline{\mathrm{Gr}}_b}{\mathrm{Re}_b^{2.7}} < 10^{-5} \qquad \text{for vertical tubes} \tag{18.43a}$$

$$\frac{\mathrm{Gr}_b}{\mathrm{Re}_b^{2}} < 10^{-3} \qquad \text{for horizontal tubes} \tag{18.43b}$$

where $\overline{\mathrm{Gr}}_b$ and Gr_b are the Grashof numbers defined in the Nomenclature section.

The criterion for the vertical case is well substantiated by experimental data on supercritical pressure fluids by Jackson et al. [63], Brassington and Cairns [77], and Fewster [78]. For the horizontal case, the criterion is consistent with the rather limited data which are available on the onset of buoyancy effects; most supercritical pressure experiments with tubes of bore large enough for circumferential variations of heat transfer to be reliably detected have exhibited strong influences of buoyancy whenever the wall temperatures have risen above T_{pc}. These data are in all cases consistent with the above criterion.

18.3.2 Empirical Correlations

As was mentioned above, there is a considerable number of experimental studies on heat transfer in the supercritical region. Due to inexact analytical methods, there are disagreements between the predicted and experimental values. A satisfactory description of the significant variations in physical properties, especially the variation of turbulent diffusivity with physical properties, and full consideration of these changes over the flow cross section, will certainly improve the accuracy of the analytical solutions.

The task of finding correlation equations capable of describing the strong influence of property variation over the entire forced convection region is hard to handle. However, some success has been achieved by using relatively simple extensions of established constant-property relationship.

Various correlations on supercritical pressure fluids are available in the literature. In general, the approach adopted in developing such equations fall into two categories. In the first type of approach, wall-to-bulk property ratios raised to suitable powers are added to the usual type of constant-property correlation equation. In the second type of approach, the properties in the Nusselt, Reynolds, and Prandtl numbers in the constant-property correlations are evaluated at some chosen reference temperature, the arithmetic mean of the wall and bulk temperatures being a value frequently used. This approach offers far less scope for correlating data over a wide range of conditions than does the property ratio method. In a number of instances, a combination of the two approaches is employed. A development of the first type, due to Krasnoshchekov and Protopopov [79], involves the use of a specific heat ratio term in which an integrated specific heat $\bar{c}_p$ is employed. Using their experimental data on CO_2 and the experimental data of other investigators for CO_2 and H_2O, they suggested the following equation:

$$\mathrm{Nu}_b = \mathrm{Nu}_{\mathrm{cp}} \left(\frac{\rho_w}{\rho_b} \right)^{0.3} \left(\frac{\bar{c}_p}{c_{p,b}} \right)^{m} \tag{18.44}$$

where $\mathrm{Nu}_{\mathrm{cp}}$ is the Nusselt number for constant physical properties given by Eq. (18.24), and $\bar{c}_p$ is the average specific heat at constant pressure defined as follows:

$$\bar{c}_p = \frac{1}{T_w - T_b} \int_{T_b}^{T_w} c_p \, dT$$

Noting that $c_p = (\partial i / \partial T)_p$ and that pressure does not vary significantly in the transverse direction in pipe flow, this can alternatively be rewritten as

$$\bar{c}_p = \frac{i_w - i_b}{T_w - T_b} \tag{18.45}$$

where i_w and i_b are the enthalpy at T_w and T_b, respectively. The exponent m in Eq. (18.44) depends on T_w/T_{pc} and T_b/T_{pc} as shown in Fig. 18.9.

Equation (18.44) predicts the experimental data with an accuracy of $\pm 15\%$ and is valid for the following ranges: $1.01 \le P/P_c \le 1.33$, $0.6 \le T_b/T_{\mathrm{pc}} \le 1.2$, $0.6 \le T_w/T_{\mathrm{pc}} \le 2.6$, $2 \times 10^4 \le \mathrm{Re}_b \le 8 \times 10^5$, $0.85 \le \mathrm{Pr}_b \le 55$, $0.09 \le \rho_w/\rho_b \le 1.0$, $0.02 \le \bar{c}_p/c_{p,b} \le 4.0$, $2.3 \times 10^4 \le q_w'' \le 2.6 \times 10^6$ W/m^2, $L/d > 15$.

Some of the earlier correlations were based on limited experimental data and consequently did not possess much generality. However, one of these early equations, the Miropolski-Shitsman correlation [80], proved to be surprisingly successful:

$$\mathrm{Nu}_b = 0.023\, \mathrm{Re}_b^{0.8} \mathrm{Pr}_{\min}^{0.8} \tag{18.46}$$

where $\mathrm{Pr}_{\min}$ is the smaller of the two Prandtl numbers calculated from the bulk temperature (Pr_b) and the wall temperature (Pr_w).

Some later correlations, such as those of Bishop et al. [81] and Swenson et al. [64], stemmed from comprehensive experimental studies of heat transfer to water at su-

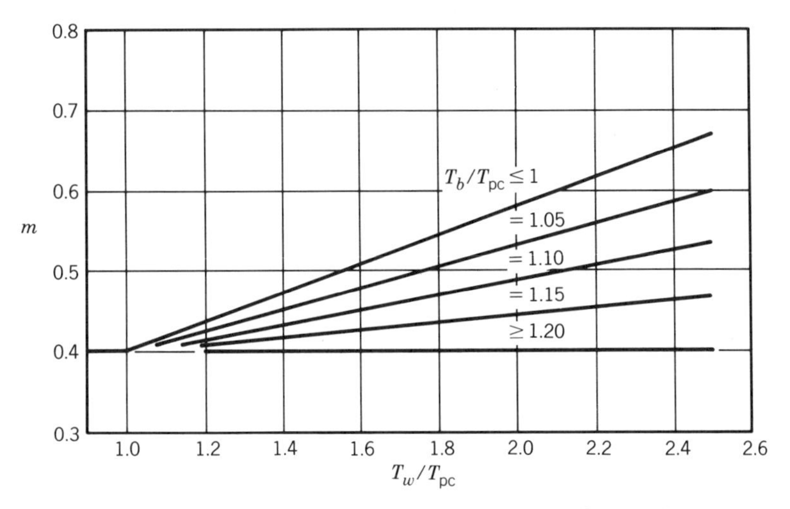

Figure 18.9. Exponent m in Eq. (18.44) vs. T_w/T_{pc} and T_b/T_{pc} [16].

percritical pressure but were developed without many data for other fluids. Swenson et al. [64] have correlated their data on the heat transfer for water by the equation

$$\mathrm{Nu}_w = 0.00459\,\mathrm{Re}_w^{0.923}\left(\overline{\mathrm{Pr}}_w\right)^{0.163}\left(\frac{\rho_w}{\rho_b}\right)^{0.231}\frac{k_w}{k_b} \tag{18.47}$$

where $\overline{\mathrm{Pr}}_w = \bar{c}_p\mu_w/k_w$ and $\bar{c}_p$ is determined from Eq. (18.45). Equation (18.47) describes the experimental data of the authors with a mean squared error of $\pm 10\%$ and covers the range of the characteristic parameters shown in Table 18.7.

The most wide-ranging and persistent attempts to correlate supercritical data have been those of Krasnoshchekov, Protopopov, Petukhov and their coworkers [79, 82, 83], correlations 6, 7, and 13 in Table 18.7. By using correlations of rather elaborate form they allowed for sufficient scope to fit a wide range of conditions. The correlation of Yamagata et al. [65] is also rather elaborate. These authors utilized three different equations, the choice of equation depending on the values of T_w and T_b in relation to T_{pc}.

A comparison of the performance of some of the earlier correlations was carried out by Hall et al. [60]. Agreement between heat transfer coefficients, calculated from various equations, was found to be poor, and it was concluded that the equations then available were inadequate for predicting heat transfer in the critical region. However, since then methods have been established for screening data to eliminate those which are influenced by buoyancy. Furthermore, a considerable body of new data has been accumulated. A further data evaluation study has been given by Jackson and Fewster [84].

The various correlations listed in Table 18.7 were used to predict heat transfer coefficients for approximately 2000 experimental conditions from the observed values of mass velocity, bulk temperature, and wall temperature. About 75% of the data points were from experiments on water, and the remainder were for carbon dioxide. They were carefully screened to exclude influences of buoyancy. The discrepancies between predicted and experimental heat transfer coefficients were tabulated for each case, and performance statistics for the correlation equations were produced [72]. The results are shown for a number of correlations.

TABLE 18.7. Turbulent Forced Convection Correlations to Fluids in Circular Ducts at Supercritical Pressure

No.	Reference	Correlation	Fluid
1	McAdams et al. [85]	$\mathrm{Nu}_b = 0.0214\,\mathrm{Re}_f^{0.8}\mathrm{Pr}_f^{0.33}(k_f/k_b)[1 + 2.3/(x/d)]$ where $T_f = (T_w + T_b)/2$, $\mathrm{Re}_f = 4\dot{m}/\pi\mu_f d$	H_2O
2	Powell [86]	$\mathrm{Nu}_b = 0.023\,\mathrm{Re}_f^{0.8}\mathrm{Pr}_f^{0.4}(k_f/k_b)(\rho_f/\rho_b)^{0.8}$	O_2
3	Humble et al. [87]	$\mathrm{Nu}_b = 0.023\,\mathrm{Re}_f^{0.8}\mathrm{Pr}_f^{0.4}(k_f/k_b)(\rho_f/\rho_b)^{0.8}$	Air
4	Bringer and Smith [88][a]	$\mathrm{Nu}_b = 0.0266\,\mathrm{Re}_{ref}^{0.77}\mathrm{Pr}_w^{0.55}(k_r/k_b)(\rho_r/\rho_b)^{0.77}$ where $\mathrm{Re}_r = 4\dot{m}/(\pi\mu_r d)$, $T_r = \begin{cases} T_b & \text{for } T_b < T_w < T_{pc} \\ T_{pc} & \text{for } T_b < T_{pc} < T_w \\ T_w & \text{for } T_{pc} < T_b < T_w \end{cases}$	CO_2
5	Miropolsky and Shitsman [80]	$\mathrm{Nu}_b = 0.023\,\mathrm{Re}_b^{0.8}\mathrm{Pr}_{min}^{0.8}$ where $\mathrm{Pr}_{min} = \min(\mathrm{Pr}_b \text{ or } \mathrm{Pr}_w)$ for $\begin{Bmatrix} H_2O \\ CO_2 \end{Bmatrix}$ when $T_w/T_{pc} < \begin{Bmatrix} 1.05 \\ 1.4 \end{Bmatrix}$	CO_2, H_2O
6	Krasnoshchekov and Protopopov [79]	$\mathrm{Nu}_b = \begin{cases} \mathrm{Nu}_{cp}(\mu_w/\mu_b)^{-0.11} & \text{for } T_b < T_w < T_{pc} \\ \mathrm{Nu}_{cp}(\mu_w/\mu_b)^{-1}(k_w/k_b)(\bar{c}_p/c_{p,b})^{0.35} & \text{for } T_b < T_{pc} < T_w \\ \mathrm{Nu}_{cp}(\mu_w/\mu_b)^{-1}(k_w/k_b)^{0.66}(\bar{c}_p/c_{p,b})^{0.35} & \text{for } T_{pc} < T_b < T_w \end{cases}$ where $\bar{c}_p = \int_{T_b}^{T_w} c_p\, dT/(T_w - T_b) = (i_w - i_b)/(T_w - T_b)$ with[b] $\mathrm{Nu}_{cp} = (f/2)\mathrm{Re}_b\mathrm{Pr}_b/[12.7(f/2)^{1/2}(\mathrm{Pr}_b^{2/3} - 1) + 1.07]$ in which[c] $f = 1/(3.64\log_{10}\mathrm{Re}_b - 3.28)^2$	CO_2, H_2O
7	Petukhov et al. [82]	$\mathrm{Nu}_b = \mathrm{Nu}_{cp}(\mu_w/\mu_b)^{-0.11}(k_w/k_b)^{0.33}(\bar{c}_p/c_{p,b})^{0.35}$ with Nu_{cp} and $\bar{c}_p$ as in No. 6	CO_2, H_2O

8	Domin [68]	$\mathrm{Nu}_b = \begin{cases} 0.036\,\mathrm{Re}_b^{0.8}\mathrm{Pr}_b^{0.4}(\mu_w/\mu_b) & \text{for } 250°\mathrm{C} < T_w < 350°C \\ 0.10\,\mathrm{Re}_f^{0.66}\mathrm{Pr}_f^{1.2} & \text{for } 350°\mathrm{C} < T_w \end{cases}$ with T_f, Re_f as in No. 1	H_2O
9	Bishop et al. [81]	$\mathrm{Nu}_b = 0.0069\,\mathrm{Re}_b^{0.9}\,\overline{\mathrm{Pr}}_b^{0.66}\,(\rho_w/\rho_b)^{0.43}[1 + 2.4/(L/d)]$ where $\overline{\mathrm{Pr}}_b = \mu_b\bar{c}_p/k_b$, with $\bar{c}_p$ as in No. 6	H_2O
10	Swenson et al. [64]	$\mathrm{Nu}_w = 0.00459\,\mathrm{Re}_w^{0.923}\overline{\mathrm{Pr}}_w^{0.613}(\rho_w/\rho_b)^{0.231}(k_w/k_b)$ where $\overline{\mathrm{Pr}}_w = \mu_w\bar{c}_p/k_w$, with $\bar{c}_p$ as in No. 6	H_2O
11	Hess and Kunz [89]	$\mathrm{Nu}_b = 0.0208\,\mathrm{Re}_f^{0.8}\mathrm{Pr}_f^{0.4}(\rho_f/\rho_b)^{0.8}(1 + 0.01457\nu_w/\nu_b)$ with T_f and Re_f as in No. 1	H_2
12	Touba and McFadden [90]	$\mathrm{Nu}_b = 0.0068\,\mathrm{Re}_b^{0.8}\,\overline{\mathrm{Pr}}_b\exp(2.19 i_b/i_{pc} - 1.75)$ with $\overline{\mathrm{Pr}}_b$ as in No. 9	H_2O
13	Krasnoshchekov and Protopopov [83]	$\mathrm{Nu}_b = \mathrm{Nu}_{cp}(\rho_w/\rho_b)^{0.3}(\bar{c}_p/c_{p,b})^n$ where[d] $n = \begin{cases} 0.4 & \text{for } T_b < T_w < T_{pc},\ 1.2T_{pc} < T_b < T_w \\ 0.4 + 0.2[(T_w/T_{pc}) - 1] & \text{for } T_b < T_{pc} < T_w \\ 0.4 + 0.2[(T_w/T_{pc}) - 1]\{1 - 5[(T_b/T_{pc}) - 1]\} & \text{for } T_{pc} < T_b < 1.2T_{pc},\ T_b < T_w \end{cases}$ with Nu_{cp} and $\bar{c}_p$ as in No. 6	CO_2, H_2O
14	Yamagata et al. [65]	$\mathrm{Nu}_b = \begin{cases} 0.024\,\mathrm{Re}_b^{0.8}\mathrm{Pr}_b^{0.8} & \text{for } T_b < T_w < T_{pc} \\ 0.0144\,\mathrm{Re}_b^{0.8}\mathrm{Pr}_b^{0.8}(\bar{c}_p/c_{p,b})^{0.66} & \text{for } T_b < T_{pc} < T_w \\ 0.024\,\mathrm{Re}_b^{0.8}\mathrm{Pr}_b^{0.8}(\bar{c}_p/c_{p,b}) & \text{for } T_{pc} < T_b < T_w \end{cases}$ with $\bar{c}_p$ as in No. 6	H_2O
15	Miropolsky and Pikus [91]	$\mathrm{Nu}_b = 0.023\,\mathrm{Re}_b^{0.8}\mathrm{Pr}_{min}^{0.8}(\rho_w/\rho_b)^{0.3}$ where $\mathrm{Pr}_{min} = \min(\mathrm{Pr}_b \text{ or } \mathrm{Pr}_w)$	H_2O

TABLE 18.7. (Continued).

No.	Reference	Correlation	Fluid
16	Lokshin et al. [92]	$\mathrm{Nu}_b = 0.021\,\mathrm{Re}_b^{0.8}\mathrm{Pr}_b^{0.4}(\alpha/\alpha_1)$ where α/α_1 = variable-property correction factor[d]	H_2O
17	Schnurr [93][e]	$\mathrm{Nu}_b = 0.0266\,\mathrm{Re}_r^{0.77}\mathrm{Pr}_w^{0.55}(k_r/k_b)(\rho_r/\rho_b)^{0.77}$ $T_r = \begin{cases} T_{pc} & \text{for } 0.025 < (T_{pc} - T_b)/(T_w - T_b) < 0.30 \\ 0.10(T_w - T_b) - 3.0(T_{pc} - T_b) + T_b & \text{for } 0 < (T_{pc} - T_b)/(T_w - T_b) < 0.025 \end{cases}$	CO_2
18	Yamagata et al. [65]	$\mathrm{Nu}_b = 0.0135\,\mathrm{Re}_b^{0.85}\mathrm{Pr}_b^{0.8}F_c$ where $F_c = \begin{cases} 1.0 & \text{for } T_b < T_w < T_{pc} \\ 0.67\,\mathrm{Pr}_{pc}^{-0.05}(\bar{c}_P/c_{p,b})^{n_1} & \text{for } T_b < T_{pc} < T_w \\ (c_p/c_{p,b})^{n_2} & \text{for } T_b < T_{pc} < T_w \end{cases}$ where $n_1 = -0.77\,(1 + 1/\mathrm{Pr}_{pc}) + 1.49$ $n_2 = 1.44\,(1 + 1/\mathrm{Pr}_{pc}) - 0.53$ Pr_{pc} = Prandtl number at pseudocritical temperature.	H_2O

[a] Equation developed from Deissler's semiempirical theory as reported in [58].

[b] Semiempirical equation for forced-convection heat transfer with uniform properties due to Petukhov and Kirillov as reported in [16].

[c] Petukhov [59] states that, in order to correlate data over a wide range of pressure, the exponent of the density ratio should take the form $0.35 - 0.05P/P_c$, which is recommended.

[d] Presented graphically in Ref. 92 as a function of i_b and $q_w'' \times 10^{-3}/G$.

[e] The equation for Nu_b was developed by Deissler [58] from semiempirical theory. Reference temperatures are the approximation of Schnurr's graphical results.

It has been concluded that the 1965 correlation of Krasnoshchekov and Protopopov [83] is the most effective. A modified form of this correlation has been tried in which the constant properties part Nu_{cp} (due to Petukhov and Kirillov; see [16]) is replaced by a simpler Dittus-Boelter form. This version gave equally good results and can be recommended for design calculations [72]. It has the following form:

$$Nu_b = 0.0183\,Re_b^{0.82}\,Pr_b^{0.5}\left(\frac{\rho_w}{\rho_b}\right)^{0.3}\left(\frac{\bar{c}_p}{c_{p,b}}\right)^{n} \tag{18.48}$$

The index n is not constant, and the equations governing its variation are given under correlation 13 (footnote d) of Table 18.7. It depends on the values of T_w and T_b, conditions for which data are available; a mean value for n was employed, and the following modified form of Eq. (18.48) is suggested [74, 84]:

$$Nu_b = 0.0183\,Re_b^{0.82}\left(\overline{Pr}_b\right)^{0.5}\left(\frac{\rho_w}{\rho_b}\right)^{0.3} \tag{18.49}$$

Jackson and Fewster [84] found that the correlation of Krasnoshchekhov and Protopopov [83] performed best, correlating 90% of the water data and 93% of the carbon dioxide data to within $\pm 20\%$. However, Eq. (18.49) correlates 78% of the water data and 79% of the carbon dioxide data to within $\pm 20\%$.

Tarasova and Leontiev (as reported in [16]) studied skin friction for the case of water flowing through smooth tubes in the supercritical region. The experiments were carried out in downward and upward flows in heated tubes over the pressure range of 2.26×10^7 to 2.65×10^7 Pa, with heat fluxes from 0.6×10^6 to 2.3×10^6 W/m^2 and in the range of Re_b from 5×10^4 to 63×10^4. The average friction factor $\bar{f}_b$ was determined in a section of length $l = 50d$ and $75d$, following a heated or unheated section of length $50d$. The experimental results are correlated by the equation

$$\frac{\bar{f}}{f_{cp}} = \left(\frac{\bar{\mu}_w}{\bar{\mu}_b}\right)^{0.22} \tag{18.50}$$

where f_{cp} is calculated from Eq. (18.23) with $Re_b = \overline{G}d/\bar{\mu}_b$; $\bar{\rho}_b$ and $\bar{\mu}_b$ are average values of ρ_b and μ_b over the tube section under consideration.

Turbulent forced-convection correlations to fluids in circular ducts at supercritical pressure are summarized in Table 18.7. The correlations given in this table cannot predict the heat transfer process with diminished heat transfer and enhanced heat transfer; these empirical equations are known for the calculation of heat transfer in the normal regime, for which the relations observed in experiments can be explained and analyzed by existing concepts of turbulent flow and heat transfer formulations for variable physical properties. In heat exchanger systems at supercritical pressures, normal heat transfer regimes usually occur, and therefore predicted results can be used.

18.4 NATURAL CONVECTION

This section summarizes the available solutions and experimental studies to describe temperature-dependent property effects as they occur in natural convection.

A characteristic common to most analytical studies of natural convection has been the neglect of all fluid-property variations, except for the essential density differences which, in the absence of mass transfer, are a consequence of temperature gradients in the fluid. This greatly simplifies analytical and experimental studies, since the number of variables which must be considered is vastly reduced.

In practice, however, experimental data usually exhibit considerable deviations from the analytical predictions, in large part because of inadequacy of the constant-property assumption. If the ratio of the absolute temperature of the wall to the ambient temperature, T_w/T_∞, is nearly unity, the constant-property assumption is valid and the most common reference temperature under the assumption of constant properties is the film temperature, $T_f = (T_w + T_\infty)/2$, which is used in the evaluation of the significant dimensionless groups, specifically the Nusselt (Nu), Grashof (Gr), Prandtl (Pr), and Rayleigh (Ra) numbers.

In theoretical natural convection analyses, the universal adoption of the Boussinesq approximation is no longer valid when the ratio of the absolute temperature of the wall to the absolute ambient temperature, T_w/T_∞, is very much different than unity. Therefore, for large temperature differences between the wall and the ambient, the adequacy of the results derived from the constant-property analysis has been in doubt.

Due to the importance of natural convection with variable fluid properties in industrial applications, there has been much analytical and experimental work directed toward determining the effects of variable properties, which are cited in Refs. 94–108. A distinction can easily be made in analytical formulation concerning which transport properties are considered variable; however, it is clearly difficult to sort out two effects in an experiment. At present, the variable-property work has been applied to only a fraction of the geometries and boundary conditions for which the constant-property solutions are available.

18.4.1 Theoretical Studies

For steady, laminar, two-dimensional (plane), vertical natural convection flow in the absence of viscous dissipation, motion pressure and volumetric energy source effects, the boundary layer equations appropriate to the variable-property situations are

$$\frac{\partial}{\partial x}(\rho u) + \frac{\partial}{\partial y}(\rho v) = 0 \tag{18.51}$$

$$\rho\left(u\frac{\partial u}{\partial x} + v\frac{\partial u}{\partial y}\right) = g(\rho_\infty - \rho) + \frac{\partial}{\partial y}\left(\mu\frac{\partial u}{\partial y}\right) \tag{18.52}$$

$$\rho c_p\left(u\frac{\partial T}{\partial x} + v\frac{\partial T}{\partial y}\right) = \frac{\partial}{\partial y}\left(k\frac{\partial T}{\partial y}\right) \tag{18.53}$$

where u and v are the vertical and horizontal velocity components, respectively, and g is taken to be in the negative x direction for heated flows. The temperature of the ambient fluid, T_∞, at large values of y is constant (Fig. 18.10).

In most of the theoretical studies, the absolute viscosity μ has been taken to be variable in the force-momentum balance, while the fluid volumetric coefficient of thermal expansion, β, the specific heat c_p, and the thermal conductivity k have been assumed to be constant. Effects of viscous dissipation, motion pressure, and volumetric energy source have also been neglected. This reduces Eqs. (18.51), (18.52), and (18.53)

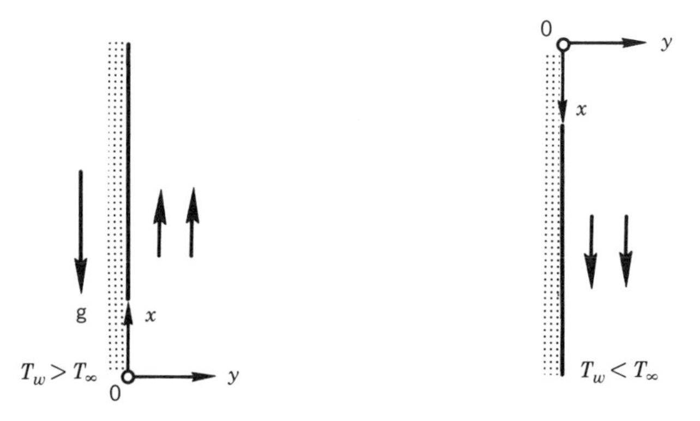

Figure 18.10. System coordinates for heating and cooling of a vertical plate.

to the following governing equations:

$$\frac{\partial u}{\partial x} + \frac{\partial u}{\partial y} = 0 \tag{18.54}$$

$$u\frac{\partial u}{\partial x} + v\frac{\partial u}{\partial y} = \frac{1}{\rho}\frac{\partial}{\partial y}\left(\mu\frac{\partial u}{\partial y}\right) + g\beta(T - T_\infty) \tag{18.55}$$

$$u\frac{\partial T}{\partial x} + v\frac{\partial T}{\partial y} = \frac{k}{\rho c_p}\frac{\partial^2 T}{\partial y^2} \tag{18.56}$$

In the above equations, the Boussinesq and boundary-layer approximations are employed.

Sparrow and Gregg [94] transformed the partial differential equations (18.52) and (18.53) to ordinary differential equations by defining a stream function and a similarity variable. They studied both idealized and real gases under the following boundary conditions (Fig. 18.10):

$$\begin{aligned} &u = 0, \quad v = 0, \quad T = T_w \qquad \text{at} \quad y = 0 \\ &u = 0, \quad T = T_\infty \qquad \text{as} \quad y \to \infty \end{aligned} \tag{18.57}$$

where T_w and T_∞ are constant wall and ambient temperatures, respectively.

Sparrow and Gregg [94] studied both idealized and real gases. The power-law variations for k and μ are commonly used approximations. Sutherland-type formulas are also used to describe the conductivity and viscosity variations; they are closer to reality than the simple power laws. The variations of c_p and the Prandtl number are also considered for air. Their heat transfer results for $p = \rho RT$, $k \propto T^{3/4}$, $\mu \propto T^{3/4}$, c_p = constant, and Pr = constant are listed in Table 18.8.

From an analytical solution for the constant-property fluid, it is found that the heat transfer (for Pr = 0.7) is

$$\frac{\mathrm{Nu}_x}{\mathrm{Gr}_x^{1/4}} = 0.353 = \frac{\overline{\mathrm{Nu}_L}}{\frac{4}{3}\mathrm{Gr}_L^{1/4}} \tag{18.58}$$

TABLE 18.8. Heat Transfer Results for an Isothermal Vertical Plate [94]

T_W/T_∞	$\mathrm{Nu}_{x,w}/\mathrm{Gr}_{x,w}^{1/4}$ (Pr = 0.7)	$\mathrm{Nu}_{L,w}/\frac{4}{3}\mathrm{Gr}_{L,w}^{1/4}$ (Pr = 0.1)
4	0.371	—
3	0.368	0.418
$\frac{5}{2}$	0.366	—
2	0.363	—
$\frac{3}{4}$	0.348	—
$\frac{1}{2}$	0.339	—
$\frac{1}{3}$	0.330	0.375
$\frac{1}{4}$	0.323	—

where the Grashof number is given by

$$\mathrm{Gr}_x = \frac{g\beta|T_w - T_\infty|x^3}{\nu^2}, \qquad \mathrm{Gr}_L = \frac{g\beta|T_w - T_\infty|L^3}{\nu^2} \tag{18.59}$$

By making some trials, Sparrow and Gregg [94] determined the temperature for evaluating k and ν in Eq. (18.58) which gives the best agreement between the constant-property result and the variable-property findings of Table 18.8. This temperature is termed the reference temperature T_r, and is found to be given by

$$T_r = T_w - 0.38(T_w - T_\infty) \tag{18.60}$$

for the entire Prandtl number range of gases. The coefficient of volumetric expansion β should be taken as $1/T_\infty$, though that is strictly correct for perfect gases only.

The error in heat transfer predictions for the constant-property result by using this reference temperature is at most 0.6% over the entire range $\frac{1}{4} \le T_w/T_\infty \le 4$ (Fig. 18.11). It is found that this method of predicting heat transfer results can be used for other gases. Sparrow and Gregg [94] also studied the variable-property problem in mercury. To facilitate numerical integrations, polynomials of the following form were fitted to the data on k, μ, c_p, and ρ (designated generically as χ):

$$\chi = \sum_{n=0}^{3} A_n T^n \tag{18.61}$$

By the numerical methods, Sparrow and Gregg [94] obtained solutions for two special cases where wall temperature, ambient temperature, Pr_w, and Pr_∞ are specified ($T_w = 1060$ K, $T_\infty = 560$ K; $T_w = 910$ K, $T_\infty = 610$ K). It is found that on evaluating k, ν, β, and Pr at

$$T_r = T_w - 0.3(T_w - T_\infty) \tag{18.62}$$

the constant-property heat transfer results coincide with those of variable-property ones.

From the findings reported in Ref. 94, it appears that natural-convection heat transfer for laminar flow under variable-property conditions can be computed using the reference-temperature relations given by Eqs. (18.60) and (18.62). Both equations

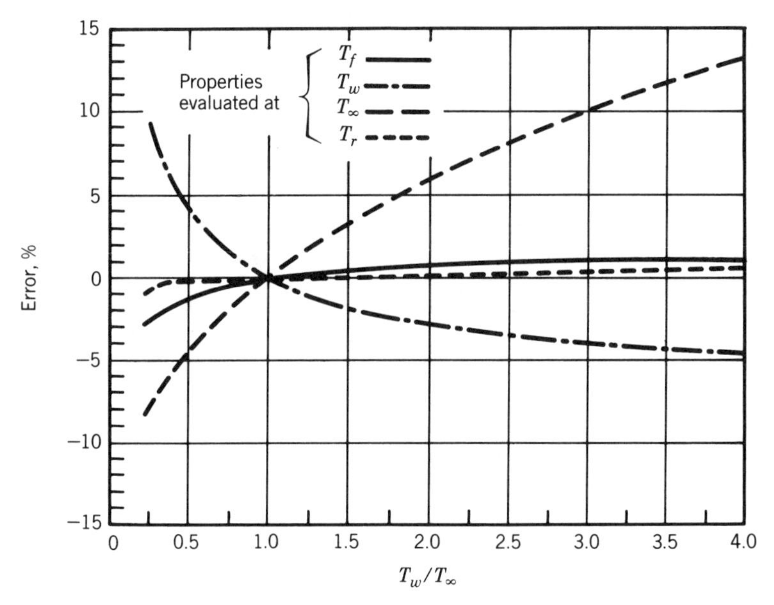

Figure 18.11. Errors made in prediction of heat transfer for a gas ($P = \rho RT$, $k \propto T^{3/4}$, $\mu \propto T^{3/4}$) by evaluating constant-property results at various temperatures [94].

include the property variation of gases and liquid mercury with temperature (with $\beta = 1/T_\infty$ for gases).

It should be remembered that variable-property relations are valid only within the range of the experimental conditions and the theorical model.

The coefficient of volumetric expansion, β, has a major bearing on natural-convection heat transfer via the buoyancy term $(\rho_\infty - \rho)g$ in the momentum equation. From its definition, the buoyancy term in the momentum equation is approximated to $\beta_\infty \rho_\infty \Delta T g$. However, β can vary appreciably with temperature for many fluids. Barrow and Sitharamaro [95] examined the effect of variable volumetric expansion coefficient β for water, and compared the result of using a constant value of β for natural convection on a constant-temperature vertical flat plate in an infinite medium. Alternately, the fluid density and its variation with temperature may be employed, but use of the coefficient β is more common and illustrates the property variation with temperature more convincingly. However, the authors ignored the temperature dependence of density due to that of β. They considered a linear variation of β with temperature as $\beta = \beta_\infty(1 + c\,\Delta T)$, where c is a constant.

By the use of integral analysis, they obtained the following expression:

$$\frac{\mathrm{Nu}(\text{variable } \beta)}{\mathrm{Nu}(\text{constant } \beta)} = \left(\frac{\beta_\infty}{\beta}\right)^{1/4}\left(1 + \frac{3c\,\Delta T}{5}\right)^{1/4} \tag{18.63}$$

Table 18.9 shows the results which have been obtained using the above equation with the following values:

$$T_\infty = 20°\mathrm{C}, \qquad T_w - T_\infty = 60°\mathrm{C}, \qquad \beta_\infty = 2.55 \times 10^{-4}\ °\mathrm{C}^{-1}$$

$$c = 2.98 \times 10^{-2}\ °\mathrm{C}^{-1}$$

TABLE 18.9. Variation of Nusselt Number with β for Water

Constant Value of	$\frac{\text{Nu(variable } \beta)}{\text{Nu(constant } \beta)}$
β_∞ at 20°C	1.20
β_f at 50°C	1.02
β_w at 80°C	0.929

As is seen, the variable-coefficient analysis results in a heat transfer coefficient value 20% greater than that based on a constant value of β_∞. The results confirm that the film value is the best one for a constant-β analysis, as would be expected.

Brown [96] also studied the effect of the temperature dependence of the coefficient of volumetric expansion on laminar free-convection heat transfer. He solved the momentum and energy boundary-layer equations by the approximate integral technique for laminar free convection with the temperature-dependent coefficient of volumetric expansion. Minkowycz and Sparrow [97] found from their computer results for steam that the reference-temperature rule of Sparrow and Gregg [94] applied, provided that the coefficient 0.38 in Eq. (18.60) is replaced by 0.46 and β is evaluated at the bulk steam temperature.

If β is evaluated as $1/T_\infty$, significant errors can arise under some conditions, particularly for temperatures close to the saturation temperature. For low and moderate steam pressures, the reference-temperature rule gave good agreement with the variable-property heat transfer within 1%. Similar accuracy was obtained at high pressures by the use of a multiplier incorporating a linear relationship in the pressure. This multiplicative correction factor has been fitted as follows [97]:

$$\mathrm{Nu} = \left(\frac{\mathrm{Gr\,Pr}}{4}\right)^{1/4} (0.4748 + 0.1251\,\mathrm{Pr} - 0.0328\,\mathrm{Pr}^2) \times \left[1 - 0.00181p\left(\frac{T_\infty}{T_w} - 1\right)\right] \tag{18.64}$$

in which β is evaluated at the bulk steam temperature, p is in atmospheres, the T's are absolute temperatures, and the other properties are at the reference temperature.

The dependence of the coefficient of volumetric expansion on temperature implies a corresponding dependence between the density and temperature, and for a linear dependence of β on T it follows that

$$\rho = \rho_\infty \exp\left[-\beta_\infty\,\Delta T\left(1 + \frac{c}{2}\,\Delta T\right)\right] \tag{18.65}$$

where $\Delta T = T_w - T_\infty$.

Using Eq. (18.65), Brown [96] integrated the integral form of the momentum and energy equations for natural convection by assuming that the thermal and boundary layers are identical and the interrelationship between ρ and T should have only secondary effects on the velocity and temperature profiles. Therefore the conventional natural convection profiles are applicable.

If one considers the case of water with $\Delta T = 60°\mathrm{C}$, $T_\infty = 20°\mathrm{C}$, $\beta_\infty\,\Delta T = 1.532 \times 10^{-2}$, and $c\,\Delta T = 2.13°\mathrm{C}$, Brown's results give

$$\frac{\mathrm{Nu}(\text{variable } \beta, \rho)}{\mathrm{Nu}(\text{constant } \beta, \rho)} = \left[\frac{\mathrm{Pr} + 0.955}{1.618(\mathrm{Pr} + 0.952)}\right]^{-1/4} \tag{18.66}$$

whereas Eq. (18.63) gives

$$\frac{\text{Nu(variable } \beta)}{\text{Nu(constant } \beta)} = 1.228 \tag{18.67}$$

For water at 20°C the Prandtl number is 6.85, and therefore, for Pr based on the bulk fluid temperature, Eq. (18.66) gives

$$\frac{\text{Nu(variable } \beta, \rho)}{\text{Nu(constant } \beta, \rho)} = 1.128 \tag{18.68}$$

Therefore, ignoring the variation of β and ρ with temperature introduced an error in heat transfer of +12.8%, whereas if the effect of variation of ρ is omitted, the suggested error is an overestimate at +22.8%.

Carey and Mollendorf [98] solved the governing equations (18.54), (18.55), and (18.56) for a vertical plate by the use of a boundary-layer similarity analysis applicable to liquid by defining the following:

$$\eta = \frac{y}{x}\sqrt[4]{\frac{\text{Gr}_x}{4}}, \qquad \Psi = 4\mu_f\sqrt[4]{\frac{\text{Gr}_x}{4}}\,\frac{f(\eta)}{\rho_f}, \qquad \phi(\eta) = \frac{T - T_\infty}{T_w - T_\infty} \tag{18.69}$$

They considered the variation of viscosity, which is important for many important fluids. For a vertical isothermal surface, similarity has been shown to exist for viscosity variation expressed as a general function of temperature. For the simpler situation of a linear variation of viscosity with temperature, calculated results have been presented for a wide range of Prandtl number. The calculations cover the range of $-1.6 \le \gamma_f \le +1.6$, where γ_f is the viscosity-variation coefficient defined as $\gamma_f = (1/\mu)(d\mu/dT)_f(T_w - T_\infty)$. An examination of fluid properties for many familiar fluids, such as glycerin, ethylene glycol, kerosene, silicone fluids, and petroleum oils, reveals that a temperature difference of 20 to 100 K, which may occur in many real flow situations, can result in values of $\gamma_f \approx 1$ and sometimes as large as $\gamma_f \approx 2$.

The local and average Nusselt numbers can be expressed as

$$\text{Nu}_x = [-\phi'(0)]\left(\frac{\text{Gr}_x}{4}\right)^{1/4} \tag{18.70}$$

and

$$\overline{\text{Nu}}_L = \tfrac{4}{3}[-\phi'(0)]\left(\frac{\text{Gr}_L}{4}\right)^{1/4} \tag{18.71}$$

The summary of heat transfer results of their numerical analysis is given in Table 18.10.

For most liquids, β is greater than zero and $(1/\mu)(d\mu/dT)$ is less than zero; the most common case for $\gamma_f < 0$ is $T_w > T_\infty$ and upward flow with $\mu_w < \mu_\infty$. The most common case for $\gamma_f > 0$ is $T_w < T_\infty$ and downward flow with $\mu_w > \mu_\infty$.

The effect of viscosity variation with temperature on the temperature and velocity profiles is seen in Fig. 18.12 for a fluid with Pr = 10, which indicates that the effect of temperature-dependent viscosity on the velocity profile is more pronounced than its effect on the temperature profile.

There are some liquids for which properties other than μ vary strongly with temperature. In particular, water and methyl alcohol exhibit strong variations of both μ and β. The analysis presented in [98] is not applicable to these liquids, since they

TABLE 18.10. Calculated Transport Parameters of Liquids with Viscosity Variations: Vertical Isothermal Flat Surface

Pr_f	γ_f	$-\phi'(0)$	Pr_f	γ_f	$-\phi'(0)$
1.0	−1.6	0.6514	100.0	−1.6	2.7168
	−0.8	0.5965		−0.8	2.3600
	0.0	0.5671		0.0	2.1914
	0.8	0.5469		0.8	2.0843
	1.6	0.5315		1.6	2.0177
10.0	−1.6	1.4076	1000	−1.6	4.9827
	−0.8	1.2476		−0.8	4.2887
	0.0	1.1693		0.0	3.9654
	0.8	1.1190		0.8	3.7602
	1.6	1.0843		1.6	3.6178

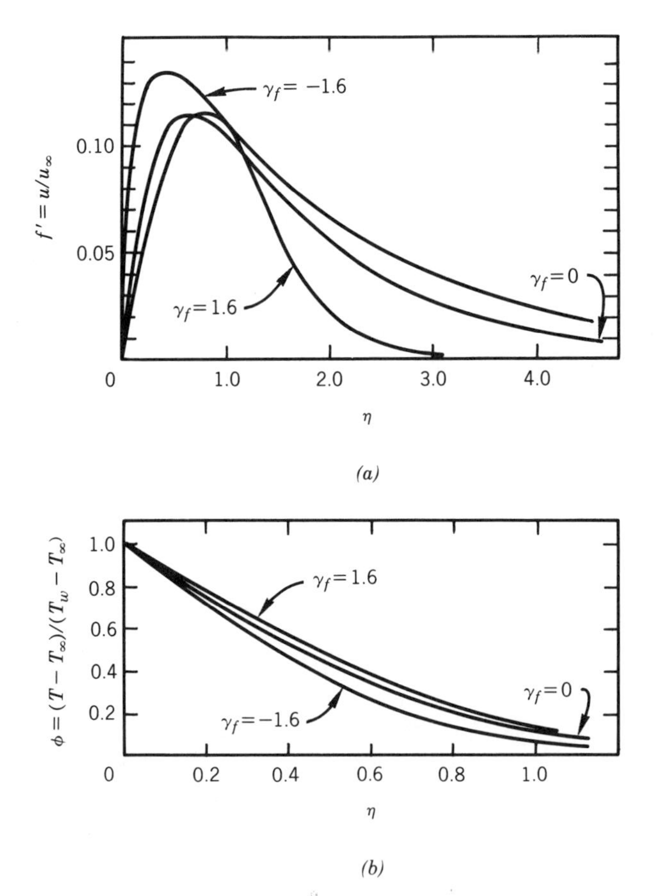

Figure 18.12. The effect of viscosity variation on (a) velocity and (b) temperature profiles for a vertical surface for Pr = 10 [98].

consider only the μ variation. Furthermore, the assumptions made in [98] are not justified for gases, since the variation of other gas properties with temperature cannot be neglected.

Carey and Mollendorf [99] presented a perturbation analysis to study the variable-viscosity effects in three laminar convection flows with liquids, namely, a freely rising plane plume, the flow above a horizontal line source on an adiabatic surface (a plane-wall plume) and the flow adjacent to a vertical uniform-flux surface. They used a perturbation method to analyze the effect of temperature-dependent viscosity on the above three vertical-plane flows. The perturbation parameter is

$$\gamma_f^* = \frac{1}{\mu}\left(\frac{d\mu}{dT}\right)_f (T_w - T_\infty)_0 \tag{18.72}$$

where γ_f^* is evaluated at the film temperature, and $(T_w - T_\infty)_0$ is the downstream temperature difference (along the x axis) which would result for $\gamma^* = 0$ (constant viscosity evaluated at the film temperature).

For the first-order linear variation of viscosity with temperature, it can be shown that

$$\gamma_f^* = \frac{\dfrac{\mu_w}{\mu_\infty} - 1}{\phi(0) + \dfrac{1}{2}\left(\dfrac{\mu_w}{\mu_\infty} - 1\right)} \tag{18.73}$$

where μ_w and μ_∞ are the viscosities at T_w and T_∞, respectively. The relationship between γ_f^* and μ_w/μ_∞ is not explicit, since $\phi(0)$ is implicitly a function of γ_f^*.

The effect of γ_f^* on heat transfer for the isothermal and uniform flux surfaces has been determined as

$$N' = \frac{\sqrt{2}\,\mathrm{Nu}_x}{(\mathrm{Gr}_x')^{1/4}} = \frac{-\phi'(0, \gamma_f^*)}{[\phi(0, \gamma_f^*)]^{5/4}} \tag{18.74}$$

where the actual physical local Grashof number is

$$\mathrm{Gr}_x' = \mathrm{Gr}_x \phi(0) \tag{18.75}$$

Figure 18.13 shows the values of the heat transfer parameter, N', predicted by the perturbation analysis for the isothermal and uniform heat flux surfaces.

Good agreement between the perturbation and similarity solution results is seen for the range of $|\gamma_f^*| \leq 0.8$. For both the isothermal and uniform heat flux surfaces, $\gamma_f^* < 0$ increases the surface heat transfer, while $\gamma_f^* > 0$ reduces it. Even for the limited range of $|\gamma_f^*| \leq 0.8$, it is found that γ^* affects the transport properties significantly for the isothermal and uniform-heat-flux surfaces. For $|\gamma_f^*| = 0.8$, the deviation from the constant-viscosity ($\gamma_f^* = 0$) result is as much as 7% for the heat transfer parameter N'. Table 18.11 shows the comparison between the heat transfer parameters predicted by the perturbation analysis [99] using the similarity solution of Carey and Mollendorf [98], and those predicted by the correlations of Fujii et al. [100] for isothermal and uniform-heat-flux conditions.

It is interesting to note that for $|\gamma_f^*| = 0.8$ the deviation from the constant-viscosity results ($\gamma_f^* = 0$) is as much as 25% for the isothermal surface and the uniform-heat-flux

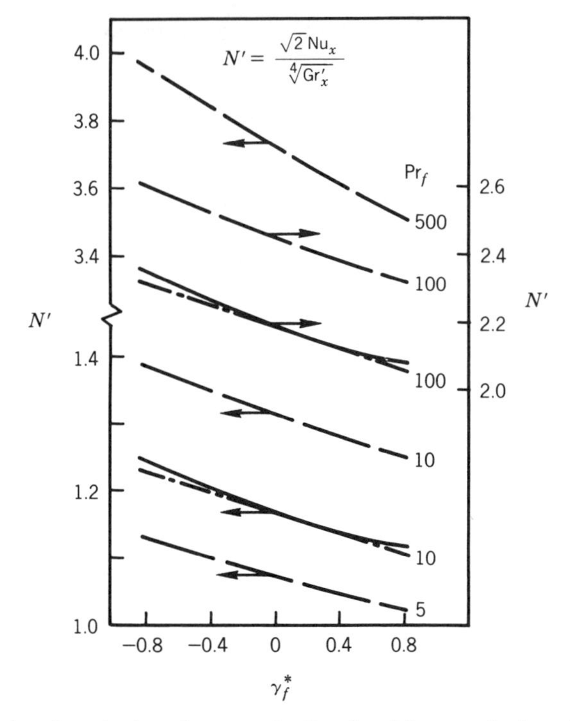

Figure 18.13. The effect of γ_f^* on heat transfer. Results of the perturbation solution are shown for the isothermal (dot-dash curves) and uniform-heat (dashed curves) surface conditions for the indicated values of the film Prandtl number. Also shown for the isothermal condition are the results of the similarity solution (solid curves) [99].

surface, while for N' the deviation is limited to 7%. This difference occurs because in perturbation analysis the reference viscosity is evaluated at T_∞, while for N' it is evaluated at T_f. This demonstrates how the choice of reference temperature influences the predicted effect of variable viscosity on heat transfer.

Zhong et al. [101] studied the variable-property effects on laminar natural convection in a square enclosure filled with air by solving the full variable-property equations including the use of the perfect-gas law for the density. Calculations have been done for a Rayleigh number range up to 10^6, and temperature differences $\theta = (T_h - T_c)/T_c$ of 0.2, 0.5, 1, and 2.0, where T_h and T_c are the hot and cold wall temperatures, respectively. They have shown that the Boussinesq approximation is adequate for $\theta < 0.10$. A correlation for the determination of a limiting value of θ under which the heat transfer data can be predicted adequately by the Boussinesq approximation has also been given. They suggested the following reference temperature for evaluating the fluid properties:

$$T_r = T_c + 0.25(T_h - T_c) \tag{18.76}$$

The correlated Nusselt number for natural convection in a square enclosure with variable properties is then given by

$$\mathrm{Nu}_r = 0.1107\,\mathrm{Ra}_r^{0.324} \tag{18.77}$$

TABLE 18.11. Comparison of the Local Heat Transfer Parameters Predicted by Numerical and Experimental Analyses for the Isothermal and Uniform Heat-Flux Conditions

(a) ISOTHERMAL

Pr_f	γ_f^*	ν_∞/ν_w	Pr_∞	$(Nu_x)_\infty/(Gr_x)_\infty^{1/4}$ [99]	$(Nu_x)_\infty/(Gr_x)_\infty^{1/4}$ [100]	Diff. (%)
100	−0.8	2.333	140	1.942	2.014	3.5
	−0.4	1.500	120	1.748	1.766	1.0
	0	1.000	100	1.550	1.550	0.0
	0.4	0.667	80	1.345	1.346	0.0
	0.8	0.429	60	1.129	1.141	1.1
100[a]	−1.6	9.000	180	2.577	2.847	9.5
	−0.8	2.333	140	1.975	2.014	1.9
	0.8	0.429	60	1.142	1.141	0.0
	1.6	0.111	20	0.636	0.653	2.6
1000[a]	−1.6	9.000	1800	4.727	5.063	6.6
	−0.8	2.333	1400	3.588	3.581	0.2
	0	1.000	1000	2.804	2.755	1.8
	0.8	0.429	600	2.060	2.030	1.5
	1.6	0.111	200	1.144	1.162	1.5

(b) UNIFORM HEAT FLUX

Pr_f	γ_f^*	ν_∞/ν_w	Pr_∞	$(Nu_x)_\infty/(Gr_x)_\infty^{1/5}$ [99]	$(Nu_x)_\infty/(Gr_x)_\infty^{1/5}$ [100]	Diff. (%)
50	−0.8	2.199	70	1.612	1.658	2.8
	−0.4	1.483	60	1.480	1.504	1.6
	0	1.000	50	1.344	1.356	0.8
	0.4	0.662	40	1.202	1.209	0.5
	0.8	0.418	30	1.048	1.055	0.7
100	−0.8	2.195	140	1.869	1.904	1.8
	−0.4	1.482	120	1.715	1.727	0.7
	0	1.000	100	1.556	1.557	0.1
	0.4	0.661	80	1.391	1.388	0.2
	0.8	0.417	60	1.211	1.212	0.1
500	−0.8	2.189	700	2.611	2.626	0.5
	−0.4	1.482	600	2.393	2.383	0.4
	0	1.000	500	2.170	2.149	1.0
	0.4	0.661	400	1.937	1.915	1.1
	0.8	0.417	300	1.686	1.672	0.8

[a] Similarity-solution results of Carey and Mollendorf [98].

in the convection-dominated region, $3 \times 10^3 < \mathrm{Ra}_r < 10^6$. The temperature difference in Ra is $T_h - T_c$, and the characteristic length is H, the height of the enclosure. It is interesting to note that this reference temperature is closer to the cold wall temperature, and it is different from that for the vertical-plate boundary-layer phenomenon as given by Sparrow and Gregg [94]. Zhong et al. [101] also correlated the variable-property results for the convection region in a square enclosure with air, without using the reference temperature, as

$$\mathrm{Nu} = \frac{\mathrm{Ra}^{0.322}}{8 + (\alpha_h/\alpha_c)^{0.84}} \qquad \text{for} \quad \mathrm{Ra}_{\mathrm{lim}} \le \mathrm{Ra} \le 10^6 \tag{18.78}$$

where $\mathrm{Ra}_{\mathrm{lim}}$ is the limiting Rayleigh number below which conduction dominates, and is given by

$$\mathrm{Ra}_{\mathrm{lim}} = 780(1 + \theta)^{2.7} \tag{18.79}$$

Further discussion of pure laminar natural convection is given in [99].

18.4.2 Experimental Studies

The number of experimental studies on the effect of the property variation is small but growing. Fujii et al. [100] studied the natural convection from the outer surface of a vertical cylinder to water, spindle oil, and Mobiltherm oil, all of which have remarkably different Prandtl numbers. Three regions—*laminar*, *transition-turbulent*, and *turbulent*—are distinguished. The local heat transfer coefficients have been correlated nondimensionally about each flow region for the cases of uniform wall temperature and uniform heat flux [100]. In each case, two kinds of experimental equations have been proposed, respectively, by using the physical properties at a reference temperature and by using the supplementary terms referred to the variation of viscosity. Nondimensional equations of average heat transfer coefficients have also been proposed for the case of uniform wall temperature.

The variation of viscosity with temperature is most conspicuous for liquids; therefore, the authors obtained nondimensional equations of local heat transfer coefficients for specified values of kinematic viscosity and neglected the variation of other properties.

In the region of transition from the laminar to the transition-turbulent region, or to the turbulent region for water, local heat transfer coefficients increase abruptly. This narrow region was termed the "transitional" region. The values of† $(\mathrm{Gr}_x\mathrm{Pr})_r$ corresponding to the transitional region vary with the type of fluid, and even for a specific fluid they are not exactly determined. The transition from the transition-turbulent to the turbulent region is continuous, and the value of $(\mathrm{Gr}_x\mathrm{Pr})_r$ corresponding to it is almost precisely defined as long as only one fluid is concerned.

The upper limits of the laminar region are $(\mathrm{Gr}_x\mathrm{Pr})_{\infty,\mathrm{cri}} = (1 \text{ to } 5) \times 10^{10}$ and $(\mathrm{Gr}_x^*\mathrm{Pr})_{\infty,\mathrm{cri}} = (0.2 \text{ to } 2.5) \times 10^{13}$; the lower limits of the transition-turbulent region of spindle oil and Mobiltherm oil, or the turbulent region of water, are $(\mathrm{Gr}_x\mathrm{Pr})_\infty = (2.5 \text{ to } 6) \times 10^{10}$ and $(\mathrm{Gr}_x^*\mathrm{Pr})_\infty = (1 \text{ to } 5) \times 10^{13}$.

† The subscript r refers to the reference temperature $T_r = T_w - \frac{1}{4}(T_w + T_\infty)$ used for evaluating fluid properties in Gr_x and Pr.

Fujii et al. [100] also expressed average Nusselt numbers for vertical cylinders with uniform wall temperature:

For water,

$$(\mathrm{Nu}_L)_\infty\left(\frac{\nu_w}{\nu_\infty}\right)^{0.21} = 0.130(\mathrm{Gr}_L\mathrm{Pr})_\infty^{1/3} - (111 \text{ to } 176)$$

$$\text{for} \quad (\mathrm{Gr}_L\mathrm{Pr})_\infty \geq (1.5 \text{ to } 4) \times 10^{10} \qquad (18.80)$$

For spindle oil,

$$(\mathrm{Nu}_L)_\infty\left(\frac{\nu_w}{\nu_\infty}\right)^{0.21} = 1.16(\mathrm{Gr}_L\mathrm{Pr})_\infty^{1/4} - (177 \text{ to } 227)$$

$$\text{for} \quad (1.5 \text{ to } 4) \times 10^{10} < (\mathrm{Gr}_L\mathrm{Pr})_\infty \leq 2.0 \times 10^{11} \qquad (18.81)$$

For Mobiltherm oil,

$$(\mathrm{Nu}_L)_\infty\left(\frac{\nu_w}{\nu_\infty}\right)^{0.21} = 0.0145(\mathrm{Gr}_L\mathrm{Pr})_\infty^{2/5} + (115 \text{ to } 65)$$

$$\text{for} \quad (\mathrm{Gr}_L\mathrm{Pr})_\infty > 2.0 \times 10^{11} \qquad (18.82)$$

The average heat transfer coefficients were observed to depend on the heights of occurrence of the transitions; therefore, the above equations involve an indefinite constant in order to cover all experimental data on each fluid. It is assumed that transitions occur discontinuously at the lower limit of $(\mathrm{Gr}_L\mathrm{Pr})_\infty$ indicated after the above equations.

The correlations of average heat transfer coefficients under the uniform-wall-temperature boundary conditions are shown in Fig. 18.14 along with the correlation curves of

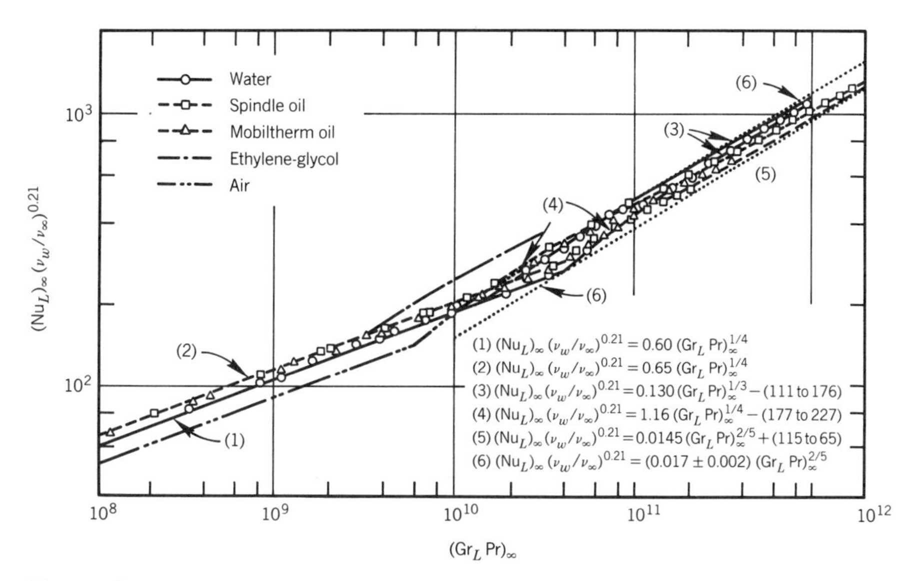

Figure 18.14. Experimental data and their correlations on average heat transfer coefficients in a vertical cylinder at constant wall temperature [100].

air by Cheesewright [102] and of ethylene glycol by Fujii [103], both derived by integration of local heat rates. Fujii et al. [100] also proposed the following correlation for water, spindle oil, Mobiltherm oil and air flow over a vertical cylinder with uniform wall temperature:

$$(\mathrm{Nu}_L)_\infty \left(\frac{\nu_w}{\nu_\infty}\right)^{0.21} = (0.017 \pm 0.002)(\mathrm{Gr}_L \mathrm{Pr})_\infty^{2/5} \qquad \text{for} \quad (\mathrm{Gr}_L \mathrm{Pr})_\infty \geq 10^{10} \quad (18.83)$$

where the properties in the nondimensional groups are evaluated at T_∞. Equation (18.83) may be useful to estimate average heat transfer coefficients approximately, but it cannot correctly correlate the data on each individual fluid, especially air.

One of the problems associated with studying the influence of variable properties experimentally is the difficulty of accurately measuring the convective heat exchange. The large characteristic lengths and large values of the ratio T_w/T_∞ usually result in radiative heat transfer rates which mask the convective heat flow. Because of this, the use of a cryogenic environment in the experimental study of natural, forced, and combined convective heat transfer is advantageous. Large property variations across the thermal boundary layer can be obtained with a relatively small temperature difference between the surface and ambient working fluid.

Clausing and Kempka [104] studied the influence of property variations in natural convection. Heat transfer from a vertical isothermal, heated cylinder surface to gaseous nitrogen upflow was experimentally investigated. For this purpose, a cylinder model of 0.28 m in height and 0.14 m in diameter was used. The ambient temperature T_∞ was varied in order to cover a large range of the Rayleigh number, including large values. The range $1 < T_w/T_\infty < 2.6$ was investigated. Experimental data are shown in Fig. 18.15. The temperature differences range from 10 to 90 K.

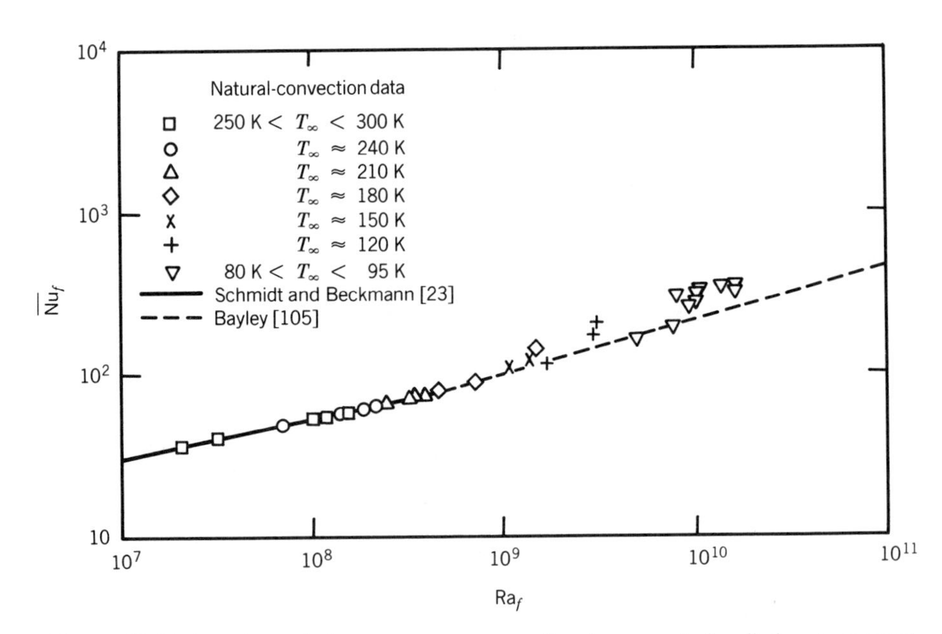

Figure 18.15. Experimental data on natural convection for a vertical cylinder at constant temperature [104].

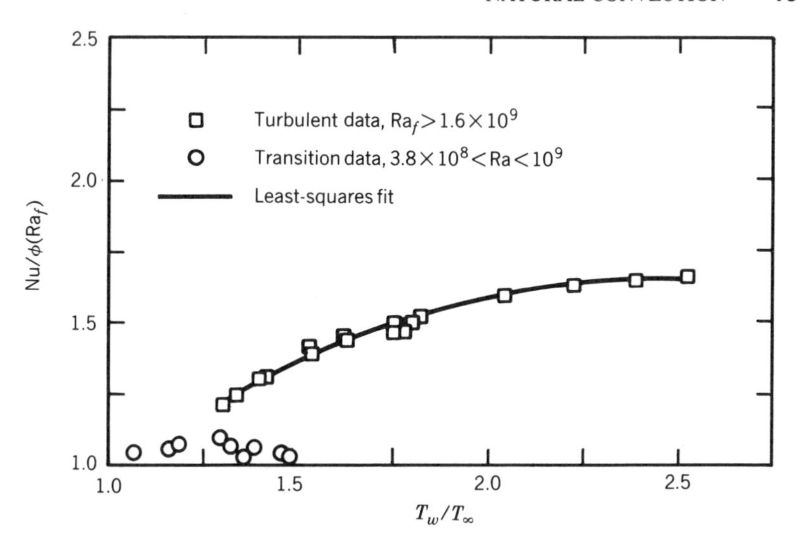

Figure 18.16. Variable-property natural-convection experimental data and correlation for an isothermal vertical cylinder [104].

In the laminar region of a vertical cylinder, for $Ra_f < 3.8 \times 10^8$, all data on the parameter T_w/T_∞ were observed to lie within 2% of Schmidt and Beckmann's correlations as reported in [23], $Nu_f = 0.52\, Ra_f^{1/4}$. On the other hand, large deviations from the Bayley correlations [105], $Nu_f = 0.10\, Ra_f^{1/3}$, were observed in the turbulent regime, $Ra_f > 1.6 \times 10^9$, as shown in Fig. 18.15. Increases in heat transfer coefficient of up to 50% from the Bayley correlation have been reported. They assumed a correlation in the following form:

$$Nu_f = \phi(Ra_f)\,\psi(T_w/T_\infty) \tag{18.84}$$

In the turbulent regime Clausing and Kempka [104] used the Bayley correlation for the constant-property correlation $\phi(Ra)$; then the authors obtained $\psi(T_w/T_\infty)$ by a least-squares fit to all data with $1.3 < T_w/T_\infty < 2.6$. The variable data with correlation $\psi(T_w/T_\infty)$ are shown in Fig. 18.16.

Clausing [106] suggested a new correlation for the constant-property portion instead of Bayley correlation by assuming the exponent 1/3, and adopting a second-order polynomial in T_w/T_∞ for the variable-property influence. By the use of a least-squares fit of the experimental data, the following correlation was obtained for variable-property turbulent natural convection over a vertical flat plate at constant temperature:

$$Nu_f = 0.082\, Ra_f^{1/3} \frac{T_w}{T_\infty} \qquad \text{for} \quad Ra > 1.6 \times 10^9 \tag{18.85}$$

where

$$\psi(T_w/T_\infty) = -0.9 + 2.4\frac{T_w}{T_\infty} - 0.5\left(\frac{T_w}{T_\infty}\right)^2 \qquad \text{for} \quad 1 < \frac{T_w}{T_\infty} < 2.6 \tag{18.86}$$

and the characteristic length is L.

The properties in Eq. (18.85) should be evaluated at the film temperature; it gives the average heat transfer coefficient for a vertical cylinder at constant wall temperature.

The *transition* regime for the vertical cylinder at constant wall temperature was determined from the experimental data to be $3.8 \times 10^8 < \text{Ra}_f < 1.6 \times 10^9$, and the following correlation has been recommended for the average Nusselt number [106]:

$$\text{Nu}_f = 1.6\,\text{Ra}_f^{0.193}\psi_{\text{tr}} \qquad \text{for} \quad 3.8 \times 10^8 < \text{Ra}_f < 1.6 \times 10^9 \tag{18.87}$$

where $\text{Nu}_f = 1.6\,\text{Ra}_f^{0.193}$ is the constant-property correlation and ψ_{tr} is the variable-property influence in the *transition* region. ψ_{tr} is given by

$$\psi_{\text{tr}} = (\psi - 1)\left(\frac{\text{Ra}_f^{1/3} - \text{Ra}_l^{1/3}}{\text{Ra}_t - \text{Ra}_l^{1/3}}\right) + 1 \tag{18.88}$$

where $\text{Ra}_l = 3.8 \times 10^8$, $R_t = 1.6 \times 10^9$, and ψ is given by Eq. (18.86).

The author obtained a good agreement in all three regimes between the correlations and the experimental data, especially in the range of parameters, 3.25 mm $\le L \le$ 7320 mm, $1.03 \le T_w/T_\infty \le 2.53$, 0.98×10^5 Pa $\le p \le 68.6 \times 10^5$ Pa, 9 K $\le \Delta T \le$ 140 K, and 82 K $\le T_\infty \le$ 305 K.

In the transition region ($3.9 \times 10^8 < Ra_f < 1.6 \times 10^9$), an average increase of 6% in heat transfer has been reported.

A detailed review of natural convection with temperature-dependent fluid properties has been given by Kakaç et al. [107].

18.5 CONCLUDING REMARKS

The heat transfer mechanisms for constant fluid properties and the influence of the variation of physical properties with temperature on heat transfer and friction coefficient in a number of important cases can be described by analytical methods. If the properties vary considerably over the flow cross section, there will be disagreements between the theoretical and experimental results which may be attributed to inexact methods of estimating the variation. This is especially true in turbulent forced convection for estimating the effect of the property variation on turbulent diffusivity.

Theoretical studies of flow and heat transfer for fluids with temperature-dependent properties are hindered by different mathematical and physical difficulties. In the case of variable physical properties, the momentum and energy equations are coupled and nonlinear; they can be solved only by the use of numerical methods. For some fluids, the variation of the properties with temperature is not the same over different ranges of the state parameters. For such a fluid it is presently impossible to describe the heat transfer and friction factor by a single relationship valid for all conditions. Theoretical findings must be verified by comparison with experimental data. However, experimental study with variable physical properties is difficult because of the need to perform experiments under large temperature differences between the surface and the fluid.

There are two commonly used schemes to account for the effects of the temperature-dependent properties on heat transfer—the reference-temperature method and the property-ratio method—which need systematic variable-property study over a range of temperature ratios and Raleigh numbers for different geometries, fluids and flow conditions. The reference temperature is a good concept for the correlation of the experimental and theoretical results, but there is no assurance that the same reference temperature is valid for all fluids in a given geometry.

Extensive literature on the influence of variable-property effects on the heat transfer and friction coefficient in internal flow and in natural convection has been reviewed in this Chapter, and the important results for internal flow are summarized in Tables 18.1, 18.4, 18.5, 18.6, and 18.7 together with the suggested constant-property heat transfer and friction factor correlations (Tables 18.2 and 18.3 and Chapter 3) which may be used in conjunction with Eqs. (18.1) and (18.2). One must be cautious when using the property ratio method to make sure that the constant-property portion is evaluated in terms of parameters specifically defined as the original author intended.

There are empirical and semiempirical theories proposed for supercritical forced convection. It is concluded that the correlations do not show sufficient agreement with experiment to justify their use except in very limited conditions. The limitations imposed on a specific correlation should be carefully studied before its use in practical applications.

Turbulent natural convection is strongly affected by property variations. Large-temperature-difference laminar experiments yielded data that were within 2% of constant-property correlations, that is within experimental errors. It is difficult if not impossible to correlate the data in the transition region. In the turbulent region, the effect of the boundary condition on the heat transfer coefficient is negligible, as in turbulent forced convection.

It is generally known that the Boussinesq approximation used in theoretical analysis is valid for small temperature differences. Therefore, the validity of this approximation should be checked if the necessary information is available, or the limits of its validity of the Boussinesq approximation should be evaluated for a given geometry and fluid flow conditions if possible [108]. This is an important problem which should be solved, since the trend in energy-related applications is to higher and higher temperatures. In addition to high temperatures, many future applications may involve high Rayleigh or Grashof numbers.

NOMENCLATURE

A	constant
c_p	specific heat at constant pressure, J/(kg · K), Btu/(lb_m · °F)
c_v	specific heat at constant volume, J/(kg · K), Btu/(lb_m · °F)
D_e	laminar equivalent diameter defined by Eq. (18.37), m, ft
D_h	hydraulic diameter = 4(minimum free flow area)/(wetted perimeter), m, ft
d	circular duct inside diameter, distance between parallel plates, m, ft
f	Fanning friction factor $= \tau_w / \frac{1}{2} u_m^2$
$f(\eta)$	defined by Eq. (18.69); $u/u_\infty = df/d\eta = f'(\eta)$
G	fluid mass velocity, kg/(m² · s), lb_m/(hr · ft²)
Gr	Grashof number $= g\beta \Delta T L^3/\nu^2$
Gr_b	Grashof number $= (\rho_b - \rho_w)\, d^3 g/\rho_b \nu^2$
$\overline{\mathrm{Gr}_b}$	Grashof number based on $\bar{\rho}$, $\overline{\mathrm{Gr}_b} = (\rho_b - \bar{\rho})\, d^3 g/\rho_b \nu^2$
Gr_L	Grashof number based on L, $= g\beta L^3 (T_w - T_\infty)/\nu^2$
Gr_x^*	local Grashof number for surface heat flux, $= g\beta q'' x^4/k\nu^2$
g	gravitational acceleration, m/s², ft/s²
H	height of an enclosure, m, ft
h	heat transfer coefficient, W/(m² · K), Btu/(hr · ft² · °F)

i	enthalpy per unit mass, J/kg, Btu/lb_m
$\bar{h}$	average heat transfer coefficient, W/($m^2 \cdot$ K), Btu/(hr $\cdot$ $ft^2 \cdot$ °F)
K_f	constant $= f$ Re
k	thermal conductivity of fluid, W/(m $\cdot$ K), Btu/(hr $\cdot$ ft $\cdot$ °F)
L	distance along the duct, or the characteristic length, m, ft
m	exponent, Eqs. (18.1b) and (18.2b)
$\dot{m}$	mass flow rate, kg/s, lb_m/s
Nu	Nusselt number $= hd/k, hL/k$
$\overline{Nu}_L$	average Nusselt number $= \bar{h}d/k$,
Nu_x	local Nusselt number $= hx/k$
n	exponent, Eqs. (18.1a) and (18.2a)
p	pressure, Pa, lb_f/ft^2
P_{cr}	thermodynamic critical pressure Pa, lb_f/ft^2
Pr	Prandtl number $= \mu c_p/k$
$\overline{Pr}$	Prandtl number based on $\bar{c}_p$, $= \mu \bar{c}_p/k$
Pr_t	turbulent Prandtl number $= \epsilon_M/\epsilon_H$
q''	heat flux, W/m^2, Btu/hr $\cdot$ ft^2
R	gas constant, J/(kg $\cdot$ K), ft $\cdot$ $lb_f/(lb_m \cdot$ °R)
Ra	Rayleigh number = Gr Pr
Re	Reynolds number $= \rho u_m d/\mu,\ \rho u_m D_h/\mu$
Re_w^*	wall Reynolds number $= u_m\, d\rho_w/\mu_w$
r	radial coordinate, m, ft
r_i	inner radius of a circular duct, m, ft
St	Stanton number = Nu/Re Pr $= h/Gc_p$
T	temperature, °C, K, °F, °R
T_f	film temperature $= (T_w + T_\infty)/2$, °C, K, °F, °R
T_{pc}	pseudocritical temperature, °C, K, °F, °R
T_r	reference temperature, °C, K, °F, °R
ΔT	temperature difference $= (T_w - T_\infty)$, °C, K, °F, °R
u_{max}	central velocity in the duct, m/s, ft/s
u	velocity component in axial direction, velocity component in x direction, m/s
u_m	mean axial velocity, m/s, ft/s
v	velocity in radial direction; velocity component in y direction, m/s, ft/s
x	axial distance, Cartesian coordinate, m, ft
y	Cartesian coordinate, distance normal to surface, m, ft
z	Cartesian coordinate, m, ft

Greek Symbols

α	thermal diffusivity $= k/\rho c_p$, m^2/s, ft^2/s
α^*	aspect ratio $= b/a$
β	coefficient of thermal expansion $= -(1/\rho)(\partial\rho/\partial T)_p$, K^{-1}, $°R^{-1}$
γ	viscosity parameter $= -(1/\mu)(d\mu/dT)$

γ_f viscosity parameter $= -(1/\mu)(d\mu/dT)_f(T_w - T_\infty)$
γ^* viscosity parameter $= -(1/\mu)(d\mu/dT)_f(T_w - T_\infty)_0$
ϵ_H thermal eddy diffusivity, m^2/s, ft^2/s
ϵ_M momentum eddy diffusivity, m^2/s, ft^2/s
η dimensionless similarity variable defined by Eq. (18.69)
θ dimensionless temperature $= (T_h - T_c)/T_c$
μ dynamic viscosity of fluid, Pa · s, $lb_m/(hr \cdot ft)$
ν kinematic viscosity of fluid, m^2/s, ft^2/s
ρ density of fluid, kg/m^3, lb_m/ft^3
$\bar{\rho}$ integrated density $= (T_w - T_b)^{-1}\int_{T_b}^{T_w} \rho \, dT$, kg/m^3, lb_m/ft^3
Φ $16/K_f$
ϕ dimensionless temperature ratio $= (T - T_\infty)/(T_w - T_\infty)$
Ψ stream function, m^2/s, ft^2/s
τ_w shear stress at the wall, Pa, lb_f/ft^2

Subscripts

a arithmetic mean
b bulk fluid condition or properties evaluated at bulk mean temperature
c critical condition, cold side
cp constant property
e equivalent
f film fluid condition or properties evaluated at film temperature
h hot side
l laminar
i inlet condition
pc pseudocritical condition
r reference condition
t turbulent
w wall condition
x local value at distance *x*
∞ free-stream condition

REFERENCES

1. R. G. Deissler, Analytical Investigation of Fully Developed Laminar Flow in Tubes with Heat Transfer with Fluid Properties Variable Along the Radius, NACA TN 2410, 1951.
2. K. T. Yang, Laminar Forced Convection of Liquids in Tubes with Variable Viscosity, *J. Heat Transfer*, Vol. 84, pp. 353–362, 1962.
3. R. L. Shannon and C. A. Depew, Forced Laminar Flow Convection in a Horizontal Tube with Variable Viscosity and Free Convection Effects, *J. Heat Transfer*, Vol. 91, pp. 251–258, 1969.
4. S. D. Joshi and A. E. Bergles, Analytical Study of Heat Transfer to Laminar In-Tube Flow of Non-Newtonian Fluids, *AIChE Symp. Ser.*, No. 199, Vol. 76, pp. 270–281, 1980.
5. E. N. Sieder and G. E. Tate, Heat Transfer and Pressure Drop of Liquids in Tubes, *Ind. Eng. Chem.*, Vol. 28, pp. 1429–1435, 1936.

6. S. Whitaker, Forced Convection Heat-Transfer Correlations for Flow in Pipes, past Flat Plates, Single Cylinders, Single Spheres, and Flow in Packed Beds and Tube Bundles, *AIChE, J.*, Vol. 18, pp. 361–371, 1972.
7. R. Oskay and S. Kakaç, Effect of Viscosity Variations on Turbulent and Laminar Forced Convection in Pipes, *METU J. Pure and Appl. Sci.*, Vol. 6, pp. 211–230, 1973.
8. V. V. Kuznetsova, Convective Heat Transfer with Flow of a Viscous Liquid in a Horizontal Tube (in Russian), *Teploenergetika*, Vol. 19, No. 5, p. 84, 1972.
9. F. L. Test, Laminar Flow Heat Transfer and Fluid Flow for Liquids with a Temperature-Dependent Viscosity, *J. Heat Transfer*, Vol. 90, pp. 385–393, 1968.
10. B. C. Sze, The Effects of Temperature-Dependent Fluid Properties on Heat Transfer and Flow Friction for Gas Flow, Ph.D. Dissertation, Stanford Univ., Stanford, Calif., 1957.
11. P. M. Worsøe-Schmidt, Heat Transfer and Friction for Laminar Flow of Helium and Carbon Dioxide in a Circular Tube at High Heating Rate, *Int. J. Heat Mass Transfer*, Vol. 9, pp. 1291–1295, 1966.
12. W. M. Kays and M. E. Crawford, *Convective Heat and Mass Transfer*, 2nd ed., McGraw-Hill, New York, 1980.
13. R. K. Shah and A. L. London, *Laminar Flow Forced Convection in Ducts*, Academic, New York, 1978.
14. S. Kakaç and Y. Yener, *Convective Heat Transfer*, Middle East Technical University Publication No. 65, Ankara, Turkey; distributed by Hemisphere, New York, 1980.
15. B. S. Petukhov and V. N. Popov, Theoretical Calculation of Heat Exchange and Frictional Resistance in Turbulent Flow in Tubes of an Incompressible Fluid with Variable Physical Properties, *High Temperature*, Vol. 1, No. 1, pp. 69–83, 1963.
16. B. S. Petukhov, Heat Transfer and Friction in Turbulent Pipe Flow with Variable Physical Properties, *Advances in Heat Transfer*, ed. J. P. Hartnett and T. F. Irvine, Academic, New York, Vol. 6, pp. 504–564, 1970.
17. R. L. Webb, A Critical Evaluation of Analytical Solutions and Reynolds Analogy Equations for Heat and Mass Transfer in Smooth Tubes, *Wärme- und Stoffübertragung*, Vol. 4, pp. 197–204, 1971.
18. C. A. Sleicher and M. W. Rouse, A Convenient Correlation for Heat Transfer to Constant and Variable Property Fluids in Turbulent Pipe Flow, *Int. J. Heat Mass Transfer*, Vol. 18, pp. 677–683, 1975.
19. V. Gnielinski, Neue Gleichungen für den Wärme- und den Stoffübergang in Turbulent Durchströmten Rohren und Kanalen, *Int. Chem. Eng.*, Vol. 16, pp. 359–368, 1976.
20. W. M. Kays and H. C. Perkins, Forced Convection, Internal Flow in Ducts, *Handbook of Heat Transfer, Fundamentals*, 2nd ed., ed. W. M. Rohsenow, J. P. Hartnett, and E. N. Ganić, Chapter 7, McGraw-Hill, New York, 1985.
21. R. K. Shah and R. S. Johnson, Correlations for Fully Developed Turbulent Flow through Circular and Noncircular Channels, *Proc. 6th National Heat and Mass Transfer Conference*, Indian Institute of Technology, Madras, pp. D75–D95, 1981.
22. D. G. Rogers, Forced Convective Heat Transfer in Single Phase Flow of a Newtonian Fluid in a Circular Pipe, CSIR Report CENG 322, Pretoria, South Africa, 1980.
23. W. H. McAdams, *Heat Transmission*, 3rd ed., McGraw-Hill, New York, 1954.
24. M. N. Özişik, *Heat Transfer—A Basic Approach*, McGraw-Hill, New York, 1985.
25. F. I. Mohammad, The Influence of Property Variations in Forced Convection, Internal Report, Dept. of Mech. Eng., Univ. of Miami, Coral Gables, Fla., 1985.
26. D. M. McElligot, P. M. Magee, and G. Leppert, Effect of Large Temperature Gradients on Convective Heat Transfer; the Downstream Region, *J. Heat Transfer*, Vol. 87, pp. 67–76, 1965.
27. H. C. Perkins and P. Worsøe-Schmidt, Turbulent Heat and Momentum Transfer for Gases in a Circular Tube at Wall to Bulk Temperature Ratios to Seven, *Int. J. Heat Mass Transfer*, Vol. 8, pp. 1011–1031, 1965.

28. V. L. Lel'chuck and B. V. Dyadyakin, Heat Transfer from a Wall to a Turbulent Current of Air, inside a Tube and the Hydraulic Resistance at Large Temperature Differentials, in Problems in Heat Transfer, AEC TR-4511, 1962.

29. A. P. Colburn, A Method of Correlating Forced Convection Heat Transfer Data and Comparison with Fluid Friction, *Trans. AIChE*, Vol. 29, pp. 174–210, 1933.

30. R. H. Norris and M. W. Sims, A Simplified Heat Transfer Correlation for Semi-turbulent Flow of Liquids in Pipes, *Trans. AIChE*, Vol. 38, 469–492, 1942.

31. J. A. Malina, and E. M. Sparrow, Variable Property, Constant Property and Entrance-Region Heat Transfer Results for Turbulent Flow of Water and Oil in a Circular Tube, *Chem. Eng. Sci.*, Vol. 19, pp. 953–962, 1964.

32. W. Hufschmidt, E. Burck, and W. Riebold, Die bestimmung Örtlicher und Wärme-übergangs-Zahlen in Rohren bei Hohen Wärmestromdichten, *Int. J. Heat Mass Transfer*, Vol. 9, pp. 539–565, 1966.

33. A. Hackl and W. Gröll, Zum Wärmeübergangsverhalten Zähflüssiger Öle, *Verfahrenstechnik*, Vol. 3, No. 4, p. 141, 1969.

34. S. Kakaç and R. Oskay, Effect of Property Variations on Forced Convection Heat Transfer in Pipe Flow, 4th International Congress CHISA, 1972, Praha, Czechoslovakia, 1972.

35. H. Hausen, Extended Equation for Heat Transfer in Tubes at Turbulent Flow, *Wärme- und Stoffübertragung*, Vol. 7, pp. 222–225, 1974.

36. R. Gregorig, The Effect of a Non-linear Temperature Dependent Prandtl Number on Heat Transfer of Fully Developed Flow of Liquids in a Straight Tube, *Wärme- und Stoffübertragung*, Vol. 9, pp. 61–72, 1976.

37. R. Oskay and S. Kakaç, Development of Two Layer Model Empirical Correlations for Variable Property Turbulent Forced Convection in Pipes, *Turbulent Forced Convection in Channels and Bundles*, ed. S. Kakaç and D. B. Spalding, Vol. 2, pp. 641–652, Hemisphere, New York, 1979.

38. L. V. Humble, W. H. Lowdermilk, and L. G. Desmon, Measurement of Average Heat Transfer and Friction Coefficients for Subsonic Flow of Air in Smooth Tubes at High Surface and Fluid Temperatures, NACA Report 1020, 1951.

39. M. F. Taylor and T. A. Kirchgessner, Measurements of Heat Transfer and Friction Coefficients for Helium Flowing in a Tube at Surface Temperatures up to 5900°R, NACA TN D-133, 1959.

40. J. R. McCarthy and H. Wolf, The Heat Transfer Characteristics of Gaseous Hydrogen and Helium, RR-60-12, Rocketdyne, Canoga Park, Calif., 1960.

41. J. F. Barnes and J. D. Jackson, Heat Transfer to Air, Carbon Dioxide and Helium Flowing through Smooth Circular Tubes under Conditions of Large Surface/Gas Temperature Ratio, *J. Mech. Eng. Sci.*, Vol. 3, No. 4, pp. 303–314, 1961.

42. W. F. Wieland, Measurement of Local Heat Transfer Coefficients for Flow of Hydrogen and Helium in a Smooth Tube at High Surface to Fluid Bulk Temperature Ratios, presented at AIChE Symposium on Nuclear Engineering Heat Transfer, Chicago, 1962.

43. M. F. Taylor, Local Heat Transfer Measurements for Forced Convection of Hydrogen and Helium at Surface Temperatures up to 5600°R, Heat Transfer and Fluid Mechanics Institute, Stanford U.P., Stanford, Calif., 1963.

44. R. G. Deissler and A. F. Prester, Computed Reference Temperature for Turbulent Variable-Property Heat-Transfer in a Tube for Several Common Gases, *Int. Div. Heat Transfer Conf.*, Boulder, Colo., pp. 579–584, 1961.

45. D. M. McEligot, Convective Heat Transfer in Internal Gas Flows with Temperature-Dependent Properties, *Adv. Transport Processes*, Vol. 4, pp. 113–200, 1985.

46. K. Rehme, A Simple Method of Predicting Friction Factors of Turbulent Flow in Non-circular Channels, *Int. J. Heat Mass Transfer*, Vol. 16, pp. 933–950, 1973.

47. J. Malák, J. Hejna, and J. Schmid, Pressure Losses and Heat Transfer in Non-circular Channels with Hydraulically Smooth Walls, *Int. J. Heat Mass Transfer*, Vol. 18, pp. 139–149, 1975.

48. E. Brundrett, Modified Hydraulic Diameter, *Turbulent Forced Convection in Channels and Bundles*, ed. S. Kakaç and D. B. Spalding, Vol. 1, pp. 361–367, Hemisphere, New York, 1979.
49. O. C. Jones, Jr., An Improvement in the Calculation of Turbulent Friction in Rectangular Ducts, *J. Fluids Eng.*, Vol. 98, pp. 173–181, 1976.
50. L. W. Carlson and T. F. Irvine, Jr., Fully Developed Pressure Drop in Triangular Shaped Ducts, *J. Heat Transfer*, Vol. 83C, pp. 441–444, 1961.
51. D. Cain and J. Duffy, An Experimental Investigation of Turbulent Flow in Elliptical Ducts, *Int. J. Mech. Sci.*, Vol. 13, pp. 451–459, 1971.
52. D. Cain, Turbulent Flow and Heat Transfer in Elliptical Ducts, Ph.D. Thesis, Liverpool Polytechnic, 1971.
53. H. Barrow and A. Roberts, *Heat Transfer 1970*, Vol. 2, Paper No. FC4.1, 1970.
54. O. C. Jones, Jr., and J. C. M. Leung, An Improvement in the Calculation of Turbulent Friction in Smooth Concentric Annuli, *J. Fluids Eng.*, Vol. 103, pp. 615–623, 1981.
55. A. Žukauskas and J. Žiugžda, *Heat Transfer in Laminar Flow of Fluid*, Institute of Physical and Technical Problems of Energetics of the Academy of Sciences of the Lithuanian SSR, Vilnius, 1969.
56. A. A. Žukauskas, *Convective Heat Transfer in Heat Exchanger* (in Russian), Nauka, Moscow, 1982.
57. L. H. Chen, Thermodynamic and Transport Properties of Gaseous Car Dioxide, *Thermodynamic and Transport Properties of Gases, Liquids, and Solids*, ASME, pp. 358–369, 1959.
58. B. S. Petukhov, Heat Transfer in a Single Phase Medium Under Supercritical Conditions, Survey Paper (in Russian), *Teplofiz. Vysokikh Temp.*, Vol. 6, No. 4, pp. 732–745, 1968.
59. S. A. Khan, Heat Transfer to Carbon Dioxide at Supercritical Pressure, Ph.D. Thesis, University of Manchester, 1965.
60. W. B. Hall, J. D. Jackson, and A. Watson, A Review of Forced Convection Heat Transfer to Fluids at Supercritical Pressures, Symposium on Heat Transfer and Fluid Dynamics of Near Critical Fluids, *Proc. Inst. Mech. Eng.*, Vol. 182, pp. 10–22, 1968.
61. W. B. Hall, Heat Transfer Near the Critical Point, *Adv. Heat Transfer*, Vol. 7, pp. 1–86, Academic, New York, 1971.
62. R. C. Hendricks, R. J. Simoneau, and R. V. Smith, Survey of Heat Transfer to Near Critical Fluids, NASA Technical Memorandum X-52612, 1969.
63. J. D. Jackson, W. B. Hall, J. Fewster, A. Watson, and M. J. Watts, Heat Transfer to Supercritical Pressure Fluids, U.K.A.E.A. A.E.R.E.-R 8158, Design Report 34, 1975.
64. H. S. Swenson, J. R. Carver, and C. R. Kakarala, Heat Transfer to Supercritical Water in Smooth-Bore Tubes, *J. Heat Transfer*, Vol. 87, pp. 477–484, 1965.
65. K. Yamagata, K. Nishikawa, S. Hasegawa, T. Fujii, and S. Yoshida, Forced Convective Heat Transfer to Supercritical Water Flowing in Tubes, *Int. J. Mass Transfer*, Vol. 15, pp. 2575–2593, 1972.
66. V. Vikhrev, D. Barulin, and A. S. Kon'kov, A Study of Heat Transfer in Vertical Tubes at Supercritical Pressures, *Thermal Eng.*, Vol. 9, pp. 116–119, 1967.
67. N. S. Kondratev, Heat Transfer and Hydraulic Resistance with Supercritical Water Flowing in Tubes (in Russian), *Teploenergetika*, Vol. 16, No. 8, pp. 49–51, 1969.
68. G. Domin, Heat Transfer to Water in Pipes in the Critical/Supercritical Region, *Brennst-Wärme-Kraft*, Vol. 15, No. 11, 1963.
69. K. R. Schmidt, Thermal Investigations with Heavily Loaded Boiler Heating Surfaces, *Mett. Verg. Gross*, No. 63, p. 391, 1959.
70. V. Vikhrev, D. Barulin, and A. S. Kon'kov, A Study of Heat Transfer in Vertical Tubes at Supercritical Pressures (in Russian), *Teploenergetika*, Vol. 14, No. 9, pp. 80–82, 1967.
71. A. P. Ornatsky, L. F. Glushchenko, E. T. Siomin, and S. I. Kalatchev, The Research of Temperature Conditions of Small Diameter Parallel Tubes Cooled by Water Under Supercritical Pressures, *Heat Transfer 1970*, Vol. 6, Paper B 8.11, 1970.

72. J. D. Jackson and W. B. Hall, Forced Convection Heat Transfer to Fluids at Supercritical Pressures, in *Turbulent Forced Convection in Channels and Bundles*, ed. S. Kakaç and D. B. Spalding, Vol. 2, pp. 563–611, Hemisphere, New York, 1979.

73. W. B. Hall and J. D. Jackson, Laminarization of a Turbulent Pipe Flow by Buoyancy Forces, ASME Paper No. 69-HT-55, 1969.

74. W. B. Hall, Forced Convective Heat Transfer to Supercritical Pressure Fluids, *Arch. Termodynamiki i Spalania*, Vol. 6, No. 3, 1975.

75. R. S. Weinberg, Experimental and Theoretical Study of Buoyancy Effects in Forced Convection to Supercritical Pressure Carbon Dioxide, Ph.D. Thesis, Mech. Eng. Dept., Manchester Univ., 1972.

76. G. A. Adebiyi, Experimental Investigation of Heat Transfer to Supercritical Pressure Carbon Dioxide in a Horizontal Pipe, Ph.D. Thesis, Mech. Eng. Dept., Manchester Univ., 1973.

77. D. J. Brassington and D. N. H. Cairns, Measurements of Forced Convective Heat Transfer to Supercritical Helium, C.E.G.B. C.E.R.L. Note RD/L/N 156, 1975.

78. J. Fewster, Heat Transfer to Supercritical Pressure Fluids, Ph.D. Thesis, Mech. Eng. Dept., Manchester Univ., 1975.

79. E. A. Krasnoshchekov and V. S. Protopopov, Heat Exchange in the Supercritical Region During the Flow of Carbon Dioxide and Water (in Russian), *Teploenergetika*, Vol. 6, No. 12, pp. 26–30, 1959.

80. Z. L. Miropolsky and M. E. Shitsman, Heat Transfer to Water and Steam at Variable Specific Heat (in Near Critical Region), *Sov. Phys.*, Vol. 2, No. 10, 1967.

81. A. A. Bishop, R. O. Sandberg, and L. S. Tong, Forced Convection Heat Transfer to Water at Near-Critical Temperatures and Supercritical Pressures, *AIChE–I. Chem. E. Symp. Ser.*, No. 2, 1965.

82. B. S. Petukhov, E. A. Krasnoshchekov, and V. S. Protopopov, An Investigation of Heat Transfer to Fluids Flowing in Pipes Under Supercritical Conditions, *Int. Dev. Heat Transfer Conf.*, Boulder, Colo., 1962.

83. E. A. Krasnoshchekov and V. S Protopopov, Experimental Study of Heat Exchange in Carbon Dioxide in the Supercritical Range at High Temperature Drops, *Teplofizika Vysokikh Temperature* (in Russian), Vol. 4, No. 3, 1966.

84. J. D. Jackson and J. Fewster, Correlation Equations for Forced Convection Heat Transfer to Fluids at Supercritical Pressure, Report, Simon Engineering Labs., Univ. of Manchester, 1980.

85. W. H. McAdams, W. E. Kennel, and J. N. Addoms, Heat Transfer to Super-heated Steam at High Pressures, *Trans. ASME*, Vol. 80, May 1950.

86. W. B. Powell, Heat Transfer to Fluid in the Region of the Critical Temperature, *Jet Propulsion*, Vol. 27, No. 27, pp. 776–783, 1957.

87. L. V. Humble, W. H. Lowdermilk, and L. G. Desmon, Measurements of Average Heat Transfer and Friction Coefficients for Subsonic Flow of Air, in Smooth Tubes at High Surface and Fluid Temperature, NACA Technical Report 1020, 1950.

88. R. P. Bringer and J. M. Smith, Heat Transfer in the Critical Region, *AIChE J.*, Vol. 3, No. 1, pp. 49–55, 1957.

89. H. L. Hess and H. R. Kunz, A Study of Forced Convection Heat Transfer to Supercritical Hydrogen, *J. Heat Transfer*, Vol. 87, pp. 41–48, 1965.

90. R. F. Touba and P. W. McFadden, Combined Turbulent Convection Heat Transfer to Near Critical Water, Purdue Research Foundation, Lafayette, Ind., Technical Report No. 18, L00-1177-18, 1966.

91. Z. L. Miropolsky and V. U. Picus, Heat Transfer in Supercritical Flows through Curvilinear Channels, *Proc. Inst. Mech. Eng.*, Vol. 183, Part 31, 1967.

92. V. A. Lokshin, I. E. Semenovker, and V. Vikhrev, Calculating the Temperature Conditions of the Radiant Heating Surfaces in Supercritical Boilers (in Russian), *Teploenergetika*, Vol. 15, No. 9, pp. 21–24, 1968.

93. N. M. Schnurr, Heat Transfer to Carbon Dioxide in the Immediate Vicinity of the Critical Point, *J. Heat Transfer*, Vol. 91, pp. 16–20, 1969.

94. E. M. Sparrow and J. L. Gregg, The Variable Fluid Property Problem in Free Convection, *Trans. ASME*, Vol. 80, pp. 869–886, 1958.

95. H. Barrow and T. L. Sitharamaro, The Effect of Variation in the Volumetric Expansion Coefficient on Free Convection Heat Transfer, *Br. Chem. Eng.*, Vol. 16, pp. 704–705, 1971.

96. A. Brown, The Effect on Laminar Free Convection Heat Transfer of the Temperature Dependence of the Coefficient of Volumetric Expansion, *J. Heat Transfer*, Vol. 97C, pp. 133–135, 1975.

97. W. J. Minkowycz and E. M. Sparrow, Free Convection Heat Transfer to Steam under Variable Property Conditions, *Int. J. Heat Mass Transfer*, Vol. 9, pp. 1145–1147, 1966.

98. V. P. Carey and J. C. Mollendorf, Natural Convection in Liquids with Temperature-Dependent Viscosity, *Heat Transfer 1978*, Vol. 2, pp. 211–217, Hemisphere, New York, 1978.

99. V. P. Carey and J. C. Mollendorf, Variable Viscosity Effects in Several Natural Convection Flows, *Int. J. Heat Mass Transfer*, Vol. 23, pp. 95–109, 1980.

100. T. Fujii, M. Takeuchi, M. Fujii, K. Suzaki, and H. Uehara, Experiments on Natural Convection Heat Transfer from the Outer Surface of a Vertical Cylinder to Liquids, *Int. J. Heat Mass Transfer*, Vol. 13, pp. 753–787, 1970.

101. Z. Y. Zhong, K. T. Yang, and S. R. Lloyd, Variable Property Effects in Laminar Natural Convection in a Square Enclosure, HTD, Vol. 26, pp. 69–75, ASME, New York, 1983.

102. R. Cheesewright, Turbulent Natural Convection from a Vertical Plane Surface, *J. Heat Transfer*, Vol. 90, pp. 1–8, 1968.

103. T. Fujii, Experiments of Free Convection Heat Transfer from a Vertical Cylinder Submerged in Liquids, *Trans. Jpn. Soc. Mech. Eng.*, Vol. 25, pp. 280–286, 1959.

104. A. M. Clausing and S. N. Kempka, The Influence of Property Variation on Natural Convection from Vertical Surfaces, *J. Heat Transfer*, Vol. 103, pp. 609–612, 1981.

105. F. J. Bayley, An Analysis of Turbulent Free Convection Heat Transfer, *Proc. Inst. Mech. Eng.*, Vol. 109, No. 20, pp. 361–370, 1955.

106. A. M. Clausing, Natural Convection Correlations for Vertical Surfaces, Including Influences of Variable Properties, *ASME J. Heat Transfer*, Vol. 105, No. 1, pp. 138–143, 1983.

107. S. Kakaç, Y. Yener and O. E. Ateşoğlu, Natural Convection with Temperature-Dependent Fluid Properties, in *Natural Convection: Fundamentals and Applications*, ed. S. Kakaç, W. Aung, and R. Viskanka, pp. 729–773, Hemisphere, New York, 1985.

108. K. T. Yang, and B. Run, eds., *Proc. Workshop on Natural Convection*, National Science Foundation, Washington, 1982.

19

INTERACTION OF RADIATION WITH CONVECTION

M. Necati Özışık

North Carolina State University
Raleigh, North Carolina

19.1 INTRODUCTION

Heat transfer by simultaneous radiation and convection has applications in numerous technological problems, including combustion, furnace design, the design of high-temperature gas-cooled nuclear reactors, nuclear-reactor safety, fluidized-bed heat exchangers, fire spreads, advanced energy conversion devices such as open-cycle coal and natural-gas-fired MHD, solar ponds, solar collectors, natural convection in cavities, turbid water bodies, photochemical reactors, and many others. When heat transfer by radiation is of the same order of magnitude as by convection, a separate calculation of radiation and convection and their superposition without considering the interaction between them can lead to significant errors in the results, because the presence of radiation in the medium alters the temperature distribution within the fluid. Therefore, in such situations, heat transfer by convection and radiation should be solved for simultaneously.

In the analysis of interaction of radiation with convection, distinction should be made between the following two classes of problems: (1) *nonparticipating media*, and (2) *participating media*. In the former, the fluid is transparent to radiation for all practical purposes; hence the coupling between radiation and convection takes place through the thermal boundary condition in the convection problem. In the latter case, the fluid is semitransparent to radiation; it may absorb, emit, and in some cases scatter radiation. Hence the coupling between radiation and convection occurs through the presence of the divergence of the radiation flux term in the energy equation for convection. In either case, the study of heat transfer by simultaneous convection and radiation necessitates some background on the role of radiation in the mathematical formulation of the problem. Therefore, in the following sections, before presenting the analysis and the results on simultaneous convection and radiation for both participating and nonparticipating media, we present an overview of the formulation and some of the recommended solution techniques for the radiation part of the problem.

Reader should consult Refs. [1–3] for comprehensive treatment of the radiation transfer in participating and nonparticipating media and the interactions with conduction and convection.

19.2 SIMULTANEOUS CONVECTION AND RADIATION IN NONPARTICIPATING MEDIA

There are many engineering applications that involve flow inside ducts or over bodies of a transparent fluid, such as air, that does not absorb, emit, or scatter radiation. If the temperature is sufficiently high or the convective heat transfer coefficient is low, the contribution of radiation to heat transfer may become important. For problems involving a prescribed heat-flux boundary condition, the effect of radiation from the surface is to alter the surface temperature, which in turn alters the heat transfer by convection. Therefore, the problems of convection and radiation are coupled through the boundary condition. On the other hand, if the problem involves a prescribed temperature boundary condition, there is no coupling through the boundary condition, because the presence of radiation does not change the surface temperature.

The complete mathematical formulation of simultaneous convection and radiation involves two parts: the convection part and the radiation part. For transparent fluids considered here, the presence of thermal radiation does not alter the standard equations of motion and energy; hence the convective part of the problem can be formulated by using the standard equations of motion and energy for nonradiating fluids well documented in heat transfer books such as Schlichting [4], Kays and Crawford [5], and Özışık [6]. To determine the effects of radiation on the boundary temperature, the problem of radiation exchange between the surfaces should be formulated and solved. Therefore, in the following subsection, we discuss the mathematical formulation of radiation exchange between surfaces pertinent to the problems of simultaneous convection and radiation, and then present the analysis and solutions of simultaneous radiation and forced convection over bodied and inside ducts.

19.2.1 Radiation Exchange among Gray Surfaces

Consider a gray enclosure whose surfaces are diffuse emitter, diffuse reflector, and divided into N zones such that radiation properties are uniform over the surface of each zone. We assume that the temperature $T_i(\boldsymbol{r})$ or the radiation heat flux $q_i^r(\boldsymbol{r})$ varies with position over the surface A_i of each zone i, which in turn implies that the *radiosity* $R_i(\boldsymbol{r})$ varies over the surfaces of the zones. Then one cannot apply the simple analysis of radiation exchange among surfaces, leading to a system of algebraic equations for the radiosities, described in numerous introductory texts on heat transfer. When the radiosity varies with position, the mathematical formulation of the radiation part of the problem leads to the following system of coupled integral equations for the radiosity functions $R_i(\boldsymbol{r})$ (Özışık [2, Chap. 5, Eqs. 5-9 and 5-10]):

$$R_i(\boldsymbol{r}_i) = \epsilon_i \bar{\sigma} T_i^4(\boldsymbol{r}_i) + \rho_i \sum_{j=1}^{N} \int_{A_j} R_j(\boldsymbol{r}_j)\, dF_{dA_i\text{-}dA_j} \tag{19.1}$$

and

$$q_i^r(\boldsymbol{r}_i) = R_i(\boldsymbol{r}_i) - \sum_{j=1}^{N} \int_{A_j} R_j(\boldsymbol{r}_j)\, dF_{dA_i\text{-}dA_j} \tag{19.2}$$

for $i = 1, 2, \ldots, N$. Here, $\boldsymbol{r}_i$ is the position vector for the coordinates at zone i, and $dF_{dA_i\text{-}dA_j}$ is the differential view factor between the elemental surfaces dA_i and dA_j at the zones A_i and A_j, respectively. The functional form of the differential view factor depends on the configurations and the geometrical arrangement of the surfaces. The methods of determination of view factors are discussed by Sparrow and Cess [1, Chaps. 3, 4] and Özışık [2, Chaps. 3, 5]. An alternative, simpler expression for the net radiation flux $q_i^r(\boldsymbol{r}_i)$ is obtained by eliminating the summation term between Eqs. (19.1) and (19.2). We find

$$q_i^r(\boldsymbol{r}_i) = \frac{1}{\rho_i}\left[\epsilon_i \bar{\sigma} T_i^4(\boldsymbol{r}_i) - (1-\rho_i) R_i(\boldsymbol{r}_i)\right], \quad \rho_i \neq 0, \qquad \text{for} \quad i = 1, 2, \ldots, N. \tag{19.3a}$$

The utility of the above set of equations for analyzing the radiation part of the problem is as follows: Suppose that temperatures $T_i(\boldsymbol{r}_i)$ are prescribed over the surfaces of each of the N zones. Then Eq. (19.1) provides N simultaneous integral equations for the determination of the N unknown radiosity functions $R_i(\boldsymbol{r}_i)$ $(i = 1, 2, \ldots, N)$. Once these radiosities are known, the radiation heat flux $q_i^r(\boldsymbol{r}_i)$ over the surface of any one of the zones A_i $(i = 1, 2, \ldots, N)$ is evaluated from Eq. (19.2) or (19.3a).

Significant simplification occurs in the formulation of the radiation part of the problem if all surfaces are assumed to be black. For black surfaces, we set $\rho_i = 0$ and

$\epsilon_i = 1$ for $i = 1,2,\ldots,N$. Then Eq. (19.1) reduces to $R_i(\mathbf{r}_i) = \bar{\sigma}T_i^4(\mathbf{r})$, and the corresponding expressions for the radiation heat flux $q_i(\mathbf{r}_i)$ are obtained from Eq. (19.2) as

$$q_i^r(\mathbf{r}_i) = \bar{\sigma}T_i^4(\mathbf{r}_i) - \sum_{j=1}^{N} \int_{A_j} \bar{\sigma}T_j^4(\mathbf{r}_j)\, dF_{dA_i\text{-}dA_j} \qquad \text{for} \quad i = 1,2,\ldots,N \quad (19.3\text{b})$$

The determination of the radiation flux $q_i^r(\mathbf{r}_i)$ from Eq. (19.3b) is a relatively straightforward matter because the analysis involves only evaluating the integrals over the zones for specified temperature distributions. On the other hand, when surfaces have reflectivity, the determination of the radiation flux $q_i^r(\mathbf{r}_i)$ from Eq. (19.2) or (19.3a) is a complicated matter because the radiosities $R_i(\mathbf{r}_i)$ are unknown and to be determined from the solution of a set of coupled integral equations (19.1). There are several methods for the solution of a system of integral equations. The reader should consult references (e.g., Özışık [2, Chap. 5]) for an overview of the methods of solving integral equations arising in radiation problems. More recently, Özışık and Yener [7] advanced a straightforward approach for solving radiation problems involving integral equations by the application of the Galerkin method.

19.2.2 Forced Convection with Radiation over a Flat Plate

Heat transfer in boundary-layer flow over surfaces is of interest in numerous engineering applications, and the solution of such problems with no radiation effects is well documented in the literature. During the past twenty years such problems have also been studied by including in the analysis the effects of radiation as coupled through the boundary condition [8–12]. Here we focus attention on the laminar boundary-layer flow of a transparent fluid over a flat plate subjected to constant applied wall heat flux at the wall surface. The mathematical formulation of the convection part of the problem can readily be done within the framework of the classical boundary-layer theory. The contribution of radiation to heat transfer is then introduced as a part of the boundary condition at the wall surface. Here our objective is to present the first order effects of radiation as a correction to the convective Nusselt number.

We consider the steady, laminar boundary-layer flow of an incompressible transparent fluid along a flat plate subjected to a constant applied wall heat flux q_w at the wall surface. Figure 19.1 illustrates the geometry and coordinate. The heat supplied to the wall is dissipated from the plate surface by convection into the fluid and by radiation to an external ambient maintained at a constant temperature T_e. The surface of the plate is opaque and gray, and has a uniform emissivity ϵ. The thermophysical properties of the fluid are constant. The external flow has a constant velocity u_∞ and constant temperature T_∞. The viscous energy dissipation is considered negligible.

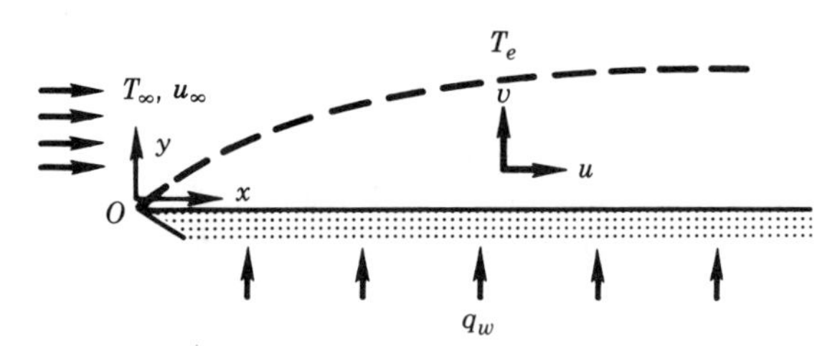

Figure 19.1. Laminar boundary-layer flow of transparent fluid over a flat plate with radiation boundary condition.

The mathematical formulation of this heat transfer problem consists of the boundary-layer continuity and momentum equations

$$\frac{\partial u}{\partial x} + \frac{\partial v}{\partial y} = 0 \tag{19.4a}$$

$$u\frac{\partial u}{\partial x} + v\frac{\partial u}{\partial y} = \nu\frac{\partial^2 u}{\partial y^2} \tag{19.4b}$$

with the boundary conditions

$$u = v = 0 \quad \text{at} \quad y = 0, \tag{19.4c}$$

$$u = u_\infty \quad \text{as} \quad y \to \infty \tag{19.4d}$$

The energy equation is given by

$$u\frac{\partial T}{\partial x} + v\frac{\partial T}{\partial y} = \alpha\frac{\partial^2 T}{\partial y^2} \tag{19.5a}$$

The boundary condition at the wall surface that allows for radiation effects and the boundary condition at far distances from the wall are taken, respectively, as

$$q_w = -k\frac{\partial T}{\partial y} + \epsilon\bar{\sigma}\left(T^4 - T_e^4\right) \quad \text{at} \quad y = 0 \tag{19.5b}$$

$$T = T_\infty \quad \text{at} \quad y \to \infty \tag{19.5c}$$

Note that radiation effects enter the analysis as a fourth-power temperature law for this simple problem; hence no radiosity is involved, because there is one surface only which is exposed to an infinite medium for radiation exchange.

The velocity part of this problem, defined by Eqs. (19.4), is uncoupled from the temperature problem and hence can be solved by the classical similarity solution of the boundary-layer equations.

To solve the energy problem defined by Eqs. (19.5), two new independent variables are defined:

$$\bar{\xi} = \frac{\epsilon\bar{\sigma}T_\infty^3}{k}\left(\frac{\nu x}{u_\infty}\right)^{1/2} \tag{19.6a}$$

$$\bar{\eta} = y\left(\frac{u_\infty}{\nu x}\right)^{1/2} \tag{19.6b}$$

Then the energy problem, Eqs. (19.5), is transformed under the transformation given by Eq. (19.6), and the velocity components obtained from the similarity solution of the velocity problem are utilized. The resulting transformed energy equation is solved by series expansion in powers of $\bar{\xi}$, utilizing the boundary condition (19.5b) to determine the expansion coefficients. The details of the analysis are well documented [2, Chap. 7]. Finally, the local Nusselt number for laminar forced convection with constant wall heat flux boundary condition is determined as

$$\frac{\text{Nu}_x}{\text{Re}_x^{1/2}} = 0.4059 - 0.620\bar{\xi} - \cdots \tag{19.7}$$

where

$$\text{Nu}_x = \frac{h_x x}{k}, \qquad \text{Re}_x = \frac{u_\infty x}{\nu} \tag{19.8}$$

and the parameter $\bar{\xi}$ which represents the radiation effects on the Nusselt number is defined by Eq. (19.6a).

The first term on the right-hand side of Eq. (19.7) is the local Nusselt number for laminar boundary-layer flow over a flat plate subjected to a constant heat flux at the wall surface. The second term represents the first order effect of radiation on the Nusselt number. Note that the radiation has no effect on the solution for the forced-convection problem with radiation having constant wall temperature boundary condition.

19.2.3 Forced Convection with Radiation inside Ducts

Heat transfer in forced convection inside ducts without the effects of radiation has been studied extensively since the original Graetz solution almost a century ago. When temperature is high or the convective heat transfer coefficient is low, radiation effects in forced convection heat transfer become important. Typical examples include, among others, the heating of a coolant in the passages of a nuclear-reactor core or inside an electrically heated tube, the heating of air in the parallel-plate channels of standard flat plate collectors. Therefore, the subject has been studied by several investigators [13–21] during the past twenty years for ducts subjected to prescribed heat flux at the wall surfaces. The effects of radiation on the duct wall temperature and the gas temperature as a function of the axial distance along the duct are quantities of practical interest. The principal difficulty in the analysis of such problems is that the convective heat transfer coefficient is not known *a priori* and must be determined as a part of the solution. To alleviate this difficulty, most investigators assumed a known convective heat transfer coefficient taken from the solution of the classical problems without the radiation. Using this coefficient in the analysis, the radiation effects on the wall and gas temperatures are determined [13–19]. Results obtained in this manner are useful to indicate the general effects of radiation on the wall surface and bulk fluid temperatures along the duct, but not so accurate as that obtained from an analysis in which the heat transfer coefficient determined as a part of the solution.

Recently, Liu and Sparrow [20] analyzed simultaneously developing laminar forced convection inside a parallel-plate channel. In this study, one wall of the channel was externally heated at a constant rate q_w and the other externally insulated. Air was considered as the heat transfer fluid. To illustrate the effects of radiation on the local Nusselt number along the directly heated wall and the wall surface and on the bulk temperature distribution, we present here the highlights of this work.

Consider laminar flow of a transparent fluid (i.e., air) inside a parallel-plate channel of length l and spacing H, as illustrated in Fig. 19.2. One wall of the channel is heated externally at a constant rate q_w while the other is externally insulated. The temperature of the fluid inlet and the inlet plenum is T_{I}, and that of the exit plenum is T_{II}. Assuming simultaneously developing laminar flow, the mathematical formulation of

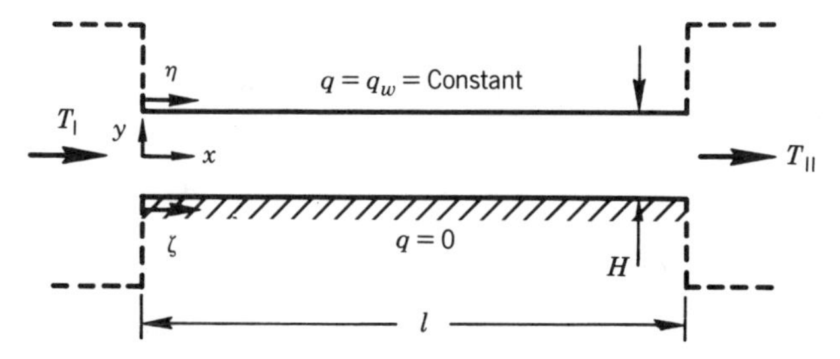

Figure 19.2. Geometry and coordinates for laminar forced convection with radiation through a parallel-plate channel.

the problem consists of the continuity, the x-momentum, and the energy equations, given in dimensionless form as

$$\frac{\partial U}{\partial X} + \frac{\partial V}{\partial Y} = 0 \tag{19.9}$$

$$U\frac{\partial U}{\partial X} + \frac{\partial U}{\partial Y} = -\frac{\partial P}{\partial X} + \frac{2}{\mathrm{Re}}\frac{\partial^2 U}{\partial Y^2} \tag{19.10}$$

$$U\frac{\partial \theta}{\partial X} + V\frac{\partial \theta}{\partial Y} = \frac{2}{\mathrm{Re\,Pr}}\frac{\partial^2 \theta}{\partial Y^2} \tag{19.11}$$

where the dimensionless variables are defined as

$$X = \frac{x}{H}, \quad Y = \frac{y}{H}, \quad U = \frac{u}{u_m}, \quad V = \frac{v}{u_m},$$
$$P = \frac{p}{\rho u_m^2}, \quad \theta = \frac{T}{T_I}, \quad \mathrm{Re} = \frac{u_m 2H}{\nu} \tag{19.12}$$

The above problem involves four unknowns: U, V, P, and θ. An additional equation is obtained from a gross mass balance through the cross section of the channel.

The velocity boundary conditions include no slip and impermeability at the walls,

$$U = V = 0 \tag{19.13}$$

a uniform velocity profile at the inlet,

$$U = 1, \qquad V = 0 \tag{19.14}$$

and the inlet fluid temperature, taken as

$$\Theta_i = 1 \tag{19.15}$$

So far the mathematical formulation of the problem is identical to the classical one for nonradiative convective heat transfer. The coupling with radiation arises through the thermal boundary condition at the walls. If q^c and q^r denote the local convective and radiative heat fluxes, we write

$$q^c(\eta) + q^r(\eta) = q_w \qquad \text{at the heat wall} \tag{19.16a}$$

$$q^c(\xi) + q^r(\xi) = 0 \qquad \text{at the insulated wall} \tag{19.16b}$$

where η and ξ denote the axial coordinates measured, respectively, along the heated and insulated walls. In Eqs. (19.16), the convective heat flux q^c is defined by

$$q^c = -k\frac{\partial T}{\partial y}\bigg|_w \tag{19.16c}$$

and the radiation heat flux q^r, assuming *black* wall surfaces, is obtained from Eq. (19.3b). Then the boundary conditions given by Eqs. (19.16a, b) are expressed in dimensionless form as follows.

At the heated surface,

$$\frac{\partial \theta}{\partial Y} = \frac{q_w H}{T_I k} - \frac{\bar{\sigma} T_I^3}{k/H}\left[\Theta^4(\eta) - \int_{\zeta=0}^{l/H} \Theta^4(\zeta)\, dF_{\eta\text{-}\zeta} - F_{\eta\text{-}\mathrm{I}}\Theta_{\mathrm{II}}^4 F_{\eta\text{-}\mathrm{II}}\right] \tag{19.17a}$$

At the insulated surface,

$$-\frac{\partial \theta}{\partial Y} = \frac{\bar{\sigma} T_I^3}{k/H}\left[-\Theta^4(\zeta) + \int_{\eta=0}^{l/H} \Theta^4(\eta)\, dF_{\zeta\text{-}\eta} + F_{\zeta\text{-}\mathrm{I}} + \Theta_{\mathrm{II}}^4 F_{\zeta\text{-}\mathrm{II}}\right] \tag{19.17b}$$

Various differential view factors are identified as discussed by Liu and Sparrow [20]. It is assumed that the inlet and exit plenum temperatures are, respectively, the same as the fluid inlet and the fluid bulk temperature at the exit.

The convective-radiative heat transfer problem defined above is solved by a finite difference procedure. We note that the boundary conditions (19.17a, b) contain two parameters

$$\frac{\bar{\sigma} T_{\mathrm{I}}^3}{k/H} \quad \text{and} \quad \frac{q_w H}{T_{\mathrm{I}} k} = \frac{q_w}{\bar{\sigma} T_{\mathrm{I}}^4}\,\frac{\bar{\sigma} T_{\mathrm{I}}^3}{k/H} \tag{19.18}$$

The first of these represents the ratio of radiation to convection, and the factor $q_w/(\bar{\sigma} T_{\mathrm{I}}^4)$ represents the relative strength of the external heating.

The independent parameters that affect the heat transfer results include Re, Pr, l/H, $\bar{\sigma} T_{\mathrm{I}}^3/(k/H)$, and $q_w/(\bar{\sigma} T_{\mathrm{I}}^4)$. Figure 19.3 shows the effect of radiation on the wall and fluid bulk mean temperature distribution along the channel for the case

$$\mathrm{Pr} = 0.7, \qquad \mathrm{Re} = 2500, \qquad \frac{\bar{\sigma} T_I^3}{k/H} = 0.5, \quad \text{and} \quad \frac{q_w}{\bar{\sigma} T_I^4} = 1$$

where all temperatures are absolute, and where the laminar flow is considered to persist up to Re = 2500 and even farther if the sources of disturbance are not situated within

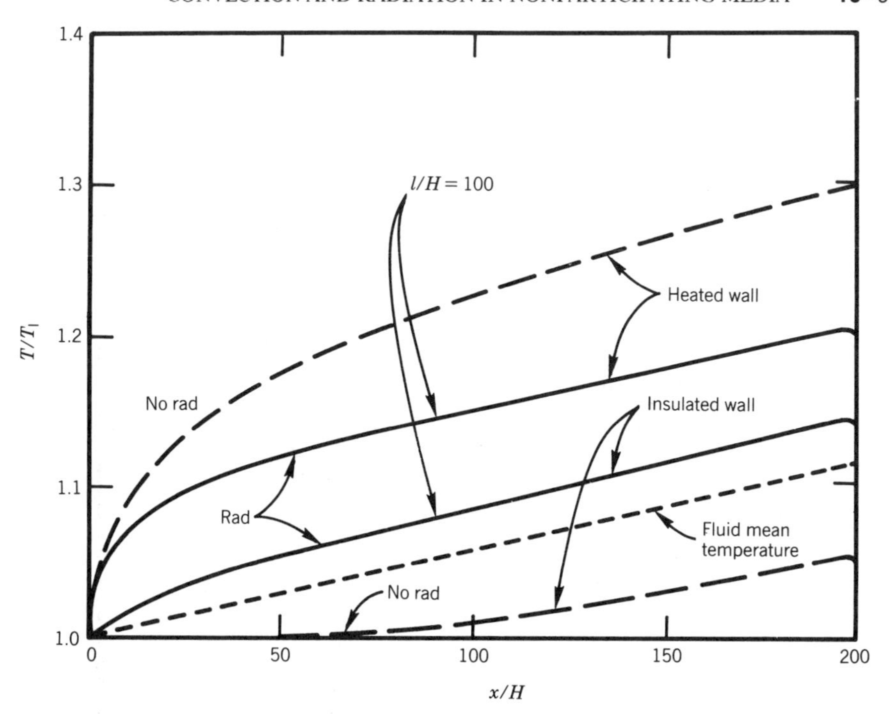

Figure 19.3. Effects of radiation on the wall and fluid bulk mean temperature for laminar flow of a transparent gas inside a parallel-plate channel with one wall at constant heat flux, the other wall adiabatic [20].

the channel. Included in this figure are the corresponding results for pure convection (i.e., neglecting radiation). An examination of this figure reveals that there is a significant difference between the results with and without radiation. The temperature of the unheated wall is significantly elevated by the effect of radiation. On the other hand, the bulk temperature lines are coincident within the scale of the figure for the with-radiation and without-radiation cases. This implies that, while the radiation process is effective in transferring heat from the heated to the insulated wall across the channel, it does not transport a significant amount of energy along the channel.

Figure 19.4 shows the effect of radiation on the wall and fluid bulk mean temperature distribution at two different Reynolds numbers. Decreasing the Reynolds number increases the wall temperature for both the heated and unheated wall. These results can be rationalized by noting that the lower the Reynolds number, the lower the convective heat transfer coefficient, and so the greater the relative importance of radiative transfer.

Figure 19.5 shows the effect of radiation on the local Nusselt number. The curve for pure convection is also included in this figure. At the region near the inlet where the boundary layer is developing, the convective heat transfer coefficient is very high; hence the importance of radiation relative to convection is insignificant. Therefore, at the inlet region the Nusselt numbers with and without radiation do not differ appreciably. At the downstream region where the convective heat transfer coefficient is low, substantial differences occur in the Nusselt number for the cases with and without radiation. In the downstream region, the Nusselt number with radiation is about 25% higher than that without radiation.

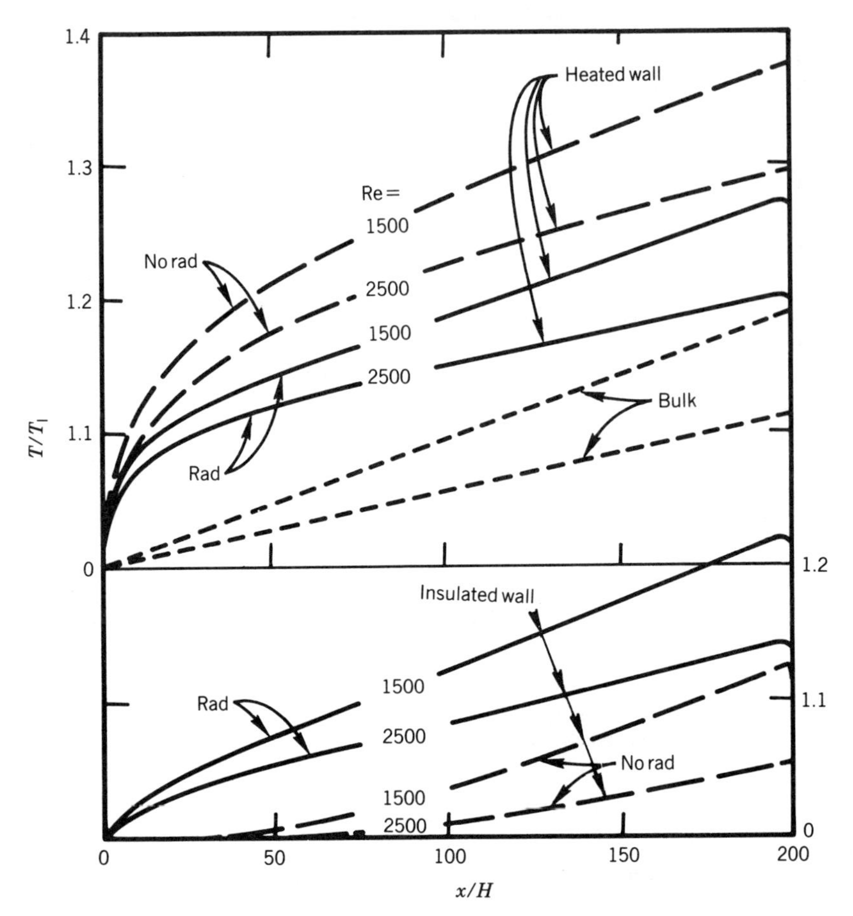

Figure 19.4. Effects of radiation on the wall and fluid bulk mean temperature for laminar flow of transparent gas inside a parallel-plate channel at different Reynolds numbers (one wall at constant heat flux, the other wall adiabatic) [20].

19.3 SIMULTANEOUS CONVECTION AND RADIATION IN PARTICIPATING MEDIA

The analysis of combined convection and radiation in participating media is significantly different from and more involved than that for a nonparticipating medium. Therefore, before introducing the subject of simultaneous convection and radiation in an absorbing, emitting, and scattering medium, we present an overview of radiation transfer in participating media, including the equation of radiative transfer, the radiation boundary conditions, and a discussion of exact and approximate methods of solution of the pure radiation problem.

19.3.1 Equation of Radiative Transfer

A fundamental quantity in the study of radiation transfer in an absorbing, emitting, and scattering media is the *spectral radiation intensity* $I_\nu(\boldsymbol{r}, \hat{\Omega})$, which represents the

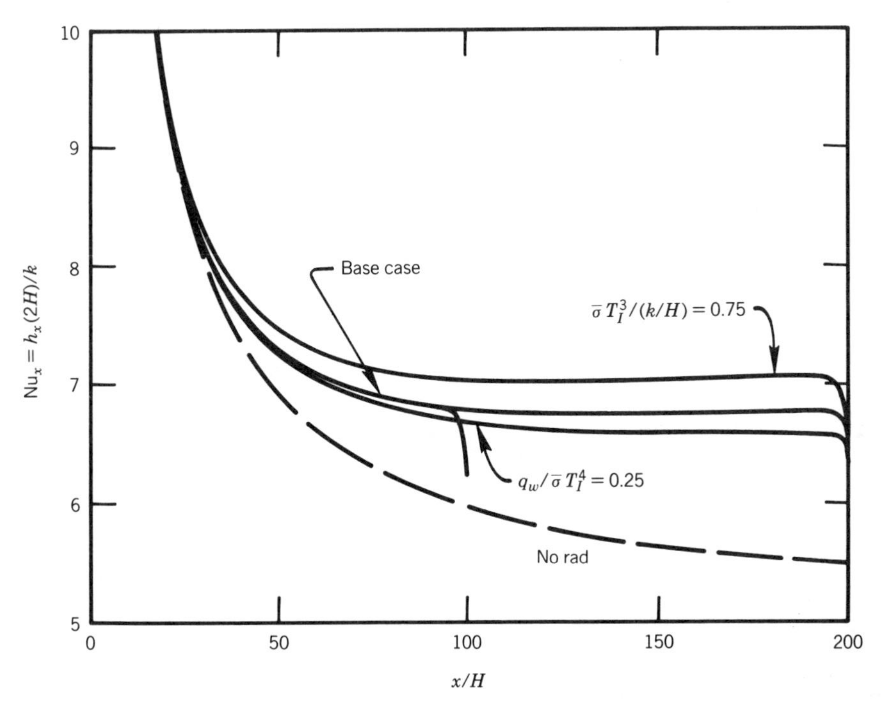

Figure 19.5. Effects of radiation on the local Nusselt number for laminar flow of a transparent gas inside a parallel-plate channel; upper plate heated, lower plate insulated [20].

amount of radiation energy streaming through a unit area perpendicular to the direction of propagation $\hat{\Omega}$, per unit solid angle about the direction $\hat{\Omega}$, per unit frequency about frequency ν, and per unit time about time t. In many engineering applications, the wavelength λ is used to characterize the spectral radiation intensity as $I_\lambda(\boldsymbol{r}, \hat{\Omega})$. The integration of the spectral intensity over all frequencies (or wavelengths) from zero to infinity yields the *radiation intensity* $I(\boldsymbol{r}, \hat{\Omega})$.

The propagation of radiation in a participating medium is governed by the *equation of radiative transfer*, the derivation of which is given in several references [1, 2, 22–24]. The absorption and scattering properties of a medium are characterized by the *absorption coefficient* κ and the *scattering coefficient* σ, both of which have dimensions of $(\text{length})^{-1}$. The sum of the absorption and scattering coefficients is called the *extinction coefficient* β (i.e., $\beta = \kappa + \sigma$). In general, these coefficients depend on the frequency or wavelength of the radiation.

We consider the propagation of radiation beam in any given direction $\hat{\Omega}$, and let s be the coordinate measured along this direction of propagation. Assuming that the radiation properties of the medium are independent of frequency, the *directional form of the equation of radiative transfer* governing the distribution of the radiation intensity $I(s, \hat{\Omega})$ is given by

$$\frac{1}{\beta}\frac{dI(s, \hat{\Omega})}{ds} + I(s, \hat{\Omega}) = (1 - \omega)\frac{n^2\bar{\sigma}T^4}{\pi} + \frac{\omega}{4\pi}\int p(\hat{\Omega}', \hat{\Omega})\, I(s, \hat{\Omega}')\, d\hat{\Omega}', \quad (19.19)$$

where

$$n = \text{refractive index}$$

$$p(\hat{\Omega}', \hat{\Omega}) = \text{phase function}$$

$$\omega = \frac{\sigma}{\beta} = \text{single-scattering albedo}$$

$$\beta = \kappa + \sigma = \text{extinction coefficient}$$

$$\bar{\sigma} = \text{Stefan-Boltzmann constant}$$

and d/ds denotes the derivative along s in the direction of propagation $\hat{\Omega}$. The phase function $p(\hat{\Omega}', \hat{\Omega})$ represents the probability that incident radiation from the direction $\hat{\Omega}'$ will be scattered into the direction $\hat{\Omega}$.

To apply Eq. (19.19) in a particular coordinate system, the directional derivative d/ds should be expressed in terms of the derivatives of the space coordinates in that particular coordinate system. The solution of the equation of radiative transfer [Eq. (19.19)] in its general form in a three-dimensional system is an extremely difficult matter.

We consider below only the one-dimensional form of Eq. (19.19) for a *plane-parallel medium*. Let Oy be the preferred coordinate axis, and θ the polar angle between the direction of propagation $\hat{\Omega}$ and the positive Oy axis as shown in Fig. 19.6. By plane-parallel we mean that the medium is stratified in planes perpendicular to the Oy axis. In addition, we assume azimuthally symmetric radiation, so that the radiation intensity is a function of the variables y and θ. For convenience in the analysis, we define

$$\mu \equiv \cos\theta \tag{19.20a}$$

$$\tau \equiv \beta y = \text{optical variable} \tag{19.20b}$$

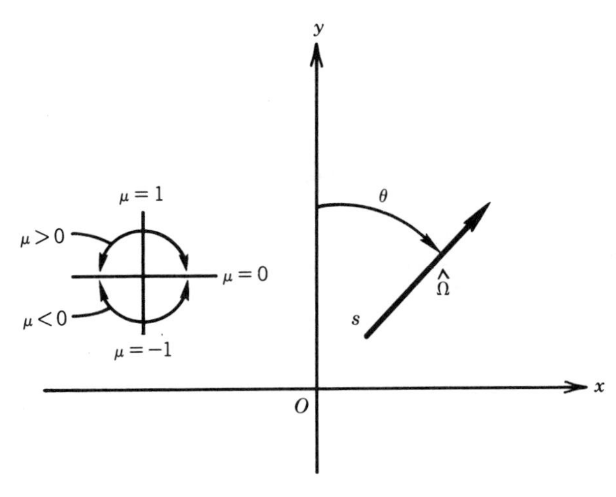

Figure 19.6. Coordinates for plane-parallel system.

Then the equation of radiation transfer, Eq. (19.19), reduces to (Özışık [2, p. 261]).

$$\mu \frac{\partial I(\tau, \mu)}{\partial \tau} + I(\tau, \mu) = (1 - \omega) \frac{n^2 \bar{\sigma} T^4(\tau)}{\pi} + \frac{\omega}{2} \int_{-1}^{1} p(\mu, \mu') I(\tau, \mu') \, d\mu' \quad (19.21a)$$

where the maximum range of the μ variable is between -1 and $+1$, and the phase function $p(\mu, \mu')$ is generally expressed in terms of the Legendre polynomials $P_n(\mu)$ as

$$p(\mu, \mu') = \sum_{n=0}^{N} a_n P_n(\mu) P_n(\mu') \qquad \text{with} \quad a_0 = 1 \quad (19.21b)$$

Some special cases of Eq. (19.21a) include:

1. *Isotropic scattering medium*. We set $p(\mu, \mu') = 1$, and Eq. (19.21a) simplifies to

$$\mu \frac{\partial I(\tau, \mu)}{\partial \tau} + I(\tau, \mu) = (1 - \omega) \frac{n^2 \bar{\sigma} T^4(\tau)}{\pi} + \frac{\omega}{2} \int_{-1}^{1} I(\tau, \mu') \, d\mu' \quad (19.22)$$

2. *Purely scattering medium*. We set $\omega = 1$, and Eq. (19.21a) takes the form

$$\mu \frac{\partial I(\tau, \mu)}{\partial \tau} + I(\tau, \mu) = \frac{1}{2} \int_{-1}^{1} I(\tau, \mu') \, d\mu' \quad (19.23)$$

which implies that there is no coupling between the equation of radiative transfer and no emission of radiation by the medium.

3. *Purely absorbing and emitting medium*. We set $\omega = 0$, and Eq. (18.21a) becomes

$$\mu \frac{\partial I(\tau, \mu)}{\partial \tau} + I(\tau, \mu) = \frac{n^2 \bar{\sigma} T^4(\tau)}{\pi} \quad (19.24)$$

We note that for the nonscattering medium, the equation of radiative transfer is a first degree ordinary differential equation and its solution is a relatively simple matter. For a scattering medium, the equation of radiative transfer is a singular integrodifferential equation, and its rigorous solution, even for the one-dimensional isotropically scattering case, is rather a complicated matter.

19.3.2 Radiation Boundary Conditions

To solve the equation of radiative transfer, appropriate radiation boundary conditions are needed. Such boundary conditions being different from the customary heat transfer boundary conditions associated with conduction and convection, we present a brief discussion of basic concepts leading to the construction of radiation boundary conditions.

We consider a plane-parallel slab of optical thickness τ_0, with τ and μ the space and angular coordinates, respectively, as illustrated in Fig. 19.7. Several different possibilities can be envisioned as boundary conditions at the surfaces $\tau = 0$ and $\tau = \tau_0$ as discussed below.

1. *Transparent boundaries*. Suppose the boundaries at $\tau = 0$ and $\tau = \tau_0$ of the slab shown in Fig. 19.7 are both transparent and subjected to externally incident

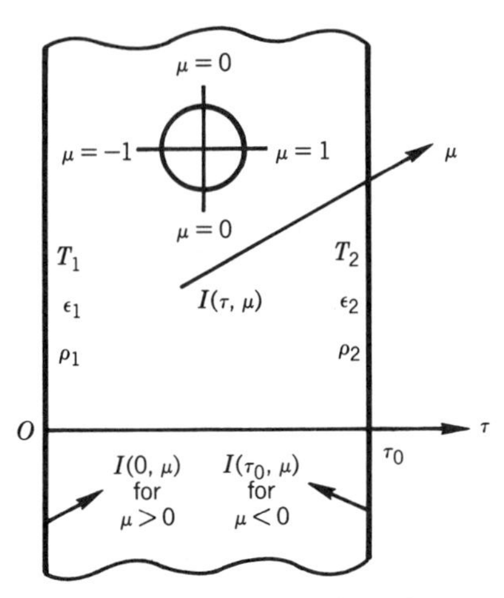

Figure 19.7. Radiation boundary conditions for a plane-parallel slab.

diffuse radiations of constant intensities f_1 and f_2, respectively. Then the boundary conditions are given by

$$I(0) = f_1 \quad \text{at } \tau = 0 \qquad \text{for } \mu > 0 \tag{19.25a}$$

$$I(\tau_0) = f_2 \quad \text{at } \tau = \tau_0 \qquad \text{for } \mu < 0 \tag{19.25b}$$

2. *Opaque, black boundaries.* Suppose the boundary surfaces at $\tau = 0$ and $\tau = \tau_0$ are opaque and black and maintained at constant temperatures T_1 and T_2, respectively. The medium has a refractive index n. The boundary conditions become

$$I(0) = \frac{n^2 \bar{\sigma} T_1^4}{\pi} \quad \text{at } \tau = 0 \qquad \text{for } \mu > 0 \tag{19.26a}$$

$$I(\tau_0) = \frac{n^2 \bar{\sigma} T_2^4}{\pi} \quad \text{at } \tau = \tau_0 \qquad \text{for } \mu < 0 \tag{19.26b}$$

3. *Diffusely reflecting, diffusely emitting boundaries.* Let the boundary surfaces at $\tau = 0$ and $\tau = \tau_0$ be opaque, diffusely emitting, diffusely reflecting surfaces, maintained at temperatures T_1 and T_2, respectively. Let ρ_1^d and ρ_2^d be the diffuse reflectivity and ϵ_1 and ϵ_2 the diffuse emissivity of the surfaces at $\tau = 0$ and $\tau = \tau_0$ respectively, and n the refractive index of the intervening medium. The boundary conditions are given by [2, p. 274]

$$I(0) = \epsilon_1 \frac{n^2 \bar{\sigma} T_1^4}{\pi} + 2\rho_1^d \int_0^1 I(0, -\mu')\mu' \, d\mu' \quad \text{at } \tau = 0 \qquad \text{for } \mu > 0 \tag{19.27a}$$

$$I(\tau_0) = \epsilon_2 \frac{n^2 \bar{\sigma} T_2^4}{\pi} + 2\rho_2^d \int_0^1 I(\tau_0, \mu')\mu' \, d\mu' \quad \text{at } \tau = \tau_0 \qquad \text{for } \mu < 0 \tag{19.27b}$$

Here the first terms on the right-hand side of Eqs. (19.27) represent the radiation emitted by the surface because of its temperature, and the second terms represent the radiation diffusely reflected from the surface.

4. *Specularly reflecting, diffusely emitting boundaries.* Let the boundary surfaces at $\tau = 0$ and $\tau = \tau_0$ be opaque, be a diffuse emitter and a specular reflector, have emissivities ϵ_1 and ϵ_2, have specular reflectivities ρ_1^s and ρ_2^s, and be maintained at temperatures T_1 and T_2 respectively. The boundary conditions become (Özışık [2, p. 275])

$$I(0, \mu) = \epsilon_1 \frac{n^2 \bar{\sigma} T_1^4}{\pi} + \rho_1^s I(0, -\mu) \quad \text{at } \tau = 0 \qquad \text{for } \mu > 0 \quad (19.27c)$$

$$I(\tau_0, \mu) = \epsilon_2 \frac{n^2 \bar{\sigma} T_2^4}{\pi} + \rho_2^s I(\tau_0, -\mu) \quad \text{at } \tau = \tau_0 \qquad \text{for } \mu < 0 \quad (19.27d)$$

Clearly, several different combinations of the above boundary conditions are possible for a radiation problem in an absorbing, emitting, and scattering slab.

19.3.3 Methods of Solving Radiation Problems in One-Dimensional Media

The mathematical techniques used in the solution of radiation problems in participating media being rather different from those commonly applied for the solution of convection problems, we present here a brief discussion of this subject in order to introduce the pertinent references and give some background information.

The methods of solving radiation transfer in participating media can be separated into three distinct categories: (1) approximate, (2) analytical, and (3) purely numerical. In each of these groups, many different techniques have been developed.

1. *Approximate methods.* The approximate methods include, among others, the Eddington approximation, exponential kernel approximation, multiflux method, low-order moment method, and low-order P_N method. Some of these approximate methods are capable of producing sufficiently accurate heat flux results under certain conditions; but in general their accuracy cannot be assessed unless they are compared with the exact solutions. A discussion of various approximate methods of analysis of radiation transfer is given in Ref. 2. However, with the recent developments in the analytical methods of calculating radiation transfer in plane-parallel media, now one can readily use an efficient analytical method just as easily.
2. *Analytical methods.* In recent years, significant advances have been made in the development of efficient analytical methods for calculating radiation transfer in absorbing, emitting, scattering plane-parallel media. Starting with Case's singular-eigenfunction technique, developed in early sixties and discussed in the Refs. 2, 25, various other approaches that followed included, among others, the Fourier-transform technique [26, 27], the source-function expansion technique [28, 29], the F_N method [30–33], and the Galerkin method [34–37]. Any of these methods is capable of producing highly accurate results for the radiation intensity and radiation flux; but for their use in the analysis of interaction problems, the efficiency and simplicity of the method as well as its compatibility with the solution of convection are important. It appears that both the F_N and

the Galerkin method are highly efficient and require very little computer time. They are also compatible with the solution of the convection problem, because radiation enters the energy equation as a source term, which can readily be calculated from the solution of the radiation problem for a specified temperature distribution in the medium. Of course, iteration is needed for the complete solution. For engineering applications, the use of the Galerkin method has the additional advantage of simplicity in the analysis.

3. *Numerical methods*. Purely numerical methods include the Monte Carlo methods [3, 38, 39, 40], the use of Gaussian quadratures [41, 42], the iterative method [43–45], the finite element method [46], and others. Such numerical methods can produce sufficiently accurate results for engineering applications, but they require large amounts of computer time and storage. Therefore, they are computationally inefficient for use in the solution of interaction problems in comparison with the analytical methods, because the complete solution requires several iterations between solutions of the convection and radiation parts of the problem.

Nonplanar Media. The solution of the equations of radiative transfer in cylindrical symmetry is of interest in the analysis of the interaction of radiation with convection for the flow of a participating fluid inside a circular tube. Some of the literature on radiation transfer in cylindrical symmetry for engineering applications has been cited in Ref. 47. Most solutions are approximate including the variational method [48], the two-flux method [49], and the differential approximation [50, 51].

19.3.4 Methods of Solving Radiation Problems in Multidimensional Media

A review of radiation transfer in multidimensional participating media by Crosbie and Linsenbart [52] reveals that the exact analysis of the subject is very limited because of the difficulties associated with the solution of the multidimensional equations of radiative transfer. Purely numerical techniques such as the Monte Carlo methods have been used to solve such problems; but they require so much computer time and memory that they are not well suited for the analysis of interaction problems. Therefore, efforts have been directed toward the development of approximate methods of analysis. For example, the P_1 and P_3 approximations have been used by Ratzel and Howell [53] to calculate radiation transfer in an isotropically scattering two-dimensional rectangular enclosure, and by Mengüc and Viskanta [54] in three-dimensional cylindrical enclosures. The discrete-ordinate method has been used by Fiveland [55] to treat two-dimensional radiation in an axisymmetric participating medium, and the source-function expansion technique has been applied by Sutton and Özışık [56] to treat radiation in an isotropically scattering two-dimensional rectangular enclosure.

An examination of these solutions reveals that approximate methods may produce acceptable results over limited ranges, but their accuracy can not be assessed unless they are compared with the exact solutions. The exact methods are rather elaborate and require much computer time.

19.3.5 Equations of Motion and Energy for Radiating Fluids

Here we present an overview of the effects of radiation on the equations of motion and energy for a participating fluid. The reader should consult Sparrow and Cess [1] and Özışık [2] for a detailed discussion of this subject.

Continuity Equation. The standard equation of continuity for a nonradiating fluid is applicable also for a radiating fluid, since the change of mass due to radiation is negligible.

Momentum Equation. When radiation is present, it exerts pressure and hence introduces an additional rate of change of momentum in the system. Hence, one envisions a radiation stress tensor analogous to the stress tensor in fluid dynamics. However, it can be shown that unless the temperature is extremely high and pressure is extremely low, the momentum equation for a radiating fluid is the same as that for a nonradiating fluid.

Energy Equation. The presence of radiation in the fluid leads to additional terms in the energy equation to account for the effects of the radiation energy density, the radiation stresses, and the radiation heat flux. However, the contribution of the radiation energy density and the radiation stress tensor in the energy equation is negligible for most practical situations encountered in engineering applications. Only the contribution of the radiation flux is important, and the energy equation for a radiating fluid can therefore be given as (Özısık [2])

$$\rho c_p \frac{DT}{Dt} = \nabla \cdot (k \nabla T) - \nabla \cdot \mathbf{q}^r + S^* + \frac{Dp}{Dt} + \tilde{\mu}\Phi \tag{19.28}$$

where $\mathbf{q}^r$ is the radiation flux vector, which should be obtained from the solution of the radiation part of the problem. The viscous energy dissipation Φ depends on the nature of the fluid; for a Newtonian fluid, it is the same as that given in the standard texts for a nonradiating fluid. Finally, S^* is an external energy input per unit volume and per unit time, p is the pressure, $\tilde{\mu}$ is the viscosity, and D/Dt is the substantial derivative.

Thus the convection and radiation parts of the problems are coupled, because the energy equation involves the divergence of the radiation flux $\nabla \cdot \mathbf{q}^r$, whereas the determination of the radiation flux $\mathbf{q}^r$ requires the solution of the equation of radiative transfer, Eq. (19.19), which involves the fourth power of the medium temperature.

Boundary-Layer Simplifications. For a boundary-layer flow of a participating fluid, the equations of motion and energy discussed above are simplified by the application of the standard order-of-magnitude analysis. Here we focus attention on two-dimensional, steady, laminar boundary-layer flow of a compressible radiating fluid over a body. Let x and y be the coordinate axes along the surface in the direction of flow and perpendicular to the surface, respectively. The boundary-layer equations are given by

continuity,
$$\frac{\partial}{\partial x}(\rho u) + \frac{\partial}{\partial y}(\rho v) = 0 \tag{19.29}$$

x momentum
$$\rho\left(u\frac{\partial u}{\partial x} + v\frac{\partial u}{\partial y}\right) = -\frac{dp}{dx} + \frac{\partial}{\partial y}\left(\tilde{\mu}\frac{\partial u}{\partial y}\right) \tag{19.30}$$

energy,
$$\rho c_p\left(u\frac{\partial T}{\partial x} + v\frac{\partial T}{\partial y}\right) = \frac{\partial}{\partial y}\left(k\frac{\partial T}{\partial y}\right) - \frac{\partial q^r}{\partial y} + u\frac{dp}{dx} + \tilde{\mu}\left(\frac{\partial u}{\partial y}\right)^2 \tag{19.31}$$

Here we neglected the conduction and radiation in the axial direction, and included the viscous energy dissipation in the energy equation.

Assuming a perfect gas, the equation of state is given by

$$p = \rho R T \tag{19.32a}$$

The viscosity and thermal conductivity are considered as prescribed functions of temperature

$$\tilde{\mu} = \tilde{\mu}(T) \quad \text{and} \quad k = k(T) \tag{19.32b}$$

and the pressure gradient term dp/dx is related to the external flow velocity $u_\infty(x)$ by

$$\frac{dp}{dx} = -\rho_\infty(x)\, u_\infty(x) \frac{du_\infty(x)}{dx} \tag{19.32c}$$

where $u_\infty(x)$ is determined from the solution of the potential flow problem.

An examination of the above system of equations reveals that the energy equation for a radiating fluid contains the divergence of the radiation flux term. The presence of this complicates the analysis significantly, because q^r or $\partial q^r/\partial y$ must be obtained from the solution of the equation of radiative transfer subject to appropriate radiation boundary conditions.

Nondimensional Parameters. When the equations of motion and energy are expressed in dimensionless form by following the standard procedures, the following well-known nondimensional parameters result:

$$E = \frac{u_0^2}{c_{p,0} T_0} = \text{Eckert number} \tag{19.33a}$$

$$\text{Pr} = \frac{c_{p,0}\mu_0}{k_0} = \text{Prandtl number} \tag{19.33b}$$

$$\text{Re} = \frac{\rho_0 u_0 L}{\mu_0} = \text{Reynolds number} \tag{19.33c}$$

and the presence of the radiation term in the energy equation gives rise to new dimensionless group

$$\text{Bo} = \frac{\rho_0 u_0 c_{p,0}}{n_0^2 \bar{\sigma} T_0^3} = \text{Boltzmann number} \tag{19.33d}$$

The Boltzmann number as defined here represents the ratio of the convective to radiative heat flux. The smaller the Boltzmann number, the larger the radiation effects.

19.3.6 Forced Convection with Radiation over a Flat Plate

Here we consider laminar boundary-layer flow of an absorbing, emitting, compressible, gray gas over a flat plate as illustrated in Fig. 19.8. It is assumed that the fluid is a perfect gas. The viscosity varies linearly with temperature, the specific heat and Prandtl number are constant, and the external flow temperature T_∞ and velocity u_∞ are uniform. The plate surface is opaque, gray, a diffuse emitter and a diffuse reflector, impervious to flow and maintained under adiabatic conditions. The flow is at high speed and gives rise to a temperature increase in the boundary layer.

The problem considered here (simultaneous convection and radiation) was studied by Taitel and Hartnett [57] for an absorbing, emitting fluid, and later on the effects of scattering of radiation were included in the analysis by Boles and Özışık [58].

The mathematical formulation of the convection part of this problem is defined by Eqs. (19.29) to (19.31); here the pressure gradient term vanishes for flow over a flat plate. These equations are transformed by the similarity transformation as discussed by Özışık [2, Chap. 13]. Then the continuity equation is identically satisfied, and the momentum and energy equations are transformed to

$$\frac{d^3f}{d\eta^3} + \tfrac{1}{2}f\frac{d^2f}{d\eta^2} = 0 \tag{19.34}$$

$$\frac{1}{\Pr}\frac{\partial^2\theta}{\partial\eta^2} + \tfrac{1}{2}f\frac{\partial\theta}{\partial\eta} = \frac{df}{d\eta}\xi^*\frac{\partial\theta}{\partial\xi^*} + \xi^*\frac{\partial Q^r}{\partial\tau} - E_\infty\left(\frac{d^2f}{d\eta^2}\right)^2 \tag{19.35}$$

where

$$\theta \equiv \theta(\xi^*, \eta) = \frac{T}{T_\infty}$$

$$E_\infty = \frac{u_\infty^2}{c_{p,\infty}T_\infty} = \text{Eckert number}$$

$$\xi^* = \frac{4n^2\bar{\sigma}T_\infty^3\kappa x}{\rho_\infty u_\infty c_{p,\infty}} .$$

$$\tau = \kappa x = \text{optical variable}$$

$$\kappa = \text{absorption coefficient}$$

$$\eta = y\left(\frac{u_\infty}{\nu x}\right)^{1/2} \tag{19.36}$$

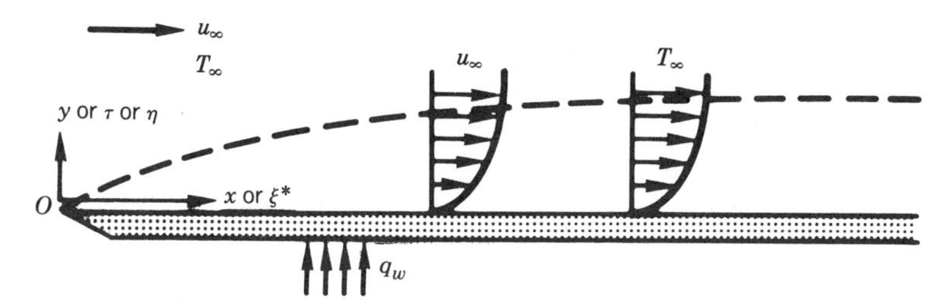

Figure 19.8. Boundary flow of radiating gas over a flat plate.

Note that the parameter ξ^* is inversely proportional to the Boltzmann number; hence, the larger ξ^*, the stronger the radiation effects. For the nonscattering medium considered here, the extinction coefficient β is replaced by the absorption coefficient κ in the definitions of τ and ξ^*.

The boundary conditions for the momentum equation (18.34) are given by

$$f = \frac{df}{d\eta} = 0 \quad \text{at} \quad \eta = 0 \tag{19.37a}$$

$$\frac{df}{d\eta} = 1 \quad \text{at} \quad \eta \to \infty \tag{19.37b}$$

which imply that the velocity components are zero at the wall and $u = u_\infty$ outside the boundary layer.

The boundary conditions for the energy equation (19.35) are taken as:

1. The sum of the conduction and radiation fluxes is zero at the wall surface.
2. The temperature is equal to the external flow temperature T_∞ outside the thermal boundary layer.
3. For $\xi^* = 0$, the solution of the energy equation is equal to the solution of the same problem for the nonradiating flow.

These requirements lead respectively to the following expressions for boundary condition:

$$-\left(\frac{N_1}{\xi^* \mathrm{Pr}}\right)^{1/2} \frac{\partial \theta}{\partial \eta} + Q^R = 0 \qquad \text{at} \quad \eta = 0 \tag{19.38a}$$

$$\theta = 1 \qquad \text{as} \quad \eta \to \infty \tag{19.38b}$$

$$\theta = \theta_0(\eta) \qquad \text{at} \quad \xi^* = 0 \tag{19.38c}$$

where

$$N_1 = \frac{k_\infty \kappa_\infty}{4n^2 \bar{\sigma} T_\infty^3} = \text{conduction-to-radiation parameter} \tag{19.39a}$$

$$Q^r = \frac{q^r}{4n^2 \bar{\sigma} T_\infty^4} = \text{dimensionless radiation heat flux} \tag{19.39b}$$

and the radiation flux q^r is related to the radiation intensity $I(\tau, \xi^*, \mu)$ by

$$q^r = 2\pi \int_{-1}^{1} I(\tau, \xi^*, \mu)\, \mu\, d\mu \tag{19.40}$$

The radiation intensity is determined from the solution of the radiation part of the problem. For an absorbing, emitting, gray medium with diffusely emitting, diffusely reflecting boundary at $\tau = 0$, the radiation part of the problem is given by

$$\mu \frac{\partial I(\tau, \xi^*, \mu)}{\partial \tau} + I(\tau, \xi^*, \mu) = \frac{n^2 \bar{\sigma} T^4(\tau, \xi^*)}{\pi} \qquad \text{for} \quad 0 < \tau < \infty, \quad -1 \le \mu \le 1 \tag{19.41a}$$

with

$$I(0,\xi^*) = \epsilon_w \frac{n^2\bar{\sigma}T_w^4(\xi^*)}{\pi} + 2(1-\epsilon_w)\int_0^1 I(0,\xi^*,-\mu^*)\,d\mu' \qquad \text{for } \mu > 0 \quad (19.41b)$$

and $I(\tau,\mu)$ remains finite as $\tau \to \infty$. Here ϵ_w is the emissivity of the wall surface and n is the refractive index of the fluid.

Clearly, the above convection and radiation problems are coupled, because the solution of the convection problem requires a knowledge of the radiation flux, whereas the radiation part of the problem involves the fourth power of the temperature in the fluid; therefore an iterative scheme is needed for the solution of the problem.

Figure 19.9 shows the temperature distribution in the boundary layer as a function of the dimensionless distance η from the surface at several values of the dimensionless axial location ξ^* for the values Pr = 1, $E_\infty = 2$, $N_1 = 0.5$, and $\epsilon_w = 1$ (i.e., a black wall). The temperature profile for $\xi^* = 0$ corresponds to nonradiating fluid. The maximum temperature in the boundary layer is highest with the nonradiating fluid. The effect of radiation within the boundary layer is to flatten the temperature profile and hence reduce the maximum temperature. Thus with increasing ξ^*, radiation effects become more dominant, which in turn flattens the temperature profile and reduces the peak. The *optically thin* and the *optically thick* limits are sometimes used to simplify the radiation part of the analysis; but the results are applicable only over limited ranges. The solutions obtained by using the optically thin and optically thick approximations are also included in this figure. Note that, for values of ξ^* of the order of 10^{-4} or less, the results obtained with the optically thin boundary-layer approximation

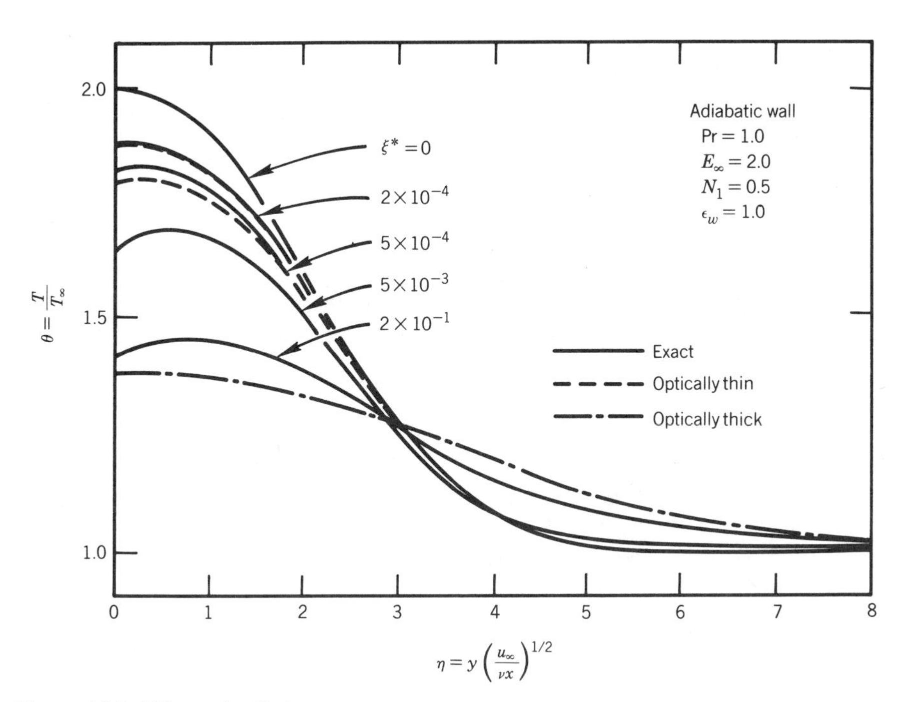

Figure 19.9. Effects of radiation on temperature profile in boundary-layer flow of a radiating gas over an adiabatic flat plate for an absorbing, emitting fluid. (From Taitel and Hartnett [57].)

seems to agree reasonably well with the exact results. For values of ξ^* of the order of unity or larger, the optically thick approximation is close to the exact results, but the slopes of the temperatures at the wall differ significantly.

With increasing radiation, a positive temperature gradient with respect to η is established at the wall, controlled by a balance between the convective heat flux to the wall and the radiative heat flux away from it.

The effects of scattering of radiation in addition to the absorption and emission by the fluid have been investigated by Boles and Özışık [58] for the problem discussed above. The scattering of radiation by the fluid reduces the effects of radiation in flattening the temperature profile in the boundary layer.

19.3.7 Forced Convection with Radiation inside Ducts

The analysis of simultaneous radiation and forced convection inside ducts is rather involved, because the energy problem for forced convection is coupled to the radiation problem. An examination of the literature on this subject reveals that much of the available work is concerned with thermally fully developed situations [59–65], or restricted to an absorbing and emitting flow [66–69]. When scattering effects are to be included, an approximate technique is employed to solve the radiation part of the problem [70–73]. More recently, heat transfer in thermally developing laminar flow has been studied [74–78] by including the scattering effects in the analysis and treating the radiation part of the problem exactly.

Of the available investigations, only a few are on the interaction of radiation and forced convection in turbulent flow inside parallel-plate channels [62–65, 72, 73, 79]. The interaction of radiation with convection for hydrodynamically fully developed turbulent and laminar flow of an absorbing and emitting gas through a black-wall circular duct with a prescribed inlet gas temperature and wall temperature or heat flux distribution has been studied by Smith and Shen [80].

We present below some of the heat transfer results under both laminar and turbulent flow conditions for the flow of an absorbing, emitting, isotropically scattering gray fluid inside a parallel-plate channel, in order to illustrate the effects of radiation on the Nusselt number.

Laminar Flow. Consider thermally developing, hydrodynamically developed steady flow of an absorbing, emitting, isotropically scattering gray fluid inside a parallel-plate channel of spacing $2L$ as illustrated in Fig. 19.10. We idealize constant thermophysical properties, and we neglect viscous dissipation and the axial conduction and radiation in

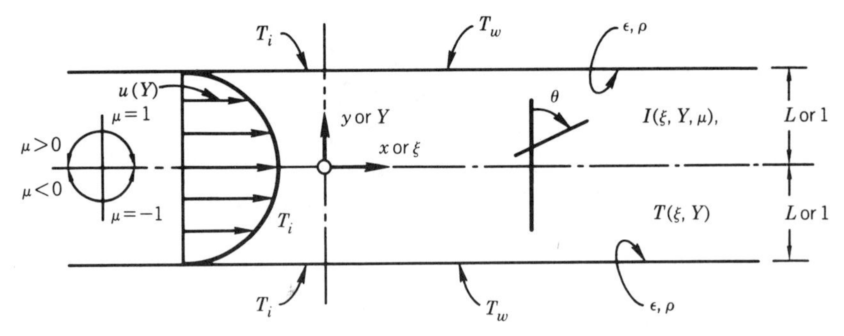

Figure 19.10. Geometry and coordinates for flow inside a parallel-plate channel.

the fluid. The mathematical formulation of the convection part of the problem is given in the dimensionless form as

$$(1 - Y^2)\frac{\partial \theta}{\partial \xi} = \frac{\partial^2 \theta}{\partial Y^2} - \frac{\tau_0}{N_2}\frac{\partial Q^r}{\partial Y} \quad \text{in} \quad 0 < Y < 1,\ \xi > 0 \tag{19.42a}$$

$$\theta(\xi, Y) = \theta_i \quad \text{at} \quad \xi = 0 \tag{19.42b}$$

$$\frac{\partial \theta(\xi, Y)}{\partial Y} = 0 \quad \text{at} \quad Y = 0 \tag{19.42c}$$

$$\theta(\xi, Y) = 1 \quad \text{at} \quad Y = 1 \tag{19.42d}$$

where the various dimensionless quantities are

$$\theta(\xi, Y) = \frac{T(\xi, y)}{T_w}, \qquad \theta_i = \frac{T_i}{T_w}, \qquad Y = \frac{y}{L}, \qquad \xi = \frac{32}{3}\frac{x/D_h}{\text{Re}\,\text{Pr}},$$

$$D_h = 4L, \qquad \text{Re} = \frac{u_m D_h}{\nu}, \qquad \text{Pr} = \frac{\nu}{\alpha}, \qquad Q^r = \frac{q^r(\xi, y)}{4n^2\bar{\sigma}T_w^4},$$

$$\tau_0 = \beta L, \qquad \beta = \kappa + \sigma, \qquad N_2 = \frac{k\beta}{4n^2\bar{\sigma}T_w^3} \tag{19.43}$$

The dimensionless radiation flux $Q^r(\xi, Y)$ is related to the dimensionless radiation intensity $\psi(\xi, Y, \mu)$ by

$$Q^r(\xi, Y) = \frac{1}{2}\int_{-1}^{1} \psi(\xi, Y, \mu')\mu'\, d\mu' \tag{19.44}$$

where

$$\psi(\xi, Y, \mu) = \frac{I(\xi, Y, \mu)}{n^2\bar{\sigma}T_w^4/\pi} \tag{19.45}$$

$I(\xi, Y, \mu)$ is the actual radiation intensity, and μ is the cosine of the angle between the radiation beam and the positive Y axis.

Here $\psi(\xi, Y, \mu)$ should be obtained from the solution of the radiation part of the problem. For an absorbing, emitting, isotropically scattering gray fluid treated as a plane-parallel medium stratified in planes perpendicular to the Y coordinate axis, $\psi(\xi, Y, \mu)$ satisfies the equation of radiative transfer given in the form

$$\frac{\mu}{\tau_0}\frac{\partial \psi}{\partial Y} + \psi(\xi, Y, \mu) = (1 - \omega)\theta^4(\xi, Y) + \frac{\omega}{2}\int_{-1}^{1} \psi(\xi, Y, \mu')\, d\mu'$$

$$\text{in} \quad 0 < \tau < \tau_0, \quad -1 \le \mu \le 1 \tag{19.46a}$$

The boundary conditions for this equation are the symmetry condition at $Y = 0$ and a diffusely emitting, diffusely reflecting, gray, opaque boundary at $Y = 1$, given by

$$\psi(\xi, 0, \mu) = \psi(\xi, 0, -\mu) \qquad \text{at} \quad Y = 0, \quad 0 < \mu < 1 \quad (19.46b)$$

$$\psi(\xi, 1, -\mu) = \epsilon_w + 2\rho^d \int_0^d \psi(\xi, 1, \mu')\mu' \, d\mu' \qquad \text{at} \quad Y = 1, \quad 0 < \mu < 1 \quad (19.46c)$$

where ϵ_w and ρ^d are the emissivity and diffuse reflectivity of the wall.

In this problem, the parameters that affect the temperature distribution in the flow, and hence the heat transfer, include the conduction-to-radiation parameter N_2; the optical thickness of the fluid, τ_0; the emissivity of the boundary surfaces, ϵ_1; and the single-scattering albedo ω.

Figure 19.11 shows the effects of the conduction-to-radiation parameter on the temperature profile at two different axial locations $\xi = 0.1$ and $\xi = 0.5$ along the channel. Note that for $N_2 \geq 5$, the temperature profiles for the cases with and without

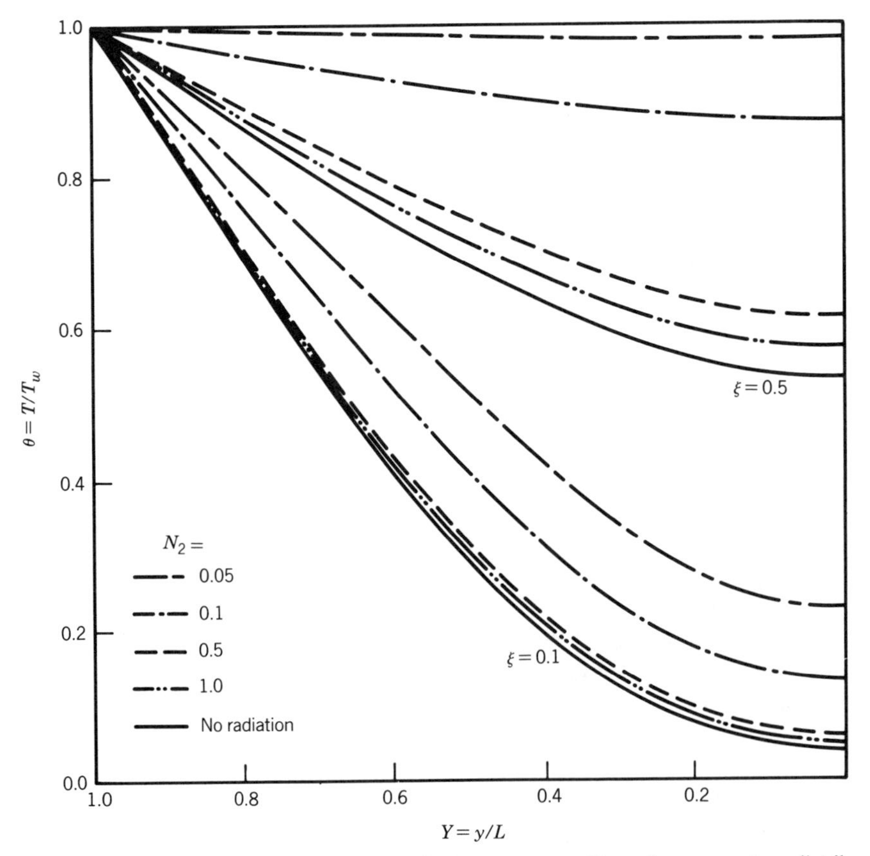

Figure 19.11. Effects of the conduction-to-radiation parameter N_2 on the temperature distribution θ for laminar flow of a participating fluid inside a parallel-plate channel, for $\tau_0 = 1.0$, $\omega = 0.1$, $\theta_i = 0.0$, $\rho^d = 0.0$. (From Mengüc et al. [77].)

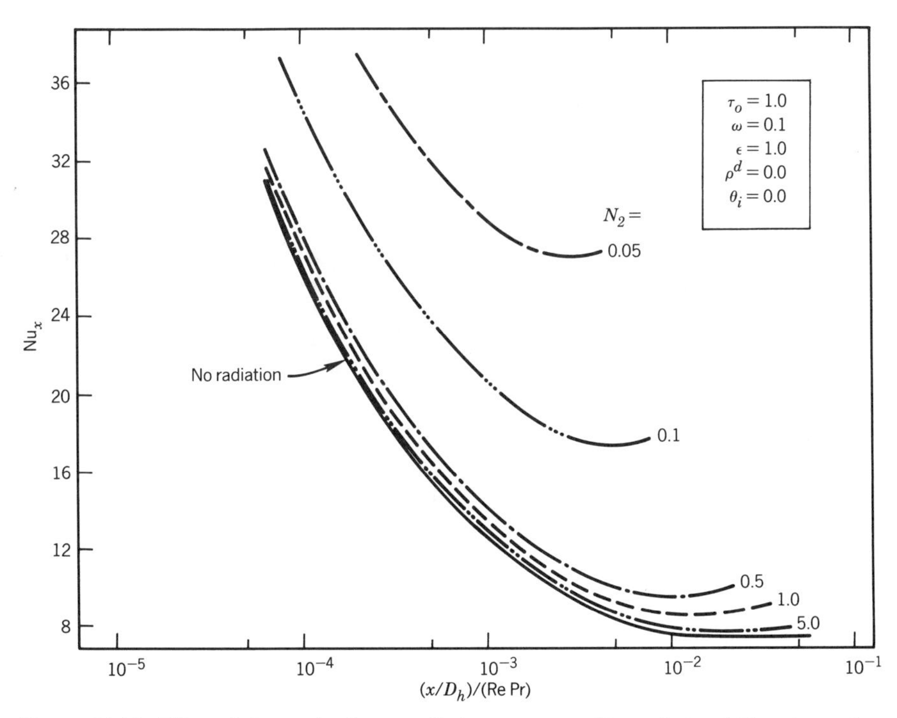

Figure 19.12. Effect of the conduction-to-radiation parameter N_2 on the local Nusselt number. (From Mengüc et al. [77].)

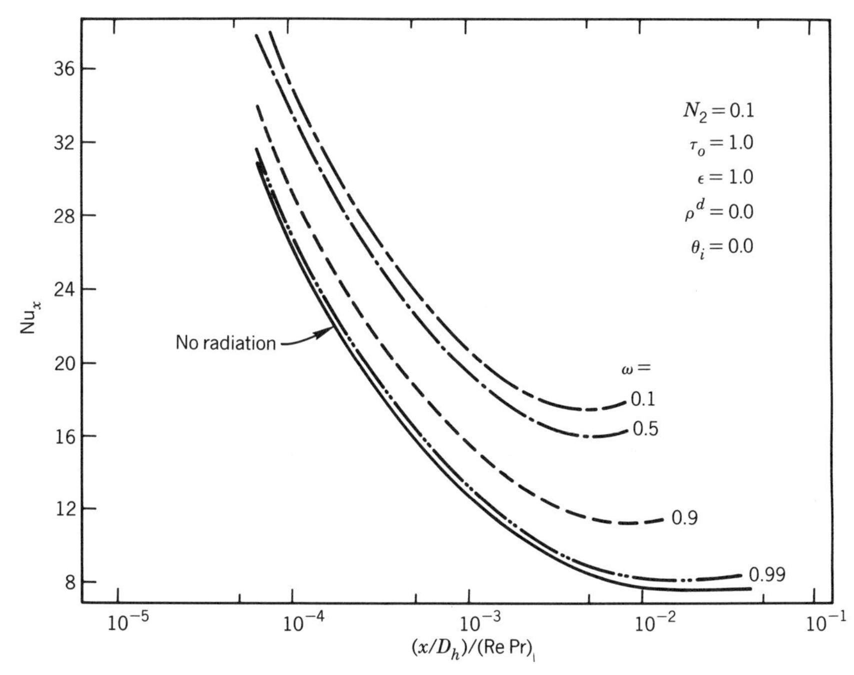

Figure 19.13. Effect of the single-scattering albedo ω on the local Nusselt number. (From Mengüc et al. [77].)

radiation are not distinguishable. For small values of N_2, the temperature profile is flattened as a result of the radiation effects.

Figure 19.12 shows the effect of the conduction-to-radiation parameter N_2 on the local Nusselt number as a function of the dimensionless distance $(x/D_h)/(\mathrm{Re\,Pr})$ along the channel. The effects of N_2 on the local Nusselt number are pronounced for smaller values of N_2, which signifies strong radiation.

Figure 19.13 shows the effect of the single-scattering albedo ω on the local Nusselt number. Clearly, as ω approaches unity, the radiation effects are weakened and the Nusselt number approaches that of forced convection without radiation. The reader should consult Mengüc et al. [77] for a discussion of the effects of other parameters on heat transfer.

Turbulent Flow. We now examine the above heat transfer problem for flow inside a parallel-plate channel for the case of turbulent flow. We consider hydrodynamically developed, thermally developing turbulent flow, idealize constant thermophysical properties, and neglect the viscous energy dissipation and the axial conduction and radiation in the flow. The coordinates and geometry are similar to that shown in Fig. 19.10 for the case of laminar flow.

The mathematical formulation of the convection part of the problem is given in the dimensionless form as

$$w(Y)\frac{\partial \theta}{\partial \chi} = \frac{\partial}{\partial Y}\left[\epsilon(Y)\frac{\partial \theta}{\partial Y}\right] - \frac{\tau_0}{N_2}\frac{\partial Q^r}{\partial Y} \qquad \text{in } 0 < Y < 1, \quad \chi > 0 \tag{19.47a}$$

$$\theta(\chi, Y) = \theta_i \qquad \text{at } \chi = 0 \tag{19.47b}$$

$$\frac{\partial \theta(\chi, Y)}{\partial Y} = 0 \qquad \text{at } Y = 0 \tag{19.47c}$$

$$\theta(\chi, Y) = 0 \qquad \text{at } \chi = 1 \tag{19.47d}$$

where various dimensionless quantities are

$$\theta(\chi, Y) = \frac{T(x, y)}{T_w}, \qquad \theta_i = \frac{T_i}{T_w}, \qquad Y = \frac{y}{L}, \qquad \chi = \frac{16}{C}\frac{x/D_h}{\mathrm{Re\,Pr}},$$

$$D_h = 4L, \qquad \mathrm{Re} = \frac{u_m D_h}{\nu}, \qquad \mathrm{Pr} = \frac{\nu}{\alpha}, \qquad w(Y) = \frac{u(y)}{u_{\max}},$$

$$C = \frac{u_{\max}}{u_m}, \qquad \epsilon(Y) = 1 + \frac{\epsilon_H}{\alpha}, \qquad \tau_0 = \beta L, \qquad N_2 = \frac{k\beta}{4n^2\bar{\sigma}T_w^3},$$

$$\beta = \kappa + \sigma \tag{19.48}$$

The dimensionless radiation heat flux $Q^r(\chi, Y)$ is related to the dimensionless radia-

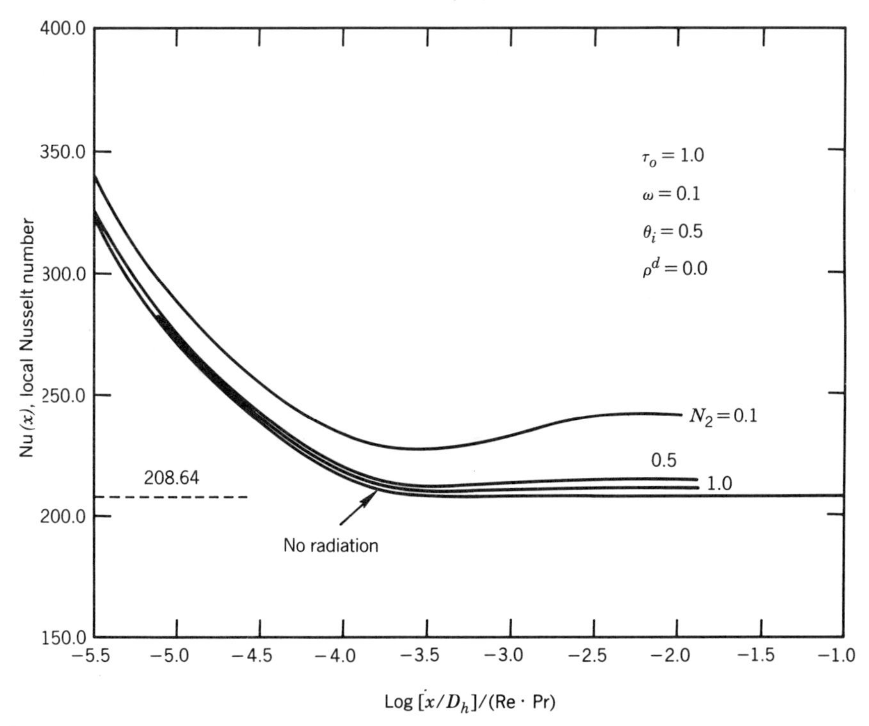

Figure 19.14. Effect of the conduction-radiation parameter N_2 on the local Nusselt number; Re = 10^5, Pr = 1. (From Yener and Özışık [79].)

tion intensity $\psi(\chi, Y)$ by

$$Q^r(\chi, Y) = \frac{1}{2}\int_{-1}^{1} \psi(\chi, Y, \mu')\mu' \, d\mu' \tag{19.49}$$

where $\psi(\chi, Y)$ is defined by Eq. (19.45) and the radiation part of the problem is given by Eqs. (19.46).

In this problem, the parameters affecting heat transfer are the same as for laminar flow, except that the results have weak dependence on the Reynolds number.

Figure 19.14 shows the effect of conduction-to-radiation parameter N_2 on the local Nusselt number for Re = 10^5, Pr = 1, black walls, and $\theta_i = 0.5$. The curve for large N_2 corresponds to the nonradiating fluid. Decreasing N_2 increases radiation effects, which in turn increases the local Nusselt number.

Figure 19.15 shows the effects of single-scattering albedo ω on the local Nusselt number. As ω approaches unity, the radiation effects are weakened, because the medium becomes highly scattering. The case $\omega = 1$ corresponds to purely scattering medium in which radiation is uncoupled from convection and the Nusselt number is the same as that for a nonradiating flow.

Figure 19.16 shows the effects of wall reflectivity on the local Nusselt number. The radiation effects are largest with black boundaries (i.e., $\rho = 0$) and diminish with increasing surface reflectivity.

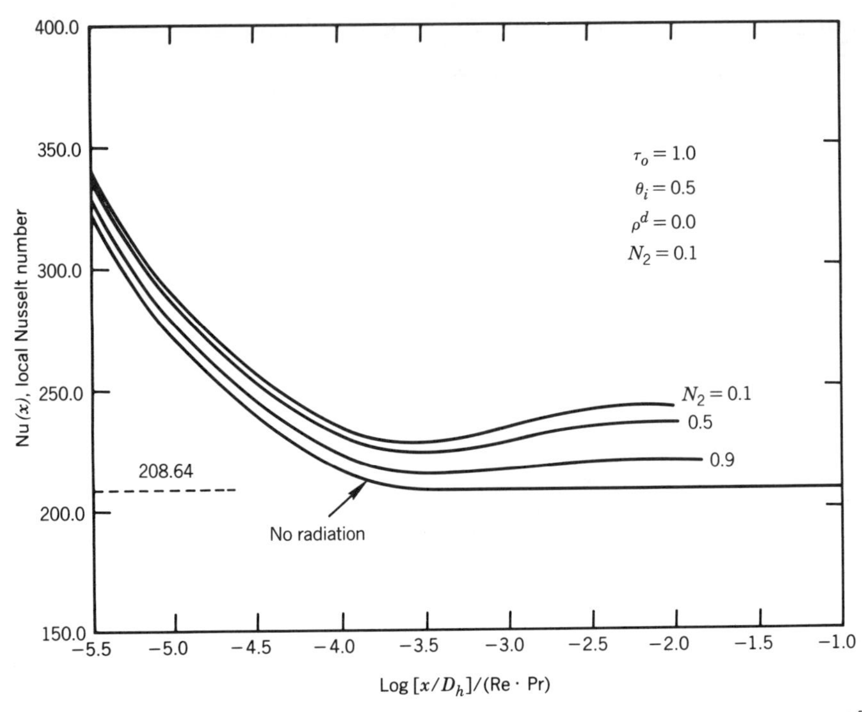

Figure 19.15. Effect of the single-scattering albedo ω on the local Nusselt number; Re = 10^5, Pr = 1. (From Yener and Ösiẓik [79].)

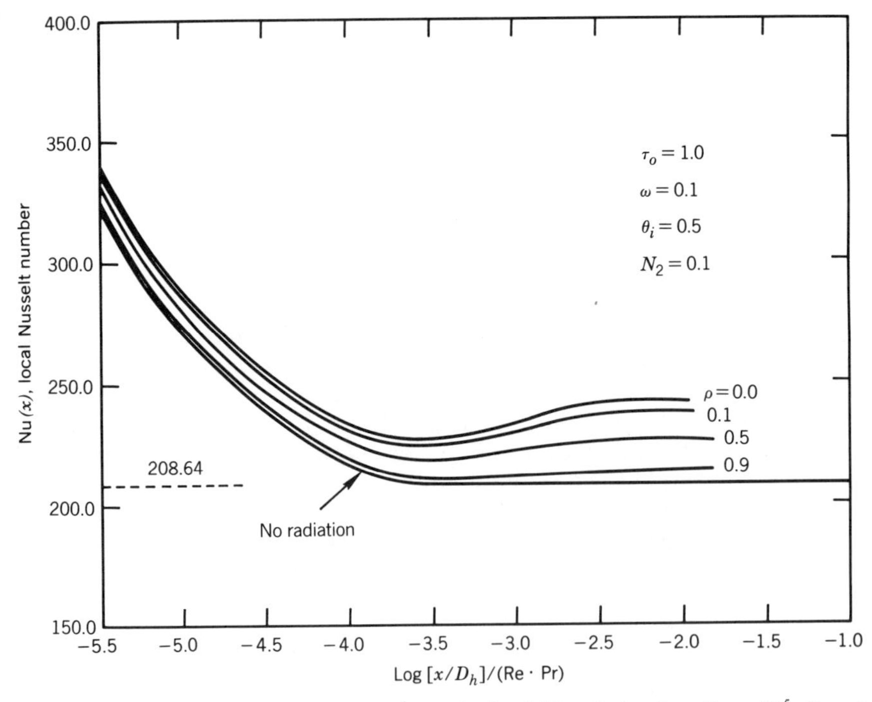

Figure 19.16. Effect of the reflectivity, ρ^d, on the local Nusselt number; Re = 10^5, Pr = 1. (From Yener and Özışık [79].)

Note that the effect of radiation on the Nusselt number is less pronounced with turbulent flow than with laminar flow, because radiation becomes less dominant in comparison with the high heat transfer rate in turbulent flow.

NOMENCLATURE

A_i	area of zone i, m^2, ft^2
Bo	Boltzmann number $= \rho_0 u_0 c_{p,0}/n_0^2 \bar{\sigma} T_0^3$
c_p, $c_{p,0}$	specific heat, J/(kg · K), Btu/(lb$_m$ · °F)
D_h	hydraulic diameter $= 4L = 2H$, m, ft
$dF_{dA_i\text{-}dA_j}$, $dF_{\zeta\text{-}\eta}$	elemental view factors
E	Eckert number $= u_0^2/c_{p,0}T_0$
f	velocity function in Eq. (19.34)
$F_{\zeta\text{-}I}$	view factor from surface ζ to surface I
H	spacing between parallel plates, m, ft
h	heat transfer coefficient, W/(m^2 · K), Btu/(hr · ft^2 · °F)
h_x	local heat transfer coefficient W(m^2 · K), Btu/(hr · ft^2 · °F)
$I(s, \Omega)$, $I(\tau, \mu)$	radiation intensity, W/(m^2 · sr), Btu/(hr · ft^2 · sr)
k	fluid thermal conductivity, W/(m · K), Btu/(hr · ft^2 · °F)
l	channel length, m, ft
$2L$	spacing between parallel plates, m, ft
n	refractive index
N_1	conduction-to-radiation parameter, Eq. (19.39a), $= k_\infty \kappa_\infty / 4n^2\bar{\sigma}T_\infty^3$
N_2	conduction-to-radiation parameter, Eq. (19.43), $= k\beta/4n^2\bar{\sigma}T_w^3$
Nu$_x$	local Nusselt number, $= h_x x/k$ for flow along a flat plate, $= h_x D_h/k$ for flow inside a parallel-plate channel
p	pressure, Pa, lb$_f$/ft^2
$p(\Omega, \Omega)$, $p(\mu, \mu')$	phase function
P	dimensionless pressure, Eq. (19.12), $= p/\rho u_m^2$
$P_n(\mu)$	Legendre polynomial
Pr	Prandtl number $= \tilde{\mu} c_p/k$
q	heat flux, W/m^2, Btu/(hr · ft^2)
q^c	convection heat flux, W/m^2, Btu/(hr · ft^2)
q^r	radiation heat flux, W/m^2, Btu/(hr · ft^2)
q_w	wall heat flux, W/m^2, Btu/(hr · ft^2)
Q^r	dimensionless radiation heat flux $= q^r/4n^2\bar{\sigma}T_\infty^4$
R	gas constant, J/(kg · K), lb$_f$ · ft/(lb$_m$ · °R)
$R_i(\mathbf{r}_i)$	radiosity at zone i, W/m^2, Btu/(hr · ft^2)
$\mathbf{r}$	position vector, m, ft
Re$_x$	local Reynolds number $= u_\infty x/\nu$
Re	Reynolds number based on hydraulic diameter, $= u_m D_h/\nu$
s	distance measured along the direction of propagation, m, ft

S^*	source term, Eq. (19.28), W/m^3, $Btu/(hr \cdot ft^3)$
T	temperature, °C, K, °F, °R
T_b	bulk fluid temperature, °C, K; °F, °R
T_I, T_{II}	plenum temperatures (see Fig. 19.2), °C, K; °F, °R
T_∞	free-stream temperature, °C, K; °F, °R
T_w	wall temperature, °C, K; °F, °R
t	time, s
U	dimensionless axial velocity $= u/u_m$
u	axial velocity, m/s, ft/s
u_m	mean velocity, m/s, ft/s
u_0	reference velocity, m/s, ft/s
u_∞	free-stream velocity, m/s, ft/s
v	normal velocity component, m/s, ft/s
V	dimensionless normal velocity component $= v/u_m$
x	axial coordinate, m, ft
X	dimensionless axial coordinate $= x/H$
y	Cartesian coordinate, m, ft
Y	dimensionless coordinate $= y/H$ or y/L

Greek Symbols

α	thermal diffusivity $= (k/\rho c_p)$, m^2/s, ft^2/s
β	extinction coefficient $= \kappa + \sigma$, m^{-1}, ft^{-1}
ϵ	emissivity
η	axial coordinate measured along the heated wall, Fig. 19.2
η	dimensionless coordinate, Eqs. (19.6a), (19.36), $= y\sqrt{u_\infty/\nu x}$
ζ	axial coordinate along the insulated wall, Fig. 19.2
θ	polar angle, rad, deg
Θ	dimensionless temperature, Eq. (19.36) or (19.12), T/T_∞ or T/T_I or T/T_w
κ	absorption coefficient, m^{-1}, ft^{-1}
μ	direction cosine $= \cos\theta$
$\tilde{\mu}$	fluid dynamic viscosity, $Pa \cdot s$, $lb_m/(ft \cdot s)$
ν	fluid kinematic viscosity, m^2/s, ft^2/s
$\tilde{\nu}$	frequency of the radiation, Hz
ξ	dimensionless axial coordinate, Eq. (19.43), $= \frac{32}{3}(x/D_h)/(\mathrm{Re\,Pr})$
ξ^*	dimensionless axial coordinate, Eq. (19.36), $= 4n^2\bar{\sigma}T_\infty^3\kappa x/\rho_\infty u_\infty c_{p,\infty}$
$\bar{\xi}$	dimensionless axial coordinate, Eq. (19.6b), $= (\epsilon\bar{\sigma}T_\infty^3/k)\sqrt{\nu x/u_\infty}$
ρ	fluid density, kg/m^3, lb_m/ft^3
ρ^d	diffuse reflectivity
ρ^s	specular reflectivity
σ	scattering coefficient, m^{-1}, ft^{-1}
$\bar{\sigma}$	Stefan-Boltzmann constant, $W/(m^2 \cdot K^4)$, $Btu/(hr \cdot ft^2 \cdot °R^4)$

τ	optical variable = βy or κy
τ_0	optical thickness
ψ	dimensionless radiation intensity, Eq. (19.45)
ω	single-scattering albedo = σ/β

Subscripts

0	reference temperature
w	wall

REFERENCES

1. E. M. Sparrow and R. D. Cess, *Radiation Heat Transfer*, Brooks/Cole, Belmont, Calif., 1970.
2. M. N. Özışık, *Radiative Transfer and Interactions with Conduction and Convection*, Wiley, New York, 1973.
3. R. Siegel and J. R. Howell, *Thermal Radiation Heat Transfer*, 2nd ed., McGraw-Hill, New York, 1981.
4. H. Schlichting, *Boundary Layer Theory*, 6th ed., McGraw-Hill, New York, 1968.
5. W. M. Kays and M. E. Crowford, *Convective Heat and Mass Transfer*, 2nd ed., McGraw-Hill, New York, 1980.
6. M. N. Özışık, *Heat Transfer*, McGraw-Hill, New York, 1985.
7. M. N. Özışık and Y. Yener, The Galerkin Method for Calculating Radiant Interchange between Surfaces, ASME paper 83-WA/HT-106, 1983.
8. R. D. Cess, The Effect of Radiation upon Forced-Convection Heat Transfer, *Appl. Sci. Res.*, Sect. A, Vol. 10, pp. 430–438, 1962.
9. E. M. Sparrow and S. H. Lin, Boundary Layers with Prescribed Heat Flux: Applications to Simultaneous Convection and Radiation, *Int. J. Heat Mass Transfer*, Vol. 8, pp. 437–448, 1965.
10. R. C. Lind and T. Cebeci, Solution of the Equations of the Compressible Laminar Boundary Layer with Surface Radiation, Report DAC-33482, Douglas Aircraft Co., Los Angeles, Calif., Dec. 1966.
11. C. C. Oliver and P. W. McFadden, The Interaction of Radiation and Convection in the Laminar Boundary Layer, *J. Heat Transfer*, Vol. 88, pp. 205–213, 1966.
12. M. S. Sohal and J. R. Howell, Determination of Plate Temperature in Case of Combined Conduction, Convection and Radiation Exchange, *Int. J. Heat Mass Transfer*, Vol. 16, pp. 2055–2066, 1973.
13. M. Perlmutter and R. Siegel, Heat Transfer by Combined Forced Convection and Thermal Radiation in a Heated Tube, *J. Heat Transfer*, Vol. 84, pp. 301–311, 1962.
14. R. Siegel and M. Perlmutter, Convective and Radiant Heat Transfer for Flow of a Transparent Gas in a Tube with a Gray Wall, *Int. J. Heat Mass Transfer*, Vol. 5, pp. 639–660, 1962.
15. B. I. Dussan and T. F. Irvine, Laminar Heat Transfer in a Round Tube with Radiating Flux at the Outer Wall, *Proc. Third Int. Heat Transfer Conf.*, Vol. 5, pp. 184–189, 1966.
16. J. C. Chen, Laminar Heat Transfer in a Tube with Nonlinear Radiant Heat-Flux Boundary Condition, *Int. J. Heat Mass Transfer*, Vol. 9, pp. 433–440, 1966.
17. E. G. Keshock and R. Siegel, Combined Radiation and Convection in an Asymptotically Heated Parallel Plate Flow Channel, *J. Heat Transfer*, Vol. 86, pp. 341–350, 1964.
18. S. T. Liu and R. S. Thorsen, Combined Forced Convection with Radiation Heat Transfer in Asymmetrically Heated Parallel Plates, *Proc. Heat Transfer and Fluid Mechanics Inst.*, Stanford U.P., Palo Alto, Calif., pp. 32–44, 1970.

19. C. M. Usiskin and R. Siegel, Thermal Radiation from a Cylindrical Enclosure with Specified Heat Flux, *J. Heat Transfer*, Vol. 82, pp. 369–374, 1960.

20. C. H. Liu and E. M. Sparrow, Convective-Radiative Interaction in a Parallel Plate Channel—Application to Air-Operated Solar Collectors, *Int. J. Heat Mass Transfer*, Vol. 23, pp. 1137–1146, 1980.

21. B. T. F. Chung and J. E. Thompson, Heat Transfer in a Gray Tube with Forced Convection, Internal Radiation and Axial Wall Conduction, ASME Paper 83-WA/HT-92, 1983.

22. V. V. Sobolev, *A Treatise on Radiative Transfer*, Van Nostrand, Princeton, N.J., 1963.

23. V. Kourganoff, *Basic Methods in Transfer Problems*, Dover, New York, 1963.

24. G. C. Pomraning, *Radiation Hydrodynamics*, Pergamon, New York, 1973.

25. K. M. Case and P. F. Zweifel, *Linear Transport Theory*, Addison-Wesley, Reading, Mass., 1967.

26. W. H. Sutton and M. N. Özışık, A Fourier Transfer Solution for Radiative Transfer in a Slab with Isotropic Scattering and Boundary Reflection, *JQSRT*, Vol. 22, pp. 55–64, 1979.

27. G. Spiga, F. Santarelli, and C. Stramigioli, Radiative Transfer in an Absorbing and Anisotropically Scattering Slab with Reflecting Boundaries, *Int. J. Heat Mass Transfer*, Vol. 23, pp. 841–852, 1980.

28. M. N. Özışık and W. H. Sutton, A Source Function Expansion in Radiative Transfer, *J. Heat Transfer*, Vol. 102C, pp. 715–718, 1980.

29. M. N. Özışık and S. M. Shouman, Source Function Expansion Method for Radiative Transfer in a Two-Layer Slab, *JQSRT*, Vol. 24, pp. 441–449, 1980.

30. C. E. Siewert and P. Benoist, The F_N Method in Neutron-Transport Theory. Part I. Theory and Application, *Nucl. Sci. Eng.*, Vol. 69, pp. 156–160, 1979.

31. P. Grandjean and C. E. Siewert, The F_N Method in Neutron-Transport Theory, Part II. Applications and Numerical Results, *Nucl. Sci. Eng.*, Vol. 69, pp. 161–168, 1979.

32. C. E. Siewert, J. R. Maiorino, and M. N. Özışık, The Use of the F_N Method for Radiative Transfer with Reflective Boundary Conditions, *JQSRT*, Vol. 23, pp. 565–579, 1980.

33. S. M. Shouman and M. N. Özışık, Radiative Transfer in an Isotropically Scattering Two-Region Slab with Reflective Boundaries, *JQSRT*, Vol. 26, pp. 1–19, 1981.

34. M. N. Özışık and Y. Yener, The Galerkin Method for Solving Radiation Transfer in Participating Plane-Parallel Media, *J. Heat Transfer*, Vol. 104, pp. 351–354, 1982.

35. Y. A. Cengel, M. N. Özışık, and Y. Yener, Radiative Transfer in a Plane-Parallel Medium with Space-Dependent Albedo, $\omega(x)$, *Int. J. Heat Mass Transfer*, Vol. 27, pp. 1919–1922, 1984.

36. Y. A. Cengel and M. N. Özışık, The Use of Galerkin Method for Radiation Transfer in an Anisotropically Scattering Slab with Reflecting Boundaries, *JQSRT*, Vol. 32, pp. 225–233, 1984.

37. Y. A. Cengel and M. N.Özışık, Radiation Transfer in an Anisotropically Scattering Plane-Parallel Medium with Space Dependent Albedo, $\omega(x)$, ASME Paper 84-HT-38, 1984.

38. J. R. Howell and M. Perlmutter, Monte Carlo Solution of Thermal Transfer Through Radiant Media Between Gray Walls, *J. Heat Transfer*, Vol. 86C, pp. 117–122, 1964.

39. M. Perlmutter and J. R. Howell, Radiant Transfer Through a Gray Gas between Concentric Cylinders Using Monte Carlo, *J. Heat Transfer*, Vol. 86C, pp. 169–179, 1964.

40. Y. S. Yang, J. R. Howell, and D. E. Klein, Radiative Heat Transfer through Randomly Packed Bed of Spheres by the Monte Carlo Method, ASME Paper 82-HT-14, 1982.

41. T. J. Love and R. J. Grosh, Radiative Heat Transfer in Absorbing, Emitting, and Scattering Media, *J. Heat Transfer*, Vol. 87C, pp. 161–166, 1965.

42. H. M. Hsia and T. J. Love, Radiative Heat Transfer between Parallel Plates Separated by a Nonisothermal Medium with Anisotropic Scattering, *J. Heat Transfer*, Vol. 89C, pp. 197–203, 1967.

43. C. V. Pao, An Interactive Method for Radiative Transfer in a Slab with Specularly Reflecting Boundary, *Appl. Math. Comput.*, Vol. 1, pp. 353–364, 1975.

44. J. A. Roux and A. M. Smith, Comparison of Three Techniques for Solving the Radiative Transport Equation, AEDC-TR-73-200, Arnold Engineering Development Center, Arnold Air Force Station, Tenn., 1974.

45. W. H. Sutton and M. N. Özışık, An Iterative Solution for Anisotropic Radiative Transfer in a Slab, *J. Heat Transfer*, Vol. 101, pp. 695–698, 1979.

46. R. T. Ackroyd, A. K. Ziver, and J. H. Goddard, A Finite Element Method for Neutron Transport, Part IV, *Ann. Nuclear Energy*, Vol. 7, pp. 335–349, 1980.

47. F. H. Azad and M. F. Modest, Evaluation of Radiative Heat Flux in Absorbing, Emitting and Linear-Anisotropically Scattering Cylindrical Media, *J. Heat Transfer*, Vol. 103, pp. 350–356, 1981.

48. S. K. Loyalka, Radiative Heat Transfer between Parallel Plates and Concentric Cylinders, *Int. J. Heat Mass Transfer*, Vol. 12, pp. 1513–1517, 1969.

49. W. W. Yuen and C. L. Tien, Approximate Solutions of Radiative Transfer in One-Dimensional Nonplanar Systems, *JQSRT*, Vol. 19, pp. 533–549, 1978.

50. A. Yücel and Y. Bayazitoglu, Radiative Transfer in Anisotropically Scattering Nonplanar Media, *AIAA J.*, Vol. 19, pp. 540–542, 1981.

51. M. F. Modest and F. H. Azad, The Differential Approximation for Radiative Transfer in an Emitting, Absorbing and Anisotropically Scattering Medium, *JQSRT*, Vol. 23, pp. 117–120, 1980.

52. A. L. Crosbie and T. L. Linsenbart, Two-Dimensional Isotropic Scattering in a Semi-infinite Medium, *JQSRT*, Vol. 19, pp. 257–284, 1978.

53. A. C. Ratzel, III and J. R. Howell, Two-Dimensional Radiation in Absorbing-Emitting Media Using the P-N Approximation, *J. Heat Transfer*, Vol. 105, pp. 333–340, 1983.

54. P. Mengüc and R. Viskanta, Radiative Transfer in Axisymmetric, Finite Cylindrical Enclosures, *Fundamentals of Thermal Radiation Heat Transfer*, ed. T. C. Min and J. L. S. Chen, ASME HTD, Vol. 40, pp. 21–28, 1984.

55. W. A. Fiveland, A Discrete Ordinates Method for Predicting Radiative Heat Transfer in Axisymmetric Enclosures, ASME Paper 82-HT-20, 1982.

56. W. H. Sutton and M. N. Özışık, An Alternative Formulation for Radiative Transfer in an Isotropically Scattering Two-Dimensional Bar Geometry and Solution Using the Source Function Expansion Technique, ASME Paper 84-HT-36, 1984.

57. Y. Taitel and J. P. Hartnett, Equilibrium Temperature in Boundary Layer Flow over a Flat Plate of Absorbing-Emitting Gas, ASME Paper 66-WA/HT-48, 1966.

58. M. Boles and M. N. Özışık, Effects of Scattering of Radiation upon Compressible Boundary Layer Flow over an Adiabatic Flat Plate, *J. Heat Transfer*, Vol. 97, pp. 311–313, 1975.

59. R. Viskanta, Interaction of Heat Transfer by Conduction, Convection and Radiation in a Radiating Fluid, *J. Heat Transfer*, Vol. 85C, pp. 318–328, 1963.

60. R. Viskanta, Heat Transfer in a Radiating Fluid with Slug Flow in a Parallel Plate Channel, *Appl. Sci. Res.*, Sect. A, Vol. 13, pp. 291–311, 1964.

61. S. De Soto and D. K. Edwards, Radiative Emission and Absorption in Non-Isothermal Nongray Gases in Tubes, *Proc. 1965 Heat Transfer and Fluid Mechanics Inst.*, Stanford U.P., Palo Alto, Calif., pp. 358–372, 1965.

62. D. K. Edwards and A. Balakrishnan, Nongray Radiative Transfer in a Turbulent Gas Layer, *Int. J. Heat Mass Transfer*, Vol. 16, pp. 1003–1015, 1973.

63. D. K. Edwards and A. Balakrishnan, Self-Absorption of Radiation in Turbulent Molecular Gases, *Combustion and Flame*, Vol. 20, pp. 401–417, 1973.

64. A. Balakrishnan and D. K. Edwards, Established Laminar and Turbulent Channel Flow of a Radiating Molecular Gas, *Heat Transfer 1974*, Vol. 1, pp. 93–97, 1974.

65. A. T. Wassel and D. K. Edwards, Molecular Radiation in a Laminar or Turbulent Pipe Flow, *J. Heat Transfer*, Vol. 98, pp. 101–107, 1976.

66. V. N. Timofeyev, F. R. Shklyar, V. M. Malkin, and A. K. H. Berland, Combined Heat Transfer in an Absorbing Stream Moving in a Flat Channel, *Heat Transfer—Sov. Res.*, Parts I, II, III, Vol. 1, No. 6, pp. 57–93, 1969.

67. Y. Kurosaki, Heat Transfer by Simultaneous Radiation and Convection in an Abosrbing and Emitting Medium in a Flow Between Parallel Plates, *Heat Transfer 1970*, Vol. 3, Paper No. R 2.5, 1970.

68. I. S. Habib and R. Greif, Heat Transfer to a Flowing Non-gray Radiating Gas: An Experimental and Theoretical Study, *Int. J. Heat Mass Transfer*, Vol. 13, pp. 1571–1582, 1970.

69. R. S. Thorsen, Combined Conduction, Convection, and Radiation Effects on Optically Thin Tube Flow, ASME Paper No. 71-HT-17, 1971.

70. J. C. Chen, Simultaneous Radiative and Convective Heat Transfer in an Absorbing, Emitting and Scattering Medium in Slug Flow between Parallel Plates, *AIChE J.*, Vol. 2, pp. 253–259, 1964.

71. J. B. Bergquam and N. S. Wang, Heat Transfer by Convection and Radiation in an Absorbing, Scattering Medium Flowing between Parallel Plates, ASME Paper No. 76-HT-50, 1976.

72. F. H. Azad and M. F. Modest, Combined Radiation and Convection in Absorbing, Emitting and Anisotropically Scattering Gas—Particulate Tube Flow, *Int. J. Heat Mass Transfer*, Vol. 24, pp. 1681–1697, 1981.

73. K. H. Im and R. K. Ahluwalia, Combined Convection and Radiation in Rectangular Ducts, ASME Paper No. 82-HT-48, 1982.

74. C. C. Lii and M. N. Özışık, Heat Transfer in an Absorbing, Emitting and Scattering Slug Flow between Parallel Plates, *J. Heat Transfer*, Vol. 95C, pp. 538–540, 1973.

75. T. C. Chawla and S. H. Chan, Spline Collocation Solution of Combined Radiation-Convection in Thermally Developing Flows with Scattering, *Numer. Heat Transfer*, Vol. 3, pp. 47–65, 1980.

76. T. C. Chawla and S. H. Chan, Combined Radiation and Convection in Thermally Developing Poiseuille Flow with Scattering, *J. Heat Transfer*, Vol. 102C, pp. 297–302, 1980.

77. M. P. Mengüc, Y. Yener, and M. N. Özışık, Interaction of Radiation and Convection in Thermally Developing Laminar Flow in a Parallel-Plate Channel, ASME Paper No. 83-HT-35, 1983.

78. C. A. Schmidt, Y. Yener, and M. N. Özışık, Simultaneous Convection and Radiation in Participating Laminar Flow between Parallel Plates, *16th Southeastern Seminar on Thermal Sci.*, Miami Beach, Fl., 1982.

79. Y. Yener and M. N. Özışık, Simultaneous Radiation and Forced Convection in Thermally Developing Turbulent Flow through a Parallel-Plate Channel, *J. Heat Transfer*, Vol. 108, pp. 985–988, 1986.

80. T. F. Smith and Z. F. Shen, Radiative and Convective Transfer in a Cylindrical Enclosure, ASME Paper No. 83-HT-53, 1983.

20

NON-NEWTONIAN FLUID FLOW AND HEAT TRANSFER

Thomas F. Irvine Jr.
Jacob Karni

State University of New York
Stony Brook, New York

20.1 INTRODUCTION

Many important industrial fluids are non-Newtonian in their flow characteristics and are referred to as rheological fluids. These include paints; various suspensions such as coal-water or coal-oil slurries, glues, inks, foods, and soap and detergent slurries; polymer solutions; and many others. Because such fluids have more complicated equations that relate the shear stresses to the velocity field than Newtonian fluids have, additional factors must be considered in examining various fluid mechanics and heat transfer phenomena.

Another important characteristic of rheological fluids, because of their large "apparent viscosities," is that they have a tendancy toward low Reynolds and Grashof numbers. Accordingly, in engineering practice laminar flow situations are encountered more often than with Newtonian fluids. Thus both laminar and turbulent flows will be considered here with equal emphasis.

It is the purpose of this chapter to discuss the definition and classification of non-Newtonian fluids, to examine different methods for measuring their rheological properties, and to consider a variety of fluid mechanics and heat transfer situations that are likely to be encountered in engineering practice. These will include both forced and free convection in internal and external flows.

Knowledge of non-Newtonian fluid mechanics and heat transfer is still in an early stage, and many aspects of the field have yet to be investigated and clarified.

20.2 DESCRIPTION AND CLASSIFICATION OF NON-NEWTONIAN FLUIDS

20.2.1 General Considerations

The most straightforward way to describe the class of non-Newtonian fluids is to define a Newtonian fluid, since all others are non-Newtonian.

Newtonian fluids possess a property called viscosity and obey a relation analogous to the Hookian relation between a stress applied to a solid and its strain, i.e., the stress is proportional to the strain rate.

Figure 20.1*a* shows a simple laminar flow situation which illustrates this relation. Shown in the figure are two parallel plates with very large dimensions relative to the distance $\hat{b}$ which separates them. The bottom plate is stationary while the top plate moves at a constant velocity u_b due to a force per unit area τ_{yx} applied to the plate. The shear stress uses the subscript convention that the first letter (here y) refers to the direction normal to the stress and the second letter (here x) refers to the direction of the stress. If the velocity distribution were measured between the plates, it would be found to vary linearly from a value zero at the bottom plate to a value u_b at the top plate, as shown in the figure.

Figure 20.1*b* shows the fluid strain or angular displacement $\Delta\gamma$ experienced by a small fluid element during a short time Δt. For small angles,

$$d\gamma = \frac{du}{dy}\, dt \tag{20.1}$$

and in the limit

$$\frac{d\gamma}{dt} = \dot{\gamma} = \frac{u_b}{\hat{b}} = \frac{du}{dy} \tag{20.2}$$

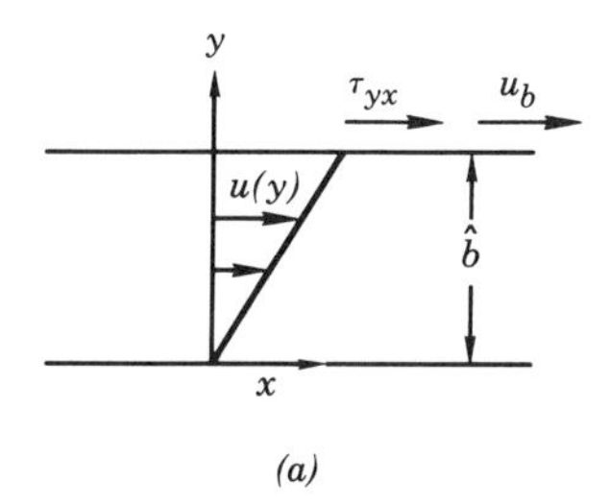

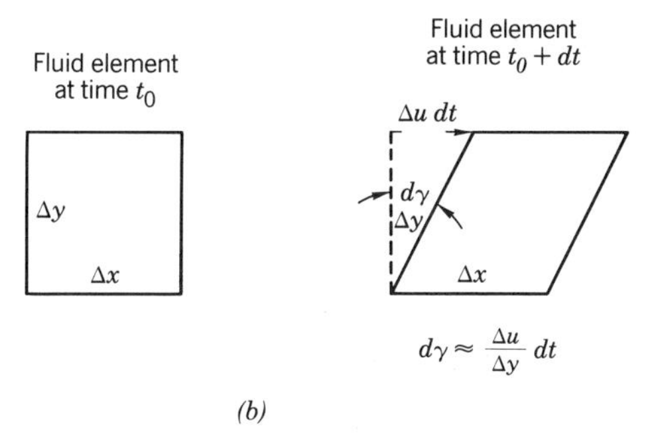

Figure 20.1. (*a*) Velocity profile caused by motion of top plate. (*b*) Angular displacement (strain) of fluid.

For a Newtonian fluid, the stress is proportional to the rate of strain, and a linear relation between the applied stress and the strain rate can be written as

$$\tau_{yx} = \mu\dot{\gamma} = \mu\frac{du}{dy} \tag{20.3}$$

where $\dot{\gamma}$ is called the strain rate or shear rate and is equal in this simple case to the velocity gradient. This reasoning can be easily applied to other one-dimensional flows where the velocity gradient is not constant. The constant of proportionality μ is called the dynamic viscosity, and for Newtonian fluids it is a transport property. In this chapter, the direction normal to the stress and the direction of the stress will usually be apparent, and stress subscripts will only be displayed when necessary to avoid confusion.

Equation (20.3), which is called a constitutive equation, indicates that if τ_{yx} is plotted against $\dot{\gamma}$, as illustrated in Fig. 20.2, a linear relation exists between the two quantities and the slope of the curve is the dynamic viscosity. Such a graph is called a flow curve and is a convenient way to examine the viscous properties of different types of fluids.

Fluids which do not obey Eq. (20.3) are non-Newtonian. As shown in Fig. 20.3, fluids can be divided into Newtonian and non-Newtonian categories. The non-Newtonian fluids can further be divided into purely viscous and viscoelastic fluids.

Purely viscous, time-independent fluids are defined as those whose shear stress depends only upon some function of the shear rate (and sometimes on an initial yield stress). For purely viscous, time-dependent fluids, the shear stress can depend upon the time history of the applied stress.

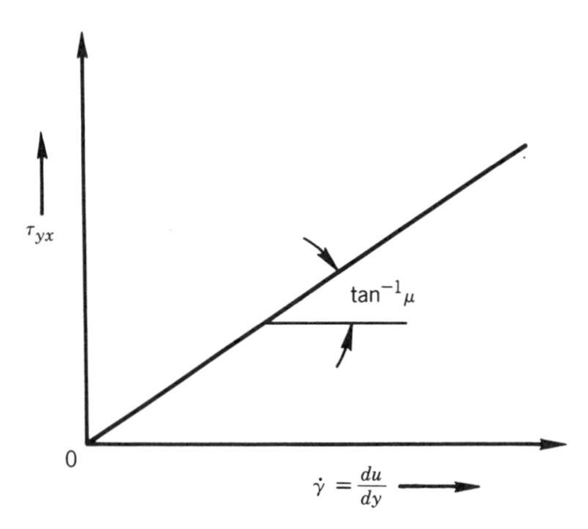

Figure 20.2. Flow curve for a Newtonian fluid.

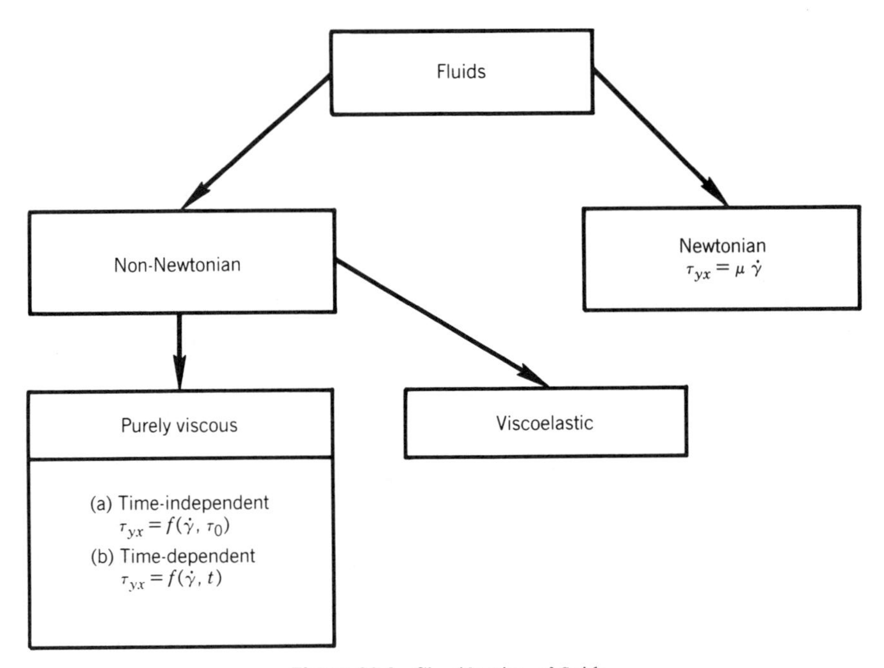

Figure 20.3. Classification of fluids.

Viscoelastic fluids are those which possess both viscous and elastic properties. They are important because of drag and heat transfer reduction in turbulent flows of such fluids. Such fluids will not be considered in this chapter, but an excellent recent review is given by Cho and Hartnett [1].

Purely viscous, time-independent fluids are illustrated in Fig. 20.4 along with the flow of a Newtonian fluid (curve c) for comparison. In the figure, curves b and d are for time-independent, purely viscous fluids where the shear stress depends only on the shear rate, but in a nonlinear way. A fluid which behaves like b is called pseudoplastic

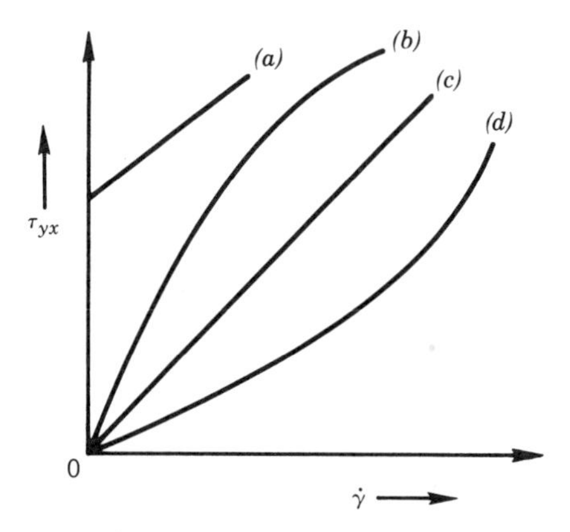

Figure 20.4. Flow curves of purely viscous, time-independent fluids: (*a*) Bingham plastic; (*b*) pseudoplastic; (*c*) Newtonian; (*d*) Dilatant.

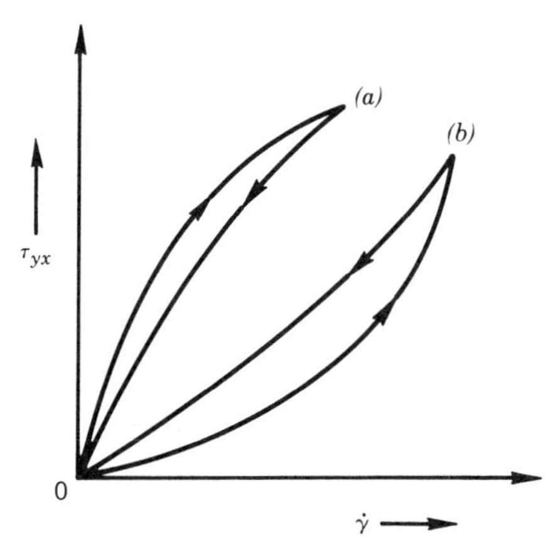

Figure 20.5. Flow curves for purely viscous, time-dependent fluids: (*a*) thixotropic, (*b*) rheopectic.

(or shear-thinning), and one which behaves like *d* is called dilatant (or shear-thickening). Curve *a* represents a fluid that possesses an initial yield stress, upon exceeding which the fluid behaves as a Newtonian fluid. Such a fluid is called a Bingham plastic fluid. Other fluids exhibit non-Newtonian behavior after the initial yield stress is exceeded.

Purely viscous, time dependent fluids are illustrated in Fig. 20.5. Here the flow curves can assume a hysteresis loop whose shape depends upon the time-dependent rate at which the shear stress is applied. Fluids which have a pseudoplastic flow curve (*a*) are called thixotropic, and those with a dilatant curve (*b*) are called rheopectic.

Time-dependent fluids are the bane of rheologists. There are so many variables to consider for these fluids that it is extremely difficult to produce any general results which can be applied to a variety of problems. Such fluids will not be considered further in this chapter.

Table 20.1 lists the classes of fluids discussed above, with examples of fluids which exhibit these characteristics.

TABLE 20.1 Examples of Rheological Fluids

Type	Examples
Newtonian	Water, air, mercury, engine oil
Pseudoplastic	Paints, glues, blood, suspensions
Dilatant	Wet sand, sugar and Borax solutions
Bingham plastic	Certain emulsions and paints, ketchup
Thixotropic	Printing inks, food materials, paints
Rheopectic	Clay suspensions
Viscoelastic	Polymer solutions (e.g., Polyox and water)

20.2.2 Apparent Viscosity

The apparent viscosity is defined as the ratio of the shear stress at some location in the flow field to the local shear rate:

$$\mu_a = \frac{\tau}{\dot{\gamma}} \tag{20.4}$$

Only for a Newtonian fluid is the apparent viscosity a true property in the sense that a property depends on the basic microstructure of the fluid. For non-Newtonian fluids, the apparent viscosity is not a fluid property but is a function of the velocity field, as will be examined in more detail in the next section. In spite of this complication, which sometimes causes confusion, the apparent viscosity is of great conceptual value and appears extensively in the rheological literature. It will be used in the present chapter along with frequent warnings about its true nature.

20.2.3 Constitutive Equations for Purely Viscous Time-Independent Fluids

A constitutive equation is one that relates the shear stress or apparent viscosity in a fluid to the shear rate through the rheological properties of the fluid. For example, Eq. (20.3) is the constitutive equation for a Newtonian fluid, with the dynamic viscosity as the rheological property. A large number of constitutive equations have been developed to describe the behavior of non-Newtonian fluids [2]. Some have as many as five rheological properties, and while they are suitable for describing in detail the relations between shear stress and stress rate for complex fluids, they are normally too cumbersome to use in engineering analyses. A selected list of constitutive equations is given in Table 20.2. Most popular in engineering work are the three property models which will be discussed next.

A convenient way to depict the constitutive equations is to plot a graph of apparent viscosity against shear rate. Figure 20.6 shows such a qualitative graph which is indicative of the behavior of many purely viscous pseudoplastic fluids. From the figure, it is seen that at low shear rates (region a) the fluid is Newtonian in behavior (constant μ_a). The viscosity μ_0 at zero shear rate is called the zero-shear-rate viscosity. At higher shear rates (region b), the apparent viscosity μ_a begins to decrease until it stabilizes into a straight line (region c). Following region c there is another transition region (d), after which, at high shear rates, the fluid becomes once again Newtonian in region e with a viscosity μ_∞. A similar dependence of the apparent viscosity on the shear rate is also seen in purely viscous dilatant fluids. However, in such fluids μ_a increases with $\dot{\gamma}$ in the transition and power-law regions (b, c, d).

TABLE 20.2 Some Constitutive Equations for Time-Independent Purely Viscous Fluids[a]

Model	Constitutive Equation	Ref.
Power law (Ostwald–de Waele)	$\mu_a = K\dot{\gamma}^{n-1}$	[3]
Ellis	$\mu_a = \dfrac{\mu_0}{1 + \left(\tau/\tau_{1/2}\right)^{\alpha-1}}$	[4]
Cross	$\mu_a = \dfrac{\mu_0}{1 + \left(\dot{\gamma}/\dot{\gamma}_{1/2}\right)^{\beta}}$	[5]
Modified power law	$\mu_a = \dfrac{\mu_0}{1 + (\mu_0/K)\dot{\gamma}^{1-n}}$	—
Sutterby	$\mu_a = \mu_0\left[\dfrac{\sinh^{-1}\lambda\dot{\gamma}}{\lambda\dot{\gamma}}\right]^{\delta}$	[6]
Carreau (A)	$\mu_a = \mu_\infty + (\mu_0 - \mu_\infty)[1 + (\lambda_c\dot{\gamma})^2]^{(n-1)/2}$	[7]

[a] See also Ref. 2.

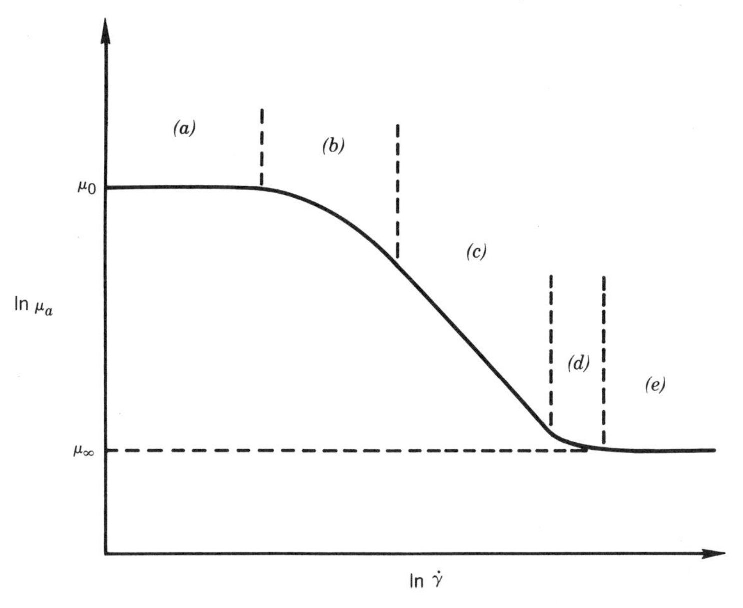

Figure 20.6. Illustrative flow curve for a pseudoplastic fluid: (a) low-shear-rate Newtonian region, (b) transition region I, (c) power-law region, (d) transition region II, (e) high-shear-rate Newtonian region.

It is evident from Fig. 20.6 that a constitutive equation which completely describes the apparent viscosity over the entire shear-rate range must contain a large number of constants (properties). Fortunately, however, for many engineering problems, only a portion or portions of the entire curve need to be considered, and therefore less constants are required.

In the following, three constitutive equations will be considered. These are the ones frequently used in engineering applications and also those most often cited in the fluid mechanics and heat transfer literature. Only regions a, b, and c of Fig. 20.6 will be considered, since many applications fall in these regions.

Power Law (Ostwald–de Waele). The constitutive equation for this model is given by [3]

$$\tau = K\dot{\gamma}^n \tag{20.5}$$

or

$$\mu_a = K\dot{\gamma}^{n-1} \tag{20.6}$$

In Eq. (20.5), which is a two-property model, K is called the fluid consistency and n the flow index. By taking the logarithm of both sides of Eq. (20.6), it can be seen that the equation represents region c in Fig. 20.6, i.e., it appears as a straight line in logarithmic coordinates with a negative slope of $n - 1$ (for $n < 1$). This is why region c in Fig. 20.6 is called the power-law region.

The power-law model is the one most frequently used in non-Newtonian fluid mechanics and heat transfer. It has proved quite successful in predicting the behavior of a large number of pseudoplastic (and dilatant) fluids in spite of the fact that it has several built-in flaws and anomalies. For example, if one considers the flow of a pseudoplastic fluid ($n < 1$) through a circular tube, then at the center of the tube, because of symmetry, the shear rate (velocity gradient) becomes zero and thus the apparent viscosity defined by Eq. (20.6) is infinite. This poses grave conceptual difficulties. In addition, if the flow field under consideration has shear rates in regions a and b of Fig. 20.6, then the power-law model cannot possibly represent the correct constitutive equation. These drawbacks must be kept in mind when power-law models are considered further.

Modified Power Law. A generalization of the power-law fluid which extends the shear-rate range is given by

$$\mu_a = \frac{\mu_0}{1 + (\mu_0/K)\dot{\gamma}^{1-n}} \tag{20.7}$$

At low shear rates, the second term in the denominator will become negligible compared to the first and thus the apparent viscosity becomes a constant equal to μ_0. This represents region a in Fig. 20.6.

Conversely, when the shear rate becomes large and the first term in the denominator becomes negligible compared to the second, then Eq. (20.7) becomes the same as the power-law fluid, or region c in Fig. 20.6. At intermediate shear rates, the equation represents the transition region b in Fig. 20.6.

An important advantage of this model is that it retains the rheological properties K and n of the power-law model plus the additional property μ_0. Thus, as will be shown later, in the flow and heat transfer equations, the same dimensionless groups (e.g., the generalized Reynolds number) will occur in both models, and one model can serve as an asymptotic solution to the other.

Ellis Model. The model proposed by Ellis (as summarized in [4]) has many of the characteristics of the modified power-law model. The Ellis equation is

$$\mu_a = \frac{\mu_0}{1 + \left(\tau/\tau_{1/2}\right)^{\alpha-1}} \tag{20.8}$$

where $\tau_{1/2}$ is the shear stress when $\mu_a = \mu_0/2$ (i.e., one-half the zero-shear-rate viscosity).

Equation (20.8) has the characteristics that at low values of the shear stress τ (which corresponds to low shear rates) the apparent viscosity becomes constant at μ_0. At large enough values of $\dot{\gamma}$, Eq. (20.8) turns into a power-law fluid where K and n are related to μ_0, $\tau_{1/2}$, and α.

The Ellis model is written as a function of shear stress rather than of shear rate as in Eqs. (20.6) and (20.7). As a result, it is simpler to obtain analytical solutions of the shear field. Its disadvantage, from an engineering point of view, is that although it is similar to the modified power-law model, it introduces a different set of properties that are related to but different from the properties in the power-law model.

As mentioned previously, there are a large number of constitutive equations which have been suggested by rheological investigators. It should be emphasized that these equations are not always based on fundamental concepts of the microstructure of rheological fluids, but they are equations that represent with differing degrees of accuracy the variation of the shear stress with shear rate.

20.3 RHEOLOGICAL PROPERTY MEASUREMENTS

20.3.1 Purely Viscous Fluids

Rheological property measurements of purely viscous fluids consist of the experimental determination of the flow curves illustrated in Fig. 20.4. Although it might appear to be a formidable task to measure shear stresses and shear rates directly, many instruments do just that. Other devices measure τ and $\dot{\gamma}$ indirectly. The choice of which instrument to use depends upon a number of factors, such as the shear-rate range, desired accuracy, instrument cost, experimental skills, etc. Instruments of both types will be considered below, and examples will be presented of property measurements made by several of them.

20.3.2 Capillary-Tube Viscometer

This is one of the most fundamental of instruments and when used carefully is capable of accuracies of better than 2% over its applicable shear-rate range. It is most appropriate for intermediate shear rates of approximately $10^2 \leq \dot{\gamma} \leq 10^5$, which corresponds to the power-law region for many non-Newtonian fluids (see Fig. 20.6). The operation of the instrument will first be illustrated by considering a Newtonian fluid.

Figure 20.7 is a schematic diagram of a capillary tube viscometer. The test fluid flows in a laminar steady manner through the capillary tube of diameter D and length L with an average velocity $\bar{u}$ under the influence of the pressure difference Δp. The tube is made long enough so that entrance effects can be neglected or a small entrance correction can be made. For fully developed laminar Newtonian constant-property flow through a circular tube, the following well-known relation applies (see Nomenclature section):

$$f\,\mathrm{Re} = 16 \tag{20.9}$$

where f is the Fanning friction factor defined as $\Delta p/L = (\rho \bar{u}^2/2)\, f/(D/4)$ and Re is the Reynolds number based on the tube diameter.

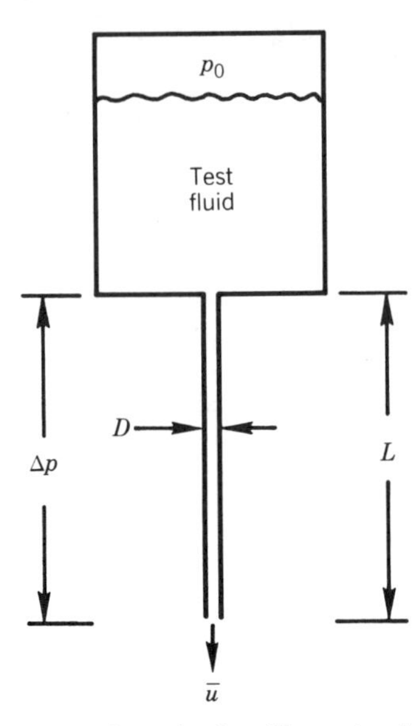

Figure 20.7. Schematic of capillary-tube viscometer.

If Eq. (20.9) is rearranged in terms of measurable quantities for a circular tube, it can be written as

$$\frac{\Delta p\, D}{4L} = \mu\left(\frac{8\bar{u}}{D}\right) \tag{20.10}$$

A comparison of Eqs. (20.3) and (20.10) shows that $\tau_w = \Delta p\, D/4L$ and $\dot{\gamma}_w = 8\bar{u}/D$. The subscript w is appended to both the shear stress and the shear rate because the left side of Eq. (20.10) can only be related to the wall shear stress through a force balance with the pressure drop. Thus both the shear stress and the shear rate are measured at wall conditions.

If logarithms are taken of both sides of Eq. (20.10), it can be written as

$$\ln\left(\frac{\Delta p\, D}{4L}\right) = \ln \mu + \ln \frac{8\bar{u}}{D} \tag{20.11}$$

Thus, the experimental measurements that are taken of Δp and $\bar{u}$ for a particular tube can be plotted on a logarithmic graph as illustrated in Fig. 20.8. From Eqs. (20.11) and (20.5), it can be seen that the slope of the logarithmic flow curve must be unity and the ordinate intercept equal to $\ln \mu$ for a Newtonian fluid.

For a power-law fluid, the experimental measurements are the same. Now, however, the relation between the friction factor and the Reynolds number is given by

$$f\,\mathrm{Re}_g = 2^{3n+1}\left(\frac{3n+1}{4n}\right)^n \tag{20.12}$$

where Re_g is a generalized Reynolds number to be discussed in Section 20.4.1 (see Nomenclature section). If Eq. (20.12) is written in terms of measurable quantities and

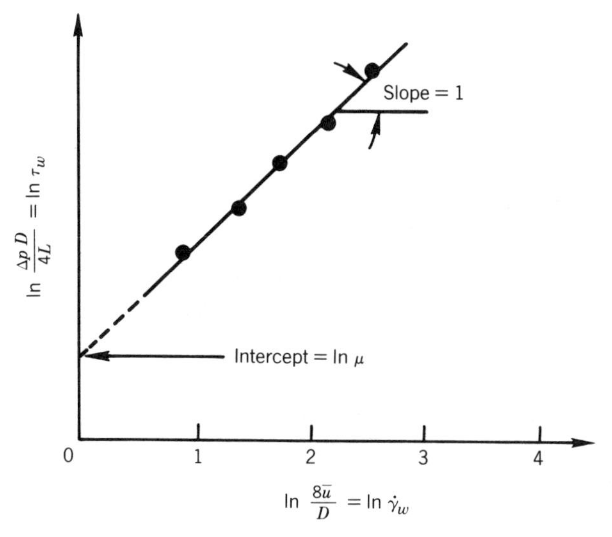

Figure 20.8. Illustrative experimental flow curve of a Newtonian fluid.

in logarithmic form for a circular tube, the result is

$$\ln \frac{\Delta p D}{4L} = \ln\left[K\left(\frac{3n+1}{4n}\right)^n\right] + n \ln\left(\frac{8\bar{u}}{D}\right) \tag{20.13}$$

When the experimental data are plotted as shown in Fig. 20.9, the flow index n can be obtained from the slope of the line and the ordinate intercept is equal to $\ln\{K[(3n+1)/4n]^n\}$, from which the consistency K can be calculated.

Figure 20.10 shows some typical data obtained with a capillary-tube viscometer [8]. The fluid was tap water with the addition of 1000 wppm (parts per million by weight)

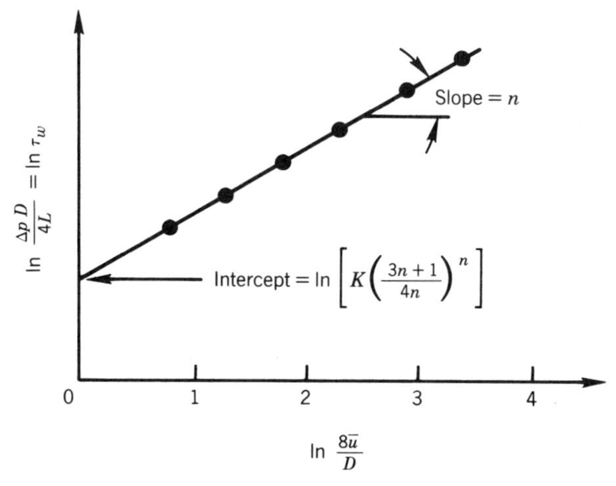

Figure 20.9. Illustrative experimental flow curve of a power-law fluid.

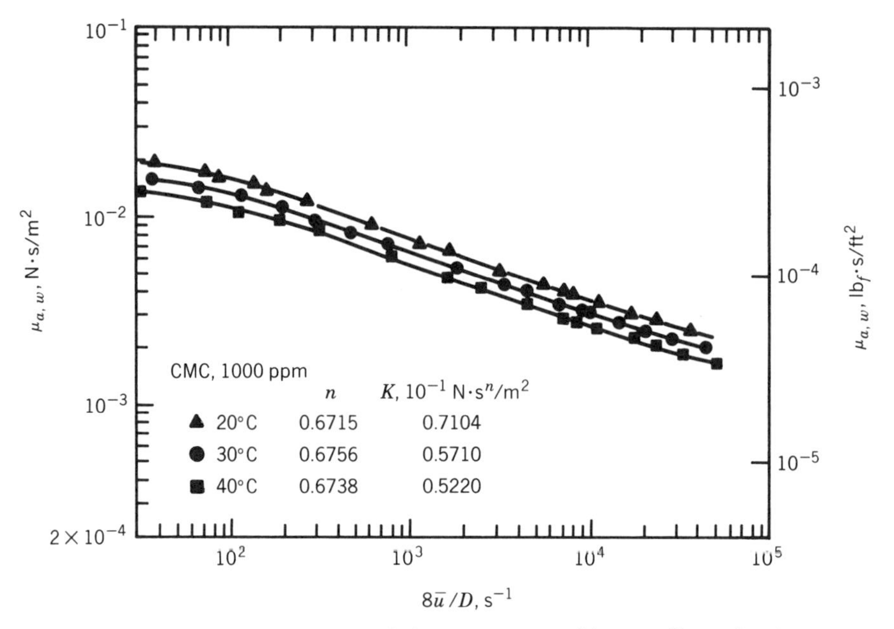

Figure 20.10. Apparent viscosity vs. wall shear rate measured by a capillary-tube viscometer.

of sodium carboxymethyl cellulose (CMC). The abscissa in Fig. 20.10 is not the logarithm of the true shear rate, as can be seen from Eq. (20.13), but is proportional to it for a fixed value of n.

The measurements illustrate an important characteristic of power-law fluids which has been alluded to in the discussion of Fig. 20.6, i.e., that purely viscous non-Newtonian fluids have power-law characteristics only over a limited shear-rate range. A careful examination of Fig. 20.10 reveals that the apparent-viscosity curves are straight lines only over the range $10^2 \leq 8\bar{u}/D < 10^4$. Above and below this range the fluid enters the transition regions shown in Fig. 20.6.

Thus, it is clear that measurements of n and K must be accompanied by information of the applicable shear-rate range. In addition, when applying these measured properties to engineering systems, the shear-rate range of the system must match that of the experimental property measurements. This rule is not always followed carefully in engineering practice.

Other variations of the classical capillary-tube viscometer have been developed. For example, Yamasaki [9] has reported on a comparative method using three capillary tubes. By measuring only flow rates, it is possible to determine n and K for a power-law fluid by comparing the flow rates of the non-Newtonian test fluid with the flow rate of a Newtonian fluid whose viscous properties are well established. No pressure measurements need be made with this method, which is a definite advantage. Other variations and descriptions of commercially available capillary tubes and other viscometers are discussed in the books by Van Wazer et al. [10], Walters [11], and Whorlow [12].

Although the capillary-tube viscometer is a straightforward and attractive instrument, it has its limitations. The most important of these is that it is only applicable over a certain shear-rate range. If the shear rate exceeds a certain value, depending upon the fluid, the flow becomes turbulent and the viscometric equations are no longer

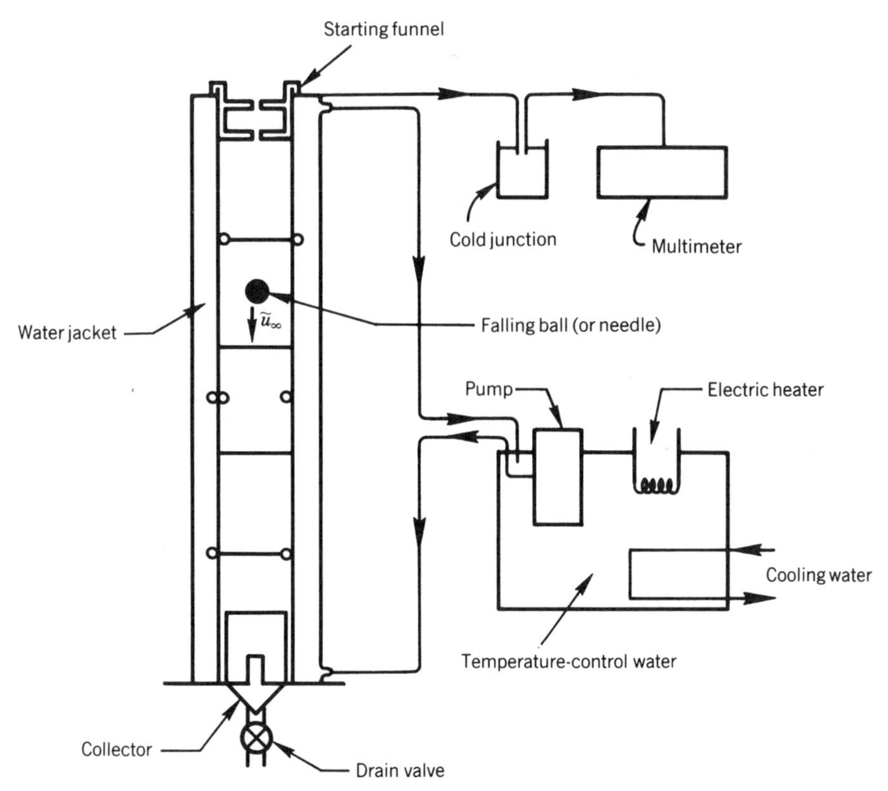

Figure 20.11. Falling-ball or falling-needle viscometer.

valid. If the shear rate becomes too low, accompanied by low flow rates, surface tension effects at the tube exit become significant relative to the pressure drop, introducing extraneous and undeterminable forces into the flow system. As a general rule, depending upon the fluid, the capillary-tube viscometer is most effective in the shear-rate range $10^2 \leq \gamma \leq 10^5$. For the shear-rate range below 10^2, the falling-ball and falling-needle viscometers discussed in the next section are more appropriate.

20.3.3 Falling-Ball Viscometer

The falling-ball viscometer is shown in schematic form in Fig. 20.11. It consists of a transparent tube which is water-jacketed for temperature control. Along the tube in the vertical direction are several horizontal lines a known distance apart which serve as reference lines for the measurement of the velocity of the falling ball.

To operate the instrument, a sphere of known diameter and weight is carefully placed in the starting funnel at the top of the tube and allowed to fall though the test liquid under the influence of gravity. After a starting transient, the weight minus the buoyancy force of the ball will equal the drag forces on the sphere, and the velocity will become a constant called the terminal velocity $\tilde{u}_\infty$. This velocity can be determined by measuring the time required for the ball to traverse the known distance between two horizontal reference marks.

If the fluid is Newtonian and the Reynolds number based on the sphere radius is less than 1.0, the flow is known as Stokes flow. For a ball falling in an infinite fluid medium, a remarkably simple expression relates the viscosity of the fluid to the terminal velocity, the sphere radius, the gravitational acceleration, and the densities of the sphere and the fluid:

$$\mu = \frac{2}{9} \frac{gR^2}{\tilde{u}_\infty} (\rho_s - \rho_l) \tag{20.14}$$

As in all experimental techniques, there are difficulties and limitations. Some of these for the falling-ball viscometer are:

1. The difficulty of obtaining spheres of the proper density so that sphere Reynolds numbers will be in the range of the Stokes solution, especially when making measurements with low-viscosity fluids.
2. The difficulty in obtaining spheres with dimensional integrity. If the sphere is not homogeneous and spherical, it may exhibit an erratic downward motion.
3. The fact that the ball is not falling in an infinite fluid medium because the fluid must be in a container. Thus a "wall correction" must be applied to Eq. (20.14) [1, p. 73].
4. The fluid must be transparent in order to observe the descent of the ball.
5. There is no unique value of the shear rate $\dot{\gamma}$, since it varies around the sphere.

In using the falling-ball viscometer for measuring the rheological properties of non-Newtonian fluids, additional difficulties arise. The most serious of these is the absence of analytical solutions for purely viscous non-Newtonian fluids similar to the Stokes solution, Eq. (20.14), for Newtonian fluids. Cho and Hartnett [1] have obtained upper- and lower-bound solutions for a power-law fluid, and Dazhi and Tanner [13] have reported a numerical solution for the same fluid. However, the wall correction factor for non-Newtonian fluids is as yet an unsolved problem.

In summary, the falling-ball viscometer is a simple device that can be used with confidence for measuring the viscosity of transparent Newtonian fluids and for measuring the low-shear-rate viscosity μ_0 of some classes of non-Newtonian fluids. For power-law fluids, at intermediate shear rates, it may also be applied, although some questions remain unresolved.

20.3.4 Falling-Needle Viscometer

The chief difficulty in obtaining solutions for the falling-ball viscometer as applied to non-Newtonian fluids is the complexity of the flow field around a sphere. Park and Irvine [14] have reported the development of a viscometer using a falling needle (cylinder), which has several advantages over the falling ball for measuring Newtonian-fluid viscosities. Park [15] describes the application of the device for the determination of the rheological properties of power-law, Cross, and Ellis fluids.

Essentially the apparatus is the same as the falling-ball viscometer (see Fig. 20.11) with the sphere replaced by a long thin cylinder (needle) with hemispherical ends which falls through the fluid in the direction of its longitudinal axis. The needle consists of a hollow glass or metal tube with wire inserts at its front end to enable the density of the needle to be varied. One advantage of this method of construction is that the falling needle is extremely stable. It is analogous to an arrow with the metal insert acting as the arrowhead and the shear stress along the sides of the needle acting as the feathers.

As is the case with the falling ball, the terminal velocity is measured from the traverse time between two horizontal reference lines. The terminal velocity must be such that the flow is Stokian around the needle ends and laminar along the needle sides.

By assuming that the needle is infinitely long (thus neglecting end effects, but taking into consideration the presence of the walls and the fluid displaced by the falling needle), the viscosity of a Newtonian fluid can be calculated from the relation

$$\mu = \frac{g\tilde{d}^2}{8\tilde{u}'_\infty}(\rho_s - \rho_l)f(\tilde{b}) \tag{20.15}$$

where $\tilde{d}$ is the needle diameter, $\tilde{b} = D/\tilde{d}$ (where D is the container diameter), and

$$f(\tilde{b}) = \frac{\tilde{b}^2[(\ln\tilde{b}) - 1] + [(\ln\tilde{b}) + 1]}{\tilde{b}^2 + 1} = (\ln\tilde{b}) - 1 \qquad \text{for} \quad \tilde{b} > 30 \tag{20.16}$$

In Eq. (20.15), $\tilde{u}'_\infty$ is not the measured terminal velocity, but the terminal velocity corrected for end effects through the aspect ratio $\tilde{a} = l/\tilde{d}$, where l is the needle length, i.e.,

$$\tilde{u}'_\infty = c_f\tilde{u}_\infty \tag{20.17}$$

$$c_f = \frac{1 + 3f(\tilde{b})/2[c_w(\tilde{a} - 1)]}{1 + 2/3(\tilde{a} - 1)} \tag{20.18}$$

$$c_w = 1 - \frac{2.04}{\tilde{b}} + \frac{2.09}{\tilde{b}^3} \tag{20.19}$$

As mentioned previously, one disadvantage of the falling-ball viscometer is that the measured results must be corrected for the presence of the container wall. In a similar way, the measured results for the falling needle must be corrected for end effects. For practical needle lengths, however, this correction is small and generally less than the wall correction for the falling-ball system. To minimize errors, Park [15] recommends the following specifications:

$$\frac{D}{\tilde{d}} \geq 30, \qquad \frac{l}{D} \geq 2.5, \qquad 0 \leq \tilde{u}_\infty \leq 10 \text{ cm/s}$$

Figure 20.12 shows measurements of the viscosity of glycerol (99.9%) by both the falling-ball and falling-needle systems [14] along with several independent measurements made in another laboratory with a rotating viscometer (Weissenberg rheogoneometer—WRG). The agreement between the falling-ball and falling-needle viscometers is excellent, and either one may be used with confidence to measure the viscosities of Newtonian fluids.

For non-Newtonian fluids, solutions have been reported by Park [15] for the falling needle in power-law, Ellis, and Cross model fluids. Since these were obtained by numerical techniques, no analytical expressions such as Eq. (20.14) or (20.15) are

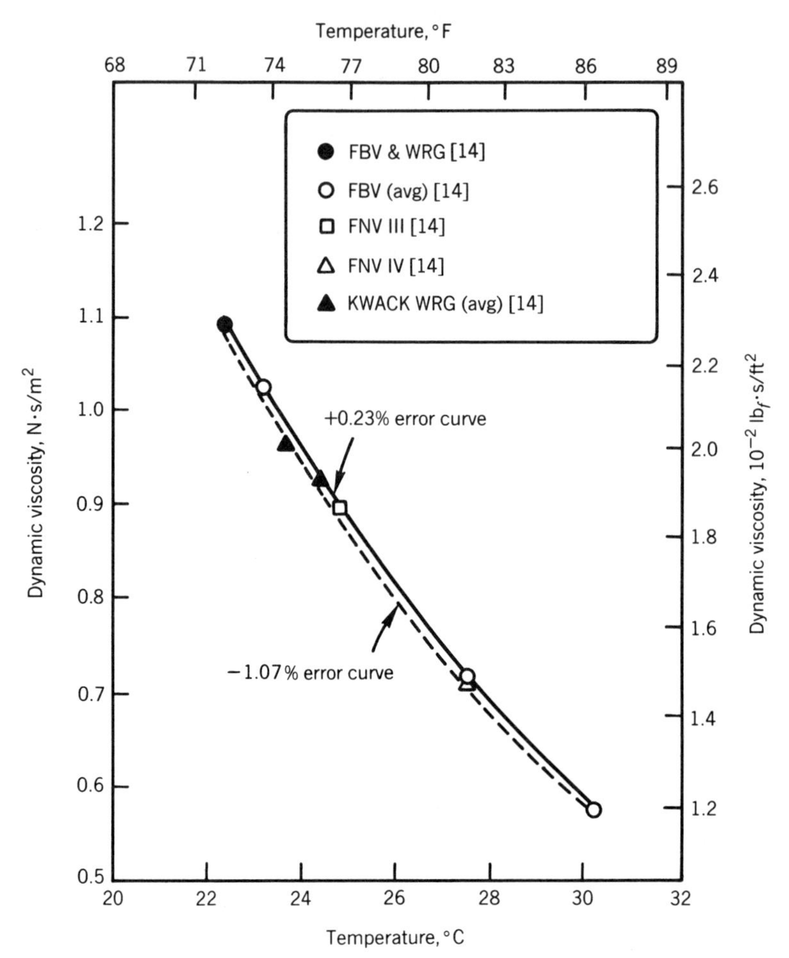

Figure 20.12. Comparison of glycerol viscosity measurements made with a falling-ball viscometer (FBV), a falling-needle viscometer (FNV), and a rotating viscometer (WRG).

available to relate the rheological properties to the measured terminal velocities. However, Park has prepared and presented data reduction tables as well as computer programs to simplify this procedure. Examples of rheological property measurements made with the falling-needle viscometer will be presented at the end of this section.

20.3.5 Rotating Viscometers

There are two primary types of rotating viscometers that are used for making absolute measurements of the constitutive equations for rheological fluids, i.e. direct measurements of the shear stress and shear rate. These are the rotating-cylinder and the cone-and-plate viscometer.

The rotating-cylinder viscometer is shown schematically in Fig. 20.13. An analysis of the system [10] for Newtonian fluids yields the working equations

$$\frac{\tilde{T}_i f(R)}{4\pi\tilde{h}} = \mu\Omega \tag{20.20}$$

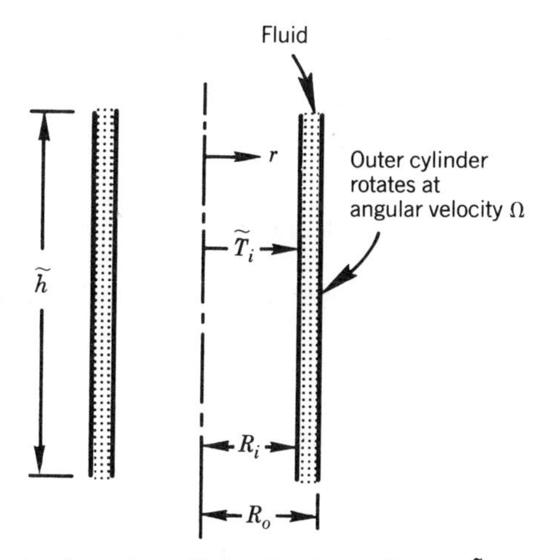

Figure 20.13. Schematic of rotating-cylinder viscometer. Torque $\tilde{T}_i$ measured on inner cylinder (radius R_i). Outer cylinder rotates with angular velocity Ω.

where

$$f(R) = \frac{1}{R_i^2} - \frac{1}{R_o^2}$$

and

$$\dot{\gamma} = \frac{2\Omega}{r^2 f(R)} \tag{20.21}$$

Equations (20.20) and (20.21) may be put into a more usable form for plotting a flow curve by noting that

$$\tau_i = \frac{\tilde{T}_i}{2\pi R_i^2 \tilde{h}} \quad \text{and} \quad \dot{\gamma}_i = \frac{2\Omega}{R_i^2 f(R)}$$

Therefore Eq. (20.20) can be written as

$$\frac{\tilde{T}_i}{2\pi R_i^2 \tilde{h}} = \mu \left[\frac{2\Omega}{R_i^2 f(R)} \right] \tag{20.22}$$

or

$$\tau_i = \mu \dot{\gamma}_i$$

By plotting $\ln(\tilde{T}_i / 2\pi R_i^2 \tilde{h})$ vs. $\ln[2\Omega / R_i^2 f(R)]$, a flow curve will be obtained whose slope is unity and whose ordinate value at $\ln[2\Omega / R_i^2 f(R)] = 0$ will be $\ln \mu$.

It should be noted that while the torque or shear stress is measured on the inner cylinder, the shear rate is measured at the outer rotating cylinder. Equation (20.22) has already transformed the shear-rate measurement to the inner cylinder through Eq. (20.21). Some viscometer manufacturers, however, do not carry out this transfer in their working equations and relate the shear stress at the inner cylinder to the shear rate at the outer cylinder. From Eq. (20.21), the ratio of the two shear rates is

$$\frac{\dot{\gamma}_i}{\dot{\gamma}_o} = \left(\frac{R_o}{R_i}\right)^2 \tag{20.23}$$

If this ratio is significant, the shear-rate transfer can be accomplished from the above equation.

Working equations may also be derived for other rheological fluids. For a power-law fluid, the analogous equations are

$$\frac{\tilde{T}_i g(R, n)}{2\pi \tilde{h}} = K\Omega^n \tag{20.24}$$

where

$$g(R, n) = \left(\frac{n}{2}\right)^n \left(\frac{1}{R_i^{2/n}} - \frac{1}{R_o^{2/n}}\right)$$

and

$$\dot{\gamma}_r = \frac{\Omega}{r^{2/n}[g(R, n)]^{1/n}} \tag{20.25}$$

Noting again that $\tau_i = \tilde{T}_i/2\pi R_i^2 \tilde{h}$ and $\dot{\gamma}_i = [\Omega/R_i^{2/n} g(R, n)^{1/n}]$, the flow curve can be plotted by using an equation obtained from Eqs. (20.24) and (20.25), and $\tau_i = K\dot{\gamma}_i^n$

$$\frac{\tilde{T}_i}{2\pi \tilde{h}} = \frac{K}{g(R, n)} \Omega^n \tag{20.26}$$

In this case, if a shear-rate correction has to be made, the appropriate relation is, from Eq. (20.25),

$$\frac{\dot{\gamma}_i}{\dot{\gamma}_o} = \left(\frac{R_o}{R_i}\right)^{2/n} \tag{20.27}$$

A schematic of the cone-and-plate rotating viscometer is shown in Fig. 20.14. An analysis [10] of the system yields the following relation between the torque $\tilde{T}_p$ measured on the plate and the angular velocity Ω of the cone, for a Newtonian fluid:

$$\tilde{T}_p = \tfrac{2}{3}\pi R^3 \mu \frac{\Omega}{\theta_0} \tag{20.28}$$

It can be seen from Eq. (20.28) that $\tau_p = \tilde{T}_p / \tfrac{2}{3}\pi R^3$ and $\dot{\gamma} = \Omega/\theta_0$. Thus the shear rate is not a function of the radius and is constant along the plate and the cone surfaces. In this case, the shear stress and shear rate are effectively measured at the same location and there is no necessity to transfer the shear rate through an equation such as Eq. (20.21), as there is for the rotating-cylinder viscometer. The Newtonian flow curve can be constructed by plotting $\ln(\tilde{T}_p / \tfrac{2}{3}\pi R^3)$ against $\ln(\Omega/\theta_0)$. The slope of the curve will be unity and the ordinate intercept at $\ln(\Omega/\theta_0) = 0$ will be $\ln \mu$.

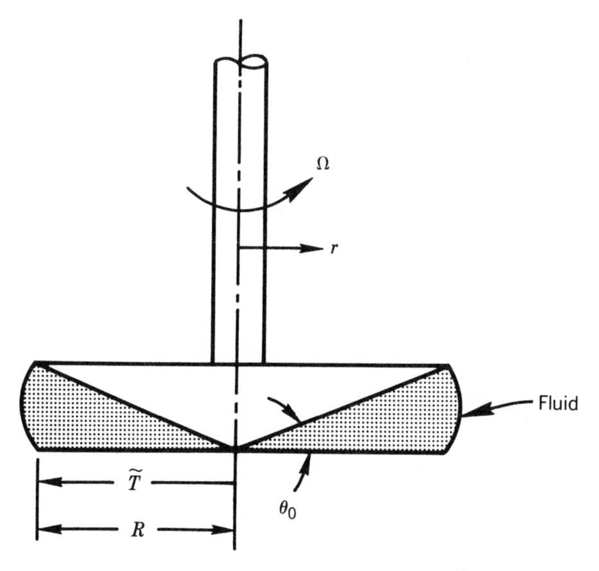

Figure 20.14. Schematic of cone-and-plate viscometer. Torque $\tilde{T}$ measured on plate; angular velocity Ω measured on rotating cone.

For a power-law fluid, the working equation of the cone-and-plate viscometer can be written as

$$\tilde{T}_p = \tfrac{2}{3}\pi R^3 K\left(\frac{\Omega}{\theta_0}\right)^n \tag{20.29}$$

and therefore if $\ln(\tilde{T}_p/\tfrac{2}{3}\pi R^3)$ is plotted against $\ln(\Omega/\theta_0)$, the slope of the line will be the flow index n and the ordinate intercept at $\ln[\Omega/\theta_0] = 0$ will be $\ln K$.

20.3.6 Representative Rheological Property Measurements

Figure 20.15 shows the measurements by Park [15] on water solutions of sodium carboxymethyl cellulose (CMC 7H4) at two different concentrations. They were made with three different viscometers: falling needle (FNV), rotating cylinder (RTN) (using four different range springs A, Aa, B, and C), and capillary tube (CTV). Thus, the different instruments can be compared in areas of data overlap, and some feeling can be acquired for the applicable shear-rate ranges of the different methods. Also included in the figure are curves of various constitutive equations which were fitted to the experimental data.

First, it is noted that the data in Fig. 20.15 cover three of the shear-rate regions illustrated in Fig. 20.6, i.e., the Newtonian, the transition I, and the power-law regions. Thus, for example, if a particular problem is analyzed by specifying power-law behavior for a CMC 10,000-wppm solution, the results are only applicable if the shear-rate range of interest is above $\dot{\gamma} = 10^2$. If the range extends below $\dot{\gamma} = 10^2$, a more general constitutive equation should be used.

In the regions of data overlap for the different instruments, it is seen that the agreement is quite satisfactory. Thus, any of the viscometers is suitable, depending

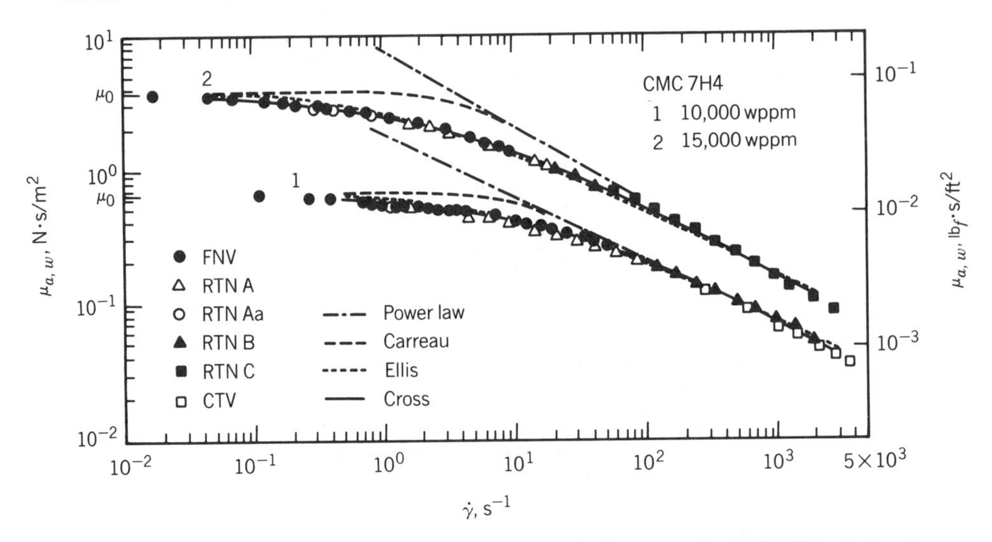

Figure 20.15. Apparent viscosity vs. shear rate and constitutive curves for CMC 7H4 solutions in water. Measurements by Park [15]. FNV falling needle; RTN, rotating-cylinder; CTV, capillary tube.

upon the shear-rate range under consideration and other factors such as instrument cost, availability of skilled technicians, fluid opacity, etc.

It is also of interest to examine the various constitutive equations which are given in Fig. 20.15 (see Table 20.2). If the different models are examined for the CMC 10,000-wppm data, it is seen that they all agree approximately in the power-law region but that there is some divergence in the transition region. It is also clear that if the power-law model is used at shear rates below its range of applicability, serious errors can occur. In this particular case, the Cross model would appear to represent the data best.

If the shear-rate ranges of the various instruments are examined in Fig. 20.15 for curve 2, it is seen that for the falling needle, $10^{-2} \leq \dot{\gamma} \leq 50$; for the rotating cylinder, $3 \times 10^{-1} \leq \dot{\gamma} \leq 2 \times 10^3$; and for the capillary tube, $3 \times 10^2 \leq \dot{\gamma} \leq 4 \times 10^3$. These ranges depend on the fluid under consideration. For polymer solutions such as those used to prepare Fig. 20.15, Fig. 20.16 gives a more general representation of the operating ranges.

20.4 PRESSURE DROPS IN NON-NEWTONIAN DUCT FLOWS

20.4.1 Introduction

When a flowing fluid enters a duct, as illustrated in Fig. 20.17, boundary layers begin to build up along the duct walls while the fluid outside the boundary layers in the core region remains inviscid and is accelerated. After the boundary layers meet, the shape of the flow (velocity) field remains constant and the fluid "forgets" that it passed through the boundary-layer region. This divides the duct into two regions: The boundary-layer or entrance region L_{hy}, where the velocity field is changing, and the fully developed region, where the velocity field remains constant. Note that the flow will only become fully developed if the duct area and fluid properties are constant and the flow is steady.

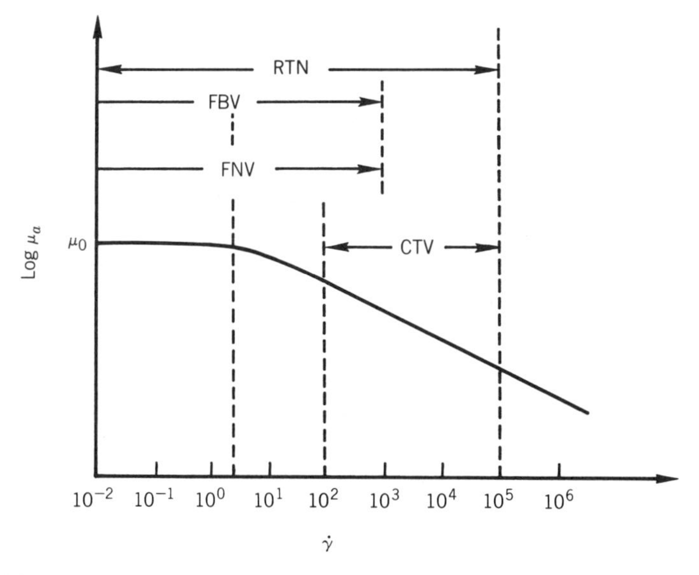

Figure 20.16. Shear-rate range (approximate) of different viscometers: rotating-cylinder viscometer (RTN), falling-ball viscometer (FBV), falling-needle viscometer (FNV) and capillary-tube viscometer (CTV).

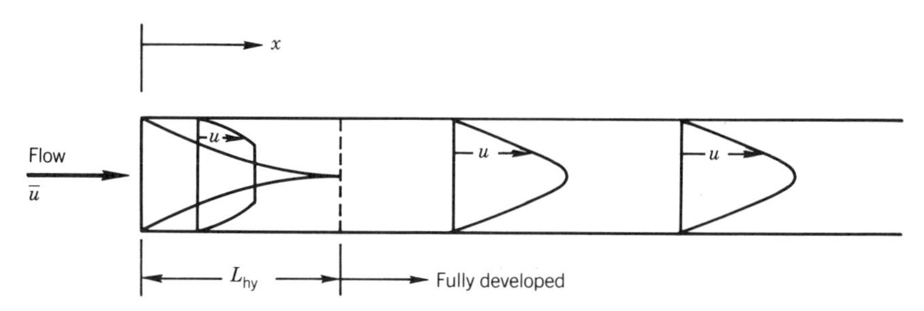

Figure 20.17. Entrance and fully developed flow regions, illustrating developing and fully developed velocity profiles.

The engineering problem in such duct flows is to predict the relation between the average flow velocity $\bar{u}$ and the pressure drop Δp over the duct length, since these two factors will determine the size of the pump required to force the fluid through the duct. This is done by defining a dimensionless pressure drop, the Fanning friction factor, and a dimensionless average velocity, the Reynolds number (see Nomenclature section):

$$\mathrm{Re} = \frac{\rho \bar{u} D_h}{\mu}, \qquad f = \frac{(dp/dx)\, D_h}{2\rho \bar{u}^2} \tag{20.30}$$

It should be noted that the above friction factor is the Fanning friction factor. The Darcy friction factor, also used in the literature, is four times the Fanning friction factor.

An analysis of the boundary-layer region will show that L_{hy} is proportional to the Reynolds number. Thus, since many non-Newtonian fluids have high effective viscosities and therefore low Reynolds numbers, the entrance lengths tend to be small and the fully developed region dominates the overall pressure drop. For these reasons, the following sections will concentrate on pressure drops in fully developed flow.

An examination of the differential equations that describe fully developed laminar duct flow will show that the friction factor is a function of the Reynolds number. For Newtonian fluids, the Reynolds number is that given in Eq. (20.30). However, for a power-law fluid the Reynolds number is defined differently and is given by

$$\mathrm{Re}_g = \frac{\rho \bar{u}^{2-n} D_h^n}{K} \tag{20.31}$$

This is called the generalized Reynolds number. When $n = 1$ (Newtonian fluid), the generalized Reynolds number becomes the same as given in Eq. (20.30). Since the definition is somewhat arbitrary, a number of generalized Reynolds numbers appear in the rheological literature. As is the case with friction factors, care must be exercised with regard to the definition. This point is discussed in greater detail by Cho and Hartnett [1]. Two other generalized Reynolds numbers in particular often appear. These are

$$\mathrm{Re}'_g = \mathrm{Re}_{\mathrm{eff}} = \frac{\mathrm{Re}_g 8^{1-n}}{[(3n+1)/4n]^n} \tag{20.32}$$

and

$$\mathrm{Re}_a = \frac{\rho \bar{u} D_h}{\mu_a} = \mathrm{Re}_g 8^{1-n} \left(\frac{3n+1}{4n} \right)^{1-n} \tag{20.33}$$

In this work, unless otherwise specifically stated, the generalized Reynolds number Re_g defined by Eq. (20.31) will be used.

20.4.2 Transition to Turbulence

For the flow of a Newtonian fluid through a circular tube, it is accepted practice to consider that when the Reynolds number, as defined by Eq. (20.30), becomes greater than approximately 2100, transition from laminar to turbulent flow occurs in a circular tube. Such a criterion is also needed for the flow of rheological fluids. While investigations to date have not produced any general agreement, there is evidence that transition occurs when Re_g exceeds approximately 2100 for a power-law fluid in a circular tube. The cautious designer, when operating between the limits $1500 \le \mathrm{Re}_g \le 3000$, will consider both cases of laminar and turbulent flow and choose a conservative design. Subsequent data which will be presented will lend credence to the above criterion.

20.4.3 Fully Developed Laminar Pressure Drops

Kozicki and his coworkers [16–19] have presented a general approximate method for calculating laminar pressure drops for rheological fluids flowing in ducts and open channels. They considered a wide variety of cross-sectional shapes and power-law, Bingham, and Ellis fluids.

For power-law fluids, the fully developed pressure drop can be calculated from

$$f \, \mathrm{Re}_g = 2^{3n+1}\left(\frac{a + bn}{n}\right)^n \tag{20.34}$$

where a and b are geometric parameters depending upon the cross-sectional shape of the duct. For circular ducts $a = 0.25$, $b = 0.75$, and for parallel-plate ducts $a = 0.5$, $b = 1.0$. Table 20.3 lists values of additional geometric parameters for a large number of different duct cross sections [16, 18]. Similar information for open-channel flow is available [17, 19]. For open-channel flow, the friction factor is defined as

$$f = \frac{D_h g \sin \theta}{2\bar{u}^2} \tag{20.35}$$

where θ is the inclination angle of the open channel to the horizontal. The generalized Reynolds number for open-channel flow is defined the same as Eq. (20.31).

The method developed by Kozicki and his coworkers is approximate. For all geometries except the circular and the parallel plate, the varying wall shear stress around the periphery is replaced by an average wall stress. Thus, one would expect the approximation to become less realistic for geometries that have widely varying peripheral wall shear stresses. Chang and Irvine [20] tested the Kozicki method experimentally, and Cheng [21] tested it numerically, by looking at $f \, \mathrm{Re}_g$ predictions in a narrow (10°) isosceles triangular duct. The results are shown in Fig. 20.18, where it is seen that the numerical and experimental results are close and the predictions of Kozicki et al. are 7% higher.

Kozicki et al., in their original paper [16], compared their method with a numerical solution by Wheeler and Wissler [22] for a rectangular duct. The results for a square duct are given in Table 20.4, where it is seen that the deviations are less than 5%.

In a subsequent paper, Kozicki and Tiu [23] presented an improved method which increases the number of geometric parameters. However, for many engineering calculations, where an accuracy of the order of 5% is suitable, the first method utilizing Eq. (20.34) would appear to be adequate.

It should once again be emphasized that the Kozicki method as applied to power-law fluids is only valid if the operating shear-rate range is in the power-law region (Region c in Fig. 20.6). If region a or b is included, then the Kozicki method as applied to an Ellis model fluid would be appropriate (see Ref. 16).

20.4.4 Fully Developed Turbulent Pressure Drops

In many engineering calculations for turbulent flow, the shear-rate range falls in region c of Fig. 20.6. Thus the power-law model is applicable. For power-law fluids flowing turbulently in circular tubes, Dodge and Metzner [24] derived the following relation between the friction factor and the generalized Reynolds number:

$$\frac{1}{f^{1/2}} = \frac{1.733}{n^{0.75}} \ln\left[\mathrm{Re}_g'(f)^{1-(n/2)}\right] - \frac{0.40}{n^{1.2}} \tag{20.36}$$

Equation (20.36) agreed quite well with their experimental data and also with subsequent measurements by Yoo [25]. Since the Reynolds number is generally an independent variable, one problem with Eq. (20.36) is that it cannot be solved explicitly for the

TABLE 20.3 Duct Flows: Geometric Constants for Use with Eq. (20.34)[a]

Geometry	α^*	a	b
Concentric annuli, $\alpha^* = \dfrac{d_i}{d_o}$	0.1	0.4455	0.9510
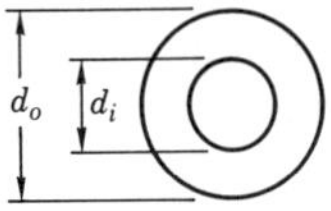	0.2	0.4693	0.9739
	0.3	0.4817	0.9847
	0.4	0.4890	0.9911
	0.5	0.4935	0.9946
	0.6	0.4965	0.9972
	0.7	0.4983	0.9987
	0.8	0.4992	0.9994
	0.9	0.4997	1.0000
	1.0[b]	0.5000	1.0000
Rectangular, $\alpha^* = \dfrac{c}{d}$	0.0	0.5000	1.0000
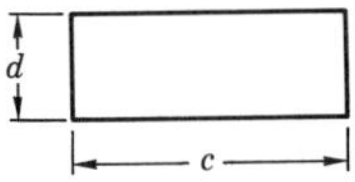	0.25	0.3212	0.8482
	0.50	0.2440	0.7276
	0.75	0.2178	0.6866
	1.00	0.2121	0.8766
Elliptical, $\alpha^* = \dfrac{c}{d}$	0.00	0.3084	0.9253
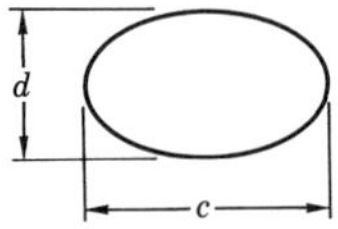	0.10	0.3018	0.9053
	0.20	0.2907	0.8720
	0.30	0.2796	0.8389
	0.40	0.2702	0.8107
	0.50	0.2629	0.7886
	0.60	0.2575	0.7725
	0.70	0.2538	0.7614
	0.80	0.2515	0.7546
	0.90	0.2504	0.7510
	1.00[c]	0.2500	0.7500
	2ϕ (deg)		
Isosceles triangular (2ϕ)	10	0.1547	0.6278
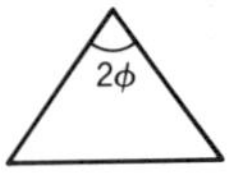	20	0.1693	0.6332
	40	0.1840	0.6422
	60	0.1875	0.6462
	80	0.1849	0.6438
	90	0.1830	0.6395
	N		
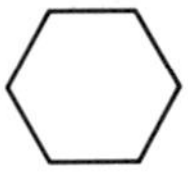 Regular polygon (N sides)	4	0.2121	0.6771
	5	0.2245	0.6966
	6	0.2316	0.7092
	8	0.2391	0.7241

[a] Data from Refs. 16–19.

[b] Parallel plates.

[c] Circle.

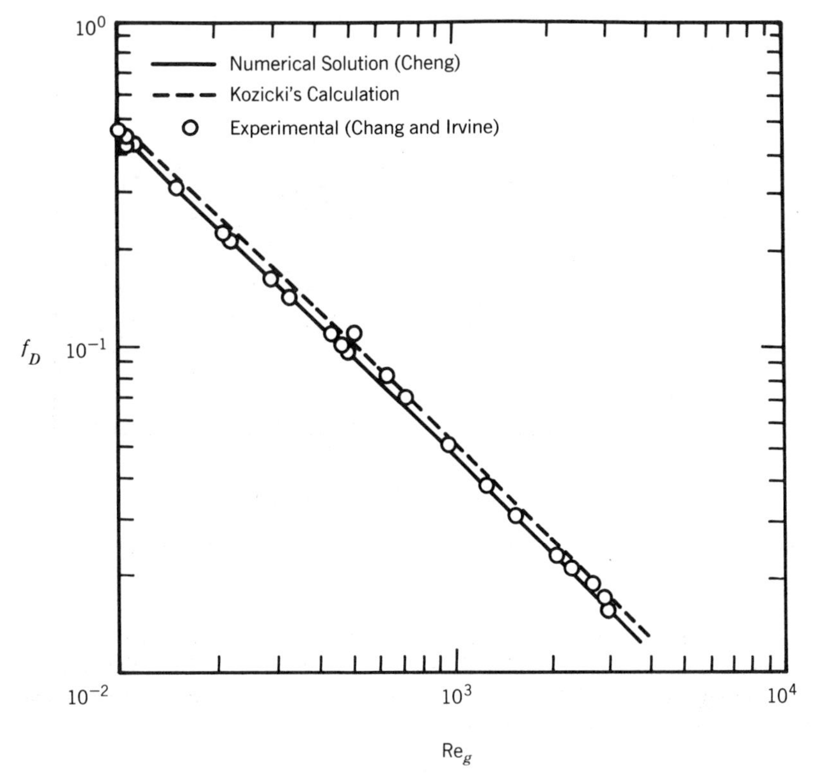

Figure 20.18. Comparison of the approximate calculation by Kozicki et al. [16] with the numerical solution by Cheng [21] and experiments by Chang and Irvine [20] for a 10° isosceles triangular duct containing a solution of 1000 wppm CMC, $n = 0.89$.

TABLE 20.4 Comparison of Numerical Solution for f_D Re_g by Wheeler and Wissler with Method of Kozicki et al. for a Power-Law Fluid in a Square Duct

	f_D Re_g		
Flow Index	Kozicki et al. [16]	Wheeler and Wissler [22]	Deviation (%)
1.0	56.88	56.91	−0.05
0.9	47.87	47.62	0.50
0.8	40.24	39.66	1.47
0.75	36.89	36.22	1.85
0.7	33.80	33.07	2.21
0.6	28.36	27.53	3.02
0.5	23.74	22.89	3.71
0.4	19.81	18.97	4.59

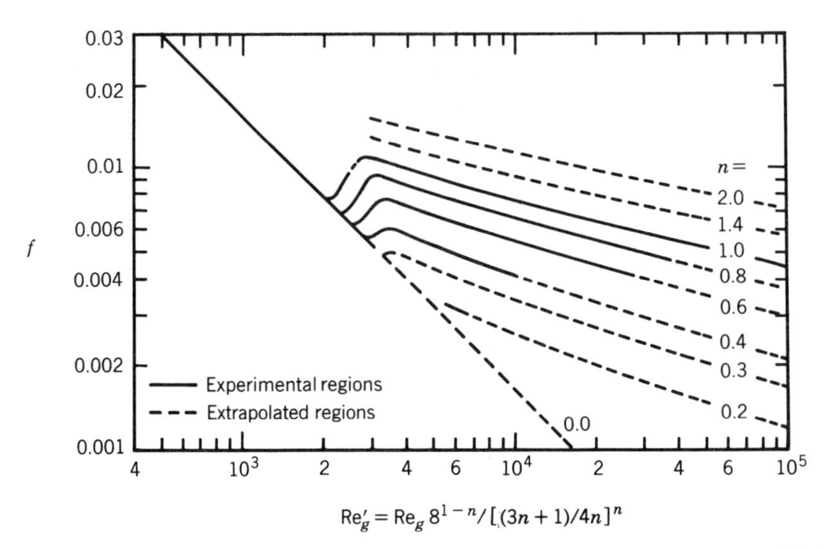

Figure 20.19. Dodge and Metzner's [24] relation between Fanning friction factor and Re'_g.

friction factor. Figure 20.19 from [24], is a graphical representation of Eq. (20.36) which makes it simpler to determine the friction factor from the Reynolds number.

A useful engineering relation where the friction factor and a Reynolds number are related explicitly has been developed by Irvine [26]:

$$f = \frac{F(n)}{\mathrm{Re}_g^{1/(3n+1)}} \tag{20.37}$$

where

$$F(n) = \frac{2(2/7^7)^{1/(3n+1)}}{[(3n+1)/4n]^n}$$

When $n = 1$ (Newtonian fluid), Eq. (20.37) reduces to the well-known Blasius equation

$$f = \frac{0.079}{\mathrm{Re}^{1/4}} \tag{20.38}$$

Figure 20.20 illustrates how the data of Dodge and Metzner [24] and Yoo [25] are correlated by the generalized Blasius relation given by Eq. (20.37).

For noncircular ducts with fully developed turbulent flow of power-law fluids, very little work has been done up to the present time. Kostic and Hartnett [27] have suggested a technique which consists of inserting the Reynolds number defined by Kozicki et al. [16] into the Dodge-Metzner equation [Eq. (20.36)] for circular ducts. In the present nomenclature, this Reynolds number is given by

$$\mathrm{Re}_{g,K} = \mathrm{Re}'_g \frac{3n+1}{4(a+bn)} \tag{20.39}$$

where a and b are the geometric constants given in Table 20.3.

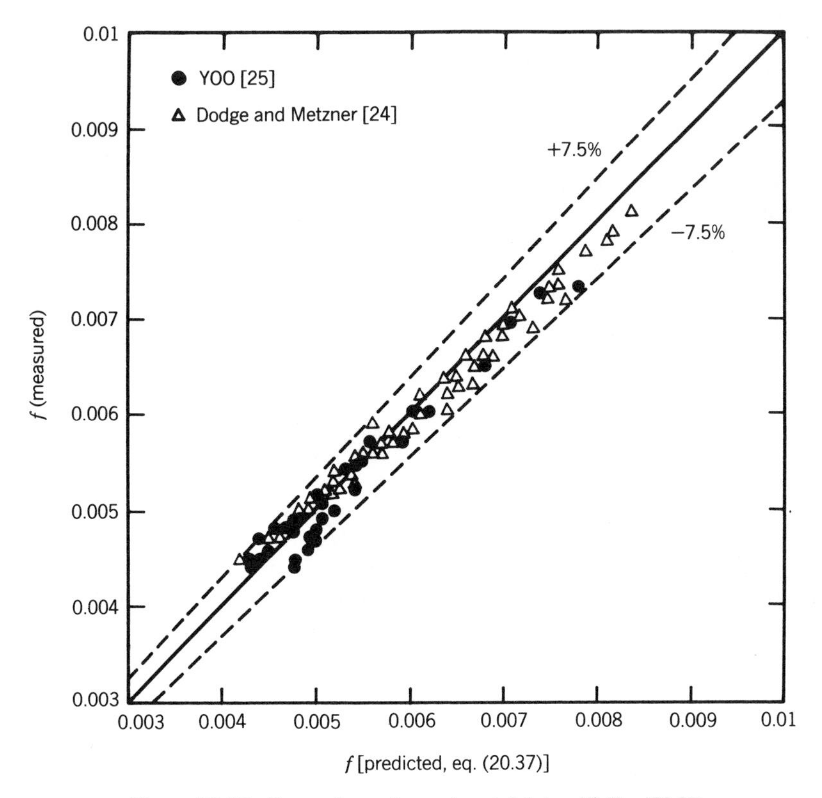

Figure 20.20. Comparison of experimental data with Eq. (20.37).

Using this technique, Kostic and Hartnett [27] found good agreement with their measurements on a square duct, and Cheng [28] also found experimental agreement in an equilateral triangular duct. Both investigators warned, however, that the technique was not valid with extreme geometries having narrow corner regions. The noncircular duct problem is also discussed in [26].

20.5 HEAT TRANSFER IN NON-NEWTONIAN DUCT FLOWS

20.5.1 Introduction

In addition to hydrodynamic boundary layers (see Fig. 20.17), thermal boundary layers are also present when a fluid entering a duct has a temperature different from the channel walls. Thus, a thermal entrance region exists, and the duct can be divided thermally into two regions: the thermal entrance region and the region of the thermally fully developed flow. Note that for thermally fully developed flow to exist, the velocity field must already be fully developed. Because rheological fluids often have high effective viscosities, the hydrodynamic entrance region is much shorter than the thermal entrance region. Therefore, in this section only fully developed and developing thermal flows will be considered. In both cases, it will be assumed that the velocity profile is already fully developed.

The engineering problem related to duct heat transfer is to be able to predict either the heat transfer q from the duct walls to the fluid for a given temperature difference, or the wall-to-fluid bulk temperature difference for a given heat flow. In either case, this can be conveniently done by defining a heat transfer coefficient h:

$$q'' = h(T_w - T_b) \tag{20.40}$$

As is the case of the pressure drop, where a dimensionless friction factor was used, a dimensionless heat transfer coefficient, called the Nusselt number, can be defined as

$$\text{Nu} = \frac{hD_h}{k} \tag{20.41}$$

As discussed previously, an examination of the equation of motion revealed that the friction factor in fully developed duct flow of a power-law fluid was a function of Re_g, the generalized Reynolds number of Eq. (20.31). Therefore the Nusselt number, which depends upon the flow field, will also usually depend upon Re_g and an additional heat transfer parameter called the generalized Prandtl number. An examination of the energy equation of the fluid will show that the Prandtl number for a power-law fluid in a circular duct has the form

$$\text{Pr}_g = \frac{Kc_p}{k}\left(\frac{\bar{u}}{D_h}\right)^{n-1} \tag{20.42}$$

Similar to the generalized Reynolds number, other generalized Prandtl numbers are also used in the rheological heat transfer literature. The two most often cited are

$$\text{Pr}'_g = \text{Pr}_{\text{eff}} = \frac{Kc_p}{k}\left(\frac{3n+1}{4n}\right)^n\left(\frac{8\bar{u}}{D_h}\right)^{n-1} \tag{20.43}$$

$$\text{Pr}_a = \frac{\mu_a c_p}{k} = \frac{Kc_p}{k}\left(\frac{3n+1}{4n}\right)^{n-1}\left(\frac{8\bar{u}}{D_h}\right)^{n-1} \tag{20.44}$$

Another consideration related to the heat transfer problem in rheological duct flow is the variation of rheological properties with temperature. For power-law fluids, which have been most often investigated, it has been found that the flow index n is essentially temperature-independent, while the fluid consistency K is a significant function of temperature. Thus, when there are large temperature differences in the duct cross section, the temperature-consistency problem must be considered. This effect will be discussed later in the appropriate sections.

An additional complication in heat transfer problems is the influence of the thermal boundary conditions on the resulting heat transfer coefficient or Nusselt number. Unlike the velocity field, which has only one general boundary condition (i.e., the fluid velocity is zero on any solid surface), a variety of thermal boundary conditions can prevail. It is now generally recognized [29, 30] that the most important and limiting thermal boundary conditions can be conveniently organized to include both circular and noncircular cross sections as follows:

	Thermal Boundary Condition	Symbol
1.	Constant wall temperature in both the flow and circumferential directions	Ⓣ
2.	Constant heat flux in the flow direction and constant temperature in the circumferential direction	(H1)
3.	Constant heat flux in the flow direction and constant heat flux in the circumferential direction	(H2)

It should be noted that because of symmetry, for circular and parallel-plate ducts, the (H1) and (H2) conditions are identical, and they will simply be referred to as Ⓗ. For all other shapes, they are distinct.

20.5.2 Fully Developed Laminar Heat Transfer

Solutions are available from Bird et al. [31] for the circular tube for thermal boundary conditions Ⓣ and Ⓗ as follows:

For the Ⓣ boundary condition:

$$\mathrm{Nu_T} = \beta_1^2 \tag{20.45}$$

An approximate method of calculating β_1 is presented in [31]. For $n = 1.0$, 0.5 and 0.33 the values of Nu_T are 3.657, 3.949 and 4.175 respectively.

For the Ⓗ boundary condition:

$$\mathrm{Nu_H} = 8\frac{(3n + 1)(5n + 1)}{31n^2 + 12n + 1} \tag{20.46}$$

In an alternative form by Grigull [32],

$$\mathrm{Nu_H} = 4.36\left(\frac{3n + 1}{4n}\right)^{1/3} \tag{20.47}$$

For noncircular shapes, only a few solutions are available. For parallel plates, the work

TABLE 20.5 Fully Developed Laminar Flow Nusselt Numbers for a Square Duct [34]

Flow behavior index n	$\mathrm{Nu_T}$	$\mathrm{Nu_{H1}}$	$\mathrm{Nu_{H2}}$
1.0	2.975	3.612	3.095
0.9	2.997	3.648	3.106
0.8	3.030	3.689	3.135
0.75	3.050	3.713	3.152
0.7	3.070	3.741	3.171
0.6	3.120	3.804	3.216
0.5	3.184	3.889	3.274

TABLE 20.6 Fully Developed Laminar Flow Nusselt Numbers for Isosceles Triangular Ducts, Calculated Numerically by Cheng [21]

Apex Angle (deg)	n	Nu_{H1}	Nu_{H2}
10	0.6	2.434	0.0846
	1.0	2.391	0.0792
	1.2	2.374	0.0756
60	0.6	3.250	1.933
	1.0	3.101	1.891
	1.2	3.060	1.882
90	0.6	3.098	1.356
	1.0	2.974	1.351
	1.2	2.946	1.350

of Yau and Tien [33] may be consulted. For a square duct, the values of Nu_T, Nu_{H1}, and Nu_{H2} have been calculated by Chandrupatla and Sastri [34] and are listed in Table 20.5. For an isosceles triangular duct, the values of Nu_{H1} and Nu_{H2} as calculated by Cheng [21] are given in Table 20.6.

For other cross-sectional shapes, an interesting correlation has been suggested by Cheng [21]. He proposed that a correction factor, such as is used in Eq. (20.47), be applied to the appropriate Newtonian Nusselt numbers to take account of the power-law effect. The correction factor he suggests is

$$\text{correction factor} = \left(\frac{(a + bn)}{(a + b)n} \right)^{1/3} \tag{20.48}$$

where a and b are the Kozicki geometric factors presented in Table 20.3. Thus, the correlation equation would be

$$\frac{Nu_{\text{power-law}}}{Nu_{\text{Newtonian}}} = \left(\frac{(a + bn)}{(a + b)n} \right)^{1/3} \tag{20.49}$$

for any thermal boundary condition.

The use of Cheng's correction factor, for example, agrees with the numerically obtained power-law Nusselt numbers in Table 20.5 for a square duct with a maximum difference of 1.5% as long as the value of $Nu_{\text{Newtonian}}$ is correct for the thermal boundary condition under consideration. For the Nusselt numbers for isosceles triangular ducts given in Table 20.6, the difference from the numerical solution is at most 4.35% and generally less than 2.5%. In the absence of numerical solutions for fully developed Nusselt numbers for power-law fluids, Eq. (20.49) should serve as a useful approximation, since the values of $Nu_{\text{Newtonian}}$ are available for a large number of cross-sectional shapes [30].

20.5.3 Laminar Heat Transfer in Thermally Developing Flow in Circular Tubes

As discussed in Sec. 20.5.1, since many rheological fluids have a large apparent viscosity, the velocity field becomes fully developed more rapidly than the temperature field. It was also mentioned that for many power-law fluids the fluid consistency has a significant variation with temperature, while the flow index is relatively independent of temperature. Thus, solutions which account for thermally developing flow plus the variation of K with temperature are of engineering interest.

Such solutions have been obtained numerically for a circular tube by Joshi and Bergles [35, 36] and compared with their experiments for fluids in the flow index range $0.2 \leq n \leq 1.0$. In their numerical solution, they considered that n was constant and that K had a temperature variation of the type

$$K = a_1 e^{-a_2 t} \tag{20.50}$$

The final equations which they recommend are given by the following expressions for the two primary boundary conditions:

For the Ⓣ boundary condition:

$$\mathrm{Nu}_{x,\mathrm{T}} = \frac{h_x D_h}{k} = 3.66\left\{\left(\frac{3n+1}{4n}\right)^{1/3}\left(\frac{K}{K_w}\right)^{0.11}\right\} \times \left\{1.0 + \left[c_1\left(\frac{K}{K_w}\right)^m\right]^{15}\right\}^{1/15} \times \left\{1.0 + [0.424\,\mathrm{Gz}^{0.34}]^3\right\}^{1/3} \tag{20.51}$$

$c_1 = d_1 \exp(-d_z \gamma \Delta T)$ $\quad \hat{\gamma}\Delta T = a_2(T_w - T_0)$

$m = f_1 \exp(-f_2 \hat{\gamma} \Delta T)$ $\quad f_1 = 2.8308 - 4.103n + 1.9327n^2$

$d_1 = 0.4230 + 0.6428n$ $\quad f_2 = 0.215 - 0.22n$

$d_2 = 0.69 - 0.1n$ $\quad \mathrm{Gz} = \dot{m}c_p/kx$

$$\hat{\gamma} = -\frac{1}{K}\frac{dK}{dT} = a_2$$

T_0 = fluid temperature when heating begins

For the Ⓗ boundary condition,

$$\mathrm{Nu}_{x,\mathrm{H}} = \frac{h_x D_h}{k} = 4.36\left(\frac{3n+1}{4n}\right)^{1/3} K^+\left\{1 + \left[(0.376x^+)^{-0.303}\right]^8\right\}^{1/8} \tag{20.52}$$

$$K^+ = \begin{cases} 1 + (0.1232 - 0.054n)\hat{\gamma}\Delta T \\ \quad -(0.0101 - 0.0068n)(\hat{\gamma}\Delta T)^2 & \text{for fully developed region} \\ \left(\dfrac{K}{K_w}\right)^{0.58-0.44n} & \text{for entrance region} \end{cases}$$

$$x^+ = 2\frac{x/D_h}{\mathrm{Re_{eff}}\,\mathrm{Pr_{eff}}} \qquad \gamma = -\frac{1}{K}\left[\frac{dK}{dT}\right]$$

$$\mathrm{Gz} = \frac{\dot{m}c_p}{kx} = \frac{\pi}{4}\frac{\mathrm{Re_{eff}}\,\mathrm{Pr_{eff}}}{x/D_h} \qquad \Delta T = \frac{q_w D_h}{2k}$$

Note that all unsubscripted fluid properties in Eqs. (20.51) and (20.52) are evaluated at the fluid bulk mean temperature.

20.5.4 Fully Developed Turbulent Heat Transfer

For heat transfer through turbulent fully developed flow in circular tubes, two approaches have been reported. Metzner and Friend [37] utilized the analogy between heat and momentum transfer to derive an equation of the form

$$\mathrm{St}_a = \frac{\mathrm{Nu}}{\mathrm{Re}_a \mathrm{Pr}_a} = \frac{f/2}{1.20 + 11.8\sqrt{f/2}\,(\mathrm{Pr}_a - 1)\mathrm{Pr}_a^{-1/3}} \tag{20.53}$$

Here f is given by Eq. (20.36). Using Eq. (20.53), they were able to correlate turbulent heat transfer data within $\pm 25\%$ for power-law fluids having a flow index range $0.39 \le n \le 1.0$.

Other investigators presented experimental correlations which are simpler to use in engineering calculations. Clapp [38] was able to represent the results of 11 experimental runs of power-law fluids within $+2\%$ and -4.5% by using a generalized form of the well known Dittus-Boelter equation for Newtonian fluids:

$$\mathrm{Nu} = 0.023(9350)^{0.8[1-(1/n'')]}\mathrm{Re}_{\mathrm{eff}}^{(0.8/n'')}\mathrm{Pr}_{\mathrm{eff}}^{0.4} \tag{20.54}$$

Yoo [25] performed experiments on power-law fluids (aqueous solutions of Carbopol and slurries of Attagel) with n ranging from 0.24 to 0.9. He was able to correlate the data with an average deviation of 2.3% with the equation

$$\mathrm{St}_a \mathrm{Pr}_a^{2/3} = 0.0152\,\mathrm{Re}_a^{-0.155} \tag{20.55}$$

Yoo recommends his correlation over the n range from 0.2 to 0.9 and an apparent Reynolds number range from 3,000 to 90,000.

It should be mentioned that none of the above turbulent heat transfer equations specifies the type of thermal boundary conditions which are applicable as was done in the section on laminar flow. This is because in turbulent flow in circular tubes, the thermal boundary conditions have much less effect—in fact usually less than the uncertainty of the data. For turbulent heat transfer in noncircular ducts where the velocity and temperature fields are not symmetrical, the thermal boundary effects may be significant.

For fully developed turbulent heat transfer in noncircular ducts for non-Newtonian fluids, no studies have yet appeared in the literature. This remains an area for future investigations.

20.6 FREE AND MIXED CONVECTION WITH NON-NEWTONIAN FLUIDS

20.6.1 Introduction

Free-convection problems introduce a new difficulty into fluid mechanics and heat transfer. This occurs because the driving forces for fluid motions are the buoyancy forces which are caused by density or temperature differences. Thus the velocity and temperature fields are coupled, and the momentum and energy equations can no longer be solved independently as they often can with forced convection.

When considering free convection with rheological fluids, these difficulties are compounded by the nonlinearity of the shear forces. For these reasons, most available solutions for free convection with non-Newtonian fluids rely on either approximate analytical or numerical techniques. Few "exact" or similarity solutions exist for laminar flows.

Mixed flows, where both free and forced convection are present, increase the complexities of the problem, since new flow parameters must be introduced. In addition, when the forced and free forces act in opposite directions, it is possible for flow instabilities to arise which invalidate many of the simple physical models which are so useful in analysis.

Also, in the case of rheological fluids a unique difficulty arises. Because the velocities in free and mixed convection are low, so are the shear rates. Hence, care must be exercised to ensure that the constitutive equations used are appropriate for the shear-rate range of the particular problem under consideration.

As an example of this concern, consider the following simple and approximate calculation. It is known from experience that for a vertical plate with a free-convection boundary layer, the range of average velocities can be of the order of 1 to 10 mm/s, while the boundary layer thicknesses are in the range of $1 \leq \delta \leq 10$ mm. If we calculate an approximate shear rate equal to $\dot{\gamma} = \bar{u}/\delta$, then the shear-rate range will be of the order of $10^{-1} \leq \dot{\gamma} \leq 10^{1}$.

Reference to Fig. 20.15 shows that for such conditions the shear-rate range falls in the transition region between the low-shear-rate Newtonian region and the power-law region for that particular polymer solution. Thus, while only a three-parameter constitutive equation would be appropriate in this case, most available solutions are for power-law fluids. It is again evident that existing solutions must be used with caution, and a match must exist between the shear-rate range of any engineering problem and that of the applicability of particular solutions.

Another aspect of free convection problems is that the effects of wall shear stresses are usually of secondary interest. For flows over external surfaces, the structural considerations relating to wall shear are normally not as important as the heat flow from the external surface to the fluid. For internal or duct flows, both the wall heat flux and the fluid velocity caused by the buoyancy forces are of interest. Consequently, these quantities will be emphasized.

In this section, a concise summary will be presented of the analytical and experimental investigations which have been reported in the literature. In addition, some useful engineering relations will be presented for several of the more common geometries which will allow estimates to be made of the heat transfer in free and mixed convection systems with rheological fluids.

20.6.2 Summary of Previous Investigations

A recent review article by Shenoy and Mashelkar [39] discusses both free and mixed convection in rheological fluids. In addition to the investigations considered in this excellent article, an additional number of studies have been reported in the literature. Table 20.7 is a summary of most of these previous investigations, both analytical and experimental.

Table 20.7 is arranged so that it is entered through the geometry under consideration. Also included are the thermal boundary conditions, the constitutive equation used in the analysis, and the existence of any experimental measurements.

Several observations can be made concerning Table 20.7. First, it is seen that almost without exception, the power-law constitutive equation was used without regard to the

TABLE 20.7 Some Previous Investigations of Steady Laminar Free and Mixed Convection with Rheological Fluids

Geometry	Thermal Boundary Conditions	Free or Mixed	Constitutive Equation	Analysis (Ref.)	Experiment (Ref.)
Vertical plate	T_w = const	Free	Power law	[40–42]	[42, 49]
	q_w = const	Free	Power law	[41, 43–45]	[43]
	T_w = const	Free	Sutterby	[46]	[47]
	q_w = const	Free	Sutterby	[46]	[47]
	q_w = const	Free	Ellis	[48]	[49]
	T_w = const	Free	Bingham plastic[a]	[50]	—
	T_w = const	Free	Power law[b]	[51]	—
Vertical plates	T_w = const	Free	Power law	[64]	—
	q_w = const	Free	Power law	[65]	—
Horiz. cylinder	q_w = const	Free	Power law	[66]	[66]
	T_w = const	Free	Power law	[40, 52]	[66]
Vertical cone	T_w = const q_w = const $T_w = Ax$	Free	Power law	[53]	—
Sphere	T_w = const	Free	Power law	[40, 54]	[54]
Vertical plate	T_w = const	Mixed	Power law	[55]	—
Horiz. tube	T_w = const	Mixed	Power law	[56, 57]	[57]
Vertical tube	T_w = const	Mixed	Power law	[58, 59, 61, 62]	—
	q_w = const	Mixed	Power law	[58, 60]	—
Vertical annulus		Mixed	Power law	[67]	—

[a] Transient solution.
[b] Turbulent flow.

shear-rate range matching problem. Next, it is obvious that there are many more analytical than experimental investigations. This is not unusual, especially since the development of computers that allow the numerical solution of nonlinear problems which may be difficult or impossible to solve by standard mathematical techniques. Nevertheless, the disparity calls attention to the fact that many more experimental investigations are needed to validate the ranges of applicability of existing analytical models or indeed to confirm the actual physical reality of the models themselves. For those wishing additional information on a particular free or mixed convection system, the appropriate reference in Table 20.7 may be consulted.

20.6.3 Engineering Relations for Free and Mixed Convection Systems

Given below are a number of useful relations, accompanied in some cases by tables and graphs. These relate to the most common geometries which might be encountered in engineering practice. Whenever available, results were chosen from previous studies which were accompanied by experimental measurements.

TABLE 20.8 Values of $T(n)$ in Eq. (20.56) as Reported by Shenoy and Mashelkar [39]

	$T(n)$		
n	[42]	[40]	[41]
0.5	0.5957	0.63	0.6098
1.0	0.6775	0.67	0.6838
1.5	0.7194	0.71	0.7229

Single Vertical Plate— T_w = Constant, Free Convection. The average Nusselt number for a power-law fluid based on the plate height is given by (analyses by Shenoy and Ulbrecht [42], experiments by Shenoy and Ulbrecht [42] and Reilly et al. [49])

$$\overline{\mathrm{Nu}}_L = T(n)\mathrm{Gr}_{L,T}^{1/(2n+2)}\mathrm{Pr}_{L,T}^{n/(3n+1)} \tag{20.56}$$

where

$$\mathrm{Gr}_{L,T} = \frac{\rho^2 L^{n+2}}{K^2}[g\beta(T_w - T_\infty)]^{2-n} \tag{20.57}$$

$$\mathrm{Pr}_{L,T} = \frac{\rho c_p}{k}\left(\frac{K}{\rho}\right)^{2/(n+1)} L^{(n-1)/(2n+2)}[g\beta(T_w - T_\infty)]^{3(n-1)/(2n+2)} \tag{20.58}$$

The function $T(n)$ in Eq. (20.56) is given in Table 20.8 as reported by Shenoy and Mashelkar [39]. The table shows the results of the analyses of Shenoy and Ulbrecht [42], Acrivos [40], and Tien [41] for comparison. It is seen that the agreement is quite satisfactory.

Single Vertical Plate— q_w = Constant, Free Convection. In external flows, the thermal boundary condition q_w = constant denotes a constant heat flux in the flow direction. For two dimensional flows, e.g., a flat plate or an infinitely long cylinder, the surface temperature will be constant in the direction transverse to the flow direction.

For the case of the thermal boundary condition q_w = constant, since the heat flux is prescribed, the quantity of interest is the wall temperature distribution, which can be obtained from the local Nusselt number. For a power-law fluid, this can be expressed by (analyses by Dale and Emery [43], Shenoy [45], and Tien [41]; experiments by Dale and Emery [43])

$$\mathrm{Nu}_x = \bar{C}\left[\mathrm{Gr}_{x,q_w}^{(3n+2)/(n+4)}\mathrm{Pr}_{x,q_w}^{n}\right]^{\bar{B}} \tag{20.59}$$

where

$$\mathrm{Gr}_{x,q_w} = \frac{\rho^2 x^4}{K^2}\left(\frac{g\beta q_w}{k}\right)^{2-n} \tag{20.60}$$

$$\mathrm{Pr}_{x,q_w} = \frac{\rho c_p}{k}\left(\frac{K}{\rho}\right)^{5/(n+4)} x^{2(n-1)/(n+4)}\left(\frac{g\beta q_w}{k}\right)^{3(n-1)/(n+4)} \tag{20.61}$$

Values of the coefficients, $\bar{C}$ and $\bar{B}$ as reported by Shenoy and Mashelkar [39] are given in Table 20.9, which compares the results of the analyses with experiment. It is

TABLE 20.9 Values of Constants in Eq. (20.59) as Reported by Shenoy and Mashelkar [39]

$\bar{C}$	$\bar{B}$	Ref.
0.600[a]	0.2101[a]	[43]
0.590	0.2144	[43]
0.692	0.2144	[41]
0.593	0.2144	[45]

[a] Experiment.

seen in the table that except for the value of $\bar{C}$ from the analysis by Tien, which is slightly high, the other values are remarkably consistent.

Single Sphere— T_w = Constant, Free Convection. For a power-law fluid, the experimental correlation of the average Nusselt number by Amato and Tien [54], which shows good agreement with the analysis of Acrivos [40], is given by

$$\overline{\mathrm{Nu}_R} = C_s Z^{\tilde{D}} \tag{20.62}$$

where

$$Z = \mathrm{Gr}_{R,T}^{1/(2n+2)} \mathrm{Pr}_{R,T}^{n/(3n+1)}$$

and

$$C_s = 0.996 + 0.120, \quad \tilde{D} = 0.682 \qquad \text{for} \quad Z < 10$$

$$C_s = 0.489 + 0.005, \quad \tilde{D} = 1.0 \qquad \text{for} \quad 10 < Z < 40$$

$\mathrm{Gr}_{R,T}$ and $\mathrm{Pr}_{R,T}$ are defined in the same way as in Eqs. (20.57) and (20.58), with the sphere radius R being used instead of L.

Horizontal Cylinder— T_w = Constant, Free Convection. Experimental correlations have been proposed for this case for both pseudoplastic ($n < 1$) and dilatant ($n > 1$) power-law fluids. For $n < 1$, Gentry and Wollersheim [52] recommend

$$\overline{\mathrm{Nu}_D} = \frac{\bar{h}D}{k} = 1.19(\mathrm{Gr}_{D,T}\mathrm{Pr}_{D,T})^{0.20} \tag{20.63}$$

and for $n > 1$, Kim and Wollersheim [66] propose

$$\overline{\mathrm{Nu}_R} = \frac{\bar{h}R}{k} = 2.816[\mathrm{Gr}_{R,T}\mathrm{Pr}_{R,T}]^{0.103} \tag{20.63a}$$

where the above generalized Grashof and Prandtl numbers are defined as in Eqs. (20.57) and (20.58) using D [Eq. (2.63)] and R [Eq. (20.63a)] instead of L.

Horizontal Cylinder— q_w = Constant, Free Convection. For this case, measurements are only available for dilatant fluids ($n > 1$). Kim and Wollersheim [66] propose the correlation equation

$$\overline{\mathrm{Nu}_R} = \frac{\bar{h}R}{k} = 3.544\left[\mathrm{Gr}_{R,q_w}\mathrm{Pr}_{R,q_w}\right]^{0.071} \tag{20.64}$$

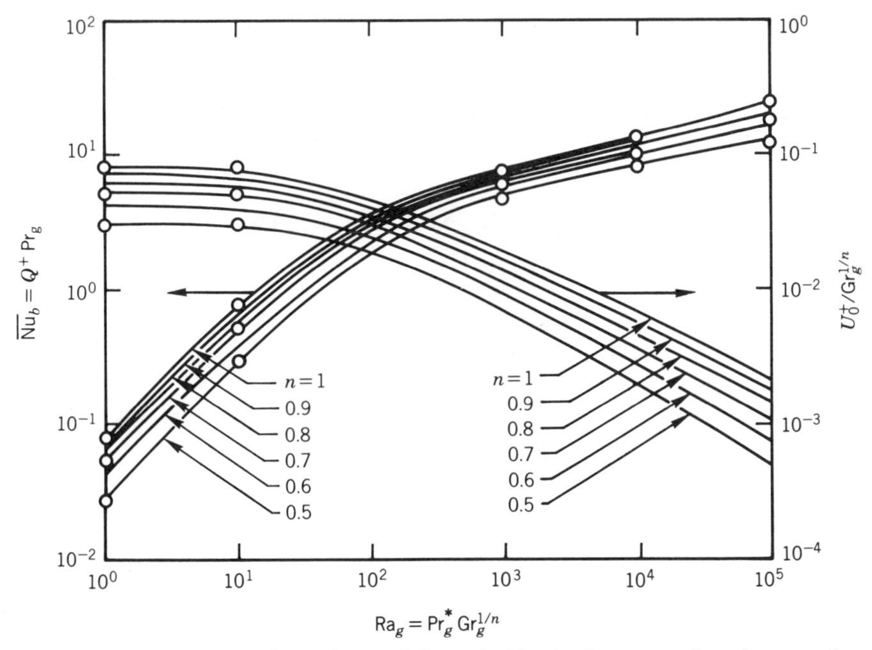

Figure 20.21. Average Nusselt numbers and flow velocities for free convection of a power-law fluid between vertical parallel plates [64]. Open circles are asymptotic solutions for large and small values of $\text{Pr}_g^* \text{Gr}_g^{1/n}$.

where the generalized Grashof and Prandtl numbers are defined as in Eqs. (20.60) and (20.61) with the cylinder radius R replacing x as the characteristic dimension.

Vertical Parallel Plates— T_w = Constant, Mixed Convection. For vertical parallel plates with height H and spacing $\hat{b}$, Irvine et al. [64] present the graphical results of a numerical solution as shown in Fig. 20.21 for a power-law fluid. Of interest are the average Nusselt number $\overline{\text{Nu}}_b$ and the dimensionless average flow velocity between the plates, U_0^+. These are shown in the figure on the right and left ordinates respectively. The dimensionless quantities used in Fig. 20.21 are defined as follows:

$$\overline{\text{Nu}}_b = \frac{\bar{h}\hat{b}}{k} \qquad U_0^+ = \frac{u_0}{u_n^*}$$

$$\text{Pr}_g^* = \frac{\rho c_p}{k}\left[\frac{\nu_K^{1/(2-n)}}{(H/\hat{b})^{(1-n)/(2-n)}\hat{b}^{2(n-1)/(2-n)}}\right] \qquad \nu_K = \frac{K}{\rho} \tag{20.65}$$

$$\text{Gr}_g = \frac{g\beta(T_w - T_\infty)\hat{b}^{(n+2)/(2-n)}}{\nu_K^{2/(2-n)}(H/\hat{b})^{n/(2-n)}} \qquad u_n^* = \frac{\nu_K^{1/(2-n)}\hat{b}^{(1-2n)/(2-n)}}{H^{(1-n)/(2-n)}}$$

For vertical parallel plates with the thermal boundary condition q_w = constant, graphical results similar to the above have been reported by Schneider and Irvine [65].

Single Vertical Plate— T_w = Constant, Mixed Convection. Shenoy [55] has pointed out that no rigorously correct solution exists for mixed laminar convection for a power-law fluid on a vertical plate. He has proposed the following correlating equation which can be used to estimate the mixed convection from a vertical plate and thus to determine from other known solutions the relative importance of the free and forced convection components:

$$\frac{\mathrm{Nu}_{x,T}}{\mathrm{Re}_{g,x}^{1/(n+1)}} = \left\{ \left(\frac{1}{0.893} \right)^3 \frac{1}{18} \left(\frac{117}{560(n+1)} \right)^{1/(n+1)} \left(\frac{2n+1}{n+1} \right) \mathrm{Pr}_{x,F} + 8 \left[\frac{1}{2} \left(\tfrac{3}{10} \right)^{1/n} g(n) \left(\frac{2n+1}{3n+1} \right) \mathrm{Pr}_{x,F} \right]^{3n/(3n+1)} \times \left(\frac{\mathrm{Gr}_{x,T}}{\mathrm{Re}_{g,x}^2} \right)^{3/(3n+1)(2-n)} \right\}^{1/3} \tag{20.66}$$

where

$$\mathrm{Re}_{g,x} = \frac{\rho u_\infty^{2-n} x^n}{K} \tag{20.66a}$$

$$\mathrm{Pr}_{x,F} = \frac{\rho c_p}{k} \left(\frac{K}{\rho} \right)^{2/(n+1)} x^{(1-n)/(1+n)} u_\infty^{3(n-1)(n+1)}. \tag{20.66b}$$

$$\mathrm{Gr}_{x,T} = \frac{\rho^2 x^{n+2}}{K^2} [g\beta(T_w - T_\infty)]^{2-n} \tag{20.66c}$$

$$g(n) = \frac{1}{15} - \frac{5}{126n} + \frac{1}{84n^2} - \frac{1}{486n^3} + \frac{1}{5103n^4} - \frac{1}{124{,}740n^5} \tag{20.66d}$$

Vertical Tube— T_w = Constant, Mixed Convection. Gori [60] has considered mixed convection for a power-law fluid flowing in a long vertical tube. It was specified that the inlet velocity distribution was the same as for fully developed forced flow. The effects of free convection on the velocity profile and the heat transfer were investigated. Both the fluid density ρ and the consistency K were taken as appropriate functions of temperature, and a numerical solution was obtained. The average Nusselt number of this numerical solution from the entrance to any axial location x can be correlated within $\pm 18\%$ by

$$\overline{\mathrm{Nu}}_T = \frac{\bar{h}D}{k} = 1.75 \left(\frac{3n+1}{4n} \right)^{1/3} \times \left[\mathrm{Gz} + 0.72 \left(2^{3n} \mathrm{Gr}_{R,T}^* \mathrm{Pr}_{R,T}^* \frac{D}{L} \right) \right]^{0.75} \left(\frac{K_x}{K_w} \right)^a \tag{20.67}$$

where

$$\overline{\mathrm{Nu}}_T = \frac{2}{\eta}\left(\frac{1 - \Theta_b}{1 + \Theta_b}\right) \qquad \eta = \frac{x/R}{\mathrm{Re}_g \mathrm{Pr}^*_{R,T}} \tag{20.67a}$$

$$\overline{\Delta T} = \frac{(T_0 - T_w) - (T_{b,x} - T_w)}{2} \tag{20.67b}$$

$$q'' = \bar{h}\overline{\Delta T}, \qquad \Theta_b = \frac{T_b - T_w}{T_0 - T_w} \tag{20.67c}$$

with

$$\mathrm{Gr}^*_{R,T} = \frac{\rho^2 \beta g (T_w - T_0) R^{2n+1} \bar{u}^{2(1-n)}}{K^2} \tag{20.67d}$$

$$\mathrm{Pr}^*_{R,T} = \frac{K c_p}{k}\left(\frac{\bar{u}}{R}\right)^{n-1} \tag{20.67e}$$

$$\mathrm{Gz} = \frac{\dot{m} c_p}{kx}, \qquad a = \frac{0.56}{[(3n+1)/n]^n}$$

T_0 = fluid temperature at $x = 0$

T_b = fluid bulk mean temperature at x

Using Eq. (20.67), the average heat flux from $x = 0$ to x, q'', can be calculated using the definitions given above as follows:

a. Known: T_0, T_w, $\bar{u}$, x, all properties (evaluated at T_w).
b. Evaluate $\mathrm{Pr}^*_{R,T}$, $\mathrm{Gr}^*_{R,T}$, and η.
c. Find Nu_T from Eq. (20.67) (assuming $K_{b,x} = K_w$).
d. Find θ_b and thus $T_{b,x}$ from Eq. (20.67c).
e. Use this $T_{b,x}$ to evaluate $K_{b,x}$ and repeat steps a–d until the iteration converges. Then find $\bar{h}$ from Eq. (20.67).
f. Determine $\overline{\Delta T}$ from Eq. (20.67b).
g. Find q'' from Eq. (20.67a).

The above calculation can be made without iteration if a constant consistency is assumed, i.e. $K_{b,x} = K_w$.

Gori [62] has also investigated the above problem numerically for the thermal boundary condition q_w = constant. Local Nusselt numbers as a function of Graetz number are presented for selected ranges of generalized Grashof and Prandtl numbers, but no correlation equation is available.

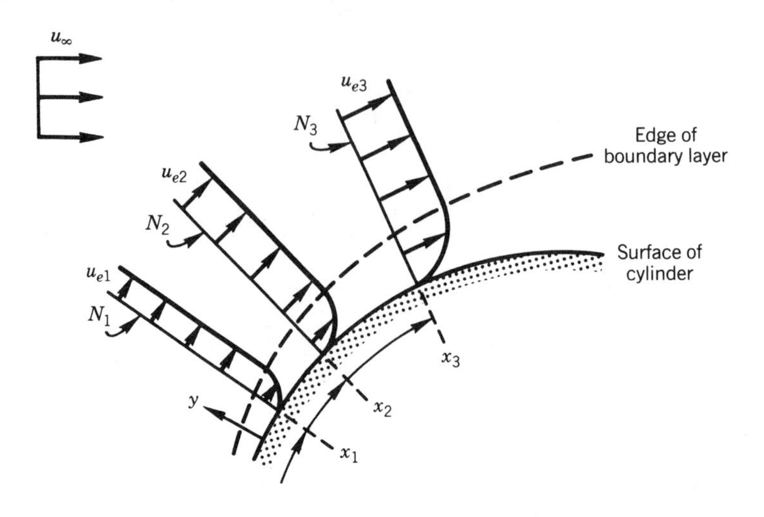

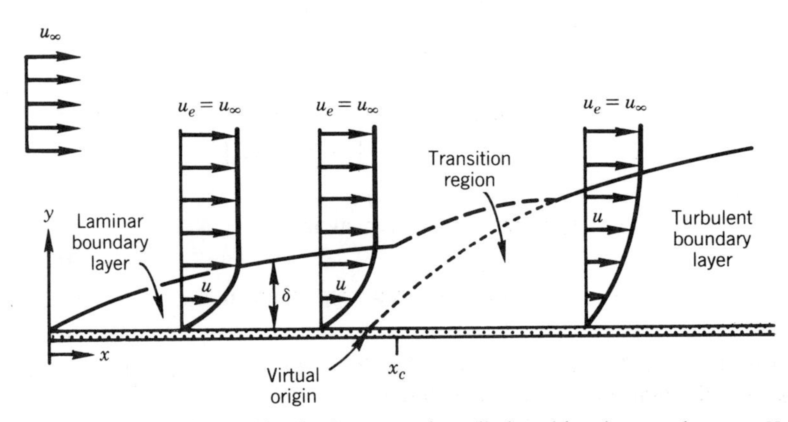

Figure 20.22. (*a*) Velocity profile of a flow around a cylinder with axis normal to u_∞. N_1, N_2, and N_3 are normals to the surface at distances x_1, x_2, and x_3 from the front stagnation point. (*b*) Boundary-layer formation on a flat plate oriented at zero incidence to an undisturbed stream with constant velocity.

20.7 FLOW DYNAMICS AND HEAT TRANSFER IN NON-NEWTONIAN EXTERNAL FLOWS

20.7.1 Introduction

A common physical situation is that involving relative motion between a solid surface and a large volume of fluid. In most cases, the relative motion between the fluid and the solid is zero at the surface. The region where the fluid velocity changes from its free-stream value to zero at the solid surface is known as the boundary layer; see Fig. 20.22. The boundary-layer thickness δ is usually defined as the normal distance from the surface to the point where the velocity is 0.99 of the external velocity u_e. Note that u_e, which is parallel to the surface, is not always equal to the free-stream velocity u_∞ (Fig. 20.22*a*).

The characteristics of the boundary layer such as its profile and thickness are functions of the distance x from the frontal stagnation point (or the leading edge), the free-stream velocity, the fluid viscosity, and some surface characteristics. Immediately downstream of the leading edge the boundary layer is laminar. As it grows with x, the flow within the boundary layer may become instable and a transition region followed by a turbulent boundary layer develop at x larger than some critical value x_c; see Fig. 20.22*b*.

An external flow, as opposed to an internal (or duct) flow, is a flow along a surface whose boundary-layer thickness is small compared to the distance to any other surface. The boundary-layer region, where fluid viscosity effects are significant, is bounded on one side by a solid surface and on the other side by the free stream, which can be assumed to behave as an inviscid (potential) flow. In an internal flow, we are usually concerned with viscous effects over the entire region between bounding surfaces, although the hydrodynamic entry region of a duct has the characteristics of an external flow; see Sec. 20.4 and Fig. 20.17.

20.7.2 Governing Principles

Conservation of Mass. The first of the governing laws of fluid motion is the conservation of mass. The mathematical expression of this principle, known as the continuity equation, is ([68], or see Chaps. 1 and 2)

$$\frac{\partial \rho}{\partial t} + \frac{\partial}{\partial x}(\rho u) + \frac{\partial}{\partial y}(\rho v) + \frac{\partial}{\partial z}(\rho w) = 0 \tag{20.68}$$

Conservation of Momentum. The application of Newton's second law to fluid motion is known as the conservation of momentum. The mathematical expression of the conservation of momentum in the x direction is ([68], or see Chaps. 1 and 2)

$$\rho\frac{\partial u}{\partial t} + \rho u\frac{\partial u}{\partial x} + \rho v\frac{\partial u}{\partial y} + \rho w\frac{\partial u}{\partial z}$$

$$= -\frac{\partial p}{\partial x} + \frac{\partial \tau_{xx}}{\partial x} + \frac{\partial \tau_{yx}}{\partial y} + \frac{\partial \tau_{zx}}{\partial z} + \rho f_x \tag{20.69}$$

where f_x is a body force (such as weight) per unit mass acting in the x direction.

Expressions similar to Eq. (20.69) are obtained in the y direction by interchanging x and y and also u and v; or in the z direction, by interchanging x and z and also u and w. These expressions, together with the continuity equation [Eq. (20.68)], govern all fluid motions. For Newtonian fluids, the stress-strain relations are relatively simple (see Sec. 20.2.1), and when they are introduced into the momentum equations the Navier-Stokes equations, are obtained. Derivations of Eq. (20.68), Eq. (20.69), and the Navier-Stokes equations are given by Schlichting [68] and Currie [69].

Conservation of Energy. The principle of conservation of energy is an application of the first law of thermodynamics to a fluid element in motion. The conservation of energy expressed in terms of the fluid temperature can be written as follows ([68], or see

Chaps. 1 and 2):

$$\rho c_p \frac{\partial T}{\partial t} + \rho c_p u \frac{\partial T}{\partial x} + \rho c_p v \frac{\partial T}{\partial y} + \rho c_p w \frac{\partial T}{\partial z}$$

$$= \left[\frac{\partial}{\partial x}\left(k \frac{\partial T}{\partial x}\right) + \frac{\partial}{\partial y}\left(k \frac{\partial T}{\partial y}\right) + \frac{\partial}{\partial z}\left(k \frac{\partial T}{\partial z}\right) \right]$$

$$+ \left[\frac{\partial P}{\partial t} + u \frac{\partial P}{\partial x} + v \frac{\partial P}{\partial y} + w \frac{\partial P}{\partial z} \right] + \tau_i \cdot \nabla \mathbf{u} + S \tag{20.70}$$

S is a heat source term; it stands for thermal energy per unit volume and time which is generated within the fluid body. τ_i is the shear stress tensor which has three vector components, one in each spatial direction. The term $\boldsymbol{\tau} \cdot \nabla \mathbf{u}$ expresses frictional heat dissipation and for simplicity is not expanded here. See Schlichting [68] and Currie [69] for derivations leading to Eq. (20.70).

Equations (20.68), (20.69) (and its y and z analogs), and (20.70) are the governing equations of momentum and energy transport in non-Newtonian and Newtonian fluids. A solution of these equations is possible only if all the relevant terms of the shear stress tensor are specified, based on the rheological properties of the fluid, in Eqs. (20.69) and (20.70).

As emphasized at the beginning of the chapter (Sec. 20.2), the present overview covers only purely viscous, time-independent flows. A review of viscoelastic flows is given by Cho and Hartnett [1].

20.7.3 Laminar Boundary-Layer Solutions

Solutions for free and mixed convection in external flows are given in Sec. 20.6 and will not be discussed here.

Laminar Flow over a Flat Plate. Consider the boundary-layer flow illustrated in the laminar portion of Fig. (20.22*b*). Assume that the free-stream velocity u_∞ is constant, the flow is steady, and all fluid properties are constant. For this simple situation, the continuity equation [Eq. (20.68)] is reduced to

$$\frac{\partial u}{\partial x} + \frac{\partial v}{\partial y} = 0 \tag{20.71}$$

The only relevant momentum equation is in the x direction [Eq. (20.69)], and it is reduced to

$$\rho u \frac{\partial u}{\partial x} + \rho v \frac{\partial u}{\partial y} = \frac{\partial \tau_{yx}}{\partial y} \tag{20.69a}$$

Acrivos et al. [70] solved Eqs. (20.71) and (20.69a) numerically for a power-law fluid [Eq. (20.5)]. Their results are expressed in the following relation:

$$\frac{\tau_w}{\rho u_\infty^2} = c(n) \mathrm{Re}_{g,x}^{-1/(n+1)} \tag{20.72}$$

where

$$\mathrm{Re}_{g,x} = \frac{\rho u_\infty^{2-n} x^n}{K} \tag{20.73}$$

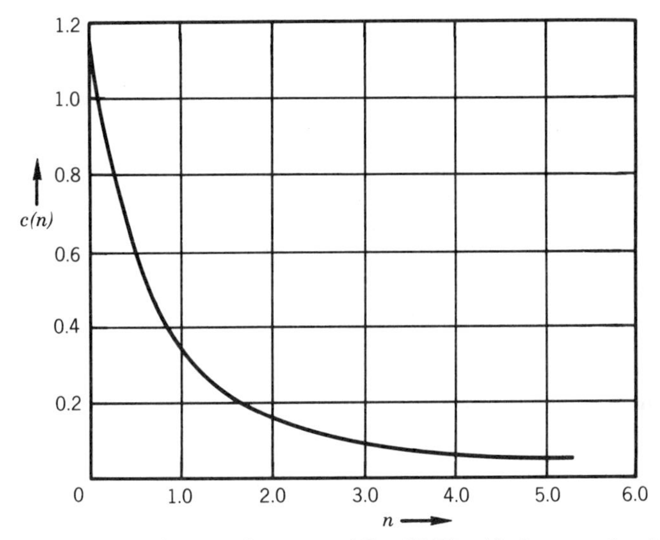

Figure 20.23. Variation of the coefficient $c(n)$ of Eq. (20.72) with the power-law index n [70].

and

$$\tau_w = K\left(\frac{\partial u}{\partial y}\right)^n\bigg|_{y=0}$$

The variation of the coefficient $c(n)$ with the power n is shown in Fig. 20.23; $c(n)$ values for selected n's are given in Table 20.10. For $n = 1$, Eq. (20.72) becomes

$$\frac{\tau_w}{\rho u_\infty^2} = \frac{0.332}{\mathrm{Re}_x^{1/2}} \tag{20.74}$$

This expression agrees perfectly with the well-known Blasius solution for a Newtonian flow over a flat plate [71].

The total frictional drag force over a distance L is obtained (assuming the boundary layer is always laminar) by integrating τ_w over L and multiplying by the width of the

TABLE 20.10 Selected $c(n)$ Values for Use with Eqs. (20.72), (20.76), (20.79b), (20.91), and (20.92)[a]

n	$c(n)$	n	$c(n)$
0.05	1.017	1.5	0.2189
0.1	0.969	2.0	0.1612
0.2	0.8725	2.5	0.1226
0.3	0.7325	3.0	0.09706
0.5	0.5755	4.0	0.06777
1.0	0.33206	5.0	0.05111

[a] Data from [70].

plate:

$$F_L = W\int_0^L \tau_w\,dx = \rho u_\infty^2 W c(n)\int_0^L \mathrm{Re}_{g,x}^{-1/(n+1)}\,dx$$
$$= \rho u_\infty^2 WL(n+1)c(n)\mathrm{Re}_{g,L}^{-1/n+1} \tag{20.75}$$

where $\mathrm{Re}_{g,L}$ is defined by Eq. (20.73) with x replaced by L.

The results expressed in Eq. (20.72) can easily be expanded to fluids with yield stress. In that case

$$\tau_w = \tau_y + \rho u_\infty^2 c(n)\mathrm{Re}_{g,x}^{-1/(n+1)} \tag{20.76}$$

and for a Bingham plastic flow

$$\tau_w = \tau_y + \rho u_\infty^2 \frac{0.332}{\mathrm{Re}_x^{1/2}} \tag{20.77}$$

Laminar Heat Transfer over a Flat Plate. The energy equation applicable for a laminar, constant-property, steady flow over a flat plate is

$$u\frac{\partial T}{\partial x} + v\frac{\partial T}{\partial y} = \frac{k}{\rho c_p}\frac{\partial^2 T}{\partial y^2} \tag{20.78}$$

The experimental results of Takahashi et al. [72] and the numerical calculations of Kim et al. [73] show that the heat transfer from such a plate to a power-law fluid can be expressed as follows:

$$\mathrm{St}_x = \frac{\tau_w}{\rho u_\infty^2}\mathrm{Pr}_{p,x}^{-2/3} \tag{20.79a}$$

or, combining with Eq. (20.72),

$$\mathrm{St}_x = c(n)\mathrm{Re}_{g,x}^{-1/(n+1)}\mathrm{Pr}_{p,x}^{-2/3} \tag{20.79b}$$

where

$$\mathrm{St}_x = \frac{h_x}{\rho c_p u_\infty} \tag{20.80}$$

is the local Stanton number, $\mathrm{Re}_{g,x}$ is defined in Eq. (20.73), and

$$\mathrm{Pr}_{p,x} = \frac{\rho c_p u_\infty x}{k}\mathrm{Re}_{g,x}^{-2/(n+1)} \tag{20.81}$$

is the generalized Prandtl number based on the distance x.

As expected, for $n = 1$ (Newtonian fluid), Eq. (20.79b) is reduced to

$$\mathrm{St}_x = 0.332\mathrm{Re}_x^{-1/2}\mathrm{Pr}^{-2/3} \tag{20.82}$$

General Solutions for Two-Dimensional Laminar Flow and Heat Transfer. Momentum and heat transfer in power-law fluid flow over arbitrarily shaped, two-dimensional, or axisymmetrical bodies were studied by Kim et al. [73] and by Lin and Chern [74]. Wedge-type flows and flows around the front portion of a circular cylinder and a sphere were examined in detail. Other experimental and theoretical studies of a power-law flow around a circular cylinder were presented by Mizushina and Usui [75], Mizushina et al. [76], and Shah et al. [77]. The interested reader is referred to the above studies.

20.7.4 Heat Transfer Correlations for a Cylinder in Non-Newtonian Crossflow

When a cylinder (or a sphere) is exposed to a crossflow, the fluid stagnates on the front point of the cylinder and then accelerates downstream symmetrically around both sides of the cylinder. A boundary layer is formed over the front portion of the cylinder; see Fig. 20.22*a*. At about 90° downstream of the front stagnation point, the boundary layer separates and a wake is created behind the cylinder. Commonly, the cylinder (or sphere) heat transfer either is expressed as a function of the angular distance from the stagnation point, or is averaged over the entire circumference.

A review of Newtonian flow and heat transfer characteristics of a cylinder is provided by Zukauskas [78]. Mizushina et al. [76] and Shah et al. [77] showed that the pressure distribution around a cylinder exposed to a pseudoplastic crossflow is similar to that obtained with Newtonian fluids.

Mizushina and Usui [75] presented an approximate solution for the heat transfer from the front portion of a cylinder in a power-law crossflow. At the front stagnation point, the heat transfer coefficient is expressed as

$$\mathrm{Nu}_d = 1.04 n^{-0.4} \mathrm{Re}_{g,d}^{1/(n+1)} \mathrm{Pr}_{p,d}^{1/3} \tag{20.83}$$

where

$$\mathrm{Nu}_d = \frac{hd}{k} \tag{20.84}$$

is the Nusselt number based on the cylinder diameter d, and

$$\mathrm{Re}_{g,d} = \frac{\rho u_\infty^{2-n} d^n}{K} \tag{20.85}$$

and

$$\mathrm{Pr}_{p,d} = \frac{\rho c_p u_\infty d}{k} \mathrm{Re}_{g,d}^{-2/(n+1)} \tag{20.86}$$

are, respectively, the generalized Reynolds and Prandtl numbers based on the cylinder diameter.

Mizushina et al. [76] obtained the following correlation for the Nusselt number averaged over the cylinder circumference:

$$\overline{\mathrm{Nu}}_d = 0.72 n^{-0.4} \mathrm{Re}_{g,d}^{1/(n+1)} \mathrm{Pr}_{p,d}^{1/3} \tag{20.87}$$

where Nu_d, $\mathrm{Re}_{g,d}$, and $\mathrm{Pr}_{p,d}$ are defined in Eqs. (20.84), (20.85), and (20.86), respectively. As shown in Ref. 76, experimental measurements in pseudoplastic flow around a circular cylinder at $n = 0.72$ to 1.0, $\mathrm{Re}_{g,d} = 43$ to 19,200, and $\mathrm{Pr}_{p,d} = 5.6$ to 40,000 agree within $\pm 20\%$ with Eq. (20.87).

20.7.5 Transition and Turbulent Boundary-Layer Flow and Heat Transfer

Estimates of the velocity and temperature fields in external turbulent flows of non-Newtonian fluids are largely based on modified Newtonian flow solutions. Only a little experimental evidence confirming the validity of these approximations is presently available.

Boundary-Layer Transition from Laminar to Turbulent. Application of viscous stability theory [79] to Newtonian fluids leads to a general criterion for transition. It states that the lowest Reynolds number for which the laminar boundary layer may become unstable (turbulent) is

$$\mathrm{Re}_{\delta_2} = 162 \tag{20.88}$$

where

$$\mathrm{Re}_{\delta_2} = \frac{\rho u_e \delta_2}{\mu}$$

The boundary-layer momentum thickness δ_2 is defined by

$$\delta_2 \equiv \int_0^\infty \frac{\rho u}{\rho_e u_e}\left(1 - \frac{u}{u_e}\right) dy \tag{20.89}$$

where y is the direction normal to the surface; see Fig. 20.22.

On a flat plate at zero incidence to a Newtonian mainstream with constant velocity ($u_\infty = u_e =$ constant), Eq. (20.88) corresponds to

$$\mathrm{Re}_x = 60{,}000 \tag{20.90}$$

Measurements on smooth surfaces show that transition to a turbulent boundary layer tends to occur at $\mathrm{Re}_x = 3 \times 10^5$ to 5×10^5. Transitions at substantially lower Re_x values were observed in flows over rough surfaces, with adverse pressure gradients, or when the mainstream was highly unstable. In other cases, the boundary layer does not become fully turbulent until Re_x reaches about 3×10^6. Based on these observations of Newtonian flows, Skelland [2] proposed a tentative criterion for transition from laminar to turbulent boundary layer flow of power-law fluid as

$$3 \times 10^5 < \left(\frac{0.33206}{c(n)}\right)^2 \mathrm{Re}_{g,x}^{2/(n+1)} < 3 \times 10^6 \tag{20.91}$$

$c(n)$ is given in Fig. 20.23 and in Table 20.10 (p. **20** • 43), and $\mathrm{Re}_{g,x}$ is defined in Eq. (20.73).

When transition of Newtonian flow would have been possible at lower Re_x, the lower limit in Eq. (20.91) should be 6×10^4 instead of 3×10^5 [79]. For all power-law flows over arbitrarily shaped surface, a safe criterion for stability in the boundary layer is

$$\left[\frac{0.33206}{c(n)}\right]^2 \mathrm{Re}_{g,\delta_2}^{2/(n+1)} < 162 \tag{20.92}$$

where Re_{g,δ_2} is defined by Eq. (20.73) with x replaced by δ_2.

Turbulent Boundary-Layer Flow over a Flat Plate. Consider the turbulent portion of the boundary-layer flow shown in Fig. 20.22*b*. Assume that the average free-stream velocity u_∞ does not vary with time and all fluid properties are constant. An approximate expression for the wall shear stress in power-law flows is given by Skelland [2] as follows:

$$\frac{\tau_w}{\rho u_\infty^2} = \frac{\psi}{\tilde{\beta}n+1}\left\{\frac{(\tilde{\beta}n+1)\tilde{\Omega}}{\psi}\left[8^{n-1}\left(\frac{3n+1}{4n}\right)^n\right]^{\tilde{\beta}} \mathrm{Re}_{g,x'}^{-\tilde{\beta}}\right\}^{1/(\tilde{\beta}n+1)} \tag{20.93}$$

where

$$\psi = \frac{2-\tilde{\beta}(2-n)}{2(1-\tilde{\beta}+\tilde{\beta}n)} - \frac{2-\tilde{\beta}(2-n)}{2-2\tilde{\beta}+3\tilde{\beta}n} \tag{20.94}$$

$$\tilde{\Omega} = \frac{\tilde{\alpha}(0.817)^{2-\tilde{\beta}(2-n)}}{2^{\tilde{\beta}n+1}} \tag{20.95}$$

and $\tilde{\alpha}$ and $\tilde{\beta}$ are given graphically in Fig. 20.24, which is based on the results of Dodge and Metzner [24].

The x' in $\mathrm{Re}_{g,x'}$ of Eq. (20.93) is the downstream distance from the *virtual origin*, which is the point where the turbulent boundary layer would have originated were it not preceded by a laminar boundary layer; see Fig. 20.22*b*. The boundary-layer thickness δ at a given distance x' from the virtual origin is

$$\delta = \left[\frac{(\tilde{\beta}n+1)\tilde{\Omega}}{\psi}\left(\frac{\tilde{\gamma}}{\rho}\right)^{\tilde{\beta}} u_\infty^{-\tilde{\beta}(2-n)}\right]^{1/(\tilde{\beta}n+1)} x'^{1/(\tilde{\beta}n+1)} \tag{20.96}$$

where

$$\tilde{\gamma} = 8^{n-1}K\left(\frac{3n+1}{4n}\right)^n \tag{20.97}$$

The location of the virtual origin can be calculated using Eq. (20.96) if δ is known at a given point on the plate.

Equations (20.93) and (20.96) are obtained assuming that the boundary-layer profile can be represented by

$$\frac{u}{u_\infty} = \left(\frac{y}{\delta}\right)^{\tilde{\beta}n/[2-\tilde{\beta}(2-n)]} \tag{20.98}$$

This expression is a relatively good approximation for pseudoplastic fluids ($n < 1$), but

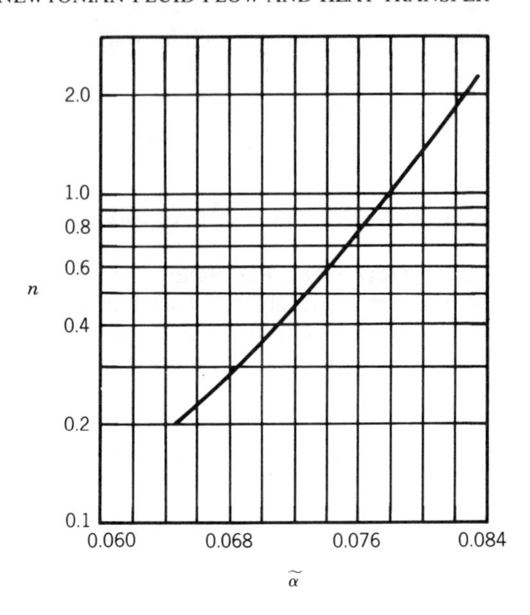

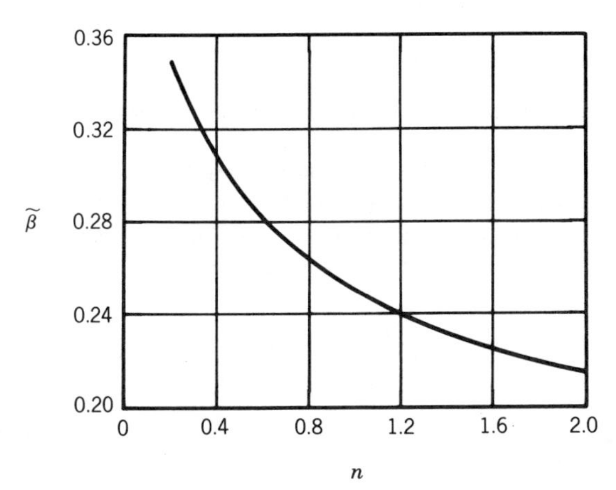

Figure 20.24. (*a*) Relation between the parameter $\tilde{\alpha}$ in Eq. (20.95) and the power-law index n [24]. (*b*) Relation between the parameter $\tilde{\beta}$ in Eq. (20.93) and the power-law index n [24].

its deviation from the real profile increases with n and thus it may introduce a significant error into estimates of dilatant flows, especially for $n > 1.3$.

An estimate of the drag force over the turbulent boundary-layer region may be obtained by integrating Eq. (20.93) from the approximate transition point x_c (measured with respect to the virtual origin) to L, and multiplying this by W. The result is the turbulent-region drag force on one side of a plate of length L and width W. It can then be added to the laminar-region drag [Eq. (20.75) with x_c in place of L] to obtain the total drag force.

Turbulent Boundary-Layer Heat Transfer over a Flat Plate. Assuming constant fluid properties and constant (time-averaged) mainstream velocity, the heat transfer from a smooth flat plate to a turbulent boundary layer of a power-law fluid flowing at zero incidence to the plate is approximated as follows:

$$\mathrm{St}_x = \frac{\tau_w}{\rho u_\infty^2} \mathrm{Pr}_{p,x'}^{-0.4} \tag{20.99}$$

where St_x and $\mathrm{Pr}_{p,x}$ are defined in Eqs. (20.80) and (20.81), respectively.

Equation (20.99) is obtained by extending the solution of Newtonian flows to include power-law fluids. It is assumed that the power index (−0.4) is independent of the rheological properties of the fluid. Thus, it is the same as that in the Newtonian-fluid expression [68]. When Eq. (20.99) is combined with Eq. (20.93), the expression for the

TABLE 20.11 Summary of Non-Newtonian Flow and Heat Transfer Topics

Section	Topic
20.2.2	Apparent viscosity
20.2.3	Constitutive equations for purely viscous time-independent fluids
20.3.1	Characteristics of purely viscous time-independent fluids
20.3.2	Capillary-tube viscometer
20.3.3	Falling-ball viscometer
20.3.4	Falling-needle viscometer
20.3.5	Rotating viscometer
20.3.6	Examples of rheological-property measurements
20.4.2	Transition to turbulent flow in ducts
20.4.3	Fully developed laminar pressure drops in ducts
20.4.4	Fully developed turbulent pressure drops in ducts
20.5.2	Fully developed laminar heat transfer in ducts
20.5.3	Laminar heat transfer in thermally developing duct flows
20.5.4	Fully developed turbulent heat transfer in ducts
20.6.2	Summary of previous investigations in free and mixed convection
20.6.3	Engineering relations for free and mixed convection systems:
	Single vertical plate—T_w = const (free)
	Single vertical plate—q_w = const (free)
	Single sphere—T_w = const (free)
	Horizontal cylinder—T_w = const (free)
	Horizontal cylinder—q_w = const (free)
	Vertical parallel plates—T_w = const (mixed)
	Single vertical plate—T_w = const (mixed)
	Vertical tube—T_w = const (mixed)
20.7.2	Governing principles of Non-Newtonian flow and heat transfer
20.7.3	External laminar boundary layer solutions:
	Laminar flow over a flat plate
	Laminar heat transfer over a flat plate
	General solutions for two-dimensional flow and heat transfer
20.7.4	Heat transfer correlations for a cylinder in non-Newtonian crossflow
20.7.5	Transition and turbulent boundary-layer flow and heat transfer:
	Boundary-layer transition from laminar to turbulent
	Turbulent boundary-layer flow over a flat plate
	Turbulent boundary-layer heat transfer over a flat plate

local Stanton number becomes

$$\mathrm{St}_x = \frac{\psi}{\tilde{\beta} n + 1} \left\{ \frac{(\tilde{\beta} n + 1)\tilde{\Omega}}{\psi} \left[8^{n-1} \left(\frac{3n+1}{4n} \right)^n \right]^{\tilde{\beta}} \mathrm{Re}_{g,x'}^{-\tilde{\beta}} \right\}^{1/(\tilde{\beta} n + 1)} \mathrm{Pr}_{p,x'}^{-0.4} \quad (20.100)$$

Note the similarity between the expressions for laminar and turbulent heat transfer over a flat plate [Eqs. (20.79a) and (20.99), respectively]. For $n = 1$, both $\mathrm{Pr}_{p,x}$ and $\mathrm{Pr}_{p,x'}$ become equal to Pr, and the well-known expressions for St in a Newtonian flow are obtained.

20.8 SUMMARY OF USEFUL RELATIONS

Table 20.11 on page **20** · 49 lists many of the flow and heat transfer problems discussed in this chapter, and the section where information on each problem is provided.

NOMENCLATURE

a	geometric parameter in Eq. (20.34) (see Table 20.3)
$\tilde{a}$	l/d = (needle length)/(needle diameter), Eq. (20.18)
$\hat{a}$	exponent in Eq. (20.67)
a_1, a_2	constants, Eq. (20.50)
b	geometric parameter in Eq. (20.34), (see Table 20.3)
$\hat{b}$	distance between parallel plates, Eqs. (20.2) and (20.65), m, ft
$\tilde{b}$	D/d = (container diameter)/(needle diameter), Eq. (20.15)
$\overline{B}$	coefficient in Eq. (20.59) (see Table 20.9)
c_p	specific heat at constant pressure, J/(kg · K), Btu/($\mathrm{lb_m}$ · °F)
C_f	correction factor, Eq. (20.17)
C_s	coefficient in Eq. (20.62)
$\overline{C}_w$	correction factor, Eq. (20.19)
C	coefficient in Eq. (20.59) (see Table 20.9)
$c(n)$	coefficient in Eq. (20.72) (see Table 20.10)
d	cylinder outer diameter in Sec. 20.7.4, m, ft
$\tilde{d}$	needle diameter in Sec. 20.3, m, ft
D	tube inside or cylinder outside diameter, m, ft
$\tilde{D}$	exponent in Eq. (20.62)
D_h	hydraulic diameter = 4 × (cross-section area)/(wetted perimeter), m, ft
f	Fanning friction factor = $\tau_w/(\rho \bar{u}^2/2) = (dp/dx) D_h/(2\rho \bar{u}^2)$
f_D	Darcy friction factor = $4f$
f_x	body force per unit mass, acting in the x direction, Eq. (20.69), N/kg, $\mathrm{lb_f/lb_m}$
$f(R)$	function of R in Eq. (20.20), 1/m², 1/ft²
$F(n)$	function of n in Eq. (20.37)

g	gravitational acceleration, m/s^2, ft/s^2
Gr_g	generalized Grashof number, Eq. (20.65)
$Gr_{L,T}$	Grashof number, Eq. (20.57); $Gr_{R,T}$ and $Gr_{D,T}$ defined the same way except for L replaced by R and D, respectively
$Gr^*_{R,T}$	modified Grashof number, Eq. (20.67e)
Gr_{x,q_w}	heat flux Grashof number, Eq. (20.60); Gr_{R,q_w} defined the same way except for x replaced by R
Gz	Graetz number = $\dot{m}c_p/kx$
$g(n)$	function of n in Eq. (20.66)
$g(R, n)$	function of R and n in Eq. (20.24), $1/m^{2/n}$, $1/ft^{2/n}$
h	convective heat transfer coefficient, $W/(m^2 \cdot °K)$, $Btu/(hr \cdot ft^2 \cdot °F)$
$\tilde{h}$	height in Eq. (20.20) and Fig. 20.13, m, ft
H	height in Eq. (20.65), m, ft
k	thermal conductivity, for fluid if no subscript, $W/(m \cdot K)$, $Btu/(hr \cdot ft \cdot °F)$
K	fluid consistency, Eq. (20.5), $N \cdot s^n/m^2$, $lb_f \cdot s^n/ft^2$
L	length or height, m, ft
L_{hy}	entrance length of a duct, Fig. 20.17, m, ft
$\dot{m}$	mass flow rate, kg/s, lb_m/s
n	flow index, Eq. (20.5)
Nu	Nusselt number = hD_h/k, for hydrodynamically and thermally fully developed internal flow if no subscript
Nu_d	Nusselt number = hd/k at the stagnation point of a circular cylinder; $\overline{Nu}_d$ is the mean Nusselt number for crossflow to a cylinder
Nu_x	axially local Nusselt number = $h_x D_h/k$ for internal flow and $h_x x/k$ for external flow
$\overline{Nu}$	axially (or along the surface) mean Nusselt number for external flow = $\bar{h}D_e/k$; $D_e = L$ for $\overline{Nu}_L$, $D_e = R$ for $\overline{Nu}_R$, and $D_e = D$ for $\overline{Nu}_D$
p	pressure, Pa, lb_f/ft^2
Pr	Prandtl number = $\mu c_p/k$
Pr_a	defined in Eq. (20.44)
Pr_{eff}	defined in Eq. (20.43)
Pr_g	generalized Prandtl number, Eq. (20.42)
Pr'_g	same as Pr_{eff}
Pr^*_g	generalized Prandtl number, Eq. (20.65)
Pr_p	generalized Prandtl number, Eqs. (20.81), (20.86).
$Pr_{L,T}$	Prandtl number, Eq. (20.58); $Pr_{R,T}$ and $Pr_{D,T}$ defined the same way except for L replaced by R and D, respectively
$Pr^*_{R,T}$	generalized Prandtl number, Eq. (20.67d); similar to Pr_g but defined with R instead of D_h
Pr_{x,q_w}	modified Prandtl number, Eq. (20.61)
q''	average heat flux from $x = 0$ to x, Eq. (20.67a), W/m^2, $Btu/(hr \cdot ft^2)$
q''_x	local heat flux at x, W/m^2, $Btu/(hr \cdot ft^2)$

r	radial distance from cylinder centerline or sphere center point, m, ft
R	radius, m, ft
Re	Reynolds number (characteristic dimension indicated by subscript)
Re_a	defined in Eq. (20.33)
$\mathrm{Re}_{\mathrm{eff}}$	defined in Eq. (20.32)
Re_g	generalized Reynolds number, defined in Eq. (20.31); $\mathrm{Re}_{g,x}$ defined in Eq. (20.73); and $\mathrm{Re}_{g,d}$ defined in Eq. (20.85)
$\mathrm{Re}_{g,K}$	generalized Reynolds number for noncircular ducts, defined in Eq. (20.39)
Re'_g	same as $\mathrm{Re}_{\mathrm{eff}}$
S	heat source in Eq. (20.70), $\mathrm{W/m^3}$, $\mathrm{Btu/(hr \cdot ft^3)}$
St_x	local Stanton number, Eq. (20.80)
t	time, s
T	temperature, °C, K, °F, °R
$\tilde{T}$	torque (moment) in Sec. 20.3.5, $\mathrm{N \cdot m}$, $\mathrm{lb}_f \cdot \mathrm{ft}$
$T(n)$	function of the flow index n in Eq. (20.56)
u	velocity component in x direction, m/s, ft/s
u_0	entrance velocity, Eq. (20.65), m/s, ft/s
u_∞	free-stream velocity of external flow, Sec. 20.7, m/s, ft/s
$\tilde{u}_\infty$	terminal velocity of sphere or needle, Sec. 20.3, m/s, ft/s
$\tilde{u}'_\infty$	correlated needle velocity, Eq. (20.15), m/s, ft/s
u^*_n	power-law shear velocity, Eq. (20.65), m/s, ft/s
$\mathbf{u}$	velocity vector, Eq. (20.70), m/s, ft/s
$\bar{u}$	average velocity, m/s, ft/s
U_0^+	dimensionless entrance velocity, Eq. (20.65)
v	velocity component in y direction, m/s, ft/s
V	volume, $\mathrm{m^3}$, $\mathrm{ft^3}$
w	velocity component in z direction, m/s, ft/s
W	plate width, m, ft
x	axial coordinate or distance measured along surface of body (in general flow direction), m, ft
x_c	critical value of x at which laminar boundary-layer flow begins to become unstable, Fig. 20.22*b*, m, ft
x'	downstream distance from virtual origin, Eqs. (20.93), (20.96), (20.100), m, ft
y	spatial coordinate or distance measured from surface of body, m, ft
z	spatial coordinate or distance measured across surface of a body, m, ft
Z	free-convection function, Eq. (20.62)

Greek Symbols

α	rheological property in Ellis fluid, Eq. (20.8).
$\tilde{\alpha}$	coefficient in Eq. (20.95), (see Fig. 20.24*a*)
α^*	aspect ratio; see Table 20.3

β	thermal expansion coefficient = $-(1/\rho)(\partial\rho/\partial T)_p$, K^{-1}, $°R^{-1}$
$\tilde{\beta}$	coefficient in Eq. (20.93), (see Fig. 20.24*b*)
γ	angular strain, Eqs. (20.1), (20.2)
$\tilde{\gamma}$	parameter in Eq. (20.96)
$\hat{\gamma}$	consistency parameter, Eq. (20.51), (20.52)
$\dot{\gamma}$	time derivative of strain (strain rate), Eq. (20.2), s^{-1}
δ	velocity boundary-layer thickness, m, ft
δ_2	boundary layer momentum thickness, defined in Eq. (20.89), m, ft
Δ	designates a difference when used as a prefix
η	parameter in Eq. (20.67)
Θ_b	dimensionless bulk temperature, Eq. (20.67a)
θ	inclination angle of open channel with the horizontal, Eq. (20.35)
θ_0	angular distance, Eqs. (20.28) and (20.29)
μ	dynamic viscosity, for Newtonian fluid without subscript, $N \cdot s/m^2$, $lb_f \cdot s/ft^2$
μ_a	apparent viscosity, Eq. (20.4), $N \cdot s/m^2$, $lb_f \cdot s/ft^2$
μ_0	apparent viscosity at zero shear rate, Eq. (20.7), Fig. 20.6, $N \cdot s/m^2$, $lb_f \cdot s/ft^2$
μ_∞	apparent viscosity at very high shear rates, Fig. 20.6, $N \cdot s/m^2$, $lb_f \cdot s/ft^2$
ν	kinematic viscosity = μ/ρ, m^2/s, ft^2/s
ν_K	K/ρ = (fluid consistency)/(fluid density), m^2/s^{2-n}, ft^2/s^{2-n}
ρ	fluid density, kg/m^3, lb_m/ft^3
τ	shear stress, Pa, lb_f/ft^2
τ_y	yield stress, Pa, lb_f/ft^2
τ_i	shear stress tensor, Eq. (20.70), Pa, lb_f/ft^2
ψ	parameter in Eq. (20.93)
Ω	angular velocity, Eq. (20.20), s^{-1}, s^{-1}
$\tilde{\Omega}$	parameter in Eq. (20.93)

Subscripts

a	apparent
b	bulk (mixed mean) flow conditions
b	in the case of parallel plates, indicates the velocity of the upper plate, Eq. (20.2)
e	flow conditions just outside boundary layer
eff	effective
F	forced convection
g	generalized
H1	thermal boundary condition of constant heat flux in the flow direction and constant temperature in the circumferential direction
H2	thermal boundary condition of constant heat flux in the flow direction and constant heat flux in the circumferential direction
i	inner surface of tube

l	liquid
o	outer surface of tube
p	plate
s	solid
T	thermal boundary condition of constant wall temperature in both flow and circumferential directions
w	wall
x	local value at section x
∞	free-stream conditions (far from any solid boundary)

REFERENCES

1. Y. I. Cho and J. P. Hartnett, Non-Newtonian Fluids in Circular Pipe Flow, *Advances in Heat Transfer*, Vol. 15, Academic, New York, pp. 59–141, 1982.
2. A. P. H. Skelland, *Non-Newtonian Flow and Heat Transfer*, Wiley, New York, 1967.
3. W. Ostwald and R. Auerbach, Über die Viskosität Kolloider hosungen im Struktur-Laminar und Turbulezgebiet, *Kolloid-Z.*, Vol. 38, pp. 261–280, 1926.
4. M. Reiner, *Deformation, Strain and Flow*, Wiley-Interscience, New York, 1960.
5. M. M. Cross, Rheology of Non-Newtonian Fluids: A New Equation for Pseudoplastic Systems, *J. Colloid. Sci.*, Vol. 20, pp. 417–437, 1965.
6. J. L. Sutterby, Laminar Converging Flow of Dilute Polymer Solutions in Conical Sections—I. Viscosity Data, New Viscosity Model, Tube Flow Solution, *AIChE J.*, Vol. 12, pp. 63–68, 1966.
7. P. J. Carreau, Rheological Equations from Molecular Network Theory, *Trans. Soc. Rheol.*, Vol. 16, pp. 99–127, 1972.
8. H. S. Lee, Rheological Fluid Properties, Measurement of CMC Water Solutions, M.S. Thesis, Mech. Eng. Dept., State Univ. of New York at Stony Brook, 1979.
9. T. Yamasaki, A New Capillary Tube Viscometer to Measure the Rheological Properties of Power Law Fluids, M.S. Thesis, Mech. Eng. Dept., State Univ. of New York at Stony Brook, 1982.
10. J. R. Van Wazer, J. W. Lyons, K. Y. Kim, and R. E. Colwell, *Viscosity and Flow Measurement*, Interscience (Wiley), New York, 1963.
11. K. Walters, *Rheometry*, Halstead (Wiley), New York, 1975.
12. R. W. Whorlow, *Rheological Techniques*, Halstead (Wiley), New York, 1980.
13. G. Dazhi and R. I. Tanner, The Drag on a Sphere in a Power Law Fluid, *J. Non-Newtonian Fluid Mech.*, Vol. 17, pp. 1–12, 1985.
14. N. A. Park and T. F. Irvine Jr., The Falling Needle Viscometer–A New Technique for Viscosity Measurements, *Wärme- und Stoffübertragung*, Vol. 18, pp. 201–206, 1984.
15. N. A. Park, Measurement of Rheological Properties of Non-Newtonian Fluids with the Falling Needle Viscometer, Ph.D. Thesis, Mech. Eng. Dept., State Univ. of New York at Stony Brook, 1984.
16. W. Kozicki, C. H. Chou, and C. Tiu, Non-Newtonian Flow in Ducts of Arbitrary Cross-Sectional Shape, *Chem. Eng. Sci.*, Vol. 21, pp. 665–679, 1966.
17. W. Kozicki and C. Tiu, Non-Newtonian Flow through Open Channels, *Can. J. Chem. Eng.*, Vol. 45, pp. 127–134, 1967.
18. C. Tiu, W. Kozicki, and T. Q. Phung, Geometric Parameters for Some Flow Channels, *Can. J. Chem. Eng.*, Vol. 46, pp. 389–393, 1968.
19. C. Tiu and W. Kozicki, Geometric Parameters for Open Circular Channels, *Can. J. Chem. Eng.*, Vol. 47, pp. 438–439, 1969.

20. T. D. Chang and T. F. Irvine Jr., Fully Developed Laminar Pressure Drop in a Triangular Duct for a Power Law Fluid, *Proc. All-Union Heat and Mass Transfer Conference*, Minsk, USSR, May, 1984.

21. J. A. Cheng, Laminar Forced Convective Heat Transfer of Power Law Fluids in Isosoceles Triangular Ducts with Peripheral Wall Conduction, Ph.D. Thesis, Mech. Eng. Dept., State Univ. of New York, Stony Brook, 1984.

22. J. A. Wheeler and E. H. Wissler, The Friction Factor–Reynolds Number Relation for the Steady Flow of Pseudoplastic Fluids through Rectangular Ducts, *AIChE J.*, Vol. 11, pp. 207–216, 1965.

23. W. Kozicki and C. Tiu, Improved Parametric Characterization of Flow Geometrics, *Can. J. Chem. Eng.*, Vol. 49, pp. 562–569, 1971.

24. D. W. Dodge and A. B. Metzner, Turbulent Flow of Non-Newtonian Systems, *AIChE J.*, Vol. 5, pp. 189–204, 1959.

25. S. S. Yoo, Heat Transfer and Friction Factors for Non-Newtonian Fluids in Turbulent Pipe Flow, Ph.D. Thesis, Univ. of Illinois at Chicago, 1974.

26. T. F. Irvine Jr., A Generalized Blasius Equation for Power Law Fluids, Internal Mechanical Engineering Dept. State Univ. New York, Stony Brook, N.Y., 1986.

27. M. Kostic and J. P. Hartnett, Predicting Turbulent Friction Factors of Non-Newtonian Fluids in Non-circular Ducts, *Int. Comm. Heat and Mass Transfer*, Vol. 11, pp. 345–352, 1984.

28. T. D. Chang, The Prediction of Fully Developed Turbulent Drops in a Triangular Duct for a Power Law Fluid, Ph.D. Thesis, Mech. Eng. Dept., State Univ. of New York, Stony Brook, 1984.

29. T. F. Irvine Jr., Non-circular Duct Heat Transfer, *Modern Developments in Heat Transfer*, W. Ibele, ed., Academic, New York, pp. 1–17, 1963.

30. R. K. Shah and A. L. London, Laminar Flow Forced Convection in Ducts, Supplement 1 to *Advances in Heat Transfer*, Academic, New York, 1978.

31. R. B. Bird, R. C. Armstrong, and O. Hassanger, *Dynamics of Polymeric Liquids*, Vol. 1, Wiley, New York, pp. 242–245, 1977.

32. U. Grigull, Wärmeübergang an Nicht-Newtonsche Flussigkeiten bei Laminarer Rohrstromung, *Chem-Ing.-Tech.*, Vol. 28, pp. 553–556, 1956.

33. J. Yau and C. Tien, Simultaneous Development of Velocity and Temperature Profiles for Laminar Flow of a Non-Newtonian Fluid in the Entrance Region of Flat Ducts, *Can. J. Chem Eng.*, Vol. 41, pp. 139–145, 1963.

34. A. R. Chandrupatla and V. M. K. Sastri, Laminar Forced Convection Heat Transfer of a Non-Newtonian Fluid in a Square Duct, *Int. J. Heat Mass Transfer*, Vol. 20, pp. 1315–1324, 1977.

35. S. D. Joshi and A. E. Bergles, Experimental Study of Laminar Heat Transfer to In-Tube Flow of Non-Newtonian Fluids, *J. Heat Transfer*, Vol. 102, pp. 397–401, 1980.

36. S. D. Joshi and A. E. Bergles, Analytical Study of Heat Transfer to Laminar In-Tube Flow of Non-Newtonian Fluids, *AIChE Symp. Ser.*, No. 199, Vol. 76, pp. 270–281, 1980.

37. A. B. Metzner and P. S. Friend, Heat Transfer to Turbulent Non-Newtonian Fluids, *Ind. Eng. Chem. J.*, Vol. 51, pp. 879–882, 1959.

38. R. M. Clapp, Turbulent Heat Transfer in Pseudoplastic Non-Newtonian Fluids, *Proc. 1961 Int. Heat Transfer Conf.*, pp. 652–661, 1963.

39. A. V. Shenoy and R. A. Mashelkar, Thermal Convection in Non-Newtonian Fluids, *Adv. Heat Transfer*, Vol. 15, pp. 143–225, Academic, New York 1982.

40. A. Acrivos, A Theoretical Analysis of Laminar Natural Convection Heat Transfer to Non-Newtonian Fluids, *AIChE J.*, Vol. 6, pp. 584–590, 1960.

41. C. Tien, Laminar Natural Convection Heat Transfer from Vertical Plate to Power Law Fluid, *Appl. Sci. Res.*, Vol. 17, pp. 233–248, 1967.

42. A. V. Shenoy and J. J. Ulbrecht, Temperature Profiles for Laminar Natural Convection Flow of Dilute Polymer Solutions Past an Isothermal Vertical Flat Plate, *Chem. Eng. Comm.*, Vol. 3, pp. 303–324, 1979.
43. J. D. Dale and A. F. Emery, The Free Convection of Heat from a Vertical Plate to Several Non-Newtonian Pseudoplastic Fluids, *J. Heat Transfer*, Vol. 94, pp. 64–72, 1972.
44. T. V. W. Chen and D. E. Wollersheim, Free Convection at a Vertical Plate with Uniform Flux Condition in Non-Newtonian Power Law Fluids, *J. Heat Transfer*, Vol. 95, pp. 123–124, 1973.
45. A. V. Shenoy, Ph.D. Thesis, Univ. of Salford, U.K., 1977.
46. T. Fujii, O. Miyatake, M. Fujii, H. Tanaka, and K. Murakami, Natural Convection Heat Transfer from a Vertical Isothermal Surface to a Non-Newtonian Sutterby Fluid, *Int. J. Heat Mass Transfer*, Vol. 16, pp. 2177–2187, 1973.
47. T. Fugii, O. Miyatake, M. Fujii, H. Tanaka, and K. Murakami, Natural Convection Heat Transfer from a Vertical Surface of Uniform Heat Flux to a Non-Newtonian Sutterby Fluid, *Int. J. Heat Mass Transfer*, Vol. 17, pp. 149–154, 1974.
48. C. Tien and H. S. Tsuei, Laminar Natural Convection Heat Transfer in Ellis Fluids, *Appl. Sci. Res.*, Vol. 20, pp. 131–147, 1969.
49. I. G. Reilly, C. Tien, and M. Adelman, Experimental Study of Natural Convective Heat Transfer from a Vertical Plate in a Non-Newtonian Fluid, *Can. J. Chem. Eng.*, Vol. 43, pp. 157–161, 1965.
50. J. Kleppe and W. J. Marner, Transient Free Convection in a Bingham Plastic Fluid on a Vertical Flat Plate, *J. Heat Transfer*, Vol. 94, pp. 371–376, 1972.
51. A. V. Shenoy and R. A. Mashelkar, Turbulent Free Convection Heat Transfer from a Flat Vertical Plate to a Power Law Fluid, *AIChE J.*, Vol. 24, pp. 344–347, 1978.
52. C. C. Gentry and D. E. Wollersheim, Local Free Convection to Non-Newtonian Fluids from a Horizontal Isothermal Cylinder, *J. Heat Transfer*, Vol. 96, pp. 3–8, 1974.
53. A. V. Shenoy and R. A. Mashelkar, Thermal Convection in Non-Newtonian Fluids, *Adv. Heat Transfer*, Vol. 15, pp. 172–173, Academic, New York 1982.
54. W. S. Amato and C. Tien, Free Convection Heat Transfer from Isothermal Spheres in Polymer Solutions, *Int. J. Heat Mass Transfer*, Vol. 19, pp. 1257–1266, 1976.
55. A. V. Shenoy, A Correlating Equation for Combined Laminar Forced and Free Convection Heat Transfer to Power Law Fluids, *AIChE J.*, Vol. 26, pp. 505–507, 1980.
56. A. B. Metzner and D. F. Gluck, Heat Transfer to Non-Newtonian Fluids under Laminar Flow Conditions, *Chem. Eng. Sci.*, Vol. 12, p. 185, 1960.
57. D. R. Oliver and V. G. Jenson, Heat Transfer to Pseudoplastic Fluids in Laminar Flow in Horizontal Tubes, *Chem. Eng. Sci.*, Vol. 19, pp. 115–129, 1964.
58. S. H. DeYoung and G. F. Scheele, Natural Convection Distorted Non-Newtonian Flow in a Vertical Pipe, *AIChE J.*, Vol. 16, pp. 712–717, 1970.
59. W. J. Marner and R. A. Rehfuss, Buoyancy Effects on Fully Developed Laminar Non-Newtonian Flow in a Vertical Tube, *Chem. Eng. J.*, Vol. 3, pp. 294–300, 1972.
60. F. Gori, Effects of Variable Physical Properties in Laminar Flow of Pseudoplastic Fluids, *Int. J. Heat Mass Transfer*, Vol. 21, pp. 247–250, 1978.
61. W. J. Marner and H. K. McMillan, Combined Free and Forced Laminar Non-Newtonian Convection in a Vertical Tube with Constant Wall Temperature, *Chem. Eng. Sci.*, Vol. 27, pp. 473–488, 1972.
62. F. Gori, Variable Physical Properties in Laminar Heating of Pseudoplastic Fluids with Constant Wall Heat Flux, *J. Heat Transfer*, Vol. 100, pp. 220–223, 1978.
63. J. R. Bodia and J. F. Osterle, The Development of Free Convection between Heated Vertical Plates, *J. Heat Transfer*, Vol. 84, pp. 40–44, 1962.
64. T. F. Irvine, K. C. Wu, and W. J. Schneider, Vertical Channel Free Convection for a Power Law Fluid, ASME paper 82-WA/HT-69, 1982.

65. W. J. Schneider and T. F. Irvine, Vertical Channel Free Convection for a Power Law Fluid with a Constant Heat Flux, ASME Paper 84-HT-16, 1984.

66. C. B. Kim and D. E. Wollersheim, Free Convection Heat Transfer to Non-Newtonian, Dilatant Fluids from a Horizontal Cylinder, *J. Heat Transfer*, Vol. 98, pp. 144–146, 1976.

67. M. Tanka and N. Mitsuishi, Non-Newtonian Laminar Heat Transfer in Concentric Annuli, *Heat Transfer—Jpn. Res.*, Vol. 4, No. 2, pp. 26–36, 1975.

68. H. Schlichting, *Boundary Layer Theory*, 7th ed., McGraw-Hill, New York, 1979.

69. I. G. Currie, *Fundamental Mechanics of Fluids*, McGraw-Hill, New York, 1974.

70. A. Acrivos, M. J. Shah, and E. E. Petersen, Momentum and Heat Transfer in Laminar Boundary Layer Flows of Non-Newtonian Fluids Past External Surfaces, *AIChE J.*, Vol. 6, p. 312, 1960.

71. H. Blasius, Grenzschichten in Flussigkeiten mit kleiner Reibung, Diss. Göttingen 1907; *Z. Math. Phys.*, Vol. 56, pp. 1–37, 1908; English translation, NACA TM 1256, 1950.

72. K. Takahashi, M. Maeda, and S. Ikai, Study on Heat Transfer to a Power-Law Non-Newtonian Fluid from a Flat Plate, *Heat Transfer Jpn. Res.*, Vol. 8, No. 3, pp. 83–91, 1979.

73. H. W. Kim, D. R. Jeng, and K. J. DeWitt, Momentum and Heat Transfer in Power-Law Fluid Flow over Two-Dimensional or Axisymmetrical Bodies, *Int. J. Heat Mass Transfer*, Vol. 26, pp. 245–259, 1983.

74. F. N. Lin and S. Y. Chern, Laminar Boundary-Layer Flow of Non-Newtonian Fluid, *Int. J. Heat Mass Transfer*, Vol. 22, pp. 1323–1329, 1979.

75. T. Mizushina and H. Usui, Approximate Solution of the Boundary Layer Equations for the Flow of a Non-Newtonian Fluid around a Circular Cylinder, *Heat Transfer—Jpn. Res.*, Vol. 7, No. 2, pp. 83–92, 1978.

76. T. Mizushina, H. Usui, K. Veno, and T. Kato, Experiments of Pseudoplastic Fluid Cross Flow Around a Circular Cylinder, *Heat Transfer—Jpn. Res.*, Vol. 7, No. 3, pp. 92–101, 1978.

77. M. J. Shah, E. E. Petersen, and A. Acrivos, Heat Transfer from a Cylinder to a Power-Law Non-Newtonian Fluid, *AIChE J.*, Vol. 8, pp. 542–549, 1962.

78. A. Žukauskas, Heat Transfer from Tubes in Crossflow, *Adv. Heat Transfer*, Vol. 8, pp. 93–160, Academic, New York, 1972.

79. S. F. Shen, Stability of Laminar Flows, in *Theory of Laminar Flows*, High Speed Aerodyn. and Jet Propulsion Ser., F. K. Moore, ed. Vol. 4, pp. 719–853, Princeton U.P., Princeton, N.J., 1964.

21

FOULING WITH CONVECTIVE HEAT TRANSFER

W. J. Marner

Jet Propulsion Laboratory
California Institute of Technology
Pasadena, California

J. W. Suitor

Jet Propulsion Laboratory
California Institute of Technology
Pasadena, California

21.1 INTRODUCTION

This chapter is concerned with the fouling of heat transfer surfaces. The treatment is limited to single-phase convective heat transfer applications; i.e., fouling with boiling or condensation and high-temperature applications where radiation heat transfer is important are given very little consideration. Fouling is a problem in a broad variety of applications involving both gases and liquids in the energy-related industries, and thus an attempt has been made to maintain a reasonable balance between gas-side and liquid-side fouling in this chapter.

The term fouling is defined, the categories of fouling are described briefly, and the deleterious effects of fouling in heat transfer equipment are considered in this section. Fouling mechanisms, or the phenomenological aspects of fouling, are discussed in Sec. 21.2, along with a treatment of the more important fouling parameters, followed by a presentation of selected fouling models. Section 21.3 deals with the design of heat transfer equipment for fouling service and makes use of fouling factors or fouling resistances. The effects of fouling on both heat transfer and pressure drop are considered, and examples are given to illustrate these effects quantitatively. Prevention, mitigation, and accommodation techniques used to combat liquid-side and gas-side fouling are discussed in some detail in Secs. 21.4 and 21.5, reflecting the practical importance of these techniques. Finally, this chapter is concluded with a brief discussion of fouling measuring devices.

The treatment of fouling here is design-oriented and relies heavily on the use of fouling factors. The effect of fouling on pressure drop is considered—an effect which is frequently overlooked because attention is focused on the degradation of heat transfer. Because of space limitations, the treatment of most topics must be brief; however, an ample supply of references is included for readers wishing to pursue topics in greater depth. Fouling references of general interest, based on recent conferences, workshops, and surveys, which reflect the current state of the art, include Somerscales and Knudsen [1], Bryers [2–4], Chenoweth and Impagliazzo [5], Marner and Webb [6], Marner and Suitor [7], Suitor and Pritchard [8], and Garrett-Price et al. [9].

21.1.1 Definition of Fouling

Fouling may be defined as the deposition of an insulating layer of material onto a heat transfer surface. The process is known as liquid-side or gas-side fouling depending on whether the fluid in contact with the surface is a liquid or a gas. In either case, the thermal-hydraulic performance can be diminished through decreased heat transfer and increased pressure drop because of the fouling deposit. A few examples illustrating the diversity of fouling include: biofouling in a condenser using seawater as the coolant, coking in heat exchangers used in the petrochemical industry, ash deposits in coal-fired boilers, chemical reaction fouling of heat transfer surfaces in the foodstuff industries, and smut deposition in low-temperature heat recovery systems.

In some cases fouling may be accompanied by either corrosion (the destruction of a metal or alloy by means of an electrochemical reaction with its environment) or erosion (the wearing away of a solid surface by the impact of a liquid stream or particulate matter in a gas stream). In this chapter, attention is focused on fouling, with erosion and corrosion treated briefly as appropriate.

21.1.2 Types of Fouling

Epstein [10, 11] has developed the following classification scheme for fouling, which has received wide acceptance:

Precipitation fouling. The precipitation of dissolved substances onto a heat transfer surface. When the dissolved substances have inverse rather than normal solubility-vs.-temperature behavior, the precipitation occurs on superheated rather than subcooled surfaces and the process is often referred to as scaling.

Particulate fouling. The accumulation of finely divided solids suspended in the fluid onto the heat transfer surface. In a minority of instances settling by gravity prevails, and the process is then referred to as sedimentation fouling.

Chemical reaction fouling. Deposits formed at the heat transfer surface by chemical reactions in which the surface material itself is not a reactant.

Corrosion fouling. The heat transfer surface itself reacts to produce corrosion products (*in situ* corrosion fouling) which thermally insulate (foul) the surface and may promote the attachment of other foulants.

Biological fouling. The attachment of macroorganisms (macrofouling) and/or microorganisms (microfouling or microbial fouling) to a heat transfer surface, along with the adherent slimes often generated by the latter.

Solidification fouling. Freezing of a liquid, or some of its higher-melting constituents, or liquid components in a gas stream onto a subcooled heat transfer surface.

Although these six categories have been listed individually, it should be emphasized that two or more fouling processes can and often do occur simultaneously and in some cases may be synergistic. For example, fouling in a condenser using seawater as the cooling medium generally is caused by the combined effects of biological and corrosion fouling. Similarly, fouling of a suspension preheater in the cement industry may involve the simultaneous action of particulate, chemical reaction, and solidification fouling.

21.1.3 Deleterious Effects of Fouling

The deleterious effects of fouling maybe broken down into the following major categories: (1) increased capital costs, (2) increased maintenance costs, (3) loss of production, and (4) energy losses.

First, increased capital costs result in part from oversurfacing the heat exchanger to compensate for reduced heat transfer due to fouling, oversizing pumps and fans to compensate for oversurfacing and the increased pressure drop arising from the reduced flow area, employing duplicate heat exchangers, and using specialty materials such as stainless steels, superalloys, and coated tubes in fouling environments. In those cases where on-line cleaning of the heat transfer surfaces is required, increased capital costs are also associated with the provision for cleaning equipment. Second, maintenance costs are primarily due to on-line and off-line cleaning and to the costs of additives. Chemical additives are used in both gas and liquid service, with gas-side additives introduced in either the fuel or the combustion gases. Third, the loss of production, due to down time or operation at reduced capacity, in industrial applications can be significant. Finally, energy losses due to fouling buildups in heat exchangers can be a major contributor to the costs of fouling. There are the obvious energy losses associated with the reduction of heat transfer and increase in pumping-power requirements. Additional energy losses are incurred when gas streams are exhausted to the environment because they contain potential foulants: the use of heat-recovery equipment is strongly discouraged if fouling is anticipated from dirty gas streams. Such occurrences are widespread in the glass, cement, and primary metal industries.

Although the basic factors which contribute to the costs of fouling are well established, quantification of these costs is extremely difficult because of the many uncertainties involved. A 1978 study estimated the annual cost of fouling in the United Kingdom to be $0.8 to 1.0 billion (Thackery [12]). More recently, the annual cost of fouling and corrosion in U.S. industries, excluding electric utilities, was placed between $3 and 10 billion in 1982 dollars [9]. It is clear that the deleterious effects of fouling are extremely costly.

21.2 PHENOMENOLOGICAL ASPECTS OF FOULING

Fouling is a combined heat, mass, and momentum transfer problem which occurs under transient conditions. Fortunately, fouling buildups usually occur sufficiently slowly that quasisteady conditions may be assumed. Even so, the many processes which can take place simultaneously during the fouling of a heat transfer surface are incredibly complicated. The various fouling mechanisms, as well as a consideration of the important fouling parameters, are treated in this section, which is concluded with a brief consideration of fouling models.

21.2.1 Basic Considerations

A simple visualization of fouling is given Fig. 21.1, which depicts fouling buildup on a plane surface at some instant in time. A fluid at temperature T_f and velocity V is flowing over a surface which is at temperature T_w. Under clean or unfouled conditions,

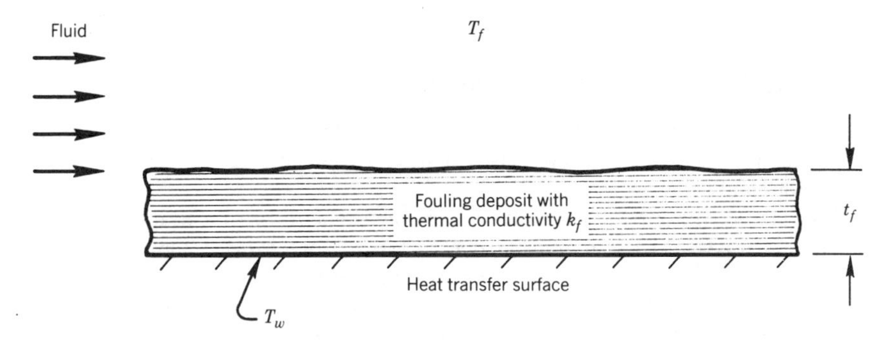

Figure 21.1. Schematic drawing of fouling deposit used to define fouling factor.

the heat transfer between the fluid and surface is given by Newton's law of cooling,

$$q''_w = \frac{T_w - T_f}{1/h} \tag{21.1}$$

Under fouled conditions, as shown in Fig. 21.1, the heat transfer is reduced by the thermal resistance of the fouling layer:

$$q''_w = \frac{T_w - T_f}{1/h + t_f/k_f} \tag{21.2}$$

In Eq. (21.2), the ratio t_f/k_f is known as the fouling factor,

$$R_f = \frac{t_f}{k_f} \tag{21.3}$$

Since information on t_f and k_f is usually not known, the fouling factor R_f is generally used in design calculations.

In some cases, it is more convenient to deal with the mass of the deposit rather than the fouling factor, using the relation

$$R_f = \frac{t_f}{k_f} = \frac{m_f}{\rho_f k_f A} \tag{21.4}$$

However, in order to relate the three different expressions in Eq. (21.4) it is apparent that a knowledge of both the deposit density and thermal conductivity is required.

Unfortunately, the situation is not as simple as described in Fig. 21.1. In general, the deposit thickness is not constant but changes with time: a variety of transport processes add material to increase the deposit thickness, while at the same time natural removal forces remove material from the surface. In addition, the composition and strength of the fouling deposit change with time. The phenomena contributing to these various fouling processes are identified and discussed in the next section.

21.2.2 Fouling Mechanisms

Fouling mechanisms—referred to as sequential events by Epstein [13]—may be grouped into the following five major categories: (1) initiation of fouling, (2) transport to surface, (3) attachment to surface, (4) removal from surface, and (5) aging of deposit.

Initiation of Fouling. The initiation of fouling appears to be a strong function of surface-related parameters such as surface material, surface finish, surface roughness, and surface coatings or films. During the initial delay—induction or incubation—period, the surface is being conditioned for the fouling which will take place later. This phenomenon, which is usually observed in crystallization or precipitation fouling and sometimes in other types of fouling, lasts on the order of hours. For example, Ritter [14] observed an induction period of about 20 hr while studying the deposition of calcium and lithium sulfate in crystalline fouling. After this delay period has been observed, the fouling resistance increases with time in some fashion.

Transport to the Surface. Of the five mechanisms listed above, transport to the surface is clearly the most studied and best understood. Transport to the surface results from a variety of processes, including: (1) diffusion, (2) electrophoresis, (3) sedimentation, (4) chemical reactions, (5) thermophoresis, (6) inertial impaction, (7) turbulent downsweeps, and (8) diffusiophoresis.

Diffusion plays a very important role in fouling, in the transport of both gaseous and particulate species. The phenomenological equation governing the transfer of mass is Fick's law,

$$J'' = -D\frac{\partial c}{\partial y} \tag{21.5}$$

A convective mass transfer coefficient h_D is defined by the relation

$$J''_w = h_D(c_f - c_w) \tag{21.6}$$

Evaluating Eq. (21.5) at the wall and equating it to Eq. (21.6) yields

$$h_D = -D\frac{(\partial c/\partial y)_w}{c_f - c_w} \tag{21.7}$$

The nondimensional convective mass transfer coefficient is the Sherwood number, defined as

$$\text{Sh} = \frac{h_D L}{D} \tag{21.8}$$

It is a function of the Reynolds number, the Schmidt number, and the geometry:

$$\text{Sh} = \text{Sh}(\text{Re}, \text{Sc}, \text{geometry}) \tag{21.9}$$

with the Schmidt number Sc defined by

$$\text{Sc} = \frac{\nu}{D} \tag{21.10}$$

In practice, h_D is determined empirically for fouling applications. For example, for turbulent flow in a circular tube, use of the Dittus-Boelter correlation [15] and the analogy between heat and mass transfer gives the Sherwood number as

$$\text{Sh} = 0.023\,\text{Re}^{0.8}\,\text{Sc}^{0.4} \tag{21.11}$$

Electrophoresis may be defined as the transport of particulate matter from a fluid stream to a solid surface due to an electric potential difference between the particles and the surface. Particle charging may occur due to field charging, diffusion charging, or charging by surface contact. The most important surface forces are the London–van der Waals forces, which are always attractive. The electrical double-layer interaction forces are attractive if the particles and wall have zeta potentials of opposite sign and repulsive if these potentials are of the same sign. In general, the smaller a particle in a fluid stream, the more significant the effect of the electrical charge it carries. In gases, electrical forces become increasingly important on charged surfaces as the particle size decreases below about 0.1 μm. For larger particles, very strong electrical fields are required to influence transport. In practice, it appears that electrophoresis is much more important in gas-side than in liquid-side fouling service. Additional details on electrophoresis may be found in Friedlander [16] and Soo [17].

Sedimentation is the transport of particulate matter in a fluid stream to a surface under the action of gravity. As might be expected, this process is important in applications where the particles are dense and the fluid velocities are low. Sedimentation is generally of importance only in liquid systems.

Material can be transported to a heat transfer surface by means of chemical reactions taking place in either the fluid stream or on the surface, but not involving the surface itself. Examples involving chemical reaction transport include a variety of applications in the foodstuffs industry, coking in the petrochemical industry, and buildups in suspension preheaters in the cement industry.

Thermophoresis is the movement of small particles in a fluid stream under the influence of a temperature gradient. Particles are bombarded by higher-energy molecules on their "hot" side, and thus tend to move toward a cold surface and away from a hot surface. Whitmore and Meisen [18] have shown that the thermophoretic velocity V_t is given by

$$V_t = -\alpha\nu\frac{\nabla T}{T} \tag{21.12}$$

where the coefficient α for gases is given by

$$\alpha = \frac{c_0}{k_p/k + 2} \tag{21.13}$$

Thermophoresis is important for particles below 5 μm in diameter and becomes dominant at about 0.1 μm. Depending on the magnitude of the temperature gradient and the other parameters in Eq. (21.12), thermophoresis can be important in both laminar and turbulent flow situations. Since the thermal conductivity of liquids tends to be much larger than that for gases, the coefficient in Eq. (21.13) is much larger for gases than for liquids; therefore, thermophoresis is much more important as a transport process in gases than in liquids. Nishio et al. [19] have recently carried out an important experimental study of thermophoresis in an engineering application of gas-side fouling.

For particles larger than about 1 μm in gases, the inertial effects of particles can become very important. For example, when a fluid flows over a circular tube in a heat exchanger, inertia can cause a particle to deviate from the fluid streamlines. Particle transport to the surface from such a process is known as inertial impaction. This transport is governed by the magnitude of the Stokes number St, defined as

$$\mathrm{St} = \frac{\rho_p d_p^2 V}{18\mu L} \tag{21.14}$$

The transport increases with St. In particular, inertial impaction becomes increasingly important as the size of the particles increases. Additional information on inertial transport can be found in Friedlander [16].

The next transport process considered here is that due to turbulent downsweeps. Based on experimental observations, Cleaver and Yates [20, 21] found that miniature tornadoes in the fluid stream are able to penetrate the laminar sublayer and transport solid material to the surface. They also observed that turbulent bursts are an effective removal mechanism.

Diffusiophoresis is the final transport process to be considered here. As water or some other vapor condenses on a surface that is below the dew-point temperature, the vapor creates a force on nearby particles in the gas stream and sweeps them toward the surface. This process is important in certain gas-side fouling situations.

Attachment to Surface. Not all of the material transported to the surface actually sticks. There is a probability between zero and one that the particle will bounce off the surface. As Pritchard [22] and others have pointed out, the reason why some particles stick and others do not is not understood very well. Certainly the forces acting on particles as they approach a surface play an important role. Also, the properties of the particles such as density, elasticity, surface condition, and state are important. Finally, the nature of the surface, including parameters such as roughness and type of material, plays an important role in the sticking probabilities of various particles.

Removal from Surface. Material can be removed from the deposit by several mechanisms, including spalling, caused by shear forces and turbulent bursts; re-solution; and erosion. The natural removal mechanism due to fluid shear stress depends on the velocity gradient at the surface, the viscosity of the fluid, and the roughness of the surface, and is reflected in the pressure drop. Turbulent bursts have already been discussed in connection with transport mechanisms to the surface. Re-solution can occur, for example, if the pH of a liquid stream is changed by additives or some other means. Finally, erosion by particulate matter or by liquid impingement can remove material from the fouling layer. In general, several of these mechanisms can occur simultaneously.

Aging of the Deposit. Once an initial deposit is placed on a heat transfer surface, it does not remain static. Initially, the deposit thickness will usually increase with time as has been discussed earlier. In addition, the mechanical strength of the deposit can change with time due to changes in the crystal structure or chemical composition of the deposit. For example, a chemical reaction taking place at the deposit surface can alter the chemical composition of the deposit and thereby change its mechanical strength. Of course, the introduction of a chemical additive into a fluid stream can cause a change in the deposit characteristics. Clearly, aging may strengthen or weaken fouling deposits.

21.2.3 Important Fouling Parameters

Fouling is a complex, transient process involving the simultaneous transport of heat, mass, and momentum and is characterized by a large number of parameters. These parameters include: (1) fluid characterization, (2) surface temperature, (3) fluid temperature, (4) fluid velocity, (5) fluid shear stress, (6) surface material and finish, (7) surface geometry, and (8) fouling-deposit strength.

Fluid characterization includes the thermodynamic and transport properties of the fluid such as its specific heat, density, thermal conductivity, viscosity, and diffusion coefficients. If the fluid contains particulate matter, the particle concentration, composition, and size distribution are also important. Condensable components and their associated dew-point temperatures are important parameters in gas-side fouling. In some cases, the chemical species in the fluid stream can react to produce new components, and it is important to know if and when such processes might take place. The importance of fluid characterization as a parameter cannot be overemphasized, because it is the constituents in the gas or liquid stream which are ultimately deposited onto heat transfer surfaces to form the fouling deposits.

The various thermodynamic and transport properties are functions of temperature, which itself can affect the transport processes taking place. For example, the diffusion coefficients for gases are very strong functions of the temperature and vary as the absolute temperature raised to the 1.8 power (e.g., Eckert and Drake [23]):

$$\frac{D}{D_{\mathrm{ref}}} = \left(\frac{T}{T_{\mathrm{ref}}}\right)^{1.8} \tag{21.15}$$

The surface temperature is the controlling parameter in those cases where species are condensing, crystallizing, or sublimating onto heat transfer surfaces. In such instances, the fluid temperature plays a very minor role in the deposition processes which take place. This phenomenon is very apparent in a classical study carried out by Brown [24], who investigated the deposition of sodium sulfate from combustion gases and found no deposition above the dew point, maximum deposition just below the dew point, and very little deposition far below the dew point. Finally, in situations involving thermophoresis, it is the temperature difference $T_s - T_f$ which is the controlling parameter. In such cases, the magnitude of either temperature is not of particular significance, except as the various thermodynamic and transport properties are affected.

The fluid velocity is important in transporting material to the surface as well as removing material from the surface. For example, in diffusion-controlled processes, the mass transfer coefficient is a reasonably strong function of the velocity. Although the fluid velocity is frequently used in removal processes, it is actually the shear stress which is the controlling parameter. The distinction between velocity and fluid shear stress depends on the geometry of the heat transfer surface. For example, for a given velocity, the shear stress at the surface will be much higher for a plate heat exchanger than for a shell-and-tube exchanger. Also, in a double-pipe exchanger, the shear stress at the outer surface of the annulus is not the same as that at the inner surface even though the average velocity of the fluid is identical. Thus, the removal forces generated by the shear stresses at the two surfaces will also be different. An additional parameter of importance in the removal process is the mechanical strength of the fouling deposit. The deposit strength is a strong function of aging and depends on the material transported to the surface, chemical reactions within the deposit, and removal of material from the deposit surface. Additives may also have a strong effect on the deposit strength.

Surface finish is especially important during the initial stages of fouling and plays a prominent role in determining the duration of the incubation period. However, once a deposit covers a surface, the finish no longer plays an important role. On the other hand, the surface material is a critical consideration in those situations where corrosion fouling takes place. In such instances, the interaction between the fouling deposit and the surface material is very important, with the surface and fluid temperatures also playing prominent roles. A variety of surface coatings, e.g., galvanizing and plastics, have been used in an attempt to overcome *in situ* corrosion fouling.

21.2.4 Fouling Models

Making use of the principle of conservation of mass, the time rate of change of the fouling resistance must equal the material deposited minus the material removed:

$$\frac{dR_f}{dt} = \phi_d - \phi_r \tag{21.16}$$

where ϕ_d and ϕ_r are the so-called deposition and removal functions, respectively. In the case of precipitation or scaling fouling, and certain other situations in which the removal function is negligible, Eq. (21.16) may easily be integrated by assuming the deposition function to be independent of time to yield

$$R_f = \phi_d t \tag{21.17}$$

Equation (21.17), first developed in 1924 by McCabe and Robinson [25], represents a linear fouling curve. Fouling in such cases will continue to increase indefinitely unless some type of cleaning is employed. Kern and Seaton [26] used Eq. (21.16), assumed the removal function to be proportional to the fouling resistance and the deposition function to be a constant, to obtain the classical relation

$$R_f = R_f^* [1 - e^{-t/\theta}] \tag{21.18}$$

Equation (21.18) is a mathematical expression for the so-called asymptotic fouling curve which is shown in Fig. 21.2, along with the linear fouling curve described by Eq. (21.17). The quantity R_f^* is known as the asymptotic fouling factor and is the value approached as the time approaches infinity. In practical situations, the time required to approach asymptotic conditions can be short, on the order of hours. The time constant θ in Eq. 21.18 indicates how quickly asymptotic fouling conditions are approached and is related to R_f^* by

$$\theta = \frac{R_f^*}{(dR_f/dt)_{t=0}} \tag{21.19}$$

Also shown in Fig. 21.2 is the falling-rate fouling curve, between the linear and asymptotic curves, which is sometimes observed in practice.

Through the years, a number of semiempirical models have been formulated to predict fouling factors under transient conditions. Epstein [10] has tabulated several

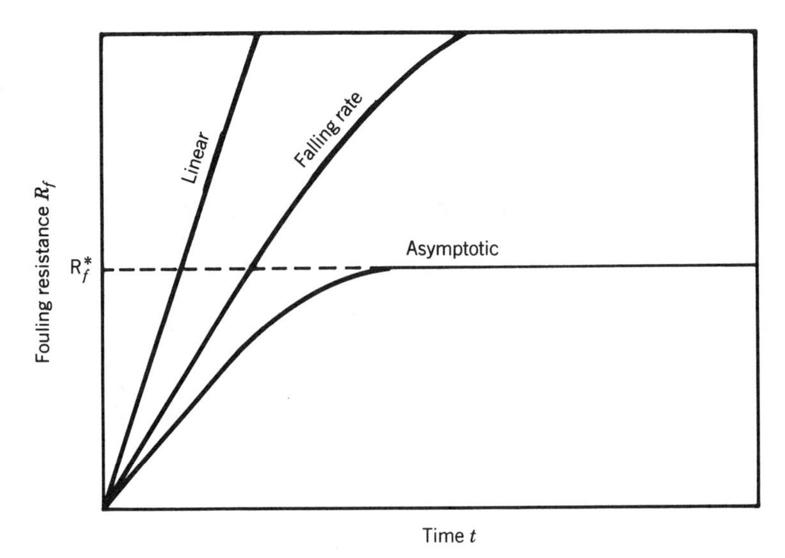

Figure 21.2. Types of fouling curves for zero incubation period.

TABLE 21.1. Some Fouling Deposition Models [10]

Investigators (Year)	Deposition Flux $\dot{m}_d$	Type of Fouling Investigated
McCabe and Robinson (1924)	$a_0 U$	Scaling of evaporators, constant ΔT
Kern and Seaton (1959)	$a_1 V c_f$	Particulate and other fouling
Parkins (1961)	$J'' c_0 \exp(-E/RT_w)$	Particulate fouling
Hasson (1962)	$a_2 U^n$	Sensible heat scaling of $CaCO_3$ solutions, $\Delta T \approx$ constant
Reitzer (1964)	$K_1(c_f - c_w)^n$	Evaporation and sensible-heat scaling of saturated solutions
Bartlett (1968)	$h_D c_f$	Convective mass transfer of depositable species
Charlesworth (1970)	$k_d c_f$	Iron oxide deposition in flow boiling
Watkinson and Epstein (1970)	$a_3 h_D (c_f - c_w) c_1 \exp(-E/RT_w)/u^{*2}$	Particulate and chemical reaction fouling
Beal (1970)	$k_d c_f = c_f/(1/h_D + 1/V_w)$	Particle deposition by eddy and Brownian diffusion, and inertial coasting
Taborek et al. (1972)	$a_4 P_v \Omega \exp(-E/RT_w)$	Cooling-water service
Galloway (1973)	$c_f/(1/h_D + x_f/\alpha_f)$	Convective mass transfer of O_2 in series with diffusion of O_2 through deposit
Ruckenstein and Prieve (1973)	$k_d c_f = c_f/(1/h_D + 1/k_r)$	Colloidal deposition across zeta potential barrier at wall; $k_r \propto \exp(-E/RT_w)$
Thomas and Grigull (1974)	$m_{d,0} \exp(-a_5 m)$	Fine magnetic deposition from aqueous suspension; autoretardation assumed

deposition and removal (reentrainment) models which have been developed, as indicated in Tables 21.1 and 21.2, respectively. Several of the transport equations are based on mass transfer models, and other approaches have also been used as indicated. The function in Tables 21.1 and 21.2 are expressed in terms of mass fluxes, which are related to ϕ_d and ϕ_r as follows:

$$\dot{m}_d = \rho_f k_f \phi_d \tag{21.20}$$

and

$$\dot{m}_r = \rho_f k_f \phi_r \tag{21.21}$$

In order to predict transient fouling factors, expressions for both the deposition and removal functions must be formulated. Of the correlations which have been developed

TABLE 21.2. Some Removal Models [10]

Investigators (Year)	Removal Flux $\dot{m}_r$	Assumed Removal Mechanism
Kern and Seaton (1959)	$b_1 \tau_w m$	Spalling
Bartlett (1968)	$b_2 \dot{m}_d$	Erosion plus bond fracture
Charlesworth (1970)	$b_0 m$	Erosion, spalling
Taborek et al. (1972)	$b_3 \tau_w m/\psi$	Spalling
Beal (1973)	$b_4 m_{\text{loose}}$	Erosion
Burrill (1977)	$b_5(c_w - c_f)$	Dissolution

to date, the greatest emphasis has been placed on cooling-water applications, using the Kern-Seaton model of asymptotic fouling as the basis. Two of the most important relations which have been proposed to date include the 1968 Watkinson-Epstein [27] equation

$$R_f^* \propto \frac{c_f \exp(-E/RT_w)}{fV^2} \tag{21.22}$$

and the 1972 Taborek et al. [28] equation

$$R_f^* \propto P_v \Omega e^{-E/RT_w} \tag{21.23}$$

where the asymptotic fouling factors have been given. Both of these models were developed for initially clean surfaces and no induction period. As can be inferred from these equations, Watkinson and Epstein extended the deposition function to be temperature-dependent, and Taborek et al. introduced a water characterization factor in the deposition function.

21.3 DESIGNING FOR FOULING SERVICE

The design of heat exchangers for fouling service must take into account the effects of fouling on both heat transfer and pressure drop. Fouling deposits reduce the effectiveness of a heat exchanger by reducing the heat transfer and by affecting the pressure drop of the exchanger, generally unfavorably but sometimes favorably. The effect of fouling on the design of heat exchangers, including both thermal and hydraulic considerations, is treated in this section.

21.3.1 Basic Considerations

The designer of a heat exchanger must determine the surface area to satisfy the required heat transfer or heat duty. The relationship between surface area and heat duty is

$$A_0 = \frac{q}{U\,\mathrm{MTD}} \tag{21.24}$$

Process conditions usually set the heat duty and temperature difference at specific values. Fouling reduces the overall heat transfer coefficient by adding thermal resistance in the heat flow path; thus the surface area becomes an adjustable parameter to account for the reduction of the overall coefficient due to fouling. The effect of fouling on the surface area as a function of the overall coefficient is shown in Fig. 21.3. It is not uncommon for the heat exchanger area to increase 100% due to fouling; in other words, fouling allowances often account for more than half of the total heat exchanger surface area.

A circular tube fouled on both the inside and the outside is shown in Fig. 21.4 with the corresponding thermal-resistance circuit. The overall coefficient is

$$\frac{1}{U} = \frac{1}{h_o} + \frac{A_o}{A_i h_i} + \frac{R_w A_o}{A_w} + R_{f,o} + \frac{R_{f,i} A_o}{A_i} \tag{21.25}$$

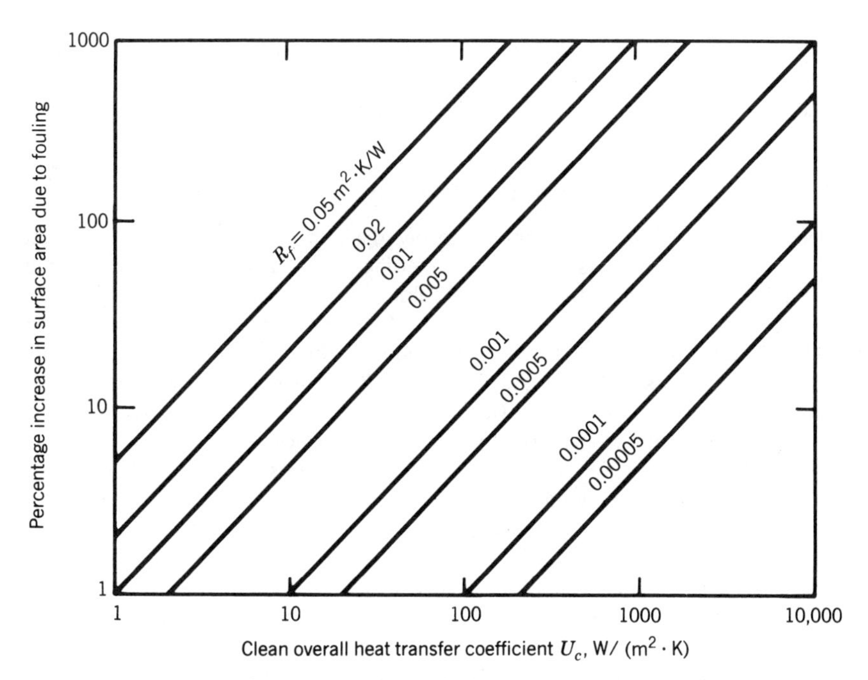

Figure 21.3. Effect of fouling on surface area.

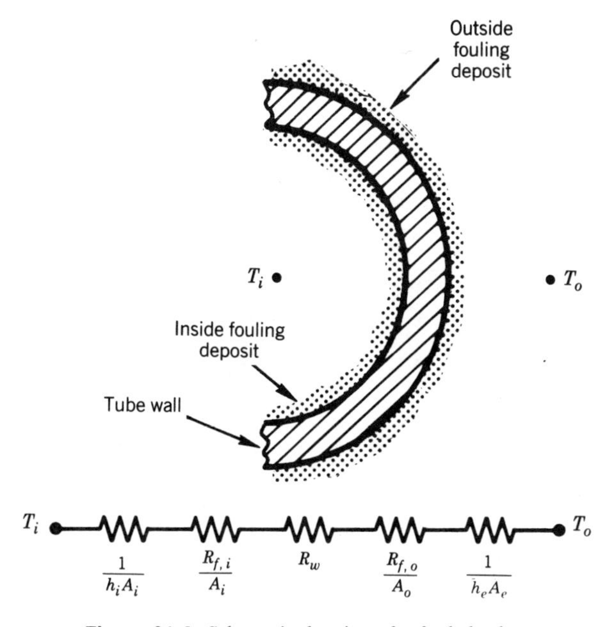

Figure 21.4. Schematic drawing of a fouled tube.

The addition of the fouling resistance lowers the overall coefficient, therefore requiring more area. The amplification of the tubeside fouling resistance by the outside-to-inside area ratio is especially important if finned tubes are used, as will be seen later.

21.3.2 Effect of Fouling on Heat Transfer

Even though the fouling factors appear to be small, they can increase the required surface area significantly. As an example, consider the industrial heat exchanger in liquid-liquid service described thermally in Table 21.3. The clean area required for this exchanger is about 59.0 m^2 (635 ft^2). However, if inside and outside fouling resistances of 0.000352 and 0.000176 $m^2 \cdot K/W$ (0.002 and 0.001 $hr \cdot ft^2 \cdot °F/Btu$) are used, the required area increases to 140 m^2 (1510 ft^2), or a 237% increase. These seemingly low fouling resistances obviously have a significant effect on the required surface area, the extent depending on the relative value of the overall coefficient.

Small fouling resistances, such as those in the previous example, will not affect gas-to-gas heat exchangers significantly, since typical gas-side heat transfer coefficients are much lower than those for liquids. Unfortunately, gas-side fouling factors are usually much larger and do significantly affect the overall heat transfer coefficient of gas-to-gas exchangers. For example, consider a heat recovery unit in a hot flue gas stream as indicated in Table 21.4. The gas-side coefficient is much lower than the liquid-side coefficient, so the overall coefficient is less. But the typical gas-side fouling factor is also larger. In this example, the clean area required is 2640 m^2 (28,400 ft^2), while the fouled required area is 4140 m^2 (44,600 ft^2), a 57% increase.

Cleanliness Factor. A term used primarily in the utility industry is the cleanliness factor, CF, which is an alternative expression for the fouling resistance and refers to the multiplying factor used with the clean overall coefficient to derate the exchanger performance. The cleanliness factor emanated from the Heat Exchange Institute [29]

TABLE 21.3. Industrial Heat Exchanger in Liquid-Liquid Service

Quantity	Value
q	5861 kW
MTD	42 K
$R_w A_o$	0.0000153 $m^2 \cdot K/W$
h_o	3974 $W/(m^2 \cdot K)$
h_i	7380 $W/(m^2 \cdot K)$
$R_{f,o}$	0.000176 $m^2 \cdot K/W$
$R_{f,i}$	0.000352 $m^2 \cdot K/W$
U_c	2365 $W/m^2 \cdot K$
U_f	996 $W/m^2 \cdot K$
A_f/A_c	2.37
A_o/A_i	1.15

CONVERSION FACTORS

$$1\ W/(m^2 \cdot K) = 0.176\ Btu/(hr \cdot ft^2 \cdot °F)$$
$$1\ m^2 \cdot K/W = 5.678\ hr \cdot ft^2 \cdot °F/Btu$$
$$1\ kW = 3412\ Btu/hr$$
$$T_F = 1.8(T_K - 273) + 32$$

TABLE 21.4. Industrial Heat Recovery Unit

Quantity	Value
q	5861 kW
MTD	42 K
$R_w A_o$	0.0000153 $m^2 \cdot K/W$
h_o	57 $W/(m^2 \cdot K)$
h_i	7380 $W/(m^2 \cdot K)$
A_o/A_i	10
$R_{f,o}$	0.0088 $m^2 \cdot K/W$
$R_{f,i}$	0.0002 $m^2 \cdot K/W$
U_c	52.9 $W/(m^2 \cdot K)$
U_f	33.7 $W/(m^2 \cdot K)$
A_f/A_c	1.57

CONVERSION FACTORS

$1\ W/(m^2 \cdot K) = 0.176\ Btu/(hr \cdot ft^2 \cdot °F)$

$1\ m^2 \cdot K/W = 5.678\ hr \cdot ft^2 \cdot °F/Btu$

$1\ kW = 3412\ Btu/hr$

$T_F = 1.8(T_K - 273) + 32$

and is used to design large surface condensers. It is a function of several variables such as velocity, tube material, and fluid temperature and is defined as

$$CF = U_f/U_c \tag{21.26}$$

The fouling factor and cleanliness factor are related by

$$CF = \frac{1}{1 + U_c R_f} \tag{21.27}$$

21.3.3 Effect of Fouling on Pressure Drop

The effect of fouling on hydraulic performance is often neglected in the design of heat exchangers, but should be considered when the fouling deposit significantly reduces the flow area. In tubular exchangers, the fouling layer roughens the surface, decreases the inside diameter, and increases the outside diameter of the tubes. A shellside deposit also influences the shellside flow patterns, in some cases actually enhancing the exchanger performance by plugging or reducing unwanted shellside fluid bypass streams. The effects on shellside performance are not easy to quantify, but the tubeside effects are, and the remainder of this discussion will center around the hydraulic effects of tubeside fouling.

Inside the tube, the fouling layer decreases the inside diameter and roughens the surface resulting in a:

Pressure drop increase due to the roughened surface.

Pressure drop increase due to the reduced flow area.

Velocity increase due to the reduced flow area.

An exaggerated schematic of the fouled surface is shown in Fig. 21.5.

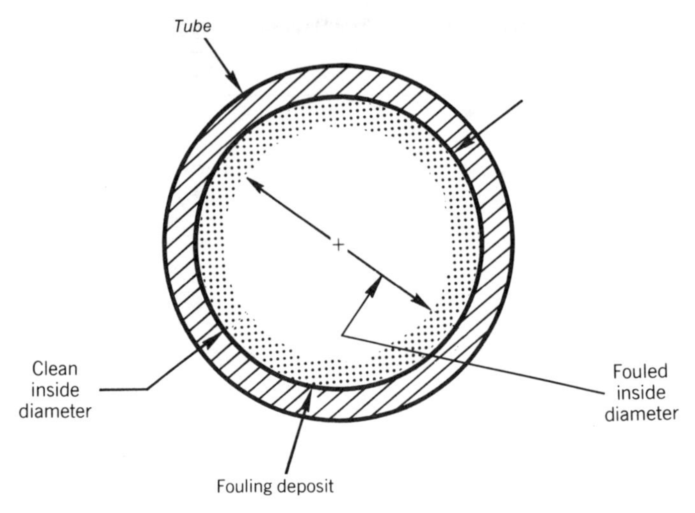

Figure 21.5. Sketch of tube cross section with fouling on inside.

To determine the effects of the fouling layer on the hydraulic performance, a computational procedure for tubeside flow is:

Step 1. Determine the clean pressure drop.
Step 2. Determine the fouling-layer thickness.
Step 3. Determine the velocity under fouled conditions assuming constant volumetric flow rate.
Step 4. Determine the pressure drop under fouled conditions.

In this discussion, turbulent flow is assumed, with the friction factors given by [30]

$$f = 0.0014 + 0.125\,\mathrm{Re}^{-0.32} \qquad \text{for smooth tubes} \tag{21.28}$$

$$f = 0.0035 + 0.264\,\mathrm{Re}^{-0.42} \qquad \text{for rough tubes} \tag{21.29}$$

and the pressure drop calculated by

$$\Delta p = \frac{2 f \rho V^2 L}{d} \tag{21.30}$$

The fouling-layer thickness is determined by assuming the fouling layer to be a cylindrical thermal resistance, much in the same way that the tube-wall resistance is determined. The fouling resistance is

$$\frac{R_f}{A_c} = \frac{\ln(d_c/d_f)}{2\pi k_f L} \tag{21.31}$$

Equation (21.31) is rearranged to express the fouled diameter as a function of the

TABLE 21.5. Fouling-Deposit Effect on Pressure Drop

	Pressure Drop, Pa/m	Change, %
Clean condition	1740	—
Fouled condition with area reduction only	2545	46
Fouled condition with area reduction and rough surface	2995	72

CONVERSION FACTOR

1 Pa/m = 0.0064 (lb_f/ft^2)/ft

fouling resistance:

$$d_f = d_c \exp\left(-\frac{2k_f R_f}{d_c}\right) \tag{21.32}$$

For most liquid-side deposits, the thermal conductivity has been found to vary between 0.20 and 10 W/(m · K) [0.116 and 5.78 Btu/(hr · ft · °F)] (see Sec. 21.3.4).

The use of this procedure and the effect on pressure drop are best illustrated by an example. Consider a shell-and-tube exchanger with 25.4 mm (1 in.) O.D., 16 BWG tubes using cooling-tower water on the tubeside. The water flows at 1.83 m/s (6 ft/s) with a tubeside fouling resistance of 0.000528 $m^2 \cdot K/W$ (0.003 $hr \cdot ft^2 \cdot °F/Btu$). Table 21.5 summarizes the pressure drop contributions due to the increase in the velocity and the roughness of the deposit as calculated by this procedure. These results illustrate that the changes in pressure drop associated with fouling deposits may be substantial and must be considered when designing a tubular heat exchanger.

In gas-side service, the effect of fouling on pressure drop may be more pronounced than that on heat transfer. For fully developed flow conditions, Shah [31] has shown that

$$h \propto \frac{1}{D_h}, \qquad \Delta p \propto \frac{1}{D_h^3} \tag{21.33}$$

For a fixed mass flow rate, there will be a slight increase in h as the fouling deposit builds up due to increases in fluid velocity and deposit-surface roughness. However, the overall heat transfer coefficient will be reduced because of the additional thermal resistance, which depends on the thickness and thermal conductivity of the deposit. Since Δp is inversely proportional to D_h^3, the pressure drop for flow inside ducts is highly sensitive to reductions in the hydraulic diameter due to fouling. However, for external flows in tubular exchangers, the hydraulic diameter depends on the tube and layout geometry and can be relatively large for heat-recovery applications. In such cases, the effect of fouling on pressure drop may not be as great as that on heat transfer.

21.3.4 Tabulated Fouling Factors

Tabulated values of both liquid-side and gas-side fouling factors are presented in this section. In addition, properties of representative fouling deposits are also given in tabular form. It should be pointed out that although fouling deposits are generally nonuniform, spatial averages are usually implied in such tabulated data.

Typical liquid-side fouling factors are shown in Table 21.6 for three standard fluid classifications. The values given are those typically encountered in applications. The seawater values are for once-through cooling such as in coastal-based utility plants. Note there are no "usual" values for hydrocarbon fluids. The general classification of hydrocarbon fluids covers a wide range of liquids, and it is difficult to establish commonly used values. Additionally, there is limited information available on hydrocarbon fouling, since the fluids and applications tend to be proprietary. Sources of fouling resistances in the open literature are very limited, in part due to the relatively recent interest in fouling research, the complexity of the problem, and the proprietary nature of many of the processes involved. In general, the designer is left with the following sources of fouling information:

Proprietary research data
Plant data
Personal or company experience
TEMA tables.

The tables found in the Tubular Exchanger Manufacturers Association (TEMA) handbook [32] are probably the most often referenced source of fouling factors used in the design of exchangers. A summary of the liquid-fouling factors from the TEMA tables is given in Table 21.7, and additional fouling factors from recent published literature in Table 21.8. It is important to understand the basis of these tables to use them properly. The tables were developed ca. 1942 and represent the combined wisdom, knowledge, and experience of individuals from industry at the time the first edition of TEMA was published. The values are not asymptotic fouling factors; instead, they are values that allow the exchanger to perform satisfactorily in the designated service for a reasonable time between cleaning. This might mean that an exchanger is sufficiently oversurfaced to provide proper performance regardless of the fouling present. The TEMA tables have not been significantly updated in later editions (up to the sixth edition at this writing). After the sixth edition was released, a commission of representatives from TEMA and Heat Transfer Research, Inc. (HTRI) was established to provide a substantial update. Release of that update has not been announced at this writing.

TABLE 21.6. Typical Values of Fouling Resistances

Fluid	Range of Values, $m^2 \cdot K/W$	Typical Value, $m^2 \cdot K/W$
Seawater	0.0000352–0.000881	0.0000881
Cooling water	0.0001760–0.00141	0.000352
Hydrocarbons	0.0000352–0.00881	None

CONVERSION FACTOR

$1\ m^2 \cdot K/W = 5.678\ hr \cdot ft^2 \cdot °F/Btu$

TABLE 21.7. TEMA Recommended Liquid-Side Fouling Factors [32]

WATER

	Fouling Factor, $m^2 \cdot K/W$			
	Heating-Medium Temp. ≤ 389 K Water Temp. ≤ 325 K		389–478 K > 325 K	
Types of Water	velocity ≤ 0.914	> 0.914 m/s	velocity ≤ 0.914	> 0.914 m/s
Seawater	0.0000881	0.0000881	0.000176	0.000176
Brackish water	0.000352	0.000176	0.000528	0.000352
Cooling tower and artificial spray pond:				
Treated makeup	0.000176	0.000176	0.000352	0.000352
Untreated	0.000528	0.000528	0.000881	0.000704
City or well water (e.g., Great Lakes)	0.000176	0.000176	0.000352	0.000352
River water:				
Minimum	0.000352	0.000176	0.000528	0.000352
Average	0.000528	0.000352	0.000704	0.000528
Muddy or silty	0.000528	0.000352	0.000704	0.000528
Hard (over 15 grains/gal)	0.000528	0.000528	0.000881	0.000881
Engine jacket	0.000176	0.000176	0.000176	0.000176
Distilled or closed-cycle condensate	0.0000881	0.0000881	0.0000881	0.0000881
Treated boiler feedwater	0.000176	0.0000881	0.000176	0.000176
Boiler blowdown	0.000352	0.000352	0.000352	0.000352

INDUSTRIAL FLUIDS

Fluid	Fouling Factor, $m^2 \cdot K/W$
Fuel oil	0.000881
Transformer oil	0.000176
Engine lube oil	0.000176
Quench oil	0.000704
Refrigerant liquids	0.000176
Hydraulic fluid	0.000176
Industrial organic heat transfer media	0.000176
Molten heat transfer salts	0.000881

CHEMICAL PROCESSING STREAMS

Fluid	Fouling Factor, $m^2 \cdot K/W$
MEA and DEA solutions	0.000352
DEG and TEG solutions	0.000352
Stable side draw and bottom product	0.000176
Caustic solutions	0.000352
Vegetable oils	0.000528

NATURAL GAS–GASOLINE PROCESSING STREAMS

Fluid	Fouling Factor, $m^2 \cdot K/W$
Lean oil	0.000352
Rich oil	0.000176
Natural gasoline and liquified petroleum gases	0.000176

TABLE 21.7. (continued)

CRUDE OIL

Temp., K	Dry			Salt		
	Velocity ≤ 0.610	0.610–1.22	≥ 1.22 m/s	Velocity ≤ 0.610	0.610–1.22	≤ 1.22 m/s
256–366	0.000528	0.000352	0.000352	0.000528	0.000352	0.000352
367–422	0.000528	0.000352	0.000352	0.000881	0.000704	0.000704
422–533	0.000704	0.000528	0.000352	0.00106	0.000881	0.000704
≥ 533	0.000881	0.000704	0.000528	0.000123	0.000106	0.000881

CRUDE AND VACUUM LIQUIDS

Fluid	Fouling Factor, $m^2 \cdot K/W$
Gasoline	0.000176
Naptha and light distillates	0.000176
Kerosene	0.000176
Light gas oil	0.000352
Heavy gas oil	0.000528
Heavy fuel oils	0.000881
Asphalt and residuum	0.00176

CRACKING AND COKING UNIT STREAMS

Fluid	Fouling Factor, $m^2 \cdot K/W$
Light crude oil	0.000352
Heavy cycle oil	0.000528
Light coker gas oil	0.000528
Heavy coker gas oil	0.000704
Bottoms slurry oil[a]	0.000528
Light liquid products	0.000352

LIGHT ENDS AND LUBE OIL PROCESSING STREAMS

Fluid	Fouling Factor, $m^2 \cdot K/W$
Liquid products	0.000176
Absorption oils	0.000352
Alkylation trace acid streams	0.000352
Reboiler streams	0.000528
Feed stock	0.000352
Solvent feed mix	0.000352
Solvent	0.000176
Extract	0.000528
Raffinate	0.000176
Asphalt	0.000881
Wax slurries	0.000528

CONVERSION FACTORS

$$1\ m^2 \cdot K/W = 5.678\ hr \cdot ft^2 \cdot {}^\circ F/Btu$$
$$1\ m/s = 3.281\ ft/s$$
$$T_F = 1.8(T_K - 273) + 32$$

[a] Minimum velocity 1.37 m/s.

TABLE 21.8. Additional Liquid-Side Fouling Factors

Fluid	Deposit	Velocity, m/s	Surface Temperature, K	Fouling Factor, $m^2 \cdot K/W$	Reference
Water	Calcium carbonate		330	0.00005–0.00025	Watkinson [33]
River water	Corrosion products			0.0015–0.008	McAllister et al. [34]
Water	Aluminum/brass corrosion products			0.009	Gutzeit [35]
Coastal seawater	Biofilm	2–8	304	0.00005–0.00035	Ritter et al. [36]
Cooling tower	Calcium phosphate	2	310	0.00025	Parry et al. [37]
Open seawater	Biofilm	1.8	298	0.0009–0.0015	Panchal et al. [38]
Coastal seawater	Biofilm	2	308	0.00026	Nosetani et al. [39]
Open seawater	Biofilm	1.8	303	0.00018	Sasscer et al. [41]
Geothermal brine	Silica	Cross flow	343	0.00002	Seki et al. [42]
Crude oil	Desalter emulsion	1.3		0.00881	Lambourn and Durrieu [43]
Cooling water with inhibitors	Zinc silicate	1–2.5	327–350	0.00001–0.0003	Knudsen et al. [44]

CONVERSION FACTORS

$1\ m^2 \cdot K/W = 5.678\ hr \cdot ft^2 \cdot °F/Btu$

$1\ m/s = 3.281\ ft/s$

$T_F = 1.8(T_K - 273) + 32$

Fouling deposits from gaseous streams can be more severe in magnitude than those from liquid streams; however, due to the lower overall coefficients, the effect on the surface area is not as great as might be expected. The TEMA tables have limited fouling factors for gas-side service as seen from Table 21.9 [32]. A study sponsored by the U.S. Department of Energy has produced additional gas-side fouling factors from industrial sources, given in Table 21.10 [7]. Weierman [45] has made recommendations regarding fouling factors, gas velocity, cleaning provisions, and fin spacing as summarized in Table 21.11.

As discussed in Sec. 21.2, the thermophysical properties of fouling deposits are needed to relate fouling resistance to the deposit mass. Properties of selected liquid-side and gas-side fouling deposits are presented in Table 21.12. The availability of such information in the open literature is very limited.

21.3.5 Heat Exchangers for Liquid-Side Fouling Service

A variety of heat exchangers are used in liquid-side fouling service. The most commonly used units include:

Shell-and-tube exchangers
Plate heat exchangers
Spiral exchangers

TABLE 21.9. TEMA Fouling Factors for Gases [32]

Type of Gas	Fouling Factor, $m^2 \cdot K/W$
INDUSTRIAL	
Manufactured gas	0.00176
Engine exhaust gas	0.00176
Steam (non-oil-bearing)	0.0000880
Exhaust steam (oil-bearing)	0.000176
Refrigerant vapors	0.000352
Compressed air	0.000325
Industrial organic heat transfer media	0.000176
CHEMICAL PROCESSING	
Acid gas	0.000176
Solvent vapors	0.000176
Stable overhead products	0.000176
PETROLEUM PROCESSING	
Atmospheric tower overhead vapors	0.000176
Light naphthas	0.000176
Vacuum overhead vapors	0.000352
Natural gas	0.000176
Overhead products	0.000176
Coke-unit overhead vapors	0.000352

CONVERSION FACTOR

$1\ m^2 \cdot K/W = 5.678\ hr \cdot ft^2 \cdot {}^\circ F/Btu$

This section focuses on designing for liquid-side fouling service using shell-and-tube exchangers, including low-finned tubes as a design option, and plate heat exchangers.

Shell-and-Tube Exchangers. Shell-and-tube heat exchangers are the most widely used ones in the process industry. Tubeside fouling deposits in shell-and-tube exchangers are reasonably uniform in thickness, with some variation from the tube inlet to outlet. The majority of the fouling data specified in sources such as the TEMA tables are related specifically to tubeside fouling.

The nonuniformity of shellside flow patterns leads to nonuniform deposits on the shellside of shell-and-tube exchangers (Fig. 21.6). Particulate material tends to settle out by sedimentation in the baffle-to-shell corners where the velocity is low and often recirculatory. Additionally, the temperature variations due to nonuniform heat transfer coefficients result in higher reaction rates for chemical-reaction fouling processes. These areas suffer from heavier deposits.

Some benefit is derived from the heavier deposition in the baffle-to-shell corners. These regions are generally not a major contributor to the total heat transfer process on the shellside, so the loss of surface area is not serious. The presence of the deposits blocks the so-called leakage or bypass streams that allow the shellside fluid to pass between the baffle and shell and through the tube-to-baffle clearances. The plugging of these bypass streams redirects the flow into the main crossflow region where most of

TABLE 21.10. Comparison of Available Gas-Side Fouling Factors [7]

	Gas-Side Fouling Factor, $m^2 \cdot K/W$				
	Weierman [45]	Zink [46]	TEMA [32]	Rogalski [47]	Henslee and Bogue [48]
CLEAN GAS					
Natural gas	0.0000881–0.000528	0.000176	—	—	—
Propane	0.000176 –0.000528	—	—	—	—
Butane	0.000176 –0.000528	—	—	—	—
Gas turbine	0.000176	—	—	—	—
AVERAGE GAS					
No. 2 oil	0.000352 –0.000704	0.000528	—	—	—
Gas turbine	0.000264	—	—	0.000528–0.00669	—
Diesel engine	0.000528	—	0.00176	—	0.0211–0.0247
DIRTY GAS					
No. 6 oil	0.000528 –0.00123	0.000881	—	—	—
Crude oil	0.000704 –0.00264	—	—	—	—
Residual oil	0.000881 –0.00352	0.00176	—	—	—
Coal	0.000881 –0.00881	—	—	—	—
MISCELLANEOUS					
Sodium-bearing waste	—	0.00528	—	—	—
Metallic oxides	—	0.00176	—	—	—
FCCU catalyst fines	—	0.00141	—	—	—

CONVERSION FACTOR

$1\ m^2 \cdot K/W = 5.678\ hr \cdot ft^2 \cdot °F/Btu$

the heat transfer occurs. The crossflow region has significant turbulence induced on the flow by the tube bundle, which tends to scrub the tubes and minimize the amount of fouling in this region.

Plate Heat Exchangers. Plate exchangers are widely used in the processing of foodstuffs when frequent and complete cleaning is required. The plates of most plate heat exchangers are corrugated and yield high heat transfer coefficients as a result of the high turbulence caused by the tortuous path of fluids passing through the exchanger. This turbulence reduces the amount of fouling in the same way as on the shellside of a shell-and-tube exchanger. Figure 21.7 shows the data obtained by Cooper et al. [59] on a plate exchanger operating in cooling tower water service. The values are very low compared to those experienced by a tubeside test unit operated simultaneously.

Velocity as a Design Parameter. The TEMA tables specify velocity ranges for the tabulated fouling factors. However, the magnitude of this velocity for a given geometry affects the specific fouling factor selected. The nominal operating velocity is about 1.83

TABLE 21.11. Design Parameters for Finned Tubes in Fossil-Fuel Exhaust Gases [45]

Type of Flue Gas	Fouling Factor, $m^2 \cdot K/W$	Minimum Spacing Between Fins, m	Maximum Gas Velocity to Avoid Erosion, m/s
CLEAN GAS (CLEANING DEVICES NOT REQUIRED)			
Natural Gas	0.0000881–0.000528	0.00127–0.003	30.5–36.6
Propane	0.000176 –0.000528	0.00178	
Butane	0.000176 –0.000528	0.00178	
Gas Turbine	0.000176		
AVERAGE GAS (PROVISIONS FOR FUTURE INSTALLATION OF CLEANING DEVICES)			
No. 2 Oil	0.000352 –0.000704	0.00305–0.00384	25.9–30.5
Gas Turbine	0.000264		
Diesel Engine	0.000528		
DIRTY GAS (CLEANING DEVICES REQUIRED)			
No. 6 Oil	0.000528 –0.00123	0.00457–0.00579	18.3–24.4
Crude Oil	0.000704 –0.00264	0.00508	
Residual Oil	0.000881 –0.00352	0.00508	
Coal	0.000881 –0.00881	0.00587–0.00864	15.2–21.3

CONVERSION FACTORS

$1\ m^2 \cdot K/W = 5.678\ hr \cdot ft^2 \cdot °F/Btu$
$1\ m = 39.37\ in.$
$1\ m/s = 3.281\ ft/s$

m/s (6 ft/s) for tubeside flow, 0.610 m/s (2 ft/s) for shellside flow, and 0.305 m/s (1 ft/s) for plate exchangers. The fouling observed in plate exchangers and the shellside of shell-and-tube exchangers is much less than that for tubeside flow. The explanation of this anomaly lies in the definition of flow area and the amount of induced turbulence for a given geometry. The flow area in tubeside flow is simply the cross-sectional area, while the flow area for shellside flow is defined as the open space between baffles at the shell diameter. The plate-exchanger flow area is based on the cross-sectional area between the gaskets, assuming smooth plates. The liquid turbulence in the latter two geometries is much higher at nominal operating velocities than for tubeside flow; therefore, the fouling factor is lower than the value found in the TEMA tables for this velocity. Hence, care must be exercised when comparing the expected performance of different geometries or when using data from one geometry and applying it to another situation.

Effects of Low-Finned Tubes on Fouling Design. External low-finned tubes are used sometimes to improve the performance of a shell-and-tube exchanger. Such tubes have finned-to-plain tube area ratios of 2 to 3 with 0.591 to 1.18 fins/mm (15 to 30 fins/in.). The fin diameter is less than the plain end diameter, which permits the retubing of heat exchangers with such tubes to increase the total available exchanger surface area. However, tubeside fouling can offset this effect.

Equation (21.25) shows that the inside fouling resistance is magnified by the outside-to-inside surface area ratio. The use of low-finned tubes therefore amplifies an

TABLE 21.12. Properties of Representative Fouling Deposits

Reference	Type of Deposit	Temperature, K	Thickness, mm	Density, kg/m^3	Thermal Conductivity, $W/(m \cdot K)$	Relative Roughness
Chow et al. [49]	MHD seed slag	700–810	2.6–9.5	—	0.33–0.40	—
Weight [50]	Coal-fired boiler	673–1373	—	—	0.1–10.0	—
Characklis [51]	Biofilm	—	0.4–5.0	—	—	0.003–0.157
Characklis [52]	Biofilm	—	—	—	0.57–0.71	—
Lister [53]	Calcium carbonate	—	1.65–2.62	—	2.6	0.0001–0.0006
Sherwood et al. [54]	Calcium carbonate	—	—	—	2.26–2.93	—
	Calcium sulfate	—	—	—	2.31	—
	Calcium phosphate	—	—	—	2.60	—
	Magnesium phosphate	—	—	—	2.16	—
	Magnetic iron oxide	—	—	—	2.88	—
	Analcite	—	—	—	1.27	—
	Biofilm	—	—	—	0.63	—
Raask [55]	Coal-fired boiler	500–1200	0–50	500	0.03–3.0	—
Pritchard [56]	Calcium carbonate	—	—	—	1.6	—
Rogalski [47]	Oil-fired diesel exhaust	—	—	—	0.0353–0.047	—
Wagoner et al. [57]	Coal-fired boiler	889	—	—	0.0520	—
Characklis [58]	Biofilm	300–301	—	10–40	0.17–1.08	—
Parry [37]	Calcium phosphate	310	0.25	—	1.0	—

CONVERSION FACTORS

$T_F = 1.8(T_K - 273) + 32$ $\quad$ $1\ kg/m^3 = 0.0624\ lb_m/ft^3$

$1\ mm = 0.00394\ in.$ $\quad$ $1\ W/(m \cdot K) = 0.578\ Btu/(hr \cdot ft \cdot °F)$

existing tubeside fouling factor, offsetting the beneficial effects of the increased surface area. The heat duty is proportional to the product UA, and although the area is increased when low-finned tubes are used, the overall heat transfer coefficient is reduced when tubeside fouling is present. This effect is illustrated in the following example.

A shellside condenser, with cooling tower water on the tubeside, is not performing satisfactorily due to tubeside fouling. There is negligible shellside fouling. The replacement of the plain tubes with low-finned tubes having 1.18 fins/mm (30 fins/in.) will increase the available surface area by nearly a factor of 3. Table 21.13 shows the

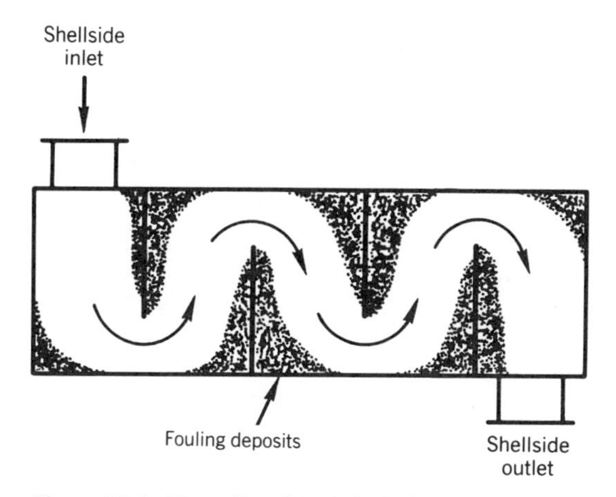

Figure 21.6. Nonuniformity of shellside fouling deposits.

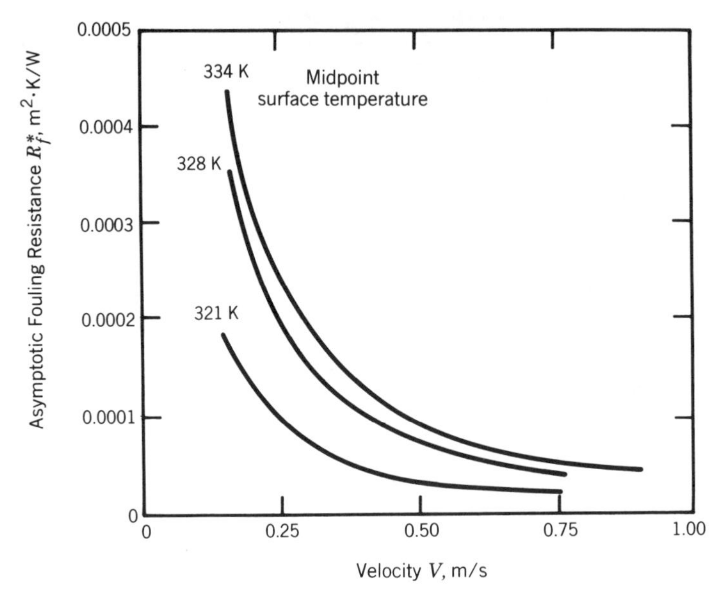

Figure 21.7. Fouling in a plate heat exchanger [59].

conditions for this example. For the plain tube, the fouled overall coefficient is 1909 W/(m² · K) [336 Btu/(hr · ft² · °F)]. For the low-finned tube, the overall coefficient (with the same tubeside fouling factor) is only 771 W/(m² · K) [136 Btu/(ht · ft² · °F)], due to the area-ratio magnification of the tubeside fouling factor. For the plain tube, the product $U_f A$ per unit length L is 153 (W/K)/m, while that for the low-finned tube is 179 (W/K)/m, an increase of only 17%, even though the surface area is increased

TABLE 21.13. Fouling in a Plain and a Low-Finned Tube

Quantity	Value	
	Plain Tube	Low-Finned Tube
d_o	25.4 mm	25.4 mm
d_i	22.9 mm	22.9 mm
$R_{f,i}$	0.00018 m² · K/W	0.00018 m² · K/W
$R_{f,o}$	0	0
h_i	8517 W/(m² · K)	8517 W/(m² · K)
h_0	8517 W/(m² · K)	8517 W/(m² · K)
k_w	17.3 W/(m · K)	17.3 W/(m · K)
Area ratio		
outside : inside	1.11	
finned : bare end		2.9
finned : inside		3.22
U_f	1909 W/(m² · K)	771 W/(m² · K)
$U_f A/L$	153 (W/K)/m	179 (W/K)/m

CONVERSION FACTORS

1 mm = 0.0394 in.
1 m² · K/W = 5.678 hr · ft² · °F/Btu
1 W/(m · K) = 0.578 Btu/(hr · ft · F)

290%. Assumptions used in the calculations include identical thermal wall resistance for both plain and low-finned tubes and 100% fin efficiency for the low-finned tubes.

21.3.6 Heat Exchangers for Gas-Side Fouling Service

The designer must consider the type of heat exchanger, materials of construction, geometric constraints, and cleaning provisions for gas-side fouling service. The basis for determining the type of exchanger must include the gas-side fouling potential, corrosion potential, erosion potential, and process requirements. Possible units that can be used include:

Recuperators
Regenerators
Fluidized bed exchangers
Direct-contact exchangers
Heat-pipe exchangers.

The placement of the dirty gas with respect to the process stream is also an important consideration.

The temperature operating regime of the exchanger will determine the materials of construction requirements. Three temperature regimes have identified [7]:

Low temperature [< 533 K (< 500°F)]. Characterized by water and acid condensation and smut deposition. In this regime, the designer can maintain the surface temperature above the dewpoint by preheating the secondary fluid; use coatings such as galvanizing, Teflon, and plastics; and use alternative materials with high corrosion resistance such as glass, stainless steel, Corten, Hastelloy, and Inconel.

Intermediate temperature [533 to 1090 K (500 to 1500°F)]. This is the operating range of most heat exchangers. Material recommendations include the use of carbon steels up to 811 K (1000°F); stainless steels; and superalloys such as Hastelloy.

High temperature [> 1090 K (> 1500°F)]. Characterized by extremely hostile environments including corrosive, erosive, and reactive gases. Material recommendations include the use of stainless steels; superalloys; and ceramics [required for surface temperatures greater than 1367 K (2000°F)].

The geometry of the exchanger must also take into account the fouling environment in which the exchanger will operate. These considerations include the possible use of:

Inline layouts to provide cleaning lanes for sootblowers
Wide pitches for dirty fuels and gases
Large fin spacing or possibly plain tubes for dirty gases
Thick fins in abrasive external environments
Sleeve inserts in abrasive internal environments
Shield tubes where erosion or corrosion is a potential problem
Enhancement on both sides when there is gas-to-gas exchange of heat.

Cleaning provisions are also dependent on the type of service and may be designed using either an on-line or off-line technique. Gas-side cleaning techniques are discussed in Sec. 21.5.4.

21.4 LIQUID-SIDE PREVENTION, MITIGATION, AND ACCOMMODATION TECHNIQUES

Although fouling is not completely avoidable, there are prevention, mitigation, and accommodation techniques to minimize the effects of fouling. Discussed in this section are the three important areas pertaining to liquid-side fouling—operational considerations, additive effects, and cleaning options.

21.4.1 Control of Operating Conditions

The amount of fouling that occurs in a heat exchanger is directly related to the operating conditions, especially the velocity of the fluid and the exchanger surface temperature. An increase in fluid velocity will generally cause a decrease in the amount of fouling deposit on the heat transfer surface, and an increase in surface temperature will increase chemical-reaction fouling deposition and will decrease the fouling deposition from biological sources (extremely low temperatures will also remove biological fouling deposits). Often an exchanger is overdesigned, with extra surface area allowed for anticipated fouling. When operation begins in the clean exchanger, the performance is better than that for which it was designed, requiring modification of the operating conditions. These modifications generally tend toward unfavorable fouling conditions, with a lower velocity and higher surface temperature. In effect, the clean exchanger is operated so that the unit will foul as predicted in the design calculations. Knudsen [60] suggested that the final fouling resistance will be even higher than estimated in the design calculations.

21.4.2 Additives

The use of chemical additives to minimize fouling is common, and a number of references are available to assist the engineer in the selection and use of such additives. Most notable of these are the reference books produced by chemical-treatment companies such as Drew Chemical Company and Betz Laboratories, Incorporated [61, 62]. Treatment of cooling water is well summarized in a recent article by Strauss and Puckorious [63].

The most commonly used liquid in heat exchangers is water, generally cooling tower water, and the best understanding of additive effects is also in water treatment. Generally, the types of fouling found in cooling waters are crystallization, particulate, biological, and corrosion. Chemical additives are used to control fouling from these sources. The following comments are summarized from Strauss and Puckorious concerning the use of additives and treatment control to minimize fouling effects:

Crystallization fouling. Minerals from the water are removed by softening, solubility of the fouling compounds is increased by the use of such chemicals as acids and polyphosphates, and crystal modification by chemical additives is used to make deposits more easily removed.

Particulate fouling. Particles are removed mechanically by filtration, flocculants are used to aid filtration, and dispersants are used to maintain particles in suspension.

Biological fouling. Chemical removal is most common, especially the use of chlorine and other biocides, although thermal backwash or temperature elevation is also used.

Corrosion fouling. Additives in the form of inhibitors or passivators are used to produce protective films on the metal surfaces, requiring the addition of chemicals to the water.

Corrosion inhibitors are required to minimize the corrosion, but they also produce fouling deposits which can reduce the effectiveness of an exchanger. Knudsen et al. [44] report that the fouling resistance increases with increasing pH and increasing inhibitor concentration.

The use of additives in other liquids is not as well documented as for water. However, Van der Wee and Tritsman [64] report effective use of steam condensate in controlling desalting-exchanger fouling in crude-oil streams, and Lambourn and Dirrieu [43] report a substantial improvement in the performance of a crude hot preheat exchanger located upstream of the furnace when a commercial dispersant is used.

21.4.3 Surface Cleaning Techniques

A decrease in the performance of a heat exchanger beyond acceptable levels requires cleaning. In some applications, the cleaning can be done on line to maintain acceptable performance without interruption of operation. At other times, off-line cleaning must be used. A summary of cleaning approaches is given by Garrett-Price et al. [9].

On-Line Cleaning. On-line cleaning generally utilizes a mechanical method designed for only tubeside and requires no disassembly. In some applications, flow reversal is required, but the interruption of the service is minimal. Chemical feeds can also be used as an on-line cleaning technique, but may upset the rest of the liquid service loop. The two mechanical techniques available for on-line tubeside cleaning are the sponge-ball and brush systems. The sponge-ball system recirculates rubber balls through a separate loop feeding into the upstream end of the exchanger. The brush system has capture cages at each end of each tube, and flow reversal is required. The main advantage of on-line cleaning is the continuity of service of the exchanger and the hope that no cleaning-mandated downtime will occur. The principal disadvantage is the added cost of a new heat exchanger installation or the large cost of retrofits. Additionally, there is no assurance that all tubes are being cleaned sufficiently.

Off-Line Chemical Cleaning. Off-line chemical cleaning is a technique that is used very frequently to clean exchangers. Some refineries and chemical plants have their own cleaning facilities for dipping bundles or recirculating cleaning solutions. However, an outside organization is often brought in to perform the cleaning during a turnaround. In general, this type of cleaning is designed to dissolve the deposit by means of a chemical reaction with the cleaning fluid.

The advantages of this approach include the cleaning of difficult-to-reach areas. Often in mechanical cleaning, there is incomplete cleaning due to regions that are difficult to reach with the cleaning tools. U-tube bundles are especially bad, since the U bends are not cleaned by brushes and lances, and there is a possibility of plugging them with the removed debris. There is no mechanical damage to the bundle from chemical cleaning, although there is a possibility of corrosion damage due to a reaction of the tube material with the cleaning fluid. This potential problem may be overcome through proper flushing of the unit.

Disadvantages of off-line chemical cleaning include corrosion damage potential, handling of hazardous chemicals, use of a complex procedure, and the disassembly required with all off-line techniques. The hazardous chemicals often used in chemical cleaning require knowledgeable personnel and probably should be left to the companies that specialize in chemical cleaning.

Off-Line Mechanical Cleaning. Off-line mechanical cleaning is a frequently used procedure and can often be done by the plant personnel. The approach is to abrade or scrape away the deposit by some mechanical means. Typical mechanical methods of cleaning include high-pressure water (20,000 psi), steam, lances, and water guns. High-pressure water works well for most deposits, but frequently a thin layer of the deposit is not removed, resulting in a greater affinity for fouling when the bundle is returned to service. Some damage from fluid impingement is also possible. High-temperature, high-pressure steam is useful for hydrocarbon deposits. Lances are scraping devices attached to long rods and sometimes include a water or steam jet for flushing and removing the deposits. Water guns are used with projectiles to clean each tube individually. The projectile is usually a stiff bristled brush, rubber plug, or metal scraper.

The advantages of off-line mechanical cleaning include excellent cleaning of each tube and good removal potential of very tenacious deposits. Disadvantages include the inability to clean U-tube bundles successfully, the usual disassembly problem, and the great labor needed.

21.5 GAS-SIDE PREVENTION, MITIGATION, AND ACCOMMODATION TECHNIQUES

A number of prevention, mitigation, and accommodation techniques are used to deal with gas-side fouling. These techniques may be grouped under the general categories of: (1) fuel cleaning techniques, (2) control of combustion conditions, (3) fuel and gas additives, (4) surface cleaning techniques, (5) quenching, (6) control of operating conditions, and (7) gas cleaning techniques.

21.5.1 Fuel Cleaning Techniques

Ash and other constituents such as sodium, sulfur, and vanadium in fuels are responsible for the deposits which form on heat transfer surfaces through the various gas-side fouling transport processes. Therefore, the removal of any of these constituents from fuels prior to combustion is highly desirable. Processes are available for both the desulfurization and de-ashing of residual fuel oils [65]. Mechanical and physical cleaning methods, including gravity separation and froth flotation, are available for removing mineral matter from coal (Essenhigh [66]). However, chemical methods for cleaning coal are still in the conceptual or experimental stage.

21.5.2 Control of Combustion Conditions

An important consideration in the prevention of gas-side fouling and corrosion is the control of combustion conditions to minimize the formation of particulate matter in exhaust gases. The type of fuel burned, type of combustion equipment, and burner design are all important factors in the formation of particulates such as soot in combustion gases. Other important combustion parameters include fuel injection pattern, fuel injection schedule, and fuel viscosity control. White [67] investigated the effects of premixing a fuel (JP-5) and oxidizer (air) on gas-side fouling downstream of a gas turbine. He found that fuel-air premixing eliminated the production of nearly all soot, and it was virtually impossible to foul the heat exchanger under these conditions.

Preheating combustion air, a technique which is widely used in practice, can improve combustion efficiency and reduce fouling, whereas varying the quality of the fuel supply can contribute to fouling problems. For example, the quality of coal supplied to industrial users tends to be much more variable than that provided for electric utilities. A final combustion-related example is the classical problem of gas-side fouling and corrosion caused by H_2SO_4 condensation onto low-temperature heat transfer surfaces. One solution to this problem is to control the amount of excess air, which in turn limits the conversion of SO_2 to SO_3, and hence the amount of H_2SO_4 produced.

21.5.3 Fuel and Gas Additives

Under some conditions, chemical additives can improve combustion efficiency, reduce emissions, and mitigate the effects of gas-side fouling and corrosion. Many proprietary additives have been marketed for use with both oil- and coal-fired systems.

In the case of residual oils, there are three chemical elements which cause fouling problems which can be alleviated with additives: sulfur, vanadium, and sodium. Although some sodium occurs naturally in crude oil, most of that found in residual oils results from contamination by seawater during ocean transport. Refineries use a desalting procedure which removes all but about 10 to 30 ppm of sodium in the oil. During the combustion of residual oils, most of the sulfur burns to become SO_2. About 1% of the SO_2 is converted to SO_3, which in turn combines with water vapor in the combustion gases to form H_2SO_4. This conversion of SO_2 to SO_3 can continue through the boiler as a result of catalytic action by iron and vanadium compounds. Meanwhile, the vanadium in the residual oil is oxidized to V_2O_5, which may then react with the sodium and other trace metals in the fuel to form metal vanadates. The various additives which have been used to attack these problems include MgO, $Mg(OH)_2$, Mg metal, Mg-Mn, and dolomite.

Battelle Columbus Laboratories, under the sponsorship of the Electric Power Research Institute, was recently involved in two activities—a literature review and a workshop—which report the state of the art for coal and oil additives [68, 69]. Although this work was directed primarily toward the utility industry, the results are generally applicable. The conclusions reached by the workshop participants regarding the use of additives in combatting residual oil fouling and corrosion include:

MgO in an oil dispersion is the additive most active in combatting corrosion from vanadium and sulfur trioxide.

Many other additives are useful, depending on circumstances.

Additive performance is site-specific; experience in one boiler is not directly transferable to another.

There are many unanswered questions regarding additive use, including the most effective test-program procedures.

Much less work has been done in the area of additives in coal-fired systems than for residual oils where MgO-oil, MgO + Mn-dry, $CaCO_3$, and boron-based additives have been used in an attempt to combat high-temperature fouling and ash modification. Only in the case of boron-based additives has there been significant experience, and even then only in the modification of ash. Regarding coal additives, the workshop participants drew the following rather general conclusions:

The use of additives developed for oil firing has been helpful in some cases, but is neither optimum nor sufficient for all cases.

Better data are needed on a range of problems and applications, including slag behavior, deposit effects, corrosion suppression, plume-visibility reduction, and coal handling.

Most additives, for both coal and oil, are magnesium-based. The following brief discussion provides some basis for the success of such additives in fuels where sodium is an important reactive constituent. The deposition of sodium sulfate, which is formed by the reaction:

$$Na_2 + SO_3 \rightarrow Na_2SO_4 \tag{21.34}$$

occurs widely in many industries. The Na_2SO_4 deposits can also react with the SO_3 in the gas stream to form sodium iron trisulfate:

$$3Na_2SO_4 + Fe_2O_3 + 3SO_3 \rightarrow 2Na_3Fe(SO_4)_3 \tag{21.35}$$

which is a tenacious, corrosive deposit. On the other hand, magnesium oxide can also react with SO_3 to form magnesium sulfate deposits:

$$MgO + SO_3 \rightarrow MgSO_4 \tag{21.36}$$

which are softer and more friable than Na_2SO_4, and hence may be removed from the surface more easily. It should be pointed out that MgO tends to be more reactive with SO_3 than Na; therefore, the reaction in Eq. (21.36) is more likely to occur than that in Eq. (21.34). In addition, the magnesium sulfate can react with sodium sulfate:

$$2MgSO_4 + Na_2SO_4 \rightarrow Na_2Mg_2(SO_4)_3 \tag{21.37}$$

to produce sodium magnesium trisulfate, $Na_2Mg_2(SO_4)_3$, which is softer and more friable than the sodium iron trisulfate discussed earlier.

21.5.4 Surface Cleaning Techniques

A number of surface cleaning techniques have been developed to remove gas-side fouling deposits, including steam and air sootblowing, sonic sootblowing, and water washing. Other approaches which are used to some extent include chemical, mechanical, and thermal cleaning. Each of these methods is discussed in this section.

The most widely used on-line technique to remove gas-side fouling deposits is sootblowing, with dry steam or air as the blowing medium. The location and spacing of the sootblowers within the heat exchanger are very important, and the frequency of sootblower operation depends on the type of fuel, amount of excess air, and operating conditions. DiCarlo [70] has presented a discussion of the two major types of sootblowers, the rotary and long retractable types. The rotary sootblower utilizes a multinozzled element and is permanently located in the tube bank to be cleaned. Rotary blowers are limited to flue gas temperatures below 811 K (1000° F), because such units are exposed to the flue gases at all times. However, the long retractable types can be used at virtually any temperature.

The frequency of sootblowing is an important consideration because of the costs of the blowing medium. In past years, a single blow on each of the three shifts during a

24-hour day was considered standard practice in many boiler operations. Today, however, more sophisticated computer-controlled methods are being employed in some installations, using either the bundle pressure drop or the outlet gas temperature as a threshold parameter to determine when cleaning is necessary.

In recent years, considerable interest has been shown in the use of both low-frequency (20 Hz) and high-frequency (220 Hz) sonic sootblowers to remove gas-side fouling deposits. Soot and other particles attached to the surface are dislodged by sound-wave vibration energy generated by the horns. The wave pressure fluidizes particles by breaking their bonds with other particles and the surface to which they are attached. Once fluidized, the particles will flow from the surface under gravitational or gas shear forces. Sonic horns work best for light, fluffy deposits and become increasingly less effective as the deposits become sticky and tenacious. The cleaning cycle is significantly different for steam sootblowers than for sonic horns. Steam sootblowers typically are operated periodically, whereas horns are blown on a more or less continuous basis. Capital costs for horns are greater than for steam sootblowers, but the operating costs are significantly lower.

Water washing is widely used to clean air preheaters as well as other types of heat transfer equipment. Water-soluble deposit accumulations formed on heat transfer surfaces may be easily removed by washing, provided that a sufficient quantity of water is used. The standard water-washing apparatus is a stationary, high-penetration, multinozzle device. The high-velocity jets produce a high fluid shear stress along with contractions caused by thermal shock to remove the deposits. The frequency of water washing depends on a number of factors, including the quality of fuel burned, nature and amount of deposit, operating conditions, and dryness of the sootblowing medium. The washing operations may be carried out under off-line or on-line conditions.

Other surface cleaning techniques may be grouped into the categories of chemical, mechanical, and thermal cleaning. Chemical cleaning is generally used as an off-line technique for gas-side fouling deposits. Thermal cleaning, also called self-cleaning or baking at elevated temperatures, consists of heating the exchanger and holding it at an elevated temperature until the deposits are baked off. Another alternative is to blow hot fluid, usually air, through a heat exchanger core or just reverse the hot and cold fluids —e.g., freezing fouling in cryogenic reversing heat exchangers. Mechanical cleaning includes the use of shot, surface vibration, mechanical brushing, and scouring with nutshells. Shaking or vibrating a tube bundle is sometimes used in applications involving dusty gases where air or steam sootblowing may not be used because the process involved is composition-sensitive—e.g., the production of H_2SO_4 from the burning of sulfur.

21.5.5 Quenching

The purpose of quenching a hot flue gas is to reduce the temperature in order to solidfy molten and soft particles in the gas, thereby preventing attachment of the particles at the cooler heat transfer surface. Suitable fluids for quenching gases include: (1) cool flue gases, (2) steam, (3) air, and (4) water spray. Of these techniques, cool flue-gas quench is probably the most desirable, and water-spray quench the least desirable.

21.5.6 Control of Operating Conditions

The importance of properly controlled operating conditions cannot be overemphasized as a technique for mitigating the effects of gas-side fouling. Severe gas-side fouling and corrosion problems can occur in heat exchangers under upset conditions or when there are frequent shutdowns and startups. Such transients should be minimized if at all possible.

One fairly standard startup procedure is to use a clean fuel during such operations. For example, if a unit is designed for operation with residual oil, it will probably be started up using natural gas. During this initial period, the combustion efficiency of a dirty fuel such as residual oil is poor and considerable quantities of unburned material might escape and deposit on heat transfer surfaces. Therefore, the low ignition temperature of natural gas makes it an ideal fuel for startup to minimize the accumulation of deposits of oil or oily soot.

Boiler units must generally be taken out of service at regular intervals for internal inspection, cleaning, and repair. If the heat transfer surfaces are to be cleaned externally, the operation should be carried out as soon as possible and should be finished by means of air lances and scrapers after the unit has cooled off [65]. If the deposits are not removed before the boiler shutdown, they will act as a sponge to collect acid and water condensate and subject the surface to corrosion.

21.5.7 Gas Cleaning Techniques

Particulate removal from gases may be carried out using electrostatic precipitators, mechanical collectors, fabric filters, and wet scrubbers. Gaseous pollutants are more difficult to remove than particulates using standard methods such as limestone addition, wet scrubbing without sulfur recovery, MgO systems with sulfur recovery, and dry sorbent systems [65]. In most applications, the particulates and gaseous pollutants are removed for environmental reasons, and this must generally be done, unfortunately, downstream of any heat transfer equipment.

21.6 FOULING MEASURING DEVICES

The practicing engineer must sometimes rely on experimental fouling factors rather than on values obtained from predictive equations or tables. Experimental fouling factors can be obtained using special fouling probes or, in some instances, existing plant heat exchangers. In addition, ancillary measurements may be carried out on test fluids and fouling deposits to obtain a better understanding of the fouling processes which are taking place. The current state of the art for liquid-side fouling measuring devices is far more advanced than that for gas-side service.

21.6.1 Liquid-Side Service

Operating heat exchangers and three important types of liquid-side fouling measuring devices are discussed in this section. Also, the use of deposit analysis and fluid characterization is given come consideration. Fischer et al. [71] and Knudsen [72] carried out comprehensive reviews of liquid-side fouling measuring devices, and these references may be consulted for additional details.

Actual Heat Exchangers As Diagnostic Tools. The use of an operating heat exchanger as a fouling measuring device is convenient, but can be inaccurate. Fouling may occur on both sides of the exchanger, making it difficult, if not impossible, to determine the fouling resistance of one of the streams. The instrumentation, particularly in older installations, may not be sufficiently accurate or complete to determine the fouling factors. Steady-state operation may not be possible because of the nature of the process. Finally, it may not be possible to determine the clean overall coefficient.

In spite of these potential shortcomings, it is possible to make measurements using installed exchangers if proper precautions are taken. The relations needed include the expression for the overall heat transfer coefficient,

$$U = \frac{q}{(A_0 \, \text{MTD})} \tag{21.38}$$

and the expression for the fouling factor in terms of the clean and fouled overall coefficients,

$$R_f = \frac{1}{U_f} - \frac{1}{U_c} \tag{21.39}$$

An important but subtle point is that the clean coefficient U_c must be adjusted for varying conditions when the fouled coefficient U_f is determined. These varying conditions may be due to variable flow rates or temperatures or both. This adjustment is made using a correction factor in the expression for fouling resistance as follows:

1. Determine U_c from temperature and flow-rate measurements at clean startup conditions.
2. Calculate U_c using an appropriate correlation (such as the Delaware method for shellside flow).
3. Compute the correction factor using

$$C = \frac{U_{c,\,\text{measured}}}{U_{c,\,\text{calculated}}} \tag{21.40}$$

4. Compute the fouling resistance using

$$R_f = \frac{1}{U_f} - \frac{1}{CU_{c,\,\text{calculated}}} \tag{21.41}$$

The use of the calculated coefficient accounts for changes in flow rates and temperatures, and the correction factor accounts for inaccuracies in the predictive methods for the particular exchanger in question.

Special Probes. Three basic types of devices have been used as liquid-side fouling probes to determine experimentally the fouling resistance of a process stream: (1) annular test section, (2) single circular tube, and (3) wire coil. In general, the annular test section has received more use than the others, especially in cooling-water studies. Single-tube units have been used in a number of applications in process stream fouling studies. The wire coil is usually used in laboratory studies with small batches of fouling fluid as single samples.

The general configuration of the annular test section is shown in Fig. 21.8. The heated surface consists of a cartridge type heating coil encased in a metal sheath. Thermocouples are embedded below the surface to measure the wall temperature, and local fouling resistances are determined using these thermocouples. The fluid flows

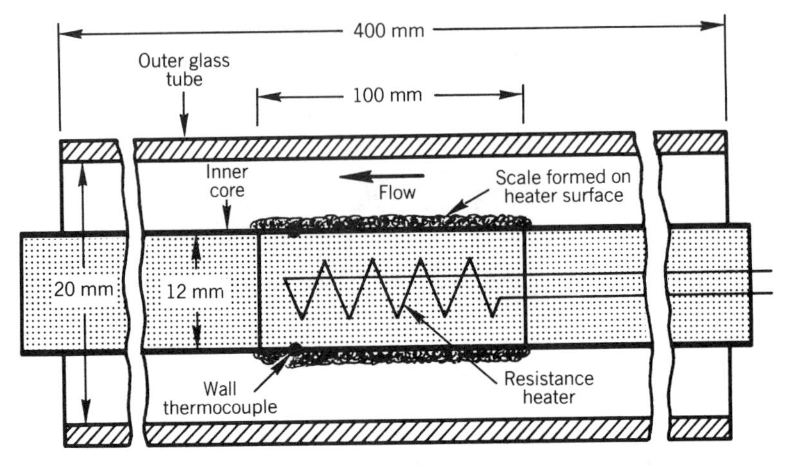

Figure 21.8. Schematic of annular fouling probe test section.

through the annulus in a direction parallel to the surface. The advantages of this unit include the fact that it is small, accurate, and removable for deposit inspection and analysis, and that a variety of test-section materials can be used. Among its limitations are the need for thermocouple calibration, sensitivity to fluctuations in the deposit layer thickness (fouling resistance), electrical-power hazard, and limitation to heating measurements.

The single circular tube test section is generally used in a double pipe exchanger. The assembly is very easy, and the materials of construction are often found in the plant shop. The fluid to be tested flows inside the tube with a utility fluid for heating or cooling flowing on the shellside. Heating is achieved with a hot liquid or a condensing vapor such as steam, while cooling tests can be conducted using a cold liquid or a boiling fluid such as a refrigerant. If a phase-change process is used, the pressure of the phase-change fluid as well as the temperature will have to be known. In cases where a condensing vapor is used, a subcooling condenser at the exit is used before measurements are made. The advantages of this system include the use of existing plant hardware, the ability to use heating or cooling, and material flexibility. Disadvantages include a large size requirement for a significant temperature difference. The heat transfer measurements are based on overall conditions rather that local measurements and require large temperature differences to reduce errors, especially when steam is used.

The wire test section has generally been employed as a laboratory test unit using small batches of material, although industrial slip streams have been used. The unit consists of a flow chamber with a straight or coiled wire perpendicular to the flow stream. The wire is heated by passing a current through it and monitoring the electrical resistance. As a deposit buildup occurs, the heat transfer is reduced, thereby increasing the wire temperature and the electrical resistance. Calibration and correlation with convective heat transfer data provide a connection between the increase in wire electrical resistance and the heat transfer coefficient. The fouling resistance can be obtained in the usual way by determining the degradation in the heat transfer performance. The advantages of this system include the compactness of the unit, quickly obtainable results, and simple construction; a major disadvantage is the questionable heat transfer data. There are large uncertainties in the heat transfer coefficient with deposit buildup, since the deposit will substantially increase the diameter of an initially small-diameter wire. Heating is the only mode of heat transfer possible, and the extrapolation of the data to tubeside or shellside flow is very difficult.

Deposit Analysis. Deposit analysis provides important information which complements the data on the fouling resistance. While the fouling resistance indicates the magnitude of the performance degradation, the deposit analysis identifies what is degrading the performance. The following types of deposit analysis are used in water-side fouling:

Atomic absorption or similar analysis determines principal elements in the deposit.

X-ray diffraction or *microprobe analysis* identifies possible compounds.

Loss on ignition (LOI) determines the mass percentage that is composed of organic material and water of hydration. The test is made by drying the deposit at 378 K (220° F) and then combusting it at 773 K (931° F).

Total carbon determines the organic content of the deposit, which includes both hydrocarbons and biological material.

Colony count determines the biological activity.

For hydrocarbon studies, a similar differential analysis can be made:

Atomic absorption or similar analysis determines principal elements in the deposit.

X-ray diffraction or *microprobe analysis* identifies possible compounds.

Pentane-soluble fraction identifies the resins and free oils.

Toluene-soluble fraction takes the residual from the pentane soluble fraction and identifies the asphaltenes.

Coke fraction is a type of LOI at 823 K (1020° F) on the toulene-soluble residue that identifies the coke content.

Ash fraction is the inorganic material left as residue from the coke-fraction test.

Fluid Characterization Measurements. A knowledge of what is on the surface is completed with the knowledge of what is in the fluid. The following points should be considered for a complete analysis:

Chemical composition of the fluid, including both cation and anion components.

Makeup chemical composition if applicable.

Treatment compounds and procedures, including schedules of when treatments are added to the system: continuous, once per week (when), once per shift (which shift), etc.

21.6.2 Gas-Side Service

In applications involving combustion gases, which are of principal interest here, the gases are always being cooled, either in direct heating applications or in heat recovery systems. Gas-side fouling measuring devices, or probes, have been used in a number of applications, including industrial and utility boilers, municipal incinerators, gas turbines, and heat recovery systems in a broad range of industries. In addition, a number of studies utilizing gas-side fouling probes have been carried out in laboratory settings.

A variety of devices, ranging from a small cylindrical disk to a full-size heat exchanger, have been used to obtain and analyze gas-side fouling deposits. Most of these probes are single, circular cylinders which utilize a controlled surface temperature maintained by air or water flowing through the inside. Although all of these probes are capable of collecting gas-side fouling deposits, very few of them are capable of measuring gas-side fouling resistances.

Gas-side fouling measuring devices come in a variety of sizes and shapes, and may be grouped in the following categories: (1) heat-flux meters, (2) mass-accumulation probes, (3) optical devices, (4) deposition probes, and (5) acid-condensation probes. The essential features of each type of probe are considered briefly. For industrial applications, current interest is focused on mass-accumulation and deposition probes. Additional details on gas-side fouling probes which have been developed and tested to date may be found in Marner and Henslee [73, 74].

A heat-flux meter measures the local heat transfer per unit area to monitor the fouling resistance. The decrease in heat flux as a function of time is thus a measure of the fouling buildup on the heat transfer surface. Most of the heat-flux work to date has been related to furnace applications, where radiation is the dominant mode of heat transfer. However, there is no conceptual reason why this approach could not be used in applications such as heat recovery systems where heat is transferred primarily by convection.

A mass-accumulation measuring device is a probe designed so that the deposit sample mass may be determined quantitatively, generally under controlled conditions of the surface temperature, gas temperature, and other parameters of interest. In a numer of cases where the interaction of fouling and corrosion occurs simultaneously, this type of probe has also been used to measure corrosion rates.

Optical fouling measuring devices are those used to determine deposition rates using optical methods. Such devices have seldom been used and to date have been restricted to relatively simple laboratory studies under closely controlled conditions.

Deposition measuring devices are probes used to collect deposits, generally under controlled conditions, on a qualitative basis. In this type of device, the actual mass of the deposit is not determined; however, quantitative analyses of the deposit itself are frequently carried out. In many cases simultaneous corrosion rates are measured using cylindrical deposition probes consisting of many segments of different materials joined together by threads, tension, or some other means.

Finally, acid-condensation probes are used to collect liquid acid which accumulates on a surface which is at a temperature below the acid dew point of the gas stream. Such probes are frequently used to measure dew point temperatures as well as acid deposition rates. A number of probes of this type are available commercially and are very important in low-temperature heat-recovery applications.

In order to measure on-line fouling resistances, the following quantities must be monitored as a function of time: (1) gas temperature T_g, (2) wall heat flux q''_w, (3) wall temperature T_w, and (4) heat transfer coefficient h. The convective heat transfer coefficient h may be expressed as

$$h = aV^b \tag{21.42}$$

where a and b are coefficients which depend on the fluid properties and the geometry. As the deposit builds up, it is possible that either a or b or both will change owing to the roughness of the fouled surface and a change in the geometry of the surface. Although the area for other than a plane surface will change as the fouling layer is deposited, it has become customary to base the calculation of the fouling resistance on the clean-surface area.

21.7 CONCLUDING REMARKS

Fouling is the undesirable accumulation of insulating material on heat transfer surfaces. Such deposits may result from either liquid or gas streams and can take place in a broad variety of energy-related applications. In general, when fouling with convective heat transfer occurs, the heat transfer decreases and the pressure drop increases. Methods have been presented in this chapter to predict these effects provided that appropriate information is available. Tabulated values of the TEMA fouling factors, additional fouling factors from the literature, and deposit thermal-conductivity values have been presented for this purpose. In cases where fouling is anticipated or cannot be avoided, the designer of heat transfer equipment can specify a number of prevention, mitigation, and accommodation techniques to deal with the problem of fouling. A variety of such techniques are available and have been discussed for both liquid-side and gas-side fouling service. As additional experimental data are collected and the phenomenological aspects of fouling are better understood, more sophisticated models will be formulated to predict the effects of fouling and more effective cleaning cycles will be employed. In the meantime, fouling will continue to plague heat exchanger operations and will cost the U.S. industrial sector on the order of 3 to 10 billion dollars annually.

NOMENCLATURE

A_c	clean surface area, m^2, ft^2
A_i	inside surface area, m^2, ft^2
A_f	fouled surface area, m^2, ft^2
A_o	outside surface area, m^2, ft^2
a_0–a_4	deposition constants in Table 21.1
b_0–b_5	removal constants in Table 21.2
C	correction factor defined by Eq. 21.40
CF	cleanliness factor
c	concentration, kg/m^3, lb_m/ft^3
c_0	sticking-probability constant in Table 21.1
c_f	concentration in fluid, kg/m^3, lb_m/ft^3
c_w	concentration at wall, kg/m^3, lb_m/ft^3
D	diffusion coefficient, m^2/s, ft^2/hr
D_h	hydraulic diameter, m, ft
d_c	clean tube diameter, m, ft
d_f	fouled tube diameter, m, ft
d_p	particle diameter, m, ft
E	activation energy, J/kg, $lb_f \cdot ft/lb_m$
f	Fanning friction factor
h	convective heat transfer coefficient, $W/(m^2 \cdot K)$, $Btu/(hr \cdot ft^2 \cdot °F)$
h_i	inside convective heat transfer coefficient, $W/(m^2 \cdot K)$, $Btu/(hr \cdot ft^2 \cdot °F)$
h_o	outside convective heat transfer coefficient, $W/(m^2 \cdot K)$, $Btu/(hr \cdot ft^2 \cdot °F)$
h_D	convective mass transfer coefficient, m^2/s, ft^2/hr
J''	fouling-species mass flux, $kg/(m^2 \cdot s)$, $lb_m/(ft^2 \cdot hr)$

K_1	coefficient in Table 21.1
k	thermal conductivity of fluid, W/(m · K), Btu/(hr · ft · °F)
k_d	deposition coefficient in Table 21.1
k_f	thermal conductivity of fouling deposit, W/(m · K), Btu/(hr · ft · °F)
k_p	thermal conductivity of particle, W/(m · K), Btu/(hr · ft · °F)
k_r	first order reaction rate constant in Table 21.1
L	tube length or characteristic length, m, ft
m	deposit mass per unit surface area, kg/m², lb$_m$/ft²
$\dot{m}_d$	deposition flux, kg/(m² · s), lb$_m$/(ft² · hr)
$m_{d,0}$	deposition flux at zero time, kg/(m² · s), lb$_m$/(ft² · hr)
m_f	mass of fouling deposit, kg, lb$_m$
m_{loose}	loose deposit mass per unit surface area, kg/m², lb$_m$/ft²
$\dot{m}_r$	removal flux, kg/(m² · s), lb$_m$/(ft² · hr)
MTD	mean temperature difference, °C, K, °F, °R
n	exponent in Table 21.1
P_v	probability function of velocity
q	heat transfer rate or heat duty, W, Btu/hr
q''_w	wall heat flux, W/m², Btu/(hr · ft²)
R	gas constant, J/(kg · K), lb$_f$ · ft/(lb$_m$ · °R)
R_f	fouling factor or fouling resistance, m² · K/W, hr · ft² · °F/Btu
$R_{f,i}$	inside fouling factor or resistance, m² · K/W, hr · ft² · °F/Btu
$R_{f,o}$	outside fouling factor or resistance, m² · K/W, hr · ft² · °F/Btu
R_f^*	asymptotic fouling factor or resistance, m² · K/W, hr · ft² · °F/Btu
R_w	wall resistance per unit area, K/W, hr · °F/Btu
Re	Reynolds number = $\rho VL/\mu$
S	sticking probability
Sc	Schmidt number = ν/D
Sh	Sherwood number = $h_D L/D$
St	Stokes number = $\rho_p d_p^2 V/18\mu L$
T	temperature, °C, K, °F, °R
T_f	fluid temperature, °C, K, °F, °R
T_s	temperature at fluid-deposit surface interface, °C, K, °F, °R
T_w	wall temperature, °C, K, °F, °R
t	time, s, hr
U	overall heat transfer coefficient, W/(m² · K), Btu/(hr · ft² · °F)
U_c	clean overall heat transfer coefficient, W/(m² · K), Btu/(hr · ft² · °F)
U_f	fouled overall heat transfer coefficient, W/(m² · K), Btu/(hr · ft² · °F)
u^*	friction velocity, m/s, ft/s
V	fluid velocity, m/s, ft/s
V_t	thermophoretic velocity, m/s, ft/s
V_w	particle velocity normal to wall, m/s, ft/s
x_f	fouling-deposit thickness, m, ft
y	spatial coordinate normal to surface, m, ft

Greek symbols

α	coefficient in Eq. (21.12)
α_f	fouling-deposit thermal diffusivity, m^2/s, ft^2/hr
ρ_f	density of fouling deposit, kg/m^3, lb_m/ft^3
ρ_p	density of particle, kg/m^3, lb_m/ft^3
μ	viscosity, $kg/(m \cdot s)$, $lb_m/(hr \cdot ft)$
ν	kinematic viscosity, m^2/s, ft^2/hr
Ω	water characterization factor
ψ	fouling deposit strength, Pa, lb_f/ft^2
τ_w	wall shear stress, Pa, lb_f/ft^2
ϕ_d	deposition function, $m^2 \cdot K/J$, $ft^2 \cdot {}^\circ F/Btu$
ϕ_r	removal function, $m^2 \cdot K/J$, $ft^2 \cdot {}^\circ F/Btu$
θ	time constant, s, hr
Δp	pressure drop, Pa, lb_f/ft^2

Subscripts

c	clean condition
f	fouled condition, fouling deposit, or fluid
d	deposition
i	inside
o	outside
p	particle
r	removal
ref	reference conditions
w	wall

REFERENCES

1. E. F. C. Somerscales and J. K. Knudsen, eds., *Fouling of Heat Transfer Equipment*, Hemisphere, Washington, 1981.
2. R. W. Bryers, ed., *Ash Deposits and Corrosion Due to Impurities in Combustion Gases*, Hemisphere, Washington, 1978.
3. R. W. Bryers, ed., *Fouling and Slagging Resulting from Impurities in Combustion Gases*, Engineering Foundation, New York, 1983.
4. R. W. Bryers, ed., *Fouling of Heat Exchanger Surfaces*, Engineering Foundation, New York, 1983.
5. J. M. Chenoweth and M. Impagliazzo, eds., *Fouling in Heat Exchange Equipment*, HTD—Vol. 17, ASME, New York, 1981.
6. W. J. Marner and R. L. Webb, eds., Workshop on an Assessment of Gas-Side Fouling in Fossil Fuel Exhaust Environments, JPL Publ. 82-67, Jet Propulsion Lab., Calif. Inst. of Technology, Pasadena, Calif., 1982.
7. W. J. Marner and J. W. Suitor, A Survey of Gas-Side Fouling in Industrial Heat-Transfer Equipment, JPL Publication 83-74, Jet Propulsion Lab., Calif. Inst. of Technology, Pasadena, Calif., 1983.

8. J. W. Suitor and A. M. Pritchard, eds., *Fouling in Heat Exchange Equipment*, HTD—Vol. 35, ASME, New York, 1984.

9. B. A. Garrett-Price, S. A. Smith, R. L. Watts, J. G. Knudsen, W. J. Marner, and J. W. Suitor, *Fouling of Heat Exchangers: Characteristics, Costs, Prevention, Control, and Removal*, Noyes Publications, Park Ridge, N.J., 1985.

10. N. Epstein, Fouling in Heat Exchangers, *Heat Transfer 1978*, Vol. 6, pp. 235–254, 1978. Hemisphere, New York.

11. N. Epstein, Fouling: Technical Aspects (Afterword to Fouling in Heat Exchangers), in *Fouling of Heat Transfer Equipment*, ed. E. F. C. Somerscales and J. G. Knudsen, pp. 31–53, Hemisphere, Washington, 1981.

12. P. A. Thackery, The Cost of Fouling in Heat Exchange Plant, *Effluent Water Treatment J.*, Vol. 20, pp. 111–115, 1980.

13. N. Epstein, Thinking about Heat Transfer Fouling: a 5 × 5 Matrix, *Heat Transfer Eng.*, Vol. 4, No. 1, pp. 43–56, 1983.

14. R. B. Ritter, Crystallization Fouling Studies, *ASME J. Heat Transfer*, Vol. 105, pp. 374–378, 1983.

15. F. W. Dittus and L. M. K. Boelter, Heat Transfer in Automobile Radiators of the Tubular Type, *Publications in Engineering*, Vol. 2, Univ. of California, Berkeley, pp. 443–461, 1930.

16. S. K. Friedlander, *Smoke, Dust and Haze*, Wiley, New York, 1977.

17. S. L. Soo, Gas-Solid Systems, *Handbook of Multiphase Systems*, G. Hetsroni, ed., pp. 3-1–3-55, Hemisphere, Washington, 1982.

18. P. J. Whitmore and A. Meisen, Estimation of Thermo- and Diffusiophoretic Particle Deposition, *Can. J. Chem. Eng.*, Vol. 55, pp. 279–285, 1977.

19. G. Nishio, S. Kitani, and K. Takahashi, Thermophoretic Deposition of Aerosol Particles in a Heat-Exchanger Pipe, *Ind. Eng. Chem. Proc. Design Dev.*, Vol. 13, pp. 408–415, 1974.

20. J. W. Cleaver and B. Yates, A Sublayer Model for the Deposition of Particles from a Turbulent Flow, *Chem. Eng. Sci.*, Vol. 30, pp. 983–992, 1975.

21. J. W. Cleaver and B. Yates, The Effect of Re-entrainment on Particle Deposition, *Chem. Eng. Sci.*, Vol. 31, pp. 147–151, 1976.

22. A. M. Pritchard, Why does It Stick?, *Fouling Prevention Digest*, Vol. 6, No. 4, p. ii, 1984.

23. E. R. G. Eckert and R. M. Drake, Jr., *Heat and Mass Transfer*, 2nd ed., McGraw-Hill, New York, 1959.

24. T. D. Brown, The Deposition of Sodium Sulphate from Combustion Gases, *J. Inst. Fuel*, Vol. 39, pp. 378–385, 1966.

25. W. L. McCabe and C. S. Robinson, Evaporator Scale Formation, *Ind. and Eng. Chem.*, Vol. 16, pp. 478–479, 1924.

26. D. Q. Kern and R. E. Seaton, A Theoretical Analysis of Thermal Surface Fouling, *Br. Chem. Eng.*, Vol. 4, pp. 258–262, 1959.

27. A. P. Watkinson and N. Epstein, Particulate Fouling of Sensible Heat Exchangers, *Heat Transfer 1970*, Vol. 1, Paper HE 1.6, 1970.

28. J. Taborek, T. Aoki, R. B. Ritter, J. W. Palen, and J. G. Knudsen, Fouling—The Major Unresolved Problem in Heat Transfer, *Chem. Eng. Prog.*, Vol. 68, No. 2, pp. 59–67; No. 7, pp. 69–78, 1972.

29. Heat Exchange Institute, Cleveland, Ohio.

30 D. Q. Kern, *Process Heat Transfer*, McGraw-Hill, New York, 1950.

31. R. K. Shah, Compact Heat Exchangers, *Handbook of Heat Transfer Applications*, 2nd ed., ed. W. M. Rohsenow, J. P. Hartnett, and E. N. Ganić, Chapter 4, Part 3, pp. 4-279–4-312, McGraw-Hill, New York, 1985.

32. *Standards of the Tubular Exchanger Manufacturers Association*, 6th ed., Tubular Exchanger Manufacturers Assoc., New York, 1978.

33. A.P. Watkinson, Water Quality Effects on Fouling from Hard Waters, *Heat Exchangers: Theory and Practice*, ed. J. Taborek, G. F. Hewitt, and N. Afgan, pp. 853–861, Hemisphere, New York, 1983.

34. R. A. McAllister, D. H. Eastham, N. A. Dougharty and M. Hollier, A Study of Scaling and Corrosion in Condenser Tubes Exposed to River Water, *Corrosion*, Vol. 17, pp. 579t–588t, 1961.

35. J. Gutzeit, Corrosion and Fouling of Admiralty, Aluminum, and Steel Tubes in Open Recirculating Cooling Water Systems, *Materials Protection*, Vol. 4, pp. 28–34, July 1965.

36. R. B. Ritter, J. W. Suitor, and G. A. Cypher, Thermal Fouling Rates of 90-10 Copper-Nickel and Titanium in Seawater Service, *Proc. OTEC Biofouling and Corrosion Symposium*, Seattle, Wash., pp. 353–365, 1977.

37. D. J. Parry, D. Hawthorn, and A. Rantell, Fouling of Power Station Condensers within the Midlands Region of the CEGB, *Fouling of Heat Transfer Equipment*, ed. E. F. C. Somerscales and J. G. Knudsen, pp. 569–586, Hemisphere, Washington, 1981.

38. C. Panchal, J. Larsen-Basse, and B. Little, Biofouling Control for Marine Heat Exchangers Using Intermittent Chlorination, *Fouling in Heat Exchange Equipment*, ed. J. W. Suitor and A. M. Pritchard, HTD—Vol. 35, ASME, New York, pp. 97–103, 1984.

39. T. Nosetani, S. Sato, K. Onda, J. Kashiwada, and K. Kawaguchi, Effect of Marine Biofouling on the Heat Transfer Performance of Titanium Condenser Tubes, *Fouling of Heat Transfer Equipment*, ed. E. F. C. Somerscales and J. G. Knudsen, pp. 345–353, Hemisphere, Washington, 1981.

40 A. P. Watkinson and O. Martinez, Scaling of Heat Exchanger Tubes by Calcium Carbonate, *ASME J. Heat Transfer*, Vol. 97, pp. 504–508, 1975.

41. D. S. Sasscer, T. O. Morgan, T. R. Tosteson and T. J. Summerson, In Situ Biofouling, Biofouling Control and Corrosion of Titanium and Aluminum Heat Exchanger Elements at Punta Tuna, Puerto Rico, *Proc. ASME-JSME Thermal Eng. Joint Conf.*, Vol. 2, pp. 291–299, ASME, New York, 1983.

42. J. Seki, C-Y Wei, S. Maezawa, and A. Tsuchida, Silica Scale Deposition on a Cylinder in Crossflow, *Proc. ASME-JSME Thermal Eng. Joint Conf.*, Vol. 2, pp. 149–153, ASME, New York, 1983.

43. G. A. Lambourn and M. Durrieu, Fouling in Crude Oil Preheat Trains, *Heat Exchangers: Theory and Practice*, ed. J. Taborek, G. F. Hewitt, and N. Afgan, pp. 841–852, Hemisphere, New York, 1983.

44. J. G. Knudsen, E. Tijitrasmoro, and X. Dun-qi, The Effect of Zink Chromate Corrosion Inhibitors on the Fouling Characteristics of Cooling Tower Water, *Fouling in Heat Exchange Equipment*, ed. J. W. Suitor and A. M. Pritchard, HTD—Vol. 35, pp. 11–18, ASME, New York, 1984.

45. R. C. Weierman, Design of Heat Transfer Equipment for Gas-Side Fouling Service, Workshop on an Assessment of Gas-side Fouling in Fossil Fuel Exhaust Environments, ed. W. J. Marner and R. L. Webb, JPL Publ. 82-67, Jet Propulsion Lab., Calif. Inst. of Technology, Pasadena, Calif., 1982.

46. Technical Information for Heat Recovery Seminar, John Zink Company, Tulsa, Oklahoma, 1981.

47. R. D. Rogalski, Fouling Effects of Turbine Exhaust Gases on Heat Exchanger Tubes for Heat Recovery Systems, *SAE Trans.*, Vol. 88, pp. 2223–2239, 1979.

48. S. P. Henslee and J. L. Bogue, Fouling of a Finned-Tube Heat Exchanger in the Exhaust of a Stationary Diesel Engine: Final Report, Report EGG-FM-6189, Idaho Nat. Eng. Lab., Idaho Falls, 1983.

49. L. S. Chow, P. F. Dunn, K. Natesan, T. R. Johnson, C. B. Reed, and B. J. Schlenger, Seed-Slag Fouling and Corrosion in an MHD Steam Bottoming Plant, *Proc. 17th IECEC*, pp. 1229–1239, IEEE, New York, 1982.

50. R. P. Weight, A Rapid Method for Determination of Thermal Diffusivity in Boiler Fireside Deposits, *Fouling and Slagging Resulting from Impurities in Combustion Gases*, ed. R. W. Bryers, pp. 319–328, Engineering Foundation, New York, 1983.
51. W. G. Characklis, A Rational Approach to Problems of Fouling Deposition, *Fouling of Heat Exchanger Surfaces*, ed. R. W. Bryers, pp. 1-31, Engineering Foundation, New York, 1983.
52. W. G. Characklis, Microbial Fouling: A Process Analysis, *Fouling of Heat Transfer Equipment*, ed. E. F. C. Somerscales and J. G. Knudsen, pp. 251–291, Hemisphere, Washington, 1981.
53. D. H. Lister, Corrosion Products in Power Generating Systems, *Fouling of Heat Transfer Equipment*, ed. E. F. C. Somerscales and J. G. Knudsen, pp. 135–200, Hemisphere, Washington, 1981.
54. T. K. Sherwood, R. L. Pigford, and C. R. Wilke, *Mass Transfer*, McGraw-Hill, New York, 1975.
55. E. Raask, *Mineral Impurities in Coal Combustion*, Hemisphere, Washington, 1985.
56. A. M. Pritchard, Fouling in Heat Exchange Plant—A British View, unpublished paper presented at the 20th ASME-AIChE National Heat Transfer Conference, Milwaukee, Wis., August, 1981.
57. C. L. Wagoner, G. Haider, J. W. Berthold, and R. A. Wessel, Measurement of Fundamental Properties Characterizing Coal Minerals and Fire-Side Deposits, Report DOE/PC/40266-3, Babcock and Wilcox, Alliance, Ohio, 1984.
58. W. G. Characklis, Biofilm Development and Destruction, Report EPRI CS-1554, Rice Univ., Houston, Tex., 1980.
59. A. Cooper, J. W. Suitor, and J. D. Usher, Cooling Water Fouling in Plate Heat Exchangers, *Heat Transfer Eng.*, Vol. 1, No. 3, pp. 50–55, 1980.
60. J. G. Knudsen, Fouling of Heat Exchangers: Are We Solving the Problem?, *Chem. Eng. Prog.*, Vol. 80, No. 7, pp. 63–69, 1984.
61. *Principles of Industrial Water Treatment*, Drew Chemical Corp., Boonton, N.J., 1977.
62. *Water Treatment Handbook*, Betz Labs. Inc., Trevose, Pa., 1980.
63. S. D. Strauss and P. R. Puckorious, Cooling-Water Treatment for Control of Scaling, Fouling, and Corrosion, *Power*, Vol. 128, No. 6, pp. S1–S24, 1984.
64. P. Van der Wee and P. A. Tritsman, What to Do about Crude Oil Preheat Exchanger Fouling, *Hydrocarbon Process.*, Vol. 45, No. 8, pp. 141–144, August 1966.
65. *Steam: Its Generation and Use*, Babcock and Wilcox, Barberton, Ohio, 1975.
66. R. H. Essenhigh, Coal Combustion, *Coal Conversion Technology*, ed. C. Y. Wen and E. S. Lee, pp. 171–312, Addison-Wesley, Reading, Mass. 1979.
67. D. J. White, Effects of Gas Turbine Combustion on Soot Deposition, Final Report, Contract N0001478-C-003, Solar Turbines International, San Diego, Calif., March 1979.
68. Electric Utility Use of Fireside Additives, EPRI Report CS-1318, Battelle Columbus Labs., Columbus, Ohio, 1980.
69. N. F. Lansing, ed., Workshop Proceedings: Applications of Fireside Additives to Utility Boilers, EPRI Report WS-80-127, Battelle Columbus Labs., Columbus, Ohio, 1981.
70. J. T. DiCarlo, Guidelines Aid in Application, Operation of Soot Blowers, *Oil and Gas J.*, Vol. 77, No. 19, pp. 88–96, 1972.
71. P. Fischer, J. W. Suitor, and R. B. Ritter, Fouling Measurement Techniques and Apparatus, *Chem. Eng. Prog.*, Vol. 71, No. 7, pp. 66–72, July 1975.
72. J. G. Knudsen, Apparatus and Techniques for Measuring of Fouling of Heat Exchangers, *Fouling of Heat Transfer Equipment*, ed. E. F. C. Somerscales and J. G. Knudsen, pp. 57–81, Hemisphere, Washington, 1981.
73. W. J. Marner and S. P. Henslee, A survey of Gas-Side Fouling Measuring Devices, JPL Publ. 84-11, Jet Propulsion Lab., Calif. Inst. of Technology, Pasadena, Calif., 1984.
74. W. J. Marner and S. P. Henslee, A Survey of Gas-Side Fouling Measuring Devices, *Industrial Heat Exchangers*, ed. A. J. Hayes, W. W. Liang, S. L. Richlen, and E. S. Tabb, pp. 209–226, ASM, Metals Park, Ohio, 1985.

22

THERMOPHYSICAL PROPERTIES

P. E. Liley

School of Mechanical Engineering and
Center for Information and Numerical Data Analysis and Synthesis (CINDAS)
Purdue University
Lafayette, Indiana

22.1 Tables

22.1 TABLES

TABLE 22.1. Fundamental Units in the SI System

Quantity	Unit	Symbol
Mass	Kilogram(me)	kg
Length	Meter (metre)	m
Time	Second	s
Temperature	Kelvin	K
Electric current	Ampere	A
Luminous intensity	Candela	cd

DEFINITIONS OF FUNDAMENTAL UNITS

Meter (metre): The meter is the length of a path traveled by light in vacuum during a time interval of 1/299,792,458 of a second.[a, b]

Kilogram(me). The kilogram is the unit of mass; it is represented by the mass of the international prototype kilogram.[a]

Second. The second is the duration of 9,192,631,770 periods of the radiation corresponding to the transition between the two hyperfine levels of the ground state of an atom of cesium 133.

Ampere. The ampere is the constant current which, maintained in two straight parallel conductors of infinite length, of negligible cross-sectional area, and placed 1 meter apart from one another in a vacuum, produces between these conductors a force equal to 2×10^{-7} mks units of force (newtons) per meter of length.

Kelvin. The thermodynamic scale of temperature is defined by means of the triple point of water as being a fixed fundamental point, and attributing to this point the temperature 273.16 kelvins exactly.

Candela. The value of the candela is such that the brilliance of total radiation, at the temperature of solidification of platinum, is 60 candela per square centimeter.

[a] "Kilogram" and "meter" are the U.S. spellings; "kilogramme" and "metre" are the international spellings.

[b] *Metrologia*, Vol. 19, pp. 163–177 (1984); *Eur. J. Phys.*, Vol. 4, pp. 190–197 (1983); *Am. J. Phys.*, Vol. 52, pp. 607 (1984).

TABLE 22.2. Prefixes for the SI Units

Prefix	Symbol[a]	Multiplier
exa	E	10^{18}
peta	P	10^{15}
tera	T	10^{12}
giga	G	10^{9}
mega	M	10^{6}
kilo	k	10^{3}
hecto	(h)	10^{2}
deca	(da)	10^{1}
deci	(d)	10^{-1}
centi	(c)	10^{-2}
milli	m	10^{-3}
micro	μ	10^{-6}
nano	n	10^{-9}
pico	p	10^{-12}
femto	f	10^{-15}
atto	a	10^{-18}

[a] Prefixes with symbols in parentheses are not recommended for use.

TABLE 22.3. Conversion Factors[a]

Density: 1 kg/m^3 = 0.06243 lb$_m$/ft^3 = 0.01002 lb$_m$/U.K. gallon = 8.3454 $\times$ 10^{-3} lb$_m$/U.S. gallon = 1.9403 $\times$ 10^{-3} slug/ft^3 = 10^{-3} g/cm^3

Energy: 1 kJ = 737.56 ft $\cdot$ lb$_f$ = 238.85 cal = 0.94783 Btu = 3.7251 $\times$ 10^{-4} hp $\cdot$ hr = 2.7778 $\times$ 10^{-4} kW $\cdot$ hr

Specific energy: 1 kJ/kg = 334.55 ft $\cdot$ lb$_f$/lb$_m$ = 0.4299 Btu/lb$_m$ = 0.2388 cal/g

Specific energy per degree: 1 kJ/(kg $\cdot$ K) = 0.23885 Btu/(lb $\cdot$ °F) = 0.23885 cal/(g $\cdot$ °C)

Heat transfer coefficient: 1 W/(m^2 $\cdot$ K) = 0.8598 kcal/(m^2 $\cdot$ hr $\cdot$ °C) = 0.1761 Btu/(ft^2 $\cdot$ hr $\cdot$ °F) = 10^{-4} W/(cm^2 $\cdot$ K) = 0.2388 $\times$ 10^{-4} cal/(cm^2 $\cdot$ s $\cdot$ °C)

Mass: 1 kg = 2.20462 lb$_m$ = 0.06852 slug = 1.1023 $\times$ 10^{-3} U.S. ton = 10^{-3} tonne = 9.8421 $\times$ 10^{-4} U.K. ton

Pressure: 1 bar = 10^5 N/m^2 = 10^5 Pa = 750.06 mm Hg at 0°C = 401.47 in. H_2O at 32°F = 29.530 in. Hg at 0°C = 14.504 lb$_f$/in.^{2} = 14.504 psia = 1.01972 kg/cm^2 = 0.98692 atm = 0.1 MPa

Temperature: T(K) = T(°C) + 273.15 = [T(°F) + 459.67]/1.8 = T(°R)/1.8

Temperature difference: $\Delta T(K) = \Delta T(°C) = \Delta T(°F)/1.8 = \Delta T(°R)/1.8$

Thermal conductivity: 1 W/(m $\cdot$ K) = 0.8604 kcal/(m $\cdot$ hr $\cdot$ °C) = 0.5782 Btu/(ft $\cdot$ hr $\cdot$ °F) = 0.01 W/(cm $\cdot$ K) = 2.390 $\times$ 10^{-3} cal/(cm $\cdot$ s $\cdot$ °C)

Thermal diffusivity: 1 m^2/s = 38750 ft^2/hr = 3600 m^2/hr = 10.764 ft^2/s

Viscosity, dynamic: 1 N $\cdot$ s/m^2 = 1 Pa $\cdot$ s = 10^7 μP = 2419.1 lb$_m$/(ft $\cdot$ hr) = 10^3 cP = 75.188 slug/(ft $\cdot$ hr) = 10 P = 0.6720 lb$_m$/(ft $\cdot$ s) = 0.02089 lb$_f$ $\cdot$ s/ft^2

Viscosity, kinematic (*see* Thermal diffusivity)

[a] E. Lange, L. F. Sokol, and V. Antoine, *Information on the Metric System and Related Fields*, 6th ed., G. C. Marshall Space Flight Center, Ala. (exhaustive bibliography); C. H. Page and P. Vigoureux, *The International System of Units*, NBS S.P. 330, Washington, 1974; E. A. Mechtly, *The International System of Units, Physical Constants and Conversion Factors*, NASA S.P. 9012, 1973. Numerous revision periodically appear; see, for example, *Pure Appl. Chem.*, Vol. 51, 1–41 (1979) and later issues. This table is only intended to enable rapid conversion to be made with moderate, (i.e., five-significant-figure) accuracy, usually acceptable in most engineering calculations. The references listed should be consulted for more exact conversions and definitions.

TABLE 22.4. Thermophysical Properties of 113 Fluids at 1 bar, 300 K[a]

Name	Formula	M	T_m, K	T_b, K	P_g, bar	ρ, kg/m^3
Acetaldehyde	C_2H_4O	44.053	149.7	293.7	2.94	
Acetic acid	$C_2H_4O_2$	60.053		391.1	0.0225	1046
Acetone	C_3H_6O	58.080	178.5	329.3	0.0318	782.5
Acetylene	C_2H_2	26.038	179.0	189.2	50.66	1.0508
Air (R 729)	mix	28.966	60.0	var	—	1.1614
Ammonia (R 717)	NH_3	17.031	195.4	239.7	10.614	
Aniline	C_6H_7N	93.129	266.8	457.5	7.206. − 4	
Argon (R 740)	Ar	39.948	83.8	87.5	—	
Benzene	C_6H_6	78.114	278.7	353.3	0.1382	871
Bromine	Br	159.81	265.4	331.5	0.310	3094
Butadiene, 1,3-	C_4H_6	54.088	164.2	268.7		
Butane, *iso*-	C_4H_{10}	58.124	113.7	261.5	3.696	
Butane, *n*-	C_4H_{10}	58.124	134.9	272.6	2.581	
Butanol	$C_4H_{10}O$	74.124	183.9	390.8	0.01064	804.3
Butylene	C_4H_8	56.108	87.8	266.9		
Carbon dioxide	CO_2	44.010	216.6	194.7	67.10	1.7734
Carbon disulfide	CS_2	76.131	161.1	319.4		1262
Carbon tetrachloride	CCl_4	153.82	250.3	349.8		1580.8
Carbon tetrafluoride	CF_4	88.005	89.5	145.2	—	
Chlorine	Cl_2	70.906	172.2	238.6	8.	
Chlorine trifluoride	ClF_3	92.449	196.8	284.9	1.8	1.820
Chlorine pentafluoride	ClF_5	130.446	178.2	260.1	4.0	1.788
Chloroform	$CHCl_3$	119.377	209.7	334.5	0.2771	1530
Cresol, *o*-	C_7H_8O	108.134	303.8	464.1		
Cresol, *m*-	C_7H_8O	108.134				
Cresol, *p*-	C_7H_8O	108.134				
Cyclobutane	C_4H_8	56.104	182.4	285.7	1.6	
Cyclohexane	C_6H_{12}	84.156	279.8	353.9		772.1
Cyclopentane	C_5H_{10}	70.130	179.3	322.5		
Cyclopropane	C_3H_6	42.081	145.5	240.3	6.	
Decane	$C_{10}H_{22}$	142.276	243.4	447.3	0.0017	724.1
Deuterium	D_2	4.028	18.7	23.7	—	
Diphenyl	$C_{12}H_{10}$	154.200	342.4	527.6		
Ethane (R 170)	C_2H_6	30.070	89.9	184.6	43.541	1.2157
Ethanol	C_2H_6O	46.069	159.0	351.5		1247
Ethyl acetate	$C_4H_8O_2$	88.106	189.4	350.3	0.1361	892.3
Ethyl bromide	C_2H_5Br	108.97	153.5	311.5		1450
Ethyl chloride (R 160)	C_2H_5Cl	64.515	134.9	285.4	1.68	
Ethyl ether	$C_4H_{10}O$	74.123	150.0	307.8		705.6
Ethyl fluoride (R 161)	C_2H_5F	48.060	130.0	236.1	9.1	
Ethyl formate	$C_3H_6O_2$	74.080	193.8	327.4		
Ethylene	C_2H_4	28.054	104.0	169.5	—	0.9035
Ethylene oxide	C_2H_4O	44.054	160.6	283.6	1.82	
Fluorine	F_2	37.997	53.5	85.1	—	1.5442
Glycerol						1284.3

c_p, kJ/(kg · K)	c_v, kJ/(kg · K)	γ	μ, 10^{-1} Pa · s	k, W/m · K	Pr	σ, N/m	$\bar{v}_s$, m/s
1.49							
2.066			11.15	0.1635	14.1		1130
2.13			3.08	0.1595	4.11	0.0229	1160
1.7032	1.380	1.234	0.1039	0.0220	0.804	—	342
1.005	0.718	1.400	0.184	0.0261	0.711	—	347.3
2.10	1.63	1.33	0.102	0.246	0.870	—	434
2.079			34.7	0.173	41.7		1615
0.522	0.313	1.667	0.2271	0.0177	0.670	—	322.6
1.73			5.88	0.1444	7.04	0.0279	1275
				0.122		0.041	
1.475			0.0876	0.0181	0.714		
			0.0760	0.0164	0.814		
1.731	1.569	1.103	0.0757	0.0160	0.811	0.0116	211
2.39			24.9	0.152	39.2	0.0241	1215
			0.0780			—	
0.845	0.657	1.288	0.150	0.0166	0.763	—	269.6
1.016				0.161			1143
0.8637			8.81	0.1026	7.42	0.0261	918
0.7071	0.6095	1.150	0.1740	0.01711	0.719	—	
0.4728			0.1371	0.00889	0.729	—	216
0.695			0.1421	0.0138	0.716		
0.745			0.1431	0.0146	0.730		
0.975			5.39	0.117			988
			70.2	0.153	96.0		
			112	0.149	174		
			126	0.144			
1.289			0.0812	0.0148	0.707		
1.865			8.69	0.123	13.2	0.0242	1230
1.81			3.97	0.132	5.46		
1.331			0.0890	0.0163	0.727	—	
2.213							
			0.126	0.141			
1.586							
1.769	1.482	1.192	0.094	0.0218	0.765	—	314
2.456			10.46	0.167	15.4	0.0222	11.37
1.947			4.37	0.143	5.95	0.0234	1120
0.880			3.63	0.101	3.16		905
0.973			0.1002	0.0126	0.774	—	
2.21			2.25	0.130	3.83		968
1.243			0.0954	0.0163	0.727		
				0.159			
1.560	1.262	1.237	0.104	0.0203	0.796	—	332
1.096			0.0879	0.0130	0.742		
0.827	0.607	1.362	0.227	0.0279	0.673	—	299
2.38			7.922	0.280	6730	0.0587	1875

TABLE 22.4. (Continued)

Name	Formula	M	T_m, K	T_b, K	P_g, bar	ρ, kg/m^3
Helium	He	4.003	—	4.3	—	0.1625
Heptane	C_7H_{16}	100.21	182.5	371.6	0.06719	678.0
Hexane	C_6H_{14}	86.178	177.8	341.9	0.06674	652.5
Hydrazine	N_2H_4	32.045	274.7	386.7		
Hydrogen, *n*-	H_2	2.016	14.0	28.4	—	0.0808
Hydrogen, *p*-	H_2	2.016				
Hydrogen bromide	HBr	80.912	186.3	206.4	25.5	
Hydrogen chloride	HCl	36.461	160.0	188.1	48.7	
Hydrogen fluoride	HF	20.006	181.8	272.7	1.3	
Hydrogen iodide	HI	127.91	222.4	237.8	8.5	
Hydrogen peroxide	H_2O_2				0.0031	1449
Hydrogen sulfide	H_2S	34.076	187.5	213.0	20.9	
Krypton	Kr	83.80	116.0	121.4	—	3.3659
Mercury	Hg	200.59	234.3	630.1		
Methane	CH_4	16.043	90.7	111.5	—	0.6443
Methanol	CH_4O	32.042	175.5	337.7	0.1860	784.9
Methyl acetate	$C_3H_6O_2$	74.080	175	330.3		
Methyl bromide (R 40 B 1)	CH_2Br	94.939	179.5	276.7	2.38	
Methyl chloride (R 40)	CH_3Cl	50.487	175.4	249.4	6.189	
Methyl fluoride (R 41)	CH_3F	34.033	131.3	194.7	39.5	
Methyl formate (R 611)	$C_2H_4O_2$	60.052	173.4	304.7		
Methylene chloride (R 30)	CH_2Cl_2	84.922	176.5	312.9		
Neon (R 720)	Ne	20.179	24.5	27.3	—	0.8091
Neopentane	C_5H_{12}	72.151	256.6	282.7		
Nitric oxide	NO	30.006		121.4	—	
Nitrogen (R 728)	N_2	28.013	63.1	77.3	—	
Nitrogen peroxide	NO_2	46.006	263	294.5		
Nitrous oxide (R 744a)	N_2O	44.013	176	184.7	57.5	
Nonane	C_9H_{20}	128.250			0.0064	712.3
Octane	C_8H_{18}	114.220	216.4	398.9	0.0207	704.2
Oxygen	O_2	31.999	54.4	90.0	—	
Pentane, *iso*-	C_5H_{12}	72.151	113.7	301.1		
Pentane, *n*-	C_5H_{12}	72.151	143.7	309.2		
Propadiene	C_3H_4	40.062	136.9	238.8	8.4	
Propane	C_3H_8	44.097	86.	231.1	9.9973	
Propanol	C_3H_8O	60.096	147.0	370.4		
Propylene	C_3H_6	42.081	87.9	225.5	12.118	
Refrigerant 11	$CFCl_3$	137.37	162.2	296.9	1.1341	
Refrigerant 12	CF_2Cl_2	120.91	115.4	243.4	6.8491	1311
Refrigerant 12 B 1	CF_2BrCl	165.37	113.7	269.2	2.8	
Refrigerant 12 B 2	CF_2Br_2	209.82	131.6	295.9	1.4	
Refrigerant 13	CF_2Cl	104.46	92.1	191.7	37.05	
Refrigerant 13 B 1	CF_3Br	148.91	105.4	215.4	16.914	
Refrigerant 21	$CHFCl_2$	102.91	138.2	282.1		
Refrigerant 22	CHF_2Cl	86.469	113.2	232.4	10.96	1194

c_p, kJ/(kg · K)	c_v, kJ/(kg · K)	γ	μ, 10^{-1} Pa · s	k, W/m · K	Pr	σ, N/m	$\bar{v}_s$, m/s
5.193	3.115	1.667	0.1976	0.165	0.668		1020
2.252			3.85	0.1242	6.96	0.0197	1120
2.270			2.87	0.123	5.30	0.0177	1130
14.27	10.18	1.405	0.0894	0.182	0.701	—	1350
14.84	10.72	1.384	0.0814	0.176	0.686	—	1310
0.360			0.1888	0.0098	0.693	—	210
0.845			0.1461	0.0170	0.728	—	310
1.456			0.1251	0.0260	0.702	—	
0.228			0.1872	0.00624	0.684	—	165
1.48			11.3	0.481	3.47		
1.00			0.1294	0.01461	0.886	—	300
0.249	0.149	1.667	0.256	0.0094	0.678	—	
0.139			15.3				1450
2.235	1.711	1.306	0.112	0.0344	0.733	—	450
2.528			5.35	0.200	6.76	0.0221	1097
			3.56	0.1555	4.89		1125
0.449			0.136	0.00832	0.734	—	905
0.808			0.1088	0.012	0.732	—	
1.100			0.1161	0.01809	0.706		
			3.26	0.186			
			4.08	0.1388	3.57		
1.030	0.618	1.667	0.317	0.0493	0.649	—	455
1.77			0.0726	0.0146	0.880		
0.996			0.1919	0.0257	0.743	—	
1.040	0.743	1.400	0.180	0.0260	0.715	—	352
0.879			0.1490	0.0166	0.790	—	275
2.22			6.53	0.129	11.2	0.022	1205
2.231			5.04	0.128	8.78	0.0210	1163
0.920	0.660	1.394	0.2072	0.0267	0.714	—	332
2.28			2.125	0.146	3.17	0.0153	976
2.41			2.24	0.152	3.55		1050
1.473			0.0844	0.0170	0.733		
1.693	1.492	1.135	0.0826	0.01844	0.762		248
2.446			19.17	0.138	34.0	0.0232	1195
1.536	1.336	1.150	0.081	0.0168	0.797		
0.573	0.509	1.126	0.1095	0.0079	0.814	—	139
0.602	0.542	1.111	0.1260	0.0097	0.781		
0.460	0.395	1.165	0.1381	0.0078	0.811		
0.367	0.316	1.160	0.1340				
0.644	0.562	1.146	0.1449	0.0122	0.766	—	
0.468	0.409	1.143	0.1575	0.0099	0.745	—	
0.594	0.504	1.179	0.115	0.0089	0.768	—	
0.647	0.544	1.190	0.1299	0.0110	0.767	—	

TABLE 22.4. (Continued)

Name	Formula	M	T_m, K	T_b, K	P_g, bar	ρ, kg/m^3
Refrigerant 23	CHF_3	70.014	118.0	191.2	—	
Refrigerant 32	CH_2F_2	52.023		221.5		
Refrigerant 113	$C_2F_3Cl_3$	187.38	238.2	320.8	0.4817	1557
Refrigerant 114	$C_2F_4Cl_2$	170.92	179.2	276.7	2.2788	
Refrigerant 115	C_2F_5Cl	154.47		234.0		
Refrigerant 116	C_2F_6	138.02				
Refrigerant 142*b*	$C_2F_2H_3Cl$	100.50	142.4	263.9	3.5825	
Refrigerant 152*a*	$C_2F_4H_4$	66.051	156	248	6.3132	
Refrigerant 216	$C_3F_6Cl_2$	220.93		308		
Refrigerant 245					4.888	
Refrigerant C318	C_4F_8	200.031	232.7	267.2	3.325	
Refrigerant 500	mix	99.303	114.3	239.7	8.081	
Refrigerant 502	mix	111.63		237	12.186	
Refrigerant 503	mix	87.267		184		
Refrigerant 504	mix	79.240		216	—	
Refrigerant 505	mix	103.43		243.6		
Refrigerant 506	mix	93.69		260.7		
Silane	SiH_4	32.12	86.8	161.8		
Sulfur dioxide	SO_2	64.059	197.8	268.4	4.168	
Sulfur hexafluoride	SF_6	146.05	222.4	209.4	—	
Toluene	C_7H_8	92.141	178.2	383.8	0.0418	859
Water	H_2O	18.015	273.2	373.2	0.0353	
Xenon	Xe	131.36	161.5	165.	—	5.291

[a] M = molecular weight; T_m = melting temperature; T_b = boiling temperature; ρ = density; c_p = specific heat at constant pressure; c_v = specific heat at constant volume; γ = specific heat ratio c_p/c_v; μ = dynamic viscosity; k = thermal conductivity; Pr = Prandtl number; σ = surface tension; $\bar{v}_s$ = velocity of sound. A blank entry means no information is available; a dash means not applicable at 300 K.

c_p, kJ/(kg · K)	c_v, kJ/(kg · K)	γ	μ, 10^{-1} Pa · s	k, W/m · K	Pr	σ, N/m	$\bar{v}_s$, m/s
0.732	0.611	1.198	0.1488	0.0148	0.736	—	
0.958			6.64	0.0747	8.52		
0.715	0.660	1.083	0.1157	0.0105	0.775	—	
0.686	0.635	1.081	0.1289	0.0119	0.745	—	
0.762	0.698	1.092	0.1453	0.0150	0.740		
				0.0117			
1.029			0.1037	0.0148	0.721		
				0.0619			
0.821	0.778	1.056	0.1187	0.0127	0.765		
						—	
				0.0117	0.769	—	
						—	
						—	
						—	
						—	
1.338			0.1184	0.0221	0.718		
0.623			0.129	0.0096	6.837	—	
0.667			0.1654	0.0141	0.784		
1.69			5.54	0.133	7.04	0.0275	1285
4.179			8.9	0.609	5.69	0.0717	1501
0.160	0.097	1.655	0.232	0.0056	0.663	—	

TABLE 22.5. Thermophysical Properties of Liquid and Saturated-Vapor Air[a]

T, K	P_l, bar	P_g, bar	v_l, m^3/kg	v_g, m^3/kg	h_l, kJ/kg	h_g, kJ/kg	s_l, kJ/(kg · K)	s_g, kJ/(kg · K)
60	0.066	0.025	0.001018	6.8752	−437.6	−243.8	2.862	6.302
65	0.159	0.076	0.001068	2.4144	−436.7	−239.0	2.872	6.057
70	0.340	0.194	0.001093	1.0205	−434.9	−234.4	2.891	5.862
75	0.658	0.424	0.001121	0.4958	−430.6	−230.1	2.950	5.701
80	1.174	0.827	0.001150	0.2677	−423.8	−226.0	3.036	5.567
85	1.955	1.473	0.001182	0.1570	−415.6	−222.2	3.134	5.455
90	3.080	2.439	0.001218	0.0981	−406.5	−218.9	3.237	5.357
95	4.627	3.808	0.001257	0.0645	−396.7	−216.2	3.341	5.270
100	6.679	5.673	0.001303	0.0440	−386.3	−214.1	3.444	5.189
105	9.317	8.128	0.001355	0.0309	−375.4	−212.9	3.548	5.113
110	12.626	11.280	0.001418	0.0220	−363.8	−212.9	3.651	5.037
115	16.687	15.242	0.001497	0.0159	−351.4	−214.3	3.756	4.959
120	21.584	20.138	0.001602	0.0115	−337.8	−217.7	3.865	4.874
125	27.41	26.10	0.001757	0.0081	−322.2	−224.3	3.985	4.773
130	34.25	33.27	0.002069	0.0054	−301.1	−237.6	4.140	4.631
132.5	38.08	37.36	0.002594	0.0041	−280.4	−250.6	4.291	4.517

T K	$c_{p,l}$, kJ/(kg · K)	$c_{p,g}$, kJ/(kg · K)	μ_l, 10^{-4}Pa · s	μ_g, 10^{-4}Pa · s	k_l, W/(m · K)	k_g, W/(m · K)	Pr_l	Pr_g
60			3.25		0.180	0.005		
65			2.64		0.171	0.006		
70			2.21	0.47	0.163	0.006	2.54	0.69
75	1.79	1.13	1.89	0.51	0.154	0.007	2.35	0.73
80	1.82	1.17	1.65	0.55	0.145	0.008	2.23	0.77
85	1.85	1.21	1.47	0.60	0.137	0.008	2.17	0.81
90	1.88	1.26	1.32	0.65	0.128	0.009	2.14	0.84
95	2.00	1.31	1.20	0.70	0.119	0.009	2.15	0.87
100	2.12	1.37	1.10	0.75	0.110	0.010	2.21	0.91
105	2.24	1.48	1.02	0.80	0.102	0.011	2.29	0.98
110	2.41	1.64	0.95	0.86	0.093	0.012	2.47	1.06
115	2.65	1.91	0.87	0.93	0.084	0.014	2.66	1.13
120	3.09	2.40	0.75	1.02	0.076	0.015	2.76	1.33
125	4.12	3.53	0.62	1.17	0.067	0.018	2.83	1.65

[a] v = specific volume; h = specific enthalpy; s = specific entropy; c_p = specific heat at constant pressure; μ = dynamic viscosity; k = thermal conductivity; Pr = Prandtl number; l = saturated liquid; g = saturated vapor. Since air is a multicomponent mixture, the dew and bubble points vary with composition and there is no unique critical point.

TABLE 22.6. Ideal-Gas Thermophysical Properties of Air[a]

T, K	v, m^3/kg	h, kJ/Kg	s, kJ/(kg · K)	c_p, kJ/(kg · K)	γ	$\bar{v}_s$, m/s	μ, 10^{-5} Pa · s	k, W/(m · K)	Pr
100	0.2783	−204.5	5.755	1.030	1.424	198.0	0.71	0.0092	0.795
110	0.3076	−194.1	5.854	1.024	1.420	208.5	0.77	0.0102	0.786
120	0.3367	−183.9	5.943	1.020	1.417	218.3	0.84	0.0111	0.778
130	0.3657	−173.7	6.024	1.016	1.415	227.6	0.91	0.0120	0.770
140	0.3946	−163.6	6.099	1.014	1.413	236.4	0.97	0.0129	0.762
150	0.4233	−153.4	6.169	1.011	1.410	244.9	1.03	0.0139	0.755
160	0.4519	−143.3	6.234	1.010	1.407	253.1	1.09	0.0147	0.749
170	0.4805	−133.3	6.296	1.009	1.404	261.1	1.15	0.0156	0.743
180	0.5091	−123.3	6.353	1.009	1.402	268.7	1.21	0.0166	0.739
190	0.5376	−113.1	6.408	1.008	1.400	276.2	1.27	0.0174	0.736
200	0.5666	−103.0	6.4591	1.008	1.398	283.3	1.33	0.0183	0.734
210	0.5949	−92.9	6.5082	1.007	1.399	290.4	1.39	0.0191	0.732
220	0.6232	−82.8	6.5550	1.006	1.399	297.3	1.44	0.0199	0.730
230	0.6516	−72.8	6.5998	1.006	1.400	304.0	1.50	0.0207	0.728
240	0.6799	−62.7	6.6425	1.005	1.400	310.5	1.55	0.0215	0.726
250	0.7082	−52.7	6.6836	1.005	1.400	317.0	1.60	0.0222	0.725
260	0.7366	−42.6	6.7230	1.005	1.400	323.3	1.65	0.0230	0.723
270	0.7649	−32.6	6.7609	1.004	1.400	329.4	1.70	0.0237	0.722
280	0.7932	−22.5	6.7974	1.004	1.400	335.5	1.75	0.0245	0.721
290	0.8216	−12.5	6.8326	1.005	1.400	341.4	1.80	0.0252	0.720
300	0.8499	−2.4	6.8667	1.005	1.400	347.2	1.85	0.0259	0.719
310	0.8782	7.6	6.8997	1.005	1.400	352.9	1.90	0.0265	0.719
320	0.9065	17.7	6.9316	1.006	1.399	358.5	1.94	0.0272	0.719
330	0.9348	27.7	6.9625	1.006	1.399	364.0	1.99	0.0279	0.719
340	0.9632	37.8	6.9926	1.007	1.399	369.5	2.04	0.0285	0.719
350	0.9916	47.9	7.0218	1.008	1.398	374.8	2.08	0.0292	0.719
360	1.0199	57.9	7.0502	1.009	1.398	380.0	2.12	0.0298	0.719
370	1.0482	68.0	7.0778	1.010	1.397	385.2	2.17	0.0304	0.719
380	1.0765	78.1	7.1048	1.011	1.397	390.3	2.21	0.0311	0.719
390	1.1049	88.3	7.1311	1.012	1.396	395.3	2.25	0.0317	0.719
400	1.1332	98.4	7.1567	1.013	1.395	400.3	2.29	0.0323	0.719
410	1.1615	108.5	7.1817	1.015	1.395	405.1	2.34	0.0330	0.719
420	1.1898	118.7	7.2062	1.016	1.394	409.9	2.38	0.0336	0.719
430	1.2181	128.8	7.2301	1.018	1.393	414.6	2.42	0.0342	0.718
440	1.2465	139.0	7.2535	1.019	1.392	419.3	2.46	0.0348	0.718
450	1.2748	149.2	7.2765	1.021	1.391	423.9	2.50	0.0355	0.718
460	1.3032	159.4	7.2989	1.022	1.390	428.3	2.53	0.0361	0.718
470	1.3315	169.7	7.3209	1.024	1.389	433.0	2.57	0.0367	0.718
480	1.3598	179.9	7.3425	1.026	1.389	437.4	2.61	0.0373	0.718
490	1.3882	190.2	7.3637	1.028	1.388	441.8	2.65	0.0379	0.718
500	1.4165	200.5	7.3845	1.030	1.387	446.1	2.69	0.0385	0.718
520	1.473	221.1	7.4249	1.034	1.385	454.6	2.76	0.0398	0.718
540	1.530	241.8	7.4640	1.038	1.382	462.9	2.83	0.0410	0.718
560	1.586	262.6	7.5018	1.042	1.380	471.0	2.91	0.0422	0.718
580	1.643	283.5	7.5385	1.047	1.378	479.0	2.98	0.0434	0.718
600	1.700	304.5	7.5740	1.051	1.376	486.8	3.04	0.0446	0.718
620	1.756	325.6	7.6086	1.056	1.374	494.4	3.11	0.0458	0.718
640	1.813	346.7	7.6422	1.060	1.371	501.9	3.18	0.0470	0.718
660	1.870	368.0	7.6749	1.065	1.369	509.3	3.25	0.0482	0.717
680	1.926	389.3	7.7067	1.070	1.367	516.5	3.32	0.0495	0.717

TABLE 22.6. (Continued)

T, K	v, m^3/kg	h, kJ/Kg	s, kJ/(kg · K)	c_p, kJ/(kg · K)	γ	$\bar{v}_s$, m/s	μ, 10^{-5} Pa · s	k, W/(m · K)	Pr
700	1.983	410.8	7.7378	1.075	1.364	523.6	3.38	0.0507	0.717
720	2.040	432.3	7.7682	1.080	1.362	530.6	3.45	0.0519	0.716
740	2.096	453.9	7.7978	1.084	1.360	537.5	3.51	0.0531	0.716
760	2.153	475.7	7.8268	1.089	1.358	544.3	3.57	0.0544	0.716
780	2.210	497.5	7.8551	1.094	1.356	551.0	3.64	0.0556	0.716
800	2.266	519.4	7.8829	1.099	1.354	557.6	3.70	0.0568	0.716
820	2.323	541.5	7.9101	1.103	1.352	564.1	3.76	0.0580	0.715
840	2.380	563.6	7.9367	1.108	1.350	570.6	3.82	0.0592	0.715
860	2.436	585.8	7.9628	1.112	1.348	576.8	3.88	0.0603	0.715
880	2.493	608.1	7.9885	1.117	1.346	583.1	3.94	0.0615	0.715
900	2.550	630.4	8.0136	1.121	1.344	589.3	4.00	0.0627	0.715
920	2.606	652.9	8.0383	1.125	1.342	595.4	4.05	0.0639	0.715
940	2.663	675.5	8.0625	1.129	1.341	601.5	4.11	0.0650	0.714
960	2.720	698.1	8.0864	1.133	1.339	607.5	4.17	0.0662	0.714
980	2.776	720.8	8.1098	1.137	1.338	613.4	4.23	0.0673	0.714
1000	2.833	743.6	8.1328	1.141	1.336	619.3	4.28	0.0684	0.714
1050	2.975	800.8	8.1887	1.150	1.333	633.8	4.42	0.0711	0.714
1100	3.116	858.5	8.2423	1.158	1.330	648.0	4.55	0.0738	0.715
1150	3.258	916.6	8.2939	1.165	1.327	661.8	4.68	0.0764	0.715
1200	3.400	975.0	8.3437	1.173	1.324	675.4	4.81	0.0789	0.715
1250	3.541	1033.8	8.3917	1.180	1.322	688.6	4.94	0.0814	0.716
1300	3.683	1093.0	8.4381	1.186	1.319	701.6	5.06	0.0839	0.716
1350	3.825	1152.3	8.4830	1.193	1.317	714.4	5.19	0.0863	0.717
1400	3.966	1212.2	8.5265	1.199	1.315	726.9	5.31	0.0887	0.717
1450	4.108	1272.3	8.5686	1.204	1.313	739.2	5.42	0.0911	0.717
1500	4.249	1332.7	8.6096	1.210	1.311	751.3	5.54	0.0934	0.718
1550	4.391	1393.3	8.6493	1.215	1.309	763.2	5.66	0.0958	0.718
1600	4.533	1454.2	8.6880	1.220	1.308	775.0	5.77	0.0981	0.717
1650	4.674	1515.3	8.7256	1.225	1.306	786.5	5.88	0.1004	0.717
1700	4.816	1576.7	8.7622	1.229	1.305	797.9	5.99	0.1027	0.717
1750	4.958	1638.2	8.7979	1.233	1.303	809.1	6.10	0.1050	0.717
1800	5.099	1700.0	8.8327	1.237	1.302	820.2	6.21	0.1072	0.717
1850	5.241	1762.0	8.8667	1.241	1.301	831.1	6.32	0.1094	0.717
1900	5.383	1824.1	8.8998	1.245	1.300	841.9	6.43	0.1116	0.717
1950	5.524	1886.4	8.9322	1.248	1.299	852.6	6.53	0.1138	0.717
2000	5.666	1948.9	8.9638	1.252	1.298	863.1	6.64	0.1159	0.717
2050	5.808	2011.6	8.9948	1.255	1.297	873.5	6.74	0.1180	0.717
2100	5.949	2074.4	9.0251	1.258	1.296	883.8	6.84	0.1200	0.717
2150	6.091	2137.3	9.0547	1.260	1.295	894.0	6.95	0.1220	0.717
2200	6.232	2200.4	9.0837	1.263	1.294	904.0	7.05	0.1240	0.718
2250	6.374	2263.6	9.1121	1.265	1.293	914.0	7.15	0.1260	0.718
2300	6.516	2327.0	9.1399	1.268	1.293	923.8	7.25	0.1279	0.718
2350	6.657	2390.5	9.1672	1.270	1.292	933.5	7.35	0.1298	0.719
2400	6.800	2454.0	9.1940	1.273	1.291	943.2	7.44	0.1317	0.719
2450	6.940	2517.7	9.2203	1.275	1.291	952.7	7.54	0.1336	0.720
2500	7.082	2581.5	9.2460	1.277	1.290	962.2	7.64	0.1354	0.720

[a] v = specific volume; h = specific enthalpy; s = specific entropy; c_p = specific heat at constant pressure; γ = specific-heat ratio c_p/c_v (dimensionless); $\bar{v}_s$ = velocity of sound; μ = dynamic viscosity; k = thermal conductivity; Pr = Prandtl number (dimensionless). Condensed from S. Gordon, Thermodynamic and Transport Combustion Properties of Hydrocarbons with Air, NASA Technical Paper 1906, 1982, Vol. 1. These properties are based on constant gaseous composition. The reader is reminded that, at the higher temperatures, the pressure can affect the composition and the thermodynamic properties.

TABLE 22.7. Thermophysical Properties of the U.S. Standard Atmosphere[a]

Z, m	H, m	T, K	P, bar	ρ, kg/m^3	g, m/s^2	$\bar{v}_s$, m/s
0	0	288.15	1.0133	1.2250	9.8067	340.3
1000	1000	281.65	0.8988	1.1117	9.8036	336.4
2000	1999	275.15	0.7950	1.0066	9.8005	332.5
3000	2999	268.66	0.7012	0.9093	9.7974	328.6
4000	3997	262.17	0.6166	0.8194	9.7943	324.6
5000	4996	255.68	0.5405	0.7364	9.7912	320.6
6000	5994	249.19	0.4722	0.6601	9.7882	316.5
7000	6992	242.70	0.4111	0.5900	9.7851	312.3
8000	7990	236.22	0.3565	0.5258	9.7820	308.1
9000	8987	229.73	0.3080	0.4671	9.7789	303.9
10000	9984	223.25	0.2650	0.4135	9.7759	299.5
11000	10981	216.77	0.2270	0.3648	9.7728	295.2
12000	11977	216.65	0.1940	0.3119	9.7697	295.1
13000	12973	216.65	0.1658	0.2667	9.7667	295.1
14000	13969	216.65	0.1417	0.2279	9.7636	295.1
15000	14965	216.65	0.1211	0.1948	9.7605	295.1
16000	15960	216.65	0.1035	0.1665	9.7575	295.1
17000	16954	216.65	0.0885	0.1423	9.7544	295.1
18000	17949	216.65	0.0756	0.1217	9.7513	295.1
19000	18943	216.65	0.0647	0.1040	9.7483	295.1
20000	19937	216.65	0.0553	0.0889	9.7452	295.1
22000	21924	218.57	0.0405	0.0645	9.7391	296.4
24000	23910	220.56	0.0297	0.0469	9.7330	297.4
26000	25894	222.54	0.0219	0.0343	9.7269	299.1
28000	27877	224.53	0.0162	0.0251	9.7208	300.4
30000	29859	226.51	0.0120	0.0184	9.7147	301.7
32000	31840	227.49	0.00889	0.01356	9.7087	303.0
34000	33819	233.74	0.00663	0.00989	9.7026	306.5
36000	35797	239.28	0.00499	0.00726	9.6965	310.1
38000	37774	244.82	0.00377	0.00537	9.6904	313.7
40000	39750	250.35	0.00287	0.00400	9.6844	317.2
42000	41724	255.88	0.00220	0.00299	9.6783	320.7
44000	43698	261.40	0.00169	0.00259	9.6723	324.1
46000	45669	266.93	0.00131	0.00171	9.6662	327.5
48000	47640	270.65	0.00102	0.00132	9.6602	329.8
50000	49610	270.65	0.00080	0.00103	9.6542	329.8

[a] Z = geometric attitude; H = geopotential attitude; ρ = density; g = acceleration of gravity; $\bar{v}_s$ = velocity of sound. Condensed and in some cases converted from *U.S. Standard Atmosphere 1976*, National Oceanic and Atmospheric Administration and National Aeronautics and Space Administration, Washington. Also available as NOAA-S/T 76-1562 and Government Printing Office Stock No. 003-017-00323-0.

TABLE 22.8. Thermophysical Properties of Condensed and Saturated-Vapor Carbon Dioxide from 200 K to the Critical Point

Temp.	Absolute Pressure	Specific Volume, m^3/kg		Specific Enthalpy, kJ/kg		Specific Entropy, kJ/(kg · K)		Specific Heat c_p kJ/(kg · K)		Thermal Conductivity, W/(m · K)		Viscosity 10^{-4} Pa · s		Prandtl Number	
T, K	P, bar	Condensed[a]	Vapor	Condensed[a]	Vapor	Consensed[a]	Vapor	Condensed[a]	Vapor	Liquid	Vapor	Liquid	Vapor	Liquid	Vapor
200	1.544	0.000644	0.2362	164.8	728.3	1.620	4.439								
205	2.277	0.000649	0.1622	171.5	730.0	1.652	4.379								
210	3.280	0.000654	0.1135	178.2	730.9	1.682	4.319								
215	4.658	0.000659	0.0804	185.0	731.3	1.721	4.264								
216.6	5.180	0.000661	0.0718	187.2	731.5	1.736	4.250								
216.6	5.180	0.000848	0.0718	386.3	731.5	2.656	4.250	1.707	0.958	0.182	0.011	2.10	0.116	1.96	0.96
220	5.996	0.000857	0.0624	392.6	733.1	2.684	4.232	1.761	0.985	0.178	0.012	1.86	0.118	1.93	0.97
225	7.357	0.000871	0.0515	401.8	735.1	2.723	4.204	1.820	1.02	0.171	0.012	1.75	0.120	1.87	0.98
230	8.935	0.000886	0.0428	411.1	736.7	2.763	4.178	1.879	1.06	0.164	0.013	1.64	0.122	1.84	0.99
235	10.75	0.000901	0.0357	420.5	737.9	2.802	4.152	1.906	1.10	0.160	0.013	1.54	0.125	1.82	1.01
240	12.83	0.000918	0.0300	430.2	738.9	2.842	4.128	1.933	1.15	0.156	0.014	1.45	0.128	1.80	1.02
245	15.19	0.000936	0.0253	440.1	739.4	2.882	4.103	1.959	1.20	0.148	0.015	1.36	0.131	1.80	1.04
250	17.86	0.000955	0.0214	450.3	739.6	2.923	4.079	1.992	1.26	0.140	0.016	1.28	0.134	1.82	1.06
255	20.85	0.000977	0.0182	460.8	739.4	2.964	4.056	2.038	1.34	0.134	0.017	1.21	0.137	1.84	1.08
260	24.19	0.001000	0.0155	471.6	738.7	3.005	4.032	2.125	1.43	0.128	0.018	1.14	0.140	1.89	1.12
265	27.89	0.001026	0.0132	482.8	737.4	3.047	4.007	2.237	1.54	0.122	0.019	1.08	0.144	1.98	1.17
270	32.03	0.001056	0.0113	494.4	735.6	3.089	3.981	2.410	1.66	0.116	0.020	1.02	0.150	2.12	1.23
275	36.59	0.001091	0.0097	506.5	732.8	3.132	3.954	2.634	1.81	0.109	0.022	0.96	0.157	2.32	1.32
280	41.60	0.001130	0.0082	519.2	729.1	3.176	3.925	2.887	2.06	0.102	0.024	0.91	0.167	2.57	1.44
285	47.10	0.001176	0.0070	532.7	723.5	3.220	3.891	3.203	2.40	0.095	0.028	0.86	0.178	2.90	1.56
290	53.15	0.001241	0.0058	547.6	716.9	3.271	3.854	3.724	2.90	0.088	0.033	0.79	0.191	3.35	1.68
295	59.83	0.001322	0.0047	562.9	706.3	3.317	3.803	4.68		0.081	0.042	0.71	0.207	4.1	1.8
300	67.10	0.001470	0.0037	585.4	690.2	3.393	3.742			0.074	0.065	0.60	0.226		
304.2[b]	73.83	0.002145	0.0021	636.6	636.6	3.558	3.558								

[a] Above the solid line, the condensed phase is solid; below the line, it is liquid.

[b] Critical point.

TABLE 22.9. Thermophysical Properties of Gaseous Carbon Dioxide at 1-Bar Pressure[a]

T, K	v, m^3/kg	h, kJ/kg	s, kJ/(kg · K)	c_p, kJ/(kg · K)	k, W/(m · K)	μ, 10^{-4} Pa · s	Pr
300	0.5639	809.3	4.860	0.852	0.0166	0.151	0.778
350	0.6595	853.1	4.996	0.898	0.0204	0.175	0.770
400	0.7543	899.1	5.118	0.941	0.0243	0.198	0.767
450	0.8494	947.1	5.231	0.980	0.0283	0.220	0.762
500	0.9439	997.0	5.337	1.014	0.0325	0.242	0.755
550	1.039	1049	5.435	1.046	0.0364	0.261	0.750
600	1.133	1102	5.527	1.075	0.0407	0.281	0.742
650	1.228	1156	5.615	1.102	0.0445	0.299	0.742
700	1.332	1212	5.697	1.126	0.0481	0.317	0.742
750	1.417	1269	5.775	1.148	0.0517	0.334	0.742
800	1.512	1327	5.850	1.168	0.0551	0.350	0.742
850	1.606	1386	5.922	1.187	0.0585	0.366	0.742
900	1.701	1445	5.990	1.205	0.0618	0.381	0.742
950	1.795	1506	6.055	1.220	0.0650	0.396	0.743
1000	1.889	1567	6.120	1.234	0.0682	0.410	0.743

[a] v = Specific volume; h = enthalpy; s = entropy; c_p = specific heat at constant pressure; k = thermal conductivity; μ = viscosity; Pr = Prandtl number.

TABLE 22.10. Thermophysical Properties of Saturated Cesium[a]

T, K	P, bar	v_l, 10^{-4} m³/kg	v_g, m³/kg	h_l, kJ/kg	h_g, kJ/kg	s_l, kJ/(kg · K)	s_g, kJ/(kg · K)
301.6	2.66. − 9	5.44		74.6		0.696	
350	1.59. − 7	5.52		86.4		0.731	
400	3.83. − 6	5.62	65400	98.5	651.9	0.765	2.148
450	4.44. − 5	5.71	6330	110.4	659.2	0.793	2.012
500	3.11. − 4	5.80	1001	122.0	666.1	0.817	1.905
600	5.65. − 3	6.00	65.5	144.9	678.4	0.859	1.748
700	4.40. − 2	6.22	9.67	167.0	688.9	0.893	1.638
800	0.2029	6.44	2.35	188.7	698.3	0.922	1.559
900	0.6622	6.69	0.796	210.6	707.3	0.948	1.500
1000	1.693	6.95	0.335	233.2	716.4	0.972	1.455
1100	3.629	7.26	0.169	256.7	725.9	0.994	1.420
1200	6.790	7.63	0.097	281.1	736.1	1.015	1.394
1300	11.41	8.04	0.061	306.2	747.1	1.035	1.374
1400	18.7	8.47	0.040	332.0	759.0	1.053	1.358
1500	27.6	8.91	0.029	358.5	771.9	1.068	1.344

T, K	$c_{p,l}$, kJ/(kg · K)	$c_{p,g}$, kJ/(kg · K)	μ_l, 10^{-4} Pa · s	μ_g, 10^{-4} Pa · s	k_l, W/(m · K)	k_g, W/(m · K)	Pr_l	Pr_g
301.6	0.245	0.158	6.82		19.7		0.0085	
350	0.243	0.164	5.23		20.0		0.0064	
400	0.240	0.170	4.25		20.3		0.0050	
450	0.236	0.182	3.62		20.4		0.0042	
500	0.232	0.198	3.18		20.5		0.0036	
600	0.224	0.234	2.54		20.5		0.0028	
700	0.219	0.265	2.15		20.1		0.0023	
800	0.217	0.282	1.86		19.4		0.0021	
900	0.222	0.288	1.71	0.21	19.0	0.0081	0.0020	
1000	0.231	0.285	1.51	0.22	17.5	0.0088	0.0020	0.713
1100	0.239	0.278	1.42	0.23	16.3	0.0094	0.0021	0.680
1200	0.248	0.269	1.32	0.24	15.0	0.0100	0.0022	0.646
1300	0.256	0.258	1.24	0.25	13.6	0.0106	0.0023	0.608
1400	0.263	0.245	1.17	0.26	12.2	0.0111	0.0025	0.572
1500	0.269	0.230	1.11	0.27	10.8	0.0115	0.0028	0.538

[a] v = specific volume; h = specific enthalpy; s = specific entropy; c_p = specific heat at constant pressure; μ = dynamic viscosity; k = thermal conductivity; Pr = Prandtl number; l = saturated liquid; g = saturated vapor. The notation 2.66. − 9 signifies 2.66×10^{-9}.

TABLE 22.11. Thermophysical Properties of Gaseous Cesium at 1-Bar Pressure[a]

T, K	943	1000	1100	1200	1300	1400	1500
v, m^3/kg	0.547	0.592	0.665	0.734	0.801	0.866	0.931
h, kJ/kg	711	727	749	769	787	804	821
s, kJ/(kg · K)	1.479	1.496	1.517	1.534	1.549	1.562	1.573
c_p, kJ/(kg · K)	0.287	0.249	0.209	0.188	0.176	0.169	0.165
c_v, kJ/(kg · K)	0.184	0.157	0.129	0.113	0.105	0.101	0.099
γ	1.566	1.634	1.653	1.666	1.672	1.674	1.672
Z	1.00	1.0005	1.0006	1.0007	1.0007	1.0008	1.0008
$\bar{v}_s$, m/s	284.0	311.1	331.6	349.7	365.9	380.9	394.6
k, W/(m · K)	0.00824	0.00828	0.00836				
μ, 10^{-4} Pa · s	0.215	0.232	0.258				
Pr	0.749	0.698	0.645				

[a] v = specific volume; h = specific enthalpy; s = specific entropy; c_p = specific heat at constant pressure; c_v = specific heat at constant volume; γ = ratio of principal specific heats = c_p/c_v; Z = compressibility factor = Pv/RT; $\bar{v}_s$ = velocity of sound; k = thermal conductivity; μ = viscosity; Pr = Prandtl number.

TABLE 22.12. Thermophysical Properties of *n*-Hydrogen at Atmospheric Pressure[a]

T, K	v, m^3/kg	h, kJ/kg	s, kJ/(kg · K)	c_p, kJ/(kg · K)	Z,	$\bar{v}_s$, m/s	k, W/(m · K)	μ, 10^{-4} Pa · s	Pr
250	10.183	3517	67.98	14.04	1.000	1209	0.162	0.079	0.685
300	12.218	4227	70.58	14.31	1.000	1319	0.187	0.089	0.685
350	14.253	4945	72.79	14.43	1.000	1423	0.210	0.099	0.685
400	16.289	5669	74.72	14.48	1.000	1520	0.230	0.109	0.684
450	18.324	6393	76.43	14.50	1.000	1611	0.250	0.118	0.684
500	20.359	7118	77.96	14.51	1.000	1698	0.269	0.127	0.684
550	22.39	7844	79.34	14.53	1.000	1780	0.287	0.135	0.684
600	24.48	8571	80.60	14.55	1.000	1859	0.305	0.143	0.684
650	26.47	9299	81.76	14.57	1.000	1934	0.323	0.151	0.684
700	28.50	10029	82.85	14.60	1.000	2006	0.340	0.159	0.684

[a] v = specific volume; h = specific enthalpy; s = specific entropy; c_p = specific heat at constant pressure; Z = compressibility factor = Pv/RT; $\bar{v}_s$ = velocity of sound; k = thermal conductivity; μ = dynamic viscosity; Pr = Prandtl number.

TABLE 22.13. Thermophysical Properties of Saturated Lithium[a]

T, K	P, bar	v_l, m^3/kg	v_g, m^3/kg	h_l, kJ/kg	h_g, kJ/kg	s_l, kJ/(kg · K)	s_g, kJ/(kg · K)
453.7	1.78. − 13	0.00191		1703	24259	6.78	56.49
500	8.21. − 12	0.00195		1905	24390	7.20	52.17
600	4.18. − 9	0.00199		2334	24674	7.98	45.22
700	3.51. − 7	0.00203		2697	24869	8.63	40.31
800	9.57. − 6	0.00207	994000	3174	25162	9.19	36.68
900	0.000124	0.00211	85540	3590	25341	9.68	33.85
1000	0.000960	0.00216	12175	4006	25477	10.12	31.59
1100	0.00509	0.00221	2494	4421	25578	10.52	29.75
1200	0.0204	0.00226	669	4835	25654	10.88	28.23
1300	0.0658	0.00231	232	5251	25717	11.21	26.95
1400	0.1794	0.00237	86	5668	25778	11.52	25.88
1500	0.4269	0.00243	38	6088	25845	11.81	24.98

T, K	$c_{p,l}$, kJ/(kg · K)	$c_{p,g}$, kJ/(kg · K)	μ_l, 10^{-4} Pa · s	μ_g, 10^{-4} Pa · s	k_l, W/(m · K)	k_g, W/(m · K)	Pr_l	Pr_g
453.7	4.30		6.00		42.8		0.0603	
500	4.34		5.31		44.3		0.0520	
600	4.23		4.26		47.6		0.0379	
700	4.19		3.58		50.9		0.0295	
800	4.16		3.10		54.1		0.0238	
900	4.16	6.96	2.75	0.116	57.2	0.0704	0.0200	1.16
1000	4.16	8.17	2.47	0.121	60.0	0.0806	0.0171	1.22
1100	4.15	9.11	2.25	0.124	62.5	0.0894	0.0149	1.26
1200	4.14	9.72	2.07	0.126	64.7	0.0968	0.0132	1.28
1300	4.16	10.02	1.92	0.129	66.5	0.1024	0.0120	1.30
1400	4.19	10.05	1.80	0.139	68.0	0.1065	0.0111	1.33
1500	4.20	9.89	1.69	0.150	69.1	0.1092	0.0103	1.35

[a] v = specific volume; h = specific enthalpy; s = specific entropy; c_p = specific heat at constant pressure; μ = dynamic viscosity; k = thermal conductivity; Pr = Prandtl number; l = saturated liquid; g = saturated vapor. The notation 1.78. − 13 signifies 1.78×10^{-13}.

TABLE 22.14. Thermophysical Properties of Lithium at 1-Bar Pressure[a]

T, K	1615	1700	1800	1900	2000
v, m^3/kg	16.91	18.77	20.40	21.90	23.32
h, kJ/kg	25934	26682	27379	27946	28425
s, kJ/(kg · K)	24.11	24.57	24.97	25.28	25.52
c_p, kJ/(kg · K)	9.561	7.794	6.234	5.173	4.469
c_v, kJ/(kg · K)	6.627	5.176	4.091	3.342	2.839
γ	1.443	1.506	1.524	1.548	1.574
Z	0.981	1.001	1.001	1.001	1.001
$\bar{v}_s$, m/s	1514	1681	1763	1841	1916
μ, 10^{-4} Pa · s	0.161	0.171	10.181	0.192	0.202
k, W/(m · K)	0.193	0.198	0.202	0.207	0.209
Pr	0.798	0.673	0.559	0.480	0.432

[a] v = specific volume; h = specific enthalpy; s = specific entropy; c_p = specific heat at constant pressure; c_v = specific heat at constant volume; γ = ratio of principal specific heats = c_p/c_v; Z = compressibility factor = Pv/RT; $\bar{v}_s$ = velocity of sound; μ = dynamic viscosity; k = thermal conductivity; Pr = Prandtl number.

TABLE 22.15. Thermophysical Properties of Saturated Mercury[a]

T, K	P, bar	v_l,[b] 10^{-5} m^3/kg	v_g, m^3/kg	h_l,[b] kJ/kg	h_g, kJ/kg	s_l,[b] kJ/(kg · K)	s_g, kJ/(kg · k)	$c_{p,l}$,[b] kJ/(kg · K)	$c_{p,g}$, kJ/(kg · K)
200	1.9. − 12			21.41	342.50	0.2740	1.8794	0.1360	0.1036
220	7.73. − 11			24.16	344.52	0.2871	1.7433	0.1399	0.1036
234.3	7.330. − 10		3.450. + 7	26.17	345.88	0.2960	1.6607	0.1420	0.1036
234.3	7.330. − 10	7.3038	3.450. + 7	37.58	345.88	0.3447	1.6607	0.1421	0.1036
240	1.668. − 9	7.3114	2.898. + 7	38.40	346.46	0.3481	1.6322	0.1419	0.1036
260	6.925. − 8	7.3381	1.801. + 6	41.22	348.53	0.3595	1.5418	0.1410	0.1036
280	5.296. − 7	7.3648	2.440. + 5	44.03	350.60	0.3699	1.4651	0.1401	0.1036
300	3.075. − 6	7.3915	4.383. + 4	46.83	352.67	0.3795	1.4200	0.1393	0.1036
320	1.428. − 5	7.4183	9.878. + 3	49.61	354.75	0.3885	1.3603	0.1386	0.1036
340	5.516. − 5	7.4451	2.679. + 3	52.38	356.82	0.3969	1.2925	0.1380	0.1036
360	1.829. − 4	7.4721	847	55.13	358.89	0.4048	1.2487	0.1375	0.1036
380	5.289. − 4	7.4991	304	57.87	360.96	0.4122	1.2099	0.1370	0.1036
400	1.394. − 3	7.5262	120	60.61	363.04	0.4192	1.1754	0.1366	0.1036
450	1.053. − 2	7.5946	18.0	67.41	368.21	0.4352	1.1037	0.1357	0.1036
500	5.261. − 2	7.6640	3.98	74.19	373.38	0.4495	1.0479	0.1353	0.1037
550	0.1949	7.7347	1.176	80.95	378.53	0.4624	1.0035	0.1352	0.1038
600	0.5776	7.8069	0.4318	87.72	383.64	0.4742	0.9674	0.1355	0.1040
650	1.4425	7.8810	0.1867	94.51	388.68	0.4850	0.9376	0.1360	0.1044
700	3.153	7.9573	0.0917	101.54	393.62	0.4951	0.9127	0.1369	0.1049
750	6.197	8.0362	0.0497	108.24	398.41	0.5046	0.8915	0.1382	0.1057
800	11.18	8.115	0.0292	115.23	403.04	0.5136	0.8734	0.1398	0.1067
850	18.82	8.203	0.0183	122.31	407.44	0.5221	0.8576	0.1417	0.1081
900	29.88	8.292	0.0121	129.53	411.61	0.5302	0.8437	0.1439	0.1099
950	45.23	8.385	0.0083	136.89	415.25	0.5381	0.8313	0.1464	0.1120
1000	65.74	8.482	0.0060	144.41	419.08	0.5456	0.8203	0.1493	0.1145

TABLE 22.15. (Continued)

T, K	μ_l, Pa · s	μ_g, Pa · s	k_l, W/(m · K)	k_g, W/(m · K)	Pr_l	Pr_g
234.3	20.51	0.217	7.64	0.0034	0.0381	0.568
240	19.85	0.222	7.72	0.0034	0.0365	0.568
260	17.91	0.240	8.00	0.0037	0.0316	0.567
280	16.39	0.259	8.27	0.0040	0.0278	0.567
300	15.21	0.277	8.54	0.0043	0.0248	0.567
320	14.24	0.295	8.80	0.0046	0.0224	0.567
340	13.44	0.314	9.06	0.0049	0.0205	0.567
360	12.77	0.332	9.31	0.0052	0.0189	0.666
380	12.19	0.350	9.56	0.0054	0.0175	0.666
400	11.73	0.370	9.80	0.0058	0.0163	0.666
450	10.74	0.419	10.38	0.0065	0.0140	0.667
500	10.08	0.470	10.93	0.0073	0.0125	0.667
550	9.51	0.523	11.45	0.0081	0.0112	0.667
600	9.10	0.577	11.94	0.0090	0.0100	0.669
650	8.76	0.631	12.39	0.0098	0.0096	0.671
700	8.49	0.687	12.82	0.0107	0.0091	0.675
750	8.26	0.742	13.21	0.0115	0.0086	0.680
800	8.08	0.798	13.57	0.0124	0.0083	0.687
850	7.92	0.854	13.90	0.0133	0.0081	0.696
900	7.78	0.909	14.20	0.0142	0.0079	0.705
950	7.65	0.964	14.46	0.0150	0.0077	0.719
1000	7.54	1.019	14.69	0.0158	0.0076	0.737

[a] v = specific volume; h = specific enthalpy; s = specific entropy; c_p = specific heat at constant pressure; μ = dynamic viscosity; k = thermal conductivity; Pr = Prandtl number; l = saturated liquid; g = saturated vapor. The notation 1.9. − 12 signifies 1.9×10^{-12}.

[b] Above the solid line, solid phase; below the line, liquid.

TABLE 22.16. Thermophysical Properties of Mercury at 1-Bar Pressure[a]

T, K	v, m^3/kg	h, kJ/kg	s, kJ/(kg · K)	c_p, kJ/(kg ·)	c_v, kJ/(kg · K)	γ	Z	$\bar{v}_s$, m/s	μ, Pa · s	k, W/(m · K)	Pr
630.1	0.26043	386.08	0.94881	0.1042	0.0623	1.672	0.9972	208.7	6.19. − 5	0.0104	0.620
650	0.26874	388.76	0.95285	0.1042	0.0623	1.671	0.9976	211.6	6.39. − 5	0.0108	0.617
700	0.28955	393.96	0.96055	0.1040	0.0623	1.670	0.9980	219.6	6.89. − 5	0.0117	0.612
750	0.31033	399.16	0.96773	0.1040	0.0623	1.670	0.9984	227.3	7.39. − 5	0.0127	0.605
800	0.33112	404.36	0.97444	0.1039	0.0622	1.669	0.9987	234.8	7.89. − 5	0.0136	0.603
850	0.35188	409.55	0.98073	0.1038	0.0622	1.668	0.9988	242.1	8.39. − 5	0.0145	0.601
900	0.37265	414.74	0.98667	0.1038	0.0622	1.668	0.9990	249.2	8.88. − 5	0.0154	0.599
950	0.39341	419.93	0.99228	0.1038	0.0622	1.668	0.9992	256.0	9.39. − 5	0.0164	0.594
1000	0.41416	425.12	0.99769	0.1038	0.0622	1.668	0.9993	263.0	9.87. − 5	0.0173	0.592
1100	0.45566	435.50	1.00742	0.1037	0.0622	1.667	0.9995	275.5	1.08. − 4	0.0189	0.594
1200	0.49715	445.87	1.01646	0.1037	0.0622	1.667	0.9996	287.8			
1300	0.53863	456.24	1.02469	0.1037	0.0622	1.667	0.9997	299.6			
1400	0.58010	466.61	1.03226	0.1037	0.0622	1.667	0.9998	311.0			
1500	0.62157	476.97	1.03956	0.1037	0.0622	1.668	0.9998	321.9			
1600	0.66304	487.34	1.04616	0.1037	0.0622	1.668	0.9999	332.5			
1800	0.74598	508.07	1.05841	0.1037	0.0622	1.668	0.9999	352.6			
2000	0.82890	528.80	1.06936	0.1037	0.0622	1.668	1.0000	371.7			

[a] v = specific volume; h = specific enthalpy; s = specific entropy; c_p = specific heat at constant pressure; c_v = specific heat at constant volume; γ = ratio of principal specific heats = c_p/c_v; Z = compressibility factor = Pv/RT; μ = dynamic viscosity; k = thermal conductivity; Pr = Prandtl number. The notation 6.19. − 5 signifies 6.19×10^{-5}.

TABLE 22.17. Thermophysical Properties of Nitrogen at Atmospheric Pressure[a]

T, K	v, m^3/kg	h, kJ/kg	s, kJ/(kg · K)	c_p, kJ/(kg · K)	Z	$\bar{v}_s$, m/s	k, W/(m · K)	μ, 10^{-4} Pa · s	Pr
250	0.7317	259.1	6.650	1.042	0.9992	322	0.0223	0.155	0.724
300	0.8786	311.2	6.840	1.041	0.9998	353	0.0259	0.178	0.715
350	1.025	363.3	7.001	1.042	0.9998	382	0.0292	0.200	0.713
400	1.171	415.4	7.140	1.045	0.9999	407	0.0324	0.220	0.710
450	1.319	467.8	7.263	1.050	1.0000	432	0.0366	0.240	0.708
500	1.465	520.4	7.374	1.056	1.0002	455	0.0386	0.258	0.706
550	1.612	573.4	7.475	1.065	1.0002	476	0.0414	0.275	0.707
600	1.758	626.9	7.568	1.075	1.0003	496	0.0442	0.291	0.708
650	1.905	681.0	7.655	1.086	1.0003	515	0.0470	0.306	0.709
700	2.051	735.6	7.736	1.098	1.0003	534	0.0496	0.321	0.711

[a] v = specific volume; h = specific enthalpy; s = specific entropy; c_p = specific heat at constant pressure; Z = compressibility factor = Pv/RT; v_s = velocity of sound; k = thermal conductivity; μ = dynamic viscosity; Pr = Prandtl number.

TABLE 22.18. Thermophysical Properties of Oxygen at Atmospheric Pressure[a]

T, K	v, m^3/kg	h, kJ/kg	s, kJ(kg · K)	c_p, kJ/(kg · K)	Z	k, W/(m · K)	μ, 10^{-4} Pa · s	Pr
250	0.6402	226.9	6.247	0.915	0.9987	0.0226	0.179	0.725
300	0.7688	272.7	6.414	0.920	0.9994	0.0266	0.207	0.716
350	0.9790	318.9	6.557	0.929	0.9996	0.0305	0.234	0.713
400	1.025	365.7	6.682	0.942	0.9998	0.0343	0.258	0.710
450	1.154	413.1	6.973	0.956	1.0000	0.0380	0.281	0.708
500	1.282	461.3	6.895	0.972	1.0000	0.0416	0.303	0.707
550	1.410	510.3	6.988	0.988	1.0001	0.0451	0.324	0.708
600	1.539	560.1	7.075	1.003	1.0002	0.0487	0.344	0.708
650	1.667	610.6	7.156	1.018	1.0002	0.0521	0.363	0.709
700	1.795	661.9	7.232	1.031	1.0002	0.0554	0.381	0.710

[a] v = specific volume; h = specific enthalpy; s = specific entropy; c_p = specific heat at constant pressure; Z = compressibility factor = Pv/RT; k = thermal conductivity; μ = dynamic viscosity; Pr = Prandtl number.

TABLE 22.19. Thermophysical Properties of Saturated Potassium[a]

T, K	P, bar	v_l, m³/kg	v_g, m³/kg	h_l, kJ/kg	h_g, kJ/kg	s_l, kJ/(kg · K)	s_g, kJ/(kg · K)	$c_{p,l}$, kJ/(kg · K)	$c_{p,g}$, kJ/(kg · K)
336.4	1.37. − 9	1.208. − 3		94		1.928		0.822	
350	3.82. − 8	1.213. − 3		105	2307	1.961	8.251	0.818	
400	1.84. − 7	1.229. − 3	4.63. + 6	146	2342	2.068	7.559	0.805	0.532
450	3.21. − 6	1.247. − 3	2.98. + 5	186	2366	2.163	7.009	0.794	0.610
500	3.13. − 5	1.266. − 3	33890	225	2390	2.246	6.576	0.785	0.670
600	9.26. − 4	1.304. − 3	1364	303	2433	2.388	5.937	0.771	0.819
700	0.0102	1.346. − 3	142	379	2468	2.506	5.490	0.762	0.965
800	0.0612	1.389. − 3	26.7	456	2498	2.608	5.161	0.761	1.066
900	0.2441	1.437. − 3	7.402	532	2525	2.697	4.912	0.769	1.116
1000	0.7322	1.488. − 3	2.691	610	2552	2.780	4.722	0.792	1.121
1100	1.864	1.546. − 3	1.143	690	2580	2.856	4.574	0.819	1.100
1200	3.963	1.605. − 3	0.584	774	2610	2.929	4.459	0.846	1.064
1300	7.304	1.672. − 3	0.335	860	2643	2.998	4.370	0.873	1.022
1400	12.44	1.742. − 3	0.207	948	2679	3.063	4.299	0.899	0.980
1500	20.0	1.816. − 3	0.136	1042	2715	3.126	4.240	0.924	0.938

T, K	μ_l, 10^{-4} Pa · s	μ_g, 10^{-4} Pa · s	k_l, W/(m · K)	k_g, W/(m · K)	$\Pr_l$	$\Pr_g$
336.4	5.57		54.8		0.00836	
350	5.18		54.2		0.00782	
400	4.13		52.0		0.00639	
450	3.46		49.9		0.00551	
500	3.01		47.9		0.00493	
600	2.38		43.9		0.00418	
700	1.98	0.121	40.4	0.0142	0.00373	0.822
800	1.71	0.134	37.1	0.0175	0.00351	0.813
900	1.51	0.148	34.0	0.0205	0.00342	0.806
1000	1.35	0.163	31.3	0.0228	0.00341	0.801
1100	1.23	0.178	28.7	0.0248	0.00351	0.790
1200	1.14	0.196	26.3	0.0266	0.00367	0.784
1300	1.05	0.212	23.9	0.0280	0.00384	0.774
1400	0.98	0.229	21.5	0.0293	0.00410	0.763
1500	0.93	0.242	19.2	0.0303	0.00448	0.750

[a] v = specific volume; h = specific enthalpy; s = specific entropy; c_p = specific heat at constant pressure; μ = dynamic viscosity; k = thermal conductivity; Pr = Prandtl number. The notation 1.37. − 9 signifies 1.37×10^{-9}.

TABLE 22.20. Thermophysical Properties of Gaseous Potassium at 1-Bar Pressure[a]

T, K	1032	1100	1200	1300	1400	1500
v, m^3/kg	1.9960	2.208	2.462	2.701	2.931	3.155
h, kJ/kg	2561	2630	2714	2785	2849	2909
s, kJ/(kg · K)	4.671	4.734	4.807	4.864	4.912	4.953
c_p, kJ/(kg · K)	1.116	0.934	0.763	0.669	0.617	0.587
c_v, kJ/(kg · K)	0.730	0.583	0.470	0.407	0.372	0.353
γ	1.529	1.600	1.625	1.645	1.658	1.665
Z	1.0006	1.0006	1.0006	1.0005	1.0005	1.0005
$\bar{v}_s$, m/s	538.3	594.4	632.5	666.6	697.0	724.7
μ, 10^{-4} Pa · s	0.167	0.184	0.207			
k, W/(m · k)	0.0234	0.0218	0.0211			
Pr	0.796	0.788	0.749			

[a] v = specific volume; h = specific enthalpy; s = specific entropy; c_p = specific heat at constant pressure; c_v = specific heat at constant volume; γ = ratio of principal specific heats = c_p/c_v; Z = compressibility factor = Pv/RT; $\bar{v}_s$ = velocity of sound; μ = dynamic viscosity; k = thermal conductivity; Pr = Prandtl number.

TABLE 22.21. Thermophysical Properties of Saturated Refrigerant 12[a]

p bar	T K	v_l, 10^{-4} m^3/kg	v_g, m^3/kg	h_l, kJ/kg	h_g, kJ/kg	s_l, kJ/(kg · K)	s_g, kJ/(kg · K)
0.10	200.1	6.217	1.365	334.8	518.1	3.724	4.640
0.15	206.3	6.282	0.936	340.1	521.0	3.750	4.627
0.20	211.1	6.332	0.716	344.1	523.2	3.769	4.618
0.25	214.9	6.374	0.582	347.4	525.0	3.785	4.611
0.30	218.2	6.411	0.491	350.2	526.5	3.798	4.606
0.4	223.5	6.437	0.376	354.9	529.1	3.819	4.598
0.5	227.9	6.525	0.306	358.8	531.2	3.836	4.592
0.6	231.7	6.570	0.254	362.1	532.9	3.850	4.588
0.8	237.9	6.648	0.198	367.6	535.8	3.874	4.581
1.0	243.0	6.719	0.160	372.1	538.2	3.893	4.576
1.5	253.0	6.859	0.110	381.2	542.9	3.929	4.568
2.0	260.6	6.970	0.0840	388.2	546.4	3.956	4.563
2.5	266.9	7.067	0.0681	394.0	549.2	3.978	4.560
3.0	272.3	7.183	0.0573	399.1	551.6	3.997	4.557
4.0	281.3	7.307	0.0435	407.6	555.6	4.027	4.553
5.0	288.8	7.444	0.0351	414.8	558.8	4.052	4.551
6.0	295.2	7.571	0.0294	421.1	561.5	4.073	4.549
8.0	306.0	7.804	0.0221	431.8	565.7	4.108	4.546
10	314.9	8.022	0.0176	440.8	569.0	4.137	4.544
15	332.6	8.548	0.0114	459.3	574.5	4.193	4.539
20	346.3	9.096	0.0082	474.8	577.5	4.237	4.534
25	357.5	9.715	0.0062	488.7	578.5	4.275	4.527
30	367.2	10.47	0.0048	502.0	577.6	4.311	4.517
35	375.7	11.49	0.0036	515.9	574.1	4.347	4.502
40	383.3	13.45	0.0025	532.7	564.1	4.389	4.471
41.2[b]	385.0	17.92	0.0018	548.3	548.3	4.429	4.429

TABLE 22.21. (Continued)

p, bar	$c_{p,l}$, kJ/(kg · K)	$c_{p,g}$, kJ/(kg · K)	μ_l, 10^{-4} Pa · s	μ_g, 10^{-5} Pa · s	k_l, W/(m · K)	k_g, W/(m · K)	Pr_l	Pr_g
0.10	0.855		6.16		0.105	0.0050	5.01	
0.15	0.861		5.61		0.103	0.0053	4.69	
0.20	0.865		5.28		0.101	0.0055	4.52	
0.25	0.868		4.99		0.099	0.0056	4.38	
0.30	0.872		4.79		0.098	0.0057	4.26	
0.4	0.876		4.48		0.097	0.0060	4.05	
0.5	0.880	0.545	4.25	1.00	0.095	0.0062	3.94	0.89
0.6	0.884	0.552	4.08	1.02	0.094	0.0063	3.84	0.88
0.8	0.889	0.564	3.81	1.04	0.091	0.0066	3.72	0.88
1.0	0.894	0.574	3.59	1.06	0.089	0.0069	3.61	0.88
1.5	0.905	0.600	3.23	1.10	0.086	0.0074	3.40	0.89
2.0	0.914	0.613	2.95	1.13	0.083	0.0077	3.25	0.90
2.5	0.922	0.626	2.78	1.15	0.081	0.0081	3.16	0.91
3.0	0.930	0.640	2.62	1.18	0.079	0.0083	3.08	0.91
4.0	0.944	0.663	2.40	1.22	0.075	0.0088	3.02	0.92
5.0	0.957	0.683	2.24	1.25	0.073	0.0092	2.94	0.93
6.0	0.969	0.702	2.13	1.28	0.070	0.0095	2.95	0.95
8.0	0.995	0.737	1.96	1.33	0.066	0.0101	2.95	0.97
10	1.023	0.769	1.88	1.38	0.063	0.0107	3.05	1.01
15	1.102	0.865	1.67	1.50	0.057	0.0117	3.23	1.11
20	1.234	0.969	1.49	1.69	0.053	0.0126	3.47	1.30
25	1.36	1.19	1.33		0.047	0.0134	3.84	
30	1.52	1.60	1.16		0.042	0.014	4.2	
35	1.73	2.5			0.037	0.016		

[a] v = specific volume; h = specific enthalpy; s = specific entropy; c_p = specific heat at constant pressure; μ = dynamic viscosity; k = thermal conductivity; Pr = Prandtl number; l = saturated liquid; g = saturated vapor.

[b] Critical point.

TABLE 22.22. Thermophysical Properties of Refrigerant 12 at 1-Bar Pressure[a]

T, K	v, m³/kg	h, kJ/kg	s, kJ/(kg · K)	μ, 10^{-5} Pa · s	c_p, kJ/(kg · K)	k, W/(m · K)	Pr
300	0.2024	572.1	4.701	1.26	0.614	0.0097	0.798
320	0.2167	584.5	4.741	1.34	0.631	0.0107	0.788
340	0.2309	597.3	4.780	1.42	0.647	0.0118	0.775
360	0.2450	610.3	4.817	1.49	0.661	0.0129	0.760
380	0.2590	623.7	4.853	1.56	0.674	0.0140	0.745
400	0.2730	637.3	4.890	1.62	0.684	0.0151	0.730
420	0.2870	651.2	4.924	1.67	0.694	0.0162	0.715
440	0.3009	665.3	4.956	1.72	0.705	0.0173	0.703
460	0.3148	697.7	4.987	1.78	0.716	0.0184	0.693
480	0.3288	694.3	5.018	1.84	0.727	0.0196	0.683
500	0.3427	709.0	5.048	1.90	0.739	0.0208	0.674

[a] v = specific volume; h = specific enthalpy; s = specific entropy; μ = dynamic viscosity; c_p = specific heat capacity at constant pressure; k = thermal conductivity; Pr = Prandtl number.

TABLE 22.23. Thermophysical Properties of Saturated Refrigerant 22[a]

T	P	v, m^3/kg		h, kJ/kg		s, kJ/(kg · K)		c_p, kJ/(kg · K)	μ, 10^{-4} Pa · s	k, W/(m · K)	σ, N/m
K	bar	Liquid	Vapor	Liquid	Vapor	Liquid	Vapor	(Liquid)	(Liquid)	(Liquid)	(Liquid)
150	0.0017	6.209. − 4[b]	83.40	268.2	547.3	3.355	5.215	1.059		0.161	
160	0.0054	6.293. − 4	28.20	278.2	552.1	3.430	5.141	1.058		0.156	
170	0.0150	6.381. − 4	10.85	288.3	557.0	3.494	5.075	1.057	0.770	0.151	
180	0.0369	6.474. − 4	4.673	298.7	561.9	3.551	5.013	1.058	0.647	0.146	
190	0.0821	6.573. − 4	2.225	308.6	566.8	3.605	4.963	1.060	0.554	0.141	
200	0.1662	6.680. − 4	1.145	318.8	571.6	3.657	4.921	1.065	0.481	0.136	0.024
210	0.3116	6.794. − 4	0.6370	329.1	576.5	3.707	4.885	1.071	0.424	0.131	0.022
220	0.5470	6.917. − 4	0.3772	339.7	581.2	3.756	4.854	1.080	0.378	0.126	0.021
230	0.9076	7.050. − 4	0.2352	350.6	585.9	3.804	4.828	1.091	0.340	0.121	0.019
240	1.4346	7.195. − 4	0.1532	361.7	590.5	3.852	4.805	1.105	0.309	0.117	0.0172
250	2.174	7.351. − 4	0.1037	373.0	594.9	3.898	4.785	1.122	0.282	0.112	0.0155
260	3.177	7.523. − 4	0.07237	384.5	599.0	3.942	4.768	1.143	0.260	0.107	0.0138
270	4.497	7.733. − 4	0.05187	396.3	603.0	3.986	4.752	1.169	0.241	0.102	0.0121
280	6.192	7.923. − 4	0.03803	408.2	606.6	4.029	4.738	1.193	0.225	0.097	0.0104
290	8.324	8.158. − 4	0.02838	420.4	610.0	4.071	4.725	1.220	0.211	0.092	0.0087
300	10.956	8.426. − 4	0.02148	432.7	612.8	4.113	4.713	1.257	0.198	0.087	0.0071
310	14.17	8.734. − 4	0.01643	445.5	615.1	4.153	4.701	1.305	0.186	0.082	0.0055
320	18.02	9.096. − 4	0.01265	458.6	616.7	4.194	4.688	1.372	0.176	0.077	0.0040
330	22.61	9.535. − 4	9.753. − 3	472.4	617.3	4.235	4.674	1.460	0.167	0.072	0.0026
340	28.03	1.010. − 3	7.479. − 3	487.2	616.5	4.278	4.658	1.573	0.151	0.067	0.0014
350	34.41	1.086. − 3	5.613. − 3	503.7	613.3	4.324	4.637	1.718	0.130	0.062	0.0008
360	41.86	1.212. − 3	4.036. − 3	523.7	605.5	4.378	4.605	1.897	0.106		
369.3	49.89	2.015. − 3	2.015. − 3	570.0	570.0	4.501	4.501	∞	—	—	0

[a] T = temperature; P = pressure; v = specific volume; h = specific enthalpy; s = specific entropy; c_p = specific heat at constant pressure; μ = dynamic viscosity; k = thermal conductivity; σ = surface tension. Sources: P, v, T, h, s interpolated and extrapolated from I. I. Perelshteyn, *Tables and Diagrams of the Thermodynamic Properties of Freons* 12, 13, 22, Moscow, 1971. c_p, μ, k interpolated and converted from *Thermophysical Properties of Refrigerants*, ASHRAE, New York, 1976. σ calculated from V. A. Gruzdev et al., *Fluid Mech. Sov. Res.*, Vol. 3, p. 172, (1974).

[b] The notation 6.209. − 4 signifies 6.209×10^{-4}.

TABLE 22.24. Thermophysical Properties of Refrigerant 22 at Atmospheric Pressure[a]

T, K	250	300	350	400	450	500
v, m^3/kg	0.2315	0.2802	0.3289	0.3773	0.4252	0.4723
h, kJ/kg	597.8	630.0	664.5	702.5	740.8	782.3
s, kJ/(kg · K)	4.8671	4.9840	5.0905	5.1892	5.2782	5.3562
c_p, kJ/(kg · K)	0.587	0.647	0.704	0.757	0.806	0.848
Z	0.976	0.984	0.990	0.994	0.995	0.996
$\bar{v}_s$, m/s	166.4	182.2	196.2	209.4	220.0	233.6
k, W/(m · K)	0.0080	0.0110	0.0140	0.0170	0.0200	0.0230
μ, 10^{-4} Pa · s	0.109	0.130	0.151	0.171	0.190	0.209
Pr	0.800	0.765	0.759	0.761	0.766	0.771

[a] v = specific volume; h = specific enthalpy; s = specific entropy; c_p = specific heat at constant pressure; Z = compressibility factor = Pv/RT; $\bar{v}_s$ = velocity of sound; k = thermal conductivity; μ = dynamic viscosity; Pr = Prandtl number.

TABLE 22.25. Thermophysical Properties of Saturated Rubidium[a]

T, K	P, bar	v_l, 10^{-4} m^3/kg	v_g, m^3/kg	h_l, kJ/kg	h_g, kJ/kg	s_l, kJ/(kg · K)	s_g, kJ/(kg · K)	$c_{p,l}$, kJ/(kg · K)	$c_{p,g}$, kJ/(kg · K)
312.7	2.46. − 9	6.752		119	1038	0.998	3.937	0.379	0.231
350	5.97. − 8	6.858		133	1046	1.038	3.647	0.377	0.251
400	1.69. − 6	6.983	230040	152	1057	1.091	3.355	0.375	0.273
450	2.23. − 5	7.102	19570	170	1068	1.135	3.130	0.372	0.301
500	1.73. − 4	7.215	2789	189	1078	1.174	2.953	0.369	0.335
600	3.66. − 3	7.463	156.6	225	1096	1.241	2.692	0.362	0.410
700	0.0317	7.728	20.75	261	1111	1.296	2.510	0.357	0.468
800	0.1584	8.013	4.662	297	1124	1.343	2.378	0.353	0.498
900	0.5746	8.319	1.487	332	1137	1.385	2.279	0.353	0.504
1000	1.467	8.651	0.605	368	1150	1.422	2.205	0.360	0.494
1100	3.295	9.009	0.291	404	1164	1.457	2.148	0.373	0.476
1200	6.466	9.398	0.159	442	1179	1.490	2.104	0.385	0.456
1300	11.43	9.823	0.097	481	1196	1.522	2.071	0.397	0.435
1400	14.7	10.29	0.081	521	1215	1.553	2.059	0.408	0.413
1500	28.5	10.80	0.045	562	1236	1.581	2.030	0.418	0.390

T, K	μ_l, 10^{-4} Pa · s	μ_g, 10^{-4} Pa · s	k_l, W/(m · K)	k_g, W/(m · K)	$\Pr_l$	$\Pr_g$
312.7	6.43		33.4		0.0073	
350	5.33		32.5		0.0062	
400	4.37		31.6		0.0052	
450	3.69		30.7		0.0045	
500	3.23		29.8		0.0040	
600	2.58	0.112	27.8	0.0073	0.0034	0.629
700	2.18	0.135	25.9	0.0089	0.0030	0.710
800	1.89	0.158	24.1	0.0103	0.0028	0.764
900	1.69	0.183	22.2	0.0115	0.0027	0.802
1000	1.53	0.208	20.3	0.0125	0.0027	0.832
1100	1.40	0.244	18.5	0.0133	0.0028	0.873
1200	1.30	0.268	16.7	0.0141	0.0030	0.867
1300	1.21	0.289	15.0	0.0149	0.0032	0.844
1400	1.14	0.314	13.6	0.0156	0.0034	0.831
1500	1.08	0.336	12.0	0.0160	0.0038	0.819

[a] v = specific volume; h = specific enthalpy; s = specific entropy; c_p = specific heat at constant pressure; μ = dynamic viscosity; k = thermal conductivity; Pr = Prandtl number; l = saturated liquid; g = saturated vapor. The notation 2.46. − 9 signifies 2.46×10^{-9}.

TABLE 22.26. Thermophysical Properties of Rubidium at 1-Bar Pressure[a]

T, K	959.2	1000	1100	1200	1300	1400	1500
v, m^3/kg	0.846	0.898	1.014	1.123	1.227	1.328	1.428
h, kJ/kg	1144	1164	1203	1236	1266	1293	1320
s, kJ/(kg · K)	2.233	2.253	2.291	2.319	2.343	2.363	2.382
c_p, kJ/(kg · K)	0.499	0.445	0.356	0.309	0.283	0.268	0.260
c_v, kJ/(kg · K)	0.324	0.277	0.218	0.187	0.170	0.161	0.156
γ	1.540	1.606	1.630	1.650	1.662	1.669	1.671
Z	1.0001	1.0005	1.0006	1.0006	1.0006	1.0006	1.0006
$\bar{v}_s$, m/s	348.5	382.3	409.3	433.3	454.6	473.8	491.7
k, W/(m · K)	0.0121	0.0118	0.0117				
μ, 10^{-4} Pa · s	0.198	0.219	0.262				
Pr	0.817	0.826	0.797				

[a] v = specific volume; h = specific enthalpy; s = specific entropy; c_p = specific heat at constant pressure; c_v = specific heat at constant volume; γ = ratio of principal specific heats = c_p/c_v; Z = compressibility factor = Pv/RT; $\bar{v}_s$ = velocity of sound; k = thermal conductivity; μ = dynamic viscosity; Pr = Prandtl number.

TABLE 22.27. Thermophysical Properties of Saturated Sodium[a]

T, K	P, bar	v_l, 10^{-3} m³/kg	v_g, m³/kg	h_l, kJ/kg	h_g, kJ/kg	s_l, kJ/(kg · K)	s_g, kJ/(kg · K)	$c_{p,l}$, kJ/(kg · K)	$c_{p,g}$, kJ/(kg · K)
380	2.55.−10	1.081	5.277.+9	219	4723	2.853	14.705	1.384	0.988
400	1.36.−9	1.086	2.187.+9	247	4740	2.924	14.158	1.374	1.023
420	6.16.−9	1.092	2.410.+8	274	4757	2.991	13.665	1.364	1.066
440	2.43.−8	1.097	6.398.+7	301	4773	3.054	13.219	1.355	1.117
460	8.49.−8	1.103	1.912.+7	328	4790	3.114	12.814	1.346	1.176
480	2.67.−7	1.109	6.341.+6	355	4806	3.171	12.443	1.338	1.243
500	7.64.−7	1.114	2.304.+6	382	4820	3.226	12.104	1.330	1.317
550	7.54.−6	1.129	2.558.+5	448	4856	3.352	11.367	1.313	1.523
600	5.05.−5	1.145	41511	513	4887	3.465	10.756	1.299	1.745
650	2.51.−4	1.160	9001	578	4915	3.569	10.241	1.287	1.963
700	9.78.−4	1.177	2449	642	4939	3.664	9.802	1.278	2.160
750	0.00322	1.194	794	705	4959	3.752	9.424	1.270	2.325
800	0.00904	1.211	301	769	4978	3.834	9.095	1.264	2.452
850	0.02241	1.229	128.1	832	4995	3.910	8.808	1.260	2.542
900	0.05010	1.247	60.17	895	5011	3.982	8.556	1.258	2.597
1000	0.1955	1.289	16.84	1021	5043	4.115	8.137	1.259	2.624
1200	1.482	1.372	2.571	1274	5109	4.346	7.542	1.281	2.515
1400	6.203	1.469	0.688	1535	5175	4.546	7.146	1.330	2.391
1600	17.98	1.581	0.258	1809	5225	4.728	6.863	1.406	2.301
1800	40.87	1.709	0.120	2102	5255	4.898	6.649	1.516	2.261
2000	78.51	1.864	0.0634	2422	5256	5.064	6.480	1.702	2.482
2200	133.5	2.076	0.0362	2794	5207	5.235	6.332	2.101	3.307
2400	207.6	2.480	0.0196	3299	5025	2.447	6.166	3.686	8.476
2500	251.9	3.323	0.0100	3850	4633	5.666	5.980		

T, K	μ_l, 10^{-4} Pa · s	μ_g, 10^{-4} Pa · s	k_l, W/(m · K)	k_g, W/(m · K)	Pr_l	Pr_g
380	6.68		85.6		0.0108	
400	6.08		84.7		0.0099	
420	5.58		83.8		0.0091	
440	5.16		82.8		0.0084	
460	4.81		81.9		0.0079	
480	4.50		80.9		0.0074	
500	4.24		80.0		0.0070	
550	3.69		77.7		0.0062	
600	3.28		75.4		0.0057	
650	2.95		73.0		0.0053	
700	2.69		70.7		0.0049	
750	2.47	0.192	68.3	0.0311	0.0046	1.44
800	2.30	0.196	65.9	0.0343	0.0044	1.40
850	2.14	0.200	63.6	0.0375	0.0042	1.36
900	2.02	0.206	61.4	0.0406	0.0041	1.32
1000	1.81	0.230	56.7	0.0455	0.0040	1.31
1200	1.51	0.275	54.5	0.0522	0.0036	1.32
1400	1.32	0.322	52.2	0.0570	0.0034	1.35
1600	1.18	0.371	49.9	0.0613	0.0033	1.39
1800	1.07					
2000	0.98					

[a] v = specific volume; h = specific enthalpy; s = specific entropy; c_p = specific heat at constant pressure; l = saturated liquid; g = saturated vapor. Converted from the tables of J. K. Fink, Argonne Nat. Lab. Rept. ANL-CEN-RSD-82-4, 205 pp. 1982. The notation 2.55.−10 signifies 2.55×10^{-10}.

TABLE 22.28. Thermophysical Properties of Sodium at 1-Bar Pressure[a]

T, K	1151	1200	1300	1400	1500	1600	1700	1800	1900	2000
v, m^3/kg	3.418	3.968	4.450	4.888	5.299	5.695	6.075	6.456	6.826	7.194
h, kJ/kg	5091	5202	5394	5544	5668	5779	5882	5979	6073	6166
s, kJ/(kg · K)	7.666	7.693	7.823	7.913	7.980	8.036	8.087	8.135	8.180	8.225
c_p, kJ/(kg · K)	2.542	2.240	1.703	1.380	1.194	1.087	1.024	0.985	0.961	0.945
c_v, kJ/(kg · K)	1.732	1.511	1.120	0.884	0.738	0.664	0.620	0.594	0.578	0.567
γ	1.468	1.536	1.559	1.590	1.618	1.638	1.651	1.659	1.663	1.666
Z	0.918	1.001	1.001	1.001	1.001	1.000	1.000	1.000	1.000	1.000
$\bar{v}_s$, m/s	710	781	833	882	926	966	1002	1035	1065	1095
μ, 10^{-5} Pa · s	0.264	0.275								
k, W/(m · K)	0.0507	0.0522								
Pr	1.324	1.180								

[a] v = specific volume; h = specific enthalpy; s = specific entropy; c_p = specific heat at constant pressure; c_v = specific heat at constant volume; γ = ratio of principal specific heats = c_p/c_v; Z = compressibility factor = Pv/RT; $\bar{v}_s$ = velocity of sound; μ = dynamic viscosity; k = thermal conductivity; Pr = Prandtl number.

TABLE 22.29. Thermophysical Properties of Saturated Ice-Water-Steam

P, bar	T, K	v_l,[b] 10^{-3} m^3/kg	v_g, m^3/kg	h_l,[b] kJ/kg	h_g, kJ/kg	μ_g, 10^{-4} Pa · s	k_l,[b] W/(m · K)	k_g, W/(m · K)	Pr_l[b]	Pr_g
0.001	252.84	1.0010	1167	−374.9	2464.1	0.0723	2.40	0.0169		
0.002	260.21	1.0010	600	−360.1	2477.4	0.0751	2.35	0.0174		
0.003	265.11	1.0010	408.5	−350.9	2486.0	0.0771	2.31	0.0177		
0.004	267.95	1.0010	309.1	−344.4	2491.9	0.0780	2.29	0.0179		
0.005	270.74	1.0010	249.6	−337.9	2497.3	0.0789	2.27	0.0180		
0.006	273.06	1.0010	209.7	−333.6	2502	0.0798	2.26	0.0182		
0.0061	273.15	1.0010	206.0	−333.5	2502	0.0802	2.26	0.0182		
0.0061	273.15	1.0002	206.0	0.0	2502	0.0802	0.566	0.0182	13.04	0.817
0.008	276.73	1.0001	159.4	21.9	2508	0.0816	0.568	0.0184	11.66	0.823
0.010	280.13	1.0001	129.2	29.4	2513.4	0.0829	0.578	0.0186	10.39	0.828
0.02	290.66	1.0013	67.00	73.5	2532.7	0.0872	0.595	0.0193	7.51	0.841
0.03	297.24	1.0028	45.66	101.1	2544.8	0.0898	0.605	0.0195	6.29	0.854
0.04	302.13	1.0041	34.80	121.4	2553.6	0.0918	0.612	0.0198	5.57	0.865
0.05	306.04	1.0053	28.19	137.8	2560.6	0.0933	0.618	0.0201	5.08	0.871
0.06	309.33	1.0065	23.74	151.5	2566.6	0.0946	0.622	0.0203	4.62	0.877
0.08	314.68	1.0085	18.10	173.9	2576.2	0.0968	0.629	0.0207	4.22	0.883
0.10	318.98	1.0103	14.67	191.9	2583.9	0.0985	0.635	0.0209	3.87	0.893
0.20	333.23	1.0172	7.65	251.5	2608.9	0.1042	0.651	0.0219	3.00	0.913
0.30	342.27	1.0222	5.23	289.3	2624.6	0.1078	0.660	0.0224	2.60	0.929
0.40	349.04	1.0264	3.99	317.7	2636.2	0.1105	0.666	0.0229	2.36	0.941
0.5	354.50	1.0299	3.24	340.6	2645.4	0.1127	0.669	0.0233	2.19	0.951
0.6	359.11	1.0331	2.73	359.9	2653.0	0.1147	0.673	0.0236	2.06	0.961
0.8	366.66	1.0385	2.09	391.7	2665.3	0.1176	0.677	0.0242	1.88	0.979
1.0	372.78	1.0434	1.6937	417.5	2675.4	0.1202	0.6805	0.0244	1.735	1.009
1.5	384.52	1.0530	1.1590	467.1	2693.4	0.1247	0.6847	0.0259	1.538	1.000
2.0	393.38	1.0608	0.8854	504.7	2706.3	0.1280	0.6866	0.0268	1.419	1.013
2.5	400.58	1.0676	0.7184	535.3	2716.4	0.1307	0.6876	0.0275	1.335	1.027
3.0	406.69	1.0735	0.6056	561.4	2724.7	0.1329	0.6879	0.0281	1.273	1.040
3.5	412.02	1.0789	0.5240	584.3	2731.6	0.1349	0.6878	0.0287	1.224	1.050

4.0	416.77	1.0839	0.4622	604.7	2737.6	0.1367	0.6875	0.0293	1.185	1.057
4.5	421.07	1.0885	0.4138	623.2	2742.9	0.1382	0.6869	0.0298	1.152	1.066
5	424.99	1.0928	0.3747	640.1	2747.5	0.1396	0.6863	0.0303	1.124	1.073
6	432.00	1.1009	0.3155	670.4	2755.5	0.1421	0.6847	0.0311	1.079	1.091
7	438.11	1.1082	0.2727	697.1	2762.0	0.1443	0.6828	0.0319	1.044	1.105
8	445.57	1.1150	0.2403	720.9	2767.5	0.1462	0.6809	0.0327	1.016	1.115
9	448.51	1.1214	0.2148	742.6	2772.1	0.1479	0.6788	0.0334	0.992	1.127
10	453.03	1.1274	0.1943	762.6	2776.1	0.1495	0.6767	0.0341	0.973	1.137
12	461.11	1.1386	0.1632	798.4	2782.7	0.1523	0.6723	0.0354	0.943	1.156
14	468.19	1.1489	0.1407	830.1	2787.8	0.1548	0.6680	0.0366	0.920	1.175
16	474.52	1.1586	0.1237	858.6	2791.8	0.1569	0.6636	0.0377	0.902	1.191
18	480.26	1.1678	0.1103	884.6	2794.8	0.1589	0.6593	0.0388	0.889	1.206
20	485.53	1.1766	0.0995	908.6	2797.2	0.1608	0.6550	0.0399	0.877	1.229
25	497.09	1.1972	0.0799	962.0	2800.9	0.1648	0.6447	0.0424	0.859	1.251
30	506.99	1.2163	0.0666	1008.4	2802.3	0.1684	0.6347	0.0449	0.849	1.278
35	515.69	1.2345	0.0570	1049.8	2802.0	0.1716	0.6250	0.0472	0.845	1.306
40	523.48	1.2521	0.0497	1087.4	2800.3	0.1746	0.6158	0.0496	0.845	1.331
45	530.56	1.2691	0.0440	1122.1	2797.7	0.1775	0.6068	0.0519	0.849	1.358
50	537.06	1.2858	0.0394	1154.5	2794.2	0.1802	0.5981	0.0542	0.855	1.386
60	548.70	1.3187	0.0324	1213.7	2785.0	0.1854	0.5813	0.0589	0.874	1.442
70	558.94	1.3515	0.0274	1267.4	2773.5	0.1904	0.5653	0.0638	0.901	1.503
80	568.12	1.3843	0.0235	1317.1	2759.9	0.1954	0.5499	0.0688	0.936	1.573
90	576.46	1.4179	0.0205	1363.7	2744.6	0.2005	0.5352	0.0741	0.978	1.651
100	584.11	1.4526	0.0180	1408.0	2727.7	0.2057	0.5209	0.0798	1.029	1.737
110	591.20	1.4887	0.0160	1450.6	2709.3	0.2110	0.5071	0.0859	1.090	1.837
120	597.80	1.5268	0.0143	1491.8	2689.2	0.2166	0.4936	0.0925	1.163	1.963
130	603.98	1.5672	0.0128	1532.0	2667.0	0.2224	0.4806	0.0998	1.252	2.126
140	609.79	1.6106	0.0115	1571.6	2642.4	0.2286	0.4678	0.1080	1.362	2.343
150	615.28	1.6579	0.0103	1611.0	2615.0	0.2373	0.4554	0.1307	1.502	2.571
160	620.48	1.7103	0.0093	1650.5	2584.9	0.2497	0.4433	0.1280	1.688	3.041
170	625.41	1.7696	0.0084	1691.7	2551.6	0.2627	0.4315	0.1404	2.098	3.344
180	630.11	1.8399	0.0075	1734.8	2513.9	0.2766	0.4200	0.1557	2.360	3.807
190	634.58	1.9260	0.0067	1778.7	2470.6	0.2920	0.4087	0.1749	2.951	8.021
200	638.85	2.0370	0.0059	1826.5	2410.4	0.3094	0.3976	0.2007	4.202	12.16

TABLE 22.29. (Continued)

P, bar	s_l,[b] kJ/(kg · K)	s_g, kJ/(kg · K)	$c_{p,l}$,[b] kJ/(kg · K)	$c_{p,g}$, kJ/(kg · K)	μ_l 10^{-4} Pa · s	$\gamma_{l'}$	γ_g	$\bar{v}_{s,l'}$, m/s	$\bar{v}_{s,g}$, m/s	σ^b N/m
0.001	−1.378	9.848	1.957							
0.002	−1.321	9.585	2.015							
0.003	−1.280	9.456	2.053							
0.004	−1.260	9.339	2.075							
0.005	−1.240	9.250	2.097	1.851						
0.006	−1.222	9.160	2.106	1.854						
0.0061	−1.221	9.159	2.116	1.854						
0.0061	0.0000	9.159	4.217	1.854	17.50					0.0756
0.008	0.0543	9.0379	4.206	1.856	15.75					0.0751
0.010	0.1059	8.9732	4.198	1.858	14.30					0.0747
0.02	0.2605	8.7212	4.183	1.865	10.67					0.0731
0.03	0.3543	8.5756	4.180	1.870	9.09					0.0721
0.04	0.4222	8.4724	4.179	1.874	8.15					0.0714
0.05	0.4761	8.3928	4.178	1.878	7.51					0.0707
0.06	0.5208	8.3283	4.178	1.881	7.03					0.0702
0.08	0.5925	8.2266	4.179	1.887	6.35					0.0693
0.10	0.6493	8.1482	4.180	1.894	5.88					0.0686
0.20	0.8321	7.9065	4.184	1.917	4.66					0.0661
0.30	0.9441	7.7670	4.189	1.935	4.09					0.0646
0.40	1.0261	7.6686	4.194	1.953	3.74					0.0634
0.5	1.0912	7.5928	4.198	1.967	3.49					0.0624
0.6	1.1454	7.5309	4.201	1.978	3.30					0.0616
0.8	1.2330	7.4338	4.209	2.015	3.03					0.0605
1.0	1.3027	7.3598	4.222	2.048	2.801	1.136	1.321	438.74	472.98	0.0589
1.5	1.4336	7.2234	4.231	2.077	2.490	1.139	1.318	445.05	478.73	0.0566
2.0	1.5301	7.1268	4.245	2.121	2.295	1.141	1.316	449.51	482.78	0.0548
2.5	1.6071	7.0520	4.258	2.161	2.156	1.142	1.314	452.92	485.88	0.0534
3.0	1.6716	6.9909	4.271	2.198	2.051	1.143	1.313	455.65	488.36	0.0521
3.5	1.7273	6.9392	4.282	2.233	1.966	1.143	1.311	457.91	490.43	0.0510
4.0	1.7764	6.8943	4.294	2.266	1.897	1.144	1.310	459.82	492.18	0.0500
4.5	1.8204	6.8547	4.305	2.298	1.838	1.144	1.309	461.46	493.69	0.0491

5	1.8604	6.8192	4.315	2.329	1.787	1.144	1.308	462.88	495.01	0.0483
6	1.9308	6.7575	4.335	2.387	1.704	1.144	1.306	465.23	497.22	0.0468
7	1.9918	6.7052	4.354	2.442	1.637	1.143	1.304	467.08	498.99	0.0455
8	2.0457	6.6596	4.372	2.495	1.581	1.142	1.303	468.57	500.55	0.0444
9	2.0941	6.6192	4.390	2.546	1.534	1.142	1.302	469.78	501.64	0.0433
10	2.1382	6.5821	4.407	2.594	1.494	1.141	1.300	470.76	502.64	0.0423
12	2.2161	6.5194	4.440	2.688	1.427	1.139	1.298	472.23	504.21	0.0405
14	2.2837	6.4651	4.472	2.777	1.373	1.137	1.296	473.18	505.33	0.0389
16	2.3436	6.4175	4.504	2.862	1.329	1.134	1.294	473.78	506.12	0.0375
18	2.3976	6.3751	4.534	2.944	1.291	1.132	1.293	474.09	506.65	0.0362
20	2.4469	6.3367	4.564	3.025	1.259	1.129	1.291	474.18	506.98	0.0350
25	2.5543	6.2536	4.640	3.219	1.193	1.123	1.288	473.71	507.16	0.0323
30	2.6455	6.1837	4.716	3.407	1.143	1.117	1.284	472.51	506.65	0.0300
35	2.7253	6.1229	4.792	3.593	1.102	1.111	1.281	470.80	505.66	0.0280
40	2.7965	6.0685	4.870	3.781	1.069	1.104	1.278	468.72	504.29	0.0261
45	2.8612	6.0191	4.951	3.972	1.040	1.097	1.275	466.31	502.68	0.0244
50	2.9206	5.9735	5.034	4.168	1.016	1.091	1.272	463.67	500.73	0.0229
60	3.0273	5.8908	5.211	4.582	0.975	1.077	1.266	457.77	496.33	0.0201
70	3.1219	5.8162	5.405	5.035	0.942	1.063	1.260	451.21	491.31	0.0177
80	3.2076	5.7471	5.621	5.588	0.915	1.048	1.254	444.12	485.80	0.0156
90	3.2867	5.6820	5.865	6.100	0.892	1.033	1.249	436.50	479.90	0.0136
100	3.3606	5.6198	6.142	6.738	0.872	1.016	1.244	428.24	473.67	0.0119
110	3.4304	5.5595	6.463	7.480	0.855	0.998	1.239	419.20	467.13	0.0103
120	3.4972	5.5002	6.838	8.384	0.840	0.978	1.236	409.38	460.25	0.0089
130	3.5616	5.4408	7.286	9.539	0.826	0.956	1.234	398.90	453.00	0.0076
140	3.6243	5.3803	7.834	11.07	0.813	0.935	1.232	388.00	445.34	0.0064
150	3.6859	5.3178	8.529	13.06	0.802	0.916	1.233	377.00	437.29	0.0053
160	3.7471	5.2531	9.456	15.59	0.792	0.901	1.235	366.24	428.89	0.0043
170	3.8197	5.1855	11.30	17.87	0.782	0.867	1.240	351.19	420.07	0.0034
180	3.8765	5.1128	12.82	21.43	0.773	0.838	1.248	336.35	410.39	0.0026
190	3.9429	5.0332	15.76	27.47	0.765	0.808	1.260	320.20	399.87	0.0018
200	4.0149	4.9412	22.05	39.31	0.758	0.756	1.280	298.10	387.81	0.0011

[a] v = specific volume; h = specific enthalpy; s = specific entropy; c_p = specific heat at constant pressure; μ = dynamic viscosity; k = thermal conductivitiy; Pr = Prandtl number; $\gamma = c_p/c_v$ ratio; v_s = velocity of sound; σ = surface tension; subscripts: l = saturated liquid; l' = wet saturated vapor; g = dry saturated vapor. Values at and above 1.0 bar rounded off from those of C. M. Tseng, T. A. Hamp, and E. O. Moeck, Atomic Energy of Canada Report AECL-5910, 1977.

[b] Above the solid line, solid phase; below the line, liquid.

TABLE 22.30. Thermophysical Properties of Steam at 1-Bar Pressure[a]

T, K	v, m^3/kg	h, kJ/kg	s, kJ/(kg · K)	c_p, kJ/(kg · K)	c_v, kJ/(kg · K)	γ	Z	$\bar{v}_s$, m/s	μ, 10^{-5} Pa · s	k, W/(m · K)	Pr
373.15	1.679	2676.2	7.356	2.029	1.510	1.344	0.9750	472.8	1.20	0.0248	0.982
400	1.827	2730.2	7.502	1.996	1.496	1.334	0.9897	490.4	1.32	0.0268	0.980
450	2.063	2829.7	7.741	1.981	1.498	1.322	0.9934	520.6	1.52	0.0311	0.968
500	2.298	2928.7	7.944	1.983	1.510	1.313	0.9959	540.3	1.73	0.0358	0.958
550	2.531	3028	8.134	2.000	1.531	1.306	0.9971	574.2	1.94	0.0410	0.946
600	2.763	3129	8.309	2.024	1.557	1.300	0.9978	598.6	2.15	0.0464	0.938
650	2.995	3231	8.472	2.054	1.589	1.293	0.9988	621.8	2.36	0.0521	0.930
700	3.227	3334	8.625	2.085	1.620	1.287	0.9989	643.9	2.57	0.0581	0.922
750	3.459	3439	8.770	2.118	1.653	1.281	9.9992	665.1	2.77	0.0646	0.913
800	3.690	3546	8.908	2.151	1.687	1.275	0.9995	685.4	2.98	0.0710	0.903
850	3.921	3654	9.039	2.185	1.722	1.269	0.9996	705.1	3.18	0.0776	0.897
900	4.152	3764	9.165	2.219	1.756	1.264	0.9996	723.9	3.39	0.0843	0.892
950	4.383	3876	9.286	2.253	1.791	1.258	0.9997	742.2	3.59	0.0912	0.886
1000	4.614	3990	9.402	2.286	1.823	1.254	0.9998	760.1	3.78	0.0981	0.881
1100	5.076	4223	9.625	2.36			0.9999	794.3	4.13	0.113	0.858
1200	5.538	4463	9.834	2.43			1.0000	826.8	4.48	0.130	0.837
1300	5.999	4711	10.032	2.51			1.0000	857.9	4.77	0.144	0.826
1400	6.461	4965	10.221	2.58			1.0000	887.9	5.06	0.160	0.816
1500	6.924	5227	10.402	2.65			1.0002	916.9	5.35	0.18	0.788
1600	7.386	5497	10.576	2.73			1.0004	945.0	5.65	0.21	0.735
1800	8.316	6068	10.912	3.02			1.0011	999.4	6.19	0.33	0.567
2000	9.263	6706	11.248	3.79			1.0036	1051.0	6.70	0.57	0.445

[a] v = specific volume; h = specific enthalpy; s = specific entropy; c_p = specific heat at constant pressure; c_v = specific heat at constant volume; γ = ratio of principal specific heats = c_p/c_v; Z = compressibility factor = Pv/RT; $\bar{v}_s$ = velocity of sound; μ = dynamic viscosity; k = thermal conductivity; Pr = Prandtl number.

TABLE 22.31. Thermophysical Properties of Water-Steam at High Pressures[a]

T, K	v, m^3/kg	h, kJ/kg	s, kJ/(kg · K)	c, kJ/(kg · K)	c, kJ/(kg · K)	γ	Z	$\bar{v}_s$, m/s	μ, Pa · s	k, W/(m · K)	Pr
					P = 10 bar						
300	1.003. − 3	113.4	0.392	4.18	4.13	1.01	0.0072	1500	8.57. − 4	0.615	5.82
350	1.027. − 3	322.5	1.037	4.19	3.89	1.08	0.0064	1552	3.70. − 4	0.668	2.32
400	1.067. − 3	533.4	1.600	4.25	3.65	1.17	0.0058	1509	2.17. − 4	0.689	1.34
450	1.123. − 3	749.0	2.109	4.39	3.44	1.28	0.0054	1399	1.51. − 4	0.677	0.981
500	0.221	2891	6.823	2.29	1.68	1.36	0.957	535.7	1.71. − 5	0.038	1.028
600	0.271	3109	7.223	2.13	1.61	1.32	0.987	592.5	2.15. − 5	0.047	0.963
800	0.367	3537	7.837	2.18	1.70	1.28	0.994	686.2	2.99. − 5	0.072	0.908
1000	0.460	3984	8.336	2.30	1.83	1.26	0.997	759.4	3.78. − 5	0.099	0.881
1500	0.692	5224	9.337	2.66			1.000	917.2	5.35. − 5	0.18	0.80
2000	0.925	6649	10.154	3.29			1.002	1050	6.70. − 5	0.39	0.57
					P = 50 bar						
300	1.001. − 3	117.1	0.391	4.16	4.11	1.01	0.0362	1508	8.55. − 4	0.618	5.76
350	1.025. − 3	325.6	1.034	4.18	3.88	1.08	0.0317	1561	3.71. − 4	0.671	2.31
400	1.064. − 3	536.0	1.596	4.24	3.64	1.16	0.0288	1519	2.18. − 4	0.691	1.34
450	1.120. − 3	751.4	2.103	4.37	3.43	1.27	0.0270	1437	1.52. − 4	0.681	0.975
500	1.200. − 3	976.1	2.575	4.64	3.25	1.43	0.0260	1246	1.19. − 4	0.645	0.856
600	0.0490	3013	6.350	2.85	1.94	1.47	0.885	560.5	2.14. − 5	0.054	1.129
800	0.0713	3496	7.049	2.31	1.74	1.32	0.966	674.5	3.03. − 5	0.075	0.929
1000	0.0911	3961	7.575	2.35	1.85	1.27	0.987	756.5	3.81. − 5	0.102	0.880
1500	0.1384	5214	8.589	2.66			1.000	918.8	5.37. − 5	0.18	0.81
2000	0.1850	6626	9.398	3.12			1.002	1053	6.70. − 5	0.33	0.64
					P = 100 bar						
300	9.99. − 4	121.8	0.390	4.15	4.09	1.01	0.0722	1516	8.52. − 4	0.622	5.69
350	1.022. − 3	329.6	1.031	4.17	3.87	1.08	0.0633	1571	3.73. − 4	0.675	2.31
400	1.061. − 3	539.6	1.590	4.23	3.64	1.16	0.0575	1532	2.20. − 4	0.694	1.34
450	1.116. − 3	754.1	2.097	4.35	3.43	1.27	0.0537	1452	1.53. − 4	0.685	0.975
500	1.193. − 3	977.3	2.567	4.60	3.24	1.42	0.0517	1269	1.21. − 4	0.651	0.853

TABLE 22.31. (Continued)

T, K	v, m^3/kg	h, kJ/kg	s, kJ/(kg · K)	c_p, kJ/(kg · K)	c_v, kJ/(kg · K)	γ	Z	$\bar{v}_s$, m/s	μ, Pa · s	k, W/(m · K)	Pr
600	0.0201	2820	5.775	5.22	2.64	1.97	0.726	502.3	2.14. − 5	0.073	1.74
800	0.0343	3442	6.685	2.52	1.82	1.38	0.929	662.4	3.08. − 5	0.081	0.960
1000	0.0449	3935	7.233	2.44	1.88	1.30	0.973	753.3	3.85. − 5	0.107	0.876
1500	0.0692	5203	8.262	2.68			1.000	921.1	5.37. − 5	0.18	0.82
2000	0.0926	6616	9.073	3.08			1.003	1057	6.70. − 5	0.31	0.67
					P = 250 bar						
300	9.93. − 4	135.3	0.385	4.12	4.06	1.02	0.1792	1542	8.48. − 4	0.634	5.50
350	1.016. − 3	341.7	1.022	4.14	3.84	1.08	0.1572	1599	3.78. − 4	0.686	2.28
400	1.053. − 3	550.1	1.578	4.20	3.62	1.16	0.1426	1568	2.24. − 4	0.704	1.33
450	1.105. − 3	762.4	2.078	4.30	3.41	1.26	0.1330	1496	1.57. − 4	0.696	0.969
500	1.175. − 3	981.9	2.541	4.50			0.1273	1331	1.24. − 4	0.666	0.838
600	1.454. − 3	1479	3.443	5.88	4.22	1.40	0.1313	896.9	8.63. − 5	0.532	0.952
800	0.0120	3261	6.086	3.41			0.813	627.3	3.29. − 5	0.109	1.03
1000	0.0173	3845	6.741	2.69	1.97	1.36	0.935	745.9	3.98. − 5	0.125	0.856
1500	0.0277	5186	7.827	2.73			1.000	929.1	5.40. − 5	0.18	0.819
2000	0.0372	6608	8.642	3.04			1.008	1068			
					P = 500 bar						
300	9.83. − 4	157.7	0.378	4.06	3.98	1.02	0.3549	1583	8.45. − 4	0.650	5.28
350	1.005. − 3	361.8	1.007	4.10	3.81	1.08	0.3112	1644	3.87. − 4	0.700	2.27
400	1.041. − 3	567.8	1.557	4.14	3.59	1.15	0.2820	1623	2.31. − 4	0.719	1.33
450	1.088. − 3	776.9	2.050	4.23	3.39	1.25	0.2618	1561	1.62. − 4	0.714	0.960
500	1.151. − 3	991.5	2.502	4.37			0.2493	1418	1.29. − 4	0.689	0.822
600	1.362. − 3	1456	3.346	5.08	3.72	1.37	0.2459	1080	9.34. − 5	0.588	0.808
800	4.576. − 3	2895	5.937	5.84	2.79	2.10	0.620	597.8	4.04. − 5	0.178	1.33
1000	8.102. − 3	3697	6.302	3.17	1.81	1.76	0.878	742.1	4.28. − 5	0.150	0.905
1500	0.0139	5157	7.484	2.82			1.004	943.6			
2000	0.0188	6595	8.310	3.04			1.018	1086			

[a] v = specific volume; h = specific enthalpy; s = specific entropy; c_p = specific heat at constant pressure; c_v = specific heat at constant volume; γ = ratio of principal specific heats = c_p/c_v; Z = compressibility factor = Pv/RT; $\bar{v}_s$ = velocity of sound; μ = dynamic viscosity; k = thermal conductivity; Pr = Prandtl number.

TABLE 22.32. Thermal Expansion Coefficient $\tilde{\alpha}$ of Water

T,[a]	$\tilde{\alpha}$, 10^{-4} K^{-1}				
K	0	2	4	6	8
270	$(-1.298)^b$	$(-8.99)^b$	−0.530	−0.185	0.137
280	0.394	0.717	0.993	1.217	1.491
290	1.723	1.944	2.157	2.361	2.558
300	2.747	2.930	3.107	3.279	3.445
310	3.607	3.764	3.917	4.067	4.213
320	4.356	4.496	4.633	4.767	4.899
330	5.029	5.157	5.287	5.407	5.530
340	5.651	5.770	5.888	6.004	6.120
350	6.234	6.347	6.459	6.570	6.681
360	6.790	6.899	7.008	7.170	7.278
370	7.385	7.492	7.600	7.707	7.814

[a] Column headings (0, 2, ...) give third digit of T.

[b] Subcooled liquid.

TABLE 22.33. Isothermal Compressibility Coefficient β_T of Water

T,[a]	β_T, 10^{-5} bar^{-1}				
K	0	2	4	6	8
270	$(5.219)^b$	$(5.135)^b$	5.057	4.986	4.922
280	4.863	4.810	4.738	4.696	4.676
290	4.640	4.607	4.577	4.551	4.527
300	4.506	4.487	4.471	4.457	4.445
310	4.435	4.428	4.422	4.418	4.415
320	4.415	4.416	4.419	4.423	4.428
330	4.436	4.444	4.454	4.465	4.478
340	4.492	4.507	4.524	4.541	4.560
350	4.580	4.602	4.624	4.648	4.673
360	4.699	4.727	4.755	4.785	4.816
370	4.848	4.882	4.916	4.953	4.992

[a] Column headings (0, 2, ...) give third digit of T.

[b] Subcooled liquid.

TABLE 22.34. Thermophysical Properties of Unused Engine Oil [a]

T, K	v_l, m^3/kg	$c_{p,l}$, kJ/(kg · K)	μ_l, Pa · s	k_l, W/(m · K)	Pr_l	α_l, m^2/s
250	1.093. − 3	1.72	32.20	0.151	367,000	9.60. − 8
260	1.101. − 3	1.76	12.23	0.149	144,500	9.32. − 8
270	1.109. − 3	1.79	4.99	0.148	60,400	9.17. − 8
280	1.116. − 3	1.83	2.17	0.146	27,200	8.90. − 8
290	1.124. − 3	1.87	1.00	0.145	12,900	8.72. − 8
300	1.131. − 3	1.91	0.486	0.144	6,450	8.53. − 8
310	1.139. − 3	1.95	0.253	0.143	3,450	8.35. − 8
320	1.147. − 3	1.99	0.141	0.141	1,990	8.13. − 8
330	1.155. − 3	2.04	0.084	0.140	1,225	7.93. − 8
340	1.163. − 3	2.08	0.053	0.139	795	7.77. − 8
350	1.171. − 3	2.12	0.036	0.138	550	7.62. − 8
360	1.179. − 3	2.16	0.025	0.137	395	7.48. − 8
370	1.188. − 3	2.20	0.019	0.136	305	7.34. − 8
380	1.196. − 3	2.25	0.014	0.136	230	7.23. − 8
390	1.205. − 3	2.29	0.011	0.135	185	7.10. − 8
400	1.214. − 3	2.34	0.009	0.134	155	6.95. − 8

[a] v = specific volume; c_p = specific heat at constant pressure; μ = dynamic viscosity; k = thermal conductivity; Pr = Prandtl number; α = thermal diffusivity, l = saturated liquid. The notation 1.093. − 3 signifies 1.093×10^{-3}.

TABLE 22.35. Functional Representations of a Thermophysical Property ψ as a Function of Absolute Temperature T

No.	Form
1	$\psi = a + bT$
2	$\psi = a + bT + cT^2$
3	$\psi = a + bT + cT^2 + dT^3$
4	$\psi = a + bT + cT^2 + dT^3 + eT^4$
5	$\ln \psi = a + \frac{b}{T}$
6	$\ln \psi = a + \frac{b}{T - c}$
7	$\ln \psi = a + \frac{b}{T} + \frac{c}{T^2}$
8	$\ln \psi = a + b \ln T$
9	$\ln \psi = a + \frac{b}{T} + cT^2 + dT^3$
10	$\psi = \frac{\sqrt{T}}{a + b/T}$
11	$\psi = \frac{\sqrt{T}}{a + b/T + c/T^2}$
12	$\psi = \frac{\sqrt{T}}{a + b/T + c/T^2 + d/T^3}$
13	$\psi = \frac{\sqrt{T}}{a + (b/T)e^{-c/T}}$
14	$\psi = \frac{\sqrt{T}}{a + bT + cT^2}$
15	$\psi = aT^s$
16	$\ln \psi = c(T - T_c)^s$

22.2 CORRELATION, ESTIMATION, AND PREDICTION

A few notes are appended which list useful equations to represent property variation with temperature, some representative sources of information on property estimation, and some representative sources of tabulated data. The information available is quite detailed, and an adequate treatment would require many times the space available here.

22.2.1. Functional Representations

Table 22.35 gives sixteen common functional representations of thermophysical properties ψ as a function of (absolute) temperature T. A survey of the use of these as applied to the specific heat at constant pressure, thermal conductivity, and viscosity of saturated liquids, saturated vapors, and dilute gases has been given.† The broadest possible recommendations of this survey are to fit saturated-liquid and saturated-vapor thermal conductivities by a linear or quadratic polynomial (form 1 or 2) for temperatures up to $0.9T_c$ (where T_c is the critical temperature), to fit saturated-liquid dynamic viscosities by form 5 or 6; to fit saturated-vapor dynamic viscosities by form 2 or 3, to fit dilute-gas dynamic viscosities and thermal conductivities by any of forms 10 to 15, and to fit the specific heat at constant pressure by any of forms 1 to 4, particularly 3. The source cited discusses critical-region effects, effects of internal degrees of freedom, etc. In addition to the forms cited for the property ψ, one can also experiment with using a reduced property ψ/ψ_c. Other types of reduction have been employed, such as using intermolecular-force parameters.

The above approach is mainly empirical, even though some of the equations cited do have a theoretical basis. Calculations based upon physical models can be made. However, except for reasonably dilute gases and/or very simply structured materials, the results obtained often differ appreciably from those of experiments. Section 22.2.2 lists a few good sources of information on the prediction of properties. Finally, Sec. 22.2.3 lists some sources of tabular data.

22.2.2. Bibliography on Estimation and Prediction

a. R. C. Reid, J. M. Prausnitz, and T. K. Sherwood, *The Properties of Gases and Liquids*, 3rd ed., McGraw-Hill, New York, 669 pp., 1977. A partial update of this material appears in *Perry's Chemical Engineers Handbook*, 6th ed., McGraw-Hill, New York, pp. 3-264–3-290, 1984.

b. S. Bretsznajder, *Prediction of Transport and Other Physical Properties of Fluids*, Pergamon, Oxford, 408 pp., 1971.

c. S. J. Walas, *Phase Equilibria in Chemical Engineering*, Butterworths, Stoneham, Mass., 671 pp., 1985.

d. T. E. Daubert, *Chemical Engineering Thermodynamics*, McGraw-Hill, New York, 469 pp., 1985.

e. W. C. Edmister and Y. I. Lee, *Applied Hydrocarbon Thermodynamics*, Vol. 1, Gulf, Houston, Tex., 1984.

f. Various journals, including *Fluid Phase Equilibria*, *Int. J. Thermophys.*, *J. Chem. Eng. Data*, A.S.M.E. Symposia on Thermophysical Properties, etc.

† P. E. Liley, in N.B.S. Spec. Publ. 590, Washington, 1980. An earlier version appeared in *ASHRAE Trans.*, Vol. 78, (No. II), p. 108, 1972.

22.2.3 Sources of Tabulated Data

a. *Perry's Chemical Engineers Handbook*, 6th ed., McGraw-Hill, New York, pp. 3-1–3-263, 1984.

b. W. M. Rosenhow, J. P. Hartnett, and E. N. Ganic, *Handbook on Heat Transfer Fundamentals*, 2nd ed., McGraw-Hill, New York, Chap. 3, 1985.

c. N. B. Vargaftik, *Tables of the Thermophysical Properties of Liquids and Gases*, 2nd ed., 758 pp., Wiley, New York, 1975.

d. *ASHRAE Handbook of Fundamentals*, 1981, ASHRAE, Atlanta, Ga., 1981.

e. W. C. Reynolds, *Thermodynamic Properties in S.I.*, Stanford Univ., Stanford, Calif., 173 pp., 1979.

NOMENCLATURE

a	constant in equation
b	constant in equation
c	constant in equation
c_p	specific heat at constant pressure, kJ/(kg · K)
$c_{p,l}$	specific heat at constant pressure of saturated liquid, kJ/(kg · K)
$c_{p,g}$	specific heat at constant pressure of saturated vapor, kJ/(kg · K)
c_v	specific heat at constant volume, kJ/(kg · K)
$c_{v,l}$	specific heat at constant volume of saturated liquid, kJ/(kg · K)
$c_{v,g}$	specific heat at constant volume of saturated vapor, kJ/(kg · K)
d	constant in equation
e	constant in equation
g	acceleration of gravity, m/s^2
h	specific enthalpy, kJ/kg
H	geopotential altitude, m
M	molecular weight
P	pressure, bar
P_l	vapor pressure of saturated liquid, bar
P_g	vapor pressure of saturated vapor, bar
Pr	Prandtl number = $\mu c_p/k$
R	gas constant, J/(kg · K)
s	temperature exponent
s	specific entropy, kJ/(kg · K)
s_l	specific entropy of saturated liquid, kJ/(kg · K)
s_g	specific entropy of saturated vapor, kJ/(kg · K)
T	temperature, K
T_b	boiling temperature, K
T_m	melting temperature, K
v	specific volume, m^3/kg
v_l	specific volume of saturated liquid, m^3/kg

v_g specific volume of saturated vapor, m^3/kg
$\bar{v}_s$ velocity of sound, m/s
Z compressibility factor = Pv/RT
Z geometric altitude, m

Greek Symbols

α thermal diffusivity = $k/\rho c_p$ m^2/s
$\tilde{\alpha}$ thermal expansion coefficient, K^{-1}
β_T isothermal compressibility coefficient, bar^{-1}
γ ratio of principal specific heats, = c_p/c_v
μ dynamic viscosity, Pa · s
μ_l dynamic viscosity of saturated liquid, Pa · s
μ_g dynamic viscosity of saturated vapor, Pa · s
ρ density, kg/m^3
σ surface tension, N/m
ψ any thermophysical property
ψ_c any thermophysical property at the critical point (such as critical temperature, pressure, etc.)

INDEX